GT	Group Technology	OCR	Optical Character Recognition
GTAW	Gas Tungsten Arc Welding	OS	Operating System
HAZ	Heat Affected Zone	PAW	Plasma Arc Welding (PAC = Cutting) (PAM = Machining)
HERF	High Energy Rate Forming		
HGVS	Human-Guided Vehicle System (fork-lift with driver)	PC	Personal Computer
		PCB	Printed Circuit Board
HIP	Hot Isostatic Pressing	PD	Pitch Diameter
IGES	Initial Graphics Exchange System	PDES	Product Design Exchange Specification
IMPSs	Integrated Manufacturing Production Systems	PLC	Programmable Logic Controller
		PROM	Programmable Read-Only Memory
I/O	Input/Output	PS	Production System
IOCS	Input/Output Control System	P/M	Powder Metallurgy
JIT	Just-In-Time	PVD	Physical Vapor Deposition
LAN	Local Area Network	QC	Quality Control
LASER	Light Amplification by Stimulated Emission of Radiation	QMS	Quality Management System
		RAM	Random Access Memory
LBM	Laser Beam Machining (LBW = Welding) (LBC = Cutting)	RIM	Reaction Injection Molding
		ROM	Read-Only Memory
LED	Light Emitting Diode	SAW	Submerged Arc Welding
LSI	Large Scale Integration	SMAW	Shielded Metal Arc Welding
M	Module	SPC	Statistical Process Control
MAP	Manufacturing Automation Protocol	SQC	Statistical Quality Control
MAPIM	Materials and Processes in Manufacturing	TCM	Thermochemical Machining
MCU	Machine Control Unit	TIG	Tungsten–Inert Gas
MDI	Manual Data Input	TOP	Technical Office Protocol
MIG	Metal–Inert Gas	TPA	Target Point Align
MPS	Manufacturing Production System	TQC	Total Quality Control
mrp	Material Requirements Planning	USM	Ultrasonic Machining (USW = Welding)
MRPII	Material Resources Planning	VA	Value Analysis
MS	Manufacturing System	WAN	Wide Area Network
N	Number	WIMS	Work-in-Manufacturing System
NC	Numerically Controlled	WIP	Work-In-Progress
NDT	NonDestructive Testing (NDE = Evaluation) NDI = Inspection)	YAG	Yttrium-Aluminum Garnet

MATERIALS AND PROCESSES IN MANUFACTURING

EIGHTH EDITION

E. Paul DeGarmo
Registered Professional Engineer, Professor of Industrial and Mechanical Engineering Emeritus, University of California-Berkeley

J T. Black
Registered Professional Engineer, Professor of Industrial Engineering, Auburn University, Auburn, Alabama

Ronald A. Kohser
Professor of Metallurgical Engineering, University of Missouri-Rolla

John Wiley & Sons, Inc.
New York • Chichester • Weinheim • Brisbane • Singapore • Toronto

Library of Congress Cataloging in Publication Data:

DeGarmo, E. Paul (Ernest Paul)
     Materials and processes in manufacturing / E. Paul DeGarmo, J.
Temple Black, Ronald A. Kohser.—8th ed.
         p.   cm.
     Includes bibliographical references and index.
     ISBN 0-471-36679-X
     1. Manufacturing processes. 2. Materials. I. Black, J. Temple.
II. Kohser, Ronald A.   III. Title
TS183.D4   1997
670.42'7—dc20   96-19151
                                                 CIP

Printed in the United States of America.

10 9 8 7 6 5 4 3 2

Earlier editions copyright 1988, 1984 by Macmillan Publishing Company, copyright 1979 and 1974 by Darvic Associates, Inc., copyright 1969 by E. Paul DeGarmo, copyright 1962 and 1957 by Macmillan Publishing Company.

CONTENTS

Preface   ix

Chapter 1   Introduction   1

1.1   Introduction   1
1.2   Manufacturing and Production Systems   5
1.3   Language of Manufacturing or Process
      Technology   14
1.4   Product Life Cycle   17
1.5   Basic Manufacturing Processes   18
1.6   Yardstick for Automation   24
1.7   Planning for Manufacturing   28
1.8   New Manufacturing System   30
Case Study: Famous Folks in Manufacturing   33

Part 1   Materials   34

Chapter 2   Properties of Materials   36

2.1   Introduction   36
2.2   Static Properties   38
2.3   Dynamic Properties   53
2.4   Physical Properties   64
Case Study: Separating Mixed Materials   67

Chapter 3   Nature of Metals and Alloys   68

3.1   Structure–Property Relationship   68
3.2   Atomic Structure   69
3.3   Atomic Bonds   70
3.4   Secondary Bonds   72
3.5   Interatomic Distances and Size of Atoms   72
3.6   Atom Arrangements in Materials   73
3.7   Crystal Structures of Metals   73
3.8   Development of a Grain Structure   76
3.9   Elastic Deformation of a Single Crystal   77
3.10 Plastic Deformation of a Single Crystal   78
3.11 Dislocation Theory of Slippage   80
3.12 Strain Hardening or Work Hardening   81
3.13 Plastic Deformation in Polycrystalline Metals   82
3.14 Grain Deformation and Anisotropic
      Properties   83
3.15 Fracture of Metals   83

3.16 Cold Working, Recrystallization,
      and Hot Working   84
3.17 Grain Growth   86
3.18 Alloys   86
3.19 Alloy Types   86
3.20 Atomic Structure and Electrical Properties   86
Case Study: Window Frame Materials and Design   89

Chapter 4   Equilibrium Diagrams and the Iron–Carbon System   90

4.1   Introduction   90
4.2   Phases   90
4.3   Equilibrium Phase Diagrams   91
4.4   Iron–Carbon Equilibrium Diagram   99
4.5   Steels and the Simplified Iron–Carbon Diagram   100
4.6   Cast Irons   102
Case Study: The Blacksmith Anvils   109

Chapter 5   Heat Treatment   110

5.1   Introduction   110
5.2   Processing Heat Treatments   111
5.3   Heat Treatments Used to Increase Strength   114
5.4   Strengthening Heat Treatments for Nonferrous
      Metals   115
5.5   Strengthening Heat Treatments for Steel   118
5.6   Surface Hardening of Steel   132
5.7   Furnaces   136
5.8   Heat Treatment and Energy   138
Case Study: A Flying Chip from a Sledgehammer   140

Chapter 6   Ferrous Metals and Alloys   141

6.1   Introduction to History-Dependent Materials   141
6.2   Iron   142
6.3   Steel   142
6.4   Stainless Steels   158
6.5   Tool Steels   161
6.6   Alloy Cast Steels and Irons   163
Case Study: Interior Tub of a Top-Loading
             Washing Machine   166

Chapter 7   Nonferrous Metals and Alloys   167

7.1	Introduction	167
7.2	Copper and Copper Alloys	168
7.3	Aluminum and Aluminum Alloys	174
7.4	Magnesium and Magnesium Alloys	182
7.5	Zinc–Based Alloys	183
7.6	Titanium and Titanium Alloys	185
7.7	Nickel–Based Alloys	186
7.8	Superalloys and Other Nonferrous Metals for High-Temperature Service	187
7.9	Lead, Tin, and Their Alloys	187
7.10	Some Less Known Metals and Alloys	190
7.11	Graphite	190
Case Study: Nonsparking Wrench		192

Chapter 8   Nonmetallic Materials: Plastics, Elastomers, Ceramics, and Composites   193

8.1	Introduction	193
8.2	Plastics	194
8.3	Elastomers	205
8.4	Ceramics	207
8.5	Composite Materials	214
Case Study: Two-Wheel Dolly Handles		222

Chapter 9   Material Selection   227

9.1	Introduction	227
9.2	Material Selection and Manufacturing	230
9.3	Design Process	231
9.4	Procedure for Material Selection	232
9.5	Additional Factors to Consider	235
9.6	Consideration of the Manufacturing Process	236
9.7	Ultimate Objective	237
9.8	Materials Substitution	237
9.9	Effect of Product Liability on Materials Selection	238
9.10	Aids to Material Selection	239
Case Study: Material Selection		242

Part 2   Measurement and Quality Assurance   243

Chapter 10   Measurement and Inspection   244

10.1	Introduction	244
10.2	Standards of Measurement	245
10.3	Allowance and Tolerance	252
10.4	Inspection Methods for Measurement	261
10.5	Measuring Instruments	263
10.6	Vision Systems for Measurement	275
10.7	Coordinate Measuring Machines	279
10.8	Angle-Measuring Instruments	282
10.9	Gages for Attributes Measuring	283
10.10	Surface Roughness Measurement	288

Chapter 11   Nondestructive Inspection and Testing   299

11.1	Destructive versus Nondestructive Testing	299
11.2	Visual Inspection	301
11.3	Liquid Penetrant Inspection	302
11.4	Magnetic Particle Inspection	303
11.5	Ultrasonic Inspection	305
11.6	Radiography	308
11.7	Eddy-Current Testing	310
11.8	Acoustic Emission Monitoring	312
11.9.	Other Methods of Nondestructive Testing and Inspection	313
11.10	Dormant versus Critical Flaws	315
Case Study: Portable Failure Analysis Kit		316

Chapter 12   Process Capability and Quality Control   317

12.1	Introduction	317
12.2	Determining Process Capability	317
12.3	Inspection and Quality Control	326
12.4	Determining Causes for Problems in Quality	334

Part 3   Casting Processes   339

Chapter 13   Fundamentals of Casting   340

13.1	Introduction to Materials Processing	340
13.2	Introduction to Casting	342
13.3	Casting Terminology	343
13.4	The Solidification Process	344
13.5	Patterns	355
13.6	Design Considerations in Casting	358
Case Study: The Cast Oil-Field Fitting		364

Chapter 14   Expendable-Mold Casting Processes   365

14.1.	Introduction	365
14.2.	Sand Casting	366
14.3	Cores and Core Making	383
14.4	Other Expendable-Mold Processes with Multiple-Use Patterns	387

14.5   Expendable-Mold Processes
       Using Single-Use Patterns   390
14.6   Summary   396
Case Study: Moveable and Fixed Jaw Pieces
       for a Heavy-Duty Bench Vise   398

Chapter 15   Multiple-Use-Mold Casting Processes   399

15.1   Introduction   399
15.2   Permanent Mold Casting   399
15.3   Die Casting   403
15.4   Squeeze Casting
       (or Liquid-Metal Forging)   408
15.5   Centrifugal Casting   408
15.6   Semicentrifugal Casting   409
15.7   Centrifuging   411
15.8   Continuous Casting   411
15.9   Electromagnetic (or Levitation) Casting 411
15.10 Melting and Pouring   412
15.11 Pouring Practice   416
15.12 Cleaning, Finishing, and Heat-Treating
       of Castings   417
15.13 Robots in Foundry Operations   419
15.14 Process Selection   419
Case Study: Baseplate for a Household Steam Iron 422

Chapter 16   Powder Metallurgy   423

16.1   Introduction   423
16.2   Basic Process   424
16.3   Powder Manufacture   424
16.4   Rapidly Solidified Powder
       (Microcrystalline and Amorphous)   426
16.5   Powder Testing and Evaluation   426
16.6   Powder Mixing and Blending   426
16.7   Compacting   427
16.8   Sintering   432
16.9   Hot Isostatic Pressing   433
16.10 Other Techniques to Produce
       High-Density P/M Products   434
16.11 Secondary Operations   434
16.12 Properties of P/M Products   436
16.13 Design of Powder Metallurgy Parts   437
16.14 Powder Metallurgy Products   438
16.15 Advantages and Disadvantages
       of Powder Metallurgy   440
Case Study: Automobile Seat Adjustment Gears 444

Part 4   Forming Processes   445

Chapter 17   Fundamentals of Metal Forming   446

17.1   Introduction   446
17.2   Forming Processes: Independent Variables 447
17.3   Dependent Variables   449
17.4   Independent–Dependent Relationships   450
17.5   General Parameters   451
17.6   Friction and Lubrication under Metalworking
       Conditions   452
17.7   Temperature Concerns   454
Case Study: Repairs to a Damaged Propeller   463

Chapter 18   Hot-Working Processes   464

18.1   Introduction   464
18.2   Classification of Deformation Processes 464
18.3   Hot-Working Processes   465
18.4   Rolling   465
18.5   Forging   472
18.6   Extrusion   484
18.7   Hot Drawing of Sheet and Plate   490
18.8   Pipe Welding   491
18.9   Piercing   491
Case Study: Outboard Motor Brackets   495

Chapter 19   Cold-Working Processes   496

19.1   Introduction   496
19.2   Squeezing Processes   497
19.3   Bending   509
19.4   Shearing Operations   518
19.5   Drawing and Sheet Metal Forming   528
19.6   Alternative Methods of Producing
       Sheet-Type Products   545
19.7   Presses   546
Case Study: The Bronze Bolt Mystery   555

Chapter 20   Fabrication of Plastics, Ceramics, and Composites   556

20.1   Introduction   556
20.2   Fabrication of Plastics   557
20.3   Processing of Rubber and Elastomers   571
20.4   Processing of Ceramics   572
20.5   Fabrication of Composite Materials   576
Case Study: Fabrication of Lavatory Wash Basins 588

Part 5   Material Removal Processes   589

Chapter 21   Fundamentals of Chip-Type Machining Processes   590
21.1   Introduction   590
21.2   Basic Chip Formation Processes   590
21.3   Understanding Chip Formation   599
21.4   Orthogonal Machining   601
21.5   Energy and Power in Machining   607
21.6   Heat and Temperature in Metal Cutting   611
21.7   Summary   612
Case Study: HSS versus Tungsten Carbide   617

Chapter 22   Cutting Tools for Machining   618
22.1   Introduction   618
22.2   Cutting Tool Materials   619
22.3   Tool Geometry   635
22.4   Tool Failure and Tool Life   637
22.5   Reconditioning Cutting Tools   643
22.6   Economics of Machining   643
22.7   Machinability   646
22.8   Cutting Fluids   647
Case Study: Indexable Drill Insert   652

Chapter 23   Turning, Boring, and Related Processes   653
23.1   Introduction   653
23.2   Fundamentals of Turning, Boring, and Facing   655
23.3   Lathe Design and Terminology   665
23.4   Types of Boring Machines   677
23.5   Cutting Tools for Lathes   682
23.6   Workholding in Lathes   687
Case Study: Break-Even Point Analysis of a Lathe Part   695

Chapter 24   Drilling and Related Hole-Making Processes   696
24.1   Introduction   696
24.2   Fundamentals of the Drilling Process   697
24.3   Types of Drills   700
24.4   Toolholders for Drills   709
24.5   Workholding Devices for Drilling   710
24.6   Machine Tools for Drilling   711
24.7   Cutting Fluids for Drilling   716
24.8   Counterboring, Countersinking, and Spot Facing   717
24.9   Drilling Practice and Problems   720
24.10 Reaming   720
Case Study: Bolt-Down Leg on a Casting   725

Chapter 25   Milling   726
25.1   Introduction   726
25.2   Fundamentals of Milling Processes   726
25.3   Milling Cutters   732
25.4   Milling Machines   737
25.5   Workholding Devices for Milling   745
25.6   Milling Tolerances   745
25.7   Milling Surface Finish   746

Chapter 26   Broaching, Sawing, Filing, Shaping, and Planing   748
26.1   Introduction to Broaching   748
26.2   Fundamentals of Broaching   749
26.3   Broaching Machines   756
26.4   Introduction to Sawing   759
26.5   Saw Blades   761
26.6   Types of Sawing Machines   764
26.7   Filing   769
26.8   Types of Files   769
26.9   Filing Machines   771
26.10 Shaping and Planing   772
26.11 Planing Machines   778
Case Study: Socket with a Triangular Hole   782

Chapter 27   Abrasive Machining Processes   783
27.1   Introduction   783
27.2   Abrasives   785
27.3   Grinding   788
27.4   Grinding Machines   798
27.5   Design Considerations in Grinding   810
27.6   Honing   811
27.7   Superfinishing   812
27.8   Lapping   813
Case Study: Aluminum Retainer Rings   815

Chapter 28   Workholding Devices   816
28.1   Introduction   816
28.2   Conventional Fixture Design   817
28.3   Design Criteria for Workholders   818
28.4   Design Steps   819

28.5 Clamping Considerations 821
28.6 Example of Jig Design 822
28.7 Chip Disposal 824
28.8 Unloading and Loading Time 824
28.9 Setup and Changeover 827
28.10 Types of Jigs 827
28.11 Examples of Conventional Fixtures 829
28.12 Clamps 832
28.13 Modular Fixturing 833
28.14 Group Jig and Fixture 834
28.15 Other Workholding Devices 838
28.16 Economic Justification of Jigs and Fixtures 840
Case Study: Overhead Crane Installation 845

Chapter 29 Numerical Control and Machining Centers 846

29.1 Introduction 846
29.2 CAD-NC versus APT 852
29.3 A(4) versus A(5) Level of Automation 852
29.4 Numerical Control versus Conventional Machine Tools 854
29.5 Basic Principles of Numerical Control 855
29.6 Machining Center Features and Trends 869
29.7 Advantages of Numerical Control 874
29.8 Economic Considerations in Numerical Control 874
Case Study: Break-Even-Point Analysis of a Lathe Part 878

Chapter 30 Thread Manufacturing 880

30.1 Introduction 880
30.2 Thread Cutting 886
30.3 Internal Thread Cutting 891
30.4 Thread Milling 899
30.5 Thread Grinding 899
30.6 Thread Rolling 900
Case Study: Vented Cap Screws 905

Chapter 31 Gear Manufacturing 906

31.1 Introduction 906
31.2 Gear Types 910
31.3 Gear Manufacturing 912
31.4 Machining of Gears 913
31.5 Gear Finishing 927
31.6 Gear Inspection 930
Case Study: Six-Second Pinion Gear 933

Chapter 32 Nontraditional Machining Processes 935

32.1 Introduction 935
32.2 Chemical NTM Processes 941
32.3 Electrochemical NTM Processes 947
32.4 Mechanical NTM Processes 952
32.5 Thermal Processes 955

Part 6 Joining Processes 965

Chapter 33 Gas Flame Processes: Welding, Cutting, and Straightening 966

33.1 Overview of Welding Processes 966
33.2 Oxyfuel Gas Welding 968
33.3 Overview of Cutting Processes 971
33.4 Oxygen Torch Cutting 972
33.5 Flame Straightening 975

Chapter 34 Arc Processes: Welding and Cutting 978

34.1 Arc Welding 978
34.2 Arc Cutting 993
34.3 Metallurgical and Heat Considerations in Thermal Cutting 996
Case Study: Repair of a Bicycle Frame 998

Chapter 35 Resistance Welding 999

35.1 Theory of Resistance Welding 999
35.2 Resistance Welding Processes 1002
35.3 Advantages and Limitations 1008

Chapter 36 Other Welding and Related Processes 1010

36.1 Introduction 1010
36.2 Solid-State Welding Processes 1010
36.3 Other Welding and Cutting Processes 1016
36.4 Welding of Plastics 1024
36.5 Welding-Related Processes 1028
Case Study: Field Repair to a Power Transformer Case 1033

Chapter 37 Brazing and Soldering 1034

37.1 Introduction 1034
37.2 Brazing 1034

37.3   Soldering   1045
Case Study: The Industrial Disposal Impeller   1050

Chapter 38   Adhesive Bonding and Mechanical Fasteners   1051

38.1   Adhesive Bonding   1051
38.2   Mechanical Fastening   1059
Case Study: Golf Club Heads   1065

Chapter 39   Manufacturing Concerns in Welding and Joining   1066

39.1   Introduction   1066
39.2   Types of Fusion Welds and Types of Joints   1066
39.3   Design Considerations   1068
39.4   Heat Effects   1070
39.5   Weldability or Joinability   1077
39.6   Summary   1077
Case Study: The Welded Frame   1080

Part 7   Processes and Techniques Related to Manufacturing   1081

Chapter 40   Surface Treatments and Finishing   1082

40.1   Introduction   1082
40.2   Mechanical Cleaning and Finishing   1083
40.3   Chemical Cleaning   1090
40.4   Burr Removal   1093
40.5   Coatings   1096
40.6   Vaporized Metal Coatings   1105
40.7   Ion Implantation   1108
40.8   Clad Materials   1109
40.9   Textured Surfaces   1109
40.10 Coil-coated Sheets   1110
40.11 Effects of Surface Processing   1110
Case Study: Burrs on Yo-gi's Collar   1116

Chapter 41   Manufacturing Systems and Automation   1117

41.1   Introduction   1117
41.2   Trends in Manufacturing Systems   1118

41.3   How the First Manufacturing System Evolved   1119
41.4   Systems Defined   1119
41.5   Manufacturing versus Production Systems   1120
41.6   Classification of Manufacturing Systems   1121
41.7   Automation   1134
41.8   Robotics   1149
41.9   Summary   1161

Chapter 42   Production Systems   1164

42.1   Introduction   1164
42.2   Typical Functional Areas   1165
42.3   Computer-Integrated Manufacturing   1189
42.4   Summary   1195
Case Study: Underground Steam Line   1198

Chapter 43   Lean Production: JIT Manufacturing Systems   1199

44.1   Introduction   1199
44.2   Ten Steps to Lean Production   1200
44.3   Design of the Linked-Cell Factory   1202
44.4   How to Design Manufacturing Cells   1207
44.5   Setup Reduction   1211
44.6   Integrated Quality Control   1214
44.7   Integrated Preventive Maintenance   1216
44.8   Leveling and Balancing the Manufacturing System   1216
44.9   Linking the Cells   1217
44.10 Reducing the Work in Process   1219
44.11 Vendor Relationships   1219
44.12 Autonomation   1221
44.13 Restructuring the Rest of the Business   1222
44.14 Benefits of Conversion   1223
44.15 Constraints to Conversion   1223
44.16 Summary   1224
Case Study: Snowmobile Accident   1227

Appendix   Additional Case Studies   1228

Bibliography   1238

Index   1241

PREFACE

In the world of manufacturing, significant changes are taking place. These changes or trends are having a profound impact on the world in which we live because, whether we like it or not, we all live in a technological society, a world of manufactured goods. Every day, we use hundreds of manufactured items, from the hardware to deliver water to kitchen faucets, telephones for communication, cars and planes for transportation, TVs, VCRs, furniture, clothing, and so on. These goods are manufactured in factories (manufacturing systems) designed (or arranged) to make these goods using manufacturing technologies. What are the trends in manufacturing systems and how do they impact manufacturing processes?

First, many manufacturing companies are redesigning their manufacturing systems and becoming lean (or Just-In-Time) producers. These companies are developing manufacturing cells to produce families of parts. This requires redesign of all the elements of the manufacturing systems—the machine tools, the workholding devices, the material-handling equipment and the retraining of the people who work in the system. The companies are trying to design the process technology and the system to be flexible, controllable, and efficient.

Second, the number and variety of products and the materials from which they are being made continues to proliferate, and production quantities have become smaller. Existing processes must be modified to be more flexible and new processes must be developed.

Third, consumers want better quality and reliability, so the processes and the people responsible for the quality must be continually improved.

Finally, the "time to market" for new products is continuing to decrease. Many companies are developing concurrent engineering efforts to bring product design and manufacturing closer to the customer. There are two key aspects here. First, products are designed to be simpler and easier to manufacture and assemble (called design for manufacture/assembly). The second aspect deals with design of the manufacturing system. If the system is fast and flexible, it can readily accommodate new products (new designs or design changes) and it can respond quickly to changing customer demands for existing products. Therefore, the company can be competitive in the global marketplace.

These trends impact the manufacturing processes which convert the materials into functionally usable goods. Processes must be selected for compatibility with the selected materials, keeping in mind that quality, cost per unit, and a necessary production rate will all play a role in the final selection.

Basically, manufacturing is a "value adding" activity, where the conversion of materials into products adds value to the original material. Thus, the objective of a company engaged in manufacturing is to add value and do so in the most efficient manner, using the least amount of time, material, money, space, and manpower. To minimize waste and maximize efficiency, the processes and operations need to be properly selected and arranged to permit smooth and controlled flow of material through the factory and also provide an optimal degree of product flexibility. Meeting these goals requires a well-designed and efficient manufacturing system.

The purpose of this book is to provide design and manufacturing engineers with basic information on materials and processing. The materials section focuses on properties and behavior. Thus, aspects of smelting and refining (or other material production processes) are presented only as they affect manufacturing and manufactured products. In terms of the processes used to manufacture items (converting materials into products), this text seeks to provide a descriptive introduction to a wide variety of options, emphasizing how each process works and its relative advantages and limitations. Our goal is to present this material in a way that can be understood by individuals seeing it for the very first time. This is not a handbook! Nor is it a graduate text where the objective is to thoroughly understand and optimize manufacturing processes. Mathematical models and analytical equations have been ignored or are used only when they enhance the basic understanding of the material.

The book also serves to introduce the "language of manufacturing." Just as there is a big difference between a "hand gun" and a "gun hand," there is also a big difference between "vertical drill" (a kind of machine tool) and "drill vertically" (a type of process). Every day English words (words like climb, bloom, allowance, chuck, coin, head, and ironing) have entirely different meanings on the factory floor, a place where misunderstandings can be very costly. Pity the engineer who has to go on the plant floor not knowing an engine lathe from a milling machine or what a press brake can do. This engineer quickly loses all credibility with the people who make the products and pay the salaries.

Materials and Processes in Manufacturing was written to provide a solid introduction into the fundamentals of manufacturing. The book begins with a survey of engineering materials, the "stuff" that manufacturing begins with, and seeks to provide the basic information that can be used to match the properties of a material to the service requirements of a component. A variety of engineering materials are presented, along with their properties and means of modifying them. This section can be used in curriculums that lack preparatory courses in metallurgy, materials science, or strength of materials, or where the student has not yet been exposed to those topics. In addition, various chapters in this section can be used as supplements to a basic materials course, providing additional information on topics such as heat treatment, plastics, composites, and material selection.

The remaining two-thirds of the text is devoted to manufacturing processes and the concepts that enable these processes to produce quality products in a competitive manner. Following the section on measurement and quality control, casting, forming, powder metallurgy, material removal, and joining are all developed as families of manufacturing processes. Each section begins with a presentation of the fundamentals on which those processes are based. This is followed by a discussion of the various process alternatives, which can be selected to operate individually or be combined into an integrated system.

Although considerable effort has been made to include many of the new developments in both materials and processes, the major emphasis is on the fundamentals of production processes, for these provide an enduring basis for understanding both existing phenomena and those that are to come. Another objective of the book is to introduce the broad spectrum of manufacturing processes to individuals who will be involved in the design and manufacture of finished products. The text material is presented in a descriptive fashion where the emphasis is on the fundamental workings of the process, its capabilities, typical applications, advantages, and limitations. Furthermore, the interactions of the material and the process are emphasized throughout the text, because if one does not understand material behavior, it is difficult to understand process behavior.

The objective of this revision and restructuring has been to produce a state-of-the art introduction to the materials and processes used in manufacturing. Approximately 20% of the present text has been revised for the Eighth Edition. Polymers, ceramics and composite materials now occupy a more prominent position in the materials section, and a separate chapter has been included to present the manufacturing processes that are unique to those materials. The section on measurements and quality control has been expanded and now includes non-destructive testing and inspection, statistical process control, and new concepts in total quality assurance. The information on manufacturing systems has been updated and restructured to reflect the many methods and strategies that are impacting the manufacturing community. The case studies have been expanded in scope, having been designed to make students aware of the great importance of properly coordinating design, material selection and manufacturing in order to produce a satisfactory and reliable product.

The text is intended for use by engineering (mechanical, manufacturing and industrial) and engineering technology students, in both 2- and 4-year undergraduate degree programs. In addition, the book has also been used by engineers and technologists in other disciplines concerned with design and manufacturing (such as aerospace). Factory personnel have found the book to be valuable reference that concisely presents the various production alternatives and the advantages and limitations of each. Individuals desiring additional, or in-depth, information on specific materials or processes are directed to the various references listed at the back of the text.

In preparing the Eighth Edition, the authors have a great many people to thank. First, Drs. Black and Kohser are forever indebted to E. Paul DeGarmo for selecting them to try to "fill his shoes" and continue the fine tradition that he has established. The authors also wish to acknowledge the multitude of assistance, information, and illustrations that have been provided by a variety of industries, professional organizations, and trade associations. The text has become known for the large number of clear and helpful photos and illustrations that have been graciously provided by a variety of sources. In some cases, equipment is photographed or depicted without safety guards so as to show important details, and personnel are not wearing certain items of safety apparel that would be worn during normal operation.

Over the many editions, there have been hundreds of reviewers and readers who have made suggestions and corrections to the text. We continue to be grateful for the time and interest they have put into this book.

Heartfelt thanks are due to our wives (Carol Black and Barb Kohser) who endured being "textbook widows" during the time when the bulk of this Eighth Edition was being written. Not only did they provide loving support, but Carol and Barb also provided hours of expert proofreading, typing, and editing as the manuscript was prepared.

A special thanks to our editor, Bill Stenquist, for putting up with two procrastinating professors who tried both his patience and his abilities as he sought to coordinate all the various activities required to produce such a text on a reasonable schedule.

History of the Text

The first edition of Materials and Processes in Manufacturing was written by E. Paul DeGarmo. It was published in 1957 and soon became the emulated standard for introductory texts in manufacturing. Second, third and fourth editions followed in 1962, 1969 and 1974. In 1977, Ronald A. Kohser wrote a letter to E. Paul DeGarmo highlighting a number of problems he found in the materials chapters. DeGarmo then invited Kohser, a professor at the University of Missouri-Rolla, to rewrite the materials section and thus Kohser became involved with the fifth edition, published in 1979.

After publication of the fifth edition, E. Paul DeGarmo expressed a desire to complete his retirement from the University of California-Berkeley by also retiring from textbook writing. It became apparent that an individual was needed who could provide expertise in areas that would complement those of Ron Kohser, and who also had the ability to present technical material in a way that first-time students could understand. After a nationwide search and screening, an offer was extended to J T. Black, who was then a professor at The Ohio State University. For the sixth edition, published in 1984, and the seventh edition, published in 1988, Ron Kohser and J T. Black have shared the responsibility for the text. The chapters on engineering materials, casting, forming, powder metallurgy, joining, and nondestructive testing have been written or revised by Ron Kohser. J T. Black has assumed the responsibility for the introduction, and chapters on material removal, metrology, quality control, and manufacturing production systems.

The eighth edition was originally slated for publication in 1995. The recent acquisition of the Macmillan Publishing Company by Prentice Hall and the associated restructuring, the untimely death of our Macmillan editor (Mr. David Johnstone), and a heart attack experienced by Ron Kohser all contributed to schedule delays. It is apparent that the same considerations involved in bringing a manufactured product to market also apply to textbooks. The "system" involves design, production planning, prototype fabrication (drafts), evaluation and quality control (editing and review), mass production, distribution, and finally, customer evaluation. Just as the best factories are managed by individuals who understand the materials, processes and systems that are used in their plants, we also benefit from the management provided through our publisher. Just as companies form product teams, we too have our team of authors, editor, reviewers, production supervisor, and others. Just as companies take pride in their products, we take pride in ours. Through all of these editions the book has been kept current with the rapidly changing developments in the field and has maintained its stature as the standard text in manufacturing processes. We hope that you will find it to be a "quality" text.

CHAPTER 1

INTRODUCTION

1.1 INTRODUCTION
Materials, Manufacturing,
    and the Standard of Living
Roles of Engineers in
    Manufacturing
Changing World Competition
1.2 MANUFACTURING AND
    PRODUCTION SYSTEMS
Production System
Manufacturing Systems
Manufacturing Processes
Job and Station
Operation
Treatments
Tools, Tooling, and
    Workholders
Tooling for Measurement
    and Inspection
Integrating Inspection into
    the Process

1.3 LANGUAGE OF
    MANUFACTURING OR
    PROCESS TECHNOLOGY
Workpiece and Its
    Configuration
Understanding What the
    Enterprise Does
1.4 PRODUCT LIFE CYCLE
1.5 BASIC
    MANUFACTURING
    PROCESSES
1.6 YARDSTICK FOR
    AUTOMATION
1.7 PLANNING FOR
    MANUFACTURE
1.8 NEW MANUFACTURING
    SYSTEM
Case Study: FAMOUS FOLKS IN
    MANUFACTURING

■ 1.1 INTRODUCTION

Materials, Manufacturing and the Standard of Living

The standard of living in any society is determined, primarily, by the *goods* and *services* that are available to its people. In most cases, materials are utilized in the form of manufactured goods. Manufactured goods are typically divided into two classes: consumer goods and producer goods. *Producer goods* are those goods manufactured for other companies to use to manufacture either producer or consumer goods. *Consumer goods* are those purchased directly by the consumer or the general public. For example, someone has to build the machine tool, a lathe, that turns the large rolls that are sold to the rolling mill to be used to roll the sheets of steel which are then formed and become the fenders of your car. Similarly, many service industries depend heavily on the use of manufactured products, just as the agricultural industry is heavily dependent on the use of large farming machines for efficient production.

Converting materials from one form to another adds value to them. The more efficiently materials can be produced and converted into the desired products that function with the prescribed quality, the greater will be the companies' productivity and the better will be the standard of living of the employees.

The history of man has been linked with his ability to work with materials, beginning with the Stone Age and ranging through the eras of copper and bronze, the Iron Age, and recently the age of steel, with our sophisticated ferrous and nonferrous materials. We are now entering the age of tailor-made materials and exotic alloys. The alloys used in the

1

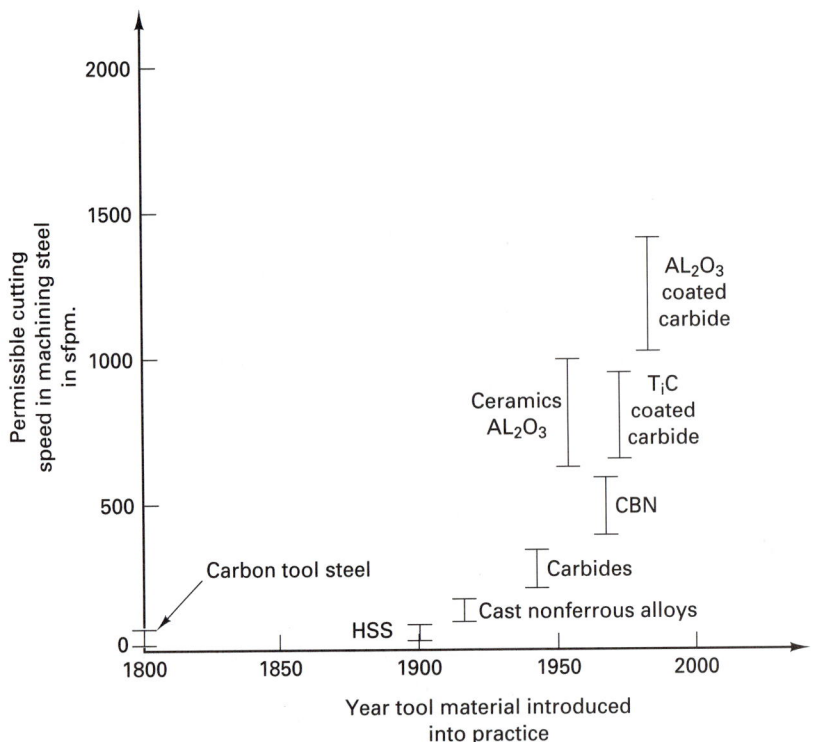

FIGURE 1-1   New cutting tool materials are continuously being introduced, resulting in faster cutting speeds (in surface feet per minute).

manufacture of jet engines have continuously changed to enhance engine performance. As shown in Figure 1-1, the materials used for cutting tools in machining processes have continuously been improved to allow for faster cutting speeds. As materials become more sophisticated, having greater strength and lighter weight, they also become more difficult to manufacture with existing manufacturing methods.

Although materials are no longer used only in their natural state or in modified forms, there is obviously an absolute limit to the amounts of many materials available here on earth. Therefore, although while the variety of man-made materials continues to increase, resources must be used efficiently and recycled whenever possible. Of course, recycling only postpones the exhaustion date.

Like materials, processes have also proliferated greatly in the last 30 years, with new processes being developed to handle the new materials more efficiently and with less waste. Advances in manufacturing technology often account for improvements in productivity. Even when the technology is proprietary, the competition often gains access to it, usually quite quickly.

Materials, men, methods, and equipment are interrelated factors in manufacturing that must be combined properly to achieve low cost, superior quality, and on-time delivery. Typically, as shown in Figure 1-2, 40% of the selling price of a product is *manufacturing cost*. Since the selling price is determined by the customer, maintaining the profit often depends on reducing manufacturing cost. Direct labor, usually the target of automation, accounts for only about 12% of manufacturing cost even though many view it as the main factor in increasing productivity. In Chapter 43, a manufacturing strategy is presented that attacks the materials cost, indirect costs, and general administration costs in addition to labor costs. The material costs include the cost of storing and handling the materials within the plant. The strategy is called *lean production.*

Referring again to Figure 1-2, of the total expenses (selling price less profit), about 68% of dollars are spent on people: about 15% for engineers; 25% for marketing, sales, and general management people; 5% for direct labor, and 10% for indirect labor (55/80 = 68.75%). The average labor cost in manufacturing in the United States was around $11 per hour for the hourly workers in 1990. Reductions in direct labor will have only marginal effects on the people costs. The optimal combination of factors for producing a small quantity of a given product may be very inefficient for a larger quantity of the same product. Consequently, a systems approach, taking all the factors into account, must be used. *This requires a sound and broad understanding of materials, processes, and equipment on the part of the decision makers, accompanied by an understanding of the manufacturing systems.* Materials and processes in manufacturing are what this book is all about.

Roles of Engineers in Manufacturing

Many engineers have as their function the designing of products. The products are brought into reality through the processing or fabrication of materials. In this capacity designers are a key factor in the material selection and manufacturing procedure. A *design engineer,* better than any other person, should know what the design is to accomplish, what assumptions can be made about service loads and requirements, what service environment the product must withstand, and what appearance the final product is to have. To meet these requirements, the material(s) to be used must be selected and specified. In most cases, to utilize the material and to enable the product to have the desired form, the designer knows that certain *manufacturing processes* will have to be employed. In many instances, the selection of a specific material may dictate what processing must be used. On the other hand, when certain processes must be used, the design may have to be modified for the process to be utilized effectively and economically. Certain dimensional sizes can dictate the processing, and some processes require certain sizes for the parts going into them. In converting the design into reality, many decisions must be made. In most instances they can be made most effectively at the design stage. It is thus apparent that design engineers are a vital factor in the manufacturing process, and it is indeed a blessing to the company

FIGURE 1-2   Manufacturing cost is the largest cost in the selling price. The largest manufacturing cost is material costs, not direct labor.

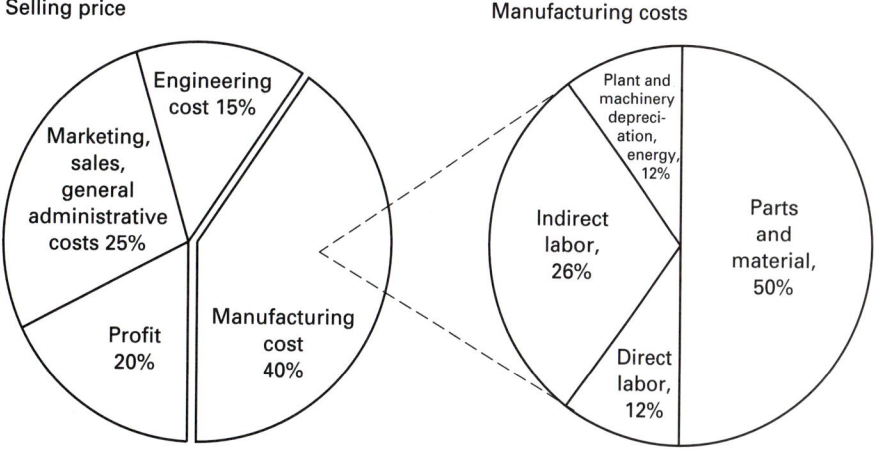

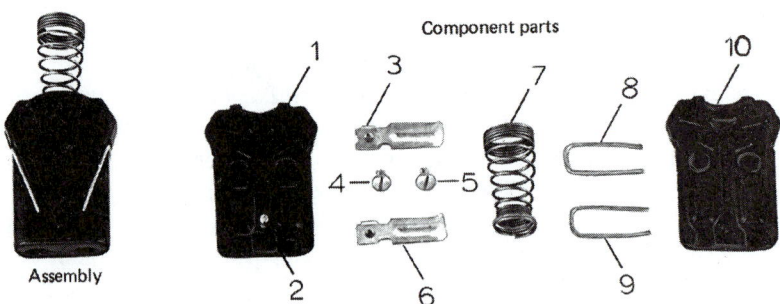

FIGURE 1-3 Component parts that make up the assembly of a low-cost appliance plug.

if they can *design for manufacturing*, that is, design the product so that it can be manufactured and/or *assembled* economically.

Manufacturing engineers select and coordinate specific processes and equipment to be used, or supervise and manage their use. Some design special tooling that is used so that standard machines can be utilized in producing specific products. These engineers must have a broad knowledge of manufacturing processes and of material behavior so that desired operations can be done effectively and efficiently without overloading or damaging machines and without adversely affecting the materials being processed. Although it is not obvious, the most hostile environment the material may ever encounter in its lifetime is the processing environment.

The machines and equipment used in manufacturing and their arrangement in the factory also comprise a design task. *Industrial* or *manufacturing engineers* who design (or lay out) factories have the same concerns of the interrelationship of design, the properties of the materials that the machines are going to process, and the interreaction of the materials and the machines.

Materials engineers devote their major efforts to developing new and better materials. They, too, must be concerned with how these materials can be processed and with the effects that the processing will have on the properties of the materials. Although their roles may be quite different, it is apparent that a large proportion of engineers must concern themselves with the interrelationships of materials and manufacturing processes.

As an example of the close interrelationship of design, materials selection, and the selection and use of manufacturing processes, consider the standard electrical appliance plug shown in Figure 1-3. Suppose that this plug is sold at the retail store for $4.49. The wholesale outlet sold the plug for $4.00 and the manufacturer probably received about $3.50 for it. As shown in Figure 1-3, it consists of 10 parts. Thus the manufacturer had to produce and assemble the 10 parts for about $1.50, an average of 15 cents per part. Only by giving a great deal of attention to design, selection of materials, selection of processes, selection of equipment used for manufacturing (tooling), and utilization of personnel could such a result be achieved.

The appliance plug is a relatively simple product, yet the problems involved in its manufacture are typical of those with which manufacturing industries must deal. The elements of design, materials, and processes are all closely related, each having its effect on the others. For example, if the two plastic shell components were to be fastened together by a rivet instead of by the two U-shaped clips, entirely different machines, processes, and assembly procedures would be required. Such a design change would have a significant impact on the entire manufacturing process and on the cost.

The performance of the clips depends on the selection of the proper material. The material had to be ductile to permit it to be bent without breaking, yet it had to be sufficiently strong and stiff to act as a spring for holding the shells together firmly. It is apparent that both the material and the processing had to be considered when the plug and the clips were designed to assure a satisfactory product that could be manufactured economically.

Changing World Competition

In recent years, major changes in the world of goods manufacturing have taken place. Three of these are:

1. Worldwide or global competition
2. Advanced technology
3. New manufacturing systems structure, strategies, and management

Worldwide (global) competition is now a fact of manufacturing life and this trend will continue in the future. The goods you buy today may have been made anywhere in the world. The second aspect, advanced manufacturing technology, usually refers to new processes or computer-aided manufacturing. The new technology is often purchased from companies who have developed the technology, so this approach is important but may not provide a unique competitive advantage in that your competitor can also buy the technology, provided that they have the capital. Some companies develop their own unique process technology and try to keep it proprietary as long as they can. A good example of unique process technology was the numerical control machine tool discussed in Chapter 29. Nowadays, computer-controlled machine tools are common to the factory floor.

The third change and perhaps the real key to success in manufacturing is to build a manufacturing system that can deliver on time to the customer, superior-quality goods at the lowest possible cost in a flexible way. This change reflects an effort to improve markedly the methodology by which goods are produced rather than simply upgrading the process technology.

■ 1.2 MANUFACTURING AND PRODUCTION SYSTEMS

Manufacturing is the economic term for making goods and services available to satisfy human wants. Manufacturing implies creating value by applying useful mental or physical labor.

The *manufacturing processes* are collected together to form a *manufacturing system* (MS). The manufacturing system takes inputs (Figure 1-4) and produces products for the customer. The production system includes the manufacturing system and services it. In this book, a *production system* will refer to the total company and will include within it the *manufacturing systems*. The following football analogy distinguishes between manufacturing systems and production systems (Figure 1-5).

FOOTBALL ANALOGY.   College football is an example of a service industry. Football players are equivalent to manufacturing machine tools. The things that they do, such as punt, pass, run, tackle, and block, are equivalent to operations. Different machines do different operations, and some machines do operations better than others.

The arrangement of machines (often called the plant layout) defines the design of the manufacturing system. In football, this arrangement is called an offensive alignment or defensive formation. Modern teams use pro "T" sets and "I" formations on offense. Factories are arranged as project shops, job shops, flow shops, and linked-cell manufacturing systems.

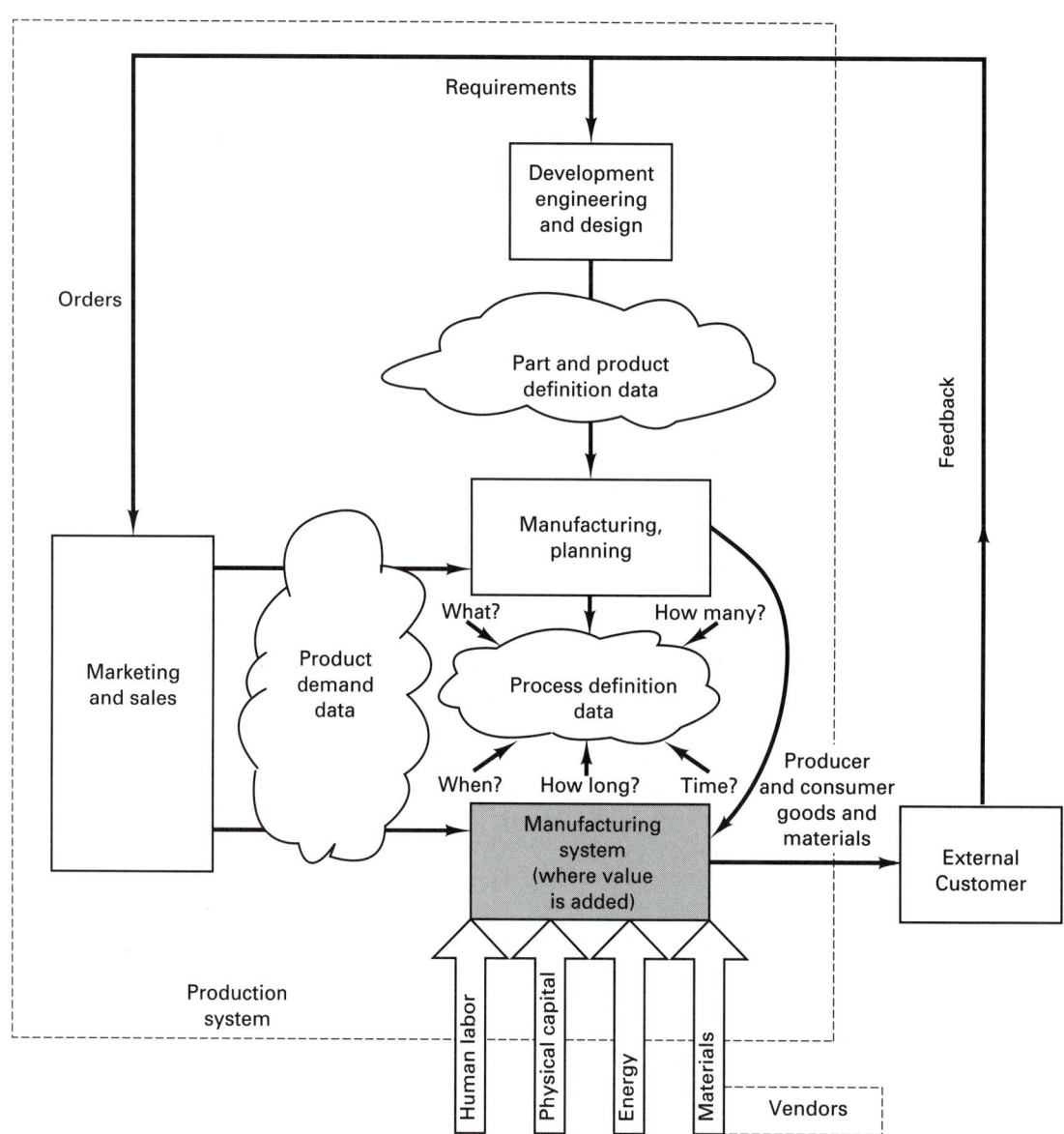

FIGURE 1-4   The production system (PS) takes product demand information and product definition data and uses them to plan the operations of manufacturing system (MS).

The job shop, with some elements of flow shop when the volume is large enough to justify some special-purpose equipment, is the most common manufacturing system. The job shop is functionally arranged; that is, similar processes are put together. The same thing is done in football, with all the linemen segregated form the backs. Coaches are equivalent to foremen, and the head coach is the supervisor.

In the football analogy, the production system would be the athletic department, which sells tickets, runs the training room (machine maintenance and repair), raises operating capital, arranges the material handling (travel), and does whatever is needed to help

Football players	=	Manufacturing processes or machine tools
Things football players do: Run, punt, pass, block, tackle, catch	=	Operations on the machine tools like turning, drilling, boring, tapping
Offensive and defensive plays Pro Tee I Back Single Wing	=	Manufacturing systems Job shop Flow job Linked-cell
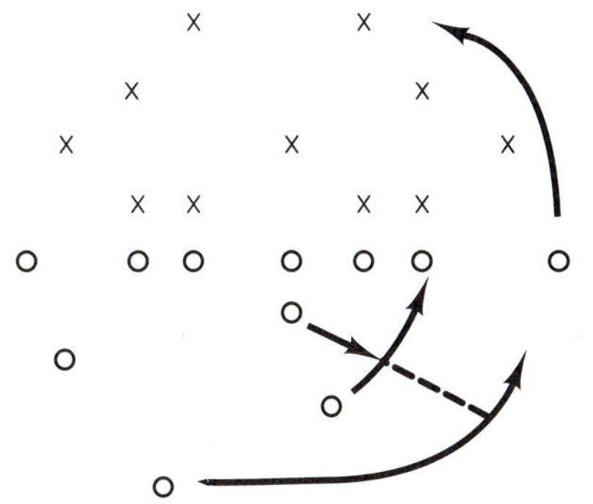	=	Layouts of manufacturing systems Functional layout Product layout Lean layout
Athletic department Maintain field Sell tickets Print programs	=	Production system Design Personnel Accounting Quality control

FIGURE 1-5 Football analogy to manufacturing production systems (MPS)

keep the manufacturing system operating but does not really do any manufacturing. Members of the athletic department never play during a game. They are all indirect labor, managerial, and staff employees. In the plant, the production system does not build parts but rather *services* the manufacturing system. The division is also known as *staff*, while people who work in manufacturing are called *line*.

As shown in Figure 1-4, the production system therefore includes the manufacturing system plus all the other functional areas of the plant for information, design, analysis, and control. These subsystems are somehow connected to each other to produce either goods or services or both. *Goods* refer to material things. *Services* are nonmaterial things that we buy to satisfy our wants, needs, or desires. *Service production systems* (SPSs) include transportation, banking, finance, savings and loan, insurance, utilities, health care, education, communication, entertainment, sporting events, and so on. They are useful labors that do not directly produce a product.

As shown in Table 1-1, production terms have a definite rank of importance somewhat like grades in the army. Confusing *system* with *section* is similar to mistaking a colonel for a corporal. In either case, knowledge or rank is necessary. The terms tend to overlap because of the inconsistencies of popular usage.

TABLE 1-1. Production Terms for Manufacturing Production Systems

Term	Meaning	Examples
Production system	All aspects of workers, machines, and information, considered collectively needed to manufacture parts or products; integration of all units of the system is critical (see Chapter 42)	Company that makes engines, assembly plant, glassmaking factory, foundry; sometimes called the enterprise or the business
Manufacturing system (sequence of operations, collection of processes)	The collection of manufacturing processes and operations resulting in specific end products; an arrangement or layout of many processes (see Chapter 41)	Rolling steel plates, manufacturing of automobiles, series of connected operations or processes, a job shop, a flow shop, a continuous process
Machine or machine tool or manufacturing process	A specific piece of equipment designed to accomplish specific processes, often called a *machine tool*; machine tools link together to make a manufacturing system.	Spot welder, milling machine, lathe, drill press, forge, drop hammer, die caster, punch press, grinder, etc.
Job (sometimes called a station)	A collection of operations done on machines or a collection of tasks performed by one worker at one location on the assembly line	Operation of machines, inspection, final assembly; forklift driver has the job of moving materials
Operation (sometimes called a process)	A specific action or treatment, the collection of which makes up the job of a worker	Drill, ream, bend, solder, turn, face, mill extrude, inspect, load
Tools or tooling	Refers to the implements used to hold, cut, shape, or deform the work materials; called *cutting tools* if referring to machining; can refer to *jigs and fixtures* in workholding and *punches and dies* in metal forming	Grinding wheel, drill bit, end milling cutter, die, mold, clamp, three-jaw chuck, fixture

An obvious problem exists here in the terminology of manufacturing and production. The same term can refer to different things. For example, *drill* can refer to the machine tool that does these kinds of operations; the operation itself, which can be done on many different kinds of machines; or the cutting tool, which exists in many different forms. It is therefore important to use modifiers whenever possible: "Use the *radial* drill *press* to drill a hole with a 1-in.-diameter spade drill." The emphasis of this book is directed toward the understanding of the processes, machines, and tools required for manufacturing and how they interact with the materials being processed. In the last three chapters, a brief introduction to the systems aspects is presented.

Production System

The highest-ranking term in the hierarchy is *production system*. A production system includes people, money, equipment, materials and supplies, markets, management, and the manufacturing system. In fact, all aspects of business and commerce (manufacturing, sales, advertising, profit, and distribution) are involved.

Much of the information given for *manufacturing production systems* (MPSs) is relevant to the *service production system* (SPS). Many *manufacturing production systems* require an SPS for proper product sales. This is particularly true in industries such as the food (restaurant) industry in which customer service is as important as quality and on-time delivery.

Manufacturing Systems

A collection of operations and processes used to obtain a desired product(s) or component(s) is called a *manufacturing system*. The manufacturing system is therefore the design or *arrangement of the manufacturing processes*. Control of a system applies to overall control of the whole, not merely of the individual processes or equipment. The entire manufacturing system must be controlled to control material movement, inventory levels, product quality, output rates, and so on.

Five manufacturing system designs can be identified: the job shop, the flow shop, the linked-cell shop, the project shop, and the continuous process. The latter system deals primarily with liquids, gases (such as an oil refinery) rather than solids or discrete parts.

The most common of these layouts is the *job shop*, characterized by large varieties of components, general-purpose machines, and a functional layout (Chapter 41). This means that machines are collected by function (all lathes together, all milling machines together) and the parts are routed around the shop in containers to the various machines. The material in the cart or container is called the *lot*.

Flow shops are characterized by larger build quantities, special-purpose machines, less variety, and more mechanization. Flow shop layouts are typically either continuous or interrupted. If *continuous*, they basically run one large-volume complex product in great quantity and nothing else. The appliance plug was made this way. A transfer line producing an engine block is another typical example. If *interrupted*, the line manufactures large quantities but is periodically "changed over" to run a similar but different component.

The *linked-cell* manufacturing system is composed of manufacturing cells connected together (linked) using a unique form of inventory and information control (kanban) (Chapters 41 and 43).

The *project shop* is characterized by the immobility of the item being manufactured. In the construction industry, bridges and roads are good examples. In the manufacture of goods, large airplanes, ships, large machine tools, and locomotives are manufactured in project shops. It is necessary that the workers, machines, and materials come to the site. The number of end items is not very large, and therefore the lot sizes of the components going into the end item are not large. Thus the job shop usually supplies parts and subassemblies to the project shop in small lots.

Naturally, there are many hybrid forms of these manufacturing systems, but the job shop with elements of the flow shop is the most common system. Because of its design, the job shop has been shown to be the least cost-efficient of all the systems. Component parts in a typical job shop spend only 5% of their time in machines and the rest of the time waiting or being moved from one functional area to the next. Once the part is on the machine, it is actually being processed (i.e., having value added to it by the changing of its shape) only about 30 to 40% of the time (Figure 1-6). The rest of the time it is being loaded, unloaded, inspected, and so on. The advent of *numerical control* machines (Figure 1-7 and Chapter 29) increased the percentage of time that the machine is making chips because tool movements are programmed and the machines can automatically change tools or load or unload parts. However, there are a number of trends that are forcing manufacturing management to consider means by which the job shop system itself can be redesigned to improve its overall efficiency. These trends have forced manufacturing companies to convert their batch-oriented job shops into linked cells. One of the ways to restructure the manufacturing system and form cells is through the use of group technology.

Group technology (GT) is a concept whereby similar parts are grouped together into part families. Parts of similar size and shape can often be processed through a *similar set of processes*. A part family based on manufacturing would have the same set or sequences of manufacturing processes. The set of processes can be collected together to form a cell. Thus, with GT, job shops can be restructured into cells, each cell specializing in a particular family of parts (Figure 1-8). The parts are handled less, machine setup time is shorter, in-process inventory is lower, and the time needed for parts to get through the manufacturing system is greatly reduced.

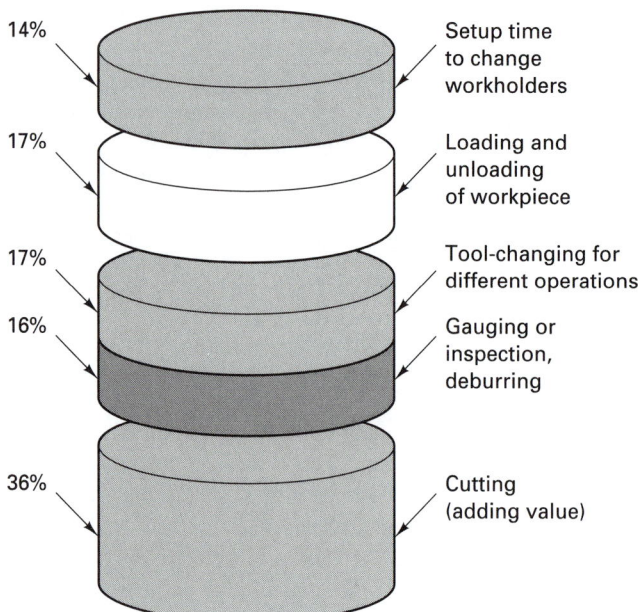

14% — Setup time to change workholders

17% — Loading and unloading of workpiece

17% — Tool-changing for different operations

16% — Gauging or inspection, deburring

36% — Cutting (adding value)

FIGURE 1-6 Typical utilization of the production time in metalturning operations with conventional tool handling, workpiece loading and unloading, setups, and inspection.

Manufacturing Processes

A *manufacturing process* converts unfinished materials to finished products often using a machine tool. For example, injection molding, die casting, progressive stamping, milling, arc welding, painting, assembling, testing, pasteurizing, homogenizing, and annealing are commonly called *processes* or *manufacturing processes.* The term *process* often implies a sequence of steps, processes, or operations for production of goods or services.

A *machine tool* is an assembly of related mechanisms on a frame or bed that together produce a desired result. Generally, the motors, controls, and auxiliary devices are included. The cutting tools and workholding devices used on the machine tool are considered separately.

A machine tool may do a single manufacturing process (cutoff saw) or multiple processes, or it may manufacture an entire component. Machine sizes vary from a tabletop drill press to a 1000-ton forging press.

Job and Station

In the classical system, a *job* is the total of the work or duties a worker performs. A station is the work area of a production line worker. (The multifunctional worker concept discussed in Chapter 44 has changed these traditional definitions.)

A job may be a group of related operations and tasks performed at one station or series of stations in cells. For example, the job at a final assembly station may consist of four tasks:

1. Attach carburetor.
2. Connect gas line.
3. Connect vacuum line.
4. Connect accelerator rod.

The job of a turret lathe (a semiautomatic machine) operator may include the following operations and tasks: load, start, index and stop, unload, inspect. The operator's job

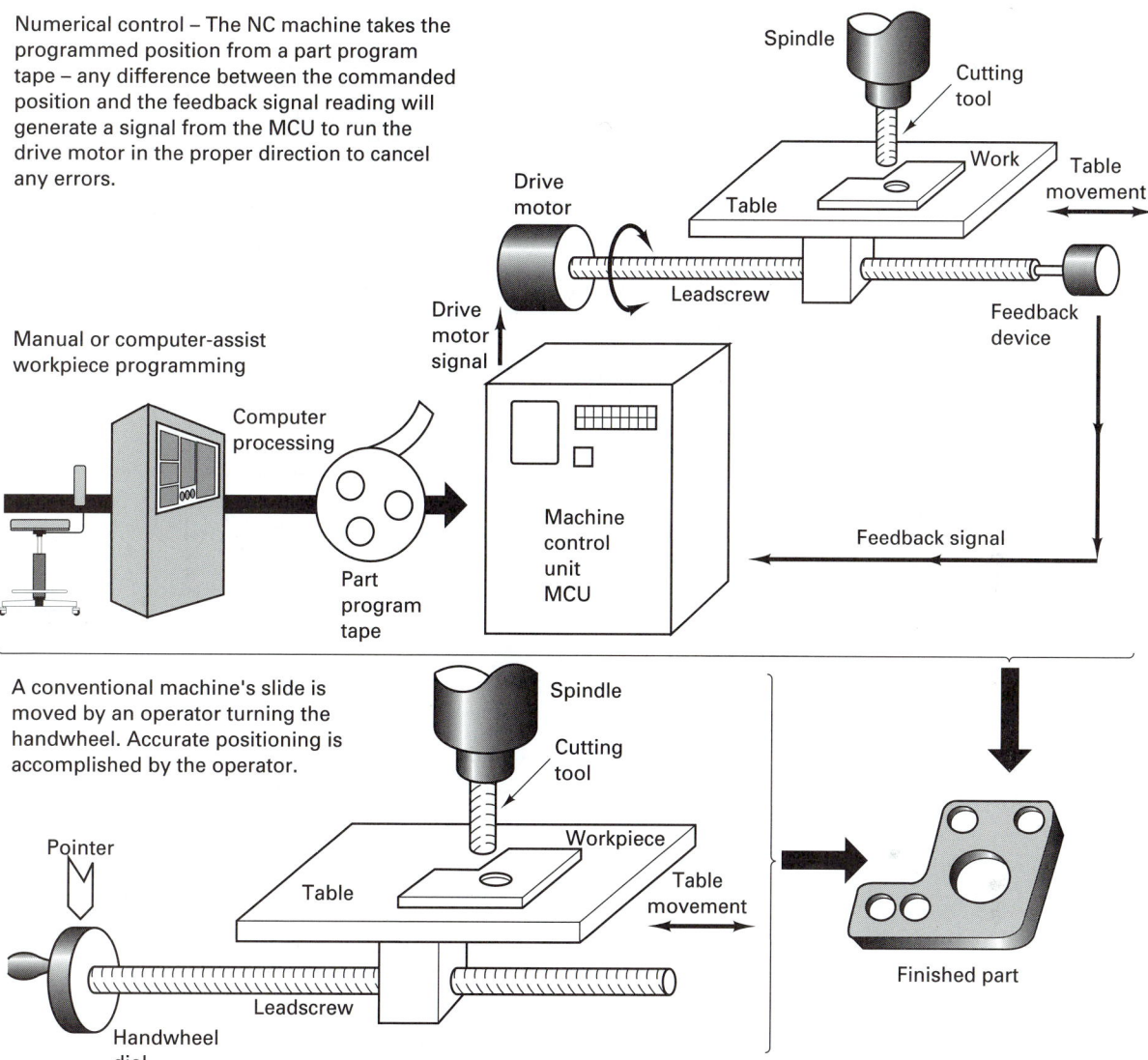

Numerical control – The NC machine takes the programmed position from a part program tape – any difference between the commanded position and the feedback signal reading will generate a signal from the MCU to run the drive motor in the proper direction to cancel any errors.

Manual or computer-assist workpiece programming

A conventional machine's slide is moved by an operator turning the handwheel. Accurate positioning is accomplished by the operator.

FIGURE 1-7   The same part can be made by NC or manual machining. The increased cost of NC can be offset by the decreased manufacturing time and improved quality.

may also include setting up the machine (i.e., getting it ready for manufacturing). Other machine operations include drilling, reaming, facing, turning, chamfering, and knurling. The operator can run more than one machine or service more than one station.

The terms *job* and *station* have been carried over to unmanned machines. A *job* is a group of related operations generally performed at one station, and a *station* is a position or location in a machine (or process) where specific operations are performed. A simple machine may have only one station. Complex machines can be composed of many stations. The job at a station often includes many simultaneous operations, such as "drill all face holes" by multiple spindle drills.

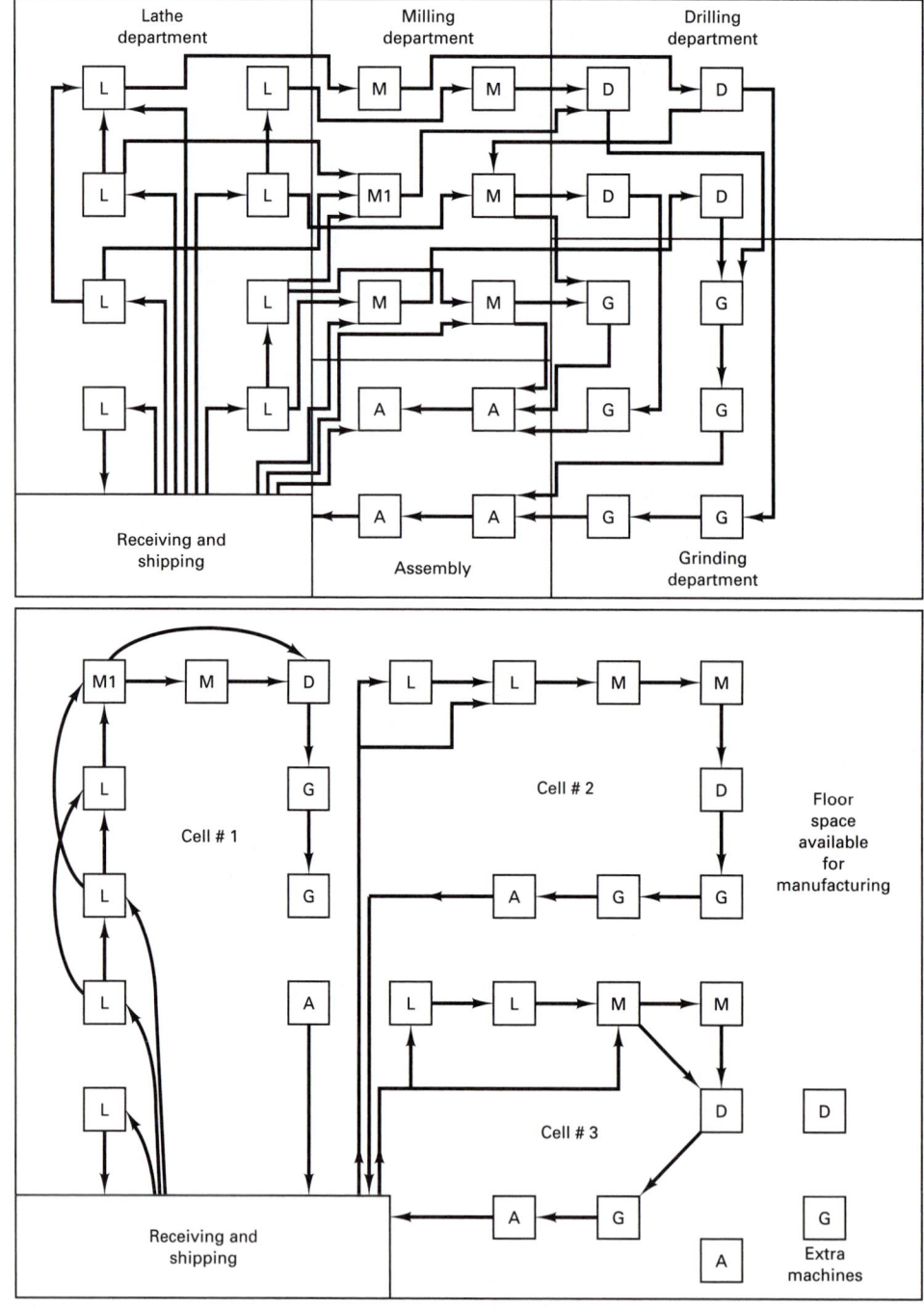

FIGURE 1-8 (above) Schematic layout of a job shop where processes are gathered functionally into areas or departments. Each square block represents a manufacturing process. (below) Schematic layout of a linked-cell manufacturing system. Group technology can be used to restructure the factory floor by grouping processes into cells to process families of parts.

Operation

An *operation* is a distinct action performed to produce a desired result or effect. Typical machine operations are loading and unloading. Operations can be divided into suboperational elements. For example, loading is made up of picking up part, placing part in jig, closing jig. However, suboperational elements will not be discussed here.

Operations categorized by function are:

1. *Materials handling and transporting*: change in position of the product
2. *Processing*: change in volume and quality, including assembly and disassembly; can include packaging
3. *Packaging*: special processing; may be temporary or permanent for shipping
4. *Inspecting and testing*: comparison to the standard or check process behavior
5. *Storing*: time lapses without further operations

These basic operations may occur more than once in some processes, or they may sometimes be omitted. *Remember, it is the manufacturing processes that change the value and quality of the materials.* Defective processes produce poor quality or scrap. Other operations may be necessary but do not, in general, add value, whereas operations performed by machines that do material processing usually do add value.

Treatments

Treatments operate continuously on the workpiece. They usually alter or modify the product-in-process without tool contact. Heat treating, curing, galvanizing, plating, finishing, (chemical) cleaning, and painting are examples of treatments. Treatments usually do add value to the part.

These processes are often difficult to include in cells because they often have long cycle times, are hazardous to workers' health, or are unpleasant to be around because of high heat or chemicals. They are often done in large tanks or furnaces or rooms. The cycle time for these processes may *dictate* the cycle times for the entire system. These operations also tend to be material specific. Many manufactured products are given decorative and protective surface treatments that control the finished appearance. A customer may not buy a new truck with a visible defect in the chrome bumper, although this defect does not alter the performance of the vehicle.

Tools, Tooling, and Workholders

The lowest mechanism in the production term rank is the *tool*. This implement is used to hold, cut, shape, or form the unfinished product. Common hand tools include the saw, hammer, screwdriver, chisel, punch, sandpaper, drill, clamp, file, torch, and grindstone.

Basically, machines are mechanized versions of such hand tools. Most tools are for cutting (drill bits, reamers, single-point turning tools, milling cutters, saw blades, broaches, and grinding wheels). Noncutting tools for forming include extrusion dies, punches, and molds.

Tools also include workholders, jigs, and fixtures. These tools and cutting tools are generally referred to as the *tooling*, which is usually considered separate from machine tools. Cutting tools wear and fail and must be replaced periodically before parts are ruined. The workholding devices must be able to locate and secure the workpieces during processing in a repeatable, mistake-proof way.

Tooling for Measurement and Inspection

Measuring tools and instruments are also important for manufacturing. Common examples of measuring tools are rulers, calipers, micrometers, and gages. Precision devices that use laser optics or vision systems coupled with sophisticated electronics are becoming commonplace. Vision systems and coordinate measuring machines are becoming critical elements for achieving superior quality.

Integrating Inspection into the Process. The integration of the *inspection* process into the manufacturing process or the manufacturing system is a critical step toward building products of superior quality. An example will help. Compare an electric typewriter with a computer that does word processing. The electric typewriter is flexible. It types whatever words are wanted in whatever order. It can type in Pica, Elite, or Orator, but the font (disk or ball that has the appropriate type size on it) has to be changed according to the size of type wanted. The computer can do all of this but can also, through its software, do italics, darken the words, vary the spacing to justify the right margin, and perform many other functions. It checks immediately for incorrect spelling and other defects, such as repeated words. The software system provides a signal to the hardware to flash the word so that the operator knows that something is wrong and can make an immediate correction. If the system were designed to prevent the typist from typing repeated words, this would be a *pokayoke*, a defect prevention. Defect prevention is better than immediate defect detection and correction. Ultimately, the system should be able to forecast the probability of a defect, correcting the problem at the source. This means that the typist would have to be removed from the process loop, perhaps by having the system type out what it is told (convert oral to written directly). Pokayoke and source inspection are keys to designing manufacturing systems that produce superior-quality products at low cost.

■ 1.3 LANGUAGE OF MANUFACTURING OR PROCESS TECHNOLOGY

In manufacturing, material things (goods) are made to satisfy human wants. *Products* result from manufacture. Manufacture also includes conversion processes such as refining, smelting, canning, and mining.

Products can be manufactured by fabricating or by processing. *Fabricating* is the manufacture of a product from pieces such as parts, components, or assemblies. Individual products or parts can also be fabricated. Separable discrete items such as tires, nails, spoons, screws, refrigerators, and hinges are fabricated.

Processing is also used to refer to the manufacture of a product by continuous means, or by a continuous series of operations, for a specific purpose. Continuous items such as steel strip, beverages, breakfast foods, tubing, chemicals, and petroleum are "processed." Many processed products are marketed as discrete items, such as bottles of beer, bolts of cloth, spools of wire, and sacks of flour.

Separable discrete products, both piece parts and assemblies, are fabricated in a *plant*, *factory*, or *mill*, for instance a textile or rolling mill. Products that *flow* (liquids, gases, grains, or powders) are processed in a *plant or refinery*. The continuous-process industries, such as petroleum and chemical plants, are sometimes called processing industries or flow industries.

To a lesser extent, the terms *fabricating industries* and *manufacturing industries* are used when referring to fabricators or manufacturers of large products composed of many parts, such as a car, a plane, or a tractor. Manufacturing often includes continuous-process treatments such as electroplating, heating, demagnetizing, and extrusion forming.

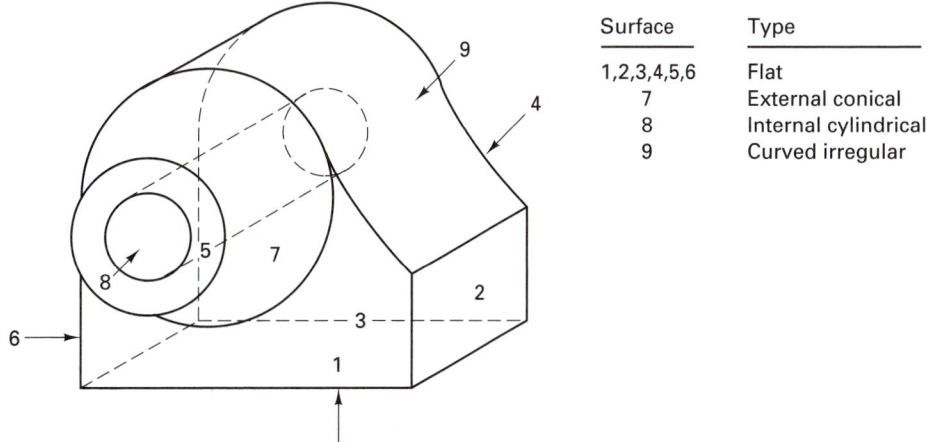

Surface	Type
1,2,3,4,5,6	Flat
7	External conical
8	Internal cylindrical
9	Curved irregular

FIGURE 1-9   Object composed of nine geometric surfaces.   Dashed lines are hidden surfaces.

Construction and *agriculture* make goods by means other than manufacturing or processing in factories.   Construction is a form of project manufacturing of useful goods. The public may not consider construction as manufacturing because the work is not usually done in a plant or factory, but it can be.   There is a company in Delaware that can build a custom house of any design in their factory, truck it to the building site, and assemble it on a foundation in two or three weeks.

Agriculture, fisheries, and commercial fishing produce real goods from useful labor. Lumbering is similar to both agriculture and mining in some respects, and mining should be considered processing.   Processes that convert the raw materials from agriculture, fishing, lumbering, and mining into other usable and consumable products are also forms of manufacturing.

Workpiece and Its Configuration

In the manufacturing of goods, the primary objective is to produce a component having a desired geometry, size, and finish.   Every component has a shape that is bounded by various types of surfaces of certain sizes that are spaced and arranged relative to each other. Consequently, a component is manufactured by producing the surfaces that bound the shape.   Surfaces may be:

1. Plane or flat
2. Cylindrical (external or internal)
3. Conical (external or internal)
4. Irregular (curved or warped)

Figure 1-9 illustrates how a shape can be analyzed and broken up into these basic bounding surfaces.   Parts are manufactured by using processes that will either (1) remove portions of a rough block of material so as to produce and leave the desired bounding surface, or (2) cause material to form into a stable configuration that has the required bounding surfaces.   Consequently, in designing an object, the designer specifies the shape, size, and arrangement of the bounding surface.   The part design must be analyzed to determine what materials will provide the desired properties, including mating to other components, and what processes can best be employed to obtain the end product at the most reasonable cost.

Understanding What the Enterprise Does

Understanding the process technology of the company is very important for everyone in the company. Manufacturing technology affects the design of the product and the manufacturing system, the way in which the manufacturing system can be controlled, the types of people employed, and the materials that can be processed. Table 1-2 outlines the factors that characterize a process technology. Take a process you are familiar with and think about these factors. One valid criticism of American companies is that their managers seem to have an aversion to understanding their companies' manufacturing technologies. Failure to understand the company business (i.e., its fundamental process technology) can lead to the failure of the company.

TABLE 1-2. Characterizing a Process Technology

Mechanics (statics and dynamics of the process)
    How does the process work?
    What are the process mechanics?
    What physically happens, and what makes it happen? (Understand the physics.)

Economics or costs
    What are the tooling costs; the engineering costs?
    Which costs are short term, which long term?
    What are the setup costs?

Time spans
    How long does it take to set up?
    How can this time be shortened?
    How long does it take to run a part, once set up?
    What process parameters affect the run time?

Constraints
    What are the process limits?
    What cannot be done?
    What constrains this process (sizes, speeds, forces, volumes, power, cost)?
    What is very hard to do within an acceptable time/cost frame?

Uncertainties and process reliability
    What can go wrong?
    How can this machine fail?
    What do people worry about with this process?
    Is this a reliable, stable process?

Skills
    What operator skills are critical?
    What is not done automatically?
    How long does it take to learn to do this process?

Flexibility
    Can this process easily do new parts of a new design or material?
    How does the process react to changes in part design and demand?
    Which changes are easy to do?

Process capability
    What are the accuracy and precision of the process?
    What tolerances does the process meet? (What is the process capability?)
    How repeatable are those tolerances?

The way to overcome technological aversion is to run the process and study the technology. Only someone who has run a drill press can understand the sensitive relationship between feed rate and drill torque and thrust. All process have these "know-how" features. Those who run the processes must be part of the decision making for the factory. The CEO who takes a vacation working on the plant floor learning the processes will be well on the way to being the head of a successful company.

■ 1.4 PRODUCT LIFE CYCLE

Manufacturing systems are dynamic and change with time. There is a general, traditional relationship between a *product's life cycle* and the kind of manufacturing system it has. Figure 1-10 simplifies the life cycle into these steps:

1. *Startup.* New product or new company, low volume, small company.
2. *Rapid growth.* Products become standardized and volume increases rapidly. Company's ability to meet demand stresses its capacity.
3. *Maturation.* Standard designs emerge. Process development is very important.

FIGURE 1-10   Relationship between product life cycle, manufacturing system development and evolution, and manufacturing cost per unit. PJS, production job shop.

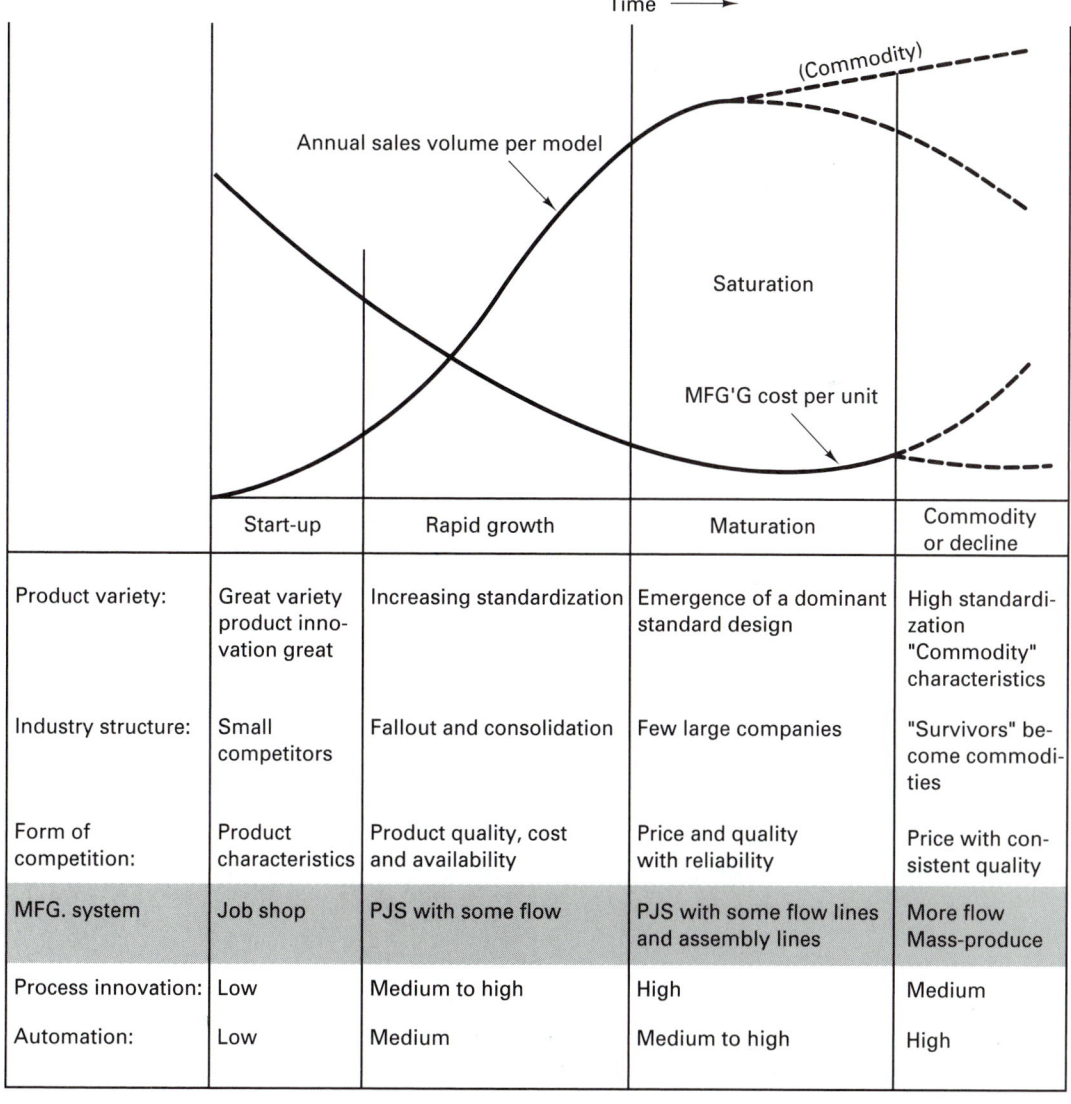

	Start-up	Rapid growth	Maturation	Commodity or decline
Product variety:	Great variety product inno-vation great	Increasing standardization	Emergence of a dominant standard design	High standardi-zation "Commodity" characteristics
Industry structure:	Small competitors	Fallout and consolidation	Few large companies	"Survivors" be-come commodi-ties
Form of competition:	Product characteristics	Product quality, cost and availability	Price and quality with reliability	Price with con-sistent quality
MFG. system	Job shop	PJS with some flow	PJS with some flow lines and assembly lines	More flow Mass-produce
Process innovation:	Low	Medium to high	High	Medium
Automation:	Low	Medium	Medium to high	High

4. *Commodity.* Long-life, standard-of-the-industry type of product.

or

Decline. Product is slowly replaced by improved products.

The maturation of a product in the marketplace generally leads to fewer competitors, with competition based more on price and on-time delivery than on unique product features. As the competitive focus shifts during the different stages of the product life cycle, the requirements placed on manufacturing—cost, quality, flexibility, and delivery dependability—also change. The stage of the product life cycle affects the product design stability, the length of the product development cycle, the frequency of engineering change orders, and the commonality of components, all of which have implications for manufacturing process technology.

In short, the product life-cycle concept provides a framework for thinking about the product's evolution through time and the kind of market segments that are likely to develop at various times. The design of the manufacturing system determines the cost per unit, which generally decreases over time with process improvements and increased volumes. The linked-cell approach discussed in Chapter 43 enables companies to decrease cost per unit significantly while maintaining flexibility and making smooth transitions from low-volume to high-volume manufacturing.

■ 1.5 BASIC MANUFACTURING PROCESSES

Manufacturing processes can be classified as:

- Casting, foundry, or molding processes
- Forming or metalworking processes
- Machining (material removal) processes
- Joining and assembly
- Surface treatments (finishing)
- Heat treating
- Other

These classifications are not mutually exclusive. For example, some finishing processes involve a small amount of metal removal or metal forming. A laser can be used either for joining or for metal removal or heat treating. Occasionally, we have a process such as shearing, which is really metal cutting but is viewed as a (sheet) metal-forming process. Assembly may involve processes other than joining. The categories of process types are far from perfect.

Casting and *molding* processes are widely used to produce parts that often require other follow-on processes, such as machining. Casting uses molten metal and a cavity. The metal retains the desired shape of the mold cavity after solidification. An important advantage of casting and molding is that, is a single step, materials can be converted from a crude form into a desired shape. In most cases, a secondary advantage is that excess or scrap material can easily be recycled. Figure 1-11 illustrates schematically some of the basic aspects of these processes.

Casting processes commonly are classified into two types: permanent mold (a mold can be used repeatedly) or nonpermanent mold (a new mold must be prepared for each casting made). Molding processes for plastics and composites are included in the chapters on forming processes.

Forming and *shearing* operations typically utilize material (metals or plastics) that previously has been cast or molded. In many cases the materials pass through a series of forming or shearing operations, so the form of the material for a specific operation may be

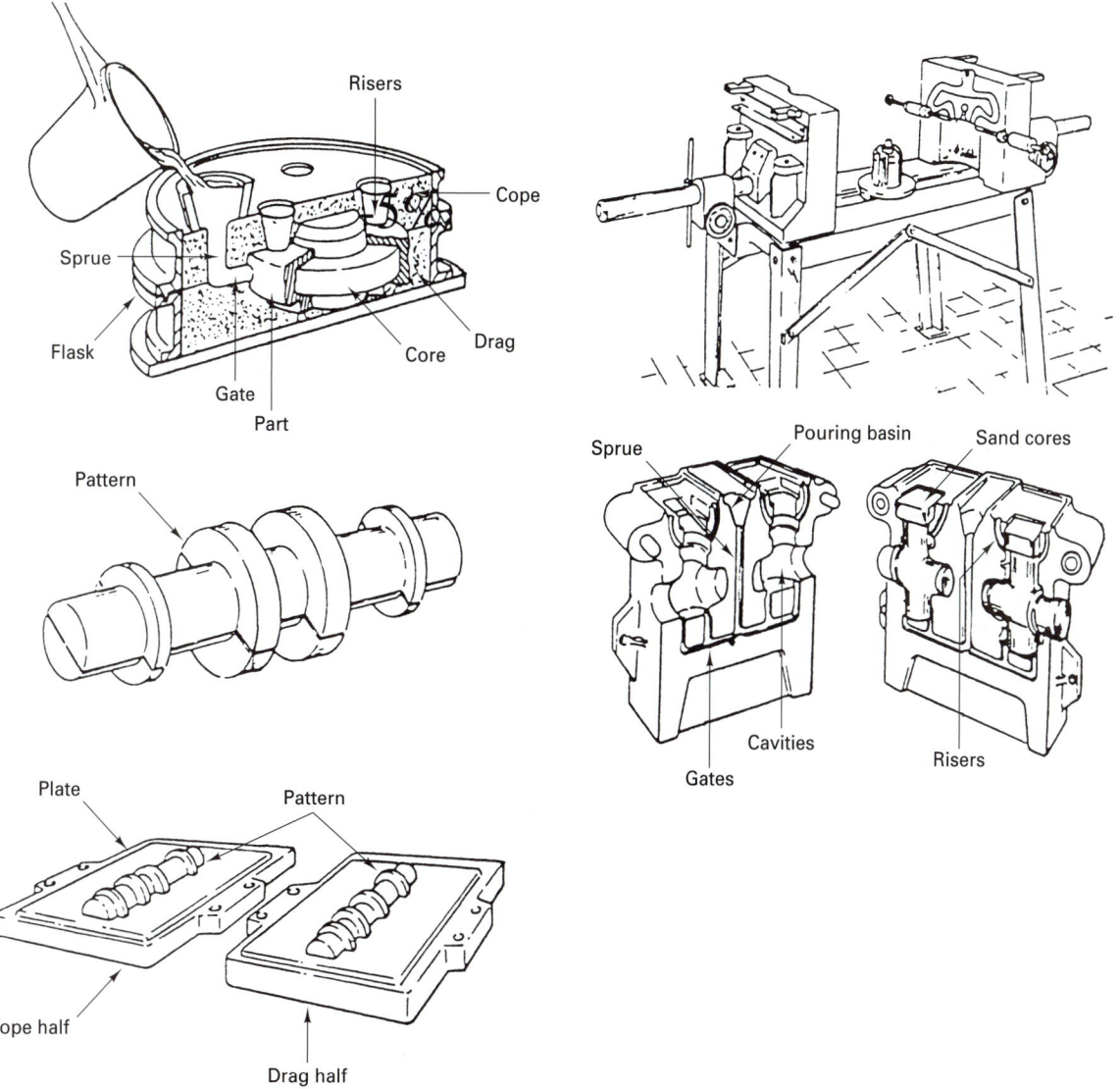

FIGURE 1-11 Casting processes: sand casting on left; Permanent mold casting above.

the result of all the prior operations. The basic purpose of forming and shearing is to modify the shape and size and/or physical properties of the material.

Some of the forming and shearing processes are shown in Figure 1-12. *Metalforming* and *shearing operations* are done both "hot" and "cold," a reference to the temperature of the material at the time it is being processed with respect to the temperature at which this material can recrystallize (i.e., grow new grain structure).

Machining or *metal removal processes* refer to the removal of certain selected areas from a part to obtain a desired shape or finish. Chips are formed by interaction of a cutting tool with the material being machined. Figure 1-13 shows a chip being formed by a single-point cutting tool.

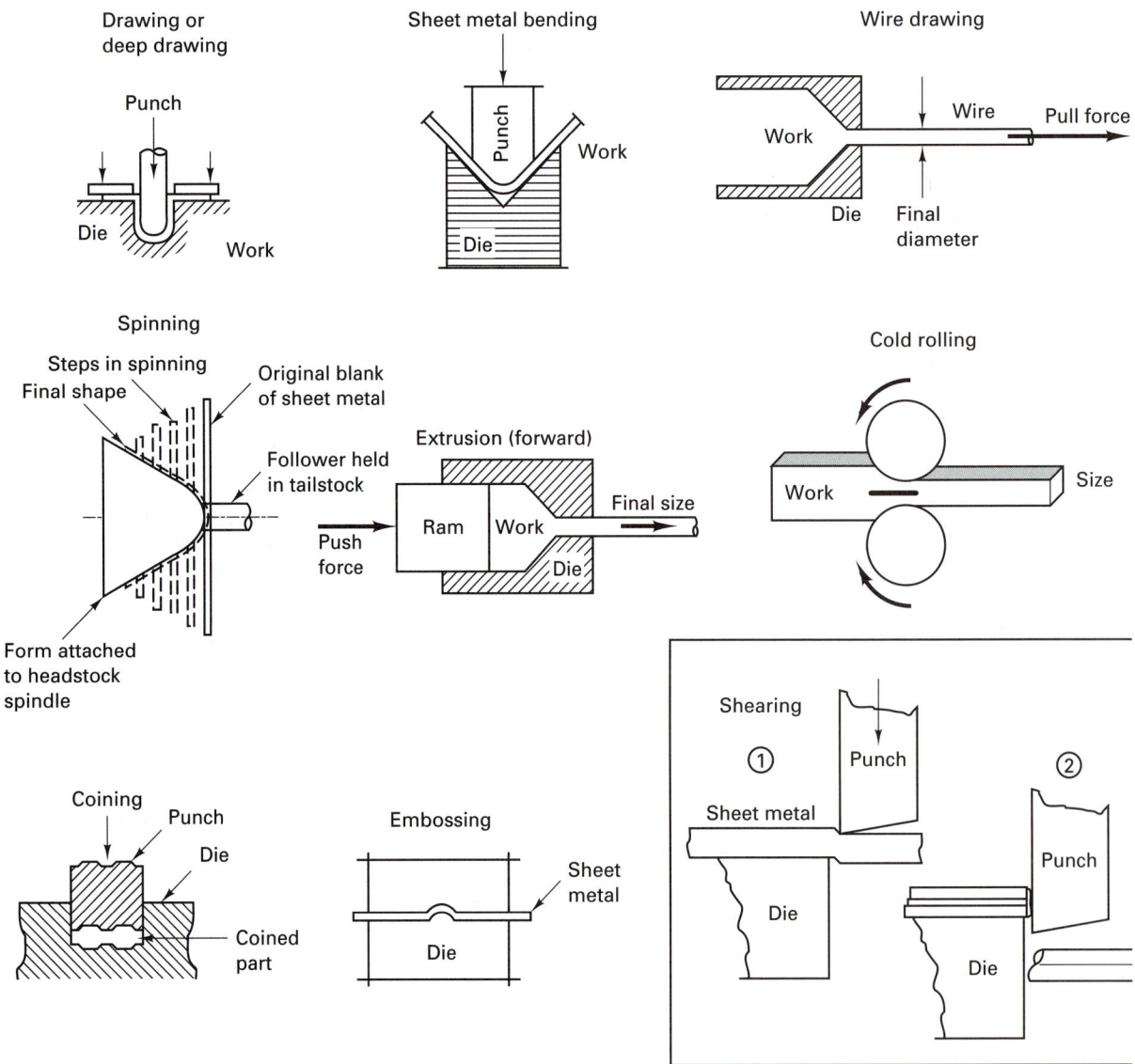

FIGURE 1-12   Common forming and shearing processes.

Cutting tools are used to perform the basic and related machining processes that are shown schematically in Figure 1-14. The cutting tools are mounted in machine tools, which provide the required movements of the tool with respect to the work (or vice versa) to accomplish the process desired. In recent years many new machining processes have been developed.

The seven basic machining processes are *shaping*, *drilling*, *turning*, *milling*, *sawing*, *broaching*, and *abrasive machining*. Historically, eight basic types of machine tools were developed to accomplish the basic processes. These are shapers (and planers), drill presses, lathes, boring machines, milling machines, saws, broaches, and grinders. Many of these machine tools are capable of performing more than one of the basic machining processes. This obvious advantage has led to the development of *machining centers* specifically

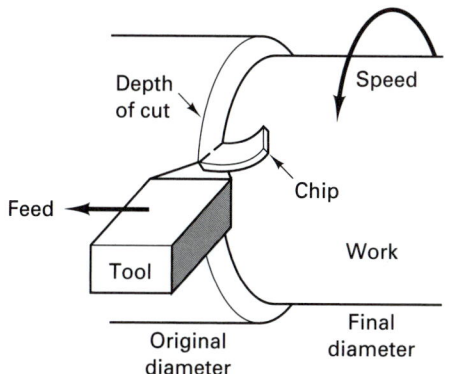

FIGURE 1-13 Single-point metalcutting process (turning) produces a chip.

designed to combine many of the basic processes, and perform other related processes, all on a single machine tool using a single workpiece setup (see Chapter 29).

Included with the machining processes are processes wherein metal is removed by chemical, electrical, electrochemical, or thermal sources. Generally speaking, these nontraditional processes have evolved to fill a specific need when conventional processes were too expensive or too slow when machining very hard materials. One of the first uses of a laser was to machine holes in ultra-high-strength metals. It is being used today to drill tiny holes in turbine blades for jet engines. Table 1-3 shows the more common chipless machining processes.

TABLE 1-3. Nontraditional Chipless Machining Processes

Process	Metal-Removal Mechanism	Examples
Chemical machining, milling, or blanking	Chemical etching	Photoengraving
Electrochemical machining (ECM) or drilling or grinding	High-intensity "reverse electroplating" using high current densities	Machine cavities in dies
Ultrasonic machining (drilling or welding)	Abrasive slurry with grits vibrated into works by ultrasonic means; actually forms chips	Tool-and-die work (in nonconductors)
Electrodischarge machining (EDM)	Spark erosion of metals by local heating and melting	Drill holes in very hard tool and die materials
Laser beam machining (LBM) or heat treating	High-energy laser melts and vaporizes metal	Drill holes in turbine blades
Electron beam machining (EBM) or welding or cutting	High-energy electron beam melts and vaporizes metal	Microhole drilling in integrated-circuit boards
Plasma jet machining or cutting or welding	Ionic plasma, very high temperature jets to melt materials	Rapid cutting of plates

Perhaps the largest collection of processes, in terms of both diversity and quantity, are the *joining* processes, which include the following:

1. Mechanical fastening
2. Soldering and brazing
3. Welding
4. Press, shrink, or snap fittings
5. Adhesive bonding
6. Assembly processes

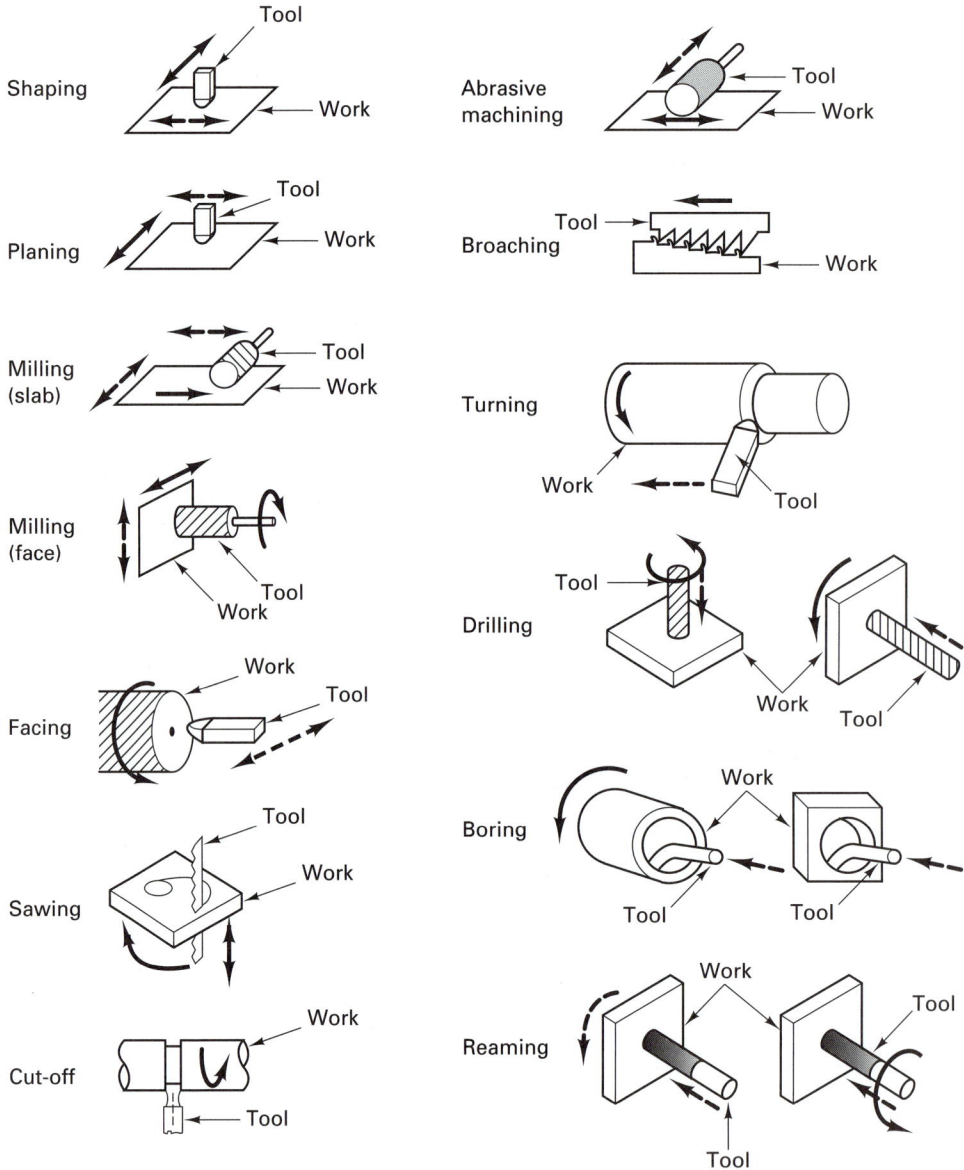

FIGURE 1-14 Schematic representation of basic machining processes.

These processes are often found in the assembly area of the plant.

Finishing processes are yet another class of processes typically employed for cleaning, removing of burrs left by machining, or providing protective and/or decorative surfaces on workpieces. Surface treatments include chemical and mechanical cleaning, deburring, painting, plating, buffing, galvanizing, and anodizing.

Heat treatment is the heating and cooling of a metal for specific purpose of altering its metallurgical and mechanical properties. Because the changing and controlling of these properties is so important in the processing and performance of metals, heat treatment is a very important manufacturing process. Each type of metal reacts differently to heat treatment. Consequently, a designer should know not only how a selected metal can be altered by heat treatment but, equally important, *how a selected metal will react, favorably or unfavorably, to any heating or cooling that may be incidental to the manufacturing processes.*

In addition to the processes, there are some other fundamental manufacturing operations that we must consider. *Inspection* determines whether the desired objectives stated by the designer in the specifications have been achieved. This activity provides feedback to design and manufacturing with regard to the process behavior. Essential to this inspection function are measurement activities.

In *testing*, a product is tried by actual function or operation or by subjection to external effects. Although a test is a form of inspection, it is often not viewed that way. In manufacturing, parts and materials are inspected for conformance to the dimensional and physical specifications, while testing may simulate the environmental or usage demands to be made on a product after it is placed in service. Complex products may require many tests during final inspection. Testing includes life-cycle tests, destructive tests, nondestructive testing to check for processing defects, wind-tunnel tests, road tests, and overload tests.

Transportation of goods in the factory is often referred to as *material handling* or *conveyance* of the goods and refers to the transporting of unfinished goods, work-in-progress in the plant, and supplies to and from, between, and during manufacturing operations. Loading, positioning, and unloading are also material-handling operations. Transportation, by truck or train, is material handling between factories. Proper manufacturing system design and mechanization can reduce material handling in countless ways.

Automatic material handling is a critical part of continuous automatic manufacturing. The word *automation* is derived from automatic material handling. Material handling, a fundamental operation done by people and by conveyors and loaders, often includes positioning the workpiece within the machine by indexing, shuttle bars, slides, and clamps. In recent years, wire-guided automated guided vehicles (AGVs) and automatic storage and retrieval systems (AS/RSs) have been developed in an attempt to replace forklift trucks on the factory floor. Another form of material handling, the mechanized removal of waste (chips, trimmings, and cutoffs), can be more difficult than handling the product. Chip removal must be done before a tangle of scrap chips damages tooling or creates defective workpieces.

Most texts on manufacturing processes do not mention *packaging*, yet the packaging is often first thing the customer sees. Also, packaging often maintains the product's quality between completion and use. (Packaging is also used in electronics manufacturing to refer to placing microelectronic chips in containers for mounting on circuit boards.) Packaging can also prepare the product for delivery to the user. It varies from filling ampules with antibiotics to steel-strapping aluminum ingots into palletized loads. A product may require several packaging operations. For example, Hershey Kisses are (1) individually wrapped in foil, (2) placed in bags, (3) put into boxes, and (4) placed in shipping cartons.

Weighing, filling, sealing, and labeling are packaging operations that are highly automated in many industries. When possible, the cartons or wrapping is formed from material on rolls in the packaging machine. Packaging is a specialty combining elements of product design (styling), materials handling, and quality control. Some packages cost more than their contents, for example, cosmetics and razor blades.

During *storage*, nothing happens intentionally to the product or part except the passage of time. Part or product deterioration on the shelf is called *shelf life*, meaning that items can rust, age, rot, spoil, embrittle, corrode, creep, and otherwise change in state or structure, while supposedly, nothing is happening to them. Storage is detrimental, wasting the company's time and money. The best strategy is to keep the product moving with as little storage as possible. Storage during processing should be *eliminated*, whenever possible, not automated or computerized. Companies should avoid investing heavily in large automated systems that do not alter the bottom line. Have the outputs improved with respect to the inputs, or has storage simply increased the costs (indirectly) without improving either the quality or the throughput time?

By not storing a product, the company avoids having to (1) retrieve it, (2) remember where it is, (3) worry about its deteriorating, or (4) pay storage costs. Storage is the biggest waste of all and must be eliminated at every opportunity.

■ 1.6 YARDSTICK FOR AUTOMATION

In 1962, Amber & Amber published their *"yardstick for automation."* The chart that they developed has been updated and is included here as Table 1-4 in a somewhat abbreviated form. The key to the chart is that each level of automation is tied to the human attribute that is being replaced (mechanized or automated) by the machine. Therefore, the A(0) level of automation, in which no human attribute was mechanized, covers the Stone Age through the Iron Age. Two of the earliest machine tools were the crude lathes the Etruscans used for making wooden bowls around 700 B.C. and the windlass-powered broach for machining of grooves into rifle barrels used over 300 years ago.

TABLE 1-4. Yardstick for Automation

Order of Automation	Human Attribute Replaced	Examples
A(0)	*None*: lever, screw, pulley, wedge	Hand tools, manual machine
A(1)	*Energy*: muscles replaced with power	Powered machines and tools, Whitney's milling machine
A(2)	*Dexterity*: Self-feeding	Single-cycle automatics
A(3)	*Diligence*: no feedback but repeats cycle automatically	Repeats cycle; open-loop numerical control or automatic screw; transfer lines
A(4)	*Judgment*: positional feedback	Closed loop; numerical control; self-measuring and adjusting
A(5)	*Evaluation*: adaptive control; deductive analysis; feedback from the process	Computer control; model of process required for analysis and optimization
A(6)	*Learning*: by experience	Limited self-programming; some artificial intelligence (AI); expert systems
A(7)	*Reasoning*: exhibits intuition; relates causes and effects	Inductive reasoning; advanced AI in control software
A(8)	*Creativeness*: performs design unaided	Originality
A(9)	*Dominance*: supermachine, commands others	Machine is master (Hal in *2001, A Space Odyssey*)

Source: G. Amber & P. Amber, *Anatomy of Automation*, Prentice Hall, Englewood Cliffs, N.J., 1962. Used by permission, modified by Black.

The first industrial revolution can be tied to the development of powered machine tools, dating from 1775, when the energetic "iron-mad" John Wilkinson, in England, constructed a horizontal boring machine for machining internal cylindrical surfaces, such as in piston-type pumps. In Wilkinson's machine, a model of which is shown in Figure 1-15, the boring bar extended through the casting to be machined and was supported at its outer end by a bearing. Modern boring machines still employ this basic design. Wilkinson reported that his machine could bore a 57-in.-diameter cylinder to such accuracy that nothing greater than an English shilling (about $\frac{1}{16}$ in. or 1.59 mm) could be inserted between the piston and the cylinder. This machine tool made Watt's steam engine a reality. At the time of his death, Wilkinson's industrial complex was the largest in the world.

FIGURE 1-15   Model of Wilkinson's horizontal boring machine. *(British Crown Copyright, Science Museum, London.)*

FIGURE 1-16   Maudsley's screw-cutting lathe *(British Crown Copyright, Science Museum, London.)*

The next machine tool was developed in 1794 by Henry Maudsley. It was an engine lathe with a practical slide tool rest. This machine tool, shown in Figure 1-16, was the forerunner of the modern engine lathe. The lead screw and change gear mechanism, which enabled threads to be cut, were added about 1800. The first planer was developed in 1817 by Richard Roberts in Manchester, England. Roberts was a student of Maudsley, who also had a hand in the career of Joseph Whitworth, the designer of screw threads. Roberts also added back gears and other improvements to the lathe. The first horizontal milling machine is credited to Eli Whitney in 1818 in New Haven, Connecticut (Figure 1-17). The development of machine tools that not only could make specific products but could also produce other machines to make other products was fundamental to the first industrial revolution.

While early work in machine tools and precision measurement was done in England, the earliest attempts at interchangeable manufacturing apparently occurred almost simultaneously in Europe and the United States with the development of *filing jigs*, with which duplicate parts could be hand-filed to substantially identical dimensions. In 1798, Eli Whitney,

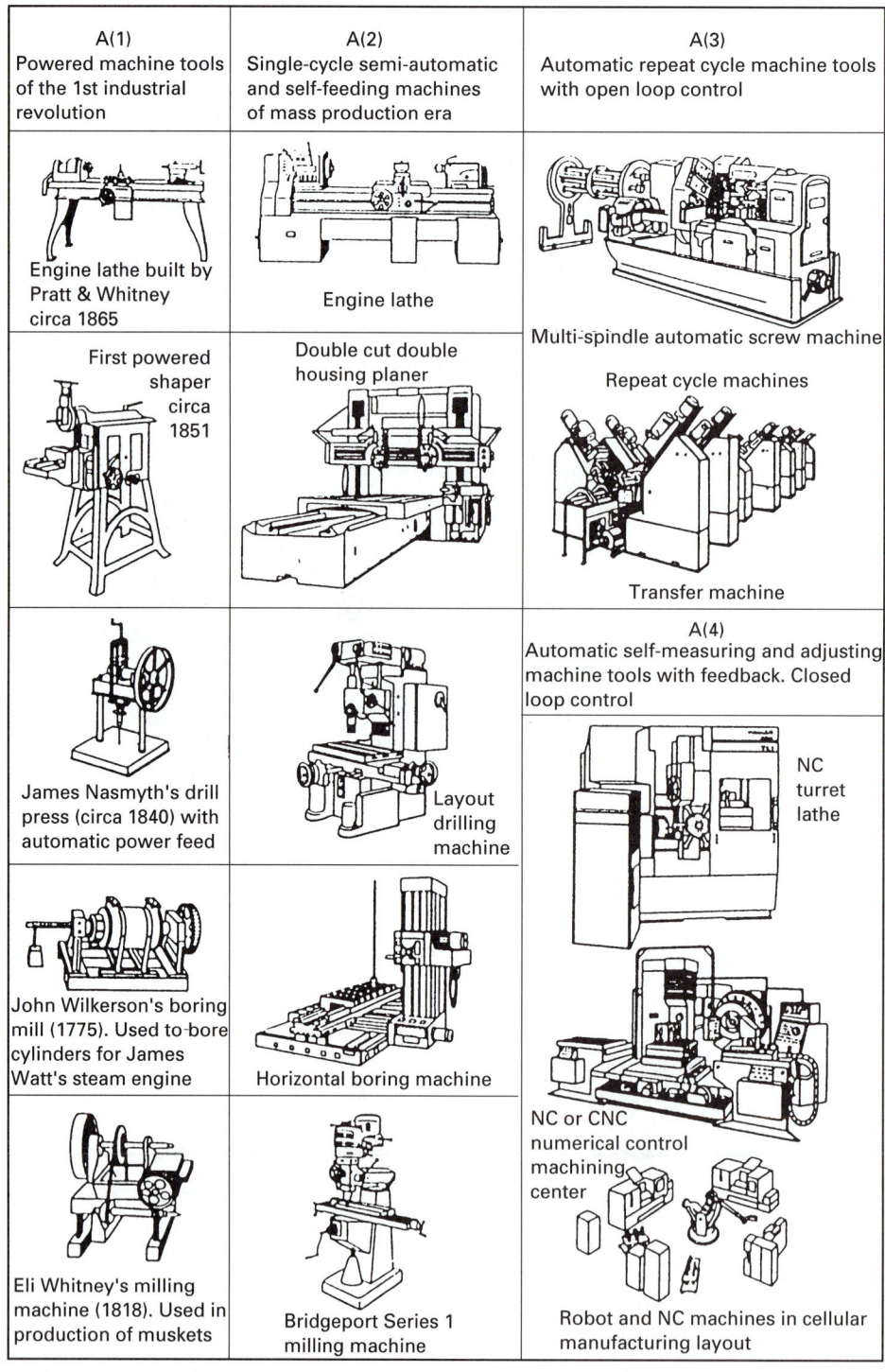

A(1) Powered machine tools of the 1st industrial revolution	A(2) Single-cycle semi-automatic and self-feeding machines of mass production era	A(3) Automatic repeat cycle machine tools with open loop control
Engine lathe built by Pratt & Whitney circa 1865	Engine lathe	Multi-spindle automatic screw machine
First powered shaper circa 1851	Double cut double housing planer	Repeat cycle machines Transfer machine
James Nasmyth's drill press (circa 1840) with automatic power feed	Layout drilling machine	A(4) Automatic self-measuring and adjusting machine tools with feedback. Closed loop control NC turret lathe
John Wilkerson's boring mill (1775). Used to bore cylinders for James Watt's steam engine	Horizontal boring machine	NC or CNC numerical control machining center
Eli Whitney's milling machine (1818). Used in production of muskets	Bridgeport Series 1 milling machine	Robot and NC machines in cellular manufacturing layout

FIGURE 1-17 Machine tools of the first industrial revolution [A(1)] and the mass production era [A(2)], and examples of levels of automation [A(3)] and [A(4)].

using this technique, was able to obtain and eventually fulfill a contact from the U. S. government to produce 10,000 army muskets, the parts of each being interchangeable. However, this truly remarkable achievement was accomplished primarily by painstaking handwork and not by specific machines.

Joseph Whitworth, starting about 1830, accelerated the use of Wilkinson's and Maudsley's machine tools by developing precision measuring methods. Later he developed a measuring machine using a large micrometer screw. Still later, he worked toward establishing thread standards and made plug and ring gages. His work was valuable because precise methods of measurement were the prerequisite for developing interchangeable parts, a requirement for later mass production.

The next significant machine tool was the drill press with automatic feed, developed by John Nasmyth, another student of Maudsley, in 1840 in Manchester. Surface-grinding machines came along about 1880 and the era was completed with the development of the bandsaw blades, which could cut metal. In total, there were eight basic machine tools in the first industrial revolution for machining: lathe, milling machine, drill press, broach, boring mill, planer (shaper), grinder, and saw. The first factories were developed so that power could be added to drive the machines. This is the A(1) level of automation.

The A(2) level of automation was clearly delineated when machine tools became single-cycle, self-feeding machines displaying *dexterity*. Many examples of this level of machine are given in Chapters 21 through 28; they exist in great numbers in many factories. The A(2) level of machine can be loaded with a part and the cycle initiated by the worker. The machine completes the cycle and stops automatically. The A(3) level requires the machine to be *diligent* or repeat cycle automatically. These machines are open loop, meaning that they do not have *feedback* and are controlled by either an internal fixed program, such as a cam, or are externally programmed with a tape or, more recently, a computer. A(3), A(4), and A(5) levels are basically superimposed on A(2)-level machines, which must be A(1) by definition. The A(3) level includes robots, numerical control (NC) machines that have no feedback, and many special-purpose machine tools.

The A(4) level of automation required that *human judgment* be replaced by a capability in the machine to measure and compare results with desired position or size and make adjustments to minimize errors. This is *feedback* or *closed-loop control* (see Figure 1-7). The first numerically controlled machine tool was developed in 1952. It had positional feedback control and is generally recognized as the first A(4) machine tool. By 1958, the first NC *machining center* (see Chapter 29) was being marketed by Kearney and Trecker. This machining center was a compilation of many machine tools capable of performing many processes: in this case, milling, drilling, tapping, and boring (see Figure 1-17). It could automatically change tools to give it greater flexibility. Almost from the start, computers were needed to help program these machines. Within 10 years, NC machine tools became computer numerical control (CNC) machine tools. Thus these machines had their own microprocessor and could be programmed directly. In any event, CNC machines are still A(4) machines.

With the advent of the NC type of machine and more recently the programmable robot, two types of automation were defined. *Hard* or *fixed automation* is exemplified by transfer machines or automatic screw machines, and *flexible* or *programmable automation* is typified by NC machines or robots that can be taught or programmed externally by means of computers. The control was in computer software rather than mechanical hardware.

The A(5) level requires that machines perform *evaluation* of the process itself. Thus the machine must be cognizant of the multiple factors on which the process performance is predicated, evaluate the current setting of the input parameters versus the outputs from the process, and then determine how to alter the inputs to optimize the process. This is called *adaptive control* (AC); see Chapters 29 and 42 for discussions.

There are very few examples of A(5) machines on the factory floor and fewer at the A(6) level, wherein the machine control has expert systems capability. *Expert systems* try to infuse the software with the deductive decision-making capability of the human brain by having the system get smarter *through experience. Artificial intelligence* (AI) carries this step higher by *infecting* the control software with programs that exhibit the ability to reason inductively. Levels A(3) and A(4) are discussed in Chapter 29. Levels A(5) and A(6) are discussed in Chapter 41. Levels A(6) and A(7) are the subjects of intensive worldwide research efforts. Levels A(8) and A(9) are left to the whims of the science fiction writer.

Automation involves machines, or integrated groups of machines, that automatically perform required machining, forming, assembly, handling, and inspection operations. Through sensing and feedback devices, these systems automatically make necessary corrective adjustments. That is, *human thinking must be automated.* There are relatively few completely automated systems, but there are numerous examples of highly mechanized machines. While the potential advantage of a completely automated plant are tremendous, in practice, step-by-step automation of individual operations is required. Therefore, it is important to have a piecewise plan to convert from the classical job shops to the simplified, integrated factory of the future that can be automated. This is discussed in more detail in Chapter 43. However, the most serious limitation in automation is available capital, as the initial investments for automation equipment and installations are large. Because proper engineering economics analysis must be employed to evaluate these investments, students who anticipate a career in manufacturing should consider a course in this area a firm requirement.

Continually reducing manufacturing costs is and always will be the primary objective. New materials are constantly being sought. The age of composites and custom-made materials is here. New processes are needed to deal with new materials that do not pollute the environment. Despite the great advances of recent years, even greater progress may be expected in the future. The consumer wants products to have high reliability and superior quality. More attention will be given to eliminating the waste of materials, reducing manufacturing time, improving quality, and considering how the product will be disposed of at the end of its useful life.

■ 1.7 PLANNING FOR MANUFACTURING

Low-cost manufacturing does not just happen. There is a close interdependent relationship between the design of a product, selection of materials, selection of processes and equipment, the design of the processes, and tooling selection and design. Each of these steps must be carefully considered, planned, and coordinated before manufacturing starts. This lead time, particularly for complicated products, may take months, even years, and the expenditure of large amounts of money may be involved. Typically, the lead time for a completely new model of an automobile or a modern aircraft may be 2 to 5 years.

Figure 1-18 shows some of the steps involved in getting one product from the original idea stage to manufacturing. The steps are closely related to each other. For example, the design of the tooling was conditioned by the design of the parts to be produced. It is often possible to simplify the tooling if certain changes are made in the design of the parts or the design of the manufacturing systems. Similarly, the material selection will affect the design of the tooling or the processes selected. Can the design be altered so that it can be produced with tooling already on hand and thus avoid the purchase of new equipment?

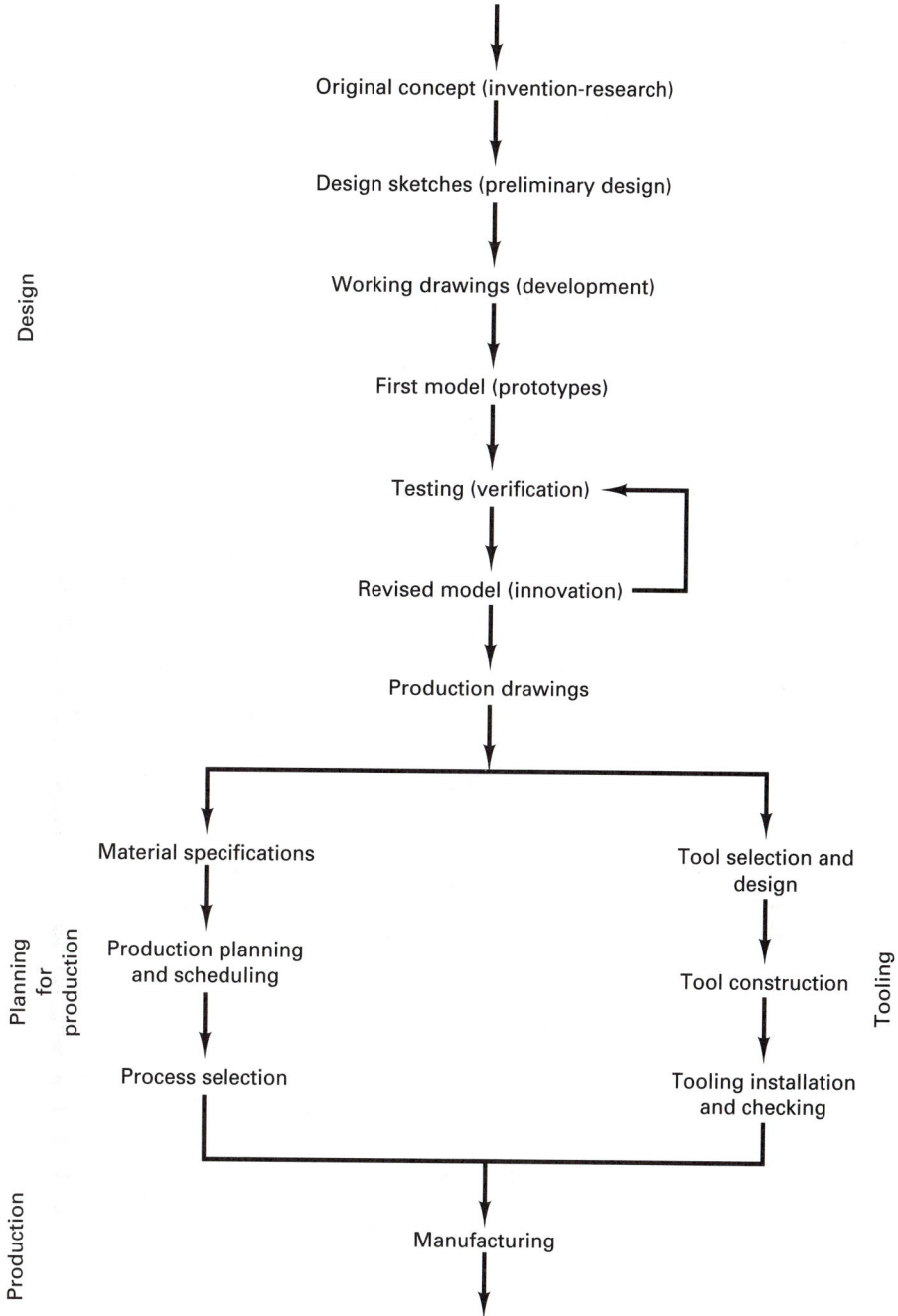

FIGURE 1-18 Traditional steps required to convert an idea into a finished product.

Close coordination of all the various phases of manufacture is essential if economy is to result. All mistakes and "bugs" should be eliminated during the preliminary phases because changes become more and more costly as work progresses.

With the advent of computers and computer-controlled machines, the integration of the design function, the planning for production function, and the manufacturing function through the computer is a reality. This is usually called CAD/CAPP/CAM (computer-aided design/computer-aided process planning/computer-aided manufacturing). The key is a common database from which detailed drawings can be made for use by both the designer, the process planner, and the manufacturing engineer and from which programs can be generated to make all the tooling. In addition, extensive computer-aided testing and inspection (CATI) of the manufactured parts is taking place. There is no doubt that this trend will continue at ever-accelerating rates as computers become cheaper and smarter.

■ 1.8 New Manufacturing System

Manufacturing systems and production systems are introduced in Chapters 41 and 42. The manufacturing process technology described herein is available worldwide. Many countries have about the same level of process development when it comes to manufacturing technology. Much of the technology existing in the world today was developed in the United States and Europe. Japan and more recently Taiwan and Korea are making great strides in becoming world industrial nations, particularly in the automotive and electronics industries. What many people have failed to recognize is that many Japanese companies have developed and promoted a totally different kind of manufacturing system. In future years, this new system, based on linked cells, will take its place with the Taylor system of scientific management and the Ford system for mass production. The original working model for this new system is the Toyota Motor Company. The system is known as the Toyota production system, just-in-time (JIT) manufacturing, or lean production.

Many American companies have successfully adopted some version of lean production. The experience of dozens of these companies is amalgamated into 10 steps, which, if followed, can make any company a factory with a future. The key steps to developing the new manufacturing system are presented in Chapter 43.

For the lean production to work, 100% good units flow rhythmically to subsequent processes without interruption. To accomplish this, an integrated quality control (IQC) program has to be developed. The responsibility for quality has been given to manufacturing. The employees are empowered to make it right the first time. There is a companywide attitude toward constant quality improvement. The goal is to make it right (perfect) every time. Make quality easy to see, stop the line when something goes wrong, and inspect things 100% if necessary to prevent defects from occurring. The results of this system are astonishing, in terms of quality, low cost, and on-time delivery of goods to the customer.

The Japanese became world-class competitors by developing superior design and process technology. Now, Japan is contentrating on new product innovation, with emphasis on high-value-added products. They are spending as much as 14% of sales dollars for research and development (R&D). The typical American company spends less than 5% for research and development. Furthermore, many Japanese firms and firms in other countries as well are now forming joint ventures with U. S. companies, so the workings of this new manufacturing system is diffusing around the world.

The most important factor in economical, and successful, manufacturing is the manner in which the resources—labor, materials, and capital—are organized and managed so as to provide effective coordination, responsibility, and control. Part of the success of the lean production system can be attributed to a different management approach. This approach is characterized by a wholistic attitude to people and includes:

1. Working in teams or groups, improving communications
2. Consensus decision making by teams
3. Decision making at the lowest possible level

4. Mutual trust, integrity, and loyalty between workers and management
5. Innovative methods of pay for all employees, including incentive pay in the form of bonuses
6. Stable (even lifetime) employment
7. Large pool of part-time temporary workers (contract workers)

There are many companies in the United States that employ some or all of these elements and, obviously, there are many different ways that a company can be organized and managed.

The real secret of successful manufacturing lies in designing a simplified system that everyone understands in terms of how it works and how it is controlled, with the decision making placed at the correct level. Today's engineers must possess a broad fundamental knowledge of design, metallurgy, processing, economics, accounting, and human relations. In the manufacturing game, low-cost mass production is the result of teamwork within an integrated manufacturing production system. This is the key to producing superior quality at less cost with on-time delivery.

■ KEY WORDS

assembling	inspection	molding
casting	job	numerical control
consumer goods	job shop	operation
continuous process	joining	producer goods
design engineer	lean production	product life cycle
fabricating	linked-cell shop	production system
feedback	machine tool	project shop
filing jig	machining	shearing
flow shop	manufacturing cost	station
forming	manufacturing engineer	tooling
group technology	manufacturing process	tools
heat treatment	manufacturing system	treatments

■ REVIEW QUESTIONS

1. What role does manufacturing play relative to the standard of living of a country?
2. Are not all goods really consumer goods, depending on how you define the customer? Discuss.
3. Explain the differences between job shop, flow shop, and project shop.
4. How does a system differ from a process? From a machine tool? From a job? From an operation?
5. Is a cutting tool the same thing as a machine tool? As tooling?
6. What are the major classifications of basic manufacturing processes?
7. Why would it be advantageous if casting could be used to produce a complex-shaped part to be made from a hard-to-machine metal?

8. How is a railroad station like a job station?
9. Since no work is being done on a part when it is in storage, it does not cost you anything. True or false? Explain!
10. List the forming processes used to make a wire coat hanger.
11. Which manufacturing system best describes your college or university? Analogize the elements in this service system to those in the manufacturing system.
12. It is acknowledged that chip-type machining, basically, is an inefficient process. Yet it probably is used more than any other to produce desired shapes. Why?
13. What is the level of automation as practiced by the surgeon during an operation in an operating room? (The surgeon is analogous to a

plumber who works on body pipes and pieces.)

14. List three purposes of packaging operations.

15. Who invented the first workholders called jigs?

16. Assembly is defined as "the putting together of all the different parts to make a complete machine." Think of (and describe) an assembly process? Is making a club sandwich an assembly process? What about building a house?

17. What kind of assembly process does a slaughterhouse have?

18. Characterize the process of squeezing toothpaste from a tube (extrusion of toothpaste) using Table 1-2 as a guideline. See Figure 1-12 for assistance.

19. What is a basic difference between mechanization and automation?

20. Give an example of a machine found in the home for as many levels of automation as you can (see Table 1-4).

21. What difficulties might result if the step "production planning and scheduling" were omitted from the procedure shown in Figure 1-18, assuming a job shop MS?

22. A company is considering making automobile bumpers from aluminum instead of from steel. List some of the factors it would have to consider in arriving at its decision.

23. Many companies are critically examining the relationship of product design to manufacturing and assembly. Discuss.

24. It has been said that low-cost products are more likely to be more carefully designed than high-priced items. Do you think this is true, and why or why not?

25. In a typical metal-cutting job using a conventional machine, which operations add no value to the part?

26. What is a proprietary process?

27. Classify the electric typewriter using the "yardstick for automation."

28. If the rolls for the cold-rolling mill that produce the sheet metal used in your car cost $300,000, how is it that your car can still cost less than $10,000?

29. Make a list of service production systems, giving an example of each and explaining the fundamental difference between an SPS and an MPS.

30. In the process of buying a calf, raising it to a cow, and disassembling it into "cuts" of meat for sale, where is the "value added"?

31. What kind of process is chrome plating?

32. In view of Figure 1-2, who really determines the selling price per unit?

33. What costs make up manufacturing cost?

■ PROBLEMS

1. In the manufacture of an $8000 car, how much do you estimate is spent on direct labor?

2. The average Japanese autoworker makes around $8 to $10 per hour and each car at Toyota has about 30 direct labor hours in it. If the average U.S. autoworker makes $20 per hour, how many worker-hours of direct labor are in the typical American car?

3. Using the data in Table 1-A, estimate the percentage of costs of direct labor in the typical medium-sized manufacturing plant. Compare to Figure 1-2.

4. What percentage of costs does power represent in the factories specified in Table 1-A?

5. Suppose that the appliance plug in Figure 1-3 were redesigned to be joined by a rivet or by a screw-and-nut assembly. (You should be able to find such a redesigned appliance plug at a local discount store). How much did it cost? How parts did it have? Make up a "new parts" list and indicate which parts would have to be redesigned and which parts would be eliminated to accommodate the new method of assembly. Estimate the manufacturing cost of the new plug assuming that manufacturing costs are 40% of the selling price. What is the disadvantage of a riveted plug as opposed to a plug held together by a screw-and-nut assembly or clips? (Assume that the male plug and wire are the same for either design.)

TABLE 1-A. What It Costs to Operate a Factory[a]

Metropolitan Area	Labor Cost (per hour)	Power Cost	Occupancy Cost (millions)	Total Cost (millions)
San Francisco	$13.34	$705,000	$3.3	$32.2
Peoria, Illinois	13.45	654,000	3.4	31.9
Cleveland	11.99	554,000	3.4	29.1
Baltimore	11.43	639,000	3.2	28.1
Minneapolis	10.96	472,000	3.5	26.9
Los Angeles	10.47	733,000	3.2	26.3
Chicago	10.45	645,000	3.7	26.2
Boston	9.93	861,000	3.6	26.1
Mobile, Alabama	10.49	501,000	2.8	25.6
Phoenix	10.18	590,000	3.1	25.4
Atlanta	10.16	500,000	2.9	25.0
Houston	10.00	590.000	3.0	24.8
Denver	10.05	531,000	3.3	24.8
Burlington, Iowa	8.78	478,000	3.2	22.2

[a] The yearly cost of operating a durable-goods manufacturing plant employing 750 hourly workers, occupying 300,000 ft^2, and making annual shipments of 33 million pounds to a national market.

Source: Boyd Co., Princeton, N.J.

*C*hapter 1 CASE STUDY

famous folks in manufacturing

Manufacturing engineering is that engineering function charged with the responsibility of interpreting product design in terms of manufacturing requirements and process capability. Specifically: the manufacturing engineer may:

- determine how the product is to be made in terms of specific manufacturing processes.
- design workholding and work transporting tooling or containers.
- select the tools (including the tool materials) that will machine or form the work materials.
- select, design, specify devices and instruments which inspect that which has been manufactured to determine its quality.
- design and evaluate the performance of the manufacturing system.
- perform all these functions (and many more) related to the actual making of the product at the most reasonable cost per unit without sacrifice of the functional requirements or the users' service life.

People who do this kind of work often have changed the course of history through what historians like to call industrial revolutions. The current design of most manufacturing systems is a combination of a job shop and a flow shop. The job shop, as a manufacturing system design, developed during the 1800's. These early factories replaced craft or cottage manufacturing and evolved as a functional design because of the method needed to drive or power the machines. That is, the first factories were built by rivers and the machines were powered by waterwheels which drove

(cont.)

C*hapter 1* CASE STUDY *(cont.)*

shafts that ran into the factory. Machines of like type were set, all in a line, underneath the appropriate power shaft, i.e., the shaft that turned at the speed needed to drive this kind of machine. So all the lathes were collected together under their own power shaft, likewise with all the milling machines, and all the presses. Belts were used to take power off the shaft to the machines. Later the waterwheel was replaced by a steam engine which allowed the factory to be built somewhere other than by a river. Eventually, the steam engine was replaced by large electric motors and then individual electric motors for each machine. But, the job shop design was replicated and the functional design held. It became known to the historians as the American Armory System. The world came to see the American Armory System and it was duplicated around the industrial world.

In the early 1900's, the first vestiges of the flow shop began to emerge. Flow line manufacturing began for small items and culminated with the moving assembly line at the Ford Motor Company. This methodology was developed by Charles Sorenson. Just as in the 1800's, the world again came to see how this system worked and this new design methodology was spread across the world, resulting in a hybrid system—a mixture of job shop and flow shop. As automation was introduced into this system, called "mass production" by the historians, it became characterized by very large lots and very long throughput times. These two eras, which resulted from the two classical Manufacturing System Designs (MSD), became known to the historians (who know nothing about MSD) as the first and secod industrial revolutions.

Now we are entering the third industrial revolution. Again it is not based on hardware or a particular process but once again on the design of the manufacturing system—the complex arrangement of physical elements characterized by measurable parameters.

In recent years, the world went to Japan to try to understand how this tiny nation has become such a giant in the global manufacturing arena. Among the Japanese manufacturers, one company has stood out as the best: **The Toyota Motor Company**. What made this company become the number one automobile manufacturer in the world? It was the development of a unique linked-cell manufacturing system, known initially as the Toyota Production System or the JIT system or World Class Manufacturing system and recently as "lean production." What was different about this system was the development of *manufacturing cells* that posses unique process technology which allows component parts to be built in synchronous fashion (one-piece flow) with final assembly. The cells are linked together in a functionally integrated, pull manufacturing methodology. The result is low unit cost (high efficiency), superior quality (no defects), on-time delivery of unique products.

Now here are some names from the past of famous and not so famous manufacturing, mechanical, and industrial engineers. Can you relate them to the development of the three manufacturing system designs and industrial revolutions?

Eli Whitney, Sam Colt, Elisha Root, Fred Taylor, Henry Ford, John Parsons, John Hall, Taiichi Ohno, Charles Sorenson, Eiji Toyoda.

PART 1

MATERIALS

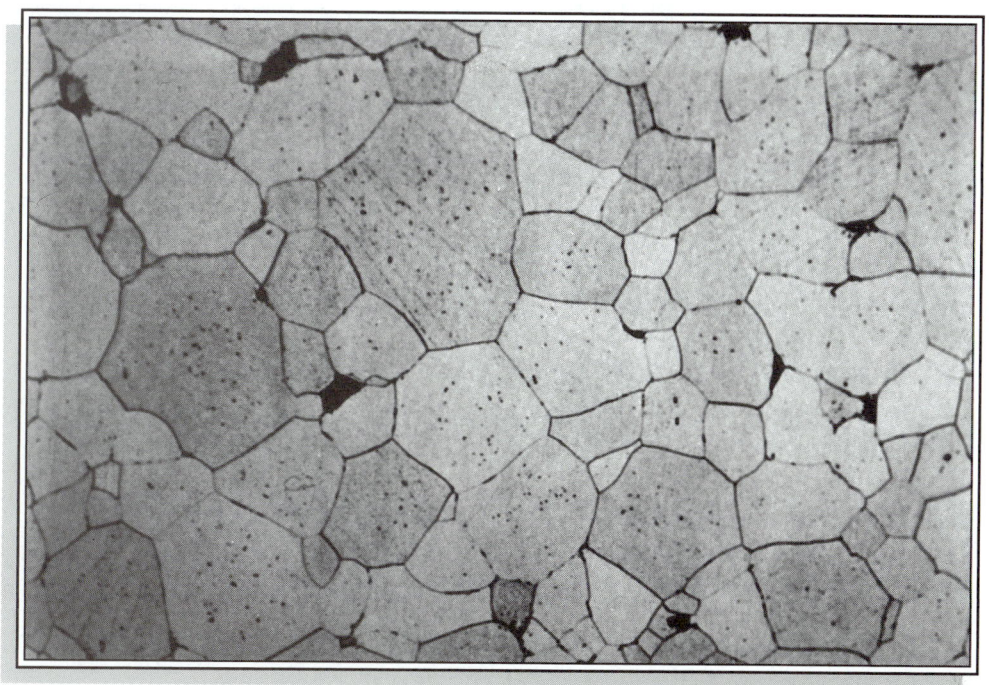

2   Properties of Materials

3   The Nature of Metals and Alloys

4   Equilibrium Diagrams and the Iron-Carbon System

5   Heat Treatment

6   Ferrous Metals and Alloys

7   Nonferrous Metals and Alloys

8   Nonmetallic Materials: Plastics, Elastomers, Ceramics, and Composites

9   Material Selection

CHAPTER 2

PROPERTIES OF MATERIALS

2.1   INTRODUCTION
    Metallic and Nonmetallic
      Materials
    Physical and Mechanical
      Properties
    Stress and Strain
2.2   STATIC PROPERTIES
    Tensile Test
    Strength Properties
    Ductility and Brittleness
    Toughness
    True Stress–True Strain
      Curves
    Repeated Loading and
      Unloading
    Strain Hardening and the
      Strain-Hardening
      Exponent
    Damping Capacity
    Hardness
    Brinell Hardness Test
    Rockwell Hardness Test
    Vickers Hardness Test

    Microhardness Tests
    Other Hardness
      Determinations
    Relationships among the
      Various Hardness Tests
    Relationship of Hardness to
      Tensile Strength
    Compression Tests
2.3   DYNAMIC PROPERTIES
    Impact Test
    Metal Fatigue and the
      Endurance Limit
    Fatigue Failures
    Temperature Effects
    Creep
    Machinability, Formability,
      and Weldability
    Fracture Toughness and the
      Fracture Mechanics
      Approach
2.4   PHYSICAL PROPERTIES
Case Study: SEPARATING MIXED
      MATERIALS

■ 2.1 INTRODUCTION

When selecting a material, a primary concern of engineers is to assure that the material properties are consistent with the operating conditions of the component. The various requirements of each part or component are first estimated or determined. These may include mechanical characteristics (strength, rigidity, resistance to fracture, or the ability to withstand vibrations or impacts), physical characteristics (weight, electrical properties, or appearance), and features relating to the service environment (ability to operate under extremes of temperature or resist corrosion). The selection of the appropriate engineering material is often based on the tabulated or recorded results of standardized tests. Although this information is usually readily available, it is important that it be used properly. The engineer must know which of the properties are significant, how the values were determined, and what restrictions or limitations should be placed on their use. Only by being familiar with the various test procedures and their capabilities and limitations can the engineer determine whether the data is applicable to a particular problem and how they should be applied.

Metallic and Nonmetallic Materials

Perhaps the most common classifying distinction among engineering materials is whether the material is metallic or nonmetallic. The common *metallic* materials include iron, copper, aluminum, magnesium, nickel, titanium, lead, tin, and zinc, as well as the alloys of these metals, such as steel, brass, and bronze. They possess the *metallic properties* of luster, high thermal conductivity, and high electrical conductivity; they are relatively ductile; and some have good magnetic properties. Some common nonmetals are wood, brick, concrete, glass, rubber, and plastics. Their properties vary widely, but they generally tend to be less ductile, weaker, and less dense than the metals and have poor electrical and thermal conductivities.

Although metals have traditionally been the more important of the two groups, the *nonmetallic* group has made great strides and new nonmetallic materials are continuously being developed. Advanced ceramics, composite materials, and engineered plastics are now receiving considerable attention. In many cases, metals and nonmetals are now competing materials for the same component. Selection is based on how well each is capable of providing the required properties. Where both can perform adequately, total cost often becomes the deciding factor, where total cost includes both the cost of the material and the cost of fabricating the desired component.

Physical and Mechanical Properties

A common means of distinguishing one material from another is by evaluating their *physical properties*. These include such characteristics as density (weight), melting point, optical properties (transparency, opaqueness, or color), the thermal properties of specific heat, coefficient of thermal expansion and thermal conductivity, electrical conductivity, and magnetic properties. In some cases, the physical properties may be of prime importance when selecting a material, and several are discussed in more detail near the end of the chapter.

More often however, the properties that describe how a material responds to applied loads (or forces) assume the dominant position in material selection. These *mechanical properties* are usually determined by subjecting prepared specimens to standard laboratory tests. In using the results, however, engineers should remember that they apply only to the specific test conditions that were employed. The actual service conditions of engineered products rarely duplicate the conditions of laboratory testing, so caution should be exercised.

Stress and Strain

When a force is applied to a component or structure, the material is deformed (*strained*), and internal reactive forces (*stresses*) are transmitted through the material. For example, if a weight, W, is suspended from a bar of uniform cross section, as in Figure 2-1, the bar will elongate by an amount ΔL. For a given weight, the magnitude of the *elongation, ΔL*, depends on the original length of the bar. The amount of elongation for each unit length, expressed as $e = \Delta L / L$, is called the *unit strain*. Although it is the ratio of a length to an-

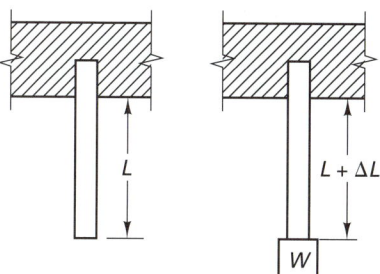

FIGURE 2-1   Tension loading and resultant elongation.

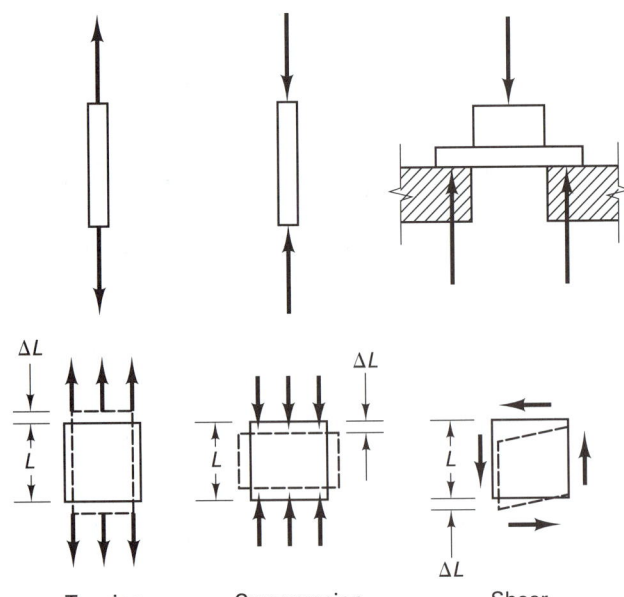

FIGURE 2-2 Examples of tension, compression, and shear loading, and their shear response.

Tension   Compression   Shear

other length and is therefore dimensionless, it is usually expressed in terms of millimeters per meter, inches per inch, or simply as a percentage.

Application of the force also produces reactive stresses within the bar, through which the load is transmitted to the supports. *Stress* is defined as the force or load being transmitted divided by the cross-sectional area transmitting the load. Thus, in Figure 2-1, the stress is $S = W/A$, where A is the cross-sectional area of the supporting bar. Stress is normally expressed in pounds per square inch (in the English system) or megapascals (in SI units).

In Figure 2-1, the weight tends to stretch or lengthen the bar, so the strain is known as a *tensile strain* and the stress as a *tensile stress*. Other types of loadings produce other types of stresses and strains (Figure 2-2). Compressive forces tend to shorten the material and produce *compressive stresses and strains*. *Shear stresses and strains* result when two forces acting on a body are offset with respect to one another.

■ 2.2 STATIC PROPERTIES

When the forces that are applied to a material are constant, or nearly so, they are said to be *static*. Since static loadings are observed in many engineering applications, it is important to characterize the behavior of materials under such conditions. Consequently, a number of standardized tests have been developed to determine the *static properties* of materials. The results of these tests can then be used to select materials, provided that the service conditions are sufficiently similar to those of testing. Even when the service conditions differ, the results of the tests can be used to qualitatively rate and compare the various materials.

Tensile Test

The most common of the static tests is the *uniaxial tensile test*, which provides information about a variety of properties. A *standard specimen* is loaded in tension in a testing machine such as the one shown in Figure 2-3. The standard specimens ensure meaningful and

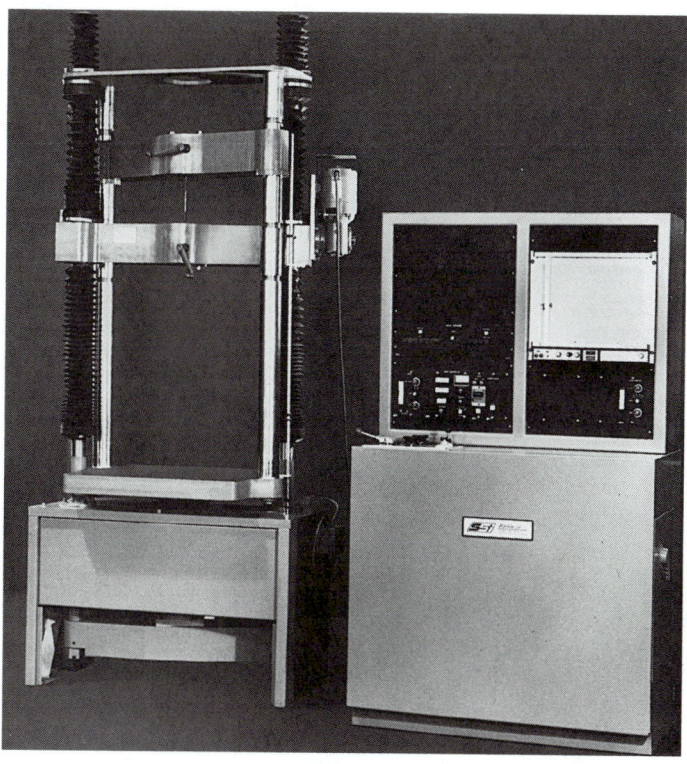

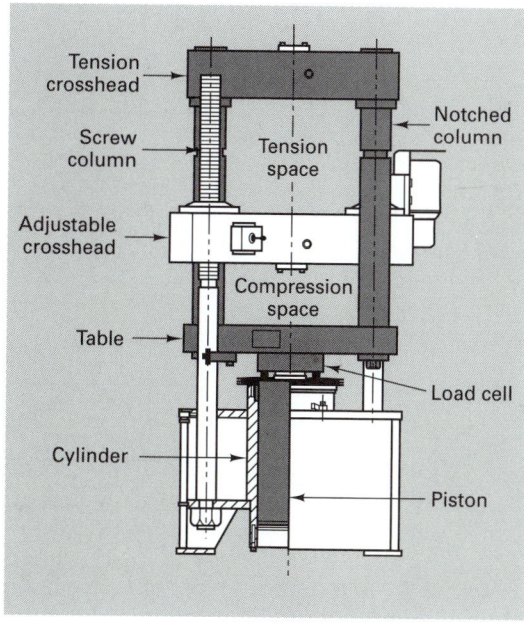

(a)

(b)

FIGURE 2-3 (a) Hydraulic tension and compression testing machine;
(b) schematic of the load frame showing how motion of the darkened yoke can
produce tension or compression with respect to the stationary (white) crosspiece.
(Courtesy of Satec Systems, Inc., Grove City, Pa.)

reproducible results and are designed to produce uniform uniaxial tension in the central portion while ensuring reduced stresses in the sections that are gripped. Figure 2-4 shows two of the most common designs.

Strength Properties.   A load, *W*, is applied and measured by the testing machine, while the elongation (ΔL) or strain over a specified gage length is simultaneously monitored. A plot of the coordinated load–elongation points produces a curve similar to that of Figure 2-5. Since the loads will differ for different-size specimens and the elongations will vary with different gage lengths, it is important to remove the size effects if we are to produce data that is characteristic of a given material and not a particular specimen. If the load is divided by the *original* cross-sectional area and the elongation is divided by the *original* gage length, the size effects are eliminated and the resulting plot becomes known as an *engineering stress-strain curve* (Figure 2-5). This is simply a load–elongation curve with the scales of both axes modified to remove the effects of specimen size.

In Figure 2-5 it can be noted that the initial response is linear—up to a certain stress, the strain and stress are directly proportional to one another. The stress at which this proportionality ceases to exist is known as the *proportional limit*. Up to this stress, the material obeys *Hooke's law*, which states that the strain is directly proportional to the stress. The proportionality constant, or ratio of stress to strain in this region, is known as *Young's*

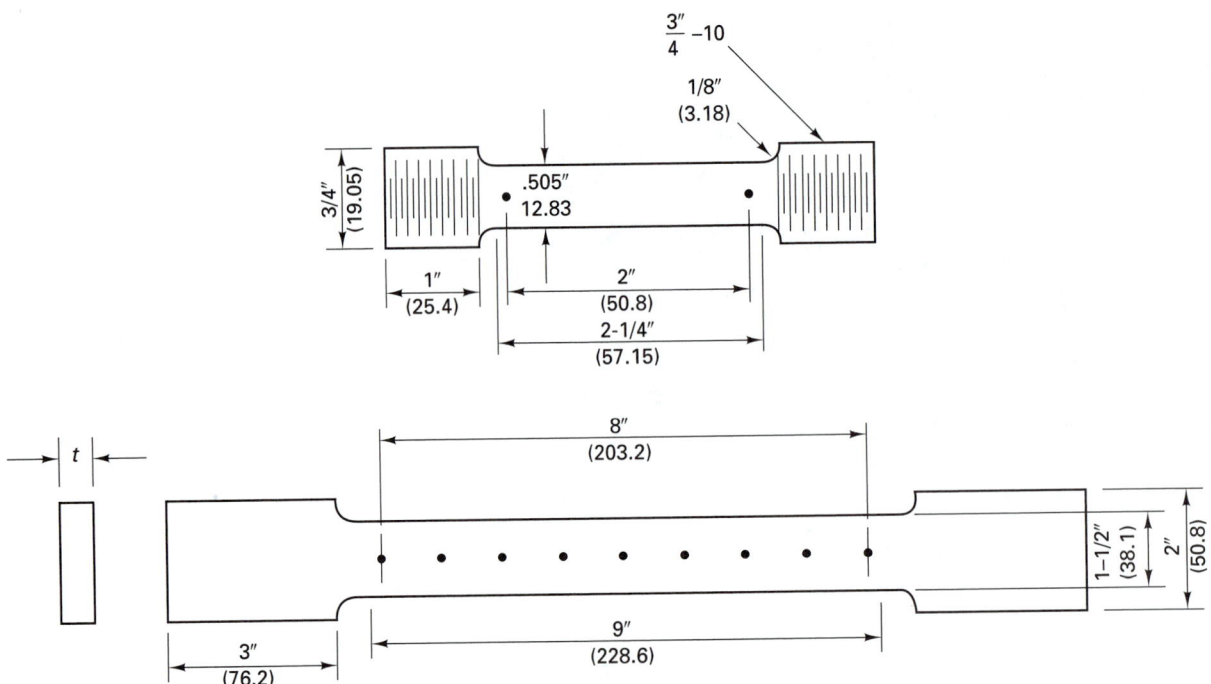

FIGURE 2-4 Two common types of standard tensile test specimens: (a) round; (b) flat. Dimensions are in inches, with millimeters in parentheses.

FIGURE 2-5 Engineering stress-strain diagram for a low-carbon steel.

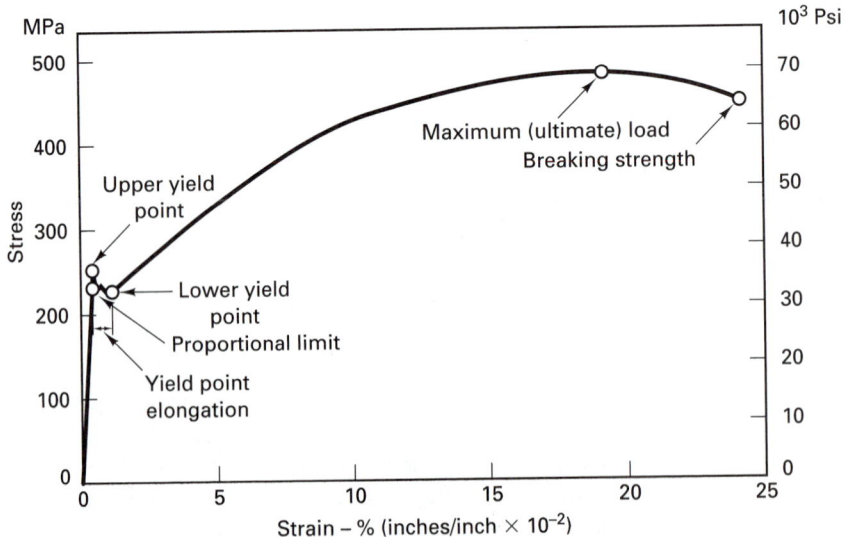

modulus or the *modulus of elasticity*. This is an inherent property of a given material and is of considerable engineering importance. As a measure of stiffness, it indicates the ability of a material to resist deflection or stretching when loaded and is commonly designated by the symbol E.

Up to a certain stress, if the load is removed, the specimen will return to its original length. For stresses between zero and this value, the response is *elastic*, and the uppermost stress is known as the *elastic limit*. For some materials the elastic limit and proportional limit are almost identical. In most cases, however, the elastic limit is slightly higher. Neither quantity, however, should be assigned great engineering significance because the values are quite dependent on the sensitivity of the test equipment.

The amount of energy that a unit volume of material can absorb while in the elastic range is called the *resilience* or the *modulus of resilience*. Because energy is the product of force times distance, the area under the load–elongation curve up to the elastic limit is equal to the energy absorbed by the specimen. By dividing the load by the original area to produce engineering stress, and the elongation by the gage length to produce engineering strain, the area under the engineering stress-strain curve becomes the energy per unit volume, or the modulus of resilience. This energy is potential energy and is released whenever the member is unloaded.

Elongation beyond the elastic limit becomes unrecoverable and is known as *plastic deformation*. When the load is removed, the specimen retains a permanent change in shape. An engineer is usually interested in either the elastic or plastic response, but rarely both. For most components, plastic flow (except for a slight amount to permit redistribution of concentrated stresses) represents failure since the dimensions will be outside allowable tolerances. In manufacturing, however, plastic deformation is often used to shape a product, and the applied stresses must be sufficient to induce plastic flow. Because of the two distinct types of response, it is important that we determine the conditions where elastic behavior transforms into plastic flow.

When the elastic limit is exceeded, increases in strain do not require proportionate increases in stress. For some materials, a stress value may be reached where additional strain occurs without further increase in stress. This point is known as the *yield point*, or *yield-point stress*. For low-carbon steels (Figure 2-5) two distinct points are significant. The highest stress preceding extensive strain is known as the *upper yield point*, and the lower, relatively constant, "run-out" value is known as the *lower yield point*. The lower value is the one that usually appears in tabulated data.

Most materials, however, do not have a well-defined yield point, but exhibit a stress-strain curve of the form shown in Figure 2-6. For these materials, the elastic-to-plastic transition is not distinct and is therefore defined through use of the *offset yield strength*, the value of the stress that will produce a given, but tolerable, amount of permanent strain. For most components, the amount of offset strain is set at 0.2%, but values of 0.1% or even 0.02% may be specified when small amounts of plastic deformation could lead to component failure. The offset yield strength is determined by drawing a line parallel to the elastic line, displaced by the offset strain, and reporting the stress where this line intersects the stress-strain curve (Figure 2-6). This value is reproducible and is independent of equipment sensitivity. However, it is meaningless unless it is reported in conjunction with the amount of offset strain used in its determination. If the applied stresses are kept below the 0.2% offset yield strength, the user can be guaranteed that any observed plastic deformation will be less than 0.2% of the original dimension.

If plastic deformation of the material is continued, the material acquires an increased ability to bear load. Since load-bearing ability is equal to material strength times cross-

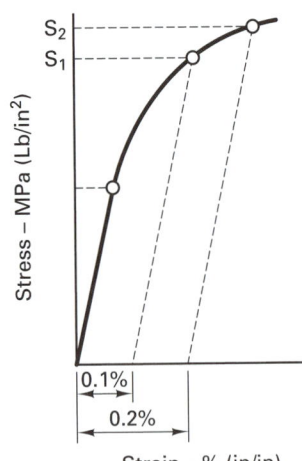

FIGURE 2-6 Stress-strain diagram for a material not having a well-defined yield point, showing the offset method for determining yield strength. S_1 is the 0.1% offset yield strength; S_2 is the 0.2% offset yield strength.

sectional area, and the area is decreasing with increased tensile stretching, one must conclude that the material is getting stronger. When the mechanism for this phenomenon is discussed in Chapter 3, we will see that the strength of a metal will continue to increase with increased deformation. As the weakest location of the specimen deforms, it becomes stronger and is no longer the weakest location. Another location then deforms, and so on, such that the specimen deforms but maintains its original cylindrical or rectangular shape. As straining progresses, however, the additional increments of strength decrease in magnitude, and a point is reached when the decrease in area cancels or dominates the increase in strength. When this occurs, the load-bearing ability peaks, and the force required to continue straining the specimen begins to diminish, as shown in Figure 2-5. The stress at which this occurs is known as the *ultimate strength*, *tensile strength*, or *ultimate tensile strength* of the material. The weakest location in the tensile bar at that time continues to be the weakest location by virtue of the decrease in area, and further deformation becomes localized. This localized reduction in cross-sectional area is known as *necking*, and is shown in Figure 2-7.

If the straining is continued far enough, the tensile specimen will ultimately fracture. The stress at which fracture occurs is known as the *breaking strength* or *fracture strength*. For relatively *ductile* materials, the breaking strength is less than the ultimate tensile strength, and necking precedes fracture. For a *brittle* material, fracture usually terminates the stress-strain curve before necking, and possibly before the onset of plastic flow.

FIGURE 2-7 Standard 0.505-in.-diameter tensile specimen showing a necked region developed prior to failure.

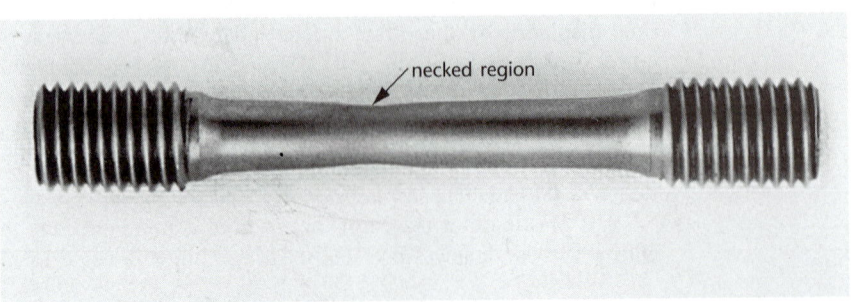

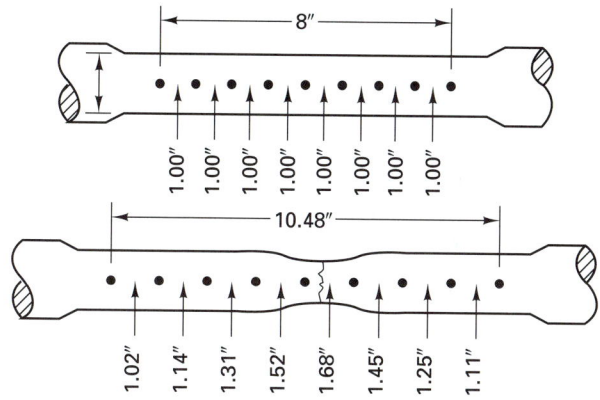

FIGURE 2-8 Final elongation in various segments of a tensile test specimen: (a) original geometry; (b) shape after fracture.

Ductility and Brittleness. The amount of *plasticity* that a material can exhibit is a significant feature when evaluating its suitability for certain manufacturing processes. For metal deformation processes, the more plastic a material is, the more it can be deformed without rupture. This ability of a material to change shape without fracture is known as *ductility.*

One of the primary ways to evaluate ductility is to determine the *percent elongation* of a tensile test specimen. As shown in Figure 2-8, however, ductile materials do not elongate uniformly when loaded beyond necking. Therefore, it has become common practice to report ductility in terms of the percent elongation of a specified gage length. The specific gage length used in the calculation can be of great significance. For example, if the entire 8-in. gage length of Figure 2-8 is used, the elongation is 31%. However, if only the center 2-in. segment is considered, the elongation becomes 60%. Quantitative comparison of materials requires that the elongations be computed from specimens with identical gage lengths.

In many cases, material "failure" is defined as the onset of localized deformation or necking. For example, consider sheet metal being formed into an automobile body panel. The operation must be performed in such a way as to maintain uniform thickness to assure uniform strength and corrosion resistance. Here, a more meaningful measure of ductility would be *uniform elongation* or *percent elongation prior to necking.* This is determined by constructing a line parallel to the elastic portion of the diagram, passing through the point of highest force or stress. The intercept where the line crosses the strain axis denotes the available uniform elongation. Since the additional deformation that occurs after necking is not considered, uniform elongation is always less than total elongation at fracture (the generally reported elongation value).

Another measure of ductility is the *percent reduction in area* that occurs in the necked region of the specimen. This is computed as

$$\frac{A_o - A_f}{A_o} \times 100\%$$

where A_o is the original cross-sectional area and A_f is the smallest area in the necked region. Percent reduction in area, therefore, can range from 0% (brittle) to 100% (extremely plastic).

When materials fail with little or no ductility, they are said to be *brittle.* Brittleness, therefore, is the opposite of ductility and should not be confused with a lack of strength. A brittle material is simply one that lacks significant ductility.

Toughness.   *Toughness*, or *modulus of toughness*, is defined as the work per unit volume required to fracture a material. The tensile test can provide one means of measuring toughness, since the total area under the stress-strain curve represents the desired value. Caution should be exercised when using toughness data, however, because the values can vary markedly with different conditions of testing. As will be seen later, variation in temperature or rate of load application can significantly alter the stress-strain curve, and hence the toughness. Toughness is commonly associated with impact or shock loadings. The values obtained from dynamic impact tests often fail to correlate with those from the static tensile test.

True Stress–True Strain Curves.   The stress-strain curve in Figure 2-5 is a plot of *engineering stress*, S, versus *engineering strain*, e, where S is computed as the applied load (W) divided by the *original* cross-sectional area (A_o) and e is the elongation, L, divided by the *original* gage length, L_o. As discussed previously, and illustrated in Figures 2-7 and 2-8, the cross section of the test specimen changes as the test proceeds, first uniformly and then in a nonuniform manner after necking begins. The actual stress within the specimen should be based on the instantaneous cross-sectional area, not the original, and will be greater than the engineering stress plotted in Figure 2-5. *True stress,* σ, can be computed by taking simultaneous readings of the load and the minimum specimen diameter. The actual area (A) can then be computed, and true stress can be determined as

$$\sigma = \frac{W}{A}$$

The determination of true strain is somewhat more complex. In place of the change in length divided by the original length that was used to compute engineering strain, true strain is defined as the summation of the incremental strains that occur throughout the test. Thus, for a specimen that has been stretched from length ℓ_o to length ℓ, the *true, natural, or logarithmic strain,* ε, would be

$$\varepsilon = \int_{\ell_o}^{\ell} \frac{dL}{L} = \ln \frac{\ell}{\ell_o} = 2 \ln \frac{D_o}{D}$$

The last equality makes use of the relationship for cylindrical specimens that

$$\frac{\ell}{\ell_o} = \frac{A_o}{A} = \frac{D_o^2}{D^2}$$

and applies only up to the onset of necking.

Figure 2-9 shows the type of curve that results when the data from a uniaxial tensile test is plotted in the form of true stress versus true strain. It should be noted that the true stress is a measure of the material strength at any point during the test, and continues to rise even after necking. Data beyond the point of necking should be used with extreme caution, however, because the geometry of the neck transforms the stress state from uniaxial tension (stretching in one direction with contractions in the other two) to triaxial tension, in which the material is being stretched or held back in all three directions. Voids or cracks (Figure 2-10) tend to open in the necked region as a precursor to final fracture. The diameter measurements no longer reflect the true load-bearing area, and the data is further distorted.

Repeated Loading and Unloading.   Figure 2-11 is an adaptation of a true stress–true strain diagram, designed to show how a ductile metal (such as steel) will behave subjected

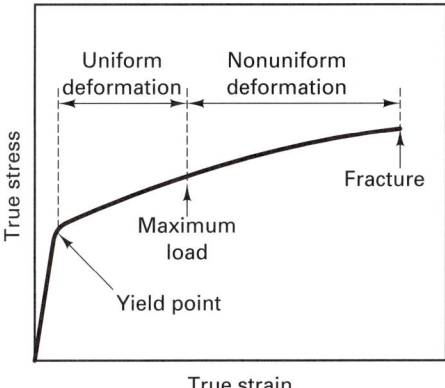

FIGURE 2-9 True stress–true strain curve for an engineering metal.

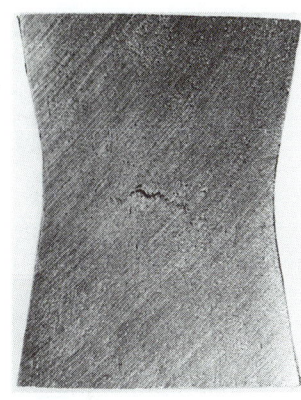

FIGURE 2-10 Section of a tensile test specimen stopped just prior to failure, showing a crack already started in the necked region. *(Courtesy of E. R. Parker.)*

to slow loading and unloading. Loading and unloading within the elastic region will result in simply cycling up and down the linear portion of the curve between points O and A. If the initial load is up to point B (in the plastic region), unloading will follow the path BeC, which is approximately parallel to the line OA. The specimen will exhibit a permanent elongation of the amount OC. Reloading from this point will follow the curve CfD, a slightly different path from that of unloading.

Strain Hardening and the Strain-Hardening Exponent. When the specimen is reloaded from point C, elastic behavior is exhibited as the stress is increased, all the way to point D. Only when the stress surpasses that of point D does further plastic deformation begins. Thus point D becomes the yield point or yield stress for the material in its present state, and the material has become stronger than it was in its initial condition. If the test were again interrupted at point E, the material would exhibit a new, even higher, yield stress. Beyond the elastic region, the true stress–true strain curve is actually the locus of the yield stress for the various amounts of strain.

When metals are plastically deformed, they *strain harden*. They become harder and stronger, and this change is a progressive one. If a stress is producing plastic deformation, an even greater stress will be required to induce further flow. In Chapter 3 we discuss the atomic-scale changes responsible for this phenomenon.

Various materials strain harden at different rates—for a given amount of deformation, different materials will exhibit different increases in strength. One method to describe this behavior is to fit the plastic region of the true stress-true strain data to the equation

$$\sigma = K\varepsilon^n$$

and determine the best-fit value of n, the *strain-hardening coefficient*.[*] As shown in Figure 2-12, a material with a high value of n strain hardens rapidly. Small amounts of deformation induce significant increases in material strength. Materials with a small n value show little change in strength with plastic deformation.

---

[*]Taking the logarithm of both sides of the equation above yields $\log \sigma = \log K + n \log \varepsilon$. This is the same form as the equation $y = mx + b$, the equation for a straight line with slope m and intercept b. Therefore, if the true stress–true strain data plotted on a log-log scale with stress on the y axis and strain on the x, the slope of the data in the plastic region will be n.

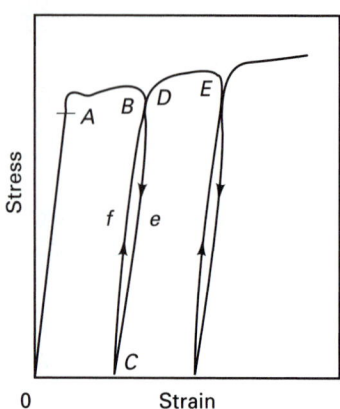

FIGURE 2-11 Stress-strain diagram obtained by unloading and reloading a specimen.

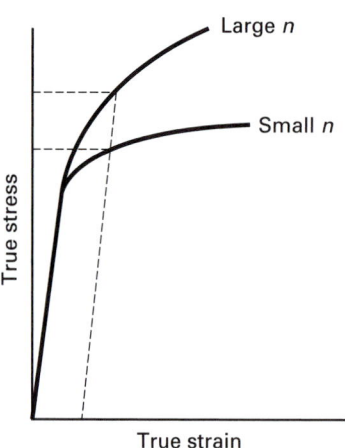

FIGURE 2-12 True stress–true strain curves for metals with large and small strain hardening. Metals with larger n values experience larger amounts of strengthening for a given strain.

Damping Capacity. In Figure 2-11 the unloading and reloading of the specimen follow slightly different paths. The area between the two curves is proportional to the amount of energy that is converted from mechanical form to heat and is therefore absorbed by the material. When this area is large, the material is said to exhibit good *damping capacity* and is able to absorb mechanical vibrations or damp them out quickly. This is an important property in applications such as crankshafts and machinery bases. Gray cast iron is used in many applications because of its high damping capacity. Materials with low damping capacity, such as brass and steel, transmit sound and vibrations.

Hardness

Hardness is a very important but hard-to-define property of materials. A number of tests have been developed around several different phenomena. The most common of the tests are based on resistance to permanent deformation (indentation) under static or dynamic loading. Other tests evaluate resistance to scratching, energy absorption under impact loading, wear resistance, or even resistance to cutting or drilling. Since these phenomena are not the same, the results of the various tests often do not correlate with one another. Caution should be exercised so that the test that is selected clearly evaluates the phenomena of interest.

Brinell Hardness Test. The *Brinell hardness test* was one of the earliest accepted methods of measuring hardness. A tungsten carbide or hardened steel ball 1 cm in diameter, is pressed into a flat surface of the material by a standard load of 500, 1500, or 3000 kg, and the load is maintained for 5 to 10 seconds to permit the full amount of plastic deformation to occur. The load and ball are then removed, and the diameter of the resulting spherical indentation (usually in the range of 2 to 5 mm) is measured using a special grid or traveling microscope. The Brinell hardness number is equal to the load divided by the spherical surface area of the indentation when the units are expressed as kilograms per square millimeter.

$$\text{Brinell hardness number (BHN)} = \frac{\text{load}}{\text{surface area of indentation}}$$

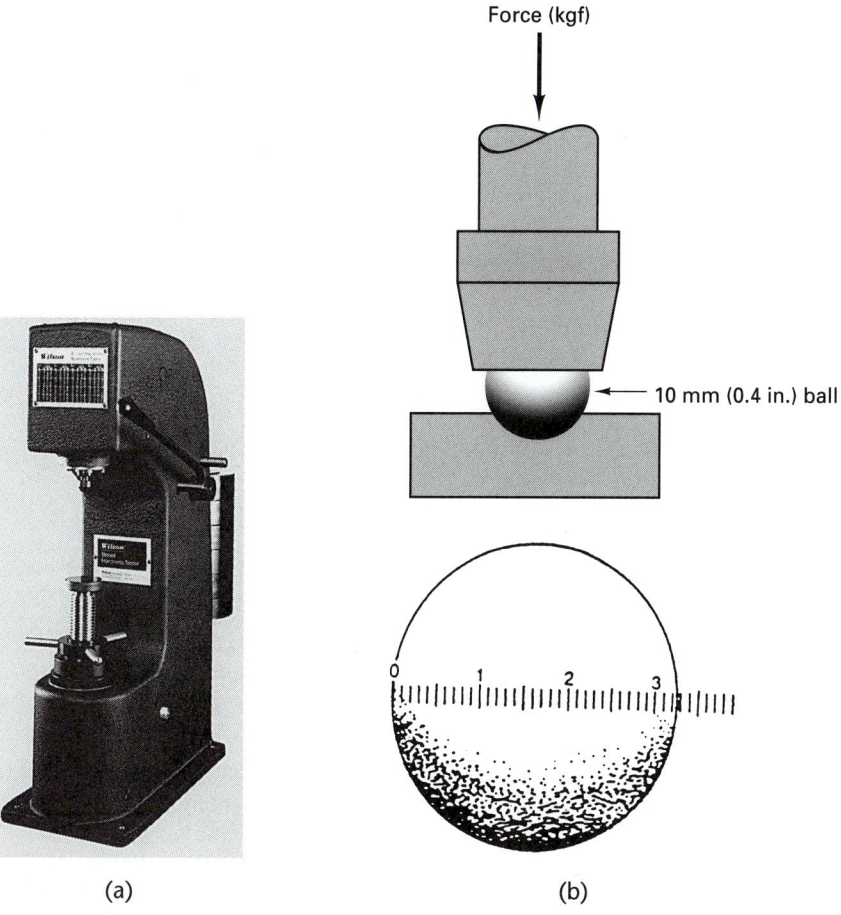

(a)   (b)

FIGURE 2-13   (a) Brinell hardness tester; (b) Brinell test sequence showing loading and measurement of the indentation under magnification with a scale calibrated in millimeters. [*(a) Courtesy of Wilson Instruments Division, Instron Corp.*]

In actual practice, the Brinell hardness number is determined from tables that correlate the Brinell number with the diameter of the indentation for the various loads. Figure 2-13 shows a typical Brinell tester along with a schematic of the testing procedure. Portable testers are also available.

The Brinell test measures the hardness over a relatively large area and is indifferent to small-scale variations in structure. In addition, it is relatively simple and easy to conduct, and is used extensively on irons and steels. On the negative side, the Brinell test has the following limitations:

1. It cannot be used on very hard or very soft materials.
2. The results may not be valid for thin specimens. It is best if the thickness of material is at least 10 times the depth of the indentation. Some standards specify the minimum hardnesses for which the tests on thin specimens will be considered valid.

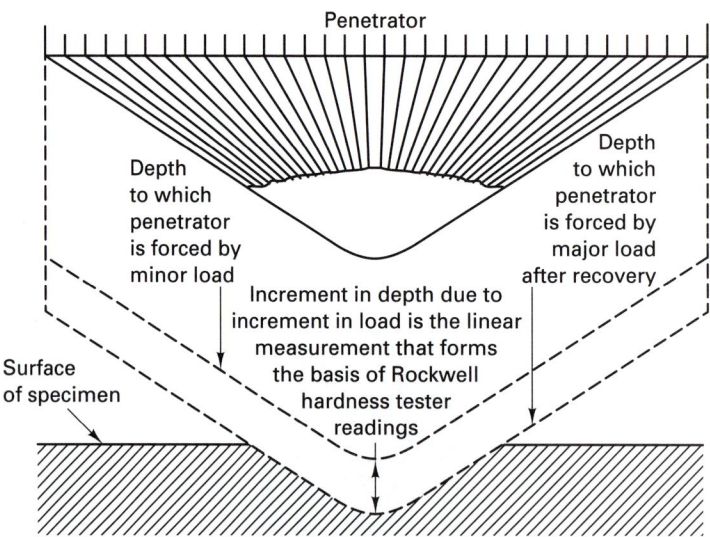

FIGURE 2-14 Operating principle of the Rockwell hardness tester. *(Courtesy of Wilson Instruments Division, Instron Corp.)*

3. The test is not valid for case-hardened surfaces.
4. The test must be conducted far enough from the edge of the material that no edge bulging can occur.
5. The substantial indentation may be objectionable on certain finished parts.
6. The edge or rim of the indentation may not be clearly defined or may be difficult to see on materials of different colors.

Rockwell Hardness Test.   The widely used *Rockwell hardness test* is similar to the Brinell test since the hardness value is also determined through an indentation produced under a static load. The exact nature of the Rockwell test and how it differs from the Brinell test can be explained through the use of Figure 2-14. A small indenter, either a small-diameter steel ball or a diamond-tipped cone called a *brale*, is first seated firmly against the material by the application of a "minor" load of 10 kg. This causes a very slight penetration into the surface, usually elastic in nature. The indicator on the dial of the tester, such as the one shown in Figure 2-15, is then set to zero, and a "major" load is then applied to the indenter to produce a deeper penetration (i.e., plastic deformation). After the indicating pointer has come to rest, the major load is removed. With the minor load still applied, the tester now indicates the appropriate Rockwell hardness number on either a dial gauge or digital display. This number is actually an indication of the *depth of plastic or permanent penetration* produced by the major load.

Different combinations of major loads and indenters are available and are used for materials with various levels of strength. Table 2-1 provides a partial listing of the Rockwell scales and typical materials for which they are used. Because of the different scales, Rockwell hardness numbers must always be accompanied by a letter indicating the particular combination of load and indenter used in the test. The notation $R_C 60$ (or Rockwell C 60) indicates that a brale indenter was used in combination with a major load of 150 kg, and a reading of 60 was obtained. The B and C scales are used more extensively than the others, B being common for copper and aluminum and C for steels.

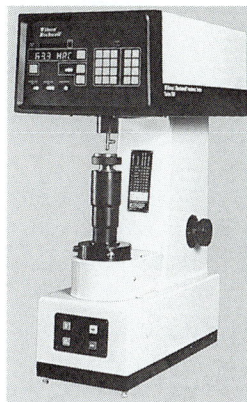

FIGURE 2-15 Rockwell hardness tester with digital readout. *(Courtesy of Wilson Instruments Division, Instron Corp.)*

TABLE 2-1.	Some Common Rockwell Hardness Tests		
Scale Symbol	Penetrator	Load (kg)	Typical Materials
A	Brale	60	Cemented carbides, thin steel, shallow case-hardened steel
B	$\frac{1}{16}$ -in. ball	100	Copper alloys, soft steels, aluminum alloys, malleable iron
C	Brale	150	Steel, hard cast irons, titanium, deep case-hardened steel
D	Brale	100	Thin steel, medium case-hardened steel
E	$\frac{1}{8}$ -in. ball	100	Cast iron, aluminum, magnesium
F	$\frac{1}{16}$ -in. ball	60	Annealed coppers, thin soft sheet metals
G	$\frac{1}{16}$ -in. ball	150	Hard copper alloys, malleable irons
H	$\frac{1}{8}$ -in. ball	60	Aluminum, zinc, lead

The Rockwell test should not be used on thin materials (generally less than $\frac{1}{16}$ -in. thick), on rough surfaces, or on materials that are not homogeneous, such as gray cast iron. Because of the small size of the indentation, variations in roughness, composition, or structure can greatly influence the results. For thin materials, or where a very shallow indentation is desired (as in the evaluation of surface treatments such as nitriding or carburizing), the *Rockwell superficial hardness test* is used. Operating on the same Rockwell principle, this test employs smaller major and minor loads and uses a more sensitive depth measuring device.

In comparison with the Brinell test, the Rockwell test offers the attractive advantage of direct readings in a single step. Because it can be conducted more rapidly, it is frequently used for monitoring the quality of products during mass production. Furthermore, it has the additional advantage of producing a smaller indentation that can be easily concealed on the finished product or easily removed in a later operation.

Vickers Hardness Test. The *Vickers hardness test* is also similar to the Brinell test, but a square-based diamond pyramid is used as the indenter. Like the Brinell test, the Vickers hardness number is defined as load divided by the surface area of the indentation expressed in units of kilograms per square millimeter. The advantage of the Vickers approach is the increased accuracy in determining the diagonal of a square impression as opposed to the diameter of a circle and the assurance that even light loads will produce plastic deformation.

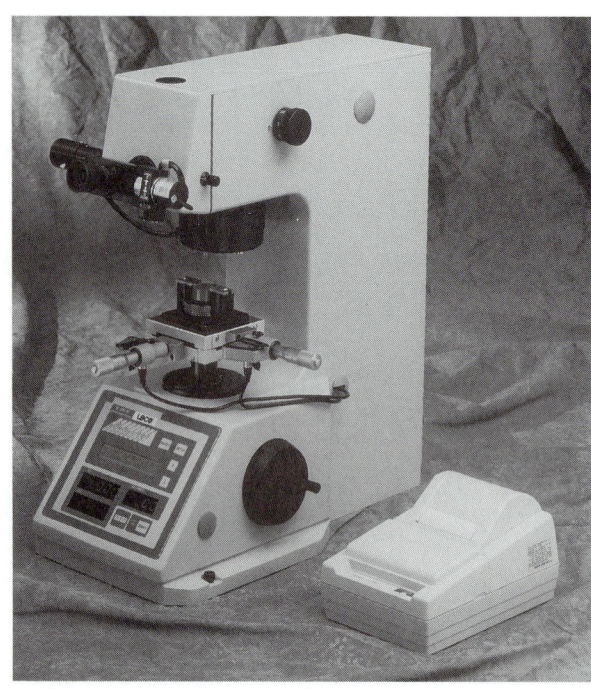

FIGURE 2-16 Microhardness tester. *(Courtesy of LECO Corporation)*

Like the other indentation hardness methods, the Vickers test offers a number of attractive features: (1) it is simple to conduct, (2) little time is involved, (3) little surface preparation is required, (4) the test can be done on location, (5) it is relatively inexpensive, and (6) it often provides results that can be used to evaluate material strength or assess product quality.

Microhardness Tests. Various microhardness tests have been developed for applications where it is necessary to determine the hardness of a very precise area of material, or where the material or surface layer is exceptionally thin. Special machines, such as the one shown in Figure 2-16, have been developed for this purpose. The location for the test is selected under high magnification. A small diamond penetrator is then loaded with a predetermined load ranging from 25 to 3600 g. In the Knoop test, an elongated diamond-shaped indenter is used and the length of the indentation is measured with the aid of a microscope (Figure 2-17). The hardness value, known as the *Knoop hardness number*, is obtained by dividing the load in kilograms by the projected area of the diamond-shaped indentation, expressed in square millimeters. The Vickers test can also be used to determine microhardness.

Other Hardness Determinations. Using a principle known as ultrasonic contact impedance (UCI) measurement, an extension of the low-load-range Vickers test has been developed to permit rapid, on-site hardness determinations without the need for measuring the size of the indentation. A Vickers diamond-tipped indenter is attached to the end of an ultrasonically oscillating rod and is pressed into the test specimen. As the diamond penetrates, the resonant frequency of the rod will change as a function of the size of the contact area. When the critical load is reached, the change in resonant frequency is noted and the corresponding Vickers hardness is displayed.

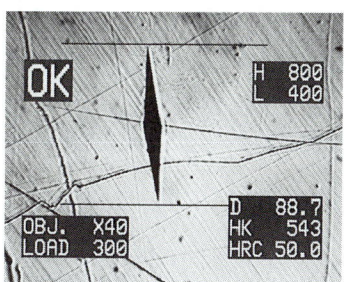

FIGURE 2-17 Knoop microhardness indentation viewed and measured under a microscope. H and L indicate the anticipated range of Knoop hardness. D is the distance between ends of the indentation, HK is the actual Knoop hardness, and HRC is a conversion to Rockwell C hardness. *(Courtesy of LECO Corporation)*

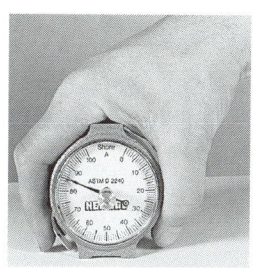

FIGURE 2-18 Durometer hardness tester. *(Courtesy of New Age Industries.)*

When testing soft, elastic materials, such as rubbers and nonrigid plastics, a *durometer* is often used. This instrument (Figure 2-18) measures the resistance of a material to elastic penetration by a spring-loaded conical steel indenter. No permanent deformation occurs. A similar test is used to evaluate the strength of molding sands used in the foundry industry.

In the *scleroscope test*, hardness is measured by the rebound of a small diamond-tipped "hammer" that is dropped from a fixed height onto the surface of the material to be tested. Obviously, this test measures the resilience of a material, and the surface on which the test is made must have a fairly high polish to yield good results. In addition, scleroscope hardness numbers are comparable only among similar materials. A comparison between steel and rubber, therefore, would not be valid.

Hardness has also been defined as the ability of a material to resist being scratched. A crude but useful test that employs this principle is the *file test*, wherein one determines whether a material can be cut by a simple metalworking file. The test can be either a pass–fail test using a single file, or a semiquantitative evaluation using a series of files that have been pretreated to various levels of known hardness.

Relationships among the Various Hardness Tests. Since the various hardness tests tend to evaluate somewhat different material phenomena, there is no simple relationship between the different types of hardness numbers that are determined. Approximate relationships have been developed, however, by testing the same material on a variety of devices. Table 2-2 presents a comparison of hardness values for plain carbon and low-alloy steels. It may be noted that for Rockwell C numbers above 20, the Brinell values are approximately 10 times the Rockwell number. Also, for Brinell hardnesses below 320, the Vickers and Brinell values agree quite closely. Since the relationships between the various tests will vary with material, mechanical processing, and heat treatment, tables such as Table 2-2 should be used with caution.

Relationship of Hardness to Tensile Strength. Table 2-2 and Figure 2-19 show a correlation between tensile strength and hardness for steel. For plain carbon and low-alloy steels, the tensile strength (in pounds per square inch) can be estimated by multiplying the Brinell hardness number by 500. Thus an inexpensive and quick hardness test can be used

TABLE 2-2. Hardness Conversion Table for Steels

Brinell Number	Vickers Number	Rockwell Number C	Rockwell Number B	Scleroscope Number	Tensile Strength ksi	Tensile Strength MPa
	940	68		97	368	2537
757[a]	860	66		92	352	2427
722[a]	800	64		88	337	2324
686[a]	745	62		84	324	2234
660[a]	700	60		81	311	2144
615[a]	655	58		78	298	2055
559[a]	595	55		73	276	1903
500	545	52		69	256	1765
475	510	50		67	247	1703
452	485	48		65	238	1641
431	459	46		62	212	1462
410	435	44		58	204	1407
390	412	42		56	196	1351
370	392	40		53	189	1303
350	370	38	110	51	176	1213
341	350	36	109	48	165	1138
321	327	34	108	45	155	1069
302	305	32	107	43	146	1007
285	287	30	105	40	138	951
277	279	28	104	39	134	924
262	263	26	103	37	128	883
248	248	24	102	36	122	841
228	240	20	98	34	116	800
210	222	17	96	32	107	738
202	213	14	94	30	99	683
192	202	12	92	29	95	655
183	192	9	90	28	91	627
174	182	7	88	26	87	600
166	175	4	86	25	83	572
159	167	2	84	24	80	552
153	162		82	23	76	524
148	156		80	22	74	510
140	148		78	22	71	490
135	142		76	21	68	469
131	137		74	20	66	455
126	132		72	20	64	441
121	121		70		62	427
112	114		66		58	

[a]Tungsten carbide ball; others standard ball.

to provide a close approximation of the tensile strength of the steel. For other materials, the relationship will be different and may exhibit too much variation to be dependable. For example, the multiplying factor for duraluminum is about 600, while for soft brass it is around 800.

Compression Tests

When a material is subjected to compressive loading, the relationships between stress and strain are similar to those for a tension test. Up to a certain value of stress, the material behaves elastically. Beyond this value, plastic flow occurs. In general, however, a com-

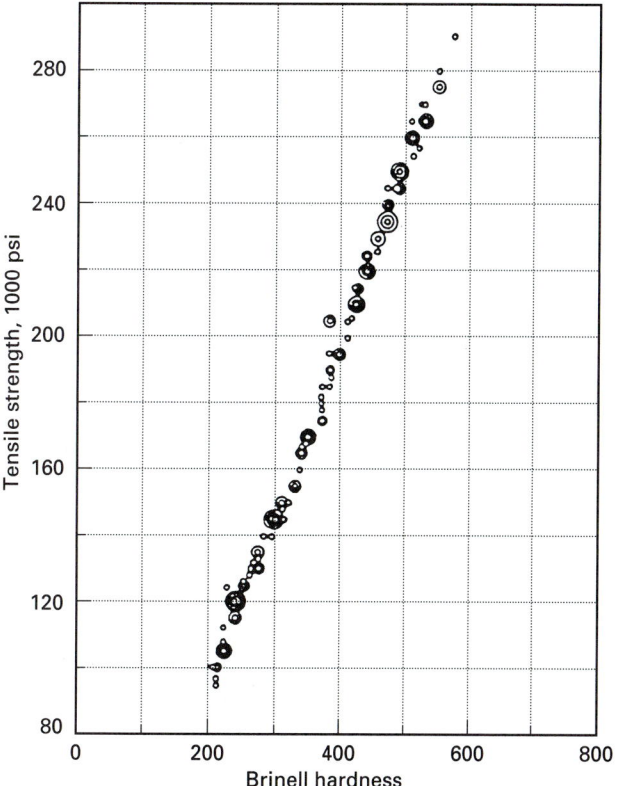

FIGURE 2-19 Relationship of hardness and tensile strength for a group of standard alloy steels. *(Courtesy of ASM International, Materials Park, Ohio.)*

pression test is more difficult to conduct than a standard tensile test. Test specimens must have larger cross-sectional areas to resist bending or buckling. As deformation proceeds, the material strengthens by strain hardening and the cross section of the specimen increases, producing a substantial increase in required load. Friction between the testing machine surfaces and the ends of the test specimen tends to alter the results if not properly considered. The selection of the tension or compression mode of testing, however, is determined largely by the type of service for which the material is intended.

■ 2.3 DYNAMIC PROPERTIES

In many engineering applications, materials are subjected to various types of dynamic loading. These may include (1) sudden loads (impacts) or loads that vary rapidly in magnitude, (2) repeated cycles of loading and unloading, or (3) frequent changes in the mode of loading, such as from tension to compression. For these operating conditions, the engineer must be concerned with properties in addition to those determined by the static tests.

Many of the dynamic tests subject standard specimens to a well-controlled set of test conditions. Since the actual application of a product will probably involve a different set of conditions, the data from the test may not be all that useful for design. Nevertheless, the dynamic tests do provide a valuable comparison of the way materials will respond to specific conditions of dynamic loading. One should always be aware of the limitations in test data.

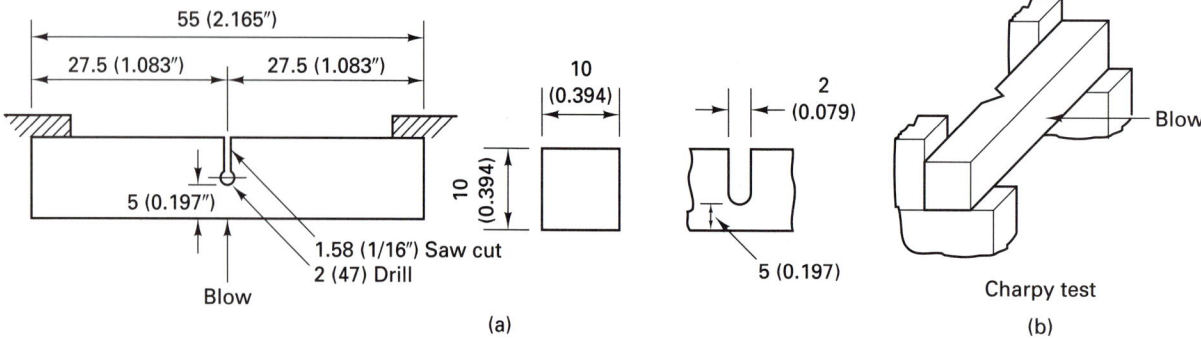

FIGURE 2-20 (a) Standard Charpy impact specimens and mode of loading. Illustrated are keyhole and U notches; dimensions are in millimeters with inches in parentheses. (b) Standard V-notch specimen showing the three-point bending type of impact.

Impact Test

Several tests have been developed to evaluate the fracture resistance of a material when subjected to rapidly applied dynamic loads, or impacts. Of those tests that have become common, two basic types have emerged: (1) bending impacts, which includes the standard Charpy and Izod tests, and (2) tension impacts.

The bending impact tests utilize specimens that are supported as beams. In the *Charpy test*, the specimen contains a V, keyhole, or U-shaped notch, the V and keyhole being the most common. As shown in Figure 2-20, the Charpy test specimen is supported on the ends, and the impact is applied to the center, behind the notch, to complete a three-point bending. The *Izod test* specimen is supported as a cantilever beam and is impacted on the end (Figure 2-21). Standard testing machines such as the one shown in Figure 2-22 apply a predetermined impact energy by means of a swinging pendulum. After breaking or deforming the specimen, the pendulum continues its upward swing with an energy equal to its original minus that absorbed by the impacted specimen. This loss of energy is measured by the angle attained by the pendulum on its upward swing.

The test specimens for bending impacts must be prepared with precision to ensure consistent and reproducible results. Notch profile, particularly the radius at the root of V-notch specimens, is extremely critical, for the test measures the energy required to both initiate

FIGURE 2-21 (a) Izod impact specimen; (b) cantilever mode of loading in the Izod test.

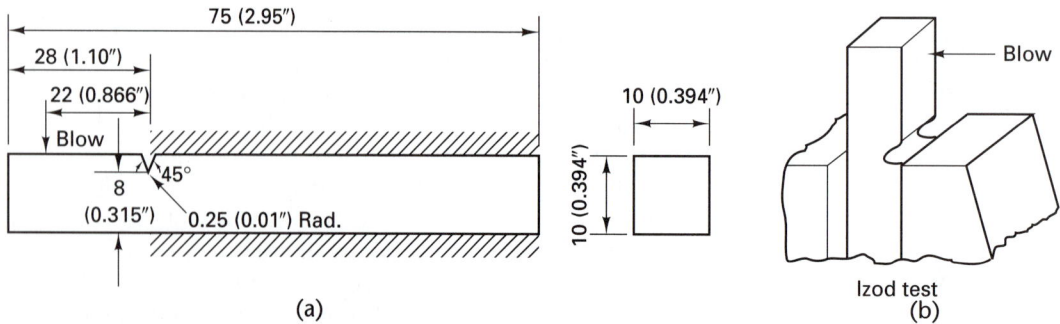

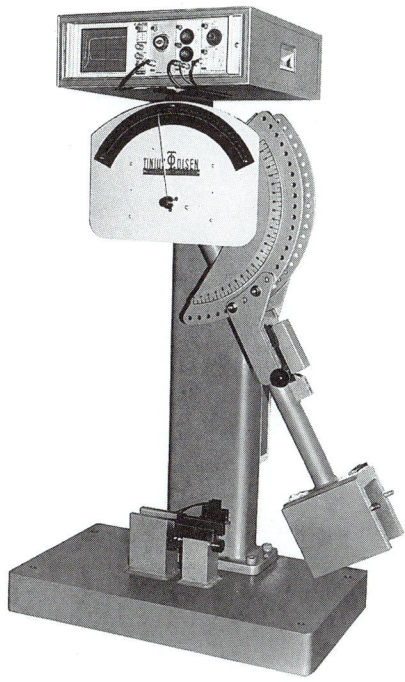

FIGURE 2-22 Impact testing machine. *(Courtesy of Tinius Olsen Testing Machine Co., Inc.)*

FIGURE 2-23 Notched and unnotched impact specimens before and after testing. Both specimens had the same cross-sectional area.

and propagate a fracture. The effect of notch profile is shown dramatically in Figure 2-23. Here two specimens have been made from the same piece of steel with the same reduced cross-sectional area. The one with the keyhole notch fractures and absorbs only 43 ft-lb of energy, whereas the other specimen resists fracture and absorbs 65 ft-lb during the impact.

Caution should also be placed on the use of impact test data for design purposes. The test results apply only to standard specimens containing a standard notch. The tests also evaluate material behavior under very specific conditions. Changes in the form of the notch, minor variations from the standard specimen geometry, or faster or slower rates of loading (speed of the pendulum) can produce significant changes in the results. Under conditions of rapid loading, wide specimens, and sharp notches, many ductile materials lose their energy-absorbing capability and fail in a brittle manner.

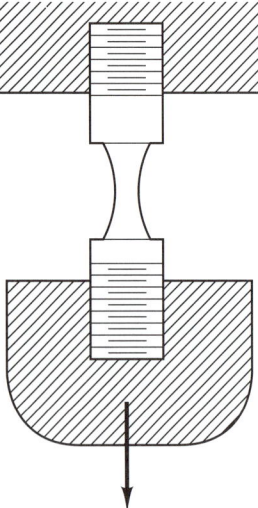

FIGURE 2-24 Tensile impact test schematic.

The results of bending impact tests, however, can be quite valuable in assessing a material's sensitivity to notches and the multiaxial stresses that exist around a notch. In addition, testing can be performed at various temperatures. As will be seen in a later section of this chapter, the evaluation of how fracture resistance varies with temperature can be valuable input when selecting engineering materials.

The *tensile impact test*, illustrated schematically in Figure 2-24, eliminates the use of a notched specimen, thereby avoiding many of the objections inherent in the Charpy and Izod tests. Specimens are subjected to uniaxial impact loadings by means of drop weights, modified pendulums, and variable-speed flywheels.

Metal Fatigue and the Endurance Limit

Metals may also fracture when subjected to repeated applications of stress, even though all of the stresses are less than the ultimate tensile strength and usually less than the yield strength as determined by a tensile specimen. This phenomenon, known as *metal fatigue,* may result from the cyclic repetition of a particular loading cycle or from entirely random variations in stress. Since fatigue failures probably account for nearly 90% of all mechanical fractures, it is important for an engineer to know how materials will respond to fatigue conditions.

Although there are an infinite number of possible repeated loadings, the periodic, sinusoidal mode is most suitable for experimental reproduction and subsequent analysis. By restricting the conditions to equal-magnitude tension–compression reversals, the test is further simplified, and data such as that presented in Figure 2-25 is generated. If this material were subjected to a standard tensile test, it would require a stress in excess of 70,000 psi to induce failure. However, under cyclic loading with a peak stress of 55,000 psi, well below the yield strength of the material, failure would occur after about 100,000 cycles. If the peak stress were reduced to 51,000 psi, the lifetime would be extended to approximately 1,000,000 cycles. If the applied stress were reduced below 49,000 psi, this steel would not fail by fatigue, regardless of the number of stress application cycles.

Curves such as that in Figure 2-25 are known as *stress versus number of cycles*, or *S–N curves*. Any point on the curve is the *fatigue strength* corresponding to the matching number of loading cycles. The limiting stress value below which the material will not fail

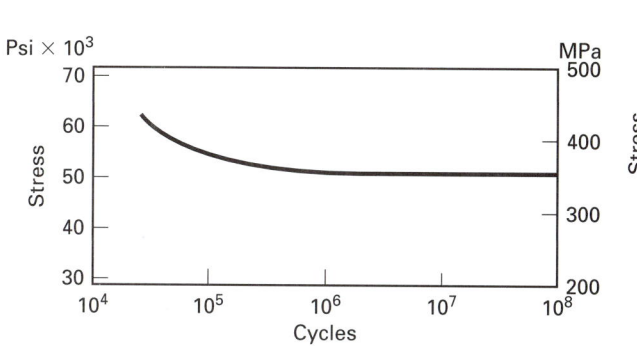

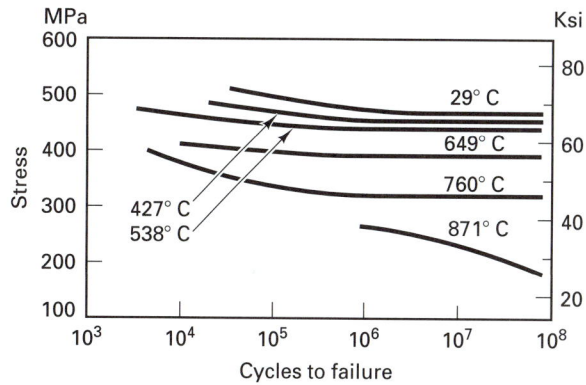

FIGURE 2-25   Typical *S–N* or endurance-limit curve for steel. Specific numbers will vary with the type of steel and treatment.

FIGURE 2-26   Fatigue strength of Inconel alloy 625 at various temperatures. *(Courtesy of Huntington Alloy Products Division, The International Nickel Company Inc.)*

regardless of the number of load cycles, known as the *endurance limit* or *endurance strength*, is an important criterion in many design applications.

A different number of loading cycles is required to determine the endurance limit for different materials. For steels, 10 million (10^7) cycles are usually sufficient. Several of the nonferrous metals require 500 million (5×10^8) cycles, and some aluminum alloys require such a great number that no endurance limit is apparent under typical test conditions.

The fatigue resistance of a material is sensitive to a number of factors. One of the most important of these is the presence of stress raisers, such as small surface cracks, machining marks, or surface gouges. Data for the reported *S–N* curves is obtained from polished specimens, and the reported lifetime is the cumulative number of cycles required to initiate a fatigue crack and propagate it to failure. If a part already contains a surface crack or flaw, the number of cycles required for crack initiation can be reduced significantly. In addition, the stress concentrator magnifies the stress applied at the tip of the crack, accelerating the rate of subsequent crack growth. Great care should be taken to eliminate stress raisers and surface flaws on parts that will be subjected to cyclic loadings. Proper design and manufacturing practices are often more important than material selection and heat treatment for fatigue applications.

Operating temperature can also shift the fatigue performance of a metal. Figure 2-26 shows the *S–N* curves for Inconel 625 (a Ni–Cr–Fe alloy) for a variety of temperatures. As temperature is increased, the fatigue strength drops significantly. Since most test data is generated at room temperature, caution should be exercised when the application involves elevated service temperatures.

Fatigue lifetime can also be altered by changes in the environment. When metals are subjected to corrosion during the cyclic loadings, the condition is known as *corrosion fatigue*, and both specimen lifetime and the endurance limit are reduced significantly. Special corrosion-resistant coatings such as zinc or cadmium may be required for these applications. The nature of the environmental attack need not be severe. Tests conducted in a vacuum have been shown to produce different results from those conducted in air, and further variations have been observed with different levels of humidity. Test results are further complicated by the selection of cycle frequency. For slower frequencies, the environment has a longer time to act between loadings; at high frequencies, the environmental effects may be somewhat masked. Direct application of test data, therefore, should be done with caution.

Residual stresses in products can also alter fatigue behavior. If the specimen surface is in a state of compression, such as that produced from shot peening, carburizing, or burnishing, it is more difficult to initiate a fatigue crack and lifetime is extended. Conversely, processes resulting in residual surface tension, such as welding or machining (Chapter 40), can significantly reduce the fatigue lifetime of a product.

If the magnitude of the load changes during service, a condition common to many components, the fatigue response of the metal becomes even more complex. Consider the wing of a commercial airplane. Many low-stress cycles (vibrations during flight) may be less damaging to the structure than a few high-stress loadings, such as those encountered during landing. In a contrary manner, however, the heavy load may stretch and blunt the fatigue crack such that many small-load cycles are required to "reinitiate" it. Evaluating how materials respond to complex patterns of loading is an area of great importance to design engineers.

Since reliable fatigue data may take considerable time to generate, it is often useful to approximate fatigue behavior from properties that can be determined more quickly. Table 2-3 shows the approximate ratio of the endurance limit to ultimate tensile strength for several engineering metals. For many steels the endurance limit can be approximated by 0.5 times the ultimate tensile strength. For the nonferrous metals, the ratio is significantly lower.

TABLE 2-3. Ratio of Endurance Limit to Tensile Strength for Various Materials

Material	Ratio
Aluminum	0.38
Beryllium copper (heat-treated)	0.29
Copper, hard	0.33
Magnesium	0.38
Steel	
AISI 1035	0.46
Screw stock	0.44
AISI 4140 normalized	0.54
Wrought iron	0.63

Fatigue Failures

Metal components that fail as a result of the repeated application of load and the fatigue phenomena are commonly called *fatigue failures*. These fractures form a major part of a larger classification known as *progressive fractures*. If the fracture surface shown in Figure 2-27 is examined closely, two points of fracture initiation can be located. These points, indicated by the two arrows, often correspond to discontinuities in the form of surface cracks, sharp corners, machining marks, or even "metallurgical notches," such as an abrupt change in metal structure. With each repeated application of load, the stress at the tip of the crack exceeds the strength of the material, and the crack grows. Crack propagation continues until the remaining section of material is no longer large enough to withstand the peak load. Sudden overload fracture then occurs through the remainder of the metal. The fracture surface tends to exhibit two distinct regions: a smooth, relatively flat region where the crack was propagating by cyclic fatigue, and a coarse, ragged region of ductile overload tearing.

The smooth areas of the fracture often contain a series of crescent-shaped ridges radiating outward from the origin of the crack. Sometimes these markings may not be visible under normal examination. They may be extremely fine; they may have been

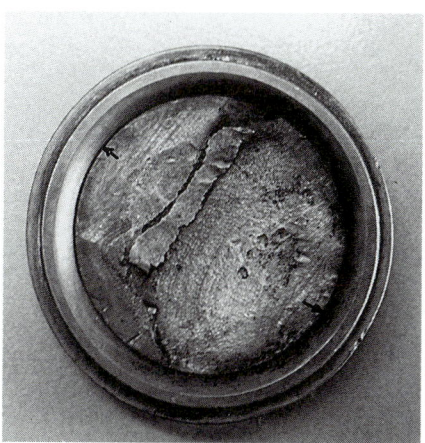

FIGURE 2-27 Progressive fracture of an axle within a ball-bearing ring, starting at two points (arrows).

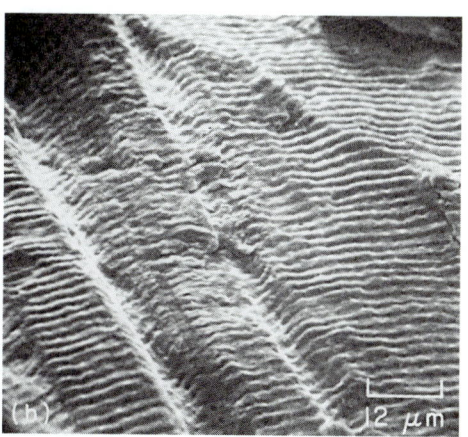

FIGURE 2-28 Fatigue fracture of AISI type 304 stainless steel viewed in a scanning electron microscope at 810×. Well-defined striations are visible. (From "Interpretation of SEM fractographs," *Metals Handbook*, Vol. 9, 8th ed., ASM International, Materials Park, Ohio, 1970, p. 70).

obliterated by a rubbing action during the compressive stage of the repeated loading; or they may be very few in number if the failure occurred after only a few cycles of loading ("low-cycle fatigue"). Electron microscope examination can frequently reveal the small parallel ridges, or *fatigue striations*, that are characteristic of fatigue failure. Figure 2-28 shows a high-magnification view of these markings.

For some fatigue fractures, the overload area may exhibit a crystalline appearance and the failure is sometimes attributed to the metal having "crystallized." As shown in Chapter 3, metals are almost always crystalline materials. The final overload fracture simply occurred in such a way that the crystalline surfaces have been revealed. The conclusion that the material crystallized is erroneous and the term is a misnomer.

Another common misnomer is to apply the term *fatigue failure* to all fractures that have the progressive failure appearance. Other progressive failure mechanisms, such as creep failure and stress–corrosion cracking, will produce the characteristic two-region fracture. In addition, the same mechanism can produce fractures with different appearances depending on the magnitude of the load, type of loading (torsion, bending, or tension), temperature, and operating environment. Correct interpretation of a metal failure takes more information than that acquired by a visual examination of a fracture surface.

A final misconception regarding fatigue failures is to assume that the failure is time dependent. The failure of materials under repeated loads below their static strengths is primarily a function of the magnitude and number of loading cycles. If the frequency of loading is increased, the time to failure should decrease proportionately. If the time does not change and appears to be the controlling feature, the failure is dominated by some environmental factor, and fatigue is a secondary component.

Temperature Effects

It cannot be overemphasized that test data used in design and engineering decisions should be obtained under conditions that best simulate those of actual service. Engineers are fre-

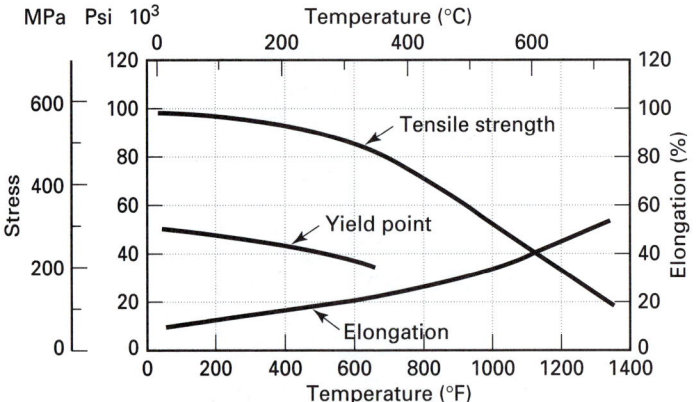

FIGURE 2-29 Some effects of temperature on the tensile properties of a medium-carbon steel.

quently confronted with the design of structures, such as aircraft, space vehicles, gas turbines, and nuclear power plants, that operate under temperatures as low as $-200°$F $(-130°$C) or as high as 2300°F (1250°C). Consequently, it is imperative that the designer know both the short- and long-range effects of temperature on the mechanical and physical properties of a material being considered for such applications. From a manufacturing viewpoint, the effects of temperature are equally important. Since numerous manufacturing processes involve the use of heat, the processing may tend to alter the properties in a favorable or unfavorable manner. Often, a material can be processed successfully, or economically, only because its properties can be changed by heating or cooling.

To a manufacturing engineer, the most important effects of temperature are those that relate to the tensile and hardness properties of a material. Figure 2-29 illustrates the changes in key properties for a medium-carbon steel. Similar effects are presented for magnesium in Figure 2-30. In general terms, an increase in temperature will induce a decrease in strength and hardness and an increase in elongation. For forming operations, heating is extremely attractive because the material is both weaker and more ductile.

Figure 2-31 shows the combined effects of temperature and strain rate (speed of testing) on the ultimate tensile strength. From this graph it can be seen clearly that the rate of

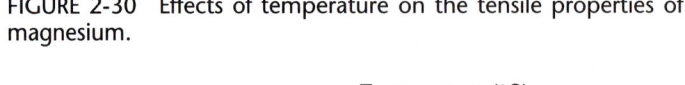

FIGURE 2-30 Effects of temperature on the tensile properties of magnesium.

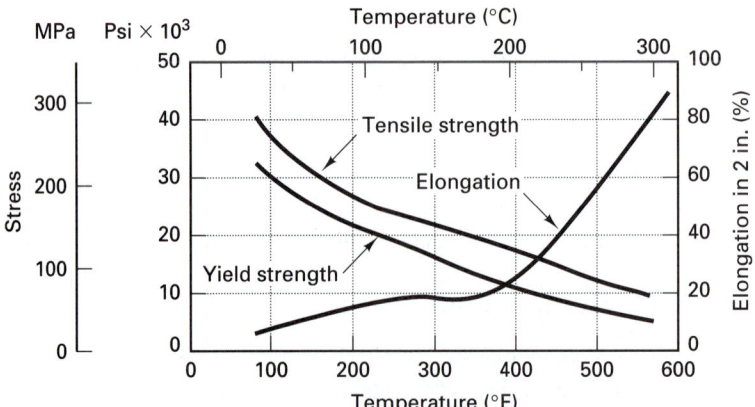

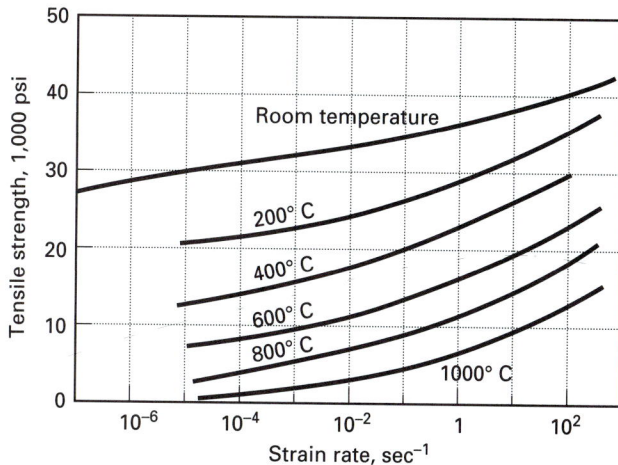

FIGURE 2-31 Effects of temperature and strain rate on the tensile strength of copper. (From A. Nadai and M.J. Manjoine, Journal of Applied Mechanics, Vol. 8, 1941, p. A82, *courtesy of ASME*)

deformation can strongly influence mechanical properties. Room-temperature standard-rate tensile test data will be of little use to an engineer concerned with the behavior of a material being hot-rolled at speeds of 5000 ft/min (1300 m/min). The effects of strain rate on the more important yield-strength value are more difficult to evaluate, but follow the same trends as tensile strength.

The effects of temperature on impact properties became the subject of intense study in the 1940s, when the increased use of welded construction led to catastrophic failures of ships, structures, and components when operating in cold environments. (*Note*: In welded construction, cracks can propagate through a weld and continue on to other segments of the structure!) Figure 2-32 shows the effect of decreasing temperature on the impact properties of two low-carbon steels. Although similar in many ways, the two steels show a distinctly different response. The steel indicated by the solid line becomes brittle at temperatures below 25°F (−4°C), while the other steel retains its fracture resistance down to −15°F (−26°C). The temperature at which the response goes from high to low energy absorption, known as the *transition temperature*, is useful in evaluating the suitability of materials for certain applications. All steels tend to exhibit the transition in impact strength when temperature is decreased, but the temperature at which it occurs varies with the material. Thus, when low temperatures are expected, special consideration should be given when selecting the material.

Creep

One of the long-term effects of elevated temperature may be a phenomenon known as *creep*. If a tensile-type specimen is subjected to a constant load at elevated temperature, it will elongate continuously until rupture occurs, even though the applied stress is below the yield strength of the material at the temperature of testing. Although the rate of elongation is small, it is sufficient to be of great importance in the design of equipment such as steam or gas turbines, power plants, and high-temperature pressure vessels that operate under loads and high temperatures for long periods of time.

If a specimen is tested under conditions of fixed load and fixed temperature, an elongation versus time plot is generated, similar to the one shown in Figure 2-33. The curve contains three distinct stages: a short-lived initial stage, a rather long second stage where the elongation rate is somewhat linear, and a short-lived third stage leading to fracture. Two significant pieces of engineering data are obtained from this curve: the rate of elongation in the second stage, or *creep rate*, and the total elapsed *time to rupture*. Tests con-

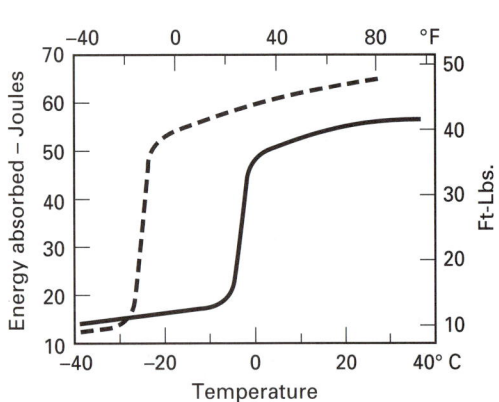

FIGURE 2-32 Effect of temperature on the impact properties of two low-carbon steels.

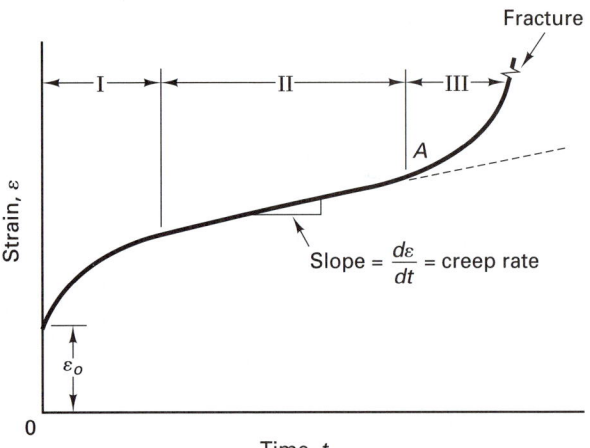

FIGURE 2-33 Creep curve for a single specimen at a fixed temperature, showing the three stages of creep and reported creep rate. Note the nonzero strain at time zero due to the initial application of the load.

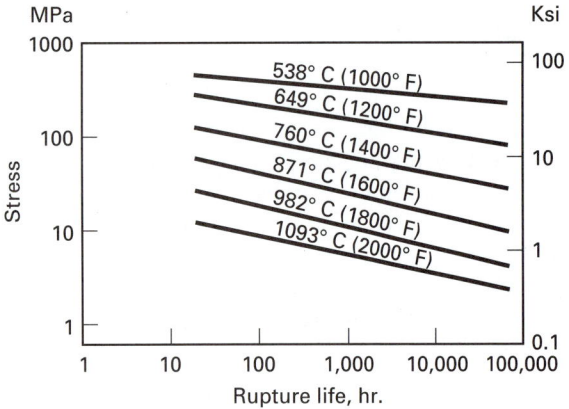

FIGURE 2-34 Stress–rupture diagram of a solution-annealed Incoloy alloy 800 (Fe–Ni–Cr alloy). *(Courtesy of Huntington Alloy Products Division, The International Nickel Company, Inc.)*

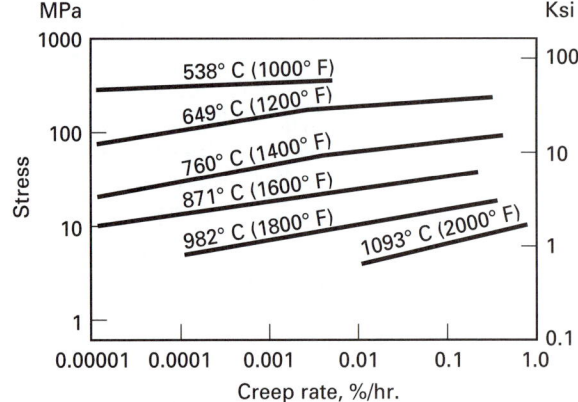

FIGURE 2-35 Creep-rate properties of a solution-annealed Incoloy alloy 800. *(Courtesy of Huntington Alloy Products Division, The International Nickel Company, Inc.)*

ducted at higher temperatures or with higher applied loads would exhibit higher creep rates and shorter rupture times.

When creep is a significant design factor, a *stress–rupture diagram* (Figure 2-34) can be a very useful engineering tool. The rupture times from a number of tests conducted at various temperatures and stresses are plotted on a single diagram to give an overall picture of material performance. In a similar manner, creep rate data can be plotted to show the effect of variations in temperature and stress. Figure 2-35 presents such a diagram.

Machinability, Formability, and Weldability

While many individuals assume that these terms refer to specific material properties, they actually refer to the way a given material responds to specific processing techniques and are therefore quite difficult to define. *Machinability*, for example, depends not only on the material being machined but also on the specific process and the aspects of the process that are of greatest interest. Machinability ratings are generally based on relative tool life data. In some cases, one may be interested in how easy or fast a metal can be cut, irrespective of the tool life or the resulting surface finish. In another application, surface finish may be of prime importance. For some processes, the formation of fine chips may be a desirable feature. Thus the term *machinability* often means different things to different people and frequently involves multiple properties of a material acting in unison.

In a similar manner, *malleability, workability,* and *formability* all refer to a material's suitability for plastic deformation processing. However, materials can behave differently at different temperatures. A material with good "hot formability" may behave poorly when deformed at room temperature. Some materials that flow nicely at low deformation speeds behave in a brittle manner when loaded at rapid rates. Thus formability needs to be evaluated for a specific combination of material, process, and process conditions, and the results cannot be extrapolated to other processes or process conditions. Similarly, the *weldability* of a material will depend on the specific welding or joining process being considered.

Fracture Toughness and the Fracture Mechanics Approach

A discussion of the mechanical properties of materials would not be complete without mention of the many tests and design concepts based on the fracture mechanics approach. Instead of treating test specimens as flaw-free materials, fracture mechanics begins with the premise that *all materials contain flaws* or defects of some given size. These may be *material defects*, such as pores, cracks, or inclusions; *manufacturing defects*, in the form of machining tool marks, arc strikes, or contact damage to external surfaces; or *design defects*, such as abrupt section changes, excessively small fillet radii, and holes. When the specimen is then loaded, the applied stresses are amplified or intensified in the vicinity of these defects, leading to accelerated failure or failure under unexpected conditions.

Fracture mechanics seeks to identify the conditions under which a defect will grow or propagate to failure and, if possible, the rate of crack or defect growth. The methods concentrate on three principal quantities: (1) the size of the largest or most critical flaw, usually denoted as a; (2) the applied stress, denoted by σ; and (3) the *fracture toughness*, a quantity that describes the resistance of a material to fracture or crack growth, which is usually denoted by K with subscripts to signify the conditions of testing. Equations have been developed that relate these three quantities (at the onset of crack growth or propagation) for various specimen geometries, flaw locations, and flaw orientations. If nondestructive testing or quality control checks (such as those described in Chapter 11) have been applied, the size of the largest flaw that could go undetected is often known. By mathematically placing this worst possible flaw in the worst possible location and orientation, and coupling this with the largest applied stress for that location, the designer can then determine the value of fracture toughness that would be necessary to prevent the flaw from propagating during service. In a reverse manner, if the material and stress conditions were defined, the size of the maximum permissible flaw could be computed. Inspection conditions could then be selected to assure that flaws greater than this magnitude are cause for product rejection. In a third approach, if a component is found to have a significant flaw and the material is known, the maximum operating stress can be determined that will assure no further growth of that flaw.

According to the philosophy of fracture mechanics, all materials contain flaws or defects, but these defects can be either dormant or growing. Dormant defects are permissible and the goal of fracture mechanics is to define the conditions under which a given defect will remain dormant. In a 1983 study, it was determined that fracture of materials costs over \$119 billion per year in the United States alone. Over 80% of this cost was attributed to excessive measures for fracture prevention, such as overdesign, excessive inspection, and the use of premium-quality materials. Another study has concluded that over 90% of all dynamic failures of materials are in some way attributable to fatigue. In contrast to the "standard" method of fatigue testing, where polished specimens are cycled and the lifetime consists of both crack initiation and crack propagation, fracture mechanics focuses on the growth rate of an already defective material (containing a crack or flaw of known size and shape). Figure 2-36 shows the crack growth rate (change in size per loading cycle denoted as *da/dN*) plotted as a function of the fracture mechanics parameter, ΔK. The results of such testing enable a far more realistic guarantee of minimum service life.

Fracture mechanics is a truly integrated blend of design (applied stresses), inspection (flaw-size determination), and materials (fracture toughness). The approach has proven valuable in many areas where fracture could be catastrophic and has shown great refinement and increased acceptance in recent years. The 1983 fracture study concluded that over 29% of the \$119 billion cost of fracture (\$34.5 billion per year in 1983 dollars) could be saved with proper application of existing technology, such as fracture mechanics.

■ 2.4 PHYSICAL PROPERTIES

For some engineering applications, the physical properties of an engineering material may be even more important than the mechanical. For this reason it is important that we become familiar with the various types of physical properties that may be considered.

In addition to the previously discussed responses of material to variations in temperature, there are three more *thermal properties* that are worthy of consideration. The *heat capacity* or *specific heat* of a material is a measure of the amount of energy that must be imparted or extracted to produce a 1° change in temperature. This would be important in processes such as casting, where heat must be extracted rapidly to promote solidification, and heat treatment, where quantities of material are heated and cooled. *Thermal conductivity* measures the rate at which heat can be transported or conducted through a material. While this is often tabulated in reference texts, it is helpful to remember that for metals, thermal conductivity is directly proportional to electrical conductivity. Thus materials such as copper, gold, and aluminum that have good electrical conductivity will also be good transporters of thermal energy. *Thermal expansion* is the final property of significance. Most materials expand upon heating and contract upon cooling, but the degree of expansion or contraction will vary with the material. For components that are machined at room temperature but put in service at elevated temperatures, or castings that solidify at elevated temperatures and then cool, corrections must be incorporated in the manufacture to compensate for the dimensional changes.

Electrical conductivity or resistivity is often an important design consideration and should be evaluated in terms of both the base material property and how that property will vary with different levels of purity or changes in temperature.

From the standpoint of *magnetic response*, materials are often classified as diamagnetic, paramagnetic, ferromagnetic, antiferromagnetic, or ferrimagnetic. These refer to the way in which the material responds to an applied magnetic field. Terms such as *saturation strength, remanence*, and *magnetic hardness or softness*, refer to the strength, duration, and nature of this response, and can be used to compare and evaluate materials.

Other physical properties of possible importance include *weight or density, melting and boiling points*, and the various *optical properties*, such as the ability to transmit, absorb, or reflect light or other electromagnetic radiation.

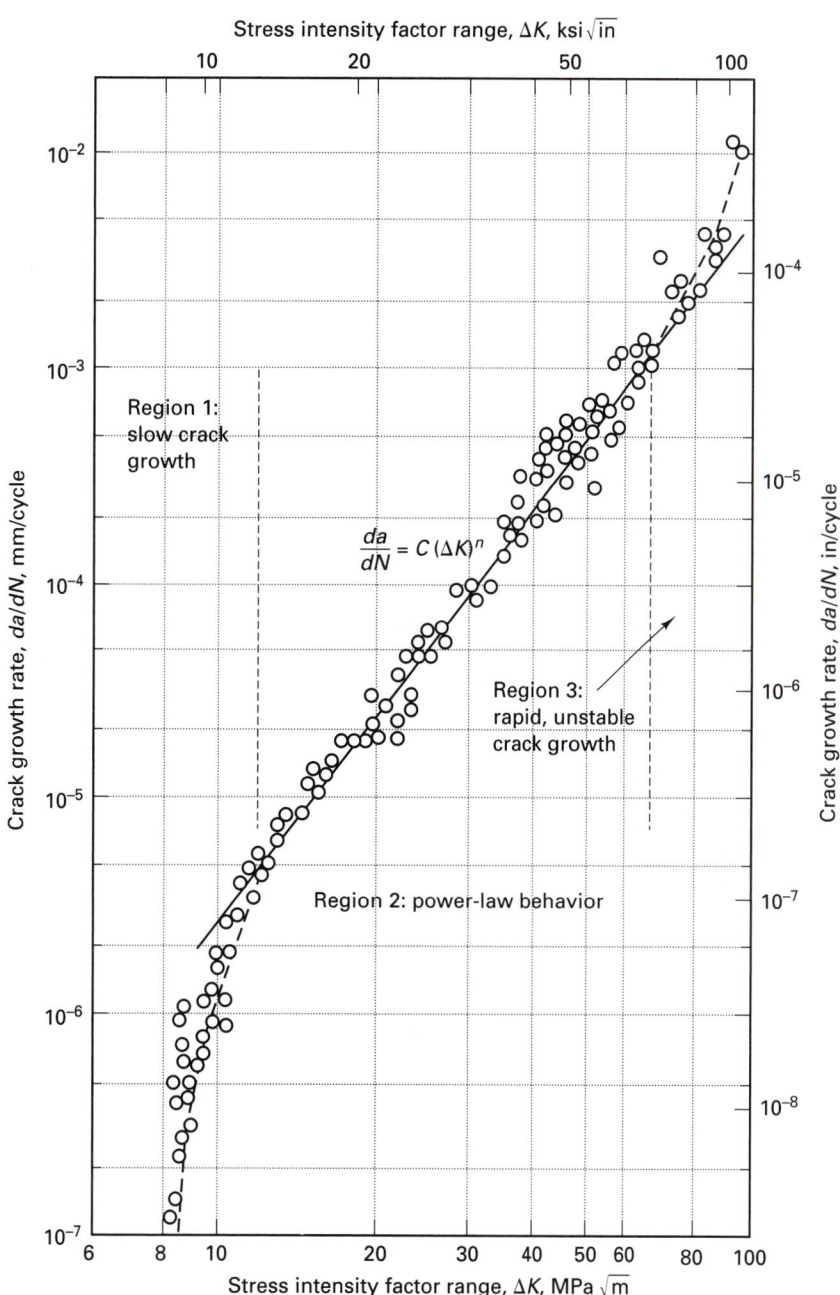

FIGURE 2-36 Plot of the fatigue crack growth rate for a typical steel using the fracture mechanics approach. Similar shape curves are obtained for most engineering metals. *(Courtesy of ASM International.)*

■ KEY WORDS

Brinell hardness
brittleness
charpy test
creep
creep rate
elongation
engineering strain
engineering stress
fatigue striations
fracture toughness
hardness
heat capacity
impact test
Knoop hardness

mechanical properties
metal fatigue
modulus of elasticity (Young's
  modulus)
offset yield strength
percent reduction in area
physical properties
Rockwell hardness
S–N curve
specific heat
static properties
strain
strain hardening
strain-hardening coefficient

stress
stress–rupture diagram
tensile test
thermal conductivity
thermal expansion
time to rupture
toughness
transition temperature
true strain
true stress
ultimate tensile strength
yield point

■ REVIEW QUESTIONS

1. What are some properties commonly associated with metallic materials?
2. What are some of the common physical properties of materials?
3. Why should caution be exercised when applying the results of the standard mechanical property tests?
4. What are the standard units of reporting or measuring stress and strain in the English system? In the metric system?
5. What are some important properties that relate to the elastic response of materials?
6. What are some tensile test properties that relate to the elastic-to-plastic transition in a material?
7. What are some tensile test properties that relate to the plastic deformation behavior of materials?
8. Describe two methods to utilize the tensile test as a means of evaluating the ductility of a material.
9. Why might the uniform elongation or percent elongation prior to necking be a more meaningful measure of useful ductility?
10. Is a brittle material necessarily weak?
11. What is the toughness of a material?
12. What is the difference between true stress and engineering stress? True strain and engineering strain?
13. What is strain hardening or work hardening? How might this phenomenon be measured or reported? How might it be used in manufacturing?
14. What are some of the various material

characteristics or responses that have been associated with the term *hardness*?
15. What are some of the limitations of the Brinell hardness test?
16. Both the Brinell and Rockwell hardness tests are penetration tests. What would you cite as the major difference between the two methods?
17. Under what conditions might one want to use a microhardness test?
18. Why might there be a lack of agreement when materials are compared using various types of hardness tests?
19. Describe the relationship between penetration hardness and the ultimate tensile strength.
20. Why is an accurate compression test more difficult to conduct than an accurate tensile test?
21. What is the difference between static and dynamic loading? Describe several different types of dynamic loading.
22. What are the two most common bending impact tests? How are the specimens supported and loaded during impact?
23. Why should a designer use extreme caution when applying impact test data for design purposes?
24. Under what conditions might a metal fracture when exposed to stresses that lie below its yield strength?
25. Fatigue strength and endurance limit are two terms that are derived from S–N diagrams. Define these terms and describe how they can be determined from a diagram.

26. What are some of the factors that can alter the fatigue lifetime or fatigue behavior of a material?
27. How might the endurance limit of a steel be approximated without requiring the long time for extensive fatigue testing?
28. What are some features that may be responsible for the initiation of a fatigue crack?
29. Why is it important for a designer or engineer to know how a material's properties will change with temperature?
30. Why should one use caution when employing a steel at low (below zero Fahrenheit) temperatures?

31. What are some evaluation tools and quantities that can be used to assess the long-term effect of elevated temperature on an engineering material?
32. During which stage of a creep test is the creep rate measured?
33. Why is there not a single standard means of assessing characteristics such as machinability, formability, or weldability?
34. What is the basic premise of the fracture mechanics approach to testing and design?
35. What three principal quantities does fracture mechanics attempt to relate?
36. What are the three primary thermal properties of a material, and what do they measure?

■ PROBLEMS

1. Select a product or component for which physical properties are more important than mechanical properties.
   (**a**) Describe the product or component and its function.
   (**b**) What are the most important properties or characteristics?
   (**c**) What are the secondary properties or characteristics that would also be desirable?
2. Repeat Problem 1 for a product or component whose dominant required properties are of a static mechanical nature.
3. Repeat Problem 1 for a product or component whose dominant requirements are dynamic mechanical properties.

Chapter 2 CASE STUDY

separating mixed materials

B ecause of the amount of handling that occurs during material production, within warehouses, and during manufacturing operations, along with the handling of loading, unloading and shipping, material mix-ups and mixed materials are not an uncommon occurrence. Mixed materials also occur when industrial scrap is collected, or when discarded products are used as new raw material through recycling.

A case of mixed materials has been identified, and you have been called upon to devise a means of separation. Assume that you have equipment to perform each of the tests described in this chapter (as well as access to the full spectrum of household and department store items and a small machine shop). For each of the material combinations below, determine a procedure that could accomplish such a separation. Use standard data-source references to help you identify distinguishable properties.

  a. Hot-rolled bars of AISI 1020 and 1040 steel
  b. Stainless steel sheets of Type 430 ferritic stainless and Type 316 austenitic stainless.
  c. 6061-T6 aluminum and AZ91 magnesium that have become mixed in a batch of machine shop scrap.
  d. Transparent bottles of polyethylene and polypropylene (both thermoplastic polymers) that have been collected for recycling.

CHAPTER 3

NATURE OF METALS AND ALLOYS

3.1	STRUCTURE–PROPERTY RELATIONSHIP	3.12	STRAIN HARDENING OR WORK HARDENING
3.2	ATOMIC STRUCTURE	3.13	PLASTIC DEFORMATION IN POLYCRYSTALLINE METALS
3.3	ATOMIC BONDS		
3.4	SECONDARY BONDS		
3.5	INTERATOMIC DISTANCES AND SIZE OF ATOMS	3.14	GRAIN DEFORMATION AND ANISOTROPIC PROPERTIES
3.6	ATOM ARRANGEMENTS IN MATERIALS	3.15	FRACTURE OF METALS
3.7	CRYSTAL STRUCTURES OF METALS	3.16	COLD WORKING, RECRYSTALLIZATION, AND HOT WORKING
3.8	DEVELOPMENT OF A GRAIN STRUCTURE	3.17	GRAIN GROWTH
		3.18	ALLOYS
3.9	ELASTIC DEFORMATION OF A SINGLE CRYSTAL	3.19	ALLOY TYPES
		3.20	ATOMIC STRUCTURE AND ELECTRICAL PROPERTIES
3.10	PLASTIC DEFORMATION OF A SINGLE CRYSTAL		
3.11	DISLOCATION THEORY OF SLIPPAGE	Case Study:	WINDOW FRAME MATERIALS AND DESIGN

■ 3.1 STRUCTURE–PROPERTY RELATIONSHIP

As discussed in Chapter 2, the success or failure of many engineering activities depends on the selection of engineering materials whose properties match the specific requirements of the application. Primitive cultures were often limited to the naturally occurring materials in their environment. If the match was not a good one, compromises were required. As civilization developed, the range of engineering materials expanded. Materials could now be processed and their properties could be altered and possibly enhanced. The alloying or heat treatment of metals and the firing of ceramics are examples of techniques that can substantially alter the properties of a material. Fewer compromises were required and enhanced design possibilities emerged. Products became more sophisticated. While the early successes in altering materials were largely the result of trial and error, we now recognize that the engineering *properties* of a material are a direct result of the *structure* of that material. Changes in the properties, therefore, are the direct result of changes in the material structure.

Since all materials are composed of the same basic components—*protons, neutrons, and electrons*—it is amazing that so many different materials exist with such widely varying properties. This variation is explained by the many possible combinations of these units in a macroscopic assembly. The subatomic particles, listed above, combine in different arrangements to form the various elemental *atoms*, each having a *nucleus* of protons and neutrons surrounded by the proper number of electrons to maintain charge neutrality.

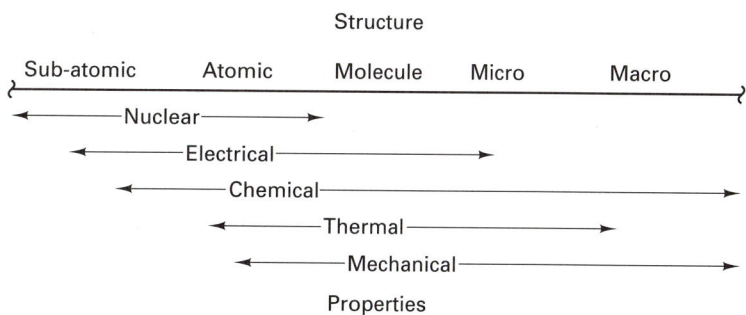

FIGURE 3-1 General relationship of the structural level to various engineering properties.

The arrangement of the electrons surrounding the nucleus affects the electrical, magnetic, thermal, and optical properties as well as how the atoms bond to one another. Atomic bonding then produces a higher level of structure, which may be in the form of a molecule, crystal, or amorphous aggregate. This structure and the imperfections that may be present have a profound effect on mechanical properties. The size, shape, and arrangement of multiple crystals, or the mixture of different structures within a material, in turn produces another level of structure: a microscopic-scale structure, or *microstructure*. Variations in microstructure also affect the material properties.

As a result of the ability to control structures through processing, and the ability to develop new structures through techniques such as composite materials, engineers now have at their disposal a wide variety of materials with an almost unlimited range of properties. The properties of these materials depend on all levels of their structure, from subatomic to macroscopic (Figure 3-1). Therefore, it is important for the engineer to understand the entire structure spectrum and the way the basic structure and changes in that structure will affect properties.

■ 3.2 ATOMIC STRUCTURE

Experiments have revealed that atoms consist of a relatively dense nucleus composed of positively charged *protons* and neutral particles of nearly identical mass, known as *neutrons*. Surrounding the nucleus are the negatively charged *electrons*, which have only 1/1839 times the mass of a neutron and appear in numbers equal to the protons, to maintain a net charge balance. Distinct groupings of these basic particles produce the known elements, ranging from the relatively simple hydrogen atom to unstable transuranium atoms over 250 times as heavy. Except for density and specific heat, however, the weight of atoms has relatively little influence on engineering properties.

The light electrons that surround the nucleus play a far more significant role in determining material properties. Again, experiments reveal that the electrons are arranged in a characteristic structure consisting of shells and subshells, each possessing a distinctive energy. Upon absorbing a small amount of energy, an electron can jump to a higher-energy shell farther from the nucleus. The reverse jump can also occur with the concurrent release of a distinct amount, or *quantum*, of energy.

Each of the various shells and subshells can contain only a limited number of electrons. The first shell, nearest the nucleus, can contain only two. The second shell can contain eight, and the third, 32. Each shell and subshell is most stable when it is completely filled. For atoms containing electrons in the third shell and beyond, however, relative stability is achieved with eight electrons in the outermost layer (subshell).

If, in its outermost layer, a normal atom has slightly less than the number of electrons required for stability, such as seven in the third shell, it will readily accept an electron from another source. It will then have one electron more than the number of protons and becomes a negatively charged atom, or *negative ion*. Depending on the number of additional electrons, ions can have negative charges of 1, 2, 3, or more. Conversely, if an atom has a slight excess of electrons beyond the number required for stability (such as sodium, with one electron in the third shell), it will readily give up the excess electron and become a *positive ion*. The remaining electrons become more strongly bound, so that the removal of electrons becomes progressively more difficult.

The number of electrons surrounding the nucleus of a neutral atom is called the *atomic number*. More important, however, are those electrons in the outermost shell or subshell, known as *valence electrons*. These are influential in determining chemical properties, electrical conductivity, some mechanical properties, the nature of interatomic bonding, atom size, and optical characteristics. Elements with similar electron configurations in their outer shells tend to have similar properties.

■ 3.3 ATOMIC BONDS

Atoms are rarely found as free and independent units; they are usually linked or bonded to other atoms in some manner as a result of interatomic forces. The electron structure of the atoms plays a strong role in determining the nature of the bond.

There are generally three types of primary bonds that are recognized, the simplest of which is the *ionic bond*. When more than one type of atom is present, the outermost electrons can break free from atoms with excesses in their valence shell, producing positive ions, and then be donated to other atoms to complete their deficient outer shell, transforming them into negative ions. The positive and negative ions have a natural attraction for each other, producing a strong bonding force. Figure 3-2 presents a crude schematic of the ionic process for sodium and chlorine. In reality, however, the atoms do not unite in simple pairs. All positively charged atoms attract all negatively charged atoms. Thus sodium ions will attempt to surround themselves with negative chlorine ions, and the chlorine ions will attempt to surround themselves with positive sodium ions. The attraction is equal in all directions and results in three-dimensional structures (Figure 3-3). For stability within

FIGURE 3-2   Ionization of sodium and chlorine, producing stable outer shells by electron transfer.

FIGURE 3-3   Three-dimensional structure of the sodium chloride crystal. Note how the various ions are surrounded by ions of the opposite charge.

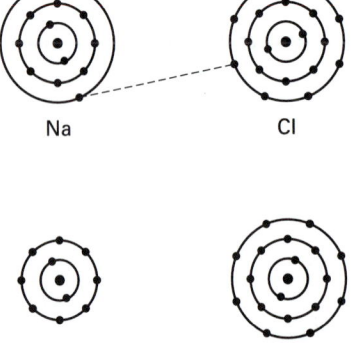

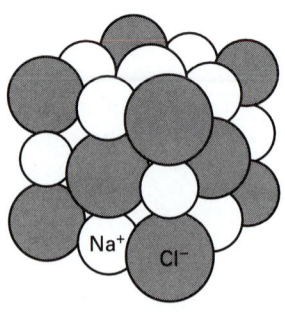

the structure, total charge neutrality must be maintained; equal numbers of positive and negative charges must be present in each neighborhood. General characteristics of materials joined by ionic bonds include moderate to high strength, high hardness, brittleness, high melting point, and low electrical conductivity (electrons are captive to atoms, so charge transport requires atom—or ion—movement).

A second type of primary bond is the *covalent* type. Here the atoms in the assembly find it impossible to produce completed shells by electron transfer but achieve the same goal through electron sharing. Adjacent atoms share outer-shell electrons so that each achieves a stable electron configuration. The shared (negatively charged) electrons locate between the positive nuclei, forming a positive–negative–positive bonding link. Figure 3-4 illustrates this type of bond for a pair of chlorine atoms, each of which contain seven electrons in the valence shell. The result is a stable two-atom *molecule*. Stable molecules can also form from the sharing of more than one electron from each atom, as in the case of nitrogen (Figure 3-5a). The atoms in the assembly need not be identical (as in HF, Figure 3-5b), the sharing does not have to be equal, and a single atom can share electrons with more than one other atom. For atoms such as carbon, with four valence electrons, one atom may share its valence electrons with each of four neighboring carbon atoms. The resulting structure then becomes a three-dimensional network of bonded atoms (Figure 3-5c) rather than a finite, well-defined molecule. Like the ionic bond, the covalent bond tends to produce materials with high strength and high melting point. Atom movement within the framework material (plastic deformation) requires the breaking of discrete bonds, thereby making the material characteristically brittle. Electrical conductivity depends on bond strength, ranging from conductive tin (weak covalent bonding), through semiconducting silicon and germanium, to insulating diamond (carbon). Engineering materials possessing ionic or covalent bonds tend to be ceramic (refractories or abrasives) or polymeric in nature.

A third type of primary bond can form when a complete outer shell cannot be formed by either electron transfer or electron sharing. This bond is known as the *metallic*

FIGURE 3-4 Formation of a chlorine molecule by a covalent bond.

FIGURE 3-5 Examples of covalent bonding in (a) N_2, (b) HF, and (c) diamond.

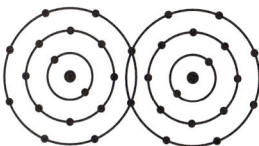

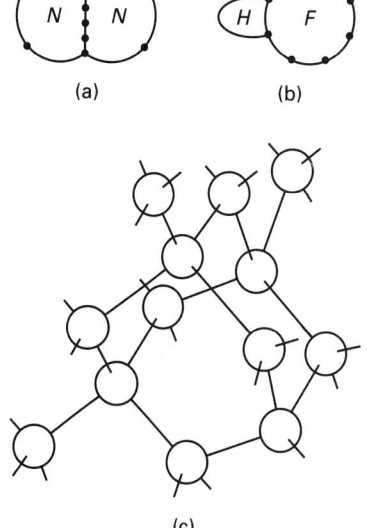

(a)   (b)

(c)

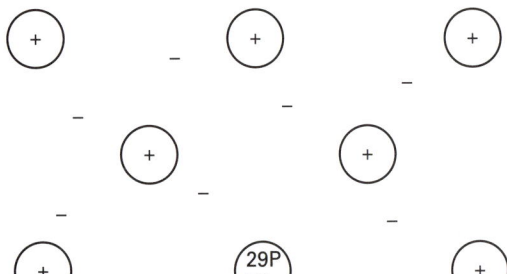

FIGURE 3-6   Schematic of the metallic bond showing the positive ions and associated electron cloud for copper. Each positive-charged ion contains a nucleus with 29 protons and stable electron shells containing 28 electrons.

bond. If there are only a few valence electrons (one, two, or three) in each of the atoms in an aggregate, these electrons can easily be removed while the remainder are held firmly to the nucleus. The resulting structure (Figure 3-6) is one of positive ions (nucleus and inner, nonvalence electrons) surrounded by a wandering array of universally-shared valence electrons, sometimes referred to as an *electron cloud* or *electron gas*. These highly-mobile, "free" electrons account for the high electrical and thermal conductivity values as well as the opaque property (free electrons can absorb the discrete energies of light radiation) observed in metals. Moreover, they provide the "cement" required for the positive–negative–positive attractions that result in bonding. Bond strength, and therefore material strength, varies over a wide range. More significant, however, is the observation that the positive ions can move within the structure without the breaking of discrete bonds. Materials bonded by metallic bonds can therefore be deformed by atom-movement mechanisms and produce a deformed material that is every bit as strong as the original. This phenomenon is the basis of metal plasticity, ductility, and many of the shaping processes used in the fabrication of metal products.

■ 3.4 SECONDARY BONDS

Weak or secondary bonds, known as *van der Waals forces*, can form between molecules that possess a nonsymmetrical distribution of charge. Some molecules, such as hydrogen fluoride and water, can be viewed as electric dipoles in that certain portions of the molecule tend to be more positive or negative than others (an effect referred to as *polarization*). The negative part of one molecule tends to attract the positive part of another, forming a weak bond. Van der Waals forces contribute to the mechanical properties of a number of molecular polymers, such as polyethylene and polyvinyl chloride (PVC).

Another form of weak bond results from the momentary polarization or charge imbalance caused by the random movement of electrons. The weak attractive force that results from random polarization is called the *dispersion effect*.

A third type of weak bond is the *hydrogen bridge*. Here, a small hydrogen nucleus is attracted to the negative electrons in two atoms simultaneously and forms a link between them. Although these bonds play a significant role in biological systems, they are rarely of engineering significance.

■ 3.5 INTERATOMIC DISTANCES AND SIZE OF ATOMS

Since the space occupied by the electron shells of an atom is far greater than the size of the actual electrons, much of the "volume" of an atom is unoccupied space. Atoms are not solid bodies, and the concept of atom size becomes somewhat nebulous. The various bonding forces tend to pull atoms together, but repelling forces exist between adjacent positive nuclei. Some equilibrium separation distance exists where the forces of attraction and repulsion are equal.

If atoms are modeled as rigid spheres, the equilibrium separation distance between neighboring atoms can be taken as the sum of the atomic radii, and atoms can be assigned a distinct size. The atomic radius is not a constant, however. Added thermal energy causes the atoms to vibrate about their equilibrium positions. Since the repulsive force as atoms approach is much greater than the attractive forces when they separate, the vibrations or oscillations are not symmetrical. The average location of the atoms is now at a larger-than-normal separation distance, and the result is the macroscopically observed thermal expansion of the material. Removing electrons from the outer shell will decrease the atomic radius, and adding electrons will increase it. Consequently, a negative ion is larger than the base atom, and a positive ion is smaller. Atomic radius also changes with the number of adjacent or nearest-neighbor atoms in a crystalline arrangement. With more neighbors, there is less attraction to any single neighbor atom, and the interatomic distance is increased. For example, iron has an *atomic radius* of 1.241 Å in a crystal structure where each atom has eight neighbors (body-centered cubic) and a radius of 1.269 Å in a structure with 12 adjacent neighbors (face-centered cubic).

■ 3.6 ATOM ARRANGEMENTS IN MATERIALS

As atoms bond together to form aggregates, we find that the particular arrangement of the atoms has a significant effect on the material properties. Depending on the manner of atomic grouping, materials are classified as having *molecular structures*, *crystal structures*, or *amorphous structures*.

Molecular structures have a distinct number of atoms that are held together by primary bonds. They have only weak attraction, however, to other similar groupings. Typical examples of molecules include O_2, H_2O, and C_2H_4 (ethylene). Each molecule is free to act more or less independently, so these materials exhibit relatively low melting and boiling points. Molecular materials tend to be weak, since the molecules can move easily with respect to one another. Upon changes of state from solid to liquid or liquid to gas, the molecules remain as distinct entities.

Solid metals and most minerals assume a crystalline-type structure. Here the atoms are arranged in a regular geometric array known as a *lattice*. Lattices are describable through a unit building block that is essentially repeated throughout space in a repetitive manner. These blocks are known as *unit cells*. Crystalline structures are discussed more fully in the coming section.

In an amorphous structure, such as glass, the atoms have a certain degree of local order (arrangement with respect to neighbors), but when viewed as an aggregate, lack the periodically ordered arrangement of atoms characteristic of a crystalline solid.

■ 3.7 CRYSTAL STRUCTURES OF METALS

From a manufacturing viewpoint, metals are an extremely important class of materials. They are frequently the material being processed, and often form the tool and machinery performing the processing. More than 50 of the known chemical elements are classified as metals, and about 40 have commercial importance. These are characterized by the metallic bond and possess the distinguishing characteristics of strength, good electrical and thermal conductivity, luster, the ability to be plastically deformed to a fair degree without fracturing, and a relatively high specific gravity (density) compared with nonmetals. The fact that some metals possess properties different from the general pattern simply expands their engineering utility.

When metals solidify, they assume a crystalline structure; that is, the atoms arrange themselves in a geometric lattice. Many metals exist in only one lattice form. Some, however, can exist in the solid state in two or more lattice forms, the particular form depending on the conditions of temperature and pressure. Such metals are said to be *allotropic* or *polymorphic*, and the change from one lattice form to another is called an *allotropic transformation*. The most notable example of such a metal is iron, where the allotropic change makes it possible for heat-treating procedures to produce a wide range of final properties. It is largely because of its allotropy that iron has become the basis of our most important alloys.

There are 14 basic types of crystal structures (lattices) that are possible. Fortunately, however, nearly all of the commercially important metals solidify into one of three types of lattice: body-centered cubic, face-centered cubic, or hexagonal close-packed. Table 3.1 lists the room temperature structure for a number of common metals. Figure 3-7 compares the common structures with one another, as well as to the easily visualized, but rarely observed, simple cubic structure.

TABLE 3-1.   Types of Lattices of Common Metals at Room Temperature

Metal	Lattice Type
Aluminum	Face-centered cubic
Copper	Face-centered cubic
Gold	Face-centered cubic
Iron	Body-centered cubic
Lead	Face-centered cubic
Magnesium	Hexagonal
Silver	Face-centered cubic
Tin	Body-centered tetragonal
Titanium	Hexagonal

As a starting point, consider the *simple cubic* structure of Figure 3-7a, which can be constructed by placing single atoms on all corners of a cube and then linking identical cube units together. Assuming that the atoms are rigid spheres with atomic radii touching one another, computation reveals that only 52% of available space is occupied. Each atom is in direct contact with only six neighbors (plus and minus directions along the *x*, *y*, and *z* axes). Both of these observations are unfavorable to the metallic bond, where atoms desire the greatest number of nearest neighbors and high-efficiency packing.

The largest region of open space is in the volumetric center of the cube, where a sphere of 0.732 times the atom diameter could be inserted.[*] If the cube is expanded somewhat to allow for the insertion of an entire atom in the center, the *body-centered-cubic* (bcc) structure results (Figure 3-7b). Each atom now has eight nearest neighbors, and 68% of the space is occupied. This structure is more favorable to metals and is observed in iron, chromium, manganese, and the other metals listed in Figure 3-7b.

---

[*]The diagonal of a cube is equal to $\sqrt{3}$ times the length of the cube edge, and the cube edge is here equal to two atomic radii or one atomic diameter. Thus the diagonal is equal to 1.732 times the atom diameter and is made up of an atomic radius, open space, and another atomic radius. Since two radii equal one diameter, the open space must be equal in size to 0.732 times the atomic diameter.

	Lattice structure	Unit cell schematic	Ping-pong ball model	Number of nearest neighbors	Packing efficiency	Typical metals
a	Simple cubic	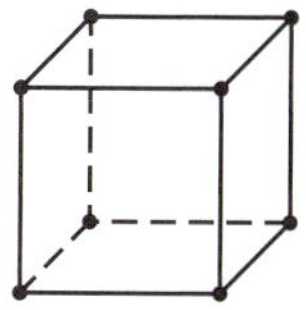	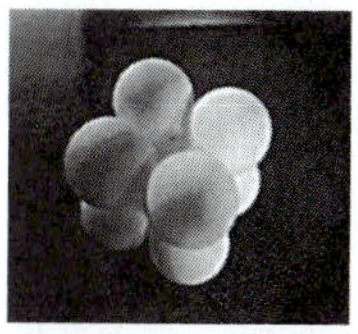	6	52%	None
b	Body-centered cubic	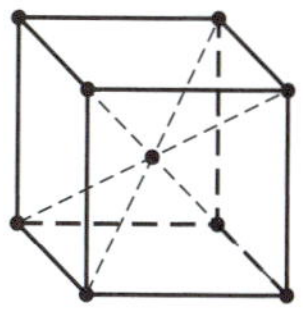		8	68%	Fe, Cr, Mn, Cb, W, Ta, Ti, V, Na, K
c	Face-centered cubic	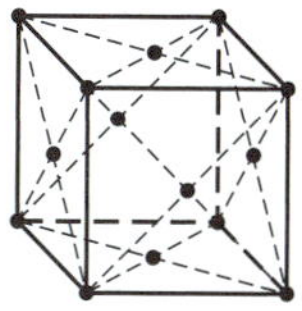		12	74%	Fe, Al, Cu, Ni, Ca, Au, Ag, Pb, Pt
d	Hexagonal close-packed	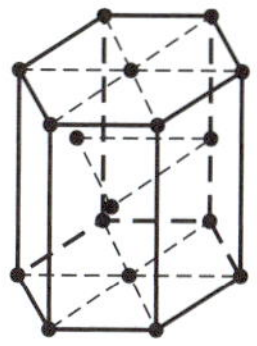	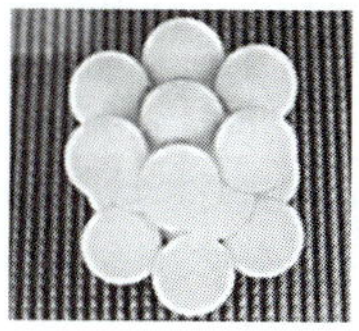	12	74%	Be, Cd, Mg, Zn, Zr

FIGURE 3-7   Comparison of crystal structures: simple cubic, body-centered cubic, face-centered cubic, and hexagonal close packed.

If Ping-Pong balls, used to simulate atoms, were placed in a box and agitated until a stable arrangement were produced, we would find that structure to consist of layered *close-packed planes*, where each plane has the configuration of Figure 3-8. Two different structures can result, depending on the sequence in which the various layers are stacked. Both, however, are identical in the number of nearest neighbors (12—six within the plane and three from each of the layers above and below) and the efficiency of occupying space (74%).

One of these sequences produces a structure that can also be viewed as an expanded cube with an atom inserted in the center of each of the six cube faces. This is the *face-centered-cubic* (fcc) structure shown in Figure 3-7c. It is the preferred structure for many of the engineering metals and tends to provide exceptionally high formability (the ability to be plastically deformed without fracture).

The other sequence of stacking produces a structure known as *hexagonal close-packed* (hcp), where the individual close-packed planes can be clearly identified (Figure 3-7d). Metals having this structure tend to have poor formability and often require special processing procedures.

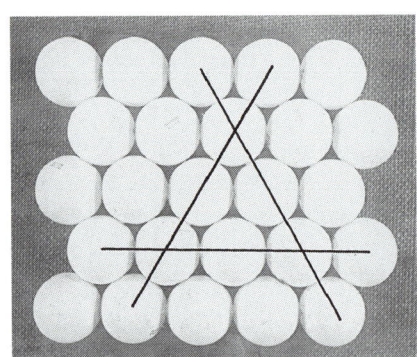

FIGURE 3-8 Close-packed atomic plane showing three directions of closest packing.

■ 3.8 DEVELOPMENT OF A GRAIN STRUCTURE

When a metal solidifies, a small particle of solid forms from the liquid with a lattice structure characteristic of the given material. This particle then acts like a *seed* or *nucleus* and grows as other atoms in the vicinity attach themselves. The basic crystalline unit is repeated throughout space, as illustrated in the body-centered cubic examples of Figure 3-9.

In actual solidification, many nuclei form independently at various locations throughout the liquid and have random orientations with respect to one another. Each then grows until it begins to interfere with its neighbors, as illustrated in two dimensions in Figure 3-10. Since adjacent lattice structures have different alignments or orientations, growth cannot produce a single continuous structure. The small, continuous volumes of solid are known as *crystals* or *grains*, and the surfaces that divide them (i.e., the surfaces of crystalline discontinuity) are known as *grain boundaries*. The process by which a grain structure is produced upon solidification is one of *nucleation and growth*.

Grains are the smallest of the structural units in a metal that are observable with ordinary light microscopy. If a piece of metal is polished to mirror finish with a series of abrasives and then exposed to an attacking chemical for a short time (etched), the grain structure is revealed. The atoms in the grain boundaries are more loosely bonded and tend

FIGURE 3-9 Growth of crystals to produce an extended lattice: (a) line schematic; (b) Ping-Pong ball model.

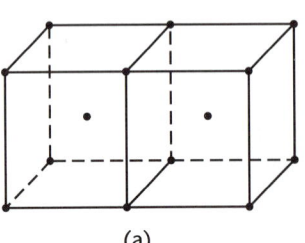

(a)

(b)

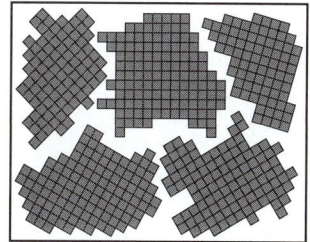 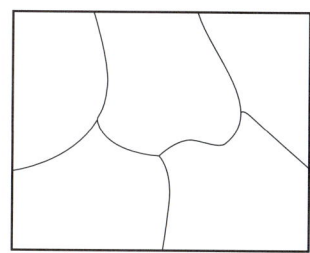

FIGURE 3-10   Schematic representation of the growth of crystals to produce a polycrystalline material.

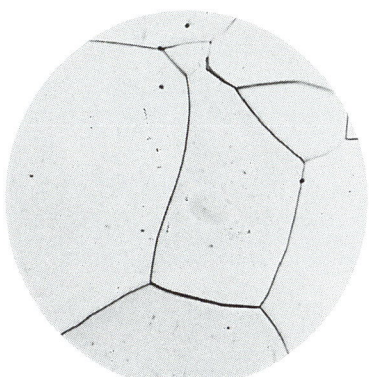

FIGURE 3-11   Photomicrograph of alpha ferrite (essentially pure iron) showing grain boundaries; 1000×. *(Courtesy of USX Corporation.)*

to react with the chemical more readily than those that are part of the grain interior. When viewed under reflected light, the attacked boundaries appear dark compared to the relatively unaffected (still flat) grains (Figure 3-11). Occasionally, grains are large enough to be seen by the unaided eye, as with some galvanized steels, but usually magnification is required.

The number and size of the grains in a metal vary with the rate of nucleation and the rate of growth. The greater the nucleation rate, the smaller the resulting grains. Conversely, the greater the rate of growth, the larger the grain. Because the resulting *grain structure* will influence certain mechanical and physical properties, it is an important property for an engineer to both control and specify. One means of specification is through the *ASTM (*American Society for Testing and Materials) *grain size number*, defined as

$$N = 2^{n-1}$$

where N is the number of grains per square inch visible in a prepared specimen at 100X and n is the ASTM grain-size number. Low ASTM numbers mean a few massive grains; high numbers refer to materials with many small grains.

■ 3.9 ELASTIC DEFORMATION OF A SINGLE CRYSTAL

To a great extent, the mechanical properties of materials are a reflection of their crystal structure. Therefore, if we are to understand mechanical behavior, we must have an understanding of the way crystals react when subjected to mechanical loads. Through the study of carefully prepared single crystals, we have learned that the mechanical behavior is dependent on (1) the type of lattice, (2) the interatomic forces (i.e., bond strength), (3) the spacing between the planes of atoms, and (4) the density of the atoms on the various planes.

If the applied loads are relatively low, the crystal responds by simply stretching or compressing the distance between atoms (Figure 3-12). The basic lattice unit does not change, and all of the atoms remain in their original position relative to one another. The applied load serves only to disrupt the force balance of the atomic bonds, and the atoms assume a new equilibrium spacing with the applied load as one of the force components.

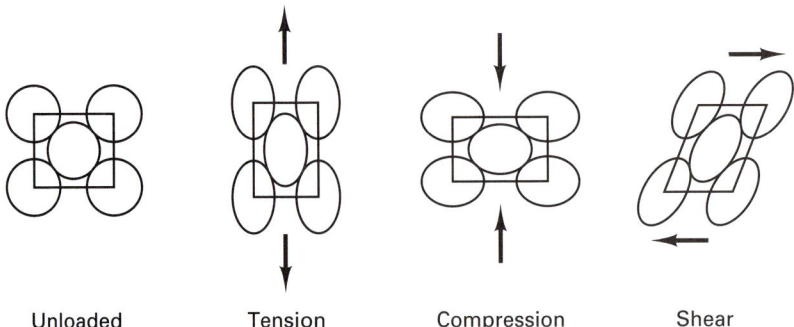

| Unloaded | Tension | Compression | Shear |

FIGURE 3-12 Distortion of a crystal lattice in response to elastic loadings.

When the load is removed, the atoms return to their original positions and the crystal resumes its original size and shape. The response to such loads is *elastic* in nature, and the amount of stretch or compression is proportional to the applied load or stress.

Elongation or compression in the direction of loading also produces an opposite change in dimensions at right angles to that direction. The ratio of lateral contraction to axial tensile strain is known as *Poisson's ratio*. This value is always less than 0.5 and is usually about 0.3

■ 3.10 PLASTIC DEFORMATION OF A SINGLE CRYSTAL

As the magnitude of applied load is increased, the distortion increases to a point where the atoms must either (1) break bonds to produce a fracture, or (2) slide over one another to produce a permanent shift of atom positions in a way that might reduce the load. For metallic materials, the second phenomenon generally requires lower loads and occurs preferentially in nature. The result is a plastic deformation where a permanent change in shape occurs without a concurrent deterioration in properties.

Investigation reveals that the mechanism of plastic deformation is the shearing of atomic planes over one another to produce a net displacement. Conceptually, this is similar to the distortion of a deck of playing cards when one card slides over another. As we shall see, the actual mechanism is really a progressive one rather than all atoms in a plane shifting simultaneously.

Recalling that a crystal structure is a regular and periodic arrangement of atoms in space, it becomes possible to link the atoms into flat planes in an almost infinite number of ways. Planes having different orientations with respect to the surfaces of the unit cell will have different atomic densities and different spacing between adjacent and parallel surfaces (Figure 3-13). Given the choice of all possibilities, plastic deformation tends to occur

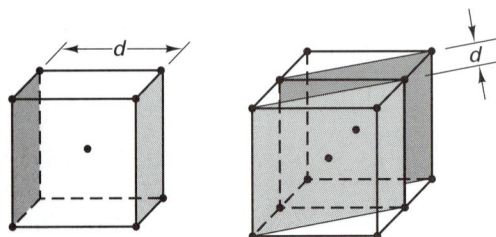

FIGURE 3-13 Schematic diagram showing crystalline planes with different atomic densities and interplanar spacings.

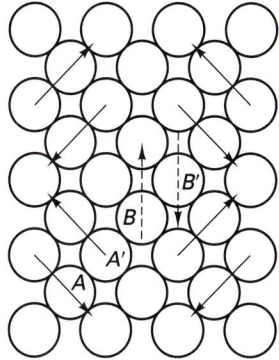

FIGURE 3-14   Planar schematic representing the greater deformation resistance of planes of lower atomic density and closer interplanar spacing.

along planes having the highest atomic density and greatest separation. The rationale for this can be seen in the simplified two-dimensional array of Figure 3-14. Planes *A* and *A'* have higher density and greater separation than planes *B* and *B'*. In visualizing relative motion, the atoms of *B* and *B'* would interfere significantly with one another, whereas planes *A* and *A'* do not experience this difficulty.

Although Figure 3-14 represents the planes of sliding as lines, the structure is actually three-dimensional. Within the preferred planes are also preferred directions. If sliding occurs in a direction that corresponds to one of the close-packed directions (shown as dark lines in Figure 3-8), atoms can simply follow one another rather than each having to negotiate its own path. Thus plastic deformation tends to occur by the preferential sliding of maximum-density planes (close-packed planes if present) in directions of closest packing. The specific combination of plane and direction is called a *slip system*, and the shear deformation is known as *slip*.

The ease with which a given metal may be deformed depends on the ease of shearing one atomic plane over an adjacent one and the favorability with which the plane is oriented with respect to the load. For example, consider that deck of playing cards. The deck does not "deform" when laid flat on the table and pressed from the top, or when stacked on edge and pressed uniformly. The cards will slide over one another only if the deck is skewed with respect to the applied load so as to induce a shear stress along the plane of sliding.

With this understanding, let us now consider the properties of the three most common crystal structures.

1. *Body-centered cubic.* In the bcc structure, there are no close-packed planes. Slip occurs on the most favorable alternatives, which are those planes with the greatest interplanar spacing (six of which are illustrated in Figure 3-15). Within these planes, slip occurs along the directions of closest packing, which are the cube diagonals. If each specific combination of plane and direction is considered as a separate slip system, we find that the bcc materials contain 48 attractive ways to slip (plastically deform). The probability that one or more of these systems will be oriented in a favorable manner is great, but the force required to produce deformation is extremely large since there are no close-packed planes. Materials with this structure generally possess high strength with moderate ductility. (Refer to the typical bcc metals in Figure 3-7).

2. *Face-centered cubic.* In the fcc structure, each unit cell contains four close-packed planes, as illustrated in Figure 3-15. Each of those planes contains three close-packed directions, or face diagonals, giving 12 possible means of slip.

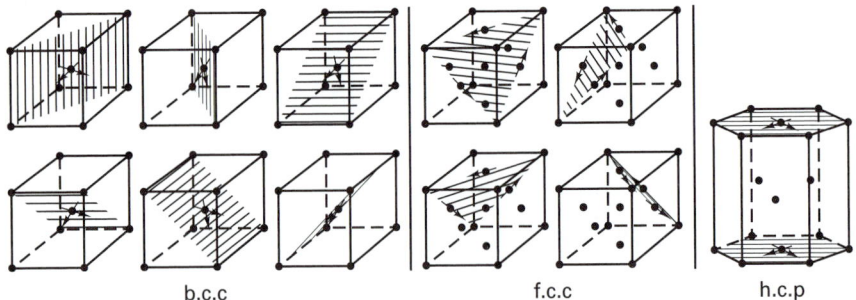

FIGURE 3-15 Slip planes of the various lattice types.

Again, the probability that one or more of these will be favorably oriented is great, and the force required to induce slip is rather low. Metals with the fcc structure are relatively weak and possess excellent ductility, as can be confirmed by a check of Figure 3-7.

3. *Hexagonal close-packed.* The hexagonal lattice also contains close-packed planes, but only one such plane exists within the lattice. Although this plane contains three close-packed directions and the force required to produce slip is rather low, the probability of favorable orientation to the applied load is rather small. As a result, metals with the hcp structure tend to have low ductility and often appear to be brittle.

■ 3.11 DISLOCATION THEORY OF SLIPPAGE

A theoretical calculation of the strength of metals based on the sliding of atomic planes over one another predicts yield strengths on the order of 3 million pounds per square inch (20,000 MPa). Observed strengths are typically 100 to 150 times less than this value, indicating a significant discrepancy between theory and reality.

An explanation is provided by the fact that plastic deformation does not occur by all the atoms in one plane slipping simultaneously over all the atoms in an adjacent plane. Instead, the motion takes place by the progressive slippage of a localized disruption known as a *dislocation*. Consider a simple analogy. A carpet has been rolled onto a floor and someone wants to move it a short distance in a given direction. One approach would be to pull on one end and try to "shear the carpet across the floor," simultaneously overcoming the frictional resistance over the entire area of contact. This would require a large force acting over a small distance. An alternative approach would be to form a wrinkle in the carpet and walk it across the floor to produce a net shift in the whole carpet—a low-force-over-large-distance approach to the same task. In the region of the wrinkle, there is an excess of carpet with respect to the floor below, and the motion of this excess is relatively easy.

Electron microscopes have revealed that metal crystals do not have all of their atoms in perfect arrangement but contain a variety of localized imperfections. Two such imperfections are the *edge dislocation* and *screw dislocation* (Figure 3-16). Edge dislocations are the edges of extra half-planes of atoms. Screw dislocations correspond to partial tearing of the crystal plane. In each case the dislocation is a disruption to the regular, symmetric arrangement of atoms and can be moved about with a rather low applied force. It is the motion of these atomic-scale dislocations under applied load that is responsible for the observed macroscopic plastic deformation.

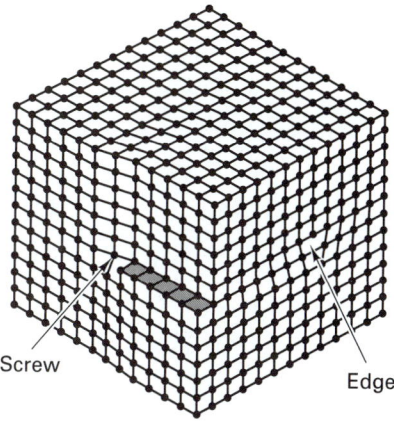

Screw   Edge   FIGURE 3-16   Schematic representation of screw and edge dislocations.

All engineering metals contain dislocations, usually in abundant quantities. The ease of deformation, therefore, depends on the ease of inducing dislocation movement. Barriers to dislocation motion tend to increase the overall strength of a metal. These barriers take the form of other crystal imperfections and may be of the point type (missing atoms or *vacancies*, extra atoms or *interstitials*, or substituting atoms of a different variety, as may occur in an alloy), line type (another dislocation), or surface type (crystal grain boundary or free surface).

■ 3.12 STRAIN HARDENING OR WORK HARDENING

As noted in our discussion of the tensile test in Chapter 2, most metals become stronger when plastically deformed, a phenomenon known as *strain hardening* or *work hardening*. Understanding of this phenomenon can now come from our knowledge of dislocations and an extension of the carpet analogy. Suppose this time that the goal is to move the carpet diagonally. The best way would be to move a wrinkle in one direction and then a second one 90° to the first. But suppose that both wrinkles were started simultaneously. We would find that wrinkle 1 would impede the motion of wrinkle 2, and vice versa. In essence, the device that makes deformation easy can also serve to impede the motion of other, similar devices. (*Note*: This can also be demonstrated with a sheet of notebook paper, but requires some coordination!)

Returning to metals, we find that plastic deformation is accomplished by the motion of dislocations. As dislocations move, they are more likely to encounter and interact with other dislocations or crystalline defects, thereby producing resistance to further motion. Moreover, mechanisms exist that markedly increase the number of dislocations in a metal during deformation, thereby enhancing the probability of interaction.

The effects of strain hardening become attractive when one considers that the mechanical working of metal is frequently performed to produce a more useful shape. Since strength can be increased substantially during deformation, a strain-hardened (deformed), inexpensive metal can often be substituted for an undeformed (such as machined to shape), costly one.

Experimental evidence strongly supports the dislocation and slippage theory of deformation. The transmission electron microscope can be used to show images of the individual dislocations in a thin metal section, and studies confirm the increase in number and interactions during deformation. Macroscopic observations also lend support. When a load is applied to a single metal crystal, deformation will begin on the slip system that is

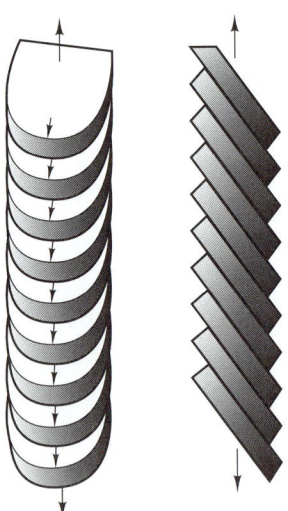

FIGURE 3-17 Schematic representation of slip and rotation resulting from deformation.

most favorably oriented. The net result is often an observable slip and rotation (Figure 3-17.) Dislocation motion on one slip system may become impeded as strain hardening produces increased resistance or rotation makes the orientation less favorable. Further deformation often occurs on alternative systems that now offer less resistance. This phenomenon known as *cross slip*, has also been observed.

■ 3.13 PLASTIC DEFORMATION IN POLYCRYSTALLINE METALS

Thus far, only the deformation of single crystals has been considered. Commercial metals are usually encountered in the form of polycrystalline aggregates. Within each crystal of the polycrystalline assembly, deformation proceeds in the manner just described. Since the various grains have different orientations, an applied load will produce different deformations in each of the crystals. This can be illustrated by Figure 3-18, where a metal

FIGURE 3-18 Slip lines in a polycrystalline material. *(From Richard Hertzberg, Deformation and Fracture Mechanics of Engineering Materials; courtesy of John Wiley & Sons, Inc.)*

has been polished and then deformed to reveal the different slip lines for each of the grains.

It may also be noted that the slip lines do not cross over from one grain to another. The grain boundaries act as barriers to dislocation motion. Thus metals with a finer grain structure—more grains per unit area—tend to exhibit greater strength and hardness, coupled with increased impact resistance. This "universal improvement in properties" is an attractive motivation for controlling grain size during processing.

■ 3.14 GRAIN DEFORMATION AND ANISOTROPIC PROPERTIES

When a metal is deformed by a substantial amount, the grains become elongated in the direction of metal flow (Figure 3-19). Concurrent with the nonsymmetry of structure is nonsymmetry or directional variation in properties. Mechanical properties, such as strength and ductility, as well as electrical and magnetic properties, may all show directional differences. Properties that vary with direction are said to be *anisotropic*. Properties that are uniform in all directions are *isotropic*.

The directional variation of properties can be useful to both the design and manufacturing engineer. By controlling the metal flow in processes such as forging, enhanced strength or fracture resistance can be imparted to certain locations. Caution should be exercised, however, for improvement of properties in one direction is usually accompanied by a decline in properties in another. Moreover, directional variation in properties may impose serious difficulties in further processing operations, as with the further forming of rolled metal sheets.

■ 3.15 FRACTURE OF METALS

If the plastic deformation of a metal is extended too far, the metal may ultimately fracture. These types of fractures are known as *ductile fractures*, noting that the initial response to the applied load was one of plastic deformation. Another possibility, however, is where fracture precedes plastic deformation, occurring in a sudden, catastrophic manner, and propagating rapidly through the material. These fractures, known as *brittle fractures*, are most common with metals having the bcc or hcp crystal structures. Whether the fracture

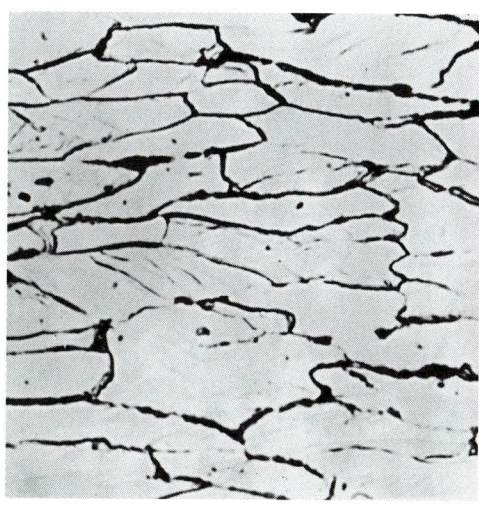

FIGURE 3-19 Deformed grains in cold-worked 1008 steel after 50% reduction by rolling; 1000×. (From *Metals Handbook*, 8th ed. ASM International, Materials Park, Ohio, 1972.)

is ductile or brittle, however, often depends on the specific conditions of material, temperature, state of stress, and rate of loading.

■ 3.16 COLD WORKING, RECRYSTALLIZATION, AND HOT WORKING

During deformation, a portion of the deformation energy becomes stored within the material in the form of additional dislocations and increased grain boundary surface area.[*] If a deformed polycrystalline metal is subsequently heated to a high-enough temperature, the material will seek to lower its energy. New, equiaxed (spherical-shaped) crystals will nucleate and grow out of the original structure (Figure 3-20). This process of reducing the internal energy through the formation of new crystals is known as *recrystallization*. The temperature at which recrystallization takes place is different for each metal and also varies with the amount of prior deformation. The greater the amount of prior deformation, the more stored energy, and the lower the recrystallization temperature. However, there is a lower limit below which recrystallization will not take place in a reasonable amount of time. Table 3-2 gives the lowest practical recrystallization temperatures for several materials. This temperature can often be estimated by taking 0.4 times the melting point of the metal when the melting point is expressed in an absolute temperature scale (Kelvin or Rankin). This is also the temperature at which atomic diffusion (atom movement within the solid) becomes significant, indicating that diffusion is an important mechanism in recrystallization.

TABLE 3-2. Lowest Recrystallization Temperature of Common Metals	
Metal	Temperature [°F (°C)]
Aluminum	300 (150)
Copper	390 (200)
Gold	390 (200)
Iron	840 (450)
Lead	Below room temperature
Magnesium	300 (150)
Nickel	1100 (590)
Silver	390 (200)
Tin	Below room temperature
Zinc	Room temperature

When metals are plastically deformed below their recrystallization temperature, the process is called *cold working*. The metal strain hardens and the structure consists of distorted grains. If the deformation is continued, the metal may fracture. Therefore, we find it common practice to recrystallize material after certain amounts of cold work. Ductility is restored, and the material is ready for further deformation. The heating process is known as a *recrystallization anneal* and enables deformation to be carried out to great lengths without the danger of fracture.

---

[*]A sphere has the least amount of surface area of any shape to contain a given volume of material. When the shape becomes altered from that of a sphere, the surface area must increase. Consider a round balloon filled with air. If the balloon is stretched or flattened into another shape, the rubber balloon is stretched further. When the applied load is removed, the balloon snaps back to its original shape, the one involving the least surface energy. Metals behave in an analogous manner. During deformation, the distortion of the crystals increases the energy of the material. Given the opportunity, the material will try to lower its energy by returning to spherical grains.

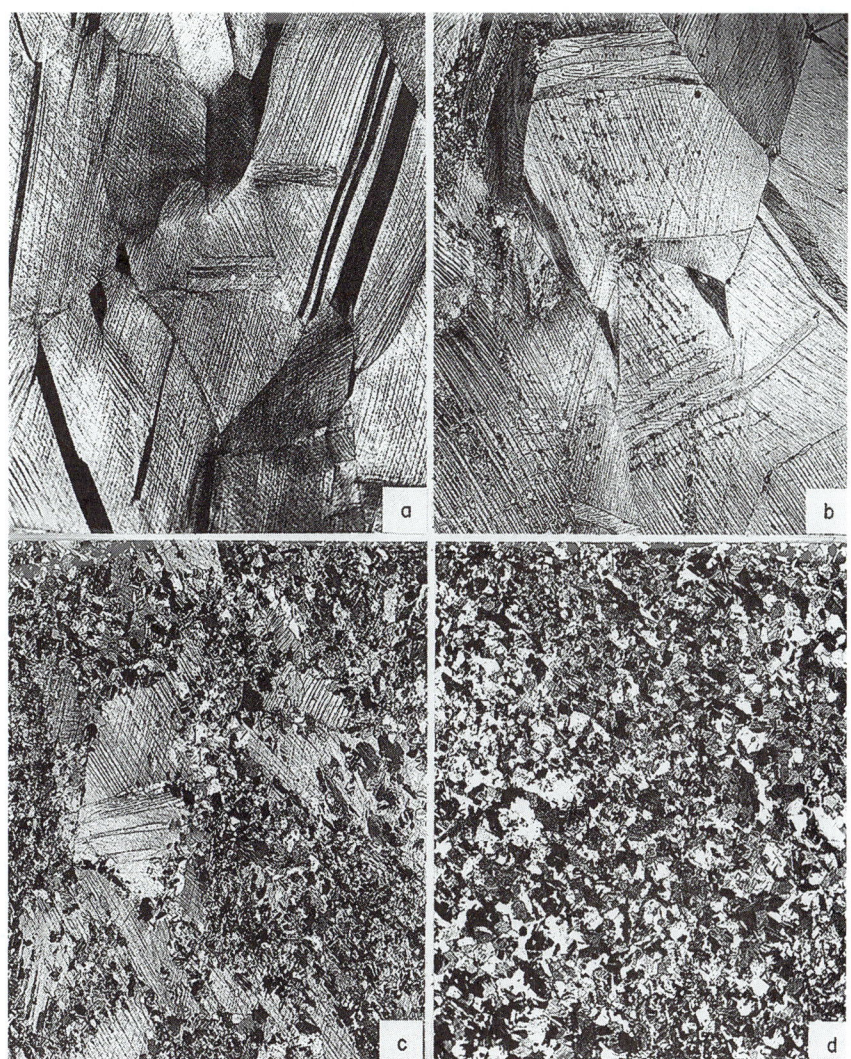

FIGURE 3-20   Recrystallization of 70-30 brass: (a) cold-worked 33%; (b) heated at 580°C (1075°F) for 3 seconds, (c) 4 seconds, and (d) 8 seconds; 45×. *(Courtesy of J.E. Burke, General Electric Company.)*

If metals are deformed at temperatures sufficiently above the recrystallization, the process is known as *hot working*. Deformation and recrystallization can take place simultaneously, and large deformations are possible. Since a recrystallized grain structure is constantly forming, the final product will not exhibit strain hardening. As shown in Table 3-2, the minimum temperature for hot working depends on the material being worked.

Having already noted the correlation of fine grain size with improved properties, we find recrystallization to be an attractive means of grain-size control. In metals that do not undergo allotropic changes, a coarse grain structure can be converted to a fine grain structure through recrystallization. The material must first be plastically deformed to store energy to provide the driving force. Control of the recrystallization process then establishes the final grain size.

■ 3.17 GRAIN GROWTH

The recrystallization process tends to produce uniform grains of comparatively small size. If the metal is held at or above the recrystallization temperature for any appreciable time, however, the grain size of the new structure will continue to increase. In effect, some of the grains will become larger at the expense of their neighbors as the material seeks to further decrease the amount of grain boundary surface area. Properties tend to diminish with increased grain size, so control of recrystallization is of prime importance. The material should be held at elevated temperature long enough to complete recrystallization. The temperature should then be decreased to prevent the further changes that would lead to grain growth.

■ 3.18 ALLOYS

Up to this point in the chapter, the discussion has been confined to the nature and behavior of pure metals. For most manufacturing applications, however, metals are not used in their pure form, but in the form of *alloys*, materials composed of two or more different elements. The formation of an alloy usually results in a change of properties. Knowledge of alloys and their properties, therefore, is important to the intelligent selection of materials.

■ 3.19 ALLOY TYPES

There are three possible ways in which a metal may respond to the addition of an alloying element. The first, and probably the simplest, response occurs when the *two materials are insoluble in one another in the solid state*. In this case the base metal and the alloying addition each maintain their individual identities, structures, and properties. The "alloy" in effect becomes a composite structure, consisting of two types of building blocks in an intimate mechanical mixture.

A second possibility occurs when the *two elements are soluble in each other in the solid state*. The two materials form a *solid solution*, where the alloy element dissolves in the base metal. The solutions can be of two distinct forms: (1) *substitutional* and (2) *interstitial*. In the substitutional solution, some atoms of the alloy element occupy lattice sites normally filled by atoms of the base metal. The replacement is totally random in nature, with the alloy atoms being dispersed throughout the base lattice. In the interstitial solution, the alloy element atoms squeeze into the open spaces between the atoms of the base metal lattice.

The third possibility occurs when the *elements combine to form intermetallic compounds*. In this case, the atoms of the alloying element combine with the atoms of the base metal in definite proportions and in definite geometric relationships. The bonding is primarily of the nonmetallic variety (i.e., ionic or covalent), and the lattice structures are often quite complex. Because of the type of bonding, intermetallic compounds tend to be hard, but brittle, high-strength materials.

Even though alloys are composed of more than one type of atom, their structure is still one of lattices and grains. Their behavior in response to applied loadings is similar to that of pure metals, with some features reflecting the increased level of structural complexity. Dislocation movement can now be impeded by the presence of unlike atoms. When neighboring grains have different chemistries and/or structures, they will show different response to the same type and magnitude of load.

■ 3.20 ATOMIC STRUCTURE AND ELECTRICAL PROPERTIES

As with the mechanical properties, the structure of a material also influences the electrical behavior. Electrical conductivity refers to the net movement of charge through the structure.

In metals, the charge carriers are the valence electrons. The more perfect the atomic arrangement, the higher the electrical conductivity. Conversely, the greater the number of lattice imperfections or irregularities, the higher the resistance to electrical conduction.

The electrical resistance of a metal depends largely on two factors: (1) lattice imperfections and (2) temperature. Vacant atomic sites, interstitial atoms, substitutional atoms, dislocations, and grain boundaries all act as disruptions to the regularity of a crystalline lattice. Thermal energy causes the atoms to vibrate about their equilibrium position. These vibrations cause the atoms to be out of position and further interferes with electron travel. For a metal, *electrical conductivity* will decrease with an increase in temperature. As the temperature is decreased, the electrical conductivity becomes primarily a function of the crystalline imperfections. The best metallic conductors, therefore, are pure metals at low temperature.

The electrical conductivity of a metal is due to the motion of the "free" electrons in the metallic bond. For covalently bonded materials, however, bonds must be broken to provide electrons for charge transport. Therefore, the electrical properties of these materials is a function of bond strength. Diamond, for instance, is a strong insulator. Silicon and germanium have weaker bonds that are more easily broken by thermal energy. These metals are known as *intrinsic semiconductors*, since moderate amounts of thermal energy will enable the materials to conduct small amounts of electricity. Tin has such weak bonding that a high number of bonds are broken at room temperature and the electrical behavior resembles that of a metal.

The electrical conductivity of the semiconductors can be substantially improved by a process known as *doping*. Both silicon and germanium have four valence electrons and form four covalent bonds. If one of the atoms is replaced with an atom containing five valence electrons, such as phosphorus or arsenic, the four covalent bonds will form, leaving an additional valence electron. This extra electron is then free to move about and provide additional conductivity. Such materials are known as *n-type extrinsic semiconductors*.

A similar effect can be created by the substitution of an atom with three valence electrons, such as aluminum. An electron will be missing from one of the bonds, creating an electron *hole*. When a voltage is applied, a nearby electron can jump into this hole, creating a hole in the spot it vacated. Movement of electron holes is equivalent to a countermovement of electrons, and thus provides conductivity. Materials containing three-electron dopants are known as *p-type semiconductors*. The control of conductivity through semiconductor devices is the functional basis of solid-state electronics and circuitry.

■ KEY WORDS

allotropic	electrical conductivity	metallic bond
alloy	extrinsic semiconductor	microstructure
amorphous structure	face-centered cubic	molecular structure
anisotropic	grain	Poisson's ratio
ASTM grain-size number	grain boundary	polymorphic
atomic number	grain growth	recrystallization
atomic radius	grain size	simple cubic
body-centered cubic	hexagonal close-packed	slip
brittle fracture	hot work	slip system
close-packed planes	intermetallic compound	solid solution
cold work	interstitial	strain hardening
covalent bond	intrinsic semiconductor	substitutional
cross slip	ion	unit cell
crystal structure	ionic bond	vacancy
dislocation	isotropic	valence electrons
ductile fracture	lattice	van der Waals forces

■ REVIEW QUESTIONS

1. Why might an engineer be concerned with controlling or altering the structure of a material?
2. What is meant by the term *microstructure*?
3. What properties or characteristics of a material are influenced by the valence electrons?
4. What are the three types of primary bonds, and what types of atoms do they unite?
5. What are some general characteristics of ionically bonded materials?
6. What are some general properties and characteristics of covalently bonded materials?
7. What are some unique property features of materials bonded by metallic bonds?
8. What causes the bonding forces in the van der Waals type of secondary bond?
9. Why does the interatomic distance increase when a material is heated (thermal expansion)?
10. What is the difference between a crystalline material and one with an amorphous structure?
11. What are some of the general characteristics of metallic materials?
12. What is an allotropic material?
13. What are the three most common crystal structures found in metals?
14. What kind of geometric symmetry is observed in a close-packed plane of atoms?
15. What is the efficiency of filling space with spheres in the simple cubic structure? Body-centered-cubic structure? Face-centered-cubic structure? Hexagonal close packed structure?
16. Describe the difference in ductility observed between the two structures formed by stacking close-packed planes.
17. What is a grain boundary?
18. What is the most common means of quantifying the grain size of a solid metal?
19. How does a metallic crystal respond to low applied loads?

20. What is plastic deformation?
21. What is a slip system in a material? What types of planes and directions tend to be preferred?
22. Based on the ease or difficulty of deformation, what is the dominant mechanical property or characteristic for each of the three most common metal crystal structures?
23. What is a dislocation? How do dislocations affect the mechanical properties of a metal?
24. What are some of the common barriers to dislocation movement that can be used to strengthen a metal?
25. What are the three major types of point defects in crystalline materials?
26. What is the mechanism (or mechanisms) responsible for the observed strain hardening of a metal?
27. Why is a fine grain size often desired in an engineering metal?
28. What is a possible cause of anisotropic properties in a metal?
29. What is the difference between brittle fracture and ductile fracture?
30. In what ways does a metal increase its internal energy during plastic deformation?
31. In what ways can recrystallization be used to enable large amounts of deformation without fear of fracture?
32. What is the major distinguishing feature between hot and cold working?
33. What types of structures can form when an alloy is added to a base metal?
34. What types of mechanical properties are typical of intermetallic compounds?
35. What features in a metal structure tend to impede or reduce electrical conductivity?
36. What is the difference between an intrinsic semiconductor and an extrinsic semiconductor?

*C*hapter 3   CASE STUDY

window frame materials and design

B ecause of your knowledge of engineering materials, a friend has asked for your assistance in evaluating various materials that are used for household window frames. He is in the process of designing a new home and wants the windows to be energy efficient and durable, and also to require low maintenance. Various suppliers have recommended wood, aluminum, and vinyl windows, each claiming their product to be superior. For each of the above materials, identify its primary assets and significant limitations. Consider ease of fabrication, strength, thermal expansion and contraction, response to moisture and humidity, durability, rigidity, ability to produce in a variety of colors, properties at low and high extremes of temperature, and any other factors that you deem important.

Based on the above considerations, combined with cost, which material would you recommend? How might your recommendation change depending on the location of the home (consider features such as temperature and humidity)? How might materials be combined to produce a window that would be superior to all of the above alternatives?

CHAPTER 4

EQUILIBRIUM DIAGRAMS AND THE IRON–CARBON SYSTEM

4.1	INTRODUCTION		Solidification of Alloy X
4.2	PHASES		Three-Phase Reactions
4.3	EQUILIBRIUM PHASE DIAGRAMS		Intermetallic Compounds
			Complex Diagrams
	Temperature–Composition Diagrams	4.4	IRON–CARBON EQUILIBRIUM DIAGRAM
	Cooling Curves	4.5	STEELS AND THE SIMPLIFIED IRON–CARBON DIAGRAM
	Solubility Studies		
	Complete Solubility in Both Liquid and Solid States		
	Partial Solid Solubility	4.6	CAST IRONS
	Insolubility	Case Study:	THE BLACKSMITH ANVILS
	Utilization of Diagrams		

■ 4.1 INTRODUCTION

As our study of engineering materials becomes more focused on specific metals and alloys, it is increasingly important that the natural characteristics and properties of the material be known. What is the basic structure of the material? Is the material uniform throughout, or is it a mixture of two or more distinct components? If there are multiple components, how much of each is present, and what are the different chemistries? Is there a component that may impart undesired properties or characteristics? What will happen if temperature is increased or decreased, pressure is changed, or chemistry is varied? The answers to these and other important questions can be obtained through the use of equilibrium phase diagrams.

■ 4.2 PHASES

Before we move to a discussion of these diagrams, it is important first to develop a working definition of the term *phase*. As a starting definition, a phase is simply a form of material possessing a characteristic structure and associated characteristic properties. Uniformity of chemistry, structure, and properties is assumed throughout a phase. More rigorously, a phase is *any physically distinct, chemically homogeneous, and mechanically separable portion of a substance*. In more common terms, a phase has a well-defined structure, uniform composition*, and distinct boundaries or interfaces.

A phase can be continuous (e.g., the air in a room) or discontinuous (e.g., grains of salt in a shaker). A phase can be solid, liquid, or gas. In addition, a phase can be a pure substance or a solution, provided that the structure and composition are uniform throughout. Alcohol and water mix in all proportions and will therefore form a single phase when combined. Oil and water tend to form isolated regions with distinct boundaries and must be regarded as two distinct phases.

---

*Within this chapter and those to follow, the term *composition* will refer not to the structural makeup of a material, but to its chemical makeup, and is usually expressed in terms of weight percent of an element.

■ 4.3 Equilibrium Phase Diagrams

The *equilibrium phase diagram* is a *graphic mapping* of the natural tendencies of a material or a material system, assuming that equilibrium has been attained for the specific conditions. There are three primary variables to be considered: *temperature*, *pressure*, and *composition*. The simplest phase diagram is a pressure–temperature (*P–T*) diagram for a fixed-composition material. Areas of the diagram are assigned to the various phases, with the boundaries indicating the equilibrium conditions of transition.

As an introduction, consider the *P–T* diagram for water presented in Figure 4-1. With composition fixed, the diagram maps the stable form of water for various conditions of temperature and pressure. If pressure is held constant and temperature is varied, the transition boundaries locate the melting and boiling points. For example, at 1 atm pressure, water melts at 0°C and boils at 100°C. Still other uses are possible. Locate a temperature where the stable phase is liquid at atmospheric pressure. Now drop the temperature until the material goes from liquid to solid (i.e., ice). Now maintain that temperature and begin to decrease the pressure. A transition is encountered where solid goes directly to gas without melting. The process described above, known as *freeze drying,* is employed in the manufacture of numerous dehydrated products. Through use of an appropriate phase diagram, process conditions could be determined that would minimize the amount of cooling and the pressure drop that would be required.

Temperature–Composition Diagrams

While the *P–T* diagram for water is an excellent introduction to phase diagrams, that and other *P–T* phase diagrams are rarely used for engineering applications. Most engineering processes are conducted at atmospheric pressure, and variations come primarily in temperature and composition. The most useful mapping, therefore, would be a *temperature–composition phase diagram* at atmospheric pressure. For the remainder of the chapter, it is this form of phase diagram that is considered.

For mapping purposes, temperature is placed on the vertical axis and composition on the horizontal. Figure 4.2 shows the form of such a mapping for the *A–B* system, where

FIGURE 4.1 Pressure–temperature diagram for water.

FIGURE 4.2 Mapping for a temperature–composition equilibrium phase diagram.

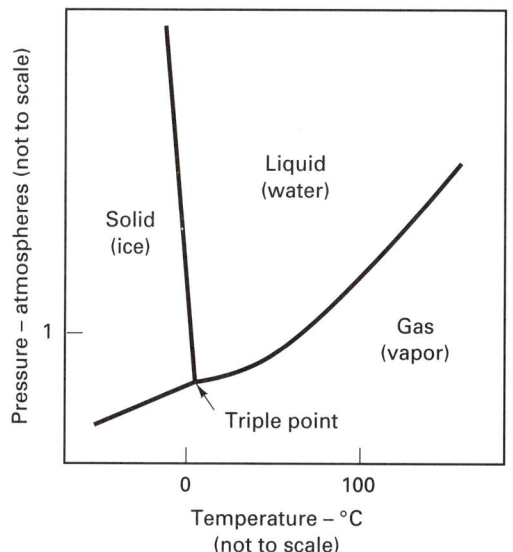

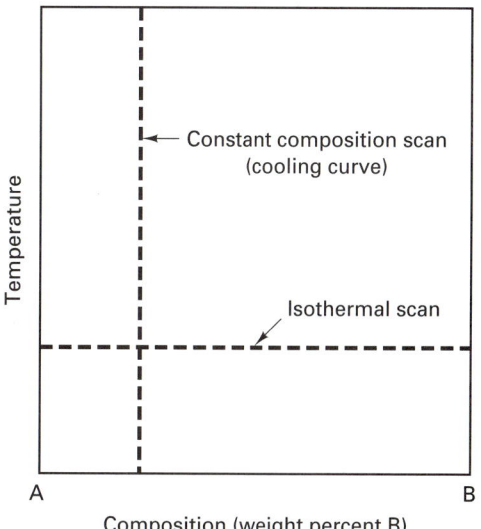

the left-hand vertical corresponds to pure material A and the percentage of B (usually expressed in weight percent) increases as we move toward pure B at the right of the diagram. The temperature range often includes only solids and liquids, since few processes involve engineering materials in the gaseous state. Experimental investigations for filling in the details of the diagram then take the form of either vertical or horizontal scans designed to locate the transitions between phases.

Cooling Curves

Considerable information can be obtained from vertical scans through the diagram in which a fixed composition material is heated and subsequently slow-cooled by removing heat at a uniformly slow rate. Transitions in structure appear as characteristic points in a temperature-versus-time plot of the cooling cycle, known as a *cooling curve*.

Consider the system composed of sodium chloride (common table salt) and water. Five different cooling curves are presented in Figure 4-3. Curve (a) is for pure water being cooled from the liquid state. A smooth continuous line is observed for the liquid where the extraction of heat produces a concurrent drop in temperature. When the freezing point is reached (point a), the material changes state and releases heat energy as part of the liquid-to-solid transition. Heat is still being extracted from the system, but its source is the change in state and not a decrease in temperature. Thus an isothermal or constant-temperature hold ($a–b$) is observed until the solidification is complete. From this point, the newly formed solid experiences a smooth drop in temperature as heat extraction continues. Such a curve is characteristic of pure metals and other substances with a distinct melting point.

Curve (b) in Figure 4-3 presents the cooling curve for a solution of 10% salt in water. The liquid region undergoes continuous cooling down to point c, where the slope

FIGURE 4.3   Cooling curves for various compositions of NaCl–H$_2$O solutions: (a) 0% NaCl; (b) 10% NaCl; (c) 23.5% NaCl; (d) 50% NaCl; (e) 100% NaCl.

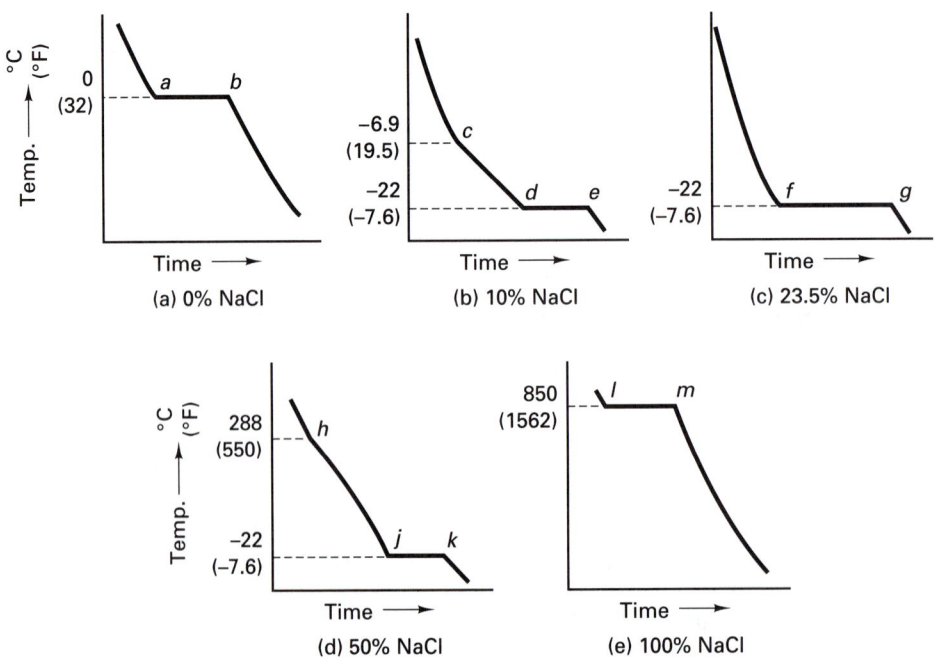

abruptly decreases. At this temperature, small particles of ice begin to form and the reduced slope is attributed to the energy released in this transition. The formation of these ice particles leaves the remaining solution richer in salt and imparts a lower freezing temperature to it. Further cooling must take place if additional solid is to form. The formation of additional solid continues to enrich the solution and further lowers the freezing point of the remaining liquid. Instead of possessing a distinct melting point or freezing point, the material is said to have a *freezing range.* When the temperature of point *d* is reached, the remaining liquid solidifies into an intimate mixture of solid salt and solid water (discussed later), and an isothermal hold is observed. Further extraction of heat produces a drop in the temperature of the solid.

For a solution of 23.5% salt in water, a distinct freezing point is again observed, as shown in curve (c). Compositions with richer salt concentration [curve (d)] show phenomena similar to those in curve (b), but with salt being the first solid to form from the liquid. Finally, pure salt [curve (e)] exhibits behavior similar to that of pure water.

The observed transition points can now be transferred to the temperature–composition diagram, with the designation letters corresponding to the transition points on the cooling curves. Figure 4-4 presents such a map, on which several key lines have been drawn. Line *a–f–l* denotes the lowest temperature at which the material is totally liquid and is known as the *liquidus line.* Line *d–f–j* denotes a particular three-phase reaction and is discussed later. Between the lines, two phases coexist, one being a liquid and the other a solid.

The cooling curve studies have enabled the determination of some key information regarding the salt-water system, including some insight into the use of salt on highways in the winter. The freezing point of water can be lowered from 0°C (32°F) to as low as −22°C (−7.6°F). The equilibrium phase diagram is a collective presentation of cooling curve data for an entire range of alloy compositions.

Solubility Studies

The observant reader will note that the ends of the diagram remain undetermined. Both pure materials have a distinct melting point, below which they appear as a single-phase solid. Can ice retain some salt in a single-phase solid solution? Can solid salt hold some water and remain a single phase? If so, how much, and does the amount vary with temperature? Completion of the diagram, therefore, requires several horizontal scans to determine any *solubility limits,* the conditions of saturation at various temperatures.

FIGURE 4.4   Partial equilibrium diagram for NaCl and H_2O derived from cooling-curve information.

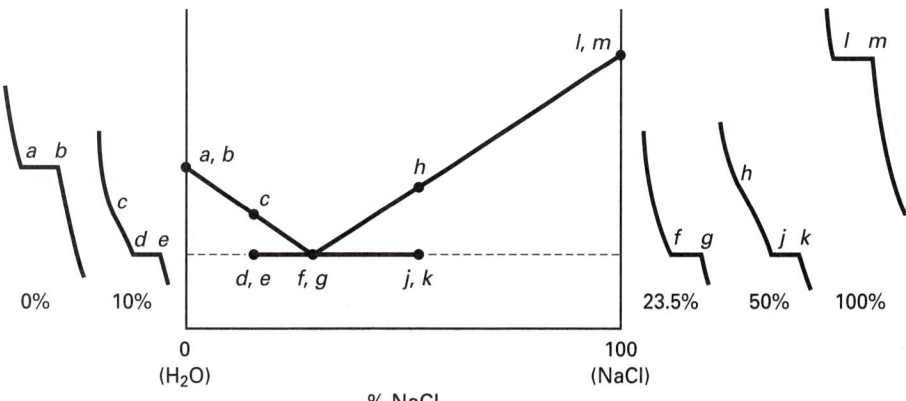

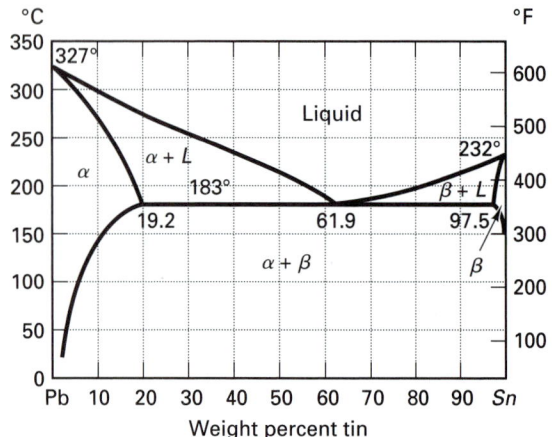

FIGURE 4.5   Lead–tin equilibrium diagram.

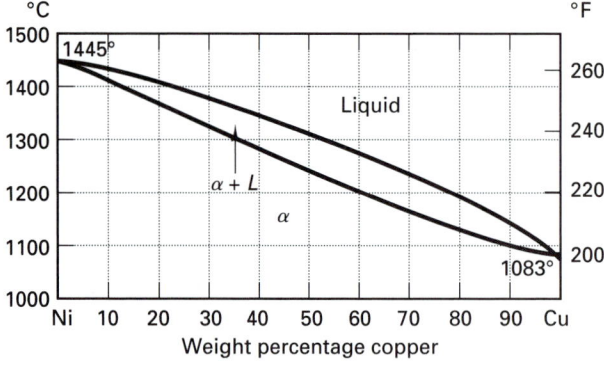

FIGURE 4.6   Copper–nickel equilibrium diagram, showing complete solubility in both liquid and solid states.

These isothermal scans usually require the preparation of specimens of various composition and their subsequent analysis by x-ray techniques, microscopy, or other methods to determine whether the structure and chemistry is uniform or a two-phase mixture. As we move away from the pure material, the first transition in the phase diagram (provided that the temperature is in the region where all compositions are solid) denotes the solubility limit and is known as a *solvus* line. Figure 4-5 presents the equilibrium phase diagram for the lead–tin system, using the conventional notation in which Greek letters are used to denote the various single-phase solids. The upper portion of the diagram closely resembles the salt-water diagram, but the solubility of one material in the other can be observed on both ends of the solid-state region.

Complete Solubility in Both Liquid and Solid States

Having developed the basic concepts of equilibrium phase diagrams, we will begin to consider various examples, moving from the simple to the more complex. If two materials are each completely soluble in the other in both the liquid and solid states, a rather simple diagram results, such as the copper–nickel diagram of Figure 4-6. The upper line is a liquidus line. At temperatures above it, the two materials form a uniform-chemistry liquid solution. The lower line denotes the highest temperature for which the material is completely solid and is known as a *solidus* line. Below the solidus, the combined materials form a solid-state solution in which the two types of atoms are uniformly distributed throughout the crystalline lattice. Between the liquidus and solidus is a two-phase region where liquid and solid solutions coexist.

Partial Solid Solubility

As might be expected, many materials fail to exhibit complete solubility in the solid state. Each is often soluble in the other up to a certain limit or saturation point which varies with temperature. Such a diagram has already been observed in the lead–tin system of Figure 4-5.

At the point of maximum solubility, 183°C, lead can hold up to 19.2 wt% tin in a single-phase solution and tin can hold up to 2.5% lead and still be single phase. If the temperature is decreased, the amount of solute that can be held in solution decreases in a continuous manner. Therefore if a saturated solution of tin in lead is cooled from 183°C,

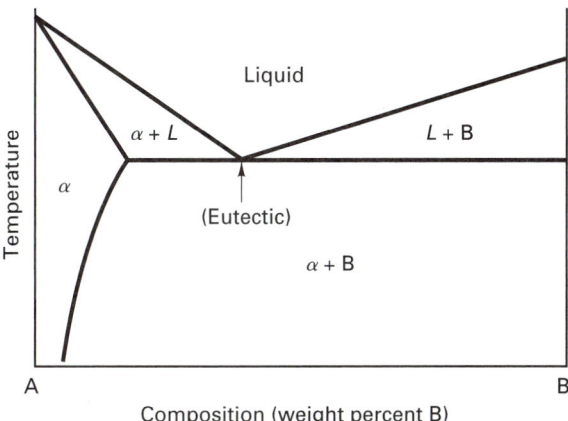

A   B

Composition (weight percent B)

FIGURE 4.7 Equilibrium diagram of two materials, one of which is completely insoluble in the other.

the material changes from a single-phase solution to a two-phase mixture. A tin-rich second phase is precipitated from solution. This change in structure can be used to alter and control the properties in a number of engineering alloys.

Insolubility

If one or both of the components is totally insoluble in the other, the diagrams also reflect this phenomenon. Figure 4-7 illustrates the case where component A is completely insoluble in component B.

Utilization of Diagrams

Before moving to the more complex diagrams, let us first consider a simple phase diagram, such as the one in Figure 4-8 and develop several tools to extract some useful information. *For each point of temperature and composition, three pieces of information can be obtained*:

1. *The phases present.* The stable phases can be determined simply by locating the point of consideration on the temperature–composition mapping and identifying the region of the diagram in which the point appears.

2. *The composition of each phase.* If the point lies in a single-phase region, the composition of the phase is the composition of the alloy being considered. If the point lies in a two-phase region, a *tie-line* is drawn. A tie-line is simply an isothermal (constant-temperature) line drawn through the point of consideration, terminating at the boundaries of the single-phase regions on either side. The compositions at which the tie-line intersects the neighboring single-phase regions is the composition of those respective phases in the two-phase mixture. For example, consider point a in Figure 4-8. The tie-line for this point runs from S_2 to L_2. The point S_2 is the intersection of the tie-line with the solid phase, and thus the solid in the two-phase mixture at point a has the composition of point S_2. Similarly, the liquid phase has the composition of point L_2.

3. *The amount of each phase present.* If the point lies in a single-phase region, there must be 100% of that phase. If the point lies in a two-phase region, the relative amounts of the two components can be determined by a *lever-law calculation* using the previously drawn tie-line. Consider the cooling of alloy X in Figure 4-8 in a manner sufficiently slow so as to preserve equilibrium. At temperatures above t_1, the

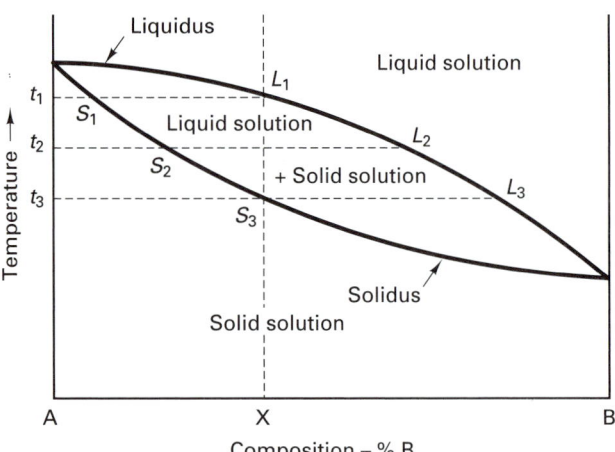

FIGURE 4.8 Equilibrium diagram showing the changes that occur during the cooling of alloy X.

material is in a single-phase liquid state. Temperature t_1 is the lowest temperature at which the alloy is 100% liquid. If we draw a tie-line at this temperature, it runs from S_1 to L_1, and lies entirely to the left of composition X. At temperature t_3, the alloy is completely solid, and the tie-line lies completely to the right of composition X. Extrapolating these observations to the intermediate temperatures, such as temperature t_2, we predict that the fraction that is liquid is equal to the fraction of the tie-line that lies to the left of point a. This fraction can be computed as

$$\frac{a - S_2}{L_2 - S_2} \times 100\%$$

In a similar manner, the fraction of solid corresponds to the fraction of the tie-line that lies to the right of point a. (*Note*: The relations above can also be derived from the conservation of A or B atoms, as the material divides into two different compositions.) Since the calculations consider the tie-line as a lever with the phases at each end and the fulcrum at the composition line, they are called lever-law calculations.

Phase diagrams can also be used to give an overall picture of an alloy system or determine the transition points for various changes in phase. The temperature required to redissolve a second phase or melt an alloy can easily be determined. The various changes that occur during the slow heating or slow cooling of a material can now be predicted. In fact, most of the questions posed at the beginning of this chapter can now be answered.

Solidification of Alloy X

Let us now use the tools we have developed and follow the solidification of alloy X in Figure 4-8. At temperature t_1, the first minute amount of solid forms with the chemistry of point S_1. As the temperature drops, more solid forms, but the chemistries of both the solid and liquid phases shift to follow the tie-line endpoints. The chemistry of the liquid follows the liquidus line and the chemistry of the solid follows the solidus. Finally, at t_3, solidification is complete, and the composition of the single-phase solid is that of alloy X, as required.

The composition of the final solid is different from that of the first solid to form. If the cooling is sufficiently slow (i.e., equilibrium is maintained or approximated), the composition of the entire mass of solid will shift as it follows the endpoint of the tie-line. This shift is made possible through the phenomenon of diffusion, whereby atoms can migrate through the crystal lattice given sufficient time at elevated temperature. If the cooling rate is too rapid, the temperature may drop before sufficient diffusion occurs, and a nonuniform material will result. The initial solid that formed will retain a chemistry that is different from the solid regions that formed later. When these variations occur on a microscopic level, the resultant structure is referred to as being *cored*.

Three-Phase Reactions

Several of the phase diagrams that were presented earlier contain a feature in which phase regions are separated by a horizontal line. These lines are further characterized by either a V intersecting from above or an inverted V intersecting from below. The intersection of the V and the line denotes the location of a *three-phase equilibrium reaction*.

One common type of three-phase reaction, known as a *eutectic*, has already been observed in Figures 4-4, 4-5, and 4-7. It is possible to understand these reactions through the use of the tie-line and lever-law concepts that we have just developed. Refer to the lead–tin diagram of Figure 4-5 and consider any alloy containing between 19.2 and 97.5 wt % tin at a temperature just above the 183°C horizontal line. Tie-line and lever-law computations reveal that the material contains either a lead-rich or tin-rich solid and remaining liquid. At this temperature, any liquid that is present will have a composition of 61.9 wt % tin, regardless of the overall composition of the alloy. If we now focus on the liquid and allow it to cool to just below 183°C, a transition occurs in which liquid of composition 61.9% tin transforms to a mixture of lead-rich solid with 19.2% tin and tin-rich solid containing 97.5% tin. The relative amounts of the two components are such as to maintain the overall chemistry of 61.9% tin. The form of this eutectic transition is similar to a chemical reaction where

$$\text{liquid} \rightarrow \text{solid}_1 + \text{solid}_2$$

Since the two solids have chemistries on either side of the intermediate liquid, a separation must have occurred within the system. Such a separation will occur in solidifying melts where the two materials are soluble in the liquid state but only partially soluble in the solid state. Separation requires atom movement, but the distances involved cannot be great. Therefore, the resulting *eutectic structure* is an intimate mixture of two single-phase solids. For a given reaction, the eutectic structure always forms from the same chemistry at the same temperature, and therefore has its own characteristic set of physical and mechanical properties. Alloys with the eutectic composition have the lowest melting point of all alloys within the system and are often used as casting alloys or filler material in soldering or brazing operations.

Figure 4-9 summarizes the variety of three-phase reactions that may occur in engineering systems. These include the *peritectic, monotectic,* and *syntectic* reactions, where the suffix *-ic* denotes that at least one of the three phases in the reaction is a liquid. If the same prefix appears with an *-oid* suffix, the reaction form is the same, but all phases involved are solids. Two such reactions are possible, the *eutectoid* and the *peritectoid*. The solid-state reactions tend to be a bit more sluggish, since all changes must occur within (usually crystalline) solids.

Intermetallic Compounds

A final feature occurs in alloy systems where the bonding attraction of the component materials is sufficiently strong so that compounds tend to form. These compounds are single-

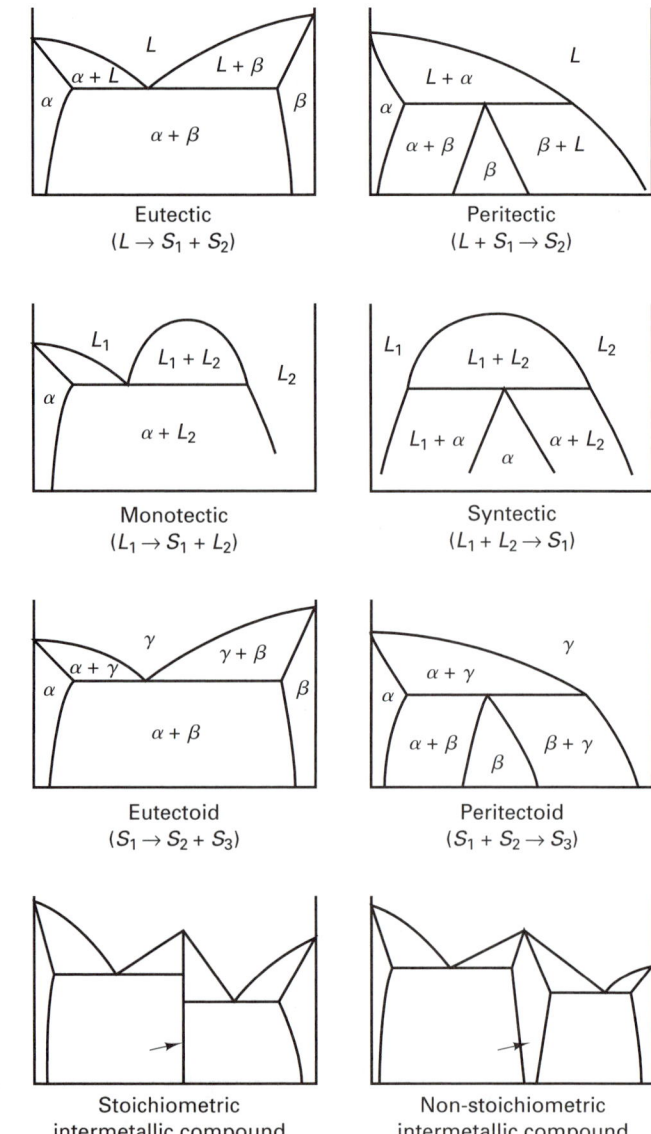

FIGURE 4.9 Schematic summary of three-phase reactions and intermetallic compounds.

phase solids and tend to break the diagram into recognizable subareas. If components A and B form a compound A_xB_y and the compound cannot tolerate any deviation from that fixed atomic ratio, the product is known as a *stoichiometric intermetallic compound* and appears as a single vertical line in the diagram. If some degree of deviation is tolerable, the vertical line expands into a single-phase region, and the compound is known as a *nonstoichiometric intermetallic compound*. Figure 4-9 shows schematic representations of both stoichiometric and nonstoichiometric compounds.

In general, intermetallic compounds tend to be hard, brittle materials, these properties being a consequence of their ionic or covalent bonding. If they are present in large quantities or lie along grain boundaries as continuous films, the overall alloy can be extremely brittle. If the same compound is dispersed throughout the alloy in the form of small discrete particles, the result can be a considerable strengthening of the base metal.

Complex Diagrams

The equilibrium diagrams of actual alloy systems consist of one of the basic types just discussed or combinations of them. In some cases the diagrams appear to be quite complex and formidable. However, by focusing on a particular composition and analyzing specific points using the tie-line and lever-law concepts, even the most complex diagram can be interpreted. If the properties of the various components are known, it is possible to predict the behavior of the resultant structures.

■ 4.4 IRON–CARBON EQUILIBRIUM DIAGRAM

Steel, composed primarily of iron and carbon, is clearly the most important engineering metal. For this reason, the iron–carbon equilibrium diagram is probably the most important for the average engineer to understand. The diagram most frequently encountered, however, is not the full iron–carbon diagram but the iron–iron carbide diagram shown in Figure 4-10. Here, a stoichiometric intermetallic compound, Fe_3C, is used to terminate the carbon range at 6.67 wt % carbon. The names of key phases and structures, and the specific notations used on the diagram, have evolved historically and will be used in their generally accepted form.

There are four single phases within the diagram. Three of these occur in pure iron, and the fourth is the carbide intermetallic at 6.67% carbon. Upon solidification, pure iron forms a body-centered-cubic solid that is stable down to 2541°F (1394°C). Known as *delta ferrite*, this phase is present only at extreme elevated temperatures and has little engineer-

FIGURE 4.10   The iron–carbon equilibrium diagram: α, ferrite; γ, austenite; δ, δ–ferrite; Fe_3C, cementite.

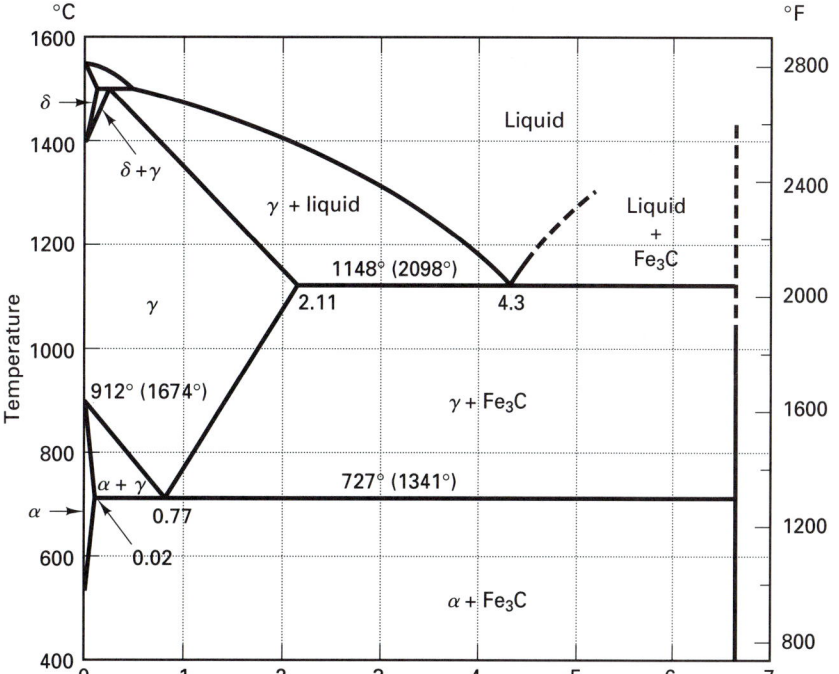

ing importance. From 2541 to 1674°F (1394 to 912°C), pure iron assumes a face-centered-cubic structure known as *austenite* (γ) in honor of the famed metallurgist Sir Robert Austen of England. Key features of austenite are the high formability that is characteristic of the face-centered-cubic structure and the high solubility of carbon (over 2% carbon can be dissolved in austenite). Hot forming of steel takes advantage of the high ductility and chemical uniformity of austenite. Most of the heat treatments of steel begin with the single-phase austenite structure. *Alpha-ferrite*, or more commonly just *ferrite*, is the stable form of iron at temperatures below 1674°F (912°C). This body-centered-cubic structure can hold only 0.02 wt % carbon in solid solution and forces the creation of a two-phase mixture in most steels. The only other change to occur upon further cooling is the non-magnetic-to-magnetic transition at the Curie point of 1418°F (770°C). Because this transition is not associated with any change in phase (but is an atomic-level transition), it does not appear on the equilibrium phase diagram.

The fourth single phase is the stoichiometric intermetallic compound, Fe_3C, which goes by the name *cementite*, or iron–carbide. Like most intermetallics, it is quite hard and brittle, and care should be exercised in controlling the structures in which it occurs. Alloys with excessive amounts of cementite, or cementite in undesirable form, tend to have brittle characteristics. Because cementite dissociates prior to melting, its exact melting point is unknown, and the liquidus line remains undetermined in the high-carbon region of the diagram.

Three distinct three-phase reactions can also be identified in the diagram. At 2723°F (1495°C), a *peritectic* occurs for alloys with a low weight percentage of carbon. Because of its high temperature and the extensive single-phase austenite region immediately below it, the peritectic reaction rarely assumes any engineering significance. A *eutectic* is observed at 2098°F (1148°C), with the eutectic composition of 4.3% carbon. All alloys containing more than 2.11% carbon will experience the eutectic reaction and are classified by the general term *cast irons*. The final three-phase reaction is a *eutectoid* at 1341°F (727°C) with the eutectoid composition of 0.77 wt % carbon. Alloys with less than 2.11% carbon miss the eutectic reaction and form a two-phase mixture when they cool through the eutectoid. These alloys are known as *steels*. Thus the point of maximum carbon solubility in iron, 2.11 wt %, forms an arbitrary separation between steels and cast irons.

■ 4.5 STEELS AND THE SIMPLIFIED IRON–CARBON DIAGRAM

If we focus on the materials normally known as steel, the diagram can be simplified considerably. Those portions near the delta phase (or peritectic) region and those with greater than 2% carbon are of little significance and can be deleted. The resulting diagram, such as the one presented as Figure 4-11, focuses on the eutectoid reaction and is quite useful in understanding the properties and processing of steel.

The key transition in this diagram is the conversion of single-phase austenite (γ) to the two-phase ferrite plus carbide mixture as the temperature drops. Control of this reaction, which arises as a result of the drastically different carbon solubilities of the face-centered and body-centered structures, enables a wide range of properties to be achieved through heat treatment.

To begin to understand these processes, consider a steel of the eutectoid composition, 0.77% carbon, being slow cooled along line *x–x'* in Figure 4-11. At the upper temperatures, only austenite is present, the 0.77% carbon being dissolved in solid solution within the face-centered structure. When the steel cools to 1341°F (727°C), several changes occur simultaneously. The iron wants to change from the face-centered-cubic austenite structure

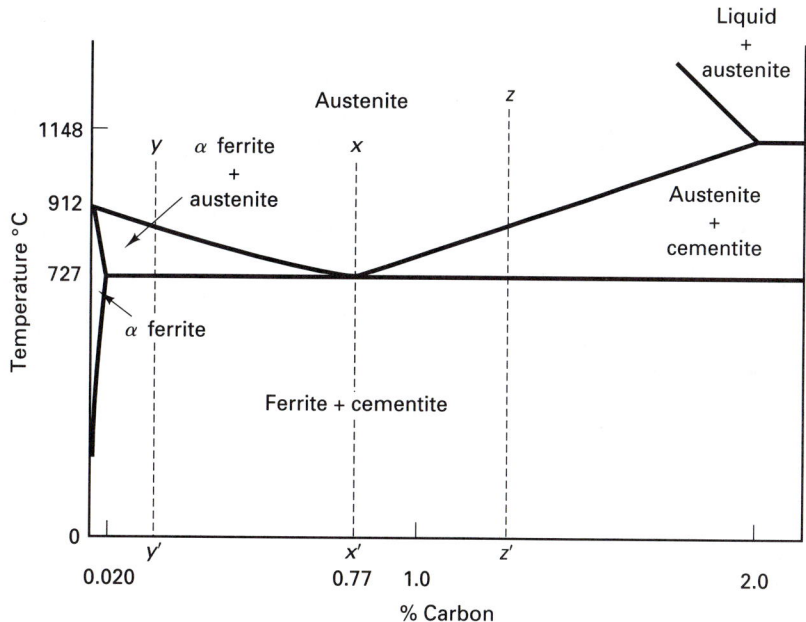

FIGURE 4.11   Simplified iron–carbon phase diagram.

to the body-centered-cubic ferrite structure, but the ferrite can only contain 0.02% carbon in solid solution. The rejected carbon forms the carbon-rich intermetallic (Fe_3C) known as cementite. The net reaction at the eutectoid, therefore, is

$$\underset{\substack{0.77\%C \\ fcc}}{\text{austenite}} \rightarrow \underset{\substack{0.02\%C \\ bcc}}{\text{ferrite}} + \underset{6.67\%C}{\text{cementite}}$$

Since the chemical separation occurs entirely within crystalline solids, the resultant structure is a fine mechanical mixture of ferrite and cementite. Specimens prepared by polishing and etching in a weak solution of nitric acid and alcohol reveal a lamellar structure composed of alternating layers or plates, as shown in Figure 4-12. Since it forms from a fixed composition at a fixed temperature, this structure has its own set of characteristic properties (even though it is composed of two distinct phases) and goes by the name *pearlite* because of its resemblance to mother-of-pearl when viewed at low magnification.

Steels having less than the eutectoid amount of carbon (less than 0.77%) are called *hypoeutectoid steels* (*hypo*- means less than). Consider now the cooling of a typical hypoeutectoid alloy along line *y–y′* in Figure 4-11. At high temperatures the material is entirely austenite. Upon cooling, however, it enters a region where the stable phases are ferrite and austenite. Tie-line and lever-law calculations show that the low-carbon ferrite nucleates and grows, leaving the remaining austenite richer in carbon. At 1341°F (727°C), the remaining austenite is of the eutectoid composition (0.77% carbon), and further cooling transforms it to pearlite. The resulting structure is a mixture of primary or proeutectoid ferrite (ferrite that forms before the eutectoid reaction) and regions of pearlite as shown in Figure 4-13.

Hypereutectoid steels (*hyper*- means greater than) are those that contain more than the eutectoid amount of carbon. When such a steel cools, as along line *z–z′* in Figure 4-11, the process is similar to the hypoeutectoid case, except that the primary or proeutectoid

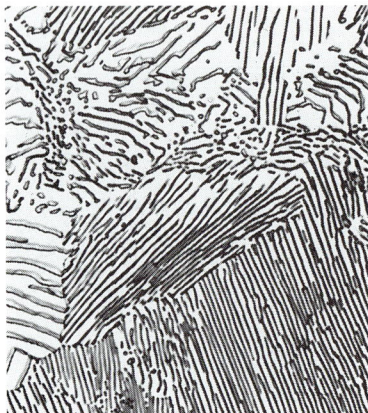

FIGURE 4.12 Pearlite; 1000×. *(Courtesy of USX Corporation.)*

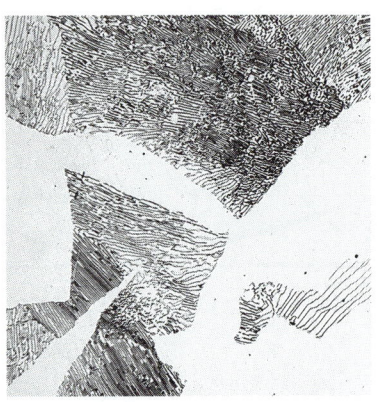

FIGURE 4.13 Photomicrograph of a hypoeutectoid steel showing regions of ferrite (white) and pearlite; 500×. *(Courtesy of USX Corporation.)*

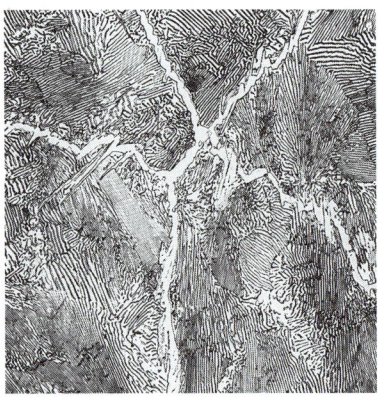

FIGURE 4.14 Photomicrograph of a hypereutectoid steel showing primary cementite along grain boundaries; 500×. *(Courtesy of USX Corporation.)*

phase is now cementite instead of ferrite. As the carbon-rich phase forms, the remaining austenite decreases in carbon content, again reaching the eutectoid composition at 1341°F (727°C). As before, this austenite transforms to pearlite upon slow cooling through the eutectoid temperature. Figure 4-14 is a photomicrograph of the resulting structure, which consists of primary cementite and ferrite. In this case the continuous network of primary cementite will cause the material to be extremely brittle.

It should be noted that the transitions just described are for equilibrium conditions, which can be approximated by slow cooling. Upon slow heating, the transitions will occur in the reverse manner. However, when the alloys are cooled rapidly, entirely different results may be obtained, since sufficient time may not be provided for the normal phase reactions to occur. In these cases, the equilibrium phase diagram is no longer a valid tool for engineering analysis. Since the rapid-cool processes are important in the heat treatment of steels and other metals, their characteristics are discussed in Chapter 5, and new tools introduced to aid our understanding.

■ 4.6 CAST IRONS

Iron–carbon alloys with more than 2.11% carbon experience the eutectic reaction during cooling and are known as *cast irons*. Being relatively inexpensive, with good fluidity and rather low liquidus temperatures, they are readily cast and occupy an important place in engineering applications.

Most commercial cast irons also contain a significant amount of silicon, the general composition being 2.0 to 4.0% carbon, 0.5 to 3.0% silicon, less than 1.0% manganese, and less than 0.2% sulfur. Silicon produces two major effects. First, it partially substitutes for carbon, so that use of the phase diagram requires replacing the weight percent carbon scale with a *carbon equivalent*. Several formulations exist to compute carbon equivalent, the simplest being the percentage of carbon plus one-third the percentage of silicon:

$$\text{carbon equivalent} = \text{wt \% carbon} + \tfrac{1}{3} \text{ wt \% silicon}$$

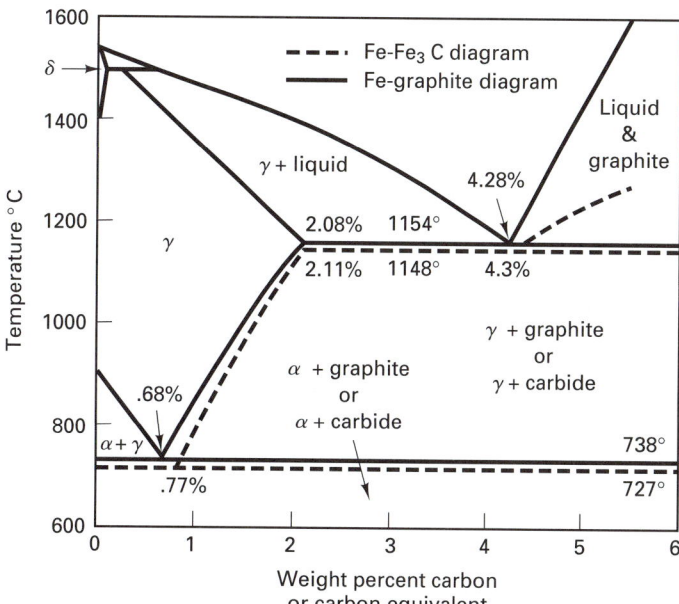

FIGURE 4.15 Iron–carbon diagram showing two possible high-carbon phases. Solid lines, iron–graphite system; dashed lines, iron–cementite (or iron-carbide).

As a second effect, silicon tends to promote the formation of *graphite* as the carbon-rich single phase instead of the Fe_3C intermetallic. Thus the eutectic reaction now has two distinct possibilities, as shown in the modified phase diagram of Figure 4-15:

$$liquid \quad \rightarrow \quad austenite + Fe_3C$$

$$liquid \quad \rightarrow \quad austenite + graphite$$

The final microstructure of cast iron has two possible extremes: (1) all of the carbon-rich phase being Fe_3C, and (2) all of the carbon-rich phase being graphite. In practice, both of these extremes can be approached by controlling both the chemistry and the other process variables. Graphite formation is promoted by slow cooling, high carbon and silicon contents, heavy section sizes, inoculation practices, and the presence of sulfur, phosphorus, aluminum, magnesium, antimony, tin, copper, nickel, and cobalt. Cementite (Fe_3C) is favored by fast cooling, low carbon and silicon levels, thin sections, and alloy additions of titanium, vanadium, zirconium, chromium, manganese, and molybdenum.

Various types of cast iron are produced, depending on the chemical composition, cooling rate, and the type and amount of inoculants that are used. *Gray cast iron,* the least expensive and most common variety, is characterized by those features that promote the formation of graphite. Typical compositions range from 2.5 to 4.0% carbon, 1.0 to 3.0% silicon, and 0.4 to 1.0% manganese. The microstructure consists of three-dimensional graphite flakes (that form during the eutectic reaction) dispersed in a matrix of ferrite, pearlite, or other iron-based structure (which forms from austenite during the eutectoid reaction). The various possibilities for the matrix structure are discussed further in Chapter 5. Figure 4-16 presents a typical section through gray cast iron, showing the graphite flakes in an unetched and etched matrix. Because the graphite flakes have no appreciable strength, they act essentially as voids in the structure. Moreover, the pointed edges of the

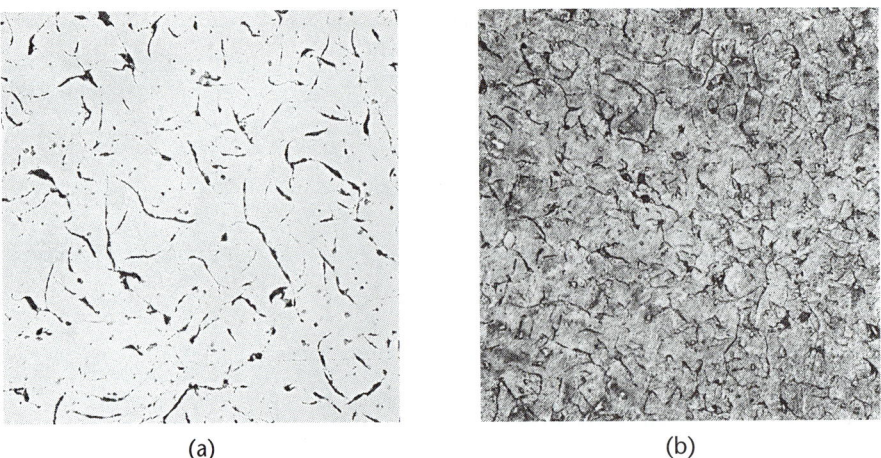

(a)   (b)

FIGURE 4.16 Photomicrographs of typical gray cast iron; 1000×: (a) unetched; (b) etched. *(Courtesy of Bethlehem Steel Corporation.)*

FIGURE 4.17 Fracture surfaces of (a) gray, (b) white, and (c) malleable cast irons. *(Courtesy of Iron Castings Society, Rocky River, Ohio.)*

flakes act as preexisting notches or crack initiation sites, giving the material its characteristic brittle nature. Since a large portion of any fracture follows the graphite flakes, the freshly exposed fracture surfaces have a characteristic gray appearance (Figure 4-17), and a graphite smudge can usually be obtained if one rubs a finger across it.

The size and shape of the graphite flakes have considerable effect on the overall properties of gray cast iron. When maximum strength is desired, small, uniformly distributed flakes are preferred with a minimum amount of intersection. A more effective means of controlling strength, however, is through control of the matrix structure. Gray cast iron is normally sold by "class," with the class number corresponding to the minimum tensile strength in thousands of pounds per square inch. Class 20 iron (minimum tensile strength of 20,000 psi) consists of high carbon-equivalent metal with a ferrite matrix. Higher strengths, up to class 40, are obtainable with lower carbon equivalents and a pearlite matrix. Above class 40, alloying is required to provide solid solution strengthening, and heat-treatment practices are often employed to modify the matrix. Gray cast irons can be obtained up through class 80, but in all cases the presence of the graphite flakes results in extremely low ductility.

Gray cast irons possess excellent compressive strengths (compressive forces do not promote crack propagation), excellent machinability (graphite acts to break up the chips and lubricate contact surfaces), good wear resistance in adhesive wear conditions (graphite flakes self-lubricate), and outstanding sound and damping characteristics (graphite flakes absorb transmitted energy). High silicon contents promote good corrosion resistance and the enhanced fluidity desired for casting operations. For these reasons, coupled with its low cost, gray cast iron is specified for a number of applications, including large equipment parts that are subjected to compressive loads and vibrations.

White cast iron has essentially all of its carbon in the form of iron carbide (Fe_3C) and receives its name from the white surface that appears when the material is fractured (Figure 4-17). Features promoting its formation are those that favor cementite over graphite: a low carbon equivalent (1.8 to 3.6% carbon, 0.5 to 1.9% silicon, and 0.25 to 0.8% manganese) and rapid cooling.

By virtue of the large amounts of iron carbide, white cast iron is very hard and brittle and finds applications where high abrasion resistance is required. For these uses it is also common to promote the formation of the hard, wear-resistant *martensite* structure as the iron-rich matrix, either directly upon solidification or by subsequent heat treatment. (*Note*: This phase is described in Chapter 5.) In this manner, both phases will contribute to the wear-resistant characteristics of the material.

Other applications involve the formation of a white cast iron surface over a substrate or base of another material. For example, mill rolls that require extreme wear resistance may have a white cast iron surface and a steel interior. Variable cooling rates produced by tapered sections or metal chill bars placed in the molding sand can be used to produce white iron surfaces or sections on an otherwise gray iron casting. Where differential cooling rates are used to produce white and gray cast irons in the same component, there is generally a transition region of mixed white and gray irons, known as the *mottled zone*.

White iron can also be put through a controlled heat treatment where the cementite dissociates into its component elements, and some or all of the carbon is converted into irregular graphite spheroids (also referred to as clump or popcorn graphite). The product, known as *malleable cast iron*, has significantly greater ductility than that exhibited by gray cast iron because the more favorable graphite shape removes the internal notches. The rapid cooling required to produce the starting white iron structure restricts the size and thickness of malleable iron products such that most weigh less than 10 lb (4.5 kg).

Two types of malleable iron can be produced, depending on the nature of the thermal cycle. If the white iron is heated and held for a prolonged time just below the melting point, the carbon in the cementite converts to graphite (first-stage graphitization). Subsequent slow cooling through the eutectoid reaction causes the carbon-containing austenite to transform to ferrite and more graphite (second-stage graphitization). The resulting product, known as *ferritic malleable cast iron,* has properties consistent with its structure of irregular graphite spheroids in a ferrite matrix: 10% elongation, 35 ksi (240 MPa) yield strength, 50 ksi (345 MPa) tensile strength, and excellent impact strength, corrosion resistance, and machinability. The heat-treatment times, however, are quite lengthy, often involving over 100 hours at elevated temperature. Figure 4-18 shows a typical resulting structure.

If the material is cooled more rapidly through the eutectoid transformation, the carbon in the austenite does not form additional graphite but is retained in a pearlite or martensite matrix. These structures are stronger than ferrite, and the resulting product, known

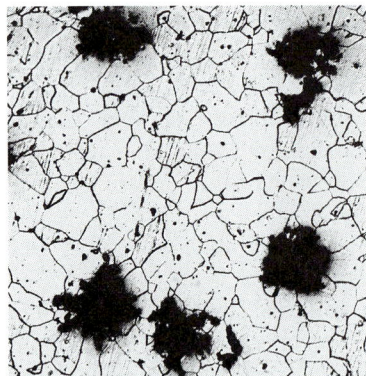

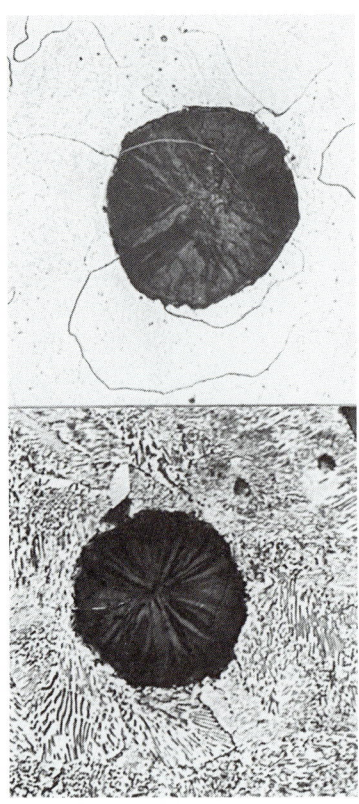

FIGURE 4.18 Photomicrograph of malleable iron showing the irregular graphite spheroids. *(Courtesy of Iron Castings Society, Rocky River, Ohio.)*

FIGURE 4.19 Ductile iron with (a) ferrite matrix and (b) pearlite matrix; 500×. Note spheroidal shape of the graphite nodule in each photo.

as *pearlitic malleable cast iron*, is characterized by higher strength and lower ductility than its ferritic counterpart. Typical properties range from 1 to 4% elongation, 45to 85 ksi (310 to 590 MPa) yield strength, and 65 to 105 ksi (450 to 725 MPa) tensile strength, with reduced machinability compared to the ferritic material.

The modified graphite structure of malleable iron provides quite an improvement in properties, but it would be even better if it could be obtained directly from solidification rather than by prolonged heat treatment at highly elevated temperatures. If a high-carbon-equivalent cast iron is sufficiently low in sulfur (either by original chemistry or by desulfurization), the addition of certain materials can promote graphite formation and change the morphology (shape) of the graphite product. Inoculation with ferrosilicon will promote the formation of graphite. If magnesium (in the form of MgFeSi or MgNi alloy) or cesium is also added to the liquid just prior to solidification, the graphite will form as smooth-surface spheres. The latter addition is known as a *nodulizer,* and the product as *ductile* or *nodular cast iron.* Subsequent control of cooling can produce a variety of matrix structures, with ferrite or pearlite being the most common (Figure 4-19). Properties span a wide range from 2 to 18% elongation, 40 to 90 ksi (275 to 620 MPa) yield strength, and 60 to 120 ksi (415 to 825 MPa) tensile strength.

The combination of good ductility, high strength, toughness, and castability makes ductile iron a rather attractive engineering material. Unfortunately, the cost of the nodulizer, higher-grade melting stock, better furnaces, and improved process control required for its manufacture all contribute to the increased cost of ductile iron. Recently, *austempered ductile iron* has emerged as a significant engineering material, offering the ability to cast intricate shapes with strength and wear-resistance properties that are similar to those of steel (with 8 to 10% lower density). To achieve these results, alloyed ductile iron is subjected to the austempering treatment described in Chapter 5. By proper alloying, ductile iron can also be made to have an austenite matrix. Known as austenitic ductile iron, this material offers good corrosion resistance, nonmagnetic behavior, and dimensional stability at elevated temperatures.

A fifth type of cast iron, *compacted graphite iron* (or vermicular graphite iron), has also begun to attract considerable attention. Produced by a method similar to that used to make ductile iron (a Mg–Ce–Ti addition is made), compacted graphite iron is characterized by a graphite structure intermediate to the flake graphite of gray iron and the nodular graphite of ductile iron and tends to possess the desirable properties and characteristics of each. Depending upon composition and section size, tensile strengths of 45 to 75 ksi (310 to 520 MPa) and yield strengths of 35 to 60 ksi (240 to 415 MPa) have been reported, with elongation values averaging 4%. Machinability, castability, damping capacity, and thermal conductivity all approach those of gray cast iron. Impact and fatigue properties approach those of ductile iron. Areas of application tend to be those requiring high strength with good machinability, thermal conductivity, and thermal shock resistance.

■ KEY WORDS

austenite	freezing range	pearlite
carbon equivalent	graphite	peritectic
cast iron	gray cast iron	peritectoid
cementite	hypereutectoid	phase
compacted graphite cast iron	hypoeutectoid	phase diagram
complete solubility	inoculation	solidus
composition	intermetallic compound	solubility limit
cooling curve	lever law	solvus
cored structure	liquidus	steel
ductile cast iron	malleable cast iron	stoichiometric
equilibrium	monotectic	syntectic
eutectic	nodular cast iron	three-phase reaction
eutectoid	nodulizer	tie-line
ferrite	nonstoichiometric	white cast iron

■ REVIEW QUESTIONS

1. What are some features that are useful in defining a phase?
2. Supplement the examples provided in the text with another example of a single phase that is each of the following: continuous, discontinuous, gaseous, and a solution.
3. What is an equilibrium phase diagram?
4. What three primary variables are generally considered in equilibrium phase diagrams?
5. Why is a pressure–temperature phase diagram not that useful for many engineering applications?
6. What is a cooling curve?
7. What features in a cooling curve indicate some form of change in a material's structure?
8. What is a solubility limit, and how might it be determined?
9. In general, how does the solubility of one material in another change as temperature is increased?

10. What types of changes occur upon crossing a liquidus line? A solidus line? A solvus line?

11. What three pieces of information can be obtained for each point in an equilibrium phase diagram?

12. What is a tie-line? For what types of phase diagram regions would it be useful?

13. What points on a tie-line are used to determine the chemistry (or composition) of the component phases?

14. What tool can be used to compute the relative amounts of the component phases in a two-phase mixture? How does this tool work?

15. Under what conditions is a cored structure produced? How is it formed?

16. What features in a phase diagram can be used to identify three-phase reactions?

17. What is the general form of a eutectic reaction?

18. Why are alloys of eutectic composition attractive for casting and as filler metals in soldering and brazing?

19. What is a stoichiometric intermetallic compound, and how would it appear in a temperature–composition phase diagram? How would a nonstoichiometric intermetallic compound appear?

20. What type of mechanical properties would be expected for intermetallic compounds?

21. In what form(s) might intermetallic compounds be undesirable in an engineering material? In what form(s) might they be attractive?

22. What are the four single phases in the iron–iron carbide diagram? Answer with both phase diagram notation and assigned name.

23. What feature in the iron–carbon diagram is used to distinguish between cast irons and steels?

24. What features of austenite make it attractive for forming operations and as a starting structure for many heat treatments?

25. Which of the three-phase reactions in the iron–carbon diagram is most important in understanding the behavior of steels? Write this reaction in terms of the interacting phases and their composition.

26. Describe the ability of iron to dissolve carbon in solution when in the form of austenite (the elevated temperature phase). When in the form of ferrite at room temperature.

27. What is pearlite? What is its structure?

28. What is a hypoeutectoid steel, and what structure will it assume upon slow cooling? What is a hypereutectoid steel and how will its structure differ from that of a hypoeutectoid?

29. In addition to iron and carbon, what other elements are generally present in a cast iron, and in what amounts? What important effects are associated with the presence of silicon?

30. What are the two possible high-carbon phases in cast irons? What features tend to favor the formation of each?

31. Describe the microstructure of gray cast iron.

32. Which of the structural units is generally altered to increase the strength of a gray cast iron.

33. What are some of the attractive engineering properties of gray cast iron?

34. What is the dominant mechanical property of white cast iron?

35. What structural feature is responsible for the increased ductility and fracture resistance of malleable cast iron?

36. What requirements of ductile iron manufacture are responsible for its increased cost over materials such as gray cast iron?

37. Compacted graphite iron has a structure and properties intermediate to what two other types of cast irons?

Chapter 4 CASE STUDY

the blacksmith anvils

You are an officer in the Western-America Blacksmith Association, and you have determined that a number of your members would like to have a modern equivalent of an 1870-vintage blacksmith anvil. Your objective is to replicate the 1870's design, but utilize the features of today's engineering materials. You ultimately want to identify a producer who will make a limited number of these items for sale and distribution through your monthly magazine. The proposed design is a large forging anvil that has a total length of 20-inches. The top surfaces must be resistant to wear, deformation and chipping.

Estimated mechanical properties call for a yield strength in excess of 70 ksi, an elongation greater than 2%, and a Brinell hardness of 200 or more on the top surface. You feel confident that you will be able to secure a minimum of 500 orders.

1. Discuss the various properties that this part must possess to adequately perform its intended task.
2. Discuss the various concerns that would influence the proposed method of fabricating the anvils.
3. Assuming that the anvils will be made from some form of ferrous metal, consider the properties of the various types of cast irons and steels with regard to this application. Which material would you recommend? Why?
4. How would you propose that a production run of 500 replica anvils be produced?

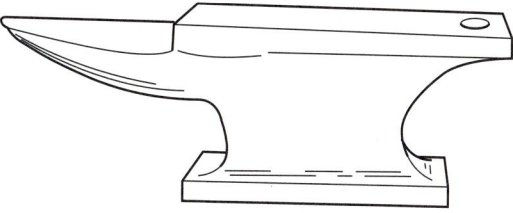

FIGURE CS4

CHAPTER 5

HEAT TREATMENT

5.1 INTRODUCTION
5.2 PROCESSING HEAT TREATMENTS
   Equilibrium Diagrams as Aids
   Processing Heat Treatments for Steel
   Heat Treatments for Nonferrous Metals
5.3 HEAT TREATMENTS USED TO INCREASE STRENGTH
5.4 STRENGTHENING HEAT TREATMENTS FOR NONFERROUS METALS
   Precipitation or Age Hardening
5.5 STRENGTHENING HEAT TREATMENTS FOR STEEL
   Isothermal Transformation Diagram
   Tempering of Martensite
   Continuous Cooling Transformations
   Jominy Test for Hardenability
   Hardenability Considerations
   Quench Media
   Design Concerns in the Heat Treatment of Steel
   Techniques to Reduce Cracking
   Ausforming
5.6 SURFACE HARDENING OF STEEL
   Selective Heating Techniques
   Techniques Involving Altered Surface Chemistry
5.7 FURNACES
   Furnace Types
   Furnace Controls
5.8 HEAT TREATMENT AND ENERGY
Case Study: A FLYING CHIP FROM A SLEDGEHAMMER

■ 5.1 INTRODUCTION

The material on cast irons presented in Chapter 4 has already shown that variations in the heating or cooling of a metal can be used to produce different structures with different properties. Many engineering materials can be characterized not just by one set of properties but by an entire spectrum that can be selected and varied at will. *Heat treatment is the controlled heating and cooling of metals for the purpose of altering their properties* and can perform this function without a concurrent change in product shape. Because both physical and mechanical properties can be altered by heat treatment, it is one of the most important and widely used manufacturing processes.

While the term *heat treatment* applies only to processes where the heating and cooling are done for the specific purpose of altering properties, heating and cooling often occur as incidental phases of other manufacturing processes, such as hot forming or welding. Properties will be altered just as though an intentional heat treatment had been performed, and the results can be beneficial or harmful. Thus the designer who selects material and the engineer who specifies its processing must be aware of the possible changes in properties during heating and cooling. Heat treatment must be understood and correlated with the other manufacturing processes if effective results are to be obtained. Both the theory of heat treatment and a survey of the various processes are presented in this chapter.

■ 5.2 PROCESSING HEAT TREATMENTS

The term *heat treatment* is often associated with those thermal processes that increase the strength of a material, but the broader definition permits inclusion of another set of processes which we will call *processing heat treatments*. These are performed as a means of preparing the material for fabrication. Specific objectives may be the improvement of machining characteristics, the reduction of forming forces, or the restoration of ductility to enable further processing. Thus heat treatment becomes even more attractive. The same metal can now be softened for ease of fabrication, and then, through another heat treatment, given a totally different set of properties, for actual service.

Equilibrium Diagrams as Aids

Most of the processing heat treatments involve rather slow cooling or extended times at elevated temperatures. These conditions tend to approximate equilibrium, and the resulting structures, therefore, can be reasonably predicted by the use of an equilibrium phase diagram. The diagrams can be used to indicate both the temperatures that must be attained to produce a desired starting structure, and the changes that will occur upon subsequent cooling. It should be noted, however, that the diagrams are for true equilibrium conditions, and departures from equilibrium may lead to substantially different results.

Processing Heat Treatments for Steel

Because many of the processing heat treatments are applied to plain-carbon and low-alloy steels, they are presented here with the simplified iron–carbon equilibrium diagram of Figure 4-11 serving as a reference guide. Figure 5-1 shows this diagram with the key transition lines labeled in standard notation. The eutectoid line is designated by the symbol A_1, and A_3 designates the boundary between austenite and ferrite + austenite.[*] The transition from austenite to austenite + cementite is designated as the A_{cm} line.

A number of process heat-treating operations are classified under the general term of *annealing*. These may be employed to reduce strength or hardness, remove residual stresses, improve toughness, restore ductility, refine grain size, reduce segregation, or alter the electrical or magnetic properties of the material. By producing a certain desired structure, characteristics can be imparted that are favorable to subsequent operations or applications. The temperature, cooling rate, and specific details of the process are determined by the material being treated and the objectives of the treatment.

In the process of *full annealing*, hypoeutectoid steels (less than 0.77% carbon) are heated to 50 to 100°F (30 to 60°C) above the A_3 temperature, held for sufficient time to convert the structure to homogeneous single-phase austenite of uniform composition and temperature, and then slowly cooled at a controlled rate to below the A_1 temperature. Cooling is usually done in the furnace by decreasing the temperature by 20 to 50°F (10 to 30°C) per hour to at least 50°F (30°C) below the A_1. At this point the metal can be removed from the furnace and air cooled to room temperature. The resulting structure is one of coarse pearlite (widely spaced lamellae) with excess ferrite in amounts predicted by the phase diagram. In this condition, the steel is quite soft and ductile.

The procedure to full-anneal a hypereutectoid alloy (greater than 0.77% carbon) is basically the same, except that the original heating is only into the austenite plus cementite

---

[*]Historically, an A_2 line once appeared between the A_1 and A_3. This line designated the magnetic property change known as the Curie point. This transition was later shown to be an atomic change, not a change in phase, and the line was deleted from the diagram without relabeling.

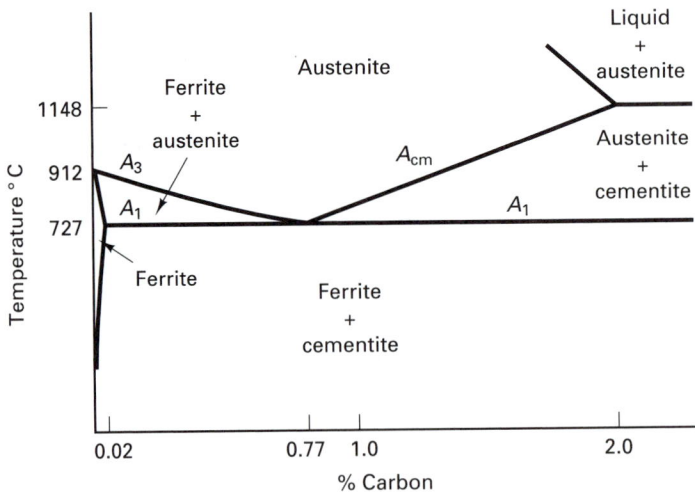

FIGURE 5.1   Simplified iron–carbon phase diagram for steels with transition lines labeled in standard notation.

region (50 to 100°F above the A_1). If the material were to be slow cooled from the all-austenite region, a continuous network of cementite may form on the grain boundaries and make the material brittle. When properly annealed, the structure of a hypereutectoid steel will be coarse pearlite with excess cementite in dispersed spheroidal form.

Full anneals are time consuming, and considerable energy is needed to maintain the elevated temperatures required during soaking and furnace cooling. When the maximum softness is not required and cost savings are desirable, *normalizing* may be specified. In this process, the steel is heated to 100°F (60°C) above the A_3 (hypoeutectoid) or A_{cm} (hypereutectoid) temperature, held at this temperature to produce uniform austenite, and then removed from the furnace and allowed to cool in still air. The resultant structures and properties depend on the subsequent cooling rate. Although wide variations are possible depending on the size and geometry of the metal, fine pearlite with excess ferrite or cementite is generally produced.

One should note a key difference between full annealing and normalizing. In the full anneal, the furnace imposes identical cooling conditions at all locations within the metal, which produces identical properties. With normalizing, the cooling will be different at different locations. Properties will vary between surface and interior, and different thickness regions will also have different properties. When subsequent processing involves a substantial amount of machining that may be automated, the added cost of a full anneal may be justified, since it produces a product with uniform machinability at all locations.

When cold working has severely strain hardened a metal, it is often desirable to restore the ductility, either for service or to permit further processing without danger of fracture. This is often achieved through the recrystallization process described in Chapter 3. When the material is a low-carbon steel (<0.25% carbon), however, the specific procedure is known as a *process anneal*. The steel is heated to a temperature slightly below the A_1, held long enough to induce recrystallization of the dominant ferrite phase, and then cooled at any desired rate (usually in still air). Since all of the temperatures remain within the same phase region, the process simply induces a change in phase morphology (size, shape, and distribution). The material is not heated to as high a temperature as in the full anneal or normalizing process, so a process anneal is somewhat cheaper and tends to produce less scaling.

A *stress-relief anneal* may be employed to reduce the residual stresses in large steel castings, welded assemblies, and cold-formed products. Parts are heated to temperatures below the A_1 (1000 to 1200°F or 550 to 650°C), held for a period of time, and then slow cooled. Times and temperatures vary with the condition of the component.

When high-carbon steels (>0.60% carbon) will undergo extensive machining or cold-forming, a process known as *spheroidization* is often employed. Here the objective is to produce a structure in which all of the cementite is in the form of small spheroids or globules dispersed throughout a ferrite matrix. This can be accomplished by a variety of techniques, including (1) prolonged heating at a temperature just below the A_1 followed by relatively slow cooling, (2) prolonged cycling between temperatures slightly above and slightly below the A_1, or (3) in the case of tool or high-alloy steels, heating to 1400 to 1500°F (750 to 800°C) or higher and holding at this temperature for several hours, followed by slow cooling.

Although the selection of a processing heat treatment often depends on the desired objectives, steel composition strongly influences the choice. Process anneals are restricted to low-carbon steels, and spheroidization is a treatment for high-carbon material. Normalizing and full annealing can be applied to all carbon contents, but even here, preferences are noted. Since different cooling rates do not produce a wide variation of properties in low-carbon steels, the air cool of a normalizing treatment often produces acceptable uniformity. For higher carbon contents, such as the 0.4 to 0.6% range, different cooling rates can produce wide property variations, and the uniform furnace cooling of a full anneal is often preferred. Figure 5-2 provides a graphical summary of the process heat treatments.

FIGURE 5.2   Graphical summary of the process heat treatments for steels on an equilibrium diagram.

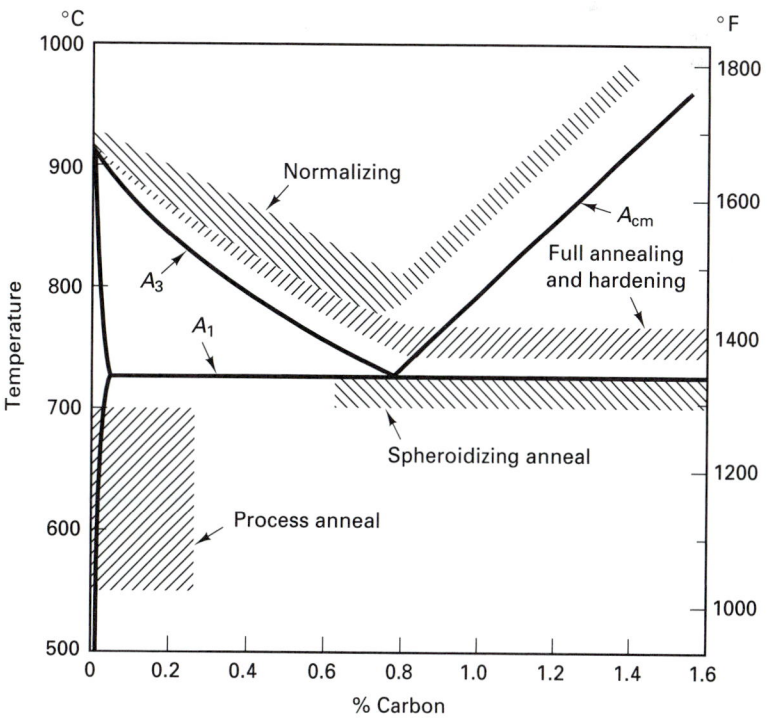

Heat Treatments for Nonferrous Metals

Most of the nonferrous metals do not have the significant phase transitions observed in the iron–carbon system, and for them, the process heat treatments do not play such a significant role. Aside from the strengthening treatment of precipitation hardening, which is discussed later, the nonferrous metals are usually heat-treated for three purposes: (1) to produce a uniform, homogeneous structure, (2) to provide stress relief, or (3) to bring about recrystallization. Castings that have been cooled too rapidly can possess a segregated solidification structure known as coring (discussed more fully in Chapter 4). *Homogenization* can be achieved by heating to moderate temperatures and then holding for a sufficient time to allow thorough diffusion to take place. Similarly, the internal stresses that are produced by forming, welding, or brazing can be reduced by heating for several hours at relatively low temperatures. Recrystallization (discussed in Chapter 3) is a function of the particular metal, the amount of prior straining, and the desired recrystallization time. In general, the more a metal has been strained, the lower the recrystallization temperature or the shorter the time. Without prior straining, however, recrystallization will not occur and heating will only produce undesirable grain growth.

■ 5.3 HEAT TREATMENTS USED TO INCREASE STRENGTH

Six major mechanisms are available to increase the strength of metals:

1. Solid-solution strengthening
2. Strain hardening
3. Grain-size refinement
4. Precipitation hardening
5. Dispersion hardening
6. Phase transformations

All of these can be induced or altered by heat treatment, but not all are applicable to any given metal.

In *solid-solution strengthening*, a base metal dissolves other atoms in solid solution, either as *substitutional solutions*, where the new atoms occupy sites in the host crystal lattice, or as *interstitial solutions*, where the new atoms squeeze into "holes" between the atoms of the base lattice. The amount of strengthening depends on the amount of dissolved solute and the size difference of the atoms involved. Since distortion of the host structure makes dislocation movement more difficult, the greater the size difference, the more effective the addition.

Strain hardening (discussed in Chapter 3) produces an increase in strength by means of plastic deformation under cold-working conditions

Because grain boundaries act as barriers to dislocation motion, a metal with small grains tends to be stronger than the same metal with larger grains. Thus *grain-size refinement* can be used to increase strength, except at elevated temperatures, where failure is by a grain-boundary diffusion-controlled creep mechanism. It is important to note that grain-size refinement is one of the few processes that can simultaneously improve both strength and ductility.

Precipitation hardening, or *age hardening*, is a method whereby strength is obtained from a nonequilibrium structure produced by a three-step heat treatment. Details of this method are provided in the following section.

Strength obtained from distinct second-phase particles dispersed throughout a base material is known as *dispersion hardening*. To be effective, the dispersed particles should be stronger than the matrix, adding strength through both their reinforcing action and the additional surface barriers presented to dislocation movement.

Phase transformation strengthening involves those alloys that can be heated to form a single phase at elevated temperature which subsequently transforms to one or more low-temperature phases upon cooling. When this feature is used to increase strength, the cooling is usually rapid and the phases that are produced are usually nonequilibrium in nature.

■ 5.4 STRENGTHENING HEAT TREATMENTS FOR NONFERROUS METALS

All six of the mechanisms just described can be used to increase the strength of nonferrous metals. Solid-solution strengthening can impart strength to single-phase materials. Strain hardening is useful if sufficient ductility is present. Eutectic-forming alloys exhibit considerable dispersion hardening. Among all of the possibilities, however, the most effective strengthening mechanism for the nonferrous metals is precipitation hardening.

Precipitation or Age Hardening

Some alloy systems, mostly nonferrous, possess a sloping solvus line, such that certain alloys can be heated to form a single-phase solid solution and, because of the decreasing solubility, will revert to two distinct phases upon cooling. If the heated single phase is cooled rapidly (quenched), it may be possible to produce a supersaturated single phase where the component that normally forms the second phase remains trapped within the parent lattice. If this substance is then heated within the two-phase region, an *aging* process can occur in which the excess solute atoms precipitate out of the supersaturated matrix to form a controllable nonequilibrium structure.

As an example, consider the aluminum-rich portion of the aluminum–copper phase diagram presented in Figures 5-3 and 5-4. Follow an alloy composed of 96% aluminum and 4% copper being slowly cooled from 1000°F. Upon crossing the solvus line at 930°F, theta phase is precipitated out of the alpha-phase solid solution because the solubility of copper in aluminum decreases from 5.65% at 1018°F to less than 0.2% at room temperature. If this same alloy were rapidly cooled from 1000°F, there would not be sufficient time for the normal two-phase structure to form and the alpha phase would be retained in a highly supersaturated state. Because this supersaturated condition is not normal, the excess copper would like to precipitate out of solution and coalesce into theta-phase particles, where theta phase consists of an intermetallic compound with the formula $CuAl_2$. Atom movement or diffusion is required, and therefore time at elevated temperature may have to be provided. If the temperature is now raised to around 350°F, high-copper clusters will begin to form. With proper control, this process can significantly increase the yield strength of the alloy.

Precipitation hardening is a three-step controlled heat treatment. The first step, known as *solution treatment*, is a heating operation designed to produce an elevated temperature single-phase solid solution. This heating should not exceed the eutectic temperature or there might be melting if a cored structure were present. After soaking to assure a uniform chemistry single phase, the alloy is *quenched* (rapidly cooled), usually in water, to prevent diffusion and produce a supersaturated solid solution. In this state, the material is often soft, and can be straightened, formed, or machined.

For the third step, precipitation-hardening materials can be divided into two types: (1) *naturally aging*, for which room temperature is capable of providing the diffusion necessary to convert the unstable supersaturated solution into the stable two-phase structure,

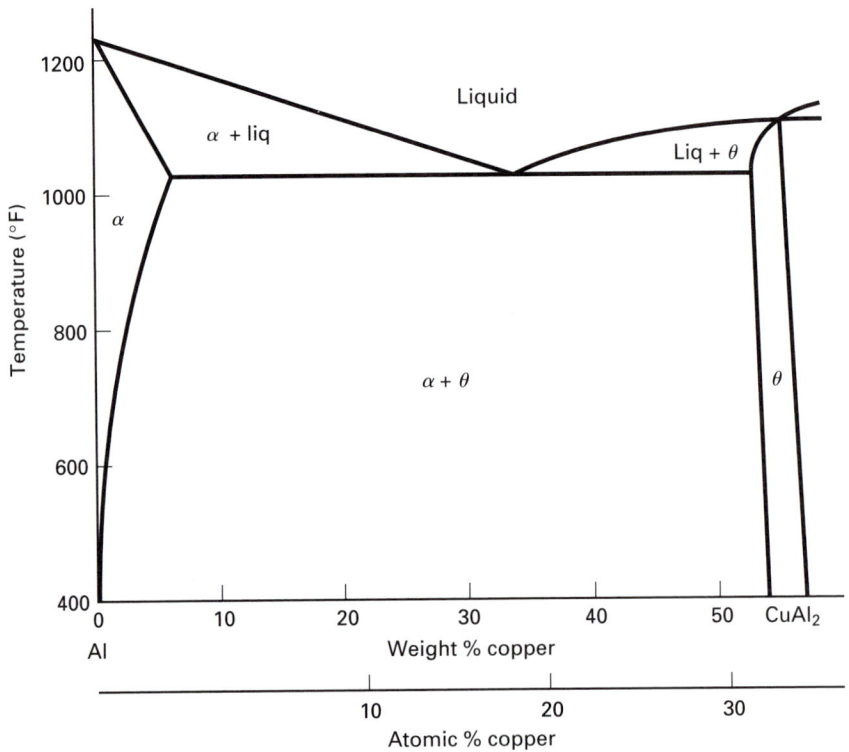

FIGURE 5.3   High-aluminum section of the aluminum–copper equilibrium phase diagram.

and (2) *artificially aging,* for which elevated temperatures are necessary to promote diffusion. With the first type, refrigeration may be required to retain the after-quench condition of softness, as with aluminum alloy rivets. Upon removal from refrigeration, they are easily headed and then progress to full strength after several days at room temperature. The properties of the second type can be readily controlled by controlling the time and temperature of the elevated temperature aging.

Aging is a continuous process which begins with the clustering of solute atoms on distinct planes within the parent lattice. Various transitions may then occur, leading ultimately to the formation of a distinct second phase with its own characteristic crystal structure.

A key concept in this sequence is that of *coherency.* If the clustered solute atoms continue to occupy lattice sites of the parent structure, the crystallographic planes remain continuous in all directions, and the solute clusters tend to distort (strain) the adjacent lattice for a sizable distance in all directions. For this reason, a small cluster appears to be much larger with respect to its ability to impede dislocation motion (i.e., impart strength). When the clusters reach a certain size, however, the associated strain becomes so great that they tend to break free of the parent structure, forming distinct second-phase particles with well-defined interphase boundaries. Coherency is lost and the mechanism of strengthening reverts to *dispersion hardening,* wherein the particles present only their physical dimensions as effective dislocation blocks. Strength and hardness begin to decrease, and the material is said to be *overaged.* Figure 5-5 presents a family of aging curves for the 4%

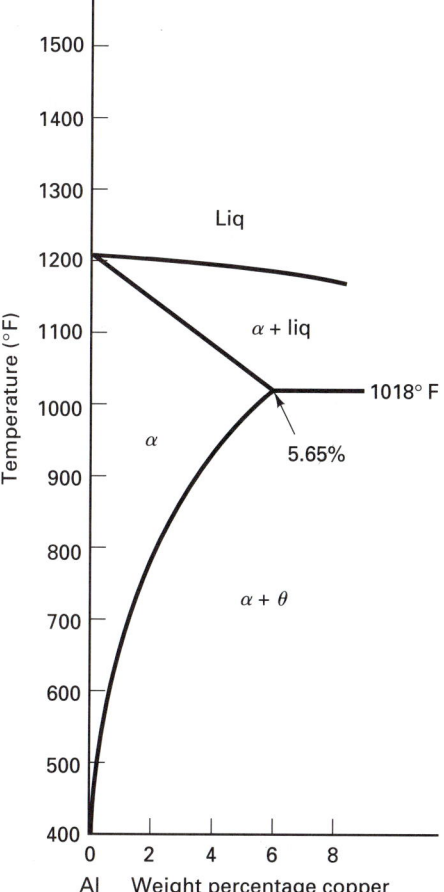

FIGURE 5.4   Enlargement of the solvus-line region of the aluminum–copper equilibrium diagram of Figure 5-3.

copper–96% aluminum alloy discussed earlier. For higher aging temperatures, the peak properties are achieved in a shorter time, but the optimum hardness (and strength) is not as great as that for lower aging temperatures. Selection of the aging conditions (temperature and time) is a decision that is made on the basis of desired strength, available equipment, and production considerations.

An attractive feature of artificial aging is the ability to stop the process at any stage by simple quenching. The structure and properties of that stage are then retained, provided that the material does not subsequently experience elevated temperature conditions that would reactivate diffusion. For example, if the alloy in Figure 5-5 were aged for one day at 375°F and then quenched, the metal would retain a hardness of 94 Vickers (and the associated strength) throughout its useful lifetime. If higher strength were desired, a lower temperature and longer time could be selected. Artificially aging alloys are quite popular because of the ability to "lock in" the peak properties that occur prior to overaging.

Precipitation hardening is an extremely effective strengthening mechanism and is responsible for the attractive engineering properties of a number of aluminum, copper, and magnesium alloys. In many cases the strength can be more than double that observed upon conventional cooling. Through special alloying, some age-hardenable ferrous alloys have also been produced.

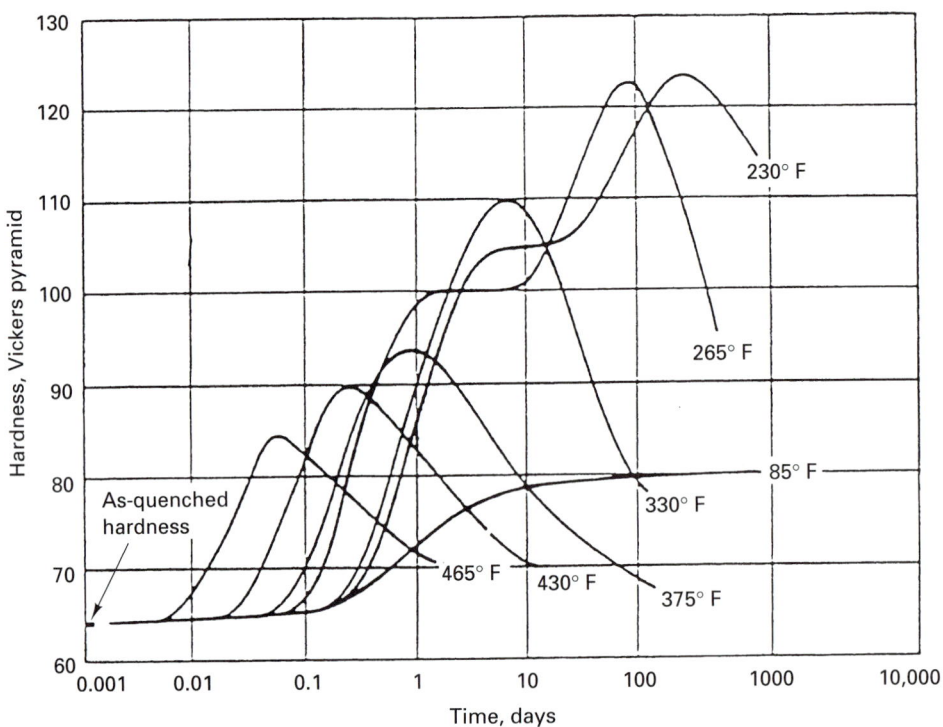

FIGURE 5.5 Aging curves for the Al–4%Cu alloy at various temperatures. (Adapted from *Journal of the Institute for Metals,* Vol. 79 p. 321, 1951.)

■ 5.5 Strengthening Heat Treatments for Steel

Iron-based metals have been heat treated for centuries, and today, over 90% of all heat-treatment operations are performed on steel. The striking changes that resulted from plunging red-hot steel into cold water or some other quenching medium were awe inspiring to the ancients. Those who performed such heat treatment in the making of swords and armor were looked upon as possessing unusual powers, and much superstition arose regarding the process. Because quality was directly related to the act of quenching, great importance was placed on the quenching medium that was used. For example, urine was thought to be a superior quenching medium, and that from a red-haired boy was deemed particularly effective, as was that from a 3-year-old goat fed only ferns.

Isothermal Transformation Diagram

It has only been within the last 100 years that the art of heat treating has begun to turn into a science. One of the major barriers to understanding was the fact that the strengthening treatments were nonequilibrium in nature. Minor variations in cooling often produced major variations in structure and properties.

One of the aids to understanding the nonequilibrium processes was the *isothermal transformation* (I-T) or *time-temperature-transformation* (T-T-T) *diagram.* The information in this diagram is obtained by heating thin specimens of a particular steel to produce uniform-chemistry austenite, "instantaneously" quenching to a temperature where austenite is not the stable phase, holding for variable periods of time, and observing the resultant products via photomicrographs.

For simplicity, consider a carbon steel of eutectoid composition (0.77% carbon) and the resulting T-T-T diagram of Figure 5-6. Above 1341°F (727°C), austenite is the stable phase. Below this temperature, the face-centered austenite would like to transform to body-centered ferrite and carbon-rich cementite. Two factors control the rate of transition: (1) the motivation or driving force for the change, and (2) the ability to form the desired products (i.e., the ability to rearrange the atoms through diffusion). The region below 1341°F in Figure 5-6 can be interpreted as follows. Zero time corresponds to a sample quenched "instantaneously" to its new, lower temperature. The structure is usually unstable austenite. As time passes (moving horizontally across the diagram), a line is encountered representing the start of transformation and a second line indicating completion of the phase change. At elevated temperatures (just below 1341°F), diffusion is rapid, but the rather sluggish driving force dominates the kinetics. At a low temperature, the driving force is high but diffusion is quite limited. The phase transformation kinetics are most rapid at a compromise intermediate temperature, resulting in the characteristic *C-curve* shape. The portion of the C that extends farthest to the left is known as the *nose* of the T-T-T diagram.

Let us now consider the products of the various phase transformations. If the transformation occurs between the A_1 temperature and the nose of the curve, the departure from equilibrium conditions is not very great. The austenite transforms into ferrite and cementite in the combined structure known as *pearlite* (as discussed in the equilibrium phase diagram description of Chapter 4). Because the diffusion rates are greater at higher temperatures, the lamellar spacing (separation distance between similar layers) will be larger for pearlite produced at higher temperatures. The pearlite formed near the A_1 temperature is known as *coarse pearlite* and the structure formed near the nose is called *fine pearlite*.

If the austenite is quenched to a temperature between the nose and the temperature designated as M_s, another product will form. The transformation is now a significant departure from equilibrium, and the diffusion required to form lamellar pearlite is no longer available. The metal still has the same goal: to change crystal structure from the face-centered austenite to the body-centered ferrite and accommodate the excess carbon in the form of cementite. The resulting structure, however, is not one of alternating plates but rather a dispersion of discrete cementite particles in a lathlike or needlelike matrix of ferrite. Electron microscopy may be required to resolve the carbides in this structure, which is known as *bainite*. Because of the fine dispersion of carbide, its strength exceeds that of fine pearlite, and ductility is retained because soft ferrite is the continuous matrix phase.

If the austenite is quenched to below the M_s temperature, a different type of transformation occurs. The steel still wants to change from the face-centered-cubic structure to body-centered cubic, but it cannot expel the required amount of carbon to form ferrite. Responding to the severe nonequilibrium conditions, it simply undergoes an instantaneous change in crystal structure with no diffusion. The excess carbon becomes trapped, distorting the structure into a body-centered tetragonal lattice (distorted body-centered cubic), with the degree of distortion being proportional to the amount of excess carbon. The new structure (Figure 5-7), is known as *martensite*, and, with sufficient carbon, is exceptionally strong, hard, and brittle. The dislocation motion necessary for metal flow is effectively blocked by the highly distorted lattice.

As shown in Figure 5-8, the hardness and strength of steel in the martensitic condition are strong functions of the carbon content. Below 0.10% carbon, martensite is not

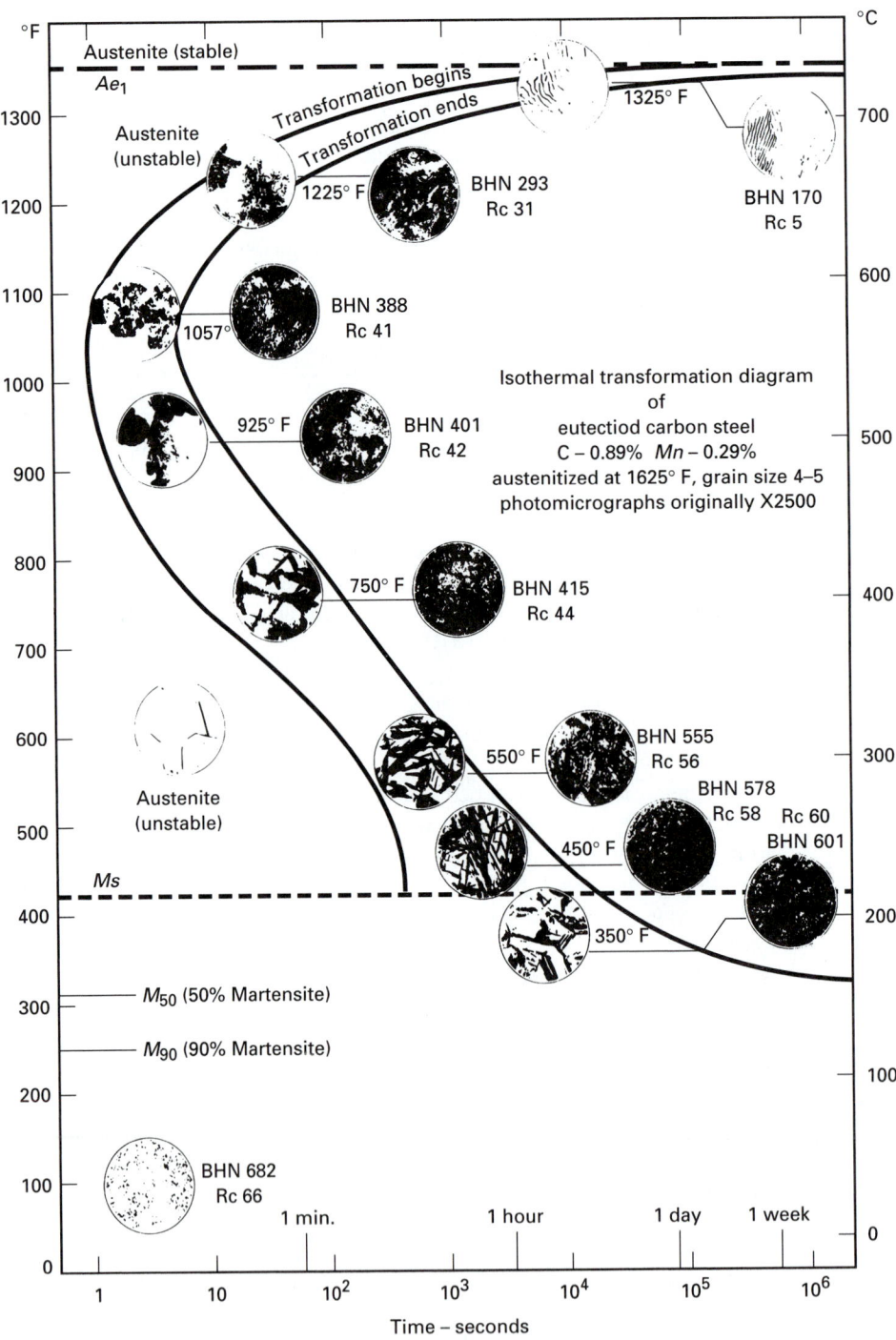

FIGURE 5.6   Isothermal transformation diagram (T-T-T diagram) for eutectoid composition steel. Structures resulting from transformation at various temperatures are shown as insets. *(Courtesy of USX Corporation.)*

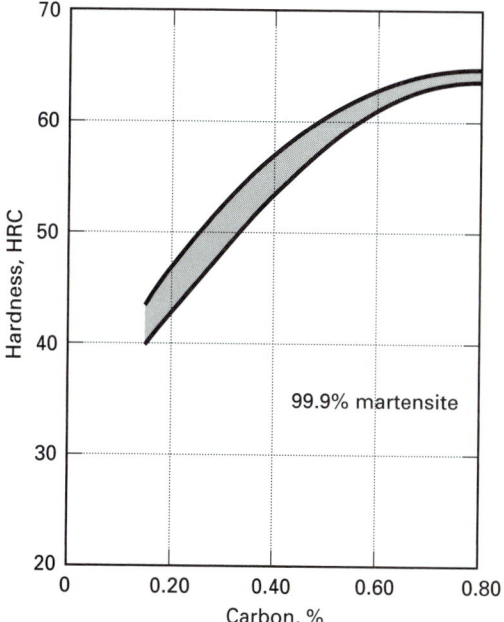

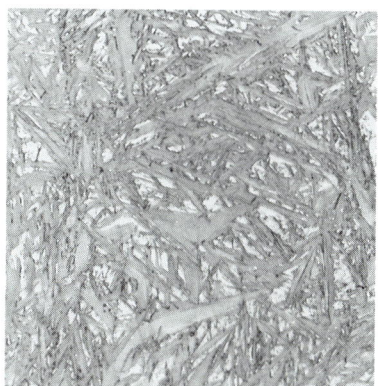

FIGURE 5.7 Photomicrograph of martensite; 1000×. *(Courtesy of USX Corporation.)*

FIGURE 5.8 Effect of carbon on the hardness of martensite.

very strong. Hardness is typically 30 to 35R$_C$ at 0.1% carbon and decreases rapidly as carbon is reduced. Since no diffusion occurs during the transformation, higher-carbon steels form higher-carbon martensite, with concurrent increase in strength and hardness and decrease in toughness and ductility. From 0.3 to 0.7% carbon, strength and hardness increase rapidly. Above 0.7% carbon, however, the rise is far less dramatic, a feature related to the presence of retained austenite.

The amount of martensite that forms upon cooling is a function of the lowest temperature that is encountered, not the time at that temperature. This feature is shown in Figure 5-9. If we return to the C curve of Figure 5-6, there is a temperature designated as M_{50}, where the structure is 50% martensite and 50% untransformed austenite. At the lower M_{90} temperature, the structure is 90% martensite. If no further cooling is performed, the untransformed austenite can remain within the structure. This *retained austenite* can cause loss of strength or hardness, dimensional instability, or cracking or brittleness. Since most quenches are to room temperature, retained austenite problems become significant when the martensite finish, or 100% martensite, temperature lies below room temperature. Higher carbon contents and alloy additions both decrease all martensite-related temperatures, and these materials may require refrigeration or a quench in liquid nitrogen to produce full hardness.

One should note that all the transformations that occur below the A_1 are one-way transitions (austenite to something). The steel is simply seeking to change its crystal structure, and the various products are the result of this change. It is impossible, therefore, to convert one product to another without first reheating to above the A_1 to again form the face-centered-cubic austenite.

T-T-T diagrams can be quite useful in determining the kinetics of transformation and the nature of the products. The left-hand curve shows the elapsed time at constant temper-

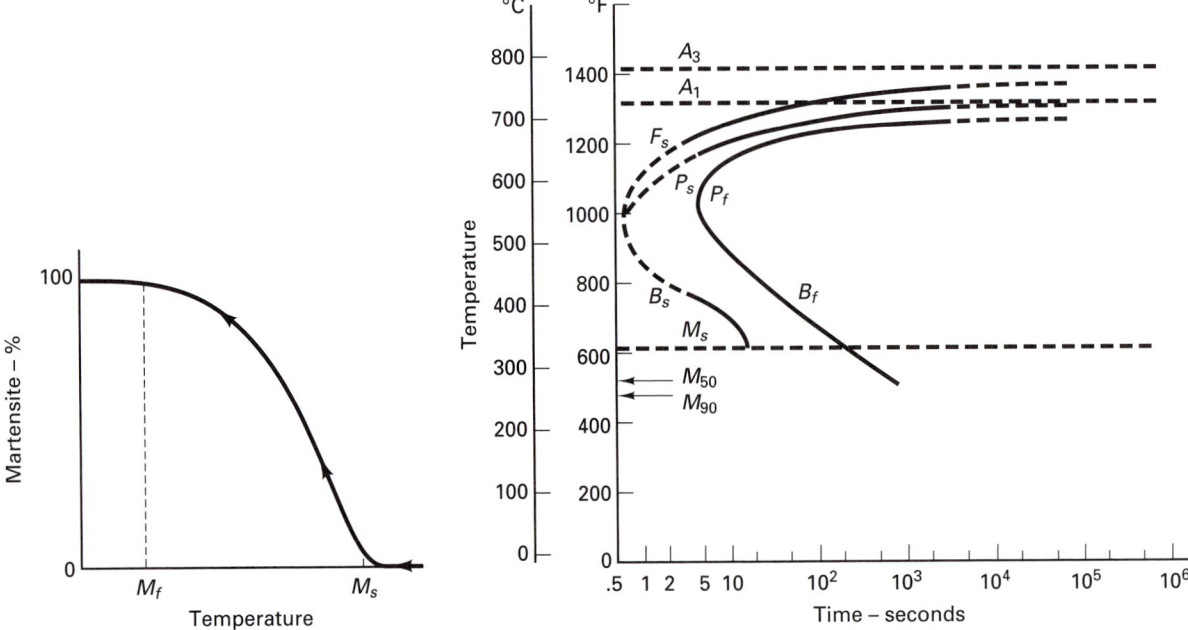

FIGURE 5.9 Schematic representation depicting the amount of martensite formed upon quenching to various temperatures from M_s through M_f.

FIGURE 5.10 Isothermal transformation diagram for a hypoeutectoid steel (1050) showing additional region for primary ferrite.

ature before the transformation begins, and the right-hand curve shows the time required to complete the transformation at that temperature. If hypo- or hypereutectoid steels were considered, additional regions would have to be added to the diagram to correspond to the primary equilibrium phases that form below the A_3 or A_{cm}. These regions would not extend below the nose, however, since the nonequilibrium bainite and martensite structures do not have to maintain a fixed amount of carbon, like the near-equilibrium pearlite. Figure 5-10 is a T-T-T curve for a hypoeutectoid steel containing an additional region for the primary ferrite.

Tempering of Martensite

Because it lacks adequate toughness and ductility, medium- or high-carbon martensite is not a useful engineering structure, despite its great strength. A subsequent heating, known as *tempering*, is usually required to impart a desired amount of toughness at the expense of a decrease in both strength and hardness.

Martensite is a supersaturated solid solution of carbon in alpha ferrite and, therefore, is a metastable structure. When heated into the range 200 to 1300°F (100 to 700°C), the excess carbon atoms are rejected from solution, and the structure moves toward a mixture of the stable ferrite and cementite phases. This decomposition of martensite into ferrite and cementite is a time- and temperature-dependent, diffusion-controlled phenomenon with a spectrum of intermediate and transitory conditions.

Table 5-1 shows a chartwise comparison of precipitation hardening and the current quench-and-temper procedure. Both are nonequilibrium heat treatments that involve three distinct stages. In both, the first step is an elevated temperature soaking designed to produce a uniform-chemistry, single-phase starting condition. Both treatments follow this soak with a rapid-cool quench. In precipitation hardening, the purpose of the quench is to prevent nucleation of the second phase, thereby producing a supersaturated solid solution. This material is usually soft, weak, and ductile, with good toughness. Subsequent aging allows the material to move toward the formation of the stable two-phase structure, and sacrifices toughness and ductility for an increase in strength. When the proper balance is achieved, the temperature is dropped, diffusion ceases, and the current structure and properties are preserved, provided that the material is never subjected to elevated temperatures that would reactivate diffusion and permit the structure to move further toward equilibrium.

TABLE 5-1.	Comparison of Age Hardening with the Quench-and-Temper Process		
Heat Treatment	Step 1	Step 2	Step 3
Age hardening	*Solution treatment.* Heat into the stable single-phase region (above the solvus) and hold to form a uniform-chemistry single-phase solid solution.	*Quench.* Rapid cool to form a nonequilibrium supersaturated single-phase solid solution (crystal structure remains unchanged, material is soft and ductile).	*Age.* A controlled reheat in the stable two-phase region (below the solvus). The material moves toward the formation of the stable two-phase structure, becoming stronger and harder. The properties can be "frozen in" by dropping the temperature to stop further diffusion.
Quench and temper for steel	*Austenize.* Heat into the stable single-phase region (above the A_3 or A_{cm}) and hold to form a uniform-chemistry single-phase solid solution (austenite).	*Quench.* Rapid cool to form a nonequilibrium supersaturated single-phase solid solution (crystal structure changes to body-centered martensite, which is hard but brittle).	*Temper.* A controlled reheat in the stable two-phase region (below the A_1). The material moves toward the formation of the stable two-phase structure, becoming weaker but tougher. The properties can be "frozen in" by dropping the temperature to stop further diffusion

For steels, the quench induces a phase transformation as the material changes from the face-centered-cubic austenite to the distorted body-centered structure known as martensite. The product is again a supersaturated, single-phase solid solution, but the associated properties are the reverse of precipitation hardening. Martensite is strong and hard but relatively brittle. When the material is tempered, strength and hardness are sacrificed for an increase in ductility and toughness. Figure 5-11 shows the properties for a steel that has been tempered at a variety of temperatures. During tempering, diffusion again permits movement *toward* the stable two-phase structure, and a drop in temperature again halts diffusion and locks in properties. Therefore, by quenching steel to form 100% martensite and then tempering it at various temperatures, an infinite range of structures and corresponding properties can be produced. This is known as the *quench-and-temper process* and the product is called *tempered martensite.*

Continuous Cooling Transformations

Although the T-T-T diagrams can provide useful information about the structures obtained through nonequilibrium thermal processing, they are not rigorously applicable to engineering applications because the assumptions of instantaneous cooling followed by constant-temperature transformation rarely match reality. Continuous cooling from elevated temperature is far more realistic, and a diagram showing the results of continuous cooling at various rates would be far more useful. What would be the result if the temperature were to be decreased at a rate of 500°F per second, 50°F per second, or 5°F per second?

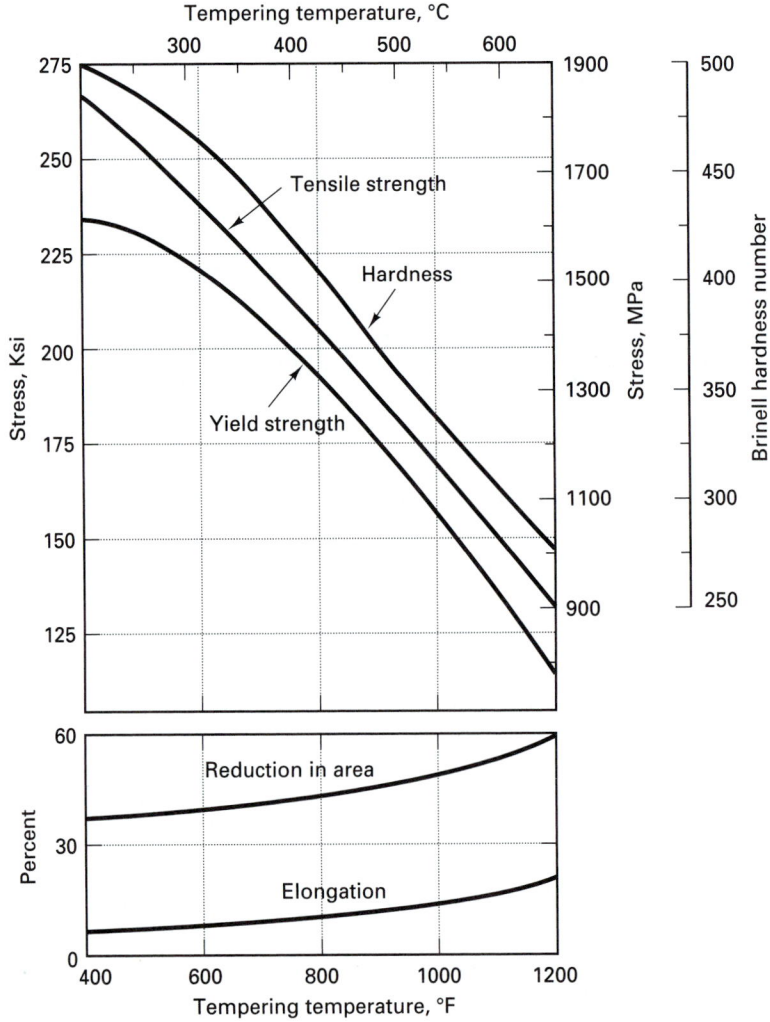

FIGURE 5.11 Properties of an AISI 4140 steel that has been austenitized, oil-quenched, and tempered at various temperatures. (Adapted from *Engineering Properties of Steel,* ASM International, Materials Park, Ohio, 1982.)

A *continuous-cooling-transformation* (C-C-T) *diagram,* such as the one shown in Figure 5-12, can provide answers to these questions and several others. The critical cooling rates required to produce various structures can easily be determined. If cooled fast enough, the structure will be martensite. A slow cool may produce coarse pearlite and some primary phase. Intermediate rates usually produce mixed structures, since the time at any one temperature is usually insufficient to complete the transformation. If each structure is regarded as providing a companion set of properties, the wide range of possibilities obtainable through the controlled heating and cooling of steel becomes even more evident.

Jominy Test for Hardenability

The *Jominy end-quench hardenability test* and associated diagrams provide another tool that is frequently used to assist in the understanding of nonequilibrium heat treatment. In

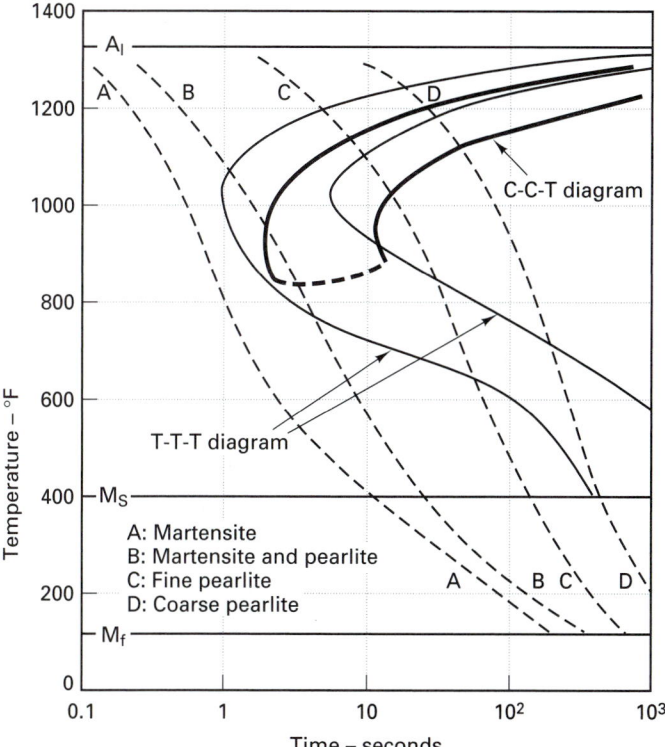

FIGURE 5.12   Schematic C-C-T diagram for a eutectoid composition steel, showing several superimposed cooling curves and the resultant structures. *(Courtesy of USX Corporation.)*

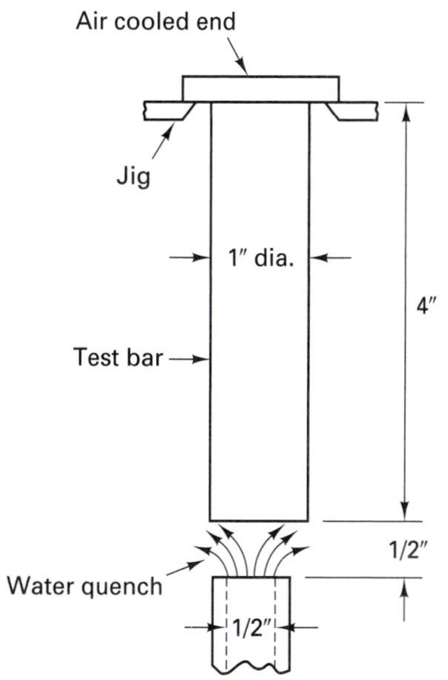

FIGURE 5.13   Schematic diagram of the Jominy hardenability test.

this test, depicted in Figure 5-13, an effort is made to reproduce an entire spectrum of cooling rates on a single specimen by quenching a heated bar from one end. The quench is standardized by specifying the specimen geometry, the quench medium (water at 75°F), internal nozzle diameter ($\frac{1}{2}$ in.), water pressure (that producing a $2\frac{1}{2}$ in. vertical fountain), and the gap between the nozzle and the specimen ($\frac{1}{2}$ in.).

After the specimen has been cooled by the quench, a flat region is ground along one side and R_C hardness readings are taken every $\frac{1}{16}$ in. along the bar. The resulting data is then plotted as shown in Figure 5-14. Since the cooling rate is known for the various locations within the bar, the resulting hardness values are correlated with position and indirectly with the cooling rate that produced them.

Application of the test assumes that identical results will be obtained if the same material undergoes the same cooling history. If the cooling rate is known for a given location within a part (from experimentation or theory), the properties at that location can be predicted to be those at the equivalent cooling-rate location on the Jominy test bar. Similarly, if specific properties are required, the necessary cooling rate can be determined for a selected material. If the cooling rates are restricted, various materials can be compared and a satisfactory alloy selected. Figure 5-15 shows the Jominy curves for several common engineering steels.

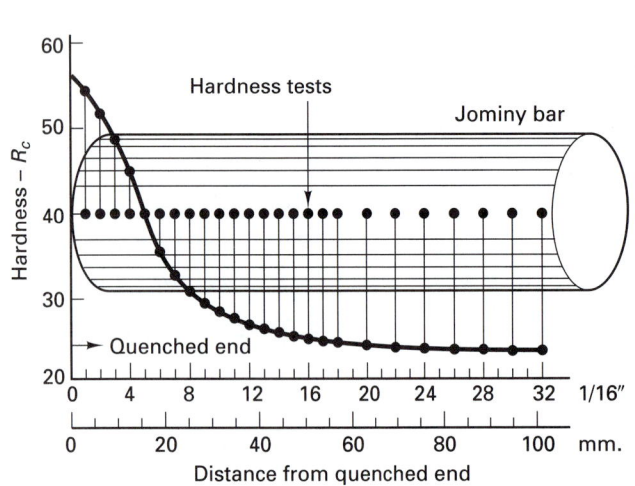

FIGURE 5.14 Typical hardness distribution along a Jominy test specimen.

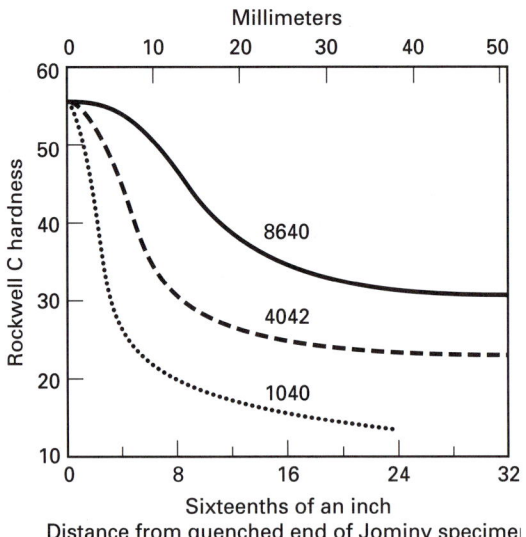

FIGURE 5.15 Jominy hardness curves for engineering steels with the same carbon content and varying types and amounts of alloy elements.

Hardenability Considerations

When one is attempting to understand the heat treatment of steel, several key effects must be considered: the effect of carbon content, the effect of alloy additions, and the effect of various quenching conditions. The first two relate to the material and the third to the heat-treatment process.

Hardness is a mechanical property related to strength and is a strong function of the carbon content of a steel. *Hardenability*, on the other hand, is a measure of the depth to which full hardness can be obtained under a normal hardening cycle and is related primarily to the amounts and types of alloying elements. In Figure 5-15, all of the steels have the same carbon content, but they differ in the type and amounts of alloy elements. Maximum hardness is the same in all cases, but the depth of hardening varies considerably. Figure 5-16 shows the results for steels containing the same alloying elements but variable amounts of carbon. Note the change in peak hardness.

The primary reason for adding alloy elements to commercial steels is to increase their hardenability, not to improve their strength. Steels with greater hardenability need not be cooled as rapidly to achieve a desired level of strength or hardness, and they can be completely hardened in thicker sections.

Materials selection for steels requires an accurate determination of need. Strength tends to be associated with carbon content, and a general rule is to select the lowest possible level that will still meet the specifications. Because heat can only be extracted from the surface of a metal, the size of the piece and depth of required hardening set the conditions for hardenability and quench. For a given quench condition, different alloys will produce different results. Because alloy additions increase the cost of a material, a general rule is to select only what is required to ensure compliance with specifications. Money is often wasted by specifying an alloy steel for an application where a plain carbon steel, or a steel with lower alloy content (less costly), would be satisfactory. Another alternative when

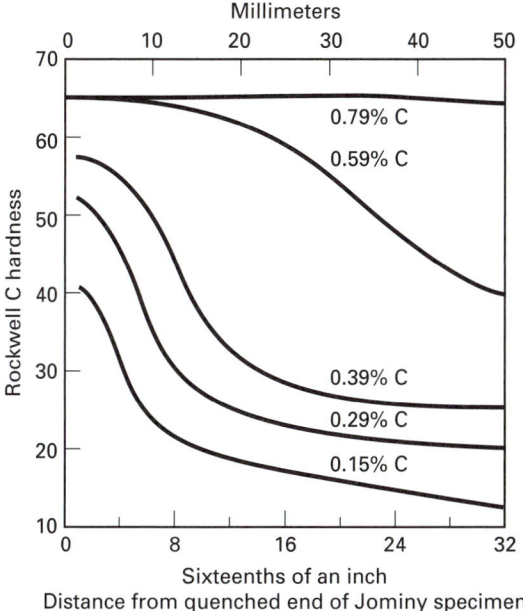

FIGURE 5.16 Jominy hardness curves for engineering steels with identical alloy conditions but variable carbon content.

greater depth of hardness is required is to modify the quench conditions so that a faster cooling rate is achieved. Quench changes may be limited, however, by cracking or warping problems, with these relating to the size, shape, complexity, and desired precision of the part being treated.

Quench Media

Quench media vary in their effectiveness, and one can best understand the variation by considering the three stages of quenching. When a piece of hot metal is first inserted into a tank of liquid quenchant, the liquid adjacent to the metal vaporizes and forms a gaseous layer between the metal and the liquid. Cooling is slow through this *vapor jacket (first stage)* since all heat transport must now be through a gas. This stage will occur if the temperature of the metal is above the boiling point of the quenchant. Soon bubbles nucleate and break the jacket; liquid again contacts the metal, vaporizes (removing its heat of vaporization from the metal), and forms another bubble; and as the bubbles are removed, the process continues. This *second stage of quenching* provides very rapid cooling as a result of the large quantities of heat removed by the vaporization mechanism. When the metal cools to below the boiling point of the quenchant, vaporization can no longer occur. Heat transfer must now take place by conduction across the solid–liquid interface, aided by convection or stirring within the liquid. This is the *third stage of quenching*.

Water is a fairly good quenching medium because of its high heat of vaporization and the fact that the second stage of quenching extends down to 212°F (100°C), usually well into the martensite range or even below. Water is also cheap, readily available, easily stored, nontoxic, nonflammable, smokeless, and easy to filter and pump. However, with a water quench, the clinging tendency of the bubbles may cause soft spots on the metal. Agitation is recommended when using a water quench. Still other problems associated with a water quench include its oxidizing nature, its corrosiveness, and the tendency for excessive distortion and possible cracking.

Brine (salt water) is a more severe quench medium than water because the salt nucleates bubbles, forcing a more rapid transition through the vapor jacket stage. Unfortunately, brine also tends to accelerate corrosion problems unless it is removed completely after the quench. Different types of salts can be used, including sodium or potassium hydroxide, and various degrees of agitation or spraying can be used to adjust the effectiveness of the quench.

When a slower cooling rate is desired, oil quenches can be employed. Various oils are available that have high flash points and different degrees of quenching effectiveness. Since the boiling points are often quite high, the transition to third-stage cooling usually precedes the martensite start temperature. The slower cooling through the M_s-to-M_f transition leads to a milder temperature gradient within the piece and reduced likelihood of cracking. Problems associated with oil quenchants include water contamination, smoke, fumes, spill and disposal problems, and fire hazard. In addition, quench oils tend to be somewhat expensive.

Quite often, there is a need for a quenchant that will cool more rapidly than the oils, but slower than water or brine. To fill this gap, a number of *polymer quench* solutions (also called *synthetic quenchants*) have been developed. Tailored quenchants can be produced by varying the concentrations of the components (such as glycol polymer and water). The polymer quenchants provide extremely uniform and reproducible results and are less corrosive than water and brine and less of a fire hazard than oils (no fires, fumes, smoke, or need for air-pollution-control apparatus). In addition, they tend to minimize distortion. Many of the polymer quench mediums, however, are extremely sensitive to concentration changes and require constant monitoring and control during use.

When slow cooling is desired, molten salt baths can be employed to provide a medium where the quench goes directly to the third stage of cooling. Still slower cooling can be obtained by cooling in still air, burying the metal in sand, or a variety of other methods.

Design Concerns in the Heat Treatment of Steel

Product design and material selection play important roles in the satisfactory and economical heat treatment of parts. Proper consideration of these factors usually leads to more simple, more economical, and more reliable products. Failure to relate design and materials to heat-treatment procedures usually produces disappointing or variable results and may lead to a variety of service failures.

From the viewpoint of heat treatment, undesirable design features include (1) nonuniform sections or thicknesses, (2) sharp interior corners, and (3) sharp exterior corners. Since these features often find their way into the design of parts, the designer should be aware of their effect on heat treatment. Undesirable results may include nonuniform structure and properties, undesirable residual stresses, cracking, warping, and dimensional changes.

Heat can only be extracted from a piece through its exposed surfaces. Therefore, if the piece to be hardened has a nonuniform cross section, any thin region will cool rapidly and may fully harden, while thick regions may harden only on the surface, if at all. These surfaces may even become tempered as a result of the heat escaping from the central mass of hot metal. The shape that might be closest to ideal from the viewpoint of quenching would be a doughnut. The uniform cross section with high exposed surface area and absence of sharp corners are quite attractive. It is recognized, however, that most shapes are designed to perform a function, and compromises are usually necessary.

Residual stresses are the often-complex stresses that are present within a body, independent of any applied load. They can be induced in a number of ways, but the complex dimensional changes that can occur during heat treatment are a primary cause. Thermal expansion during heating and contraction during cooling is a well-understood phenomenon, but nonuniform heating or cooling can produce complex results. In addition, the various phases and structures that form are usually characterized by different densities. Therefore, a volume expansion or contraction will accompany any phase transformation. When austenite transforms to martensite, for example, there is a volume expansion of up to 4%. Austenite transforming to pearlite also experiences a volume expansion but of a smaller magnitude.

If all the temperature changes occurred uniformly throughout a part, all of the associated dimensional changes would occur simultaneously and the resultant product would be free of residual stresses. However, most of the parts being heat-treated experience nonuniform temperatures during the cooling or quenching operation. Consider a plate of hot aluminum being cooled by water sprays from top and bottom. For simplicity, let us model the plate as a three-layer sandwich (Figure 5-17). At the start of the quench, all layers are uniformly hot. However, as the water spray begins, the surface layers cool and contract. The center layer, however, does not experience the quench and remains hot. Since the part is actually one piece, the various layers must accommodate each other. The contracting surface layers exert compressive forces on the hot, weak interior and cause it to contract through plastic deformation. As time passes, the interior cools and wants to contract, but finds itself sandwiched between cold, strong surface layers. It pulls on the surface layers, placing them in compression, and the surface layers hold it back, creating tension in the interior. While the net force is zero (since there is no applied load), counterbalancing tension and compression stresses exist within the product.

Now consider the same situation, but change the material to steel. When heated, all three layers are hot face-centered-cubic austenite. Upon rapid cooling, the surface layers transform to martensite (the structure changes to body-centered tetragonal) and *expand*!

FIGURE 5.17   Three-layer model of a plate undergoing cooling: (a) material such as aluminum that contracts upon cooling; (b) situation for steel, which expands during the phase transformation.

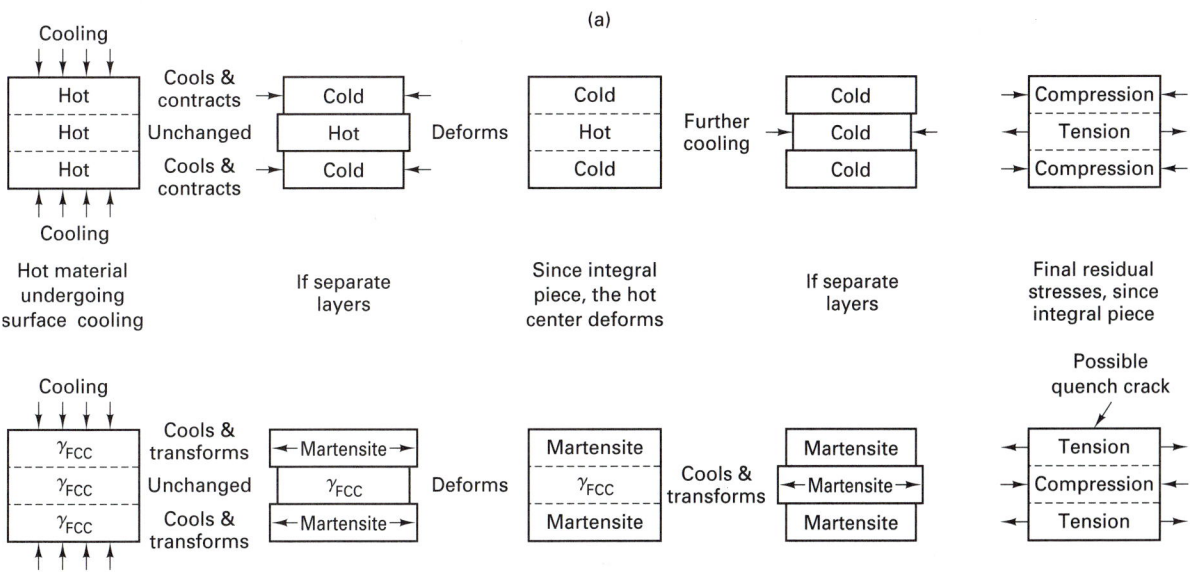

The expanding surfaces deform the soft, weak, austenite center, which then cools and wants to expand as it undergoes the crystal structure change to martensite or one of the other transformation products. The hard, strong surface layers hold it back, placing the center in compression, and the expanding center tries to stretch the surface layers, producing surface tension. If the tension at the surface becomes great enough, cracking can result, a phenomenon known as *quench cracking*. One should note that for the conditions just described, the aluminum will not quench crack, because the surface is in compression, but the steel might, because the residual stresses are reversed. If the cooling were not symmetrical from top and bottom, there might be more contraction or expansion on one side and the plate might warp. For more complex shapes or nonuniform modes of cooling, the residual stresses induced by heat treatment can be quite complex.

The problems associated with residual stresses can be minimized if the cross sections are sufficiently uniform that the temperature differences are minimized and are not concentrated at any specific location. If this is not possible, slower cooling may be required, coupled with a material that will produce the desired properties with an oil or air quench. Because materials with greater hardenability are invariably more expensive, the design alternative clearly has advantages.

When temperature differences and the resultant residual stresses become severe or localized, cracking or distortion problems can be expected. Figure 5-18 shows an example in which a sharp interior corner is placed at a change in cross section. Upon quenching, stresses will concentrate along line *A–B*, and a crack is almost certain to result. When changes in cross section or other transitions must be made, they should be gradual, as in the redesigned version of Figure 5-18a. Generous fillets at interior corners, radiused exterior corners, and smooth transitions all reduce problems. The use of a more hardenable material or a less severe quench will also help to reduce heat-treating problems.

Figure 5-18b shows the cross section of a blanking die that cracked consistently during hardening. By rounding the sharp corners and adding additional holes to produce a more uniform cross section during quenching, the problem was eliminated.

One of the ominous features of poorly designed heat-treated parts is the fact that the residual stresses may not produce immediate failure but may contribute to a failure at a later time. Applied stresses are added to the residual stresses already present within the part. Therefore, it is possible for applied stresses that are well within the "safe" designed limit to couple with residual stresses to produce a value sufficient to induce failure. Corrosion reactions can be accelerated substantially in the presence of residual stresses. Dimensional changes or warping can occur when the equilibrium balance of the residual stresses is upset by subsequent machining or grinding operations. After the removal of some material, the remaining piece adjusts its shape to produce a new balance. Considering all the possible difficulties, it is apparent that considerable time and money could be

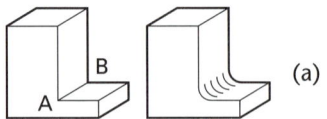

(a)

FIGURE 5.18   (a) Shape containing nonuniform sections joined by a sharp interior corner that may crack during quenching. This is improved by using a large radius to join the sections. (b) Original design containing holes, which can be modified to produce more uniform sections.

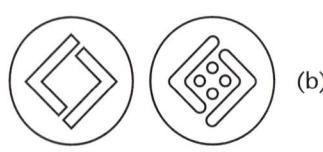

(b)

saved if good design, material selection, and heat-treatment practices are employed. *If used properly*, heat treatment can enable better results to be produced with less costly materials.

Techniques to Reduce Cracking

Figure 5-19a shows the cause of quench cracking on a T-T-T diagram (a misuse of the diagram, but helpful for visualization). Follow the cooling curve for the surface, and note that when the surface is transforming to martensite and expanding, the center is still hot, untransformed austenite. When the center then cools to the martensite transformation and wants to expand, the surface is already cold and hard.

Two variations of rapid quenching have been developed to produce strong structures but reduce the likelihood of cracking. The rapid cool is still required to prevent transformation to the weaker pearlitic structure, but instead of quenching through the martensite transformation, the component is rapidly quenched into a liquid medium that is several degrees above the martensite start (M_s) temperature. Holding for a period of time in this bath allows the piece to come to a nearly uniform temperature. If the material is held at this temperature for long enough time, the austenite will transform to bainite, and the process is known as *austempering*. If the material is brought to a uniform temperature and then slow cooled through the martensite transformation, the process is known as *martempering*, or *marquenching*. Here the product is martensite, which must be tempered the same as the martensite that forms upon rapid quenching. Bainite usually has sufficient toughness that a temper is not required. Figure 5-19b illustrates the modified processes and shows that the transformations (and related volume expansions) of the surface and center occur at the same time, thereby eliminating the residual stresses and tendency to crack.

FIGURE 5.19   (a) Schematic representation of the cooling paths of surface and center during a direct quench; (b) modified cooling paths experienced during the austempering and martempering processes.

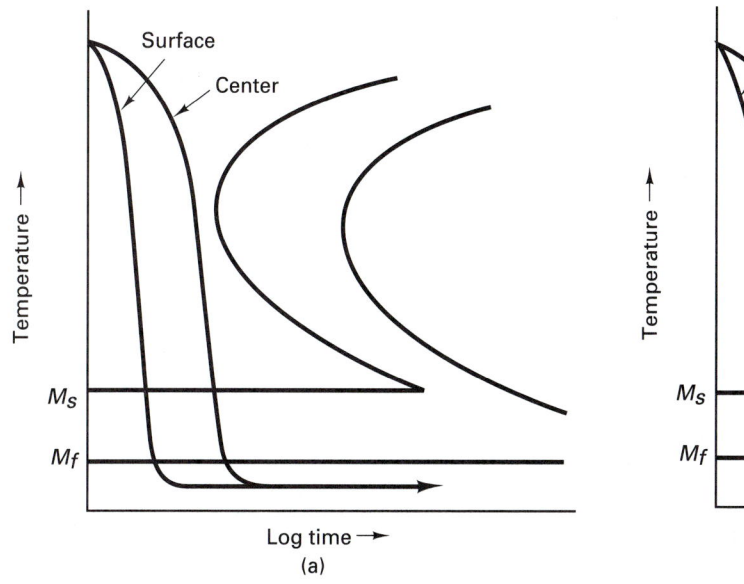

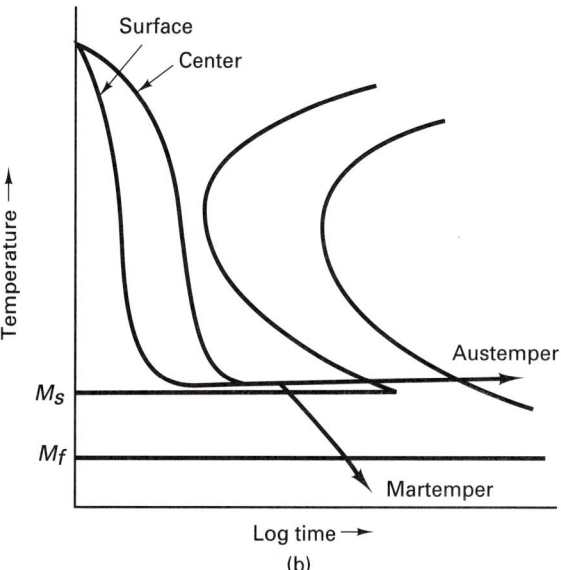

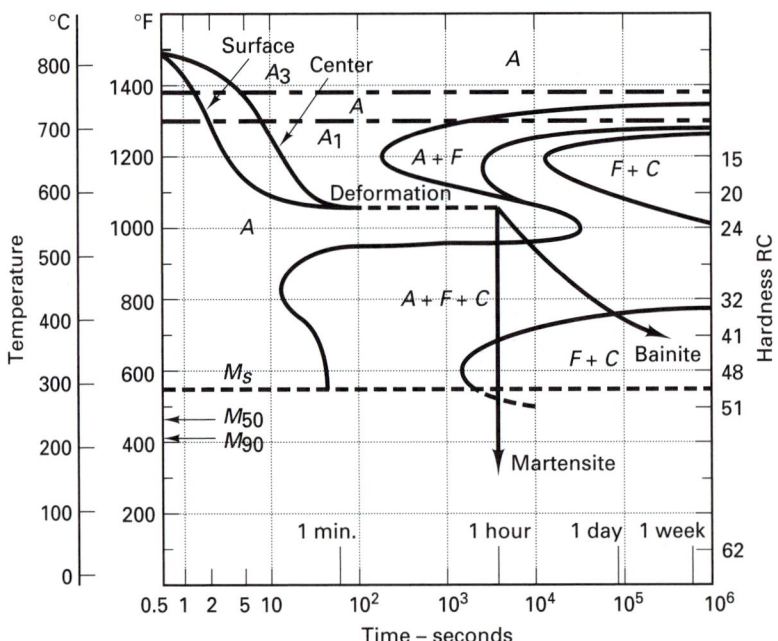

FIGURE 5.20   T-T-T diagram for 4340 steel, showing the bay and a schematic of the ausforming process. *(Courtesy of USX Corporation.)*

Ausforming

A process that is often confused with austempering is *ausforming*. Certain alloys tend to retard the pearlite transformation far more than the bainite reaction and produce a T-T-T curve of the shape shown in Figure 5-20. If this material is heated to form austenite and then quenched to the temperature of the "bay" between the pearlite and bainite reactions, it can retain its austenite structure for a useful period of time. Deformation can be performed on an austenite structure at a temperature where it technically should not exist. Benefits include the increased ductility of the face-centered-cubic crystal structure, the finer grain size that forms upon recrystallization at the lower temperature, and the possibility of some degree of strain hardening. Following the deformation, the metal can be slowly cooled to produce bainite or rapidly quenched to martensite, which must then be tempered. The resulting product has exceptional strength and ductility, coupled with good toughness, creep resistance, and fatigue life—properties that are far superior to those produced if the deformation and transformation processes were conducted in their normal sequence. Ausforming is an example of a growing class of *thermomechanical processes* in which deformation and heat treatment are intimately combined.

■ 5.6 SURFACE HARDENING OF STEEL

Many products require different properties at different locations. Quite frequently, this variation takes the form of a hard, wear-resistant surface coupled with a tough, fracture-resistant core. The methods developed to produce such properties can generally be classified into three basic groups: selective heating of the surface, altered surface chemistry, and deposition of an additional surface layer.

Selective Heating Techniques

If a steel has sufficient carbon to attain the desired surface hardness, generally greater than 0.3%, the different properties can be obtained simply by varying the thermal histories of the various regions. Maximum hardness depends on the carbon content of the material, while the depth of that hardness depends on the depth of heating and the material's hardenability.

Flame hardening uses a high-intensity oxyacetylene flame to raise the surface temperature high enough to reform austenite. This region is then water quenched[*] and tempered to the desired level of toughness. Heat input is quite rapid and is concentrated on the surface. Slow heat transfer and short heating times leave the interior at low temperature and therefore free from any significant change.

Considerable flexibility is provided since the rate and depth of heating can easily be varied. Depth of hardening can range from thin skins to over $\frac{1}{4}$-in. Flame hardening is often used on large objects, since alternative methods tend to be limited by both size and shape. Equipment varies from crude, hand-held torches to fully automated and computerized units.

In induction hardening the steel part is placed inside a conductor coil, which is then energized with alternating current. The changing magnetic field induces surface currents in the steel, which heat by electrical resistance. The heating rates can be extremely rapid and efficiency is high. *Induction heating* is particularly well suited to surface hardening since the rate and depth of heating can be controlled directly through the amperage and frequency of the generator.

Induction hardening is ideal for round bars and cylindrical parts but can also be adapted to more complex geometries. The process offers high quality, good reproducibility, and the possibility of automation. Figure 5-21 shows a cross section of an induction-hardened gear, where hardening has been applied to those areas expected to see high wear. Distortion during hardening is negligible since the dark areas remain cool and rigid throughout the entire process.

Laser-beam hardening has been used to produce hardened surfaces on a wide variety of geometries. An absorptive coating such as zinc or manganese phosphate is often applied to the steel to improve the efficiency of converting light energy into heat. The surface is then scanned with the laser, where beam size, beam intensity, and scanning speed (often as high as 100 in./min) have been selected to obtain the desired amount of heat input and depth of heating. It is possible that the heat can be effectively removed through transfer into the cool, underlying metal (autoquenching), but a water or oil quench is often used.

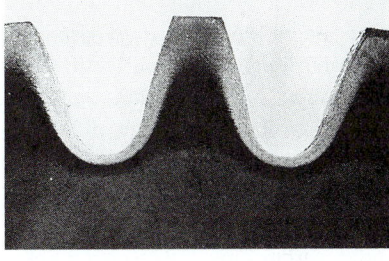

FIGURE 5.21 Section of gear teeth showing induction-hardened surfaces. *(Courtesy of TOCCO Division, Park-Ohio Industries, Inc.)*

---

[*]There is no real danger of surface cracking during the water quench. When the surface is reaustenitized, the soft austenite adjusts to the underlying material. Upon quenching, the surface austenite expands during transformation. The interior is cold and restrains the expansion, producing a surface in compression and no tendency toward cracking. This is just the opposite of the conditions that occur during the through-hardening of a furnace-soaked workpiece!

Through laser beam hardening, 0.4% carbon steel can attain surface hardnesses as high as Rockwell C 65. The process operates at high speeds, produces little distortion, induces residual compressive stresses on the surface, and can be used to harden selected surface areas while leaving the remaining surfaces unaffected. Computer software and automation can be used to control the process parameters, and conventional mirrors and optics can be used to shape and manipulate the beam.

Electron-beam hardening is very similar to laser-beam hardening. Here the heat source is a beam of high-energy electrons rather than a beam of light, and the charged particles can be focused and directed by means of electromagnetic controls. Like laser-beam treating, the process can readily be automated and production equipment can perform a variety of operations with efficiencies often greater than 90%. Electrons cannot travel in air, however, so the entire operation must be performed in a hard vacuum, and this provides the major limitation. More information on laser and electron beam techniques, as well as other means of heating material, is provided in the chapters on welding. Still other selective heating techniques employ immersion in a *lead pot* or *salt bath* as the means of heating the surface.

Techniques Involving Altered Surface Chemistry

When the steels contain insufficient carbon to achieve the desired surface properties through selective heating, an alternative approach is to alter the surface chemistry. *Carburizing,* the most common technique within this category, involves the addition of carbon by diffusion from a high-carbon source.

In the *pack carburizing process*, components are packed in a high-carbon solid medium (such as carbon powder or cast iron turnings) and heated in a furnace for 6 to 72 hours at roughly 1650°F (900°C). The hot carburizing compound produces CO gas, which reacts with the metal, releasing carbon, which is readily absorbed by the hot austenite. When sufficient carbon has diffused to the desired depth, the boxes are removed from the furnace and the parts are thermally processed. Direct quenching from the carburization treatment can often produce the different surface and core properties due to the different carbon contents of the steel at these locations and the different cooling rates. A slow cool from carburizing, reaustenitizing, and quench or a duplex process involving separate surface heat treatment are alternative thermal processes which can produce improved product properties. The carbon content of the surface usually varies from 0.7 to 1.2% depending on the details of the process. Case depth may range from a few thousandths of an inch to over $\frac{3}{8}$-in., but cases over 0.06 in. (1.5 mm) are seldom employed.

Several problems are encountered in the pack carburizing process. Heating is inefficient, temperature uniformity is questionable, handling is often difficult, and the process is not readily adaptable to continuous operation. *Gas carburizing* overcomes many of these difficulties by replacing the solid carburizing compound with a carbon-containing gas, usually containing an excess of C. Although the mechanisms and processing are the same, the operation is faster and more easily controlled. Accuracy and uniformity are increased, and continuous operation is possible. Special types of furnaces are required, however, to safely contain the CO-containing gas.

In *liquid carburizing* the steel parts are simply immersed in a molten carbon-containing bath. Historically, most liquid carburizing baths contained cyanide, which supplied both carbon and nitrogen to the surface. Safety and environmental concerns associated with the toxic cyanide fumes have severely limited the process, but noncyanide liquid compounds have been developed to preserve the technique. Most applications involve the production of thin cases on small parts.

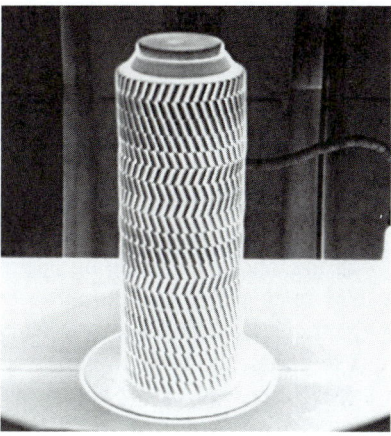

FIGURE 5.22 Stack of gears undergoing ionitriding. *(Courtesy of Abar Ipsen Industries.)*

Nitriding hardens the surface by producing alloy nitrides in special steels that contain nitride-forming elements such as aluminum, chromium, molybdenum, or vanadium. The parts are heat-treated and tempered at 1000 to 1250°F (525 to 675°C) prior to nitriding. After cleaning and removal of any decarburized surface material, they are heated in an atmosphere containing dissociated ammonia (nitrogen and hydrogen) for 10 to 40 hours at 950 to 1150°F (500 top 625°C). Nitrogen diffusing into the steel then forms alloy nitrides, hardening the metal to a depth of about 0.025 in. (0.65 mm). Very hard cases are formed and distortion is low. No subsequent thermal processing is required. In fact, subsequent heating should be avoided because the differential thermal expansions and contractions will crack the hard, nitrided case. Finish grinding should also be avoided, if possible, because of the exceptionally thin nitrided layer. Thus, while the surface hardness is higher than for most other hardening methods, the long times at elevated temperatures, coupled with the exceptionally thin case, restrict the application of nitriding to the production of high-quality surfaces.

Ionitriding is a plasma process that has emerged as an attractive alternative to the conventional methods. Parts to be treated are placed in an evacuated "furnace" and a direct-current potential of 500 to 1000 V is applied between the parts and the furnace walls. Low-pressure nitrogen gas is introduced into the chamber and becomes ionized. The ions are accelerated toward the product surface, where they impact and generate sufficient heat to promote inward diffusion. This is the only heat associated with the process; the "furnace" acts only as a vacuum container and electrode. Figure 5-22 shows the ionized plasma surrounding a stack of gears. Advantages of the process include shorter cycle times, reduced consumption of gases, significantly reduced energy costs, reduced space requirements and the possibility of total automation. Product quality is improved over that of conventional nitriding and the process is applicable to a wider range of materials.

In *ion carburizing*, a low-pressure methane plasma is created, producing atomic carbon which is transferred to the surface. Production time is generally 30% less than with gas carburizing and the product is oxide-free due to the near-vacuum conditions. *Ion plating* and *ion implantation* are other new technologies in surface modification. In Chapter 40 we expand on the surface treatment of materials and compare the processes discussed above to various platings, coatings, and other techniques.

■ 5.7 FURNACES

Furnace Types

To facilitate production heat treatment, many styles of furnaces have been developed in a wide range of sizes. These furnaces are generally classified as *batch* or *continuous* type. Batch furnaces are those in which the workpiece remains stationary throughout its treatment. Continuous furnaces move the components through the heat-treat operation at rates selected to be compatible with the other manufacturing operations.

Horizontal batch furnaces are often called *box furnaces* because of their overall shape, and generally use gas or electricity as their source of heat. As shown in Figure 5-23, a door is provided on one end to permit the work to be inserted and removed. When large or very long workpieces are to be heated, a *car-bottom* box furnace may be employed, like the one shown in Figure 5-24. Here the work is loaded onto a refractory-topped flatcar, which can be rolled into and out of the furnace on railway rails.

Another batch furnace design is the *bell furnace*. Here the heating elements are contained within a bottomless "bell" that is lowered over the work. An airtight inner shell is often placed around the workpieces to contain a protective atmosphere during the heating and cooling operations, thereby reducing tarnish and oxidation. After the work is heated, the furnace unit can be lifted off and transferred to another batch, while the inner shell maintains the protective atmosphere during cooling. If extremely slow cooling is desired, an insulated cover can be placed over the heated shell.

The *elevator furnace* is an interesting modification of the bell design where the bell remains stationary and the workpieces are raised into it on a movable platform that forms the bottom of the furnace. In one design, the platform can occupy three vertical positions.

FIGURE 5.23 Box electric heat-treating furnace. *(Courtesy of Lindberg, A Unit of General Signal.)*

FIGURE 5.24 Car-bottom box furnace. *(Courtesy of Hevi Duty Electric Company.)*

The middle position is for loading and unloading. When raised into the upper position, the work is in the furnace, and in the lower position, it is immersed in a quench tank. These furnaces are attractive for applications where the work must be quenched as soon as possible after being removed from the furnace.

All of the furnaces described above are horizontal designs. Horizontal furnaces are easy to construct in a variety of sizes and easy to use, but offer little resistance to the sagging or warping of long, slender products when they are laid on their side. For these types of workpieces, a *vertical pit furnace* is preferred. These are usually cylindrical chambers sunk into the floor with a door on top that can be swung aside to permit the work to be lowered into the furnace. Suspended in this manner, the long, slender workpieces are far less likely to warp. These furnaces can also be used to heat large quantities of small parts, which can be loaded into wire-mesh baskets and stacked within the column.

Continuous furnaces are used for large production runs of the same or similar parts that undergo the same thermal processing. Here a steady flow of workpieces is moved through the furnace by some type of conveyor or push mechanism, and the furnace often ends in such a way that the workpieces fall into a quench tank to complete the treatment. Complex cycles of heating, holding, and quenching or cooling can be conducted in an exact and repeatable manner with low labor cost. Circular continuous furnaces, where the workpieces move on a rotating hearth, are convenient when a single workstation is used to load and unload the furnace.

While all of the furnaces described above can heat in air, most commercial furnaces can also employ artificial gas atmospheres. These are selected to prevent scaling or tarnishing, prevent decarburization, or even to provide carbon or nitrogen for surface modification. Many of the artificial atmospheres are generated from natural gas, but nitrogen-based atmospheres frequently offer reduced cost, energy savings, increased safety, and environmental attractiveness.

When a liquid heating medium is preferred over gas, *salt bath furnaces* are a popular choice. Electrically conductive salt can be heated by passing a current between two electrodes suspended in the bath. These electrical currents also cause the bath to circulate and thus maintain uniform temperature. Nonconductive salt baths can be heated by some form of immersion heater, or the containment vessels can be externally fired. The molten salt not only serves as a uniform source of heat but can also be selected to prevent scaling or decarburization. The *lead pot* is a similar furnace, where molten lead replaces salt as the heat transfer medium.

The heating rates of gas atmosphere furnaces can be made comparable to those of liquid baths by incorporating the *fluidized-bed* concept. These furnaces consist of a bed of dry, inert particles, such as aluminum oxide, which are heated (by elements in the furnace walls) and fluidized (suspended) in a stream of flowing gas. Products introduced into the bed become engulfed in the particles, which then radiate uniform heat. Temperature and atmosphere can be altered quickly and high heat transfer rates, high thermal efficiency, and low fuel consumption have been observed. Since atmosphere changes can be performed in minutes, one furnace can be used for nitriding, stress relieving, carburizing, carbonitriding, annealing, and hardening.

Electrical induction heating is another popular means of heating conductive materials, such as metal. Small parts can be through-heated and hardened as in the other methods. Long products can be heated and quenched in a continuous manner by passing them through a stationary heating coil or by having a moving coil traverse a stationary part. Localized or selective heating can also be performed at rapid production rates. As an additional attractive feature, a standard induction unit can be adapted to a wide variety of products simply by changing the induction coil and adjusting the equipment settings.

Furnace Controls

All thermal processing operations should be carried out with rigid control if the desired results are to be obtained in a consistent fashion. Most furnaces are equipped with some form of temperature-indicating thermocouple or pyrometer. These can be coupled with a controller or computer control to regulate the temperature and rate of heating or cooling. It should be remembered, however, that it is the temperature of the workpiece, not the temperature of the furnace, that controls the result, and it is this temperature that should be monitored if at all possible. In addition, time should be allotted to allow the workpieces to attain the uniform temperature of the furnace.

■ 5.8 HEAT TREATMENT AND ENERGY

From a consideration of temperatures and times, it is apparent that heat treatments can consume considerable amounts of energy. However, if one considers the broader picture, heat treatment may actually prove to be an energy conservation measure. Through its use, we can manufacture higher-quality, more durable products, which eliminates the need for frequent replacements. Higher strengths may also permit the use of less material in the manufacture of a product, thereby saving additional energy.

Further savings can often be obtained by integrating the manufacturing operations. For example, a direct quench and temper from hot forging may be used to replace the conventional air cool, reheat, quench, and temper sequence. One should note, however, that the integrated procedure would have greater variability in temperature, uniformity of that temperature, and austenite grain size going into the quench. If this variation is too great, additional energy may be required to conduct the reheat and soak.

■ KEY WORDS

A_1
A_3
A_{cm}
age hardening
aging
annealing
ausforming
austempering
bainite
batch furnaces
C-C-T diagram
carburizing
coherency
continuous furnaces
dispersion hardening
flame hardening
fluidized-bed furnaces

full anneal
grain-size refinement
hardenability
heat treatment
homogenization
induction hardening
ionitriding
isothermal transformation diagram
Jominy test
martempering
martensite
nitriding
normalize
overaged
pearlite
phase transformation strengthening
polymer quench

precipitation hardening
process anneal
quench cracking
quenching
residual stresses
retained austenite
solid-solution strengthening
solution treatment
spheroidization
stress-relief anneal
surface hardening
T-T-T diagram
tempered martensite
tempering
thermomechanical processing

■ REVIEW QUESTIONS

1. What is heat treatment?
2. Why should people performing hot forming or welding be aware of the effects of heat treatment?
3. What is the major goal of the processing heat treatments? Cite some of the specific objectives that may be sought.
4. Why might equilibrium phase diagrams be

useful aids in designing and understanding the processing heat treatments?

5. What are some of the possible objectives of annealing operations?

6. While full anneals often produce the softest and most ductile structures, what may be some of the objections or undesirable features of these treatments?

7. Why are the hypereutectoid steels not furnace-cooled from the all-austenite region?

8. What might be cited as the major process difference between full annealing and normalizing?

9. What are some of the process heat treatments that can be performed without reaustenitizing the material (heating above the A_1 temperature)?

10. What types of steel would be candidates for a process anneal? Spheroidization?

11. Why is the recrystallization temperature of a metal not a well-defined temperature?

12. What are the six major mechanisms that can be used to increase the strength of a metal?

13. What is the most effective strengthening mechanism for the nonferrous metals?

14. What are the three steps in an age-hardening treatment?

15. What is the difference between natural and artificial aging? Which offers more flexibility? Over which does the engineer have more control?

16. What is the difference between a coherent precipitate and a distinct second-phase particle?

17. What types of heating and cooling conditions are imposed in the I-T or T-T-T diagram? Is this realistic for the processing of commercial items?

18. For steels below the A_1 temperature, what are the stable equilibrium phases as predicted by the equilibrium phase diagram?

19. Considering the details of the T-T-T diagram, what are some nonequilibrium structures that may be produced in heat-treated steels?

20. Which of the structures in heat-treated steel is the result of a diffusionless phase change?

21. What is the major factor that influences the strength and hardness of steel in the martensitic structure?

22. Why is retained austenite an undesirable structure in heat-treated steels?

23. Why are martensitic structures usually tempered before being put into use? What properties increase during tempering? Which ones decrease?

24. In what ways is the quench-and-temper heat treatment similar to age hardening? How are the property changes different in the two processes?

25. What is a C-C-T diagram?

26. How do the various locations of a Jominy test specimen correlate with the various cooling rates?

27. What conditions are used to standardize the quench in the Jominy test?

28. What is the assumption that enables the data from a Jominy test to be used to predict the properties of various locations on a manufactured product?

29. What is hardenability?

30. What are some of the ways in which a steel product might be hardened to a greater depth?

31. What are the three stages of liquid quenching?

32. What are some of the major advantages and disadvantages of a water quench?

33. Why is an oil quench less likely to produce quench cracks than water or brine?

34. What are some of the attractive qualities of a polymer of synthetic quench?

35. What are some undesirable design features in parts that are to be heat treated?

36. Why are the residual stresses in steel often different from the residual stresses in an identically processed aluminum part?

37. What are some of the undesirable effects of residual stresses?

38. Describe several techniques that reduce residual stresses by enabling the volumetric changes to occur simultaneously throughout the part.

39. What is thermomechanical processing?

40. What are some of the methods that can be used to selectively heat treat the surface of metal parts?

41. What are some of the attractive features of surface hardening with a laser beam?

42. Why might gas carburizing be more attractive than pack carburizing for high-volume part production?

43. In what ways might ionitriding be more attractive than conventional nitriding or carburizing?

44. For what type of products or product mixes might a batch furnace be preferred to a continuous furnace?

45. What are some of the possible functions of artificial atmospheres used during heat treating?

46. How are parts heated in a fluidized-bed furnace? What are some of their attractive features?

47. In what ways might the heat treatment of metals actually be an energy conservation measure?

■ PROBLEMS

1. A number of heat treatments have been devised to harden the surfaces of steel and other engineering metals. Consider the processes of: (1) Flame hardening (2) Induction hardening (3) Carburizing and (4) Nitriding. For each of the above processes, provide information relating to: a) A basic description of how the process works, b) Typical materials on which the process is performed, c) Type of equipment required, d) Typical times, temperatures and atmospheres required, e) Typical depth of hardening and reasonable limits, f) Hardness achievable, g) Subsequent treatments or processes that might be required, h) Information relating to distortion and/or stresses, i) Ability to use the process to harden selective areas.

2. This chapter has presented four processing-type heat treatments whose primary objective is to soften, weaken, enhance ductility, or promote machinability. Consider each of the following processes as they are applied to steels. (1) Full annealing (2) Normalizing (3) Process annealing and (4) Spheriodizing. Provide information relating to: a) A basic description of how the process works and what are its primary objectives, b) Typical materials on which the process is performed, c)Type of equipment used, d) Typical times, temperatures and atmospheres required, e) Recommended rates of heating and cooling, f) Typical properties achieved.

Chapter 5 CASE STUDY

a flying chip from a sledgehammer

Industrial sledgehammers are used throughout JCL Industries, most having a 15-pound head made of AISI 1060 steel. To reduce tool replacement costs, the company machine shop periodically gathers hammers with heavily-deformed (mushroomed) heads and grinds off the deformed segment.

A reground hammer was returned to use. Upon striking a metal plate, a chip flew from a corner of the hammer head and lodged in the eye of a nearby worker. A lawsuit resulted.

Subsequent investigation revealed that the head of new hammers had a bulk hardness between R_C44 and 55. The chip, however, had a hardness of approximately R_C65 on the surface where it fractured from the hammer. Inspection of the other hammers in the shop revealed numerous chipped regions. All of the chips, however, were on redressed hammers.

What do you suspect might be the problem? How would you alter the procedures or policies of JCL Industries to eliminate a possible recurrence, yet minimize expense?

CHAPTER 6

FERROUS METALS AND ALLOYS

6.1 INTRODUCTION TO
       HISTORY-DEPENDENT
       MATERIALS
6.2 IRON
6.3 STEEL
       Solidification Concerns
       Deoxidation and
          Degassification
       Plain-Carbon Steel
       Alloy Steels
       Effects of the Various
          Alloying Elements
       AISI-SAE Classification
          System
       Selecting Balanced Alloy
          Steels
       High-Strength Low-Alloy
          Structural Steels

       Microalloyed Steels in
          Manufactured Products
       Free-Machining Steels
       Bake-Hardenable Steel Sheet
       Precoated Steel Sheet
       Steels for Electrical and
          Magnetic Applications
       Maraging Steels
       Steels for High-Temperature
          Service
6.4 STAINLESS STEELS
6.5 TOOL STEELS
6.6 ALLOY CAST STEELS
       AND IRONS
Case Study: INTERIOR TUB OF A
             TOP-LOADING WASHING
             MACHINE

■ 6.1 INTRODUCTION TO HISTORY-DEPENDENT MATERIALS

Engineering materials are available with a wide range of useful properties and characteristics. Some of these are inherent to the particular material, but many others can be varied by controlling the manner of production and the details of processing. Metals are classic examples of such history-dependent materials; their final properties are clearly affected by their past processing history. The particular details of the smelting and refining process control the resulting purity and the type and nature of any influential contaminants. The initial solidification process imparts structural features that may be transmitted to the final product. Preliminary operations such as the rolling of sheet or plate often impart directional variations to properties, and their impact should be considered during subsequent processing and use. Thus, while it is easy to take the attitude that "metals come from warehouses," it is important to recognize that aspects of prior processing can significantly influence further operations as well as the final properties of the product. The breadth of this book does not permit coverage of the processes and methods involved in the production of engineering metals, but certain aspects will be presented because of their role in affecting subsequent performance.

In the present chapter we introduce the major ferrous (iron-based) metals and alloys. These materials have been the backbone of civilization and numerous varieties have been developed over the years to meet the specific needs of various industries. In recent years, there has been a quiet revolution. Over half of the steel used in a mid-1990s automobile did not exist as little as 10 years ago.

The availability of so many alternatives, both old and new, has often led to poor materials selection. Money can be wasted in the unnecessary selection of an expensive alloy or one that is difficult to fabricate. At other times, such materials may be absolutely necessary, and selection of a cheaper alloy would mean certain failure. Thus it is the responsibility of

141

the design and manufacturing engineer to be knowledgeable in the area of engineering materials and be able to make the best selection from among the available alternatives.

■ 6.2 IRON

Iron is the fourth most plentiful element in the earth's crust and for centuries has been the most important of the basic engineering metals. The variety of metals and alloys derived from iron has played a central role in the development of civilization and will probably continue in this role in the foreseeable future. Even now, new advances in the technology of iron and iron alloys continue to expand their utility in a myriad of engineering applications.

Iron is rarely found in the metallic state but occurs in a variety of mineral compounds, known as ores. The most attractive of these ores are iron oxides coupled with companion impurities. These are first processed in a manner that breaks the iron–oxygen bonds to produce metallic iron (chemical reducing reactions). Ore, limestone, coke (carbon), and air are continuously introduced into specifically designed furnaces and molten metal is periodically withdrawn.

In the environment of the furnace, other impurity oxides are also reduced. All of the phosphorus and most of the manganese enter the iron. Oxides of silicon and sulfur are partially reduced, and these elements also become part of the resulting metal. Other contaminant elements, such as calcium, magnesium, and aluminum, are collected in the limestone-based slag and are removed from the system. Thus the resulting *pig iron* tends to have roughly the following composition:

Carbon	3.0–4.5%
Manganese	0.15–2.5%
Phosphorus	0.1–2.0%
Silicon	1.0–3.0%
Sulfur	0.05–0.1%

A small portion of this iron is cast directly into final shape and is classified as cast iron. Most commercial cast iron however, is produced by remelting scrap iron and steel, with the possible addition of some pig iron. The metallurgical properties of cast iron have been presented in Chapter 4, and melting and utilization in the casting process are developed in Chapters 13 through 15. Most pig iron is simply transferred in the molten state and further processed into steel.

■ 6.3 STEEL

The manufacture of *steel* is essentially an oxidation process that decreases the amount of carbon, silicon, manganese, phosphorus and sulfur in a mixture of molten pig iron and steel scrap. In 1856, the *Kelly–Bessemer process* opened up the industry by enabling the manufacture of commercial quantities of steel. The *open-hearth process* surpassed the Bessemer process in tonnage produced in 1908 and was producing over 90% of all steel in 1960. Currently, *oxygen furnaces* of a variety of types and *electric arc furnaces* produce most of our commercial steels.

In many of these processes, air or oxygen passes over or through the molten metal to drive a variety of exothermic refining reactions. Carbon oxidizes to form gaseous CO or CO_2, which then leaves the melt. Other elements, such as silicon and phosphorus, are similarly oxidized and rise to be collected in a removable slag. At the same time, oxygen and other elements from the reaction gases dissolve in the molten metal and may later become a cause for concern.

Solidification Concerns

Regardless of the method by which the steel is made, it must undergo a change of state from liquid to solid before it can become a usable product. The liquid can be converted directly into finish-shape steel castings or solidified into a form suitable for further processing. In most cases, the latter option is exercised through either *continuous casting* or the forming of discrete *ingots*, both of which can serve as the feedstock material for subsequent forging or rolling operations.

When solidifying the steel, the desire is to obtain a product that is as free of flaws as possible. Most steelmaking furnaces first pour metal into containment vessels, known as *ladles*. Historically, these served simply as transfer and pouring containers, but they have recently emerged as the site for additional processing. *Ladle metallurgy* refers to a variety of processes designed to provide final purification and to fine-tune both the chemistry and temperature of the melt. Alloy additions can be made, carbon can be further reduced, dissolved gases can be reduced or removed, and steps can be taken to control subsequent grain size, limit inclusion content, reduce sulfur, and control the shape of any included sulfides. Stirring, degassing, reheating, and various injection procedures can be performed to increase the cleanliness of the steel and provide for tighter control of the chemistry and properties. Figure 6-1 shows a schematic of a ladle metallurgy station that incorporates injection of powdered alloys, injection of cored wire, argon purging, and argon stirring.

The processed metal is then poured from these ladles into ingot molds or continuous casters, usually through some form of bottom-pouring process such as the one shown schematically in Figure 6-2. By extracting the metal from the bottom of the ladle, slag and floating matter are not transferred, and a cleaner product results.

When we extend the observation that most nonmetallic contaminants have lower density than the molten metal, we find that the highest-quality ingots are also formed by a

FIGURE 6-1   Schematic of a ladle metallurgy station with powder and wire injection, argon purging, and argon stirring. *(Courtesy of Quanex MacSteel.)*

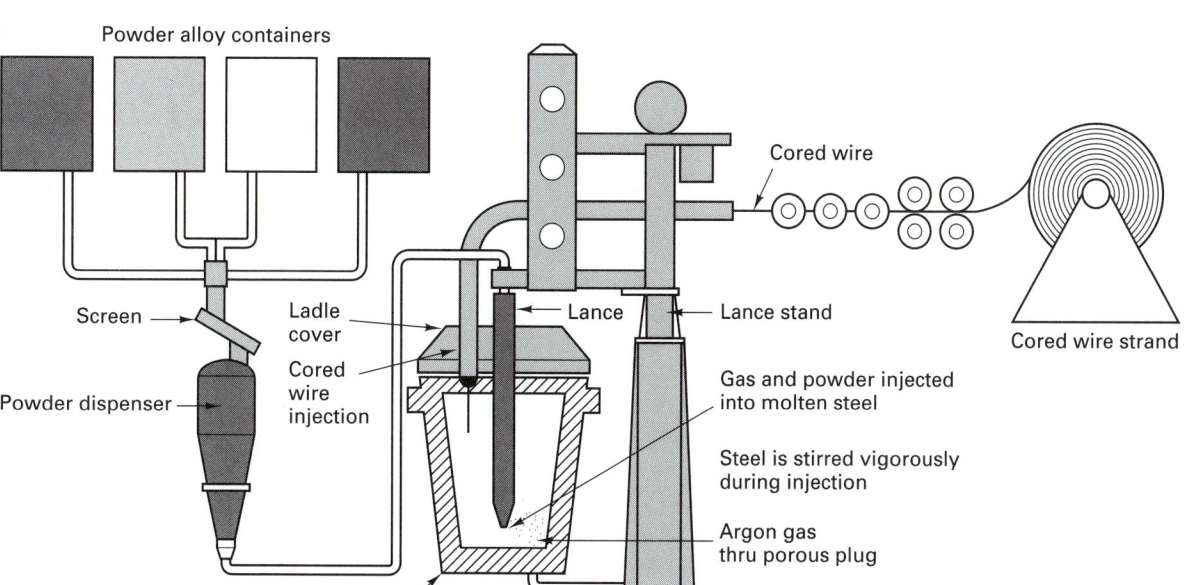

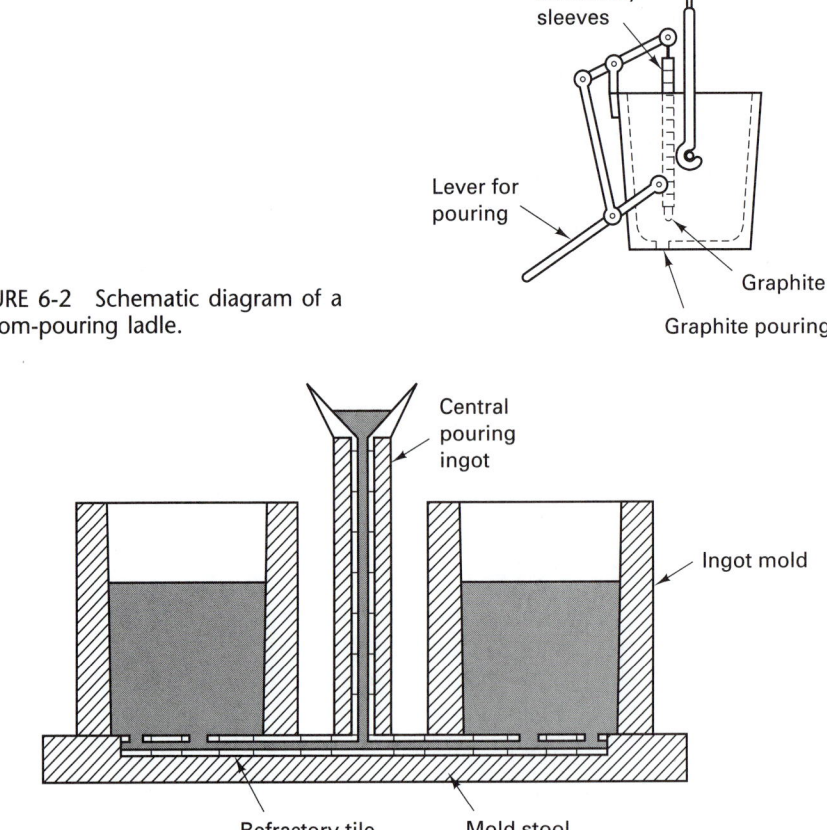

FIGURE 6-2   Schematic diagram of a bottom-pouring ladle.

FIGURE 6-3   Pouring of ingots by the bottom-pouring process. The bottom of the center mold is connected to the bottom of the remaining molds in the cluster by ceramic channels.

method that fills the ingot mold from the bottom. As illustrated in Figure 6-3, the bottom of several ingot molds is connected to a central pouring ingot by ceramic tile tunnels. Hot metal is poured into the center ingot and is conveyed through the tunnels to fill the outer molds. Any contaminants in the pour stream rise to the top of the central pouring ingot. In addition, the surrounding molds fill smoothly, avoiding any turbulence or splashing and the additional contamination that might result as the hot material reacts with air.

Figure 6-4 schematically shows the dimensions of a metal undergoing cooling and solidification. The large discontinuity at the melting point is known as *solidification shrinkage* and reflects the difference in density of the liquid and solid states of the metal. Thus when metals solidify, it is quite likely that a shrinkage void will form in the region of the last material to be liquid. In ingots, solidification proceeds inward from the mold walls and upward from the bottom. Shrinkage takes the form of a *pipe* coming in from the top (Figure 6-5). Since the pipe surface has been exposed to the atmosphere while at elevated temperature, oxides and surface contaminants form that would prevent the metal from welding back together during subsequent processing. That portion of the ingot containing the pipe is usually removed and recycled as scrap.

Although the amount of shrinkage cannot be changed, the shape and location can be greatly controlled. For example, the amount of pipe can be reduced by placing a ceramic

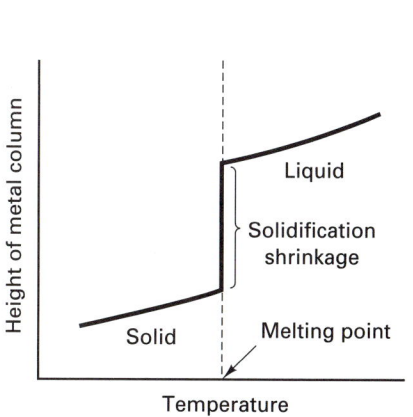

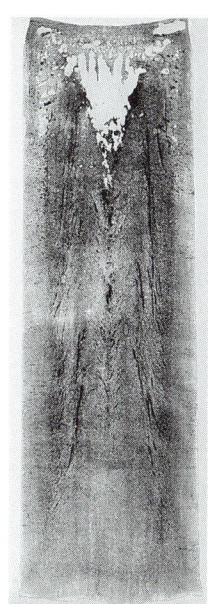

FIGURE 6-4   Height of a column of metal as a function of temperature, showing the significant shrinkage that occurs upon solidification.

FIGURE 6-5   Section of an ingot showing a pipe (top) and segregation (dark areas). *(Courtesy of Bethlehem Steel Corporation.)*

hot top or *exothermic material* on the top of the ingot. With additional heat at the top, the liquid reservoir at the end of solidification is more of a uniform layer, and the depth of pipe penetration is reduced. Variation in the size and shape of the mold can further control both the geometry of solidification and the final structure of the ingot. Tapered ingots with the big end up are commonly used to produce the greatest soundness.

Continuous casting processes have been developed to overcome a number of ingot-related difficulties, such as piping, entrapped slag, and structure variation along the length of the product. Figure 6-6 illustrates a typical continuous casting procedure, in which molten metal flows from a ladle, through a tundish, into a bottomless, water-cooled mold, usually made of copper. Cooling is controlled so that the outside has solidified before the metal exits the mold. The metal is then cooled by direct water sprays to produce complete solidification. The cast solid is still hot and is either bent and fed horizontally through a short reheat furnace and on to rolling, or is simply cut to desired lengths. Mold shape, and thus the shape of the cast product, may vary so that products may be cast with cross sections closer to the desired final shape.

In terms of product quality, continuous casting virtually eliminates the problems of piping and mold spatter. From a production viewpoint, it eliminates pouring into molds, stripping the molds from the solidified metal, and handling and reheating the ingots prior to rolling. By producing a continuous shape, hot and ready for rolling, cost, energy, and scrap are all reduced significantly. In addition, the products have improved surfaces, more uniform chemical composition, and fewer oxide inclusions.

Deoxidation and Degassification

During the oxidation process of steelmaking, considerable amounts of oxygen can dissolve in the molten metal. When the metal is then cooled to induce solidification, the oxygen

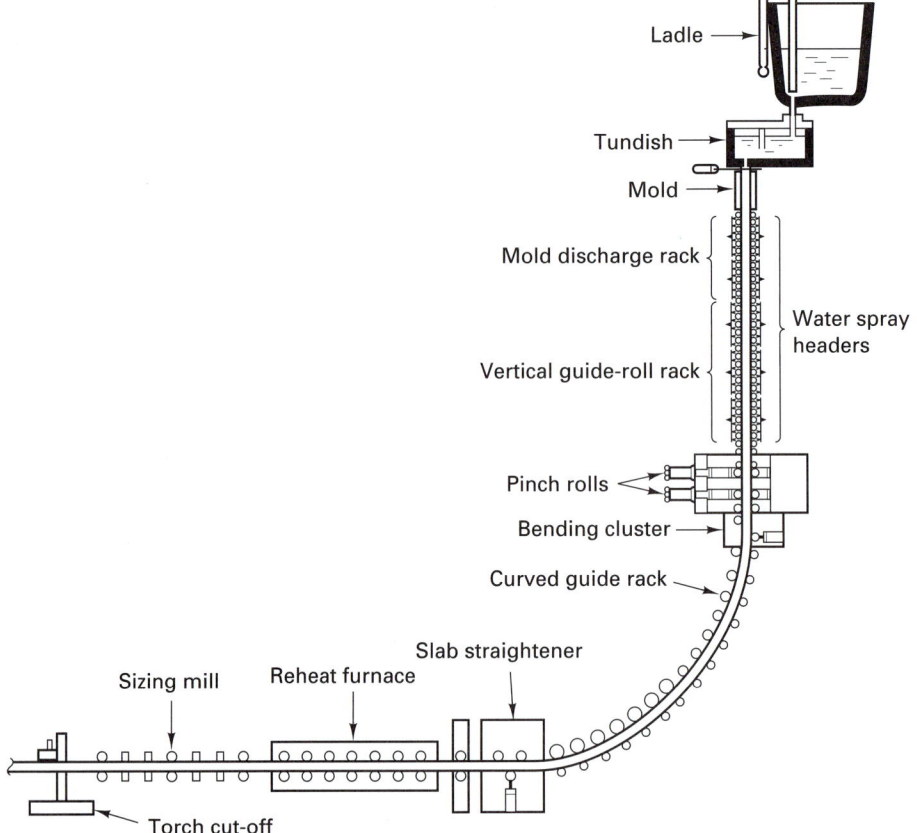

FIGURE 6-6 Schematic representation of the continuous casting process for producing billets, slabs, and bars. *(Courtesy of Materials Engineering.)*

and other gases are rejected as the solubility levels decrease markedly (Figure 6-7). The rejected oxygen frequently links with atomic carbon to produce carbon monoxide gas. It is possible that this gas can escape through the liquid, but quite frequently the bubbles become trapped and a porous structure results. The bubble-induced porosity may take the form of small, dispersed voids or large blowholes. Those pores that remain within the material can often be welded shut during subsequent hot forming (if sufficient deformation is performed). Some of the voids may not be fully closed, however, and others may have become sufficiently contaminated that they do not weld upon closure. Cracks and internal voids become an inherent part of the finished product.

In many cases it is desirable to avoid potential porosity problems by either removing the oxygen or rendering it nongaseous prior to solidification. When high-quality steel is desired, or where subsequent deformation might not be sufficient to weld the pores, a steel is often *deoxidized* (or *killed*). Aluminum, ferromanganese, or ferrosilicon is added to the molten steel to provide a material whose affinity for oxygen is higher than carbon's. The rejected oxygen reacts with the addition to produce solid metal oxides dispersed throughout the structure. High-carbon steels are often fully killed.

For steels with lower carbon contents, a partial deoxidation may be employed to produce a *semikilled* steel. Enough deoxidant is added to partially suppress bubble evolution but not enough to completely eliminate the effect of oxygen. Some pores still form in the

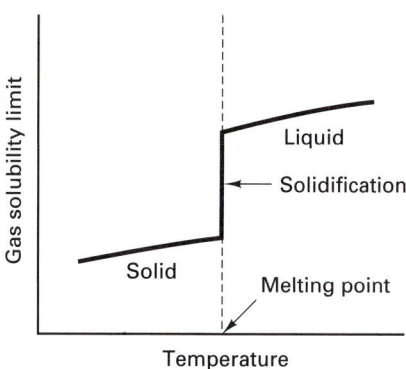

FIGURE 6-7 Solubility of gas in a metal as a function of temperature.

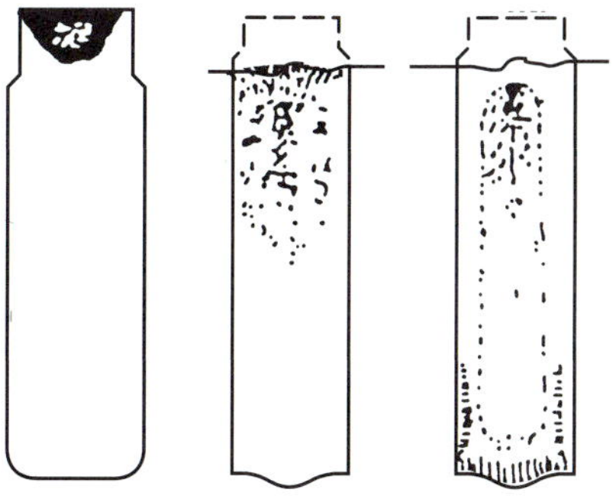

FIGURE 6-8 Schematic comparison of killed, semikilled, and rimmed ingot structures. *(Courtesy of American Iron and Steel Institute, Washington, D.C.)*

center of the ingot, their volume serving to cancel some of the solidification shrinkage, thereby reducing the extent of piping and scrap generation.

For steels with sufficiently low carbon, (usually less than 0.2%), a process known as *rimming* may be employed. The steel is only partially deoxidized before solidification. When the material is poured into the mold, the first metal to solidify is almost pure iron, being very low in carbon, oxygen, sulfur, and phosphorus. These elements are rejected from the solid into the remaining liquid. When the concentration of dissolved gases exceeds the solubility limit, a layer of CO bubbles evolves. Further solidification produces additional regions of porosity in the inner portions of the ingot. If properly controlled, the rimming action can produce a porosity level that will approximately compensate for the solidification shrinkage.

The outside of the rimmed ingot is clean, low-carbon metal that forms an excellent, blemish-free surface when the ingot is rolled into flat strips or similar product. The holes on the inside have bright, clean surfaces because they have not been exposed to air. When sufficient amounts of hot deformation are performed, the surfaces weld to produce a pore-free product. Figure 6-8 shows a schematic comparison of killed, semikilled, and rimmed ingots.

While the methods described above can effectively address the presence of dissolved oxygen, small amounts of other dissolved gases, particularly hydrogen and nitrogen, can have deleterious effects on the performance of steels. This is particularly important for alloy steels because several of the major alloying elements, such as vanadium, niobium, and chromium, tend to increase the solubility of these gases. As a result, *degassing* processes have been devised to reduce the amounts of all dissolved gases. Figure 6-9 illustrates the process of *vacuum degassing*, one of the most widely used degassing operations. An ingot mold is placed in an evacuated chamber, and the metal stream passes through the vacuum during pouring. By creating a large amount of exposed surface during the pouring, the vacuum is able to extract much of the dissolved gas.

An alternative to vacuum degassing is the *consumable-electrode remelting* process. Here an already solidified metal electrode replaces the molten metal ladle. When this electrode is remelted by an electric arc, the molten droplets pass through the vacuum and the extremely high surface area provides an effective means of gas removal. When the melting is done by electric arc, the process is known as *vacuum arc remelting* (VAR). If induction heating replaces the electric arc, the process is known as *vacuum induction melting* (VIM).

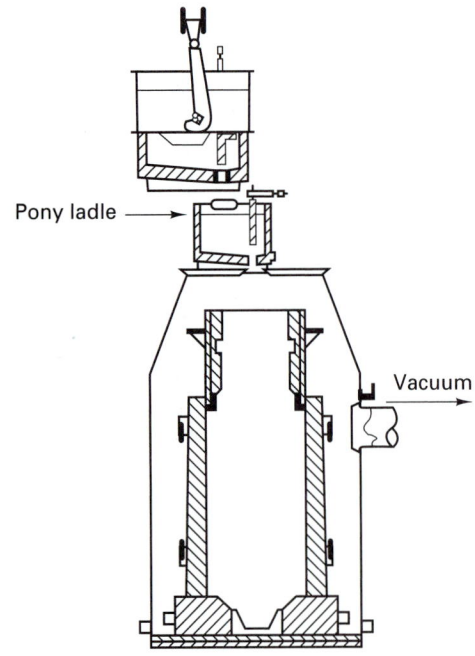

Pony ladle

Vacuum

FIGURE 6-9   Method of vacuum degassing
steel while pouring ingots.

Both are highly effective in removing dissolved gases but are relatively ineffective in removing nonmetallic impurities from the metal.

When extremely clean, gas-free metal is required; it can be obtained by means of the *electroslag remelting process* (ESR; Figure 6-10). Here the solid electrode is again melted and recast by means of an electric current, but the surface of the recast liquid is now covered with a blanket of molten flux. Nonmetallic impurities float and are collected in the flux, leaving beneath a newly solidified structure with improved quality. No vacuum is required since all molten material exists below the shroud of molten flux. (*Note*: This process is simply a large-scale version of the electroslag welding process discussed in Chapter 36.)

Plain-Carbon Steel

Commercial steel, while theoretically an alloy of iron and carbon, actually contains manganese, phosphorus, sulfur, and silicon in significant and detectable quantities. When these four additional elements are present in their normal percentages, the product is referred to as *plain-carbon steel*. Its strength is primarily a function of its carbon content, increasing with increasing carbon. Unfortunately, the ductility of plain-carbon steels decreases as the carbon content is increased and its hardenability is quite low. In addition, the properties of ordinary carbon steels are impaired by both high and low temperatures (loss of strength and embrittlement, respectively), and they are subject to corrosion in most environments.

Plain-carbon steels are generally classed into three subgroups based on their carbon content. *Low-carbon steels* have less than 0.30% carbon and possess good formability and weldability, but lack sufficient hardenability to be hardened to any significant depth. Their structures are usually ferrite and pearlite, and the material is generally used as it comes from the hot-forming or cold-forming processes, or in the as-welded condition. *Medium-carbon steels* have between 0.30 and 0.80% carbon, and they can be quenched to form martensite or bainite if the section size is small and a severe water or brine quench is used. The best balance of properties is obtained at these carbon levels where the high toughness

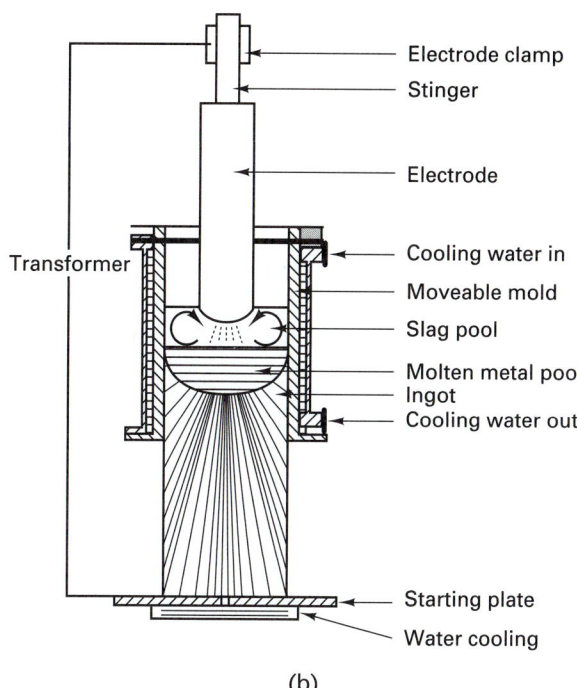

FIGURE 6-10   (a) Production of an ingot by the electroslag remelting process; (b) schematic representation of this process. *(Courtesy of Carpenter Technology Corporation.)*

and ductility of the low-carbon material is in good compromise with the strength and hardness that comes with higher carbon contents. These steels are extremely popular and find numerous mechanical applications. *High-carbon steels* have more than 0.80% carbon. Toughness and formability are quite low, but hardness and wear resistance are high. Severe quenches can form martensite, but hardenability is still poor. Quench cracking is often a problem when the material is pushed to its limit.

Plain-carbon steels are the lowest-cost steel material and should be given first consideration for many applications. Often, however, their limitations become restrictive. When improved performance is required, these steels can often be upgraded by the addition of one or more alloying elements.

Alloy Steels

The differentiation between plain-carbon and alloy steel is often somewhat arbitrary. Both contain carbon, manganese, and usually silicon. Copper and boron are possible additions to both classes. Steels containing more than 1.65% manganese, 0.60% silicon, or 0.60% copper are usually designated as *alloy steels*. Also, a steel is considered to be an alloy steel if a definite or minimum amount of other alloying element is specified. The most common alloy elements are chromium, nickel, molybdenum, vanadium, tungsten, cobalt, boron, and copper, as well as manganese, silicon, phosphorus, and sulfur in amounts greater than are normally present.

Effects of the Various Alloying Elements

In general, alloying elements are added to steels in small percentages (usually less than 5%) to improve strength or hardenability, or in much larger amounts (often up to 20%) to produce special properties such as corrosion resistance or stability at high or low temperatures. Certain additions may be made during the steelmaking process to remove dissolved oxygen from the melt. Manganese, silicon, and aluminum are frequently used for this deoxidation. Aluminum, and to a lesser extent, vanadium, niobium, and titanium, are used to control austenite grain size. Other elements may be added to improve the strength or toughness properties of the product metal. Manganese, silicon, nickel, and copper add strength by forming solid solutions in ferrite. Chromium, vanadium, molybdenum, tungsten, and other elements increase strength by forming dispersed second-phase carbides. Niobium, vanadium, and zirconium can be used for ferrite grain-size control. Nickel and copper are added in small amounts to improve corrosion resistance.

For the constructional alloy steels, the primary reason for the alloying elements is to increase hardenability. The most common elements for this purpose (in order of decreasing effectiveness) are manganese, molybdenum, chromium, silicon, and nickel. Small quantities of vanadium are quite effective, but the response drops off as the quantity is increased. Boron is also extremely effective in steels with less than 0.65% carbon.

The following information describes the distinct properties and characteristics imparted by the various alloy elements used in steel:

1. *Manganese* combines with sulfur to produce soft manganese sulfides, thereby preventing the formation of iron sulfide, which would coat the grain boundaries and impart brittleness. In somewhat larger amounts, usually greater than 1%, manganese increases hardenability, slightly strengthens ferrite, and lowers the martensite transformation temperatures. When manganese is added in large amounts (11 to 14%), an austenitic alloy is produced. Its high hardness, coupled with good ductility, high strain hardening capacity, and excellent wear resistance, make it ideal for impact-resisting tools and similar applications.

2. *Sulfur* is usually not desired in steel because of the embrittling effect of iron sulfide. If the product is manganese sulfide, however, the sulfur is not harmful (provided that the sulfides are not in large quantities and are well dispersed). If large quantities of manganese sulfide are present in the proper form, they can actually impact desirable machining properties. Some *free-machining steels* contain as much as 0.08 to 0.15% sulfur in combination with an increased manganese content.

3. *Nickel* is often added to increase toughness and impact resistance, primarily at low temperature. It is generally used in amounts from 2 to 5%, often in combination with other alloying elements, such as chromium and molybdenum. Alloys with 12 to 20% nickel and low amounts of carbon possess outstanding corrosion resistance. An iron alloy with 36% nickel, commonly known as *Invar*, has a near-zero thermal expansion coefficient and is used for sensitive measuring devices.

4. Although large percentages of *chromium* can impart corrosion resistance and heat resistance, in the amounts used in low-alloy steels, these effects are relatively minor. For these alloys, chromium, in amounts that are usually less than 2%, serves primarily to increase hardenability and strength. In many of these alloys, chromium and nickel are used together in a ratio of about one part chromium to two parts nickel. When combined with carbon, the chromium carbides can enhance wear resistance.

5. *Molybdenum* is used in alloy steels in amounts less than 0.3% to improve hardenability and increase strength properties, particularly under dynamic and high-temperature conditions. Resistance to temper embrittlement is also attributed to the presence of molybdenum. Molybdenum carbides are extremely stable at elevated temperatures and are used in alloys to retain fine grain size and provide strength and creep resistance at elevated temperature. Molybdenum carbides are also used in hot-work tool steels, such as those used in forging dies, to impart hardness that will persist at red heat.

6. *Vanadium* is another alloying element that forms strong carbides that persist at elevated temperature. Thus 0.03 to 0.25% vanadium in the form of carbides can effectively inhibit grain growth and increase strength properties (most notably elastic limit, yield point, and impact strength) with almost no loss of ductility.

7. Another element that forms stable carbides, *tungsten*, is also used as a primary alloying element in tool steels that must maintain their hardness at high elevated temperatures.

8. *Copper* has been known to resist atmospheric corrosion for centuries, but only recently has it been used as an addition to steel (in amounts from 0.10 to 0.50%) to provide this property. Low-carbon steel sheet and structural steels often contain a copper addition to enhance corrosion resistance, but surface quality and hot-working behavior tend to deteriorate somewhat.

9. In small percentages, *silicon* has an effect that is similar to nickel, increasing the strength properties with little companion loss of ductility. It is an important alloying element in certain high-yield-strength structural steels and is also used in spring steels (in amounts of about 2%) and to promote the large grain size that is desirable in steels used for magnetic applications.

10. *Boron* is a very powerful hardenability agent, being from 250 to 750 times as effective as nickel, 75 to 125 times as effective as molybdenum, and about 100 times as powerful as chromium. Only a few thousandths of a percent are sufficient to produce the desired effect in low-carbon steels, but the results diminish rapidly with increasing carbon content. Since no carbide formation or ferrite strengthening is produced, improved machinability and cold-forming capability often result from the use of boron in place of other hardenability additions.

11. In addition to its use as a deoxidizer, *aluminum* may be added to steels in amounts of 0.95 to 1.30% to produce aluminum nitrides during a nitriding operation. *Titanium* and *niobium* (columbium) are additional carbide formers. Steels with 0.15 to 0.35% *lead* show substantially improved machinability, but the environmental effects of lead may limit its acceptability. *Bismuth* is another alloy that promotes machinability, as well as *selenium* and *tellurium*. *Zirconium, cerium,* and *calcium* can control the shape of inclusions and thereby promote toughness.

Table 6-1 summarizes the primary effects of the common alloying elements in steel. A working knowledge of the information contained in this table may be useful to the design engineer in selecting an alloy steel to meet a given set of requirements. Alloying elements are often used in combination, however, resulting in the immense variety of alloy steels that are commercially available. To simplify this situation, a classification system has been developed and has achieved general acceptance in a variety of industries.

AISI–SAE Classification System

The most important group of alloy steels is undoubtedly those that are incorporated into the AISI identification system. This system, which classifies alloys by chemistry, was started by the Society of Automotive Engineers (SAE) to provide some standardization for

TABLE 6-1. Principal Effects of Major Alloying Elements in Steel		
Element	Percentage	Primary Function
Aluminum	0.95–1.30	Alloying element in nitriding steels
Bismuth	—	Improves machinability
Boron	0.001–0.003	Powerful hardenability agent
Chromium	0.5–2	Increase of hardenability
	4–18	Corrosion resistance
Copper	0.1–0.4	Corrosion resistance
Lead	—	Improved machinability
Manganese	0.25–0.40	Combines with sulfur to prevent brittleness
	>1	Increases hardenability by lowering transformation points and causing transformations to be sluggish
Molybdenum	0.2–5	Stable carbides; inhibits grain growth
Nickel	2–5	Toughener
	12–20	Corrosion resistance
Silicon	0.2–0.7	Increases strength
	2	Spring steels
	Higher percentages	Improves magnetic properties
Sulfur	0.08–0.15	Free-machining properties
Titanium	—	Fixes carbon in inert particles
		Reduces martensitic hardness in chromium steels
Tungsten		Hardness at high temperatures
Vanadium	0.15	Stable carbides; increases strength while retaining ductility; promotes fine grain structure

the steels used in the automotive industry. It was later adopted and expanded by the American Iron and Steel Institute (AISI) and has been incorporated into the Universal Numbering System that was developed to include all engineering metals. Both plain-carbon and low-alloy steels are identified by a four-digit number, with the first number indicating the major alloying elements and the second number designating a subgrouping within the major alloy system. The meaning of these first two digits can be summarized by a list such as that presented in Table 6-2. The last two digits indicate the approximate amount of carbon, expressed as "points" of carbon, where one point is equal to 0.01% carbon. Thus a 1080 steel would be a plain carbon steel with 0.80% carbon. Similarly, a 4340 steel would be a Mo–Cr–Ni alloy with 0.40% carbon.[*]

A letter prefix may be used to designate the process used to produce the steel, such as the letter *E* for electric furnace. An *X* prefix is used to indicate permissible variations in the range of manganese, sulfur, or chromium. The letter *B* between the second and third digits indicates that the base metal has been supplemented by the addition of boron. Similarly, a letter *L* in this position indicates a lead addition for enhanced machinability.

The *H grade of AISI steels* is designated by the letter *H* placed as a suffix to the standard designation. These steels are specified when hardenability is a major requirement. The chemistry specifications are somewhat less stringent, but the steel must also comply with a hardenability standard. The hardness values at specific distances from the quenched end of a Jominy specimen (see Chapter 5) must all lie within a predetermined band.

Other designation organizations, such as the American Society for Testing and Materials (ASTM) and the U.S. government (MIL and federal), base their specification systems

---

[*]The four digits are generally read as two-digit numbers. For example, a 4340 steel is designated as a "forty-three forty"; the 43 refers to the alloy family, and the 40, to the amount of carbon.

TABLE 6-2. Some AISI-SAE Standard Steel Designations

AISI Number	Type	Alloying Elements (%)					
		Mn	Ni	Cr	Mo	V	Other
1xxx	Carbon steels						
10xx	Plain carbon						
11xx	Free cutting (S)						
12xx	Free cutting (S) and (P)						
15xx	High manganese						
13xx	High manganese	1.60–1.90					
2xxx	Nickel steels		3.5–5.0				
3xxx	Nickel–chromium		1.0–3.5	0.5–1.75			
4xxx	Molybdenum						
40xx	Mo				0.15–0.30		
41xx	Mo, Cr			0.40–1.10	0.08–0.35		
43xx	Mo, Cr, Ni		1.65–2.00	0.40–0.90	0.20–0.30		
44xx	Mo				0.35–0.60		
46xx	Mo, Ni (low)		0.70–2.00		0.15–0.30		
47xx	Mo, Cr, Ni		0.90–1.20	0.35–0.55	0.15–0.40		
48xx	Mo, Ni (high)		3.25–3.75		0.20–0.30		
5xxx	Chromium						
50xx				0.20–0.60			
51xx				0.70–1.15			
6xxx	Chromium–vanadium						
61xx				0.50–1.10		0.10–0.15	
8xxx	Ni, Cr, Mo						
81xx			0.20–0.40	0.30–0.55	0.08–0.15		
86xx			0.40–0.70	0.40–0.60	0.15–0.25		
87xx			0.40–0.70	0.40–0.60	0.20–0.30		
88xx			0.40–0.70	0.40–0.60	0.30–0.40		
9xxx	Other						
92xx	High silicon						1.20–2.20 Si
93xx	Ni, Cr, Mo		3.00–3.50	1.00–1.40	0.08–0.15		
94xx	Ni, Cr, Mo		0.30–0.60	0.30–0.50	0.08–0.15		

more on specific applications. Acceptance into a given classification is generally based on physical or mechanical properties rather than the chemistry of the metal. Many low-carbon and structural steels are specified by their ASTM designations.

Selecting Balanced Alloy Steels

From the previous discussion it is apparent that two or more alloying elements can often produce similar effects. Thus steels with substantially different chemical compositions can possess almost identical mechanical properties. Figure 6-11 clearly demonstrates that steels of quite different composition (chemistry) can have almost identical property ratios *when properly heat treated.* This fact should be kept in mind when selecting and specifying alloy steels. It is particularly important when one realizes that some alloying elements can be very costly, and others may be in short supply due to emergencies or political constraints. Overspecification has often been employed to guarantee success despite sloppy manufacturing and heat-treatment practice. The correct steel, however, is usually the least expensive one that can be consistently processed to achieve the desired properties. This usually involves taking advantage of the effects provided by *all* of the alloy elements. Steels with "balanced" compositions can avoid needlessly large amounts of expensive alloy elements.

In selecting alloy steels, it is important to consider both use and fabrication. For one

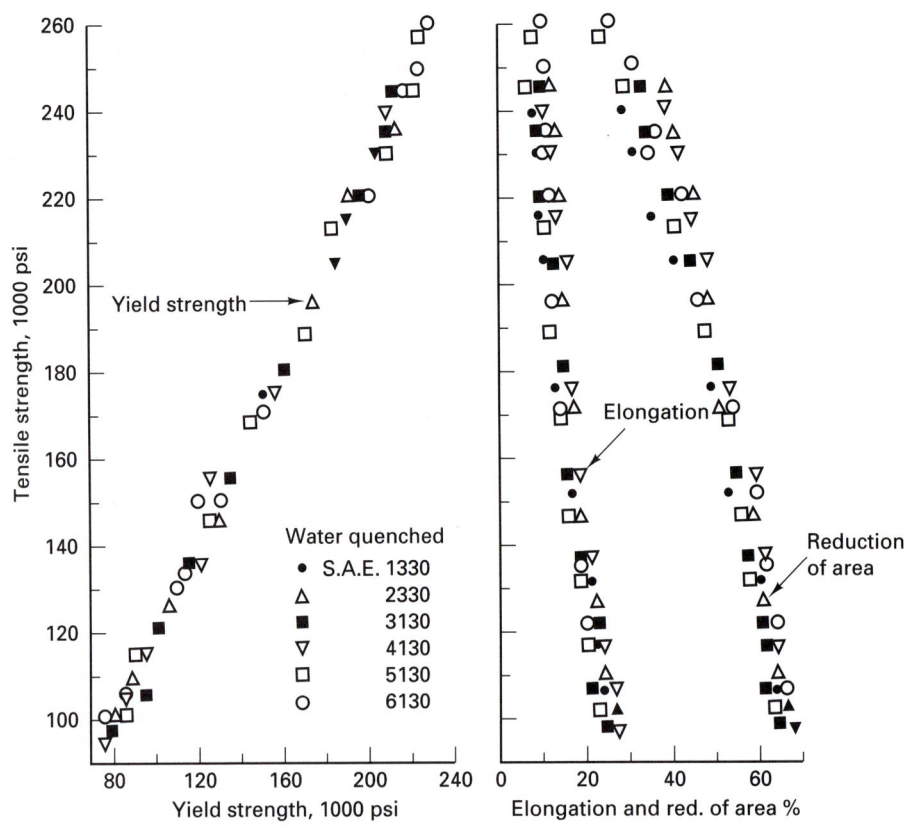

FIGURE 6-11 Straight-line relationship between the mechanical properties of properly heat-treated SAE alloy steels. *(Courtesy of ASM International, Materials Park, Ohio.)*

product, it might be permissible to increase the carbon content to obtain greater strength. For another application, such as one involving assembly by welding, it might be best to keep the carbon content low and use a balanced amount of alloy elements to obtain the desired strength without risking the possibility of cracking due to high-carbon martensite. Steel selection, therefore, often involves defining the required properties, determining the best microstructure to provide those properties, and selecting the steel with the best carbon content and hardenability characteristics to achieve that goal. One approach is to purchase material on the basis of properties rather than exact chemical composition. The supplier or producer is then free to provide any material that will possess the desired properties, and substantial cost savings can result. To assure success, however, it is important that *all* of the necessary properties be specified.

High-Strength Low-Alloy Structural Steels

There are two general categories of alloy steels: (1) the constructional alloy steels, where the desired properties are typically developed by thermal treatment and the alloys are often selected for their effect on hardenability, and (2) the *high-strength low-alloy* (HSLA) or microalloyed types, which rely largely on chemical composition to develop the desired mechanical properties in the as-rolled or normalized condition.

The latter type often finds application in large, welded structures, where the size of the product would preclude subsequent heat treatment. The dominant property require-

TABLE 6-3.   Typical Compositions and Strength Properties of Several Groups of High-Strength Low-Alloy Structural Steels

Group	Chemical Compositions[a] (%)					Strength Properties				
	C	Mn	Si	Cb	V	Yield		Tensile		Elongation in 2 in. (%)
						ksi	MPa	ksi	MPa	
Columbium or vanadium	0.20	1.25	0.30	0.01	0.01	55	379	70	483	20
Low manganese–vanadium	0.10	0.50	0.10		0.02	40	276	60	414	35
Manganese–copper	0.25	1.20	0.30			50	345	75	517	20
Manganese–vanadium–copper	0.22	1.25	0.30		0.02	50	345	70	483	22

[a]All have 0.04% P, 0.05% S, and 0.20% Cu.

ments for these applications are high yield strength, good weldability, and acceptable corrosion resistance. Only a limited ductility is required, and hardenability is no longer a consideration.

The low-alloy structural steels often exhibit yield strengths that are as much as twice that of the plain-carbon structurals. This increase in strength, coupled with the resistance to martensite formation in the weld zone, is obtained by controlling the amounts of carbon, manganese, and silicon and adding low amounts niobium, vanadium, or other alloys. About 0.2% copper can be added to improve corrosion resistance. Many types of high-strength low-alloy steels have been developed by combining the alloy elements in various combinations and amounts. Table 6-3 presents the chemistries and properties of several of the more common types.

The HSLA steels have significantly changed the use of structural materials. Because of their higher yield strength, weight savings of 20 to 30% can often be achieved with no sacrifice to strength or safety. Rolled and welded HSLA steels are now being used in automobiles, trains, bridges, and buildings. Because of their low alloy content and high-volume application, their cost is often little more than that of the ordinary plain-carbon structurals.

Microalloyed Steels in Manufactured Products

Steels containing small amounts of alloying elements, such as niobium, vanadium, titanium, zirconium, boron, rare earth elements, or combinations thereof are also being used increasingly as substitutes for heat-treated steels in the manufacture of small-to-medium-sized discrete parts. These *microalloyed steels* derive their name from the fact that the alloying elements are only added in amounts ranging from 0.05 to 0.15%, where their primary effect is to provide grain refinement and/or precipitation strengthening. These are effective strengthening mechanisms, and yield strengths between 70 and 110 ksi can be obtained without heat treatment, without interfering with the material processing, and without significantly compromising toughness. Weldability can be retained or even improved if the carbon content is decreased simultaneously. In essence, these steels offer maximum strength with minimum carbon, while simultaneously preserving weldability, machinability, and formability.

Cold-formed microalloyed steels require less cold work to achieve a desired level of strength, so they tend to have greater residual ductility. By means of accurate temperature control and controlled-rate cooling, hot-formed products, such as forgings, can often be used in the air-cooled condition. Many mechanical properties match those of quenched-and-tempered material. Machinability can actually be enhanced because of the more uniform hardness and the fact that the ferrite–pearlite structure of the microalloyed steel is often more machinable than the ferrite–carbide structure of the quenched-and-tempered variety. Fatigue life and wear resistance can be superior to those of the heat-treated counterparts. Thus, in applications where the properties are adequate, microalloyed steels can often provide attractive cost savings. Energy savings can be substantial, straightening or stress relieving after heat treatment is no longer necessary, and quench cracking is not a problem. Due to the increase in material strength, the size and weight of finished products can often be reduced. As a result, the cost of a finished forging could be reduced by 5 to 25%.

Certain precautions must be observed, however, if these materials are to attain their optimum properties. During the elevated-temperature segments of processing, the material must be heated high enough to place all the alloys into solution. After forming, the products should be rapidly air cooled to 1000 to 1100°F before dropping into collector boxes. Microalloyed steels tend to through-harden upon air cooling, so products fail to exhibit the lower-strength, higher-toughness interiors that are typical with the quenched-and-tempered materials.

Free-Machining Steels

The increased use of high-speed machining, particularly with automated machine tools, has spurred the use and development of several varieties of *free-machining steels*. These steels machine readily and form small chips when cut. The smaller chips reduce the length of contact between the chip and cutting tool, thereby reducing the associated friction and heat (and therefore, power and tool wear). The formation of small chips also reduces the likelihood of chip entanglement in the machine and makes chip removal much easier.

Free-machining steels are basically carbon steels that have been modified by an alloy addition to enhance machinability. Sulfur, lead, bismuth, selenium, tellurium, and phosphorus have all been added to enhance machinability. Sulfur (0.08 to 0.33%) combines with manganese (0.7 to 1.6%) to form soft manganese sulfide inclusions. These act as discontinuities in the structure which serve as sites to form broken chips. The inclusions also provide a built-in lubricant that prevents formation of a built-up edge on the cutting tool and imparts an altered cutting geometry (see Chapter 18). Insoluble lead particles serve the same purpose.

The bismuth free-machining steels have become an attractive alternative to the previous varieties. Compared with lead, bismuth is a better free-machining agent. It is more environmentally acceptable, has less tendency to form stringers, and can be more uniformly dispersed since its density is closer to that of iron. Machinability is improved because the cutting action generates enough heat to form a thin film of liquid bismuth that lasts for only fractions of a microsecond. Tool life is noticeably extended and the machined product is still weldable.

Use of free-machining steels is not without compromise, however. Ductility and impact properties are somewhat reduced compared to the unmodified steels. Copper-based braze joints tend to embrittle when used to join bismuth free-machining steels, and the machining additions reduce the strength of shrink-fit assemblies. If these compromises are objectionable, other methods may be employed to enhance machinability. For example, the machinability of steels can be improved by cold working the metal. As the strength and hardness of the metal increase, the metal loses ductility, and subsequent machining produces chips that tear away more readily and fracture into smaller segments.

Bake-Hardenable Steel Sheet

The somewhat recent development of *bake-hardenable sheet* has helped the steel industry maintain its dominance in automotive sheet applications. These low-carbon steels are processed in such a way that they are aging resistant during normal storage but begin to age during forming. The subsequent exposure to heat during the paint baking operation completes the aging process and adds an additional 5 to 10 ksi, raising the final yield strength to approximately 40 ksi. Since the increase in strength occurs after forming, the material offers a combination of good formability and improved dent resistance in the final product. In addition, it allows weight savings to be achieved without compromising the attractive features of steel sheet, which include good crash energy absorption, low cost, and easy recyclability through magnetic separation.

Precoated Steel Sheet

Traditional sheet metal fabrication has involved the fabrication of components from bare steel, followed by the finishing of these products on a piece-by-piece basis. In this system it is not uncommon for the finishing processes to be the most expensive and time-consuming stages of manufacture.

An alternative to this procedure is to purchase mill-coated sheet material with the coating having been applied by the steel supplier when the material was in the form of a long, continuous strip. Numerous coatings can be specified, including the entire spectrum of dipped and plated metals (including aluminum, zinc, and chromium), vinyls, paints, and others. Extra caution must be exercised during fabrication to prevent damage to the coating, but this additional effort and expense often add up to far less than the cost of finishing individual pieces.

Steels for Electrical and Magnetic Applications

Several types of steels are widely used in the electrical industry. *Silicon steels*, which contain 0.5 to 5.0% silicon, have noticeably increased electrical resistivity and magnetic permeability. Increased resistivity decreases eddy-current losses, while increased permeability decreases hysteresis losses. When such steels are used in electrical motors, generators, and transformers, the power losses and associated heat problems are reduced. For this reason, silicon steel is often specified for use in the magnetic circuits of electrical equipment. Since silicon causes the steel to become brittle, the amount of silicon should be kept as low as possible for applications where the component is subjected to dynamic forces.

Cobalt increases the magnetic saturation of a steel when it is added in amounts up to 36%. For this reason, *cobalt alloy steels* are often used in electrical equipment where high magnetic densities must exist. Many permanent magnets are made from high-cobalt alloys.

More recently, amorphous metals have shown considerable promise in this area. Since the material has no grains or grain boundaries: (1) the magnetic domains can move freely in response to magnetic fields, (2) the properties are the same in all directions, and (3) corrosion resistance is improved. The high magnetic strength and low hysteresis losses offer the possibility of smaller, lighter-weight magnets. When used to replace silicon steel in power transformer cores, this material has the potential of reducing core losses by as much as 50%. One estimate cites a potential savings to the United States alone of about $1 billion per year.

Maraging Steels

When super-high strength is required from a steel, one possible alternative is often the maraging grades. These alloys contain between 15 And 25% nickel, plus significant amounts of cobalt, molybdenum, and titanium, all added to a very low carbon steel. A typical composition of a maraging steel is:

0.03% C	0.10% Al
18.5% Ni	0.003% B
7.5% Co	0.10% Si maximum
4.8% Mo	0.10% Mn maximum
0.40% Ti	0.01% S maximum
0.01% Zr	0.01% P maximum

This steel can be hot worked at temperatures between 1400 and 2300°F (760 to 1260°C). When air cooled from 1500°F (815°C), it has a hardness of about $30R_C$ and a structure of soft, tough, low-carbon martensite. It is easily machined and, because of its low work-hardening rate, can be cold worked to a high degree. Aging at 900°F (480°C) for 3 to 6 hours, followed by air cooling, raises the hardness to about $52R_C$, with a yield strength in excess of 250 ksi (1725 MPa) and elongation in excess of 11%.

Maraging alloys are very useful in applications where ultrahigh strength and good toughness are important. They can be welded, if welding is followed by the full solution and aging treatment. As might be expected from the large amount of alloy additions (over 30%) and multistep thermal processing, maraging steels are quite expensive and should be specified only when their outstanding properties are required.

Steels for High-Temperature Service

Continued developments in missiles, jet aircraft, and nuclear power have increased the demand for metals that have good strength characteristics, corrosion resistance, and, particularly, creep resistance at high operating temperatures. Much work has been done to produce both ferrous and nonferrous alloys that have useful properties at temperatures in excess of 1000°F (550°C).

The ferrous alloys, which are normally restricted to use below 1400°F (760°C), tend to be low-carbon materials with less than 0.1% carbon. One such alloy is a modified 18-8 stainless steel, stabilized with either niobium or titanium. A 1000-hour rupture stress of 6 to 7 ksi (40 to 50 MPa) is observed at 1400°F (760°C), with considerably higher strengths at lower temperatures. Iron is also a major component of other high-temperature alloys, but when the amounts become less than 50%, the metal can hardly be classified as ferrous in nature. High strength at high temperature usually requires the more expensive nonferrous materials (discussed in Chapter 7).

■ 6.4 STAINLESS STEELS

Corrosion-resistant or stainless steels contain sufficient amounts of chromium that they can no longer be considered low-alloy steels. The corrosion resistance is imparted by the formation of a strongly adherent chromium oxide on the surface of the metal. Good resistance to many of the corrosive media encountered in the chemical industry can be obtained by the addition of 4 to 6% chromium to low-carbon steel. When improved corrosion resistance and outstanding appearance are required, materials should be specified that utilize a superior chromium oxide that forms when the amount of atomic chromium in solution (excluding chromium carbides and other forms where the chromium is unavailable to react with oxygen) exceeds 12%. This category forms what has been commonly called the *true stainless steels*.

Several classification schemes have been devised to categorize these alloys. The American Iron and Steel Institute (AISI) groups the metals by chemistry and assigns a three-digit number that identifies the basic family and the particular alloy within that family. The material in this book, however, will group these alloys by microstructural families, for it is the basic structure that controls the engineering properties of the metal. Table 6-4 presents the AISI designation scheme for stainless steels and correlates it with the microstructural families.

TABLE 6-4. AISI Designation Scheme for Stainless Steels

Series	Alloys	Structure
200	Chromium, nickel, manganese, or nitrogen	Austenitic
300	Chromium and nickel	Austenitic
400	Chromium only	Ferritic or martensitic
500	Low chromium (<12%)	Martensitic

Chromium is a ferrite stabilizer, the addition of chromium tending to increase the temperature range over which ferrite is the stable structure. If sufficient chromium is added to the iron, and carbon is kept low, an alloy can be produced that is ferrite at all temperatures below solidification. These alloys are known as *ferritic stainless steels*. They possess rather poor ductility or formability because of the bcc crystal structure, but they are readily weldable. No martensite can form in the welds because there is no possibility of forming austenite that can then transform during cooling. Since the ferritic alloys are the cheapest type of stainless steel, they should be given first consideration when a stainless alloy is required.

If increased strength is needed, the *martensitic stainless steels* should be considered. For these alloys, insufficient chromium is added to fully eliminate the austenite region, resulting in a corrosion-resistant material that can be austenite at high temperature and ferrite at low. As with the standard steels, carbon can be dissolved in the face-centered-cubic austenite, which can then be quenched to form a martensitic structure. Variable carbon contents are available, providing a range of possible strengths and hardnesses. However, caution should be taken to assure more than 12% chromium in solution. Slow cools may allow the carbon and chromium to react and form chromium carbides. When this occurs, the chromium is not available to react with oxygen and form the protective oxide. Thus, the martensitic stainless steels may only be "stainless" when in the martensitic condition (when the chromium is trapped in atomic solution), and may be susceptible to red rust when annealed or normalized for machining or fabrication. The martensitic stainless steels cost about $1\frac{1}{2}$ times as much as the ferritic alloys, part of this being due to the additional heat treatment, which generally consists of an austenitization, quench, stress relief, and temper.

Nickel is an austenite stabilizer, and with sufficient amounts of both chromium and nickel, it is possible to produce a stainless steel in which austenite is the stable structure at room temperature. Known as *austenitic stainless steels*, these alloys may cost twice as much as the ferritic variety, with the added expense being attributed to the cost of the alloying nickel and chromium. Manganese and nitrogen are also austenite stabilizers and may be substituted for some of the nickel to produce a lower-cost, somewhat lower-quality austenitic stainless steel.

Austenitic stainless steels are nonmagnetic and are highly resistant to corrosion in almost all media except hydrochloric acid and other halide acids and salts. In addition, they may be polished to a mirror finish and thus combine attractive appearance and corrosion resistance. Formability is outstanding (characteristic of the fcc crystal structure), and these steels strengthen significantly when cold worked. The response of the popular 304 alloy (also known as 18-8 because of the composition of 18% chromium and 8% nickel) to a small amount of cold work is as follows:

	Water Quench	Cold Rolled 15%
Yield strength [ksi (MPa)]	38 (260)	117 (805)
Tensile strength [ksi (MPa)]	90 (620)	140 (965)
Elongation in 2 in. (%)	68	11

These materials are often used in the water-quenched condition, where the water quench serves to retain the alloys in solid solution. No phase transformations occur during the quench since austenite is the stable phase for all temperatures involved.

Austenitic stainless steels are costly materials and should not be specified where the less expensive ferritic or martensitic alloys would be adequate or where a true stainless steel is not required. Figure 6-12 lists some of the popular alloys from each of the three major classifications and schematically denotes their key properties. Table 6-5 shows the typical alloy compositions for the three basic families of stainless steels.

TABLE 6-5. Typical Compositions (in wt. %) of the Ferritic, Martensitic, and Austenitic Stainless Steels

Element	Ferritic	Martensitic	Austenitic[a]
Carbon	0.08–0.20	0.15–1.2	0.03–0.25
Chromium	11–27	11.5–18	16–26
Manganese	1–1.5	1	2 (5.5–10)
Molybdenum			Some cases
Nickel			3.5–22
Phosphorus and sulfur			Normal (0)
Silicon	1	1	1–2 (0)
Titanium			Some cases

[a]Values in parentheses are for one type.

A fourth and special class of stainless steels is the *precipitation-hardening* variety. These alloys are basically martensitic or austenitic types, modified by the addition of alloying elements such as aluminum that permit age hardening at relatively low temperatures. By adding age hardening to the existing strengthening mechanisms, these materials are capable of attaining properties such as a 260-ksi (1790-PMa) yield strength and 265-ksi (1825-MPa) tensile strength with a 2% elongation. However, the additional alloys and extra processing make the precipitation-hardening alloys some of the most expensive stainless steels, and they should be used only when absolutely required.

Duplex stainless steels contain between 21 and 25% chromium and 5 to 7% nickel and are water quenched from a hot-working temperature that is between 1830 and 1920°F to produce a microstructure that is approximately half ferrite and half austenite. This struc-

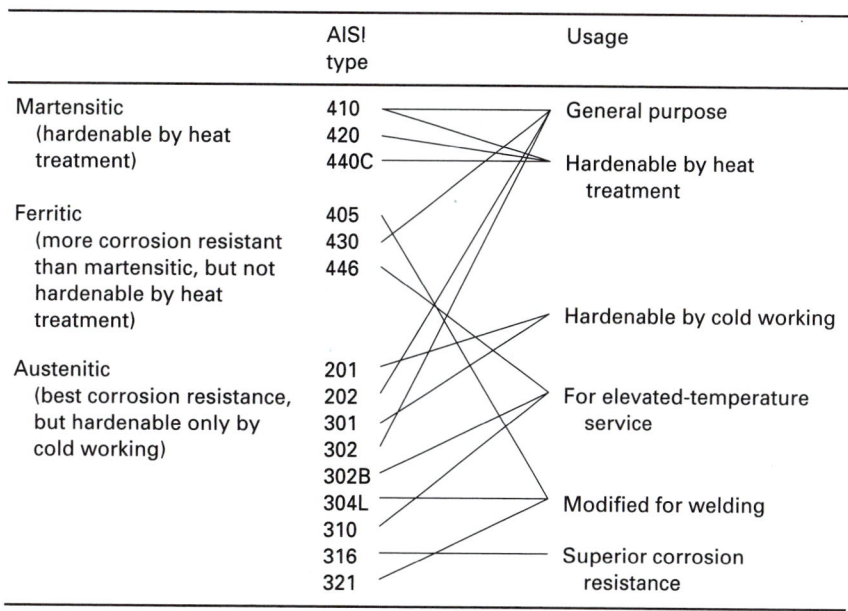

FIGURE 6-12 Popular alloys and key properties for the three primary types of stainless steels.

ture offers a higher yield strength and greater resistance to stress corrosion cracking than either the austenitic or ferritic grades.

Still other stainless alloys have been developed to meet special needs. Ordinary stainless steels are difficult to machine because of their work-hardening properties and their tendency to seize during cutting. Special *free-machining alloys* have been produced within each family, with additions of sulfur or selenium raising the machinability to approximately that of a medium-carbon steel.

Problems with stainless steels are often due to the loss of corrosion resistance (*sensitization*) when the amount of chromium in solution drops below 12%. Since chromium depletion is usually caused by the formation of chromium carbides along grain boundaries, and these carbides form at elevated temperatures, various means have been developed to prevent their formation. One approach is to keep the carbon content of stainless steels as low as possible, usually less than 0.10%. Another method is to tie up the carbon with small amounts of "stabilizing" elements, such as titanium or niobium, that have a stronger affinity for carbon than does chromium. Rapidly cooling of these metals through the carbide-forming range of 900 to 1500°F (480 to 820°C) also works to prevent carbide formation.

Another problem with high-chromium stainless steels is an embrittlement that occurs after long times at elevated temperatures. This is attributed to the formation of *sigma phase*, a brittle compound that forms at elevated temperature and coats grain boundaries, thereby producing a brittle crack path through the metal. Stainless steels used in high-temperature service should be checked periodically to detect sigma-phase formation.

■ 6.5 TOOL STEELS

Tool steels are ferrous alloys that have been designed to provide wear resistance and toughness combined with high strength. They are basically high-carbon steels, modified by alloy additions to provide the desired balance of toughness and wear.

Several classification systems have been applied to tool steels, some using chemistry as a basis and others employing hardening method and major mechanical property. The AISI–SAE system uses a letter designation to identify basic features such as quenching method, primary application, special characteristic, or specific industry involved. Table 6-6 lists the seven basic families of tool steels, the corresponding AISI–SAE letter grades, and associated feature or characteristic. Individual alloys within the letter grades are then listed numerically to produce a letter–number identification system.

TABLE 6-6. Basic Types of Tool Steel and Corresponding AISI–SAE Grades

Type	AISI–SAE Grade	Significant Characteristic
1. Water-hardening	W	
2. Cold-work	O	Oil-hardening
	A	Air-hardening medium alloy
	D	High-carbon–high-chromium
3. Shock-resisting	S	
4. High-speed	T	Tungsten base
	M	Molybdenum base
5. Hot-work	H	H1–H19: chromium base
		H20–H39: tungsten base
		H40–H59: molybdenum base
6. Plastic-mold	P	
7. Special-purpose	L	Low alloy
	F	Carbon–tungsten

Water-hardening tool steels (W-grade) are essentially high-carbon plain-carbon steels, and account for a large percentage of all of the tool steels used. They are the least expensive variety and are used for a wide range of parts that are usually quite small and not subject to severe usage or elevated temperature. Because strength and hardness are functions of the carbon content, a wide range of properties can be achieved through composition variation. These steels must be quenched in water to attain high hardness. Because their hardenability is low, they can be used only for relatively thin sections if the full depth of hardness is desired. They are also rather brittle, particularly at higher hardness.

Typical uses of the various plain-carbon steels are as follows:

- *0.60–0.75% carbon*: machine parts, chisels, setscrews, and similar products where medium hardness is required coupled with good toughness and shock resistance
- *0.75–0.90% carbon*: forging dies, hammers, and sledges
- *0.90–1.10% carbon*: general-purpose tooling applications that require good balance of wear resistance and toughness, such as drills, cutters, shear blades, and other heavy-duty cutting edges
- *1.10–1.30% carbon*: small drills, lathe tools, razor blades, and other light-duty applications in which extreme hardness is required without great toughness

In applications where improved toughness is desired, small amounts of manganese, silicon, and molybdenum are often added. Vanadium additions of about 0.20% are used to form strong, stable carbides that retain fine grain size during heat treating. One of the main weaknesses of the plain-carbon tool steels is their loss of hardness at elevated temperature. Prolonged exposure to temperatures over 300°F (150°C) usually results in undesired softening.

When larger parts must be hardened or distortion must be minimized, *oil-* or *air-hardening grades* (O and A designations, respectively) are often preferred. These metals contain alloy additions, and their higher hardenability permits hardening by less severe quenches. Tighter dimensional tolerances can be maintained during heat treatment, and the cracking tendency is reduced.

Manganese tool and die steels form one segment of this group. These typically contain 0.75 to 1.0% carbon and 1.0 to 2.0% manganese and are moderate in cost. When greater toughness is desired, the manganese can be reduced and chromium, silicon, or nickel is added. These modified alloys are not as hard as the plain manganese types, but they have less tendency to crack because of their greater hardenability and less severe quench requirement.

The low-chromium tool and die steels are also members of this class. With additions of 0.5 to 5.0% chromium, these alloys behave in much the same way as the plain-carbon steels, with the chromium addition providing increased hardenability and toughness. The high-chromium tool steels are designated by a different letter, D, and contain between 10 and 18% chromium. These steels generally require an annealing treatment before they can be machined. After machining, they are hardened and tempered and can usually retain full hardness at temperatures up to 800°F (425°C). Forging dies, die-casting die blocks, and drawing dies are often made from high-chromium tool steels.

Shock-resisting tool steels (S designation) have been developed for both hot and cold impact applications. Low carbon content (approximately 0.5% carbon) is usually specified to provide the necessary toughness, with carbide-forming alloys providing the necessary abrasion resistance, hardenability, and hot-work characteristics.

High-speed tool steels are used for cutting tools and other applications where strength and hardness must be retained at temperatures up to or exceeding red-heat (about 1400°F/760°C). One popular member is the T1 alloy, also known as 18–4–1 because the 0.7% carbon steel is alloyed with 18% tungsten, 4% chromium, and 1% vanadium. It offers a balanced combination of shock resistance and abrasion resistance, and is used for a wide variety of cutting applications. Other high-speed tool steels have cobalt added to improve the hardness at elevated temperature. Toughness diminishes, however, and forming is more difficult.

The *molybdenum high-speed steels* (M designation) were developed to reduce the amount of tungsten and chromium required to produce the high-speed properties. The M2 variety is now the most widely used high-speed steel, its higher carbon content and balanced analysis producing properties that are applicable to a number of high-speed applications.

Hot-work tool steels (H designation) have been developed to provide strength and hardness during prolonged exposure to elevated temperature. All employ substantial additions of carbide-forming alloys. H1 to H19 are chromium-based alloys with about 5.0% chromium; H20 to H39 are tungsten-based types with 9 to 18% tungsten coupled with 3 to 4% chromium; and H20 to H59 are molybdenum-based.

Other types of tool steels include (1) the *plastic mold steels* (P designation), designed to meet the requirements of zinc die casting and plastic injection molding dies; (2) the *low-alloy special-purpose tool steels* (L designation), such as the L6 extreme toughness variety; and (3) the *carbon-tungsten type* of special-purpose tool steels (F designation), which are water hardening but substantially more wear-resistant than the plain-carbon tool steels.

■ 6.6 ALLOY CAST STEELS AND IRONS

The effects of alloying elements are the same regardless of the process used to produce the final shape. Some alloys, however, can also enhance process-specific features, such as fluidity and as-solidified properties, when the shape is to be made by casting. If the material contains less than about 2.0% carbon, it is considered to be a cast steel. Alloys with more than 2% carbon are cast irons.

Cast steels are usually heat-treated to produce a quenched-and-tempered structure, and the alloy additions are selected to provide the desired hardenability and balance of properties. Similarly, if a cast iron is to be quenched, chromium, molybdenum, and nickel are frequently added to improve hardenability. The chromium tends to offset the undesirable quenching effects of the high carbon, and the molybdenum and nickel compensate for the high silicon.

Most cast irons, however, are used in the as-cast condition, with the only heat treatment being a stress relief or annealing. The alloy elements are now selected for their ability to alter properties by (1) affecting the formation of graphite or cementite, (2) modifying the morphology of the carbon-rich phase, (3) strengthening the matrix material, or (4) enhancing wear resistance through the formation of alloy carbides. Nickel promotes graphite formation and tends to promote a finer graphite structure. Chromium, on the other hand, retards graphite formation and stabilizes cementite. They are frequently used together in a ratio of two or three parts of nickel to one part of chromium.

If the silicon content of a gray cast iron is lowered, the strength is increased considerably. If the decrease in silicon is carried too far, however, a white cast iron tends to form. The addition of about 2% nickel can be used to suppress the formation of white iron. Thus, by adjusting the silicon content and adding small amounts of nickel, a cast iron can be produced with good strength and adequate machinability.

Between 0.5 and 1.0% molybdenum is often added to gray cast iron to impart additional strength and form alloy carbides. In addition, it helps to control the size of the graphite flakes.

High-alloy cast irons are often designed to provide enhanced corrosion resistance, particularly at elevated temperatures such as those encountered in the chemical industry. Within this family, the austenitic gray cast irons have become quite popular. These contain about 14% nickel, 5% copper, and 2.5% chromium. They offer good corrosion resistance to many acids and alkalis at temperatures up to about 1500°F (800°C). Alloy cast irons and steels are usually specified through their ASTM designation number.

■ KEY WORDS

AISI–SAE designation
alloy steel
austenitic stainless steel
bake-hardenable steel
cast steel
continuous casting
degassification
deoxidation
duplex stainless steel
electroslag remelting
ferritic stainless steel
free-machining steel
high-strength low-alloy steel

ingot
iron
killed
ladle metallurgy
maraging steel
martensitic stainless steel
microalloyed steel
pig iron
plain-carbon steel
precipitation-hardenable stainless
  steel
precoated steel
rimmed

semikilled
sensitization
sigma phase
solidification shrinkage
stainless steel
steel
tool steel
vacuum arc remelting
vacuum degassing
vacuum induction melting

■ REVIEW QUESTIONS

1. Why might it be important to know the prior processing history of an engineering material?
2. What is a ferrous material?
3. When iron ore is reduced to metallic iron, what other elements are generally present in the metal?
4. How does steel differ from pig iron?
5. What are some of the changes that can be made to a steel by ladle metallurgy operations?
6. What is the advantage of pouring molten metal from the bottom of a ladle?
7. What is solidification shrinkage?
8. What are some of the techniques that can be used to control the shape and location of solidification shrinkage in an ingot?
9. What are some of the attractive economic and processing advantages of continuous casting?
10. What are some of the techniques used to reduce the amount of dissolved oxygen in molten steel? How might other gases, such as nitrogen and hydrogen, be reduced?
11. What are some of the attractive features of electroslag remelting?
12. What is plain-carbon steel?
13. What properties account for the high-volume use of medium-carbon steels?
14. Why should plain-carbon steels be given first consideration for applications requiring steel?
15. What are some of the most common alloy elements added to steel?
16. What are some of the different reasons that alloying elements might be added to steel?
17. What alloys are particularly effective in increasing the hardenability of steel?
18. While chromium imparts corrosion resistance to the stainless steels, it does not perform this function for most of the alloy steels. What features does a chromium addition impart?
19. What are some of the alloy elements that tend to form stable carbides within a steel?
20. What is the significance of the last two digits in a typical four-digit AISI–SAE steel designation?
21. Why should fabrication processes enter into the considerations when selecting a steel?
22. How are the final properties usually obtained in the constructional alloy steels? In the HSLA steels?
23. What are microalloyed steels?
24. What are some of the potential benefits that may be obtained through the use of microalloyed steels?
25. What are some of the various alloy additions that have been used to improve the machinability of steels?
26. What are some of the compromises associated with the use of free-machining steels?
27. What are the attractive characteristics of the bake-hardenable steels?
28. What economic factors might justify the use of precoated steel sheet?
29. Why have the amorphous metals attracted attention as potential materials for magnetic applications?
30. Under what conditions might maraging steels be required?
31. What structural feature is responsible for the observed corrosion resistance of stainless steels?
32. Why should ferritic stainless steels be given first consideration when selecting a stainless steel?
33. Which of the major types of stainless steel is likely to contain significant amounts of carbon? Why?
34. Why might a martensitic stainless steel "rust" when exposed to a hostile environment?
35. What are some of the unique properties of austenitic stainless steels?
36. What is a duplex stainless steel?
37. What is sensitization of a stainless steel, and how can it be prevented?
38. What is a tool steel?
39. How does the AISI–SAE designation system for tool steels differ from that for plain-carbon and alloy steels?
40. For what types of applications might an air-hardenable tool steel be attractive?
41. What alloying elements are used to produce the hot-work tool steels?
42. What are some of the reasons that alloy additions are made to cast irons that are not to undergo heat treatment?

Chapter 6 CASE STUDY

interior tub of a top-loading washing machine

The interior tub of a washing machine is the container that holds the clothes and solutions during the washing and rinsing cycles, but also contains the perforations that permit removal of the water by draining and spinning. The component will see mechanical loadings from the weight of the clothes and water, and also the dynamic action of spinning water-laden fabrics. There will be exposure to a wide range of water quality, as well as the full spectrum of soaps, detergents, bleaches, and other laundry additives. The surfaces should also be resistant to the impact and abrasion of buttons, zippers and snaps.

This part has traditionally been manufactured by the deep drawing, perforating and trimming of metal sheet, followed by some form of surface coating treatment. For a long time, the standard material was "enameling iron"—a steel sheet with less than 0.03% carbon—which was then coated with a fired porcelain enamel. Due to the difficulties of producing ultra-low carbon material in today's steelmaking operations, enameling iron became increasingly scarce, and manufacturers were forced to substitute the lowest carbon, readily available material, namely 1008 steel. This substitution further required modification of the enameling process to prevent CO and CO_2 blistering.

Your employer, Carefree Appliances, is presently manufacturing these tubs from 1008 steel sheet and subsequently coating them with fired porcelain enamel. Your marketing staff reports that consumers tend to view a stainless steel tub to be of higher quality. As a result, your supervisor has asked you to evaluate the merits of converting to this material. You should familiarize yourself with the current product (the material, the forming process, and the porcelain enameling), and then determine what might be involved in converting to stainless steel. Consider the following specific questions:

1. What are the obvious pros and cons of the present product and process? Where would you expect most problems to occur in the manufacturing process? Which aspects of fabrication are likely to be the most costly?

2. What would be the pros and cons of converting to stainless steel? In what ways would the product be superior? Are there any assets or liabilities associated with product fabrication?

3. Which stainless steel would you recommend. Begin by considering the basic types (ferritic, austenitic, and martensitic) and then refine your relection to a specific alloy if possible. Discuss the rationale for your selection.

4. Since deep drawing is a metal deformation process, we could use cold work (strain hardening) as a strengthening mechanism. Would you find this to be attractive, or would you prefer to use a recrystallization anneal after drawing and prior to use? Why? If you elect to use cold work, might you want to at least perform a stress relief heat treatment prior to use? Could this be done and still preserve the deformation strengthening? In deep drawing, the deformation is not uniform (increasing as we move up the sidewalls of the container), and the bottom of the tub will simply retain the properties of the starting sheet. In order to assure a minimum amount of strength at all locations, it may be desirable to begin the drawing with a partially cold-rolled sheet. Do you find this suggestion to be desirable? Why or why not?

5. After drawing and perforating, the residual drawing lubricant is removed from the part. Do you feel that any additional surface treatment would be required? What would be your recommendation?.

CHAPTER 7

NONFERROUS METALS AND ALLOYS

7.1 INTRODUCTION

7.2 COPPER AND COPPER ALLOYS

General Properties and Characteristics

Commercially Pure Copper

Copper-Based Alloys

Copper–Zinc Alloys

Copper–Tin Alloys

Copper–Nickel Alloys

Other Copper-Based Alloys

7.3 ALUMINUM AND ALUMINUM ALLOYS

General Properties and Characteristics

Commercially Pure Aluminum

Aluminums for Mechanical Applications

Corrosion Resistance of Aluminum and Its Alloys

Classification System

Wrought Aluminum Alloys

Aluminum Casting Alloys

Aluminum–Lithium Alloys

7.4 MAGNESIUM AND MAGNESIUM ALLOYS

General Properties and Characteristics

Magnesium Alloys and Their Fabrication

7.5 ZINC–BASED ALLOYS

7.6 TITANIUM AND TITANIUM ALLOYS

7.7 NICKEL–BASED ALLOYS

7.8 SUPERALLOYS AND OTHER NONFERROUS METALS FOR HIGH-TEMPERATURE SERVICE

7.9 LEAD, TIN, AND THEIR ALLOYS

7.10 SOME LESS KNOWN METALS AND ALLOYS

7.11 GRAPHITE

Case Study: NONSPARKING WRENCH

■ 7.1 INTRODUCTION

Nonferrous metals and alloys are playing increasingly important roles in modern technology. Because of their number and the fact that their properties vary widely, they provide an almost limitless range of properties for the design engineer. While they are often more costly than iron or steel, these metals often possess certain properties or combinations of properties that are not available in ferrous metals:

1. Resistance to corrosion
2. Ease of fabrication
3. High electrical and thermal conductivity
4. Light weight
5 Strength at elevated temperatures
6. Color

Although it is true that corrosion resistance can be obtained in certain ferrous alloys, several nonferrous metals possess this property without requiring special and expensive alloying elements. Nearly all of the nonferrous alloys possess at least two of the qualities listed, and some possess nearly all. For many applications, specific combinations of these

167

properties are highly desirable, and the availability of materials that provide them directly is a strong motivation for the use of the nonferrous alloys.

In most cases the strength of the nonferrous alloys is inferior to that of steel. Also, the modulus of elasticity may be considerably lower, a fact that places them at a distinct disadvantage when stiffness is a necessary characteristic. Fabrication, however, is usually easier than for steel. Those alloys with low melting points are often easy to cast, either in sand molds, permanent molds, or dies. Many alloys have high ductility coupled with low yield points, the ideal combination for easy cold work and high formability. Good machinability is characteristic of many nonferrous alloys. The savings obtained through ease of fabrication can often overcome the higher cost of the nonferrous material and favor its use in place of steel. The one fabrication area in which the nonferrous alloys are somewhat inferior to steel is weldability. Due to a number of developments and improvements in welding, however, it is generally possible to produce satisfactory weldments in all of the nonferrous metals.

■ 7.2 COPPER AND COPPER ALLOYS

General Properties and Characteristics

Copper is an important engineering metal that has been in use for over 6000 years. Used in its pure state, copper is the backbone of the electrical industry. It is also the major metal in a number of highly important engineering alloys, namely the brasses and bronzes.

The wide use of copper is based, primarily, on three important properties: its high electrical and thermal *conductivity,* useful strength with high *ductility,* and *corrosion resistance.* Obviously, its excellent conductivity accounts for its importance to the electrical industry, and about one-third of all copper produced is used in some form of electrical application, such as the commutators shown in Figure 7-1. The better grades of conductor copper now have a conductivity rating of about 102% IACS, reflecting metallurgical improvements made since 1913, when the International Annealed Copper Standard was established and the conductivity of pure copper was set at 100% IACS.

Pure copper in its annealed state has a tensile strength of only about 30,000 psi (200 MPa), with an elongation of nearly 60%. By cold working, however, the tensile strength can be raised to over 65,000 psi (450 MPa), with a decrease in elongation to about 5%. Its relatively low strength and high ductility make copper a very desirable metal for applications where extensive forming is required. Furthermore, the hardening effects of cold working can easily be removed, since the recrystallization temperature for copper is less than 500°F (260°C). The copper alloys also lend themselves nicely to the whole spectrum of fabrication processes, including casting, machining, and welding.

Unfortunately, copper is heavier than iron. Although the strengths can be quite high, the strength/weight ratio for the copper alloys is usually less than that for the weaker aluminum and magnesium alloys. In addition, several problems occur when copper is used at elevated temperature. If copper is stressed for a long period of time at high temperature, it is subject to intercrystalline failure at about half of its normal room-temperature strength. Material containing more than 0.3% oxygen is also subject to hydrogen embrittlement when it is exposed to reducing gases above 750°F (400°C).

Commercially Pure Copper

Refined copper containing between 0.02 and 0.05% oxygen (the principal impurity in copper) is called *electrolytic tough-pitch (ETP) copper.* It is often used as a base for copper

FIGURE 7-1   Copper and copper alloys are used for a variety of electrical applications, such as these electrical commutators. *(Courtesy of The Electric Materials Company.)*

alloys and may be used for electrical applications such as wire and cable, where the highest conductivity is not required. For superior conductivity, additional refining is required to reduce the oxygen content and produce *oxygen-free high-conductivity (OFHC) copper.*

Copper-Based Alloys

Copper, as a pure metal, is not used extensively in manufactured products, except in electrical applications, and even here, alloy additions such as silver, arsenic, cadmium and zirconium, are used to enhance various properties without significantly impairing conductivity. More often, copper is the base metal for some alloy, to which it imparts its good ductility, corrosion resistance, and electrical and thermal conductivity.

Copper-based alloys are commonly identified through a system of numbers standardized by the Copper Development Association (CDA). Table 7-1 presents a breakdown of this system, which has been adopted by the American Society for Testing and Materials (ASTM), Society of Automotive Engineers (SAE), and the U.S. government. Alloys numbered from 100 to 190 are mostly copper with less than 2% alloy addition. Numbers 200 to 799 are other wrought alloys. The 800 and 900 series are all casting alloys.

TABLE 7-1. Standard Designations for Copper and Copper Alloys (CDA System)

	Wrought Alloys		Cast Alloys
100–155	Commercial coppers	833–838	Red brasses and leaded red brasses
162–199	High–copper alloys	842–848	Semired brasses and leaded semired brasses
200–299	Copper–zinc alloys (brasses)	852–858	Yellow brasses and leaded yellow brasses
300–399	Copper–zinc–lead alloys (leaded brasses)	861–868	Manganese and leaded manganese bronzes
400–499	Copper–zinc–tin alloys (tin brasses)	872–879	Silicon bronzes and silicon brasses
500–529	Copper–tin alloys (phosphor bronzes)	902–917	Tin bronzes
532–548	Copper–tin–lead alloys (leaded phosphor bronzes)	922–929	Leaded tin bronzes
600–642	Copper–aluminum alloys (aluminum bronzes)	932–945	High-leaded tin bronzes
647–661	Copper–silicon alloys (silicon bronzes)	947–949	Nickel-tin bronzes
667–699	Miscellaneous copper–zinc alloys	952–958	Aluminum bronzes
700–725	Copper–nickel alloys	962–966	Copper nickels
732–799	Copper–nickel–zinc alloys (nickel silvers)	973–978	Leaded nickel bronzes

Copper–Zinc Alloys

Zinc is by far the most popular alloying addition to copper, with the resulting alloys being known as *brasses*. If the copper content is not over 36%, the brass is a single-phase solid solution. Since this structure is identified as the alpha phase, these alloys are often called *alpha brasses*. They are quire ductile and formable, these characteristics increasing with the zinc content up to about 36%. Cartridge brass, the 70% copper–30% zinc alloy, offers the best combination of strength and ductility, and is quite popular for sheet-forming operations such as deep drawing.

Above 36% zinc, the copper–zinc alloys enter a two-phase region involving the brittle beta phase, and ductility drops markedly. Cold-working properties are rather poor for these high-zinc brasses, but deformation is rather easy when performed hot. Like copper, brass is hardenable by cold working and is commercially available in various degrees of hardness.

Table 7-2 lists some of the most common copper–zinc alloys and their composition, properties, and typical uses. Brasses range in color from copper to nearly white, with the lower-zinc brasses being more coppery than those with more zinc. The addition of a third element can significantly affect the color, however.

TABLE 7-2. Composition, Properties, and Uses of Some Common Copper–Zinc Alloys

| CDA Number | Common Name | Composition (%) | | | | | Condition | Tensile Strength | | Elongation in 2 in. (%) | Typical Uses |
		Cu	Zn	Sn	Pb	Mn		ksi	MPa		
200	Commercial indent bronze	90	10				Soft sheet	38	262	45	Screen wire, hardware, screws, jewelry
							Hard sheet	64	441	4	
							Spring	73	503	3	
240	Low brass	80	20				Annealed sheet	47	324	47	Drawing, architectural work, ornamental
							Hard	75	517	7	
							Spring	91	627	3	
260	Cartridge brass	70	30				Annealed sheet	53	365	54	Munitions, hardware, musical instruments, tubing
							Hard	76	524	7	
							Spring	92	634	3	
270	Yellow brass	65	35				Annealed sheet	46	317	64	Cold forming, radiator cores, springs, screws
							Hard	76	524	7	
280	Muntz metal	60	40				Hot-rolled	54	372	45	Architectural work; condenser tube
							Cold-rolled	80	551	5	
443–445	Admiralty metal	71	28	1			Soft	45	310	60	Condenser tube (salt water), heat exchangers
							Hard	95	655	5	
360	Free-cutting brass	61.5	35.3		3		Soft	47	324	60	Screw-machine parts
							Hard	62	427	20	
675	Manganese bronze	58.5	39	1		0.1	Soft	65	448	33	Clutch disks, pump rods, valve stems, high-strength propellers
							Bars, half hard	84	579	19	

Most brasses have good corrosion resistance. In the 0 to 40% zinc region, the addition of a small amount of tin imparts improved resistance to seawater corrosion. Cartridge brass with tin becomes admiralty brass; Muntz metal with tin is called naval brass. Brasses with 20 to 36% zinc are subject to a selective corrosion, known as *dezincification,* when exposed to acidic or salt solutions. This can be suppressed through an addition of arsenic. Brasses with more than 15% zinc often experience season cracking or *stress-corrosion cracking*. Both stress and exposure to corrosive media are required for this failure to occur (but residual stresses and atmospheric moisture may be sufficient!). As a result, cold-worked brass is usually stress relieved (to remove the residual stresses) before being placed in service.

Many uses of brass relate to the high electrical and thermal conductivities coupled with useful engineering strength. Plating characteristics are excellent and make the material an excellent base for decorative chrome or similar coatings. A unique property of alpha brass is its ability to have rubber vulcanized to it without any special treatment except thorough cleaning. As a result, it is widely used in mechanical rubber goods.

When high machinability is required, as with automatic screw machine stock, 2 to 3% lead is added to the brass to ensure the formation of free-breaking chips. Brass casting alloys are quite popular for use in plumbing fixtures and fittings, low-pressure valves, and a variety of decorative hardware. An alloy containing between 50 and 55% copper and the remainder zinc is often used as a filler metal in brazing. It is an effective material for joining steel, cast iron, brasses, and copper, and produces joints that are nearly as strong as those obtained by welding.

Copper–Tin Alloys

Alloys of copper and tin, commonly called *tin bronzes*, are usually specified when they offer some form of special property. The term *bronze* is often confusing since it is used to designate any copper alloy where the major alloy addition is neither zinc nor nickel. To provide clarification, the major alloy addition is usually included in the designation.

The true tin bronzes usually contain less than 12% tin. Strength continues to increase as tin is added, up to about 20%, but the high-tin alloys tend to be brittle. The copper–tin bronzes offer good strength, toughness, wear resistance, and corrosion resistance. They are often used for bearings, gears, and fittings that are subjected to heavy compressive loads. When the copper–tin alloys are used for bearing applications, up to 10% lead is often added.

The most popular wrought[*] bronze is phosphor bronze, which usually contains from 1 to 11% tin. Alloy 521 (CDA) is typical of this class and contains 92% copper, 8% tin, and 0.15% phosphorus. Hard sheet of this material has a tensile strength of 110 ksi (760 MPa) and an elongation of 3%. Soft sheet has a tensile strength of 55 ksi (380 MPa) and 65% elongation. The material is specified for pump parts, gears, springs, and bearings.

Alloy 905 is a commonly used bronze casting alloy containing 88% copper, 10% tin, and 2% zinc. In the cast condition, the tensile strength is about 45 ksi (310 MPa), with an elongation of 45%. It has very good resistance to seawater corrosion and is used on ships for pipe fittings, gears, pump parts, bushings, and bearings.

Bronzes can also be made by mixing powders of copper and tin and processing by powder metallurgy (described in Chapter 16). The resulting porous product can be used as filters for high-temperature or corrosive media, or can be infiltrated with oil to produce self-lubricating bearings.

---

[*]The term *wrought* means "shaped or fabricated in the solid state," as opposed to *cast*, which is shaped as a liquid.

Copper–Nickel Alloys

Copper and nickel exhibit complete solubility, as seen in Figure 4-6, and a wide range of useful alloys has been developed. Key features include high thermal conductivity and high-temperature strength, coupled with corrosion resistance to a range of materials, including seawater. Coupled with a high resistance to stress-corrosion cracking, these properties make the copper–nickel alloys a good choice for heat exchangers, cookware, desalination apparatus, and a wide variety of coinage. *Cupronickels* contain 2 to 30% nickel. *Nickel silvers* contain no silver, but 10 to 30% nickel and at least 5% zinc. Their bright silvery luster make them attractive for ornamental applications, and they are also used for musical instruments. An alloy with 45% nickel is known as *constantan*, and the 67% nickel alloy is called *Monel*.

Other Copper-Based Alloys

The copper alloys discussed previously acquire their strength primarily through solid-solution strengthening and cold work. In the copper alloy family, three alloying elements can produce materials that are precipitation hardenable: aluminum, silicon, and beryllium.

Aluminum bronze alloys are best known for their combination of high strength and excellent corrosion resistance, and are often considered as cost-effective alternatives to stainless steel and nickel-based alloys. The wrought alloys can be strengthened through solid-solution strengthening, cold work, and the precipitation of iron- or nickel-rich phases. With less than 8% aluminum, the alloys are very ductile. When aluminum exceeds 9%, however, the ductility drops and the hardness approaches that of steel. Still higher aluminum contents result in brittle but very wear resistant materials. By varying the aluminum content and heat treatment, the tensile strength can range from about 60 to 145 ksi (415 to 1000 MPa). Typical applications include marine hardware, power shafts, sleeve bearings, and pump and valve components for handling seawater, sour mine water, and various industrial fluids. Cast alloys are available for applications where casting is the preferred means of manufacture. Since aluminum bronze exhibits large amounts of solidification shrinkage, castings made of this material should be designed with this in mind.

Silicon bronzes contain up to 4% silicon and 1.5% zinc (higher zinc contents may be used when the material is to be cast). Strength, formability, machinability, and corrosion resistance are quite good. Tensile strengths range from a soft condition of about 55 ksi (380 MPa) through a maximum that approaches 130 ksi (900 MPa). Uses include boiler, tank, and stove applications which require a combination of weldability, high strength, and corrosion resistance.

Copper–Beryllium alloys can be age hardened to produce the highest strengths of the copper-based metals. They ordinarily contain less than 2% beryllium but are quite expensive to use. When annealed the material has a yield strength of 25 ksi (170 MPa), tensile strength of 70 ksi (480 MPa), and an elongation of 50%. After heat treatment, these properties can rise to 160 ksi (1100 MPa), 180 ksi (1250 MPa), and 5%, respectively. Cold work coupled with age hardening can produce even stronger material. The modulus of elasticity is about 18×10^6 psi (125,000 MPa) and the endurance limit is around 40 ksi (275 MPa). These properties make the material an excellent choice for electrical contact springs, but cost limits application to small components requiring long life and high reliability. Other applications, such as spark-resistant safety tools and spot welding electrodes, utilize the unique combination of properties: the material has the strength of steel and is nonsparking, nonmagnetic, and conductive. Recent concerns over the toxicity of beryllium have somewhat reduced the popularity of this material, however, and have created a demand for substitute alloys with similar properties.

■ 7.3 ALUMINUM AND ALUMINUM ALLOYS

General Properties and Characteristics

Although aluminum has been a commercial metal for just a little over 100 years, it now ranks second to steel in both worldwide quantity and expenditure and is clearly the most important of the nonferrous metals. It has achieved importance in virtually all segments of the world economy, with principal uses in transportation, construction, electrical applications, containers, consumer durables, and mechanical equipment.

A number of unique and attractive properties account for the engineering significance of aluminum. These include its *workability*, *light weight*, *corrosion resistance*, and good *electrical and thermal conductivity*. Aluminum has a specific gravity of 2.7 compared to 7.85 for steel, making aluminum about one-third the weight of steel for an equivalent volume. Cost comparisons are often made on the basis of cost per pound, where aluminum is at a distinct disadvantage, but there are a number of applications where the more appropriate comparison would be based on cost per unit volume. Since a pound of aluminum would produce three times as many same-size parts as a pound of steel, the cost difference becomes markedly less.

Probably the most serious weakness of aluminum from an engineering viewpoint is its relatively *low modulus of elasticity*, also about one-third that of steel. Under identical loadings, an aluminum component will deflect three times as much as a steel component of the same design. Since the modulus of elasticity cannot be significantly altered by alloying or heat treatment, it is usually necessary to provide stiffness through design features such as ribs or corrugations. These can be incorporated with relative ease, however, because aluminum adapts quite readily to the full spectrum of fabrication processes.

Commercially Pure Aluminum

In its pure state aluminum is soft, ductile, and not very strong. In the annealed condition, pure aluminum has only about one-fifth the strength of hot-rolled structural steel. Thus commercially pure aluminum is used primarily for its physical rather than its mechanical properties.

Electrical-conductor-grade aluminum is used in large quantities and has replaced copper in many applications, such as electrical transmission lines. Commonly designated by the letters EC, this grade contains a minimum of 99.45% aluminum and has an electrical conductivity that is 62% that of copper for the same size wire and 200% that of copper on an equal-weight basis.

Aluminums for Mechanical Applications

For nonelectrical applications, most aluminum is used in the form of alloys. These have much greater strength than pure aluminum, yet retain the advantages of light weight, good conductivity, and corrosion resistance. While usually weaker than steel, some alloys are now available that have tensile properties (except for ductility) that are superior to those of the HSLA structural grades. Since alloys can be as much as 30 times stronger than pure aluminum, designers can frequently optimize their design and then tailor the material to their specific requirements.

On a strength-to-weight basis, most of the aluminum alloys are superior to steel and other structural metals, but wear, creep, and fatigue properties are generally rather poor. Since aluminums often lack an endurance limit, fatigue failures often occur, even at rather low stresses. Because of their low melting point, aluminum alloys rapidly lose their

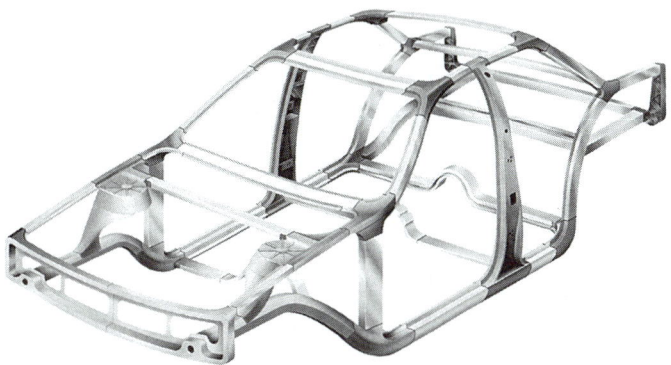

FIGURE 7-2 Automotive space frame constructed from aluminum extrusions and die-cast aluminum connecting nodes. *(Courtesy of Aluminum Company of America)*

strength as temperature is increased and should not be considered for applications involving service temperatures much above 300°F (150°C).

The selection of steel or aluminum for any given component is often a matter of cost, although in many cases the advantages of reduced weight or corrosion resistance may be used to justify additional expense. Aluminum generally replaces steel or cast iron where there is a strong need for light weight, corrosion resistance, low maintenance expense, or high thermal or electrical conductivity. In modern motor vehicles, aluminum is seeing increased use in body panels, engine blocks, manifolds, and transmission cases, where the reduced weight serves to increase fuel economy. The aluminum space frame shown in Figure 7-2 can reduce the primary body structure of an automobile from up to 300 spot-welded steel stampings to a welded assembly of fewer than 100 formed extrusions and cast nodes, while decreasing its weight by 35% or more.

Corrosion Resistance of Aluminum and Its Alloys

Pure aluminum is very reactive and forms a tight, adherent oxide coating on the surface as soon as it is exposed to air. This oxide is resistant to many corrosive media and serves as a corrosion-resistant barrier to protect the underlying metal. Thus, like stainless steels, the corrosion resistance of the metal is actually a property of the surface oxide. As alloys are added to the aluminum, oxide formation is somewhat retarded. Aluminum alloys, in general, do not have quite the superior corrosion resistance of pure aluminum.

The oxide coating on aluminum alloys also causes some difficulty in relation to its weldability. For resistance welding of consistent quality, it is usually necessary to remove the oxide immediately before welding. During fusion welding, the aluminum would oxidize so readily that special fluxes or protective inert-gas atmospheres must be employed. Although welding aluminum may be more difficult than steel, suitable techniques have been developed so that high-quality, cost-effective welds can be produced by most of the welding processes.

Classification System

The aluminum alloys can be divided into two major groups, wrought alloys and casting alloys, based on the method of fabrication. Wrought alloys are those that are shaped as solids by plastic deformation, and are therefore designed to have attractive forming characteristics, such as low yield strength, high ductility, good fracture resistance, and good strain hardening. Attractive features for the casting alloys, on the other hand, include low melting point, high fluidity, and attractive as-solidified structures and properties. Clearly, these properties are distinctly different, and the alloys that have been designed to meet them are also different. As a result, different classification systems exist for the wrought and cast aluminum alloys.

Wrought Aluminum Alloys

The wrought aluminum alloys can be further divided into two basic types: those that achieve *strength by solid-solution strengthening and cold working* and those that are *strengthened by heat treatment* (age hardening). Table 7-3 lists some of the common wrought aluminum alloys in each family, using the standard four-digit designation system for aluminums. The first digit indicates the major alloy grouping as follows:

Major Alloying Element	
Aluminum, 99.00% and greater	1xxx
Copper	2xxx
Manganese	3xxx
Silicon	4xxx
Magnesium	5xxx
Magnesium and silicon	6xxx
Zinc	7xxx
Other element	8xxx

The second digit is usually zero. Nonzero numbers are used to indicate some modification to the original alloy. The last two digits simply indicate the particular alloy within the family. For example, 2024 simply means alloy number twenty-four within the 2xxx, or aluminum–copper, system. For the 1xxx series, the last three digits are used to denote the purity of the aluminum.

The first four digits of a wrought aluminum designation identify the chemistry of the alloy. Clarification of the alloy condition is then made through a *temper designation*, in the form of a letter–number suffix using the following system:

-F: as fabricated
-H: strain-hardened
  -H1: strain-hardened by working to desired dimensions; a second digit, 1 through 9, indicates the degree of hardness, 8 being commercially full-hard and 9 extrahard
  -H2: strain-hardened by cold working, followed by partial annealing, second-digit numbers 2 through 8, as above
  -H3: strain-hardened and stabilized
-O: annealed
-T: thermally treated (heat treated)
  -T1: cooled from hot working and naturally aged
  -T2: cooled from hot working, cold-worked, and naturally aged
  -T3: solution-heat-treated, cold-worked, and naturally aged
  -T4: solution-heat-treated and naturally aged
  -T5: cooled from hot working and artificially aged
  -T6: solution-heat-treated and artificially aged

-T7:   solution-heat-treated and stabilized
-T8:   solution-heat-treated, cold-worked, and artificially aged
-T9:   solution-heat-treated, artificially aged, and cold-worked
-T10:   cooled from hot working, cold-worked, and artificially aged
-W: solution-heat-treated only

It can be noted from Table 7-3 that the work-hardenable alloys (those that cannot be age-hardened) are primarily those in the 1xxx (pure aluminum), 3xxx (aluminum–manganese), and 5xxx (aluminum–magnesium) series. Within these series the 1100, 3003, and 5052 alloys have become the most popular.

Because of their higher strengths, the precipitation-hardenable alloys are somewhat more numerous. These are found primarily in the 2xxx, 6xxx, and 7xxx series. Alloy 2017, the original *duralumin*, is probably the oldest age-hardenable aluminum alloy. The 2024 alloy is stronger than 2017 and has seen considerable use in aircraft applications. An attractive feature of the 2xxx series is the fact that ductility does not significantly decrease during the strengthening heat treatment. Within the 7xxx series are some newer alloys with strengths that approach or exceed those of the high-strength structural steels. Ductility, however, is less than that of steel, and fabrication is more difficult than for the 2024-type alloys. Nevertheless, the 7xxx series alloys have also found wide use in aircraft applications. The age-hardenable alloys should not be used at temperatures over 350°F (175°C) in their age-hardened condition.

Because of their two-phase structure, the heat-treatable alloys tend to have poorer corrosion resistance than either pure aluminum or the single-phase work-hardenable alloys. Thus where both high strength and superior corrosion resistant are desired, the wrought aluminum is often produced as *Alclad* material. A thin layer of corrosion-resistant aluminum is bonded to one or both surfaces of the high-strength alloy during rolling and the material is further processed as a composite.

Because only moderate temperatures are required to lower the strength of aluminum alloys, extrusions and forgings are relatively easy to produce and are manufactured in large quantities. Deep drawing and other sheet-metal-forming operations can be carried out quite easily. In general, the high ductility and low yield strength of the aluminum alloys make them appropriate for almost all forming operations. Good dimensional tolerances and fairly intricate shapes can be produced with relative ease.

The machinability of aluminum-based alloys, however, can vary greatly, and special tools and techniques may be desirable if large amounts of machining are required. Free-machining alloys, such as 2011, have been developed for screw-machine work. These can be machined at very high speeds and have replaced brass screw-machine stock in many applications.

Aluminum Casting Alloys

Although its low melting temperature tends to make it suitable for casting, pure aluminum is seldom cast. Its high shrinkage and susceptibility to hot cracking cause considerable difficulty, and scrap is high. By adding small amounts of alloying elements, however, very suitable casting characteristics are obtained and strength is increased. Aluminum alloys are cast in considerable quantity, and many of the most popular contain enough silicon to produce the eutectic reaction, giving the materials low melting points, good fluidity, and high as-cast strength. Copper, zinc, and magnesium are other popular alloy additions that permit the formation of age-hardening precipitates.

Table 7-4 list some of the commercial aluminum casting alloys and employs the designation system of the Aluminum Association. The first digit indicates the alloy group as follows:

TABLE 7-3. Composition, Typical Properties, and Designations of Some Wrought Aluminum Alloys

Designation[a]	Composition (%) Aluminum = Balance					Form Tested	Tensile Strength		Yield Strength[b]		Elongation in 2 in. (%)	Brinell Hardness	Uses and Characteristics
	Cu	Si	Mn	Mg	Others		ksi	MPa	ksi	MPa			
						Work-Hardening Alloys—Not Heat-Treatable							
1100–0	0.12				99 Al	-in. sheet	13	90	5	34	35	23	Commercial Al: good forming properties
1100–H14						-in. sheet	16	110	14	97	9	32	Good corrosion resistance, low yield strength
110–H18						-in. sheet	24	165	21	145	5	44	Cooking utensils; sheet and tubing
3003–0	0.12		1.2			-in. sheet	16	110	6	41	30	28	Similar to 1100
3003–H14						-in. sheet	22	152	21	145	8	40	Slightly stronger and less ductile
3003–H18						-in. sheet	29	200	27	186	4	55	Cooking utensils; sheetmetal work
5052–0				2.5	0.25 Cr	-in. sheet	28	193	13	90	25	45	Strongest work-hardening alloy
5052–H32						-in. sheet	33	228	28	193	12	60	High yield strength and fatigue limit
5052–H36						-in. sheet	40	276	35	241	8	73	Highly stressed sheetmetal products
						Precipitation-Hardening Alloys—Heat-Treatable							
2017–0	4.0	0.5	0.7	0.6		-in. sheet	26	179	10	69	20	45	Duralumin, original strong alloy
2017–T4						-in. sheet	62	428	40	276	20	105	Hardened by quenching and aging

Aluminum Association number and temper[a]	Cu	Si	Mn	Mg	Other	Product form	Tensile strength, kpsi	Tensile strength, MPa	Yield strength,[b] kpsi	Yield strength,[b] MPa	Elongation, %	Brinell hardness	Characteristics and uses
2024-0	4.4		0.6	1.5		$\frac{1}{16}$-in. sheet	27	186	11	76	20	42	Stronger than 2017
2024-T4						$\frac{1}{16}$-in. sheet	64	441	45	290	19	120	Used widely in aircraft construction
2014-0	4.4	0.8	0.8	0.5		$\frac{1}{2}$-in. extruded shapes	27	186	14	97	12	45	Strong alloy for extruded shapes
2014-T6						Forgings	65	448	55	379	10	125	Strong forging alloy
2014-T6						$\frac{1}{16}$-in. sheet	70	483	60	413	8		Higher yield strength than Alclad 2024
Alclad 2014-T6	4.5	1.0	0.8	0.4		$\frac{1}{16}$-in. sheet	63	434	56	386	7		Clad with heat-treatable alloy[c]
7075-0	1.6	0.2		2.5	{ 0.3 Cr · 5.6 Zn }	$\frac{1}{16}$-in. sheet	33	228	15	103	17	60	Alloy of highest strength
7075-T6						$\frac{1}{16}$-in. sheet	76	524	67	462	11	150	Lower ductility than 2024
Alclad 7075-T6						$\frac{1}{16}$-in. sheet	76	524	67	462	11		Strongest Alclad product
7075-T6						$\frac{1}{2}$-in. extruded shapes	80	552	70	483	6		Strongest alloy for extrusions
6061-T6	0.28	0.6		1.0	0.20 Cr	$\frac{1}{2}$-in. extruded shapes	42	290	40	276	12	95	Strong, corrosion resistant
6063-T6		0.4		0.7		$\frac{1}{2}$-in. rod extruded	35	241	31	214	12	80	Good forming properties and corrosion resistance
6151-T6		0.9		0.6	0.25 Cr	Forgings	48	331	43	297	17	90	For intricate forgings
2025-T6	4.5	0.8	0.8			Forgings	55	379	30	207	18	100	Good forgeability, lower cost
2018-T6	4			0.7	2 Ni	Forgings	55	379	40	276	10	100	Strong at elevated temperatures; forged pistons
4032-T6	0.9	12.2		1.1	0.9 Ni	Forgings	55	379	46	317	9	115	Forged aircraft pistons
2011-T3	5.5			(0.5 Bi)	0.5 Pb	$\frac{1}{2}$-in. rod	55	379	43	297	15	95	Free cutting, screw-machine products

[a]O, annealed; T, quenched and aged; H, cold-rolled to hard temper.

[b]Yield strength taken at 0.2% permanent set.

[c]Cladding alloy: 1.0 Mg, 0.7 Si, 0.5 Mn.

TABLE 7-4. Composition, Properties, and Designations of Some Aluminum Casting Alloys

Alloy Designation[a]	Process[b]	Composition (%) (Major Alloys > 1%)						Temper	Tensile Strength		Elongation in 2 in. (%)	Uses and Characteristics
		Cu	Si	Mg	Zn	Fe	Other		ksi[c]	MPa		
208	S	4.0	3.0		1.0	1.2		F	19	131	1.5	General-purpose sand castings, can be heat treated
242	S, P	4.0		1.6		1.0	2.0 Ni	T61	40	276	—	Withstands elevated temperatures
295	S	4.5	1.0			1.0		T6	32	221	3.0	Structural castings, heat-treatable
296	P	4.5	2.5			1.2		T6	35	241	2.0	Permanent-mold version of 295
308	P	4.5	5.5		1.0	1.0		F	24	166	—	General-purpose permanent mold
319	S, P	3.5	6.0		1.0	1.0		T6	31	214	1.5	Superior casting characteristics
354	P	1.8	9.0					T6	—	—	—	High-strength, aircraft
355	S, P	1.3	5.0		1.0			T6	32	221	2.0	High strength and pressure tightness
C355	S, P	1.3	5.0					T61	40	276	3.0	Stronger and more ductile than 355
356	S, P		7.0					T6	30	207	3.0	Excellent castability and impact strength
A356	S, P		7.0					T61	37	255	5.0	Stronger and more ductile than 356
357	S, P		7.0					T6	45	310	3.0	High-strength-to-weight castings
359	S, P		9.0					—	—	—	—	High-strength aircraft usage
360	D		9.5			2.0		F	44[d]	303	2.5[d]	Good corrosion resistance and strength
A360	D		9.5			2.0		F	46[d]	317	3.5[d]	Similar to 360
380	D	3.5	8.5		3.0	2.0		F	46[d]	317	2.5[d]	High strength and hardness
A380	D	3.5	8.5		3.0	1.3		F	47[d]	324	3.5[d]	Similar to 380
383	D	1.5	10.5		3.0	1.3		F	45[d]	310	3.5[d]	High strength and hardness
384	D	3.75	11.3		1.0	1.3		F	48[d]	331	2.5	High strength and hardness
413	D	1.0	12.0			2.0		F	43[d]	297	2.5[d]	General purpose, good castability
A413	D	1.0	12.0			1.3		F	42[d]	290	3.5[d]	Similar to 413
443	D	5.25				2.0		F	33[d]	228	9.0[d]	General purpose, good castability
B443	S, P	5.25				2.0		F	17	117	3.0	General-purpose casting alloy
514	S			4.0				F	22	152	6.0	High corrosion resistance
518	D			8.0		1.8		F	45[d]	310	5.0[d]	Good corrosion resistance, strength, and toughness
520	S			10.0				T4	42	290	12.0	High strength with good ductility
535	S			6.9				F	35	241	9.0	Good corrosion resistance and machinability
712	S				5.8			F	34	234	4.0	Good properties without heat treatment
713	S, P				7.5	1.1		F	32	221	3.0	Similar to 712
771	S				7.0			T6	42	290	5.0	Aircraft and computer components
850	S, P	1.0					6.3 Sn, 1.0 Ni	T5	16	110	5.0	Bearing alloy

[a] Aluminum Association.
[b] S, sand-cast; P, permanent-mold-cast; D, die cast.
[c] Minimum figures unless noted.
[d] Typical values.

Major Alloying Element

Major Alloying Element	
Aluminum, 99.00% and greater	1xx.x
Copper	2xx.x
Silicon with Cu and/or Mg	3xx.x
Silicon	4xx.x
Magnesium	5xx.x
Zinc	7xx.x
Tin	8xx.x
Other elements	9xx.x

The second and third digits identify the particular alloy or aluminum purity, and the last digit, separated by a decimal point, indicates the product form (e.g., casting or ingot). A modification of the original alloy is indicated by a letter before the numerical designation.

Casting alloys are designed for both properties and process. When the strength requirements are low, as-cast properties are usually adequate. High-strength castings usually require the use of alloys that can subsequently be heat treated. Sand casting has the fewest process restrictions. The aluminum alloys used for permanent mold casting are designed to have lower coefficients of thermal expansion (or contraction) because the molds offer restraint to the dimensional changes that occur upon cooling. Die-casting alloys require high degrees of fluidity because they are often cast in thin sections. Many die-casting alloys are also designed to produce high "as-cast" strength without heat treatment, using the rapid cooling conditions of the die-casting process to promote a fine grain size and fine eutectic structure. Several of the aluminum permanent-mold and die-casting alloys offer tensile strengths in excess of 40 ksi (275 MPa).

Aluminum–Lithium Alloys

In the search for aluminum alloys with higher strength, greater stiffness, and lighter weight, *aluminum–lithium alloys* are emerging as an attractive aerospace material. Each percent of lithium (up to 4%) reduces the overall weight by 3% and increases stiffness by 6%. Alloys have already been developed that have 8 to 10% lower density, 15 to 20% greater stiffness, strengths comparable to those of existing alloys, and good resistance to fatigue crack propagation, but fracture toughness, ductility, and stress-corrosion resistance are usually poorer than for conventional alloys. These materials are available in both wrought and cast forms and can be fabricated just like other aluminum alloys. They are highly machinable, can be welded, and are readily adaptable to forming by forging or extrusion. Some sheet material can even be fabricated by superplastic forming. Thus significant weight savings can be achieved without the major overhaul of manufacturing equipment that would be associated with a switch to advanced composites.

In recent years, conventional aluminum alloys have accounted for up to 80% of the weight of commercial aircraft. Through material substitution, redesign of components to take advantage of greater stiffness, and reduction in the size of related parts (since the entire structure will be lighter), a 10 to 15% weight savings would appear to be possible in an aircraft structure. On a more graphic basis, the weight of a Boeing 747 could be reduced by about 14,000 lb. Fuel savings over the life of the airplane would more than compensate for any additional manufacturing expense.

■ 7.4 MAGNESIUM AND MAGNESIUM ALLOYS

General Properties and Characteristics

Magnesium is the lightest of the commercially important metals, having a specific gravity of about 1.74 (compared to 2.7 for aluminum and 7.9 for iron or steel). Like aluminum, magnesium is relatively weak in the pure state and for engineering purposes is almost always used as an alloy. Even in alloy form, however, the metal can be characterized by poor wear, creep, and fatigue properties. Strength drops rapidly when the temperature exceeds 200°F (100°C), so magnesium should not be considered for elevated-temperature service. Its modulus of elasticity is even less than that of aluminum, being between one-fourth and one-fifth that of steel. Thick sections are required to provide adequate stiffness, but the alloy is so light that it is often possible to use thicker sections for the required rigidity and still have a lighter structure than can be obtained with any other metal. Cost per unit volume is low, so the use of thick sections is generally not prohibitive. Moreover, since a large portion of magnesium components are cast, the thicker sections actually become a desirable feature.

Magnesium alloys also possess limited ductility, a characteristic of their hcp crystal structure. Designers should be aware that brittle fracture is a probable mode of failure when components are loaded beyond the designed conditions. Consider the lightweight magnesium automobile wheels. Although they are quite attractive for racetrack applications, most passenger car manufacturers prefer aluminum wheels, so that impacts with high curbing results only in a denting deformation and not fracture.

On the more positive side, magnesium alloys have a relatively high strength-to-weight ratio with some commercial alloys attaining strengths as high as 55 ksi (380 MPa). While many magnesium alloys require enamel or lacquer finishes to impart adequate corrosion resistance, this property has been improved markedly with the development of higher-purity alloys. In the absence of unfavorable galvanic couples, these materials resist corrosion better than steel and aluminum and have paved the way for applications in many areas, including the automotive market. The numerous limitations, however, generally restrict the use of magnesium to applications where light weight is a dominant concern. While aluminum alloys are often used for the load-bearing members of mechanical structures, magnesium alloys are best suited for those applications where lightness is the primary consideration and strength is a secondary requirement.

Magnesium Alloys and Their Fabrication

The designation system for magnesium alloys is not as well standardized as in the case of steels or aluminums, but most producers follow a system using one or two prefix letters, two or three numerals, and a possible suffix letter. The prefix letters designate the two largest alloying metals according to the following format presented in ASTM specification B93:

A	aluminum	H	thorium	Q	silver
B	bismuth	K	zirconium	R	chromium
C	copper	L	beryllium	S	silicon
D	cadmium	M	manganese	T	tin
E	rare earth	N	nickel	Z	zinc
F	iron	P	lead		

Aluminum, zinc, zirconium, and thorium promote precipitation hardening; manganese improves corrosion resistance; and tin improves castability. Aluminum is the most common alloying element. The numerals in the designation correspond to the rounded-off whole number percentages of the two main alloy elements and are arranged in the same order as the letters. Thus the AM60 alloy would contain approximately 6% aluminum and less than 0.5% manganese. The suffix letter is used to denote variations of the same base alloy, where A is the original material. A temper designation system is also employed that is much the same as that used with the aluminum alloys. Table 7-5 lists some of the more common magnesium alloys together with their properties and uses.

Sand, permanent mold, and die casting are all well developed for magnesium alloys and take advantage of the low melting point of the alloys. Die casting is clearly the most popular manufacturing process for magnesium. Although the magnesium alloys cost about twice as much as aluminum, its hot chamber die casting process is easier, more economical, and 40 to 50% faster than the cold chamber process generally required for aluminum. Magnesium die castings often replace plastic injection-molded components when improved stiffness or dimensional stability is required.

Forming behavior is poor at room temperature, but most conventional processes can be performed when the material is heated to temperatures between 450 and 700°F (230 to 370°C). Since these temperatures are easily attained and generally do not require a protective atmosphere, many formed and drawn magnesium products are manufactured.

The machinability of magnesium alloys is the best of any commercial metal and, in many applications, the savings in machining costs, achieved through high cutting speeds and long tool life, more than compensate for the increased cost of the material. It is necessary, however, to keep the tools sharp and provide adequate cooling for the chips.

Magnesium alloys can be spot welded nearly as easily as aluminum, but scratch brushing or chemical cleaning is necessary before forming the weld. Fusion welding is carried out most easily by processes using an inert shielding atmosphere of argon or helium gas.

Considerable misinformation exists regarding the fire hazards when processing or using magnesium alloys. It is true that magnesium alloys are highly combustible when in a finely divided form, such as powder or fine chips, and this hazard should never be ignored. When the metal is heated above 800°F (425°C), a noncombustible, oxygen free atmosphere is required to suppress burning. Casting operations often require additional precautions due to the reactivity of magnesium with sand and water. In the form of sheet, bar, extruded product, or finished castings, however, magnesium alloys present no real fire hazard.

■ 7.5 ZINC-BASED ALLOYS

Properties and Applications

As a pure metal, the only major use of zinc is the *galvanizing* of iron and steel. In this process the iron-based material is coated with a layer of zinc by one of a variety of processes that include direct immersion in a bath of molten metal (hot dipping) and electrolytic plating. The resultant coating provides excellent corrosion resistance, even when the surface is badly scratched or marred. Moreover, the corrosion resistance will persist until all of the sacrificial zinc has been depleted. Galvanizing accounts for about 35% of all zinc used.

Zinc is also used as the base metal for a variety of die-casting alloys. For this purpose, zinc offers low cost, a low melting point (only 715°F or 380°C), and the attractive property of not adversely affecting steel dies when in molten metal contact. Unfortunately, pure zinc is almost as heavy as steel and is also rather weak and brittle. Therefore, when

TABLE 7-5. Composition, Properties, and Uses of Common Magnesium Alloys

| Alloy | Temper | Composition (%) | | | | | | Tensile Strength[a] | | Yield Strength[a] | | Elongation in 2 in. (%) | Uses and Characteristics |
		Al	Rare Earths	Mn	Th	Zn	Zr	ksi	MPa	ksi	MPa		
AM60A	F	6.0		0.13				30	207	17	117	6	Die castings
AM100A	T4	10.0		0.1				34	234	10	69	6	Sand and permanent-mold castings
AZ31B	F	3.0				1.0		32	221	15	103	6	Sheet, plate, extrusions, forgings
AZ61A	F	6.5				1.0		36	248	16	110	7	Sheet, plate, extrusions, forgings
AZ63A	T5	6.0				3.0		34	234	11	76	7	Sand and permanent-mold castings
AZ80A	T5	8.5				0.5		34	234	22	152	2	High-strength forgings, extrusions
AZ81A	T4	7.6				0.7		34	234	11	76	7	Sand and permanent-mold castings
AZ91A	F	9.0				0.7		34	234	23	159	3	Die castings
AZ92A	T4	9.0				2.0		34	234	11	76	6	High-strength sand and permanent-mold castings
EZ33A	T5		3.2			2.6	0.7	20	138	14	97	2	Sand and permanent-mold castings
HK31A	H24				3.2		0.7	33	228	24	166	4	Sheet and plates; castings in T6 temper
HM21A	T5			0.8	2.0			33	228	25	172	3	High-temperature (800°F) sheets, plates, forgings
HZ32A	T5				3.2	2.1	0.7	27	186	13	90	4	Sand and permanent-mold castings
ZH62A	T5				1.8	5.7	0.7	35	241	22	152	5	Sand and permanent-mold castings
ZK51A	T5					4.6	0.7	34	234	20	138	5	Sand and permanent-mold castings
ZK60A	T5					5.5	0.45	38	262	20	138	7	Extrusions, forgings

[a]Properties are minimums for the designated temper.

alloys are designed for die casting, the alloy elements are usually selected for their ability to increase strength and toughness in the as-cast condition while retaining the low melting point. High fluidity and good dimensional stability are additional considerations.

Two of the most popular zinc die-casting alloys are characterized in Table 7-6. Alloy AG40A (also known as alloy 903 or Zamak 3) is widely used because of its excellent dimensional stability. Alloy AC41A offers higher strength and better corrosion resistance. As a whole, the zinc die-casting alloys offer a strength greater than that of all other die-cast metals, with the exception of the copper-based alloys. They can be cast to close dimensional limits with extremely thin sections and offer good tensile strength coupled with exceptional impact resistance. Moreover, these alloys are then machinable at a minimum of cost. Resistance to surface corrosion is adequate for a number of applications. While prolonged contact with moisture will result in the formation of a white corrosion product, this can easily be suppressed by inexpensive forms of surface treatment.

TABLE 7-6. Characteristics of Two Zinc Die-Casting Alloys

	ASTM AG40A (SAE 903) (Zamak 3)	ASTM AC41A (SAE 925) (Zamak 5)
Composition (%)		
Aluminum	3.5–4.3	3.5–4.3
Cadmium, maximum	0.004	0.004
Copper	0.25	0.75–1.25
Iron, maximum	0.1	0.1
Lead, maximum	0.005	0.005
Magnesium	0.02–0.05	0.03–0.08
Tin, maximum	0.003	0.003
Zinc	Remainder	Remainder
Properties		
Tensile strength, as cast (ksi)	41	47.6
Tensile strength, 10 years of aging	35	39.3
Elongation in 2 in., as cast (%)	10	7
Charpy impact strength, as cast (ft-lb)	43	48
Melting point (°F)	717	717

The attractiveness of zinc die casting has been further enhanced by the development of several zinc–aluminum casting alloys with rather large amounts of aluminum (ZA8, ZA12, and ZA27, with 8, 12 and 27% aluminum, respectively). Initially developed for sand permanent mold, and graphite mold casting, these alloys can also be die cast to achieve higher performance characteristics than obtained with any of the conventional alloys. Strength, hardness, and wear resistance are all improved, and several of the alloys have demonstrated excellent bearing properties. As a result of their lower melting and casting costs, these alloys are becoming attractive alternatives to the conventional aluminum, brass, and bronze casting alloys, as well as cast iron.

■ 7.6 TITANIUM AND TITANIUM ALLOYS

Titanium is a strong, lightweight, corrosion-resistant metal that has been of commercial importance only since 1948. Because its properties are between those of steel and aluminum, its importance has been increasing rapidly. Its yield strength is about 60 ksi (415 MPa), but this can be raised to 190 ksi (1300 MPa) by alloying and heat treatment, a strength comparable to that of many alloy steels. Density, on the other hand, is only 56% that of steel, and the modulus of elasticity ratio is also about one-half. Good mechanical properties are retained up to temperatures of 900°F (480°C), so the metal is often considered as a high-temperature engineering material. From the negative side, titanium and its alloys suffer from high cost, fabrication difficulties, a high energy content (they require about 10 times as much energy to produce as steel), and a high reactivity at elevated temperatures (above 900°F).

Titanium alloys are generally grouped into three classes based on their microstructural features. These classes are known as alpha-, beta-, and alpha-beta-titanium alloys, the terms denoting the stable phase or phases at room temperature. Fabrication can be by casting, forging, rolling, or extruding, provided that special process modifications and controls are implemented. Advanced processing methods include powder metallurgy, mechanical alloying, rapid-solidification processing (RSP), superplastic forming, and hot-isostatic pressing (HIP).

The uses of titanium relate primarily to its *high strength-to-weight ratio*, its *good stiffness*, its *corrosion resistance*, and its *retention of mechanical properties at elevated temperatures*. Aerospace applications tend to dominate, but titanium alloys are also used in chemical and electrochemical processing equipment, marine implements, and ordinance equipment. They are often used in place of steel where weight savings is desired and to replace aluminums where high-temperature performance is necessary. Some bonding applications utilize the unique property that titanium wets glass and some ceramics. The titanium–6% aluminum–4% vanadium alloy is the most popular titanium alloy, accounting for nearly 50% of all titanium usage worldwide.

■ 7.7 NICKEL-BASED ALLOYS

Nickel-based alloys are most noted for their outstanding strength and corrosion resistance, particularly at high temperatures. *Monel* metal, containing about 67% nickel and 30% copper, has been used for years in the chemical and food-processing industries because of its outstanding corrosion characteristics. In fact, Monel probably has better corrosion resistance to more media than does any other commercial alloy. It is particularly resistant to salt water, sulfuric acid, and even high-velocity, high-temperature steam. For the latter reason, Monel has been used for steam turbine blades. It can be polished to have an excellent appearance, similar to that of stainless steel, and is often used in ornamental trim and household ware. In its most common form, Monel has a tensile strength ranging from 70 to 170 ksi (480 to 1170 MPa), with a companion elongation of 50 to 2%.

There are three special grades of Monel that contain small amounts of added alloying elements. K-Monel contains about 3% aluminum and can be precipitation hardened to a tensile strength of 160 to 180 ksi (1100 to 1240 MPa). H-Monel has 3% silicon added, and S-Monel has 4% silicon. These are used for casting applications and can also be precipitation hardened. To improve the machining characteristics, a special free-machining variety, known as R-Monel, has been produced with about 0.35% sulfur.

Nickel-based alloys have also been used for electrical resistors and heating elements. These materials are primarily nickel–chromium alloys and are known by the trade name *Nichrome*. One popular alloy contains 80% nickel and 20% chromium. Another has 60% nickel, 16% chromium, and 24% iron. They have excellent resistance to oxidation while retaining useful strength at red heats.

When the nickel-based alloys are designed to provide good mechanical properties at extremely high temperatures, the materials are classified as superalloys. These alloys will be discussed along with other, similar materials in the following section.

Nickel-based materials are generally difficult to cast, but they can be forged and hot-worked. Heating operations, however, are generally performed in controlled atmospheres to avoid intercrystalline embrittlement. Welding operations can be performed with little difficulty.

■ 7.8 SUPERALLOYS AND OTHER NONFERROUS METALS FOR HIGH-TEMPERATURE SERVICE

Titanium and titanium alloys have already been cited as being useful in providing strength at elevated temperatures, but the maximum temperature for these materials is approximately 900°F (480°F). Jet engine, gas turbine, rocket, and nuclear applications often require materials that possess high strength, creep resistance, oxidation and corrosion resistance, and fatigue resistance at temperatures up to and in excess of 2000°F (1100°C). One class of materials offering these properties is the *superalloys*, first developed extensively in the 1940s for use in turbojet aircraft. Table 7-7 lists some of the more common superalloys. It should be noted that nickel, iron and nickel, or cobalt forms the base metal for these alloys. Most are precipitation hardenable, and yield strengths above 100 ksi (690 MPa) are readily attained. The nickel-based alloys tend to have higher strengths at room temperature, with yield strengths up to 175 ksi (1200 MPa) and ultimate tensile strengths as high as 210 ksi (1450 MPa). This is in comparison with 115 ksi (790 MPa) and 170 ksi (1170 MPa), respectively, for the cobalt-based alloys. The 1000-hour rupture strengths of the nickel-based alloys at 1500°F (815°C) are also higher than those of the cobalt-based material, up to 65 ksi (450 MPa) versus 33 ksi (228 MPa). Unfortunately, the density of the superalloys is much greater than iron, so their use is often at the expense of additional weight.

Most of the superalloys are very difficult to form or machine, so methods such as electrodischarge, electrochemical, or ultrasonic machining are often utilized, or the products are made to final shape as investment castings. Powder metallurgy techniques are also used extensively in the manufacture of superalloy components. Because of their ingredients, all of these alloys are quite expensive, and this limits their use to small or critical parts where the cost is not the determining factor.

Some engineering applications have already exceeded the temperature limits of these alloys, and still others await the development of materials that would make them feasible. One source, for example, estimates the exhaust temperatures of future jet engines to be in excess of 2600°F (1425°C). Materials such as TD-nickel (a nickel alloy containing 2% dispersed thoria) can operate at service temperatures somewhat above 2000°F (1100°C). The *refractory metals* can be used to temperatures as high as 3000°F (1650°C), provided that they are protected by coatings that isolate them from gases in their operating environment. These metals, along with their melting points, are *niobium* (4480°F/2470°C), *molybdenum* (4730°F/2610°C), *tantalum* (5430°F/3000°C), *rhenium* (5740°F/3170°C), and *tungsten* (6170°F/3410°C). Table 7-8 presents key properties for several of the refractory metals. All are heavier than steel, and several are significantly heavier. Figure 7-3 illustrates one high-temperature application. Other materials and technologies that offer promise for high-temperature service include directionally solidified eutectics, single crystals, intermetallic compounds, engineered ceramics, and advanced coating systems.

■ 7.9 LEAD, TIN, AND THEIR ALLOYS

The principal uses of *lead* as a pure metal include storage batteries, cable cladding, and sound-dampening shields. These consume about 60% of the annual production of the metal. *Tin* is used as a coating on steel to impart corrosion resistance.

TABLE 7-7. Some Nonferrous Alloys for High-Temperature Service

Alloy	Composition (%)														
	C	Mn	Si	Cr	Ni	Co	Mo	W	Cb	Ti	Al	B	Zr	Fe	Other
Nickel base															
Hastelloy X	0.1	1.0	1.0	21.8	Balance	2.5	9.0	0.6	—	—	—	—	—	18.5	—
IN-100	0.18	—	—	10.0	Balance	15.0	3.0	—	—	4.7	5.5	0.014	0.06	—	1.0 V
Inconel 601	0.05	0.5	0.25	23.0	Balance	—	—	—	—	—	1.4	—	—	14.1	0.2 Cu
Inconel 718	0.04	0.2	0.2	19.0	Balance	—	3.0	—	5.0	0.9	0.5	0.005	—	18.5	0.2 Cu
M-252	0.15	0.5	0.5	19.0	Balance	10.0	10.0	—	—	2.6	1.0	0.01	—	—	—
Rene 41	0.09	—	—	19.0	balance	11.0	10.0	—	—	3.1	1.5	0.01	—	—	—
Rene 80	0.17	—	—	14.0	Balance	9.5	4.0	4.0	—	5.0	3.0	0.015	0.03	—	—
Rene 95	0.15	—	—	14.0	Balance	8.0	3.5	3.5	3.5	2.5	3.5	0.01	0.05	—	—
Udimet 500	0.08	—	—	19.0	Balance	18.0	4	—	—	3.0	3.0	0.005	—	0.5	—
Udimet 700	0.07	—	—	15.0	Balance	18.5	5.0	—	—	3.5	4.4	0.025	—	0.5	—
Waspaloy B	0.07	0.75	0.75	19.5	Balance	13.5	4.3	—	—	3.0	1.4	0.006	0.07	2.0	0.1 Cu
Iron–nickel base															
Illium P	0.20	—	—	28.0	8.0	—	2.0	—	—	—	—	—	—	Balance	3.0 Cu
Incoloy 825	0.03	0.5	0.2	21.5	42.0	—	3.0	—	—	0.9	0.1	—	—	30	2.2 Cu
Incoloy 901	0.05	0.4	0.4	13.5	42.7	—	6.2	—	—	2.5	0.2	—	—	34	—
16-25-6	0.08	1.35	0.7	16.0	25.0	—	6.0	—	—	—	—	—	—	Balance	0.15 N
Cobalt base															
Haynes 150	0.08	0.65	0.75	28.0	—	Balance	—	—	—	—	—	—	—	20.0	—
MAR-M322	1.00	0.10	0.1	21.5	—	Balance	—	9.0	—	0.75	—	—	2.25	—	4.5 Ta
S-816	0.38	1.20	0.4	20.0	20.0	Balance	4.0	4.0	4.0	—	—	—	—	4.0	—
WI-52	0.45	0.5	0.5	21.0	1.0	Balance	—	11.0	2.0	—	—	—	—	2.0	—

| TABLE 7-8. | Properties of Some Refractory Metals | | | | | | |

| Metal | Melting Temperature [°F(°C)] | Density (g/cm³) | Room Temperature | | | Elevated Temperature [1832°F (1000°C)] | |
			Yield Strength (ksi)	Tensile Strength (ksi)	Elonga-tion (%)	Yield Strength (ksi)	Tensile Strength (ksi)
Molybdenum	4730 (2610)	10.22	80	120	10	30	50
Niobium	4480 (2470)	8.66	20	45	25	8	17
Tantalum	5430 (3000)	16.6	35	50	35	24	27
Tungsten	6170 (3410)	19.25	220	300	3	15	66

FIGURE 7-3   Niobium coated with fused silicide is used in the afterburner components of the Pratt & Whitney F100 jet pictured. *(Courtesy of Pratt & Whitney.)*

When in the form of lead- or tin-based alloys, the two metals are almost always used together. Bearing material and solder are the two most important uses. One of the oldest and best bearing materials is an alloy of 84% tin 8% copper, and 8% antimony, known as genuine or tin *babbitt*. Because of the high cost of tin, however, lead babbitt, composed of 85% lead, 5% tin, 10% antimony, and 0.5% copper, is a more widely used bearing material. For slow speeds and moderate loads, the lead-based babbitts have proven to be quite adequate.

Figure 7-4 shows a photomicrograph of a lead babbitt. The antimony combines with the tin to form hard particles within the softer lead matrix, a multiphase structure that is typical of many bearing metals. The shaft rides on the harder particles with little friction while the softer matrix acts as a cushion that can distort sufficiently to compensate for mis-alignment and assure a proper fit between the two surfaces.

Soft solders are basically lead–tin alloys with a chemistry near the eutectic composition of 61.9% tin (see Figure 4-5). While the eutectic alloy has the lowest melting temperature, the high cost of tin has forced many users to specify solders with a lower-than-optimum tin content. A variety of compositions are available, each with its own characteristic melting range. Additional information on solders and soldering is provided in Chapter 37.

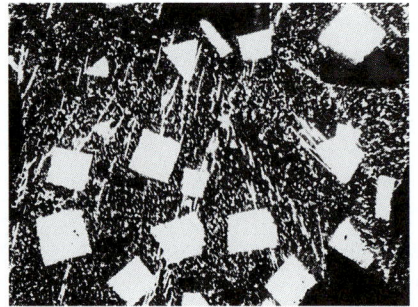

FIGURE 7-4 Photomicrograph of lead babbitt metal; 75X. Square or triangular white masses are SbSn; small white particles are CuSn.

■ 7.10 SOME LESS-KNOWN METALS AND ALLOYS

Several of the less known metals and alloys have achieved importance in modern technology as a result of their somewhat unique physical and mechanical properties. *Hafnium*, *thorium*, and *beryllium* are used in nuclear reactors because of their low neutron-absorption characteristics. Depleted *uranium*, because of its very high density (19.1 g/cm^3), is useful in special applications where maximum weight must be put into a limited space, as in counterweights and flywheels. *Cobalt*, in addition to its use as a base metal for superalloys, is used as a binder in various powder-based components and sintered carbides, where it provides good high-temperature strength.

Zirconium is used for its outstanding corrosion resistance to most acids, chlorides, and organic acids. It offers high strength, good weldability and fatigue resistance, and attractive neutron absorption characteristics. High electrical conductivity can be obtained by alloying with copper. An alloy containing a small percentage of hafnium offers a yield strength of 84 ksi (580 MPa) and a tensile strength of 90 ksi (620 MPa). Some applications of zirconium alloys include heat exchangers, protective claddings, heat shields, and engine components.

Rare-earth metals have been incorporated into magnets that offer increased strength compared to the standard ferrite variety. Neodymium–iron–boron and samarium–cobalt are two common varieties. Current efforts in aerospace, nuclear, high-temperature, and electronic applications will undoubtedly bring about increased research and development of these and other uncommon metals.

■ 7.11 GRAPHITE

While technically not a metal, *graphite* is an engineering material that has considerable potential. It possesses the unique property of increasing in strength when the temperature is elevated. Recrystallized polycrystralline graphites have mechanical strengths up to 10 ksi (70MPa) at room temperature, which double at 4500°F (2500°C).

Large quantities of graphite are used as electrodes in arc furnaces, but other uses are developing rapidly. The addition of small amounts of borides, carbides, nitrides, and silicides greatly lowers the oxidation rate at elevated temperatures and improves the mechanical strength. This makes the material highly suitable for use as rocket-nozzle inserts and permanent molds for casting various metals. It can be machined quite readily to excellent surface finishes. Graphite fibers have also found extensive use in composite materials. This application is discussed in Chapter 8.

■ KEY WORDS

aluminum
brass
bronze
cast
copper
dezincification
galvanizing
graphite

lead
magnesium
Monel
nickel
nonferrous
refractory metals
solder
stress-corrosion cracking

superalloys
temper designation
tin
titanium
wrought
zinc

■ REVIEW QUESTIONS

1. What types of properties do nonferrous metals possess that are not available in the ferrous metals?
2. In what respects are the nonferrous metals usually inferior to steel?
3. For what type of fabrication processes might the low-melting-point alloys be attractive?
4. What are the three primary properties of copper and copper alloys that account for many uses and applications?
5. What properties make copper attractive for cold-working processes?
6. What are some of the limiting properties of copper that might restrict its area of application?
7. How are wrought alloys distinguished from casting alloys in the copper designation system?
8. What are some of the attractive engineering properties that account for the wide use of the copper–zinc alpha-brasses?
9. Why might the term *bronze* be potentially confusing when used in reference to a copper-based alloy?
10. What are some attractive engineering properties of copper–nickel alloys?
11. Describe the somewhat unique property combination that exists in heat-treated copper–beryllium alloys. What has limited its use in recent years?
12. What are some of the attractive engineering properties of aluminum and aluminum alloys?
13. How does aluminum compare to copper in terms of electrical conductivity?
14. What features might limit the uses and applications of aluminum and aluminum alloys?
15. How is the corrosion-resistance mechanism of aluminum and aluminum alloys similar to that observed in stainless steels?

16. How are the wrought alloys distinguished from the cast alloys in the aluminum designation system?
17. What feature in the wrought aluminum designation scheme is used to denote the condition or structure of a given alloy?
18. What is the primary mechanism used to provide strength to the high-strength "aircraft-quality" aluminum alloys?
19. What specific material qualities or features might make an aluminum casting alloy particularly attractive for permanent mold casting? For die-casting?
20. What features of aluminum–lithium alloys make them particularly attractive for aerospace applications?
21. What are some limiting properties of magnesium and magnesium alloys?
22. What particular design feature or requirement is generally necessary to justify the use of magnesium alloys?
23. In what way can ductility be imparted to magnesium alloys so that they can be formed by conventional processes?
24. Under what conditions should magnesium be considered to be a flammable or explosive material?
25. What is the primary application of pure zinc? Of the zinc-based engineering alloys?
26. What are some of the attractive engineering properties of titanium and titanium alloys?
27. What temperature is generally considered to be the upper limit for which titanium alloys retain their useful engineering properties?
28. What property of Monel alloys accounts for a majority of its applications?
29. What metals or combinations of metals form the bases of the superalloys?

30. What class of metals or alloys must be used when the operating temperatures exceed the limits of the superalloys?

31. Which metals are classified as refractory metals?

32. When selecting a material for a solder application, why might the lead content be higher than the lowest-melting-point eutectic composition?

33. What is one of the unique properties of graphite that makes it attractive for elevated-temperature applications?

*C*hapter 7   CASE STUDY

nonsparking wrench

The Terrific Tool Company is considering an expansion of its line of conventional hand tools to provide a set of safety tools, capable of being used in areas such as gas leaks where the potential of explosion or fire exists. Conventional irons and steels are pyrophoric (i.e., small slivers or fragments can burn in air, forming sparks if dropped or impacted on a hard surface).

You are asked to specify the materials and processes to manufacture a non-sparking pipe wrench. The new product must retain the same shape and size as a conventional wrench as well as all of the characteristic properties (strength in the handle, hardness in the teeth, fracture resistance, corrosion resistance, etc.). In addition, the new safety wrench must be non-sparking (or non-pyrophoric).

You restrict your initial search to the nonferrous metals, such as aluminum- and copper-based alloys. Both casting and forging appear to be viable manufacturing options. You note that copper-2% beryllium can be age-hardened to provide the strength and hardness properties equivalent to the steel that is currently being used for the jaws of the wrench. However, the cost of this material is also quite high. You feel that the wrench must be made in the most economical manner possible if the new line of safety tools is to be accepted by potential customers.

Suggest some alternative manufacturing systems (materials and methods of fabrication) that could be used to produce the desired wrench. What might be the advantages and disadvantages of each? Which of them would you recommend to your supervisor?

CHAPTER 8

NONMETALLIC MATERIALS: PLASTICS, ELASTOMERS, CERAMICS, AND COMPOSITES

8.1 INTRODUCTION

8.2 PLASTICS

Molecular Structure of Plastics

Isomers

Forming Molecules by Polymerization

Thermosetting and Thermoplastic Materials

Properties and Applications

Common Types or Families of Plastics

Additive Agents in Plastics

Plastics as Adhesives

Oriented Plastics

Plastics for Tooling

Engineering Plastics

Plastics versus Other Materials

Recycling of Plastics

8.3 ELASTOMERS

Rubber

Artificial Elastomers

Elastomers for Tooling Applications

8.4 CERAMICS

Nature and Structure of Ceramics

Clay and Whiteware Products

Refractory Materials

Abrasives

Ceramics for Electrical and Magnetic Applications

Glasses

Glass Ceramics

Cermets

Ceramics for Mechanical Applications

Advanced Ceramics as Cutting Tools

8.5 COMPOSITE MATERIALS

Laminar or Layered Composites

Particulate Composites

Fiber-Reinforced Composites

Advanced Fiber-Reinforced Composites

Hybrid Composites

Design and Fabrication

Assets and Limitations

Areas of Application

Case Study: TWO-WHEEL DOLLY HANDLES

■ 8.1 INTRODUCTION

Because of their wide range of attractive properties, the nonmetallic materials have always played a significant role in manufacturing. Wood has been a key engineering material down through the centuries, and craftspersons have learned to select and utilize the various types and grades to manufacture a broad spectrum of quality products. Stone and rock continue to be key construction materials, and clay products can be traced to antiquity. Even leather has been a construction material, and was used for fenders in early automobiles.

More recently, however, the *nonmetallic materials* family has expanded greatly from the natural materials described above and now includes an extensive listing of plastics

193

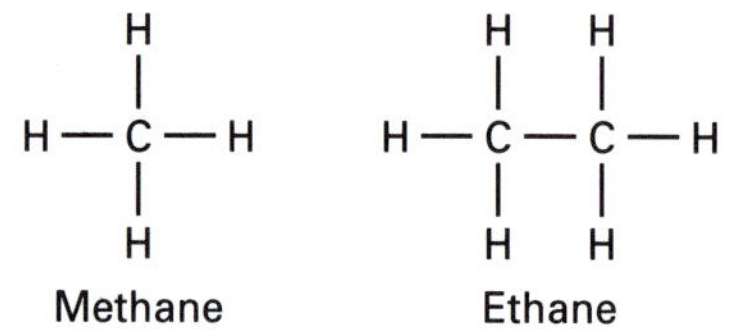

FIGURE 8-1 Linking of hydrogen and carbon in methane and ethane molecules.

(polymers), elastomers, ceramics, and composites. Most of these are manufactured, so a wide variety of properties and characteristics can be obtained. New materials and variations are being created on a continuous basis, and their uses and applications are expanding rapidly. Many observers now refer to a *materials revolution* as these new materials are developed, and research and development efforts bring their cost into competition with steel and the other engineering metals. New products have been designed to utilize the new properties, and existing products are continually reevaluated for the possibility of material substitution. As the design requirements of engineering products continue to push the limits of the traditional materials, the role of manufactured nonmetallic materials will continue to expand.

Because of the breadth and number of nonmetallic engineering materials, we do not attempt to provide information about all of them. Instead, the emphasis will be on the basic nature and properties of the various families so that the reader will be able to determine if they may be reasonable candidates for specific products and applications. For detailed information about specific materials within these families, more extensive and dedicated texts, handbooks, and compilations should be consulted.

■ 8.2 PLASTICS

It is difficult to provide a precise definition of the term *plastics*. From a technical viewpoint, the term is applied to a group of engineered materials characterized by large molecules that are built up by the joining of smaller molecules. On a more practical level, these materials are natural or synthetic resins, or their compounds, that can be molded, extruded, cast, or used as thin films or coatings. They offer low density, low tooling costs, good corrosion resistance, cost reduction, and design versatility. From a chemical viewpoint, most are organic substances containing hydrogen, oxygen, carbon, and nitrogen.

Molecular Structure of Plastics

It is helpful to have an understanding of the molecular structure of plastics. Many are based on paraffin-type hydrocarbons, in which carbon and hydrogen combine in the relationship $C_n H_{2n+2}$. Theoretically, the atoms can link together indefinitely to form very large molecules, extending the series depicted in Figure 8-1. The bonds between the various atoms are all single pairs of shared (covalent) electrons. Bonding within the molecule, therefore, is quite strong, but the attractive forces between adjacent molecules are much weaker. Because there is no provision for additional atoms to be added to the chain, such molecules are said to be *saturated*.

Carbon and hydrogen can also form molecules in which the carbon atoms are held together by double or triple covalent bonds. Ethylene and acetylene are examples (Figure 8-2). Because such molecules do not have the maximum possible number of hydrogen atoms, they are said to be *unsaturated* and are important in the *polymerization* process, where small molecules link to form large ones with the same constituent atoms.

In each of the organic compounds, four electron pairs surround each carbon atom and one electron pair is shared with each hydrogen atom. Other atoms or structures can

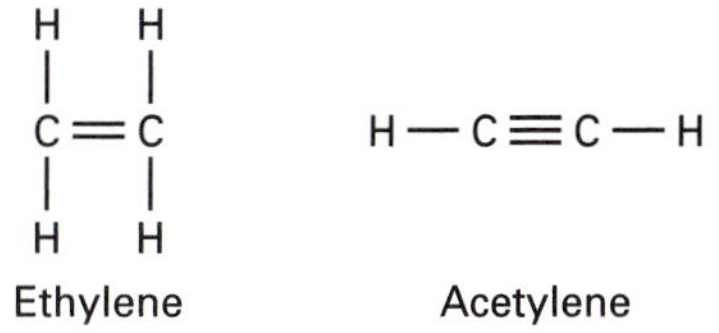

H H
| |
C = C   H — C ≡ C — H
| |
H H
Ethylene   Acetylene

FIGURE 8-2 Covalent bonds in unsaturated ethylene and acetylene molecules.

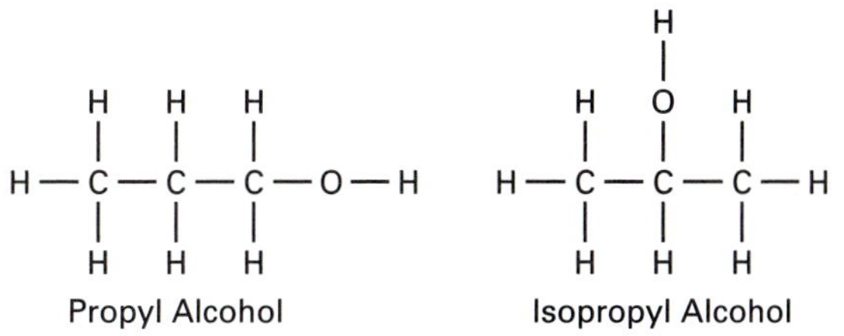

H H H
| | |
H — C — C — C — O — H
| | |
H H H
Propyl Alcohol

H
|
H O H
| | |
H — C — C — C — H
| | |
H H H
Isopropyl Alcohol

FIGURE 8-3 Linking of eight hydrogen, one oxygen, and three carbon atoms to form two isomers, propyl and isopropyl alcohol.

be substituted for carbon and hydrogen, however. Chlorine, fluorine, or even a benzene ring can take the place of hydrogen. Oxygen, silicon, sulfur, or nitrogen can take the place of carbon. Because of these substitutions, a wide range of organic compounds can be created.

Isomers

The same kind and number of atoms can also unite in different structural arrangements, known as *isomers*, where these behave as different compounds with different engineering properties. Figure 8-3 shows an example of this feature, involving propyl and isopropyl alcohol. Isomers can be considered analogous to allotropism or polymorphism in crystalline materials, where the same material possesses different properties because of the different crystal structures.

Forming Molecules by Polymerization

The polymerization process takes place by either an *addition* or *condensation* mechanism. Figure 8-4 illustrates polymerization by addition, where a number of basic units (*monomers*) are added together to form a large molecule (*polymer*) in which there is a repeated unit (*mer*). Activators or catalysts, such as benzoyl peroxide, initiate and terminate the chain. Thus, the amount of activator relative to the amount of monomer determines the av-

FIGURE 8-4 Polymerization by addition: the uniting of identical monomers.

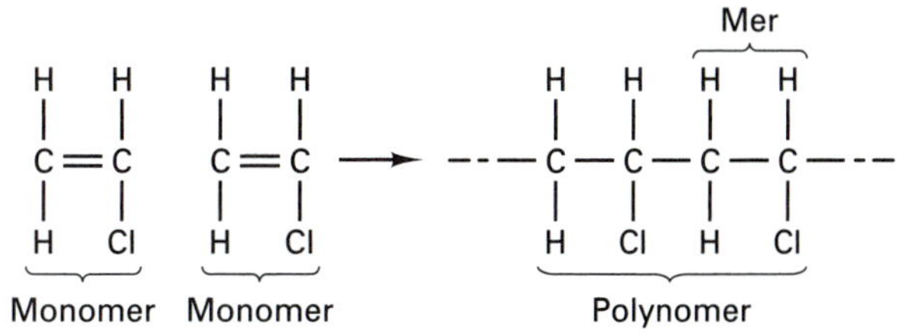

Mer

H H   H H   H H H H
| | | | | | | |
C = C C = C → - - - C — C — C — C — - - -
| | | | | | | |
H Cl H Cl   H Cl H Cl
Monomer Monomer   Polynomer

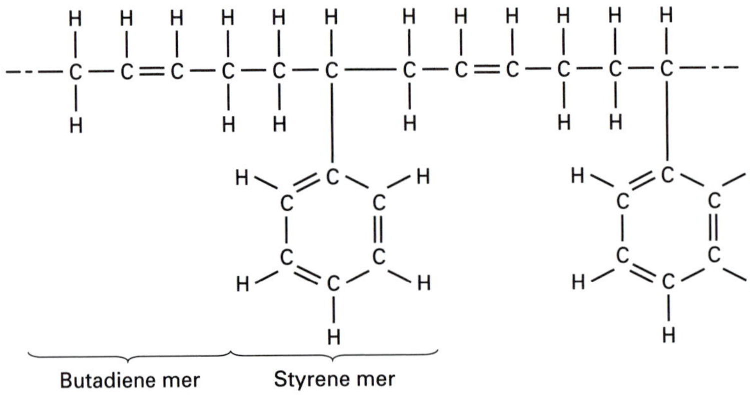

Butadiene mer   Styrene mer

FIGURE 8-5 Polymerization by the addition of two kinds of mers: copolymerization.

erage molecular weight or length of the chain. The average number of mers in the poly-mer, known as the *degree of polymerization*, ranges from 75 to 750 for most commercial plastics. *Copolymers* are a special category of polymer where two different types of mers are combined into the same addition chain. The formation of copolymers (Figure 8-5) greatly expands the possibilities of creating new types of plastics with improved physical and mechanical properties. *Terpolymers* extend the combinations to include three different monomers.

In contrast to polymerization by addition, where all of the component atoms appear in the product molecule, condensation polymerization occurs as reactive molecules com-bine with one another to produce a polymer plus small, by-product molecules, such as water. Heat, pressure, and catalysts are often required to drive the reaction. Figure 8-6 il-lustrates the reaction between phenol and formaldehyde to form Bakelite, first performed in 1910. The structure of condensation polymers can be either linear chains or a three-dimen-sional framework in which all atoms are linked by strong, primary bonds.

FIGURE 8-6 Formation of phenol–formaldehyde (Bakelite) by condensation polymerization.

Formaldehyde   Phenol   Phenol   Phenol Formaldehyde   Water

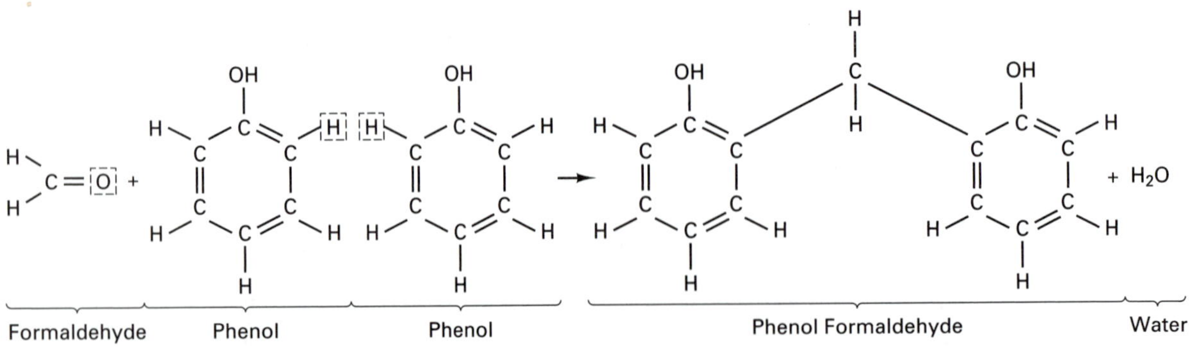

Thermosetting and Thermoplastic Materials

The terms *thermosetting* and *thermoplastic* refer to the material's response to elevated temperature. Addition polymers can be viewed as long chains of tightly bonded carbon atoms with strongly attached pendants of hydrogen, fluorine, chlorine, or benzene rings. All of the bonds within the molecules are strong primary bonds. The attraction between neighboring molecules, however, is only by the much weaker van der Waals forces. For these materials, the mechanical and physical properties are largely determined by the intermolecular forces. In general, they tend to be flexible and tough. Because the secondary bonds are weakened by elevated temperature, plastics of this type soften with increasing temperature and become harder and stronger when cooled. The softening and hardening of these *thermoplastic* materials can be repeated as often as desired and no chemical change is involved. Because they contain molecules of different lengths, thermoplastic materials do not have a definite melting temperature, but instead, soften over a range of temperatures.

Above the melting temperature, the bonds between the molecular chains are so weak that an applied force will produce flow with no elastic strain. The material can be poured and cast, or formed by injection molding. Below the melting temperature, the material can retain its amorphous structure, but with companion properties that are somewhat rubbery. The application of a force produces both elastic and plastic deformation. Large amounts of permanent deformation are available and make this range attractive for molding and extrusion. At still lower temperatures, the bonds become stronger and the polymer is stiffer and somewhat leathery. Many commercial polymers (e.g., polyethylene) have useful strength in this condition. When further cooled below the *glass transition temperature*, the linear polymer retains its amorphous structure but becomes hard, brittle, and glasslike.

Many thermoplastics can partially crystallize when cooled below the melting temperature. During this process the chains closely align over appreciable distances, with a companion increase in density. In addition, the polymer becomes stiffer, harder, less ductile, and more resistant to solvents and heat. The ability of a polymer to crystallize depends on the complexity of its molecules, the degree of polymerization (length of the chains), the cooling rate, and the amount of deformation during cooling.

The mechanical behavior of an amorphous thermoplastic can be modeled by a common cotton ball. The individual molecules are bonded within by strong covalent bonds and are analogous to the individual fibers of cotton. The bonding forces between molecules are much weaker and are similar to the friction forces between the strands of cotton in the assembly. When pulled or stretched, plastic deformation occurs by slippage between adjacent fibers or molecular chains. Methods to increase the strength of thermoplastics, therefore, focus on restricting intermolecular slippage. Longer chains have less freedom of movement and are therefore stronger. Connecting adjacent chains with primary bond cross-links, as with the sulfur links when vulcanizing rubber, can also impede deformation. Since the strength of the secondary bonds is inversely related to the separation distance of the molecules, processes such as deformation or "crystallization" can be used to produce a tight parallel alignment of adjacent molecules and a concurrent increase in strength, stiffness, and density. Polymers with large side structures, such as chlorine atoms or benzene rings, may be stronger or weaker than those with just hydrogen, depending on whether the dominant effect is the impediment to slippage or the increased separation distance. Branched polymers, where the chains divide in a "Y" with primary bonds linking all segments of the chain, are often weaker since branching reduces the density and close packing of the chains.

Thermosetting plastics, on the other hand, usually have a highly *cross-linked* or three dimensional framework structure in which all atoms are connected by strong, covalent bonds. These materials are generally produced by condensation polymerization where elevated temperature promotes the irreversible reaction, hence the term *thermosetting*. Once set, additional heatings do not produce softening. Instead, the materials maintain their mechanical properties up to the temperature at which they char or burn. Deformation requires the breaking of primary bonds, so these plastics tend to be strong but brittle. The thermosetting polymers, therefore, are significantly stronger and more rigid than the thermoplastics, but have a lower ductility and poorer impact properties.

Whether a polymer is thermosetting or thermoplastic is of great importance in determining how it will perform in service. In addition, consideration should be given to how it will be processed. For example, thermoplastics are easily molded. After the hot, soft, material has been formed to the desired shape, the mold must be cooled so that the plastic will harden and be able to retain its shape when removed. In contrast, when a part is produced from thermosetting materials, the mold can remain at an elevated temperature throughout the entire process, but the time per part is often longer to permit complete "setting" or curing of the resins. The material hardens as a result of the reaction and has strength and rigidity even when hot. Product removal can be performed without cooling the mold.

Properties and Applications

Because there are so many varieties of plastics, and new ones are being developed almost continuously, it is helpful to have a knowledge of the general properties of plastics as well as the specific properties of the basic families. General properties of plastics include:

1. *Light weight.* Most plastics have specific gravities between 1.1 and 1.6, compared with about 1.75 for magnesium. Thus they are among the lightest of the engineering materials.
2. *Corrosion resistance.* Many plastics perform well in hostile, corrosive environments. A number are notably resistant to acid corrosion.
3. *Electrical resistance.* Plastics are widely used as insulating materials.
4. *Low thermal conductivity.* Plastics are relatively good thermal insulators.
5. *Variety of optical properties.* Many plastics have an almost unlimited color range and the color goes throughout, not just on the surface. Both transparent and opaque materials exist.
6. *Formability.* Objects can frequently be produced from plastics in a single operation. Raw material can be converted to final shape through such processes as casting, extrusion, and molding. Relatively low temperatures are required for the forming of plastics.
7. *Surface finish.* Excellent surface finishes are often obtained by the same processes that produce the shape. Additional surface finishing operations may not be required.
8. *Comparatively low cost.* The low cost of plastics generally applies to both the material itself and the manufacturing process. Plastics frequently offer reduced tool costs and high rates of production.
9. *Low energy content.*

While the attractive features of plastics tend to be in the area of physical properties, the inferior features often relate to mechanical strength. None of the plastics possess strength properties that approach those of the engineering metals, but their low density allows them to compete effectively on a strength-to-weight (or specific strength) basis. Many have low impact strength, although several (such as ABS, high-density polyethylene, polycarbonate, and cellulose propionate) are exceptions to this rule. The dimensional stability of plastics tends to be greatly inferior to that of metals, and the coefficient of thermal expansion is rather high. The thermoplastics are quite sensitive to heat, and the strength of many drops rapidly as temperatures increase above normal environmental conditions. Thermosetting materials offer good strength retention at elevated temperature but have an upper limit of about 500°F (250°C). While the corrosion resistance of plastics is generally good, they often absorb moisture, and this, in turn, decreases strength. Some thermoplastics can exhibit a 50% drop in tensile strength as the humidity increases from 0 to 100%. Finally, radiation, both ultraviolet and particulate, can markedly alter the properties. Many plastics used in an outdoor environment have ultimately failed due to the cumulative effect of ultraviolet radiation.

Table 8-1 summarizes the properties of a number of common plastics. By considering the information in this table along with the discussion of general properties above, it becomes apparent that plastics are best used in applications that require materials with low to moderate strength, light weight, low electrical and/or thermal conductivity, a wide range of available colors, and ease of fabrication into finished products. No other family of materials can offer this combination of properties. Because of their light weight, attractiveness, and ease of fabrication, plastics have been selected for many "packaging" or container applications. This classification includes such items as stereo cabinets, clock cases, and household appliance housings, which serve, primarily, as containers for the interior mechanisms. Applications such as insulation on electrical wires and handles for hot articles capitalize on the low electrical and thermal conductivities. Soft, pliable, foamed plastics are used extensively as cushioning material. Rigid foams are used inside sheet metal structures to provide compressive strength. Some plastics are used as adhesive or bonding agents in the assembly of products, an application discussed in Chapter 37. Others can provide inexpensive tooling for applications where pressures, temperatures, and wear requirements are not extreme.

There are many applications where only one or two of the properties of plastics are sufficient to dictate their use. In other cases, special characteristics are desired that are not normally found in the commercial plastics, such as high directional strength. To meet these demands, *composite materials* can be designed that incorporate a fabric or fiber reinforcement within a plastic resin. These are discussed in some detail later in the chapter.

Common Types or Families of Plastics

The following is descriptive information about several of the types of plastics listed in Table 8-1.

THERMOPLASTICS

- *ABS:* contains acrylonitrile, butadiene, and styrene; low weight, good strength, and very tough; resists heat, weather, and chemicals quite well; dimensionally stable but flammable
- *Acrylics:* highest optical clarity, transmitting over 90% of light; common trade names include *Lucite* and *Plexiglas*; high impact, flexural, tensile, and dielectric strengths; available in a wide range of colors; resist weathering; stretch rather easily
- *Cellulose acetate:* wide range of colors; good insulating qualities; easily molded; high moisture absorption in most grades and affected by alcohols and alkalies
- *Cellulose acetate butyrate:* higher impact strength and moisture resistance than cellulose acetate; will withstand rougher usage
- *Ethyl cellulose:* high electrical resistance and impact strength: retains toughness at low temperatures

TABLE 8-1. Properties and Major Characteristics of Common Types of Plastics

Material	Specific Gravity	Tensile Strength (1000 lb/in²)	Impact Strength Izod (ft-lb/in. of Notch)	Top Working Temperature [°F (°C)]	Dielectric Strength[b] (V/mil)	24-Hour Water Absorption (%)	Weatherability	Colorability	Optical Clarity	Chemical Resistance	Injection Molding	Extrusions	Formable Sheet	Film	Fiber	Compression or Transfer Moldings	Castings	Reinforced Plastics	Moldings	Industrial Thermosetting Laminates	Foam
Thermoplastics																					
ABS material	1.02–1.06	4–8	1.3–10.0	250 (121)	300–400	0.2–0.3	0	x		0	•	•	•								
Acetal	1.4	10	1.5	200 (93)	1200	0.22	0	x		0	•	•									
Acrylics	1.12–1.19	5.5–10	0.2–2.3	200 (93)	400–530	0.2–0.4	x		x	0	•	•	•		•		•				
Cellulose acetate	1.25–1.50	3–8	0.75–4.0	260 (127)	300–600	2.0–6.0		x	x		•	•	•	•	•						
Cellulose acetate butyrate	1.18–1.24	2–6	0.6–3.2	130 (54)	250–350	1.8–2.1	x	x	x		•	•	•								
Cellulose propionate	1.19–1.24	1–5	0.8–9	140 (60)	300	1.8–2.1		x	x		•	•	•	•							
Chlorinated polyether	1.4	6	3.3	300 (149)	400	0.01				x	•	•	•								
Ethyl cellulose	1.16	3–6	1.8–4.0	150 (66)	350	1.6–2.2		x	x		•	•	•								
TFE-fluorocarbon	2.1–2.3	1.5–3	2.5–4.0	500 (260)	450	0	x			x			•	•	•	•					
CFE-fluorocarbon	2.1–2.15	4.5–6	3.5–3.6	390 (199)	550	0	x			x			•	•	•						
Nylon	1.1–1.2	8–10	2	250 (121)	385–470	0.4–5.5		0	0	0	•	•	•	•	•						
Polycarbonate	1.2	9.5	14	250 (121)	400	0.15	x	0	0	x	•	•	•	•							
Polyethylene	0.96	4	10	200 (93)	440	0.003		0	x	x	•	•	•	•	•						
Polypropylene	0.9–1.27	3.4–5.3	1.02	230 (110)	520–800	0.03		0	x	x	•	•	•		•						
Polystyrene	1.05–1.15	5–9	0.3–0.6	190 (88)	400–600	<0.2		x	x	0	•	•	•	•							•
Modified polystyrene	1.0–1.1	2.5–6	0.25–11.0	212 (100)	300–600	0.03–0.2		x	x	x	•	•	•								
Vinyl	1.16–1.55	1–5.9	0.25–2.0	220 (104)	25–500	0.2–1		x	x	x	•	•	•	•	•						•
Thermosetting plastics																					
Epoxy	1.1–1.7	4–13	0.4–1.5	325 (163)	500	0.1–0.5	x	x		x							•	•		•	•
Melamine	1.76–1.98	5–8		350 (177)	460	0.1		x		0						•	•	•		•	
Phenolic	1.2–1.45	5–9	0.25–5	300 (149)	100–500	0.2–0.6				0						•	•	•		•	•
Polyester (other than molding compounds)	1.06–1.46	4–10	0.18–0.4	300 (149)	340–570	0.5				0				•					•		
Polyester (alkyd, DAP)	1.6–1.75	3.2–8	3.6–8			0.16–0.67		x						•			•	•			
Silicone	2.0	3–5	0.2–3.0	550 (288)	250–350	0.4–0.5			x							•		•		•	•
Urea	1.41–1.80	4–8.5	0.2–0.5	185 (85)	300–600	1–3	x	x	0							•	•	•		•	•

[a] x denotes a principal reason for its use; 0 indicates a secondary reason.

[b] Short-time ASTM test.

- *Fluorocarbons:* inert to most chemicals; high temperature resistance; very low co-efficients of friction (Teflon), used for nonlubricated bearings and nonstick coatings for cooking utensils and electrical irons
- *Nylon (polyamides):* low coefficient of friction; good strength, abrasion resistance, and toughness; excellent dimensional stability; used for bearings and as monofilaments for textiles, fishing line, and ropes
- *Polycarbonates:* high strength and outstanding toughness; good dimensional stability
- *Polyethylenes:* inexpensive, tough, good chemical resistance and high electrical resistance; low strength; subject to weathering via ultraviolet light; flammable; used for grocery bags, milk jugs, tubes, pipes, sheeting, and electrical wire insulation
- *Polystyrenes:* high dimensional stability and low water absorption; best all-around dielectric; clear, hard, and brittle; often used for rigid packaging, can be foamed to produce expanded polystyrene; burns readily; trade name of Styrofoam
- *Vinyls:* wide range of types, from thin, rubbery films to rigid forms; tear resistant; good aging properties; good dimensional stability and water resistance in rigid forms; used for floor and wall covering, upholstery fabrics, and lightweight water hose; common trade names include Saran and Tygon

THERMOSETS

- *Epoxies:* good strength, toughness, elasticity, chemical resistance, moisture resistance, and dimensional stability; easily compounded to cure at room temperature; used as adhesives, bonding agents, coatings, and in fiber laminates
- *Melamines:* excellent resistance to heat, water, and many chemicals; full range of translucent and opaque colors; excellent electric-arc resistance; tableware (but stained by coffee); used extensively in treating paper and cloth to impart water-repellent properties
- *Phenolics:* oldest of the plastics but still widely used; hard, strong, low cost, and easily molded, but rather brittle; resistant to heat and moisture; dimensionally stable; opaque, but with a wide color range; wide variety of forms: sheet, rod, tube, and laminate
- *Polyesters:* strong and resist environmental influences well; uses include boat and car bodies, pipes, vents and ducts, textiles, adhesives, coatings, and laminates
- *Silicones:* Semiorganic spine molecules with alternating silicon and oxygen atoms; heat and weather resistant; low moisture absorption; chemically inert; high dielectric properties; excellent sealants
- *Urea–formaldehyde:* properties similar to those of phenolics but available in lighter colors; useful in containers and housings but not outdoors; used in lighting fixtures because of translucence in thin sections

Additive Agents in Plastics

For most uses, additional materials are incorporated into plastics to (1) improve their properties, (2) reduce their cost, (3) improve their moldability, and/or (4) impart color. These additive constituents are usually classified as *fillers*, *plasticizers*, *lubricants*, *coloring agents*, *stabilizers*, *antioxidants*, and *flame retardants*.

Ordinarily, fillers comprise a large percentage of the total volume of a molded plastic product. Their primary purpose is to improve strength, stiffness, or toughness; reduce shrinkage; reduce weight; or simply serve as an "extender," providing cost-saving bulk (often at the expense of reduced moldability). To a large degree, they determine the general properties of a molded plastic. Selection tends to favor materials that are much less expensive than the plastic resin, but the various fillers can impart different properties and characteristics. Some of the most common fillers and their properties are:

1. *Wood flour* (fine sawdust): a general-purpose filler; low cost with fair strength; good moldability
2. *Cloth fibers:* improved impact strength; fair moldability
3. *Macerated cloth:* high impact strength; limited moldability
4. *Glass fibers:* high strength; dimensional stability; translucence
5. *Mica:* excellent electrical properties and low moisture absorption
6. *Calcium carbonate, silica, talc, and clay:* serve primarily as extenders

When fillers are used with a plastic resin, the resin acts as a binder, surrounding the filler material and holding the mass together. The surface of a molded part, therefore, will be almost pure resin with no exposed filler.

Coloring agents may be either *dyes*, which are soluble in the resins, or insoluble *pigments*, which impart color simply by their presence. In general, dyes are used for transparent plastics and pigments for the opaque ones. Since most fillers do not produce attractive colors, coloring agents are usually needed.

Plasticizers can be added in small amounts to improve the flow of the plastic during molding or to increase the flexibility of the thermoplastic products by reducing the intermolecular contact and strength of the secondary bonds between the polymer chains. When used for molding purposes, the amount used is governed by the intricacy of the mold. In general, it should be kept to a minimum because it is likely to affect the stability of the finished product through a gradual aging loss. When used for flexibility, plasticizers should be selected with minimum volatility, so as to impart the desired property for as long as possible.

Lubricants such as waxes, stearates, and soaps can be added to improve the moldability of plastics and to facilitate removal of parts from the mold. They are also used to keep thin polymer sheets from sticking to each other when stacked or rolled. Only a minimum amount should be used, however, because the lubricants adversely affect most engineering properties.

Heat, light (especially ultraviolet), and oxidation tend to degrade polymers. Stabilizers and antioxidants can be added to retard these effects. Flame retardants can be added when nonflammability is important. Antistatic agents may be incorporated into plastics used for applications such as electronics packaging. Antimicrobial additives can provide long-term protection from both fungus (such as mildew) and bacteria. Table 8-2 summarizes the purposes of the various additives.

TABLE 8-2. Additive Agents in Plastics

Type	Purpose
Fillers	Enhance mechanical properties, reduce shrinkage, reduce weight, or provide bulk
Plasticizer	Increase flexibility, improve flow during molding, reduce elastic modulus
Lubricant	Improve moldability and extraction from molds
Coloring agents (dyes and pigments)	Impart color
Stabilizers	Retard degradation due to heat or light
Antioxidants	Retard degradation due to oxidation
Flame retardants	Reduce flammability

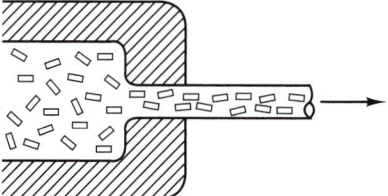

FIGURE 8-7   Schematic representation of the aligning of plastic molecules in the orienting process.

Plastics as Adhesives

Structural adhesives capable of withstanding high stresses are used in many industrial applications. They are quite attractive for the bonding of dissimilar materials, such as metals to nonmetals, and have even been used to replace welding or riveting. A wide range of mechanical properties is available through variations in composition and additives, and a variety of curing mechanisms can be used. The various features of adhesive bonding are discussed in greater detail in Chapter 37.

Oriented Plastics

Because the intermolecular bond strength increases with reduced separation distance, processing that aligns the molecules parallel to the applied load can be used to give the long-chain thermoplastics high strength in a given direction. This *orientation* process can be accomplished by stretching, rolling, or extrusion, as illustrated in Figure 8-7. The material is usually heated prior to the orienting process to aid in overcoming the intermolecular forces and is cooled immediately afterward to "freeze" the molecules in the desired orientation.

Orienting may increase the tensile strength by more than 50%, but a 25% increase is more typical. In addition, the elongation may be increased by several hundred percent. If the oriented plastics are reheated, they tend to deform back toward their original shape, due to the phenomenon of *viscoelastic memory*. The various shrink-wrap materials are examples of this effect.

Plastics for Tooling

Because of their wide range of properties, their ease of conversion into desired shapes, and their excellent properties when loaded in compression, plastics are widely used in tooling to make jigs, fixtures, and forming-die components. Both thermoplastics and thermosets (particularly the cold-setting types) are used for drill and trim jigs, routing and assembly jigs and fixtures, and form blocks. Thermoplastic materials can also be used for punch-press punches. Thermosetting materials have been used for stretch-press dies and tube-bending mandrels. The use of plastics in these applications usually results in less costly tooling, permitting smaller quantities of products to be produced more economically. In addition, the tooling can often be produced in a much shorter time, so that production can begin at an earlier date.

Engineering Plastics

The commercial, or standard, polymers are lightweight, corrosion-resistant materials with low strength and low stiffness. They are relatively inexpensive and are readily formed into a wide range of useful shapes, but are not suitable for use at elevated temperatures.

In contrast, a group of true engineering plastics has been developed with improved thermal properties (up to 650°F/350°C), first-rate impact and stress resistance, high rigidity, superior electrical characteristics, excellent processing properties, and little dimensional change with varying temperature and humidity. These materials include the polyamides, polyacetals, polyarylates, polycarbonates, modified polyphenylene oxides, polybutylene terepthalates, polyketones, polysulfones, polyetherimides, and liquid crystal polymers. While the conventional plastics can be upgraded by stabilizers, fibrous reinforcements, and particulate fillers, there is usually an accompanying reduction in other properties. The engineering plastics, on the other hand, offer a more balanced set of properties. They are usually produced in small quantities and are often quite expensive.

Using advanced technology, materials producers have also developed electroconductive polymers with tailored electrical and electronic properties and high-crystalline polymers with properties comparable to metals.

Plastics versus Other Materials

Polymeric materials now compete with traditional materials in a number of areas. Plastics have taken over about 20% of the glass market, especially in the area of containers and flat glass. PVC pipe and fittings have replaced copper and brass in a number of plumbing applications. Plastics have even replaced ceramics in areas as diverse as sewer pipe and lavatory facilities.

While plastics and metals are often viewed as competing materials, their engineering properties are often considerably different. Many of the attractive features of plastics have already been discussed. In addition to these, one may wish to note that the higher cost of the plastic material is often offset by (1) the ability to be fabricated with lower tooling costs; (2) the ability to be molded at the same rate as product assembly, thereby reducing inventory; (3) a possible reduction in assembly operations and easier assembly through snap fits, friction welds, or the use of self-tapping fasteners; (4) the ability to reuse manufacturing scrap; and (5) reduced finishing costs.

Metals, on the other hand, are often cheaper and offer faster fabrication speeds and greater impact resistance. They are considerably stronger and more rigid and can withstand traditional paint cure temperatures. In addition, resistance to flames, acids, and various solvents is significantly better. Table 8-3 compares the cost per pound and elastic modulus of five engineering plastics with wood, steel, and aluminum. It should be noted, however, that cost per cubic inch may be a more useful cost comparison when size is fixed. If this standard is used, the plastics become more attractive because of their low density.

TABLE 8-3. Comparison of Materials (Modulus and Cost[a])			
Material	Modulus ($\times 10^6$ psi)	$/lb	$/in^3
Aluminum	10.0	0.85	0.086
Steel	30.0	0.27	0.077
Wood (oak)	1.8	0.36	0.0078
Glass/epoxy	3.0	0.92	0.066
Nylon	0.1	1.85	0.076
Polycarbonate	0.35	1.85	0.080
Polypropylene	0.2	0.57	0.018
Polystyrene	0.3	0.57	0.022

[a] Cost figures are as of 1994.

The automotive industry is a good indication of the expanding use of plastics. Polymeric materials accounted for 243 lb of a typical 1992 vehicle, as opposed to only 195 lb of a heavier, 2800-lb (1270-kg) car in 1980. In addition to the traditional application areas of dashboards, interiors, body panels, and trim, plastics are now being used for intake manifolds, valve covers, fuel tanks, and fuel lines and fittings.

Recycling of Plastics

Because of the wide variety of types and compositions with similar physical properties, the recycling of mixed plastics is far more difficult than the recycling of mixed metals. Not only should these materials be sorted on the basis of resin type, but type of filler and color are additional considerations. Therefore, when thermoplastics and thermosets are mixed in variable amounts, the material may have more value as an alternative fuel (competing with coal and oil) than as a resource for recycling into quality engineering material. Plastic "lumber" is another possible use, where the material offers weather and insect resistance and a reduction in required maintenance, but at higher cost than that of traditional material.

If the various types of resins can be identified and kept separate, however, many of the thermoplastic materials can be readily recycled into useful products. Packaging is the largest single market for plastics, and there is currently a well-established network to collect and recycle PET (the polyester used in soft-drink bottles), and high-density polyethylene (the plastic used in milk, juice, and water jugs). In general, however, the properties deteriorate with recycling, so the applications must often be downgraded with multiple reuse. PET is being recycled into new bottles, fiber-fill insulation, and carpeting. Recycled polyethylene is used for new containers, plastic bags, and recycling bins. Polystyrene has been recycled into cafeteria trays and videocassette cases. In contrast to the thermoplastics, the thermosetting resins cannot easily be reprocessed.

■ 8.3 ELASTOMERS

Elastomers are a special class of linear polymers that display an exceptionally large amount of elastic deformation when a force is applied. Many can be stretched to several times their original length. Upon release of the force, the deformation can be completely recovered, as the material quickly returns to its original shape. In addition, the cycle can be repeated numerous times with identical results, as with the stretching of a rubber band.

The elastic properties of most engineering materials are the result of a change in the distance between adjacent atoms (i.e., bond length) when loads are applied. Hooke's law is commonly obeyed, wherein twice the force will produce twice the stretch. When the applied load is removed, the interatomic forces return all of the atoms to their original position and the elastic deformation is recovered completely.

In the elastomeric polymers, the linear chain-type molecules are twisted or curled, much like a coil spring. When a force is applied, the polymer stretches by uncoiling. When the load is removed, the molecules recoil and the material returns to its original size and shape. The relationship between force and stretch does not, however, follow Hooke's Law.

In reality, the behavior of elastomers is a bit more complex. While the chains indeed uncoil when placed under load, they also tend to slide over one another to produce a small degree of viscous deformation. When the load is removed, the molecules recoil, but the viscous deformation is not recovered and the elastomer retains some permanent change in shape.

By cross-linking the coiled molecules, however, it is possible to restrict the viscous deformation while retaining the large elastic response. The elasticity or rigidity of the product can be determined by controlling the number of cross-links within the material. Small amounts of cross-linking leave the elastomer soft and flexible, as in a rubber band. Additional cross-linking restricts some of the uncoiling, and the material becomes harder, stiffer, and more brittle, like the rubber used in bowling balls. Thus engineering elastomers can be tailored to possess a wide range of properties and stress-strain characteristics.

If placed under constant strain, however, even highly cross-linked material will exhibit some viscous flow over time. Consider a rubber band stretched between two nails. While the dimensions remain fixed, the force or stress being applied to the nails will continually decrease. This phenomenon is known as *stress relaxation*. The rate of this relaxation depends on the material, the force, and the temperature.

Rubber

Natural rubber, the oldest commercial elastomer, is made from the processed sap of a tropical tree. In its crude form it is an excellent adhesive and many cements can be made by dissolving it in suitable solvents. Its use as an engineering material dates from 1839, when Charles Goodyear discovered that it could be vulcanized (cross-linked) by the addition of about 30% sulfur followed by heating to a suitable temperature. The cross-linking restricts the movement of the molecular chains and imparts strength. Subsequent research found that the properties could be further improved by various additives (e.g., carbon black) which act as stiffeners, tougheners, and antioxidants. Accelerators have been found that speed up the vulcanization process. These have enabled a reduction in the amount of sulfur such that most rubber compounds now contain less than 3% sulfur. Softeners can be added to facilitate processing, and fillers can be used to add bulk.

Rubber can now be compounded to provide a wide range of characteristics, ranging from soft and gummy to extremely hard. When additional strength is required, textile cords or fabrics can be coated with rubber. The fibers carry the load and the rubber serves as a matrix to join the cords while isolating them from one another to preventing chafing. For severe service, steel wires can be used as the load-bearing medium. Vehicle tires and heavy-duty conveyor belts are examples of this technology.

Natural rubber compounds are outstanding for their flexibility, good electrical insulation, low internal friction, and resistance to most inorganic acids, salts, and alkalies. However, they have poor resistance to petroleum products, such as oil, gasoline, and naphtha. In addition, they lose their strength at elevated temperatures, so it is advisable that they not be used at temperatures above 175°F (80°C). They also deteriorate fairly rapidly in direct sunlight unless specially compounded.

Artificial Elastomers

Seeking to overcome some of these limitations as well as the uncertainty in the supply and price of natural rubber, a number of synthetic or artificial elastomers have been developed and have come to assume great commercial importance. While some are a bit inferior to natural rubber, others offer distinctly different and, frequently, superior properties. Polyisoprene offers the same molecular structure as that of natural rubber, with equal or superior properties. Silicone rubbers are based on a linear chain of silicon and oxygen atoms and can operate in service environments as hot as 450°F (230°C). Various mixes and blends now offer retention of physical properties at elevated temperatures; flexibility at low temperatures; resistance to acids, bases, and other aqueous and organic fluids; resistance to flex fatigue; ability to absorb energy and provide damping; good weatherability; ozone resistance; and availability in a variety of different hardnesses. Thus elastomeric materials can now be selected and used for a wide range of engineering applications.

Table 8-4 lists some of the more common artificial elastomers, along with natural rubber for comparison, and gives their properties and some typical uses. It should be remembered, however, that the properties of these materials can vary widely, depending on the details of their compounding and processing.

Elastomers for Tooling Applications

When an elastomer is confined, it acts like a fluid, transmitting force uniformly in all directions. For this reason, elastomers can be substituted for one half of a die set in sheet-metal-forming operations. Elastomers are also used to perform bulging and form reentrant sections that would be impossible to form with rigid dies except through the use of costly multipiece tooling. Because they can be compounded to range from very soft to very hard, hold up well under compressive loading, and can quickly and economically be made into a desired shape, the engineering elastomers have become increasingly popular as tool materials. The urethanes are currently the most popular for these applications.

■ 8.4 CERAMICS

Ceramic materials have had a long history in the electrical industry because of their high electrical resistance, but have also assumed considerable importance in a variety of other engineering applications. Most of these utilize their outstanding physical properties, including their ability to withstand high temperatures (refractories and refractory coatings), provide a variety of electrical properties (solid-state electronics), and resist wear (coated cutting tools). In general, ceramics are hard, brittle, high-melting-point materials with low electrical and thermal conductivity, low thermal expansion, good chemical and thermal stability, good creep resistance, high elastic modulus, and high compressive strengths. More recently, a family of structural ceramics has emerged, and these materials now provide enhanced mechanical properties that make them attractive for a number of load-bearing applications.

In terms of 1993 sales, glass and glass products accounted for 53% of the ceramic materials market. Advanced ceramic materials (incorporating the structural ceramics, electrical and magnetic ceramics, and fiber optic material) comprised 20%. Whiteware products and porcelain enameled products (primarily household appliances) each accounted for 9%, refractories for 6%, and structural clay products for 2%.

Nature and Structure of Ceramics

Ceramic materials are compounds of metallic and nonmetallic elements (often in the form of oxides, carbides, and nitrides) and exist in a wide variety of compositions and forms. Most have crystalline structures, but unlike metals, the bonding electrons are generally captive in strong ionic or covalent bonds. The absence of free electrons makes the ceramic materials poor electrical conductors and results in many being transparent in thin sections. Because of the strength of the primary bonds, most ceramics have high melting temperatures.

The crystal structures of ceramic materials can be quite different from those observed in metals. In many ceramics, atoms that differ greatly in size must be accommodated within the same structure (as with the sodium chloride crystal of Figure 3-3), and the interstitial sites become extremely important. Charge neutrality must be maintained throughout the ionic structures. Covalent materials must have structures with a limited number of nearest neighbors, set by the number of shared-electron bonds. These features often dictate a less efficient packing, and hence lower densities, than those observed for metallic mate-

TABLE 8-4. Properties and Uses of Common Elastomers

Elastomer	Specific Gravity	Durometer Hardness	Tensile Strength (psi)		Elongation (%)		Service Temperature [°F (°C)]		Resistance to:[a]			Typical Application
			Pure Gum	Black	Pure Gum	Black	Min.	Max.	Oil	Water Swell	Tear	
Natural rubber	0.93	20–100	2500	4000	750	650	−65 (−54)	180 (82)	P	G	G	Tires, gaskets, hose
Polyacrylate	1.10	40–100	350	2500	600	400	0 (−18)	300 (149)	G	P	F	Oil hose, O-rings
EDPM (ethylene propylene)	0.85	30–100	1	3		500	−40 (−40)	300 (149)	P	G	G	Electric insulation, footwear, hose, belts
Chlorosulfonated polyethylene	1.10	50–90	4	2		400	−65 (−54)	250 (121)	G	E	G	Tank lining, chemical hose, shoe, soles and heels
Polychloroprene (neoprene)	1.23	20–90	3500	4000	800	550	−50 (−46)	225 (107)	G	G	G	Wire insulation, belts, hose, gaskets, seals, linings
Polybutadiene	1.93	30–100	1000	3000	800	550	−80 (−62)	212 (100)	P	P	G	Tires, soles and heels, gaskets, seals
Polyisoprene	0.94	20–100	3000	4000		600	−65 (−54)	180 (82)	P	G	G	Same as natural rubber
Polysulfide	1.34	20–80	350	1000	600	400	−65 (−54)	180 (82)	E	G	G	Seals, gaskets, diaphragms, valve disks
SBR (styrene–butadiene)	0.94	40–100	2			1200	−65 (−54)	225 (107)	P	G	G	Molded mechanical goods, disposable pharmaceutical items
Silicone	1.1	25–90		1200		450	−120 (−84)	450 (232)	F	E	P	Electric insulation, seals, gaskets, O-rings
Epichlorohydrin	1.27	40–90	2			325	−50 (−46)	250 (121)	G	G	G	Diaphragms, seals, molded goods, low-temperature parts
Urethane	0.85	62–95	5000		700		−54 (−65)	212 (100)	E	F	E	Caster wheels, heels, foam padding
Fluoroelastomers	1.65	60–90	1	3		400	−40 (−40)	450 (232)	E	E	F	O-rings, seals, gaskets, roll coverings

[a] P, poor; F, fair; G, good; E, excellent.

rials. As in metals, the same chemistry material can often exist in more than one structural arrangement (polymorphism). Silica (SiO_2), for example, can exist in three forms—quartz, tridymite, and crystobalite—depending on the conditions of temperature and pressure.

Ceramic materials can also exist in the form of chains, similar to the linear molecules in plastics. As with the polymeric materials having this structure, the bonds between the chains are not as strong as those within the chains. Consequently, when forces are applied, cleavage occurs between the chains. In other ceramics, the atoms bond in the form of sheets and produce layered structures. Relatively weak bonds exist between the sheets, and these surfaces become the preferred sites for fracture. Mica is an example of such a material.

A noncrystalline structure is also possible in solid ceramics. This amorphous condition is referred to as the *glassy state*, and the materials are known as *glasses*.

Clay and Whiteware Products

Many ceramic products are based on clay, to which various amounts of quartz and feldspar have been added. Selected proportions are mixed with water, shaped, dried, and fired to produce the structural clay products of brick, roof and structural tiles, and drainage and sewer pipe, as well as the whiteware products of sanitary ware (toilets, sinks, and bathtubs), dinnerware, china, decorative floor and wall tile, pottery, and other artware.

Refractory Materials

Refractory materials are ceramics that have been designed to provide acceptable mechanical or chemical properties while at high temperatures. They may take the form of bricks and shaped products, bulk refractory materials (often used as coatings), and insulating ceramic fibers. Most are based on stable oxide compounds, where the coarse oxide particles are bonded by finer refractory material. Various carbides, nitrides, and borides can also be used in refractory applications.

Three distinct classes of refractories can be identified: acidic, basic, and neutral. Common acidic refractories are based on silica (SiO_2) and alumina (Al_2O_3) and can be compounded to provide high-temperature resistance along with high hardness and good mechanical properties. (*Note:* Machinable silica ceramics have been used as insulating tiles on the U.S. space shuttle.) Magnesium oxide (MgO) is the core material for most basic refractories. These are generally more expensive than the acidic materials but are often required in metal-processing applications to provide compatibility with the metal. Neutral refractories, containing chromite (Cr_2O_3), are often used to separate the acidic and basic materials since they tend to attack one another. The combination is often attractive, however, when a basic refractory is necessary on the surface for chemical reasons and the cheaper, acidic material is used beneath to provide strength and insulation. Table 8-5 presents the compositions of some typical refractories. Figure 8-8 shows a variety of high-strength alumina components.

Abrasives

Because of their high hardness, ceramic materials, such as silicon carbide and aluminum oxide (alumina), are often used for abrasive applications, such as grinding. Materials such as manufactured diamond and cubic boron nitride have such phenomenal properties that they are often termed *superabrasives*. Materials used for abrasive applications are discussed in greater detail in Chapter 27.

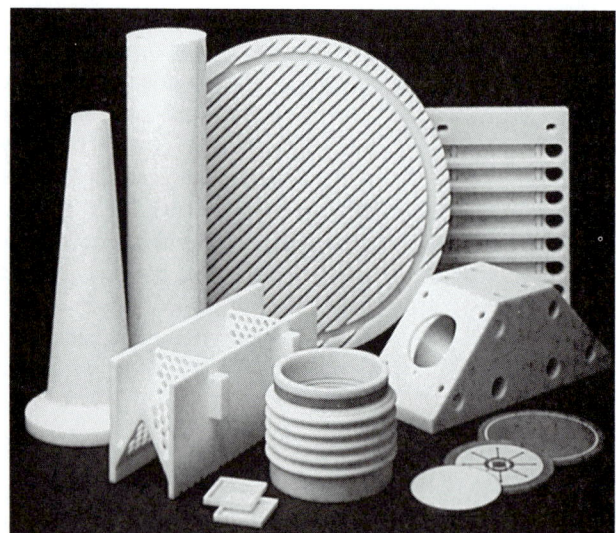

FIGURE 8-8   Variety of high-strength alumina (acid refractory) components, including a filter for molten metal. *(Courtesy of Wesgo Division, GTE.)*

TABLE 8-5.   Composition of Some Typical Refractories

Refractory	SiO_2	Al_2O_3	MgO	Fe_2O_3	Cr_2O_3
Acidic					
Silica	95-97				
Superduty fire brick	51-53	43-44			
High-alumina fire brick	10-45	50-80			
Basic					
Magnesite			83-93	2-7	
Olivine	43		57		
Neutral					
Chromite	3-13	12-30	10-20	12-25	30-50
Chromite–magnesite	2-8	20-24	30-39	9-12	30-50

Source: Ceramic Data Book, Cahners Publishing Co., Westport, Conn., 1982.

Ceramics for Electrical and Magnetic Applications

Ceramic materials also have a variety of useful electrical and magnetic properties. Some ceramics, such as silicon carbide, are used as resistors and heating elements for electric furnaces. Others have semiconducting properties and are used for thermistors and rectifiers. Dielectric, piezoelectric, and ferroelectric behavior can be utilized in a number of applications. Barium titanate, for example, is used in capacitors and transducers. High-density clay-based ceramics and aluminum oxide make excellent high-voltage insulators. Considerable attention has also been directed toward the ceramic superconductors.

Glasses

As with the polymers, glass is an *amorphous* solid that has been cooled to a rigid condition without crystallizing. Most commercial glasses are based on silica (SiO_2) with additives to alter the structure or reduce the melting point. Various chemistries can be used to optimize optical properties, thermal stability, and resistance to thermal shock. Applications range from the traditional window glass, containers, light bulbs, TV and computer display tubes, and fiberglass insulation, to glass cookware and the special fibers used for fiber-optic communication.

Glass Ceramics

These materials are first shaped as a glass and then heat treated to promote *devitrification* or recrystallization of the material, resulting in a structure that contains large amounts of crystalline material within the amorphous base. The crystalline phase helps to retard creep at high temperatures. Strength is greater than the traditional glasses, and the thermal expansion coefficient is near zero, thereby providing good resistance to thermal shock. The white Pyroceram (trade name) material commonly found in Corningware is an example of a common glass ceramic.

Cermets

Cermets are combinations of metals and ceramics (usually oxides, carbides, nitrides, or carbonitrides), bonded together by the procedures of powder metallurgy. This usually involves pressing the powders in molds at pressures ranging from 10 to 40 ksi (70 to 280 MPa) followed by sintering in a controlled-atmosphere furnace at about 3000°F (1650°C). Cermets combine the high hardness and refractory characteristics of ceramics with the toughness and thermal shock resistance of metals. They are used as crucibles, jet engine nozzles, aircraft brakes, and in other applications requiring hardness, strength, and toughness at elevated temperature.

Ceramics for Mechanical Applications: Structural Ceramics

Because of their strong interatomic bonding and high shear resistance, ceramic materials tend to have low ductility and high compressive strength. Theoretically, ceramics could also have high tensile strengths. Because of their high melting points and lack of ductility, most ceramics are processed in the solid state and products are made from powdered material. As a result, small cracks, pores, and impurities are an integral part of most ceramics and act as stress concentrators. When loads are applied, the effect of these flaws cannot be reduced through plastic flow, and the result is often brittle fracture. Applying the principles of fracture mechanics, we find that ceramics are sensitive to very small flaws. Failures typically occur at tensile stress values between 3 and 30 ksi (20 to 210 MPa), more than an order of magnitude less than their compressive strength.

Since the size, number, shape, and location of the flaws is likely to differ from part to part, ceramic parts produced from identical material by identical methods often fail at very different applied loads. Thus the mechanical properties of ceramic products tend to follow a statistical spread that is much less predictable than for metals. This feature tends to limit the use of ceramics in critical high-strength applications. If the various flaws and defects could be eliminated or reduced to very small size, however, high and consistent tensile strengths could be obtained. Hardness, wear resistance, and strength at elevated temperatures would be attractive properties, along with light weight (specific gravities of 2.3 to 3.85), high stiffness, dimensional stability, low thermal conductivity, corrosion resistance, and chemical inertness. Reliability might still be low, however, and failure would still occur by brittle fracture, with little, if any, prior warning. Because of the poor thermal conductivity, thermal shock may be a problem. The cost of these "flaw-free" materials is also rather high, and it is extremely difficult to form reliable bonds with other engineering

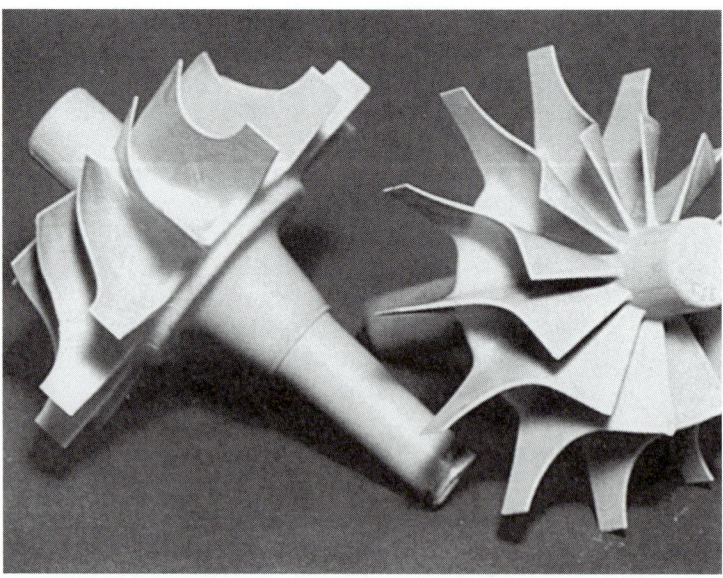

FIGURE 8-9   Gas turbine rotors made of silicon nitride. This lightweight material (one-half the weight of stainless steel) offers strength at elevated temperature as well as excellent resistance to corrosion and thermal shock. *(Courtesy of Wesgo Division, GTE.)*

materials. Machining is also difficult, so products must be fabricated through the use of net-shape processing.

The structural ceramic materials include silicon nitride, silicon carbide, partially stabilized zirconia (limited to temperatures below 1300°F due to a rapid drop in strength), transformation-toughened zirconia, alumina, sialons, boron carbide, boron nitride, titanium diboride, and ceramic composites (such as ceramic fibers in a glass, glass–ceramic, or ceramic matrix). Silicon carbide and silicon nitride offer excellent strength and wear resistance with moderate toughness. They work well in high-stress, high-temperature applications such as turbine blades, and may well replace nickel- or cobalt-based superalloys. Figure 8-9 shows gas-turbine rotors made from injection-molded silicon nitride. They are designed to operate at 2300°F (1250°C), where the material retains over half of its room-temperature strength, and do not require external cooling. Figure 8-10 shows some additional silicon nitride products. Table 8-6 presents the mechanical properties of some of the structural ceramics.

TABLE 8-6.   Properties of Some Structural Ceramics

Material	Density (g/cm^3)	Tensile Strength (ksi)	Compressive Strength (ksi)	Modulus of Elasticity (10^6 psi)	Fracture Toughness (ksi $\sqrt{\text{in.}}$)
Al_2O_3	3.98	30	400	56	5
Sialon	3.25	60	500	45	9
SiC	3.1	25	560	60	4
Z_rO_2 Partially stabilized	5.8	65	270	30	10
Z_rO_2 Transformation toughened	5.8	50	250	29	11
Si_3N_4 (hot pressed)	3.2	80	500	45	5

FIGURE 8-10   Variety of components manufactured from silicon nitride, including an exhaust valve and turbine blade. *(Courtesy of Wesgo Division, GTE.)*

Partially stabilized zirconia combines the resistance to thermal shock, wear, and corrosion, the low thermal conductivity, and the low friction coefficient of zirconia with the enhanced strength and toughness brought about by doping the material with oxides of calcium, yttrium, or magnesium. Transformation-toughened zirconia has even greater toughness as a result of dispersed second phases throughout the ceramic matrix. When a crack approaches the metastable phase, it transforms to a more stable structure, expanding in volume, thereby compressing and stopping the crack.

Sialon (a *si*licon–*al*uminum–*o*xygen–*n*itrogen structural ceramic) is stronger than steel, extremely hard, and as light as aluminum. It has good resistance to corrosion, wear, and thermal shock, is an electrical insulator, and retains good tensile and compressive strength up to 2550°F (1400°C). It has excellent dimensional stability, with a coefficient of thermal expansion that is only one-third that of steel and one-tenth that of plastic. When overloaded, however, it will fail by brittle fracture.

A ceramic engine block continues to be an area of considerable interest. If perfected, it would allow higher operating temperatures with a companion increase in engine efficiency. In addition, it would lower sliding friction and permit the elimination of radiators, fan belts, cooling system pumps, coolant lines, and coolant. The net result would be reduced weight and a more compact design. Estimated fuel savings could amount to 30% or more.

The high cost of the structural ceramics continues to be a barrier to their widespread acceptance. High-grade ceramics are currently about eight times more expensive than their metal counterparts. Factoring in enhanced lifetime and improved performance, there is still a need to reduce cost. Research continues toward the production of a low-cost ($3 to $10 per pound), high-strength, high-toughness ceramic with a useful temperature range up to 2350°F (1300°C). Parallel research is under way in the area of nondestructive evaluation, where flaws in the range of 10 to 50 m must be readily detectable. If this work is successful, potential applications include engines, turbochargers, gas turbines, bearings, pump and valve seals, and other uses involving high-temperature, high-stress environments. These are not traditional applications for ceramics but are areas where tool steels, powdered metals, coated materials, and tungsten carbide are now being used.

Advanced Ceramics as Cutting Tools

Cutting tools have improved significantly through the application of ceramic technology. Silicon carbide is a common abrasive in many grinding wheels. Cobalt-bonded tungsten

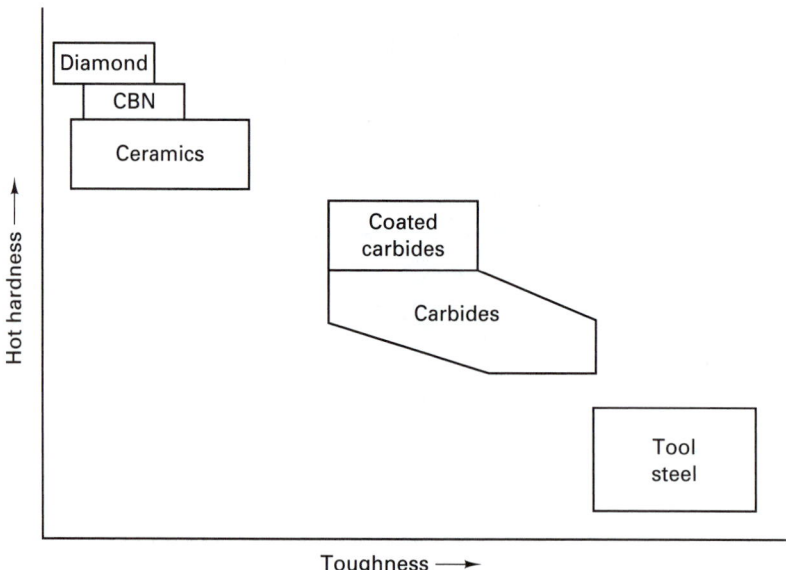

FIGURE 8-11   Graphical comparison of toughness and hardness for a variety of cutting-tool materials.

carbide is a popular alternative to high-speed tool steels for many cutting applications. Many of these are now enhanced by a variety of vapor-deposited ceramic coatings. Thin layers of titanium carbide, titanium nitride, and aluminum oxide can significantly reduce wear and friction and enable faster rates of cutting. Silicon nitride, cubic boron nitride, and polycrystalline diamond cutting tools now offer even greater tool life, higher cutting speeds, and reduced machine downtime. Aluminum oxide reinforced with 35% silicon carbide whiskers can machine nickel-based superalloys more than 10 times faster than standard carbide materials. Cutting speeds can be increased from 200 to nearly 5000 ft/min. Use of these ultrahigh-speed materials, however, will require companion developments in the area of machine tools. High-speed spindles must be perfectly balanced, and workholding devices must withstand the high centrifugal forces. Chip removal methods must be developed that can remove the chips as fast as they are being formed. Yet another trend with ceramic tools is the direct machining of material that once required grinding. Figure 8-11 shows a comparison of toughness and hardness for a variety of cutting-tool materials.

■ 8.5 COMPOSITE MATERIALS

A *composite material* is a heterogeneous solid consisting of two or more different materials that are mechanically or metallurgically bonded together. Each of the various components retains its identity in the composite and maintains its characteristic structure and properties. There are recognizable interfaces between the materials. The composite material, however, generally possesses characteristic properties, such as stiffness, strength, weight, high-temperature performance, corrosion resistance, hardness, or conductivity, that are not possible with the individual components by themselves. Analysis of these properties shows that they depend on (1) the properties of the individual components; (2) the relative amounts of the components; (3) the size, shape, and distribution of the discontinuous components; (4) the degree of bonding between components; and (5) the orientation of the various compo-

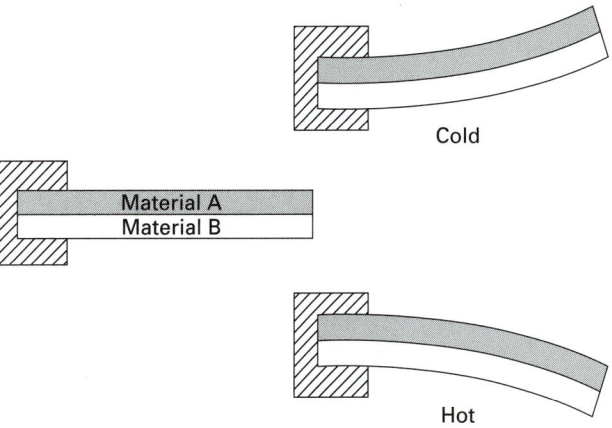

FIGURE 8-12   Schematic of a bimetallic strip where material A has the greater coefficient of thermal expansion.

nents. The various materials involved can be organics, metals, or ceramics. Hence a wide range of freedom exists, and composite materials can often be designed to meet a desired set of engineering properties and characteristics.

There are many types of composite materials and several methods of classifying them. One such method is based on geometry and consists of three distinct families: laminar or layered composites, particulate composites, and fiber-reinforced composites.

Laminar or Layered Composites

Laminar composites are those having distinct layers of material bonded together in some manner and include thin coatings, thicker protective surfaces, claddings, bimetallics, laminates, sandwiches, and others. Plywood is probably the most common engineering material in this category and is an example of a laminate material. Layers of wood veneer are adhesively bonded with their grain orientations at various angles to one another. Strength and fracture resistance are improved, properties are somewhat uniform within the plane of the sheet, swelling and shrinkage tendencies are minimized, and large pieces are now available at reasonable cost. Safety glass is another laminate in which a layer of polymeric adhesive is placed between two pieces of glass and serves to retain the fragments when the glass is broken. *A*ramid–*Al*uminum *L*aminates (Arall) consist of thin sheets of aluminum bonded with adhesive-impregnated aramid fibers. The combination offers light weight coupled with high fracture, impact, and fatigue resistance. Other laminates are used in decorative items, such as Formica countertops and furniture.

Bimetallic strip consists of two metals with different coefficients of thermal expansion bonded together in a laminate. Changes in temperature produce flexing or curvature in the product, which may be employed in thermostat and other heat-sensing applications. Figure 8-12 illustrates this effect. Still other laminar composites are designed to provide enhanced surface characteristics while retaining a low-cost, high-strength, or lightweight core. Many clad materials fit this description. Alclad metal, for example, consists of high-strength, heat-treatable aluminum with an exterior cladding of one of the more corrosion-resistant, non-heat-treatable alloys. U.S. coinage is another laminate, designed to conserve the more costly, high-nickel-content material while providing a lustrous, corrosion-resistant surface. Other laminates have surface layers that have been selected primarily for enhanced wear resistance or improved appearance.

Sandwich material is a laminar structure composed of a thick, low-density core placed between thin, high-density surfaces. Corrugated cardboard is an example of a sandwich structure. Other engineering sandwiches incorporate cores of a polymer foam or honeycomb structure, to produce a lightweight, high-strength, high-rigidity composite.

Particulate Composites

Particulate composites consist of discrete particles of one material surrounded by a matrix of another material. Concrete is a classic example, consisting of sand and gravel particles surrounded by cement. Asphalt consists of similar aggregate in a matrix of bitumin, a thermoplastic polymer. In both of these examples, the particles are rather coarse. Other particulate composites involve extremely fine particles and include many of the multicomponent powder metallurgy products, specifically those where the dispersed particles do not diffuse into the matrix material.

Dispersion-strengthened materials are particulate composites where a small amount of hard, brittle, small particles (typically, oxides or carbides) are dispersed throughout a softer, more ductile matrix. Pronounced strengthening can be induced, which decreases only gradually as temperature is increased. Creep resistance is, therefore, improved significantly. Examples of dispersion-strengthened materials include sintered aluminum powder (SAP), which consists of an aluminum matrix strengthened by up to 14% aluminum oxide, and TD-nickel, a nickel alloy containing 1 to 2 wt% thoria (ThO_2).

Other types of particulate composites contain large amounts of rather coarse particles and are usually designed to produce some desired combination of properties rather than increased strength. Cemented carbides, for example, consists of hard ceramic particles, such as tungsten carbide, tantalum carbide, or titanium carbide, embedded in a metal matrix, generally cobalt. While the hard, stiff carbide could withstand the high temperatures and pressure of cutting, it is also extremely brittle. Toughness is imparted by combining the carbide particles with cobalt powder, pressing the material into the desired shape, heating to melt the cobalt, and then resolidifying the compacted material. Varying levels of toughness can be imparted by varying the amount of cobalt in the composite.

Grinding and cutting wheels are often formed by bonding abrasives, such as alumina (Al_2O_3), silicon carbide (SiC), cubic boron nitride (BN), or diamond, in a matrix of glass or polymeric material. As the hard particles wear, they fracture or pull out of the matrix, exposing new cutting edges. By combining tungsten powder and powdered silver or copper, electrical contacts can be produced that offer both high conductivity and resistance to wear and arc erosion. Foundry molds and cores are often made from sand (particles) and an organic or inorganic binder (matrix). Particulate-type metal-matrix composites have been made by introducing a variety of ceramic or glass particles into aluminum or magnesium matrices. Particulate-toughened ceramics using zirconia and alumina matrices are being used as bearings, bushings, valve seats, die inserts, and cutting-tool inserts. In addition, many plastics might be considered to be particulate composites because the additive fillers and extenders can be viewed as dispersed particles. Designation as a composite, however, is usually reserved for those polymers where the particles are added for the primary purpose of property modification. One such example would be the combination of granite particles in an epoxy matrix currently being used in some machine tool bases. This unique material offers high strength and a vibration damping capacity that exceeds that of gray cast iron.

Due to their unique geometry, the properties of particulate composites are often *isotropic*, that is, uniform in all directions. This may be particularly important in engineering applications.

Fiber-Reinforced Composites

The most popular type of composite material is the *fiber-reinforced composite* geometry, where continuous or discontinuous thin fibers of one material are embedded in a matrix of another. The matrix supports and transmits forces to the fibers, protects them from environments and handling, and provides ductility and toughness, while the fibers carry most of the load and impart enhanced stiffness. Wood and bamboo are two naturally occurring fiber composites, consisting of cellulose fibers in a lignin matrix. Bricks of straw and mud may well have been the first human-made material of this variety, dating back to near 800 B.C. Automobile tires now use fibers of nylon, rayon, aramid (Kevlar), or steel in various numbers and orientations to reinforce the rubber and provide added strength and durability. Steel-reinforced concrete is actually a double composite, consisting of a particulate matrix reinforced with steel "fibers."

Glass-fiber reinforced resins, the first of the modern fibrous composites, were developed shortly after World War II in an attempt to produce lightweight materials with high strength and high stiffness. Glass fibers about 10 μm in diameter are bonded in a variety of polymers, generally epoxy or polyester resins. Between 30 and 60% by volume is made up of fibers of either E-type borosilicate glass or the stronger, stiffer, S-type magnesia–alumina–silicate glass. Glass fibers are still the most widely used reinforcement, primarily because of their lower cost and adequate properties for many applications. Limitations are generally related to stiffness, since the glass fibers have a modulus of only 10 to 13×10^6 psi (70,000 to 90,000 MPa).

Efforts to improve the strength and modulus focused on the development of improved fibers. Boron–tungsten fibers (boron deposited on a tungsten core) offer an elastic modulus of 55×10^6 psi (380,000 MPa) with tensile strengths in excess of 400 ksi (2750 MPa). Silicon carbide filaments (SiC on tungsten) have an even higher modulus of elasticity. These high-strength, high-stiffness materials generally fail in a sudden, catastrophic manner due to the propagation of flaws. When used as fibers, however, the propagation of a flaw only leads to the failure of an individual fiber, not the entire assemblage as with a monolithic block of the same material.

Graphite and aramid (Kevlar) are popular as reinforcing fibers. Graphite fibers can be either the PAN type, produced by the thermal pyrolysis of synthetic organic fibers, such as viscose rayon or polyacrylonitrile, or pitch type, made from petroleum pitch. They have low density and a range of high tensile strengths and elastic moduli. Graphite's negative thermal expansion coefficient can also be used to offset the positive values of most matrix materials. Kevlar is an organic aramid fiber with 450 ksi (3100 MPa) tensile strength, 19×10^6 psi (131,000 MPa) elastic modulus, a density approximately one-half that of aluminum, good toughness, and a negative thermal expansion coefficient. In addition, it is flame retardant and transparent to radio signals, making it attractive for a number of military and aerospace applications where the service temperature is not excessive. Table 8-7 lists some of the key engineering properties for several of the common reinforcing fibers.

Within the composite, the reinforcing fibers can be arranged in a variety of orientations. Fiberglass, for example, contains short, randomly oriented fibers. Unidirectional fibers can be used to produce highly directional properties, with the fiber directions being tailored to the direction of loading. Woven fabrics or tapes can be produced and then layered in various orientations to produce a plywoodlike product. The layered materials can be stitched together to add a third dimension to the weave, and complex three-dimensional shapes can be woven from fibers and later injected with a matrix material.

The properties of fiber-reinforced composites depend strongly on several characteristics: (1) the properties of the fiber material, (2) the volume fraction of fibers, (3) the aspect

TABLE 8-7. Properties and Characteristics of Some Common Reinforcing Fibers

Fiber Material	Specific Strength[a] (10^6 in.)	Specific Stiffness[b] (10^6 in.)	Density (lb/in.^{3})	Melting Temperature[c] (°F)
Al_2O_3 whiskers	21.0	434	0.142	3600
Boron	4.7	647	0.085	3690
Ceramic fiber (Mullite)	1.1	200	0.110	5430
E-type glass	5.6	114	0.092	<3140
High-strength graphite	7.4	742	0.054	6690
High-modulus graphite	5.0	1430	0.054	6690
Kevlar	10.1	347	0.052	—
SiC whiskers	26.2	608	0.114	4890

[a]Strength divided by density.

[b]Elastic modulus divided by density.

[c]Or maximum temperature of use.

ratio of the fibers, i.e., the length-to-diameter ratio, (4) the orientation of the fibers, (5) the degree of bonding between the fiber and the matrix, and (6) the properties of the matrix. The matrix materials should be strong, tough, and ductile, so they can transmit the loads to the fibers and prevent cracks from propagating through the composite. Both thermosetting and thermoplastic resins have been used. Popular thermosets include epoxies, polyesters, bismaleimides, and polyimides. From a manufacturing viewpoint it is often easier and faster to heat and cool a thermoplastic than to cure a thermoset, and the thermoplastics are tougher and more tolerant to damage. Attractive thermoplastic matrix materials are usually those with improved high-temperature properties, which includes the thermoplastic polyimides, polyphenylene sulfide (PPS), polyether ether ketone (PEEK), and the liquid-crystal polymers. The matrix material should be selected to match the temperature of operation. Polymeric materials can only be used for temperatures below 600°F (315°C). Above this temperature, metal or ceramic matrices should be considered.

Advanced Fiber-Reinforced Composites

Advanced composites are materials that have been developed for applications requiring exceptional combinations of strength, stiffness, and light weight. Fiber content generally exceeds 50% (by weight) and the modulus of elasticity is typically greater than 16×10^6 (psi). Superior creep and fatigue resistance, low thermal expansion, low friction and wear, vibration damping characteristics, and environmental stability are other properties that may be required in these materials.

There are four basic types of advanced composites where the matrix material is matched to the fiber and application:

1. The advanced *organic or resin-matrix composites* frequently use high-strength, high-modulus fibers of graphite, aramid (Kevlar), or boron. Properties can be put in desired locations or orientations at about one-half the weight of aluminum (or one-sixth that of steel). Thermal expansions can be designed to be low, or even negative. Unfortunately, these materials have a maximum service temperature of about 600°F (315°C) because the polymeric matrix loses strength when heated. Table 8-8 compares the properties of some of the common resin-matrix

TABLE 8-8. Comparison of the Properties of Fiber Composites (in the Fiber Direction) with Those of Lightweight or Low-Thermal-Expansion Structural Metals

Material	Specific Strength[a] (10^6 in.)	Specific Stiffness[b] (10^6 in.)	Density (lb/in^3)	Thermal Expansion Coefficient [in./(in.-°F)]	Thermal Conductivity [Btu/(hr-ft-°F)]
Boron–epoxy	3.3	457	0.07	2.2	1.1
Glass–epoxy (woven cloth)	0.7	45	0.065	6	0.1
Graphite–epoxy: high modulus (unidirectional)	2.1	700	0.063	−0.5	75
Graphite–epoxy: high strength (unidirection)	5.4	400	0.056	−0.3	3
Kevlar–epoxy (woven cloth)	1	80	0.5	1	0.5
Aluminum	0.7	100	0.10	13	100
Beryllium	1.1	700	0.07	7.5	120
Invar[c]	0.2	70	0.29	1	6
Titanium	0.8	100	0.16	5	4

[a]Strength divided by density.

[b]Elastic modulus divided by density.

[c]A low-expansion metal containing 36% Ni and 64% Fe.

composites with those of several of the lightweight or low-thermal-expansion metals. Typical applications include sporting equipment (tennis rackets, skis, golf clubs, and fishing poles), lightweight armor plate, and a myriad of low-temperature aerospace components.

2. *Metal-matrix composites* (MMCs) can be used for operating temperatures up to 2300°F (1250°C), where the conditions require high strength coupled with ductility and toughness. The ductile matrix material can be aluminum, copper, magnesium, titanium, nickel, superalloy, or even intermetallic compound, and the reinforcing fibers may be graphite, boron carbide, alumina, or silicon carbide. Fine whiskers (tiny needlelike single crystals of 1 to 10 μm in diameter) of sapphire, silicon carbide, and silicon nitride have also been used as the reinforcement, as well as wires of titanium, tungsten, molybdenum, beryllium, and stainless steel.

Compared to the engineering metals, these composites offer higher stiffness and strength (especially at elevated temperatures), a lower coefficient of thermal expansion, and enhanced resistance to fatigue, abrasion, and wear. Compared to the organic matrix materials, they offer higher heat resistance, as well as improved electrical and thermal conductivity. They are nonflammable and do not absorb water or gases. Unfortunately, these materials are quite expensive, the vastly different thermal expansions of the components may lead to debonding, and the assemblies may be prone to galvanic corrosion.

Graphite-reinforced aluminum can be designed to have near-zero thermal expansion in the fiber direction. Aluminum oxide-reinforced aluminum has been used in automotive connecting rods to provide stiffness and fatigue resistance with lighter weight. Aluminum reinforced with silicon carbide whiskers has been fabricated into aircraft wing panels, providing a 20 to 40% weight savings. In the future, fiber-reinforced superalloys may well become a preferred material for applications such as turbine blades.

3. *Carbon–carbon composites* (graphite fibers in a carbon matrix) offer the possibility of a heat resistant material that could operate at temperatures up to 6000°F (3300°C), with a strength that is 20 times that of conventional graphite and a den-

sity that is 30% lighter (1.38 g/cm^3). Not only does this material withstand high temperatures, it actually gets stronger when heated. For temperatures over 1000°F (540°C), however, the composite requires some form of coating to protect it from oxidizing. Various coatings can be used for different temperature ranges. Current applications include: the nose cone and leading edge of the space shuttle, racing car disc brakes (become stronger when hotter), aerospace turbines and jet engine components, rocket nozzles, and surgical implants.

4. *Ceramic-matrix composites* (CMCs) offer light weight, high-temperature strength, and good dimensional and environmental stability. The matrix provides high-temperature resistance. Glass matrices can operate at temperatures as high as 2700°F (1500°C). The crystalline ceramics, usually based on alumina, silicon carbide, silicon nitride, boron nitride, or zirconia, can be used for even hotter conditions. The fibers add directional strength, increase fracture toughness, improve thermal shock resistance, and can be incorporated in unwoven, woven, knitted, and braided form. Typical reinforcements include carbon fiber, glass fiber, fibers of the various matrix materials, and ceramic whiskers. A silicon nitride material reinforced with silicon carbide whiskers has 40% more resistance to internal cracking and 25% more resistance to complete fracture than does unreinforced silicon nitride. Ceramic composites using continuous fiber reinforcement typically fail in a non-catastrophic manner. Unfortunately, the cost of ceramic–ceramic composites ranges from about $10 per pound to as much as $1000 per pound.

Hybrid Composites

Hybrid composites involve two or more types of fibers set in a common matrix. The particular combination of fibers is usually selected to balance strength and stiffness, provide dimensional stability, reduce cost, reduce weight, or improve fatigue and fracture resistance. Types of hybrid composites include (1) interply (alternating layers of fibers), (2) intraply (mixed strands in the same layer), (3) interply–intraply, (4) selected placement (where the more costly material is used only where needed), and (5) interply knitting (where plys of one fiber are stitched together with fibers of another type).

Design and Fabrication

The design of composite materials involves the selection of components; the determination of size, shape, distribution, and orientation of the components; and the selection of an appropriate fabrication method. Many of the possible fabrication methods have been specifically developed for use with composite materials. For example, fibrous composites can be manufactured into useful shapes through simple compression molding, filament winding, pultrusion (where bundles of coated fibers are drawn through a heated die), cloth laminations, and autoclave curing (where pressure and elevated temperature are applied simultaneously). Fiber-reinforced plastics have also been injection molded to produce products that compete with zinc die castings. Sheets of fiber-reinforced plastic (sheet molding compounds) can be press formed to provide lightweight, corrosion-resistant products that are similar to those made from sheet metal. A more complete description of fabrication methods that have been developed for composite materials is provided in Chapter 20. With the wide variety of materials, geometries, and processes, designers can now tailor a composite material specifically for a given application.

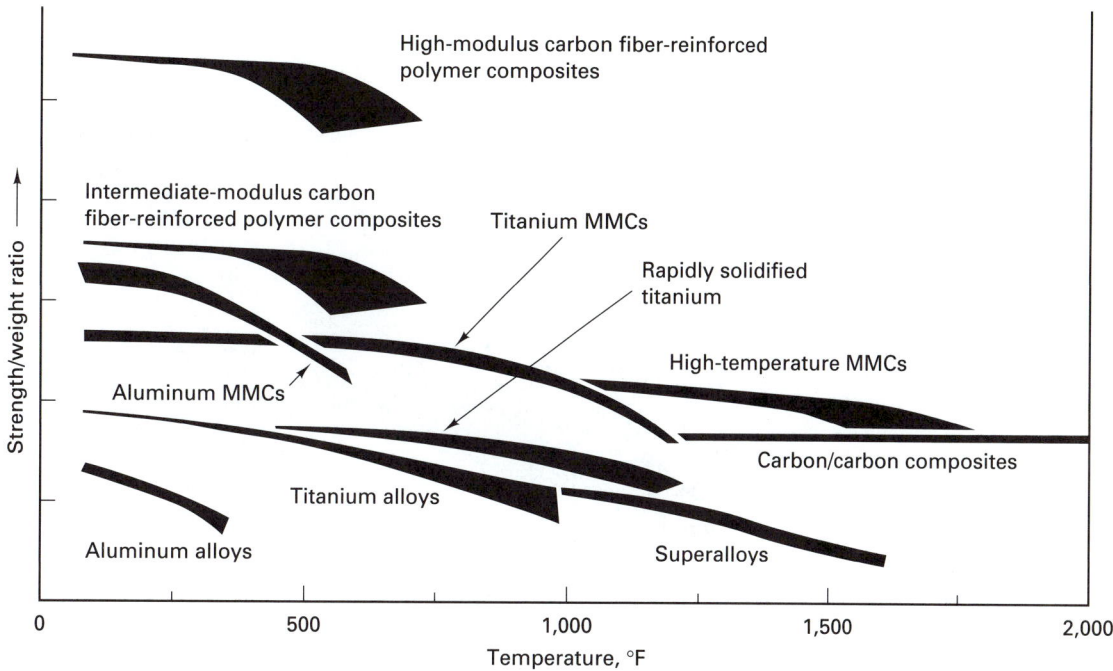

FIGURE 8-13 Graphical depiction of the strength/weight ratio of various aerospace materials as a function of temperature. (*Adapted with permission of DuPont Company.*)

Assets and Limitations

Figure 8-13 presents the strength-to-weight ratios of various aerospace materials as a function of temperature. The superiority of the various advanced composites over the aerospace metals is clearly evident. The weight of a graphite–epoxy composite I-beam is less than one-fifth that of steel, one-third that of titanium, and one-half that of aluminum. Its ultimate tensile strength equals or exceeds that of the other three materials, and it possesses an almost infinite fatigue life. The greatest limitations to the use of this and other composites are their relative brittleness and high cost. Graphite fibers, which cost between $400 and $500 per pound in the late 1960s, are now available for about $10 per pound, but even that cost is considerably higher than the cost of many alternative materials. Metal-matrix composites now cost between $10 and $200 per pound. Ceramic-matrix composites range from $20 per pound to over $1000 per pound for specialty materials such as the space shuttle tiles.

While the field has definitely matured, manufacturing with composites is still quite labor intensive and there is still a lack of trained designers, established design guidelines and data, information about fabrication costs, and well-developed methods of quality control and inspection. It is often difficult to predict the interfacial bond strength, the strength of the composite, its response to impacts, and probable modes of failure. Defects can involve delaminations, voids, missing layers, contamination, fiber breakage, and hard-to-detect improperly cured resin. There is often some concern about heat resistance, and many composites with polymeric matrices are sensitive to moisture, acids, chlorides, organic solvents, oils, and ultraviolet radiation, and tend to cure forever, causing continually changing properties. In addition, most composites have limited ability to be repaired if damaged,

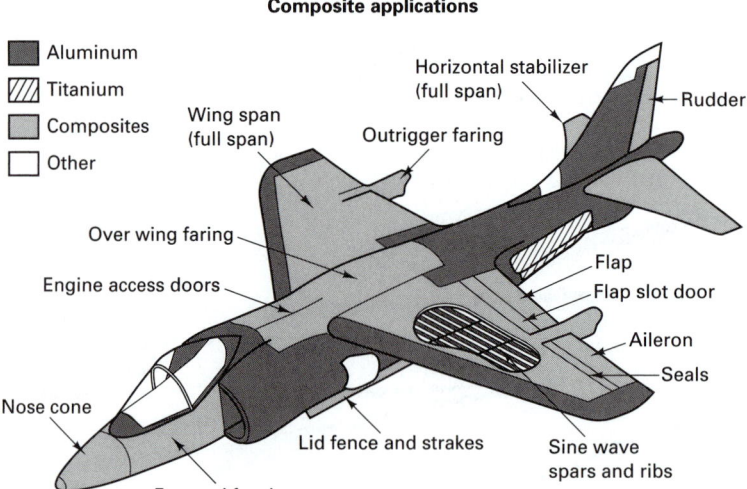

FIGURE 8-14 Schematic diagram of the U.S. Marine Corps' AV-8B Harrier II fighter plane (built by McDonnell Douglas Corp.) depicting the extensive use of composite materials. Composites account for 26% of the plane's weight and are primarily graphite fiber-reinforced epoxy. (Reprinted with permission of *Metalworking News,* copyright 1987, Fairchild Publications, a Capital Cities/ABC Company.)

preventive maintenance procedures are not well established, and recycling is often extremely difficult. Assembly operations with composites generally require the use of industrial adhesives.

On the positive side, the availability of a corrosion-resistant material with strength and stiffness greater than those of steel at only one-fifth the weight may be sufficient to justify some engineering compromises. In addition, products can often be designed to significantly reduce the number of parts, number of fasteners, assembly time, and cost.

Areas of Application

Many composite materials are stronger than steel, lighter than aluminum, and stiffer than titanium. They offer low thermal conductivity, good heat resistance, good fatigue life, low corrosion rates, and adequate wear resistance. For these reasons they have become well established in a number of major areas. Aerospace applications frequently require light weight, stiffness, and fatigue resistance, and composites may well account for 65% of the weight of an airplane in turn-of-the-century subsonic designs. Figure 8-14 shows a schematic of the U.S. Marine Corps' AV-8B Harrier II (built by McDonnell Douglas Corp.). This plane, which can take off on a 1500-ft runway and make vertical landings, is 26% composite material (by weight), primarily graphite–epoxy. The forward fuselage, which is 67% composite, weighs 25% less than an equivalent metal design. The composite wing assembly is 330 lb lighter than a metal wing. An added benefit to military aircraft is reduced detectability by radar.

Sporting equipment such as golf club shafts, baseball bats, fishing rods, tennis rackets, bicycle frames, and skis are now available in a variety of fibrous composites. Figure 8-15 shows several of these applications. Potential automotive uses, in addition to body

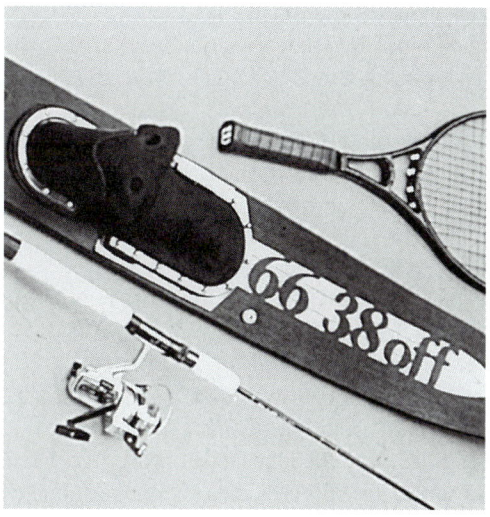

FIGURE 8-15   Composite materials are often used in sporting applications to provide light weight, high stiffness, strength, and attractive styling.   (*Courtesy of Fiberite Corporation.*)

panels, include drive shafts, springs, and bumpers. Weight savings compared to existing parts is generally 20 to 25%. North American truck manufacturers now use fiber-reinforced composites for such parts as cab shells and bodies, oil pans, fan shrouds, instrument panels, and engine covers. One prediction estimated that the demand for composite materials will increase from 4 billion pounds per year to over 16 billion pounds per year by the year 2000.

■ KEY WORDS

addition polymerization	elastomer	plastic
additive agents	fiber-reinforced composite	plasticizer
amorphous	fillers	polymer
ceramic-matrix composite	glass transition temperature	refractory material
cermet	hybrid composite	saturated monomer
composite	isomer	thermoplastic
condensation polymerization	laminar composite	thermosetting
copolymer	metal-matrix composite	unsaturated monomer
cross-linked	oriented plastics	
degree of polymerization	particulate composite	

■ REVIEW QUESTIONS

1. What are some naturally occurring nonmetallic materials that have been used for engineering applications?
2. What are some material families that would be classified under the general term *nonmetallic engineering materials*?
3. How might plastics be defined from the viewpoints of chemistry, structure, and fabrication processing?
4. What is the difference between a saturated and an unsaturated molecule?
5. Describe and differentiate the two means of forming polymers: addition and condensation.
6. Describe and differentiate thermoplastic and thermosetting plastics.
7. Describe the mechanical property behaviors that may be observed as a thermoplastic is cooled from above the melting temperature.
8. What are some of the ways that a thermoplastic polymer can be made stronger?
9. Why are thermosetting polymers characteristically brittle?
10. How do thermosetting polymers respond to subsequent heating?
11. Describe how thermoplastic or thermosetting characteristics affect productivity during the fabrication of a molded part.
12. What are some attractive engineering properties of polymeric materials?
13. What are some limiting properties of plastics, and in what general area do they fall?
14. What are some environmental conditions that might adversely affect the engineering properties plastics?
15. What are some reasons that additive agents are incorporated into plastics?
16. What are some functions of a filler material?
17. What is the primary engineering benefit of an oriented plastic?
18. What are some properties and characteristics of the "true engineering plastics"?
19. What are some products for which plastics have captured segments of markets traditionally held by other materials?
20. Why is the recycling of mixed plastics more difficult than the recycling of mixed metals?
21. What is the unique mechanical property of elastomeric materials, and what structural feature is responsible for it?
22. How can cross-linking be used to control the engineering properties of elastomers?
23. What are some attractive engineering properties of natural and artificial elastomers?
24. What are some outstanding physical properties of ceramic materials?
25. Why are the crystal structures of ceramics frequently more complex than those observed for metals?
26. What is the dominant property of refractory ceramics?
27. What is the dominant property of ceramic abrasives?
28. What are cermets, and what properties or combination of properties do they offer?
29. Why do most ceramic materials fail to possess their theoretically high tensile strength?
30. Why do the mechanical properties of ceramics generally show a wider statistical spread than the same properties of metals?
31. If all significant flaws or defects could be eliminated from the structural ceramics, what features might still limit their possible applications?
32. What are some specific materials that are classified as structural ceramics?
33. What are some attractive and limiting properties of sialon (one of the structural ceramics)?
34. What are some attractive possibilities that could accompany the development of a ceramic engine block?
35. What are some ceramic materials that are currently being used for cutting-tool applications, and what features or properties make them attractive?
36. What is a composite material?
37. What are the basic structural features of a composite material which influence and determine its properties?
38. What are the three primary geometries of composite materials?
39. What feature in a bimetallic strip makes its shape sensitive to temperature?
40. What is the attractive aspect to the strength that is induced by the particles in a dispersion-strengthened material?

41. Which of the three primary composite geometries is most likely to possess isotropic properties?
42. What is the primary role of the matrix in a fiber-reinforced composite?   Of the fibers?
43. What are some possible fiber orientations or arrangements in a fiber-reinforced composite material?
44. What are some features that influence the properties of the fiber-reinforced composites?
45. In what ways are metal-matrix composites superior to straight engineering metals? To organic-matrix composites?
46. What features might be imparted by the fibers in a ceramic matrix composite?
47. What are some major limitations to the extensive use of composite materials in engineering applications?
48. What are some properties of composite materials that make them attractive for aerospace applications?

■ PROBLEMS

1. **(a)** One of Leonardo da Vinci's sketchbooks contains a crude sketch of an underwater boat (or submarine). Leonardo did not attempt to develop or refine this sketch further, possibly because he recognized that the engineering materials of his day (wood, stone, and leather) were inadequate for the task. What properties would be required for the body of a submersible vehicle? What materials might you consider?

   (b) Another of Leonardo's sketches bears a crude resemblance to a helicopter—a flying machine. What properties would be desirable in a material that would be used for this type of application?

   (c) Try to identify a possible engineering product that would require a material with properties that do not exist among today's engineering materials. For your application, what are the demanding features or requirements? If a material were to be developed for this application, from what family or group do you think it would emerge? Why?

2. Select a product (or component of a product) that can reasonably be made from materials from two or more of the basic materials families (metals, polymers, ceramics, and composites).

   (a) Describe briefly the function of the product or component.

   (b) What properties would be required for this product or component to perform its function?

   (c) What two materials groups might provide reasonable candidates for your product?

   (d) Select a candidate material from the first of your two families and describe its characteristics. In what ways does it meet your requirements? How might it fall short of the needs?

   (e) Repeat part (d) for a candidate material from the second material family.

   (f) Compare the two materials to one another. Which of the two would you prefer? Why?

*C*hapter 8 CASE STUDY

two-wheel dolly handles

T he items illustrated in Figure CS-8 are the handle grips for an industrial-quality, pneumatic-tire, two-wheel dolly. They are designed to be bolted onto box-channel tubular sections using four bolt-holes, which are sized to accomodate 3/8-inch diameter bolts. The major service requirements are strength, durability, fracture resistance, reasonable appearance, and possibly weight.

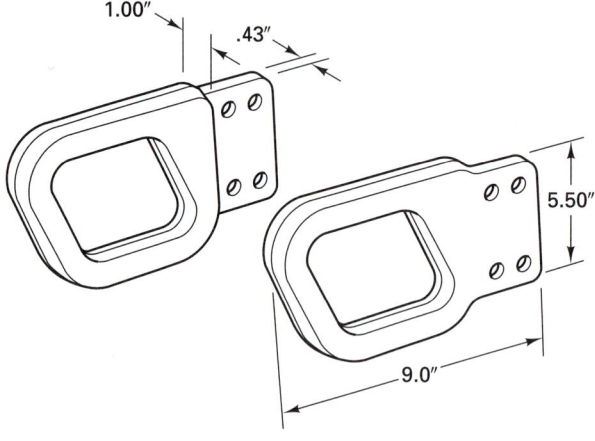

1.00″   .43″   5.50″   9.0″

FIGURE CS-8

Your employer, Industrial Equipment Company, is currently marketing such a dolly using handles that are made as permanent-mold aluminum castings. The firm is in the process of updating its line, and is reevaluating the design and manufacture of each of its products. You have been assigned the dolly and have been asked specifically to determine whether the handles might be replaced by a less expensive material, such as a polymer or low-cost composite.

Investigate the properties and cost* of alternative materials, including means of fabricating the desired shape, and make your recommendation. Since the existing design was for cast metal, might you want to make minor modifications? Would the alternative materials possess adequate properties in the bolt-hole region, or would you want to use some form of reinforcement?

---

*Since the size of the part will remain relatively unchanged, material costs should be compared on the basis of $/in. and not $/lb.

CHAPTER 9

MATERIAL SELECTION

9.1	INTRODUCTION	9.7	ULTIMATE OBJECTIVE
9.2	MATERIAL SELECTION AND MANUFACTURING	9.8	MATERIALS SUBSTITUTION
9.3	DESIGN PROCESS	9.9	EFFECT OF PRODUCT LIABILITY ON MATERIALS SELECTION
9.4	PROCEDURE FOR MATERIAL SELECTION		
9.5	ADDITIONAL FACTORS TO CONSIDER	9.10	AIDS TO MATERIAL SELECTION
9.6	CONSIDERATION OF THE MANUFACTURING PROCESS	Case Study:	MATERIAL SELECTION

■ 9.1 INTRODUCTION

The objective of any practical work dealing with the manufacture of products is to produce components that will adequately perform their designated tasks. Meeting this objective implies the manufacture of components from *selected engineering material*s with the *required geometrical shape and precision* and with *companion material structures* that are optimized for the service environment that the component must withstand. The ideal design is one that will just meet all requirements. Anything better tends to waste money and material. Anything worse, and we have failed to manufacture an adequate product.

It has not been all that long ago that each of the materials groups had their own well-defined uses and markets. Metals were specified when strength, toughness, and durability were the primary requirements. Ceramics were generally limited to low-value applications where heat or chemical resistance was required and any loadings were compressive. Glass was used for its optical transparency, and plastics were relegated to low-value applications where low cost and light weight were attractive features and performance properties were secondary.

Such clear delineations no longer exist. Many of the metal alloys in use today did not exist as little as 30 years ago, and the common alloys that have been in use for a century or more have been much improved due to advances in metallurgy and production processes. New on the scene are amorphous metals, oxide dispersion-strengthened alloys produced by powder metallurgy, mechanical alloying, and directionally solidified materials. Ceramics, polymers, and composites are now available with specific properties that often transcend the traditional limits and boundaries. Advanced structural materials offer higher strength and stiffness, strength at elevated temperature, light weight, and resistance to corrosion, creep, and fatigue. Other materials have enhanced thermal, electrical, optical, magnetic, and chemical properties.

As a result of these dynamic changes, the selection of engineering materials has become extremely important and the process now requires constant reevaluation. New materials are continually being developed, others may no longer be available, and prices are always subject to change. Concerns regarding environmental pollution, recycling, and worker health and safety have imposed new constraints. New desires for weight reduction, energy savings, or improved corrosion resistance may well motivate a change in engineering material. Pressures from domestic and foreign competition, increased demand for quality and serviceability, and negative customer feedback can all prompt a reevaluation.

227

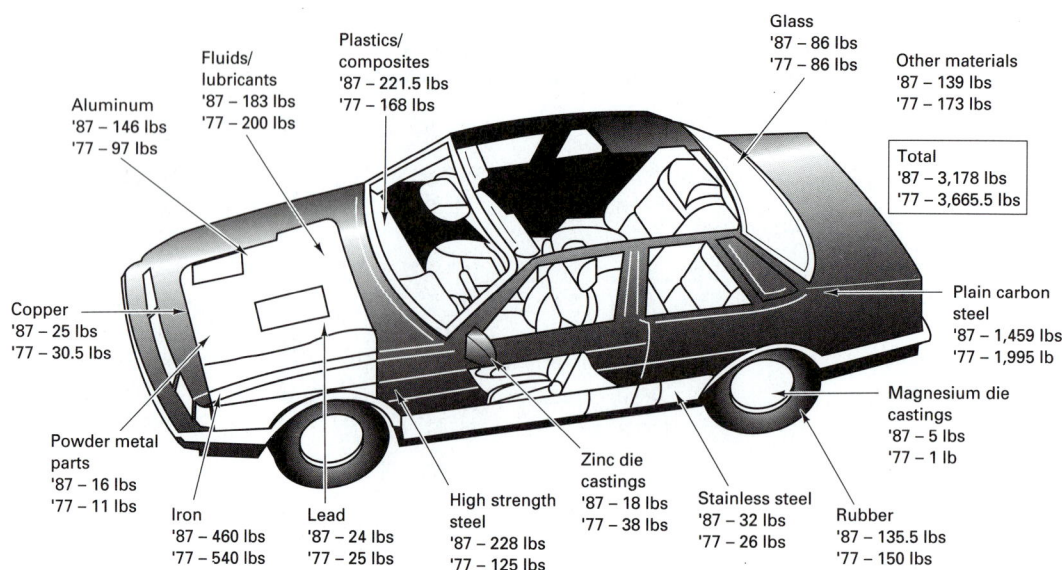

FIGURE 9-1   Changes in the material makeup of a typical U.S. automobile between 1977 and 1987. (Reprinted with the permission of *Metalworking News,* copyright 1987, Fairchild Publications, a Capital Cities/ABC Company.)

Finally, the proliferation of product liability actions, many of which are the result of improper material use, has further emphasized the need for constant reevaluation of the engineering materials in a product.

The automotive industry alone consumes approximately 60 million metric tons of engineering materials worldwide every year—primarily steel, cast iron, aluminum, copper, glass, lead, polymers, rubber, and zinc. Figure 9-1 shows the dramatic change in the material content of a typical American car between 1977 and 1987. Notice the increased use of lighter-weight materials and high-strength steels, as well as plastics and composites. In the four years between 1987 and 1991, the use of aluminum increased another 45 lb, to a total of 191 lb, an amount that is projected to remain constant through the year 2000. U.S. automakers predict that the average automobile in the year 2000 will weigh 2992 lb and be composed of 53% steel, 12% cast iron, 10% polymers, 6% aluminum, and 19% other materials.

An additional $\frac{1}{2}$ million metric tons of engineering materials go into aerospace applications. Here the principal materials are aluminum, magnesium, titanium, superalloys, polymers, rubber, steel, metal-matrix composites, and polymer-matrix composites. Competition is intense, and materials substitutions are frequent. The use of advanced composite materials in aircraft construction has risen from less than 2% in 1970 to the point where they will account for over 30% of the structure on the U.S. Air Force's new YF-22 Advanced Tactical Fighter (which also employs titanium for the exterior skins surrounding the engines, as well as the engine frames). The Beech Starship (Figure 9-2) uses no metal in the airframe construction. The frame is made entirely of molded, fiber-reinforced composites.

Looking to the future, consider the materials requirements of advanced hypersonic structures such as the proposed X-30 National Aerospace Plane (NASP). Building a plane that can take off from a runway, fly directly into space (achieving speeds in excess of 25 times the speed of sound), and return to land on another runway may seem logical enough, but the engineering problems are immense. Jet engines cannot reach the required speeds to escape the atmosphere, and rocket engines require too much heavy fuel. If the propul-

FIGURE 9-2   The lightweight Beech Starship aircraft utilizes an airframe constructed entirely of molded, fiber-reinforced composites. *(Courtesy of Beech Aircraft Corp.)*

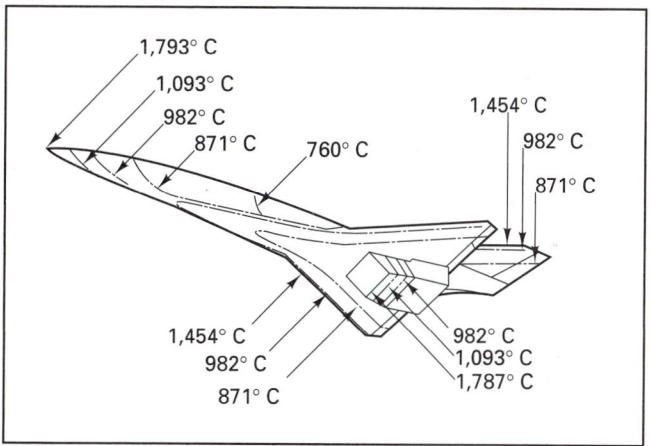

FIGURE 9-3   Predicted maximum temperatures of a transatmospheric hypersonic aircraft, such as the X-30 national aerospace plane.

sion problems can be solved, the incredible speeds will generate elevated temperatures far in excess of those encountered by present aircraft. Figure 9-3 shows that the maximum temperatures at leading edges of the body will be near 1500°C (2700°F). Other components could reach temperatures as high as 3300°C (6000°F). Radically different aerospace materials will be required to provide the properties of high strength, light weight, and stability at these highly elevated temperatures.

At one time, bicycle frames were constructed almost exclusively from welded steel tubing. Now, companies offer frames in a wide range of engineering materials, including aluminum alloys, titanium, and various fiber-reinforced composites. One top-of-the-line carbon fiber frame now weighs only 2.5 lb!

The vacuum cleaner assembly shown in Figure 9-4 is typical of many engineering products, with nine different materials being used in the parts that are illustrated. Table 9-1

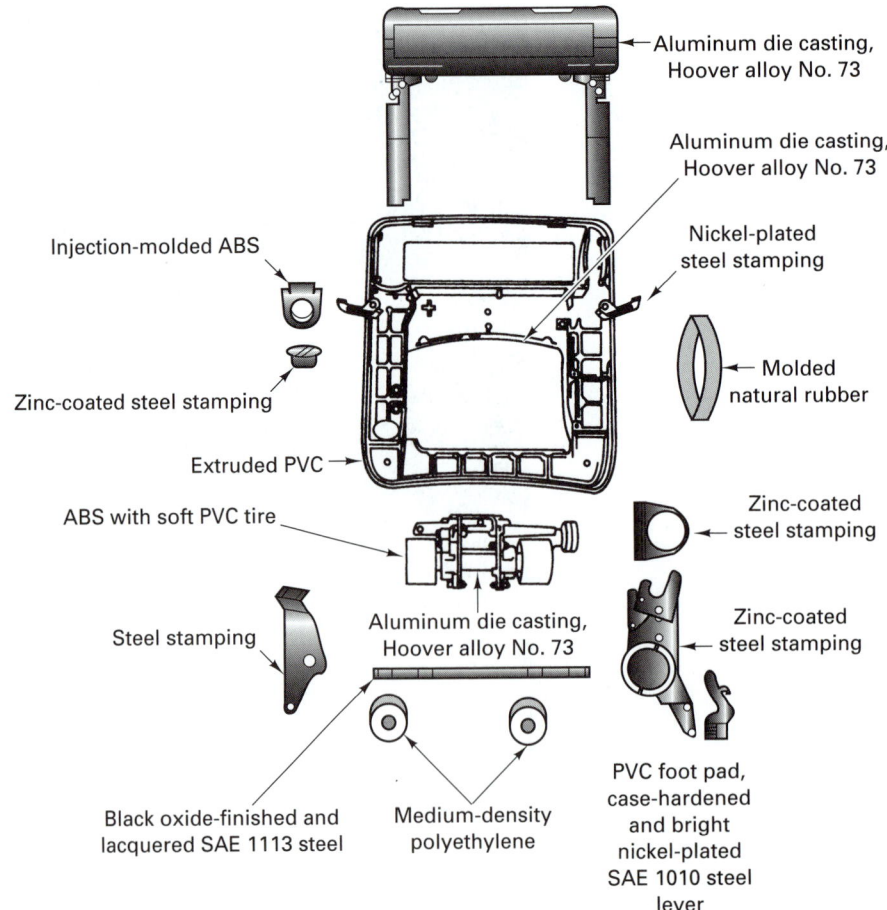

FIGURE 9-4 Materials used in various parts of a vacuum cleaner assembly. *(Courtesy of Metals Progress.)*

lists the material changes that were recommended in just one revision of the appliance. The materials for 12 components were changed completely, and that for a thirteenth was modified. Eleven different reasons were given for the changes.

The list of *engineering materials* now includes metals and alloys, ceramics, glasses, plastics, elastomers, concrete, composite materials, and others. It is not surprising, therefore, that a single person might have difficulty making the necessary decisions concerning the materials in even a simple manufactured product. The design engineer will, therefore, frequently work in conjunction with various materials specialists to select the materials that will be used to convert the designs into reality.

■ 9.2 MATERIAL SELECTION AND MANUFACTURING

The interdependence between materials and their processing must also be recognized. New processes are frequently associated with new materials, and their implementation can often cut production costs and improve product quality. Therefore, a change in material may well require a change in the manufacturing process, and improvements in processes will often lead to a reevaluation of the materials being processed. Improper processing of a

TABLE 9-1. Examples of Material Selection and Substitution in a Modern Vacuum Cleaner			
Part	Former Material	New Material	Benefits
Bottom plate	Assembly of steel stampings	One-piece aluminum die casting	More convenient servicing
Wheels (carrier and caster)	Molded phenolic	Molded medium-density polyethylene	Reduced noise
Wheel mounting	Screw-machine parts	Preassembled with a cold-headed steel shaft	Simplified replacement, more economical
Agitator brush	Horsehair bristles in a die-cast zinc or aluminum brush back	Nylon bristles stapled to a polyethylene brush back	Nylon bristles last seven times longer and are now cheaper than horsehair
Switch toggle	Bakelite molding	Molded ABS	Breakage eliminated
Handle tube	AISI 1010 lock-seam tubing	Electric seam-welded tubing	Less expensive, better dimensional control
Handle bail	Steel stamping	Die-cast aluminum	Better appearance, allowed lower profile for cleaning under furniture
Motor hood	Molded cellulose acetate (replaced Bakelite)	Molded ABS	Reasonable cost, equal impact strength, much improved heat and moisture resistance: eliminated warpage problems
Extension-tube spring latch	Nickel-plated spring steel, extruded PVC cover	Molded acetal resin	More economical
Crevice tool	Wrapped fiber paper	Molded polyethylene	More flexibility
Rug nozzle	Molded ABS	High-impact styrene	Reduced costs
Hose	PVC-coated wire with a single-ply PVC extruded covering	PVC-coated wire with a two-ply PVC extruded covering separated by a nylon reinforcement	More durability, lower cost
Bellows, cleaning-tool nozzles, cord insulation, bumper strips	Rubber	PVC	More economical, better aging and color, less marking

Source: Metal Progress, by permission.

well-chosen material can often result in a defective product. Thus, if satisfactory results are to be achieved, considerable care must be exercised in selecting both the engineering materials and the manufacturing processes used to produce the product.

■ 9.3 DESIGN PROCESS

The first step in the manufacture of any product is design, which usually takes place in several distinct stages: (1) conceptual, (2) functional, and (3) production. During the *conceptual-design* stage, the designer is concerned primarily with the functions that the product is to fulfill. Several concepts are often considered, and a determination is made that the concept is either not practical, or is sound and should be developed further. Here the only concern about materials is that materials exist that can provide the desired properties. If such materials are not available, consideration is given to whether there is a reasonable prospect that new ones could be developed within the limitations of cost and time.

At the *functional-* or *engineering-design* stage, a workable design is developed, including a detailed plan for manufacturing. Geometric features are determined and dimensions are specified, along with allowable tolerances. Specific materials are selected for each component. Consideration is given to appearance, cost, reliability, producibility, and serviceability, in addition to the various functional factors. Often, a *prototype* or working model is constructed to permit a full evaluation of the product. It is possible that the prototype evaluation will show that some changes have to be made in either the design or material before the product can be advanced to production. This should not be taken, however, as an excuse for not doing a thorough job of material selection. There is much merit to the practice that all prototypes be built with the same materials that will be used in production and, where possible, with the same manufacturing techniques. It is of little value to have a perfectly functioning prototype that cannot be manufactured economically in the desired volume, or one that is substantially different from what the production units will be like.

It is important to have a complete understanding of the functions and performance requirements of each component and to follow this with a thorough materials analysis, selection, and specification. If these decisions are deferred, they may end up being made by individuals who are less knowledgeable about all of the functional aspects of the product.

In the *production-design* stage, the materials concerns should be directed to whether the specified materials are compatible with the manufacturing processes and equipment. Can they be processed economically, and are they available in the necessary quantities and quality?

As actual manufacturing begins, changes in both the materials and processes may be suggested. In most cases, however, changes made at this stage (after the tooling and machinery are placed in production) tend to be quite costly. Good material selection and thorough product evaluation can do much to eliminate the need for change.

The subsequent availability of new materials and new processes, however, may present possibilities for cost reduction or improved performance. Before adopting new materials, however, they should be evaluated very carefully to assure that all of their characteristics are well established. Remember that it is indeed rare that as much is known about the properties and reliability of a new material as about an established one. Numerous product failures and product liability cases have resulted from new materials being substituted before their long-term properties were fully known.

■ 9.4 PROCEDURE FOR MATERIAL SELECTION

The selection of an appropriate material and its subsequent conversion into a useful product with desired shape and properties can be a rather complex process. Nearly every engineered item goes through the following sequence of activities: design → material selection → process selection → production → evaluation → and possible redesign or modification. Numerous engineering decisions are made along the way.

Several methods have been developed for approaching a design and selection problem. The *case-history method* assumes that if something has worked successfully before, a similar component could be made with the same engineering material and method of manufacture. This approach is quite useful, and many manufacturers evaluate their competitors' products for just this purpose. Nevertheless, minor variations in service requirements may well require different materials or different manufacturing operations. Moreover, this approach tends to preclude the use of new technology, new materials, and other manufacturing advances that have occurred since the formulation of the previous solution. It is equally unwise, however, to totally ignore the benefits of past experience.

Other design and selection activities involve the *modification of an existing product*, generally in an effort to reduce cost or to improve quality. Efforts here can begin with an evaluation of the current product and its present method of manufacture. The most frequent pitfall is to overlook one of the original design requirements and recommend a change that in some way compromises the total performance of the product. Examples of such oversights are provided in Section 9.8.

The safest and most thorough method for design and selection is to approach the task as though it were the *development of an entirely new product.* Here the full sequence of design, material selection, and process selection will be followed prior to production.

The *first step* in any material selection problem is to *define the needs of the product.* Without prior biases about material or method of fabrication, the engineer should develop a clear picture of all the characteristics necessary for this part to adequately perform its intended function. These requirements will fall into three major areas: (1) shape or geometry considerations, (2) property requirements, and (3) manufacturing concerns.

The *shape considerations* primarily influence decisions relating to the proposed method or methods of fabrication. Although the features of part geometry may be somewhat obvious, geometric considerations are often more complex than first imagined. Typical questions include:

1. What is the relative size of the component?
2. How complex is its shape? Are there any axes or planes of symmetry? Are there any uniform cross sections? Could the component be divided into several simpler shapes that might be easier to manufacture?
3. How many dimensions must be specified?
4. How precise must these dimensions be? Are all precise? How many are restrictive, and which ones?
5. How does this component interact geometrically with other components? Are there any restrictions imposed by this interaction?
6. What are the surface finish requirements? Must all surfaces be finished? Which ones do not?
7. How much can a dimension change by wear or corrosion and the part still function adequately?
8. Could a minor change in part geometry increase the ease of manufacture or improve the performance (fracture resistance, fatigue resistance, etc.) of the part?

The various *property requirements* strongly influence the selection of both the material and the method of fabrication. Some of the aspects that should be considered include:

MECHANICAL PROPERTIES

1. How much static strength is required?
2. If the part is overloaded, it might fail by either plastic deformation or fracture. Is one method preferred or unacceptable?
3. Do you anticipate any impact loadings? If so, of what type and magnitude?
4. Can you envision cyclic loadings? If so, of what type, magnitude, and frequency?
5. Is wear resistance needed? Where? How much? How deep?
6. Over what temperature range must these properties be present? Which properties are needed at the lowest extreme? At the highest?
7. How much can the material bend, stretch, or compress and still function properly?

PHYSICAL PROPERTIES

1. Are there any electrical requirements? Conductivity? Resistivity?
2. Are any magnetic properties desired?
3. Are thermal properties significant? Thermal conductivity? Changes in dimension with change in temperature?
4. Are there any optical requirements?
5. Is weight a significant factor?
6. How important is appearance? Is there a preferred color, texture, or "feel"?

Another important area to evaluate is the *service environment* of the product throughout its lifetime:

1. What are the lowest, highest, and normal operating temperatures for the component? Will temperature changes be cyclic? How fast will temperature changes occur?
2. What is the most severe environment that is anticipated as far as corrosion or deterioration of material properties is concerned?
3. What is the desired service lifetime for the product?
4. What is the anticipated level of inspection and maintenance?
5. What is the potential liability if the product should fail?
6. Should the product be manufactured with disassembly or recyclability in mind? Are there any disposal concerns?

A final area of concern is a variety of factors that will directly influence the method of manufacture. Some of these *manufacturing concerns* are:

1. How many of the components are to be made? At what rate?
2. Have standard sizes and shapes been specified wherever possible (both as finished shapes and as starting raw material)?
3. Has the design addressed the requirements that will facilitate ease of manufacture? (machinability, castability, formability, weldability, hardenability)
4. What are the largest and smallest section thicknesses?
5. What is the desired level of quality compared to similar products on the market?
6. What are the quality control and inspection requirements?
7. Are there any assembly (or disassembly) concerns?

Although there is a tendency to want to jump directly to "the answer," time spent determining the various requirements will be well rewarded. It is important that *all* factors be listed and *all* service conditions and uses be considered. Many failures and product liability claims have resulted from simple engineering oversights or failure to consider the entire spectrum of conditions that a product might experience in its lifetime. As an example, consider the failure of several large electric power transformers. Fatigue cracks formed around several horizontal cooling fins that had been welded to the exterior of the casing. The loss of cooling oil through the cracks led to overheating and failure of the transformer coils. Since transformers operate under static conditions, fatigue had not been considered in the original design and material selection. When the horizontal fins were left unsupported *during shipping*, however, the resulting vibrations were sufficient to induce the fatal cracks. It is also not uncommon for the most severe corrosion environment to be experienced during shipping or storage as opposed to normal operation. Products also encounter unexpected service conditions. Numerous parts failed on earthmoving equipment used during construction of the Alaskan pipeline. This equipment had not been designed to operate in the extreme subzero temperatures that were encountered in this new environment.

Assuming that a complete and thorough evaluation has been made of required properties, it is often helpful to assign a relative importance to the various needs. Some requirements may be *absolutes*, while others may be *relative*. Absolutes are those properties where there can be no compromise. For example, if ductility is a "must," gray cast iron would have to be ruled out. If the product must possess good electrical conductivity, plastics would not be appropriate. On the basis of absolutes, many materials can be quickly eliminated from consideration.

On rare occasions, a single engineering material will emerge as the obvious choice. Most likely, several materials will meet the specific requirements for the part but with differing degrees of compatibility. Compromise and opinion now enter into the decision making, and it is important that the final choice be the "best" alternative and not overlook a major requirement. Listing and ranking of required properties will go a long way to ensure that the person making the selection has considered all the necessary factors. If no material meets the requirements or the compromises appear to be too severe, it may be necessary to redesign the product.

■ 9.5 ADDITIONAL FACTORS TO CONSIDER

When attempting to evaluate candidate materials, the engineer is often forced to utilize handbook-type data obtained through standardized materials characterization tests. It is important to note the conditions of these tests in comparison with those of the proposed application. Significant variations in factors such as temperature, rates of loading, or surface finish can lead to major changes in a material's behavior. In addition, one should keep in mind that the handbook values often represent an average or mean and that actual material properties will vary to either side of that value. Where vital information is missing or the data may not be applicable to the proposed use, one is advised to consult with the various materials producers or qualified materials engineers.

At this point in our procedure it is appropriate to introduce *cost* as an additional selection factor. Because of competition and marketing pressures, economic considerations are often as important as technological considerations. However, we have adopted the philosophy that cost should not be considered until a material has been shown to meet the necessary property requirements. If acceptable candidates have been identified, cost becomes an important part of the selection process, and both material cost and the cost of fabrication should be considered. Often, the final decision involves a compromise between cost, ease of fabrication, and performance or quality. Numerous questions might be asked, such as:

1. Is the material too expensive to meet the marketing objectives?
2. Should a more expensive material be used if it offers improved performance?
3. How much additional expense might be justified to gain ease of fabrication?

In addition, it is important that the appropriate cost figures be considered. Most often, material costs are reported in the form of dollars per pound, or some other form of cost per unit weight. If the product has fixed size, however, comparison should probably be based on cost per unit volume. For example, aluminum has a density about one-third that of steel. For products where the size is fixed, 1 lb of aluminum can be used to produce three times as many parts as 1 lb of steel. As long as the "per pound" cost of aluminum is less than three times that of steel, aluminum will be the cheaper material. Table 9-2 presents the cost per pound and cost per cubic inch for several of the more common engineering metals. When the density of materials is quite different, as with magnesium and stainless steel, the ranking on the two different cost scales may differ greatly.

Material availability is another important consideration. The material selected may not be available in the size, quantity, or shape desired, or may not be available in any form at all. The reliability of supply may be another consideration, especially when the material or a major component of it is imported and political events may drastically alter availability. In these cases one should be prepared to recommend alternative materials, provided that they, too, are feasible candidates for the specific use.

Still other factors to be considered when making material selections include:

TABLE 9-2. Cost Comparison of Some Common Engineering Metals Showing Both Cost Per Unit Weight and Cost Per Unit Volume[*]

Material	Cost		Density (lb/in^3)
	$/lb	$/in^3	
Aluminum	0.90–1.25	0.09–0.12	0.100
Cast iron	0.20–0.25	0.05–0.06	0.250
Magnesium	2.00–2.25	0.13–0.15	0.066
Stainless steel type 304	2.00	0.57	0.287
Steel			
1015	0.30	0.09	0.285
4140 Q&T	1.25	0.36	0.285
Titanium	6.00–6.25	1.02–1.06	0.170

[*]All cost values are estimates as of 1994.

1. Are there possible misuses of the product that should be considered? If the product is to be used by the general public, one should definitely anticipate the worst. Screwdrivers are routinely used as chisels and prybars (different forms of loading than that of the torsional twist intended). Scissors may be used as wire cutters. Other products are similarly misused.

2. Have there been any failures of this or similar products? If so, what were the identified causes? Failure analysis results should definitely be made available to the designers, who can directly benefit from them.

3. Has the material (or class of materials) selected established a favorable or unfavorable performance record? Under what conditions was unfavorable performance noted?

4. Has an attempt been made to benefit from material standardization, whereby multiple components are manufactured from the same material or use the same manufacturing process? Although function, reliability, and appearance should not be sacrificed, one should not overlook the potential for savings and simplification that standardization has to offer.

■ 9.6 CONSIDERATION OF THE MANUFACTURING PROCESS

The overall value of an engineering material depends not only on its physical and mechanical properties but also on our ability to shape it into useful objects in an economical and timely manner. Without the necessary shape, parts cannot perform, and without economical production, the material will be limited to a few high-value applications. Therefore, it is now appropriate for our selection to be further refined by considering the possible fabrication processes and the suitability of each "prescreened" material to each process. Familiarity with the various manufacturing alternatives is a necessity, together with a knowledge of the associated limitations, economics, product quality, surface finish, precision, and so on. All processes are not compatible with all materials. Steel, for example, cannot be fabricated by die casting. Titanium can be forged successfully by isothermal techniques but generally not by conventional drop hammers. Wrought alloys cannot be cast, and casting alloys are not attractive for forming.

Certain fabrication processes have distinct ranges of product size, shape, and thickness, and these should be compared with the requirements of the product. Each has its characteristic precision and surface finish. Some may require prior heating or subsequent heat treatment. Still other considerations include production rate, production volume, level of desired automation, and the amount of labor required, especially if it is skilled labor. All of these concerns will be reflected in the cost of fabrication. There may also be additional constraints, such as the need to design a product so that it can be produced with existing equipment or facilities.

It is not uncommon for a certain process to be implied by the geometric details of a component design, such as the presence of cored holes, the magnitude of draft allowances, or the recommended surface finish. A good engineer, however, will consider all possible methods of manufacture and, if necessary, will recommend a change in design to accommodate a more attractive means of production.

■ 9.7 ULTIMATE OBJECTIVE

The real objective of this entire activity is to arrive at a combination of material and process (or sequence of processes) that is the "best" solution for the product. Numerous decisions will be made along the way in a manner analogous to the purchase of a new product. Consider, for example, the purchase of a new toaster at a discount department store. Is the two-slice model adequate, or do you want or need the four-slice model? Is the wide-slice feature (to accommodate waffles and bagels) desirable or necessary? How important is appearance? What level of quality is preferred? If your specific needs can be met by models that are "good," "better," and "best", is the added expense of "better" or "best" a worthwhile investment? Various heating-element designs may be used, ranging from widely spaced heavy-duty wire wrapped around an insulator to closer-spaced wrappings or even fine coils. Do you feel that this is related to quality of performance? If so, do you have a preference? Ultimately, how much are you willing to spend, and which do you want to buy? The many decisions related to the purchase of a new car can be used to provide a similar analogy.

While the various considerations in developing an optimized *manufacturing system* have been presented in a definite, sequential manner, one should be aware that they are often rearranged and are definitely interrelated. For example, part geometry may dictate a specific method of manufacture, and material is then selected for compatibility with that process. Decisions in one area frequently impose restrictions or limitations in another. Production requirements, such as formability, weldability, or machinability, tend to restrict the candidate materials. Similarly, specific materials can limit the number of fabrication possibilities. Each fabrication method imparts characteristic properties to the material, and all of these may not be beneficial. Processes designed to alter certain properties (such as heat treatment) may affect others adversely. Economics, environment, energy, efficiency, recycling, inspection, and serviceability all tend to influence decisions.

Engineers working in this area must be capable of exercising sound judgment. They must understand the product, the materials, the manufacturing processes, and all of the various interrelations. The development of new materials, technological advances in processing methods, increased restrictions in environment and energy, and the demand for enhanced performance of existing products provide continuing challenge to these individuals and ensure their value to employers.

■ 9.8 MATERIALS SUBSTITUTION

As new technology is developed or market pressures arise, it is not uncommon for new materials to be substituted into an existing design or manufacturing system. Quite often, the substitution brings about improved quality, reduced cost, ease of manufacturing, simplified assembly, or enhanced performance. When making a material substitution, however, it is also possible to overlook certain requirements and cause more harm than good.

Consider the efforts related to the production of a lighter-weight, more-fuel-efficient automobile. The development of high-strength low-alloy steel sheets (HSLA) provided the opportunity to match the strength of the traditional body panels with thinner-gage material. Having overcome some of the early fabrication problems, the substitution appeared to be a natural one. However, it is important that the engineer consider the total picture and be aware of any possible compromises. While strength was increased, corrosion resistance and elastic stiffness (rigidity) were essentially unaltered. Thinner sheets would corrode in a shorter time and undesirable vibrations could become a significant problem. Design modifications would probably be necessary to accommodate the new material.

Aluminum castings might be considered as a possible replacement for cast iron transmission housings. Corrosion resistance would be enhanced and weight savings would be substantial. However, the mechanical properties must be assured to be adequate, and consideration would have to be given to the area of noise and vibration. Cast iron is a damping material and eliminates these undesirable features. Aluminum transmits noise and vibration, and its use would probably require the addition of some form of sound isolation material. Cast iron engine blocks do an excellent job of damping out vibrations. When aluminum was used for this application, vibrations became a problem, and a companion redesign of the motor mounts was required.

Polymeric materials have been used successfully for body panels, bumpers, fuel tanks, pumps, and housings. Composite-material driveshafts have been used in place of metal. Cast crankshafts have replaced forgings. Cast metal, powder metallurgy, and composite materials have all been used for connecting rods. Ceramic and reinforced plastic components have been used on engines. Numerous other examples could be cited. When making a materials substitution, the responsible engineer should first consider *all* of the design requirements. Approaching a design or material modification as thoroughly as one approaches a new problem may well avoid costly errors.

■ 9.9 EFFECT OF PRODUCT LIABILITY ON MATERIALS SELECTION

Product liability actions, court awards, and rising insurance costs have made it imperative that designers and companies employ the very best procedures in selecting materials. Although many people would agree that the situation has grown to absurd proportions, there have also been many unfortunate instances where sound procedures were not used in selecting materials and methods of manufacture. No designer or engineer can afford to be so negligent.

The five most common causes of product liability losses have been:

1. Failure to know and use the latest and best information about the materials being specified
2. Failure to foresee, and account for, *all reasonable uses* of the product (*Note*: The designer is also advised to consider the entire spectrum of possible misuses, since a number of product liability cases have been lodged where the claimant was injured during misuse of a product)
3. Use of materials for which there were insufficient or uncertain data, particularly with regard to long-term properties
4. Inadequate, and unverified, quality control procedures
5. Material selection made by people who were completely unqualified

An examination of the faults listed above will lead one to conclude that there is no good reason for them to exist. Consideration of each of them, however, is good practice when seeking to assure the production of a good-quality product and can greatly reduce the number and magnitude of product liability claims.

■ 9.10 Aids to Material Selection

From the discussion in this chapter, it is apparent that those who select materials should have a broad, basic understanding of the nature and properties of materials and their processing characteristics. Providing this background is a primary purpose of this book. However, the number of engineering materials is so great, and the mass of information that is both available and useful is so large, that a single book of this type and size cannot be expected to furnish all that is required. Anyone who does much work in material selection needs to have ready access to many sources of data.

One very useful reference is the "Materials Selector" issue of *Materials Engineering*. This annual issue of the monthly magazine provides tabulated data about a number of engineering materials, as well as typical uses, process capabilities, and the compatibility between materials and processes.

Another "must" is the materials-related volumes of the *Metals Handbook*, published by the ASM International. The various volumes provide information on classes of engineering metals or manufacturing processes. The one-volume *Metals Handbook Desk Edition* (1984) provides similar information in a less voluminous, more concise format. The ASM *Engineered Materials Handbook* series provides similar information for the nonmetallic materials. Various volumes address composites, plastics, adhesives, and ceramics.

ASM also offers a one-volume *ASM Metals Reference Book* (3rd edition, 1993) that provides extensive data about metals and metalworking in tabular or graphic form. The *ASM Engineered Materials Reference Book* (2nd edition, 1993) provides similar material for ceramics, plastics, composites, and electronic materials. Additional handbooks often focus on specific materials, such as titanium alloys, stainless steels, and tool steels. Various technical magazines often provide special issues that serve as databooks. Some of these include *Modern Plastics, Industrial Ceramics*, and ASM's *Advanced Materials and Processes*.

In addition to these references, persons selecting materials should have available several of the handbooks published by various technical societies and trade associations. They may be material related (such as the Aluminum Association's *Aluminum Standards and Data* and the Copper Development Association's *Standards Handbook: Copper, Brass, and Bronze*), process-related (such as the *Steel Castings Handbook* by the Steel Founder's Society of America and the *Heat Treater's Guide* by the American Society for Metals), or profession related (such as the *SAE Handbook* by the Society of Automotive Engineers, *ASME Handbook* by the American Society for Mechanical Engineers, and the *Tool and Manufacturing Engineers Handbook* by the Society for Manufacturing Engineers). These may be supplemented further by a variety of supplier-published references. Although the latter are excellent, low-cost references, the user should recognize that they are clearly focused on the products of the publisher and may not provide a truly objective viewpoint.

It is also important to have accurate information on the cost of various materials. Since these tend to fluctuate, it may be necessary to consult a daily or weekly publication such as the *American Metal Market* newspaper. Costs associated with various processing operations are more difficult to obtain and can vary greatly from one company to another. These costs may be available from within the firm or may have to be estimated from outside sources. One valuable reference is the *American Machinist Cost Estimator* or the associated computer software.

Rating chart for selecting materials

Material	Go-No-Go** screening			Relative rating number (†rating number x *weighting factor)									Material rating number
	Corrosion	Weldability	Brazability	Strength (5)*	Toughness (5)	Stiffness (5)	Stability (5)	Fatigue (4)	As-welded strength (4)	Thermal stresses (3)	Cost (1)		Σ rel rating no. / Σ rating factors

*Weighting factor (range = 1 lowest to 5 most important)
† Range = 1 poorest to 5 best
**Code = S = satisfactory
          U = unsatisfactory

FIGURE 9-5   Rating chart that may be used for comparing materials for a specific application.

In addition to the various information sources, tools have been developed to assist in identifying the best candidates for a given application. One simple tool is a comparison chart, such as the one shown in Figure 9-5. The various desired properties are weighted as to their significance, and candidate materials are evaluated on a scale such as 1 to 5 or 1 to 10 with regard to their ability to provide that property. A *rating number* is then computed by multiplying the property rating by its weighted significance and summing the results. Potential materials can then be compared in a uniform, unbiased manner, and the best candidates can often be identified. In addition, by placing all the requirements on a single sheet of paper, the designer is less likely to overlook a major requirement. Finalist materials should then be reevaluated to assure that no key requirement has been overlooked or excessively compromised.

With the development of high-speed computers with large amounts of storage, computerized materials selection is now a reality. Programs such as ASM's Metal Selector and a variety of on-line databases are now readily available. The various property requirements can be set (including compatibility with the desired processing methods) and the entire spectrum of engineering materials can be searched to identify possible candidates. Search parameters can then be tightened or relaxed so as to produce a desired number of finalist materials. In addition, a much broader range of materials can be considered than would be possible by an individual performing a manual selection. Nevertheless, it is still important for the final materials to be reevaluated to assure full compliance with the needs of the product.

■ KEY WORDS

absolute requirement
case history
conceptual design
functional design
material availability

material selection
material substitution
product liability
production design
prototype

recyclability
relative requirements
service environment

■ REVIEW QUESTIONS

1. In a manufacturing environment, why should the selection and use of engineering materials be a matter of constant reevaluation?
2. What are some of the recent shifts in the materials used in a family automobile? How do the materials in use today compare to those used in cars 50 or 75 years ago?
3. How have some of the new demands on aerospace vehicles prompted a change in materials and the development of new ones?
4. Discuss the interrelation between engineering material and the fabrication processes used to produce the desired shape and properties.
5. What are the three primary phases of product design, and how does the consideration of materials differ in each?
6. What is the benefit of requiring prototype products to be manufactured from the same materials that will be used in production and by the same manufacturing techniques?
7. What precautions should be exercised when considering a new material or process for the manufacture of a product?
8. What are some of the possible pitfalls in the case-history approach to materials selection?
9. What is the most frequent problem when seeking to improve an existing product?
10. What should be the first step in any materials selection problem?
11. What are some of the important aspects of the

service environment to be considered when selecting an engineering material?
12. Why is it important to resist jumping to "the answer" and first perform a thorough evaluation of product needs and requirements?
13. What is the difference between an absolute and relative requirement?
14. What are some possible pitfalls when using handbook data to assist in materials selection?
15. Why might it be appropriate to defer cost considerations until after evaluating the performance capabilities of various engineering materials?
16. Give an example of a product or component where material cost should be compared on a cost per pound basis. Give a contrasting example where cost per unit volume would be more appropriate.
17. In what way might failure analysis data be useful in a material selection decision?
18. Why should consideration of the various fabrication process possibilities be included in material selection?
19. Why is it important to consider all of the design and service requirements when considering material substitution in an existing product?
20. What are some of the most common causes of product liability losses?
21. Why might a single-page comparative rating chart be a useful tool in materials selection?

■ PROBLEMS

1. Three materials, X, Y, and Z, are available for a certain use. Any material selected must have good weldability. Tensile strength, stiffness, stability, and fatigue strength have also been identified as key requirements. Fatigue strength is considered the most important of these requirements, and stiffness is least important. The three materials can be rated as follows:

	X	Y	Z
Weldability	Excellent	Poor	Good
Tensile strength	Good	Excellent	Fair
Stiffness	Good	Good	Good
Stability	Good	Excellent	Good
Fatigue strength	Fair	Good	Excellent

Use a rating chart such as the that in Figure 9-5 to determine which material you would recommend.

*C*hapter 9 CASE STUDY

material selection

T his study is designed to get you to question why parts are made from a particular material and how they could be fabricated to their final shape. For one or more of the products listed below, write a brief evaluation which addresses the following questions.

QUESTIONS:

1. What are the normal use or uses of this product or component? What are its normal operating conditions in terms of temperatures, loadings, impacts, corrosive media, etc.?
2. What are the major properties or characteristics that the material must possess in order for the product to function?
3. What material (or materials) would you suggest and why?
4. How might you propose to fabricate this product?
5. Would the product require heat treatment For what purpose? What kind of treatment?
6. Would this product require any surface treatment or coating? For what purpose? What would you recommend?
7. Would there be any concerns relating to environment? Recycling? Product liability?

PRODUCTS:

A. The head of a carpenter's claw hammer
B. The lid for a top-loading washing machine
C. Residential interior door knob
D. A paper clip
E. A thumb tack
F. A pair of scissors
G. A moderate-to-high-quality household cookpot
H. Case for a jeweler-quality wrist watch
I. Jet engine turbine blade to operate in the exhaust region of the engine
J. Standard open-end wrench
K. A socket-wrench socket to install and remove spark plugs
L. The frame of a 10-speed bicycle
M. Interior panels of a microwave oven
N. Handle segments of a retractable blade utility knife with internal storage for additional blades
O. The outer skin of an automobile muffler
P. The interior crank handle for an automobile window
Q. The basket section of a grocery-store shopping cart
R. The body of a child's toy wagon
S. Decorative handle for a kitchen cabinet
T. Automobile engine block
U. The motor housing for a chain saw
V. Nails for the installation of aluminum siding on homes
W. Household dinnerware (knife, fork and spoon)
X. The blades on a high-quality cutlery set
Y. A shut-off valve for a 1/2-inch household water line
Z. The base plate (with heating element) for an electric steam iron

PART 2

MEASUREMENT AND QUALITY ASSURANCE

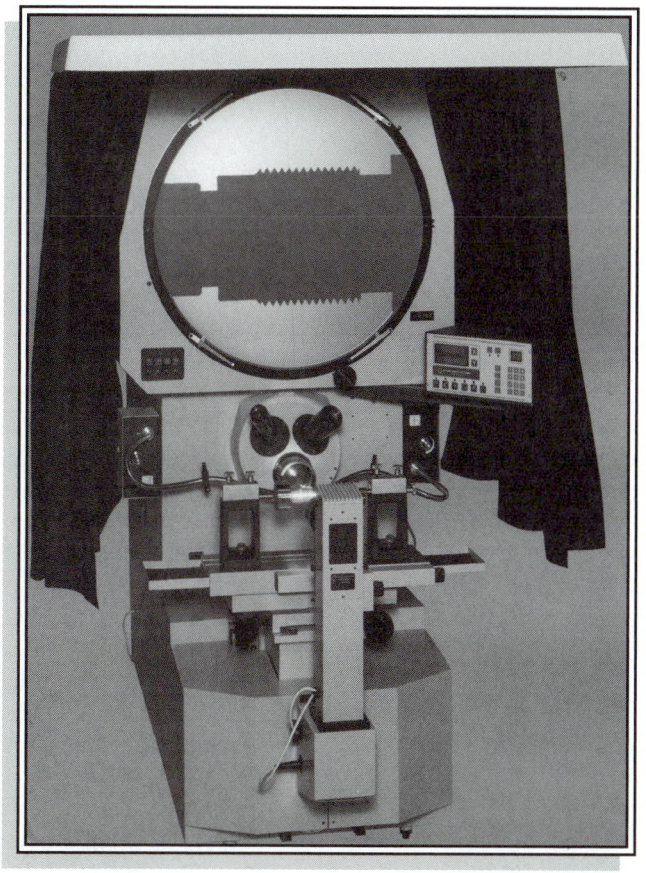

10 Measurement and Inspection

11 Nondestructive Measurement and Testing

12 Process Capability and Quality Control

CHAPTER 10

MEASUREMENT AND INSPECTION

10.1 INTRODUCTION
Attributes versus Variables

10.2 STANDARDS OF MEASUREMENT
Linear Standards
Length Standards in Industry
Standard Measuring Temperature

10.3 ALLOWANCE AND TOLERANCE
Specifying Tolerance and Allowance
Accuracy versus Precision in Processes
Geometric Tolerances

10.4 INSPECTION METHODS FOR MEASUREMENT
Factors in Selecting Inspection Equipment

10.5 MEASURING INSTRUMENTS
Linear Measuring Instruments
Measuring with Light and Lasers

10.6 VISION SYSTEMS FOR MEASUREMENT

10.7 COORDINATE MEASURING MACHINES

10.8 ANGLE-MEASURING INSTRUMENTS

10.9 GAGES FOR ATTRIBUTES MEASURING
Fixed-Type Gages
Deviation Gages

10.10 SURFACE ROUGHNESS MEASUREMENT

■ 10.1 INTRODUCTION

Measurement, the act of measuring or being measured, is the fundamental activity of inspection. The intent of *inspection* is to ensure that what is being manufactured will conform to the specifications of the product. Most products are manufactured to standard sizes and shapes. For example, the base of a 60-W light bulb has been standardized so that when one bulb burns out, the next will also fit the socket in the lamp. The socket in the lamp has also been designed and made to accept the standard bulb size. Christmas tree light bulbs are made to a different standard size. Standardization is a necessity for interchangeable parts and is also important for economic reasons. A 69-W light bulb cannot be purchased because that is not a standard wattage. Light bulbs are manufactured only in standard wattages so that they can be mass produced in large volumes by high-speed automated equipment. This results in a low unit cost.

Large-scale manufacturing based on the principles of standardization of sizes and interchangeable parts became common practice early in the twentieth century. Size control must be built into machine tools and workholding devices through the precision manufacture of these machines and their tooling. The output of the machines must then be checked carefully (1) to determine the capability of specific machines and (2) for the control and maintenance of the quality of the product. A designer who specifies the dimensions and tolerances of a part quite often does so to enhance the function of the product, but the designer is also determining the machines and processes needed to make the part. Quite often, it is necessary for the design engineer to alter the design or the specifications to make the product easier or less costly to manufacture, assemble, or inspect (or all of these). Designers should always be prepared to do this provided that they are not sacrificing functionality, product reliability, or performance.

Attributes versus Variables

The examination of the product either during or after manufacture, either manually or automatically, falls in the province of *inspection*. Basically, inspection of items or products can be done in two ways:

1. By attributes, with the use of gages to determine if the product is good or bad, resulting in a yes or no, go or not-go decision
2. By variables, with the use of calibrated instruments to determine the actual dimensions of the product for comparison with the size desired

In an automobile, a speedometer and oil pressure gage are variable types of measuring instruments, and an oil pressure light is an attributes type of gage. As is typical of an attributes gage, the driver does not know *what* the pressure actually is if the light goes on, only that it is not good.

Measurement is the generally accepted industrial term for inspection by variables. *Gaging* (or *gauging*) is the term for determining whether the dimension or characteristic is large or smaller than the established standard or range of acceptability. Variable types of inspection generally take more time and are more expensive than attribute inspection, but they give more information because the magnitude of the characteristic is known in some standard unit of measurement.

■ 10.2 STANDARDS OF MEASUREMENT

The four fundamental measures on which all others depend are *length, time, mass,* and *temperature*. Three of these basic measures are defined in terms of material constants, as shown in Table 10-1, along with the original definitions. These four measures, along with

TABLE 10-1. International System of Units, Founded on Seven Base Quantities on Which All Others Depend

Quantity	Name of Base	Symbol	Definition or Comment
Length	Meter (or metre)	m	Original: 1/10,000,000 of quadrant of earth's meridian passing through Barcelona and Dunkirk
			Present: 1,650,763.73 wavelengths in vacuum of transition between energy levels $2p_{10}$ and $5d_5$ of krypton-86 atoms, excited at triple point of nitrogen ($-210°C$).
Mass	Kilogram	kg	Original: Mass of 1 cubic decimeter (1000 cubic centimeters) of water at its maximum density ($4°C$).
			Present: Mass of Prototype Kilogram No. 1 kept at International Bureau of Weights and Measures at Sèvres, France
Time	Second	s	Original: 1/86,400 of mean solar day.
			Present: 9,192,631,770 cycles of frequency associated with transition between two hyperfine levels of isotope cesium-133.
Electric current	Ampere	A	Present: The rate of motion of charge in a circuit is called the *current*. The unit of current is the *ampere*. One ampere exists when the charge flows at a rate of 1 coulomb per second.
Thermodynamic temperature	Degree Celsius (Kelvin)	°C (K)	Present: 1/273.16 of thermodynamic temperature of the triple point of water ($0.01°C$).
Amount of substance	Mole	mol	Present: A mole is an artificially chosen number ($N_0 = 6.02 \times 10^{23}$) that measures the number of molecules.
Luminous intensity	Candle	cd	Present: One lumen per square foot is a footcandle or $I = F/4$.

the *ampere* and the *candela*, provide the basis for all other units of measurement, as shown in Figure 10-1. Most mechanical measurements involve combinations of units of mass, length, and time. Thus the newton, a unit of force, is derived from Newtons's second law of motion ($f = ma$) and is defined as the force that gives an acceleration of 1 meter per second to a mass of 1 kilogram. A figure and table for metric–English conversions are provided in this chapter (see Table 10-2 and the conversion scales below.)

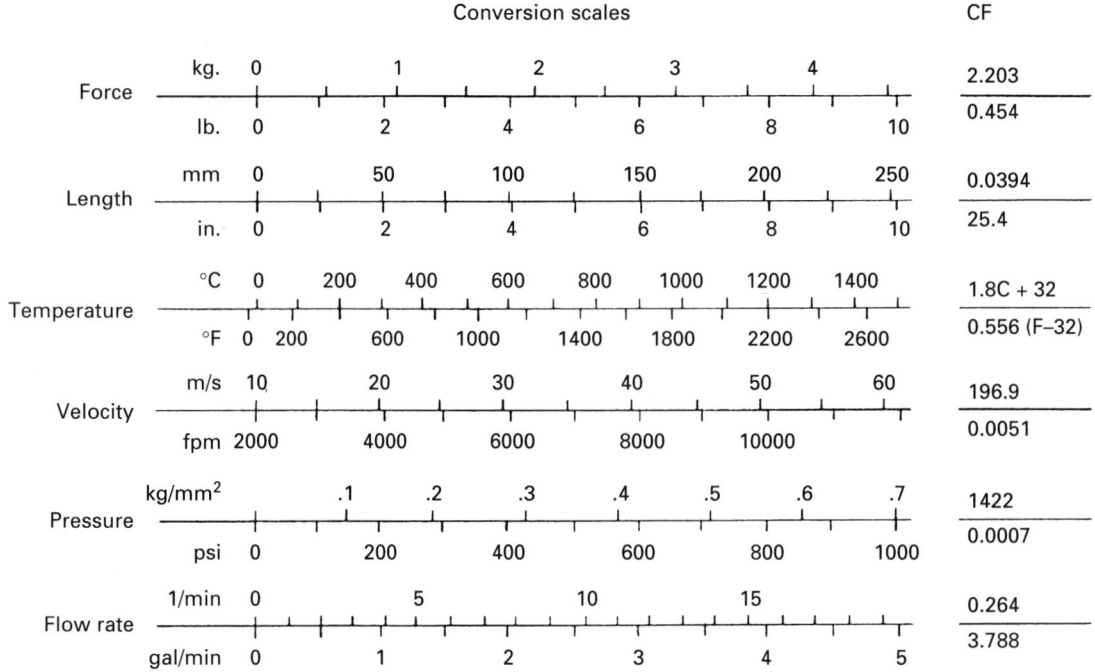

CF = conversion factor; example; 10 kg = (2.203) (10) lb., 10 in = 2.54 (10) mm.
1 Newton ≈ 0.1 kg ≈ 0.22 lb; 1 micron = 40 microinches

Set of surface-roughness standards being used in a drafting room. *(Courtesy of Surface Checking Gage Co.)*

Linear Standards

When people first sought a unit of length, they adopted parts of the human body, mainly the hands, arms, or feet. Such tools were not very satisfactory because they were not universally standard in size. Satisfactory measurement and gaging must be based on a reliable, preferably universal, standard or standards. These have not always existed. For example, although the musket parts made in Eli Whitney's shop were interchangeable, they were not interchangeable with parts made by another contemporary gun maker *from the same drawings* because the two gunsmiths *had different foot rulers.* Today, the entire industrialized world has adopted the *international meter* as the standard of linear measurement. The inch, used by both the United States and Great Britain, has been defined officially as 2.54 centimeters. Thus the U.S. standard inch is 41,929.399 wavelengths of the orange-red light from krypton-86.

Although *officially* the United States is committed to conversion to the metric (SI) system of measurement, which uses millimeters for virtually all linear measurements in

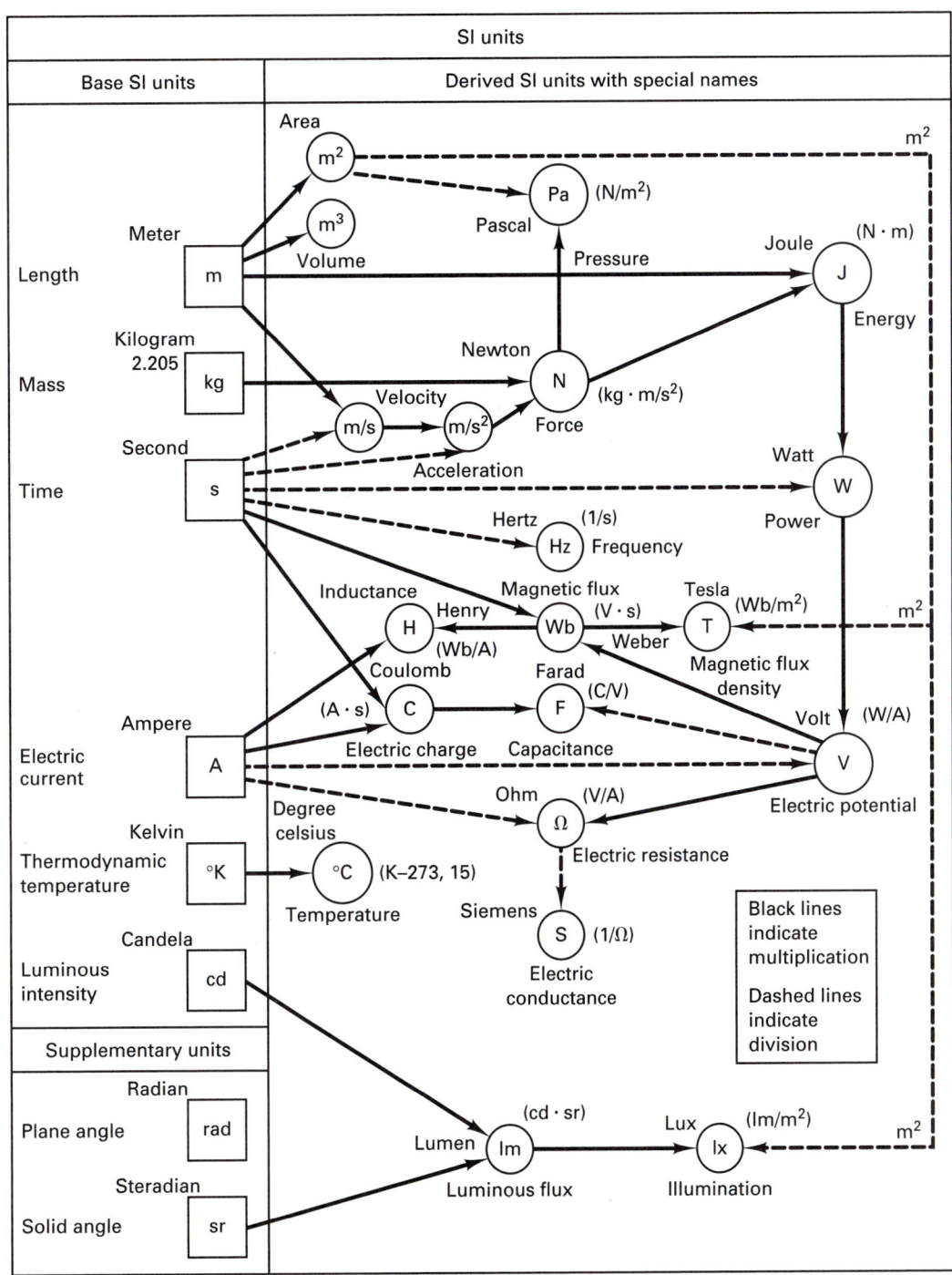

FIGURE 10-1   Relationship of secondary physical quantities to basic SI units. Solid lines, multiplication; dashed lines, division.

TABLE 10-2. Metric-English Conversions[a]

Measurement	Metric Symbol	Metric Unit	English Conversion
Linear dimensions	m	metre	1 in. = 0.0254 m[a]
			1 ft = 0.348 m[a]
	cm	centimetre	1 in. = 2.54 cm[a]
	mm	millimetre	1 in. = 25.4 mm[a]
	μm	micrometre	1 μin. = 0.0254 μm[a]
Area	m^2	square metre	1 ft^2 = 0.093 m^2[b]
	cm^2	square centimetre	1 $in.^2$ = 6.45 cm^2[b]
	mm^2	square millimetre	1 $in.^2$ = 645.16 mm^2[a]
Volume and capacity	m^3	cubic metre	1 ft^3 = 0.028 m^3[b]
	cm^3	cubic centimetre	1 in^3 = 16.39 cm^3[b]
	mm^3	cubic millimetre	1 $in.^3$ = 16.387 mm^3[b]
	L	litre	1 ft^3 = 28.32 L[b]
			1 U.S. gal = 3.79 L[b]
Velocity, acceleration, and flow	m/s	metres per second	1 ft/s = 0.3048 m/s[a]
	m/min	metres per minute	1 ft/min = 0.3048 m/mm[a]
	m/s^2	metres per second squared	1 $in./s^2$ = 0.0254 m/s^2[a]
	L/mm	litres per minute	1 ft^3/min = 28.3 L/min[b]
			1 gallon (U.S. liquid)/min = 3.758L/min[b]
Mass	g	gram	1 oz = 28.36g[b]
	kg	kilogram	1 lb = 0.45 kg[b]
	tonne	metric ton	1 short ton (2000 lb) = 0.7072 tonne[b]
Force	N	newton	1 lb-force = 4.448 N[b]
	kN	kilonewton	1 short ton-force (2000 lb) = 8.896 kN[b]
Bending moment or torque	N·m	newton-metre	1 oz-force/in. = 0.007 N·m[b]
			1 lb-force/in. = 0.113 N·m[b]
			1 lb-force/ft = 1.3558 N·m[b]
Pressure	Pa	pascal	1 lb/ft^2 = 47.88 Pa[b]
	kPa	kilopascal	1 $lb/in.^2$ = 6.895 kPa[b]
Energy, work, or quantity of heat	J	joule	1 Btu[c] = 1055.056[b]
	kJ	kilojoule	1 Btu[c] = 1.055 kJ[b]
Power	W	watt	1 hp (550 ft-lb/s) = 745.7 W[b]
			1 hp (electric) = 746 W[a]
	kW	kilowatt	1 hp (550 ft-lb/s) = 0.7457 kW[b]
Temperature	C	Celsius	degrees C = $\dfrac{\text{degrees F} - 32}{1.8}$
Frequency	Hz	hertz	1 cycle per second = 1 Hz
	kHz	kilohertz	1000 cycles per second = 1 kHz
	MHz	megahertz	1000000 cycles per second = 1 MHz

[a]Comments on the metric system:

　　The *re spelling of* metre conforms with such standards as ANZI Z210.1 ISO, Standard 1000, and recommendations of the Society of Manufacturing Engineers Metric Advisory Committee. The use of metre also avoids confusion with meter and micrometer measuring instruments.

　　Metric symbols are usually presented the same way in singular and in plural (1 mm, 100 mm), and periods are not used after symbols, except at the end of sentences. Degree (instead of radian) continues to be used for plane angles, but angles are expressed with decimal subdivisions rather than minutes and seconds. Surface finishes are specified in micrometres.

　　Accuracy of conversion. Multiplying an English measurement by an exact metric conversion factor often provides an accuracy not intended by the original value. In general, use one less significant digit to the right of the decimal point than was given in the original value; 0.032 in. (0.81) in. (0.008 mm). etc. Fractions of an inch are converted to the nearest tenth of a millimetre: $\frac{1}{8}$ in. (3.2 mm).

　　Weight, mass, and force. Confusion exists in the use of the term *weight* as a quantity to mean either force or mass. In nontechnical circles, the term *weight* nearly always means mass, and this use will probably persist. Weight is a force generated by mass under the influence of gravity. Mass is expressed in gram (g), kilogram (kg), or metric ton (t) units, and force in newton (N) or kilonewton (kN) units.

　　Pressure. Kilopascal (kPa) is the recommended unit for fluid pressure. Absolute pressure is specified either by using the identifying phrase "absolute pressure" or by adding the word *absolute* after the unit symbol, separating the two by a comma or a space.

[b]Exact.

[c]Approximate.

[d]International Table.

FIGURE 10-2   Standard set of rectangular gage blocks with ±0.000050-in. accuracy, and three individual blocks.

manufacturing, the English system of feet and inches is still being used by many manufacturing plants, and its use will probably continue for some time.

Length Standards in Industry

Gage blocks provide industry with linear standards of high accuracy that are necessary for everyday use in manufacturing plants. The blocks are small, rectangular, square, or round in cross section and are made from steel or carbide with two very flat and parallel surfaces that are certain specified distances apart (Figure 10-2). These gage blocks were first conceived by Carl E. Johansson in Sweden just before 1900. By 1911 he was able to produce sets of such blocks on a very limited scale, and they came into limited but significant use during World War I. Shortly after the war, Henry Ford recognized the importance of having such gage blocks generally available. He arranged for Johansson to come to the United States, and through facilities provided by the Ford Motor Company, methods were devised for the large-scale production of gage blocks sets. Today, gage block sets of excellent quality are produced by a number of companies in this county and abroad.

Gage blocks are made of alloy steel, hardened to $65R_C$ and carefully heat treated (*seasoned*) to relieve internal stresses and to minimize subsequent dimensional change. Carbide gage blocks provide extra wear resistance. The measuring surfaces of each block are surface ground to approximately the required dimension and are then lapped and mirror polished to bring the block to the final dimension and to produce a very flat and smooth surface. The surface finish is 0.4 millionths of an inch (0.0l μm). (Surface finish is discussed in Section 10.10.)

Gage blocks are commonly made to conform to National Bureau of Standards No. 731/222 131, having Grades 1, 2, and 3 with tolerances as follows (ANSI/ASME B89.1.OM-1984):

Grade		Inches	Millimeters	Recalibration Period
0.5	(laboratory)	±0.000001	±0.00003	Annually
1	(laboratory)	±0.000002	±0.00005	Annually
2	(precision)	+0.000004	+0.00010	Monthly to semiannually
		−0.000002	−0.00005	
3	(working)	+0.000008	+0.00020	Monthly to quarterly
		−0.000004	−0.00010	

Blocks up to 1 in. in length have absolute accuracies as stated, whereas the tolerances are per inch of length for blocks larger than 1 in. Some companies supply blocks in AA quality (which corresponds to Grade 1) and in A+ quality (which corresponds to Grade 2).

Grade 0.5 (grand-master) blocks are used as a basic reference standard in calibration laboratories. Grade 1 (laboratory-grade) blocks are used for checking and calibrating other grades of gage blocks. Grade 2 (precision-grade) blocks are used for checking Grade 3 blocks and master gages. Grade 3 (B or working-grade) blocks are used to calibrate or check routine measuring devices such as micrometers, or in actual gaging operations.

The dimensions of individual blocks are established by light-beam interferometry, with which it is possible to calibrate these blocks routinely with an uncertainty as low as 1 part per million.

Gage blocks usually come in sets containing various numbers of blocks of various sizes, such as those shown in Figure 10-2. By wringing the blocks together in various combinations, as shown in Figure 10-3, any desired dimension can be obtained. For example, here is the breakdown of blocks that one would have in an 81-block set of English No. 2 gage blocks.

9 blocks	0.1001 through 0.1009 in.	in steps of 0.0001 in.
49 blocks	0.101 through 0.149 in.	in steps of 0.001 in.
19 blocks	0.050 through 0.950 in.	in steps of 0.050 in.
4 blocks	1.000 through 4.000 in.	in steps of 1.000 in.

Gage blocks are wrung together by sliding one past another using hand pressure. They will adhere to one another with considerable force and must not be left in contact for extended periods of time. Gage blocks are available in different shapes (squares, angles, rounds, and pins), so that standards of high accuracy can be obtained to fill almost any need. In addition, various auxiliary clamping, scribing, and base block attachments are available that make it possible to form very accurate gaging devices, such as the device shown in Figure 10-4.

Standard Measuring Temperature

Because all the commonly used metals are affected dimensionally by temperature, a standard measuring temperature of 68°F (20°C) has been adopted for precision-measuring work. All gage blocks, gages, and other precision-measuring instruments are calibrated at

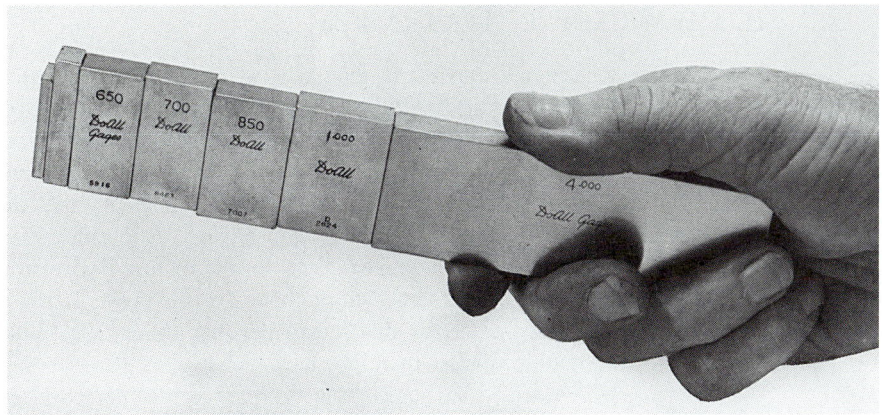

FIGURE 10-3   Seven gage blocks wrung together to build up a desired dimension. *(Courtesy of DoALL Company.)*

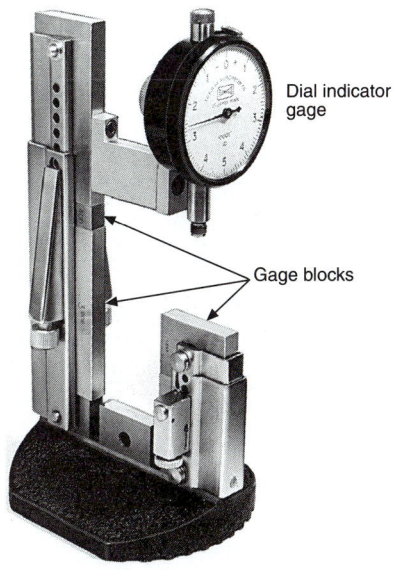

Dial indicator gage

Gage blocks

FIGURE 10-4   Wrung-together gage blocks in a special holder, used with a dial gage to form an accurate comparator. *(Courtesy of DoALL Company.)*

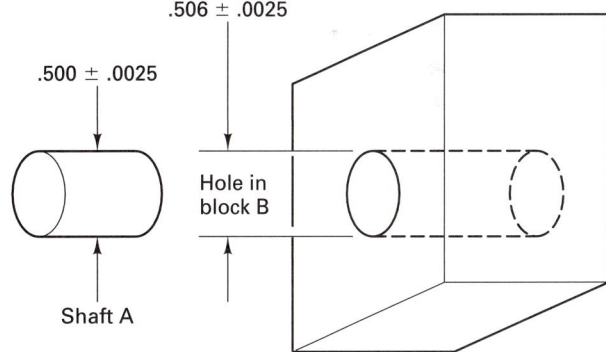

.506 ± .0025

.500 ± .0025

Hole in block B

Shaft A

FIGURE 10-5   When mating parts are designed, each shaft must be smaller than each hole for a clearance fit.

this temperature. Consequently, when measurements are to be made to accuracies greater than 0.0001 in. (0.0025 mm), the work should be done in a room in which the temperature is controlled at standard. Although it is true that to some extent both the workpiece and the measuring or gaging device *may* be affected to about the same extent by temperature variations, one should not rely on this. Measurements to even 0.0001 in. (0.0025 mm) should not be relied on if the temperature is very far from 68°F (20°C).

■ 10.3 ALLOWANCE AND TOLERANCE

Two factors, allowance and tolerance, must be specified if one is to obtain the desired fit between mating parts. *Allowance* is the intentional, desired difference between the dimensions of two mating parts. It is the difference between the dimension of the largest interior-fitting part (shaft) and that of the smallest exterior-fitting part (hole). Figure 10-5 shows Shaft A designed to fit into the hole in block B. This difference thus determines the condition of *tightest* fit between mating parts. Allowance may be specified so that either *clearance* or *interference* exists between the mating parts. In the case of a shaft and mating hole, it is the difference in diameters of the largest shaft and the smallest hole. With clearance fits, the largest shaft is smaller than the smallest hole, whereas with interference fits, the hole is smaller than the shaft.

Tolerance is an undesirable but permissible deviation from a desired dimension in recognition that no part can be made *exactly* to a specified dimension, except by chance, and that such exactness is neither necessary nor economical. Consequently, it is necessary to permit the actual dimension to deviate from the desired theoretical dimension (called the *nominal*) and to control the degree of deviation so that satisfactory functioning of the mating parts will still be ensured.

Now we can see that the objective of *inspection*, by means of measurement techniques, is to provide feedback information on the actual size of the parts with reference to the size specified by the designer on the part drawing.

The processes that make the shaft are different from those that make the hole, but both the hole and the shaft are subject to deviations in size because of variability in the processes and the materials. Thus, while the designer wishes ideally that all the shafts would be exactly 0.500 and all the holes 0.506, the reality of processing is that there will be deviations in size around these nominal or ideal sizes.

Most manufacturing processes result in products whose measurements of the geometrical features and sizes are distributed normally (Figure 10-6). That is, most of the measurements are clustered around the average dimension, $\overline{X}$. $\overline{X}$ will be equal to the nominal dimension only if the process is accurate, i.e. perfectly centered. Parts will be distributed on either side of the average. In normal distributions, 99.73% of the measurements will fall within plus or minus 3 standard deviations ($\pm 3\sigma$); 95.46% in Figure 10-6 will be within $\pm 2\sigma$, and 68.26% within $\pm 1\sigma$, where

$$\sigma = \sqrt{\frac{\sum_{i=1}^{n}(X_i - \overline{X})^2}{n}} \tag{10-1}$$

and

$$\overline{X} = \frac{\sum_{i=1}^{n} X_i}{n} \quad \text{for } n \text{ items} \tag{10-2}$$

In general, the designer applies nominal values to the mating parts according to the desired fit between the parts. Tolerances are added to those nominal values in recognition of the fact that all processes have some natural amount of variability.

Assume that the distributions for both the hole and the shaft are normally distributed. As shown in Figure 10-7, the average fit of two mating parts is equal to the difference between the mean of the shaft distribution and the mean of the hole distribution. The *range*

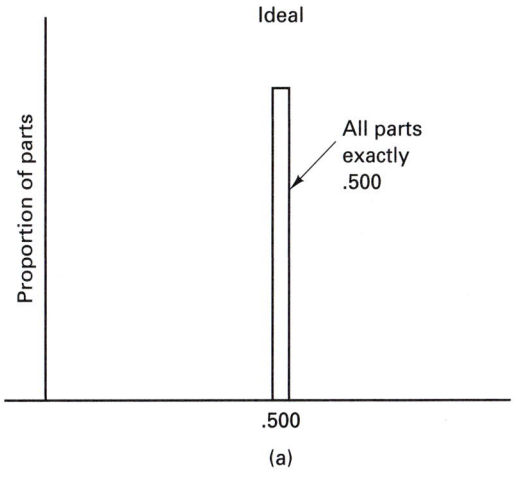

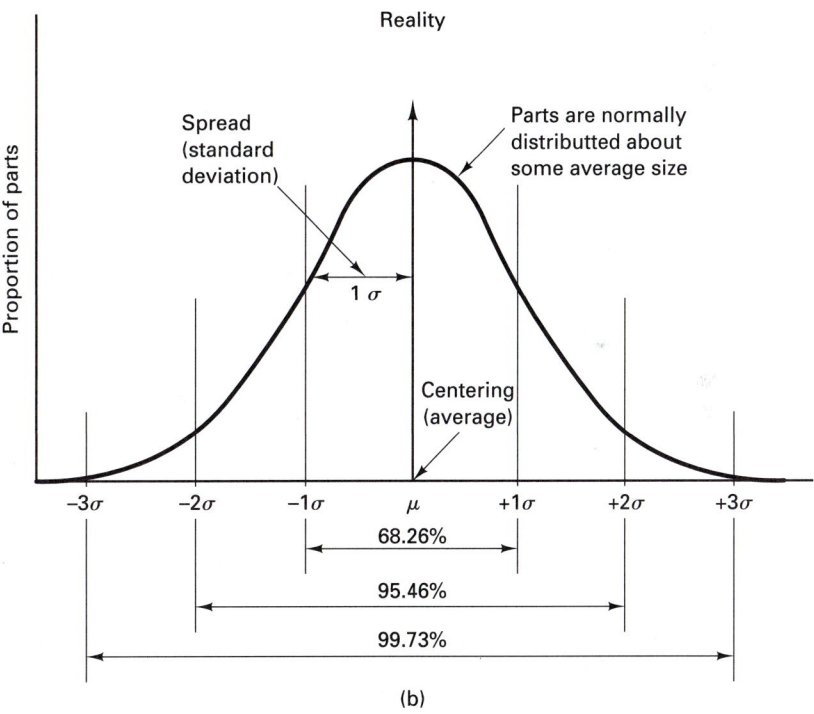

FIGURE 10-6   In the ideal situation, the process would make all parts exactly the same size. In the real world of manufacturing, parts have variability in size. The distribution of sizes is often normal.

of fit will be the difference between the minimum diameter shaft and the maximum diameter hole. The minimum *clearance* would be the difference between the smallest hole and the largest shaft. The tighter the distributions (i.e., the more precise the process), the better fit between the parts.

   If tool wear is considered, the diameter of the shaft will tend to get larger as the tool wears. On the other hand, the diameter of the hole in part B decreases in size with tool

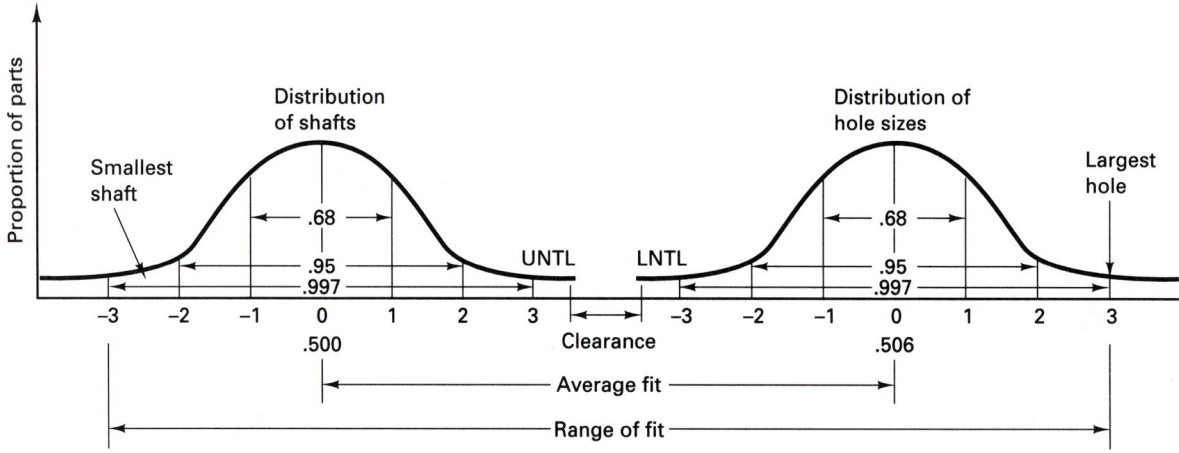

FIGURE 10-7 The manner in which the distributions of the two mating parts interact determines the fit. UNTL, upper natural tolerance limit = $\mu + 3\sigma$; LNTL, lower natural tolerance limit = $\mu + 3\sigma$.

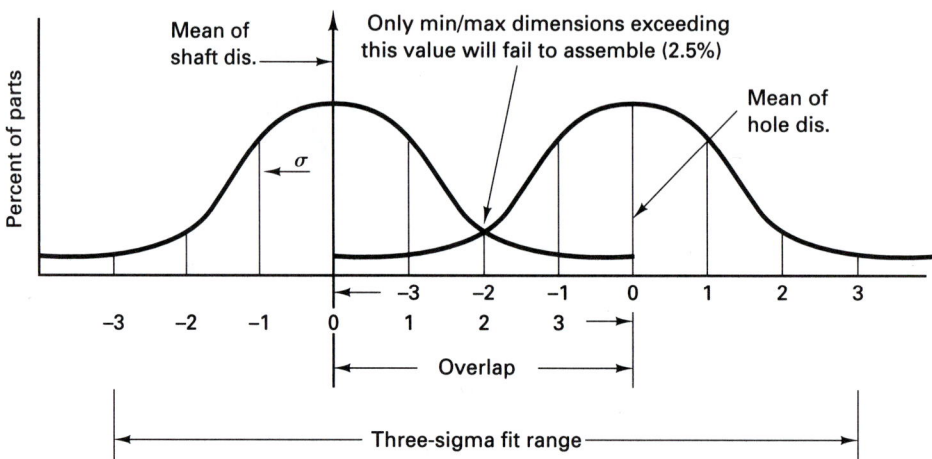

FIGURE 10-8 Shifting the means of the distributions toward each other results in some interface fits.

wear. If no corrective action is taken, the means of the distributions will shift toward each other, and the fit will become increasingly tight (the clearance will be decreased). If the hole and the shaft distributions overlap by 2 standard deviations, as shown in Figure 10-8, 6 parts out of every 10,000 could not be assembled. As the shaft and the hole distributions move closer together, more interference between parts will occur, and the fit will become tighter, eventually becoming an interference fit.

The designer must specify the tolerances according to the function of the mating parts. Suppose that the mating parts are the cap and the body of an ink pen. The cap must fit snugly but must be easily removed by hand. A snug fit would be too tight for a dead bolt in a door lock. A snug fit is also too tight for a high-speed bearing for rotational parts but is not tight enough for permanently mounting a wheel on an axle. In the next section we introduce the manner in which tolerances and allowances are specified, but design engineers are expected to have a deeper understanding of this topic.

Specifying Tolerance and Allowance

Tolerance can be specified in three ways: bilateral, unilateral, and limits. *Bilateral* tolerance is specified as a plus or minus deviation from the nominal size, such as 2.000 ± 0.002 in. More modern practice uses the *unilateral* system, where the deviation is in one direction from the basic size, such as

$$2.000 \text{ in.} \quad \begin{matrix} +0.004 \text{ in.} \\ -0.000 \text{ in.} \end{matrix} \quad \text{or} \quad 50.8 \text{ mm} \quad \begin{matrix} +0.1 \text{ mm} \\ -0.0 \text{ mm} \end{matrix} \quad \text{in metric}$$

In the first case, that of bilateral tolerance, the dimension of the part could vary between 1.998 and 2.002 in., a total tolerance of 0.004 in. For the example of unilateral tolerance, the dimension could vary between 2.000 and 2.004 in., again a tolerance of 0.004 in. Obviously, to obtain the same maximum and minimum dimensions with the two systems, different basic sizes must be used. The maximum and minimum dimensions that result from application of the designated tolerance are called *limit dimensions*, or *limits*.

There can be no rigid rules for the amount of clearance that should be provided between mating parts; the decision must be made by the designer, who considers how the parts are to function. The American National Standards Institute, Inc. (ANSI) has established eight classes of fits that serve as a useful guide in specifying the allowance and tolerance for typical applications and that permit the amount of allowance and tolerance to be determined merely by specifying a particular class of fit. These classes are as follows:

- CLASS 1. *Loose fit:* large allowance. Accuracy is not essential.
- CLASS 2. *Free fit:* liberal allowance. For running fits where speeds are above 600 rpm and pressures are 600 psi (4.1 MPa) or above.
- CLASS 3. *Medium fit:* medium allowance. For running fits below 600 rpm and pressures below 600 psi (4.1 MPa) and for sliding fits.
- CLASS 4. *Snug fit:* zero allowance. No movement under load is intended, and no shaking is wanted. This is the tightest fit that can be assembled by hand.
- CLASS 5. *Wringing fit:* zero to negative allowance. Assemblies are selective and not interchangeable.
- CLASS 6. *Tight fit:* slight negative allowance. An interference fit for parts that must not come apart in service and are not to be disassembled or are to be disassembled only seldom. Light pressure is required for assembly. Not to be used to withstand other than very light loads.
- CLASS 7. *Medium force fit:* an interference fit requiring considerable pressure to assemble; ordinarily assembled by heating the external member or cooling the internal member to provide expansion or shrinkage. Used for fastening wheels, crank disks, and the like to shafting. The tightest fit that should be used on cast iron external members.
- CLASS 8. *Heavy force and shrink fits:* considerable negative allowance. Used for permanent shrink on steel members.

The allowances and tolerances that are associated with the ANSI classes of fits are determined according to the theoretical relationship shown in Table 10-3. The actual

TABLE 10-3.	ANSI Recommended Allowances and Tolerances			
Class of Fit	Allowance	Average Interference	Hole Tolerance	Shaft Tolerance
1	$0.0025 \sqrt[3]{d^2}$	—	$+ 0.0025 \sqrt[3]{d}$	$- 0.0025 \sqrt[3]{d}$
2	0.0014	—	+ 0.0013	− 0.0013
3	0.0009	—	+ 0.0008	− 0.0008
4	0	—	+ 0.0006	− 0.0004
5	—	0	+ 0.0006	+ 0.0004
6	—	0.00025d	+ 0.0006	+ 0.0006
7	—	0.0005d	+ 0.0006	+ 0.0006
8	—	0.001d	+ 0.0006	+ 0.0006

resulting dimensional values for a wide range of basic sizes can be found in tabulations in drafting and machine design books.

In the ANSI system, the hole size is always considered basic, because the majority of holes are produced through the use of standard-size drills and reamers. The internal member, the shaft, can be made to any one dimension as readily as to another. The allowance and tolerances are applied to the basic hole size to determine the limit dimensions of the mating parts. For example, for a basic hole size of 2 in. and a class 3 fit, the dimensions would be:

Allowance	0.0014 in.
Tolerance	0.0010 in.
Hole	
Maximum	2.0010 in.
Minimum	2.0000 in.
Shaft	
Maximum	1.9986 in.
Minimum	1.9976 in.

It should be noted that for both clearance and interference fits, the permissible tolerances tend to result in a looser fit.

The *ISO System of Limits and Fits* (Figure 10-9) is widely used in a number of leading metric countries. This system is considerably more complex than the ANSI system just discussed. In this system each part has a *basic size*. Each limit of size of a part, high and low, is defined by its *deviation* from the basic size, the magnitude and sign being obtained by subtracting the basic size from the limit in question. The difference between the two limits of size of a part is called *tolerance*, an absolute amount without sign.

There are three classes of fits: (1) *clearance fits*; (2) *transition fits* (the assembly may have either clearance or interference); and (3) *interference fits*. Either a *shaft-* or *hole-basis system* may be used (Figure 10-10). For any given basic size, a range of tolerances and deviations may be specified with respect to the line of zero deviation, called the *zero line*. The tolerance is a function of the basic size and is designated by a number symbol, called the *grade* (e.g., the *tolerance grade*). The *position* of the tolerance with respect to the zero line, also a function of the basic size, is indicated by a letter symbol (or two letters), a

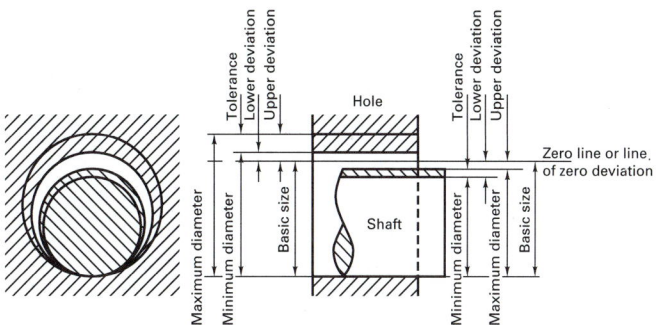

FIGURE 10-9   Basic size, deviation, and tolerance in the ISO system. (By permission from Recommendations R286-1962, *System of Limits and Fits,* copyright 1962, American Standards Institute, New York.)

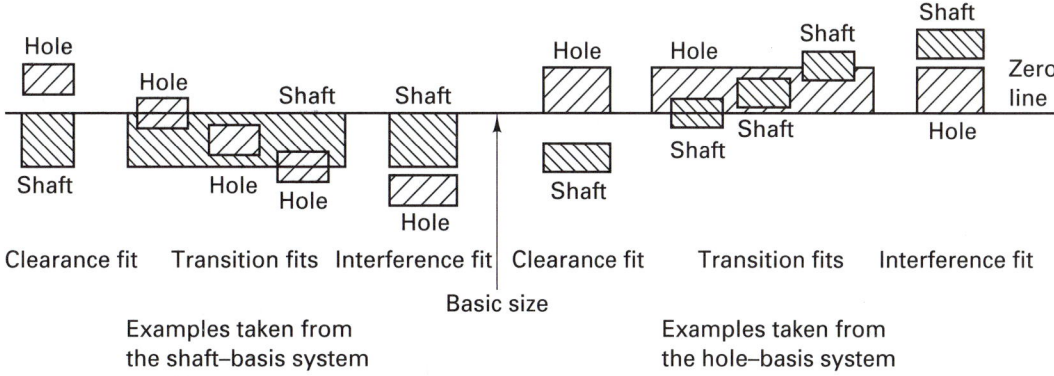

FIGURE 10-10   Shaft-basis and hole-basis system for specifying fits in the ISO system. (By permission from Recommendations R286-1962, *System of Limits and Fits,* copyright 1962, American Standards Institute, New York.)

capital letter for holes, and a lowercase letter for shafts (Figure 10-11).[*] Thus the specification for a hole and a shaft having a basic size of 45 mm might be 45 H8/g7.

Eighteen standard grades of tolerances are provided, called IT 01, IT 0, and IT 1 through IT 16, providing numerical values for each nominal diameter, in arbitrary steps up to 500 mm (e.g., 0–3, 3–6, 6–10, …, 400–500 mm). The value of the tolerance unit, i, for grades 5 through 16 would be

$$i = 0.45 \sqrt[3]{d} + 0.001\, D$$

where i is in micrometers and D in millimeters.

Standard shaft and hole deviations are provided by similar sets of formulas. However, for practical applications, both tolerances and deviations are provided in three sets of rather complex tables. Additional tables give the values for basic sizes above 500 mm, and

---

[*] It will be recognized that the "position" in the ISO system essentially provides the "allowance" of the ANSI system.

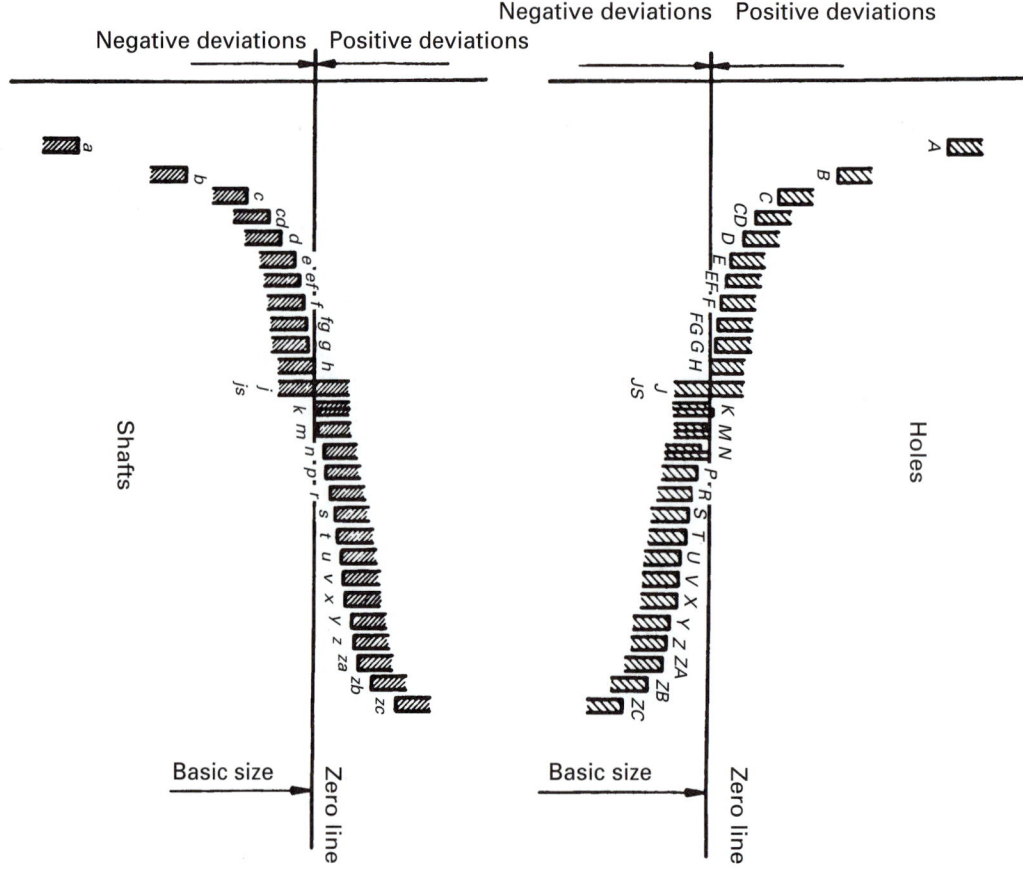

FIGURE 10-11 Position of the various tolerance zones for a given diameter in the ISO system. (By permission from Recommendations R286-1962, *System of Limits and Fits*, copyright 1962, American Standards Institute, New York.)

for "commonly used shafts and holes" in two categories: "general purpose" and "fine mechanisms and horology" (horology is the art of making timepieces).

Accuracy versus Precision in Processes

It is vitally important that the difference between accuracy and precision be understood. *Accuracy* refers to the ability to hit what is aimed at (the bull's-eye), and *precision* refers to the repeatability of the process. Suppose that five sets of five shots are fired at a target from the same gun. Figure 10-12 shows some of the possible outcomes. In Figure 10-12a, inspection of the target shows that this is a good process—accurate and precise. Figure 10-12b shows precision (repeatability) but poor accuracy. The agreement with a standard is not good. In Figure 10-12c the process is, "on the average," quite accurate, as the "X" is right in the middle of the bull's-eye, but the process has too much scatter or variability; it does not repeat. Finally, in Figure 10-12d, a failure to repeat accuracy between samples with respect to time is observed; the process is not stable. These four outcomes are typical but not all-inclusive of what may be observed.

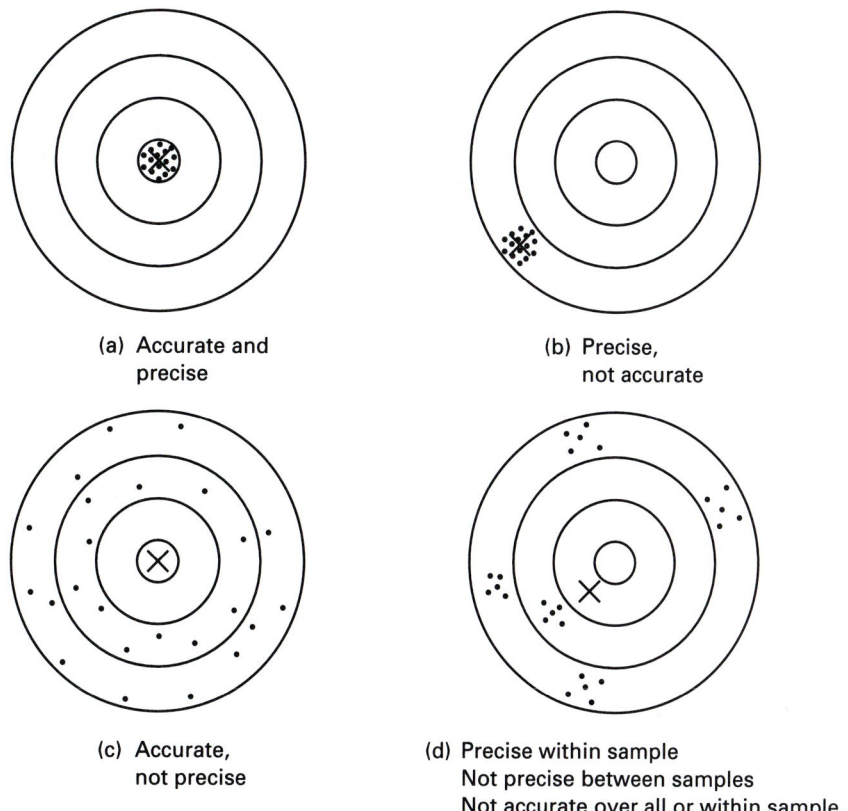

(a) Accurate and
precise

(b) Precise,
not accurate

(c) Accurate,
not precise

(d) Precise within sample
Not precise between samples
Not accurate over all or within sample

FIGURE 10-12   Accuracy versus precision. Dots in targets represent location of shots. Cross (X) represents the location of the average position of all shots.

Geometric Tolerances

Geometric tolerances state the maximum allowable deviation of a form or a position from the perfect geometry implied by a drawing. These tolerances specify the diameter or the width of a tolerance zone necessary for a part to meet its required accuracy. Figure 10-13a shows the various symbols used to specify the required geometric characteristics of dimensioned drawings. A modifier is used to specify the limits of size of a part when applying geometric tolerances. The *maximum material condition* (MMC) indicates that a part is made with the largest amount of material allowable (i.e., a hole at its smallest permitted diameter or a shaft at its largest permitted diameter). The least material condition (LMC) is the converse of the maximum material condition. *Regardless of feature size* (RFS) indicates that tolerances apply to a geometric feature for any size it may be. Many geometric tolerances or *feature control symbols* are stated with respect to a particular datum or reference surface. Up to three datum surfaces can be given to specify a tolerance. Datum surfaces are generally designated by a letter symbol. Figure 10-13b gives examples of the symbols used for datum planes and feature control symbols.

There are four tolerances that specify the permitted variability of forms: *flatness, straightness, roundness,* and *cylindricity.* Form tolerances describe how an actual feature may vary from a geometrically ideal feature.

The surface of a part is ideally flat if all its elements are coplanar. The *flatness* specification describes the tolerance zone formed by two parallel planes that contain all the

	Tolerance	Characteristic	Symbol
Individual features	Form	Straightness	——
		Flatness	▱
		Circularity	○
		Cylindricity	/○/
Individual or related features	Profile	Line	⌒
		Surface	⌓
Related features	Orientation	Angularity	∠
		Perpendicularity	⊥
		Parallelism	//
	Location	Position	⊕
		Concentricity	○
	Runout	Circular runout	↗
		Total runout	↗↗

Notes	⌀	Ⓜ	Ⓛ	Ⓢ
	DIA	MMC	LMC	RFS

(a)

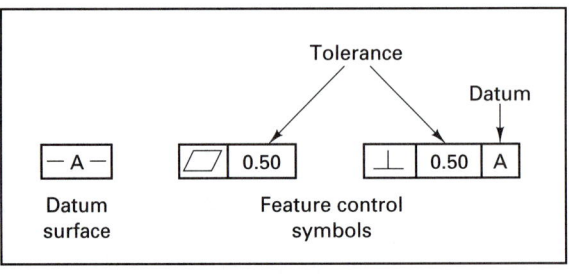

(b)

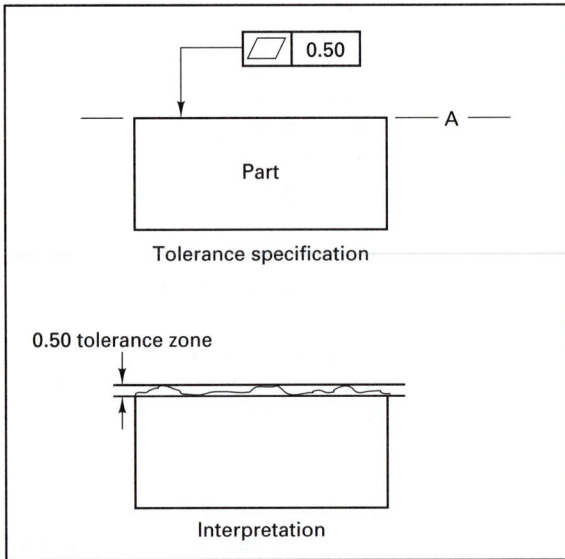

(c)

FIGURE 10-13 (a) Geometric tolerancing symbols; (b) feature control symbols for part drawings; (c) example of use of geometric tolerancing: form tolerancing for flatness.

elements on a surface. A 0.5-mm tolerance zone is described by the feature control symbol in Figure 10-13c. The distance between the highest point on the surface to the lowest point on the surface may not be greater than 0.5 mm.

In Section 10.7, on coordinate measuring machines, additional examples of geometric tolerances can be found.

■ 10.4 INSPECTION METHODS FOR MEASUREMENT

The field of *metrology*, even limited to geometrical or dimensional measurements, is far too large to cover here. The remainder of this chapter concentrates on basic linear measurements and the attributes and variables devices most commonly found in a company's metrology or quality control facility. At a minimum, such labs would typically contain optical flats: one or two granite measuring tables; an assortment of indicators, calipers, micrometers, and height gages; an optical comparator; a set or two of Grade 1 gage blocks; a coordinate measuring machine; a laser scanning device; a laser interferometer; a toolmaker's microscope; and pieces of equipment specially designed to accommodate the company's products.

Table 10-4 provides a summary of inspection methods, listing five basic kinds of devices; air, light optical and electron optical, electronic, and mechanical. The variety seems to be endless, but digital electronic readouts connected to any of the measuring devices are becoming the preferred method.

The discrete digital readout on a clear liquid-crystal display (LCD) eliminates reading interpretations associated with analog scales and can be entered directly into dedicated microprocessors or computers for permanent record and analysis. The added speed and ease of use for this type of equipment have allowed it to be used routinely on the plant floor instead of in the metrology lab. In summary, the trend toward tighter tolerances (greater precision) and accuracy associated with the need for superior quality and reliability has greatly enhanced the need for improved measurement methods.

Factors in Selecting Inspection Equipment

Many inspection devices use electronic output to communicate directly to microprocessors. Inspection devices are being built into the processes themselves and are often computer-aided. In-process inspection generates feedback sensory data from the process or its output to the computer control of the machine, which is the first step in making the processes responsive to changes (Adaptive Control). In addition to in-process inspection, many other quality checks and measurements of parts and assemblies are needed. In general, six factors should be considered when selecting equipment for an inspection job by measurement techniques.

1. *Gage Capability.* The measurement device (or working gage) should be 10 times more precise than the tolerance to be measured. This is known on the factory floor as "the rule of 10". The rule actually applies to all stages in the inspection sequence, as shown in Figure 10-14. The master gage should be ten times more precise than that of the inspection device. The reference standard used to check the master gage should be 10 times more precise than the master gage. The application of the rule greatly reduces the probability of rejecting good parts or accepting bad components and performing additional work on them.
2. *Linearity.* This factor refers to the calibration accuracy of the device over its full working range. Is it linear? What is its degree of nonlinearity? Where does it become nonlinear, and what, therefore, is its real linear working region?

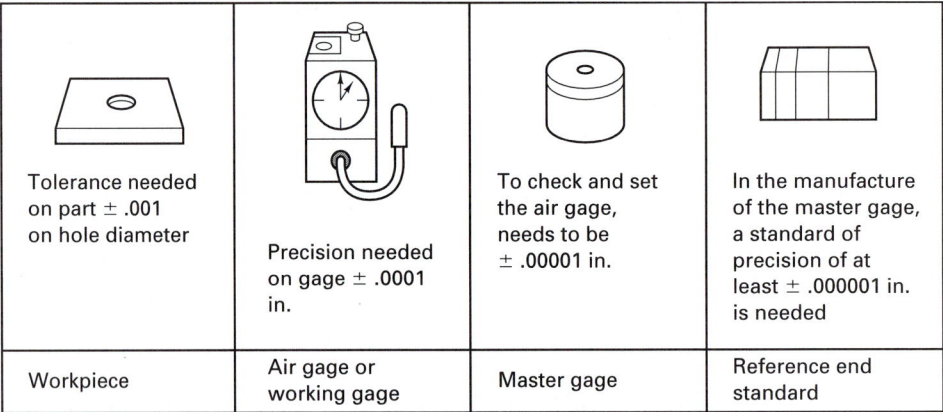

Tolerance needed on part ± .001 on hole diameter	Precision needed on gage ± .0001 in.	To check and set the air gage, needs to be ± .00001 in.	In the manufacture of the master gage, a standard of precision of at least ± .000001 in. is needed
Workpiece	Air gage or working gage	Master gage	Reference end standard

FIGURE 10-14   The rule of 10 states that for reliable measurements each successive step in the inspection sequence should have 10 times the *precision* of the preceding step.

TABLE 10-4.   Five Basic Kinds of Inspection Method

Method	Typical Accuracy	Major Applications	Comments
Air	0.5–10 μin. or 2 to 3% of scale range	Gaging holes and shafts using a calibrated difference in air pressure or airflow, with magnifications of 20,000–40,000 to 1; also used for machine control, sorting, and classifying.	High precision and flexibility; can measure out-of-round, taper, concentricity, camber, squareness, parallelism, and clearance between mating parts; noncontact principle good for delicate parts.
Optical light energy	0.2–2 μin. or better with laser interferometry 0.5–1 second of arc in autocollimation optical comparators	Interferometry; checking flatness and size of gage blocks; finding surface flaws; measuring spherical shapes, flatness of surface plates, accuracy of rotary index tables; includes all light microscopes and devices common on plant floor	Largest variety of measuring equipment; autocollimators are used for making precision angular measurements; lasers are used to make precision inprocess measurements; laser scanning.
Optical electron energy	100 Å	Precision measurement in scanning electron microscopes of microelectronic circuits and other small precision parts.	Part size restricted by vacuum chamber size; electron beam can be used for processing and part testing of electronic circuits.
Electronics	0.5–10 μin.	Widely used for machine control, on-line inspection, sorting, and classification; ODs, IDs, height, surface, and geometrical relationships, profile tracing for roundness, surface roughness, contours, etc.; most devices are comparators with movement of stylus or spindle producing an electronic signal that is amplified electronically; commonly connected to microprocessors and minicomputers for process adjustment.	Electronic gages come in many forms but usually have a sensory head or detector combined with an amplifier; capable of high magnification with resolution limited by size or geometry for sensory head; readouts commonly have multiple magnification steps; solid-state digital electronics make these devices small, portable, stable, and extremely flexible, with extremely fast response time.
Mechanical	1–10 μin.	Large variety of external and internal measurements using dial indicators, micrometers, calipers, and the like; commonly used for bench comparators for gage calibration work.	Moderate cost and ease of use make many of these devices and workhorses of the shop floor; highly dependent on workers' skills and often subject to problems of linkages.

3. *Repeat accuracy.* How repeatable is the device in taking the same reading over and over on a given reference standard?

4. *Stability.* How well does this device retain its calibration over a period of time? Stability is also called *drift.* As devices become more accurate, they often lose stability and become more sensitive to small changes in temperature and humidity.

5. *Magnification.* The amplification of the output portion of the device over the actual input dimension. The more accurate the device, the greater must be its magnification factor, so that the required measurement can be read out (or observed) and compared with the desired standard. Magnification is often confused with resolution, but they are not the same thing.

6. *Resolution.* Sometimes called *sensitivity,* resolution refers to the smallest unit of scale or dimensional input that the device can detect or distinguish. The greater the resolution of the device, the smaller will be the things it can resolve and the greater will be the magnification required to expand these measurements up to the point where they can be observed by the naked eye.

Some other factors of importance in selecting inspection devices include the type of measurement information desired, the range or the span of sizes the device can handle versus the size and geometry of the workpieces, the environment, the cost of the device, and the cost of installing and using the device. The last factor depends on the speed of measurement, the degree to which it can be automated, and the functional life of the device in service.

■ 10.5 Measuring Instruments

Because of the great importance of measuring in manufacturing, a great variety of instruments are available that permit measurements to be made routinely, ranging in accuracy from $\frac{1}{64}$ to 0.00001 in. and from 0.5 to 0.0003 mm. Machine-mounted measuring devices (probes and lasers) for automatically inspecting the workpiece during manufacturing are beginning to compete with postprocess gaging and inspection, in which the part is inspected, automatically or manually, after it has come off the machine. In-process inspection for automatic size control has been used for some years in grinding to compensate for the relatively rapid wear of the grinding wheel. Today, touch trigger probes, with built-in automatic measuring systems, are being used on machine tools to determine cutting tool offsets and tool wear. These systems are discussed in Chapter 29.

For manually operated analog instruments, the ease of use, precision, and accuracy of measurements can be affected by (1) the least count of the subdivisions on the instrument, (2) line matching, and (3) the parallax in reading the instrument. Elastic deformation of the instrument and workpiece and temperature effects must be considered. Some instruments are more subject to these factors than others. In addition, the skill of the person making the measurements is very important. Digital readout devices in measuring instruments lessen or eliminate the effect of most of these factors, simplify many measuring problems, and lessen the chance of making a math error.

Linear Measuring Instruments

Linear measuring instruments are of two types: direct reading and indirect reading. *Direct-reading instruments* contain a line-graduated scale so that the size of the object being measured can be read directly on this scale. *Indirect-reading instruments* do not contain line graduations and are used to transfer the size of the dimension being measured to a direct-reading scale, thus obtaining the desired size information indirectly.

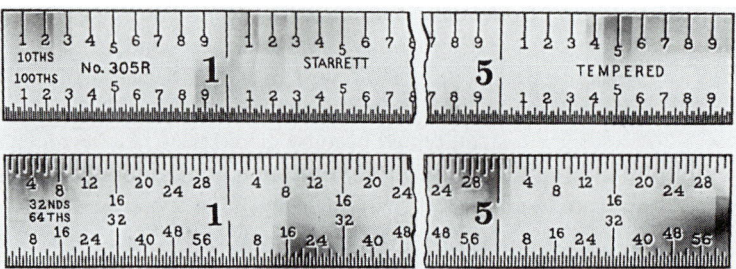

FIGURE 10-15   Machinist's rules: (Top) metric; (Center and bottom) inch graduations; 10ths and 100ths on one side, 32nd and 64ths on the opposite side. *(Courtesy of L.S. Starret Company.)*

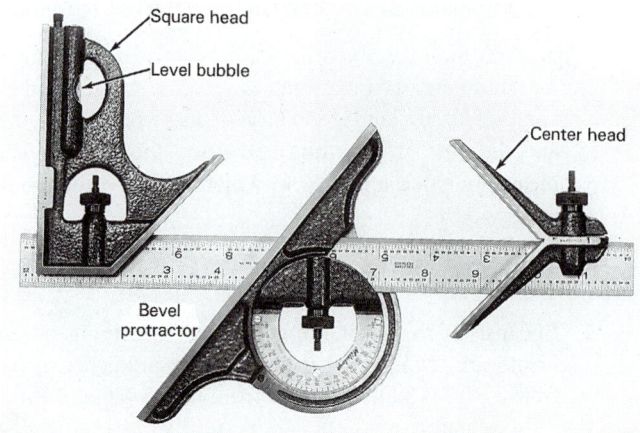

FIGURE 10-16   Combination set. *(Courtesy of MTI Corporation.)*

The simplest and most common direct-reading linear measuring instrument is the *machinist's rule* (Figure 10-15). Metric rules usually have two sets of line graduations on each side; $\frac{1}{2}$- and l-mm divisions; English rules have four sets; $\frac{1}{16}$, $\frac{1}{32}$, $\frac{1}{64}$, and $\frac{1}{100}$-in. divisions. Other combinations can be obtained in each type.

The machinist's rule is an end- or line-matching device. For the desired reading to be obtained, an end and a line, or two lines, must be aligned with the extremities of the object or the distance being measured. Thus the accuracy of the resulting reading is a function of the alignment and the magnitude of the smallest scale division. Such scales ordinarily are not used for accuracies greater than $\frac{1}{64}$ in. (0.01 in.) or about $\frac{1}{2}$ mm.

Several attachments can be added to a machinist's rule to extend its usefulness. The *square head* (Figure 10-16) can be used as a miter or trisquare or to hold the rule in an upright position on a flat surface for making height measurements. It also contains a small bubble-type level so that it can be used by itself as a level. The *bevel protractor* permits the measurement or layout of angles. The *center head* permits the center of cylindrical work to be determined.

The *vernier caliper* (Figure 10-17) is an end-measuring instrument, available in various sizes, that can be used to make both outside and inside measurements to theoretical accuracies of 0.001 in. or 0.01 mm. End-measuring instruments are more accurate and somewhat easier to use than line-matching types because their jaws are placed against either end of the object being measured, so that any difficulty in aligning edges or lines is avoided. However, the difficulty remains in obtaining uniform contact pressure, or "feel," between the legs of the instrument and the object being measured.

A major feature of the vernier caliper is the auxiliary scale (Figure 10-18). The caliper shown has a graduated beam with a metric scale on the top, a metric vernier plate, and an English scale on the bottom with an English vernier. The manner in which readings are made is explained in the figure. Figure 10-17 also shows a vernier depth gage for measuring the depth of holes or the length of shoulders on parts and a vernier height gage for making height measurements. Figure 10-19 shows calipers that have a dial indicator or a digital readout that replace the vernier. The latter two calipers are capable of making inside and outside measurements as well as depth measurements, as shown in Figure 10-20, which shows typical applications for this end-measuring device. The digital device can be zeroed at any position, as shown in the left-hand sketch of applications A through D, which greatly speeds the inspection processes and improves reading accuracy (eliminates math errors).

The *micrometer caliper*, more commonly called a *micrometer*, is one of the most widely used measuring devices. Until recently, the type shown in Figure 10-21 was virtually standard. It consists of a fixed anvil and a movable spindle. When the thimble is rotated on the end of the caliper, the spindle is moved away from the anvil by means of an accurate screw thread. On English types, this thread has a lead of 0.025 in., and one revolution of the thimble moves the spindle this distance. The barrel, or sleeve, is calibrated in 0.025-in. divisions, with each $\frac{1}{10}$ of an inch being numbered. The circumference at the edge of the thimble is graduated into 25 divisions, each representing 0.001 in.

A major difficulty with this type of micrometer is making the reading of the dimension shown on the instrument. To read the instrument, the division on the thimble that coincides with the longitudinal line on the barrel is added to the largest reading exposed on the barrel.

Micrometers graduated in ten-thousandths of an inch are the same as those graduated in thousandths, except that an additional vernier scale is placed on the sleeve so that a reading of ten-thousandths is obtained and added to the thousandths reading. The vernier

FIGURE 10-17   Variations in the vernier caliper design result in other basic gages.

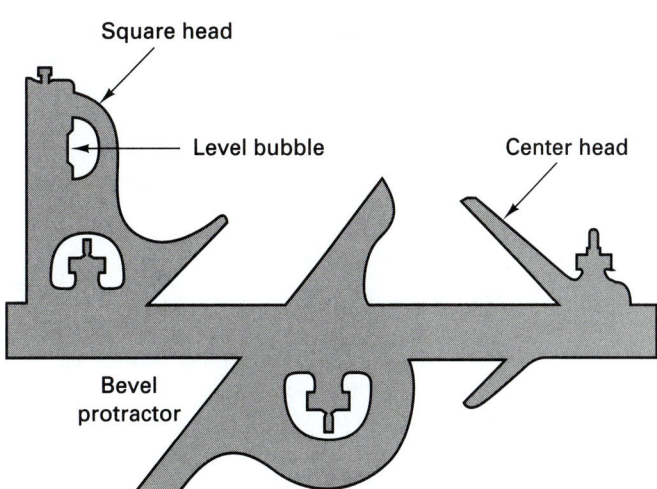

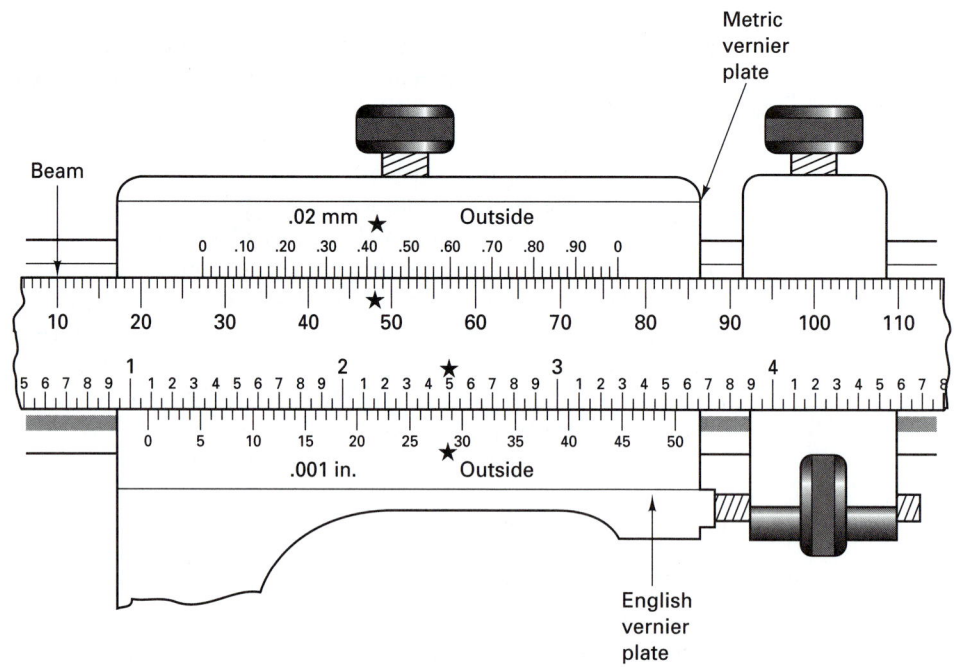

FIGURE 10-18 Vernier caliper graduated for English and metric (direct) reading. The metric reading is 27 + 0.42 = 27.42 mm.

consists of 10 divisions on the sleeve, shown in B, which occupy the same space as 9 divisions on the thimble. Therefore, the difference between the width of 1 of the 10 spaces on the vernier and 1 of the 9 spaces on the thimble is one-tenth of a division on the thimble, or one-tenth of one-thousandth, which is one ten-thousandth. To read a ten-thousandths micrometer, first obtain the thousandths reading, then see which of the lines on the vernier coincides with a line on the thimble. If it is the line marked "1," add one ten-thousandths; if it is the line marked "2," add two ten-thousandths; and so on.

> EXAMPLE: REFER TO INSETS A AND B IN FIGURE 10-21.
>
> | | |
> |---|---:|
> | The "2" line on sleeve is visible, representing | 0.200 in. |
> | Two additional lines, each representing 0.025 in. | 2 x 0.025 in. = 0.050 in. |
> | Line "0" on the thimble coincides with the reading line on the sleeve, representing | 0.000 in. |
> | The "0" line on the vernier coincide with lines on the thimble, rerpresenting | 0.000 in. |
> | The micrometer reading is | 0.2500 in. |

Now you try to read inset C.

> | | |
> |---|---:|
> | The "2" line on sleeve is visible, representing | 0.200 in. |
> | Two additional lines, each representing 0.025 in. | 2 x 0.025 in. = 0.050 in. |
> | The reading line on the sleeve lies between the "0" and "1" on the thimble, so nothing is to be added as read from the thimble. | |
> | The "7" line on the vernier coincides with a line on the thimble, rerpresenting 7 x 0.0001 in. = 0.0007 in. | |
> | The micrometer reading is | 0.2507 in. |

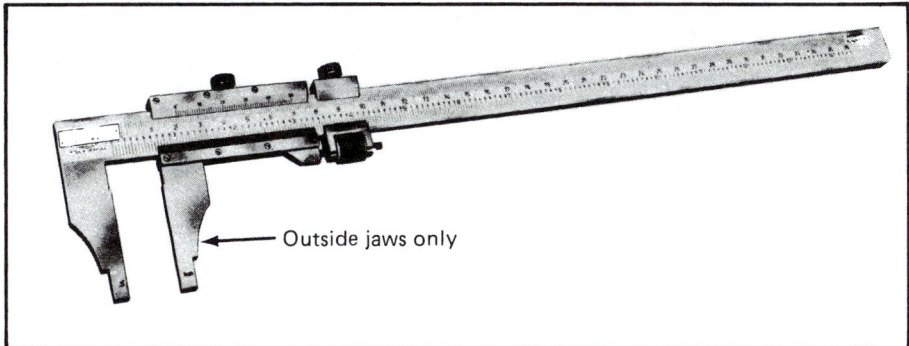

Vernier caliper with inch or metric scales and .001'' accuracy

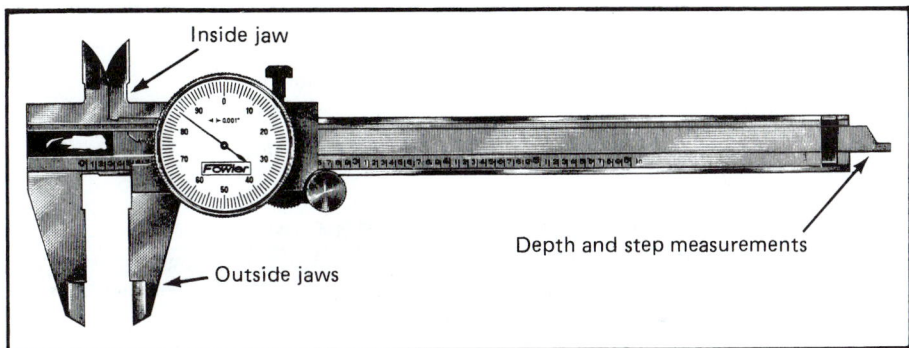

Dial caliper with .001'' accuracy

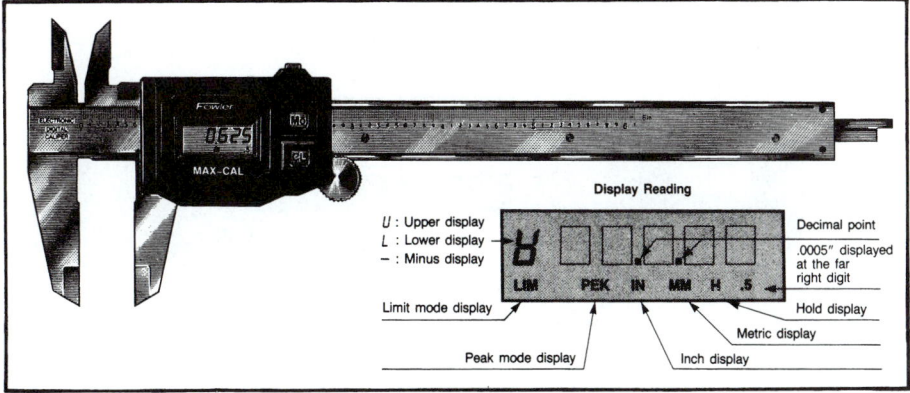

FIGURE 10-19 Three styles of calipers in common use today.

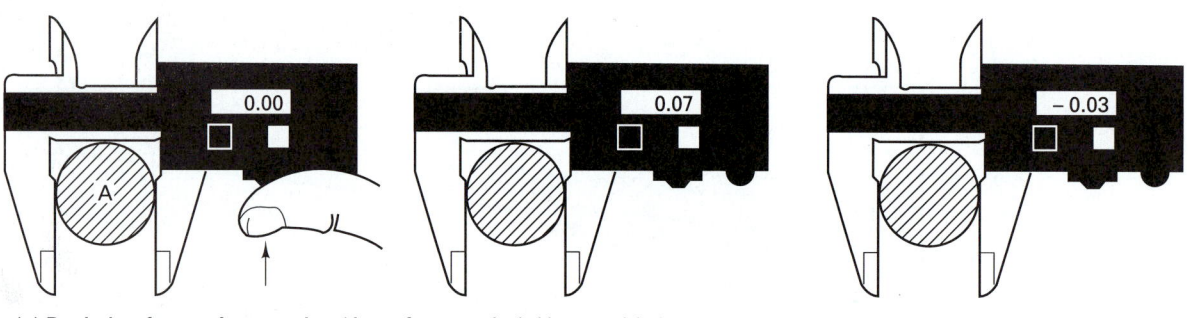

(a) Deviation from reference size. (A = reference size). Use outside jaws.

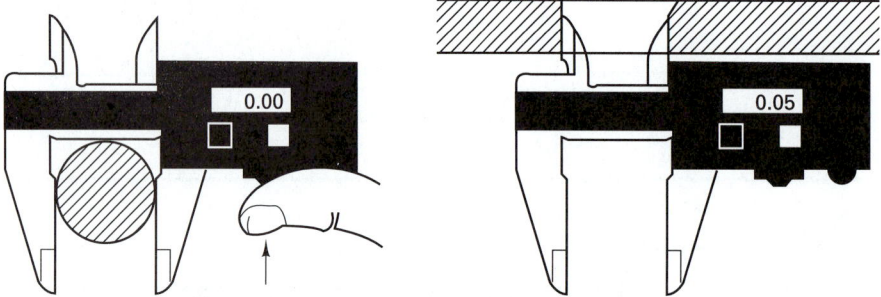

(b) Comparison, e.g. between plug and hole. Use outside + inside jaws.

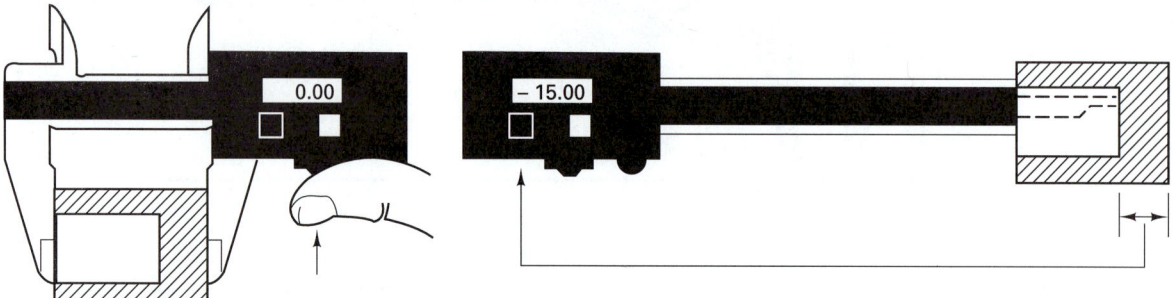

(c) Measurement of wall thickness. Use outside jaws + depth rod.

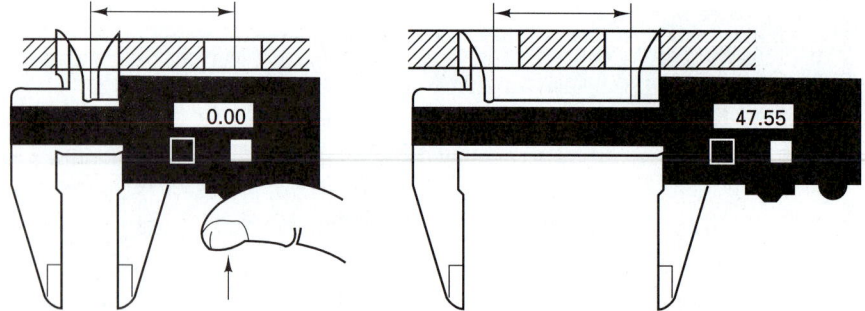

(d) Measurement of center of distance between two identical holes. Use inside jaws.

FIGURE 10-20 Four typical applications of a digital caliper. Reading reset to zero (0.000) in each left-hand figure.

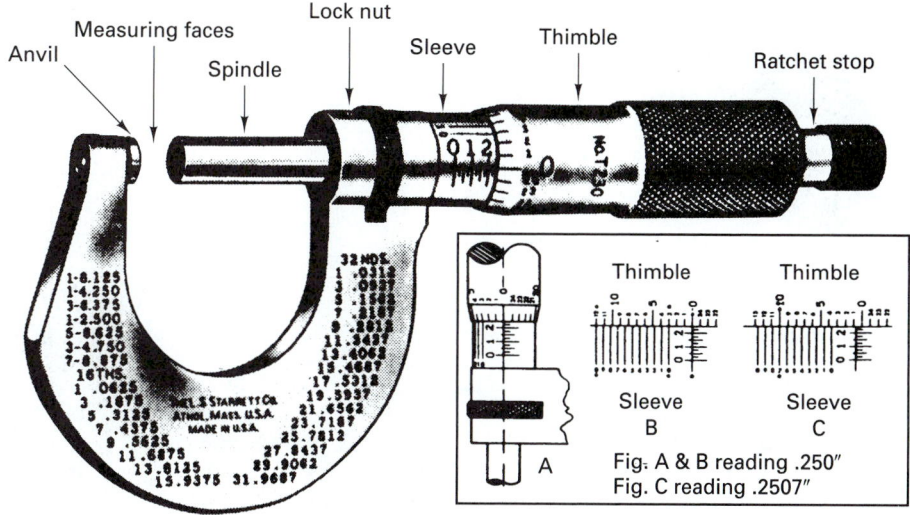

FIGURE 10-21   Micrometer caliper graduated in ten-thousandths of an inch. *(Courtesy of L.S. Starrett Company.)*

FIGURE 10-22   Digital micrometer for measurements from 0 to 1in., in 0.0001-in. graduations.

However, owing to the lack of pressure control, micrometers can seldom be relied on for accuracy beyond 0.0005 in., and such vernier scales are not used extensively. On metric micrometers the graduations on the sleeve and thimble are usually 0.5 and 0.0l mm, respectively (see the problems at the end of the chapter).

Many errors have resulted from the ordinary micrometer being misread, the error being ±0.025 in. or ±0.5 mm. Consequently, direct-reading micrometers have been developed. Figure 10-22 shows a digital outside micrometer that reads to 0.001 in. on the digit counter and 0.0001 in. on the vernier on the sleeve. The range of a micrometer is limited to 1 in. Thus a number of micrometers of various sizes are required to cover a wide range of dimensions. To control the pressure between the anvil, the spindle, and the piece being measured, most micrometers are equipped with a ratchet or a friction device, as shown in Figures 10-21 and 10-22. Calipers that do not have this device may be overtightened and sprung by several thousandths by applying excess torque to the thimble. Micrometer calipers should not usually be relied on for measurements of greater accuracy than 0.01 mm or 0.001 in., unless they are of the new digital design.

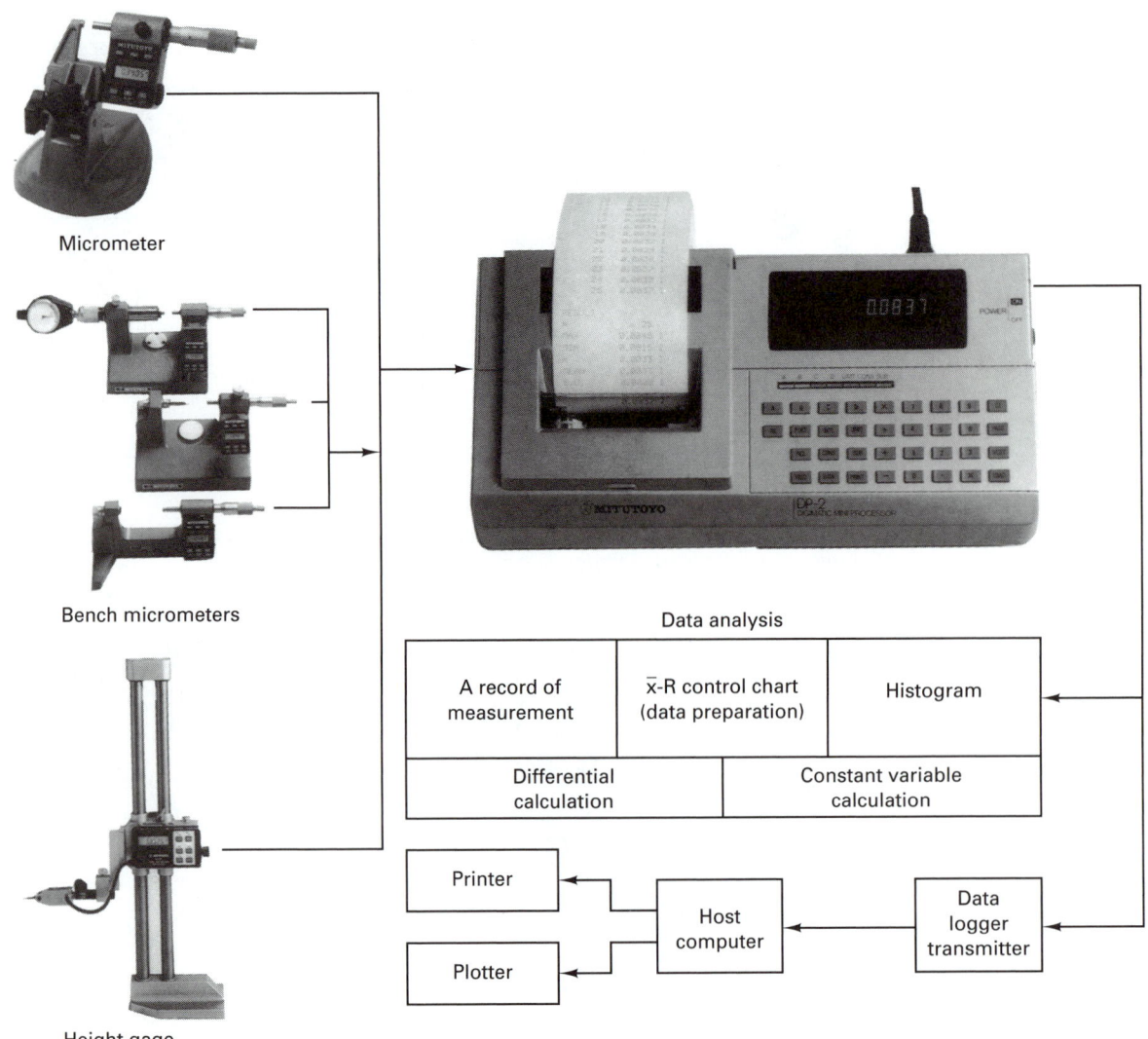

FIGURE 10-23   Direct gaging system for process control and statistical analysis of inspection data. *(Courtesy of MITUTOYO.)*

Micrometer calipers are available with a variety of specially shaped anvils and/or spindles, such as point, balls, and disks, for measuring special shapes, including screw threads. Micrometers are also available for inside measurements, and the micrometer principles is also incorporated into a *micrometer depth gage.*

Digital bench micrometers with direct readout to data processors are becoming standard inspection devices on the plant floor (Figure 10-23). The data processor provides a record of measurement as well as control charts and histograms. Direct gaging height gages, calipers, indicators, and micrometers are also available with statistical analysis capability. See Chapter 12 for more discussion on control charts and process capability.

Larger versions of micrometers, called *supermicrometers*, are capable of measuring 0.0001 in. when equipped with an indicator that shows that a selected pressure between the

FIGURE 10-24   Pratt & Whitney super micrometer modified with a digital electronic readout with accuracy to ± 0.00005 in. *(Courtesy of CDI.)*

anvils has been obtained. The addition of a digital readout permits the device to measure to ±0.00005 in. (0.001 mm) directly when it is used in a controlled-temperature environment (Figure 10-24).

The *toolmaker's microscope* (Figure 10-25) is a versatile instrument that measures by optical means; no pressure is involved. Thus it is very useful for making accurate measurements on small or delicate parts. The base, on which the special microscope is mounted, has a table that can be moved in two mutually perpendicular, horizontal directions (*X* and *Y*) by means of accurate micrometer screws that can be read to 0.0001 in., or, if so equipped, by means of the digital readout. Parts to be measured are mounted on the table, the microscope is focused, and one end of the desired part feature is aligned with the cross-line in the microscope. The reading is then noted, and the table is moved until the other extremity of the part coincides with the cross-line. From the final reading, the desired measurement can be determined. In addition to a wide variety of linear measurements, accurate angular measurements can be made by means of a special protractor eyepiece. These microscopes are available with digital readouts.

The *optical projector* or *comparator* (Figure 10-26) is a large optical device on which both linear and angular measurements can be made. As with the toolmaker's microscope, the part to be measured is mounted on a table can be moved in the *X* and *Y* directions by accurate micrometer screws. The optical system projects the image of the part on a screen, magnifying it from 5 to more than 100 times. Measurements can be made directly by means of the micrometer dials, digital readouts, or dial indicators, or on the magnified image on the screen by means of an accurate rule. A very common use for this type of instrument is the checking of parts, such as dies and screws. A template is drawn to an enlarged scale and is placed on the screen. The projected contour of the part is compared to the desired contour on the screen. Some projectors also function as low-power microscopes by providing surface illumination.

Measuring with Light and Lasers

One of the earliest and most common metrological uses of low-power lasers has been in interferometry. The interferometer is a device that can determine distance and thickness by measuring wavelengths (Figure 10-27). First, a beam splitter divides a beam of light into

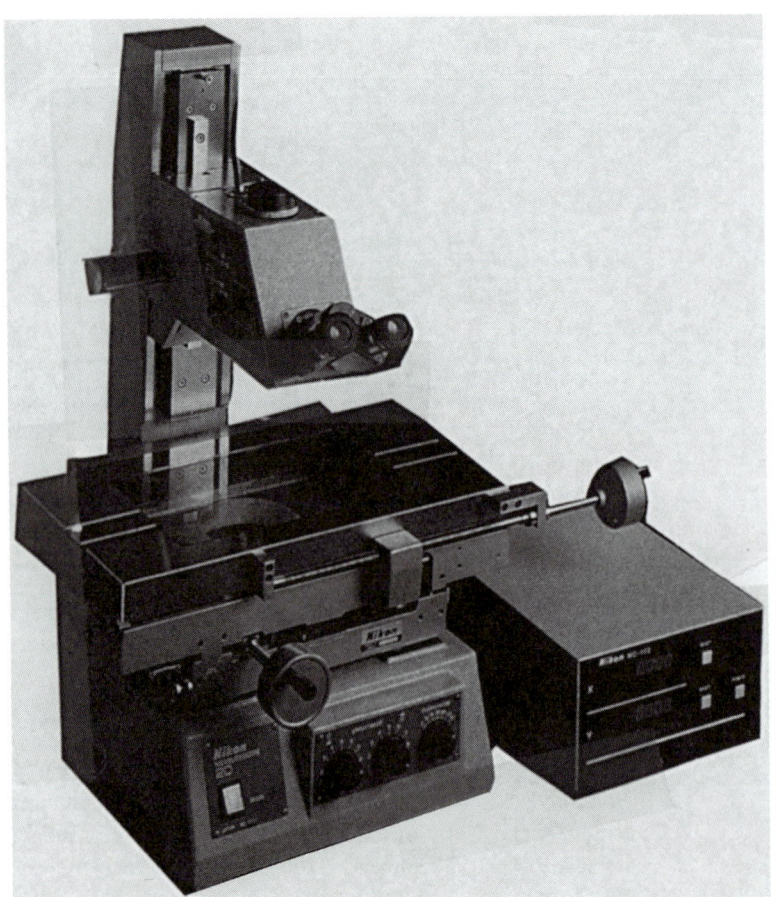

FIGURE 10-25   Toolmakers microscope with digital readouts for *X* and *Y* table movements *(Courtesy of Nikon.)*

a measurement beam and a reference beam. The measurement beam travels to a reflector (optical glass plate A) resting on the part whose distance is to be measured, while the reference beam is directed at fixed reflector B. Both beams are reflected back through the beam splitter, where they are recombined into a single beam before traveling to the observer. This recombined beam produces interference fringes, depending on whether the waves of the two returning beams are in phase (called *constructive interference*) or out of phase (termed *destructive interference*). In-phase waves produce a series of bright bands, and out-of-phase waves produce dark bands. The number of fringes can be related to the size of the object, measured in terms of light waves of a given frequency. The following example explains the basics of the method.

To determine the size of object *U* in Figure 10-28, a calibrated reference standard *S*, plus an optical flat and a toolmaker's flat, are needed along with a monochromatic light source. *Optical flats* are quartz or special glass disks from 2 to 10 in. (50 to 250 mm) in diameter and about $\frac{1}{2}$ to 1 in. (12 to 25 mm) thick whose surfaces are very nearly true planes and nearly parallel. Flats can be obtained with the surfaces within 0.000001 in. (0.00003 mm) of true flatness. It is not essential that both surfaces be accurate or that they be exactly parallel, but one must be certain that only the accurate surface is used in making

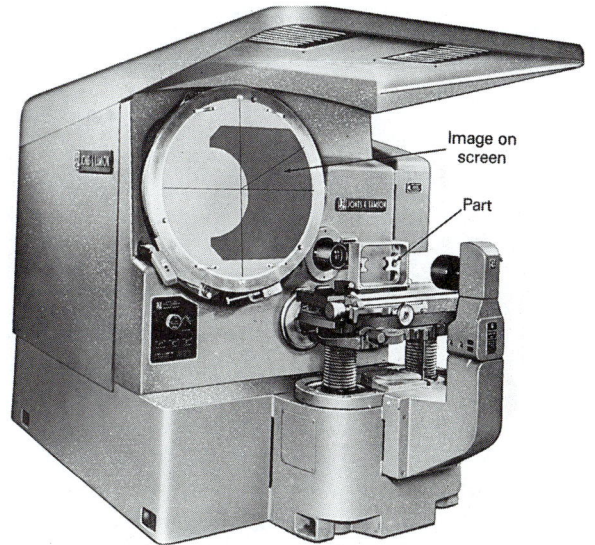

FIGURE 10-26   Optical comparator, measuring the contour on a workpiece. Digital indicators with in./mm conversions add to the utility of optical comparators.

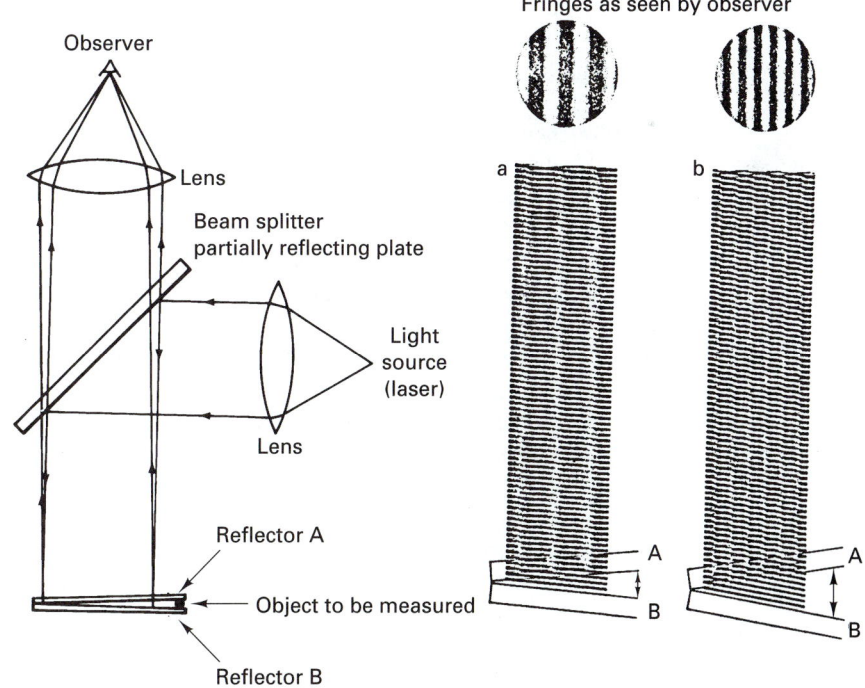

FIGURE 10-27   Interference bands can be used to measure the size of objects to great accuracy *(based on Michelson interferometer invented in 1882).*

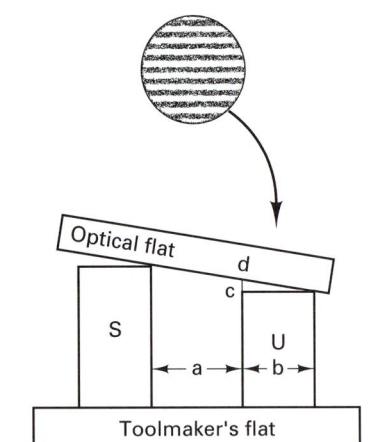

FIGURE 10-28   Method of calibrating gage block by light-wave interference

measurements. A *toolmaker's flat* is similar to an optical flat but is made of steel and usually has only one surface that is accurate. A *monochromatic light source*, light of a single wavelength, must be used. Selenium, helium, or cadmium sources are commonly used along with helium–neon lasers.

The block to be measured is *U* and the calibrated block is *S*. Distances *a* and *b* must be known but do not have to be measured with great accuracy. By counting the number of *interference bands* shown on the surface of block *U*, the distance c–d can be determined. Because the difference in the distances between the optical flat and the surface of *U* is one-half wavelength, each dark band indicates a change of one-half wavelength in the elevation. If a monochromatic light source having a wavelength of 23.2 μin. (0.589 μm) is used, each interference band represents 11.6 μin (0.295 μm). Then, by simple geometry, the difference in the heights of the two blocks can be computed. The same method is applicable for making precise measurements of other objects by comparing them with a known gage block.

Light interference also makes it possible to determine easily whether a surface is exactly flat. The achievement of interference fringes is largely dependent on the coherence of the light used. The availability of highly coherent *laser* light (in-phase light of a single frequency) has made interferometry practical in far less restrictive environments than in the past. The sometimes arduous task of extracting usable data from a close-packed series of interference fringes has been taken over by microprocessors. The laser light has less tendency to diverge (spread out) and is also monochromatic (of the same wavelength). A process that has been largely confined to the optical industry and the metrology lab is now suitable for the factory, where its extremely precise distance-measuring capabilities have been applied to the alignment and calibration of machine tools.

Accurate measurement of distances greater than a few inches was very difficult until the development of laser interferometry, which permits accuracies of ±0.5 part per million over a distance of 6.1 m with 0.01-μm resolution. Such equipment is particularly useful in checking the movement of machine tool tables, aligning and checking large assembly jigs, and making measurements of intricate machined parts such as tire-tread molds.

The Hewlett-Packard *laser interferometer* (Figure 10-29) uses a helium–neon laser beam split into two beams, each of different frequency and polarized. When the beams are recombined, any relative motion between the optics creates a Doppler shift in the frequency. This shift is then converted into a distance measurement.

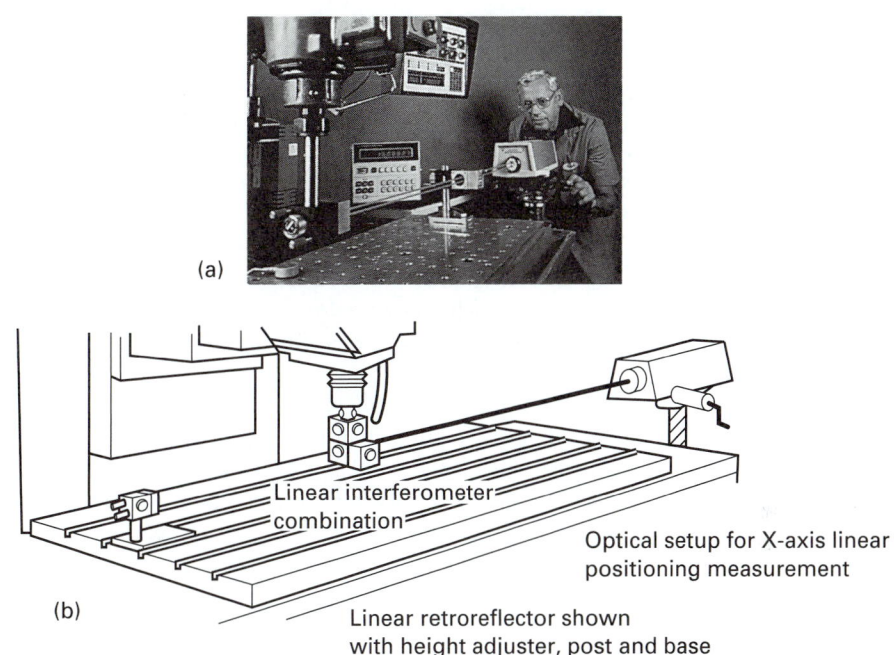

(a)

(b)

Linear interferometer combination

Optical setup for X-axis linear positioning measurement

Linear retroreflector shown with height adjuster, post and base

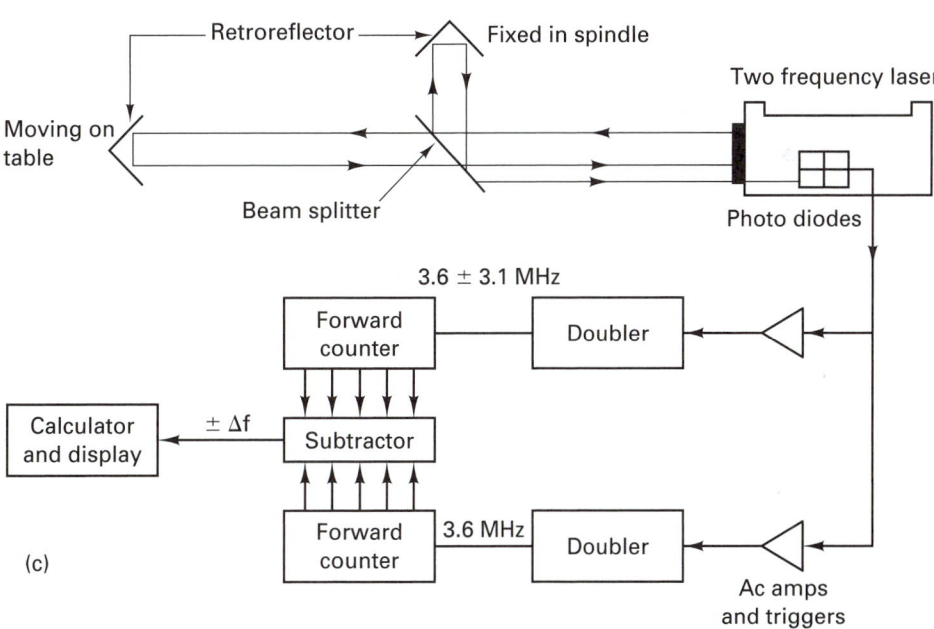

(c)

Retroreflector — Fixed in spindle

Two frequency laser

Moving on table

Beam splitter

Photo diodes

3.6 ± 3.1 MHz

Forward counter

Doubler

Calculator and display

± Δf

Subtractor

Forward counter

3.6 MHz

Doubler

Ac amps and triggers

FIGURE 10-29   (Top) Calibrating the *X*-axis linear table displacement of a vertical spindle milling machine; (Middle) schematic of optical setup; (Bottom) schematic of components of a two frequency laser interferometer. *(Courtesy of Hewlett-Packard.)*

The company's first two-frequency interferometer calibration system was introduced in 1970 to overcome workplace contamination by thermal gradients, air turbulence, oil mist, and so on, which affect the intensity of light. Doppler laser interferometers are relatively insensitive to such problems. The system can be used to measure linear distances, velocities, angles, flatness, straightness, squareness, and parallelism in machine tools.

Lasers provide for accurate machine tool alignment. Large, modern machine tools can move out of alignment in a matter of months, causing production problems often attributed to the cutting tools, the workholders, the machining conditions, or the numerical control part program.

The most widely used laser technique for inspection and in-process gaging is known as *laser scanning*. At its most basic level, the process consists of placing an object between the source of the laser beam and a receiver containing a photodiode. A microprocessor then computes the object's dimensions based on the shadow that the object casts (Figure 10-30).

The noncontact nature of laser scanning makes it well suited to in-process measurement, including such difficult tasks as the inspection of hot-rolled, or extruded material, and its comparative simplicity has led to the development of highly portable systems. The bench gage versions can measure to resolutions of 0.0001 mm.

FIGURE 10-30   Scanning laser measuring system.

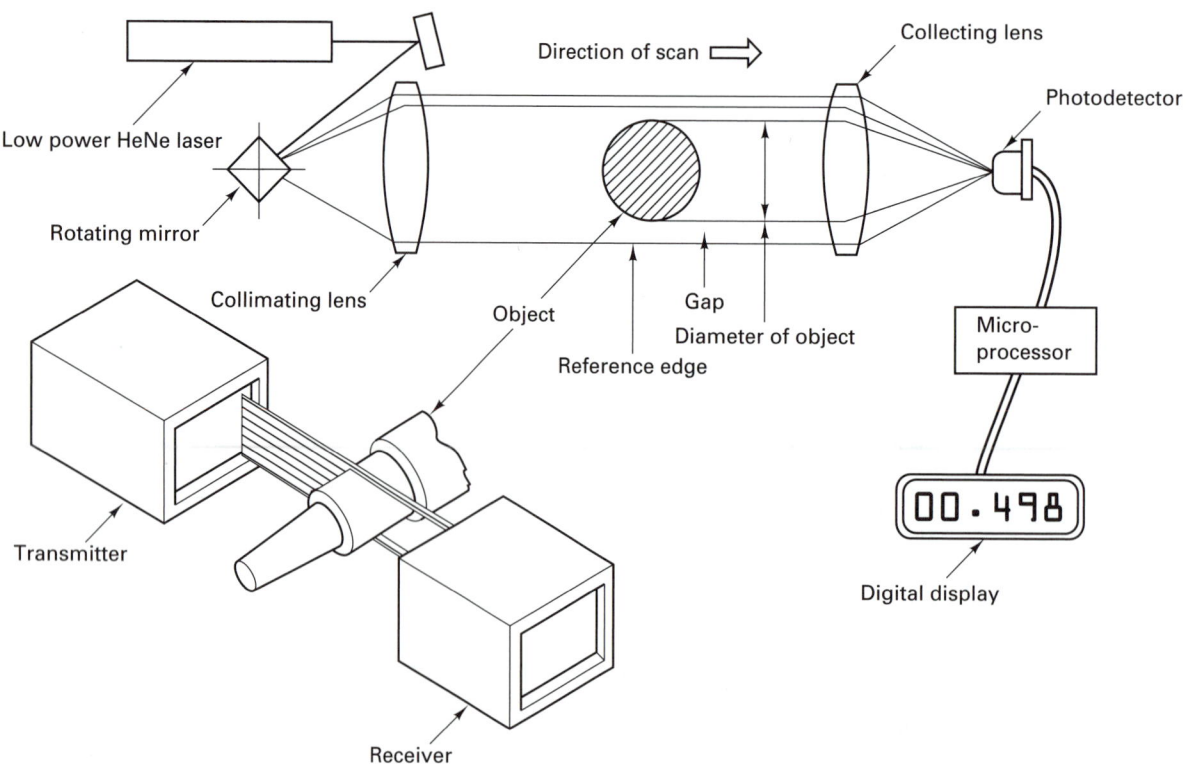

■ 10.6 VISION SYSTEMS FOR MEASUREMENT

If a picture is worth a thousand words, vision systems are the tome of inspection methods (Figure 10-31). Machine vision is used for visual inspection, for guidance and control, or for both. Normal TV image formation on photosensitive surfaces or arrays is used, and the video signals are analyzed information about the object. Each picture frame represents the object at a brief interval of time. Each frame must be dissected into picture elements (called *pixels*). Each pixel is *digitized* (has binary numbers assigned to it) by fixing the brightness or gray level of each pixel to produce a *bit-map* of the object (Figure 10-32). That is, each pixel is assigned a numerical value based on its shade of gray. Image preprocessing improves the quality of the image data by removing unwanted detail. The bit map is stored in a buffer memory. By analyzing and processing the digitized and stored bit map, the patterns are extracted, edges located, and dimensions determined.

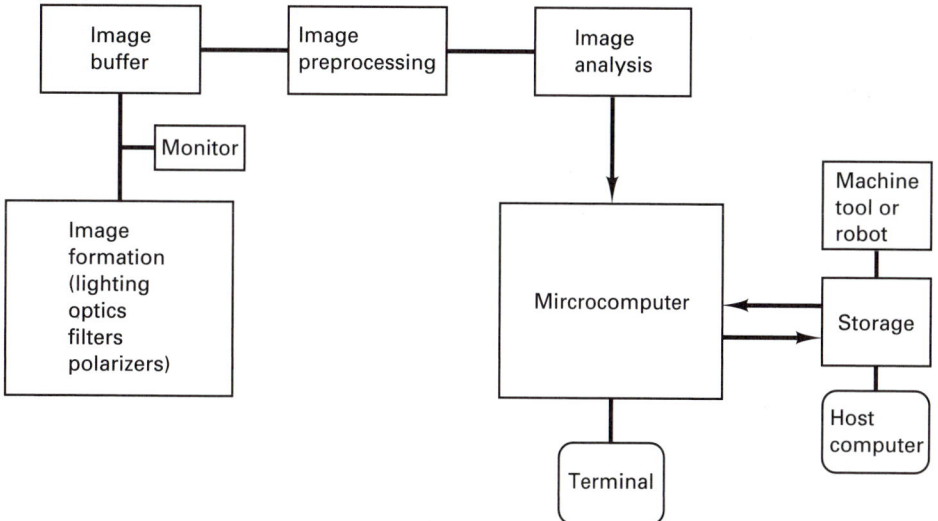

FIGURE 10-31   Schematic of element of a machine vision system

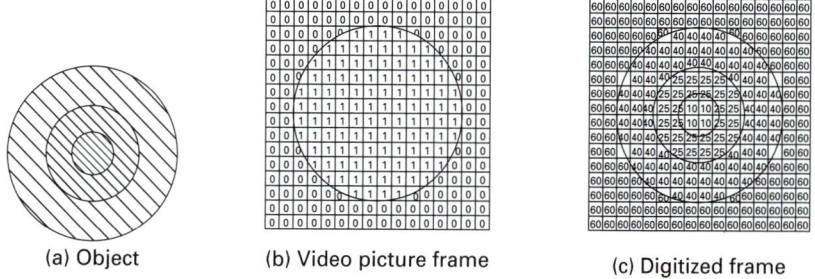

(a) Object   (b) Video picture frame   (c) Digitized frame

FIGURE 10-32   Vision systems use a gray scale to identify objects: (a) object with three different gray values; (b) one frame of object (pixels); (c) each pixel assigned a gray scale number.

Sophisticated computer algorithms using artificial intelligence have greatly reduced the computer operations needed to achieve a result, but even the most powerful video-based systems currently require 1 to 2 seconds to achieve a measurement. This may be too long for many on-line production applications. Table 10-5 provides a comparison of vision systems to laser scanning. Additional material on vision systems is provided in Chapters 29 and 42.

With the recent emphasis on quality and 100% inspection, applications for inspection by machine vision have increased markedly. Vision systems can check hundreds of parts per hour for multiple dimensions. Resolutions of ± 0.01 in. have been demonstrated, but 0.02 in. is more typical for part location. Machine vision is useful for robot guidance in material handling, welding, and assembly, but nonrobotic inspection and part location applications are still more typical. The use of vision systems in inspection, quality control, sorting, and machining tool monitoring will continue to expand. Systems can cost $100,000 or more to install and must be justified on the basis of improved quality rather than labor replacement.

TABLE 10-5. Laser Scanning versus Vision Systems

Variable	Laser-Scanning Systems	Video-Based Systems
Ambient lighting	Independent	Dependent
Object motion	Object usually stationary	Multiple cameras or strobe lighting may be required
Adaptability to robot systems	Readily adapted; some limitations on robot motion speed or overall system operation	Readily adapted; image-processing delays may delay system operation
Signal processing	Simple; computers often not required	Requires relatively powerful computers with sophisticated software
Cycle time	Very fast	Seconds of computer time may be needed
Applicability to simple tasks	Readily handled; edges and features produce sharp transitions in signal	Requires extensive use of sophisticated software algorithms to identify edges
Sizing capability	Can size an object in a single scan per axis	Can size on horizontal axis in one scan; other dimensions require full-frame processing
Three-dimensional capability	Limited three dimensionality; needs ranging capability	Uses two views of two cameras with sophisticated software or structured light
Accuracy and precision	Submicrometer 0.001 to 0.0001 in. or better accuracy; highly repeatable	Depends on resolution of cameras and distance between camera and object; systems with 0.004-in. precision and 0.006-in accuracy are typical

■ 10.7 COORDINATE MEASURING MACHINES

Precision measurements in three-dimensional Cartesian coordinate space can be made with *coordinate measuring machines* (CMMs) of the design shown in Figure 10-33. The vertical column rides on a bridge beam and carries a touch-trigger probe. Such machines use digital readouts, air bearings, computer controls, and granite tables to achieve accuracies on the order of 0.0002 to 0.0004 in. over spans of 10 to 30 in. These systems may have computer routines that give the best fit to feature measurements and that provide the means of establishing geometric tolerances discussed earlier in the chapter. Figure 10-34 and 10-35 give a partial listing of the results one can achieve with these machines.

■ 10.8 ANGLE-MEASURING INSTRUMENTS

Accurate angle measurements are usually more difficult to make than linear measurements. Angles are measured in degrees (a degree is 1/360 part of a circle) and decimal subdivisions of a degree (or in minutes and seconds of arc). The SI system calls for measurements of plane angles in radians, but degrees are permissible. The use of degrees will continue in manufacturing, but with minutes and seconds of arc possibly being replaced by decimal portions of a degree.

The bevel protractor (Figure 10-36) is the most general angle-measuring instrument. The two movable blades are brought into contact with the sides of the angular part, and the angle can be read on the vernier scale to 5 minutes of arc. A clamping device is provided to lock the blades in any desired position so that the instrument can be used for both direct measurement and layout work. As indicated previously, an angle attachment on the combination set can also be used to measure angles, similar to the way that a bevel protractor is used but usually with somewhat less accuracy.

FIGURE 10-33   Coordinate measuring machine with inset showing probe and a part being measured.

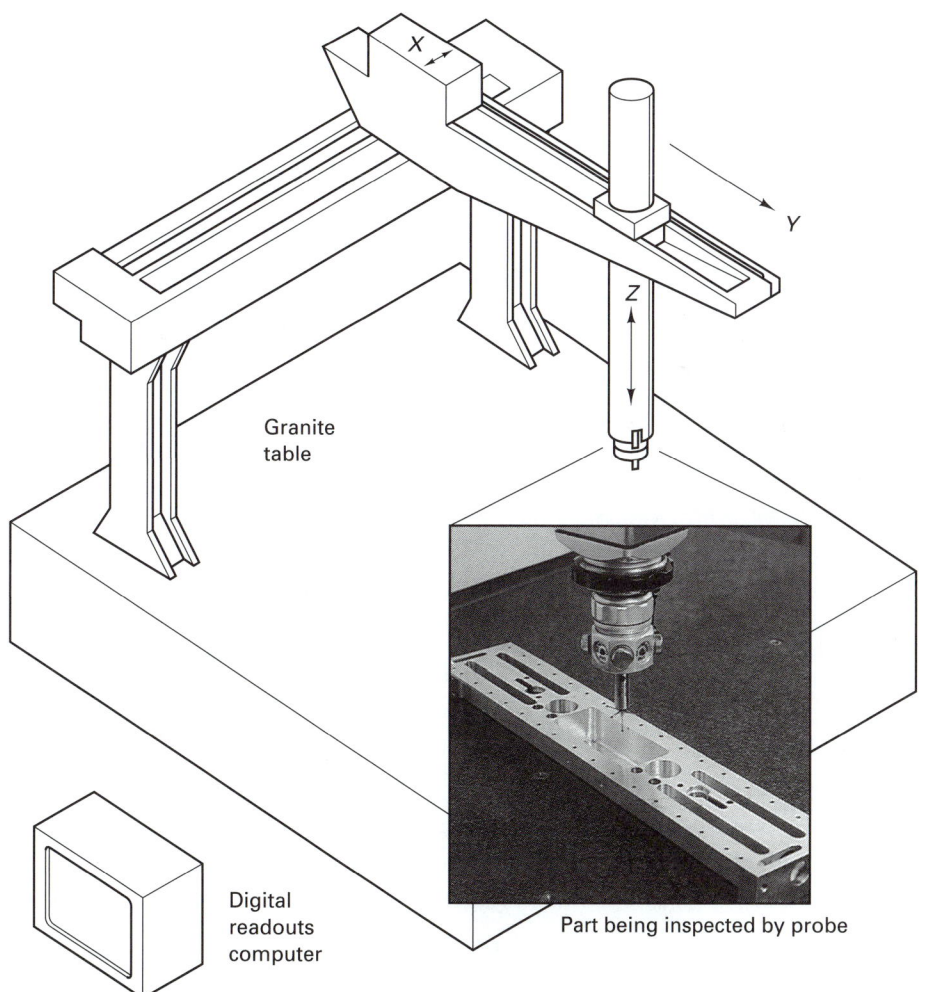

Straightness 	Straightness Measured or previously calculated points may be used to determine a 'best fit' line. The form routine establishes two reference lines which are parallel to the 'best fit' line, and which just contain all of the measured or calculated points. Straightness is defined as the distance D between these two reference lines.
Flatness 	Flatness Measured or previously calculated points may be used to determine the 'best fit' plane. The form routine establishes two reference planes which are parallel to the 'best fit' plane, and which just contain all of the measured or calculated points. Flatness is defined as the distance D between these two reference planes.
Roundness 	Roundness Measured or previously calculated points may be used to determine the 'best fit' circle. The form routine establishes two reference circles which are concentric with the 'best fit' circle, and which just contain all of the measured or calculated points. Roundness is defined as the difference D in radius of these two reference circles.
Cylindricity 	Cylindricity Measured or previously calculated points may be used to determine the 'best fit' cylinder. The form routine establishes two reference cylinders which are co-axial to the 'best fit' cylinder, and which just contain all of the measured or calculated points. Cylindricity is the difference D in radius of these two reference cylinders. Also applicable to stepped cylinders.
Conicity 	Conicity Measured or previously calculated points may be used to determine the 'best fit' cone. The form routine establishes two reference cones which are co-axial with and similar to the 'best fit' cone, and which just contain all of the measured or calculated points. Conicity is defined as the distance D between the side of these two reference cones.

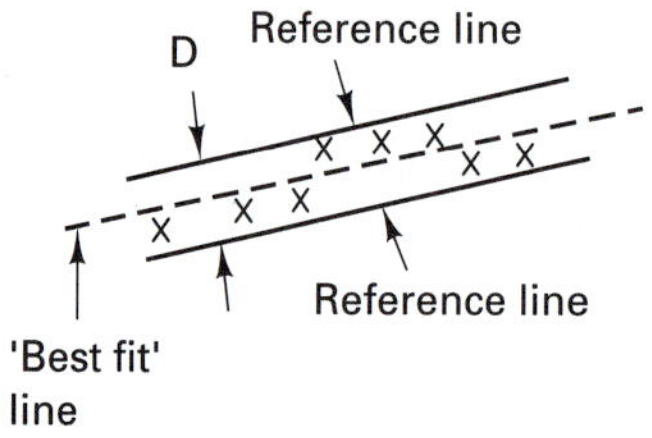

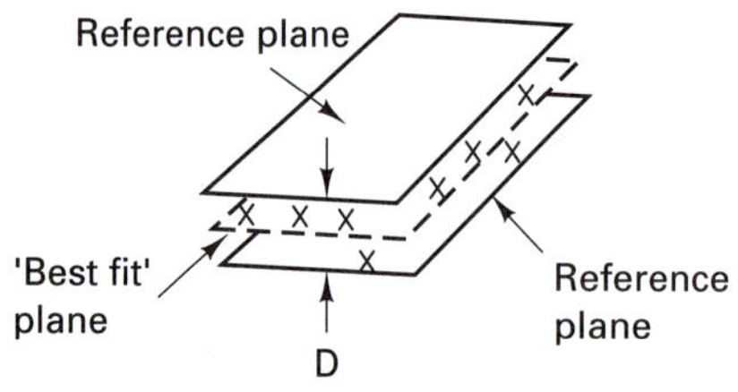

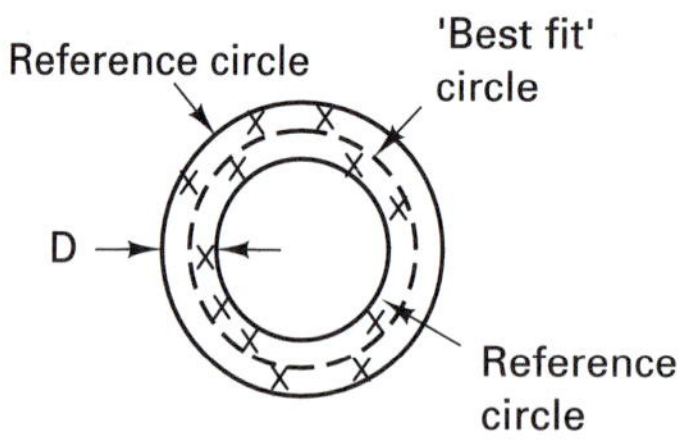

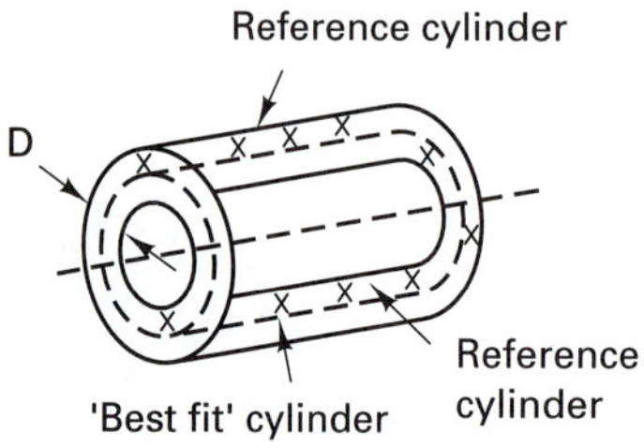

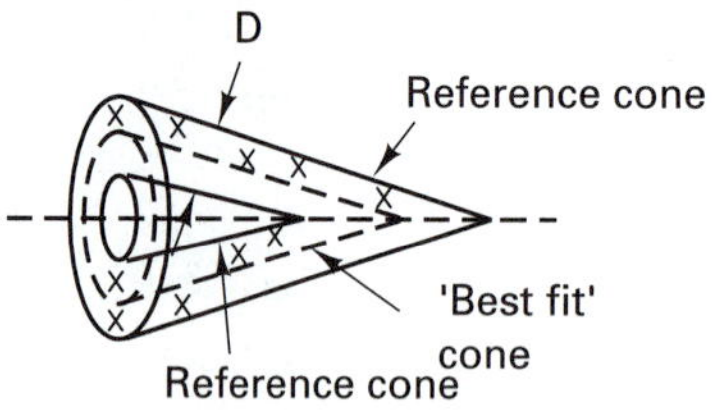

FIGURE 10-34   Examples of geometric form tolerances developed by probing surface with a CMM.

Parallelism

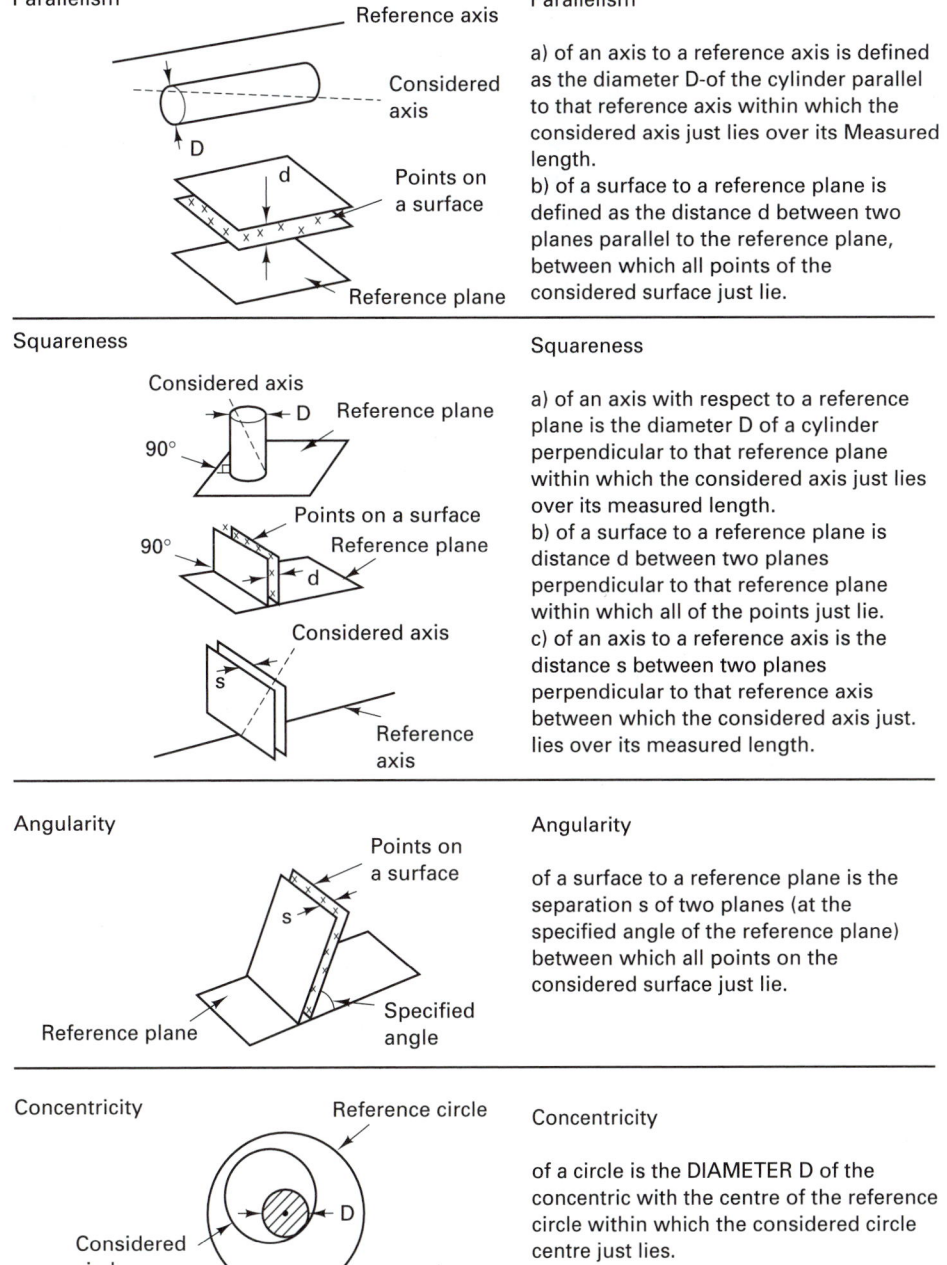

Parallelism

a) of an axis to a reference axis is defined as the diameter D-of the cylinder parallel to that reference axis within which the considered axis just lies over its Measured length.
b) of a surface to a reference plane is defined as the distance d between two planes parallel to the reference plane, between which all points of the considered surface just lie.

Squareness

Squareness

a) of an axis with respect to a reference plane is the diameter D of a cylinder perpendicular to that reference plane within which the considered axis just lies over its measured length.
b) of a surface to a reference plane is distance d between two planes perpendicular to that reference plane within which all of the points just lie.
c) of an axis to a reference axis is the distance s between two planes perpendicular to that reference axis between which the considered axis just. lies over its measured length.

Angularity

Angularity

of a surface to a reference plane is the separation s of two planes (at the specified angle of the reference plane) between which all points on the considered surface just lie.

Concentricity

Concentricity

of a circle is the DIAMETER D of the concentric with the centre of the reference circle within which the considered circle centre just lies.

FIGURE 10-35   Geometric tolerances established by CMM

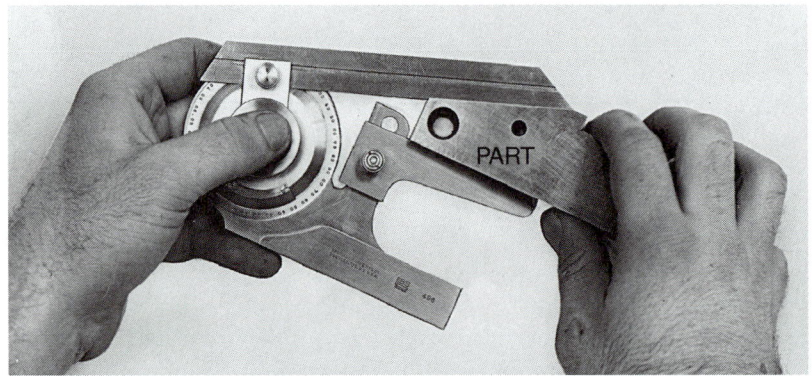

FIGURE 10-36   Measuring an angle on a part with a bevel protractor.
(Courtesy of Brown & Sharpe Mfg. Co.)

The toolmaker's microscope is very satisfactory for making angle measurements, but its use is restricted to small parts. The accuracy obtainable is 5 minutes of arc. Similarly, angles can be measured on the optical contour projector. Angular measurements can also be made by means of an angular interferometer with the laser system.

A *sine bar* may be used to obtain accurate angle measurements if the physical conditions permit. This device (Figure 10-37) consists of an accurately ground bar on which two accurately ground pins of the same diameter are mounted an exact distance apart. The distances used are usually either 5 or 10 in., and the resulting instrument is called a 5- or 10-inch sine bar. Sine bars also are available with millimeter dimensions. Measurements are made by using the principle that the sine of a given angle is the ratio of the opposite side to the hypotenuse of the right triangle.

The object being measured is attached to the sine bar, and the inclination of the assembly is raised until the top surface is exactly parallel with the surface plate. If a stack of gage blocks is used to elevate one end of the sine bar, as shown in Figure 10-37, the height of the stack directly determines the difference in height of the two pins. The difference in height of the pins can also be determined by a dial indicator gage or any other type of gage. The difference in elevation is then equal to either 5 or 10 times the sine of the angle being measured, depending on whether a 5- or 10-in. bar is being used. Tabulated values of the angles corresponding to any measured elevation difference for 5- or 10-in. sine bars are available in various handbooks. Several types of sine bars are available to suit various requirements.

Accurate measurements of angles to 1 second of arc can be made by means of *angle gage blocks*. These come in sets of 16 blocks that can be assembled in desired combinations. Angle measurements can also be made to $\pm 0.001°$ on rotary indexing tables having suitable numerical control.

■ 10.9 GAGES FOR ATTRIBUTES MEASURING

In manufacturing, particularly in mass production, it may not be necessary to know the exact dimensions of a part, only that it is within previously established limits. Limits can often be determined more easily than specific dimensions by the use of attributes-type instruments called *gages*. They may be of either fixed type or deviation type, may be used

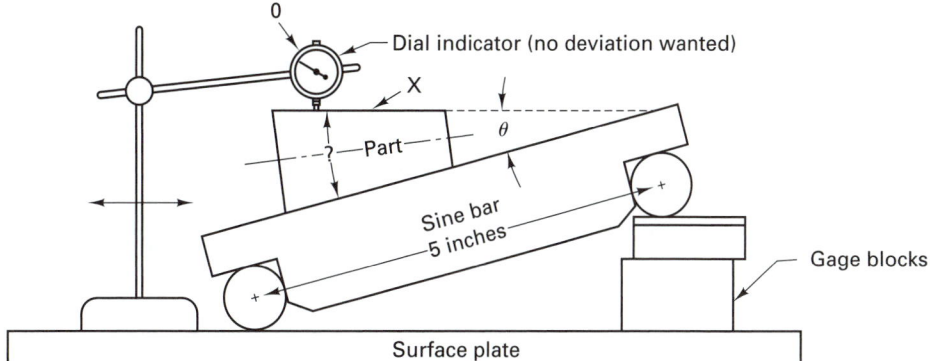

FIGURE 10-37 Setup to measure an angle on a part using a sine bar. The dial indicator is used to determine when the part surface *X* is parallel to the surface plate.

for both linear and angular dimensions, and may be used manually or mechanically (automatically).

Fixed-Type Gages

Fixed-type gages are designed to gage only one dimension and to indicate whether it is larger or smaller than the previously established standard. They do not determine how much larger or smaller the measured dimension is than the standard. Because such gages fulfill a simple and limited function, they are relatively inexpensive and are easy to use.

Gages of this type are ordinarily made of hardened steel of proper composition and are heat treated to produce dimensional stability. Hardness is essential to minimize wear and maintain accuracy. Because steels of high hardness tend to be dimensionally unstable, some fixed gages are made of softer steel with a hard chrome plating on the surface to provide surface hardness. Chrome plating can also be used for reclaiming some worn gages. Where gages are to be subjected to extensive use, they may be made of tungsten carbide at the wear points.

The *plug gage* is one of the most common types of fixed gages. As shown in Figure 10-38, plug gages are accurately ground cylinders used to gage internal dimensions, such as holes. The gaging element of a *plain plug gage* has a single diameter. To control the minimum and maximum limits of a given hole, two plug gages are required. The smaller, or *go gage* controls the minimum because it must go (slide) into any hole that is larger than the required minimum. The larger, or *not-go gage* controls the maximum dimension because it will not go into any hole unless that hole is over the maximum permissible size. As shown in Figure 10-38, the go and not-go plugs are often designed with two gages on a single handle for convenience in use. The not-go plug is usually much shorter than the go plug; it is subjected to little wear because it seldom slides into any holes. *Step-type go*, *go/not-go gages* have the go and not-go diameters on the same end of a single plug, the go portion being the outer end. The question is simply, does the gage go or not go into the hole? Such gages require careful use and should never be forced into (or onto) the part.

In designing plug and snap ring gages, the key principle is: *It is better to reject a good part than declare a bad part to be within specifications.* All gage design decisions are made with this principle in mind. Gages must have tolerances like any manufactured components. All gages are made with gage and wear tolerances. Gage tolerance allows for

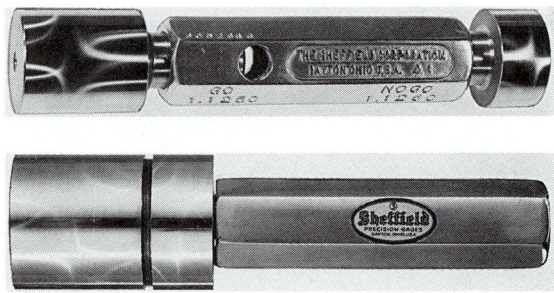

FIGURE 10-38 (Top) Plain plug gage having go member on one end and not-go member on the other; (Bottom) plug gage with stepped go and not-go member. *(Courtesy of Automation and Measurement Division, Bendix Corporation).*

the permissible variation in the manufacture of the gage. They are typically 5 to 20% (depending on the industry) of the tolerance on the dimension being gaged. Wear tolerances compensate for the wear of the gage surface as a result of repeated use. Wear tolerance is applied only to the go side of the gage since the not-go side should seldom see contact with a part surface. It is typically 5 to 20% of the dimensional tolerance.

Plug-type gages are also made for gaging shapes other than cylindrical holes. Three common types are *taper plug gages, thread plug gages*, *and spline gages*. Taper plug gages gage both the angle of the taper and its size. Any deviation from the correct angle is indicated by looseness between the plug and the tapered hole. The size is indicated by the depth to which the plug fits into the hole, the correct depth being denoted by a mark on the plug. Thread plug gages come in go and not-go types. The go gage must screw into the threaded holes, and the not-go gage must not enter.

Ring gages are used to gage shaft or other external round members. These are also made in go and not-go types (Figure 10-39). Go ring gages have plain knurled exteriors, whereas not-go ring gages have a circumferential groove in the knurling so that they can easily be distinguished. *Ring thread gages* are made to be slightly adjustable because it is almost impossible to make them exactly to the desired size. Thus they are adjusted to exact, final size after the final grinding and polishing have been completed.

Snap gages are the most common type of fixed gage for measuring external dimensions. As shown in Figure 10-40, they have a rigid, U-shaped frame on which are two or three gaging surfaces, usually made of hardened steel or tungsten carbide. In the

FIGURE 10-39   Go and not-go ring gages for checking a shaft. *(Courtesy of Automation and Measurement Division, Bendix Corporation.)*

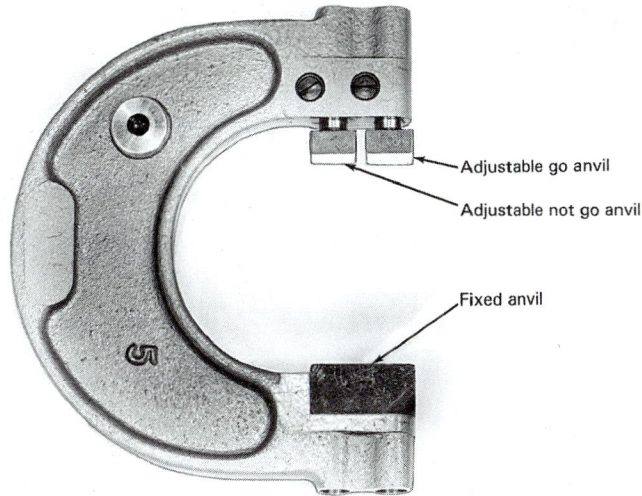

FIGURE 10-40   Adjustable go and not-go snap gage.
(Courtesy of Automation and Measurement Division, Bendix Corporation.)

adjustable type shown, one gaging surface is fixed and the other(s) may be adjusted over a small range and locked at the desired position(s). Because in most cases one wishes to control both the maximum and the minimum dimensions, the *progressive* or *step-type snap gage* (Figure 10-40) is used most frequently. These gages have one fixed anvil and two adjustable surfaces to form the outer go and the inner not-go openings, thus eliminating the use of separate go and not-go gages.

Snap gages are available in several types and wide range of sizes. The gaging surfaces may be round or rectangular. They are set to the desired dimensions with the aid of gage blocks.

Many types of special gages are available or can be constructed for special applications. The *flush-pin gage* (Figure 10-41) is an example for gaging the depth of a shoulder. The main section is placed on the higher of the two surfaces, with the movable step pin resting on the lower surface. If the depth between the two surfaces is sufficient but not too great, the top of the pin, but not the lower step, will be slightly above the top surface of the gage body. If the depth is too great, the top of the pin will be below the surface. Similarly, if the depth is not great enough, the lower step on the top of the pin will be above the surface of the gage body. When a finger, or fingernail, is run across the top of the pin, the pin's position with respect to the surface of the gage body can readily be determined.

Several types of *form gages* are available for use in checking the *profile* of various objects. Two of the most common types are *radius gages* (Figure 10-42) and *screw-thread pitch gages* (Figure 10-43).

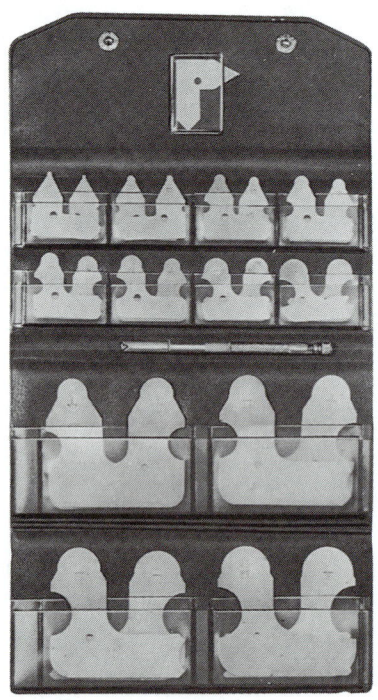

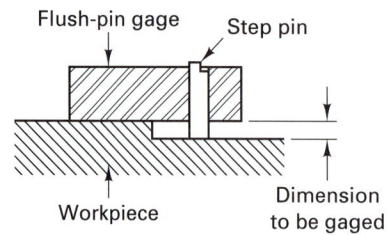

FIGURE 10-41   Flush-pin gage being used to check height of step.

FIGURE 10-42   Set of radius gages, showing how they are used. *(Courtesy of MTI Corporation.)*

Deviation-Type Gages

A large amount of gaging, and some measurement, is done through the use of *deviation-type gages*, which determine the amount by which a measured part deviates, plus or minus, from a standard dimension to which the instrument has been set. In most cases the deviation is indicated directly in units of measurement, but in some cases the gage shows only whether the deviation is within a permissible range. A good example of a deviation-type gage is a flashlight battery checker, which shows if the battery is good (green), bad (red), or borderline (yellow), but not how much voltage or current is generated. Such gages use mechanical, electrical, or fluidic amplification techniques so that very small linear deviations can be detected. Most are quite rugged, and they are available in a variety of designs, amplifications, and sizes.

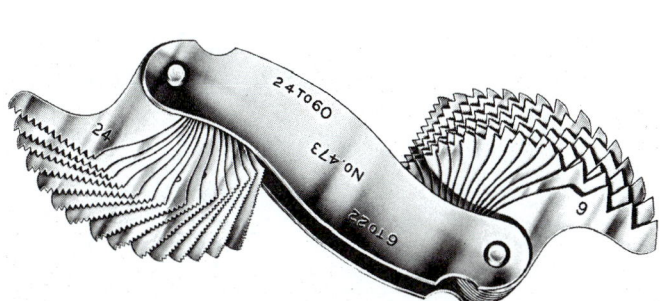

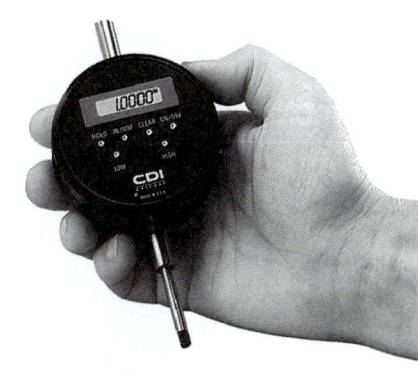

FIGURE 10-43   Thread pitch gages. *(Courtesy of L.S. Starret Company.)*

FIGURE 10-44   Digital dial indicator with 1-in. range and 0.0001-in. accuracy. *(Courtesy of CDI.)*

Dial indicators Figure 10-4 are a widely used form of deviation-type gage. Movement of the gaging spindle is amplified mechanically through a rack and pinion and a gear train and is indicated by a pointer on a graduated dial. Most dial indicators have a spindle travel equal to about $2\frac{1}{2}$ revolutions of the indicating pointer and are read in either 0.001 or 0.0001 in. or 0.02 or 0.002 mm.

The dial can be rotated by means of the knurled bezel ring to align the zero point with any position of the pointer. The indicator is often mounted on an adjustable arm to permit its being brought into proper relationship with the work. It is important that the axis of the spindle be aligned exactly with the dimension being gaged if accuracy is to be achieved. Digital dial indicators are also readily available (Figure 10-44).

Dial indicators should be checked occasionally to determine if their gage capability has been lost through wear in the gear train. Also, it should be remembered that the pressure of the spindle on the work varies because of spring pressure as the spindle moves into the gage. This spring pressure normally causes no difficulty unless the spindles are used on soft or flexible parts.

In comparators using electronic magnification (Figure 10-45), the gaging head is small and readily portable and can be mounted in many ways. The end of the sensing lever is shaped so as to compensate automatically for misalignment in the measuring plane up to about ±15 degrees. The indicator may use either a pointer and graduated scale or a digital readout. Accuracies up to ±0.0001 in. (0.0003 mm) are available, and several ranges can usually be selected merely by turning a knob.

Linear variable differential transformers (LVDTs) are used as sensory elements in many electronic gages, usually with a solid-state diode display or in automatic inspection setups. These devices can frequently be combined into multiple units for the simultaneous gaging of several dimensions. Ranges and resolutions down to 0.0005 and 0.00001 in. (0.013 and 0.00025 mm, respectively) are available.

Air gages have special characteristics that make them especially suitable for gaging holes or the internal dimensions of various shapes. A typical gage of this type, shown earlier in Figure 10-14, indicates the clearance between the gaging head and the hole by measuring either the volume of air that escapes or the pressure drop resulting from the airflow. The gage is calibrated directly in 0.0001-in. or 0.02-mm divisions. Air gages have an advantage over mechanical or electronic gages for this purpose in that they detect not only linear size deviations but also out-of-round conditions. Also, they are subject to very

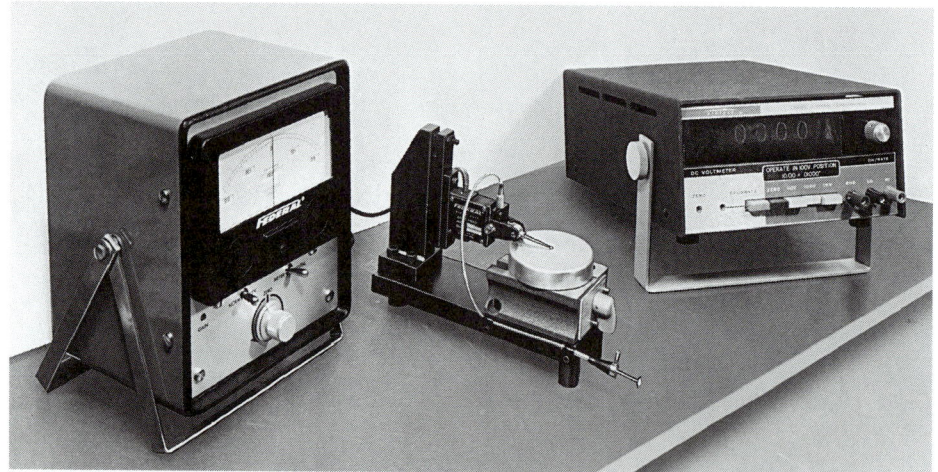

FIGURE 10-45   Electronic-magnification gage being used to gage a computer memory core 0.76 mm (0.030 in.) in diameter and 0.15 mm (0.006 in.) thick. *(Courtesy of Federal Products Corporation.)*

little wear because the gaging member is always slightly smaller than the hole and the air-flow minimizes rubbing. Special types of air gages can be used for external gaging.

◼ 10.10 SURFACE ROUGHNESS MEASUREMENT

The machining processes discussed in Chapters 21 through 32 generate a wide variety of surface patterns. *Lay* is the term used to designate the direction of the predominate surface pattern produced by the machining process. In addition, certain other terms and symbols have been developed and standardized for specifying the surface quality. The most important terms are *surface roughness*, *waviness*, and *lay* (Figure 10-46). *Roughness* refers to the finely spaced surface irregularities. It results from machining operations in the case of machined surfaces. *Waviness* is surface irregularity of greater spacing than in roughness. It may be the result of warping, vibration, or the work being deflected during machining.

Roughness is measured by the size or height of the irregularities with respect to an average line as depicted in Figure 10-47. These measurements are usually expressed in micrometers or microinches. In most cases, the arithmetical average (AA) is used. In terms of the measurements indicated in Figure 10-47, the AA should be as follows:

$$AA = \frac{\sum_{i=1}^{n} y_i}{n}$$

Cutoff refers to the sampling length used for the calculation of the roughness height. When it is not specified, a value of 0.030 in. (0.8 mm) is assumed. So in the preceding equation, y_i is a vertical distance from the centerline and n is the total number of vertical measurements taken within a specified cutoff distance. This average roughness value is also called R_a. Occasionally used is the *root-mean-square* (rms) value, which is defined as

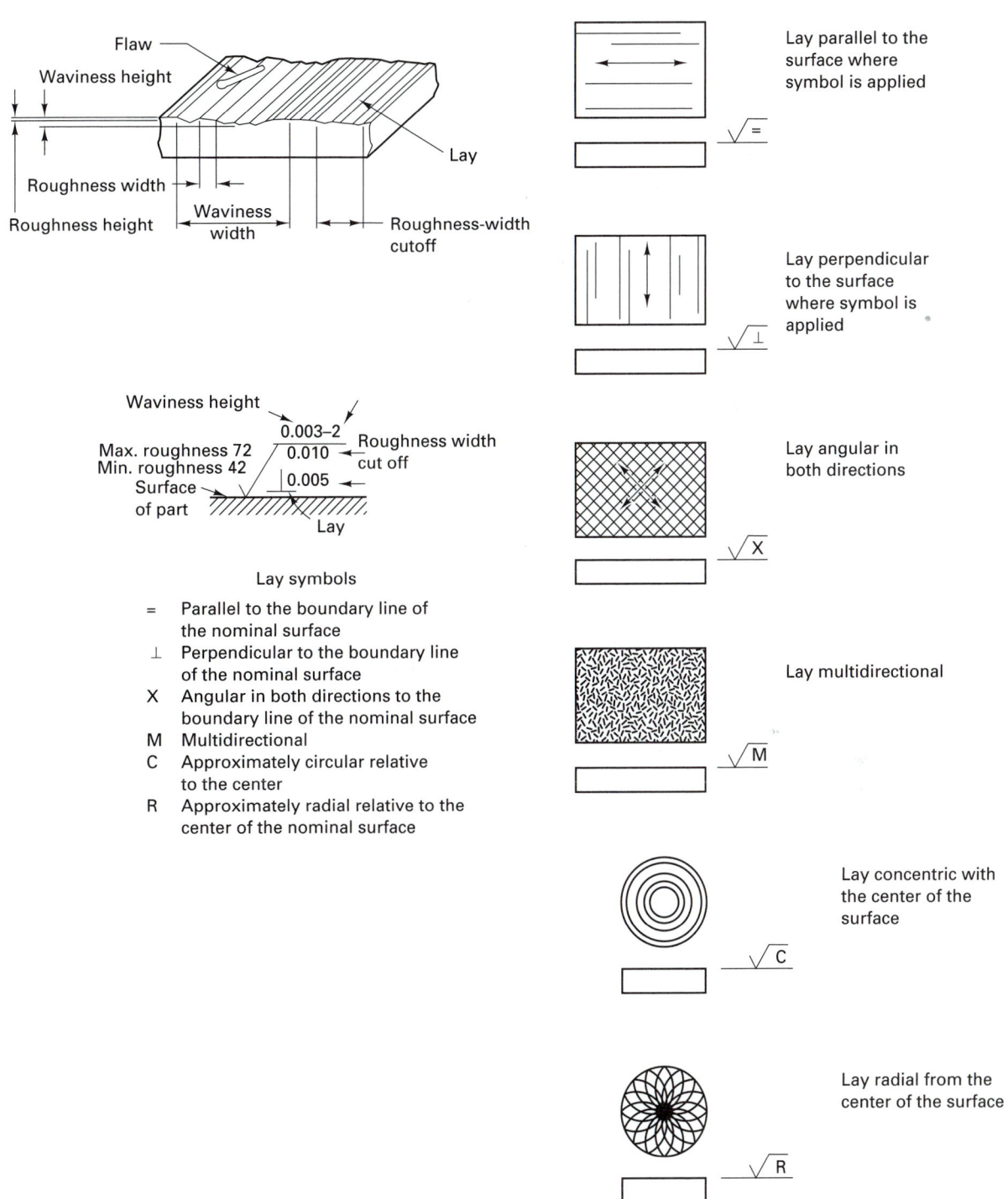

FIGURE 10-46 (Above left) Terminology used in specifying and measuring surface quality; (Above middle) symbols used on drawing by part designers, with definitions of symbols; (Above bottom) lay symbols; (Right) lay symbols applied on drawings.

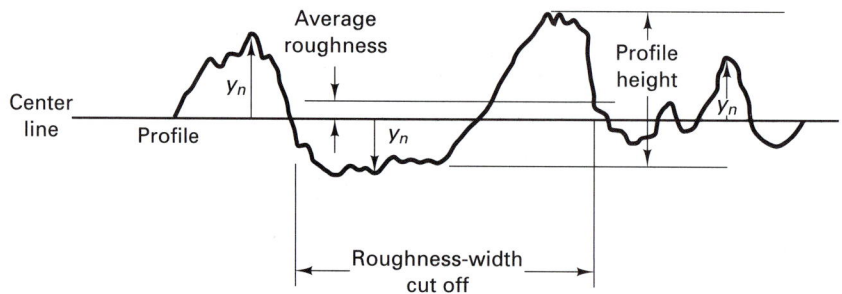

FIGURE 10-47 Schematic of surface profile as produced by a stylus device showing some typical *y* values with respect to the center line.

$$\text{rms} = \sqrt{\frac{\sum_{i=1}^{n} y_i^2}{n}}$$

A variety of instruments are available for measuring surface roughness and surface profiles. The majority of these devices use a diamond stylus that is moved at a constant rate across the surface, perpendicular to the lay pattern. The rise and fall of the stylus is detected electronically (often by an LVDT device), is amplified and recorded on a strip-chart, or is processed electronically to produce AA or rms readings for a meter (Figure 10-48.) The unit containing the stylus and the driving motor may be hand-held or supported by the workpiece or some other supporting surface.

The instrument shown in Figure 10-49 is capable of making a series of parallel offset traces on the surface, providing the two-dimensional profile maps shown in the bottom of the figure. Notice how the different machining processes create different roughness profiles. Areas of from 0.005 × 0.005 in. (0.13 × 0.13 mm) up to 2 × 2 in. (50.8 × 50.8 mm), depending on the magnification selected, can be profiled.

FIGURE 10-48 Schematic of stylus profile device for measuring surface roughness and surface profile with two readout devices shown: a meter for AA or rms values and a strip-chart recorder for surface profile.

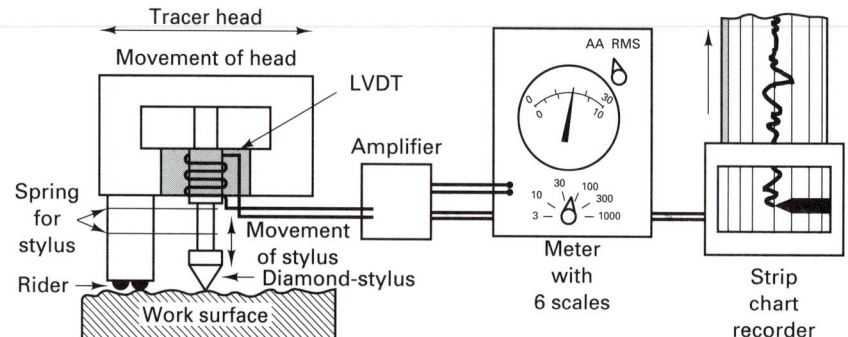

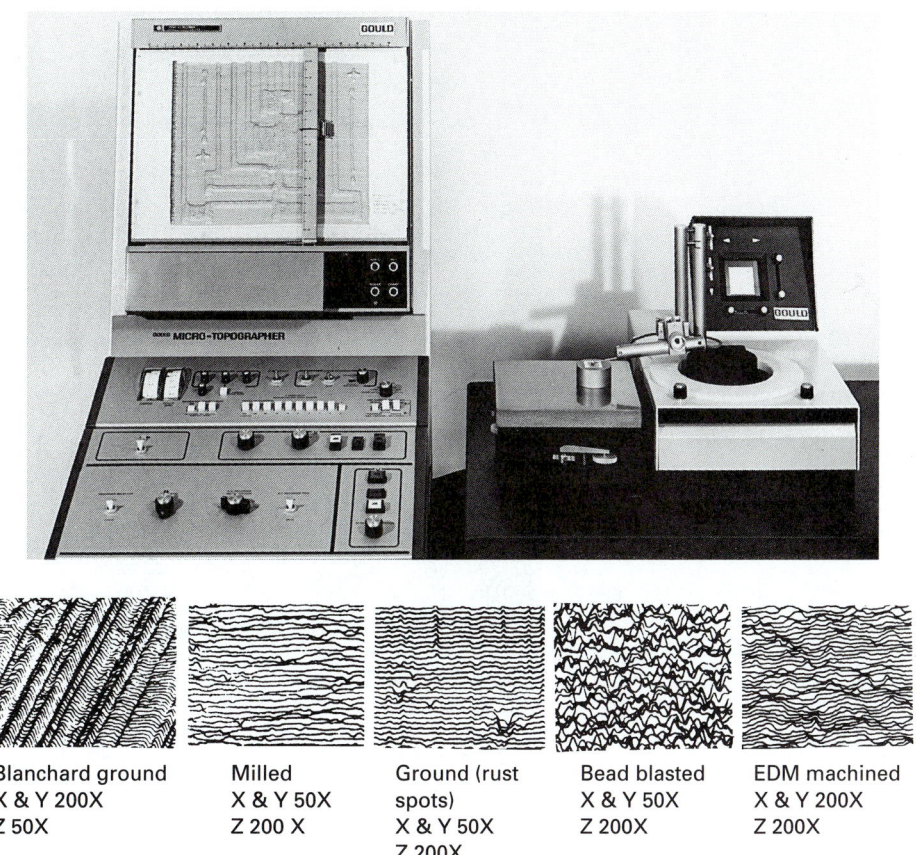

Blanchard ground	Milled	Ground (rust	Bead blasted	EDM machined
X & Y 200X	X & Y 50X	spots)	X & Y 50X	X & Y 200X
Z 50X	Z 200 X	X & Y 50X	Z 200X	Z 200X
		Z 200X		

FIGURE 10-49   (Top) Microtopographer, a stylus profile device used to measure and depict surface roughness and character (surface profile); (Bottom) Typical surface-roughness profiles. *(Courtesy of Measurement System Division, Gould Inc.)*

The resolution of these devices is determined by the radius or the diameter of the tip of the stylus. When the magnitude of the geometric features begins to approach the magnitude of the tip of the stylus, great caution should be used in interpreting the output from these devices. As a case in point, Figure 10-50 shows a scanning electron micrograph of a face-milled surface on which has been superimposed (photographically) a scanning electron micrograph of the tip of a diamond stylus (tip radius of 0.0005 in). Both micrographs have the same final magnification. Surface flaws of the same general size as the roughness created by the machining process are difficult to resolve with the stylus-type device, where both these features are about the same size as the stylus tip.

This example points out the difference between resolution and detection. Stylus tracing devices often can detect the presence of a surface crack, step, or ridge on a part but cannot resolve the geometry of the defect when the defect is of the same order of magnitude as the stylus tip or smaller.

Another problem with these devices is that they produce a reading (a line on the chart) where the stylus tip is not touching the surface, as demonstrated in Figure 10-51a,

FIGURE 10-50 Typical machined steel surface as created by face milling and examined in the SEM. A micrograph (same magnification) of a 0.0000.5-in. stylus tip has been superimposed at the top.

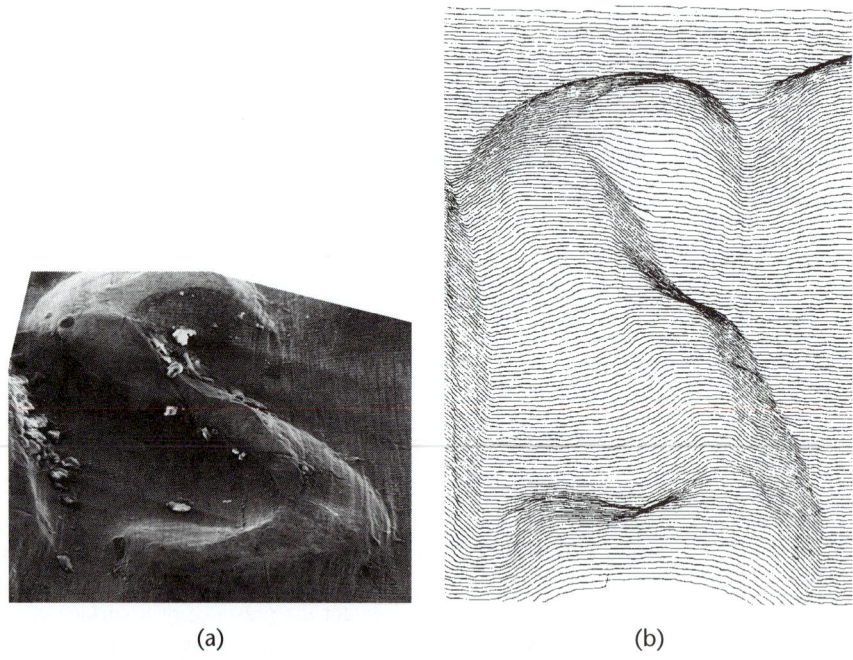

(a)   (b)

FIGURE 10-51 (a) SEM micrograph of a U.S. dime, showing the "S" in the word TRUST after the region has been traced by a stylus-type machine; (b) topographical map of the "S" region of the word TRUST from a U.S. dime (compare to part (a)).

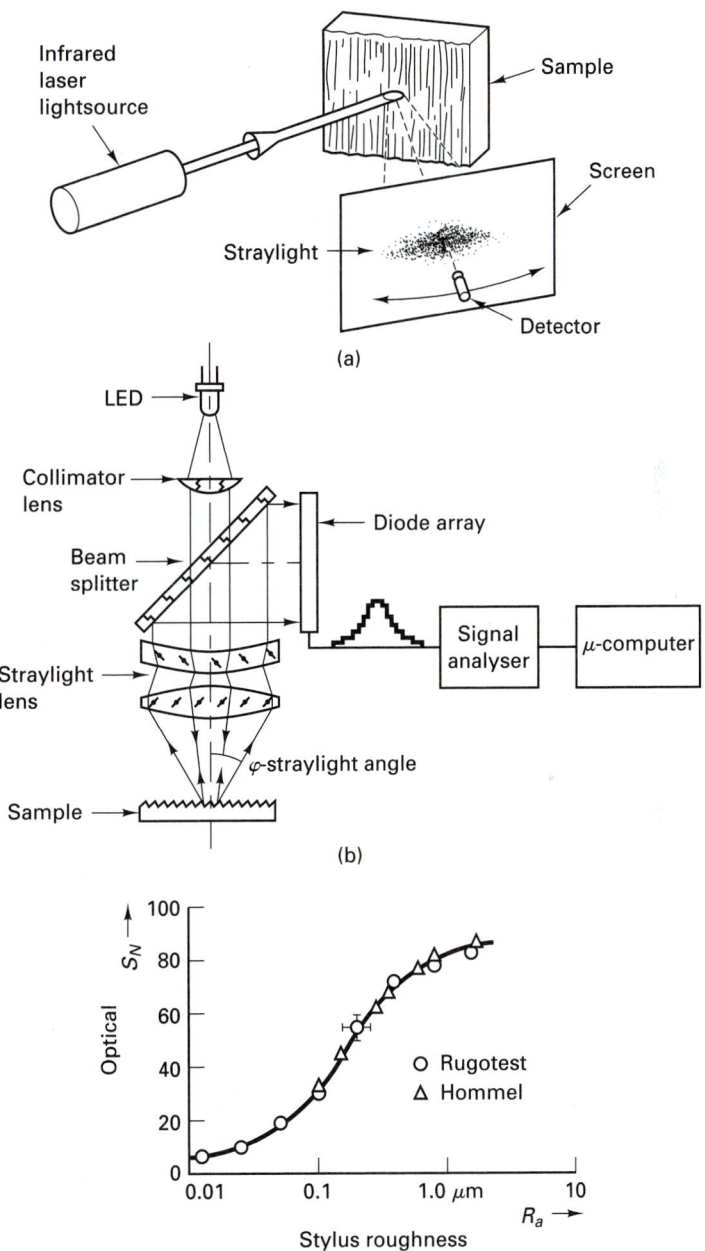

FIGURE 10-52   (a) Laser-light scattering on rough surface; (b) schematic of roughness measuring system; (c) good correlation between optical and stylus roughness in the low rms ranges. *(Courtesy of Rodenstock.)*

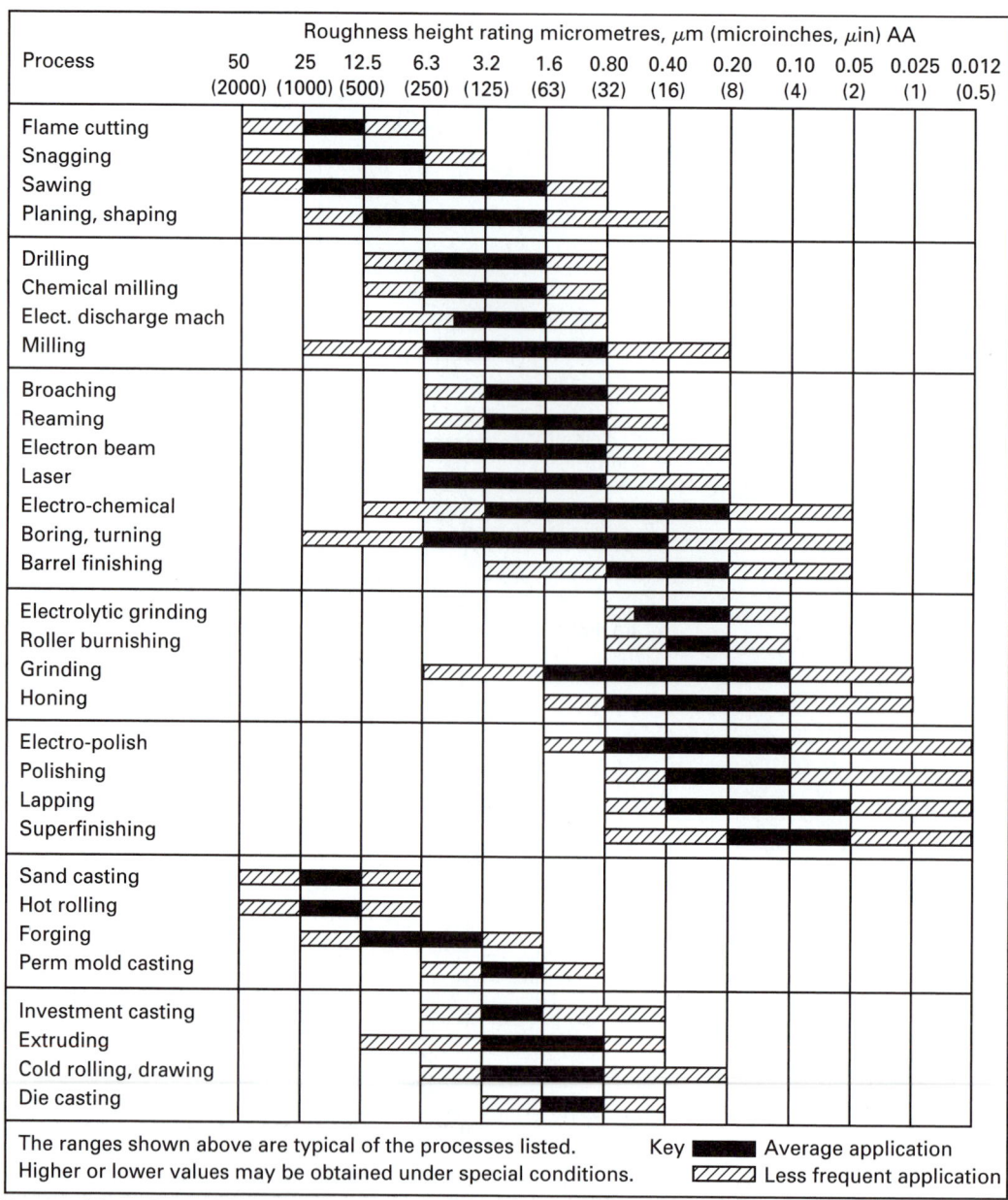

FIGURE 10-53   Comparison of surface roughness produced by common production processes. *(Courtesy of American Machinist.)*

FIGURE 10-54   Set of surface-roughness standards being used in a drafting room. *(Courtesy of Surface Checking Gage Co.)*

which shows the "*s*" from the word *trust* on a U.S. dime. The SEM micrograph was made after the topographical map of Figure 10-51b had been made. Both figures are at about the same magnification. The tracks produced by the stylus tip are easily seen in the micrograph. Notice the difference between the features shown in the micrograph and the trace, indicating that the stylus tip was not in contact with the surface many times during its passage over the surface (left no track in the surface), yet the trace itself is continuous.

Figure 10-52 shows schematics for a new laser-based instrument capable of measuring surface roughness. This instrument has the advantage of not needing to contact the surface to obtain a reading. The parameter s_n is a mean value from the scattered light distribution and has an entirely different meaning from R_a and is not (yet) an established standard. The noncontact feature, coupled with the measurement speed (20 readings per second) and the ability to read small parts easily are strong advantages. Clearly, such systems will be incorporated into machines in the future for in-process control of surface roughness.

The range of surface roughnesses that are typically produced by various manufacturing processes are indicated in Figure 10-53. For the designer's use, sets of *surface-finish blocks*, such as those shown in Figure 10-54, are very useful. It is difficult to visualize a surface having a given roughness, inasmuch as the same value of roughness may reflect different surface characteristics when produced by different processes.

■ Key Words

accuracy	length	snap gage
allowance	linearity	stability
ampere	machinist's rule	supermicrometer
attributes	magnification	surface roughness
candela	mass	temperature
clearance fit	metrology	time
coordinate measuring machine	micrometer caliper	tolerence
drift	optical comparator	toolmaker's flat
gage blocks	plug gage	toolmaker's microscope
geometric tolerances	precision	variables
interference bands	resolution	vernier caliper
interference fit	ring gage	vision system
laser interferometer	rule of 10	waviness
lay	sine bar	

■ Review Questions

1. What are some of the advantages to the consumer of standardization and of interchangeable parts?
2. Why is it important to interface the manufacturing engineering requirements with the design phase as early as possible?
3. Explain the difference between attributes and variables inspection.
4. Why have so many variable-type devices in autos been replaced with attribute-type devices?
5. What are the four basic measures upon which all others depend?
6. What is a pascal, and how is it made up of the basic measures?
7. What are the different grades of gage blocks, and why do they come in sets?
8. What keeps gage blocks together when they are "wrung together"?
9. What is the difference between tolerance and allowance?
10. What type of fit would describe the following situations?
    (a) The cap of a ball-point pen
    (b) The lead in a mechanical lead pencil, at the tip
    (c) The bullet in a barrel of a gun
11. List the items you can think of which are assembled with a medium-force fit.
12. Why might you use a shrink fit rather than welding to join two steel parts? What does the word *shrink* imply?
13. Explain the difference between accuracy and precision.

14. Into which of the five basic kinds of inspection does interferometry fall?
15. What factors should be considered in selecting measurement equipment?
16. Explain how you could determine if your ordinary bathroom scale is linear and has good repeat accuracy, assuming that the scale is analog.
17. Design and describe a simple experiment that demonstrates the difference between magnification and resolution?
18. Explain what is meant by the statement that usable magnification is limited by the resolution of the device.
19. What is parallax?
20. What is the rule of 10?
21. What is the principle of vernier calipers?
22. What are the two most likely sources of error in using micrometer calipers?
23. What is the major disadvantage of a micrometer caliper as compared with a vernier caliper? The advantages?
24. What would be the major difficulty in obtaining an accurate measurement with a micrometer depth gage if it were not equipped with a ratchet or friction device for turning the thimble?
25. Suppose that you had a 2-ft steel bar in your supermicrometer. Could you detect a length change if the temperature of the bar changed by 20°F?
26. Why is the toolmaker's microscope particularly useful for making measurement on delicate parts?
27. In what two ways can linear measurements be made using an optical projector?

28. What type of instrument would you select for checking the accuracy of the linear movement of a machine tool table through a distance of 50 in.?
29. What are the chief disadvantages of using a vision system for measurement compared to laser scanning?
30. What is a coordinate measuring machine?
31. Upon what principle is a sine bar based?
32. How can the not-go member of a plug gage be easily distinguished from the go member?
33. What is the primary precaution that should be observed in using a dial gage?
34. What tolerances are added to gages when they are being designed?
35. Explain how a go/not-go ring gage works for check a shaft.
36. Why are air gages particularly well suited for gaging the diameter of a hole?

37. Explain the principle of measurement by light-wave interference.
38. How does a toolmaker's flat differ from an optical flat?
39. Why may two surfaces that have the same microinch roughness be quite different in appearance?
40. Why are surface-finish blocks often used for specifying surface finish rather than just microinch values?
41. What limits the resolution of a stylus type surface-measuring device in finding profiles?
42. What is the general relationship between surface roughness and tolerance? Between tolerance and cost to produce the surface and/or tolerance?
43. What is the main disadvantage of the laser-based instrument for surface measurement?

■ PROBLEMS

1. Read the 25-division vernier graduated in English (Figure 10-A).

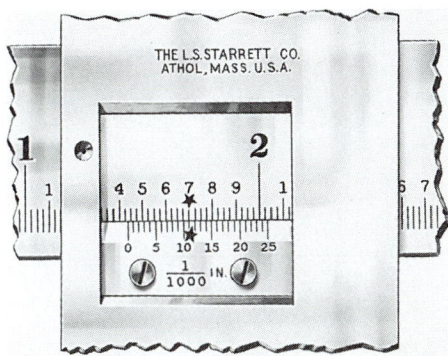

FIGURE 10-A

2. Read the 25-division vernier graduated in metric (direct reading) (Figure 10-B).
3. Convert the larger of the two readings in Figures 10-A and 10-B to units of the smaller and subtract.
4. In Figure 10-37, suppose that the height of the gage blocks is 3.2500 in. What is the angle θ assuming that the dial indicator is reading 0.0 ± 0.001 in.?
5. What is the estimated error in this measurement, given that Grade 3 working blocks are being used?
6. In Figure 10-C, the sleeve-thimble regions of

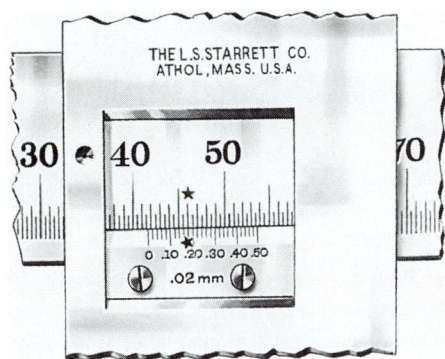

FIGURE 10-B

three micrometers graduated in thousandths of an inch are shown. What are the readings for these three micrometers? (*Hint*: Think of the various units as if you were making change from a $10 bill. Count the figures on the sleeve as dollars, the vertical lines on the sleeve as quarters, and the divisions on the thimble as cents. Add up your change, and put a decimal point instead of a dollar sign in front of the figures.)

7. Figures 10-D shows the sleeve-thimble region of two micrometers graduate in thousandths of an inch with a vernier for an additional ten-thousandths. What are the readings?
8. Two examples of metric vernier micrometer are shown in Figure 10-E. The micrometer is

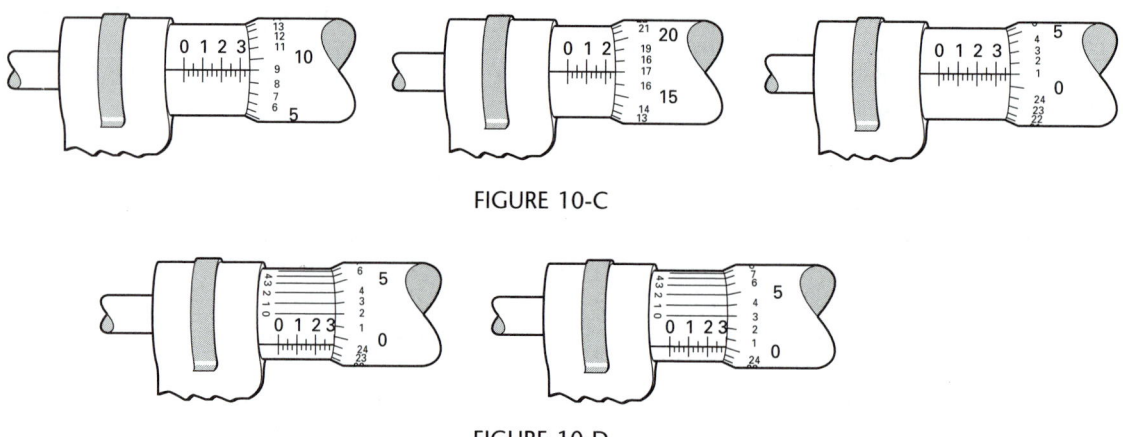

FIGURE 10-C

FIGURE 10-D

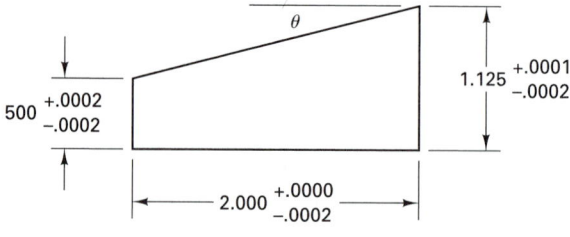

Thimble

Thimble

Sleeve

Sleeve

FIGURE 10-E

graduated in hundredths of a millimeter (0.01 mm), and an additional reading in two-thousandths of a millimeter (0.002 mm) is obtained from the vernier on the sleeve. What are the readings?

9. In checking a 1-in gage block by means of a helium light source, five dark bands were observed. There was a 2-in. distance between the front edges of the two blocks. What was the difference in height between the two blocks?

10. In Figure 10-F, the angle θ on the part needs to be inspected. The setup used is shown in Figure 10-G. (No sine plate was available.) Determine the angle θ from the part drawing and the value of X for the stack of gage blocks.

FIGURE 10-F

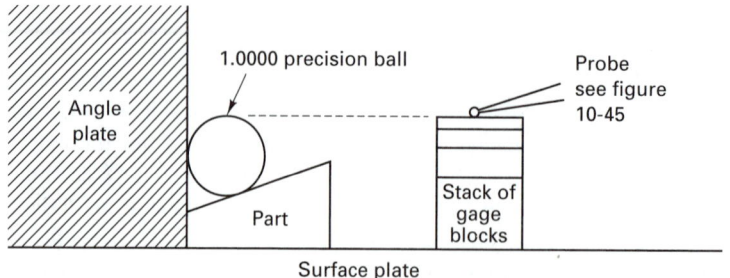

1.0000 precision ball

Angle
plate

Part

Stack of
gage
blocks

Probe
see figure
10-45

Surface plate

FIGURE 10-G

CHAPTER 11

NONDESTRUCTIVE INSPECTION AND TESTING

11.1 DESTRUCTIVE VERSUS NONDESTRUCTIVE TESTING
11.2 VISUAL INSPECTION
11.3 LIQUID PENETRANT INSPECTION
11.4 MAGNETIC PARTICLE INSPECTION
11.5 ULTRASONIC INSPECTION
11.6 RADIOGRAPHY
11.7 EDDY-CURRENT TESTING
11.8 ACOUSTIC EMISSION MONITORING
11.9. OTHER METHODS OF NONDESTRUCTIVE TESTING AND INSPECTION
   Leak Testing
   Thermal Methods
   Strain Sensing
   Advanced Optical Methods
   Resistivity Methods
   Computed Tomography
   Chemical Analysis and Topography of Surfaces
11.10 DORMANT VERSUS CRITICAL FLAWS
Case Study: PORTABLE FAILURE ANALYSIS KIT

■ 11.1 DESTRUCTIVE VERSUS NONDESTRUCTIVE TESTING

A major emphasis in any manufacturing operation is the production of a high-quality product, and such a product implies the absence of defects that may cause poor performance or product failure. Care in product design, material selection, fabrication of the desired shape, heat treatment, and surface treatment, as well as consideration of all possible service conditions, can do much to assure the manufacture of a high-quality product. However, it is important that we confirm that our efforts have been successful—that the product is indeed free from any harmful defects or flaws.

A variety of tests can be conducted to evaluate product quality and assure the absence of any flaws that might impair performance. *Destructive testing* provides one approach. Here sample components are subjected to test conditions that intentionally induce failure. By determining the specific conditions of product failure, insight can be gained into the performance characteristics of the larger production lot. Statistical methods are then used to determine the probability that the remaining products are good, based on the number of representative specimens that prove to be satisfactory. For example, let's assume that 100 parts are produced and one is tested to failure with satisfactory results. Is it safe to assume that the remaining 99 are acceptable? The satisfactory test of another randomly selected part would serve to increase our confidence in the remaining 98. Additional tests would further increase confidence, but the cost of destroying each of the products tested must be borne by the remaining quantity. Regardless of the degree of testing, however, there will still be some degree of uncertainty about the quality of each of the remaining products, because they have never been subjected to direct inspection.

Another means of assuring quality is through *proof testing*. Here a product is subjected to loads or pressure of a determined magnitude (generally equal to or greater than

299

the designed capacity). If the part remains intact, there is reason to believe that it will perform adequately, provided it is not subjected to abuse or loads in excess of its rated level. Proof tests can be conducted under laboratory conditions or at the site of installation, as with large manufactured assemblies such as pressure vessels.

In some situations, *hardness tests* can be used to provide insight into the quality of a product. Proper material and heat treatment can be reasonably ensured by requiring that all test results fall within a desirable range. (Abnormal results usually correlate with some form of manufacturing error, such as improper material or missed operations.) Hardness tests can be performed quickly and the surface indentations are often small enough that they can be concealed or easily removed from a product. The results, however, relate only to the surface strength of the product and bear no correlation to such defects as cracks or voids.

Table 11-1 provides a summary of the advantages and limitations of destructive testing and compares it with the nondestructive testing approach. In *nondestructive testing*, the product is examined in a manner that will not render it useless for future service. Tests can be performed on parts during or after manufacture, or even on parts that are already in

---

TABLE 11-1. Advantages and Limitations of Destructive and Nondestructive Testing

Destructive Testing

ADVANTAGES
1. Provides a direct and reliable measurement of how a material or component will respond to service conditions.
2. Provides quantitative results, useful for design.
3. Does not require interpretation of results by skilled operators.
4. Usually find agreement as to meaning and significance of test results.

DISADVANTAGES
1. Applied only to a sample; must show that the sample is representative of the group.
2. Tested parts are destroyed during testing.
3. Usually cannot repeat a test on the same item or use the same specimen for multiple tests.
4. May be restricted for costly or few-in-number parts.
5. Hard to predict cumulative effect of service usage.
6. Difficult to apply to parts in use; if done, testing terminates their useful life.
7. Extensive machining or preparation of test specimens is often required.
8. Capital equipment and labor costs are often high.

Nondestructive Testing

ADVANTAGES
1. Can be performed directly on production items without regard to cost or quantity available.
2. Can be performed on 100% of production lot (when high variability is observed) or a representative sample (if sufficient similarity is noted).
3. Different tests can be applied to the same item, and a test can be repeated on the same specimen.
4. Can be performed on parts that are in service; the cumulative effects of service life can be monitored on a single part.
5. Little or no specimen preparation is required.
6. The test equipment is often portable.
7. Labor costs are usually low.

DISADVANTAGES
1. Results often require interpretation by skilled operators.
2. Different observers may interpret the test results differently.
3. Properties are measured indirectly and results are often qualitative or comparative.
4. Some test equipment requires a large capital investment.

service. An entire production lot can be inspected, or representative samples can be taken. Different tests can be applied to the same item, either simultaneously or sequentially, and the same test can be repeated on the same specimen for additional verification. Little or no specimen preparation is required and the equipment is often portable, permitting on-site testing in most locations.

Nondestructive tests can have a variety of objectives, including the detection of internal or surface flaws, the measurement of dimensions, the determination of a material's structure or chemistry, or the evaluation of a material's physical or mechanical properties. In general, nondestructive tests incorporate the following aspects: (1) a probing medium that can be applied to the product; (2) a means by which a flaw, defect, material property, or specimen feature interacts with or modifies the probing medium; (3) a sensor to detect the response; (4) a device to indicate or record the response; and (5) a means of interpretation and evaluation of quality.

How you look at a material or product generally depends on what you are looking at, what you wish to see, and to what detail you wish to see it. Each of the various inspection processes has characteristic advantages and limitations. Some can be performed only on certain types of materials (such as electrical conductors or ferromagnetic materials). Each is limited in the type, size, and orientation of flaws that it can detect. There may be geometric restrictions as to part size, part complexity, or the degree of accessibility required for critical surfaces or locations. The availability of required equipment, the cost of operation, the need for a skilled operator or technician, and the possibility of producing a permanent test record are additional considerations when selecting a test procedure.

Regardless of the specific method, nondestructive testing can be a vital element in good manufacturing practice, and its potential value is becoming widely recognized as productivity demands increase, consumers demand higher-quality products, and product liability continues to be a concern. Rather than being an added manufacturing cost, its practice can actually expand profit by ensuring product reliability and customer satisfaction. In addition to its contribution to quality control, nondestructive testing can be used to aid in product design, provide on-line control of a manufacturing process, and reduce overall manufacturing costs.

The remainder of this chapter consists of an overview of the various nondestructive test methods. Each is presented along with its underlying principle, associated advantages and limitations, compatible materials, and typical applications.

■ 11.2 VISUAL INSPECTION

Probably the simplest and most widely used nondestructive testing method is visual inspection (Table 11-2). The human eye is a very discerning instrument and, with training, the brain can readily interpret the signals. Optical aids such as mirrors, magnifying glasses, and microscopes can expand the capabilities of the system. Video cameras and computer

---

TABLE 11-2. Visual Inspection

Principle: Illuminate the test specimen and observe the surface. Use of optical aids or assists is permitted.
  While most inspection is by human eye, video cameras and computer-vision systems can be employed.
Advantages: Simple, easy to use, relatively inexpensive.
Limitations: Depend on skill and knowledge of inspector. Limited to detection of surface flaws.
Material limitations: None.
Geometrical limitations: Any size or shape providing viewing accessibility of surfaces to be inspected.
Permanent record: Photographs or videotapes are possible. Inspectors' reports also provide valuable records.

systems, such as digital image analyzers, can be used to automate the inspection and perform quantitative geometrical evaluations. Borescopes and similar tools can provide accessibility to otherwise inaccessible locations. Only the surfaces of the product can be examined, however, and this forms the primary limitation of the method.

■ 11.3 LIQUID PENETRANT INSPECTION

Liquid penetrant testing (Table 11-3) is an effective method of detecting surface defects in metals and other nonporous material surfaces and is shown schematically in Figure 11-1. The piece to be tested is first subjected to a thorough cleaning and is dried prior to the test. Then a *penetrant*, a liquid material capable of wetting the entire surface and being drawn into fine openings, is applied to the surface of the workpiece by dipping, spraying, or brushing. Sufficient time is given for capillary action to draw the penetrant into the surface discontinuities, and the excess penetrant liquid is then removed. The surface is then coated with a thin film of "developer," an absorbent material capable of drawing traces of penetrant from the defects back onto the surface. Brightly colored dyes or fluorescent materials that radiate in ultraviolet light are generally added to the penetrant to make these traces more visible, and the developer is often selected to provide a contrasting background. Ra-

TABLE 11-3. Liquid Penetrant Inspection
Principle: A liquid penetrant is drawn into surface flaws by capillary action and subsequently revealed by developer material in conjunction with visual inspection.
Advantages: Simple, inexpensive, versatile, portable, easily interpreted, and applicable to complex shapes.
Limitations: Can only detect flaws that are open to the surface; surfaces must be cleaned before and after inspection; deformed surfaces and surface coatings may prevent detection; and the penetrant may be wiped or washed out of large defects.
Material limitations: Must have nonporous surface.
Geometrical limitations: Any size or shape permitting accessibility of surfaces to be inspected.
Permanent record: Photographs, videotapes, and inspectors' reports provide the most common records.

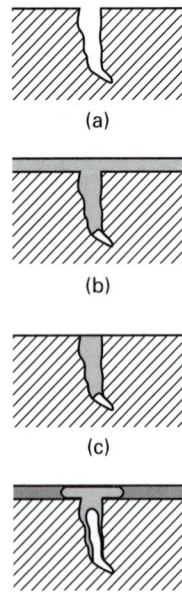

(a)

(b)

(c)

FIGURE 11-1 Liquid-penetrant testing: (a) initial surface with open crack; (b) penetrant is applied and is pulled into the crack by capillary action; (c) excess penetrant is removed; (d) developer is applied, some penetrant is extracted, and the product inspected.

(d)

dioactive tracers can be added and used in conjunction with photographic paper to produce a permanent image of the defects. Cracks, laps, seams, lack of bonding, pinholes, gouges, and tool marks can all be detected. After inspection, the developer and residual penetrant are removed by a second cleaning operation.

The detection of surface defects must be correlated with the manufacturing operations. If previous processing involves techniques that could have induced the flow of surface material, such as shot peening, honing, burnishing, machining, or various forms of cold working, a chemical etching may first be required to remove material that might be covering critical flaws. An alternative procedure is to penetrant test before the final surface-finishing operation, when significant defects are still open and available for detection. Penetrant inspection systems can range from aerosol spray cans of cleaner, penetrant, and developer (for portable applications), to automated, mass-production equipment using sophisticated computer vision systems.

■ 11.4 MAGNETIC PARTICLE INSPECTION

Magnetic particle inspection (Table 11-4) is based on the principle that ferromagnetic materials (such as iron, steel, nickel, and cobalt alloys), when magnetized, will have distorted magnetic fields in the vicinity of material defects (Figure 11-2). Surface and subsurface flaws, such as cracks and inclusions, can produce magnetic anomalies that can be mapped with the aid of magnetic particles on the specimen surface.

As with the previous methods, the specimen must be cleaned prior to inspection. A suitable magnetic field is then established in the part. As shown in Figure 11-3, orientation considerations are quite important. For a flaw to be detected it must produce a significant disturbance of the magnetic field at or near the surface. If a bar of steel is placed within

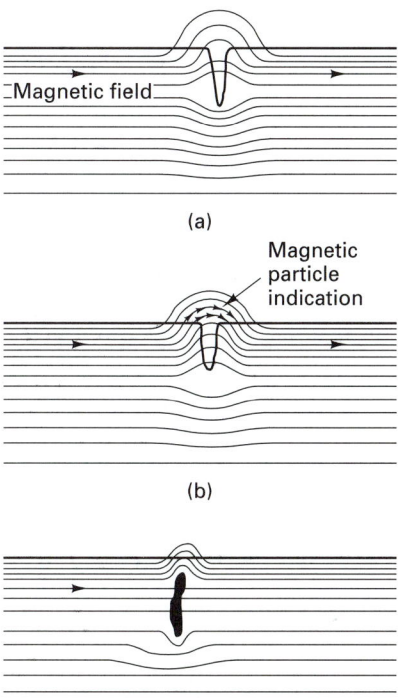

FIGURE 11-2   (a) Magnetic field showing disruption by a surface crack; (b) magnetic particles are applied and are preferentially attracted to field leakage; (c) subsurface defects can also produce surface-detectable disruptions if they are sufficiently close to the surface.

TABLE 11-4.   Magnetic Particle Inspection

Principle: When magnetized, ferromagnetic materials will have a distorted magnetic field in the vicinity of flaws and defects. Magnetic particles will be strongly attracted to surface regions where the flux is concentrated.

Advantages: Relatively simple, fast, easy-to-interpret portable units exist; can reveal both surface and subsurface flaws and inclusions (as much as $\frac{1}{4}$ in. deep) and small, tight cracks.

Limitations: Parts must be relatively clean; alignment of the flaw and the field affects the sensitivity so that multiple inspections with different magnetizations may be required; can only detect defects at or near surfaces; must demagnetize part after test; high current source is required; some surface processes can mask defects; postcleaning may be required.

Material limitations: Must be ferromagnetic; nonferrous metals such as aluminum, magnesium, copper, lead, tin, and titanium, and the ferrous (but not ferromagnetic) austenitic stainless steels cannot be inspected.

Geometrical limitations: Size and shape are almost unlimited, most restrictions relate to the ability to induce uniform magnetic fields within the piece.

Permanent record: Photographs, videotapes, and inspectors' reports are most common. In addition, the defect pattern can be preserved on the specimen by an application of transparent lacquer, or transferred to a piece of transparent tape that has been applied to the specimen and peeled off.

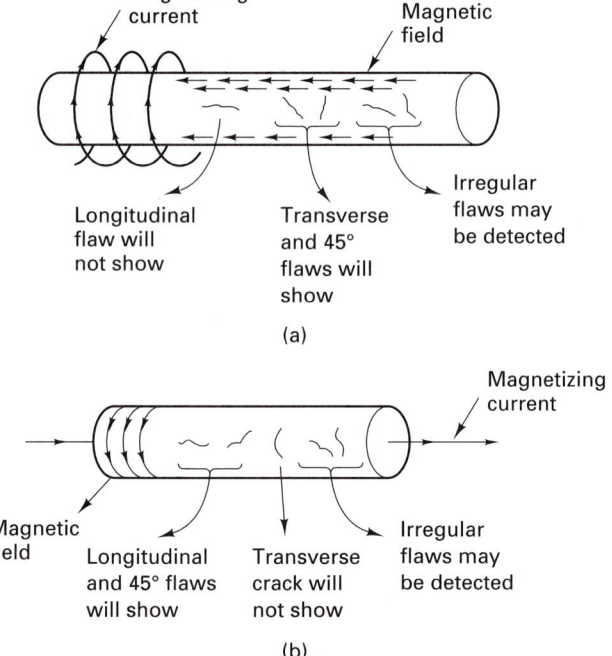

FIGURE 11-3   (a) A bar placed within a magnetizing coil will have an axial magnetic field. Defects parallel to this field may go unnoticed while those that disrupt the field and are sufficiently close to a surface are likely to be detected. (b) When magnetized by a current passing through it, the bar has a circumferential magnetic field and the geometries of detectable flaws are reversed.

an energized coil, a magnetic field will be produced whose lines of flux travel along the axis of the bar. Any defect perpendicular to this axis will alter the field significantly. If the perturbation is sufficiently large and close enough to the surface, the flaw can be detected by the inspection methods. However, if the flaw is in the form of a crack running along the specimen axis, there will be little perturbation of the lines of flux and the flaw is likely to go undetected.

If the same sample is then magnetized by passing a current through it, a circumferential magnetic field will be produced. The axial defect now becomes a significant perturbation and the defect perpendicular to the axis may go unnoticed. Thus a series of inspections utilizing different forms of magnetization may be required to fully inspect a

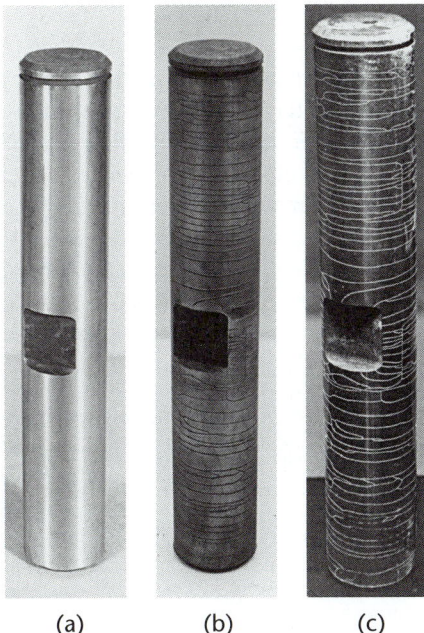

FIGURE 11-4 Front-axle king pin for a truck. (a) as manufactured and apparently sound; (b) inspected under conventional magnetic particle inspection to reveal numerous grinding-induced cracks.; (c) fluorescent particles and ultraviolet light make the cracks even more visible. *(Courtesy of Magnaflux Corporation.)*

(a)　　　　(b)　　　　(c)

product. Passing a current between various points of contact is a popular means of inducing the desired fields. Electromagnetic coils of various shapes and sizes are also used. Alternating-current methods are most sensitive to surface flaws, while direct-current methods are more capable of detecting subsurface defects, such as nonmetallic inclusions.

Once the specimen is magnetized, magnetic particles are applied to the surface either as a dry powder or a suspension in a liquid carrier. The particles are often made in an elongated form to better reveal the orientation of the lines of flux. In addition, they can be treated with a fluorescent material to enhance observation under ultraviolet light, or they can be coated with a lubricant to prevent oxidation and enhance their mobility. The resultant distribution of particles is then examined and any anomalies are interpreted. Figure 11-4 shows a component of a truck front-axle assembly: as manufactured, under straight magnetic particle inspection, and under ultraviolet light with fluorescent particles.

Some residual magnetization will be retained by all parts subjected to magnetic particle inspection. Therefore, it is usually necessary to demagnetize them before further processing or before placing them in use. One common means of demagnetization is to place the parts inside a coil powered by alternating current and then gradually reduce the current to zero. A final cleaning operation completes the process.

In addition to in-process and final inspection of parts during manufacture, magnetic particle inspection is used extensively during the maintenance and overhaul of equipment and machinery. The testing equipment ranges from small, portable units to complex automated systems.

■ 11.5 ULTRASONIC INSPECTION

Sound has long been used to provide an indication of product quality. A cracked bell will not ring true, but a fine crystal goblet will have a clear ring when tapped lightly. Striking an object and listening to the characteristic "ring" is an ancient art but is limited to the detection of large defects because the wavelength of audible sound is rather large compared to the size of most defects. By reducing the wavelength of the signal to the ultrasonic range, typically between 100,000 and 25 million hertz, ultrasonic inspection can detect rather small defects and flaws.

Ultrasonic inspection, therefore, involves sending high-frequency vibrations through a material and observing what happens (Table 11-5). Within a specimen, sound waves are affected by voids, impurities, changes in density, delaminations, interfaces with materials having a different speed of sound, and other imperfections. At an interface, part of the ultrasonic wave will be reflected and part will be transmitted. If the incident beam is at an angle to an interface where materials change, the transmitted portion of the beam will be bent to a new angle by the phenomenon of refraction. By receiving and interpreting the transmitted or reflected signals, ultrasonic inspection can be used to detect flaws within the material, measure thickness from only one side, or characterize metallurgical structure.

TABLE 11-5.   Ultrasonic Inspection

Principle: Sound waves are propagated through a test specimen and the transmitted or reflected signal is monitored and interpreted.

Advantage: Can reveal internal defects; high sensitivity to most cracks and flaws; high-speed test with immediate results; can be automated and recorded; portable; high penetration in most important materials (up to 60 ft in steel); indicates flaw size and location; access to only one side is required; can also be used to measure thickness, Poisson's ratio, or elastic modulus; presents no radiation or safety hazard.

Limitations: Difficult to use with complex shapes; external surfaces and defect orientation can affect the test (may need dual transducer or multiple inspections); a couplant is required; the area of coverage is small (inspection of large areas requires scanning); trained, experienced, and motivated technicians may be required.

Material limitations: Few—can be used on metals, plastics, ceramics, glass, rubber, graphite, and concrete, as well as joints and interfaces between materials.

Geometric limitations: Small, thin, or complex-shape parts or parts with rough surfaces and nonhomogeneous structure pose the greatest difficulty.

Permanent record: Ultrasonic signals can be recorded for subsequent playback and analysis. Strip charts can also be used.

An ultrasonic inspection system begins with a pulsed oscillator and transducer, a device that transforms electrical energy into mechanical vibrations. The pulsed oscillator generates a burst of alternating voltage, with a characteristic principal frequency, duration, profile, and repetition rate. This burst is then applied to a sending transducer, which uses a piezoelectric crystal to convert the electrical oscillations into mechanical vibrations. Because air is a poor transmitter of ultrasonic waves, an acoustic *coupling medium*—generally a liquid such as oil or water—is required to link the *transducer* to the piece to be inspected and transmit the vibrations into the part. The pulsed vibrations then propagate through the part with a velocity that depends on the density and elasticity of the test material. A receiving transducer is then used to convert the transmitted or reflected vibrations back into electrical signals. The receiving transducer is often identical to the sending unit, and the same transducer can actually perform both functions. A receiving unit then amplifies, filters, and processes the signal for display on an oscilloscope, possible recording on some form of electromechanical recorder, and final interpretation. An electronic clock is generally integrated into the system to time the responses and provide reference signals for comparison purposes.

Depending on the test objectives and part geometry, several different inspection methods can be employed:

1. In the *pulse-echo* technique, an ultrasonic pulse is introduced into the piece to be inspected and the echoes from the opposing surfaces and any intervening flaws are monitored by the receiver. The time interval between the initial emitted pulse and the various echoes is displayed on the horizontal axis of an oscilloscope screen. Defects

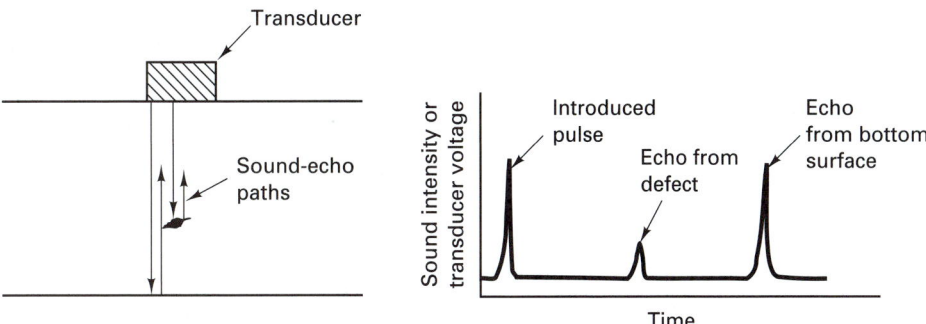

FIGURE 11-5 (a) Ultrasonic inspection of a flat plate with a single transducer; (b) plot of sound intensity or transducer voltage versus time showing the initial pulse and echoes from the bottom surface and intervening defect.

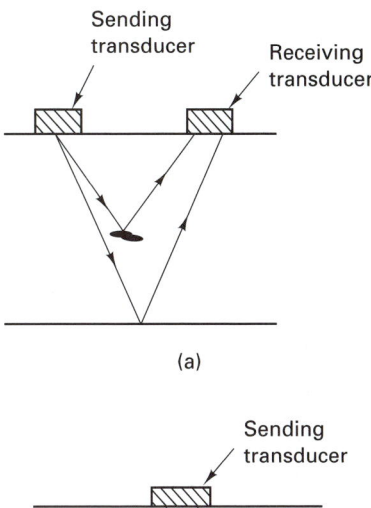

(a)

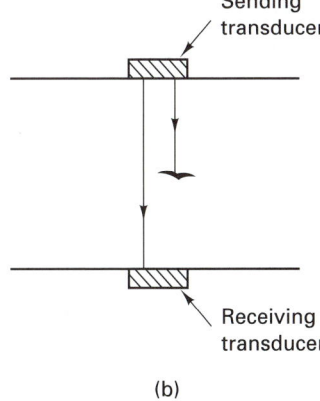

(b)

FIGURE 11-6 (a) Dual transducer ultrasonic inspection in the pulse-echo mode; (b) dual transducers in through-transmission configuration.

are identified by the position and amplitude of the various echoes. Figure 11-5 shows a schematic of a single-transducer pulse-echo inspection and the companion signal as it would appear on an oscilloscope. Figure 11-6a depicts a dual-transducer pulse-echo examination. Both cases require access to only one side of the specimen. **2.** The *through-transmission* technique requires separate sending and receiving transducers. As shown in Figure 11-6b, a pulse is emitted by the sending transducer and detected by the receiver. Flaws in the material decrease the amplitude of the transmitted signal because of back-reflection and scattering.

3. *Resonance testing* can be used to determine the thickness of a plate or sheet from one side of the material. Input pulses of varying frequency are fed into the material. When resonance is detected by an increase in energy at the transducer, the thickness can be calculated from the speed of sound in the material and the time of traverse. Ultrasonic thickness gages can be calibrated to provide direct digital readout of the thickness of a material.

Reference standards—specimens of known thickness or containing various types and sizes of machined "flaws"—are often used to ensure consistent results and aid in interpreting any indications of internal discontinuities.

■ 11.6 Radiography

Radiographic inspection (Table 11-6) employs the same principles and techniques as those of medical X-rays. A shadow pattern is created when certain types of radiation (X-rays, gamma rays, or neutron beams) penetrate an object and are differentially absorbed due to variations in thickness, density, or chemistry, or the presence of defects in the specimen. The transmitted radiation is registered on a photographic film that provides a permanent record and a means of analyzing the component. Fluorescent screens can provide direct conversion of radiation into visible light and enable fast and inexpensive viewing without the need for film processing. The fluorescent image, however, usually does not offer the sensitivity of the photographic methods.

TABLE 11-6. Radiography

Principle: Some form of radiation is passed through the sample and is differentially absorbed depending on the thickness, type of material, and the presence of flaws or defects.

Advantages: Probes the internal regions of a material; provides a permanent record of the inspection; can be used to determine the thickness of a material; very sensitive to density changes.

Limitations: Most costly of the NDT methods (involves expensive equipment, film, and processing); radiation precautions are necessary (potentially dangerous to human health); the defect must be at least 2% of the total section thickness to be detected (thin cracks can be missed if oriented perpendicular to the beam); film processing requires time, facilities, and care; the image is a two-dimensional projection of a three-dimensional object, so the location of an internal defect requires a second inspection at a different angle; complex shapes can present problems; a high degree of operator training is required.

Material limitations: Applicable to most engineering materials.

Geometric limitations: Complex shapes can present problems in setting exposure conditions and obtaining proper orientation of source, specimen, and film. Two-side accessibility is required.

Permanent record: A photographic image is part of the standard test procedure.

Various types of radiation can be used for inspection. *X-rays* are an extremely short wavelength form of electromagnetic radiation that are capable of penetrating many materials that reflect or absorb visible light. They are generated by high-voltage electrical apparatus—the higher the voltage, the shorter the X-ray wavelength and the greater the energy and penetrating power of the beam. *Gamma rays* are also electromagnetic radiation but are emitted during the disintegration of radioactive nuclei. Various radioactive isotopes can be selected as the radiation source. *Neutron beams* for radiography are generally derived from nuclear reactors, nuclear accelerators, or radioisotopes. For most applications it is necessary to moderate the energy and collimate the beam before use.

The absorption of X-rays and gamma rays depends on the thickness, density, and atomic structure of the material being inspected. The higher the atomic number, the greater

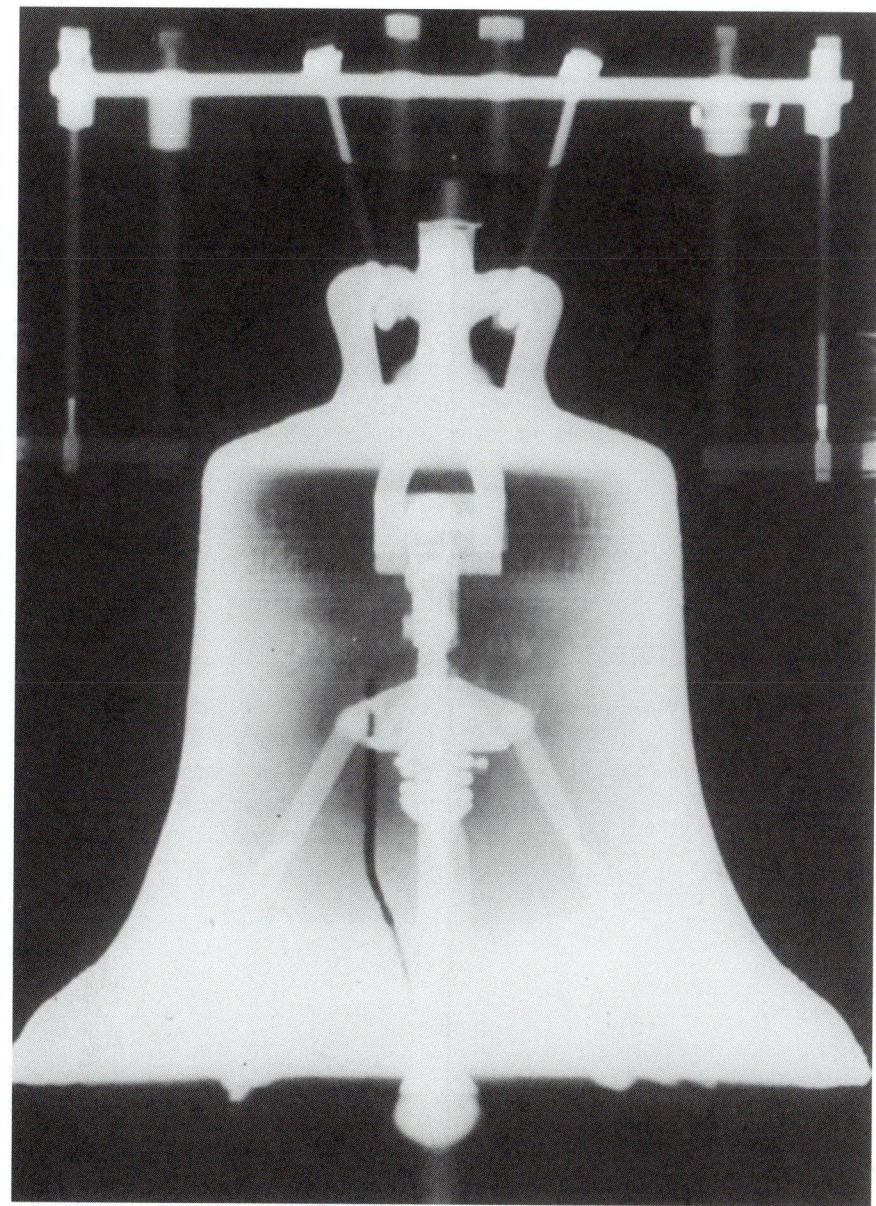

FIGURE 11-7 Full-size radiograph of the Liberty Bell. The photo reveals the famous crack, as well as the iron spider installed in 1915 to support the clapper and the steel beam and supports which were set into the yoke in 1929. *(Courtesy of Eastman Kodak company.)*

the attenuation of the beam. Figure 11-7 shows a radiograph of the historic Liberty Bell. The famous crack is clearly visible, along with the internal spider (installed to support the clapper in 1915) and the steel beam and supports installed in the wooden yoke in 1929. Other radiographs disclosed shrinkage separations and additional cracks that were not known to exist as well as a crack in the bell's clapper.

In contrast to X-ray absorption, neutron absorption varies widely from atom to atom, with no pattern in terms of atomic number. Unusual contrasts can be obtained that would be impossible with other inspection methods. For example, hydrogen has a high neutron absorption, so the presence of water in a product can easily be detected by neutron radiography. X-rays, on the other hand, are readily transmitted through water.

When a radiation beam is passed through an object, part of the radiation is scattered in all directions. This scatter produces an overall "fogging" of the radiograph and a reduction in contrast and sharpness of the image. The thicker the material, the more troublesome the scattered radiation becomes. Photographic considerations of exposure time and development also affect the quality of the radiographic image.

It is a common practice to include a standard test piece, or *penetrameter,* in a radiographic exposure. Penetrameters are made of the same or similar material as the specimen and contain features with known dimensions. The image of the penetrameter can then be compared to the image of the product being inspected. Regions of similar intensity are attributed to be of similar thickness. Subtle but important variations in photographic density can often be revealed by use of image-enhancing computer software.

Radiography is not inexpensive. Therefore, many users recommend extensive use during the development of a product or process, followed by spot-checks and statistical methods during subsequent production.

■ 11.7 EDDY-CURRENT TESTING

When an electrically conductive material is exposed to an alternating magnetic field such as that generated by a coil of wire carrying an alternating current, small electric currents are induced on or near the surface of the material (Figure 11-8). These induced *eddy currents*, in turn, generate their own opposing magnetic field, which then reduces the strength of the field from the coil. This change in magnetic field causes a change in the impedance of the coil, which in turn changes the magnitude of the current flowing through it. By

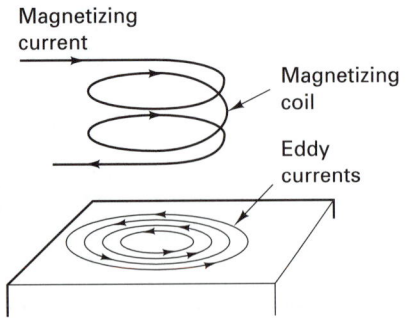

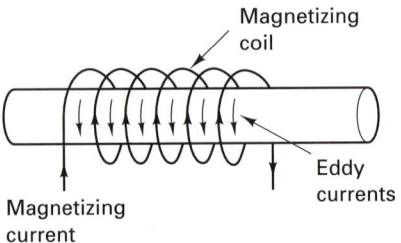

FIGURE 11-8 Relation of the magnetizing coil, magnetizing current, and induced eddy currents. The magnetizing current is actually an alternating current, producing a magnetic field that forms, collapses, and reforms in the opposite direction. This dynamic magnetic field induces the eddy currents and the changes in the eddy currents produce a secondary magnetic field which interacts with the sensor coil or probe.

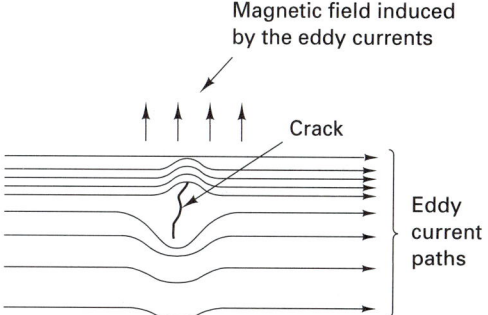

Magnetic field induced
by the eddy currents

Crack

Eddy
current
paths

FIGURE 11-9   Eddy currents are constrained to travel within the conductive material, but the magnitude and path of the currents will be affected by defects and changes in material properties. By focusing on the magnitude of the eddy currents, features such as differences in heat treatment can be detected.

monitoring the *impedance* of the exciting coil, or a separate indicating coil, eddy-current testing can be used to detect any condition that would affect the current-carrying conditions (or conductivity) of the test specimen. Figure 11-9 shows how the eddy-current paths would be forced to alter around a crack, thereby changing the characteristics of the induced magnetic field in that vicinity.

Eddy-current testing (Table 11-7) can be used to detect surface and near-surface flaws, such as cracks, voids, inclusions, and seams. Stress concentrations, differences in metal chemistry, or variation in heat treatment will affect the magnetic permeability and conductivity of a metal and will therefore alter the eddy-current characteristics. Material mix-ups and processing errors can be detected. Specimens can be sorted by hardness, case depth, residual stresses, or any other structure-related property. Thicknesses or variation in thickness of platings, coatings, or even corrosion can be detected or measured.

TABLE 11-7.   Eddy-Current Testing

Principle: When an electrically conductive material is brought near an alternating-current coil that produces an alternating magnetic field, surface currents (eddy currents) are generated in the material. These surface currents generate their own magnetic field, which interacts with the original, modifying the impedance of the originating coil. Various material properties and/or defects can affect the magnitude and direction of the induced eddy currents and can be detected by the electronics.

Advantages: Can detect both surface and near-surface irregularities; applicable to both ferrous and nonferrous metals; versatile—can detect flaws, variations in alloy or heat treatment, variations in plating or coating thickness, wall thickness, and crack depth; intimate contact with the specimen is not required; can be automated; electrical circuitry can be adjusted to select sensitivity and function; pass/fail inspection is easily conducted; high speed; low cost; no final cleanup is required.

Limitations: Response is sensitive to a number of variables, so interpretation may be difficult; sensitivity varies with depth, and depth of inspection depends on the test frequency; reference standards are needed for comparison; trained operators are generally required.

Material limitations: Only applicable to conductive materials, such as metals; some difficulties may be encountered with ferromagnetic materials.

Geometric limitations: Depth of penetration is limited; must have accessibility of coil or probe; constant separation distance between coils and specimen is required for good results.

Permanent record: Electronic signals can be recorded using devices such as strip-chart recorders.

Eddy-current test equipment can range from simple, portable units with hand-held probes to fully automated systems with computer control and analysis. Each system, however, includes:

1. A source of magnetic field capable of inducing eddy currents in the part being tested. This source generally takes the form of a coil (or coil-containing probe) carrying alternating current. Various coil geometries are used for different-shaped specimens.

2. A means of sensing the field changes caused by the interaction of the eddy currents with the original magnetic field. Either the exciting coil itself or a secondary sensing coil can be used to detect the impedance changes. Differential testing can be performed using two oppositely wound coils wired in series. In this method, only differences in the signals between the two coils are detected as one or both coils are scanned over the specimen.

3. A means of measuring and interpreting the resulting impedance changes. The simplest method is to measure the induced voltage of the sensing coil, a reading that evaluates the cumulative effect of all variables affecting the eddy-current field. Phase analysis can be used to determine the magnitude and direction of the induced eddy-current field. Familiarity with characteristic impedance responses can then be used to identify selected features in the specimen.

When comparing alternative techniques, eddy current is usually not as sensitive as penetrant testing in detecting small, open flaws, but it requires none of the cleanup operations and is noticeably faster. In a similar manner, it is not as sensitive as magnetic particle inspection to small subsurface flaws but can be applied to all metals (ferromagnetic and nonferromagnetic alike). In addition, eddy-current testing offers capabilities that cannot be duplicated by the other methods, such as the ability to differentiate between various chemistries and heat treatments.

■ 11.8 ACOUSTIC EMISSION MONITORING

Materials undergoing stressing, deformation, or fracture emit sound waves in frequencies as high as 1 MHz. While these sounds are inaudible to the human ear, they are detectable through the use of sophisticated electronics. Transducers, amplifiers, filters, counters, and microcomputers can be used to isolate and analyze the sonic emissions of a cracking or deforming material. Much like the warning sound of ice cracking underneath boots or skates, the acoustic emissions of materials can be used to provide a warning of impending danger. They can detect deformations as small as 10^{-12} in./in. (that occur in short intervals of time), initiation or propagation of cracks (including stress-corrosion cracking), delamination of layered materials, and fiber failure in composites. By using multiple sensors, it is possible to accurately pinpoint the source of these sounds by a triangulation method similar to that used to locate seismic sources (earthquakes) in the earth.

Acoustic emission monitoring (Table 11-8) involves listening for indications of failure. Temporary monitoring can be used to detect the formation of cracks in materials during production operations, such as welding and subsequent cooling of the weld region. Monitoring can also be employed to assure the absence of plastic deformation during pre-service proof testing. Continuous surveillance may be justified when the product or component is particularly critical, as with bridges and nuclear reactor pressure vessels. The sensing electronics can be coupled to an alarm and safety system to protect and maintain the integrity of the structure.

In contrast to the previous inspection methods, acoustic emission cannot detect an existing defect in a static product. Instead, it is a monitoring technique designed to detect a dynamic change in the material, such as the formation or growth of a crack or defect, or the onset of plastic deformation.

TABLE 11-8. Acoustic Emission Monitoring

Principle: Almost all materials will emit high-frequency sound (acoustic emissions) when stressed, deformed, or undergoing structural changes, such as the formation or growth of a crack or defect. These emissions can now be detected and provide an indication of dynamic change within the material.

Advantages: The entire structure can be monitored with near-instantaneous detection and response; only "active" flaws are detected; defects inaccessible to other methods can be detected; inspection can be in harsh environments; and the location of the emission source can be determined.

Limitations: Only growing flaws can be detected (the mere presence of defects is not detectable); background signals may cause difficulty; there is no indication of the size or shape of the flaw; and experience is required to interpret the signals.

Material limitations: Virtually unlimited, provided that they are capable of transmitting sound.

Geometric limitations: Requires continuous sound-transmitting path between the source and the detector. Size and shape of the component affects the strength of the emission signals that reach the detector.

■ 11.9 OTHER METHODS OF NONDESTRUCTIVE TESTING AND INSPECTION

Leak Testing

Leak testing is a form of nondestructive testing designed to determine the existence or absence of leak sites and the rate of material loss through the leaks. Various testing methods have been developed, ranging from the rather crude bubble-emission test (pressurize, immerse, and look for bubbles) to advanced techniques involving tracers, detectors, and sophisticated apparatus. Each has its characteristic advantages, limitations, and sensitivity. Selection should be on the basis of cost, sensitivity, reliability, and compatibility with the specific product to be tested.

Thermal Methods

Temperature-sensing devices (including thermometers, thermocouples, pyrometers, temperature-sensitive paints and coatings, liquid crystals, infrared scanners, infrared film, and others) can also be used to evaluate the soundness of engineering materials and components. Parts can be heated and then inspected to detect abnormal temperature distributions that are the result of faults or flaws. The locations of "hot spots" on an operating component can be a valuable means of defect detection and can provide advanced warning of impending failure. Faulty electrical components tend to be hotter than defect-free counterparts. Composite materials can be subjected to brief pulses of intense heat and then inspected to reveal the pattern of thermal conductivity. Thermal anomalies tend to appear in areas where the bonding between the components is poor or incomplete.

Strain Sensing

Although used primarily during product development, strain sensing techniques can also be used to provide valuable insight into the stresses and stress distribution within a part. Brittle coatings, photoelastic coatings, or electrical resistance strain gages can be applied to the external surfaces of a part, which is then subjected to an applied stress. The extent and nature of cracking, the photoelastic pattern produced, or the electrical resistance changes then provide insight into the strain at various locations. X-ray diffraction methods and extensiometers have also been used.

Advanced Optical Methods

Although visual inspection is often the simplest and least expensive of the nondestructive inspection methods, there are also a number of advanced optical methods. Monochromatic laser light can be used to detect differences in the backscattered pattern from a part and a "master." The presence or absence of geometrical features such as holes or gear teeth is readily detected. Holograms can provide three-dimensional images of an object, and holographic interferometry can detect minute changes in the shape of an object under stress.

Resistivity Methods

The *electrical resistivity* of a conductive material is a function of its chemistry, processing history, and structural soundness. Measurement of resistivity can, therefore, be used for alloy identification, flaw detection, or the assurance of proper processing. Tests can be developed to evaluate the effects of heat treatment, the amount of cold work, the integrity of welds, or the depth of case hardening. The development of sensitive micro-ohm-meters (or microhmeters) has greatly expanded the possibilities in this area.

Computed Tomography

While X-ray radiography provides a single image of the X-ray intensity being transmitted through an object, X-ray computed tomography (CT) is an inspection technique that provides a cross-sectional view of the interior of an object along a plane parallel to the X-ray beam. This is the same technology that has revolutionized medical diagnostic imaging (CAT scans), with the process parameters (such as the energy of the X-ray source) being adapted to permit the nondestructive probing of industrial products. Basic systems include an X-ray source, an array of detectors, a mechanical system to move and rotate the test object, and a dedicated computer system. The intensity of the received signal is recorded at each of the numerous detectors with the part in a variety of orientations. Complex numerical algorithms are then used to construct an image of the interior of the component. Internal boundaries and surfaces can be determined clearly, enabling inspection and dimensional analysis of a product's interior. The presence of cracks, voids, or inclusions can be detected and their precise location can be determined.

CT inspections are both slow and costly, so they are currently used when the component is critical and the other inspection methods prove to be inadequate. Complex shapes are difficult to inspect by alternative methods, and the resolution of detail is often limited. In addition, the video images of the CT technique permit easy visualization and interpretation.

Chemical Analysis and Topography of Surfaces

While nondestructive inspection is usually associated with the detection of flaws and defects, various nondestructive techniques can also be employed to determine the chemical and elemental analysis of surface and near-surface material. These techniques include Auger electron spectroscopy (AES), energy-dispersive X-ray analysis (EDX), electron spectroscopy for chemical analysis (ESCA), and various forms of secondary-ion mass spectroscopy (SIMS). Because of its large depth of focus, the scanning electron microscope has become an extremely useful tool for observing the surfaces of materials. More recently, the atomic-force microscope and scanning tunneling microscope have extended this capability and can be used to provide information about surface topography with resolution to the atomic scale.

■ 11.10 DORMANT VERSUS CRITICAL FLAWS

There was a time when the detection of a flaw was considered to be sufficient cause for rejecting a material or component, and material specifications often contained the term *flaw-free*. Such a criterion, however, is no longer practical, because the sensitivity of detection methods has increased dramatically. If materials were rejected upon detection of a flaw, we would find ourselves rejecting nearly all commercial engineering materials.

If a defect is sufficiently small, it is possible for it to remain dormant throughout the useful lifetime of a product, never changing in size or shape. Such a defect is clearly allowable. Larger defects, however, may grow or propagate under the same conditions of loading, often causing sudden or catastrophic failure. These flaws would be clearly unacceptable. The objective (or challenge), therefore, is to determine the threshold below which a flaw remains dormant and above which it becomes critical, and a cause for rejection. This issue is addressed in the section "Fracture Toughness and the Fracture Mechanics Approach" in Chapter 2.

■ KEY WORDS

acoustic emission	impedance	radiography
computed tomography	leak testing	resonance
coupling medium	nondestructive testing (*also*	tomography
critical flaw	nondestructive inspection)	transducer
dormant flaw	penetrameter	ultrasonic
eddy current	penetrant	visual inspection
electrical resistivity	proof test	
hardness testing	pulse-echo method	

■ REVIEW QUESTIONS

1. What are some unattractive features of destructive testing methods?
2. What is a proof test, and what assurance does it provide?
3. What quality-related features can a hardness test reasonably ensure?
4. What exactly is nondestructive testing, and what are some attractive features of the approach?
5. What are some possible objectives of nondestructive testing?
6. What are some factors that should be considered when selecting a nondestructive testing method?
7. How might the costs of nondestructive testing actually be considered as an asset rather than a liability?
8. What is the primary limitation of a visual inspection?
9. What types of defects can be detected in a liquid penetrant test?
10. What is the primary materials-related limitation of magnetic particle inspection?
11. Describe how the orientation of a flaw with respect to a magnetic field can affect its detectability during magnetic particle inspection.
12. What is the major limitation of sonic testing, where one listens to the characteristic "ring" of a product in an attempt to detect defects?
13. What types of defects can be detected by ultrasonic inspection?
14. What are three types of ultrasonic inspection methods?
15. What types of radiation can be used in radiographic inspection of manufactured products?
16. What are penetrameters, and how are they used in radiographic inspection?
17. Why would we not expect eddy-current examination to be useful with ceramics or polymeric materials?
18. What types of defects or product features can be detected by eddy-current inspection?
19. Why can't acoustic emission methods be used to detect the presence of an existing but static defect?
20. How can acoustic emission be used to determine the location of a flaw or defect?
21. How can temperature be used to reveal defects?
22. What kinds of product features can be evaluated

by electrical resistivity methods?

23. What type of information can be obtained through computed tomography?

24. Why is it important to set the size of "allowable flaws," as opposed to rejecting any material that contains a detectable flaw?

■ PROBLEMS

1. A manufacturing company routinely specifies X-ray radiography to assure the absence of cracks in its products. The primary reason for selecting radiography is the availability of a hard-copy record of each inspection for use in any possible liability litigation. Discuss the pros and cons of their selection. What other processes might you want to consider? If some form of permanent record is desirable, discuss how these might be obtained for the various alternative processes?

2. For each of the inspection methods listed below, cite one major limitation to its use.
   (a) Visual inspection
   (b) Liquid penetrant inspection
   (c) Magnetic particle inspection
   (d) Ultrasonic inspection
   (e) Radiography
   (f) Eddy-current testing
   (g) Acoustic emission monitoring

3. Which of the major nondestructive inspection methods might you want to consider to inspect products made from the following materials?
   (a) Ceramic
   (b) Polymeric
   (c) Composite with (1) polymer matrix and (2) metal matrix

4. Discuss the application of nondestructive inspection methods to powder metallurgy (metallic) products with low, average, and high density.

*C*hapter 11 CASE STUDY

portable failure analysis kit

You are a member of a corporate failure analysis group, working out of the home office of Big-One Petroleum Corporation, a large petroleum company with production sites throughout North America. These sites generally contain such equipment as pumps, valves, pipelines, and storage tanks, which occasionally experience failure under both normal and abnormal production conditions. Much of this equipment is large and cumbersome. Therefore, when a failure occurs, you must frequently travel to the site of the failure to begin your investigation. Your primary objectives are to collect information, inspect the site, and acquire specimens and samples that can be taken back to your base laboratory for further investigation and study.

Your present assignment is to design and equip a portable failure analysis kit for on-site investigation. This should take the form of a suitcase that could be stored in your laboratory. When a problem occurs, you could simply pick up the kit and be on the next flight to that location.

1. Your container is restricted to the size of a typical suitcase and must be carried by a single individual (suggested container weight of less than 50 pounds). Moreover, you are being asked to purchase and equip the case for less than $1000.00. (It is hoped that your case can be duplicated and distributed to other individuals throughout the corporation). With these constraints, develop a list of the items that you would include in your case. For each item, discuss briefly: a description of the item, its intended use or uses, and justification of the cost (if over $50.00), the weight (if over five pounds), and the size (if it occupies over 20% of the suitcase).

2. If additional funds were subsequently provided to upgrade your suitcase, but the size and weight restrictions were maintained, what additional or upgraded items would you want to include? Discuss each in the manner used in part 1.

CHAPTER 12

PROCESS CAPABILITY AND QUALITY CONTROL

12.1 INTRODUCTION
12.2 DETERMINING PROCESS CAPABILITY
Making PC Studies
What PC Studies Tell about the Process
Taguchi Methods
12.3 INSPECTION AND QUALITY CONTROL

Statistical Process Control
Sampling Errors
Quality Control Charts
12.4 DETERMINING CAUSES FOR PROBLEMS IN QUALITY
Total Quality Control

■ 12.1 INTRODUCTION

All manufacturing processes display some level of variability, referred to as inherent capability or inherent uniformity. For example, suppose that we view "shooting at a metal target" as a "process" for putting holes in a piece of metal. I hand you the gun and tell you to take five shots at the bull's-eye. Thus you are the operator of the process. To measure the *process capability* (PC), that is, your ability to hit consistently what you are aiming at, the bull's-eye, the target is inspected after you have finished shooting. So the capability of manufacturing processes is determined by measuring the output of the process. In *quality control* (QC), the product is examined to determine whether or not the processing accomplished was what was specified in the design.

PC studies use statistical and analytical tools, as does QC, except that the results are directed at the machines used in the processing rather than the output or products from the processes. Going back to our example, a PC study would be directed toward quantifying the inherent accuracy and precision in the shooting process. The QC program would be designed to root out problems that can cause defective products during production. Traditionally, the objective of QC studies has been to find the defects in the process. The more progressive point of view is to inspect to prevent defects from occurring (Figure 12-1).

■ 12.2 DETERMINING PROCESS CAPABILITY

The *nature of the process* refers to both the *variability* (or inherent uniformity) and the aim of the process. Thus in the target-shooting example, a perfect process would be capable of placing five shots right in the middle of the bull's-eye, one right on top of the other. The process would display no variability with perfect *accuracy*. Such performance would be very unusual in a real industrial process. The variability may have assignable causes and may be correctable if the cause can be found and eliminated. That variability to which no cause can be assigned and which cannot be eliminated is inherent in the process and is therefore its "nature."

Some examples of assignable causes of variation in processes include multiple machines for the same components, operator blunders, defective materials, or progressive wear in the tools during machining. Sources of inherent variability in the process include

317

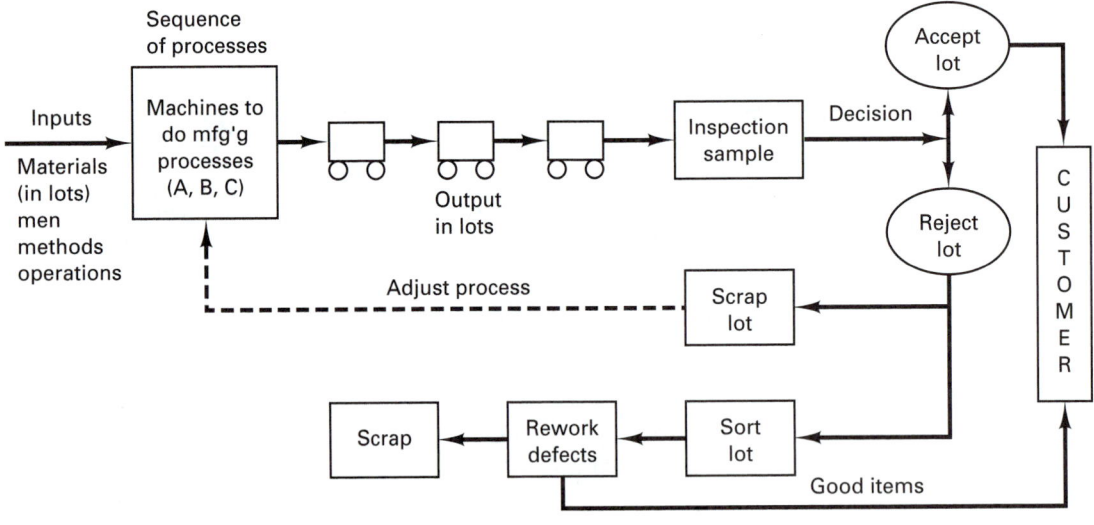

(a) Inspection that finds defects

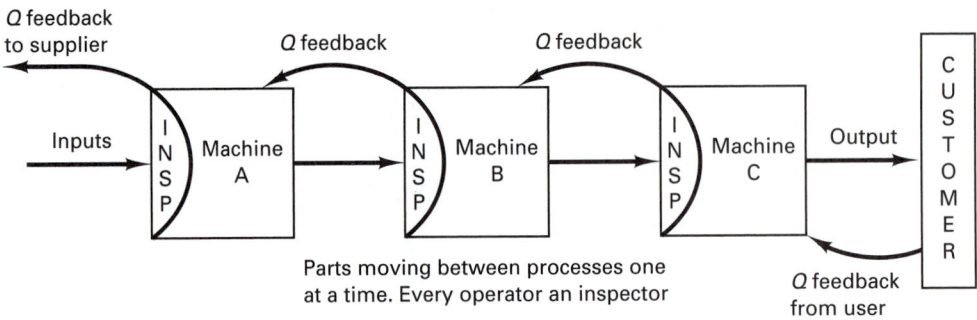

(b) Inspection to prevent defects

FIGURE 12-1   Don't inspect to find defects; inspect to prevent the defect from happening.

variation in material properties, operator variability, vibrations and chatter, and the wear of the sliding components in the machine, perhaps resulting in poorer operation of the machine. These kinds of variations, which occur naturally in processes, usually display a random nature and often cannot be eliminated. In QC terms, these are referred to as *chance causes*. Sometimes, the causes of assignable variation cannot be eliminated because of cost. Almost every process has multiple causes of variability occurring simultaneously, so it is extremely difficult to separate the effects of the different sources of variability during the analysis.

Making PC Studies

The object of the PC study is to determine the inherent nature of the process as compared to the desired specifications. The output of the process must be examined under normal conditions, or what is typically called *hands-off conditions*. The inputs (e.g., materials, set-ups, cycle times, temperature, pressure, and operator) are fixed or standardized. The process is allowed to run without tinkering or adjusting while the output (i.e., the product or units or components) is documented carefully with respect to (1) time, (2) source, and (3) order of production. A sufficient amount of data has to be taken to ensure confidence in the statistical analysis of the data. The precision of the measurement system should exceed the process capability by at least one order of magnitude. (See the discussion of the rule of 10 in Chapter 10.)

Prior to data collection, these steps must be taken:

1. Design the PC experiment.
   A. Use normal or hands-off process conditions; specify machine settings for speed, feed, volume, pressure, material, temperature, operator, and so on. (This is the *standard* PC study *approach.*)
   Alternative to A. Use specified combinations of all of the input parameters, at various levels, that are believed to influence the quality characteristics being measured. These combinations should be run with the objective of selecting the best. For example, speed levels may be high, normal, and low; and operators may be fast or slow. (This is the *Taguchi* or *factorial approach.*)
2. Define the inspection method and the inspection means (the procedure and the instrumentation).
3. Decide how many items will be needed to perform the statistical analysis. (For a Taguchi approach, this means deciding how many replications are desired at each level of combination.)
4. For a standard PC study, use homogeneous input material, and try to contrast it with normal (more variable) input material. (For a Taguchi approach, material *is* an input variable specified at different levels: normal, homogeneous, and highly variable. If a material is not controllable, it is considered a noise factor.)
5. Data sheets must be designed to record date, time, source, order of production, and all the process parameters being used (or measured) while the data are being gathered. (For a Taguchi method, this is usually a simplified experimental design, called an orthogonal array.)
6. Assuming that the standard PC study approach is being used, the process is run and the parts are made and measured.

Let us assume that the designer specified the part to be 1.001 ± 0.005 in. After 70 units have been manufactured without any adjustment of the process, each unit is measured and the data are recorded on the data sheet. A frequency distribution, or *histogram* (Figure 12-2), is developed. This histogram shows the raw data and the desired value, along with the upper and lower *specification limits*, where LSL represents the lower specification limit and USL the upper specification limit. [These are also called the (lower and upper) tolerance limits.] The statistical data are used to estimate the mean and the standard deviation of this distribution.

The mechanics of this statistical analysis are shown in Figure 12-3. The true mean of the distribution, designated $\bar{\bar{X}}'$ ("X bar prime") is to be compared with the nominal value. The estimate of the true *standard deviation*, designated $\hat{\sigma}'$ ("sigma prime"), is used to determine how the process compares with the desired tolerance. *The purpose of the analysis is to obtain estimates of $\bar{X}'$ and σ' values, the true process parameters, as they are not known.* A sample size of 5 was used in this example, so $n = 5$. Fourteen groups of samples were drawn from the process, so $k = 14$. For each sample the *sample mean* $\bar{X}$ and *sample range* R are computed. For large samples ($n > 12$), the standard deviation of each sample should be computed rather than the range. Next, the average of the sample averages, $\bar{\bar{X}}$, is

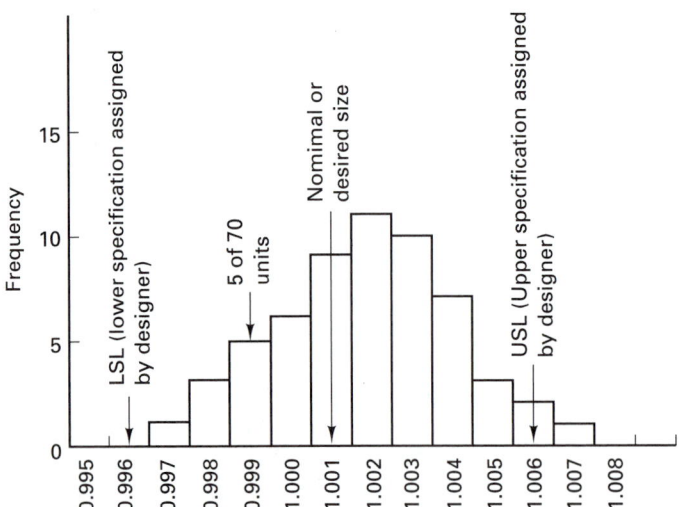

FIGURE 12-2   Histogram of 70 measurements of a parameter.
The design specification was 1.001 ± 0.005 in.

computed. Sometimes called the *grand average*. This is used to estimate the mean of the process, $\overline{X}'$. The standard deviation of the process, which is a measure of the spread or variability of the process, is estimated from either the average of the sample ranges, $\overline{R}$, or the average of the sample standard deviations, $\overline{\sigma}$, using either $\overline{R}/d_2$ or $\overline{\sigma}/c_2$. The factors d_2 and c_2 depend on the sample size n and are given in Table 12.1. The process capability is defined by $\pm 3\hat{\sigma}'$ or $6\hat{\sigma}'$. Thus $\overline{X}' \pm 3\hat{\sigma}'$ defines the natural capability limits of the process, assuming that the process is approximately normally distributed.

Note that a distinction is made between a sample and a population. A sample is of a specified, limited size and is drawn from the population. The population is the large source of items, which can include all the items the process will ever produce under the conditions specified. Our calculations assume that this population was normal or bell-shaped. Figure 10-6 shows a typical normal curve and the areas under the curve as defined by the standard deviation. Other distributions are possible, but the histogram in Figure 12-2 clearly suggested that this process can best be described by a normal probability distribution. Now it remains for the process engineer and the operator to combine their knowledge of the process with the results from the analysis to draw conclusions about the ability of this process to meet specifications.

What PC Studies Tell about the Process

PC studies provide answers to these questions:

1. Does the process have the ability to meet specifications? To answer this question, a process capability ratio, C_p, is often computed, where

$$C_p = \frac{\text{tolerance spread}}{6\hat{\sigma}'} = \frac{\text{USL} - \text{LSL}}{6\hat{\sigma}'} \tag{12-1}$$

A value of $C_p \geq 1.33$ is considered good.

2. Is the process well centered with respect to the desired nominal specification? To answer this question, the amount of mismatch, D, is computed, where

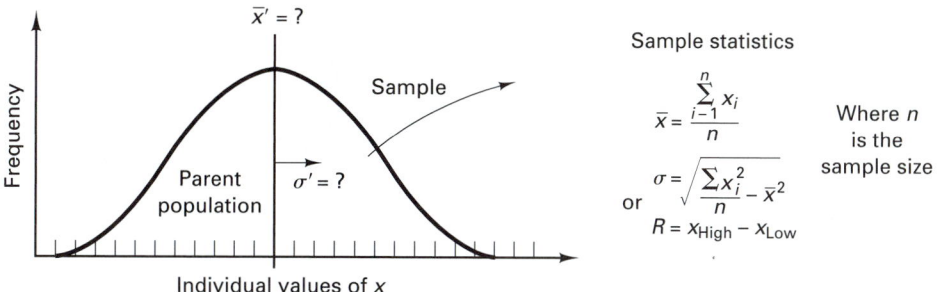

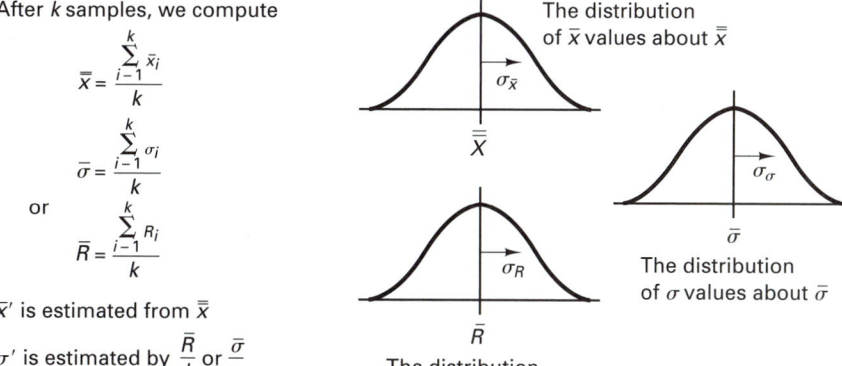

Each distribution of averages of samples, or standard deviation of samples or ranges of samples must have its own average and standard deviation.

FIGURE 12-3   Calculations needed to obtain estimates of the mean ($\overline{X}'$) and the standard deviation (σ')of the parent population, as in a process capability study.

$$D = \frac{\text{estimated process mean} - \text{nominal}}{\frac{1}{2}\,(\text{tolerance spread})} = \frac{\overline{X}' - \text{nominal}}{\frac{1}{2}\,(\text{USL} - \text{LSL})} \qquad (12\text{-}2)$$

The condition for not producing rejects is

$$\overline{X}' - \text{nominal} + 3\hat{\sigma}' < \tfrac{1}{2}(\text{USL} - \text{LSL}) \qquad (12\text{-}3)$$

The capability index combines both factors and is defined as

$$C_{pk} = \frac{\min\{|\text{USL} - \overline{X}'|, |\text{LSL} - \overline{X}'|\}}{3\hat{\sigma}'} \qquad (12\text{-}4)$$

In response to the first question, the width of the histogram is compared with the specifications (Figure 12-4). The natural spread of the process, $6\hat{\sigma}'$, is computed and then compared with the upper and lower tolerance limits. Three situations can exist:

1. $6\hat{\sigma}' < \text{USL} - \text{LSL}$ or $C_p > 1$, or process variability less than tolerance spread (Figure 12-4a).
2. $6\hat{\sigma}' = \text{USL} - \text{LSL}$ or $C_p = 1$, or process variability equal to tolerance spread (Figure 12-4c).

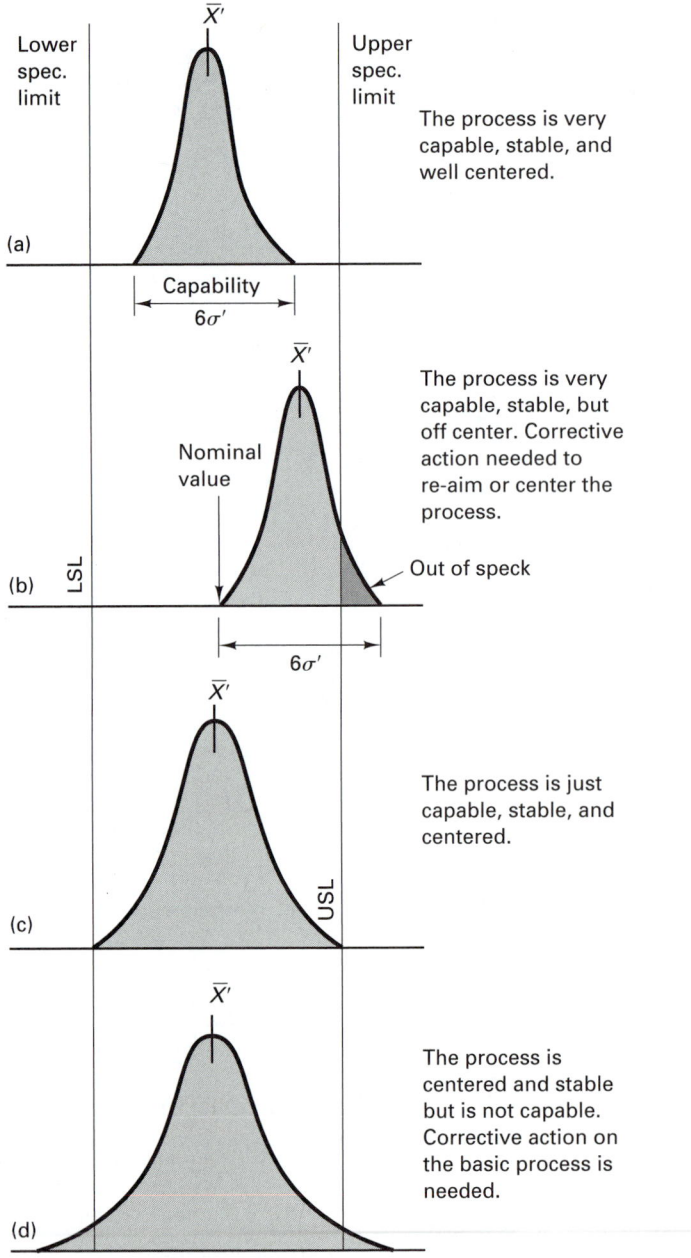

FIGURE 12-4 What PC studies tell about the process.

3. $6\hat{\sigma}' >$ USL $-$ LSL or $C_p < 1$, or process variability greater than tolerance spread (Figure 12-4d).

In situation 1 the machine is capable of meeting the tolerances applied by the designer. Generally speaking, if process capability is on the order of two-thirds to three-fourths of the design tolerance, there is a high probability that the process will produce all good parts over a long period of time. If the PC is on the order of one-half or less of the

TABLE 12.1. Factors for Estimating the Standard Deviation (σ') of a Parent Population for Sample-Data for the Average Range ($\bar{R}$) or the Average Sample Standard Deviation ($\bar{\sigma}$)

Sample Size or the Number of Observations in Subgroup n	Range Factor for Estimate from $\bar{R}$, $d_2 \equiv \bar{R}/\sigma'$	Standard Deviation Factor for Estimate from $\bar{\sigma}$, $c_2 \equiv \bar{\sigma}/\sigma'$	Factor That Relates σ_R with σ' (σ_R = Standard Deviation of Range Distribution), $d_3 = \sigma_R/\sigma'$
2	1.128	0.5642	0.8525
3	1.693	0.7236	0.8884
4	2.059	0.7979	0.8798
5	2.326	0.8407	0.8641
6	2.534	0.8686	0.8480
7	2.704	0.8882	0.8330
8	2.847	0.9027	0.8200
9	2.970	0.9139	0.8084
10	3.078	0.9227	0.7970
11	3.173	0.9300	0.7870
12	3.258	0.9359	0.7780
13	3.336	0.9410	For $n > 12$, C_2 factors
14	3.407	0.9453	should be used with $\bar{\sigma}$
15	3.472	0.9490	
16	3.532	0.9523	
17	3.588	0.9551	
18	3.640	0.9576	
19	3.689	0.9599	
20	3.735	0.9619	
25	3.931	0.9696	
30	4.086	0.9748	

Source: 1950 Manual on Quality Control of Materials, copyright ASTM, 1916 Race Street, Philadelphia, Pa. 19103. Adapted, with permission.

design tolerance, it may be that the selected process is "'too good"; that is, the company is producing ball bearings when what is called for is marbles. In this case it may be possible to trade off some precision in this process for looser specifications elsewhere, resulting in an overall economic gain. Quality in well-behaved processes can be maintained by checking the first and last parts of a lot or production run. If these parts are good, the lot is certain to be good. This is called $N = 2$. Naturally, if the lot size is 2 or less, this is 100% inspection. Sampling and control charts are also used under these conditions to maintain the process aim and variability.

In situation 3 the process is not capable of meeting the design specifications. There are a variety of alternatives here, including:

a. Shifting this job to another machine with greater process capability.
b. Getting a review of the specifications to see if they may be relaxed.
c. Sorting the product, to separate the good from the bad. This entails 100% inspection of the product, which may not be a feasible economic alternative unless it can be done automatically. Automatic sorting of the product on a 100% basis can ensure near-perfect quality of all the accepted parts. The automated station shown in Figure 12-5 checks parts for the proper diameter with the aid of a linear variable differential transformer (LVDT). As a part approaches the inspection station on a motor-driven conveyor system, a computer-based controller activates a clamping device. Embedded in the clamp is an LVDT position sensor with which the control computer can measure the diameter of the part. Once the measurement has been

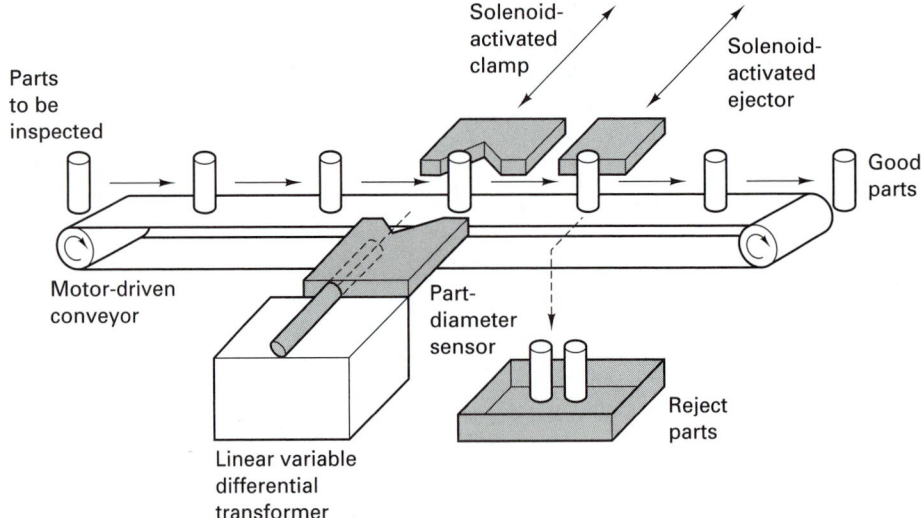

FIGURE 12-5   A linear variable differential transformer (LVDT) is a key element in an inspection station checking part diameters.   Momentarily clamped into the sensor fixture, a part pushed the LVDT armature into the device winding.   The LVDT output is proportional to the displacement of the armature.   The transformer makes highly accurate measurements over a small displacement range.

made, the computer releases the clamp, allowing the part to be carried away.   If the diameter of the part is within a given tolerance, a solenoid-actuated gate operated by the computer lets the part pass.   Otherwise, the part is ejected into a bin. With the fast-responding LVDT, 100% of manufactured parts can be automatically sorted quickly and economically.

Automated sorting does not determine what caused the defects, so this example is an "automated defect finder."   How would one change this inspection system to make it "inspect to prevent" the defect from occurring?

d. Determining whether the *precision* of the process can be improved by:
(1) Switching cutting tools, workholding devices, or materials.
(2) Overhauling the existing process and/or developing a preventive maintenance program.
(3) Finding and eliminating the causes of variability, or
(4) Combinations of (1), (2), and (3).
(5) Using a factorial (Taguchi) procedure, that is, designing experiments to reduce the variability of the process.

In Figure 12-4c the process capability is almost exactly equal to the assigned tolerance spread, so if the process in not perfectly centered, defective products will always result.   Thus situation 2 should be treated like situation 3 unless the process can be perfectly centered and maintained.   Tool wear, which causes the distribution to shift, must be negligible.   Then the situation can be treated as in 1, particularly if a small percentage of parts just outside tolerance is acceptable.

The second question deals with the ability of the process to maintain centering so that the average of the distribution comes as close as possible to the nominal value desired (see Figure 12-4b).   This process needs to be recentered so that the mean of the process distribution is at or near the nominal value. Most processes can be reamed. Poor accuracy is often due to assignable causes, which can be eliminated.

In addition to direct information about the accuracy and precision of the process, PC studies can also tell the manufacturing engineer how pilot processes compare with production processes, and vice versa. If the source and time of each product are carefully recorded, information about the instantaneous reproducibility can be found and compared with the repeatability of the process with respect to time (time-to-time variability). More important, since almost all processes are duplicated, PC studies generate information about machine-to-machine variability. Going back to our target-shooting example, suppose that five different guns were used, all of the same make and type. The results would have been different, just as having five marksmen use the same gun would have resulted in yet another outcome. Thus PC studies generate information about the homogeneity and the differences in multiple machines and operators.

It is quite often the case in such studies that one variable dominates the process. Target shooting viewed as a process is "operator"-dominated in that the outcome is highly dependent on the skill (the capability) of the "worker." Processes that are not well engineered nor highly automated, or in which the worker is viewed as "highly skilled," are usually operator dominated. Processes that change or shift uniformly with time but that have good repeatability in the short run are often machine dominated. For example, the mean of a process ($\overline{X}'$) will usually shift after a tool change, but the variability may decrease or remain unchanged. Machines tend to become more precise (to have less variability within a sample) after they have been broken in (i.e., the rough contact surfaces have smoothed out because of wear) but will later become less precise (will have less repeatability) due to poor fits between moving elements (called *backlash*) of the machine under varying loads. Other variables that can dominate processes are setup, input components, and even information.

In many machining processes in use today, the task of tool setting has been replaced by an automatic tool-positioning capability (see Chapter 29), which means that one source of variability in the process has been eliminated and the process becomes more repeatable. In the same light, it will be very important in the future for manufacturing engineers to know the process capability of robots they want to use in the workplace.

The discussion to this point has assumed that the *parent population* is normally distributed, that is, has the classic bell-shaped distribution in which the percentages shown in Figure 10-6 are dictated by the number of standard deviations from the central value or mean. The shape of the histogram may reveal the nature of the process to be skewed to the left or the right (unsymmetrical), often indicating some natural limit in the process. Drilled holes exhibit such a trend as the drill tends to make the hole oversize. Another possibility is a bimodal distribution (two distinct peaks), often caused by two processes being mixed together. The possibilities are endless and require a careful recording of all the sources of the data to track down the factors that result in loss of precision and accuracy in the process. Rapid feedback on quality is perhaps the most important factor.

Taguchi Methods

The alternative approach to making a PC study may involve the Taguchi method, which requires a more sophisticated analysis. The Taguchi approach uses a truncated experimental design (called an *orthogonal array*) to determine which process inputs have the greatest effect on process variability (i.e., precision) and which have the least. Those inputs that have the greatest influence are set at levels that minimize their effect on process variability. As shown in Figure 12-6, factors A, B, C, and D all have an effect on process variable V. By

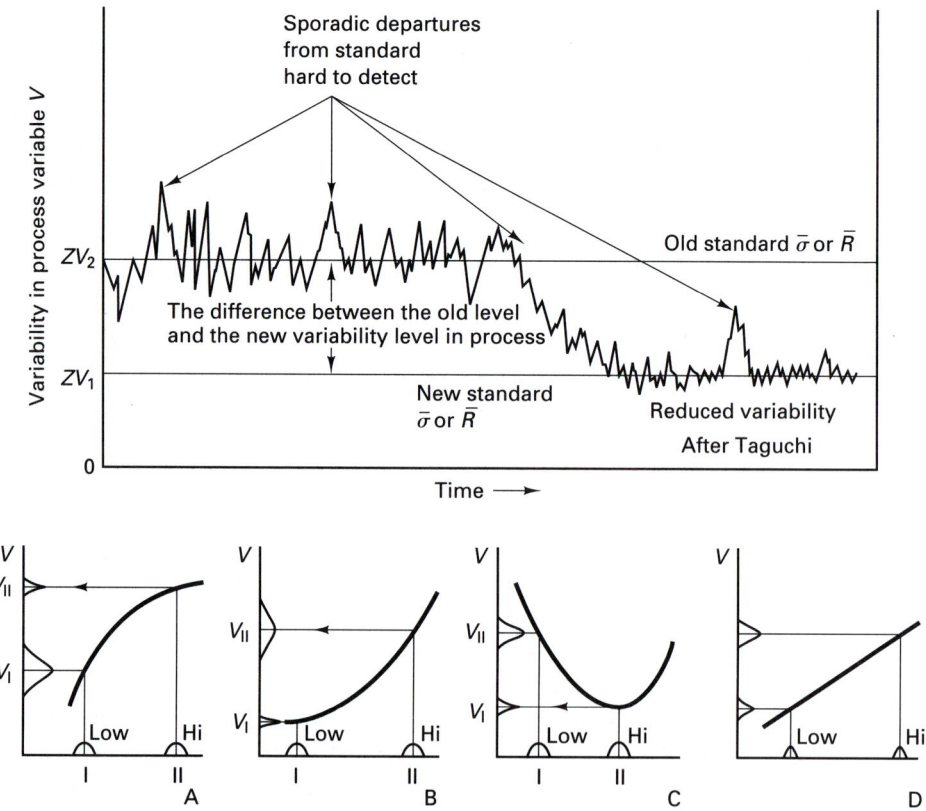

FIGURE 12-6   The use of Taguchi methods can reduce the inherent process variability as shown in the upper figure; factors *A, B, C,* and *D* versus process variable *V* shown in lower figure.

selecting a high level of *A* and low levels of *B, C,* and *D,* the inherent variability of the process can be reduced. Those factors that have little effect on the process variable *V* are used to adjust or recenter the process aim. In other words, Taguchi methods seek to minimize or dampen the effect of the causes of variability and thus to reduce the total process variability.

In Chapter 29 process capability is addressed again in the context of machining centers, programmable (NC) machines. In machine tools, accuracy and precision in processing are affected by machine alignment, the setup of the workholder, the design and rigidity (accuracy) of the workholder, the accuracy of the cutting tools, the design of the product, the temperature, and the operating parameters. The Taguchi methods provide a means of determining which of the input parameters are most influential in product quality.

■ 12.3 INSPECTION AND QUALITY CONTROL

In virtually all manufacturing, it is extremely important that the dimensions and quality of individual parts be known and maintained. This is of particular importance where large quantities of parts, often made in widely separated plants, must be capable of interchangeable assembly. Otherwise, difficulty may be experienced in subsequent assembly or in service, and costly delays and failures may result. In recent years, defective products resulting in death or injury to the user have resulted in expensive litigation and damage awards against manufacturers. Inspection is that function which controls the quality (e.g., the dimensions, the performance, and the color)—manually, by using operators or inspectors or automatically with machines, as discussed previously.

The economics-based question "How much should be inspected?" has three possible answers.

1. *Inspect every item being made.* 100% Inspect every item being made; 100% checking with prompt execution of feedback and immediate corrective action can ensure perfect quality.
2. *Sample.* Inspect some of the product by sampling and make decisions about the quality of the process based on the sample.
3. *None.* Assume that everything made is acceptable or that the product is inspected by the consumer, who will exchange it if it is defective. (This procedure is not recommended.)

The reasons for not inspecting all of the product (i.e., for sampling) include:

1. Everything has not yet been manufactured—the process is continuing to make the item—so we look at some before we are done with all.
2. The test is destructive.
3. There is too much product for all of it to be inspected.
4. The testing takes too much time or is too complex or too expensive.
5. It is not economically feasible to inspect everything even though the test is simple, cheap, and quick.

Some characteristics are nondissectible; that is, they cannot be measured during the manufacturing process because they do not exist until after a whole series of operations have taken place. The final edge geometry of a razor blade is a good example, as is the yield strength of a rolled bar of steel.

Sampling (looking at some percentage of the whole) requires the use of statistical techniques that permit decisions about the acceptability of the whole based on the quality found in the sample. This is known as *statistical process control* (SPC).

Statistical Process Control

Looking at some (sampling) and deciding about the behavior of the whole (the parent population) is common in industrial inspection operations. The basic SPC techniques are the histogram and control charts. The histogram (Figure 12-7) could be used to classify the range of shaft diameters after turning. The X axis represents the diameter measurement of the turned part and the Y axis represents the frequency with which these measurements were found. How many parts were measured? The natural tolerance limits of the distribution can be compared with the engineering specifications to determine if the process is centered at the print *nominal* value and to estimate if diameters are being produced within tolerance values (or product specifications).

Control charts for variables are used to monitor the output of a process by sampling (looking at some), by measuring selected quality characteristics, by plotting the sample data on the chart, and then by making decisions about the performance of the process.

Figure 12-8 shows the basic structure of three control charts commonly used for variables types of measurements. *The data plotted on these charts are sample values, not individual values.* The $\overline{X}$ chart tracks the aim (accuracy) of the process. The R chart (or σ chart) tracks the precision or variability of the process. Usually, only the $\overline{X}$ chart and the

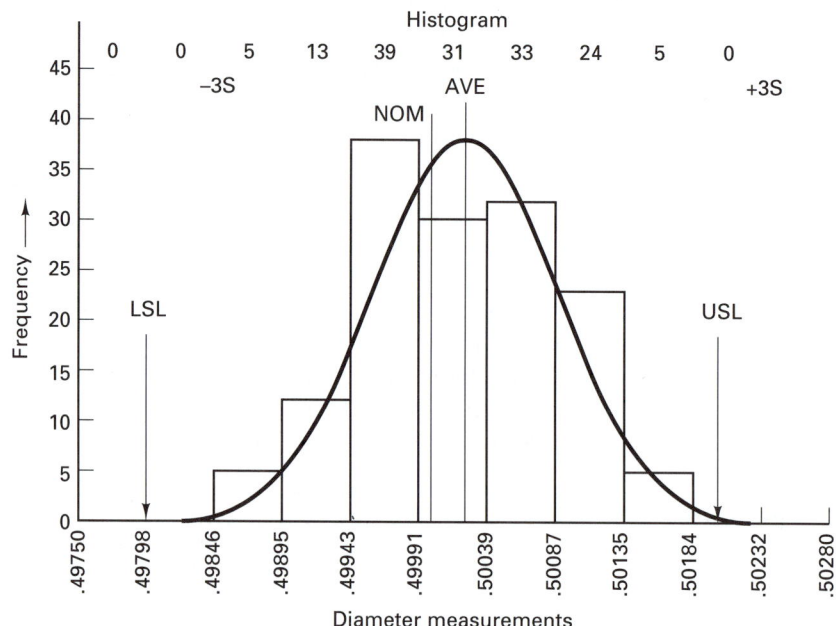

FIGURE 12-7   Histogram of a process to compare measured sizes with designer specifications.

R chart are used unless the sample size is large, and then σ charts are used in place of R charts. Because some sample statistics tend to be normally distributed about their own mean (see Figure 12-3), $\overline{X}$ values are normally distributed about $\overline{\overline{X}}$, R values are normally distributed about $\overline{R}$, and σ values are normally distributed about $\overline{\sigma}$.

Sampling Errors

It is important to understand that in sampling, two kinds of decision errors are always possible. Suppose that the process is running perfectly but the sample data indicate that something is wrong. You decide to stop the process to make adjustments. This is a Type I error. Suppose that the process was not running perfectly and was making defective products. However, the sample data did not indicate that anything was wrong. You did not stop the process and set it right. This is a type II error. Both types of errors are possible in sampling. For a given sample size, reducing the chance of one type of error will increase the chance of the other. Increasing the sample size or the frequency of sampling reduces the probability of errors but increases the cost of the inspection. Many companies determine the size of the errors they are willing to accept according to the overall cost of making the errors plus the cost of inspection. If, for example, a type II error is very expensive in terms of product recalls or legal suits, the company may be willing to make more type I errors, to sample more, or even to go to 100% inspection on very critical items to ensure that the company is not accepting defective materials as good and passing them on to the customer. The inspection should take place immediately after the processing.

As mentioned earlier, in any continuing manufacturing process, variations from established standards are of two types: (1) *assignable cause variations*, such as those due to malfunctioning equipment or personnel, or to defective material, or to a worn or broken tool; and (2) *normal chance variations*, resulting from the inherent nonuniformities that exist in

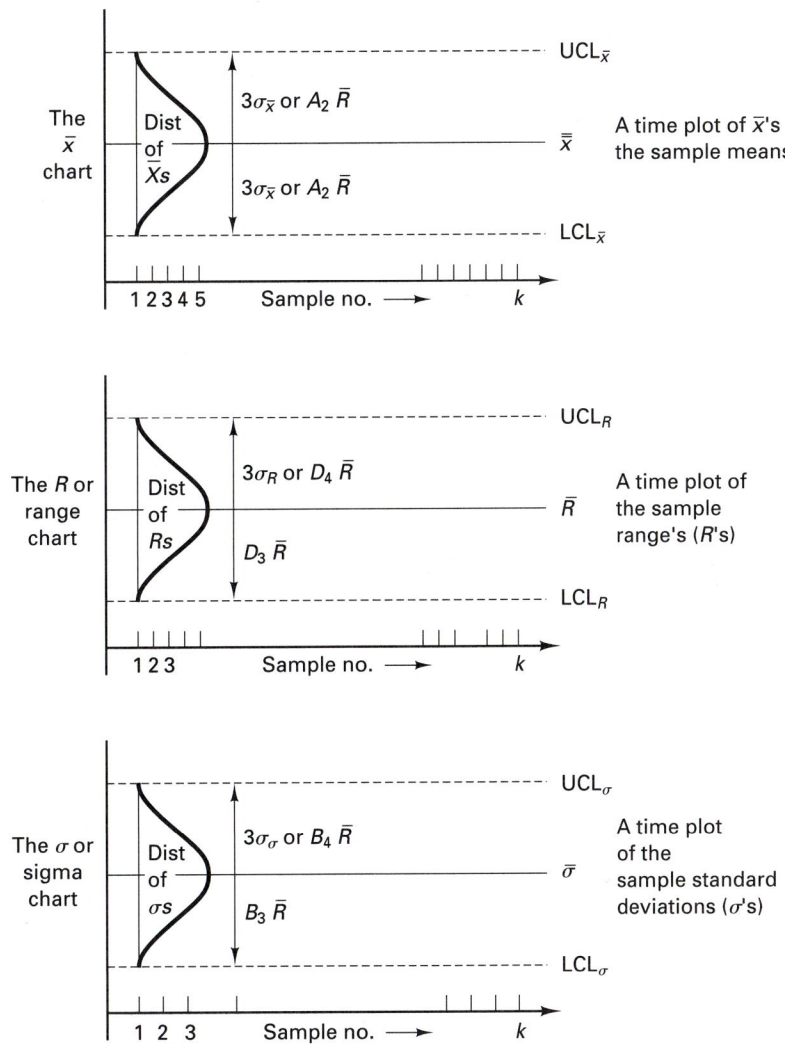

FIGURE 12-8   Basic design of the $\overline{X}$ chart, R chart, and σ chart used in SPC. Values for A_2, D_3, and D_4 given in Table 12-2; see also Figure 12-3.

materials and in machine motions and operations. Deviations due to assignable causes may vary greatly. Their magnitude and occurrence are unpredictable and one thus wishes to prevent their occurrence. On the other hand, if the assignable causes of variation are removed from a given operation, the magnitude and frequency of the chance variations can be predicted with great accuracy. Thus if one can be assured that only chance variations will occur, the quality of the product will be better known and manufacturing can proceed with assurance about the results. By using statistical process control procedures, one may detect the presence of an assignable-cause variation and remove the cause, often before it causes quality to become unacceptable. Also, the astute application of statistical experimental design methods (Taguchi experiments) can help to identify some assignable causes.

To sum up, use PC analysis and process improvements to get the process in the condition described in Figure 12-4a. Use control charts to keep the process centered ($\overline{X}$ chart) with no change in variability (R chart).

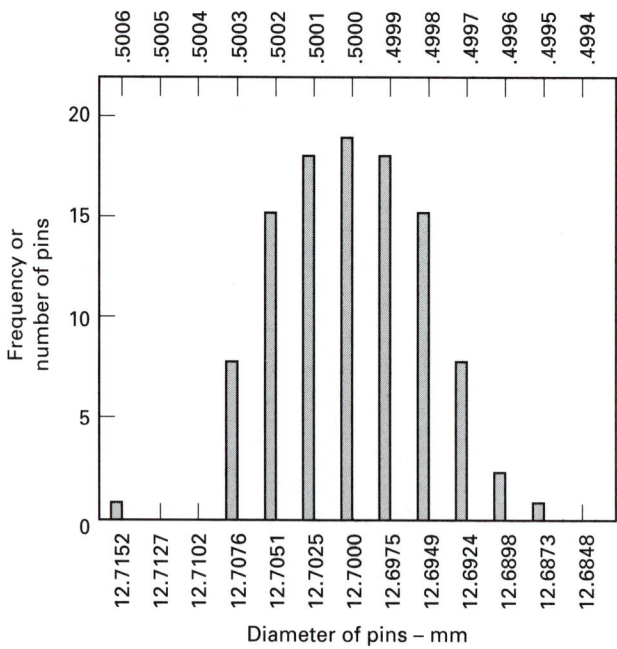

FIGURE 12-9   Frequency distribution of 100 ground pins having a nominal diameter of 1.700 mm (0.5000 in.).

As an example, examine the data in Figure 12-9. This is the frequency distribution of measurements of the diameters of 100 ground pins, which represent the entire population in this case. These pins are supposed to have a nominal diameter of 0.500 in. (12.700 mm) and this was the size appearing most frequently (there were 18 of these). The population is assumed to be normal and only chance variations are occurring. Even so, no pins between 0.5003 and 0.5006 in. were found. The mean for the entire population, μ, is obtained by

$$\mu = \frac{\sum\limits_{i=1}^{n} X_i}{n} = \frac{\sum\limits_{i=1}^{100} X_i}{100} \tag{12-5}$$

where X_i is an individual measurement and n is the number of items or measurements. The mean is a measure of the central tendency around which the individual measurements tend to group. The variability of the individual measurements about the average may be indicated by the standard deviation, σ, where

$$\sigma = \sqrt{\frac{\sum\limits_{i=1}^{n} (X_i - \mu)^2}{n}} \tag{12-6}$$

The standard deviation is of particular interest. For normal distributions, 68.26% of all measurement values will lie within the $\pm 1\sigma$ range from the average, 95.46% within the

$\pm 2\sigma$ range, and 99.73% within the $\pm 3\sigma$ range, refer to Figure 10-6. If the standard deviation for a population of items is known, not over 0.07% of the items will fall outside $\mu \pm 3\sigma$ *as long as only chance variations occur.* For the pin-grinding operation the average diameter, μ, was 0.5000 in. (12.7000 mm), and the standard deviation, σ, was 0.00019 in. (0.0048 mm), making $3\sigma = 0.0006$ in. (0.0144 mm). The manufacturing specifications could be set at (0.5006 and 0.4994) with the assurance that fewer than 3 parts in 1000 would fall outside those limits due to chance causes of variation. More likely, a process would be selected that produced a distribution of parts whose $6\sigma'$ value is less than the total part tolerance.

Note that these are not the same values as were obtained earlier in the PC study. Here we are finding the mean and standard deviation of 100 items, the entire population. In the PC study, the true mean and standard deviation of the population were being *estimated* from sample data, which is more typical of the situation one has to deal with.

Quality Control Charts

Quality control charts are widely used as aids in maintaining quality and in achieving the objective of detecting trends in quality variation before defective parts are actually produced. These charts are based on the previously discussed concept that if only chance causes of variation are present, the deviation from the specified dimension or attribute will fall within predetermined limits.

When sampling inspection is used, the typical sample sizes are from 3 to about 12 units. Figure 12-10 shows an example of two control charts. The $\overline{X}$ chart tracks the sample averages ($\overline{X}$ values), and the R chart plots the range values (R values). Let us assume that the data shown in Figure 12-9 now represent the results of 20 samples of size 5 taken from a large number of ground pins rather than the entire parent population. The sample data will be used to prepare the control charts shown in Figure 12-10.

The centerline of the $\overline{X}$ chart was computed prior to actual usage of the charts in control work,

$$\overline{\overline{X}} = \frac{\sum\limits_{i=1}^{k} \overline{X}i}{k} \tag{12-7}$$

where $\overline{X}_i$ is the sample average and k is the number of sample averages. The horizontal axis for the charts is time; thus indicating *when* the sample was taken, $\overline{\overline{X}}$ serves as an estimate for $\overline{X}'$, the true center of the process distribution and the centerline of the $\overline{X}$ chart. The upper and lower control limits are commonly based on three standard error units, $3\sigma_{\overline{X}}$ (i.e., the standard deviations for the distribution of the $\overline{X}'$ values about $\overline{\overline{X}}$).
Thus

$$\text{UCL}_{\overline{X}} = \text{upper control limit, } \overline{X} \text{ chart} = \overline{\overline{X}} + 3\sigma_{\overline{X}} = \overline{X}' + A_2\overline{R}$$

$$\text{LCL}_{\overline{X}} = \text{lower control limit, } \overline{X} \text{ chart} = \overline{\overline{X}} - 3\sigma_{\overline{X}} = \overline{X}' - A_2\overline{R} \tag{12-8}$$

where, $\sigma_{\overline{X}} = \sigma'/\sqrt{n}$ (see Table 12-2 for A_2). The standard deviation for the distribution of sample averages, $\sigma_{\overline{X}}$, is determined directly from the estimate of the standard deviation of the parent population, σ'. The $\overline{X}$ chart is used to track the central tendency (aim) of the process. In this example, assume that samples are being taken hourly and that the average of *each sample* (not individual values) is plotted.

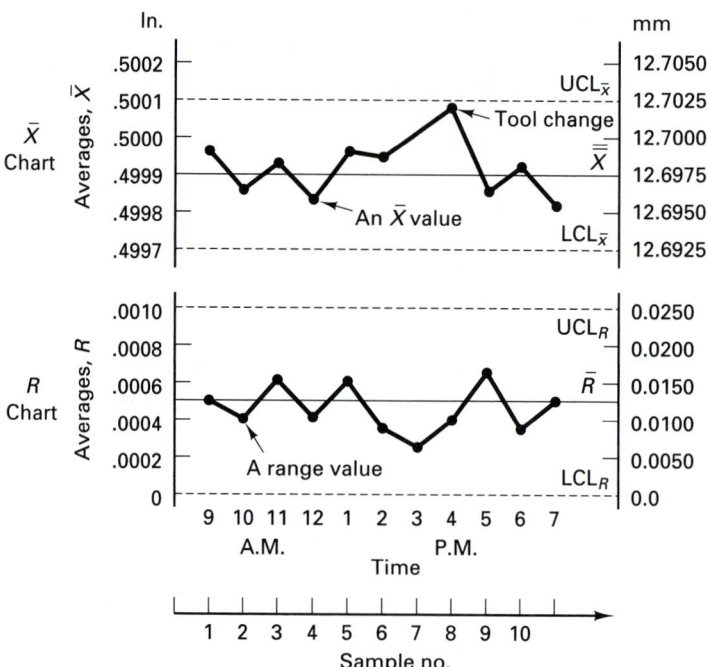

FIGURE 12-10   Statistical quality control charts for 12.700-mm (0.5000-in.)-diameter pins. Note the trend in the $\bar{X}$ curve before the tool change, indicating that the mean was shifting due to tool wear.

The R chart is used to track the variability or dispersion of the process. A σ chart could also be used. R is computed for each sample, where $R = (X_{\text{high}} - X_{\text{low}})$. The value of $\bar{R}$ was determined previously in PC study from

$$\bar{R} = \frac{\sum\limits_{i=1}^{k} R}{k} \qquad (12\text{-}9)$$

where $\bar{R}$ represents the average range of k range values. The range values may also be normally distributed about $\bar{R}$, with standard deviation σ_R. To determine the upper and lower control limits for the charts, the following relationships are used:

$\text{UCL}_R = \text{upper control limit } R \text{ chart, } \bar{R} + 3\sigma_R = D_2\bar{R} = 2.11\bar{R} \quad \text{for} \quad n = 5$

$$(12\text{-}10)$$

$\text{UCL}_R = \text{lower control limit, } \bar{R} - 3\sigma_R = D_3\bar{R} = 0$

where D_4 and D_3 are constants and are given in Table 12-2. For small values of n, the distance between centerline R and LCL_R is less than $3\sigma_R$, but LCL_R cannot be negative, as negative range values are not allowed, by definition. Hence $D_3 = 0$ for values of n up to 6.

After control charts have been established, and the average and range values have been plotted for each sample group, the chart acts as a control indicator for the process. If the process is operating under chance cause conditions, the data will appear random (will have no trends or pattern; Figure 12-11). If $\bar{X}$ or R values fall outside the control limits or

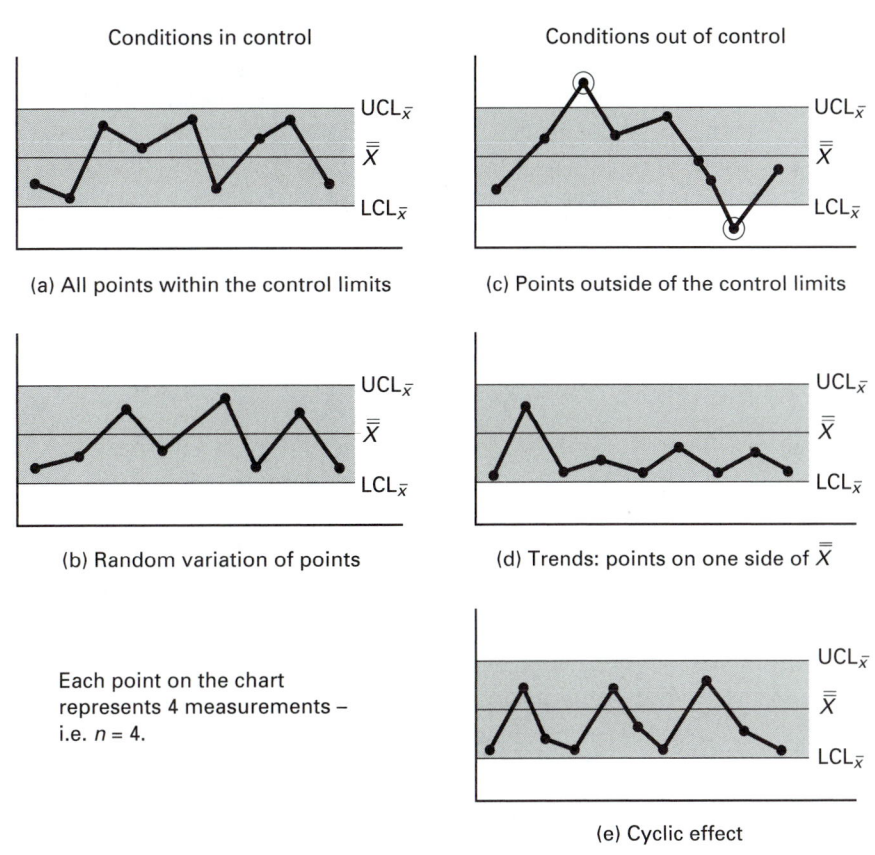

FIGURE 12-11 Examples of $\overline{X}$ control charts.

if nonrandom trends (run or cyclic effects) appear, an assignable cause or change may have occurred, and some action should be taken to correct the problem.

As shown in Figures 12-10 and 12-11, trends in the charts often indicate the existence of an assignable cause factor before the process actually produces a point outside the control limit. Thus $\overline{X}$ values for samples 5, 6, 7, and 8 show a trend toward oversized parts. In this grinding operation, the wheel has worn down (become undersized), so now the parts are becoming oversized and corrective action should be taken. (Redress and reset wheel or replace with new wheel.) Note that defective parts can be produced even if the points on the charts are in control. That is, it is possible for something to change in the process, causing defective parts to be made and the sample point still to be within the control limits. Since no corrective action was suggested by the charts, a type II error was made. Subsequent operations will then involve performing additional work on products already defective. Thus the effectiveness of the SPC approach in improving quality is often deterred by the lag in time between the discovery of an abnormality and the corrective action.

With regard to control charts in general, it should be kept in mind that the charts are only capable of indicating that something has happened, not what happened, and that a certain amount of detective work will be necessary to find out what has occurred to cause a break from the random, normal pattern of sample points on the charts. Keeping careful track of when and where the sample was taken will be very helpful in such investigations,

TABLE 12-2.   Factors for Determining $\overline{X}$ and R Chart Control Limits, 3 Standard Deviations Assumed

Number of Observations in Subgroup n	Factors for $\overline{X}$ Chart A_2	Factors for R Chart	
		Lower Control Limit D_3	Upper Control Limit D_4
2	1.88	0	3.27
3	1.02	0	2.57
4	0.73	0	2.28
5	0.58	0	2.11
6	0.48	0	2.00
7	0.42	0.08	1.92
8	0.37	0.14	1.86
9	0.34	0.18	1.82
10	0.31	0.22	1.78
11	0.29	0.26	1.74
12	0.27	0.28	1.72
13	0.25	0.31	1.69
14	0.24	0.33	1.67
15	0.22	0.35	1.65
16	0.21	0.36	1.64
17	0.20	0.38	1.62
18	0.19	0.39	1.61
19	0.19	0.40	1.60
20	0.18	0.41	1.59

Upper control limit for $\overline{X} = \text{UCL}_x = \overline{X}' + A_2\overline{R}$
Lower control limit for $\overline{X} = \text{LCL}_x = \overline{X}' - A_2\overline{R}$
Upper control limit for $R = \text{UCL}_R = D_4\overline{R}$
Lower control limit for $R = \text{LCL}_R = D_3\overline{R}$

Note that since $\sigma_{\overline{X}} = \sigma'/\sqrt{n}$, $3\sigma_{\overline{X}} = 3\,(\sigma'/\sqrt{n}) = 3\overline{R}/\sqrt{n}\,d_2 = A_2\overline{R}$ where $A_2 = 3/\sqrt{n}\,d_2$. Also, $\text{UCL}_R = \overline{R} + 3\sigma_R = d_2\sigma' + 3d_3\sigma'$ (see Table 12-1). Thus $\text{UCL}_R = (d_2 + 3d_3)\sigma' = (1 + 3d_3/d_2)$.

Source: 1950 Manual on Quality Control of Materials, copyright ASTM, 1916 Race Street, Philadelphia, Pa., 19103. Adapted, with permission.

but the best procedure is to have the operator do the inspection and run the chart. In this way, quality feedback is very rapid and the causes of defects readily found.

■ 12.4 DETERMINING CAUSES FOR PROBLEMS IN QUALITY

The best way to isolate quality problems quickly is to make everyone an inspector. This means that every worker, foreman, supervisor, engineer, manager, and so on, is responsible for making it right the first time and every time. One very helpful tool in this effort is the *fishbone* diagram (Figure 12-12). The problem (bad welds) can have many causes. Generally, the cause will lie in the process, operators, materials, or method (i.e., the four main branches on the chart). Every time a quality problem is caused by one of these events, it is noted by the observer and corrective action taken. As before, experimental design procedures can help identify causes that effect performance.

In summary, the data gathered to develop the PC study can be used to prepare the initial control charts from the process after the removal of all assignable causes for variability and proper setting of the process average. After the charts have been in place for some time and a large amount of data has been obtained from the process output, the PC study

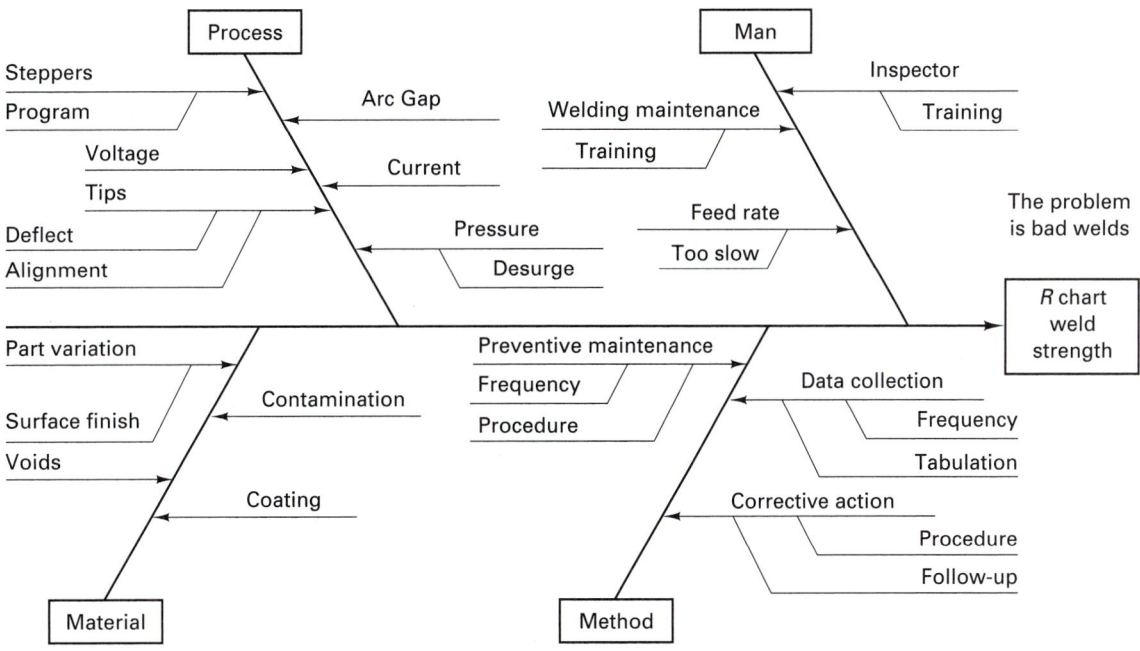

FIGURE 12-12 Fishbone diagram used to determine the causes of quality problems.

can be redone to obtain better estimates of the natural spread of the process during actual production. On-line or in-process methods such as SPC and off-line (Taguchi) methods are key elements in a total quality control program.

Total Quality Control

The phrase *total quality control* (TQC) was first used by A. V. Feigenbaum in *Industrial Quality Control* in May 1957, the idea being that all departments of a company must participate in quality control. Quality control is the responsibility of workers at every level in every department, all of whom have studied quality control. It begins at the product design stage and carries through the manufacturing system, where the emphasis is on making it right the first time. Few companies have embarked on implementing Taguchi methods in order to reduce the inherent variability in their processes. A change in company culture is often needed to give the responsibility for quality to the worker, along with the authority to stop the process when something goes wrong. An attitude of defect prevention and a habit of constant improvement of quality are fundamental to lean production (see Chapter 43). Companies such as Toyota have accomplished TQC (they call it companywide quality control) by extensive education of the workers, giving them the analysis tools they need (control charts with cause-and-effect diagrams) to find and expose the problems. Workers are encouraged to correct their own errors, and 100% inspection (often done automatically) is the rule. Passing defective products on to the next process is not allowed. The goal is perfection. *Quality circles*, now popular in the United States, are just one of the methods used by Japanese industries to achieve perfection.

The designer of the product must have quality in mind during the design phase, seeking the least costly means to assure the quality of the product having the desired functional characteristics. Major factors that can be handled during the early stages of the product design cycle include temperature, humidity, power variations, and deterioration of materials, tools, and so on. Compensation for these factors is difficult or even impossible to implement after the product is in production. The distinction between superior- and poor-quality products can be in their variability in the face of internal and external causes. This is where Taguchi parameter design methods are critical.

The secret to successful process control is putting the control of quality in the hands of the workers. Many companies in this country are currently engaged in SQC (statistical quality control) but they are still inspecting to *find* defects. The number of defects will not be reduced merely by making the inspection stage better or faster or automated. You are simply more efficient at discovering defects. The trick is to *inspect* to prevent defects. How can this be done? Here are some basic ideas. Use source inspection techniques that control quality at the stage where defects originate. Use 100% inspection with immediate feedback rather than sampling. Make every worker an inspector. Minimize the time it takes to carry out corrective action. Remember that people are human and not infallible. Devise methods and devices that prevent them from making errors. Such devices and methods are called pokayoke devices. Can you think of such a device? Does your car have a procedure that prevents you from locking the ignition keys inside the car? That is a pokayoke.

Concentrate on making the system more efficient, not simply on making the operators and operations more proficient. Continuous improvement requires that you redesign the manufacturing continuously, reducing the time required for products to move through the system (i.e., the throughput time). This approach seems to have been the American stumbling block. Industrial engineers can do operations improvement work and need to do more systems improvement work. Too often we devise fancy, complex computerized solutions to solve complex manufacturing process problems. Why not simplify the manufacturing system so that the need for complex solutions disappears? For additional reading on this topic, see Chapters 41 and 43.

■ KEY WORDS

accuracy	precision	standard deviation
control chart	process capability	statistical process control
fishbone diagram	quality control	Taguchi methods
histogram	sample mean	variability
nominal	sample range	
parent population	specification limit	

■ REVIEW QUESTIONS

1. Define a process capability study in terms of accuracy or precision.
2. To what does *nature of the process* refer?
3. Suppose you had a process that was accurate and precise, as shown in Figure 12-1a. What might the hole patterns in the target look like if, occasionally while shooting, a sharp gust of wind blew left to right?
4. Review the steps required to making a PC study of a process.
5. Why don't standard tables exist detailing the

natural variability of a given process, such as rolling, extruding, or turning?
6. What are Taguchi or factorial experiments, and how are they related to experimental designs?
7. How does the Taguchi approach differ from the typical experimental approach?
8. Why are Taguchi experiments so important compared to classical experiments?
9. Here are some common, everyday processes with which you are familiar. What variable do you think dominates these processes?

(a) Baking a cake (from scratch; from a cake mix)

(b) Mowing the lawn

(c) Washing dishes in a dishwasher

10. Explain why the diameter measurements for holes produced by the process of drilling could have a skewed rather than a normal distribution.

11. What are some common manufactured items that may not receive any final inspection?

12. What are common reasons for sampling inspection rather than 100% inspection?

13. Fill in this table with one of the following statements: type I, type II, no error

		In reality, if we looked at everything the process made, we would know that it had:	
		Changed	Not changed
The sample suggested that the process had:	Changed		
	Not changed		

14. Why, when we sample, can we not avoid making type I and type II errors?

■ PROBLEMS

1. For the items listed below, obtain a quantity of 48. Measure the indicated characteristics and determine the process mean and standard deviation. Use a sample size of 4, so that 12 samples are produced.

Item	Characteristic(s) You Can Measure
Flat washer	Weight, width, diameter of hole, outside diameter
Paper clip	Length, diameter of wire
Coin (penny, dime)	Diameter, thickness at point, weight
Your choice	Your choice

2. Perform a process capability study to determine the PC of the process that makes M&M candy. You will need to decide what characteristics you want to measure, (weight, diameter, thickness, etc.), how you will measure it (use the rule of

15. Which error can lead to legal action from the consumer for a defective product that caused bodily injury?

16. Define and explain the difference between σ', $\sigma_{\overline{X}}$, and σ_R.

17. What is C_p, and why is a value of 0.80 not good? How about value of 1.00? 1.3?

18. Explain what is measured by the mismatch factor, D?

19. What are some of the alternatives available to you when you have the situation where $6\sigma' > USL - LSL$?

20. C_{pk} is also a process capability index. How does it differ form C_p?

21. In a sigma chart, are σ values for the samples normally distributed about $\overline{\sigma}$? Why or why not?

22. What is an assignable cause, and how is it different from a chance cause?

23. Why is the range used to measure variability when the standard deviation is really a better statistic?

24. How is the standard deviation of a distribution of a sample means related to the distribution from which the samples were drawn?

25. What is the real secret to superior quality control?

26. Think of (or design) a pokayoke device for use around the house.

10), and what kind of M&M's you want to inspect (how many bags of M&M's you wish to sample). Take a sample size of 4 ($n = 4$). Make a histogram of the individual data and estimate $\overline{X}'$ and σ' as outlined in Figure 12-3. If you decide to measure the weight characteristics, you can check your estimate of $\overline{X}'$ by weighing all the M&M's together and dividing by the total number of M&M's.

3. For the data given in Figure 12-2, compute C_p, D, and C_{pk}, making any assumptions needed to perform the calculations.

4. For the data given in Figure 12-7, compute C_p, D, and C_{pk}.

5. Calculate $\overline{\overline{X}}$ and $\overline{R}$ and the control limits for the $\overline{X}$ and R control charts shown in Figure 12-A. The sample mean, $\overline{X}$, and range for the first nine subgroups and the data for each sample is

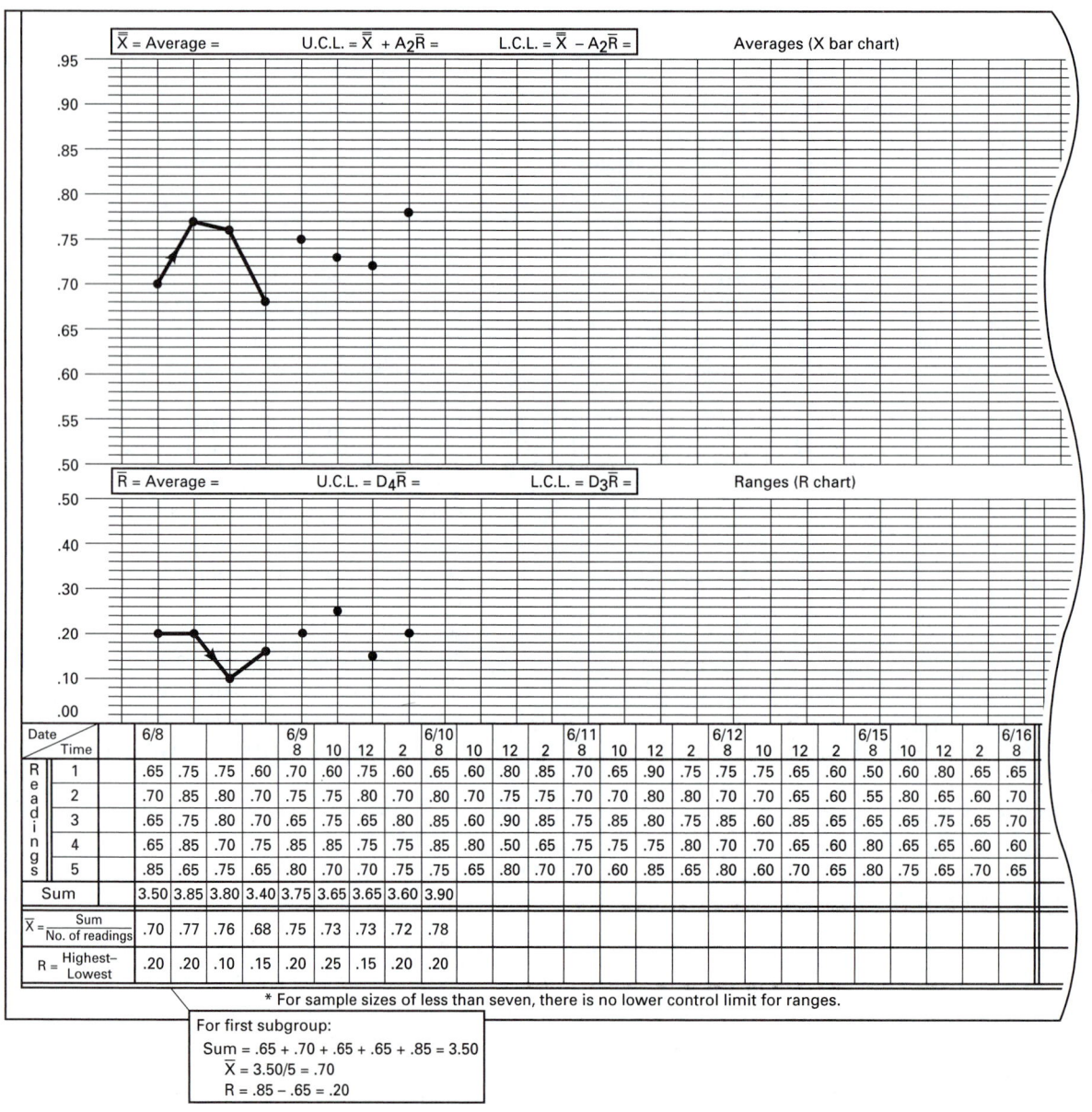

FIGURE 12-A

given at the bottom of the figure. There are 25 samples of size 5. Therefore, $k = 25$, $n = 5$. Complete the bottom part of the table and then compute the control limits for both charts. Construct the charts plotting $\overline{X}$ and $\overline{R}$ as solid lines and control limits as dashed lines. The first eight data points have been plotted. Are they all correctly plotted? The first four data points have been connected. Plot the rest of the

data on the charts and comment on your findings.

6. For the data given in Figure 12-A, estimate the mean and standard deviation for the process from which these samples were drawn (i.e., the parent population) and discuss the process capability. The USL and LSL for this dimension are 0.9 and 0.5, respectively, and the nominal is 0.7.

PART 3

CASTING PROCESSES

13 Fundamentals of Casting

14 Expendable-Mold Casting Processes

15 Multiple-Use-Mold Casting Processes

16 Power Metallurgy

CHAPTER 13

FUNDAMENTALS OF CASTING

13.1 INTRODUCTION TO
       MATERIALS PROCESSING
13.2 INTRODUCTION TO
       CASTING
       Basic Requirements of
         Casting Processes
13.3 CASTING TERMINOLOGY
13.4 THE SOLIDIFICATION
       PROCESS
       Cooling Curves
       Prediction of Solidification
         Time
       The Cast Structure
       Molten Metal Problems

       Fluidity
       Pouring Temperature
       Gating System
       Solidification Shrinkage
       Risers and Riser Design
       Risering Aids
13.5 PATTERNS
13.6 DESIGN
       CONSIDERATIONS IN
       CASTING
Case Study: THE CAST OIL-FIELD
             FITTING

■ 13.1 INTRODUCTION TO MATERIALS PROCESSING

Materials processing has been defined as the science and technology by which a material is converted into a useful shape with a structure and properties that are optimized for the proposed service environment. More loosely, materials processing is "all that is done to convert stuff into things."

Numerous operations and processes are involved in the manufacture of products and components, and many of them are associated with the production of a desired shape. These processes are often grouped into four basic "families," as indicated in Figure 13-1. *Casting processes* exploit the fluidity of a liquid as it assumes the shape of a prepared container and solidifies upon cooling. The *material removal processes* begin with an oversize piece and remove material to leave a desired shape. While these processes have often been referred to as *machining*, that term is generally used to describe the mechanical cutting of materials. The more general term *material removal* is used to incorporate material removal by all means, including chemical, thermal, and physical processes. *Deformation processes* exploit the ductility of certain materials, most notably metals, and produce the desired shape by mechanical rearrangement, or plasticity. *Consolidation processes* are those processes that put pieces together, and include welding, brazing, soldering, adhesive bonding, and mechanical fasteners. *Powder metallurgy* is the manufacture of the desired shape from particulate material and involves aspects of casting, forming, and consolidation.

Each of the various families has distinctive advantages and limitations, and the various processes within the families have unique characteristics. Cast products can have extremely complex shapes but also have structures that are produced by solidification and are subject to associated defects, such as shrinkage and porosity. Material removal processes are capable of outstanding dimensional precision but produce scrap as material is cut away to produce the desired shape. Deformation processes can have high rates of production but generally require powerful equipment and dedicated tools or dies. Complex products can often be assembled from simple shapes, but the joint areas are often modified by the joining process and may possess characteristics different from the base material.

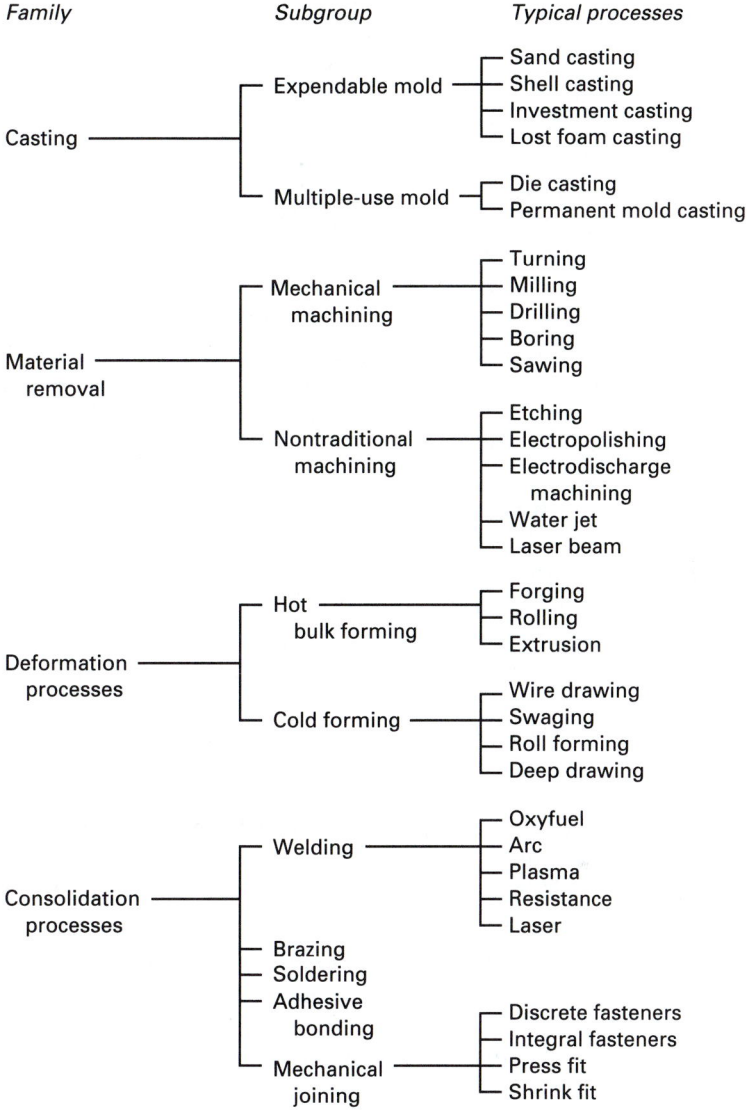

FIGURE 13-1   Materials processing families, subgroups, and typical processes.

The manufacture of any product or component involves not only design and material selection but also the selection of the process or processes involved in obtaining the desired shape and achieving the desired properties. Decisions should be made with the knowledge of all available alternatives and their associated assets and limitations. A large portion of this book is dedicated to the presentation of the various processes applied to engineering materials. They are grouped according to the four basic categories, with powder metallurgy included at the end of this section. Emphasis is on process fundamentals, descriptions of the various alternatives, and an assessment of associated assets and limitations. We begin with casting.

■ 13.2 INTRODUCTION TO CASTING

In *casting* processes, a solid material is first melted, heated to proper temperature, and sometimes treated to modify its chemical composition. The molten material, generally metal, is then poured into a cavity or mold that contains it in the desired shape during solidification. Thus, *in a single step*, simple or complex shapes can be made from any material that can be melted. The resulting product can have virtually any configuration the designer desires. In addition, the resistance to working stresses can be optimized, directional properties can be controlled, and a pleasing appearance can be produced.

Cast parts range in size from a fraction of an inch and a fraction of an ounce (such as the individual teeth on a zipper), to over 30 ft (10 m) and many tons (such as the huge propellors and stern frames of ocean liners). Moreover the casting processes have distinct advantages in the production of complex shapes, parts having hollow sections or internal cavities, parts that contain irregular curved surfaces (except those made from thin sheet metal), very large parts, and parts made from metals that are difficult to machine. Because of these features, casting is one of the most important of the manufacturing processes.

It is almost impossible to design a part that cannot be cast by one or more of the commercial casting processes. However, as with all manufacturing techniques, the best results and lowest cost are achieved only if the designer understands the various options and tailors the design to use the most appropriate process in the most efficient manner. The various processes are distinguished primarily by the mold material (whether sand, metal, or other material) and pouring method (gravity, vacuum, low pressure, or high pressure). All of them share the requirement that the material should solidify in a manner that will maximize the properties and avoid the formation of defects, such as shrinkage voids, gas porosity, and trapped inclusions.

Basic Requirements of Casting Processes

Six basic requirements are associated with most casting processes:

1. A *mold cavity*, having the desired shape and size, must be produced with due allowance for shrinkage of the solidifying material. Any geometrical feature desired in the finished casting must exist in the cavity. Consequently, the mold material must be able to reproduce the desired detail and also have a refractory character so that it will not contaminate the molten material that it will contain. Either a new mold must be prepared for each casting (*single-use molds*) or the mold must be made from a material that can withstand repeated use. The latter type, known as *multiple-use molds*, are generally made of metal or graphite. Since they tend to be quite costly, their use is generally restricted to large production runs. The more economical single-use molds are usually preferred for the production of smaller quantities.

2. A *melting process* must be capable of providing molten material not only at the proper temperature, but also in the desired quantity, with acceptable quality, and within a reasonable cost.

3. A *pouring technique* must be devised to introduce the molten metal into the mold. Provision should be made for the escape of all air or gases present in the cavity prior to pouring, as well as those generated by the introduction of the hot metal. The molten material is then free to fill the cavity, producing a high-quality casting that is fully dense and free of defects.

4. The *solidification process* should be properly designed and controlled. Castings should be designed so that solidification and solidification shrinkage can occur

without producing internal porosity or voids. In addition, the molds should not provide excessive restraint to the shrinkage that accompanies cooling. If they do, the casting may crack when it is still hot and its strength is low.

5. It must be possible to remove the casting from the mold (i.e., *mold removal*). With single-use molds that are broken apart and destroyed after each casting, there is no serious difficulty. With multiple-use molds, however, the removal of a complex-shaped casting may present a major design problem.

6. After the casting is removed from the mold, various *cleaning, finishing, and inspection* operations may be required. Extraneous material is usually attached where the metal entered the cavity, excess material may be present along mold parting lines, and mold material often adheres to the casting surface. All of these must be removed from the finished casting.

Each of these requirements is considered in more detail as we move through the chapter. The fundamentals of solidification, pattern design, gating, and risering are developed. Various defects are considered, together with their causes and cures.

■ 13.3 CASTING TERMINOLOGY

Before proceeding with the process fundamentals, it is helpful if we first become familiar with a variety of casting terms. Figure 13-2 shows the cross section of a two-part sand mold and incorporates many features of a typical casting process. The process starts with the construction of a *pattern*, an approximate duplicate of the final casting. The molding material is then packed around the pattern and the pattern is removed to produce a mold cavity. The *flask* is the box that contains the molding aggregate. In a two-part mold, the *cope* is the name given to the top half of the pattern, flask, mold, or core. The *drag* refers to the bottom half of any of these features. A *core* is a sand shape that is inserted into the mold to produce the internal features of a casting, such as holes or passages for water cooling. A *core print* is that region added to the pattern, core, or mold that is used to locate and support the core within the mold. The mold material and the core then combine to form the *mold cavity*, the shaped hole into which the molten metal is poured and solidified to produce the desired casting. A *riser* is an extra void created in the mold that will also fill with molten metal. It provides a reservoir of material that can flow into the mold cavity to compensate for any shrinkage that occurs during solidification. If the riser contains the last material to solidify, shrinkage voids should be located in the riser and not the final casting.

The *gating system* is the network of channels used to deliver the molten metal to the mold cavity. The *pouring cup* (or pouring basin) is the portion of the gating system that initially receives the molten metal from the pouring vessel and controls its delivery to the rest of the mold. From the pouring cup, the metal travels down a *sprue* (the vertical portion of the gating system), then along horizontal channels, called *runners*, and finally through controlled entrances, or *gates*, into the mold cavity. Additional channels, known as *vents*, may be included to provide an escape for the gases that are generated within the mold.

The *parting line* or *parting surface* is the interface that separates the cope and drag halves of a mold, flask, or pattern and also the halves of a core in some core-making processes. *Draft* is the taper on a pattern or casting that permits it to be withdrawn from the mold. The mold or die used to produce casting cores is known as a *core box*. Finally, the term *casting* is used to describe both the process and the product when molten metal is poured and solidified in a mold.

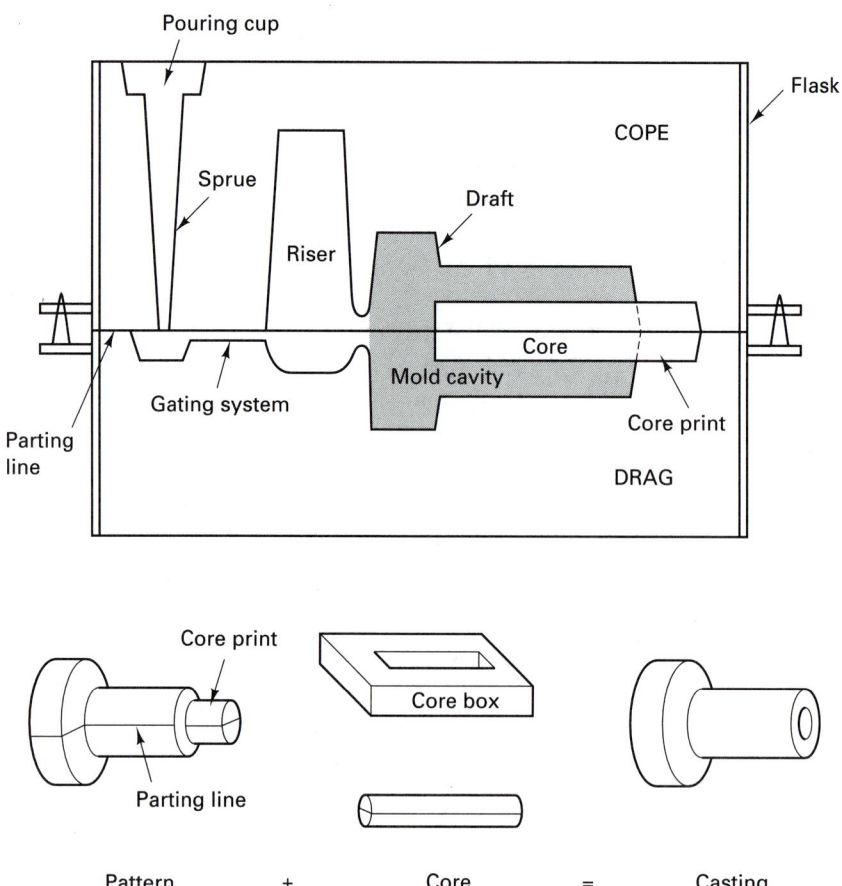

FIGURE 13-2 Cross section of a typical two-part sand mold, indicating various mold components and terminology.

■ 13.4 THE SOLIDIFICATION PROCESS

Casting is a solidification process where the molten material is poured into a mold and then allowed to freeze into the desired final shape. Many of the structural features that ultimately control product properties are set during solidification. Furthermore, many casting defects, such as gas porosity and solidification shrinkage, are solidification phenomena, and they can be reduced or eliminated by controlling the solidification process.

Solidification occurs in two stages, nucleation and growth, and its is important to control both of these processes. *Nucleation* occurs when a particle of a stable solid forms from within the molten liquid. As the material changes state, its internal energy is reduced since at lower temperatures the solid phase is more stable than the liquid. At the same time, however, interface surfaces are created between the new solid and the parent liquid. Formation of these surfaces requires a positive contribution of energy. As a result, nucleation generally occurs at a temperature somewhat below the equilibrium melting point (the temperature where the internal energies of the liquid and solid are equal). The difference between the melting point and the temperature of nucleation is known as the amount of *undercooling*.

In most practical situations, the nucleation process utilizes existing surfaces where solidification can begin without the need to create a full, surrounding interface. These surfaces are usually present in the form of mold or container walls, or solid impurity particles within the molten liquid.

Each nucleation event produces a crystal or grain in the final casting. Since fine-grained materials (many small grains) possess enhanced mechanical properties, efforts to promote nucleation tend to be beneficial to the final product. It is not uncommon to intentionally introduce impurities into the liquid before pouring into the mold. These small particles of solid provide numerous sites for nucleation and promote formation of a uniform, fine-grained product. The practice of intentionally introducing impurities is known as *inoculation* or *grain refinement*.

The second step in the solidification process is *growth*, which occurs as the evolved heat of fusion is continually extracted from the liquid material. The direction, rate, and type of growth can be controlled by the way in which the heat is extracted. *Directional solidification*, in which the solidification interface sweeps continuously through the material, can be used to assure the production of a sound casting. The molten material on the liquid side of the interface can flow into the mold to continuously compensate for the shrinkage that occurs as the material changes from liquid to solid. Faster rates of cooling generally produce products with finer grain size and superior mechanical properties.

Cooling Curves

Cooling curves, such as those introduced in Chapter 4, can provide one of the most useful tools for studying the solidification process. By inserting thermocouples into a casting and monitoring the temperature versus time, one can obtain valuable insight into what is happening in the various regions.

Figure 13-3 shows a typical cooling curve for a pure or eutectic-composition material and is useful for depicting many of the principal features and terms. The *pouring temperature* is the temperature of the liquid metal when it first enters the mold cavity. *Superheat* is the difference between the pouring temperature and the freezing temperature of the material. The higher the superheat, the more time allowed for the material to flow into the intricate details of the mold cavity before it begins to freeze. The *cooling rate* is the rate at which the liquid or solid is cooling and can be viewed as the slope of the cooling curve at any given point. A *thermal arrest* is the plateau in the cooling curve that occurs during the solidification of a material with fixed melting point. At this temperature, the heat being removed from the mold comes from the latent heat of fusion being released during the solidification process and does not require a decrease in the material's temperature. The time from the start of pouring to the end of solidification is known as the *total solidification time*. The time from the start of solidification to the end of solidification is called the *local solidification time*.

If an alloy is used that does not have a distinct melting point, such as the one shown in Figure 13-4, the difference between the liquidus and solidus temperatures is known as the *freezing range*. The onset and termination of solidification appear as slope changes in the cooling curve. If undercooling is required to induce the initial nucleation, the subsequent solidification may release enough heat to cause an increase in temperature back to the melting point. This increase in temperature, known as *recalescence*, is shown in Figure 13-5.

The specific form of a cooling curve depends on the type of material being poured, the nature of the nucleation process, and the rate and means of heat removal from the mold. Analysis of experimental cooling curves can provide valuable insight into the process and the product. Fast cooling rates and short solidification times lead to finer structures and improved mechanical properties.

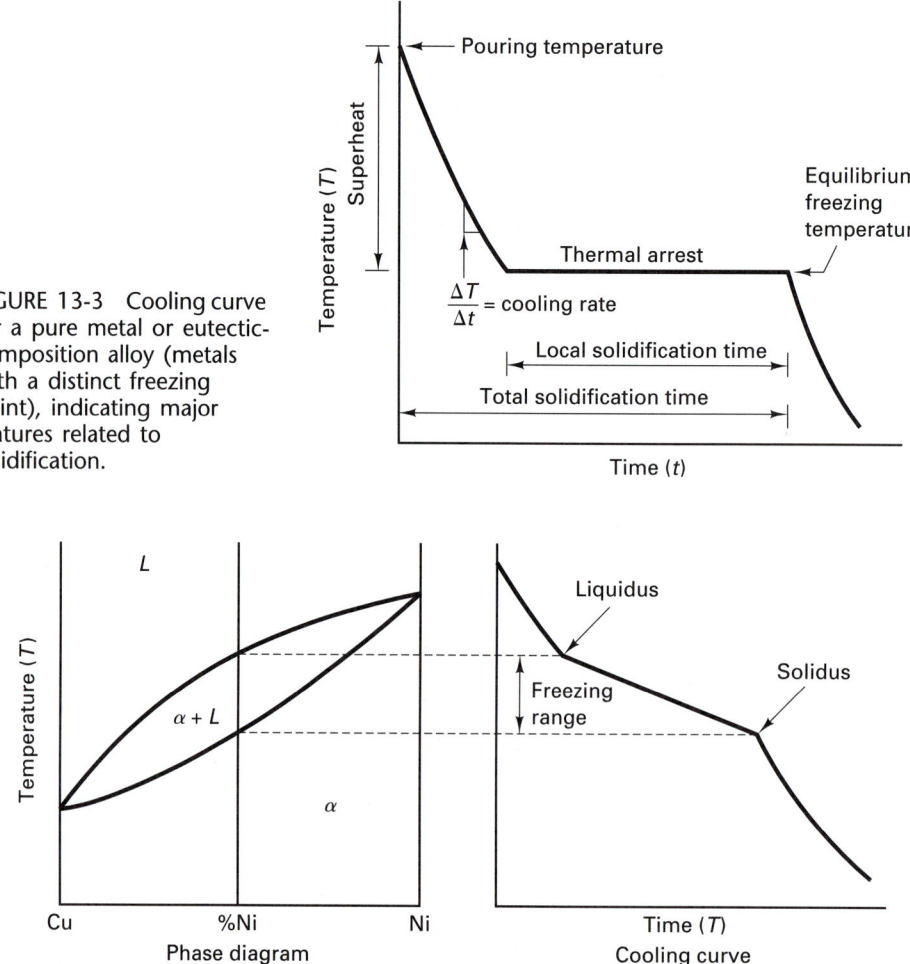

FIGURE 13-3 Cooling curve for a pure metal or eutectic-composition alloy (metals with a distinct freezing point), indicating major features related to solidification.

FIGURE 13-4 Cooling curve of an alloy that has a freezing range. Slope changes indicate the onset and termination of solidification.

Prediction of Solidification Time: Chvorinov's Rule

The amount of heat that must be removed from a casting to cause it to solidify is directly proportional to the amount of superheating and the amount of metal in the casting, or the casting volume. Conversely, the ability to remove heat from a casting is directly related to the amount of exposed surface area through which the heat can be extracted and the environment surrounding the molten material (i.e., the mold and mold surroundings). These observations are reflected in *Chvorinov's rule*[*], which states that t_s, the total solidification time, can be computed by

$$t_s = B(V/A)^n \qquad \text{where } n = 1.5 \text{ to } 2.0$$

---

[*] N. Chvorinov, *Proc. Inst. Brit. Found.*, Vol. 32, p. 229 (1938–1939).

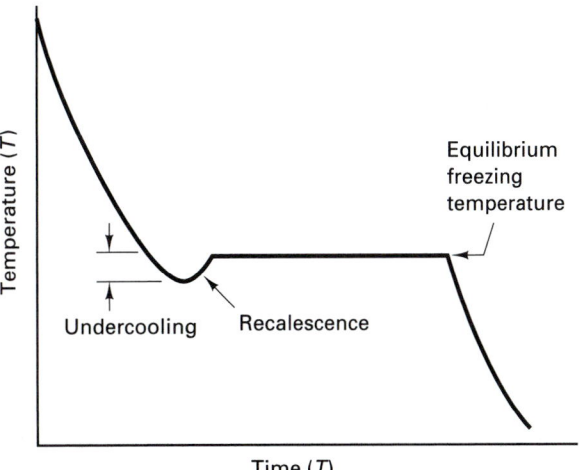

FIGURE 13-5 Cooling curve depicting undercooling and subsequent recalescence.

The total solidification time, t_s, is the time from pouring to the completion of solidification; V is the volume of the casting; A is the surface area; and B is the *mold constant*, which incorporates the characteristics of the metal being cast (its density, heat capacity, and heat of fusion), the mold material (its density, thermal conductivity, and heat capacity), the mold thickness, and the amount of superheat.

Test specimens can be cast to determine B for a given mold material, casting material, and condition of casting. This value can then be used to compute the solidification times for other castings made under the same conditions. Since a riser and casting are both within the same mold and fill with the same metal under the same conditions, Chvorinov's rule can be used to compare the solidification times of each and thereby assure that the casting will solidify before the riser. This condition is absolutely essential if the liquid within the riser is to effectively feed the casting and compensate for solidification shrinkage. Aspects of riser design are developed later in this chapter.

Different cooling rates and solidification times can produce substantial variation in the structure and properties of the resulting casting. For instance, the die casting process uses metal molds, and the faster cooling produces higher-strength castings than sand casting, which uses a more insulating mold material. In addition, the various types of sands can produce different cooling rates. Sands with high moisture contents extract heat faster than ones with low moisture.

The Cast Structure

The structure that results when molten metals are poured into molds and permitted to solidify may have as many as three distinct regions or zones. The *chill zone* is a narrow band of randomly oriented crystals that forms on the surface of a casting. Rapid nucleation occurs here due to the presence of the mold walls and the relatively rapid surface cooling. As additional heat is removed from the surfaces, the grains of the chill zone begin to grow inward, and the rate of heat extraction and solidification decreases. Since most crystals have directions of rapid growth, a selection process occurs. Those crystals whose rapid-growth direction is perpendicular to the casting surface grow fast and shut off adjacent

FIGURE 13.6   Cross-selectional structure of a cast metal bar showing the chill zone at the periphery, columnar grains growing toward the center, and central shrinkage cavity.

grains whose rapid-growth direction is at some intersecting angle. The favorably oriented crystals can then grow until all of the liquid has solidified, producing long, thin columnar grains. This region is known as the *columnar zone*. The resulting properties are highly directional, since the selection process has converted the purely random structure of the surface into one of aligned parallel crystals. Figure 13-6 shows a cast structure containing both chill and columnar zones.

In many materials, new crystals can nucleate in the interior of the casting and then grow to produce another region of spherical, randomly oriented crystals, known as the *equiaxed zone*. Low pouring temperatures, alloy additions, or the addition of inoculants can be used to promote the formation of this region, which is far more desirable than columnar grains. Isotropic properties (uniform in all directions) are observed in this region of the casting.

Molten Metal Problems

Castings begin with molten metal, and many of the reactions that occur between molten metal and its surroundings can lead to defects in the casting. Oxygen and molten metal often react to produce metal oxides (i.e., ceramic material), which can then be carried with the molten metal during pouring and filling of the mold. Known as *dross* or *slag*, this material can become trapped in the casting and impair surface finish, machinability, and mechanical properties.

Dross and slag can be controlled by using special precautions during melting and pouring and by good design practice. Fluxes can be used to cover and protect molten metal during melting, or melting and pouring can be performed under a vacuum or protec-

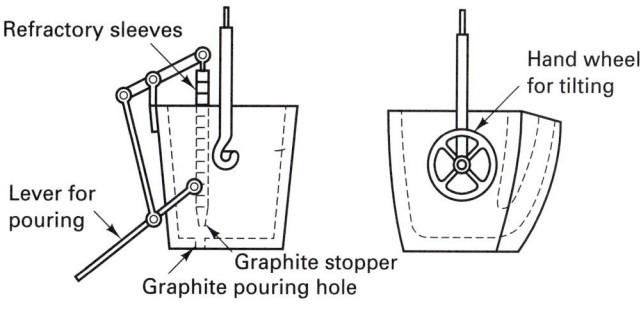

FIGURE 13.7 Two types of ladles used in the pouring of castings. Note how each avoids pouring the impure material from the top of the molten pool.

tive atmosphere. Measures can be taken to agglomerate the dross and cause it to float to the surface of the metal, where it can be skimmed off prior to pouring. Special ladles can be used which pour from beneath the surface. Figure 13-7 presents two popular designs. Gating systems can often be designed to trap any dross that might enter the mold and keep it from flowing into the mold cavity. In addition, filters can be inserted into the feeder channels of the mold.

Significant amounts of gas can dissolve in many liquid metals. When these metals then solidify, the solid structure cannot accommodate the gas, and the rejected atoms often form bubbles or *gas porosity* within the casting. Figure 13-8 shows the maximum solubility of hydrogen in aluminum as a function of temperature. Note the large decrease that occurs as the material goes from liquid to solid.

Several different techniques can be used to prevent the formation of gas porosity. One approach is to prevent the gas from initially dissolving in the molten metal. Melting can be performed under vacuum, in an environment of low-solubility gases, or under a protective flux which excludes contact with the air. Superheat temperatures can be kept low to minimize the solubility. In addition, careful handling and pouring can do much to streamline the flow of molten metal and minimize the turbulence that brings air and molten metal into contact.

Other techniques attempt to remove the gas from the molten metal before pouring. *Vacuum degassing* subjects the molten metal to a low-pressure environment. Under these conditions, the amount of dissolved gas reduces as the system seeks to establish an equilibrium with its new surroundings. (See a discussion of Sievert's law in any basic chemistry text.) Passing small bubbles of inert or reactive gas through the melt, known as *gas flushing*, is also effective. In seeking equilibrium, the dissolved gas enters the flushing gas and is carried away. Bubbles of nitrogen or chlorine, for example, are particularly effective in removing hydrogen from molten aluminum.

Still another approach is to react the dissolved gas with something to produce a low-density compound. These compounds then float to the surface and can be removed with the dross or slag. Oxygen can be removed from copper by the addition of phosphorus. Steels can be deoxidized with additions of aluminum or silicon. The resulting phosphorus, aluminum, or silicon oxides are then removed by skimming, or left on the top of the container as the remaining high-quality metal is poured from beneath the surface.

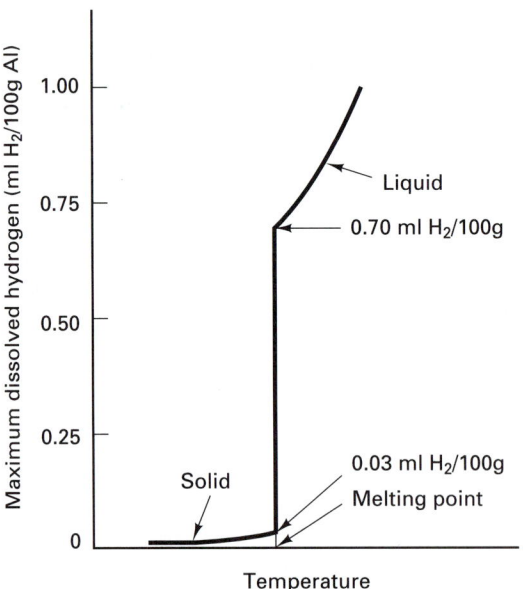

FIGURE 13-8 Maximum solubility of hydrogen in aluminum as a function of temperature.

Fluidity

When molten metal is poured to produce a casting, it should first *flow* into all regions of the mold cavity and then *freeze* into this new shape. It is vitally important that these two functions occur in the proper sequence. If the metal begins to freeze before it has filled the mold completely, defects known as *misruns* and *cold shuts* are produced.

The ability of a metal to flow and fill a mold is known as *fluidity*. Although no single method has been accepted as a means of measuring fluidity, various "standard molds" have been developed where the results are sensitive to metal flow. One popular approach produces castings in the form of a long, thin spiral that progresses outward from a central sprue. The length of the final casting provides the measure of fluidity.

Pouring Temperature

Fluidity is dependent on the composition, freezing temperature, and freezing range of the metal or alloy, but the most important controlling factor is usually the pouring temperature or the amount of superheat. The higher the pouring temperature, the higher the fluidity. Excessive temperatures should be avoided, however. At high pouring temperatures, metal–mold reactions are accelerated and the fluidity may be so great as to permit *penetration*. Penetration is a defect where the metal not only fills the mold cavity but also fills the small voids between the sand particles in a sand mold. The product surface then contains small particles of embedded sand.

Gating System

As molten metal is poured into a mold, the gating system serves to deliver it to all sections of the mold cavity. The speed or rate of metal movement is important as well as the degree of cooling that occurs while it is flowing. Slow filling and high loss of heat can result

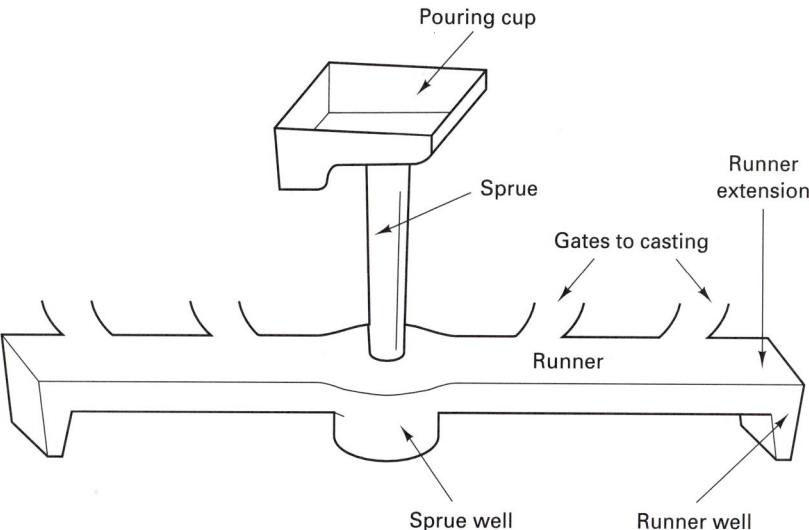

FIGURE 13-9   Typical gating system for a horizontal parting plane mold, showing key components involved in controlling the flow of metal into the mold cavity.

in misruns and cold shuts. Rapid rates or filling, on the other hand, can produce erosion of the gating system and mold cavity and might result in the entrapment of mold material in the final casting. The cross-sectional areas of the various channels can be selected to regulate flow. In addition, the shape and length of the channels are influential in controlling temperature loss. When heat loss is to be minimized, short channels with round or square cross sections are the most desirable. Multiple gates and runners are often used to introduce metal to more than one point of a large casting.

Another function of the gating system is to minimize *turbulent flow*, which tends to promote excessive absorption of gases and oxidation of the metal and accelerates mold erosion. Figure 13-9 shows a typical gating system for a mold with a horizontal parting line and can be used to identify some of the key components that can be optimized to promote the smooth flow of molten metal. Short sprues are desirable, since they minimize the distance that the metal must fall when entering the mold. Rectangular pouring cups prevent a vortex of funnel from forming, which tends to suck gas and oxides into the sprue. Tapered sprues also prevent vortex formation. A large sprue well can be used to dissipate the kinetic energy of the falling stream. Finally, the *choke*, the smallest cross-sectional area which controls the rate of flow, should be located at the base of the sprue. If the choke were moved to the gates, the metal would enter the mold cavity with a fountain effect, a very turbulent mode of flow.

The gating system can also be designed to trap dross and sand particles and keep them from entering the mold cavity. Given sufficient time, the lower-density particles will rise to the top of the molten metal. Long, flat runners can be beneficial (but these promote cooling of the metal), as well as gates that exit from the lower portion of the runners. In addition, the first metal to enter the mold is most likely to contain the foreign matter (dross from the top of the pouring ladle and loose particles washed from the walls of the gating system). Runner extensions and wells (see Figure 13-9) can be used to catch and trap this first metal and keep it form entering the mold cavity. These are particularly effective with aluminum castings since aluminum oxide has approximately the same density as molten aluminum. Screens or ceramic filters can also be inserted into the gating system to trap the foreign material.

The specific details of a gating system often depend on the metal being cast. Turbulent-sensitive metals (such as aluminum and magnesium) and alloys with low melting points generally employ gating systems that concentrate on eliminating turbulence and trapping dross. Turbulent-insensitive alloys (such as steel, cast iron, and most copper alloys), and alloys with a high melting point, generally use short, open gating systems that provide for quick filling of the mold cavity.

Solidification Shrinkage

Once they are in the mold cavity and begin to cool, most metals and alloys undergo a noticeable contraction. There are three principal stages of *shrinkage*: (1) shrinkage of the liquid, (2) *solidification shrinkage* as the liquid turns to solid, and (3) solid metal contraction as the solidified material cools to room temperature. The amount of liquid metal contraction depends on the coefficient of thermal contraction (a property of the metal being cast) and the amount of superheat. Liquid contraction, however, is rarely a problem in casting production because the metal in the gating system continues to flow into the mold cavity as the metal in the cavity cools and contracts.

As the metal cools between the liquidus and solidus temperatures and changes state from liquid to solid, significant amounts of shrinkage can occur, as indicated by the data in Table 13-1. As indicated by this table, however, not all metals contract upon solidification. Some actually expand, such as gray cast iron, where the low-density graphite flakes form in the solid structure.

TABLE 13-1. Solidification Shrinkage (Percent) of Some Common Engineering Metals	
Aluminum	6.6
Copper	4.9
Magnesium	4.0
Zinc	3.7
Low-carbon steel	2.5–3.0
High-carbon steel	4.0
White cast iron	4.0–5.5
Gray cast iron	−1.9

When solidification shrinkage does occur, it is important to control the form of the resulting void. Metals and alloys with short freezing ranges, such as pure metals and eutectic alloys, tend to form large cavities or pipes. These can be avoided by designing the casting to have directional solidification. Here, the solidification begins farthest away from the feed gate or riser and moves progressively toward it. As the metal solidifies and shrinks, the shrinkage void is continually filled with liquid metal. The final shrinkage void, therefore, is located in the riser or gating system.

Alloys with large freezing ranges have a period of time when the material is in a mushy (liquid plus solid) condition. As the material cools, the amount of solid increases and tends to isolate small pockets of remaining liquid. It is almost impossible for additional liquid to feed into the shrinkage areas, and the resultant structure contains small but numerous shrinkage pores dispersed throughout. This type of shrinkage is far more difficult to prevent by means of gating and risering, and a porous product may be inevitable.

If a gas- or liquid-tight product is desired, the castings can be impregnated (the pores filled with a resinous material or lower-melting-temperature metal) in a subsequent operation. Castings with dispersed porosity tend to have poor ductility, toughness, and fatigue life.

After solidification is complete, a casting will contract further as it cools to room temperature. Compensation should be made for these dimensional changes when the mold cavity or pattern is designed. Additional concern arises, however, when the casting is produced in a rigid mold, such as the metal molds used in die casting. If the mold provides constraint during the time of contraction, tensile forces can be generated within the casting and cracking can occur. It may be desirable to eject the castings as soon as solidification is complete.

Risers and Riser Design

Risers are added reservoirs designed to feed liquid metal to the solidifying casting as a means of compensating for solidification shrinkage. To perform this function, the risers must solidify after the casting. If the reverse were true, liquid metal would flow from the casting into the solidifying riser and the casting shrinkage would be even greater. Hence the casting should be designed to produce directional solidification which sweeps from the extremities of the mold cavity toward the riser. In this way, the riser can feed molten metal continuously and will compensate for the solidification shrinkage of the entire mold cavity. If this type of solidification is not possible, multiple risers may be necessary with various sections of the casting solidifying toward their respective risers.

The risers should also be designed to conserve metal. If we define the yield of a casting as the casting weight divided by the total weight of metal poured (sprue, gates, risers, and casting), it is clear that there is a motivation to make the risers as small as possible, yet still able to perform their task. This is usually done through proper consideration of riser size, shape, and location, as well as the type of connection between the riser and casting.

According to Chvorinov's rule, a good shape for a riser would be one that has a long freezing time (i.e., a small surface area per unit volume). While a sphere would make the most efficient riser, this shape presents considerable difficulty to both the patternmaker and the moldmaker, who must remove the pattern from the mold. As a result, the most popular shape for a riser is a cylinder, where the height/diameter ratio is varied depending upon the nature of the alloy, location of the riser, the size of the flask, and other variables.

Risers should be located so that directional solidification occurs from the extremities of the mold cavity back toward the riser. Since the thickest regions of a casting will be the last to freeze, the risers should feed directly into these locations. Various types of risers are possible. A *top riser* is one that sits on top of a casting. Because of their location, top risers have shorter feeding distances and occupy less space within the flask, thereby providing more freedom for the layout of the pattern and gating system. *Side risers* are located adjacent to the mold cavity, displaced horizontally along the parting line. Figure 13-10 depicts both a top and a side riser. If the riser is contained entirely within the mold, it is known as a *blind riser*; if it is open to the atmosphere, it is called an *open riser*. Blind risers are usually larger than open risers because of the additional solidification and heat loss that occurs where the top of the riser is in contract with mold material.

Live risers (also known as hot risers) receive the last hot metal that enters the mold and generally do so at a time when the metal in the mold cavity has already begun to cool and solidify. Thus they can be smaller than *dead (or cold) risers*, which fill with metal that has already flowed through the mold cavity. As shown in Figure 13-10, top risers are almost always dead risers. Risers that are part of the gating system are generally live risers.

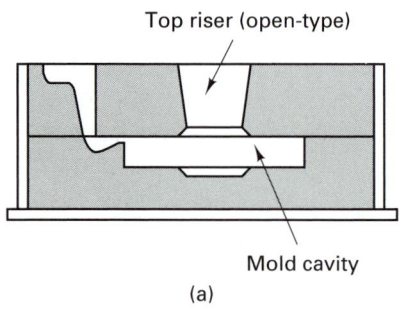

Top riser (open-type)

Mold cavity

(a)

Side riser (blind-type)

Mold cavity

(b)

FIGURE 13-10   Schematic of a sand casting mold, showing an open-type top riser (*left*) and a blind-type side riser (*right*). The side riser is a live riser, receiving the last hot metal to enter the mold. The top riser is a dead riser, receiving metal that has flowed through the mold cavity.

The minimum size of a riser can be calculated from Chvorinov's rule by setting the total solidification time for the riser to be greater than the total solidification time for the casting. Since both cavities receive the same metal and are in the same mold, the mold constant, B, will be the same for both regions. Assuming that $n = 2$, and a safe difference in solidification time is 25% (the riser takes 25% longer to solidify than the casting), we can write this condition as

$$t_{\text{riser}} = 1.25\ t_{\text{casting}} \tag{13-1}$$

or

$$(V/A)^2\ \text{riser} = 1.25\ (V/A)^2\ \text{casting} \tag{13-2}$$

Calculation of the riser size then requires selection of a riser geometry, which is generally cylindrical. For a cylinder of diameter D and height H, the volume and surface area can be written as

$$V = \pi D^2 H / 4$$

$$A = \pi D H + 2(\pi D^2 / 4)$$

Specifying the riser height as a function of the riser diameter then enables the V/A ratio for the riser to be written as a simple expression with one unknown, D. The V/A ratio for the casting can be calculated from its particular geometry. Substitution of this information into equation (13-2) produces a relation with one unknown, the size of riser required. Note, however, that if the riser and casting share a surface, as with a blind top riser, the common surface area should be subtracted from both components since it will not be a surface of heat loss to either. There are actually a wide variety of methods to calculate riser size, but the Chvorinov's rule method is the only one presented in this book.

A final aspect of riser design is the connection between the riser and the casting. Since the riser will ultimately be separated from the casting, it is desirable that the connection area be as small as possible. On the other hand, the connection area must be large enough that the link does not freeze before solidification of the casting is complete. Risers should be placed close to the casting with relatively short connections. Short connections are most desirable. The mold material surrounding the link will then receive heat from both the casting and the riser. It will heat rapidly and remain hot throughout the cast, thereby preventing solidification of the metal in the channel.

Risering Aids

Various methods have been developed to assist the risers in performing their job. Some are intended to promote directional solidification, while others seek to reduce the number and size of the risers, thereby increasing the yield of a casting. These techniques often work by either speeding the solidification of the casting or retarding the solidification of the riser.

External chills are masses of high-heat-capacity, high-thermal-conductivity material that are placed in the mold (adjacent to the casting) to accelerate the cooling of various regions. Chills can effectively promote directional solidification or increase the effective feeding distance of a riser. They can often be used to reduce the number of risers required for a casting.

Internal chills are pieces of metal that are placed within the mold cavity to absorb heat and promote more rapid solidification. Since some of this metal will melt during the operation, it will absorb not only the heat-capacity energy, but also some heat of fusion. Since they ultimately become part of the final casting, internal chills must be made from the same alloy as that being cast.

Ways to slow the cooling of risers include (1) switching from a blind riser to an open riser, (2) placing insulating sleeves and toppings around the riser, and (3) surrounding the sides or top of the riser with exothermic material that supplies added heat to just the riser segment of the mold. These techniques generally seek to reduce the riser size rather than promote directional solidification.

Finally, it is important to note that risers are not always necessary. For alloys with large freezing ranges, the risers would not be particularly effective, and one generally accepts the fine, dispersed porosity. For processes such as die casting, low-pressure permanent molding, and centrifugal casting, the positive pressures provide the feeding action that is required to compensate for solidification shrinkage.

■ 13.5 PATTERNS

Casting processes can be divided into two basic categories: those for which a new mold must be created for each casting (the *expendable-mold processes*) and those that use a permanent, reusable mold. Almost all of the expendable mold processes begin with a permanent, reusable *pattern*, a duplicate of the part to be cast, modified to reflect the casting process and the material being cast.

The modifications that are incorporated into a pattern are called *allowances*, and the most important of these is the *shrinkage allowance*. Following solidification, a casting continues to contract as it cools, the amount of this contraction being as much as 2% or $\frac{1}{4}$ in./ft. To produce the desired final dimensions, the pattern must be slightly larger than the casting. The exact amount of this compensation depends on the metal that is being cast. Typical allowances for some common engineering metals are:

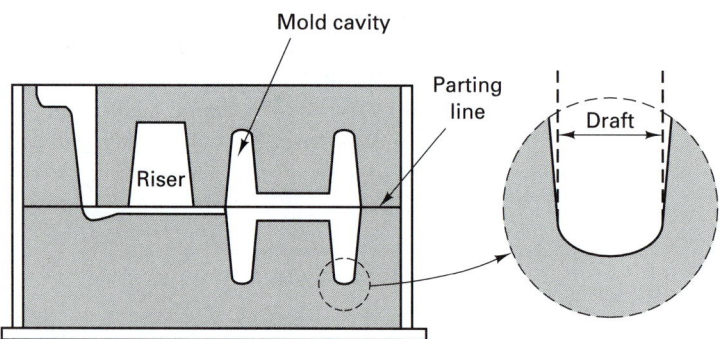

FIGURE 13-11 Two-part mold showing the parting line and the incorporation of a draft allowance on vertical surfaces.

Cast iron	0.8–1.0%	($\frac{1}{10} - \frac{1}{8}$ in./ft.)
Steel	1.5–2.0%	($\frac{3}{16} - \frac{1}{4}$ in./ft.)
Aluminum	1.0–1.3%	($\frac{1}{8} - \frac{5}{32}$ in./ft.)
Magnesium	1.0–1.3%	($\frac{1}{8} - \frac{5}{32}$ in./ft.)
Brass	1.5%	($\frac{3}{16}$ in./ft.)

Shrinkage allowances are often incorporated into a pattern through use of special *shrink rules*, measuring devices that are larger than a standard rule by the desired shrink allowance. For example, a shrink rule for brass would designate 1 ft at a length that is actually 1 ft $\frac{3}{16}$ in. A pattern made to shrink rule dimensions would produce a proper-size casting after cooling.

Some caution should be exercised when using shrink rules, however, for thermal contraction may not be the only factor affecting the final dimensions. The various phase transformations discussed in Chapter 4 are often accompanied by significant expansions or contractions. Examples include eutectoid reactions, martensitic reactions, and graphitization.

In many casting processes, mold material is formed around the pattern and the pattern is then removed to create the necessary cavity. To facilitate pattern removal, molds are often made in two or more sections. Consideration must then be given to the location of the *parting line*, the surface where one section of the mold mates with the other section or sections. If the pattern contains surfaces that are perpendicular to the parting line (parallel to the direction of pattern withdrawal), the friction between the pattern and the mold, or any horizontal movement of the pattern during extraction, would tend to damage the mold. This damage would be particularly severe at the corners where the mold cavity intersects the parting surface. By incorporating a slight taper, or *draft*, on all surfaces parallel to the direction of withdrawal, this difficulty can be minimized. As soon as the pattern is withdrawn a slight amount, it is free from the sand on all surfaces, and it can be withdrawn further without damaging the mold. Figure 13-11 illustrates the use of parting lines and draft.

The amount of draft is determined by the size and shape of the pattern, the depth of the cavity, the method used to withdraw the pattern, the pattern material, the mold material, and the molding procedure. Draft is seldom less than 1° or $\frac{1}{8}$ in./ft, with a minimum of about $\frac{1}{16}$ in. on any surface. On interior surfaces where the opening is small, such as a hole in the center of a hub, the draft should be increased to about $\frac{1}{2}$ in./ft. However, since draft allowances tend to increase the size of a pattern and thus the size and weight of a casting, it is generally desirable to keep them to the minimum that will permit satisfactory pattern removal. Modern molding procedures, which provide higher strength to the molding material before the pattern is withdrawn, and the use of molding machines that incorporate mechanical pattern withdrawal, have permitted substantial reductions in draft allowances. These improvements have enabled the production of lighter castings with thinner sections, thereby saving both weight and machining.

When machined surfaces must be provided on castings, it is often necessary to provide a *machining allowance*, or *finish allowance,* on the pattern. This allowance depends to a great extent on the casting process and the mold material. Ordinary sand castings have rougher surfaces than those of shell-mold castings. Die castings are sufficiently smooth so that very little or no metal has to be removed, and investment castings frequently require no additional machining. Consequently, the designer should relate the finishing allowance to the casting process and also remember that draft may provide part or all of the extra metal needed for machining.

If a core is to be used to form a hole or interior cavity, it too must be made oversized to compensate for shrinkage (all of the metal surrounding the hole will contract, making the hole smaller). However, if a machining allowance is to be included, it should be *subtracted* from the core dimensions because machining will increase the size of the hole. If the casting is made directly into a metal mold, all of the "pattern allowances" should be incorporated into the mold cavity. In addition, the change in mold dimensions caused by the heating of the mold from room temperature to its elevated operating temperature should be included as an additional correction.

Some casting shapes still require an additional allowance for *distortion*. For example, the arms of a U-shaped section may be restrained by the mold, while the base of the U is free to shrink. This restraint will result in a final casting with outwardly sloping arms. However, by designing the arms to originally slope inward, they will distort to a straight shape upon cooling. Long, horizontal sections tend to sag in the center unless adequate support is provided by suitable ribbing. Distortion depends greatly on the particular configuration of the casting, and the designer must use experience and judgment to provide the required distortion allowance.

Figure 13-12 illustrates the manner in which the various allowances are incorporated into a pattern. Since allowances increase both the weight of a casting and the amount of metal that has to be removed by machining, efforts are generally made to reduce them to the lowest value possible.

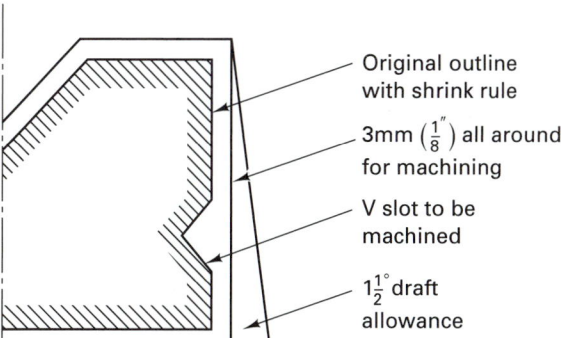

Original outline with shrink rule

3mm ($\frac{1}{8}''$) all around for machining

V slot to be machined

$1\frac{1}{2}°$ draft allowance

FIGURE 13-12   Various allowances incorporated into a casting pattern.

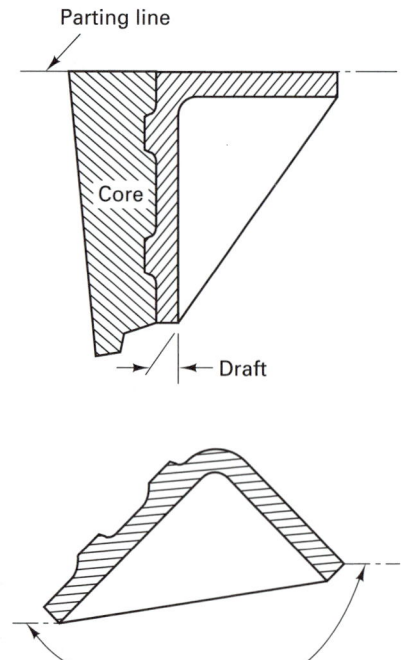

FIGURE 13-13   Elimination of a core by changing the location of the parting plane.

■ 13.6 DESIGN CONSIDERATIONS IN CASTINGS

To produce the best-quality product at the lowest possible cost, it is important that the designer of castings give careful attention to several process requirements and, if possible, work closely with the producing foundry. It is not uncommon for minor and readily permissible changes in design to greatly facilitate and simplify the casting of a component and also reduce the number and severity of defects.

One of the first features that must be considered by a designer is the *location of the parting plane*, an important part of all processes that use segmented or separable molds. The location of the parting plane can affect each of the following: (1) the number of cores, (2) the use of effective and economical gating, (3) the weight of the final casting, (4) the method of supporting the cores, (5) the final dimensional accuracy, and (6) the ease of molding.

In general, it is desirable to minimize the use of cores. Often, a change in the location or orientation of the parting plane can assist in this objective, as illustrated in Figure 13-13. Note that the change also reduces the weight of the casting by eliminating the need for draft. Figure 13-14 shows another example of how a simple design change can eliminate the need for a core.

The location of the parting plane can also be dictated by certain design features. Figure 13-15 shows how the specification of round edges can restrict the location of the parting plane. The specification of draft can also fix the parting plane, as indicated in Figure 13-16. This figure also shows that considerable freedom can be provided simply by noting the need to provide for a draft or by letting it be an option of the foundry. Since mold closure may not always be consistent, consideration should also be given to the fact that dimensions across the parting plane are subject to more variation than those that lie within a given segment of the mold.

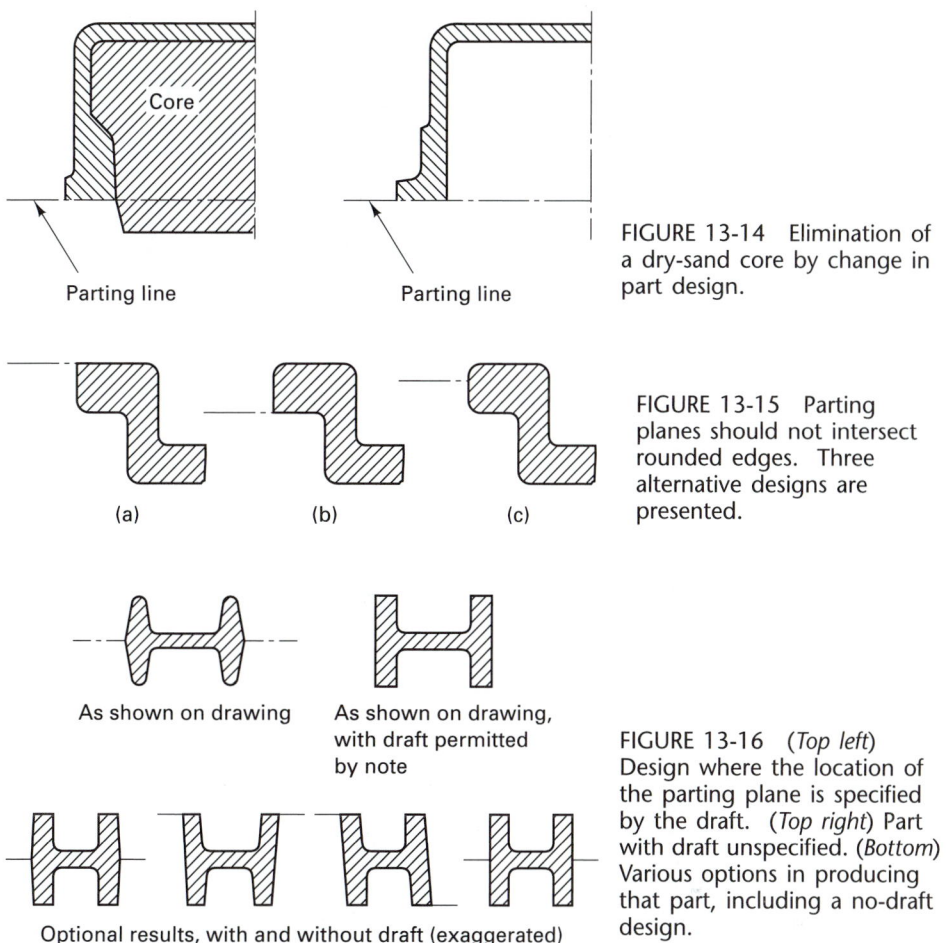

FIGURE 13-14 Elimination of a dry-sand core by change in part design.

Parting line

Parting line

FIGURE 13-15 Parting planes should not intersect rounded edges. Three alternative designs are presented.

(a) (b) (c)

As shown on drawing

As shown on drawing, with draft permitted by note

Optional results, with and without draft (exaggerated)

FIGURE 13-16 (*Top left*) Design where the location of the parting plane is specified by the draft. (*Top right*) Part with draft unspecified. (*Bottom*) Various options in producing that part, including a no-draft design.

Controlling the solidification process is of prime importance in obtaining quality castings, and this control is also related to design. Those portions of a casting that have a high ratio of surface area to volume will experience more rapid cooling and will be stronger and harder than the other regions. Heavier sections will cool more slowly and, unless special precautions are observed, may contain shrinkage cavities and porosity or have large grain-size structures. Ideally, a casting should have uniform thickness in all directions. In most cases, however, this is not possible. When the section thickness must change, it is best if these changes be gradual, as indicated in the recommendations of Figure 13-17.

When sections of castings intersect, two problems can arise. The first of these is stress concentration. This problem can be minimized by providing generous fillets (inside radii) at all interior corners. Excessive fillets, however, can augment the second problem, known as *hot spots*. Figure 13-18 shows that localized thick sections tend to exist where sections of castings intersect. These thick sections cool more slowly than the others and tend to be sites of localized, abnormal shrinkage. When the differences in section are large, like those illustrated in Figure 13-19, the hot-spot areas are likely to contain serious defects, such as porosity or shrinkage cavities.

Defects, such as voids, porosity, and cracks, can be sites of subsequent failure and should be prevented if at all possible. Sometimes cored holes, like those illustrated in

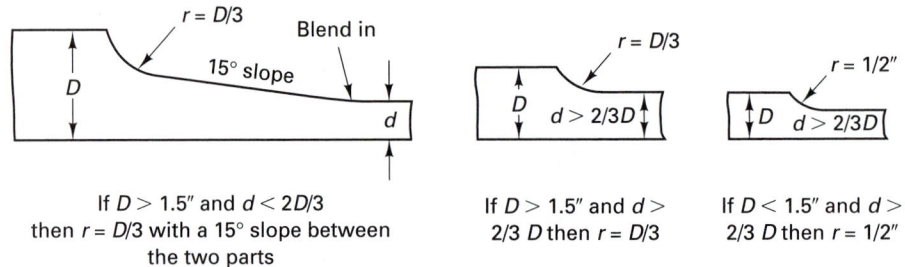

If $D > 1.5''$ and $d < 2D/3$
then $r = D/3$ with a 15° slope between
the two parts

If $D > 1.5''$ and $d > 2/3 D$ then $r = D/3$

If $D < 1.5''$ and $d > 2/3 D$ then $r = 1/2''$

FIGURE 13-17   Guidelines for section changes in castings.

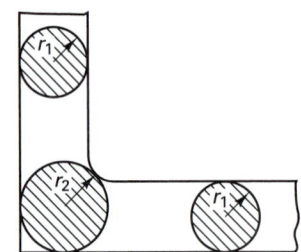

FIGURE 13-18   "Hot spot" at section r_2 caused by intersecting sections.

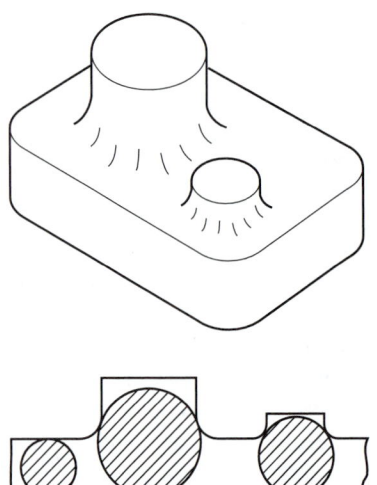

FIGURE 13-19   Hot spots resulting from intersecting sections of various thickness.

FIGURE 13-20   Method of eliminating unsound metal at the center of heavy sections in castings by using cored holes.

Figure 13-20, can be used to avoid hot spots. Where heavy sections must exist, an adjacent riser is often used to feed the section during solidification and shrinkage. If the riser is designed properly, the shrinkage cavity will lie totally within the riser, as illustrated in Figure 13-21, and can be removed when the riser is cut off.

Intersecting ribs can cause shrinkage problems and should be given special consideration by the designer. Where sections intersect to form continuous ribs, contraction occurs in opposite directions as the various ribs contract. As a consequence, cracking frequently

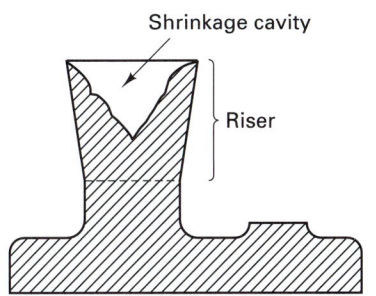

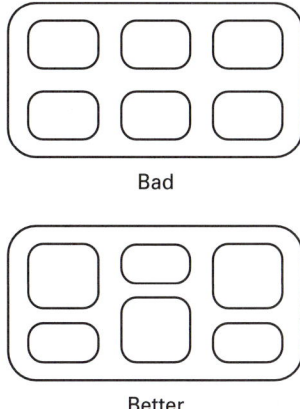

FIGURE 13-21 Use of a riser to keep the shrinkage cavity out of a casting.

FIGURE 13-22 Method of using staggered ribs to prevent cracking during cooling.

occurs at the intersections. By staggering the ribs, as shown in Figure 13-22, there is opportunity for distortion to occur, providing relaxation to the high residual stresses that would otherwise induce cracking.

Large unsupported areas should be avoided in all types of casting, since these regions tend to warp during cooling. The warpage then disrupts the good, smooth surface appearance that is so often desired. Another appearance consideration is the location of the parting line. Some small amount of fin, or flash, is often present at this location. When the flash is removed, or if it is considered small enough to leave in place, a region of surface imperfection will be present. If this is in the middle of a flat surface, it will be clearly visible. However, if the parting line is placed to coincide with a corner, the "defect" line will go largely unnoticed.

When designing castings, minimum section thickness should also be considered. Specific values are rarely given, however, because they tend to vary with the shape and size of the casting, the type of metal, the method of casting, and the practice of the individual foundry. Table 13-2 presents average values of the minimum and desirable section thicknesses for some common foundry materials and compatible casting processes.

TABLE 13-2. Recommended Minimum Section Thicknesses for Various Engineering Metals and Casting Processes

Material	Minimum		Desirable		Casting Process
	mm	in.	mm	in.	
Steel	4.76	$\frac{3}{16}$	6.35	$\frac{1}{4}$	Sand
Gray iron	3.18	$\frac{1}{8}$	4.76	$\frac{3}{16}$	Sand
Malleable iron	3.18	$\frac{1}{8}$	4.76	$\frac{3}{16}$	Sand
Aluminum	3.18	$\frac{1}{8}$	4.76	$\frac{3}{16}$	Sand
Magnesium	4.76	$\frac{3}{16}$	6.35	$\frac{1}{4}$	Sand
Zinc alloys	0.51	0.020	0.76	0.030	Die
Aluminum alloys	1.27	0.050	1.52	0.060	Die
Magnesium alloys	1.27	0.050	1.52	0.060	Die

The design considerations associated with the production of a high-quality casting are usually quite complex and interrelated. For these reasons, the input of a skilled and knowledgeable foundry worker is usually recommended.

■ KEY WORDS

allowance	external chill	pattern
blind riser	flask	penetration
casting	fluidity	pouring cup
chill zone	freezing range	pouring temperature
choke	gas flushing	recalescence
Chvorinov's rule	gas porosity	riser
cold shut	gate	runner
columnar zone	grain refinement	shrink rule
cooling curve	growth	shrinkage
cooling rate	hot spot	side riser
cope	inoculation	single-use mold
core	internal chill	solidification shrinkage
core box	live riser	sprue
core print	local solidification time	superheat
dead riser	misruns	thermal arrest
directional solidification	mold cavity	top riser
draft	mold constant	total solidification time
drag	multiple-use mold	turbulant flow
dross	nucleation	undercooling
equiaxed zone	open riser	vacuum degassing
expendable-mold process	parting line (parting surface)	

■ REVIEW QUESTIONS

1. What is materials processing?
2. What are the four basic families of shape-production processes? Cite one advantage and one limitation of each family.
3. Describe the capabilities of the casting process in terms of size and shape of the product.
4. How might the desired production quantity influence the selection of a single- or multiple-use molding process?
5. Why is it important to provide a means of venting gases from the mold cavity?
6. What types of problems can arise if the mold material provides too much restraint to the solidifying and cooling metal?
7. What is a casting pattern? Flask? Core? Mold cavity? Riser?
8. What are some or the components that combine to make up the gating system of a mold?
9. What is a parting line or parting surface?
10. What are the two stages of solidification, and what occurs during each?
11. Why is it that most solidification does not begin until the temperature falls somewhat below the equilibrium melting temperature (i.e., undercooling is required).
12. Why might it be desirable to promote nucleation in a casting?
13. Heterogeneous nucleation begins at preferred sites within a mold. What are some probable sites for heterogeneous nucleation?
14. Why might directional solidification be desirable in the production of a cast product?
15. Describe some of the key features observed in the cooling curve of a pure metal, such as the one depicted in Figure 13-3.
16. What is the freezing range for a metal or alloy?
17. Discuss the roles of casting volume and surface area as they relate to the total solidification time and Chvorinov's rule.
18. What characteristics of a specific casting process are incorporated into the mold constant, B, of Chvorinov's rule?
19. What is the chill zone of a casting, and why does it form?

20. Which of the three regions of a cast structure is least desirable? Why are its properties highly directional?
21. What is dross or slag, and how can it be prevented from becoming part of a finished casting?
22. What are some of the possible approaches that can be taken to prevent the formation of gas porosity in a metal casting?
23. What is fluidity, and how can it be measured?
24. Why is it important to design the geometry of the gating system to control the rate of metal flow from the pouring ladle into the mold cavity?
25. What are some of the undesirable consequences that could result from turbulence of the metal in the gating system and mold cavity?
26. What features can be incorporated into the gating system to aid in trapping dross and mold material that is flowing with the molten metal?
27. What features of the metal being cast tend to influence whether the gating system is designed to minimize turbulence and reduce dross, or promote rapid filling to minimize temperature loss?
28. What are the three stages of contraction or shrinkage as a liquid is converted into a finished casting?
29. Why is it more difficult to prevent shrinkage

30. voids from forming in metals or alloys with large freezing ranges?
31. What type of flaws or defects are associated with the further shrinkage of a casting after it has solidified?
31. Why is it desirable to design a casting to have directional solidification sweeping from the extremities of the mold to the riser?
32. Based on Chvorinov's rule, what would be an ideal shape for a casting riser? A desirable shape from a practical perspective?
33. Define the following riser-related terms: top riser; side riser; open riser; blind riser; live riser; dead riser.
34. What assumptions were made when using Chvorinov's rule to calculate the size of a riser in the manner presented in the text?
35. What are the two common objectives of risering aids, and how might they be achieved?
36. What types of modifications or allowances are generally incorporated into a casting pattern?
37. What is a shrink rule, and how does it work?
38. What is the purpose of a draft or taper on pattern surfaces?
39. Why is it desirable to make the pattern allowances as small as possible?
40. Fillets are used to reduce the magnitude of stress raisers at intersecting sections. What can occur if too large a fillet is used?

■ PROBLEMS

1. Using Chvorinov's rule as presented in the text with $n = 2$, calculate the dimensions of an effective riser for a casting that is a 2 in. by 4 in. by 6 in. rectangular plate. Assume that the casting and riser are not connected, except through a gate and runner, and that the riser is a cylinder of height/diameter ratio $H/D = 1.5$. The finished casting is what fraction of the combined weight of the riser and casting?
2. Reposition the riser in Problem 2 so that it sits

directly on top of the flat rectangle, with its bottom circular surface being part of the surface of the casting, and recompute the size and yield fraction. Which approach is more efficient?

3. A rectangular casting having the dimensions 3 in. by 5 in. by 10 in. solidifies completely in 11.5 minutes. Using $n = 2$ in Chvorinov's rule, calculate the mold constant B. Then compute the solidification time of a 0.5 in. by 8 in. by 8 in. casting poured under the same conditions.

Chapter 13 CASE STUDY

the cast oil-field fitting

A cast iron, T-type fitting is being produced for the oil drilling industry, using an air-set or no-bake sand for both the mold and the core. A silica sand has been used in combination with a catalyzed alkyd-oil/urethane binder. Figure CS-13 shows a cross section of the mold with the core in place (part a), and a cross section of the finished casting (part b). Note that there are several significant defects. Gas bubbles are observed at one location in the base of the tee. A penetration defect is observed near the bottom of the inside diameter, and there is an enlargement of the casting at location "C".

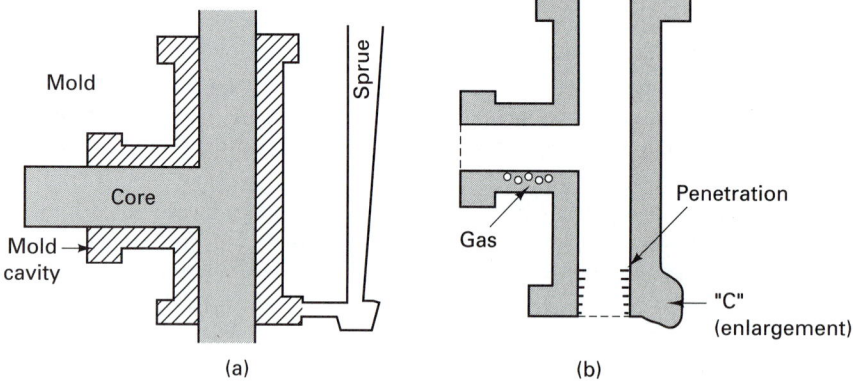

FIGURE CS-13

1. What is the most likely source of the gas bubbles? Why are they present only at the location noted? What might you recommend as a solution?

2. What factors may have caused the penetration defect? Why is the defect present on the inside of the casting, but not on the outside? Why is the defect near the bottom of the casting, but not near the top?

3. What factors led to the enlargement of the casting at point "C"? What would you recommend to correct this problem?

4. Another producer has noted penetration defects on all surfaces of the casting, both interior and exterior. What would be some possible causes? What could you recommend as possible cures?

5. Could these molds and cores be reclaimed after breakout? Discuss.

CHAPTER 14

EXPENDABLE-MOLD CASTING PROCESSES

14.1. INTRODUCTION
14.2. SAND CASTING
Patterns and Pattern
  Materials
Types of Patterns
Sand Conditioning
Sand Testing
Sand Properties and Sand-
  Related Defects
The Making of Sand Molds
Green-Sand, Dry-Sand, and
  Skin-Dried Molds
Sodium Silicate–CO_2
  Molding
No-Bake, Air-Set, or
  Chemically Bonded Sands
Shell Molding
Other Sand-Based Molding
  Methods
14.3 CORES AND CORE
  MAKING

14.4 OTHER EXPENDABLE-
  MOLD PROCESSES WITH
  MULTIPLE-USE
  PATTERNS
Plaster Mold Casting
Ceramic Mold Casting
Expendable Graphite Molds
Rubber-Mold Casting
14.5 EXPENDABLE-MOLD
  PROCESSES USING
  SINGLE-USE PATTERNS
Investment Casting
Full-Mold and Lost-Foam
  Casting
14.6 SUMMARY
Case Study: MOVEABLE AND FIXED
  JAW PIECES FOR A
  HEAVY-DUTY BENCH
  VISE

■ 14.1 INTRODUCTION

Because of the number and variety of casting processes used to produce manufactured items, it is helpful to have some form of process classification. One approach focuses on the molds and patterns and utilizes the following three categories:

1. Single-use molds with multiple-use patterns
2. Single-use molds with single-use patterns
3. Multiple-use molds

Categories 1 and 2 are often combined under the more general heading "expendable-mold casting processes," and these are presented in the present chapter. Sand, plaster, ceramics, and other refractory materials are combined with binders to form the mold material. Those processes where a mold can be used multiple times are presented in Chapter 15; here the molds are usually made from metal.

Although some nonmetals are cast, the casting processes are associated primarily with the production of metal products, and the emphasis of these chapters will be on metals casting. The metals most frequently cast are iron, steel, aluminum, brass, bronze, magnesium, certain zinc alloys, and nickel-based superalloys. Of these, cast iron is the dominant casting material, primarily because of its low cost, good fluidity, low shrinkage, ease of control, and wide range of properties, including useful strength and rigidity. In Chapter 20 we discuss the processes used to fabricate products from polymers, ceramics (including glass), and composites.

■ 14.2. SAND CASTING

Sand casting, by far the most popular of the casting processes, uses ordinary sand as the primary mold material. The sand grains are mixed with small amounts of other materials, such as clay and water, to improve moldability and cohesive strength, and are then packed around a pattern that has the shape of the desired casting. Because the grains will pack into thin sections and can be used economically in large quantities, products spanning a wide range of sizes and detail can be made by this method. If the pattern must be removed before pouring, the mold is usually made in two or more pieces. An opening called a *sprue hole* is cut from the top of the mold through the sand and connected to a system of channels called *runners*. The molten metal is poured into the sprue hole, flows through the runners, and enters the mold cavity through an opening called a *gate*. Gravity flow is the most common means of introducing the metal into the mold. After solidification, the mold is broken and the finished casting is removed. Because the mold is destroyed, a new mold must be made for each casting. Figure 14-1 shows the essential steps and basic components of a sand casting process.

Patterns and Pattern Materials

The first step in making a sand casting is the design and construction of a *pattern*. This is a duplicate of the part to be cast, modified in accordance with the requirements of the casting process, metal being cast, and particular molding technique that is being used. The pattern material is determined primarily by the number of castings to be made but is also influenced by the size and shape of the casting, the desired dimensional precision, and the molding process. Wood patterns are relatively easy to make and are frequently used when small quantities of castings are required. Wood, however, is not very dimensionally stable. It may warp or swell with changes in humidity, and it tends to wear with repeated use. Metal patterns are more expensive but are more dimensionally stable and more durable. Hard plastics, such as urethanes, offer another alternative, and are often preferred with processes that use strong, organically bonded sands that tend to stick to other pattern materials. In the full-mold process, *expanded polystyrene* (EPS) is used, and investment casting uses wax patterns. In the later processes, each pattern can be used only once.

Types of Patterns

Since many types of patterns are used in the foundry industry, selection is usually based on the number of duplicate castings required and the complexity of the part.

One-piece or *solid patterns*, such as the one shown in Figure 14-2, are the simplest and often the least expensive type to make. This type of pattern is essentially a duplicate of the part to be cast, modified only by the various allowances discussed in Chapter 13 and by the possible addition of core prints. They are relatively cheap to construct, but the molding process is usually slow. One-piece patterns, therefore, are generally used when the shape is relatively simple and the number of duplicate castings is rather small.

If a one-piece pattern is simple in shape and contains a flat surface, it can be placed directly on a follow board. The entire mold cavity will be in one segment of the mold, and the follow board will create the parting surface. If the shape is more complex or the parting plane is to be more centrally located, it may be necessary for the molder to hand-cut an irregular parting surface. This is a time-consuming process and requires a skilled worker. More often, special follow boards are used with inset cavities designed so that the one-piece pattern is positioned at the depth of the desired parting line. Here, the follow board again determines the parting surface, as illustrated in Figure 14-3.

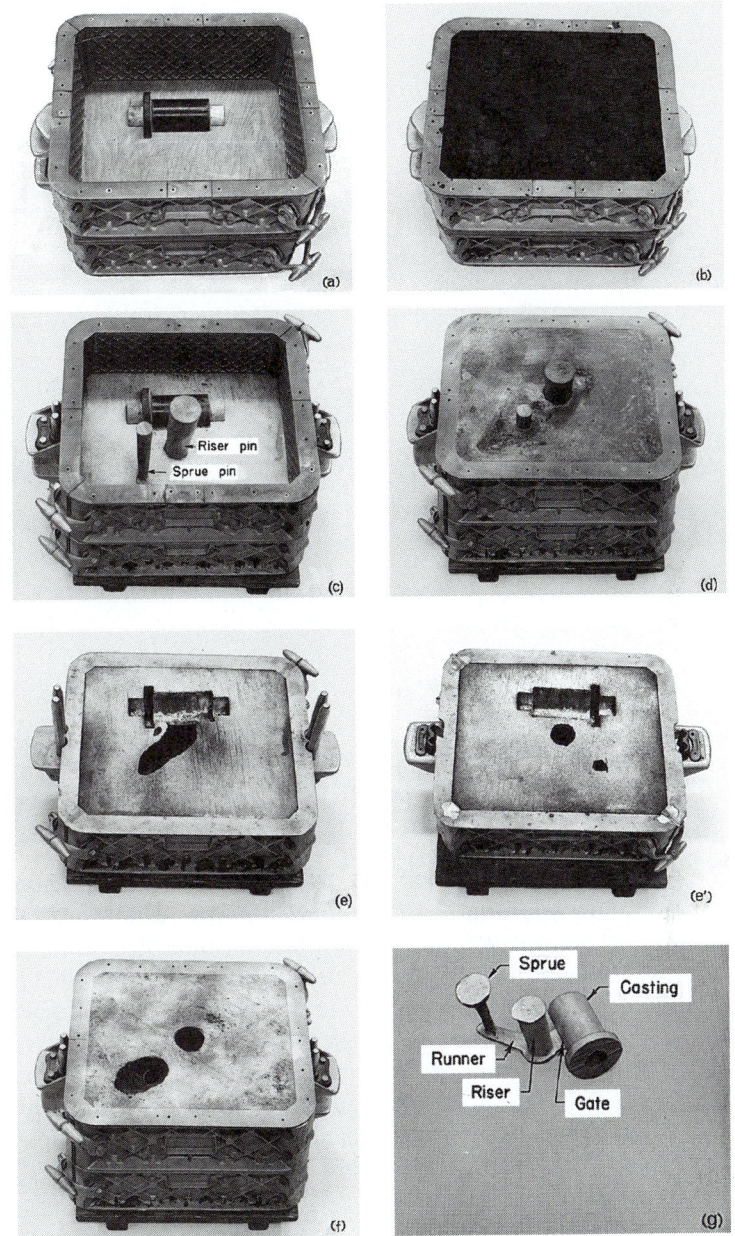

FIGURE 14-1 Sequential steps in making a sand casting. (a) A pattern board is placed between the bottom (drag) and top (cope) halves of a flask, with the bottom side up. (b) Sand is then packed into the drag half of the mold. (c) A bottom board is positioned on top of the packed sand, and the mold is turned over, showing the top (cope) half of pattern with sprue and riser pins in place. (d) The cope half of the mold is then packed with sand. (e) The mold is opened, the pattern board is drawn (removed), and the runner and gate are cut into the surface of the sand. (e') The parting surface of the cope half of the mold is shown with the pattern and pins removed. (f) The mold is reassembled with the pattern board removed, and molten metal is poured through the sprue. (g) The contents are shaken from the flask and the metal segment is separated from the sand, ready for further processing.

FIGURE 14-2   Single-piece pattern for a pinion gear.

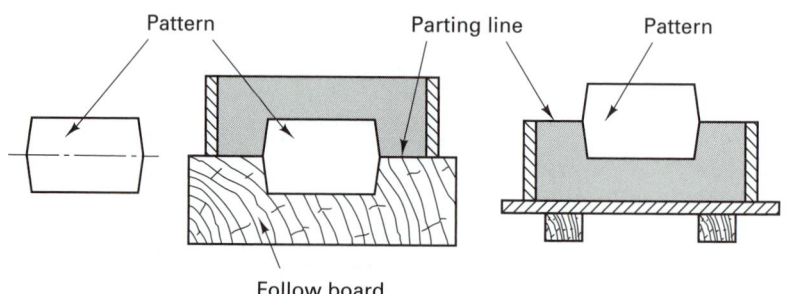

FIGURE 14-3   Method of using a follow board to position a pattern and locate a parting surface.

Split patterns are used when moderate quantities of duplicate castings are to be made. The pattern is divided into two segments along a single parting plane, which will correspond to the parting plane of the mold. The bottom segment of the pattern is positioned in a flask and the lower (drag) portion of the mold is formed. This flask is then inverted, and the upper segment of the pattern is attached. Tapered pins in the cope half of the pattern align with holes in the drag segment to assure proper positioning. With the full pattern in place, mold material is then packed into the upper (cope) flask. The flasks are separated and the pattern is removed, completing the mold cavity. Sprues and runners are cut and the mold is poured. Figure 14-4 shows a split pattern with several core prints.

Match-plate patterns, such as the one in Figure 14-5, further simplify the process and can be coupled with modern molding machines to produce large quantities of duplicate castings. The cope and drag segments of a split pattern are permanently fastened to opposite sides of a wood or metal *match plate*. Using holes that align with pins on the flask, the match plate is positioned between the upper and lower segments of the flask. Mold material is then packed around the pattern to complete the cope and drag segments of a mold. The mold sections are then separated and the match-plate pattern is removed. When the mold is then reassembled using the pins and guide holes, the cavities in the cope and drag join in proper alignment.

In most cases, the necessary gates and runners are incorporated on the match plate. This eliminates the need for hand cutting and guarantees that these features will be uniform and of the proper size in each mold, thereby reducing the possibility of defects. The gate and runner system can be seen as the dark center section on the cope side of the match-

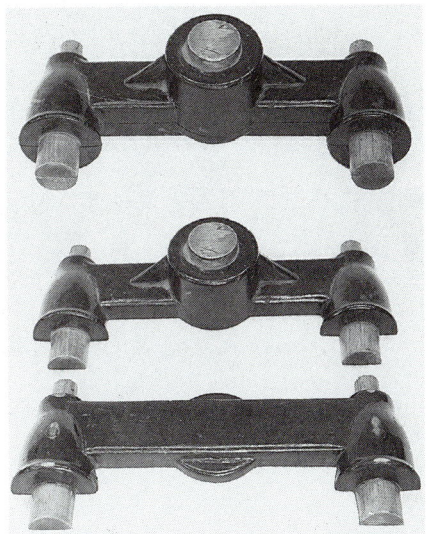

FIGURE 14-4 Split pattern, showing the two sections together and separated. Light-colored portions are core prints.

FIGURE 14-5 Match-plate pattern to produce two identical parts in a single flask. (*Left*) Cope side; (*right*) drag side.

plate pattern in Figure 14-5 and on the cope section in Figure 14-6. These figures also include core prints and risers, and further illustrate the common practice of including more than one pattern on a match plate.

When large quantities are to be produced, or when the casting is large, it may be desirable to have the cope and drag halves of split patterns attached to separate match plates to produce mating *cope-and-drag patterns*. This enables independent molding of the cope and drag segments of a mold. Large molds can be handled more easily in separate segments, and small molds can be made at a faster rate. Figure 14-6 shows the mating pieces of a cope-and-drag pattern.

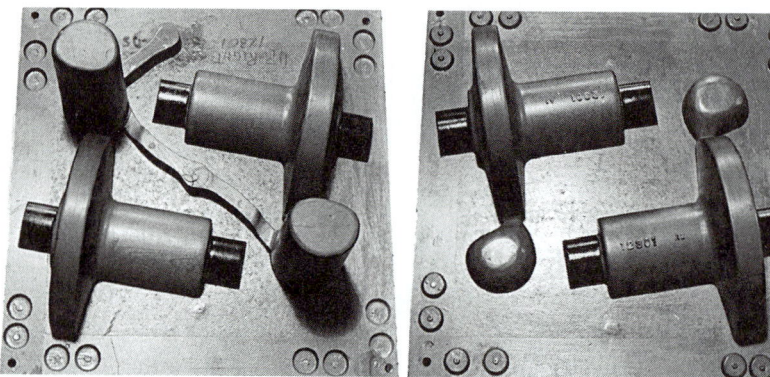

FIGURE 14-6 Cope-and-drag pattern for molding two heavy parts. (*Left*) Cope section; (*right*) drag section.

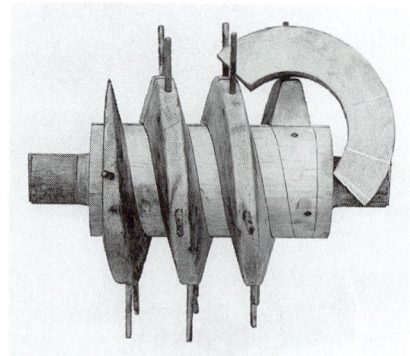

FIGURE 14-7 Loose piece pattern for molding a large worm. After sufficient sand is packed around the pattern halves to hold the pieces in position, the wooden pins are withdrawn. The mold is completed and the pieces of the pattern are removed in sequence.

When the product has protruding sections arranged such that a one-piece or split pattern could not be removed from the molding sand, a *loose-piece pattern* can sometimes be developed. Loose pieces are held to the primary segment of the pattern by beveled grooves or pins (Figure 14-7). After molding, the primary segment of the pattern is removed by direct withdrawal. The hole that is created then permits the remaining segments to be moved in the direction necessary for their extraction. In some loose-piece patterns, a single sliding pin is used to hold all the segments in place. After the sand is compacted around the pattern, the pin is removed and the individual segments are withdrawn. Loose-piece patterns are expensive, they require careful maintenance, slow the molding process, and increase molding costs. They do, however, enable the sand casting of complex shapes that would otherwise require the full-mold or investment processes. Whenever possible, however, design changes should be considered that would eliminate the need for split-piece patterns.

In all foundry patterns it is important that internal corners contain a small radius, called a *fillet*, rather than meet in a distinct line. Fillets avoid shrinkage cracks at the intersections and reduce the stress concentrations in the finished product. Designers should provide generous fillets whenever possible, with radii of $\frac{1}{4}$ and $\frac{1}{8}$ in. (6.35 and 3.18 mm) being most common. Fillets are often added to patterns through the use of pre-radiused wax, leather, or plastic strips that are glued to the pattern or pressed into place with a heated fillet tool.

Sand Conditioning

The sand used to make molds must be carefully prepared if it is to provide satisfactory and uniform results. Ordinary silica (SiO_2), zircon, or olivine (forsterite and fayalite) sands are compounded with additives to meet four requirements:

1. *Refractoriness:* the ability to withstand high temperatures
2. *Cohesiveness:* (also referred to as *bond*): the ability to retain a given shape when packed into a mold
3. *Permeability:* the ability to permit gases to escape through it
4. *Collapsibility:* the ability to permit the metal to shrink after it solidifies and ultimately to free the casting by disintegration of the surrounding mold

Refractoriness is provided by the basic nature of the sand. Cohesiveness, bond, or strength is obtained by coating the sand grains with clays, such as bentonite, kaolinite, or illite, that become cohesive when moistened. Collapsibility is sometimes obtained by adding cereals or other organic material, such as cellulose, that burn out when they come in contact with the hot metal. The combustion reduces both the volume and strength of the restraining sand. Permeability is a function of the size of the sand particles, the amount and type of clay or bonding agent, the moisture content, and the compacting pressure.

Good molding sand always represents a compromise between conflicting factors. The size of the sand particles, the amount of bonding agent (such as clay), the moisture content, and the organic matter are all selected to obtain an acceptable compromise of the four requirements. The composition must be carefully controlled to assure satisfactory and consistent results. (A typical green-sand mixture contains about 88% silica sand, 9% clay, and 3% water.) Since molding material is often reclaimed and recycled, the temperature of the mold during pouring and solidification is also important. If organic materials have been incorporated into the mix to provide collapsibility, a portion will burn during the pour. Some of the mold material may have to be discarded and replaced with new.

It is also important for each grain of the sand to be coated uniformly with the additive agents. This is achieved by putting the ingredients through a *muller,* a device that kneads, rolls, and stirs the sand. Figure 14-8 shows both a continuous and batch-type muller, which use blades and wheels to produce the mixing. After mixing, the sand is often discharged through an *aerator,* which fluffs it so that it does not pack too hard during handling.

Sand Testing

Maintaining consistent sand quality is of little concern to the casting designer, but it is a significant matter to the foundry worker, who is expected to deliver consistent, high-quality products. Standard tests and procedures have been developed to evaluate *grain size, moisture content, clay content, and compactability,* as well as *mold hardness, permeability, and strength.*

Grain size is determined by shaking a known amount of clean, dry sand downward through a set of 11 standard sieves of decreasing mesh size. After shaking for 15 minutes, the amount remaining on each sieve is weighed, and the weights are converted into an AFS (American Foundrymen's Society) grain fineness number.

Moisture content is usually determined by a special device that measures the electrical conductivity of a small sample of sand that is compressed between two prongs. Another method is to measure the weight lost from a 50-g sample after it has been subjected to a temperature of about 230°F (110°C) for sufficient time to drive off all the water.

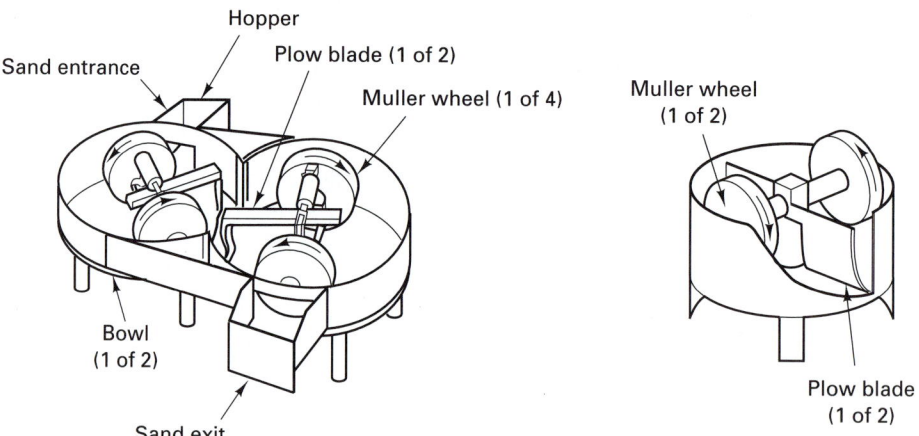

FIGURE 14-8 Schematic diagram of a continuous (*left*) and batch-type (*right*) sand muller. Plow blades move the sand and the muller wheels mix the components. *(Courtesy of ASM International.)*

Clay content can be determined by washing the clay from a 50-g sample of molding sand in water that contains sufficient sodium hydroxide to make it alkaline. Several cycles of agitation and washing may be required to fully remove the clay. The remaining sand is then dried and weighed to determine the amount of clay in the original sample.

Permeability and strength tests are conducted on a *standard rammed specimen*. A sufficient amount of sand is placed into a 2-in.-diameter steel tube so that after a 14-lb weight is dropped three times from a height of 2 in., the final height of the specimen is within $\frac{1}{32}$ in. of 2 in.

Permeability is a measure of how easily gases can pass through the narrow voids between the sand grains. Air in the mold before pouring (plus the steam that is produced when the hot metal contacts the moisture in the sand) must be allowed to escape, rather than be trapped in the casting as porosity or blow holes. During the permeability test, the sample tube containing the rammed specimen is placed on a device like that pictured in Figure 14-9, and subjected to an air pressure of 10 g/cm^2. By means of either a flow rate determination or measurement of the pressure between the orifice and the sand, an *AFS permeability number*[*] is determined. Most test devices are calibrated to provide a direct readout of the permeability number.

The compressive strength of the sand is determined by removing the rammed specimen from the tube and placing it in a mechanical testing device. A compressive load is then applied until the specimen breaks, usually in the range of 10 to 30 psi (0.07 to 0.2 MPa). When there is too little moisture in the sand, the grains are poorly bonded and strength is poor. When there is excess moisture, the extra water acts as a lubricant and strength is again poor. Thus there is a maximum strength and an optimum water content that will vary with the content of other materials in the mix. A similar optimum also ap-

---

[*] The AFS permeability number is defined as follows:

$$\text{AFS number} = V \times H / P \times A \times T$$

where V is the volume of air (2000 cm^2), H the height of the specimen (5.08 cm), P the pressure (10 g/cm^2), A the cross section of the specimen (20.268 cm^3), and T the time in seconds to pass a flow of 2000 cm^3. Substituting the constants, the permeability number is equal to 3000.2/ T, where T is the time required (in seconds) to pass 2000 cm^3 of air through the specimen.

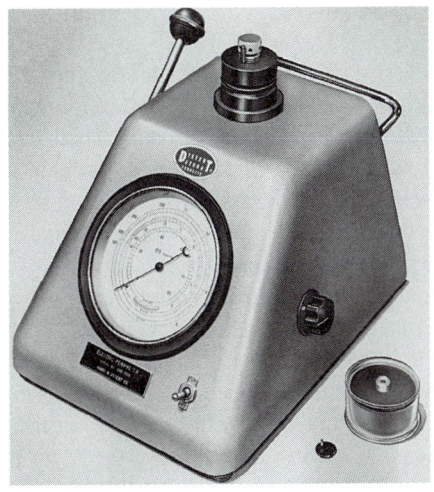 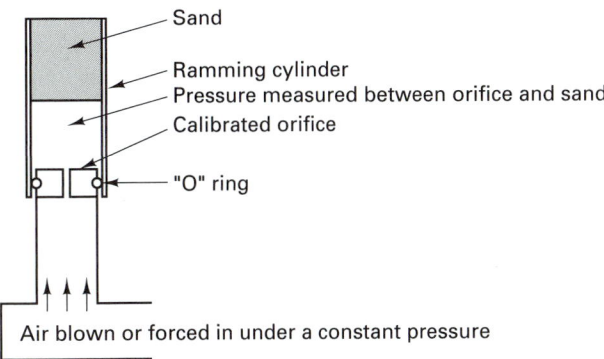

Sand
Ramming cylinder
Pressure measured between orifice and sand
Calibrated orifice
"O" ring
Air blown or forced in under a constant pressure

FIGURE 14-9 (*Left*) Permeability tester for foundry sand. A standard sample in a sleeve is sealed by an o-ring on top of the unit. (*Right*) Schematic of the permeability tester in operation. (*Courtesy of Harry W. Dietert Company.*)

FIGURE 14-10 Mold hardness tester. (*Courtesy of Harry W. Dietert Company.*)

plies to permeability, since unwetted clay blocks passages, as does excess water. Sand coated with a uniform thin film of moist clay provides the best molding properties. A ratio of 1 part water to 3 parts clay (by weight) is often a good starting point.

The hardness of the compacted sand can provide a quick indication of mold strength and give additional insight into the strength–permeability characteristics. It can be measured by an instrument like the one shown in Figure 14-10, which determines the resistance of the sand to penetration by a 0.2-in. (5.08-mm)-diameter spring-loaded steel ball.

Compactability is determined by sifting sand into a steel cylinder, leveling off the column, striking it three times with a standard weight (as in the permeability test), and then measuring the final height. The percent compactability is the change in height divided by the original height, times 100%. The value can often be correlated with the moisture content of the sand, with a compactability around 45% indicating a proper level of moisture. A low compactability correlates with too little moisture.

Sand Properties and Sand-Related Defects

The characteristics of the sand granules can be very influential in determining the properties of the molding material. Round grains give good permeability and minimize the amount of clay required because of their low surface area. Angular sands give better green strength because of the mechanical interlocking of the grains. Large grains provide good permeability and better resistance to high temperature melting and expansion, while fine-grained sands produce a better surface finish on the final casting. Uniform-size sands give good permeability, while a wide distribution of sizes enhances surface finish.

When hot metal is poured into a silica sand mold, the sand becomes hot, undergoes one or more phase transformations, and has a substantial expansion in volume. Because sand is a poor thermal conductor, only the sand that is adjacent to the mold cavity becomes hot and expands. The remaining material stays fairly cool, does not expand, and provides a high degree of mechanical restraint. Because of this uneven heating, the sand at the surface of the mold cavity may buckle or fold. Castings having large, flat surfaces are more prone to *sand expansion defects* since a considerable amount of expansion must occur in a single, fixed direction.

Sand expansion defects can be minimized in a number of ways. Certain sand geometries permit the grains to slide over one another, thereby relieving the expansion stresses. Excess clay can be added to absorb the sand expansion, or volatile additives, such as cellulose, can be added to the mix. As the sand becomes hot, the cellulose burns, creating voids that can accommodate the sand expansion. Finally, olivine sands can be used in place of silica. Since olivine sand does not undergo phase transformations upon heating, its expansion is only about half that of silica sand. Unfortunately, it is much more expensive.

Castings can also contain voids that form where the molten metal is held back by trapped or evolved gas. These are usually attributed to low sand permeability and/or large amounts of gas evolution caused by high moisture or excessive amounts of volatiles. If adjustments to the mold composition are not sufficient to eliminate the voids, vent passages may have to be cut, a procedure that adds significantly to the mold-making cost.

The molten metal can also penetrate between the sand grains, causing the mold material to become embedded in the surface of the casting. *Penetration* can be the result of high pouring temperatures (excess fluidity), high metal pressure (possibly due to excessive cope height or pouring from too high an elevation above the mold), or the use of high-permeability sands with coarse, uniform particles. Fine-grained materials, such as silica flour, can be used to fill the voids, but this reduces permeability and increases the likelihood of gas and expansion defects.

Hot tears or cracks can form in castings made from alloys with a large freezing range. During solidification, the metal tries to contract but may find itself restrained by a strong mold or core. Tensile stresses can develop while the metal is still partially liquid, and if they become great enough, the casting will crack. Hot tears can often be attributed to a lack of *collapsibility*, the ability of the sand to break down and crumble after the casting has been poured and solidified. Sand additives, such as cellulose, can be particularly helpful in providing collapsibility. Table 14-1 summarizes some of the desirable properties of a sand-based molding material.

The Making of Sand Molds

Hand ramming may be the preferred method of mold making when only a few castings are to be made from any given design, and some small foundries still make their molds by this

TABLE 14-1. Desirable Properties of a Sand-Based Molding Material

1. Is inexpensive in bulk quantities
2. Retains properties through transportation and storage
3. Uniformly fills a flask or container
4. Can be compacted or set by simple methods
5. Has sufficient elasticity to remain undamaged during pattern withdrawal
6. Can withstand high temperatures and maintains its dimensions until the metal has solidified
7. Is sufficiently permeable to allow the escape of gases
8. Is sufficiently dense to prevent metal penetration
9. Is sufficiently cohesive to prevent wash-out of mold material into the pour steam
10. Is chemically inert to the metal being cast
11. Can yield to solidification and thermal shrinkage, thereby preventing hot tears and cracks
12. Has good collapsibility to permit easy removal and separation of the casting
13. Can be recycled

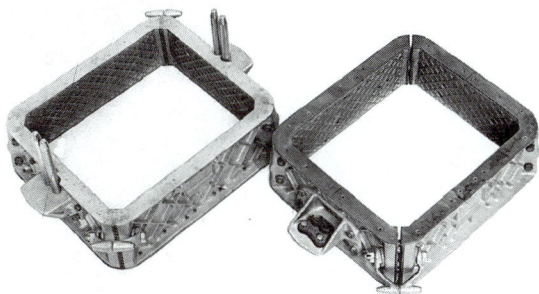

FIGURE 14-11   Halves of a tapered cam-latch snap flask, closed at left and open at right.

method. In most cases, however, sand molds are made by specially designed molding machines. The various methods differ in the type of flask required, the way sand is packed within the flask, whether mechanical assistance is provided to turn or handle the mold, and whether a flask is even required. In all cases, however, the molding machines greatly reduce the labor and skill required, and lead to castings with better dimensional accuracy and consistency.

Molding usually begins with a pattern, such as the match-plate pattern discussed earlier, and a *flask*. The flasks may be either straight-walled containers with guide pins or removable jackets, and they are generally constructed of aluminum or magnesium. Figure 14-11 shows a snap flask that can open slightly to permit withdrawal after the mold packing is complete.

The sand is generally packed in the flask by one or more basic techniques. In a method known as *jolting,* sand is placed on top of the pattern, and the pattern, flask, and sand are then lifted and dropped several times, as shown in Figure 14-12. The kinetic energy of the sand produces optimum packing around the pattern. Jolting machines can be used on the first half of a match-plate pattern or on both halves of a cope-and-drag operation.

Squeezing machines use either an air-operated squeeze head, a flexible diaphragm, or small individually activated squeeze heads to compact the sand. *Squeezing* provides firm packing near the squeeze head, but the density diminishes as you move farther into the mold. High-pressure machines with a flexible diaphragm, commonly called *Taccone machines,* can produce a more uniform density around all parts of an irregular pattern. Figure 14-13 illustrates the squeezing process, and Figure 14-14 compares squeezing with a flat plate and squeezing with a flexible diaphragm.

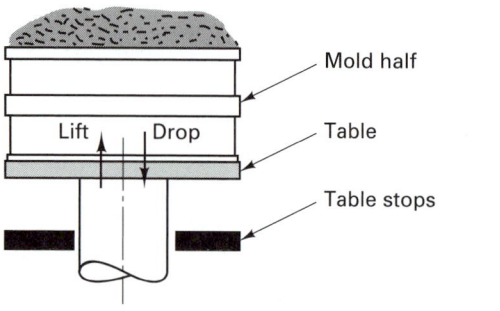

FIGURE 14-12   Jolting a mold half.

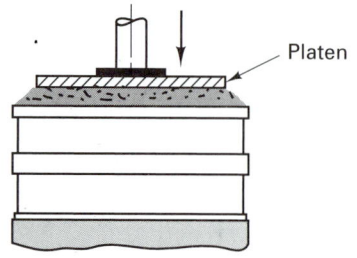

FIGURE 14-13   Squeezing operation

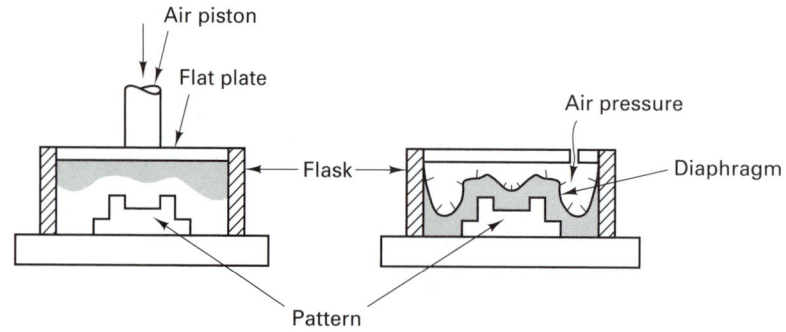

FIGURE 14-14   Schematic diagram showing relative sand densities obtained by flat-plate squeezing (*left*) and by flexible-diaphragm squeezing (*right*).

A combination of jolting and squeezing is often used to produce a more uniform density throughout the mold. Here a match-plate pattern is positioned between the cope and drag sections of a flask, and the assembly is placed upside down on the molding machine. A parting compound is sprinkled on the pattern, and the top section of the flask is filled with sand. The entire assembly is then jolted a specified number of times to pack the sand around the pattern. A squeeze head is then swung into place, and pressure is applied to complete the upper portion of the mold. The flask can be inverted and operations repeated on the cope half, or the cope and drag can be made on separate machines using cope-and-drag patterns. Unless the molds are very small, the molding machines usually provide mechanical assistance for inverting the heavy molds.

While some patterns include the sprue hole, it may also be cut by hand. (Hand cutting is performed before removal of the pattern, so loose sand does not fall into the mold cavities.) The pouring basin may also be hand cut, or it may be formed by a protruding shape on the squeeze board. The gates and runners are usually included on the pattern. Because of the cost and time required for hand labor, the growing trend is to design patterns that will minimize the amount of hand working.

After the mold is completed, the tapered flask may be removed to prevent possible damage to the flask during the pour. A *slip jacket,* an inexpensive metal band, may be positioned around the mold to hold the sand in place. Heavy metal weights are often placed on top of the molds to prevent the sections from separating as the hydrostatic pressure of the molten metal presses upward on the cope. This slip jackets and added weights are needed only during pouring and for a few minutes afterward, while solidification occurs.

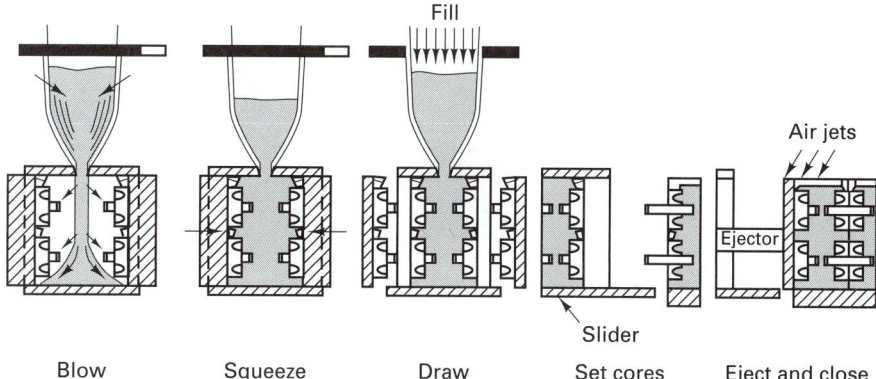

FIGURE 14-15   Vertically parted flaskless molding. *(Courtesy of Belens Corporation.)*

They can then be removed and placed on other molds, thereby reducing the amount of equipment needed in the operation.

For mass-production molding, a number of automatic mold-making devices have been developed. These include automatic match-plate machines, automatic cope-and-drag machines, and machines that produce some form of stacked segments. Figure 14-15 depicts a *vertically parted flaskless molding machine*, where the cope-and-drag patterns are incorporated into opposing sides of the mold. Sand is deposited between the patterns and squeezed with a horizontal motion. The patterns are withdrawn, cores are set, and the mold block is joined to those that were previously molded. Since each block contains the right-hand cavity of one mold and the left-hand cavity of another, an entire mold is made with each cycle of the machine. (Previous techniques required two separate molding operations to produce the individual cope and drag segments of a mold.) A vertical gating system is usually included on the pattern. Vertically parted molds can be poured individually, or a common runner and sprue system can be used to connect a number of mold segments. The latter method is known as the *H-process*.

In *stack molding*, sections containing both a cope and a drag impression are piled vertically on top of one another. Metal is poured down a common sprue, which is connected to a gating system at each of the parting planes. For molds that are too large to be made by either hand ramming or with one of the many types of molding machines, large flasks can be place on the foundry floor. Various types of mechanical aids are then used to add and pack the sand. One such device, a *sand slinger*, uses impeller blades to fling the sand into the flask at reasonably high velocity. If this is done skillfully, large molds can be made with uniform compaction throughout. Additional tamping can be done with a pneumatic rammer.

Extremely large molds can be made in sunken pits. Because of the size, the complexity, and the need for strength, pit molds are frequently assembled using smaller sections of baked or dried sand. Added binders may be required to provide the necessary strength.

Green-Sand, Dry-Sand, and Skin-Dried Molds.

In *green-sand molding*, the mold material is composed of sand with a binder of clay, water, and additives. Tooling costs are low, and the entire process is quite inexpensive. Almost

FIGURE 14-16   Some aluminum products produced by conventional sand casting. *(Courtesy of Bodine Aluminum Inc.)*

any metal can be cast, and there are few limits on the size, shape, weight, and complexity of the products. Design limitations are usually related to the rough surface finish, poor dimensional accuracy, and the need for subsequent machining. Still other problems can be attributed to the low strength of the mold material and the moisture that is present in the binder. Table 14-2 summarizes the process of green-sand molding. Figure 14-16 shows a variety of parts produced by conventional sand casting.

The problems of the green-sand process can be reduced by heating the mold to a temperature of 300°F or higher, and baking until most of the moisture is driven off. This strengthens the mold and reduces the amount of gases generated when the hot metal enters the cavity. These *dry-sand molds* are not very popular, however, because of the long times required for drying, the added cost of that operation, and the availability of practical alternatives. An attractive compromise is to produce a *skin-dried mold*, drying only the sand that is adjacent to the mold cavity. Torches are often used to perform the drying, and the water is usually removed to a depth of about one-half inch.

TABLE 14-2.	Green-Sand Casting
Process:	Sand, bonded with clay and water, is packed around a wood or metal pattern. The pattern is removed and molten metal is poured into the cavity. When the metal has solidified, the mold is broken and the casting is removed.
Advantages:	Almost no limit on size, shape, weight or complexity; low cost; almost any metal can be cast.
Limitations:	Tolerances and surface finish are poorer than in other casting processes; some machining is often required; relatively slow production rate.
Common metals:	Cast iron, steel, stainless steel, and casting alloys of aluminum, copper, magnesium, and nickel
Size limits:	1 oz to 6000lb
Thickness limits	As thin as $\frac{3}{32}$ in., with no maximum
Typical tolerances:	$\frac{1}{32}$ in. for first 6 in., 0.003 in. for each additional inch; additional increment for dimensions across the parting line
Draft allowance:	1–3°
Surface finish:	100-1000 μin. rms

Molds used for the casting of steel are almost always skin-dried, because the pouring temperatures are significantly higher than those for cast iron. These molds may also be given a high-silica wash prior to drying to increase the refractoriness of the surface, or the more-stable zircon sand can be used as a facing. Additional binders, such as molasses, linseed oil, or corn flour, may be added to the facing sand to provide additional strength to the skin-dried segment.

Sodium Silicate–CO_2 Molding

Molds (and cores) can also be made from a sand that receives its strength from the addition of 3 to 4% sodium silicate, a liquid inorganic binder that is also known as *water glass*. The sand can be mixed with the liquid sodium silicate in a standard muller and can be packed into flasks by any of the methods discussed previously. It remains soft and moldable until it is exposed to a flow of CO_2 gas, when it hardens in a matter of seconds by the reaction

$$Na_2SiO_3 + CO_2 \rightarrow Na_2CO_3 + SiO_2 \text{ (colloidal)}$$

The CO_2 gas is nontoxic and odorless, and no heating is required to drive the reaction. The hardened sands, however, have poor collapsibility, making shakeout and core removal difficult. Unlike most other sands, the heating that occurs as a result of the pour makes the mold even stronger (a phenomenon similar to the firing of a ceramic material). Additives that will burn out during the pour are often used to enhance the collapsibility of sodium silicate molds. In addition, care must be taken to prevent the carbon dioxide in the air from hardening the sand before the mold-making process is complete.

A modification of the CO_2 process can be used when certain portions of a mold require higher strength, better accuracy, thinner sections, or deeper draws than can be achieved with ordinary molding sand. Sand mixed with sodium silicate is packed around a metal pattern to a depth of about 1 in., followed by regular molding sand as a backing material. After the mold is fully rammed, CO_2 is introduced through vents in the metal pattern. This hardens the adjacent sand, and the pattern can now be withdrawn with less possibility of damaging the mold.

No-Bake, Air-Set, or Chemically Bonded Sands

An alternative to the sodium silicate process involves the use of organic resin binders that cure by chemical reactions that occur at room temperature. Two or more binder components are mixed with the sand just prior to the molding operation, and the curing reactions begin immediately. Because the mix is workable for only a short period of time, the molds (or cores) must be made in a reasonably rapid fashion. After a few minutes to a few hours (depending on the specific binder and curing agent), the sands harden enough to be removed from the pattern and are ready to pour.

Various *no-bake sand* systems are available, with selection being based on the metal being poured and the specific sand performance characteristics that are desired. Each system is based on organic resin binders, curing agents or catalysts, and various additives and modifiers. Like the sodium silicate molds, no-bake offers high dimensional accuracy, good hot strength, and high resistance to mold-related casting defects. Patterns can incorporate thinner sections and deeper draws. In contrast to the sodium silicate material, however, the no-bake molds decompose readily after the metal has solidified, providing excellent shakeout characteristics.

Shell Molding

Many molds are now being made by the *shell-molding process*, which offers better surface finish than can be obtained with ordinary sand molding, better dimensional accuracy, and a higher production rate with reduced labor requirements. In many cases, the process can be completely mechanized and adapted for mass production.

Figure 14-17 illustrates the six basic steps of the shell process:

1. A mixture of sand and thermosetting plastic binder is dumped onto a metal pattern that has been heated to 300 to 450°F (150 to 230°C). It is allowed to stand for a few minutes while the heat from the pattern partially cures a layer of the sand–plastic mixture. This forms a strong, solid-bonded region about $\frac{1}{8}$ in. (3.5 mm) thick adjacent to the pattern. The actual thickness depends on the pattern temperature and the time of contact.
2. The pattern and sand mixture are then inverted. All of the excess sand drops free, leaving only the layer of partially cured material that adhered to the pattern.
3. The pattern and partially cured "shell" are then placed in an oven for a few minutes to complete the curing process.
4. The hardened shell is then stripped from the pattern.
5. Two or more shells are then clamped or glued together to produce a mold.
6. The bonded shells are often placed in a pouring jacket and surrounded with metal shot or sand to provide extra support during the pour.

Because the sand is compounded for almost no shrinkage and a metal pattern is used, the shell has excellent dimensional accuracy. Tolerances of 0.003 to 0.005 in. (0.08 to 0.13 mm) are quite common. Shell-mold sand is finer than ordinary foundry sand and, in combination with the plastic resin, produces a very smooth shell and casting surface. Cleaning, machining, and other finishing costs can be reduced significantly. In addition, the shell-mold process offers a product consistency that is superior to that of green-sand casting.

Figure 14-18 shows a set of patterns, the two shells before clamping, and the resulting shell-mold casting. Machines for making shell molds vary from simple ones for small operations, to large, completely automated devices for mass production. The cost of a metal pattern is often rather high and its design must include the gate and runner system since these cannot be cut after molding. Fairly large amounts of expensive binder are re-

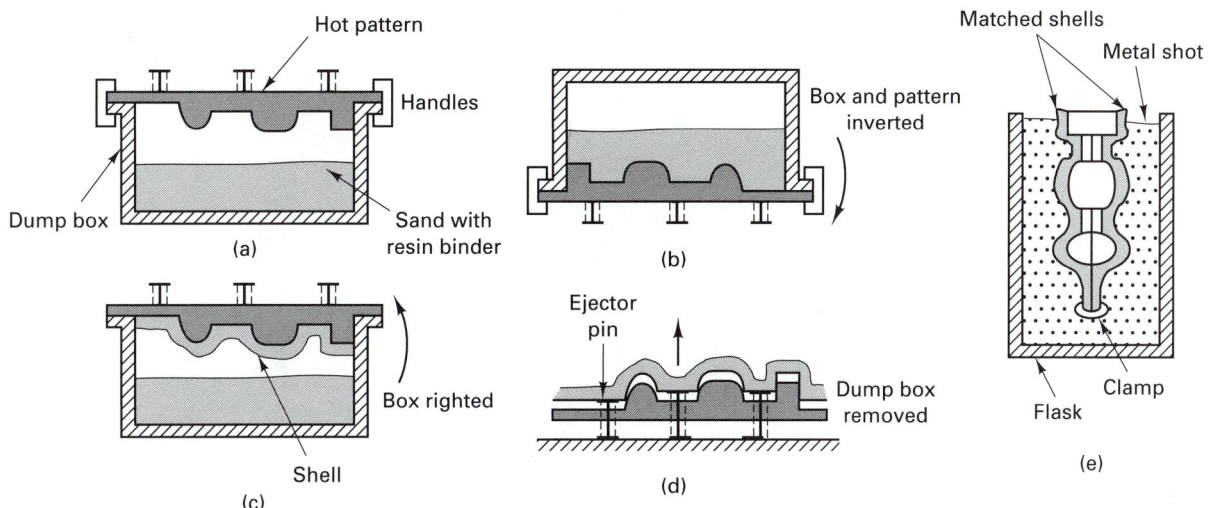

FIGURE 14-17 Schematic of the shell-molding process. A heated pattern is placed over a dump box containing a sand and resign mixture. The box is inverted and a shell partially cures around the pattern. The box is righted, the top is removed, and the shell is further cured and is finally stripped from the pattern. Matched shells are then joined and supported in a flask ready for pouring.

quired, but the amount of material required to form a thin shell (usually between 0.2 and 0.4 in. thick) is not that great. High productivity, low labor costs, smooth surfaces, and a level of precision that reduces that amount of required machining all combine to make the process economical for even moderate quantities. The thin shell provides for the easy escape of the gases that evolve during the pour. When the shell becomes hot, some of the resin binder burns out, providing excellent collapsibility and shakeout characteristics. Table 14-3 summarizes the features of shell molding.

TABLE 14-3.	Shell-Mold Casting
Process:	Sand coated with a thermosetting plastic resin is dropped onto a heated metal pattern, which cures the resin. The shell segments are stripped from the pattern and assembled. When the poured metal solidifies, the shell is broken away from the finished casting.
Advantages:	Faster production rate than sand molding; high dimensional accuracy with smooth surfaces.
Limitations:	Requires expensive metal patterns. Plastic resin adds to cost; part size is limited.
Common metals:	Cast irons and casting alloys of aluminum and copper
Size limits:	1 oz minimum; usually less than 25 lb; mold area usually less than 500 in^2
Thickness limits:	Minimums range from $\frac{1}{16}$ to $\frac{1}{4}$ in., depending on material
Typical tolerances:	Approximately 0.005 in./in.
Draft allowance:	$\frac{1}{4}-\frac{1}{2}$ °
Surface finish:	50–150 μin. rms

Other Sand-Based Molding Methods

Over the years, various processes have been proposed to overcome some of the limitations of the more traditional methods. While few have become commercially significant, several are included here to illustrate the nature of these efforts.

FIGURE 14-18 (*Top*) Two halves of a shell-mold pattern. (*Bottom*) The two shells before clamping, and the final shell-mold casting. (*Courtesy of Shalco Systems, an Acme-Cleveland Company.*)

In one method, known as the *V-process* or *vacuum molding*, a vacuum is used in place of a sand binder. Figure 14-19 depicts the production sequence, which begins by draping a thin sheet of plastic over a special pattern, which is then drawn tightly to the pattern surface by a pattern vacuum. A vacuum flask is then placed over the pattern, the flask is filled with sand, a sprue and pouring cup are formed, and a second sheet of plastic is placed over the mold. A second vacuum is then drawn on the flask itself, compacting the sand and providing the necessary strength and hardness. The pattern vacuum is released, the pattern is withdrawn, and the mold halves are assembled. The mold is poured while maintaining a vacuum in both the cope and drag segments of the flask.

Advantages of the vacuum process include the total absence of moisture-related defects. Since no binder is used, binder cost is eliminated and the sand is completely reusable. No fumes (binders burning up) are generated during the pouring operation. Shakeout

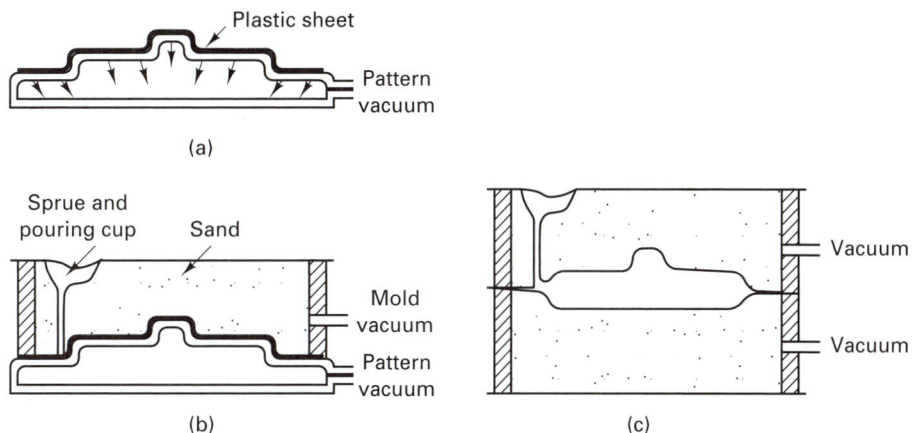

FIGURE 14-19   Schematic of the V-process or vacuum molding. A vacuum is pulled on a pattern, drawing a plastic sheet tightly against it. A vacuum flask is placed over the pattern, filled with sand, a second sheet placed on top, and a mold vacuum is drawn. The pattern vacuum is then broken and the pattern is withdrawn. The mold halves are then assembled, and the molten metal is poured.

characteristics are exceptional; the mold virtually collapses when the vacuum is released. Unfortunately, the process is relatively slow because of the additional steps and the time required to pull a sufficient vacuum.

In the *Eff-set process*, sand with a small amount of clay and quite a bit of water is first packed around a pattern. The pattern is removed and liquid nitrogen is sprayed onto the mold surface. The ice that forms becomes the binder, and the mold is then poured while it is in its frozen condition. As with the V-process, binder cost is low and shakeout is excellent.

■ 14.3 Cores And Core Making

Casting processes are unique in their ability to incorporate internal cavities or reentrant sections with relative ease. To produce these features, however, it is often necessary to use *cores* as part of the mold. Figure 14-20 shows an example of a product that could not be made by any process other than casting with cores. While these cores constitute an added cost, they do much to expand the capabilities of the process, and good design practice can often facilitate and simplify their use.

Consider the simple belt pulley shown schematically in Figure 14-21. Various methods of fabrication are suggested in the four sketches, beginning with the casting of a solid form and the subsequent machining of the through-hole for the drive shaft. A large volume of metal would have to be removed through a substantial amount of costly machining. A more economical approach would be to make the pulley with a cast-in hole of the approximate final size. Figure 14-21b depicts an approach where each half of the pattern includes a tapered hole, which receives the same green sand that is used for the remainder of mold. While these protruding sections are an integral part of the mold, they are also known as *green-sand cores*. Unfortunately, green-sand cores have a relatively low level of strength. If the protrusions are narrow or long, it might be difficult to withdraw the pattern without breaking them, or they may not have enough strength to even support their own weight. For long cores, a considerable amount of machining may be required to remove the draft

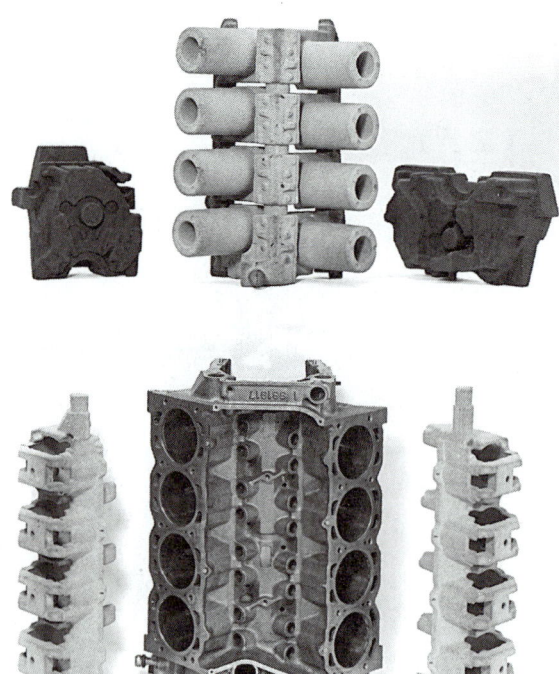

FIGURE 14-20   V-8 engine block (*bottom center*) and the five dry-sand cores that are used in its construction. *(Courtesy of Central Foundry Division of General Motors Corporation.)*

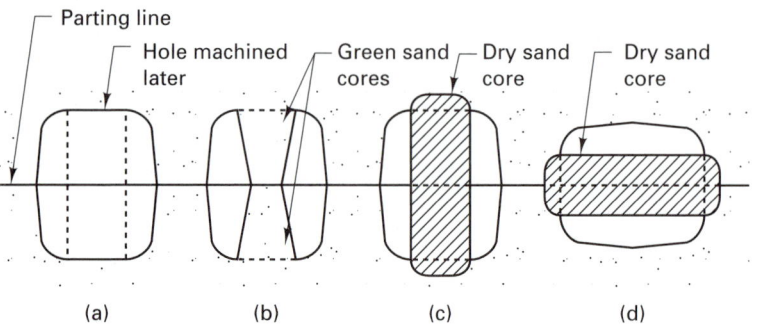

FIGURE 14-21   Four methods of making a hole in a cast pulley.

that must be provided on the pattern, and the advantage of the core is substantially reduced. For more complex shapes, it might be impossible to withdraw the pattern, and green-sand cores could not be used.

Dry-sand cores can be used to overcome some of the cited difficulties. These cores are made independent from the remainder of the mold and are then inserted into core prints to hold them in position. The remaining sketches in Figure 14-21 show dry-sand cores in the vertical and horizontal positions. Dry-sand cores can be made in several ways. In each, the sand, mixed with some form of binder, is packed into a wood or metal core box that contains a cavity of the desired shape. A *dump core box* such as the one shown in Figure 14-22 offers a very simple approach. Sand is packed into the box and scraped level with the top surface (which acts like the parting line in a traditional mold). A wood or metal plate is then placed on top of the box, and the box is turned over and lifted upward,

FIGURE 14-22   (*Clockwise from upper right*) Core box, two core halves ready for baking, and the completed core made by gluing the two halves together.

leaving the molded sand resting on the plate. After baking or hardening, the core segments are joined with hot-melt glue or some other bonding agent. Rough spots along the parting line are removed with files or sanding belts, and the final product may be given a thin coating to provide a smoother surface or greater resistance to heat. Graphite, silica, or mica can be sprayed or brushed onto the surface.

Single-piece cores are often made in a *split core box*. Two halves are clamped together, and an opening is provided in one or both ends through which sand is introduced and rammed. After the sand is compacted, the halves of the box are separated to permit removal of the core. Cores that have a uniform cross section can be made by a core-extruding machine that is similar to the familiar meat grinder. These cores are cut to length as they come from the machine and are placed in core supports for hardening. More complex cores can be made in core-blowing machines that use separating dies and receive the sand in a manner similar to injection molding or die casting.

Since cores are often the most fragile part of a mold assembly, special binders are often required. Various processes have been developed for the production of cores and core sands. The oldest of these processes, the *core-oil process*, uses a vegetable or synthetic oil as a binder, and water with cereal or clay to develop green strength.[*] The wet sand is blown or rammed at room temperature to produce the desired shape, and the uncured cores are supported on flat plates or special supports during oven drying. Forced hot air at 400 to 500°F (200 to 260°C) is the primary means of curing. The heat causes the binder to cross link or polymerize, producing a strong organic bond between the grains of sand. While the process is simple and the materials are inexpensive, the dimensional accuracy of the resultant cores is often difficult to maintain.

In the *hot-box method*, sand containing a liquid thermosetting binder and catalyst is blown into a core box that has been heated to around 450°F (230°C). When the sand contacts the hot surface, the catalyst is heated, and curing occurs within 10 to 30 seconds. The core can then be removed from the pattern and will hold its shape during further handling. For some materials, the cure will complete through an exothermic curing reaction. Others require further baking to complete the process.

Room-temperature curing is the characteristic feature of the *cold-box process*. Binder-coated sand is first blown into the core box, which is then sealed and a gas is

---

[*]Green strength refers to strength after molding but before curing.

passed through the core to polymerize the resin. Hollow cores can be produced by introducing the curing gas through holes in the core-box pattern. The uncured sand in the center can be dumped free and reused. In some cases the gases used can be extremely toxic, and special handling of both incoming and exhaust gas is required.

Room-temperature cores can also be made with the air-set or no-bake sands. These systems use an organic resin as the binder and a catalyst to promote curing, eliminating the gassing operation that is required for the cold-box materials. Shell molding is another core-making alternative, producing cores with excellent permeability since they are usually hollow.

Selection of the actual method of core production is based on numerous considerations, including production quantity, production rate, required precision, required surface finish, and the metal being poured. Certain metals may be sensitive to gases that are emitted from the cores when they come into contact with the hot metal. Other materials may have low pouring temperatures that are inadequate to break down the binder and permit collapsibility and removal from the final casting.

To function properly, cores must have the following characteristics:

1. Sufficient hardness and strength (after baking or hardening) to withstand handling and the forces of the molten metal. Compressive strength should be between 100 and 500 psi.
2. Sufficient strength before hardening to permit handling in that condition.
3. Adequate permeability to permit the escape of gases. Since cores are largely surrounded by molten metal, they should possess exceptionally good permeability.
4. Collapsibility. After pouring, the cores must be weak enough to permit shrinkage of the solidified casting as it cools, thereby preventing cracking. In addition, they must be easily removable from the interior of the finished product via shakeout.
5. Adequate refractoriness. Since the cores are largely surrounded by hot metal, they can become quite a bit hotter than the adjacent mold material.
6 A smooth surface
7. Minimum generation of gases when heated during the pour.

Various techniques have been developed to enhance the natural properties of cores and core materials. Internal wires or rods can be used to impart additional strength. Collapsibility can be enhanced by making the cores hollow or by placing a material such as straw in the center. Enhanced collapsibility is particularly important in steel castings, where a large amount of shrinkage is observed. All but the smallest of cores must be vented to permit the escape of trapped and evolved gases. Vent holes can be made by pushing small wires into the core. Coke or cinders are sometimes placed in the center of large cores to improve permeability.

Since the core material must be removed from the finished casting, the cores must be connected to the outer surfaces of the mold cavity. Recesses at these connection points, known as *core prints*, are used to support the cores and hold them in proper position during mold filling. Figure 14-23 presents a cross section of molds where core prints are used to hold a vertical and a horizontal core.

In some cases the cores do not pass completely through the casting, and additional measures must be taken to support the weight of the core and keep it from being moved or floated by the molten metal. Small metal supports, called *chaplets*, can be positioned between the core and the surfaces of the mold cavity, as shown in Figure 14-24. The use of chaplets should be minimized, however, because they become an integral part of the finished casting and may cause defects or be a location of weakness. Chaplets should be of the same, or at least comparable, composition as the casting material. They should be

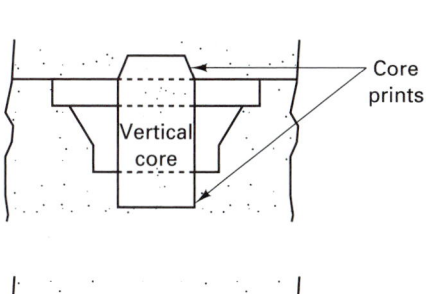

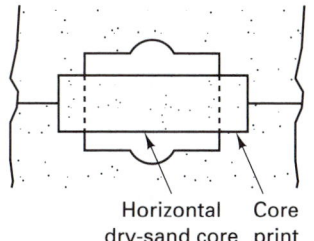

FIGURE 14-23 Molds cavities containing core prints to hold and position cores. (*Left*) Vertical core; (*right*) horizontal core.

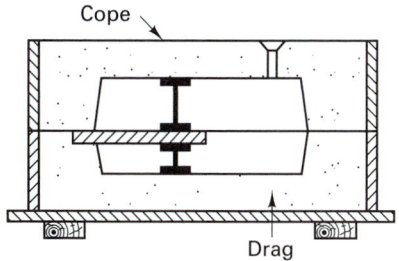

FIGURE 14-24 (*Top*) Typical chaplets. (*Bottom*) Method of supporting cores by use of chaplets (relative size of the chaplets is exaggerated).

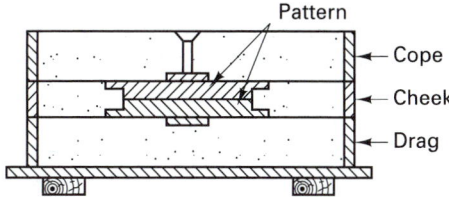

FIGURE 14-25 Method of making a reentrant angle using a three-piece flask.

large enough that they do not completely melt and permit the core to float, but small enough that their surface melts and fuses with the metal being cast.

Core sections can also be used to facilitate the molding of shapes that contain reentrant angles or sections. Consider a round pulley with a recessed groove around its perimeter. Figure 14-25 produces the part by using a third segment of flask, called a *cheek*. By adding a second parting plane, the entire mold can be made by green-sand molding around

FIGURE 14-26 Molding a reentrant section using a dry-sand core.

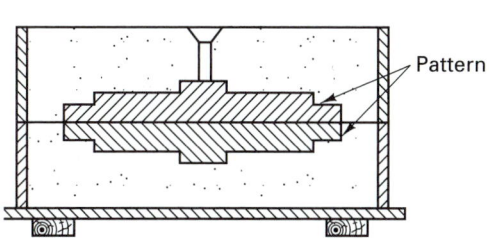

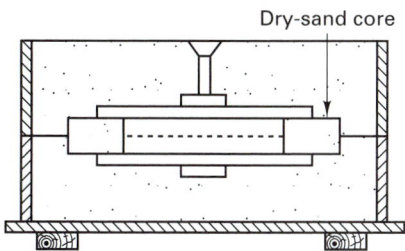

withdrawable patterns. Although additional molding operations are required, this approach may be attractive when only a few molds are to be made, since it eliminates the need for a special core box.

If a substantial quantity of the pulley were required, a simple green-sand mold can be used in conjunction with a ring-shaped core, as shown in Figure 14-26. Rapid machine molding can be used, with the green sand pattern being modified to provide a seat for the separately manufactured core. Molding time is reduced at the expense of a core box and a separate core-making operation.

■ 14.4 OTHER EXPENDABLE-MOLD PROCESSES WITH MULTIPLE-USE PATTERNS

Plaster Mold Casting

In plaster molding the mold material is plaster of paris (also known as calcium sulfate or gypsum), with various additions to improve green strength, dry strength, permeability, and castability. Talc or magnesium oxide helps to prevent cracking and reduces the setting time, lime or cement helps to control expansion during baking, and glass fibers can be added to improve strength. Sand is often used as a filler.

The mold material is first mixed with water, and the resultant slurry is immediately poured over a pattern and allowed to set. Hydration of the plaster produces a hard solid that can then be stripped from the pattern, which is usually made from metal. (*Note*: When complex angular surfaces or reentrant angles are required, flexible rubber patterns can be used. Because of the strength of the plaster, the rubber patterns can be removed without damage to the mold cavity.) After removal from the pattern, the mold is then baked to remove the excess water, assembled, and poured.

The metal patterns, coupled with the plaster mold material, provide an excellent surface finish and good dimensional accuracy. Because of the low heat capacity and low thermal conductivity of the mold, cooling is slow, which enables the metal to flow into and replicate thin sections and fine detail. Unfortunately, only the lower-melting-temperature nonferrous alloys (such as aluminum, copper, magnesium, and zinc) can be cast in plaster molds. At the high temperatures of ferrous casting, the plaster would first undergo a phase transformation and then melt. Table 14-4 summarizes the features of plaster casting.

Plaster Casting

Process:	A slurry of plaster, water, and various additives is poured over a pattern and allowed to set. The pattern is removed and the mold is baked to remove excess water. After pouring and solidification, the mold is broken and the casting is removed.
Advantages:	High dimensional accuracy and smooth surface finish; can reproduce thin sections and intricate detail to make net- or near-net-shaped parts.
Limitations:	Lower-temperature nonferrous metals only; long molding time restricts production volume or requires multiple patterns; mold material is not reusable; maximum size is limited.
Common metals:	Primarily aluminum and copper
Size limits:	As small as 1 oz but usually less than 15 lb.
Thickness limits:	Section thickness as small as 0.025 in.
Typical tolerances:	0.005 in. on first 2 in., 0.002 in. per additional inch
Draft allowance:	$\frac{1}{2}$–1°
Surface finish:	50–125 μin. rms

A variation of plaster molding known as the *Antioch process* uses a mold material consisting of about 50% sand and 50% plaster. Advantages, compared to the conventional plaster method, include increased permeability and more rapid cooling of the casting (i.e., higher-strength products). The major limitation is the long processing time required to make a mold.

Ceramic Mold Casting

Ceramic mold casting (summarized in Table 14-5) is similar to plaster mold casting, except that the mold can now withstand the higher-melting-point metals. Cope-and-drag molds are formed around withdrawable patterns (which can be metal or wood), using a ceramic slurry as the mold material. As with the plaster process, ceramic molding can produce thin sections, fine detail, and smooth surfaces and eliminate a considerable amount of finish machining. These advantages, however, must be weighed against the greater cost of the mold material. For large molds, the ceramic material can be used to produce a facing around the pattern, which is then backed up by a material such as fireclay. Not only is the back up material less expensive, it can be reused.

TABLE 14-5.	Ceramic Mold Casting
Process:	Stable ceramic powders are combined with binders and gelling agents to produce the mold material.
Advantages:	Intricate detail, close tolerances, and smooth finishes.
Limitations:	Mold material is costly and not reusable
Common metals:	Ferrous and high-temperature nonferrous metals are most common; can also be used with alloys of aluminum, copper, magnesium, titanium, and zinc.
Size limits:	Several ounces to several tons
Thickness limits:	As thin as 0.050 in.; no maximum
Typical tolerances:	0.005 in. on the first inch, 0.003 in. per each additional inch
Draft allowance:	1° preferred.
Surface finish:	75–150 μin. rms

One of the most popular of the ceramic molding techniques is the *Shaw process*. A reusable pattern is placed inside a slightly tapered flask, and a slurry like mixture of refractory aggregate, hydrolyzed ethyl silicate, alcohol, and a gelling agent is poured on top. The mixture sets to a rubbery state that permits removal of the pattern and the flask, and the mold surface is then ignited with a torch. During burn-off, most of the volatiles are consumed, and a three-dimensional network of microscopic cracks (*microcrazing*) forms in the ceramic. The gaps are small enough to prevent metal penetration but large enough to provide venting of air and gas (permeability) and to accommodate the thermal expansion of the ceramic particles during the pour and subsequent shrinkage of the solidified metal (i.e., provide collapsibility). A subsequent baking operation removes all of the remaining volatiles, making the mold hard and rigid. Before pouring, the ceramic molds are often preheated to ensure proper filling and to control the solidification characteristics of the metal. Figure 14-27 shows a set of intricate cutters that were produced by this process. Figure 14-28 shows how the Shaw process can be combined with the lost-wax process (discussed in Section 14.5) to produce even more complex products. Cores can also be used in this technique in much the same way as with sand casting.

FIGURE 14-27 Group of cutters produced by ceramic mold casting. *(Courtesy of Avnet Shaw Division of Avnet, Inc.)*

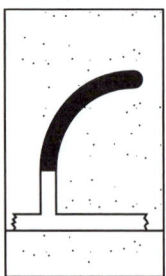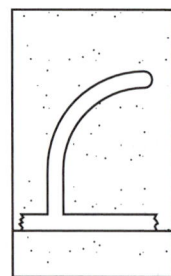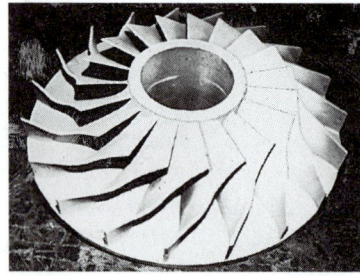

FIGURE 14-28 Method of combining ceramic mold casting and wax pattern casting to produce the complex vanes of an impeller. A wax pattern is added to a metal pattern. After the metal pattern is withdrawn *(center)*, the wax pattern is melted out, leaving a cavity as shown *(right)*. *(Courtesy of Avnet Shaw Division of Avnet, Inc.)*

Expendable Graphite Molds

For metals such as titanium which tend to react with many of the more common mold materials, powdered graphite can be combined with cement, starch, and water and compacted around a pattern. The pattern is then removed and the mold is fired at 1800°F (1000°C) to consolidate the graphite. After pouring, the mold is broken to remove the casting.

Rubber-Mold Casting

Several types of artificial elastomers are available that can be compounded in liquid form and then poured over a pattern to form a semi-rigid mold. The molds are sufficiently flexible to permit stripping from an intricate pattern or shapes with reentrant surfaces. Unfortunately, rubber molds are suitable only for small castings of low-melting-point materials. Wax patterns for investment casting can be made as well as finished castings of plastics and metals which can be poured at temperatures below 500°F (260°C).

■ 14.5 EXPENDABLE-MOLD PROCESSES USING SINGLE-USE PATTERNS

Investment Casting

While *investment casting* is actually a very old process and has been performed by dentists and jewelers for a number of years, it was not until the end of World War II that it attained any degree of industrial importance. Developments and demands in the aerospace industry, such as rocket components and jet engine turbine blades, required high-precision complex shapes from high-melting-point metals that are not readily machinable. Investment casting offers almost unlimited freedom in both the complexity of shapes and types of materials that can be cast. As experience and capabilities developed, so did the market. At present, millions of investment castings are produced each year.

Investment casting uses the same type of binder and molding aggregate as the ceramic molding process (Shaw process) and typically involves the following steps:

1. *Produce a master pattern.* The pattern is a replica of the desired product made from metal, wood, plastic, or some other easily worked material.

2. *From the master pattern, produce a master die.* This can be made from low-melting-point metal, steel, or possibly even wood. If low-melting-point metal is used, the die can often be cast directly from the master pattern. Steel dies may be machined directly, eliminating the need for step 1. Rubber molds can also be made from the master pattern.

3. *Produce wax patterns.* Patterns are made by pouring molten wax into the master die, or injecting it under pressure, and allowing it to harden. Plastic and frozen mercury have also been used as pattern material.

4. *Assemble the wax patterns onto a common wax sprue.* The individual wax patterns are attached to a central sprue and runners by means of heated tools and melted wax. In some cases, several pattern pieces may first be united to form a complex, single pattern that if made in one piece, could not be with drawn from a master die. The result of this step is a pattern cluster, or *tree*.

5. *Coat the cluster with a thin layer of investment material.* This step is usually accomplished by dipping the cluster into a watery slurry of finely ground refractory. A thin but very smooth layer of investment material is deposited onto the wax pattern, ensuring a smooth surface and good detail in the final product.

6. *Produce the final investment around the coated cluster.* After the initial layer is formed, the cluster can be redipped, but this time the wet ceramic is coated with a layer of sand. This process can be repeated until the investment coating is the desired thickness, or the single-coated cluster can be placed upside down in a flask and liquid investment material poured around it.

7. *Vibrate the flask to remove entrapped air and settle the investment material around the cluster.* This step is performed when the investment material is poured around the cluster.

8. *Allow the investment to harden.*

9. *Melt or dissolve the wax pattern to remove it from the mold.* This is generally accomplished by placing the molds upside down in an oven, where the wax melts and runs out, and any residue subsequently vaporizes. This step is the most distinctive feature of the process, because it enables a complex pattern to be removed from a single-piece mold. Extremely complex shapes with reentrant sections can readily be cast.

    In the early years of the process, only small parts were cast. When the molds were placed in the oven, the molten wax was absorbed into the porous investment. Because the wax "disappeared," the process was called the *lost-wax process*, and the name is still used occasionally.

10. *Preheat the mold in preparation for pouring.* Heating to 1000 to 2000°F (550 to 1100°C) ensures complete removal of the mold wax, cures the mold to give added strength, and allows the molten metal to retain its heat and flow more readily to all thin sections. It also gives better dimensional control because the mold and the metal will shrink together during cooling.

11. *Pour the molten metal.* Various methods, beyond simple pouring, can be used to ensure complete filling of the mold, especially when complex, thin sections are involved. Among these methods are the use of positive air pressure, evacuation of the air from the mold, and a centrifugal process.

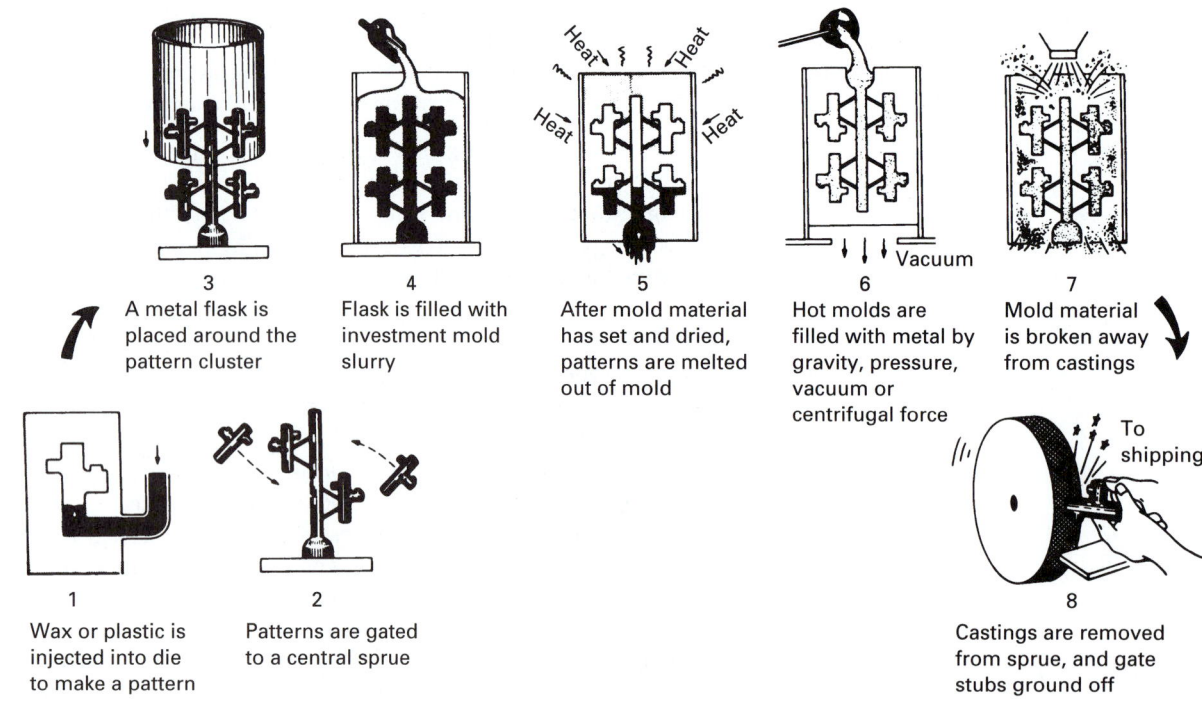

3
A metal flask is placed around the pattern cluster

4
Flask is filled with investment mold slurry

5
After mold material has set and dried, patterns are melted out of mold

6
Hot molds are filled with metal by gravity, pressure, vacuum or centrifugal force

7
Mold material is broken away from castings

To shipping

1
Wax or plastic is injected into die to make a pattern

2
Patterns are gated to a central sprue

8
Castings are removed from sprue, and gate stubs ground off

FIGURE 14-29   Investment flask-casting procedure.   *(Courtesy of Investment Casting Institute, Dallas, Texas.)*

12. *Remove the casting from the mold.* This is accomplished by breaking the mold away from the casting. Techniques include mechanical vibration and high-pressure water.

Figure 14-29 depicts the investment procedure where the investment material fills the entire flask. Figure 14-30 shows the shell-investment method. Table 14-6 summarizes the features of the investment casting process.

Investment casting is a complex process and tends to be rather expensive. Its unique advantages, however, often justify its use, and many of the steps can be automated. Extremely complex shapes can be cast as a single piece. Thin sections, down to 0.015 in.

TABLE 14-6.	Investment Casting
Process:	A refractory slurry is formed around a wax or plastic pattern and allowed to harden. The pattern is then melted out and the mold is baked. Molten metal is poured into the mold and solidifies. The mold is then broken away from the casting.
Advantages:	Excellent surface finish; high dimensional accuracy; almost unlimited intricacy; almost any metal can be cast; no flash or parting line concerns.
Limitations:	Costly patterns and molds; labor costs can be high; limited size.
Common metals:	Aluminum, copper, and steel dominate; also performed with stainless steel, nickel, magnesium, and the precious metals.
Size limits:	As small as $\frac{1}{10}$ oz but usually less than 10 lb
Thickness limits:	As thin as 0.025 in., but less than 3.0 in.
Typical tolerances:	0.005 in. for the first inch and 0.002 in. for each additional inch.
Draft allowance:	None required
Surface finish:	50 to 125 μin. rms

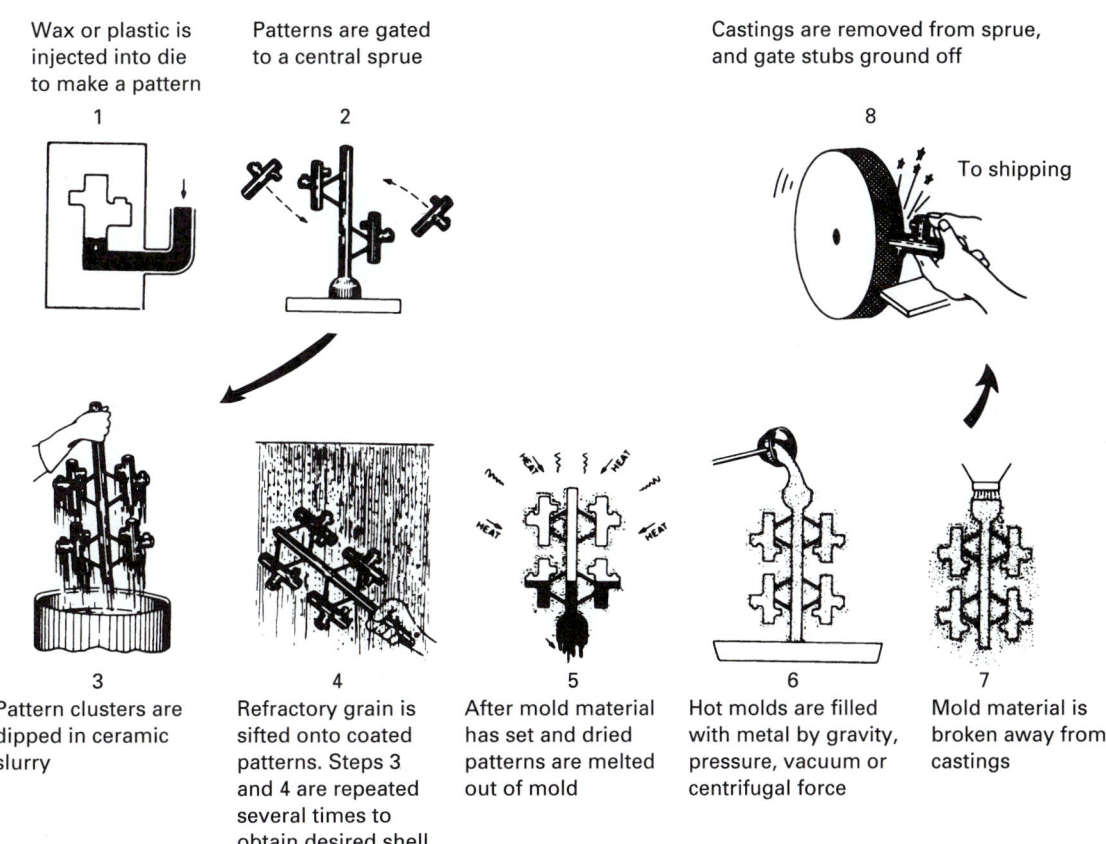

1 Wax or plastic is injected into die to make a pattern

2 Patterns are gated to a central sprue

8 Castings are removed from sprue, and gate stubs ground off

To shipping

3 Pattern clusters are dipped in ceramic slurry

4 Refractory grain is sifted onto coated patterns. Steps 3 and 4 are repeated several times to obtain desired shell

5 After mold material has set and dried patterns are melted out of mold

6 Hot molds are filled with metal by gravity, pressure, vacuum or centrifugal force

7 Mold material is broken away from castings

FIGURE 14-30 Investment shell-casting procedure. *(Courtesy of Investment Casting Institute, Dallas, Texas.)*

(0.40 mm), can be produced. Dimensional tolerances of 0.005 in./in. are routinely obtained in combination with very smooth surfaces. Machining can often be completely eliminated or greatly reduced. Where machining is required, allowances of 0.015 to 0.040 in. (0.4 to 1 mm) are usually ample. These advantages are especially attractive when difficult-to-machine metals are involved, since the investment casting process can be applied to a wide spectrum of metals.

Although most investment castings are less than 3 in. in size and weigh less than 1 pound castings up to 36 in. and 80 pounds have been produced. Some typical investment castings are shown in Figure 14-31. It should be noted that a high degree of shape complexity is their most common characteristic.

Full-Mold and Lost-Foam Casting

Several limitations are common to most of the casting processes that have been presented. Some form of pattern is usually required, and this pattern may be costly to design and fabricate, especially when the number of identical castings is rather small. In addition, the pattern must be withdrawn from the mold, and this withdrawal often requires some form of design modification or compromise, a complex pattern, or special molding procedures. In-

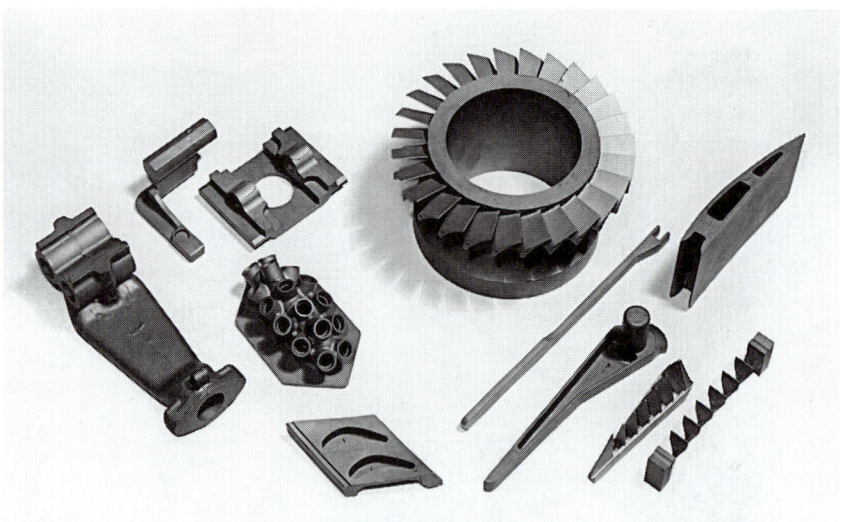

FIGURE 14-31   Typical parts produced by investment casting. *(Courtesy of Haynes Stellite Company.)*

vestment casting overcomes the withdrawal limitations through the use of patterns that can be removed by melting and vaporization. Unfortunately, this process also has its set of limitations, including a large number of individual operations and the need to remove the finished castings from the investment material.

In the full-mold and lost-foam processes, the pattern is made of expanded polystyrene, which *remains in the mold during the pouring of the metal.* When the molten metal is poured, the polystyrene melts and burns, and the metal fills the space that was occupied by the pattern.

When small quantities are required, the pattern can be hand cut or machined from pieces of foamed polystyrene. This material is extremely light in weight (1.2 lb/ft^3) and can be cut by a number of methods, including ones as simple as an electrically heated wire. Foamed material in the form of a pouring basin, sprue, runner segments, and risers can be glued on to form a complete pattern assembly. Small products can be assembled into clusters, similar to investment casting.

When larger quantities are desired, hard beads of polystyrene are first steam expanded and dried. The expanded beads are then injected into a heated metal mold at low pressure, where they further expand, fill the die, and fuse to form a pattern. The dies can be very complex and multiple patterns can be produced accurately and rapidly. When size or complexity is great, the pattern can be divided into separate segments which are assembled by gluing.

There are various options for the completion of the mold. In the *full-mold process,* shown schematically in Figure 14-32, green sand or some type of chemically bonded sand is compacted around the pattern, taking care not to crush or distort it. In the *lost foam* technique, shown in Figure 14-33, the polystyrene assembly is first dipped into a water-based ceramic that wets the surface and forms a coating about 0.005 in. thick, rigid enough to prevent mold collapse during pouring, but thin and porous enough to permit the escape of the molten and gaseous material. The coated pattern is then positioned in a flask and surrounded by coarse unbonded sand that is vibrated into place. During the pour, the coating contains the metal, isolating it from the loose, unbonded sand. Figure 14-34 presents a complex part that was made from expanded polystyrene patterns.

The full-mold and lost-foam processes can be used for castings of any size, both ferrous and nonferrous (although nonferrous is most common). Because of the reduced pat-

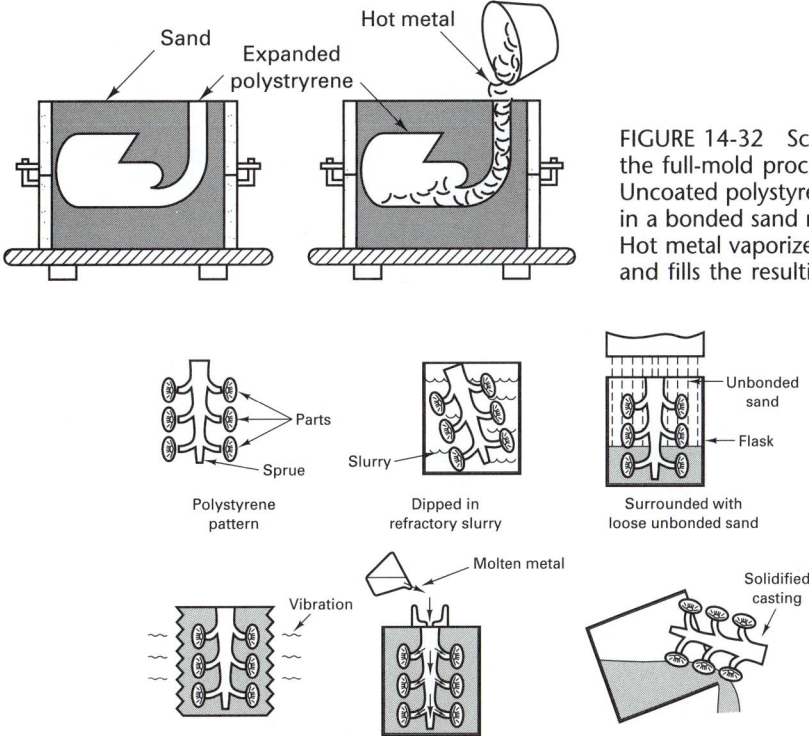

FIGURE 14-32 Schematic of the full-mold process. (*Left*) Uncoated polystyrene pattern is in a bonded sand mold. (*Right*) Hot metal vaporizes the pattern and fills the resulting cavity.

FIGURE 14-33 Schematic of the lost-foam casting process. In this process, the polystyrene pattern is dipped in a ceramic slurry, and the coated pattern is surrounded with loose, unbonded sand.

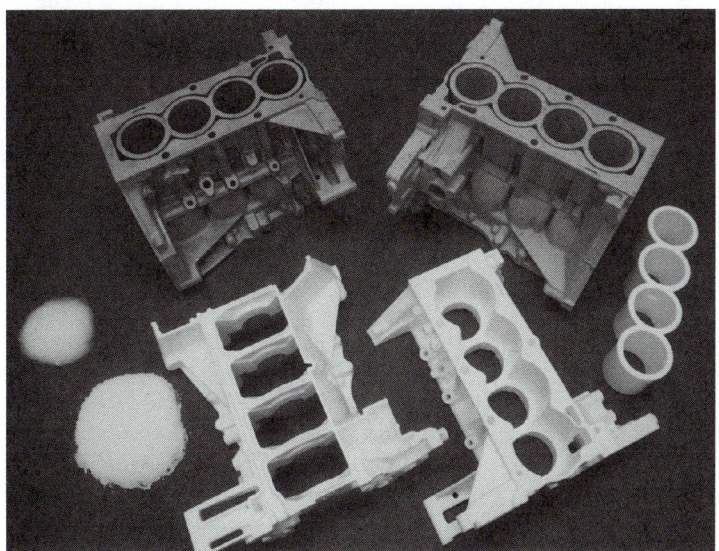

FIGURE 14-34 The stages of full-mold casting, counterclockwise from the left: polystyrene beads, expanded polystyrene pellets, foam pattern segments, assembled and dipped polystyrene pattern, and the finished metal casting. (*Courtesy of Saturn Corporation, Spring Hill, TN.*)

TABLE 14-7.	Full-Mold Casting
Process:	A Pattern containing a sprue, runners, and risers is made from single or multiple pieces of foamed plastic, such as polystyrene. It is dipped in a ceramic material, dried, and positioned in a flask, where it is surrounded by loose sand. Molten metal is poured directly onto the pattern, which vaporizes and is vented through the sand.
Advantages:	Almost no limits on shape and size; most metals can be cast; no draft is required and no flash is present (no parting lines).
Limitations:	Pattern cost can be high for small quantities; patterns are easily damaged or distorted because of their low strength.
Common metals:	Aluminum, iron, steel, and nickel alloys; also performed with copper and stainless steel.
Size limits:	1 lb to several tons
Thickness limits:	As small as 0.1 in. with no upper limit
Typical tolerances:	$\frac{1}{32}$ in./ft or less
Draft allowance:	None required
Surface finish:	100–1000 μin. rms

tern cost, small quantities can often be produced in an economical manner. Since the pattern need not be withdrawn, no draft is required in the design, and the processes are attractive for complex shapes that would ordinarily require cores, loose-piece patterns, or extensive finish machining. Precision and surface finish are sufficiently good that many machining and finishing operations can be reduced or eliminated. Cores and parting lines are not required, metal yield (product weight versus the weight of poured metal) is high, and the back-up sand is often directly reusable. For many castings, risers are not required. As the metal moves into the pattern, it loses heat due to the melting and volatilizing of the foam. Thus the material farthest from the gate is the coolest, and solidification proceeds in a directional manner back to the gate (thereby eliminating the need for a riser). For these and other reasons, full-mold casting is growing rapidly in popularity and use. Table 14-7 summarizes the process and its capabilities.

■ 14.6 Summary

A number of processes have been developed to utilize the fluidity of a liquid and its subsequent solidification as a means of producing a desired shape. Each has its own characteristic set of capabilities, advantages, and limitations, and the selection of the best method for a given application requires an understanding of all possible alternatives. In the present chapter we presented those processes that utilize a single-use (expendable) mold. Chapter 15 supplements this knowledge with a survey of the multiple-use mold processes.

■ Key Words

ceramic mold
chaplets
cheek
cohesiveness
cold-box process
collapsibility
compactability
core
core-oil process
core prints
dry-sand mold
expanded polystyrene
expendable mold
flask

full-mold casting
graphite mold
green-sand molding
hand ramming
hot-box method
investment casting
jolting
lost-foam casting
lost-wax process
muller
no-bake sand
pattern
penetration
permeability

plaster mold
refractoriness
rubber mold
sand expansion defects
sand slinger
Shaw process
shell molding
skin-dried mold
slip jacket
sodium silicate–CO_2 molding
squeezing
stack molding

■ REVIEW QUESTIONS

1. What is a casting pattern?
2. What are some of the materials used in making casting patterns? What features should be considered when selecting a pattern material?
3. What is the benefit of a split pattern over a one-piece or solid pattern? How are the two halves maintained in proper alignment?
4. How is a cope-and-drag pattern different from a matchplate pattern? Why might this be attractive?
5. For what types of products might a loose-piece pattern be required?
6. What are the four primary requirements of a molding sand? How is each provided by the sand and additive aggregate?
7. In what ways might a molding sand be a compromise material?
8. What is a muller, and what function does it perform?
9. What are some of the properties or characteristics of foundry sands that are evaluated by standard tests?
10. What is a standard rammed specimen for evaluating foundry sands, and how is it produced?
11. What is permeability, and why is it important in molding sands?
12. How does the size and shape of the sand grains relate to molding sand properties?
13. What is the cause of sand expansion defects?
14. What features can cause the penetration of molten metal between the grains of the molding sand?
15. Describe the distribution of sand density after compaction by jolting; squeezing; and a jolt–squeeze combination.
16. What is the purpose of heavy metal weights that are placed on top of mold sections prior to pouring?
17. How can the use of vertically parted flaskless molding reduce the number mold sections required to produce a series of castings?
18. How might extremely large molds be made?
19. What are the two major limitations of green sand as a mold material?
20. What restricts the popularity of dry sand molding?
21. What are some of the advantages and limitations of the sodium silicate–CO_2 process?
22. What is the primary feature of no-bake sands?
23. What material serves as the binder in the shell-molding process, and how is it cured?
24. What is the sand binder in the V-process? The Eff-set process?
25. What types of geometric features might require the use of cores?
26. What is the primary limitation of green-sand cores?
27. What is the sand binder in the core-oil process, and how is it cured?
28. What is the binder in the hot-box core-making process?
29. What is the primary attraction of the cold-box coremaking process? The use of no-bake or air-set sands for cores?
30. What is an attractive feature of shell-molded cores?
31. Why is it common for greater permeability and collapsibility to be required of cores than for the base molding sand?
32. Why is it important to provide adequate permeability and venting to cores?
33. Why is it important that chaplets not completely melt during the pouring and solidification of a casting?
34. Why are plaster molds only suitable for the lower-melting-temperature nonferrous metals and alloys?
35. How can permeability be imparted to ceramic molds?
36. What are some of the attractive features of ceramic and plaster molding?
37. For what materials might a graphite mold be required?
38. What is the major limitation of rubber-mold casting?
39. What materials are used to produce the patterns for investment casting?
40. Why are investment casting molds generally preheated prior to pouring?
41. What are some of the benefits of not having to remove the pattern from the mold (as in investment casting and the full-mold process)?
42. What are some of the attractive features of the full-mold process?

Chapter 14 CASE STUDY

moveable and fixed jaw pieces for a heavy-duty bench vise

F igure CS-14 presents a cut-away sketch of the moveable and fixed jaw pieces of a heavy-duty vise that might see use in vocational schools, factories and machine shops. The vise is intended to have a rated maximum clamping force of 15 tons. The slide of the moving jaw has been designed to be a $2\frac{1}{4}$ inch box channel. The jaw width is 5 inches, the maximum jaw opening is 6 inches, and the depth of the throat is $4\frac{1}{2}$ inches. The designer has elected to use replaceable, serrated jaws, and suggests that the material used for the receiving jaw pieces have a yield strength in excess of 35 ksi, with at least 15% elongation in a uniaxial tensile test (to assure that an overload or hammer impact would not produce brittle fracture).

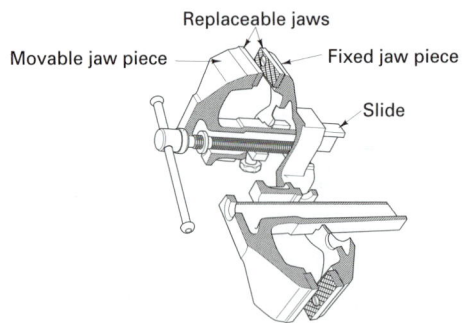

Note: shaded surfaces have been produced by cross-sectional cuts

FIGURE CS-14

1. Determine some possible combinations of material and process that could fabricate the desired shapes with the required properties. Of the alternatives presented, which would you prefer and why?

2. Would the components require some form of heat treatment? Consider the possibilities of stress relief, homogenization, or the establishment of desired final properties. What would you recommend?

3. One of your colleagues has suggested that the slides be finished with a coat of paint. Do you think a surface treatment is necessary or desirable? If so, what would you recommend?

CHAPTER 15

MULTIPLE-USE-MOLD CASTING PROCESSES

15.1 INTRODUCTION
15.2 PERMANENT MOLD CASTING
Slush Casting
Corthias Casting
Low-Pressure Permanent Mold Casting
Vacuum Permanent Mold Casting
15.3 DIE CASTING
15.4 SQUEEZE CASTING (OR LIQUID-METAL FORGING)
15.5 CENTRIFUGAL CASTING
15.6 SEMICENTRIFUGAL CASTING
15.7 CENTRIFUGING
15.8 CONTINUOUS CASTING
15.9 ELECTROMAGNETIC (OR LEVITATION) CASTING

15.10 MELTING AND POURING
Cupolas
Indirect Fuel-Fired Furnaces (or Crucible Furnaces)
Air Furnaces or Direct Fuel-Fired Furnaces
Arc Furnaces
Induction Furnaces
15.11 POURING PRACTICE
15.12 CLEANING, FINISHING, AND HEAT-TREATING OF CASTINGS
Cleaning and Finishing
Heat Treatment and Inspection of Castings
15.13 ROBOTS IN FOUNDRY OPERATIONS
15.14 PROCESS SELECTION
Case Study: BASEPLATE FOR A HOUSEHOLD STEAM IRON

■ 15.1 INTRODUCTION

With the expendable-mold casting processes discussed in Chapter 14, a separate mold must be created for each casting. Variations in mold consistency, mold strength, moisture content, pattern removal, and other factors contribute to dimensional and property variation from casting to casting. In addition, the need to create and then destroy a separate mold for each pour results in rather low rates of production.

The multiple-use-mold casting processes overcome many of these limitations, but they, in turn, have their own assets and liabilities. Since the molds are generally made from metal, many of the processes are restricted to the casting of the lower-melting-point nonferrous metals and alloys. Part size is often limited, and the dies or molds can be rather costly.

■ 15.2 PERMANENT MOLD CASTING

In the *permanent mold casting* process, a reusable mold is machined from gray cast iron, steel, bronze, graphite, or other material, and the mold segments are usually hinged to permit rapid and accurate opening and closing. The mold is preheated and clamped shut, and molten metal is poured in using simple gravity flow. After solidification, the mold is opened and the product is removed. The mold is then reclosed, and since the heat from

399

the previous cast is usually sufficient to maintain mold temperature, another casting can be poured. Aluminum-, magnesium-, and copper-based alloys are the metals most frequently cast. If graphite is used as the mold material, iron and steel castings can also be made by the permanent mold process.

Numerous advantages can be cited for the permanent mold process, which is also known as *gravity die casting*. The mold is reusable, and a good surface finish is obtained if the mold is in good condition. Dimensional accuracy can often be held to within 0.010 in. (0.25 mm). Directional solidification can be promoted by selectively heating or chilling various portions of the mold or by varying the thickness of the mold wall. The result is a sound, relatively defect-free casting with good mechanical properties. The faster cooling rates of the metal mold produce stronger products than would result from a sand casting process. Expendable sand cores or *retractable metal cores* can be used to increase the complexity of the casting. Multiple cavities can often be included in a single mold.

On the negative side, the process is generally limited to the lower-melting-point alloys. When applied to steels or cast irons, the mold life tends to be extremely short. Even for the low-temperature metals, the mold life is limited because of erosion by the molten metal and thermal fatigue. The actual mold life varies with:

1. *Alloy being cast.* The higher the melting point, the shorter the mold life.
2. *Mold material.* Gray cast iron has about the best resistance to thermal fatigue and machines easily. Thus it is used most frequently for permanent molds.
3. *Pouring temperature.* Higher pouring temperatures reduce mold life, increase shrinkage problems, and induce longer cycle times.
4. *Mold temperature.* If the temperature is too low, misruns are produced and high temperature differences form in the mold. If the temperature is too high, excessive cycle times result and mold erosion is aggravated.
5. *Mold configuration.* Differences in section sizes of either the mold or the casting can produce temperature differences and reduce the life of the mold.

Mold complexity is often restricted because the rigid cavity offers no collapsibility to compensate for shrinkage of the casting. Therefore, as a best alternative, it is common practice to open the mold and remove the casting immediately after solidification. This prevents the formation of hot tears that may form if the product is restrained during cooldown.

Permanent molds are usually heated at the beginning of a run and maintained at a fairly uniform elevated temperature. This minimizes the degree of thermal fatigue, facilitates metal flow, and controls the cooling rate of the metal being cast. Since the mold temperature rises when a casting is produced, it may be necessary to provide a cooldown delay before the cycle is repeated. Refractory washes or graphite coatings can be applied to the mold walls to prevent the casting from sticking and prolong the mold life. When pouring cast iron, an effective approach is to use an acetylene torch and apply a coating of carbon black to the mold.

Since the molds are not permeable, special provision must be made for venting. This is usually accomplished through the slight cracks between mold halves or by very small vent holes that permit the escape of trapped air but not the passage of molten metal. Since gravity is the only means of inducing metal flow, risers must still be employed to compensate for shrinkage, and yields are generally less than 60%. Both sand and retractable metal cores can be used to increase part complexity.

Mold costs are generally high, so high-volume production is usually required to justify the expense. Automated machines are used to coat the mold, pour the metal, and remove the casting. Figure 15-1 shows a variety of automobile and truck pistons that were manufactured by the permanent mold process, which is summarized in Table 15-1.

FIGURE 15-1   Truck and car pistons, mass-produced by the millions using permanent mold casting. *(Courtesy of Central Foundry Division of General Motors Corporation.)*

TABLE 15-1.	Permanent Mold Casting
Process:	Mold cavities are machined into mating metal die blocks, which are then preheated and clamped together. Molten metal is then poured into the mold and enters the cavity by gravity flow. After solidification, the mold is opened and the casting is removed.
Advantages:	Good surface finish and dimensional accuracy; metal mold gives rapid cooling and fine-grain structure; multiple-use molds (up to 25,000 uses).
Limitations:	High initial mold cost; shape, size, and complexity are limited; yield rate rarely exceeds 60%, but runners and risers can be directly recycled; mold life is very limited with high-melting-point metals such as steel.
Common metals:	Alloys of aluminum, magnesium, and copper are most frequently cast; irons and steels can be cast into graphite molds; alloys of lead, tin, and zinc are also cast.
Size limits:	Several ounces to about 150 lb
Thickness limits:	Minimum depends on material but generally greater than $\frac{1}{8}$ in.; maximum thickness about 2.0 in.
Typical tolerances:	0.015 in. for the first inch and 0.002 in. for each additional inch; 0.01 in. added if the dimension crosses a parting line
Draft allowance:	2–3°
Surface finish:	100 to 250 μin. rms

Slush Casting

Slush casting is a variation of the permanent mold process in which the metal is permitted to remain in the mold only until a shell of the desired thickness has formed. The mold is then inverted and the remaining liquid is poured out. When the mold halves are separated, the resulting casting is a hollow shape with good surface detail but variable wall thickness. This process is frequently used to cast low-melting-temperature metals into ornamental objects such as candlesticks, lamp bases, and statuary.

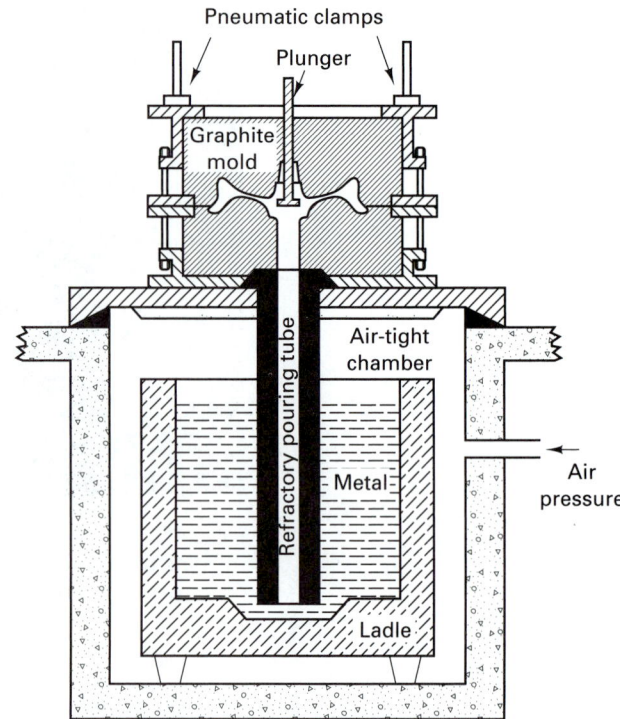

FIGURE 15-2   Schematic of the low-pressure permanent mold process. *(Courtesy of Amsted Industries.)*

Corthias Casting

Another variation of the permanent mold process, *Corthias casting,* utilizes a plunger that is pushed down into the pouring cup, sealing the sprue hole and displacing some of the molten metal into the outer portions of the mold cavity. The positive pressure produces additional detail and permits thinner sections to be cast successfully.

Low-Pressure Permanent Mold Casting

In contrast to the gravity pouring of conventional permanent mold casting, *low-pressure permanent mold* (LPPM) casting forces the molten metal up through a vertical tube by applying low pressure gas at 5 to 15 psi to a molten bath below. Figure 15-2 provides an illustration of this process, which can be applied to a wide variety of metals, including steel and cast iron.

Clean metal from the center of the melt is fed directly into the mold (a distance of only 3 to 4 in.), never passing through the atmosphere. This is particularly attractive for maintaining the cleanliness of rapidly oxidizing metals such as aluminum. By controlling the pressure, the mold can be filled in a controlled, nonturbulent manner, which further minimizes gas porosity and dross. Mold cooling is designed to promote directional solidification from the top down. The applied pressure continually feeds molten metal to compensate for shrinkage, and the unused metal in the feed tube simply drops back into the crucible after the pressure is released. Since no risers are used (the pressure is maintained until solidification is complete and the pressurized feed tube acts as a riser) and the molten metal in the feed tube can be reused immediately, yields are generally greater than 85%. Mechanical properties are typically 5% superior to those of conventional permanent mold castings, but cycle times are somewhat longer.

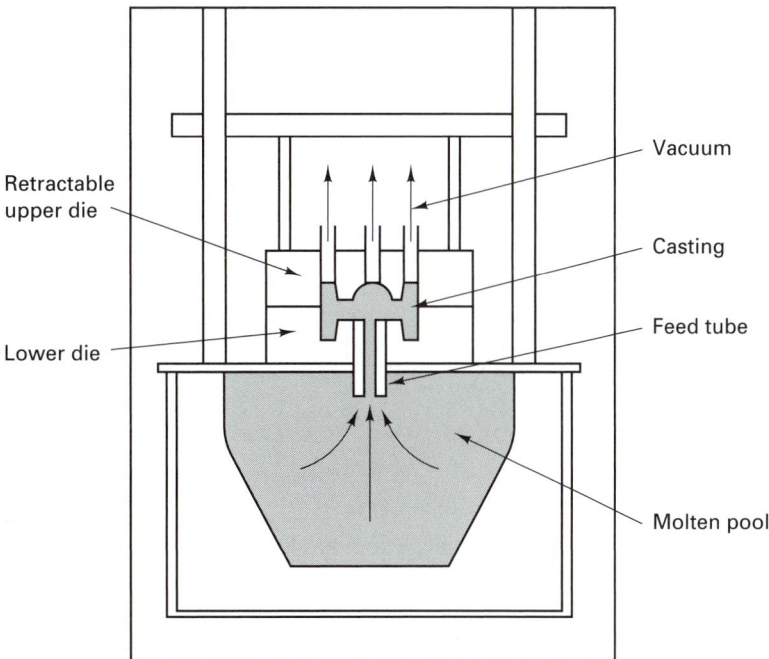

FIGURE 15-3   Schematic illustration of vacuum permanent mold casting.

Vacuum Permanent Mold Casting

Figure 15-3 shows yet another variation of permanent mold casting, where a vacuum is used to draw material into the mold cavity. All of the benefits and features of the low-pressure process are retained, including the subsurface extraction of molten metal from the melt, the bottom feed to the mold, the minimal metal disturbance during pouring, the self-risering action, and the downward directional solidification. Thin-walled castings can be produced with high metal yield and excellent surface quality. Because of the vacuum, the cleanliness of the metal and dissolved gas content are superior even to that of the low-pressure process. Final castings typically range from 0.4 to 10 lb, and have mechanical properties that are 10 to 15% superior to those of conventional permanent mold products. The process can be further modified by (1) performing both melting and casting under vacuum conditions, or (2) replacing the permanent mold atop the feed tube with expendable shell or sand molds (the *Hitchiner process*).

■ 15.3 DIE CASTING

Die casting differs from permanent mold casting in that the molten metal is forced into the molds by pressures of thousands of pounds per square inch and held under this pressure during solidification. Most die castings are made from nonferrous metals and alloys, but ferrous die castings are now being produced. Because of the combination of metal molds or dies and high pressure, fine sections and excellent detail can be achieved, together with long mold life. The attractiveness of the process has been further expanded by the development of special zinc-, copper-, and aluminum-based alloys that have excellent properties when die cast.

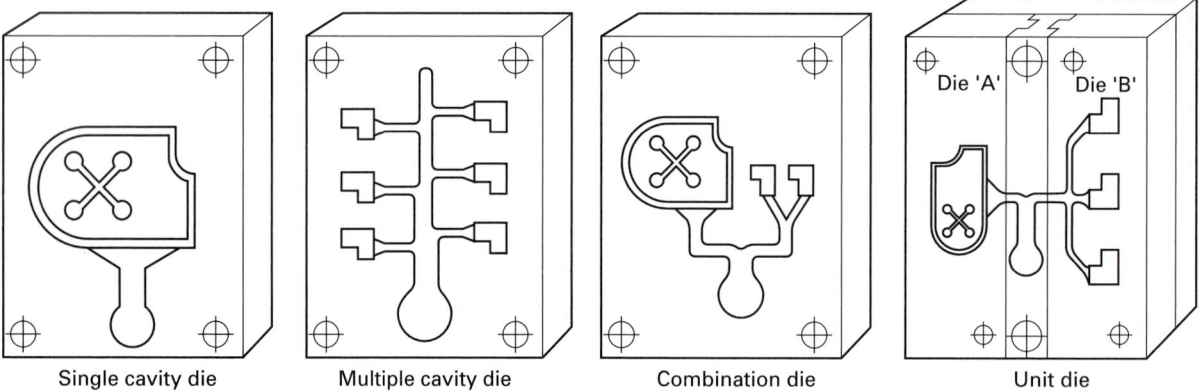

FIGURE 15-4   Some common die-casting designs. *(Courtesy of American Die Casting Institute, Inc., Des Plaines, Illinois.)*

Because die-casting dies are usually made from hardened hot-worked tool steels (cast iron cannot withstand the pressures), they tend to be rather expensive. As shown in Figure 15-4, the dies may be relatively simple, containing only one or two mold cavities, or they may be quite complex, containing multiple cavities of the same or different products. To permit removal of the casting, the dies must open, separating into at least two pieces. It is not uncommon, however, for dies to be far more complicated, containing multiple sections that move in several different directions. In addition, the separate die sections often contain water cooling passages, retractable cores, and knock-out pins, which eject the finished casting. Consequently, die casting dies often cost in excess of $5000, and $10,000 is not considered excessive.

Die life is usually limited by wear (or erosion), which is strongly dependent on the temperature of the molten metal. Surface cracking of the die surfaces as a result of the cyclic heating and cooling (*heat checking* or *thermal fatigue*) can also be a problem. The economical use of die casting, therefore, is closely related to its high production rates, the good strength properties of its products, the excellent dimensional precision and surface qualities, and the almost complete elimination of subsequent machining. The size and weight of die castings are increasing constantly, and parts weighing up to 20 lb (10 kg) and measuring up to 24 in. (600 mm) can now be made routinely.

In the die-casting process, the water-cooled dies are usually lubricated and then clamped tightly together as molten metal is injected under pressure. Because high injection pressures may cause turbulence and air entrapment, the amount of pressure and time of application varies considerably. A recent trend is toward the use of larger gates and lower injection pressure, followed by higher pressure after the mold has been filled completely and the metal has started to solidify. This tends to improve the density and reduce porosity in the finished casting. After solidification is complete, the dies separate and ejector pins extract the finished casting, along with its attached runners and sprues.

There are two basic types of die-casting machines. Figure 15-5 schematically illustrates the *hot-chamber* (or gooseneck) design. A "gooseneck" is partially submerged in a reservoir of molten metal. To begin the cycle, a port opens, which allows the molten metal to fill the gooseneck. A mechanical plunger or air injection system then forces the metal out of the gooseneck and into the die, where it rapidly solidifies. The *hot-chamber machine* offers fast cycling times (up to about 15 cycles per minute), and the added advantage that the molten metal is injected from the same chamber in which it is melted (i.e., there is no han-

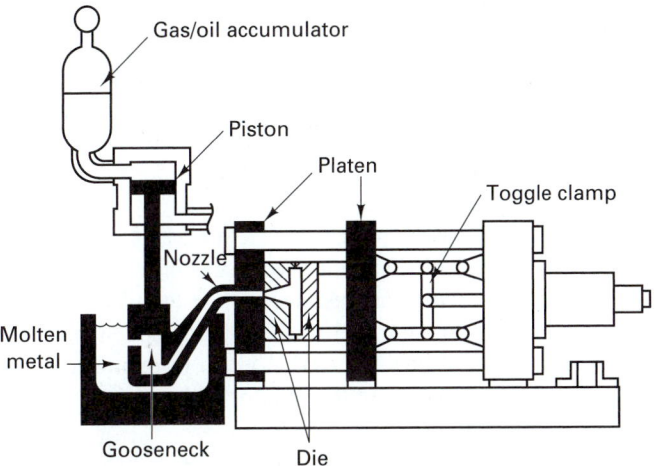

FIGURE 15-5 Principal components of a hot-chamber die-casting machine. *(Courtesy of Noranda Sales Corp. Ltd.)*

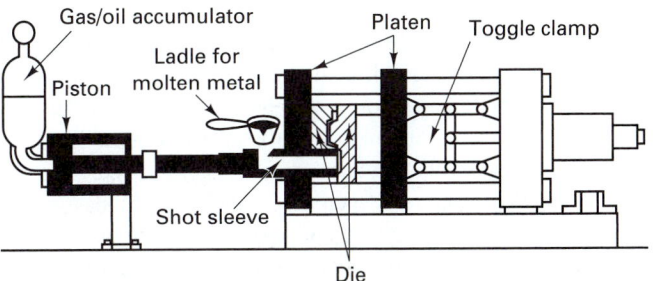

FIGURE 15-6 Principal components of a cold chamber die-casting machine. *(Courtesy of Noranda Sales Corp. Ltd.)*

dling or transfer of molten metal). Unfortunately, the hot chamber design cannot be used for the higher-melting-point metals, and there is a tendency for molten aluminum to pick up some iron as a result of the extended time in contact with the casting equipment. Therefore, hot-chamber machines see primary use with zinc-, tin-, and lead-based alloys.

Cold-chamber machines are usually employed for the die casting of materials that are not suitable for the hot-chamber design. These include alloys of aluminum, magnesium, and copper. As illustrated in Figure 15-6, the metal is melted in a separate furnace, then transported to the die-casting machine, where a measured quantity is fed into an unheated shot chamber (or injection cylinder) and subsequently driven into the die by a hydraulic or mechanical plunger. The pressure is then maintained or increased during the solidification process. Since molten metal must be transferred to the chamber for each cycle, the cold-chamber process has a longer operating cycle than that of the hot-chamber machines. Nevertheless, productivity is still high.

A double plunger is used in a modification of the cold-chamber process, known as the *Acurad process.* After the molten metal has been injected into the die and begins to solidify and shrink, a second plunger advances to force more metal into the die cavity and assure production of a better casting.

Die-casting dies fill with metal so fast that there is little time for air in the mold cavity to escape, and the metal molds offer no permeability. Trapped air causes problems in the form of blow holes, porosity, or misruns. To minimize these, it is crucial that the dies be properly vented, usually by wide, thin (0.005 in. or 0.13 mm) vents along the parting line. Metal protruding into these vents must be trimmed off after the casting has been removed from the mold. This can be done with special trimming dies that also serve to remove the sprues and runners.

No risers are used in the die-casting process since the high pressures help feed metal from the gating system into the casting. However, die castings frequently contain some porosity due to entrapped air and the turbulent mode of die filling. The porosity tends to be confined to the center of the casting, and smooth metal flow, good venting, and proper application of pressure can do much to minimize its extent. The rapidly solidified surface is usually harder than the interior and is usually sound and suitable for plating or decorative applications.

In the *pore-free casting process*, air problems are eliminated by introducing oxygen into the mold before each die-casting shot. When the molten metal is injected, it reacts with the oxygen to form fine, dispersed oxide particles, virtually eliminating gas porosity. The products can be welded and heat-treated and possess greater strength than do conventional die castings. Pore-free casting has been performed on aluminum, zinc, and lead alloys.

Another recent innovation is the *heated-manifold direct-injection die casting* of zinc (also known as direct-injection zinc die casting or runnerless zinc die casting). Molten zinc is forced through a heated manifold and heated mininozzles directly into the die cavity, thereby eliminating the need for sprues, gates, and runners. Less cooling of the metal occurs, so the product surface is improved. Energy is conserved, scrap is reduced, and product quality is increased. Existing die-casting machines can be used with the addition of a heated manifold and modification of the various dies.

Sand cores cannot be used in die casting because the high pressures and flow rates cause the cores to either disintegrate or have excessive metal penetration. As a result, metal cores are required, and provisions must be made for retracting them, usually before the die is opened for removal of the casting. A close fit must be maintained between the halves of the die and between the moving cores and die sections if metal is to be prevented from flowing into the gap. Core retraction motions must be either in a straight line or a circular arc. As an alternative, loose core pieces can be inserted into the die at the beginning of each cycle and then removed from the casting after it has ejected from the die. This procedure permits complex shapes to be cast, such as holes with internal threads, but production rate is slowed and costs increase.

The die-casting process also permits the incorporation of cast-in *inserts*. These include prethreaded bosses, electrical heating elements, threaded studs, and high-strength bearing surfaces that are positioned in the die before the metal is injected. Suitable recesses must be provided in the die for positioning and support, and the casting cycle tends to be slowed by the additional operations.

Smooth surfaces and excellent dimensional accuracy are attractive features of die casting. For aluminum-, magnesium-, zinc-, and copper-based alloys, linear tolerances of 0.003 in./in. (3 mm/m) are not uncommon. Minimum section thickness and draft depend on the type of metal, as follows:

FIGURE 15-7   Variety of aluminum and zinc die castings.   *(Courtesy of Yoder Die Casting Corporation.)*

Metal	Minimum Section	Minimum Draft
Aluminum alloys	0.035 in. (0.89 mm)	1:100 (0.010 in./in.)
Brass and bronze	0.050 in. (1.27 mm)	1:80 (0.015 in./in.)
Magnesium alloys	0.050 in. (1.27 mm)	1:100 (0.010 in./in.)
Zinc alloys	0.025 in. (0.63 mm)	1:200 (0.005 in./in.)

Because of the features above, most die castings require no finish machining except for the removal of the small amount of excess metal fin, or flash, around the parting line and the possible drilling or tapping of holes. Production rates are high and a set of dies can produce many thousands of castings without significant change in dimensions. Table 15-2 summarizes the features of the die-casting process, and Figure 15-7 presents a variety of aluminum and zinc die castings.

TABLE 15-2.	Die Casting
Process:	Molten metal is injected into closed metal dies under pressures ranging from 1500 to 25,000 psi. Pressure is maintained during solidification, after which the dies separate and the casting is ejected along with its attached sprues and runners. Cores must be simple and retractable and take the form of moving metal segments
Advantages:	Extremely smooth surfaces and excellent dimensional accuracy; rapid production rate.
Limitations:	High initial die cost; limited to high-fluidity nonferrous metals; part size is limited; porosity may be a problem; some scrap in sprues, runners, and flash, but this can be directly recycled
Common metals:	Alloys of aluminum, zinc, magnesium, and lead; also possible with alloys of copper and tin
Size limits:	Less than 1 oz up through about 15 lb most common
Thickness limits:	As thin as 0.03 in., but generally less than $\frac{1}{2}$ in.
Typical tolerances:	Varies with metal being cast; typically 0.005 in. for the first inch and 0.002 in. for each additional inch
Draft allowances:	2°
Surface finish:	40–100 μin. rms.

■ 15.4 SQUEEZE CASTING (OR LIQUID-METAL FORGING)

The *squeeze casting* process is actually a combination of casting and forging. A precise amount of molten metal is poured into the bottom half of a preheated die set and allowed to partially solidify. An upper die then descends, applying pressure throughout the duration of solidification. Intricate shapes can be produced at pressures that are far less than would normally be required for hot or cold forging. Both retractable and disposable cores can be used to create holes and internal passages. Gas and shrinkage porosity are substantially reduced, and mechanical properties are enhanced. The process can be applied to both ferrous and nonferrous alloys, and both wrought and cast alloys can be processed.

An adaptation of the process can be used to produce metal–matrix composites by forcing the pressurized liquid around foamed or fiber reinforcements that have been positioned in the mold. Another modification involves the use of thixotropic semisolid material. Here, the need to introduce a precise amount of molten metal into the die is eliminated by starting with chunks of solid metal that have been heated into the semisolid (liquid plus solid) range. *Thixotropic material* can be handled mechanically, like a solid, but shaped at low pressures because it flows like a liquid when agitated or squeezed. The absence of turbulent flow minimizes gas pickup and entrapment. Since the material is already partially solid, solidification shrinkage and related porosity is reduced. Cooling while under pressure completes the solidification, while simultaneously producing high-quality intricate parts with good finish and precision.

■ 15.5 CENTRIFUGAL CASTING

In *centrifugal casting* (a category that includes true centrifugal casting, semicentrifugal casting, and centrifuging), the inertial forces of rotation act to distribute the molten metal into the mold cavity or cavities. In *true centrifugal casting*, a dry-sand, graphite, or metal mold rotates about either a horizontal or vertical axis at speeds of 300 to 3000 rpm. As the molten metal is introduced, it is flung to the surface of the mold, where it solidifies into some form of hollow product. The exterior profile is usually round (as with pipes and gun barrels), but hexagons and other symmetrical shapes are also possible.

No core or mold is needed to shape the interior, which will always have a round profile because the molten metal is uniformly distributed by the centrifugal forces. When rotation is about the horizontal axis, as illustrated in Figure 15-8, the inner surface is always cylindrical. If the vertical axis is used, gravitational forces cause the inner surface to become a section of a parabola. As shown in Figure 15-9, the exact shape is a function of

FIGURE 15-8 Schematic representation of a horizontal centrifugal casting machine. *(Courtesy of American Cast Iron Pipe Company.)*

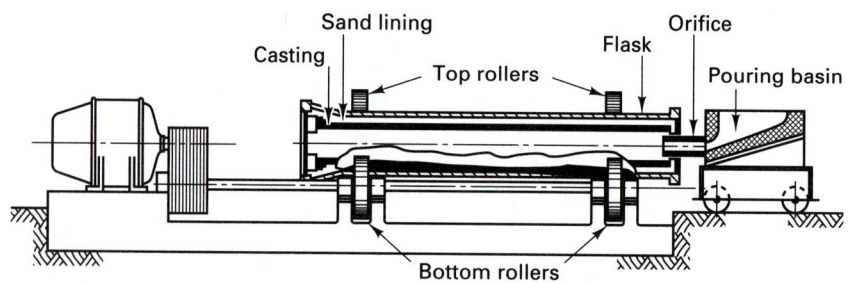

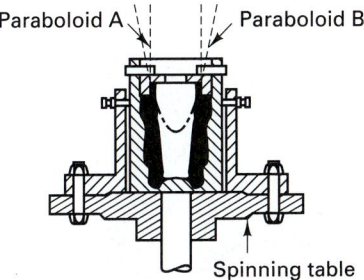

Paraboloid A Paraboloid B

Spinning table

FIGURE 15-9 Vertical centrifugal casting, showing the effect of rotational speed on the shape of the inner surface. Paraboloid A results from fast spinning, paraboloid B from slow spinning.

the speed of rotation. Wall thickness can be controlled by varying the amount of metal that is introduced into the mold.

In true centrifugal casting, the metal is forced against the outer walls of the mold with considerable force, and solidification begins at the outer surface. Centrifugal force continues to feed molten metal, compensating for shrinkage, so no risers are required. The final product has a strong, dense exterior with all the lighter impurities (including dross and pieces of the refractory mold coating) collecting on the inner surface of the casting. For some applications, this surface is then removed by a light machining cut.

Pipe, pressure vessels, cylinder liners, brake drums, and the starting material for bearing rings and parts such as those of Figure 15-10 can all be manufactured by centrifugal castings. The equipment is rather specialized and can be quite expensive for large castings. The permanent molds can also be quite costly, but they do offer a long service life, especially when coated with some form of refractory dust or wash. Since no sprues, gates, or risers are required, yields can be greater than 90%. In addition, the process can be used to produce composite products by the centrifugal casting of a second material onto the inside surface of another alloy. Table 15-3 summarizes the features of centrifugal casting.

TABLE 15-3.	Centrifugal Casting
Process:	Molten metal is introduced into a rotating sand, metal, or graphite mold, and held against the mold wall by centrifugal force until it is solidified
Advantages:	Can produce a wide range of cylindrical parts, including ones of large size; good dimensional accuracy, soundness, and cleanliness
Limitations:	Shape is limited; spinning equipment can be expensive
Common metals:	Iron, steel, stainless steel, and alloys of aluminum, copper, and nickel
Size limits:	Up to 10 ft in diameter and 50 ft in length
Thickness limits:	Wall thickness 0.1–5 in.
Typical tolerances:	O.D. to within 0.1 in.; I.D. to about 0.15 in.
Draft allowance:	$\frac{1}{8}$ in./ft.
Surface finish:	100–500 μin. rms.

■ 15.6 SEMICENTRIFUGAL CASTING

In *semicentrifugal casting* (Figure 15-11) the centrifugal force assists the flow of metal from a central reservoir to the extremities of a rotating symmetrical mold, which may be either expendable or multiple-use. The rotational speeds are usually lower than for true centrifugal casting, and it is not uncommon for several molds to be stacked on top of one another, being fed by a common pouring basin.

The central reservoir acts as a riser and must be large enough to assure that it will be the last material to freeze. Since the lighter impurities concentrate in the center, the proc-

FIGURE 15-10 Electrical products (collector rings, slip rings, and rotor end rings) that have been centrifugally cast from aluminum and copper. *(Courtesy of The Electric Materials Company.)*

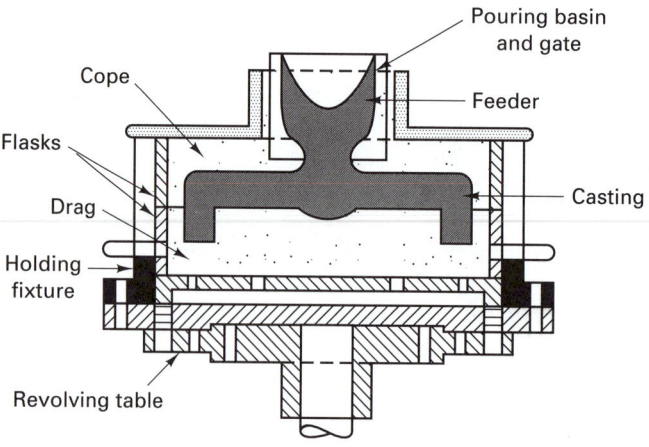

FIGURE 15-11 Semicentrifugal casting process.

ess is best used for castings where the central region will ultimately be hollow. Cores can be used to increase the complexity of the product.

■ 15.7 CENTRIFUGING

Centrifuging (Figure 15-12) uses centrifugal action to force the metal from a central pouring reservoir into separate mold cavities that are offset from the axis of rotation. Relatively low rotational speeds are required to produce sound castings with thin walls and intricate shapes. Centrifuging is often used to assist in the pouring of investment casting trees.

In an adaptation of the process, pewter, zinc, or wax can be cast into spinning rubber molds to produce products with close tolerances, smooth surfaces, and excellent detail. These can be finished products or patterns for subsequent assembly onto the investment casting trees.

■ 15.8 CONTINUOUS CASTING

As discussed in Chapter 6 and depicted in Figure 6-6, *continuous casting* is usually employed in the solidification of basic shapes which become the feedstock for deformation processes such as rolling and forging. By producing a special mold, continuous casting can also be used to produce long lengths of complex cross-section product, such as the one depicted in Figure 15-13. Since each product is simply a cutoff section of the continuous strand, a single mold is all that is required to produce a large number of pieces. Quality is high as well, since the metal can be protected from contamination during melting and pouring, and only a minimum of handling is required.

■ 15.9 ELECTROMAGNETIC (OR LEVITATION) CASTING

In *electromagnetic casting* the metal is contained and solidified in an electromagnetic field designed to counteract the gravitational forces and hydrostatic pressure of the metal column. Since there is no container, the surfaces of the molten metal are completely exposed and open to the direct impingement of water jets or spray. The intense stirring induced by the electromagnetic field helps to create a final structure that is homogeneous, equiaxed,

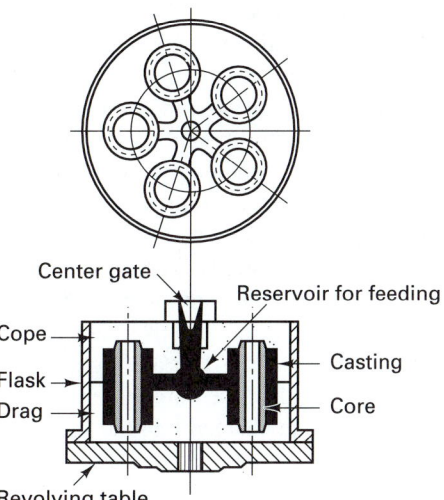

FIGURE 15-12   Method of casting by the centrifuging process. Metal is poured into the central pouring sprue and spun into the various mold cavities. *(Courtesy of American Cast Iron Pipe Company.)*

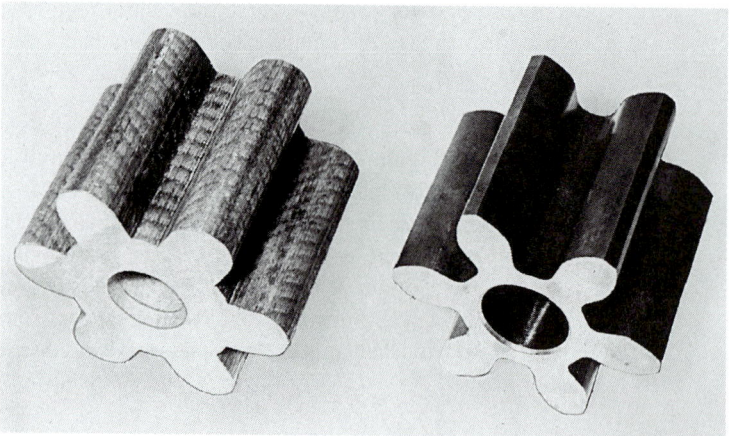

FIGURE 15-13   Gear produced by continuous casting. (*Left*) As-cast material; (*right*) after machining. (*Courtesy of American Smelting and Refining Company.*)

and fine grained. The exterior surface is quite smooth, and the process can be automated and adapted for continuous casting.

■ 15.10 MELTING AND POURING

All casting processes require a source of molten metal. Ideally, this source should be able to provide an adequate amount of molten material, at the desired temperature, with the desired chemistry and minimum contamination. It should be capable of holding material for an extended period of time without deterioration of quality, be economical to operate, and be capable of being operated without contributing to the pollution of the environment. Except for experimental or very small operations, virtually all foundries use cupolas, air furnaces, (also known as direct fuel-fired furnaces), electric-arc furnaces, or electric-induction furnaces. In locations such as fully integrated steel mills, molten metal may be taken directly from a steelmaking furnace and poured into casting molds. This practice is usually reserved for exceptionally large castings. For small operations, gas-fired crucible furnaces are often used, but these have rather limited capacities.

Selection of the most appropriate melting procedure depends on such factors as (1) the temperature needed to melt and superheat the metal, (2) the alloy being melted and the form of available charge material, (3) the desired melting rate or quantity of metal, (4) the desired quality of the metal, (5) the availability and cost of various fuels, (6) the variety of metals or alloys to be melted, (7) whether melting is to be batch or continuous, (8) the required level of emission control, and (9) the various capital and operating costs.

The feedstock to the melting furnace may take several forms. While prealloyed ingot may be purchased for remelt, it is not uncommon for the starting material to be a mix of commercially pure primary metal and commercial scrap, which also includes recycled gates, runners, sprues, and risers, as well as defective castings. Alloy elements are then added as either pure materials or master alloys that are high in a particular element but are designed to have a lower melting point than the pure material and a density that allows for good mixing. Preheating the metal being charged into the various types of furnaces can increase the melting rate by as much as 30%.

Cupolas

A large amount of gray, nodular, and white cast iron is still melted in *cupolas*, although many foundries have now converted to electric induction furnaces. Basically, a cupola is a refractory-lined, vertical steel shell into which alternating layers of coke, iron (pig iron and/or scrap), limestone or other flux, and possible alloy additions are charged and melted under forced-air draft. Operation is similar to that of a blast furnace, with the molten metal collecting at the bottom of the cupola to be tapped off at periodic intervals.

Cupolas are simple and economical, can be obtained in a wide range of capacities, and can produce cast iron of excellent quality if the proper raw materials are used and good control is practiced. They are used exclusively for the melting of cast iron and can be operated either continuously or as a batch-type melter. Temperature and chemistry control is somewhat difficult, however. The nature of the charged materials and the reaction that occur within the cupola can all affect the product chemistry. Moreover, by the time the final chemistry is determined through analysis of the tapped product, a substantial charge of material is already working its way through the furnace. Final chemistry adjustments are often performed in the ladle, using the various techniques of ladle metallurgy discussed in Chapter 6.

Various methods can be used to increase the melting rate and improve the economy of operation. In a *hot-blast cupola*, the stack gases are put through a heat exchanger, enabling the incoming air to be preheated to temperatures as high as 1200°F (650°C). Oxygen-enriched blasts can be used to further increase the temperature and accelerate the rate of melting. Plasma torches can also be used to melt the iron scrap in a cupola. Melting rates can be quite high, such that it is not uncommon for a continuously operating cupola to produce as much as 120 tons of hot metal per hour.

If the air currents within the stack can be reduced sufficiently, iron turnings can be incorporated into the feedstock, further reducing the operating costs. Water cooling may be provided to the shell as a means of prolonging the life of the cupola. With sufficient cooling, the refractory lining may not be necessary, and its elimination permits a wider variety of slags to be used. Environmental considerations require a large volume of stack gases to be passed through dust collectors and various types of pollution-control equipment. Consequently, a modern cupola installation can be quite complex, as indicated by the schematic of Figure 15-14.

Indirect Fuel-Fired Furnaces (or Crucible Furnaces)

While limited in size and melting rate, *indirect fuel-fired furnaces* are often used to melt small batches of nonferrous metal. They usually take the form of crucibles or holding pots whose outer surface is heated by an external flame. Stirring action, temperature control, and chemistry control are poor, but these furnaces offer low capital cost. The containment crucibles are generally made from clay and graphite, silicon carbide, cast iron, or steel.

Air Furnaces or Direct Fuel-Fired Furnaces

Direct fuel-fired furnaces, also known as *air furnaces*, are similar to small open-hearth furnaces but are less sophisticated. As illustrated in Figure 15-15, the surface of the metal is heated directly by the burning fuel. Capacity is much larger than the crucible furnace, but the operation is still limited to the batch melting of nonferrous metals and the holding of cast iron that has been melted previously in a cupola. The rate of heating and melting and the temperature and composition of the molten metal are all easily controlled.

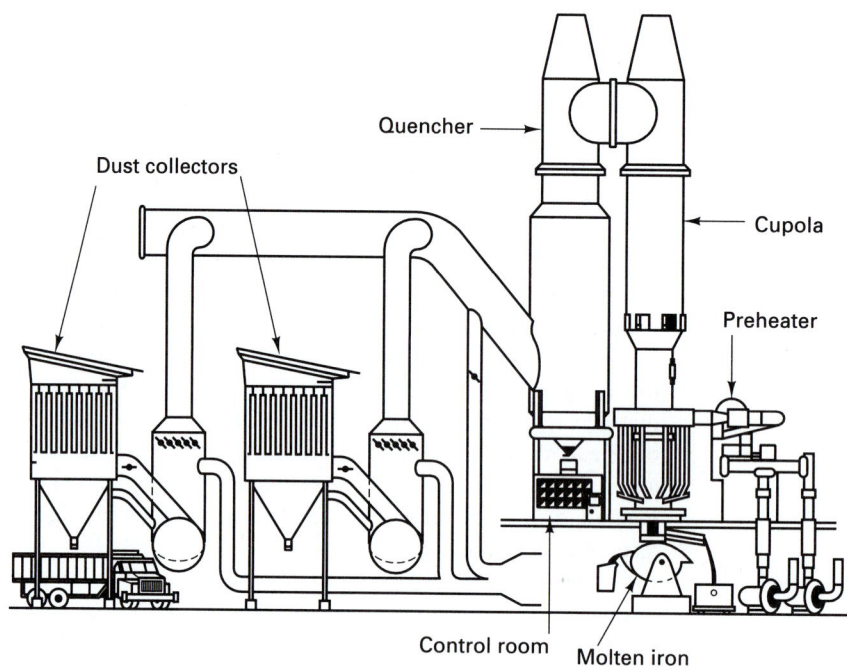

FIGURE 15-14   Cupola installation showing associated pollution control equipment.   *(Courtesy of Foundry Management and Technology.)*

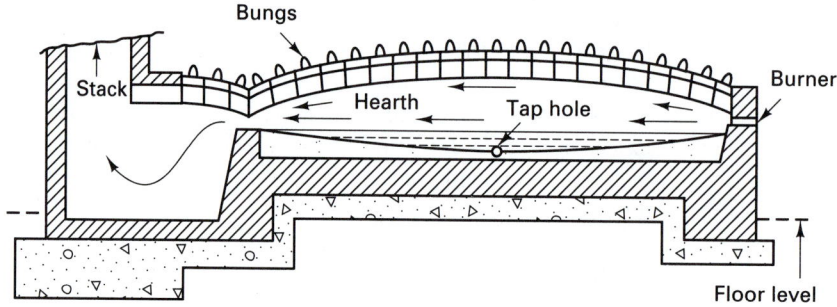

FIGURE 15-15   Cross section of an air furnace.

Arc Furnaces

Arc furnaces have become the preferred method of melting in many foundries because of (1) their rapid melting rates, (2) their ability to hold the molten metal for any desired period of time, and (3) the greater ease of incorporating pollution control equipment.

The basic features and operating cycle of a *direct-arc furnace* can be described with the aid of Figure 15-16. The top is first lifted or swung aside to permit the introduction of charge material. The top is then replaced, and the electrodes are lowered to create an arc between the electrodes and the metal charge. The path of the heating current is usually through one electrode, across an arc, through the metal charge, and back through another arc to another electrode.

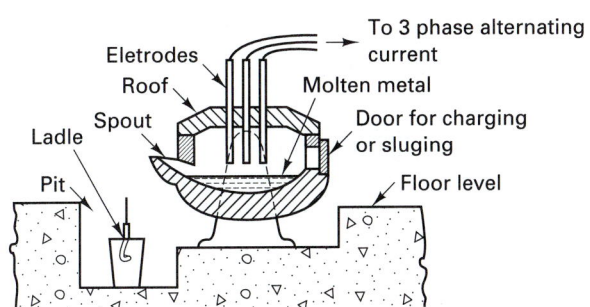

FIGURE 15-16   Schematic diagram of a three-phase electric arc furnace.

FIGURE 15-17   Electric arc furnace, tilted for pouring. *(Courtesy of Pittsburgh Lectromelt Furnace Corporation.)*

Fluxing materials are usually added to provide a protective cover to the pool of molten metal. Because the metal is covered and can be maintained at a given temperature for long periods of time, arc furnaces can be used to produce high-quality metal of almost any desired composition. They are available in sizes up to about 200 tons (but capacities of 25 tons or less are most common), and up to 50 tons per hour can be melted conveniently in batch operations. Arc furnaces are generally used with ferrous alloys, especially steel, and provide good mixing and homogeneity to the molten bath. Unfortunately, the noise and level of particle emissions can be rather high, and the consumption of electrodes, refractories, and power results in high operating costs. Figure 15-17 shows the pouring of an electric arc furnace.

Induction Furnaces

Because of their very rapid melting rates and the relative ease of controlling pollution, electric *induction furnaces* have become a very popular means of melting metal. There are two basic types of induction furnaces. The *high-frequency*, or *coreless* units, shown schematically in Figure 15-18, consist of a crucible surrounded by a water-cooled coil of copper tubing. A high-frequency electrical current passes through the coil, creating an alternating magnetic field. The varying magnetic field induces secondary currents in the metal being melted, which bring about a rapid rate of heating.

Coreless induction furnaces are used for virtually all common alloys, the maximum temperature being limited only by the refractory and the ability to insulate against heat loss. They provide good control of temperature and composition and are available in a range of capacities up to about 65 tons. Because there is no contamination from the heat source, they produce very pure metal. Operation is generally on a batch basis.

Low-frequency or *channel-type* induction furnaces are also seeing increased use. As shown in Figure 15-19, only a small channel is surrounded by the primary (current-carry-

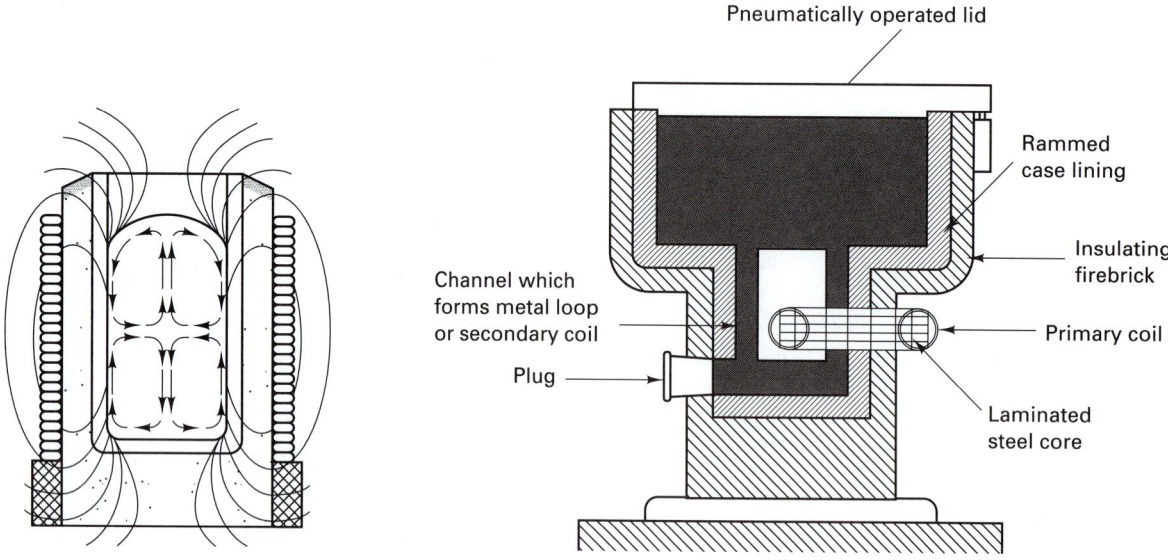

FIGURE 15-18 Schematic showing the basic principle of a coreless induction furnace.

FIGURE 15-19 Cross section showing the principle of the low-frequency or channel-type induction furnace.

ing) coil. A secondary coil is formed by a loop, or channel, of molten metal, and all the metal is free to circulate through the loop and gain heat. Enough molten metal must be placed into the furnace to fill the secondary coil, with the remainder of the charge taking a variety of forms. The heating rate is very high and the temperature can be accurately controlled. As a result, channel-type furnaces are often preferred as holding furnaces, where the molten metal is maintained at a constant temperature for an extended period of time. Capacities can be quite large, up to about 250 tons.

■ 15.11 POURING PRACTICE

To transfer the metal from the melting furnace to the molds, some type of pouring device, or ladle, is required. The primary considerations for this operation are (1) to maintain the metal at the proper temperature for pouring, and (2) to assure that only high-quality metal is introduced into the molds. The specific type of *pouring ladle* is determined largely by the size and number of castings to be poured. In small foundries, a hand-held, shank-type ladle such as that shown in Figure 15-20 is extremely common. In larger foundries, either bottom-pour or teapot-type ladles are used, like the ones illustrated in Figure 13-7. These are often used in conjunction with a conveyor line which moves the molds past the pouring station. By extracting metal from beneath the surface, slag and other impurities that float on top of the melt are not permitted to enter the mold.

High-volume, mass-production foundries often use automatic pouring systems, such as the one shown in Figure 15-21. Molten metal is transferred from the main melting furnace to a holding furnace by overhead crane. A programmed amount of molten metal is further transferred into individual pouring ladles and is then poured automatically into the corresponding molds. Laser-based control units position the pouring ladle over the sprue and control the flow rate into the sprue cup.

FIGURE 15-20 Pouring a mold using a shank-type ladle.

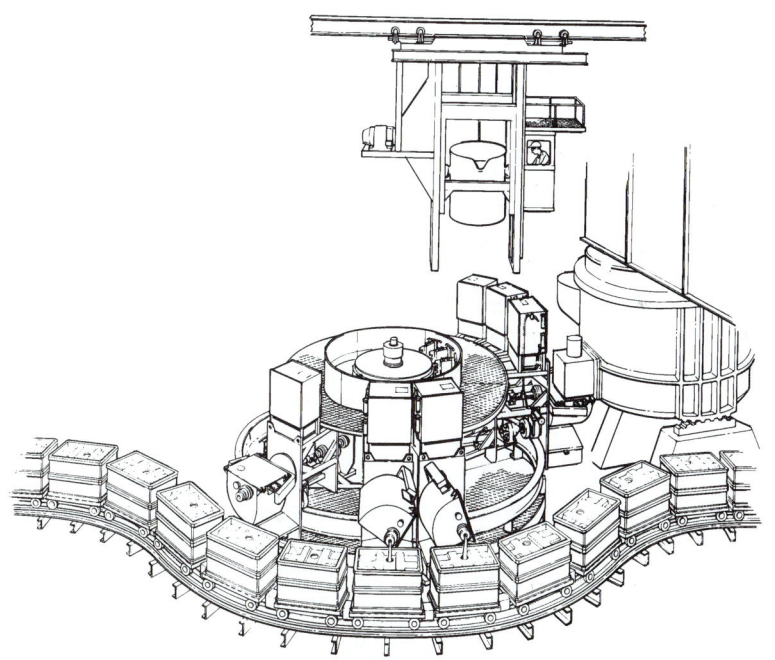

FIGURE 15-21 Machine for automatic pouring of molds on a conveyor line. *(Courtesy of Roberts Corporation.)*

■ 15.12 CLEANING, FINISHING, AND HEAT TREATING OF CASTINGS

Cleaning and Finishing

After solidification and removal from molds, most castings require some additional cleaning and finishing. Specific operations may include all or several of the following:

1. Removing cores
2. Removing gates and risers
3. Removing fins, flash, and rough spots from the surface
4. Cleaning the surface
5. Repairing any defects

Cleaning and finishing operations can be quite expensive, so consideration should be given to their minimization when designing the product and selecting the specific method of casting. In addition, consideration should also be directed toward the possibility of automating the cleaning and finishing.

Sand cores can usually be removed by mechanical shaking. At times, however, they must be removed by chemically dissolving the core binder. On small castings, gates and risers can sometimes be knocked off. For larger castings, a cutting operation is usually required. Most nonferrous metals and cast irons can be cut with an abrasive cutoff wheel, power hacksaw, or bandsaw. Steel castings frequently require an oxyacetylene torch. Plasma arc cutting can also be used to remove sprues, gates, and risers.

The specific method of cleaning often depends on the size and complexity of the casting. After the gates and risers have been removed, small castings are often processed through tumbling barrels to remove fins, flash, and sand that have adhered to the surface. Tumbling may also be done to remove cores and, in some cases, gates and risers. Metal shot or abrasive material is often added to the barrel to aid in the cleaning. Conveyors can be used to pass larger castings through special cleaning chambers were they are subjected to blasts of abrasive or cleaning material. Extremely large castings usually require manual finishing, using pneumatic chisels, portable grinders, and manually directed blast hoses.

While defect-free castings are always desired, flaws such as cracks, voids, and laps are not uncommon. In some cases, especially when the part is large and the production quantity is small, it may be more attractive to repair the part rather than change the pattern, die, or process. If the material is weldable, repairs are often made by removing the defective region (usually by chipping or grinding) and filling the created void with deposited weld metal. Porosity that is at or connected to free surfaces can be filled with resinous material, such as polyester, by a process known as *impregnation*. (See Chapter 16 for a further discussion of this process).

Heat Treatment and Inspection of Castings

Heat treatment is an attractive means of altering properties while retaining the shape of the product. Steel castings are frequently given a full anneal to reduce the hardness (and brittleness) of rapidly cooled, thin sections, and to reduce the internal stresses that result from uneven cooling. Nonferrous castings are often heat treated to provide stress relief and prepare them for subsequent machining. For final properties, virtually all of the treatments discussed in Chapter 5 can be applied. Ferrous-metal castings often undergo a quench-and-temper treatment, and many nonferrous castings are age hardened to impart additional strength. The variety of heat treatments is largely responsible for the wide range of properties and characteristics available in cast metal products.

Virtually all of the nondestructive *inspection techniques* that were presented in Chapter 11 can be applied to cast metal products. X-ray radiography, liquid penetrant inspection, and magnetic particle inspection are extremely common.

■ 15.13 ROBOTS IN FOUNDRY OPERATIONS

Many of the operations that are performed in a foundry are ideally suited for robotic *automation* since they tend to be dirty, dangerous, and dull. Robots can dry molds, coat cores, vent molds, and clean or lubricate dies. They can tend stationary, cyclic equipment, such as die casting machines, and if the machines are properly grouped, one robot can often service two or three machines. In the finishing room, robots can be equipped with plasma cutters or torches to remove sprues, gates, and runners. They can perform grinding and blasting operations, as well as various functions involved in the heat treatment of castings. In the investment casting process, robots can be used to dip the wax patterns into the refractory slurry and produce the desired molds. In a similar manner, they have been used to dip full-mold Styrofoam patterns in their refractory coating and hang them on conveyors to dry. In a fully automized operation, robots could be used to position the pattern, fill the flask with sand, pour the metal, and use a torch to remove the sprue.

■ 15.14 PROCESS SELECTION

As shown in the individual process summaries that have been included throughout Chapters 14 and 15, each of the casting processes has its own characteristic set of capabilities, assets, and limitations. The requirements of the particular product (such as size, complexity, required dimensional precision, desired surface finish, total quantity to be made, and desired rate of production) often limit the number of processes that should be considered as production candidates. Further selection is often based on cost.

Some aspects of product cost, such as the cost of the material and the energy required to melt it, are somewhat independent of specific process. The cost of other features, such as patterns, molds, dies, melting and pouring equipment, scrap material, cleaning, inspection, and all related labor, can vary markedly and be quite dependent on the process. For example, pattern and mold costs for sand casting are quite a bit less than the cost of die-casting dies. Die casting, on the other hand, offers high production rates and a high degree of automation. When a small quantity of parts is desired, the cost of the die or tooling must be distributed over the total number of parts, and unit cost (or cost per casting) is high. When the total quantity is large, the tooling cost is distributed over many parts, and cost per piece decreases.

Figure 15-22 shows the relationship between unit cost and production quantity for both sand and die casting. Sand casting is an expendable mold process. Since an individual mold is made for each pour, increasing quantity does not lead to a significant drop in unit cost. Die casting involves a multiple-use mold, and the cost of the die can be distributed over the total number of parts. As shown, sand casting is often less expensive for small production runs, and processes such as die casting are preferred for large quantities. [*Note:* While the die-casting curve in Figure 15-22 is a smooth line, it is not uncommon for an actual curve to contain abrupt discontinuities. If the lifetime of a set of tooling is 50,000 casts, the cost per part for 45,000 pieces (one set of tooling) would actually be less than 60,000, since the latter would require a second set of dies.]

In most cases, more than one or two processes are possible candidates for production, and curves for all of the options should be included. The final selection is then based on a combination of economic and technical considerations.

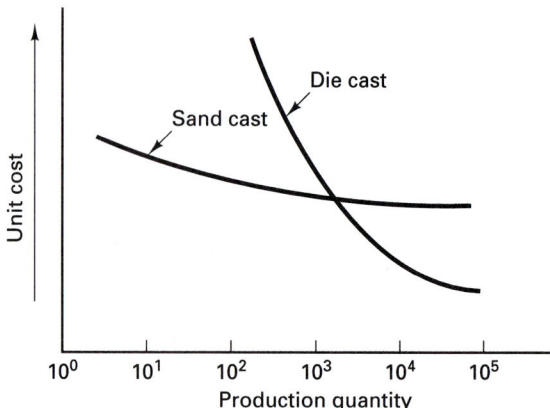

FIGURE 15-22   Typical unit cost of castings for both sand casting and die casting.

■ KEY WORDS

air furnace
arc furnace
automation
centrifugal casting
centrifuging
cold-chamber machines
continuous casting
core removal
cupola
die casting
direct fuel-fired furnace

electromagnetic casting
heat treatment
heated-manifold direct-injection die
 casting
hot-chamber machines
indirect fuel-fired furnace
induction furnace
inserts
inspection
low-pressure permanent mold casting
permanent mold casting

pouring ladle
retractable metal cores
semicentrifugal casting
slush casting
squeeze casting
thermal fatigue
thixotropic material
true centrifugal casting
vacuum permanent mold casting

■ REVIEW QUESTIONS

1. What are some of the major disadvantages of the expendable-mold casting processes?
2. What are some possible limitations to the use of multiple-use molds?
3. What are some common mold materials for permanent mold casting? What are some of the metals more commonly cast?
4. Describe some of the process advantages of permanent mold casting.
5. Why are permanent mold castings generally removed from the mold immediately after solidification has been completed?
6. How is venting provided in the permanent mold process? Are risers required? Can sand cores be used?
7. What types of products would be possible candidates for manufacture by slush casting?
8. How does low-pressure permanent mold casting differ from the traditional gravity-pour process?

9. What are some of the attractive features of the low-pressure permanent mold process?
10. What are some additional advantages of vacuum permanent mold casting over the low-pressure process?
11. Contrast the feeding pressures on the molten metal in low-pressure permanent molding and die casting.
12. Contrast the materials used to make dies for gravity permanent mold casting and die casting. Why is there a notable difference?
13. Why might the pressure on the molten metal vary during the die casting cycle?
14. For what types of materials would a hot-chamber die casting machine be appropriate?
15. How does the air in the mold cavity escape in the die-casting process?
16. Are risers employed in die casting? Can sand cores be used?

17. What are some of the claimed advantages of the heated-manifold direct-injection die casting of zinc?

18. What are some of the more attractive features of the die casting process compared to alternative casting methods?

19. Describe the steps in the squeeze casting process?

20. What is a thixotropic material? How does it provide an attractive alternative to squeeze casting?

21. Contrast the structure and properties of the outer and inner surfaces of a centrifugal casting.

22. What is the difference between semicentrifugal casting and centrifuging?

23. What are some of the attractive features of electromagnetic casting?

24. What are some of the factors that influence the selection of a furnace type or melt procedure in a casting operation?

25. What types of metals are commonly melted in cupolas?

26. In cupola-melting operations, why are many of the chemistry control operations performed in the ladle via ladle metallurgy?

27. What are some of the pros and cons of indirect fuel-fired furnaces?

28. What are some of the attractive features of arc furnaces in foundry applications?

29. Why are channel induction furnaces attractive for metal holding applications where molten metal must be held at a specified temperature for long periods of time?

30. What are the primary functions of a pouring ladle?

31. What are some of the typical cleaning and finishing operations that are performed on castings?

32. How might defective castings be repaired to permit successful use in their intended applications?

33. What are some of the ways that industrial robots can be employed in metal casting operations?

34. Describe some of the features that affect the cost of a cast product. Why might the cost vary significantly with the quantity to be produced?

Chapter 15 CASE STUDY

baseplate for a household steam iron

The item depicted in Figure CS-15 is the baseplate of a high-quality household steam iron. It is rated for operation at up to 1200 watts, and is designed to provide both steady steam and burst of steam features. Incorporated into the design is an integral electrical resistance heating "horseshoe" that must be thermally coupled to the baseplate, but remain electrically insulated. (This component often takes the form of a resistance heating wire, surrounded by ceramic insulation, all encased in a metal tube). There is a complex series of channels designed to receive the steam and distribute it out evenly through a number of small vent holes in the base, each about an 1/16-inch in diameter. There are about a dozen larger threaded recesses, about 1/8-inch in diameter, that are used in assembling the various components.

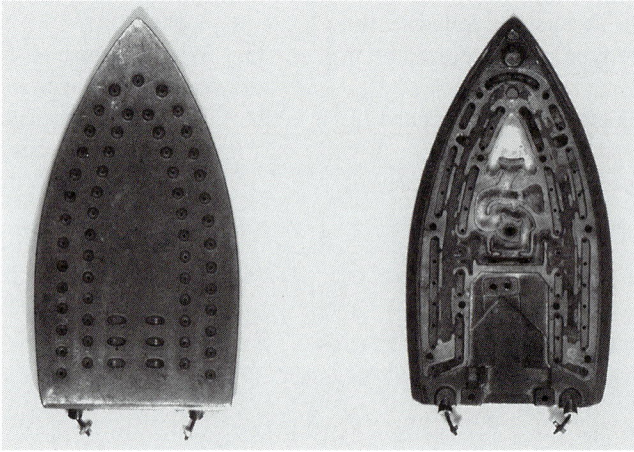

FIGURE CS-15

1. Discuss the various features that this component must possess in order to function in an adequate fashion. Consider strength, impact resistance, thermal conductivity, corrosion resistance, weight, and other factors.

2. What material or materials would appear to be strong candidates?

3. What are some possible means of producing the desired shape? Which would you prefer? Could the heating element assembly be incorporated during manufacture, or does it have to be added as a secondary operation? What are the major advantages of the method you propose?

4. Could all of the design features (holes, webs, and recesses) be incorporated in the initial manufacturing operation, or would secondary processing be required? If secondary processing is required, for what features, and how would you recommend that they be produced?

5. Some commercial irons have baseplates that are finished by a simple buff and polish, while others have a teflon coating or have been anodized. If your desire is to produce a recognized high-quality product, what form of surface finishing would you recommend?

CHAPTER 16

POWDER METALLURGY

16.1 INTRODUCTION
16.2 BASIC PROCESS
16.3 POWDER MANUFACTURE
16.4 RAPIDLY SOLIDIFIED POWDER
16.5 POWDER TESTING AND EVALUATION
16.6 POWDER MIXING AND BLENDING
16.7 COMPACTING
   Compaction Tooling (Punches and Dies)
   P/M Injection Molding
16.8 SINTERING
16.9 HOT ISOSTATIC PRESSING

16.10 OTHER TECHNIQUES TO PRODUCE HIGH-DENSITY P/M PRODUCTS
16.11 SECONDARY OPERATIONS
16.12 PROPERTIES OF P/M PRODUCTS
16.13 DESIGN OF POWDER METALLURGY PARTS
16.14 POWDER METALLURGY PRODUCTS
16.15 ADVANTAGES AND DISADVANTAGES OF POWDER METALLURGY
Case Study: AUTOMOBILE SEAT ADJUSTMENT GEARS

■ 16.1 INTRODUCTION

Powder metallurgy is the name given to the process by which fine powdered materials are blended, pressed into a desired shape (compacted), and then heated (sintered) in a controlled atmosphere to bond the contacting surfaces of the particles and establish the desired properties. The process, commonly designated as P/M, readily lends itself to the mass production of small, intricate parts of high precision, often eliminating the need for additional machining or finishing. There is little material waste; unusual materials or mixtures can be utilized; and controlled degrees of porosity or permeability can be produced. Major areas of application tend to be those for which the P/M process has strong economical advantage or where the desired properties and characteristics would be difficult to obtain by any other method.

While a crude form of powder metallurgy appears to have existed in Egypt as early as 3000 B.C., mass manufacturing of P/M products did not begin until the mid- or late nineteenth century. As this time, powder metallurgy was used to produce copper coins and medallions, platinum ingots, and tungsten wires, the primary material for light bulb filaments early in the twentieth century. By the 1920s tungsten carbide cutting-tool tips and nonferrous bushings were being produced. Self-lubricating bearings and metallic filters were other early products. A period of rapid technological development occurred after World War II based primarily on automotive applications, and iron and steel replaced copper as the dominant P/M material. Aerospace and nuclear developments created accelerated demand for refractory and reactive materials where P/M processing is quite attractive. Full-density products emerged in the 1960s, and high-performance superalloy components, such as aircraft turbine engine parts, were a highlight of the 1970s. Developments in the 1980s included the commercialization of rapidly-solidified and amorphous powders and the development of P/M injection molding technology.

Recent years have been ones of rapid development and growth for the P/M industry. From 1960 to 1980, the consumption of iron powder increased tenfold. A similar increase occurred between 1980 and 1990, and this exponential growth continues. P/M has become

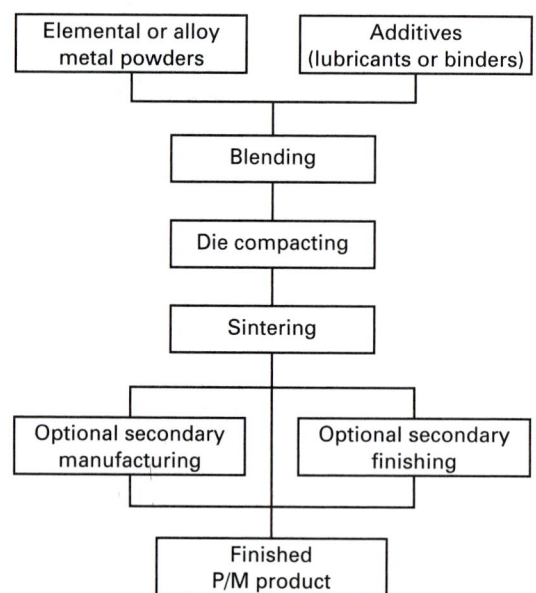

FIGURE 16-1 Simplified flowchart of the powder metallurgy process.

a proven method of parts production and is now considered as an alternative in the manufacture of many components. While most products are still under 2 in. (50 mm) in size, some have been produced with weights up to 100 lb (45 kg) and linear dimensions up to 20 in. (500 mm). Automotive applications account for nearly 75% of the powder metallurgy market. In 1990, the average U.S. automobile contained about 21 lb (10 kg) of P/M parts. By 1992, use increased to 26 lb (12 kg), and by 1995, new car models contained over 30 lb (13.6 kg) of P/M parts. Other areas where powder metallurgy products are used extensively include household appliances, recreational equipment, hand tools, hardware items, business machines, industrial motors, and hydraulics. Areas of rapid growth include aerospace applications, advanced composites, electronic components, magnetic materials, metalworking tools, and a variety of biomedical and dental applications. Iron and low-alloy steels now account for 85% of all P/M usage, with copper and copper-based powders comprising 6 to 7%. Stainless steel, high-strength and high-alloy steels, and aluminum and aluminum alloys are also high-volume materials.

■ 16.2 BASIC PROCESS

The powder metallurgy process generally consists of four basic steps: (1) powder manufacture, (2) mixing or blending, (3) compacting, and (4) sintering. Optional secondary processing often follows to obtain special properties or enhanced precision. Figure 16-1 presents a simplified flowchart of the P/M process.

■ 16.3 POWDER MANUFACTURE

The properties of powder metallurgy products are highly dependent on the characteristics of the metal (or material) powders that are used. Some important properties and characteristics include chemistry and purity, *particle size, size distribution, particle shape*, and the surface texture of the particles. Several processes can be used to produce powdered material, with each imparting distinct properties and characteristics to the powder and hence to the final product.

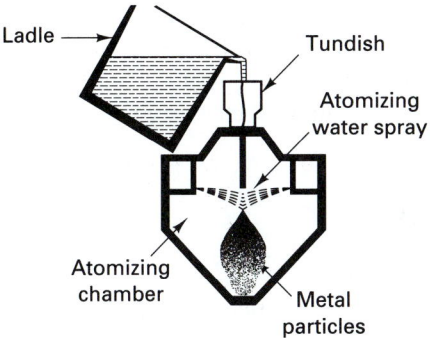

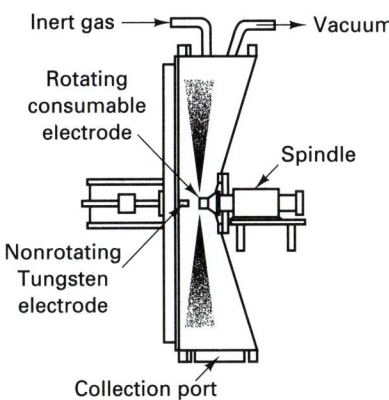

FIGURE 16-2   Two methods for producing metal powders: (a) melt atomization; (b) atomization from a rotating consumable electrode.

Over 80% of all commercial powder is produced by some form of melt *atomization*, where a liquid is fragmented into molten droplets which then solidify into particles. Various forms of energy have been used to form the droplets. In the method illustrated in Figure 16-2a, molten metal is atomized by a stream of impinging gas or liquid as it emerges from an orifice. In Figure 16-2b, an electric arc impinges on a rapidly rotating electrode (all contained within a chamber purged with inert gas), with centrifugal force causing the molten droplets to fly from the surface of the electrode. Regardless of the particular form, melt atomization processes are an extremely useful means of producing prealloyed powders. By starting with an alloyed melt or electrode, each powder particle is of the desired alloy composition. Powders of aluminum alloys, copper alloys, stainless steel, nickel-based alloys (such as Monel), titanium alloys, cobalt-based alloys, and various low-alloy steels have all been commercially produced. The size and shape of the powder particles can be varied and depend on process features such as the velocity and media of the atomizing jets or the speed of electrode rotation, the starting temperature of the liquid (which affects the time that surface tension can act on the individual droplets prior to solidification), and the environment provided for cooling. When cooling is slow (such as in gas atomization) and surface tension is high, spherical shapes can form before solidification. With more rapid cooling, such as with water atomization, irregular shapes tend to be produced.

Other methods of powder manufacture include *chemical reduction of particulate compounds* (generally crushed oxides or ores), *electrolytic deposition from solutions or fused salts*, *pulverization or grinding of brittle materials (comminution)*, *thermal decomposition of hydrides or carbonyls*, *precipitation from solution*, and *condensation of metal vapors*.

Almost any metal, metal alloy, or nonmetal (ceramic, polymer, or wax or graphite lubricant) can be converted into powder form by one or more of the methods noted above. Some methods can produce only elemental powder (often of high purity), while others can produce prealloyed particles. Operations such as drying or heat treatment may be required prior to further processing.

■ 16.4 RAPIDLY SOLIDIFIED POWDER (MICROCRYSTALLINE AND AMORPHOUS)

Increasing the cooling rate of liquid material can result in the formation of an ultrafine or microcrystalline grain size. In these materials, a large percentage of the atoms are located in grain boundary regions, giving unusual properties (such as high diffusivity), expanded alloy possibilities, and good formability. If the cooling rate approaches or exceeds 10^6 °C/sec, metals can solidify without becoming crystalline. These amorphous or glassy metals can also exhibit unusual or unique properties, which include high strength, improved corrosion resistance, and reduced energy to induce and reverse a magnetization. Amorphous metal transformer cores lose 60 to 70% less energy in magnetization than conventional silicon steels. As a result, it is estimated that over half of all new power distribution transformers purchased in the United States will utilize amorphous metal cores.

Production of *amorphous* material, however, requires immensely high cooling rates and hence ultra-s426426mall dimensions. Atomization with rapid cooling and the "splat quenching" of a metal stream onto a cool surface to produce a continuous ribbon are two prominent methods. Since much of the ribbon material is further fragmented into powder, powder metallurgy thus becomes the primary means of fabricating useful products.

■ 16.5 POWDER TESTING AND EVALUATION

In addition to evaluation of bulk chemistry, surface chemistry, particle size and size distribution, particle shape, surface texture, and internal structure, metal powders must also be evaluated for their suitability for further processing. *Flow rate* is a measure of the ease by which powder can be fed and distributed into a die. Poor flow characteristics can result in nonuniform die filling and in nonuniform density and properties in a product.

Associated with the flow characteristics is the *apparent density*, a measure of the powder's ability to fill available space without the application of external pressure. A low apparent density means that there is a high percentage of unfilled space in the loose-fill powder. *Compressibility tests* evaluate the effectiveness of applied pressure in raising the density of the powder, and *green strength* is used to describe the strength of the pressed powder immediately after compacting. It is well established that higher product density results in superior mechanical properties, such as strength and fracture resistance. Good green strength is required to maintain smooth surfaces, sharp corners, and intricate details during ejection from the compacting die or tooling and the subsequent transfer to the sintering operation.

■ 16.6 POWDER MIXING AND BLENDING

It is rare that a single powder will possess all of the characteristics desired in a given process and product. Most likely, the starting material will be a mixture of various grades or sizes of powder, or powders of different compositions, with additions of *lubricants* or *binders*.

The final product chemistry is often obtained by combining pure metal or nonmetal powders, rather than using prealloyed material. Sufficient diffusion must then occur during the sintering operation to produce a uniform chemistry and structure in the final product. Unique composites can also be produced, such as the distribution of an immiscible reinforcement material in a matrix, or the combination of metals and nonmetals in a single product such as a tungsten carbide–cobalt matrix cutting tool for high-temperature service.

Some powders, such as graphite, can even play a dual role, serving as a lubricant during compacting and a source of carbon as it alloys with iron during sintering to produce steel. Lubricants such as graphite or stearic acid improve the flow characteristics and compressibility at the expense of reduced green strength. Binders produce the reverse effect. Most lubricants or binders are not wanted in the final product and are removed (volatilized or burned off) in the early stages of sintering, leaving holes that are reduced in size or closed during subsequent heating.

Blending or mixing operations can be done either dry or wet, where water or other solvent is used to improve mixing, reduce dusting, and lessen explosion hazards. Large lots of powder can be homogenized with respect to both chemistry and distribution of components, sizes, and shapes. Quantities up to 35,000 lb (16,000 kg) have been blended in single lots to ensure uniform behavior during processing and the production of a consistent product.

■ 16.7 COMPACTING

One of the most critical steps in the P/M process is *compacting*. Loose powder is compressed and densified into a shape known as a *green compact*, usually at room temperature. High product density and the uniformity of that density throughout the compact are generally desired characteristics. In addition, the compacts should possess sufficient green strength for in-process handling and transport to the sintering furnace.

Most compacting is done with mechanical presses and rigid tools, but hydraulic and hybrid (combinations of mechanical, hydraulic, and pneumatic) presses can also be used. Figure 16-3 shows a typical mechanical press for compacting powders and a removable set of compaction tooling. The removable die sets allow the time-consuming alignment and synchronization of tool movements to be set up while the press is producing parts with another die set. Compacting pressures generally range between 3 and 120 tons/in² (40 to 1650 MPa) depending on material and application (see Table 16-1), with 10 to 30 tons/in² (140 to 415 MPa) being the most common. While most P/M presses have total capacities of less than 100 tons (9×10^5N), increasing numbers are being purchased with higher capacity. As a result, powder metallurgy products are often limited to cross sections of less than 3 in² (2000 mm²). With increased press capacity, sections up to 10 in² (6500 mm²) have become more common. Some P/M presses now have capacities up to 3000 tons (27 MN) and 100 in² (65,000 mm²). Where larger products are desired, compaction can be performed by dynamic methods, where an explosively induced shock wave provides the compaction pressure. Metal-forming processes, such as rolling, forging, extrusion, and swagging, have also been adapted to compact powders.

TABLE 16-1. Typical Compacting Pressures for Various Applications

	Compaction Pressures	
Application	tons/in²	MPa
Porous metals and filters	3–5	40–70
Refractory metals and carbides	5–15	70–200
Porous bearings	10–25	146–350
Machine parts	20–50	275–690
High-density iron and steel parts	50–120	690–1650

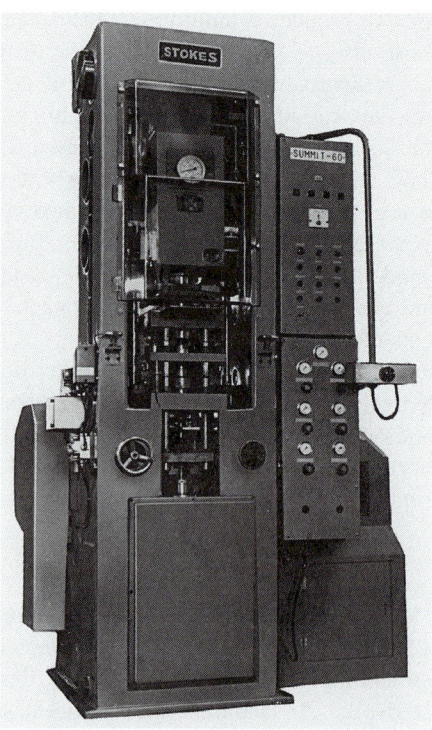

FIGURE 16-3 (*Left*) Typical press for the compacting of metal powders. The removable die set (*right*) allows the machine to be producing parts with one die set while another is being fitted to produce a second product. *(Courtesy of Sharples-Stokes Division, Pennwalt Corporation.)*

Figure 16-4 shows the typical compaction sequence for a mechanical press. With the bottom punch in its fully raised position, a feed shoe moves into position over the die. The feed shoe is an inverted container filled with powder, connected to the powder supply by a flexible feed tube. With the feed shoe in position, the bottom punch descends to a preset fill depth, and the shoe retracts, leveling the powder. The upper punch then descends and compacts the powder as it penetrates the die. The upper punch retracts and the bottom punch raises to eject the green compact. As the die shoe advances for the next cycle, its forward edge clears the compact from the press, and the cycle repeats.

During compacting, the powder particles move primarily in the direction of the applied force. The powder does not flow like a liquid but simply compresses until an equal and opposing force is created. This opposing force is probably a combination of (1) resistance by the bottom punch, and (2) friction between the particles and the die surfaces. As illustrated in Figure 16-5, when the pressure is applied by only one punch, maximum density occurs below the punch and decreases as one moves down the column. It is very difficult to transmit uniform pressures and produce uniform density throughout a compact. By use of a double-action press (Figure 16-6), more uniform density can be obtained and thicker products can be compacted. Since sidewall friction is a key factor in compaction, the resulting density is a strong function of both the thickness and width of the part being pressed. For good, uniform compaction, the ratio of thickness/width should be kept below 2.0 whenever possible. Products with ratios above 2.0 tend to exhibit considerable variation in density.

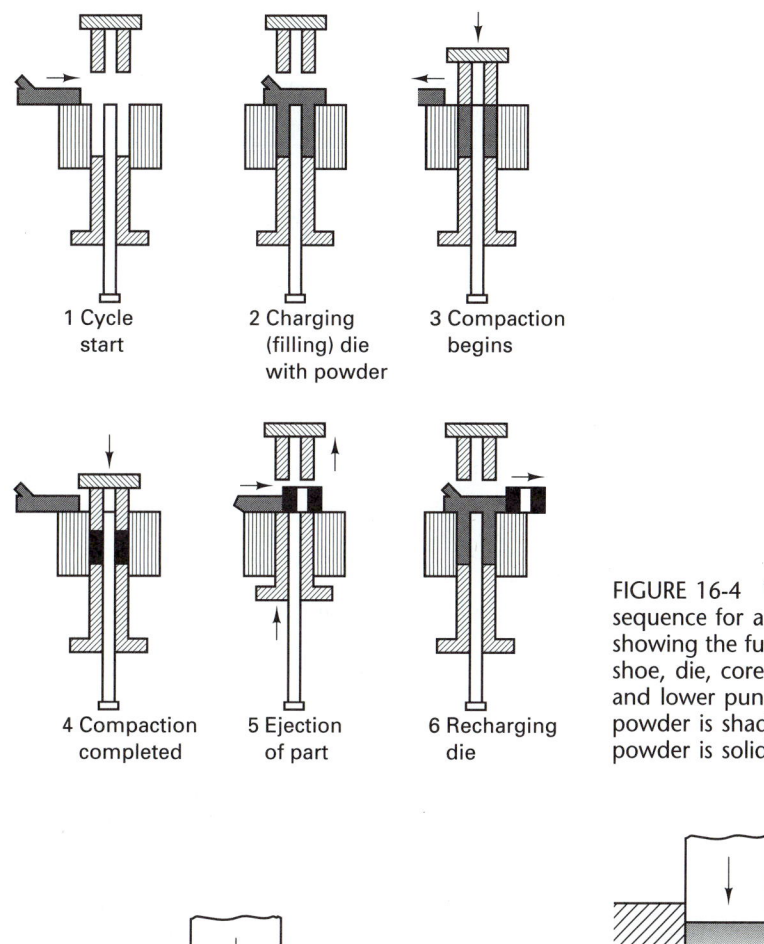

1 Cycle
start

2 Charging
(filling) die
with powder

3 Compaction
begins

4 Compaction
completed

5 Ejection
of part

6 Recharging
die

FIGURE 16-4   Typical compaction sequence for a single-level part, showing the functions of the feed shoe, die, core rod, and upper and lower punches. Loose powder is shaded; compacted powder is solid black.

FIGURE 16-5   Compaction with a single punch, showing the resultant nonuniform density.

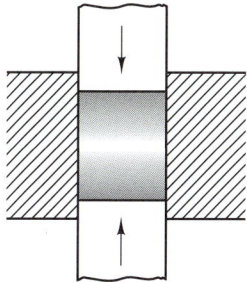

FIGURE 16-6   Density distribution obtained with a double-acting press and two moving punches.

As shown in Figure 16-7, the average density of the compact depends on the amount of pressure that is applied, with the particular response being a function of the powder being compressed (its size, shape, surface texture, mechanical properties, etc.). The final density may be reported as either an absolute density in units such as grams per cubic centimeter, or as a percentage of the theoretical density, where the difference between this number and 100% is the amount of void space left within the compact.

Figure 16-8 shows that a single displacement will produce different degrees of compaction in different thicknesses of powder. Therefore, it is impossible for a single punch to produce uniform density in a multi-thickness part. When the product requires nonuni-

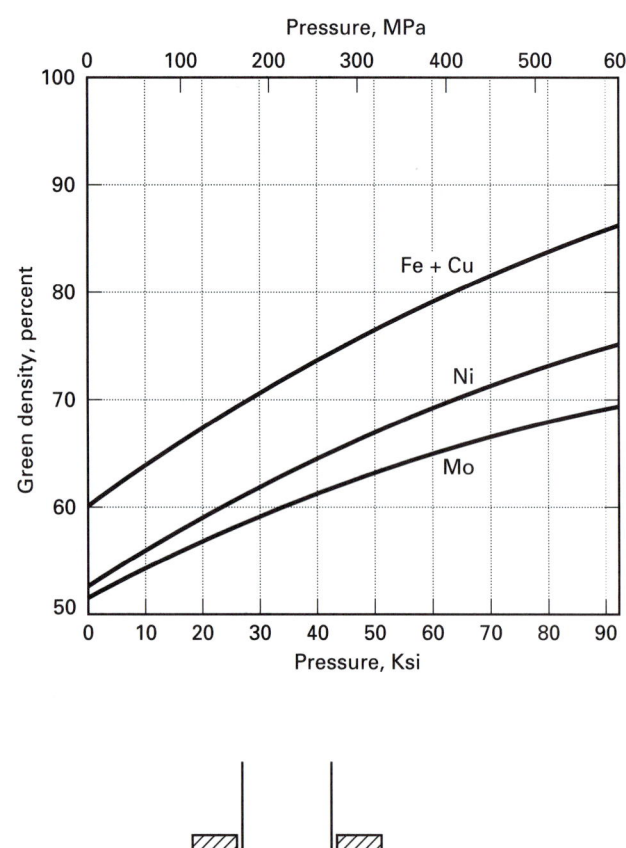

FIGURE 16-7 Effect of compacting pressure on green density (after compaction but before sintering) for several commercial powders.

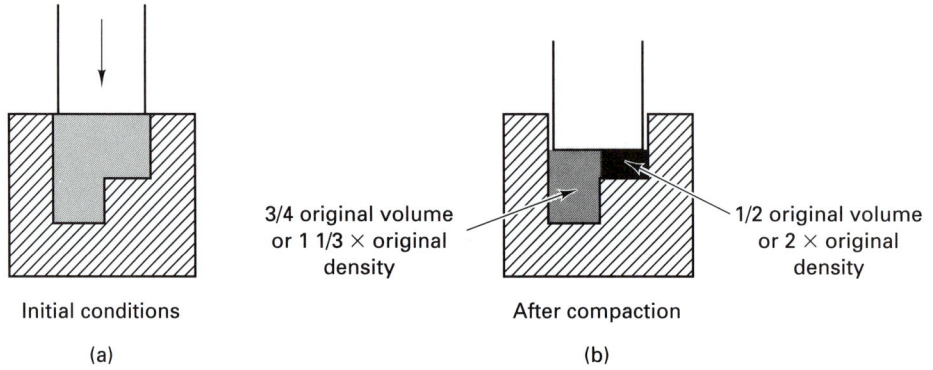

FIGURE 16-8 Two-thickness part with only one moving punch. (*Left*) Initial conditions; (*right*) after compaction by the upper punch, showing drastic difference in compacted density.

form thickness, more complicated presses or compaction methods must be employed. Figure 16-9 illustrates two methods to compact a dual-thickness part. By providing different amounts of motion to the various punches and synchronizing these movements to provide simultaneous compaction, a uniformly compacted product can be produced. When extremely complex shapes are desired, the powder is generally encapsulated in a flexible mold and immersed in a pressurized gas or liquid, a process known as *isostatic* (uniform pressure) *compaction*. Production rates in this process are extremely low, but parts up to several hundred pounds can be compacted effectively.

Another means of promoting uniform compaction is to increase the amount of lubricant in the powder. This reduces the friction between the powder and the die wall and enhances

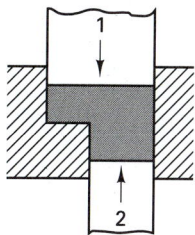

Single lower punch

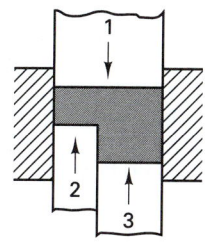

Double lower punch

FIGURE 16-9 Two methods of compacting two thickness parts to near-uniform density. Both involve the controlled movement of two or more punches.

the transmission of compaction pressure through the powder. Unfortunately, additional lubricant may reduce the green strength to the point where it is insufficient for part ejection and handling, or reduce final properties to the point where they are unacceptable.

Pressing rates vary widely, but small mechanical presses can typically produce up to 100 pieces per minute. By means of mechanisms such as bulk movement of particles, deformation of individual particles, and particle fracture or fragmentation, mechanical compaction can raise the density of loose powder to about 80% of an equivalent cast or forged metal. Sufficient strength can be imparted to hold the shape and withstand a reasonable amount of handling. In addition, the process sets the nature and uniformity of the porosity remaining in the product.

Compaction Tooling (Punches and Dies)

Because powder particles tend to be somewhat abrasive and high pressures are involved during compaction, wear of the tool components is a major concern. Consequently, compaction tools are usually made of hardened tool steel. For particularly abrasive powders, or for high-volume production, cemented carbides may be employed. Die surfaces should be highly polished and the dies should be heavy enough to withstand the high pressing pressures. Lubricants are also used to reduce die wear.

P/M Injection Molding

Small, complex-shaped components have been fabricated from plastic for many years by means of *injection molding*. A thermoplastic resin is heated to impart the necessary degree of fluidity and is then pressure injected into a die, where it cools and hardens. Die casting is a similar process for metals but is restricted to alloys with relatively low melting temperature (lead-, zinc-, aluminum-, and copper-based materials). Small, complex-shaped products of the higher-melting-point metals are generally made by more costly processes, such as investment casting, machining directly from metal stock, or conventional powder metallurgy.

Injection molding of P/M products is a recently developed alternative to conventional powder metallurgy compaction. While powdered material does not flow like a fluid, complex shapes can be produced by mixing ultrafine (usually less than 10 μm) metal, ceramic, or carbide powder with a thermoplastic/wax material (up to 50% by volume), heating it to a pastelike consistency (about 500°F), and injecting it into a mold cavity under conditions of pressure (about 10,000 psi) and temperature selected to assure die filling. After ejection, the binder material is removed by either solvent extraction or controlled heating to above the volatilization temperature. The parts then undergo conventional sintering, during which they shrink 15 to 25% and obtain the desired density (up to 98% of ideal) and properties.

Removing the binder is currently the most expensive and time consuming part of the process. Heating rates, temperatures, and debinding times must be carefully controlled and adjusted for part thickness. Up to 3 days may be required for certain thick-walled sections. A water-soluble methylcellulose binder (used in the Rivers process) is one attractive alternative to the thermoplastics. Water from the binder evaporates early during heating and the methylcellulose burns off during sintering, reducing the processing time significantly. Other variations of P/M injection molding use different binders or fluidizers and different means of removal. While the size of conventional P/M products is generally limited by press capacity, the size of P/M injection moldings is more limited by economics (cost of the fine powders) and binder removal. The best candidates for injection molding are parts with thicknesses of less than $\frac{1}{2}$ in. and weights under 2 oz.

Despite the high cost of ultrafine powders and extensive processing, P/M injection molding has become an attractive method for producing small, intricate parts with thin walls or delicate cross sections that would be impossible to compact in a conventional press or might otherwise require significant amounts of secondary machining. Die design and manufacture are sufficiently costly that large production volumes (more than 2000 to 5000 identical parts) are generally required to justify the process. However, the relatively high final density (94 to 98% of ideal compared to 75 to 90% for conventional P/M parts), uniformity of that density, and close tolerances (0.3 to 0.5%) make the process attractive for many applications. The final properties are generally equivalent to wrought or cast metals.

■ 16.8 SINTERING

In the *sintering* operation, the pressed-powder compacts are heated in a controlled-atmosphere environment to a temperature below the melting point but high enough to permit solid-state diffusion, and held for sufficient time to permit bonding of the particles. Most metals are sintered at temperatures of 70 to 80% of their melting point, while certain refractory materials may require temperatures near 90%. When the product is composed of more than one material, the sintering temperature may even be above the melting temperature of one or more components. The lower-melting-point materials then melt and flow into the voids between the remaining particles.

Most sintering operations involve three stages, and many sintering furnaces employ three corresponding zones. The first operation, the *burn-off* or *purge*, is designed to combust any air, volatilize and remove lubricants or binders that would interfere with good bonding, and slowly raise the temperature of the compacts in a controlled manner. Rapid heating would produce high internal pressure from air entrapped in closed pores or volatilizing lubricants and would result in swelling or fracture of the compacts. When the compacts contain appreciable quantities of volatile materials, their removal increases the *porosity* and *permeability* of the pressed shape. The manufacture of products such as metal filters is designed to take advantage of this feature. When the products are load-bearing components, however, high amounts of porosity are undesirable, and the amount of volatilizing lubricant is kept to an optimized minimum. The second, or *high-temperature stage* is where the desired solid-state diffusion and bonding between the powder particles take place. The time must be sufficient to produce the desired density and final properties, and usually varies from 10 minutes to several hours. Finally, a *cooling period* is required to lower the temperature of the products while maintaining them in a controlled atmosphere. This feature serves to prevent both oxidation that would occur upon direct discharge into air and possible thermal shock from rapid cooling. Both batch and continuous furnaces can be used.

All three stages of sintering must be conducted in a protective atmosphere. This is critical because the compacted shapes have residual porosity and internal voids that are connected to exposed surfaces. At elevated temperatures, rapid oxidation would occur and significantly impair the quality of interparticle bonding. Reducing atmospheres, commonly based on hydrogen, dissociated ammonia, or cracked hydrocarbons, are preferred since they can reduce any oxide already present on the particle surfaces and combust harmful gases that are liberated during the sintering. Inert gases cannot reduce existing oxides but will prevent the formation of any additional contaminants. Vacuum sintering is frequently employed with stainless steel, titanium, and the refractory metals. Nitrogen atmospheres are also common.

During the sintering operation, a number of changes occur in the compact. Metallurgical bonds form between the powder particles as a result of solid-state atomic diffusion, and strength, ductility, toughness, and electrical and thermal conductivities all increase. If different chemistry powders were blended, diffusion may also promote the formation of alloys or intermetallic phases. In addition, the bond formation results in an increase in density and a contraction in product dimensions. This dimensional shrinkage will have to be compensated through the design of oversized compaction dies. Not all porosity is removed, however, and conventional pressed-and-sintered P/M products generally contain between 5 and 25% residual porosity.

■ 16.9 HOT ISOSTATIC PRESSING

Hot isostatic pressing (HIP) combines powder compaction and sintering into a single operation. Here the powder is sealed in a flexible, airtight, evacuated container and then subjected to a high-temperature, high-pressure environment. Conditions for processing irons and steels are generally around 10,000 to 15,000 psi (70 to 100 MPa) and about 2300°F (1250°C), with the nickel-based superalloys, refractory metals, and ceramic powders requiring equipment capable of 45,000 psi (310 MPa) and 2750°F (1500°C). Products emerge at full density with uniform, isotropic properties that are often superior to those of other processes. Near-net shapes are possible, thereby eliminating material waste and costly machining operations. Since the powder is totally isolated and compaction and sintering occur simultaneously, the process is attractive for reactive and brittle materials such as beryllium, uranium, zirconium, and titanium. The HIP process has also been employed to densify existing parts (such as those that have been conventionally pressed and sintered), heal internal porosity in castings, and seal internal cracks in various products.

Several aspects of the HIP process make it expensive and unattractive for high-volume production. The first is the high cost of "canning" the powder in a flexible isolating medium that can resist the subsequent temperatures and pressures, and then later removing this material from the product ("decanning"). Sheet metal containers are most common, but glass and even ceramic molds have been used. The second problem involves the relatively long time for the HIP cycle. While new advances have reduced cycle times from 24 hours to 6 to 8 hours, production is still limited to several loads a day, and the number of parts per load is limited by the ability to produce and maintain uniform temperature throughout the chamber.

The *sinter-HIP process* and *pressure-assisted sintering* are techniques that can produce full-density products without the expense of canning and decanning. Conventionally compacted P/M parts are placed in a pressurizable chamber and first sintered (heated) under vacuum for sufficient time to close and isolate all remaining porosity (i.e., no internal pores remain connected to external surfaces). While maintaining the elevated temperature, the vacuum is broken and high pressure is applied for the remainder of the process. The sealed surface produced during vacuum sintering acts as the isolating can during the high-pressure stage. By starting with as-compacted powder parts, these processes eliminate the additional heating and cooling cycle that would be required if parts were first compacted and sintered in the conventional manner and then subjected to the HIP process for final densification.

■ 16.10 Other Techniques to Produce High-Density P/M Products

High-density P/M parts can also be produced by using high-temperature forming processes. Sheets of sintered powder can be reduced in thickness and further densified by rolling. Rods, wires, and small billets can be produced by the hot extrusion of encapsulated powder or pressed-and-sintered slugs. Similarly, forging can be applied to form complex shapes from canned powder or sintered preforms. These processes offer the combined benefits of powder metallurgy and the respective forming process, such as the production of fabricated shapes with uniform fine grain size and uniform chemistry or unusual alloy composition.

The *Ceracon process* is another method of raising conventional pressed-and-sintered P/M products to full density without the need for encapsulation or canning. A heated preform is totally surrounded by hot granular material (usually ceramic) capable of transmitting pressure in a pseudouniform manner, and the entire assembly is compacted in a conventional hydraulic press. Since no gas or liquid is involved, the porous or permeable preform does not require encapsulation. Cycle times are on the order of several seconds, the part and the pressurizing medium separate freely, and the pressure-transmitting granules can be reheated and reused.

Another means of producing a high-density shape from fine particles is *in-situ compaction* or *spray forming* (also known as the *Osprey process*). Consider an atomizer similar to that of Figure 16-2a, in which jets of inert or harmless gas (nitrogen or carbon dioxide) propel molten droplets down into a collecting container. If the droplets solidify before impact, the container fills with loose powder. If the droplets remain liquid, the container fills with molten metal and then solidifies into a conventional casting. However, if the cooling of the droplets is controlled so that they are semisolid upon impact (and computers can provide the necessary process control), they tend to flatten and solidify into a uniform chemistry, uniformly fine grain size, relatively high-density (in excess of 98% of theoretical) solid. Depending on the shape of the collecting container, this product can be a finished part, a strip or plate, a deposited coating, or a preform for subsequent operations, such as forging. Both ferrous and nonferrous products can be produced with deposition rates as high as 400 lb (180 kg) per minute. Simultaneous deposition of two or more materials, injecting secondary particles into the stream, and promoting in-stream reactions can all be used to produce unique composites.

■ 16.11 Secondary Operations

For numerous applications, P/M parts are ready to use after they have emerged from the sintering furnace. Many products, however, utilize one or more secondary operations to provide enhanced precision, improved properties, or special characteristics.

During sintering, product dimensions shrink due to densification. Warping or distortion may occur during cooldown from elevated temperature. Thus a second pressing operation, known as *repressing, coining, or sizing*, may be required to restore or improve dimensional precision. The part is placed in a die and subjected to pressures equal to or greater than the initial pressing pressure. A small amount of plastic flow takes place,

FIGURE 16-10 Comparison of conventional forging and the forging of a powder metallurgy preform to produce a gear blank (or gear). Moving left to right, the top sequence shows the sheared stock, upset section, forged blank, and exterior and interior scrap associated with conventional forging. The finished gear is generally machined from the blank. The bottom pieces are the powder metallurgy preform and forged gear produced without scrap by P/M forging. (Courtesy of GKN Forging Limited.)

producing high dimensional accuracy, sharp detail, and improved surface finish. In addition, the associated cold working may increase part strength by 25 to 50%.

If massive metal deformation takes place in the second pressing, the operation is known as *P/M forging*. Conventional powder metallurgy is used to produce a preform, which is one forging operation removed from the finished shape. The normal forging sequence of billet or bloom production, shearing, reheating, and sequential deformation to the desired shape is replaced by the manufacture of a comparatively simple-shaped powder metallurgy preform followed by a final hot-forging operation. Here, forging produces the complex shape, adds precision, and provides the benefits of metal flow. In addition, forging increases the density of the preform, often up to 99% of theoretical density, and significantly improves mechanical properties. While *protective atmospheres* or coatings must be used to prevent oxidation during heating and hot forging, the process can significantly reduce scrap or waste (by controlling preform weight to within 0.5%, flash-free forging can often be performed) and enables powder metallurgy to be used to produce products with improved properties, increased size, and increased complexity. The forged P/M products also have no segregation, a uniform grain size, and can be made from novel alloys or composites. The tolerance requirements of cams, splines, and gears can usually be met without subsequent machining. Figure 16-10 shows an example of a part made by conventional forging and then by the P/M forge method. The substantial reduction in scrap is clearly illustrated. Figure 16-11 shows several P/M forged connecting rods. More than 25 million such rods have been produced and are being used by a number of automotive manufacturers.

Low-density P/M products have interconnected porosity, or permeability, and this opens up the possibilities of additional processing by impregnation or infiltration. *Impregnation* refers to the forcing of oil or other liquid into the porous network. This can be done by immersing the part in a bath and applying pressure or by a combination vacuum-pressure process. The most common application is that of oil-impregnated bearings. After impregnation, the bearing material contains from 10 to 40% oil by volume, which will provide lubrication over an extended lifetime of operation. In a similar manner, P/M parts can

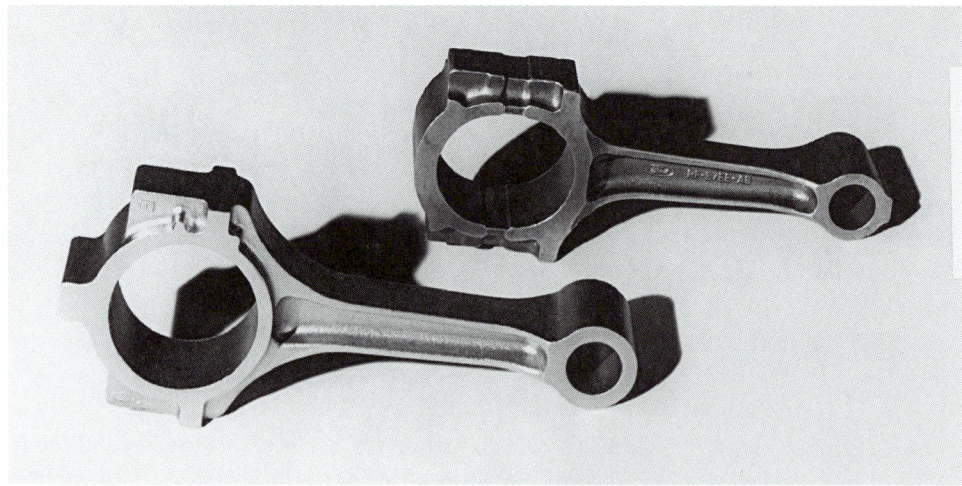

FIGURE 16-11   P/M forged connecting rods. *(Courtesy of Metal Powder Industries Federation.)*

be impregnated with fluorocarbon resin (such as Teflon) to produce products offering a combination of high strength and low friction.

When the presence of pores is undesirable, P/M products may be subjected to metal *infiltration*. In this process a molten metal with a melting point lower than the P/M constituent is forced into the product under pressure or is absorbed by capillary action. After infiltration, engineering properties such as strength and toughness are generally comparable to those of solid metal products. Infiltration may also be used to seal pores prior to plating, improve machinability or corrosion resistance, or make the components gas- or liquidtight.

Powder metallurgy products can also be subjected to the more conventional finishing operations such as heat treatment, machining, or surface treatment. If the part is of high density or has been metal impregnated, conventional processing is employed. Special precautions must be taken, however, when processing low-density P/M products. During heat treatment, protective atmospheres must again be used and certain liquid quenchants should be avoided. When machining, speeds and feeds must be adjusted and care taken to avoid pickup of lubricant or coolant. When large amounts of machining are required, special machinability-enhancing additions may be incorporated into the initial powder blend. Nearly all common methods of surface finishing can be applied, including platings and coatings, diffusion treatments, surface hardening, and steam treatment (which is used to produce a hard, corrosion-resistant oxide on ferrous parts). As with the other secondary processes, however, some process modifications may be required for porous or low-density parts.

■ 16.12 PROPERTIES OF P/M PRODUCTS

Because the properties of powder metallurgy products depend on so many variables—type and size of powder, amount and type of lubricant, pressing pressure, sintering temperature and time, finishing treatments, an so on—it is difficult to provide generalized information. Products can range all the way from low-density, highly porous parts with tensile strengths as low as 10 ksi (70 MPa) up to high-density pieces with tensile strengths of 180 ksi (1250 MPa) or more.

In general, most mechanical properties show a strong dependence on product density, with the fracture-limited properties of toughness, ductility, and fatigue life being more sensitive than strength and hardness. The voids in the P/M part act as stress concentrators and assist in starting and propagating fractures. The yield strength of P/M products made from the weaker metals is often equivalent to the same material in wrought form. If higher-strength materials are used or the fracture-related tensile strength is specified, the P/M properties tend to fall below those of wrought equivalents by varying but usually substantial amounts. Table 16-2 shows the properties of a few powder metallurgy materials compared with those of wrought material of similar composition. When larger presses or processes such as P/M forging or HIP are employed to produce higher density, the strength of the P/M products approaches that of the wrought material. With full density and fine grain size, P/M parts often have properties that exceed their wrought or cast equivalents. Since the mechanical properties of powder metallurgy products are so dependent upon density, it is important that P/M products be designed and materials selected so that the final properties will be achieved with the anticipated amount of final porosity.

TABLE 16-2.   Comparison of Properties of Powder Metallurgy Materials and Equivalent Wrought Metals

Material[a]	Form and Composition	Condition[b]	Theoretical Density (%)	Tensile Strength		Elongation in 2 in. (%)
				10^3 psi	MPa	
Iron	Wrought	HR	—	48	331	30
	P/M—49% Fe min	As sintered	89	30	207	9
	P/M—99% Fe min	As sintered	94	40	276	15
Steel	Wrought AISI 1025	HR	—	85	586	25
	P/M—0.25%C, 99.75% Fe	As sintered	84	34	234	2
Stainless steel	Wrought Type 303	Annealed	—	90	621	50
	P/M Type 303	As sintered	82	52	358	2
Aluminum	Wrought 2014	T6	—	70	483	20
	P/M 201 AB	T6	94	48	331	2
	Wrought 6061	T6	—	45	310	15
	P/M 601 AB	T6	94	36.5	252	2
Copper	Wrought OFHC	Annealed	—	34	234	50
	P/M Copper	As sintered	89	23	159	8
		Repressed	96	35	241	18
Brass	Wrought 260	Annealed	—	44	303	65
	P/M 70% Cu-30% Zn	As sintered	89	37	255	26

[a]Equivalent wrought metal shown for comparison.

[b]HR, hot rolled; T6, age hardened.

Physical properties can also be affected by *porosity*. Corrosion resistance tends to be reduced due to the presence of entrapment pockets and fissures. Electrical, thermal, and magnetic properties all vary with density. However, porosity actually promotes good sound and vibration damping, and many P/M parts are designed to take advantage of this feature.

■ 16.13 DESIGN OF POWDER METALLURGY PARTS

Powder metallurgy is a manufacturing system whose ultimate objective is to economically produce products for specific engineering applications. Success begins with good design and follows with good material and proper processing. In designing parts that are to be made by a powder metallurgy, it must be remembered that P/M is a special manufacturing process and provision should be made for a number of unique factors. Products that are converted from other manufacturing processes without modification in design rarely perform as well as parts designed specifically for manufacture by powder metallurgy. Some basic rules for the design of P/M parts are:

1. The shape of the part must permit ejection from the die. Perpendicular sidewalls are preferred, and holes or recesses should be uniform in size and parallel to the axis of punch travel.
2. The shape of the part should be such that powder is not required to flow into small cavities such as thin walls, narrow splines, or sharp corners.
3. The shape of the part should permit the construction of strong tooling.
4. The shape of the part should be within the thickness range for which P/M parts can be adequately compacted.
5. The part should be designed with as few changes in section thickness as possible.
6. Parts can be designed to take advantage of the fact that certain forms and properties can be produced by P/M which are impossible, impractical, or uneconomical to obtain by any other method.
7. If necessary, the design should be consistent with available equipment. Pressing areas should match press capability, and the number of thicknesses should be consistent with the number of available press actions.
8. Consideration should also be made for product tolerances. Higher precision and repeatability is observed for dimensions in the radial direction (set by the die) than for those in the axial or pressing direction (set by punch movement).
9. Finally, design should consider and compensate for the dimensional changes that will occur after pressing, such as the shrinkage that occurs during sintering.

The ideal powder metallurgy part has a uniform cross section and a single thickness that is small compared to the cross-sectional width or diameter. More complex shapes are indeed possible, but it should be remembered that uniform strength and properties require uniform density. Designs can easily accommodate holes that are parallel to the direction of pressing. Holes at angles to this direction must be made by secondary processing. Multiple-stepped diameters, reentrant holes, grooves, and undercuts should be eliminated whenever possible. Abrupt changes in section, narrow deep flutes, and internal angles without generous fillets should be avoided. Straight serrations can be molded readily, but diamond knurls cannot. Punches should be designed to eliminate sharp points or thin sections that could easily wear or fracture. Figure 16-12 illustrates some of these points.

■ 16.14 Powder Metallurgy Products

The products that are commonly produced by powder metallurgy can generally be classified into five groups.

1. *Porous or permeable products, such as bearings, filters, and pressure or flow regulators.* Oil-impregnated bearings, made from either iron or copper alloys, constitute a large volume of P/M products. They are widely used in home appliance and automotive applications since they require no lubrication or maintenance during their service life. P/M filters can be made with pores of almost any size, some as small as 0.0001 in. (0.0025 mm). Unlike many alternative filters, they can withstand conditions of elevated temperature, high applied stresses, and corrosive environments.

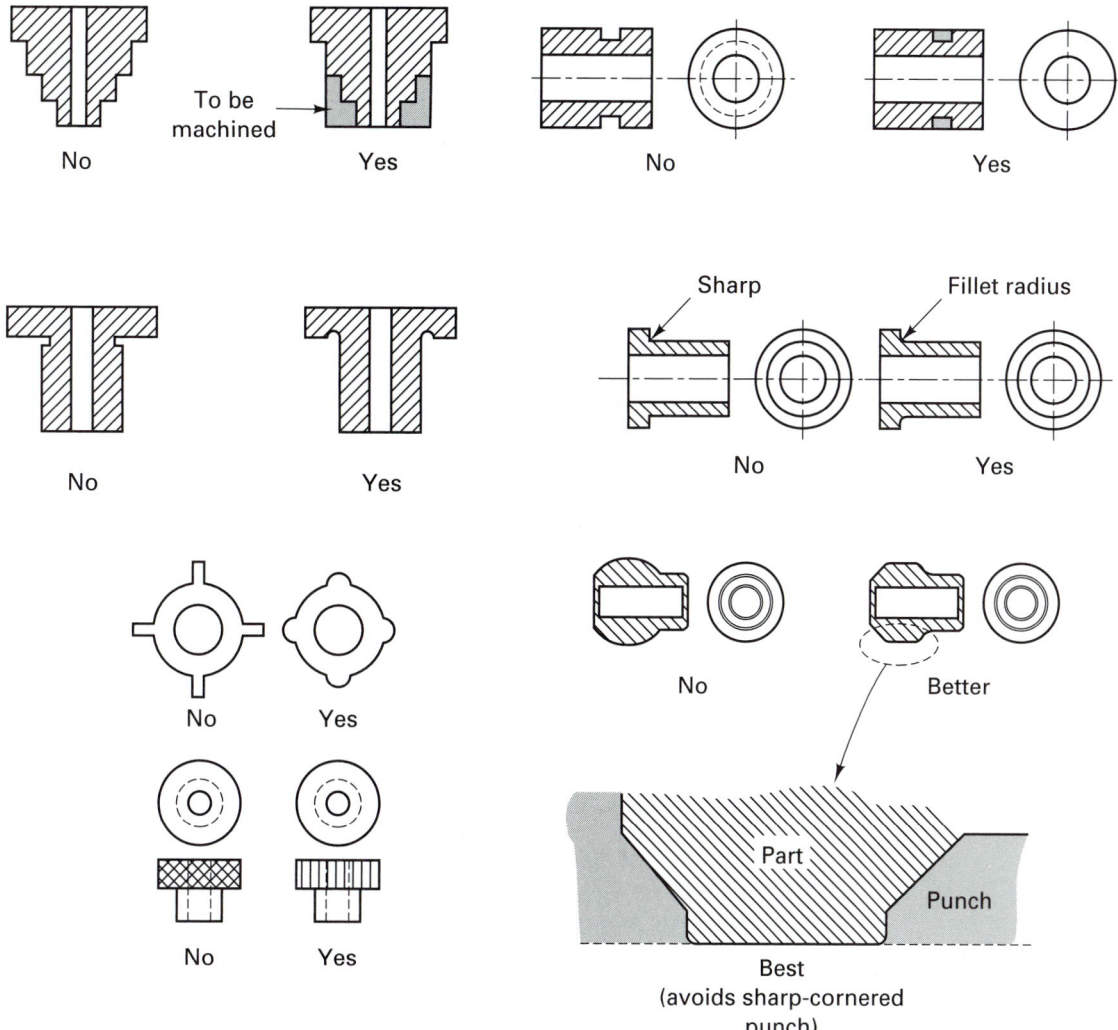

FIGURE 16-12   Examples of poor and good design details for use in powder metallurgy parts.

2. *Products of complex shapes that would require considerable machining when made by other processes.* Because of the accuracy and fine finish characteristic of the P/M process, many parts require no further processing and others require only a small amount of finish machining. Large numbers of small gears are made by the powder metallurgy process. Other complex shapes, such as pawls, cams, and small activating levers, can be made quite economically.

3. *Products made from materials that are difficult to machine or with high melting points.* Some of the first modern uses of powder metallurgy were the production of tungsten lamp filaments and tungsten carbide cutting tools.

FIGURE 16-13   Typical parts produced by the powder metallurgy process.
(Courtesy of PTX-Pentronix, Inc.)

4. *Products where the combined properties of two or more metals (or both metals and nonmetals) are desired.* This unique capability of the powder metallurgy process is applied to a number of products. In the electrical industry, copper and graphite are frequently combined in such applications as motor or generator brushes, copper providing the current-carrying capacity, with graphite providing lubrication. Similarly, bearings have been made of graphite combined with iron or copper, or of mixtures of two metals, such as tin and copper, where the softer metal is placed in a harder metal matrix. Electrical contacts often combine copper or silver with tungsten, nickel, or molybdenum. The copper or silver provides high conductivity, while the material with high melting temperature provides resistance to fusion during the conditions of arcing and subsequent closure.

5. *Products where the powder metallurgy process produces clearly superior properties.* The development of processes that produce full density has resulted in P/M products that are clearly superior to those produced by competing techniques. In areas of critical importance such as aerospace applications, the additional cost of the processing may be justified by the enhanced properties of the product. In the production of P/M magnets, a magnetic field can be used to align the particles prior to sintering, thereby producing a high flux density in the product.

Figure 16-13 shows an array of typical powder metallurgy products.

■ 16.15 ADVANTAGES AND DISADVANTAGES OF POWDER METALLURGY

Like all other manufacturing processes, powder metallurgy has distinct advantages and disadvantages that should be considered if the technique is to be employed economically and successfully. Among the important advantages are:

1. *Elimination or reduction of machining.* The dimensional accuracy and surface finish of P/M products are such that subsequent machining operations can be totally eliminated for many applications. If unusual dimensional accuracy is required, simple coining or sizing operations can often give accuracies equivalent to those of most production machining.

2. *High production rates.* All steps in the P/M process are simple and readily automated. Labor requirements are low, and product uniformity and reproducibility are among the highest in manufacturing.

3. *Complex shapes can be produced.* Subject to the limitations discussed previously, complex shapes can be produced, such as combination gears, cams, and internal keys. It is often possible to produce parts by powder metallurgy that cannot be machined or cast economically.

4. *Wide variations in compositions are possible.* Parts of very high purity can readily be produced. Metals and ceramics can be intimately mixed. Immiscible materials can be combined, and solubility limits can be exceeded. In most cases the chemical homogeneity of the product exceeds that of all competing techniques.

5. *Wide variations in properties are available.* Products can range from low-density parts with controlled permeability to high-density parts with properties that equal or exceed those of equivalent wrought counterparts. Damping of noise and vibration can be tailored into a P/M product. Magnetic properties, wear properties, and others can all be designed to match the needs of a specific application.

6. *Scrap is eliminated or reduced.* Powder metallurgy is the only common manufacturing process in which no material is wasted. In casting, machining, and press forming the scrap can often exceed 50% of the starting material. This is particularly important where expensive materials are involved and may make it possible to use more costly materials without increasing the overall cost of the product. An example of such a product would be the rare-earth magnets.

The major disadvantages of the powder metallurgy process are:

1. *Inferior strength properties.* Because of the residual porosity, powder metallurgy parts generally have mechanical properties that are inferior to wrought or cast products of the same material.* Their use may be limited when high stresses are involved. However, if the additional expense is justified, the required strength and fracture resistance can often be obtained by using different materials or by employing alternate or secondary processing techniques.

2. *Relatively high die cost.* Because of the high pressures and severe abrasion involved in the process, the P/M dies must be made of expensive materials and be relatively massive. Because of the need for part-specific tooling, production volumes of less than 10,000 identical parts are normally not practical.

3. *High material cost.* On a unit weight basis, powered metals are considerably more expensive than wrought or cast stock. However, the absence of scrap and the elimination of machining can often offset the higher cost of the starting mate-

---

*It is not uncommon to compare processes by specifying the material, size, and shape of a product and then compare the quality of the parts made by various techniques. Because of their residual porosity, P/M products tend to be mechanically inferior to their wrought or cast counterparts. A more valid comparison of processes, however, might be to specify the general size and shape and the required mechanical properties. Materials and geometric details could then be optimized to the process. P/M could then utilize unique materials, such as the blends of iron and copper that cannot be processed by the other methods. Since all products would meet the mechanical requirements, the final comparison might be on the basis of something such as total production cost.

rial. In addition, powder metallurgy is usually employed for rather small parts where the material cost per part is not very great.

4. *Design limitations.* The powder metallurgy process is simply not feasible for many shapes. Parts must be able to be ejected from the die. The thickness/diameter (or thickness/width) ratio is limited. Thin vertical sections are difficult, and the overall size must be within the capacity of available presses. Few parts exceed 25 in^2 in pressing area.

5. *Density variations produce property variations.* The nonuniform product density that is frequently produced in compacting operations generally results in property variations throughout the part. For some products, these variations would be unacceptable.

6. *Health and safety hazards.* Many metals, such as aluminum, titanium, magnesium, and iron, are pyrophoric—they can ignite or explode when in particle form with large surface/volume ratios. Fine particles can also remain airborne for long times and can be inhaled by workers. To minimize the health and safety hazards, the handling of metal powders frequently requires the use of inert atmospheres, dry boxes, and hoods, as well as special cleanliness of the working environment.

■ KEY WORDS

amorphous	impregnation	porosity
apparent density	infiltration	powder metallurgy
atomization	injection molding	protective atmosphere
binder	isostatic compaction	rapidly solidified powder
blending	lubricant	repressing
compacting	mixing	sintering
compressibility tests	P/M forging	size distribution
flow rate	particle shape	spray forming (Osprey)
green strength	particle size	
hot isostatic pressing	permeability	

■ REVIEW QUESTIONS

1. What were some of the earliest powder metallurgy products?
2. What are some of the primary market areas for P/M products?
3. What are the four basic steps that are usually involved in making products by powder metallurgy?
4. What are some of the important properties and characteristics of metal powders to be used in powder metallurgy?
5. What is the most common method of producing metal powders?
6. What are some of the other techniques that can be employed to produce particulate material?
7. Which of the powder manufacturing processes can be used to produce prealloyed particles?
8. Why is powder metallurgy a key process in producing products from rapidly solidified powders?
9. What is apparent density, and how is it related to the final density of a P/M product?
10. What is green strength, and why is it important to the manufacture of high-quality P/M products?
11. What are some of the objectives of powder mixing or blending?
12. How does the addition of a lubricant affect compressibility? Green strength?
13. How might the use of a graphite lubricant be fundamentally different from the use of wax or stearates?
14. What are some of the objectives of the compacting operation?
15. What limits the cross-sectional size of most P/M parts to several square inches or less?

16. Describe the movement of powder particles during compaction. What feature is responsible for the fact that powder does not flow and transmit pressure like a liquid?

17. Why might double-action pressing be more attractive than compaction with a single moving punch?

18. In what ways might the final density of a P/M product be reported?

19. What is isostatic compaction? For what product shapes might it be preferred?

20. What is the role of the thermoplastic material in P/M injection molding?

21. Why are P/M injection molding products injection molded to sizes that are considerably larger than the desired product?

22. For what types of parts is P/M injection molding an attractive manufacturing process?

23. What are the three stages associated with most P/M sintering operations?

24. How is the sintering temperature usually related to the melting temperature of the material being sintered?

25. Why is it necessary to raise the temperature of P/M compacts slowly to the temperature of sintering?

26. Why is a protective atmosphere required during sintering?

27. What are some of the changes that occur to the compact during sintering?

28. What are some of the attractive properties of HIPped products?

29. What are some of the major limitations of the HIP process, and how does the sinter-HIP process eliminate or minimize them?

30. What are some of the other methods that can produce high-density P/M products?

31. What is the purpose of repressing, coining, or sizing operations?

32. Why can the original compaction tooling not be used to shape the product during repressing?

33. What is the difference between repressing and P/M forging?

34. What is the difference between impregnation and infiltration? How are they similar?

35. The properties of P/M products are strongly tied to density. Which properties show the strongest dependence?

36. What advice would you want to give to a person who is planning to convert the manufacture of a component from die casting to powder metallurgy?

37. What is the shape of an "ideal" powder metallurgy product?

38. Give an example of a product where two or more materials are mixed to produce a composite P/M product with a unique set of properties.

39. How might you respond to the criticism that P/M parts have inferior strength?

40. Why is P/M not attractive for parts with low production quantities?

41. What features of the P/M process often compensate for the higher cost of the starting material?

■ PROBLEMS

1. When specifying the starting material for casting processes, the primary variables are chemistry and purity. Any structure of the starting material will be erased by the melting. For forming processes, the material remains in the solid state, so the principal concerns relating to the starting material are: chemistry and purity, ductility, yield strength, strain hardening characteristics, grain size, and so on. What are some of the characteristics that should be specified for the starting powder to assure the success of a powder metallurgy process? In what ways are these similar or different from those mentioned for casting and forming processes?

2. In conventional powder metallurgy manufacture, the material is compacted with applied pressure at room temperature and then sintered by elevated temperature at atmospheric pressure. With P/M hot pressing, the loose powder is subjected to pressure at elevated temperature. It would appear, therefore, that hot pressing could produce a finished part in a single operation and would be a more economical and attractive manufacturing process. What features have been overlooked in this argument that would tend to favor the press-and-sinter sequence for conventional manufacture?

*C*hapter 16 CASE STUDY

automobile seat adjustment gears

T he components pictured in Figure CS-16 include the two gears used to adjust the inclination of a bucket seat in an automobile. The smaller gear is approximately 1 inch in diameter, and the large, flat gear has a diameter of 2.25 in. Both have been machined from steel (by labor-intensive hand machining) to produce prototypes for testing a new design. Based on that prototype, your company, Progressive Seating, has been awarded a contract to provide the seats for a major automotive manufacturer. You now need to move into mass production. Based on the prototype and laboratory testing, your design team has specified a minimum tensile strength of 110 ksi., with a surface hardness in excess of Rockwell-C 25.

FIGURE CS-16

While the prototype gears were machined from round steel bars, you are free to specify any material or method of fabrication that will provide the required properties. While it appears desirable to retain some form of ferrous metal, your design team is suggesting manufacture by powder metallurgy. What features of the parts tend to favor powder metallurgy manufacture? What would be some of the alternative processes that could also produce the gears? Consult a powder metallurgy reference to determine whether the desired properties could be achieved by a commercial ferrous powder, with the densities commonly achieved by a normal press- and sinter method of fabrication. Could the required surface hardness be achieved without secondary heat treatment? If not, what would be necessary? How would you recommend that these products be produced.

PART 4

FORMING PROCESSES

17 Fundamentals of Metal
      Forming

18 Hot-Working Processes

19 Cold-Working Processes

20 Fabrication of Plastics,
      Ceramics, and Composites

CHAPTER 17

FUNDAMENTALS OF METAL FORMING

17.1   INTRODUCTION

17.2   FORMING PROCESSES: INDEPENDENT VARIABLES

17.3   DEPENDENT VARIABLES

17.4   INDEPENDENT–DEPENDENT RELATIONSHIPS

17.5   GENERAL PARAMETERS

17.6   FRICTION AND LUBRICATION UNDER METALWORKING CONDITIONS

17.7   TEMPERATURE CONCERNS

Hot Working
    Structure and Property Modification by Hot Working
    Temperature Variations
Cold Working
    Metal Properties and Cold Working
    Preparing Metals for Cold Working
Warm Working
Isothermal Forming

Case Study: REPAIRS TO A DAMAGED PROPELLER

■ 17.1 INTRODUCTION

The casting processes described in Chapters 13 through 15, and the powder metallurgy techniques presented in Chapter 16, have included a variety of methods for producing a desired shape from an engineering material. Each method was shown to have its own characteristic set of capabilities, advantages, and limitations. Selection of the best method to make a given product, however, requires considerable information about the part to be made and a complete understanding of *all* of the available techniques for shape production and their related characteristics. Therefore, we will continue our survey of manufacturing processes by considering a new family, the deformation processes.

The deformation processes have been designed to exploit a remarkable property of some engineering materials (most notably metals)—*plasticity*, the ability to flow as solids without deterioration of their properties. Since all processing is done in the solid state, there is no need to handle molten material or deal with the complexities of solidification. Since the material is simply moved (or rearranged) to produce the shape, as opposed to the cutting away of unwanted regions, the amount of waste is reduced substantially. Unfortunately, the forces required are often high. Machinery and tooling can be quite expensive, and large production quantities may be necessary to justify the approach.

The overall usefulness of metals is due largely to the ease by which they can be formed into useful shapes. Nearly all metal products undergo metal deformation at some stage of their manufacture. By rolling, cast ingots, strands, and slabs are reduced in size and converted into basic forms such as sheets, rods, and plates. These forms then undergo further deformation to produce wire, or the myriad of finished products formed by processes such as forging, extrusion, sheet metal forming, and others. The deformation may be bulk flow in three dimensions, simple shearing, simple or compound bending, or complex combinations of these. The stresses producing these deformations can be tension, compression,

446

TABLE 17-1. Classification of States of Stress

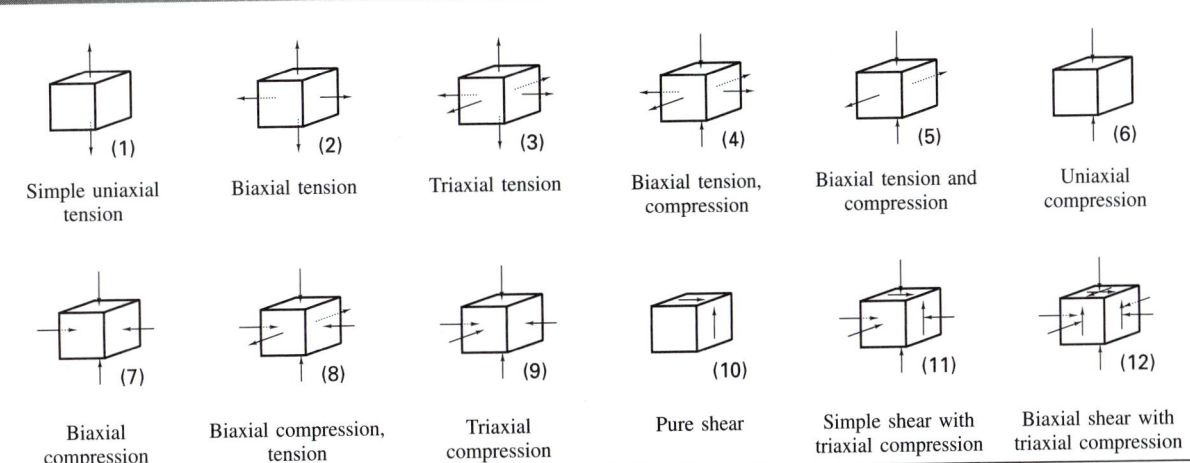

(1)	(2)	(3)	(4)	(5)	(6)
Simple uniaxial tension	Biaxial tension	Triaxial tension	Biaxial tension, compression	Biaxial tension and compression	Uniaxial compression
(7)	(8)	(9)	(10)	(11)	(12)
Biaxial compression	Biaxial compression, tension	Triaxial compression	Pure shear	Simple shear with triaxial compression	Biaxial shear with triaxial compression

shear, or any of the other varieties included in Table 17-1. As shown in Table 17-2, the specific processes are numerous and varied. A wide range of speeds, temperatures, tolerances, surface finishes, and deformation amounts are possible.

■ 17.2 FORMING PROCESSES: INDEPENDENT VARIABLES

Forming processes tend to be complex systems consisting of independent variables, dependent variables, and independent-dependent interrelations. *Independent variables* are those aspects of the process over which the engineer has direct control, and they are generally selected or specified when setting up the process. Consider some of the independent variables in a typical forming process:

1. *Starting material.* The engineer is often free to specify the chemistry and condition, and hence the properties and characteristics, of the material to be deformed. These may be chosen for ease in fabrication, or they may be restricted by the final properties that are desired for the product.
2. *Starting geometry of the workpiece.* This may be dictated by previous processing, or it may be selected by the engineer from a variety of available shapes. Economics often influence this decision.
3. *Tool or die geometry.* This is an area of major significance and has many aspects, such as the diameter of a rolling mill roll, the bend radius in a sheet-forming operation, the die angle in wire drawing or extrusion, and the cavity details when forging. Since the tooling will produce and control the metal flow, success or failure of a process often depends on tool geometry.
4. *Lubrication.* It is not uncommon for friction to account for more than 50% of the power supplied to a deformation process. Since lubricants also act as coolants, thermal barriers, corrosion inhibitors, and parting compounds, their selection is an aspect of great importance. Specification includes type of lubricant, amount to be applied, and the method of application.
5. *Starting temperature.* Temperature is one of the most influential of the process variables. Many material properties vary greatly with temperature, so its selection and control may well dictate the success or failure of an operation. Specification

TABLE 17-2. Classification of Some Forming Operations

Number	Process	Schematic Diagram	State of Stress in Main Part During Forming [a]
1	Rolling		7
2	Forging		9
3	Extrusion		9
4	Shear spinning		12
5	Tube spinning		9
6	Swaging or kneading		7
7	Deep drawing		In flange of blank, 5 In wall of cup, 1
8	Wire and tube drawing		8
9	Stretching		2
10	Straight bending		At bend, 2 and 7
11	Contoured flanging	(a) Convex	At outer flange, 6 At bend, 2 and 7
		(a) Concave	At outer flange, 1 At bend, 2 and 7

[a] See Table 17-1 for key.

448

may well include the starting temperature of both the workpiece and the tooling.

6. *Speed of operation.* Most deformation processing equipment can be operated over a range of speeds. Since speed can directly influence the lubricant effectiveness, the forces required for deformation (see Figure 2-31), and the time available for heat transfer, it is obvious that its selection would be significant in a forming operation.

7. *Amount of deformation.* While some processes control this variable through die design, others, such as rolling, permit its selection at the discretion of the engineer.

■ 17.3 DEPENDENT VARIABLES

After the engineer specifies the independent variables, the process then determines the nature and values for a second set of variables. Known as *dependent variables*, these, in essence, are the consequences of the dependent variable selection. Examples of dependent variables include:

1. *Force or power requirements.* To convert a selected material from a starting shape to a final shape, with a specified lubricant, tooling geometry, speed, and starting temperature, will require a certain amount of *force* or *power*. A change in any of the independent variables will bring about a change in the force or power required, but the effect is indirect. Engineers cannot directly specify the force or power; they can only specify the independent variables and then experience the consequences of the selection. The ability to predict the forces or powers, however, is extremely important, for only by having this knowledge will the engineer be able to specify or select the equipment for the process, select appropriate tool or die materials, compare various die designs or deformation methods, and ultimately optimize the process.

2. *Material properties of the product.* While the engineer can specify the properties of the starting material, the combined effects of the deformation and temperature experienced during the process will certainly change them. The customer is not interested in the starting properties but is concerned with our ability to produce the desired final shape with the desired *final* properties. Thus, while it may be desirable to select the starting properties based on compatibility with the process, it is also necessary to know or be able to predict how the process will alter them.

3. *Exit (or final) temperature.* Deformation generates heat. Hot workpieces cool in cold tooling. Lubricants can break down or decompose when overheated. Engineering properties can be altered by both the mechanical and thermal history of the material. Thus it is important to know and control the temperature of the material throughout the process. (*Note*: The fact that temperature may vary from location to location within the deforming and finished product adds further to the complexity of this variable.)

4. *Surface finish and precision.* Both are characteristics of the resultant product that are dependent on the specific details of the process.

5. *Nature of the material flow.* Deformation processes generally exert external constraints on the material through control and movement of its surfaces. How it flows or deforms internally depends on the specifics of the material and the process. Since properties depend on deformation history, control here is vital. The customer is satisfied only if the desired geometric shape is produced with the right set of companion properties, and without surface or internal defects.

■ 17.4 INDEPENDENT–DEPENDENT RELATIONSHIPS

Figure 17-1 serves to illustrate a major problem facing the metal-forming engineer. On one side are the independent variables, those aspects of the process for which control is direct and immediate. On the other are the dependent variables, those aspects for which control is totally indirect. It is the dependent variables that we want to control, but the *dependent variables are determined by the process*, as consequences of the independent variable selection. If we want to change a dependent variable, we must determine which independent variable (or variables) is to be changed, in what manner, and by how much. Thus it is important for us to develop a knowledge of the *independent variable–dependent variable interrelations*.

The link between independent and dependent variables is the most important area of knowledge for a person in metal forming. Unfortunately, such links are often difficult to obtain. Metal-forming processes are complex systems composed of the material being deformed, the tooling performing the deformation, lubrication at surfaces and interfaces, and various other process parameters. The number of different forming processes (and variations thereof) is quite large. In addition, various materials often behave differently in the same process. Multitudes of different lubricants exist, and some processes are sufficiently complex that they have 15 or more interacting independent variables.

The ability to predict and control dependent variables can be obtained in three distinct ways:

1. *Experience*. Unfortunately, this requires long-time exposure to the process and is generally limited to the specific materials, equipment, and products encountered in the realm of past contact.
2. *Experiment*. While possibly the least likely to be in error, direct experiment is both time consuming and costly. Size and speed of deformation are often reduced when conducting laboratory studies. Lubricant performance and heat transfer behave differently at different speeds and sizes, and their effects are generally altered. The most valid experiment, therefore, is one conducted under full-size and full-speed production conditions—generally too costly to consider to any great degree. Laboratory experiments can provide valuable insight, but caution should be exercised when extrapolating their results to more realistic production conditions.

FIGURE 17-1   Schematic of the metal-forming system showing independent variables, dependent variables, and the various means of relating the two.

Independent variables		Dependent variables
Starting material		Force or power requirements
Starting geometry	-Experience-	
Tool geometry		Product properties
Lubrication	-Experiment-	Exit temperature
Starting temperature		Surface finish
Speed of deformation	-Modeling-	Dimensional precision
Amount of deformation		Material flow details

3. *Process modeling.* Here one approaches the problem with a high-speed computer and one or more mathematical models of the process. Numerical values are provided for the various independent variables and the models are used to compute predictions for the dependent variables. Most techniques rely on the applied theory of plasticity with various simplifying assumptions. Alternatives vary from crude, first-order approximations, such as slab equilibrium or uniform deformation energy calculations, to sophisticated, computer-based solutions, such as the finite element and finite difference methods. Solutions may be algebraic equations which describe the process and reveal trends between the variables, or simply a numerical answer based on the specific input values.

Although the recent trend strongly favors the mathematical models, both for process design and computer control, it is important to note that the accuracy of the models can be no better than that of the input variables. For example, the mechanical properties of the deforming material (i.e., yield strength, ductility, etc.) must be known for the specific conditions of temperature, strain (amount of prior deformation), and strain rate (speed of deformation) being considered. The mathematical descriptions of material behavior under various process conditions are known as *constitutive relations*, and their development is receiving considerable attention. The task is not an easy one, however, because the same material may respond differently to the same conditions if its microstructure is different. A 1040 steel that has been annealed (ferrite and pearlite) will not respond the same as material that has been quenched and tempered (tempered martensite). Microstructure and its effects are difficult to describe in quantitative terms that can be handled by a computer.

Another rather elusive variable is the friction between the tool and the workpiece. While studies have shown friction to depend on contact pressure, area, surface finish, lubricant, speed, and the two contacting materials (and we know that these parameters often vary from location to location within a given process and change with time during a process), many models tend to account for its effect with a single variable of constant magnitude. The variations with time and location are simply ignored in favor of mathematical simplicity.

At first glance, these problems appear to offer a significant barrier to the theoretical approach. However, it must be noted that the same lack of knowledge hinders the person trying to document, characterize, and extrapolate the results of experience or experiments. Process modeling often reveals features that might otherwise go unnoticed and can be quite useful when attempting to prevent or eliminate defects, optimize performance, or extend a process into a previously unknown area.

■ 17.5   GENERAL PARAMETERS

While much metal-forming knowledge is specific to a given process, there are certain features that are common to all processes, and these will be presented here.

It is extremely important to characterize the *material being deformed*. What is its strength or resistance to deformation at the relevant conditions of temperature, speed of deformation, and amount of prior straining? What are its formability and fracture resistance? What is the effect of temperature or variations in temperature? To what extent does the material strain harden? What are the recrystallization kinetics? Does the material react with various environments or lubricants? These and many other questions must be answered to assess the suitability of a material to a given deformation process. Since the properties of engineering materials vary widely, the details will not be presented at this time. The reader is referred to the various chapters on engineering materials, as well as the more in-depth references cited in Chapter 9 and the Appendix.

Another general parameter is the *speed of deformation* and the various related effects that are possible. Some rate-sensitive materials may shatter or crack if impact loaded, but will deform plastically when subjected to slow-speed loading. Other materials appear stronger when deformed at a faster speed. For them, more energy is required to produce the same result if we wish to do it faster. Mechanical data obtained from slow-speed tensile tests may be totally useless if the deformation process operates at a significantly greater rate of deformation. It has also been observed that speed sensitivity (the degree to which behavior varies with speed) is generally greater when the material is at elevated temperature, as is often the case in metal-forming operations.

In addition to the changes in mechanical properties, faster speeds tend to promote improved lubricant efficiency. They also reduce the time for heat transfer and cooling. During hot working, the workpiece stays hotter and less heat is transferred to the tools.

Other general parameters include *friction and lubrication* and *temperature*. Both are of sufficient importance that they will be discussed in some detail.

■ 17.6 FRICTION AND LUBRICATION UNDER METALWORKING CONDITIONS

An important consideration in metal deformation processes is the friction developed between the tool and the workpiece. For some processes, more than 50% of the input energy is spent in overcoming friction. The surface finish and dimensional precision of the product are often directly related to friction. Changes in lubrication can alter the mode of material flow during forming and, in so doing, create or eliminate defects, or modify the properties of the final product. Production rates, tool design, tool wear, and process optimization all depend on the ability to determine and control process friction.

In most cases, our desire is to economically reduce the effects of friction. Some processes, however, can operate only when sufficient friction is present, as is the case with rolling. Regardless of the process, friction effects are hard to measure. Moreover, since they depend on variables such as contact area, speed, and temperature, friction effects are difficult to scale down for laboratory testing or extrapolate up to production conditions.

It should be noted that friction under metalworking conditions is significantly different from the friction encountered in most mechanical devices. The friction conditions of gears, bearings, journals, and similar components generally involve (1) two surfaces of similar material and strength, (2) under elastic loads such that neither body undergoes permanent change in shape, (3) with wear-in cycles that produce surface compatibility, and (4) generally low-to-moderate temperatures. Metal-forming operations, on the other hand, involve a hard, nondeforming tool interacting with a soft workpiece at pressures sufficient to cause plastic flow in the weaker material. Only a single pass is involved as the tool induces deformation, and the workpiece is often at elevated temperature.

Figure 17-2 shows the relationship between frictional resistance and contact pressure. For light, elastic loads, friction is directly proportional to the pressure, with the proportionality constant (often denoted by μ) being known as the *coefficient of friction*. At high pressures, friction becomes independent of contact pressure and is more dependent on the strength of the weaker material.

An understanding of these results can be obtained from modern friction theory, whose primary premise is that "flat surfaces are not flat" but have some degree of roughness. When two irregular surfaces interact, sufficient contact is established to support the applied load. At the lightest of loads, only three points of contact may be necessary to support a plane. As the load is increased, the contact area increases, initially in a linear

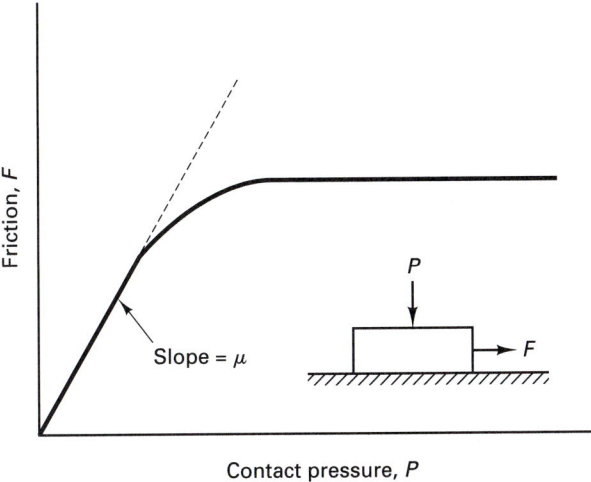

FIGURE 17-2 Effect of contact pressure on the frictional resistance between two surfaces.

fashion. At some reasonably high load, the entire surface is in physical contact. Additional loads can no longer bring additional area into contact, and friction remains constant.

Friction is the resistance to sliding motion along an interface. From a mechanics viewpoint, this resistance can be attributed to (1) the force necessary to plow the peaks of the harder material through the softer one, and/or (2) the force necessary to rip apart microscopic weldments that form between the two materials. Since the tears generally occur in the weaker of the two materials, the resistance attributed to both components is proportional to the strength of the weaker material and to the actual area of metal-to-metal contact. The relation shown in Figure 17-2, therefore, can be viewed as a plot of actual contact area at the interface versus contact pressure. Unfortunately, Figure 17-2 and the associated theory applies only to unlubricated metal-to-metal contact. The addition of a lubricant, as well as variation in its type or amount, can alter the response significantly.

Wear behavior is a related concern. Since the workpiece only interacts with the tooling during a single contact, any wear experienced by the workpiece is usually not objectionable. In fact, the shiny, fresh-metal surface produced by wear is often viewed as desirable. Manufacturers who fail to produce enough wear, and thereby retain some of the original dull finish, may be accused of selling old or substandard products. Wear on the tooling, however, is quite the reverse. Tooling is expensive and it is expected to shape many workpieces. Wear here generally means that the dimensions of the workpiece will change. Tolerance control is lost, and at some point, the tools will have to be replaced. Other consequences of tool wear include increased frictional resistance (increased required power and decreased process efficiency), poor surface finish on the product, and loss of production during tool changes.

Lubrication is a key factor in metal-forming operations. While lubricants are selected primarily for their ability to reduce friction and suppress tool wear, secondary considerations may include their ability to act as a thermal barrier, keeping heat in the workpiece and away from the tooling; their ability to act as a coolant, removing heat from the tools; and their ability to retard corrosion if left on the formed product. Other influencing factors include ease of application and removal; lack of toxicity, odor, and flammability; reactivity or lack of reactivity with material surfaces; adaptability over a useful range of pressure, temperature, and velocity; surface wetting characteristics; cost; availability; and their ability to flow or thin and still function. Selection is further complicated by the fact that lubricant performance may change with any variation of the interface conditions. The exact response will depend on such factors as the finish of both surfaces, the area of contact, the applied load, the speed, the temperature, and the amount of lubricant.

Selection of an appropriate lubricant can often be the key factor in determining whether a process is successful or unsuccessful, efficient or inefficient. For example, if one can prevent mechanical contact between the tool and the workpiece (full-fluid separation), the forces and power required may decrease by 30 to 40%, and tool wear becomes almost nonexistent. Considerable effort, therefore, has been directed to the study of friction and lubrication (a subject known as *tribology*) as it applies to both general metalworking conditions and specific metal-forming processes. A science base has been established that can aid in optimizing the use of lubricants in metalworking.

■ 17.7 TEMPERATURE CONCERNS

In metalworking operations, temperature effects are every bit as important as lubrication. The role of temperature in altering the properties of the workpiece material has been discussed in Chapter 2. In general, an increase in temperature brings about a decrease in strength, an increase in ductility, and a decrease in the rate of strain hardening—all effects that would tend to promote ease of deformation.

Forming processes tend to be classified as hot working, cold working, or warm working based on both the temperature and the material being formed. In *hot working*, the deformation is performed under conditions of temperature and strain rate where *recrystallization* occurs simultaneously with the deformation. To achieve this, the temperature of deformation is usually in excess of 0.6 times the melting point of the material on an absolute temperature scale (Kelvin or Rankine). *Cold working* is deformation under conditions where the recovery processes are not active. Here the working temperatures are usually less than 0.3 times the workpiece melting temperature. *Warm working* is deformation under the conditions of transition (i.e., a working temperature between 0.3 and 0.6 times the melting point).

Hot Working

Hot working is defined as the plastic deformation of metals above their recrystallization temperature. It is important to note, however, that the recrystallization temperature varies greatly with different materials. Tin is near hot-working conditions at room temperature; steels require temperatures near 2000°F; and tungsten does not enter the hot-working regime until about 4000°F. Thus the term *hot working* does not necessarily imply high or elevated temperature, although such is often the case.

As shown in Figures 2-29 and 2-30, elevated temperatures bring about a decrease in the yield strength of a metal and an increase in ductility. At the temperatures of hot working, recrystallization eliminates the effects of strain hardening, so there is no significant increase in yield strength or hardness, or corresponding decrease in ductility. The true stress–true strain curve is essentially flat for strains above the yield point, and deformation can be used to drastically alter the shape of a metal without fear of fracture and without the requirement of excessively high forces. In addition, the elevated temperatures promote diffusion that can remove or reduce chemical inhomogeneities; pores can be welded shut or reduced in size during the deformation; and the metallurgical structure can often be altered through recrystallization to improve the final properties. An added benefit is observed for steels, where hot working involves the deformation of weak, ductile, face-centered-cubic austenite, as opposed to the stronger body-centered-cubic ferrite that is stable at lower temperatures.

From a negative viewpoint, the high temperatures of hot working may promote undesirable reactions between the metal and its surroundings. Tolerances are poorer due to thermal contractions and possible nonuniform cooling. The metallurgical structure may also be nonuniform, since the final grain size depends on the reduction, temperature at last deformation, cooling history after the deformation, and other factors, all of which may vary throughout a workpiece.

Structure and Property Modification by Hot Working.   When metals solidify, particularly in the large sections that are typical of ingots or continuously cast strands, coarse structures tend to form with a certain amount of chemical segregation. The size of the grains is usually not uniform, and undesirable grain shapes can be quite common, such as the columnar grains that are revealed in Figure 17-3. Small gas cavities or shrinkage porosity can also form during solidification.

If a metal is reheated without prior deformation, it will simply experience grain growth and a concurrent decrease in properties. However, if the metal has experienced sufficient deformation, the distorted structure is rapidly replaced by new strain-free grains. The recrystallization is then followed by either (1) grain growth, (2) additional deformation and recrystallization, or (3) a drop in temperature to terminate diffusion and "freeze in" the recrystallized structure. The structure in the final product is that formed by the last recrystallization and the thermal history that follows. By replacing the starting structure with one of fine, spherical-shaped grains, one can produce an increase not only in strength but also in ductility and toughness.

Properties can also be improved through the reorientation of inclusions or impurity particles that are present in the metal. With normal melting and cooling, many impurities tend to locate along grain boundary interfaces. If these are unfavorably oriented, they can initiate a crack or assist its propagation through a metal. When a metal is plastically deformed, the impurities often flow along with the base metal, or fracture into rows of fragments (*stringers*) that are aligned in the direction of working. These nonmetallic impurities do not recrystallize with the base metal but retain their distorted shape and orientation. The product exhibits a *fiber structure*, like the one shown in Figure 17-4, and properties tend to vary in different directions. Through proper design, the impurities can often be reoriented into a "crack-arrestor" configuration where they are perpendicular to the direction

FIGURE 17-3   Cross section of a cast copper bar (4 in. in diameter) showing the as-cast grain structure.

FIGURE 17-4   "Fiber" structure of a hot-formed (forged) transmission gear blank. *(Courtesy of Bethlehem Steel Corporation.)*

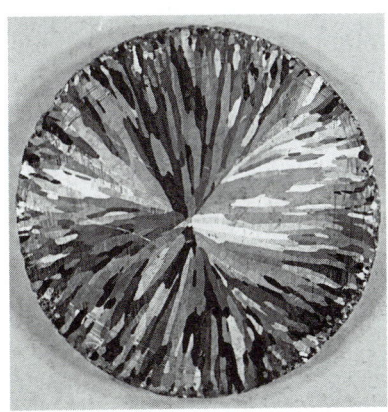

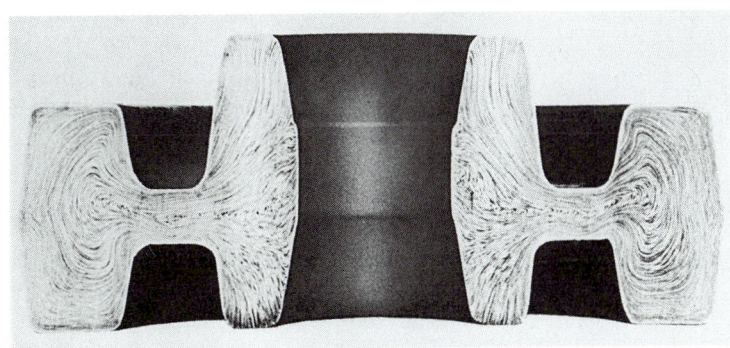

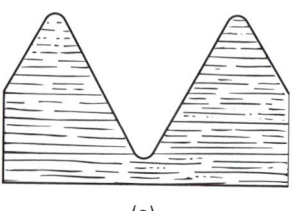

FIGURE 17-5 Schematic comparison of the grain flow characteristics in a machined thread (a) and a rolled thread (b). The rolling operation further deforms the axial structure produced by the previous wire- or rod-forming operations, while machining simply cuts through it.

of crack propagation. The outer lobe of the forging in Figure 17-4 has excellent fracture resistance since all flow lines are parallel to the free surface. The impurities are oriented as crack initiators only at the top and bottom of the inner lobe, which hopefully is a low-stress or noncritical location.

Figure 17-5 schematically compares a machined thread and a rolled thread in a threaded fastener. By reorienting the axial defects to be parallel to the thread surfaces, the rolled thread offers improved strength and fracture resistance.

Temperature Variations. The success or failure of a hot deformation process often depends on the ability to control the temperatures within the workpiece. Over 90% of the energy imparted to a deforming workpiece will be converted into heat. If the deformation process is sufficiently rapid, the temperature of the workpiece may actually increase. More common, however, is the cooling of the workpiece in its lower-temperature environment. Heat is lost through the workpiece surfaces, with the majority of the loss occurring where the workpiece is in direct contact with lower-temperature tooling. Nonuniform temperatures are produced, and flow of the hotter, weaker, interior may well result in cracking of the colder, less ductile, surfaces. Thin sections cool faster than thick sections, and this may further complicate the flow behavior.

To minimize problems, it is desirable to keep the workpiece temperatures as uniform as possible. Heated dies can reduce the rate of heat transfer, but die life tends to be compromised. For example, dies are frequently heated to 600 to 800°F (320 to 420°C) when used in the hot forming of steel. Tolerances could be improved and contact times could be increased if the tool temperatures could be raised to 1000 to 1200°F (540 to 650°C), but tool life drops so rapidly that these conditions become quite unattractive.

A final concern is the cool-down from the temperatures of hot working. Nonuniform cooling can introduce significant amounts of residual stress in hot-worked products.

Cold Working

Plastic deformation of metals below the recrystallization temperature is known as *cold working*. The process is usually performed at room temperature, but mildly elevated temperatures may be used to provide increased ductility and reduced strength. From a manufacturing viewpoint, cold working has a number of distinct advantages, and the various cold-working processes have become extremely important. Recent advances have added to their attractiveness, and the trend toward increased cold working appears likely to continue.

When compared to hot working, the advantages of cold working include:

1. No heating is required.
2. Better surface finish is obtained.
3. Superior dimensional control is achieved, so little, if any, secondary machining is required.
4. Products possess better reproducibility and interchangeability.
5. Strength, fatigue, and wear properties are improved through *strain hardening*.
6. Directional properties can be imparted.
7. Contamination problems are minimized.

Some disadvantages associated with cold-working processes include:

1. Higher forces are required to initiate and complete the deformation.
2. Heavier and more powerful equipment is required.
3. Less ductility is available.
4. Metal surfaces must be clean and scale-free.
5. Intermediate anneals may be required to compensate for the loss of ductility that accompanies strain hardening.
6. The imparted directional properties may be detrimental.
7. Undesirable residual stresses may be produced.

The strength levels induced by strain hardening are often comparable to those produced by the strengthening heat treatments. Even when the precision and surface finish of cold working is not required, it may be cheaper to produce a product by cold working a less expensive alloy (achieving the strength by strain hardening) than by heat-treating parts that have been hot formed from an alloy that will respond to the strengthening heat treatment. Because the cold-forming processes require powerful equipment and product-specific tools or dies, they are best suited for large-volume production of precision parts, where the quantity of products can justify the cost of the equipment and tooling. Considerable effort has been devoted to developing and improving cold-forming machinery. In addition, better and more ductile metals and an improved understanding of plastic flow have done much to reduce the difficulties experienced in earlier years. As an added benefit, most cold-working processes eliminate or minimize the production of waste material and the need for subsequent machining. With increasing emphasis on conservation and materials recycling, this feature can be quite significant.

Although the cold-forming processes tend to be better suited to large-scale manufacturing, much effort has been directed to developing methods that enable these processes and their associated equipment to be used economically for quite modest production quantities. By grouping products made from the same starting material and using quick-change *tooling*, cold-forming processes can often be adapted to small-quantity or just-in-time manufacture.

Metal Properties and Cold Working.   The suitability of a metal for cold working is determined primarily by its tensile properties, and these are a direct consequence of its metallurgical structure. Cold working will then change the structure, thereby altering the tensile properties of the resulting product. It is important for both of these relationships to be considered by the designer when selecting metals that are to be processed by cold working.

Figure 17-6 presents the true stress–true strain curves for both a low- and a high-carbon steel. Focusing on the low-carbon material, we note that plastic deformation cannot

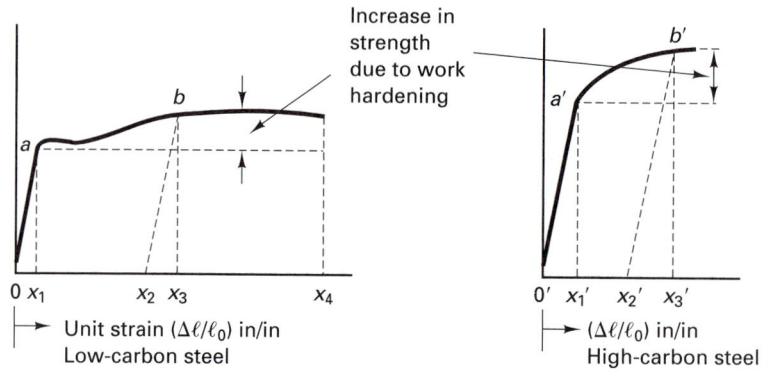

FIGURE 17-6   Use of true stress–true strain diagrams to reveal the tensile properties of two metals and assess their suitability for cold working.

occur until the strain exceeds X_1, the strain associated with the elastic limit, point a, on the stress–strain curve. Plastic deformation continues until the strain reaches the value X_4, where the metal ruptures. From the viewpoint of cold working, two features are significant: (1) the magnitude of the yield-point stress, which determines the force required to initiate permanent deformation, and (2) the extent of the strain region from X_1 to X_4, which indicates the amount of plastic deformation (or ductility) that can be achieved without fracture. If a considerable amount of deformation is desired, a material like the low-carbon steel is more desirable than the high-carbon variety. Greater ductility would be available and less force would be required to initiate and continue the deformation. The curve on the right, however, has a higher strain-hardening coefficient (see Chapter 2 for discussion). If strain hardening is being used to impart strength, this material would have a greater increase in strength for the same amount of cold work. In addition, the material on the right would be more attractive for shearing operations and would be easier to machine (see Chapter 21).

Springback is another cold-working phenomenon that can be explained with the aid of a stress-strain diagram. When a metal is deformed by the application of a load, part of the resulting deformation is elastic. For example, if a metal is stretched to point X_1 in Figure 17-6 and the load is removed, it will return to its original size and shape because all of the deformation is elastic. If, on the other hand, the metal is stretched by an amount X_3, corresponding to point b on the stress-strain curve, the total strain is made up of two parts, a portion that is elastic and another that is plastic. When the deforming load is removed, the stress relaxation will follow line bX_2, and the final strain will only be X_2. The decrease in strain, X_3–X_2, is known as *elastic springback*.

In cold-working processes, springback can be extremely important. If a desired size is to be achieved, the deformation must be extended beyond that point by an amount equal to the springback. Since different materials have different elastic moduli, the amount of springback from a given load will change from one material to another. A substitution in material may well require adjustments in the forming process. Fortunately, springback is a predictable phenomenon, and most difficulties can be prevented by proper design procedures.

Preparing Metals for Cold Working.   The success or failure of a cold-working operation often depends on the quality of the starting material. To obtain a good surface finish and maintain dimensional precision, the starting material must be clean and free of oxide or scale that might cause abrasion and damage to the dies or rolls. Scale can be removed by *pickling*, in which the metal is dipped in acid and then washed. In addition, sheet metal and plate is sometimes given a light cold rolling prior to the major deformation. The rolling operation not only assures uniform starting thickness but also produces a smooth starting surface.

The light cold-rolling pass can also serve to remove the yield-point phenomenon and the associated problems of nonuniform deformation and surface irregularities in the product. Figure 17-7 presents a blow-up of the left-hand region of Figure 2-5, a stress-strain curve that is typical of many low-carbon steels. After loading to the upper yield point, the material exhibits a *yield-point runout* wherein the material can strain up to several percent with no additional force being required. Consider a piece of sheet metal that is to be formed into an automotive body panel. If a segment of that panel were to receive a total stretch less than the yield-point value, a stress equal to the yield-point stress would have to be applied. Under this stress, the material is free to not deform at all, to deform the entire amount of the yield-point runout, or to select some point in between. It is not uncommon for some regions to deform the entire amount and thin correspondingly while adjacent regions resist deformation and retain the original thickness. The resulting ridges and valleys, shown in Figure 17-7, are referred to as *Luders bands* or *stretcher strains* and are very difficult to remove or conceal. By first cold rolling the material to a strain near or past the yield-point runout, subsequent forming occurs in a region of the curve where a well-defined strain corresponds to each value of stress. When the body panel is shaped, the material deforms and thins uniformly.

Annealing is another treatment that may be given to a metal prior to cold working. If the amount of deformation is large, it may be desirable to have as much starting ductility as possible. In other cases, annealing may be performed after the workpiece has been partially shaped. These anneals, called *intermediate anneals*, serve to restore sufficient ductility to enable further processing without danger of fracture. If the last anneal is properly positioned in the deformation cycle, the desired shape can be produced with the mechanical properties that accompany the amount of cold work imparted since the anneal. For example, if the last anneal were moved earlier in a deformation sequence, a product that used to contain 10% cold work may now contain 40% cold work and be stronger but less ductile. In all annealing operations, however, care should be exercised to control the grain size of the resulting material. Grain sizes that are too large or too small can both be detrimental.

Warm Forming

Deformation produced at temperatures intermediate to hot and cold forming is known as *warm forming*. Compared to cold forming, warm forming offers the advantages of reduced loads on the tooling and equipment, increased material ductility, and a possible reduction in the number of anneals due to a reduction in the amount of strain hardening. The use of higher forming temperatures can often expand the range of materials and geometries that can be formed by a given process or piece of equipment. High-carbon steels can often be formed without a prior spheroidization treatment. Compared to hot forming, the lower temperatures of warm working produce less scaling and decarburization, and enable production of products with better dimensional precision and smoother surfaces. Finish machining is reduced and less material is converted into scrap. Because of the finer structures and the presence of some strain hardening, the as-formed properties may be adequate for many applications, enabling the elimination of final heat treatment operations. The warm regime generally requires less

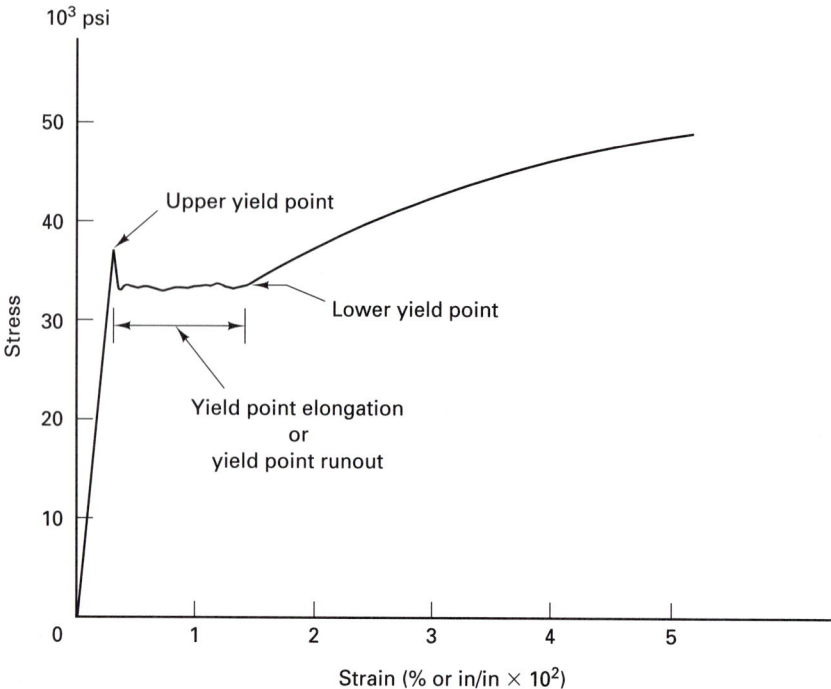

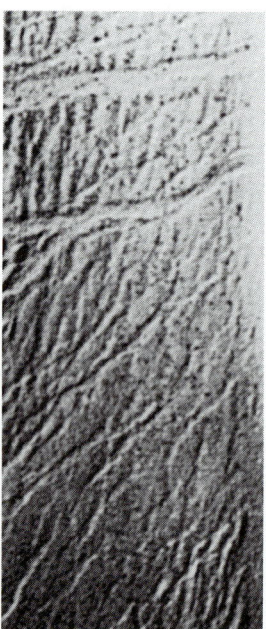

FIGURE 17-7 (*Left*) Stress-strain curve for a low-carbon steel showing the commonly observed yield-point runout; (*Right*) Luders bands or stretcher strains that form when this material is stretched to an amount less than the yield-point runout.

energy than hot working due to the decreased energy in heating the workpiece (lower temperature), energy saved through higher precision (less material being heated), and the possible elimination of postforming heat treatments. Tools last longer, for while they must exert 25 to 60% higher forces, there is less thermal shock and thermal fatigue.

Prior to the high cost of energy, metal forming was usually conducted in either the hot- or cold-working regimes, and warm working was largely ignored. As a result, material behavior is less well characterized at these less-used temperatures [the warm-working temperatures for steel are between 1000 and 1500°F (550 to 800°C)]. Lubricants have not been fully developed for the newer conditions of temperature and pressure, and die design technology is not well established. Nevertheless, the pressures of energy conservation and material conservation, coupled with the other cited benefits, strongly favor the continued development of warm working. Cold forming is still the preferred method for fabricating small components, but warm forming is considered to be attractive for larger parts (up to about 10 pounds) and steels with more than 0.35% carbon and/or high alloy content.

Isothermal Forming

Figure 17-8 shows the relationship between yield strength (or forging pressure) and temperature for several engineering metals. The 1020 and 4340 steels show a moderate increase in strength with decreasing temperature. In contrast, the strength of the titanium alloy and the A-286 nickel-based superalloy shows a much stronger dependence on temperature. Within the realm of typical hot-working temperatures, cooling of as little as

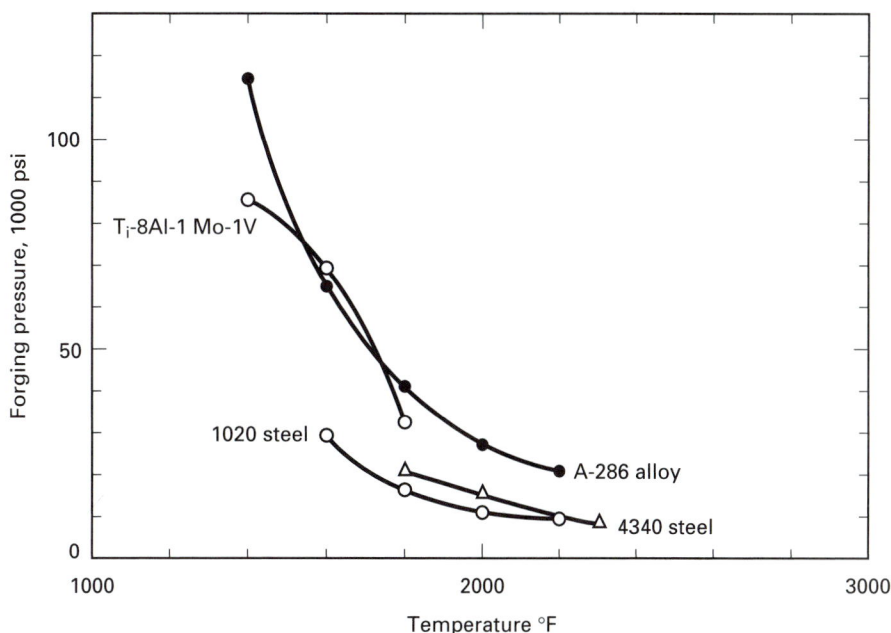

FIGURE 17-8   Variation of material strength (as indicated by pressure required to forge a standard specimen) with temperature. Materials with steep curves may have to be isothermally formed. *(From* A Study of Forging Variables, *ML-TDR-64-95, March 1964; courtesy of Battelle Columbus Laboratories.)*

200°F (100°C) can produce a doubling in strength. During hot forming, cooler surfaces surround a hotter interior, and the variations in strength can result in nonuniform deformation and cracking of the surface.

To adequately deform temperature-sensitive materials, deformation may have to be performed under isothermal conditions. The dies or tooling must be heated to the workpiece temperature, sacrificing die life for product quality. Deformation speeds must be slowed so that any heat generated by deformation can be removed in a manner that would maintain a uniform and constant temperature. Inert atmospheres may be required because of the long times at elevated temperature. Although such methods are indeed costly, they are often the only means of producing satisfactory products from certain materials. Because of their forming conditions, isothermally formed components generally exhibit close tolerances, low residual stresses, and fairly uniform metal flow.

■ KEY WORDS

coefficient of friction	isothermal forming	speed sensitivity
cold working	lubrication	strain hardening
constitutive relation	Luders bands	stretcher strains
dependent variable	oriented structure	stringers
elastic springback	pickling	surface finish
exit temperature	plasticity	tooling
fiber structure	recrystallization	tribology
hot working	required force	warm working
independent variable	required power	workpiece
intermediate anneal		

■ REVIEW QUESTIONS

1. What are some of the general assets of the metal deformation processes? Some general liabilities?
2. Why might large production quantities be necessary to justify metal deformation as a means of manufacture?
3. What is an independent variable in a metal-forming process?
4. What is the significance of tool and die geometry in designing a successful metal-forming process?
5. Why is lubrication often a major concern in metal forming?
6. What are some of the secondary effects that may occur when the speed of a metal-forming process is varied?
7. What is a dependent variable in a metal-forming process?
8. Why is it important to be able to predict the forces or powers required to perform specific forming processes?
9. Why is it important to know and control the thermal history of a metal as it undergoes deformation?
10. Why is it often difficult to determine the specific relationships between independent and dependent variables?
11. What are the three distinct ways of determining the interrelation of independent and dependent variables?
12. What features limit the value of laboratory experiments in modeling metal-forming processes?
13. What feature may serve to limit the accuracy of a mathematical model?
14. What is a constitutive relation for an engineering material?
15. What simplifying assumptions are often made regarding friction between the tool and workpiece?
16. What type of information about the material being deformed may be particularly significant to a metal-forming engineer?
17. Why are friction conditions difficult to model in laboratory experiments?
18. What are several ways in which the friction conditions during metalworking differ from the friction conditions found in most mechanical equipment?
19. According to modern friction theory, frictional resistance is equal to the product of what two physical variables?
20. Discuss the significance of wear in metal forming: wear on the workpiece and wear on the tooling.
21. Lubricants are often selected for properties in addition to their ability to reduce friction. What are some of these additional properties?
22. What are some of the benefits that can be obtained by fully separating a tool and workpiece by an intervening layer of lubricant?
23. If the temperature of a material is increased, what changes in properties might occur that would promote the ease of deformation?
24. Define the various regimes of cold working, warm working, and hot working in terms of the melting point of the material being formed.
25. What is an acceptable definition of *hot working*?
26. What are some of the attractive manufacturing and metallurgical features of hot-working processes?
27. What are some of the negative aspects of hot working?
28. Describe how hot working can be used to improve the grain structure of a metal.
29. If the deformed grains recrystallize during hot working, how can the process impart an oriented or fiber structure (and directionally dependent properties) to the product?
30. Why are heated dies or tools often employed in hot-working processes?
31. What generally restricts the upper temperature to which dies or tooling is heated?
32. Compared to hot working, what are some of the advantages of cold-working processes?
33. What are some of the disadvantages of cold-forming processes?
34. How can the tensile test properties of a metal be used to assess its suitability for cold forming?
35. Why is elastic springback an important consideration in cold-forming processes?
36. What are Luders bands or stretcher strains, and what causes them to form?
37. What are some of the advantages of warm forming compared to cold forming? Compared to hot forming?
38. For what types of materials might isothermal forming be required?

■ PROBLEMS

1. **(a)** List and discuss the various economic factors that should be considered when evaluating a possible switch from cold forming to warm forming.

   (b) Repeat part (a) for a possible conversion from hot forming to warm forming.

2. An advertisement for automobile spark plugs cited the superiority of rolled threads over machined threads. Figure 17-5 shows such a comparison for hot forming. The spark plug threads, however, were cold rolled. Discuss the assets and liabilities of the cold rolling of threads compared to thread formation by conventional machining.

*C*hapter 17 **CASE STUDY**

repairs to a damaged propeller

The propeller of a moderately large pleasure boat has been cast from a nickel-aluminum-bronze alloy that contains 82% Cu, 9% Al, 4% Ni, 4% Fe, and 1% Mn. It is approximately 13 inches in diameter with three 10-pitch blades, and has been designed for both fresh- and saltwater usage.

1. One of the blades has struck a rock and is badly bent. A replacement propeller is quite expensive and cannot be obtained for several weeks. An attractive alternative, therefore, may be to repair the existing piece. What would you recommend? Can it simply be hammered back into shape? Would you recommend any additional processing, either before or after the repair? How would you proceed? What is the rationale for your recommendations?

2. A second propeller, identical to the one above, has also been damaged by an impact. This time the damage is in the form of a crack at the base of one of the blades, as shown in Figure CS-17. Since the crack does not penetrate into the hub, it is proposed that a repair be made using some form of welding or brazing process. Would you recommend such a repair? If so, how would you suggest that the repair be made? Explain the rationale for your recommendations, and outline the procedure that should be followed to perform the repair. Would there be any sacrifice in quality or performance with the repaired propeller?

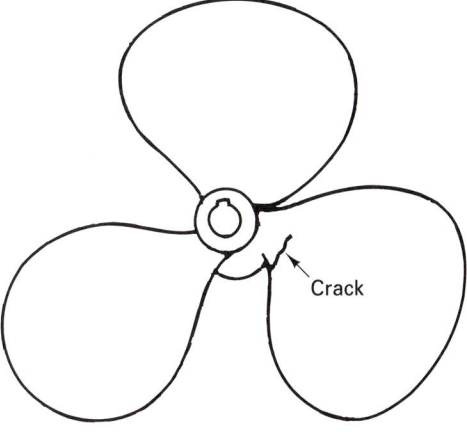

FIGURE CS-17

CHAPTER 18

HOT-WORKING PROCESSES

18.1   INTRODUCTION
18.2   CLASSIFICATION OF DEFORMATION PROCESSES
18.3   HOT-WORKING PROCESSES
18.4   ROLLING
     Basic Rolling Process
     Rolling Temperatures
     Rolling Mill Configurations
     Continuous Rolling Mills
     Ring Rolling
     Characteristics, Quality, and Precision of Hot-Rolled Products
     Flatness Control and Rolling Defects
     Thermomechanical Processing and Controlled Rolling
18.5   FORGING
     Open-Die Drop-Hammer Forging

     Impression-Die Drop-Hammer Forging
     Design of Impression-Die Forgings
     Press Forging
     Upset Forging
     Automatic Hot Forging
     Roll Forging
     Swaging
     Net-Shape and Near-Net-Shape Forging
18.6   EXTRUSION
     Extrusion Methods
     Extrusion of Hollow Shapes
     Metal Flow in Extrusion
18.7   HOT DRAWING OF SHEET AND PLATE
18.8   PIPE WELDING
     Butt-Welded Pipe
     Lap-Welded Pipe
18.9   PIERCING
Case Study: OUTBOARD MOTOR BRACKETS

◼ 18.1 INTRODUCTION

The shaping of metal by deformation is as old as recorded history. The Bible, in the fourth chapter of Genesis, introduces Tubal-cain and cites his ability as a worker of metal. While we do not know of his equipment, it is well established that metal forging was practiced before written records. Processes such as rolling and wire drawing were common in the Middle Ages and probably date back much further. In North America, by the 1680 the Saugus Iron Works near Boston had an operating drop forge, rolling mill, and slitting mill.

Although the basic concepts of many forming processes have remained largely unchanged throughout history, the details and equipment have evolved considerably. Manual processes were converted to machine processes during the industrial revolution. The machinery then became bigger, faster, and more powerful. Waterwheel power was replaced by steam and then electricity. More recently, computer-controlled, automated operations have emerged.

◼ 18.2 CLASSIFICATION OF DEFORMATION PROCESSES

A wide variety of processes have been developed to mechanically shape material, and a number of methods have been proposed to provide classification. One approach divides the processes into primary and secondary. Primary processes reduce a cast material into inter-

464

mediate shapes, such as slabs, plates, or billets. Secondary processes further convert these shapes into finished or semifinished products. Unfortunately, some processes clearly fit both categories, depending on the particular product being made.

A more useful division focuses on the size and shape of the workpiece and how that size and shape is changed. *Bulk deformation processes* are those where the surface area of the workpiece changes significantly. Thicknesses or cross sections are reduced or shapes are changed. Since the volume of the material remains constant, other dimensions must change in proportion. Thus the enveloping surface area is altered, usually increasing as the product lengthens or the shape becomes more complex. In contrast, *sheet-forming* operations involve the deformation of a material where the thickness and surface area remain relatively constant. Even here, the division is not without confusion. Coining, for example, begins with sheet material but alters the thickness in a complex manner that is essentially bulk deformation.

In Chapters 18 and 19 we present a survey of metal deformation processes, where the division of processes is based on the workpiece temperature. Processes that are normally performed "hot" are presented in Chapter 18, and processes normally performed "cold" are deferred to Chapter 19. Even here, the division is somewhat blurred, especially in view of the increased emphasis on energy conservation, the growth of "warm working," and new advances in technology. Hot-working processes are often performed cold, and cold-forming processes can often be aided by some degree of heating. Since sheet material has such a large surface-to-volume ratio, it tends to lose heat rapidly. Therefore, most sheet-forming operations are performed cold.

■ 18.3 Hot-Working Processes

An obvious reason for the popularity of the hot-working processes is that they often provide an attractive means of producing a desired shape. At elevated temperatures, metals weaken and become more ductile. With continual recrystallization, massive deformation can take place without exhausting material plasticity. In steels, hot forming involves the deformation of the weaker, austenite structure, which then cools to the stronger, room-temperature, ferrite, or much stronger nonequilibrium structures.

Some of the hot-working processes that are of major importance in modern manufacturing are:

1. Rolling
2. Forging
3. Extrusion
4. Hot drawing
5. Pipe welding
6. Piercing

■ 18.4 Rolling

As shown in Figure 18-1, *rolling* is usually the first process that is used to convert material into a finished wrought product. Thick starting stock can be rolled into blooms, billets, or slabs, or these shapes can be obtained directly from continuous casting. A *bloom* has a square or rectangular cross section, with a thickness greater than 6 inches and a width no greater than twice the thickness. A *billet* is usually smaller than a bloom and has a square or circular cross section. Billets are usually produced by some form of deformation process, such as rolling or extrusion. A *slab* is a rectangular solid where the width is greater than twice the thickness. Slabs can be further rolled to produce *plate*, *sheet*, and *strip*.

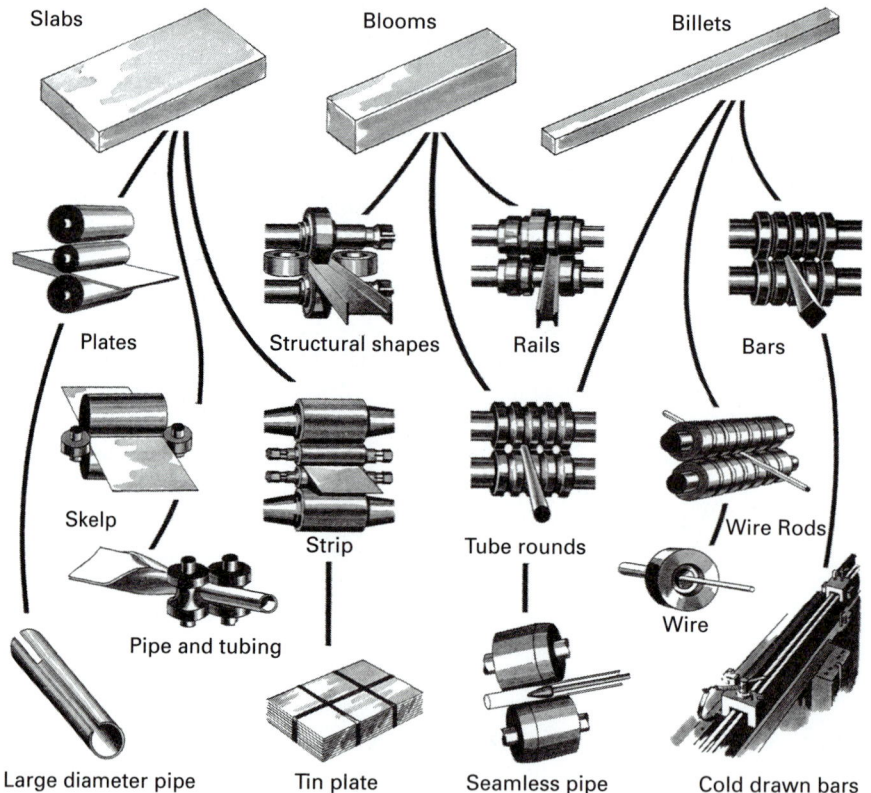

FIGURE 18-1 Schematic flowchart for the production of various finished and semifinished steel shapes. Note the abundance of rolling operations. *(Courtesy of American Iron and Steel Institute, Washington, D.C.)*

These hot-worked products often form the starting material for subsequent processing using techniques such as cold forming or machining. Sheet and strip can be fabricated into products or further cold rolled into thinner, stronger material, or even into foil. Blooms and billets can be further rolled into finished products, such as *structural shapes* or railroad rail, or they can be processed into semifinished shapes, such as *bar*, *rod*, *tube*, or *pipe*.

From a tonnage viewpoint, hot rolling is clearly predominant among all manufacturing processes, and hot-rolling equipment and practices are sufficiently advanced that standardized, uniform-quality products can be produced at relatively low cost. Because shaped rolls are both massive and costly, hot-rolled products can normally be obtained only in standard shapes and sizes for which there is sufficient demand to permit economical production.

Basic Rolling Process

As shown in Figure 18-2, heated metal is passed between two rolls that rotate in opposite directions, the gap between the rolls being somewhat less than the thickness of the entering metal. Because the rolls rotate with a surface velocity that exceeds the speed of the incoming metal, friction along the contact interface acts to propel the metal forward. The metal is squeezed and elongates to compensate for the decrease in cross-sectional area. The amount of deformation that can be achieved in a single pass between a given pair of rolls

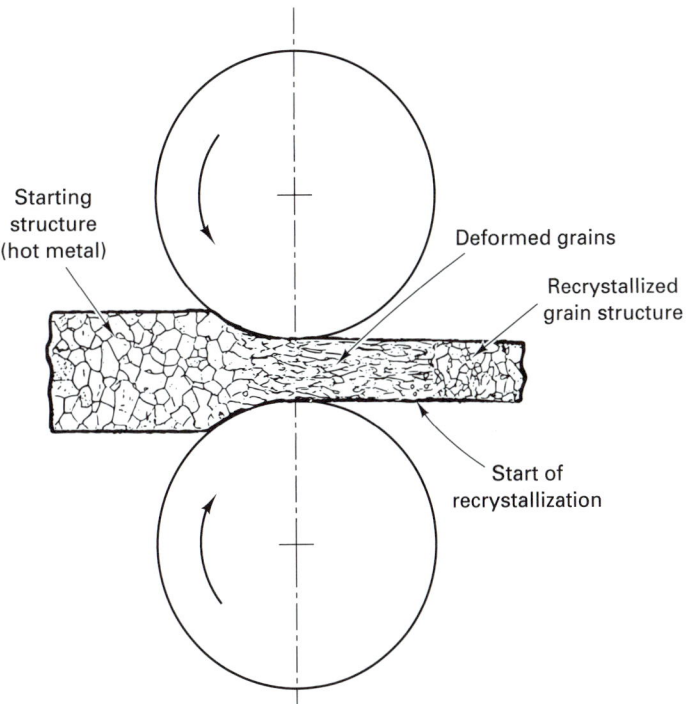

FIGURE 18-2 Schematic representation of the hot-rolling process, showing the deformation and recrystallization of the metal being rolled.

depends on the friction conditions along the interface. If too much is demanded, the rolls cannot advance the material and simply skid over its surface. Too little deformation per pass, however, results in excessive production cost.

Rolling Temperatures

In hot rolling, as with all hot-working processes, temperature control is crucial to the success of the process. Ideally, the starting material will be heated to a uniform elevated temperature. If the temperature is not uniform, the subsequent deformation will not be uniform. For example, if the material was being reheated prior to rolling, and insufficient time was provided, the hotter exterior will flow in preference to the cooler, stronger interior. If the part is removed from a furnace and cools prior to working, or has cooled during previous operations, the cooler surfaces will tend to resist deformation. Cracking and tearing of the surface may result as the hotter, weaker interior tries to deform.

It is not uncommon for high-volume producers to begin with continuous-cast feedstock. The cooling from solidification is controlled so as to enable direct insertion into a hot-rolling operation without additional handling or reheating. For smaller operations or secondary processing, the starting material is often a room-temperature solid, such as an ingot, slab, or bloom. These must first be brought to the desired rolling temperature, usually in gas- or oil-fired soaking pits or furnaces. For plain-carbon and low-alloy steels, the soaking temperature is usually about 2200°F (1200°C). For smaller cross sections, induction coils may be used to heat the material for rolling.

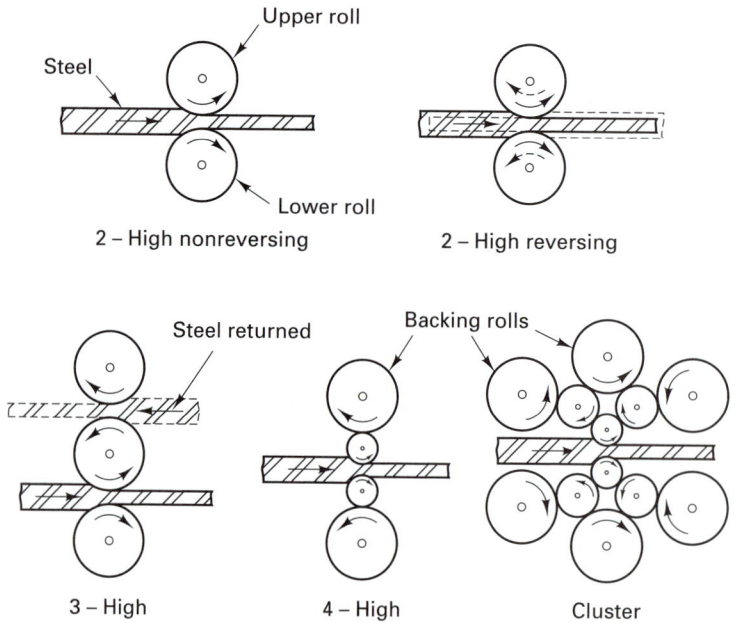

FIGURE 18-3 Various roll configurations used in rolling operations.

Hot rolling is usually terminated when the temperature falls to about 100 to 200°F (50 to 100°C) above the recrystallization temperature of the material. Such a *finishing temperature* assures the production of a uniform fine grain size and prevents the possibility of unwanted strain hardening. Before additional deformation can be performed, a period of reheating is required to reestablish desirable hot-working conditions.

Rolling Mill Configurations

As illustrated in Figure 18-3, rolling mill stands are available in a variety of roll configurations. Early reductions, often called primary roughing or breakdown passes, usually employ a two- or three-high configuration with 24- to 55-in. (600- to 1400-mm) diameter rolls. The two-high nonreversing mill is the simplest design, but the material can only pass through the mill in one direction. The two-high reversing mill permits back-and-forth rolling, but the rolls must be stopped, reversed, and brought back to rolling speed between each pass. The three-high mill eliminates the need for roll reversal but requires some form of elevator on each side of the mill to raise or lower the material and mechanical manipulators to turn or shift the product between passes.

As shown in Figure 18-4, smaller-diameter rolls produce less length of contact for a given reduction and therefore require lower force and less energy to produce a given change in shape. The smaller cross section, however, provides reduced stiffness, and the rolls are prone to flex elastically since they are supported on the ends and pressed apart by the metal passing through the middle. Four-high and cluster arrangements use backup rolls to support the smaller work rolls. These configurations are used in the hot rolling of wide plate and sheets, and in cold rolling, where even small deflections in the roll would result in an unacceptable variation in product thickness. Foil is almost always rolled on *cluster mills* since the small thickness requires small-diameter rolls. In a cluster mill, the roll in contact with the work can be as small as $\frac{1}{4}$ in. in diameter. To counter the need for even

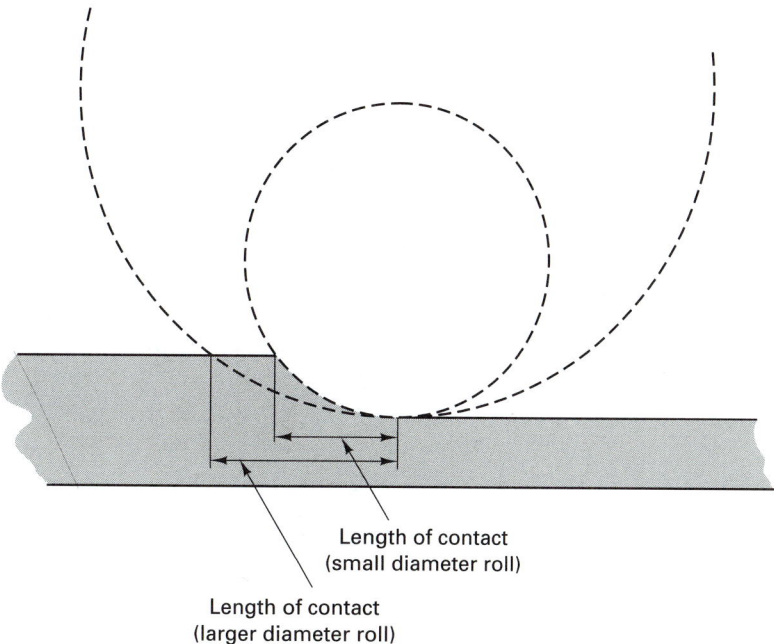

Length of contact
(small diameter roll)

Length of contact
(larger diameter roll)

FIGURE 18-4 Schematic showing the effect of roll diameter on length of contact for a given reduction.

smaller rolls, some foils are produced by *pack rolling*, a process where two or more layers of metal are rolled simultaneously as a means of providing a thicker input material. Household aluminum foil is usually pack rolled, as evidenced by the one shiny side (in contact with the roll) and one dull side (in contact with the other piece of foil).

In the rolling of nonflat or shaped products, such as structural shapes and railroad rail, the sets of rolls contain contoured grooves that sequentially form the desired shape and cross section and control the metal flow. Figure 18-5 shows some typical roll-pass sequences used in the production of structural shapes.

Continuous Rolling Mills

When the volume of a product justifies the investment, it may be rolled on a continuous rolling mill. Billets, blooms, or slabs are heated and fed through an integrated series of nonreversing stands. Continuous mills for the hot rolling of steel strip, for example, often consist of a roughing train of approximately four four-high mill stands and a finishing train of six or seven additional four-high stands. In a continuous structural mill, the rolls in each stand contain only one set of shaped grooves, in contrast to the multigrooved rolls used when the product is produced by back-and-forth passes through a single stand.

In a continuous rolling mill, the same amount of material must pass through each stand in a given period of time. If the cross section is reduced, the speed must be increased proportionately. Thus the rolls of each successive stand must turn faster than those of the preceding one by an amount equal to the change in cross-sectional area. If this synchronization is not maintained, material may accumulate between stands, or the demand for incoming material may place the material under excessive tension, and cause a tearing or rupture.

The synchronization of six or seven mill stands is not an easy task, especially when key variables such as temperature and lubrication may change during a single run and the

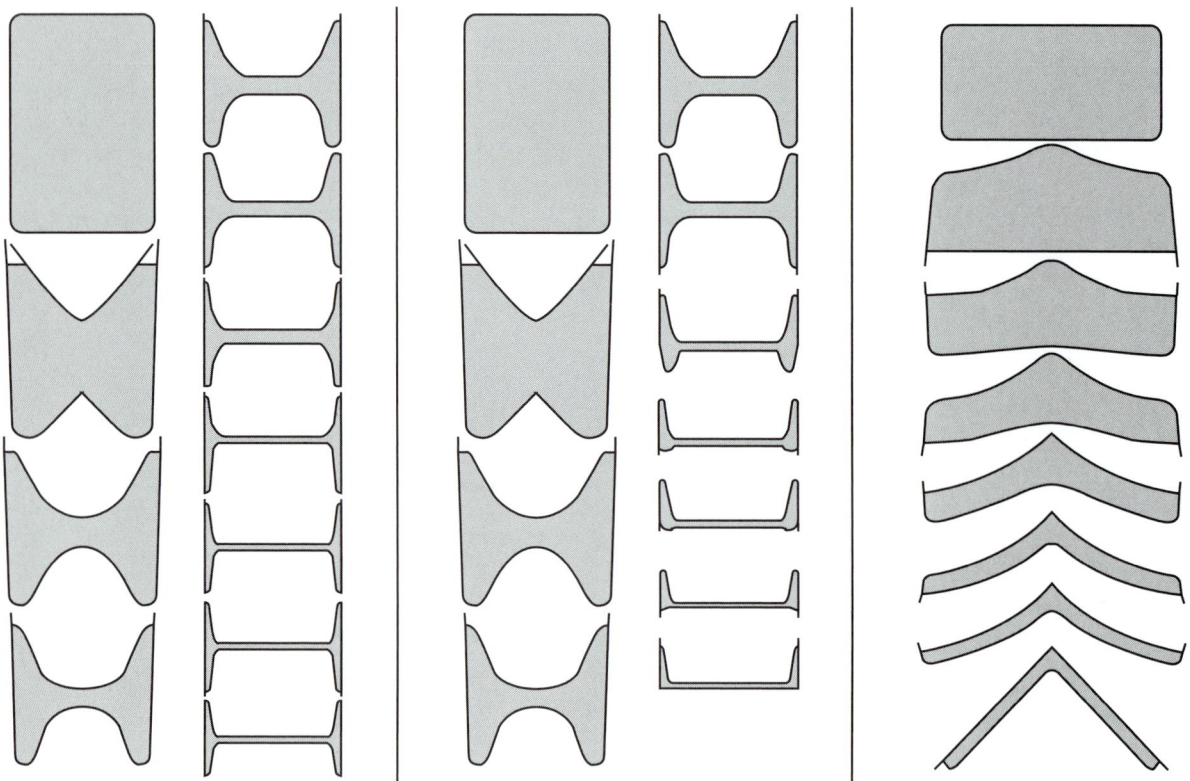

FIGURE 18-5 Typical roll-pass sequences used in producing various structural shapes.

product may be exiting the final stand at speeds in excess of 70 miles per hour (110 kilometers per hour). Computer control is basic to successful rolling, and modern mills are equipped with numerous sensors to provide the needed information. When continuous casting units feed directly into continuous rolling mills, the time lapse from final solidification to finished rolled product is often a matter of a few minutes or less.

Ring Rolling

In the ring rolling process, one roll is placed through the hole of a thick-walled ring, and a second roll presses in from the outside (Figure 18-6). As the rolls squeeze and rotate, the wall thickness is reduced and the diameter of the ring increases. Shaped rolls can be used to produce a wide variety of cross-section profiles. The resulting seamless rings find application in products such as rockets, turbines, airplanes, pipelines, and pressure vessels.

Characteristics, Quality, and Precision of Hot-Rolled Products

Because they are rolled and finished above the recrystallization temperature, hot-rolled products have little directionality in their properties and are relatively free of deformation-induced residual stresses. These characteristics may vary, however, depending on the thickness of the product and the presence of complex sections. Nonmetallic inclusions do not recrystallize, so they may impart some degree of directionality. Substantial residual stresses can be induced during nonuniform cooling from the temperatures of hot working.

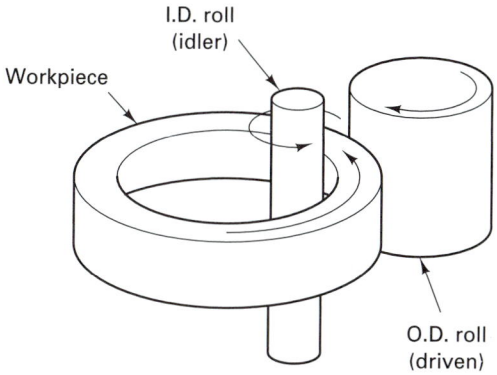

FIGURE 18-6 Schematic of a horizontal ring rolling operation. As the thickness of the ring is reduced, its diameter will increase.

Thin sheets often show some definite directional characteristics, whereas thicker plate (such as that above 0.8 in. or 20 mm) will usually have very little. Because of the high residual stresses in the rapidly cooled edges, a complex shape, such as an I- or H-beam, may warp noticeably if a portion of one flange is cut away.

As a result of the hot deformation and the good control that is maintained during processing, hot-rolled products are normally of uniform and dependable quality, and considerable reliance can be placed on them. It is quite unusual to find any voids, seams, or laminations when these products are produced by reliable manufacturers. Of course, the surfaces of hot-rolled products are usually a bit rough and are originally covered with a tenacious high-temperature oxide, known as mill scale. This is usually removed by an acid pickling operation, resulting in a surprisingly smooth surface finish.

The dimensional tolerances of hot-rolled products vary with the kind of metal and the size of the product. For most products produced in reasonably large tonnages, the tolerances are within 2 to 5% of the specified dimension (either height or width).

Flatness Control and Rolling Defects

The rolling of flat material with uniform thickness requires that the gap between the rolls be a uniform one. Attaining such an objective, however, may be difficult. Consider the upper roll in a set that is rolling sheet or plate. The material presses upward in the middle of the roll, and the roll is held in place by bearings that are mounted to either end and supported in the mill frame. Thus the roll is loaded in three-point bending and tends to flex in a manner that produces a thicker center and thinner edge. If the roll is always used to roll the same material at the same temperature, the forces and deflections can be predicted, and the roll can be designed to have a specified amount of crowning (i.e., a barrel-shape deviation from a cylindrical profile). When the roll is subjected to the specified load, it will "deflect into flatness."

If the applied load is not of the designed magnitude, the profile will not be flat and defects may result. For example, if the correction is insufficient, the edges will still be thinner than the center. The thinner material will try to become longer but must remain attached to the thicker, and therefore shorter, center. The result may be wavy edges or fractures in the center. If the correction is excessive, the center becomes thinner and longer, and the result can be a wavy center or cracking of the edges.

Thermomechanical Processing and Controlled Rolling

As with most deformation processes, rolling is generally viewed as being a means of changing the shape of a material. While heat may be used to reduce forces and promote plasticity, thermal processes to produce or control mechanical properties (heat treatments) are usually performed as subsequent operations. *Thermomechanical processing*, of which *controlled rolling* is an example, consists of simultaneously performing both deformation and controlled thermal processing so as to produce the desired levels of strength and toughness directly in the as-worked product. The heat for the property modification is the same heat used in the rolling operation, and subsequent heat treatment is thereby eliminated.

A successful operation begins with process design. The starting material must be specified and the composition closely maintained. Then a time–temperature–deformation system must be developed to achieve the desired objective. Possible goals include production of a uniform fine grain size; controlling the nature, size, and distribution of the various transformation products (such as ferrite, pearlite, bainite, and martensite in steels); controlling the reactions that produce solid solution strengthening or precipitation hardening; and producing a desired level of toughness. Starting structure (controlled by composition and prior thermal treatments), deformation details, temperature during the various stages of deformation, and the cooldown from the working temperature must all be specified and controlled. Moreover, the attainment of uniform properties requires uniform temperatures and deformations throughout the product. Computer-controlled rolling facilities are almost a necessity if thermomechanical processing is to be successfully performed.

Possible benefits of thermomechanical processing include improved product properties; substantial energy savings (by eliminating subsequent heat treatment); and the possible substitution of a cheaper, less-alloyed metal for a highly alloyed one that responds to heat treatment. For these reasons it is expected that processes such as controlled rolling will assume increased significance in the years to come.

■ 18.5 FORGING

Forging is the term applied to a family of processes where the deformation is induced by localized compressive forces. The equipment can be manual or power hammers, presses, or special forging machines. While the deformation can be done in the hot, cold, warm, or isothermal mode, the term *forging* usually implies hot forging done above the recrystallization temperature.

Forging is the oldest known metalworking process. From the days when prehistoric peoples discovered that they could heat sponge iron and beat it into a useful implement by hammering with a stone, forging has been an effective method of producing many useful shapes. Modern forging has developed from the ancient art practiced by the armor makers and the immortalized village blacksmith. High-powered hammers and mechanical presses have replaced the strong arm, the hammer, and the anvil, and modern metallurgical knowledge supplements the art and skill of the craftsman in controlling the heating and handling of the metal.

A variety of forging processes have been developed that offer a wide range of capabilities. A single piece can be economically fashioned by some methods, while others can mass produce thousands of identical parts. The metal may be (1) *drawn out* to increase its length and decrease its cross section, (2) *upset* to decrease the length and increase the cross section, or (3) *squeezed in closed impression dies* to produce multidirectional flow. As indicated in Table 17-2, the state of stress in the work is primarily uniaxial or multiaxial compression.

Common forging processes include:

1. Open-die drop-hammer forging
2. Impression-die drop forging

3. Press forging
4. Upset forging
5. Automatic hot forging
6. Roll forging
7. Swaging

Open-Die Drop-Hammer Forging

In concept, *open-die hammer forging* is the same type of forging done by the blacksmith of old, but massive mechanical equipment is now used to impart the repeated blows. The metal is first heated to the proper temperature by gas, oil, or electric furnaces or by electrical induction heating. The impact is then delivered by some type of mechanical hammer, the simplest type being a gravity drop or *board hammer*. Here the hammer is attached to the lower end of a hardwood board, which is raised by being gripped between two rotating, roughened rollers, which then separate to release it for free-fall. Few of these are still in use. *Steam or air hammers*, which use pressure to both raise and propel the hammer, are far more common. These give higher striking velocities, more control of striking force, easier automation, and the ability to shape pieces up to several tons. *Programmable, computer-controlled hammers* can provide blows of differing impact speed (energy) for each of the various stages of an operation. Their use can greatly increase the efficiency of the process and also minimize the amount of noise and vibration, which are the most common outlets for the excess energy not absorbed in the deformation of the workpiece. Figure 18-7 shows a large double-frame hammer as well as a simple schematic of its basic tooling.

Open-die forging does not confine the flow of metal. The operator must obtain the desired shape by orienting and positioning the workpiece between blows. The hammer may contact the workpiece directly, or specially-shaped tools can be inserted to assist in making simple shapes (such as round, concave, or convex), forming holes, or performing a cutoff operation. Manipulators may be used to position the larger workpieces, which may weigh several tons. While some finished parts can be made by this technique, open-die forging is usually employed to preshape the metal in preparation for further operations. For example, consider such massive parts as turbine rotors and generator shafts, which may be as much as 70 ft in length and up to 3 ft in diameter. Open-die forging is used to both induce oriented plastic flow and minimize the amount of subsequent machining. Figure 18-8 schematically depicts the unrestricted flow of material and the formation of both a multi-diameter cylindrical shaft and a seamless ring by open-die forging.

Impression-Die Drop-Hammer Forging

Open-die hammer forging (or smith forging, as it has been called) is a simple and flexible process, but it is not practical for large-scale production. It is a slow operation and the size and shape of the resulting workpiece are dependent on the skill of the operator. *Impression-die* or *closed-die forging* overcomes these difficulties by using shaped dies to control the flow of metal (Figure 18-9). Figure 18-10 shows a typical set of dies, one half of which attaches to the hammer and the other half to the anvil. The heated metal is positioned in the lower cavity and struck one or more blows by the upper die. This hammering causes the metal to flow so as to completely fill the die cavity. Excess metal is squeezed out around the periphery of the cavity to form a *flash*. When the final forging is completed, the flash is trimmed off by a trimming die.

In *flashless forging*, also known as true closed-die forging, the metal is deformed in a cavity that provides total confinement. Accurate workpiece sizing is required since

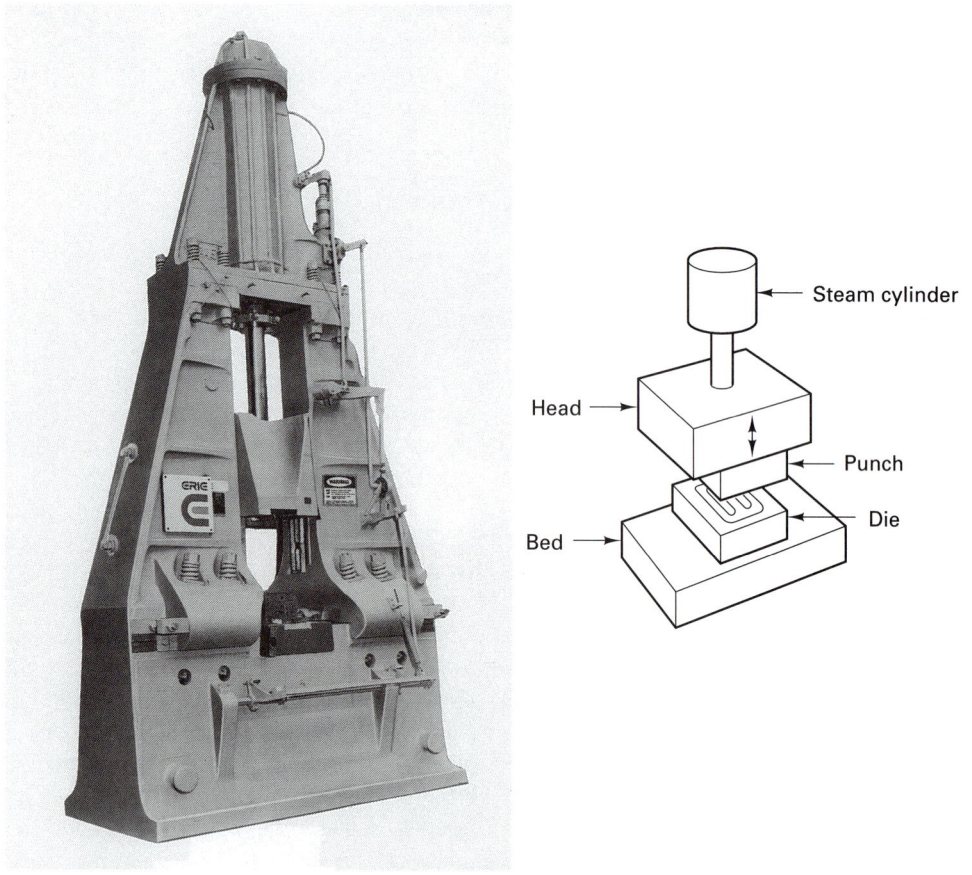

FIGURE 18-7 Double-frame drop hammer and a schematic of the basic tooling.
(Courtesy of Erie Press Systems, Erie, Pa.)

complete filling of the cavity must be assured with no excess material. Accurate workpiece positioning is also necessary, along with good die design and control of lubrication. The major advantage is the elimination of the scrap generated during flash formation, an amount that is typically between 20 and 45% of the starting material.

Most conventional forgings are impression-die with flash and are produced in dies with a series of cavities. The first impression is often an *edging*, *fullering*, or *bending* impression to distribute the metal roughly in accordance with the requirements of the later cavities. Intermediate impressions are for *blocking* the metal to approximately its final shape, with generous corner and fillet radii. For small production lots, the cost of further cavities may not be justified, and the blocker-type forgings are finished by machining. Usually, however, the final shape and size are set by additional forging in the *final or finisher impression*. These steps and the shape of a part at the conclusion of each step are shown in Figure 18-10. Because each product is shaped in the same die cavities, each is a close duplicate of all the others.

The shape of the various cavities forces the metal to flow in the desired direction. This flow, in turn, imparts the oriented structure discussed in Chapter 17. In addition, forging permits the cross section to be varied, so the metal can be distributed as needed to resist the applied loads. These factors, coupled with a fine recrystallized grain structure (hot working) and the absence of voids (compressive forming stresses), make it possible to

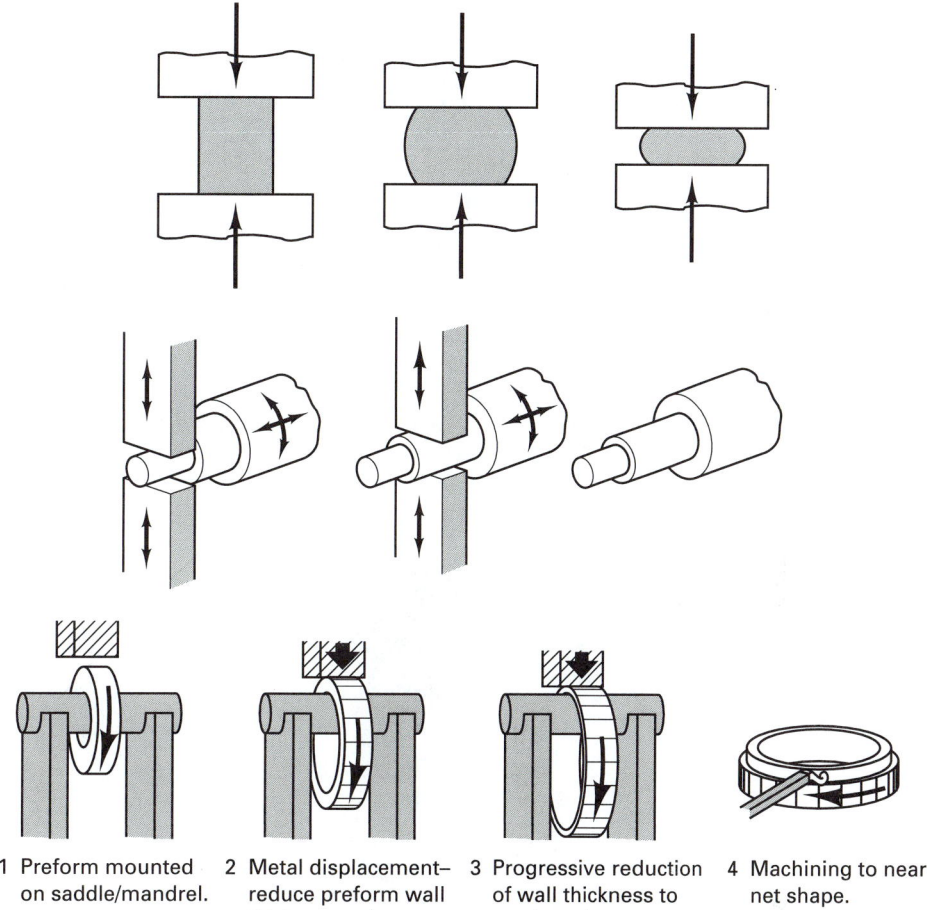

1 Preform mounted on saddle/mandrel.

2 Metal displacement–reduce preform wall thickness to increase diameter.

3 Progressive reduction of wall thickness to produce ring dimensions.

4 Machining to near net shape.

FIGURE 18-8 (*Top*) Schematic of the unrestrained flow of material in open-die forging. Note the barrel shape that forms due to friction between the die and material. (*Middle*) Open-die forging of a multidiameter shaft. (*Bottom*) Forging of a seamless ring by the open-die method. (*Courtesy of Forging Industry Association, Cleveland, Ohio.*)

obtain about 20% higher strength/weight ratios with forgings than with cast or machined parts of the same material.

Board hammers, steam hammers, and air hammers are all used in impression die forging. An alternative to the hammer and anvil arrangement is the *counterblow machine*, or *impactor* (Figure 18-11). These machines have two horizontal hammers that simultaneously impact a workpiece that is positioned between them. Excess energy simply becomes recoil, in contrast to the hammer and anvil arrangement, where energy is lost to the machine foundation and a heavy machine base is required. The impactors operate more quietly and with less vibration. Each of the processes described above can be automated further through the use of induction heating, mechanical feeding, positioning and manipulation, and the direct heat treatment of parts as they emerge from the machine.

Conventional closed-die forging begins with a simple hot-rolled shape and utilizes reheating and progressive working to convert it into a more complex geometry whose metallurgical structure and properties are superior to cast equivalents. Several alternative

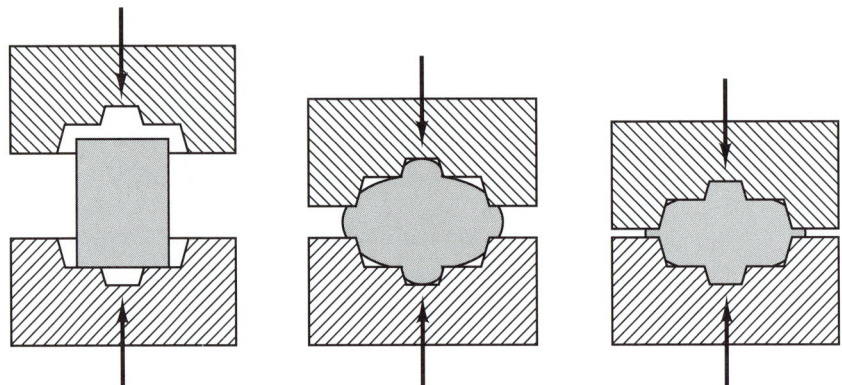

FIGURE 18-9   Schematic of the impression-die forging process showing partial die filling and the beginning of flash formation in the center sketch, and the final shape with flash in the right-hand sketch.

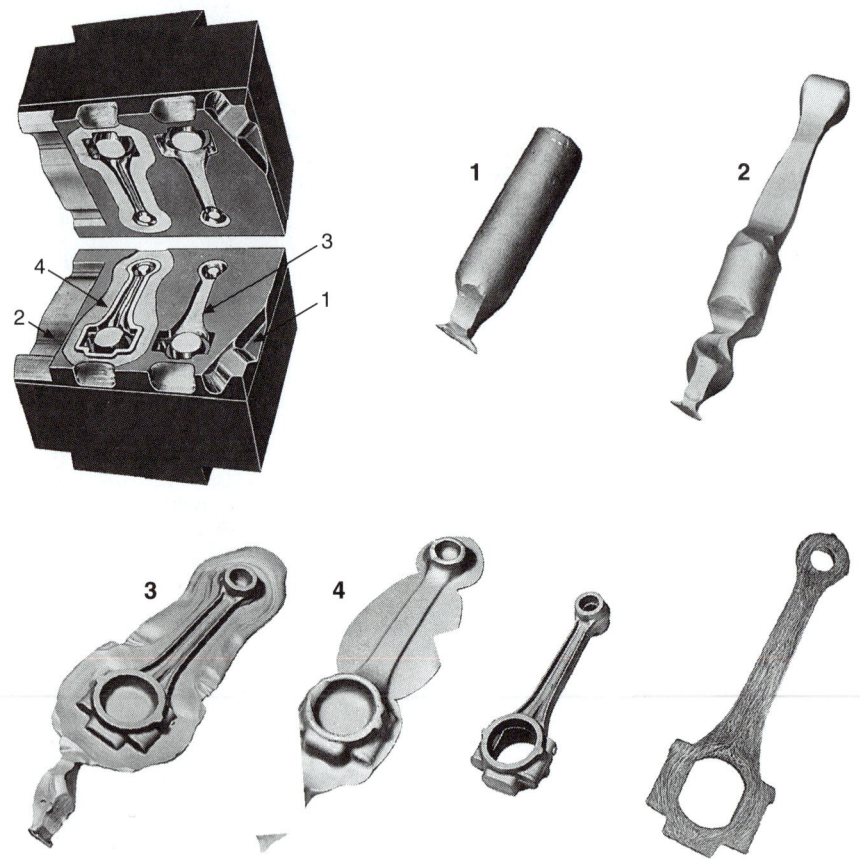

FIGURE 18-10   Impression drop-forging dies and the product resulting from each impression. The flash is trimmed from the finished connecting rod in a separate trimming die. The sectional view shows the grain fiber resulting from the forging process. *(Courtesy of Forging Industry Association, Cleveland, Ohio.)*

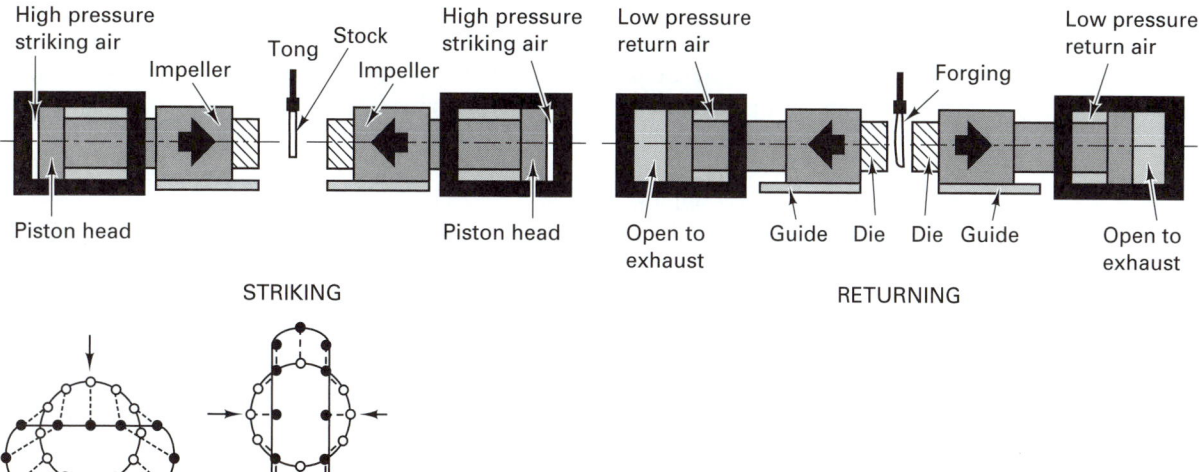

High pressure striking air
Tong
Stock
Impeller
Impeller
High pressure striking air
Piston head
Piston head

STRIKING

Low pressure return air
Forging
Low pressure return air
Open to exhaust
Guide Die Die Guide
Open to exhaust

RETURNING

(a)

Conventional forged disc with paths of flow

(b)

Disc formed by impacter with paths of flow

FIGURE 18-11 Schematic diagram of an impactor in the striking and returning modes, along with a comparison of the metal flow in conventional forging and impacting. *(Courtesy of Chambersburg Engineering Company.)*

processes have been developed to conserve energy while simultaneously producing a product that is somewhere between a conventional forging and a conventional casting. In the *Autoforge* process, used primarily for nonferrous metals, a cast preform is removed from the mold while hot and is finish forged in a single-cavity die. After forging, the flash is trimmed and the part is quenched to room temperature. Forging preforms can also be produced by the spraying of atomized droplets into shaped collectors, as with the *Osprey process* described in Chapter 16. The preforms are then removed from the mold, and the final shape and properties are imparted by forging. Yet another approach is squeeze casting (liquid-metal forging) or semi-solid forging, discussed in Chapter 15.

Design of Impression-Die Forgings

Forging dies are usually made of high-alloy or tool steel and can be quite costly to design and construct. Impact resistance, wear resistance, strength at elevated temperature, and the ability to withstand cycles of rapid heating and cooling must all be outstanding. In addition, considerable care is required to produce and maintain a smooth and accurate cavity and parting plane. Better and more economical results will be obtained if the following rules are observed:

1. The dies should part along a single, flat plane if at all possible. If not, the parting plane should follow the contour of the part.
2. The parting surface should be a plane through the center of the forging and not near an upper or lower edge.
3. Adequate draft should be provided—at least 3° for aluminum and 5 to 7° for steel.
4. Generous fillets and radii should be provided.
5. Ribs should be low and wide.
6. The various sections should be balanced to avoid extreme differences in metal flow.
7. Full advantage should be taken of fiber flow lines.
8. Dimensional tolerances should not be closer than necessary.

The various design details, such as the number of intermediate steps, the shape of each, the amount of excess metal required to assure die filling, and the dimensions of the flash at each step, are often a matter of experience. Each component is a new design entity and brings its own unique challenges. Computer-aided design has made notable advances, and the development and accessibility of high-speed, expanded-memory computers has enabled accurate modeling of complex shapes. Nevertheless, good forging design is still a combination of science and art.

Good dimensional accuracy is one motivation for using impression-die forging. It must be remembered, however, that the dimensions across the parting plane are set by closure of the dies and are dependent on die wear and the thickness of the final flash. With reasonable care, these dimensions (for steel products) can be maintained within the tolerances shown in Table 18-1. Dimensions contained entirely within a single segment of the die can be maintained at a greater level of accuracy. Draft angles can sometimes be reduced, occasionally approaching zero, but this is not recommended for general practice.

TABLE 18-1.   Thickness Tolerances for Steel Drop Forgings

Mass of Forging		Minus		Plus	
lb	kg	in.	mm	in.	mm
1	0.45	0.006	0.15	0.018	0.48
2	0.91	0.008	0.20	0.024	0.61
5	2.27	0.010	0.25	0.030	0.76
10	4.54	0.011	0.28	0.033	0.84
20	9.07	0.013	0.33	0.039	0.99
50	22.68	0.019	0.48	0.057	1.45
100	45.36	0.029	0.74	0.087	2.21

Selection of a lubricant is also critical to successful forging. The lubricant not only affects the friction and wear and associated metal flow, but it may also be expected to act as a thermal barrier (restricting heat flow from the workpiece to the dies) and a parting compound (preventing the part from sticking in the cavities).

Press Forging

In hammer or impact forging, the metal flows in response to the energy in the hammer–workpiece collision. Speeds are high, so the forming time is short and there is little cooling of the workpiece. However, if all of the energy can be dissipated by deformation in the surface of the metal, coupled with absorption by the anvil and foundation, the interior regions of the workpiece can remain undeformed. (For example, consider the deformation of a metal wedge after being struck repeatedly by a sledge hammer.) When larger pieces or thicker products must be formed, *press forging* is often required. Here deformation is analyzed in terms of forces or pressures (rather than energy), and the slow squeezing action penetrates completely through the metal, producing a more uniform deformation and flow. Problems can arise, however, because of the longer time of contact between the dies and the workpiece. As the surface of the workpiece cools, it becomes stronger and less ductile and may crack if deformation is continued. Heated dies are generally used to reduce heat loss, promote surface flow, and enable the production of finer details and closer tolerances. Periodic reheating of the workpiece may also be required.

Forging presses are of two basic types, mechanical and hydraulic, and are usually quite massive. *Mechanical presses* use means such as cams, cranks, or toggles to produce

FIGURE 18-12 A 35,000-ton (310-MN) forging press. In the foreground is a 262-lb, 121-in. aluminum part that was forged on this press. *(Courtesy of Wyman-Gordon Company.)*

a preset and reproducible stroke. Because of their mechanical drives, production presses are capable of up to 50 strokes per minute. Hydraulic presses, on the other hand, move in response to oil pressure in a piston, and are slower, more massive, and more costly to operate. On the positive side, they usually are much more flexible and have greater capacity. Since motion is in response to pressure and flow of the drive fluids, they can be programmed to have different strokes for different operations and even different speeds within a stroke. *Hydraulic presses* with capacities up to 50,000 tons (445 MN) are currently in operation in the United States. Figure 18-12 shows a large hydraulic press that is capable of forging large sections of aircraft structures as a single piece.

Press forgings usually require less draft than drop forgings and have higher dimensional accuracy. In addition, press forgings can often be completed in a single closing of the dies, and the process can be readily automated.

Upset Forging

Upset forging involves increasing the diameter of a material by compressing its length. In terms of the number of pieces produced, it is the most widely used of all forging processes, and its use has increased greatly in recent years. Parts can be upset forged both hot and

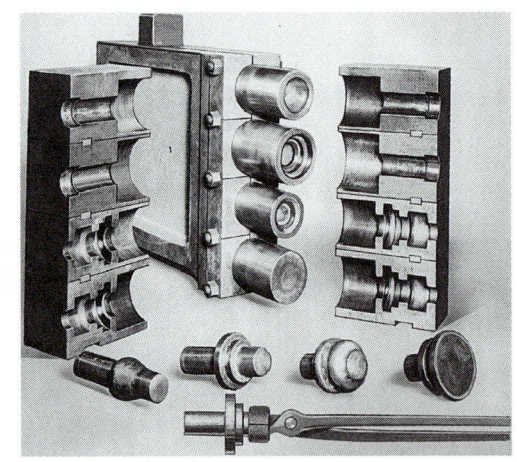

FIGURE 18-13 Set of upset forging dies and punches. The product resulting from each of the four positions is shown along the bottom. *(Courtesy of Ajax Manufacturing Company.)*

cold on special high-speed machines where the workpiece is rapidly moved from station to station. The starting stock is usually wire or rod, but some machines can forge bars up to 10 in. (250 mm) in diameter.

Upset forging generally employs split dies that contain multiple positions or cavities, such as the set shown in Figure 18-13. The dies separate enough for the heated bar to advance between them and move into position. They are then clamped together and a heading tool or ram moves longitudinally against the bar, upsetting it into the cavity. Separation of the dies then permits transfer to the next position or removal of the product. By starting a new piece with each die separation and performing an operation in each cavity simultaneously, a finished product can be made with each cycle of the machine. If a shearing operation is performed as the initial piece moves into position, the process can operate with continuous coil or long-length rod as its feedstock.

Upset-forging machines are often used to form heads on bolts and other fasteners and to shape valves, couplings, and many other small components. The following three rules, illustrated in Figure 18-14, should be followed in designing parts that are to be upset-forged:

1. The length of unsupported metal that can be gathered or upset in one blow without injurious buckling should be limited to three times the diameter of the bar.
2. Lengths of stock greater than three times the diameter may be upset successfully provided that the diameter of the cavity is not more than $1\frac{1}{2}$ times the diameter of the bar.
3. In an upset requiring stock length greater than three times the bar diameter and where the diameter of the cavity is not more than $1\frac{1}{2}$ times the diameter of the bar (the conditions of rule 2), the length of unsupported metal beyond the face of the die must not exceed the diameter of the bar.

Figure 18-15 illustrates the variety of parts that can be produced by upsetting and subsequent piercing, trimming, and machining operations.

Automatic Hot Forging

Several equipment manufacturers now offer highly automated upset equipment in which mill-length steel bars (typically, 24 ft long) are fed into one end at room temperature and hot-forged products emerge from the other end at rates of up to 180 parts per minute

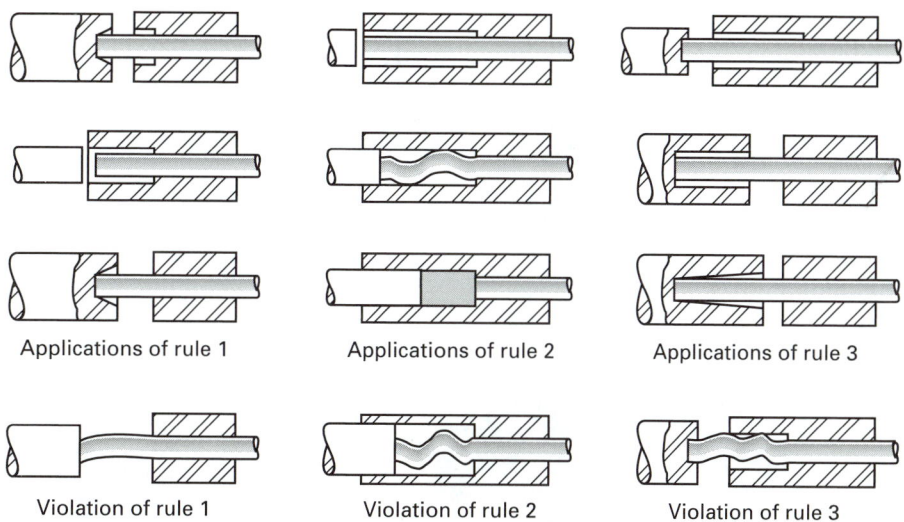

Applications of rule 1   Applications of rule 2   Applications of rule 3

Violation of rule 1   Violation of rule 2   Violation of rule 3

FIGURE 18-14   Schematics illustrating the rules governing upset forging. *(Courtesy of National Machinery Company.)*

FIGURE 18-15   Typical parts made by upsetting and related operations. *(Courtesy of National Machinery Company.)*

(86,400 parts per 8-hour shift). These parts can be solid or hollow, round or symmetrical, up to 12 lb (6 kg) in weight, and up to 7 in. (180 mm) in diameter.

The process begins with the lowest-cost steel bar stock: hot-rolled and air-cooled carbon or alloy steel. The bar is first heated to 2200 to 2350°F (1200 to 1300°C) in under 60 seconds as it passes through high-power induction coils. It is then descaled by rolls, sheared into individual blanks, and transferred through several successive forming stages, during which it is upset, preformed, final-forged, and finally, pierced (if necessary). Small parts can be produced at up to 180 parts per minute, with rates for larger pieces on the order of 90 parts per minute. Figure 18-16 shows a typical sequence and a variety of ferrous products.

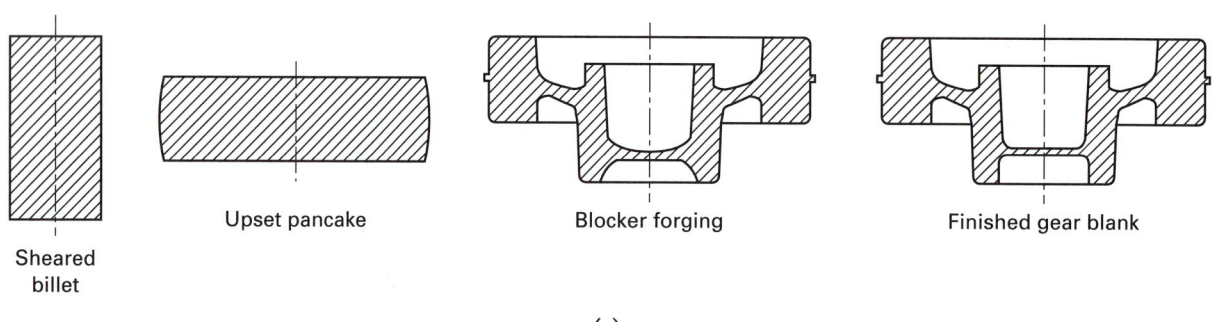

| Sheared billet | Upset pancake | Blocker forging | Finished gear blank |

(a)

(b)

FIGURE 18-16 *(Top)* Typical four-step sequence to produce a spur-gear forging. The sheared billet is progressively shaped into an upset pancake, blocker forging, and finished gear blank. *(Bottom)* Samples of ferrous parts produced by automatic hot forging at rates between 90 and 180 parts per minute. *(Courtesy of National Machinery Company.)*

The process has a number of attractive features. Low-cost input material and high production speeds have already been cited. Minimum labor is required, and since no flash is produced, material savings can be as much as 20 to 30% over conventional forging. With a consistent finishing temperature near 1900°F (1050°C), an air cool can often produce a structure suitable for machining, eliminating the need for an additional anneal or normalizing treatment. Tolerances are generally ±0.012 in., surfaces are clean, and draft

 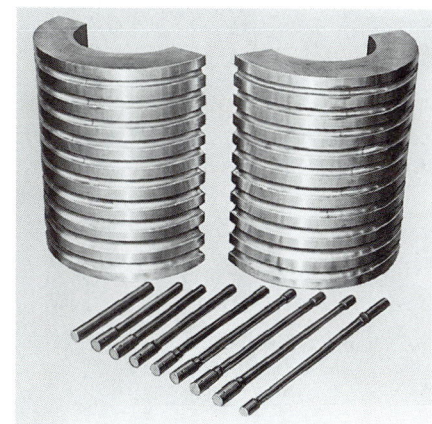

FIGURE 18-17   (*Left*) Roll forging machine in operation. (*Right*) Rolls from a roll forging machine and the various stages in roll forging a part. *(Courtesy of Ajax Manufacturing Company.)*

angles need only be $\frac{1}{2}$ to $1°$ (as opposed to the conventional 3 to $5°$). Tool life is nearly double that of conventional forging because the contact times are only on the order of 6/100 of a second.

To justify such an operation, however, large quantities of a given product must be required. A single installation of automatic hot-forging equipment may well require an initial investment in excess of $10 million.

Automatic hot formers can also be coupled with high-rate cold-forming operations. Preform shapes can be hot-formed at rates that approach 180 parts per minute. These products are then cold-formed to final shape on machines that operate near 90 parts per minute. The benefits of the combined operations include high-volume production at low cost, coupled with the precision, surface finish, and strain hardening that are characteristic of a cold-finished material.

Roll Forging

In *roll forging*, round or flat bar stock is reduced in thickness and increased in length to produce such products as axles, tapered levers, and leaf springs. As illustrated in Figure 18-17, roll forging is performed on machines that have two cylindrical or semicylindrical rolls, each containing one or more shaped grooves. A heated bar is inserted between the rolls when they are in the open position. When the bar encounters a stop, the rolls rotate, and the bar is progressively shaped as it is rolled out toward the operator. The piece can be reinserted between the next set of grooves (or rotated and reinserted) and the process repeated to produce the desired size and shape. Figure 18-18 schematically presents the roll forging process.

Swaging

Swaging generally involves the hammering of a rod or tube to reduce its diameter where the die itself acts as the hammer. Repeated blows are delivered from various angles,

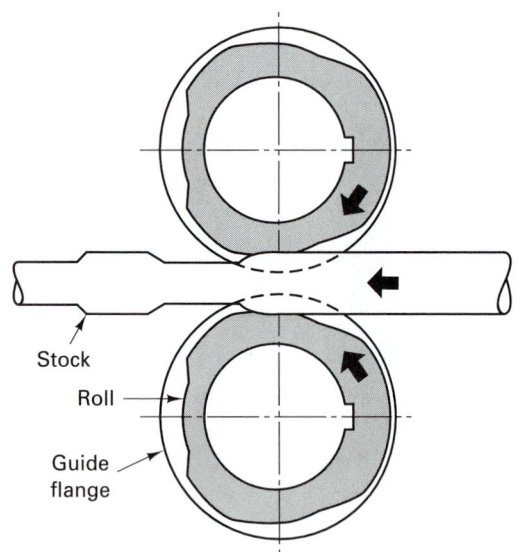

FIGURE 18-18   Schematic of the roll forging process showing the two shaped rolls and the stock being formed. *(Courtesy of Forging Industry Association, Cleveland, Ohio.)*

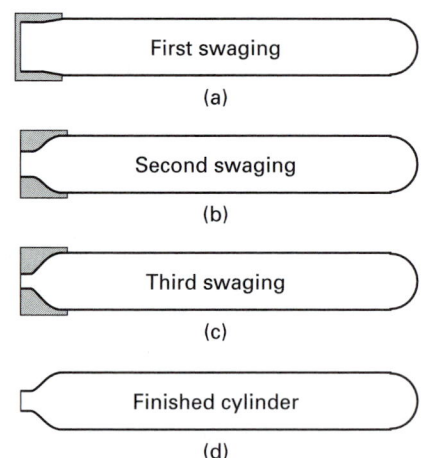

FIGURE 18-19   Steps in swaging a tube to form the neck of a gas cylinder. *(Courtesy of USX Corporation.)*

causing the metal to flow inward and assume the shape of the die. Since this operation is usually performed cold, it will be discussed further in Chapter 19.

The term *swaging* is also applied to processes where material is forced into a confining die to reduce its diameter. Figure 18-19 illustrates the application of hot swaging to reduce and form the end of a gas cylinder.

Net-Shape and Near-Net-Shape Forging

As much as 80% of the cost of a forged gear can be incurred during machining operations, and a finished aerospace wing spar may contain as little as 4% of the original billet weight (the remaining 96% being lost as scrap in the forging and subsequent machining operations). To minimize both the expense and waste, considerable effort has been made to develop processes that can form parts close enough to final dimensions that little or no final machining is required. These are known as *net-shape*, or *near-net-shape*, operations. Cost savings often result from the reduction or elimination of secondary machining (and the associated handling, positioning, and fixturing), a reduction in scrap, and a decrease in the energy required to produce the product.

Precision or near-net-shape forgings can now be produced with draft angles of less than 1° (or even zero draft). Complex shapes can be forged with such close tolerances that little or no finish machining is required. At present, the design and implementation of net-shape processing can be rather expensive. As a result, application is usually reserved for parts where a significant cost reduction can be achieved.

■ 18.6 EXTRUSION

In the *extrusion* process, metal is compressed and forced to flow through a suitably shaped die to form a product with reduced but constant cross section. Although extrusion may be

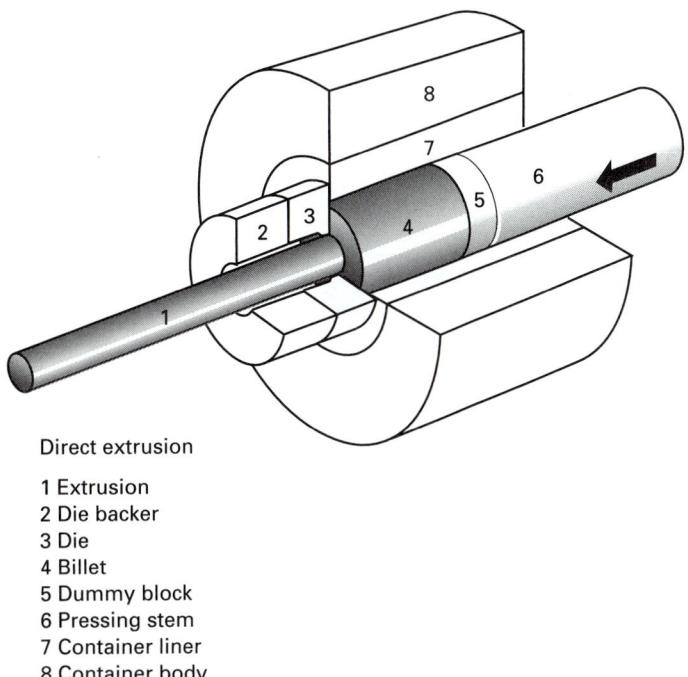

Direct extrusion

1 Extrusion
2 Die backer
3 Die
4 Billet
5 Dummy block
6 Pressing stem
7 Container liner
8 Container body

FIGURE 18-20 Direct extrusion schematic showing the various equipment components. *(Courtesy of Wean United, Inc., Hydraulic Machinery Division.)*

performed either hot or cold, hot extrusion is commonly employed for many metals to reduce the forces required, eliminate cold-working effects, and reduce directional properties. Basically, the extrusion process is like squeezing toothpaste out of a tube. In the case of metals, a common arrangement is to have a heated billet placed inside a confining chamber. A ram advances from one end, first causing the billet to upset and conform to the confining chamber. As the ram continues to advance, the pressure builds until the material flows plastically through the die, as illustrated in Figure 18-20. The stress state within the material is one of triaxial compression.

Lead, copper, aluminum, magnesium, and alloys of these metals are commonly extruded, taking advantage of the relatively low yield strengths and low hot-working temperatures. Steels, stainless steels, and nickel-based alloys are far more difficult to extrude. Their yield strengths are high and the metal has the tendency to weld to the walls of the die and confining chamber under the required conditions of temperature and pressure. With the development and use of phosphate-based and molten glass lubricants, however, substantial quantities of extrusions are now produced from these high-strength, high-temperature metals. These lubricants are able to withstand the required temperatures and adhere to the billet, flowing and thinning in a way that prevents metal-to-metal contact throughout the process.

As shown in the left-hand segment of Figure 18-21, almost any cross-sectional shape can be extruded from the nonferrous metals. Size limitations are few because presses are now available that can extrude any shape that can be enclosed within a 30-in. (750-mm) circle. In

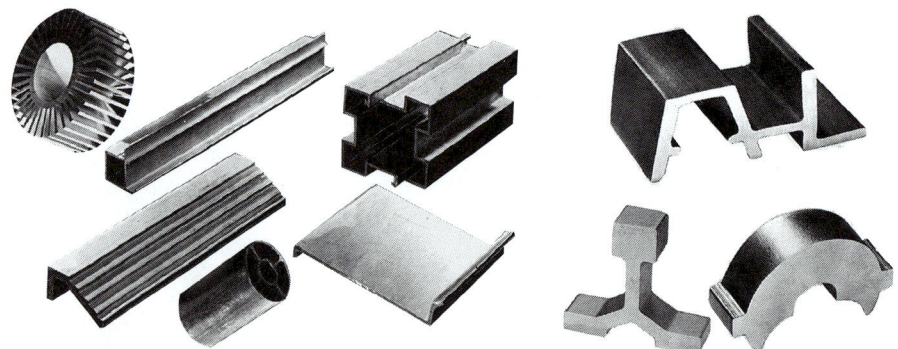

FIGURE 18-21 Typical shapes produced by extrusion. (*Left*) Aluminum products. (*Courtesy of Aluminum Company of America.*) (*Right*) Steel products. (*Courtesy of Allegheny Ludlum Steel Corporation.*)

the case of steels and the other high-strength metals, the shapes and sizes are more limited, but, as the right-hand segment of Figure 18-21 shows, considerable freedom still exists.

Extrusion has a number of attractive features. Many shapes can be produced as extrusions that are not possible by rolling, such as ones containing reentrant angles or longitudinal holes. No draft is required, so extrusions can offer savings in both metal and weight. Since the deformation is compressive, the amount of reduction in a single step is limited only by the capacity of the equipment. Billet-to-product cross-sectional area ratios can be in excess of 100:1 for the weaker metals. In addition, extrusion dies can be relatively inexpensive, and one die may be all that is required to produce a given product. Conversion from one product to another requires only a single die change, so small quantities of a desired shape can be produced economically. The major limitation of the process is the requirement that the cross section be uniform for the entire length of the product.

Extruded products have good dimensional precision. For most shapes, tolerances of ±0.003 in./in., with a minimum of ±0.003 in. are easily attainable. Grain structure is typical of other hot-worked metals, but strong directional properties (longitudinal versus transverse) are usually observed. Standard product lengths are about 20 to 24 ft, but lengths in excess of 40 feet have been produced.

Extrusion Methods

Extrusions can be produced by various techniques and equipment configurations. Hot extrusion is usually done by either the direct or indirect method, both of which are illustrated in Figure 18-22. In *direct extrusion*, a solid ram drives the entire billet to and through a stationary die and must provide additional power to overcome the frictional resistance between the surface of the moving billet and the confining chamber. In *indirect extrusion*, a hollow ram drives the die back through a stationary, confined billet. Since there is no relative motion, friction between the billet and the chamber is eliminated. The required force is lower and longer billets can be used with no penalty in power or efficiency. Figure 18-23 shows the ram force versus ram position for both direct and indirect extrusion. The areas below the curves have the units of foot-pounds or inch-pounds and are proportional to work or power. The area between the two curves is the power associated with overcoming billet–chamber friction during direct extrusion, an amount that can be saved by converting to indirect extrusion. Unfortunately, the added complexity of the indirect process

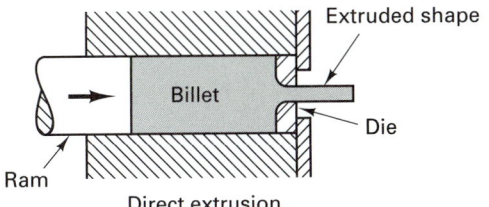

Direct extrusion

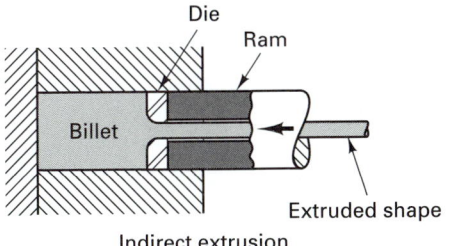

Indirect extrusion

FIGURE 18-22 Direct and indirect extrusion. In direct extrusion, friction with the chamber opposes forward motion of the billet. For indirect extrusion, there is no friction, since there is no relative motion.

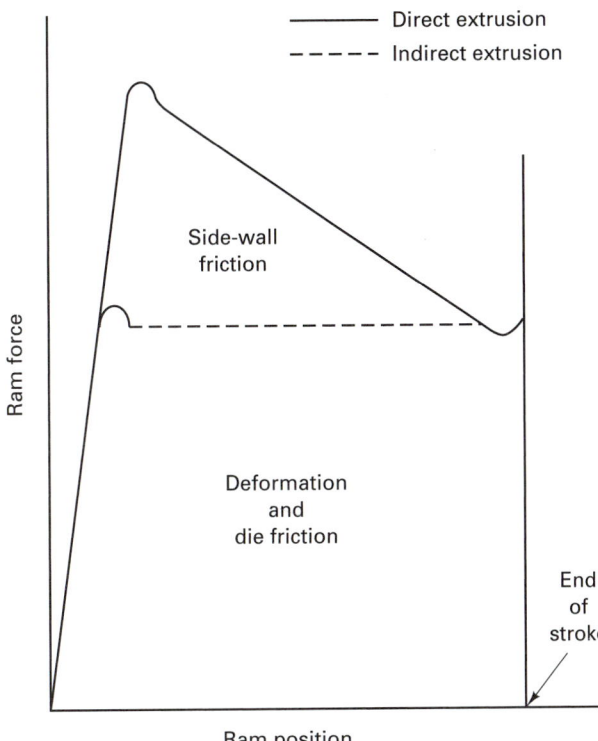

FIGURE 18-23 Diagram of the ram force versus ram position for both direct and indirect extrusion of the same product. The area under the curve corresponds to the amount of work (force × distance) performed. The difference between the two curves is attributed to billet-chamber friction.

(applying force through a hollow ram, extracting product through the hollow, and removing residual billet material at the end of the stroke) serves to increase both the purchase price and maintenance cost of the required equipment.

Regardless of the process, the speeds of hot extrusion are often rather fast, to minimize the cooling of the billet within the chamber. The speed may be restricted, however, by the large amounts of heat that are generated by the massive deformation and the associated rise in temperature. Sensors are often used to monitor the temperature of the emerging product

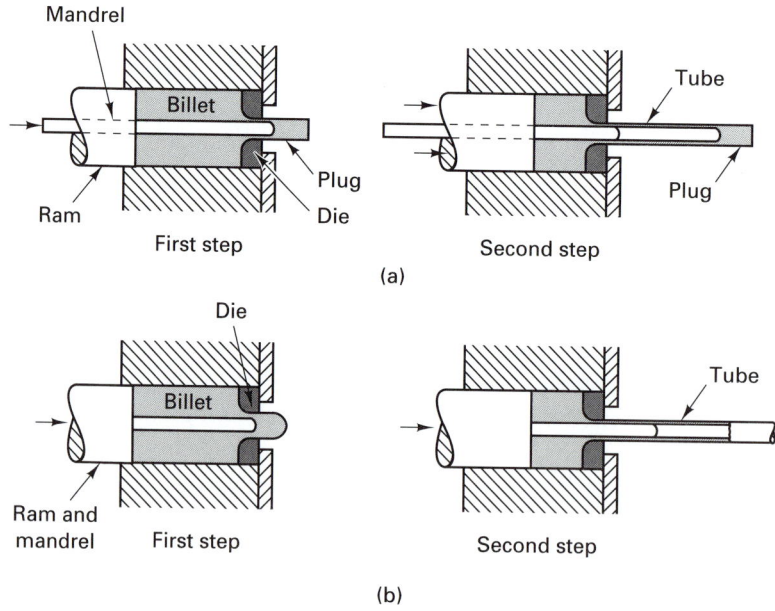

FIGURE 18-24   Two methods of extruding hollow shapes using internal mandrels. In the upper schematic the mandrel and ram have independent motions; in the lower sequence they move as a single unit.

and feed this information to a control system. For materials whose properties are not sensitive to strain rate, ram speed is often maintained at the highest level that will keep the product temperature below a predetermined maximum.

Lubrication is another primary concern. If the reduction ratio (cross section of billet to cross section of product) is 100, the product is 100 times longer than the starting billet. If the product has a complex cross section, its perimeter can be significantly greater than a circle of equivalent area. Thus the lubricant that is applied to the starting billet must thin considerably as the material passes through the die. At all stages of the process, it will be expected to function both as a lubricant and a barrier to heat transfer.

Impact extrusion, *hydrostatic extrusion*, and *continuous extrusion* are usually performed cold and are discussed in Chapter 19.

Extrusion of Hollow Shapes

Hollow shapes, and shapes with more than one longitudinal cavity, can be extruded by several methods. For tubular products, the stationary or moving mandrel processes of Figure 18-24 are quite common. For products with multiple or more-complex cavities, a *spider-mandrel die* (also known as a porthole, bridge, or torpedo die) may be required. As illustrated in Figure 18-25, the hot metal flows around the arms of the spider, where further reduction forces the seams to close and weld back together. Since the metal has never been exposed to contamination, perfect welds result. Unfortunately, lubricants cannot be used, since they will contaminate the surfaces to be welded. The process is therefore limited to materials that can be extruded without lubrication and that can easily be pressure welded. Since additional tooling is required, hollow extrusions will obviously cost more than solid ones, but a wide variety of shapes can be produced that cannot be made economically by any other process.

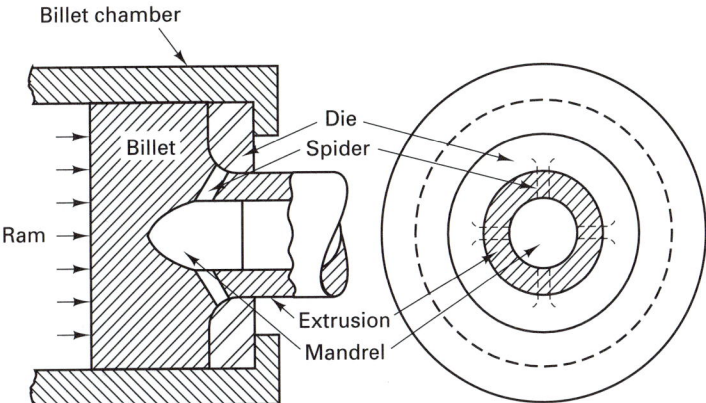

FIGURE 18-25   Hot extrusion of a hollow shape using a spider-mandrel die. Note the four arms connecting the die and the mandrel.

Metal Flow in Extrusion

The flow of metal during extrusion is often quite complex and some care must be exercised to prevent surface cracks, interior cracks, and other flow-related defects. Metal near the center of the chamber can often pass through the die with little distortion, while metal near the surface undergoes considerable shearing. In direct extrusion, friction between the forward-moving billet and the stationary chamber and die serves to further impede surface flow. The result is often a deformation pattern similar to the one shown in Figure 18-26. If the surface regions of the billet undergo excessive cooling, the deformation is further impeded and cracks tend to form on the product surface. If quality is to be maintained, process control must be exercised in the areas of design, lubrication, extrusion speed, and temperature.

FIGURE 18-26   Grid pattern showing the metal flow in a direct extrusion. The billet was sectioned and the grid pattern was engraved prior to extrusion.

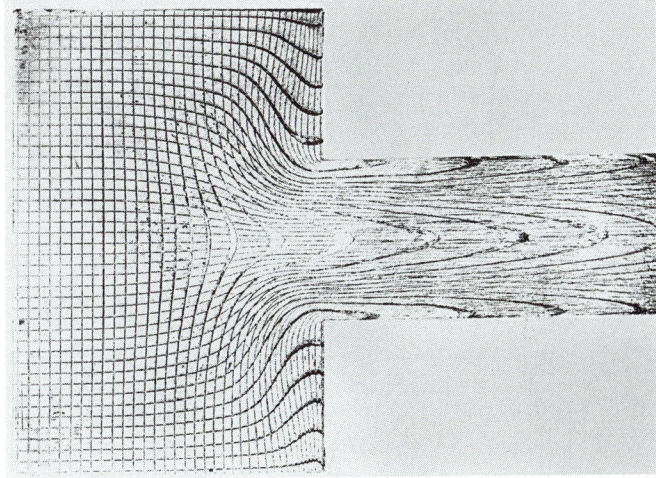

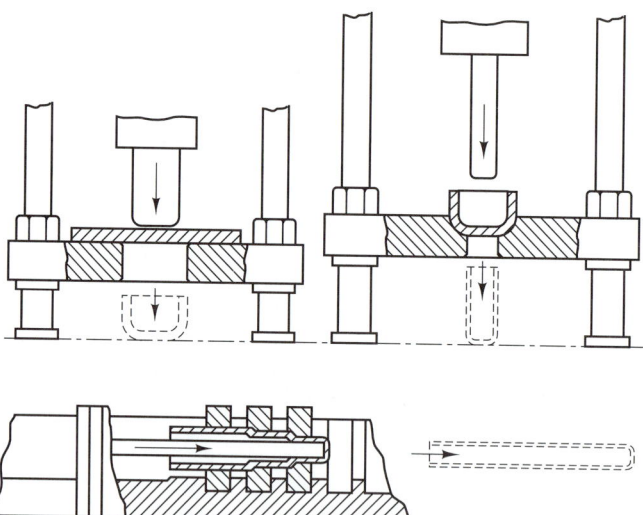

FIGURE 18-27 Methods of cup forming or hot drawing.
(*Upper left*) First draw. (*Upper right*) Redraw operation.
(*Lower*) Multiple-die drawing. (*Courtesy of USX Corporation.*)

■ 18.7 HOT DRAWING OF SHEET AND PLATE

Drawing is a plastic deformation process in which a flat sheet or plate is formed into a recessed, three-dimensional part with a depth more than several times the thickness of the metal. As a punch descends into a mating die (or the die moves upward over a mating punch), the metal assumes the desired configuration. Hot drawing is used for forming relatively thick-walled parts of simple geometries, usually cylindrical. Because the material is hot, there is often considerable thinning as it passes through the dies. In contrast, cold drawing uses relatively thin metal, changes the thickness very little or not at all, and produces parts in a wide variety of shapes.

Hot drawing is illustrated in the upper left-hand schematic of Figure 18-27. A heated sheet or plate is positioned over a female die. A punch then descends, pushing the metal through the die, converting the circular blank to a cylindrical cup. The height of the cup walls is determined by the difference between the diameter of the original blank and the diameter of the punch. This dimension is limited by the formation of several defects. Wrinkles can appear in the cup walls as the circumference is reduced, or the punch can act as a piercing tool, tearing the blank around the punch perimeter.

There are several means of producing parts with taller walls. If the gap between the punch and the die is less than the thickness of the incoming material, the cup wall is thinned and elongated simultaneously (a process often called *ironing* or *wall ironing*). If this thinning is objectionable, an intermediate shape can be produced, with the further reduction in diameter (and concurrent increase in wall height) being taken in a subsequent redrawing with a smaller punch and die, as shown in the upper right-hand segment of Figure 18-27. The lower segment of this figure illustrates still another alternative, where the cup is pushed through a series of dies with a single punch.

Some drawn products are designed to utilize part of the original disk as a flange around the lip of the cup. In this case the punch does not push the material completely through the die, but descends to a predetermined depth and then retracts. The partially drawn product is then ejected upward, and the perimeter of the remaining flange is trimmed to the desired size and shape.

■ 18.8 Pipe Welding

Large quantities of steel pipe are made by two processes that use hot forming of steel strip coupled with the deformation-induced welding of its free edges. Both of these processes, *butt welding* of pipe, and *lap welding* of pipe, utilize steel in the form of *skelp*—long strips with specified width, thickness, and edge configuration. Because the skelp has been hot rolled previously and the welding process produces further compressive working and recrystallization, pipe welded by these processes tends to be very uniform in quality.

Butt-Welded Pipe

In the butt-welding process for making pipe, steel skelp is heated to a specified hot-working temperature by passing it through a furnace. Upon exiting the furnace, it is pulled through forming rolls that shape it into a cylinder and bring the free ends into contact. The pressure exerted between the edges of the skelp is sufficient to upset the metal and produce a welded seam. Additional sets of rollers then size and shape the pipe and it is cut to standard, preset lengths. Product diameters range from $\frac{1}{8}$ in. (3 mm) to 3 in. (75 mm), and speeds can approach 500 ft/min.

Lap-Welded Pipe

The lap-welding process for making pipe differs from the butt-welding technique in that the skelp now has beveled edges and the rolls form the weld by forcing the lapped edges down against a supported mandrel. This process is used primarily for larger sizes of pipe, from about 2 in. (50 mm) to 14 in. (400 mm) in diameter. Because the product is driven over a supported mandrel, product length is limited to about 20 to 25 feet.

■ 18.9 Piercing

Thick-walled seamless tubing can be made by *rotary piercing*, a process illustrated in Figure 18-28. A heated billet is fed longitudinally into the gap between two large, convex-tapered rolls. These rolls are powered to rotate in the same direction, but the axes of the rolls are offset from the axis of the billet by about 6°, one to the right and the other to the left. The clearance between the rolls is preset at a value less than the diameter of the billet. As the billet is caught by the rolls, it is simultaneously rotated and driven forward. The reduced clearance between the rolls forces the billet to deform into a rotating ellipse. As shown in the right-hand segment of Figure 18-28, rotation of the elliptical section causes the metal to shear about the major axis. A crack tends to form down the center axis of the billet, and the cracked billet is then forced over a pointed mandrel that enlarges and shapes the opening to form a seamless tube. The result is a short length of thick-walled seamless tubing, which can then be passed through a reeler and sizing rolls to straighten it and reduce the diameter and/or wall thickness. Seamless tubes can also be expanded in diameter by passing them over a larger mandrel. As the diameter and circumference increase, the walls correspondingly thin.

The *Mannesmann mills* commonly used in hot piercing can be used to produce tubing up to 12 in. (300 mm) in diameter. Larger-diameter tubes can be produced on *Stiefel mills*, which use the same principle but replace the convex rolls of the Mannesmann mill with larger-diameter conical disks.

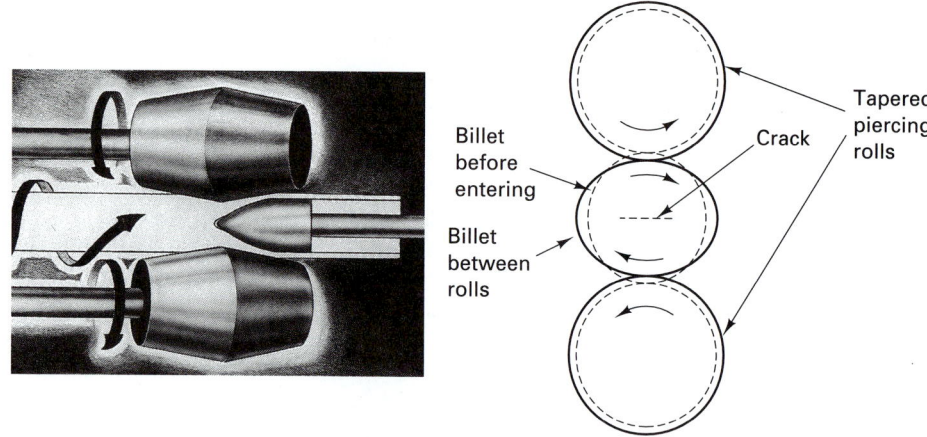

FIGURE 18-28 *(Left)* Principle of the Mannesmann process of producing seamless tubing. *(Courtesy of American Brass Company.)* *(Right)* Mechanism of crack formation in the Mannesmann process.

■ KEY WORDS

automatic hot forging	flashless forging	pipe welding
billet	forging	press forging
bloom	hydraulic press	roll forging
bulk deformation processes	impression-die forging	rolling
butt-welding	indirect extrusion	rotary piercing
cluster mill	ironing	sheet forming
controlled rolling	lap-welding	skelp
counterblow machine	Mannesmann mill	slab
direct extrusion	mechanical press	spider-mandrel die
drawing	near-net-shape forging	swaging
drop-hammer forging	net-shape forging	thermomechanical processing
extrusion	open-die forging	upset
finishing temperature	pack rolling	upset forging
flash	piercing	

■ REVIEW QUESTIONS

1. Briefly describe the evolution of forming equipment from ancient to modern.
2. What are some of the possible means of classifying metal deformation processes?
3. Why is the division of forming processes into hot working and cold working a somewhat nebulous classification?
4. What are some features of metals at elevated temperature that make hot forming an attractive technique to alter shape?
5. Why are hot-rolled products generally limited to standard shapes and sizes?
6. Why is it undesirable to minimize friction between the workpiece and tooling in a rolling operation?

7. Discuss the relative advantages and typical uses of two-high rolling mills with large-diameter rolls, three-high mills, and four-high mills.
8. Why is speed synchronization of the various rolls so vitally important in a continuous or multistand rolling mill?
9. What types of products are produced by ring rolling?
10. Discuss the problems in maintaining uniform thickness in a rolled product and some of the associated defects.

11. What is thermomechanical processing, and what are some of its possible advantages?

12. Why are steam or air hammers more attractive than board hammers for hammer forging?

13. What is the difference between open-die and impression-die forging?

14. Why is open-die forging not a practical technique for large-scale production of identical products?

15. What additional controls must be exercised to perform flashless forging satisfactorily?

16. What attractive features are offered by counterblow forging equipment, or impactors?

17. What are some of the attractive features of the Autoforge or Osprey processes compared to conventional forging?

18. Why are different tolerances usually applied to dimensions contained within a single die cavity and dimensions across the parting plane?

19. For what types of forging products or conditions might a press be preferred over a hammer?

20. Why are heated dies generally employed in hot press forging operations?

21. Describe some of the primary differences between hammers, mechanical presses, and hydraulic presses.

22. What is upset forging?

23. What are some of the typical products produced by upset-forging operations?

24. What are some of the attractive features of automatic hot forging? What is a major limitation?

25. How does roll forging differ from a conventional rolling operation?

26. What are some possible objectives of near-net-shape forging?

27. What are some of the attractive features of the extrusion process?

28. What is the primary shape limitation of the extrusion process?

29. If the area under the indirect extrusion curve of Figure 18-23 is proportional to the power required to extrude a product without billet-chamber frictional resistance, how could the relative regions of the direct extrusion curve be used to determine a crude measure of the mechanical "efficiency" of direct extrusion?

30. What property of a lubricant is critical in extrusion that might not be required for processes such as forging?

31. Why can lubricants not be used in conjunction with a spider-mandrel extrusion die?

32. How can tall, thin cups be produced by hot drawing? (*Note*: The wall height is greater than can be produced in a single drawing operation.)

33. What is meant by the term *ironing* in hot-drawing operations?

34. What two hot-forming operations can be used to produce pipe from steel strip?

35. What limits the length of seamless pipe that can be produced by hot-piercing operations?

■ PROBLEMS

1. Some snack foods, such as rectangular corn chips, are often formed by a rolling-type operation and are subject to the same types of defects common to rolled sheet and strip. Obtain a bag of such a snack and examine the chips to identify examples of rolling-related defects such as those discussed in Section 18.4 and shown in Figure 18-29.

2. Consider the extrusion of a cylindrical billet, and compute the following.
   (a) Assume the starting billet to have a length of 1 ft and a diameter of 5 in. This is extruded into a cylindrical product that is 1 in. in diameter and 25 ft long (a reduction ratio of 25). Neglecting the areas on the two ends, compute the ratio between the

FIGURE 18-29   Some typical defects that occur during flat rolling: wavy edges, edge cracking, and center cracking.

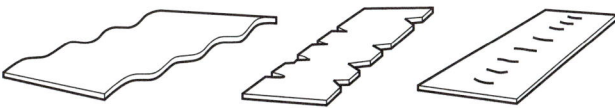

product surface area (wraparound cylinder) and the surface area of the starting billet.

(b) How would this ratio change if the product were a square with the same cross-sectional area as that of the 1-in. diameter circle?

(c) Consider a cylinder-to-cylinder extrusion with a reduction ratio of R. Derive a general expression of the relative surface areas of product to billet as a function of R. (*Hint*: Start with a cylinder with length and diameter both equal to 1 unit. Since the final area will be $1/R$ times the original, the final length will be R units and the final diameter will be proportional to $1/\sqrt{R}$.)

3. The force required to compress a cylindrical solid between flat parallel dies has been estimated (by the theory of plasticity analysis) to be

$$\text{force} = \pi R^2 \, \sigma_0 \, \frac{1 + 2mR}{3\sqrt{3}\,T}$$

where

R = radius of the cylinder
T = thickness of the cylinder
σ_0 = yield strength
m = friction factor

An engineering student is attempting to impress his date by demonstrating some of the neat aspects of metal forming. He places a shiny copper penny between the platens of a 60,000-lb capacity press and proceeds to apply pressure. Assume that the coin has a $\frac{3}{4}$ in. diameter and is $\frac{1}{16}$ in. thick. The yield strength is estimated as 50,000 psi, and since no lubricant is applied, friction is that of complete sticking, or $m = 1.0$.

(a) Compute the force required to induce plastic deformation.

(b) If this force is greater than the capacity of the press (60,000 lb), compute the pressure when the full-capacity force is applied.

(c) If the press surfaces are made from thick plates of a material with a yield strength of 120,000 psi, describe the results of the demonstration.

4. Mathematical analysis of the rolling of flat strip reveals that the roll-separation force (the squeezing force required to deform the strip) is directly proportional to the term

$$1 + \frac{K_1 \, mL}{t_{av}}$$

where

K_1 = geometric constant
m = friction factor
L = length of contact
t_{av} = average thickness of the strip in the roll bite

Since L is proportional to the roll radius, R, $K_1 L$ can be replaced by $K_2 R$, so the force becomes proportional to the term:

$$1 + \frac{K_2 \, mR}{t_{av}}$$

If K_2 and m are both positive numbers, how will the roll-separation force change as the strip becomes thinner? How can this effect be minimized? Relate your observations to the types of rolling mills used for various thicknesses of product.

Chapter 18 **CASE STUDY**

outboard motor brackets

The components depicted in Figure CS-18 are two styles of mounting brackets used to attach outboard motors to the stern plates of small boats. The sketches show the parts as unfinished castings or forgings. Finish machining will add a threaded hole through the bottom of the short leg to accommodate a tightening turn screw. Holes will also be drilled through the long leg to permit attachment of the motor.

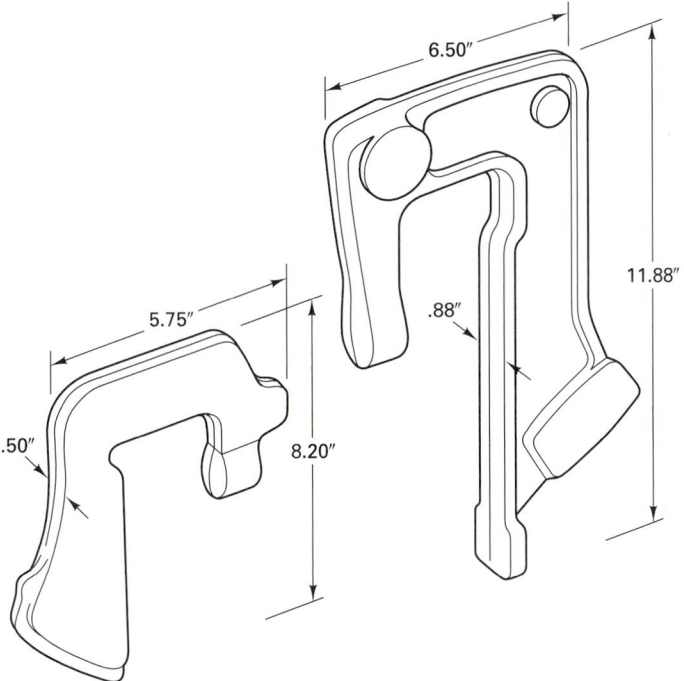

FIGURE CS-18

Since outboard motors are frequently carried to the boat, weight is a definite concern. The bracket will also operate in a wet environment, possibly even salt water, and will be subjected to impacts and vibration, in addition to its steady-state force. The designer has recommended a minimum yield strength of 50 ksi, coupled with a 5% minimum elongation in a uniaxial tensile test to assure adequate resistance to brittle fracture.

1. Briefly discuss the major service requirements for these components. Which of the various engineering materials would you recommend for this application?

2. For your recommended material, what are some of the possible means of producing the desired shapes? Which of the possible alternatives would you recommend? Why?

3. Would heat treatment be necessary to establish the required properties? If so, what type of treatment should be used?

4. For your recommended material, would some form of surface treatment be desirable? What type of treatment would you recommend?

CHAPTER 19

COLD-WORKING PROCESSES

19.1 INTRODUCTION
19.2 SQUEEZING PROCESSES
    Cold Rolling
    Cold-Rolled Shapes
    Thread Rolling
    Swaging
    Cold Forging
    Cold Extrusion (Also Known
        as Impact Extrusion)
    Hydrostatic Extrusion
    Continuous Extrusion
    Roll Extrusion
        Sizing
        Riveting
        Staking
        Coining
        Hubbing
        Surface Improvement by
        Cold Working
19.3 BENDING
    Angle Bending
    Design for Bending
    Air-Bend Versus Bottoming
        Dies
    Roll Bending
    Draw Bending and
        Compression Bending
    Cold-Roll Forming
    Seaming and Flanging
    Straightening
19.4 SHEARING
    Simple Shearing
    Slitting
    Piercing and Blanking
    Tools and Dies for Piercing
        and Blanking
    Design for Piercing and
        Blanking
19.5 DRAWING AND SHEET
    METAL FORMING
    Rod, Bar, and Tube Drawing
    Wire Drawing
    Spinning
    Shear Forming
    Stretch Forming
    Sheet Metal Drawing (Deep
        Drawing, Shell Drawing,
        and Shallow Drawing)
    Forming with Rubber
        Tooling or Fluid Pressure
    Drawing on a Drop Hammer
    High-Energy-Rate Forming
        Ironing
        Embossing
        Superplastic Sheet
        Forming
        Design Aids for Sheet
        Metal Forming
19.6 ALTERNATIVE
    METHODS OF
    PRODUCING SHEET-TYPE
    PRODUCTS
    Electroforming
    Plasma Spray Forming
19.7 PRESSES
    Classification of Presses
    Types of Press Frame
    Special Types of Presses
    Press-Feeding Devices
Case Study: THE BRONZE BOLT
    MYSTERY

■ 19.1 INTRODUCTION

A number of attractive features can be associated with the cold deformation of a metal. Strain hardening acts to increase the strength of the material, so cheaper, weaker starting materials may be selected, or costly heat treatment of the finished product may be eliminated. By properly positioning the last anneal, a final product can be produced with a specified amount of cold work, and properties can be tailored to meet specific needs.

Surface finish and dimensional precision are quite good. In addition, substantial energy savings are possible. While greater forces are required to induce deformation, heating is not required to bring the workpiece to the forming temperature.

As a result of these features, a number of cold-working processes have been developed to perform a wide variety of deformations. These processes can be classified into four basic categories: *squeezing operations*, *bending operations*, *shearing operations*, and *drawing operations*. Table 19-1 lists some of the cold-working processes according to their primary form of deformation.

TABLE 19-1.   Classification of the Major Cold-Working Operations

Squeezing		Bending	
1. Rolling	7. Staking	1. Angle	5. Seaming
2. Swaging	8. Coining	2. Roll	6. Flanging
3. Cold forging	9. Peening	3. Draw and	7. Straightening
4. Extrusion	10. Burnishing	compression	
5. Sizing	11. Hubbing	4. Roll-forming	
6. Riveting	12. Thread rolling		

Shearing		Drawing	
1. Shearing; slitting	4. Notching;	1. Bar and tube	5. Stretch forming
2. Blanking	Nibbling	drawing	6. Sheet metal drawing
3. Piercing; lancing;	5. Shaving	2. Wire drawing	7. Ironing
perforating	6. Trimming	3. Spinning	8. Superplastic forming
	7. Cutoff	4. Embossing	
	8. Dinking		

■ 19.2 SQUEEZING PROCESSES

Most of the squeezing-type cold-working processes have identical hot-working counterparts or are extensions of hot-working processes. The primary reason for deforming cold is to obtain better dimensional accuracy and improved surface finish. In many cases, the equipment is basically the same, except that it must be more powerful to deform the higher-strength starting material and overcome the additional resistance caused by strain hardening. If power is limited, compromise may have to be made in either the size of the workpiece or the amount of deformation.

Cold Rolling

Cold rolling is clearly the dominant cold-working process in terms of product tonnage. Sheets, strips, bars, and rods are cold-rolled to obtain products that have smooth surfaces and accurate dimensions. Because of the size and strength of the material, most cold rolling is performed on four-high or cluster-type rolling mills.

Cold-rolled sheet and strip can be obtained in various conditions, including *skin-rolled*, *quarter-hard*, *half-hard*, and *full hard*. Skin-rolled metal is given only a $\frac{1}{2}$ to 1% reduction to produce a smooth surface and uniform thickness, and to remove the yield-point phenomenon (i.e., prevent formation of Luders bands upon further forming). This material is well suited for subsequent cold-working operations where good ductility is required. Quarter-hard, half-hard, and full-hard sheet and strip have experienced greater amounts of cold reduction, up to 50%. Consequently, their yield points have been increased, properties have become directional, and ductility has decreased. Quarter-hard steel

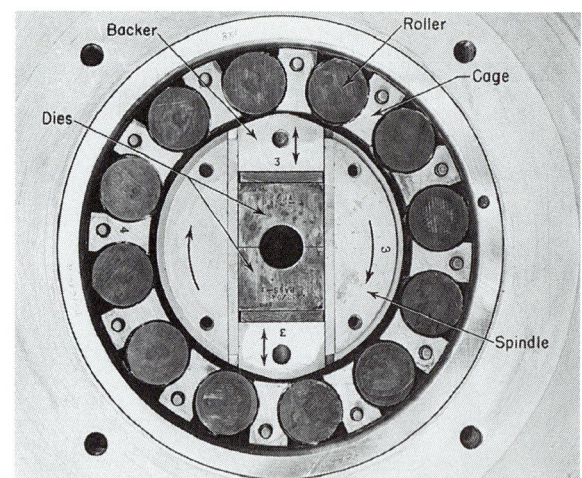

FIGURE 19-1 Basic components and motions of a rotary swaging machine. *Note:* The cover plate has been removed to reveal the interior workings. *(Courtesy of Torrington Company.)*

can be bent back on itself across the grain without breaking. Half-hard and full-hard can be bent back 90° and 45°, respectively, about a radius equal to the material thickness.

Cold-Rolled Shapes

If a product has a uniform (or nearly uniform) cross section and relatively small transverse dimensions (less than about 2 in.), cold rolling of rod or bar may be an attractive alternative to extrusion or machining. Strain hardening can be employed to provide up to 20% additional strength to the material, and the process offers smooth surfaces and high dimensional precision. Like the hot forming of structural shapes discussed in Chapter 18, the cold rolling of shapes generally requires a series of shaping operations. The production of a given part may require separate passes (and roll grooves) for sizing, breakdown, roughing, semiroughing, semifinishing, and finishing. A minimum order of several tons of product may be required to justify the expense of tooling and make the process economical.

Thread Rolling

A more specialized cold-rolling operation is thread rolling, which is discussed in some detail in Chapter 30.

Swaging

Swaging (also known as rotary swaging or radial forging) is a process that reduces the diameter, tapers, or points round bars or tubes by external hammering. If a shaped mandrel is inserted into a tube before swaging, the metal can be collapsed around it to shape and size simultaneously both the interior and exterior of a product.

Cold swaging is performed by a rotary machine like the one shown in Figures 19-1 and 19-2. High-speed rotation of the spindle generates centrifugal force, which causes the matching die segments and backer blocks to separate. As the spindle rotates, however, the backer blocks encounter opposing rollers that are mounted in a massive machine housing. To pass beneath these rollers, the dies must be squeezed tightly together. Once cleared, the dies can again separate and the cycle repeats. The power to sustain the rotary motion is usually supplied by an external motor connected to a large, massive flywheel.

FIGURE 19-2 Tube being reduced in a rotary swaging machine. *Note:* The cover plate has been reinstalled. *(Courtesy of Torrington Company.)*

The operator simply inserts a rod or tube between the dies and advances it during the periods of die separation. The repeated closures squeeze the workpiece from a variety of angles (since the parting line of the dies changes as the spindle rotates), reducing the diameter and increasing the length. When the desired length has been swaged, the product is extracted from the machine as the spindle continues to rotate.

Swaging operations can also be used to form products with internal shapes, usually of constant cross section. A tubular or closed-end workpiece is placed over a shaped mandrel and the assembly is inserted between the rotating dies. As the dies reciprocate and rotate, they compress the exterior of the workpiece and simultaneously force the interior to conform to the shape of the mandrel. When a long product is desired, the workpiece can be fed over a short, stationary mandrel that is positioned between the dies. Swaging over a mandrel can be used to form parts with internal gears, splines, recesses, and sockets. Figure 19-3 shows a variety of swaged products which contain shaped holes or recesses.

Cold Forging

Large quantities of products are now being made by *cold forging*, a family of processes in which slugs of material are squeezed into shaped die cavities to produce finished parts of precise shape and size. *Cold heading*, illustrated schematically in Figure 19-4, is used for making enlarged sections on the ends of rod or wire, such as the heads of nails, bolts, rivets, or other fasteners. Two variations of the process are common. In the sequence illustrated, a piece of rod is first sheared to a preset length and then transferred to a holder-ejector assembly. Heading punches then strike one or more blows on the exposed end to perform the upsetting. If intermediate shapes are required, the piece is transferred from station to station, or the various heading punches sequentially rotate into position. When the heading is completed, the ejector stop advances to expel the product.

In the second variation, a continuous rod (or wire) is fed forward to produce a preset extension, clamped, and the head is formed. The rod is then advanced to a second preset length and sheared. The cycle then repeats. This procedure is particularly attractive for producing nails, since the point can be formed in the cutoff operation.

Enlarged sections can be produced at locations other than the ends of a rod or wire by means of the *upsetting process* illustrated in Figure 19-5. To produce the deformation and control final dimensions, ejector stops must be provided in both the punch and the die.

FIGURE 19-3   A variety of swaged parts containing internal details. *(Courtesy of Cincinnati Milacron, Inc.)*

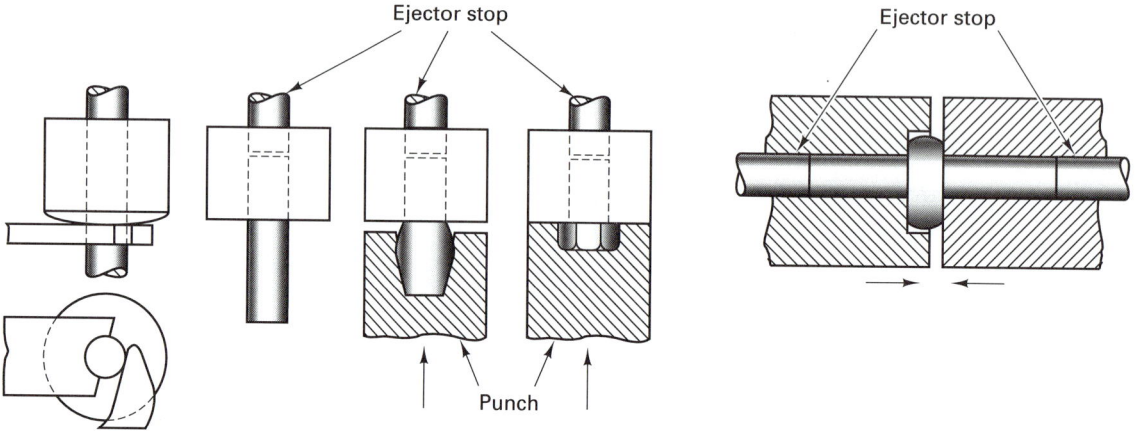

FIGURE 19-4   Typical steps in a shearing and cold-heading operation.

FIGURE 19-5   Method of upsetting the center portion of a rod. The stock is supported in both dies during upsetting.

By using various types of dies and combining high-speed operations such as heading, upsetting, extrusion, bending, coining, thread rolling, and knurling, a wide variety of relatively complex parts can be cold formed to close tolerances. Figure 18-15 has already presented a display of typical products. The larger ones are hot formed and machined, and the smaller ones are cold formed. Since cold forming is a chipless manufacturing process, producing parts that would otherwise be machined from bar stock or hot forgings, the material is used more efficiently and waste is reduced.

Figure 19-6 compares the manufacture of a spark-plug body by machining from hexagonal bar stock to manufacture by cold forming. Material is saved, machining time and cost are reduced, and the product is stronger, due to cold work, and tougher, as illustrated

CUTTING (74% WASTE)   COLD FORMING (6% WASTE)

FIGURE 19-6   Manufacture of a spark plug body: (*left*) by machining from hexagonal bar stock; (*right*) by cold forming. *(Courtesy of National Machinery Co.)*

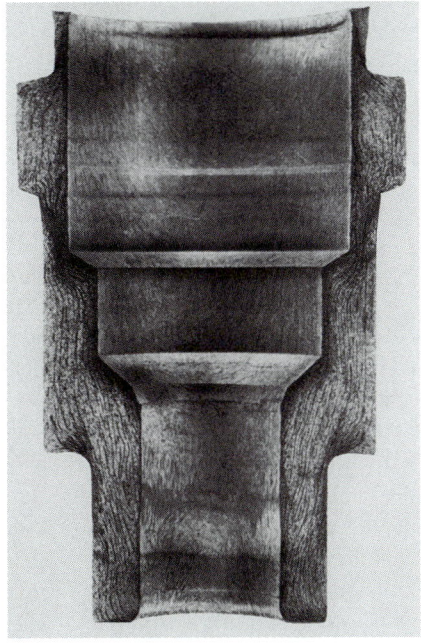

FIGURE 19-7   Section of the cold-formed spark-plug body of Figure 19-6, etched to reveal the flow lines. The cold-formed structure produces an 18% increase in strength over the machined product. *(Courtesy of National Machinery Co.)*

by the flow lines in Figure 19-7. In another example, a manufacturer of cruise-control housings converted from screw machining to cold forming, and reduced material usage by 65% while increasing production rate by a factor of 5.

Historically, cold forming has been associated with small parts made from the weaker nonferrous metals, but the process is now used extensively on steel, and parts can be up to 100 lb in weight and 7 in. in diameter. At the other end of the scale, microformers are cold forming small electronic components with dimensional accuracies within 0.0002 in. Shapes are usually axisymmetric or those with relatively small departures from symmetry. As with the other cold-forming processes, production rates are high, dimensional tolerances and surface finish are excellent, and machining can be reduced (material

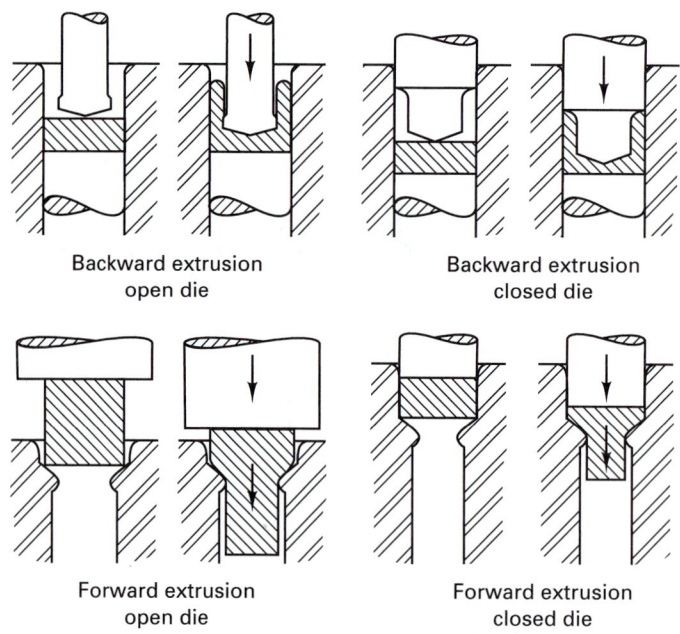

Backward extrusion
open die

Backward extrusion
closed die

Forward extrusion
open die

Forward extrusion
closed die

FIGURE 19-8 Various methods of cold extrusion.

savings). Strain hardening can provide additional strength, and favorable grain flow can enhance toughness and fatigue life. Unfortunately, the cost of the required tooling, coupled with the high production speed, generally requires large-volume production.

Cold Extrusion (Also Known as Impact Extrusion)

Chapter 18 has already presented the processes of direct and indirect extrusion, where the workpiece is usually deformed hot. Great advances have also been made in *cold extrusion*, a family of processes that also go by the name *impact extrusion*. Figure 19-8 illustrates the basic principles of *forward* and *backward* cold extrusion using open and closed dies. These processes were first used with low-strength metals such as lead, tin, zinc, and aluminum to produce such items as collapsible tubes for toothpaste, medications, and other creams; small "cans" for shielding electronic components; zinc cases for flashlight batteries; and larger cans for food and beverages.

In recent years, cold extrusion has also been used for forming mild steel parts, often in combination with cold heading, as shown in the example of Figure 19-9. When heading alone is used, there is a definite limit to the ratio of the head and stock diameters (as presented in Figure 18-14 and related discussion). The combination of forward extrusion and cold heading overcomes this limitation by using an intermediate starting diameter. The shank diameter is reduced by forward extrusion and the head is increased by upsetting. Figure 19-10 illustrates still another cold-forming operation, this time involving two extrusions and a central upset. In both cases, considerable metal is saved compared to the machining of parts from larger-diameter stock.

Hydrostatic Extrusion

Another type of cold extrusion is known as *hydrostatic extrusion*, illustrated schematically in Figure 19-11. Here high-pressure fluid applies the force necessary to extrude

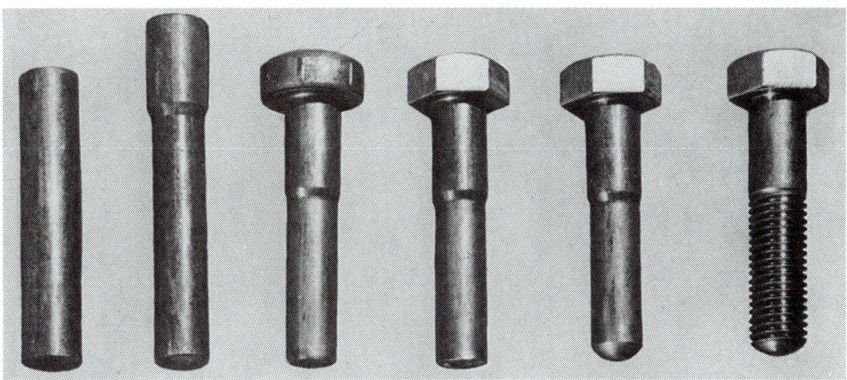

FIGURE 19-9   Steps in the forming of a bolt by cold extrusion, cold heading, and thread rolling. *(Courtesy of National Machinery Co.)*

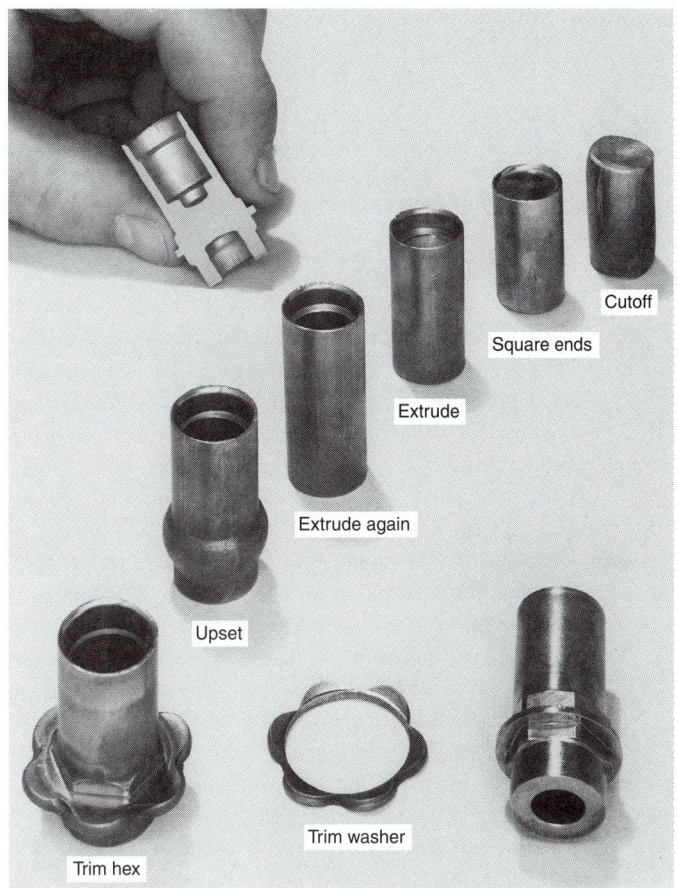

FIGURE 19-10   Cold-forming sequence involving cutoff, squaring, two extrusions, an upset, and a trimming operation. Also shown are the finished part and the trimmed scrap. *(Courtesy of National Machinery Co.)*

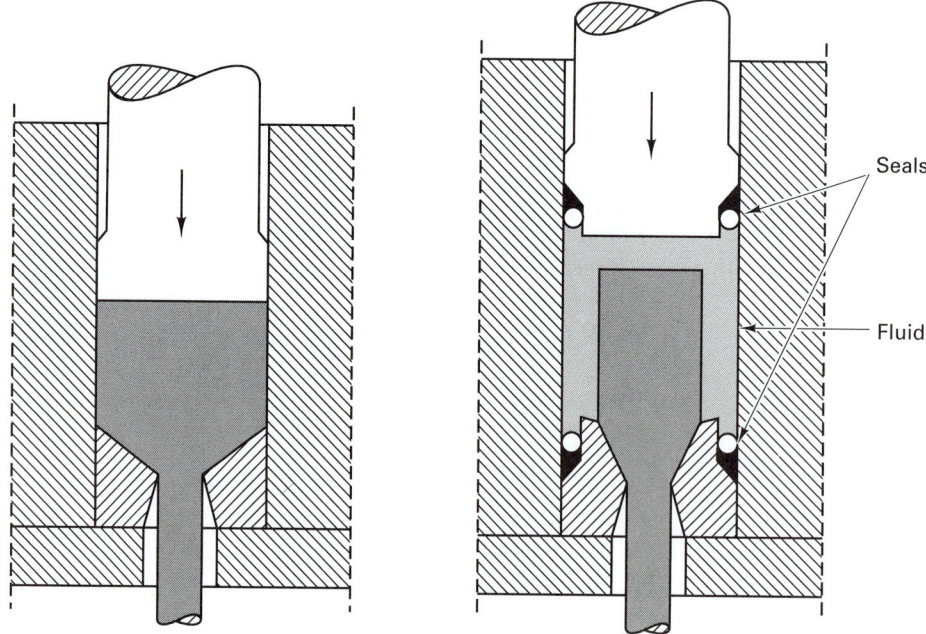

FIGURE 19-11 Comparison of conventional (*left*) and hydrostatic (*right*) extrusion. Note the addition of the pressurizing fluid and the O-ring and miter-ring seals on both the die and ram.

the workpiece through a die, with the product emerging either into atmospheric pressure or a lower-pressure fluid-filled chamber. The process resembles direct extrusion, but the fluid pressure surrounding the billet prevents upsetting. Billet-chamber friction is eliminated, and the pressurized fluid acts as a lubricant between the billet and the die. While process efficiency is greater than most other extrusion processes, there are problems associated with the fluid and high fluid pressures, which are typically in the range 125 to 250 ksi. Temperatures are limited since the fluid acts as a heat sink and the common fluids (light hydrocarbons and oils) burn or decompose at moderately low temperatures. Seals must be designed to contain the pressurized fluid without leaking, and measures must be taken to prevent the complete ejection of the product, often referred to as *blowout*. Because of these features, hydrostatic extrusion is usually employed only where the process offers unique advantages that cannot be duplicated by the more conventional methods.

One such advantage is the pressure-to-pressure process, in which the product emerges into a second high-pressure chamber. In effect, the metal deformation is performed in a high-compression environment. Crack formation is suppressed, leading to a phenomenon known as *pressure-induced ductility*. Relatively brittle materials such as molybdenum, beryllium, tungsten, and various intermetallic compounds can be plastically deformed without fracture, and materials with limited ductility become highly plastic. Thus products can be made that could not otherwise be produced, and materials can be considered that would have been rejected because of their limited ductility at room temperature and atmospheric pressure.

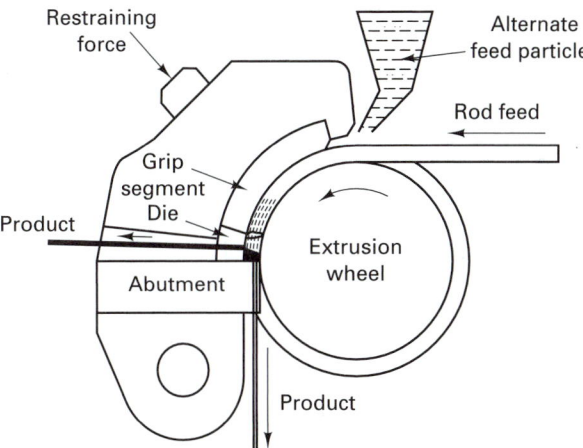

FIGURE 19-12   Cross-sectional schematic of the Conform continuous extrusion process showing various types of feeds and alternate locations of product withdrawal. *(Courtesy of United Kingdom Atomic Energy Authority, Springfields Nuclear Laboratories.)*

Continuous Extrusion

Conventional extrusion is a discontinuous process, converting finite-length billets into finite-length products. If the pushing force could be applied to the periphery of the feed-stock, rather than the back, continuous feedstock could be converted to continuous product, and the process could become one of *continuous extrusion*. The first continuous extrusion of metal feedstock was performed in 1970. Since then a number of techniques have been proposed, and these have met with varying degrees of success. In terms of commercial manufacturing, the most significant is probably the *Conform process*, illustrated schematically in Figure 19-12. Continuous feedstock is inserted into a grooved wheel and is friction propelled into a mating die shoe. Here it is upset to conform to the chamber, pressurized, and finally extruded through a die opening.

Since surface friction is the propulsion force, the feedstock can be solid rod, metal powder, punchouts, or chips from machining operations. Metallic and nonmetallic powders can be intimately mixed and extruded. Rapidly solidified materials can be formed to shape without exposure to the elevated temperatures that would harm their properties. Polymeric materials and even fiber-reinforced plastics have been extruded. The primary feed, however, is nonferrous metal rodstock, generally based on aluminum or copper.

Continuous extrusion complements and competes with wire drawing and shape rolling as a means of producing nonferrous products with small, but uniform, cross sections. It is particularly attractive for complex profiles and cross sections that contain one or more holes, since extrusion operations can perform massive reductions through a single die. One Conform extrusion can produce an amount of deformation equivalent to 10 conventional drawing or cold-rolling passes. In addition, sufficient heat can be generated that the product can emerge in an annealed condition, ready for further processing without intermediate heat treatment.

Roll Extrusion

Thin-walled cylinders can be produced from thicker-wall material by the *roll-extrusion process*. As depicted in Figure 19-13a, internal rollers increase the internal diameter as they squeeze material against an external confining ring. The tube elongates as the wall thickness is reduced. In Figure 19-13b, the internal diameter is maintained as external rollers squeeze the material against an internal mandrel. Although cylinders from 0.75 to

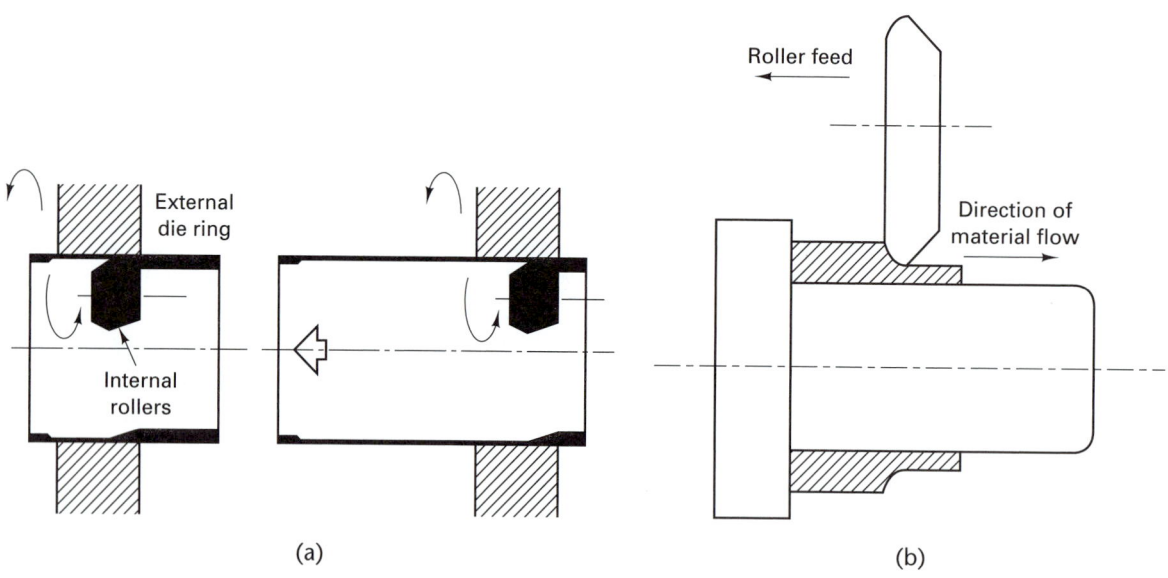

FIGURE 19-13 The roll extrusion process: (a) with internal rollers expanding the inner diameter; (b) with external rollers reducing the outer diameter.

156 in. in diameter have been made by this process, it is most commonly used for products with diameters between 3 and 20 in.

Sizing

Sizing involves squeezing all or selected areas of forgings, ductile castings, or powder metallurgy products, to achieve a desired thickness or precision. By using this process, designers can make the initial tolerances of a part more liberal, enabling the use of less-costly production methods. Those dimensions that must be precise are then set by one or more sizing operations that are usually performed on simple, mechanically driven presses.

Riveting

In *riveting*, a head is formed on the shank end of a fastener to permanently join sheets or plates of material. Although riveting is usually done hot in structural applications, in manufacturing, it is almost always done cold. Where there is access to both sides of the work, the method illustrated in Figure 19-14 is commonly used. The shaped punch may be held and advanced by a press or contained in a special, hand-held riveting hammer. When a press is used, the rivet is usually headed in a single squeezing action, although the heading punch may rotate to shape the head progressively, as in orbital forming. Special riveting machines, such as those used in aircraft assembly, can punch the hole, place the rivet in position, and perform the heading operation, all in about 1 second.

It is often desirable to use riveting in situations where there is access to only one side of the assembly. Special types of rivets are used for these "blind" applications, such as those depicted in Figure 19-15. The explosive type is activated by touching a heated tool to the rivet head. The heat causes the charge to explode, expanding the shank into a retaining head. In the pull type, or pop-rivet, the shank is mechanically expanded, after which the pull pin breaks or is cut off flush with the head.

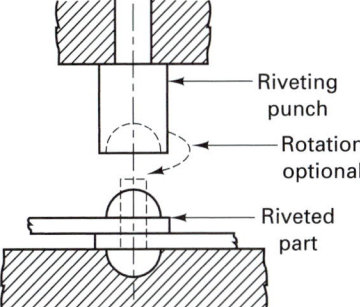

FIGURE 19-14 Joining components by riveting.

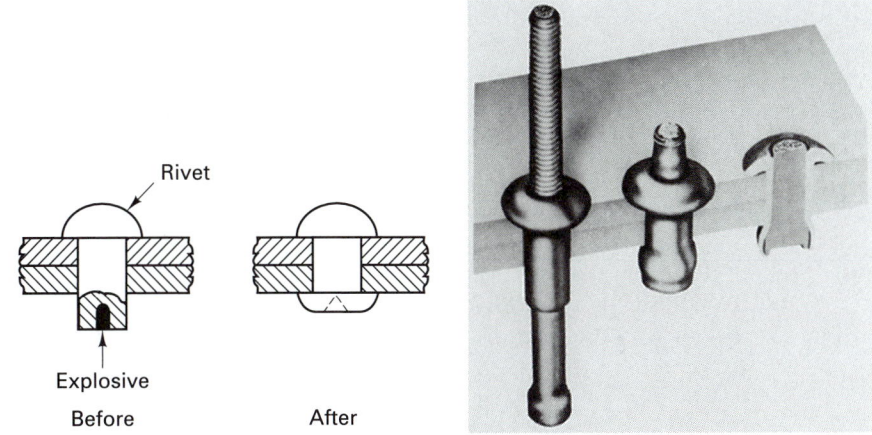

FIGURE 19-15 Rivets for use in "blind" riveting: (*left*) explosive type; (*right*) shank-type pull-up. *(Courtesy of Huck Manufacturing Company.)*

Staking

Staking is a method of permanently joining parts together when one part protrudes through a hole in the other. As shown in Figure 19-16, a shaped punch is driven into the end of the protruding piece. The deformation causes radial expansion, mechanically locking the two pieces together. Because the tooling is simple and the operation can be completed with a single stroke of a press, staking is a convenient and economical method of fastening when permanence is desired and the appearance of the punch mark is not objectionable. Figure 19-16 includes some of the decorative punch designs that are commonly used.

Coining

The term *coining* refers to the cold squeezing of metal while all of the surfaces are confined within a set of dies. The process, illustrated schematically in Figure 19-17, is used to produce coins, medals, and other products where exact size and fine detail are required, and thickness varies about a well-defined average. Because of the total confinement, there is no possibility for excess metal to flow from the die, and pressures as high as 200 ksi are often required. Accurate volumetric measurement of the metal is also essential to avoid breakage of the dies or press.

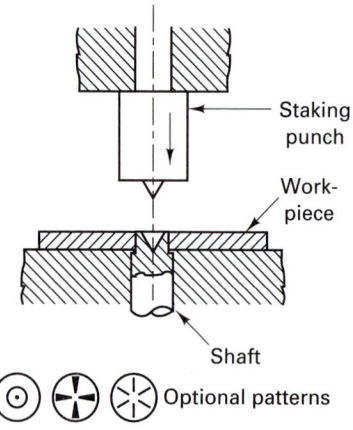

FIGURE 19-16   Fastening by staking.

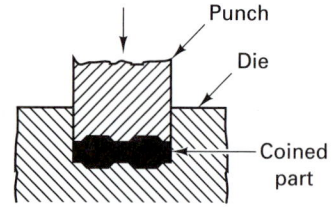

FIGURE 19-17   The coining process.

Hubbing

Hubbing[*] is a cold-working process that is used to form recessed cavities in various types of female dies. As shown in Figure 19-18, a male *hub* (or master) is made with the profile of the desired cavity. The hub is then hardened and pressed into an annealed die block (usually by a hydraulic press) until the desired impression is produced. Production of the cavity can often be aided by machining away some of the metal where large amounts of

FIGURE 19-18   Hubbing a die block in a hydraulic press. Inset shows close-up of the hardened hub and the impression in the die block. The die block is contained in a reinforcing ring. The upper surface of the die block is then machined flat to remove the bulged metal.

---

[*]This process should not be confused with hobbing, a machining process used for cutting gears (see Chapter 31).

plastic flow would occur. The die block is usually round and is reinforced by a heavy steel ring during the hubbing operation. After the cavity has been formed, the die is separated from the reinforcing ring, the displaced metal is removed by a facing-type machining cut, and the piece is hardened by heat treatment.

Hubbing is frequently more economical than conventional die sinking (machining the cavity). One hub can be used to form a number of identical cavities, and it is generally easier to machine a male profile than a female cavity.

Surface Improvement by Cold Working

Cold-working processes can also be used to improve or alter the surfaces of metal products. *Peening* is the mechanical working of surfaces by repeated blows of impelled shot or a round-nose tool. In most manufacturing operations, peening is produced by shot impellers. The highly localized impacts flatten and broaden the metal surface, but the broadening is resisted by the underlying material, resulting in a surface that is loaded in residual compression. Since the residual compression is subtracted from any applied tensile loadings, peening tends to enhance fracture resistance under a wide range of conditions, including fatigue. For this reason, shafting, crankshafts, connecting rods, gear teeth, and other cyclic-loaded components are frequently peened.

Due to solidification shrinkage and thermal contraction, the surfaces of most weldments are in residual tension. Peening can be used to reduce or cancel this effect, reducing any associated distortion and preventing cracking. For this type of peening, manual or pneumatic hammers are commonly used.

Burnishing involves rubbing a smooth, hard object (under considerable pressure) over the minute surface irregularities that are produced during machining or shearing. The edges of sheet metal stampings can be burnished by pushing the stamped parts through a slightly tapered die having its entrance end a little larger than the workpiece and its exit slightly smaller. As the part rubs along the sides of the die, the pressure is sufficient to smooth the slightly rough edges that are produced during the blanking operation (see Figure 19-35).

Roller burnishing, illustrated in Figure 19-19, can be used to improve the size and finish of cylindrical and conical surfaces, both internal and external. The hardened rolls of the tool press against the surface and deform the protrusions to a more-nearly-flat geometry. Since the surfaces are cold worked and in residual compression, they possess improved wear and fatigue resistance.

■ 19.3 BENDING

As we approach the subject of bending, it is important to first establish a few definitions. *Bending* is the plastic deformation of metals about a linear axis with little or no change in the surface area. Multiple bends can be made simultaneously, but to be classified as true bending, and treatable by simple bending theory, each axis must be linear and independent of the others. When multiple bends are made with a single die, the process is sometimes called *forming*. If the axes of deformation are not linear, or are not independent, the process becomes one of drawing and/or stretching, not bending. These operations are treated later in the chapter.

As shown in Figure 19-20, bending causes the metal on the outside to be stretched while that on the inside is compressed. The location that is neither stretched nor compressed is known as the *neutral axis* of the bend. Since the yield strength of metals in compression is somewhat higher than the yield strength in tension, the metal on the outer side yields first, and the neutral axis is displaced from the center of the two surfaces. The neutral axis

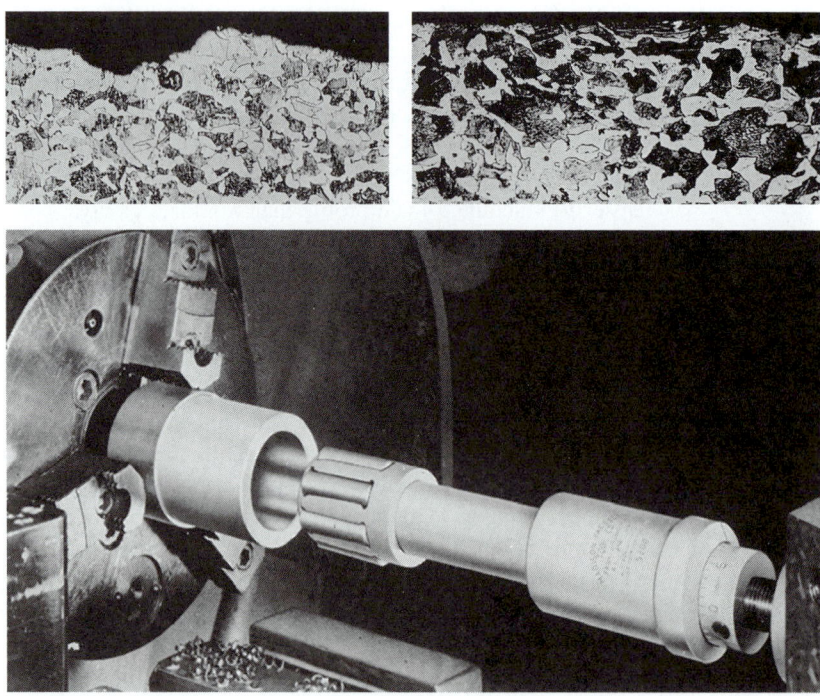

FIGURE 19-19   Tool for roller burnishing: (*top left*) surface structure before burnishing; (*top right*) surface structure after burnishing. The burnishing rolls are moved outward by means of a taper. (*Courtesy of Madison Industries, Inc.*)

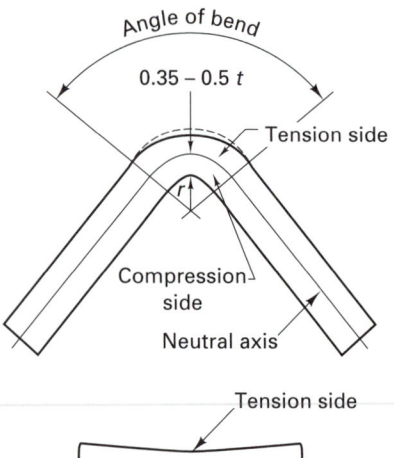

FIGURE 19-20   (*Top*) Nature of a bend in sheet metal showing tension on the outside and compression on the inside; (*bottom*) upper portion of the bend region, viewed from the side, showing how the center portion thins more than the edges.

generally locates between one-third and one-half of the way from the inner surface, depending on the bend radius and the material being bent. Because of the preferred tensile deformation, the metal is thinned somewhat at the bend. This thinning is most pronounced in the center of the sheet, where the material cannot compensate by the sideways contraction of the free edges. Both of these effects are shown schematically in Figure 19-20.

Moving to the inner side of the bend, it is possible for the compressive forces to induce upsetting. While the associated thickening somewhat offsets the thinning of the outer

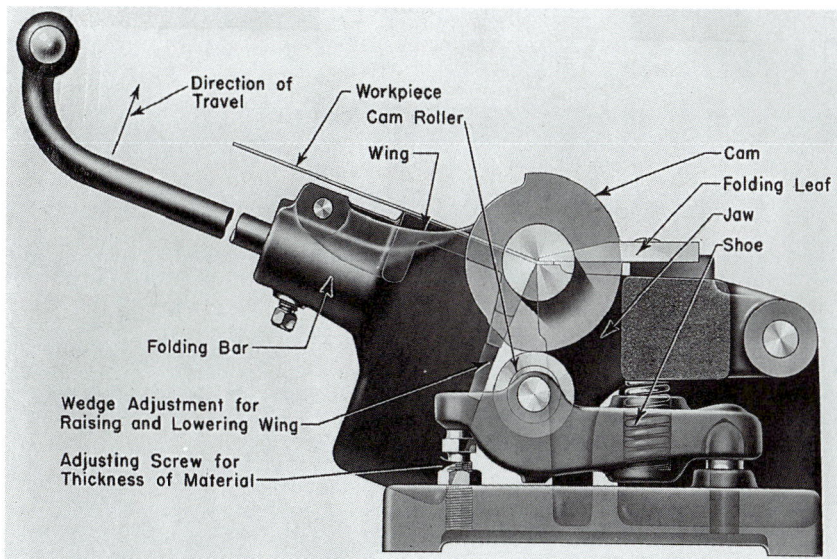

FIGURE 19-21 Phantom section of a bar folder, showing position and operation of internal components. *(Courtesy of Niagara Machine and Tool Works.)*

section, the upsetting can also induce an outward expansion of the free edges. The distortion of the edge surfaces can become quite pronounced in the bending of thick, narrow plates.

Still another consequence of the combined tension and compression is the elastic recovery that occurs during unloading. This manifests itself as a small amount of "unbending," usually referred to as *springback*. Thus, if a specified angle is desired, the metal must be slightly overbent to compensate for the unbending action of springback.

Angle Bending

Machines like the *bar folder*, shown in Figure 19-21, can be used to make angle bends up to 150° in sheet metal under $\frac{1}{16}$ in. (1.5 mm) thick. The workpiece is inserted under the folding leaf and aligned in the proper position. Raising the handle then actuates a cam, causing the leaf to clamp the sheet. Further motion of the handle bends the metal to the desired angle. Bar folders are manually operated and produce linear bends up to about 12 ft. in length.

Bends in heavier sheet, or more complex bends in thin material, are generally made on *press brakes*, like the one shown in Figure 19-22. These are mechanically or hydraulically driven presses with a long, narrow bed and short, adjustable strokes. The metal is bent between interchangeable dies that are attached to both the bed and the ram. As illustrated in Figures 19-22 and 19-23, different dies can be used to produce many types of bends. The metal can be repositioned between strokes to produce complex contours or repeated bends, such as corrugations. Figure 19-24 shows how a roll bead can be formed with repeated strokes, repositioning, and more than one set of tooling. While seaming, embossing, punching, and other operations can also be performed in press brakes, these operations can usually be done more efficiently on other types of equipment.

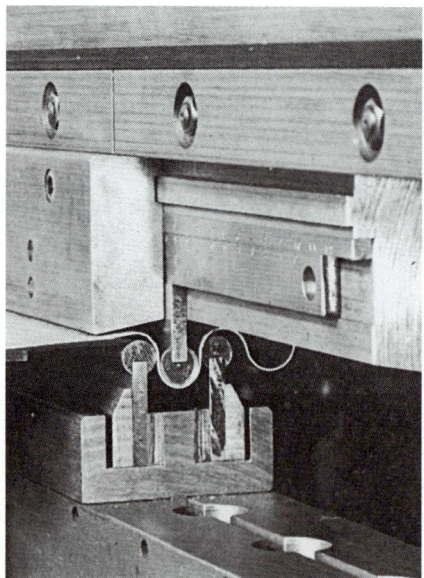

FIGURE 19-22 (*Left*) Press brake with CNC gauging system. (*Courtesy of DiAcro Division, Houdaille Industries, Inc.*) (*Right*) Close-up view of press brake dies. (*Courtesy of Cincinnati Incorporated.*)

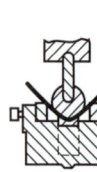
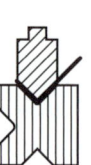
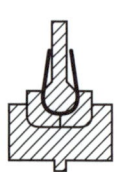

FIGURE 19-23 Press brake dies used to form various angles and rounds. (*Courtesy of Cincinnati Incorporated.*)

Design for Bending

Several factors must be considered when designing parts that are to be shaped by bending. One of the primary concerns is determining the smallest bend radius that can be formed without metal cracking (i.e., the *minimum bend radius*). This is directly dependent on the ductility of the metal and can be related to the percent reduction in area observed in a standard tensile test. Figure 19-25 shows how the ratio of the minimum bend radius R to the thickness of the material t varies with material ductility. As can be noted, it requires an extremely ductile material to produce a bend with radius less than the thickness of the metal. In general, bends should be designed with the largest possible radii. This permits easier forming and allows the designer to select from a wider variety of engineering materials.

If the metal has previously been cold worked or has marked directional properties, these features should be considered in the bending operation. Whenever possible, it is best to make the bend axis perpendicular to the direction of previous working. If two perpendicular bends are involved, it is preferable to orient the bend axes at angles of 45° to the direction of prior rolling.

  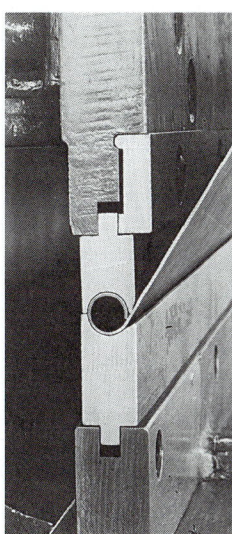

FIGURE 19-24   Dies and stages used in the press brake forming of a roll bead. *(Courtesy of Cincinnati Incorporated.)*

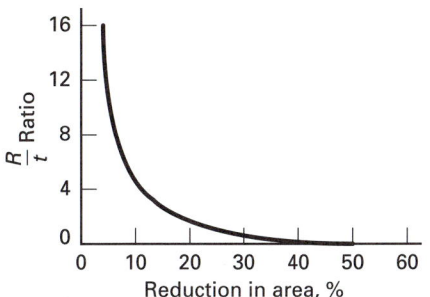

FIGURE 19-25   Relationship between the minimum bend radius (relative to thickness) and the ductility of the metal being bent (as measured by the reduction in area in a uniaxial tensile test).

Another design concern is determining the dimensions of a flat blank that will produce a bent part of the desired precision. The fact that the neutral axis is not at the centerline of the metal causes metal to thin and lengthen during bending. The amount of lengthening is a function of both the stock thickness and the bend radius. Figure 19-26 illustrates one method that has been found to give satisfactory results in determining the blank length for bent products. It may also be important to determine the minimum length of a leg that can be successfully bent. In most cases, the length of a protruding leg should be at least $1\frac{1}{2}$ times the thickness of the metal plus the bend radius.

Whenever possible, the tolerance on bent parts should not be less than $\frac{1}{32}$ in. Severe bends (90° or greater) should not be specified without first determining whether the material and bending method will permit them. Parts with multiple bends should be designed with most (or all) of them to be of the same bend radius. This will reduce setup time and tooling costs. Consideration should also be given to providing regions for adequate clamping or support during manufacture. Bending near the edge of a material will distort the edge. If an undistorted edge is required, additional material must be included and a trimming operation performed after bending.

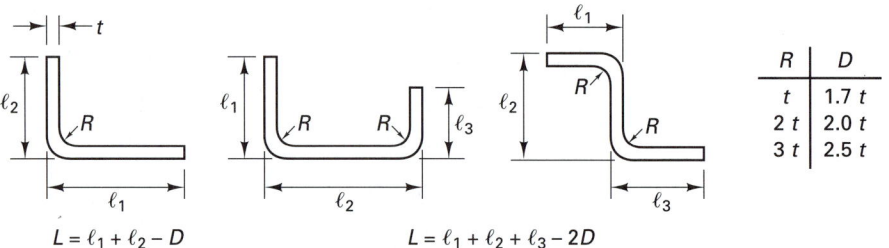

FIGURE 19-26 One method of determining the blank size for several bending operations. Due to thinning, the product will lengthen during forming.

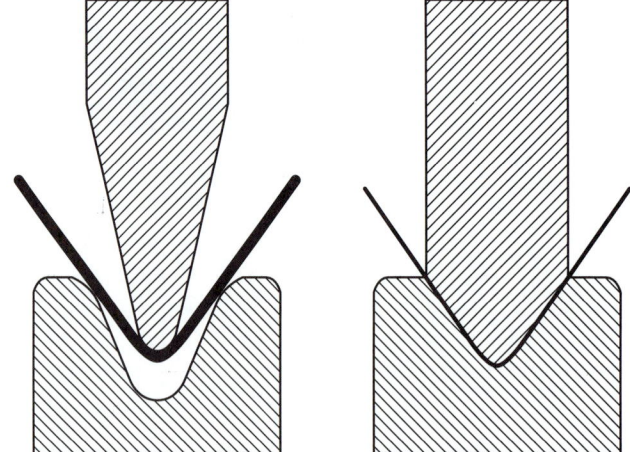

FIGURE 19-27 Comparison of air-bend (*left*) and bottoming (*right*) press brake dies. With the air-bend die, the amount of bend is controlled by the bottoming position of the upper die.

Air-Bend Versus Bottoming Dies

A decision relating to equipment and tooling involves the use of air-bend or bottoming dies. As shown in Figure 19-27, *bottoming dies* compress the full area within the tooling. The angle of the resulting bend is set by the geometry of the tooling, as modified by springback. If the results are outside specifications, or the material is changed and produces a different springback, the geometry of the tooling will have to be modified. Once the geometry is set, however, reproducibility of the bend geometry is excellent.

In contrast, *air-bend dies* produce the geometry by simple three-point bending, and the resulting angle is controlled by setting the bottoming position of the upper die. The bend angle is adjustable, and reproducibility and process control must be maintained through control of the press stroke. Adaptive control and on-the-fly corrections are possible with air-bend tooling.

Roll Bending

A continuous form of three-point bending is *roll bending*, where plates, sheets, and rolled shapes can be bent to a desired curvature on forming rolls, like those shown in Figure 19-28. These machines usually have three rolls in the form of a pyramid, with the two lower rolls being driven and the position of the upper roll being adjustable to control the degree of curvature. When long rolls are supported by a frame on each end, one of the supports can often be swung clear to permit the removal of closed shapes from the rolls.

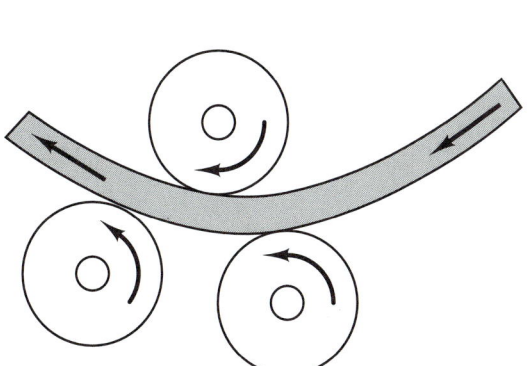

FIGURE 19-28   (*Left*) Schematic of the roll-bending process; (*right*) the roll bending of an I-beam section. Note how the material is continuously subjected to three-point bending. (*Courtesy of Buffalo Forge Company.*)

Roll bending machines are available in a wide range of sizes, some being capable of bending plate up to 6 in. thick.

Draw Bending and Compression Bending

Bending machines can also utilize a clamp and pressure tool to produce bending about a form block. *Draw bending*, illustrated in Figure 19-29, is a particularly versatile and accurate means of bending. The workpiece is clamped against a bending form and the entire assembly rotates to draw the workpiece across a stationary pressure tool. In *compression bending*, also shown in Figure 19-29, the bending form remains stationary and the pressure tool moves along the workpiece.

Cold-Roll Forming

The *roll forming* of flat strip into complex sections has become a highly developed forming technique that competes directly with press brake forming, extrusion, and stamping. As depicted in Figures 19-30 and 19-31, the process involves the progressive bending of metal strip as it passes through a series of forming rolls at speeds up to 100 ft/min. Any material that can be bent can be roll formed, including cold rolled, hot rolled, polished, prepainted, coated, and plated metals in thicknesses ranging from 0.005 through $\frac{3}{4}$ in. Various moldings, channeling, gutters and downspouts, automobile beams and bumpers, and other shapes of uniform wall thickness and uniform cross section are now being formed.

   By changing the rolls, a single machine can produce a wide variety of different shapes. However, changeover, setup, and adjustment may take several hours, so a production run of at least 10,000 ft of a given product is usually required. When tubes or pipe are desired, a resistance welding unit or seaming operation can be combined with the roll forming.

Seaming and Flanging

Seaming is a bending operation that can be used to join the ends of sheet metal to form containers such as cans, pails, and drums. The seams are formed by a series of small rollers

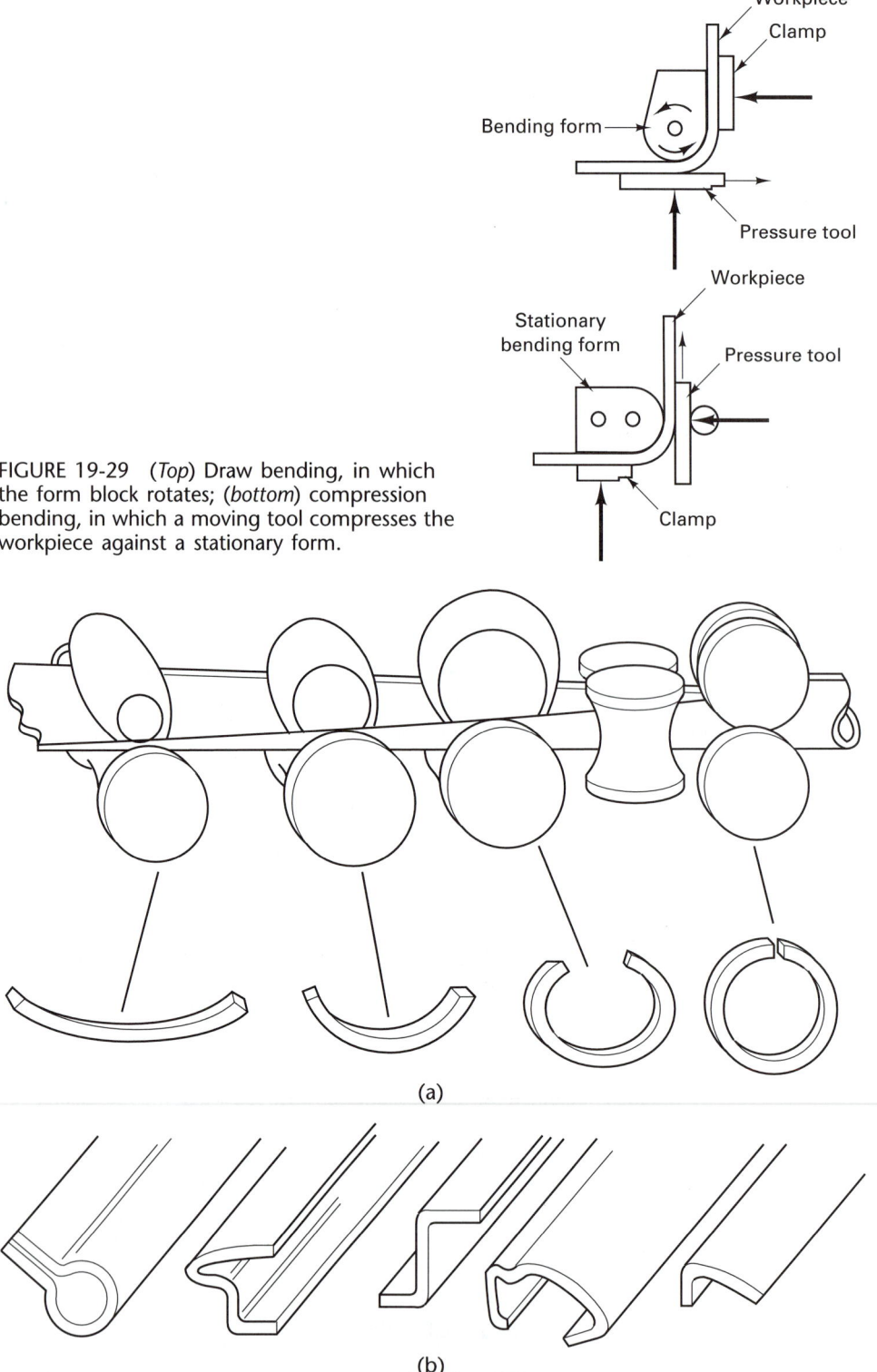

FIGURE 19-29   (*Top*) Draw bending, in which the form block rotates; (*bottom*) compression bending, in which a moving tool compresses the workpiece against a stationary form.

(a)

(b)

FIGURE 19-30   (a) Schematic representation of the cold roll-forming process being used to convert sheet or plate into tube. (b) Some typical shapes produced by roll forming.

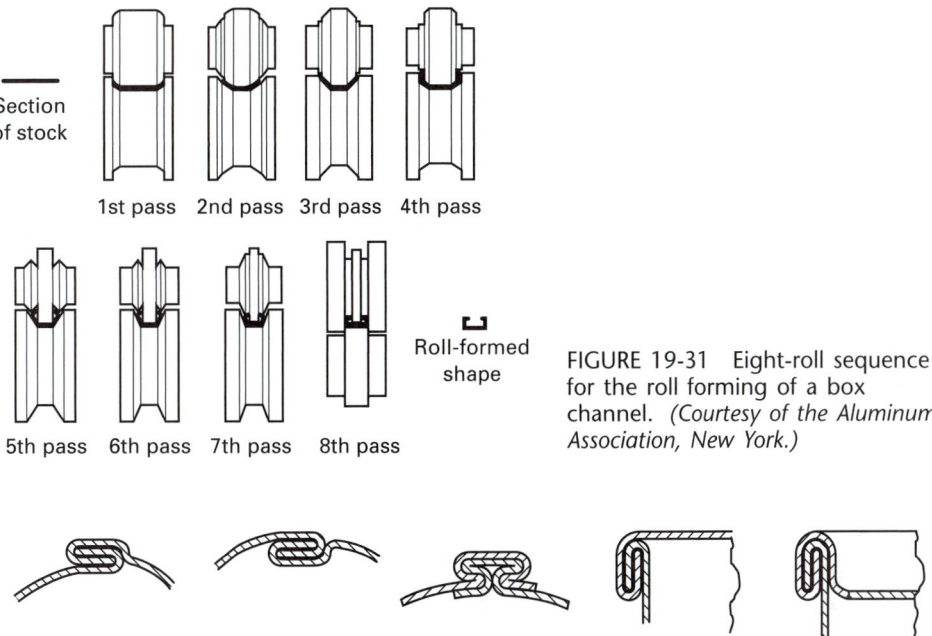

Section of stock

1st pass   2nd pass   3rd pass   4th pass

5th pass   6th pass   7th pass   8th pass

Roll-formed shape

FIGURE 19-31   Eight-roll sequence for the roll forming of a box channel. *(Courtesy of the Aluminum Association, New York.)*

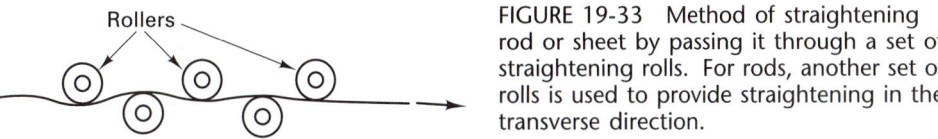

FIGURE 19-32   Various types of seams used on sheet metal.

Rollers

FIGURE 19-33   Method of straightening rod or sheet by passing it through a set of straightening rolls. For rods, another set of rolls is used to provide straightening in the transverse direction.

on seaming machines that range from small hand-operated types to large automatic units capable of producing hundreds of seams per minute. Figure 19-32 shows several of the more common seam designs.

Flanges can be rolled on sheet metal in essentially the same manner as seaming. In many cases, however, the forming of flanges and seams is a drawing operation, since the bending occurs along a curved axis.

Straightening

The objective of *straightening* or *flattening* is the opposite of bending, and these operations are often performed before subsequent cold forming to assure the use of flat or straight material. *Roll straightening* or *roller leveling*, illustrated in Figure 19-33, subjects the material to a series of reverse bends. The rod, sheet, or wire is passed through a series of rolls with progressively decreased offsets from a straight line. As the metal is bent back and forth, it is stressed slightly beyond its elastic limit, thereby replacing any permanent set with a flat or straight profile.

Sheet may also be straightened by a process called *stretcher leveling*. Here the sheets are gripped mechanically and stretched slightly beyond the elastic limit to produce the desired flatness.

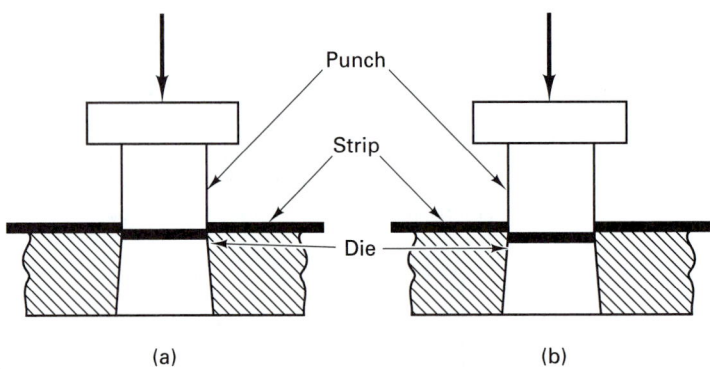

FIGURE 19-34 Simple blanking with a punch and die.

■ 19.4 SHEARING OPERATIONS

Shearing is the mechanical cutting of materials without the formation of chips or the use of burning or melting. When the two cutting blades are straight, the process is called shearing. When the blades are curved, the processes have special names, such as *blanking*, *piercing*, *notching*, and *trimming*. In terms of tool design and material behavior, all are shearing-type operations.

A simple type of shearing operation is illustrated in Figure 19-34. As the punch (or upper blade) pushes the workpiece, the metal responds by plastically flowing into the die (or over the lower blade). Because the clearance between the two tools is small, usually between 5 and 10% of the thickness of the metal being cut, the deformation is in the form of highly localized shear. The punch pushes downward on the metal, the material flows into the die, and the opposite surface bulges slightly. When the penetration is between 15 and 60% of the metal thickness, the actual amount depending on the material ductility and strength, an instability arises. The applied stress exceeds the shear strength and the metal tears or ruptures through the remainder of its thickness. These two stages of the shearing process, deformation and fracture, are often visible on the edges of sheared parts, as shown in Figure 19-35.

Because of the normal nonhomogeneities in a metal and the possibility of nonuniform clearance between the shear blades, the final shearing does not occur in a uniform manner. Fracture and tearing begins at the weakest point and proceeds progressively and intermittently to the next-weakest location. The result is usually a rough and ragged edge.

If the punch and die (or shearing blades) have proper clearance and are maintained in good condition, sheared edges may be produced that are sufficiently smooth that they may be used without further finishing. The quality of the sheared edge can be further improved if the strip stock is firmly clamped against the die (from above), the punch and die are maintained with proper clearance and alignment, and the movement of the piece through the die is restrained by an opposing plunger or rubber die cushion that applies pressure from below the workpiece. These measures cause the shearing to take place more uniformly around the perimeter of the cut.

If the shearing is performed in a compressive environment, fracture is suppressed and the relative amount of smooth edge (produced by deformation) is increased. Above a certain pressure, no fracture occurs and the entire edge is smooth. Figure 19-36 schematically depicts one method of generating this compressive environment, a process known as *fineblanking*. A V-shaped protrusion is incorporated into the hold-down or pressure plate at

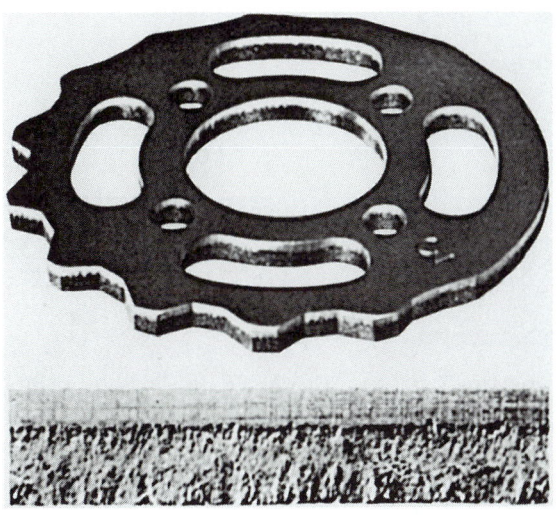

FIGURE 19-35   Conventionally sheared surface showing the distinct regions of deformation and fracture. *(Courtesy of American Feintool Inc.)*

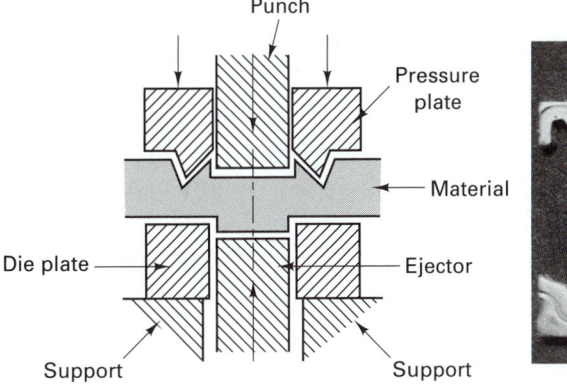
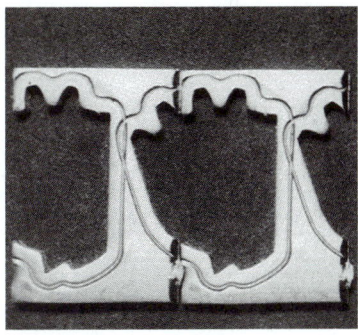

FIGURE 19-36   *(Left)* Method of obtaining a smooth edge in shearing by using a shaped pressure plate to put the metal into localized compression and a squeezing ejector descending in unison with the punch. *(Courtesy of Metal Progress.)* *(Right)* Stock skeleton after shearing, showing the compressed indentation. *(Courtesy of Clark Metal Products Company.)*

a location slightly external to the contour of the cut. As the hold-down is pressed against the material to be sheared, the protrusion penetrates the surface and places the region to be cut into a localized state of compression. Matching upper and lower punches then grip the material and descend in unison to extract the desired segment with a smooth and square sheared edge, as shown in Figure 19-37. A reduced clearance is used between the punch and the die, and a triple action press is generally required. Fineblanked parts are usually less than $\frac{1}{4}$ in. (6 mm) in thickness and typically possess complex-shaped perimeters. Dimensional tolerances are quite small, and the production of holes, slots, bends, and semipierced projections can often be incorporated into the fineblanking operation.

Another means of shearing under compression is illustrated in Figure 19-38. Incoming bar stock is pressed against the closed end of a feed hole, putting the stock in a state of compression. A transverse punch then shears the material into burr-free slugs, ready for further processing.

FIGURE 19-37 Fine-blanked surface of the same component as shown in Figure 19-35. *(Courtesy of American Feintool, Inc.)*

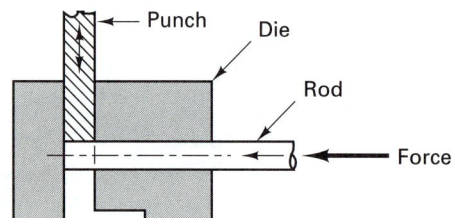

FIGURE 19-38 Method of smooth shearing rod by putting it into compression during shearing.

FIGURE 19-39 A 10-ft (3-m) power shear for $\frac{1}{4}$-in. (6.5-mm) steel. *(Courtesy of Cincinnati Incorporated.)*

Simple Shearing

When sheets of metal are to be sheared along a straight line, *squaring shears*, like the one shown in Figure 19-39, are frequently used. As the upper ram descends, a clamping bar or set of clamping fingers presses the sheet of metal against the machine table to hold it firmly in position. The moving blade then comes across the fixed blade and shears the metal. On larger shears, the moving blade is often set at an angle or "rocks" as it descends, making the cut in a progressive fashion from one side of the material to the other. This action significantly reduces the amount of cutting force required, although the total energy expended is still the same. (Since work is force × distance, a low force-long stroke operation can often be used in place of a high force and short stroke.)

Slitting

Slitting is the shearing process used to cut rolls of sheet metal into several rolls of narrower width. Here the shearing blades take the form of circumferential mating grooves on

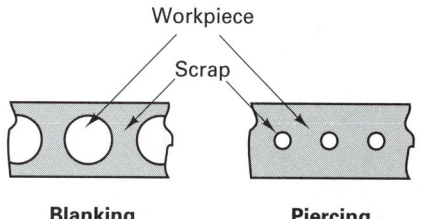

Blanking   **Piercing**

FIGURE 19-40   Schematic showing the difference between piercing and blanking.

FIGURE 19-41   (*Left to right*) Piercing and lancing, and blanking precede the forming of the final ashtray. The small round holes assist positioning and alignment.

cylindrical rolls, the raised ribs of one roll matching the recessed grooves on the other. The process is continuous and can be performed rapidly and economically. Moreover, since the distance between adjacent sets of shearing edges is fixed, the resultant strips have accurate and constant width, more consistent than that obtained from alternative cutting processes.

Piercing and Blanking

Piercing and *blanking* are shearing operations where the shear blades are closed, curved lines along the edges of a punch and die. Both involve the same basic cutting action, the primary difference being one of definition. Figure 19-40 shows that in blanking, the piece being punched out becomes the workpiece and any major burrs or undesirable features should be left on the remaining strip. In piercing, the punch-out is the scrap and the remaining strip is the workpiece. Piercing and blanking are usually done on some form of mechanical press.

There are a number of variations of piercing and blanking and some have come to acquire specific names. *Lancing* is a piercing operation that forms either a line cut (slit) or an actual hole in the metal, like those shown in the left-hand portion of Figure 19-41. The purpose of lancing is to permit the adjacent metal to flow more readily in subsequent forming operations. In the case illustrated, the lancing makes it easier to form the radial grooves, which were shaped before the ashtray was blanked from the strip stock and shallow drawn to final shape.

Perforating consists of piercing a large number of closely spaced holes.

Notching is essentially the same as piercing, except that the edge of the strip or blank forms part of the punch-out perimeter. It is used to remove segments from along the edge of an existing product.

In *nibbling*, a contour is cut by producing a series of overlapping slits or notches, as shown in Figure 19-42. In this manner, simple tools can be used to cut a complex shape

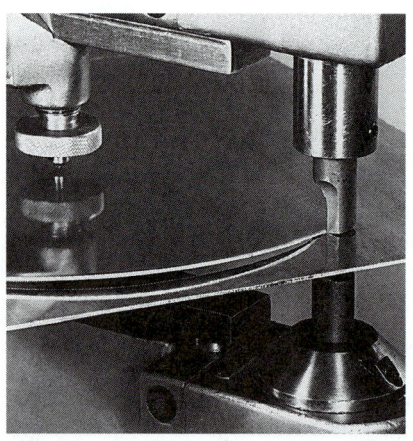

FIGURE 19-42 Shearing operations being performed on a nibbling machine. *(Courtesy of Tech-Pacific.)*

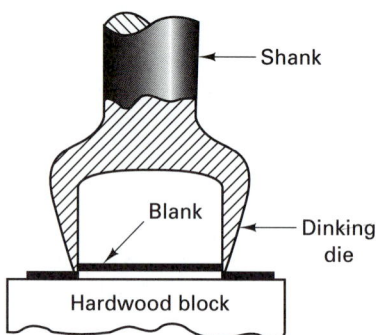

FIGURE 19-43 The dinking process.

from sheets of metal up to $\frac{1}{4}$ in. thick. Interior segments can be removed by starting the cut from a punched or drilled hole.

Shaving is actually a finishing operation in which a small amount of metal is sheared away from the edge of an already blanked part. Its primary use is to obtain greater dimensional accuracy, but it may also be employed to produce a squared or smoother edge. Because only a small amount of metal is removed, the punches and dies must be made with very little clearance. Blanked parts, such as small gears, can be shaved to produce dimensional accuracies within 0.001 in.

A *cutoff* is a punch and die operation used to separate a stamping or other product from a strip of stock. Frequently, the contour of the cutoff completes the periphery of the workpiece. Cutoff operations are quite common in progressive die sequences, like several to be presented shortly.

Dinking is a modified shearing operation that is used to blank shapes from low-strength materials, such as rubber, fiber, or cloth. As illustrated in Figure 19-43, the shank of a die is either struck with a hammer or mallet or the entire die is driven downward by some form of mechanical press.

Tools and Dies for Piercing and Blanking

As shown in Figure 19-44, the basic components of a piercing and blanking die set are a *punch*, a *die*, and a *stripper plate*, which is attached above the die to keep the strip from

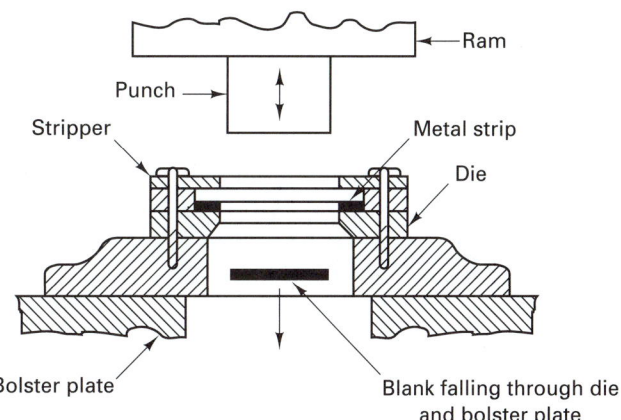

FIGURE 19-44   The basic components of piercing and blanking dies.

ascending with the retracting punch. The position of the stripper plate and the size of its hole should be such that it does not interfere with either the horizontal motion of the strip as it feeds into position or the vertical motion of the punch.

Theoretically, the punch should just fit within the die with a uniform clearance that approaches zero. On its downward stroke, it should not enter the die but should stop just as its base aligns with the top surface of the die. In general practice, the clearance is between 5 to 7% of the stock thickness and the punch enters slightly into the die cavity.

If the face of the punch is normal to the axis of motion, the entire perimeter is cut simultaneously. By tilting the punch face on angle, a feature known as *shear*, the cutting force can be reduced substantially. As shown in Figure 19-45, the periphery is now cut in a progressive fashion, similar to the action of a pair of scissors or the opening of a "pop-top" beverage can. Variation in the shear angle controls the length of cut that is being made at any given time and the total stroke that is necessary to complete the operation. Adding shear to a punch reduces the force but increases the stroke and may be necessary if existing equipment is expected to cut thicker or stronger material.

It is also important that the punch and die be in proper alignment so that a uniform clearance is maintained around the entire periphery. The die is usually attached to the bolster plate of the press, which, in turn, is attached to the main press frame. The punch is attached to the movable ram, enabling motion in and out of the die with each stroke of the press. The punch and die can also be mounted on a *punch holder* and *die shoe*, like the one shown in Figure 19-46, to create an independent die set. The holder and shoe are permanently aligned and guided by two or more guide pins. Once a punch and die are aligned and fastened to the die set, the entire unit can be inserted into a press without having to set or check the tool alignment. This can significantly reduce the amount of production time lost during tool change. Moreover, when a given punch and die are no longer needed, they can be removed and new tools attached to the shoe and holder assembly.

In most cases the punch holder is attached directly to the ram of the press, and ram motion acts to both raise and lower the punch. On small die sets, springs can be incorporated to provide the upward motion. The press ram contacts the top of the punch holder and forces it downward. As the ram retracts, the springs cause the punch to return to its starting position. This form of construction makes the die set fully self-contained. It can be inserted and removed from a press quite rapidly, thereby reducing setup time.

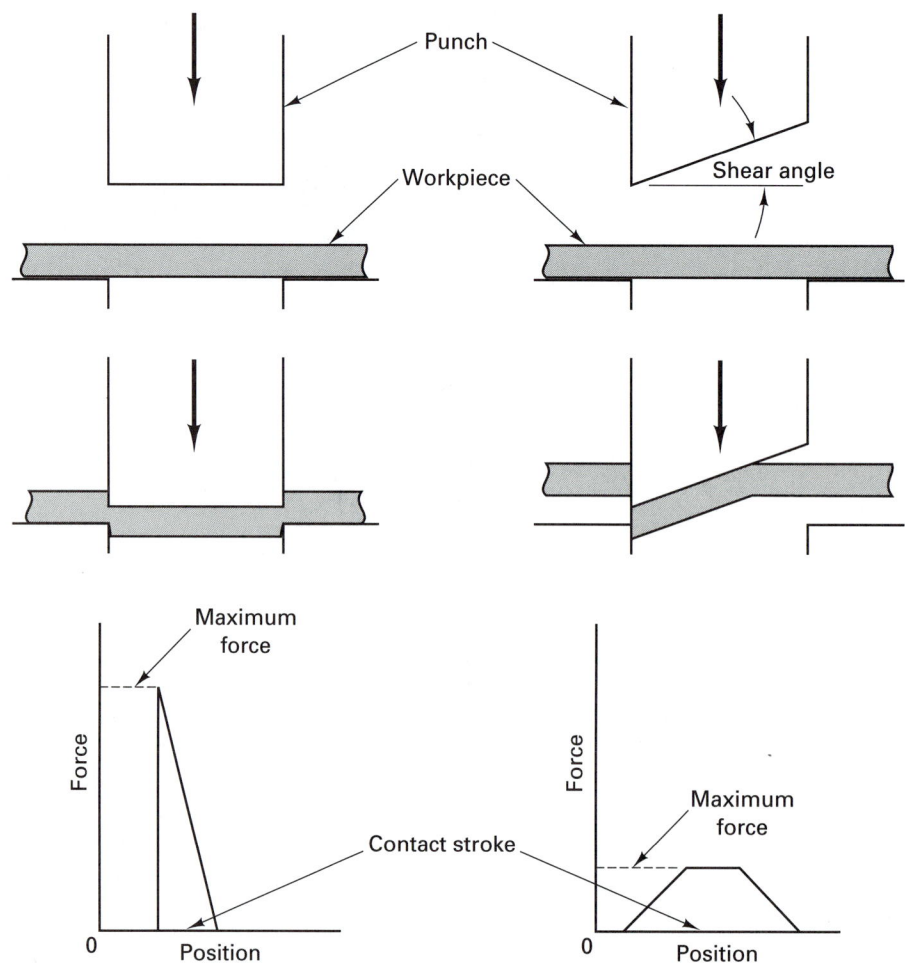

FIGURE 19-45   Blanking with a square-faced punch (*left*) and one containing angular shear (*right*). Note the difference in maximum force and contact stroke. The total work (the area under the curve) is the same for both processes.

A wide variety of standardized, self-contained die sets have been developed. Known as *subpress dies*, these can often be assembled and combined on the bed of a press to pierce or blank large parts that would otherwise require large and costly complex die sets. Figure 19-47 shows an assembly of subpress dies.

Punches and dies are usually made of nondeforming, or air-hardening, tool steel so they can be hardened after machining without danger of warpage. The die profile is maintained for a depth of about $\frac{1}{8}$ in. from the upper face, beyond which an angular clearance or back relief is provided (see Figure 19-44) to reduce friction between the part and the die and to permit the part to fall freely from the die after being sheared. The $\frac{1}{8}$ in. depth provides adequate strength and sufficient metal so that the die can be resharpened by grinding a few thousandths of an inch from its face.

Dies may be made in a single piece, or they can be made in component sections that are assembled on the die shoe. The latter procedure simplifies production and enables the

FIGURE 19-46 Typical die set having two alignment guideposts. *(Courtesy of Danly Machine Specialties, Inc.)*

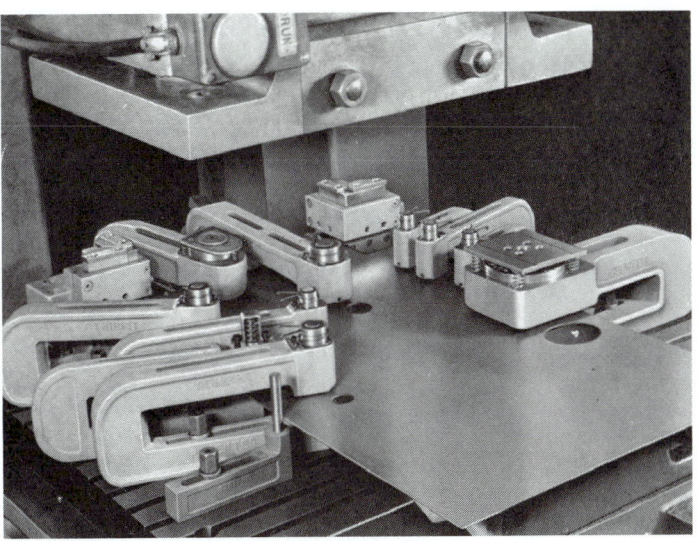

FIGURE 19-47 A piercing and blanking setup using self-contained subpress tool units. *(Courtesy of Strippit Division, Houdaille Industries, Inc.)*

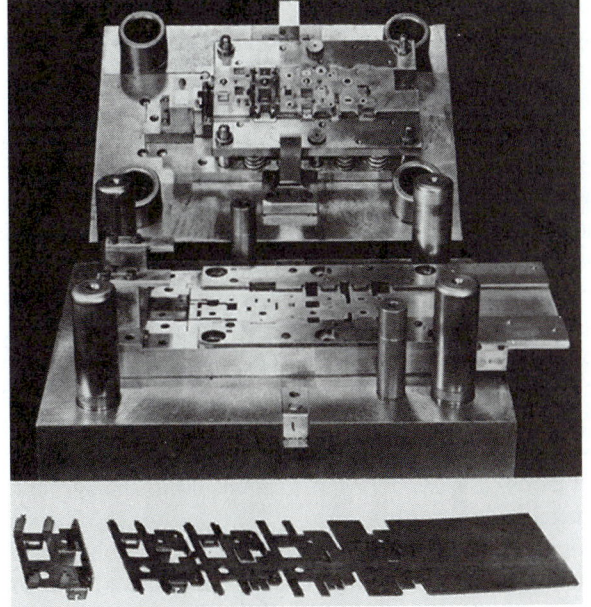

FIGURE 19-48 A progressive piercing, forming, and cutoff die set built up mostly from standard components. The part produced is shown at the bottom. *(Courtesy of Oak Manufacturing Company.)*

replacement of single sections in the event of wear or fracture. Complex dies like the one shown in Figure 19-48 can often be assembled from the many standardized punch and die components that are available. Substantial savings can often be achieved by modifying the design of parts so that standard die components can be utilized. An added advantage of this approach is that when the die set is no longer needed, the components can be removed and used to construct tooling for another product.

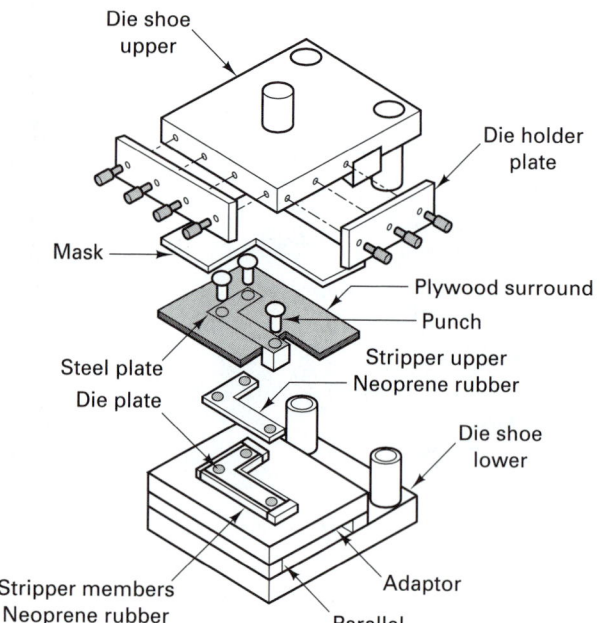

FIGURE 19-49   Construction of a steel-rule die set. *(Courtesy of J. J. Raphael.)*

Another technique that can be used to cut metal and a wide variety of softer materials is the "steel-rule" or "cookie-cutter" die, like the one illustrated in Figure 19-49. Here the cutting die is fashioned from hardened steel strips, known as steel rule, that are mounted on edge and held in position by some means, such as saw-cut grooves in a piece of plywood. The mating piece of tooling may be either a flat piece of hardwood or steel, a male shape that conforms to the part profile, or a set of matching grooves into which the protruding rule can descend. Neoprene rubber pads are usually inserted between the strips to replace the stripper plate. During the compression stroke, the rubber compresses and allows the cutting action to proceed. As the ram ascends, the rubber then expands to push the blank free of the steel-rule cavity. Steel-rule dies are less expensive to construct than solid dies and are therefore more attractive for producing small quantities of parts.

Most of the tooling that has been presented can perform only simple piercing or simple blanking. Many parts, however, require multiple cutting-type operations, and it is often desirable to perform these with a single cycle of a press. Multiple operations can be accomplished with the types of dies shown in Figures 19-50 and 19-51. For simplicity, their operation will be discussed in terms of manufacturing simple, flat washers from a continuous strip of metal.

The *progressive die set*, depicted in Figure 19-50, is the simpler of the two types. Basically, it consists of two or more sets of punches and dies mounted in tandem. The strip stock is fed into the first die, where a hole is pierced as the ram descends on its first stroke. When the ram raises, the stock is advanced and positioned under the blanking punch. As the ram descends on the second stroke, a pilot on the bottom of the blanking punch enters the hole that was pierced on the previous stroke to ensure accurate alignment. Further descent of the punch blanks the completed washer from the strip, and at the same time, the first punch pierces the hole for the next washer. As the process continues, a finished part is completed with each stroke of the press.

Progressive dies can be used for many combinations of piercing, blanking, forming, lancing, and drawing, as shown by the examples in Figures 19-41 and 19-52. They are

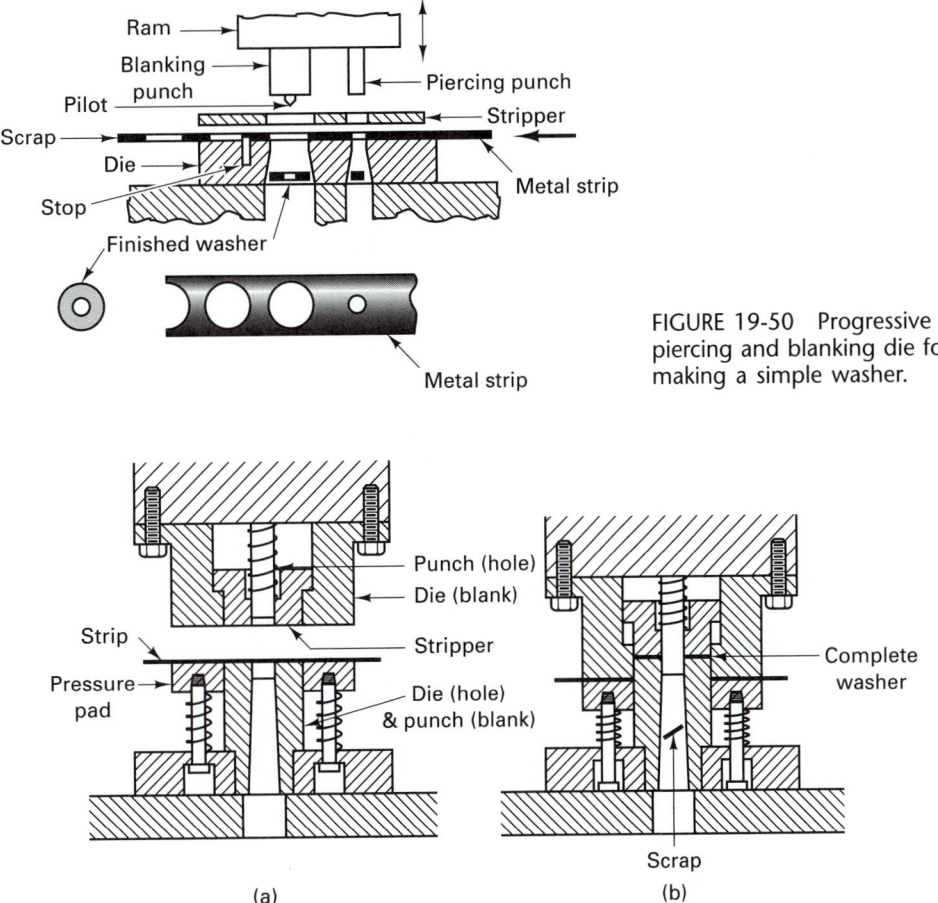

FIGURE 19-50 Progressive piercing and blanking die for making a simple washer.

FIGURE 19-51 Method for making a simple washer in a compound piercing and blanking die. Part is blanked (a) and subsequently pierced (b). The blanking punch contains the die for piercing.

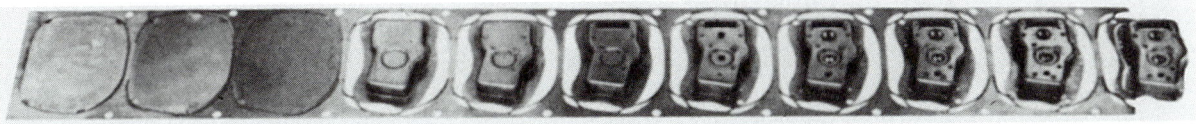

FIGURE 19-52 The various stages of an 11-station progressive die. *(Courtesy of the Minster Machine Company.)*

relatively simple to construct and are economical to maintain and repair, since a defective punch or die does not require replacement of the entire die set. However, if accurate alignment of the various operations is required, they are not as precise as compound dies.

In *compound dies* like the one shown schematically in Figure 19-51, piercing and blanking, or other combinations of operations, occur sequentially during a single stroke of the ram. While dies of this type are more precise, they are usually more expensive to construct and also more susceptible to breakage.

When many holes of varying sizes and shapes are to be placed in sheet components, numerically controlled turret-type punch presses may be specified. In these machines, as many as 60 separate punches and dies are contained within a turret that can quickly be rotated to provide the specific tooling required for an operation. Between operations, the workpiece is repositioned through numerically controlled X–Y movements of the worktable.

Design for Piercing and Drawing

The construction, operation, and maintenance of piercing and blanking dies can be greatly facilitated if designers of the parts to be fabricated keep a few simple rules in mind:

1. Diameters of pierced holes should not be less than the thickness of the metal, with a minimum of 0.025 in. Smaller holes can be made, but with difficulty.
2. The minimum distance between holes, or between a hole and the edge of the stock, should be at least equal to the metal thickness.
3. The width of any projection or slot should be at least $1\frac{1}{2}$ times the metal thickness and never less than $\frac{3}{32}$ in.
4. Keep tolerances as large as possible. Tolerances below about ±0.003 in. mean that shaving will be required.
5. Arrange the pattern of parts on the strip to minimize scrap.

■ 19.5 DRAWING AND SHEET METAL FORMING

Cold drawing is a term that can refer to two somewhat different operations. If the starting stock is sheet metal, cold drawing refers to the forming of parts where plastic flow occurs over a curved axis. This is one of the most important of all cold-working operations because a wide range of shapes, from small cups to large automobile body panels, can be readily fabricated. Cold drawing is similar to hot drawing (discussed briefly in Chapter 18), but the higher deformation forces, thinner metal, limited ductility, and closer dimensional tolerances create some distinctive problems.

If the starting stock is wire, rod, or tubing, cold drawing refers to the process of reducing the cross section of the material by pulling it through a die. These processes are similar to extrusion, but the forces are tensile, pulling on the product rather than pushing on the workpiece.

Rod, Bar, and Tube Drawing

One of the simplest cold-drawing operations is *rod* or *bar drawing*, illustrated schematically in Figure 19-53. One end of a rod is reduced or pointed, so that it can pass through a die of somewhat smaller cross section. The protruding material is then placed in grips and pulled in tension, drawing the remainder of the rod through the die. The rods reduce in section, elongate, and become stronger (strain harden). If the product cannot be bent or coiled conveniently, straight-pull *draw benches* are employed on finite-length stock. Hydraulic cylinders can be used to provide the pull for short-length products, while chain drives, such as the one shown schematically in Figure 19-54, can draw products up to 100 ft in length.

The reduction in area is usually restricted to between 20 and 50%, since higher values require higher pulling forces that tend to exceed the tensile strength of the reduced product. To produce a desired size or shape, multiple draws may be required through a series of progressively smaller dies. Intermediate anneals may also be required to restore ductility and enable further deformation.

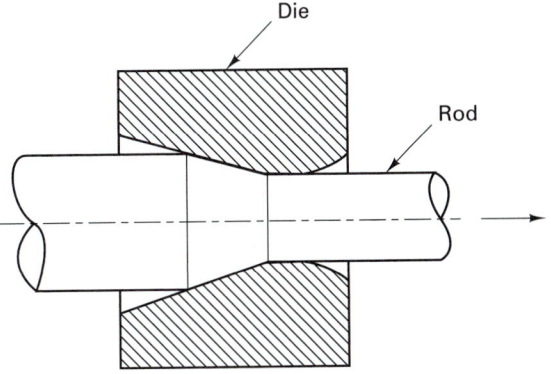

FIGURE 19-53 Schematic diagram of the rod or bar-drawing process.

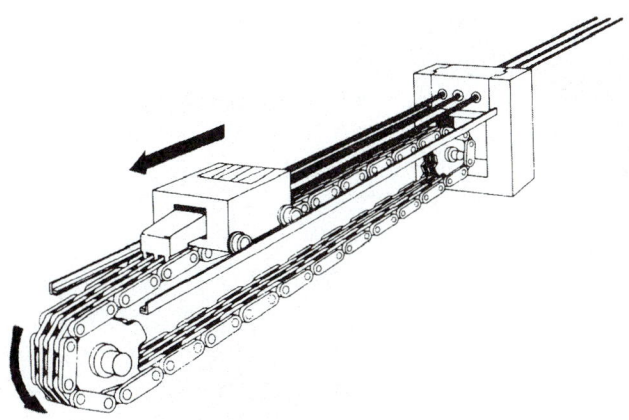

FIGURE 19-54 Schematic diagram of a chain-driven multiple-die draw bench used to produce finite lengths of straight rod or tube. *(Courtesy of Wean United, Inc.)*

Tube drawing can be used to produce high-quality tubing where the product requires the smooth surfaces, thin walls, accurate dimensions, and added strength (from the strain hardening) that are characteristic of cold forming. Internal mandrels are often used to control the inside diameter of tubes from about $\frac{1}{2}$ to 10 in. in diameter. As shown in Figure 19-55, these mandrels are inserted through the incoming stock and are held in place during the draw.

Thick-walled tubes and those less than $\frac{1}{2}$ in. in diameter are often drawn without a mandrel in a process known as *tube sinking*. Precise control of the inner diameter is sacrificed in exchange for process simplicity and the ability to draw long lengths of product. If a controlled internal diameter must be produced in a long-length product, the manufacturer may utilize a *floating plug*, like the one shown in Figure 19-56. This plug must be designed for the specific conditions of material, reduction, and friction, since it must assume a stable position within the die. If the friction on the plug surface is too great, it will be pulled too far forward by the flowing tube, pinching off or fracturing the tube wall. If the amount of friction is insufficient, the plug will chatter or vibrate within the tube and will not assume a stable position. If properly designed, the floating plug will size the internal diameter while the external die shapes and sizes the outside of the tube.

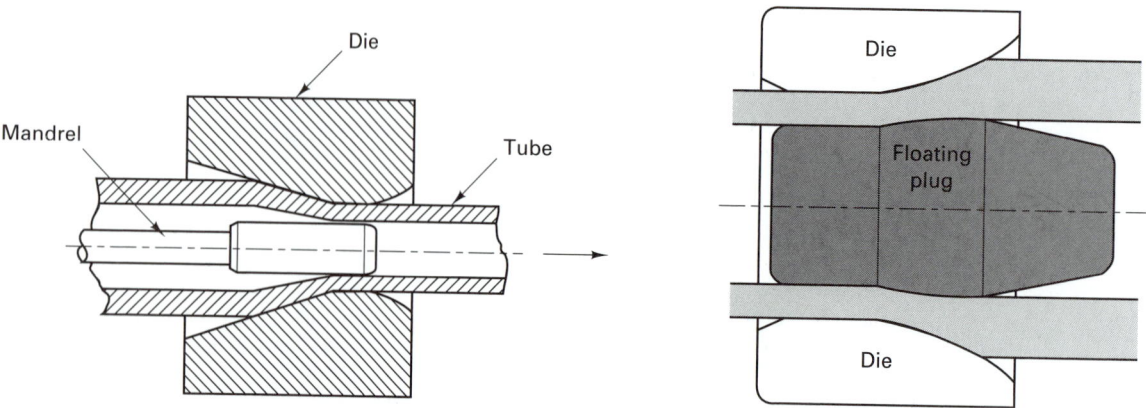

FIGURE 19-55 Cold-drawing smaller tubing from larger tubing. The die sets the outer dimension while the stationary mandrel sizes the inner diameter.

FIGURE 19-56 Tube drawing with a floating plug.

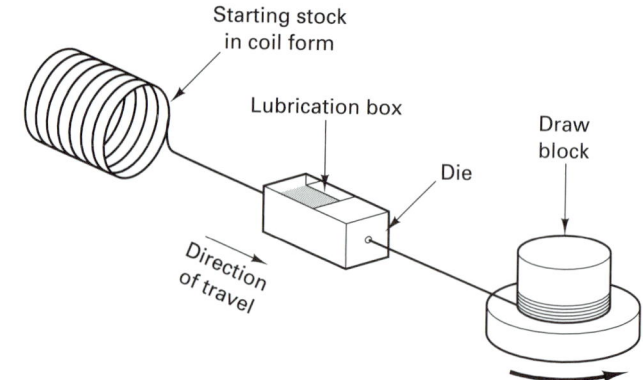

FIGURE 19-57 Schematic of wire drawing with a rotating draw block. The rotating motor on the draw block provides a continuous pull on the incoming wire.

Products with shaped cross section can also be produced by the drawing of bar stock. By using cold drawing instead of hot extrusion, precise dimensions can be produced. Inexpensive materials strengthened by strain hardening can often replace stronger alloys or ones that would require additional heat treatment. An economical means of producing small parts with complex but constant cross sections might be to cold draw shaped bars and then section them into the individual products. Steels, copper alloys, and aluminum alloys have all been cold drawn into shaped bars.

Wire Drawing

Wire drawing is essentially the same process as bar drawing except that it involves smaller-diameter material. Because the material can be coiled, the process can now be conducted in a somewhat continuous manner on draw blocks, like the one illustrated schematically in Figure 19-57. Wire drawing usually begins with large coils of hot-rolled material approximately $\frac{3}{8}$ in. (9 mm) in diameter. After descaling or other forms of surface preparation, one end of the coil is pointed, fed through a die, gripped, and the drawing process begins.

Wire dies generally have a configuration similar to the one shown in Figure 19-58. The contact regions are usually made of wear-resistant tungsten carbide or polycrystalline,

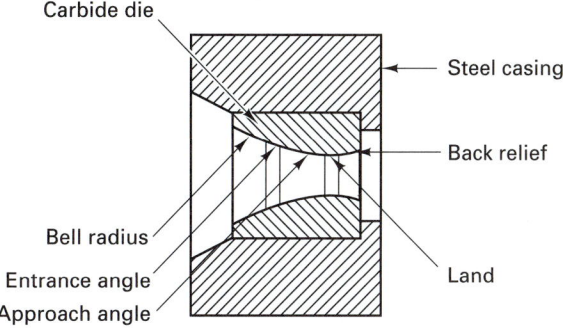

FIGURE 19-58 Cross section through a typical carbide wire drawing die showing the characteristic regions of the contour.

manufactured diamond. Single-crystal diamonds can be used for the drawing of very fine wire, and wear-resistant and low-friction coatings can be applied to the various die material substrates. Lubrication boxes often precede the individual dies to help reduce friction drag and prevent wear of the dies.

Because of the limited reductions associated with the process, multiple draws are usually required to affect any significant change in size. To convert hot-rolled rod stock to the fine wire that is used in household telephone lines requires passes through as many as 20 or 30 individual dies. To minimize handling and labor, these operations are usually performed on tandem machines, like the one shown schematically in Figure 19-59. Between 3 and 12 dies are mounted in a single machine, and the material moves continuously from one station to another in a synchronized manner that prevents any localized accumulation or tension that might induce fracture.

After passing through all the dies in a tandem machine, the material usually requires an intermediate anneal before it can be subjected to further deformation. By controlling the placement of the last anneal in the process cycle (i.e., the amount of subsequent deformation), wires can be made with a wide range of strengths (or tempers). For maximum ductility and conductivity, the wire should be annealed in controlled-atmosphere furnaces after the final draw.

Spinning

Spinning is a cold-forming operation in which a rotating disk of sheet metal is shaped over a male form, or mandrel. Localized pressure is applied through a simple round-ended wooden or metal tool or small roller, which traverses the entire surface of the part, as shown in the progressive schematic of Figure 19-60.

FIGURE 19-59 Schematic of a multistation synchronized wire drawing machine. To prevent accumulation or breakage, it is necessary to assure that the same volume of material passes through each station in a given time. The loops around the sheaves between the stations use wire tensions and feedback electronics to provide the necessary speed control.

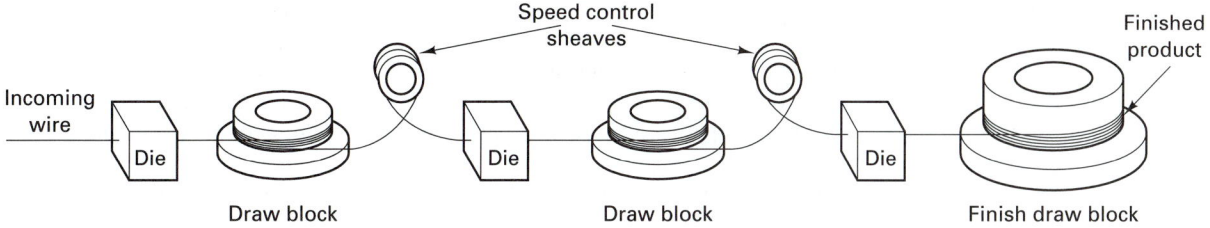

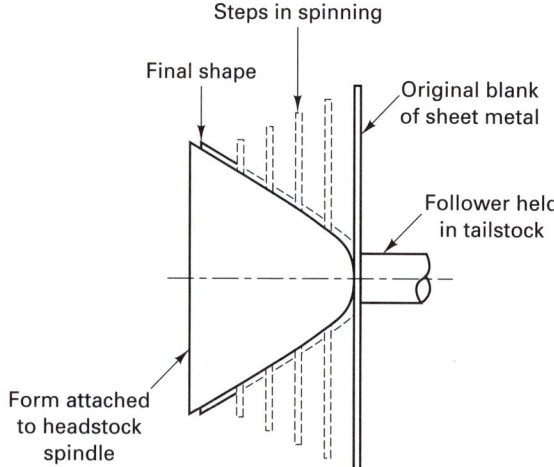

FIGURE 19-60   Progressive stages in the spinning of a sheet metal product.

FIGURE 19-61   Two stages in the spinning of a metal reflector. *(Courtesy of Spincraft, Inc.)*

The form block is attached to a rotating spindle, which is often the drive section of a simple lathe. A disk of metal is centered on the small end of the form and held in place by a follower attached to the tailstock of the lathe. As the disk and form rotate, the operator applies localized pressure against the metal, causing it to flow progressively against the form, as shown in the two stages of Figure 19-61. Because the final diameter of the part is less than that of the initial disk, the circumferential length decreases. This decrease must be compensated by either an increase in thickness, a radial elongation, or circumferential buckling. Control of the process is highly dependent on the skill of the operator.

Since the form block sees only localized compression, and the metal is not pulled across it under pressure, it can often be made of hardwood or even plastic. Its major requirement is to replicate the shape with a smooth surface, since any irregularities will transfer to the finished part. Since the tooling cost can be extremely low, spinning can be an attractive process for making small quantities of a single product. If it is sufficiently automated, however, spinning can also be used to mass-produce such high-volume items as

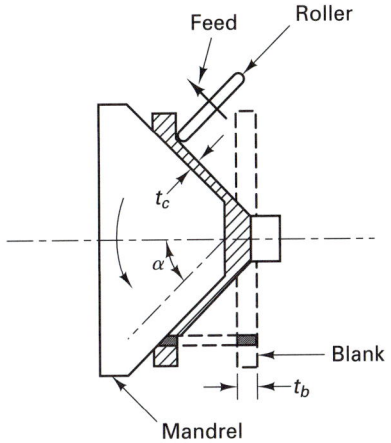

FIGURE 19-62   Schematic representation of the basic shear-forming process.

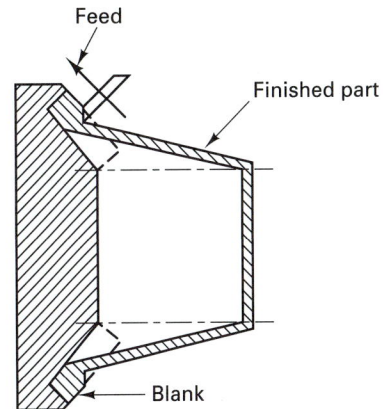

FIGURE 19-63   Forming a conical part by reverse shear forming.

lamp reflectors, cooking utensils, bowls, and the bells of some musical instruments. When large quantities are being produced, a metal form block is generally preferred.

Spinning is usually considered for simple shapes that can be withdrawn directly from a one-piece form. More complex shapes, such as those with reentrant angles, can often be spun over multipiece or offset forms. Form blocks can also be made from frozen water, which is simply melted out after spinning.

Shear Forming

Cones, hemispheres, and similar shapes are often formed by *shear forming* or *flow turning*, a modification of the spinning process in which each element of the blank maintains its distance from the axis of rotation. Because there is no circumferential shrinkage, the metal flow is entirely in shear and no compensating stretch has to occur. As shown in Figure 19-62, the wall thickness of the product, t_c, will vary with the angle of the particular region according to the relationship: $t_c = t_b \sin \alpha$, where t_b is the thickness of the starting blank. If α is less than 30°, it may be necessary to complete the forming in two stages with an intermediate anneal in between. Reductions in wall thickness as high as 8:1 are possible, but the limit is usually set at about 5:1 or 80%.

Conical shapes are usually shear formed by the *direct process* depicted in Figure 19-62. They can also be formed, however, by the *reverse process* illustrated in Figure 19-63. By controlling the position and feed of the forming rollers, the reverse process can be used to shape concave, convex, or conical parts without a matching form block or mandrel.

Cylinders can be shear formed by both the direct and reverse processes. As shown in Figure 19-64, the direct process restricts the length of the product to the length of the mandrel. No schematic is provided for the reverse process, because it is essentially the same as the roll extrusion process, depicted in Figure 19-13.

To produce a long, thin-walled cylinder, one might consider the "flo-reform" process, illustrated in Figure 19-65. This process is a sequential combination of shear forming and conventional spinning. The first steps shear the blank into a cone with no change in workpiece diameter (shear forming), while the latter stages pull the cone into a cylinder (spinning).

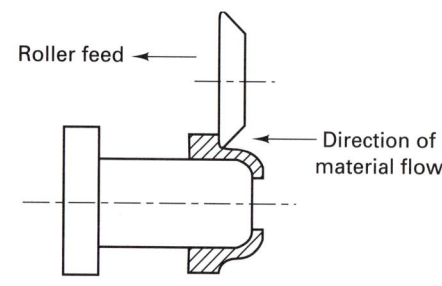

FIGURE 19-64   Shear forming a cylinder by the direct process

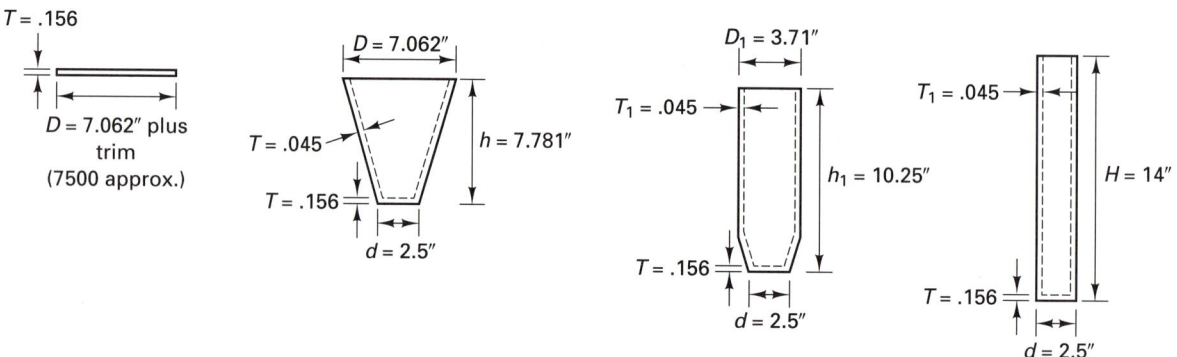

FIGURE 19-65   Forming a long, thin-walled cylinder from a flat blank by the Flo-reform process. Schematics show the starting, final and two intermediate shapes. *(Courtesy of Lodge and Shipley Company.)*

Stretch Forming

Stretch forming, illustrated in principle in Figure 19-66, is an attractive means of producing large sheet metal parts in low or limited quantities. A sheet of metal is gripped by two or more sets of jaws that stretch it and wrap it around a single form block. Various combinations of stretching, wrapping, and motion of the block or grips can be employed, depending on the shape of the part.

Because most of the deformation is induced by the tensile stretching, the forces on the form block are far less than those normally encountered in bending or forming. Consequently, there is very little springback, and the workpiece conforms very closely to the shape of the tool. Because the forces are so low, the form blocks can often be made of wood, low-melting-point metal, or even plastic.

Stretch forming, or *stretch-wrap forming* as it is often called, is quite popular in the aircraft industry and is frequently used to form aluminum and stainless steel into cowlings, wing tips, scoops, and other large panels. Low-carbon steel can be stretch formed to produce large panels for the automotive and truck industry.

Occasionally, mating male and female dies are used to shape the metal while it is being stretched. If this is done, the process is known as *stretch-draw forming*.

Sheet Metal Drawing (Deep Drawing, Shell Drawing, and Shallow Drawing)

The drawing of closed-bottom cylindrical or rectangular containers from metal sheet is one of the most important and widely used manufacturing processes. Because one of the

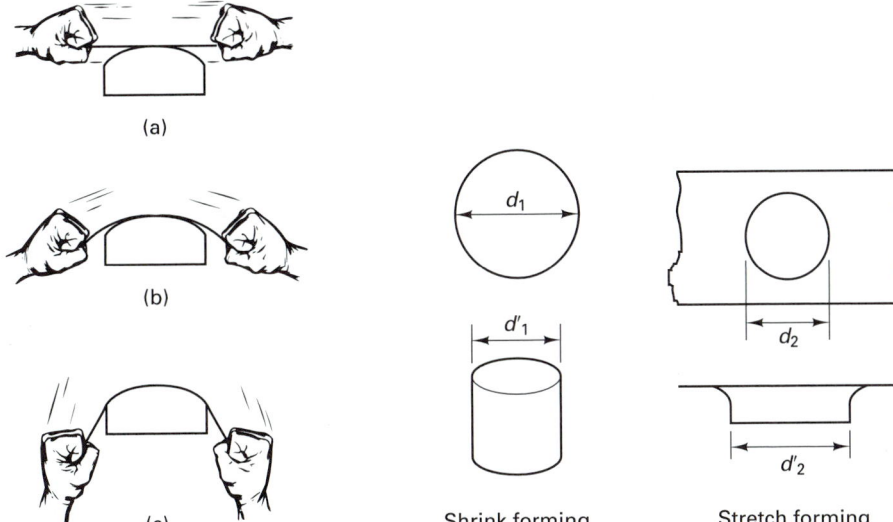

(a)

(b)

(c)

FIGURE 19-66   Schematic representation of the motions and steps involved in stretch-wrap forming.

Shrink forming   Stretch forming

FIGURE 19-67   The two basic types of flow during drawing.

earliest uses was the manufacture of artillery shells and cartridge cases, the process is sometimes called *shell drawing*. When the depth of the product is less than its diameter (or the smallest dimension of its opening), the process is considered to be *shallow drawing*. When the depth is greater than the diameter, it is known as *deep drawing*.

As illustrated in Figure 19-67, there are two basic types of drawing, both meeting the definition of multiaxis or curved-axis bending. In *shrink forming*, the circumference decreases as the blank diameter is reduced from d_1 to the cup wall diameter of d'_1. Since the volume of material must remain constant, the decrease in circumferential dimension must be compensated by an increase in another dimension, such as thickness. The material is thin, however, and an alternative response is to relieve the circumferential compression by a buckling or wrinkling. To suppress the wrinkles, it is often necessary to compress the sheet between the die and a hold-down surface during forming. This hold-down action forces an increase in either radial length or thickness.

In *stretch forming* the diameter and the circumference both increase, with a corresponding decrease in thickness that can be the cause of tearing during forming, or subsequent service failures such as premature failure in a corrosive environment.

Since many drawn products contain regions of both shrink and stretch forming, effective design and process control can be a complex problem. In general, it is best if the surface area and thickness remain relatively constant. Regions with large amounts of shrink or stretch will probably cause some degree of difficulty because of the associated buckling or thinning.

Computer models are now available to assist with the design of complex-shaped drawn products. Their success, however, requires accurate modeling and subsequent control of the drawn material, lubricant, dies and die surfaces, press speed, and numerous other variables. Nevertheless, the results are generally quite useful, but successful drawing is still a combination of science, experience, empirical data, and experimentation.

To draw a part without wrinkling, tearing, or undesirable variation in thickness, the flow of metal must be controlled throughout the operation. This is usually accomplished

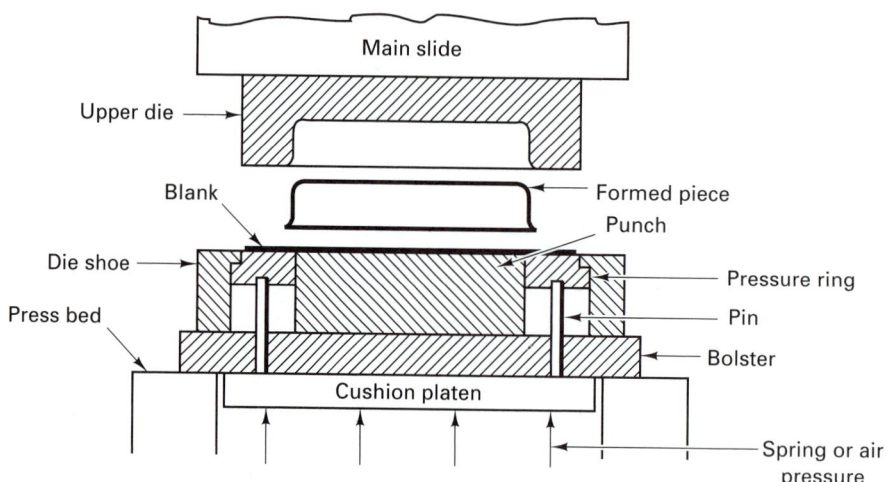

FIGURE 19-68   Deep drawing with a single-action press. Secondary springs or air pressure provides the hold-down force.

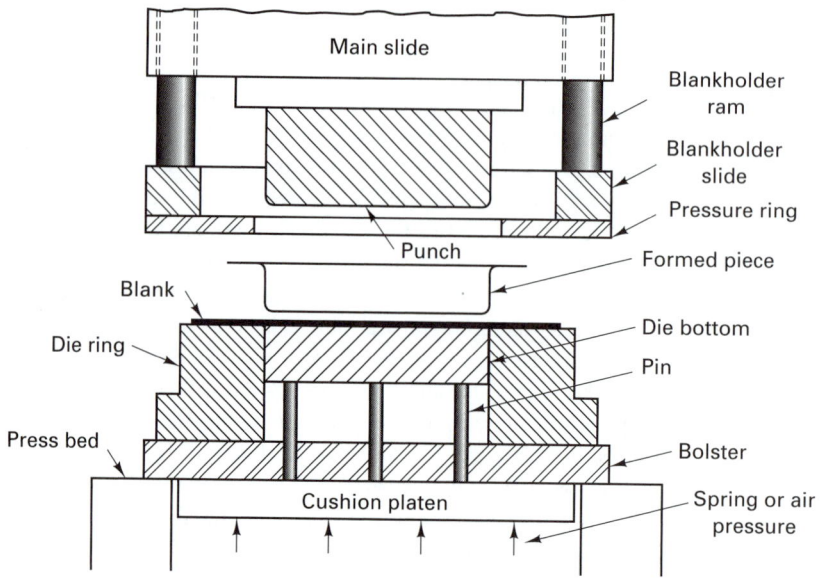

FIGURE 19-69   Drawing on a double-action press, where the blank holder uses the second press action.

through some form of pressure ring or pad. In single-action presses, there is only one movement that is available. As shown in Figure 19-68, springs or air pressure are often used to clamp the metal between the upper die and pressure ring. Multiple-action presses offer two or more independent motions. As shown in Figure 19-69, the *hold-down force* can be applied to the pressure ring in a manner that is independent of the main slide position. Hold-down pressure can be applied and varied as needed during the drawing operation. For this reason, multiple-action presses are usually specified for the drawing of more complex parts, while single-action presses can be used for the simpler operations.

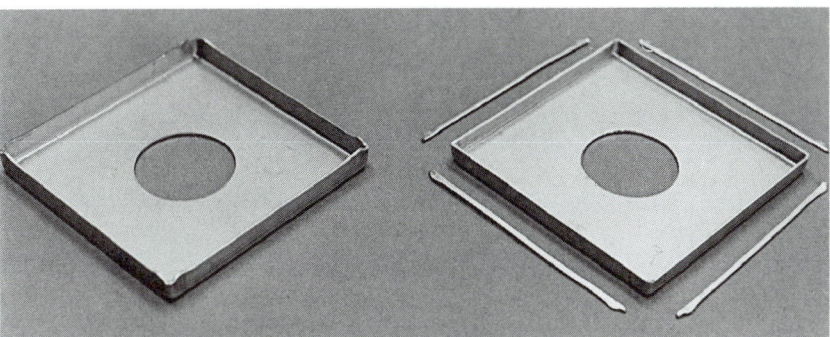

FIGURE 19-70   Pierced, blanked, and drawn part before and after trimming.

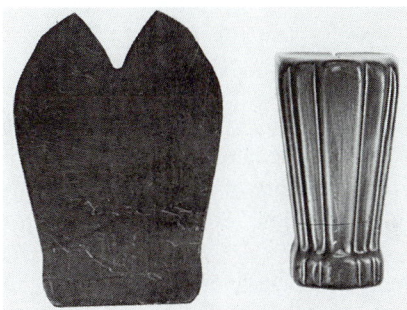

FIGURE 19-71   Irregularly shaped blank and finished drawn stove leg. No final trimming is necessary.

Because of prior rolling and other metallurgical and process variables, the flow of metal is generally not uniform in a drawing operation. Excess material may be required to assure final dimensions, and trimming may be required to establish both the size and uniformity of the final part. Figure 19-70 shows a shallow-drawn part before and after trimming. Obviously, such trimming adds to the production cost because it not only converts some of the starting material to scrap and but also adds another operation to the manufacturing process. Since the shape has already been produced, trimming requires accurate positioning or manipulation of the workpiece or the construction of a separate trimming die.

Trimming can often be eliminated through the use of complex-shaped blanks and improved metallurgical and process control. Figure 19-71 illustrates the use of a complex blank to produce a finished part without trimming. A special blanking die is required and scrap will be generated during the blanking operation. However, the approach eliminates the handling, positioning, processing, and scrap associated with trimming. The choice of method is usually dictated by process economics and depends on such factors as the complexity of the part and the quantity to be produced.

Forming with Rubber Tooling or Fluid Pressure

Blanking and drawing operations usually require mating male and female, or upper and lower, die sets, and such tooling can be quite expensive. In addition, process setup requires that the various tool components be precisely positioned and aligned. Consequently, numerous methods have been developed to reduce tooling cost and/or setup time and expense. Still other approaches seek to extend the amount of deformation that can be performed with a single set of tools, thereby eliminating both the intermediate operations and

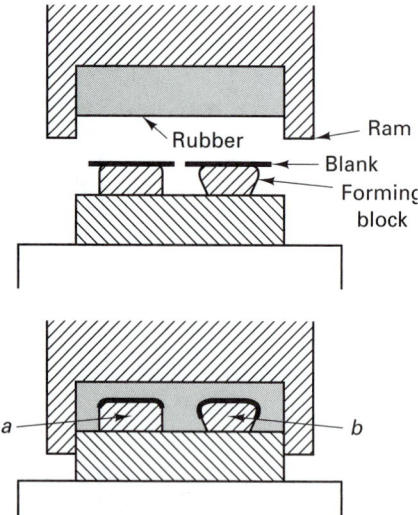

FIGURE 19-72 The Guerin process for forming sheet metal products.

associated tooling, as well as any intermediate anneals. Although most of these methods have distinct limitations, such as complexity of shape or types of metal that can be formed, they also have definite areas of application. Some alternative processes will now be presented.

Several methods of forming eliminate either the male or female member of the die set by employing rubber or fluid pressure to produce the deformation. The *Guerin process*, depicted in Figure 19-72, is based on the phenomenon that rubber of the proper consistency, *when totally confined*, acts as a fluid and transmits pressure uniformly in all directions. Blanks of sheet metal are placed on top of form blocks, which can be made of wood, Bakelite, polyurethane, epoxy, or low-melting-point metal. The upper ram contains a pad of rubber 8 to 10 in. thick mounted within a steel container. As the ram descends, the rubber pad becomes confined and transmits force to the metal, causing it to bend to the desired shape. Since no female die is used and inexpensive form blocks replace the male die, the total tooling cost is quite low. There are no mating tools to align, process flexibility is quite high (different shapes can even be formed at the same time), wear on the material and tooling is low, and the surface quality of the workpiece is easily maintained. When reentrant sections are produced (as in Figure 19-72b), it must be possible to slide the parts lengthwise from the form blocks or to disassemble a multipiece form from within the product.

The Guerin process was developed by the aircraft industry, where the production of small numbers of duplicate parts clearly favors the low cost of tooling. It can be used on aluminum sheet up to $\frac{1}{8}$ in. thick and on stainless steel up to $\frac{1}{16}$ in. Magnesium can also be formed if it is heated and shaped over heated form blocks.

Most forming done with the Guerin process is multiple-axis bending, but some shallow drawing can also be performed. The process can also be used to pierce or blank thin gages of aluminum, as illustrated in Figure 19-73. For this application the blanking blocks are shaped the same as the desired workpiece, with a face, or edge, of hardened steel. Round-edge supporting blocks are positioned a short distance from the blanking blocks to support the scrap skeleton and permit the metal to bend away from the shearing edges.

In the *high-pressure flexible-die process* (also known as "rubber bag forming," *Hydroforming*, or Flexforming), the rubber pad is replaced by a flexible rubber diaphragm backed by controlled hydraulic pressures at values between 20,000 and 30,000 psi (140 to

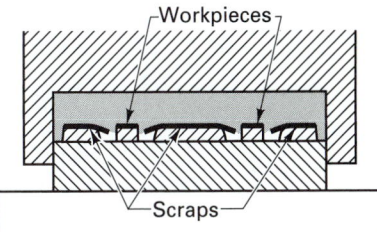

FIGURE 19-73   Method of blanking sheet metal using the Guerin process.

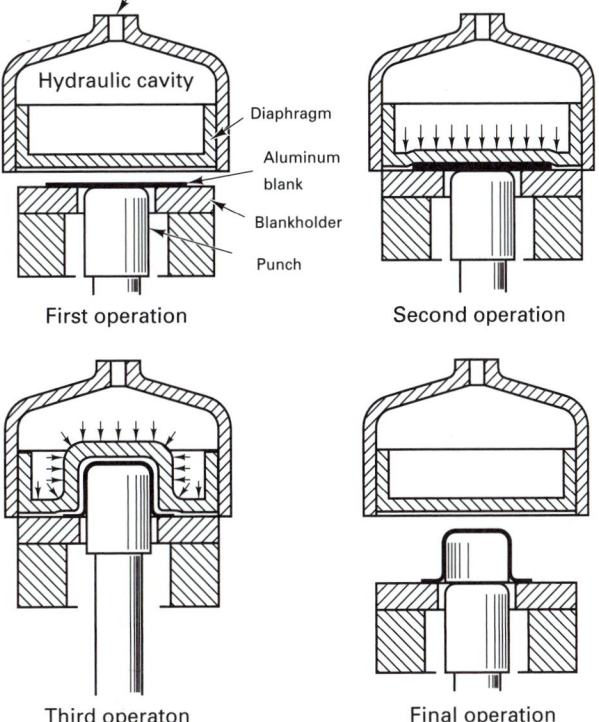

FIGURE 19-74   High-pressure flexible-die forming, showing (1) the blank in place with no pressure in the cavity; (2) press closed and cavity pressurized; (3) ram advanced with cavity maintaining fluid pressure; and (4) pressure released and ram retracted. *(Courtesy of Aluminum Association, New York.)*

200 MPa). Deeper parts can be formed with truly uniform fluid pressure, as illustrated by the cycle depicted in Figure 19-74. Cycle times are on the order of 1 to 3 minutes, but the reduced tool costs make the process attractive for prototype manufacturing and low-volume production (up to about 10,000 identical parts).

In *bulging*, fluid or rubber is used to transmit the pressure required to expand a metal blank or tube outward against a split female mold or die. For simple shapes, rubber tooling can be inserted, compressed, and then easily removed, as shown in Figure 19-75. For more complicated shapes, fluid pressure is used to form the bulge. More complex equipment is now required since pressurized seals must be formed and maintained while still enabling the easy insertion and removal of material that is required for mass production.

Drawing on a Drop Hammer

When small quantities of shallow-drawn parts are required, they can often be made most economically through the use of low-melting-point metal dies (zinc alloys are common)

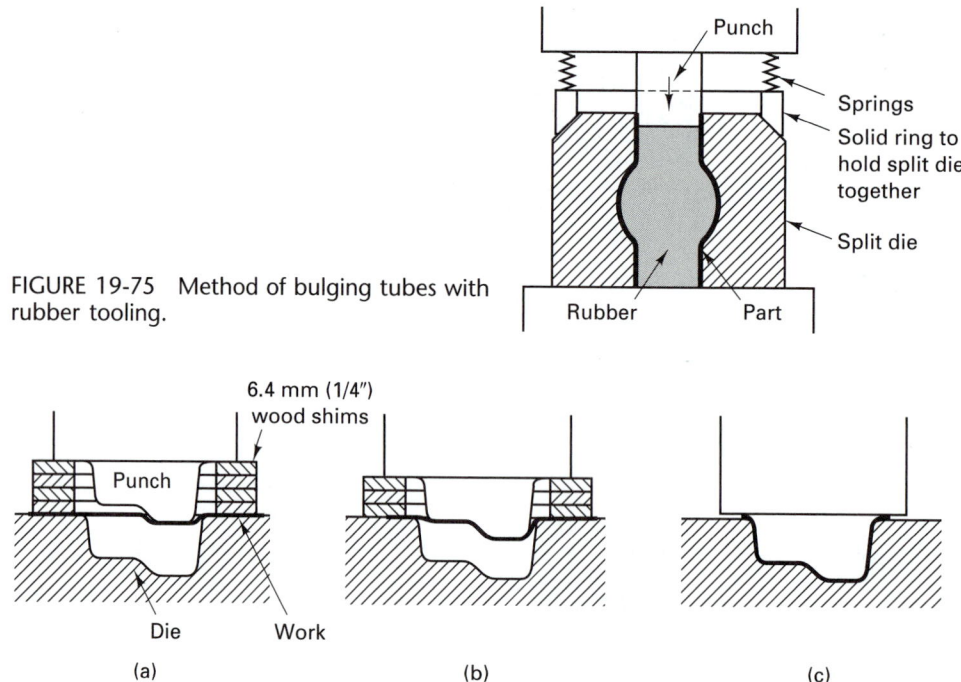

FIGURE 19-75   Method of bulging tubes with rubber tooling.

FIGURE 19-76   Method of deep drawing on a drop hammer using inexpensive dies and a stack of shims.

and a drop hammer. The dies can be directly cast, eliminating the expense of machining operations, and they can be remelted and cast into other shapes when no longer needed. Often, a stack of shims, made of thin plywood, is placed on top of the sheet of metal, as shown in Figure 19-76, to restrict the descent of the ram. These shims also act as a pressure pad or hold-down to restrict and control the flow of metal and thus inhibit wrinkling. One shim is withdrawn after each stroke of the ram, permitting the ram to fall farther and deepen the draw. Wrinkles that form can be hammered out between strokes with a hand mallet.

Drawing on a drop hammer is considerably cruder than drawing with steel dies, but it is often the most economical method for producing small quantities. It is most suitable for aluminum alloys, but carbon and stainless steel sheets, if sufficiently thin, can also be drawn successfully.

High-Energy-Rate Forming

A number of methods have been developed to form metals through the application of large amounts of energy in a very short time interval. These are known as *high-energy-rate forming processes* and often go by the abbreviation HERF. Many metals tend to deform more readily under the ultrarapid load application rates used in these processes. As a consequence, HERF makes it possible to form large workpieces and difficult-to-form metals with less-expensive equipment and tooling than would otherwise be required.

Another advantage of HERF is that there is less difficulty related to springback. This is probably associated with two factors: (1) high compressive stresses are set up in the metal when it is forced against the die, and (2) some slight elastic deformation of the die

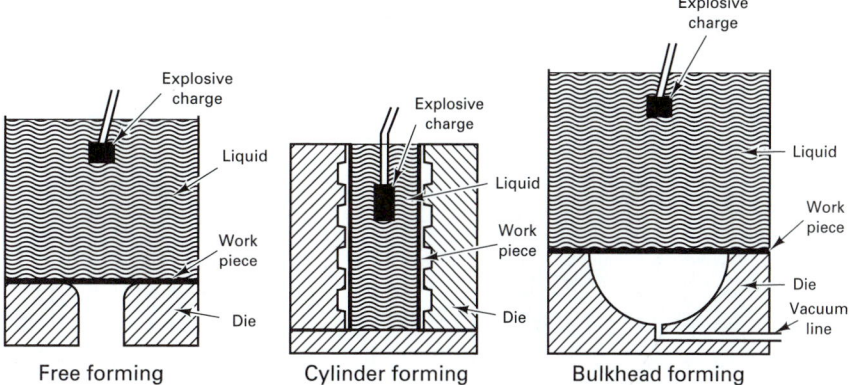

FIGURE 19-77 Three methods of high-energy-rate-forming with explosive charges. *(Courtesy of Materials Engineering.)*

occurs under the ultrahigh pressure. The latter results in a slight overforming of the workpiece and the appearance that no springback has occurred.

High energy-release rates can be obtained by five distinct methods: (1) underwater explosions, (2) underwater spark discharge (electrohydraulic techniques), (3) pneumatic–mechanical means, (4) internal combustion of gaseous mixtures, and (5) the use of rapidly formed magnetic fields (electromagnetic techniques). Specific processes have been developed around each of these approaches.

Figure 19-77 illustrates three commonly used procedures involving the use of *explosive charges*: free forming, cylinder forming, and bulkhead forming. While these procedures can be used for a wide range of products, they are particularly suited for parts of thick material like the 10-ft-diameter elliptical dome shown in Figure 19-78. Only a tank of water in the ground is required, with about 6 ft of water above the workpiece. The female die can be made of an inexpensive material such as wood, plastic, or low-melting-temperature metal.

The *spark-discharge method*, shown in Figure 19-79, uses the energy of an electrical discharge to shape the metal. Electrical energy is stored in large capacitor banks and is then released in a controlled discharge, either between two electrodes or across an exploding bridgewire. High-energy shockwaves propagate through a pressure-transmitting medium and deform the workpiece material. The initiating wire can be preshaped, and shockwave reflectors can be used to adapt the process to a variety of components. The space between the workpiece and the die is usually evacuated before the discharge occurs, to prevent the possibility of puckering due to entrapped air.

The spark-discharge methods are most often used for bulging operations in small parts, but parts up to 50 in. (1.3 m) in diameter have been formed. Compared to explosive forming, the discharge techniques are easier and safer, use smaller tanks, and do not have to be performed in remote areas.

The *pneumatic-mechanical* and *internal combustion* techniques are preferred when the HERF methods are applied to mass production within a plant. In a pneumatic-mechanical press, one portion of the forming die is attached to the stationary bolster of the press bed and the other to a movable piston. Low-pressure gas acts on the entire bottom area of the piston, holding it up against a small-area seal. High-pressure gas is then applied to the other side of the seal and the pressure is steadily increased. When the force pushing down (high pressure on a small area) exceeds that pushing up (low pressure on a large area), the

FIGURE 19-78   Explosively formed elliptical dome 10 ft. in diameter being removed from a forming die. *(Courtesy of NASA.)*

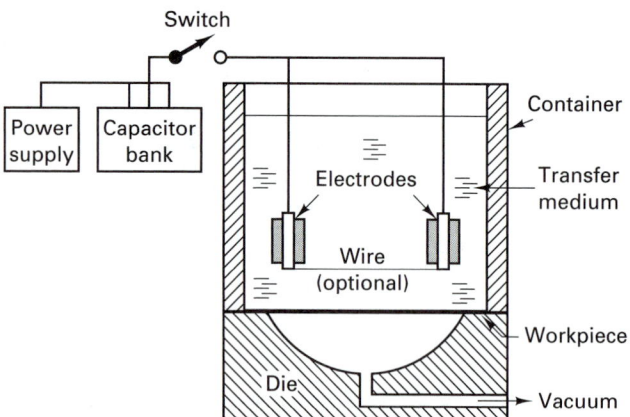

FIGURE 19-79   Components in the spark-discharge method of high-energy-rate forming.

seal is broken and the entire area of the piston is exposed to high-pressure gas. The piston moves rapidly downward, bringing the dies into contact. Figure 19-80 shows some typical parts made by this method.

Internal combustion presses operate on the same principle as that of an automobile engine. A gaseous mixture is exploded within a cylinder, causing a piston to be driven downward in a rapid fashion. The upper segment of the forming die is attached to the bottom of the piston. Internal combustion presses can produce die velocities up to 50 ft per second and cycle rates up to 60 strokes per minute. Either single or repeated blows can be used to form a part.

Electromagnetic forming is based on the principle that the electromagnetic field of an induced current always opposes the electromagnetic field of the inducing current. A large capacitor bank is discharged, producing a current surge through a coiled conductor. If the coil has been placed within a conductive cylinder, around a cylinder, or adjacent to a flat sheet of metal, the discharge induces a secondary current in the workpiece, causing it to be repelled from the coil and conformed to a die or mating workpiece. The process is very rapid and is used primarily to expand or contract tubing, or to permanently assemble component parts. Coining, forming, and swaging can also be performed with electromagnetic forces. Figure 19-81 shows some typical electromagnetically formed components.

Ironing

Ironing is the name given to the process of thinning the walls of a drawn cylinder by passing it between a punch and die whose separation is less than the original wall thickness. As shown in Figure 19-82, the walls are thinned and lengthened, while the thickness of the base remains unchanged. Common examples of ironed products include brass cartridge cases and the thin-walled beverage can.

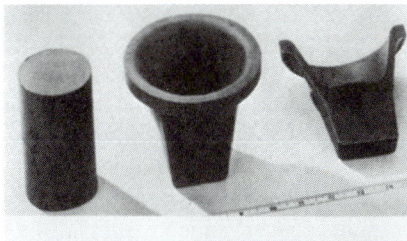

FIGURE 19-80 Typical parts manufactured by the pneumatic–mechanical high-energy-rate forming technique. *(Courtesy of Interstate Drop Forge Inc.)*

FIGURE 19-81 Two parts shaped by electromagnetic forming. Both are approximately 5 in. (125 mm) in diameter. *(Courtesy of General Atomic Division of General Dynamics Corporation.)*

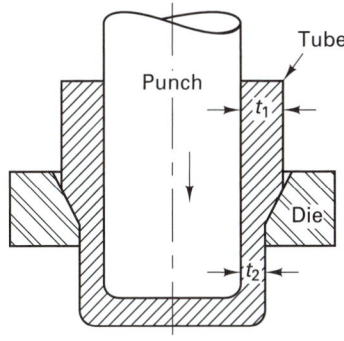

FIGURE 19-82 Schematic of the ironing process.

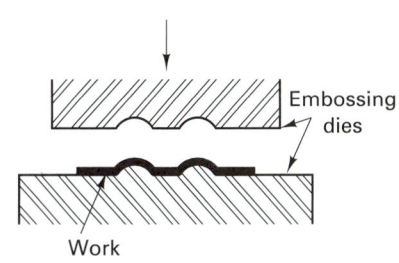

FIGURE 19-83 Embossing.

Embossing

Embossing, shown in Figure 19-83, is a pressworking process in which raised lettering or other designs are impressed in sheet material. Basically, it is a very shallow drawing operation where the depth of the draw is limited to one to three times the thickness of the metal, and the material thickness remains largely unchanged.

Superplastic Sheet Forming

Conventional metals and alloys exhibit tensile elongations in the range of 10 to 30%. In contrast, the commercial development of ultrafine-grain-size superplastic metal sheet (materials whose elongation at selected strain rates and temperatures exceeds 100% and may be as high as 2000 to 3000%) has made possible the economical production of large, complex-shaped products with compound curves. Deep or complex shapes can now be made as one-piece, single-operation pressings rather than multistep conventional pressings or multipiece assemblies. The elevated temperatures required to promote superplasticity (about 1650°F for titanium, between 840 and 970°F for aluminum, and generally above one-half of the absolute melting point) can reduce the required forces to the level where many of the forming techniques are adaptations of processes used to form thermoplastics (discussed in Chapter 20), and the tooling becomes relatively inexpensive. Because of the low forming pressures, form blocks can often be used in place of die sets. In thermoforming, a vacuum or pneumatic pressure causes the sheet to conform to a heated male or female die. Blow forming, vacuum forming, deep drawing, and combined superplastic forming and diffusion bonding are other possibilities. Precision is excellent and fine details or surface textures can be reproduced accurately. Springback and residual stresses are almost nonexistent, and the products have a fine, uniform grain size.

The major limitation to superplastic forming is the low forming rate that is required to maintain superplastic behavior. Cycle times may range from 2 minutes to as much as 2 hours per part, compared to the several seconds that is typical of conventional presswork. As a result, applications tend to be limited to low-volume products such as those common to the aerospace industry. By making the products larger and eliminating assembly operations, the weight of products can be reduced, there are fewer fastener holes to initiate fatigue cracks, tooling and fabrication costs are reduced, and there is a shorter production lead time.

Design Aids for Sheet Metal Forming

A majority of sheet metal failures occur due to thinning or fracture, and both are the result of excessive deformation in a given region. A quick and economical means of evaluating the severity of deformation in a formed part is to use *strain analysis* and a *forming limit diagram*. A pattern or grid, such as the one in Figure 19-84, is placed on the surface of a sheet by scribing, printing, or etching. The sheet is then deformed and the distorted pattern is measured and evaluated. Regions where the enclosed area has expanded are locations of sheet thinning and possible failure. Regions where the area has contracted have undergone sheet thickening and may be sites of possible buckling or wrinkles.

The major strains (strain in the direction of the largest radius or diameter) and the associated minor strains (strain 90° from the major) can be determined for a variety of locations, and the values can be plotted on a forming-limit diagram such as the one shown in Figure 19-84. If both major and minor strains are positive (right-hand side of the diagram), the deformation is known as *stretching*, and the sheet metal will definitely decrease in thickness. If the minor strain is negative, the positive stretching in the direction of the major strain will be partially or wholly compensated by the minor strain contraction. Such deformation is known as *drawing*, and the thickness may decrease, increase, or stay the same, depending on the relative magnitudes of the two strains. Regions where both strains are negative do not appear on the diagram, since its purpose is to reveal locations of possible fracture, and fracture occurs only in a tensile environment.

Those strains that fall above the forming limit line indicate regions of probable fracture. Possible corrective actions include modification of the lubricant, change in the

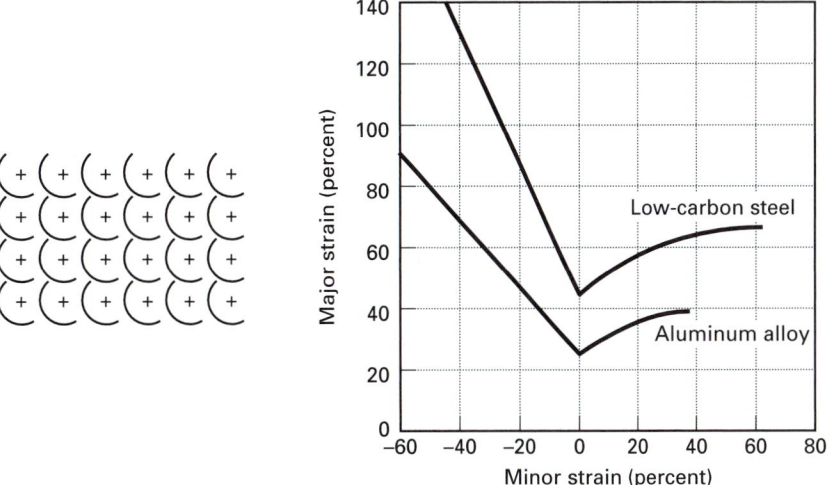

FIGURE 19-84 (*Left*) Typical pattern for sheet metal deformation analysis; (*right*) forming limit diagram used to determine whether a metal can be shaped without risk of fracture. Fracture is expected when strains fall above the lines.

die design, or variation in the clamping or hold-down pressure. Strain analysis can also be used for determining the best orientation of blanks relative to the rolling direction, comparing the effectiveness of various lubricants, or assisting in the design of dies for complex-shaped products.

■ 19.6 ALTERNATIVE METHODS OF PRODUCING SHEET-TYPE PRODUCTS

Electroforming

Several manufacturing processes have been developed to produce sheet-type products by directly depositing metal onto preshaped forms or mandrels. In a process known as *electroforming*, the metal is electroplated (by the process discussed in Chapter 40) onto a preshaped pattern or mandrel that has been fashioned from a material such as plastic, glass, Pyrex, or other metal, such as aluminum or stainless steel. If the pattern material is nonconductive, a conductive coating is first applied. Then, metals, such as nickel, iron, copper, or silver, are plated in thicknesses up to $\frac{5}{8}$ in. When the desired thickness has been attained, the workpiece is then stripped from the mandrel.

A wide variety of sizes and shapes can be made by electroforming, the principal limitation being the need to strip the product from the mandrel. Dimensional tolerances are good, and surface finishes of 2 μin. can be obtained on the replicated interior. For most applications, such as the production of multiple molds from a single master pattern, the interior surface is the critical one, and the wall thickness serves only to provide the necessary strength. Irregularities in the exterior are not critical, and various types of backup material can be employed to provide additional support. For other applications, the exterior dimensions are also important, and uniform deposition is desired. Since the fabrication of a product requires only a single pattern or mandrel, low production quantities can be made in an economical fashion.

Plasma Spray Forming

Similar parts can also be formed by injecting powder into an intensely hot stream of ionized gas (a plasma with temperatures of up to 20,000°F or 11,000°C), where it melts and is propelled onto a shaped form or mandrel. Upon contact, the droplets flatten and undergo rapid solidification to produce a dense, fine-grained product. Multiple layers can be deposited to build up a desired size, shape, and thickness. Because of the high adhesion, the mandrel or form is usually removed by machining or chemical etching. Most applications of plasma spray forming involve the fabrication of specialized products from difficult-to-form or ultrahigh-melting-point materials.

■ 19.7 PRESSES

Classification of Presses

Many types of presses have been developed to perform the various cold-working operations discussed in this chapter. When selecting a press for a given application, consideration should be given to the capacity required, the type of power (*manual, mechanical, or hydraulic*), the number of slides or drives, the type of drive, the stroke length for each drive, and the type of frame or construction. Table 19-2 lists some of the major types of presses, and Figure 19-85 illustrates some of the more important drive mechanisms. In general, *mechanical drives* provide faster motion and more positive control of displacement. Once built, however, the flexibility of a mechanical press is limited, since the length of the stroke is set by the design of the drive. The available force usually varies with position, so mechanical presses are preferred for operations that require the maximum pressure near the bottom of the stroke, such as cutting, shallow forming and drawing (up to about 4 in.), and progressive and transfer die operations. Typical capacities range up to about 6000 tons.

TABLE 19-2. Classification of the Various Drive Mechanisms of Commercial Presses		
Manual	Mechanical	Hydraulic
Kick presses	Crank	Single-slide
	Single	Multiple-slide
	Double	
	Eccentric	
	Cam	
	Knuckle joint	
	Toggle	
	Screw	
	Rack and pinion	

In contrast, *hydraulic presses* produce motion as the result of piston movement, and longer or variable-length strokes can be programmed within the limitations of the cylinder (which may be as long as 100 in.). Forces and pressures are more accurately controlled and do not depend on slide position. Speeds can be programmed to vary or remain constant throughout a stroke. Since position is varied through fluid displacement, the reproducibility of position will have greater variation than a mechanical press. Hydraulic presses are available in capacities exceeding 50,000 tons and are preferred for operations requiring a steady pressure throughout a substantial stroke (such as deep drawing), operations requiring wide

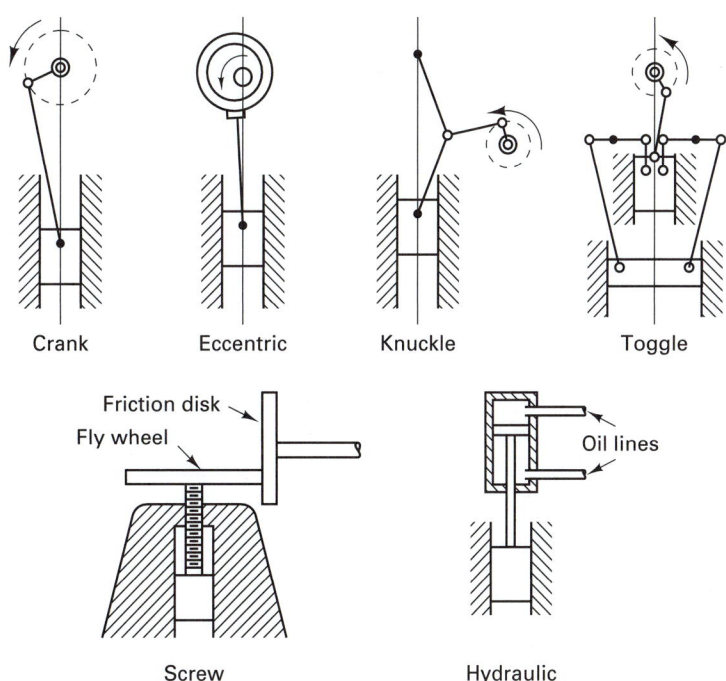

FIGURE 19-85   Schematic representation of the various types of press drive mechanisms.

variation in stroke length, and operations requiring high or widely variable forces. In general, hydraulic presses tend to be slower than the mechanical variety, but some are available that can provide up to 600 strokes per minute in a high-speed blanking operation. By using several hydraulic cylinders, programmed loads can be applied to the main ram, while a separate force and timing are used on the blank holder.

Manually operated presses such as foot-operated or *kick presses* are generally used for very light work such as shearing small sheets. *Crank-driven presses* are the most common type because of their simplicity. They are used for most piercing and blanking operations and for simple drawing. Double-crank presses offer a means of actuating blank holders or operating multiple-action dies. *Eccentric* or *cam drives* are used where only a short ram stroke is required. Cam action can also provide a dwell at the bottom of the stroke and is often the preferred method of actuating the blank holder in deep-drawing processes. *Knuckle-joint drives* provide a very high mechanical advantage along with fast action. They are often preferred for coining, sizing, and Guerin forming. *Toggle mechanisms* are used principally in drawing presses to actuate the blank holder, and screw-type drives offer great mechanical advantage coupled with an action that resembles a drop hammer (but slower and with less impact). For this reason, *screw presses* have become quite popular in the forging industry.

Types of Press Frame

Another matter of importance in selecting a press is determining the type of frame, because this often imposes limitations on the size and type of work that can be accommodated, how that work is fed and unloaded, and the overall stiffness of the machine. Table 19-3 presents a classification of basic frame types.

TABLE 19-3. Classification of Presses According to Type of Frame		
Arch	Gap	Straight-Sided
Crank or eccentric	Foot	Many variations, but all with straight-sided frames
Percussion	Bench	
	Vertical	
	Inclinable	
	Inclinable	
	Open back	
	Horn	
	Turret	

Presses that have their frames in the shape of an arch (arch-frame presses) are seldom used today, except with screw drives for coining operations. *Gap-frame presses*, whose frames are in the shape of the letter C, are among the most versatile and commonly preferred presses. They provide unobstructed access to the dies from three directions and permit large workpieces to be fed into the press. Gap-frame presses are available in a wide variety of sizes, from small bench types of about 1 ton up to about 300 tons, or more.

Several popular design features include open back, inclinability, adjustable bed, and sliding bolster. Open-back presses allow for ejection of the products or scrap through an opening in the back side of the press frame. Inclinable presses can be tilted, so that ejection can be assisted by gravity or compressed air jets. As a result of these features, the open-back inclinable (OBI) presses are the most common form of gap-frame press. The addition of an adjustable bed allows the base of the machine to be varied in position as a means of accommodating different size workpieces. Finally, the OBI press shown in Figure 19-86 contains a sliding bolster, a feature that permits a second die to be set up on the press while another is in operation. Die changeover then requires only a few minutes to unclamp the punch segment of the active die, move the second die set into position, clamp the new set to the press ram, and resume operation.

A *horn press* is a special type of upright open-back press where a heavy cylindrical shaft or "horn" appears in place of the usual bed. Curved or cylindrical workpieces can be placed over the horn for such operations as seaming, punching, and riveting, as illustrated in Figure 19-87. On some presses, both a horn and a bed are provided, with provision for swinging the horn aside when not needed.

Turret presses are especially useful in the production of sheet metal parts with numerous holes or slots that vary in size and shape. They usually employ a modified gap-frame structure, and add upper and lower turrets that carry a number of punches and dies. The two turrets are geared together so that any desired tool set can be quickly rotated into position. Another type uses a single turret that carries a number of subpress dies.

Straight-sided presses have frames that consist of a crown, two uprights, a base or bed, and one or more moving slides. Accessibility is generally from the front and rear, but openings are often provided in the side uprights to permit feeding and unloading of workpieces. Straight-sided presses are available in a wide variety of sizes and designs and are the preferred design for most hydraulic, large-capacity, or specialized mechanical-drive presses. As an added benefit, any elastic deflections tend to be uniform across the working surface, as opposed to the angular deflections that are typical of the gap-frame design. Figure 19-88 shows a typical straight-sided press.

Special Types of Presses

A number of presses have been designed to perform specific types of operations. In the *transfer press*, a single long slide enables multiple operations to be performed simultaneously

FIGURE 19-86 Inclinable gap-frame press with sliding bolster to accommodate two die sets for rapid change of tooling. *(Courtesy of Niagara Machine & Tool Works.)*

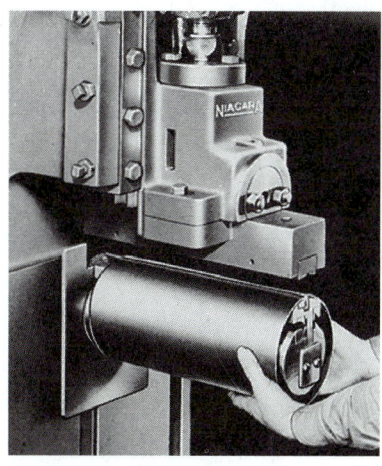

FIGURE 19-87 Making a seam on a horn press. Note the protruding "horn" that replaces the lower press bed. *(Courtesy of Niagara Machine & Tool Works.)*

FIGURE 19-88 A 200-ton (1800-kN) straight-side press. *(Courtesy of Rousselle Corporation.)*

in a single machine. Multiple die sets are mounted side by side along the slide. After the completion of each stroke, a continuous coil or individual workpieces are automatically and progressively advanced (transferred) to the next station by a mechanism like the one shown in Figure 19-89. Transfer presses can be used to perform blanking, piercing, forming, trimming, drawing, flanging, embossing, and coining. Figure 19-90 illustrates the production of a single part that illustrates a variety of these operations.

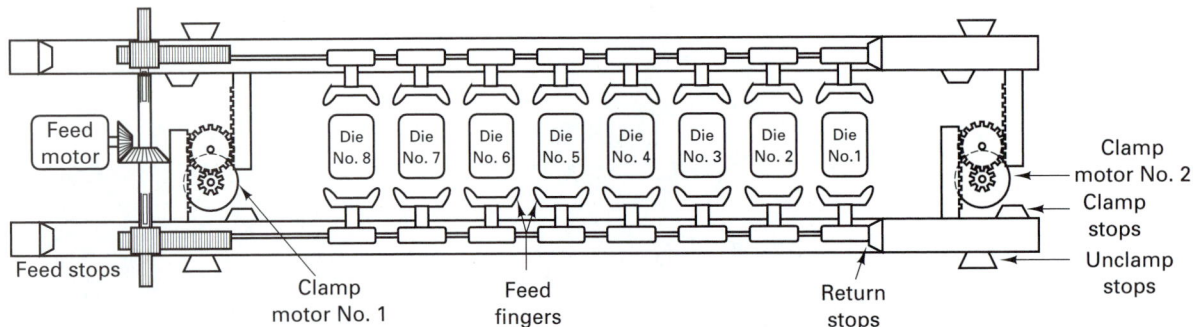

FIGURE 19-89   Schematic showing the arrangement of dies and the transfer mechanism used in transfer presses. *(Courtesy of Verson Allsteel Press Company.)*

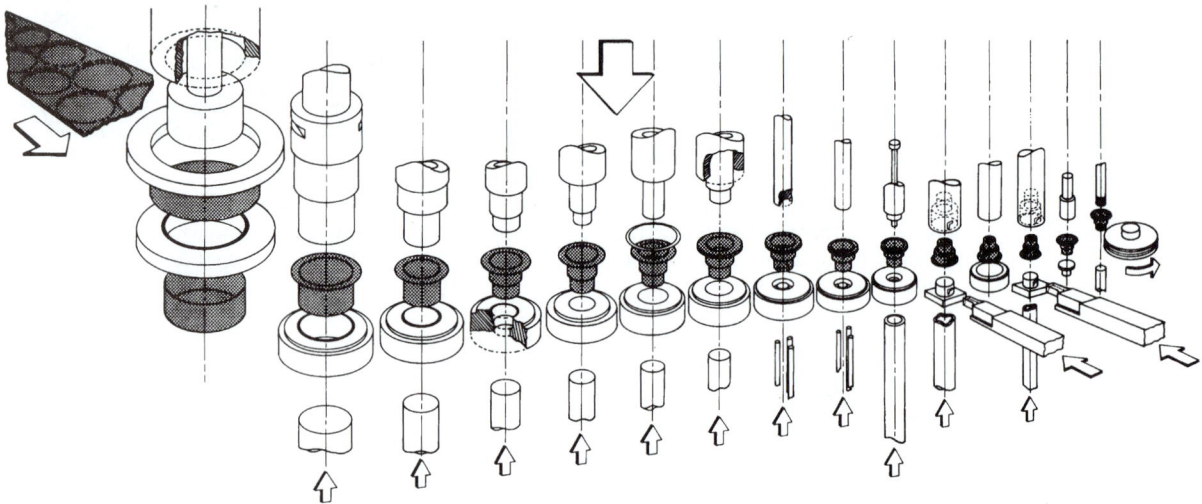

FIGURE 19-90   Various operations that can be performed during the production of stamped and drawn parts on a transfer press. *(Courtesy of U.S. Baird Corporation, Stratford, Conn.)*

By using a single press to perform multiple operations, transfer presses offer high production rates, high flexibility, and reduced costs (associated with reduced labor, floor space, energy, and maintenance). Since production is usually between 500 and 1500 parts per hour, process economics usually restricts these machines to operations where 4000 or more identical parts are required daily, each involving three or more separate operations. A total production run of 30,000 or more identical parts is generally desired between major changes in tooling. As a result, transfer presses are used primarily in industries such as automotive and appliances, where large numbers of identical products are being produced.

Four-slide or *multislide* machines like the one shown in Figure 19-91 are extremely versatile presses that are designed to produce small, intricately shaped parts from wire or coil feeds. The basic machine has four power-driven slides set 90° apart. The attached tooling is controlled by cams and designed to operate in a progressive cycle. In sheet metal forming, coiled strip stock is fed into the machine, where it is straightened and progressively pierced, notched, bent, and cut off at the various slide stations. Figure 19-92

FIGURE 19-91   Multislide machine with guards and covers removed. *(Courtesy of U.S. Baird Corporation, Stratford, Conn.)*

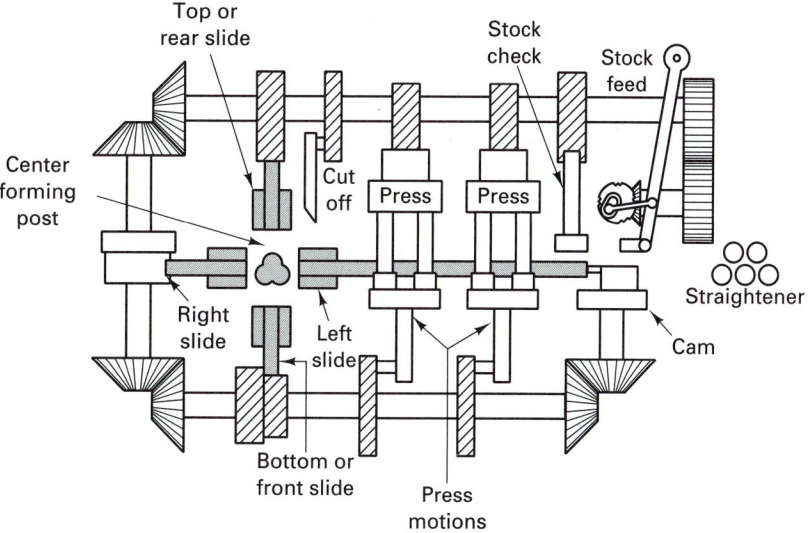

FIGURE 19-92   Schematic of the operating mechanism of a multislide machine. The material enters on the right and progresses toward the left as operations are performed. *(Courtesy of U.S. Baird Corporation, Stratford, Conn.)*

schematically presents the operating mechanism of one such machine. As the material moves from right to left, it undergoes a straightening, two successive pressing operations, various operations from all four directions, and a final cutoff. Figure 19-93 shows the carrier strip and successive operations as flat strip is pierced, blanked, and formed into a

FIGURE 19-93   Example of the piercing, blanking, and forming operations performed on a multislide machine. *(Courtesy of U.S. Baird Corporation, Stratford, Conn.)*

folded sheet metal product. The strip stock may be up to 3 in. (75 mm) wide and $\frac{3}{32}$ in. (2.5 mm) thick, and wires up to about $\frac{1}{8}$ in. (3 mm) in diameter are commonly processed. Products such as hinges, links, clips, and razor blades can be formed at very high rates on these machines. Setup times are long, so large production runs are preferred.

Press-Feeding Devices

Although hand feeding may still be used in some press operations, operator safety and the desire to increase productivity have motivated a strong shift to feeding by some form of mechanical device. When continuous strip is used it can be fed automatically by double-roll feeds mounted on the side of the press. Discrete products can be moved and positioned in a wide variety of ways. Dial-feed mechanisms enable an operator to insert workpieces into the front holes of a dial. The dial then indexes with each stroke of the press to move the parts progressively into proper position between the punch and die. Lightweight parts can be fed by suction-cup mechanisms, vibratory-bed feeders, and similar devices. Robots are being used in increasing numbers to place parts into presses and remove them after forming. Technology and equipment now exists to replace almost all manual feeding if such a transition is truly desired.

■ KEY WORDS

air-bend die	forming limit diagram	roll forming
backward extrusion	forward extrusion	roll straightening
bending	Guerin process	rubber tooling
blanking	high-energy-rate forming	seaming
bottoming die	hold-down force	shear forming
bulging	hubbing	shearing
burnishing	hydraulic press	sizing
coining	hydroform process	slitting
cold forging	hydrostatic extrusion	spinning
cold heading	impact extrusion	springback
cold rolling	ironing	staking
cold working	mechanical press	strain analysis
compound die	minimum bend radius	stretch forming
continuous extrusion	multislide press	stretcher leveling
draw bench	notching	subpress die
draw bending	peening	superplastic forming
drawing	piercing	swaging
electroforming	plasma spray forming	transfer press
embossing	press brake	trimming
fineblanking	progressive die	tube drawing
flanging	riveting	upsetting
floating plug	roll bending	wire drawing
forming	roll extrusion	

■ Review Questions

1. What are some of the attractive features of cold-working processes?
2. Why might cold-working equipment have to be more powerful than that used for hot working?
3. What benefits can be obtained by subjecting sheet metal to a skin-rolling pass?
4. Why would the cold rolling of shaped product not be an attractive means to produce a small amount of prototype product?
5. How can the swaging process impart different sizes and shapes to an interior cavity and the exterior of a product?
6. What is the primary difference between heading and upsetting?
7. How might cold forging be used to substantially reduce material waste?
8. What are some of the attractive properties or characteristics of cold forging or cold-forged products?
9. If a product contains a large-diameter head and a small-diameter shank, how can the processes of cold extrusion and cold heading be combined to save metal?
10. What are some of the unique capabilities and special limitations of hydrostatic extrusion?
11. How is the feedstock pushed through the die in continuous extrusion processes?
12. What type of products are made by the roll extrusion process?
13. What types of rivets can be used when there is access to only one side of a joint?
14. Why might hubbing be an attractive way to produce a number of identical die cavities?
15. How might a peening operation increase the fracture resistance of a product?
16. What is burnishing?
17. When making bends in sheet metal, what is the distinction between bending, forming, and drawing?
18. Why does a metal usually become thinner in the region of a bend?
19. What is springback, and why is it a concern during bending?
20. What types of operations can be performed on a press brake?
21. Why is it generally desirable to specify the largest possible bend radius?
22. If a right-angle bend is to be made in a cold-rolled sheet, should it be made with the bend lying along or perpendicular to the direction of previous rolling?
23. From a manufacturing viewpoint, why is it desirable for all bends in a product (or component) to have the same radius?
24. What is the difference between air-bend and bottoming dies? Which is more flexible? Which produces more reproducible bends?
25. What type of product geometry can be produced by cold roll forming? Is the process appropriate for making short lengths of specialized products?
26. What are two methods of straightening or flattening rod or sheet?
27. What is a definition of shearing?
28. Why are sheared or blanked edges generally not smooth?
29. What measures can be employed to improve the quality of a sheared edge?
30. Why are fineblanking presses more complex than those used in conventional blanking?
31. What are the similarities and differences between piercing and blanking?
32. What are some types of blanking or piercing operations that have come to acquire specific names?
33. What is the purpose of incorporating an angle, or shear, on the surface of a punch?
34. Why is it important that a blanking punch and die be in proper alignment?
35. What is the major benefit of mounting punches and dies on independent die sets?
36. What is the major benefit of assembling a complex die set from standard subpress dies?
37. What is a progressive die set?
38. How do compound dies differ from progressive dies?
39. What two distinctly different processes are often referred to as cold drawing?
40. What is the difference between tube drawing and tube sinking?
41. For what types of products might a floating plug be employed?
42. Why are rods generally drawn on draw benches, while wire is drawn on draw block machines?
43. Why are multiple passes usually required in wire-drawing operations?

44. Why is the tooling cost for a spinning operation relatively low?

45. How is shear forming different from spinning?

46. For what types of products would stretch forming be an appropriate manufacturing technique?

47. What types of defects or failures are generally associated with a shrink-forming mode of deformation? With the stretch forming mode?

48. What is the function of the pressure ring or hold-down in a deep-drawing operation?

49. Why is a trimming operation often included in a deep-drawing manufacturing sequence?

50. How does the Guerin process reduce the cost of tooling? What approach is used in the Hydroform process?

51. What are some of the basic methods that have been used to achieve the high energy-release rates needed in the HERF processes?

52. Why is springback rather minimal in high-energy-rate forming?

53. What are some well-known products that have been produced by processes including ironing? By embossing?

54. What is the major limitation of the superplastic forming of sheet metal? What are some of the more attractive features?

55. How can strain analysis be used to determine locations of possible defects or failure in sheet metal components?

56. What is a forming limit diagram?

57. Describe two alternative methods of producing complex-shaped thin products without requiring sheet metal deformation techniques.

58. What are the primary assets and limitations of mechanical press drives? Of hydraulic drives?

59. What is the purpose of inclining or tilting a press?

60. Describe how multiple operations are preformed simultaneously in a transfer press.

■ PROBLEM

1. Compare the forming processes of wire drawing, conventional extrusion, and continuous extrusion with respect to continuity, reduction in area possible in a single operation, possible materials, speeds, typical temperatures, and other important processing variables.

*C*hapter 19 CASE STUDY

the bronze bolt mystery

Y̶ou are employed by the Mountainous Irrigation Company, which had 10,000 special "oversized-head" bolts made from phosphor bronze (10% tin) for use in its various pumping plants. Figure CS-19 shows the details of the bolt design with associated dimensions. Within a few weeks after a number of these bolts were installed, the heads broke off several of them, the fracture being along the dashed line shown in Figure CS-19. The broken bolts were replaced, with special precautions being taken to assure that they were not overloaded. Again, after only a few days, several of the replacement bolts broke in the same manner. You have been assigned the task of determining the cause of the failures. Upon examining several unused bolts, you discover that many of them have fine cracks at the intersection of the head and body. Upon checking with the manufacturer, you learn that the bolts were made by a conventional upset-forging bolt manufacturing process.

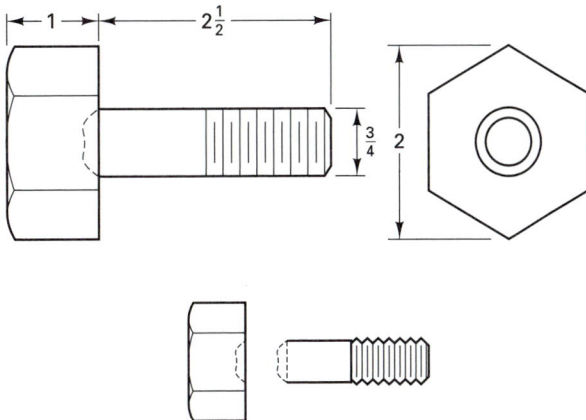

FIGURE CS-19

Describe some possible causes for the failures. How might you confirm the actual cause? What tests might you want to conduct? What changes might you want to make to avoid future difficulties? Consider the material, method of fabrication, and procedures for installation.

CHAPTER 20

FABRICATION OF PLASTICS, CERAMICS, AND COMPOSITES

20.1 INTRODUCTION

20.2 FABRICATION OF PLASTICS

Casting

Blow Molding

Compression Molding or Hot-Compression Molding

Transfer Molding

Cold Molding

Injection Molding

Reaction Injection Molding

Extrusion

Thermoforming

Rotational Molding

Foam Molding

Other Plastic-Forming Processes

Machining of Plastics

Other Finishing and Assembly Operations

Designing for Fabrication

Inserts

Design Factors Related to Finishing

20.3 PROCESSING OF RUBBER AND ELASTOMERS

20.4 PROCESSING OF CERAMICS

Fabrication Techniques for Glasses

Fabrication of Crystalline Ceramics

Firing of Ceramic Products

Machining of Ceramics

Joining of Ceramics

Design of Ceramic Components

20.5 FABRICATION OF COMPOSITE MATERIALS

Fabrication of Particulate Composites

Fabrication of Laminar Composites

Fabrication of Fiber-Reinforced Composites

Production of Reinforcing Fibers

Processes Designed to Combine Fibers and a Matrix

Fabrication of Final Shapes from Fiber-Reinforced Composites

Pultrusion

Filament Winding

Lamination and Lamination-Type Processes

Spray Molding

Sheet Stamping

Injection Molding

Braiding, Three-Dimensional Knitting, and Three-Dimensional Weaving

Fabrication of Fiber-Reinforced Metal–Matrix Composites

Fabrication of Fiber-Reinforced Ceramic–Matrix Composites

Secondary Processing and Finishing of Fiber-Reinforced Composites

Case Study: FABRICATION OF LAVATORY WASH BASINS

■ 20.1 INTRODUCTION

As was shown in the materials chapters (most notably Chapters 6, 7, and 8), *plastics, ceramics*, and *composites* are substantially different from metals in both structure and properties. As a result, the principles of material selection and product design, and the processes of product fabrication, will also tend to be different. Some distinct similarities will certainly exist. For example, the specific material will still be selected for its ability to provide the required properties, and the fabrication process, for its ability to produce the desired shape in an economical and practical manner.

There are also noteworthy differences. Plastics, ceramics, and composites tend to be used closer to their design limits, and many of the fabrication processes are capable of converting raw material into a finished product in a single operation. This is in distinct contrast to the sequence of activities usually encountered with metals. Large, complex shapes can often be formed as a single unit, eliminating the need for multipart assembly operations. If they are needed, the joining and fastening operations tend to be quite different from those used with metals. In addition, plastics, ceramics, and composites can often provide integral and variable color, and the processes used to manufacture the shape can frequently produce the desired finish and precision. As a result, finishing operations are often unnecessary. This is an attractive feature, because for many of these materials, altering the final dimensions or surface by machining-type operations would be both difficult and costly.

As with metals, however, the properties of the product are usually dependent on the process used to fabricate the shape. Thus fabrication of an acceptable product involves the selection of both (1) an appropriate material and (2) a companion method of processing, such that the resulting combination can provide the desired shape, properties, precision, and finish.

■ 20.2 FABRICATION OF PLASTICS

Chapter 8 has already presented the wide variety of specific plastics or polymers that are currently used as engineering materials. As we move to the fabrication of products, we find that there is also a large number of processes by which plastics can be shaped or formed. Which method or methods are preferred depends to a large extent on the nature of the polymer, specifically whether it is thermoplastic or thermosetting. Thermoplastic polymers can be heated to produce either a formable solid or a liquid. The material can then be cast, injected into a mold, or forced into or through dies to produce a desired shape. Thermosetting polymers have far fewer options, because once the polymerization has occurred, the framework structure is established and no further deformation can occur. Thus the polymerization reaction and the shape-forming process are usually accomplished simultaneously.

Casting, blow molding, compression molding, transfer molding, cold molding, injection molding, reaction injection molding, extrusion, thermoforming, rotational molding, and *foam molding* are all used extensively to shape polymers. Each process has certain advantages and limitations that relate to part design, compatible materials, and production cost. Since it is desirable to convert raw material directly into a finished product in a single operation, it is important that the process be capable of producing both the shape and the desired properties.

Casting

Casting is the simplest of the shape forming processes because no fillers are used and no pressure is required, as the liquid polymer (or any resin that will polymerize at low temperatures and atmospheric pressure) is simply poured into a container having the shape of the desired part. Several variations of the process have been developed, and while not all plastics can be cast, there are a number of castable thermoplastics (acrylics, nylons, urethanes, and

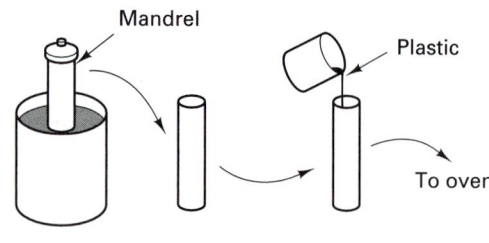

FIGURE 20-1   Steps in the casting of plastic parts.

PVC plastisols) and thermosets (phenolics, polyesters, epoxies, silicones, and urethanes). Plate glass can be used as a mold to cast thick plastic sheets. Continuous sheets and films can be produced by introducing the liquid polymer between two moving belts of highly polished stainless steel, the width and thickness being set by resilient gasket strips on either end of the gap. In another process, continuous thin sheets are made by ejecting the molten liquid from a gap-slot die onto a temperature-controlled chill roll. The molten plastic can also be spun against a rotating mold wall (centrifugal casting) to produce hollow or tubular shapes. Small products can be cast in shaped molds.

Because of the need for curing, the casting of thermoset resins often involves additional processing. Figure 20-1 depicts a process where a steel pattern is dipped into molten lead, withdrawn, and allowed to cool. A thin lead sheath is produced when the pattern is removed, and this becomes the mold for the plastic resin. The resin is held in the mold and is cured, either at room temperature or by heating for long times at temperatures in the range of 150 to 200°F (65 to 95°C). After curing, the product is removed, and the lead sheaths are reused.

Since cast plastics contain no fillers, they have a distinct lustrous appearance, and a wide range of transparent and translucent colors are available. Since the product is shaped as a liquid, fiber or particulate reinforcement can easily be incorporated. The process is relatively inexpensive because of the comparative lack of costly dies, equipment, and controls. Typical products include sheets, plates, films, rods, and tubes, as well as small objects, such as jewelry, ornamental shapes, gears, and lenses. While dimensional precision can be quite high, quality problems can occur because of inadequate mixing, air entrapment, gas evolution, and shrinkage.

Blow Molding

A variety of *blow molding* processes have been developed, the most common being used to convert thermoplastic polyethylene, polyvinyl chloride (PVC), polypropylene, and PEEK resins into bottles and other hollow-shape containers. A round, solid-bottom, hollow tube, known as a *preform* or *parison*, is made from the heated plastic by either extrusion or injection molding. The preform is then positioned between the halves of a split mold, the mold closes, and the preform is expanded against the mold by air or gas pressure. The mold is then cooled, the halves separated, and the product is removed. Any flash is then trimmed for direct recycling. Figure 20-2 depicts a schematic of this process, which has currently been expanded to include the engineering thermoplastics and to produce products as varied as automotive fuel tanks, seat backs, ductwork, and bumper beams.

Variations of blow molding have been designed to provide both axial and radial expansion of the plastic (for enhanced strength) as well as to produce multilayered products. In yet another modification, sheets of heated plastic are placed between two cavities, with

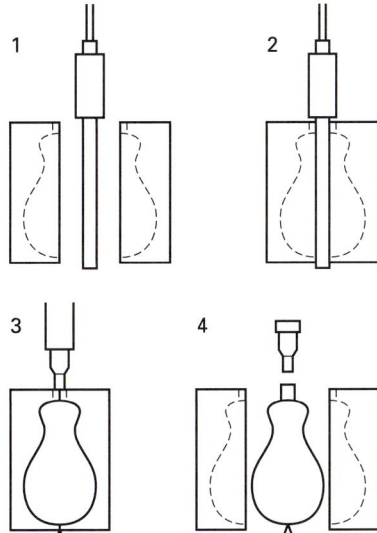

FIGURE 20-2 Steps in blow molding plastic parts: (1) a tube of heated plastic is placed in the open mold, (2) the mold closes over the tube, (3) air forces the tube against the sides of the mold, and (4) the mold opens to release the product.

the bottom one having the shape of the product. Both cavities are then pressurized to 300 to 600 psi with a nonreactive gas such as argon. When the pressure in the lower segment is vented, the gas in the upper segment "blows" the material into the lower die cavity.

The molds for this process consist of cast or machined blocks that contain the desired cavity as well as a cooling system, venting system, flash pockets, and one or more pinch-offs. The mold material must provide thermal conductivity and durability while being inexpensive and compatible with the resins being processed. Beryllium copper, aluminum, tool steels, and stainless steels are all popular.

Compression Molding or Hot-Compression Molding

In *compression molding*, as illustrated schematically in Figure 20-3, solid granules or preformed tablets of unpolymerized plastic are introduced into an open, heated cavity. A heated plunger then descends to close the cavity and apply pressure. As the plastic melts and becomes fluid, it is driven into all portions of the cavity. The heat and pressure are maintained until the material has "set" (i.e., cured or polymerized). The mold is then opened and the part is removed. A wide variety of heating systems and mold materials are used, and multiple cavities can be placed within a mold to produce more than one part in a single pressing. The process is simple and is used primarily for the *thermosetting polymers*, although recent developments permit the shaping of *thermoplastic polymers* and composites. Cycle times are set by the rate of heat transfer and the reaction or curing rate of the polymer. They typically range from under 1 minute to as much as 20 minutes or more.

Tool and machinery costs are often lower than for competing processes, and the dimensional precision and surface finish are high, thereby reducing or eliminating secondary operations. Compression molding is most economical when it is applied to small production runs of parts requiring close tolerances, high impact strength, and low mold shrinkage. It is a poor choice when the part contains thick sections (the cure times become quite long) or when large quantities are desired. Most products have relatively simple shapes because the flow of material is rather limited. Typical compression-molded parts include gaskets, seals, exterior automotive panels, aircraft fairings, and a wide variety of interior panels.

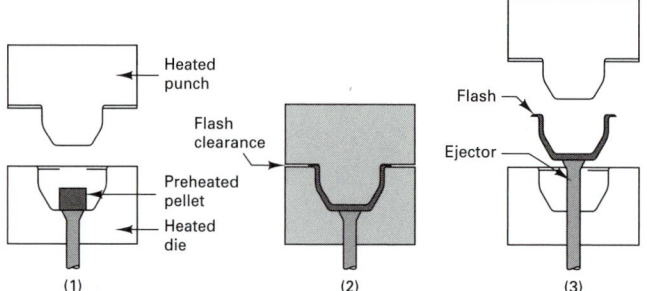

FIGURE 20-3   Schematic representation of the hot-compression molding process: (1) solid granules or a preform pellet is placed in a heated die, (2) a heated punch descends and applies pressure, and (3) after curing (thermosets) or cooling (thermoplastics), the mold is opened and the part is removed.

More recently, compression molding has become a means of forming fiber-reinforced plastics, both thermoplastic and thermoset, producing parts with properties that rival the engineering metals. In the thermoset family, polyesters, epoxies, and phenolics can be used as the base of fiber-containing sheet-molding compound, bulk-molding compound, or sprayed-up reinforcement mats. These are introduced into the mold and shaped and cured in the normal manner. Cycle times range from about 1 to 5 minutes per part, and typical products include wash basins, bathtubs, equipment housings, and various electrical components.

If the starting material is a fiber-containing *thermoplastic*, precut blanks are first heated in an infrared oven until the resin melts, producing a soft, pliable material. They are then transferred to the press, where they are shaped and cooled in specially designed dies. Compared to the thermosets, cycle times are reduced and the scrap is often recyclable. In addition, the products can be joined or assembled using the thermal "welding" processes applied to plastics.

Compression molding equipment is usually rather simple, typically consisting of a hydraulic or pneumatic press with parallel platens that apply the heat and pressure. Pressing areas range from 6 inches square to as much as 8 feet square, and the force capacities range from 6 to 10,000 tons. The molds are usually made of tool steel and are polished or chrome plated to improve material flow and product quality. Mold temperatures typically run between 300 and 400°F but can go as high as 1200°F. They are heated by a variety of means, including electric heaters, steam, oil, and gas.

Transfer Molding

Transfer molding is sometimes used to reduce the turbulence and uneven flow that often results from the high pressures of hot-compression molding. As shown schematically in Figure 20-4, the unpolymerized raw material is placed in a plunger cavity, where it is heated until molten. The plunger then descends, forcing the molten plastic through channels or runners into adjoining die cavities. Temperature and pressure are maintained until the thermosetting resin has cured completely. The charge material is frequently preheated to shorten the cycle and minimize erosion of the cavity, plunger, runner, and gates.

Because the material enters the die cavities as a liquid, there is little pressure until the cavity is completely filled. Thin sections, excellent detail, and good tolerances and finish are all characteristics of the process. In addition, transfer molding can be used when

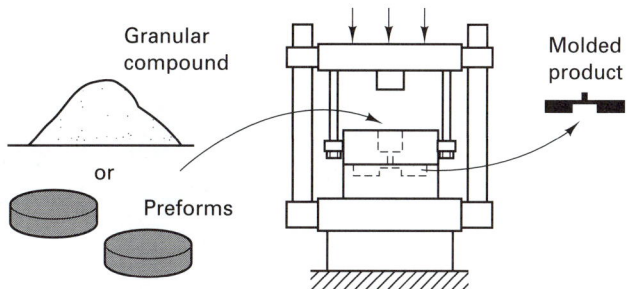

FIGURE 20-4 Schematic diagram of the transfer molding process where a plunger drives molten material into an adjacent die.

inserts are to be incorporated into the product. These are positioned within the cavity and are maintained in place as the liquid resin is introduced around them. In essence, transfer molding combines elements of both compression and injection molding and enables some of the advantages of injection molding to be utilized with thermosetting polymers. The resins can be reinforced with fillers, such as cellulose, glass, silica, alumina, or mica, to improve the mechanical or electrical properties and reduce shrinkage or warping. Common products include electrical switchgear and wiring devices, parts of household appliances that require heat resistance, structural parts that require hardness and rigidity under load, under-hood automotive parts, and parts that require good resistance to chemical attack.

Cold Molding

In the *cold-molding* process, the raw thermosetting material is pressed to shape while cold and is then removed from the mold and cured in a separate oven. While the process is quite economical, the resulting products do not have very good surface finish or dimensional precision.

Injection Molding

Injection molding is the most widely used process for high-volume production of thermoplastic resin parts. Figure 20-5 illustrates one approach to the process, where granules of raw material are fed by gravity from a hopper into a cavity that lies ahead of a moving plunger. As the plunger advances, the material is forced into a heated chamber, where it is preheated. From there it is forced through a *torpedo* section, where it is melted and superheated to 400 to 600°F (200 to 300°C). The material exits the torpedo section through a nozzle that seats against a mold. Other types of melt and injection units use screws that both rotate and reciprocate axially, or combinations of screws and plungers, to control the flow of material and to provide the pressure. Alternative means of heating include heated barrels, as well as heat generation by the shearing action of the screws.

As with die casting, sprues and runners then channel the molten material to the various closed-die cavities. Since the dies remain cool, the plastic solidifies almost as soon as the mold is filled. Premature solidification would cause defective parts, so the material is rapidly forced into the mold cavities by pressures in the range 5 to 20 ksi (35 to 140 MPa). The mold halves must clamp tightly together during molding and then be easily separated for part ejection. In addition, impact forces should be minimized during die

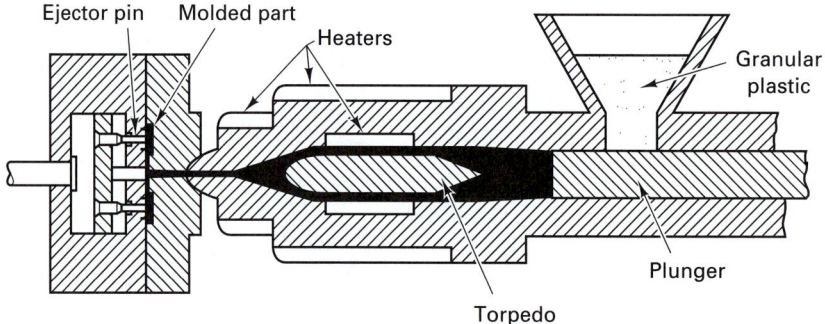

FIGURE 20-5   Schematic diagram of the injection molding process. This particular design uses a heated manifold and heated torpedo.

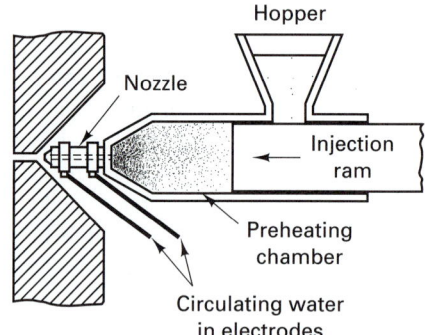

FIGURE 20-6   Modification of the injection molding process for the forming of thermosetting plastics.

closure, since they can adversely affect die life. Various types of clamping designs have been developed, including toggle, hydraulic, and hydromechanical.

Complex control systems have been developed to coordinate all of the functions of the process, including the time required for cooling within the mold. By heating the material for the next part as the mold is separating for part ejection, a complete molding cycle can be completed in 1 to 30 seconds. The entire process is very similar to the die casting of molten metals, and the properties are also somewhat similar, provided that the lower rigidity of the polymer is acceptable. The resulting form is usually a finished product needing no other work before assembly or use.

If injection molding is applied to the thermosetting materials, the process must be modified to provide the temperature and pressure required for curing. Figure 20-6 shows one such modification where the polymer is preheated in the feed chamber to about 200°F (95°C) and then heated further to the temperature of polymerization as it passes through a nozzle. Additional time in a heated mold is sufficient to complete the curing process. A major concern is to prevent the material from curing while it is still in the nozzle. Water cooling is introduced to this area as soon as the cavity is nearly filled. This cools the material and retards completion of the hardening reaction. The major deterrent to the injection molding of the thermosets, however, is the relatively long cycle times that are required.

Some injection molding machines have been modified to incorporate a hot runner distribution system to transfer the material from the injection nozzle to the mold cavities. The material in a cold runner solidifies with each cycle and needs to be ejected and reprocessed or disposed. With hot runners, the thermoplastic material is kept in a liquid state

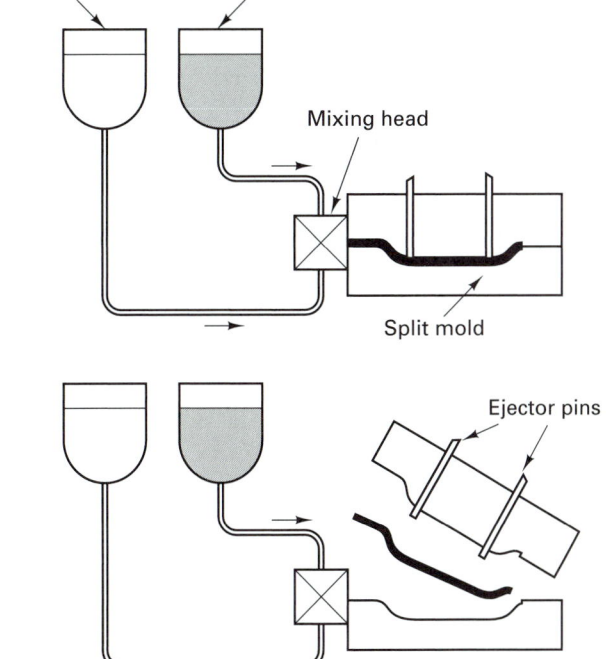

FIGURE 20-7 The reaction injection molding process. (*Left*) When the mixing head opens, the reactants mix and enter the mold at high speed. The mixing head closes when the mold is full. (*Right*) After curing, the mold is opened and the component is ejected.

until it reaches the gate. By utilizing the material in the runners in the subsequent shot, shot size is reduced, along with cycle time, since less material must be heated. Quality is improved, since all material enters the mold at the same temperature, recycled sprues and runners are not incorporated into the charge, and some degree of pressure is maintained throughout the charge since material is not being injected into empty runners. Hot runners do add an additional degree of complexity to the design as well as operation and control of the system, and the additional cost must be weighed against the benefits cited.

Reaction Injection Molding

Figure 20-7 depicts the *reaction injection molding* process, in which two or more liquid monomers are metered into a unit where they are intimately mixed by the impingement of liquid streams that have been pressurized to between 2000 and 3000 psi. The pressure is then reduced and the combined stream exits the mixhead directly into a mold. There a chemical reaction takes place between the two components, resulting in polymerization. Heating is not required; rather, the chemical reaction gives off energy that must be removed. Production rates are set primarily by the curing time of the polymer, which is often less than 1 minute. Molds are made from steel, aluminum, or nickel shell, with selection being made on the basis of number of parts to be made and the desired quality. They are generally clamped in low-tonnage presses.

At present, the dominant materials for reaction injection molding are polyurethanes, polyamides, and composites containing short fibers or flakes. Properties can span a wide range, depending on the combination and percentage of base chemicals and the additives that are used. Different formulations result in elastomeric or flexible, structural foam (foam core with a hard, solid outer skin), solid (no foam core), or composite products. Part size can range from 1 to 100 pounds, shapes can be quite complex (with variable wall

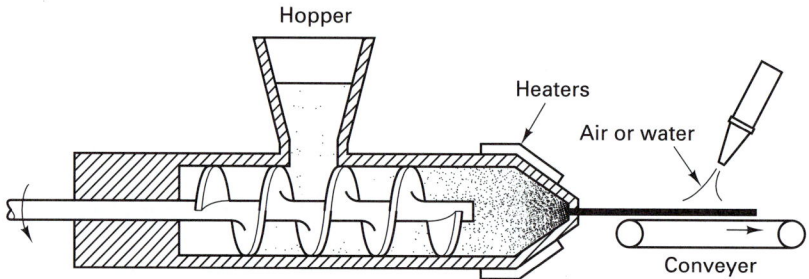

FIGURE 20-8   Extrusion process for producing plastic parts.

thickness), and surface finish is excellent. Automotive applications include steering wheels, airbag covers, instrument panels, door panels, armrests, headliners, and center consoles, as well as body panels, bumpers, and wheel covers. Rigid polyurethanes are also used in such products as household refrigerators, water skis, hot-water heaters, and picnic coolers.

The low processing temperatures and low injection pressures make the process attractive for the molding of large parts, and the large size can often enable parts consolidation. Thermoset parts can be fabricated with less energy than the injection molding of thermoplastics, with similar cycle times and similar degree of automation. While the metering and mixing equipment and related controls tend to be quite sophisticated and costly, the lower temperatures and pressures enable the use of cheaper molds, which can be quite large and easily changed.

Extrusion

Long plastic products with uniform cross sections are readily produced by the *extrusion* process depicted in Figure 20-8. Thermoplastic pellets or powders are fed through a hopper into the barrel chamber of a screw extruder. A rotating screw propels the material through a preheating section, where it is heated, homogenized, and compressed, and then forces it through a heated die and onto a conveyor belt. As the plastic passes onto the belt, it is cooled by jets of air or sprays of water which harden it sufficiently to preserve its newly imparted shape. It continues to cool as it passes along the belt and is then either cut into lengths or coiled, depending on whether the material is rigid or flexible and the desires of the customer. The process is continuous and provides a cheap and rapid method of molding. Common production shapes include a wide variety of solid forms, as well as tubes, pipes, and even coated wires and cables. If an emerging tube is expanded by air pressure, allowed to cool, and then rolled, the product can be a double layer of sheet or film.

Thermoforming

In the *thermoforming* process, thermoplastic sheet material is heated to a working temperature and then formed into a finished shape by heat, pressure, or vacuum. The starting material can be either discrete sheets or a continuous roll of material. If continuous material is used, it is usually heated by passing through an oven or other heating device. The material emerges between the upper and lower halves of a mold and is formed either up or down by the application of vacuum and/or pressure. Cooling occurs upon contact with the mold, and the product hardens in its new shape. Ejection and trimming follow to complete the process, and the remaining strip material is rewound in preparation for recycling.

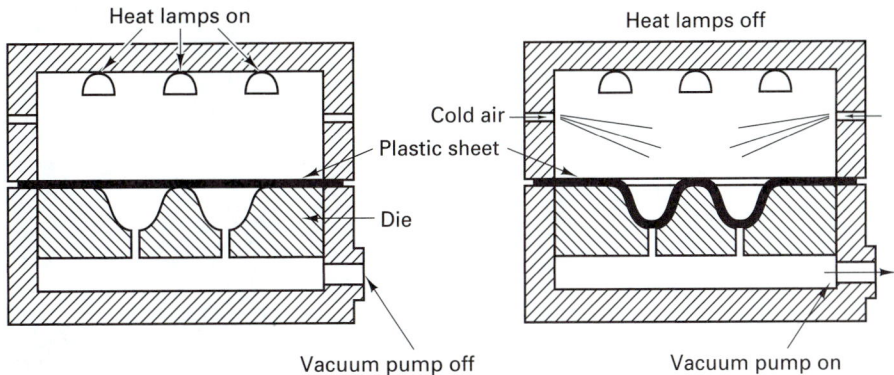

FIGURE 20-9   One method of molding plastic sheets by heat and vacuum (thermoforming).

Figure 20-9 shows the process using a female mold cavity and discrete sheets of material. Here the material is placed directly over the die or pattern and heated in place. Pressure and/or vacuum is then applied, causing the material to draw into the cavity. Male form blocks and mating male and female dies can also be used. After sufficient cooling, the part is removed from the die. An entire cycle requires only a few minutes. Products range from panels for light fixtures to pages of Braille text for the blind.

Rotational Molding

Rotational molding is used to produce hollow, seamless products of a wide variety of sizes and shapes, including storage tanks, bins and refuse containers, doll parts, footballs, helmets, and even boat hulls. The process begins with closed molds or cavities that have been filled with a premeasured amount of thermoplastic powder or liquid. The molds are either preheated or are placed in a heated oven, and are then rotated simultaneously about two perpendicular axes. Other designs rotate the molds about one axis while tilting or rocking about another. In either case, the resin melts and is distributed in the form of a uniform-thickness coating over all of the surfaces of the mold. The molds are then transferred to a cooling chamber, where the motion is continued and air or water is used to slowly drop the temperature. When the material has solidified, the molds are opened and the hollow product is removed. The lightweight rotational molds are frequently made from cast aluminum, but sheet metal is common for larger parts, and electroformed or vaporformed nickel can be used to reproduce fine detail.

Foam Molding

Foamed plastics have become an important and widely used form of polymer product. In *foam molding*, a foaming agent is mixed with the plastic resin and releases gas or volatilizes when the combination is heated during molding. The materials expand to 2 to 50 times their original size, resulting in products with densities ranging from 2 to 40 lb/ft^3 (32 to 641 kg/m^3). *Open-cell foams* have interconnected pores that permit the permeability of gas or liquid. *Closed-cell foams* have the property of being gas- or liquid-tight.

Both rigid and flexible foams have been produced, using both thermoplastic and thermosetting material. The rigid type is useful for structural applications (including housings for computers and business machines), packaging and shipping containers, as patterns for the full-mold casting process (see Chapter 14), and for injection into the interiors of thin-

FIGURE 20-10   Plastic gear having a solid outer skin and a rigid foam core. *(From American Machinist.)*

skinned metal components, such as aircraft fins and stabilizers. Flexible foams are used primarily for cushioning.

Variations of the conventional processes are used to produce the desired shapes. One unique modification involves the use of multiple injection devices and a single mold to make products with a solid outer skin and a rigid foam core. Figure 20-10 shows the cross section of a plastic gear that has this type of dual structure.

Other Plastic-Forming Processes

In the *calendering* process, molten thermoplastic is forced between two or more counter-rotating rolls to produce thin sheets or films of polymer, which are then cooled to induce hardening. Product thicknesses generally range between $\frac{1}{16}$ and $\frac{1}{100}$ in. (1.0 to 0.3 mm) but can be reduced further to as low as 0.002 in. (0.05 mm) by subsequent stretching. Embossed designs can be incorporated into the rolls to produce products with textures or patterns.

Conventional *drawing* and *rolling* can be performed to produce fibers or change the shape of plastic extrusions. In addition to changing the product dimensions, these processes can also serve to induce crystallization or produce a preferred orientation to the thermoplastic polymer chains.

Filaments, fibers, and yarns can be produced by *spinning*, a modified form of extrusion. Molten thermoplastic polymer is forced through a die containing many small holes. Where multistrand yarns or cables are desired, the dies can rotate or spin to produce the twists and wraps.

Machining of Plastics

Plastics can be milled, sawed, drilled, and threaded much like metals, but their properties are so variable that it is impossible to give instructions that are exactly correct for all. It is important, however, to remember some of the general characteristics of plastics that will affect their machinability. Since plastics tend to be poor thermal conductors, little of the heat that results from chip formation will be conducted into the material or be carried away in the chips. As a result, tools that are machining plastic run very hot and may fail more rapidly than when cutting metal. Carbide tools may be preferred over high-speed tool steels if the cuts are of moderate duration or if high-speed cutting is performed.

In response to the high temperatures that develop at the point of cutting, thermoplastics tend to soften and swell, and occasionally bind or clog the cutting tool. In addition,

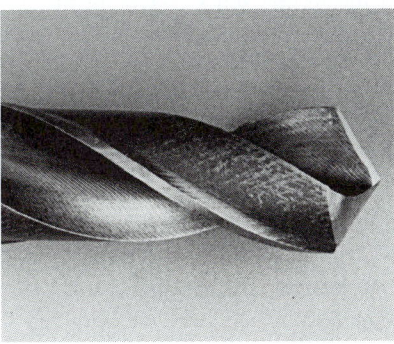

FIGURE 20-11 Straight-flute drill (*top*) and "dubbed" drill (*bottom*) used for drilling plastics.

considerable elastic flexing can occur, and this, coupled with material softening, reduces the precision of final dimensions. Because of their higher rigidity and reduced softening, the thermosetting materials are easier to machine to precise dimensions.

The tools that are used to machine plastics should be kept sharp at all times. Drilling is best done by means of straight-flute drills or by "dubbing" the cutting edge of a regular twist drill to produce a zero rake angle. Such configurations are shown in Figure 20-11. Rotary files, saws, and milling cutters should be run at high speeds so as to improve cooling but with the feed carefully adjusted to avoid clogging the cutter. Coolants can often be used advantageously if they do not discolor the plastic or induce gumming. Water, soluble oil and water, and weak solutions of sodium silicate have been used effectively. Filled and laminated plastics can be quite abrasive, and the fine dust that is produced during machining can be a health hazard.

Laser machining can be an attractive process for plastics. By vaporizing the material instead of forming chips, precise cuts can be achieved. Minute holes can be drilled, such as those in the nozzles of aerosol cans.

Other Finishing and Assembly Operations

Other finishing processes that can be applied to plastics include printing, hot stamping, vacuum metallizing, electroplating, and painting. Some of these are developed in Chapter 40.

Many plastics can be joined by heating the relevant surfaces or regions. Heat can be applied through a stream of hot gases or by a tool like a soldering iron. Joining heat can also be generated through ultrasonic vibrations. The welding techniques that are applied to plastics are presented in Chapter 36. Adhesive bonding, another popular means of joining plastic, is presented in Chapter 38.

Designing for Fabrication

The primary objective of any manufacturing activity is the production of satisfactory components or products, and this involves the selection of an appropriate material or materials. When polymers are selected as the material of construction, it is usually as a result of one or more of their somewhat unique properties, which include light weight, corrosion resistance, good thermal and electrical insulation, and the possibility of integral color. While these properties are indeed attractive, one should also be aware of the more common limitations, such as softening or burning at elevated temperatures, poor dimensional stability, or the deterioration of properties with age. The basic properties and characteristics of polymeric materials are developed in Chapter 8.

A second area of manufacturing concern is selecting the process or processes to be used in producing the shape and establishing the desired properties. Each of the wide variety of fabrication processes has distinct advantages and limitations, and efforts should be made to utilize the unique features. In addition, the production of quality products also requires an awareness of all of the various aspects of a given process. For example, consider a molding process in which a fluid or semifluid polymer is introduced into a mold cavity and allowed to harden. The proper amount of material must be introduced and caused to flow in such a way as to completely fill the cavity. Air that originally occupied the cavity needs to be vented and removed. Shrinkage will occur during solidification and/or cooling and may not occur in a uniform manner. Heat transfer must be provided to control the cooling and/or solidification. Finally, a means must be provided for part removal or ejection from the mold. Surface finish and appearance, the resultant engineering properties, and the ultimate cost of production are all dependent on the good design and execution of the molding process.

In all molded products, it is important to provide adequate fillets between adjacent sections to assure smooth flow of the plastic into all sections of the mold and to eliminate stress concentrations at sharp interior corners. These fillets also make the mold less expensive to produce and reduce the danger of mold fracture during use. Even the exterior edges should be rounded where possible. A radius of 0.010 to 0.015 in. (0.25 to 0.40 mm) is scarcely noticeable but will do much to prevent an edge from chipping. Sharp corners should also be avoided in products that will be used for electrical applications, since they tend to increase voltage gradients, which can lead to product failure.

Wall or section thickness is also very important, since the hardening or curing time is determined by the thickest section. If possible, sections should be kept nearly uniform in thickness, since nonuniformity can lead to serious warpage and dimensional control problems. As a general rule, one should use the minimum thickness that will provide satisfactory end-use performance. The specific value will be determined primarily by the size of the part and, to some extent, the process and the type of plastic being used. Recommended minimum thicknesses for molded plastics are as follows:

Small parts	0.050 in. (1.25 mm)
Average-sized parts	0.085 in. (2.15 mm)
Large parts	0.125 in. (3.20 mm)

Thick corners should be avoided because they can lead to gas pockets, undercuring, or cracking. When extra strength is needed in a corner, it can usually be provided by incorporating ribs into the design.

Economical production is also facilitated by appropriate dimensional tolerances. A minimum tolerance of ±0.003 in. (0.08 mm) should be allowed in directions that are

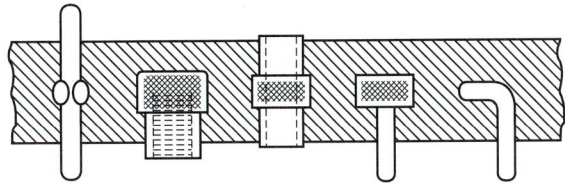

FIGURE 20-12 Typical metal inserts used to provide threaded cavities, holes, and alignment pins in plastic parts.

parallel to the parting line of the mold or contained within a mold segment. In directions that cross a parting surface, a minimum tolerance of ±0.010 in. (0.25 mm) is desirable. In both cases, increasing these values by about 50% can simultaneously reduce both manufacturing difficulty as well as cost.

Since most molds are reusable, careful attention should be given to the removal of the part. Rigid, metal molds should be designed so that they can be easily opened and closed. A small amount of unidirectional taper should be provided to facilitate part withdrawal. Undercuts should be avoided whenever possible, since they will prevent part removal unless additional mold sections are used. These must move independently of the major segments of the mold, adding to the costs of mold production and maintenance and slowing the rate of production.

Inserts

Metal *inserts* are often incorporated into plastic products to provide enhanced performance or unique features. Since molded threads are difficult to produce, machined threads require additional processing, and both types tend to chip or deform, threaded inserts are frequently used when assemblies require considerable strength or when frequent disassembly and reassembly is anticipated. Figure 20-12 depicts one form of threaded insert, along with other types that serve to provide alignment or mounting pins and holes. These metal inserts are usually made from brass or steel.

The successful use of inserts requires careful attention to design since they are generally held in place by only a mechanical bond. Knurling or grooving is often required to provide suitable sites for gripping. A medium or coarse knurl is usually adequate to resist torsional loads and moderate axial forces. Circumferential grooves are excellent for axial loads but offer little resistance to torsional rotation. Consideration should also be given to minimizing the stresses that occur around a cast-in insert as a result of the differential contraction during cooling. Preheating the inserts can be beneficial, the plastic can be selected to minimize shrinking, and nonmetallic inserts should be considered.

If an insert is to act as a boss for mounting or serve as an electrical terminal, it should protrude slightly above the surface of the plastic. This permits a firm connection to be made without creating an axial load that would tend to pull the insert from its surroundings. If the insert serves to hold two mating parts together, it should be flush with the surface. In this way, the parts can be held snugly together without danger of loosening the insert. In all cases, the wall thickness of the surrounding plastic must be sufficient to support any load that may be transmitted by the insert. For small inserts, the wall thickness should be at least half the diameter of the insert. For inserts larger than $\frac{1}{2}$ in. (13 mm) in diameter, the wall thickness should be at least $\frac{1}{4}$ in. (6.5 mm).

Design Factors Related to Finishing

Because plastics are frequently used where consumer acceptance is of great importance, special attention should be given to finish and appearance. In many cases, plastic parts can

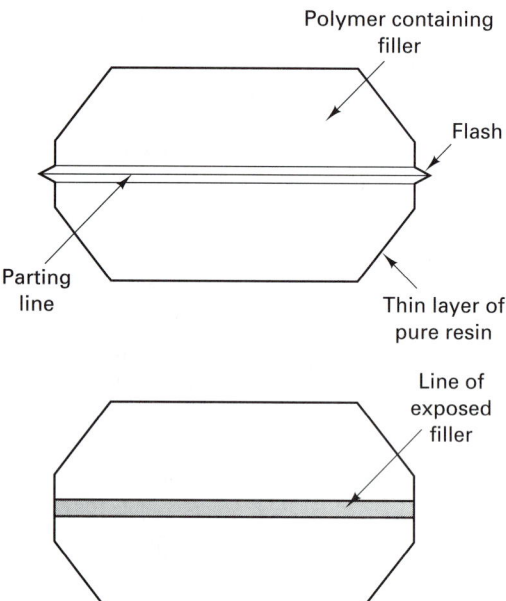

FIGURE 20-13 Trimming the flash from a plastic part creates a line of exposed filler.

be designed to require very little finishing or decorative treatment. Fins and rough spots can often be removed by a barrel tumbling with suitable abrasives or polishing agents. Smoothing and polishing occur in the same operation. By etching the surfaces of a mold, decorations or letters can be produced that protrude approximately 0.004 in. (0.01 mm) above the surface of the plastic. When higher relief is required, the mold can be engraved, but this adds significantly to mold cost.

Whenever possible, depressed letters or designs should be avoided. These features, when transferred to the mold, become raised above the surrounding surface. Mold making then requires a considerable amount of intricate machining as the surrounding material is cut away from the design or letters. When recessed features are required, manufacturing cost can be reduced if they can be incorporated into a small area that is raised above the primary surface.

A prime objective in the design of plastic parts is often the elimination of secondary machining, especially on surfaces that would be exposed to the customer. Even when fillers are used (as they are in most plastics), the surfaces of molded parts have a thin film of pure resin. This film provides the high luster that is characteristic of polymeric products. Machining cuts through the surface, exposing the underlying filler. The result is a poor surface appearance, as well as a site for the absorption of moisture.

One location that frequently requires machining is the parting line that is produced where the mold segments come together. Perfect mating is difficult to achieve, especially when the parting surfaces are not flat. As a result, a small fin or "flash" is produced around the part perimeter, as illustrated in Figure 20-13. When the flash is trimmed off, the resulting line of exposed filler may be objectionable. By locating the parting line along a sharp corner, it is easier to maintain satisfactory mating of the mold sections, and the exposed filler that is created by flash removal will be confined to a corner, where it is less noticeable.

Because plastics have a low modulus of elasticity, large flat areas are not rigid and should be avoided whenever possible. Ribbing or doming, like that illustrated in Figure

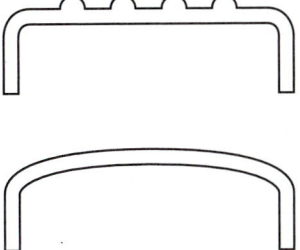

FIGURE 20-14 Method of providing stiffness to large surfaces of plastic parts through the use of ribbing or doming.

20-14, can be used to provide the required stiffness. In addition, flat surfaces tend to reveal flow marks from the molding operation, as well as scratches that occur during service. External ribbing then serves the dual function of increasing strength and rigidity while masking the surface flaws. Dimpled or textured surfaces can also be used to provide a pleasing appearance and conceal scratches.

Holes that are formed by pins that protrude from the mold often require special consideration. In compression molding, these pins can be subjected to considerable bending during mold closure and filling. When these pins are supported only at one end, their length should not exceed twice the diameter. In processes with reduced filling pressures, the length can be as much as five times the diameter without excessive problems.

Holes that are to be threaded or will be used for self-tapping screws, should be countersunk. This not only assists in starting the tap or screw but also reduces chipping at the outer edge of the hole. If the threaded hole is less than $\frac{1}{4}$ in. (6.35 mm) in diameter, it is best to cut the threads after molding, using some form of thread tap. For diameters greater than $\frac{1}{4}$ inch, it is usually better to mold the thread or use an insert. If the threads are molded, however, special provisions must be made to remove the part from the mold. These additional operations are generally uneconomical, since they extend the molding time and reduce productivity.

■ 20.3 PROCESSING OF RUBBER AND ELASTOMERS

Rubber and elastomeric products can also be produced by a variety of fabrication processes. Probably the simplest of these is *dipping*, a process used to produce relatively thin parts with uniform wall thickness, such as boots, gloves, and fairings. A master form is first produced, usually from some type of metal. This form is then immersed into a liquid preparation or compound (usually based on natural rubber, neoprene, or silicone), then removed and allowed to dry. With each dip, a certain amount of the liquid adheres to the surface, and repeated dips are used to produce a final desired thickness. After vulcanization, usually in steam, the products can be stripped from the molds.

Dipping can be accelerated by using electrostatic charges. A negative charge is introduced to the latex particles, and the form or mold is charged positively, either through an applied voltage or by a coagulant coating that releases positive ions when dipped into the solution. The attraction and neutralization of the charges causes the elastomeric particles to be deposited on the form at a faster rate and in thicker layers. With electrostatic deposition, many products can be made in a single immersion.

When products are to be made of solid elastomers, the first step is the compounding of the elastomers, vulcanizers, fillers, antioxidants, accelerators, and other pigments. This is usually done in some form of mixer, which breaks down the elastomer and permits the addition of the other components to form a homogeneous mass. Adaptations of the processes previously discussed for plastics are frequently used to produce the shapes. Injection,

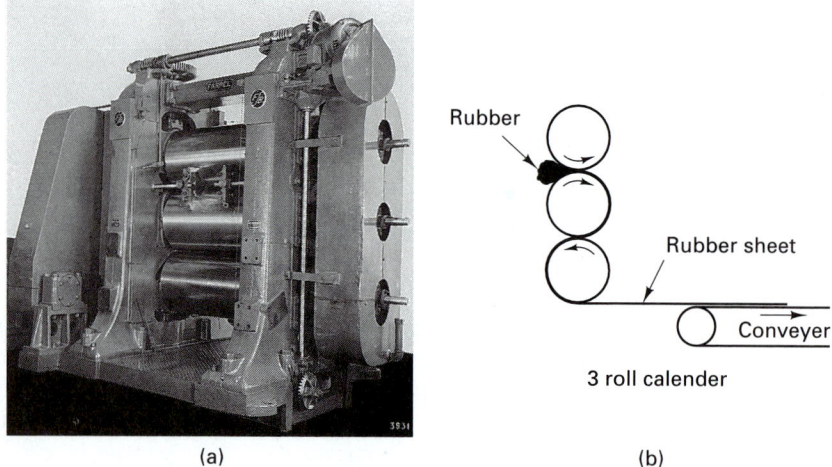

(a)   (b)

FIGURE 20-15   (*Left*) Three-roll calender used for producing rubber or plastic sheet. *(Courtesy of Farrel-Birmingham Company, Inc.)* (*Right*) Schematic diagram showing the method of making sheets of rubber with a three-roll calender.

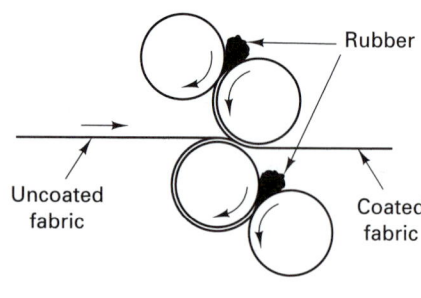

FIGURE 20-16   Arrangement of the rolls, fabric, and coating material for coating both sides of a fabric in a four-roll calender.

compression, and transfer molding are used, along with special techniques for foaming. Urethanes and silicones can be cast to shape.

Rubber compounds can be made into sheets using *calenders*, like that shown in Figure 20-15. The sheet coming from the calender is often rolled with a fabric liner to prevent the material from sticking. Three- or four-roll calenders can also be used to place a rubber or elastomer covering over cord or woven fabric. In the three-roll geometry, only one side of the fabric can be coated in each pass. The four-roll arrangement, shown schematically in Figure 20-16, enables both sides to be coated simultaneously.

Products such as inner tubes, garden hose, tubing, and strip moldings are produced by the extrusion process. The compounded material is forced through a die by a screw device similar to that described for plastics.

Adhesives have been developed that enable the bonding of rubber or artificial elastomers to metal, usually brass or steel. Only moderate pressures and temperatures are required to obtain excellent adhesion.

■ 20.4 PROCESSING OF CERAMICS

The fabrication processes applied to ceramic materials generally fall into two distinct classes, based on the properties of the material. *Glasses* can be manufactured into useful

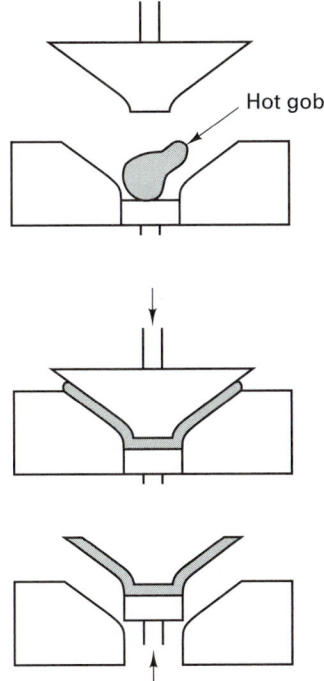

Hot gob

FIGURE 20-17 The pressing of viscous glass with mating male and female die members.

articles by first heating the material to produce a molten or viscous state, shaping the material by means of *viscous flow*, and then cooling the material to produce a solid product. *Crystalline ceramics* have a characteristically brittle behavior and are normally manufactured into useful components by pressing moist aggregates or powder into a shape, followed by drying, and then bonding by one of a variety of mechanisms, which include chemical reaction, *vitrification* (cementing with a liquefied material), and sintering (solid-state diffusion).

Fabrication Techniques for Glasses

Glass is generally shaped at elevated temperatures where the viscosity can be controlled. A number of the processes begin with material in the liquid or molten condition. Sheet and plate glass is formed by processes such as rolling through water-cooled rolls or floating on a bath of molten tin. Other glass shapes can be produced by pouring the molten material directly into a mold. The cooling rate is then controlled, usually as slow as possible, to minimize residual stresses and the tendency for cracking. Glass *fibers* can be produced by forcing liquid glass through multiple openings in a metal die.

Other processes begin with viscous masses and use mating male and female die members to press the material into the desired shape, as illustrated in Figure 20-17. A process similar to the blowmolding of plastics can be used to expand cup-shaped pieces of viscous material against the outside of heated dies. This process is illustrated in Figure 20-18.

Special heat treatments can also be applied to glass material. By applying forced cooling to the exposed surfaces, a residual stress pattern of surface compression can be induced. (The surface layers cool and contract. This is followed by the cooling of the interior, which then tries to contract but is restrained by the surface, creating the surface compression.) Such glass is stronger and more fracture resistant, since cracks tend to

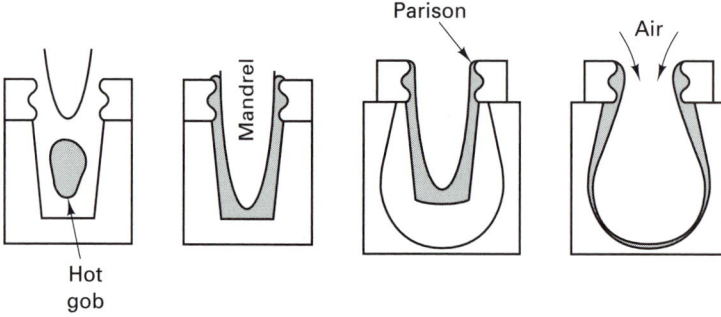

FIGURE 20-18 Forming a thin-walled glass shape by a combination of pressing and blow molding.

initiate on free surfaces. Annealing can be used to reduce unfavorable residual stresses that might lead to cracking. Heating might also be used to promote devitrification, the precipitation of a crystalline phase from within the glass.

Glass-ceramics form a unique class of materials that are part crystalline and part glass. They are fabricated into shape as a glass, and are then subjected to a special heat treatment that controls the nucleation and growth of the crystalline component. The final properties include good strength and toughness, along with low thermal expansion. Typical products include cookware (such as the white Corning products), ceramic stove tops, and materials used in electrical and computer components.

Fabrication of Crystalline Ceramics

Crystalline ceramics are hard, brittle materials with high melting points. As a result, they cannot be formed by techniques requiring either plasticity (i.e., forming methods) or melting (i.e., casting methods). Instead, these materials are generally processed in the solid state, by techniques that utilize particles or aggregates and resemble those used in powder metallurgy. Dry powders can be pressed into useful shapes either at environmental or elevated temperatures. Dry pressing, isostatic pressing, and hot-isostatic pressing (HIP) are all common techniques and exhibit features and limitations similar to those discussed in Chapter 16.

In the *slip casting* process, ceramic powder is mixed with a liquid to form a slurry, which is then cast into a mold containing very fine pores. Capillary action pulls the liquid from the slurry, allowing the ceramic particles to arrange into a "green" body with sufficient strength for subsequent handling. Hollow shapes can be produced by pouring out the remaining slurry once a desired thickness of solid has formed on the mold walls.

Clay products are based on special types of ceramics blended with water and various additives to produce a material that can be shaped by most of the traditional forming methods. *Plastic forming* can also be applied to other ceramics if the ceramic particles are combined with additives, such as plastic, which impart plasticity when subjected to pressure and heat. Processes such as injection molding can be used to form complex, three-dimensional shapes. Extrusion can produce products with constant cross sections. After the shape is produced, the additive material is usually removed by dissolving in a solvent or controlled heating, and the remaining ceramic is fused together by a firing operation.

Table 20-1 summarizes the primary processes used to fabricate shapes from crystalline ceramics.

TABLE 20-1.	Processes Used to Form Crystalline Ceramics		
Process	Starting material	Advantages	Limitations
Dry axial pressing	Dry powder	Low cost; can be automated	Limited cross sections; density gradients
Isostatic processing	Dry powder	Uniform density; variable cross sections; can be automated	Long cycle times Small number of products per cycle
Slip casting	Slurry	Large sizes; complex shapes; low tooling cost	Long cycle times; labor-intensive
Injection molding	Ceramic–plastic blend	Complex cross sections; fast; can be automated; high volume	Binder must be removed; high tool cost
Forming processes (e.g., extrusion)	Ceramic–binder blend	Low cost; variable shapes (such as long lengths)	Binder must be removed; particles oriented by flow
Clay products	Clay, water, and additives	Easily shaped by forming methods; wide range of size and shape	Requires controlled drying

Firing of Ceramic Products

The processes described above can all be used to produce useful shapes from ceramic materials, but useful strength usually requires a subsequent heating operation, known as *firing* or sintering. Slurry-type materials must first be dried in a controlled manner, designed to control dimensional changes and minimize stresses, distortion, and cracking. The material is then heated and diffusion processes act to fuse the particles together and impart the desired mechanical and physical properties. The temperature and time are selected to control the resulting grain size, pore size, and pore shape. In some firing operations, surface melting or component reactions can produce a substantial amount of liquid material (a process known as vitrification). The liquid then flows to produce a glassy bond between the ceramic particles.

An alternative to elevated temperature firing is *cementation.* A liquid binder material is used to coat the ceramic particles, and a subsequent chemical reaction converts the liquid to a solid, forming strong, rigid bonds.

Machining of Ceramics

Most ceramic materials are brittle, and the techniques used to cut metals would often produce uncontrolled or catastrophic cracks. In addition, ceramic materials are generally hard. Since ceramics are often used as abrasives or coatings on cutting tools, the tools needed to cut them would have to be even harder.

When material removal operations are performed on ceramic materials, one may have the option of machining before or after the final firing. Before firing, the material is often rather weak and fragile. While fracture is always a concern, an additional consideration might be the dimensional changes that will occur upon firing. Since the shrinkage may be as much as 30%, it may be difficult to achieve or maintain close tolerances. For this reason, machining before firing, known as green machining, is usually rough machining designed to reduce final finishing. When machining is performed after firing, the processes are generally ones we might consider to be nonconventional. Grinding, lapping, and polishing with diamond abrasives, drilling with diamond-tipped tooling, cutting with diamond saws, ultrasonic machining, laser and electron beam machining, and chemical etching have all been used. When mechanical forces are applied, material support is quite critical (since the material is usually brittle). Since ceramics are hard materials, the tools must be quite rigid. Selection and use of coolants are also important issues.

The materials producers have worked to develop "machinable" ceramics, and we now have engineering ceramics that lend themselves to precision shaping by the more-traditional machining operations. It should be noted, however, that these are indeed special materials.

Joining of Ceramics

Once again, the unique properties of ceramics introduce fabrication limitations. Brittle ceramics cannot be joined by fusion welding or deformation bonding, and threaded assemblies should be avoided whenever possible. Therefore, most joining utilizes some form of adhesive bonding, brazing, diffusion bonding, or special cements. Even with these methods, the stresses that develop on the surfaces can lead to premature failure. As a result, most ceramic products are designed to be monolithic (single-piece) structures, rather than multipart assemblies.

Design of Ceramic Components

Since ceramics are brittle materials, special care should be taken to minimize bending and tensile loading as well as design stress raisers. Sharp corners and edges should be avoided where possible. Outside corners should be chamfered to reduce the possibility of edge chips. Inside corners should have fillets of sufficient radius to minimize crack initiation. Undercuts are difficult to produce and should be avoided. Specifications should generally use the largest possible tolerances, since these can often be met with products in the "as-fired" condition. Extremely precise dimensions usually require hand grinding, and costs can escalate significantly. In addition, consideration should be given to any surface finish requirements, since grinding, polishing, and lapping operations increase production cost substantially.

■ 20.5 FABRICATION OF COMPOSITE MATERIALS

As shown in Chapter 8, composite materials can be designed to offer a number of attractive properties. In some market areas, such as aerospace and sporting goods, their acceptance and growth have been phenomenal. Usage can only occur, however, if the material can be converted into useful shapes at an acceptable cost and rate of production. Many of the manufacturing processes designed for composites are slow, and some require extensive amounts of hand labor. There is often a high amount of variability between nominally "identical" products, and inspection and quality control methods are not well developed. While these limitations may be acceptable for certain applications, they clearly restrict the use of composites for high-volume, mass-produced items. Faster production speeds, increased use of automation, reduced variability, and integrated quality control will all be important in the expanded use of composite materials.

In Chapter 8 we classified composite materials by their basic geometry as particulate, laminar, and fiber-reinforced. Since the various fabrication processes are often unique to a specific type of composite, they will also be classified according to the three basic geometries.

Fabrication of Particulate Composites

These materials usually consist of a metallic or polymeric matrix and dispersed particles of a second material. Their fabrication, however, rarely requires processes unique to the composite. Instead, the particles are simply dispersed in the matrix by introduction into a liquid melt or slurry, or by blending the various components as solids, using powder metallurgy methods. Subsequent processing generally follows the conventional methods of casting or forming, or utilizes techniques that are common to powder metallurgy. These have been presented elsewhere in the book and will not be repeated here.

Fabrication of Laminar Composites

Laminar composites include coatings and protective surfaces, claddings, bimetallics, laminates, and a host of other materials. Their production generally involves processes designed to form a high-quality bond between distinct layers of different materials. When the layers are metallic, as in claddings and bimetallics, the composites can be produced by hot or cold *roll bonding*. Sheets of the various materials are passed simultaneously through the rolls of a conventional rolling mill. If the amount of deformation is great enough, surface oxides and contaminants are broken up and dispersed, metal-to-metal contact is established, and the two surfaces become joined by a solid-state bond. U.S. coinage is a common example of a roll-bonded material.

   Explosive bonding is another practical means of bonding layers of metal. A sheet of explosive material progressively detonates above the layers to be joined, causing a pressure wave to sweep across the interface. A small open angle is maintained between the two surfaces. As the pressure wave propagates, any surface films are liquefied or scarfed off and are jetted out the open interface. Clean metal surfaces are then forced together at high pressures, forming a solid-state bond with a characteristically wavy configuration at the interface. Wide plates (too wide to roll bond conveniently) and dissimilar materials with large differences in mechanical properties are attractive candidates for explosive bonding.

   Adhesive bonding is another attractive means of joining the various plies and can be applied to both metallic and nonmetallic materials. The *lamination* of polymer matrix composites (to be discussed later in this chapter when each ply is a fiber-reinforced or woven layer) often utilizes films of unpolymerized resin which are placed between the layers. Pressing at elevated temperature cures the resin and completes the bond. *Brazing* can be employed when the layers are metallic and can withstand the moderate elevated temperatures.

   In *sandwich structures* such as corrugated cardboard or the honeycomb shown in Figure 20-19, thin layers of facing material are bonded to a lightweight filler material. Special fabrication methods may be employed to produce the filler foam, corrugated layer, or honeycomb material.

Fabrication of Fiber-Reinforced Composites

A number of processes have been developed to produce and shape the fiber-reinforced composites. Variations are based primarily on the orientation of the fibers, the length of continuous filaments, and the geometry of the final product. Each seeks to embed the fibers in a selected matrix with the proper alignment and spacing necessary to produce the desired properties. Discontinuous fibers can be combined with the matrix to provide either a random or a preferred orientation. Continuous fibers are normally aligned in a unidirectional fashion in rods or tapes, woven into fabric layers, wound around a mandrel, or woven into a three-dimensional shape.

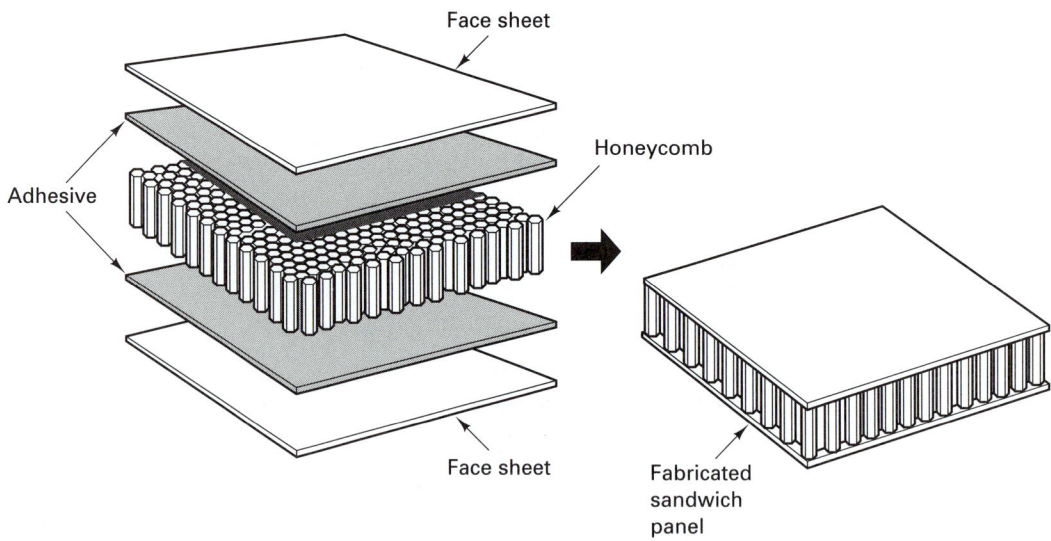

FIGURE 20-19 Fabrication of a honeycomb sandwich structure using adhesive bonding to join the facing sheets to the lightweight filler. *(Courtesy of ASM International)*

Production of Reinforcing Fibers. A number of processes have been developed to produce the various types of reinforcement fibers used in composites. Metallic fibers, glass fibers, and many polymeric fibers (including the popular Kevlar) are produced by variations of conventional wire drawing and extrusion. Boron, carbon, and ceramic fibers such as silicon carbide are too brittle to be produced by the deformation methods. Boron fibers are produced by chemical vapor deposition around a tungsten filament. Carbon (graphite) fibers can be made by carbonizing (decomposing) an organic material that is more easily formed to the desired shape.

The fine filaments are often bundled into *yarns* (twisted assemblies of filaments), *tows* (untwisted assemblies of fibers), and *rovings* (untwisted assemblies of yarns or tows). Fibers can also be chopped into short lengths, usually $\frac{1}{2}$ in. or less, for incorporation into the various sheet or bulk molding compounds. In these materials the fibers usually assume a random orientation.

Processes Designed to Combine Fibers and a Matrix. A variety of processes have been developed to combine the fiber and the matrix into a unified material suitable for further processing. If the matrix material can be liquefied and the temperature is not harmful to the fibers, *casting-type processes* can be an attractive means of coating the reinforcement. The pouring of concrete around steel reinforcing rod is a crude example of this method. In the case of the high-tech fiber-reinforced plastics and metals, the liquid can be introduced between the fibers by means of capillary action, vacuum infiltration, or pressure casting. Another alternative is to draw the fibers through a bath of molten material, and combine them into aligned bundles before the liquid solidifies.

Prepregs involve the formation of a woven fabric which is then infiltrated with the matrix material. *Mats* are sheets of nonwoven, randomly oriented fibers in a matrix material. When the matrix is a polymeric material, the infiltration is usually conducted under conditions where the resin undergoes only a partial cure. Later fabrication then involves the stacking of layers and the application of heat and pressure to further cure the resin into a continuous solid matrix. Prepreg layers can be combined in various orientations to provide the desired directional properties.

Fibers can also be wound around a mandrel with a specified spacing, prepregged with matrix material, and then removed to produce *tapes* that contain continuous, unidirectionally aligned filaments. These tapes are generally one fiber diameter in thickness and can be up to 48 in. wide.

When the temperatures of the molten matrix become objectionable, the matrix can often be bonded to the fibers by means of either diffusion or deformation bonding (hot pressing or rolling). A common arrangement is to position aligned or woven fibers between sheets of foil material. Loosely woven fibers can also be infiltrated with a particulate matrix, which is then compacted at high pressures and sintered to form a solid mass.

Individual filaments can be coated with the matrix by drawing through a molten bath, plasma spraying, vapor deposition, electrodeposition, and other techniques. These coated fibers can then be used, either individually or in the assemblies noted above.

Sheet-molding compounds (SMC) are composed of chopped fibers and partially cured resin, along with fillers, pigments, and other additives, in sheets approximately 0.1 in. thick. With strengths in the range of 5 to 10 ksi (35 to 70 MPa) and the ability to be press-formed in heated dies, these materials offer a feasible alternative to sheet metal in applications where light weight, corrosion resistance, and integral color are attractive features. Final curing can often be completed in less than 60 seconds.

Bulk-molding compounds (BMC) are fiber-reinforced thermoset molding materials, where the short fibers are distributed in random orientation. The starting material is usually a bulk material with the consistency of putty or modeling clay, although pellets and granules are also possible. The final shape is usually produced by compression molding in heated dies, but transfer molding and injection molding are other possibilities.

Fabrication of Final Shapes from Fiber-Reinforced Composites.

A number of processes have been developed for the production of finished products from fiber-reinforced material. Many are simply extensions or adaptations of processes that are used to shape the matrix material (usually metals or polymers). Others are unique to the family of fiber-reinforced composites. The dominant techniques are discussed individually in the sections to follow.

PULTRUSION. *Pultrusion* is a continuous process that is used to produce relatively simple shapes of uniform cross section, such as round, rectangular, tabular, plate, sheet, and structural products. As shown in Figure 20-20, bundles of continuous reinforcing fibers are drawn through a bath of thermoset polymer resin, and the impregnated material is then gathered to produce a desired cross-sectional shape. This material is then pulled through one or more heated dies, which further shape the product and cure the resin. Upon emergence from the heated dies, the product is cooled by air or water, cut to length, and further fabricated into products such as fishing poles, golf club shafts, and ski poles. Extremely high strengths are possible (since the reinforcement can be as much as 75% of the final structure), with densities about 20% that of steel or 60% that of aluminum. Cross sections can be as much as 60 in. wide and 12 in. thick.

FILAMENT WINDING. The availability of resin-coated or resin-impregnated, high-strength, continuous filaments, bundles, or tapes made of materials such as glass, graphite, or boron has made it possible to produce container-type shapes that have exceptional strength-to-weight ratios. The filaments are wound over a form, using longitudinal, circumferential, or helical patterns, or a combination of these, designed to take advantage of their highly directional strength properties. By adjusting the density of the filaments in various locations and selecting the orientation of the wraps, products can be designed to

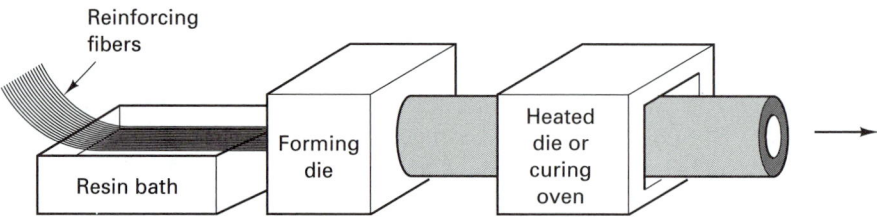

FIGURE 20-20   Schematic diagram of the pultrusion process.

FIGURE 20-21   A large tank being made by filament winding. *(Courtesy of Rohr Corporation.)*

have strength where needed and lighter weight in the less critical regions. After the resin has been cured, the product can be stripped from the form. The matrix, often an epoxy-type polymer, binds the structure together and transmits the stresses to the fibers.

Figure 20-21 shows a large tank being made by this process. Products such as pressure tanks and rocket motor casings can be made in virtually any size, some as large as 15 feet in diameter and 65 feet long. Moderate production quantities are feasible, and because the process can be highly mechanized, uniform quality can be maintained. The special tooling for a new size or design (i.e., new form block) is relatively inexpensive, so the process offers tremendous potential for cost savings and high flexibility.

LAMINATION AND LAMINATION-TYPE PROCESSES. In the lamination process, pre-pregs, mats, or tapes are stacked to produce a desired thickness and cured under pressure and heat. The resulting products possess unusually high strength properties as a result of the integral fiber reinforcement. Because the surface is a thin layer of pure resin, laminates usually possess a smooth, attractive appearance. If the resin is transparent, the fiber material is visible and can impart a variety of decorative effects.

Laminated materials can be produced as sheets, tubes, and rods. Flat sheets can be made using the method illustrated in Figure 20-22. Figure 20-23 depicts the technique used to produce rods or tubes. For tubing, the impregnated stock is wound around a mandrel of the desired internal diameter. Solid rods are made by using a small-diameter mandrel, which is removed prior to curing, or by wrapping the material tightly about itself. In all cases, the final operation is a curing, usually performed under both heat and pressure.

Because of their excellent strength properties, plastic laminates find a wide variety of uses. Some sheets can be easily blanked and punched. Gears machined from thick laminated sheets have unusually quiet operating characteristics when matched with metal gears.

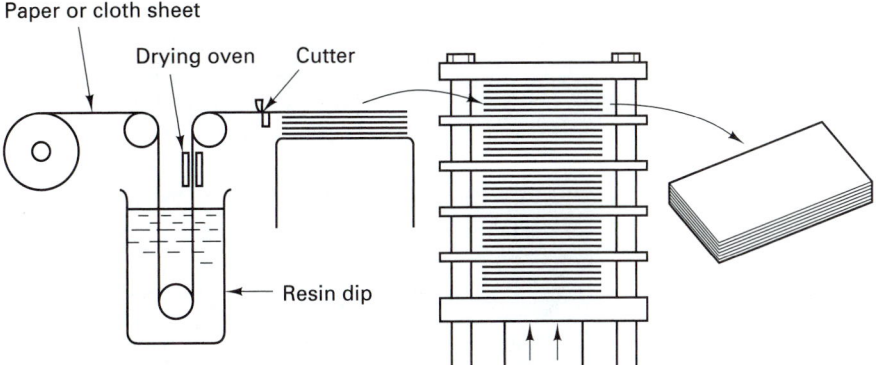

FIGURE 20-22   Method of producing laminated plastic sheets.

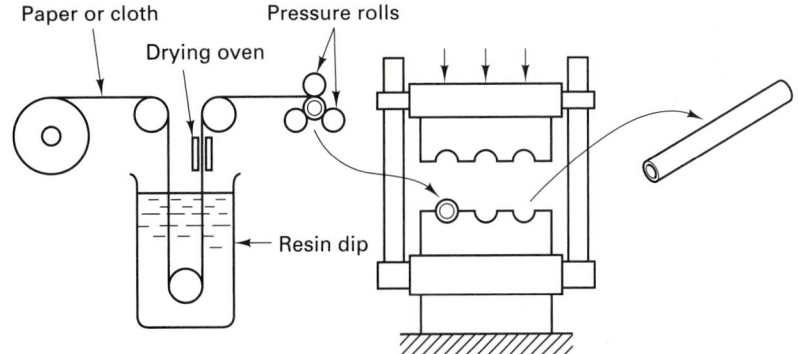

FIGURE 20-23   Method of producing laminated plastic tubing.

Many laminated products are not flat but contain relatively simple curves and contours. Boats, automobile body panels, aerospace panels, safety helmets, and similar products can be manufactured by processes that require zero to moderate pressure and relatively low curing temperatures. In one technique, only a female mold is required and this can be made from metal, hard wood, or even particleboard. The layers of prepreg or resin-dipped fabric are stacked in various orientations until the desired thickness is obtained. Caution must be taken to avoid the entrapment of air bubbles and assure that no impurities (such as oil, dirt, or other contaminants) are introduced between the layers. The entire mold assembly is then placed in a nonadhering, flexible bag, and the contained air is evacuated. During this *vacuum-bag molding* process, air pressure holds the laminate against the mold during the curing operation, which may occur at room temperature or require exposure to heat, often using a medium such as live steam. In a variation known as *pressure-bag molding*, a flexible membrane is positioned over a female mold cavity and pressurized to force the individual plies together and drive out entrapped air and excess resin. Pressures usually range from 30 to 50 psi but can be as high as 250 psi. The various pressure-molding processes have been used to produce extremely large components, such as the skins of military aircraft, the large air-deflectors for tractor-trailers, and body panels for trucks.

FIGURE 20-24 Aerodynamic styling and smooth surfaces characterize the hood and fender assembly for Ford Motor Company's AeroMax truck. This panel was produced as a resin transfer molding by Rockwell International. *(Courtesy of ASM International.)*

Higher heats and pressures can be used when the part is cured in an *autoclave*. The supporting molds and bagged layups are placed inside a heated pressure vessel where curing occurs at elevated temperatures and pressures in the range 50 to 100 psi. Denser void-free moldings are produced, and the properties can be further enhanced through the use of matrix resins that require higher-temperature cures. The size of the autoclave limits the size of the product.

When production quantities are large, the parts are sufficiently small, and quality needs to be high, matched metal dies can be substituted for the mold and bag. The process then becomes a modification of *compression molding*. The dies are heated and curing occurs during the compression operation.

Resin-transfer molding is a low-pressure process that is intermediate to the slow, labor-intensive layup processes and the faster compression molding or injection molding processes, which generally require higher-cost tooling. Continuous fiber mat or woven material (usually employing glass fiber) is positioned dry in the bottom half of a matching mold, which is then closed and clamped. Because of the low pressures employed in the process, the mold tooling does not need to be steel but can be electroformed nickel shells, epoxy composite, or aluminum. In addition, low-capacity presses can be used to clamp the mold segments, and inflatable bags can be used to produce simple holes or hollows, in much the same way that cores are used in conventional casting. A low-viscosity catalyzed resin is injected into the mold, where it permeates the reinforcement and cures at low temperatures. The products can have excellent surfaces on both sides, since both mold surfaces can be coated with a pigmented gel. Large parts can often be made as a single unit with a relatively low capital investment. Cycle times range from a few minutes to a few hours, depending on the part size and the resin system being used. The aerodynamic hood and fender assembly for Ford Motor Company's new AeroMax heavy-duty truck (Figure 20-24) is an example of a large resin transfer molding.

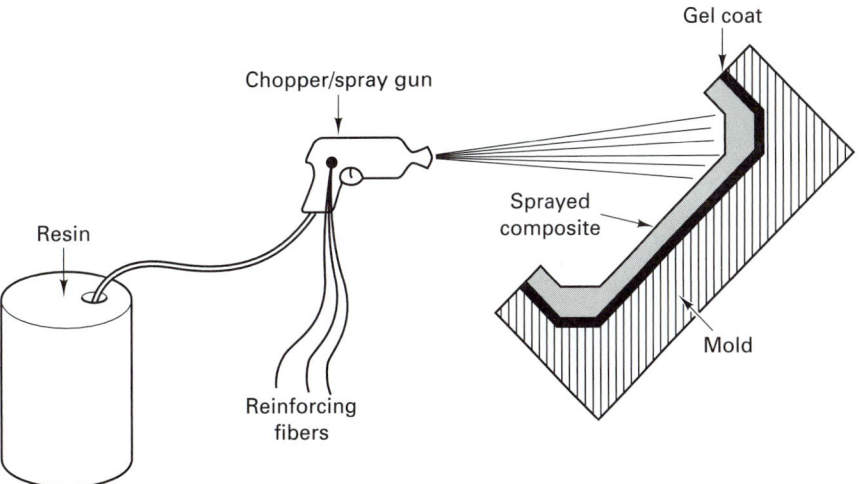

FIGURE 20-25   Schematic diagram of the spray forming of fiber-reinforced composites.

When the quality demands are not as great and the reinforcement-to-resin ratio is not exceptionally high, pressing operations can often be eliminated. The layers of pliable resin-coated cloth are simply placed in the mold or draped over a form in a process known as *hand layup* or *open mold processing*. Squeegees or rollers are used to manually assure good contact and the absence of entrapped air, and the assembly is then allowed to cure. If prepreg layers are not used, a layer of mat, cloth, or woven roving can be put in place and a layer of resin brushed, sprayed, or poured on. The process can then be repeated to build the desired thickness. While the hand layup process is slow and labor intensive and has part-to-part and operator-to-operator variability, the tooling costs are sufficiently low that single items or small quantities become economically feasible. Molds or forms can be made from wood, plaster, plastics, aluminum, or steel, so design changes and the associated tool modifications are rather inexpensive, and manufacturing lead time can be quite short. In addition, large parts can be produced as single units, significantly reducing the amount of assembly, and various types of reinforcement can be incorporated into a single product, expanding design options. High-quality surfaces can be produced by applying a pigmented gel coat to the mold before the layup.

SPRAY MOLDING.   When continuous or woven fibers are not required to produce the desired properties, sheet-type parts can be produced by mixing chopped fibers and catalyzed resin and spraying the combination into or onto a mold form as shown in Figure 20-25. Rollers or squeegees are often used to remove entrapped air and work the resin into the reinforcement. Room-temperature curing is usually preferred, but elevated temperatures are sometimes used to accelerate the cure. As with the hand layup process, an initial gel coat can be used to produce a smooth, pigmented surface.

SHEET STAMPING.   Thermoplastic sheets that have been reinforced with nonwoven fiber can often be heated and press-formed in a manner similar to conventional sheet metal forming. Precut blanks are heated and placed between the halves of a matched metal mold that is mounted in a vertical press. Ribs, bosses, and contours can be formed in parts with essentially uniform thickness. Cycle times range from 25 to 50 seconds for most parts.

INJECTION MOLDING. The injection molding of fiber-reinforced plastics is a process designed to compete with metal die castings and offers comparable properties at considerably reduced weight. In the simplest variation, chopped or continuous fibers are placed in a mold cavity which is then closed and injected with resin. An improved method utilizes chopped fibers, up to $\frac{1}{4}$ in. in length, which are premixed with the heated thermoplastic (often nylon) prior to injection. Another variation uses a feedstock of discrete pellets that have been manufactured by slicing continuous-fiber pultruded rods. Benefits of adding fiber reinforcement (compared to conventional plastic molding) include increased rigidity and impact strength, reduced possibility of brittle failure during impact, better dimensional stability at elevated temperatures and in humid environments, improved abrasion resistance, and better surface finish, due to the reduced dimensional contraction and absence of related sink marks.

BRAIDING, THREE-DIMENSIONAL KNITTING, AND THREE-DIMENSIONAL WEAVING. In these processes the high-strength reinforcing fibers are interwoven into a three-dimensional preform. Resin is injected into the assembly and the resultant product is cured for use. Complex shapes can be produced with the fiber orientations selected for optimum properties. Moreover, because of the three-dimensional interweaving of the fibers, products from these processes lack the delamination (layer-separation) planes that are present in many of the lamination-type composites. Computers can be used to design and control the weaving, making the process less expensive than many of the more labor-intensive techniques.

Fabrication of Fiber-Reinforced Metal–Matrix Composites.

Continuous-fiber *metal–matrix composites* can be produced by variations of filament winding, extrusion, and pultrusion. Fiber-reinforced sheets can be produced by electroplating, plasma spray deposition coating, or vapor deposition of metal onto a fabric or mesh, which are then shaped and bonded. Diffusion bonding of foil–fabric sandwiches, roll bonding, and coextrusion are other means of producing fiber-reinforced metal products. Casting processes have been developed that force liquid metal around the fibers by means of capillary action, pressure casting, or vacuum infiltration. Products that incorporate discontinuous fibers can also be produced by powder metallurgy or spray forming techniques and further fabricated by hot pressing, superplastic forming, forging, or some types of casting. In general, efforts are made to reduce or eliminate the need for finish machining, which would require the use of diamond or carbide tools, or methods such as EDM.

In terms of properties, graphite-reinforced aluminum has been shown to be twice as stiff as steel, one-third to one-fourth the weight, with practically zero thermal expansion. Aluminum reinforced with silicon carbide exhibits increased strength (tension, compression, and shear at both room and elevated temperature), hardness, fatigue strength, and elastic modulus. Thermal creep and thermal expansion are both reduced, but ductility, thermal conductivity, and electrical conductivity are also decreased. Magnesium, copper, and titanium alloys have also been used as the metal matrix in fiber-reinforced composites.

Fabrication of Fiber-Reinforced Ceramic–Matrix Composites.

Unlike polymeric– or metal–matrix composites, where failures originate in or along the reinforcement fibers, *ceramic–matrix composites* often fail due to flaws in the matrix. If the reinforcement is bonded strongly to the matrix, a matrix crack will propagate right through to the fibers. To impart toughness, therefore, it is often desirable to promote a weaker bond between the fiber and matrix. Cracks are then deflected along the fiber–matrix interface rather than through the fiber.

Fabrication techniques for the ceramic–matrix composites are often quite different from the other composite families. Common techniques include the chemical vapor deposition or chemical vapor infiltration of a coated fiber base. The coating serves to weaken an otherwise strong bond. Silicon nitride matrices can be formed by reaction bonding. The reinforcing fibers are dispersed in silicon powder, which is then reacted with nitrogen. Hot pressing techniques can also be used with the various ceramic matrices. When the matrix is a glass, the heated material behaves much like a polymer, and similar processing methods are used.

Secondary Processing and Finishing of Fiber-Reinforced Composites. The various fiber-reinforced composites can often be processed further with conventional equipment (sawed, drilled, routed, tapped, threaded, turned, milled, sanded, and sheared), but special considerations should be exercised. Cutting some materials may be like cutting multilayer cloth, and precautions should be used to prevent the formation of splinters and cracks as well as frayed or delaminated edges. Sharp tools, high speeds, and low feeds are generally required. Cutting debris should be removed quickly to prevent the cutters from becoming clogged.

In addition, many of the reinforcing fibers are extremely abrasive in nature and quickly dull most conventional cutting tools. Diamond or polycrystalline diamond tooling may be required to achieve realistic tool life. Abrasive slurrys can be used in conjunction with rigid tooling to assure the production of smooth surfaces. Lasers and water jets are alternative cutting tools. However, lasers may burn or carbonize the material or produce undesirable heat-affected zones. Water jets can create moisture problems with some plastic resins.

When fiber-reinforced materials must be joined, the major concern is the lack of continuity of the fibers in the joint area. Thermoplastics can be softened and welded, by applying pressure with heated tools, combining pressure and ultrasonic vibration, or using pressure and induction heating. Thermoset materials generally require the use of mechanical joints or adhesives, with each method having its characteristic advantages and limitations. Metals are often brazed.

■ KEY WORDS

adhesive bonding
autoclave
blow molding
braiding
bulk-molding compound
calendering
casting
cementation
ceramics
ceramic–matrix composites
cold molding
composites
compression molding
crystalline ceramics
dipping
explosive bonding
extrusion
fibers

filament winding
firing
foam molding
glasses
hand layup
injection molding
inserts
lamination
mats
metal–matrix composites
plastics
prepregs
pressure-bag molding
pultrusion
reaction injection molding
resin-transfer molding
roll bonding
rotational molding

rovings
sandwich structures
sheet-molding compound
slip casting
spinning
spray molding
tapes
thermoforming
thermoplastic polymer
thermosetting polymer
tows
transfer molding
vacuum-bag molding
viscous flow
vitrification
yarns

■ REVIEW QUESTIONS

1. Why are the fabrication processes applied to plastics, ceramics, and composites often different from those applied to metals? What are some of the key differences?
2. How does the fabrication of a thermoplastic polymer differ from the processing of a thermosetting polymer?
3. What are some of the ways that plastic sheet, plate, and tubing can be cast?
4. Why do cast plastic resins typically have a lustrous appearance?
5. What types of polymers are most commonly blow molded?
6. For what types of parts and production volumes would compression molding be an appropriate process?
7. What are typical mold temperatures for compression molding? What is the most common mold material?
8. What are some of the attractive features of the transfer molding process?
9. What is the most widely used process for the fabrication of thermoplastic materials (in terms of number of parts produced)?
10. In what ways is injection molding of plastic similar to the die casting of metal?
11. Why is the cycle time for the injection molding of thermosetting polymers significantly longer than that for the thermoplastics?
12. What are some of the benefits of using a hot runner distribution system when injection molding a thermoplastic?
13. What are some of the distinctive features of the reaction injection molding process?
14. What are some of the typical production shapes that are produced by the extrusion of plastics?
15. For what types of materials and products might thermoforming be considered to be appropriate?
16. What types of products are produced by rotational molding?
17. What is the difference between open-cell and closed-cell foamed plastics?
18. What are some typical applications for rigid-type foamed plastics?
19. What are some of the general properties of plastics that affect their machinability?
20. What are some of the attractive properties of plastics? What are some of the major limitations?
21. Why should adequate fillets be included between adjacent sections of a mold? What is a major benefit of rounding exterior corners?
22. Why is it most desirable to have uniform wall thickness in plastic products?
23. Why are product dimensions less precise when they cross a mold parting line?
24. Why might threaded inserts be preferred over other means of producing threaded holes in a plastic component?
25. When designing a decorative surface (design or lettering) on a plastic product, why is it desirable that the details be raised on the product rather than depressed.
26. What are some of the typical products that can be produced from elastomeric materials by the dipping process?
27. What are the two basic classes of ceramic materials, and how does their processing differ?
28. What are some of the special heat treatment operations performed on glass products?
29. What are some of the techniques that can be used to impart some degree of plasticity to crystalline ceramic materials?
30. What is the purpose of the firing or sintering operations in the processing of crystalline ceramic products?
31. What are the various benefits and limitations of machining ceramic materials before firing versus after firing?
32. Why are joining operations usually avoided when fabricating products from ceramic materials?
33. What are some common or typical limitations when fabricating products from composite materials?
34. What are some of the processes that can be used to produce a high-quality bond between the layers of a laminar composite?
35. What is a prepreg?

36. What are sheet molding compounds (SMC)? Bulk-molding compounds (BMC)?

37. In what way is pultrusion similar to wire drawing or the extrusion of metal?

38. What are some typical products that are made by filament winding? Why might this process be attractive for the production of small quantities of large parts?

39. What are some of the various molding processes that can be used to shape products from laminated sheets of woven fibers?

40. What are some of the more common defects or problems associated with the lamination-type processes?

41. What type of reinforcing fibers can be incorporated in the spray molding process? Sheet stamping? Injection molding?

42. Discuss some of the ways in which reinforcing fibers and resin can be combined for shaping by injection molding.

43. Describe some of the ways in which a metal matrix can be introduced into a fiber-reinforced composite.

44. Why might it be desirable to "weaken" the bond between a reinforcing fiber and a ceramic matrix material?

45. What is the major design concern when considering the joining of fiber-reinforced composites?

■ PROBLEM

1. Consider some of the more prominent sporting goods that are fabricated from composite materials, such as skis, tennis rackets, golf club shafts, and body panels for racing cars. For two products, identify both appropriate composite materials and companion shape-producing fabrication methods.

*C*hapter 20 CASE STUDY

fabrication of lavatory wash basins

L avatory wash basins (bathroom sinks) have been successfully made from a variety of engineering materials, including cast iron, steel, stainless steel, ceramics, and polymers (such as melamine). Your company, Diversified Household Products, Inc., is considering a possible entrance into this market and has assigned you the tasks of (1) assessing the competition, and (2) recommending the "best" approach toward producing this product.

1. For each of the above materials (or families of materials), describe the material properties that are attractive for a washbasin application. What are the primary limitations or disadvantages?

2. For each of the above materials (or families of materials), describe possible means of fabricating lavatory wash basins. Consider sheet metal forming, casting, molding, joining, and other types of fabrication processes. If multiple options exist, which one do you consider to be most attractive? Comment on the attractive features of the proposed system (materials and process) as well as the relative quality and cost.

3. Wash basins generally require a surface that is nonporous and stain resistant, scratch resistant, corrosion resistant, and attractive (and possibly available in a variety of colors). One approach to providing these properties is through a coating of porcelain enamel. For each of the "systems" discussed in Question 2, discuss the need for additional surface treatment. What type of treatment would you recommend?

4. If your company were to consider producing lavatory wash basins on a competitive basis, which of the alternative manufacturing systems (material and manufacturing process) would you recommend? What features make it the most attractive?

PART 5

MATERIAL REMOVAL PROCESSES

21 Fundamentals of Chip-Type Machining Processes

22 Cutting Tools for Machining

23 Turning, Boring, and Related Processes

24 Drilling and Related Hole-making Processes

25 Milling

26 Broaching, Sawing, Filing, Shaping and Planing

27 Abrasive Machining Processes

28 Workholding Devices

29 Numerical Control and Machining Centers

30 Thread Manufacturing

31 Gear Manufacturing

32 Nontraditional Machining Processes

CHAPTER 21

FUNDAMENTALS OF CHIP-TYPE MACHINING PROCESSES

21.1   INTRODUCTION
21.2   BASIC CHIP FORMATION PROCESSES
21.3   UNDERSTANDING CHIP FORMATION
21.4   ORTHOGONAL MACHINING
      Effects of Work Material Properties

      Mechanics of Machining
21.5   ENERGY AND POWER IN MACHINING
21.6   HEAT AND TEMPERATURE IN METAL CUTTING
21.7   SUMMARY
Case Study: HSS VERSUS TUNGSTEN CARBIDE

■ 21.1 INTRODUCTION

Machining is the process of removing unwanted material from a workpiece in the form of chips. If the workpiece is a metal, the process is often called *metal cutting* or *metal removal.* U. S. industries annually spend $60 billion to perform metal removal operations because the vast majority of manufactured products require machining at some stage in their production, ranging from relatively rough or nonprecision work, such as cleanup of castings or forgings, to high-precision work involving tolerances of 0.000l in. or less and high-quality finishes. Thus machining undoubtedly is the most important of the basic manufacturing processes.

Over the past 80 years, the process has been the object of considerable research and experimentation that have led to improved understanding of the nature of both the process itself and the surfaces produced by it. Although this research effort has led to improvements in machining productivity, the complexity of the process has resulted in slow progress in obtaining a complete theory of chip formation.

What makes this process so unique and difficult to analyze?

- Different materials behave differently.
- The process is asymmetrical and unconstrained, bounded only by the cutting tool.
- The level of strain is very large.
- The strain rate is very high.
- The process is sensitive to variations in tool geometry, tool material, temperature, environment (cutting fluids) and process dynamics (chatter and vibration).

The objective of this chapter is to put all this in perspective for the practicing engineer.

■ 21.2 BASIC CHIP FORMATION PROCESSES

There are seven basic chip formation processes: *shaping, turning, milling, drilling, sawing, broaching,* and *grinding (abrasive machining).* Grinding is treated separately in Chapter 27. For all metal-cutting processes, it is necessary to distinguish between speed, feed, and

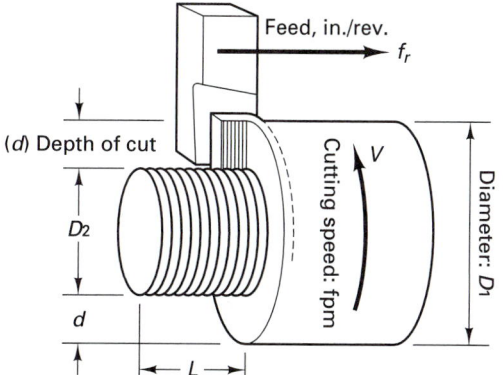

Feed, in./rev. f_r

(d) Depth of cut

D_2

d

Cutting speed: fpm

V

Diameter: D_1

L

Speed, stated in surface feet per minute, (sfpm) is the peripheral speed at the cutting edge. Feed per per revolution in turning is a linear motion of the tool parallel to the rotating axis of the workpiece. The depth of cut reflects the third dimension.
L = length of cut

FIGURE 21-1   Relationship of speed, feed, and depth of cut in turning

depth of cut. The turning process will be used to introduce these terms (see Figure 21-1). In general, *speed* (*V*) is the primary cutting motion, which relates the velocity of the rotating *workpiece* with respect to the stationary cutting tool. It is generally given in units of surface feet per minute (sfpm) or inches per minute (in./min), or meters per minute (m/m) or meters per second (m/s). Speed (*V*) is shown in Figure 21-1 with the heavy dark arrow. *Feed* (f_r) is the amount of material removed per revolution or per pass of the tool over the workpiece. In turning, feed is in inches/revolution and the tools feeds parallel to the rotational axis of the workpiece. Feed units are inches per revolution, inches per cycle, inches per minute, or inches per tooth, depending on the process. Feed is shown with dashed arrows. The *depth of cut* (DOC), *d*, represents the third dimension. In turning, it is the distance the tool is plunged into the surface. It is half the difference in the diameters D_1, the initial diameter, and D_2, the final diameter:

$$DOC = \frac{D_1 - D_2}{2} = d \qquad (21\text{-}1)$$

The surface speed of the rotating part is related to the outer diameter of the workpiece:

$$V = \frac{\pi D_1 N_s}{12} \qquad (21\text{-}2)$$

where D_1 is in inches, *V* the speed in surface feet per minute, and N_s the revolutions per minute (rpm) of the workpiece.

Figure 21-2 shows a typical *machine tool* for the turning process, a lathe. Workpieces are held in *workholding* devices. See Chapter 28 for details on the design of workholders. In this example, a three-jaw chuck is used to hold the workpiece and rotate it against the tool. The chuck is attached to the spindle, which is driven through gears by the motor. The *cutting tool* is used to machine the workpiece and is the most critical component. Cutting tools are discussed in detail in Chapter 22. Cutting tool material and geometry must be selected before speed and feed can be determined. Machine tools, cutting tools, and workholding devices are usually manufactured by separate companies and the tooling can cost as much or more than the machine tool.

To process different metals, the input parameters to the machine tools must be determined. For the lathe, the input parameters are DOC, the feed, and the rpm value of the

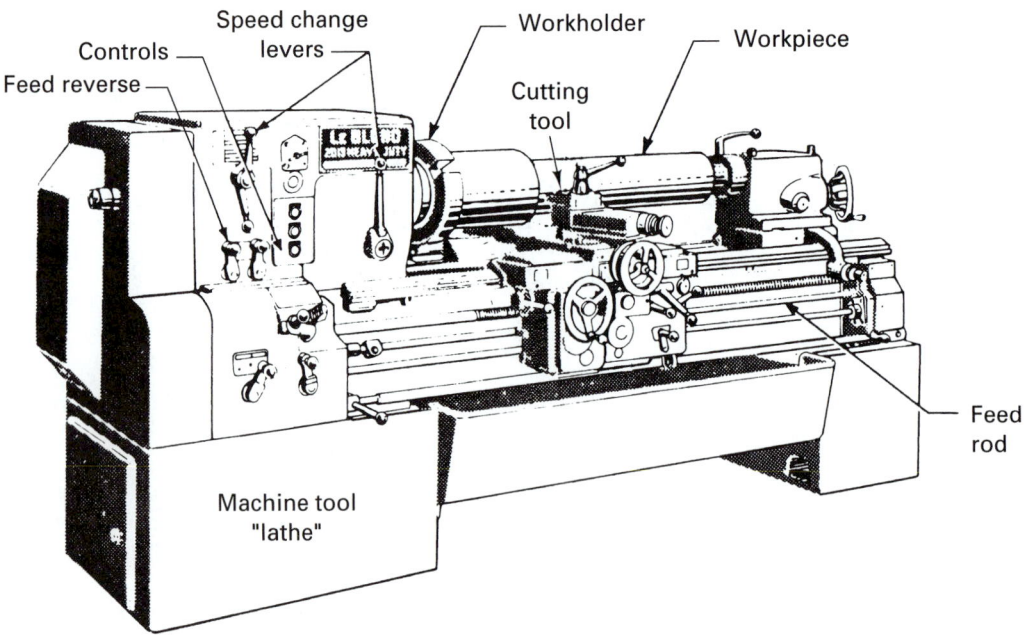

FIGURE 21-2 The basic machine tool for turning is called a lathe. This figure shows an engine lathe.

spindle. The rpm value depends on the selection of the cutting speed, V. Cutting speed, feed, and DOC selection depend on many factors, and a great deal of experience and experimentation is required to find the best combinations. A good place to begin is by consulting tables of recommended values, as shown in Table 21-1. Most tables are arranged according to the process being used, the material being machined, the hardness, and the cutting-tool material. The table given is simply a sample, to be used for solving turning problems in this book, and should not be used for industrial calculations. Standard references listed in the appendix should be consulted.

This table, extracted from the *Tool and Manufacturing Engineers' Handbook* is for turning processes only. The amount of metal removed per pass determines the DOC. In practice, roughing cuts are heavier than finishing cuts in terms of feed and DOC and are run at a lower surface speed.

Some tables provide recommendations of V and f in both English and metric units based on the DOC value needed to perform the job. Table values are usually conservative and should be considered starting points for determining the operational parameters for a process.

Once cutting speed, V, has been selected, equation (21-1) is used to determine the spindle rpm, N_s. The speed and feed can be used with the DOC to estimate the metal removal rate for the process, MRR. For turning the MRR in in^3/min., is

$$MRR \simeq 12 \, Vf_r d \qquad (21\text{-}3)$$

This is an approximate equation for MRR. For turning MRR values can range from 0.1 to 600 in^3/min. The MRR can be used to estimate the horsepower needed to perform a cut,

TABLE 21-1. Suggested Machining Parameters for Turning Various Materials with Carbide Tools

Work Material	Cutting Speed [sfpm (m/min)]		Feed Rate [in./rev (mm/rev)]		Depth of Cut [in. (mm)]	
	Roughing	Finishing	Roughing	Finishing	Roughing	Finishing
Free-machining carbon steels: AISI 1100, 1200 series, 140–190 BHN	205–1100 (76–335)	1000–2000 (305–610)	0.010–0.085 (0.25–2.16)	0.005–0.015 (0.13–0.38)	0.125–0.675 and up (3.18–17.15)	Up to 0.180 (4.57)
Plain carbon steels: AISI 1000 series, 185–240 BHN	200–800 (61–244)	700–1600 (213–488)	0.010–0.085 (0.25–2.16)	0.005–0.015 (0.13–0.38)	0.125–0.675 and up (3.18–17.15)	Up to 0.180 (4.57)
Alloy steels: AISI 1300, 4000, 5000, 8000, and 9000 series; 190–240 BHN	175–600 (53–183)	550–1200 (168–366)	0.010–0.085 (0.25–2.16)	0.005–0.015 (0.13–0.38)	0.125–0.675 and up (3.18–17.15)	Up to 0.180 (4.57)
Cast irons: gray, nodular, and malleable, 150–210 BHN	200–1200 (61–366)	200–750 (61–229)	0.010–0.055 (0.25–1.40)	0.005–0.015 (0.13–0.38)	0.125–0.675 and up (3.18–17.15)	Up to 0.180 (4.57)
Martensitic stainless steels: wrought 400 and 500 series and PH types, 175–210 BHN	175–450 (53–137)	450–850 (137–259)	0.010–0.040 (0.025–1.02)	0.005–0.015 (90.13–0.38)	0.125–0.500 (3.18–12.70)	Up to 0.180 (4.57)
Austentic stainless steels: wrought 200 and 300 series, 140–190 BHN	125–425 (38–130)	425–650 (130–198)	0.010–0.040 (0.25–1.02)	0.005–0.015 (0.13–0.38)	0.125–0.500 (3.18–12.70)	Up to 0.180 (4.57)
Superalloys: iron, nickel, titanium, and cobalt alloys, 240–300 BHN	30–150 (9–46)	150–400 (46–122)	0.010–0.025 (0.25–1.02)	0.005–0.015 (90.13–0.38)	0.100–0.300 (2.54–7.62)	Up to 0.100 (4.57)
Tool steels, wrought high-speed, shock resistant, and hot and cold work, 210–240 BHN	100–300 (30–91)	275–750 (84–229)	0.010–0.065 (0.25–1.65)	0.005–0.015 (0.13–0.38)	0.125–0.675 and up (3.18–17.15)	Up to 0.180 (4.57)
Nonferrous free-machining alloys: aluminum, copper, zinc, and brass alloys, 80–120 BHN	400–1200 (122–366)	1000–2000 (305–610)	0.010–0.085 (0.25–2.16)	0.005–0.015 (0.13–0.38)	0.125–0.675 and up (3.18–17.15)	Up to 0.10 (4.57)
Nonmetallics: nylons, acrylics, and phenolic resins	350–800 (107–244)	800–1500 (244–457)	0.010–0.040 (0.25–1.02)	0.005–0.015 (0.13–0.38)	0.125–0.500 (3.18–12.70)	Up to 0.180 (4.57)

Source: Tool and Manufacturing Engineers' Handbook, Vol. 1, 4th edition, Kennametal, Inc.

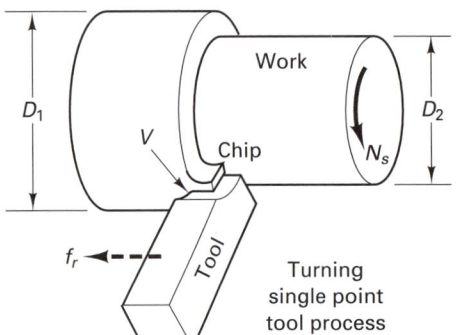

The rpm of the rotating workpiece is N_s. It established the cutting speed V, at the tool, according to

$$V = \pi D_1 N_s/12$$

The depth of cut, d, is equal to $(D_1 - D_2)/2$. The length of cut is the distance the tool travels parallel to the axis, L.

FIGURE 21-3 Basics of the turning process. Derivations for equations for cutting time (CT) and metal-removal rate (MRR) are given in Chapter 23.

as will be shown later. For most processes, the MRR equation can be viewed as the volume of metal removed divided by the time needed to remove it:

$$MRR = \frac{\text{volume of cut}}{CT}$$

where CT is the cutting time in minutes. For turning, the cutting time depends on the length of cut, L, divided by the rate of traverse of the cutting tool past the rotating workpiece, $f_r N_s$. Therefore,

$$CT = \frac{L + \text{Allowance}}{f_r N_s} \tag{21-4}$$

An allowance is usually added to the L term to allow for the tool to enter and exit the cut.

Turning is an example of a single-point tool process, as is shaping. Milling and drilling are examples of multiple-point tool processes. Figures 21-3 through 21-9 show the basic processes schematically. Speed (V) is shown in these figures with a dark heavy arrow. Feed (f) is the amount of material removed per pass of the tool over the workpiece and is shown as a dashed arrow.

For many of the basic process, the equations for CT and MRR are given. These equations are commonly referred to as *shop equations* and are as fundamental as the processes themselves, so the student should be as familiar with them as with the basic processes. If one keeps track of the units and visualizes the process, the equations are, for the most part, straightforward.

In addition to turning, other operations can also be performed on the lathe. For example, as shown in Figure 21-4, a flat surface on the rotating part can be produced by facing or a cutoff operation.

The process of milling requires two figures because it takes different forms depending on the selection of the machine tool and the cutting tool. Milling, a multiple-tooth process, has two feeds: the amount of metal an individual tooth removes, called the feed per tooth, f_t, and the rate at which the table translates past the rotating tool, called the table feed, f_m, in inches per minute:

$$f_m = f_t n N_s \tag{21-5}$$

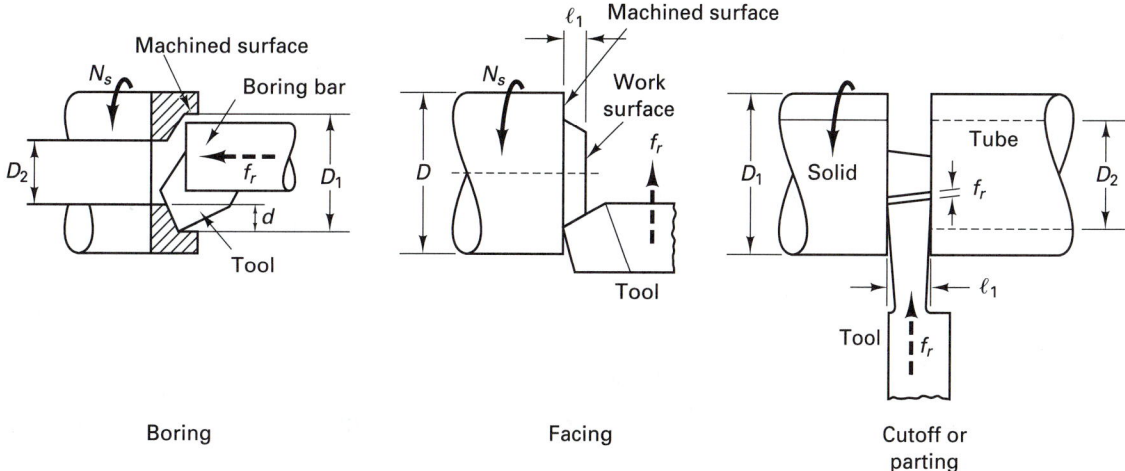

FIGURE 21-4   Basics of boring, facing, and cutoff processes. Equations for cutting time and metal removal given in Chapter 23.

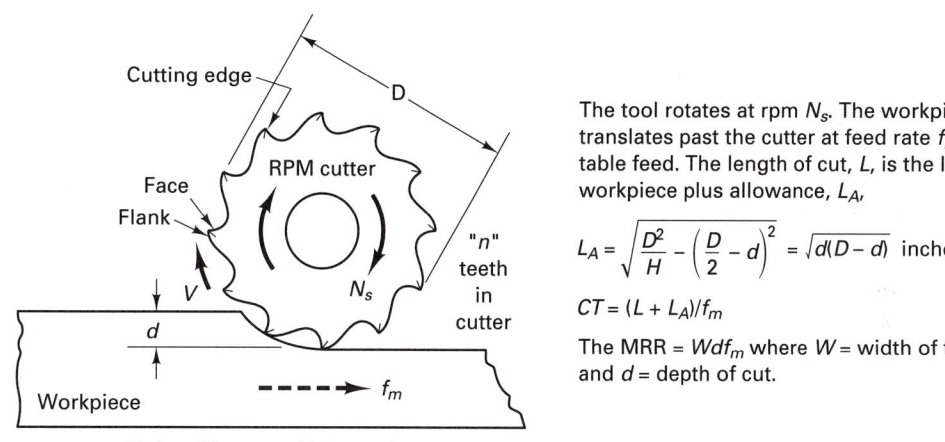

The tool rotates at rpm N_s. The workpiece translates past the cutter at feed rate f_m, the table feed. The length of cut, L, is the length of workpiece plus allowance, L_A,

$$L_A = \sqrt{\frac{D^2}{H} - \left(\frac{D}{2} - d\right)^2} = \sqrt{d(D - d)} \text{ inches}$$

$$CT = (L + L_A)/f_m$$

The MRR $= W d f_m$ where $W =$ width of the cut and $d =$ depth of cut.

FIGURE 21-5   Basics of the milling process (slab milling) as usually performed in a horizontal milling machine. Equations for CT and MRR, derived in Chapter 25.

where n is the number of teeth in a cutter and N_s is the rpm value of the cutter. Standard tables of speeds and feeds for milling provide values for the feed per tooth, f_t.

Table 21-2 provides a summary of the basic machining process parameters. Milling has pretty much replaced shaping and planing, although gear shaping is still a viable process. Milling combined with other rotational multiple-edge tool processes (drilling or reaming) is often performed in machining centers rather than on milling machines. The turret lathe has been replaced by CNC turning centers with multiple turrets (Machining centers are discussed in Chapter 29).

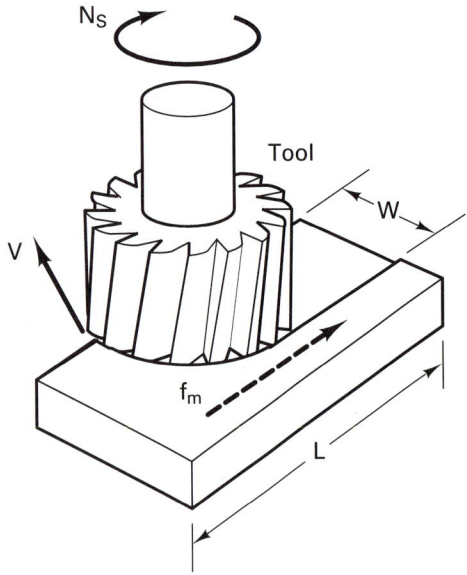

Given a selected cutting speed V and a feed per tooth f_t, the rpm of the cutter is $N_s = 12V/\pi D$ for a cutting of diameter D. The table feed rate is $f_m = f_t n N_s$ for a cutter with n teeth. The cutting time, $CT = (L + L_A + L_o)/f_m$
where $L_o = L_A = \sqrt{W(D - W)}$ for $W < D/2$
or $L_o = L_A = D/2$ for $W \geqslant D/2$.
The MRR $= Wdf_m$ where $d =$ depth of cut.

Face milling
Multiple tooth cutting

FIGURE 21-6   Basics of the milling process (face and end milling) as performed on a vertical spindle machine, including equations for cutting time and metal-removal rate.

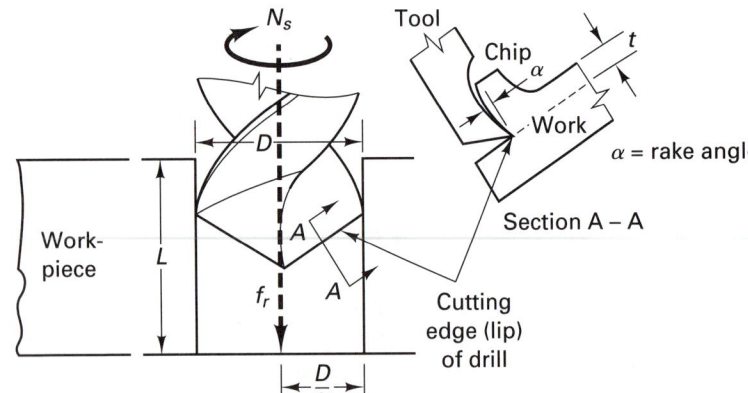

In drill, $D =$ diameter of the drill which rotates 2 cutting edges at rpm N_s. $V =$ velocity of outer edge of the lip of the drill.
$N_s = 12V/\pi D$
$CT = L + A/f_r N_s$ where f_r is the feed rate in in. per rev. The allowance $A = D/2$.
The MRR $= (\pi D^2/4)f_r N_s$ in.^{3} /min which is approximately $3DVf_r$.

Drilling-multiple edge tool

FIGURE 21-7   Basics of the drilling (hole-making) process, including equations for cutting time (CT) and metal-removal rate (MRR).

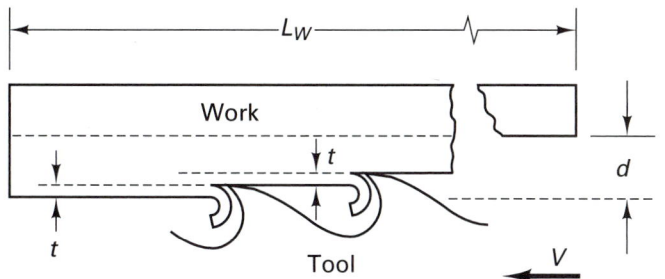

L_W
Work
t
d
t
Tool
V
The CT for broaching is $CT = L/12V$. The MRR (per tooth) is $12tWV$ in^3/min where V = cutting velocity in fpm, W is the width of cut, t = rise per tooth.

FIGURE 21-8 Process basics of broaching. Equations for cutting time and metal-removal rate, developed in Chapter 26.

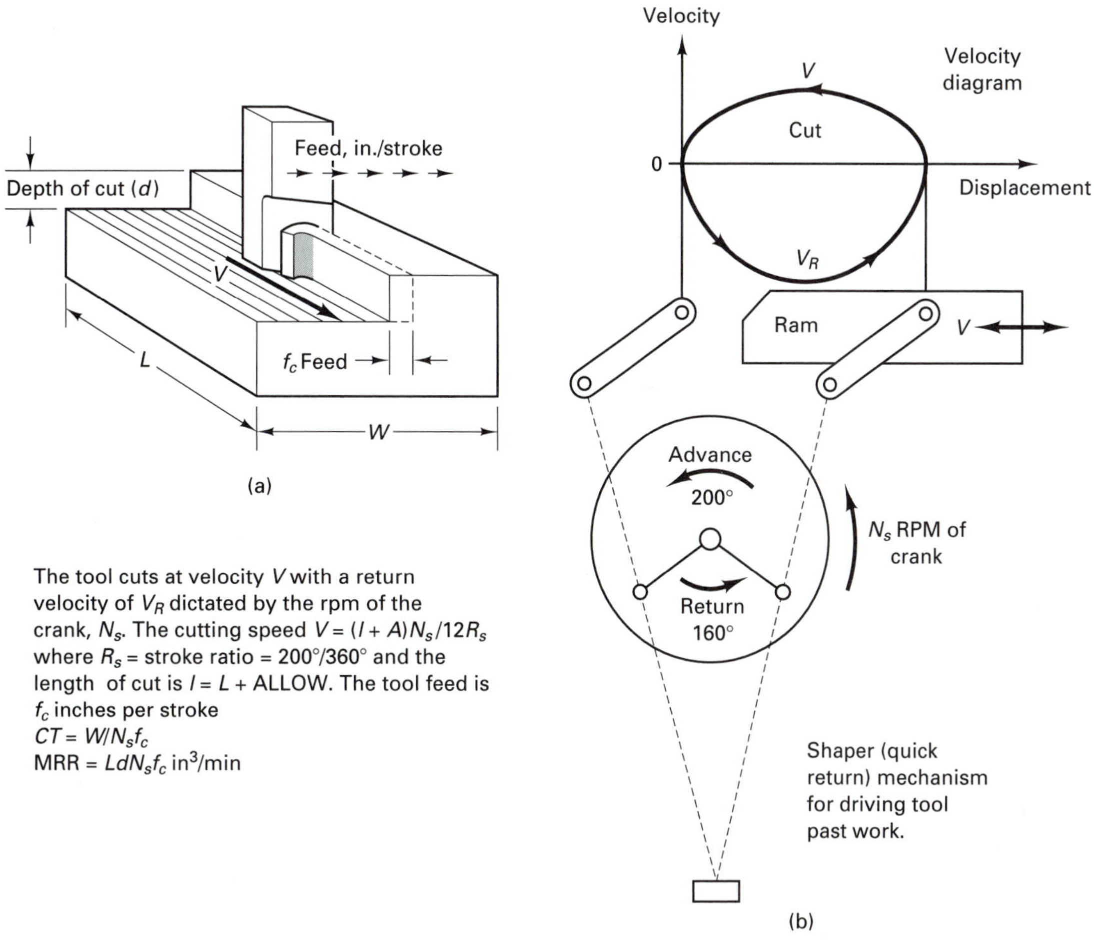

Depth of cut (d)
Feed, in./stroke
V
L
f_c Feed
W
(a)
The tool cuts at velocity V with a return velocity of V_R dictated by the rpm of the crank, N_s. The cutting speed $V = (l + A)N_s/12R_s$ where R_s = stroke ratio = 200°/360° and the length of cut is $l = L$ + ALLOW. The tool feed is f_c inches per stroke
$CT = W/N_s f_c$
MRR = $LdN_s f_c$ in^3/min
Velocity
V
Velocity diagram
Cut
0
Displacement
V_R
Ram
V
Advance
200°
Return
160°
N_s RPM of crank
Shaper (quick return) mechanism for driving tool past work.
(b)

FIGURE 21-9 Basics of the shaping process, including equations for cutting time (CT) and metal-removal rate (MRR).

TABLE 21-2. Summary of Basic Machining Processes

Applicable Process	Raw Material Form	Size		Typical Production Rate	Material Choice	Typical Tolerance	Typical Surface Roughness
		Maximum	Minimum				
Turning (engine lathes)	Cylinders Preforms Castings Forgings	78 in. diameter × 73 in. long	$\frac{1}{64}$ in. typical	1–10 parts/hr	All ferrous and nonferrous material considered machinable	±0.002 in. on diameter common; ±0.001 in. obtainable	125–250
Turning (CNC)	Bar Rod Tube Preforms	36 in. diameter × 93 in. long	$\frac{1}{64}$ in. diameter	1–2 parts/min to 1–4 parts/hr	Any material with good machinability rate	±0.001 in. on diameter where needed; ±0.0005 in. possible	63 or better
Turning (automatic screw machine)	Bar Rod	Generally 2 in. dia. × 6 in. long	$\frac{1}{16}$ in. diameter and less, weight less than 1 oz	10–30 parts/min	Any material with good machinability rating	±0.0005 in. possible; ±0.001 – ±0.003 in. common	63 average
Turning (Swiss automatic machining)	Rod	Collets adapt to $\frac{1}{2}$ in. diameter	Collets adapt to less than $\frac{1}{2}$ in.	12–30 parts/min	Any material with good machinability rating	±0.0002 to ±0.001 in. common	63 and better
Boring (vertical)	Casting Preforms	98 in. × 72 in.	2 in. × 12 in.	2–20 hr/piece	All ferrous and nonferrous	±0.0005 in.	90–250
Milling	Bar Plate Rod Tube	4–6 ft long	Limited usually by ability to hold part	1–100 parts/hr	Any material with good machinability rating	±0.0005 in. possible; ±0.001 in. common	63–250
Hobbing (milling gears)	Blanks Preforms Rods	10-ft-diameter gears, 14-in. face width	0.100 in. diameter	1 part/min	Any material with good machinability rating	±0.001 in. or better	63
Drilling	Plate Bar Preforms	$3\frac{1}{2}$ -in.-diameter drills (1 in. diameter normal)	0.002-in. drill diameter	2–20 sec/hole after setup	Any unhardened material; carbides needed for some case-hardened parts	±0.002–±0.010 in. common; ± 0.001 in. possible	63–250
Sawing	Bar Plate Sheet	2-in. armor plate ($\frac{1}{2}$ in. is preferred)	0.010 in. thick	3–30 parts/hr	Any nonhardened material	±0.015 in. possible	250–1000
Broaching	Tube Rod Bar Plate	74 in. long	1 in.	300–400 parts/min	Any material with good machinability rating	±0.0005–±0.001 in.	32–125
Grinding	Plate Rod Bars	36 in. wide × 7 in. diameter	0.020 in. diameter	1–1000 pieces/hr	Nearly all metallic materials plus many nonmetallic	0.0001 in. and less	16
Shaping	Bar Plate Casing	3 ft × 6 ft	Limited usually by ability to hold part	1–4 parts/hr	Low- to medium-carbon steels and nonferrous metals best; no hardened parts	±0.001–±0.002 in. (larger parts), ±0.0001–±0.0005 in. (small–medium parts)	63–250
Planing	Bar Plate Casting	42 ft wide × 18 ft high × 76 ft long	Parts too large for shaper work	1 part/hr	Low- to medium-carbon steels or nonferrous materials best	±0.001–±0.005 in.	63–125
Gear shaping	Blanks	120-in.-diameter gears 6-in. face width	1 in. diameter	1–60 parts/hr	Any material with good machinability rating	±0.001 in. or better at 200 D.P. to 0.0065 in. at 30 D.P.	63

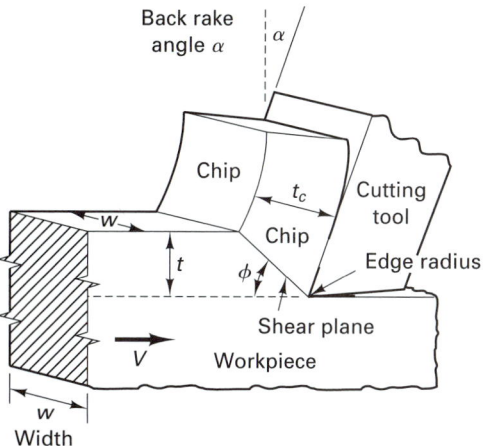

Back rake
angle α

Chip

t_c

Cutting
tool

Chip

w

Edge radius

t

ϕ

Shear plane

V Workpiece

w
Width

FIGURE 21-10   Schematic of orthogonal machining. The cutting edge of the tool is perpendicular to the direction of motion (V). The back rake angle is α. The shear angle is ϕ.

■ 21.3 UNDERSTANDING CHIP FORMATION

In order to understand this complex process, the tool geometry is simplified from the three-dimensional (oblique) geometry, which typifies most processes, to a two-dimensional (orthogonal) geometry. The workpiece is a flat plate (Figure 21-10). The workpiece is moving past the tool at velocity V. The feed of the tool is now called t, the uncut chip thickness. The DOC is the width of the plate, w. The cutting edge of the tool is perpendicular to the direction of motion, V. The angle that the tool makes with respect to a vertical from the workpiece is called the *back rake angle*, α. A positive angle is shown in the schematic. The chip is formed by *shearing*. The shear plane angle, with respect to the horizontal, is ϕ. This model is sufficient to allow us to consider the behavior of the work material during chip formation, the influence of the most critical elements of the tool geometry (the edge radius of the cutting tool and the back rake angle α), and the interactions that occur between the tool and the freshly generated surfaces of the chip against the rake face and the new surface as rubbed by the flank of the tool.

Basically, the chip is formed by a localized shear process that takes place over a very narrow region. This large-strain, high-strain-rate plastic deformation evolves out of a radial compression zone that travels ahead of the tool as it passes over the workpiece (Figure 21-11). This radial compression zone has, like all plastic deformations, an elastic compression region that changes into a plastic compression region as the yield strength of the material is exceeded. The plastic compression generates dense dislocation tangles and networks in annealed metals. When this work hardening reaches a saturated condition (fully work hardened), the material has no recourse but to shear. The onset of the shear process takes place along the lower boundary of the shear zone defined by the shear angle ϕ. The shear lamella (microscopic shear planes) lie at the angle ψ to the shear plane. This can be seen in Figure 21-12, a videograph and Figure 21-13, a schematic made from the videograph. The videograph was made by videotaping the orthogonal machining of an aluminum plate at over 100X with a highspeed videotaping machine capable of 1000 frames per second. By machining at low speeds (V= 8.125 ipm), the behavior of the process could be captured and then observed in playback at very slow frame rates. The uncut chip thickness was $t = 0.020$ in. The termination of the shear process as defined by ψ cannot be observed in the still videograph but can easily be seen in the videos.

If the work material has hard second phase particles dispersed in it, they can act as barriers to the shear front dislocations, which cannot penetrate the particle. The dislocations

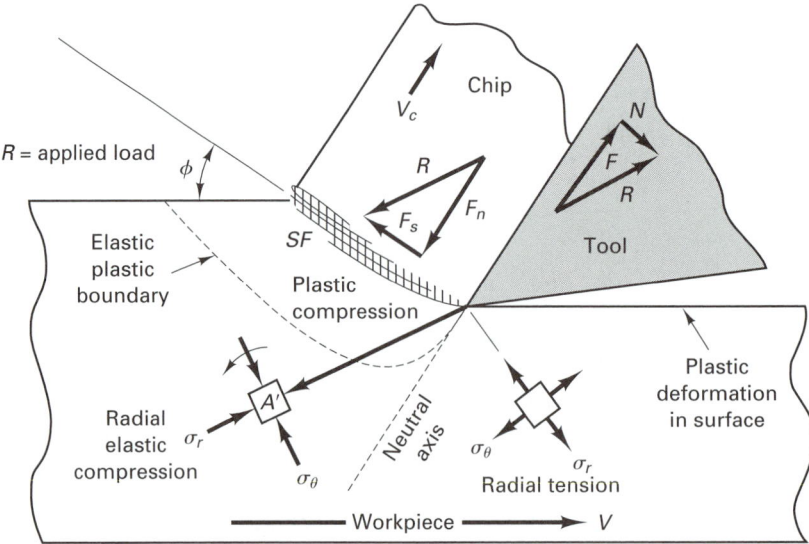

FIGURE 21-11 The machining process produces a radial compression ahead of the shear process. The stress reverses from compression to tension across the neutral axis (NA).

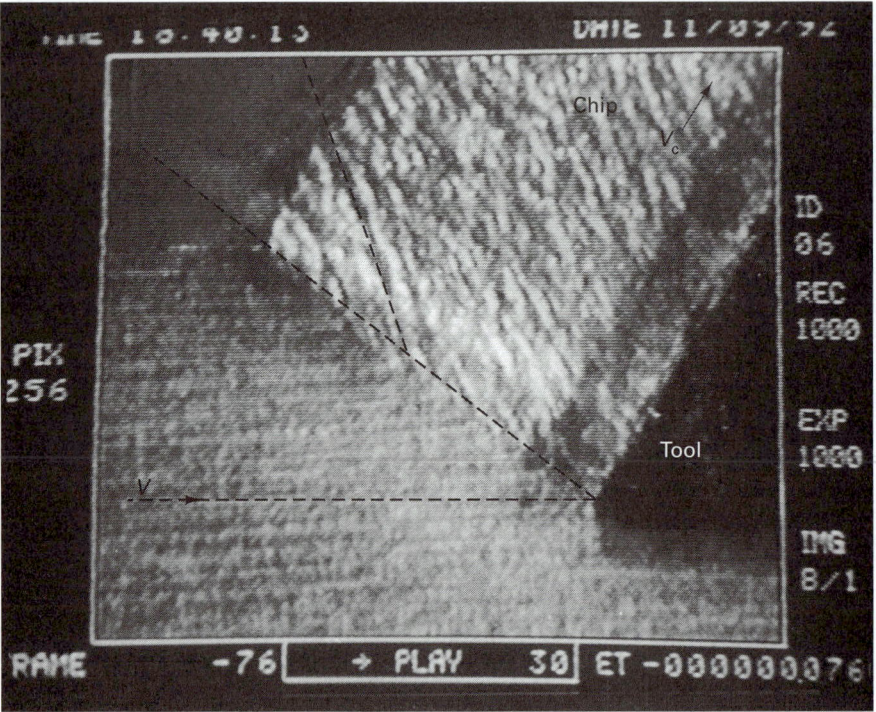

FIGURE 21-12 Videograph of the orthogonal machining process.

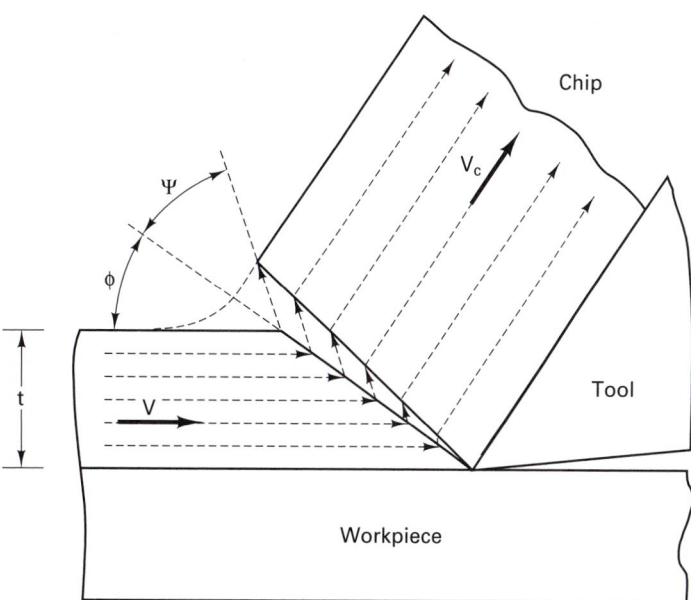

FIGURE 21-13   Schematic representation of the material flow, i.e., the chip forming shear process.

create voids around the particles. If there are enough particles of the right size and shape, the chip will fracture through the shear zone, forming segmented chips. *Free-machining steels*, which have small percentages of hard second-phase particles added to them, use this metallurgical phenomena to break up the chips for easier chip handling.

■ 21.4 ORTHOGONAL MACHINING

Orthogonal machining is done to test machining mechanics and theory. Orthogonal machining (measuring two forces) can be obtained in practice by:

1. Machining a plate as shown in Figure 21-10, and Figure 21-12.
2. End cutting a tube wall in a turning setup.

In *oblique machining*, as in shaping, drilling, milling, and single-point turning, the cutting edge and the cutting motion usually are not perpendicular to each other. In the orthogonal case, the cutting velocity vector and the cutting edge are perpendicular. Since the orthogonal case is more easily modeled, it will be used here to describe the process further.

For the purpose of modeling chip formation, assume that the shear process takes place on a single narrow plane rather than on a set of shear fronts which actually comprise a narrow shear zone. Further, assume that the tool cutting edge is perfectly sharp and that no contact is being made between the flank of the tool and the new surface. The workpiece passes the tool with velocity V, the cutting speed (Figure 21-14a). The uncut chip thickness is t. Ignoring the plastic compression, chips having thickness t_c are formed by the shear process. The chip has velocity V_c. The shear process then has velocity V_s and occurs at shear angle ϕ. The tool geometry is given by the back rake angle α and the clearance

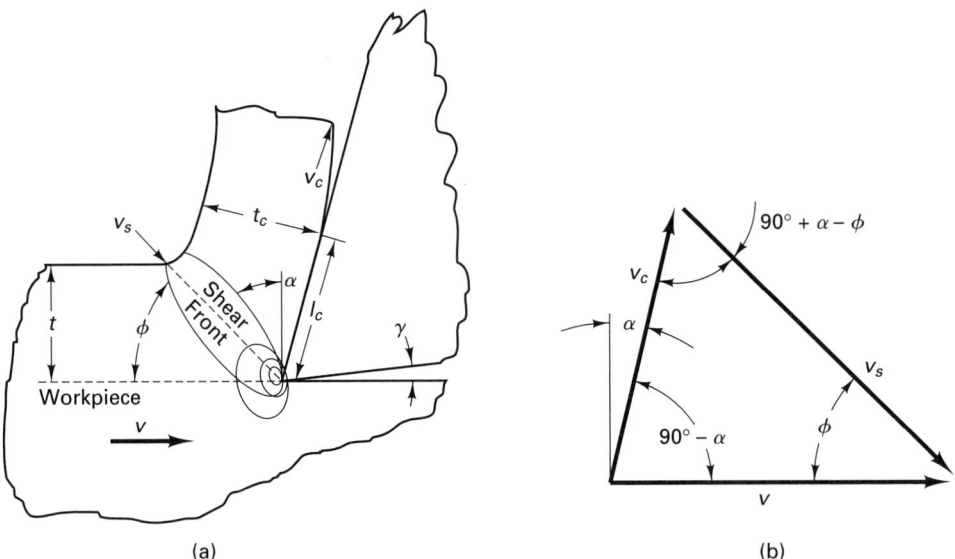

FIGURE 21-14 (a) Schematic of orthogonal machining process; (b) velocity diagram associated with orthogonal machining.

angle γ. The velocity triangle for V, V_c, and V_s is shown in Figure 21-14b. The chip makes contact with the rake face of the tool over length l_c. The plate thickness is called w. To compute the shear angle, the equation for *chip thickness ratio*, r_c, defined as t/t_c, can be derived:

$$r_c = \frac{t}{t_c} = \frac{AB \sin\phi}{AB \cos (\phi - \alpha)} \tag{21-6}$$

where AB is the length of the shear plane from the tool tip to the free surface.

Equation (21-6) may be solved for the *shear angle* ϕ as a function of the measurable chip thickness ratio by expanding the cosine term and simplifying:

$$\tan\phi = \frac{r_c \cos \alpha}{1 - r_c \sin \alpha} \tag{21-7}$$

There are numerous other ways to measure chip ratios and obtain shear angles both during (dynamically) and after (statically) the cutting process. For example, the ratio of the length of the chip, L_c, to the length of the cut, L, can be used to determine r_c. Many researchers use the chip compression ratio, which is the reciprocal of r_c, as a parameter. (See Problem 2 at the end of the chapter for another method.) The shear angle can be measured statically by instantaneously interrupting the cut through the use of *quick-stop devices*. These devices disengage the cutting tool from the workpiece while cutting is in progress, leaving the chip attached to the workpiece. Optical and scanning electron microscopy is then used to observe the direction of shear. Figure 21-15 was made using a quick-stop device. High-speed motion pictures and high-speed video graphic systems have also been used to observe the process at frame rates as high as 30,000 frames per second. Figure 21-12 is a high-speed videograph. Machining stages have been built that allow the process to be performed inside a scanning electron microscope and recorded on videotapes for high-resolution, high-magnification examination of the deformation process. Using sophisticated

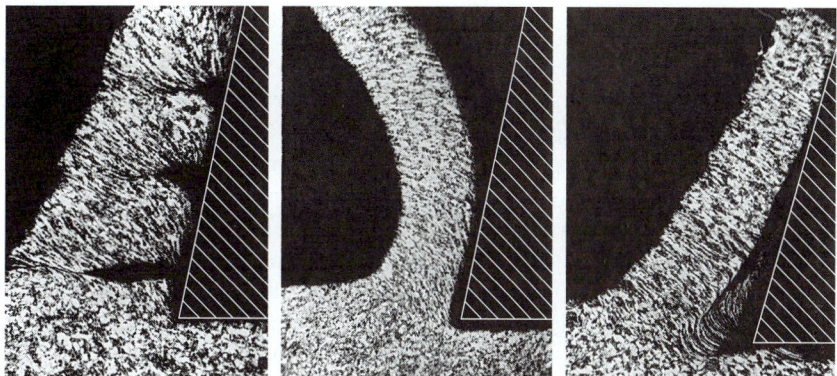

FIGURE 21-15   Three characteristic types of chips.   (Left to right) discontinuous, continuous, and continuous with built-up edge.   Chip samples produced by quick-stop techniques.   (*Courtesy of Cincinnati Milacron, Inc.*)

electronics and slow-motion playback, this technique can be used to measure (for the first time) the *shear velocity.* The vector sum of V_s and V_c equals V.

For consistency of volume, we observe that

$$r_c = \frac{t}{t_c} = \frac{\sin \phi}{\cos (\phi - \alpha)} = \frac{V_c}{V} \tag{21-8}$$

indicating that the chip ratio (and therefore the shear angle) can be determined dynamically if a reliable means to measure V_c can be found.

The ratio of V_s to V is

$$\frac{V_s}{V} = \frac{\cos \phi}{\cos (\phi - \alpha)} \tag{21-9}$$

These velocities are important in power calculations, heat determinations, and vibration analysis associated with chip formation.

The shear front angle ψ can be shown to be

$$\psi = 45° - \phi + \frac{\alpha}{2} \tag{21-10}$$

During the cutting, the chip undergoes a *shear strain* of

$$\varepsilon = \frac{2 \cos \alpha}{1 + \sin \alpha} \tag{21-11}$$

which shows that the *shear* strain is dependent only on the rake angle α. Generally speaking, metal-cutting strains are quite large compared to other plastic deformation processes, being on the order of 1 to 2 in./in.

This large strain occurs, however, over very narrow regions resulting in extremely high shear strain rates, $\dot{\varepsilon}$, typically in the range of 10^4 to 10^8 in./in. per second. It is this combination of large strains and high strain rates operating within a process constrained only by the rake face of the tool that results in great difficulties in theoretical analysis of this process.

Effects of Work Material Properties

As noted previously, the properties of the work material are important in chip formation. High-strength materials require larger forces than materials of lower strength, causing greater tool and work deflection and increased friction, heat generation, and operating temperatures, and requiring greater work input. The structure and composition also influence metal cutting. Hard or abrasive constituents such as carbides in steel accelerate tool wear.

Work material *ductility* is an important factor. Highly ductile materials not only permit extensive plastic deformation of the chip during cutting, which increases work, heat generation, and temperature, but also result in longer, "continuous" chips that remain in contact longer with the tool face, thus causing more frictional heat. Chips of this type are severely deformed and have a characteristic curl. On the other hand, some materials, such as gray cast iron, lack the ductility necessary for appreciable plastic deformation. Consequently, the compressed material ahead of the tool fails in brittle fracture, sometimes along the shear front, producing small fragments. Such chips are termed *discontinuous* or *segmented* (see Figure 21-15).

A variation of the continuous chip, often encountered in machining ductile materials, is associated with a *built-up edge* (BUE) formation on the cutting tool. The local high temperature and extreme pressure in the cutting zone cause the work material to adhere or pressure weld to the cutting edge of the tool forming the built-up edge, rather like a dead metal zone in the extrusion process. Although this material protects the cutting edge from wear, it modifies the geometry of the tool. BUEs are not stable and will break off periodically, adhering to the chip or passing under the tool and remaining on the machined surface. Built-up edge formation can be eliminated or minimized by reducing the depth of cut, altering the cutting speed, using positive rake tools, applying a coolant, or changing cutting tool materials.

Mechanics of Machining

Orthogonal machining has been defined as a two-force system. Consider Figure 21-16, which shows a free-body diagram of a chip that has been separated at a shear plane. It is assumed that the resultant force R acting on the back of the chip is equal and opposite to the resultant force R' acting on the shear plane. The resultant R is composed of the *friction force F* and the normal force N acting on the tool/chip interface contact area. The resultant force R' is composed of a *shear force* F_s and normal force F_n acting on the shear plane area A_s. Since neither of these two sets of forces can usually be measured, a third set is needed which can be measured using a dynamometer (force transducer) mounted either in the workholder or the tool holder. Note that this set has resultant R''', which is equal in magnitude to all the other resultant forces in the diagram. The resultant force R''' is composed of a *cutting force* F_c and a tangential (normal) force F_t. Now it is necessary to express the desired forces (F_s, F_n, F, N) in terms of the measured dynamometer components, F_c and F_t, and appropriate angles. To do this, a circular force diagram is developed in which all six forces are collected in the same force circle (Figure 21-17). The only symbol in this figure as yet undefined is β, which is the angle between the normal force N and the resultant R. The friction angle, β, is used to describe the friction coefficient μ on the tool–chip interface area, which is defined as F/N, so that

$$\beta = \tan^{-1}\mu = \tan^{-1}\frac{F}{N} \tag{21-12}$$

The friction force F and its normal N can be shown to be

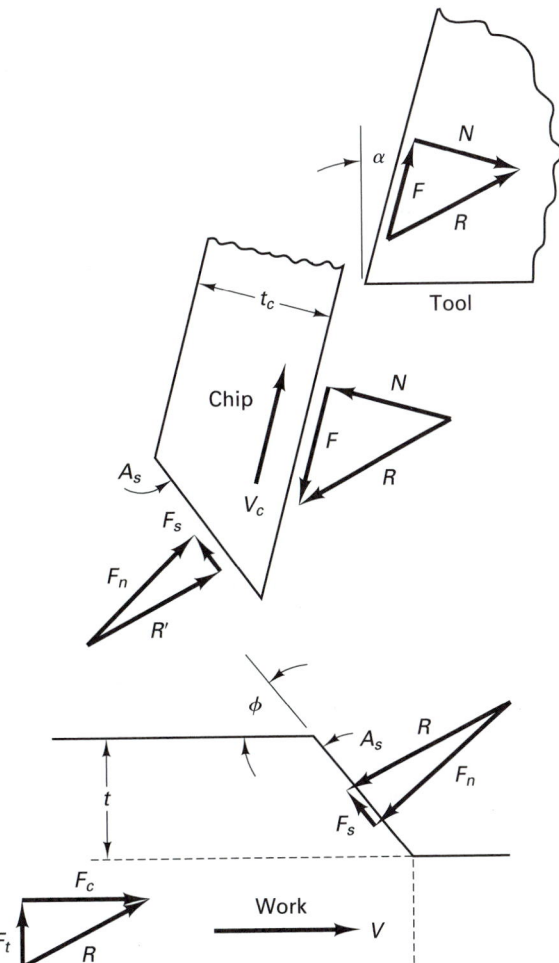

FIGURE 21-16 Free-body diagram of orthogonal chip formation process, showing equilibrium condition between resultant forces R and R'.

$$F = F_c \sin \alpha + F_t \cos \alpha \qquad (21\text{-}13)$$

$$N = F_c \cos \alpha - F_t \sin \alpha \qquad (21\text{-}14)$$

and the resulting R is

$$R = \sqrt{F_c^2 + F_t^2} \qquad (21\text{-}15)$$

Notice that in the special situation where the back rake angle is zero, $F = F_t$ and $N = F_c$, so that in this orientation, the friction force and its normal can be measured directly by the dynamometer.

The forces parallel and perpendicular to the shear plane can be shown from the force circle diagram (Figure 21-17) to be

$$F_s = F_c \cos \phi - F_t \sin \phi \qquad (21\text{-}16)$$

$$F_n = F_c \sin \phi - F_t \cos \phi \qquad (21\text{-}17)$$

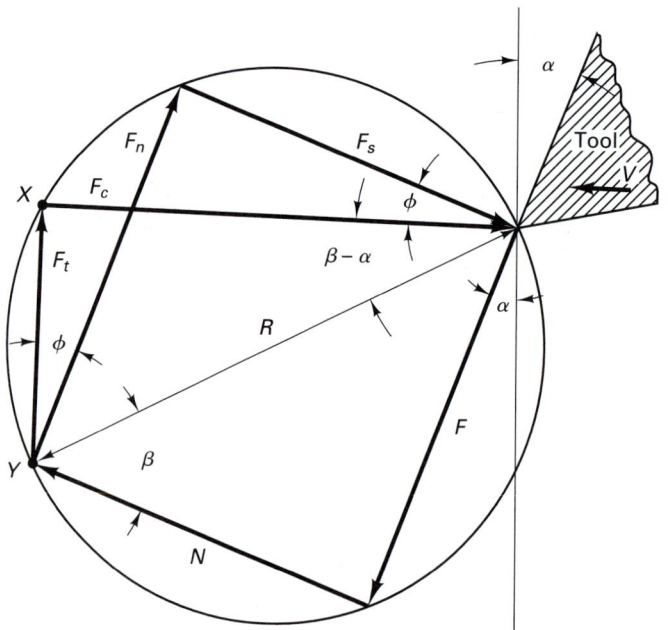

FIGURE 21-17   Circular force diagram used to derive equations A for F_s, F_n, F, and N as functions of F_c, F_t, ϕ, α, and β.

F_s is of particular interest, as it is used to compute the shear stress on the shear plane. This shear stress is defined as

$$\tau_s = \frac{F_s}{A_s} \tag{21-18}$$

where

$$A_s = \frac{tw}{\sin\phi} \tag{21-19}$$

recalling that t was the uncut chip thickness and w was the width of the workpiece. The *shear stress (flow stress)* is, therefore,

$$\tau_s = \frac{F_c \sin\phi \cos\phi - F_t \sin^2\phi}{tw} \text{ psi} \tag{21-20}$$

For a given polycrystalline metal, the shear stress is a material constant, not sensitive to variations in cutting parameters, tool material, or the cutting environment. Figure 21-18. gives some typical values for the flow stress for a variety of metals, plotted against hardness

It is the objective of some metal-cutting researchers to be able to derive (predict) the shear stress τ_s and the shear direction ϕ, from dislocation theory. Correlations of the shear stress with metallurgical measures such as hardness or dislocation stacking fault energy have been useful in these efforts. Equation (21-20) is useful for estimating the cutting force, F_c. One simply measures the thickness of a chip and computes ϕ from r_c. Then one assumes that $F_t = F_c/2$ and rewrites equation (21-20) to solve for F_c.

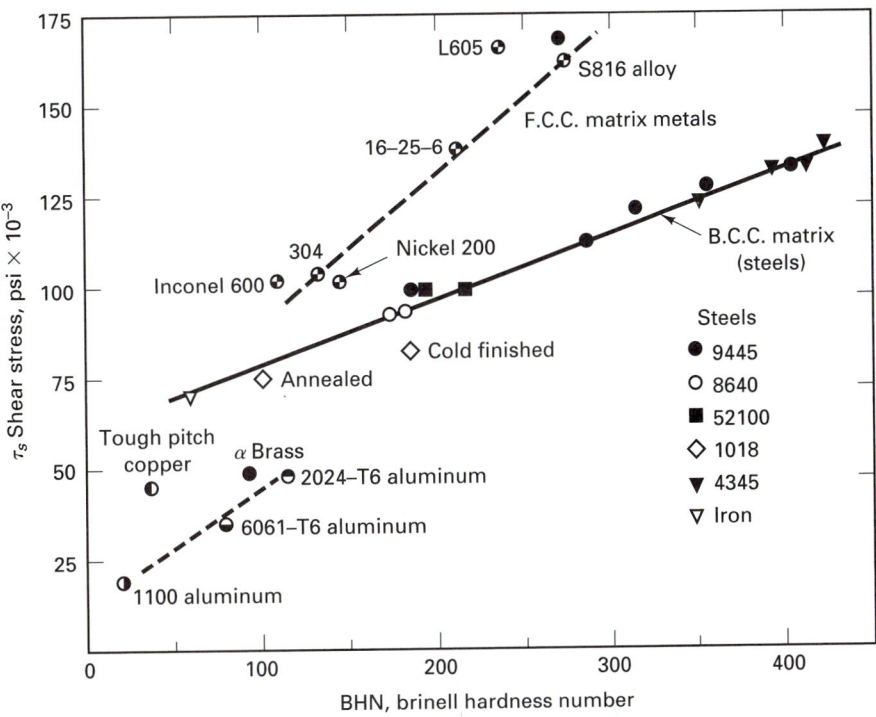

FIGURE 21-18   Shear stress τ_s variation with the Brinell hardness number for a group of steels and aerospace alloys. Data of some selected FCC metals are also included. (Adapted with permission from S. Ramalingham and K. J. Trigger, *Advances in Machine Tool Design and Research*, Pergamon Press, Elmsford, N.Y., 1971.)

■ 21.5 ENERGY AND POWER IN MACHINING

The cutting force system in a conventional, oblique chip formation process is shown schematically in Figure 21-19. Oblique cutting has three components:

1. F_c: primary cutting force acting in the direction of the cutting velocity vector. This force is generally the largest force and accounts for 99% of the power required by the process.
2. F_f: feed force acting in the direction of the tool feed. This force is usually about 50% of F_c but accounts for only a small percentage of the power required because feed rates are usually small compared to cutting speeds.
3. F_r: radial or thrust force acting perpendicular to the machined surface. This force is typically about 50% of F_f and contributes very little to power requirements because velocity in the radial direction is negligible.

Figure 21-3 shows the oblique geometry and Figure 21-19 also shows the general relationship between these forces and speed, feed, and depth of cut. These figures cannot be used to determine forces for a specific process, as the following discussion will explain.

The power required for cutting is

$$P = F_c V \quad \text{ft–lb/min} \tag{21-21}$$

The horsepower at the spindle of the machine is therefore

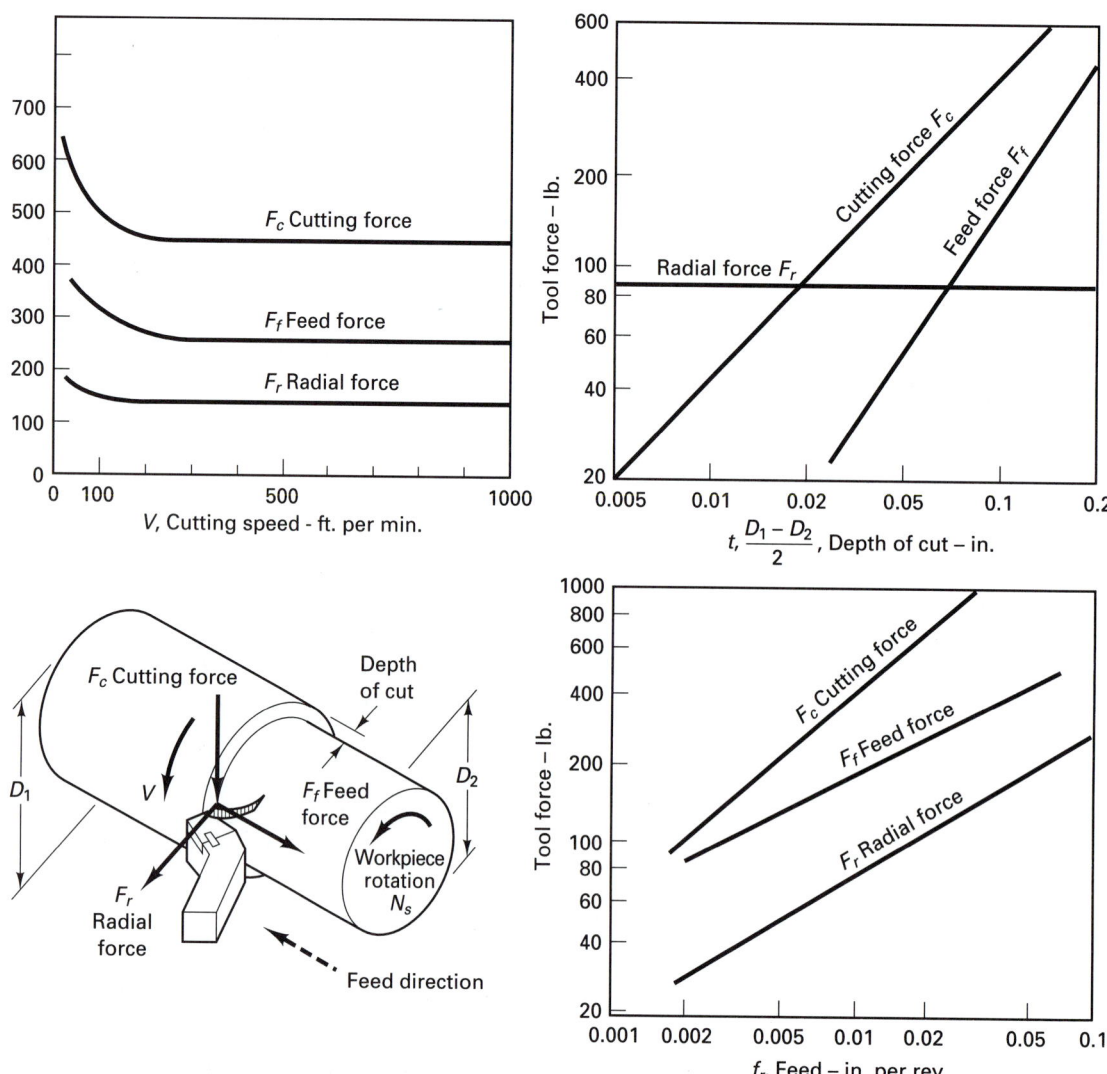

FIGURE 21-19 Three measurable components of forces acting on a single-point tool (oblique machining) with effects on the forces of varying speed, depth of cut, and feed.

$$HP = \frac{F_c V}{33,000} \qquad (21\text{-}22)$$

In metal cutting a very useful parameter is called the unit or *specific horsepower* HP_s, which is defined as

$$HP_s = \frac{HP}{MRR} \quad (\text{hp/in}^3/\text{min}) \qquad (21\text{-}23)$$

In turning, for example, where MRR $\simeq 12\ Vf_r d$,

$$HP_s = \frac{F_c}{396,000 \, f_r \, d} \tag{21-24}$$

Thus this term represents the approximate power needed at the spindle to remove a 1 in³ of metal per minute. Specific horsepower factors for some common materials are given in Table 21-3. Specific horsepower is related to and correlates well with shear stress for a given metal. Notice the similarity between equations (21-20) and (21-24). Unit horsepower is sensitive to material properties (e.g., hardness) and rake angle, depth of cut, and feed. τ_s is sensitive to material properties only.

TABLE 21-3. Values for Specific or Unit HP$_s$ for Various Metals during Metal Removal

Material	Hardness (BHN or R)	HP$_s$ hp/in³/min	HP$_s$ kW/cm³/min	Comments
Steels, including plain carbon, alloy, tool, hot or cold rolled, or cast	85–200 35–40R$_c$ 40–50R$_c$ 50–55R$_c$ 55–58R$_c$	1.1 1.4 1.5 2.0 3.4	0.050 0.064 0.068 0.091 0.155	Values assume normal feed ranges and sharp tools. Multiply value by 1.25 for a dull tool.
Cast iron	100–190 190–300	0.7–1.0 1.4–1.6	0.03–0.045 0.05–0.07	Add 10% for milling to table value.
Stainless steels	150–450	1.2–1.4	0.05–0.068	
Iron-based alloys	180–320	1.2–1.6	0.055–0.073	High-temperature alloys.
Nickel alloys	80–360	1.8–2.0	0.82–0.091	
Nickel–cobalt-based alloys	200–360	2.0–2.5	0.09–0.11	High-temperature materials.
Aluminum 2014-T6, 2017-T4 6064-T6 Pure-108 Hard (rolled)	30–150 at 500 kg 55 Hard	0.25 – 0.34 0.16 0.33	0.014 – 0.016 0.007 0.015	
Magnesium alloys	40–90 at 500 kg	0.16	0.007	
Copper	50R$_B$	0.9–1.0	0.041–0.046	
Copper alloys	10–80R$_B$ 80–100R$_B$	0.5–0.6 0.8–1.0	0.022–0.030 0.036–0.046	
Titanium	250–375	1.8–2.0	0.82–0.091	
Tungsten, tantalum	210–320	2.6–2.8	0.12–0.13	

Specific power can be used in a number of ways. First, it can be used to estimate the motor horsepower required to perform a machining operation for a given material. HP$_s$ values from the table are multiplied by the approximate MRR value for the process. The motor horsepower, HP$_m$, is then

$$HP_m = \frac{HP_s \times MRR \times CF}{E} \tag{21-25}$$

where E is the efficiency of the machine. This factor accounts for the power needed to overcome friction and inertia in the machine and drive moving parts. Usually, 80% is used. Correction factors (CFs) may also be used to account for variations in cutting speed, feed, and rake angle. There is usually a tool wear correction factor of 1.25 used to account for the fact that dull tools use more power than sharp tools.

The primary cutting force F_c can be roughly estimated according to

$$F_c \simeq \frac{\text{HP}_s \times \text{MRR} \times 33{,}000}{V} \tag{21-26}$$

This type of estimate of the major force F_c is useful in analysis of deflection and vibration problems in machining and in the proper design of workholding devices, as these devices must be able to resist movement and deflection of the part during the process.

In general, increasing the speed, feed, or depth of cut will increase the power requirement. Doubling the speed doubles the HP directly. Doubling the feed or the depth of cut doubles the cutting force F_c. In general, increasing the speed does not increase the cutting force F_c, which has always been a puzzle. However, recent research has shown that the periodicity or spacing of the shear fronts remains constant for a given material regardless of the cutting speed. Doubling the velocity essentially doubles the number of shear fronts produced in a given amount of time. Put another way, if speed is doubled, chip length is doubled for the same amount of cutting time. For constant shear front spacing (constant lamella size) this means twice as many shear fronts; therefore, energy is doubled. Cutting force F_c, on the other hand, reflects a change in F_s, where $F_s = \tau / A_s$. Neither of these two quantities were changed by the change in speed. Thus F_c remains constant. Since a change in feed or depth of cut changes area A_s directly, changes in f_r or d affect F_c directly. However, speed has a strong effect on tool life because most of the input energy is converted into heat, which raises the temperature of the chip, the work, and the tool, to the latter's detriment.

Equation (21-27) can be used to estimate the maximum depth of cut, d, for a process as limited by the available power:

$$d_{\max} = \frac{\text{HP}_M \cdot E}{12\,\text{HP}_s \cdot V f_r (\text{CF})} \tag{21-27}$$

It is interesting to compute the total energy in the process and determine how it is distributed between the primary shear and the secondary tool/chip interface shear/friction. It is safe to assume that the majority of the input energy is consumed by these two regions. Therefore,

$$U = U_s + U_f \tag{21-28}$$

where total energy is

$$U = \frac{F_c V}{f_r d V} = \frac{F_c}{f_r d} \tag{21-29}$$

and specific shear energy is

$$U_s = \frac{F_s V_s}{f_r d V} = \frac{F_s \cos \alpha}{f_r V \cos (\phi - \alpha)} = \tau_s \gamma \tag{21-30}$$

Specific friction energy is

$$U_f = \frac{F V_c}{f_r d V} = \frac{F r_c}{f_r d} \tag{21-31}$$

Usually, 30 to 40% of the total energy goes into friction and 60 to 70% into the shear process.

■ 21.6 HEAT AND TEMPERATURE IN METAL CUTTING

In metal cutting, the power put into the process ($F_c V$) is largely converted to heat, elevating the temperatures of the chip, the workpiece, and the tool. These three elements of the process, along with the environment (which includes the cutting fluid), act as the heat sinks. Figure 21-20 shows the distribution of the heat to these three sinks as a function of cutting speed.

There are three main *sources* of heat. Listed in order of their heat-generating capacity, they are:

1. The shear process itself, where plastic deformation results in the major heat source. Most of this heat stays in the chip.
2. The tool–chip interface contact region, where additional plastic deformation takes place in the chip and there is considerable heat generated due to sliding friction.
3. The flank of the tool, where the freshly produced workpiece surface rubs the tool.

There have been numerous experimental techniques developed to measure cutting temperatures and some excellent theoretical analysis of this "moving" multiple-heat-source problem. Space does not permit us to explore this problem in depth. Figure 21-21a shows the effect of cutting speed on the tool/chip interface temperature. The rate of wear of the tool at the interface can be shown to be directly related to temperature (Figure 21-21b).

FIGURE 21-20 Distribution of heat generated in machining to the chip, tool, and workpiece. Heat going to the environment is not shown.

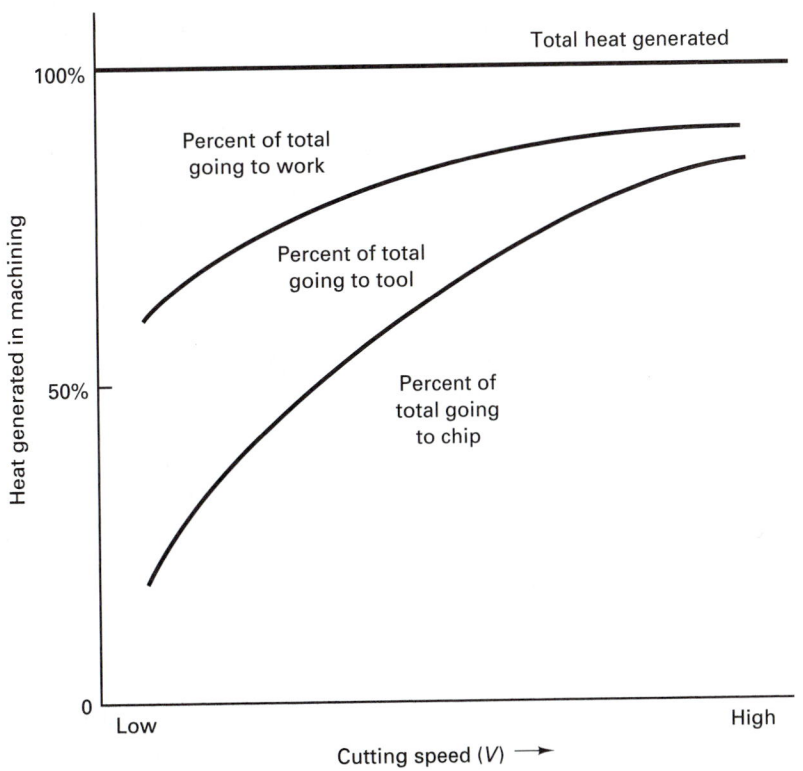

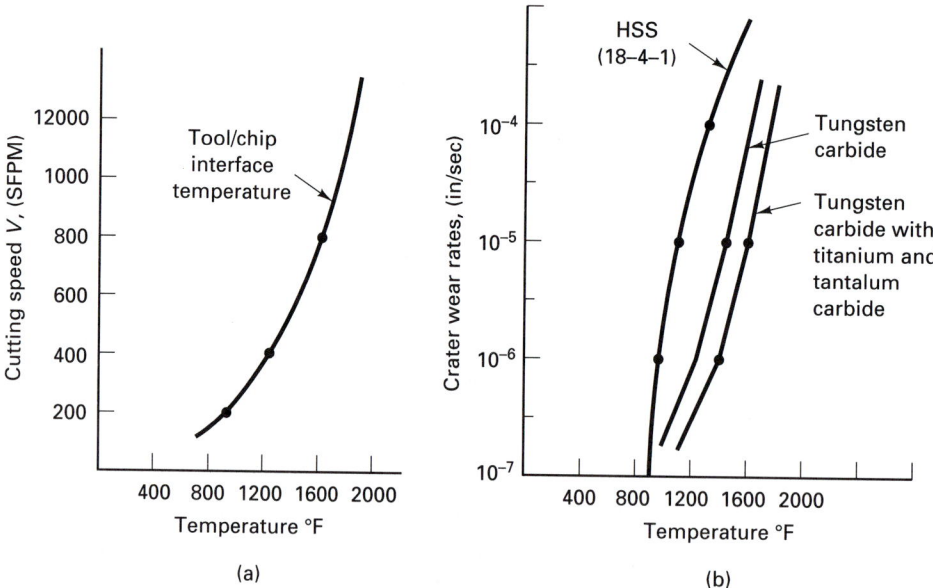

FIGURE 21-21   Typical relationships of temperature to cutting speed and crater wear for machining steels.

Because cutting forces are concentrated in small areas near the cutting edge, these forces produce large pressures. The tool material must be hard to resist wear and tough to resist cracking and chipping. Tools used in interrupted cutting, such as milling must be able to resist impact loading as well. Tool materials must sustain their hardness at elevated temperatures as shown in Figure 21-22, which relates the hardness of various tool materials to temperature.

The challenge to manufacturers of cutting tools has always been to find materials that satisfy these severe conditions. Cutting tool materials that do not lose hardness at the high temperatures associated with high speeds are said to have "hot hardness". Obtaining this property usually requires a trade-off in toughness, as hardness and toughness are generally opposing properties. In Chapter 22 cutting tool materials will be addressed in more depth.

■ 21.7 SUMMARY

In this chapter the basics of the machining processes have been presented. Chapter 23 through 30 provide additional information on the various operations and machine tools. Modern machine tools can often perform several basic operations, as shown in Figure 21-23. As described in Chapters 1 and 29, machining centers are now widely used. These chapters must be studied carefully. The relationships between the basic processes and the machine tools that can be used to perform these processes has been presented. Generally speaking, these machines will be of the A(2) or A(3) level of automation. Machining centers, which are numerical control machines, are A(4). Machining centers have automatic tool change capability and are usually capable of milling, drilling, boring, reaming, tapping (hole threading), and other minor machining processes. For particular machines, you will need to become familiar with new terminology, but in general all machining processes will need inputs concerning rpm (given that you selected the cutting speed), feeds, and depths of cut. Note also from these tables that the same process can be performed on two or more

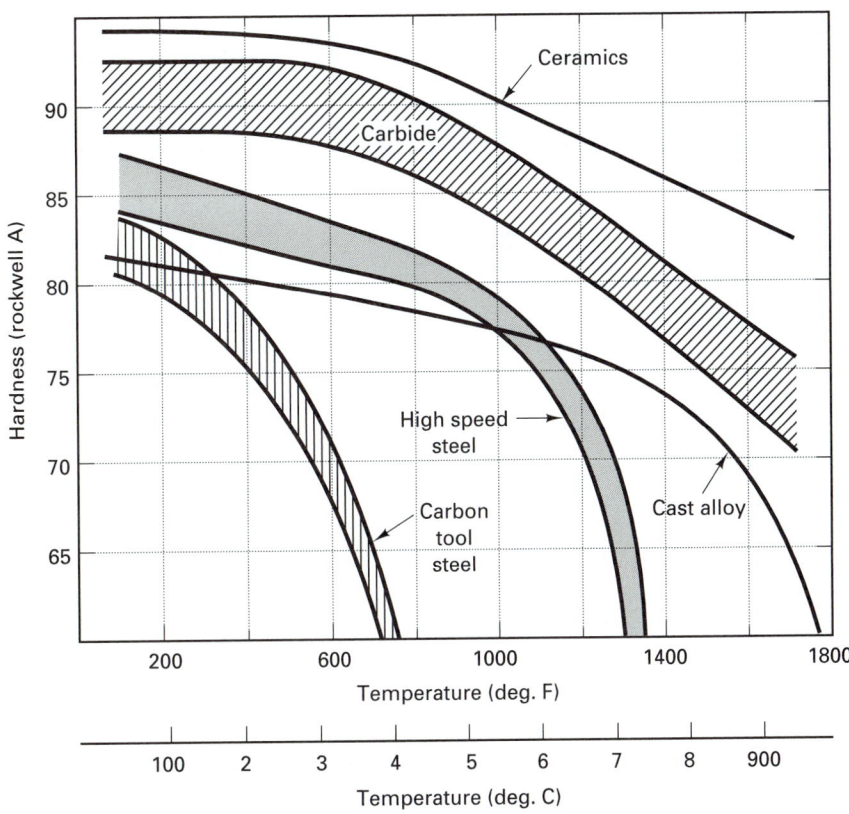

FIGURE 21-22   Hardness of tool materials decreasing with increasing temperature. Some material display a more rapid drop in hardness above some temperatures. *(From Metal Cutting Principles, 2nd ed.; courtesy of Ingersoll Cutting Tool Company.)*

different machine tools. There are many ways to produce flat surfaces, internal and external cylindrical surfaces, and special geometries in parts. Generally, the quantity to be made is the driving factor in the selection of processes, as we shall see in later discussions.

■ KEY WORDS

back rake angle	friction force	shear force
broaching	grinding (abrasive machining)	shear strain
built-up edge	machine tool	shear velocity
chip thickness ratio	metal cutting	shear stress (flow stress)
chip velocity	milling	specific horsepower
cutting force	oblique machining	speed
cutting tool	orthogonal machining	turning
depth of cut	sawing	workholding device
drilling	shaping	workpiece
feed	shear angle	

Operation	Block diagram	Most commonly used machines	Machines less frequently used	Machines seldom used
Turning	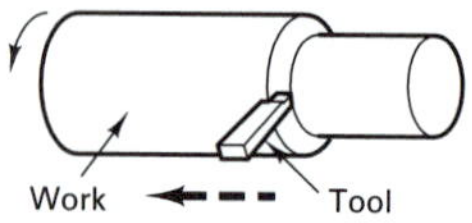	Lathe turning control	Boring mill	
Grinding	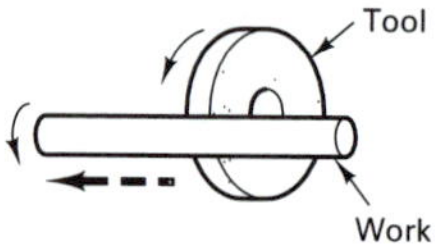	Cylindrical grinder		Lathe (with special attachment)
Sawing (of plates)	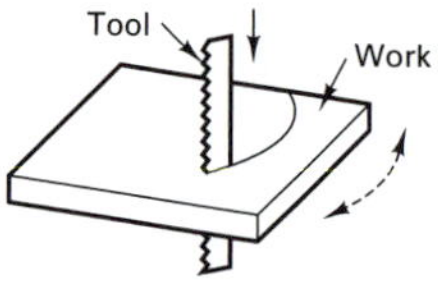	Contour or band saw	Laser Flame cutting Plasma arc	
Drilling	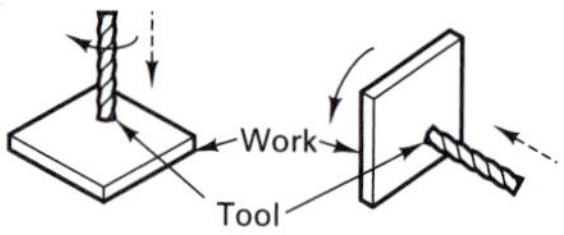	Drill press Machining center Vert. milling machine	Lathe Horizontal boring machine	Horizontal milling machine Boring mill
Boring	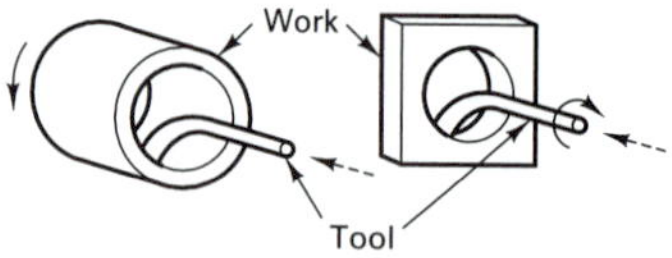	Lathe Boring mill Horizontal boring machine Machining center		Milling machine Drill press
Reaming	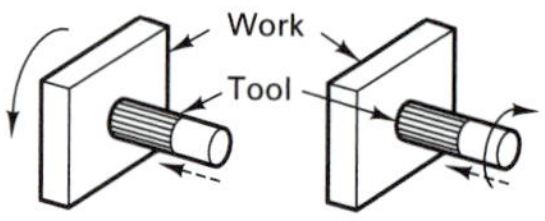	Lathe Drill press Boring mill Horizontal boring machine Machining center	Milling machine	
Grinding	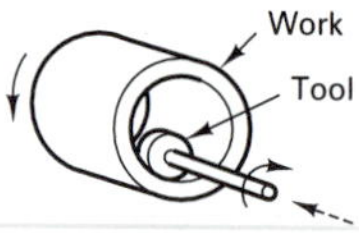	Cylindrical grinder		Lathe (with special attachement)
Sawing	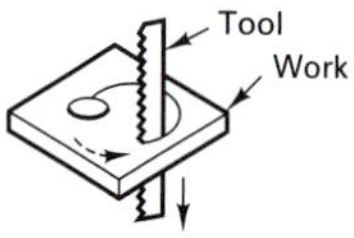	Contour or band saw		
Broaching	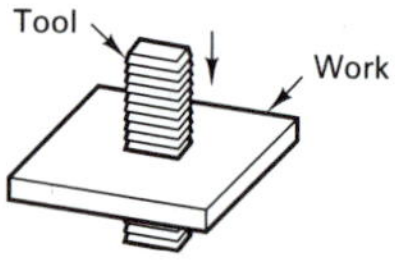	Broaching machine	Arbor press (keyway broaching)	

FIGURE 21-23   Operations and machines for machining cylindrical surfaces.

Operation	Block diagram	Most commonly used machines	Machines less frequently used	Machines seldom used
Facing	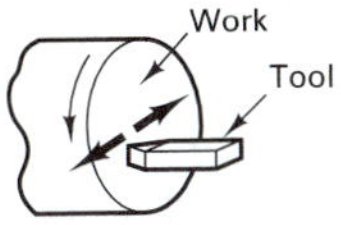	Lathe	Boring mill	
Broaching	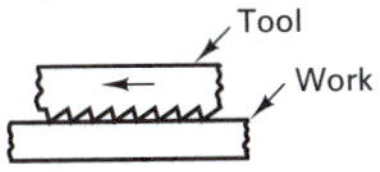	Broaching machine		Turret broach
Grinding	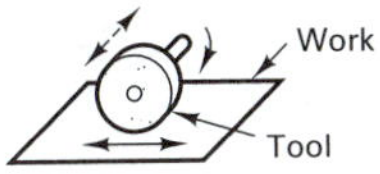	Surface grinder		Lathe (with special attachment)
Sawing	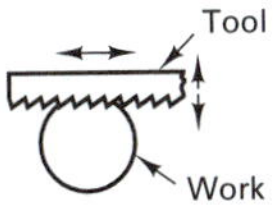	Cutoff saw	Contour saw	
Shaping	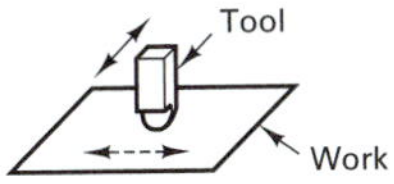	Horizontal shaper	Vertical shaper	
Planing	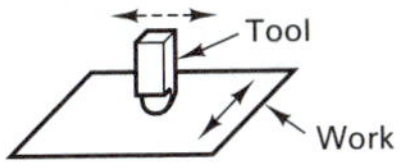	Planer		
Milling	slab milling 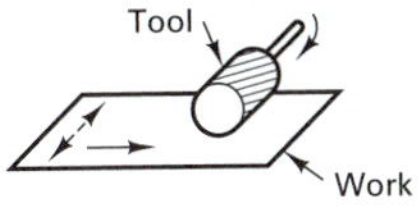	Milling machine	Lathe with special milling tools	
	face milling 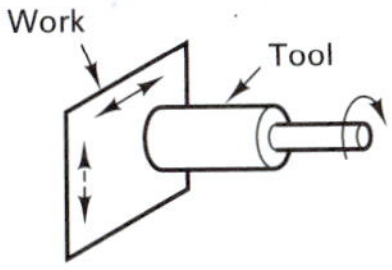	Milling machine Machining center	Lathe with special milling tools	Drill press (light cuts)

FIGURE 21-23 (cont.)

■ REVIEW QUESTIONS

1. Why has the metal-cutting process resisted theoretical solution for so many years?
2. What variables must be considered in understanding a machining process?
3. Which of the basic chip formation processes are single point, and which are multiple point?
4. How is feed related to speed in machining operations such as turning?
5. Milling has two feeds. What are they, and which one is an input parameter to the machine tool?
6. What is the fundamental mechanism of chip formation?
7. What are the implications of Figure 21-12, given that this videograph was made at a very low cutting speed?
8. What is the difference between oblique machining and orthogonal machining?

9. Note that the units for the approximate equation for MRR for turning are not correct. When is the approximate equation not very good (yields a large error in MRR values)?

10. For orthogonal machining, the cutting edge radius is assumed to be small compared to the uncut chip thickness. Why?

11. How does the magnitude of the strain and strain rate values of metal cutting compare to tensile testing?

12. Why is titanium such a difficult metal to machine? (Note its high value of HP_s).

13. Explain why you get segmented or discontinuous chips when you machine cast iron.

14. Why is shear stress so important?

15. Which of the three cutting forces in oblique cutting consumes most of the power?

16. How is the energy in a machining process typically consumed?

17. Where does the energy consumed ultimately go?

18. State two ways of estimating the primary force, F_c.

19. How is cutting speed related to tool wear?

20. What is the relationship between hardness and temperature in metal cutting tool materials?

■ PROBLEMS

1. Suppose that you have the following data obtained from a metal cutting experiment (orthogonal machining). Compute the shear angle, the shear stress, the coefficient of friction at the tool/chip interface, and the specific horsepower. How do your HP_s and τ_s values compare with the values found in Chapter 21?

Machining data for 1020 steel, as-received, in air with a K3H carbide tool, orthogonally (tube cutting on lathe) with tube OD = 2.875 in. The cutting speed was 530 ft/min. The tube wall thickness was 0.200 in. The back rake angle was zero for all cuts.

Run Number	Forces (lb) F_c	Forces (lb) F_t	Feed (in./rev. × 1/1000)	Chip Ratio, r_c	ϕ	τ_s	μ	HP_s
1	330	295	4.89	0.331				
2	308	280	4.89	0.381				
3	410	330	7.35	0.426				
4	420	340	7.35	0.426				
5	510	350	9.81	0.458				
6	560	395	9.81	0.453				

2. Suppose that you weighted a short chip fragment. You measured the length of the short chip fragment, L_c. Assume that it weighs W. The density of this metal is ρ, lb/in³. Can you obtain the chip thickness ratio, r_c?

3. For the data in Problem 1, determine the specific shear energy and the specific friction energy.

4. Derive equations for F and N using the circular force diagram. (*Hint*: Make a copy of the diagram. Extend a line from point X intersecting force F perpendicularly. Extend a line from point Y intersecting the previous line perpendicularly. Find the angle α made by these constructions.)

5. Derive equations for F_s and F_n using the circular force diagram. (*Hint*: Construct a line through X parallel to vector F_n. Extend vector F_s to intersect this line. Construct a line from X perpendicular to F_n. Construct a line through point Y perpendicular to the line through X.)

6. For the data in problem 1, calculate the shear strain and compare it to $1/r_c$. Comment on the comparison. $1/r_c = t_c/t = \dfrac{L}{L_c}$ assuming that $W = W_c$.

7. A manufacturing engineer needs an estimate of the cutting force, F_c, in order to estimate the loss of accuracy of a machining process due to deflection. The material being machined is Inconal 600 with a BHN value of 100. The

cutting speed was 250 ft/min, the feed was 0.020 in./rev, and the depth of cut was 0.250 in. The chip from the process measured 0.080 in. thick. Estimate the cutting force F_c, assuming that $F_t = F_c/2$.

8. Using Table 21-1, determine the maximum and minimum MRR values for rough machining (turning) a 1020 carbon steel with a BHN value of 200. Repeat for finish machining assuming a DOC value equal to 10% of the roughing DOC.

9. Estimate the horsepower needed to remove metal at 550 in^3/min with a feed of 0.005 in/rev at a DOC value of 0.675 in. The cutting force, F_c, was measured at 10,000 lb. Comment on these values.

10. For a turning process, the horsepower required was 24 hp. The metal removal rate was 550 in^3/min. Estimate the specific horsepower and compare to published values for 1020 steel at 200 BHN.

Chapter 21 CASE STUDY

hss versus tungsten carbide

The Tonto Machine Works, which does job shop machining, has received an order to make 40 duplicate pieces, made of AISI 4140 steel, which will require 1 hour per piece of actual cutting time if an ordinary high-speed-steel milling cutter is used. Joseph Chen, a new machinist, says the cutting time could be reduced to not over 25 minutes per piece if the company would purchase a suitable tungsten carbide milling cutter. Hugh Fellows, the foreman for the milling area, says he does not believe that Joseph's estimate is realistic, and he is not going to spend $450 of the company's money on a carbide cutter that probably would not be used again. The machine-hour rate, including labor, is $40 per hour. Joseph and Hugh have come to you, the supervisor of the shop, for a decision on whether or not to buy the cutter, which is readily available from a local supplier.

What factors should you consider in this situation? Who do you think is right, Joseph or Hugh?

CHAPTER 22

CUTTING TOOLS FOR MACHINING

22.1	INTRODUCTION		Polycrystalline Cubic Boron Nitrides
22.2	CUTTING TOOL MATERIALS		
	Tool Steels	22.3	TOOL GEOMETRY
	High Speed Steels	22.4	TOOL FAILURE AND TOOL LIFE
	Cast Cobalt Alloys		
	Carbides or Sintered Carbides		Taylor's Tool Life Model
		22.5	RECONDITIONING CUTTING TOOLS
	Ceramics	22.6	ECONOMICS OF MACHINING
	Coated Carbide Tools		
	TiN-Coated High-Speed Steel	22.7	MACHINABILITY
		22.8	CUTTING FLUIDS
	Cermets	Case Study:	INDEXABLE DRILL INSERT
	Diamonds		

■ 22.1 INTRODUCTION

Success in metal cutting depends on selection of the proper cutting tool (material and geometry) for a given work material. A wide range of cutting tool materials is available with a variety of properties, performance capabilities, and cost. These include high carbon steels and low/medium alloy steels, high-speed steels, cast cobalt alloys, cemented carbides, cast carbides, coated carbides, coated high speed steels, ceramics, cermets, whisker-reinforced ceramics, sialons, sintered polycrystalline cubic boron nitride (CBN), sintered polycrystalline diamond, and single-crystal natural diamond. Figure 22-1 presents a chronological listing of common cutting-tool materials. The tool materials are ranked by the maximum cutting speed needed to machine a unit volume of steel materials, assuming equal tool lives. As the speed increases, so does the metal removal rate. The time required to remove a given unit volume of material therefore decreases. Notice the fivefold increase in speed that the $TiC/Al_2O_3/TiN$-coated carbide has over the WC/Co tool ($250 \rightarrow 1200$ sfpm). Today, approximately 85% of carbide tools are coated, almost exclusively by the *chemical vapor deposition* (CVD) process. The cutting tool is the most critical part of the machining system.

The cutting tool material, cutting parameters, and tool geometry selected directly influence the productivity of the machining operation. Figure 22-2 outlines the input variables that influence the tool material selection decision. The elements that influence the decision are:

- Work material characteristics (chemical and metallurgical state)
- Part characteristics (geometry, accuracy, finish, and surface-integrity requirements)
- Machine tool characteristics, including the workholders (adequate rigidity with high horsepower, and wide speed and feed ranges)

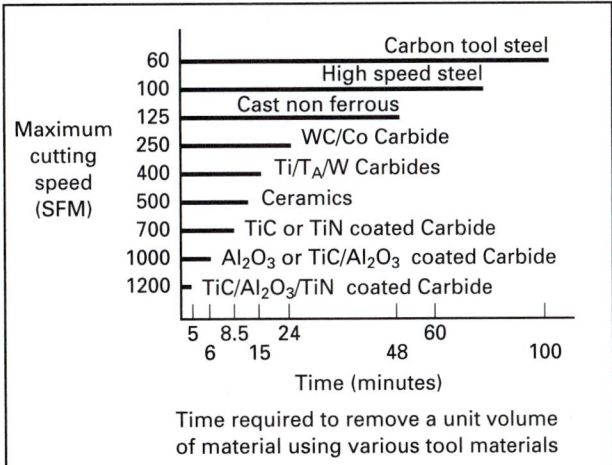

FIGURE 22-1
Improvements in cutting tool materials have reduced machining time.

- Support systems (operator's ability, sensors, controls, method of lubrication and chip removal)

Tool material technology is advancing rapidly, enabling many difficult-to-machine materials to be machined at higher removal rates and/or cutting speeds with greater performance reliability. Higher speed and/or removal rates usually improve productivity. Predictable tool performance is essential when machine tools are computer controlled and have minimal operator interaction. Long tool life is desirable when machines are placed in cellular manufacturing systems.

The cutting tool is subjected to severe conditions. Tool temperatures of 1000°C, severe friction, and high local stresses require that the tool have these characteristics.

1. High hardness (Figure 22-3)
2. Resistance to abrasion, wear, chipping of the cutting edge
3. High toughness (impact strength)
4. High *hot hardness* (refer to Figure 21-22)
5. Strength to resist bulk deformation
6. Good chemical stability (inertness or negligible affinity with the work material)
7. Adequate thermal properties
8. High elastic modulus (stiffness)
9. Consistent tool life
10. Correct geometry and surface finish

Table 22-1 compares some of these properties for various cutting tool materials. Figure 22-3 compares various tool materials on the basis of hardness, the most critical characteristic. Naturally, it would be most convenient if these materials were also easy to fabricate, readily available, and inexpensive, since cutting tools are routinely placed. Obviously, many of the requirements conflict and therefore tool selection will always require trade-offs.

■ 22.2 CUTTING TOOL MATERIALS

In nearly all machining operations, cutting speed and feed are limited by the capability of the tool material. Speeds and feeds must be kept low enough to provide for an acceptable tool life. If not, the time lost changing tools may outweigh the productivity gains from increased cutting speed.

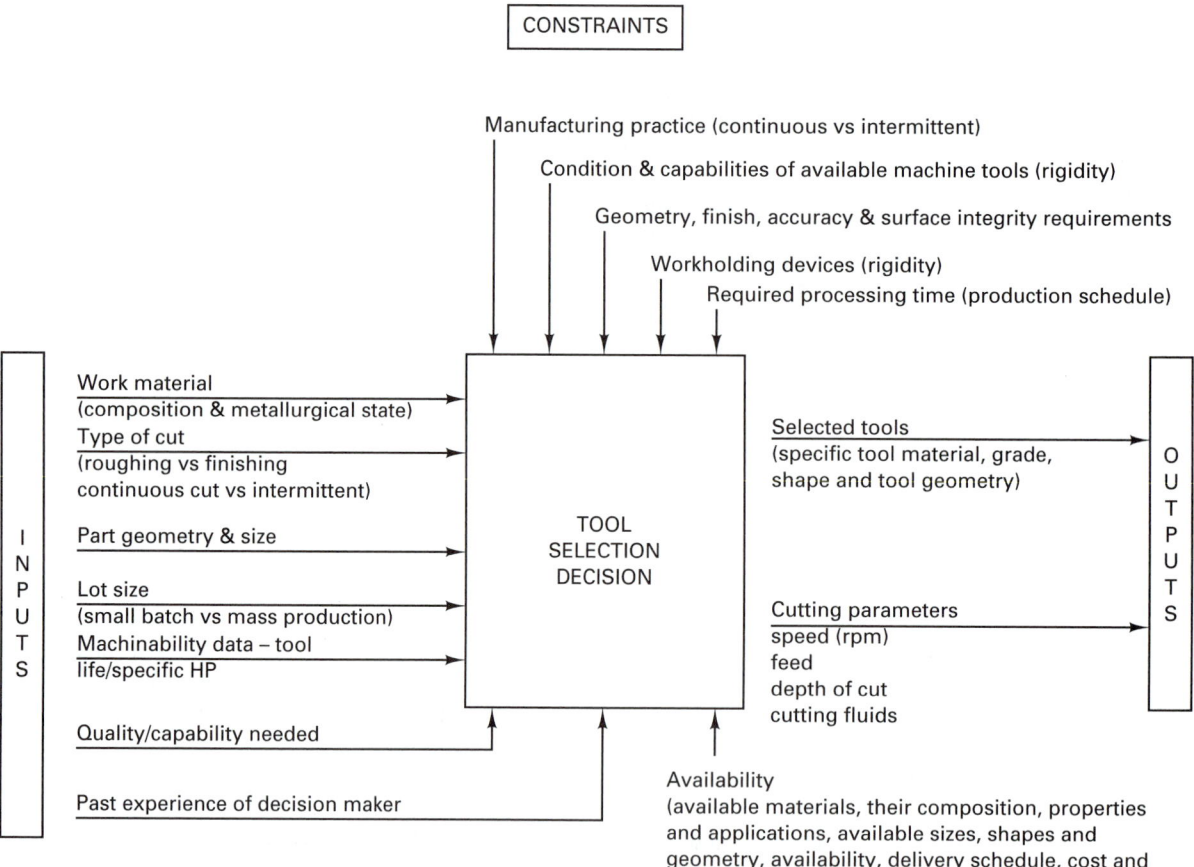

FIGURE 22-2   The selection of the cutting tool material and geometry and the cutting conditions for a given application depends on many variables.

Coated and uncoated high-speed steels and carbides are currently the most extensively used tool materials. Coated tools cost only about 15 to 20% more than uncoated tools, so even a modest improvement in performance can justify the added cost. About 15 to 20% of all tool steels are coated, mostly by the *physical vapor deposition* (PDV) processes. Diamond and CBN are used for applications in which, despite higher cost, their use is justified. Cast cobalt alloys are being phased out because of the high raw material cost and the increasing availability of alternative tool materials. New ceramic materials called *cermets* (ceramic material in a metal binder) are being introduced that will have significant impact on future manufacturing productivity.

Tool requirements for other processes that use noncontacting tools [as in electrodischarge machining (EDM) and electrochemical machining (ECM)] or *no tools at all* (as in laser machining) are discussed in Chapter 32. Grinding abrasives are discussed in Chapter 27.

Tool Steels

Carbon steels and low/medium-alloy steels, called *tool steels*, were once the most common cutting-tool materials. Plain-carbon steels of 0.90 to 1.30% carbon when hardened and tempered have good hardness and strength and adequate toughness and can be given a keen

TABLE 22-1. Salient Properties of Cutting Tool Materials[a]

	Carbon and Low/Medium-Alloy Steels	High-Speed Steels	Sintered (Cemented) Carbides	Coated HSS	Coated Carbides	Ceramics	Polycrystalline CBN	Diamond
Toughness	◄————————————— Decreasing —————————————►							
Hot hardness	◄————————————— Increasing —————————————►							
Impact strength	◄————————————— Decreasing —————————————►							
Wear resistance	◄————————————— Increasing —————————————►							
Chipping resistance	◄————————————— Decreasing —————————————►							
Cutting speed	◄————————————— Increasing —————————————►							
Depth of cut	Light to medium	Light to heavy	Light to heavy	Light to heavy	Light to heavy	Light to heavy	Light to heavy	Very light for single-crystal diamond
Finish obtainable	Rough	Rough	Good	Good	Good	Very good	Very good	Excellent
Method of manufacture	Wrought	Wrought cast, HIP sintering	Cold pressing and sintering, PM	PVD[b] after forming	CVD[c]	Cold pressing and sintering or HIP sintering	High-pressure–high-temperature sintering	High-pressure–high-temperature sintering
Fabrication	Machining and grinding	Machining and grinding	Grinding	Machining and grinding, coating	Grinding before coating	Grinding	Grinding and polishing	Grinding and polishing
Thermal shock resistance	◄————————————— increasing —————————————►							
Tool material cost	◄————————————— increasing —————————————►							

[a] Overlapping characteristics exist in many cases. Exceptions to the rule are very common. In many classes of tool materials a wide range of composition and properties is obtainable.
[b] Physical vapor deposition.
[c] Chemical vapor disposition.

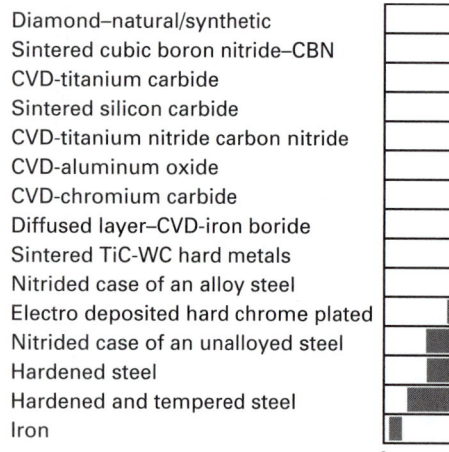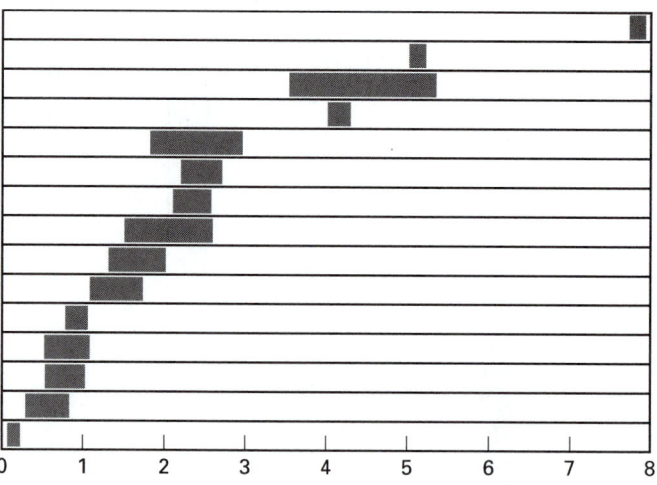

Diamond–natural/synthetic
Sintered cubic boron nitride–CBN
CVD-titanium carbide
Sintered silicon carbide
CVD-titanium nitride carbon nitride
CVD-aluminum oxide
CVD-chromium carbide
Diffused layer–CVD-iron boride
Sintered TiC-WC hard metals
Nitrided case of an alloy steel
Electro deposited hard chrome plated
Nitrided case of an unalloyed steel
Hardened steel
Hardened and tempered steel
Iron

Knoop hardness scale – 1,000 Kp/mm$_2$

FIGURE 22-3   Vickers hardness ranges for various cutting tool materials.

cutting edge. However, tool steels lose hardness at temperatures above 400°F because of tempering (refer to Figure 21-22) and have largely been replaced by other materials for metal cutting.

Low/medium-alloy steels have alloying elements such as Mo and Cr, which improve hardenability, and W and Mo, which improve wear resistance. These tool materials also lose their hardness rapidly when heated to about their tempering temperature of 300 to 650°F and they have limited abrasion resistance. Consequently, low/medium-alloy steels are used in relatively inexpensive cutting tools (e.g., drills, taps, dies, reamers, broaches, and chasers) for certain low-speed cutting applications when the heat generated is not high enough to reduce their hardness significantly. High-speed steels, cemented carbides, and coated tools are also used extensively to make these kinds of cutting tools. Although more expensive, they have longer tool life and improved performance.

High Speed Steels

First introduced in 1900 by Taylor and White, high-alloy steel was superior to tool steel in that it retained its cutting ability at temperatures up to 1100°F, exhibiting good "red hardness." Compared with carbon steel, it could operate at about double the cutting speed with equal life, resulting in its name *high-speed steel* (HSS).

Today's high-speed steels contain significant amounts of W, Mo, Co, V, and Cr besides Fe and C. W, Mo, Cr, and Co in the ferrite as a solid solution provide strengthening of the matrix beyond the tempering temperature, thus increasing the hot hardness. Vanadium (V), along with W, Mo, and Cr, improves hardness and wear resistance. Extensive solid solutioning of the matrix also ensures good hardenability of these steels.

Although many formulations are used (Table 22-2) a typical composition is that of the 18-4-1 type (tungsten 18%, chromium 4%, vanadium 1%), called T1. Comparable performance can also be obtained by the substitution of approximately 8% molybdenum for the tungsten, referred to as a tungsten equivalent (W_{eq}). High-speed steel is still widely used for drills and many types of general-purpose milling cutters and in single-point tools used in general machining. For high-production machining it has been replaced almost completely by carbides, coated carbides, and coated HSS.

TABLE 22-2. Nominal Percentage of Chemical Compositions of High-Speed Steels

AISI Tool Steel Type	Percentage of Chemical Composition[a]							
	C	Cr	V	W	Mo	Co	Cb	W_{eq}[b]
Tungsten High-Speed Steel								
T1[c]	0.70	4.0	1.0	18.0				18.0
T2[c]	0.85	4.0	2.0	18.0				18.0
T3	1.0	4.0	3.0	18.0	0.60			18.8
T4	0.75	4.0	1.0	18.0	0.60	5.0		19.2
T5	0.80	4.25	1.0	18.0	0.90	8.0		19.8
T6	0.80	4.25	1.5	2.0	0.90	12.0		21.8
T7	0.80	4.0	2.0	14.0				14.0
T8	0.80	4.0	2.0	14.0	0.90	5.0		15.8
T9	1.20	4.0	4.0	18.0				18.0
T15	1.55	4.50	5.0	12.0	0.60	5.0		13.2
Molybdenum High-Speed Steel								
M1[c]	0.80	4.0	1.00	1.5	8.0			17.5
M2[c]	0.85	4.0	2.00	6.0	5.0			16.0
M3	1.00	4.0	2.75	6.0	5.0			16.0
M4	1.30	4.0	4.00	5.5	4.5			14.5
M6	0.80	4.0	1.50	4.0	5.0	12.0		14.0
M7[c]	1.00	4.0	2.00	1.75	8.75			19.25
M8	0.80	4.0	1.50	5.0	5.0		1.25	15.0
M10	0.85	4.0	2.00		8.0			16.0
High-Hardness (Molydbenum-Based) Cobalt High-Speed Steels								
M30	0.85	4.0	1.25	2.0	8.0	5.0		18.0
M34	0.85	4.0	2.00	2.0	8.0	8.0		18.0
M35	0.85	4.0	2.00	6.0	5.0	5.0		16.0
M36	0.85	4.0	2.00	6.0	5.0	8.0		16.0
M41	1.10	4.25	2.00	6.75	3.75	5.0		20.5
M42	1.10	3.75	1.15	1.50	9.50	8.25		14.25
M43	1.20	3.75	1.60	2.75	8.00	8.25		18.75
M44	1.55	4.25	2.00	5.25	6.50	12.00		18.25
M45	1.25	4.25	1.60	8.25	5.0	5.50		18.25
M46	1.25	4.00	3.20	2.00	8.25	8.25		18.0

[a]Normal ranges of manganese, silicon, phosphorus, and sulfur are assumed. Balance Fe in all cases.

[b]$W_{eq} = 2 \, (\%Mo) + \%W$.

[c]Widely available.

High-speed-steel tools are fabricated by three methods: cast, wrought, and sintered (using the powder metallurgy technique). Improper processing of cast and wrought products can result in carbide segregation, formation of large carbide particles and significant variation of carbide size, and nonuniform distribution of carbides in the matrix. The material will be difficult to grind to shape and will cause wide fluctuations of properties, inconsistent tool performance, distortion, and cracking.

To overcome some of these problems, a powder metallurgy technique has been developed that uses the hot isostatic pressing (HIP) process on atomized, prealloyed tool steel mixtures. Because the various constituents of the PM alloys are "locked" in place by the compacting procedure, the end product is a more homogeneous alloy. PM high-speed-steel cutting tools exhibit better grindability, greater toughness, better wear resistance, and higher red (or hot) hardness, and they perform more consistently. They are about double the cost of regular HSS.

Cast Cobalt Alloys

Cast cobalt alloys, popularly known as *stellite tools*, are cobalt-rich, chromium–tungsten–carbon cast alloys having properties and applications in the intermediate range between high-speed steel and cemented carbides. Although comparable in room-temperature hardness to high-speed steel tools, cast cobalt alloy tools retain their hardness to a much higher temperature. Consequently, they can be used at higher cutting speeds (25% higher) than HSS tools. Cast cobalt alloys are hard as cast and cannot be softened or heat treated.

Cast cobalt alloys contain a primary phase of Co-rich solid solution strengthened by Cr and W and dispersion hardened by complex hard, refractory carbides of W and Cr. Other elements added include V, B, Ni, and Ta. The casting provides a tough core and elongated grains normal to the surface. The structure is not, however, homogeneous.

Tools of cast cobalt alloys are generally cast to shape and finished to size by grinding. They are available only in simple shapes, such as single-point tools and saw blades, because of limitations in the casting process and expense involved in the final shaping (grinding). The high cost of fabrication is due primarily to the high hardness of the material in the as-cast condition. Materials machinable with this tool material include plain-carbon steels, alloy steels, nonferrous alloys, and cast iron.

Cast cobalt alloys are currently being phased out for cutting-tool applications because of increasing costs, shortages of strategic raw materials (Co, W, and Cr), and the development of other, superior tool materials at lower cost.

Carbides or Sintered Carbides

Carbides, which are nonferrous alloys, are also called *sintered* (or cemented) carbides because they are manufactured by powder metallurgy techniques. See Figure 22-4 and Chapter 16 for details on powder metallurgy process. These materials became popular during World War II, as they afforded a four- or fivefold increase in cutting speeds. The early versions had tungsten carbide as the major constituent, with a cobalt binder in amounts of 3 to 13%. Most carbide tools in use today are either straight WC or multicarbides of W–Ti or W–Ti–Ta, depending on the work material to be machined. Cobalt is the binder. These tool materials are much harder, are chemically more stable, have better hot hardness, high stiffness, and lower friction, and operate at higher cutting speeds than do HSS. They are more brittle and more expensive and use strategic metals (W, Ta, Co) more extensively.

Cemented carbide tool materials based on TiC have been developed primarily for auto industry applications using predominantly Ni and Mo as a binder. These are used for higher-speed (>1000 ft/min) finish machining of steels and some malleable cast irons.

Cemented carbide tools are available in insert form in many different shapes; squares, triangles, diamonds, and rounds. They can be either brazed or clamped mechanically onto the tool shank. Mechanical clamping (Figure 22-5) is more popular because when one edge or corner becomes dull, the insert is rotated or turned over to expose a new cutting edge. Mechanical inserts can be purchased in the "as-pressed" state or the insert can be ground to closer tolerances. Naturally, precision-ground inserts cost more. Any part tolerance of less than ±0.003 in. normally cannot be manufactured without radial adjustment of the cutting tool, even with ground inserts. If no radial adjustment is performed, precision-ground inserts should be used only when the part tolerance is between ±0.003 and ±0.006 in. Pressed inserts have an application advantage, as the cutting edge is unground and thus does not leave grinding marks on the part after machining. Ground inserts can break under heavy cutting loads because the grinding marks on the insert produce stress concentrations that result in brittle fracture. Diamond grinding is used to finish carbide tools. Abusive grinding

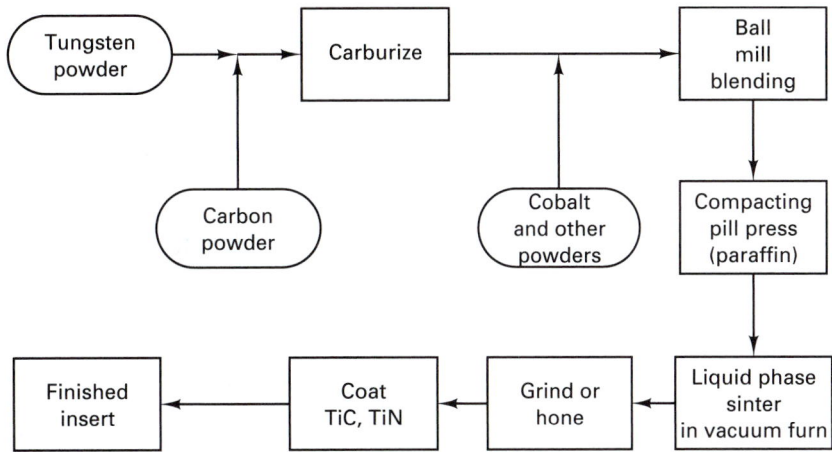

Tungsten is carburized in a high-temperature furnace, mixed with cobolt and blended in large ball mills. After ball milling, the powder is screened and dried. Paraffin is added to hold the mixture together for compacting. Carbide inserts are compacted using a pill press. The compacted powder is sintered in a high-temperature vacuum furnace. The solid cobalt dissolves some tungsten carbide, then melts and fills the space between adjacent tungsten carbide grains. As the mixture is cooled, most of the dissolved tungsten carbide precipitates onto the surface of existing grains. After cooling, inserts are finish ground and honed or used in the pressed condition.

FIGURE 22-4   PM process for making cemented carbide insert tools.

can lead to thermal cracks and premature (early) failure of the tool. Brazed tools have the carbide insert brazed to the steel tool shank. These tools will have a more accurate geometry than the mechanical insert tools but are more expensive.

Since cemented carbide tools are relatively brittle, a 90° corner angle at the cutting edge is desired. To strengthen the edge and prevent edge chipping, it is rounded off by honing, or an appropriate chamfer or a negative land on the rake face is provided. The preparation of the cutting edge can affect tool life. The sharper the edge (smaller edge radius), the more likely it is to chip or break. Increasing the edge radius will increase the cutting forces, so a trade-off is required. Typical edge radius values are 0.001 to 0.003 in.

A *chip grove* (section A–A of Figure 22-5) with a positive rake angle at the tool tip may also be used to reduce cutting forces without reducing the overall strength of the insert significantly. The groove also breaks up the chips for easier disposal by causing them to curl tightly.

For very low speed cutting operations, the chips tend to weld to the tool face and cause subsequent microchipping of the cutting edge. Cutting speeds are generally in the range 150 to 600 ft/min. Higher speeds (>1000 ft/min) are recommended for certain less difficult-to-machine materials (such as aluminum alloys), and much lower speeds (100 ft/min) for more difficult-to-machine materials (such as titanium alloys). In interrupted cutting applications, it is important to prevent edge chipping by choosing the appropriate cutter geometry and cutter position with respect to the workpiece. For interrupted cutting, finer grain size and higher cobalt content improve toughness in straight WC–Co grades.

After use, carbide inserts (called *disposable* or *throwaway inserts*) are generally recycled by separating Ta, WC, and Co. This recycling not only conserves strategic materials but also reduces costs. A new trend is to regrind these tools for future use where the actual size of the insert is not of critical concern.

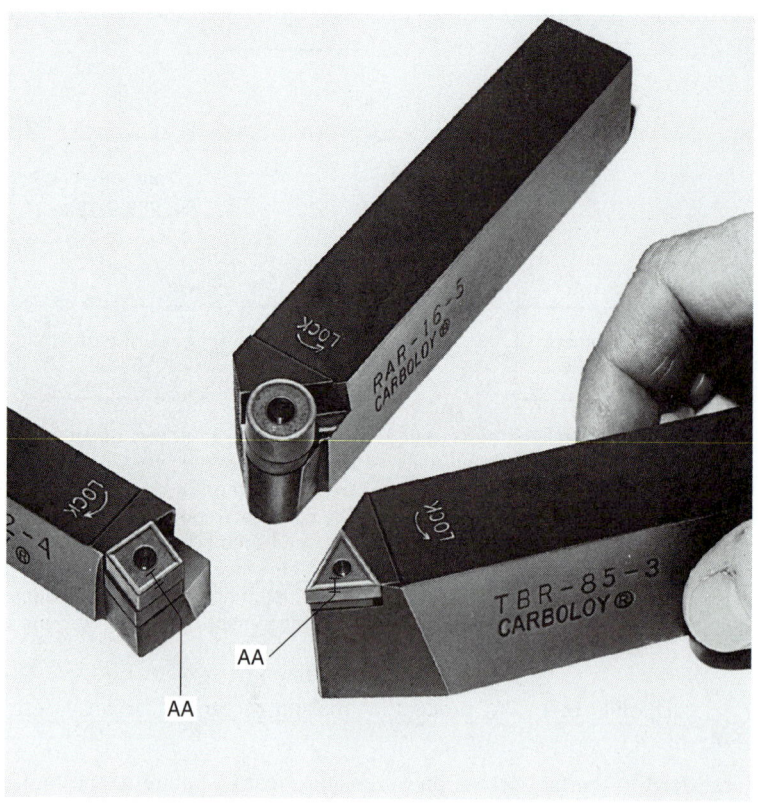

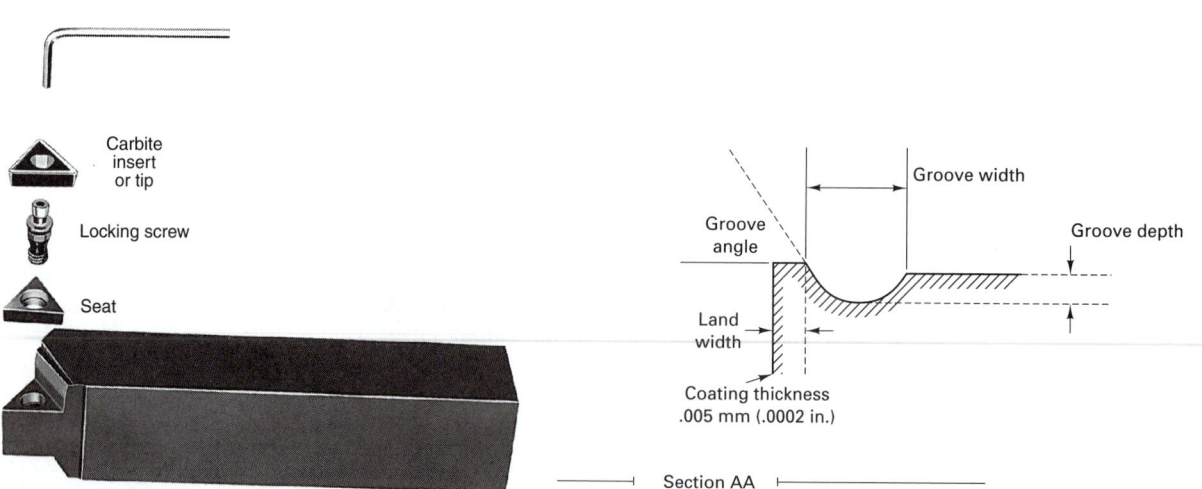

FIGURE 22-5 (Top) Examples of throwaway carbide cutting tool tip with chipbreaker grooves on rake face; (Left) components of a typical mounting holder; (Section AA) groove design on coated tool to reduce forces and breakup chips. *(Courtesy of General Electric.)*

Ceramics

Ceramics are made of pure *aluminum oxide*, Al_2O_3. Very fine particles are PM formed into cutting tips under a pressure of 20 to 28 tons/in^2 (267 to 386 MPa) and sintered at about 1800°F (1000°C). Unlike the case with ordinary ceramics, sintering occurs without a vitreous phase.

Ceramics usually are in the form of disposable tips. They can be operated at from two to three times the cutting speeds of tungsten carbide, almost completely resist cratering, usually require no coolant, and have about the same tool life at these higher speeds as tungsten carbide does at lower speeds. As shown in Table 22-3, ceramics are usually as hard as carbides but are more brittle (lower bend strength) and therefore require more rigid toolholders and machine tools to take advantage of their capabilities. Their hardness and chemical inertness make ceramics a good material for high-speed finishing and/or high-removal-rate machining applications of superalloys, hard-chill cast iron, and high-strength steels. Because ceramics have poor thermal and mechanical shock resistance, interrupted cuts and interrupted application of coolants can lead to premature tool failure. Edge chipping is usually the dominant mode of tool failure. Ceramics are not suitable for aluminum, titanium, and other materials that react chemically with alumina-based ceramics. Recently, whisker-reinforced ceramic materials that have greater transverse rupture strength have been developed. The whiskers are made from silicon carbide.

TABLE 22-3. Properties of Cutting-Tool Materials Compared for Carbides, Cast Cobalt Ceramics, and HSS[a]

	Hardness, Rockwell A or C	Transverse Rupture (Bend) Strength ($\times 10^3$ psi)	Compressive Strength ($\times 10^3$ psi)	Modulus of Elasticity, E ($\times 10^6$ psi)
Carbide C1–C4	90–95R$_A$	250–320	750–860	89–93
Carbide C5–C8	91–93R$_A$	100–250	710–840	66–81
Cast cobalt	46–62R$_C$	80–120	220–335	40
Ceramic (oxide)	92–94R$_A$	100–125	400–650	50–60
High-speed steel	86R$_A$	600	600–650	30

[a]Exact properties depend upon materials, grain size, bonder content, and volume.

Coated Carbide Tools

Coated tools are becoming the norm in the metalworking industry because coating can consistently improve tool life 200 or 300% or more. In cutting tools, material requirements at the surface of the tool need to be abrasion resistant, hard, and chemically inert to prevent the tool and the work material from interacting chemically with each other during cutting. A thin, chemically stable, hard refractory coating of TiC, TiN, or Al_2O_3 accomplishes this objective. The bulk of the tool is a tough, shock-resistant carbide that can withstand high-temperature plastic deformation and resist breakage. The result is a composite tool as shown in Figure 22-6.

To be effective, the coatings should be hard, refractory, chemically stable, and chemically inert to shield the constituents of the tool and the workpiece from interacting chemically under cutting conditions. The coatings must be fine grained, free of binders and porosity. Naturally, the coatings must be metallurgically bonded to the substrate. Interface coatings are graded to match the properties of the coating and the substrate. The coatings must be thick enough to prolong tool life but thin enough to prevent brittleness.

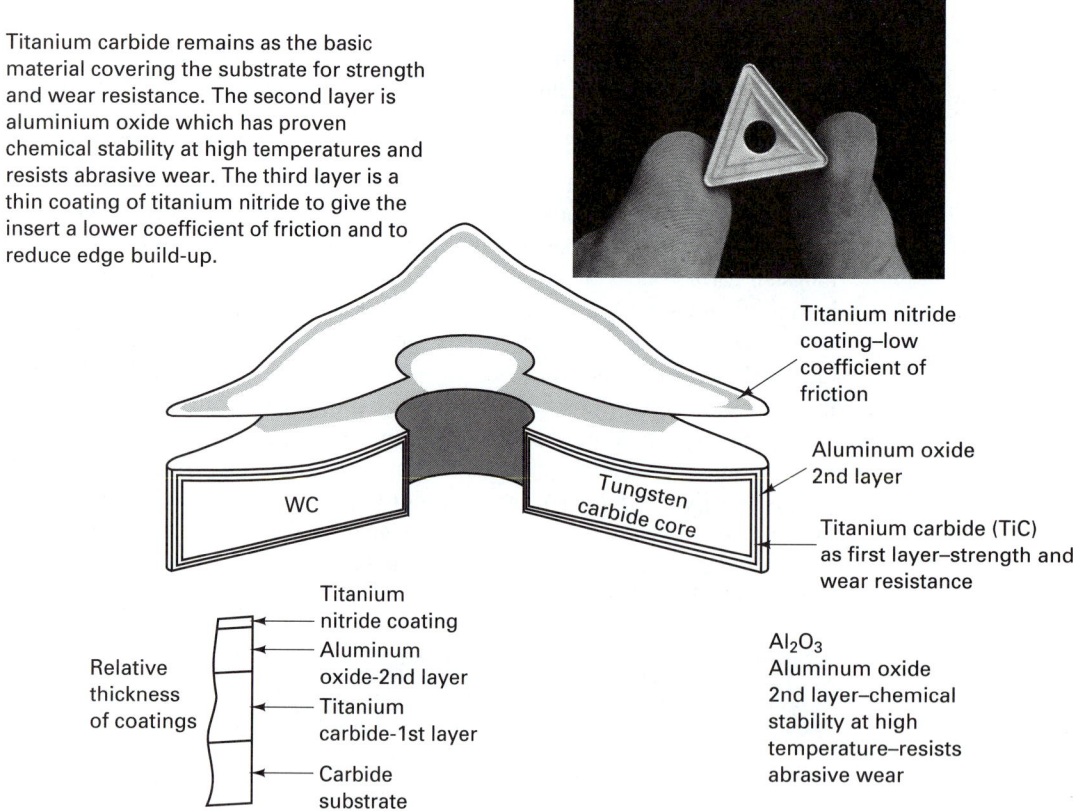

Titanium carbide remains as the basic material covering the substrate for strength and wear resistance. The second layer is aluminium oxide which has proven chemical stability at high temperatures and resists abrasive wear. The third layer is a thin coating of titanium nitride to give the insert a lower coefficient of friction and to reduce edge build-up.

Titanium nitride coating–low coefficient of friction

Aluminum oxide 2nd layer

Titanium carbide (TiC) as first layer–strength and wear resistance

WC

Tungsten carbide core

Al_2O_3 Aluminum oxide 2nd layer–chemical stability at high temperature–resists abrasive wear

Relative thickness of coatings

Titanium nitride coating

Aluminum oxide-2nd layer

Titanium carbide-1st layer

Carbide substrate

FIGURE 22-6   Triple-coated carbide tools provide resistance to wear and plastic deformation in machining of steel, abrasive wear in cast iron, and built-up edge formation.

Coatings should have a low coefficient of friction so that the chips do not adhere to the rake face. TiC-coated tools were introduced in 1969. Coating materials now include single coatings of TiC, TiN, Al_2O_3, HfN, or HfC. Multiple coatings are used, with each layer imparting its own characteristic to the tool. The most successful combinations are TiN/TiC/TiCN/TiN and TiN/TiC/Al_2O_3. Chemical vapor deposition (CVD) is the technique used to coat carbides. (See Figure 22-7 and Table 22-4 for additional details on coating processes.) The coatings are formed by chemical reactions that take place only on or near the substrate. Like electroplating, chemical vapor deposition is a process in which the deposit is built up atom by atom. It is therefore capable of producing deposits of maximum density and of closely reproducing fine detail on the substrate surface. Figure 22-7 illustrates a typical TiC coating apparatus. A mixture of hydrogen, methane, and titanium tetrachloride gases is formed in the gas-mixing chamber. The gas mixture is routed into the coating chamber, where the carbide tool blanks are heated to a temperature of about 1000°C. The tungsten carbide catalyzes the reaction that produces the TiC coating:

$$TiCl_4 + CH_4 \rightarrow TiC \times 4HCl \uparrow \qquad (21\text{-}1)$$

By substituting nitrogen for methane, a *titanium nitride* (TiN) coating is produced.

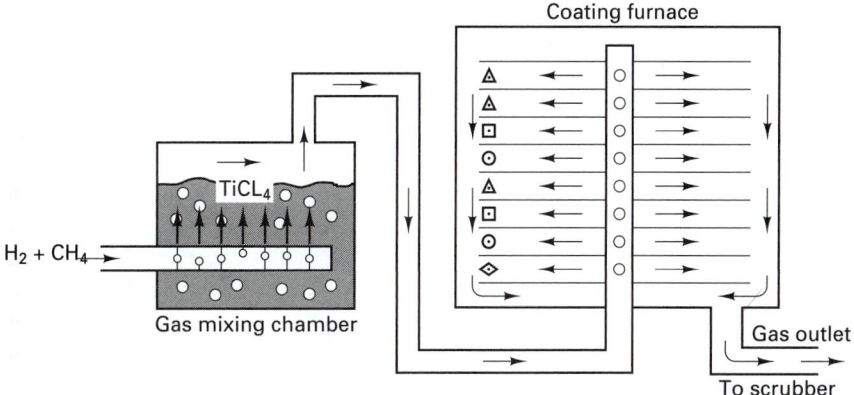

FIGURE 22-7   Chemical vapor deposition (CVD) is used to apply coatings (TiC, TiN, etc.) to carbide cutting tools.

Control of critical variables such as temperature, gas concentration, and flow pattern is required to assure adhesion of the coating to the substrate. The coating-to-substrate adhesion must be better for cutting tool inserts than for most other coatings applications to survive the cutting pressure and temperature conditions without flaking off. Grain size and shape are controlled by varying temperature and/or pressure.

The purpose of multiple coatings is to tailor the coating thickness for prolonged tool life. Multiple coatings allow a stronger metallurgical bond between the coating and the substrate and provide a variety of protection processes for machining different work materials, thus offering a more general-purpose tool material grade. A very thin (5 μm) final TiN coating can effectively reduce crater formation on the tool face by one to two orders of magnitude relative to uncoated tools.

Coated inserts of carbides are finding wide acceptance in many metal-cutting applications. Coated tools have two or three times the wear resistance of the best uncoated tools with the same breakage resistance. This results in a 50 to 100% increase in speed for the same tool life. Because most coated inserts cover a broader application range, fewer grades are needed and thus inventory costs are lower. Aluminum oxide coatings have demonstrated excellent crater wear resistance by providing a chemical/diffusion reaction barrier at the tool/chip interface, permitting a 90% increase in cutting speeds in machining some steels. Coated carbide tools have progressed to the place where in the United States about 80 to 90% of the carbide tools used in metalworking are coated.

TiN-Coated High-Speed Steel

Coated high-speed steel (HSS) does not routinely provide as dramatic improvements in cutting speeds as do coated carbides, with increases of 10 to 20% being typical. First introduced in 1980 for gear cutters (hobs) and in 1981 for drills, TiN-coated HSS tools have demonstrated their ability to more than pay for the extra cost of the coating process.

In addition to hobs, gear-shaper cutters, and drills, HSS tooling coated by TiN now includes reamers, taps, chasers, spade-drill blades, broaches, bandsaw and circular saw blades, insert tooling, form tools, end mills, and an assortment of other milling cutters.

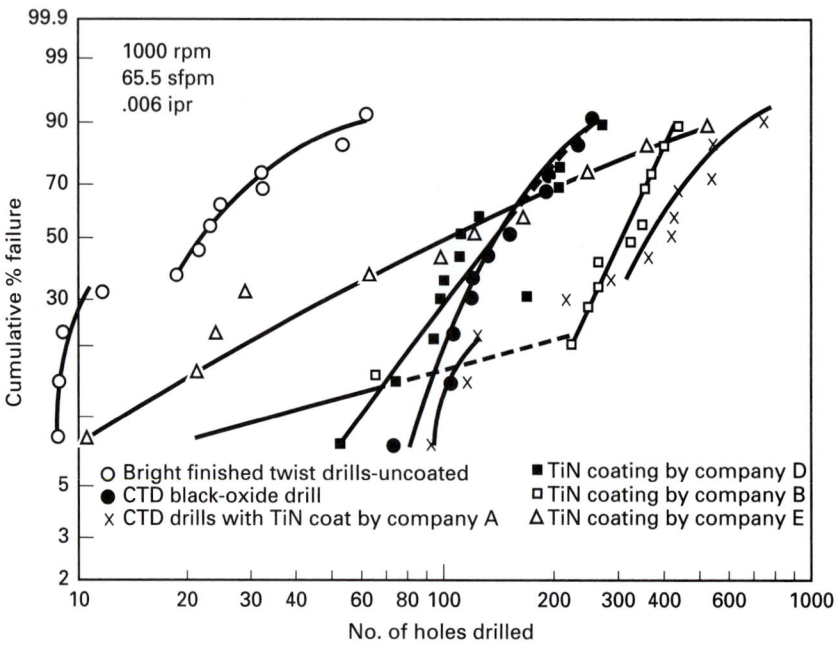

Drill performance based on the number of holes drilled with 1/4 inch diam. drills in T–1 structural steel.

FIGURE 22-8   Tool life data for various coated drills.   Notice how TiN-coated HSS drills outperform uncoated drills.

Physical vapor deposition (PVD) has proved to be the best process for coating HSS, primarily because it is a relatively low-temperature process that does not exceed the tempering point of HSS.   Therefore, no subsequent heat treatment of the cutting tool is required.   Films 0.0001 to 0.0002 in. in thickness adhere well and withstand minor elastic, plastic, and thermal loads.   Thicker coatings tend to fracture under the typical thermomechanical stresses of machining.

There are many variations to the physical vapor deposition (PVD) process, as outlined in Table 22-4.   The process usually depends on gas pressure and is performed in a vacuum chamber.   PVD processes are carried out with the workpieces heated to temperatures in the range 400 to 900°F.   Substrate heating enhances coating adhesion and film structure.

Because surface pretreatment is critical in PVD processing, tools to be coated are subject to a vigorous cleaning process.   Precleaning methods typically involve degreasing, ultrasonic cleaning, and Freon drying.   Deburring, honing, and more active cleaning methods are also used.

The advantage of TiN-coated HSS tooling is reduced tool wear.   Less tool wear results in less stock removal during tool regrinding, thus allowing individual tools to be reground more times.   For example, a TiN hob can cut 300 gears per sharpening; the uncoated tool would cut only 75 parts per sharpening.   Therefore, the cost per gear is reduced from 20 cents to 2 cents.   Naturally, reduced tool wear means longer tool life, as shown in Figure 22-8.   Tenfold increases in tool life have been reported with MRR values doubled.

TABLE 22-4.   Surface Treatments for Cutting Tools

Process	Method	Hardness[a] and Depth	Advantages	Limitations
Black oxide	HSS cutting tools are oxidized in a steam atmosphere at 1000°F	No change in prior steel hardness	Prevents built-up edge formations in machining of steel	Strictly for HSS tools
Nitriding case hardening	Steel surface is coated with nitride layer by use of cyanide salt at 900 to 1600°F, or ammonia, gas, or N_2 ions.	To $72R_C$; case depth: 0.0001–0.100 in.	High production rates with bulk handling; high surface hardness; diffuses into the steel surfaces; simulates strain hardening.	Can only be applied to steel; process has embrittling effect because of greater hardness; postheat treatment needed for some alloys.
Electrolytic electroplating	The part is the cathode in a chromic acid solution; anode is lead. Hard chrome plating is the most common process for wear resistance.	70–$72R_C$; 0.0002–0.100 in.	Low friction coefficient, antigalling; corrosion resistance; high hardness	Moderate production; pieces must be fixtured; part must be very clean; coating does not diffuse into surface, which can affect impact properties
Vapor deposition, chemical vapor deposition (CVD)	Deposition of coating material by chemical reactions in the gaseous phase. Reactive gases replace a protective atmosphere in a vacuum chamber. At temperatures of 1800–1200°F, a thin diffusion zone is created between the base metal and the coating.	To $84R_C$; 0.0002–0.0004 in.	Large quantities per batch; short reaction times reduce substrate stresses; excellent adhesion, recommended for forming tools; multiple coatings can be applied (TiN, TiC, Al203); line of sight not a problem.	High temperatures can affect substrate metallurgy, requiring post heat treatment, which can cause dimensional distortion (except when coating somtered carbodes); necessary to reduce effects of hydrogen chloride on material properties, such as impact strength; usually not diffused; tolerances of +0.001 in. required for HSS tools
Physical vapor deposition (PVD sputtering)	Plasma is generated in a vacuum chamber by ion bombardment to dislodge particles from a target made of the coating material. Metal is evaporated and is condensed or attracted to substrate surfaces.	To $84R_C$; to 0.0002 in. thick	A useful experimental procedure for developing wear surfaces; can coat substrates with metals, alloys, compounds, and refractories; applicable for all tooling.	Not a high production method; requires care in cleaning; usually not diffused
PVD (electron beam)	A plasma is generated in vacuum by evaporation from a molten pool, which is heated by an electron-beam gun.	To $84R_C$; to 0.0002 in. thick	Can coat reasonable quantities per batch cycle; coating materials are metals, compounds, alloys, and refractories; substrate metallurgy is preserved; very good adhesion; fine particle deposition; applicable for all tooling.	Parts require fixturing and orientation in line-of-sight process; ultracleanliness required.
PVD/ARC	Titanium is evaporated in a vacuum and reacted with nitrogen gas. Resulting titanium nitride plasma is ionized and electrically attracted to the substrate surface. A high-energy process with multiple plasma guns.	To $85R_C$; to 0.0002 in. thick	Process at 900 °F preserves substrate metallurgy; excellent coating adhesion; controllable deposition of grain size and growth; dimensions, surface finish, and sharp edges are preserved; can coat all high speed steels without distortion.	Parts must be fixtured for line-of-sight process; parts must be very clean; no by-products formed in reaction; usually only minor diffusion

[a] Rockwell hardness values above 68 are estimates.

Higher hardness, with typical values for the thin coatings, "equivalent" to $R_C 80$ to 85, versus $R_c 65$ to 70 for hardened HSS, means reduced abrasion wear. Relative inertness (i.e., TiN does not react significantly with most workpiece materials) results in greater tool life via a reduction in adhesion. TiN coatings have a low coefficient of friction at the tool/chip interface. This results in an increase in the shear angle, which reduces the cutting forces, spindle power, and heat generated by the deformation processes. PVD coatings generally fail in high-stress applications such as cold extrusion, piercing, roughing, and high speed machining.

Cermets

Cermets are a new class of tool materials best suited for finishing. Cermets are ceramic materials in a metal binder. TiC, nickel, cobalt, tantalum nitrides, TiN, and other carbides are used for binders. Cermets have higher hot hardness and oxidation resistance than do cemented carbides. The better surface finish imparted by a cermet is due to its low level of chemical reaction with iron [less cratering and *built-up edge* (BUE)]. Compared to carbide, the cermet has less toughness, lower thermal conductivity, and greater thermal expansion, so thermal cracking can be a problem during interrupted cuts.

Cermets are usually cold pressed, and proper processing techniques are required to prevent insert cracking. New cermets are designed to resist thermal shocking during milling by using high nitrogen content in the titanium carbonitride phase (produces finer grain size) and adding WC and TaC to improve shock resistance.

Figure 22-9 shows a comparison of speed/feed coverage of typical cermets compared to ceramics, carbides, and coated carbides. The values illustrate that cermets can clearly cover a wide range of important metalcutting applications.

Diamonds

Diamond is the hardest material known. Industrial diamonds are now available in the form of polycrystalline compacts, which are finding industrial application in the machining of aluminum, bronze, and plastics, greatly reducing the cutting forces compared to carbides. Diamond machining is done at high speeds with fine feeds for finishing and produces excellent surface finishes. Recently, *single-crystal* diamonds, with a cutting-edge radius of 100 Å or less, have been used for precision machining of large curved mirrors. However, single-crystal diamonds have been used for years to machine brass watch faces, thus eliminating polishing. They have also been used to slice biological materials into thin films for viewing in transmission electron microscopes. (This process, known as ultramicrotomy, is one of the few industrial versions of orthogonal machining in common practice.)

The salient features of diamond tools include high hardness, good thermal conductivity, the ability to form a sharp edge of cleavage (single-crystal, natural diamond), very low friction, nonadherence to most materials, the ability to maintain a sharp edge for a long period of time, especially in machining soft materials such as copper and aluminum, and good wear resistance.

To be weighed against these advantages are some shortcomings, which include a tendency to interact chemically with elements of group IVB to group VIII of the periodic table. In addition, diamond wears rapidly when machining or grinding mild steel. It wears less rapidly with high-carbon-alloy steels than with low-carbon steel, and has occasionally machined gray cast iron (which has a high carbon content) with long life. Diamond has a tendency to revert at high temperatures (700°C) to graphite and/or to oxidize in air. Diamond is very brittle and is difficult and costly to shape into cutting tools.

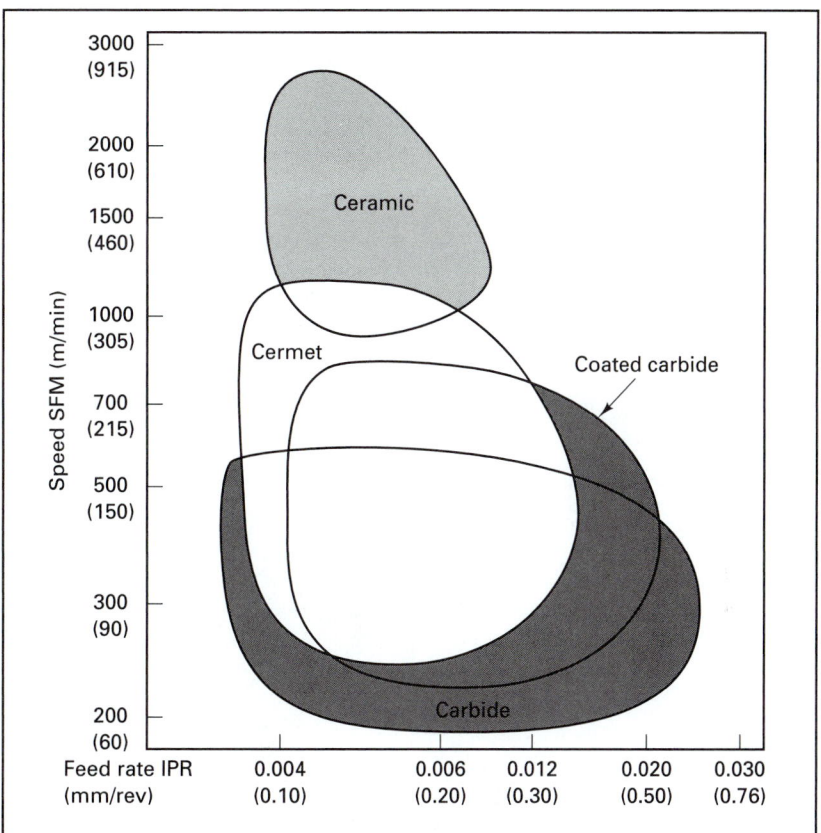

FIGURE 22-9   Cermets compared to other tool materials in terms of possible cutting speed and feed rates. SFM, surface feet per minute; ipr, inches per revolution.

Limited supply, increasing demand, and high cost of natural diamonds led to the ultrahigh-pressure (50 kbar), high-temperature (1500°C) synthesis of diamond from graphite at the General Electrical Company in the mid-1950s, and the subsequent development of *polycrystalline* sintered diamond tools in the late 1960s.

Polycrystalline diamond (PCD) tools consist of a thin layer (0.5 to 1.5 mm) of fine-grain-size diamond particles sintered together and metallurgically bonded to a cemented carbide substrate. A high-temperature/high-pressure process, using conditions close to those used for the initial synthesis of diamond, is needed. Fine diamond powder (1 to 30 μm) is first packed on a support base of cemented carbide in the press. At the appropriate sintering conditions of pressure and temperature in the diamond stable region, complete consolidation and extensive diamond-to-diamond bonding take place. Sintered diamond tools are then finished to shape, size, and accuracy by laser cutting and grinding (Figure 22-10). The cemented carbide provides the necessary elastic support for the hard and brittle diamond layer above it. The main advantages of sintered polycrystalline tools over natural single-crystal tools are better quality, greater toughness, and improved wear resistance, resulting from the random orientation of the diamond grains and the lack of large cleavage planes.

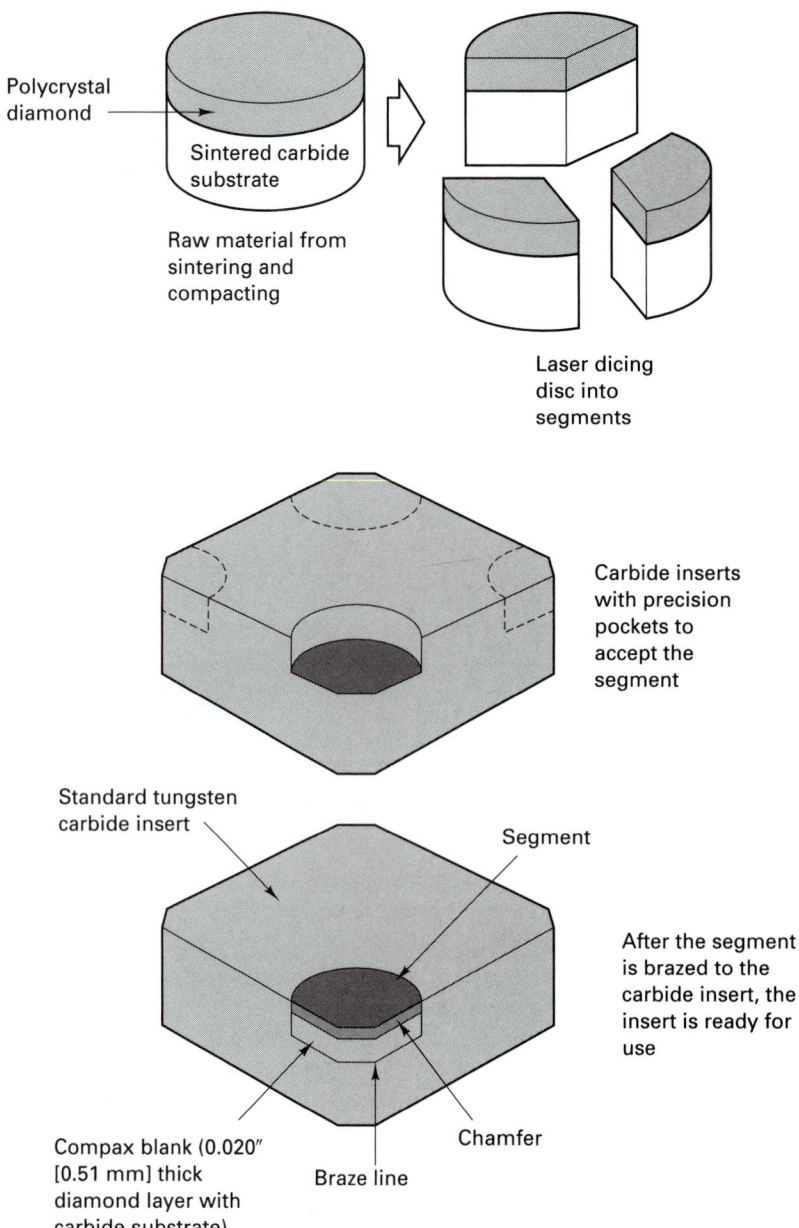

Polycrystal diamond

Sintered carbide substrate

Raw material from sintering and compacting

Laser dicing disc into segments

Carbide inserts with precision pockets to accept the segment

Standard tungsten carbide insert

Segment

After the segment is brazed to the carbide insert, the insert is ready for use

Compax blank (0.020″ [0.51 mm] thick diamond layer with carbide substrate)

Braze line

Chamfer

FIGURE 22-10   Polycrystalline diamond tools are carbides with diamond inserts. They are restricted to simple geometries.

Sintered polycrystalline diamond tools are more expensive (an order of magnitude or more) than conventional cemented carbides or ceramic tools because of the high cost of both the processing technique and the finishing methods. This is also the case with sintered polycrystalline cubic boron nitride, described in the next section. Despite the higher cost, many applications have been found for these tools because they increase productivity, provide long tool life, and can be economical on the basis of overall cost per part.

Diamond tools offer dramatic performance improvements over carbides. Tool life is often greatly improved, as is control over part size, finish, and surface integrity.

Positive rake tooling is recommended for the vast majority of diamond tooling applications. If BUE is a problem, increasing cutting speed and the use of more positive rake angles may eliminate it. If edge breakage and chipping are problems, one can reduce the feed rate. Coolants are not generally used in diamond machining unless, as in the machining of plastics, it is necessary to reduce airborne dust particles. Diamond tools can be reground.

There is much commercial interest in being able to coat HSS and carbides directly with diamond, but getting the diamond coating to adhere reliably has been difficult. Diamond-coated inserts would deliver roughly the same performance as PCD tooling when cutting nonferrous materials but could be given more complex geometries and chip breakers while reducing the cost per cutting edge.

Polycrystalline Cubic Boron Nitrides

Polycrystalline cubic boron nitride (PCBN) is a man-made tool material widely used in the automotive industry for machining hardened steels and superalloys. It is made in a compact form for tools by a process quite similar to that used for sintered polycrystalline diamonds. It retains its hardness at elevated temperatures (Knoop 4700 at 20°C, 4000 at 1000°C) and has low chemical reactivity at the tool/chip interface. This material can be used to machine hard aerospace materials such as Inconel 718 and René 95 as well as chilled cast iron.

Although not as hard as diamond, PCBN is less reactive with such materials as hardened steels, hard-chill cast iron, and nickel- and cobalt-based superalloys. PCBN can be used efficiently and economically to machine these difficult-to-machine materials at higher speeds (fivefold) and with a higher removal rate (fivefold) than cemented carbide, and with superior accuracy, finish, and surface integrity. PCBN tools are available in basically the same sizes and shapes as sintered diamond and are made by the same process. The cost of an insert is somewhat higher than either cemented carbide or ceramic tools, but the tool life may be five to seven times that of a ceramic tool. Therefore, to see the economy of using PCBN tools, it is necessary to consider all the factors. (See Section 22.6 and the problems at the end of the chapter.)

The two predominant wear modes of PCBN tools are notching at the *depth-of-cut line* (DCL) and *microchipping*. In some cases, the tool will exhibit flank wear of the cutting edge. These tools have been used successfully for heavy interrupted cutting and for milling white cast iron and hardened steels using negative lands and honed cutting edges.

Since diamond and PCBN are extremely hard but brittle materials, new demands are being placed on the machine tools and on machining practice to take full advantage of the potential of these tool materials. These demands include:

- Use of more rigid machine tools, and machining practices involving gentle entry and exit of the cut to prevent microchipping
- Use of high-precision machine tools, because these tools are capable of producing high finish and greater accuracy
- Use of machine tools with higher power, because these tools are capable of higher removal rates and faster spindle speeds

■ 22.3 TOOL GEOMETRY

For cutting tools, geometry depends mainly on the properties of the tool material and the work material. The standard terminology is shown in Figure 22-11 for single-point tools. The most important angles are the rake angles and the end and side relief angles.

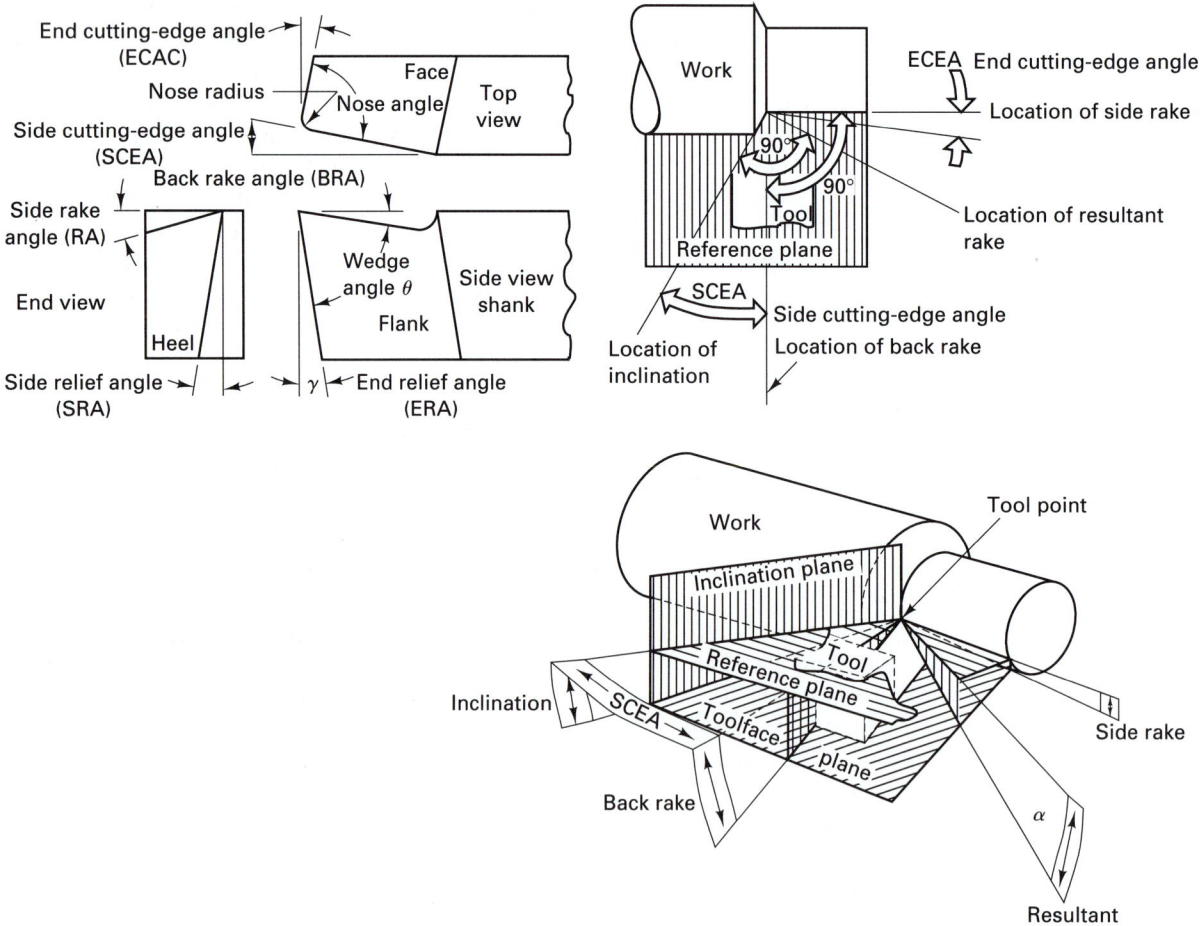

FIGURE 22-11   Standard terminology to describe the geometry of single-point tools.

The *back rake angle* affects the ability of the tool to shear the work material and form the chip. It can be positive or negative. Positive rake angles reduce the cutting forces, resulting in smaller deflections of the workpiece, toolholder, and machine. In machining hard work materials, the back rake angle must be small, even negative for carbide and diamond tools. Generally speaking, the higher the hardness, the smaller the back rake angle. For high-speed steels, back rake angle is normally chosen in the positive range, depending on the type of tool (turning, planing, end milling, face milling, drilling, and so on) and the work material.

For carbide tools, inserts for different work materials and toolholders can be supplied with several standard values of back rake angle; -6° to +6°. The side rake angle and the back rake angle combine to form the effective rake angle (See Figure 22-11). This is also called the true rake angle or resultant rake angle of the tool.

True rake inclination of a cutting tool has a major effect in determining the amount of chip compression and the shear angle. A small rake angle causes high compression, tool forces, and friction, resulting in a thick, highly deformed, hot chip. Increased rake angle reduces the compression, the forces, and the friction, yielding a thinner, less-deformed and cooler chip. Unfortunately, it is difficult to take much advantage of these desirable effects of larger positive rake angles, since they are offset by the reduced strength of the cutting tool, due to the reduced tool section, and by its greatly reduced capacity to conduct heat away from the cutting edge.

To provide greater strength at the cutting edge and better heat conductivity, zero or negative rake angles commonly are employed on carbide, ceramic, polydiamond, and PCBN cutting tools. These materials tend to be brittle, but their ability to hold their superior hardness at high temperatures results in their selection for high-speed and continuous machining operations. The negative rake angle increases tool forces, but this is necessary to provide the added support to the cutting edge. This is particularly important in making intermittent cuts and in absorbing the impact during the initial engagement of the tool and work.

In general, the power consumption is reduced by approximately 1% for each 1° in α. γ is the end relief angle. The wedge angle θ determines the strength of the tool and its capacity to conduct heat and depends on the values of α and γ. The relief angles mainly affect the tool life and the surface quality of the workpiece. To reduce the deflections of the tool and the workpiece and to provide good surface quality, larger relief values are required. For high-speed steel, relief angles in the range 5 to 10° are normal, with smaller values being for the harder work materials. For carbides, the relief angles are lower to give added strength to the tool.

The side and end-cutting edge angles define the nose angle and characterize the tool design. The nose radius has a major influence on surface finish. Increasing the nose radius usually decreases tool wear and improves surface finish.

Tool nomenclature varies with different cutting tools, manufacturers, and users. Many terms are still not standard because of all this variety. The most common tool terms will be used in later chapters to describe specific cutting tools.

The introduction of coated tools has spurred the development of improved tool geometries. Specifically, *low-force groove* (LFG) geometries have been developed that reduce the total energy consumed and break up the chips into shorter segments (see Figure 22-5). The effect of these grooves is effectively to increase the rake angle, which increases the shear angle and lowers the cutting force and power. This means that higher cutting speeds or lower cutting temperatures (and better tool lives) are possible.

As a chip breaker, the groove deflects the chip at a sharp angle and causes it to break into short pieces that are easier to remove and are not so likely to become tangled in the machine and possibly cause damage to personnel. This is particularly important on high-speed, mass-production machines.

The shapes of cutting tools used for various operations and materials are compromises, resulting from experience and research so as to provide good overall performance. Table 22-5 gives representative rake angles and suggested cutting speeds. For coated tools, edge strength is an important consideration. A thin coat enables the edge to retain high strength, but a thicker coat exhibits better wear resistance. Normally, tools for turning have a thickness of 6 to 12 μm. Edge strength is higher for multilayer coated tools. The radius of the edge should be 0.0005 to 0.005 in.

■ 22.4 TOOL FAILURE AND TOOL LIFE

In metal cutting, the failure of the cutting tool can be classified into two broad categories according to the failure mechanisms that caused the tool to die (or fail):

TABLE 22-5. Representative Machining Conditions for Various Work and Tool–Material Combinations[a]

Work Material	Tool	Rake Angles (deg) Back	Rake Angles (deg) Side	Cutting Speed m/min	Cutting Speed ft/min
B112 steel	HSS	16	22	69	225
	WC	0	3	168	550
	Ceramic	−5	−5	427	1400
4140 steel	HSS	12	14	40	130
	WC	0	3	91	300
	Ceramic	−5	−5	274	900
8620 steel	HSS		Uncoated		100
	WC		Uncoated		400
	WC		Coated with TiC		600
	WC		Coated with Al_2O_3		1100
	WC, Al_2O_3, with LFG				1300
18-8 steel (stainless)	HSS	8	14	27	90
	WC	4	8	84	275
	Ceramic	−5	−5	152	500
Gray cast iron (medium)	HSS	5	12	34	100
	WC	0–4	2–4	69	225
	Ceramic	−5	−5	244	800
Brass (free machining)	HSS	0	0	76	250
	WC	0	4	221	725
Aluminum alloys	HSS	35	15	91	300+
	WC	10–20	10–20	122	400+
Magnesium alloys	HSS	0	10	91	300+
	WC	10	10	213	700+
Titanium (turning)	WC	0	5	46	150

[a]Table value typical for
lathe turning operation,
single-point tool. Feed: 0.38 mm/rev (0.025 in./rev); depth: 3.18 mm (0.125 in.).

1. *Slow-death mechanisms:* gradual tool wear on the flank(s) of the tool or on the rake face of tool (called *crater wear*) or both
2. *Sudden-death mechanisms:* rapid, usually unpredictable and often catastrophic failures resulting from abrupt, premature death of a tool

Figure 22-12 shows a sketch of a "worn" tool, showing *crater wear* and *flank wear*, along with wear of the tool nose radius and an outer diameter groove (the DCL groove). As the tool wears, its geometry changes. This geometry change will influence the cutting forces, the power being consumed, the surface finish obtained, the dimensional accuracy, and even the dynamic stability of the process, as worn tools often chatter in processes that are usually relatively free of vibration. The actual wear mechanisms active in this high-temperature environment are abrasive, adhesion, diffusion, or chemical interactions. It appears that in metal cutting, any or all of these mechanisms may be operative at a given time in a given process.

The sudden-death mechanisms are more straightforward but less predictable. These mechanisms are categorized as plastic deformation, brittle fracture, fatigue fracture, or edge chipping. Here again it is difficult to predict which mechanism will dominate and result in a tool failure in a particular situation. What can be said is that tools, like people, die (or fail) from a great variety of causes under widely varying conditions. Therefore, tool life should be treated as a random variable, or probabilistically, and not as a deterministic quantity.

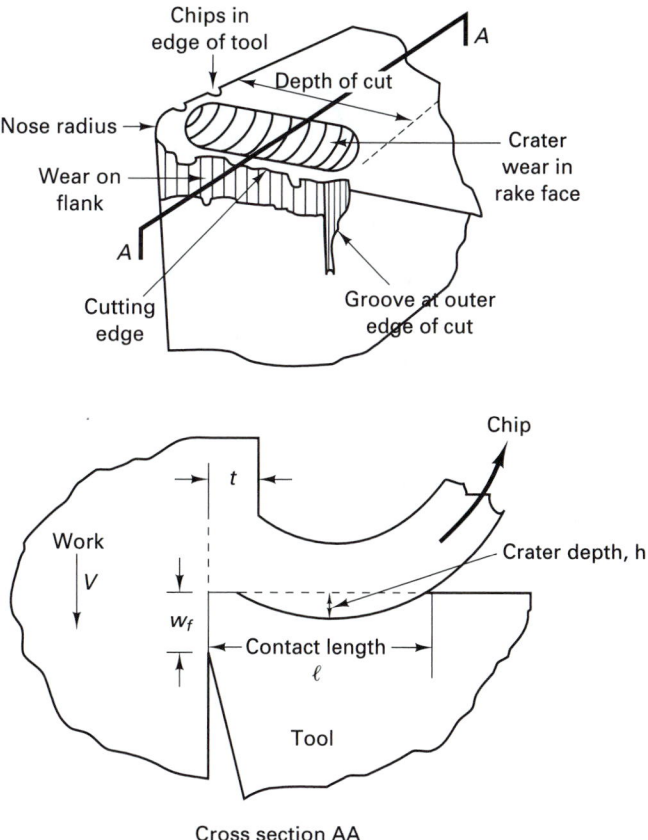

FIGURE 22-12   Sketch of a worn tool showing various wear elements resulting during oblique cutting: W_f, flank wear land length; t, uncut chip thickness.

During machining, the tool is performing in a hostile environment wherein high contact stresses and high temperatures are commonplace, and therefore tool wear is always an unavoidable consequence. In Figure 22-13, four characteristic tools wear curves (average values) are shown for four different cutting speeds, V_1 through V_4. V_4 is the fastest cutting speed and therefore generates the highest interface temperatures and the fastest wear rates. Such curves often have three general regions, as shown in the figure. The central region is a steady-state region (or the region of secondary wear). This is the normal operating region for the tool. Such curves are typical for both flank wear and crater wear. When the amount of wear reaches the value W_f, the permissible tool wear on the flank, the tool is said to be worn out. W_f is typically 0.025 to 0.030 in. For crater wear, the depth of the crater is used to determine tool failure.

Suppose that the tool wear experiment were to be repeated 15 times without changing any of the input parameters. The result would look like Figure 22-14, which depicts the variable nature of tool wear and shows why tool wear must be treated as a random variable. In Figure 22-14 the average time is denoted as μ_t and the standard deviation as σ_t, where the wear limit criterion was 0.025 in. At a given time during the test, 35 minutes, the tool displayed flank wear ranging from 0.013 to 0.021 in. with an average of $\mu_w = 0.0175$ in. with standard deviation σ_w.

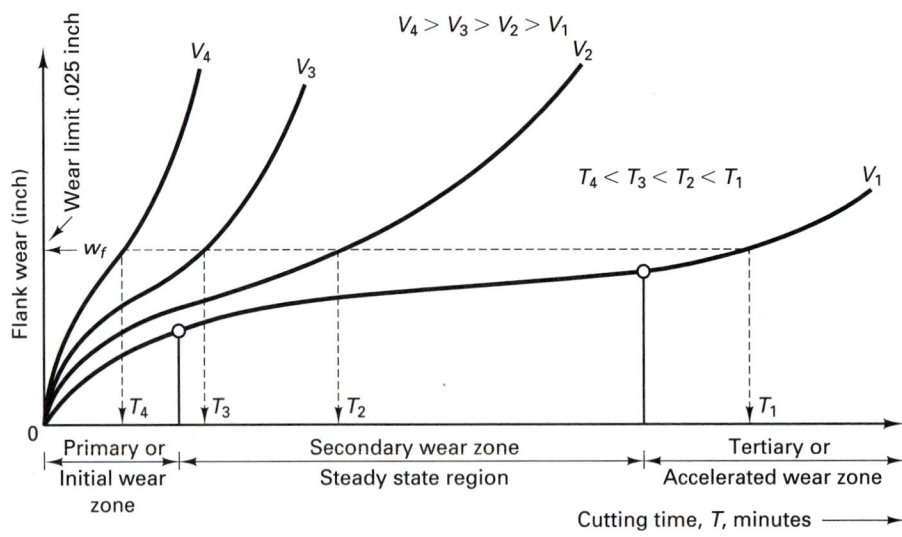

FIGURE 22-13   Typical tool wear curves for flank wear at different velocities.

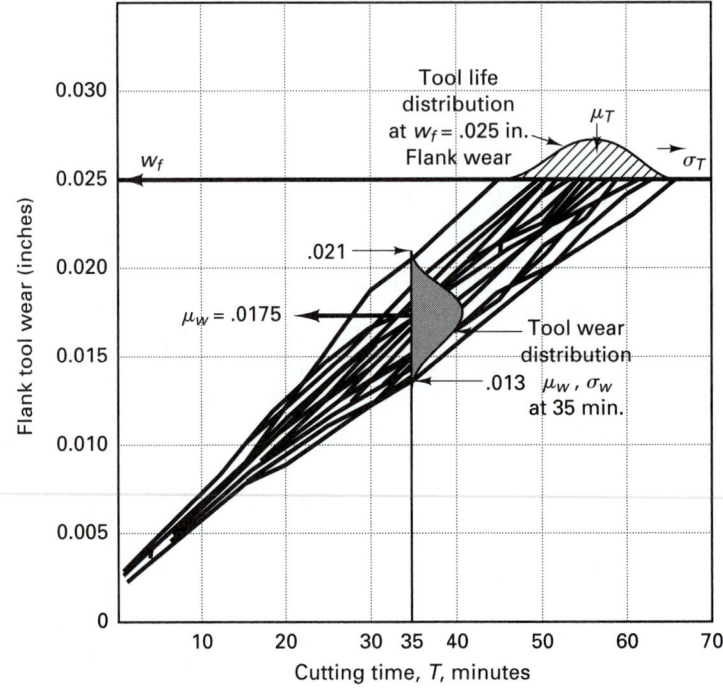

FIGURE 22-14   Tool wear on the flank displays a random nature, as does tool life.

Other criteria that can be used to define tool death in addition to wear limits are surface finish, failure to conform to size (tolerances), increases in cutting forces and power, time required to drill a hole under a given load, or complete failure of the tool. In automated processes, it is very beneficial to be able to monitor the tool wear on-line, so that the tool can be replaced prior to failure, wherein defective products may also result. The feed force has been shown to be a good, indirect measure of tool wear. That is, as the tool wears and dulls, the feed force increases more than the cutting force increases.

Once criteria for failure have been established, tool life is that time elapsed between start and finish of the cut, in minutes. Other ways to express tool life, other than time, include:

1. Volume of metal removed between regrinds or replacement of tool
2. Number of holes drilled with a given tool (see Figure 22-8)
3. Number of pieces machined per tool

Taylor's Tool Life Model

In 1907, F. W. Taylor published his now-famous *Taylor tool life equation*, wherein tool life (T) was related to cutting speed (V) and feed (f). This equation had the form

$$T = \frac{\text{constant}}{f_x V_y}$$

which over the years took the more widely published form,

$$VT^n = C \tag{22-2}$$

where n is an exponent that depends primarily on tool material but is affected by work material, cutting conditions, and environment, and C is a constant that depends on all the input parameters, including feed. This equation was obtained from empirical log-log plots, such as Figure 22-15, which used the values of T (time in minutes) associated with V (cutting speed) for a given amount of tool wear, W_f. (See the dashed-line construction in Figure 22-13.) Figure 22-16 shows typical tool life curves for one tool material and three work materials. Notice that all three plots have about the same slope, n. Typical values for n are 0.14 to 0.16 for HSS, 0.21 to 0.25 for uncoated carbides, 0.30 for TiC insets, 0.33 for polydiamonds, 0.35 for TiN inserts, and 0.40 for ceramic-coated inserts.

It takes a great deal of experimental effort to obtain the constants for the Taylor equation, as each combination of tool and work material will have different constants. Note that for a tool life of 1 minute, $C = V$, or the cutting speed that yields about 1 minute of tool life for this tool. A great deal of research has gone into developing more sophisticated versions of the Taylor equation, wherein constants for other input parameters (typically feed, depth of cut, and work material hardness) are experimentally determined: for example,

$$VT^n f^m d^p = K' \tag{22-3}$$

where n, m, and p are exponents and K' is a constant. Equations of this form are also deterministic.

The problem has been approached probabilistically in the following way. Since T depends on speed, feed, materials, and so on, one writes

$$T = \frac{C^{1/n}}{V^{1/n}} = \frac{K}{V^m} \tag{22-4}$$

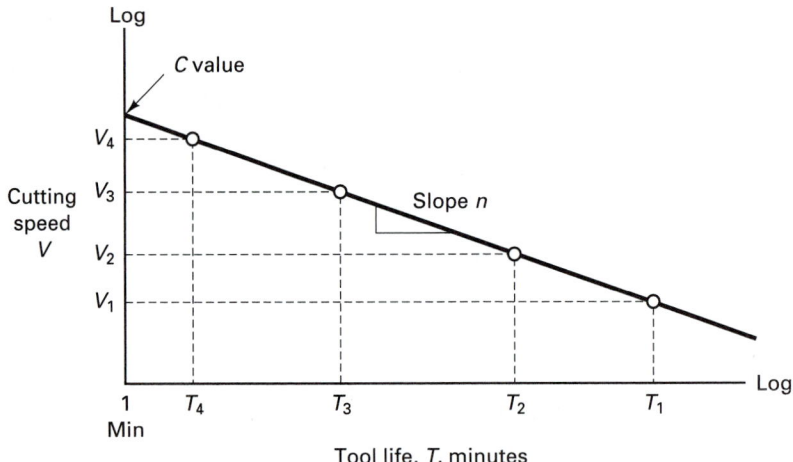

FIGURE 22-15 Construction of the Taylor tool life curve using data from deterministic tool wear plots like those of Figure 22-13

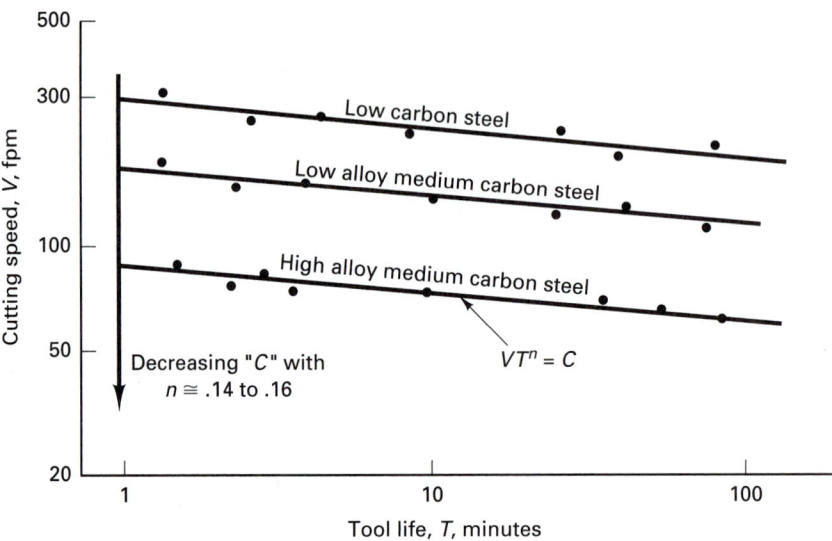

FIGURE 22-16 Log-log tool life plots for three steel work materials cut with HSS tool material.

where K is now a random variable that represents the effects of all unmeasured factors and V^m is an input variable.

The sources of tool life variability include such factors as:

1. Variation in work material hardness (from part to part and within a part)
2. Variability in cutting-tool materials, geometry, and preparation
3. Vibrations in machine tool, including rigidity of work and tool-holding devices
4. Changing surface characteristics of workpieces

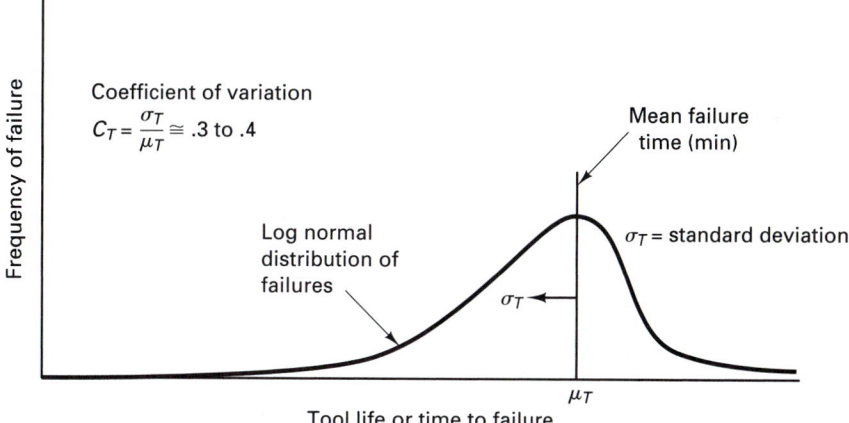

FIGURE 22-17 Tool life viewed as random variable has a log normal distribution with a large coefficient of variation.

The examination of the data from a large number of tool life studies wherein a variety of steels were machined has shown that regardless of the tool material or process, tool life distributions are usually log normal and typically have a large standard deviation. As shown in Figure 22-17, tool life distribution has a large coefficient of variation C_V so tool life is not very predictable.

■ 22.5 RECONDITIONING CUTTING TOOLS

In the reconditioning of tools by sharpening and recoating, care must be taken in grinding a tool's surfaces. The following guidelines should be observed:

1. Resharpen to original tool geometry specifications. Restoring the original tool geometry will help the tool achieve consistent results on subsequent uses. Computer numerical control (CNC) grinding machines for tool resharpening have made it easier to restore a tool's original geometry.
2. Grind cutting edges and surfaces to a fine finish. Rough finishes left by poor and abusive regrinding hinder the performance of resharpened tools. For coated tools, tops of ridges left by rough grinding will break away in early tool use, leaving uncoated and unprotected surfaces that will cause premature tool failure.
3. Remove all burrs on resharpened cutting edges. If a tool with a burr is coated, premature failure can occur because the burr will break away in the first cut, leaving an uncoated surface exposed to wear.
4. Avoid resharpening practices that overheat and burn or melt (called *glazed over*) the tool surfaces, as this will cause problems in coating adhesion. Polishing or wire brushing of tools causes similar problems.

The cost of each recoating is about one-fifth the cost of purchasing a new tool. By recoating, the tooling cost per workpiece can be cut by between 20 and 30%, depending on the number of parts being machined.

■ 22.6 ECONOMICS OF MACHINING

The cutting speed has such a great influence on the tool life compared to the feed or the depth of cut that it greatly influences the overall economics of the machining process. For

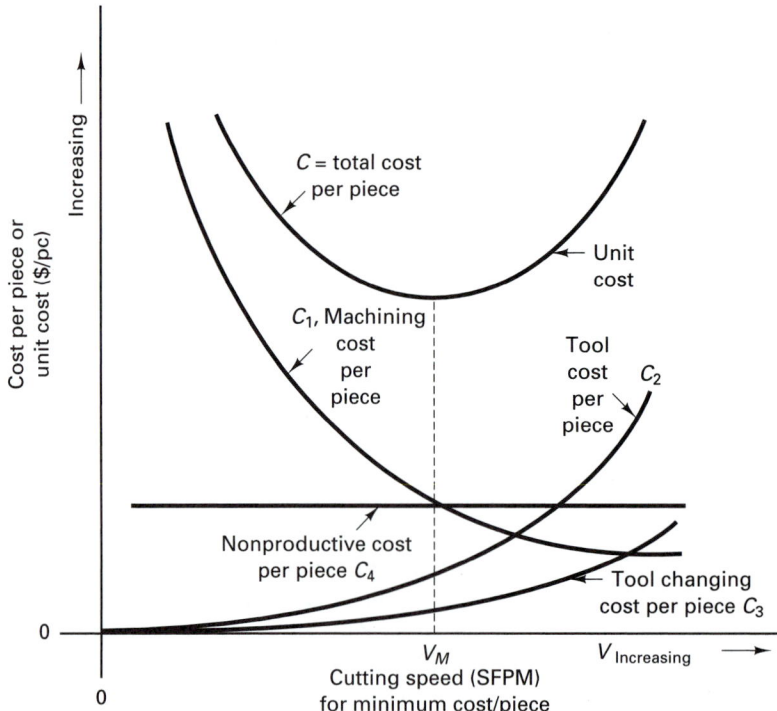

FIGURE 22-18   Cost per unit for a machining process versus cutting speed.

a given combination of work material and tool material, a 50% increase in speed results in a 90% decrease in tool life, while a 50% increase in feed results in a 60% decrease in tool life. A 50% increase in depth of cut procedures only a 15% decrease in tool life. This says that in limited-horsepower situations, depth of cut and then feed should be maximized while speed is held constant and horsepower consumed is maintained within limits. As cutting speed is increased, the machining time decreases but the tools wear out faster and must be changed more often. In terms of costs, the situation is as shown in Figure 22-18, which shows the effect of cutting speed on the cost per piece.

The total cost per operation is comprised of four individual costs: machining costs, tool costs, tool-changing costs, and handling costs. The machining cost is observed to decrease with increased cutting speed because the cutting time decreases. Cutting time is proportional to the machining costs. Both the tool costs and the tool changing costs increase with increases in cutting speeds. The handling costs are independent of cutting speed. Adding up each of the individual costs results in a total unit cost curve which is observed to go through a minimum point. Therefore, as shown in Figure 22-18, there is some minimum cost related to a specific cutting speed. To find the cutting velocity that will yield the minimum cost per piece, a total cost equation is written that is comprised of the machining cost, a tool cost, and tool-changing cost. All the terms in the equation are written as a function of cutting velocity, and then the equation is differentiated and set equal to zero to find the velocity V_M which yields the minimum cost per piece. Following is the equation for V_M:

$$V_M = \frac{C}{\{[(1 / n) - 1] (t_c)\}^n} \tag{22-5}$$

where C and n were constants from equation (22-2) and t_c is the tool-changing time in minutes.

A word of caution is appropriate here. This equation is totally dependent on the Taylor tool life equation. Such data may not be available for new tool materials because it is expensive and time consuming to obtain. Even when the tool life data is available, this procedure assumes that the tool fails only by whichever wear mechanism (flank or crater) was described by this equation and by no other failure mechanism. Recall that tool life had a very large coefficient of variation and was probabilistic in nature, but this derivation also assumes that for a given V, there is one T. The model also assumes that the workpiece material is homogeneous, the tool geometry is preselected, the depth of cut and feed rate are known and remain unchanged during the entire process, sufficient horsepower is available for the cut at the economic cutting conditions, and the cost of operating time is the same whether the machine is cutting or not cutting.

Table 22-6 shows a more realistic example of the analysis of tooling economics, where a comparison is being made between two tool materials (insert tools). A manufacturer of diesel engines is producing an in-line six-cylinder engine block which is machined on a transfer line. Each cylinder hole must be bored to accept a sleeve liner. This operation has a depth of cut of 0.062 in. per side, for a total of 0.125 in. stock removal. The tolerance on this bore is ±0.001 in. and the spindle is operating at 2000 sfpm. Ceramic inserts are used on this operation, but with these inserts, wear was severe enough to require indexing after only 35 pieces. The ceramic insert was replaced with PCBN inserts made of a high-content CBN. Both inserts had a 0.001 to 0.002 in. radius hone for edge preparation.

Table 22-6 is a cost comparison between the ceramic and PCBN insert. The PCBN insert used in the application is a full-top PCBN insert, meaning that the entire top of the insert is a layer of PCBN material. At first glance the PCBN tool appears to be extremely

TABLE 22-6. Cost Comparison for Machining Liner Bores in 1500 Engine Blocks[a]

	Ceramic TNG-433	PCBN BTNG-433
Cost per insert	$14.90	$208.00
Edges per insert	6	3
Cost per edge	$2.48	$69.33
Time per index (six tools)	0.25 hr	0.25 hr
Time per index at $45 per hour	$11.25	$11.25
Indexes per 1500 blocks	43	3
Indexing cost (indexes × $11.25)	$483.75	$33.75
Insert cost for six spindles	$638.34	$1248.00
Labor and tool cost	$1122.09	$1281.00
Cost per bore	$0.125	$0.142
Total number of tool changes	43	3
Downtime for 1500 blocks	$\frac{\times\ 0.25\ \text{hr}}{10.75\ \text{hr}}$	$\frac{\times\ 0.25\ \text{hr}}{10.75\ \text{hr}}$

[a]To see the economy of using PCBN cutting tools, it is important to consider all factors of the operation, especially downtime for tool changing.

expensive. The insert cost $208.00 and provides only three usable edges, whereas the ceramic insert costs $14.90 and provided six usable edges.

The ceramic tools must be indexed every 35 pieces, where the PCBN tool is indexed every 500 pieces. The cost per bore, including insert cost and the cost of labor to perform indexing, comes to $0.125 per bore for the ceramic tool and $0.142 per bore for the PCBN tool. This appears to make the ceramic tool more cost-effective, but downtime for indexing has not been accounted for. The ceramic insert required 10.75 hours of downtime for indexing, whereas the PCBN tool required only 0.75 hour of downtime for indexing. Use of the PCBN cutting tool will significantly reduce the total cost per piece by eliminating 10 hours of downtime and increasing the productivity of the machine.

■ 22.7 Machinability

Machinability is a much maligned term which has many different meanings but generally refers to the ease with which a metal can be machined to an acceptable surface finish. The principal definitions of the term are entirely different: the first based on material properties, the second based on tool life, and the third based on cutting speed.

1. Machinability is defined by the ease or difficulty with which the metal can be machined. In this light, specific energy, specific horsepower, or shear stress are used as measures, and in general, the larger the shear stress or specific power values, the more difficult the material is to machine, requiring greater forces and lower speeds. In this definition, the material is the key.
2. Machinability is defined by the relative cutting speed for a given tool life of the tool cutting some material, compared to a standard material cut with the same tool material. As shown in Figure 22-19, tool-life curves are used to develop machinability ratings. The material chosen for the standard material was B1112 steel, which has a tool life of 60 minutes at a cutting speed of 100 sfpm. Material X has a 70% rating, which implies that steel X has a cutting speed of 70% of B1112 for equal tool lives. Note that this definition assumes that the tool fails when machining X by whatever mechanism dominated the tool failure when machining the B1112. There is no guarantee that this will be the case. ISO Standard 3685 has machinability index numbers based on 30 minutes of tool life with flank wear of 0.33 mm.
3. Cutting speed is measured by the maximum speed at which a tool can provide satisfactory performance for a specified time under specified conditions. See ASTM Standard E 618-81: "Evaluating machining performance of ferrous metals using an automatic screw/bar machine."
4. Other definitions of machinability are based on the ease of removal of the chips (chip disposal), the quality of the surface finish of the part itself, the dimensional stability of the process, or the cost to remove a given volume of metal.

Further definitions are being developed based on the probabilistic nature of the tool failure, in which machinability is defined by a tool reliability index. Using such indexes, various tool replacement strategies can be examined and optimum cutting rates obtained. These approaches account for the tool life variability by developing coefficients of variation for common cutting tool/work material combinations. The coefficient of variation is the ratio of the standard deviation of the tool-life distribution divided by the mean of that distribution (see Figure 22-17).

The results to date are very promising. One thing is clear, however, from this sort of research: Although many manufacturers of tools have worked at developing materials that have greater tool life at higher speeds, few have worked to develop tools that have less

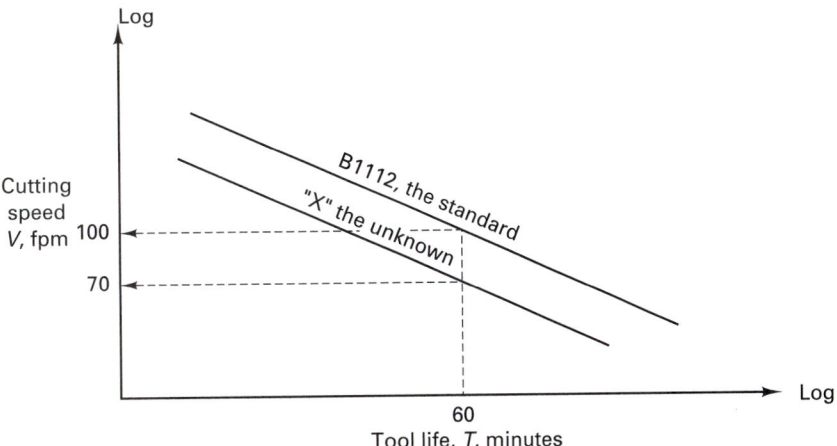

FIGURE 22-19 Machinability ratings defined by deterministic tool life curves.

variability in tool life at all speeds. The reduction in variability is fundamental to achieving smaller coefficients of variation, which typically are of the order of 0.3 to 0.4. This means that a tool with a 100-minute average tool life has a standard deviation of 30 to 40 minutes, so there is a good probability that the tool will fail early. In automated equipment, where early, unpredicted tool failures are extremely costly, reduction of the tool life variability will pay great benefits in improved productivity and reduced costs.

■ 22.8 CUTTING FLUIDS

From the day that Frederick W. Taylor demonstrated that a heavy stream of water flowing directly on the cutting process allowed the cutting speeds to be doubled or tripled, *cutting fluids* have flourished in use and variety and are employed in virtually every machining process. The cutting fluid acts primarily as a coolant and secondly as a lubricant, reducing the friction effects at the tool/chip interface and the work/flank regions. The cutting fluids also carry away the chips and provide friction (and force) reductions in regions where the bodies of the tools rub against the workpiece. Thus in such processes as drilling, sawing, tapping, and reaming, portions of the tool apart from the cutting edges come in contact with the work, and these (sliding friction) contacts greatly increase the power needed to perform the process, unless properly lubricated.

The reduction in temperature greatly aids in retaining the hardness of the tool, thereby extending the tool life or permitting increased cutting speed with equal tool life. In addition, the removal of heat from the cutting zone reduces thermal distortion of the work and permits better dimensional control. Coolant effectiveness is closely related to the thermal capacity and conductivity of the fluid used. Water is very effective in this respect but presents a rust hazard to both the work and tools and is not effective as a lubricant. Oils offer less effective coolant capacity but do not cause rust and have some lubricant value. In practice, straight cutting oils or emulsion combinations of oil and water or wax and water are frequently used. Various chemicals can also be added to serve as wetting agents or detergents, rust inhibitors, or polarizing agents to promote formation of a protective oil film on the work. The extent to which the flow of a cutting fluid washes the very hot chips away from the cutting area is an important factor in heat removal. Thus the application of a coolant should be copious and of some velocity.

The possibility that a cutting fluid provides lubrication between the chip and the tool face is an attractive one. An effective lubricant can modify the geometry of chip formation so as to yield a thinner, less deformed, and cooler chip. Such action could discourage the formation of a built-up edge on the tool and thus promote improved surface finish. However, the extreme pressure at the tool/chip interface and the rapid movement of the chip away from the cutting edge make it virtually impossible to maintain a conventional hydrodynamic lubricating film at the tool/chip interface. Consequently, any lubrication action is associated primarily with the formation of solid chemical compounds of low shear strength on the freshly cut chip face, thereby reducing chip/tool shear forces or friction. For example, carbon tetrachloride is very effective in reducing friction in machining several different metals and yet would hardly be classified as a good lubricant in the usual sense. Chemically active compounds, such as chlorinated or sulfurized oils, can be added to cutting fluids to achieve such a lubrication effect. Extreme-pressure lubricants are especially valuable in severe operations, such as internal threading (tapping), where the extensive tool/work contact results in high friction with limited access for a fluid. In addition to functional effectiveness as coolant and lubricant, cutting fluids should be stable in use and storage, noncorrosive to work and machines, and nontoxic to operating personnel. The cutting fluid should also be restorable by using a closed recycling system that will purify the used coolant and cutting oils. Cutting fluids become contaminated in three ways (Table 22-7). All these contaminants can be eliminated by filtering, hydrocycloning, pasteurizing, and centrifuging. Coolant restoration eliminates 99% of the cost of disposal and 80% or more of new fluid purchases. See Figure 22-20 for a schematic of a coolant recycling system.

FIGURE 22-20   A well-designed recycling system for coolants will return more than 99% of the fluid for reuse.

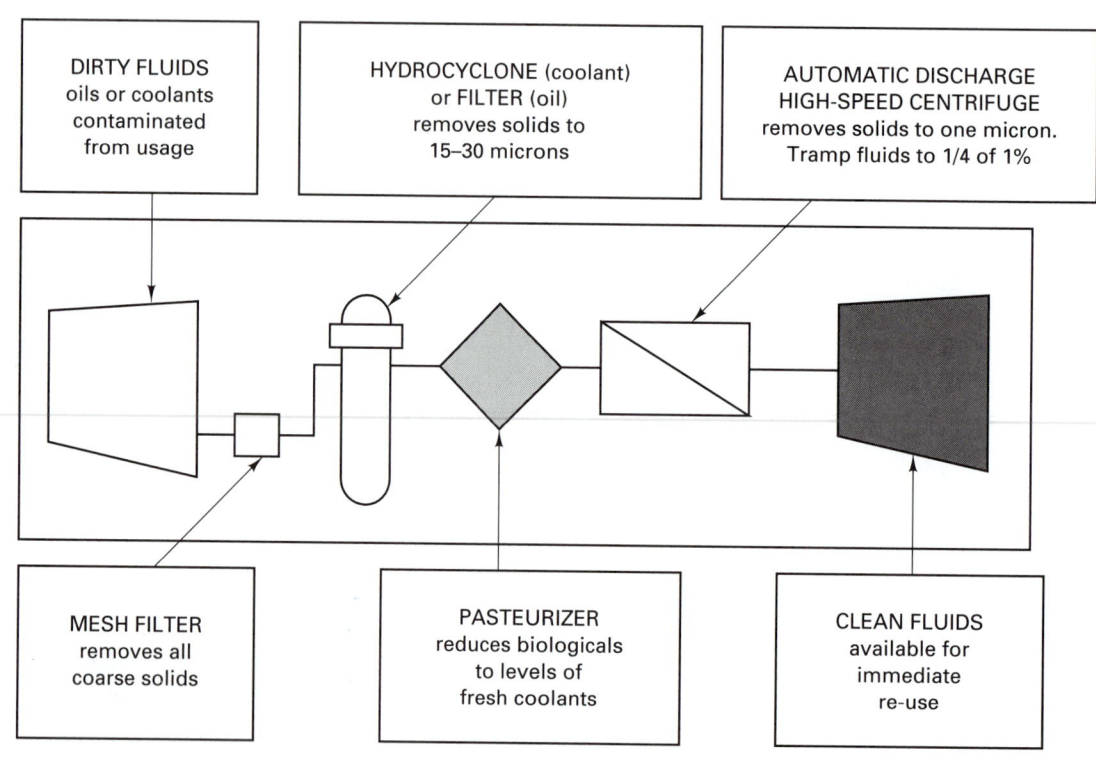

TABLE 22-7. Cutting Fluid Contaminants		
Category	Contaminants	Effects
Solids	Metallic fines and chips, grease and sludge, debris and trash	Scratch product's surface, plug coolant lines, wear on tools and machines
Tramp fluids	Hydraulic oils (coolant), water (oils)	Decrease cooling efficiency, cause smoking, clog paper filters, grow bacteria faster
Biologicals (coolants)	Bacteria, fungi, mold	Acidify coolant; break down emulsions; cause rancidity, dermatitis; require toxic biocides

■ KEY WORDS

aluminum oxide
built-up edge
carbides
cast cobalt alloy
ceramics
cermets
chemical vapor deposition
coated tools

crater wear
cubic boron nitride
cutting fluids
cutting-tool materials
depth-of-cut line
diamonds
hot hardness
machinability

physical vapor deposition
polycrystalline
single crystal
sintered
stellite
titanium nitride
wear land

■ REVIEW QUESTIONS

1. For metal-cutting tools, what is the most important material property (ie., the most critical characteristic)? Why?
2. What is hot hardness?
3. What is impact strength, and how is it measured?
4. Why is impact strength an important property in cutting tools?
5. What is HIP, and how is it used for tool fabrication?
6. What are the primary considerations in tool selection?
7. What is the general strategy behind coated tools?
8. What is a cermet?
9. How is a CBN tool manufactured?
10. F. W. Taylor was one of the discoverers of high-speed steel. For what else is he well known?
11. What casting process do you think was used to fabricate cast cobalt alloys?
12. What does the sintering step do in the manufacture of carbides?
13. What does *cemented* mean in the manufacture of carbides?
14. What advantage do ground carbide inserts have over pressed carbide inserts?

15. What is a chip groove?
16. What is the DCL?
17. Suppose that you made four beams out of carbide, HSS, ceramic, and cobalt. The beams are identical in size and shape, differing only in material. Assuming you applied the same load to each beam, which beam would:
    (a) Deflect the most?
    (b) Resist penetration the most?
    (c) Bend the farthest without breaking?
    (d) Support the greatest compressive load?
18. Why are multiple coats or layers put on the carbide base for coated tools?
19. What are the most common surface treatments for
    (a) High-speed steels?
    (b) Carbides?
    (c) Ceramics?
20. What makes the process that makes TiC coatings for tools a problem? [See equation (22-1)].
21. Why does machining with a TiN-coated tool consume less power than an uncoated HSS under exactly the same cutting conditions?
22. Why is CBN better than diamond for machining steel?

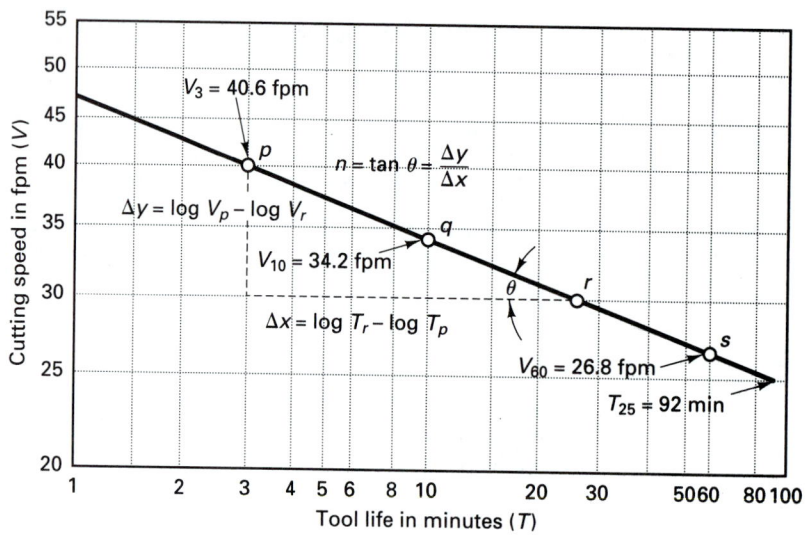

FIGURE 22-A

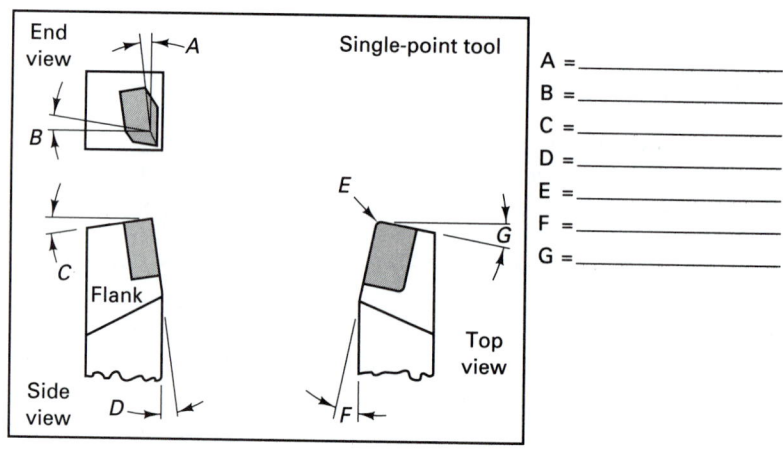

FIGURE 22-B

23. What is a coefficient of variation?
24. What is meant by the statement: "Tool life is a random variable"?
25. What is the typical value of a coefficient of variation in metal-cutting tool life distributions?
26. How is machinability defined?
27. What are the chief functions of cutting fluids?

■ PROBLEMS

1. In Figure 22-A are data for cutting speed and tool life. Determine the constants for the Taylor tool life equation for these data. What do you think the tool material might have been?

2. Suppose that you have a turning operation using a tool with zero back rake and 5° end relief.

The insert flank has a wear land of 0.020 in. How much has the diameter of the workpiece grown (increased) due to this flank wear, assuming that the tool has not been reset to compensate for the flank wear?

3. In Figure 22-B, a single-point tool is shown.

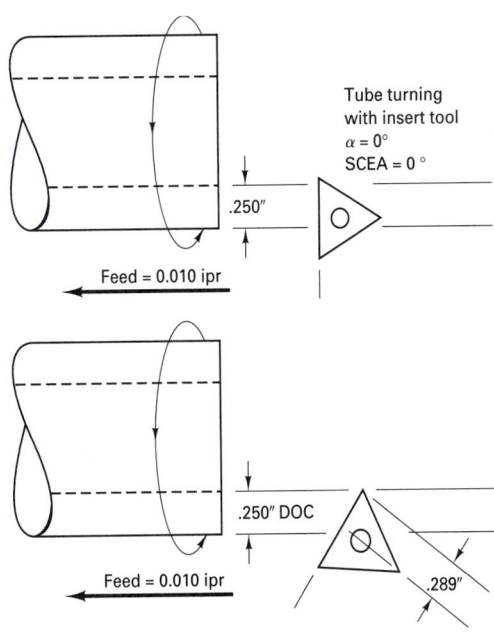

Tube turning
with insert tool
$\alpha = 0°$
SCEA = 0°

.250″

Feed = 0.010 ipr

.250″ DOC

.289″

Feed = 0.010 ipr

FIGURE 22-C

(a) Fill in the blanks for the tool nomenclature:

A = _____   F = _____
B = _____   G = _____
C = _____   E = _____
D = _____

(b) Is the back rake angle positive or negative?
(c) What kind of single-point tool is this?

4. The following data has been obtained for machining AA390 aluminum, a Si-Al alloy.

Workpiece Material	Tool Material	Cutting Speed (m/min) for 20 min	Tool Life (m/min) of: 30 min	Tool Life (m/min) of: 60 min
Sand casting	Diamond polycrystal	731	642	514
Permanent mold casting	Diamond polycrystal	591	517	411
PMC with flood cooling	Diamond polycrystal	608	554	472
Sand casting	WC-K-20	175	161	139

Compute the C and n values for the Taylor tool life equation. How do these n values compare to the typical values?

5. In Figure 22-C, a tube is being turned using a triangular insert tool. The insert at the top is set with a zero side-cutting-edge angle. The insert at the bottom is set so that the edge contact length is increased from a 0.250-in. depth of cut to 0.289 in. The feed was 0.010 in./rev.

(a) Determine the side-cutting-edge angle of the offset tool.
(b) What is the uncut chip thickness in the offset position?
(c) What effect will this have on the process?

6. Tool cost is often used as the major criterion for justifying tool selection. Either silicon nitride or PCBN insert tips can be used to machine (bore) a cylinder block on a transfer line at a rate of 312,000 parts/yr (material: gray cast iron). The operation required 12 inserts (two per tool), as six bores are machined simultaneously. The machine was run at 2600 sfpm with a feed of 0.014 in./rev at 0.005 DOC for finishing. Here is some additional data.

	SiN	PCBN
Tips in use per part	12	12
Tool life (parts per tool)	200	4700
Cost per tip	$1.25	$28.50

(a) Which tool material would you recommend?
(b) On what basis?

*C*hapter 22 CASE STUDY

indexable drill insert

The Linus Drilling Company is trying to decide what kind of inserts to use in their indexable drills. These drills do not cut at dead center but rather form a thin center slug that is pushed out of the way by the drill body. These indexable drills often provide a marked improvement in productivity over conventional HSS drills, particularly when they are combined with coated insert tools. The company is trying to determine which type of insert to use in the drill for machining some hot-rolled 8620 steel shafts using triangular inserts. The operating cost of the machine tool is $60.00 per hour. It takes about 3 minutes to change the inserts and about 30 seconds to unload a finished part and load a new part in the machine. The company is currently using uncoated inserts at the following operating conditions: 400 sfpm and 0.020 in./rev. These speeds and feeds resulted in each cutting edge producing about 40 pieces before the tool's cutting edge became dull. The tool was then indexed. Since it was a triangular tool, each tool had six cutting edges available before it had to be replaced. At these speeds and feeds, the drilling time was 4.8 minutes and the production rate was 11 parts per hour, while the machining cost per piece was $4.80 ($1.00 per minute × 4.8 minutes per piece). The manufacturing engineer on the job, Brian Paul, has found three new tool materials being used in triangular insert tools. They are listed in Table CS-22 along with the data for the uncoated tool. The new materials are TiC-coated carbide, Al_2O_3-coated carbide, and a ceramic-coated insert with a single-side, low-force-groove geometry. The expected cutting conditions, speeds, and feeds are given in the table along with Brian's estimates of the production rates in pieces per hour for each of these new tool materials. The low-force-groove geometry tools can only be used three times (they cannot be flipped over), so only three cutting edges are available per insert before it has to be replaced. Brian has argued that even though the ceramic-coated inserts cost more, they result in a lower cost per piece, considering all the costs. Determine the machining cost per piece, the tool-changing cost per piece, and the tool cost per piece, which make up the total cost per piece, and verify Brian's belief that the coated tools will provide some cost savings.

TABLE CS-22. Cost Comparison of Four Tool Materials, Based on Equal Tool Life

	Uncoated	Tic-Coated	Al_2O_3-Coated	Al_2O_3 LFG
Cutting speed (sfpm)	400	640	1100	1320
Feed (in./rev)	0.020	0.022	0.024	0.028
Cutting edges available per insert	6	6	6	3
Cost of an insert ($/insert)	4.80	5.52	6.72	6.72
Tool life (pieces/cutting edge)	192	108	60	40
Tool change time per piece (min)	0.075	0.075	0.075	0.075
Nonproductive cost per piece ($/piece)	0.50	0.50	0.50	0.50
Machining time per piece (min/piece)	4.8	2.7	1.50	1.00
Machining cost per piece ($/piece)	4.80			
Tool change cost per piece ($/piece)	0.08			
Cutting tool cost per piece ($/piece)	0.02			
Total cost per piece ($)	5.40			
Production rate (pieces/hr)	11	18	29	38
Improvement in productivity based on pieces/hr (%)	0	64	164	245

Source: Data from T. E. Hale et al., "High productivity approaches to metal removal," *Materials Technology*, Spring 1980, p. 25.

CHAPTER 23

TURNING, BORING, AND RELATED PROCESSES

23.1 INTRODUCTION
23.2 FUNDAMENTALS OF TURNING, BORING, AND FACING
Turning
Boring
Facing
Parting
Drilling
Reaming
Knurling
Turning and Boring Tapers
Special Attachments
Dimensional Accuracy
23.3 LATHE DESIGN AND TERMINOLOGY
Lathe Design
Size Designation of Lathes
Types of Lathes
    Turret Lathes
    Vertical Turret Lathes
    Automatic Turret Lathes
    Single-Spindle Automatic Screw Machines
    Multiple-Spindle Automatic Screw Machines

23.4 TYPES OF BORING MACHINES
Vertical Boring and Turning Machines
Jig Borers
Horizontal Boring Machines
Mass-Production Boring Machines
Boring Machine Precision
23.5 CUTTING TOOLS FOR LATHES
Lathe Cutting Tools
Form Tools
Turret Lathe Tools
23.6 WORKHOLDING IN LATHES
Workholding Devices for Lathes
Lathe Centers
Mandrels
Lathe Chucks
Collets
Faceplates
Mounting Work on the Carriage
Steady and Follow Rests
Case Study: BREAK-EVEN POINT ANALYSIS OF A LATHE PART

■ 23.1 INTRODUCTION

Turning is the process of machining external cylindrical and conical surfaces. It is usually performed on a lathe. Photographs and schematics of lathes are shown later in this chapter. As indicated in Figure 23-1, relatively simple work and tool movements are involved in turning a cylindrical surface. The workpiece is rotated into a longitudinally fed, single-point cutting tool. If the tool is fed at an angle to the axis of rotation, an external conical surface results. This is called *taper turning*. If the tool is fed at 90° to the axis of rotation, using a tool that is wider than the width of the cut, the operation is called *facing*, and a flat surface is produced.

By using a tool having a specific form or shape and feeding it radially or inward against the work, external cylindrical, conical, and irregular surfaces of limited length can

653

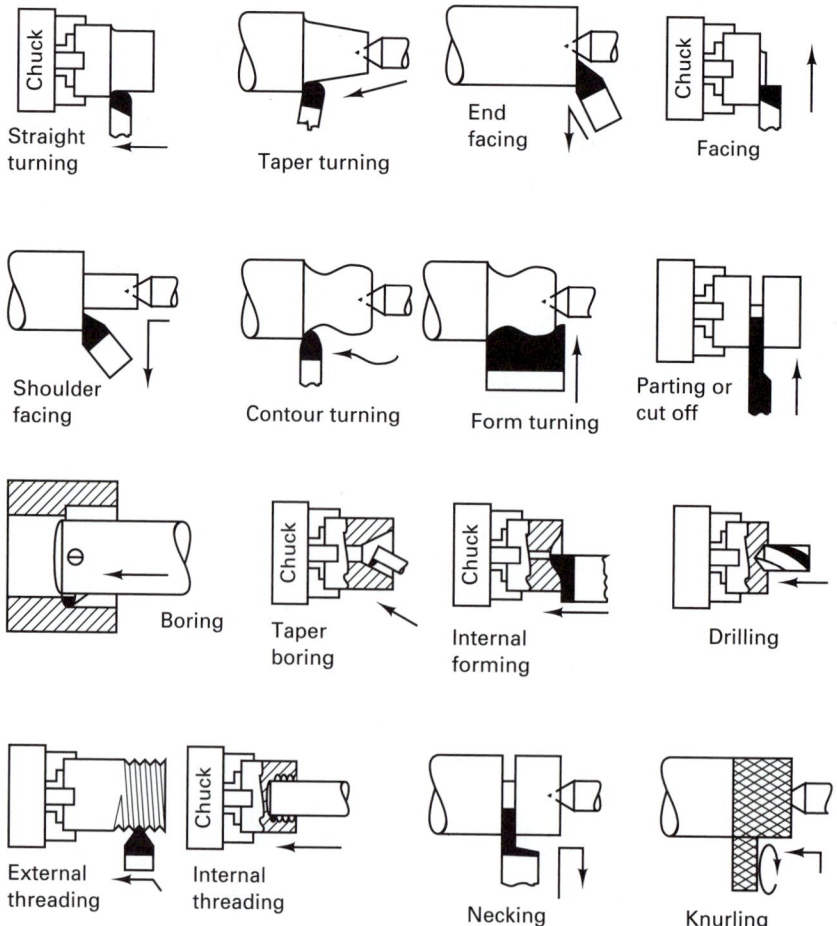

FIGURE 23-1   Turning, facing, boring, and related processes typically done on a lathe. The arrows indicate the motion of the tool relative to the work.

also be turned. The shape of the resulting surface is determined by the shape and size of the cutting tool. Such machining is called *form turning*. If the tool is fed all the way to the axis of the workpiece, it will be cut in two. This is called *parting* or *cutoff* and a simple, thin tool is used. A similar tool is used for *necking* or *partial cutoff*.

Boring is a variation of turning. Essentially, boring is internal turning. Boring can use single-point cutting tools to produce internal cylindrical or conical surfaces. It does not create the hole but rather, machines or opens the hole up to a specific size. Boring can be done on most machine tools that can do turning. However, boring also can be done using a rotating tool with the workpiece remaining stationary. Also, specialized machine tools have been developed that will do boring, drilling, and reaming but will not do turning. Other operations, such as *threading* and *knurling*, can be done on machines used for turning. In addition, drilling, reaming, and tapping can be done on the rotation axis of the work.

■ 23.2 FUNDAMENTALS OF TURNING, BORING AND FACING

Turning

Turning constitutes the majority of lathe work. The cutting forces, resulting from feeding the tool from right to left, should be directed toward the headstock to force the workpiece against the workholder and thus provide better work support.

If good finish and accurate size are desired, one or more roughing cuts usually are followed by one or more finishing cuts. Roughing cuts may be as heavy as proper chip thickness, tool life, lathe horsepower, and the workpiece permit. Large *depths of cut* and smaller feeds are preferred to the reverse procedure, because fewer cuts are required and less time is lost in reversing the carriage and resetting the tool for the following cut.

On workpieces that have a hard surface, such as castings or hot-rolled materials containing mill scale, the initial roughing cut should be deep enough to penetrate the hard materials. Otherwise, the entire cutting edge operates in hard, abrasive material throughout the cut, and the tool will dull rapidly. If the surface is unusually hard, the cutting speed on the first roughing cut should be reduced accordingly.

Finishing cuts are light, usually being less than 0.015 in. in depth, with the feed as fine as necessary to give the desired finish. Sometimes a special finishing tool is used, but often the same tool is used for both roughing and finishing cuts. In most cases one finishing cut is all that is required. However, where exceptional accuracy is required, two finishing cuts may be made. If the diameter is controlled manually, a short finishing cut ($\frac{1}{4}$ in. long) is made and the diameter checked before the cut is completed. Because the previous micrometer measurements were made on a rougher surface, it may be necessary to reset the tool in order to have the final measurement, made on a smoother surface, check exactly.

In turning, the primary cutting motion is rotational with the tool feeding parallel to the axis of rotation (Figure 23-2). The rpm value of the rotating workpiece, N, establishes the cutting velocity, V, at the cutting tool. The feed, f_r, is given in inches per revolution (in./rev). The depth of cut is d, where

$$d = \text{DOC} = \frac{D_1 - D_2}{2} \quad \text{inches} \tag{23-1}$$

The length of cut is the distance traveled parallel to the axis, L, plus some allowance or overrun, A, to allow the tool to enter and/or exit the cut.

FIGURE 23-2   Basics of the turning process, normally performed in a lathe.

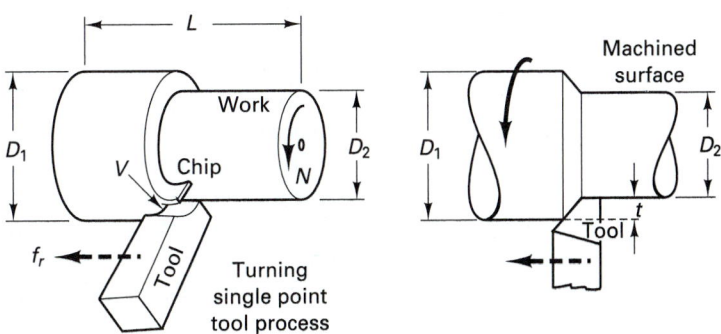

Once the cutting speed, feed, and depth of cut have been selected for a given material being cut with a tool of known cutting-tool material, the rpm value for the machine tool can be determined.

$$N = \frac{12V}{\pi D_1} \tag{23-2}$$

(using the larger diameter); where the value 12 converts feet to inches. The cutting time

$$CT = \frac{L + A}{f_r N} \tag{23-3}$$

where A is the overrun allowance. The *metal removal rate*

$$MRR = \frac{\text{volume removed}}{\text{time}} = \frac{(\pi D_1^2 - \pi D_2^2)L}{4L/f_r N}$$

(omitting the allowance term). By rearranging and subbing for N, an exact expression for MRR is obtained:

$$MRR = 12\, Vf_r\, \frac{D_1^2 - D_2^2}{4D_1} \tag{23-4}$$

Since

$$\frac{D_1^2 - D_2^2}{4D_1} = \frac{D_1 - D_2}{2} \cdot \frac{D_1 + D_2}{2D_1}$$

Since

$$d = \frac{D_1 - D_2}{2} \text{ and } \frac{D_1 + D_2}{2D_1} \simeq 1$$

so

$$MRR \simeq 12\, Vf_r d \quad \text{in}^3/\text{min} \tag{23-5}$$

Note that equation (23-5) is an approximate equation which assumes that the depth of cut, d, is small compared to the uncut diameter D_i (see Problem 7).

Boring

Boring always involves the enlarging of an existing hole, which may have been made by a drill or may be the result of a core in a casting. An equally important and concurrent purpose of boring may be to make the hole concentric with the axis of rotation of the workpiece and thus correct any eccentricity that may have resulted from the drill drifting off the centerline. Concentricity is an important attribute of bored holes.

When boring is done in a lathe, the work usually is held in a chuck or on a faceplate. Holes may be bored straight, tapered, or to irregular contours. Figure 23-3a shows the relationship of the tool and the workpiece for boring. Boring is essentially internal turning while feeding the tool parallel to the rotation axis of the workpiece.

Given V and f_r, for a cut of length, L, the cutting time is

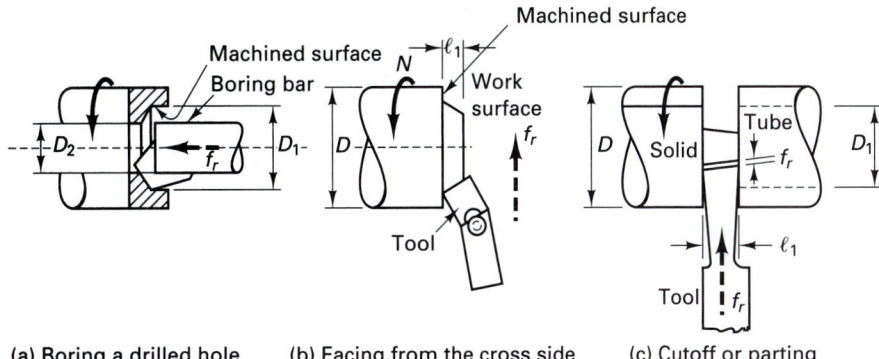

(a) Boring a drilled hole   (b) Facing from the cross side   (c) Cutoff or parting

FIGURE 23-3 Basic movements of boring, facing, and cutoff (or parting) processes.

$$CT = \frac{L+A}{f_r N} \tag{23-6}$$

where $N = 12\ V/\pi D_1$ for D_1 the diameter of the bore and A the overrun allowance.

$$MRR = \frac{L(\pi D_1^2 - \pi D_2^2)/4}{L/f_r N}$$

where D_2 is the original hole diameter.

$$MRR \simeq 12 V f_r d \tag{23-7}$$

(omitting the allowance term), where d is the depth of cut.

In most respects, the same principles are used for boring as for turning. Again, the tool should be set exactly at the same height as the axis of rotation. Slightly larger end clearance angles sometimes have to be used to prevent the heel of the tool from rubbing on the inner surface of the hole. Because the tool overhang will be greater, feeds and depths of cut may be somewhat less than for turning to prevent tool vibration and chatter. In some cases, the boring bar may be made of tungsten carbide because of this material's greater stiffness.

Facing

Facing is the producing of a flat surface as the result of the tool being fed across the end of the rotating workpiece, as shown in Figure 23-3b. Unless the work is held on a mandrel, if both ends of the work are to be faced, it must be turned end for end after the first end is completed and the facing operation repeated.

The cutting speed should be determined from the largest diameter of the surface to be faced. Facing may be done either from the outside inward or from the center outward. In either case, the point of the tool must be set exactly at the height of the center of rotation. Because the cutting force tends to push the tool away from the work, it is usually desirable to clamp the carriage to the lathe bed during each facing cut to prevent it from moving slightly and thus producing a surface that is not flat.

In the facing of castings or other materials that have a hard surface, the depth of the first cut should be sufficient to penetrate the hard material to avoid excessive tool wear.

In facing, the tool feeds perpendicular to the axis of the rotating workpiece. Because the rpm value is constant, the speed is continually decreasing as the axis is approached. The length of cut, L, is $D/2$ or $(D - D_1)/2$ for a tube.

$$\text{Cutting time} = \text{CT} = \frac{L + A}{f_r N} \quad \text{minutes}$$

$$= \frac{\dfrac{D}{2} + A}{f_r N}$$

$$\text{MRR} = \frac{\text{VOL}}{\text{CT}} = \frac{\pi D^2 l_1 f_r N}{4L} = 6\, V f_r l_1 \quad \text{in}^3/\text{min} \tag{23-8}$$

where l_1 is the depth of cut and $L = D/2$ is the length of cut.

Parting

Parting is the operation by which one section of a workpiece is severed from the remainder by means of a cutoff tool as shown in Figure 23-3c. Because parting tools are quite thin and must have considerable overhang, this process is more difficult to perform accurately. The tool should be set exactly at the height of the axis of rotation, be kept sharp, have proper clearance angles, and be fed into the workpiece at a proper and uniform feed rate.

In parting or cutoff work, the tool is fed (plunged) perpendicular to the rotational axis, as it was in facing. The length of cut for solid bars is $D/2$. For tubes,

$$L = \frac{D - D_1}{2}$$

In cutoff operations, the width of the tool is l_1 in inches. The equations for CT and MRR are then basically the same as for facing.

In boring, facing, and cutoff operations, the speeds and feeds selected are generally less than those recommended for turning because of the large overhang of the tool often needed to complete the cuts. Recall the basic equation for deflection of a cantilever beam; modified for machining,

$$\delta = \frac{Pl^3}{3EI} = \frac{F_c l^3}{3EI} \tag{23-9}$$

The overhang of the tool, l, greatly affects the deflection δ, so it should be minimized whenever possible. The reduction of the feed or the depth of cut reduces the forces operating on the tools, and the reduction of the speed usually reduces the probability of chatter and vibration.

Any imbalance in the cutting forces will deflect the tool to the side, resulting in loss of accuracy in cutoff lengths. At the outset the forces will be balanced if there is no side rake on the tool. As the cutoff tool reaches the axis of the rotating part, the tool will be deflected away from the spindle, resulting in a change in the length of the part.

Sometimes bored holes are slightly bell-mouthed because the tool deflects away from the work as it progresses into the hole. This usually can be corrected by repeating the cut with the same tool setting. Alternatively, a more robust setup can be used. Large holes may be precision-bored using the lathe setup shown in Figure 23-4, where a pilot bushing

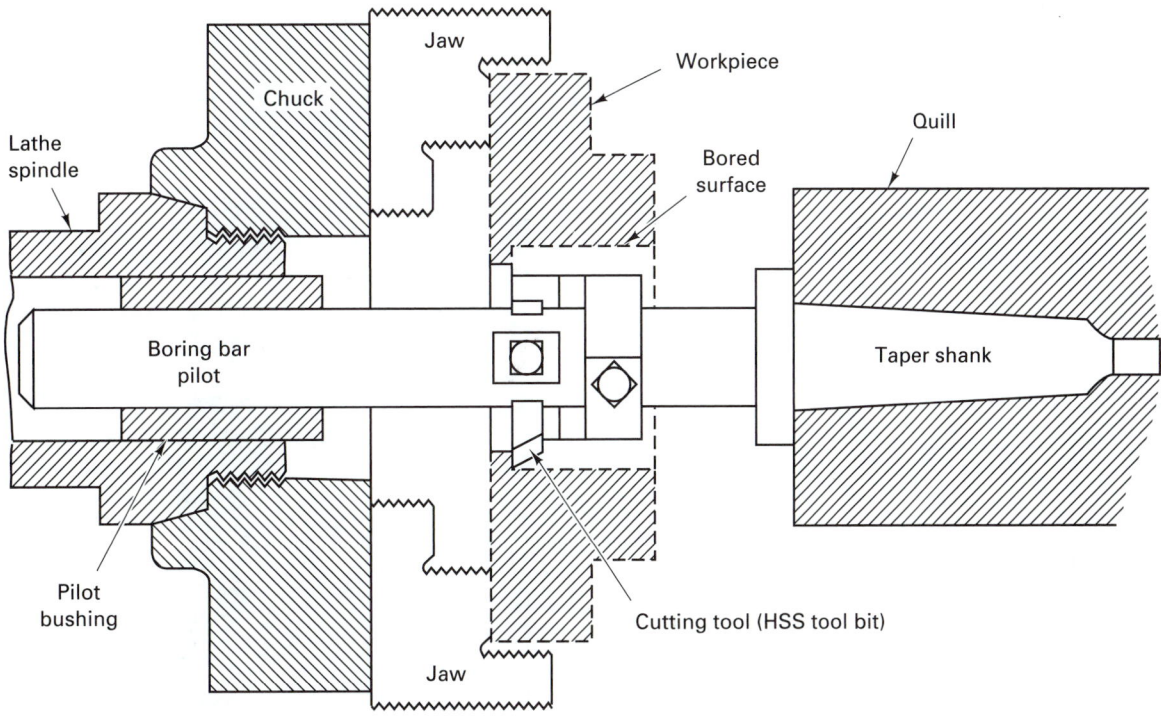

FIGURE 23-4   Pilot boring bar mounted in tailstock of lathe for precision boring large hole in casting. The size of the hole is controlled by the rotation diameter of the cutting tool.

is placed in the spindle to mate with the hardened ground pilot of the boring bar. This setup eliminates the cantilever problems common to boring.

Because the rotational relationship between the work and the tool is a simple one and is employed on several types of machine tools, such as lathes, drilling machines, and milling machines, boring very frequently is done on such machines. However, several machine tools have been developed primarily for boring, especially in cases involving large workpieces or for large-volume boring of smaller parts. Such machines as these are also capable of performing other operations, such as milling and turning. Because boring frequently follows drilling, many boring machines also can do drilling, permitting both operations to be done with a single setup of the work.

Drilling

Drilling on lathes is done with the drill mounted in the tailstock quill of engine lathes and fed by hand against a rotating workpiece. Straight-shank drills can be held in Jacobs chucks as shown in Figure 23-5. Drills with taper shanks are mounted directly in the quill hole. Drills can also be mounted in the turrets of modern slant-bed lathes and fed automatically on the rotational axis of the workpiece. It also is possible to drill on a lathe with the drill bit mounted and rotated in the spindle while the work remains stationary, supported on the tailstock or the carriage of the lathe.

Usual speeds are used for drilling in a lathe. Because the feed is manual, care must be exercised, particularly in drilling small holes. Coolants should be used where required.

Drilling on a lathe

Boring the drilled hole

Face plate

FIGURE 23-5   Drilling and boring in a lathe using fixture (fix) mounted on a face plate (fp) to hold the workpiece (w).

In drilling deep holes, the drill should be withdrawn occasionally to clear chips from the hole and to aid in getting coolant to the cutting edges. See Chapter 24 for further discussion on drilling.

Reaming

Reaming on a lathe involves no special precautions. Reamers are held in the tailstock quill, taper-shank types being mounted directly and straight-shank types by means of a drill chuck. Rose-chucking reamers usually are used (see Chapter 24). Fluted-chucking reamers also may be used, but these should be held in some type of holder that will permit the reamer to float.

Knurling

Knurling produces a regularly shaped, roughened surface on a workpiece. Although knurling also can be done on other machine tools, even on flat surfaces, in most cases it is done on external cylindrical surfaces using lathes. Knurling is a chipless, cold-forming process, using a tool of the type shown in Figure 23-6. The two hardened rolls are pressed against the rotating workpiece with sufficient force to cause a slight outward and lateral displacement of the metal so as to form the knurl, a raised, diamond pattern. Another type of knurling tool produces the knurled pattern by cutting chips. Because it involves less pressure and thus does not tend to bend the workpiece, this method is often preferred for workpieces of small diameter and for use on automatic or semiautomatic machines.

Turning and Boring Tapers

The turning and boring of uniform tapers are common lathe operations. Such tapers can be specified either in degrees of included angle between the sides or as the change in diameter per unit of length—millimeters per millimeter or inches per foot.

Four methods are available for turning external tapers on a lathe, and three for boring internal tapers. The simplest method employs the compound rest. (Refer ahead to Figures 23-10 and 23-11 for details and terminology of an engine lathe.) This method is suitable for both external and internal tapers. See Figure 23-1 for a taper turning schematic. However, because the length of travel of the compound rest is quite limited—seldom over a few inches—only short tapers can be turned or bored by this method. This method is partially useful for steep tapers. The compound rest is swiveled to the desired angle and locked in position. The compound slide is then fed manually to produce the desired taper. The tool should be set at exactly the height of the axis of rotation of the workpiece in all taper turning and boring.

Because the graduated scale on the base of the compound rest usually is calibrated only to 1° divisions, it is difficult to make the angle setting with accuracy. If accuracy is required, tapers made by this method are checked by means of plug or ring gages, and the setting of the compound rest readjusted until the gage fits perfectly. Also, the compound rest cannot be set directly to the correct angle if the taper is dimensioned in millimeters per millimeter or inches per foot.

Both external and internal tapers can be made on a lathe by using a *taper attachment* (Figure 23-7). In this device there is an extension bolted to the rear of the carriage. When the carriage is moved, the cross slide is caused to move transversely.

A raised *guide bar* is pivoted to any angle desired (within its limits). A *guide shoe* slides on the guide bar, so that when the carriage is moved longitudinally along the ways of the lathe, the guide follows the guide bar and moves the cross slide and tool post transversely to provide the proper taper angle.

Graduations of taper in mm/mm or in./ft are provided at one end (degrees at the other end), so the attachment can be set to the taper desired. While taper attachments provide an excellent and convenient method of cutting tapers, they ordinarily can be used only for tapers of less than 0.5 mm/mm or 6 in./ft.

External tapers also can be turned on workpieces that are mounted between centers by *setting over the tailstock*. This method is illustrated in Figure 23-8. The tailstock is

FIGURE 23-6 (a) Knurling in a lathe, using a forming-type tool, and showing the resulting pattern on the workpiece; (b) knurling tool with forming rolls. *(Courtesy of Armstrong Brothers Tool Company.)*

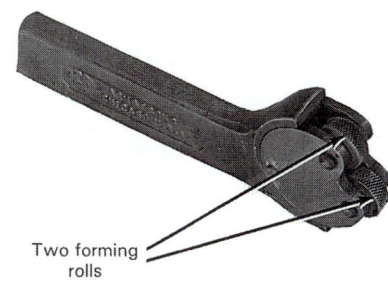

Two forming rolls

(a)   (b)

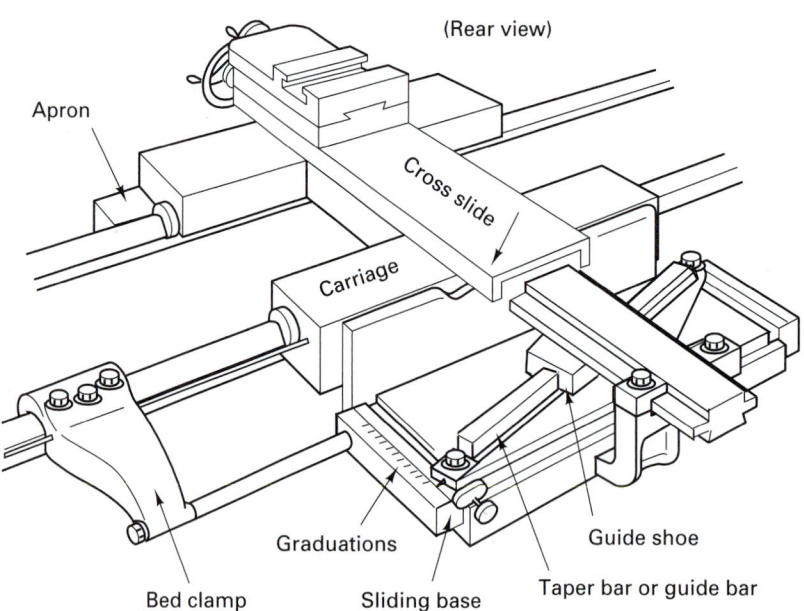

FIGURE 23-7   Rear view of carriage of an engine lathe. Taper attachment moves cross slide transversely when carriage moves, but only if the bed clamp is fastened.

moved out of line with the headstock spindle. The set-off distance from the centerline is given by the formula in the figure. This method is limited to small tapers and is seldom used.

In specifying tapers or drawings, one must remember that it is difficult for the machinist to measure the smaller diameter of a taper accurately if it is the end of a workpiece.

Both internal and external tapers can be programmed on a numerically controlled (NC) lathe. The movement of two perpendicular axes can be programmed, resulting in a tapered surface. See Chapter 29 for a discussion of interpolation and NC equipment.

Special Attachments

Milling can be done on a lathe but requires a special attachment for engine lathes. The *milling attachment* is a special vise that attaches to the cross slide to hold work. The milling cutter is mounted and rotated by the spindle. The work is fed by means of the cross-slide screw. *Tool-post grinders* are often used to permit grinding to be done on a lathe.

Duplicating attachments are available that, guided by a template, will automatically control the tool movements for turning irregularly shaped parts. In some cases the first piece, produced in the normal manner, may serve as the template for duplicate parts. To a large extent, duplicating lathes using templates have been replaced by numerically controlled lathes.

Dimensional Accuracy

Dimensional accuracy in turning operations is controlled by many factors, including the wear at the nose of the tool (Figure 23-9). Precision is influenced by deflection due to the cutting forces and surface roughness. Tool wear causes the workpiece dimension to change from the initial diameter when the tool is sharp to the diameter obtained after the tool has

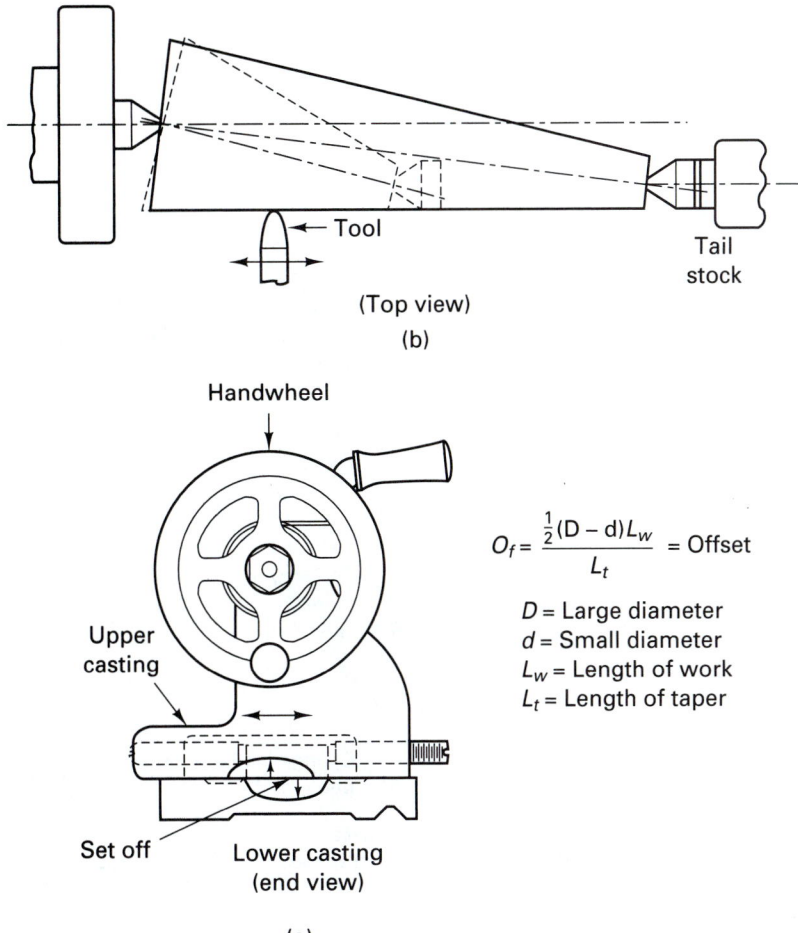

(Top view)
(b)

$O_f = \dfrac{\frac{1}{2}(D-d)L_w}{L_t}$ = Offset

D = Large diameter
d = Small diameter
L_w = Length of work
L_t = Length of taper

(end view)

(a)

FIGURE 23-8 Method of turning tapers by offsetting the tailstock.

worn. The cutting forces increase as the tool wears, resulting in increased deflection between the workpiece and the cutting tool. Built-up edge (BUE) may form at the tip of the cutting tool. BUE has the tendency to change the actual diameter of the workpiece. Thus, to hold close tolerances, the size of the wear land, the magnitude of the radial (thrust) force, and elimination of the BUE should be taken into account.

Dimensional accuracy will also be influenced by the workpiece shape, the material, the rigidity of all elements, the surface finish, and vibrations. For example, holding the dimension accuracy of a boring operation on a deep hole is a problem, due to the deflection (rigidity) of the boring bar.

Turned surfaces display characteristic turning grooves that are produced by the feed and the tool tip corner radius, as shown in Figure 23-9. The roughness resulting from feed marks from a round-nosed tool can be approximated by the formula

$$y = CR - \frac{\sqrt{CR^2 - f_r^2}}{4} \simeq \frac{f_r^2}{8CR} \qquad (23\text{-}10)$$

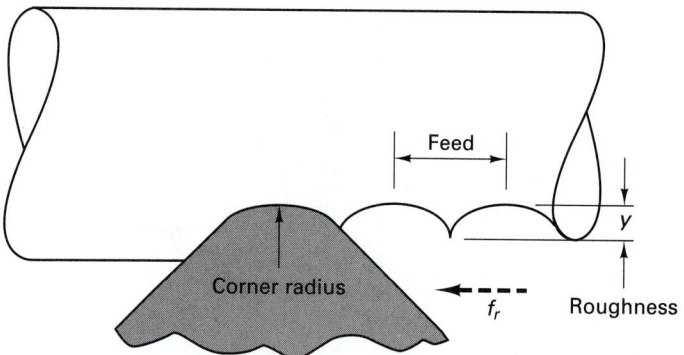

The feed and the corner radius of the cutting tool influences the surface roughness

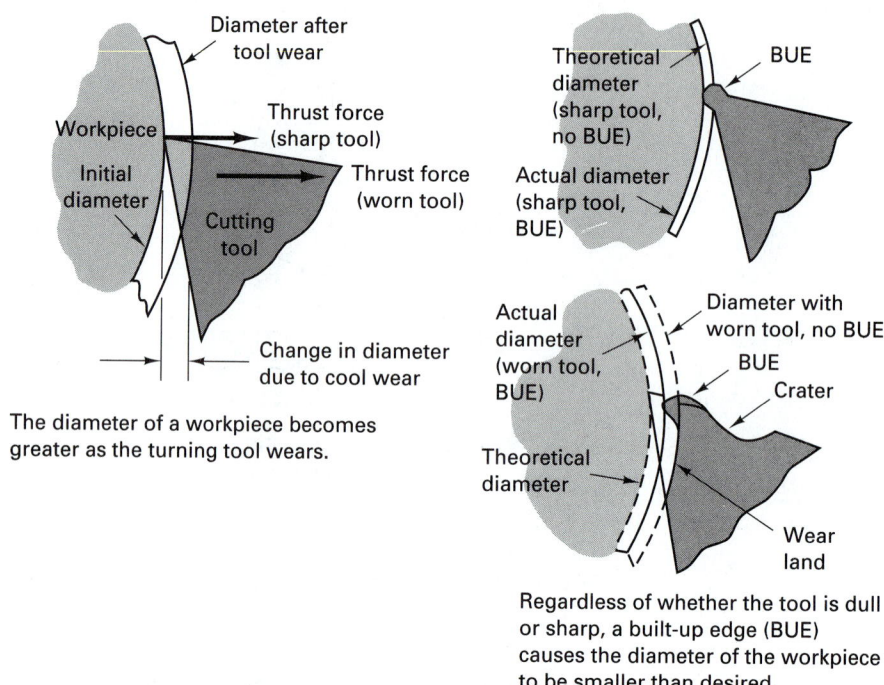

The diameter of a workpiece becomes greater as the turning tool wears.

Regardless of whether the tool is dull or sharp, a built-up edge (BUE) causes the diameter of the workpiece to be smaller than desired.

FIGURE 23-9   Accuracy and precision in turning is a function of many factors.

where y is the roughness height, CR the corner radius of insert, and f_r the feed rate (in./rev). To improve the surface finish, reduce feed and increase the corner radius.

Other factors, including built-up edge formations, cutting-edge sharpness, and tool-wear grooves in the flank wear area, also affect the surface finish in turning. Flank wear and BUE can combine to affect both surface finish and accuracy, as shown in Figure 23-9. Wear on the corner radius may cause grooves and nicks, which produce additional surface roughness on the finish-turned surfaces. Thus, to hold the surface roughness within specified limits, minimize tool wear and use small feeds and large-corner-radius tools. To minimize BUE formation, employ cutting speeds higher than those used in rough turning operations.

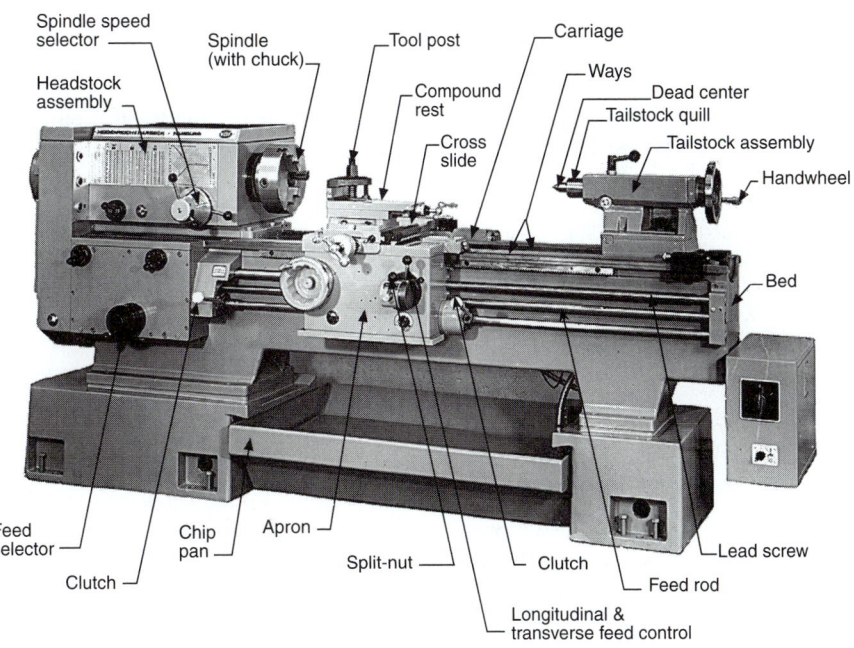

FIGURE 23-10   Modern engine lathe, with the principal parts named. *(Courtesy of Heidenreich & Harbeck.)*

■ 23.3 LATHE DESIGN AND TERMINOLOGY

Knowing the terminology of a machine tool is fundamental to understanding how it performs the basic processes, how the workholding devices are interchanged, and how the cutting tools are mounted and interfaced to the work. *Lathes* are machine tools designed primarily to do turning, facing, and boring. Very little turning is done on other types of machine tools, and none can do it with equal facility. Because lathes also can do facing, drilling, and reaming, their versatility permits several operations to be done with a single setup of the workpiece. Consequently, the lathe is the most common machine tool.

Lathes in various forms have existed for more than 2000 years, but modern lathes date from about 1797, when Henry Maudsley developed one with a leadscrew, providing controlled, mechanical feed on the tool. This ingenious Englishman also developed a change-gear system that could connect the motions of the spindle and leadscrew and thus enable threads to be cut (see Chapter 1).

Lathe Design

The essential components of an *engine lathe* (Figure 23-10) are the bed, headstock assembly, tailstock assembly, carriage assembly, quick-change gearbox, and the leadscrew and feed rod. The *bed* is the base and backbone of a lathe. It usually is made of well-normalized or aged gray or nodular cast iron and provides a heavy, rigid frame on which all the other basic components are mounted. Two sets of parallel, longitudinal *ways*, inner and outer, are contained on the bed. On modern lathes, the ways are surface-hardened and precision-machined, and care should be taken to assure that the ways are not damaged. Any inaccuracy in them usually means that the accuracy of the entire lathe is destroyed.

The *headstock*, mounted in a fixed position on the inner ways, provides powered means to rotate the work at various rpm values. Essentially, it consists of a hollow spindle, mounted in accurate bearings, and a set of transmission gears—similar to a truck transmission—through which the spindle can be rotated at a number of speeds. Most lathes provide from 8 to 18 choices of rpm. On modern lathes all the rpm rates can be obtained merely by moving from two to four levers. An increasing trend is to provide a continuously variable spindle rpm through electrical or mechanical drives.

The accuracy of a lathe is greatly dependent on the *spindle*. It carries the workholders and is mounted in heavy bearings, usually preloaded tapered roller or ball types. The spindle has a hole extending through its length, through which long bar stock can be fed. The size of this hole is an important dimension of a lathe because it determines the maximum size of bar stock that can be machined when the materials must be fed through the spindle.

The spindle protrudes from the gearbox and contains means for mounting various types of workholding devices (chucks; face and dog plates; collets). Power is supplied to the spindle from an electric motor through a V-belt or silent-chain drive. Most modern lathes have motors of from 5 to 25 hp to provide adequate power for carbide and ceramic tools at the higher cutting speeds.

For the classic engine lathe, the *tailstock* assembly consists, essentially, of three parts. A lower casting fits in the inner ways of the bed, can slide longitudinally, and can be clamped in any desired location. An upper casting fits on the lower one and can be moved transversely upon it, on some type of keyed ways, to permit aligning the tailstock and headstock spindles (for turning tapers). The third major component of the assembly is the *tailstock quill*. This is a hollow steel cylinder, usually about 2 to 3 in. in diameter, that can be moved longitudinally in and out of the upper casting by means of a handwheel and screw. The open end of the quill hole has a Morse taper. Cutting tools or a *lathe center* (see Figure 23-33) are held in the quill. A graduated scale usually is engraved on the outside of the quill to aid in controlling its motion in and out of the upper casting. A locking device permits clamping the quill in any desired position. In recent years, dual-spindle NC turning centers have emerged, where a subspindle replaces the tailstock assembly. Parts can be automatically transferred from the spindle to the subspindle for turning the back end of the part. See Chapter 29 for a discussion of NC turning centers.

The *carriage assembly* (Figure 23-11), together with the apron, provides the means for mounting and moving cutting tools. The *carriage*, a relatively flat H-shaped casting, rides on the outer set of ways on the bed. The *cross slide* is mounted on the carriage and can be moved by means of a feed screw that is controlled by a small handwheel and a graduated dial. The cross slide thus provides a means for moving the lathe tool in the facing or cutoff direction.

On most lathes, the tool post is mounted on a *compound rest*. The compound rest can rotate and translate with respect to the cross slide, permitting further positioning of the tool with respect to the work. The *apron*, attached to the front of the carriage, has the controls for providing manual and powered motion for the carriage and powered motion for the cross slide. Figure 23-11 shows front and rear views of a typical apron. The carriage is moved parallel to the ways by turning a handwheel on the front of the apron, which is geared to a pinion on the back side. The pinion engages a rack that is attached beneath the upper front edge of the bed in an inverted position.

Powered movement of the carriage and cross slide is provided by a rotating *feed rod*, shown in Figure 23-10. The feed rod, which contains a keyway, passes through two reversing bevel pinions (Figure 23-11) and is keyed to them. Either pinion can be activated

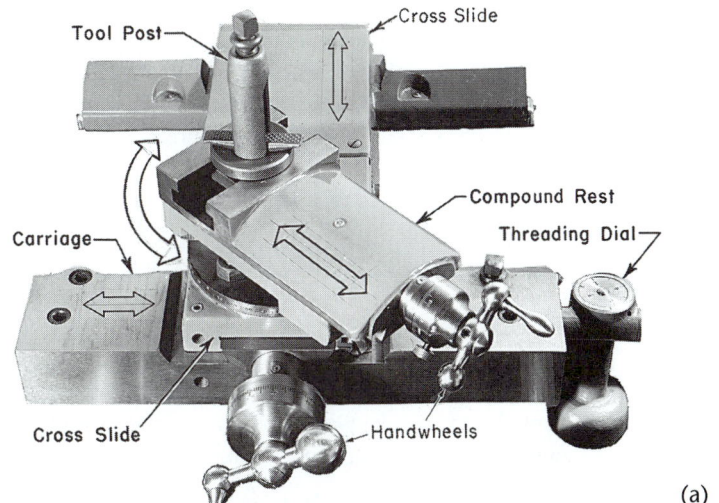

(a)

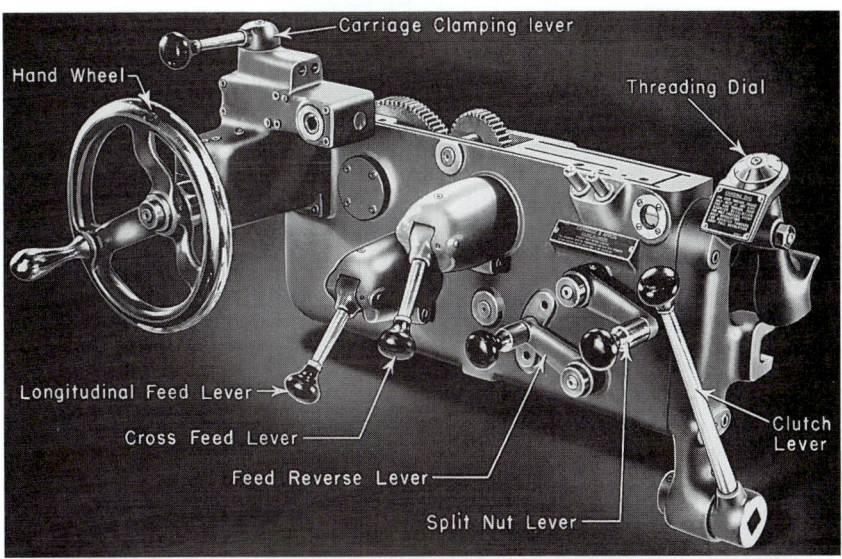

(b)

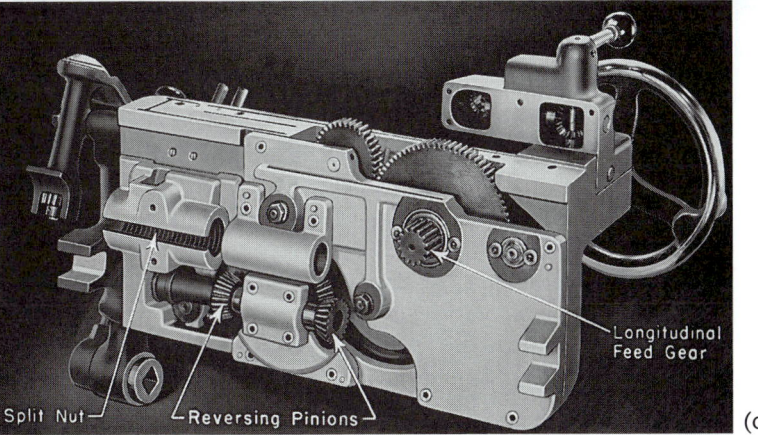

(c)

FIGURE 23-11 The carriage, cross slide, and apron assembly for an engine lathe.

by means of the feed reverse lever, thus providing "forward" or "reverse" power to the carriage. Suitable clutches connect either the rack pinion or the cross-slide screw to provide longitudinal motion of the carriage or transverse motion of the cross slide.

For cutting threads, a *leadscrew* (Figure 23-10) is used. When a friction clutch is used to drive the carriage, motion through the leadscrew is by a direct, mechanical connection between the apron and the leadscrew. A *split nut* (Figure 23-11) is closed around the leadscrew by means of a lever on the front of the apron directly driving the carriage without any slippage.

Modern lathes have *quick-change gearboxes*, driven by the spindle, which connect the feed rod and leadscrew. The associated gearing, the leadscrew, and feed rod connect the carriage to the spindle, so the cutting tool can be made to move a specific distance, either longitudinally or transversely, for each revolution of the spindle. The calculations for turning rpm and feed in inches per revolution are "mechanically related." Typical lathes may provide as many as 48 feeds, ranging from 0.002 to 0.118 in. (0.05 to 3 mm) per revolution of the spindle, and through the leadscrew, leads from $1\frac{1}{2}$ to 92 threads per inch.

Size Designation of Lathes

The size of a lathe is designated by two dimensions. The first is known as the *swing*. This is the maximum diameter of work that can be rotated on a lathe. Swing is approximately twice the distance between the line connecting the lathe centers and the nearest point on the ways. The maximum diameter of a workpiece that can be mounted between centers is somewhat less than the swing diameter because the workpiece must clear the carriage assembly as well as the ways. The second size dimension is the *maximum distance between centers*. The swing thus indicates the maximum workpiece diameter that can be turned in the lathe, while the distance between centers indicates the maximum length of workpiece that can be mounted between centers.

Types of Lathes

Lathes used in manufacturing can be classified as speed, engine, toolroom, turret, automatics, tracer, and numerical control turning centers, the latter discussed in Chapter 29. Speed lathes usually have only a headstock, a tailstock, and a simple tool post mounted on a light bed. They ordinarily have only three or four speeds and are used primarily for wood turning, polishing, or metal spinning. Spindle speeds up to about 4000 rpm are common.

Engine lathes are the type most frequently used in manufacturing. Figure 23-10 is an example of this type. They are heavy-duty machine tools with all the components described previously and have power drive for all tool movements except on the compound rest. They commonly range in size from 12 to 24 in. swing and from 24 to 48 in. center distances, but swings up to 50 in. and center distances up to 12 ft are not uncommon. Binns Machinery, in Cincinnati, Ohio, builds very large lathes (36- to 60-ft-long beds) which are extremely rigid and therefore capable of performing roughing cuts in iron and steel at depths of cut of $\frac{1}{2}$ to 2 in., cutting speed 50 to 200 sfpm with WC tools run at feed rates of 0.010 to 0.100 in./rev. To perform such heavy cuts requires rigidity in the machine tool, the cutting tools, the workholder, and the workpiece (using steady rests and other supports) and high horsepower values (50 to 100 hp).

Most engine lathes are equipped with chip pans and a built-in coolant circulating system. Smaller engine lathes, with swings usually not over 13 in., also are available in *bench type*, designed for the bed to be mounted on a bench or table.

Toolroom lathes have somewhat greater accuracy and, usually, a wider range of speeds and feeds than ordinary engine lathes. Designed to have greater versatility to meet the requirements of tool and die work, they often have a continuously variable spindle speed range and shorter beds than ordinary engine lathes of comparable swing, since they are generally used for machining relatively small parts. They may be either bench or pedestal type.

Several types of special-purpose lathes are made to accommodate specific types of work. On a *gap-bed lathe*, for example, a section of the bed, adjacent to the headstock, can be removed to permit work of unusually large diameter to be swung. Another example is the *wheel lathe*, which is designed to permit the turning of railroad-car wheel-and-axle assemblies.

Although engine lathes are versatile and very useful, the time required for changing and setting tools and for making measurements on the workpiece is often a large percentage of the cycle time. Often, the actual chip-production time is less than 30% of the total cycle time. Methods to reduce setup and tool change time sharply are discussed in Chapters 28 and 29. Much of the operator's time is consumed by simple, repetitious adjustments and in watching chips being made. The placement of machines into cells as discussed in Chapter 41 greatly increases the productivity of the workers because they can run more than one machine. Turret lathes, screw machines, and other types of semiautomatic and automatic lathes have been highly developed and are widely used in manufacturing as another means to improve cutting productivity.

Turret Lathes. The basic components of a *turret lathe* are depicted in Figure 23-12 and 23-13. Basically, a longitudinally feedable, hexagon turret replaces the tailstock. The turret, on which six tools can be mounted, can be rotated about a vertical axis to bring each tool into operating position, and the entire unit can be translated parallel to the ways, either manually or by power, to provide feed for the tools. When the turret assembly is backed away from the spindle by means of a capstan wheel, the turret indexes automatically at the end of its movement, thus bringing each of the six tools into operating position in sequence.

The square turret on the cross slide can be rotated manually about a vertical axis to bring each of the four tools into operating position. On most machines, the turret can be moved transversely, either manually or by power, by means of the cross slide, and longitudinally through power or manual operation of the carriage. In most cases, a fixed toolholder also is added to the back end of the cross slide; this often carries a parting tool.

Through these basic features of a turret lathe, a number of tools can be set up on the machine and then quickly be brought successively into working position so that a complete part can be machined without the necessity for further adjusting, changing tools, or making measurements.

The two basic types of turret lathes are the ram-type turret lathe and the saddle-type turret lathe. The *saddle-type turret lathe* provides a more rugged mounting for the hexagon turret than can be obtained by the ram-type mounting (Figure 23-14). In the *ram-type turret lathe*, the ram and the turret are moved up to the cutting position by means of the capstan wheel, and the power feed is then engaged. As the ram is moved toward the headstock, the turret is automatically locked into position so that rigid tool support is obtained. A set of rotary stopscrews, such as those shown in Figure 23-13, control the forward travel of the ram, one stop being provided for each face on the turret. The proper stop is brought into operating position automatically when the turret is indexed. A similar set of stops usually is provided to limit movement of the cross slide. In *saddle-type lathes*, the main turret is mounted directly on the saddle, and the entire saddle and turret assembly reciprocates. Larger turret lathes usually have this type of mounting. However, because the saddle-turret assembly is rather heavy, this type of mounting provides less rapid turret reciprocation. When such lathes are used with heavy tooling for making heavy or multiple

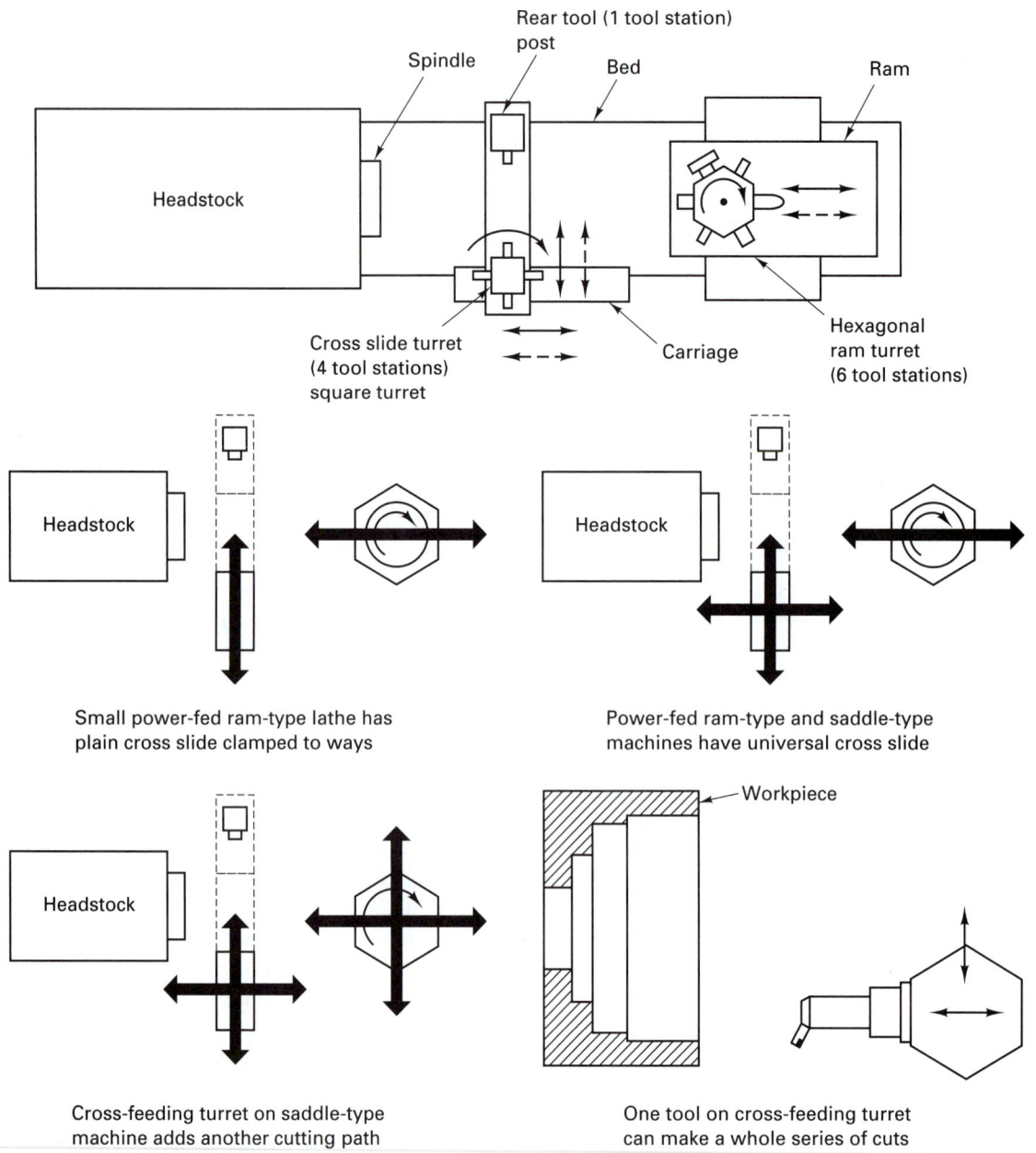

FIGURE 23-12   Block diagrams (top views) of a ram-type turret lathe.

cuts, a *pilot arm* attached to the headstock engages a pilot hole attached to one or more faces of the turret to give additional rigidity. Such a device is shown in Figure 23-14. Turret lathe headstocks can shift rapidly between spindle speeds and brake rapidly to stop the spindle very quickly. They also have automatic stock feeding for feeding bar stock through the spindle hole. If the work is to be held in a chuck, some type of air-operated chuck, or special clamping fixture is often employed to reduce the time required for part loading and unloading.

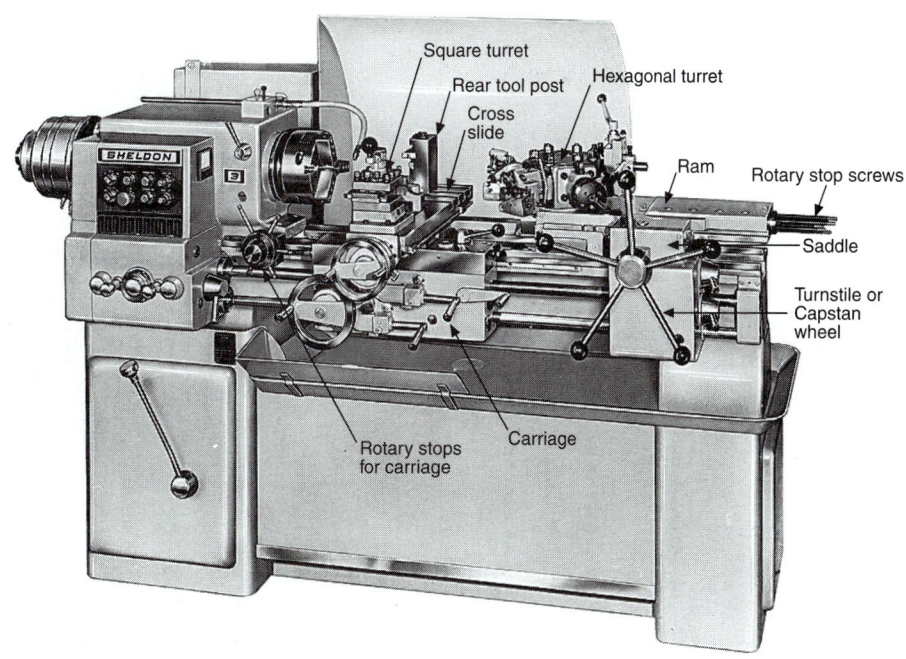

FIGURE 23-13   Ram-type turret lathe. *(Courtesy of Sheldon Machine Company, Incorporated.)*

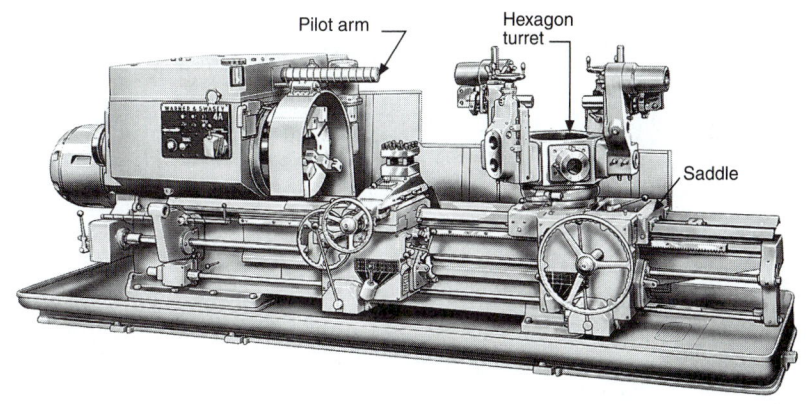

FIGURE 23-14   Saddle-type turret lathe with a side-hung carriage. *(Courtesy of Warner & Swassey Company.)*

Vertical Turret Lathes.   In machining large and/or heavy parts, such as pinions, couplings, and ring gears, vertical turret lathes are used. These are essentially regular turret lathes turned on end. Their rotary work tables commonly range from 24 to 48 in. in diameter and are equipped with both removable chuck jaws and T-slots for clamping the work. The saddle carries one or two five- or six-sided turrets. Usually, each motion of successive tools can be controlled by means of stops so that duplicate workpieces can be machined with one tooling setup. Figure 23-15 shows a twin-spindle vertical turret lathe with the guards, covers, and tools removed.

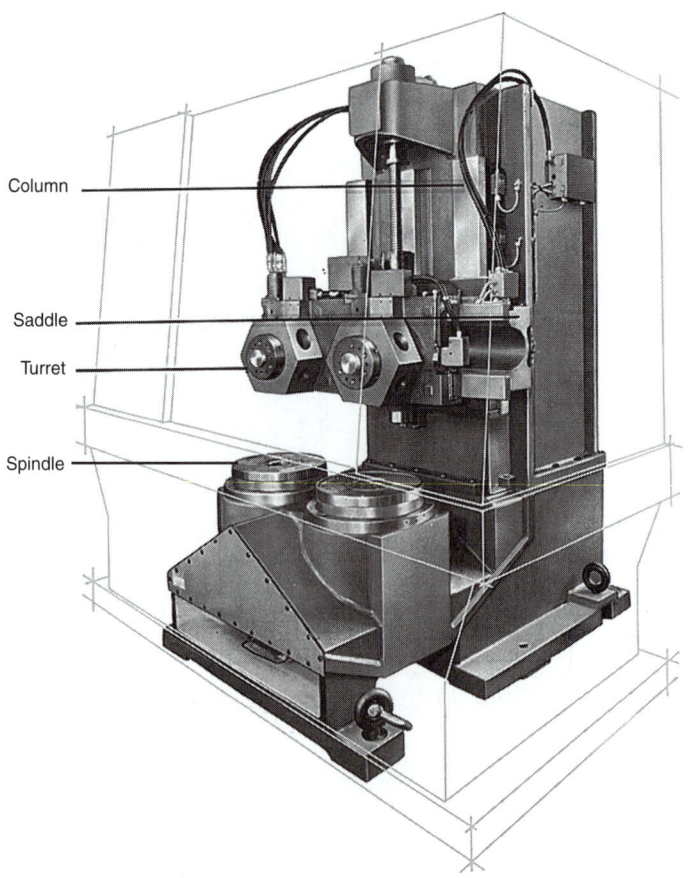

FIGURE 23-15 Basic structure of a twin-spindle vertical turret lathe, typically NC or CNC controlled. *(Courtesy of Bullard Aku-Turn.)*

Automatic Turret Lathes. After a turret lathe is tooled, the skill required of the operator is very low, and the motions are simple and repetitive. As a result, several types of automatic turret lathes have been developed that require no operator. One type uses buttons and knobs on a control panel to define (program) machine motions. A second type has a turret, the movement of which is controlled by setting trip blocks and pins. Ordinary turret lathes use the 10-station tooling setups for complete machining of a piece and minimize machine-controlling time. However, an operator is required to control the machine and to feed the work into machining position. Automatic turret lathes can eliminate these last two functions. They usually have provision for manual operation and are not quite as productive as *screw machines*, which are lathes designed for completely automatic operation. Screw machines originally were designed for machining small parts, such as screws, bolts, bushings, and so on, from bar stock—hence the name "screw machines." Now they are used for producing a wide variety of parts, covering a considerable range of sizes, and are even used for some chucking-type work.

Single-Spindle Automatic Screw Machines. There are two common types of *single-spindle screw machines.* One, an American development and commonly called the turret

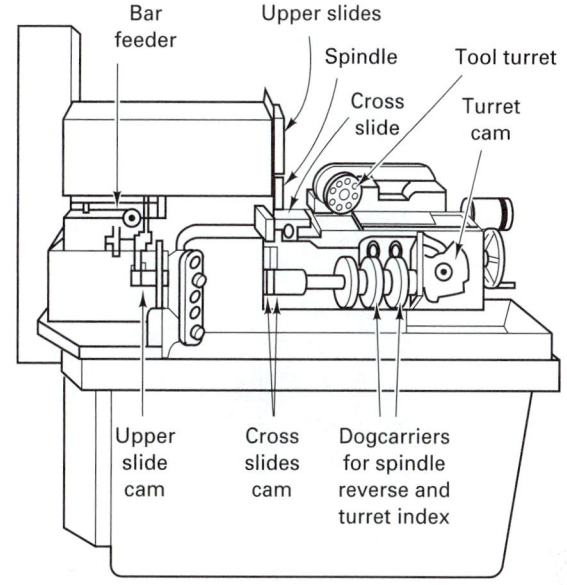

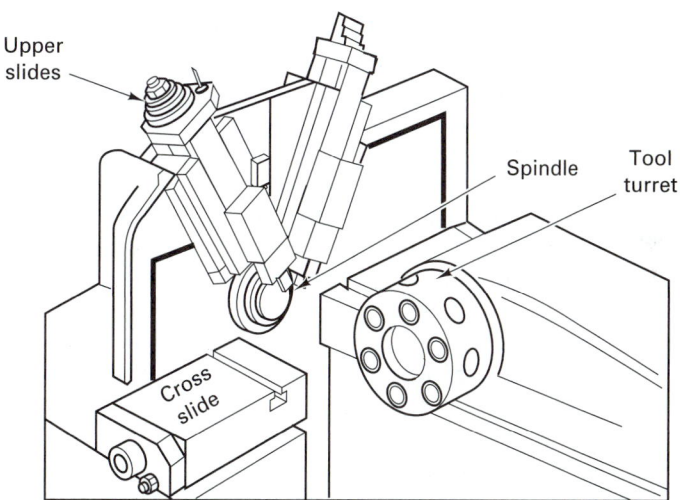

FIGURE 23-16   On the turret-type single-spindle automatic, the tools must take turns to make cuts.

type (Brown & Sharpe), is shown in Figure 23-16. The other is of Swiss origin and is referred to as the Swiss type. The *Brown & Sharpe screw machine* is essentially a small automatic turret lathe, designed for bar stock, with the main turret mounted in a vertical plane on a ram. Front and rear toolholders can be mounted on the cross slide. All motions of the turret, cross slide, spindle, chuck, and stock-feed mechanism are controlled by cams. The turret cam is essentially a program that defines the movement of the turret during a cycle. These machines usually are equipped with an automatic rod-feeding magazine that feeds a new length of bar stock into the collet (the workholding device) as soon as one rod is completely used.

FIGURE 23-17   Close-up view of a Swiss-type screw machine, showing the tooling and radial tool sides, actuated by rocker arms. *(Courtesy of George Gorton Machine Corporation.)*

Often, screw machines of the Brown & Sharpe type are equipped with a transfer or "picking" attachment. This device picks up the workpiece from the spindle as it is cut off and carries it to a position where a secondary operation is performed by a small, auxiliary power head. In this manner screwdriver slots are put in screw heads, small flats are milled parallel with the axis of the workpiece, or holes are drilled normal to the axis.

On the *Swiss-type automatic screw machine*, the cutting tools are held and moved in radial slides (Figure 23-17). Disk cams move the tools into cutting position and provide feed into the work in a radial direction only; they provide any required longitudinal feed by reciprocating the headstock.

Most machining on Swiss-type screw machines is done with single-point cutting tools. Because they are located close to the spindle collet, the workpiece is not subjected to much deflection. Consequently, these machines are particularly well suited for machining very small parts.

Both types of single-spindle screw machines can produce work to close tolerances, the Swiss-type probably being somewhat superior for very small work. Precision of 0.0002 to 0.0005 in. are not uncommon. The time required for setting up the machine is usually an hour or two and can be much less. One person can tend many machines, once they are properly tooled. They have short cycle times, frequently less than 30 seconds per piece.

Multiple-Spindle Automatic Screw Machines.   Single-spindle screw machines utilize only one or two tooling positions at any given time. Thus the total cycle time per work-piece is the sum of the individual machining and tool-positioning times. On *multiple-*

spindle screw machines, sufficient spindles, usually four, six, or eight, are provided so that all tools cut simultaneously. Thus the cycle time per piece is equal to the maximum cutting time of a single tool position plus the time required to index the spindles from one position to the next.

The two distinctive features of multiple-spindle screw machines are shown in Figure 23-18. First, the six spindles are carried in a rotatable drum that indexes in order to bring each spindle into a different working position. Second, a nonrotating tool slide contains the same number of toolholders as there are spindles and thus provides and positions a cutting tool (or tools) for each spindle. Tools are fed by longitudinal reciprocating motion. Most machines have a cross slide at each spindle position so that an additional tool can be fed from the side for facing, grooving, knurling, beveling, and cutoff operations. These slides also are shown in Figure 23-18. All motions are controlled automatically.

Study the steps on the tooling sheet for making a part shown in Figure 23-18. With a tool position available on the end tool slide for each spindle (except for a stock-feed stop at position 6), when the slide moves forward, these tools cut essentially simultaneously. At the same time, the tools in the cross slides move inward and make their cuts. When the forward cutting motion of the end tool slide is completed, it moves away from the work, accompanied by the outward movement of the radial slides. The spindles are indexed one position, by rotation of the spindle carrier, to position each part for the next operation to be performed. At spindle position 5, finished pieces are cut off. Bar stock is fed to the correct length for the beginning of the next operation. Thus a piece is completed each time the tool slide moves forward and back.

Multiple-spindle screw machines are made in a considerable range of sizes, determined by the diameter of the stock that can be accommodated in the spindles. There may be four, five, six, or eight spindles. The operating cycle of the end tool slide is determined by the operation that requires the longest time.

Once a multiple-spindle screw machine is setup, it requires only that the bar stock feed rack be supplied and the finished products checked periodically to make sure that they are within desired tolerances. One operator usually services many machines.

Most multiple-spindle screw machines use cams to control the motions. Setting up the cams and the tooling for a given job may require from 2 to 20 hours. However, once such a machine is set up, the processing time per part is very short. Often, a piece may be completed every 10 seconds. Typically, a minimum of 2000 to 5000 parts are required in a lot to justify setting up and tooling a multiple-spindle automatic screw machine. The precision of multiple-spindle screw machines is good, but seldom as good as that of single-spindle machines. However, tolerances from 0.0005 to 0.001 in. on the diameter are typical.

Although screw machines and automatic turret lathes are automatic *types* of lathes, the term *automatic lathe* generally is applied to a lathe that is semiautomatic and makes simultaneous cuts using "massed" tooling (Figure 23-19) but does not use turret or screw-machine principles. The tools are fed and retracted automatically by means of cam-controlled mechanisms. In most cases an operator is required to load and unload the machine, so they do not repeat cycles automatically. The majority of automatic lathes have only a single spindle, but some specialized multispindle machines are also used.

In a typical *single-spindle automatic lathe*, the "massed" cutting tools are held in *tool blocks*, or *slides*, which are power actuated and controlled to move the tools into position and feed them along the work. Sometimes the front block provides only radial motion for facing, form cutting, and cutoff operations. The rear block has both radial and longitudinal motions, which are controlled by a plate cam, as illustrated in Figure 23-19, which shows the

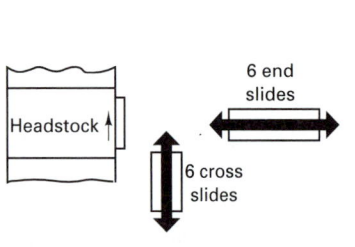

All spindles on multiple-spindle automatic have the same tool path

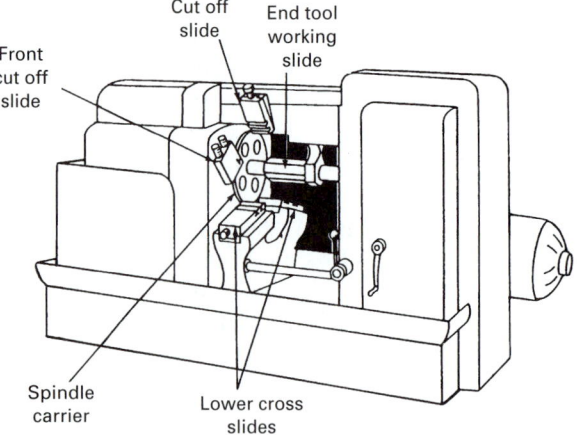

The six spindle automatic

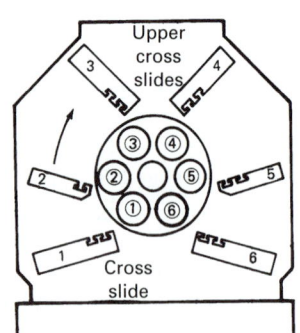

Spindle arrangement for 6 spindle automatic. The shaded circle shows the position where the barstock is usually fed. The cutoff position is the one preceding the bar feed position.

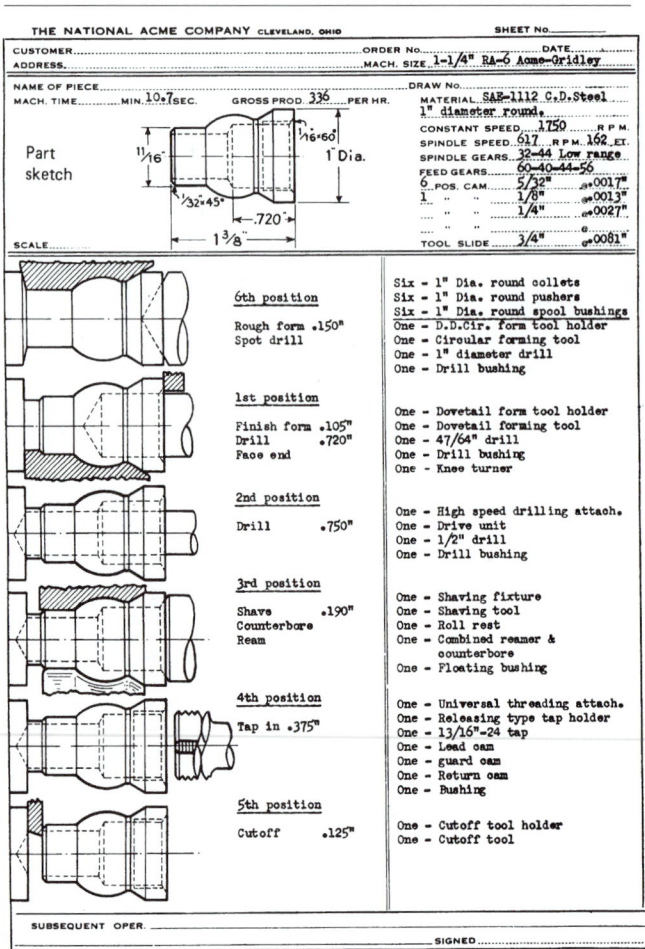

Tooling sheet for making a part on a six-spindle.

FIGURE 23-18 The multiple spindle automatic makes all cuts simultaneously and then performs the noncutting functions (tool withdrawal, index, bar feed) at high speed.

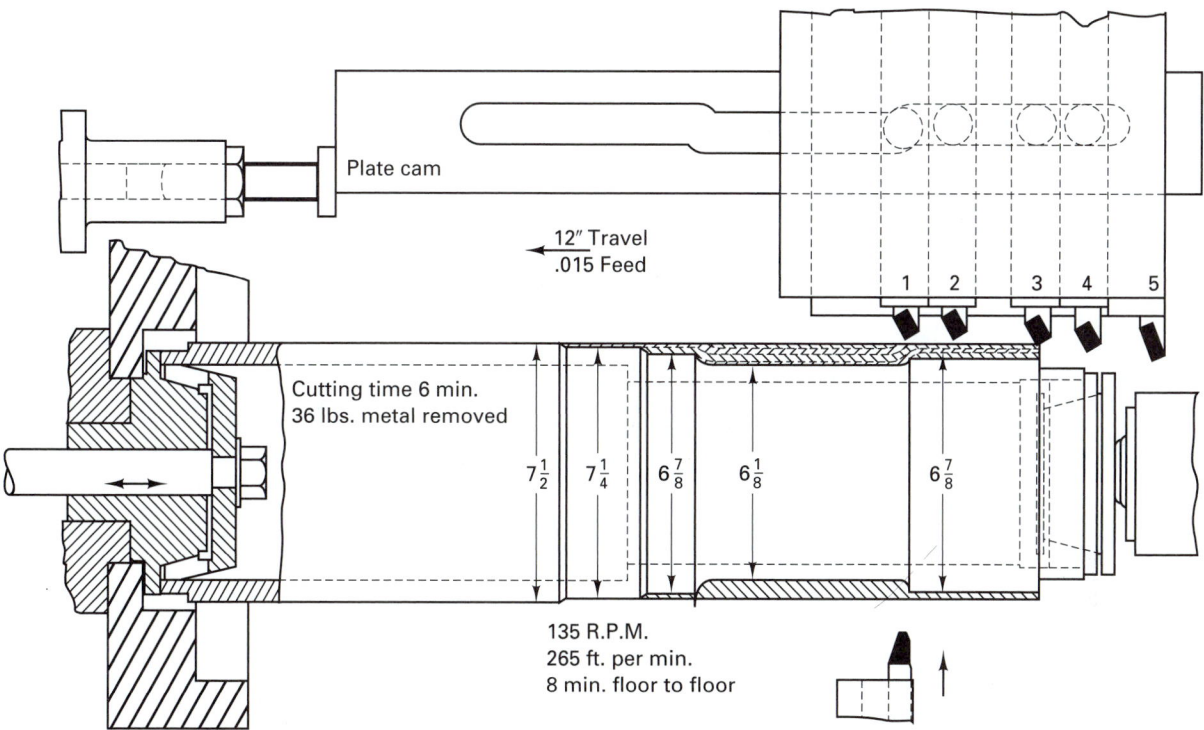

Plate cam

12″ Travel
.015 Feed

Cutting time 6 min.
36 lbs. metal removed

$7\frac{1}{2}$ $7\frac{1}{4}$ $6\frac{7}{8}$ $6\frac{1}{8}$ $6\frac{7}{8}$

135 R.P.M.
265 ft. per min.
8 min. floor to floor

1 2 3 4 5

FIGURE 23-19 Movements of the tool blocks for the machining process in a single-spindle automatic lathe, with the metal to be removed by each tool indicated. *(Courtesy of Gisholt Corporation.)*

motions of the tool blocks and the portion of the metal removed by each tool. On some machines a third overhead tool block is also provided.

In most cases the work is held between centers, utilizing various types of power-actuated chucks, collets, and tailstocks so that the work-handling time is minimized. In some cases the work is fed into the machine from a hopper-type feeding device and clamped and discharged automatically.

The total machining cycle on an automatic lathe is usually very short, often less than 1 minute. Sometimes a part is put successively into two to four automatic lathes to complete its machining. Because they are fairly flexible, quite a variety of shapes and sizes can be handled in one lathe by changing the tooling setup. Tracer and NC lathes are discussed in Chapter 29.

■ 23.4 TYPES OF BORING MACHINES

Vertical Boring and Turning Machines

Figure 23-20 shows the basic elements of a vertical boring and turning machine. These machine tools are structurally similar to double-housing planers except that the table rotates instead of reciprocating. Functionally, a vertical boring machine essentially is the same as a vertical turret lathe, but it usually has two main toolheads instead of a turret. Thus turning, facing, and usually boring (but not milling) are done on vertical boring machines.

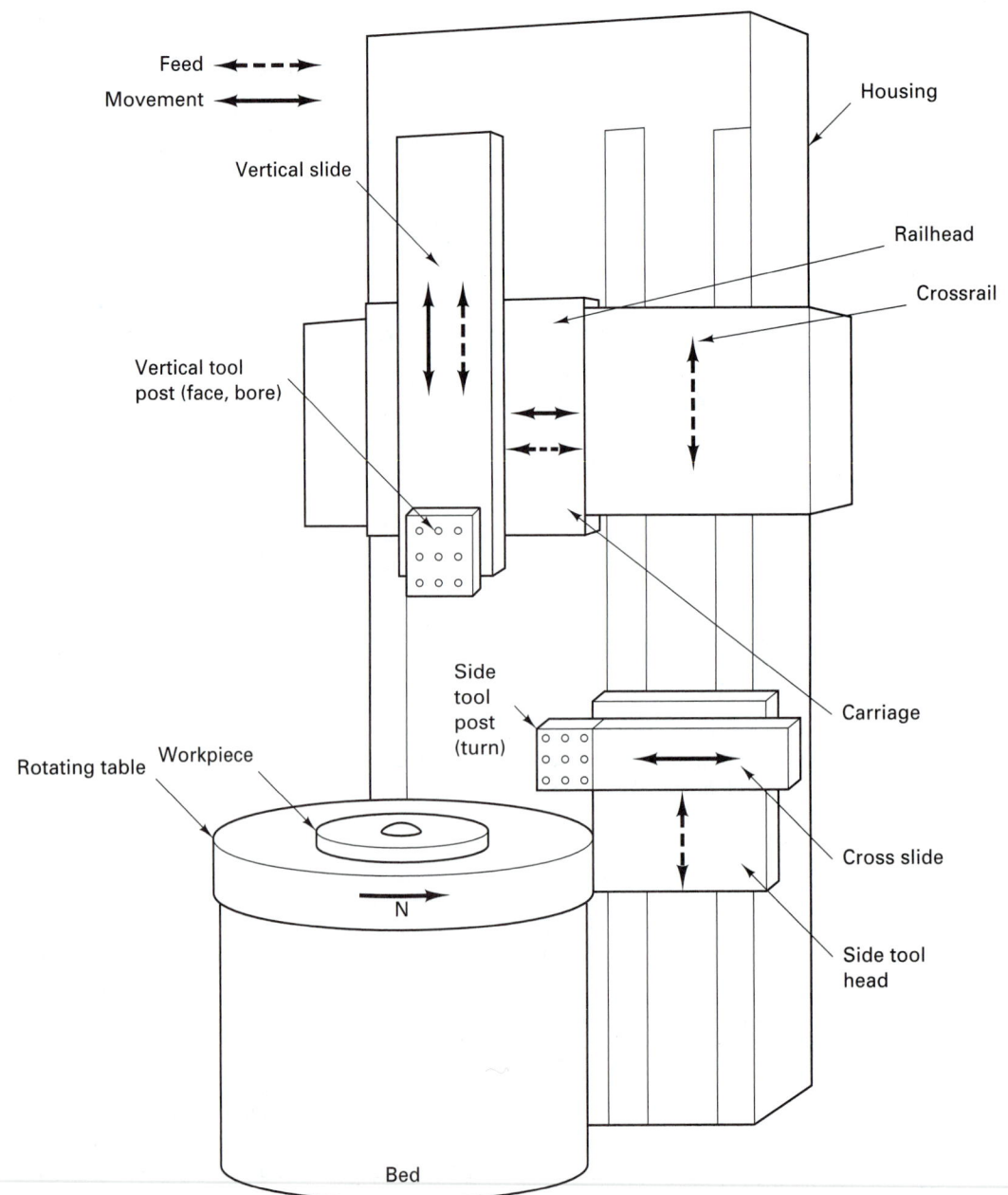

Feed

Movement

Housing

Vertical slide

Railhead

Crossrail

Vertical tool
post (face, bore)

Carriage

Side
tool
post
(turn)

Rotating table   Workpiece

Cross slide

N

Side tool
head

Bed

FIGURE 23-20   Block diagram of a vertical boring and turning machine.

Vertical boring machines come with tables ranging from about 3 to 40 ft in diameter. Toolheads have both horizontal and vertical feed and therefore can be used for boring and for facing cuts. Usually, one or both can also be swiveled about a horizontal axis to permit boring at an angle. Most machines also have a side toolhead, sometimes provided with a four-sided turret. This toolhead has vertical and horizontal feed and is used primarily for turning. Single-point tools customarily are used, for turning, facing, and boring operations.

Many modern boring machines are numerically controlled (see Chapter 29). This permits the operator to make tool settings and adjustments merely by the input of numerical data. The tool can be moved very quickly to the proper position for the next cut, reducing the amount of machine-controlling time and increasing the productivity of these large and costly machines.

Jig Borers

Jig borers are very precise vertical-type boring machines designed for use in making *jigs* and *fixtures*. From the viewpoint of boring operations, they contain no unusual features, except that the spindle and spindle bearings are constructed with very high precision. Their unique features are in the design of the worktable controls, which permits very precise movement and control, thus making them especially useful in layout work. Modern NC machining centers are capable of doing similar work, often eliminating the need for jig borers.

Horizontal Boring (Drilling, and Milling) Machines

Horizontal boring machines are very versatile and thus particularly useful in machining large parts. The basic components of these machines are indicated in Figure 23-21. The essential features are as follows:

1. A rotating spindle that can be fed horizontally (tool rotates)
2. A table that can be moved and fed in two directions in a horizontal plane
3. A headstock that can be moved vertically
4. An outboard bearing support for a long boring bar

The spindle will accept both drills and milling cutters in addition to boring bars. A wide range of rpm values is provided, and heavy bearings are incorporated that will absorb thrust in all directions. The spindle is also provided with longitudinal power feed so that drilling and boring can be done over a considerable distance without moving the table.

Boring on this type of machine is done by means of a rotating single-point tool. The tool can be mounted in either a stub-type bar, held only in the spindle, as shown in Figures 23-22 and 23-23, or in a long line-type bar that has its outer end supported in a bearing on the outboard column, as shown in Figure 23-21. The outboard bearing provides rigid support for the boring bar and permits very accurate work to be done. However, because of the flexibility inherent in long boring bar and offset tool holder, horizontal boring machines are used primarily for boring holes less than 12 in. in diameter, for long holes, or for a series of in-line holes. Unless they are very long or the shape of the workpiece prevents it, larger holes usually are bored on a vertical boring mill.

Mass Production Boring Machines

Special boring machines are built for machining specific parts in mass production. The workpiece usually remains stationary, and boring is done by one or more rotating boring tools, typically carried in a reciprocating powerhead, such as is shown in Figure 23-24. In most cases the operation is automatic once the workpiece is placed in the workholding device. Such machines usually are very accurate and often are equipped with automatic gaging and sizing controls. (See the discussion of transfer lines in Chapter 42.)

With rotating workpieces, the size of the hole is controlled by transverse movement of the toolholder. When boring is done with a rotating tool, size is controlled by changing the offset radius of the cutting-tool tip with respect to the axis of rotation. A general-

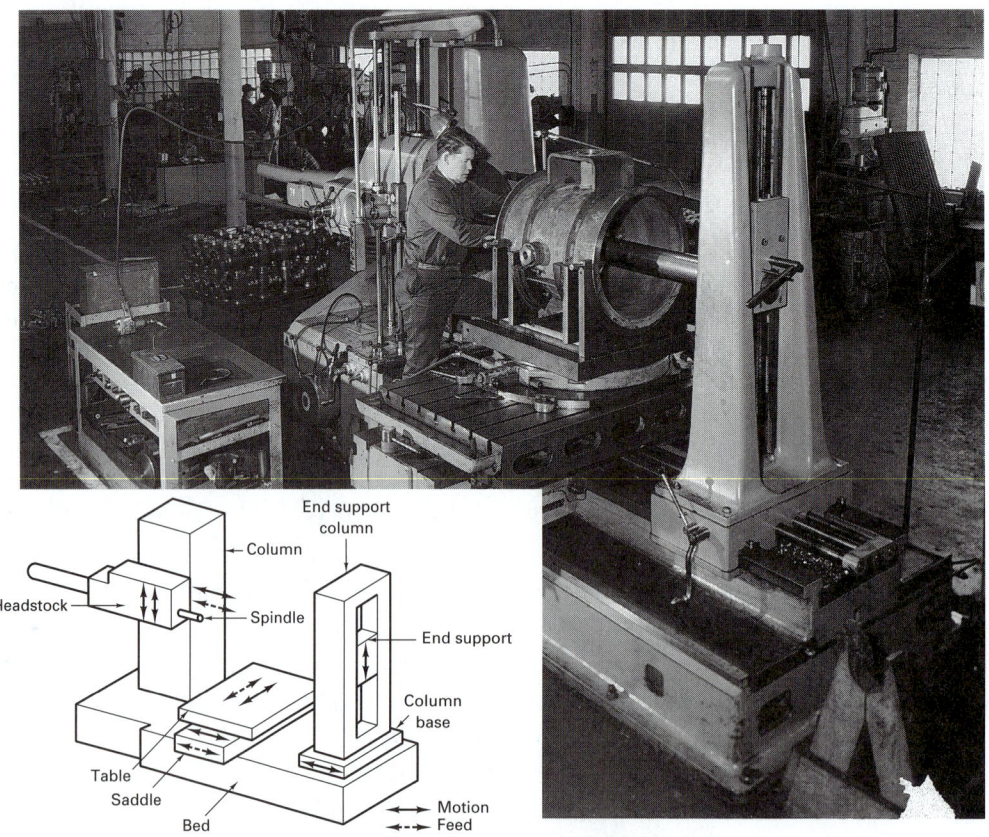

FIGURE 23-21 Boring a weldment on horizontal boring, drilling, and milling machine. A line-type boring bar is being used with an outboard bearing support. *(Courtesy of Lucas Machine Division, The New Britain Machine Company.)*

FIGURE 23-22 Adjusting boring bar, using the offset-radius principle.

purpose type of adjustable boring bar is shown in Figure 23-22. The type shown in Figure 23-23 has more precise control and is used on larger-scale manufacturing. Two or more adjustable cutting tools can be built into a single bar, thus permitting more than one diameter to be bored simultaneously. For boring relatively long holes, the type of boring bar shown in Figure 23-25 has a special advantage. As shown, the smaller, forward bit corrects misalignment of the original hole and provides a guide hole for the nose cone. The

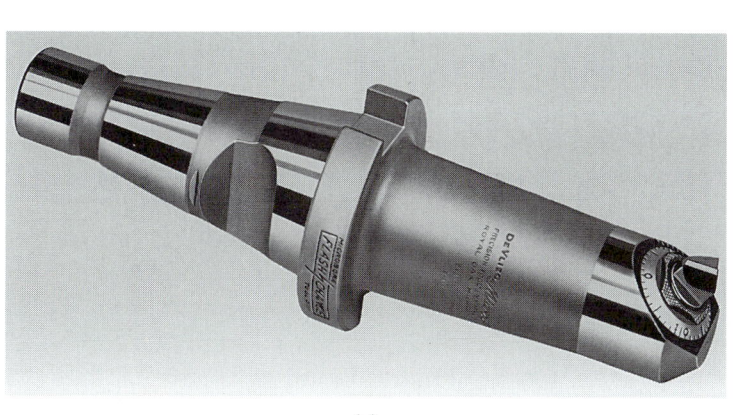
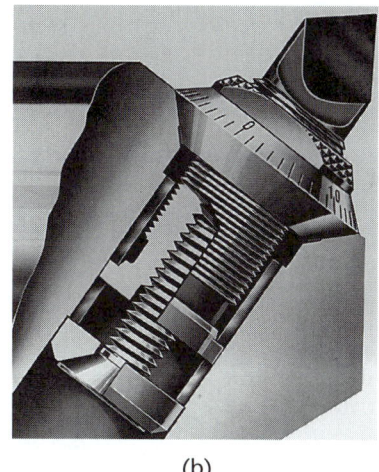

(a)   (b)

FIGURE 23-23   Adjustable boring tool. Extension of the single-point tool from the bar is adjustable, as shown in sectional view. *(Courtesy of DeVlieg Machine Company.)*

(a)   (b)

FIGURE 23-24   (a) Production-type boring machine, having multiple heads, that completes a part in 51 seconds. (b) Close-up view of one multiple-spindle boring head on a production-type machine. *(Courtesy of Healt Machine Company).*

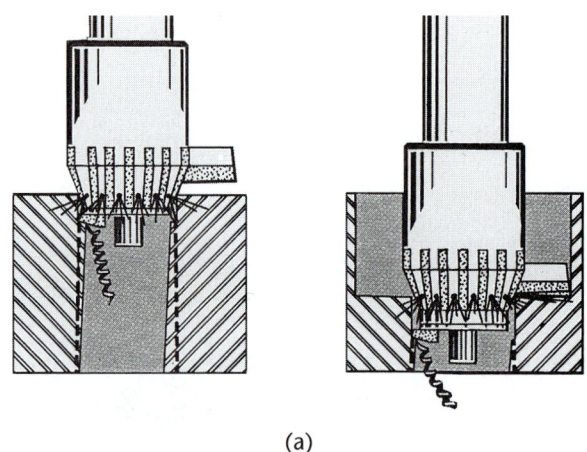

(a)

(b)

FIGURE 23-25   Boring tool employing a centering tool and conical guide, for boring large holes in a single operation. *(Courtesy of Vernon Devices, Incorporated.)*

nose cone then provides good alignment and support for the rear bit, which bores the final hole to size.

Boring Machine Precision

As with turning, the precision obtainable in boring depends considerably on the rigidity of the tool support. On specialized, production-type boring machines, tolerances are readily held to within 0.0005 in. on small diameters, whereas on general-purpose machines tolerances of 0.001 in. are typical unless the boring bar overhang becomes excessive.

■ 23.5 CUTTING TOOLS FOR LATHES

Lathe Cutting Tools

Most lathe operations are done using single-point *cutting tools*, such as those illustrated in Figure 23-26. On right-hand and left-hand turning and facing tools, the cutting takes place on the side of the tool; therefore, the side rake angle is of primary importance, particularly

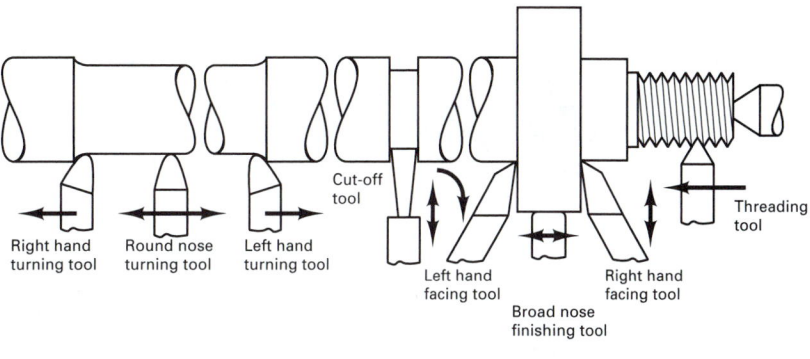

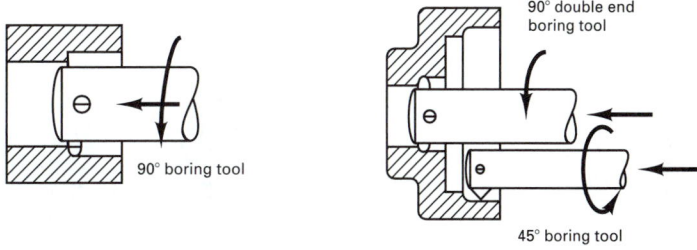

FIGURE 23-26   Schematic of common single-point lathe tools for showing how they can be used.

when deep cuts are being made. On the round-nose turning tools, cutoff tools, finishing tools, and some threading tools, cutting takes place on or near the end of the tool, and the back rake is therefore of importance. Such tools are used with relatively light depths of cut.

Because tool materials are expensive, it is desirable to use as little as possible. At the same time, it is essential that the cutting tool be supported in a strong, rigid manner to minimize deflection and possible vibration. Consequently, lathe tools are supported in various types of heavy, forged steel toolholders, as shown in Figure 23-27. The high-speed steel (HSS) tool bit should be clamped in the toolholder with minimum overhang. Otherwise, tool chatter and a poor surface finish may result.

In the use of carbide, ceramic, or coated carbides for mass production work, throw-away inserts are used that can be purchased in a great variety of shapes, geometrics (nose radius, tool angles, and groove geometry), and sizes (See Figure 23-28 for some examples).

When several different operations on a lathe are performed repeatedly in sequence, the time required for changing and setting tools may constitute as much as 50% of the total cycle time. Quick-change toolholders (Figure 23-29) are used to reduce manual tool-changing time. The individual tools, preset in their holders, can be interchanged in the special tool post in a few seconds. With some systems, a second tool may be set in the tool post while a cut is being made with the first tool and can then be brought into proper position by rotating the post.

In lathe work, the nose of the tool should be set exactly at the same height as the axis of rotation of the work. However, because any setting below the axis causes the work to tend to "climb" up on the tool, most machinists set their tools a few thousandths of an inch above the axis, except for cutoff, threading, and some facing operations.

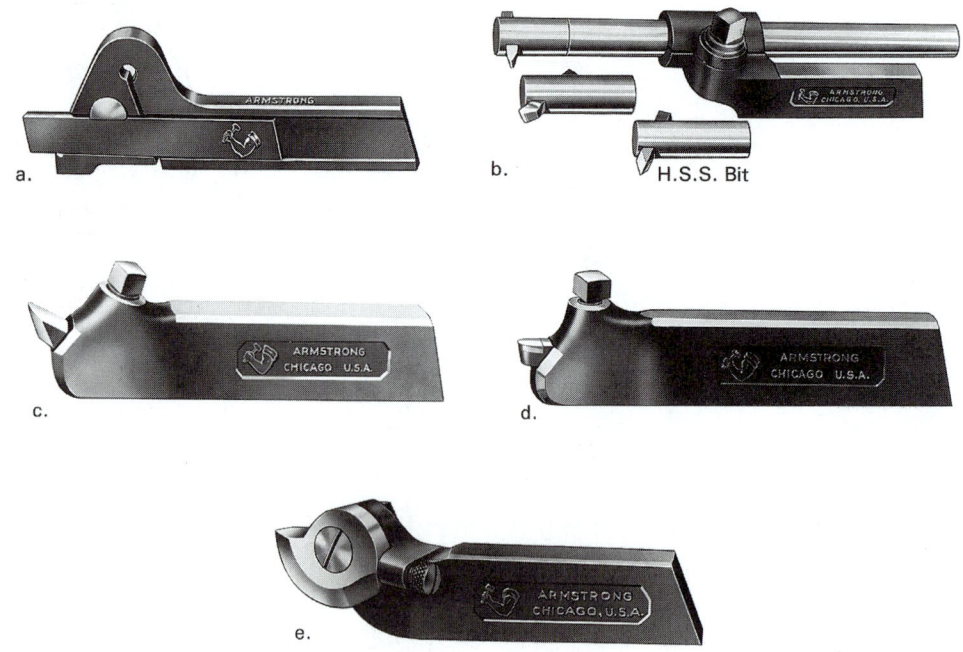

FIGURE 23-27 Common types of forged tool holders: (a) cutoff; (b) boring bars; (c) right-hand facing; (d) left-hand turning; (e) threading. *(Courtesy of Armstrong Brothers Tool Company.)*

Insert shape	Available cutting edges	Typical insert holder
Round ◯	4–10 on a side 8–20 total	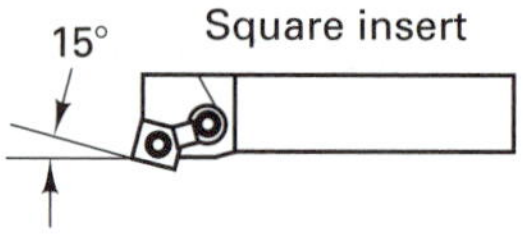 15° Square insert
80°/100° diamond ◇	4 on a side 8 total	
Square ▢	4 on a side 8 total	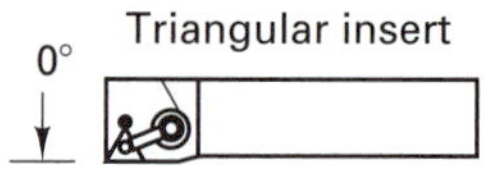 0° Triangular insert
Triangle △	3 on a side 6 total	
55° diamond ◇	2 on a side 4 total	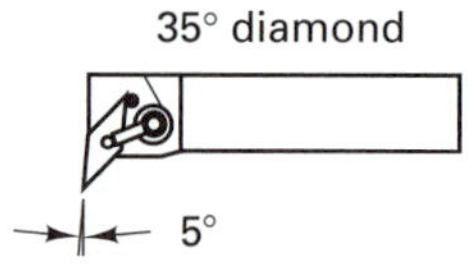 35° diamond
35° diamond ◇	2 on a side 4 total	5°

FIGURE 23-28 Typical insert shapes, available cutting edges per insert and insert holders for throwaway insert cutting tools. *(Adapted from* Turning Handbook of High Efficiency Metal Cutting, *courtesy of General Electric Company.)*

QUICK CHANGE
TOOL POST

TURNING, FACING AND
BORING TOOL HOLDER
V-Slot holds Round
Boring Bars as well as
Square Tool Bits

TURNING AND FACING
TOOL HOLDER
Takes Turning and
Facing Tool Bits

KNURLING TOOL
HOLDER
Revolving head, self-
centering.
3 pairs of knurls

FIGURE 23-29   Quick-change tool post and accompanying toolholders. *(Courtesy of Armstrong Brothers Tool Company.)*

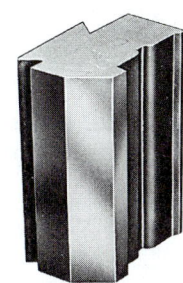

FIGURE 23-30   Circular and block types of form tools. *(Courtesy of Speedi Tool Company, Incorporated.)*

Form Tools

In Figure 23-18, the use of form tools was shown. Form tools are made by grinding the inverse of the desired work contour into a block of HSS or tool steel. A threading tool is often a form tool. Although form tools are relatively expensive to manufacture, they make it possible to machine a fairly complex surface with a single inward feeding of one tool. For mass-production work, adjustable form tools of either flat or rotary types (Figure 23-30) are used. These, of course, are expensive to make initially but can be resharpened merely by grinding a small amount off the face and then raising or rotating the cutting edge to the correct position.

The use of form tools is limited by the difficulty of grinding adequate rake angles for all points along the cutting edge. A rigid setup is needed to resist the large cutting forces that develop with these tools.

Turret Lathe Tools

In turret lathes, the work is generally held in collets and correct amount of bar stock is fed into the machine to make one part. The tools are arranged in sequence at the tool stations with depths of cut all preset. The following factors should be considered when setting up a turret lathe.

1. *Setup time:* time required to install and set the tooling and set the stops. Standard tool-holders and tools should be used as much as possible to minimize setup time. Setup time can be greatly reduced by eliminating adjustment in the setup (see Chapter 43.)
2. *Workholding time:* time to load and unload parts and/or stock.
3. *Machine-controlling time:* time required to manipulate the turrets. Can be reduced by combining operations where possible. Dependent on the sequence of operations established by the design of the setup.
4. *Cutting time:* time during which chips are being produced. Should be as short as is economically practical and represent the greatest percentage of the total cycle time possible.
5. *Cost:* cost of the tool, setup labor cost, lathe operator labor cost, and the number of pieces to be made.

There are essentially 11 tooling stations, as shown in Figure 23-31, with six in the turret, four in the indexable tool post, and one in the rear tool post. The tooling is more

FIGURE 23-31   Turret lathe tooling setup for producing part shown. Numbers in circles indicate the sequence of operation from 1 to 9. Operation 3 is a combined operation. The roll turner is turning surface F while tool 3 on the square post is turning surface B.

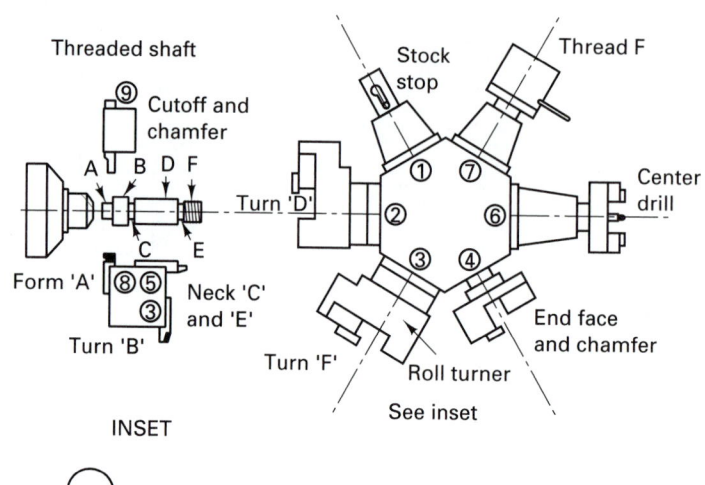

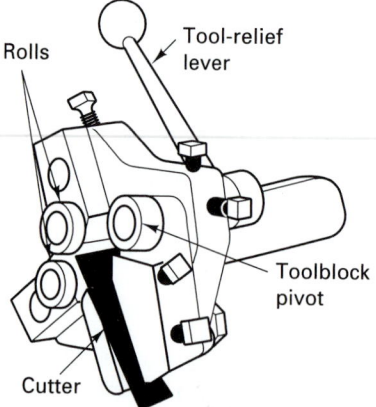

Roller turner has rolls to support the work against the cutting forces

rugged in turret lathes because heavy, simultaneous cuts are often made. Tools mounted in the hex turret that are used for turning are often equipped with pressure rollers set on the opposite side of the rotating workpiece from the tool to counter the cutting forces.

Turret lathes are most economical in producing lots too large for engine lathes but too small for automatic screw machines or automatic lathes. However, in recent years much of this work has been assumed by numerical control lathes (discussed in Chapter 29). Numerical control lathes represent a higher level of automation, A(4), than turret lathes (see the discussion of the Yardstick for Automation in Section 1.6).

■ 23.6 WORKHOLDING IN LATHES

Workholding Devices for Lathes

Five methods are commonly used for supporting workpieces in lathes:

1. Held between centers
2. Held in a chuck
3. Held in a collet
4. Mounted on a face plate
5. Mounted on the carriage

In the first four of these methods the workpiece is rotated during machining. In the fifth method, which is not used extensively, the tool rotates while the workpiece is fed into the tool.

Lathe Centers

Workpieces that are relatively long with respect to their diameters usually are machined between centers (Figure 23-32). Two *lathe centers* are used, one in the spindle hole and the other in the hole in the tailstock quill. Two types are used, called dead and live. *Dead centers* are *solid*, that is, made of hardened steel with a Morse taper on one end so that they will fit into the spindle holes (Figure 23-33). The other end is ground to a 60° taper.

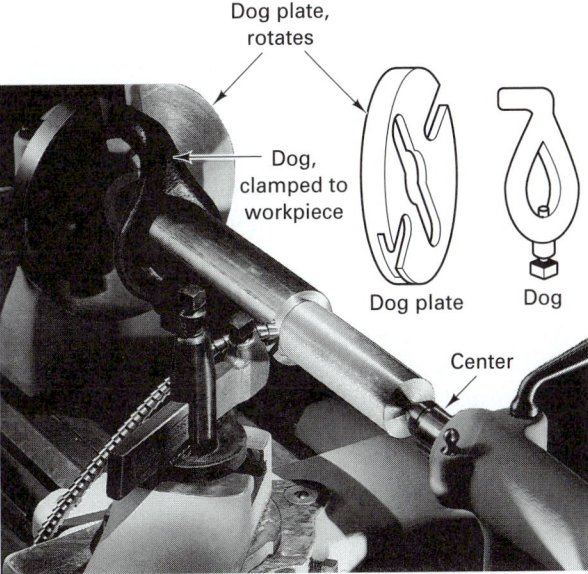

FIGURE 23-32   Work being turned between centers in a lathe, showing the use of a dog and dog place. *(Courtesy of South Bend Lathe.)*

FIGURE 23-33 Solid or "dead" lathe center. *(Courtesy of Chicago-Latrobe Twist Drill Works.)*

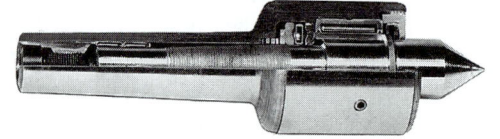

FIGURE 23-34 "Live" lathe center. *(Courtesy of Motor Tool Manufacturing Company.)*

Sometimes the tip of this taper is made of tungsten carbide to provide better wear resistance. Before a center is placed in position, the spindle hole should be carefully wiped clean. The presence of foreign material will prevent the center from seating properly and it will not be aligned accurately.

Live centers of the type shown in Figure 23-34 are designed so that the end that fits into the workpiece is mounted on ball or roller bearings. It is free to rotate. No lubrication is required. Live centers may not be as accurate as the solid type and therefore are not often used for precision work.

Before a workpiece can be mounted between lathe centers, a 60° center hole must be drilled in each end. This can be done in a drill press or in a lathe by holding the work in a chuck. A combination center drill and countersink ordinarily is used, with care taken that the center hole is deep enough so that it will not be machined away in any facing operation and yet is not drilled to the full depth of the tapered portion of the center drill (see Figure 24-9).

Because the work and the center of the headstock end rotate together, no lubricant is needed in the center hole at this end. The center in the tailstock quill does not rotate; adequate lubrication must be provided. A mixture of white lead and oil is often used. Failure to provide proper lubrication at all times will result in scoring of the workpiece center hole and the center, and inaccuracy and serious damage may occur. Live centers are often used in the tailstock to overcome these problems.

The workpiece must rotate freely, yet no looseness should exist. Looseness usually will be manifested in "chattering" of the workpiece during cutting. The setting of the centers should be checked after cutting for a short time. Heating and thermal expansion of the workpiece will reduce the clearances in the setup.

A mechanical connection must be provided between the spindle and the workpiece to cause it to rotate. This is accomplished by some type of *lathe dog* and *dog plate*. The dog is clamped to the work. The *tail* of the dog enters a slot in the dog plate, which is attached to the lathe spindle in the same manner as a lathe chuck. For work that has a finished surface, a piece of soft metal, such as copper or aluminum, can be placed between the work and the dog setscrew clamp to avoid marring.

Mandrels

Workpieces that must be machined on both ends or are disk-shaped are often mounted on *mandrels* for turning between centers. Three common types of mandrels are shown in Figure 23-35. *Solid mandrels* usually vary from 4 to 12 in. in length and are accurately ground with a 1:2000 taper (0.006 in./ft). After the workpiece is drilled and/or bored, it is pressed on the mandrel. The mandrel should be mounted between centers so that the cutting force tends to tighten the work on the mandrel taper. Solid mandrels permit the work to be machined on both ends as well as on the cylindrical surface. They are available in stock sizes but can be made to any size desired.

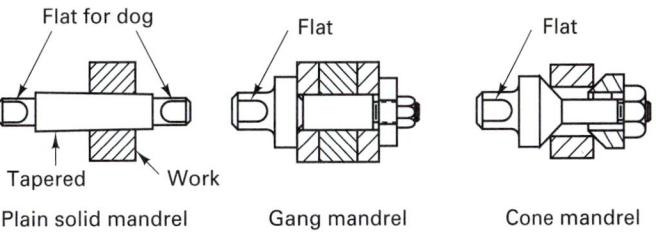

FIGURE 23-35 Three types of mandrels.

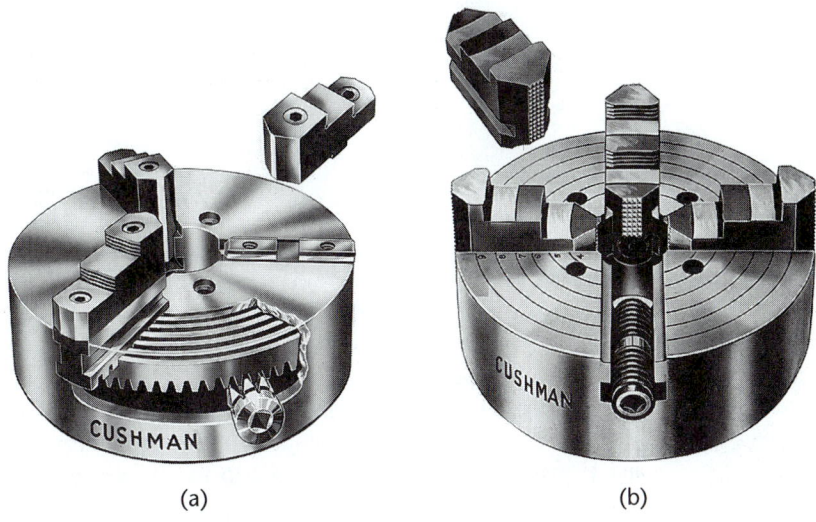

FIGURE 23-36 Lathe chucks. (*Left*) four-jaw independent. (*Right*) three-jaw, self-centering. (*Courtesy of Cushman Industries, Incorporated.*)

Gang (or *disk*) *mandrels* are used for production work because the workpieces do not have to be pressed on and thus can be put in position and removed more rapidly. However, only the cylindrical surface of the workpiece can be machined when this type of mandrel is used. *Cone mandrels* have the advantage that they can be used to center workpieces having a range of hole sizes.

Lathe Chucks

Lathe *chucks* are used to support a wider variety of workpiece shapes and to permit more operations to be performed than can be accomplished when the work is held between centers. Two basic types of chucks are used (Figure 23-36).

Three-jaw, self-centering chucks are used for work that has a round or hexagonal cross section. The three jaws are moved inward or outward simultaneously by the rotation of a spiral cam, which is operated by means of a special wrench through a bevel gear. If they are not abused, these chucks will provide automatic centering to within about 0.001 in. However, they can be damaged through use and will then be considerably less accurate.

Each jaw in a *four-jaw independent chuck* can be moved inward and outward independent of the others by means of a chuck wrench. Thus they can be used to support a

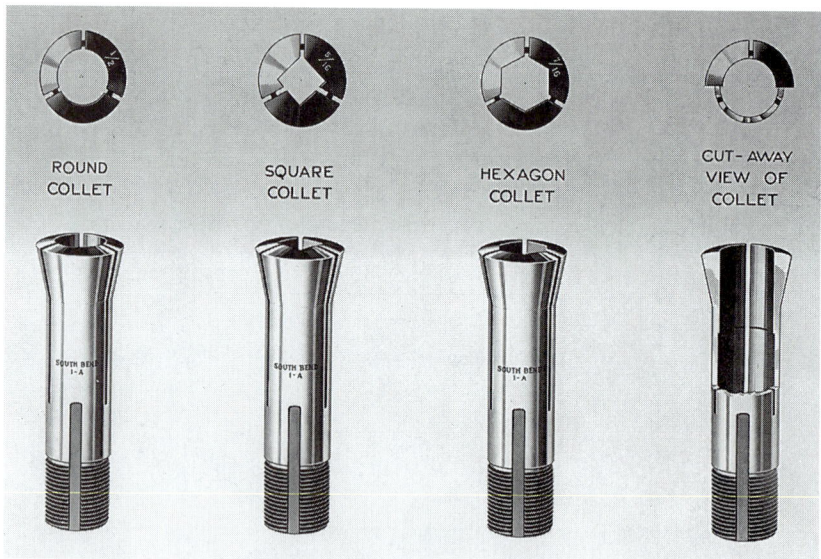

FIGURE 23-37   Several types of lathe collets. *(Courtesy of South Bend Lathe.)*

wide variety of work shapes. A series of concentric circles engraved on the chuck face aid in adjusting the jaws to fit a given workpiece. Four-jaw chucks are heavier and more rugged than the three-jaw type, and because undue pressure on one jaw does not destroy the accuracy of the chuck, they should be used for all heavy work. The jaws on both three- and four-jaw chucks can be reversed to facilitate gripping either the inside or outside of workpieces.

Combination four-jaw chucks are available in which each jaw can be moved independently or can be moved simultaneously by means of a spiral cam. Two-jaw chucks are also available. For mass-production work, special chucks often are used in which the jaws are actuated by air or hydraulic pressure, permitting very rapid clamping of the work. The rapid exchange of chuck jaws and chucks are discussed in Chapters 28 and 43.

Collets

Collets are used to hold smooth cold-rolled bar stock or machined workpieces more accurately than with regular chucks. As shown in Figure 23-37, collets are relatively thin tubular steel bushings that are split into three longitudinal segments over about two-thirds of their length. At the split end, the smooth internal surface is shaped to fit the piece of stock that is to be held. The external surface of the collet is a taper that mates with an internal taper of a collet sleeve. When the collet is pulled inward into the spindle (by means of the draw bar), the action of the two mating tapers squeezes the collet segments together, causing them to grip the workpiece (Figure 23-38).

Collets are made to fit a variety of symmetrical shapes. If the stock surface is smooth and accurate, good collets will provide very accurate centering, with runout less than 0.0005 in. However, the work should be no more than 0.002 in. larger or 0.005 in. smaller than the nominal size of the collet. Consequently, collets are used only on drill-rod, cold-rolled, extruded, or previously machined stock.

Collets that can open automatically and feed bar stock forward to a stop mechanism are commonly used on automatic lathes and turret lathes. Another type of collet similar to

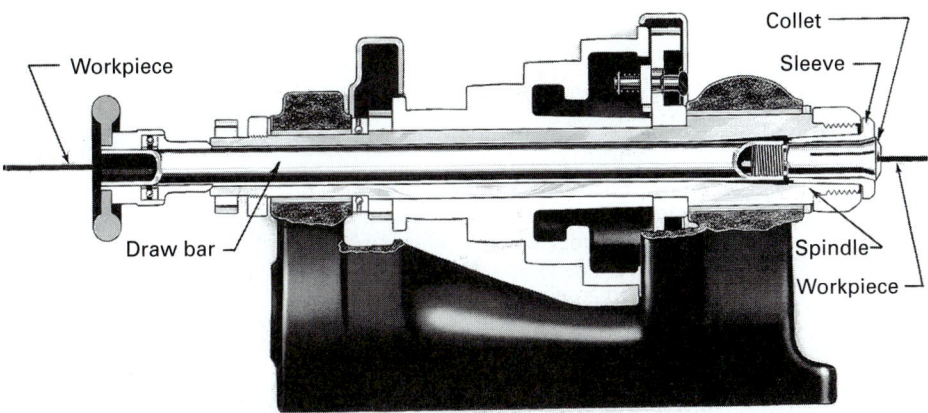

FIGURE 23-38   Method of using a draw-in collet in a lathe spindle. *(Courtesy of South Bend Lathe.)*

a Jacobs drill chuck has a greater size range than ordinary collets, therefore fewer are required.

Faceplates

Faceplates are used to support irregularly shaped work that cannot be gripped easily in chucks or collets. The work can be bolted or clamped directly on the faceplate or can be supported on an auxiliary fixture that is attached to the faceplate. The latter procedure is timesaving when identical pieces are to be machined (see Figure 23-5).

Mounting Work on the Carriage

When no other means is available, boring occasionally is done on a lathe by mounting the work on the carriage, with the boring bar mounted between centers and driven by means of a dog.

Steady and Follow Rests

If one attempts to turn a long, slender piece between centers, the radial force exerted by the cutting tool, or the weight of the workpiece itself, may cause it to be deflected out of line. *Steady rests* and *follow rests* (Figure 23-39) provide means for supporting such work between the headstock and the tailstock. The steady rest is clamped to the lathe ways and has three movable fingers that are adjusted to contact the work and align it. A light cut should be taken before adjusting the fingers to provide a smooth contact-surface area.

A steady rest also can be used in place of the tailstock as a means of supporting the end of long pieces, pieces having too large an internal hole to permit using a regular dead center, or work where the end must be open for boring. In such cases the headstock end of the work must be held in a chuck to prevent longitudinal movement. Tool feed should be toward the headstock.

The follow rest is bolted to the lathe carriage. It has two contact fingers that are adjusted to bear against the workpiece, opposite the cutting tool, to prevent the work from being deflected away from the cutting tool by the cutting forces.

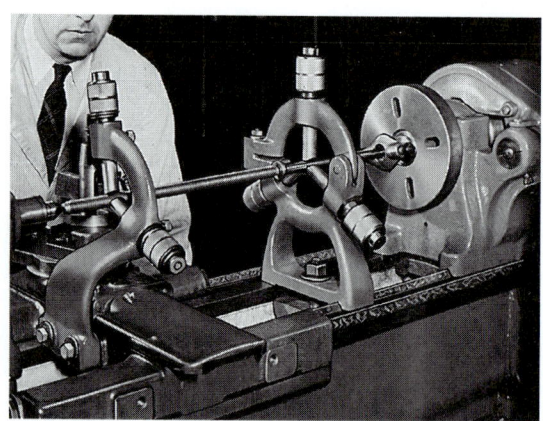

FIGURE 23-39   Cutting a thread on a long, slender workpiece, using a follow rest (left) and a steady rest (right) on an engine lathe. Note the use of a dog and face plate to drive the workpiece. *(Courtesy of South Bend Lathe.)*

■ KEY WORDS

apron	engine lathe	parting
automatic lathe	faceplates	quill
bed	facing	reaming
boring	feed	screw machine
carriage	follow rest	steady rest
chucks	headstock	tailstock
collets	knurling	taper turning
cutoff	lathe centers	turning
cutting tools	mandrels	turret lathe
depth of cut	metal removal rate	workholding
drilling	milling	

■ REVIEW QUESTIONS

1. What is the tool–work relationship in turning?
2. What different kinds of surfaces can be produced by turning?
3. How does form turning differ from ordinary turning?
4. What is the basic difference between a facing and a cutoff operation?
5. Which machining operations shown in Figure 23-1 does not use chip formation?
6. Why is it difficult to make heavy cuts if a form turning tool is complex in shape?
7. What is the swing of a lathe?
8. Why is the spindle of the lathe hollow?
9. What function does a lathe carriage have?
10. How is feed specified on a lathe?
11. What function is provided by the leadscrew on a lathe that is not provided by the feed rod?
12. How can work be held and supported in a lathe?
13. How is a workpiece that is mounted between centers on a lathe driven (rotated)?

14. What will happen to the workpiece when turned if held between centers with the centers not exactly in line?
15. Why is it not advisable to hold hot-rolled steel stock in a collet?
16. How does a steady rest differ from a follow rest?
17. What are the advantages and disadvantages of a four-jaw independent chuck as contrasted with a three-jaw chuck?
18. Why should the distance that a lathe tool projects from the toolholder be minimized?
19. If a lathe tool is set below the centerline of the workpiece in turning, what might occur?
20. How can a tapered part be turned on a lathe?
21. Why is it desirable to use a heavy depth of cut and a light feed in turning rather than the opposite?
22. On what diameter is the rpm value based for a facing cut, assuming given work and tool materials?

23. Why is it usually necessary to take relatively light feeds and depths of cut when boring on a lathe?

24. How does the corner radius of the tool influence the surface roughness?

25. What effect does a BUE have on the diameter of the workpiece in turning?

26. What important factor must be kept in mind in tooling multiple-spindle screw machines that does not have to be considered in single-spindle machines?

27. What is the objective of boring?

28. Why does boring assure concentricity between the hole axis and the axis of rotation of the workpiece (for boring tool), whereas drilling does not?

29. Why are vertical boring mills better suited than a lathe for machining large workpieces?

30. What is the principal advantage of a horizontal boring machine over a vertical boring machine for large workpieces?

31. Where in Chapter 23 is a workpiece being held in a fixture mounted on a faceplate?

32. How is the workpiece in Figure 23-6 being held?

33. In what three figures is a dead center shown in Chapter 23?

34. In what two figures is a live center shown in Chapter 23?

35. In which figures showing setups do you find the following being used as a workholding device?
    (a) Three-jaw chuck
    (b) Collet
    (c) Faceplate
    (d) Four-jaw chuck

36. How many form tools are being utilized in the process shown in Figure 23-18 to machine the part?

■ PROBLEMS

1. A cutting speed of 200 ft/min has been selected. At what rpm value should a 3-in.-diameter bar be rotated?

2. Assume that the workpiece in Problem 1 is 8 in. (203.2 mm) long and a feed of 0.020 in. (0.51 mm) per revolution is used. How long will a cut across its entire length require? Don't forget to add an allowance.

3. If the depth of cut in Problem 2 is $\frac{1}{8}$ in., what is the metal removal rate (MRR)?

4. The following data apply for machining a part on a turret lathe and on an engine lathe:

	Engine Lathe	Turret Lathe
Times to machine part	30 min	5 min
Cost of special tooling	0	300
Time to setup	30 min	3 hr
Labor rates	$ 8 /hr	$ 8 /hr
Machine rates	$10 /hr	$12 /hr

(a) How many pieces would have to be made for the cost of the engine lathe to just equal the cost of the turret lathe? This is the BEQ.

(b) What is the cost per unit at the BEQ?

5. A hole 89 mm in diameter is to be drilled and bored through a piece of 1340 steel that is 200 mm long, using a horizontal boring, drilling, and milling machine. High-speed tools will be used. The job will be done by center drilling, drilling with an 18-mm drill, followed by a 76-mm drill, then bored to size in one cut, using a feed of 0.50 mm/rev. Drilling feeds will be 0.25 mm/rev for the smaller drill and 0.64 mm/rev for the larger drill. The center drilling operation requires 0.5 minute. To set or change any given tool and set the proper machine speed and feed requires 1 minute. Select the initial cutting speeds, and compute the total time required for doing the job. (Neglect the setup time for the workpiece.)

6. In Figure 23-A three plots of unit production cost ($/unit) versus production volume (Q = build quantity) are shown. Note that this plot is made on log-log paper. The cost per unit for a particular process decreases with increased volume. For a particular process there is no minimum cost but rather, production volumes within which particular processes are most economical.

(a) Each of these curves is a plot of the equation for total cost per unit, which means that each is the sum of the fixed cost per unit (setup, tooling, overhead) and the variable costs per unit (direct labor, direct material). Estimate these costs for the NC lathe from the data in curves.

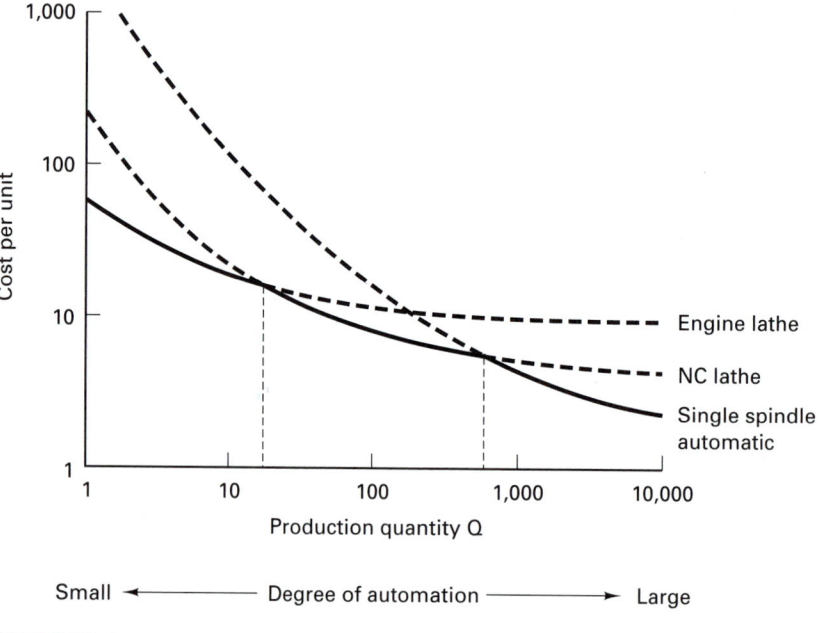

FIGURE 23-A

(b) For what build quantities is the NC lathe most economical (approximately)?

(c) What cost per unit does the NC lathe approach as the build quantity becomes very large?

(d) What happens to these plots if you plot them on regular Cartesian coordinates? Try it and comment on what you find.

(e) Many Japanese manufacturers have found innovative ways to eliminate setup time in many of their processes. What is the impact of this on these kinds of plots, on cost per unit economics, and on job shop inventories?

7. The derivation of the approximate equation (23-5) for the MRR for turning process requires an assumption regarding the diameters of the parts being turned. Determine the error in the equation for Problem 3. What is the assumption?

Chapter 23 CASE STUDY

break-even point analysis of a lathe part

You have received the part drawing for a typical lathe part that will require turning, facing, grooving, boring, and threading as it is machined from a casting. Unfortunately, you do not yet know what the quantity will be. To be prepared, you have developed some cost data for the manufacture of the part by four different lathe processes (Table CS-23). Complete the table by determining the run cost per batch, the cost per unit at the various quantities, and the total cost per batch.

Answer the following questions regarding this situation.

1. Of the four costs listed for each process, which costs are fixed and which are variable?
2. For which of these costs would you have to estimate the machining time per piece and the cycle time per piece, including the time to change parts and cutting tools?
3. How would you go about estimating this time? What time elements might be included in the cycle time in addition to the machining time?
4. How would you use this estimate of time in the cost table? Show the calculation.
5. Make a plot of cost in dollars versus quantity, with all four methods on one plot.
6. Make a plot of cost per unit versus quantity, again with all four methods on one plot. Find the break-even quantities. (*Hint:* Did you plot the data on log paper?)
7. Discuss these plots and the break-even quantities that you found in question 5 versus question 6.
8. When would you use the turret lathe? When would you use the turret lathe if you have no NC lathe?

TABLE CS-23. Cost Data

	Make Quantity				
	10,000 Units	1,000 Units	100 Units	10 Units	1 Units
Cost to Produce on Six-Spindle Automatic					
Total cost of batch	—	—	—	—	—
Engineering 2.5 hr at $20/hr	50.00	50.00	50.00	50.00	50.00
Tooling (Cutting tools and workholders)	600.00	600.00	600.00	600.00	600.00
Setup 8 hr at $15/hr	120.00	120.00	120.00	120.00	120.00
Run cost per batch: 50 cents per piece	—	—	—	—	—
Cost per unit	—	—	—	—	—
Cost to Produce on Turret Lathe					
Total cost of batch	—	—	—	—	—
Engineering 2 hr at $20/hr	40.00	40.00	40.00	40.00	40.00
Tooling	150.00	150.00	150.00	150.00	150.00
Setup 4 hr at $12/hr	48.00	48.00	48.00	48.00	48.00
Run cost per batch: $8 per piece	—	—	—	—	—
Cost per unit	—	—	—	—	—
Cost to Produce on Engine Lathe					
Total cost of batch	—	—	—	—	—
Engineering 1 hr at $20/hr	20.00	20.00	20.00	20.00	20.00
Tooling	—	—	—	—	—
Setup 2 hr at $12/hr	24.00	24.00	24.00	24.00	24.00
Run cost per batch: $12 per piece	—	—	—	—	—
Cost per unit	—	—	—	—	—
Cost to Produce on NC Lathe					
Total cost of batch	—	—	—	—	—
Engineering and programming	150.00	150.00	150.00	150.00	150.00
Tooling	100.00	100.00	100.00	100.00	100.00
Setup 1 hr at $20/hr	20.00	20.00	20.00	20.00	20.00
Run cost per batch: $2 per piece	—	—	—	—	—
Cost per unit	—	—	—	—	—

DRILLING AND RELATED HOLE-MAKING PROCESSES

24.1	INTRODUCTION	24.8	COUNTERBORING, COUNTERSINKING, AND SPOT FACING
24.2	FUNDAMENTALS OF THE DRILLING PROCESS		
24.3	TYPES OF DRILLS	24.9	DRILLING PRACTICE AND PROBLEMS
24.4	TOOLHOLDERS FOR DRILLS	24.10	REAMING
24.5	WORKHOLDING DEVICES FOR DRILLING		Types of Reamers
			Reaming Practice
24.6	MACHINE TOOLS FOR DRILLING	Case Study:	BOLT-DOWN LEG ON A CASTING
24.7	CUTTING FLUIDS FOR DRILLING		

■ 24.1 INTRODUCTION

In manufacturing it is probable that more holes are produced than any other shape, and a large proportion of these are made by *drilling*. Of all the machining processes performed, drilling makes up about 25%. Consequently, drilling is a very important process. Although drilling appears to be a relatively simple process, it is really a complex process. Most drilling is done with a tool having two cutting edges. These edges are at the end of a relatively flexible tool. Cutting action takes place inside the workpiece. The only exit for the chips is the hole that is filled by the drill. Friction results in heat that is additional to that due to chip formation. The counterflow of the chips makes lubrication and cooling difficult. There are four major actions taking place at the point of a drill:

1. A small hole is formed by the web—chips are not cut here in the normal sense.
2. Chips are formed by the rotating *lips*.
3. Chips are removed from the hole by the screw action of the helical flutes.
4. The drill is guided by the margin's rubbing against the walls of the hole.

In recent years, new drill point geometries and TiN coatings have resulted in improved hole accuracy, longer life, self-centering action, and increased-feed-rate capabilities. However, the great majority of drills manufactured are twist drills, having the conventional point and geometry shown in Figure 24-1. One estimate has U. S. manufacturing companies consuming 250 million twist drills per year.

When HSS drills wear out, the drill is reground. If regrinding is not done properly, the original drill geometry may be lost and so will drill accuracy and precision. Drill performance also depends on the drilling machine tool, the workholding device, the drill holder, and the surface of the workpiece. Poor surface conditions (sand pockets and/or chilled hard spots on castings, hard oxide scale on hot-rolled metal) can accelerate early tool failure and degrade the hole-drilling process.

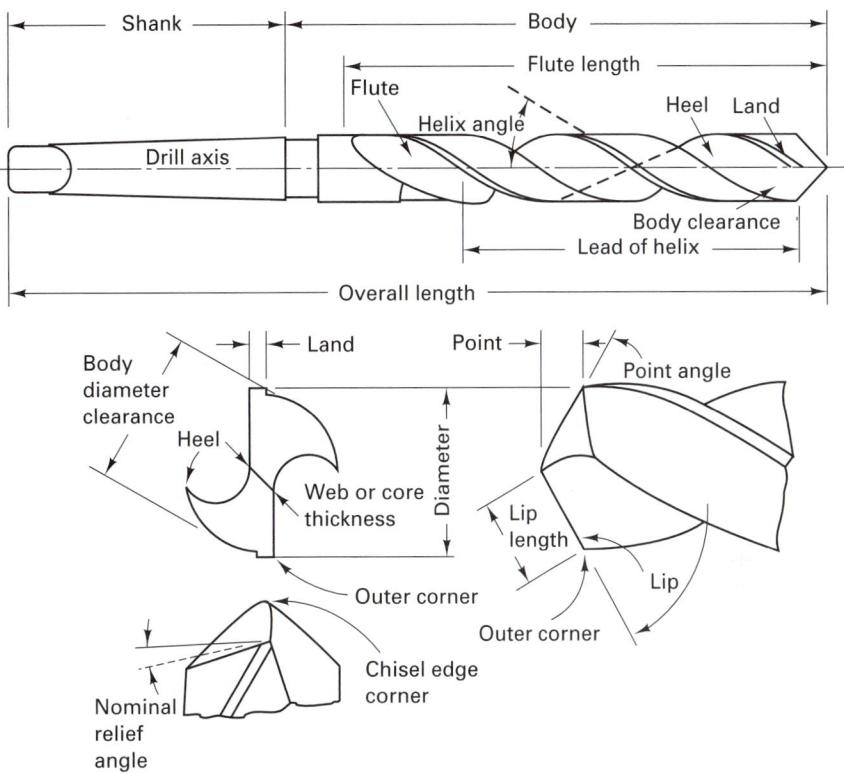

FIGURE 24-1 Nomenclature and geometry of conventional twist drill.

■ 24.2 FUNDAMENTALS OF THE DRILLING PROCESS

The process of drilling creates two chips. A conventional two-flute drill, with drill of diameter D, has two principal cutting edges rotating at an rpm rate of N and feeding axially. The rpm rate of the drill is established by the selected cutting velocity or cutting speed

$$N = \frac{12V}{\pi D} \qquad (24\text{-}1)$$

with V in surface feet per minute and D in inches. This equation assumes that V is the cutting speed at the outer edges of the cutting lip (point X in Figure 24-2).

The feed, f_r, is given in inches per revolution. The depth of cut in drilling is equal to half the feed rate, or $t = f_r/2$ (see section A–A in Figure 24-2). The feed rate in inches per minute, f_m, is $f_r N$. The length of cut in drilling equals the depth of the hole, L, plus an allowance for approach and for the tip of drill, usually A = $D/2$. As in turning, the speed and feed used in drilling depend on the material being machined and the cutting tool material. Table 24-1 gives some typical values for V and f_r for carbide indexable-insert drills, a type of drill shown later in the chapter.

After selecting the cutting speed and feed for drilling the hole, the rpm value of the spindle of the machine is determined from equation (24-1), the maximum velocity occurring at the extreme ends of the drill lips. The velocity is very small near the center of the chisel end of the drill. For drilling, the cutting time is

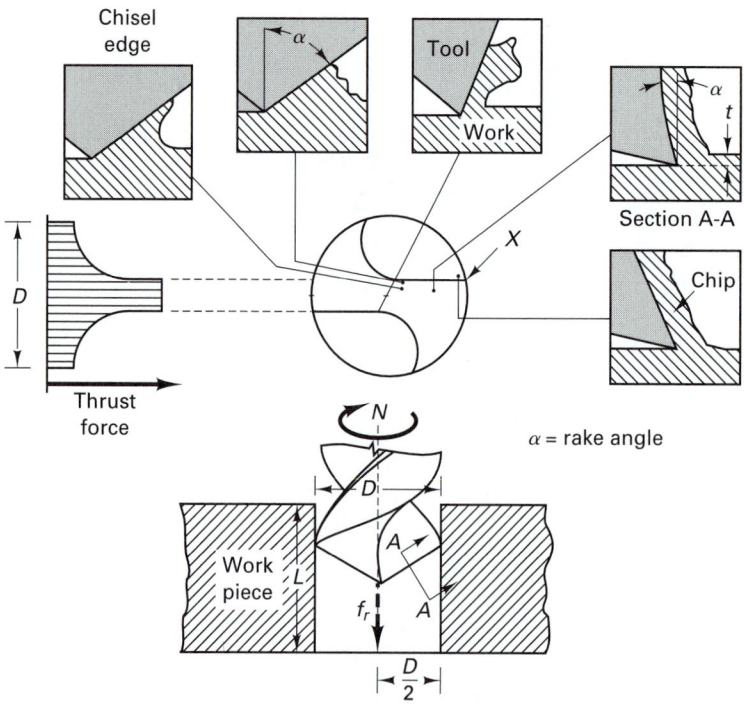

FIGURE 24-2 Conventional drill geometry and symbols with cutting cross sections and trust force.

$$CT = \frac{L+A}{f_r N} = \frac{L+A}{f_m} \tag{24-2}$$

The metal removal rate is

$$MRR = \frac{\text{volume}}{CT} \tag{24-3}$$

$$= \frac{\pi D^2 L / 4}{L / f_r N} \quad \text{(omitting allowances)}$$

which reduces to

$$MRR = (\pi D^2 / 4) f_r N \quad \text{in}^3/\text{min} \tag{24-4}$$

Substituting for N with equation (24-1), we obtain the approximate form

$$MRR \simeq 3 D V f_r \tag{24-5}$$

Note that the units here are not consistent. A summary of drilling equations and parameters is given in Table 24-2, including some simple cost equations. In line 8, the "number of holes drilled" is the measure of tool life for drills (see Figure 22-9 for examples).

TABLE 24-1.   Recommended Speeds and Feeds for Indexable-Insert Drills[a]

Material Group	Size Range	Cutting Speed (sfpm)	Feed Rate (in./rev)
Cast iron: modular, ductile, or malleable	$\frac{13}{16}-1\frac{1}{8}$	165–300	0.004–0.008
	$1-1\frac{3}{8}$	165–300	0.005–0.010
	$1\frac{1}{4}-1\frac{5}{8}$	165–300	0.006–0.012
	$1\frac{1}{2}-2\frac{1}{2}$	165–300	0.008–0.014
	$2\frac{3}{8}-3\frac{1}{2}$	165–300	0.010–0.015
1000 Series steels like 1081, 1020, etc.	$\frac{13}{16}-1\frac{1}{8}$	300–400	0.003–0.005
	$1-1\frac{3}{8}$	350–450	0.003–0.006
	$1\frac{1}{4}-1\frac{5}{8}$	400–550	0.004–0.007
	$1\frac{1}{2}-2\frac{1}{2}$	450–600	0.004–0.007
	$2\frac{3}{8}-3\frac{1}{2}$	500–700	0.005–0.009
Low-carbon unalloyed case-hardening steels	$\frac{13}{16}-1\frac{1}{8}$	200–300	0.003–0.005
	$1-1\frac{3}{8}$	250–350	0.004–0.006
	$1\frac{1}{4}-1\frac{5}{8}$	300–425	0.005–0.007
	$1\frac{1}{2}-2\frac{1}{2}$	330–490	0.005–0.008
	$2\frac{3}{8}-3\frac{1}{2}$	350–550	0.006–0.010
High-carbon alloyed heat-treated steels	$\frac{13}{16}-1\frac{1}{8}$	200–300	0.003–0.005
	$1-1\frac{3}{8}$	250–325	0.004–0.006
	$1\frac{1}{4}-1\frac{5}{8}$	300–400	0.005–0.008
	$1\frac{1}{2}-2\frac{1}{2}$	335–450	0.005–0.008
	$2\frac{3}{8}-3\frac{1}{2}$	350–500	0.006–0.010
High-tensile steels	$\frac{13}{16}-1\frac{1}{8}$	165–250	0.004–0.005
	$1-1\frac{3}{8}$	195–300	0.004–0.006
	$1\frac{1}{4}-1\frac{5}{8}$	230–300	0.005–0.007
	$1\frac{1}{2}-2\frac{1}{2}$	265–390	0.006–0.008
	$2\frac{3}{8}-3\frac{1}{2}$	265–425	0.006–0.009
Stainless steels	$\frac{13}{16}-1\frac{1}{8}$	230–280	0.003–0.004
	$1-1\frac{3}{8}$	265–300	0.004–0.005
	$1\frac{1}{4}-1\frac{5}{8}$	280–345	0.004–0.005
	$1\frac{1}{2}-2\frac{1}{2}$	295–395	0.004–0.005
	$2\frac{3}{8}-3\frac{1}{2}$	300–400	0.004–0.006
Titanium steels	$\frac{13}{16}-1\frac{1}{8}$	100–135	0.003–0.004
	$1-1\frac{3}{8}$	100–150	0.004–0.007
	$1\frac{1}{4}-1\frac{5}{8}$	115–165	0.005–0.008
	$1\frac{1}{2}-2\frac{1}{2}$	130–175	0.006–0.009
	$2\frac{3}{8}-3\frac{1}{2}$	135–190	0.006–0.010

[a]Ultimate speeds and feeds may differ from recommended speeds and feeds depending on materials, rigidity of machine and setup, workpiece, and depth of cut.

TABLE 24-2.   Summary of Drilling Parameters

1. Cutting speed at drill outside diameter = $(\pi/12) \times$ (drill diameter) $\times$ (rpm)
2. Feed rate (in./min) = feed (in./rev) $\times$ (rpm)
3. Maximum chip load = feed (in./rev)/2....for two-fluted drill
4. Material removal rate (in^3/min) = $(\pi/4) \times$ (drill diameter)$^2 \times$ feet rate (in./min)
5. Drilling time/hole = $\dfrac{\text{length drilled} + \text{air cut (allowance)}}{\text{feed rate (in./min)}}$

   $+ \dfrac{\text{rapid traverse length, including withdrawl}}{\text{rapid traverse rate}}$

   $+$ prorated downtime to change drill per hole
6. Cost/hole = (drilling time/hole) $\times$ (labor $+$ machine rate)

   $+$ prorated cost of purchasing and regrinding drill/hole
7. Prorated downtime to change drill/hole = $\dfrac{\text{drill change downtime}}{\text{holes drilled per drill regrind}}$
8. Prorated cost of purchasing and regrinding drill/hole

   $= \dfrac{(\text{purchase cost} + \text{cost per regrind}) \times (\text{number of regrinds})}{\text{number of holes drilled}}$

■ 24.3 Types of Drills

The most common type of drills are *twist drills*. These have three basic parts: the *body*, the *point*, and the *shank*, shown in Figure 24-1. The body contains two or more spiral or helical grooves, called *flutes*, separated by *lands*. To reduce the friction between the drill and the hole, each land is reduced in diameter except at the leading edge, leaving a narrow *margin* of full diameter to aid in supporting and guiding the drill and thus aiding in obtaining an accurate hole. The lands terminate in the point, with the leading edge of each land forming a cutting edge. The flutes serve as channels through which the chips are withdrawn from the hole and coolant gets to the cutting edges. Although most drills have two flutes, some, as shown in Figure 24-3, have three, and some have only one.

The principal rake angles behind the cutting edges are formed by the relation of the flute *helix angle* to the work. This means that the rake angle of a drill varies along the cutting edges (or lips), being negative close to the point and equal to the helix angle out at the lip. Because the helix angle is built into the drill, the primary rake angle cannot be changed by normal grinding. The helix angle of most drills is 24°, but drills with larger helix angles—often above 30°—are used for materials that can be drilled very rapidly, resulting in a large volume of chips. Helix angles ranging from 0 to 20° are used for soft materials, such as plastics and copper. Straight-flute drills (zero helix and rake angles) are also used for drilling thin sheets of soft materials. It is possible to change the rake angle adjacent to the cutting edge by a special grinding procedure called *dubbing*.

The cone-shaped point on a drill contains the cutting edges and the various clearance angles. This cone angle affects the direction of flow of the chips across the tool face and into the flute. The 118° cone angle that is used most often has been found to provide good cutting conditions and reasonable tool life when drilling mild steel, thus making it suitable for much general-purpose drilling. Smaller cone angles—from 90 to 118°—are sometimes used for drilling more brittle materials, such as gray cast iron and magnesium alloys. Cone angles from 118 to 135° often are used for the more ductile materials, such as aluminum alloys. Cone angles less than 90° frequently are used for drilling plastics. Many methods of grinding drills have been developed that produce point angles other than 118°.

FIGURE 24-3   Types of twist drills and shanks. *Bottom to top*: Straight-shank, three-flute core drill; taper-shank; bit-shank; straight-shank, high-helix-angle; straight-shank, straight-flute; taper-shank, subland drill.

The drill produces a thrust force T and a torque M. Drill torque increases with feed (in./rev) and drill diameter, while the thrust force is influenced greatly by the web or chisel end design, as shown in Figures 24-2 an 24-4.

The relatively thin *web* between the flutes forms a metal column or backbone. If a plain conical point is ground on the drill, the intersection of the web and the cone produces a straight-line *chisel end*, which can be seen in the end view of Figure 24-2. The chisel point, which also must act as a cutting edge, forms a 56° negative rake angle with the conical surface. Such a large negative rake angle does not cut efficiently, causing excessive deformation of the metal. This results in high thrust forces and excessive heat being developed at the point. In addition, the cutting speed at the drill center is low, approaching

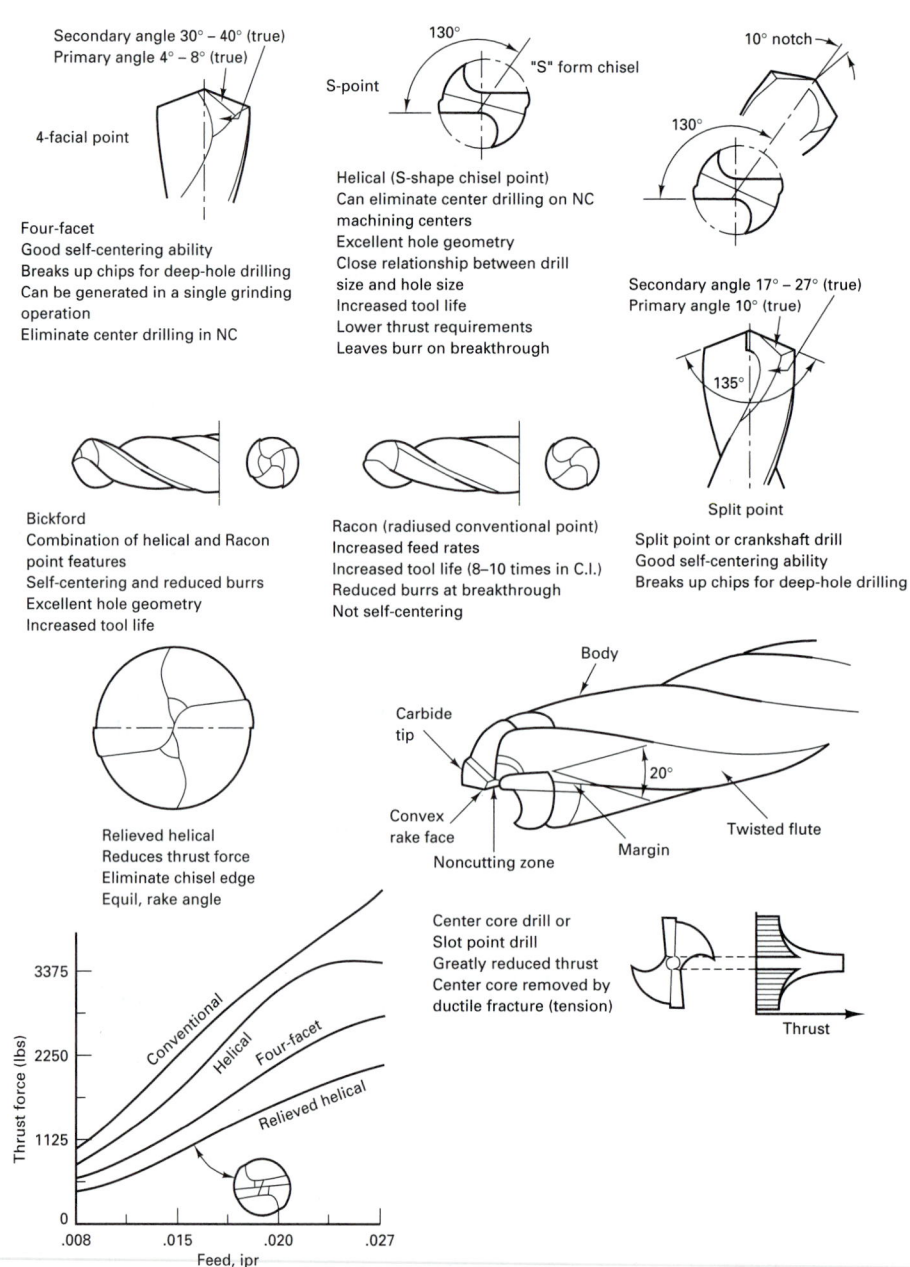

Secondary angle 30° – 40° (true)
Primary angle 4° – 8° (true)

4-facial point

Four-facet
Good self-centering ability
Breaks up chips for deep-hole drilling
Can be generated in a single grinding
operation
Eliminate center drilling in NC

S-point

130°

"S" form chisel

Helical (S-shape chisel point)
Can eliminate center drilling on NC
machining centers
Excellent hole geometry
Close relationship between drill
size and hole size
Increased tool life
Lower thrust requirements
Leaves burr on breakthrough

10° notch

130°

Secondary angle 17° – 27° (true)
Primary angle 10° (true)

135°

Split point

Split point or crankshaft drill
Good self-centering ability
Breaks up chips for deep-hole drilling

Bickford
Combination of helical and Racon
point features
Self-centering and reduced burrs
Excellent hole geometry
Increased tool life

Racon (radiused conventional point)
Increased feed rates
Increased tool life (8–10 times in C.I.)
Reduced burrs at breakthrough
Not self-centering

Body

Carbide
tip

Convex
rake face

Noncutting zone

20°

Twisted flute

Margin

Relieved helical
Reduces thrust force
Eliminate chisel edge
Equil, rake angle

Center core drill or
Slot point drill
Greatly reduced thrust
Center core removed by
ductile fracture (tension)

Thrust

Thrust force (lbs)

3375

2250

1125

0

.008   .015   .020   .027

Feed, ipr

Conventional

Helical

Four-facet

Relieved helical

FIGURE 24-4   Variations of the conventional drill geometry are usually aimed at reducing the thrust force at the chisel end of the drill.

zero. As a consequence, drill failure on a standard drill occurs at the center, where the cutting speed is lowest, and at the outer tips of the cutting edges, where the speed is highest.

When the rotating, straight-line chisel point comes in contact with the workpiece, it has a tendency to slide or "walk" along the surface, thus moving the drill away from the desired location. The conventional point drill, when used on machining centers or high-speed

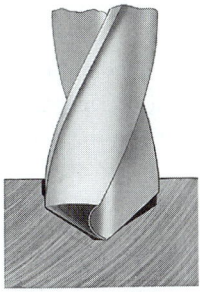 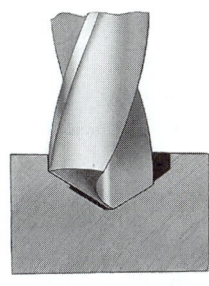

FIGURE 24-5   Effects of improper drill grinding. (Left) angles of two tips are different; (Right) lengths of the lips are not equal. *(Courtesy of Cleveland Twist Drill Company.)*

automatics, will require additional supporting operations such as center drilling, burr removal, and tool change, all of which increase total production time and reduce productivity.

Many special methods of grinding drill points have been developed to eliminate or minimize the difficulties caused by the chisel point and to obtain better cutting action and tool life (see Figure 24-4 for examples).

The center core or slot point drill is relatively new to the marketplace. This drill has twin carbide tips brazed on a steel shank and a hole (or slot) in the center. The work material in the slot is not machined but rather, fractured away. The center core drill has a self-centering action and greatly relieves the thrust force produced by the chisel edge of conventional twist drills. This drill operates at about 30 to 50% less thrust than that of conventional drills. All rake angles of the cutting edge are positive, which further reduces the cutting force.

The conventional point also has a tendency to produce a burr on the exit side of a hole. Some type of chip breaker often is incorporated into drills. One procedure is to grind a small groove in the rake face, parallel with and a short distance back from the cutting edge. Drills with a special chip-breaker rib as an integral part of the flute are available. The rib interrupts the flow of the chip, causing it to break into short lengths.

The split-point drill is a form of *web thinning* to shorten the chisel edge. This design reduces thrust and allows for higher feed rates. Web thinning uses a narrow grinding wheel to remove a portion of the web near the point of the drill. Such methods have had varying degrees of success, and they require special drill-grinding equipment.

Also shown in Figure 24-4 is a four-facet self-centering point that works well in tougher materials. The facets refer to the number of edges on the clearance surfaces exposed to the cutting action.

The original drill point produced by the manufacturer lasts only until the first regrind; thereafter performance and life depend on the quality of regrind. Proper regrinding (reconditioning) of a drill is a complex and important operation. If satisfactory cutting and hole size are to be achieved, it is essential that the point angle, lip clearance, lip length, and web thinning be correct. As illustrated in Figure 24-5, incorrect sharpening often results in unbalanced cutting forces at the tip, causing misalignment and oversized holes. Drills, even small drills, should always be machine ground, never hand ground. Drill grinders, often computer controlled, should be used to assure exact reproduction of the geometry established by the manufacturer of the drill. This is extremely important when drills are used on mass-production or numerically controlled machines. Companies invest huge sums in NC machining centers but overlook the value of a top-quality drill-grinding machine.

Drill shanks are made in several types. The two most common types are the straight and the taper. *Straight-shank* drills are usually used for sizes up to $\frac{1}{2}$ in. and must be held in some type of drill chuck. *Taper shanks* are available on drills from $\frac{1}{8}$ to $\frac{1}{2}$ in. and are common on drills above $\frac{1}{2}$ in. Morse tapers, having a taper of approximately $\frac{5}{8}$ in./ft, are used on taper-shank drills, ranging from a number 1 taper on $\frac{1}{8}$ in. drills to a number 6 on a $3\frac{1}{2}$ in. drill.

Taper-shank drills are held in a female taper in the end of the machine tool spindle. If the taper on the drill is different from the spindle taper, adapter sleeves are available. The taper assures the drill's being accurately centered in the spindle. The *tang* at the end of the taper shank fits loosely in a slot at the end of the tapered hole in the spindle. The drill may be loosened for removal by driving a metal wedge, called a *drift*, through a hole in the side of the spindle and against the end of the tang. It also acts as a safety device to prevent the drill from rotating in the spindle hole under heavy loads. However, if the tapers on the drill and in the spindle are proper, no slipping should occur. The driving force to the drill is carried by the friction between the two tapered members.

Standard drills are available in four size series, the size indicating the diameter of the drill body.

- *Millimeter series:* 0.01- to 0.50-mm increments, according to size, in diameters from 0.015 mm
- *Numerical series:* No. 80 to No. 1 (0.0135 to 0.228 in.)
- *Lettered series:* A to Z (0.234 to 0.413 in.)
- *Fractional series:* $\frac{1}{64}$ to 4 in. (and over) by 64ths

TiN coating of conventional drills greatly improves drilling performance. The increase in tool life of TiN-coated drills over uncoated drills in machining steels is over 200 to 1000%.

The bores of rifle barrels were once drilled using conventional drills. Today, *deep-hole drills*, or *gun drills*, are used when deep holes are to be drilled. These drills have special names, such as *BTA* (Boring Trepanning Association) *drills* and *ejector drills*. A deep hole is one in which the length (or depth) of the hole is three or more times the diameter. Coolants can be fed internally through these drills to the cutting edges. See Figure 24-21 for schematic of an ejector drill and the machine tool used for gun drilling. The coolants flush the chips out the flutes. The special design of these drills reduces the tendency of the drill to drift, thus producing a more accurately aligned hole. Figure 24-6 shows some typical BTA deep-hole drilling tools, designed for single-lip end cutting of a hole in a single pass. Solid-deep hole drills have alloy-steel shanks with a carbide-edged tip that is fixed to it mechanically. The cutting edge cuts through the center on one side of the hole, leaving no area of material to be extruded. The cutting is done by the outer and inner cutting angles, which meet at a point. Theoretically, the depth of the hole has no limit, but practically, it is restricted by the torsional rigidity of the shank.

Gun drills have a single-lip cutting action. Bearing areas and lifting forces generated by the coolant pressure counteract the radial and tangential loads. The single-lip construction forces the edge to cut in a true circular pattern. The lip thus follows the direction of its own axis. The *trepanning gun drill* leaves a solid core.

Hole straightness is affected by variables such as diameter, depth, uniformity of the workpiece material, condition of the machine, sharpness of cutting edges, feeds and speeds used, and whether the tool or the workpiece is rotated or counterrotated.

Two-flute drills are available that have holes extending throughout the length of each land to permit coolant to be supplied, under pressure, to the point adjacent to each cutting edge. These are helpful in providing cooling and also in promoting chip removal from the hole in drilling to moderate depths. They require special fittings through which the coolant can be supplied to the rotating drill and are used primarily on automatic and semiautomatic machines. See Table 24-3 for a comparison of drilling processes.

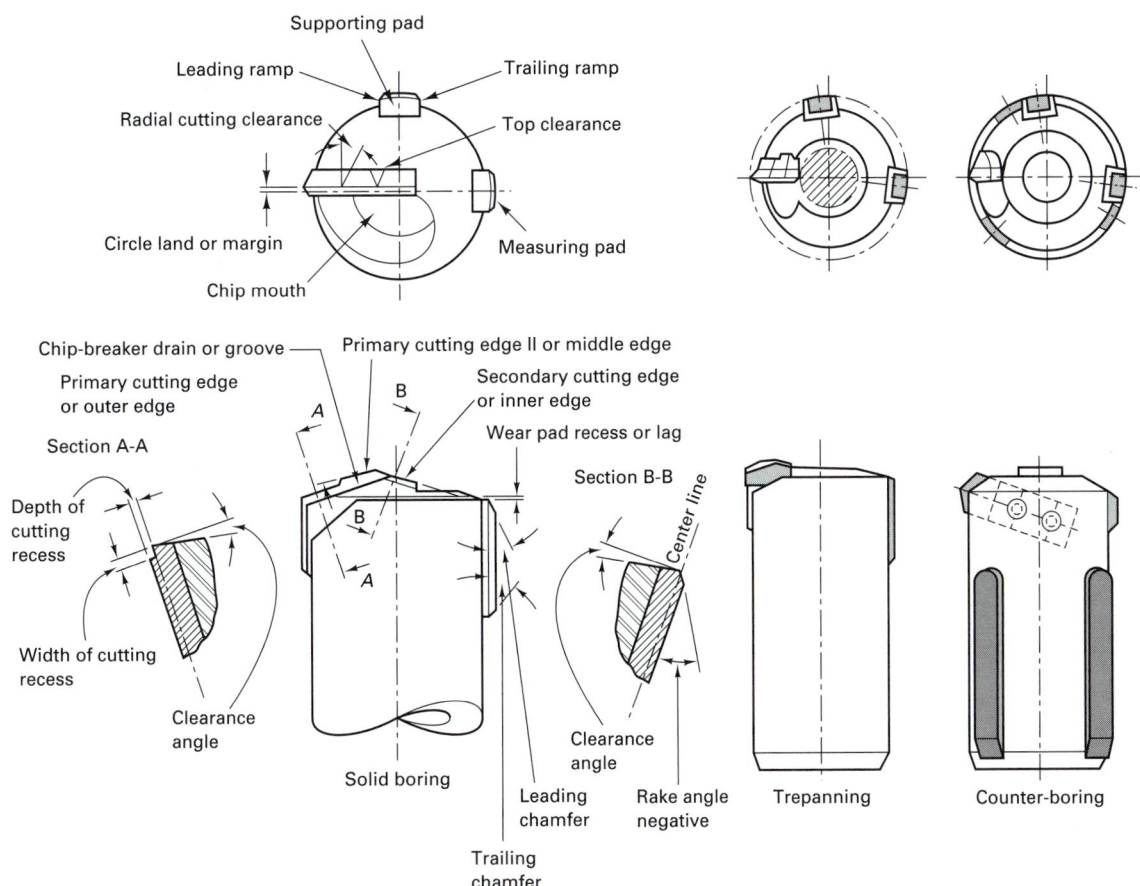

FIGURE 24-6   BTA drills for boring trepanning, counterboring, and drilling deep holes.

TABLE 24-3.	Comparison of Drilling Processes					
	Type of Drill					
	Twist Drill	Gun Drill	Spade Drill	Ejector Drill	Trepanning	Solid Coring Drill
Typical size range [diameter (mm)]	0.5–50	1.4–25	25–150	18–60	42–250	12–200
Hole depth/diameter ratio minimum practical	No minimum	<1	No minimum	1	10	1
Common maximum	5–10	100	>40	50	100	100
Ultimate	>50	200	10 (vertical), >100 (horizontal)	>50	>100	>100

Larger holes in thin material may be made with a *hole cutter* (Figure 24-7), whereby the main hole is produced by the thin-walled, multiple-tooth cutter with saw teeth. Hole cutters are often called *hole saws*.

When starting to drill a hole, a drill can deflect rather easily because of the "walking" action of the chisel point. Hole location accuracy is lost. Consequently, to assure that

FIGURE 24-7 Hole cutter used for thin sheets. *(Courtesy of Armstrong-Blum Manufacturing Company.)*

FIGURE 24-8 Combination center drill and countersink. *(Courtesy of Chicago-Latrobe Twist Drill Works.)*

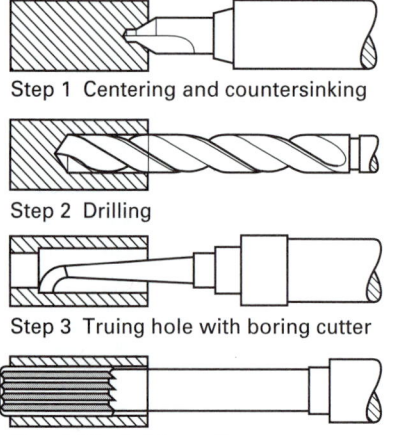

Step 1 Centering and countersinking

Step 2 Drilling

Step 3 Truing hole with boring cutter

Step 4 Final sizing with reamer

FIGURE 24-9 Sequence of operations required to obtain a hole that is accurate as to size and aligned on center.

a hole is started accurately, a *center drill* and *countersink* (Figure 24-8) is used prior to a regular chisel-point twist drill. The center drill and countersink has a short, straight drill section extending beyond a 60° taper portion. The heavy, short body provides rigidity so that a hole can be started with little possibility of tool deflection. The hole should be drilled only partway up on the tapered section of countersink. The conical portion of the hole serves to guide the drill being used to make the main hole. Combination center drills are made in four sizes to provide the proper-size starting hole for any drill. If the drill is sufficiently large in diameter, or if it is sufficiently short, satisfactory accuracy often may be obtained without center drilling. Special drill holders are available that permit drills to be held with only a very short length protruding.

Because of its flexibility, a drill may drift off centerline during drilling. The use of a center (start) drill helps to assure that a drill starts drilling at the desired location. Non-homogeneities in the workpiece and imperfect drill geometries may also cause the hole to be oversize or off-line. If accuracy in these respects is desired, it is necessary to follow center drilling and drilling by boring and reaming (Figure 24-9). Boring corrects the hole alignment; reaming brings the hole to accurate size and improves the surface finish.

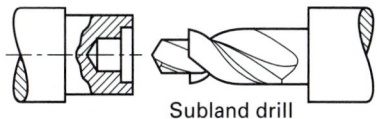

Subland drill

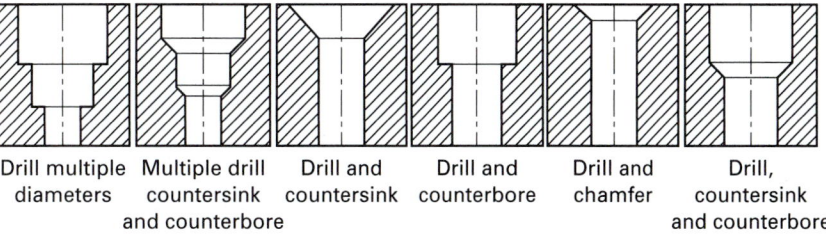

Drill multiple diameters | Multiple drill countersink and counterbore | Drill and countersink | Drill and counterbore | Drill and chamfer | Drill, countersink and counterbore

FIGURE 24-10   Special-purpose subland drill (*above*), and some of the operations possible with such drills (*below*).

Special *combination drills* are made that can drill two or more diameters, or drill and countersink and/or counterbore, in a single operation. Some of these are shown in Figure 24-10. The operations called countersinking and counterboring usually follow drilling. These operations are described in more detail later in this chapter. A *step drill* has a single set of flutes and is ground to two or more diameters. *Subland drills* have a separate set of flutes, on a single body, for each diameter or operation; they provide better chip flow, and the cutting edges can be ground to give proper cutting conditions for each operation. Combination drills are expensive and may be difficult to regrind but can be economical for production-type operations if they reduce work handling, setups, or separate machines and operations.

Spade drills (Figure 24-11) are widely used for making holes 1 in. or larger in diameter at low speeds/high feeds (Tables 24.4 and 24.5). The workpiece usually has an existing hole, but a spade drill can drill deep holes in solids of stacked materials. Spade drills

TABLE 24-4. Recommended Surface Cutting Speeds with Spade Drills for Various Materials	
Material	Surface Speed (ft/min)
Mild machinery steel 0.2 and 0.3 carbon	65–110
Steel, annealed 0.4 to 0.5 carbon	55–80
Tool steel, 1.2 carbon	45–60
Steel forging	35–50
Alloy steel	45–70
Stainless steel, free machining	50–70
Stainless steel, hard	25–40
Cast iron, soft	80–150
Cast iron, medium hard	55–100
Cast iron, hard, chilled	25–40
Malleable iron	79–90
Brass and bronze, ordinary	200–300
Bronze, high tensile	70–150
Monel metal	35–50
Aluminum and its alloys	200–300
Magnesium and its alloy	250–400

TABLE 24-5. Feeds for Spade Drilling (in./rev)

Drill Size (in.)	Cast Iron Malleable Iron Brass Bronze	Medium Steel Stainless Steel Monel Metal Drop-Forged Alloys Tool Steel (annealed)	Tough Steel Drop Forging Aluminum
$1-1\frac{1}{4}$	0.010–0.020	0.008–0.014	0.006–0.012
$1\frac{1}{4}-\frac{3}{4}$	0.010–0.024	0.008–0.018	0.008–0.012
$1\frac{3}{4}-2\frac{1}{2}$	0.010–0.030	0.010–0.024	0.010–0.017
$2\frac{1}{2}-4$	0.012–0.032	0.012–0.030	0.010–0.017
$4-6$	0.012–0.032	0.010–0.024	0.008–0.017

Source: Waukesha Cutting Tools, Inc.

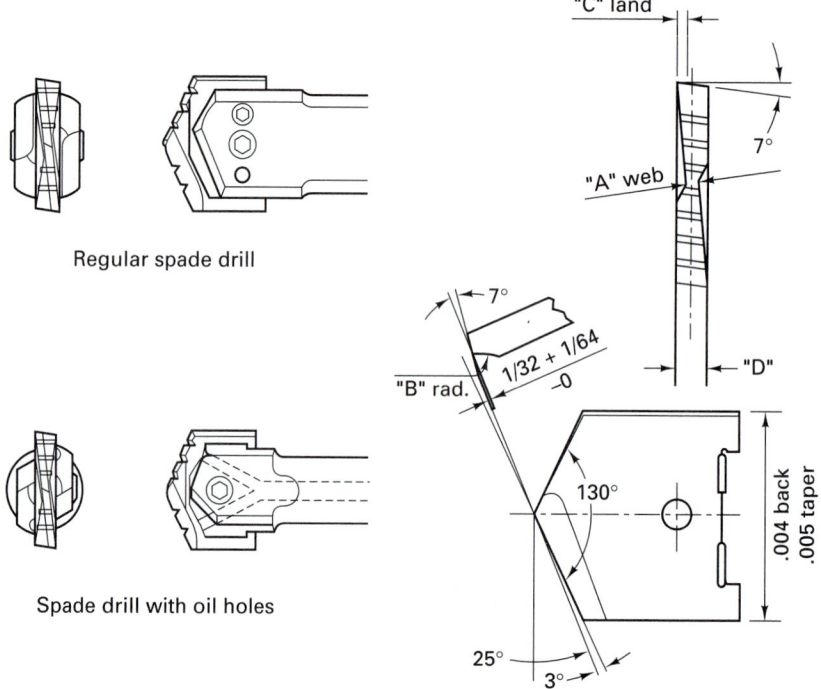

FIGURE 24-11 (*Top*) Regular spade drill; (*Middle*) spade drill with oil holes; (*Bottom*) spade drill geometry, nomenclature.

are less expensive because the long supporting bar can be made of ordinary steel. The drill point can be ground with a minimum chisel point. The main body can be made more rigid because no flutes are required. The main body can be provided with a central hole through which a fluid can be circulated to aid in cooling and in chip removal. The cutting blade is easier to sharpen. Only the blades need to be TiN-coated.

Spade drills are often used to machine a shallow locating cone for a subsequent smaller drill and at the same time to provide a small bevel around the hole to facilitate later tapping or assembly operations. Such a bevel also frequently eliminates the need for deburring. The practice is particularly useful on mass-production and numerically controlled machines.

Insert

FIGURE 24-12 One-
and two-lipped insert
drills. *(Courtesy of
Waukesha.)*

Carbide-tipped drills and drills with indexable inserts are also available (Figure
24-12) with one- and two-piece inserts for drilling shallow holes in solid workpieces. *Indexable insert drills* can produce a hole four times faster than a spade drill because they
run at high speeds/low feeds and are really more of a boring operation than a drilling process.

The diameter of the hole (*D*) and the length/diameter ratio usually determine what
kind of drill to use. Figure 24-13 shows the drilling areas of major types of drills. As
shown, different kinds of tools are used depending on the depth and diameter. Section A
shows the drilling area of relatively shallow holes and small diameters. About half of all
the drilling processes fall within the category of this section. It is the section for which the
majority of the work is done by twist drills and very few cemented carbide drills. Section
B is the drilling of deep holes for which cemented carbide gun drills are used. Section C
is that of shallow holes having large diameters, for which spade drills are used. Section D
is that of deep holes having large diameters, for which BTA tools are used.

■ 24.4 TOOLHOLDERS FOR DRILLS

Straight-shank drills must be held in some type of drill *chuck* (Figure 24-14). Chucks are
adjustable over a considerable size range and have radial steel fingers. When the chuck is
tightened by means of a chuck key, these fingers are forced inward against the drill. On
smaller drill presses, the chuck often is permanently attached to the machine spindle,
whereas on larger drilling machines the chucks have a tapered shank that fits into the female Morse taper of the machine spindle. Special types of chucks in semiautomatic or
fully automatic machines permit quite a wide range of sizes of drills to be held in a single
chuck (Figure 24-15).

Chucks using chuck keys require that the machine spindle be stopped to change a
drill. To reduce the downtime when drills must be changed frequently, *quick-change
chucks* are used. Each drill is fastened in a simple round collet that can be inserted into

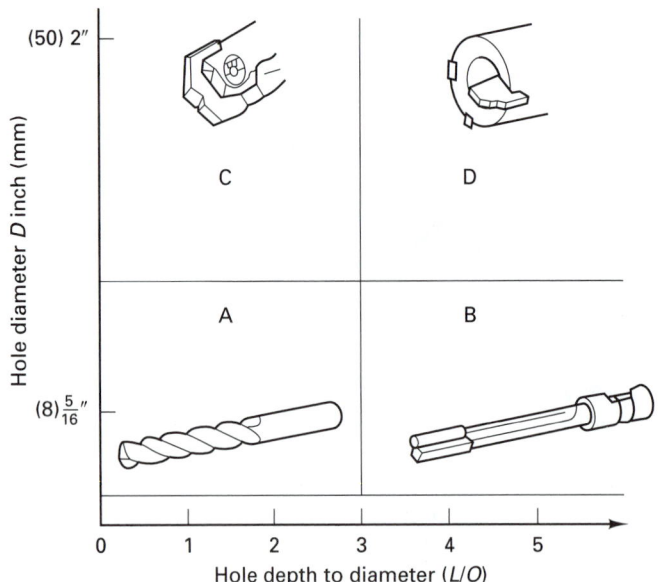

Sector	Typical drill types
A	Twist drill (HSS) Center core drill
B	Twist drill Gun drill BTA Ejector drill
C	Twist drill Indexable insert drill Spade drill Center core drill
D	BTA Ejectror drill

FIGURE 24-13   Drill selection depends on hole diameter and hole depth.

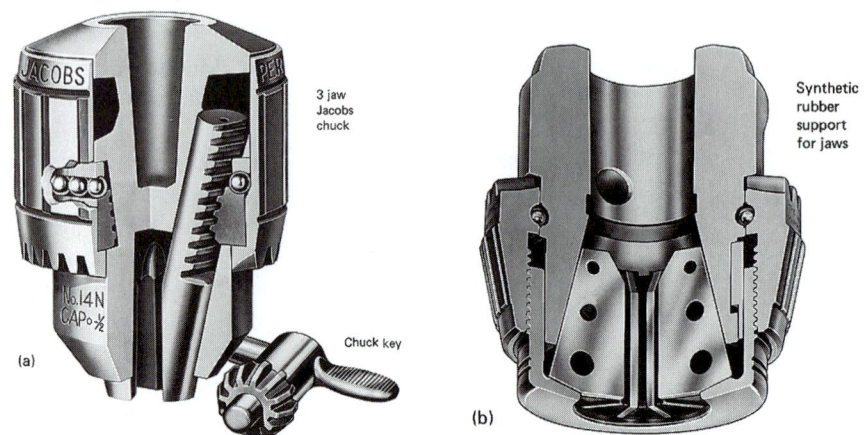

FIGURE 24-14   Two of the most commonly used types of drill chucks. *(Courtesy of Jacobs Manufacturing Company.)*

the chuck hole while it is turning merely by raising and lowering a ring on the chuck body. With the use of this type of chuck, center drills, drills, counterbores, reamers, and so on, can be manually changed in quick succession.

■ 24.5 WORKHOLDING DEVICES FOR DRILLING

Work that is to be drilled is ordinarily held in a vise or in specially designed workholders called *jigs*. Workholding devices are the subject of Chapter 28, where the design of workholding devices is discussed. Many examples of drill jigs are shown.

FIGURE 24-15   Three types of drill chucks used on automatic equipment: (*left*) universal type. *(Courtesy of Brookfield Tool Company.)* (*center*) collet type; (*right*) quick-change drill chuck. *(Courtesy of Consolidated Machine Tool Division, Farrel Birmingham Company, Inc.)*

Even in light drilling operations, the work should not be held on the table by hand unless adequate leverage is available. This is a dangerous practice and can lead to serious accidents, because the drill has a tendency to catch on the workpiece and cause it to rotate. Work that is too large to be held in a jig can be clamped directly to the machine table using suitable bolts and clamps and the slots or holes in the table.

■ 24.6 MACHINE TOOLS FOR DRILLING

The basic work and tool motions required for drilling—relative rotation between the workpiece and the tool, with relative longitudinal feeding—also occurs in a number of other machining operations. Thus drilling can be done on a variety of machine tools, such as lathes, vertical milling machines, boring machines, and machining centers. In this chapter only those machines that are designed, constructed, and used primarily for drilling are considered.

The common name for the machine tool used for drilling is the *drill press*. Drill presses consist of a *base*, a *column* that supports a *powerhead*, a *spindle*, and a *worktable*. On small machines, the base rests on a workbench, whereas on larger machines it rests on the floor (Figure 24-16). The column may be either round or of box-type construction, the latter being used on larger, heavy-duty machines, except in radial types. The powerhead contains an electric motor and means for driving the spindle in rotation at several speeds. On small drilling machines this may be accomplished by shifting a belt on a step-cone pulley, but on larger machines a geared transmission is used.

The heart of any drilling machine is its spindle. To drill satisfactorily, the spindle must rotate accurately and also resist whatever side forces result from the drilling process. In virtually all machines the spindle rotates in preloaded ball or taper-roller bearings. In

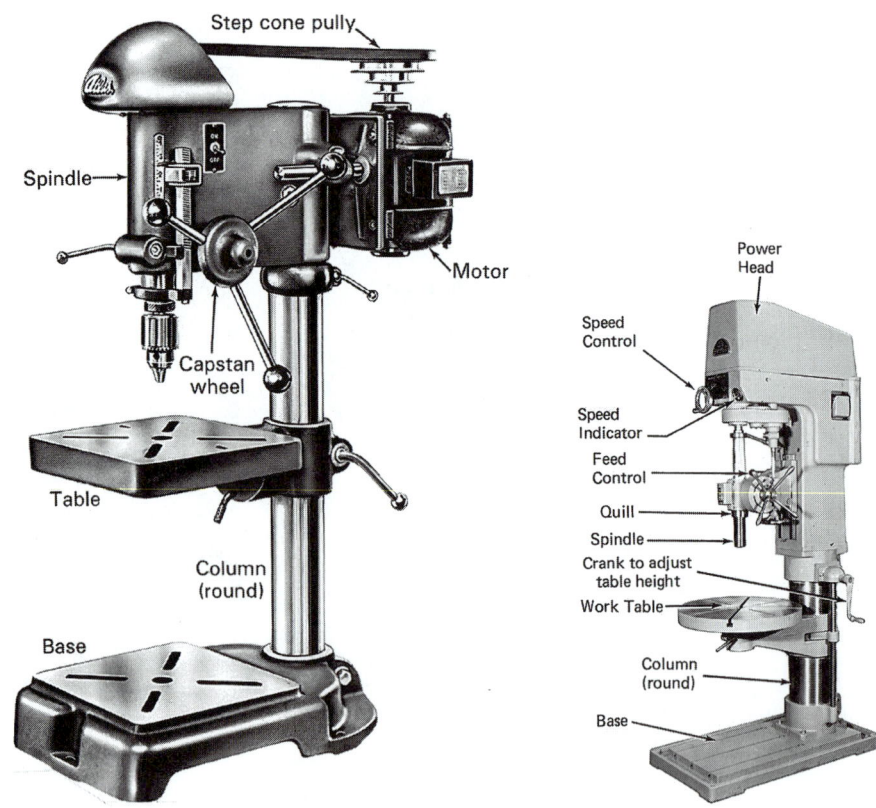

FIGURE 24-16   (*left*) Fifteen-inch bench-type drill press, usually set on a table. (*Courtesy of Atlas Press Company*); (*right*) upright drilling machine. (*Courtesy of Buffalo Forge Company.*)

addition to powered rotation, provision is made so that the spindle can be moved axially to feed the drill into the work. On small machines the spindle is fed by hand, using the handles extending from the capstan wheel, whereas on larger machines power feed is provided. Except on some small bench types, the spindle contains a hole with a Morse taper in its lower end into which taper-shank drills or drill chucks can be inserted.

The worktables on drilling machines may be moved up and down on the column to accommodate work of various sizes. On round-column machines the table usually can also be rotated out of the way so that workpieces can be mounted directly on the base. On some box-column machines the table is mounted on a subbase so that it can be moved in two directions in a horizontal plane by means of feed screws.

Figure 24-17 shows some block diagrams and schematics for four common types of drilling machines used in production environments. Drilling machines usually are classified in the following manner:

    1. Bench (Figure 24-16)
       a. Plain
       b. Sensitive
    2. Upright (Figure 24-16)
       a. Single-spindle (Figure 24-17)

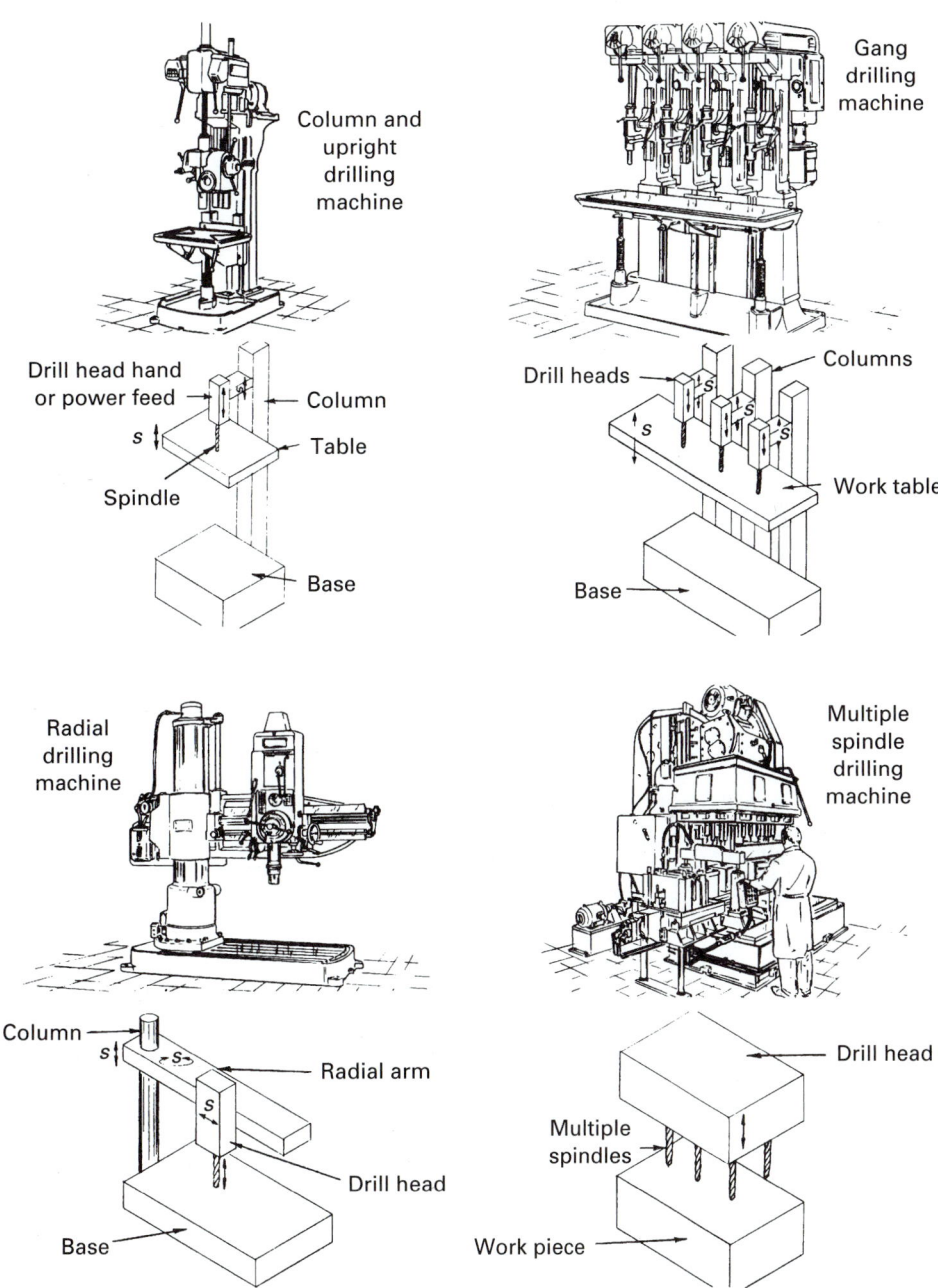

FIGURE 24-17 Four principal types of drilling machines. *(From Manufacturing Producibility Handbook, courtesy of General Electric Company.)*

    b. Turret
    c. NC turret (Figure 24-18)
3. Radial (Figure 24-17)
    a. Plain
    b. Semiuniversal
    c. Universal
4. Gang (Figure 24-17).
5. Multispindle (Figure 24-17)
6. Deep-hole (Figure 24-21)
    a. Vertical
    b. Horizontal
7. Transfer

With bench drill presses, holes up to $\frac{1}{2}$ in. in diameter can be drilled. The same type of machine can be obtained with a long column so that it can stand on the floor. The size of bench and upright drilling machines is designated by *twice* the distance from the center-line of the spindle to the nearest point on the column, this being an indication of the max-imum size of work that can be drilled in the machines. For example, a 15-in. drill press will permit a hole to be drilled at the center of a workpiece 15 in. in diameter.

Sensitive drilling machines are essentially smaller plain bench-type machines with more accurate spindles and bearings. They are capable of operating at high speeds, up to 30,000 rpm. Very sensitive hand-operated feeding mechanisms are provided for use in drilling small holes. Such machines are used for tool and die work and for drilling very small holes, often less than a few thousandth of an inch in diameter, when high spindle speeds are necessary to obtain proper cutting speed and sensitive feel to provide delicate feeding to avoid breakage of the very small drills.

Upright drilling machines usually have spindle speeds ranges from 60 to 3500 rpm and power feed rates, in from 4 to 12 steps, from about 0.004 to 0.025 in./rev. Most mod-ern machines use a single-speed motor and a geared transmission to provide the range of speeds and feeds. The feed clutch disengages automatically when the spindle reaches a preset depth.

Worktables on most upright drilling machines contain holes and slots for use in clamp-ing work and nearly always have a channel around the edges to collect cutting fluid, when it is used. On box-column machines, the table is mounted on vertical ways on the front of the column and can be raised or lowered by means of a crank-operated elevating screw.

In mass production, *gang-drilling machines* are often used when several related op-erations, such as holes of different sizes, reaming, or counterboring, must be done on a sin-gle part. These consist essentially of several independent columns, heads, and spindles mounted on a common base and having a single table. The work can be slid into position for the operation at each spindle. They are available with or without power feed. One or several operators may be used. This machine would be an example of a simple small cell except that the machines are usually not single-cycle automatics.

Turret-type, upright drilling machines (Figure 24-18) are used when a series of holes of different size, or a series of operations (such as center drilling, drilling, reaming, and spot facing), must be done repeatedly in succession. The tools selected are mounted in the turret. Each tool can quickly be brought into position merely by rotation of the turret. These machines automatically provide individual feed rates for each spindle and are often numerically controlled.

Radial drilling machines tools are used on large workpieces that cannot easily be handled manually. As shown in Figure 24-17, these machines have a large, heavy, round,

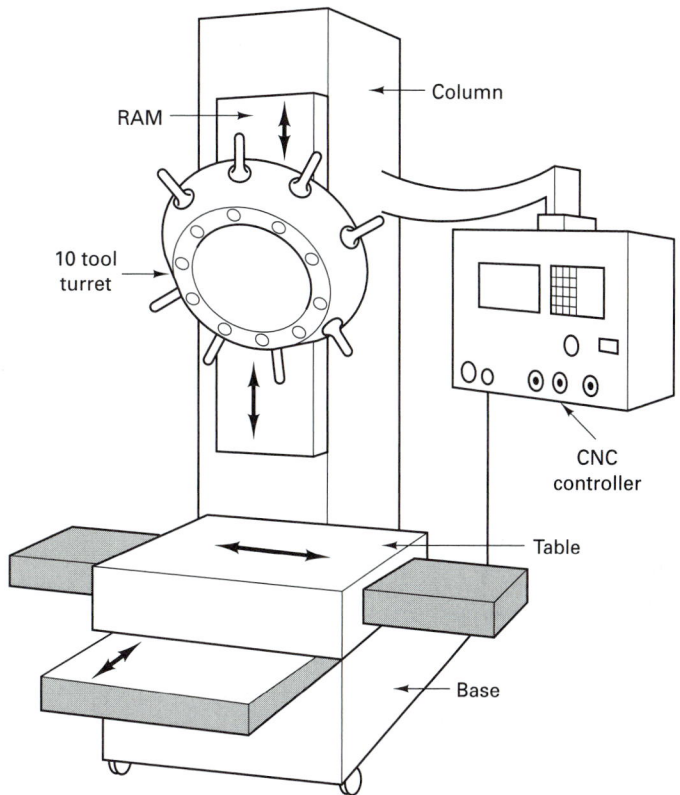

FIGURE 24-18   NC turret drilling machines have largely replaced the gang drilling machines in many production facilities.

vertical column supported on a large base. The column supports a radial arm that can be raised and lowered by power and rotated over the base. The spindle head, with its speed- and feed-changing mechanism, is mounted on the radial arm. It can be moved horizontally to any desired position on the arm. Thus the spindle can quickly be properly positioned for drilling holes at any desired point on a large workpiece mounted either on the base of the machine or even sitting on the floor.

Plain radial drilling machines provide only a vertical spindle motion. On *semiuniversal machines*, the spindle head can be pivoted at an angle to a vertical plane. On *universal machines*, the radial arm is rotated about a horizontal axis to permit drilling at any angle.

Radial drilling machines are designated by the radius of the largest disk in which a center hole can be drilled when the spindle head is at its outermost position. Sizes from 3 to 12 ft are available. Radial drilling machines have a wide range of speeds and feeds, can do boring, and include provisions for tapping (internal threading) (see Chapter 30).

Multiple-spindle drilling machines (Figure 24-17) are mass-production machines with as many as 50 spindles driven by a single powerhead and fed simultaneously into the work. Figure 24-19 shows an adjustable multiple-spindle head that can be mounted on a regular single-spindle drill press. Figure 24-20 shows the methods of driving and positioning the spindles, which permits them to be adjusted so that holes can be drilled at any location within the overall capacity of the head. Special drill jigs are often designed and built for each job to provide accurate guidance for each drill. Although such machines and workholders are quite costly, they can be cost-justified when the quantity to be produced will justify the setup cost and the cost of the jig. Reducing setup on these machines is

FIGURE 24-19 Multiple-spindle drill head attachment head, showing six spindles. *(Courtesy of Thriftmaster Products Incorporated.)*

difficult (see Chapter 44). Numerically controlled drill presses other than turret drill presses are not common because drilling and all its related processes can be done on vertical or horizontal NC machining centers equipped with automatic tool changers (see Chapter 29).

Special machines are used for drilling long (deep) holes, such as those in rifle barrels, connecting rods, and long spindles. High cutting speeds, very light feeds, and a copious flow of cutting fluid assure rapid chip removal. Adequate support for the long, slender drills is required. In most cases horizontal machines are used (Figure 24-21). The work is rotated in a chuck with steady rests providing support along its length, as required. The drill does not rotate and is fed into the work. Vertical machines are also available for shorter workpieces. Notice the similarity between this process and boring.

■ 24.7 CUTTING FLUIDS FOR DRILLING

For shallow holes, the general rules relating to cutting fluids, as given in Chapter 22, are applicable. When the depth of the hole exceeds one diameter, it is desirable to increase the lubricating quality of the fluid because of the rubbing between the drill margins and the wall of the hole. The effectiveness of cutting a fluid as a coolant is quite variable in drilling. While the rapid exit of the chips is a primary factor in heat removal, this action also tends to restrict entry of the cutting fluid. This is of particular importance in drilling materials that have poor heat conductivity. Recommendations for cutting fluids for drilling are given in Table 24-6.

If the hole depth exceeds two or three diameters, it is usually advantageous to withdraw the drill each time it has drilled about one diameter of depth, to clear chips from the hole. Some machines are equipped to provide this "pecking" action automatically.

Where cooling is desired, the fluid should be applied copiously. For severe conditions, drills containing coolant holes have a considerable advantage. Not only is the fluid supplied near the cutting edges, but the coolant flow aids in flushing the chips from the hole. Where feasible, drilling horizontally has distinct advantages over drilling vertically downward.

Adjustable drill head
Spindle: 6 Production: 50 pieces

Geared drill head
Spindle: 8 Production: 80,000 pieces

Gearless drill head
Spindle: 16 Production: 30,000 pieces

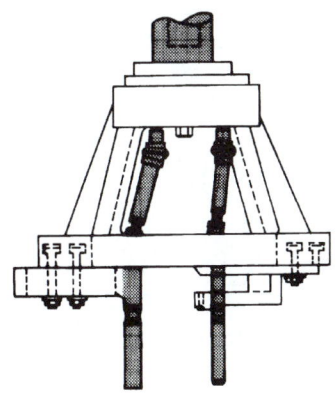
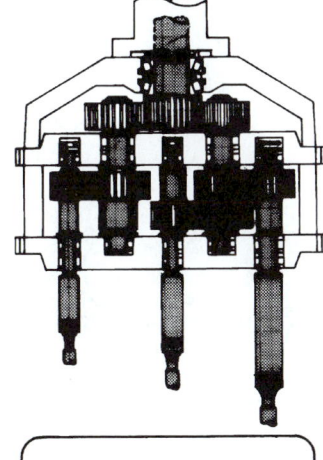
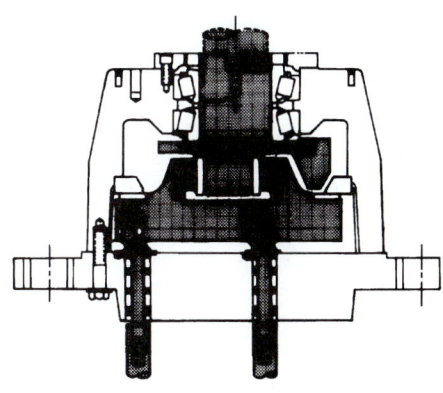

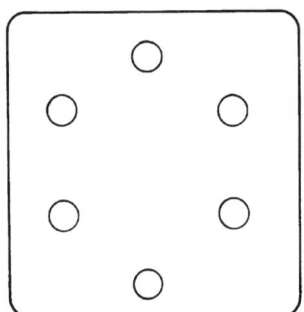
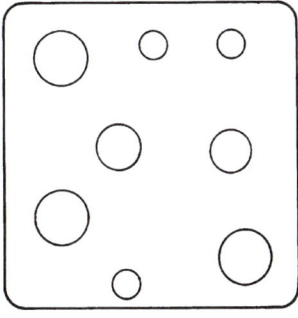
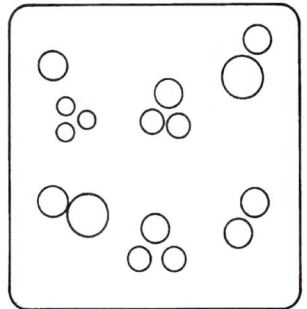

An adjustable drill head should be considered for low production jobs. However, many short-run jobs such as this would be required to justify a multiple-spindle head.

A geared drill head is most appropriate in this situation where there is a large difference in sizes and a high daily production.

Only a gearless head can perform this operation in one pass, due to the close proximity of the spindle centers.

FIGURE 24-20   Three basic types of multiple-spindle drill heads: (*left*) adjustable; (*middle*) geared; (*right*) gearless.   *(Courtesy of Zagar Incorporated.)*

■ 24.8 COUNTERBORING, COUNTERSINKING, AND SPOT FACING

Drilling often is followed by *counterboring*, *countersinking*, or *spot facing*. As shown in Figure 24-22, each provides a bearing surface at one end of a drilled hole. They are usually done with a special tool having from three to six cutting edges.

Counterboring provides an enlarged cylindrical hole with a flat bottom so that a bolt head, or a nut, will have a smooth bearing surface that is normal to the axis of the hole; the depth may be sufficient so that the entire bolt head or nut will be below the surface of the part. The pilot on the end of the tool fits into the drilled hole and helps to assure concentricity with the original hole. Two or more diameters may be produced in a single counterboring operation. Counterboring also can be done with a single-point tool, although

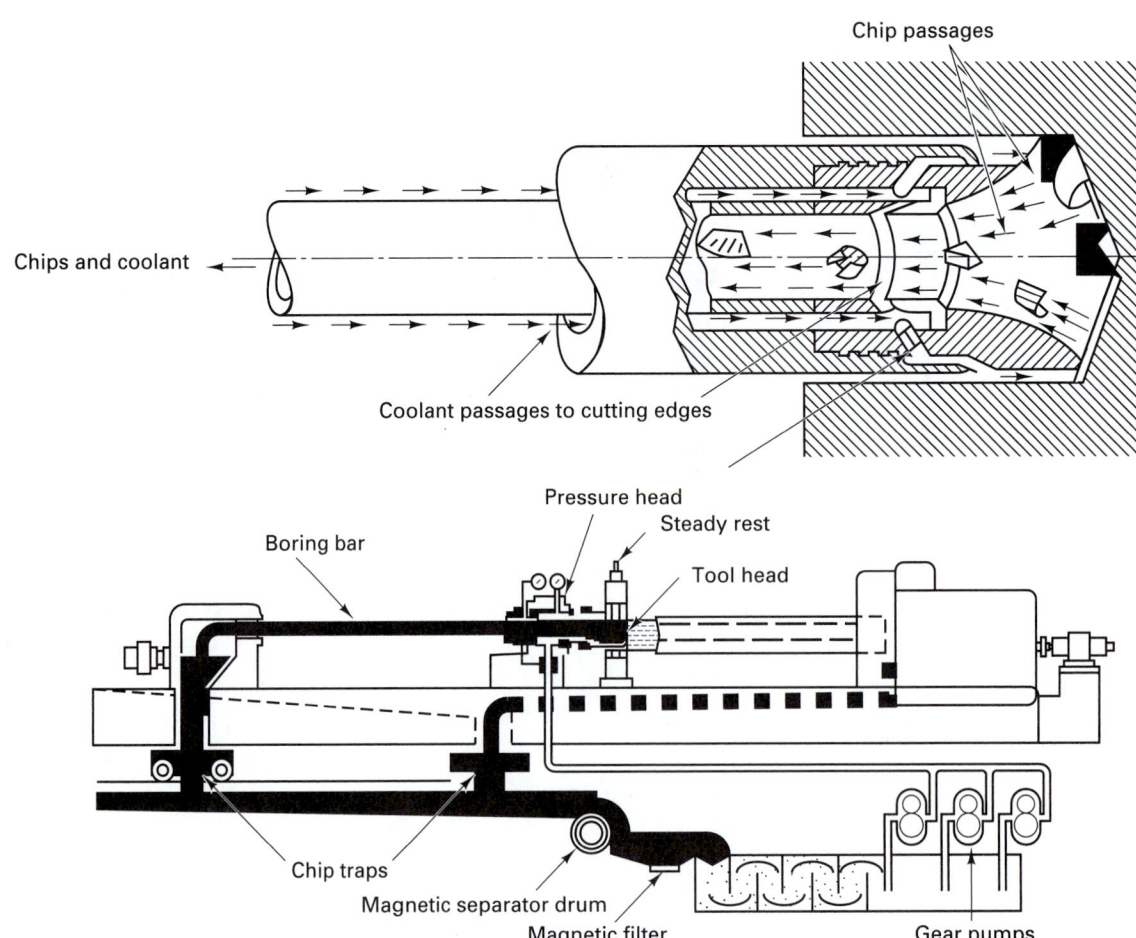

FIGURE 24-21   Horizontal deep-hole drilling machine (below) with enlargement (above) of the cutting tool, called an ejector drill. *(From S. Azad and S. Chandrashekar, Mechanical Engineering, Sept. 1985, pp. 62, 63.)*

TABLE 24-6.	Cutting Fluids for Drilling
Work Material	Cutting Fluid
Aluminum and its alloys	Soluble oil, kerosene, and lard-oil compounds; light, nonviscous neutral oil; kerosene and soluble oil mixtures
Brass	Dry or a soluble oil; kerosene and lard-oil compounds; light, nonviscous neutral oil
Cast iron	Dry or with a jet of compressed air for cooling
Copper	Soluble oil, strained lard oil, oleic-acid compounds
Malleable iron	Soluble oil, nonviscous neutral oil
Monel metal	Soluble oil, sulfurized mineral oil
Stainless steel	Soluble oil, sulfurized mineral oil
Steel, ordinary	Soluble oil, sulfurized oil, high-extreme-pressure-value mineral oil
Steel, very hard	Soluble oil, sulfurized oil, turpentine
Wrought iron	Soluble oil, sulfurized oil, mineral-animal oil compound

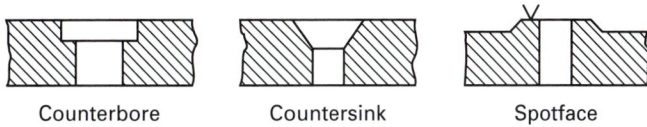

Counterbore   Countersink   Spotface

FIGURE 24-22   Surfaces produced by counterboring, countersinking, and spot facing.

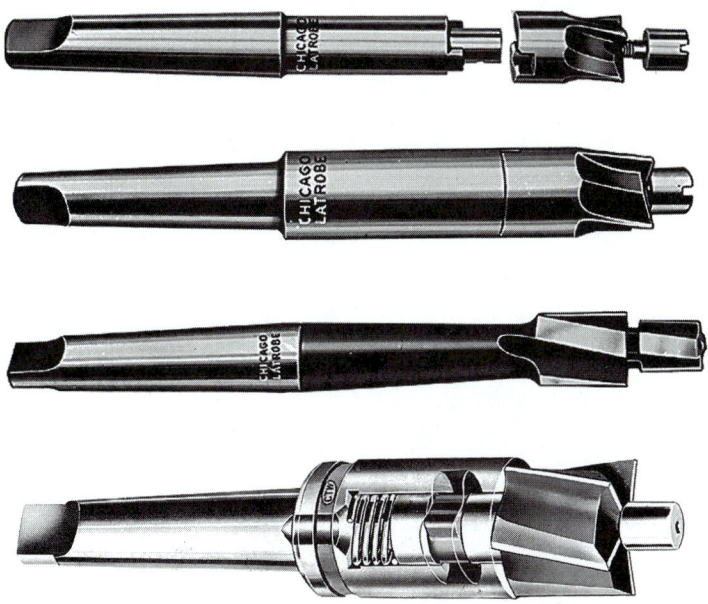

FIGURE 24-23   Counterboring tools. (Bottom to Top) Interchangeable counterbore; solid, taper-shank counterbore with integral pilot; replaceable counterbore and pilot; replaceable counterbore, disassembled. (*Courtesy of Ex-Cell-O Corporation and Chicago Latrobe Twist Drill Works.*)

this method ordinarily is used only on large holes and essentially is a boring operation. Some counterboring tools are shown in Figure 24-23.

Countersinking makes a beveled section at the end of a drilled hole to provide a proper seat for a flat-head screw or rivet. The most common angles are 60, 82, and 90°. Countersinking tools are similar to counterboring tools except that the cutting edges are elements of a cone, and they usually do not have a pilot because the bevel of the tool causes them to be self-centering.

Spot facing is done to provide a smooth bearing area on an otherwise rough surface at the opening of a hole and normal to its axis. Machining is limited to minimum depth that will provide a smooth, uniform surface. Spot faces thus are somewhat easier and more economical to produce than counterbores. They usually are made with a multiedged end-cutting tool that does not have a pilot, although counterboring tools frequently are used.

■ 24.9 DRILLING PRACTICE AND PROBLEMS

Table 24-7 lists some typical drilling problems together with their probable causes and cures. One very simple way to check drill performance is to examine the chips. They should be the same from both flutes. If the chips are not of the same size, the point has probably been improperly ground so that one lip is doing most of the cutting.

TABLE 24-7. Causes of Drilling Problems	
Problem	Cause
Outer corners of drill break down	Cutting speed too high; hard spot in material; no cutting fluid at tips; flutes clogged with chips; insufficient feed in materials such as stainless steel
Chipping in cutting edge	Feed too high; lip relief too large
Checks or cracks in cutting edges	Overheating; poor cooling; drill cooled too rapidly while sharpening or drilling
Chipped drill margin	Wrong-size drill bushing; oversize, allowing vibration; undersize, causing interference
Broken drill body	Point improperly ground; feed too heavy for work material; backlash or looseness in spindle, workholder, or workpiece; dulling of drill, increasing cutting forces; flutes clogged with chips
Broken tang	Bad fit between taper shank and socket caused by chips, burrs, dirt, wear, etc.
Drill splits up center	Lip relief angles too small or feed too large
Drill will not enter	Bad (unequal) point geometry; poor lubrication at point; wrong cutting fluid; feed too large; insufficient rigidity in workholder or spindle
Hole oversize	Unequal rake angles on cutting edges; one edge duller than the other; unequal edge length; loose spindle

The drilling machine should be rigid and have sufficient strength and power to withstand the cutting forces. A lack of rigidity in the tool, the workpiece, or the machine permits the affected members to deflect under cutting forces and cause chatter. The load builds up, the material being cut is sheared, the load is suddenly released, and the deflected member springs back to its normal position. The cycle repeats itself rapidly, with the result that the cutting lips have a vibrating action against the work. As part of the routine preventive maintenance effort, backlash in the feed mechanism should be minimized to reduce strain on the drill when it breaks through the bottom of the hole. Jigs and workholding devices on indexing machines must be free of play and firmly seated. The tapers in the machine spindle, drill sleeves, adapters, and shanks should be free of burrs and scratches that may cause drill runout. Spindle overhang should be minimized. Spindle bearings should be in good condition to prevent runout and end play.

■ 24.10 REAMING

Reaming removes a small amount of material from the surface of holes. It is done for two purposes: to bring holes to a more exact size and to improve the finish of an existing hole. Multiedge cutting tools are used as shown in Figure 24-24. No special machines are built for reaming. The same machine that was employed for drilling the hole can be used for reaming by changing the cutting tool.

To obtain proper results, only a minimum amount of materials should be left for removal by reaming. As little as 0.005 in. is desirable, and in no case should the amount exceed 0.015 in. A properly reamed hole will be within 0.001 in. of the correct size and have a fine finish.

Chucking reamer

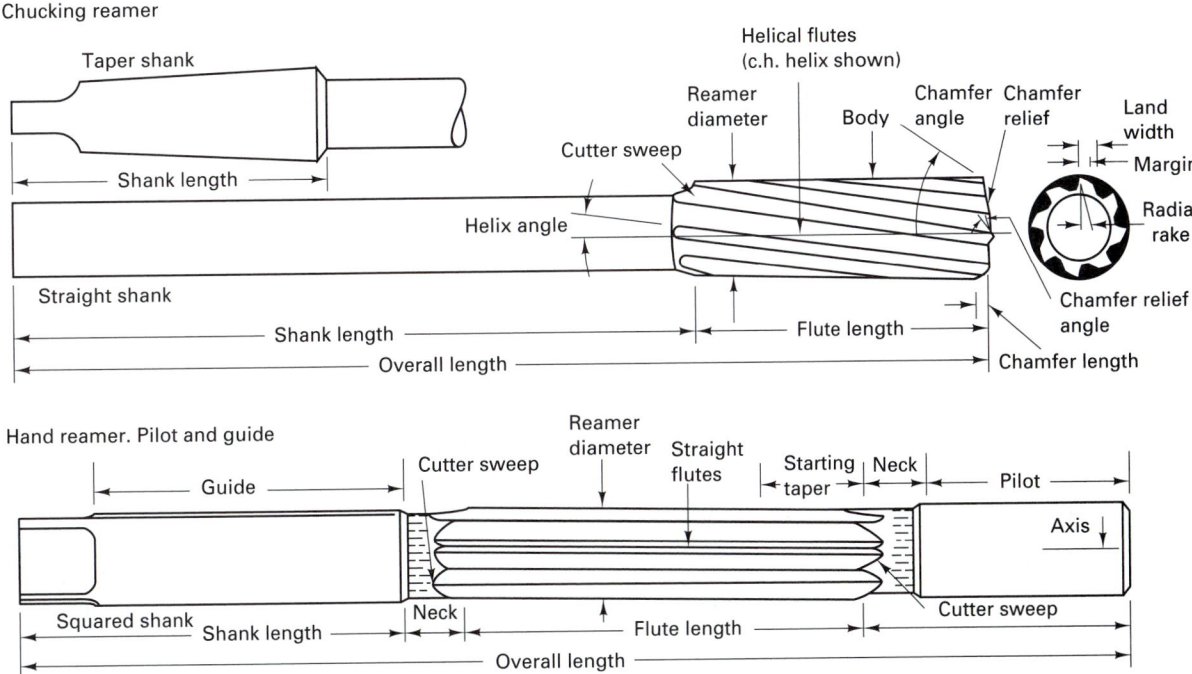

FIGURE 24-24 Standard nomenclature for hand and chucking reamers.

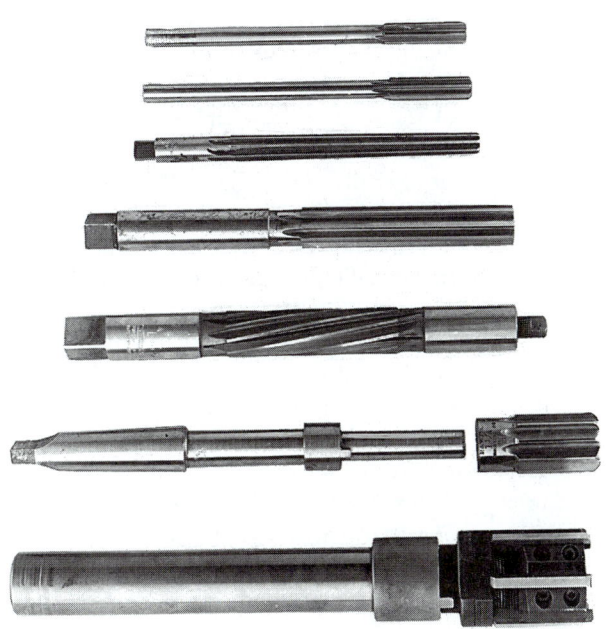

FIGURE 24-25 Types of reamers. *Top to bottom*: Straight-fluted rose reamer; straight-fluted chucking reamer; straight-fluted taper reamer; straight-fluted hand reamer; expansion reamer; shell reamer; adjustable insert-blade reamer.

Types of Reamers

The principal types of reamers, shown in Figures 24-24 and 24-25, are:

1. Hand reamers
   a. Straight
   b. Taper
2. Machine or chucking reamers
   a. Rose
   b. Fluted
3. Shell reamers
4. Expansion reamers
5. Adjustable reamers

Hand reamers are intended to be turned and fed by hand and to remove only a few thousandths of an inch of metal. They have a straight shank with a square tang for a wrench. They can have straight or spiral flutes and be solid or expandable. The teeth have relief along their edges and thus may cut along their entire length. However, the reamer is tapered from 0.005 to 0.010 in. in the first third of its length to assist in starting it in the hole, and most of the cutting therefore takes place in this portion.

Machine or *chucking reamers* are for use with various machine tools at slow speeds. The best feed is usually two to three times the drilling feed. Machine reamers have chamfers on the front end of the cutting edges. The chamfer causes the reamer to seat firmly and concentrically in the drilled hole, allowing the reamer to cut at full diameter. The longitudinal cutting edges do little or no cutting. Chamfer angles are usually 45°. Reamers have straight or tapered shanks and either straight or spiral flutes. *Rose chucking reamers* are ground cylindrical and have no relief behind the outer edges of the teeth. All cutting is done on the beveled ends of the teeth. *Fluted chucking reamers*, on the other hand, have relief behind the edges of the teeth as well as beveled ends. They can therefore cut on all portions of the teeth. Their flutes are relatively short and they are intended for light finishing cuts. For best results they should not be held rigidly but permitted to float and be aligned by the hole.

Shell reamers often are used for sizes over $\frac{3}{4}$ in. to save cutting-tool material. The shell, made of tool steel for smaller sizes and with carbide edges for larger sizes or for mass-production work, is held on an arbor that is made of ordinary steel. One arbor may be used with any number of shells. Only the shell is subject to wear and need be replaced when worn. They may be ground as rose or fluted reamers.

Expansion reamers can be adjusted over a few thousandths of an inch to compensate for wear or to permit some variation in hole size to be obtained. They are available in both hand and machine types.

Adjustable reamers have cutting edges in the form of blades that are locked in a body. The blades can be adjusted over a greater range than expansion reamers. This permits adjustment for size and to compensate for regrinding. When the blades become too small from regrinding, they can be replaced. Both tool steel and carbide blades are used.

Taper reamers are used for finishing holes to an exact taper. They may have up to eight straight or spiral flutes. Standard tapers, such as Morse, Jarno, or Brown & Shape, come in sets of two. The *roughing reamer* has nicks along the cutting edge to break up the heavy chips that result as a cylindrical hole is cut to a taper. The *finishing reamer* has smooth cutting edges.

Reaming Practice

If the material to be removed is free-cutting, reamers of fairly light construction will give satisfactory results. However, if the material is hard, tough, solid-type reamers are recommended, even for fairly large holes.

To meet quality requirements, including both finish and accuracy (tolerances on diameter, roundness, straightness, and absence of bell-mouth at ends of holes), reamers must have adequate support for the cutting edges, and reamer deflection must be minimal. Reaming speed is usually two-thirds the speed for drilling the same materials. However, for close tolerances and fine finish, speeds should be slower.

Feeds are usually much higher than those for drilling and depend upon material. A feed of between 0.0015 and 0.004 in. per flute is recommended as a starting point. Use the highest feed that will produce the required finish and accuracy. Recommended cutting fluids are the same as those for drilling.

Reamers, like drills, should not be allowed to become dull. The chamfer must be reground long before it exhibits excessive wear. Sharpening is usually restricted to the starting taper or chamfer. Each flute must be ground exactly evenly or the tool will cut oversize.

Reamers tend to chatter when not held securely, when the work or workholder is loose, or when the reamer is not properly ground. Irregularly spaced teeth may help reduce chatter. Other cures for chatter in reaming are to reduce the speed, vary the feed rate, chamfer the hole opening, use a piloted reamer, reduce the relief angle on the chamfer, or change the cutting fluid. Any misalignment between the workpiece and the reamer will cause chatter and improper reaming.

■ KEY WORDS

center drill	gang-drilling machine	reaming, hand
chisel end	gun drill	reaming, machine
chuck	helix angle	shell reamer
counterboring	indexable insert drill	spade drill
countersinking	jig	spot facing
drill press	lip	turret drilling machine
drilling	multiple-spindle drilling machine	twist drill
flute	radial drilling machine	web

■ REVIEW QUESTIONS

1. What functions are performed by the flutes on a drill?
2. What determines the rake angle of a drill? (See Figure 24-2).
3. Basically, what determines what helix angle a drill should have?
4. When a large-diameter hole is to be drilled, why is a small-diameter hole often drilled first?
5. Equation (24-4) for the MRR for drilling can be thought of as _____ times _____, where $f_r N$ is the feed rate of the drill bit.
6. Are the recommended surface speeds given in Table 24-4 typically higher or lower that those recommended for twist drills? How about the feeds?
7. What can happen when an improperly ground drill is used to drill a hole?
8. Why are most drilled holes oversize with respect to the nominally specified diameter?
9. What are the two primary functions of a combination center drill?
10. What is the function of the margins on a twist drill?
11. What factors tend to cause a drill to "drift" off the centerline of a hole?
12. For what types of holes are drills having coolant passages in the flutes advantageous?
13. In drilling, the deeper the hole, the greater the torque. Why?
14. Why do cutting fluids for drilling usually have more lubricating qualities than those for most other machining operations?
15. Find the drills shown in this chapter that have oil holes. (*Hint:* There are three).

16. How does a gang-drilling machine differ from a multiple-spindle drilling machine?
17. How does the thrust force vary with feed?
18. What may result from holding the workpiece by hand when drilling?
19. What is the rationale behind the operation sequence shown in Figure 24-9?
20. In terms of thrust, what is unusual about the slot-point drill compared to other drills?
21. What is the purpose of spot facing?
22. How does the purpose of counterboring differ from that of spot facing?
23. What are the primary purposes of reaming?
24. What are the advantages of shell reamers?

25. A drill that operated satisfactorily for drilling cast iron gave very short life when used for drilling a plastic. What might be the reason for this?
26. What precautionary procedures should be used when drilling a deep, vertical hole in mild steel when using an ordinary twist drill?
27. What is the advantage of a spade drill? Is it really a drill?
28. What is a "pecking" action in drilling?
29. Why does drill feed increase with drill size?
30. Suppose you specified a feed that was too large. What kinds of problems do you think this might cause? See Table 24-7 for help.

■ PROBLEMS

1. Suppose that you wanted to drill a 1.5-in.-diameter hole through a piece of 1020 cold-rolled steel that is 2 in. thick, using an indexable-insert drill? Values for feed and cutting speed need to be specified along with an appropriate allowance. Is this the correct tool? What other drill types could be used?
2. How much time will be required to drill this hole?
3. What is the metal-removal rate when a $1\frac{1}{2}$-in.-diameter hole, 2 in. deep, is drilled in 1020 steel at a cutting speed of 200 ft/min with a feed of 0.010 in./rev?
4. If the specific horsepower for the steel in Problem 3 is 0.9, what horsepower would be required?
5. If the specific power of AISI steel is 0.7, and 75% of the output of the 1.5-kW motor of a drilling machine is available at the tool, what is the maximum feed that can be used in drilling a 2-in.-diameter hole with a carbide drill? (Use the cutting speed suggested in Problem 3).
6. Show how the approximate equation (24-5) for MRR in drilling was obtained. What assumption was needed?
7. In Table 24-2, is the equation for MRR, line 4, the same as equation (24-4)? Explain.
8. In Table 24-2, what is the prorated downtime or prorated cost of purchasing? How are these prorated values calculated?
9. An indexable-insert drill can produce a hole four times faster than a spade drill can but may cost (with inserts) 50 to 75% more than the

equivalent spade blade and holder. For making only a couple of holes, the extra cost is not justified.

Manufacturers' charts will help determine the best feed and speed to run the drills. For example, a $1\frac{1}{2}$-inch hole is to be drilled in 4140 steel annealed to Bhn 275. For the spade drill, speed is 80 sfpm, feed 0.009 in./rev, and spindle rotation 204 rpm. For the indexable-insert drill, speed is 358 sfpm, feed 0.007 in./rev, and spindle rotation 891 rpm. Determine the number of holes needed to justify the extra cost of the indexable-insert drill. Some additional cost data are given below.

Ignore tool life and assume that the blades and the indexable drills make about the same number of holes. (Why is this a reasonable assumption?) The holes are 3 in. deep, with no allowance needed. Cost of drills:

SPADE BLADE	INDEXABLE-INSERT DRILL
$139.00 holder	$273.00 drill
21.90 per blade	12.80 per two inserts
$160.90	$285.80

Assume, for this example, that a $45/hr machine rate includes the cost of labor and machine burden.

10. Let us assume that you are drilling eight holes, equally spaced in a bolt-hole circle. That is, there would be holes at 12, 3, 6, and 9 o'clock and four additional holes equally spaced between them. The diameter of the bolt hole

circle is 6 in. The designer says that the holes must be 45 ± 1° from each other around the circle.

(a) Compute the tolerance between hole centers.

(b) Do you think a typical multiple-spindle drill

setup could be used to make this bolt circle, using eight drills all at once? Why or why not?

(c) Do you think that the use of a jig may help improve the situation?

*C*hapter 24 CASE STUDY

bolt-down leg on a casting

Figure CS-24 shows the design of one of four legs on a casting made by the Fulkerson Company. These legs are used to attach the device to the floor. The section drawing to the right shows the typical loading to which the leg is subjected. The company is currently drilling the bolt hole and then counterboring the land but manufacturing has experienced some difficulty in machining these legs. They report a lot of a drill breakage. Quality control reports that distances between the four holes are frequently too large. Sales has recently reported that a substantial number of in-service failures have occurred with these legs. You have obtained a sketch from sales showing where the legs typically fail. This casting is manufactured from gray cast iron using the sand casting process.

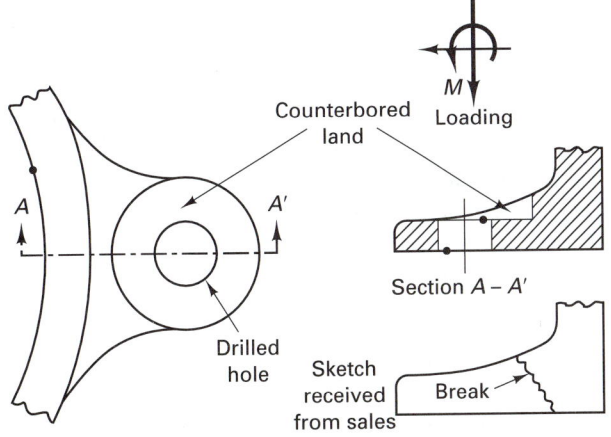

FIGURE CS-24

1. What machining difficulties would you expect this leg to have?
2. What do you think has caused the failures?
3. What do you recommend for solving this problem in the future in terms of materials, design, and manufacture?
4. What do you recommend be done with the units in the field to stop the failures?

CHAPTER 25

MILLING

25.1 INTRODUCTION
25.2 FUNDAMENTALS OF MILLING PROCESSES
25.3 MILLING CUTTERS
  Types of Milling Cutters
25.4 MILLING MACHINES
  Basic Milling Machine Construction
  Universal Column-and-Knee Milling Machines
  Turret Milling Machines
  Bed Milling Machines

Planer Milling Machines
Rotary-Table Milling Machines
Profilers and Duplicators
Accessories for Milling Machines
25.5 WORKHOLDING DEVICES FOR MILLING
25.6 MILLING TOLERANCES
25.7 MILLING SURFACE FINISH

■ 25.1 INTRODUCTION

Milling is a basic machining process by which a surface is generated by progressive chip removal. The workpiece is fed into a rotating cutting tool. Sometimes the workpiece remains stationary, and the cutter is fed to the work. In nearly all cases, a multiple-tooth cutter is used so that the material removal rate is high. Often the desired surface is obtained in a single pass of the cutter or work, and because very good surface finish can be obtained, milling is particularly well suited and widely used for mass-production work. Many types of milling machines are used, ranging from relatively simple and versatile machines that are used for general-purpose machining in job shops and tool-and-die work to highly specialized machines for mass production. Unquestionably, more flat surfaces are produced by milling than by any other machining process.

The cutting tool used in milling is known as a *milling cutter.* Equally spaced peripheral teeth will intermittently engage and machine the workpiece. This is called *interrupted cutting*.

■ 25.2 FUNDAMENTALS OF MILLING PROCESSES

Milling operations can be classified into two broad categories called peripheral milling and face milling. Each has many variations. In *peripheral milling* the surface is generated by teeth located on the periphery of the cutter body (Figure 25-1). The surface is parallel with the axis of rotation of the cutter. Both flat and formed surfaces can be produced by this method, the cross section on the resulting surface corresponding to the axial contour of the cutter. This process, often called *slab milling*, is usually performed on horizontal-spindle milling machines. In slab milling, the tool rotates (mills) at some rpm value (N) while the work feeds past the tool at a table feed rate f_m in inches per minute.

As in the other processes, the cutting speed V and feed per tooth are "selected" by the engineer or the machine tool operator. As before, these variables depend on the work material, the tool material, and the specific process. The cutting velocity is that which occurs at the cutting edges of the teeth in the milling center. The rpm of the spindle is determined from the surface cutting speed, where D is the cutter of diameter in inches according to

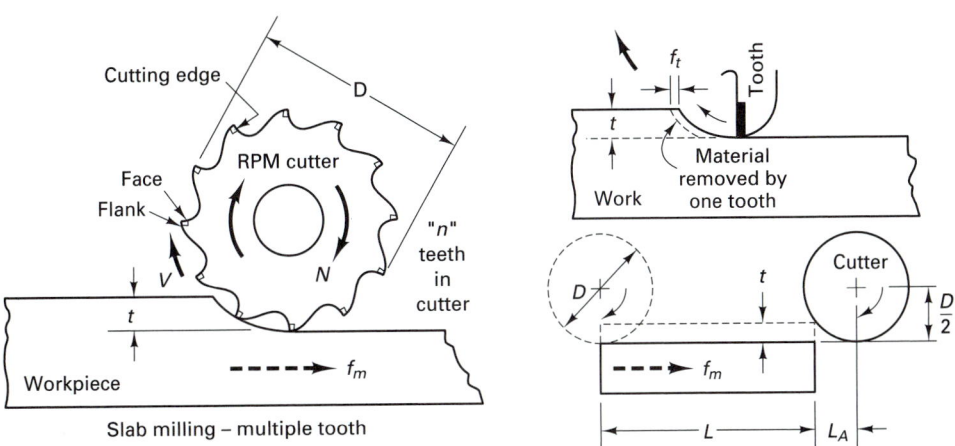

FIGURE 25-1   Basics of peripheral or slab milling process (see also Figure 25-8).

$$N = \frac{12V}{\pi D} \tag{25-1}$$

The depth of cut, called t in Figure 25-1, is simply the distance between the old and new machined surfaces.

The width of cut is the width of the cutter or the work, in inches, and is given the symbol W. The length of the cut, L, is the length of the work plus some allowance, L_A, for approach and overtravel. The feed rate of the table, f_m, in inches per minute is related to the amount of metal each tooth removes during a revolution (this is called the feed per tooth), f_t, according to

$$f_m = f_t N n \tag{25-2}$$

where n is the number of teeth in the cutter (teeth/rev). The *cutting time*

$$CT = \frac{L + L_A}{f_m} \qquad \text{min} \tag{25-3}$$

where

$$\text{length of approach} = L_A = \sqrt{\frac{D^2}{4} - \left(\frac{D}{2} - t\right)^2} = \sqrt{t(D - t)} \tag{25-4}$$

$$MRR = \frac{\text{vol.}}{CT} = \frac{LWt}{CT} = Wtf_m \qquad \text{in}^3/\text{min} \tag{25-5}$$

ignoring L_A. Values for f_t are given in Table 25-1 along with recommended cutting speeds in feet per minute.

In *face milling* and *end milling*, the surface generated is at right angles to the cutter axis (Figure 25-2). Most of the cutting is done by the peripheral portions of the teeth, with the face portions providing some finishing action. Face milling is done on both horizontal- and vertical-spindle machines.

The tool rotates (face mills) at some rpm rate (N) while the work feeds past the tool. The rpm value is related to the surface cutting speed, where the cutting diameter is D, according to equation (25-1). The depth of cut is t in inches, as shown in Figure 25-3. The

TABLE 25-1. Suggested Starting Feeds and Speeds Using High-Speed Steel and Carbide Cutters[a]

Material	Feed (in./tooth) / Speed (ft/min)	Carbide Cutters — Face Mills	Slab Mills	End Mills	Full and Half Side Mills	Saws	Form Mills	High-speed Steel Cutters — Face Mills	Slab Mills	End Mills	Full and Half Side Mills	Saws	Form Mills
Malleable iron, soft/hard	Feed	.005–.015	.005–.015	.005–.010	.005–.010	.003–.004	.005–.010	.005–.015	.005–.015	.003–.015	.006–.012	.003–.006	.005–.010
	Speed	200–300	200–300	200–350	200–300	200–350	175–275	60–100	60–90	60–100	60–100	60–100	60–80
Cast steel, soft/hard	Feed	.008–.015	.005–.015	.003–.010	.005–.010	.002–.004	.005–.010	.010–.015	.010–.015	.005–.010	.005–.010	.002–.005	.008–.012
	Speed	150–350	150–350	150–350	150–350	150–300	150–300	40–60	40–60	40–60	40–60	40–60	40–60
BHN Steel 100–150	Feed	.010–.015	.008–.015	.005–.010	.008–.012	.003–.006	.004–.010	.015–.030	.008–.015	.003–.010	.010–.020	.003–.006	.008–.010
	Speed	450–800	450–600	450–600	450–800	350–600	350–600	80–130	80–130	80–140	80–130	70–100	70–100
150–250	Feed	.010–.015	.008–.015	.005–.010	.007–.012	.003–.006	.004–.010	.010–.020	.008–.015	.003–.010	.010–.015	.003–.006	.006–.010
	Speed	300–450	300–450	300–450	300–450	300–450	300–450	50–70	50–70	60–80	50–70	50–70	50–70
250–350	Feed	.008–.015	.007–.012	.005–.010	.005–.012	.002–.005	.003–.008	.005–.010	.005–.010	.003–.010	.005–.010	.002–.005	.005–.010
	Speed	180–300	150–300	150–300	160–300	150–300	150–300	35–60	35–50	40–60	35–50	35–50	35–50
350–450	Feed	.008–.015	.007–.012	.004–.008	.005–.012	.001–.004	.003–.008	.003–.008	.005–.008	.003–.101	.003–.008	.001–.004	.003–.008
	Speed	125–180	100–150	100–150	125–180	100–150	100–150	20–35	20–35	20–40	20–35	20–35	20–35
Cast Iron Hard, BHN 180–225	Feed	.005–.010	.005–.010	.003–.008	.003–.010	.002–.003	.005–.010	.005–.012	.005–.010	.003–.008	.005–.010	.002–.004	.005–.010
	Speed	125–200	100–175	125–200	125–200	125–200	100–175	40–60	35–50	40–60	40–60	35–60	35–50
Medium, BHN 180–225	Feed	.008–.015	.008–.015	.005–.010	.005–.012	.003–.004	.006–.012	.010–.020	.008–.015	.003–.010	.008–.015	.003–.005	.008–.012
	Speed	200–275	175–250	200–275	200–275	200–250	175–250	60–80	50–70	60–90	60–80	60–70	50–60
Soft, BHN 150–180	Feed	.015–.025	.010–.020	.005–.012	.008–.015	.003–.004	.008–.015	.015–.030	.010–.025	.004–.010	.010–.020	.002–.005	.008–.010
	Speed	275–400	250–350	275–400	275–400	250–350	250–350	80–120	70–110	80–120	80–120	70–110	70–100
Bronze, soft/hard	Feed	.010–.020	.010–.020	.005–.010	.008–.012	.003–.004	.008–.015	.010–.025	.008–.020	.003–.010	.008–.015	.003–.005	.008–.015
	Speed	300–1000	300–800	300–1000	300–1000	300–1000	200–800	50–225	50–200	50–250	50–225	50–250	50–200
Brass, soft/hard	Feed	.010–.020	.010–.020	.005–.010	.008–.012	.003–.004	.008–.015	.010–.025	.008–.020	.005–.015	.008–.015	.003–.005	.008–.015
	Speed	500–1500	500–1500	500–1500	500–1500	500–1500	500–1500	150–300	100–300	150–350	150–350	150–300	100–300
Aluminum alloy, soft/hard	Feed	.010–.040	.010–.030	.003–.015	.008–.025	.003–.006	.008–.015	.010–.040	.015–.040	.015–.040	.010–.030	.004–.008	.010–.020
	Speed	2000 up	2000 up	2000 up	2000 up	2000 up	2000 up	300–1200	300–1200	300–1200	300–1200	300–1000	300–1200

[a]Generally, lower end of range used for inserted blade cutters, higher end of range for indexable insert cutters.

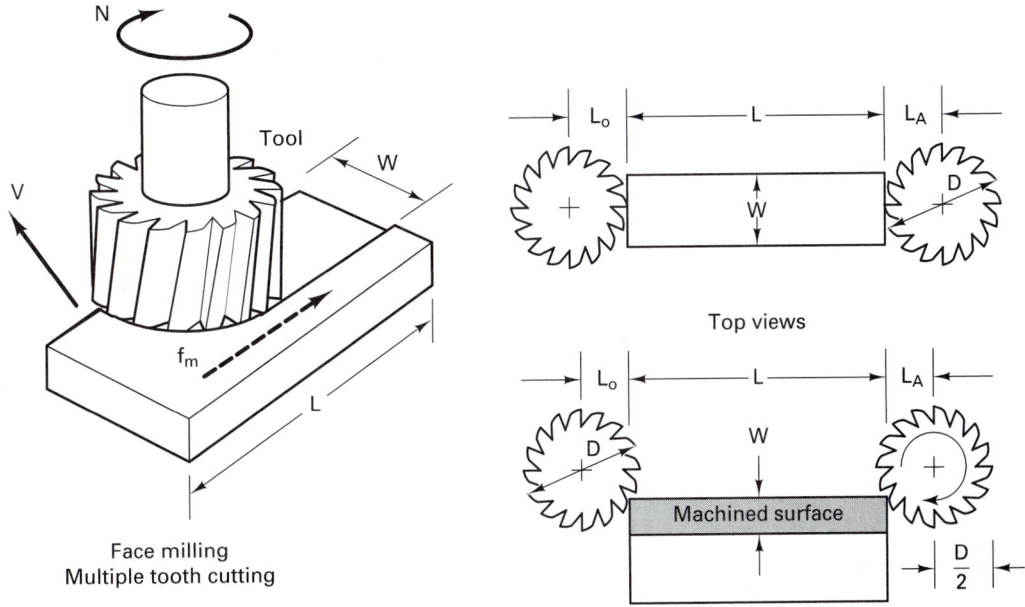

Top views

Machined surface

FIGURE 25-2 Basics of face and end milling process (see also Figure 25-19).

width of cut is W in inches and may be the width of the workpiece or the width of the cutter, depending on the setup. The length of cut is the length of the workpiece, L, plus an allowance for approach, L_A, and overtravel, L_O, in inches. The feed rate of the table, f_m, in inches per minute is related to the amount of metal each tooth removes during a pass over the work, and this is called the feed per tooth, f_t, where $f_m = f_t N n$. The number of teeth in the cutter is n. The cutting time

$$\text{CT} = \frac{L + L_A + L_O}{f_m} \qquad \text{min} \qquad (25\text{-}6)$$

The *metal removal rate*

$$\text{MRR} = \frac{\text{vol.}}{\text{CT}} = \frac{LWt}{\text{CT}} = Wtf_m \qquad \text{in}^3/\text{min}$$

(ignore L_O and L_A). The length of approach is usually equal to the length of overtravel, which usually equals $D/2$ inches. For a setup where the tool does not completely pass over the workpiece,

$$L_O = L_A = \begin{cases} \sqrt{W(D - W)} & \text{for } W < \dfrac{D}{2} \qquad (25\text{-}7) \\[2ex] \dfrac{D}{2} & \text{for } W \geq \dfrac{D}{2} \qquad (25\text{-}8) \end{cases}$$

For either slab or face milling, surfaces can be generated by two distinctly different methods (Figure 25-4). *Up milling* is the traditional way to mill and is called *conventional milling*. The cutter rotates against the direction of feed rate of the workpiece. In *climb* or *down milling*, the cutter rotation is in the same direction as the feed rate. The method of chip formation is completely different in the two cases.

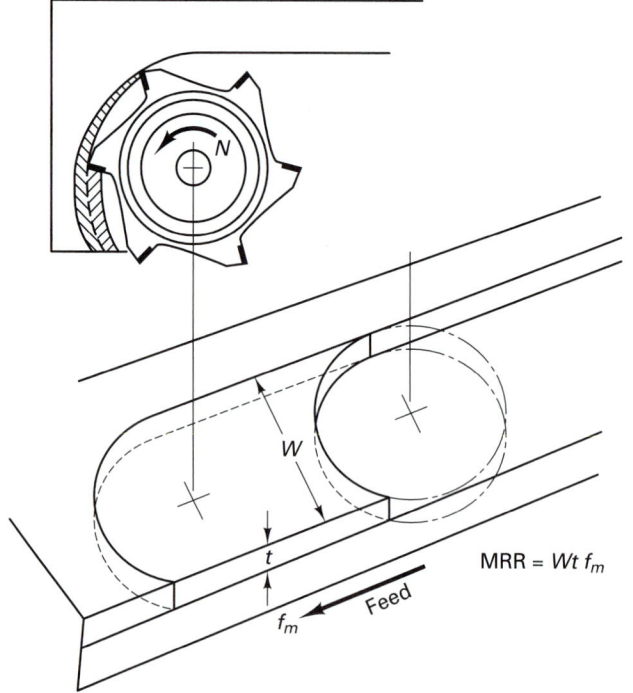

FIGURE 25-3   Face milling showing relationship of MRR parameters (w = width, t = depth) and feed per tooth versus feed rate in inches per minute of table.

$$MRR = Wt\, f_m$$

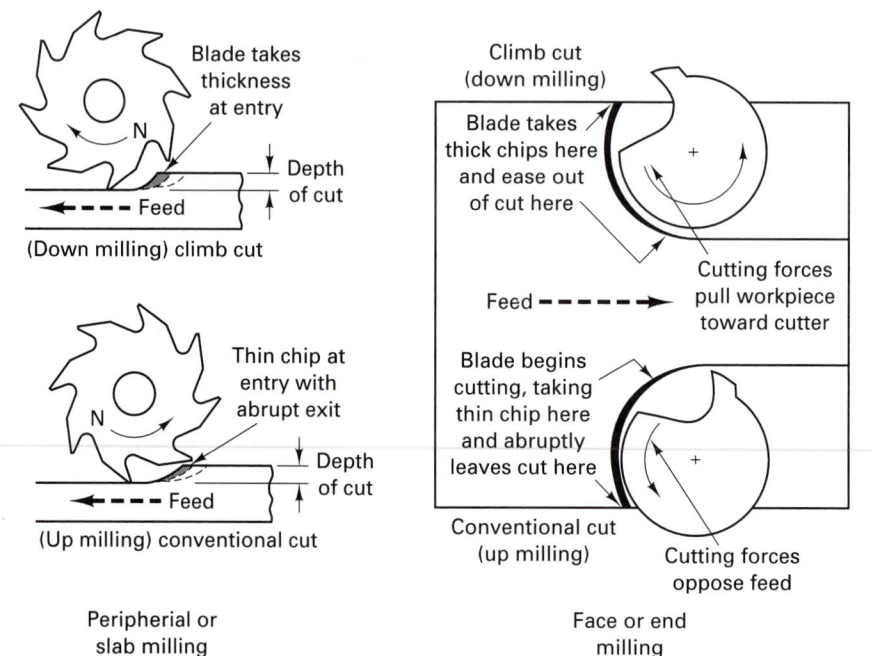

FIGURE 25-4   Climb cut (or down) milling versus conventional cut (or up) milling for slab or end milling.

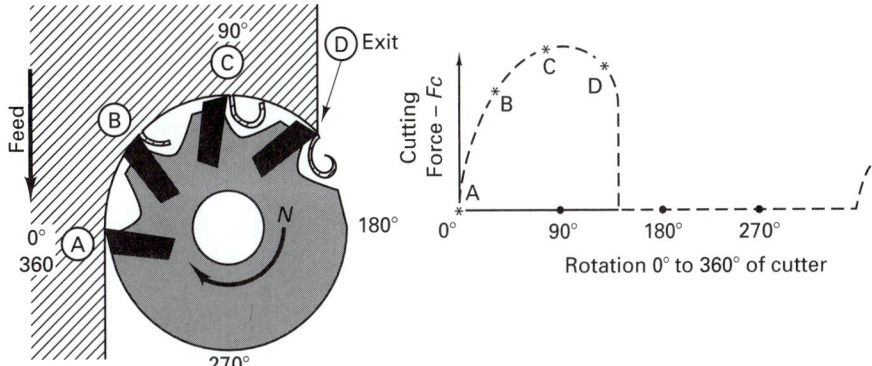

FIGURE 25-5 Conventional face milling (*left*) with cutting force diagram for F_c, (*right*) showing the interrupted nature of the process. (*From* Metal Cutting Principles, *2nd ed., Ingersoll Cutting Tool Company*)

In up milling, the chip is very thin at the beginning, where the tooth first contacts the work, and increases in thickness, becoming a maximum where the tooth leaves the work. The cutter tends to push the work along and lift it upward from the table. This action tends to eliminate any effect of looseness in the feed screw and nut of the milling machine table and results in a smooth cut. However, the action also tends to loosen the work from the clamping device; therefore, greater clamping forces must be employed. In addition, the smoothness of the generated surface depends greatly on the sharpness of the cutting edges. In up milling, chips can be carried into the newly machined surface, causing the surface finish to be poorer (rougher) than in down milling.

In down milling, maximum chip thickness occurs close to the point at which the tooth contacts the work. Because the relative motion tends to pull the workpiece into the cutter, any possibility of looseness in the table feed screw must be eliminated if down milling is to be used. It should never be attempted on machines that are not designed for this type of milling. Virtually all modern milling machines are capable of doing down milling. Because the material yields in approximately a tangential direction at the end of the tooth engagement, there is less tendency (than when up milling is used) for the machined surface to show toothmarks, and the cutting process is smoother with less chatter. Another advantage of down milling is that the cutting force tends to hold the work against the machine table, permitting lower clamping forces. However, the fact that the cutter teeth strike against the surface of the work at the beginning of each chip can be a disadvantage if the workpiece has a hard surface, as castings sometimes do. This may cause the teeth to dull rapidly. Metals that readily work harden should be climb milled.

Milling is an interrupted cutting process wherein entering and leaving the cut subjects the tool to impact loading, cyclic heating, and cycle cutting forces. As shown in Figure 25-5, the cutting force, F_c, builds rapidly as the tool enters the work at Ⓐ and progresses to Ⓑ, peaks as the blade crosses the direction of feed at Ⓒ, decreases to Ⓓ, and then drops to zero abruptly upon exit. The diagram does not indicate the impulse loads caused by impacts. The interrupted-cut phenomenon explains in large part why milling cutter teeth are designed to have small positive or negative rakes, particularly when the tool material is carbide or ceramic. These brittle materials tend to be very strong in compression, and negative rake results in the cutting edges being placed in compression by the cutting forces rather than tension. Cutters made from HSS are made with positive rakes, in the main, but must be run at lower speeds. Positive rake tends to lift the workpiece, while negative rakes compress the workpiece and allow heavier cuts to be made. Table 25-2 summarizes some additional milling problems.

TABLE 25-2. Probable Causes of Milling Problems

Problem	Probable Cause	Cures
Chatter (vibration)	1. Lack of rigidity in machine, fixtures, arbor, or workpiece	Use larger arbors.
	2. Cutting load too great	Decrease feed per tooth or number of teeth in contact with work.
	3. Dull cutter	Sharpen or replace inserts.
	4. Poor lubrication	Flood coolant.
	5. Straight-tooth cutter	Use helical cutter.
	6. Radial relief too great	
	7. Rubbing, insufficient clearance	Check tool angles.
Loss of accuracy (cannot hold size)	1. High cutting load causing deflection	Decrease number of teeth in contact with work or feed per tooth.
	2. Chip packing, between teeth	Adjust cutting fluid to wash chips out of teeth.
	3. Chips not cleaned away before mounting new piece of work	
Cutter rapidly dulls	1. Cutting load too great	Decrease feed per tooth or number of teeth in contact.
	2. Insufficient coolant	Add blending oil to coolant.
Poor surface finish	1. Feed too high	Check to see if all teeth are set at same height.
	2. Tool dull	
	3. Speed too low	
	4. Not enough cutter teeth	
Cutter digs in (hogs into work)	1. Radial relief too great	Check to see that workpiece is not deflecting and is clamped securely.
	2. Rake angle too large	
	3. Improper speed	
Work burnishing	1. Cut is too light	Enlarge feed per tooth.
	2. Tool edge worn	Sharpen cutter.
	3. Insufficient radial relief	
	4. Land too wide	
Cutter burns	1. Not enough lubricant	Add sulfur-based oil.
	2. Speed too high	Reduce cutting speed.
		Flood coolant.
Teeth breaking	1. Feed too high	Decrease feed per tooth.
		Use cutter with more teeth.
		Reduce table feed rate.

■ 25.3 MILLING CUTTERS

Milling cutters can be classified according to the way the cutter is mounted in the machine tool. *Arbor cutters* have a center hole so that they can be mounted on an arbor. *Shank cutters* have either a tapered or a straight integral shank. Those with tapered shanks can be mounted directly in the milling machine spindle, whereas straight-shank cutters are held in a chuck. *Facing cutters* usually are bolted to the end of a stub arbor. Common types of milling cutters classified in this manner are as follows:

Arbor Cutters	Shank Cutters
Plain	End mills
Side	Solids
Staggered-tooth	Insert-tooth
Slitting saws	Shell
Angle	Hollow
Inserted-tooth	T-slot
Form	Woodruff key seat
Fly	Fly

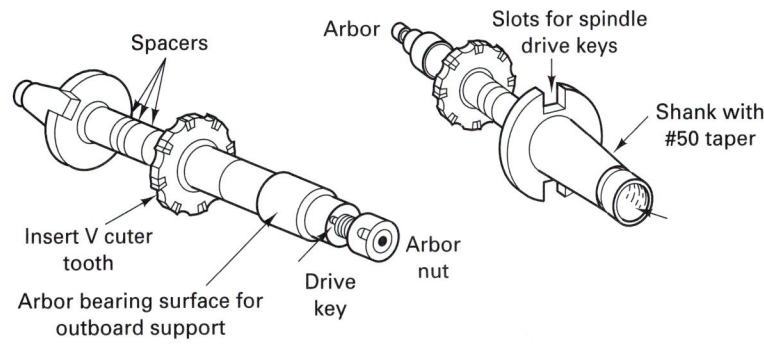

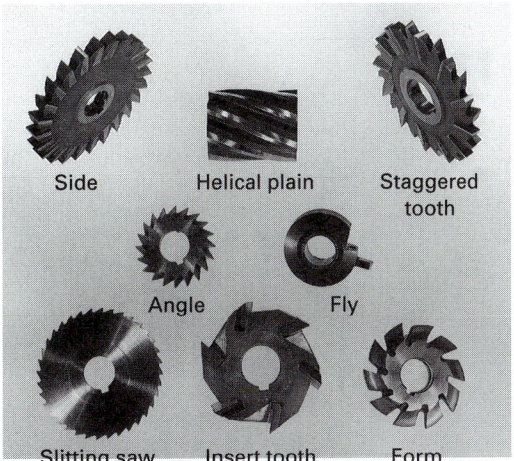

FIGURE 25-6   Sketch of an arbor (two views) used on a horizontal spindle milling machine and photos of some typical arbor-type milling cutters.

Figures 25-6 and 25-7 show several types of arbor and shank milling cutters, respectively.
Another method of classification applies only to face and end-mill cutters and relates to the direction of rotation. A *right-hand cutter* must rotate counterclockwise when viewed from the front end of the machine spindle. Similarly, a *left-hand cutter* must rotate clockwise. All other cutters can be reversed on the arbor to change them from one hand to the other. Positive rake angles are used on general-purpose HSS milling cutters. Negative rake angles are commonly used on carbide- and ceramic-tipped cutters employed in mass-production milling to obtain the greater strength and cooling capacity which they provide. TiN coating of these tools is quite common, resulting in significant increases in tool life.

Types of Milling Cutters

Plain milling cutters used for plain or slab milling have straight or helical teeth on the periphery and are used for milling flat surfaces. *Helical mills* (Figure 25-8) engage the work gradually, and usually more than one tooth cuts at a given time. This reduces shock and chattering tendencies and promotes a smoother surface. Consequently, this type of cutter usually is preferred over one with straight teeth.

FIGURE 25-7 Shank-type milling cutters. Left to right: T-slot, shell end mill, Woodruff key seater, hollow end mill, solid end mill.

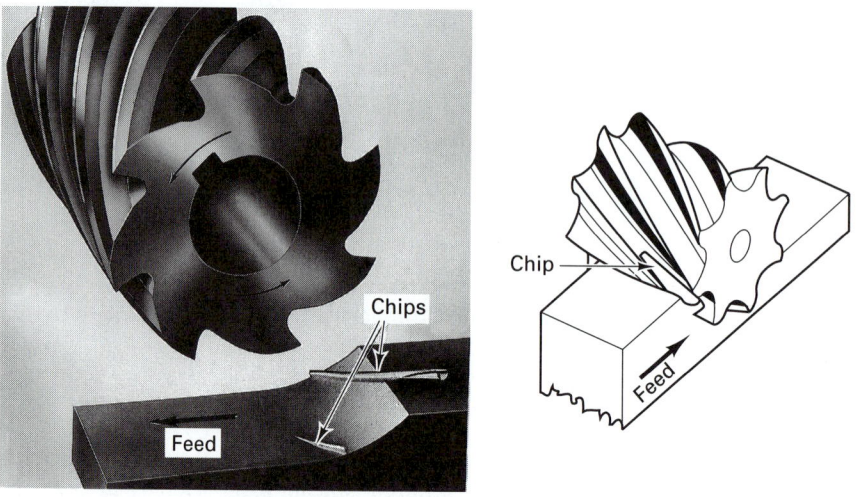

FIGURE 25-8 Manner in which chips are formed progressively by the teeth of a plain helical-tooth milling cutter in up milling. *(Courtesy of Cincinnati Milacron, Inc.)*

Side milling cutters are similar to plain milling cutters except that the teeth extend radially partway across one or both ends of the cylinder toward the center. See Figure 25-6 for an example. The teeth may be either straight or helical. Frequently, these cutters are relatively narrow, being disklike in shape. Two or more side milling cutters often are spaced on an arbor to straddle the workpiece (called *straddle milling*) and two parallel surfaces are machined at once.

Interlocking slotting cutters consist of two cutters similar to side mills but made to operate as a unit for milling slots. The two cutters are adjusted to the desired width by inserting shims between them.

Staggered-tooth milling cutters are narrow cylindrical cutters having staggered teeth, and with alternate teeth having opposite helix angles. They are ground to cut only on the

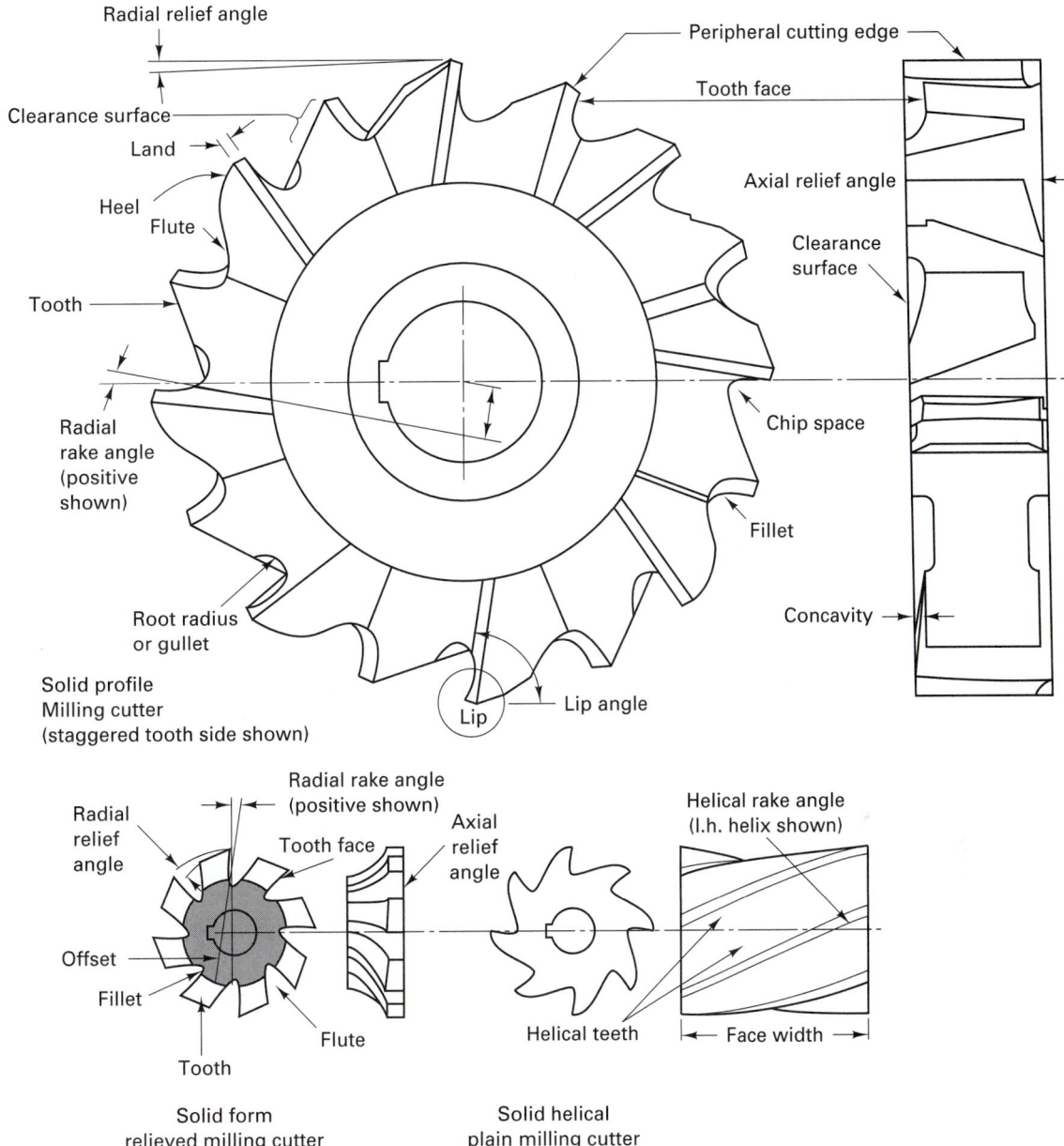

FIGURE 25-9 Geometrical features of three solid (HSS) arbor cutters for milling.

periphery, but each tooth also has chip clearance ground on the protruding side. These cutters have a free cutting action that makes them particularly effective in milling deep slots (Figures 25-6 and 25-9).

Slitting saws are thin, plain milling cutters, usually from $\frac{1}{32}$ to $\frac{3}{16}$ in. thick, which have their sides slightly "dished" to provide clearance and prevent binding. They usually have more teeth per unit of diameter than ordinary plain milling cutters and are used for milling deep narrow slots and cutting-off operations.

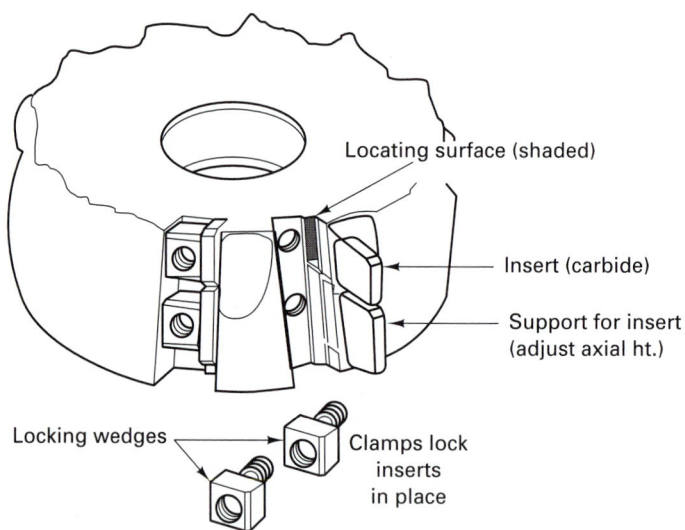

FIGURE 25-10   Insert-tooth face milling cutter.

Angle milling cutters are made in two types: single-angle and double-angle. Angle cutters are used for milling slots of various angles or for milling the edges of workpieces to a desired angle. *Single-angle cutters* have teeth on the conical surface, usually at an angle of 45 to 60° to the plane face. *Double-angle cutters* have V-shaped teeth, with both conical surfaces at an angle to the end faces but not necessarily at the same angle. The V angle usually is 45, 60, or 90°.

Form milling cutters have the teeth ground to a special shape—usually an irregular contour—to produce a surface having a desired transverse contour. They must be sharpened by grinding only the tooth face, thereby retaining the original contour as long as the plane of the face remains unchanged with respect to the axis of rotation. Convex, concave, corner-rounding, and gear-tooth cutters are common examples (Figure 25-9).

Most larger milling cutters are of the *insert-tooth type*. The cutter body is made of steel, with the teeth made of high-speed steel, carbides, or TiN carbides, fastened to the body by various methods. An insert tooth cutter using indexable carbide or ceramic inserts is shown in Figure 25-10. This type of construction reduces the amount of costly material that is required and can be used for any type of cutter but most often is used with face mills.

End mills are shank cutters having teeth on the circumferential surface and one end. They thus can be used for facing, profiling, and end milling. The teeth may be either straight or helical, but the latter is more common. Small end mills have straight shanks, whereas taper shanks are used on larger sizes.

Plain end mills have multiple teeth that extend only about halfway toward the center on the end. They are used in milling slots, profiling, and facing narrow surfaces. *Two-lip mills* have two straight or helical teeth that extend to the center. Thus they may be sunk into material, like a drill, and then fed lengthwise to form a groove, a slot, or a pocket.

Shell end mills are solid multiple-tooth cutters, similar to plain end mills but without a shank. The center of the face is recessed to receive a screw head or nut for mounting the cutter on a separate shank or a stub arbor. One shank can hold any of several cutters and thus provides great economy for larger end mills.

Hollow end mills are tubular in cross section, with teeth only on the end but having internal clearance. They are used primarily on automatic screw machines for sizing cylindrical stock, producing a short cylindrical surface of accurate diameter (Figure 25-7).

T-slot cutters are integral-shank cutters with teeth on the periphery and *both* sides. They are used for milling the wide groove of a T-slot. To use them, the vertical groove must first be made with a slotting mill or an end mill to provide clearance for the shank. Because the T-slot cutter cuts on five surfaces simultaneously, it must be fed with care (Figure 25-7).

Woodruff keyseat cutters are made for the single purpose of milling the semicylindrical seats required in shafts for Woodruff keys. They come in standard sizes corresponding to Woodruff key sizes. Those below 2 in. in diameter have integral shanks; the larger sizes may be arbor mounted (Figure 25-7).

Occasionally, *fly cutters* may be used for face milling or boring. Both operations may be done with a single tool at one setup. A single-point cutting tool is attached to a special shank, usually with provision for adjusting the effective radius of the cutting tool with respect to the axis of rotation. The cutting edge can be made in any desired shape and, because it is a single-point tool, is very easy to grind.

■ 25.4 MILLING MACHINES

Because the milling process is versatile and highly productive, a variety of machines have been developed to employ the milling principle. The basic general-purpose milling machines provide a high degree of flexibility. Another type, *duplicators*, are used exclusively for reproducing parts from templates or patterns. A third type encompasses *special-purpose machines* that are used in mass-production manufacturing. Machines of the fourth type are numerically controlled and do other basic machining operations in addition to milling. These machines can change tools and are called *machining centers* (see Chapter 29).

Milling machines have an accurate, rugged, rotating spindle. They can also be used for other machining operations, such as drilling and boring. The common types may be classified according to their general characteristics as follows:

1. Column-and-knee type (general purpose)
    a. Plain or horizontal
        (1) Power table feed
        (2) Hand table feed
    b. Universal
    c. Vertical
    d. Turret-type universal
2. Bed type (manufacturing)
    a. Simplex
    b. Universal
    c. Triplex
3. Planer type (large work only)
4. Special
    a. Rotary table
    b. Drum type
    c. Profilers
    d. Duplicators

Basic Milling Machine Construction

Most basic milling machines are of *column-and-knee* construction, employing the components and motions shown in Figure 25-11. The column, mounted on the base, is the main supporting frame for all the other parts and contains the spindle with its driving mechanism. This construction provides controlled motion of the worktable in three mutually perpendicular directions: (1) through the *knee* moving vertically on ways on the front of the column, (2) through the *saddle* moving transversely on ways on the knee, and (3) through the *table* moving longitudinally on ways on the saddle. All these motions can be imparted either by manual or powered means. In most cases, a powered rapid traverse is provided in addition to the regular feed rates for use in setting up work and in returning the table at the end of a cut.

Milling machines having only the three mutually perpendicular table motions are called *plain column-and-knee* type. These are available with both horizontal and vertical spindles (Figure 25-11). On the horizontal-spindle type, an adjustable over-arm provides an outboard bearing support for the end of the cutter arbor which was shown in Figure 25-6. These machines are well suited for slab, side, or straddle milling.

In some vertical-spindle machines the spindle can be fed up and down, either by power or by hand. Vertical-spindle machines are especially well suited for face- and end-milling operations. They also are very useful for drilling and boring, particularly where holes must be spaced accurately in a horizontal plane, because of the controlled table motion.

Universal Column-and-Knee Milling Machines

Universal column-and-knee milling machines differ from plain column-and-knee machines in that the table is mounted on a housing that can be swiveled in a horizontal plane, thereby increasing its flexibility. Helices, as found in twist drills, milling cutters, and helical gear teeth, can be milled on universal machines.

Turret Milling Machines

Turret column-and-knee milling machines (Figure 25-12) have dual heads that can be swiveled about a horizontal axis on the end of a horizontal adjustable ram. The permits milling to be done horizontally, vertically, or at any angle. This added flexibility is advantageous when a variety of work has to be done, as in tool-and-die or experimental shops. They are available with either plain or universal tables.

Bed Milling Machines

In production manufacturing operations, ruggedness and the capability of making heavy cuts are of more importance than versatility. *Bed milling machines* (Figure 25-12) are made for these conditions. The table is mounted directly on the bed and has only longitudinal motion. The spindle head can be moved vertically to set up the machine for a given operation. Normally, once the setup is completed, the spindle head is clamped in position and no further motion of it occurs during machining. However, on some machines vertical motion of the spindle occurs during each cycle.

After such milling machines are set up, little skill is required to operate them, permitting faster learning time for the operators. Some machines of this type are equipped with automatic controls so that all the operator has to do is load and unload workpieces into the fixture and set the machine into operation. For stand-alone machines, a fixture can be located at each end of the table so that one workpiece can be loaded while another is being machined.

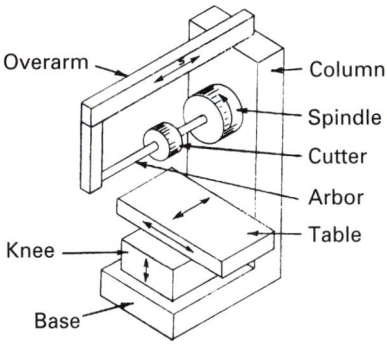

Plain horizontal knee type milling machine

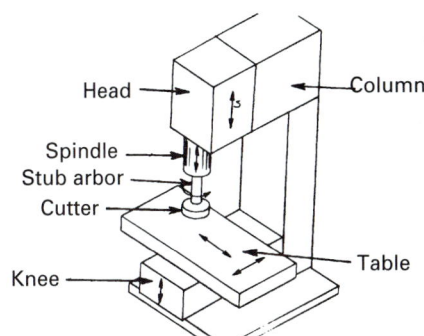

Veritcal knee type milling machine

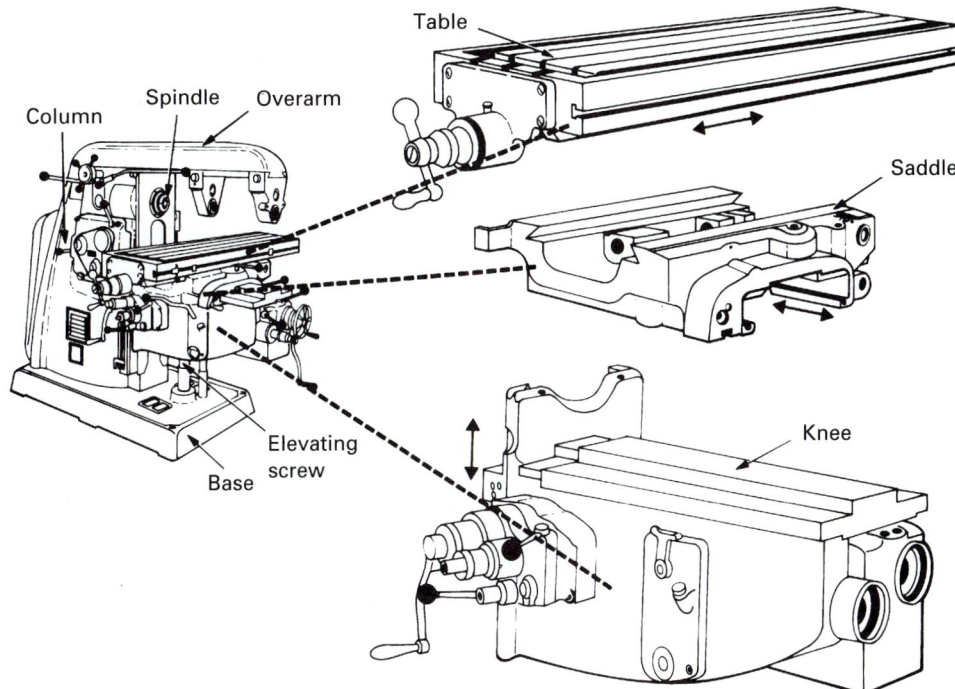

FIGURE 25-11   Major components of a plain column-and-knee-type milling machine, which can have horizontal spindle or vertical spindle. *(Top, from Manufacturing Producibility Handbook: courtesy of General Electric Company, bottom, courtesy of Cincinnati Milacron, Inc.)*

Bed milling machines with single spindles sometimes are called *simplex milling machines*; they are made with both horizontal and vertical spindles. Bed-type machines also are made in *duplex* and *triplex* types, having two or three spindles, respectively, permitting the simultaneous milling of two or three surfaces at a single pass (Figure 25-12).

Planer Milling Machines

Planner milling machines (Figure 25-13) utilize several milling heads, which can remove large amounts of metal while permitting the table and workpiece to feed quite slowly.

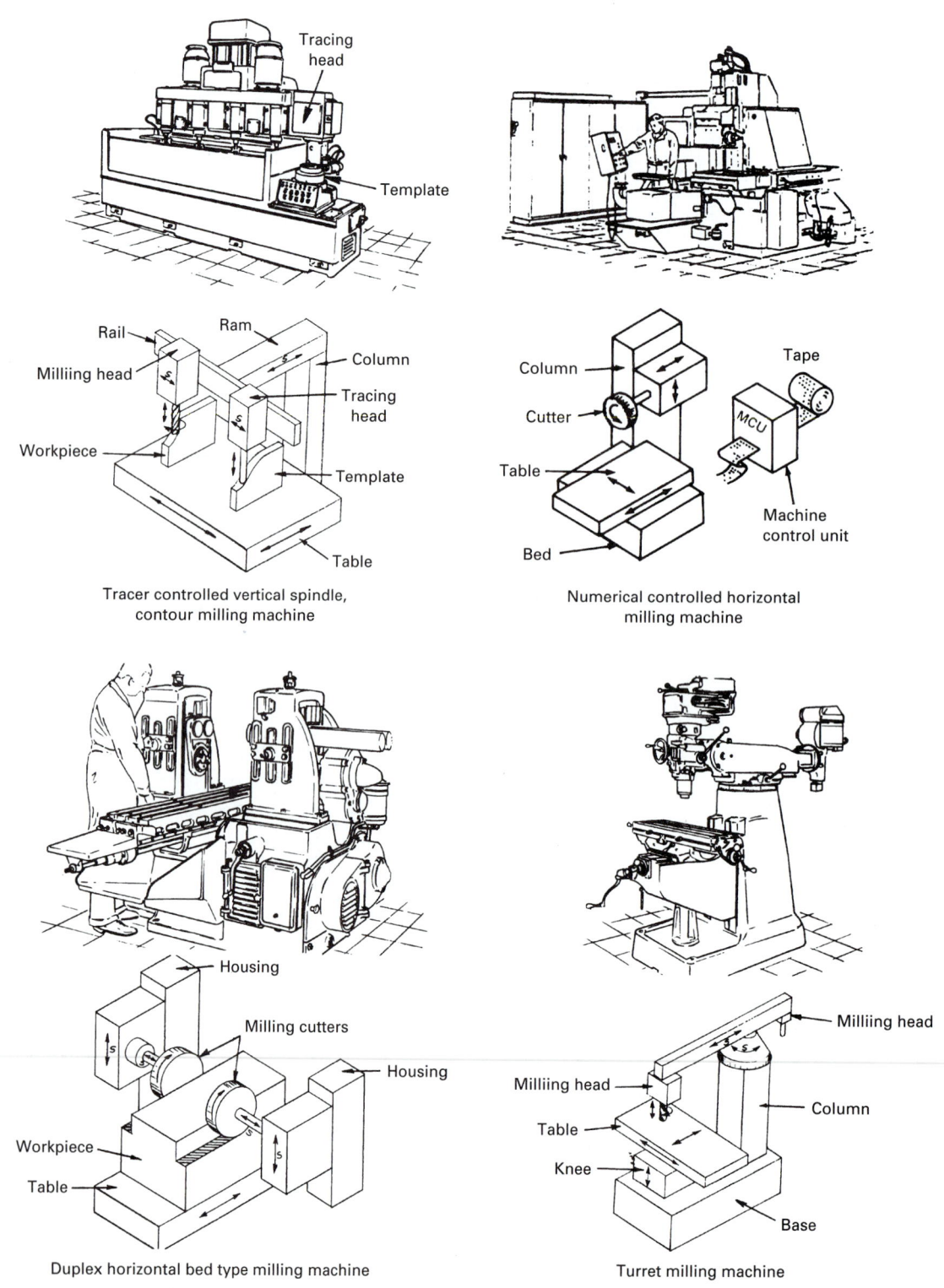

Tracer controlled vertical spindle,
contour milling machine

Numerical controlled horizontal
milling machine

Duplex horizontal bed type milling machine

Turret milling machine

FIGURE 25-12   Four types of milling machine tools.

FIGURE 25-13 Large planer-type milling machine. Inset shows 90° head being used. *(Courtesy of Cosa Corporation.)*

Often, a single pass of the workpiece past the cutters is required. Through the use of different types of milling heads and cutters, a wide variety of surfaces can be machined with a single setup of the workpiece. This is an advantage when heavy workpieces are involved.

Rotary-Table Milling Machines

Some types of face milling in mass-production manufacturing are often done on *rotary-table milling machines*. Roughing and finishing cuts can be made in succession as the workpieces are moved past the several milling cutters while held in fixtures on the rotating table. The operator can load and unload the work without stopping the machine.

Profilers and Duplicators

Milling machines that can duplicate external or integral geometries in two dimensions are called *profilers* or tracer-controlled machines. As shown in Figure 25-12, a tracing probe follows a two-dimensional template and, through electronic or air-actuated mechanisms, controls the cutting spindles in two mutually perpendicular directions (Figure 25-14).

Hydraulic tracer control is not new, but a great variety of control systems for contouring and straight-line milling are now available. These range from simple single-axis control systems for die sinking to control for profiling and three-axis contouring.

Hydraulic tracers have transformed standard and special milling machines into production tools with virtually unlimited capabilities for contouring and straight-line milling. They can produce shapes that are impossible to produce manually, such as free-form dies and molds.

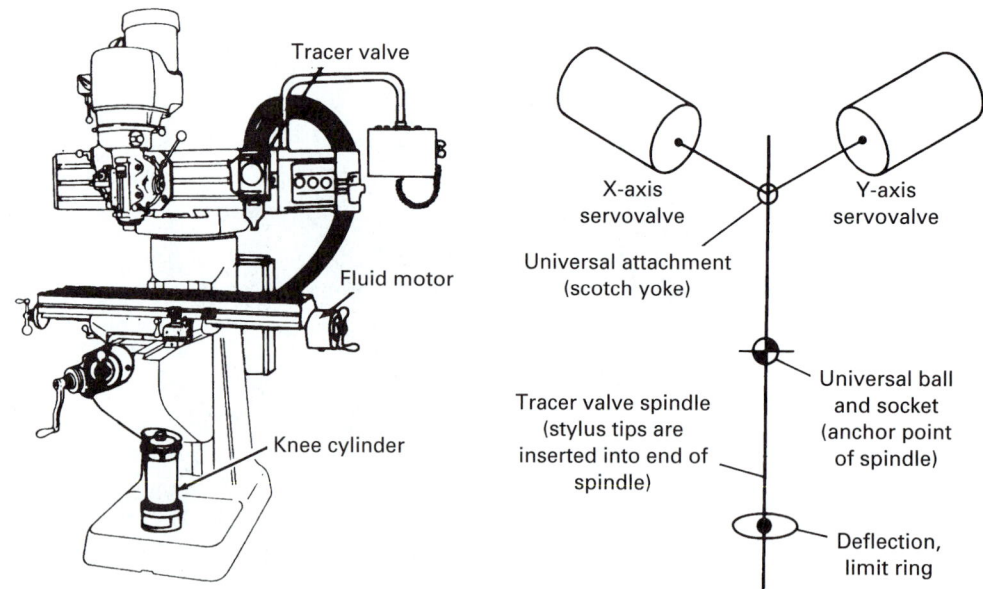

FIGURE 25-14   Small knee-type vertical spindle milling machine modified with tracer control system for X and Y control.

All hydraulic tracers work basically the same way, in that they utilize a stylus connected to a precision servomechanism for each axis of control. Figure 25-14 illustrates a stylus that is linked mechanically to the servos in a two-axis tracer valve. The servos are connected to hydraulic actuators on the machine slides. As the stylus traces a template, the servos control the motion of the slides so that the milling cutter duplicates the template shape onto the workpiece. See Figure 25-15 for details of the principle of operation.

As the tracing stylus is tilted toward the pattern, the servovalve is shifted off null, resulting in Y slide motion opposite to that of the stylus direction. When the stylus is in contact with the pattern, the servovalve is at its null position and Y slide motion is stopped.

Duplicators produce forms in three dimensions. A tracing probe follows a three-dimensional master. Often, the probe does not actually contact the master, a variation in the length of an electric spark between the probe and the master controlling the drives to the quill and the table, thereby avoiding wear on the master or possible deflection of the probe. On some machines, the ratio between the movements of the probe and cutter can be varied.

Duplicators are widely used to machine molds and dies and sometimes are called *die-sinking machines*. They are used extensively in the aerospace industry to machine parts from wrought plate or bar stock as substitutes for forgings when the small number of parts required would make the cost of forging dies uneconomical.

Accessories for Milling Machines

The usefulness of ordinary milling machines is greatly extended by employing various accessories or attachments. Here are some examples.

The *vertical milling attachment* (Figure 25-16) is used on a horizontal milling machine to permit vertical milling to be done. Ordinarily, heavy cuts cannot be made with such an attachment.

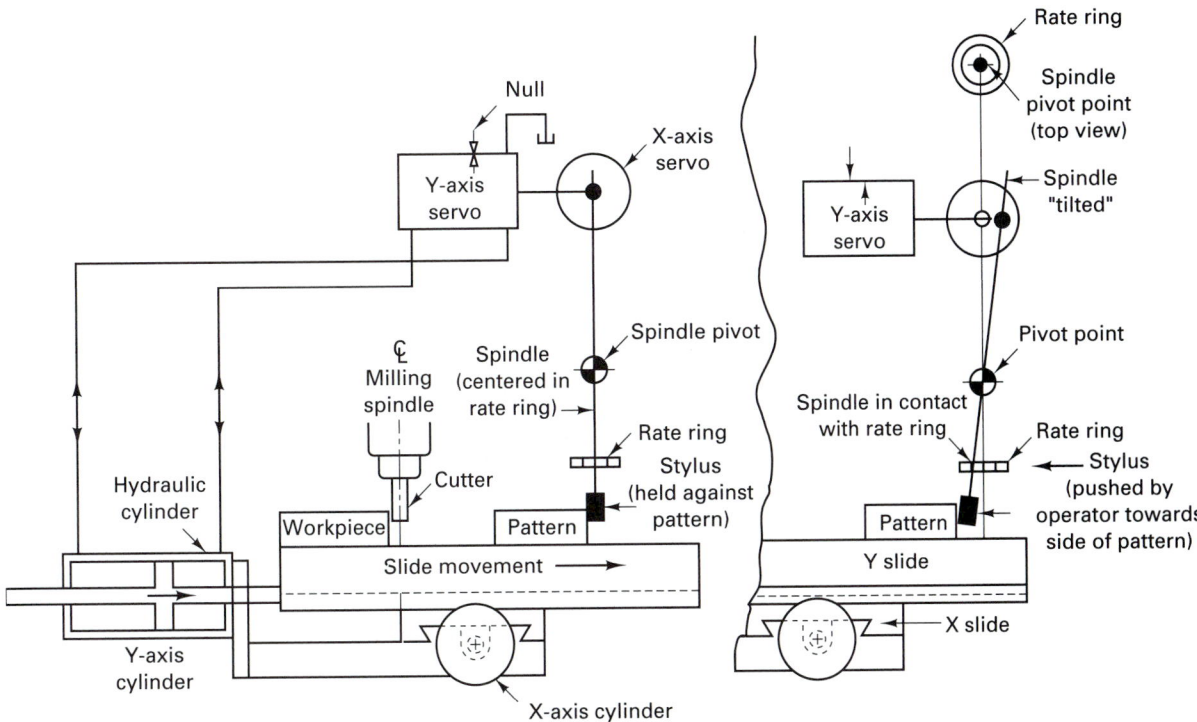

FIGURE 25-15   Principle of operation of a hydraulic tracer.

FIGURE 25-16   Vertical milling attachment for horizontal milling machine. *(Courtesy of Brown & Sharpe Manufacturing Company.)*

FIGURE 25-17   End milling a helical groove on a horizontal spindle milling machine using a universal dividing head and a universal milling attachment. *(Courtesy of Cincinnati Milacron, Inc.)*

The *universal milling attachment* (Figure 25-17) is similar to the vertical attachment but can be swiveled about both the axis of the milling machine spindle and a second, perpendicular axis to permit milling to be done at any angle.

The *universal dividing head* is by far the most widely used milling machine accessory, providing a means for holding and indexing work through any desired arc of rotation. The work may be mounted between centers (Figure 25-17) or held in a chuck that is mounted in the spindle hole of the dividing heat. The spindle can be tiled from about 5° below horizontal to beyond the vertical position.

Basically, a dividing head is a rugged, accurate, 40:1 worm-gear reduction unit. The spindle of the dividing head is rotated one revolution by turning the input crank 40 turns. An index plate mounted beneath the crank contains a number of holes, arranged in concentric circles and equally spaced, with each circle having a different number of holes. A plunger pin on the crank handle can be adjusted to engage the holes of any circle. This permits the crank to be turned an accurate, fractional part of a complete circle as represented by the increment between any two holes of a given circle on the index plate. Utilizing the 40:1 gear ratio and the proper hole circle on the index plate, the spindle can be rotated a precise amount by the application of either of the following rules:

$$\text{number of turns of crank} = \frac{40}{\text{cuts per revolution of work}}$$

$$\text{holes to be indexed} = \frac{40 \times \text{holes in index circle}}{\text{cuts per revolution of work}}$$

If the first rule is used, an index circle must be selected that has the proper number of holes to be divisible by the denominator of any resulting fractional portion of a turn of the crank. In using the second rule, the number of holes in the index circle must be such that the numerator of the fraction is an even multiple of the denominator. For example, if 24 cuts are to be taken about the circumference of a workpiece, the number of turns of the crank required would be $1\frac{2}{3}$. An index circle having 12 holes could be used with one full turn plus eight additional holes. The second rule would give the same result. Adjustable *sector arms* are provided on the index plate that can be set to a desired number of holes, less than a full turn, so that fractional turns can be made readily without the necessity for counting holes each time. Dividing heads are made having ratios other than 40:1. The ratio should be checked before using.

Because each full turn of the crank on a standard dividing head represents 360/40, or 9° of rotation of the spindle, indexing to a fraction of a degree can be obtained. For example, the space between two adjacent holes on a 36-hole circle represents $\frac{1}{4}°$. Indexing can be done in three ways. *Plain indexing* is done solely by the use of the 40:1 ratio in the dividing head. In *compound indexing* the index plate is moved forward or backward a number of hole spaces each time the crank handle is advanced. For *differential indexing* the spindle and the index plate are connected by suitable gearing so that as the spindle is turned by means of the crank, the index plate is rotated a proportional amount.

The dividing head also can be connected to the feed screw of the milling-machine table by means of gearing. This procedure is used to provide a definite rotation of the workpiece with respect to the longitudinal movement of the table, as in cutting helical gears. This procedure is illustrated in Chapter 31.

■ 25.5 WORKHOLDING DEVICES FOR MILLING

Although T slots are provided on milling machine tables so that workpieces can be clamped directly to the table, more often various workholding devices, called *vises* or *fixtures*, are utilized. Smaller workpieces usually are held in a vise mounted on the table. A universal vise (Figure 25-18) is particularly useful in tool-and-die work. Fixtures are used for larger volumes (Figure 25-19). Fixtures reduce the time it takes to put the part in the machine and assure proper part location with respect to the cutting tools. Fixtures provide clamping forces that counteract the cutting forces. The design of workholding devices is discussed in Chapter 28.

■ 25.6 MILLING TOLERANCES

As with all tooling applications, the tolerances that can be maintained in milling are dependent on the rigidity of the workpiece, the accuracy and rigidity of the machine spindle, the precision and accuracy of the workholding device, and the quality of the cutting tool itself. Milling produces forces that contribute to chatter and vibration because of the intermittent cutting action. Soft materials tend to adhere to the cutter teeth and make it more difficult to hold tolerances. Materials, such as cast iron and aluminum, are easy to mill.

Within these criteria, properly maintained cutters used in rigid spindles on properly fixtured workpieces can expect to machine within tolerances ±0.0005 in. with surface flatness tolerances of 0.001 in./ft. Such tolerances are also possible on "slotting" operations with milling cutters, but +0.001 to +0.002 in. is more probable. Flatness specifications are more difficult to maintain in steel, easier to maintain in some types of aluminum, cast iron, and other nonferrous materials.

FIGURE 25-18   Universal vise on a horizontal spindle milling machine. *(Courtesy of Cincinnati Milacron, Inc.)*

FIGURE 25-19   End milling a circular slot using a fixture mounted on a circular-milling attachment in a vertical spindle milling machine. *(Courtesy of Cincinnati Milacron, Inc.)*

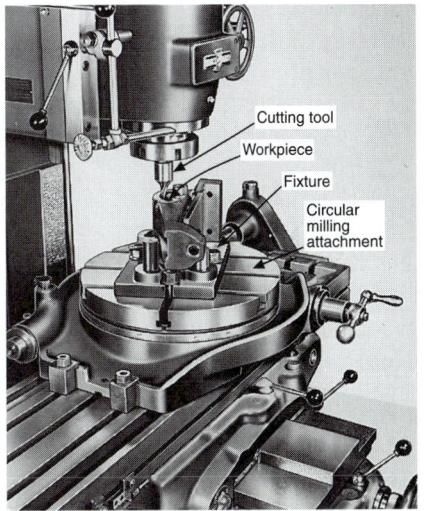

■ 25.7 MILLING SURFACE FINISH

The average surface finishes that can be expected on free-machining materials range from 60 to 150 μin. Conditions can exist, however, that can produce wide variations on either side of these ranges.

■ KEY WORDS

climb (down) milling
column-and-knee milling machine
conventional (up) milling
cutting time
duplicator
end milling
face milling
fixtures

insert-tooth milling cutter
interrupted cutting
machining center
metal removal rate
milling
milling cutters
milling machines
peripheral milling

profiler
saddle
slab milling
staggered-tooth milling cutter
straddle milling
Woodruff keyseater

■ REVIEW QUESTIONS

1. Why is milling better suited than shaping for producing flat surfaces in mass-production machining?
2. How does face milling differ basically from peripheral milling?
3. Which type of milling (up or down) do you think uses the least power? Explain.
4. Why may down milling dull the cutter more rapidly than up milling in machining sand castings?
5. What are two common ways of classifying milling cutters?
6. What kind of milling do you have if the diameter D of the cutter shown in Figure 25-2 is less than the width W of the workpiece?
7. What is the advantage of a helical-tooth cutter over a straight-tooth cutter for slab milling?
8. What would the cutting force diagram for F_c look like if all eight teeth are considered? (See Figure 25-5b.)
9. Explain what steps are required to produce a T slot by milling.
10. In a typical solid arbor milling cutter shown in Figure 25-9, why are the teeth staggered?
11. Why would a plain column-and-knee milling machine not be suitable for milling the flutes on a large twist drill?
12. Why is a turret-type milling machine often preferred for tool-and-die work?
13. What is the distinctive feature of a duplex, horizontal-bed type of milling machine?
14. Explain how controlled movements of the work

in three mutually perpendicular directions are obtained in column-and-knee milling machines.
15. What is the function of the rate ring in the tracer milling machine?
16. Why are bed milling machines preferred over column-and-knee types for production milling?
17. How does a duplicator differ from a profiler?
18. Why have planer milling machines replaced ordinary planers?
19. The Bridgeport vertical-spindle milling machine is perhaps the single most popular machine tool. Virtually every factory (or shop) that does machining has one or more of these machines. What made it so popular and widely copied?
20. What is the basic principle of a universal dividing head?
21. The input end of a universal dividing head can be connected to the feed screw of the milling machine table. For what purpose?
22. What is the purpose of the hole-circle plate on a universal dividing head?
23. Explain how a standard universal dividing head, having hole circles of 21, 24, 27, 30, and 32 holes, would be operated to cut an 18-tooth gear.
24. Explain how table feed rate (in./min) and spindle rpm are specified or computed for a milling machine.
25. Why must the number of teeth on the cutter be known when calculating milling machine table feed rate, in inches per minute.
26. In Figure 25-12, duplex milling is shown. Is the

cutter performing up or down milling, and why?

27. Why is the question of up or down milling more critical in horizontal slab milling than in vertical-spindle (end or face) milling?

■ PROBLEMS

1. In Question 28 you were asked to select a feed per tooth and a cutting speed for a milling process. Reasonable values for feed and speed are 0.005 to 0.010 in. per tooth and 200 sfpm. Compute the input values for the machine tool. Use the information in Problem 6 as input when needed.

2. How much time will be required to face mill an AISI 1020 steel surface that is 12 in. long and 5 in. wide, using a 6-in.-diameter eight-tooth tungsten carbide inserted-tooth cutter? Select values of feed per tooth and cutting speed from Table 25-1.

3. If the depth of cut is 0.35 in., what is the metal-removal rate in Problem 2.

4. Estimate the power required for the operation of Problem 3. Do not forget to consider Figure 25-5.

5. If all the flat surfaces on the bearing block shown in Figure 28-5 must be machined, would you machine them in the same sequence on a shaper as on a milling machine? Explain why or why not.

6. A gray cast iron surface 6 in. wide and 18 in. long may be machined on either a vertical milling machine, using an 8-in.-diameter face mill having 10 inserted HSS teeth, or on a hydraulic shaper with a HSS tool. If milling is used, the feed per tooth is 0.01 in. For shaping, a feed of 0.015 in. per stroke would be used. Setup time for the milling machine would be 60 minutes and for the shaper would be 10

28. Suppose that you wanted to machine a cast iron with BHN of 275. The process to be used was face milling and a HSS cutter was going to be used. What feed and speed values would you select?

minutes. The time required to load and unload either machine is 2 minutes. Machine-hour charges for the milling machine and shaper would be $24.50 and $16.50, respectively. Labor cost would be $8.75 per hour in each case. Which machine would be more economical for this job?

7. In Problem 6, what percentage of the total time for the job is consumed in non-metal-removal activities (setup time and part loading/unloading) assuming that 10 parts are to be made?

8. In Figure 21-5, the feed per tooth is 0.006 in. per tooth. The cutter is rotating at an rpm rate that will produce the desired surface cutting speed of 125 sfpm. The cutter diameter is 5 in. The depth of cut is 0.5 in. The block is 2 in. wide. What is the feed rate, in inches per minute, of the milling machine table?

9. What is the MRR for the situation described in Problem 8? Is up milling or down milling depicted?

10. Suppose that you want to do the job described in Problem 6 by slab milling. You have selected a 6-in.-diameter cutter with an eight-tooth TiN-coated HSS cutter. The cutting speed will be 500 sfpm and the feed per tooth will be 0.010 in./tooth. Determine the input parameters for the machine (rpm of arbor and table feed), then calculate the CT and MRR. Compare these answers with what you got for face milling or shaping the block.

CHAPTER 26

BROACHING, SAWING, FILING, SHAPING, AND PLANING

26.1	INTRODUCTION TO BROACHING	26.4	INTRODUCTION TO SAWING	
26.2	FUNDAMENTALS OF BROACHING	26.5	SAW BLADES	
	Advantages and Limitations of Broaching	26.6	TYPES OF SAWING MACHINES	
	Broach Design		Power Hacksaws	
	Broaching Speeds		Bandsawing Machines	
	Broaching Materials and Construction		Cutting Fluids	
	Sharpening Broaches		Feeds and Speeds	
26.3	BROACHING MACHINES		Circular-Blade Sawing Machines	
	Broaching Presses	26.7	FILING	
	Vertical Pull-Down Machines	26.8	TYPES OF FILES	
	Vertical Pull-Up Machines	26.9	FILING MACHINES	
	Vertical Surface-Broaching Machines	26.10	SHAPING AND PLANING	
	Horizontal Broaching Machines		Workholding Devices for Shapers	
	Continuous Surface-Broaching Machines	26.11	PLANING MACHINES	
	Rotary Broaching Machines		Workholding and Setup for Planers	
		Case Study:	SOCKET WITH A TRIANGULAR HOLE	

■ 26.1 INTRODUCTION TO BROACHING

The process of *broaching* is one of the most productive of the basic machining processes. The machine tool is called a *broaching machine* and the cutting tool (Figures 26-1 and 26-2) is called the *broach*. Broaching competes economically with milling and boring and is capable of producing precision-machined surfaces. The broach finishes an entire surface in a single pass. Broaches are used in production to finish holes, splines, and flat surfaces.

The feed per tooth in broaching is the change in height of successive teeth. This is called the rise per tooth. Broaching is similar to sawing except that the saw makes many passes through the cut, whereas the broach produces a finished part in one pass. The heart of this process lies in the broaching tool, in which roughing, semifinishing, and finishing teeth are combined into one tool. Broaching is unique in that it is the only one of the basic machining processes in which *feed*, which determines the chip thickness, is built into the cutting tool. The machined surface is always the inverse of the profile of the broach, and in most cases, it is produced with a single linear stroke of the tool across the workpiece (or the workpiece across the broach).

A broach is composed of a series of teeth, each tooth standing slightly higher than the last. This rise per tooth (RPT), also known as *step* or the feed per tooth, determines

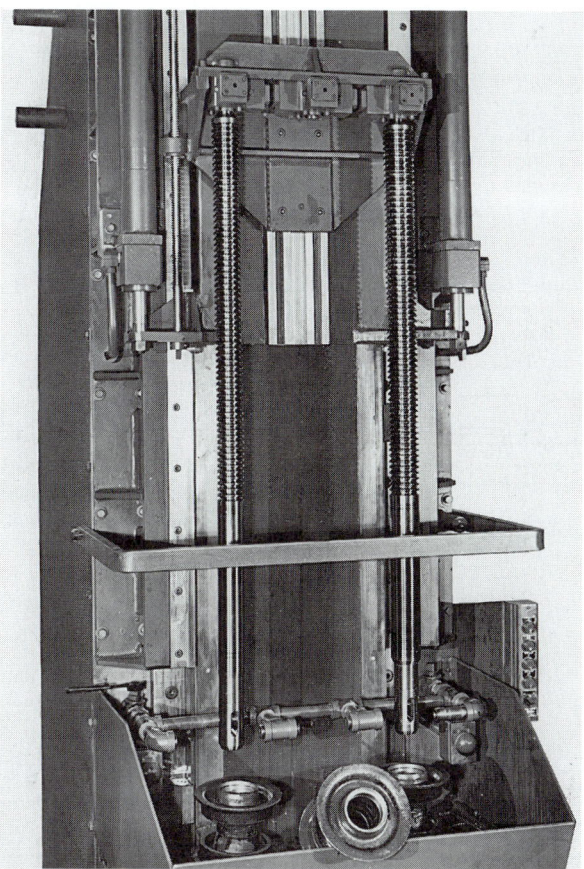

FIGURE 26-1   Vertical pull-down broaching machine shown with parts in position ready for the two broaches to be inserted. An extra part is shown lying at the front of the machine.

the amount of material removed. There is no feeding of the broaching tool required. The frontal contour of the teeth determines the shape of the resulting machined surface. As the result of these conditions built into the tool, no complex motion of the tool relative to the workpiece is required and the need for highly skilled machine operators is minimized.

Figures 26-1 and 26-2 show a *pull broach* in a vertical pull-down broaching machine. The pull end of the broach is passed through the part and a key mates to the slot. The broach is pulled through the part. The broach is retracted (pulled up) out of the part. The part is transferred from the left fixture to the right fixture. One finished part is completed in every time cycle.

■ 26.2 FUNDAMENTALS OF BROACHING

In broaching, the tool (or work) is translated past the work (or tool) with a single stroke of velocity V. The feed is provided by a gradual increase in height of successive teeth. The rise per tooth varies depending on whether the tooth is for roughing (t_r), semifinishing (t_s), or final sizing or finishing (t_f). In a typical broach there are three to five semifinishing and finishing teeth specified. The number of roughing teeth must be determined so that broach length, which is needed to estimate the cutting time, can be calculated. Other lengths needed for a typical pull broach are shown in Figure 26-2. The chip breakers in the first section of roughing teeth may be extended to more teeth if the cut is heavy or material difficult to machine. The distance between the teeth, called the pitch, P, is important because

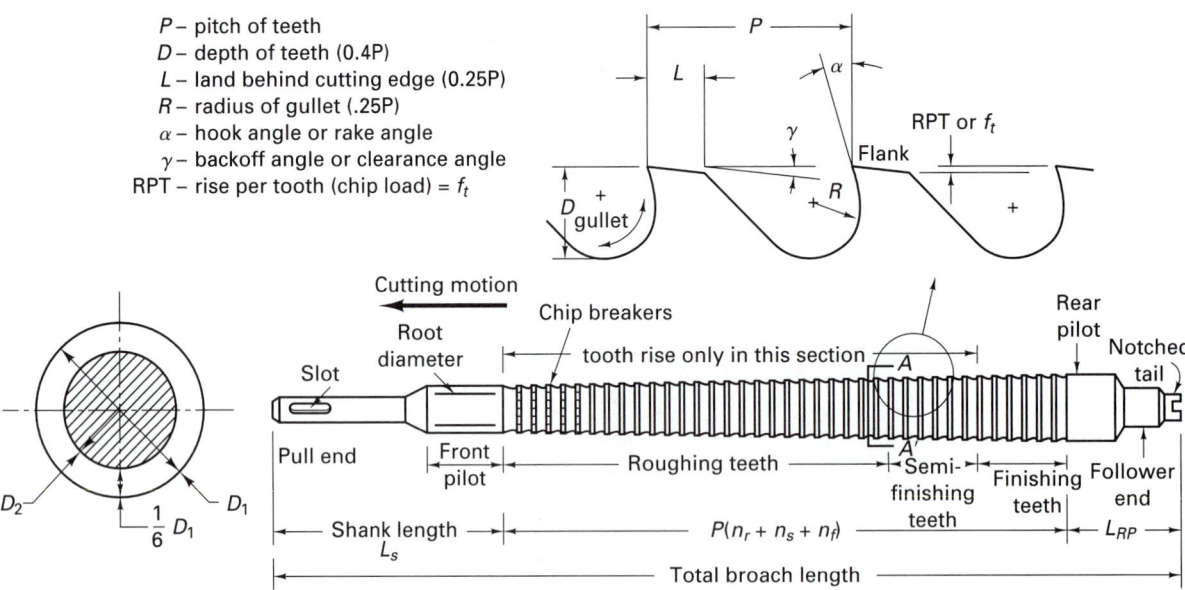

P – pitch of teeth
D – depth of teeth (0.4P)
L – land behind cutting edge (0.25P)
R – radius of gullet (.25P)
α – hook angle or rake angle
γ – backoff angle or clearance angle
RPT – rise per tooth (chip load) = f_t

FIGURE 26-2   Basic shape and nomenclature for a conventional pull (hole) broach. Section A–A shows the cross section of a tooth.

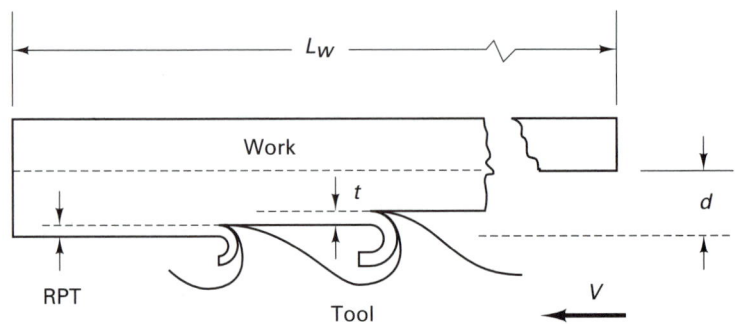

FIGURE 26-3   The feed in broaching depends on the rise per tooth (RPT). The sum of the RPT gives the depth of cut, d.

it determines the tooth construction and strengths and the number of teeth actually cutting at a given instant. It is preferable that at least two teeth are in contact at any instant.

The pitch or distance between teeth is

$$P \simeq 0.35\sqrt{L_w} \qquad (26\text{-}1)$$

where length of cut usually equals L_w as shown in Figure 26-3. The number of roughing teeth is

$$n_r = \frac{d - n_f t_f - n_s t_s}{t_r}$$

where d is the total amount of metal to be removed and t_r is the rise per tooth.

The overall length of the broach, $L_B = (n_r + n_s + n_f)P + L_s + L_{RP}$ for a pull broach.

The length of stroke $L = L_B - L_w$, in inches, if the broach moves past the work, or $L_w + L_B$ if the work moves past the broach. The cutting time

$$CT = \frac{L}{12V} \tag{26-2}$$

where V is the cutting speed, in surface feet per minute.

The metal removal rate depends on the number of teeth (roughing) contacting the work.

$$MRR \text{ (per tooth)} = 12t_r WV \qquad \text{in}^3/\text{min per roughing tooth} \tag{26-3}$$

where W is the width of the broach tooth.

The number of roughing teeth in contact with part $n \simeq L_w/P$ for a broach larger than the part.

$$MRR \text{ (for process)} = 12t_r WVn \qquad \text{in}^3/\text{min}$$

where n is usually rounded off to the next-largest whole number.

The pull broach must be strong enough so that it will not be pulled apart. The strength of a pull broach is determined by its minimum cross section, which occurs either at the root of the first tooth or at the pull end:

$$allowable \; pull = \frac{area \; of \; minimum \; section \times Y.S. \; of \; broach \; material}{factor \; of \; safety} \tag{26-4}$$

The *push broach* must be strong enough so that it will not buckle. If the length-to-diameter ratio, L/D_r, is greater than 25, the broach must be considered a long column which can buckle if overloaded. Let

$$L = \text{length from push end to first tooth}$$

$$D_r = \text{root diameter at } \tfrac{1}{2}L$$

$$S = \text{factor of safety}$$

$$allowance \; load = \frac{13.5 \times 10^6 \times D_r^4}{SL^2} \tag{26-5}$$

For L/D_r less than 25, the normal broach loads are not critical.

Calculations of the total push or pull load depend on the number of teeth engaged, n, estimated from L_w/P; the width of the cut, W; the RPT per engaged tooth, t_r; and the shear strength of the metal being machined.

The force necessary to operate a broach depends on the material being broached, the conditions of the tool, and the nature of the process. An empirical constant is required in the force calculation to account for the large amount of rubbing (friction) between the tool, the chips captured in the tooth gullet, and the workpiece.

Let F_{CB} be the broach pull force in pounds:

$$F_{CB} \simeq 5\tau_s n t_r W \tag{26-6}$$

τ_s is found from Figure 21-15 depending on the BHN for the metal. This force estimate can be used to estimate the horsepower needed for the broaching machine.

Advantages and Limitations of Broaching

Because of the features built into a broach, it is a simple and rapid method of machining. There is a close relationship between the contour of the surface to be produced, the amount of material that must be removed, and the design of the broach. For example, the total depth of the material to be removed cannot exceed the total step provided in the broach, and the step of each tooth must be sufficient to provide proper chip thickness for the type of material to be machined. Consequently, either a special broach must be made for each job, or the workpiece must be designed so that a standard broach can be used. Broaching is widely used and particularly well suited for mass production because the volume can easily justify the cost of the broaching tool, which can easily be $15,000 to $30,000 per tool. It is also used for certain simple and standardized shapes, such as keyways, where inexpensive standard broaches can be used.

Broaching originally was developed for machining internal keyways. However, its obvious advantages quickly led to its development for mass-production machining of various surfaces, such as flat, interior or exterior, cylindrical and semicylindrical, and many irregular surfaces. Because there are few limitations as to the contour form that broach teeth may have, there is almost no limitation in the shape of surfaces that can be produced by broaching. The only physical limitations are that there must be no obstruction to interfere with the passage of the entire tool over the surface to be machined and that the workpiece must be strong enough to withstand the forces involved. In internal broaching, a hole must exist in the workpiece into which the broach may enter. Such a hole can be made by drilling, boring, or coring.

Broaching usually produces better accuracy and finish than can be obtained by drilling, boring, or reaming. Although the relative motion between the broaching tool and the work usually is a single linear one, a rotational motion can be added to permit the broaching of spiral splines or gun-barrel rifling.

Broach Design

Broaches commonly are classified by design features as follows:

Purpose	Motion	Construction	Function
Single	Push	Solid	Roughing
Combination	Pull	Built-up	Sizing
	Stationary		Burnishing

Figure 26-2 shows the principal components of a pull broach and the shape and arrangement of the teeth. Each tooth is essentially a single-edge cutting tool, arranged much like the teeth on a saw except for the step, which determines the depth cut by each tooth. The rise per tooth, which determines the chip load, varies from about 0.006 in. for roughing teeth in machining free-cutting steel to a minimum 0.001 in. for finishing teeth. Typically the RPT is 0.003 to 0.006 in. in surface broaching and 0.0012 to 0.0025 in. on the diameter for internal broaching. The exact amount depends on several factors. Too-large cuts impose undue stresses on the teeth and the work; too-small cuts result in rubbing rather than cutting action. The strength and ductility of the metal being cut are the primary factors.

Where it is desirable for each tooth to take a deep cut, as in broaching castings or forgings that have a hard, abrasive surface layer, *rotor-cut* or *jump-cut* tooth design may be used (Figure 26-4). In this design, two or three teeth in succession have the same diameter, or height, but each tooth of the group is notched or cut away so that it cuts only

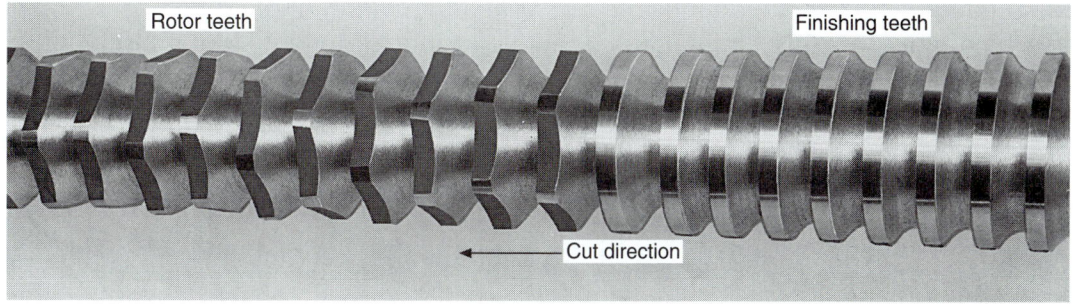

FIGURE 26-4   Rotor or jump tooth broach design.

FIGURE 26-5   Round, push-type broach with chip-breaker grooves on alternate teeth except at the finishing end. *(Courtesy of duMont Corp.)*

a portion of the circumference or width. This permits deeper but narrower cuts by each tooth without increasing the total load per tooth. This tooth design also reduces the forces and the power requirements. Chip-breaker notches are also used on round broaches to break up the chips (Figure 26-5).

A similar idea can be used for flat surfaces. Tooth loads and cutting forces also can be reduced by using the *double-cut* construction shown in Figure 26-6. Four consecutive teeth get progressively wider. The teeth remove metal over only a portion of their width until the fourth tooth completes the cut.

A third technique for reducing tooth loads utilizes the principle illustrated in Figure 26-7. Employed primarily for broaching wide, flat surfaces, the first few teeth in *progressive* broaches completely machine the center, while succeeding teeth are offset in two groups to complete the remainder of the surface. Rotor, double-cut, and progressive designs require the broach to be made longer than if normal teeth were used, and they therefore can be used only on a machine having adequate stroke length.

The cutting edges of the teeth on surface broaches may be either normal to the direction of motion or at an angle of from 5 to 20°. The latter, *shear-cut* broaches provide smoother cutting action with less tendency to vibrate. Other shapes that can be broached are shown in Figure 26-8 along with push- or pull-type broaches used for the job. The pitch of the teeth and the gullet between them must be sufficient to provide ample room for chip clearance. All chips produced by a given tooth during its passage over the full length of the workpiece must be contained in the space between successive teeth. At the same time, it is desirable to have the pitch sufficiently small so that at least two or three teeth are cutting at all times (Figure 26-9).

The *hook* determines the primary rake angle and is a function of the material being cut. It is 15 to 20° for steel and 6 to 8° for cast iron. *Back-off* or end clearance angles are from 1 to 3° to prevent rubbing.

Most of the metal removal is done by the *roughing teeth. Semifinishing teeth* provide surface smoothness, whereas *finishing teeth* produce exact size. On a new broach all the

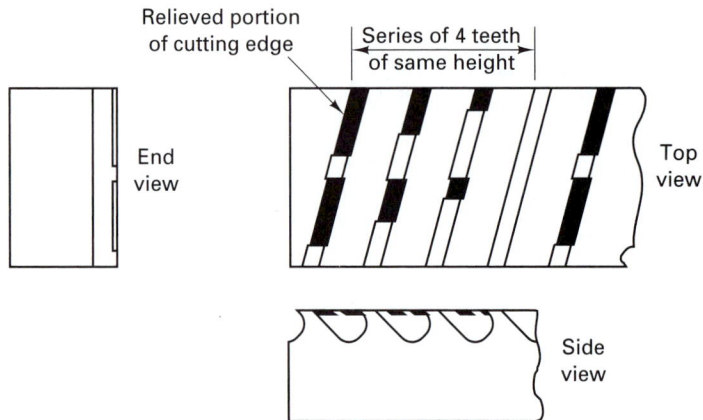

FIGURE 26-6 Overlapping successive teeth permits large RPT without increasing the load per tooth.

FIGURE 26-7 Progressive surface broach. *(Courtesy of Detroit Broach & Machine Company.)*

finishing teeth usually are made the same size. As the first finishing teeth become worn, those behind continue the sizing function. On some round broaches, *burnishing teeth* are provided for finishing. These have no cutting edges but are rounded disks that are from 0.001 to 0.003 in. larger than the size of the hole. The resulting rubbing action smooths and sizes the hole. They are used primarily on cast iron and nonferrous metals.

The *pull end* of a broach provides a means of quickly attaching the broach to the pulling mechanism. The *front pilot* aligns the broach in the hole before it begins to cut, and the *rear pilot* keeps the tool square with the finished hole as it leaves the workpiece. *Shank length* must be sufficient to permit the broach to pass through the workpiece and be attached to the puller before the first roughing tooth engages the work. If a broach is to be used on a vertical machine that has a tool-handling mechanism, a *tail* is necessary.

A broach should not be used to remove a greater depth of metal than that for which it is designed—the sum of the steps of all the teeth. In designing workpieces, a minimum of 0.020 in. should be provided on surfaces that are to be broached, and about 0.025 in. is the practical maximum.

Broaching Speeds

Cutting speeds for broaching are relatively low (25 to 20 sfpm), seldom exceeding 50 ft/min. However, because a surface usually is completed in a single stroke, the productivity is high. A complete cycle usually requires only from 5 to 30 seconds, with most of that time being taken up by the return stroke, broach handling, and workpiece loading and unloading. Such cutting conditions facilitate cooling and lubrication and result in very low tool wear rates,

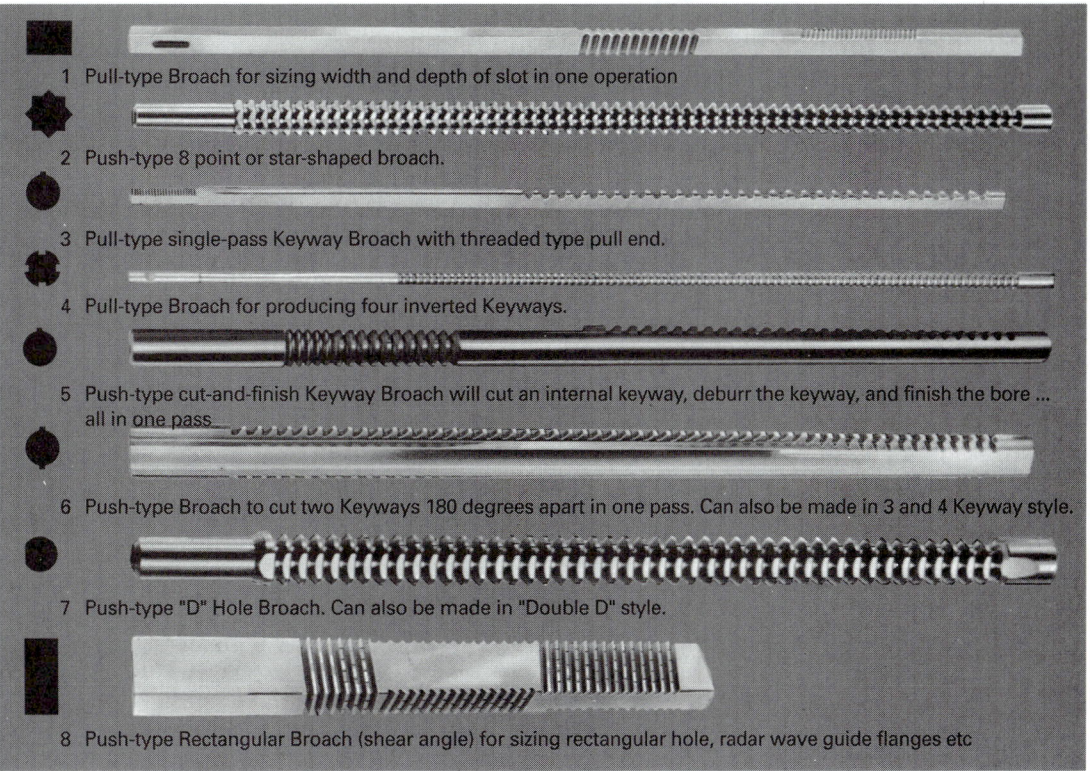

1 Pull-type Broach for sizing width and depth of slot in one operation

2 Push-type 8 point or star-shaped broach.

3 Pull-type single-pass Keyway Broach with threaded type pull end.

4 Pull-type Broach for producing four inverted Keyways.

5 Push-type cut-and-finish Keyway Broach will cut an internal keyway, deburr the keyway, and finish the bore ... all in one pass

6 Push-type Broach to cut two Keyways 180 degrees apart in one pass. Can also be made in 3 and 4 Keyway style.

7 Push-type "D" Hole Broach. Can also be made in "Double D" style.

8 Push-type Rectangular Broach (shear angle) for sizing rectangular hole, radar wave guide flanges etc

FIGURE 26-8   Examples of push-and-pull-type broaches. *(Courtesy of DuMont Corp.)*

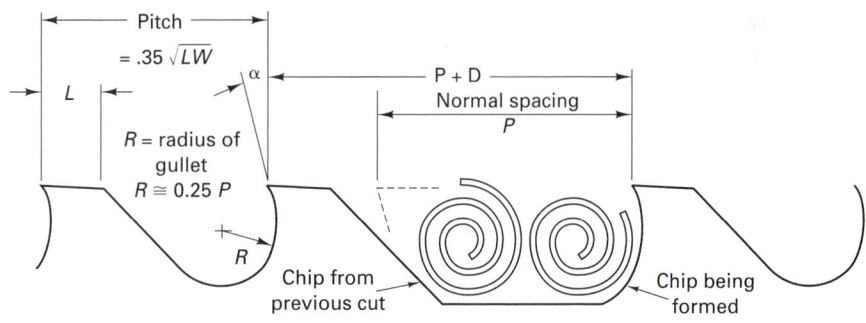

Extra-wide spacing may be used when chip disposal is a problem.
Chip adhering to broach tooth will be displaced by the next chip formed.

FIGURE 26-9   The gullet area provides room for the chips.

which reduces the necessity for frequent resharpening and prolongs the life of the expensive broaching tool.

For a given cutting speed and material, the force required to pull or push a broach is a function of the tooth width, the step, and the number of teeth cutting. Consequently, it is necessary to design or specify a broach within the stroke length and power limitations of the machine on which it is to be used.

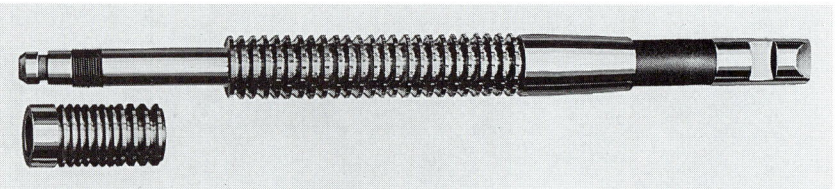

FIGURE 26-10   Shell construction for a pull broach.

Broaching Materials and Construction

Because of the low cutting speeds employed, most broaches are made of alloy or high-speed tool steel. Carbide-tipped broaches are seldom used for machining steel parts or forgings, as the cutting edges tend to chip on the first stroke, probably due to a lack of rigidity in the machine tool/cutting tool combination. TiN coating of HSS broaches is becoming more common, greatly prolonging the life of broaches. When used in continuous mass-production machines, particularly in surface broaching of cast iron, tungsten carbide teeth may be used, permitting the broach to be used for long periods of time without resharpening.

Internal broaches are of solid construction, or quite often they are made of *shells* mounted on an arbor (Figure 26-10). When the broach (or a section of it) is subject to rapid wear, a single shell can be replaced. This will be much cheaper than replacing an entire solid broach. Shell construction, however, is initially more expensive than a solid broach of comparable size.

Small surface broaches may be of solid construction, but larger ones are usually built up in sections (Figure 26-11). Sectional construction makes the broach easier and cheaper to construct and sharpen. It also often provides some degree of interchangeability of the sections.

Sharpening Broaches

Most broaches are resharpened by grinding the hook faces of the teeth. The lands of internal broaches must not be reground because this would change the size of the broach. Lands of flat surface broaches sometimes are ground, in which case all of them must be ground to maintain their proper relationship.

■ 26.3 BROACHING MACHINES

Because all the factors that determine the shape of the machined surface and that determine all cutting conditions except speed are built into the broaching tool, broaching machines are relatively simple. Their primary functions are to impart plain reciprocating motion to the broach and to provide a means for handling the broach automatically.

Most broaching machines are driven hydraulically, although mechanical drive is used in a few special types. The major classification relates to whether the motion of the broach is vertical or horizontal, as follows:

Vertical	Horizontal	Rotary
Broaching presses (push broaching)	Pull	Special types
Pull down (Figure 26-1)	Surface	
Pull up	Continuous	
Surface		

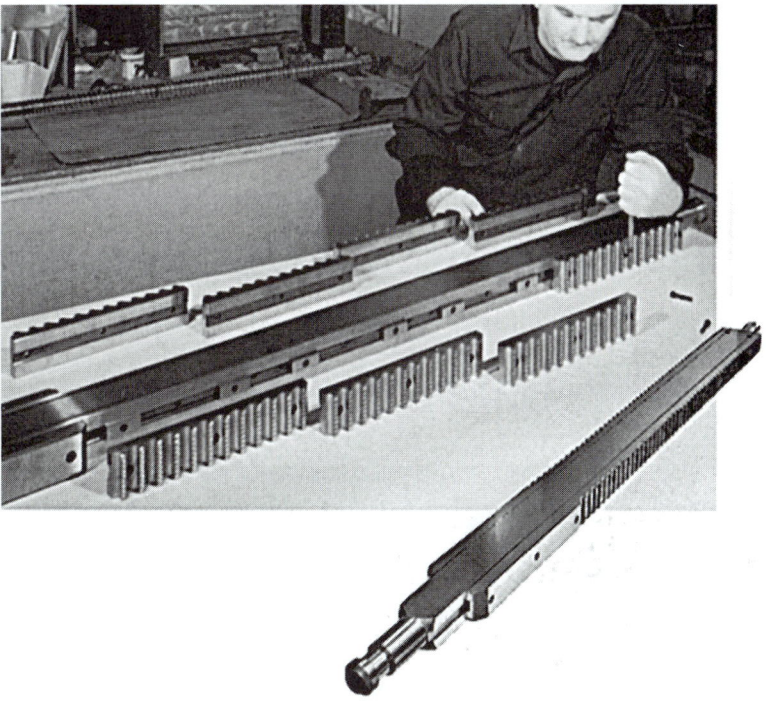

FIGURE 26-11   An 80-in.-long broach constructed from inserts is cheaper to build and sharpen in sections.

The choice between vertical and horizontal machines is determined primarily by the length of the stroke required and the available floor space. Vertical machines seldom have strokes greater than 60 in. because of height limitations. Horizontal machines can have almost any length of stroke, but they require greater floor space.

Broaching Presses

As shown in Figure 26-12, *broaching presses* essentially are arbor presses with a guided ram. They are used with push broaches, have a capacity of from 5 to 50 tons, and are used only for internal broaching. The forward guide of the broach is inserted through the hole in the workpiece as it rests on the press table, often in a fixture. As the ram descends, it engages the upper end of the broach and pushes it through the work.

Broaching presses are relatively slow in comparison with other broaching machines, but they are inexpensive, flexible, and can be used for other types of operations, such as bending and staking.

Vertical Pull-Down Machines

The major components of vertical pull-down machines are a worktable, usually having a spherically-seated workholder, a broach elevator above the table, and a pulling mechanism below the table. As shown in Figure 26-1, when the elevator raises the broach above the table, the work can be placed into position. The elevator then lowers the pilot end of the broach through the hole in the workpiece, where it is engaged by the puller. The elevator then releases the upper end of the broach, and it is pulled through the workpiece. The workpieces are removed from the table, and the broach is raised upward to be engaged by

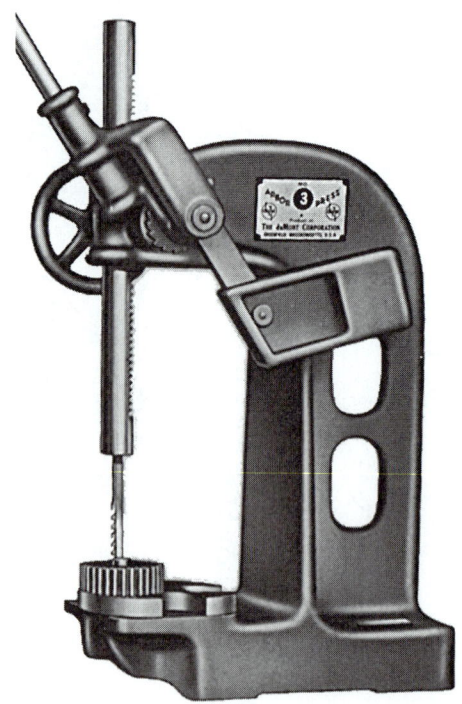

FIGURE 26-12   Arbor press used to broach keyway in a gear.

the elevator mechanism. In some cases wherein machines have two rams, they are arranged so that one broach is being pulled down while the work is being unloaded and the broach raised at the other station. In Figure 26-1, the part is being broached in two passes, first on the left, then on the right.

Vertical Pull-Up Machines

In *vertical pull-up machines*, the pulling ram is above the worktable and the broach-handling mechanism below it. The work is placed in position, above the pilot, while the broach is lowered. The handling mechanism then raises the broach until it engages the puller head. As the broach is pulled upward, the work comes to rest against the underside of the table, where it is held until the broach has been pulled through. The work then falls free, often sliding down a chute into a tote bin.

Pull-up machines may have up to eight rams. Because the workpieces need only be placed in the machines, and the broach handling and work removal are automatic, they are highly productive. For certain types of work, automatic feeding can be provided.

Vertical Surface-Broaching Machines

On *vertical surface-broaching machines* the broaches usually are mounted on guided slides to provide support against lateral thrust. Because there is no need for handling the broach, they are simpler but much heavier than pull- or push-broaching machines. Many have two or more slides so that work can be loaded at one while another part is machined at the other. The operating cycle is very short, as there is no handling of the broach. Slide or rotary-indexing fixtures are usually used to hold the work. This reduces the work-handling time and minimizes the total cycle time.

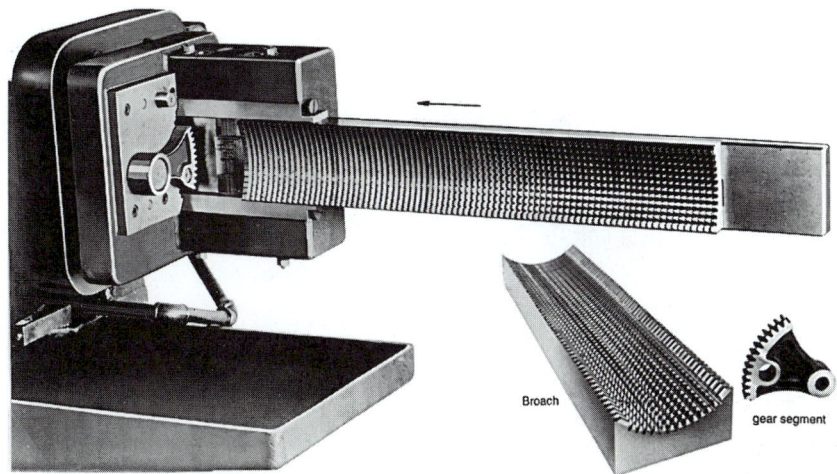

Broach

gear segment

FIGURE 26-13 (a) Broaching the teeth in a gear segment by horizontal surface broaching in one pass; (b) broach and gear segment. *(Courtesy of Apex Broach and Machine Company.)*

Horizontal Broaching Machines

The primary reason for employing a *horizontal* configuration for pull- and surface-broaching machines is to make possible longer strokes and the use of longer broaches than can conveniently be accommodated in vertical machines. Horizontal pull-broaching machines are vertical machines turned on their side (Figure 26-13). When internal surfaces are to be broached, such as holes, the broach must have a diameter-to-length ratio large enough to make it self-supporting without appreciable deflection. Consequently, horizontal machines are seldom used for small holes. In surface broaching, the broach is always supported in guides, and therefore no such limitation is encountered (Figure 26-14). Broaching that requires rotation of the broach, as in rifling and spiral splines, usually is done on horizontal machines.

Continuous Surface-Broaching Machines

In *continuous surface-broaching machines*, the broaches usually are stationary, and the work is pulled past the cutters by means of a conveyor. Fixtures are usually attached to the conveyor chain so that the workpieces can be placed in them at one end of the machine and removed at the other, sometimes automatically. Such machines are being used increasingly in mass production (Figure 26-14).

Rotary Broaching Machines

In *rotary broaching machines*, occasionally used in mass production, the broaches are stationary and the work is passed beneath or between them. The work is held in fixtures on a rotary table. The advantage of these machines is that there is no lost time due to non-cutting, reciprocating strokes.

■ 26.4 INTRODUCTION TO SAWING

Sawing is a basic machining process in which chips are produced by a succession of small cutting edges, or *teeth*, arranged in a narrow line on a saw "blade." As shown in Figure 26-15,

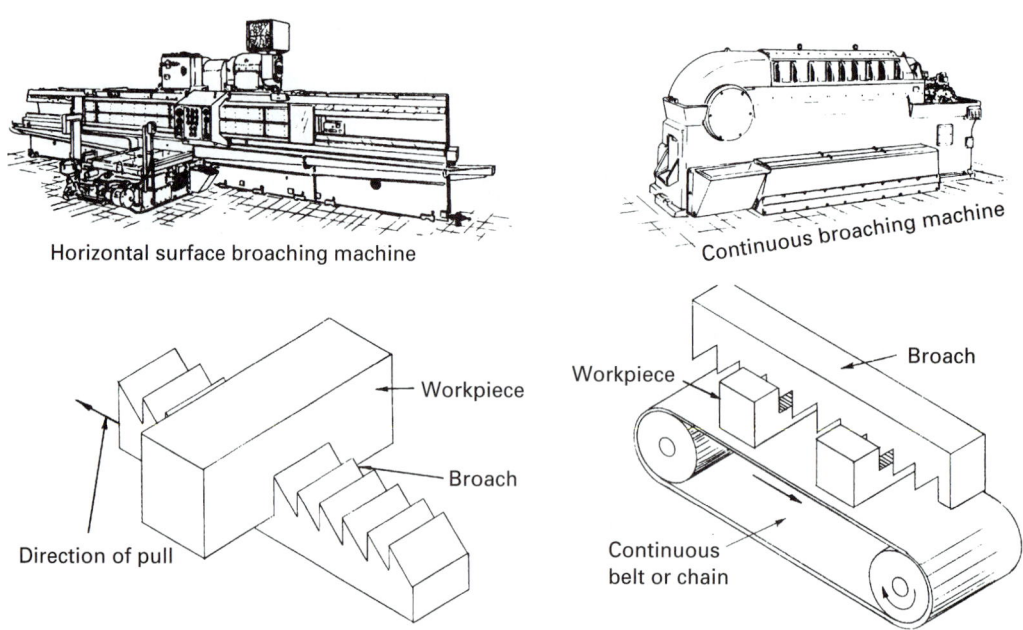

FIGURE 26-14   Large horizontal surface broaching machine on left and continuous horizontal, surface broacher on right.   *(Courtesy of General Electric Company.)*

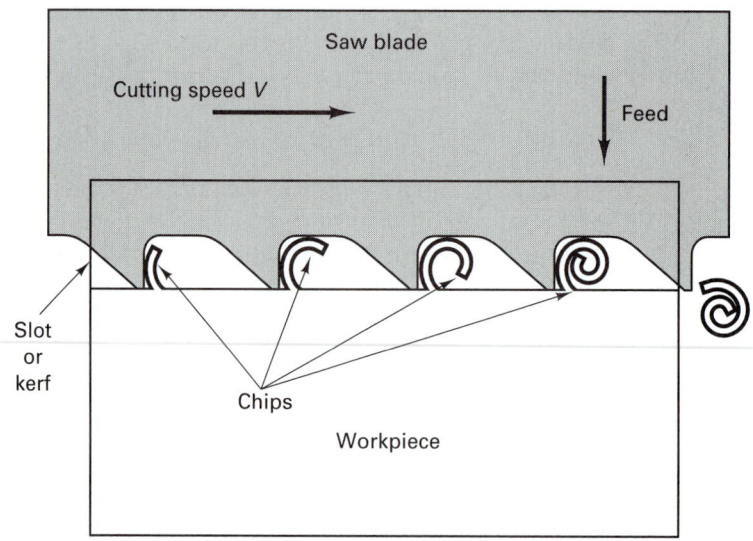

FIGURE 26-15   Formation of chips in sawing.   *(Courtesy of DoALL Company.)*

each tooth forms a chip progressively as it passes through the workpiece, and the chip is contained within the space between two successive teeth until these teeth pass from the work. Because sections of considerable size can be severed from the workpiece with the removal of only a small amount of the material in the form of chips, sawing is probably the most economical of the basic machining processes with respect to the waste of material and power consumption, and in many cases with respect to labor.

In recent years vast improvements have been made in saw blades and sawing machines, resulting in improved accuracy and precision of the process. Most sawing is done to sever bar stock and shapes into desired lengths for use in other operations. There are many cases in which sawing is used to produce desired shapes. Frequently, and especially for producing only a few parts, contour sawing may be more economical than any other machining process.

■ 26.5 SAW BLADES

Figure 26-16 gives the standard nomenclature for a saw blade. Saw blades are made in three basic configurations. The first, commonly called a *hacksaw blade*, is straight, relatively rigid, and of limited length with teeth on one edge. The second is sufficiently flexible so that a long length can be formed into a continuous band with teeth on one edge; these are known as *bandsaw blades*. The third form is a rigid disk having teeth on the periphery; these are called *circular saws* or *cold saws*.

All saw blades have certain common and basic features: (1) material, (2) tooth form, (3) tooth spacing, (4) tooth set, and (5) blade thickness or gage. Small hacksaw blades usually are made entirely of tungsten or molybdenum high-speed steel. Blades for power-operated hacksaws often are made with teeth cut from a strip of high-speed steel that has been electron-beam welded to the heavy main portion of the blade, which is made from a tougher and cheaper alloy steel. Bandsaw blades frequently are made with this same type of construction (Figure 26-17) but with the main portion of the blade made of relatively thin, high-tensile-strength alloy steel to provide the required flexibility. Bandsaw blades also are available with tungsten carbide teeth and TiN coatings. Three common *tooth forms* are shown in Figure 26-18.

Tooth spacing is very important in all sawing because it determines three factors. First, it controls the size of the teeth. From the viewpoint of strength, large teeth are desirable. Second, tooth spacing determines the space (*gullet*) available to contain the chip that is formed. As shown in Figure 26-15, the chip cannot drop from this space until it emerges from the slot cut in the workpiece, called the *kerf*. The space must be such that there is no crowding of the chip and no tendency for chips to become wedged between the teeth and not drop out when the saw emerges from the cut. Third, tooth spacing determines how many teeth will bear against the work. This is very important in cutting thin material, such as tubing. At least two teeth should be in contact with the work at all times. If the teeth are too coarse, only one tooth rests on the work at a given time, permitting the saw to rock, and the teeth may be stripped from the saw. Hand hacksaw blades have 14 to 32 teeth per inch. To make it easier to start a cut, some hand hacksaw blades are made with a short section at the forward end having teeth of a special form with negative rake angles. Tooth spacing for power hacksaw blades ranges from 4 to 18 teeth per inch.

Tooth set (Figures 26-16 and 26-18) refers to the manner in which the teeth are offset from the centerline to make the kerf wider than the gage, the thickness of the back portion of the blade. This permits the saw to move more freely in the kerf and reduces friction and heating. *Raker-tooth saws* are used in cutting most steel and iron. *Straight-set teeth* are

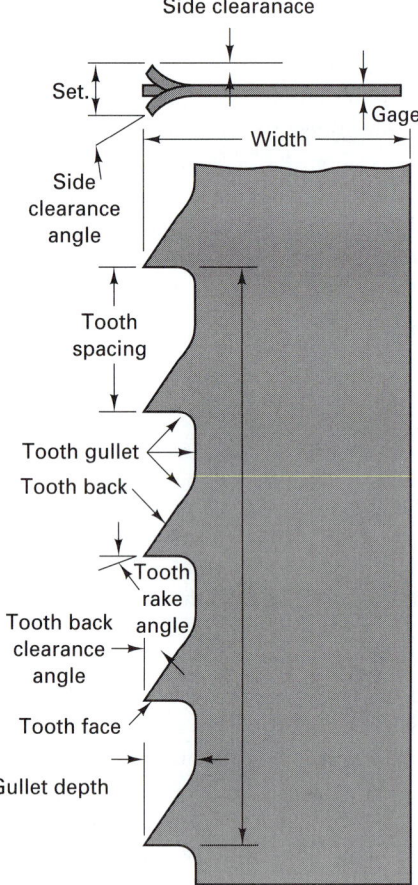

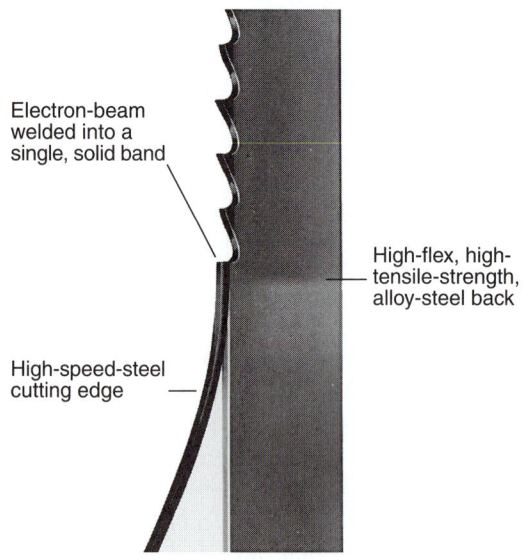

FIGURE 26-16 Standard nomenclature for a saw blade. Note how teeth are offset to provide tooth-set to the blade.

FIGURE 26-17 Method of providing HSS teeth on a softer steel band. (*Courtesy of DoALL Company.*)

used for sawing brass, copper, and plastics. Saws with *wave-set teeth* are used primarily for cutting thin sheets and thin-walled tubing.

The gage or *blade thickness* of nearly all hand hacksaw blades is 0.025 in. Saw blades for power hacksaws vary in thickness from 0.050 to 0.100 in. Hand hacksaw blades come in two standard lengths, 10 and 12 in. All are $\frac{1}{2}$ in. wide. Blades for power hacksaws vary in length from 12 to 24 in. and in width from 1 to 2 in. Wider and thicker blades are desirable for heavy-duty work. The blade should be at least twice as long as the maximum length of cut that is to be made.

Bandsaw blades are available in straight, raker, wave, or combination sets. To reduce the noise from high-speed bandsawing, it is becoming increasingly common to use blades that have more than one pitch, size of teeth, and type of set. Blade width is very important in bandsawing because it determines the minimum radius that can be cut. This relationship is illustrated in Figure 26-19. The most common widths are from $\frac{1}{16}$ to $\frac{1}{2}$ in., although wider blades can be obtained. Because wider blades are stronger, as wide a blade as possible should be used. Consequently, cutting small radii requires a narrower and weaker blade and is more time consuming. Bandsaw blades come in tooth spacings from 2 to 32 teeth per inch.

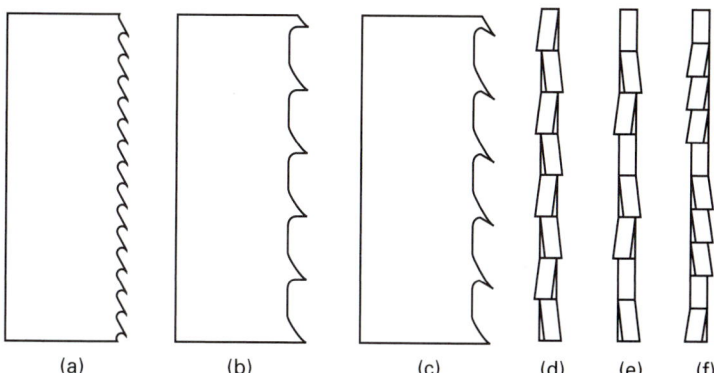

(a) Regular or standard tooth for ferrous metal, general purpose cutting
(b) Skip tooth with large gullets for machining softer, nonferrous metals
(c) Hook (10° positive rake) tooth for harder nonferrous alloys
(d) Symmetrical tooth straight set saw for brass & plastic
(e) Raker tooth set for general purpose sawing, uniform thickness
(f) Wavy tooth set for thin, or non uniform thickness of stock.

FIGURE 26-18 Bandsaw-blade tooth geometries.

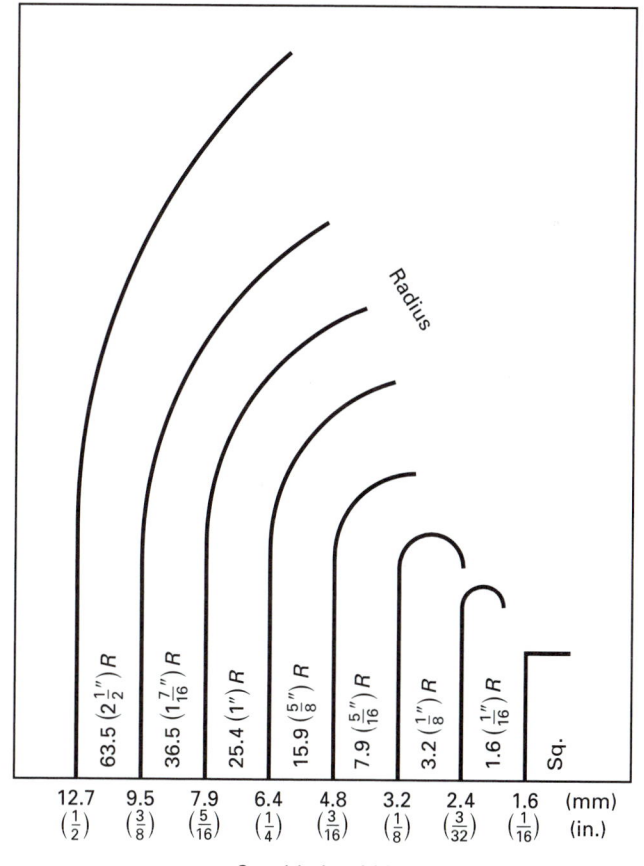

FIGURE 26-19 Relationship of bandsaw width to the minimum radius that can be cut.

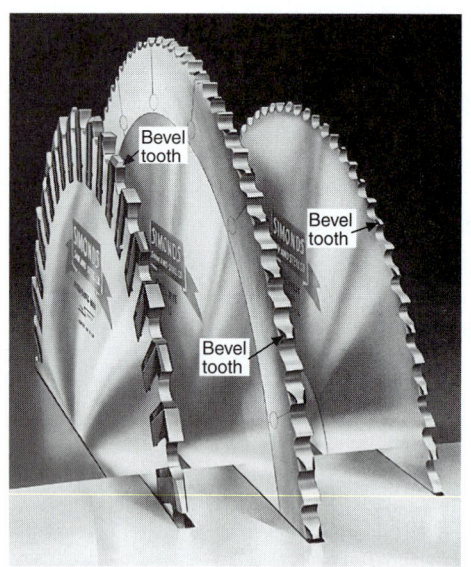

FIGURE 26-20   *Left to right:* Inserted tooth, segmental-tooth, and integral-tooth forms of saw constructions.   *(Courtesy of Simonds Saw and Steel Company.)*

Circular saws for cutting metal are often called *cold saws*, to distinguish them from friction-type disk saws, which heat the metal to the melting temperature at the point of metal removal.   Cold saws cut rapidly and produce surfaces that are comparable in smoothness and accuracy with surfaces made by slitting saws in a milling machine or by a cutoff tool in a lathe.

Disk or *circular saws* necessarily differ somewhat from straight blade forms.   Because they must be relatively large in comparison with the work, only the sizes up to about 18 in. in diameter have teeth that are cut into the disk (Figure 26-20).   Larger saws use either *segmented* or *inserted* teeth.   The teeth are made of high-speed steel or tungsten carbide.   The remainder of the disk is made of ordinary, less expensive, and tougher steel. *Segmental* blades are composed of segments mounted around the periphery of the disk, usually fitted with a tongue and groove and fastened by means of screws or rivets.   Each segment contains several teeth.   If a single tooth is broken, only one segment need be replaced to restore the saw to operating conditions.

Figure 26-20 shows a common tooth form used in circular saws, in which every other tooth is beveled on both sides.   Sometimes the first tooth is beveled on the left side, the second tooth on both sides, the third tooth on the right side, the fourth tooth on the left side, and so on.   Another method is to bevel the opposite sides of successive teeth.   Beveling is done to produce a smoother cut.   Precision circular saws made from carbide are becoming available which are very thin (0.03 in.) and have high cutting-off accuracy, around ±0.0008 in., with negligible burrs.

■ 26.6 TYPES OF SAWING MACHINES

Metal-sawing machines may be classified as follows:

    1. Reciprocating saw
        a. Manual hacksaw
        b. Power hacksaw
    2. Bandsaw
        a. Vertical cutoff

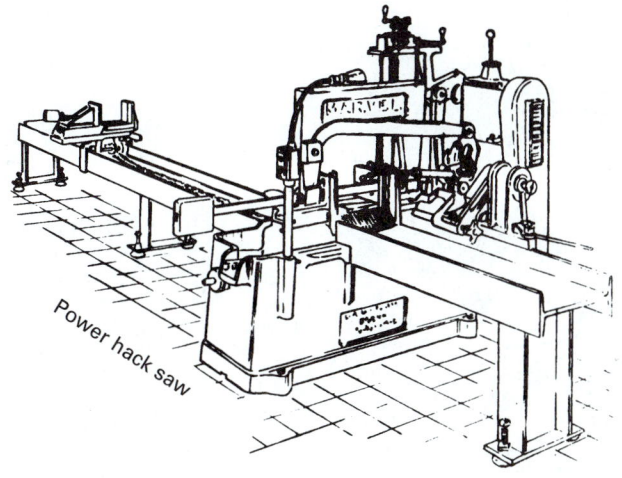

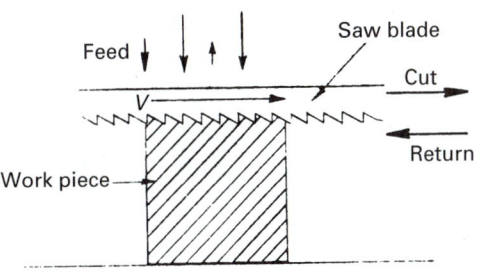

FIGURE 26-21   Power hacksaw with reciprocating blade. (From *Manufacturing Producibility Handbook, courtesy of General Electric Company.*)

   b. Horizontal cutoff
   c. Combination cutoff and contour
   d. Friction
   3. Circular saw
   a. Cold saw
   b. Steel friction disk
   c. Abrasive disk

Power Hacksaws

As the name implies, power hacksaws are machines that mechanically reciprocate a large hacksaw blade. As shown in Figure 26-21, they consist of a bed, a workholding frame, a power mechanism for *reciprocating* the saw frame, and some type of feeding mechanism. Because of the inherent inefficiency of cutting in only one stroke direction, they have often been replaced by more efficient, horizontal bandsawing machines.

Bandsawing Machines

The earliest metal-cutting bandsawing machines were direct adaptations from wood-cutting bandsaws. Modern machines of this type are much more sophisticated and versatile and have been developed specifically for metal cutting. To a large degree they were made possible by the development of vastly better and more flexible bandsaw blades and simple flash-welding equipment, which can weld the two ends of a strip of bandsaw blade together to form a band of any desired length. Three basic types of bandsawing machines are in common use.

FIGURE 26-22   Cutting a pipe at a 45° angle on an upright cutoff bandsawing machine. *(Courtesy of Armstrong-Blum Mfg. Co.)*

Upright, cutoff, bandsawing machines (Figure 26-22) are designed primarily for cut-off work on single stationary workpieces that can be held on a table. On many machines the blade mechanism can be tilted to about 45°, as shown, to permit cutting at an angle. They usually have automatic power feed of the blade into the work, automatic stops, and provision for supplying coolant.

Horizontal metal-cutting bandsawing machines were developed to combine the flexibility of reciprocating power hacksaws and the continuous cutting action of vertical bandsaws. These heavy-duty automatic bandsaws feed the saw vertically by a hydraulic mechanism and have automatic stock feed that can be set to feed the stock laterally any desired distance after a cut is completed and clamp it automatically for the next cut. Such machines can be arranged to hold, clamp, and cut several bars of material simultaneously.

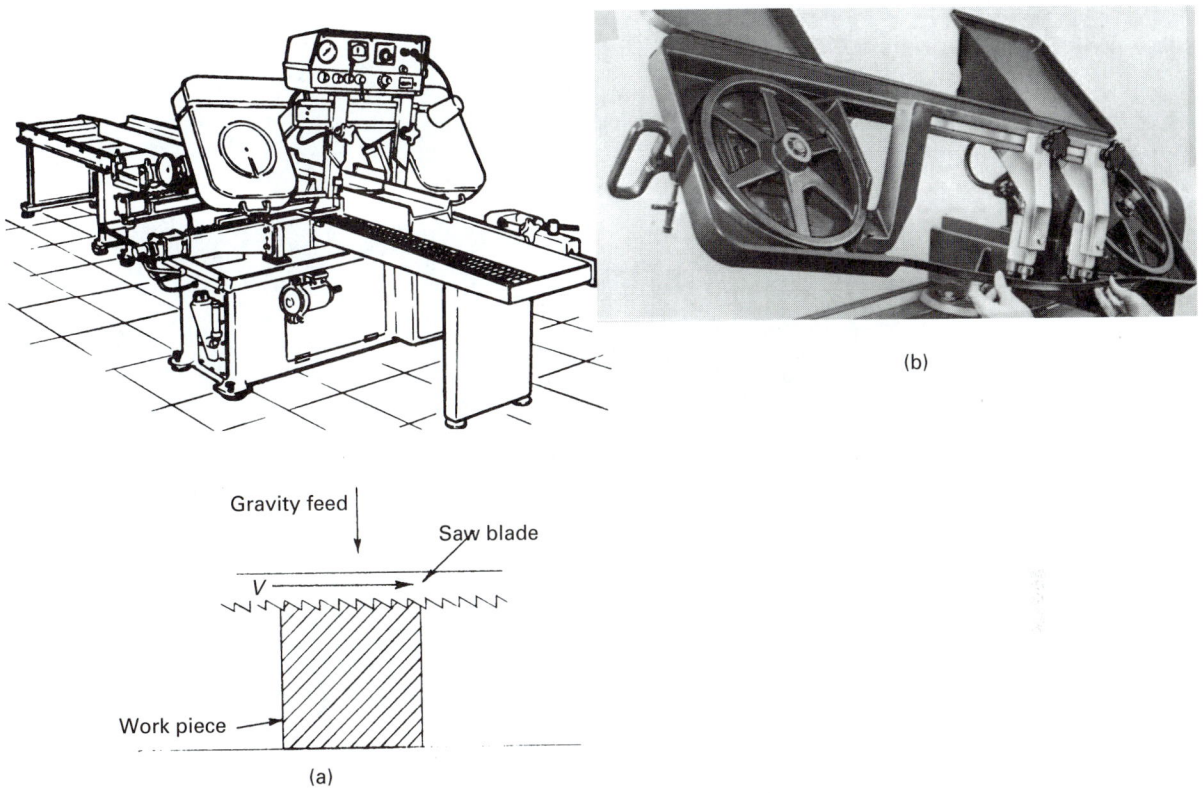

(b)

Gravity feed

Saw blade

V

Work piece

(a)

FIGURE 26-23   Horizontal bandsawing machine (*left*) sawing an I-beam; (*right*) horizontal bandsaw with machine blade guards removed for blade installation.

CNC bandsaws are available with automatic storage and retrieval systems for the bar stock. Smaller and less expensive types have swing-frame construction, with the bandsaw head mounted in a pivot on the rear of the machine. Feed is accomplished by gravity through rotation of the head about the pivot point. Because of their continuous cutting action, horizontal bandsawing machines are very efficient (Figure 26-23).

Combination cutoff and contour bandsawing machines (Figure 26-24) can be used not only for cutoff work but also for contour sawing. They are widely used for cutting irregular shapes in connection with making dies and the production of small numbers of parts and are often equipped with rotary tables. Additional features on these machines include a table that pivots so that it can be tilted to any angle up to 45°. Usually, these machines have a small flash butt welder on the vertical column, so that a straight length of bandsaw blade can be welded quickly into a continuous band. A small grinding wheel is located beneath the welder so that the flash can be ground from the weld to provide a smooth joint that will pass through the saw guides. This welding and grinding unit makes it possible to cut internal openings in a part by first drilling a hole, inserting one end of the saw blade through the hole, and then butt-welding the two ends together. When the cut is finished, the band is again cut apart and removed from the opening. The cutting speed of the saw blade can be varied continuously over a wide range to provide correct operating conditions for any material. A method of power feeding the work is provided, sometimes gravity actuated.

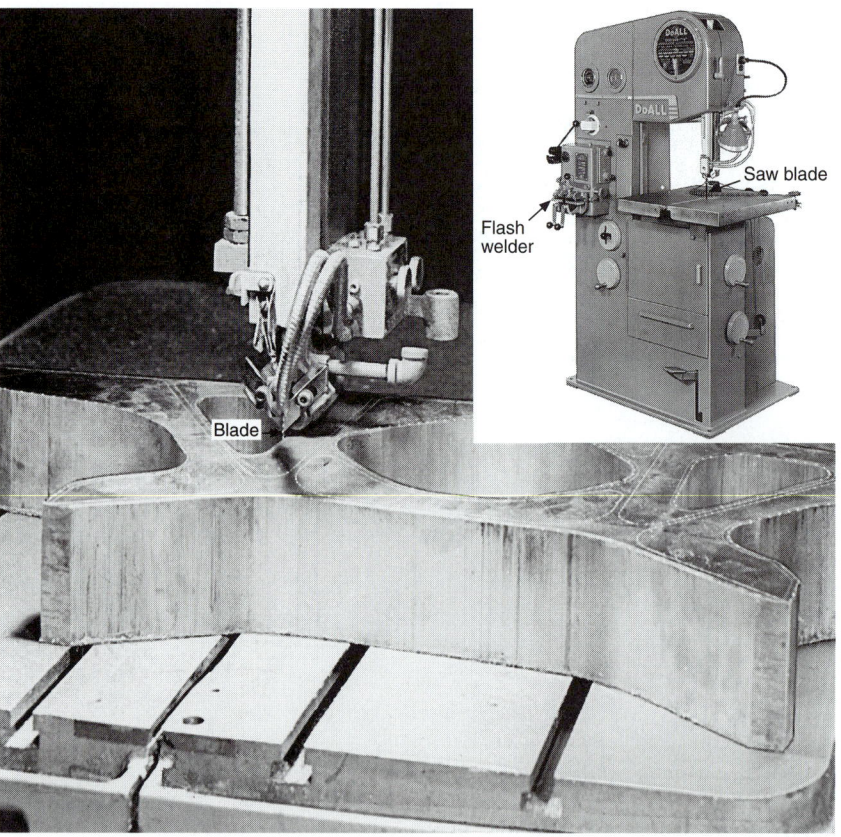

FIGURE 26-24   Contour bandsawing on vertical bandsawing machine. Bandsaw machine shown in inset.

Contour-sawing machines are made in a wide range of sizes, the principal size dimension being the throat depth. Sizes from 12 to 72 in. are available. The speeds available on most machines range from about 50 to 2000 ft/min. Modern horizontal bandsaws are accurate to ±0.002 in. per vertical inch of cut but have feeding accuracy of only ±0.005 in., subject to the size of the stock and the feed rate. Repeatability from one feed to the next may be ±0.010 to ±0.020 in.

CNC controlled sawing centers with microprocessor controls have opened up new automation aspects for sawing. Such control systems can improve accuracy to within ±0.005 in. over entire cuts by controlling saw speed, blade feed pressure, and feed rate.

Special bandsawing machines are available with very high speed ranges, up to 14,000 ft/min. These are known as *friction bandsawing machines*. Material is not cut by chip formation. Instead, the friction between the rapidly moving saw blade and the work is sufficient to raise the temperature of the material at the end of the kerf to or just below the melting point, where its strength is very low. The saw blade then pulls the molten, or weakened, material out of the kerf. Consequently, the blades do not need to be sharp; they frequently have no teeth, only occasional notches in the blade to aid in removing the metal.

Almost any material, including ceramics, can be cut by friction sawing. Because only a small portion of the blade is in contact with the work for an instant and then is cooled by its passage through the air, it remains cool. Usually, the major portion of the work, away from contact with the saw blade, also remains quite cool. The metal adjacent to the kerf is heat-affected, recast, and sometimes harder than the bulk metal. It is also a very rapid method for trimming the flash from sheet-metal parts and castings.

Cutting Fluids

Cutting fluids should be used for all bandsawing, with the exception that cast iron is always cut dry. Commercially available oils or light cutting oils will give good results in cutting ferrous materials. Beeswax or paraffin are common lubricants for cutting aluminum and aluminum alloys.

Feeds and Speeds

Because of the many different types of feed involved in bandsawing, it is not practical to provide tabular feed or pressure data. Under general conditions, however, an even pressure, without forcing the work, gives best results. A nicely curled chip usually indicates an ideal feed pressure. Burned or discolored chips indicate excessive pressure, which can cause tooth breakage and premature wear.

Most bandsaws provide recommended cutting speed information right on the machine, depending on the material being sawed. In general, HSS blades are run at 200 to 300 ft/min when cutting 1-in.-thick low- and medium-carbon steels. For high-carbon steels, alloy steels, and tool-and-die steels, the range is from 150 to 225 ft/min, and most stainless steels are cut at 100 to 125 ft/min.

Circular-Blade Sawing Machines

Machines employing rotating circular or cold saw blades are used exclusively for cutoff work. A cold saw is shown in Figure 26-25. These range from small, simple types, in which the saw is fed manually, to very large saws having power feed and built-in coolant systems, commonly used for cutting off hot-rolled shapes as they come from a rolling mill. In some cases friction saws are used for this purpose, having disks up to 6 ft in diameter and operating at surface speeds up to 25,000 ft/min. Steel sections up to 24 in. can be cut in less than 1 minute by this technique.

Although technically not a sawing operation, cutoff work up to about 6 in. is often done utilizing thin *abrasive* disks. The equipment used is the same as for sawing. It has the advantage that very hard materials that would be very difficult to saw can be cut readily. A thin rubber- or resinoid-bonded abrasive wheel is used. Usually, a somewhat smoother surface is produced.

■ 26.7 FILING

Basically, the metal-removing action in *filing* is the same as in sawing, in that chips are removed by cutting teeth that are arranged in succession along the same plane on the surface of a tool, called a *file*. There are two differences: (1) the chips are very small, and therefore the cutting action is slow and easily controlled, and (2) the cutting teeth are much wider. Consequently, fine and accurate work can be done.

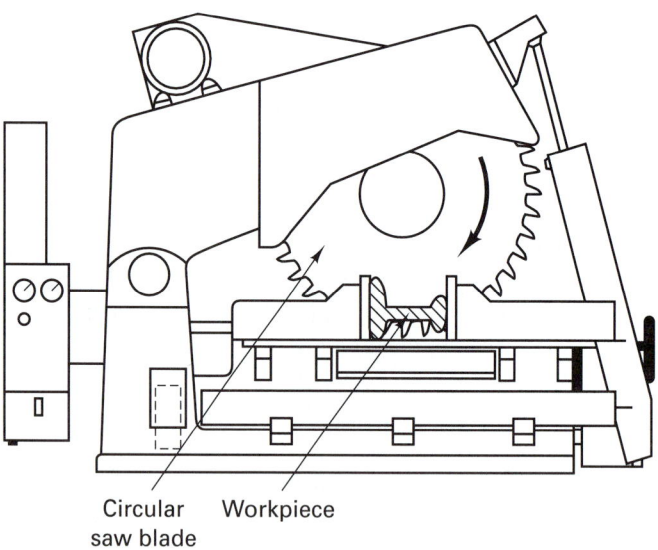

FIGURE 26-25 Circular or cold sawing a structural shape.

Circular saw blade   Workpiece

■ 26.8 TYPES OF FILES

Files are classified according to the following:

1. The type, or *cut*, of the teeth
2. The degree of coarseness of the teeth
3. Construction
   a. Single solid units for hand use or in die-filing machines
   b. Band segments, for use in band-filing machines
   c. Disks, for use in disk-filing machines

Four types of *cuts* are available. *Single-cut files* have rows of parallel teeth that extend across the entire width of the file at the angle of from 65 to 85°. *Double-cut files* have two series of parallel teeth that extend across the width of the file. One series is cut at an angle of 40 to 45°. The other series is coarser and is cut at an opposite angle that varies from about 10 to 80°. A *vixen-cut file* has a series of parallel curved teeth, each extending across the file face. On a *rasp-cut file*, each tooth is short and is raised out of the surface by means of a punch. These four types of cuts are shown in Figure 26-26.

The coarseness of files is designated by the following terms, arranged in order of increasing coarseness: *dead smooth*, *smooth*, *second cut*, *bastard*, *coarse*, and *rough*. There is also a series of finer Swiss pattern files, designated by numbers from 00 to 8.

Files are available in a number of cross-sectional shapes: *flat*, *round*, *square*, *triangular*, and *half-round*. Flat files can be obtained with no teeth on one or both narrow edges, known as *safe edges*. Safe edges prevent material from being removed from a surface that is normal to the one being filed. Most files for hand filing are from 10 to 14 in. in length and have a pointed *tang* at one end on which a wood or metal handle can be fitted for easy grasping.

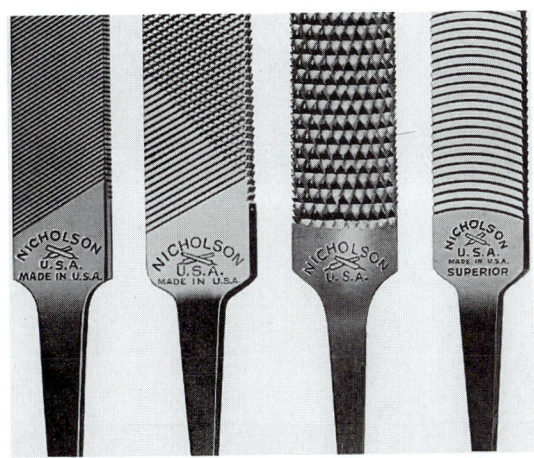

FIGURE 26-26 Four cuts of files. *Left to right:* Single, double, rasp, and curved (Vixen). *(Courtesy of Nicholson File Company.).*

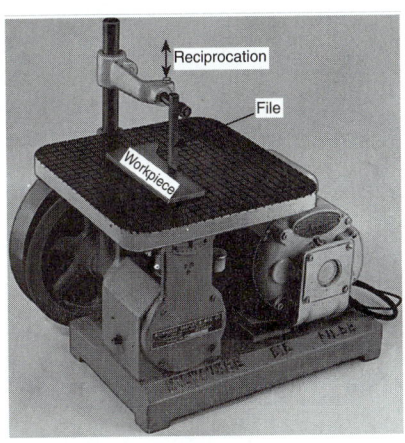

FIGURE 26-27 Die-filing machine.

■ 26.9 FILING MACHINES

Although an experienced machinist can do very accurate work by hand filing, it is a slow and tiresome task. Consequently, three types of filing machines have been developed that permit quite accurate results to be obtained rapidly and with much less effort. *Die filing machines* (Figure 26-27) hold and reciprocate a file that extends upward through the worktable. The file rides against a roller guide at its upper end, and cutting occurs on the downward stroke; therefore, the cutting force tends to hold the work against the table. The table can be tilted to any desired angle. Such machines operate at from 300 to 500 strokes per minute, and the resulting surface tends to be at a uniform angle with respect to the table. Quite accurate work can be done. Because of the reciprocating action, approximately 50% of the operating time is nonproductive.

Band-filing machines provide continuous cutting action. Most band filing is done on contour bandsawing machines by means of a special band file that is substituted for the usual bandsaw blade. The principle of a band file is shown in Figure 26-28. Rigid, straight file segments about 3 in. long are riveted to a flexible steel band near their leading ends. One end of the steel band contains a slot that can be hooked over a pin in the other end to form a continuous band. As the band passes over the drive and idler wheels of the machine, it flexes so that the ends of adjacent file segments move apart. When the band becomes straight, the ends of adjacent segments move together and interlock to form a continuous straight file. Where the file passes through the worktable, it is guided and supported by a grooved guide, which provides the necessary support to resist the pressure of the work against the file. Band files are available in most of the standard cuts and in several widths and shapes. Operating speeds range from about 50 to 250 ft/min.

Although band filing is considerably more rapid than can be done on a die-filing machine, it usually is not quite as accurate. Frequently, band filing may be followed by some finish filing on a die-filing machine.

Some *disk-filing machines* are used, having files in the form of disks (Figure 26-29). These are even simpler then die-filing machines and provide continuous cutting action. However, it is difficult to obtain accurate results by their use.

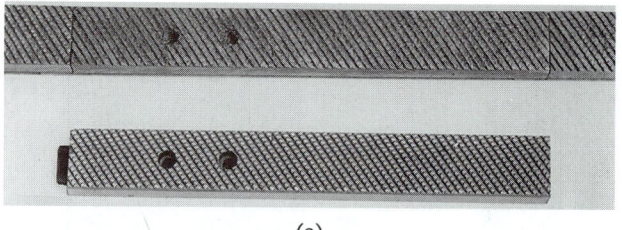

(a)

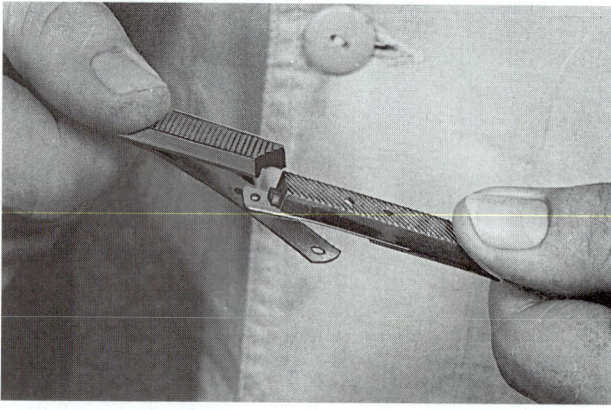

(b)

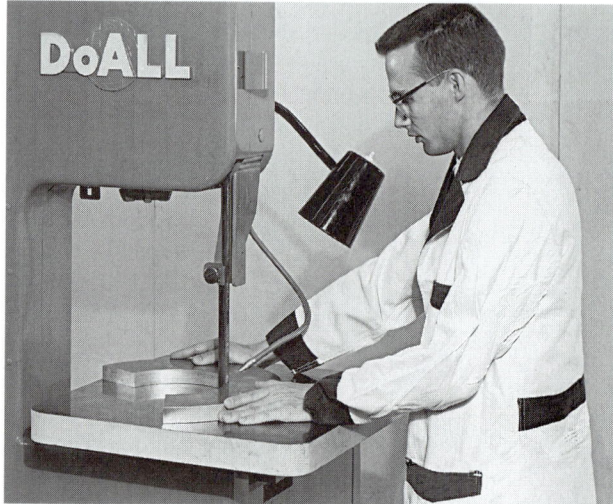

(c)

FIGURE 26-28   Band file segments (a) are joined together to form a continuous band (b) which runs on a band filing machine (c). *(Courtesy of DoALL Co.)*

■ 26.10 SHAPING AND PLANING

The process of *shaping* and *planing* are among the oldest single-point machining processes. They have largely been replaced by milling and broaching, as production processes. From a consideration of the relative motions between the tool and the workpiece, shaping and planing use a straight-line cutting motion with a single-point cutting tool to generate a flat surface. In shaping, the workpiece is fed at right angles to the cutting motion between successive strokes of the tool, as shown in Figure 26-30, where f_c is the feed

FIGURE 26-29 *(Above)* Disk-type filing machine; *(left)* some of the available types of disk files. *(Courtesy of Jersey Manufacturing Company.)*

per stroke, V is the cutting speed, and t is the DOC. (In planing, discussed next, the workpiece is reciprocated and the tool is fed at right angles to the cutting motion.) For either shaping or planing, the tool is held in a clapper box which prevents the cutting edge from being damaged on the return stroke of the tool.

In addition to plain flat surfaces, the shapes most commonly produced on the shaper and planer are those illustrated in Figure 26-31. Relatively skilled workers are required to operate shapers and planers, and most of the shapes that can be produced on them also can be made by much more productive processes, such as milling, broaching, or grinding. Consequently, except for certain special types, planers that will do only planing have become obsolete, and shapers are used mainly in tool-and-die work, in very low volume production, or in the manufacture of gear teeth (see Chapter 31).

In shaping, the cutting tool is held in the tool post located in the ram, which reciprocates over the work with a forward stroke, cutting at velocity V and a quick return stroke at velocity V_R. The rpm rate of the drive crank (N_s) drives the ram and determines the velocity of the operation (Figure 26-32). The stroke ratio

$$R_s = \frac{\text{cutting stroke angle}}{360°} = \frac{200°}{360°} = \frac{5}{9}$$

The tool is advancing 55% of the time. The number of strokes per minute is N_s, determined by the rpm rate of the drive crank. Feed, f_c, is in inches per stroke and is at right angles to the cutting direction.

The length of stroke, l, must be greater than the length of the workpiece (or length of cut), L, since velocity is position variant. Let l = twice the length of the block being cut, or $2L$. The cutting velocity, V, is assumed to be twice the average forward velocity, V, of the ram. The general relationship between cutting speed and rpm is

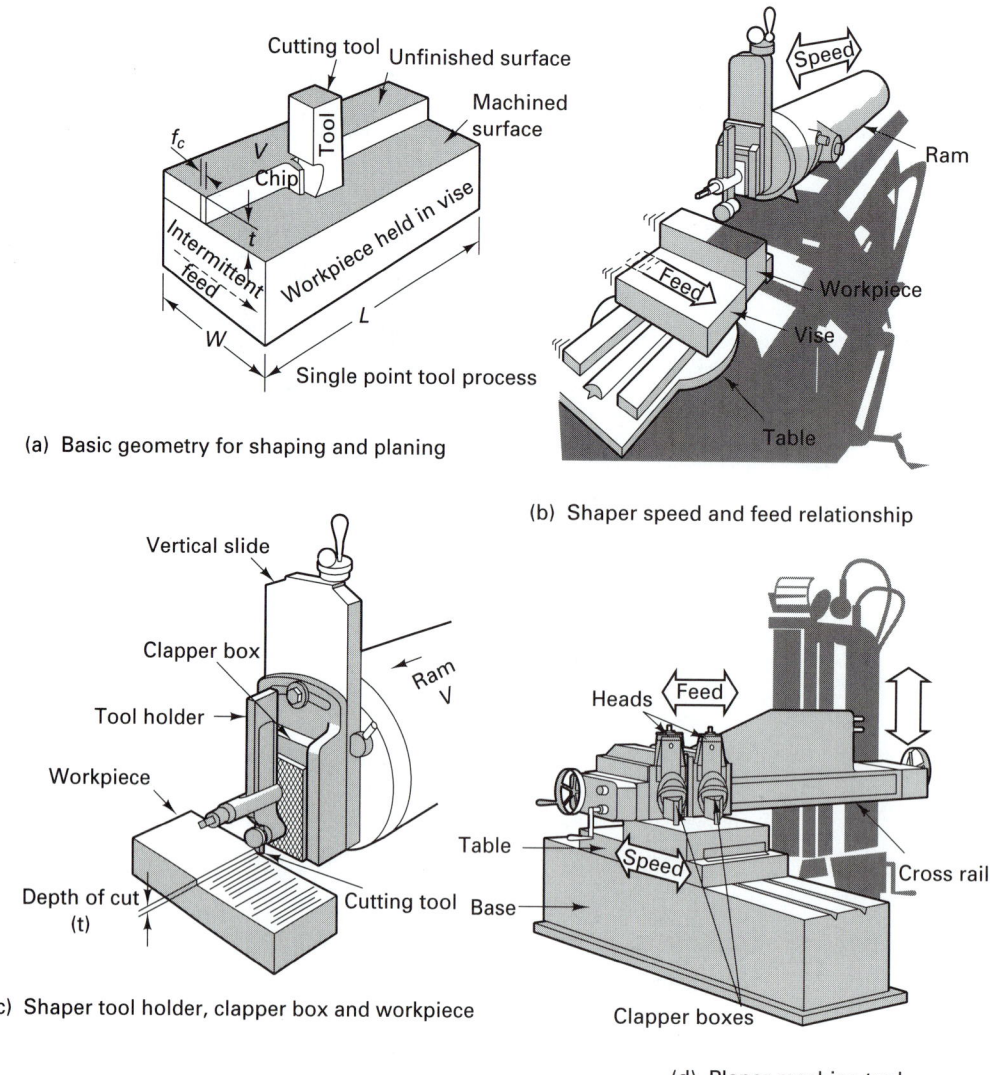

(a) Basic geometry for shaping and planing

(b) Shaper speed and feed relationship

(c) Shaper tool holder, clapper box and workpiece

(d) Planer machine tool

FIGURE 26-30   Basics of shaping and planing.

$$V = \frac{\pi D N_s}{12} \qquad \text{ft/min} \tag{26-7}$$

where D is the diameter (of rotational member) in inches. For shaping, the cutting speed

$$V = \frac{2lN_s}{12R_s} \qquad \text{ft/min} \tag{26-8}$$

Once a cutting speed (V) is selected, the rpm rate, N_s, of the machine can be calculated. Tables for suggested feed values, f_c, in inches per stroke (or cycle) and recommended depths of cut are also available. The maximum depth of cut is based on the

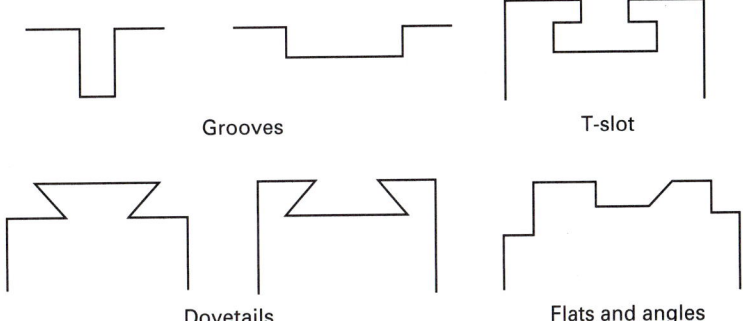

FIGURE 26-31 Types of surfaces commonly machined by shaping and planing.

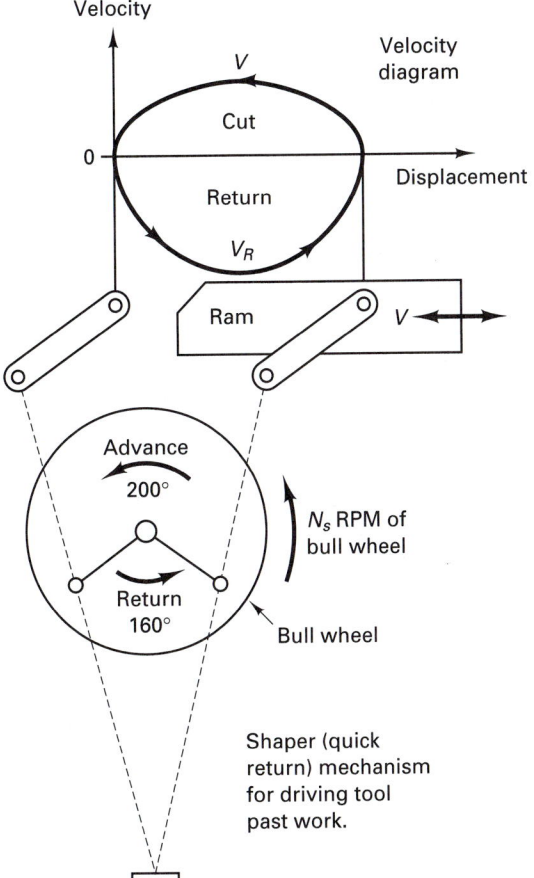

Shaper (quick return) mechanism for driving tool past work.

FIGURE 26-32 The ram of the shaper carries the cutting tool at velocity V and reciprocated at cutting velocity V_R by the rotation of a bull wheel turning at rpm N_s.

horsepower available to form the chips. This calculation requires that the metal removal rate (MRR) be known. The MRR is the volume of metal removed per unit time:

$$\text{MRR} = \frac{LWt}{CT} \qquad \text{in}^3/\text{min} \tag{26-9}$$

where W is the width of block being cut and L is the length of block being cut, so volume of cut $= WLt$. t is the depth of cut and CT is the time in minutes to cut that volume.

In general, CT is the total length of the cut divided by the feed rate. For shaping, it is the width of block divided by the feed rate of the tool across the width. Thus, for shaping,

$$CT = \frac{W}{N_s \times f_c} \qquad \text{min} \qquad \qquad (26\text{-}10)$$

Also,

$$CT = \frac{S}{N_s}$$

where the number of strokes for the job is $S = W/f_c$ for a surface of width W.

Shapers, as machine tools, usually are classified according to their general design features as follows.

1. Horizontal
   a. Push-cut
   b. Pull-cut or draw cut shaper
2. Vertical
   a. Regular or slotters
   b. Keyseaters
3. Special

They are also classified as to the type of drive employed: *mechanical drive* or *hydraulic drive*. Most shapers are of the *horizontal push-cut* type (see Figures 26-30 and 26-33), where cutting occurs as the ram *pushes* the tool across the work.

Workholding Devices for Shapers

On horizontal push-cut shapers, the work is usually held in a heavy vise mounted on the top side of the table. Shaper vises have a very heavy movable jaw, because the vise must often be turned so that the cutting forces are directed against this jaw.

In clamping the workpiece in a shaper vise, care must be exercised to make sure that it rests solidly against the bottom of the vise (on parallel bars) so that it will not be deflected by the cutting force and that it is held securely yet not distorted by the clamping pressure.

Figure 26-34 illustrates the use of parallel bars for raising work to the proper height in the vise jaws and also several methods of clamping rough and irregularly shaped work. The latter procedures prevent the clamping action from tilting the work in the vise. Work that cannot be held conveniently in a vise can be clamped directly to the top or sides of the table, using T-slots in the table.

Most shaping is done with simple high-speed-steel or carbide-tipped cutting-tool bits held in a heavy, forged toolholder, as shown in Figure 26-30c. Although shapers are versatile tools, the precision of the work done on them is greatly dependent on the operator. Feed dials on shapers nearly always are graduated in 0.001-in. divisions, and work is seldom done to greater precision than this. A tolerance of 0.002 to 0.003 in. is desirable on parts that are to be machined on a shaper, because this gives some provision for variations due to clamping, possible looseness or deflection of the table, and deflection of the tool and ram during cutting.

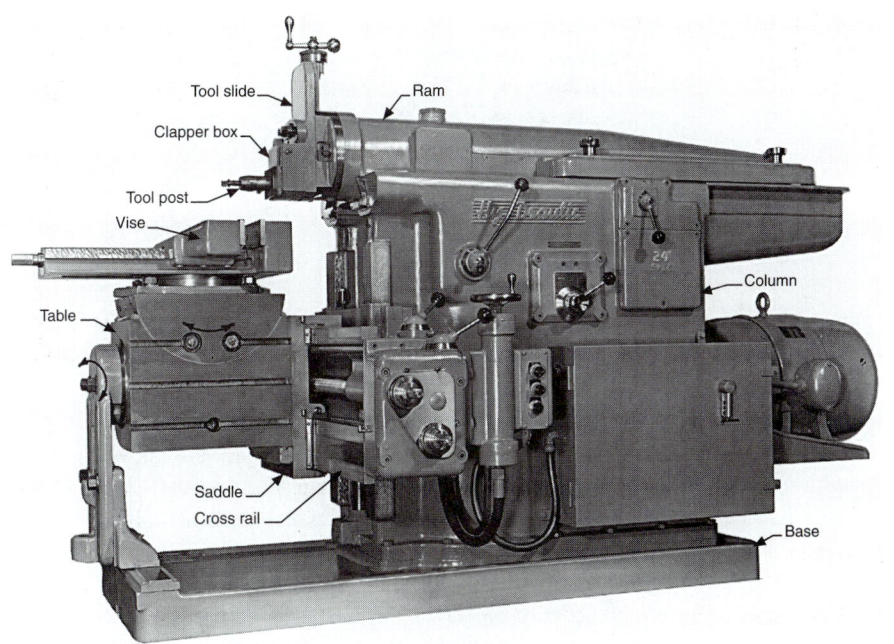

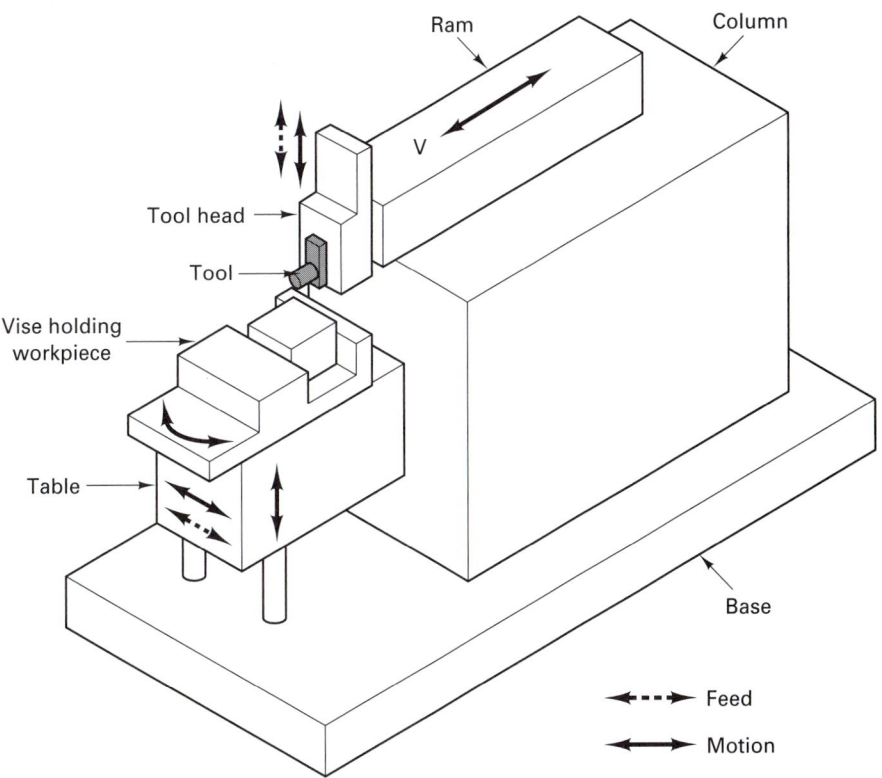

FIGURE 26-33 Details of a horizontal, push cut shaper.

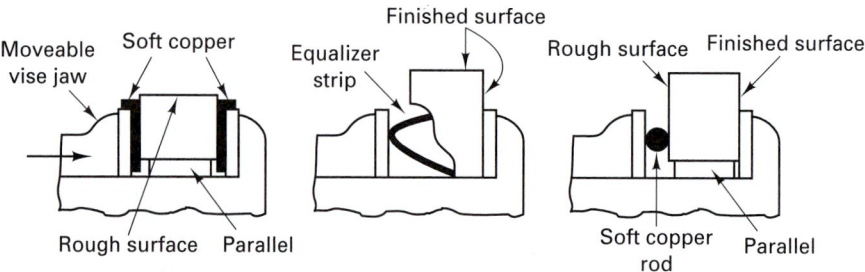

FIGURE 26-34   Methods of clamping workpieces in a shaper vise.

■ 26.11 PLANING MACHINES

Planing can be used to produce horizontal, vertical, or inclined flat surfaces on workpieces that are too large to be accommodated on shapers. However, planing is much less efficient than other basic machining processes, such as milling, that will produce such surfaces. Consequently, planing and planers have largely been replaced by planer milling machines or machines that can do both milling and planing.

Figure 26-35 shows the basic components and motions of planers. In most planing, the action is opposite to that of shaping. The work is moved past one or more stationary single-point cutting tools. Because a large and heavy workpiece and table must be reciprocated at relatively low speeds, several tool heads are provided, often with multiple tools in each head. In addition, many planers are provided with tool heads arranged so that cuts occur on both directions of the table movement. However, because only single-point cutting tools are used and the cutting speeds are quite low, planers are low in productivity compared with some other types of machine tools.

Planers are made in four basic types. Figures 26-35 depicts the most common, double-housing type. It has a closed housing structure, spanning the reciprocating worktable, with a cross rail supported at each end on a vertical column and carrying two tool heads. An additional tool head usually is mounted on each column, so that four tools (or four sets) can cut during each stroke of the table. Obviously, the closed-frame structure of this type of planer limits the size of the work that can be machined. Open-side planers have the cross rail supported on a single column. This provides unrestricted access to one side of the table to permit wider workpieces to be accommodated. Some open-side planers are convertible, in that a second column can be attached to the bed when desired so as to provide added support for the cross rail.

Workholding and Setup for Planers

Workpieces in planers are usually large and heavy. They must be securely clamped to resist large cutting forces and the high-inertia forces that result from the rapid velocity changes at the ends of the strokes. Special stops are provided at each end of the workpiece to prevent it from shifting.

Considerable time usually is required to set up the planer, thus reducing the time the machine is available for producing chips. Sometimes special setup plates are used for quick setup of the workpiece. Another procedure is to use two tables. Work is set up on one table while another workpiece is being machined on the other. The tables can be fastened together for machining long workpieces.

The large workpieces can usually support heavy cutting forces, so large depths of cut are recommended, which decreases the cutting time. Consequently, planer tools usually

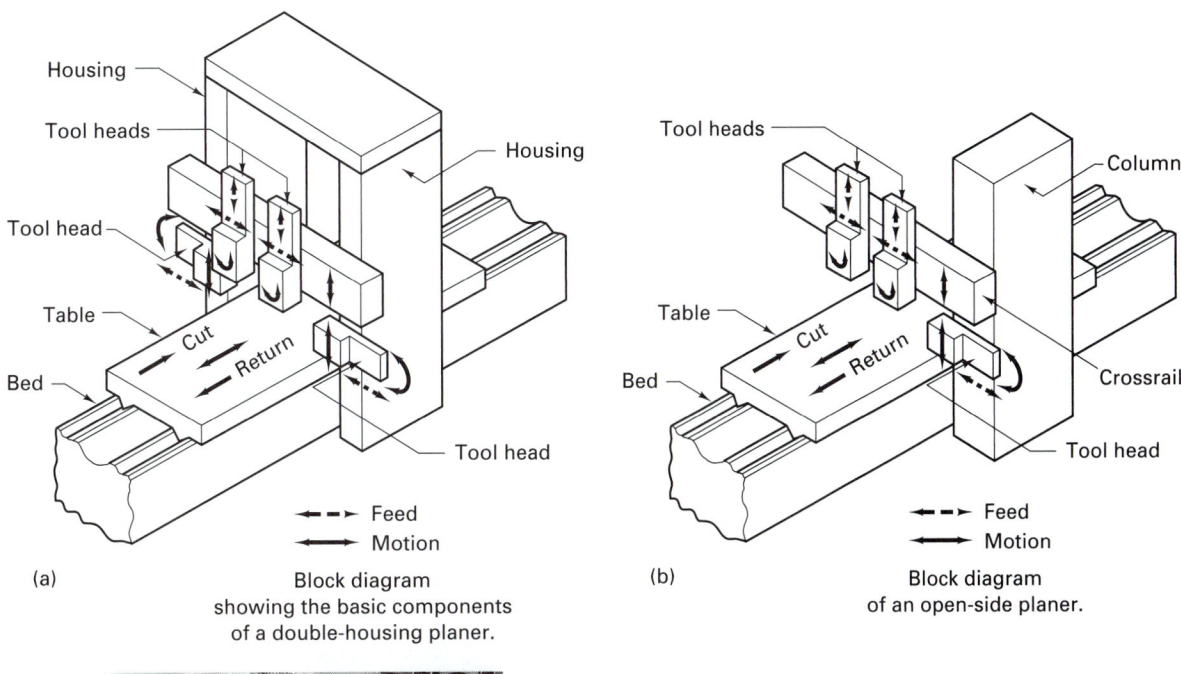

(a) Block diagram showing the basic components of a double-housing planer.

(b) Block diagram of an open-side planer.

(c)

FIGURE 26-35 Schematic of planers. (a) double-housing planer; (b) open-sided planer; (c) interchangeable multiple tool holder for use in planers. *(Photograph courtesy Gebr. Boehringer GMBH.)*

are quite massive and can sustain the large cutting forces. Usually, the main shank of the tools is made of plain-carbon steel, with tips made from high-speed steel or carbide. Chip breakers should be used to avoid long and dangerous chips in ductile materials.

Theoretically, planers have about the same precision as shapers. The feed and other dimension-controlling dials usually are graduated in 0.001-in. divisions. However, because larger and heavier workpieces are usually involved, and much longer beds and tables, the working tolerances for planer work should be somewhat greater than for shaping.

■ KEYS WORDS

band-filing machine	filing	rise per tooth
bandsaw	hacksaw	sawing
broach	kerf	shaping
broaching	planing	surface broach
burnishing teeth	pull broach	tooth set
circular saw	push broach	
cold saw	reciprocating saw	

■ REVIEW QUESTIONS

1. What is unique about broaching compared with the other basic machining processes?
2. Can a thick saw blade be used as a broach? Why or why not?
3. Broaching machines are simpler in a basic design than most other machine tools. Why is this?
4. Why is broaching particularly well suited for mass production?
5. In designing a broach, what would be the first thing you have to calculate?
6. Why is it necessary to relate the design of a broach to the specific workpiece that is to be machined?
7. What two methods can be utilized to reduce the force and power requirements for a particular broaching cut?
8. For a given job, how would a broach having rotor-tooth design compare in length with one having regular full-width teeth?
9. Why are the pitch and radius of the gullet between teeth on a broach of importance?
10. Why are broaching speeds usually relatively low compared with other machining operations?
11. What are the advantages of shell-type broach construction?
12. Why are most broaches made from alloy or high-speed steel rather than from tungsten carbide?
13. What are the advantages of TiN-coated broaching tools?
14. For mass-production operations, which process is preferred, pull-up broaching or pull-down broaching?
15. Can continuous broaching machines be used for broaching holes? Why or why not?
16. The sides of a square blind hole must be machined all the way to the bottom. The hole is drilled to full depth and the bottom end milled flat. Is it possible to machine the hole square by broaching? Why or why not?
17. The interior, flat surfaces of socket wrenches, which have one "closed" end, often are finished to size by broaching. By examining one of these, determine what design modification was incorporated to make this operation possible.
18. Why is sawing one of the most efficient of the chip-forming processes?
19. Explain why tooth spacing is important in sawing.
20. What is the tooth gullet for a saw blade?
21. Explain what is meant by the *set* of the teeth on a saw blade.
22. Why can a bandsaw blade not be hardened throughout the entire width of the band?
23. What is the general relationship between the width of a bandsaw blade and the radius of a cut that can be made with it?
24. What are the advantages and disadvantages of circular saws?
25. Why have bandsawing machines largely replaced those using reciprocating saws?
26. Explain how the hole in Figure 26-24 is made on a contour bandsawing machine.
27. How does friction sawing differ from ordinary bandsawing?
28. What is the disadvantage of using gravity to feed a saw in cutting round bar stock?
29. To what extent is filing different from sawing?
30. What is a safe edge on a file?
31. Why is an end-filing machine more efficient than a die-filing machine?
32. How does a rasp-cut file differ from other types of files?
33. How does the process of shaping differ from planing?

34. How is feed per stroke in shaping related to feed per tooth in milling?

35. What are some ways to improve the efficiency of a planer? Do any of these apply to the shaper?

■ PROBLEMS

1. A surface 12 in. long is to be machined with a flat, solid broach that has a rise per tooth of 0.0047 in. What is the minimum cross-sectional area that must be provided in the chip gullet between adjacent teeth?

2. The pitch of the teeth on a simple surface broach can be determined by equation (26-1). If a broach is to remove 0.25 in. of material from a gray iron casting that is 3 in. wide and 17.75 in. long, and if each tooth has a rise per tooth of 0.004 in., what will be the length of the roughing section of the broach?

3. Estimate the (approximate maximum) horsepower needed to accomplish the operation described in Problem 2 at a cutting speed of 10 m/min. (*Hint:* First find the HP used per tooth and determine the maximum number of teeth engaged at any time. What are those units?)

4. Estimate the approximate force acting in the forward direction during cutting for the conditions stated in Problems 2 and 3.

5. In making a 6-in. cut in a piece of AISI 1020 CR steel that is 1 in. thick, the material is fed to a bandsaw blade with teeth having a pitch of 1.27 mm (20 pitch) at the rate of 0.0001 in. per tooth. Estimate the cutting time for the cut.

6. The strength of a pull broach is determined by its minimum cross section, which usually occurs either at the root of the first tooth or at the pull end (see Figure 26-2). Suppose that the minimum root diameter is D_r, the pull end diameter is D_p, and the width of the pull slot is W. Write an equation for the allowable pull, in pounds, using 200,000 psi as the yield strength for the broach material.

7. Suppose that you want to shape a block of metal 7 in. wide and 4 in. long, using a shaper as set up in Figure 26-30. You have determined for this metal that the cutting speed should be 25 sfpm, the depth of cut needed here for roughing is 0.25 in., and the feed will be 0.1 in. per stroke. Determine the approximate crank rpm and then estimate the cutting time and the MRR.

8. Could you have saved any time in Problem 7 by cutting the block in the 7-in. direction? Redo with $L = 7$ and $W = 4$ in.

9. Derive equation (26-8), the shaping cutting speed.

10. How many strokes per minute would be required to obtain a cutting speed of 36.6 m (120 ft) per minute on a typical mechanical-drive shaper if a 254-mm (10-in.) stroke is used?

11. How much time would be required to shape a flat surface 254 mm (10 in.) wide and 203 mm (8 in.) long on a hydraulic drive shaper using a cutting speed of 45.7 m (150 ft) per minute, a feed of 0.51 mm (0.020 in.) per stroke, and an overrun of 12.7 mm ($\frac{1}{2}$ in.) at each end of the cut?

12. What is the metal-removal rate in Problem 11 if the depth of cut is 6.35 mm ($\frac{1}{4}$ in.)?

13. If the work material in Problem 11 is No. 6 gray cast iron, what will be the power required?

14. A hydraulic-drive shaper has a 5.6-kW ($7\frac{1}{2}$-hp) motor, and 75% of the motor output is available at the cutting tool. The specific power for cutting a certain metal is 0.03 W/mm³ (0.67 hp/in³) per minute. What is the maximum depth of cut that can be taken in shaping a surface in this material if the surface is 305 × 305 mm (12 × 12 in.), the feed is 0.64 mm (0.025 in.) per stroke, and the cutting speed is 54.9 mm (180 ft) per minute?

Chapter 26 CASE STUDY

socket with a triangular hole

The Cochran Company estimated that its annual requirements for the socket component shown in Figure CS-26 will be at least 50,000 units and that this volume will continue for at least 5 years. Consequently, it wants to consider all practical methods for making the component and has assigned you the task of determining these methods and of recommending which should be explored in detail to determine the most effective and economical process. The specifications call for a lightweight metal, such as an aluminum alloy, that must have a tensile strength of at least 138 MPa and an elongation of at least 3%.

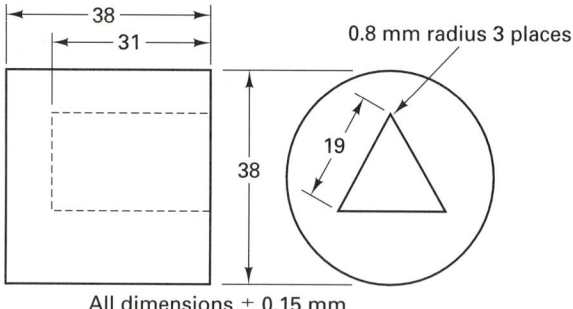

All dimensions ± 0.15 mm

FIGURE CS-26

Determine at least five practicable methods for producing the part. Suggest which two appear most likely to be economical and thus should be investigated more fully. (Give reasons for your selections.)

CHAPTER 27

ABRASIVE MACHINING PROCESSES

27.1 INTRODUCTION
27.2 ABRASIVES
    Abrasive Grain Size and Geometry
27.3 GRINDING
    Grinding Wheel Structure and Grade
    G Ratio
    Bonding Materials for Grinding Wheels
    Abrasive Machining
    Snagging
    Low-Stress Grinding
    Dressing and Truing
    Grinding Wheel Identification
    Grinding Wheel Geometry
    Balancing Grinding Wheels
    Safety in Grinding
    Use of Cutting Fluids in Grinding

27.4 GRINDING MACHINES
    Cylindrical Grinding
    Centerless Grinding
    Tool-Post Grinders
    Surface Grinding Machines
    Creep Feed Grinding Machines
    Disk Grinding Machines
    Tool and Cutter Grinders
    Mounted Wheels and Points
    Coated Abrasives
27.5 DESIGN CONSIDERATIONS IN GRINDING
27.6 HONING
    Honing Stones
27.7 SUPERFINISHING
27.8 LAPPING
Case Study: ALUMINUM RETAINER RINGS

■ 27.1 INTRODUCTION

Abrasive machining is a material-removal process that involves the interaction of abrasive grits with the workpiece at high speeds and shallow penetration depths. The chips that are formed resemble those formed by other machining processes. Unquestionably, abrasive machining is the oldest of the basic machining processes. Museums abound with examples of utensils, tools, and weapons that ancient peoples produced by rubbing hard stones against softer materials to abrade away unwanted portions, leaving desired shapes. For centuries, only natural abrasives were available, and other, more modern, basic machining, processes, were developed using superior cutting materials. However, the development of manufactured abrasives and a better fundamental understanding of the abrasive machining process has resulted in placing abrasive machining and its variations among the most important of all basic machining processes.

    The results that can be obtained by abrasive machining range from the finest and smoothest surfaces produced by any machining process, in which very little material is removed, to rough, coarse surfaces that accompany high material-removal rates. The abrasive particles may be (1) free (see Chapter 32); (2) mounted in resin on a belt (called *coated product*); or (3) close packed into wheels or stones, with abrasives held together by bonding material (called *bonded product* or a grinding wheel). See Figure 27-1,

783

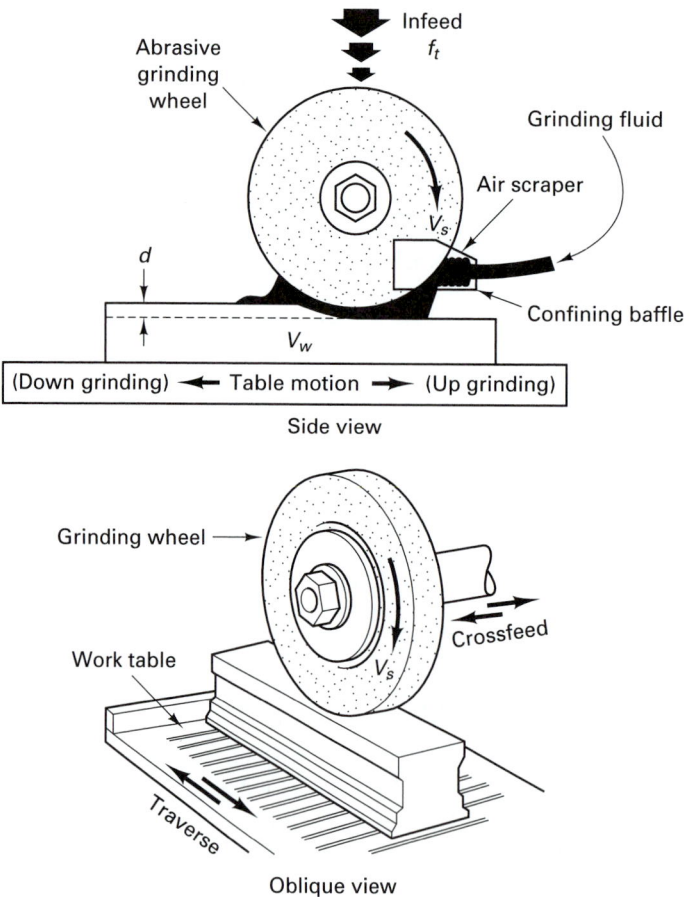

FIGURE 27-1 Schematic of a surface grinding, showing infeed and cross-feed motions along with cutting speeds, V_S, and workpiece velocity, V_W.

which shows a surface grinding process using a grinding wheel. The depth of cut, d, is determined by the infeed and is usually very small, 0.002 to 0.005 in., so the arc of contact (and the chips) are small. The table reciprocates back and forth beneath the rotating wheel. The work feeds into the wheel (see crossfeed direction) after the work is clear of the wheel.

The metal-removal process is basically the same in all abrasive machining processes but with important differences due to spacing of active grains (grains in contact with work) and the rigidity and degree of fixation of the grains. Table 27-1 summarizes the primary abrasive processes.

Abrasive machining processes have two unique characteristics. First, each cutting edge is very small, and many of these edges can cut simultaneously. When suitable machines are employed, very fine cuts are possible, and fine surfaces and close dimensional control can be obtained. Second, because extremely hard abrasive grits can be produced, very hard materials, such as hardened steel, glass, carbides, and ceramics, can readily be machined. As a result, the abrasive machining processes are not only important as manufacturing processes, they are indeed essential. Many of our modern products, such as

TABLE 27-1.	Abrasive Machining Processes	
Process	Particle Mounting	Features
Grinding	Bonded	Uses wheels, accurate sizing, finishing, low MRR; can be done at high speeds (over 12,000 sfpm)
Creep feed grinding	Bonded open, soft	Uses wheels with long cutting arc, very slow feed rate and large depth of cut
Abrasive machining[a]	Bonded	High MRR, to obtain desired shapes and approximate sizes
Snagging	Bonded belted	High MRR, rough rapid technique to clean up and deburr castings, forgings
Honing	Bonded	"Stones" containing fine abrasives; primarily a hole-finishing process
Lapping	Free	Fine particles embedded in soft metal or cloth; primarily a surface-finishing process
Abrasive waterjet	Free in jet	Waterjets with velocities up to 3000 ft/sec carry abrasive particles (silica and garnet; see Chapter 32)

[a]The term *abrasive machining* applied to one particular form of the grinding process is unfortunate, because all these processes are machining with abrasives.

modern machine tools, automobiles, space vehicles, and aircraft, could not be manufactured without these processes.

■ 27.2 ABRASIVES

An *abrasive* is a hard material that can cut or abrade other substances. Natural abrasives have existed from the earliest times. For example, sand stone was used by ancient peoples to sharpen tools and weapons. Early grinding wheels were cut from slabs of sandstone, but because they were not uniform in structure throughout, they wore unevenly and did not produce consistent results. *Emery*, a mixture of alumina (Al_2O_3) and magnetite (Fe_3O_4), is a natural abrasive used on coated paper and cloth. *Corundum* (natural Al_2O_3) and diamonds are other naturally occurring abrasive materials. As outlined in Chapter 22, these cutting materials have certain important properties. Today, the only natural abrasives that have commercial importance are quartz, sand, garnets, and diamonds. *Quartz* sand is used primarily in coated abrasives and in air blasting (discussed in Chapter 40), but artificial abrasives are also making inroads in these applications. The development of artificial abrasives having known uniform properties has permitted abrasive processes to become a precision manufacturing process.

Hardness, the ability to resist penetration, is the key property for an abrasive. Table 27-2 lists the primary abrasives and their approximate Knoop hardness (kg/mm^2). The particles must be able to decompose at elevated temperatures. Two other properties are significant in abrasive grits: attrition and friability. *Attrition* refers to the abrasive wear action of the grits, resulting in dulled edges, grit flattening, and wheel glazing. *Friability* refers to the fracture of the grits and is the opposite of toughness. In grinding, it is important that grits be able to fracture to expose new, sharp edges.

Diamonds are the hardest of all materials. Those that are used for abrasives are either natural, off-color stones (called *garnets*) that are not suitable for gems, or small, synthetic stones that are produced specifically for abrasive purposes. Manufactured stones appear to be somewhat more friable and thus tend to cut faster and cooler. They do not perform as satisfactorily in metal-bonded wheels. Diamond abrasive wheels are used extensively for sharpening carbide and ceramic cutting tools. Diamonds also are used for truing and dressing other types of abrasive wheels. Diamonds are usually used only when cheaper abrasives

TABLE 27-2.	Knoop Hardness Values of Common Abrasives			
Abrasive Material	Year of Discovery	Hardness (Knoop)	Temperature of Decomposition in Oxygen (°C)	Comments and Uses
Quartz (SiO$_2$)	?	320		Sand blasting
Aluminum oxide	1893	1600–2100	1700–2400	Softer and tougher than silicon carbide; used on steel, iron, brass silicon
Carbide	1891	2200–2800	1500–2000	Used for brass, bronze, aluminum, and stainless and cast iron
Borazon [cubic boron nitride (CBN)]	1957	4200–4700	1200–1400	For grinding hard, tough tool steels, stainless steel, cobalt and nickel based superalloys, and hard coatings
Diamond (synthetic)	1955	6000–9000	700–800	Used to grind nonferrous materials, tungsten carbide and ceramics

FIGURE 27-2   Loose abrasive grains at high magnification, showing their irregular, sharp cutting edges. *(Courtesy of Norton Company.)*

will not produce the desired results. Garnets are used primarily in the form of very finely crushed and graded powders for fine polishing.

Artificial abrasives date from 1891, when E. G. Acheson, while attempting to produce precious gems, discovered how to make *silicon carbide* (SiC). Silicon carbide is made by charging an electric furnace with silica sand, petroleum coke, salt, and sawdust. By passing large amounts of current through the charge, a temperature of over 4000°F is maintained for several hours, and a solid mass of silicon carbide crystals results. After the furnace has cooled, the mass of crystals is removed, crushed, and graded (sorted) into various desired sizes. As can be seen in Figure 27-2, the resulting grits, or grains, are irregular in shape, with cutting edges having every possible rake angle. Silicon carbide crystals are very hard, friable, and rather brittle. This limits their use. Silicon carbide is sold under the trade names Carborundum and Crystolon.

Aluminum oxide (Al$_2$O$_3$), is the most widely used artificial abrasive. Also produced in an arc furnace, from bauxite, iron filings, and small amounts of coke, it contains aluminum hydroxide, ferric oxide, silica, and other impurities. The mass of aluminum oxide that is formed is crushed, and the particles are graded to size. Common trade names for aluminum oxide abrasives are Alundum and Aloxite. Although aluminum oxide is softer than silicon carbide, it is considerably tougher. Consequently, it is a better general-purpose abrasive.

Cubic boron nitride (CBN) is not found in nature. It is produced by a combination of intensive heat and pressure in the presence of a catalyst. CBN is extremely hard, registering at 4700 on the Knoop scale. It is the second hardest substance created by nature or manufactured and is often referred to, along with diamonds, as a superabrasive. Hardness, however, is not everything.

CBN far surpasses diamond in the important characteristic of thermal resistance. At temperatures of 650°C, at which diamond may begin to revert to plain carbon dioxide, CBN continues to maintain its hardness and chemical integrity. When the temperature of 1400°C is reached, CBN changes from its cubic form to a hexagonal form and loses hardness. CBN can be used successfully in grinding iron, steel, alloys of iron, Ni-based alloys, and other materials. CBN works very effectively (long wheel life, high G-ratio, good surface quality, no burn or chatter, low scrap rate, and overall increase in parts/shift) on hardened materials (R_C50 or higher). It can also be used for soft steel under selected situations. CBN does well at conventional grinding speeds (6000 to 12,000 ft/min), resulting in lower total grinding cost/piece in conventional equipment. CBN can also perform well a high grinding speeds (12,000 ft/min and higher) and will enhance the benefits from future machine tools. CBN can solve difficult-to-grind jobs, but it also generates cost benefits in many production grinding operations despite its higher cost. CBN is manufactured by the General Electric Company under the trade name of Borazon.

Abrasive Grain Size and Geometry

To enhance the process capability of grinding, abrasive grains are sorted into sizes by mechanical sieving machines. The number of openings per linear inch in a sieve (or screen) through which most of the particles of a particular size can pass determines the grain size (Figure 27-3).

A No. 24 grit would pass through a standard screen having 24 openings per inch but would not pass through one having 30 openings per inch. These numbers have since been specified in terms of millimeters and micrometers (see ANSI B74.12 for details). Commercial practice commonly designates grain sizes from 4 to 24, inclusive, as *coarse*; 30 to 60, inclusive, as *medium*; and 70 to 600, inclusive, as *fine*. Silicon carbide is obtainable in grit sizes ranging from 2 to 240 and aluminum oxide in sizes from 4 to 240. Superabrasive grit sizes normally range from 120 grit for CBN to 400 grit for diamond. Sizes from 240 to 600 are designated as *flour* sizes. These are used primarily for lapping or in fine honing stones.

FIGURE 27-3   Typical screens for sifting abrasives into sizes. The larger the screen number (of openings per linear inch), the smaller the grain size. *(Courtesy of Carborundum Company).*

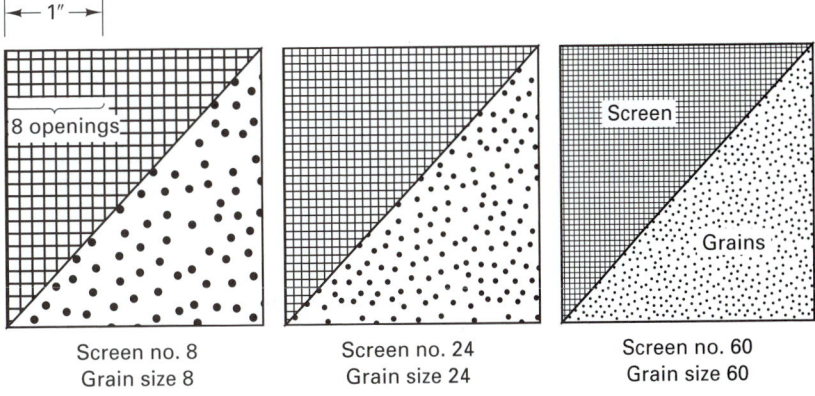

| Screen no. 8 | Screen no. 24 | Screen no. 60 |
| Grain size 8 | Grain size 24 | Grain size 60 |

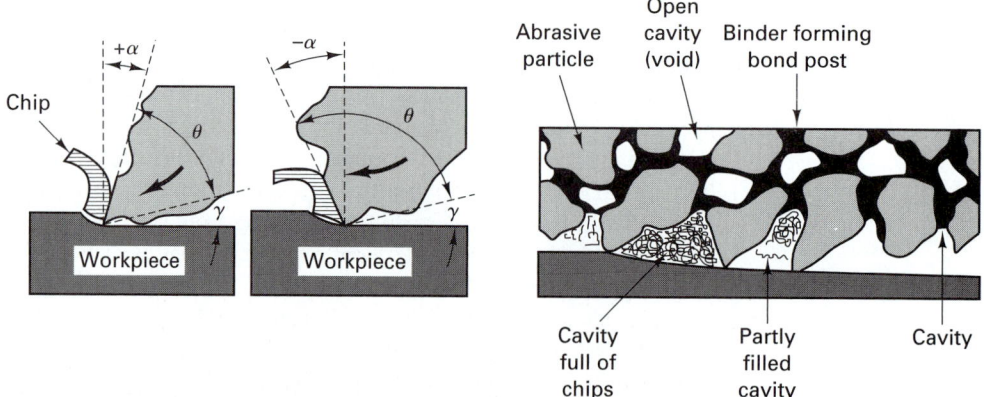

FIGURE 27-4   The rake angle of abrasive particles can be positive, zero, or negative. The cavity or voids between the grains must be large enough to hold all the chips during the cut.

The grain diameter can be estimated from the screen number (S), which corresponds to the number of openings per inch. The mean diameter of the grain (g) is related to the screen number by $g \simeq 0.7/S$.

Regardless of the size of the grain, only a small percentage (2 to 5%) of the surface of the grain is operative at any one time. That is, the depth of cut for an individual grain (the actual feed per grit) with respect to the grain diameter is very small. Thus the chips are small. As the grain diameter decreases, the number of active grains per unit area increases. The cuts become finer because grain size is the controlling factor for surface finish (roughness). Of course, the MRR also decreases.

The grain shape is also important, because it determines the tool geometry—that is, the back rake angle and the clearance angle at the cutting edge of the grit (Figure 27-4). In the figure, γ is the clearance angle, θ is the wedge angle, and α is the rake angle. The cavities between the grits provide space for the chips. The volume of the cavities must be greater than the volume of the chips generated during the cut.

Obviously, there is no specific rake angle but rather a distribution of angles. Thus a grinding wheel can present to the surface rake angles ranging from $+45°$ to $-60°$ or greater. Grits with large negative rake angles or rounded cutting edges do not form chips but will rub or *plow* a groove in the surface (Figures 27-5 and 27-6). Thus abrasive machining is a mixture of *cutting*, *plowing*, and *rubbing* with the percentage of each being highly dependent on the geometry of the grit. As the grits are continuously abraded, fractured, or dislodged from the bond, new grits are exposed and the mixture of cutting, plowing, and rubbing is changing continuously. A high percentage of the energy used for rubbing and plowing goes into the workpiece. In cutting, 95 to 98% of the energy (the heat) goes into the chip. Figure 27-6 shows a scanning electron microscope (SEM) micrograph of a ground surface with a plowing track.

■ 27.3 GRINDING

Grinding, wherein the abrasives are bonded together into a wheel, is the most common abrasive machining process. The performance of grinding wheels is greatly affected by the bonding material and the spatial arrangement of the grits.

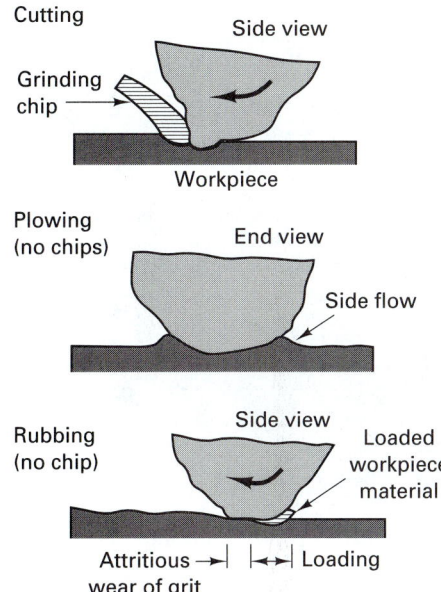

Cutting

Side view

Grinding chip →

Workpiece

Plowing (no chips)

End view

Side flow

Rubbing (no chip)

Side view

Loaded workpiece material

Attritious → ‖ ‖←→‖ Loading
wear of grit

FIGURE 27-5   The grits interact with the surface three ways: cutting, plowing, and rubbing.

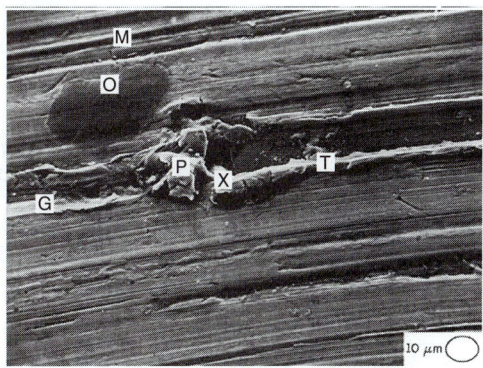

FIGURE 27-6   SEM micrograph of a ground steel surface showing a plowed track (T) in the middle and a machined track (M) above. The grit fractured, leaving a portion of the grit in the surface (X), a prow formation, (P), and a groove (G) where the fractured portion was pushed farther across the surface. The area marked (O) is an oil deposit.

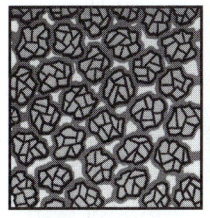
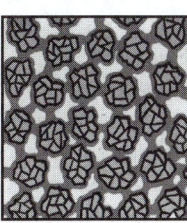

Dense spacing   Medium spacing   Open spacing

FIGURE 27-7   Meaning of grinding wheel "structure". *(Courtesy of Carborundum Company.)*

Grinding Wheel Structure and Grade

The spacing of the abrasive particles with respect to each other is called *structure*. Close-packed grains have dense structure; open structure means widely spaced grains. Open-structure wheels have larger chip cavities but fewer cutting edges per unit area (Figure 27-7).

In grinding, the chips are small but are formed by the same basic mechanism of compression and shear as discussed in Chapter 21 for regular metal cutting. Figure 27-8 shows steel chips from a grinding process, at high magnification. They show the same structure as chips from other machining processes. Chips from the grinding process often have sufficient heat energy to burn or melt in the atmosphere. Burning chips are the sparks observed during grinding with no cutting fluid. The feeds and depths of cut in grinding are small while the cutting speeds are high, resulting in high specific horsepower numbers. Because cutting is obviously more efficient than plowing or rubbing, grain fracture and

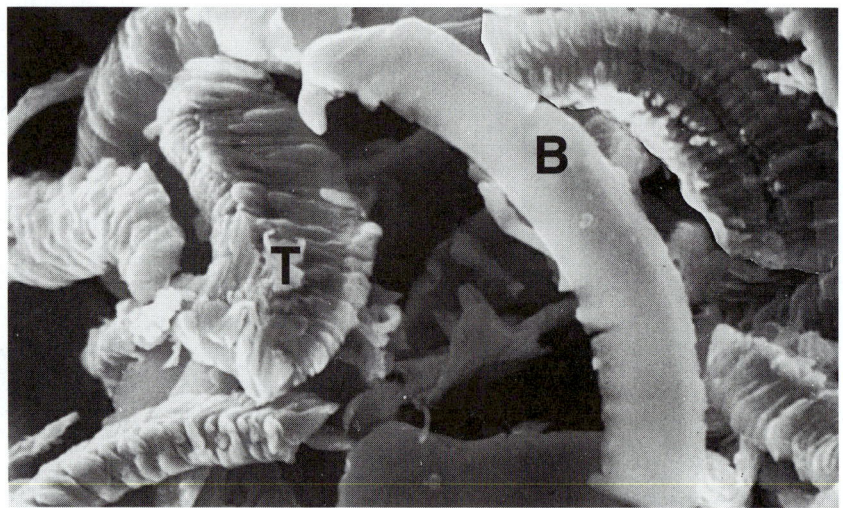

FIGURE 27-8   SEM micrograph of stainless steel chips from a grinding process. The tops (T) of the chips have the typical shear-front-lamella structure while the bottoms (B) are smooth; 4800×.

FIGURE 27-9   The grade of a grinding wheel depends on the amount of bonding agent on the holding of abrasive grains in the wheel. *(Courtesy of Carborundum Company.)*

Weak "Posts"   Medium Strength"Posts"   Strength"Posts"

grain pullout are natural phenomena used to keep the grains sharp. As the grains become dull, cutting forces increase and there is an increased tendency for the grains to fracture or break free from the bonding material. The latter action can be controlled by varying the strength of the bond, known as the *grade*. Grade thus is a measure of how strongly the grains are held in the wheel. It really is dependent on two factors: the strength of the bonding materials and the amount of the bonding agent connecting the grains. The latter factor is illustrated in Figure 27-9. Abrasive wheels are usually porous. The grains are held together with "posts" of bonding material. If these posts are large in cross section, the force required to break a grain free from the wheel is greater than when the posts are small. If a high dislodging force is required, the bond is said to be *hard*. If only a small force is required, the bond is said to be *soft*. Wheels commonly are referred to as hard or soft, referring to the net strength of the bond, resulting from both the strength of the bonding material and its disposition between the grains.

G Ratio

The loss of grains from the wheel means that the wheel is changing size. The grinding ratio or *G ratio* is defined as the cubic inches of stock removed divided by the cubic inches of wheel lost. In conventional grinding, the *G* ratio is in the range 20:1 to 80:1. The *G* ratio is a measure of grinding production and reflects the amount of work a wheel can do during its useful life. As the wheel losses material, it must be reset or repositioned to maintain workpiece size.

A typical vitrified grinding wheel will consist of 50 vol % abrasive particles, 10 vol % bond, and 40 vol % cavities; that is, the wheels have porosity. The manner in which the wheel performs is influenced by the following factors.

1. The mean force required to dislodge a grain from the surface (the grade of the wheel)
2. The cavity size and distribution or the porosity (the structure)
3. The mean spacing of active grains in the wheel surface (grain size and structure)
4. The properties of the grain (hardness, attrition, and friability)
5. The geometry of the cutting edges of the grains (rake angles and cutting edge radius compared to depth of cut)
6. The process parameters (speeds, feeds, cutting fluids) and type of grinding (surface, cylindrical)

It is easy to see why grinding is a complex process, difficult to control.

Bonding Materials for Grinding Wheels

Bonding material is a very important factor to be considered in selecting a grinding wheel. It determines the strength of the wheel, thus establishing the maximum operating speed. It determines the elastic behavior or deflection of the grits in the wheel during grinding. The wheel can be hard or rigid, or it can be flexible. Finally, the bond determines the force required to dislodge an abrasive particle from the wheel and thus plays a major role in the cutting action. Bond materials are formulated so that the ratio of bond wear matches the rate of wear of the abrasive grits. Bonding materials in common use are:

1. *Vitrified bonds.* They are composed of clays and other ceramic substances. The abrasive particles are mixed with the wet clays so that each grain is coated. Wheels are formed from the mix, usually by pressing, and then dried. They are then fired in a kiln, which results in the bonding material's becoming hard and strong, having properties similar to glass. Vitrified wheels are porous, strong, rigid, and unaffected by oils, water, or temperature over the ranges usually encountered in metal cutting. The operating speed range in most cases is 5500 to 6500 ft/min, but some wheels now operate at surface speeds up to 16,000 ft/min.
2. *Resinoid*, or phenolic resins. Because plastics can be compounded to have a wide range of properties, such wheels can be obtained to cover a variety of work conditions. They have, to a considerable extent, replaced shellac and rubber wheels. Composite materials are being used in rubber-bonded or resinoid-bonded wheels that are to have some degree of flexibility or are to receive considerable abuse and side loading. Various natural and synthetic fabrics and fibers, glass fibers, and nonferrous wire mesh are used for this purpose.
3. *Silicate* wheels use silicate of soda (waterglass) as the bond material. The wheels are formed and then baked at about 500°F for a day or more. Because they are more brittle and not so strong as vitrified wheels, the abrasive grains are released more readily. Consequently, they machine at lower surface temperatures than vitrified wheels and are useful in grinding tools when heat must be kept to a minimum.
4. *Shellac-bonded* wheels are made by mixing the abrasive grains with shellac in a heated mixture, pressing or rolling into the desired shapes, and baking for several

hours at about 300°F. This type of bond is used primarily for strong, thin wheels having some elasticity. They tend to produce a high polish and thus have been used in grinding such parts as camshafts and mill rolls.

5. *Rubber* bonding is used to produce wheels that can operate at high speeds but must have a considerable degree of flexibility so as to resist side thrust. Rubber, sulfur, and other vulcanizing agents are mixed with the abrasive grains. The mixture then is rolled out into sheets of the desired thickness, and the wheels are cut from these sheets and vulcanized. Rubber-bonded wheels can be operated at speeds up to 16,000 ft/min. They commonly are used for snagging work in foundries and for thin cutoff wheels.

6. Superabrasive wheels are either electroplated (single layer of superabrasive plated to OD of a steel blank) or impregnated. The later type of wheels can use resin, metal, or vitrified bonding. Selection of bond grade and structure (also called abrasive concentration) is critical. Table 27-3 details some aspects of superabrasive wheels.

TABLE 27-3.   Types of Superabrasive Wheels

Type of Wheel	Bonding System	Method of Manufacturing
Electroplated	Nickel plating	Steel blank accurately ground or turned on periphery. Single layer of CBN or diamond plated to OD of steel blank.
Impregnated		Superabrasives mixed with bonding media and molded (or
Low porosity	Resin or metal	performed and sintered) into a ring. Ring is mounted on
Medium porosity	Metal or vitrified	split steel body. Porosity varied (to alter structure) by
High porosity	Vitrified	varying preform pressure or using "pore forming" additives to the bond material which are vaporized during the sintering cycle.

Abrasive Machining

The condition wherein very rapid metal removal can be achieved by grinding is the one to which some have applied the term *abrasive machining*. The metal-removal rates are compared with, or exceed, those obtainable by milling or turning or broaching, and the size tolerances are comparable. It obviously is just a special type of grinding, using abrasive grains as cutting tools, as do all other types of abrasive machining. Abrasive grinding will produce sufficient localized plastic deformation and heat in the surface so as to develop tensile residual stresses, layers of overtempered martensite (in steels), and even microcracks, because this process is quite abusive (Figure 27-10).

Snagging

Snagging is a type of rough manual grinding that is done to remove fins, gates, risers, and rough spots from castings or flash from forgings, preparatory to further machining. The primary objective is to remove substantial amounts of metal rapidly without much regard for accuracy so this is a form of abrasive machining except that pedestal-type or *swing grinders* ordinarily are used. Portable electric or hand air grinders also are used for this purpose and for miscellaneous grinding in connection with welding.

Low-Stress Grinding

Conventional grinding should be replaced by procedures that develop lower surface stresses in those applications where service failures due to fatigue or stress corrosion are possible

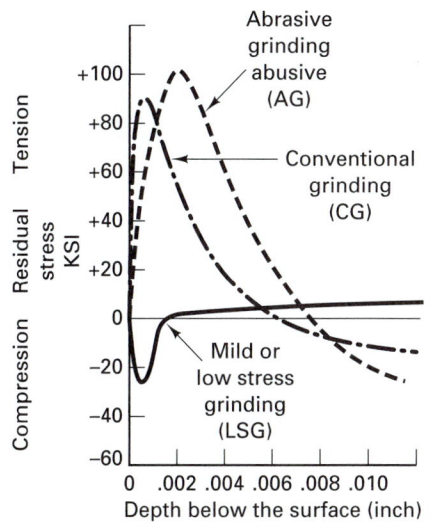

Grinding conditions	Abusive AG	Conventional CG	Low stress LSG
Wheel	A46MV	A46KV	A46HV or A60IV
Wheel speed ft/min.	6,000 to 18,000	4,500 to 6,500	2500-3000
Down feed in/pass	.002 to .004	.001 to .003	.0002 to .005
Cross feed in/pass	.040 .060	.040 .060	.040-.060
Table speed ft/min	40 to 100	40 to 100	40 to 100
Fluid	Dry	Sol oil (1 : 20)	Sulfurized oil

FIGURE 27-10 Typical residual stress distributions produced by surface grinding with different grinding conditions for abrasive, conventional, and low stress grinding. Material 4340 steel. (*From M. Field and W. P. Koster, "Surface integrity in grinding," in* New Developments in Grinding, *Carnegie-Mellon University Press, Pittsburgh, 1972, p. 666.*)

(Figure 27-10). This is accomplished by employing softer grades of grinding wheels, reducing the grinding speeds and infeed rates, using chemically active cutting fluids (e.g., highly sulfurized oil or KNO_2 in water), as outlined in the table of grinding conditions in Figure 27-10. These procedures may require the addition of a variable-speed drive to the grinding machine. Generally, only about 0.005 to 0.010 in. of surface stock needs to be finish ground in this way, as the depth of the surface damage due to conventional grinding or abrasive grinding is 0.005 to 0.007 in. High-strength steels, high-temperature nickel, and cobalt-based alloys and titanium alloys are particularly sensitive to surface deformation and cracking problems from grinding. Other postprocessing processes, such as polishing, honing, and chemical milling plus peening, can be used to remove the deformed layers in critically stressed parts. It is strongly recommended, however, that testing programs be used along with service experience on critical parts before these procedures are employed in production. See the discussion of residual stresses in Chapter 40.

Dressing and Truing

As the wheel is used, there is a tendency for the wheel to become *loaded* (metal chips become lodged in the cavities between the grains) and the grains dull or glaze (grits wear, flatten, and polish). Unless the wheel is cleaned and sharpened (or dressed), the wheel will not cut as well and will tend to plow and rub more. Figure 27-11 shows an arrangement for stick *dressing* a grinding wheel. The dulled grains cause the cutting forces on the grains to increase, ideally resulting in the grains' fracturing or being pulled out of the bond, thus providing a continuous exposure of sharp cutting edges. Such a continuous action ordinarily will not occur for light feeds and depths of cut. For heavier cuts, grinding wheels do become somewhat self-dressing, but the workpiece may become overheated and turn a bluish temper color (this is called *burn*) before the wheel reaches a fully dressed condition. A burned surface, the consequence of an oxide layer formation, results in the scrapping of several workpieces before parts of good quality are ground.

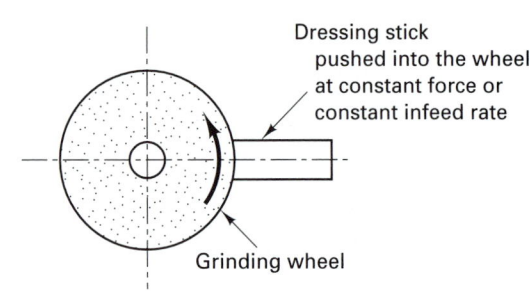

Dressing stick pushed into the wheel at constant force or constant infeed rate

Grinding wheel

FIGURE 27-11 Schematic arrangement of stick dressing.

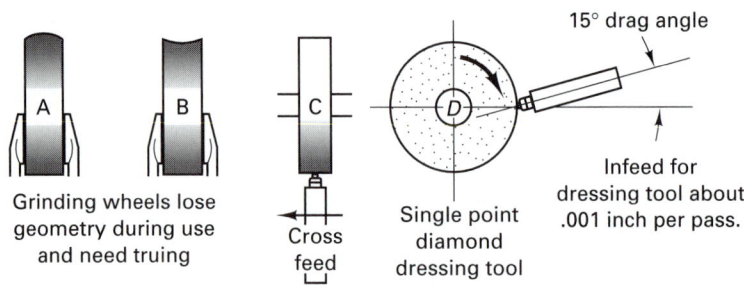

A   B

Grinding wheels lose geometry during use and need truing

C

Cross feed

D

Single point diamond dressing tool

15° drag angle

Infeed for dressing tool about .001 inch per pass.

FIGURE 27-12 Diamond nibs may be used for truing wheels in batch operations.

Grinding wheels lose their geometry during use. *Truing* restores the original shape. A single-point diamond tool can be used to *true* the wheel while fracturing abrasive grains to expose new grains and new cutting edges on worn, glazed grains (Figure 27-12). Truing can also be accomplished by grinding the grinding wheel with controlled-path or powered rotary devices using conventional abrasive wheels. The precision in generating a trued wheel surface by these methods is poorer than by the method described earlier.

Resin-bonded wheels can be trued by grinding with hard ceramics such as tungsten carbide. The procedure for truing and dressing a CBN wheel in a surface grinder might be as follows: Use 0.0002-in. downfeed per pass and cross-feed slightly more than half the wheel thickness at moderate table speeds. The wheel speed is the same as the grinding speed. The grinding power will gradually increase, as the wheel is getting dull, while being trued. When the power exceeds normal power drawn during workpiece grinding, stop the truing operation. Dress the wheel face open using a J-grade stick, with abrasive one grit size smaller than CBN. Continue the truing. Repeat this cycle until the wheel is completely trued.

Modern grinding machines are equipped so that the wheel can be dressed and/or trued continuously or intermittently while grinding continues. A common way to do this is by *crush dressing*. Crush dressing consists of forcing a hard roll (WC or HSS) having the same contour as the part to be ground against the grinding wheel while it is revolving—usually quite slowly. A water-based coolant is used to flood the dressing zone at 5 to 10 gal/min. The crushing action fractures and dislodges some of the abrasive grains, exposing fresh sharp edges, allowing free cutting for faster infeed rates. This procedure usually is employed to produce and maintain a special contour to the abrasive wheel. This is also called wheel profiling. Crush dressing is a very rapid method of dressing grinding wheels, and because it fractures abrasive grains, results in free cutting and somewhat

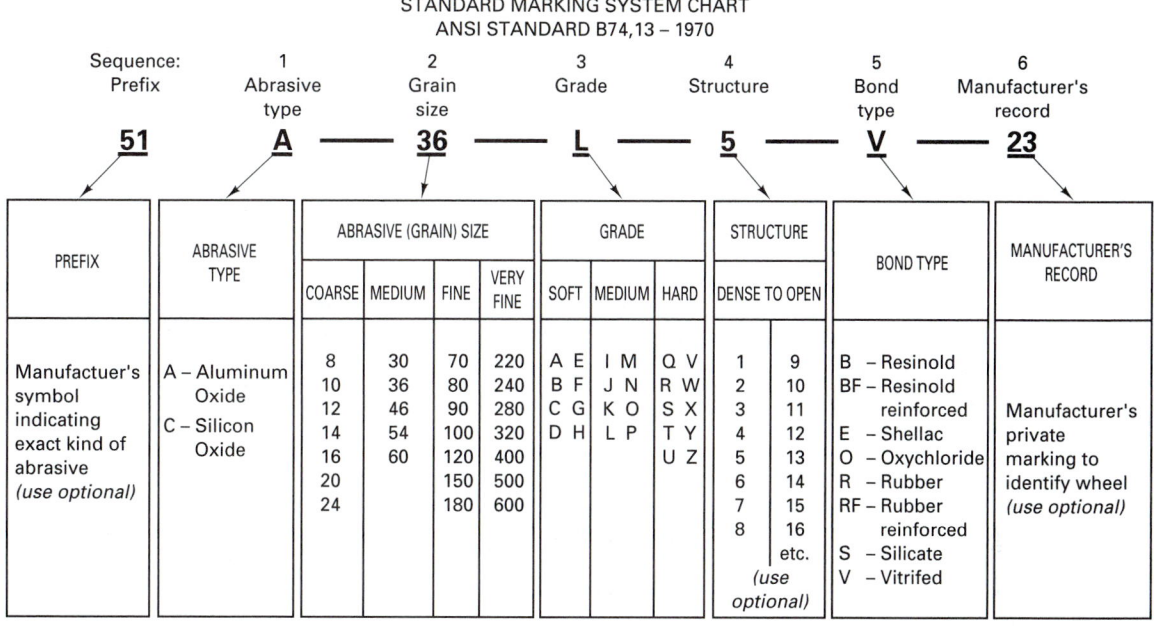

FIGURE 27-13 Standard marking system for grinding wheels (ANSI standard B74. 13-1970).

cooler grinding. The resulting surfaces may be slightly rougher than when diamond dressing is used (see Figure 27-16).

Grinding Wheel Identification

Most grinding wheels are identified by a standard marking system that has been established by the American National Standards Institute, Inc. This system is illustrated and explained in Figure 27-13. The first and last symbols in the marking are left to the discretion of the manufacturer.

Grinding Wheel Geometry

The shape and size of the wheel are critical selection factors. Obviously, the shape must permit proper contact between the wheel and all of the surface that must be ground. Grinding wheel shapes have been standardized and eight of the most commonly used types are shown in Figure 27-14. Types 1, 2, and 5 are used primarily for grinding external or internal cylindrical surfaces and for plain surface grinding. Type 2 can be mounted for grinding either on the periphery or the side of the wheel. Type 4 is used with tapered safety flanges so that if the wheel breaks during rough grinding, such as snagging, these flanges will prevent the pieces of the wheel from flying and causing damage. Type 6, the straight cup, is used primarily for surface grinding but can also be used for certain types of offhand grinding. The flaring-cup type of wheel is used for tool grinding. Dish-type wheels are used for grinding tools and saws. Type 1, the straight grinding wheels, can be obtained with a variety of standard faces. Some of these are shown in Figure 27-15.

The size of the wheel to be used is determined, primarily, by the spindle rpm values available on the grinding machine and the proper cutting speed for the wheel, as dictated

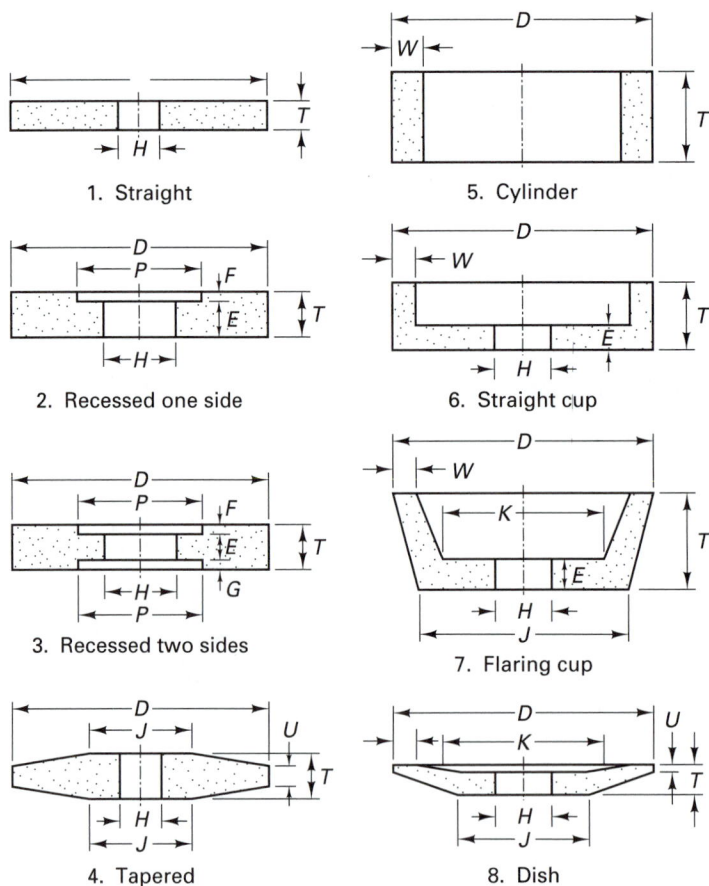

FIGURE 27-14   Standard grinding-wheel shapes commonly used. *(Courtesy of Carborundum Company.)*

by the type of bond. For most grinding operations the cutting speed is about 2500 to 6500 ft/min. Different types and grades of bond often justify considerable deviation from these speeds. For certain types of work using special wheels and machines, as in thread grinding and "abrasive machining," much higher speeds are used.

The operation for which the abrasive wheel is intended will also influence the wheel shape and size. The major use categories are:

1. *Cutting off:* for slicing and slotting parts; use thin wheel organic bond
2. *Cylindrical between centers:* grinding outside diameters of cylindrical workpieces
3. *Cylindrical, centerless:* grinding outside diameters with work rotated by regulating wheel
4. *Internal cylindrical:* grinding bores and large holes
5. *Snagging:* removing large amounts of metal without regard to surface finish or tolerances
6. *Surface grinding:* grinding flat workpieces
7. *Tool grinding:* for grinding cutting edges on tools such as drills, milling cutters, taps, reamers, and single-point high-speed-steel tools
8. *Off-hand grinding:* work or the grinding tool is hand-held

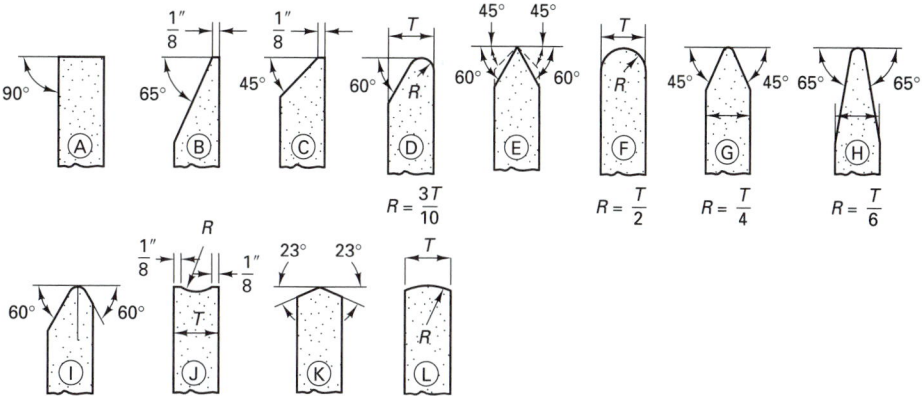

FIGURE 27-15   Standard face contours for straight grinding wheels. *(Courtesy of Carborundum Company.)*

In many cases, the classification of processes coincides with the classification of machines that do the process. Other factors that will influence the choice of wheel to be selected include the workpiece material, the amount of stock to be removed, the shape of the workpiece, and the accuracy and surface finish desired. Workpiece material has a great impact on choice of the wheel. Hard, high-strength metals (tool steels, alloy steels) are generally ground with aluminum oxide wheels or cubic boron nitride wheels. Silicon carbide and CBN are employed in grinding brittle materials (cast iron and ceramics) as well as softer, low-strength metals such as aluminum, brass, copper, and bronze. Diamonds have taken over the cutting of tungsten carbides, and CBN is used for precision grinding of tool and die steel, alloy steels, stainless, and other very hard materials. There are so many factors that affect the cutting action that there are no hard and fast rules with regard to abrasive selection.

Selection of grain size is determined by whether course or fine cutting and finish are desired. Course grains take larger depths of cut and cut more rapidly. Hard wheels with fine grains leave smaller tracks and therefore usually are selected for finishing cuts. If there is a tendency for the work material to load the wheel, larger grains with more open structure may be used for finishing.

Balancing Grinding Wheels

Because of the high rotation speeds involved, grinding wheels must never be used unless they are in good balance. A slight imbalance will produce vibrations that will cause waviness in the work surface. It may cause a wheel to break, with the probability of serious damage and injury. The wheel should be mounted with proper bushings so that it fits snugly on the spindle of the machine. Rings of blotting paper should be placed between the wheel and the flanges to assure that the clamping pressure is evenly distributed. Most grinding wheels will run in good balance if they are mounted properly and trued. Most machines have provision for compensating for a small amount of wheel imbalance by attaching weights to one mounting flange. Some have provision for semiautomatic balancing with weights that are permanently attached to the machine spindle.

Safety in Grinding

Because the rotational speeds are quite high, and the strength of grinding wheels usually is much less than the materials being ground, serious accidents occur much too frequently in connection with the use of grinding wheels. Virtually all such accidents could be avoided and are due to one or a combination of four causes. First, grinding wheels occasionally are operated at unsafe and improper speeds. All grinding wheels are clearly marked with the maximum rpm value at which they should be rotated. They are all tested to considerably above the designated rpm and are safe at the specified speed *unless abused. They should never, under any condition, be operated above the rated speed.* Second, a most common form of abuse, frequently accidental, is dropping the wheel or striking it against a hard object. This can cause a crack (which may not be readily visible), resulting in subsequent failure of the wheel while rotating at high speed under load. If a wheel is dropped or struck against a hard object, it should be discarded and never used unless tested at above the rated speed in a properly designed test stand. A third common cause of grinding wheel failure is improper use, such as grinding against the side of a wheel that was designed for grinding only on its periphery. The fourth and most common cause of injury from grinding is the absence of a proper safety guard over the wheel and/or over the eyes or face of the operator. The frequency with which operators will remove safety guards from grinding equipment or fail to use safety goggles or face shields is amazing and inexcusable.

Use of Cutting Fluids in Grinding

Because grinding involves cutting, the selection and use of a cutting fluid is governed by the basic principles discussed in Chapter 22. If a fluid is used, it should be applied in sufficient quantities and in a manner that will assure that the chips are washed away and not trapped between the wheel and the work. This is of particular importance in grinding horizontal surfaces. In hardened steel, the use of a fluid can help to prevent fine microcracks that result from highly localized heating. The air scraper shown in Figure 27-1 permits the cutting fluid (lubricant) to get onto the face of the wheel.

Much snagging and off-hand grinding is done dry. On some types of material, dry grinding produces a better finish than can be obtained by wet grinding. Grinding fluids strongly influence the performance of CBN wheels. Straight, sulfurized, or sulfochlorinated oils can enhance performance considerably.

■ 27.4 Grinding Machines

Grinding machines commonly are classified according to the type of surface they produce. Table 27-4 presents such a classification, with further subdivision to indicate characteristic features of different types of machines within each classification. Grinding on all machines is done in three ways. In the first, the depth of cut (d_t) is obtained by *infeed*—moving the wheel down into the work or the work up into the wheel (see Figure 27-1). The desired surface in then produced by traversing the wheel across (cross feed) the workpiece, or vice versa (see Figure 27-22). In the second method, known as *plunge-cut* grinding, the basic movement is of the wheel being fed radially into the work while the latter revolves on centers. It is similar to form cutting on a lathe; usually, a formed grinding wheel is used (Figure 27-16). In the third method, the work is fed very slowly past the wheel and the total downfeed or depth (d) is accomplished in a single pass (Figure 27-17). This is called *creep feed grinding* (FG). See Table 27-5, which compares CFG to conventional and high-speed grinding for CBN applications.

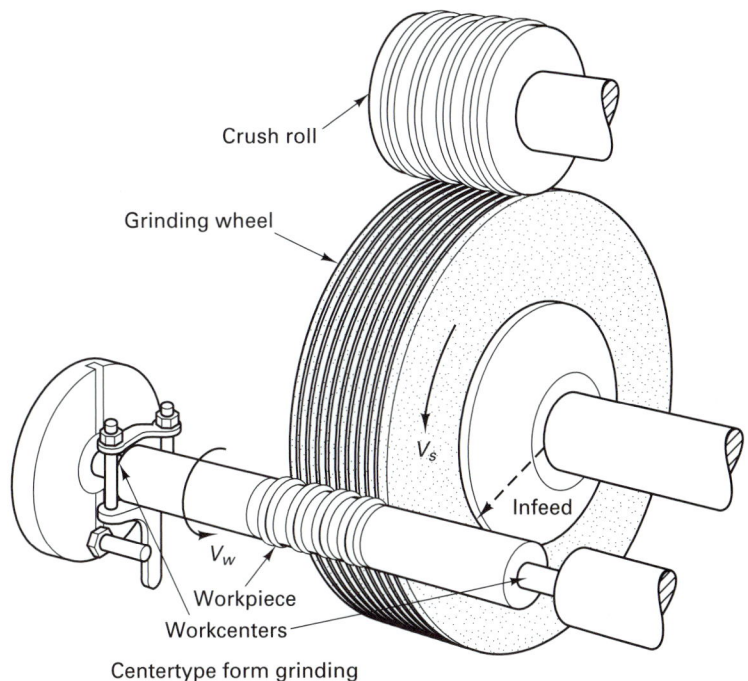

Crush roll

Grinding wheel

V_s

Infeed

V_w

Workpiece

Workcenters

Centertype form grinding

FIGURE 27-16 Plunge cut grinding of cylinder held between centers. Note that crush roll dressing is shown.

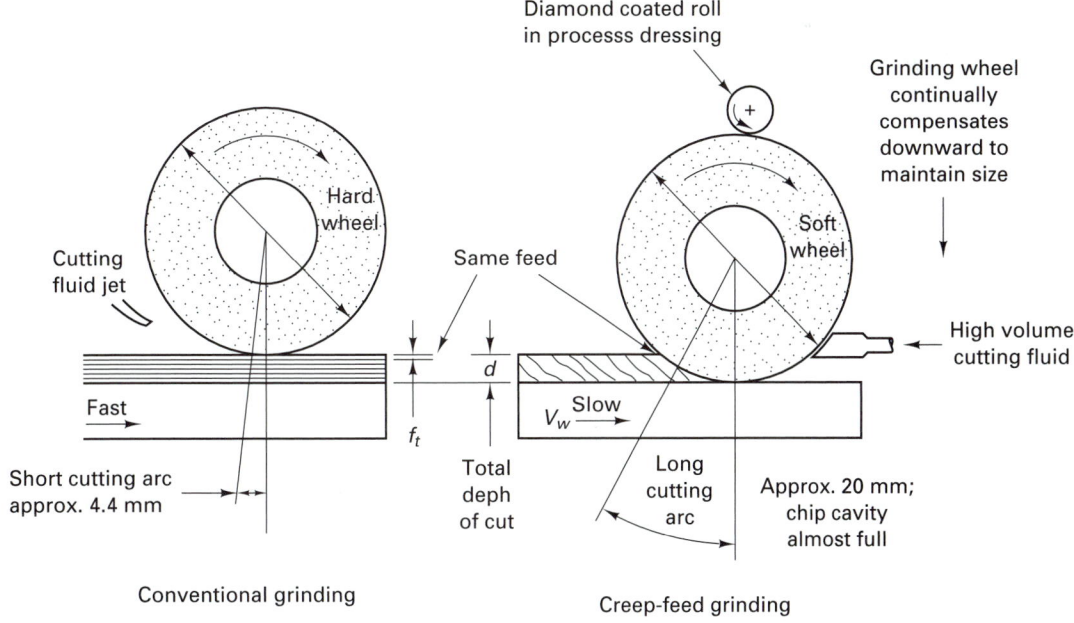

Diamond coated roll in processs dressing

Grinding wheel continually compensates downward to maintain size

Hard wheel

Cutting fluid jet

Same feed

Soft wheel

High volume cutting fluid

Fast

f_t

d

V_w Slow

Short cutting arc approx. 4.4 mm

Total deph of cut

Long cutting arc

Approx. 20 mm; chip cavity almost full

Conventional grinding

Creep-feed grinding

FIGURE 27-17 Conventional grinding contrasted to creep feed grinding. Note that crush roll dressing is indicated.

TABLE 27-4.   Grinding Machines

Type of Machine	Type of Surface	Specific Types or Features
Cylindrical External	External surface on rotating, usually cylindrical parts	Work rotated between centers Centerless Chucking Tool post Crankshaft, cam, etc.
Internal	Internal diameters of holes	Chucking Planetary (work stationary) Centerless
Surface Conventional	Flat surfaces	Reciprocating table or rotating table Horizontal or vertical spindle
Creep feed	Deep slots, profiles in hard steels, carbides and ceramics using CBN and diamond	Rigid, chatter-free, creep feed rate Continuous dressing Heaving coolant flows NC or CNC control Variable speed wheel
Tool grinders	Tool angles and geometries	Universal Special
Other	Special or any of the above	Disk, contour, thread, flexible shaft, swing frame, snag, pedestal, bench

TABLE 27-5.   Starting Conditions for CBN Grinding

Grinding Variable	Conventional Grinding	Creep Feed Grinding	High-Speed Grinding
Wheel speed (ft/min)	5500–9500 versus 4500–6500 vitrified	5000–9000 versus 3000–5000	12,000–25,000
Table speed, V_w (ft/min)	80–150	0.5–5	5–20
Feed, f_t (in./pass)	0.0005–0.0015	0.100–0.250	0.250–0.500
Grinding fluids	10% heavy-duty soluble oil or 3–5% light-duty soluble for light feeds	Sulfurized or sulfochlorinated straight grinding oil applied at 80–100 gal/min at 100 psi or more	

Grinding machines that are used for precision work have certain important characteristics that permit them to produce parts having close dimensional tolerances. They are constructed very accurately, with heavy, rigid frames to assure permanency of alignment. Rotating parts are accurately balanced to avoid vibration. Spindles are mounted in very accurate bearings, usually of the preloaded ball-bearing type. Controls are provided so that all movements that determine dimensions of the workpiece can be made with accuracy, usually to 0.001 or .0001 in.

The abrasive dust that results from grinding must be prevented from entering between moving parts. All ways and bearings must be fully covered or protected by seals. If this is not done, the abrasive dust between moving parts becomes embedded in the softer of the two, causing it to act as a lap and abrade the harder of the two surfaces, resulting in permanent loss of accuracy.

These special characteristics add considerably to the cost of these machines and require that they be operated by trained personnel. Production-type grinders are more fully automated and have higher metal-removal rates and excellent dimensional accuracy. Fine surface finish can be obtained very economically.

Cylindrical Grinding

Center-type cylindrical grinding is commonly used for producing external cylindrical surfaces. Figures 27-16 and 27-18 show the basic principles and motions of this process. The grinding wheel revolves at an ordinary cutting speed, and the workpiece rotates on centers at a much slower speed, usually from 75 to 125 ft/min. The grinding wheel and the workpiece move in opposite directions at their point of contact. The depth of cut is determined by infeed of the wheel or workpiece. Because this motion also determines the finished diameter of the workpiece, accurate control of this movement is required. Provision is made to traverse the workpiece with the wheel or the work can be reciprocated past the wheel. In very large grinders, the wheel is reciprocated because of the massiveness of the work. For form or plunge grinding, the detail of the wheel is maintained by periodic crush roll dressing.

A *plain center-type cylindrical grinder* is shown in Figure 27-18. On this type the work is mounted between headstock and tailstock centers. Solid dead centers are always used in the tailstock, and provision usually is made so that the headstock center can be operated either dead or alive. High-precision work usually is ground with a dead headstock center, because this eliminates any possibility that the workpiece will run out of round due to any eccentricity in the headstock.

The table assembly can be reciprocated, in most cases, by using a hydraulic drive. The speed can be varied, and the length of the movement can be controlled by means of adjustable trip dogs.

Infeed is provided by movement of the wheelhead at right angles to the longitudinal axis of the table. The spindle is driven by an electric motor that is also mounted on the wheelhead. If the infeed movement is controlled manually by some type of vernier drive to provide control to 0.001 in. or less, the machine is usually equipped with digital readout equipment to show the exact size being produced. Most production-type grinders have automatic infeed with retraction when the desired size has been obtained. Such machines are usually equipped with an automatic diamond wheel-truing device that dresses the wheel and resets the measuring element before grinding is started on each piece.

The longitudinal traverse should be about one-fourth to three-fourths of the wheel width for each revolution of the work. For light machines and fine finishes, it should be held to the smaller end of this range. The depth of cut (infeed) varies with the purpose of the grinding operation and the finish desired. When grinding is done to obtain accurate size, infeeds of 0.002 to 0.004 in. commonly are used for roughing cuts. For finishing, the infeed is reduced to 0.00025 to 0.0005 in. The design allowance for grinding should be from 0.005 to 0.010 in. on short parts and on parts that are not to be hardened. On long or large parts and on work that is to be hardened, a grinding allowance of from 0.015 to 0.030 in. is desirable. When grinding is used primarily for metal removal (called abrasive machining), infeeds are much higher, 0.020 to 0.040 in. being common. Continuous downfeed often is used, rates up to 0.100 in./min being common.

Grinding machines are available in which the workpiece is held in a chuck for grinding both external and internal cylindrical surfaces. *Chucking-type external grinders* are production-type machines for use in rapid grinding of relatively short parts, such as ball-bearing races. Both chucks and collets are used for holding the work, the means dictated by the shape of the workpiece and rapid loading and removal.

In chucking-type internal grinding machines, the chuck-held workpiece revolves, and a relatively small, high-speed grinding wheel is rotated on a spindle arranged so that it can be reciprocated in and out of the workpiece. Infeed movement of the wheelhead is normal to the axis of rotation of the work (see Figure 27-18).

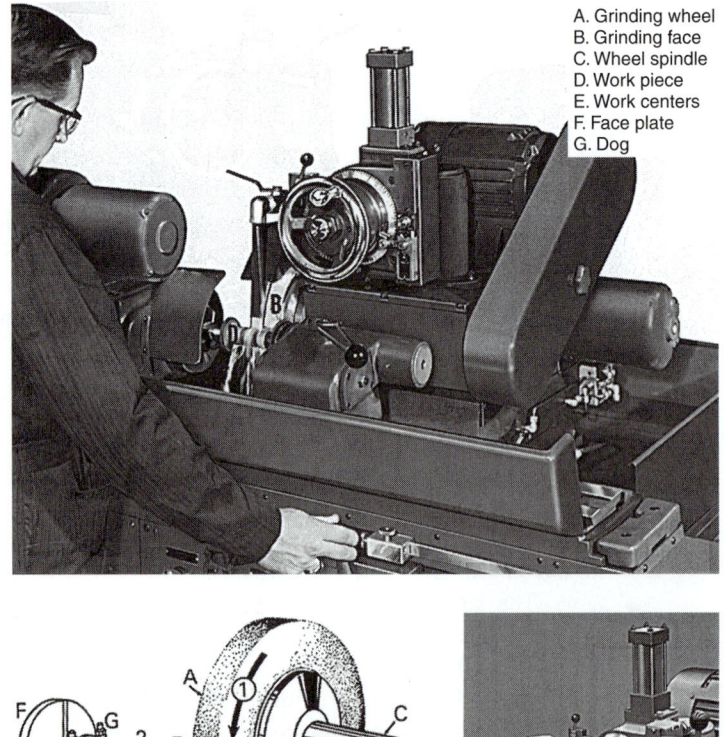

A. Grinding wheel
B. Grinding face
C. Wheel spindle
D. Work piece
E. Work centers
F. Face plate
G. Dog

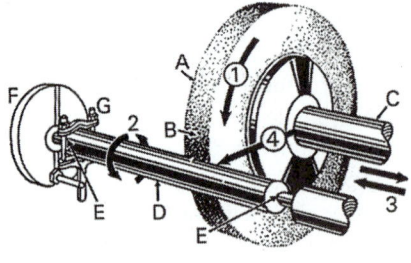

Movements

1. Wheel	2. Work (rotates)
3. Traverse	4. Infeed

FIGURE 27-18 Cylindrical grinding between centers, *lower right:* Internal cylindrical grinding on same machine.

Centerless Grinding

Centerless grinding makes it possible to grind both external and internal cylindrical surfaces without requiring the workpiece to be mounted between centers or in a chuck. This eliminates the requirement of center holes in some workpieces and the necessity for mounting the workpiece, thereby reducing the cycle time.

The principle of *centerless external grinding* is illustrated in Figure 27-19. Two wheels are used. The larger one operates at regular grinding speeds and does the actual grinding. The smaller wheel is the *regulating* wheel. It is mounted at an angle to the plane of the grinding wheel. Revolving at a much slower surface speed—usually 50 to 200 ft/min—the regulating wheel controls the rotation and longitudinal motion of the workpiece and usually is a plastic- or rubber-bonded wheel with a fairly wide face.

The workpiece is held against the work-rest blade by the cutting forces exerted by the grinding wheel and rotates at approximately the same surface speed as that of the

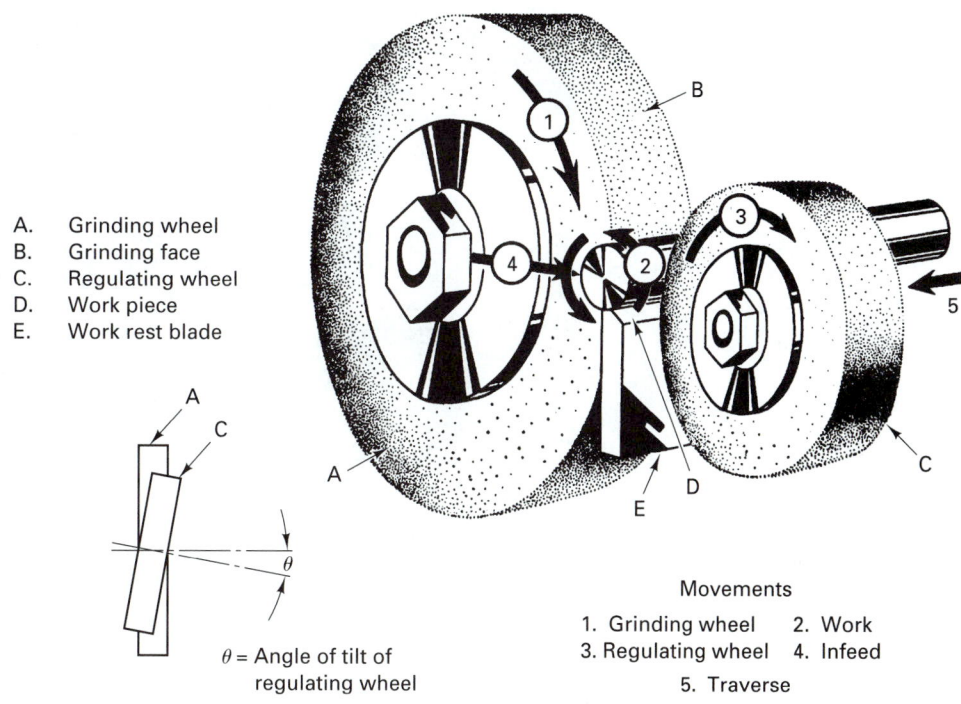

A. Grinding wheel
B. Grinding face
C. Regulating wheel
D. Work piece
E. Work rest blade

θ = Angle of tilt of
regulating wheel

Movements
1. Grinding wheel   2. Work
3. Regulating wheel   4. Infeed
5. Traverse

FIGURE 27-19   Centerless grinding showing the relationship between the grinding wheel, the regulating wheel, and the workpiece in centerless method. *(Courtesy of Carborundum Company.)*

regulating wheel. This axial feed is calculated approximately by the equation

$$F = dN \sin\theta \qquad (27\text{-}1)$$

where

F = feed (mm/min or in./min)
d = diameter of the regulating wheel (min or in.)
N = revolutions per minute of the regulating wheel
θ = angle of inclination of the regulating wheel

Centerless grinding has several important advantages:

1. It is very rapid; infeed centerless grinding is almost continuous.
2. Very little skill is required of the operator.
3. It can often be made automatic.
4. Where the cutting occurs, the work is fully supported by the work rest and the regulating wheel. This permits heavy cuts to be made.
5. Because there is no distortion of the workpiece, accurate size control is easily achieved.
6. Large grinding wheels can be used, thereby minimizing wheel wear.

Thus centerless grinding is ideally suited to certain types of mass-production operations. The major disadvantages are as follows:

1. Special machines are required that can do no other type of work.
2. The work must be round—no flats, such as keyways, can be present.

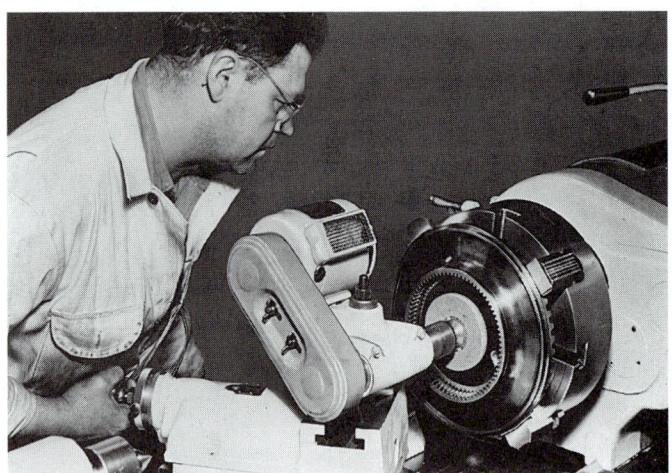

FIGURE 27-20 Grinding a bearing seat by means of a tool-post grinder on a lathe. *(Courtesy of Dunore Company.)*

3. Its use on work having more than one diameter or on curved parts is limited.

4. In grinding tubes, there is no guarantee that the OD and ID are concentric.

Special centerless grinding machines are available for grinding balls and tapered workpieces. The centerless grinding principle can also be applied to internal grinding, but the external surface of the cylinder must be finished accurately before the internal operation is started. However, it assures that the internal and external surfaces will be concentric. The operation is easily mechanized for many applications.

Tool-Post Grinders

Tool-post grinders (Figure 27-20) are used on lathes, occasionally, to grind cylindrical parts. The wheelhead is either a high-speed electric or air motor with the grinding wheel often mounted directly on the motor shaft. The entire mechanism is mounted either on the tool post or on the compound rest. The lathe spindle provides rotation for the workpiece, and the lathe carriage is used to reciprocate the wheelhead. Although tool-post grinders are versatile and useful, care should be taken to cover the ways of the lathe with a closely woven cloth to provide protection from the abrasive dust that can become entrapped between the moving parts.

Surface Grinding Machines

Surface grinding machines are used primarily to grind flat surfaces. However formed, irregular surfaces can be produced on some types of surface grinders by use of a formed wheel. There are four basic types of surface grinding machines, differing in the movement of their tables and the orientation of the grinding wheel spindles:

1. Horizontal spindle and reciprocating table
2. Vertical spindle and reciprocating table
3. Horizontal spindle and rotary table
4. Vertical spindle and rotary table

These machines are illustrated in Figure 27-21.

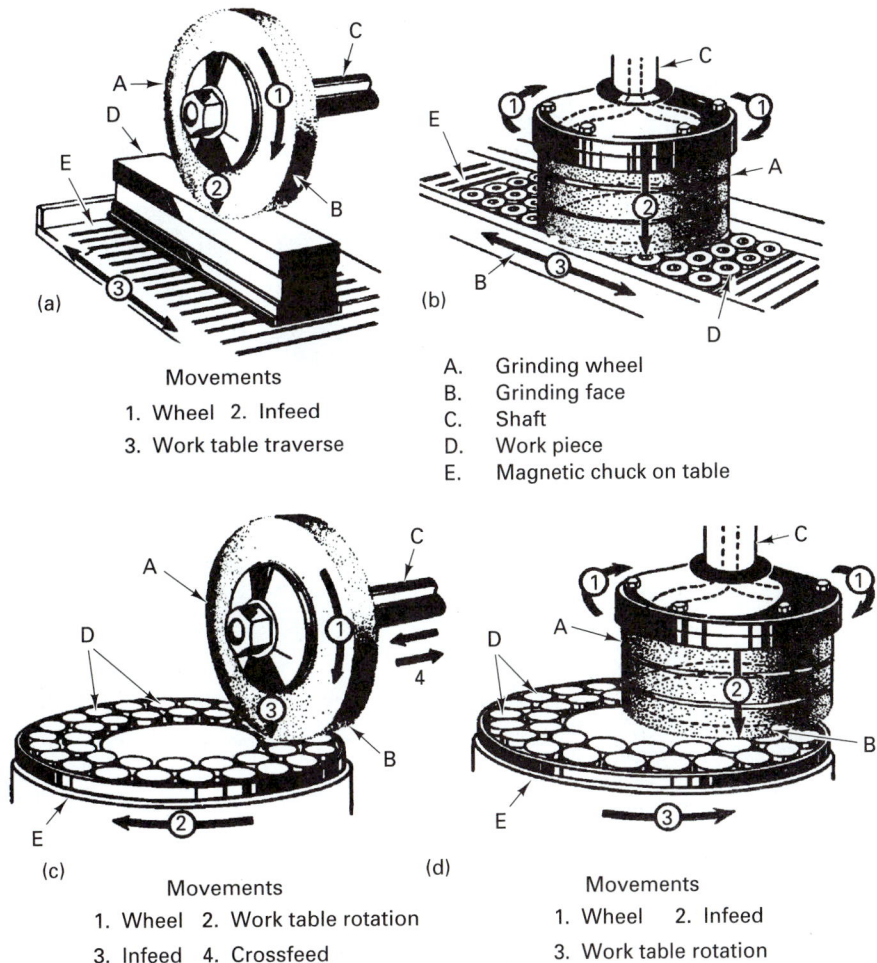

Movements

1. Wheel 2. Infeed
3. Work table traverse

A. Grinding wheel
B. Grinding face
C. Shaft
D. Work piece
E. Magnetic chuck on table

Movements

1. Wheel 2. Work table rotation
3. Infeed 4. Crossfeed

Movements

1. Wheel 2. Infeed
3. Work table rotation

FIGURE 27-21 Surface grinding: (a) horizontal surface grinding with reciprocating table; (b) vertical spindle with reciprocating table; (c) and (d) both horizontal and vertical spindle machines can have rotary tables. *(Courtesy of Carborundum Company.)*

The most common type of surface grinding machine has a reciprocating table and horizontal spindle (Figure 27-22). The table can be reciprocated longitudinally either by handwheel or by hydraulic power. The wheelhead is given transverse (crossfeed) motion at the end of each table motion, again either by handwheel or by hydraulic power feed. Both the longitudinal and transverse motions can be controlled by limit switches. Infeed or downfeed on such grinders is controlled by handwheels or automatically. The size of such machines is designated by the size of the surface that can be ground.

In using such machines, the wheel should overtravel the work at both ends of the table reciprocation, so as to prevent the wheel from grinding in one spot while the table is being reversed. The transverse or crossfeed motion should be one-fourth to three-fourths of the wheel width between each stroke.

Vertical-spindle reciprocating-table surface grinders differ basically from those with horizontal spindles only in that their spindles are vertical and that the wheel diameter must

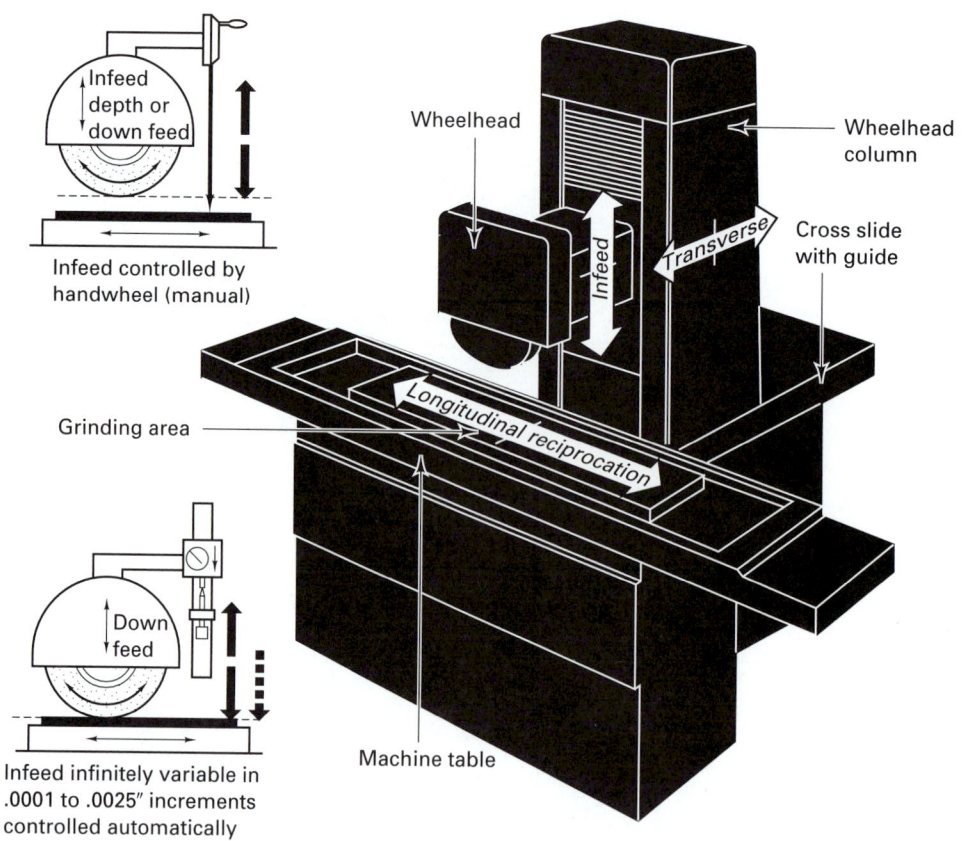

FIGURE 27-22   Horizontal spindle surface grinder, with insets showing movements of wheelhead.

exceed the width of the surface to be ground. Usually, no traverse motion of either the table or the wheelhead is provided. Such machines can produce very flat surfaces.

Rotary-table surface grinders can have either vertical or horizontal spindles, but those with horizontal spindles are limited in the type of work they will accommodate and therefore are not used to a great extent. *Vertical-spindle rotary-table surface grinders* are primarily production-type machines. They frequently have two or more grinding heads, and therefore both rough grinding and finish grinding are accomplished in one rotation of the workpiece. The work can be held either on a magnetic chuck or in special fixtures attached to the table.

By using special rotary feeding mechanisms, machines of this type often are made automatic. Parts are dumped on the rotary feeding table and fed automatically onto workholding devices and moved past the grinding wheels. After they pass the last grinding head, they are automatically unloaded.

Creep Feed Grinding Machines

This is a grinding method, often done in the surface grinding mode, that is markedly different from conventional surface grinding. As shown in Figure 27-17, in contrast to conventional techniques, the depth of cut is increased 1000 to 10,000 times and the work feed is decreased in the same proportion; hence the name *creep feed grinding* (see Table 27.5). The long arc of contact between the wheel and the work increases the cutting forces and the power required. Therefore, the machine tools to perform this type of grinding must be specially designed with high static and dynamic stability, stick-slip free ways, adequate damping, increased horsepower, infinitely variable spindle speed, variable but extremely consistent table feed (especially in the low ranges), high-pressure cooling systems, integrated devices for dressing the grinding wheels, and specially designed (soft with open structure) grinding wheels. The process is mainly being applied to grinding deep slots with straight parallel sides or to grinding complex profiles in difficult-to-grind materials. The process is capable of producing extreme precision at relatively high metal removal rates. Because the process can operate at relatively low surface temperatures, the surface integrity of the metals being ground is good.

However, in CFG, the grinding wheels must maintain their initial profile much longer. This process is greatly enhanced by continuous dressing (form-truing and dressing the grinding wheel throughout the process rather than between cycles). Continuous crush dressing results in higher MRRs, improved dimensional accuracy and form tolerance, reduced grinding forces (and power), and reduced thermal effects while sacrificing wheel wear. Creep feed grinding eliminates preparatory operations such as milling or broaching since profiles are ground into the solid workpiece. This can result in significant savings in unit part costs.

Disc Grinding Machines

Disc grinders have relatively large side-mounted abrasive discs. The work is held against one side of the disc for grinding. Both single and double disc grinders are used; in the latter type the work is passed between the two discs and is ground on both sides simultaneously. On these machines, the work is always held and fed automatically. On small, single-disc grinders the work can be held and fed by hand while resting on a supporting table. Although manual disc grinding is not very precise, flat surfaces can be obtained quite rapidly with little or no tooling cost. On specialized, production-type machines, excellent accuracy can be obtained very economically.

Tool and Cutter Grinders

Simple, single-point tools often are sharpened by hand on bench or pedestal grinders (*off-hand grinding*). More complex tools, such as milling cutters, reamers, hobs, and single-point tools for production-type operations, require more sophisticated grinding machines, commonly called *universal tool and cutter grinders*. These machines are similar to small universal cylindrical center-type grinders, but they differ in four important respects:

1. The headstock is not motorized.
2. The headstock can be swiveled about a horizontal as well as a vertical axis.
3. The wheelhead can be raised and lowered and can be swiveled through at 360° rotation about a vertical axis.
4. All table motions are manual. No power feeds being provided.

Specific rake and clearance angles must be created, often repeatedly, on a given tool or on duplicate tools. Tool and cutter grinders have a high degree of flexibility built into them so that the required relationships between the tool and the grinding wheel can be established for almost any type of tool. Although setting up such a grinder is quite complicated and requires a highly skilled worker, after the setup is made for a particular job, the

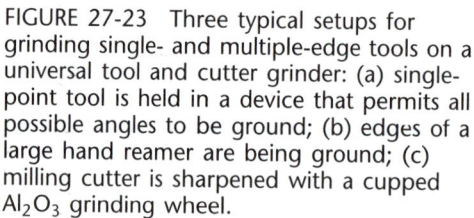

FIGURE 27-23 Three typical setups for grinding single- and multiple-edge tools on a universal tool and cutter grinder: (a) single-point tool is held in a device that permits all possible angles to be ground; (b) edges of a large hand reamer are being ground; (c) milling cutter is sharpened with a cupped Al_2O_3 grinding wheel.

actual grinding is accomplished rather easily. Figure 27-23 shows several typical setups on a tool and cutter grinder.

Hand-ground cutting tools are not accurate enough for automated machining processes. Many NC machine tools have been sold on the premise that they can position work to very close tolerances—within ±0.0001 to 0.0002 in.—only to have the initial workpieces produced by those machines out of tolerance by as much as 0.015 to 0.020 in. In most instances, the culprit was a poorly ground tool. For example, a twist drill with a point ground 0.005 in. off center can "walk" as much as 0.015 in., thus causing poor hole location. Many companies are turning to CNC grinders to handle the regrinding of their cutting tools. A six-axis CNC grinder is capable of restoring the proper tool angles (rake and clearance), concentricity, cutting edges, and dimensional size.

Mounted Wheels and Points

Mounted wheels and points are small grinding wheels of various shapes that are permanently attached to metal shanks that can be inserted in the chucks of portable, high-speed electric or air motors. They are operated at speeds up to 100,000 rpm, depending on their diameters, and are used primarily for deburring and finishing in mold and die work. Several types are shown in Figure 27-24.

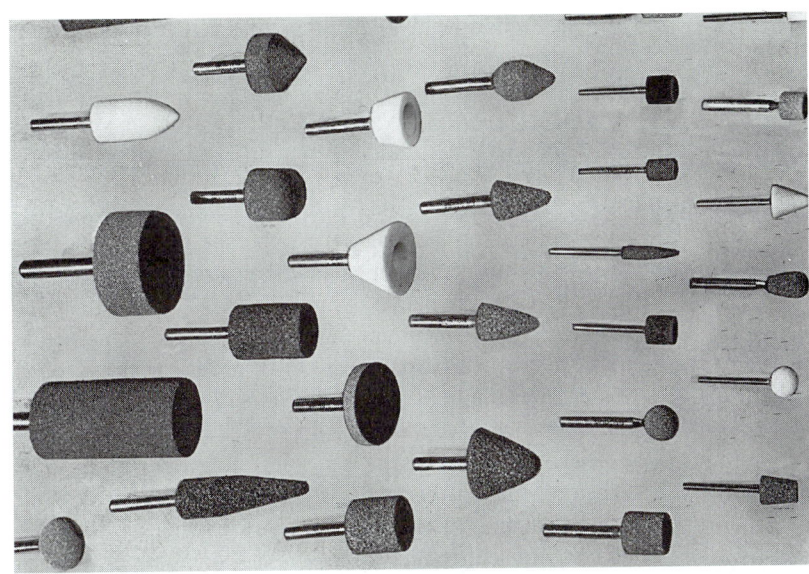

FIGURE 27-24   Examples of mounted abrasive wheels and points. *(Courtesy of Norton Company.)*

FIGURE 27-25   Examples of coated abrasives, belts and discs. *(Courtesy of Carborundum Company.)*

Coated Abrasives

Coated abrasives are being used increasingly in finishing both metal and nonmetal products. These are made by gluing abrasive grains onto a cloth or paper backing. Synthetic abrasives—Carborundum and Alundum—are used most commonly, but some natural abrasives—sand, flint, garnet, and emery—also are employed. Various types of glues are utilized to attach the abrasive grains to the backing, usually compounded to allow the finished product to have some flexibility.

Coated abrasives are available in sheets, rolls, endless belts, and discs of various sizes. Some of the available forms are shown in Figure 27-25. Although the cutting action of coated abrasives basically is the same as with grinding wheels, there is one major difference: they have little tendency to be self-sharpened when dull grains are pulled from the backing. Consequently, when the abrasive particles become dull or the belt loaded, the belt must be replaced.

■ 27.5 DESIGN CONSIDERATIONS IN GRINDING

Almost any shape and size of work can be finished on modern grinding equipment, including flat surfaces, straight or tapered cylinders, irregular external and internal surfaces, cams, antifriction bearing races, threads, and gears. For example, the most accurate threads are formed from solid cylindrical blanks on special thread-grinding machines. Gears that must operate without play are hardened and then finish ground to close tolerances. Two important design recommendations are to reduce the area to be ground and to keep all surfaces that are to be ground in the same or parallel planes (Figure 27-26). This is an example of *design for manufacturing* (DFM).

Abrasive machining can remove scale as well as parent metal. Large allowances of material, needed to permit conventional metal-cutting tools to cut below hard or abrasive inclusions, are not necessary for abrasive machining. An allowance of 0.015 in. is adequate, assuming, of course, that the part is not warped or out of round. This small allowance requirement results in savings in machining time, in material (often 60% less metal is removed), and in shipping of unfinished parts.

FIGURE 27-26   Reducing area to be ground and keeping all surfaces to be ground in the same or parallel planes are two important design recommendations. *(From Machine Design, June 1, 1972, p. 87.)*

Original design of base plate

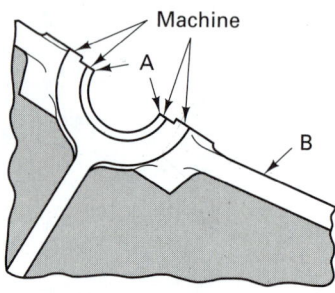

Original design of crankshaft bearing bracket

Redesigned to reduce weight and grinding time

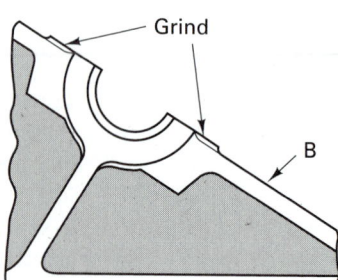

Redesign eliminated shoulders and made part suitable for grinding in single setup

■ 27.6 HONING

Honing is a stock-removal process that uses fine abrasive stones to remove very small amounts of metal. Cutting speed is much lower than that of grinding. The process is used to size and finish bored holes, remove common errors left by boring (taper, waviness, and tool marks) or remove the tool marks left by grinding. The amount of metal removed is typically about 0.005 in. or less. Although honing occasionally is done by hand, as in finishing the face of a cutting tool, it usually is done with special equipment. Most honing is done on internal cylindrical surfaces, such as automobile cylinder walls. The honing stones usually are held in a honing head, with the stones being held against the work with controlled light pressure. The honing head is not guided externally but, instead, *floats* in the hole, being guided by the work surface (Figure 27-27).

The stones are given a complex motion so as to prevent a single grit from repeating its path over the work surface. Rotation is combined with an oscillatory axial motion. For external and flat surfaces, varying oscillatory motions are used. The length of the motions should be such that the stones extend beyond the work surface at the end of each stroke. A cutting fluid is used in virtually all honing operations. The critical process parameters are rotational speed, V_r, oscillation speed, V_o, the length and position of stroke, and the honing stick pressure. V_c and the inclination angle are both products of V_o and V_r.

Honing Stones

Virtually all honing is done with stones made by bonding together various fine artificial abrasives. *Honing stones* differ from grinding wheels in that additional materials, such as sulfur, resin, or wax, are often added to the bonding agent to modify the cutting action. The abrasive grains range in size from 80 to 600 grit. The stones are equally spaced about the periphery of the tool. Table 27-6 gives some reference values for V_c and honing stick pressure, P_s, for various abrasives.

FIGURE 27-27   Schematic of honing head showing the manner in which the stones are held. The rotary and oscillatory motions combine to produce a crosshatched lay pattern.

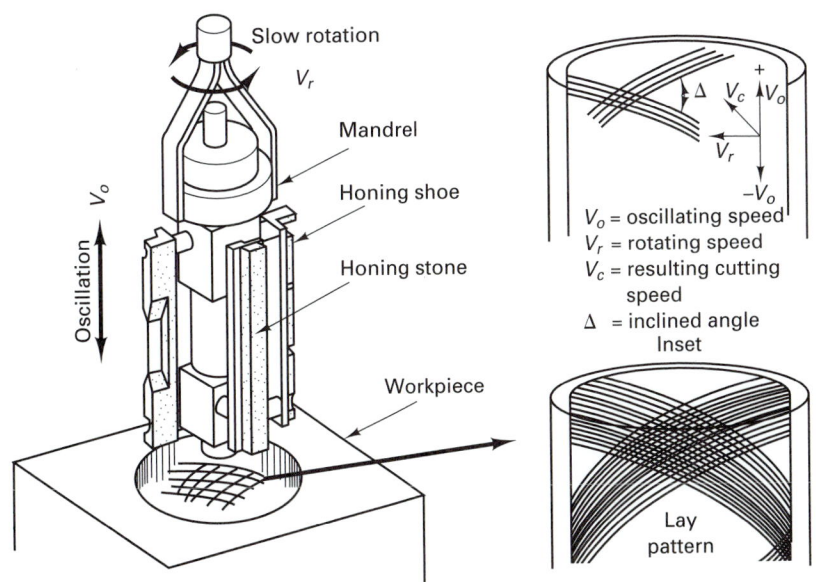

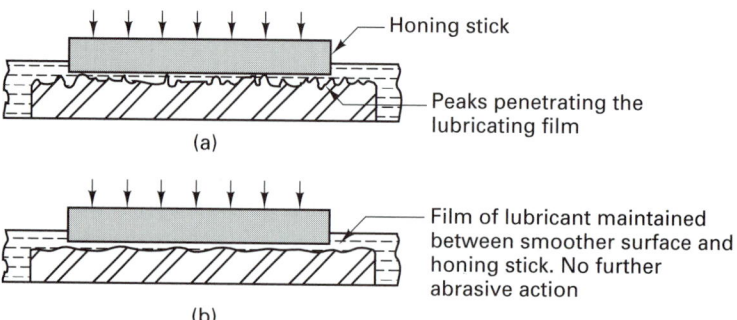

FIGURE 27-28   Manner in which a film of lubricant is established between the work and the abrasive stone in superfinishing as the work becomes smoother.

TABLE 27-6.   Honing Reference Values

For:	Honing Parameters	Conventional Abrasives	Diamonds	CBN
High MRR	V_c (m/min)	20–30	40–70	35–90
	P_s (N/mm²)	1–2	2–8	2–4
Best quality service	V_c (m/min)	5–30	40–70	20–60
	P_s (N/mm²)	0.5–1.5	1.0–3.0	1.0–2.0

Single- and multiple-spindle honing machines are available in both horizontal and vertical types. Some are equipped with special, sensitive measuring devices that collapse the honing head when the desired size has been reached.

For honing single, small, internal cylindrical surfaces, a procedure is often used wherein the workpiece is manually held and reciprocated over a rotating hone. If the volume of work is sufficient, honing is a fairly inexpensive process. A complete honing cycle, including loading and unloading the work, is often less than 1 minute. Size control within 0.0003 in. is achieved routinely.

■ 27.7 SUPERFINISHING

Superfinishing is a variation of honing that is typically used on flat surfaces. The process is:

1. Very light, controlled pressure, 10 to 40 psi
2. Rapid (over 400 cycles per minute), short strokes—less than $\frac{1}{4}$ in.
3. Stroke paths controlled so that a single grit never traverses the same path twice
4. Copious amounts of low-viscosity lubricant–coolant flooded over the work surface

This procedure, illustrated in Figure 27-28, results in surfaces of very uniform, repeatable smoothness.

Superfinishing is based on the phenomenon that a lubricant of a given viscosity will establish and maintain a separating, lubricating film between two mating surfaces if their roughness does not exceed a certain value and if a certain critical pressure, holding them apart, is not exceeded. Consequently, as the minute peaks on a surface are cut away by the honing stone, applied with a controlled pressure, a certain degree of smoothness is achieved. The lubricant establishes a continuous film between the stone and the workpiece and separates them so that no further cutting action occurs. Thus, with a given pressure, lubricant, and honing stone, each workpiece is honed to the same degree of smoothness.

Superfinishing is applied to both cylindrical and plane surfaces. The amount of metal removed usually is less than 0.002 in., most of it being the peaks of the surface roughness. Copious amounts of lubricant-coolant maintain the work at a uniform temperature and wash away all abraded metal particles to prevent scratching.

■ 27.8 LAPPING

Lapping is an abrasive surface-finishing process wherein fine abrasive particles are *charged* (caused to become embedded) into a soft material, called a *lap*. The material of the lap may range from cloth to cast iron or copper, but it is always softer than the material to be finished, being only a holder for the hard abrasive particles. Lapping is applied to both metals and nonmetals.

As the charged lap is rubbed against a surface, the abrasive particles in the surface of the lap remove small amounts of material from the surface to be machined. Thus the abrasive does the cutting, and the soft lap is not worn away because the abrasive particles become embedded in its surface instead of moving across it. This action always occurs when two materials rub together in the presence of a fine abrasive: the softer one forms a lap, and the harder one is abraded away.

In lapping, the abrasive is usually carried between the lap and the work surface in some sort of a vehicle, such as a grease, oil, or water. The abrasive particles are from 120 grit up to the finest powder sizes. As a result, only very small amounts of metal are removed, usually considerably less than 0.001 in. Because it is such a slow metal-removing process, lapping is used only to remove scratch marks left by grinding or honing, or to obtain very flat or smooth surfaces, such as are required on gage blocks or for liquidtight seals where high pressures are involved.

Materials of almost any hardness can be lapped. However, it is difficult to lap soft materials because the abrasive tends to become embedded. The most common lap material is fine-grained cast iron. Copper is used quite often and is the common material for lapping diamonds. For lapping hardened metals for metallographic examination, cloth laps are used.

Lapping can be done either by hand or by special machines. In hand lapping, the lap is flat, similar to a surface plate. Grooves usually are cut across the surface of a lap to collect the excess abrasive and chips. The work is moved across the surface of the lap, using an irregular, rotary motion, and is turned frequently to obtain a uniform cutting action.

In lapping machines for obtaining flat surfaces, workpieces are placed loosely in holders and are held against the rotating lap by means of floating heads. The holders, rotating slowly, move the workpieces in an irregular path. When two parallel surfaces are to be produced, two laps may be employed, one rotating below and the other above the workpieces.

Various types of lapping machines are available for lapping round surfaces. A special type of centerless lapping machine is used for lapping small cylindrical parts, such as piston pins and ball-bearing races.

Because the demand for surfaces having only a few micrometers of roughness on hardened materials has become quite common, the use of lapping has increased greatly. However, it is a very slow method of removing metal, obviously costly compared with other methods, and should not be specified unless such a surface is absolutely necessary.

■ KEY WORDS

abrasive machining	dressing	rubber bond
aluminum oxide	emery	shellac bond
attrition	friability	silicate bond
cubic boron nitride	*G* ratio	silicon carbide
centerless grinding	garnet	snagging
coated abrasive	grade	surface grinding
corundum	grinding	structure
creep-feed grinding	honing	truing
crush dressing	lapping	vitrified bond
cylindrical grinding	quartz	
diamond	resinoid bond	

■ REVIEW QUESTIONS

1. What machining processes use abrasive particles for cutting tools?
2. What is attrition in an abrasive grit?
3. Why is friability an important grit property?
4. What is the relationship between grit size and surface finish?
5. Why is aluminum oxide used more frequently than silicon carbide as an abrasive?
6. Why is CBN superior to silicon carbide as an abrasive in some applications?
7. What materials commonly are used as bonding agents in grinding wheels?
8. Why is the grade of a bond in a grinding wheel important?
9. How does grade differ from structure?
10. What is crush dressing?
11. How does loading differ from glazing?
12. What is meant by the statement that grinding is a mixture of processes?
13. What is accomplished in dressing a grinding wheel?
14. How does abrasive machining differ from ordinary grinding?
15. What is a grinding ratio or *G* ratio?
16. How is the feed of the workpiece controlled in centerless grinding?
17. Why is grain spacing important in grinding wheels?
18. How is the machining time in surface grinding calculated?
19. Why should a cutting fluid be used in copious quantities when doing wet grinding?
20. How does plunge-cut grinding compare to cylindrical grinding? Compare Figures 27-16 and 27-18.
21. If grinding machines are placed among other machine tools, what precautions must be taken?
22. How is the machining time for creep feed surface grinding calculated?
23. What is the purpose of low-stress grinding?
24. How is low-stress grinding done?
25. The number of grains per square inch that actively contact and cut a surface decreases with increasing grain diameter. Why is this so?
26. Why are centerless grinders so popular in industry compared to center-type grinders?
27. How is the through-feed rate varied in a centerless grinder? Could you change the feed rate while the process is in operation?
28. Why are vacuum chucks used in surface grinding and not in milling? See Chapter 28 for a discussion of workholding devices.
29. How does creep feed grinding differ from conventional surface grinding?
30. Why does a lap not wear, since it is softer than the material being lapped?
31. How do honing stones differ from grinding wheels?
32. What is meant by "charging" a lap?
33. Why is a honing head permitted to float in a hole that is being honed?
34. In what respect does a coated abrasive differ from an abrasive wheel?
35. What is the relationship of wheel wear to metal-removal rate?
36. What is the inclined angle in honing, and what determines it?
37. What are the common causes of grinding accidents?
38. What other machine tool does a surface grinder resemble?

■ PROBLEMS

1. Perhaps you have observed the following wear phenomena: A set of marble or wooden stairs shows wear on the treads in the regions where people step when they climb (or descend) the stairs. The higher up the stairs, the less the wear on the tread. Given that soles of shoes (leather, rubber) are far softer than marble or granite, explain:
   (a) Why the stairs wear.
   (b) Why the lower stairs are more worn than the upper stairs.

2. Explain why it is that a small particle of a material can be used to abrade a surface made of the same material (i.e., why does the small particle act harder or stronger than the bulk material)?

3. In grinding, both the wheel and workpiece are moving (or rotating). Using the data in Figure 27-10 and assuming that you are doing surface grinding (see Figure 27-1), what are some typical MRR values? How do these compare to MRR values for other machining processes, such as milling? What is the significance of this?

*C*hapter 27 CASE STUDY

aluminum retainer rings

T he Briggs Corporation has to make 5000 retainer rings, as shown in Figure CS-27. It is essential that the surfaces be smooth with no sharp corners on the circumferential edges. Determine the most economical method for manufacturing these rings.

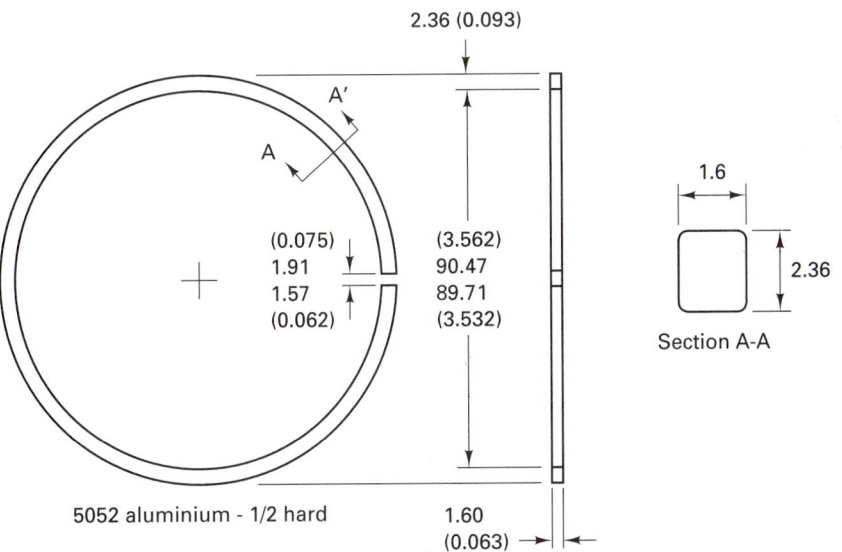

FIGURE CS-27   Aluminum snap-in retainer ring.

CHAPTER 28

WORKHOLDING DEVICES

28.1	INTRODUCTION		28.8	UNLOADING AND LOADING TIME
28.2	CONVENTIONAL FIXTURE DESIGN		28.9	SETUP AND CHANGEOVER
28.3	DESIGN CRITERIA FOR WORKHOLDERS		28.10	TYPES OF JIGS
	Positive Location		28.11	EXAMPLES OF CONVENTIONAL FIXTURES
	Repeatability			
	Adequate Clamping Forces		28.12	CLAMPS
	Reliability		28.13	MODULAR FIXTURING
	Ruggedness		28.14	GROUP JIG AND FIXTURE
	Design and Construction Ease		28.15	OTHER WORKHOLDING DEVICES
	Low Profile			Assembly Jigs
	Workpiece Accommodation			Magnetic Workholders
	Ergonomics and Safety			Electrostatic Workholders
	Freedom from Part Distortion			Vacuum Chucks
	Flexibility		28.16	ECONOMIC JUSTIFICATION OF JIGS AND FIXTURES
28.4	DESIGN STEPS			
28.5	CLAMPING CONSIDERATIONS		Case Study:	OVERHEAD CRANE INSTALLATION
28.6	EXAMPLE OF JIG DESIGN			
28.7	CHIP DISPOSAL			

■ 28.1 INTRODUCTION

In previous chapters attention has repeatedly been directed to the manner in which workpieces are mounted and held in the various machine tools. Workholding devices—that is, jigs and fixtures—are critical components in the manufacturing of interchangeable parts. *Workholders* hold (locate) the work in the machine tool with respect to the cutting tool. See Figures 29-2 and 29-22 in the next chapter for more examples of fixtures in relation to the machine tool and the cutting tool. With workholders, process capability (repeatability) can be achieved which otherwise would be impossible with a given combination of cutting tools and machine tools. In this chapter, workholding devices (jigs and fixtures) will be considered as important production tools or adjuncts, with primary attention being directed toward their functional characteristics, their relationship to the machine tools, and the manufacturing processes.

In recent years, designing the workholding device to be more flexible (i.e., able to accommodate more than one part or able to be quickly exchanged) has become very important. Flexible workholders are a critical element in manufacturing cells, both manned and unmanned. For the cell to be flexible, workholding devices should be able to accommodate all the parts within the *family of parts* (see Chapters 41 and 43). This design requirement has added significantly to the complexity of conventional jig and fixture design. Let's begin with a discussion of the basics of jig and fixture design.

■ 28.2 CONVENTIONAL FIXTURE DESIGN

In the conventional method of fixture design, tool designers rely on their experience and intuition to design single-purpose fixtures for specific machining operations, often using a trial-and-error method until the workholders perform satisfactorily. Of course, these designers should calculate the clamping forces or stress distributions in the fixturing elements to determine the loads that will deform the fixtures or the workpieces elastically or plastically. In the design of the workholding devices, two primary functions must be considered: locating and clamping. *Locating* refers to orienting and positioning the part in the machine tool with respect to the cutting tools to achieve the required specifications. *Clamping* refers to holding or maintaining the part in that location during the operations.

Jigs and *fixtures* are specially designed and built workholding devices that hold the work during machining or assembly operations. In addition, a jig determines a location dimension that is produced by machining or fastening. For example, location dimensions determine the position of a hole on a plate (Figure 28-1). Consider the subject of dimensioning as used in drafting practice. Dimensions are of two types: size and location. *Size* dimensions denote the size of geometrical shapes—holes, cubes, parallelepipeds, and so on—of which objects are composed. *Location* dimensions, on the other hand, determine the position or location of these geometrical shapes *with respect to each other.* Thus *a* and *c* in Figure 28-1 are location dimensions, whereas *e* and *g* are size dimensions. With location dimensions in mind, one can precisely define a jig as follows: *A jig is a special workholding device that, through built-in features, determines location dimensions that are produced by machining or fastening operations.* The key requirements of a jig is that it determines a location dimension. Thus jigs accomplish layout automatically.

In establishing location dimensions, jigs may do a number of other things. They frequently guide tools, as in drill jigs, and thus determine the location of a component geometrical shape. However, they do not always guide tools. In the case of welding jigs,

FIGURE 28-1   Drawing of a plate showing locating dimension versus sizing
dimensions.

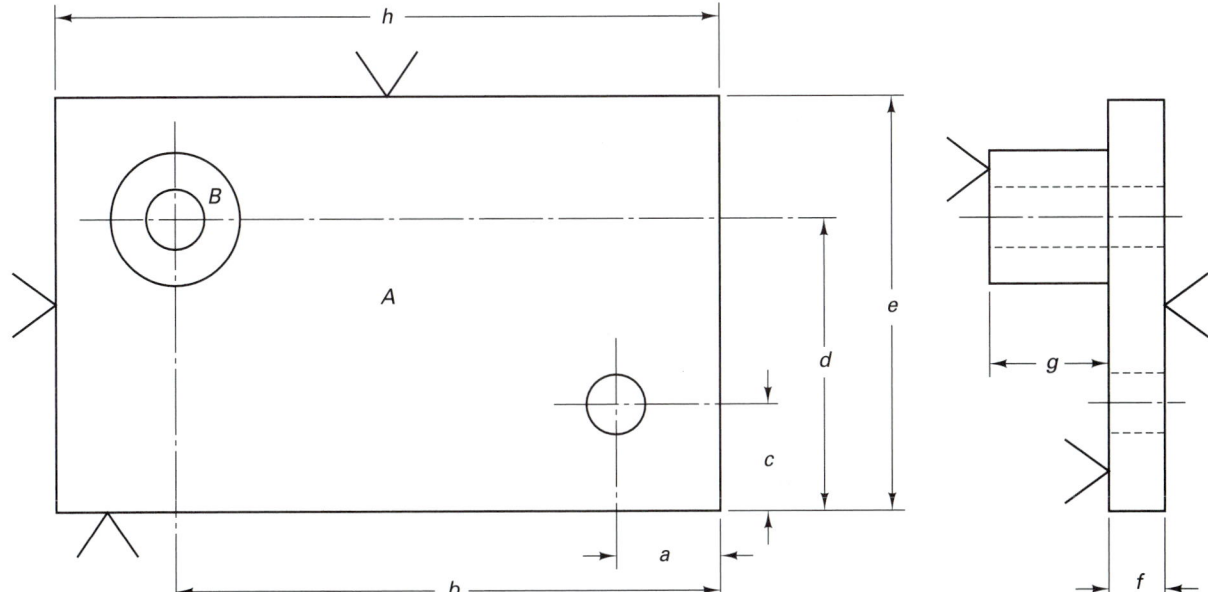

component parts are held (located) in a desired relationship with respect to each other while an unguided tool accomplishes the fastening. The guiding of a tool is not a necessary requirement of a jig.

Similarly, jigs usually hold the work that is to be machined, fastened, or assembled. However, in certain cases, the work actually supports the jig. Thus, although a jig *may* incidentally perform other functions, the basic requirement is that, through qualities that are built into it, certain critical dimensions of the workpiece are determined.

A *fixture* is *a special workholding device that holds work during machining or assembly operations and establishes size dimensions.* The key characteristic is that it is a *special* workholding device, designed and constructed for a particular part or shape. A general-purpose device, such as a vise, chuck, or clamp, is usually not considered to be a fixture. Thus a fixture has as its specific objective the facilitating of *setup,* or making the part holding easier. Because many jigs hold the work while determining critical location dimensions, they usually meet all the requirements of a fixture. Alternatively, many fixtures are used in NC machines holding parts where holes are located and drilled according to a program. (See the discussion of machining centers in Chapter 29.) So the strict definition of jigs and fixtures has been blurred by the changes in technology.

■ 28.3 Design Criteria for Workholders

To meet all the design criteria for workholders is impossible. Compromise is inevitable. Still, it is useful to know the optimal design objectives to illustrate the positioning, holding, and supporting functions that fixtures must fulfill.

Positive Location

A fixture must, above all else, hold the workpiece precisely in space to prevent each of 12 kinds of degrees of freedom-linear movement in either direction along the *X*, *Y*, and *Z* axes and rotational movement in either direction about each axis. (See the discussion of the 3-2-1 principle in Section 28.4.)

Repeatability

Identical workpieces should be located by the workholder in precisely the same space on repeated loading and unloading cycles. It should be impossible to load the workpiece incorrectly. This is called "fool proofing" the jig or fixture.

Adequate Clamping Forces

The workholder must hold the workpiece immobile against the forces of gravity, centrifugal forces, inertial forces, and cutting forces. Milling and broaching operations, in particular, tend to pull the workpiece out of the fixture, and the designer must calculate these machining forces against the fixture's holding capacity. The device must be rigid.

Reliability

The clamping forces must be maintained during machine operation every time the device is used. The mechanism must be easy to maintain and lubricate.

Ruggedness

Workholders usually receive more punishment during the loading and unloading cycle than during the machining operation. The device must endure impact and abrasion for at least the life of the job. Elements of a device that are subject to damage and wear should be easily replaceable.

Design and Construction Ease

Workholders should use standard elements as much as possible to allow the engineer to concentrate on function rather than on construction details. Modular fixtures epitomize this design rule as the entire workholder is made from standard elements, permitting a bolt-together approach for substantial time and cost savings over custom workholders.

Low Profile

Workholder elements must be clear of the cutting tool path. Designing lugs on the part for clamping can simplify the fixture and allow proper tool clearance.

Workpiece Accommodation

Surface contours of castings or forgings vary from one part to the next. The device should tolerate these variations without sacrificing positive location or other design objectives.

Ergonomics and Safety

Clamps should be selected and positioned to eliminate pinch points and facilitate ease of operation. The workholder elements should not obstruct the loading or unloading of workpieces. In manual operations, the operator should not have to reach past the tool to load or unload parts. A rule sometimes used is that the operator can repeatedly exert a force of 30 to 40 lb to open or close a clamp but greater forces than this can cause ergonomic problems.

Freedom from Part Distortion

Parts being machined can be distorted by gravity, the machining forces, or the clamping forces. Once clamped into the device, the part must be unstressed or, at least, undistorted. Otherwise, the newly machined surfaces take on any distortions caused by the clamping forces.

Flexibility

The workholding device should be designed so that it can be quickly exchanged and/or so that it can locate and restrain more than one type (design) of part. Many different schemes are being proposed to provide workholder flexibility: Modular vise fixturing, programmable clamps using air-activated plungers, part encapsulation with a low-melting-point alloy, and NC-controlled clamping machines are some of the more recently developed systems. Despite their flexibilities, these clamping systems have some significant drawbacks. They are expensive, and the individual systems may not integrated well into individual machine tools. [See the discussions of intermediate jig concept (Chapter 43) and group jigs (Chapter 41) for additional thoughts on flexibility.]

■ 28.4 DESIGN STEPS

The classical design of a workholder (e.g., a drill jig) involves the following steps:

1. Analyze the drawing of the workpiece and determine (visualize) the machining operations required to machine it. Note the critical (size and location) dimensions and tolerances.

2. Determine the orientations of the workpiece in relation to the cutting tools and their movements.
3. Perform an analysis to estimate the magnitude and direction of the cutting forces (see Chapter 21).
4. Study the standard devices available for workholders and for the clamping functions. Can an off-the-shelf device be modified? What standard elements can be used?
5. Form a mental picture of the workpiece in position in the workholder in the machine tool with the cutting tools performing the required operation(s). See the figures in Chapters 28 and 29 for examples.
6. Make a three-dimensional sketch of the workpiece in the workholder in its required position to determine the location of all the elements: clamps, locator buttons, bushings, and so on.
7. Make a sketch of the workholder and workpiece in the machine tool to show the orientation of these elements with respect to the cutting tool in the machine tool.

After determining the orientation of the workpiece in the workholder, the next step is to locate it in that position. This location is also used for all similar workpieces. The designer must select or design locating devices (supports) *which ensure that every workpiece placed in the device occupies the same position with respect to the cutting tools.* Thus, when the machining operation is performed, the workpieces are processed identically. This is, of course, the key to making interchangeable parts. In locating the workpiece, the basic *3-2-1 principle* of location is used (Figure 28-2). For positive location, the fixture must position the workpiece in each of three perpendicular planes. Positioning processes can vary greatly, but workholder design always begins by defining the first plane of reference with three points. Once the object is defined in a single plane, supported at three points (like a three-legged stool on a floor), a second plane can be assigned that is perpendicular to the first. To do this, the object is brought up against any two points in the second plane. To continue the example, the stool is slid along the floor until two legs touch a wall.

FIGURE 28-2 Workpiece location is based on the 3-2-1 principle.

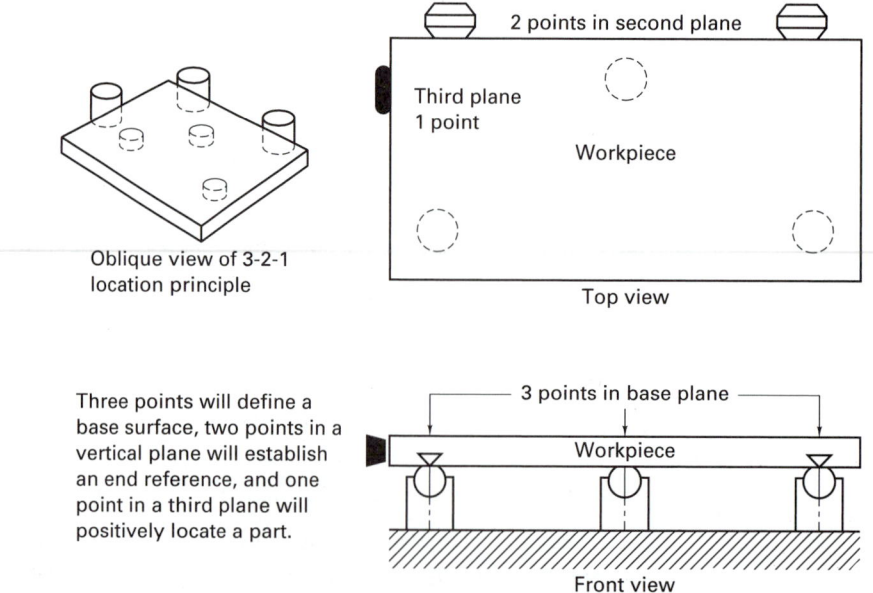

Oblique view of 3-2-1 location principle

Top view

2 points in second plane

Third plane 1 point

Workpiece

Three points will define a base surface, two points in a vertical plane will establish an end reference, and one point in a third plane will positively locate a part.

3 points in base plane

Workpiece

Front view

A third plane, perpendicular to each of the other two, is then defined by designating one point on it. As long as an object is in contact with three points on the first plane, two points on the second, and a single point on the third, it is positively located in space. The location points within each plane should be selected as far apart as possible for maximum stability.

In practice, it is often necessary to support a workpiece on more points than this 3-2-1 formula dictates. The machining of a large rectangular plate, for example, typically requires support at four or more points. However, any extra points must be established carefully to support the workpiece in a plane defined by three—and only three—points.

Appropriate clamping devices are selected so that the clamping forces hold the workpiece in the proper location and resist the effects of the cutting forces, centrifugal forces, and vibrations. If possible, the machining forces should act into the location points, not into the clamps, so that smaller clamps can be used. In reality, the worker often determines clamping force when loading the part into the workholder.

Fixtures are usually fastened to the table of the machine tool. Although used primarily on milling and broaching machines, fixtures are also designed and used to hold workpieces for various operations on most of the standard machine tools and machining centers. Some additional design rules for fixture design are given in Table 28-1.

TABLE 28-1. Twenty Principles of Jig and Fixture Design

1. Determine the critical surfaces or points for the part.
2. Decide on locating points and clamping arrangements.
3. For mating parts, use corresponding locating points or surfaces to ensure proper alignment when assembled.
4. Try to use 3-2-1 location, with 3 assigned to largest surface. Additional points should be adjustable.
5. Locating points should be visible so that the operator can see if they are clean. Can they be replaced if worn?
6. Provide clamps that are as quick acting and easy to use as is economically justifiable.
7. Clamps should not require undue effort by the operator to close or to open, nor should they harm hands or fingers during use.
8. Clamps should be integral parts of device. Avoid loose parts that can get lost.
9. Avoid complicated clamping arrangements or combinations that can wear out or malfunction.
10. Locate clamps opposite locators (if possible) to avoid deflection/distortion during machining and springback afterward.
11. Take the thrust of the cutting forces on the locators (if possible) and not on the clamps.
12. Arrange the workholder so that the workpiece can easily be loaded and unloaded from the device and so that it can be loaded only in the correct manner (mistake-proof) and in such a way that the location can be found quickly.
13. Consistent with strength and rigidity, make the workholder as light as possible.
14. Provide ample room for chip clearance and removal.
15. Provide accessibility for cleaning.
16. Provide for entrance and exit of cutting fluid (which may carry off chips) if one is to be used.
17. Provide four feet on all movable workholders.
18. Provide hold-down lugs on all fixed workholders.
19. Provide keys to align fixtures on machine tables.
20. Do not sacrifice safety for production.

■ 28.5 CLAMPING CONSIDERATIONS

Clamping of the work is closely related to support of the work. Any clamping, of course, induces some stresses into the part that can cause some distortion of the workpiece, usually elastic. If this distortion is measurable, it will cause some inaccuracy in final dimensions, as illustrated in an exaggerated manner in Figure 28-3. The obvious solution is to spread

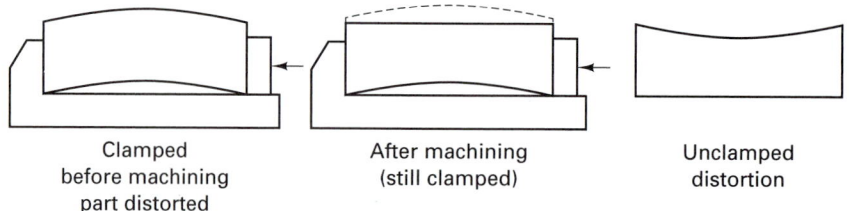

Clamped
before machining
part distorted

After machining
(still clamped)

Unclamped
distortion

FIGURE 28-3   Exaggerated illustration of the manner in which excessive clamping forces can affect the final dimensions of a workpiece.

the clamping forces over a sufficient area to reduce the stresses to a level that will not produce appreciable distortion.

The clamping forces should direct the work against the points of location and work support. Clamped surfaces often have some irregularities that may produce force components in an undesired direction. Consequently, clamping forces should be applied in directions that will assure that the work will remain in the desired position.

Whenever possible, jigs and fixtures should be designed so that the forces induced by the cutting process act to hold the workpiece in position against the supports. These forces are predictable, and proper utilization of them can materially aid in reducing the magnitude of the required clamping stresses. In addition to locating the work properly, the stops or work-supporting areas must be arranged so as to provide adequate support against the cutting forces. As shown in Figure 28-4a, having the cutting force act against a fixed portion of the jig or fixture and not against a movable section permits lower clamping forces to be used. Figure 28-4b illustrates the principle of keeping the points of clamping as nearly as possible in line with the action forces of the cutting tool so as to reduce their tendency to pull the work from the clamping jaws. Compliance with this principle results both in lower clamping stresses and less massive clamping devices. The location points should be as far apart as possible but positioned so as not to allow the cutting force to distort the work. The cutting forces may distort the work, with resulting inaccuracy or broken tools. These design suggestions materially reduce vibration and chatter during the cutting process.

As many operations as possible and practical should be performed with each clamping of the workpiece. This principle has both physical and economic aspects. Because some stresses result from each clamping, with the possibility of accompanying distortion, greater accuracy is achieved if multiple operations are performed with each clamping. From the economic viewpoint, if the number of jigs or fixtures is reduced, less capital will be required and less time will be spent handling the workpiece loading and unloading.

■ 28.6 EXAMPLE OF JIG DESIGN

Some of these principles of work location and tool guidance are illustrated in Figure 28-5. The two mounting holes in the base of the bearing block are to be located and drilled. The dimensions, A, B, and C, are determined by the jig. There also is one other location dimension that must be controlled. The axes of the mounting holes must be at right angles to the bottom surface of the block.

The way in which the part dimensions are obtained in the finished workpiece is as follows. The surfaces marked with a V are reference (or location) surfaces and are finished (machined) prior to insertion of the part into the drill jig. The part rests in the jig on four buttons marked X in Figure 28-5. These buttons, made of hardened steel, are set into the

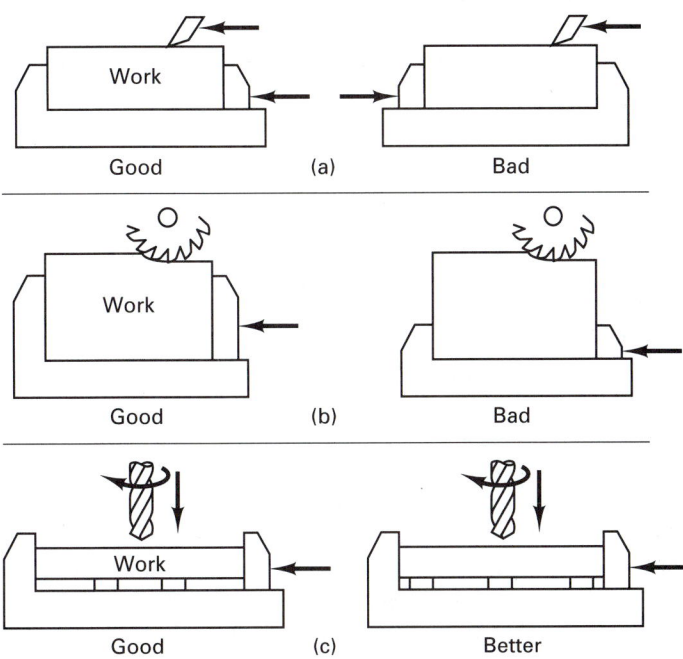

FIGURE 28-4   Proper work support to resist the forces imposed by cutting tools. In (c), three buttons form triangle for work to rest on.

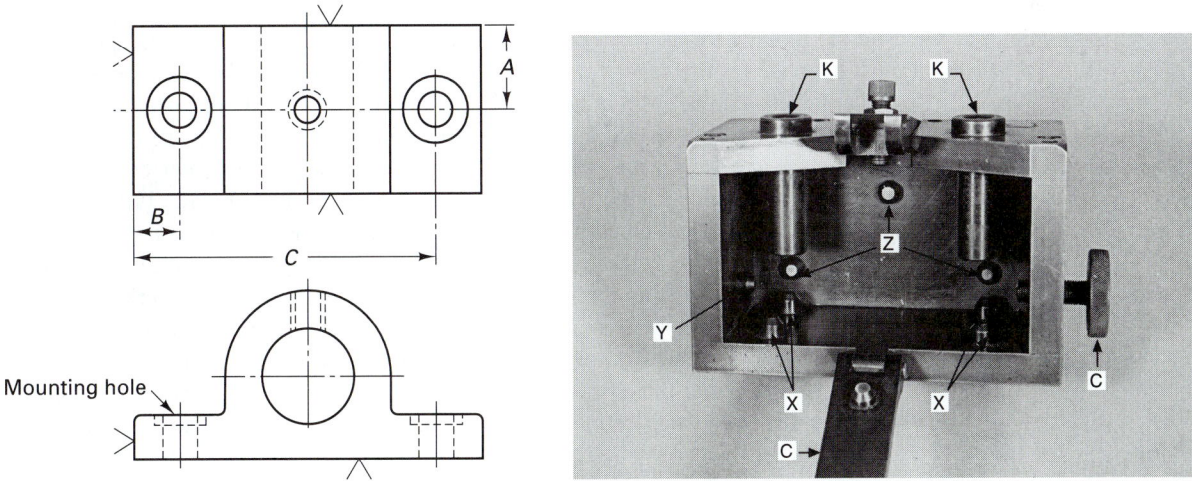

FIGURE 28-5   (*left*) Bearing block, with two mounting holes. (*right*) Box drill jig for drilling the holes.

bottom plate of the jig and are accurately ground so that their surfaces are in a single plane. The left-hand end of the part is held against another button Y in Figure 28-5. This locating button is build into the jig so that its surface is at right angles to the plane of the X buttons. When the block is placed in the jig, its rear surface rests against three more buttons marked Z. These buttons are located and ground so that their surfaces lie in a plane that is at right angles to the planes of both the X and Y buttons. The part is held in its located position by the two clamps marked C.

The use of four buttons on the bottom of this jig (X buttons) appears not to adhere to the 3-2-1 principle stated previously. However, although only two X buttons would have been required for complete location, the use of only two buttons would not have provided adequate support during drilling. The thrust from the drills would have dislodged the part from the locators. Thus the 3-2-1 principle is a *minimum* concept and often must be exceeded.

To assure that the mounting holes are drilled in their proper locations, the drill must be located and then guided during the drilling process. This is accomplished by the two drill bushings marked K in Figure 28-5. Such drill bushings are accurately made of hardened steel with their inner and outer cylindrical surfaces concentric. The inner diameter is made slightly larger than the drill—usually 0.0005 to 0.002 in.—so that the drill can turn freely but not shift appreciably. The bushings are accurately mounted in the upper plate of the jig and positioned so that their axes are exactly perpendicular to the plane of the X buttons, at a distance A from the Z buttons and at distances B and C, respectively, from the plane of the Y button. Note that the bushings are sufficiently long that the drill is guided close to the surface where it will start drilling. Consequently, when the workpiece is properly placed and clamped in the jig, the drill will be located and guided by the bushings so that the critical dimensions on the workpiece will be correct. Figure 28-6 shows the right hole being drilled in a vertical spindle drill press (not running). The box construction is rigid but open for chip removal.

■ 28.7 CHIP DISPOSAL

When jigs or fixtures are used in connection with chip-making operations, adequate provision must be made for the easy removal of the chips. This is essential for several reasons. First, if chips become packed around the tool, heat will not be carried away and tool life can be decreased. Figure 28-7 illustrates how insufficient clearance between the end of a drill bushing and the workpiece can prevent the chips from escaping, whereas too much clearance may not provide accurate drill guidance and can result in broken drills.

A second reason why chips must be removed is so that they do not interfere with proper seating of the work in the jig or fixture (Figure 28-8). Even though chips and dirt always have to be cleaned from the locating and supporting surfaces by a worker or by automatic means, such as an air blast, the design details should be such that chips and other debris will not readily adhere to, or be caught in or on, the locating surfaces, corners, or overhanging elements and thereby prevent the work from seating properly. Such a condition results in distortion, high clamping stresses, and incorrect workpiece dimensions.

■ 28.8 UNLOADING AND LOADING TIME

The cost of the workholders must be justified by the quantities of production involved, and their primary purpose is to increase productivity and quality. While work is being put into or being taken out of jigs and fixtures, the machines with which they are used are not making chips. The loading and unloading time plus the machining time (also called the run

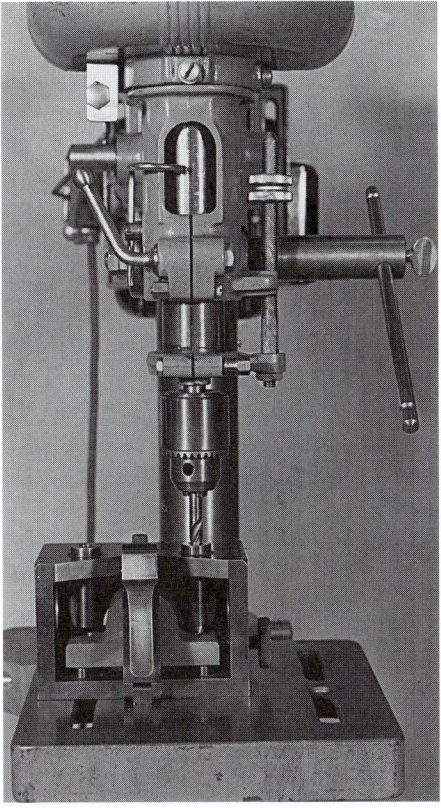

FIGURE 28-6 (*left*) Drilling the mounting holes in a bearing block, using the drill jig shown in Figure 28-5. (*right*) Close-up view, showing the block in the jig, resting on the location buttons, and the drill being guided by the drill bushing.

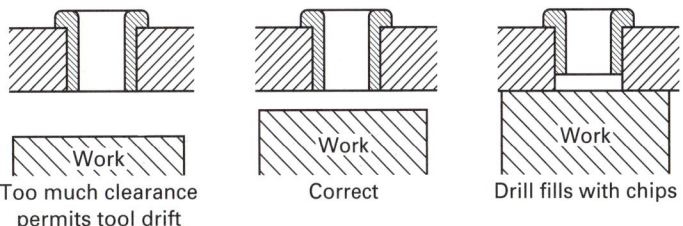

Too much clearance permits tool drift

Correct

Drill fills with chips

FIGURE 28-7 Proper clearance between drill bushing and workpiece.

time) plus any delay times equals the cycle time for a part. The loading and unloading time is greatly influenced by the choice of clamps.

In Table 28-2, the effect of reducing unloading and loading time by 1 minute is shown for two conditions: one in which the machining time is 10 minutes and the other in which it is only 5 minutes. In the first case, *A*, an operator receives $10.00 per hour, and the cost for the machine is $23.50 per hour, which includes all overhead and interest costs. In the second case, *B*, the operator is paid $8.00 per hour, but this results in a cost for the machine of $26.00 per hour. The results of decreasing the loading and unloading time by

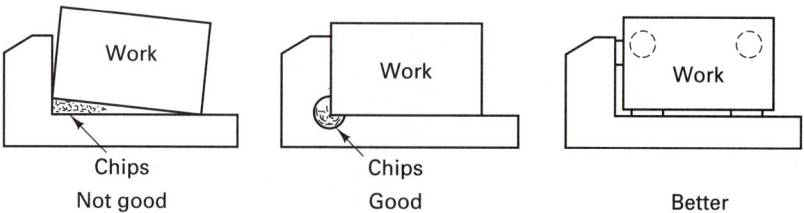

FIGURE 28-8 Methods of providing chip clearance to assure proper seating in the work.

a constant amount are very evident. Although the cost saving in each case happens to be the same ($0.57 per unit), the percentage savings is much greater in the case of the more productive machine. Many machines are equipped with matched (or multiple) fixtures so one part can be loaded while another part is being machined. Figure 29-30 shows a CNC machining center with a pallet exchange system. One part is being loaded while the previous part is being machined. The initial cost for the workholders is increased and the savings must come from decreased loading and unloading time.

TABLE 28-2.	Effects of Reducing Loading and Unloading Time				
	Loading and Unloading (min)	Machine Time (min)	Total Cycle Time (min)	Cost per Unit	Cost Savings per Unit (%)
A. Labor: $10.00/hr	3	10	13	$7.25	7.8
Machine: $23.50/hr	2	10	12	6.69	
B. Labor: $8.00/hr	3	5	8	4.53	12.6
Machine: $26.00/hr	2	5	7	9.96	

There are several ways in which jigs and fixtures can be made easier to load and unload. Some clamping methods can be operated more readily than others. For example, in the drill jig shown in Figure 28-5, a *knurled clamping screw* is used to hold the block against the buttons at the end of the jig. To clamp or unclamp the block in this direction requires several motions. On the other hand, a *cam latch* is used to close the jig and hold the workpiece against the rear locating buttons. This type of latch can be operated with a single motion.

Certainly, the device should be designed so that the part cannot be loaded incorrectly. Defect prevention is often accomplished by the clamping device. If the part is not loaded properly, it cannot be clamped. Ease of operation of workholders not only directly increases the productivity of such equipment but also results indirectly in better quality and fewer lost-time accidents.

As will be pointed out in Chapter 29, the workholder is as critical as the machine tool and the cutting tool to the final quality of the part (see Figure 29-21). The use of the workholder eliminates manual layout of the desired features of the part on the raw material. Manual layout requires a highly skilled worker and is very time consuming. The workholder permits a lower-skilled person to achieve quality and repeatable production with far greater efficiency.

■ 28.9 SETUP AND CHANGEOVER

Every part coming out of the workholder should be the same, resulting in interchangeable parts. But what about the first part? What about the initial setup of the workholder into the machine? In many cases this setup operation takes hours and the machine is not producing anything during this time. Rapid exchanges in tooling will be discussed as a key technique in integrated pull manufacturing systems. See Chapter 43 for a discussion of the elimination of setup and SMED and the intermediate jig concept. Reducing setup times permits shorter production runs (smaller lot sizes). Do not confuse initial setup (of workholders) with part loading and unloading or tool changing. The trick with initial setup is to do it quickly and to get the first part out of the process as a good part, with no adjustment of the machine, the tooling, or the workholder. Quick tool-and-die exchange is a critical component in the strategy for the factory with a future.

■ 28.10 TYPES OF JIGS

Jigs are made in several basic forms and carry names that are descriptive of their general configurations or predominant features. Several of these are illustrated in Figure 28-9.

A *plate jig* is one of the simplest types, consisting only of a plate that contains the drill bushings and a simple means of clamping the work in the jig or the jig to the work. In the latter case, wherein the jig is clamped to the work, the device is sometimes called a *clamp-on jig*. Such jigs frequently are used on large parts, where it is necessary to drill one or more holes that must be spaced accurately with respect to each other or to a corner of the part, but that need not have an exact relationship with other portions of the work.

Channel jigs also are simple and derive their name from the cross-sectional shape of the main member. They can be used only with parts having fairly simple shapes.

Ring jigs are used only for drilling round parts, such as pipe flanges. The clamping force must be sufficient to prevent the part from rotating in the jig.

Diameter jigs provide a means of locating a drilled hole exactly on a diameter of a cylindrical or spherical piece.

Leaf jigs derive their name from the hinged leaf or cover that can be swung open to permit the workpiece to be inserted and then closed to clamp the work in position. Drill bushings may be located in the leaf as well as in the body of the jig to permit locating and drilling holes on more than one side of the workpiece. Such jigs are called *rollover jigs* or *tumble jigs* when they require turning to permit drilling from more than one side.

Box jigs are very common, deriving their name from their boxlike construction. They have five fixed sides and a hinged cover or leaf, or, as shown in Figure 28-9, a cam that locks the workpiece in place. Usually, the drill bushings are located in the fixed sides to assure retention of their accuracy. The fixed sides of the box usually are fastened by means of dowel pins and screws so that they can be taken apart and reassembled without loss of accuracy. Because of their more complex construction, box jigs are costly, but their inherent accuracy and strength can be justified when there is sufficient volume of production. They have two obvious disadvantages: (1) it usually is more difficult to put work into them than into simpler types, and (2) there is a greater tendency for chips to accumulate within them. Figure 28-5 shows a box-type jig.

Because jigs must be constructed very accurately and be made sufficiently rugged so as to maintain their accuracy despite the use (and abuse) to which they inevitably are subjected, they are expensive. Consequently, several methods have been devised to aid in lowering the cost of manufacturing jigs. One way to reduce this cost is to use simple, standardized plate and clamping mechanisms called *universal jigs* (Figure 28-10). These

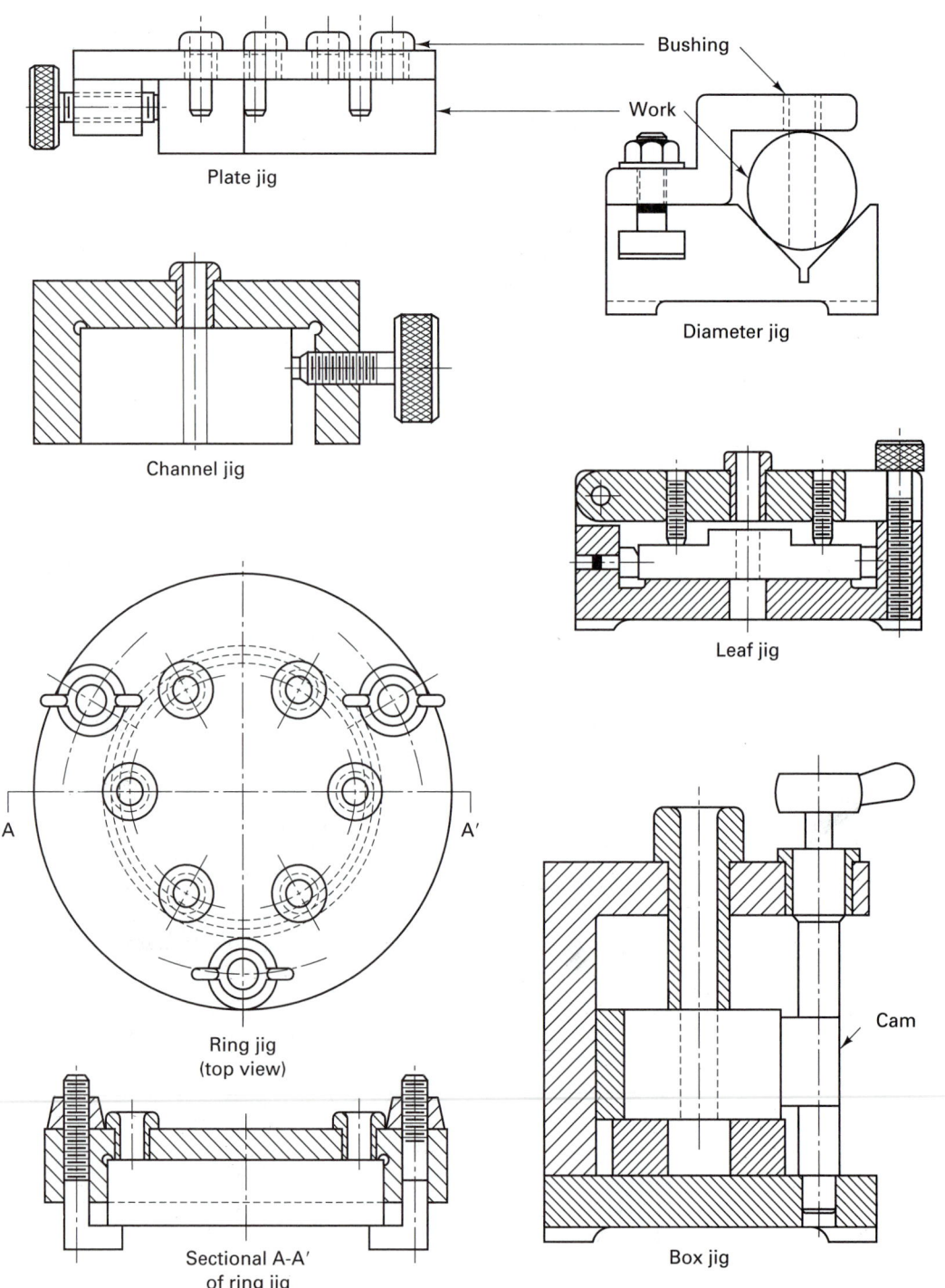

FIGURE 28-9 Examples of some common types of workholders—jigs.

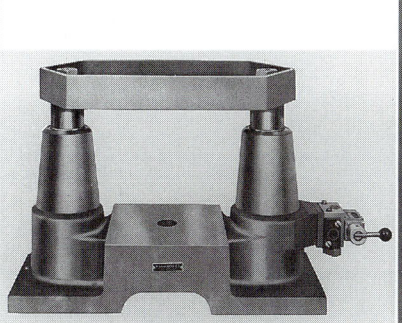
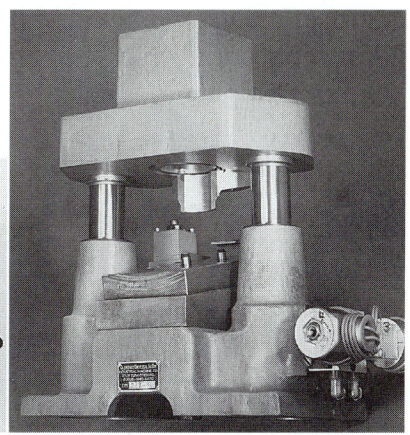

FIGURE 28-10   Two types of universal jigs: Manual (*left*) and power-actuated (*center*). (*right*) Completed jig, made from unit shown at center. (*Courtesy of Cleveland Universal Jig Division, The Industrial Machine Company.*)

can easily be equipped with suitable locating buttons and drill bushings to construct a jig for a particular job. Such universal jigs are available in a variety of configurations and sizes, and because they can be produced in quantities, their cost is relatively low. However, the variety of work that can be accommodated by such jigs obviously is limited.

■ 28.11 EXAMPLES OF CONVENTIONAL FIXTURES

Many examples of conventional fixtures have appeared in the text. Production milling, broaching, and boring processes as performed on NC machines, conventional equipment, or machining centers routinely use fixtures to locate and hold the part properly with respect to the cutting tools on the machine tool. Like cutting tools, fixtures are sold separately and are not usually supplied by the machine tool builder. Traditionally, beginning with Eli Whitney, manufacturers have designed and built custom-made, dedicated fixtures (see Figures 25-19 and 26-13 for two examples). Because of the pressure of shorter production runs and smaller lot sizes, many companies are turning to modular fixturing approaches. The greatest advantage of these systems is that the fixture can be constructed quickly.

Perhaps the most common fixture uses the vise as its base element. Figure 28-11 shows a schematic and photo of a typical commercially available vise that can be adapted for use as a fixture. As shown, the vise jaws are readily modified to conform to the 3-2-1 location principle and provide adequate clamping forces for most every machining operation. The four vises shown in Figure 28-11 are mounted on a subplate for rapid insertion and location in the machine.

The chucks used in lathes are really general-purpose fixtures for rotational parts. Newer chuck designs have greatly improved their flexibility (the range of diameters the chuck can accommodate in a given setup and speed of setup). Figure 28-12 shows a complete change of top jaws for a three-jaw chuck being done in less than 5 minutes. The normal time for this part of the setup might exceed 15 minutes. New quick-change insert top jaws may even snap in by hand with no jaw nuts, keys, screws, or tools. Jaws that can be exchanged by robots can also be designed.

Figure 28-13 shows an example of the *intermediate jig concept*, also discussed in Chapter 43, applied to lathes and chucks. An adapter or intermediate fixture is bolted to

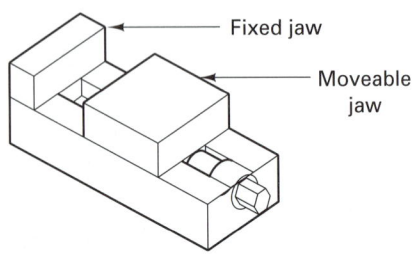

Fixed jaw

Moveable jaw

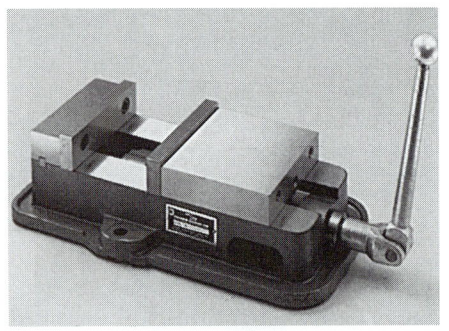

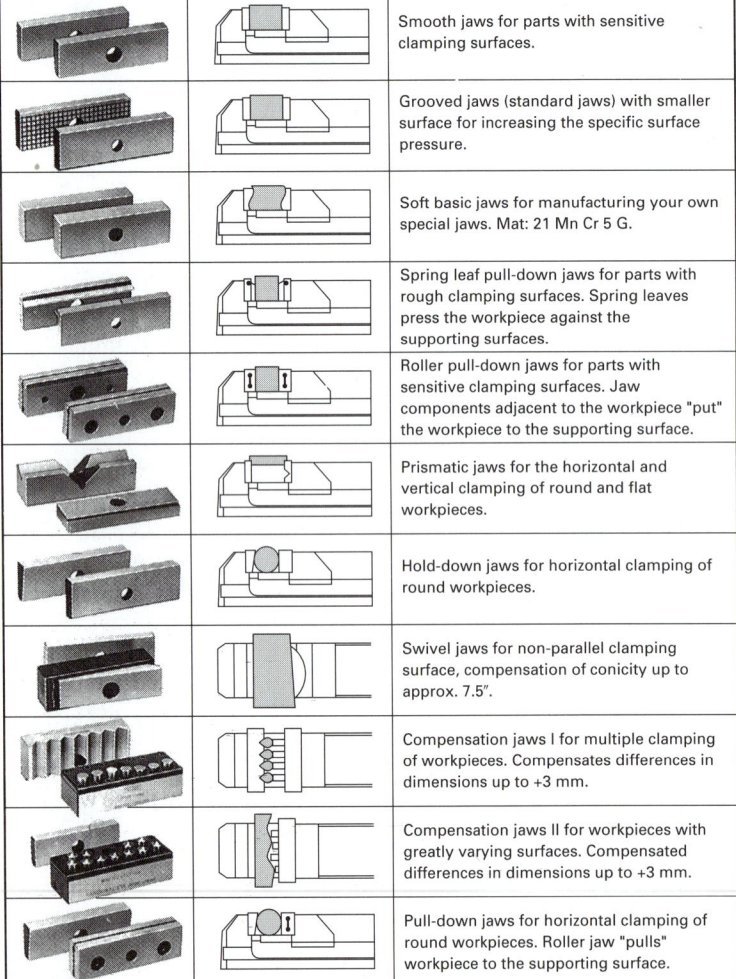

		Smooth jaws for parts with sensitive clamping surfaces.
		Grooved jaws (standard jaws) with smaller surface for increasing the specific surface pressure.
		Soft basic jaws for manufacturing your own special jaws. Mat: 21 Mn Cr 5 G.
		Spring leaf pull-down jaws for parts with rough clamping surfaces. Spring leaves press the workpiece against the supporting surfaces.
		Roller pull-down jaws for parts with sensitive clamping surfaces. Jaw components adjacent to the workpiece "put" the workpiece to the supporting surface.
		Prismatic jaws for the horizontal and vertical clamping of round and flat workpieces.
		Hold-down jaws for horizontal clamping of round workpieces.
		Swivel jaws for non-parallel clamping surface, compensation of conicity up to approx. 7.5″.
		Compensation jaws I for multiple clamping of workpieces. Compensates differences in dimensions up to +3 mm.
		Compensation jaws II for workpieces with greatly varying surfaces. Compensated differences in dimensions up to +3 mm.
		Pull-down jaws for horizontal clamping of round workpieces. Roller jaw "pulls" workpiece to the supporting surface.

FIGURE 28-11 (*above left*) Schematic of a conventional vise; (*above right*) standard vise with removable jaw plates; (*below right*) four vises with modified jaws used as a milling fixture; (*below left*) various designs of vise jaws available for standard vise.

1. Pre-assembled Mini-System and top jaws.

2. Assembly being inserted into the Master Jaw.

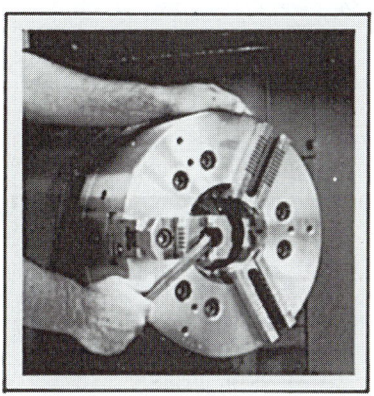

3. Quickly retighten cap screw.

4. All 3 jaws changed in 5 minutes or less.

FIGURE 28-12 Quick-changing of the top jaws on a three-jaw chuck. *(Courtesy of Huron Machine Products).*

the machine spindle and is a permanent part of the machine tool. The intermediate fixture will accept mating chucks that have been preset for the workpiece prior to insertion. Different chuck designs mount interchangeably on the common actuator. This method greatly reduces setup time and permits the operator to perform chuck maintenance and retooling (setup) while the machine is running. These chucks can be exchanged automatically.

Most producers of chucks use some variation of equation (28-1) to compute the maximum rpm rate at which the chuck can run:

$$S_m^2 = \frac{F_m}{3 \times (2.84 \times 10^{-5}) \times W \times D} \tag{28-1}$$

where

S_m = maximum rpm value at which gripping force equals $\frac{1}{3} F_m$
F_m = maximum rated gripping force, at rest (lb)
W = combined weight of jaws (lb)
D = distance from spindle centerline to center of jaw mass (in.)

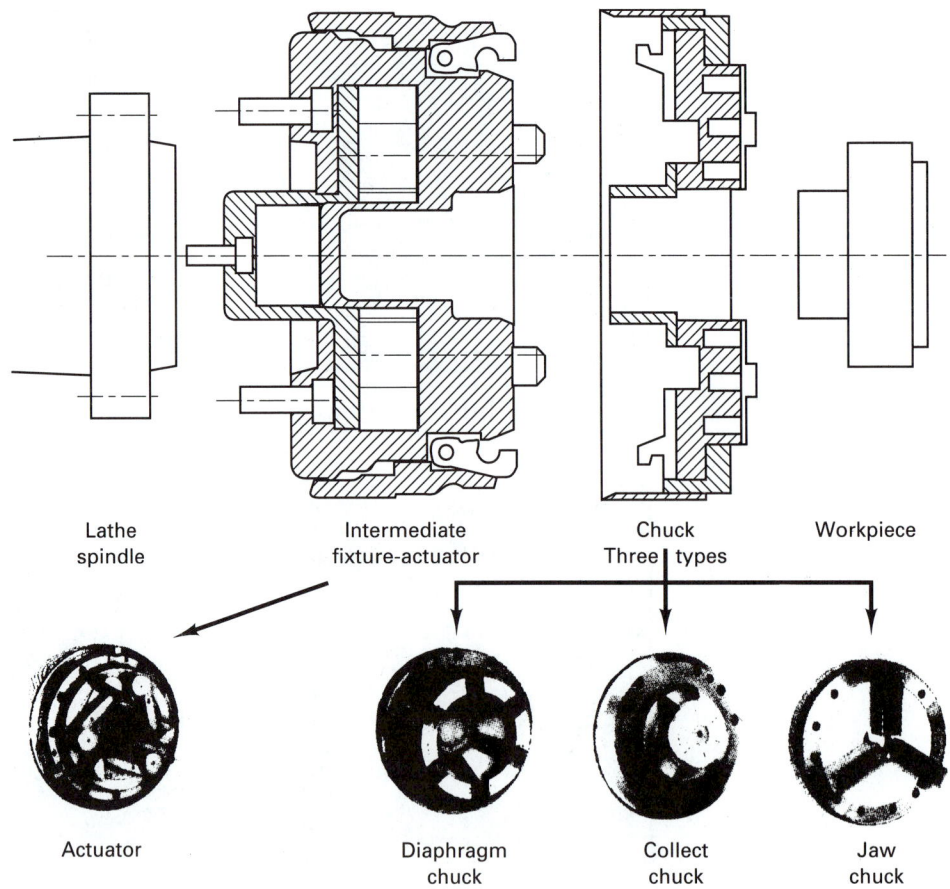

| Lathe
spindle | Intermediate
fixture-actuator | Chuck
Three types | Workpiece |

| Actuator | Diaphragm
chuck | Collect
chuck | Jaw
chuck |

FIGURE 28-13   Example of the intermediate jig concept applied to lathe chucks.
(Courtesy of Sheffer Collet Company).

Thus, with this equation, a 10-in. power chuck with a published rating, F_m, of 13,200 lb would retain one-third of its initial gripping force at 2507 rpm. (Check this calculation using $W = 8$ lb, $D = 3.1$ in.) The higher the rpm value, the greater the centrifugal force factor. This is an important factor in high-speed machining operations in which the part is rotating.

■ 28.12 CLAMPS

Manual clamps include screw, strap, swing, edge, cam, toggle, and C-clamps, each with certain strengths and weaknesses.

In Figure 28-14 typical types of clamps that are used in fixtures are shown. The *strap clamp* comes in many forms and sizes and is simple, low cost, and flexible. The force can be applied by a hand knob, a cam, or a wrench turning down a nut. A conventional *toggle clamp* accommodates only a small thickness variation from part to part, yet provides an excellent, consistent clamping force.

Figure 28-15 shows some examples of power-actuated clamps. Power clamping provides more consistent clamping forces than do manual clamps, especially in applications

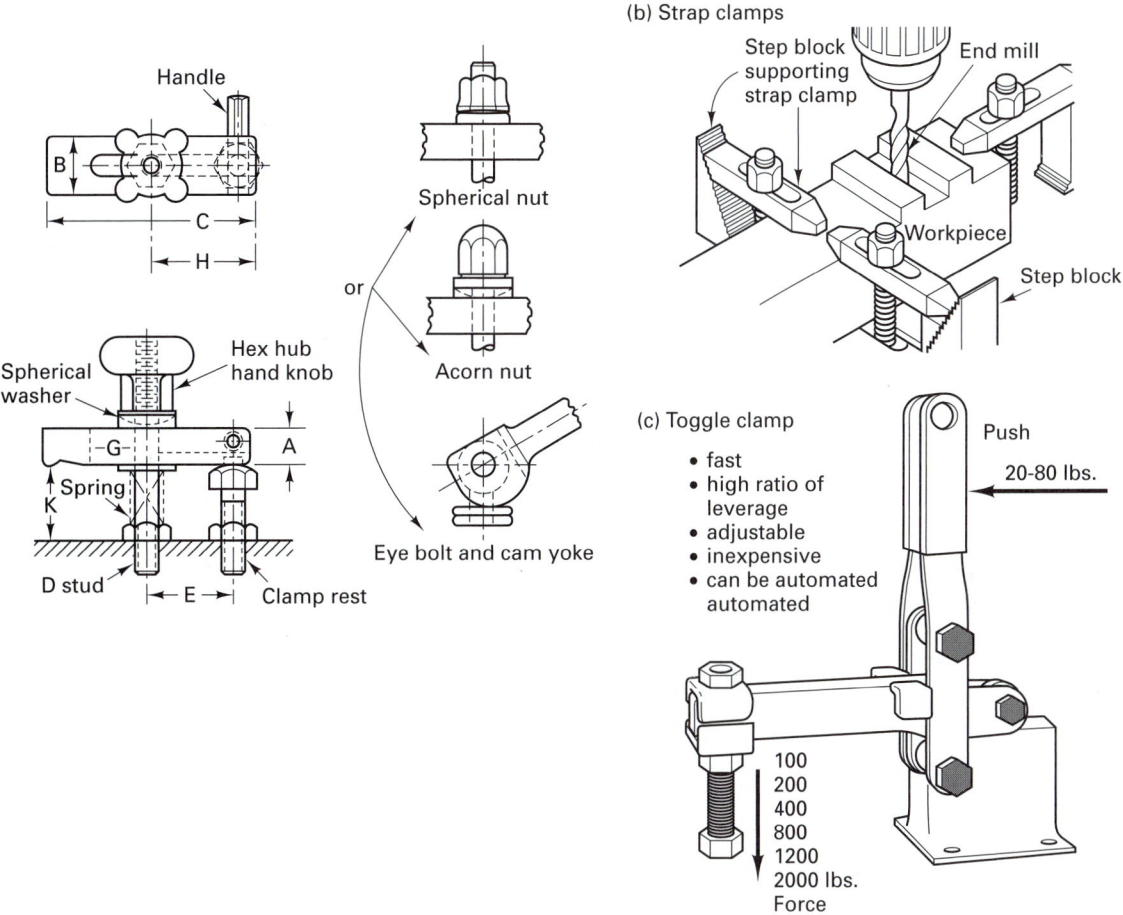

FIGURE 28-14 Examples of common types of clamps used in workholders.

that promote operator fatigue. The higher cost must be weighed against the capability for consistent and repeatable operation, automatic adjustment of holding forces, remote actuations, and automating sequencing of clamping actions. *Extending clamps* operate in a manner similar to that of a manual clamp-strap assembly. They extend forward horizontally, then clamp down. *Edge clamps* have a very low profile. They clamp down and forward simultaneously.

■ 28.13 MODULAR FIXTURING

Modular fixtures have all the same design criteria as those of conventional fixtures, plus one more, *versatility*. Modular fixture elements must be useful for a variety of machining applications and easily adaptable to different workpiece geometries. Individual fixture designs can be photographed or entered into a CAD library for future reference. After the job is done, the fixture itself can be dismantled and the elements returned to the toolroom.

The erector-set approach uses either T slot or dowel-pin designs. Figures 28-16 and 28-17 show two examples of modular fixturing. The designs begin with base plates. Elements for locating and clamping are added to the subplate. Rectangular, square, and round are the typical patterns for the subplates. Also shown are the typical components for modular fixturing systems used for mounting points, locators, attachments, and so on. The

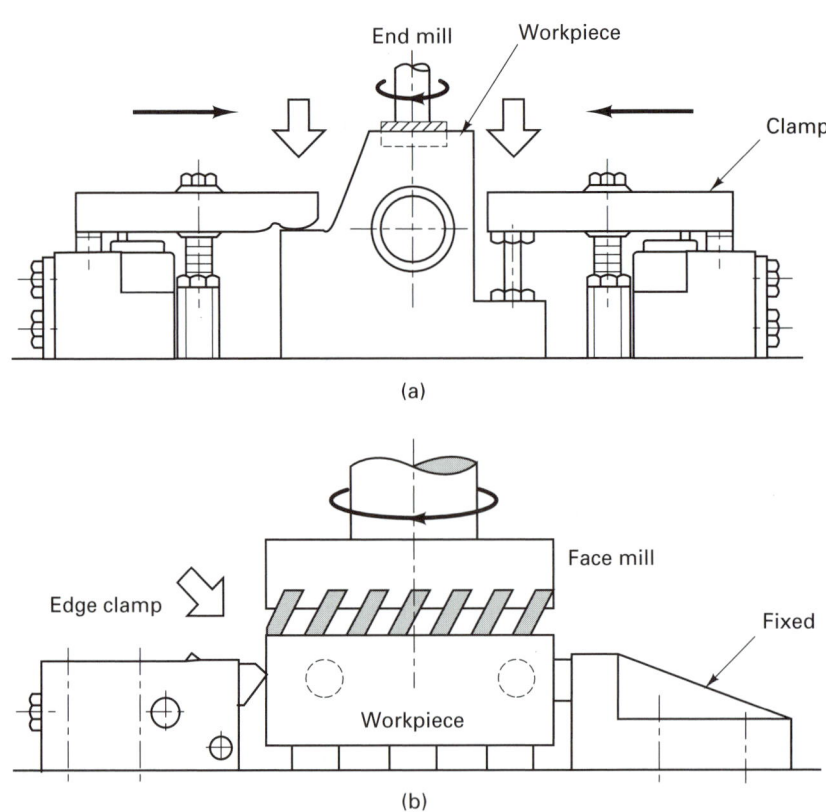

FIGURE 28-15   Examples of power clamping devices: (a) extending clamp; (b) edge clamp.

standard elements needed to construct the fixture include riser blocks, vee blocks, angle plates, cubes, box parallels, and the like. Smaller elements such as locator pins, supports, pads, and clamps are added to the subplate on the larger structural elements. Mechanical clamping devices are shown, but power-assisted clamps are available. The base and fixturing elements are made to tolerances of ±0.0002 to 0.0004 in. in flatness, parallelism, and size. Figure 28-17 shows a part in a dedicated fixture compared to a modular fixture. The dedicated fixture represents a capital investment that must be absorbed by the job and must be maintained after the job is complete. The modular fixture is disassembled and the elements reused later in fixtures for other parts. Modular fixtures are commonly used for prototype tooling and small batch production runs. They are being incorporated more frequently into regular production as users gain confidence in this approach.

■ 28.14 THE GROUP JIG AND FIXTURE

The concept of group jigs and fixtures originated from the technology (GT) concept. GT is discussed in Chapter 41 as a method to form cellular manufacturing systems by determining a family of parts. For a cell to be flexible, the workholding devices should be able to accommodate all the parts within the parts family. For unmanned cells, the workholders will also have to compensate for variation in cutting forces, centrifugal forces, and so on. Group workholders are designed to accept every part family member, with adapters that

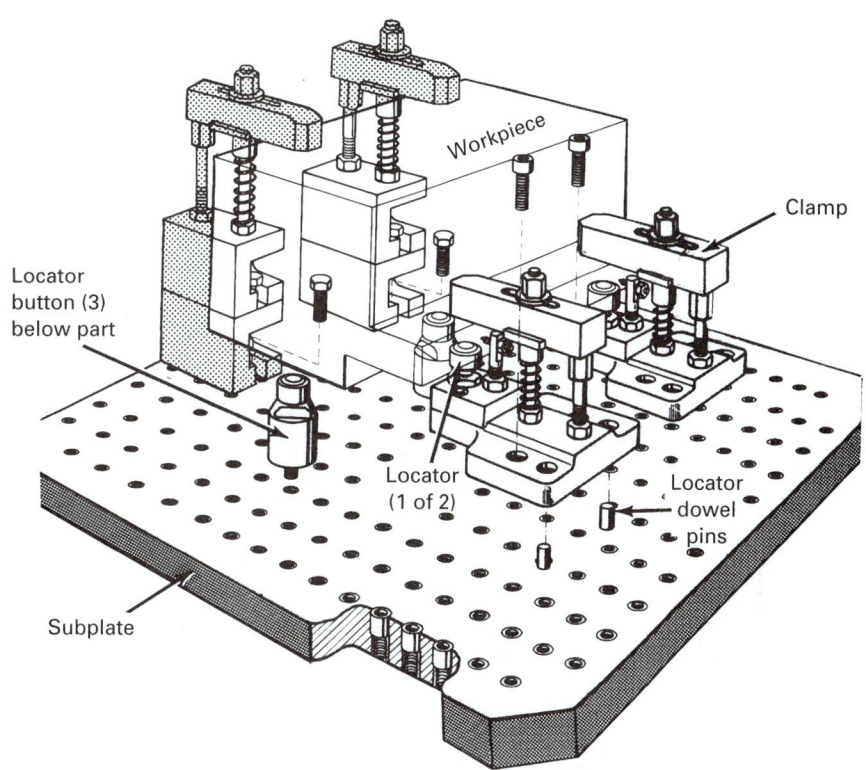

FIGURE 28-16 Modular fixturing begins with a subplate (grid base) and adds locators and clamps.

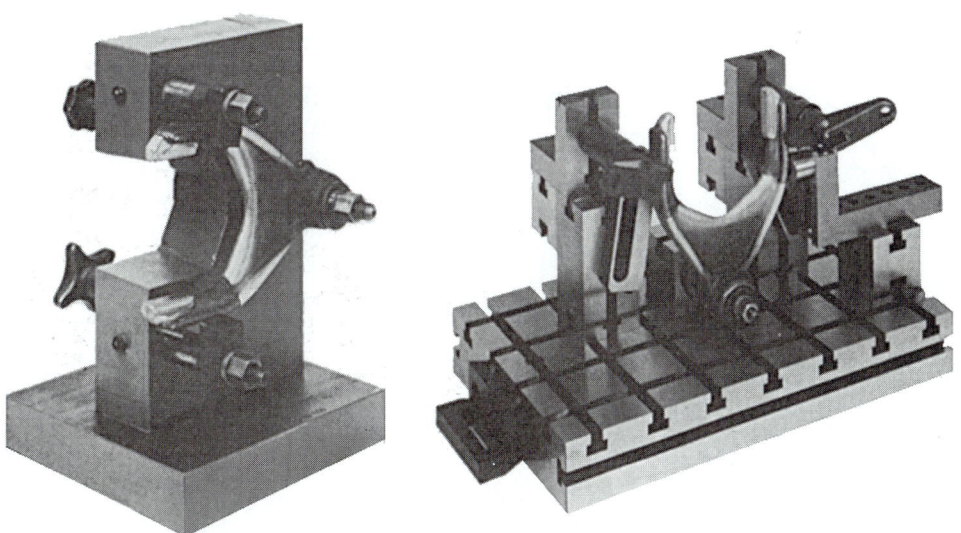

FIGURE 28-17 Dedicated fixture on left versus modular fixture on the right. (From *Manufacturing Engineering,* Jan. 1984.)

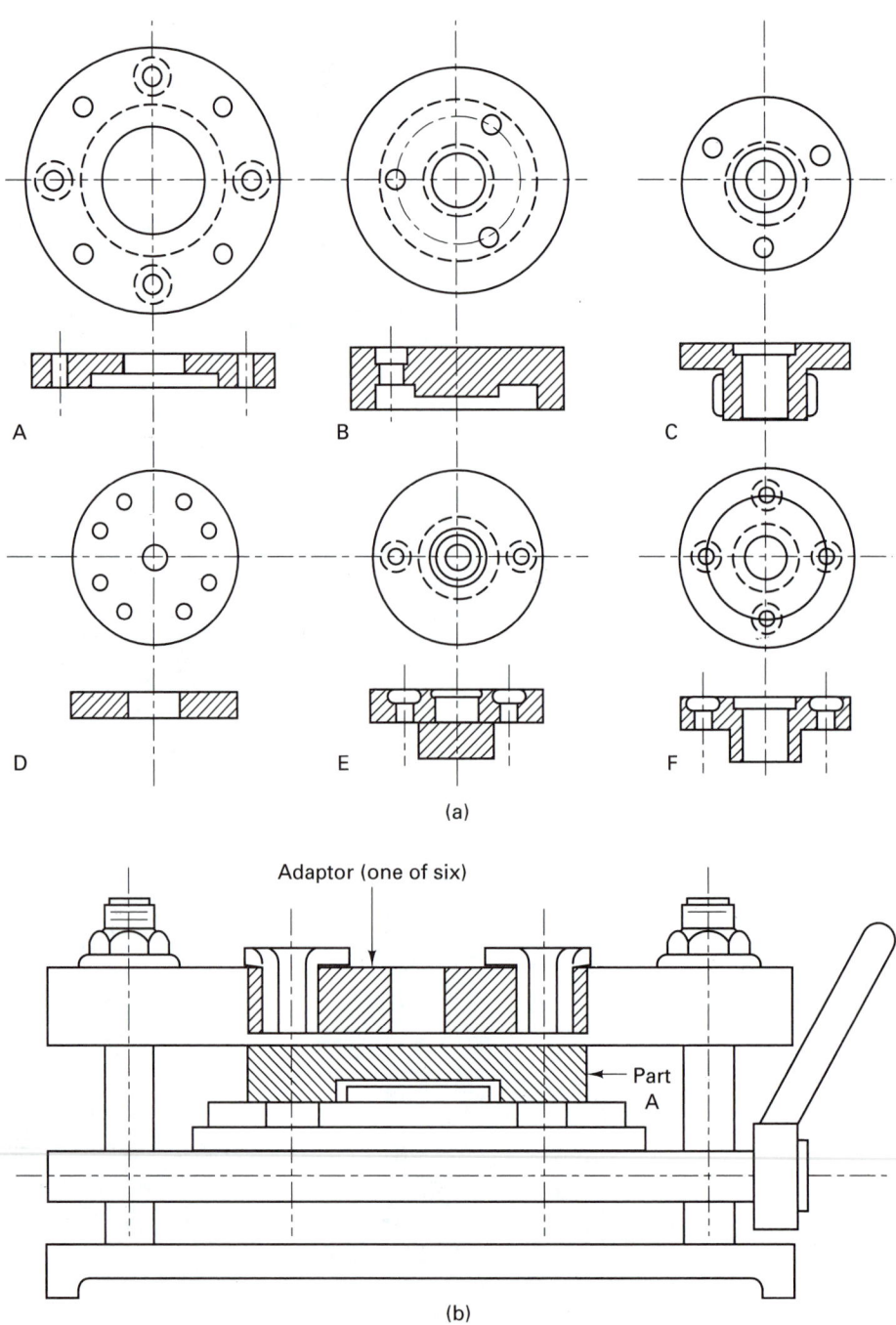

FIGURE 28-18   Master jig designed for part family: (a) part family of round plates (six parts); (b) group jig for drilling, showing adapter and part A.

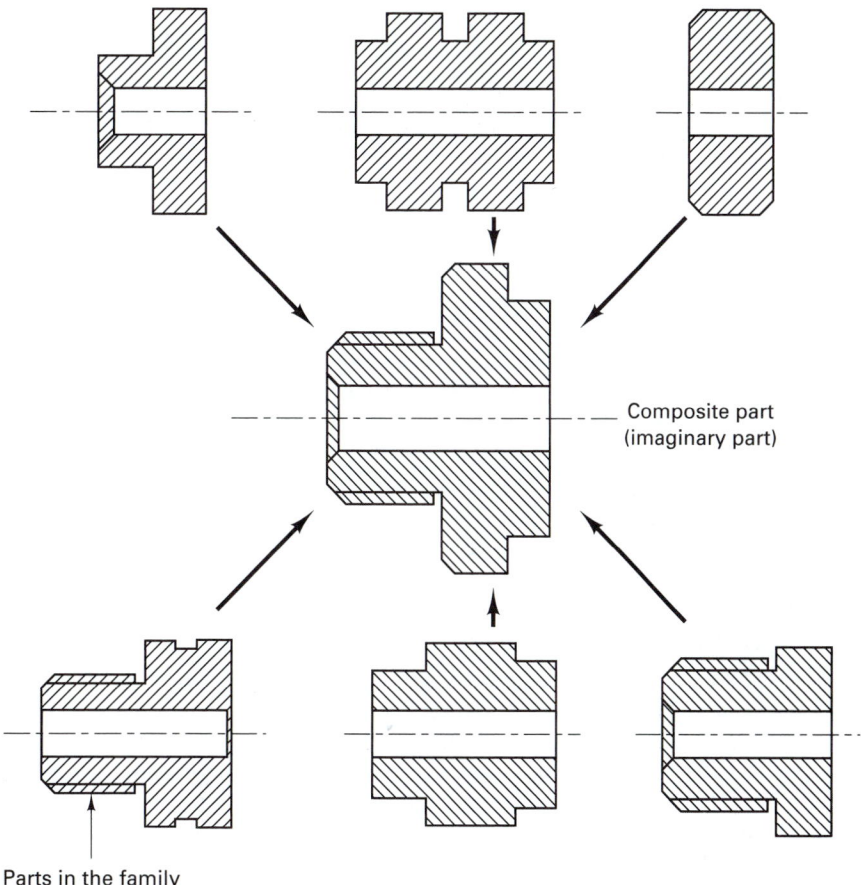

Composite part
(imaginary part)

Parts in the family

FIGURE 28-19   Example of composite part for a part family with six parts.

accommodate minor part variations. Figure 28-18 shows a master jig for processing a family of six parts. Drilling the holes of different parts in this part family requires only one group jig. This jig has six different auxiliary adapters to accommodate differences in hole sizes, numbers, and locations, and in part sizes and shapes. Therefore, instead of designing, fabricating, and using individual drill jigs as is done in a conventional production method, only one group jig with adapters is required. The jig never leaves the machine and therefore never gets lost.

Within group technology is the composite part concept, which is used solely as an aid in designing group fixtures and jigs and group tooling setups. As Figure 28-19 shows, a group of similar parts (i.e., similar in design) is represented by the composite part that possesses all the part family's shape characteristics and machining features. In other words, the composite part is an envelope, the shape of which encompasses the shapes of all the parts in the family. The theory is that if the tooling is designed for the composite part, any part that fits within the envelope could be machined without any tooling changes.

These GT approaches have been shown to be effective in the reduction of setup time.

■ 28.15 OTHER WORKHOLDING DEVICES

Assembly Jigs

Because *assembly jigs* usually must provide for the introduction of several component parts and the use of some type of fastening equipment, such as welding or riveting, they commonly are of the open-frame type. Such jigs are widely used in the automobile and aircraft industries. Large jigs of this type are shown in Figure 28-20 for the assembly of aircraft wings. This jig is constructed mainly of reinforced concrete.

Magnetic Workholders

Because of the light cuts and low cutting forces, workpieces can be held in a different manner on surface grinders than on other machine tools. *Magnetic chucks* are used for ferromagnetic materials. To obtain high accuracy, it is desirable to reduce clamping forces and distribute them over the entire area of the workpiece. Also, grinding is frequently done on quite thin or relatively delicate workpieces, which would be difficult to clamp by normal methods. In addition, there often is the problem of grinding a number of small, duplicate workpieces. Magnetic chucks solve all these problems very satisfactorily. Magnetic chucks are available in disk or rectangular shapes. Dry-disk rectifiers are used to provide the necessary direct-current power. Some magnetic chucks utilize permanent magnets as shown in Figure 28-21 and can be tilted so that angles can be ground.

Magnetic chucks provide an excellent means of holding workpieces provided that the cutting or inertial forces are not too great. The holding force is distributed over the entire contact surface of the work, the clamping stresses are low, and therefore there is little tendency for the work to be distorted. Consequently, pieces can be held and ground accurately. Also, a number of small pieces can be mounted on a chuck and ground at the same time. Magnetic chucks provide great part-to-part repeatability because the same holding power from one part to the next is the same. Initial setup is usually fast, simple, and relatively inexpensive. Parts loading and unloading is also relatively easy.

It often is necessary to demagnetize work that has been held on a magnetic chuck. Some electrically powered chucks provide satisfactory demagnetization by reversing the direct current briefly when the power is shut off.

FIGURE 28-20   Example of large assembly jig for an airplane wing.

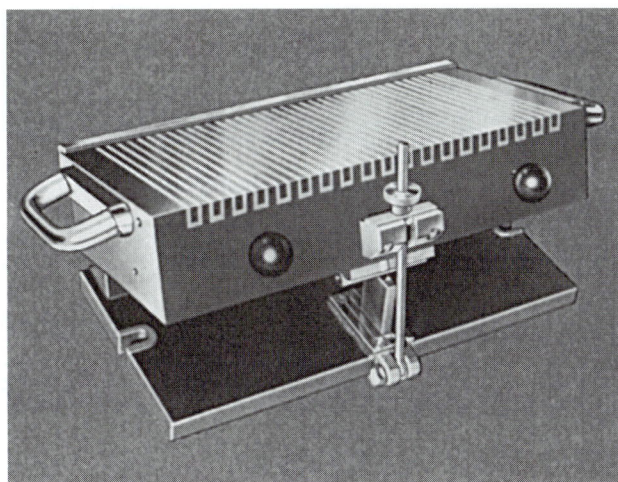

FIGURE 28-21 Example of magnetic chuck. *(Courtesy of O. S. Walker.)*

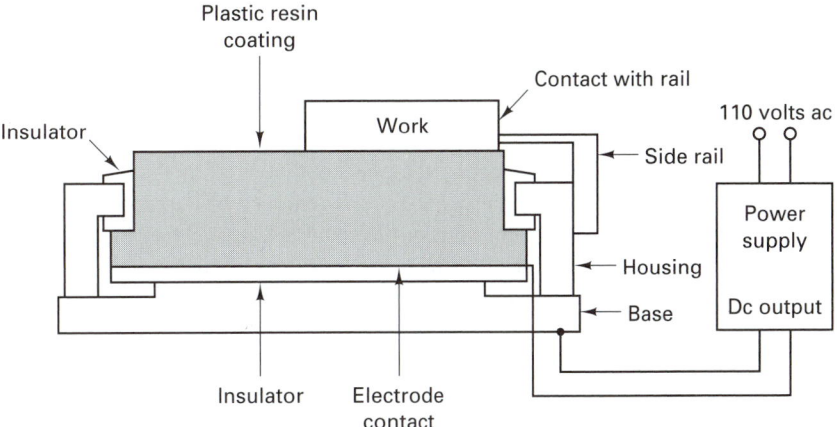

FIGURE 28-22 Principle of electrostatic chuck.

Electrostatic Workholders

Magnetic chucks can be used only with ferromagnetic materials. Electrostatic chucks can be used with any electrically conductive material. This principle (Figure 28-22) directs that work be held by mutually attracting electrostatic fields in the chuck and the workpiece. These provide a holding force of up to 20 psi (21,000 Pa). Nonmetal parts usually can be held if they are flashed (i.e., coated) with a thin layer of metal. These chucks have the added advantage of not inducing residual magnetism in the work.

Vacuum Chucks

Vacuum chucks are also available. In one type, illustrated in Figure 28-23, the holes in the work plate are connected to a vacuum pump and can be opened or closed by means of valve screws. The valves are opened in the area on which the work is to rest. The other type has a porous plate on which the work rests. The workpiece and plate are covered with a polyethylene sheet. When the vacuum is turned on, the film forms around the work piece,

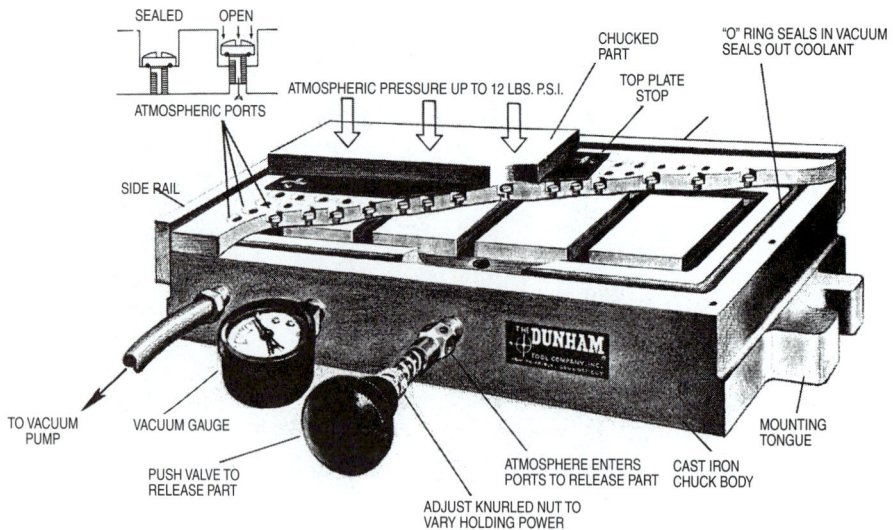

FIGURE 28-23 Cutaway view of a vacuum chuck. *(Courtesy of Dunham Tool Company, Inc.)*

covering and sealing the holes not covered by the workpiece and thus producing a seal. The film covering the workpiece is removed or the first cut removes the film covering the workpiece. Vacuum chucks have the advantage that they can be used on both nonmetals and metals and can provide an easily variable force. Magnetic, electrostatic, and vacuum chucks are used for some light milling and turning operations.

■ 28.16 ECONOMIC JUSTIFICATION OF JIGS AND FIXTURES

As discussed previously, workholders are expensive, even when designed and constructed by using standard components. Obviously, their cost is a part of the total cost of production, and one must determine whether they can be justified economically by the savings in labor and machine cost and improvements in quality that will result from their use. Often, it is only through the use of such devices that the design specifications can be met and sustained from part to part.

To determine the economic justification of any special tooling, the following factors must be considered:

1. The cost of the tooling
2. Interest or profit charges on the tooling cost
3. The savings resulting from the use of the tooling; can result from reduced cycle times or improved quality or lower-cost labor
4. The savings in machine cost due to increased productivity
5. The number of units that will be produced using the tooling

The economic relationship between these factors can be expressed in the following manner:

savings per piece (exclusive of tooling costs) $\geq$ additional cost per piece

$$\underbrace{\begin{array}{l} \text{total cost per piece} \\ \text{without tooling} \end{array} - \begin{array}{l} \text{total cost per} \\ \text{piece using tooling} \\ \text{(exclusive of tooling cost)} \end{array}}_{} \geq \text{tooling cost per piece}$$

$$\underbrace{\begin{array}{l} \text{labor cost} \\ \text{per piece} \\ \text{without} \\ \text{tooling} \end{array} + \begin{array}{l} \text{machine and} \\ \text{overhead cost} \\ \text{per piece} \\ \text{without tooling} \end{array}}_{} - \underbrace{\begin{array}{l} \text{labor cost} \\ \text{per piece} \\ \text{with} \\ \text{tooling} \end{array} + \begin{array}{l} \text{machine and} \\ \text{overhead cost} \\ \text{per piece} \\ \text{with tooling} \end{array}}_{} \geq \begin{array}{l} \text{cost} \\ \text{of} \\ \text{tooling} \end{array} + \begin{array}{l} \text{interest on} \\ \text{tooling cost} \end{array}$$

$$[(R)\,(t) + (R_m)\,(t)] - [(R_t)\,(t_t) + (R_m)\,(t_t)] \geq \frac{C_t + (C_t/2)\,(n)\,(i)}{N} \tag{28-2}$$

where

$R =$ labor rate per hour, without tooling
$R_t =$ labor rate per hour, using tooling
$t =$ hours per piece, without tooling
$t_t =$ hours per piece, using tooling
$R_m =$ machine cost per hour, including all overhead
$C_t =$ cost of the special tooling
$n =$ number of years tooling will be used
$i =$ interest rate (or what invested capital is worth)
$N =$ number of pieces that will be produced with the tooling

Equation (28-2) can be expressed in a simpler form:

$$(R + R_m)t - (R_t + R_m)t_t \geq \frac{C_t}{N}\left(1 + \frac{n \times i}{2}\right) \tag{28-3}$$

This equation assumes straight-line depreciation and computes interest on the average amount of capital invested throughout the life of the tooling.[*] When the time over which the tooling is to be used is less than one year, companies often do not include an interest cost. If this factor is neglected, the right-hand term of equation (28-3) reduces to C_t/N.

The equations assume that the material cost will be the same regardless of whether or not special tooling is used. This is not always true. Although these equations are not completely accurate for all cases, they are satisfactory for determining tooling justification in most cases, because the life of such tooling seldom exceeds five years and more frequently does not exceed two years. The equation does not include the cost of poor quality. This can be included by estimating the decrease in the number of defective parts when the workholder is used versus when it is not used.[†]

The following example illustrates the use of equation (28-3) to determine tooling justification. In drilling a series of holes on a radial drill, the use of a drill jig will reduce the time from $\frac{1}{2}$ hour per piece to 12 minutes per piece. If a jig is not used, a machinist,

---

[*]For the use of more sophisticated economic analysis, see E. P. DeGarmo, J. R. Canada, and W. G. Sullivan, *Engineering Economy*, 6th ed., Macmillan Publishing Company, Inc., New York, 1979.

[†]For a discussion of new measures and methods on cost accounting, see C. S. Park, "Counting the costs," *Mechanical Engineering*, Jan. 1987, p. 66.

whose hourly rate is $8.00, must be used. If the jig is used, the job can be done by a machinist whose rate is $6.50 per hour. The hourly rate for the radial drill is $8.75 per hour.

The cost of making the jig would include $250 for design, $75 for material, and 50 hours of toolmaker's labor, which is charged at the rate of $12.00 per hour to include all machine and overhead costs in the toolmaking department. Investment capital is worth 16% to the company. It is estimated that the jig would last 3 years and that it would be used for the production of 300 parts over this period. Is the jig justified?

The cost of the jig, C_t, is estimated to be $925.00.

$$C_t = \$250 + \$75 + \$12 \times \$50 = \$925$$

Substituting the values given in equation (28-3) yields

$$(8.00 + 8.75)\,0.5 - \left(\frac{6}{50} + 8.87\right) 0.2 \geq \frac{925}{300}\left(1 + \frac{3 \times 0.16}{2}\right)$$

or

$$5.33 \geq 3.82$$

Thus the use of the jig is justified.

One could also ask how many parts would have to be produced with the jig to break even (i.e., increased costs just equal savings). By omitting the value 300 in the solution above and solving for N, it is found that at least 216 pieces would have to be produced with the jig for it to break even.

It should be noted that equations (28-2) and (28-3) assume that the time (of the people and machines) saved by the use of the special tooling can be used for other operations. If this is not the case, the cost analysis should be altered to take this important fact into account. Otherwise, the tooling justification may be substantially in error.

The concepts of group technology and NC machines, discussed in Chapters 29 and 41, may eliminate the need for designing and building a new jig or fixture every time a new part is designed. New measures of manufacturing performance that include terms for quality and flexibility are being developed.

■ KEY WORDS

3-2-1 principle	fixture	plate jig
assembly jig	intermediate jig concept	ring jig
box jig	jig	strap clamp
channel jigs	leaf jig	toggle clamp
clamping	location	vacuum chuck
electrostatic workholder	magnetic chuck	workholder

■ REVIEW QUESTIONS

1. What are the two primary functions of a workholding device?
2. What distinguishes a jig from a fixture?
3. An early treatise defined a jig as "a device that holds the work and guides a tool." Why was this definition incorrect?
4. Why would an ordinary vise not be considered to be a fixture?
5. What basic criteria should be considered in designing jigs and fixtures?
6. What are the critical surfaces of a part (i.e., what makes a part surface critical)? (This question requires an understanding of basic part drawings.)
7. What difficulties can result from not keeping clamping stresses low in designing jigs and fixtures?

8. Explain the 3-2-1 concept for workpiece location.

9. Which of the basic design principles relating to jigs and fixtures would most likely be in conflict with the 3-2-1 location concept?

10. What are two reasons for not having drill bushings actually touching the workpiece? Notice in Figure 28-9 how many of the designs violate this rule. It is not uncommon to have conflicts and trade-offs in fixture design situations.

11. Why does the use of down milling often make it easier to design a milling fixture than if up milling were used?

12. Explain what is meant by the expression "a jig is a skill-transfer unit."

13. A large assembly jig for an airplane-wing component gave difficulty when it rested on four-point support. The assembled wing components were not consistent in shape. The jig was satisfactory when only three supporting points were used. Why?

14. Explain why the use of a given fixture may not be economical when used with one machine tool but may be economical when used in conjunction with another machine tool.

15. What are roll-over jigs, and what advantages do they offer?

16. In the clamps shown in Figure 28-14, what is the purpose of the spherical washer?

17. What are other common types of clamps?

18. What is the purpose of dimensioning the strap clamp assembly in Figure 28-14 with letters?

19. Figure 28-2 showed three types of locator buttons. Identify them.

20. In Figure 28-5, why aren't there three points put on the X plane, two points on the Z plane, and one point on the Y plane?

21. Which set of locators in Figure 28-5 establishes the A dimension?

22. To prepare the workpiece shown in Figure 28-5, which surface would you have milled first, the bottom, the back, or the front?

23. For the part shown in Figure 28-5, why don't you drill the holes first, then mill? Why mill at all?

24. The holes are countersunk after they are drilled. How must the jig be designed to put the countersinks on the mounting holes while the part is in the jig?

■ PROBLEMS

1. Using the following values, determine the number of pieces that would have to be made to justify the use of a jig costing $3000.

$$R = \$5.75$$
$$R_t = \$4.50$$
$$t_t = 1\tfrac{1}{4}$$
$$t = 2\tfrac{1}{2}$$
$$i = 10\%$$
$$R_m = \$4.50$$
$$n = 3$$

2. Suppose in the sample problem in Section 28.16 that modular fixturing is used, which reduces the toolmaker's labor to 4 hours and the design cost to $100 (4 hours at $25 per hour). The material cost (modular elements) for the subplate structural elements, clamps, and so on, was $600. What is the break-even quantity for a modular fixture? (*Note:* The modular fixture is used for the job, then disassembled and returned to the tool room. The parts are reused in other workholders.)

3. Here is a typical problem in clamping force analysis (Figure 28-A). A milling cutter is face milling the top surface of a 1500-lb casting. The cutting force is 1800 lb, and the thrust force is 900 lb. The feed force is assumed to be zero for this analysis. The unknown forces are as follows:

F_R = total force from all clamps on right side

F_L = total force from all clamps on left side

R_1 = horizontal reaction force from the fixed stop

R_2 = vertical reaction force from the fixed stop

R_3 = vertical reaction force from the right side

N = normal force ($F_L + F_R + 1500$ lb + 900 lb)

μ = coefficient of friction (0.19)

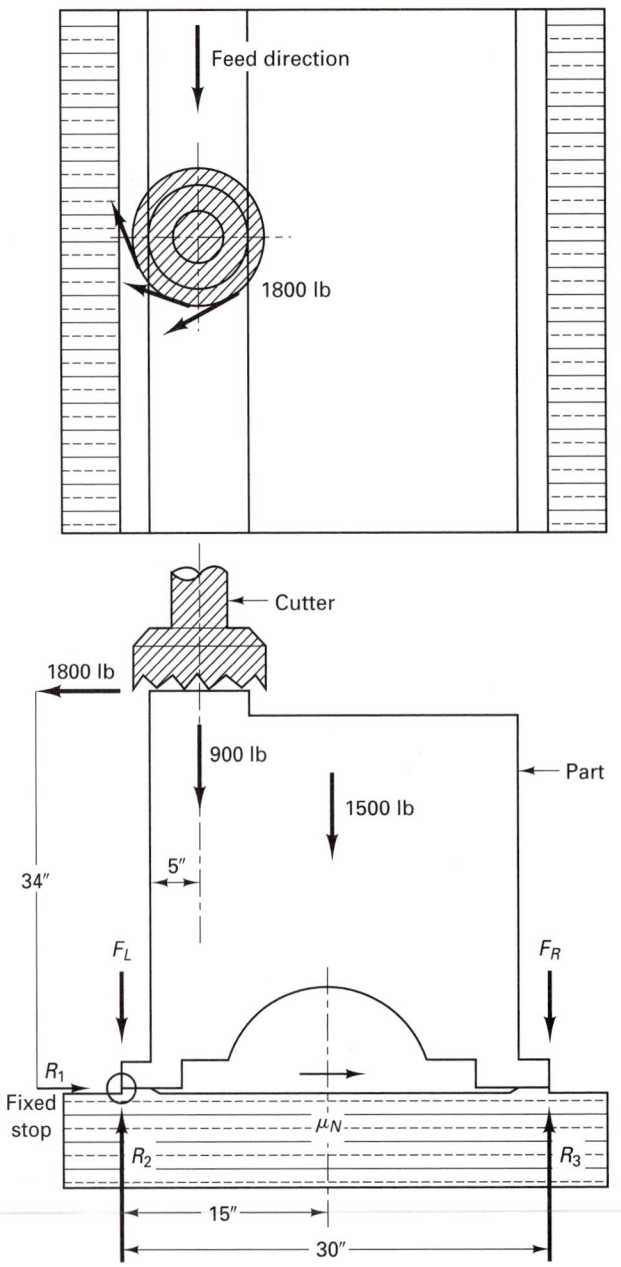

FIGURE 28-A Forces on a
1500-lb workpiece produced
by face milling.

What are the required clamping forces, F_R and
F_L? (*Hint:* For a static condition, the sum of
the X direction (horizontal) is zero, the sum of
the Y direction (vertical) is zero, and the sum of

moments around any point is zero. To solve the
equations, you will have to assume that some of
the forces are zero.)

*C*hapter 28 **CASE STUDY**

overhead crane installation

To install an overhead crane (Figure CS-28) in one bay of the Crawford Manufacturing Company, an assembly plant, brackets for the rails of the crane are to be mounted on eight columns, four on each side of the bay area facing each other. The rails for the crane will span four columns. Each bracket on each column will need six holes in a circular pattern. The holes must be accurately spaced within 5 minutes of arc of each other. The axis of the holes must be parallel and normal to the face of the columns. The center of the bolt-hole circle must be at a height of at least 20 ft from the floor, but the centers of all eight bolt-hole circles must be on the same parallel plane, so that the rails for the crane are level and parallel with each other. Four of the columns along the wall have their faces flush with the wall surface so that mechanical clamping or attachments cannot be used. The building code will permit no welding of anything to these columns.

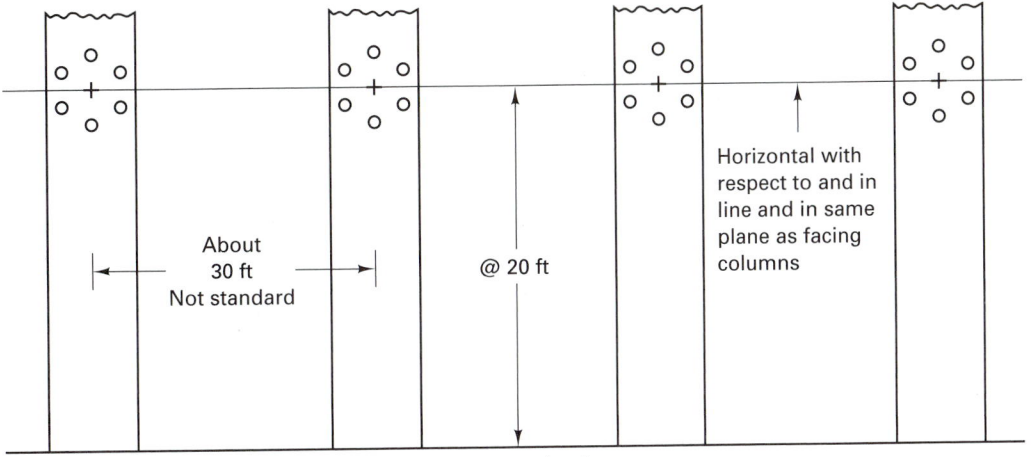

FIGURE CS-28

1. How would you proceed to get the bolt holes located in the right position on the beams?
2. How would you get the hole patterns located properly with respect to each other on all eight beams?
3. List the equipment you will need.
4. Make a sketch of any special tool you recommend.

(*Hint:* Review Chapter 28 for drill jig designs and ask your favorite civil engineer for suggestions.)

CHAPTER 29

NUMERICAL CONTROL AND MACHINING CENTERS

29.1 INTRODUCTION
Brief History of Numerical Control
29.2 CAD-NC versus APT
29.3 A(4) versus A(5) LEVEL OF AUTOMATION
29.4 NUMERICAL CONTROL versus CONVENTIONAL MACHINE TOOLS
29.5 BASIC PRINCIPLES OF NUMERICAL CONTROL
Part Programming
Contouring in NC and CNC Machines

Computer Languages for Numerical Control
29.6 MACHINING CENTER FEATURES AND TRENDS
CNC Turning Centers
29.7 ADVANTAGES OF NUMERICAL CONTROL
29.8 ECONOMIC CONSIDERATIONS IN NUMERICAL CONTROL
Case Study: BREAK-EVEN-POINT ANALYSIS OF A LATHE PART

■ 29.1 INTRODUCTION

Machining, as a fundamental method of material removal, has evolved from individual machines that performed individual processes to machines capable of performing many processes. The processes of turning (boring), drilling (and related processes of reaming and tapping), milling, shaping (and planing), broaching, and grinding have been discussed in earlier chapters. Processes are what this book is all about. However, in the early 1950s, it became possible to control the movement of machine tools by numbers, and soon *numerically controlled* (NC) milling machines, NC drilling machines, and NC boring machines were commercially available. In 1968, Kearney & Trecker marketed a NC machine that could automatically change tools so that many different processes could be done on one machine. Such a machine became known as a *machining center*, a machine that can perform a variety of processes and change tools automatically while under programmable control [see Figure 29-1, which depicts the evolution of the computer numerically controlled (CNC) machining center]. Programmable control means that the workpiece (or the tool) can be *moved*, using software, to specific locations and can even machine multiple surfaces.

So the study of *machining centers* begins with the history of numerical control. NC is programmable automation in which certain functions of the machine tools are controlled by numbers or letters. The program of instruction defines how a particular part is to be made. If the part design or the manufacturing method change, it is necessary to change the program instructions. Nowadays, writing software programs for instruction and control of machines is fairly common, but in 1955, before the advent of NC machines, it was unheard of.

Brief History of Numerical Control

The advent and wide-scale adoption of numerically (tape- and computer-) controlled machine tools has been the most significant development in manufacturing in the past 40

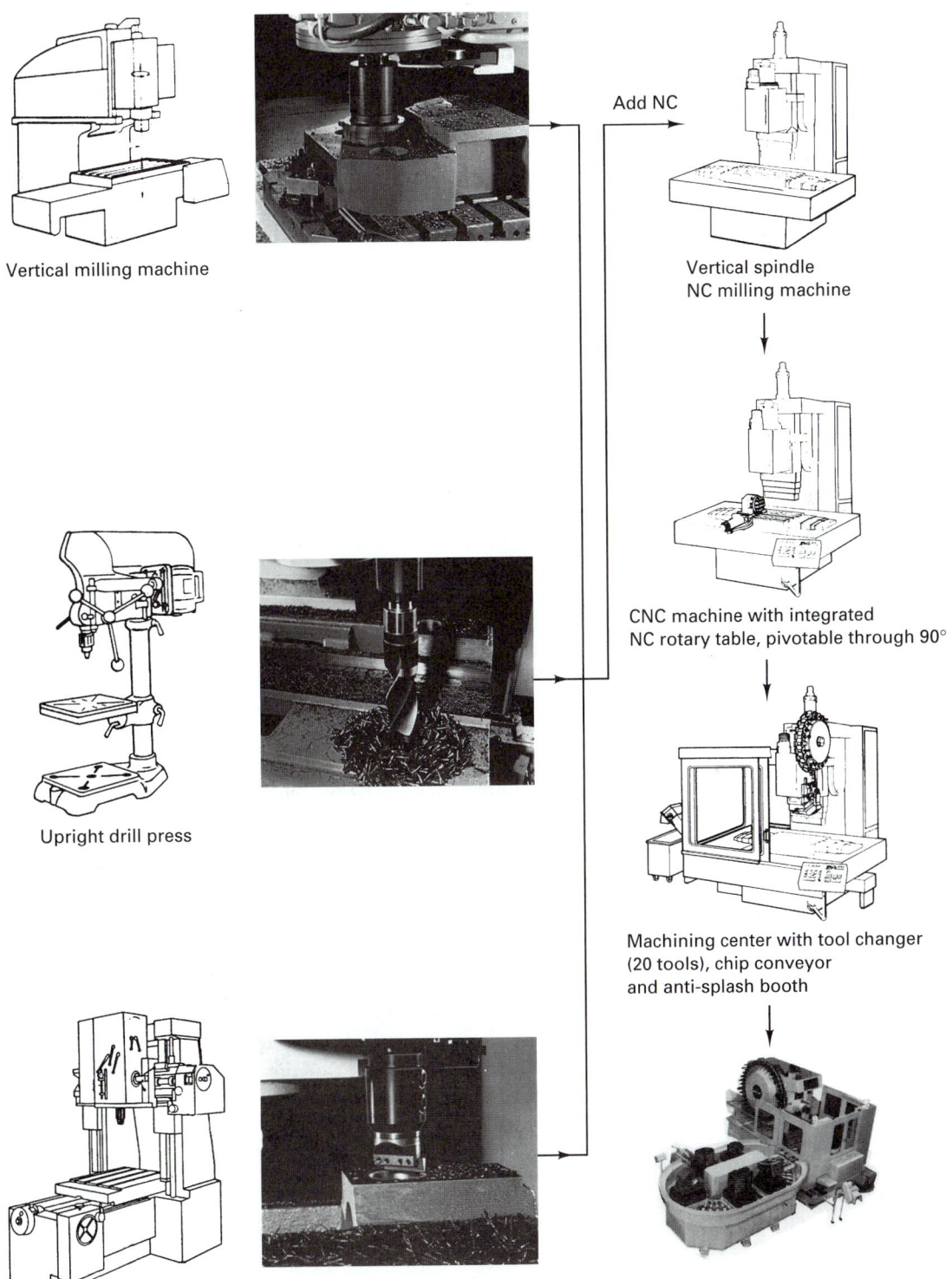

Vertical milling machine

Upright drill press

Add NC

Vertical spindle
NC milling machine

CNC machine with integrated
NC rotary table, pivotable through 90°

Machining center with tool changer
(20 tools), chip conveyor
and anti-splash booth

FIGURE 29-1 Evolution of the machining center.

years. These machines raised automation to a new level [level A(4)] by providing positional feedback as well as programmable flexibility. Numerical control of machine tools created entirely new concepts in manufacturing. Certain operations are now routine that previously were very difficult if not impossible to accomplish. In earlier years, highly trained NC programmers were required. The development of low-cost, solid-state microprocessing chips has resulted in machines that can be programmed in a very short time, by personnel having only a few hours of training, using only simple machine shop language. As a consequence, there are few manufacturing facilities today, from the largest down to the smallest job shops, that do not have, and routinely use, one or more numerically controlled machine tools.

NC came into being to fill a need. The U.S. Air Force (USAF) and the airframe industry were seeking a means to manufacture complex contoured aircraft components to close tolerances on a highly repeatable basis. John Parsons of the Parsons Corporation of Traverse, Michigan, had been working on a project for developing equipment that would machine templates to be used for inspecting helicopter blades. He conceived of a machine controlled by numerical data to make these templates and took his proposal to the USAF. Parsons convinced the USAF to fund the development of a machine. The Massachusetts Institute of Technology (MIT) was subcontracted to build a prototype machine. The prototype was a conventional two-axis tracer mill retrofitted with servomechanisms. As luck would have it, the servomechanism lab was located next to a lab where one of the very first digital computers (Whirlwind) was being developed. This computer generated the digital numerical data for the servomechanisms.

By 1962, NC machines accounted for about 10% of total dollar shipments in machine tools. Today about three-fourths of the $35 billion spent for machine tools (drill presses, milling machines, lathes, and machining centers) goes for NC equipment.

Early on, NC machines were continuous-path or contouring machines where the entire path of the tool was controlled with close accuracy in regard to position and velocity. Today, milling machines, machining centers, and lathes are the most popular applications of continuous-path control requiring feedback control. Next, point-to-point machines were produced where the path taken between operations is relatively unimportant and therefore not monitored continuously. Point-to-point machines are used primarily for drilling, milling straight cuts, cutoff, and punching. Automatic tool changers, which require that the tools be precisely set to a given length prior to installation in the machines, permitted the merging of many processes into one machine (Figures 29-1 and 29-2). Machines are often equipped with two pallets so that one can be setup while the other is working. The two- (or four-sided) "tombstone" fixture shown has multiple mounting and locating holes for attaching part-dedicated fixture plates, which greatly extends the utility of a horizontal machining center.

During the early days of NC, the machine tools had to be programmed using a machine control language, a difficult, time-consuming task, prone to error. Many companies developed computer languages to aid the NC programming task, each developing a different language to describe tool geometry, tool movements, and machining instructions. Confusion reigned.

The USAF sponsored the placement of many NC machines in the major aerospace companies. The companies soon concluded that a universally accepted NC programming language was needed. Under the auspices of the Aerospace Industries Association (AIA), these companies, with the USAF, sponsored additional research at MIT to develop a computer language that would use simple English-like statements to produce an output that would control the NC machines. This language, called *APT* (automatically programmed

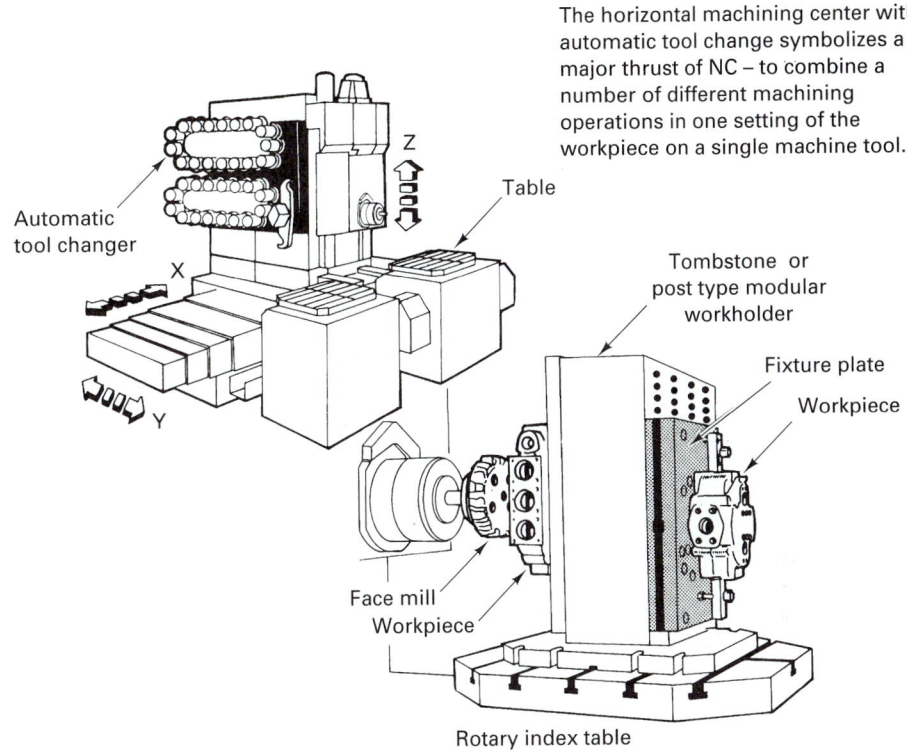

The horizontal machining center with automatic tool change symbolizes a major thrust of NC – to combine a number of different machining operations in one setting of the workpiece on a single machine tool.

FIGURE 29-2   Horizontal machining center with three axes of simultaneous control typifies the state of the art in machining tools. *(Adapted from* 1987 NC/CIM Guide Book, Modern Machine Shop.)

tools), was announced by MIT in 1959 when computer technology was in its adolescence. Hundreds of thousands of parts have been programmed using APT, initially on large mainframe computers (Figure 29-3). *Postprocessing* takes the output from the APT program and converts it to input for a particular machine. Traditional postprocessing yields NC workpiece programs that are not exchangeable. To machine an identical workpiece on another machine, the program must be postprocessed again unless the machine and control are exactly the same.

With the arrival of high-resolution *computer-aided design* (CAD) graphics, many people think that APT will be phased out. But for parts where complex tool control is required, APT is still the best choice. The chief problems in NC programming were tool radius compensation and tool path interpolation (discussed later). Computer software with the capability to perform linear, circular, parabolic, and other kinds of interpolations had to be developed. The capabilities were included in APT and each software is now routinely available on CNC machines.

In the 1960s, it was envisioned that a large computer could be used directly to control, in real time, a number of machines. A limited number of *direct numerical control* (DNC) systems were developed, with the idea that the programs were to be sent directly to the machines (eliminating paper tape handling). The mainframe computer would be shared on a real-time basis by many machine tools. The machine operator would have access to the main computer through a remote terminal at the machine, while management would

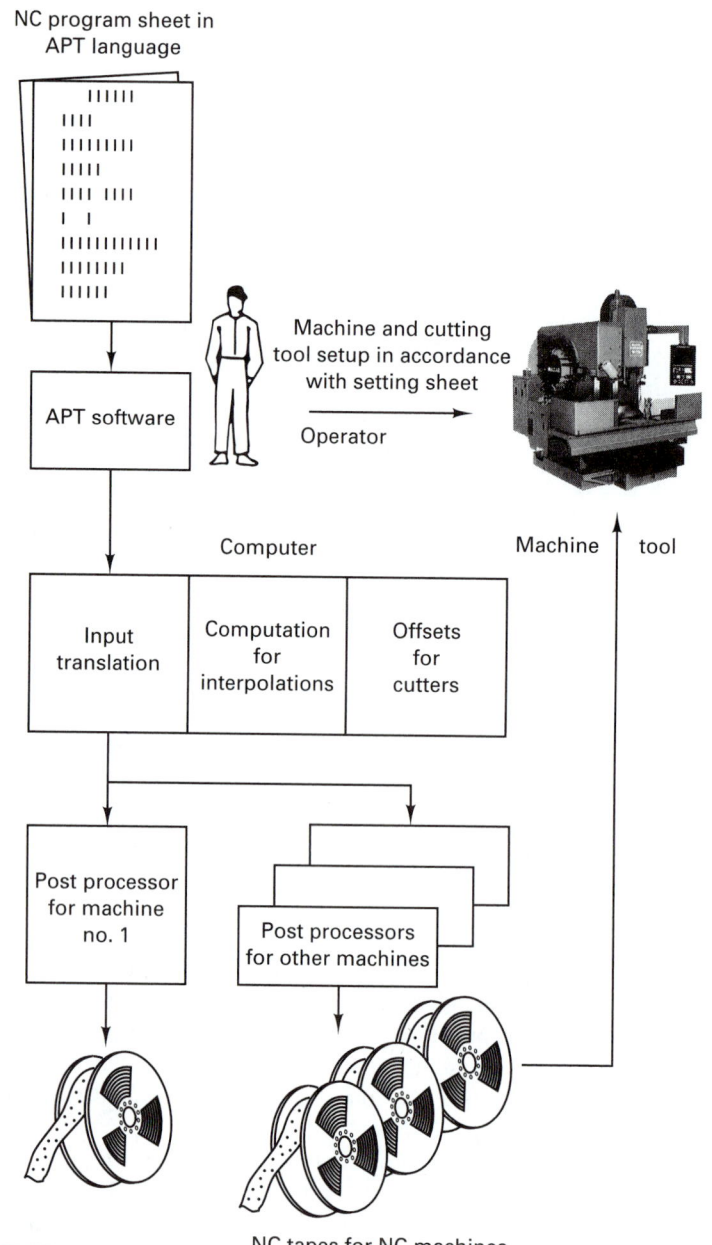

FIGURE 29-3 Steps in NC tape preparation, using APT computer software.

have up-to-the-minute data on production status and machine utilization. This version of DNC had very few takers. Instead, NC machines became *computer numerical control* (CNC) machines through the development of small inexpensive computers, microprocessors with large memories, and programmable computers (Figure 29-4). Now functions such as program storage, tool offset and tool compensation, program-editing capability, various degrees of computation, and the ability to send and receive data from a variety of

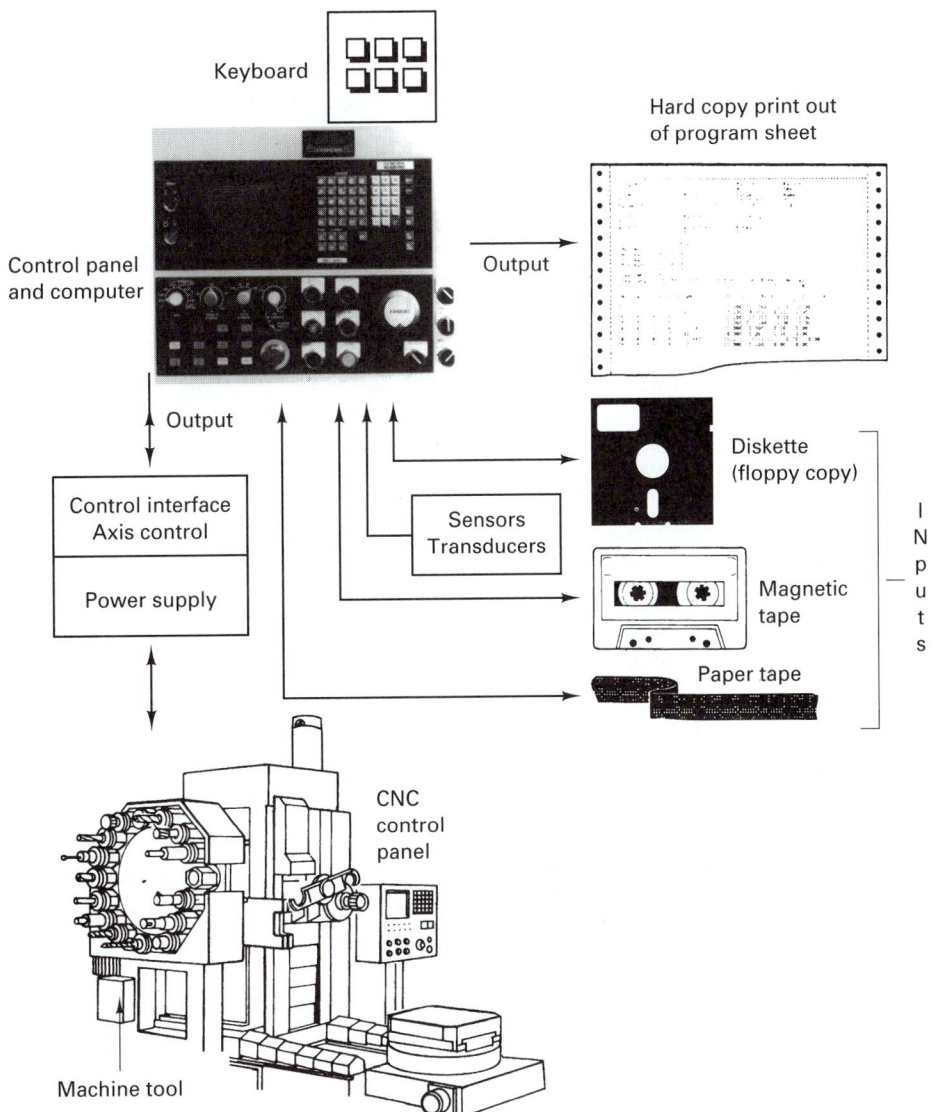

FIGURE 29-4 Inputs to the CNC control panel and onboard computer provide output to the machine tool.

sources, including remote locations, are routinely available through the onboard computer. The computer can store multiple-part programs, recalling them as needed for different parts. Immediately, it was found that the machine tool operator could readily learn how to program these machines (manually) for many component parts, often eliminating the need for a part programmer.

In recent years the DNC concept has been revived with the small computers at the machines being networked to a larger computer to provide enhanced memory and computer computational capacity. Minicomputers, supervising and controlling a *flexible manufacturing system* (FMS) or manufacturing cell, are networked to a large central computer. This

DNC now means *distributed numerical control*, with the distribution of NC programs by a central computer to individual CNC units. The FMS is a special type of manufacturing system discussed in detail in Chapter 41. Historically speaking, the first examples of FMS systems appeared in the late 1960s, but few companies adopted them because of their high initial cost.

■ 29.2 CAD-NC VERSUS APT

The development of computer-aided design in the 1980s led to an attractive alternative method for NC programming. However, virtually all CAD graphics systems were developed as part (or product) design tools, with NC capability added as an afterthought. CAD programs depict the final part and do not really deal with how the part is "shaped" from the raw material. The major advantage of NC graphics is visualization of a part; You can see how to machine it. Even so, it takes considerable training (months) to achieve proficiency in the use of CAD-NC, or as it is commonly called, CAD-CAM. (CAM stands for *computer-aided manufacturing*).

APT was designed in exactly the opposite way to CAD-NC. APT starts with the cutter and proceeds to "machine" the raw material into the final part shape. However, APT users must be able to visualize or imagine the part and "see" it being machined in their mind's eye. APT also requires considerable training to become expert. For many CAD-CAM applications for complex parts, the NC programming functions may be too limited to generate the necessary NC program. That is, the CAD capability to design a complex part far exceeds the NC capability of the CAD system. It is estimated that 5 to 10% of the parts in the aerospace and defense industries still require APT programming.

■ 29.3 A(4) VERSUS A(5) LEVEL OF AUTOMATION

In Chapter 1, a yardstick for automation was presented. The NC or CNC machine represents the A(4) level in these scheme. The next level of automation, the A(5) level, requires that the control system perform an evaluation function of the process. Figure 29-5 shows diagrams for A(3), *open-loop control*, and three schemes for A(4), *closed-loop control*. The closer the feedback device is to the tool, the better the control scheme, but usually, the more expensive. Figure 29-6 compares block diagrams for A(4) closed-loop control (with encoder) to A(5) adaptive control. A(5) requires feedback from the process or the product and seeks to optimize the process. CNC machines, with their onboard computers, are potentially capable of the A(5) level of automation. In the standard versions of NC and CNC machines in use today, speed and feed are fixed in the program unless the operator overrides them at the control panel of the machine. If either speed or feed is too high, the result can be rapid tool failure, poor surface quality, or damaged parts. If the speed and feed are too low, production time is greater than desired for best productivity. An *adaptive control* (AC) system that can sense deflection, force, heat, torque, and the like will use these measurements to make decisions about how the input parameters might be altered to *optimize* the process. This means that the computer must have *mathematical models* in its software that describe how this process behaves and mathematical functions that state what is to be optimized (cost per piece, surface quality, MRR, power consumed, and so on). The models require theory, and herein lies the problem. The theory for the mechanics of shape-generating processes (metal cutting and metal forming) is segregated, incomplete, and empirical. With incomplete theory, the models become suspect and the systems unreliable. In addition, current versions of AC systems have been very costly to develop, suffer from inadequate data banks, and produce

Open loop A(3)

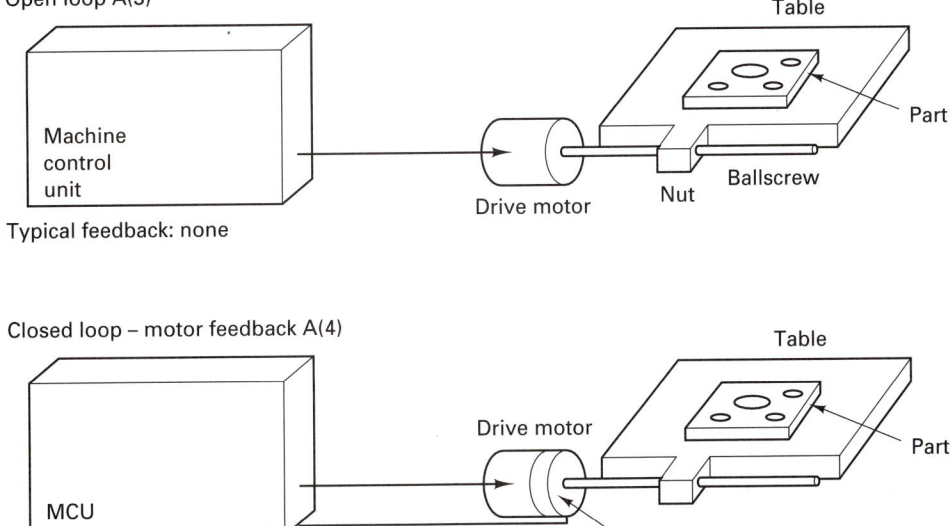

Typical feedback: none

Closed loop – motor feedback A(4)

Typical feedback: motor has turned 1.003 revolutions

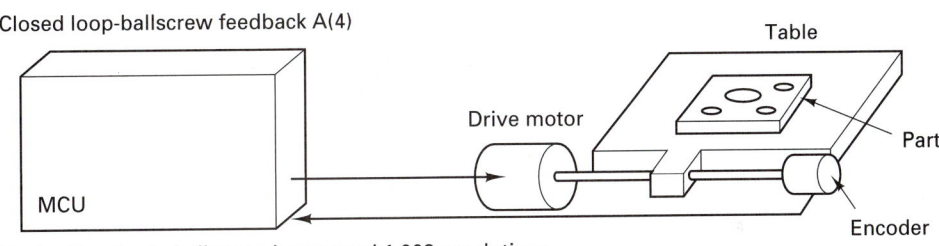

Closed loop-ballscrew feedback A(4)

Typical feedback: ballscrew has turned 1.003 revolutions

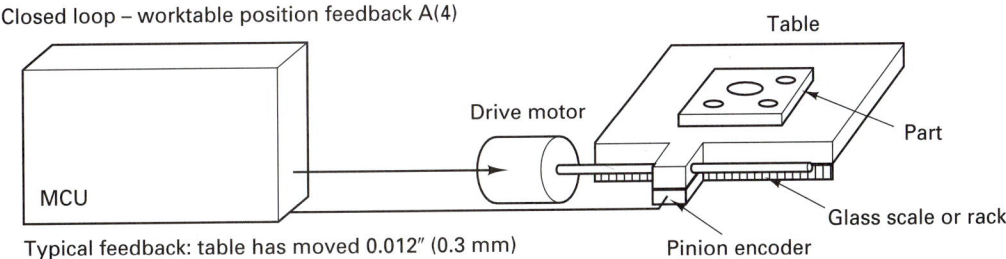

Closed loop – worktable position feedback A(4)

Typical feedback: table has moved 0.012″ (0.3 mm)

FIGURE 29-5   Open-loop NC versus three position control schemes for NC and CNC machine tools.

variable cycle times. However, AC systems have the potential of markedly improving productivity, quality, and machine utilization. Reaching this level of automation on a routine shop floor basis requires great expenditures of time and money. Most AC systems monitor such parameters as spindle deflection, horsepower, or cutting forces and override feed rates while trying to optimize metal-removal rates.

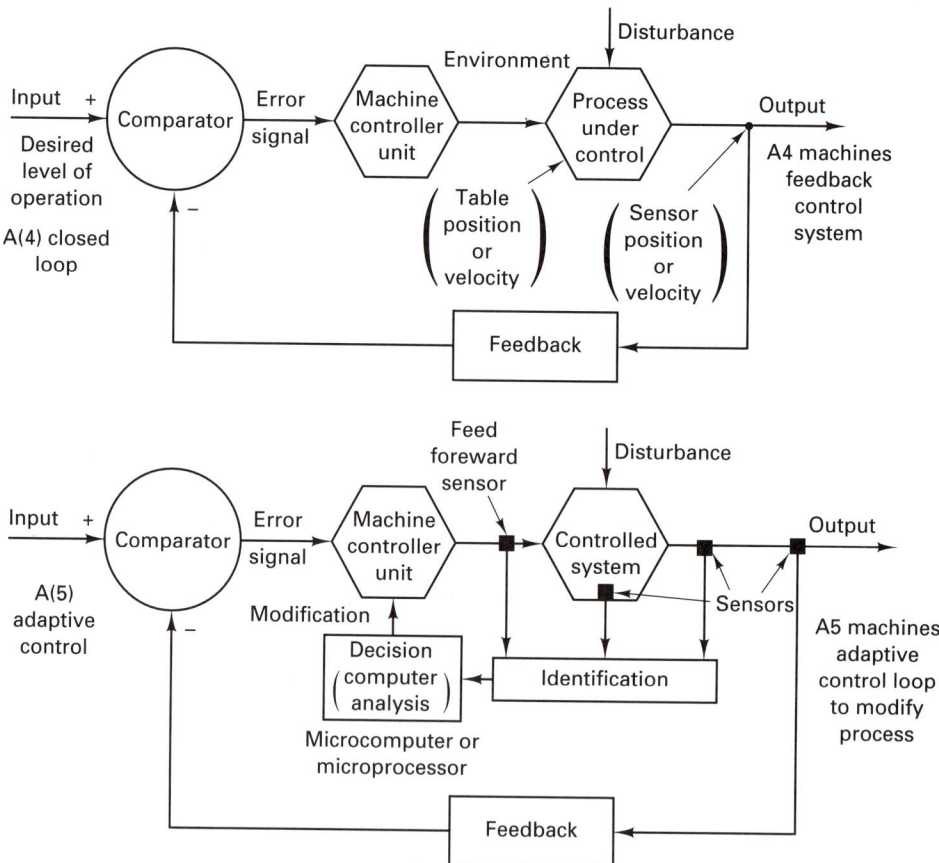

FIGURE 29-6   Block diagrams of A(4) and A(5) level of control: A(4), closed-loop NC (see Figure 29-10 for details); A(5), closed-loop NC with adaptive control loop.

■ 29.4 Numerical Control versus Conventional Machine Tools

NC uses a processing language to control the movement of the cutting tool or workpiece or both. The programs contain information about the machine tool and cutting tool geometry, the part dimensions (from rough to finish size), and the machining parameters (speeds and feeds). Thus using today's NC machines (either tape or computer controlled), consecutive parts can be duplicated, and a part made at a later date will be the same as one made today. Thus repeatability and quality are improved. Workholding devices can be made more universal and setup time reduced, along with tool-change time, thus making programmable machines economical for producing small lots or even a single piece. When combined with the managerial and organizational strategies of group technology (GT) and cellular manufacturing, programmable machines lead to tremendous improvements in productivity (see Chapters 41 and 43). GT basically leads to the creation of families of parts made in machining cells containing flexible/programmable machines. The compatibility of the components (similarly in process and sequences of processes) greatly enhances the productivity (utility) of the programmable equipment.

A side result has been the decrease in the non-chip-producing time of machine tools. The operator was relieved of the jobs of changing speeds and feeds and locating the tool relative to the work. Even simple forms of NC and digital readout equipment have provided both greater productivity and increased accuracy. Most early NC machine tools were developed for special types of work where accuracies of as much as 0.00005 in. were required, and many NC machines are built to provide accuracies of at least 0.0001 in., whether it was needed or not. This forced many machine tool builders to redesign their machines and improve their quality because the operator was not available to compensate for the machine (positioning) error. While most NC machine tools today will provide greater accuracy than is required for most jobs, the tendency is toward greater accuracy and precision (i.e., better quality) but at no increase in cost. Therefore, NC machines will continue to be the very backbone of the machine tool business. Hopefully, all builders will one day arrive at a common language, so that if one can program one machine, one can program them all.

■ 29.5 Basic Principles of Numerical Control

As the name implies, *numerical control* is a method of controlling the motion of machine components by means of numbers or coded instructions. Assume that three 1-in. holes in the part shown in Figure 29-7 are to be drilled and bored on the vertical spindle machining center shown in Figure 29-8. The centers of these holes must be located relative to each other and with respect to the left-hand edge of the workpiece (*X* direction) and the bottom edge (*Y* direction). The depth of the hole will be controlled by the *Z* (or *W*) axis.

The holes will be produced by center drilling, hole drilling, boring, reaming, and counter boring (five tool changes). If this were done conventionally or in manned cells, three or four different machines might be required. On the NC machining center, it only requires changing the tool. The movements of the table are controlled by coordinate systems. The accurate positioning of the cutting tool with respect to the work is established by zero points. The center of the table or a point along the edge of the traverse range is commonly used (Figure 29-8). The workpiece is positioned on the table with respect to this zero point on the table. For our example, the lower left-hand corner of the part is placed at the zero point on the table.

A program is written that instructs the table to move with respect to the axis of the spindle to bring the holes to the correct location for machining. The machine shown in Figure 29-8 is called a five-axis machine because it has five movements (shown by the dark arrows) under numerical control.

NC and CNC machines can be subdivided into three types (Figure 29-9). In *point-to-point machines*, the tool path is not controlled. *Straight line* controls permit axially parallel traverses at desired table feed rates, but only one axis drive is operated at a time. *Contouring* permits two or three axes to be controlled simultaneously, permitting two- or three-dimensional geometries to be generated. Another term for contouring is *continuous path*. Most machining centers have contouring capability and are closed loop. The point-to-point machines are often open loop rather than closed loop, as defined in Figure 29-6.

The *X* axis of the five-axis vertical spindle CNC machine tool shown in Figure 29-8 will be used to explain how the closed-loop positional control works. The CNC control system, shown schematically in Figure 29-10, uses a resolver or encoder to provide axis-position feedback to the *machine control unit* (MCU). A closed-loop control requires a transducer or sensing device to detect machine table position (and velocity for contouring) and transmit that information back to the MCU to compare the current status with the

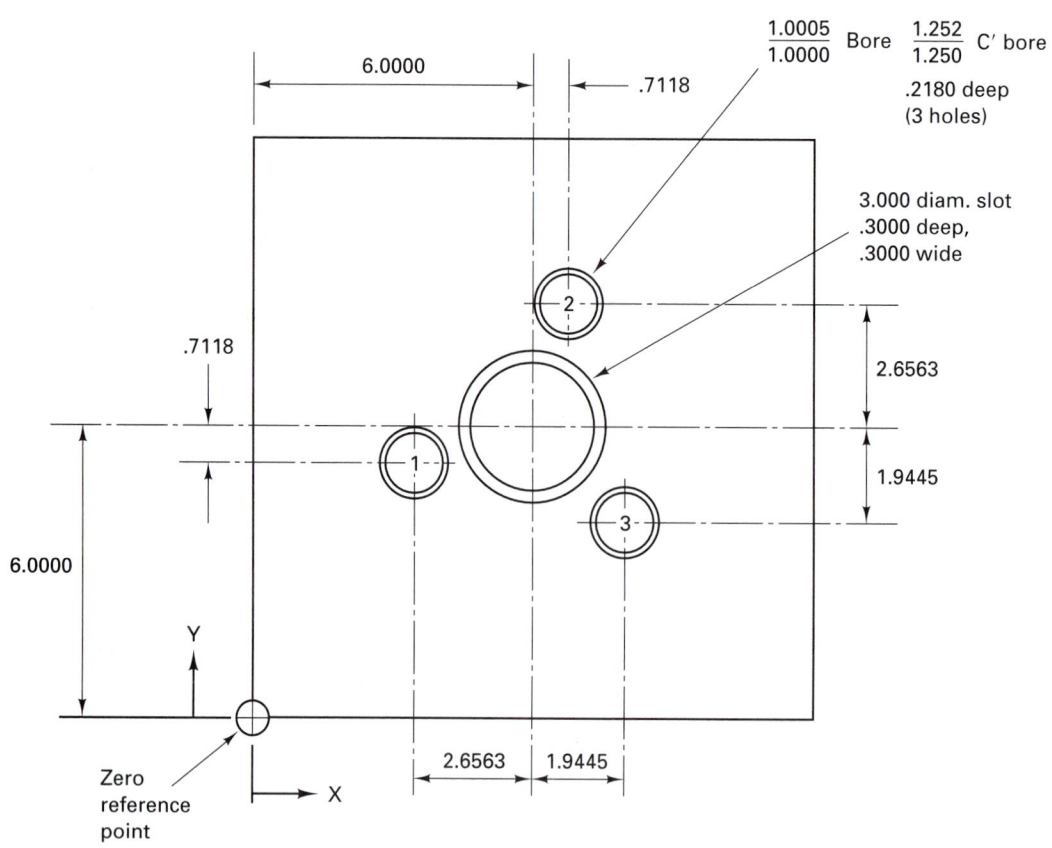

FIGURE 29-7   Part to be machined on a numerically controlled point-to-point machine tool.

desired state. If they are different, the control unit produces a signal to the drive motors to move the table, reducing the error signal and ultimately moving the table to the desired position at the desired velocity. For most NC controls, the feedback signals are supplied by transducers actuated either by the feed screw or by the actual movement of the component. The transducers may provide either *digital* or *analog* information (signals). The resolver (Figure 29-11) measures (indirectly) the movement of the table by the rotation of the ball screw. A pulse disc on the end of the ball screw converts analog movement back into digital pulses, which are used to calculate table movement. The tachometer on the drive motor measures the table velocity.

Two basic types of digital transducers are used. One supplies *incremental* information and tells how much motion of the input shaft or table has occurred since the last time. The information supplied is similar to telling a newspaper carrier that papers are to be delivered to the first, fourth, and eighth houses from a given corner on one side of a block. To follow the instructions, the carrier would need a means of counting the houses (pulses) as he passed them. He would deliver papers as he has counted 1, 4, and 8. The second type of digital information is *absolute* in character, with each pulse corresponding to a specific location of the machine components. To continue the carrier analogy, this would correspond to telling the carrier to deliver papers to the houses having house numbers 2400, 2406, and 2414. In this case it would only be necessary for the carrier (machine component) to be able to read

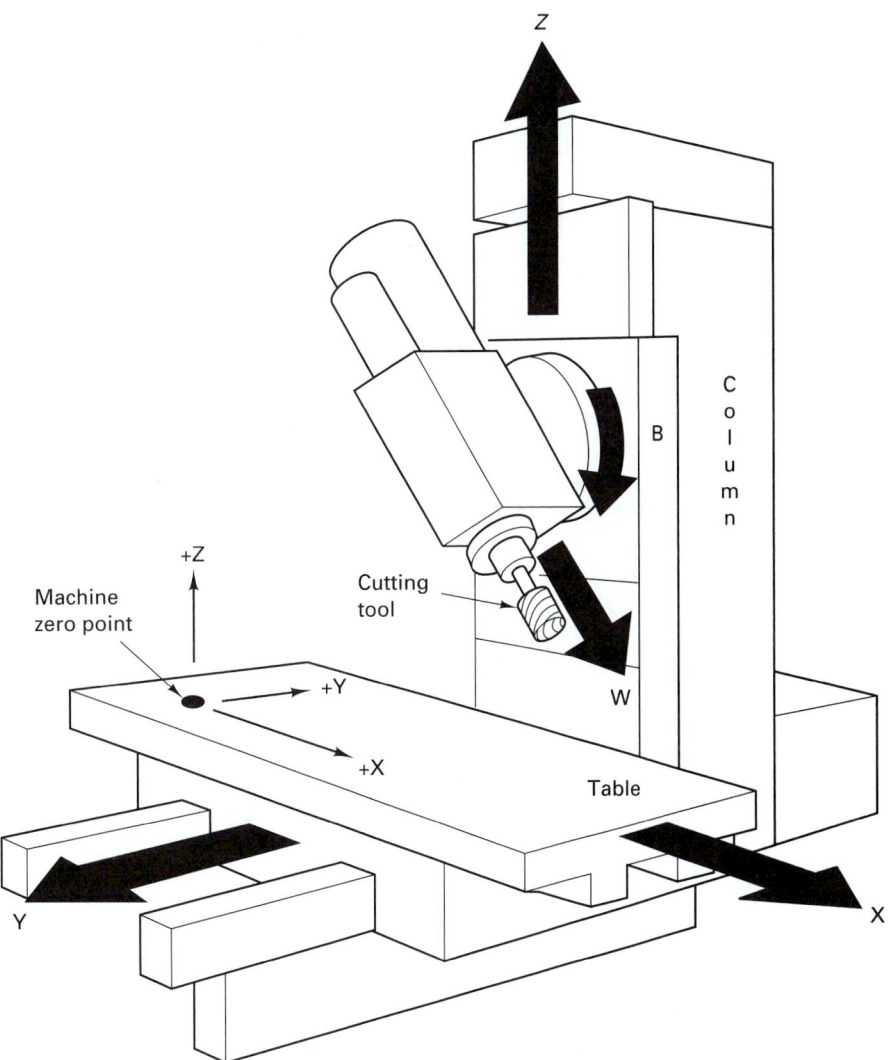

FIGURE 29-8   Five-axis vertical spindle machining center showing location of machine zero point on table.

the house numbers (addresses) and stop and deliver a paper when arriving at a proper address. This "address" system is a common one in numerical-control systems, because it provides absolute location information relative to a machine zero point.

When analog information is used, the signal is usually in the form of an electric voltage that varies as the input shaft is rotated or the machine component is moved, the variable output being a function of movement. The movement is evaluated by measuring, or matching, the voltage, or by measuring the ratios between the applied and feedback voltages; this eliminates the effect of supply-voltage variations.

The input information (i.e., the location of the holes) is given in binary form to the machine control unit in the form of a punched tape, magnetic tape in a cassette, or a floppy disc, as shown in Figure 29-4. The data can be input directly via the control panel. This

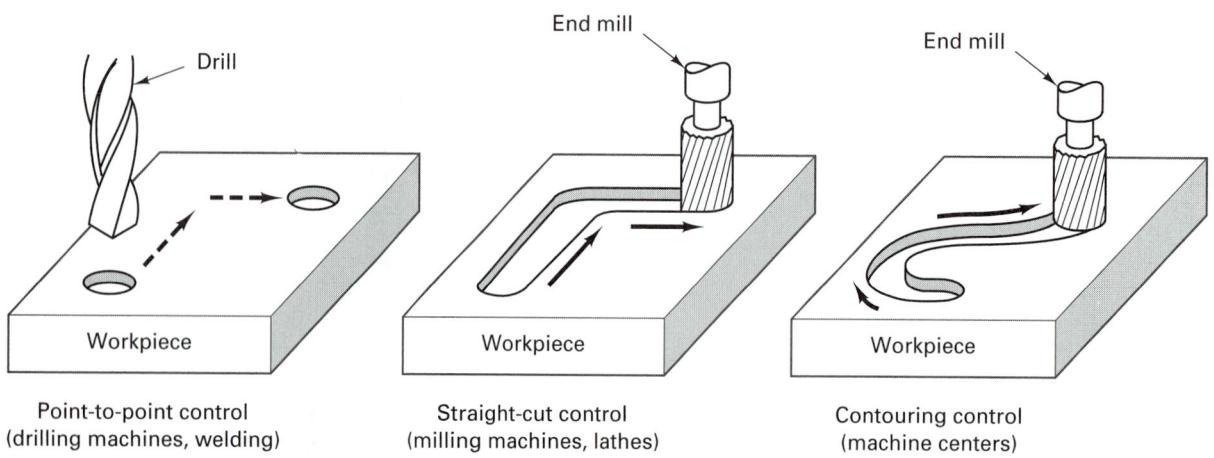

Point-to-point control
(drilling machines, welding)

Straight-cut control
(milling machines, lathes)

Contouring control
(machine centers)

FIGURE 29-9   NC and CNC systems are subdivided into three basic categories:
point-to-point controls, straight-cut controls, and contouring controls.

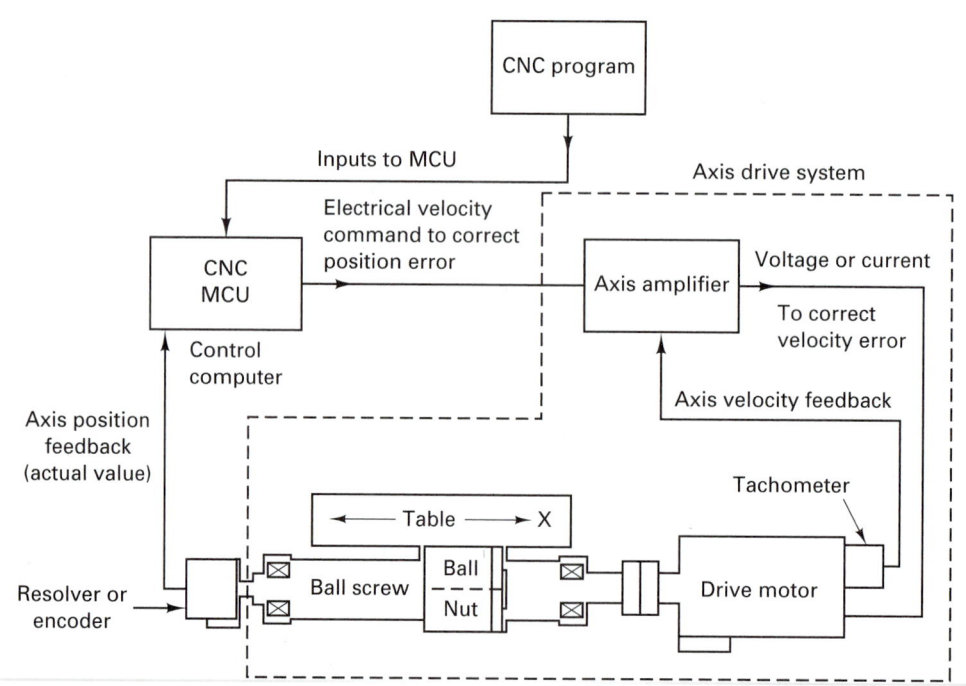

FIGURE 29-10   A CNC control system includes a velocity loop within an axis drive
system, and a position loop external to the axis drive system.

command signal is converted into pulses by the machine control unit (MCU), which in turn
drives the servomotor or stepper motor.

Ac (alternating-current) servomotors are rapidly replacing dc (direct-current) motors
on new CNC machine tools. The reasons for change are better reliability, better performance-
to-weight ratio, and lower power consumption (Figure 29-12). The magnets are mounted
on the rotor of an ac servomotor, and the windings are located in the stator. On a dc servomotor, the magnets are mounted in the stator and the windings are located in the rotor.

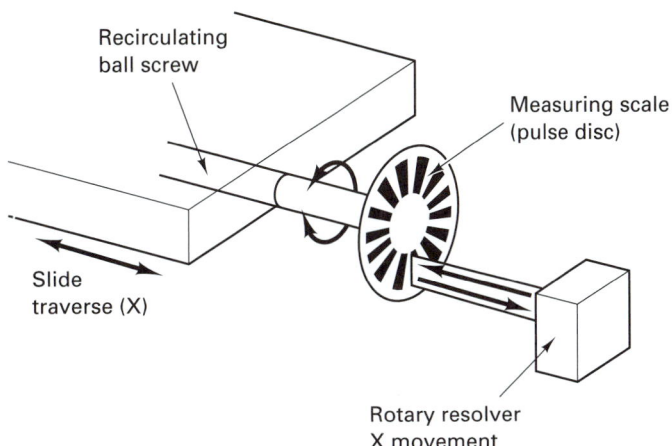

FIGURE 29-11   Table positions can be measured using a resolver.

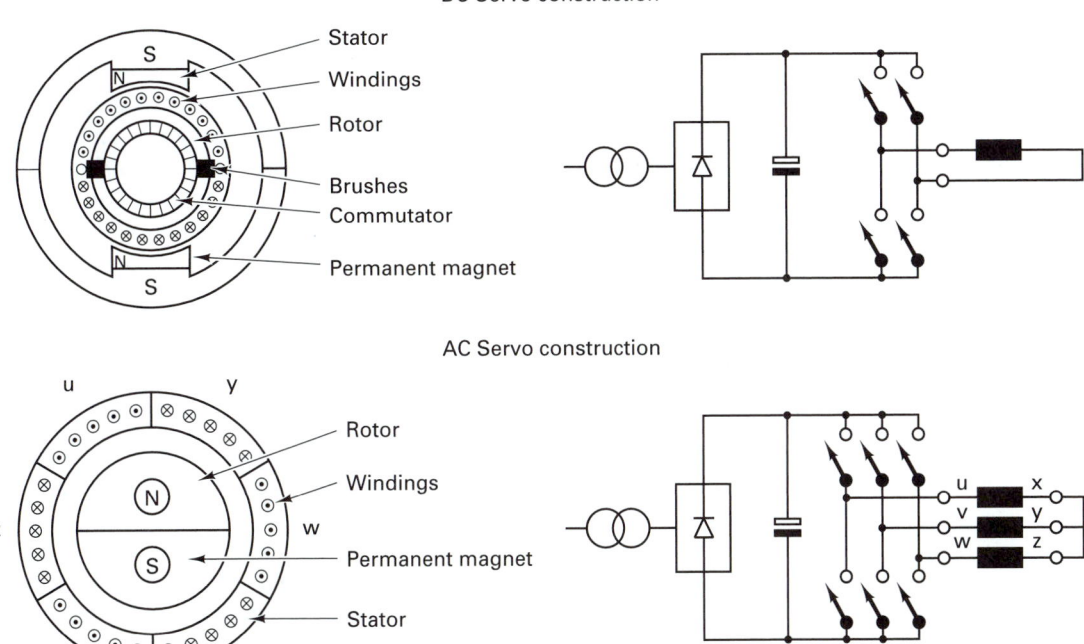

FIGURE 29-12   DC servomotor construction versus AC servo construction. (Modern Machine Shop, *July 1983, p. 72.)*

With an ac servomotor, the armature current is synchronized to the magnetic field by sensing the position of the magnetic poles through electrical commutation and control of the current in the stator winding. Electronic communication avoids the wear caused by spark formation between the dc motor's brushes and commutator bar and by friction of the brushes. It also avoids spot overheating of the commutator during long standstill periods under load.

The ac servomotor's high-gain, closed-loop positioning system also has attained positioning and holding accuracy equivalent to that of the dc motor. These benefits, along with the higher speed capabilities and torque-to-inertia ratio, are available without increasing the complexity of the system with additional electronics. When the command counter reaches zero, the correct number of pulses has been sent to move the table to the desired position. In a closed-loop system, a comparator is used to compare feedback pulses with the original value, generating an error signal. Thus, when the machine control unit receives a signal to execute this command, the table is moved to the specified location, with the actual position being monitored by the feedback transducer. Table motion ceases when the error signal has been reduced to zero and the function (drilling a hole) takes place. Closed-loop systems tend to have greater accuracy and respond faster to input signals but may exhibit stability problems (oscillating about a desired value instead of achieving it) not found in open-loop machines.

If the system is point-to-point (or positioning), the control system disregards the paths between points. Some positioning systems provide for control of straight cuts along the machine axes and produce diagonal paths at 45° to the axes by maintaining one-to-one relationships between the motions of perpendicular axes. Contouring systems generate paths between points by interpolating intermediate coordinate positions. As many of these systems as desired can be combined to provide control in several axes—two- and three-axis controls are most common, but some machines have as many as seven. In many, conversion to either English or metric measurement is available merely by throwing a switch.

The components required for such a numerical control system now are standardized items of hardware. In most cases the drive motor is electric, but hydraulic systems are also used. They are usually capable of moving the machine elements, such as tables, at high rates of speed, up to 200 in./min being common. Thus exact positioning can be achieved more rapidly than by manual means. As shown in Figure 29-5, the transducer can be placed on the drive motor or connected directly to the lead screw, with special precautions being taken, such as the use of extra-large screws and ball nuts, to avoid backlash and to assure accuracy. In other systems, the encoder is attached to the machine table, providing direct measurement of the table position. Various degrees of accuracy are obtainable. Guaranteed positioning accuracies of 0.001 or 0.0001 in. are common, but greater accuracies can be obtained at higher cost. Most NC systems are built into the machines, but they can be retrofitted to some machine tools.

Initially, NC machines provided only tool change, tool setting, and speed, feed, and depth-of-cut setting positioning the work relative to the tool, with the remaining functions controlled by the operator. Gradually, the functions were incorporated into the control system so that the machines could change the tools automatically, change the speeds and feeds as needed for different operations, position the work relative to the tools, control the cutter path and velocity, reposition the tool rapidly between operations, and start and stop the sequence as needed.

Contouring requires directional changes at controlled velocity. This is illustrated in Figure 29-13, where a cutter is milling a circular slot for the part in Figure 29-7. Any lost motion in the system will distort the shape of the circle. The positional feedback transducer for the X axis is placed on the lead screw of the table. The velocity transducer (a tachometer) is placed on the rear of the motor. This arrangement provides for minimal distortion (flats on the circle at the points of axis velocity reversal) in the contoured circle.

Recirculating ball screw drives, of the type shown in Figure 29-14, greatly reduce the backlash in the drive systems, helping to eliminate problems of servoloop oscillation and machine instability. Using such hardware, NC machines are manufactured with greater

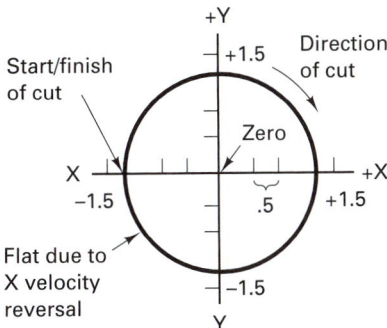

X and Y position to machine a circle

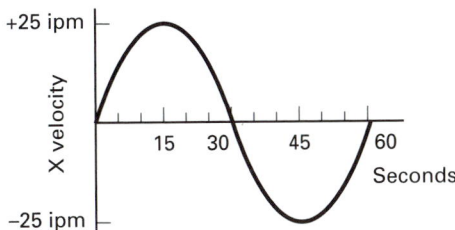

X axis velocity versus time at approximately 25 ipm

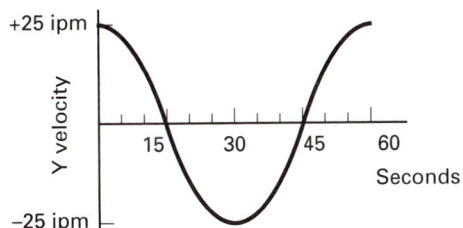

Y axis velocity versus time at approximately 25 ipm

FIGURE 29-13   Contouring requires velocity feedback as well as positional feedback.

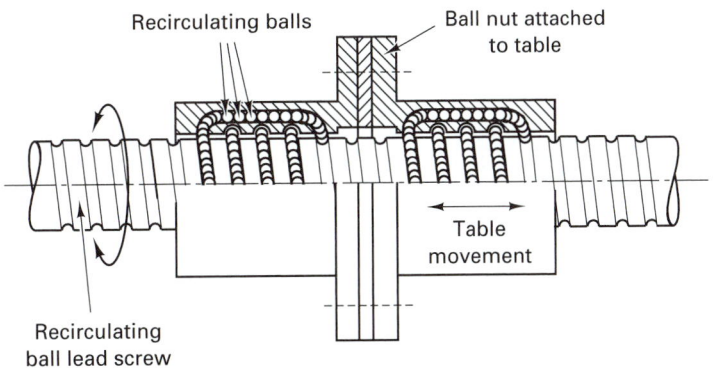

FIGURE 29-14   The ball lead screw adds to the accuracy and precision of NC and CNC machines.

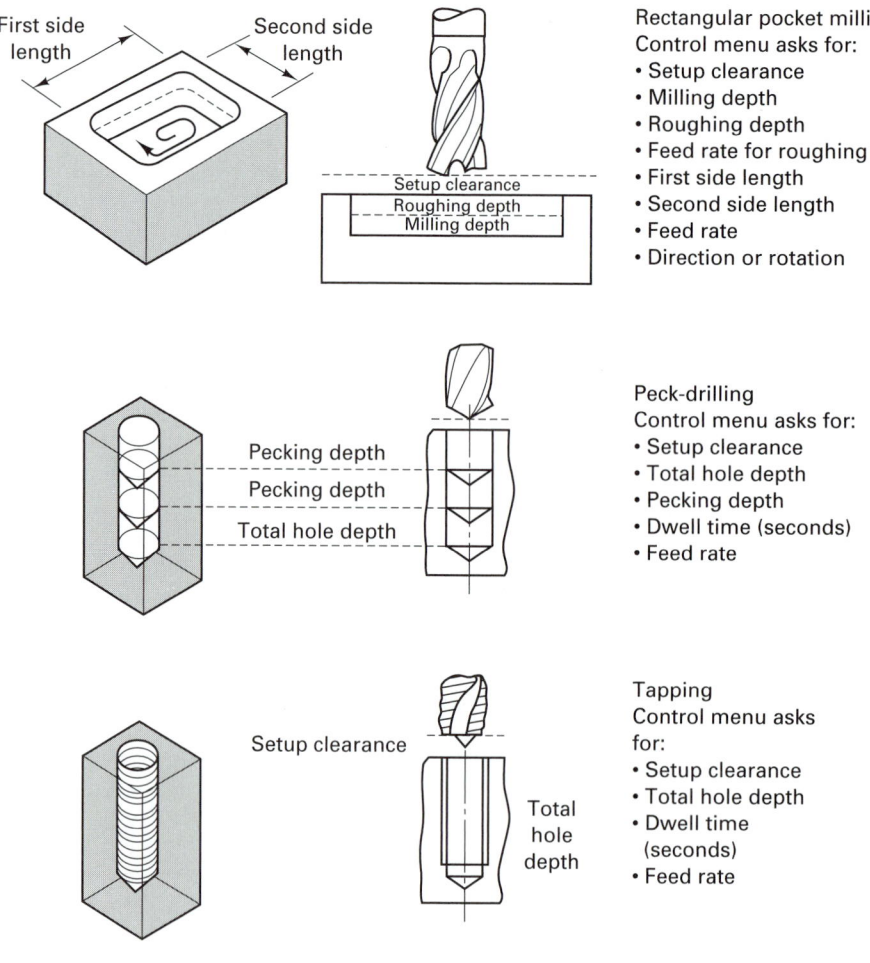

Source: Heidenhain Corp. (Elk Grove Village, III.)

FIGURE 29-15 Canned or preprogrammed machining routines greatly simplify programming CNC machines. *(Courtesy of Heidenkain Corp., Elk Grove Village, Ill.)*

accuracy and repeatability and more rapid table movements than is possible for conventional machines.

Some of the functions in programmable machines require feedforward or preset loops. The machine must know in advance the rough dimensions of a casting of a forging so that it can determine how many roughing cuts are needed prior to the finishing cut. However, for most CNC work, the operator (or part programmer) still plans the sequence of operations, selects the cutting tools and workholding devices, and selects the speeds, feeds, and depths of cut. Common machining routines such as pocket milling or peck drilling have been preprogrammed into many CNC machines. These are called *canned cycles*. As shown in Figure 29-15, the operator merely supplies the information requested by the control menu and the machine fills in the necessary data for the previously written program to perform the desired machining routines.

To ensure accurate machining of a workpiece on a CNC machine, the control system has to "know" certain dimensions of the tools. These *tool dimensions* are referenced to a

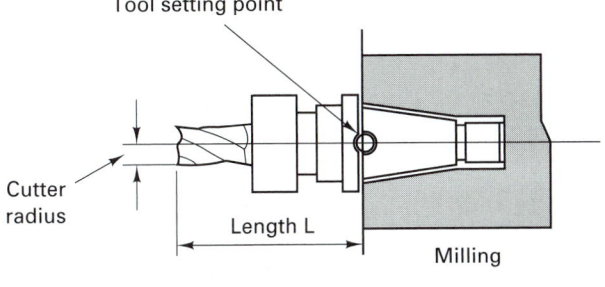

Tool setting point

Cutter radius

Length L

Milling

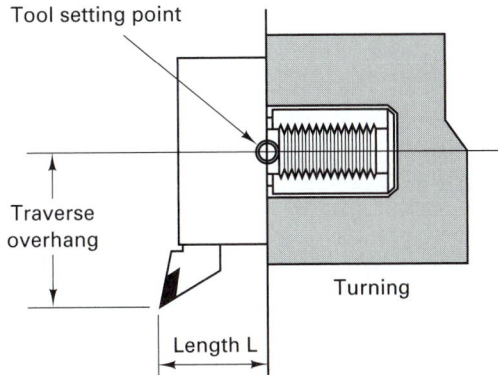

Tool setting point

Traverse overhang

Turning

Length L

FIGURE 29-16 In NC machines, the tool dimensions must be accurately set and known to the control system.

fixed *setting point* on the toolholder (Figure 29-16). For the milling cutter the dimensions are length L and cutter radius. For the turning tool the dimensions are length L and transverse overhang. These dimensions are part of the information given to the operator on the setting sheet (see Figure 29-3). The program assumes that the tools will have the specified dimensions.

Part Programming

Obviously, the preparation of the tape (or other input media) for use in NC is a critical step. In most cases, simple standard languages and programs have been developed. For older machines, tapes are manually prepared on a typewriter-like machine or on devices designed specifically for punching tapes. See Figure 29-17 for examples of NC tapes.

The basic steps can be illustrated by reference to the part shown in Figure 29-7. The first step is to modify the drawing to establish the zero reference axes: the X, Y, and Z directions. The zero reference point was the lower left-hand corner of the part. The hole labeled 1 is located $+3.3437$ in the X direction and $+5.2882$ in the Y direction with respect to the zero point. The part should be dimensioned with respect to the reference zero. This may require redrawing and redimensioning the part. Obviously, this step can be avoided if the original drawing is made in the desired form. The setup instructions given to the operator establish the position of the workpiece properly on the machine table with respect to the tool.

The second step is to make a *part program*. The program (1) defines the sequence of operations required to fabricate the part; (2) gives the X, Y, and Z coordinate positions of the operations; and (3) specifies the spindle traverse that determines the depth of the cut, the spindle speed, and feed, and determines also whether the same tool can continue the

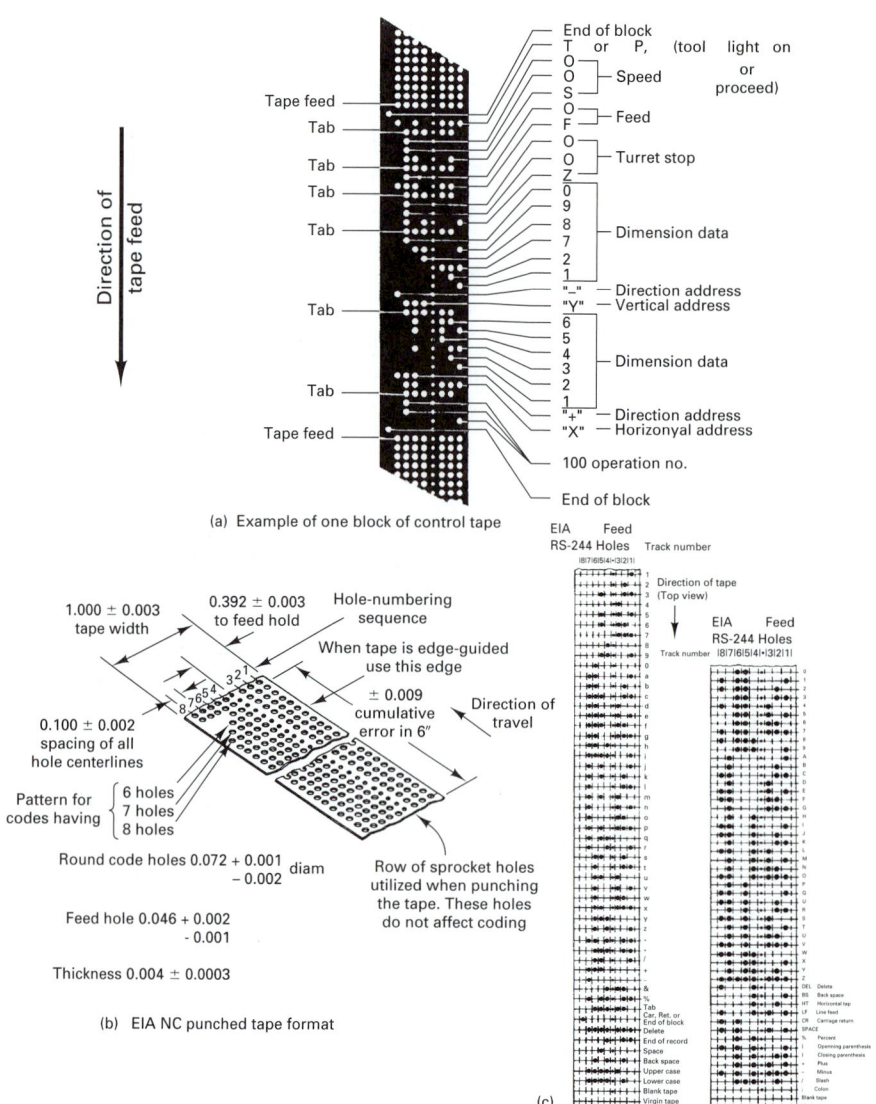

FIGURE 29-17   (a) The *X, Y, Z,* positions and other instructions comprise a block of machine information; (b) standard NC tape format defined by Electronics Industries Association; (c) original standard for tape coding as defined by EIA was RS-244. The ASCII subset is now EIA standard RS-358. Both codes are used.

next operation or whether a tool change is required. The last four items are specified by code symbols or what are called *NC* words (see Table 29-1 and Figure 29-17). The NC words are put together in a specified order to define a *block* of information needed to execute an operation. By convention, the data are usually arranged in blocks in the sequential order shown in the table.

After the program has been prepared, it is used to prepare the tape or may be directly entered into the control panel of a CNC machine. Usually, longer programs are entered by tape or disk and short programs entered manually. The tape uses a binary code to convert the numerical and instructional information into commands for the machine. Table 29-2

NC Word	Use
	TABLE 29-1. Definitions of Common NC Words
NC Word	Use
N	*Sequence number:* identifies the block of information
G	*Preparatory function:* requests different control functions, including preprogrammed machining routines
X, Y, Z, A, B, C	*Dimensional coordinate data:* linear and angular motion commands for the axis of the machine
F	*Feed function:* set feed rate for this operation
S	*Speed function:* set cutting speed for this operation
T	*Tool function:* tells the machine the location of the tool in the toolholder or tool turret
M	*Miscellaneous function:* turn coolant on or off, open spindle, reverse spindle, tool change, etc.
EOB	*End of block:* indicates to the MCU that a full block of information has been transmitted and the block can be executed

shows the Arabic versus the binary numbers. Binary numbers facilitate the rapid and accurate operation of control systems and computers. The conversion from Arabic to binary is done by the control system or computer, not by the programmer.

Punched tape is not a recent idea. The old player-piano roll was a form of tape control, and punched cards had been used for many years for controlling complicated weaving and business machines. Thus tape control of machine tools is an extension of an existing basic concept in which holes, representing information that has been punched into the tape, are "read" by sensing devices and used to actuate the devices that control various electrical or mechanical mechanisms.

Tape-controlled machine tools use a 1-in.-wide paper or Mylar tape containing eight information channels. Figure 29-17 shows an example of a block of such tape, punched with the information required for one operation on a turret-type drilling machine. Either of two codes are used, EIA244A code or ASCII, which is used in computer and telecommunications work as well as in NC. The codes are not interchangeable even though both are based on binary numbering. The binary base 2 system is used because electronic circuitry responds to either of two conditions, on or off or zero or one. Thus all the numbers, symbols, and letters that are needed to control the machine are communicated to the machine by the presence or absence of holes in the eight tracks of the tape.

Historically, four basic types of tape format are used for NC input to communicate dimensional and nondimensional information: *fixed-sequential format*, *work-address format*, *tab-sequential format*, and *word-address format*. Most new NC or CNC systems use the word-address format, which allows the words to be presented in any order and is the most flexible.

After the program is converted to tape or disk, but before it is used, it is *verified,* or checked, to make sure that it is correct. The verification step can use a special NC plotting machine that will trace out all the toolwork paths as they would occur on the machine tool. A sample part is machined in plastic or wax for checking the part specifications from the drawing against the real part. Today, CNC machines permit the user to program the interface between the machine tool and the control, greatly reducing the number of machining system components and interconnections. Current CNCs have extensive self-diagnostics and performance-monitoring systems. The greatest advances in CNC technology, however, are in part programming, where easy-to-use, menu-driven software makes programming almost as simple as setting up the machine manually. In NC machines, the operator may override the tape when necessary but cannot reprogram the machine unless a new tape is prepared. The CNC machine has the capability of reading a program into its computer

TABLE 29-2. Comparison of Arabic and Binary Numbers		
Arabic	Binary	Power of 2
0	0	
1	1	2^0
2	10	2^1
3	11	
4	100	2^2
5	101	
6	110	
7	111	
8	1000	2^3
9	1001	
10	1010	
11	1011	
12	1100	
13	1101	
14	1110	
15	1111	
16	10000	2^4
17	10001	
18	10010	
19	10011	
20	10100	
21	10101	
22	10110	
23	10111	
24	11000	
25	11001	
26	11010	
27	11011	
28	11100	
29	11101	
30	11110	
31	11111	
32	1000000	2^5
64	1000000	2^6
128	10000000	2^7

Note: Any number raised to the zero power equals 1.

Examples:

Value of 182 expressed in Arabic numbers

$$1 \times 10^2 = 100$$
$$8 \times 10^1 = 80$$
$$2 \times 10^0 = 2$$

1 8 2   182

Value of 182 expressed in binary numbers

$$1 \times 2^7 = 128$$
$$0 \times 2^6 = 0$$
$$1 \times 2^5 = 32$$
$$1 \times 2^4 = 6$$
$$0 \times 2^3 = 0$$
$$1 \times 2^2 = 4$$
$$1 \times 2^1 = 2$$
$$0 \times 2^0 = 0$$

1 0 1 1 0 1 1 0   182

memory, and the program can be modified at the machine like any other computer program.

On CNC machines, the machine tool operator may perform all the programming steps right at the console of the machine, programming the processing steps for the part directly into the computer memory. The program can be saved by having the machine print out a copy of the program, which can be used later for reorders of the same part. Features such as program edit, canned routines, program storage, diagnostics, constant surface speed, and tape punch are common on today's CNC machines.

As more and more design work is done on the computer (CAD) using databases and software that are compatible to the machine tools, there will be less dependency on tape for program storage and more utilization of floppy disks, hard disks, and other typical computer storage means. For example, a machining cell composed of NC machine tools designed for a family of 10 component parts may be able to do the 10 different parts without needing retooling or refixturing, but it will still have 10 different programs for these parts for each machine. If the programs are computer stored, they can be readily accessed, but if they are stored on tape, delays will occur in dumping the different programs in or out of the control computer.

Contouring in NC and CNC Machines

Although the majority of NC and tape-controlled machine tools do not provide for machining contoured surfaces, many do. Most CNC and DNC machines provide this feature. The required curves and contours are generated approximately by a series of very short, straight lines or segments of some type of regular curves, such as hyperbolas. This is called *interpolation*. The program fed to the machine is arranged to approximate the required curve within the desired accuracy. Figure 29-18 illustrates how a desired straight line or curved surface can be approximated by means of short segments. Interpolation refers to the fact that curved surfaces as generated by machine tools must be approximated by a series of very short, straight-line movements in the X, Y, and Z directions. The length of the segment must be varied in accordance with the deviation permitted. Most machine tools with contouring capability will produce a surface that is within 0.001 in. of the one desired, and many will provide considerably better performance. Most contouring machines have either two- or three-axis capability, but a good many have up to five-axis capability.

Obviously, contour machining (and interpolation) requires that complex information be entered into the MCU because the number of straight-line or curved segments may be quite large. Manual programming of the tape can be quite laborious. Computer programming can translate simple commands into the complex information required by the machine.

Computer Languages for Numerical Control

The control mechanisms on NC machines do not understand ordinary shop language that people use to describe what machining, or machine-work movements, must take place. In addition, the effects of various sizes and types of cutting tools, requiring different offsets as illustrated in Figure 29-18, must be accounted for. Tool radius offset accounts for the fact that the center of tool rotation (as held in the spindle) must be offset from the workpiece surface to be generated. The path of the cutter center line will have dimensions different from those of the surface. The capabilities of the machine tool regarding available power, speeds, feeds, table travel, and so on, must be taken into account. This barrier has been diminished by the use of CAD-based NC programming, which has reduced the usage of APT, a special programming language that the computer can understand and convert into the commands re-

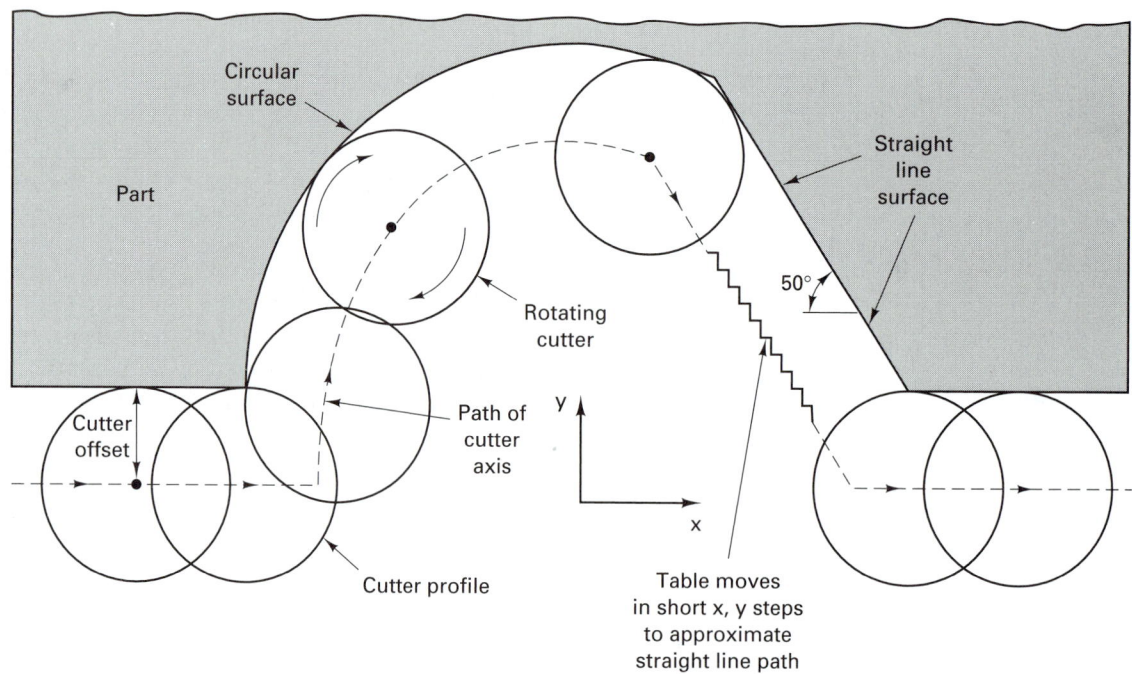

FIGURE 29-18   Two classic problems in NC programming are the determination of cutter offset and interpolation of cutter paths.

quired by NC machine controls. APT is the most widely used language in the United States and is used for both positioning and continuous-path programming. APT is designed to run on mainframe computers with large memories. ADAPT, a language developed by IBM under an Air Force contract, has many of the features of APT but is designed for smaller computers. Another language, AUTOSPOT, was developed by IBM for point-to-point positioning, but today's version can be used for contouring. In 1974 the APT standard ANSI × 3.J7-1974 was approved. Revision 3 of the standard was approved in 1989. Today it is possible to implement APT on personal computers, but unlike the CAD-NC programs, the APT system does not have visual aids or colorful graphics. The basic sequence of operations for computer-assisted part programming is as follows (see Figure 29-3).

The part programmer prepares a manuscript that specifies the part geometry, the tool path, and the operations needed for their sequence, using English-like statements. Auxiliary machine functions, such as "coolant on," are also programmed. The program is entered into the computer. The computer performs an input translation, followed by the required arithmetic calculations, including such things as cutter offset computations, to obtain the coordinate points the cutter must follow. The individualities of the specific machine tools, relative to the available speeds, feeds, accelerations, and so on, are provided for by a postprocessor program. The output from the postprocessor is the required NC tape (or disk or direct program), which is fed to the machine tool.

The workpiece, no matter how complex, can be conceived of as a compilation of points, straight lines, planes, circles, cylinders, and other mathematically defined geometries. Historically, the part programmer's job has been to translate the component parts into basic geometric elements, defining each geometric element in terms of workpiece dimensions. In recent years, much of this work has become routine in CAD systems. However, NC

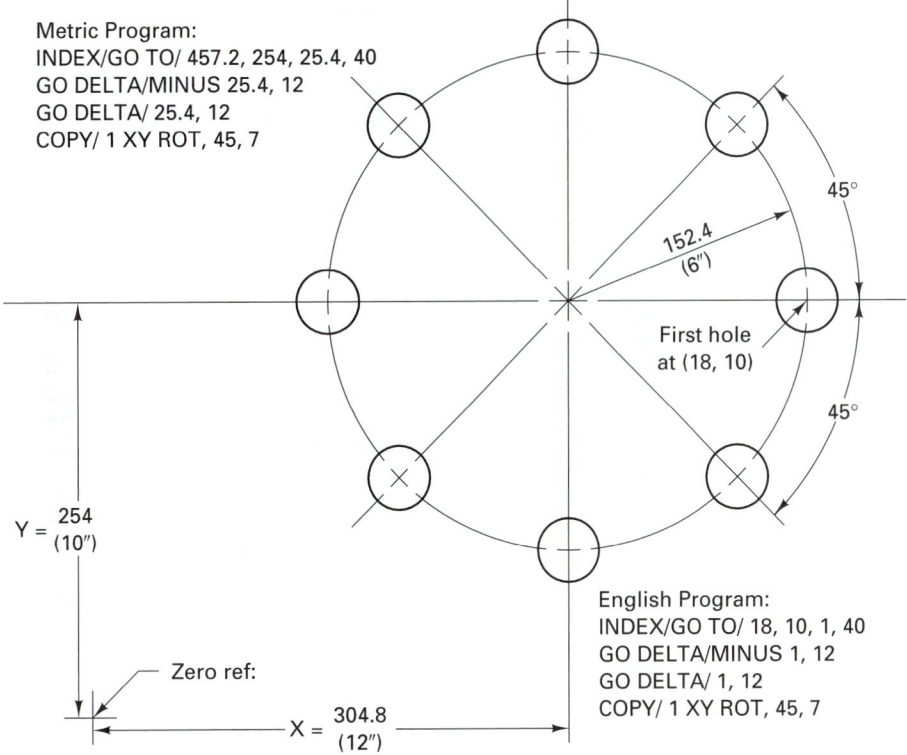

Metric Program:
INDEX/GO TO/ 457.2, 254, 25.4, 40
GO DELTA/MINUS 25.4, 12
GO DELTA/ 25.4, 12
COPY/ 1 XY ROT, 45, 7

45°

152.4
(6")

First hole
at (18, 10)

45°

$Y = \dfrac{254}{(10")}$

Zero ref:

English Program:
INDEX/GO TO/ 18, 10, 1, 40
GO DELTA/MINUS 1, 12
GO DELTA/ 1, 12
COPY/ 1 XY ROT, 45, 7

$X = \dfrac{304.8}{(12")}$

FIGURE 29-19   Bolt hole circle and APT programs for machining the eight holes.

programming languages and CAD software are often incompatible, which has hindered the CAD-to-CAM step.

An example of how the APT language would be used on a part is shown in Figure 29-19. Essentially, it is necessary to tell the computer the location of a center of the bolt hole circle or one of the holes in the circle with respect to a zero point, the number of holes in the circle, and the radius of the circle. Numerals 18, 10, and 1 (457.2, 254, 25.4) are X, Y, and Z coordinates for table and tool. Numeral 40 is table movement rate. Numeral 12 is feed rate for drill. Numerical 45 is 45° of rotation. Numeral 7 is the instruction for seven duplicate holes to be drilled. The trend to small computers, programmable controllers, and CNC machines will continue to lessen the need for and use of mainframe and minicomputer APT. Many machines tools, NC and conventional, are being retrofitted to CNC at a cost of $15,000 to $40,000.

■ 29.6 MACHINING CENTER FEATURES AND TRENDS

Computer and numerical control is used on a wide variety of machines. These range from single-spindle drilling machines, which often have only two-axis control and can be obtained for about $10,000, to machining centers, such as shown in Figure 29-20. The machining center can do drilling, boring, milling, tapping, and so on, with four-axis control. It can automatically select and change 32 preset tools. The table can move left/right or in/out and

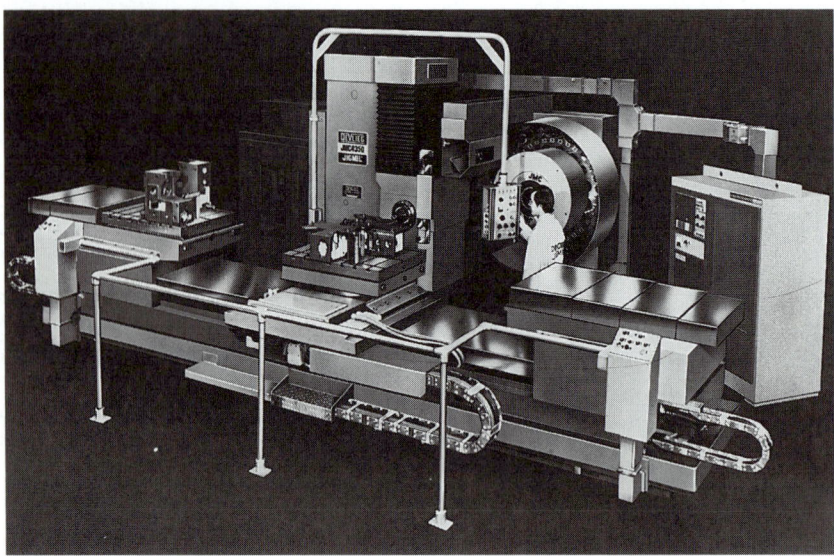

FIGURE 29-20 Horizontal spindle, four-axis CNC machining center. *(Courtesy of DeVlieg Machine Company.)*

the spindle can move up/down or in/out, with positioning accuracy in the range 0.0003 in. over 40 in. of travel. The machine has automatic tool change and automatic work transfer capability, so that workpieces can be loaded and unloaded while machining is in process. Such a machine can cost over $200,000. Between these extremes are numerous machine tools that do less varied work than the highly sophisticated machining centers but which combine high output and minimum setup time in changing from one job to another, and remarkable flexibility because of the number of tool motions that are provided.

CNC Turning Centers

The modern lathe has CNC control and tools mounted in turrets on slanted beds. The tailstock has been replaced by a live, powered spindle and chuck. On some lathes the concept of automatic tool changing has been implemented (Figure 29-21). The tools are held on a rotating tool magazine and a gantry-type tool changer is used to change the tools. Each magazine holds one type of cutting tool. This is an example of the trend of providing greater versatility along with high productivity in lathes. The versatility is being further increased by combining both rotary-work and rotary-tool operations—turning and milling in a single machine. Live, powered or "driven" tools replace regular tools in the turrets and perform milling and drilling operations when the spindle is stopped. There are numerous tape-controlled machines that provide four- and five-axis contouring capability. Tools are changed in 6 seconds or less. It is also common to provide two or more worktables, permitting work to be set up while machining is done on the workpiece in the machine, with the tables being interchanged automatically. Consequently, the productivity of such machines can be very high, with the chip-producing time often approaching 50% of the total.

Numerical control has been applied to a wide variety of other production processes. NC turret punches with *X-Y* control on the table, CNC wire EDM machines, laser welders and machining, flame cutters, and many other machines are readily available.

Two new trends are observed in the development of machining centers. One is the growing interest in smaller, more compact machining centers, and the other is the emphasis

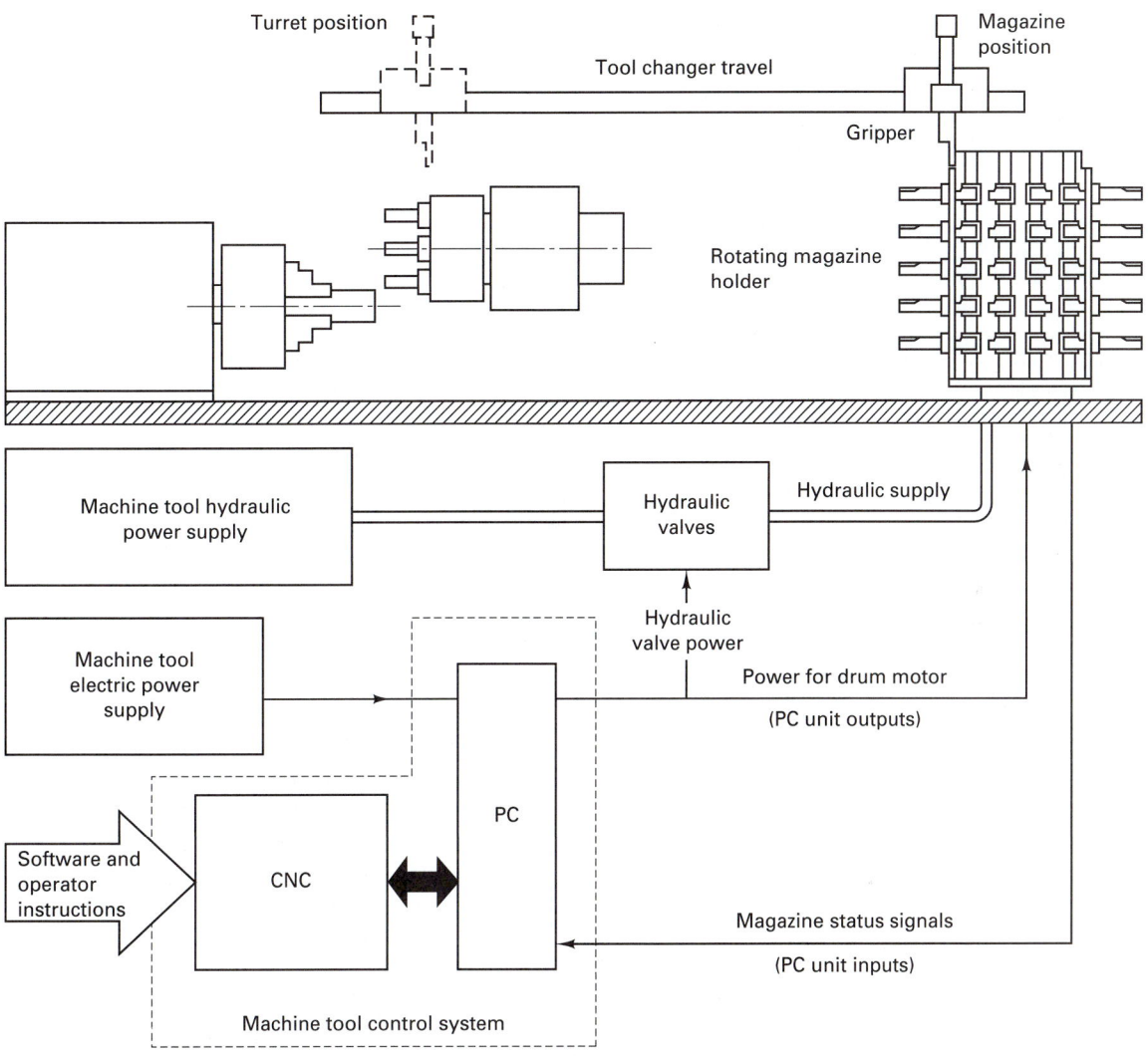

FIGURE 29-21 The concept of automatic tool changing has been extended to lathes. *(Courtesy of Sandvik.)*

on extended-shift or even unmanned operations. Modern machining centers have contributed significantly to improved productivity in many companies. They have eliminated the time lost in moving workpieces from machine to machine and the time needed for workpiece loading and unloading for separate operations. In addition, they have minimized the time lost in changing tools, carrying out gaging operations, and aligning workpieces on the machine.

The latest generation of machining centers is aimed at further improving utilization by reducing the time when machines are stopped, either during pauses in a shift or between shifts. Delays are caused by tool breakage, unforeseen tool wear, limited number of tools, or an inadequate number of available workpieces. Machines are fitted with tool breakage monitors, tool wear compensating devices, and means for increasing the number of tools and workpieces available.

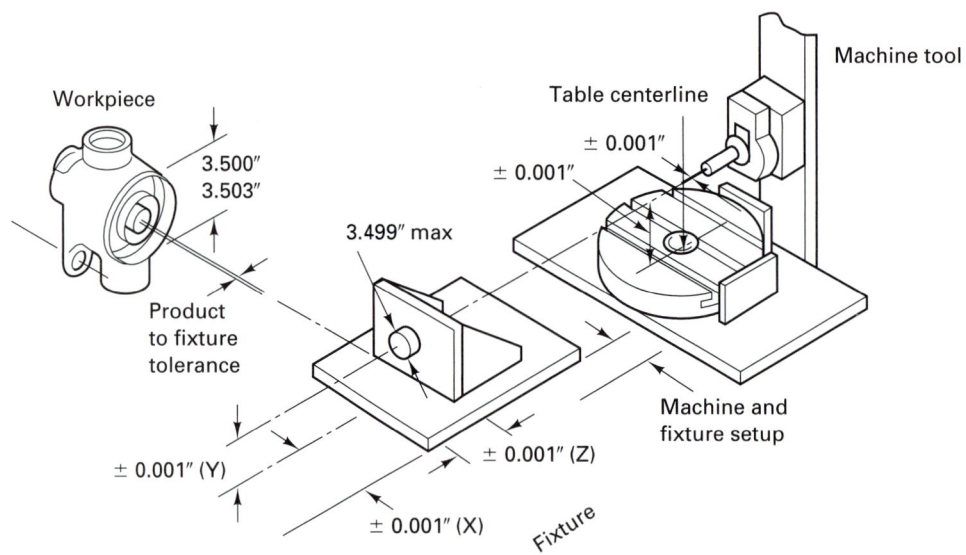

FIGURE 29-22 Process capability in NC machines is affected by many factors.

Probes on NC machines can greatly improve the process capability of the machine tool. There is a big difference between the claimed program resolution for a NC machine and the accuracy and precision (the process capability) in the actual parts. As shown in Figure 29-22, true positioning accuracy and precision are affected by machine alignment, machine and fixture setup, variations in the workholding device, raw material variations, workpiece location in the fixture variations, and cutting tool tolerances. Thus the finished workpiece may be unacceptable even though the machine is more than capable of producing the part to the design specifications. The part program has no assurance that the part is properly located in the fixture or that the fixture is properly located on the table of the machine. A probe, carried in the tool storage magazine and mounted when needed in the spindle like a cutting tool, can establish the location of the surface features relative to each other and to the spindle axis within 0.0005 in. (Figure 29-23). The machine controller, using the probe data, will then shift the program reference data accordingly. The probe can be used to determine the amount of material on a rough casting, locate a corner of a part, define the center of a hole, or check for the presence or absence of a feature. All of the variability described in Figure 29-23 can be compensated for except for variations in the cutting-tool geometry. These can be handled by a probe mounted on the machine tool. The CNC turret lathe shown in Figure 29-24 has a retractable probe mounted in the headstock. This probe can be used to check the tool location, tool dimensions (see Figure 29-16), and even tool wear, automatically updating tool-offset data in the control computer. A second probe is mounted on the tool turret for checking part setup and alignment as well as in-process inspection. Thus the machine tool can function like a coordinate measuring machine. By comparing the actual touched location with the programmed location, the measuring routine determines appropriate compensation.

Many of the new machining centers are equipped with novel automatic pallet changers and workpiece loading and unloading devices. Robots are increasingly being applied for workpiece handling in machine groups, including manufacturing cells. In some cases the robot is also used for tool-changing functions (see Chapter 41).

FIGURE 29-23 (a) Probe carried in the tool changer can be mounted in the spindle (b) for checking the location of part features accurately.

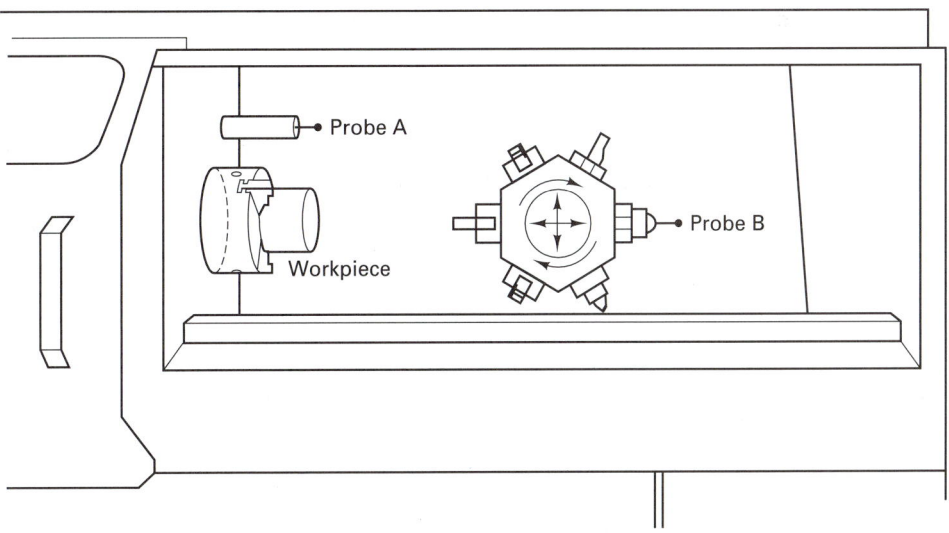

FIGURE 29-24 CNC turret lathe with retractable probe A in the headstock and probe B mounted on the turret.

■ 29.7 ADVANTAGES OF NUMERICAL CONTROL

The advantages of programmable machine tools can be summarized as follows:

1. *Flexibility.* It is easy to change from one part to another or to change part design.
2. *Superior process capability.* Greater accuracy and precision is built into the machines, resulting in better quality and a high order of repeatability.
3. *Higher production rates.* Optimum feeds and speeds can be determined for each operation, with less time spent in noncutting functions.
4. *Lower tooling costs.* Expensive jigs and templates may not be needed. More general-purpose workholders can be used. (See Chapter 28 for some examples.)
5. *Less lead time.* Programs can be prepared in less time than conventional jigs and templates. Less setup time is required.
6. *Fewer setups per workpiece.* More operations can be done at each setup of the workpiece.
7. *Better machine utilization.* There may be less machine idle time, owing to more efficient table or tool movement between successive operations and fewer setups. Cycle time is easily altered.
8. *Reduced inventory.* The overall inventory level may be reduced if parts can be run economically in smaller quantities. (See the Case Study at the end of this chapter.)
9. *Reduction in space required.* Smaller lot sizes may reduce the space required for the machines, especially if manufacturing cells are used.
10. *Less scrap.* Operator errors are reduced substantially. The first part off the machine can be a good part. Follow-on runs of parts previously manufactured can readily be duplicated.
11. *Less skill required of the operator.* Program planning in preparing tapes reduces the necessity for operator decisions.
12. *Manufacture of unique geometries.*

Through the use of computer programming of NC or CNC machines, one can generate surfaces and geometrical configurations that are not possible to make by any other method, at least not economically. Such a part is seen in Figure 29-25. This is the copper base of a cooling device. The spiral groove has constant width but constantly varying depth from start to finish. The APT programming language was used and the part machined on a three-axis CNC continuous-path vertical milling machine.

■ 29.8 ECONOMIC CONSIDERATIONS IN NUMERICAL CONTROL

The major disadvantage of NC machine tools is their high initial cost. This means that they must be justified from an economic viewpoint. Modern aggressive approaches to engineering economic analysis are strongly recommended. The return of investment with NC machines comes from many more and different areas than the return generated by conventional equipment. For starters, one needs to examine the impact on design, manufacturing engineering, setup, reduced lot sizes, and quality to find all the sources of savings. Considerations beyond savings in machining time are difficult to express in dollars but must be considered when trying to justify NC equipment.

The control equipment now is virtually all made of solid-state modules, and reliability is excellent. Programming has been greatly simplified but a manufacturing engineer may be needed for programming complex parts. Like all manufacturing tools, NC and

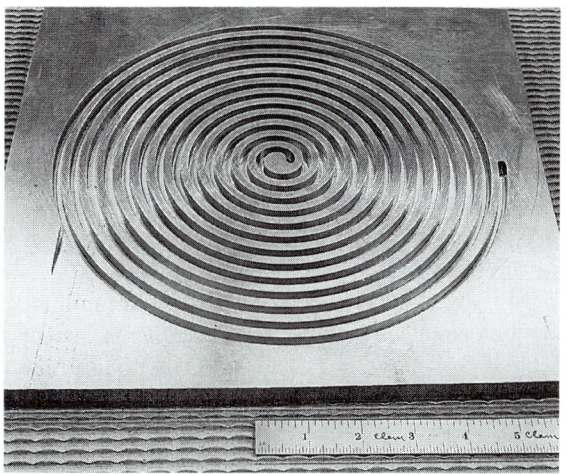

FIGURE 29-25   Spiral groove with variable depth machined in copper plate. *(Courtesy of Robert Tidmore, Fabrication Division, NASA Marshall Space Flight Center Test Laboratory.)*

CNC machines do require maintenance. Overall, NC machines have provided a much needed solution for small- and medium-quantity production, and it is easy to see why they have been so widely adopted.

NC and CNC machines are costly, but their use can generally be justified economically in from one to three years, primarily through substantial savings in setup and machining time (particularly when the parts are being processed in a manufacturing cell), accompanied by significant improvements in quality. The NC machine may be the key machine in a manned manufacturing cell. CNC level machining centers have been used in unmanned manufacturing cells with parts being loaded and unloaded robotically. CNCs are used routinely in FMS's where parts are usually exchanged with transfer pallets (see Chapter 41 for additional discussion).

■ KEY WORDS

APT (automatically programmed tools)	CNC (complete numerical control)	numerical control
canned cycles	contouring	open-loop control
closed-loop control	DNC (direct numerical control)	part program
CAD (computer-aided design)	DNC (distributed numerical control)	point-to-point machines
CAM (computer-aided manufacturing)	FMS (flexible manufacturing system)	
	MCU (machine control unit)	
	machining center	

■ REVIEW QUESTIONS

1. How is a machining center different from a vertical-spindle milling machine?
2. What role did John Parsons play in the development of NC?
3. What was DNC as first practiced, and what does DNC usually mean today?
4. How does an adaptive control system differ from a numerical control system?
5. In the decision analysis portion of the AC system, optimization of the process is required. What does this imply about the process?
6. How does feedforward differ from feedback in these systems?
7. What are some examples of everyday devices or machines that employ feedback in their control systems?
8. Why was it necessary for machine tool builders to improve the feed screws on their machines when they made them into NC machines?

9. Explain the problems of cutter offset and interpolation in NC programming.
10. Some of the functions performed by the operator in piece-part manufacturing are very difficult to automate completely. Name the functions and explain why.
11. The first NC machines were closed-loop control. Later some machines were open loop. What change did this require on the part of machine tool builders?
12. Can a continuous-path NC machine be open loop? Why or why not?
13. Why are there no NC shapers or broaches?
14. Why isn't manual programming used for continuous-path NC? What about for straight-cut or point-to-point programming?
15. What is the name of the modular workholders shown in Chapter 29? (Modular workholders were described in Chapter 28.)

16. What are the three basic closed-loop feedback [A(4)] schemes?
17. What is the difference between the zero reference point and the machine zero point?
18. What is an encoder?
19. Why does contouring require both velocity and positional feedback?
20. What is a peck-drilling subroutine for a CNC machine?
21. What is pocket milling, and what kind of milling cutter is usually used to perform it?
22. What is the tool setting point on a NC or CNC machine?
23. How are probes used in CNC machines to improve process capability?
24. What are G words used for in NC?
25. How can an NC program be verified?
26. What is APT?

■ PROBLEMS

1. What are the X and Y dimensions for the center position of holes 2 and 3 from the part shown in Figure 29-7?
2. Configurations obtained from continuous-path machining are the result of a series of straight-line, parabolic span, or higher-order curves. The degree to which curved surfaces correspond to their design depends on how many lines or spans are used. Four equal chords in a circle describe a square (Figure 29-A). Six make a hexagon, and as the number increases the lines themselves come closer to a perfect circle. The number of lines needed is determined by a maximum tolerance allowed between the design of the curved section and the actual chord programmed. The program for a parabolic span control unit requires enough spans for any deviation to stay within an acceptable tolerance. For a tolerance of $T = 0.001$ in., how long should the span be for a curve with a 5-in. radius? Assume that the arc is part of a circle. What is the span angle here, in degrees?
3. In Problem 2, suppose that the acceptable tolerance was 0.0001 in. Determine the span angle.

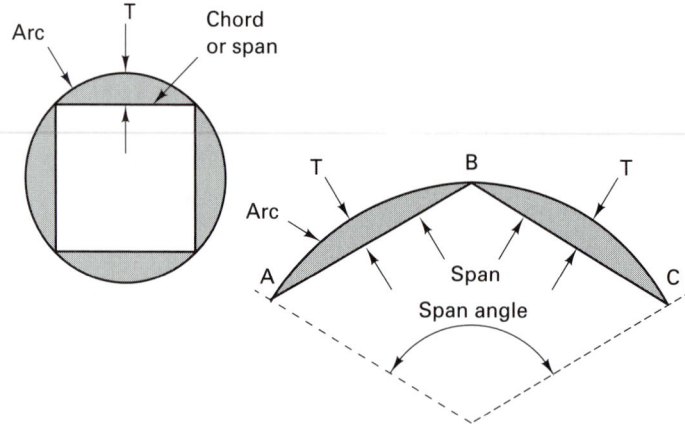

FIGURE 29-A

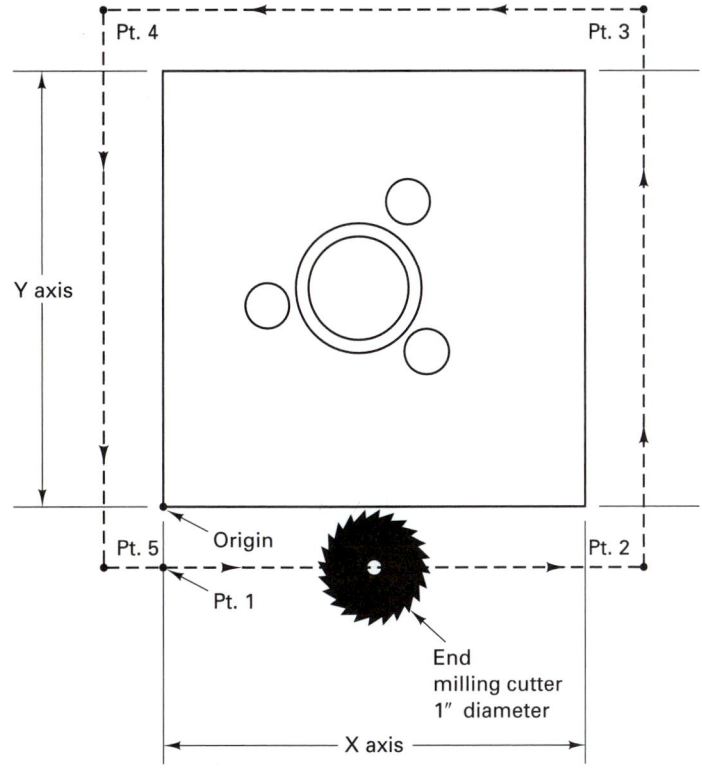

FIGURE 29-B

4. For the bolt hole circle shown in Figure 29-19, compute the X and Y dimensions for the hole 45° above the first hole.

5. Write the dimension for the first hole at (18,10) in Figure 29-19 in binary-coded decimals.

6. Suppose that the plate shown in Figure 29-7 was to be profile milled around the periphery with a 1-in.-diameter end-milling cutter, as shown in Figure 29-B. The dashed line is the cutter path. The programmer must calculate an offset path to allow for cutter diameter. Since the programmed points are followed by the cutter centerline and the profile is made at the tool's periphery, the programmer called for a $\frac{1}{2}$-in. cutter offset. If working with computer assistance, the programmer would describe the part profile to be machined and specify the cutter. The computer would generate the cutter path. Complete the table below to specify the cutter path, starting with the origin at the zero reference point. Move the tool around the plate counterclockwise.

Programmed Point Locations		
PT	X	Y
1		
2		
3		
4		
5		
1		

7. Suppose that surface finish is very important for the profile milling job described in Problem 6. Thus downmilling is going to be used. Rewrite the NC program points to accommodate this requirement. Show the new path on a sketch such as Figure 29-B.

8. The circular slot in the plate in Figure 29-7 can be end milled. This is a contouring cut (see Figure 29-13). Compute the cutter's X and Y velocity components as it is passing the number 3 hole, 225° around the slot from the start of the cut, using 25 in./min as the peak velocity value.

Chapter 29 CASE STUDY

break-even-point analysis of a lathe part

Y ou have received the part drawing (Figure CS-29) for a typical lathe part that will require turning, facing, grooving, boring, and threading as it is machined from a casting. Unfortunately, you do not yet know what the quantity will be. To be prepared, you have developed some cost data for the manufacture of the part by four different lathe processes (Table CS-29). Complete the table by determining the run cost per batch, the cost per unit at the various quantities, and the total cost per batch.

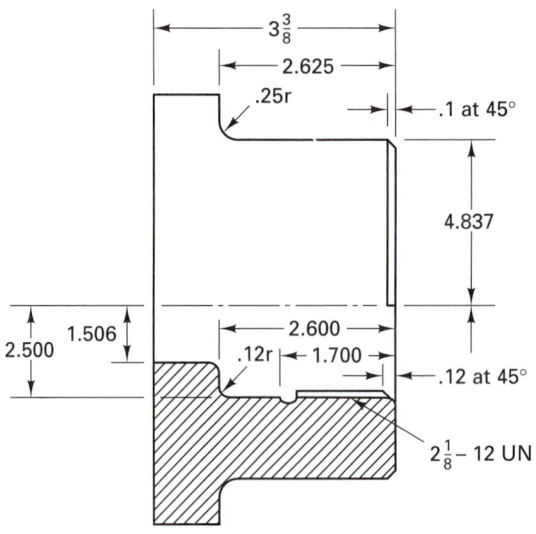

FIGURE CS-29

Answer the following questions regarding this situation:

1. Of the four costs listed for each process, which costs are fixed and which are variable?
2. For which of these costs would you have to estimate the machining time per piece and the cycle time per piece, including the time to change parts and setups?
3. How would you go about estimating this time, and what time elements might be included in the cycle time in addition to the machining time?
4. How would you use this estimate of time in the cost table? Show the calculation.
5. Make a plot of cost in dollars versus quantity, with all four methods on one plot.
6. Make a plot of cost per unit versus quantity, again with all four methods on one plot. Find the break-even quantities. (*Hint:* Did you plot the data on log paper?)
7. Discuss these plots and the break-even quantities that you found in part 5 versus part 6. (They should be the same.)
8. When would you use the NC lathe? The turret lathe? If you had no NC lathe, when would you use the turret lathe?

Chapter 29 CASE STUDY *(cont.)*

TABLE CS-29.   Cost Data for Lathe Processes

	Make Quantity				
	10,000 Units	1,000 Units	100 Units	10 Units	1 Units
Cost to produce on six-spindle automatic					
Total cost of batch	—	—	—	—	—
Engineering 2.5 hr at $40/hr	50.00	50.00	50.00	50.00	50.00
Tooling (cutting tools and workholders)	600.00	600.00	600.00	600.00	600.00
Setup 8 hr at $15/hr	120.00	120.00	120.00	120.00	120.00
Run cost per batch: 50 cents per piece	—	—	—	—	—
Cost each	—	—	—	—	—
Cost to produce on turret lathe					
Total cost of batch	—	—	—	—	—
Engineering 2 hr at $20/hr	40.00	40.00	40.00	40.00	40.00
Tooling	150.00	150.00	150.00	150.00	150.00
Setup 4 hr at $20/hr	48.00	48.00	48.00	48.00	48.00
Run cost per batch: $8 per piece	—	—	—	—	—
Cost each	—	—	—	—	—
Cost to produce on engine lathe					
Total cost of batch	—	—	—	—	—
Engineering 1 hr at $20/hr	20.00	20.00	20.00	20.00	20.00
Tooling	—	—	—	—	—
Setup 2 hr at $12/hr	24.00	24.00	24.00	24.00	24.00
Run cost per batch: $12	—	—	—	—	—
Cost each	—	—	—	—	—
Cost to produce on NC lathe					
Total cost of batch	—	—	—	—	—
Engineering and programming	150.00	150.00	150.00	150.00	150.00
Tooling	100.00	100.00	100.00	100.00	100.00
Setup 1 hr at $20/hr	20.00	20.00	20.00	20.00	20.00
Run cost per batch: $2 per piece	—	—	—	—	—
Cost each	—	—	—	—	—

CHAPTER 30

THREAD MANUFACTURING

30.1   INTRODUCTION
Screw-Thread
    Standardization and
    Nomenclature
Types of Screw Threads
Thread Classes
Thread Designation
30.2   THREAD CUTTING
Cutting Threads on a Lathe
Cutting Threads on a CNC
    Lathe
Cutting Threads with Dies
Self-Opening Die Heads
30.3   INTERNAL THREAD
        CUTTING
Collapsing Taps

Hole Preparation
Machine Tapping
Tapping Cutting Time
Special Threading and
    Tapping Machines
Common Tapping Problems
Tapping High-Strength
    Materials
Cutting Fluids for Tapping
30.4   THREAD MILLING
30.5   THREAD GRINDING
30.6   THREAD ROLLING
Chipless Tapping
Machining versus Rolling
    Threads
Case Study: VENTED CAP SCREWS

■ 30.1 INTRODUCTION

Screw threads probably are the most important of all the machine elements. *Threading*, *thread cutting*, *or thread rolling* refers to the manufacture of threads on external diameters. *Tapping* refers to machining threads in (drilled) holes. Without these processes our present technological society would come to a grinding halt. More screw threads are made each year than any other machined element. They range in size from those used in small watches to threaded shafts 10 in. in diameter. They are made in quantities ranging from one to several million duplicate threads. Their precision varies from that of inexpensive hardware screws to that of lead screws for the most precise machine tools. Consequently, it is not surprising that many very different procedures have been developed for making screw threads and that the production cost by the various methods varies greatly. Fortunately, some of the most economical methods can provide very accurate results. However, as in the design of most products, the designer can greatly affect the ease and cost of producing specified screw threads. Thus understanding thread-making processes permits the designer to specify and incorporate screw threads into designs while avoiding needless and excessive cost.

A screw thread is a ridge of uniform section in the form of a helix on the external or internal surface of a cylinder or in the form of a conical spiral on the external or internal surface of a frustrum of a cone. These are called *straight* or *tapered* threads, respectively. Tapered threads are used on pipe joints or other applications where liquidtight joints are required. Straight threads, on the other hand, are used in a wide variety of applications, most commonly on fastening devices, such as bolts, screws, and nuts, and as integral elements on parts that are to be fastened together. But as mentioned previously, they find very important applications in transmitting controlled motion, as in lead screws and precision measuring equipment.

880

Three basic methods are used to produce threads: *cutting, rolling,* and *casting.* Although both external and internal threads can be cast, relatively few are made in this manner, primarily in connection with die casting, investment casting, or the molding of plastics. Today by far the largest number of threads are made by rolling. Both external and internal threads can be made by rolling, but the material must be ductile. Because rolling is a less flexible process than thread cutting, it essentially is restricted to standardized and simple parts. Consequently, large numbers of threads still are, and will continue to be, made by cutting processes, including grinding.

Screw-Thread Standardization and Nomenclature

Starting with Sir Joseph Whitworth in England in 1841 and William Sellers in the United States in 1864, much time and effort have been devoted to screw-thread standardization. In 1948, representatives of the United States, Canada, and Great Britain adopted the Unified and American Screw Thread Standards, based on the form shown in Figure 30-1. In 1968 the International Organization for Standards (ISO) recommended the adoption of a set of metric standards, based on the basic thread profile shown in Figure 30-2. It appears likely that both types of threads will continue to be used for some time to come. The symbol P, the pitch, refers to the distance from a point on one screw thread to the corresponding point on the next thread, measured parallel to the length axis of the part.

The standard nomenclature for screw-thread components is illustrated in Figure 30-3. In both the *Unified* and *ISO systems,* the crests of external threads may be flat or rounded. The root usually is made rounded to minimize stress concentrations at this critical area. The internal thread has a flat crest in order to mate with either a rounded or V-root of the external thread. A small round is used at the root to provide clearance for the flat crest of the external thread.

In the metric system, the *pitch* always is expressed in millimeters, whereas in the American (Unified) system, it is a fraction having as the numerator 1 and as the denominator the number of threads per inch. Thus a $\frac{1}{16}$ pitch is $\frac{1}{16}$ of an inch. Consequently, in the Unified system threads more commonly are described in terms of threads per inch rather than by the pitch.

While all elements of the thread form are based on the *pitch diameter,* screw-thread sizes are expressed in terms of the *outside,* or *major* diameter and the *pitch* or *number of threads per inch.* In threaded elements, *lead* refers to the axial advance of the element during one revolution; therefore, lead equals pitch on a single-thread screw.

Types of Screw Threads

Eleven types, or series, of threads are of commercial importance, several having equivalent series in the metric system and Unified systems:

1. *Coarse-thread series* (UNC and NC). For general use where not subjected to vibration.
2. *Fine-thread series* (UNF and NF). For most automotive and aircraft work.
3. *Extra-fine-thread series* (UNEF and NEF). For use with thin-walled material or where a maximum number of threads are required in a given length.
4. *Eight-thread series* (8UN and 8N). Eight threads per inch for all diameters from 1 to 6 in. Used primarily for bolts on pipe flanges and cylinder-head studs where an initial tension must be set up to resist steam or air pressures.
5. *Twelve-thread series* (12UN and 12N). Twelve threads per inch for diameters from $\frac{1}{2}$ through 6 in. Not used extensively.

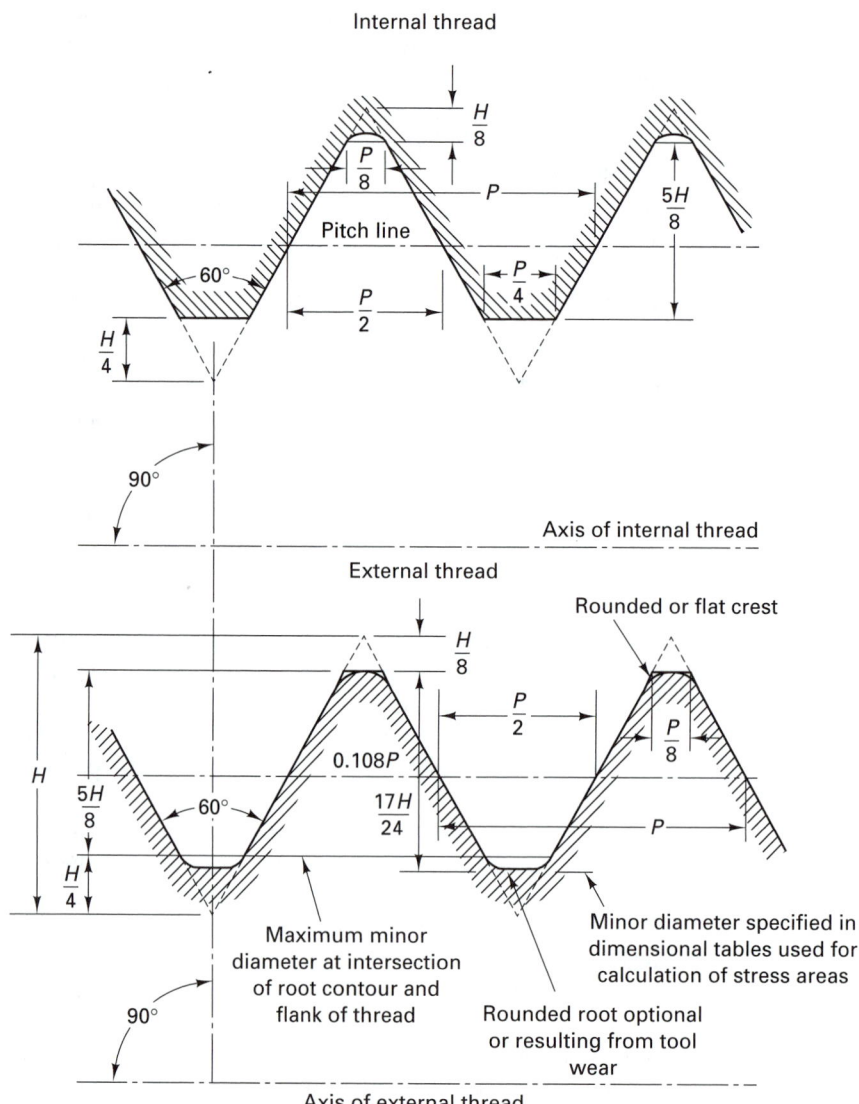

FIGURE 30-1 Unified and American screw-thread form for internal and external threads.

6. *Sixteen-thread series* (16UN and 16N). Sixteen threads per inch for diameters from $\frac{3}{4}$ through 6 in. Used for a wide variety of applications that require a fine thread.

7. *American Acme thread.* See Figure 30-4. This thread and the following three are used primarily in transmitting power and motion.

8. *Buttress thread.* See Figure 30-4 for details.

9. *Square thread.* See Figure 30-4 for details.

10. *29° Worm thread.*

11. *American, standard pipe thread.* This thread, also shown in Figure 30-4, is the standard tapered thread used on pipe joints in this country. The taper on all pipe threads is $\frac{3}{4}$ in./ft.

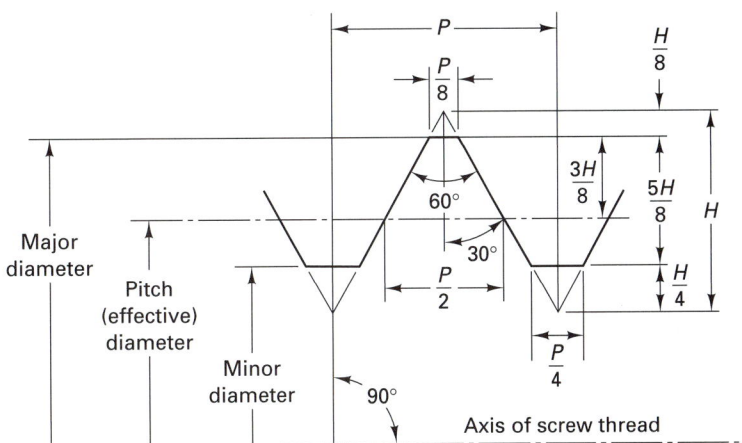

D = Major diameter of thread of nut ⎫ Normal
d = Major diameter of screw ⎭ diameter
D_1 = Minor diameter of thread of nut
d_1 = Minor diameter of screw
D_2 = Pitch diameter of thread of nut
d_2 = Pitch diameter of screw
H = Height of the complete theoretical thread profile
H_1 = Engagement
P = Pitch

$$H = \frac{\sqrt{3}}{2} P$$

FIGURE 30-2   Basic profile of metric general-purpose screw thread per 1S0/R68-1969.

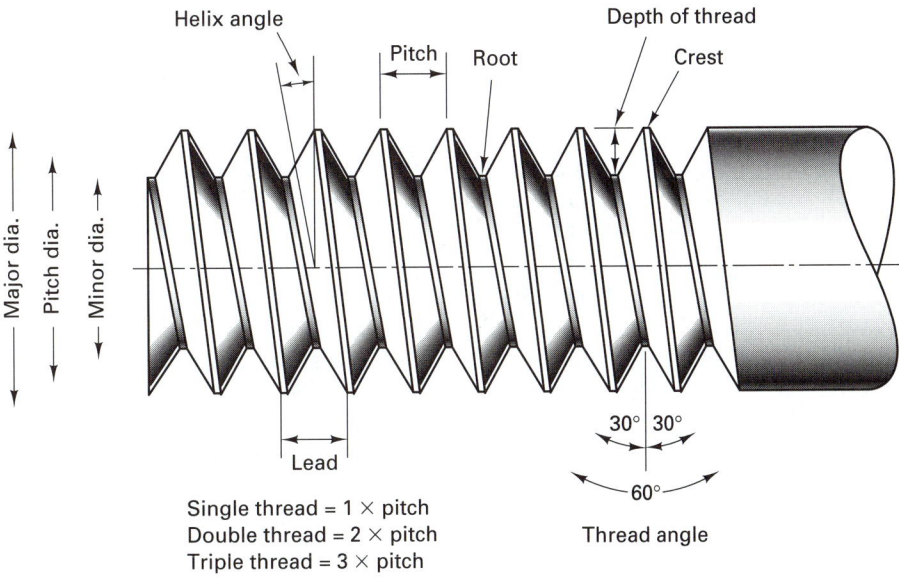

Single thread = 1 × pitch
Double thread = 2 × pitch
Triple thread = 3 × pitch

FIGURE 30-3   Standard screw-thread nomenclature.

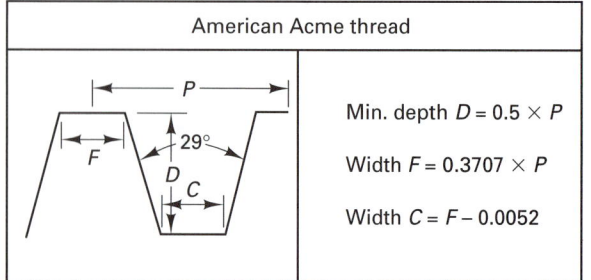

American Acme thread

Min. depth $D = 0.5 \times P$

Width $F = 0.3707 \times P$

Width $C = F - 0.0052$

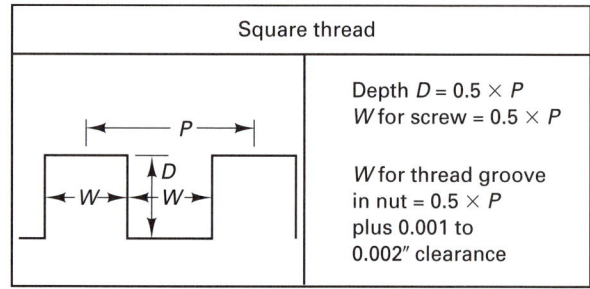

Square thread

Depth $D = 0.5 \times P$
W for screw $= 0.5 \times P$

W for thread groove
in nut $= 0.5 \times P$
plus 0.001 to
0.002" clearance

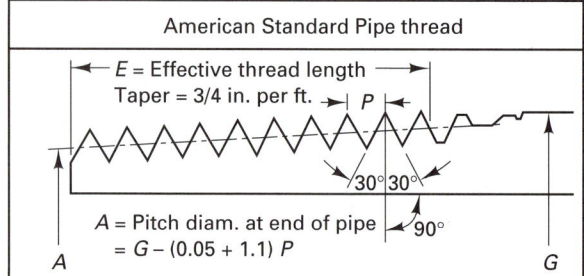

American Standard Pipe thread

E = Effective thread length
Taper = 3/4 in. per ft.

A = Pitch diam. at end of pipe
$= G - (0.05 + 1.1) P$

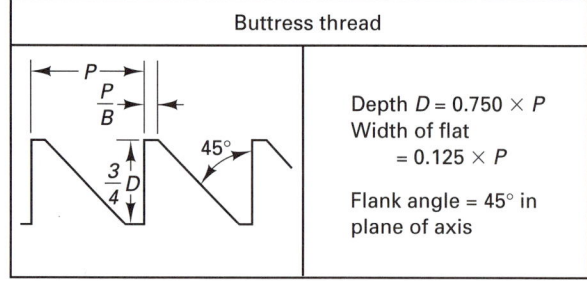

Buttress thread

Depth $D = 0.750 \times P$
Width of flat
$= 0.125 \times P$

Flank angle $= 45°$ in
plane of axis

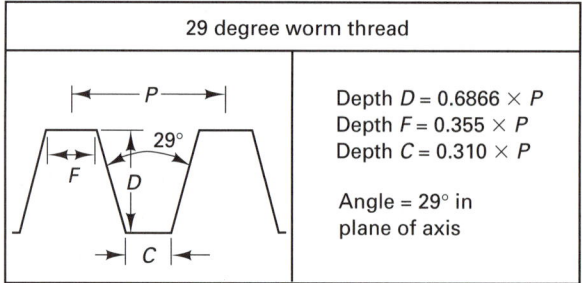

29 degree worm thread

Depth $D = 0.6866 \times P$
Depth $F = 0.355 \times P$
Depth $C = 0.310 \times P$

Angle $= 29°$ in
plane of axis

FIGURE 30-4 Special thread forms.

As has been indicated, the Unified threads are available in coarse (UNC and NC), fine (UNF and NF), extrafine (UNEF and NEF), and three-"pitch" (8, 12, and 16) series, the number of threads per inch being according to an arbitrary determination based on the major diameter.

Many nations have now adopted ISO threads into their national standards. Besides metric ISO threads, there are also inch-based ISO threads: namely, the UN series with which people in the United States, Canada, and Great Britain are familiar. ISO offers a wide range of metric sizes. Individual countries have the choice of accepting all or a selection of the ISO offerings.

The size listings of metric threads start with "M" and continue with the outside diameter in millimeters. Most ISO metric thread sizes come in coarse, medium, and fine pitches. When a coarse thread is designated, it is not necessary to spell out the pitch. *Example:* A coarse 10-mm-OD thread is called out as "M 12." This thread has a pitch of 1.75 mm, but the pitch may be omitted from the callout. A fine 12-mm-OD thread is available. It has a 1.25-mm pitch and must be designated "M 12 × 1.25." An extrafine

12-mm-OD thread having a 0.75-mm pitch would receive the designation "M 12 × 0.75." See Table 30-1 for a comparison of Unified and ISO thread sizes.

TABLE 30-1. Comparison Between Selection Unified and ISO Threads

Diameter		Unified (threads/in.)		ISO (threads/in.)		
Number, inches	mm	UNC	UNF	Coarse		Fine
No. 2	2.18	56	64	M2 × 0.4	(63.5)	
No. 4	2.84	40	48	M2 × 0.45	(56.4)	
No. 8	4.17	32	36	M4 × 0.7	(36.3)	
No. 10	4.82	24	28	M5 × 0.8	(31.8)	
$\frac{1}{4}$ in.	6.35	20	28	M6 × 1.0	(25.4)	
$\frac{1}{2}$ in.	12.7	13	20	M12 × 1.75	(14.5)	M12 × 1.25 (20.3)
$\frac{3}{4}$ in.	10.05	10	16	M20 × 2.5	(10.2)	M29 × 1.5 (16.9)
1 in.	25.4	8	14	M24 × 3	(8.47)	M24 × 2 (12.7)

The sign "×" is not employed as a multiplication symbol in metric practice but is used to relate these two attributes of the threads. The full description of a thread fastener obviously includes information beyond the thread specification. Head type, length, length of thread, design of end, thread runout, heat treatment, applied finishes, and other data may be needed to fully specify a bolt in addition to the designation of the thread. The "×" sign should not be used to separate any of the other characteristics.

The availability of fasteners, particularly nuts, containing plastic inserts to make them self-locking and thus able to resist loosening due to vibration, and the use of special coatings that serve the same purpose, have resulted in less use of finer-thread-series fasteners in mass production. Coarser-thread fasteners are easier to assemble and less subject to cross-threading (binding).

Thread Classes

In the Unified system, manufacturing tolerances are specified by three classes. Class 1 is for ordnance and other special applications, class 2 threads are the normal production grade, and class 3 threads have minimum tolerances where tight fits are required. The letters A and B are added after the class numerals to indicate external and internal threads, respectively.

In the ISO system, tolerances are applied to "positions" and "grades." Tolerance positions denote the limits of pitch and crest diameters, using "e" (large), "g" (small), and "H" (no allowance) for internal threads. The grade is expressed by numerals 3 through 9. Grade 6 is roughly equivalent to U.S. grades 2A and B, medium-quality, general-purpose threads. Below grade 6 is fine quality and/or short engagement. Above grade 6 is coarse quality and/or long length of engagement.

Thread Designation

In the Unified system, screw threads are designated by symbols as follows:

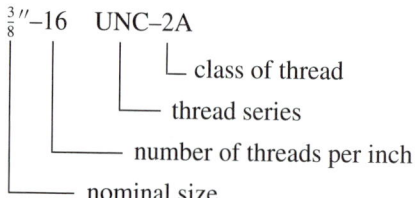

This type of designation applies to right-hand threads. For left-hand threads, the letters LH are added after the class of thread symbol.

In the ISO system, threads are designated as follows:

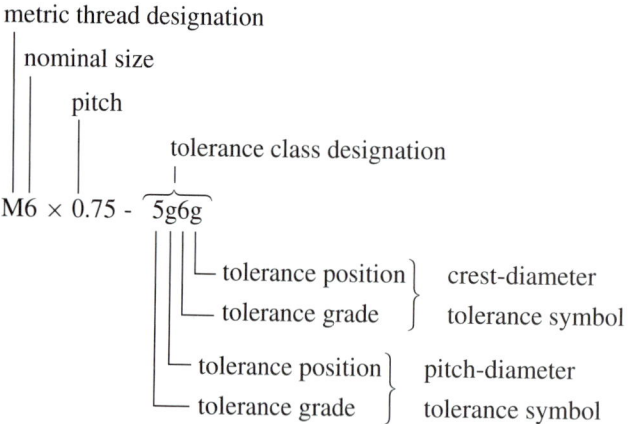

■ 30.2 THREAD CUTTING

The methods by which threads can be cut are summarized in Table 30-2.

TABLE 30-2. Thread Cutting Methods

External	Internal
Threading on an engine lathe	Threading (on an engine lathe or NC lathe)
Threading on a NC lathe	
With a die held in a stock (manual)	With a tap and holder (manual NC, machine, semiautomatic, or automatic)
With an automatic die (turret lathe or screw machine) or NC lathe	With a collapsible tap (turret lathe, screw machine, or special threading machine)
By milling	By milling
By grinding	

Cutting Threads on a Lathe

Lathes provided the first method for cutting threads by machine. Although most threads are now produced by other methods, lathes still provide the most versatile and fundamentally simple method. Consequently, they often are used for cutting threads on special workpieces where the configuration or nonstandard size does not permit them to be made by less costly methods.

FIGURE 30-5   Cutting a screw thread on a lathe, showing the method of
supporting the work and the relationship of the tool to the work. Inset shows face
of threading dial. *(Courtesy of South Bend Lathe.)*

There are two basic requirements for thread cutting on a lathe. First, an accurately
shaped and properly mounted tool is needed because thread cutting is a form-cutting operation. The resulting thread profile is determined by the shape of the tool and its position
relative to the workpiece. Second, the tool must move longitudinally in a specific relationship to the rotation of the workpiece, because this determines the *lead* of the thread. This
requirement is met through the use of the *lead screw* and the *split nut*, which provide
positive motion of the carriage relative to the rotation of the spindle.

External threads can be cut with the work mounted either between centers (Figure
30-5) or held in a chuck. For internal threads, the work must be held in a chuck. The cutting tool usually is checked for shape and alignment by means of a thread template (Figure
30-6). Figure 30-7 illustrates two methods of feeding the tool into the work. If the tool is
fed radially, cutting takes place simultaneously on both sides of the tool. With this true
form-cutting procedure, no rake should be ground on the tool, and the top of the tool must
be horizontal and be set exactly in line with the axis of rotation of the work (Figure 30-8);
otherwise, the resulting thread profile will not be correct. An obvious disadvantage of this
method is that the absence of side and back rake results in poor cutting (except on cast iron
or brass). The surface finish on steel usually will be poor. Consequently, the second
method commonly is used, with the compound swiveled 20°. The cutting then occurs
primarily on the left-hand edge of the tool, and some side rake can be provided.

Proper speed ratio between the spindle and the lead screw is set by means of the
gear-change box. Modern industrial lathes have ranges of ratios available so that nearly all
standard threads can be cut merely by setting the proper levers on the quick-change
gearbox.

To cut a thread, it also is essential that a constant positional relationship be maintained between the workpiece, the cutting tool, and the lead screw. If this is not done, the

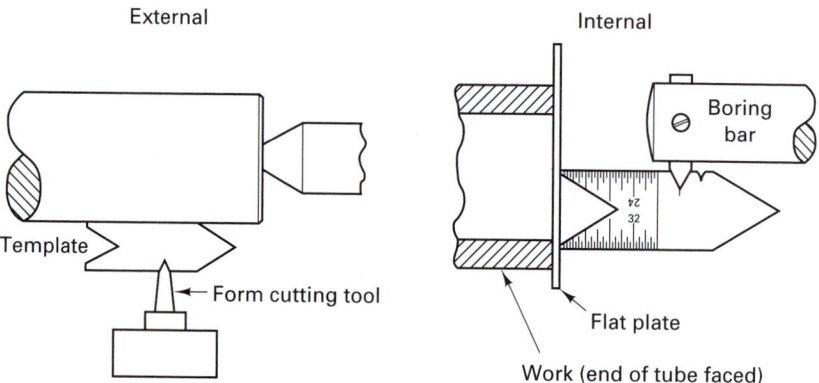

FIGURE 30-6 Methods of checking the form and setting of the cutting tool for thread cutting by means of a template. *(Courtesy of South Bend Lathe.)*

tool will not be positioned correctly in the thread space on successive cuts. Correct relationship is obtained by means of a *threading dial* (Figure 30-5), which is driven directly by the lead screw through a worm gear. Because the workpiece and the lead screw are directly connected, the threading dial provides a means for establishing the desired positional relationship between the workpiece and the cutting tool. The threading dial is graduated into an even number of major and half divisions. If the feed mechanism is engaged in accordance with the following rules, correct positioning of the tool will result:

1. *For even-numbered threads:* at any line on the dial
2. *For odd-numbered threads:* at any numbered line on the dial
3. *For threads involving $\frac{1}{2}$ numbers:* at any odd-numbered line on the dial
4. *For $\frac{1}{4}$ or $\frac{1}{8}$ threads:* return to the original starting line on the dial

To start cutting a thread, the tool usually is fed inward until it just scratches the work, and the cross-slide dial reading is then noted or set at zero. The split nut is engaged and the tool permitted to run over the desired thread length. When the tool reaches the end of the thread, it is quickly withdrawn by means of the cross-slide control. The split nut is then disengaged and the carriage returned to the starting position, where the tool is clear of the workpiece. At this point the future thread will be indicated by a fine scratch line. This permits the operator to check the thread lead by means of a scale or thread gage to assure that all settings have been made correctly.

Next, the tool is returned to its initial zero depth position by returning the cross slide to the zero setting. By using the compound rest, the tool can be moved inward the proper depth for the first cut. A depth of 0.010 to 0.025 in. usually is used for the first cut and smaller amounts on each successive cut, until the final cut is made with a depth of only 0.001 to 0.003 in. to produce a good finish. When the thread has been cut nearly to its full depth, it is checked for size by means of a mating nut or thread gage. Cutting is continued until a proper fit is obtained.

To cut right-hand threads, the tool is moved from right to left. For left-hand threads the tool must be moved from left to right. Otherwise, the procedure is essentially the same. Internal threads are cut in the same basic manner except that the tool is held in a boring bar. Tapered threads can be cut either with a taper attachment or by setting the tailstock off center. It should be remembered that in cutting a tapered thread, the tool must be set normal to the axis of rotation of the workpiece, not the tapered surface.

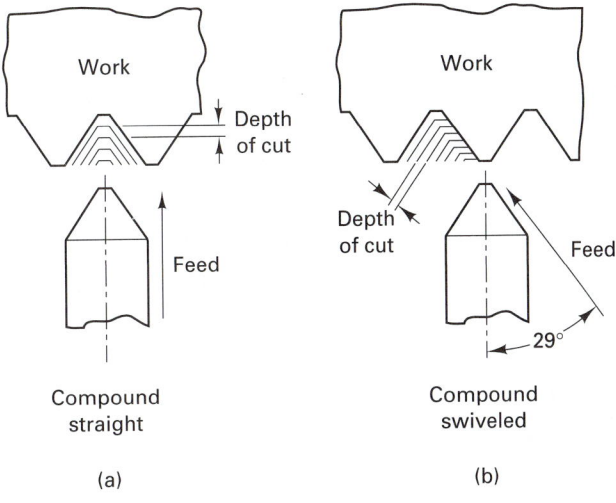

FIGURE 30-7   Top view of two methods of feeding the tool into the work in cutting threads on a lathe: (a) radial feed, compound straight; (b) half thread angle feed, compound swiveled.

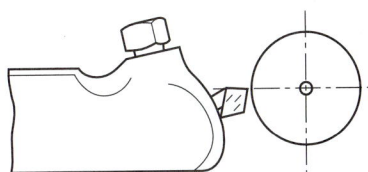

FIGURE 30-8   Proper relationship of the thread-cutting tool to the workpiece centerline. *(Courtesy of South Bend Lathe.)*

Cutting screw threads on a lathe is a slow, repetitious process that requires considerable operator skill. The cutting speeds usually employed are from one-third to one-half of regular speeds to enable the operator to have time to manipulate the controls and to ensure better cutting. The cost per part can be high, which explains why other methods are used whenever possible.

Cutting Threads on a CNC Lathe

CNC lathes and turning centers can be programmed to machine straight, tapered, or scroll threads. Threads are machined using the same type of special tool have the thread shape shown in Figure 30-7. The tool is positioned at a specific starting distance from the end of the work (Figure 30-9). This distance will vary from machine to machine. Its value can be found in the machine's programming manual. The CNC software will have a set of pre-programmed machining routines (called G codes) specifically for threading. Beginning at the start point, the tool accelerates to the feed rate required to cut the threads. The tool creates the thread shape by repeatedly following the same path as axial infeed is applied. For standard vee threads, the infeed can be applied along a $0°$ or $29°$ angle. The depth of cut for the first pass is the largest. The cutting depth is then decreased for each successive pass until the required thread depth is achieved. A final finishing pass is then made with the tool set at the thread depth.

Cutting Threads with Dies

Straight and tapered external threads up to about $1\frac{1}{2}$ in. in diameter can be cut quickly manually by means of threading dies (Figure 30-10a). Basically, these dies are similar to hardened, threaded nuts with multiple cutting edges. The cutting edges at the starting end

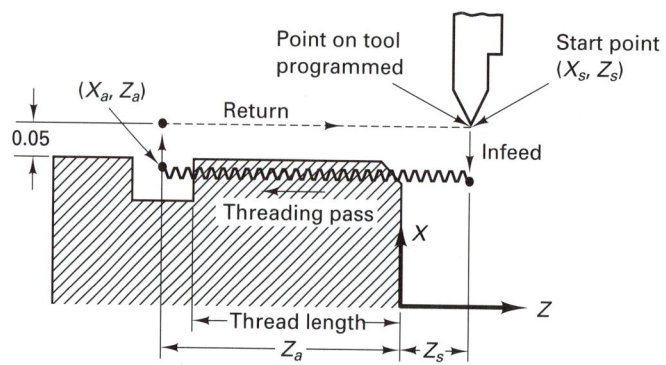

X_a Specifies the absolute X coordinate of the tool after axial infeed.
G32   Initiates the single-pass threading cycle.
Z_a Specifies the absolute Z coordinate of the tool after the threading pass.
Fn   n specifies the feed rate
$X_s Z_s$ Specifies the absolute X and Z coordinates of the start point.

FIGURE 30-9   Canned subroutines called G codes are used on CNC lathes to produce threads.

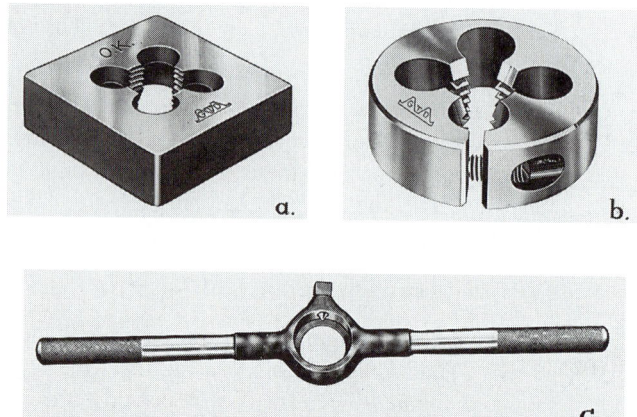

FIGURE 30-10   (a) Solid threading die; (b) solid-adjustable threading die; (c) threading-die stock for round die (die removed). *(Courtesy of TRW-Greenfield Tap & Die.)*

are beveled to aid in starting the dies on the workpiece. As a consequence, a few threads at the inner end of the workpiece are not cut to full depth. Such threading dies are made of carbon or high-speed tool steel.

Solid dies are seldom used in manufacturing because they have no provision for compensating for wear. The solid-adjustable type (Figure 30-10b) is split and can be adjusted over a small range by means of a screw to compensate for wear or to provide a variation in the fit of the resulting screw thread. These types of threading dies usually are held in a *stock* for hand rotation. A suitable lubricant is desirable to produce a smoother thread and to prolong the life of the die, since there is extensive friction during the cutting process.

Self-Opening Die Heads

A major disadvantage of solid threading dies is that they must be unscrewed from the workpiece to remove them. They are therefore not suitable for use on high-speed, production machines, and *self-opening die heads* are used instead on turret lathes, screw machines, NC lathes, and special threading machines for cutting external threads.

There are three types of self-opening die heads, all having four sets of adjustable multiple-point cutters that can be removed for sharpening or for interchanging for different thread sizes. This permits one head to be used for a range of thread sizes (Figure 30-11). The cutters can be positioned radially or tangentially, resulting in less tool flank contact and friction rubbing. In some self-opening die heads, the cutters are circular, with an interruption in the circular form to provide an easily sharpened cutting face. The cutters are mounted on the holder at an angle equal to the helix angle of the thread.

As the name implies, the cutters in self-opening die heads are arranged to open automatically when the thread has been cut to the desired length, thereby permitting the die head to be withdrawn quickly from the workpiece. On die heads used on turret lathes, the operator usually must reset the cutters in the closed position before making the next thread. The die heads used on screw machines and automatic threading machines are provided with a mechanism that automatically closes the cutters after the heads are withdrawn.

Cutting threads by means of self-opening die heads frequently is called *thread chasing*. However, some people apply this term to other methods of thread cutting, even to cutting a thread in a lathe.

■ 30.3 INTERNAL THREAD CUTTING

The cutting of an internal thread by means of a multiple-point tool is called *thread tapping*, and the tool is called a *tap*. A hole of diameter slightly larger than the minor diameter of the thread must already exist, made by drilling/reaming, boring, or die casting. For small holes, solid *hand taps* (Figure 30-12) are generally used. The flutes create cutting edges on the thread profile and provide space for the chips and the passage of cutting fluid. Such taps are made of either carbon or high-speed steel and are now routinely coated with TiN. The flutes can be either straight, helical, or spiral.

Hand taps (Figure 30-12) have square shanks and are usually made in sets of three. The *taper tap* has a tapered end that will enter the hole a sufficient distance to help align the tap. In addition, the threads increase gradually to full depth, and therefore this type of tap requires less torque to use. However, only a through hole can be threaded completely with a taper tap because it cuts to full depth only behind the tapered portion. A blind hole can be threaded to the bottom using three types of taps in succession. After the taper tap has the thread started in proper alignment, a *plug tap*, which has only a few tapered threads to provide gradual cutting of the threads to depth, is used to cut the threads as deep into the hole as its shape will permit. A *bottoming tap*, having no tapered threads, is used to finish the few remaining threads at the bottom of the hole to full depth. Obviously, producing threads to the full depth of a blind hole is time consuming, and it also frequently results in broken taps and defective workpieces. Such configurations usually can be avoided if designers will give reasonable thought to the matter.

Taps operate under very severe conditions because of the heavy friction (high torque) involved and the difficulty of chip removal. Also, taps are relatively fragile. *Spiral-fluted*

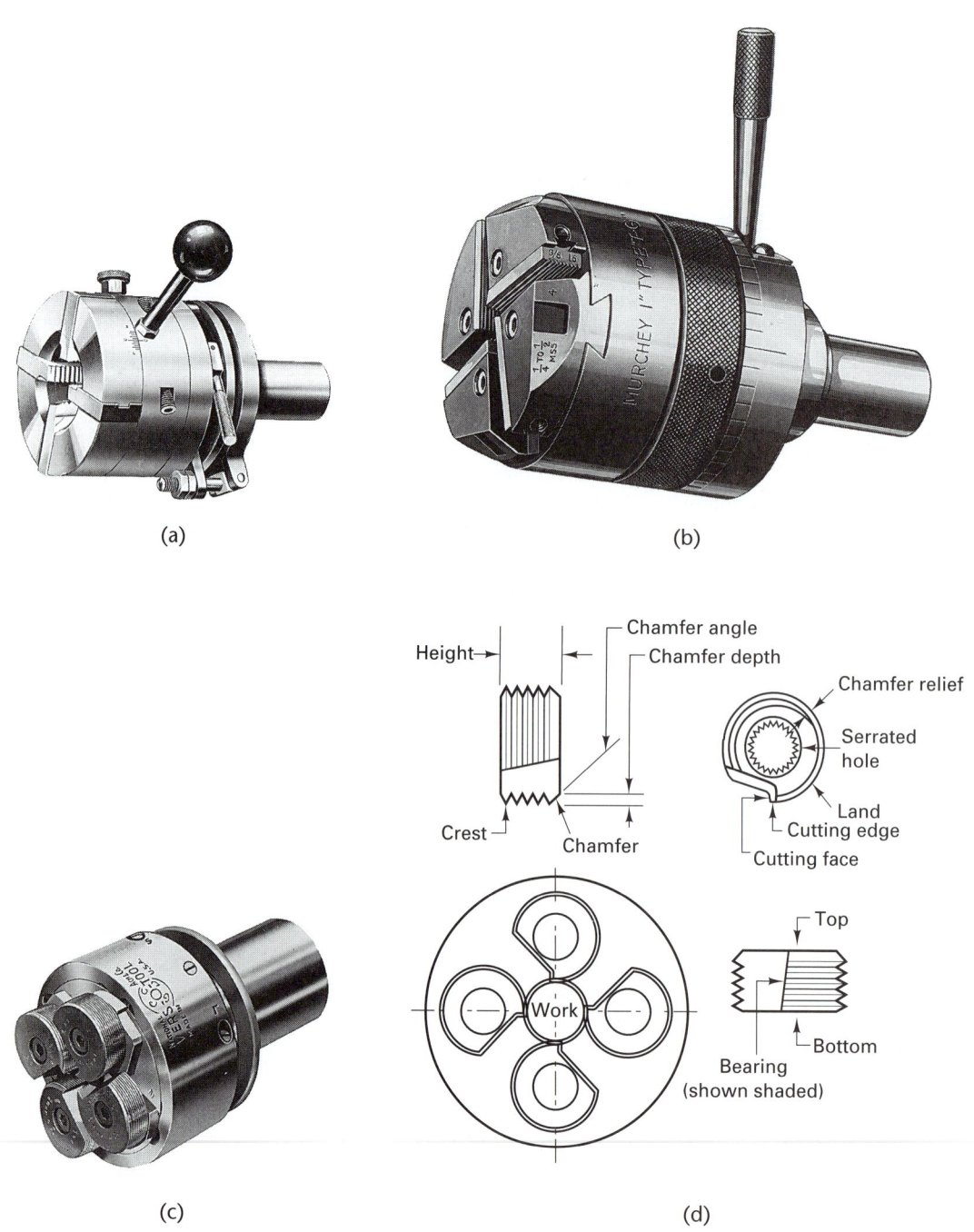

(a)

(b)

(c)

(d)

FIGURE 30-11 Self-opening die heads, with (a) radial cutter, (b) tangential cutters, and (c) circular cutters. (d) Terminology of circular chasers and their relation to the work. *(Courtesy of Geometric Tool Co., Warner & Swasey Co., National Acme Co., and TRW-Greenfield Tap & Die, respectively.)*

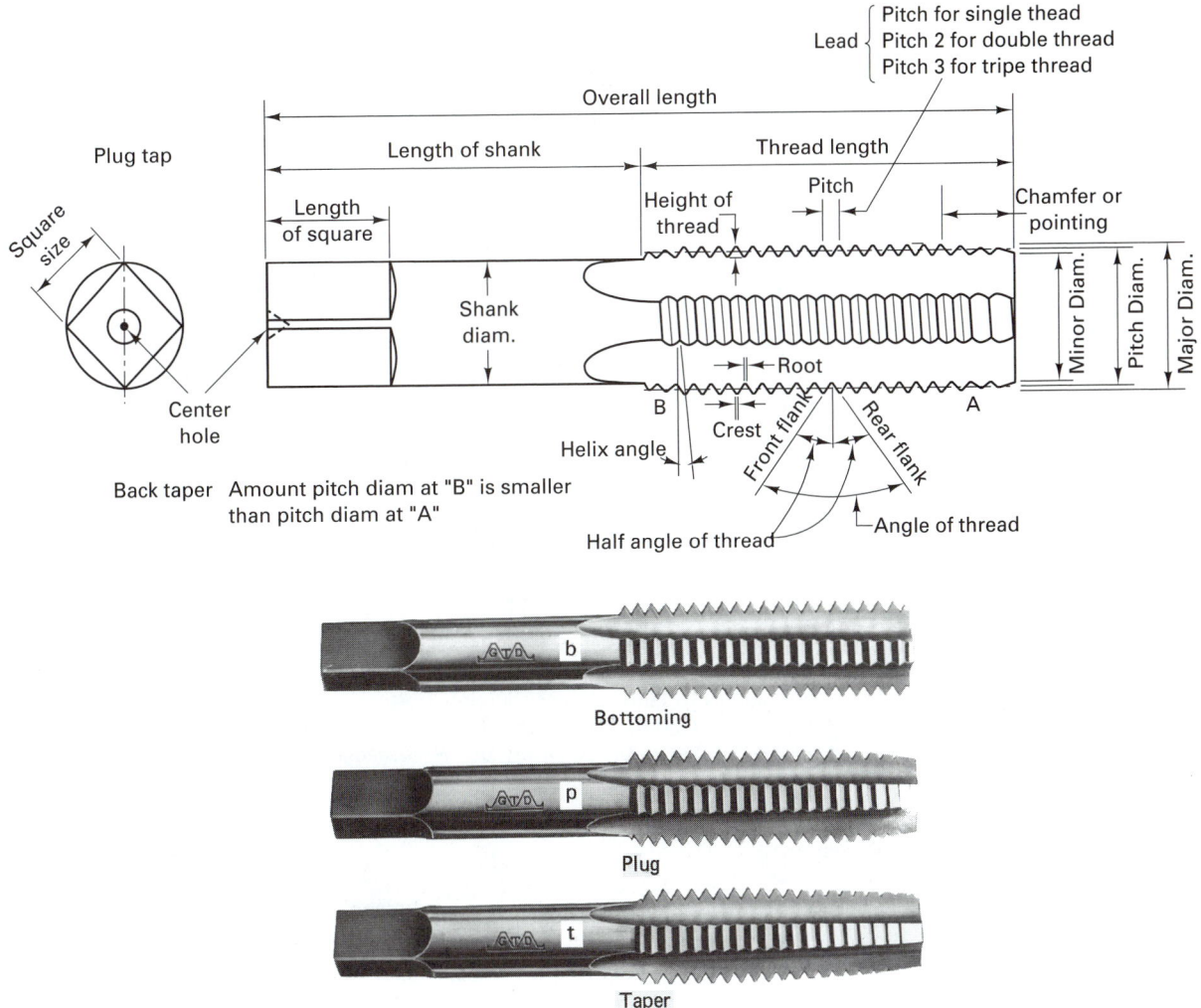

FIGURE 30-12   Terminology for a plug tap with photographs of taper (t), plug (p), and bottoming (b) taps which are used serially in threading holes. *(Courtesy of TRW-Greenfield Tap & Die.)*

taps (Figure 30-13) provide better removal of chips from a hole, particularly in tapping materials that produce long, curling chips. They also are helpful in tapping holes where the cutting action is interrupted by slots or keyways. The *spiral point* cuts the thread with a shearing action that pushes the chips ahead of the tap so that they do not interfere with the cutting action and the flow of cutting fluid into the hole.

Collapsing Taps

Collapsing taps are similar to self-opening die heads in that the cutting elements collapse inward automatically when the thread is completed. This permits withdrawing the tap from the workpiece without the necessity of unscrewing it from the thread. They can either be self-setting, for use on automatic machines, or require manual setting for each cycle. Figure 30-14 shows some of the types available.

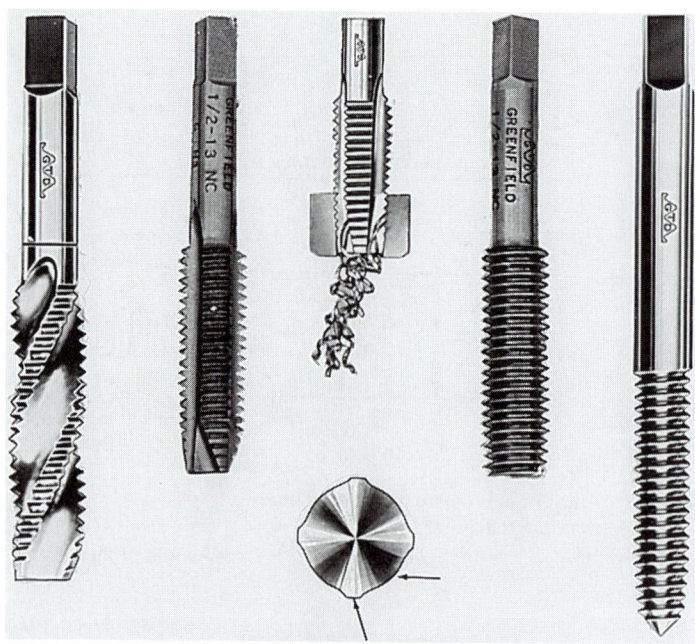

FIGURE 30-13 *Left to right:* Spiral fluted tap; spiral point tap; spiral point tap cutting chips; fluteless bottoming tap and fluteless plug tap for cold-forming internal threads. *Inset:* Cross section of fluteless forming tap. *(Courtesy of TRW-Greenfield Tap & Die.)*

Hole Preparation

Drilling is the most common method of preparing holes for tapping, and when close control over size is required, reaming may also be necessary. The drill size determines the final thread contour and the drilling torque. Unless otherwise specified, the tap drill size for most materials should produce approximately 75% thread, that is, 75% of full thread depth.

Machine Tapping

Solid taps also are used in tapping operations on machine tools, such as lathes, drill presses, and special tapping machines. In tapping on a drill press, a tapping attachment often is used. These devices rotate the tap slowly when the drill press spindle is fed downward against the work. When the tapping is completed and the spindle raised, the tap is automatically driven in the reverse direction at a higher speed to reduce the time required to back the tap out of the hole. Some modern machine tools provide for extremely fast spindle reversal for backing taps out of holes.

When solid taps are used on a screw machine or turret, the tap is prevented from turning while it is being fed into the work. As the tap reaches the end of the hole, the tap is free to rotate with the work. The work is then reversed and the tap, again prevented from rotating, is backed out of the hole.

The machine should have adequate power, rigidity, speed and feed ranges, cutting fluid supply, and positive drive action. Chucks, tap holders, and collets should be checked regularly for signs of wear or damage. Accurate alignment of the tap holder, machine spindle, and workpiece is vital to avoid broken taps or bell-mouthed, tapered, or oversized holes.

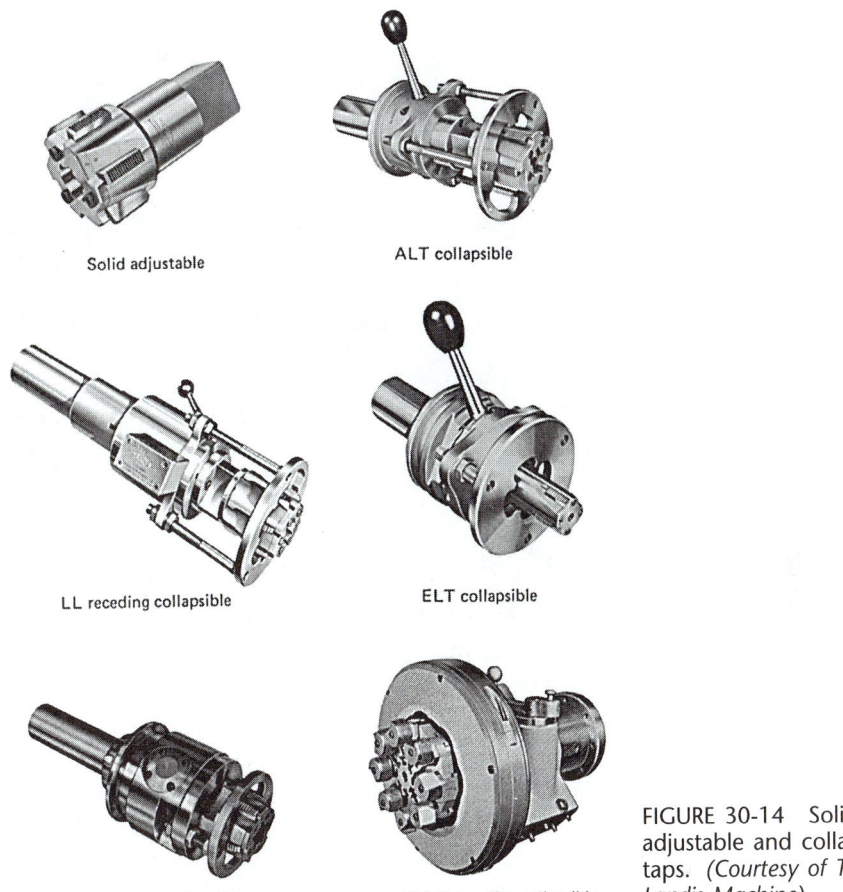

Solid adjustable

ALT collapsible

LL receding collapsible

ELT collapsible

2LLS receding collapsible

CBLM receding collapsible

FIGURE 30-14 Solid adjustable and collapsible taps. *(Courtesy of Teledyne Landis Machine).*

Tapping Cutting Time

The equation to calculate the cutting for tapping is

$$CT \cong \frac{\pi D L n}{8V} \tag{30-1}$$

where

CT = cutting time (min)

D = tap diameter (in.)

L = depth of tapped hole (in.)

n = number of threads per inch (tpi)

V = cutting speed (sfpm)

Special Threading and Tapping Machines

Special machines are available for production threading and tapping. Threading machines usually have one or more spindles on which a self-opening die head is mounted, with

FIGURE 30-15 Two-spindle automatic threading machine.
(Courtesy of Landis Machine Company.)

suitable means for clamping and feeding the workpiece. A typical machine of this type is shown in Figure 30-15. Some special tapping machines are similar in construction, with self-collapsing taps substituted for the threading dies. More commonly, tapping machines resemble drill presses, modified to provide spindle feeds both upward and downward, with the speed and feed more rapid on the upward motion.

Common Tapping Problems

Tap overloading is often caused by poor lubrication, lands that are too wide, chips packed in the flutes, or tap wear. Surface roughness in the threads has many causes. A negative grind on the heel may prevent the tap from tearing threads when backing out.

When a tap loses speed or needs more power, it generally indicates that the tap is dull (or improperly ground) or that the chips are packed in flutes (loaded). The flutes may be too shallow or the lands too deep. When tapping soft, ductile metals, loading can usually be overcome by polishing the tap before usage.

Improper hole size due to drill wear increases the percentage of threads being cut. Dull tools can also produce a rough finish or work-harden the hole surface and cause the tap to dull more quickly. Check to see that the axis of the hole and tap are aligned. If the tap cuts when backing out, check to see if the hole is oversize.

Tapping High-Strength Materials

High-strength, thermal-resistant materials, sometimes called *exotics*, cause special problems in tapping. A variety of materials are classified as exotics: stainless steel, precipitation-hardened stainless steel, high-alloy steels, iron-based superalloys, titanium, Inconel, Hastelloy, Monel, and Waspalloy. Their most important attribute is their high strength-to-weight ratio.

Each material presents different problems to efficient tapping, but they all share certain similarities. Toughness and general abrasiveness top the list. It is also difficult to impart a good surface finish to exotics, heat tends to localize in the shear zone, and exotics tend to workharden and grab the tool.

A tap's chamfer and first full thread do virtually all the cutting. The remaining ground threads serve merely as chasers. Because of this, taps to thread exotics are increasingly manufactured with short threads and reduced necks. Such taps diminish problems caused by material closure and provide more space for coolant and chip ejection.

When tapping exotics, the largest tap core diameter possible should be applied. Cutting 75 or 65% threads places less stress on the taps, lengthening tool life and reducing breakage. To cut threads in exotic alloys successfully, taps must combine geometries specifically tailored for those materials and be made of premium tool steels subjected to precisely controlled heat-treatment processes (see Table 30-3).

Stainless steel is known for its tendency to workharden and slow heat-dissipation characteristics and requires a tap geometry with a positive 6 to 9° rake, preventing workhardening and reducing torque. Grinding an appropriate eccentric thread and back-taper relief onto the tap will reduce friction. A surface treatment promotes lubricity. That is, the tool should be made from high-vanadium, high-cobalt tool steel and have a surface treatment so that coolant adheres to it. Stainless steel generates long chips and requires a tap with a 38° helix angle and adequate flute depth to promote chip evacuation. Proper hook and radial relief guarantee accurate thread-hole size and long tool life.

When tapping a titanium alloy, the material's tendency to concentrate heat in a small contact area must be considered. Concentrated heat often leads to excessive cutting-edge wear. Titanium generates average-to-short chips, is abrasive, and is prone to chip welding and high friction. These characteristics degrade tap performance and shorten tool life.

Taps for threading titanium are constructed of premium tool steel, nitrided for hardness. Titanium nitride (TiN) coatings cannot be used because they react chemically with the workpiece material, causing rapid tap failure. For tapping through-holes, a 3 to 5° rake and short thread design with high eccentric relief will promote efficient chip evacuation.

Nickel-based Inconel, Monel, Waspalloy, and Hastelloy present severe tapping problems. Among their machining characteristics are toughness, workhardening, heat retention, and built-up edge (BUE). Taps designed to thread these materials need tremendous stability and a strong cross-sectional construction. The most popular tap materials for nickel alloys are high-vanadium, high-cobalt tool steel or powdered metal (PM) tool steel. Tapping blind holes in these alloys requires a 3 to 5° rake to shear and deflect the cutting forces downward toward the tap's root, its strongest area. A 26° helix angle promotes chip evacuation, and a nitride or TiN coating reduces friction and tool wear.

Because of nickel alloys' toughness, taps to cut them should have the longest taper possible. This allows the cutting edges to gain thread height progressively before the first full thread begins its cut, distributing the load over a wider area. Spiral-pointed, straight-flute taps have a four- to five-thread taper. For tapping blind holes, the first two or three threads—more if possible—should be tapered.

Cutting Fluids for Tapping

Cutting fluids should be kept as clean as possible and should be supplied in copious quantities to reduce heat and friction and to aid in chip removal. Long tap life has been reported to result from routing high-pressure coolants through the tap to flush out the chips and cool the cutting edges. Recommended cutting fluids are listed in Table 30-4.

TABLE 30-3. High Performance Taps

COOLANTS AND LUBRICANTS
S = Sulphur-Chlorinated Oil
E = Water Soluble, Emulsion

TYPE OF LEAD
t = 4-5 tapered threads
p = 2-3 tapered threads

TOOL STEEL
F = High Vanadium
G = High-Vanadium/High-Cobalt
P = ASP-30* (PM tool steel)

TYPE OF HOLE
□ = Through Hole
● = Blind Hole
◙ = Through or Blind Hole

Main Material Group	Material Sub-Groups	Hardness (R°)	SFM	Chip Form
1. Steel	1.1 Magnetic soft steel	—	45-60	X Long
	1.2 Structural & case carbonizing steel	≤ 14	45-60	Mid/Long
	1.3 Plain Carbon steel	≤ 23	45-60	Long
	1.4 Alloy Steel	≤ 23	30-45	Long
	1.5 Alloy, hardened & tempered steel	> 24, ≤ 37	20-33	Long
	1.6 Alloy, hardened & tempered steel	> 38	5-15	Long
2. Stainless Steel	2.1 Free machining stainless steel	≤ 23	15-25	Middle
	2.2 Austenitic	≤ 23	12-15	Long
	2.3 Ferritic, Fer./Austenitic, Martensitic	≤ 30	5-15	Long
3. Titanium	3.1 Titanium, unalloyed	≤ 14	32-45	X Long
	3.2 Titanium, alloyed	≤ 26	25-40	Mid/Short
	3.3 Titanium, alloyed	> 26, ≤ 36	12-15	Mid/Short
4. Nickel	4.1 Nickel, unalloyed	—	25-40	X Long
	4.2 Nickel, alloyed	≤ 26	5-15	Long
	4.3 Nickel, alloyed	> 26, ≤ 36	5-15	Long

Tap specification columns (read across the top of the matrix):

Type of Lead	p	p	p	p	p	p	p	p	t	t	t	t
Coolant	S	S	S	S	S	S	E, S	E, S	S	S	E, S	E, S
Type of Hole	◙	□	□	◙	●	●	●	●	◙	◙	◙	◙
Thread Depth	1.5	1.5	1.5	2.5	1.5	1.5	2.5	2.5	1.5	1.5	1.5	1.5
Helix Angle (°)	—	30	—	15	26	15	38	45	15	—	—	—
Rake Angle (°)	6-8	6-8	3-5	6-8	14-17	11-13	7-9	3-5	6-8	3-5	6-8	6-8
Tool Steel	F	G	GP	GP	F	F	G	FG	GP	GP	G	F

*P - ASP stands for Powder metallurgy high speed steel made by the anti-segregation process (ASP).

TABLE 30-4. Cutting Fluids for Tapping (HSS Tools)	
Work Material	Cutting Fluid
Aluminum	Kerosene and lard oil, kerosene and light-base oil
Brass	Soluble oil or light-base oil
Naval brass	Mineral oil with lard or light-base oil
Manganese bronze	Mineral oil with lard or light-base oil
Phosphor bronze	Mineral oil with lard or light-base oil
Copper	Mineral oil with lard or light-base oil
Iron	
Cast	Dry or soluble oil
Malleable	Soluble oil or sulfur-based oil
Magnesium	Light-base oil diluted with kerosene
Monel metal	Sulfur-based oil
Steels	
Up to 0.25 carbon	Sulfur-based or soluble oil
Free machining	Sulfur-based or soluble oil
0.30–0.60 carbon, annealed	Sulfur-based oil
0.30–0.60 carbon, heat treated	Chlorinated sulfur-based oil
Tool, high-carbon, HSS	Chlorinated sulfur-based oil
Stainless	Chlorinated sulfur-based oil
Titanium	Chlorinated sulfur-based oil
Zinc die castings	Kerosene and lard oil

■ 30.4 THREAD MILLING

Highly accurate threads, particularly in larger sizes, are often form-milled. Either a single- or a multiple-form cutter may be used. A single-form cutter has a single annular row of teeth (Figure 30-16). The milling cutter is tilted at an angle equal to the helix angle of the thread and is fed inward radially to full depth while the work is stationary. The workpiece then is rotated slowly, and the cutter simultaneously is moved longitudinally, parallel with the axis of the work (or vice versa), by means of a lead screw, until the thread is completed. The thread can be completed in a single cut, or roughing and finish cuts can be used. This process is used primarily for large- or multiple-lead threads.

Some threads can be milled more quickly by using a multiple-form cutter having multiple rows of teeth set perpendicular to the cutter axis (the rows having no lead). The cutter must be slightly longer than the thread to be cut. It is set parallel with the axis of the workpiece and fed inward to full-thread depth while the work is stationary. The work then is rotated slowly for a little over one revolution, and the rotating cutter is simultaneously moved longitudinally with respect to the workpiece (or vice versa) according to the thread lead. When the work has revolved one revolution, the thread is complete. This process cannot be used on threads having a helix angle greater than about 3°, because clearance between the sides of the threads and the cutter depends on the cutter diameter's being substantially less than that of the workpiece. Thus, although the process is rapid, its use is restricted to threads of substantial diameter and not more than about 2 in. long.

■ 30.5 THREAD GRINDING

Grinding can produce very accurate threads, and it also permits threads to be produced in hardened materials. Three basic methods are used. *Center-type grinding with axial feed* is the most common method, being similar to cutting a thread on a lathe. A shaped grinding

FIGURE 30-16   Checking a large thread that was milled with a single-form cutter. The cutter can be seen behind the thread. *(Courtesy of Lees-Bradner Company.)*

wheel replaces the single-point tool. Usually, a single-ribbed grinding wheel is employed, but multiple-ribbed wheels are used occasionally. The grinding wheels are shaped by special diamond dressers or by crush dressing and must be inclined to the helix angle of the thread. Wheel speeds are in the high range. Several passes usually are required to complete the thread.

Center-type infeed thread grinding is similar to multiple-form milling in that a multiple-ribbed wheel, as wide as the length of the desired thread, is used. The wheel is fed inward radially to full thread depth, and the thread blank is then turned through about $1\frac{1}{2}$ turns as the grinding wheel is fed axially a little more than the width of one thread.

Centerless thread grinding (Figure 30-17) is used for making headless setscrews. The blanks are hopper fed to position *A*. The regulating wheel causes them to traverse the grinding wheel face, from which they emerge at position *B* in completed form. A production rate of 60 to 70 screws of $\frac{1}{2}$-in. length per minute is possible.

■ 30.6 THREAD ROLLING

Thread rolling is used to produce threads in substantial quantities. This is a cold-forming process operation in which the threads are formed by rolling a thread blank between hardened dies that cause the metal to flow radially into the desired shape. Because no metal is removed in the form of chips, less material is required, resulting in substantial savings. In addition, because of cold working, the threads have greater strength than cut threads, and a smoother, harder, and more wear-resistant surface is obtained. In addition, the process is fast, with production rates of one per second being common. The quality of cold-rolled

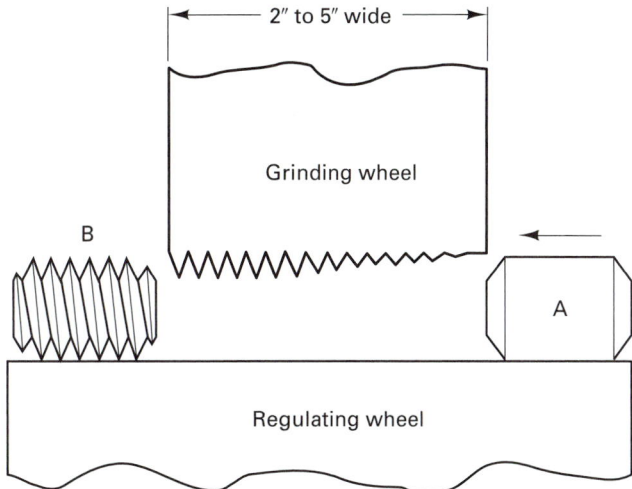

FIGURE 30-17 Principle of centerless thread grinding.

products is consistently good. Chipless operations are cleaner and there is a savings in material (15 to 20% savings in blank stock weight is typical).

Thread rolling is done by four basic methods. The simplest of these employs one fixed and one movable flat rolling die (Figure 30-18). After the blank is placed in position on the stationary die, movement of the moving die causes the blank to be rolled between the two dies and the metal in the blank is displaced to form the threads. As the blank rolls, it moves across the die parallel with its longitudinal axis. Prior to the end of the stroke of the moving die, the blank rolls off the end of the stationary die, its thread being completed.

One obvious characteristic of a rolled thread is that its major diameter always is greater than the diameter of the blank. When an accurate class of fit is desired, the diameter of the blank is made about 0.002 in. larger than the thread-pitch diameter. If it is desired to have the body of a bolt larger than the outside diameter of the rolled thread, the blank for the thread is made smaller than the body.

Thread rolling can be done with cylindrical dies. Figure 30-19 illustrates the three-roll method commonly employed on turret lathes and screw machines. Two variations are used. In one, the rolls are retracted while the blank is placed in position. The rolls then move inward radially, while rotating, to form the thread. More commonly the three rolls are contained in a self-opening die head similar to the conventional type used for cutting external threads. The die head is fed onto the blank longitudinally and forms the thread progressively as the blank rotates. With this procedure, as in the case of cut threads, the innermost $1\frac{1}{2}$ to 2 threads are not formed to full depth because of the progressive action of the rollers.

The two-roll method is commonly employed for automatically producing large quantities of externally threaded parts up to 6 in. in diameter and 20 in. in length. A planetary machine is for mass production of rolled threads on diameters up to 1 in.

Not only is thread rolling very economical but the threads are excellent as to form and strength. The cold working contributes to increased strength, particularly at the critical root areas. There is less likelihood of surface defects (produced by machining), which can act as stress raisers.

Large numbers of threads are rolled on thin, tubular products. In this case external and internal rolls are used. The threads on electric lamp bases and sockets are examples of this type of thread.

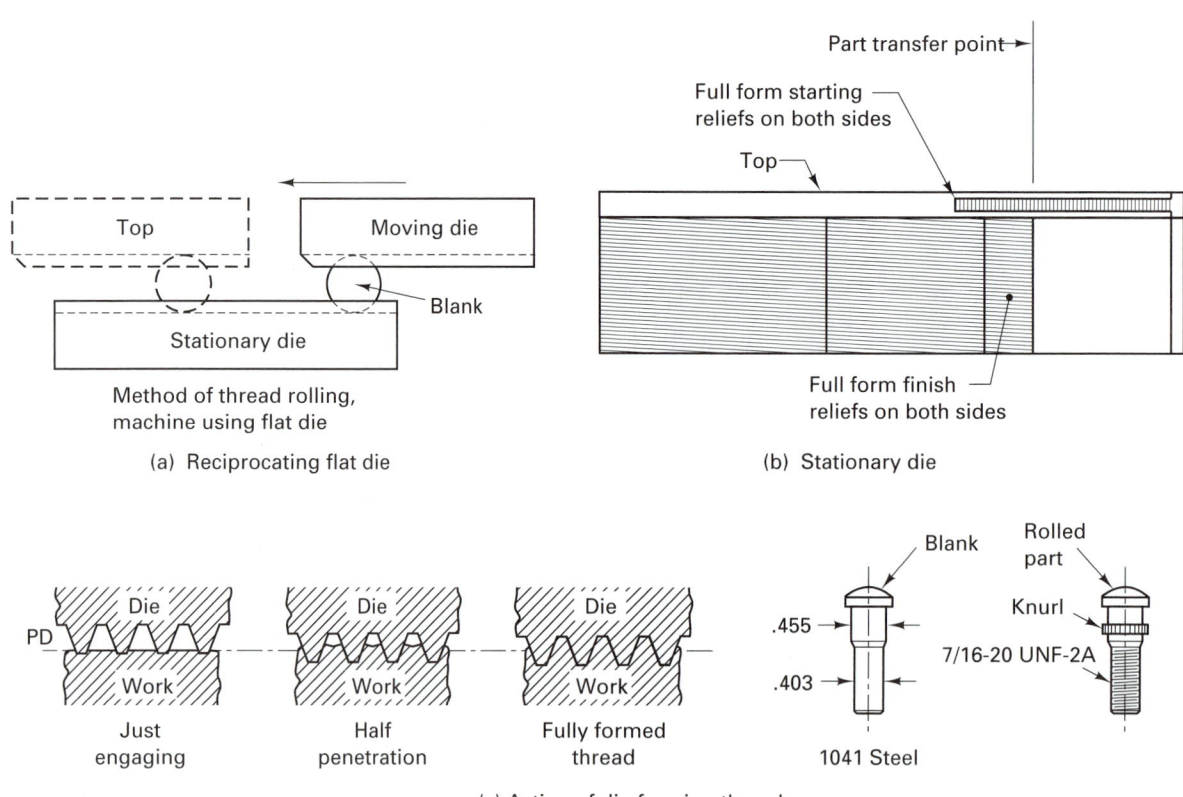

FIGURE 30-18 Combination thread rolling and knurling of wheel bolt at 70 per minute by flat die roll threading.

Chipless Tapping

Unfortunately, most internal threads cannot be made by rolling; there is insufficient space within the hole to permit the required rolls to be arranged and supported, and the required forces are too high. However, many internal threads, up to about $\frac{1}{2}$ in. in diameter, are cold formed in holes in ductile metals by means of *fluteless taps*. Such a tap and its special cross section are shown in Figure 30-13. The forming action is essentially the same as in rolling external threads. Because of the forming involved and the high friction, the torque required is about double that for cutting taps. Also, the hole diameter must be controlled carefully to obtain full thread depth without excessive torque. However, fluteless taps produce somewhat better accuracy than cutting taps and tap life is often greater than that of HSS machine taps. A lubricating fluid should be used, water-soluble oils being quite effective. Fluteless taps are especially suitable for forming threads in dead-end holes because no chips are produced. They come in both plug and bottoming types.

Machining versus Rolling Threads

Threads are machined or cut when full thread depth is needed (more than one pass necessary), for short production runs; when the blanks are not very accurate; when proximity to the shoulder in end threading is needed; for tapered threads; or when the workpiece material is not adaptable for rolling.

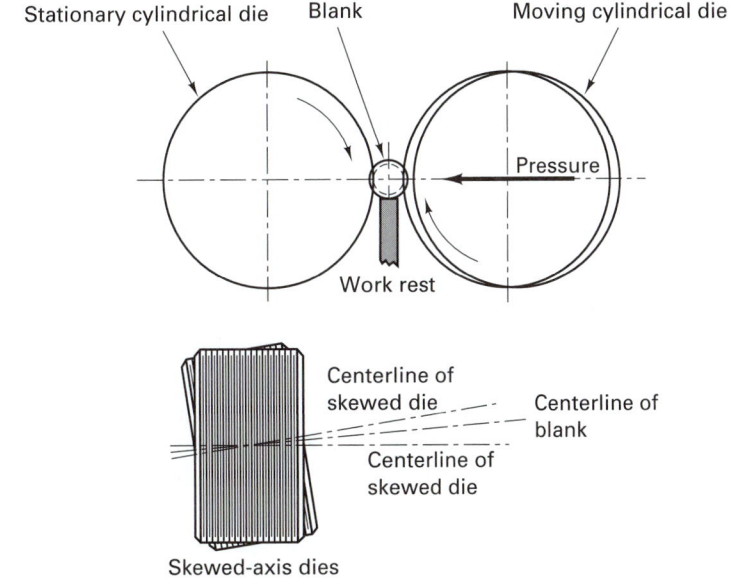

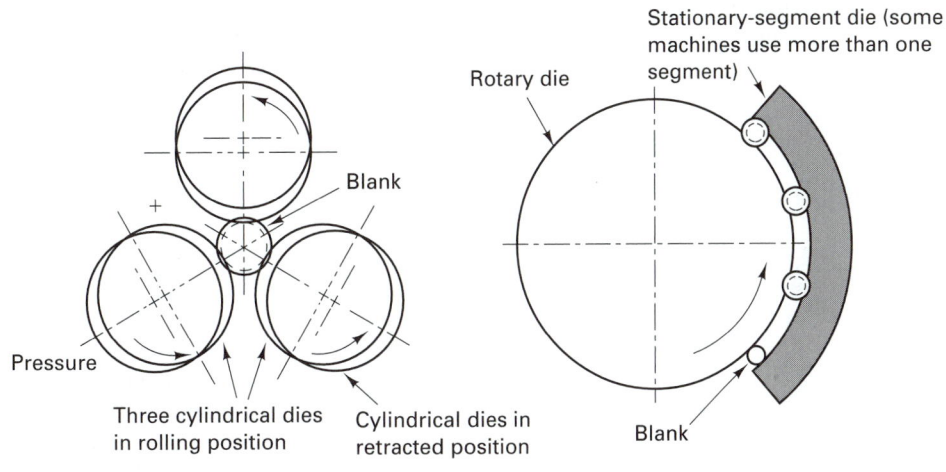

FIGURE 30-19   Methods for roll forming threads using cylindrical dies.

■ KEY WORDS

bottoming tap	plug tap	thread rolling
collapsible tap	self-opening die head	thread tapping
fluteless tap	taper tap	threading
pitch	thread cutting	

■ REVIEW QUESTIONS

1. How does the pitch diameter differ from the major diameter for a standard screw thread?
2. For what types of threads are the pitch and the lead the same?
3. What is the helix angle of a screw thread?
4. Why are pipe threads tapered?
5. By what three basic methods can external threads be produced?
6. Explain the meaning of $\frac{1}{4}$"-20 UNC-3A.
7. What is meant by the designation M20 × 2.5-6g6g? (What does the "×" mean?)
8. What are two reasons that fine-series threads are being used less now than in former years?
9. In cutting a thread on a lathe, how is the pitch controlled?
10. What is the function of a threading dial on a lathe?
11. In Chapter 23, what figure(s) showed a threading dial?
12. What controls the lead of a thread when it is cut by a threading die?
13. What is the basic purpose of a self-opening die head?
14. What is the reason for using a taper tap before a plug tap in tapping a hole?
15. What difficulties are encountered if full threads are specified to the bottom of the dead-end hole?
16. Can a fluteless tap be used for threading a hole in gray cast iron? Why or why not?
17. What provisions should a designer make so that a dead-end hole can be threaded?
18. What is the major advantage of a spiral-point tap?
19. Can a fluteless tap be used for threading to the bottom of a dead-end hole? Why or why not?
20. Is it desirable for a tapping fluid to have lubricating qualities? Why?
21. How does thread milling differ from thread turning?
22. What are the advantages of making threads by grinding?
23. Why has thread rolling become the most commonly used method for making threads?
24. How may you determine whether a thread has been produced by rolling rather than by cutting?
25. What is a fluteless tap?

■ PROBLEMS

1. Calculate the cutting needed to tap a $\frac{3}{4}$ × 2 in. hole using a cutting speed of 30 sfpm. The tap has 10 threads/in. (tpi).
2. The new manufacturing manager has recommended to you that chipless tapping be adopted for tapping holes in the deep, dead-end holes on the two-cylinder engine blocks that the company makes. Chipless tapping can run at twice the speed of the conventional tapping process, works well on deep holes, and provides better quality and finish with longer tool life. The tapping process is the bottleneck process in machining all the cast iron engine blocks; also, about 10% of the blocks have to be scrapped due to broken taps. What do you recommend?

Chapter 30 CASE STUDY

vented cap screws

T he machine shop at the E. D. Fly Space Laboratories received an order from their engineering department for 100 of the vented cap screws shown in Figure CS-30. The machine shop supervisor returned the order, stating that there was no practical way to make these other than by hand. The design engineer insisted that the vent slot, as designed, was essential to assure no pressure buildup around the threads of the screw body in the intended application, and that he knew they could be made because he had seen such cap screws.

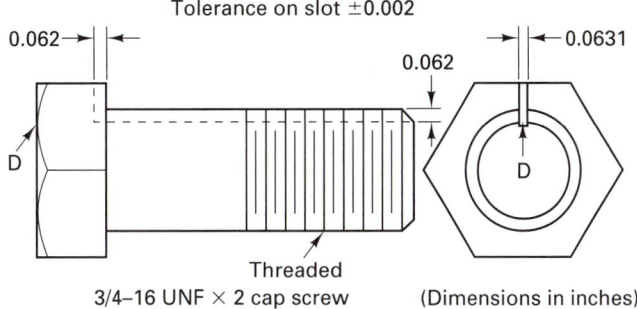

Tolerance on slot ±0.002
0.062
0.062
0.0631
D
D
Threaded
3/4–16 UNF × 2 cap screw
(Dimensions in inches)

FIGURE CS-30

To accommodate the shop, however, the design engineer redesigned the part, now specifying that a vent hole be drilled parallel to the axis of the cap at location *D*. The small hole, with cross-sectional area equal to the previous 0.062 in. × 0.031 in. slot, would be drilled $\frac{1}{2}$ in. deep through the cap to connect to the longitudinal slot. This would eliminate the right-angle slot geometry.

1. What do you think of this solution? What problems will be incurred in the manufacture of the vented cap screw?
2. As the manufacturing engineer, working with the design engineer, can you suggest an alternative redesign that would make a slot practical to manufacture? Show a sketch of your redesigned cap screw.

GEAR MANUFACTURING

31.1 INTRODUCTION
Gear Theory and
Terminology
Physical Requirements
of Gears
31.2 GEAR TYPES
31.3 GEAR MANUFACTURING
31.4 MACHINING OF GEARS
Form Milling
Broaching

Shaping
Hobbing
Bevel Gear Generation
Cold-Roll Forming of Gears
Other Gear-Making
Processes
31.5 GEAR FINISHING
31.6 GEAR INSPECTION
Case Study: SIX-SECOND PINION
GEAR

■ 31.1 INTRODUCTION

Gears transmit power or motion mechanically between parallel, intersecting, or nonintersecting shafts. Although usually hidden from sight, gears are one of the most important mechanical elements in our civilization, possibly even surpassing the wheel, since most wheels would not be turning were power not being applied to them through gears. They operate at almost unlimited speeds under a wide variety of conditions. Millions are produced each year in sizes from a few millimeters up to more than 30 ft in diameter. Often the requirements that must be, and are routinely, met in their manufacture are amazingly precise. Consequently, the machines and processes that have been developed for producing gears are among the most ingenious we have. To understand the functional requirements of these machines and processes, it is helpful briefly to consider the basic theory of gears and their operation.

Gear Theory and Terminology

Basically, gears are modifications of wheels, with *gear teeth* added to prevent slipping and to assure that their relative motions are constant. However, it should be noted that the relative surface velocities of the wheels (and shafts) are determined by the diameters of the wheels.

Although wooden teeth or pegs were attached to disks to make gears in ancient times, the teeth of modern gears are produced by machining or forming teeth on the outer portion of the wheel. The *pitch circle* (Figures 31-1 and Figure 31-2) corresponds to the diameter of the wheel. Thus the angular velocity of a gear is determined by the diameter of this imaginary pitch circle. All design calculations relating to gear performance are based on the pitch-circle diameter or, more simply, the *pitch diameter* (PD).

For two gears to operate properly, their pitch circles must be tangential to each other. The point at which the two pitch circles are tangent, at which they intersect the centerline connecting their centers of rotation, is called the *pitch point*. The common normal at the point of contact of mating teeth must pass through the pitch point. This condition is illustrated in Figure 31-2.

To provide uniform pressure and motion and to minimize friction and wear, gears are designed to have rolling motion between mating teeth rather than sliding motion. To

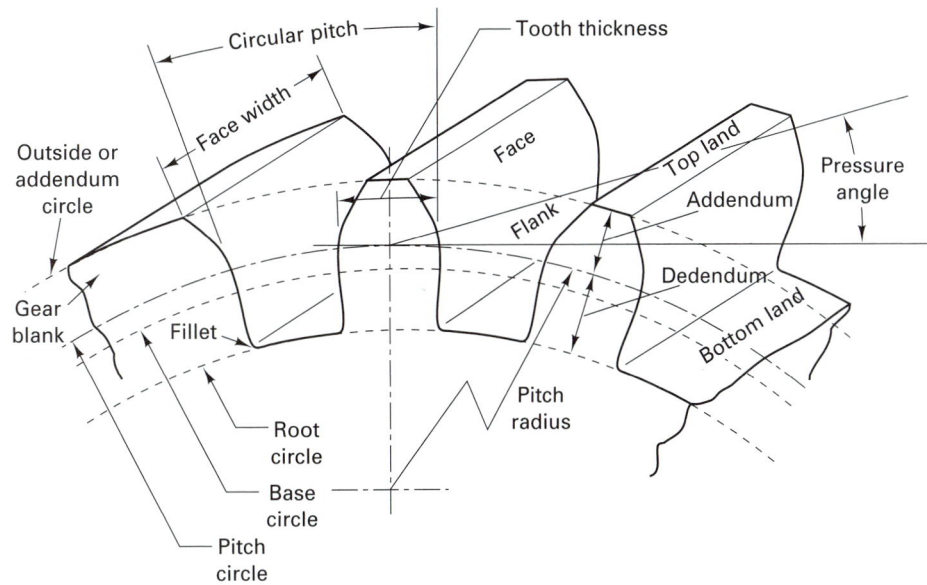

FIGURE 31-1   Gear-tooth nomenclature.

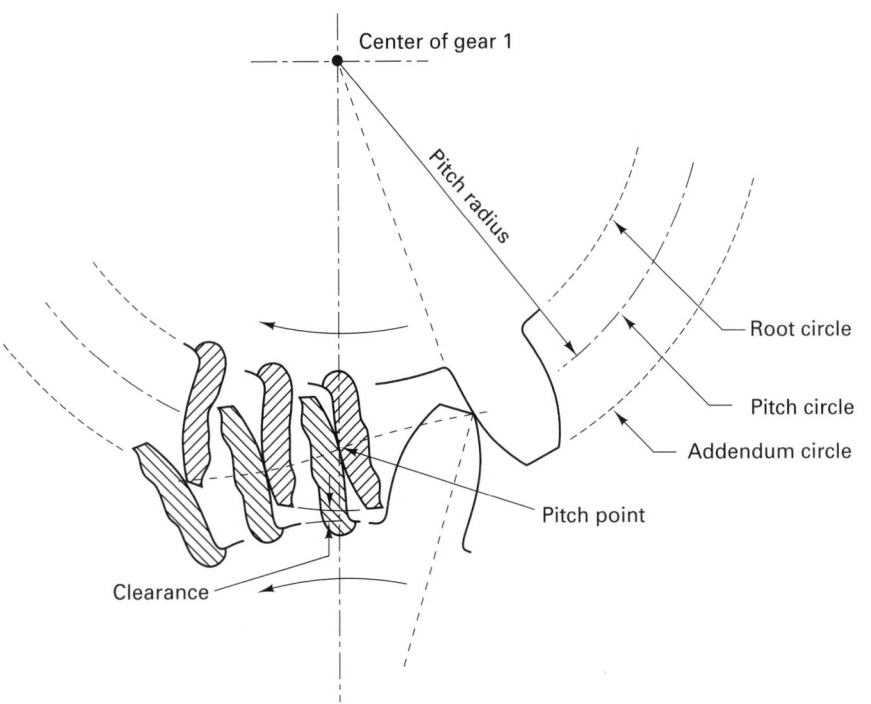

FIGURE 31-2   Tangent pitch circles between two gears produce a pitch point.

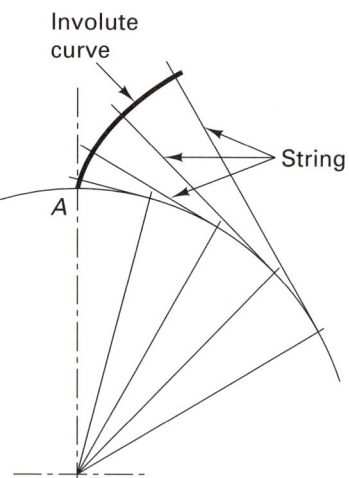

FIGURE 31-3 Method of generating an involute curve by unwinding a string from a cylinder.

achieve this condition, most gears utilize a tooth form that is based on an *involute curve*. This is the curve that is generated by a *point* on a straight line when the line rolls around a *base circle*. A somewhat simpler method of developing an involute curve is that shown in Figure 31-3, by unwinding a tautly held string from a base circle; point A generates an involute curve.

There are three other reasons for using the involute form for gear teeth. First, such a tooth form provides the desired pure rolling action. Second, even if a pair of involute gears is operated with the distance between the centers slightly too large or too small, the common normal at the point of contact between mating teeth will always pass through the pitch point. Obviously, the theoretical pitch circles in such cases will be increased or decreased slightly. Third, the *line of action*, or *path of contact*, that is, the locus of the points of contact of mating teeth, is a straight line that passes through the pitch point and is tangent to the base circles of the two gears.

Cutting an involute shape in gear blanks can be done by simple form cutting (i.e., milling the shape into the workpiece) or by generating. Generating involves relative motion between the workpiece and the cutting tool. True involute tooth form can be produced by a cutting tool that has straight-sided teeth. This permits a very accurate involute tooth profile to be obtained through the use of a simple and easily made cutting tool. The straight-sided teeth are given a rolling motion relative to the workpiece to create the curved gear-tooth face.

The basic size of gear teeth may be expressed in two ways. The common practice, especially in the United States and England, is to express the dimensions as a function of the *diametral pitch* (DP). DP *is the number of teeth (N) per unit of pitch diameter* (PD); thus DP = N/PD. Dimensionally, DP involves inches in the English system and millimeters in the SI system, and it is a measure of tooth size. Metric gears use the module system *(M)*, defined as *the pitch diameter divided by the number of teeth*, or M = PD/N. It thus is the reciprocal of diametrical pitch and is expressed in millimeters. Any two gears having the same diametral pitch or module will mesh properly if they are mounted so as to have the correct distances and relationship.

The important tooth elements can be specified in terms of the diametral pitch or the module and are as follows:

1. *Addendum:* the radial distance from the pitch circle to the outside diameter.

$$\text{Addendum} = \frac{1}{\text{DP}} \quad \text{inches}$$

2. *Dedendum:* the radial distance from the pitch circle to the root circle. It is equal to the addendum plus the *clearance*, which is provided to prevent the outer corner of a tooth from touching against the bottom of the tooth space.
3. *Circular pitch:* the distance between corresponding points of adjacent teeth, measured along the pitch circle. It is numerically equal to

$$\pi/\text{diametrical pitch}$$

4. *Tooth thickness:* the thickness of a tooth, measured along the pitch circle. When tooth thickness and the corresponding *tooth space* are equal, no *backlash* exists in a pair of mating gears.
5. *Face width:* the length of the gear teeth in an axial plane.
6. *Tooth face:* the mating surface between the pitch circle and the addendum circle.
7. *Tooth flank:* the mating surface between the pitch circle and the root circle.
8. *Pressure angle:* the angle between a tangent to the tooth profile and a line perpendicular to the pitch surface.

Four shapes of involute gear teeth are used in the United States:

1. $14\frac{1}{2}°$ pressure angle, full depth (used most frequently)
2. $14\frac{1}{2}°$ pressure angle, composite (seldom used)
3. $20°$ pressure angle, full depth (seldom used)
4. $20°$ pressure angle, stub tooth (second most common)

In the $14\frac{1}{2}°$ full-depth system, the tooth profile outside the base circle is an involute curve. Inward from the base circle the profile is a straight radial line that is joined with the bottom land by a small fillet. With this system, the teeth of the basic rack have straight sides.

The $14\frac{1}{2}°$ composite system and the $20°$ full-depth system provide somewhat stronger teeth. However, with the $20°$ full-depth system, considerable undercutting occurs in the dedendum area; therefore, stub teeth often are used. The addendum is shortened by 20%, thus permitting the dedendum to be shortened a similar amount. This results in very strong teeth without undercutting. Table 31-1 gives the formulas for computing the dimensions of gear teeth in the $14\frac{1}{2}°$ full-depth and $20°$ stub–tooth systems.

Physical Requirements of Gears

A consideration of gear theory leads to five requirements that must be met in order for gears to operate satisfactorily:

1. The actual tooth profile must be the same as the theoretical profile.
2. Tooth spacing must be uniform and correct.
3. The *actual* and theoretical pitch circles must be coincident and be concentric with the axis of rotation of the gear.

TABLE 31-1.   Standard Dimensions for Involute Gear Teeth

	$14\frac{1}{2}°$, Full Depth	$20°$, Stub Tooth
Pitch diameter	$\dfrac{N}{DP}$	$\dfrac{N}{DP}$
Addendum	$\dfrac{1}{DP}$	$\dfrac{0.8}{DP}$
Dedendum	$\dfrac{1.157}{DP}$	$\dfrac{1}{DP}$
Outside diameter	$\dfrac{N+2}{DP}$	$\dfrac{N+1.6}{DP}$
Clearance	$\dfrac{0.157}{DP}$	$\dfrac{0.2}{DP}$
Tooth thickness	$\dfrac{1.5708}{DP}$	$\dfrac{1.5708}{DP}$

4. The face and flank surfaces must be smooth and sufficiently hard to resist wear and prevent noisy operation.
5. Adequate shafts and bearings must be provided so that desired center-to-center distances are retained under operational loads.

The first four of these requirements are determined by the material selection and manufacturing process. The various methods of manufacture that are used represent attempts to meet these requirements to varying degrees with minimum cost, and their effectiveness must be measured in terms of the extent to which the resulting gears embody these requirements. Before looking at the ways to manufacture gears, let's look at some examples of gears.

■ 31.2 GEAR TYPES

The more common types of gears are shown in Figure 31-4. *Spur gears* have straight teeth and are used to connect parallel shafts. They are the most easily made and the cheapest of all types.

The teeth on *helical gears* lie along a helix, the angle of the helix being the angle between the helix and a pitch cylinder element parallel with the gear shaft. Helical gears can connect either parallel or nonparallel nonintersecting shafts. Such gears are stronger and quieter than spur gears because the contact between mating teeth increases more gradually and more teeth are in contact at a given time. Although they usually are slightly more expensive to make than spur gears, they can be manufactured in several ways and are produced in large numbers.

Helical gears have one disadvantage. When they are in use, a side thrust is created that must be absorbed in the bearings. *Herringbone gears* neutralize this side thrust by having, in effect, two helical-gear halves, one having a right-hand and the other a left-hand helix. The *continuous* herring-bone type is rather difficult to machine but is very strong. A modified herringbone type is made by machining a groove, or gap, around the gear blank where the two sets of teeth would come together. This provides a runout space for the cutting tool in making each set of teeth.

A *rack* is a gear with infinite radius, having teeth that lie on a straight line on a plane. The teeth may be normal to the axis of the rack or helical so as to mate with spur or helical gears, respectively.

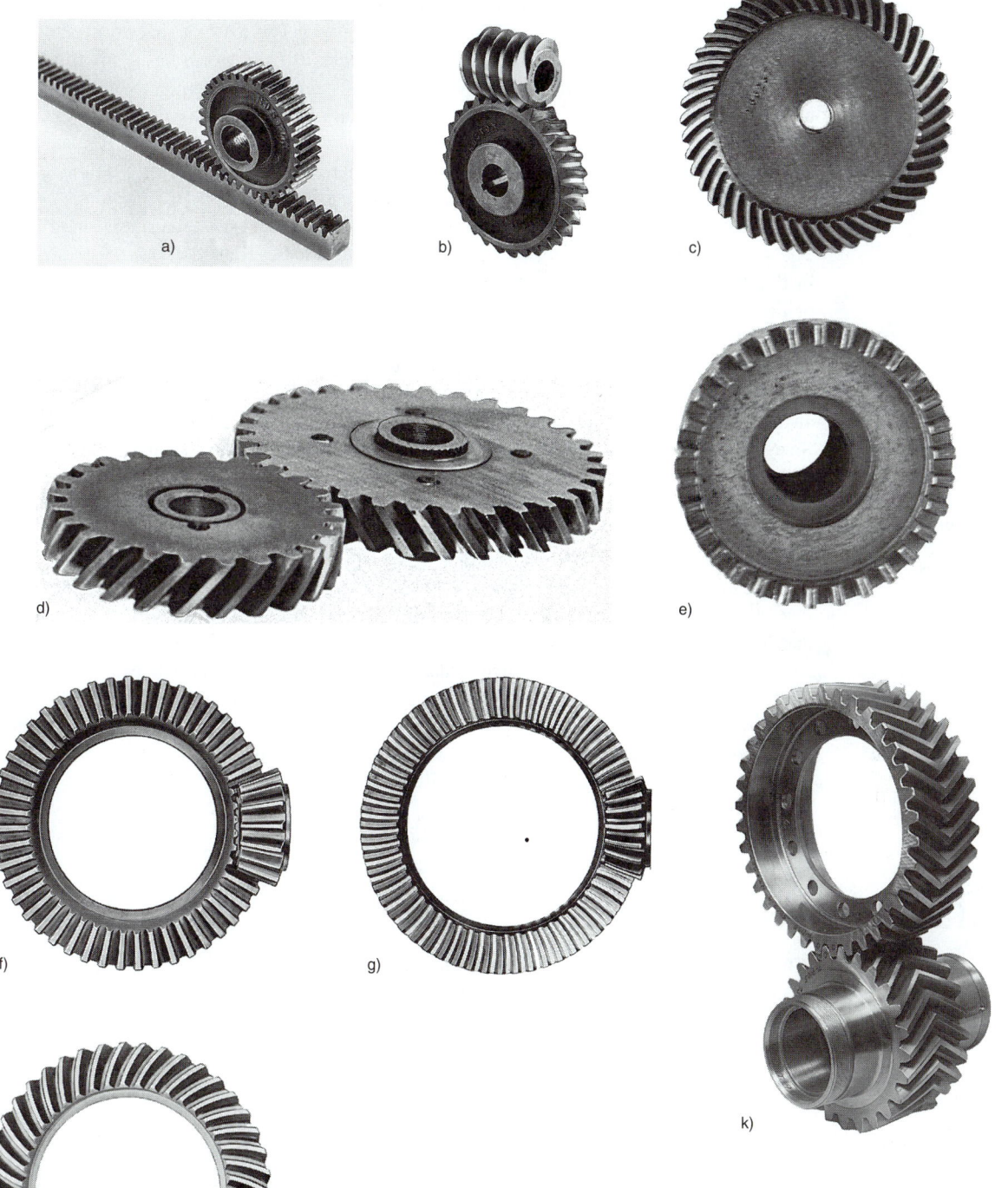

FIGURE 31-4   Types of gears: a) Spur gear and rack; b) worm and worm gear; c) spiral bevel gear; d) helical gears; crown gear; f) straight bevel gears; g) Zerol bevel gears; h) hypoid bevel gears; k) continuous herringbone gears. *(Courtesy of Gleason Works).*

A *worm* is similar to a screw. It may have one or more threads, the multiple-thread type being very common. Worms usually are used in conjunction with a *worm gear*. High gear ratios are easily obtainable with this combination. The axes of the worm and worm gear are nonintersecting and usually are at right angles. If the worm has a small helix angle, it cannot be driven by the mating worm gear. This principle frequently is employed to obtain nonreversible drives. Worm gears usually are made with the top land concave to permit greater area of contact between the worm and the gear. A similar effect can be achieved by using a *conical worm*, in which the helical teeth are cut on a double-conical blank, thus producing a worm that has an hourglass shape.

Bevel gears, teeth on a cone, are used to transmit motion between intersecting shafts. The teeth are cut on the surface of a truncated cone. Several types of bevel gears are made, the types varying as to whether the teeth are straight or curved and whether the axes of the mating gears intersect. On *straight-tooth* bevel gears the teeth are straight, and if extended all would pass through a common apex. *Spiral-tooth* bevel gears have teeth that are segments of spirals. Like helical gears, this design provides tooth overlap so that more teeth are engaged at a given time and the engagement is progressive. *Hypoid* bevel gears also have a curved-tooth shape but are designed to operate with nonintersecting axes. Rear-drive automobiles used hypoid gears in the rear axle so that the drive shaft axis can be below the axis of the axle and thus permit a lower floor height. *Zerol* bevel gears have teeth that are circular arcs, providing somewhat stronger teeth than can be obtained in a comparable straight-tooth gear. They are not used extensively. When a pair of bevel gears are the same size and have their shafts at right angles, they are termed *miter gears*.

A *crown gear* is a special form of bevel gear having a 180° cone apex angle. In effect, it is a disk with the teeth on the side of the disk. It also may be thought of as a rack that has been bent into a circle so that its teeth lie in a plane. The teeth may be straight or curved. On straight-tooth crown gears the teeth are radial. Crown gears seldom are used, but they have the important quality that they will mesh properly with a bevel gear of any cone angle, provided that the bevel gear has the same tooth form and diametral pitch. This important principle is incorporated in the design and operation of two very important types of gear-generating machines that will be discussed later.

Most gears are of the external type, the teeth forming the outer periphery of the gear. Internal gears have the teeth on the inside of a solid ring, pointing toward the center of the gear.

■ 31.3 GEAR MANUFACTURING

Whether produced in large or small quantities, in cells, or job shop batches, the sequence of processes for gear manufacturing requires four sets of operations,

1. Blanking
2. Gear cutting
3. Heat treatment
4. Grinding

Blanking refers to the initial forming or machining operations that produce a semifinished part ready for gear cutting, starting from a piece of raw material. Turning on chuckers or lathes, face and centering of shafts, milling, and sometimes grinding, fall into this category of operations. Good-quality blanks are essential in precision gear manufacturing.

Hobbing, shaping, and shaving machines are the most frequently used machines for gear cutting, producing gears for automotive, truck, agricultural, and construction equipment. Other processes used in industrial gear production include broaching, rolling, grinding, milling, and shiving. The process selected depends on finding a cost-effective application based on quality specification, production volumes, and economic conditions.

The gear cutting or machining operations can be divided into operations executed prior to heat treatment, when the material is still soft and easily machinable and after heat treatment, performed on parts that have acquired high hardness and strength.

Heat treatment gives the material the strength and durability to withstand high loads and wear but results in a reduction in dimensional and geometrical accuracy. The metallurgical transformations that occur during hardening, quenching, and tempering cause a general quality deterioration in the gears. Therefore, precision grinding operations are used on external and internal bearing diameters, critical length dimensions, and fine surface finishes after heat treatment. Cylindrical grinders, angle-head grinders, internal grinders, and surface grinders are commonly used.

Gears are made in very large numbers by cold-roll forming, and in addition, significant quantities are made by extrusion, by blanking, by casting, and some by powder metallurgy and by a forging process. However, it is only by machining that all types of gears can be made in all sizes, and although roll-formed gears can be made with accuracy sufficient for most applications, even for automobile transmissions, machining still is unsurpassed for gears that must have very high accuracy. Also, roll forming can be used only on ductile metals.

■ 31.4 MACHINING OF GEARS

Form Milling

Form cutting or *form milling* is illustrated in Figure 31-5. The cutter has the same form as the *space* between adjacent teeth. Usually, a multiple-tooth form cutter (Figure 31-6) is used, as shown in Figure 31-5. The tool is fed radially toward the center of the gear blank to the desired tooth depth, then across the tooth face to obtain the required tooth width. When one tooth space has been completed, the tool is withdrawn, the gear blank is indexed using a dividing head, and the next tooth space is cut. Basically, form cutting is a simple and flexible method of machining gears. The equipment and cutters required are relatively simple, and standard machine tools (milling machines) often are used. However, in most cases the procedure is quite slow, and considerable care is required on the part of the operator; therefore, it usually is employed where only one or a few gears are to be made. In machining gears by the form-cutting process, the form cutter is mounted on the machine spindle, and the gear blank is mounted on a mandrel held between the centers of some type of indexing device. Figure 31-7 shows the arrangement that is employed when, as is often the case, the work is done on a universal milling machine; the cutter is mounted on an arbor, and a dividing head is used to index the gear blank. When a helical gear is to be cut, as in the case shown, the table must be set at an angle equal to the helix angle, and the dividing head is geared to the longitudinal feed screw of the table so that the gear blank will rotate as it moves longitudinally.

Standard cutters usually are employed in form-cutting gears. In the United States, these come in eight sizes for each diametral pitch and will cut gears having the number of teeth indicated in Table 31-2. A single cutter will not produce a theoretically perfect tooth profile for all sizes of gears in the range for which it is intended. However, the change in tooth profile over the range covered by each cutter is very slight, and most of the time, satisfactory results can be achieved. When greater accuracy is required, half-number cutters (such as $3\frac{1}{2}$) can be obtained. Typical cutters are shown in Figures 31-6, 31-7 and 31-8. Cutters are available for all common diametral pitches and $14\frac{1}{2}$ and $20°$ pressure angles. If the amount of

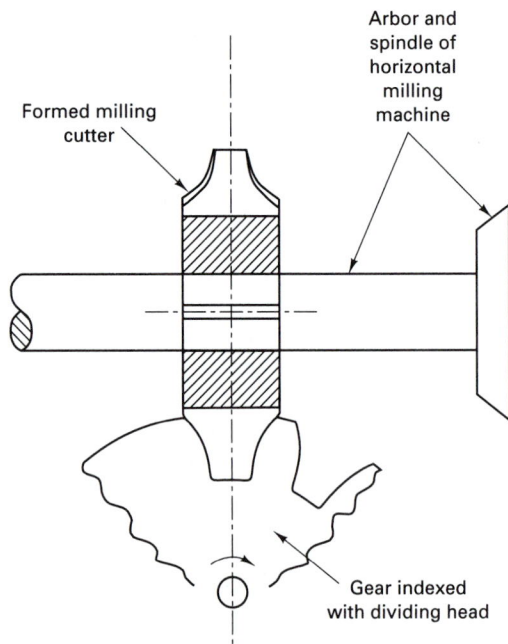

FIGURE 31-5 Basic method of machining a gear by form milling or form cutting.

FIGURE 31-6 Form cutter (*left*) and stocking cutter for machining gear teeth. See Figures 31-7 and 31-8 for setups. *(Courtesy of Brown and Sharpe Manufacturing Company.)*

metal that must be removed to form a tooth space is large, roughing cuts may be taken with a *stocking cutter.* The stepped sides of the stocking cutter remove most of the metal and leave only a small amount to be removed subsequently by the regular form cutter in a finish cut.

Straight-tooth bevel gears can be form cut on a milling machine, but this is seldom done. Because the tooth profile in bevel gears varies from one end of the tooth to the other, after one cut is taken to form the correct tooth profile at the smaller end, the relationship between the cutter and the blank must be altered. Shaving cuts then are taken on the side of each tooth to form the correct profile throughout the entire tooth length.

FIGURE 31-7 Form cutting a helical gear on a universal milling machine, using an indexing head. Inset shows closeup of gear and cutter.

TABLE 31-2.	Standard Cutter Sizes
Cutter Number	Gear Tooth Range (teeth to rack)
1	135
2	55–134
3	35–54
4	26–34
5	21–25
6	17–20
7	14–16
8	12–13

Although the form cutting of gears on a milling machine is a flexible process and is suitable for gears that are not to be operated at high speeds or that need not operate with extreme quietness, the process is slow and requires skilled labor. Semiautomatic machines are available for making gears by the form-cutting process. Such a machine is shown in Figure 31-8. The procedure utilized is essentially the same as on a milling machine, except that, after setup, the various operations are completed automatically. Gears made on such machines are no more accurate than those produced on a milling machine, but the possibility of error is less, and they are much cheaper because of reduced labor requirements. For large quantities, however, form cutting is not used.

FIGURE 31-8   Cutting a gear on a semiautomatic gear-cutting machine with cutters making simultaneous roughing and finish cuts. *(Courtesy of Brown & Sharpe Manufacturing Company.)*

FIGURE 31-9   Blind gear spline broaching machine for producing internal gears. The machine has automated pick-and-place arms for load/unload. *(Courtesy of Apex Broach and Machine Company.)*

Broaching

Broaching is another way to form cut teeth. All the tooth spaces are cut simultaneously and the tooth is formed progressively. The circular table in Figure 31-9 holds 10 sets of progressive tooling. The table rotates moving one set of tooling at a time under two work-pieces. The arms load and unload a set of parts every 15 seconds. Excellent gears can be made by broaching. However, a separate broach must be provided for each size of gear. The tooling tends to be expensive, a restriction for this method, even though the cycle time can be very fast.

Shaping

Most high-quality gears that are made by machining are made by the *generating process.* This process is based on the principle that any two involute gears, or any gear and a rack, of the same diametral pitch will mesh together properly. Utilizing this principle, one of the gears (or the rack) is made into a cutter by proper sharpening. It can be used to cut into a mating gear blank and thus generate teeth on the blank. The two principal methods for gear generating are *shaping* and *hobbing.*

To carry out the shaping process, the cutter and the gear blank must be attached rigidly to their respective shafts, and the two shafts must be interconnected by suitable gearing so that the cutter and the blank rotate positively with respect to each other and have the same pitch-line velocities. To start cutting the gear, the cutter is reciprocated and is fed radially into the blank between successive strokes. When the desired tooth depth has been obtained, the cutter and blank are then slightly indexed after each cutting stroke. The resulting generating action is indicated schematically in the upper diagram of Figure 31-10a and shown in the cutting of an actual gear tooth in Figure 31-10b.

FIGURE 31-10 *(Top)* Generating action of a Fellows gear-shaper cutter; *(bottom)* series of photographs showing various stages in generating one tooth in a gear by means of a gear shaper, action taking place from right to left, corresponding to a diagram above. One tooth of the cutter was painted white. *(Top Courtesy of Fellows Gear Shaper Co.)*

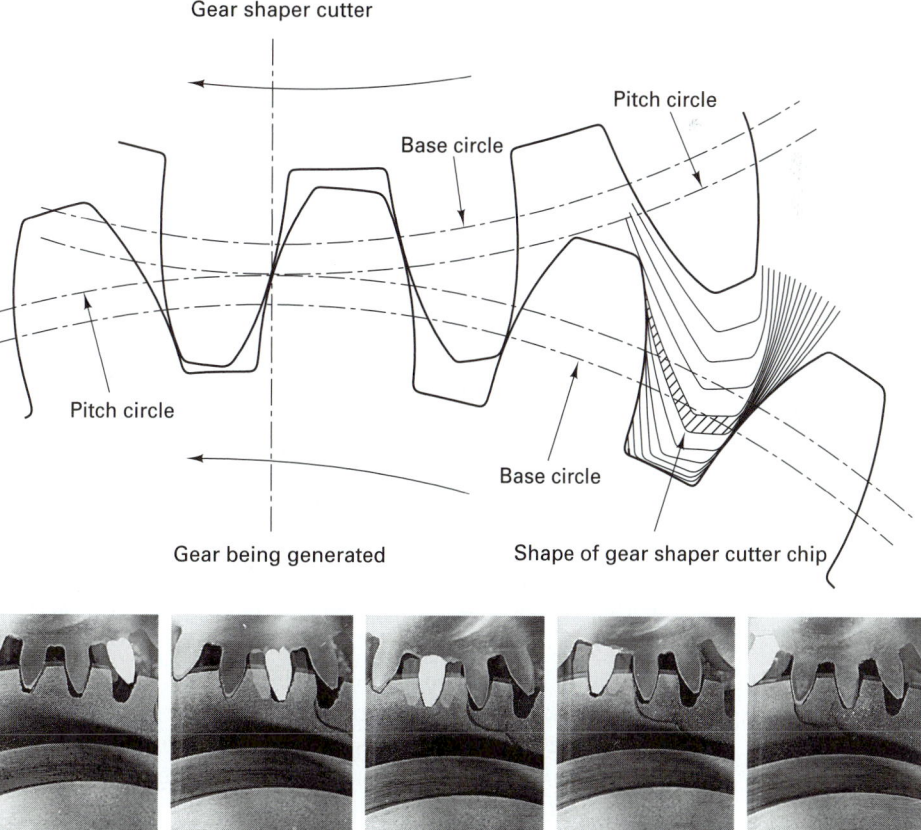

FIGURE 31-11 Fellows gear shaper, showing the motions of the gear blank and the cutter. Inset shows a close-up view of the cutter and teeth. *(Courtesy of Fellows Gear Shaper Company.)*

Figure 31-11 shows a machine called a gear shaper. *Gear shapers* generate gears by a reciprocating tool motion. The gear blank is mounted on the rotating table (or vertical spindle) and the cutter on the end of a vertical, reciprocating spindle. The spindle and the table are connected by means of gears so that the cutter and gear blank revolve with the same pitch-line velocity. Cutting occurs on the downstroke (sometimes on the upstroke). At the end of each cutting stroke, the spindle carrying the blank retracts slightly to provide clearance between work and tool on the return stroke. Because of the reciprocating action of the cutter, these machines are commonly called *gear shapers*. Details of the cutter are shown in Figure 31-12.

The conventional cutter for gear shaping is made from high-speed steel which can be coated to improve wear life with a superhard layer of titanium nitride (TiN) using the physical vapor deposition (PVD) process. Recently, new throwaway disk-shape insert tools have been developed for gear shaping, eliminating regrinding and recoating operations on the conventional tools. Regrinding the conventional cutter required two adjustments (resetting operations) on the machine tool. Because the throwaway blades are all sized the same, machine resetting after cutting tool changeover is eliminated.

To start cutting a gear, the cutter is fed radially inward before each cutting stroke, as it and the blank rotate. When the proper depth is reached, the inward feed stops and the cutter and blank continue their rotation until all the teeth have been machined by the generation process.

Either straight- or helical-tooth gears can be cut on gear shapers. To cut helical teeth, both the cutter and the blank are given an oscillating rotational motion during each stroke of the cutter, turning in one direction during the cutting stroke and in the opposite direction during the return stroke. Because the cutting stroke can be adjusted to end at any desired point, gear shapers are particularly useful for cutting cluster gears. Some machines

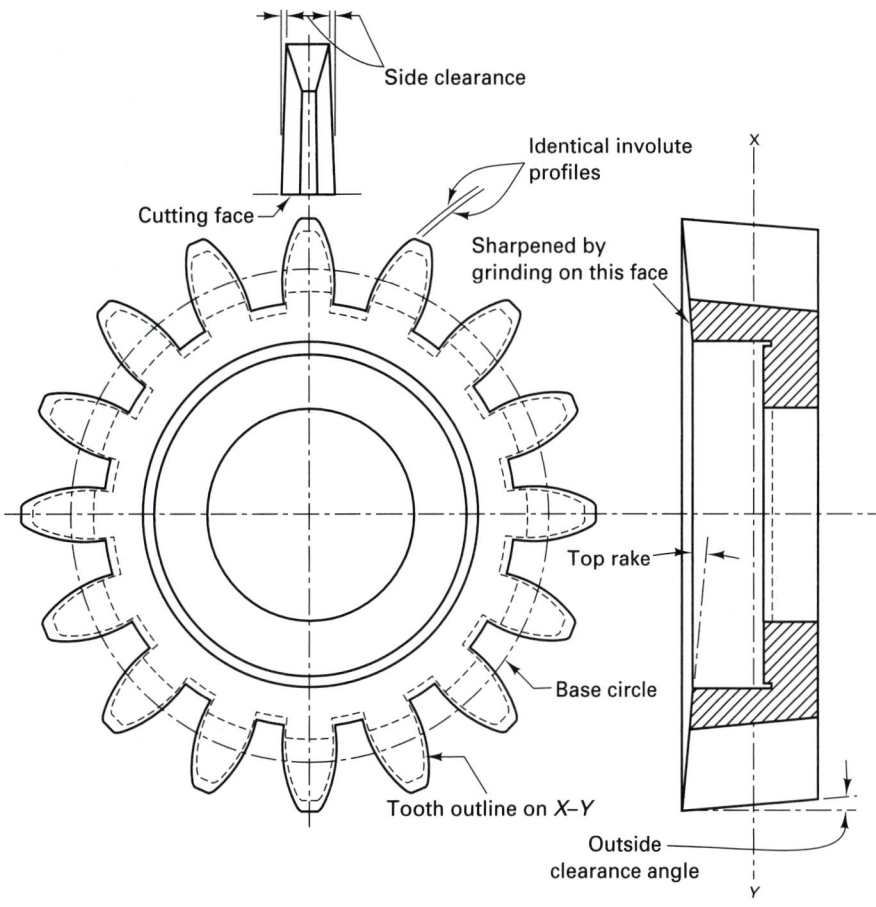

FIGURE 31-12 Details of the cutter used on the Fellows gear shaper. *(Courtesy of Fellows Gear Shaper Company.)*

can be equipped with two cutters simultaneously to cut two gears, often of different diameters. Gear shapers also can be adapted for cutting internal gears.

Special types of gear shapers have been developed for mass-production purposes. The *rotary gear shaper* essentially is 10 shaper units mounted on a rotating base and having a single drive mechanism. Nine gears are cut simultaneously while a finished gear is removed and a new blank is put in place on the tenth unit. *Planetary gear shapers* holding six gear blanks move in planetary motion about a large, central gear cutter. The cutter has no teeth in one portion to provide a space where the gear can be removed and a new blank placed on the empty spindle.

CNC gear shapers are now available with hydromechanical stroking systems which produce a uniform cutting velocity during the cutting portion of the downstroke. These machines can operate at 500 to 1700 rpm and use TiN-coated cutters to enhance tool life.

Vertical shapers for gear generating (Figure 31-13) have a vertical ram and a round table that can be rotated in a horizontal plane by either manual or power feed. These machine tools are sometimes called *slotters*. Usually, the ram is pivoted near the top so that it can be swung outward from the column through an arc of about 10°.

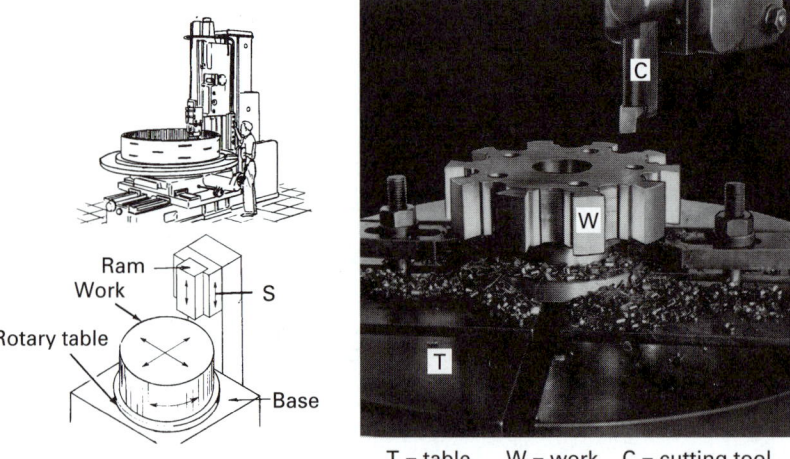

T = table   W = work   C = cutting tool

FIGURE 31-13   (*Left*) Schematic of vertical shaper. (*From* Manufacturing Producibility Handbook, *courtesy of General Electric Company*); (*right*) typical job setup in vertical shaper. (*Courtesy of Colt Industries.*)

Because one circular and two straight-line motions and feeds are available, vertical shapers are very versatile tools and thus find considerable use in one-of-a-kind manufacturing. Not only can vertical and inclined flat surfaces be machined, but external and internal cylindrical surfaces can be generated by circular feeding of the table between strokes. This may be cheaper than turning or boring for very small lot sizes. A vertical shaper can be used for generating gears or machining curved surfaces, interior surfaces, and arcs by using a stationary tool and rotating the workpiece. A *keyseater* is a special type of vertical shaper designed and used exclusively for machining keyways on the inside of wheel and gear hubs. For machining continuous herringbone gears, a *Sykes gear-generating machine* is used.

Hobbing

Involute gear teeth could be generated by a cutter that has the form of a rack. Such a cutter would be simple to make but has two major disadvantages. First, the cutter (or the blank) would have to reciprocate, with cutting occurring only during one stroke direction. Second, because the rack would have to move longitudinally as the blank rotated, the rack would need to be very long (or the gear very small) or the two would not be in mesh after a few teeth were cut. A *hob* overcomes the preceding two difficulties. As shown in Figure 31-14, a hob can be thought of, basically, as one long rack tooth that has been wrapped around a cylinder in the form of a helix and fluted at intervals to provide a number of cutting edges. Relief is provided behind each of the teeth. The cross section of each tooth, normal to the helix, is the same as that of a rack tooth. (A hob can also be thought of as a gashed worm gear.)

The action of a *hobbing* machine cutting a spur gear is illustrated in Figure 31-15. To cut a spur gear, the axis of the hob must be set off from the normal to the rotational axis of the blank by the helix angle of the hob. In cutting helical gears, the hob must be set over an additional amount equal to the helix angle of the gear. The cutting of a gear by means of a hob is a continuous action. The hob and the blank are connected by proper

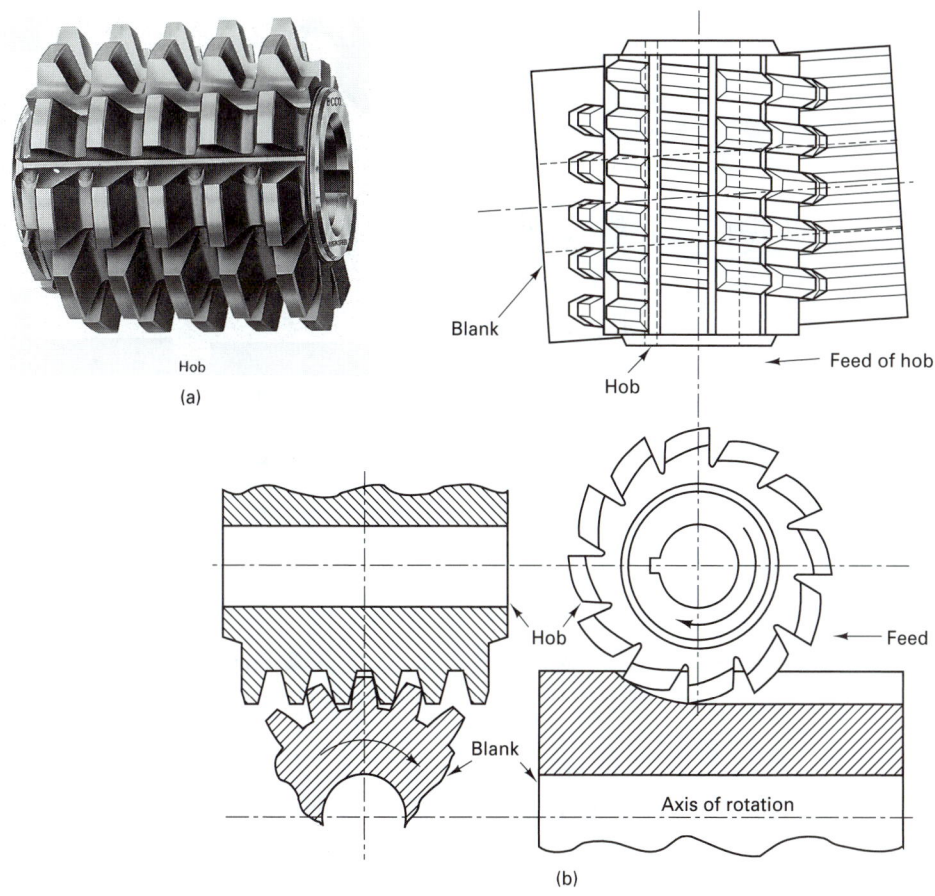

Hob
(a)

Blank

Hob   Feed of hob

Blank

Hob   Feed

Blank

Axis of rotation

(b)

FIGURE 31-14   Hob and the relationship of the hob to the gear blank in machining a spur gear by hobbing.

gearing so that they rotate in mesh. To start cutting a gear, the rotating hob is fed inward until the proper setting for tooth depth is obtained. The hob is then fed in a direction parallel with the axis of rotation of the blank. As the gear blank rotates, the teeth are generated and the feed of the hob across the face of the blank extends the teeth to the desired tooth face width.

Hobbing is rapid and economical. More gears are cut by this process than by any other. The process produces excellent gears and can also be used for splines and sprockets. Single-, double-, and triple-thread hobs are used. Multiple-thread types increase the production rate but do not produce accuracy as high as single-thread hobs.

Gear-hobbing machines are made in a wide range of sizes. Machines for cutting accurate large gears frequently are housed in temperature-controlled rooms, and the temperature of the cutting fluid is controlled to avoid dimensional change due to variations in temperature.

Bevel Gear-Generating

The machines used to generate the teeth on various types of bevel gears are among the most ingenious of all machine tools. Two basic types utilize the principle that a crown

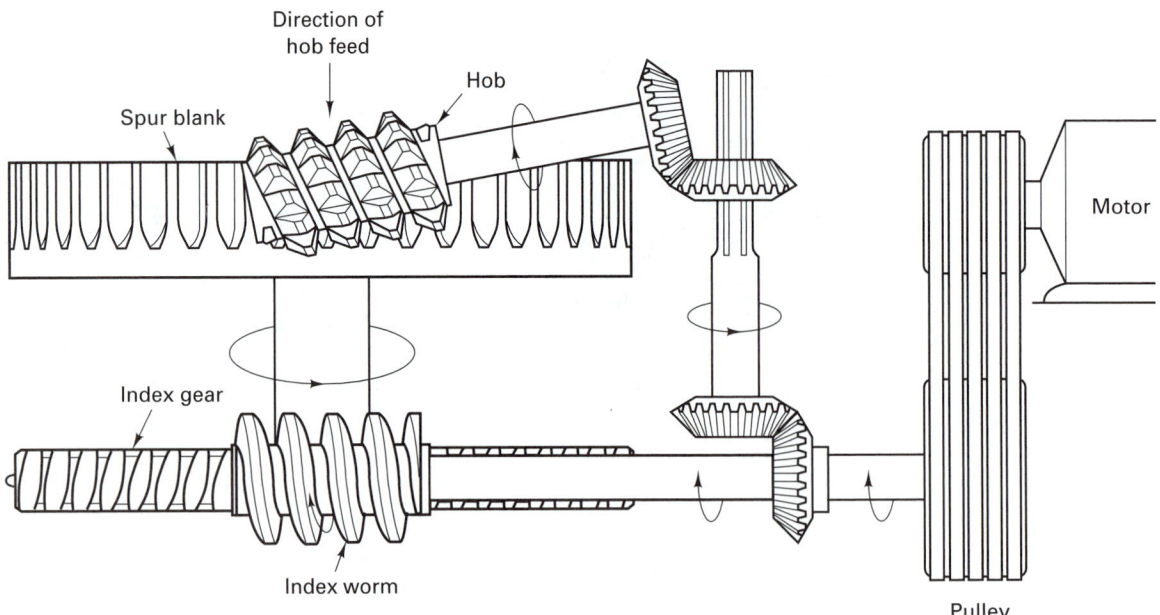

FIGURE 31-15   Schematic of gear hobbing machine for hobbing a spur gear.

gear will mate properly with any bevel gear having the same diametral pitch and tooth form. One is used for cutting straight-tooth bevel gears and the other for cutting curved-tooth bevel gears.

The basic principle of these machines is indicated in Figure 31-16. The blank and a connecting gear, having the same cone angle and diametral pitch, are mounted on a common shaft so that the connecting gear meshes with the regular teeth on the crown gear, which also has a reciprocating cutter to generate the teeth on the gear blank. As the connecting gear rolls on the crown gear, the reciprocating cutter generates a tooth in the gear blank. Obviously, with only one cutter tooth, only a single tooth space would be cut in the gear blank. However, this limitation is overcome in actual machines by two modifications. First, by indexing the connecting gear and blank unit, the process is repeated as often as required to cut all the teeth in the blank. Second, instead of a single reciprocating cutter, two half-tooth cutters are approximating the inner sides of two adjacent teeth so that they cut both faces of one tooth on the blank.

The application of this modified basic principle in actual machines is shown in Figures 31-17 and 31-18. The cradle acts as a crown gear and carries the two cutters, which reciprocate in slides simultaneously in opposite directions. It can be noted in Figure 31-18 that straight lines through the faces of the cutters, the axis of rotation of the gear blank, and elements of the pitch cone of the gear blank all meet at a common point that is in the plane of the imaginary crown gear and on the axis of the cradle, thus fulfilling the requirements shown in Figure 31-16.

In operation, while the cradle is at the extreme upward position, the gear blank is fed inward toward the cutter until the position for nearly full tooth depth is reached. The cradle then starts to roll downward, and a tooth is generated during the downward roll. When the cradle reaches the full-down position, the blank is automatically fed inward a small amount, usually 0.020 to 0.015 in. and a finish cut is made during the upward roll of the

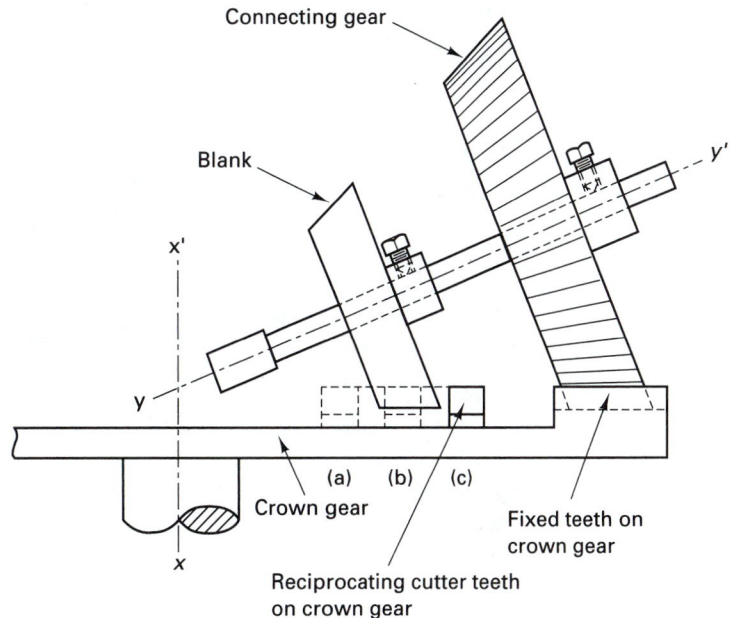

FIGURE 31-16   Principle of bevel gear generating machines.

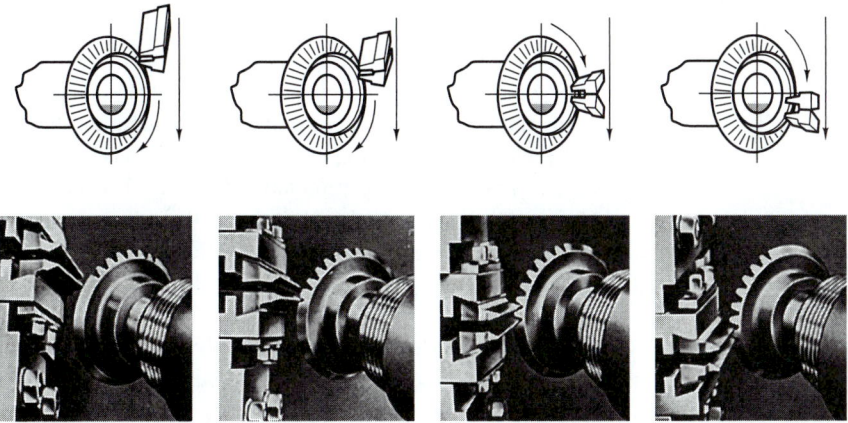

FIGURE 31-17   (*Top*) Schematic showing the roll of the gear blank and cutters during the cutting of one tooth on a bevel-gear generating machine, as seen from the front; (*bottom*) photographs showing the same roll action, as seen from the back of the machine. *(Courtesy of Heidenreich & Harbeck.)*

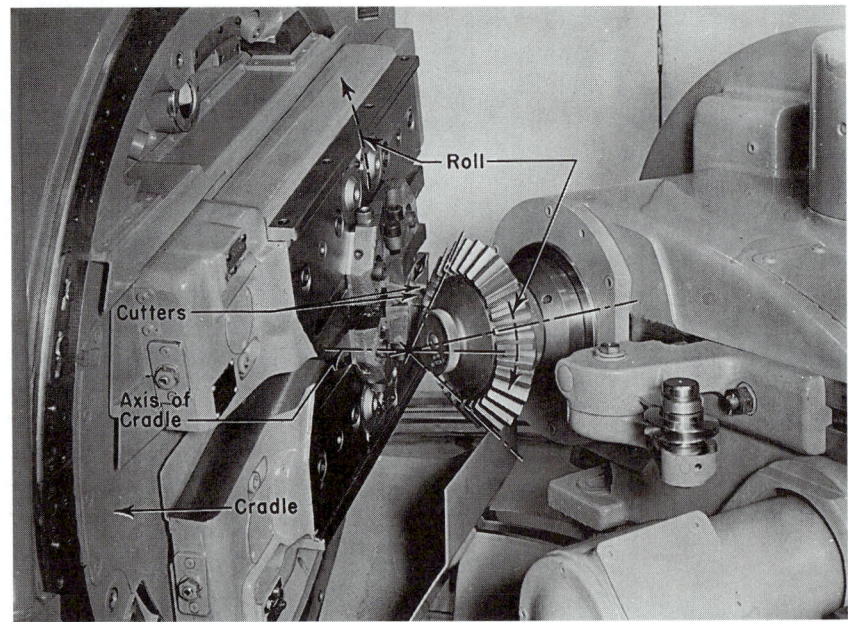

FIGURE 31-18 Cutters and gear blank on a Gleason straight-tooth gear-generating machine, showing the relationship between the axis of rotation of the blank and the axis of the cradle, which, with the cutter, simulates a crown gear. *(Courtesy of Gleason Works.)*

cradle. At the completion of the upward roll, the blank is withdrawn, indexed, and moved inward automatically, ready to start the next tooth. These machines can cut bevel gears of different cone angles.

Provision is made so that the spindle on which the gear blank is mounted can be swung through an angle to accommodate bevel gears of different cone angles. This spindle is connected to the cradle by means of suitable internal change gears to provide the required positive rolling motion between the two.

For cutting small straight-tooth bevel gears, a machine employing two revolving disk-type cutters is used. These cutters reciprocate on the cradle slides as they rotate. Because of their multiple cutting edges and continuous rotary rather than reciprocating cutting action, they are much more productive (Figure 31-19).

Generating machines for straight-tooth bevel gears often are provided with a mechanism that produces a slight crown on the teeth. Such teeth are slightly thicker at the middle than at the ends, to avoid having applied loads concentrated at the tooth ends where they are the weakest.

Machines used for generating spiral, zerol, and hypoid bevel gears employ the same basic crown-gear principle, but a multitooth rotating cutter is used. This cutter has its axis of rotation parallel with the axis of roll of the cradle. Figure 31-20 shows the relationship of the gear blank to the axis of the theoretical crown gear (the cradle) and of the cutter teeth to the theoretical mating bevel gear.

Cold-Roll Forming

The manufacture of gears by *cold-roll forming* has been highly developed and widely adopted in recent years. Currently, millions of high-quality gears are produced annually by

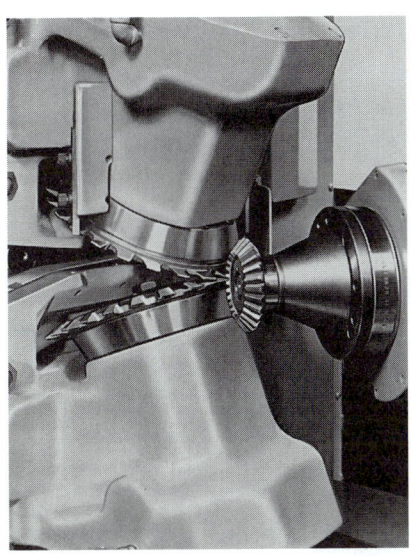

FIGURE 31-19   Bevel-gear generating machine, utilizing two disk-type cutters. *(Courtesy of Gleason Works.)*

FIGURE 31-20   Relationship of the gear blank to the axis of the theoretical crown gear (the cradle) and of the cutter teeth to the theoretical mating bevel gear.   *(Courtesy of Gleason Works.)*

this process; many of the gears in automobile transmissions are made this way. As indicated in Figure 31-21, the process is basically the same as that by which screw threads are roll formed, except that in most cases the teeth cannot be formed in a single rotation of the forming rolls; the rolls are fed inward gradually during several revolutions.

Because of the metal flow that occurs, the top lands of roll-formed teeth are not smooth and perfect in shape; a depressed line between two slight protrusions can often be seen, as shown encircled in Figure 31-22. However, because the top land plays no part in gear-tooth action, if there is sufficient clearance in the mating gear, this causes no difficulty. Where desired, a light turning cut is used to provide a smooth top land and correct addendum diameter.

The hardened forming rolls are very accurately made, and the roll-formed gear teeth usually have excellent accuracy. In addition, because the severe cold working produces tooth faces that are much smoother and harder than those on ordinary machined gears, they seldom require hardening or further finishing, and they have excellent wear characteristics.

The process is rapid (up to 50 times faster than gear machining) and easily mechanized. No chips are made and thus less material is needed. Less skilled labor is required. Small gears often are made by rolling a length of shaft and then slicing off the individual gear blanks. Usually, soft steel is required and 4 to 5 in. in diameter is about the limit, with fewer than six teeth, coarser than 12 diametral pitch, and no pressure angle less than 20°.

Other Gear-Making Processes

Gears can be made by the various casting processes. *Sand-cast gears* have rough surfaces and are not accurate dimensionally. They are used only for services where the gear moves

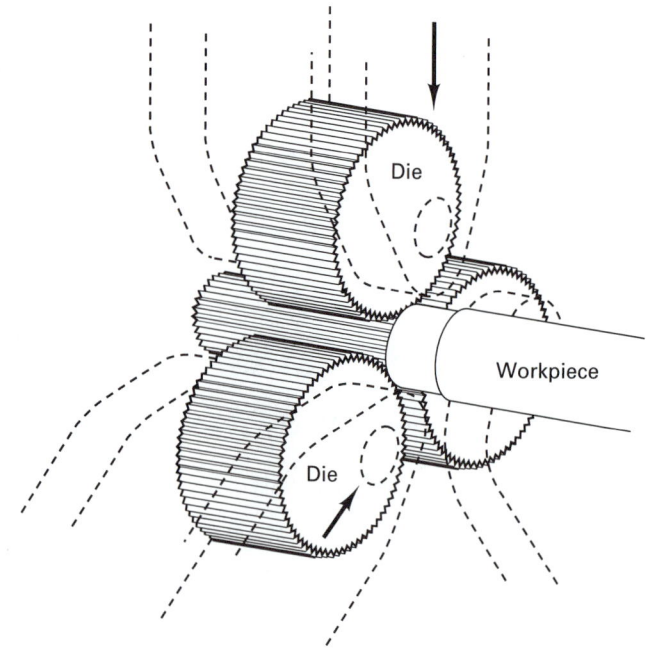

FIGURE 31-21 Method for forming gear teeth and spline by cold forming.

FIGURE 31-22 (*Top*) Worm gear being rolled by means of rotating rolling tools; (*bottom left*) typical worm made by rolling, with enlarged view of end of one tooth; (*right*) gear made by rolling. (*Courtesy of Landis Machine Company.*)

slowly and where noise and inaccuracy of motion can be tolerated. Gears made by *die casting* are fairly accurate and have fair surface finish. They can be used to transmit light loads at moderate speeds. Gears made by *investment casting* may be accurate and have good surface characteristics. They can be made of strong materials to permit their use in transmitting heavy loads. In many instances, gears that are to be finished by machining are made from cast blanks, and in some larger gears the teeth can be cast to approximate shape to reduce the amount of machining.

Large quantities of gears are produced by *blanking* in a punch press. The thickness of such gears usually does not exceed about $\frac{1}{16}$ in. By shaving the gears after they are blanked, excellent accuracy can be achieved. Such gears are used in clocks, watches, meters, and calculating machines. *Fine blanking* is also used to produce thin, flat gears of good quality.

High-quality gears, both as to dimensional accuracy and surface quality, can be made by the *powder metallurgy process*. Usually, this process is employed only for small sizes, ordinarily less than 1 in. in diameter. However, larger and excellent gears are made by forging powder metallurgy preforms. This results in a product of much greater density and strength than usually can be obtained by ordinary powder metallurgy methods, and the resulting gears give excellent service at reduced cost. Gears made by this process often require little or no finishing.

Large quantities of plastic gears are made by *plastic molding*. The quality of such gears is only fair, and they are suitable only for light loads. Accurate gears suitable for heavy loads frequently are machined out of laminated plastic materials. When such gears are mated with metal gears, they have the quality of reducing noise.

Quite accurate small gears can be made by the *extrusion* process. Typically, long lengths of rod, having the cross section of the desired gear, are extruded. The individual gears are then sliced from this rod. Materials suitable for this process are brass, bronze, aluminum alloys, magnesium alloys, and occasionally, steel.

Flame machining (oxyacetylene cutting) can be used to produce gears that are to be used for slow-moving applications wherein accuracy is not required.

A few gears are made by the hot roll-forming process. In this process a cold master gear is pressed into a hot blank as the two are rolled together.

■ 31.5 GEAR FINISHING

To operate efficiently and have satisfactory life, gears must have accurate tooth profiles and the faces of the teeth must be smooth and hard. These qualities are particularly important when gear must operate quietly at high speeds. When they are produced rapidly and economically by most processes except cold-roll forming, the tooth profiles may not be as accurate as desired, and the surfaces are somewhat rough and subject to rapid wear. Also, it is difficult to cut gear teeth in a hardened gear blank, and therefore, economy dictates that the gear be cut in a relatively soft blank and subsequently be heat treated to obtain greater hardness. Such heat treatment usually results in some slight distortion and surface roughness. Although most roll-formed gears have sufficiently accurate profiles, and the tooth faces are adequately smooth and frequently have sufficient hardness, this process is feasible only for relatively small gears. Consequently, a large proportion of high-quality gears are given some type of finishing operation after they have received primary machining or after heat treatment. Most of these finishing operations can be done quite economically because only minute amounts of metal are removed.

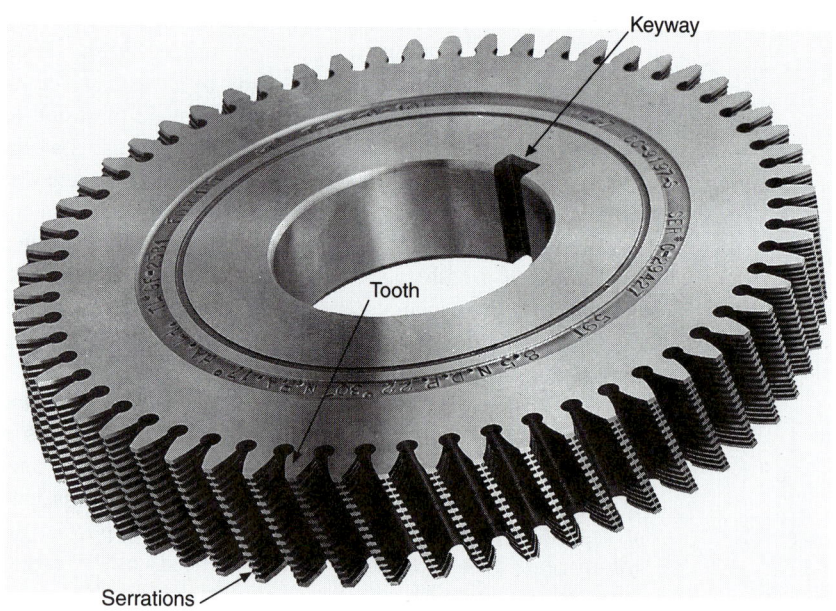

FIGURE 31-23   Rotary gear shaving cutter (see Figure 31-24).

Gear shaving is the most commonly used method for finishing spur and helical gear teeth prior to hardening. The gear is run, at high speed, in contact with a shaving tool, usually of the type shown in Figure 31-23. Such a tool is a very accurate, hardened, and ground gear that contains a number of peripheral serrations, thus forming a series of sharp cutting edges on each tooth. The gear and shaving cutter are run in mesh with their axes crossed at a small angle, usually about 10° (Figure 31-24). As they rotate, the gear is reciprocated longitudinally across the shaving tool (or vice versa). During this action, which usually requires less than 1 minute, very fine chips are *shaved* from the gear-tooth faces, thus eliminating any high spots and producing a very accurate tooth profile.

Rack shaving cutters sometimes are used for shaving small gears, the cutter reciprocating lengthwise, causing the gear to roll along it, as it is moved sideways across the cutter and fed inward.

Although shaving cutters are costly, they have a relatively long life because only a very small amount of metal is removed, usually 0.001 to 0.004 in. Some gear-shaving machines produce a slight crown on the gear teeth during shaving. Most gears are not hardened prior to shaving, although it is possible to remove very small amounts of metal from hardened gears if they are not too hard. However, modern heat-treating equipment makes it possible to harden gears after shaving without harmful effects, and therefore this practice is usually followed.

Roll finishing is a cold-forming process that is used to finish helical gears. The unhardened gear is rolled with two hardened, accurately formed rolling dies. The center distance between the dies is reduced to cold work the surfaces and produce highly accurate tooth forms. High points on the unhardened gear are plastically deformed so that a smoother surface and more accurate tooth form are achieved. Because the operation is one of localized cold working, some undesirable effects may accrue, such as localized residual

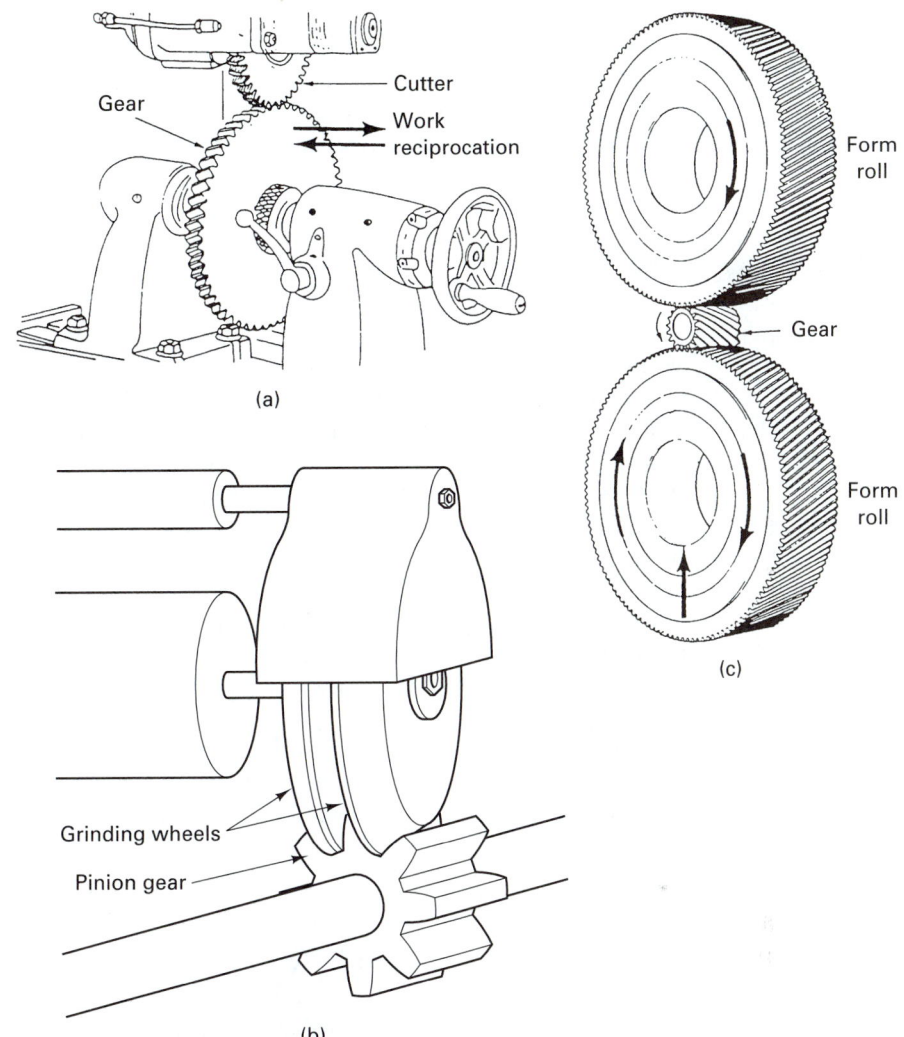

FIGURE 31-24 Methods for gear finishing by shaving, grinding, and roll forming or finishing.

stresses and nonuniform surface characteristics. Surface finishes of 6 to 8 μin. have been achieved. If roll finishing is to be used, attention must be paid to the prerolled geometry. Designers should consult the manufacturers of gear-rolling machines for specific recommendations.

Grinding is used to obtain very accurate teeth on hardened gears. Two methods are used. One employs a formed grinding wheel that is trued to the exact form of a tooth by means of diamonds mounted on a special holder and guided by a large template. The other method is involute-generation grinding, which uses straight-sided grinding wheels which simulate one side of a rack tooth. The surface of the gear tooth is ground as the gear rolls (and reciprocates) past the grinding wheels. Grinding produces very accurate gears, but because it is slow and expensive, it is used only on the highest-quality, hardened gears.

Lapping can also be used for finishing hardened gears. The gear to be finished is run in contact with one or more cast-iron lapping gears under a flow of very fine abrasive in oil. Because lapping removes only a very small amount of metal, it usually is employed on gears that previously have been shaved and hardened. This combination of processes produces gears that are nearly equal to ground gears in quality but at considerably lower cost.

■ 31.6 GEAR INSPECTION

As with all manufactured products, gears must be checked to determine whether the resulting product meets the design specifications and requirements. Because of their irregular shape and the number of factors that must be measured, inspection of gears is somewhat difficult. Among the factors to be checked are linear tooth dimensions such as thickness, spacing, depth, and so on; tooth profile; surface roughness; and noise. Several special devices, most of them automatic or semiautomatic, are used for such inspection.

Gear-tooth vernier calipers can be used to measure the thickness of gear teeth on the pitch circle (Figure 31-25). CNC gear inspection machines (Figure 31-26) can quickly check several factors, including variations in circular pitch, involute profile, lead, tooth spacing, and variations in pressure angle. The gear usually is mounted between centers. The probe is moved to the gear through X, Y, and Z translations. The gear is be rotated between measurements. The inset in Figure 31-26 shows a typical display for involute profile.

Because noise level is important in many applications, not only from the viewpoint of noise pollution but also as an indicator of probable gear life, special equipment for its measurement is quite widely used, sometimes integrated into mass-production assembly lines.

FIGURE 31-25 Using gear-tooth vernier calipers to check the tooth thickness at the pitch circle.

FIGURE 31-26 CNC gear inspection machine. *(Courtesy of Illitron.)*

■ KEY WORDS

addendum
bevel gear
blanking
broaching
circular pitch
cold roll-forming
crown gear
dedendum
diametral pitch
form milling

gear hobbing
gear shaping
gear finishing
gear grinding
gear teeth
helical gear
herringbone gear
hob
hobbing
hypoid

hypoid bevel gear
involute curve
pitch circle
pitch
pitch diameter
planetary gear
rack
spur gear
worm gear

■ REVIEW QUESTIONS

1. Why can the relative angular velocities of two mating spur gears not be determined by their outside diameters alone?

2. Why is the involute form used for gear teeth?
3. What is the diametral pitch of a gear?
4. What is the relationship between the

diametral pitch and the module of a gear?

5. On a sketch of a gear, indicate the pitch circle, addendum circle, dedendum circle, and the circular pitch.

6. What five requirements must be met for gears to operate satisfactorily? Which of these are determined by the manufacturing process?

7. What are the advantages of helical gears compared with spur gears?

8. What is the principal disadvantage of helical gears?

9. A gear that has a pitch diameter of 6 in. and a diametral pitch of 4. What number of form cutter would be used in cutting it?

10. What difficulty would be encountered in hobbing a herringbone gear?

11. What is the only type of machine on which full-herringbone gears can be cut?

12. What modification in design is made to herringbone gears to permit them to be cut by hobbing?

13. Why are not more gears made by broaching?

14. What is the most important property of a crown gear?

15. What are three basic processes for machining gears?

16. Which basic gear-machining process is utilized in a Fellows gear shaper?

17. Can a helical gear be machined on a plain milling machine? Why or why not?

18. When a helical gear is machined on a milling machine, the table lead screw and the universal dividing head are connected by a gear train. Why?

19. What is the relationship between a crown gear and Gleason gear generators?

20. Why is a gear-hobbing machine much more productive than a gear shaper?

21. In gear shaping and gear hobbing, the tooth profiles are generated. What does that mean?

22. What are the advantages of cold-roll forming for making gears?

23. Assume that 10,000 spur gears, $1\frac{1}{8}$ in. in diameter and 1 in. thick are to be made of 70-30 brass. What manufacturing methods would you consider?

24. If only three gears described in Question 23 were to be made, what process would you select?

25. Why is cold-roll forming not suitable for making gray cast iron gears?

26. Under what conditions can shaving not be used for finishing gears?

27. What inherent property accrues from cold-roll forming of gears that may result in improved gear life?

28. Can lapping be used to finish cast iron gears?

29. What factors usually are checked in inspecting gears?

30. What are the basic methods for gear finishing?

■ PROBLEMS

1. A hob that has a pitch diameter of 76.2 mm is used to cut a gear having 36 teeth. If a cutting speed of 27.4 m/min is used, what will be the rpm value of the gear blank?

2. If the gear in Problem 1 has a face width of 76.2 mm, a feed of 1.9 mm/rev of the workpiece is used, and the approach and overtravel distances of the hob are 38 mm, how much time will be required to hob the gear?

3. In Figure 31-5 a form-milling operation is shown. The cutter is shown in Figure 31-6. The gear is to be made from 4340 steel, $R_C 50$ prior to heat treat and final grind. Select the proper speeds and feeds for the job (the cutter is 4 in. in diameter) and compute the cutting time to mill this gear. Did you use up milling or down milling?

4. Compare the cutting time from Problem 3 with that of gear shaping.

5. A gear broaching machine of the type shown in Figure 31-7 could do the gear in 15 seconds (about 240 parts per hour). How many additional gears per year are needed to cover the broaching tooling cost if each broach on the machine costs $250. Do you think the broach tool life is sufficient to handle that number of parts? What about TiN coating the broaches (cost = $10.00 per broach)?

*C*hapter 31 CASE STUDY

six-second pinion gear

The Strom Company, manufacturers of a new type of gear-making machine, invites you, an independent consultant, to attend a meeting at company headquarters to examine a problem they are having with their new gear-cutting process. They claim that their process generates finished helical pinion gears (the kind used in automatic transmissions in cars) 10 times faster than conventional processes (see Chapter 31). In their system a rotating pinion blank feeds between two cutter heads that turn in opposite directions, as shown in Figure CS-31a. Each of the two cutter heads contains 90 radially mounted carbide insert tools. As shown in Figure CS-31b, the cutting tools make a series of roughing and finishing cuts while the pinion blank rotates. As shown in Figure CS-31c, this combination of the linear motion of the carbide inserts, along with the rotation of he gear blank, generates the helical profile. This machine was designed to produce 600 gears per hour. The carbide insert teeth are clamped in the cutter bodies, as shown in Figure CS-31d. Pockets ground into the steel cutter body provide accurate insert location for the carbide tools.

At the meeting, the company describes the nature of their problem. They have purchased carbide tools from two vendors. These tools have very long tool life at the speeds and feeds that are to be used to machine the gears. At the meeting are representatives from the two carbide tool manufacturing companies who supplied the tools. Both provide data confirming that the tools should last 40 hours or more without appreciable flank or crater wear. They present results of turning tests, cutting the same steel to be used in the pinions with the same structure and properties (hardness). These results confirm the expected good performance. In addition, the geometry of the inserts provided by both companies is identical.

The gear machine company has conducted tests on the machine using HSS (high-speed-steel) tools and aluminum gear blanks. The machine ran continuously for over 100 hours, producing over 600 gears per hour with no problems. (They now have a room full of aluminum pinion gears.) When they switched to steel pinion gears and carbide tools, however, they experienced rapid tool failure and could not machine more than six pinions before they had to stop the machine, even though the same cutting fluids were used on the machine as had been employed during the favorable tool life tests. The machine operator observed that "the thing spit out carbide teeth like a machine gun."

Also present at the meeting was another consultant who had performed a complete vibration analysis of the machine. His results, however, could neither confirm nor disprove that chatter was the cause of the rapid tool failure.

You request that the following test be run. The machine is to be setup to cut the gears at a rate of *one per minute* instead of 10 per minute. This requires some gear changes and adjustments to the machine. Again, the machine performed perfectly for the HSS tools and aluminum gears. When the carbide cutters were inserted in place of the HSS tools, the carbides again experienced rapid tool failure. Everyone now knew the cause of the problem. Explain the real reason why this machine failed to meet the performance specifications and the cause of the rapid tool failure.

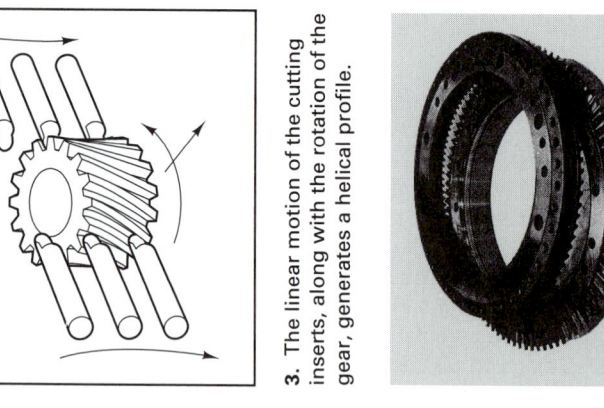

3. The linear motion of the cutting inserts, along with the rotation of the gear, generates a helical profile.

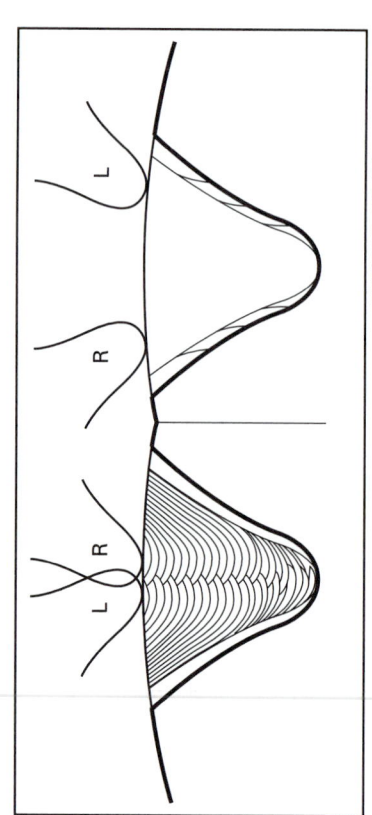

4. Machine uses cutter heads of large diameter with a series of small radially clamped carbide inserts.

1. Two cutter heads, each studded with 90 carbide inserts, turn in opposite directions to cut teeth in a rotating blank as it passes through, and again as it returns, generating a finished pinion.

2. Tools shown together here are actually on opposite sides of the gear. The effective cutting pattern for the roughing cut appears on the left; and the pattern for the finish cut, on the return stroke of the spindle slide, is on the right.

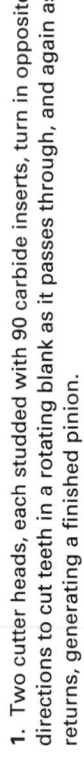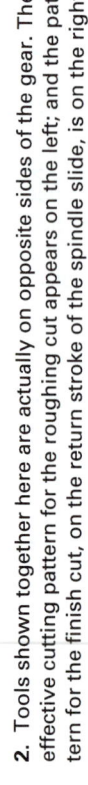

FIGURE CS-31

CHAPTER 32

NONTRADITIONAL MACHINING PROCESSES

32.1   INTRODUCTION
32.2   CHEMICAL NTM PROCESSES
        Chemical Machining
            Chemical Milling or Blanking
            Use of Scribed Maskants
            Chemical Machining to Multiple Depths
            Design Factors in Chemical Machining
            Advantages and Disadvantages of Chemical Machining
        Thermochemical Machining
32.3   ELECTROCHEMICAL NTM PROCESSES
        Electrochemical Machining
        Electrochemical Hole Machining

        Electrochemical Grinding
        Electrochemical Deburring
32.4   MECHANICAL NTM PROCESSES
        Ultrasonic Machining
        Waterjet Machining
        Abrasive Waterjet Machining
        Abrasive Flow Machining
32.5   THERMAL PROCESSES
        Electrical Discharge Machining
            Advantages and Disadvantages of Electrical Discharge Machining
        Electron Beam Machining
        Laser Beam Machining
        Plasma Arc Cutting

■ 32.1 INTRODUCTION

Machining processes that involve compression-shear chip formation have a number of inherently adverse characteristics and limitations. Although often necessary, chip formation can be an expensive, difficult process. Large amounts of energy are utilized in producing an unwanted product—chips. Further expenditure of energy and money is required to remove these chips from the machines and to dispose of, or recycle, them. A large amount of energy ends up as undesirable heat that often produces problems of distortion and surface cracking. Cutting forces create problems in holding the work and sometimes cause distortion. Undesirable deformation and residual stresses in the workpiece often require further processing to remove the effects. Finally, there are definite limitations in regard to the delicacy of the work that can be machined. For example, the production of the semiconductor device, also called a chip, shown in Figure 32-1, would not be possible with any of the traditional machining processes. In view of these adverse and limiting characteristics, it is not surprising that in recent years substantial effort has been devoted to developing and perfecting material-removal processes that replace conventional machining. *Nontraditional machining* (NTM) processes is one designation for this diverse family of unconventional processes, which are generally nonmechanical, do not produce chips or a lay pattern in the surface and often involve new energy modes.

For the purposes of our discussion, these NTM processes can be divided into four basic groups:

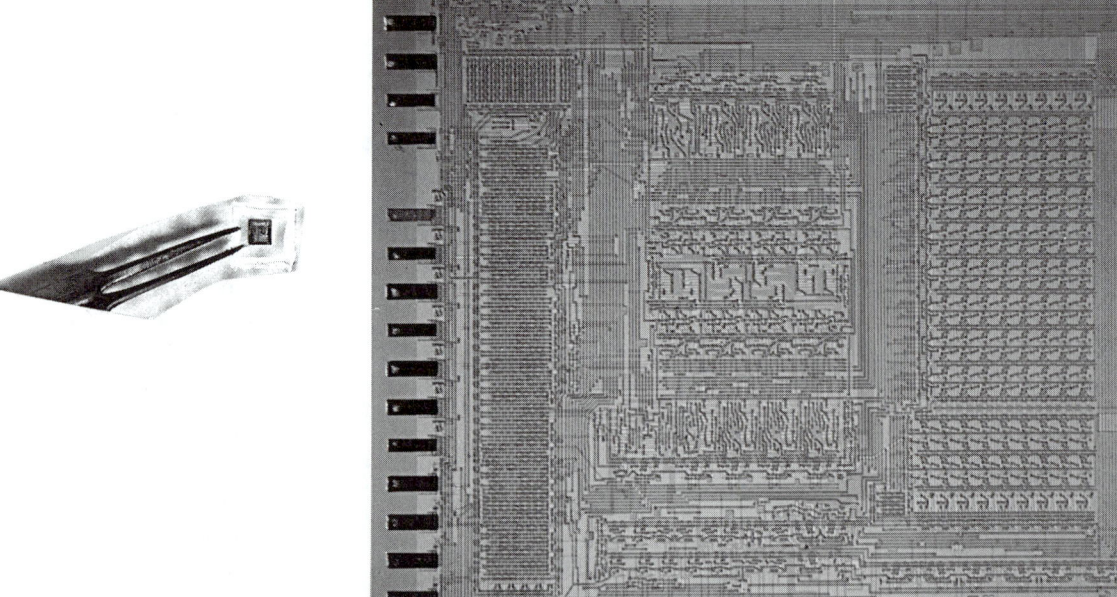

FIGURE 32-1   (*Right*) Enlarged view of one portion of a microprocessor chip.   This chip, shown to the left in actual size, measures only 5 mm on a side and contains over 3000 transistors. *(Courtesy of Bell Laboratories.)*

1. *Chemical.* Chemical reaction, sometimes enhanced by electrical or thermal energy, is the dominant mode of material removal.
2. *Electrochemical.* Electrolytic dissolution dominates the material removal process.
3. *Mechanical.* Multipoint cutting or erosion dominates the removal process.
4. *Thermal.* High temperatures in very localized regions to melt and vaporize material dominate the removal process.

Tables 32-1 through 32-4 give a summary of the characteristics for these procedures. Not included in these tables are the new forming processes, such as hot isostatic pressing or HERF, discussed in earlier chapters, which do not remove material.   Another grouping of processes, called rapid prototyping processes, are not included as they do not in general remove material.   When examining these tables, recall that conventional turning has these typical values: surface finish, 32 to 250 μin. AA; MRR, 10 to 200 in^3/min; HP$_s$, 0.5 to 2 HP/in^3-min; *V*, 100 to 1000 ft/min; accuracy $\simeq$0.002 in. NTM processes typically have low metal-removal rates compared to machining and very high specific horsepowers.   They typically have better accuracy, usually at slow rates of processing, which often results in

TABLE 32-1. Summary of Chemical NTM Processes

Process	Typical Surface Finish AA (μin.)	Typical Metal Removal Rate	Typical Specific Horsepower (HP/in³-min)	Typical Penetration Rate (ipm) or Cutting Speed (sfpm)	Typical Accuracy (in.)	Comments
Chemical machining	63–250, but can go as low as 8	30 in³/min	Chemical energy	0.001–0.002 ipm	0.001–0.006; material and process dependent	Almost all materials possible; depth of cut limited to $\frac{1}{2}$ in.; no burrs; no surface stresses; tooling low cost
Electropolishing	4–32, but can go as low as 2 or 1 or better	Very slow	50–200 A/ft²	0.0005–0.0015 ipm	NA[a]; process used to obtain finish	High quality, no stress surface; removes residual stresses; makes corrosion-resistant surfaces; may be considered to be an electrochemical process
Photochemical machining (blanking)	63–250, but can go as low as 8	Same as chemical milling	Dc power	0.0004–0.0020 ipm	10% of sheet thickness or 0.001–0.002 in.	Limited to thin material; burr-free blanking of brittle material; tooling low cost; used in microelectronics
Thermochemical machining (combustion machining for deburring)	Burr-free	Minute with rapid cycle time	NA[a]	15- to 50-sec cycle times	NA[a]	Vaporizes burrs and fins on cast or machined parts; deburrs steel gears automatically

[a]NA, not applicable for this process.

TABLE 32-2. Summary of Electrochemical NTM Processes

Process	Typical Surface Finish AA (µin.)	Typical Removal Rates (in³/1000 A-min)	Typical Specific Horsepower (HP/in³-min)	Typical Penetration Rate (in./min) or Cutting Speed (sfpm)	Typical Accuracy (in.)	Comments
Electrochemical machining (ECM)	16–63	0.06 in W, Mo 0.16 in Cl 0.13 in steel, Al 0.60 in Cu	160	0.1 to 0.5 ipm	0.0005–0.005 ≃ 0.002 in cavities	Stress-free metal removal in hard to machine metals; tool design expensive; disposal of chemicals a problem; MRR independent of hardness; deep cuts will have tapered walls
Electrochemical grinding (ECC)[a]	8–32	0.010 in chromium 0.126 Al 0.268 Cu 0.135 Fe, Ti 0.060 W	High	Cutting rates about same as grinding; wheel speed, 4000–6000 sfpm	0.001–0.0005	Special form of ECM; grinding with ECM assist; good for grinding hard conductive materials like tungsten carbide tool bits; no heat-damage burrs or residual stresses
Electrolytic hole machining (Electrostream)[b]	16–63	NA[c]	NA[c]	0.060–0.120 ipm	≃ 0.001 or 5% of diameter hole	Special version of ECM for hole drilling small round or shaped holes; multiple-hole drilling; typical holes 0.004 to 0.03 in. in diameter with depth/diameter ratio of 50:1

[z]Honing can also be done with EC assistance, which can quadruple the MRR over convention honing and yield a 2-µin. AA finish.

[b]Trademark of General Electric Company.

[c]NA, not applicable for this process.

TABLE 32-3. Summary of Mechanical NTM Processes

Process	Surface Finish AA (μin.)	Typical Metal Removal Rate (in³/min)	Typical Specific Horsepower (HP/in³-min)	Typical Penetration Rate (ipm) or Cutting Speed (sfpm)	Typical Accuracy (in.)	Comments
Abrasive flow machining	30–300; can go as low as 2	Low	NA[a]	Low	0.001–0.002	Typically used to finish inaccessible internal passages; often used to remove recast layer produced by EDM; used burr removal (cannot do blind holes)
Fluid or water jet machining	50–100	Very low	NA[a]	Depends on materials	±0.010 at 3 to 4 in. standoff	Used on wood, nonmetals; pressures of 55,000 psi and jet velocity of 1700 to 3000 ft/sec
Abrasive waterjet machining	50–75 for 0.003 to 0.020 in. diameter stream	Very low; fine finishing process, 0.001	NA[a]	Very low, 0.6–100 ipm	$\simeq$ 0.005 typical ±0.020 in. cut-line tolerance	Use in heat-sensitive or brittle materials, glass, titanium, and composites and nonmetals; produces tapered walls in deep cuts; burrless
Hydrodynamic machining	Generally 30–100	Depends on material	NA[a]	Depends on material	0.001 possible	Used for soft nonmetallic slitting; no heat-affected zone; produces narrow kerfs (0.001–0.020 in.); high noise levels
Ultrasonic machining (impact grinding)	16–63; as low as 6 to 10 with 9 μm abrasive	Slow, 0.05 typical	200	0.020–0.150 ipm	0.001–0.0005	Most effective in hard materials, $R_C > 40$; tool wear and taper limit hole depth to width at 2.5 to 1; tool also wears

[a]NA, not applicable for this process.

939

TABLE 32-4. Summary of Thermal NTM Processes

Process	Surface Finish AA (μin.)	Typical Metal Removal Rate (in^3/min)	Typical Specific Horsepower (HP/in^3·min)	Typical Penetration Rate (ipm) or Cutting Speed (sfpm)	Typical Accuracy (in.)	Comments
Electron beam machining (EBM)	32–250	0.005 maximum; extremely low	10,000	200 sfpm; 6 ipm	0.001–0.0002	Micromachining of thin materials and hole-drilling minute holes with 100:1 depth/diameter ratios; work must be placed in vacuum but suitable for automatic control; beam can be used for processing and inspection; used widely in microelectronics
Laser beam machining (LBM)	32–250	0.0003; extremely low	60,000	0.1–4 ipm hole drilling	0.005–0.0005	Can drill 0.005- to 0.050-in.-diameter holes in materials 0.100 in. thick in seconds; same equipment can weld, surface heat-treat, engrave, trim, blank, etc.; has heat-affected zone and recast layers which may need to be removed
Electrical discharge machining (EDM)	32–105	0.3	40	0.5 ipm	0.002–0.00015	Oldest of NTM processes; widely used and automated; tools and dies expensive; cuts any conductive material regardless of hardness; delicate, burr-free parts possible; always forms recast layer
Electrical discharge wire cutting	32–64	0.10–0.3	40	4–10 ipm using molybdenum wire	≃0.0002–±0.0001	Special form of EDM using traveling wire; cuts straight narrow kerfs in metals 0.001 to 3 in. thick; wire diameters of 0.002 to 0.010 in. used; NC machines allow for complex shapes
Plasma arc machining (PBM)	25–500	10	20	50 sfpm; 10 ipm; 120 ipm in steel	± 0.02 to ±0.125	Clean rapid cuts and profiles in almost all plates up to 8 in. thick with 5 to 10° taper; other names; plasma beam cutting

less subsurface damage than conventional processing. These processes are usually employed when conventional machining or grinding cannot be used, often because the materials are too hard. There are numerous hybrid forms of all these processes, generally developed for special applications. Only the main ones are described here because of space limitations.

■ 32.2 CHEMICAL NTM PROCESSES

Chemical Machining

Basically, *chemical machining* (CHM) is the simplest and oldest of the chipless machining processes. It has been used for many years in the production of engraved plates for printing and in making small name plates. However, in its use as a machining process it is applied to parts ranging from very small electronic circuits, such as that shown in Figure 32-1, to very large parts up to 15 m (50 ft) long.

In chemical machining, material is removed from selected areas of a workpiece by immersing it in a chemical reagent. Material is removed by microscopic electrochemical cell action, as occurs in corrosion or chemical dissolution of a metal; no external electrical circuit is involved. This controlled chemical dissolution will simultaneously etch all exposed surfaces, which enhances the productivity even though the penetration rates of the etch may be only 0.0005 to 0.0030 in./min. The basic process takes many forms: *chemical milling*, for pockets, contours, and overall metal removal, so named because in its earliest use it replaced milling; *chemical blanking*, for etching through thin sheets; *photochemical machining*, for etching photosensitive resists in microelectronics; *gel milling*, using reagent in gel form; *chemical or electrochemical polishing*, where weak chemical reagents are used, sometimes with remote electric assist, for polishing or deburring; *chemical jet machining*, using a single chemically active jet.

Chemical Milling or Blanking. The material-removal processing steps in chemical milling and chemical blanking are:

1. *Prepare*. Degrease, clean, rinse, pickle, or preclean the surface to provide for good adhesion for masking material.
2. *Mask*. Coat or protect areas not to be etched.
3. *Etch*. Chemically dissolve by spray or dip followed by rinse.
4. *Remove mask*. Strip or demask, clean and desmut as necessary.
5. *Finish*. Posttreatments and finish inspection.

After cleaning and masking, the part is immersed or sprayed with the proper etchant and is permitted to remain in the reagent. With the spray technique, this is not required; therefore, this procedure usually is preferred when the size and shape of the workpiece permit. The major complexity in the process involves providing a maskant on the surface of the workpiece so that etching will occur only on desired areas. Usually, the entire surface is covered with maskant, portions of which are manually removed from those areas to be etched. A variation of CHM is *photochemical machining*, wherein photographic techniques are used to apply the maskant. Figure 32-2 shows the steps that are involved when chemical machining is done through the use of photosensitive resists. These are as follows:

1. Clean the workpiece.
2. Coat the workpiece with a light-sensitive emulsion, usually by dipping or spraying. The emulsion, or resist, is then dried, usually in an oven.

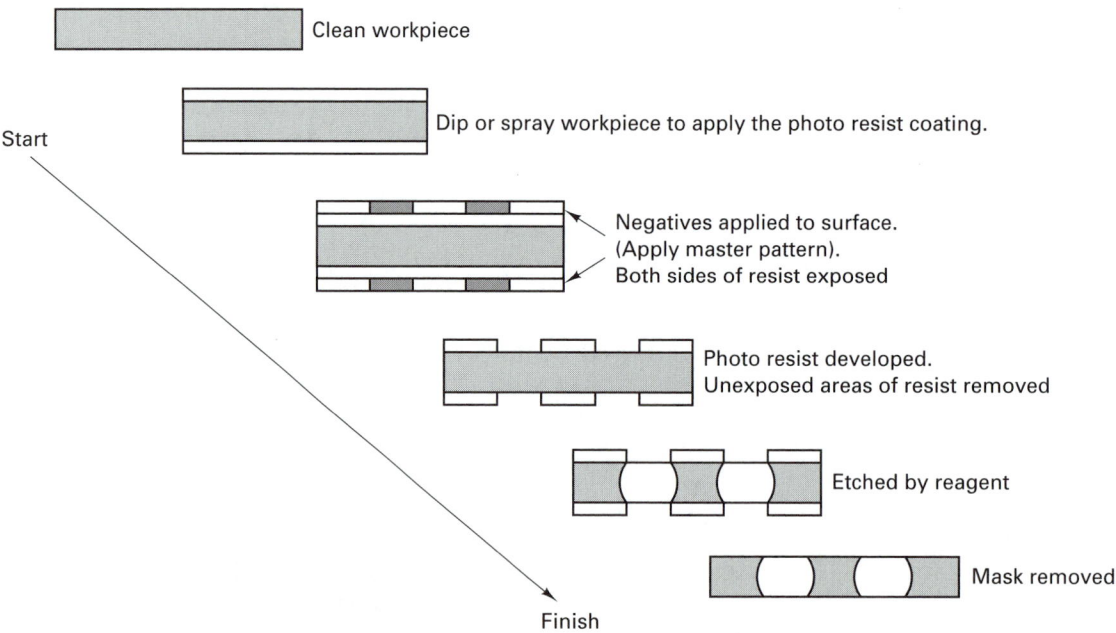

FIGURE 32-2 Basic steps in photochemical machining (PCM).

3. Prepare the "artwork." An accurate drawing of the workpiece is made, usually on polyester drafting film or glass and up to 50 or more times the size of the final part. With such magnifications, an accuracy of 0.025 in. in the original drawing will permit 0.0005 in. to be achieved in the workpiece. By special procedures, lines of 2 μm can be made.

4. Reduce the original drawing by photographic means to obtain a negative that is the master pattern, exactly the size of the finished part. This reduction may require several steps, using industrial photographic equipment.

5. Apply the master pattern to the workpiece using a vacuum frame to ensure good contact. Precise registry of duplicate negatives on each side of the workpiece is essential for accurately blanked parts. Expose to blue light, passing the light through the negative. Mercury-vapor lamps commonly are employed as the light source. Exposure to the light hardens the selected areas of the resist so that it will not be washed away in the subsequent developing.

6. Remove the negatives and develop the workpiece. This removes, or dissolves away, the unexposed areas of the resist, thereby exposing the areas of the workpiece that are to be acted on by the chemical reagent. The final developing step is to rinse away all residual material.

7. Spray the workpiece with (or immerse it in) the reagent.

8. Remove the remaining maskant.

Chemical machining with the aid of photosensitive resists has been widely used for the production of small, complex parts such as electronic circuit boards, and very thin parts that are too small or too thin to be blanked by ordinary blanking dies (Figure 32-3). Chemical machining is often followed by plating, sputtering, and vacuum deposition (see Chapter 40) to deposit metallic films of controlled thickness. The process has been the basic technology of the microelectronics industry.

FIGURE 32-3 Typical parts produced by chemical machining or blanking. *(Courtesy of Chemcut Corporation.)*

Use of Scribed Maskants. Although photosensitive resists are used in the majority of chemical machining operations, there are some cases in which scribed-and-peeled maskants are employed: (1) when the workpiece is not flat, (2) when it is very large, and (3) for low-volume work when the several steps required in using photosensitive resists are not economically justified. In this procedure the maskant is applied to the entire surface of the workpiece, usually by dipping or spraying. It is then removed from those areas where metal removal is desired by scribing through the maskant with a knife and peeling away the desired portions. When volume permits, scribing templates can be used.

Chemical Machining to Multiple Depths. If all areas are to be machined to the same depth, only a single masking, or resist application sequence, and immersing are required. Machining to two or more depths, called *step machining*, can be accomplished by removing the maskant from additional areas after the original immersion. Figure 32-4 illustrates the steps required for stepped chemical machining.

Parts having either uniformly or variably tapered cross sections can be produced by chemical machining by withdrawing them from the etch bath at controlled rates, in the vertical position. In this way, different areas are exposed to the chemical action for differing amounts of time.

Design Factors in Chemical Machining. When designing parts that are to be made by chemical machining, several unique factors related to the process must be kept in mind. First, dimensional variations can occur through size changes in the artwork due to temperature and humidity changes. These usually can be eliminated or controlled by drawing the artwork on thicker-polyester films or on glass. If very accurate dimensions must be held, the room temperature and humidity should be controlled. The photographic film used in making the master negative also can be affected to some degree by temperature and humidity, but control of handling and processing conditions can eliminate this difficulty.

The second item that must be considered is the *etch factor* or *etch radius*, which describes the undercutting of the maskant. The etchant acts on whatever surface is exposed. Areas that are exposed longer will have more metal removed from them. Consequently, as the depth of etch increases, there is a tendency to undercut or etch under the maskant (Figure 32-5). When the etch depth is only a few hundredths of a millimeter, as is often the case, this causes little or no difficulty. But when the depth is substantial, whether etching

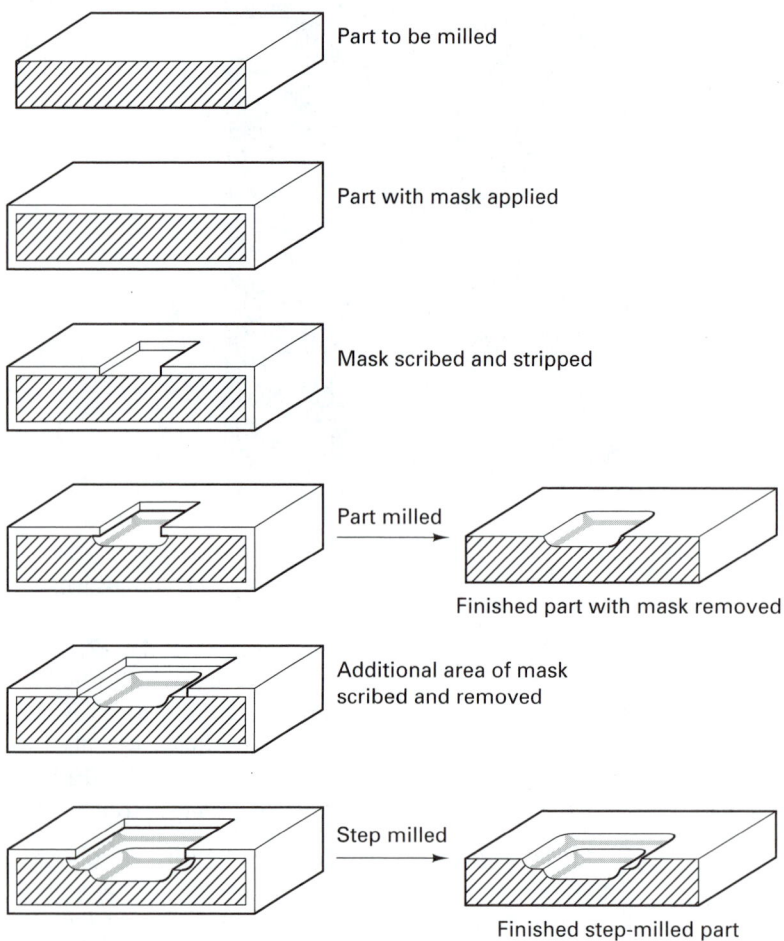

FIGURE 32-4 Steps required to produce a stepped contour by chemical machining.

from only one or both sides, and when doing chemical blanking, the conditions shown in Figure 32-6 result. In making grooves, the width of the opening in the maskant must be reduced by an amount sufficient to compensate for the etch radius. This radius varies from about one-fourth to three-fourths of the *depth* of the etch, depending on the type of material and to some extent on the depth of the etch. Consequently, it is difficult to produce narrow grooves except when the etch depth is quite small.

An allowance for the etch factor must be taken into account in designing the part and the original artwork or scribing template. The values indicated in Figure 32-6 are *minimum* values; it has been found that results will vary between etching machines, and actual etch allowances will have to be somewhat greater and adapted to the specific conditions.

In chemical blanking, with etching occurring from both sides, a sharp edge remains along the line at which breakthrough occurs, as in Figure 32-6c. Because such an edge is usually objectionable, etching ordinarily is continued to produce the straight-wall condition shown in Figure 32-6d. Etching from both sides, of course, requires the preparation of two maskant patterns and careful registration of them on the two sides of the workpiece.

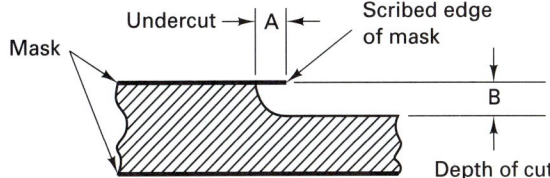

$$\text{Etch factor} = \frac{\text{Depth of cut}}{\text{Undercut}} = \frac{B}{A}$$

FIGURE 32-5 Etch factor in photochemical machining. (*Note:* Inverse of that used in chemical machining.)

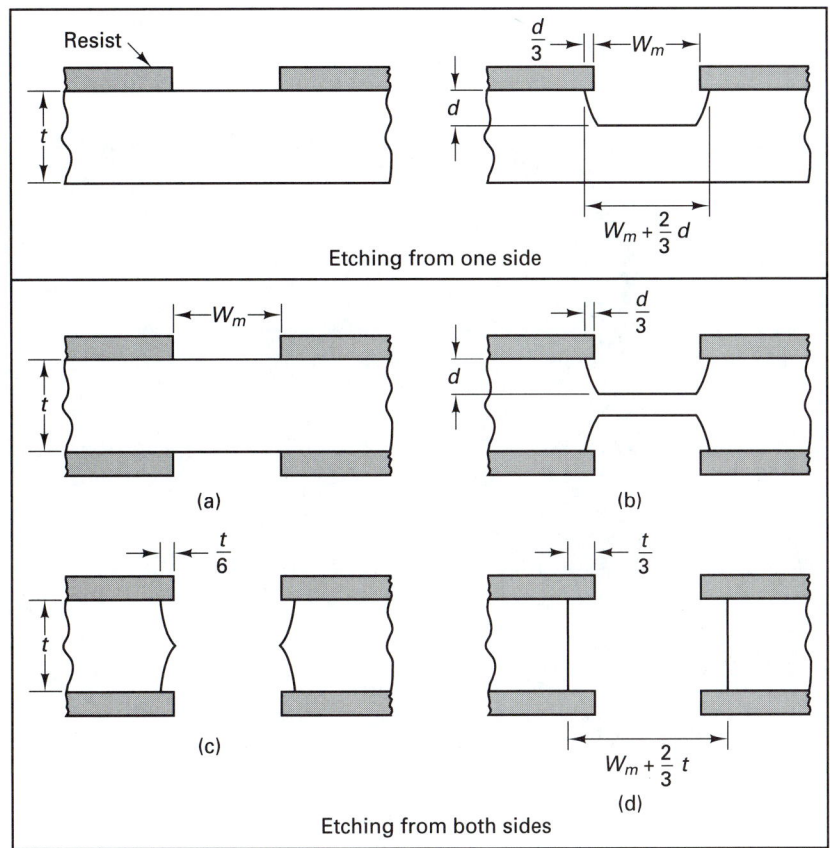

FIGURE 32-6 Effect of the "etch factor" in chemical machining from one side and both sides of a plate.

If the bath is not agitated properly, the "overhang" condition depicted in Figure 32-7 may result, particularly on deep cuts. Not only is the resulting dimension of the opening incorrect, but a very sharp edge may be produced. Other common defects, such as *islands* and *dishing*, and processing problems such as selective etching at the grain boundaries, microcracks, and pitting, caused by selective etching (etching at unequal etch rates) of the workpiece, are to be expected.

Advantages and Disadvantages of Chemical Machining.

Chemical machining has a number of distinct advantages. Except for the preparation of the artwork and master

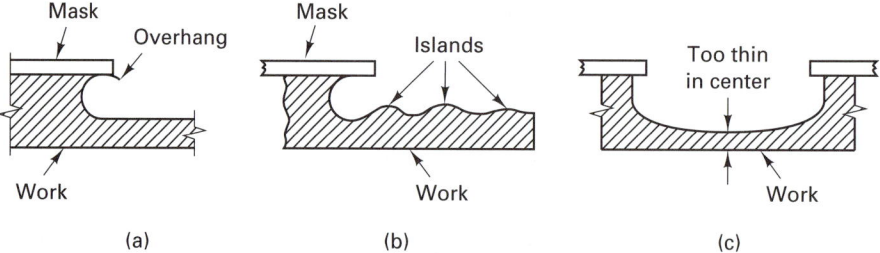

FIGURE 32-7 Typical chemical milling defects: (a) overhang: deep cuts with improper agitation; (b) islands: isolated high spots from dirt, residual maskant or work material inhomegeneity; (c) dishing: thinning in center due to improper agitation or stacking of parts in tank.

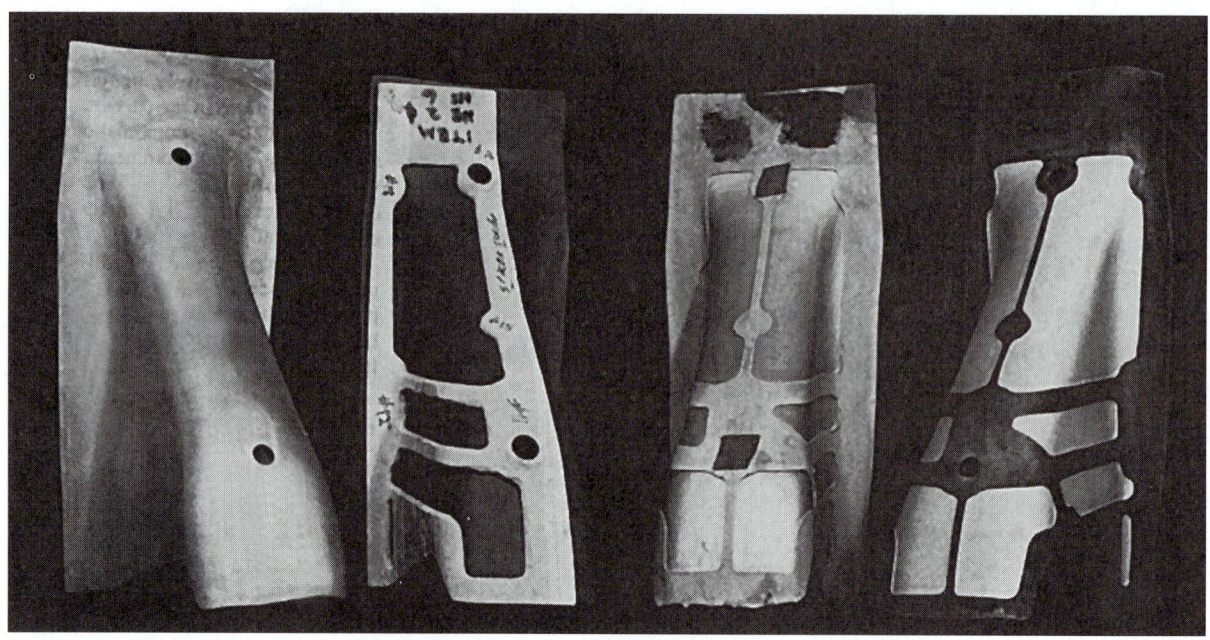

FIGURE 32-8 Iconel 718 aircraft engine parts. These sheet metal parts for a jet fighter engine, are chemically milled to remove weight. (*Left*) As-formed workpiece; (*middle left*) workpiece coated with liquid rubber, fiberglass scribing template in place; (*middle right*) scribed workpiece; (*right*) finished part. About 0.035 in. of stock is removed from the 0.070-in.-thick workpieces. Tolerances are held to ±0.004 in. (*From* Manufacturing Engineering, *Aug. 1983, p. 55.*)

negative, or a scribing template, the process is relatively simple, does not require highly skilled labor, induces no stresses or cold working in the metal, and can be applied to almost any metal—aluminum, magnesium, titanium, and steel being most common. Large areas can be machined; tanks for parts up to 12 by 50 ft are available. Machining can be done on parts of virtually any shape. Thin sections, such as honeycomb, can be machined because there are no mechanical forces involved. Consequently, chemical machining is very useful and economical for weight reduction. Figure 32-8 shows a typical large, thin part that has been chemically milled to reduce the weight of the part.

The tolerances expected with chemical machining range from ±0.0005 in. with care on small etch depth to ±0.004 in. in routine production involving substantial depths. The surface finish is generally good, as shown in Table 32-1.

In using chemical machining, some disadvantages and limitations should be kept in mind. The metal removal rate is slow *in terms of unit area exposed*, being about 0.2 to 0.04 lb/min per square foot exposed in the case of steel. However, because large areas can be exposed all at once, the overall removal rate may compare favorably with other metal-removal processes, particularly when the workpiece metal is thin and unable to sustain large cutting forces.

The soundness and homogeneity of the metal are very important. Wrought materials should be uniformly heat treated and stress relieved prior to processing. Although chemical machining induces no stresses, it may release existing residual stresses in the metal and thus cause warping. Castings can be chemically machined provided that they are not porous and have uniform grain size. Lack of the latter can cause difficulty. Because of the different grain structures that exist near welds, weldments usually are not suitable for chemical machining.

Chemical milling and conventional machining often can be combined advantageously for producing parts, a fact that often is overlooked. The tolerance in chemical milling increases with the depth of cut and with faster etch rates and varies for different metals (Figure 32-9).

Thermochemical Machining

Thermochemical machining (TCM) has been developed for the removal of burrs and fins by exposing the workpiece to hot corrosive gases for a shorter period of time (Figure 32-10). The workpiece remains unaffected and relatively cool because of its low surface-to-mass ratio and the short exposure time. Alternatively, fine burrs can be removed quickly by exposing the parts to a suitable chemical spray and at much less cost than if it were done by hand. Of course, smaller amounts of metal are removed from all exposed surfaces, and this must be permissible if the process is to be used. Consequently, the procedure usually can be used only for removing small burrs.

The hot gases are formed by detonating explosive mixtures of oxygen, hydrogen, and natural gas in a chamber with the parts. A thermal shock wave vaporizes the burrs found on gears, die castings, valves, and so on, in a few milliseconds. The process has been automated and cycle times of 15 to 50 seconds are typical.

TCM will remove burrs or fins from a wide range of materials, but it is particularly effective with materials of low thermal conductivity. It will deburr thermosetting plastics, but not thermoplastic materials. Any workpiece of modest size requiring manual deburring or flash removal should be considered a candidate for thermal deburring. Die castings, gears, valves, rifle bolts, and similar small parts are deburred readily, including blind, internal, and intersecting holes in inaccessible locations. Carburetor parts are processed in automated equipment. Maximum burr thickness should be about $\frac{1}{15}$ of the thinnest feature on the workpiece. Uniformity of results and greater quality assurance over hand deburring is a special advantage of TCM. See Chapter 40 for further discussion of deburring.

■ 32.3 ELECTROCHEMICAL NTM PROCESSES

Electrochemical Machining

Electrochemical machining, commonly designated ECM, removes material by anodic dissolution with a rapidly flowing electrolyte. It is basically a *deplating* process in which the

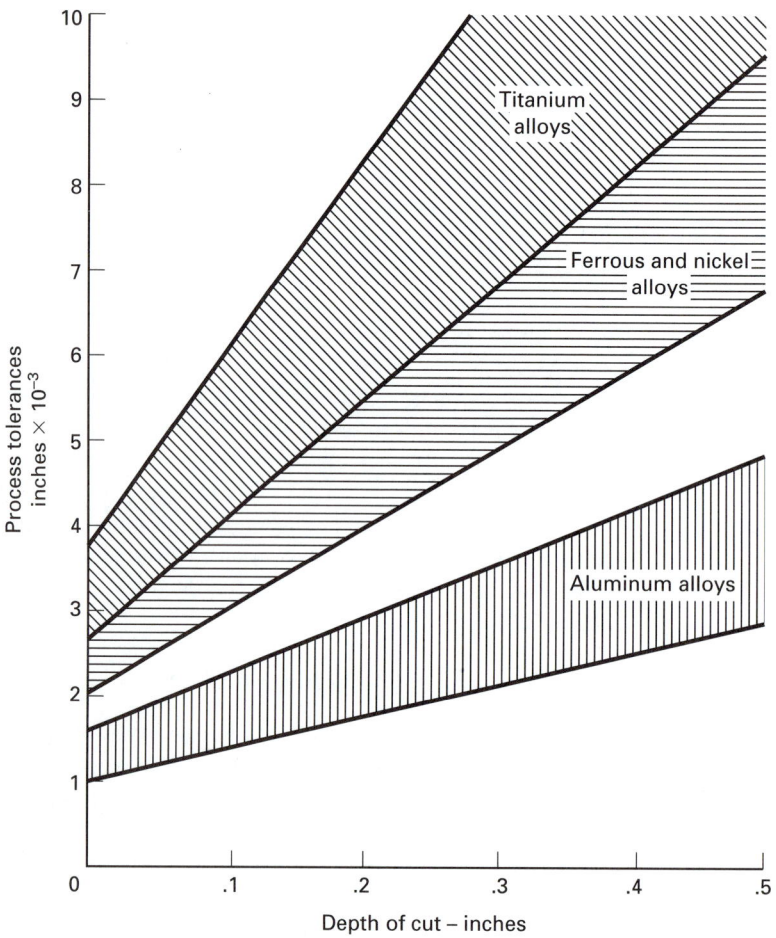

FIGURE 32-9 Chemical milling tolerance bands showing variation with respect to depth of cut and different metals. *(Adapted from* Chemical Machining: Production with Chemistry, *MDC 82-1-2, Metcut Research Associates, Inc., Machinability Data Center, Cincinnati, Ohio, 1982.)*

tool is a cathode and the workpiece is the anode; both must be electrically conductive. The electrolyte, which can be pumped rapidly through or around the tool, sweeps away the waste product (sludge) and captures it by settling in filters. The shape of the cavity is the mirror image of the tool, which is advanced by means of a servomechanism that controls the gap (0.003 to 0.030 in., with 0.010 in. typical) between the electrodes. The tool advances into the work at a constant feed rate that matches the dissolution and deplating rates of the workpiece. The electrolytes are highly conductive solutions of inorganic salts, usually NaCl, KCl, and $NaNO_3$ (or other proprietary mixtures) and are operated at about 90 to 125° with flow rates ranging from 50 to 200 ft/sec (fps). Tools are usually made of copper or brass and sometimes stainless steel. The process is shown schematically in Figure 32-11.

As shown in Figure 32-12, the metal-removal rate, in terms of penetration of the tool into the workpiece, is primarily a function of the current density. Current densities from 1500 to 2000 A/in^2 are used and, in suitable applications, ECM provides metal-removal rates on the order of 0.1 in^3/min per 1000 A. The cutting rate is solely a function of the

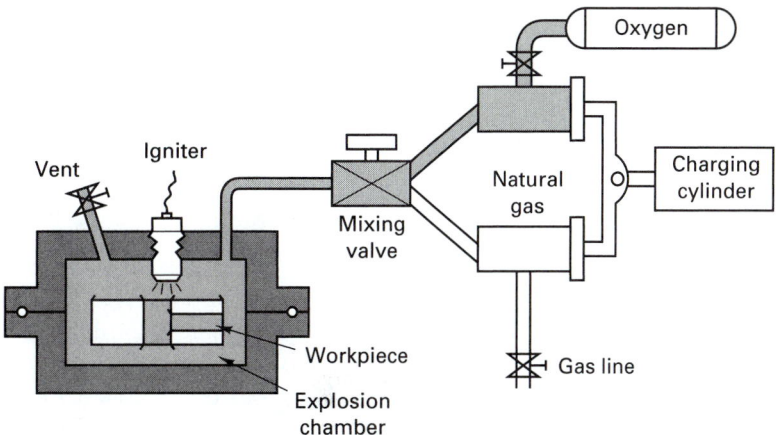

FIGURE 32-10 Thermochemical machining process for the removal of burrs and fins. *(From* Machining Data Handbook, *Vol. 2.)*

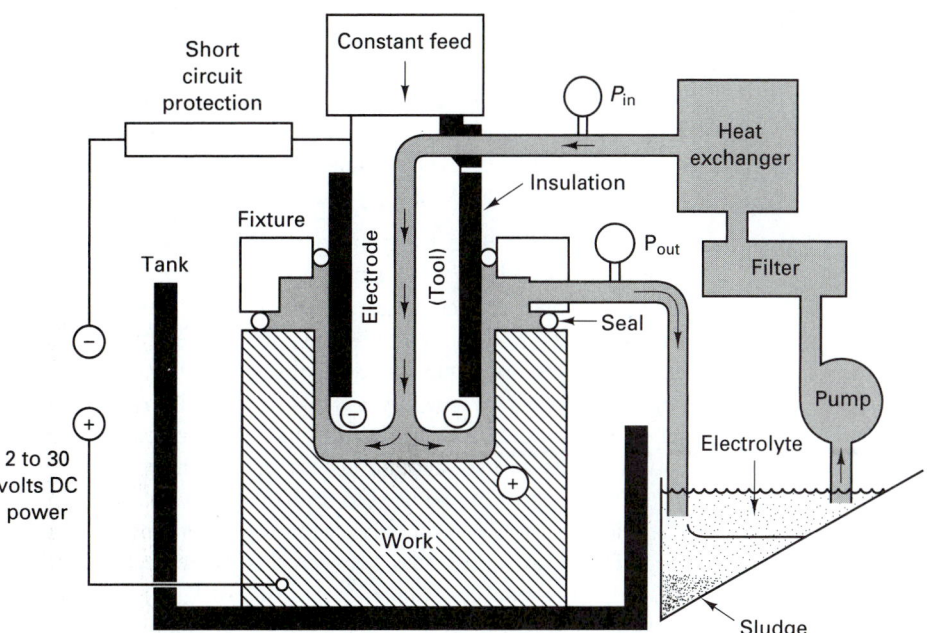

FIGURE 32-11 Schematic diagram of electrochemical machining process (ECM).

ion-exchange rate and is not affected by the hardness or toughness of the work material. Cutting rates up to 0.1 in. of depth per minute are obtained routinely in Waspalloy, a very hard metal alloy.

ECM is well suited for mass production of complex shapes in difficult-to-machine but conductive materials. The principal tooling cost is for the preparation of the tool electrode, which can be time consuming and costly, requiring several cut-and-try efforts, except for simple shapes. There is no wear of the tool during actual cutting as the tool is protected cathodically. The process produces a stress-free surface. The ability to cut the

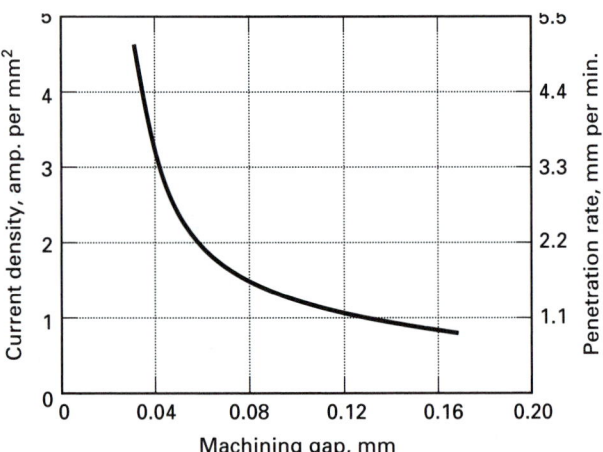

FIGURE 32-12 Relationship of current density, penetration rate, and machining gap in electrochemical machining.

entire cavity simultaneously is very productive. Process control must be exact to obtain tight tolerances, and the tools must be designed to compensate for the variable current densities produced by electrode geometries or electrolyte variations. For example, corners in cavities are automatically rounded because of the concentration of the current density at the edge of the tool.

Electrochemical polishing is a modification of the ECM process that operates essentially the same as ECM, but the feed is halted. Lower current densities and slower electrolyte flow rates greatly reduce the metal removal rates so that the surface develops a fine finish, 10 to 12 μin. AA being typical.

Electrochemical Hole Machining. *Electrochemical hole-drilling* processes have been developed for drilling very small holes using high voltages and acid electrolytes. The process is sometimes called *electrostream*. The tool is a drawn glass nozzle with an internal electrode. Multiple sets of glass tubes are employed and over 50 holes per stroke can be done. This technique was developed to drill the cooling holes in turbine blades for jet engines. Stress-free holes from 0.004 to 0.030 in. in diameter with 50:1 depth-to-diameter ratios are routinely accomplished in nickel and cobalt alloys. Acid is used so that the dissolved metals go into solution instead of forming a sludge.

The process can be used to drill shaped holes in difficult-to-machine, conductive metals. Holes up to 24 in. in depth with diameters ranging from 0.020 to 0.050 in. are possible. The major differences between this process and the hole-drilling process described above are the reduced voltage levels (5 to 10 V dc) and special electrodes, which are long, straight, acid-resistant tubes coated with an enamel insulation. The acid is pressure-fed through the tube and returns through the gap (0.001 to 0.002 in.) between the insulated tube wall and the hole wall.

Electrochemical Grinding. *Electrochemical grinding*, commonly designated ECG, is a variant of electrochemical machining. In ECG, the tool electrode is a rotating, metal-bonded, diamond grit grinding wheel. The setup shown in Figure 32-13 for grinding a cutting tool is typical. As the electric current flows between the workpiece and the wheel, through the electrolyte, the surface metal is changed to a metal oxide, which is ground away by the abrasives. As the oxide film is removed, new surface metal is oxidized and removed. ECG is a low-voltage high-current electrical process. The MRR is dependent on many variables. Table 32-5 gives typical values. The metal bond of the wheel is the cathode.

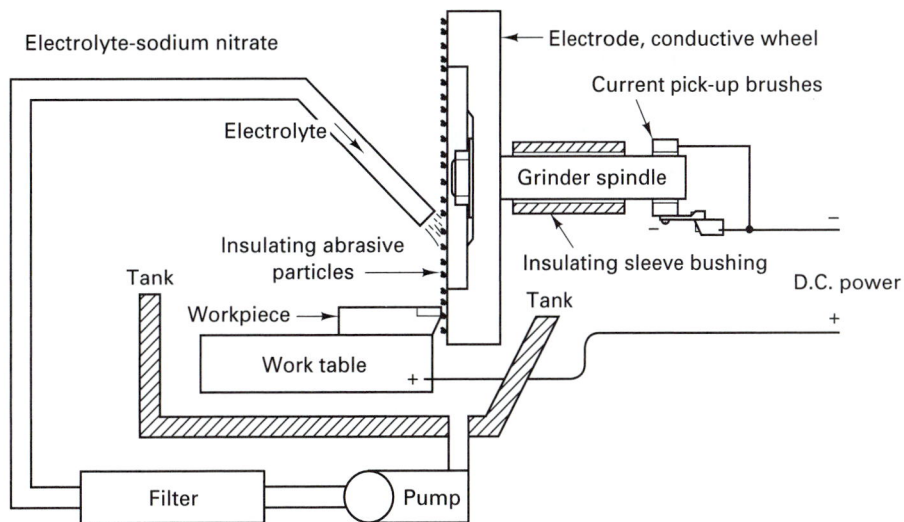

FIGURE 32-13 Equipment setup and electrical circuit for electrochemical grinding.

The wheels used in ECG must be electrically conductive abrasive wheels. For most metals, resin-bond aluminum oxide wheels are recommended. The resin bond is loaded with copper to provide for electrical conductivity. Electrical resistance must be negligible. The wheels are dressable, using a variety of wheel-dressing measures, and can then be used for precision form-grinding operations.

The purpose of the abrasive is to increase the efficiency of the ECG process and permit the continuance of the process. The abrasive particles are always nonconductive material such as aluminum oxide, diamond, or borazon (CBN). Thus they act as an insulating spacer maintaining a separation of from 0.0005 to 0.003 in. (0.012 to 0.050 mm) between the electrodes. A dead short would result if the insulating spacer were absent. The particles also serve to wipe away the residue and to cut chips if the wheel should contact the workpiece, particularly in the event of a power failure. With proper operation, less than 5% of the material is removed by normal chip forming. The process is used for shaping and sharpening carbide cutting tools, which cause high wear rates on expensive diamond wheels in normal grinding. Electrochemical grinding greatly reduces this wheel wear. Fragile parts (honeycomb structures), surgical needles, and tips of assembled turbine blades have been ECG-processed successfully. The lack of heat damage, burrs, and residual stresses is very beneficial, particularly when coupled with MRRs that are competitive with conventional grinding but with far less wheel wear.

Electrochemical Deburring. Limited use is made of the ECM principle for removing burrs from parts. The work is put into a rotating, electrically insulated drum that contains two current-carrying electrodes that are insulated from the drum. Small graphite spheres, added to the electrolyte, receive an inductive charge from the electrodes. The potential gradient across the sphere-to-workpiece gap is sufficient to cause electrochemical machining to occur as the spheres move randomly over the workpiece. Because the current density is higher at the protrusions of the burrs than at smooth areas on the workpiece, they are preferentially removed. As in chemical deburring, there is a slight dimensional change throughout the workpiece, in this case due to the general ECM action and to the natural abrasive character of the graphite spheres.

TABLE 32-5. Metal Removal Rates for ECG for Various Metals

Metal	Valency	Density		Metal-Removal Rate at 1000 A		
		lb/in^3	g/cm^3	lb/hr	in^3/min	cm^3/min
Aluminum	3	0.098	2.67	0.74	0.126	2.06
Beryllium	2	0.067	1.85	0.37	0.092	1.50
Chromium	2	0.260	7.19	2.14	0.137	2.25
	3			1.43	0.092	1.51
	6			0.71	0.046	0.75
Cobalt	2	0.322	8.85	2.42	0.125	2.05
	3			1.62	0.084	1.38
Niobium	3	0.310	8.57	2.55	0.132	2.16
(columbium)	4			1.92	0.103	1.69
	5			1.53	0.082	1.34
Copper	1	0.324	8.96	5.22	0.268	4.39
	2			2.61	0.134	2.20
Iron	2	0.284	7.86	2.30	0.135	2.21
	3			1.53	0.090	1.47
Magnesium	2	0.063	1.74	1.00	0.265	4.34
Manganese	2	0.270	7.43	2.26	0.139	2.28
	4			1.13	0.070	1.15
	7			0.65	0.040	0.66
Molybdenum	3	0.369	10.22	2.63	0.119	1.95
	4			1.97	0.090	1.47
	6			1.32	0.060	0.98
Nickel	2	0.322	8.90	2.41	0.129	2.11
	3			1.61	0.083	1.36
Silicon	4	0.084	2.33	0.58	0.114	1.87
Silver	1	0.379	10.49	8.87	0.390	6.39
Tin	2	0.264	7.30	4.88	0.308	5.05
	4			2.44	0.154	2.52
Titanium	3	0.163	4.51	1.31	0.134	2.19
	4			0.99	0.101	1.65
Tungsten	6	0.697	19.3	2.52	0.060	0.98
	8			1.89	0.045	0.74
Uranium	4	0.689	19.1	4.90	0.117	1.92
	6			3.27	0.078	1.29
Vanadium	3	0.220	6.1	1.40	0.106	1.74
	5			0.84	0.064	1.05
Zinc	2	0.258	7.13	2.69	0.174	2.85

Source: 1985 SCTE Conference Proceedings, ASM, Metals Park, Ohio, 1986.

■ 32.4 MECHANICAL NTM PROCESSES

Ultrasonic Machining

Ultrasonic machining, sometimes called *impact grinding*, employs an ultrasonically vibrating tool to impel the abrasive in a slurry against the workpiece. The tool forms a reverse image in the workpiece as the abrasive-loaded slurry abrades (machines) the material. Boron carbide, aluminum oxide, and silicon carbide are the most commonly used grit materials. The process can cut virtually any material but is most effective on materials with hardness greater than $R_C 40$. Figure 32-14 shows a simple schematic of this process.

Ultrasonic machining uses a transducer to impart high-frequency vibrations to the toolholder. Abrasive particles in the slurry are accelerated to great speed by the vibrating

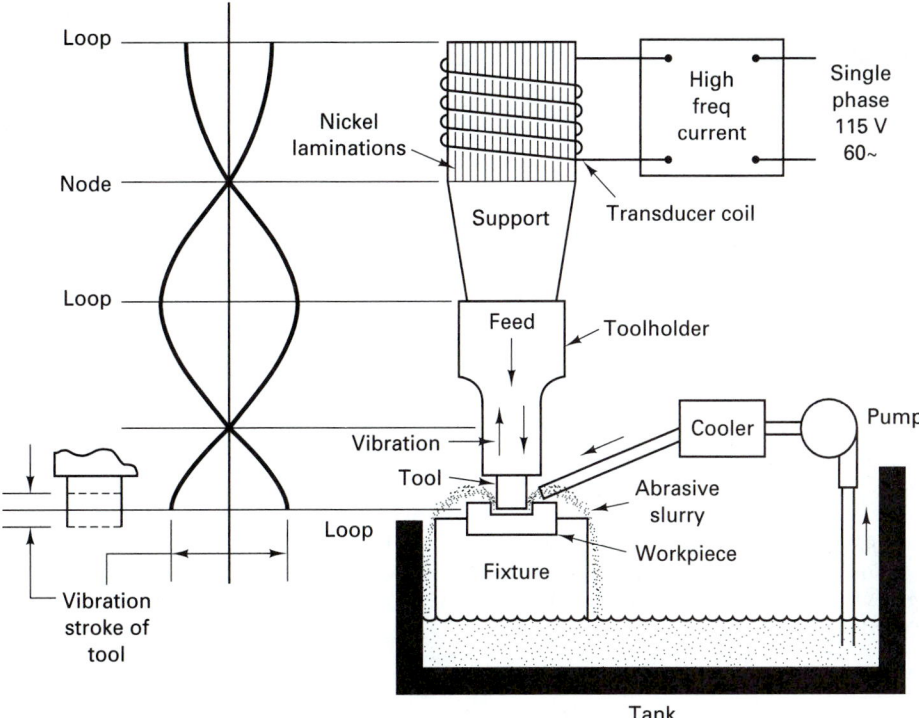

FIGURE 32-14 Sinking a hole in a workpiece by ultrasonic machining.

tool and perform the actual cutting. The tool materials are usually brass, carbide, mild steel, or tool steel and will vary in tool wear depending on their hardness. Wear ratios of 1:1 or 100:1 (material removed versus tool lost to wear) are possible. The tool must be strong enough to resist fatigue failure.

The cut will be oversize by about twice the size of the abrasive grit being used, and holes will be tapered, usually limiting the hole depth to a diameter ratio of about 3:1. Surface roughness is controlled by the size of the grit (finer finish with smaller grits). Holes, slots, or shaped cavities can be readily eroded in any material, conductive or non-conductive, metallic, ceramic, or composite.

A variation of ultrasonic machining, *rotary ultrasonic machining*, uses a rotating diamond tool vibrating at 20,000 Hz during rotation, to drill, thread, grind, or mill hard, brittle materials. It does not use an abrasive slurry.

Ultrasonics have also been employed for coining, lapping, deburring, and broaching. Plastics can be welded using ultrasonics (see Chapter 36). Three other mechanical NTM processes are listed in Table 32-3.

Waterjet Machining

Waterjet machining uses a high-velocity fluid jet (Mach 2) impinging the workpiece, performing a slitting operation. The process is also called *hydrodynamic machining* (Figure 32-15). A long-chain polymer may be added to the water of the jet to make the jet coherent (i.e., not to come out of the nozzle as a mist) under the 10,000- to 60,000-psi nozzle pressure. Alternative fluids (alcohol, glycerine, cooking oils) have been used in

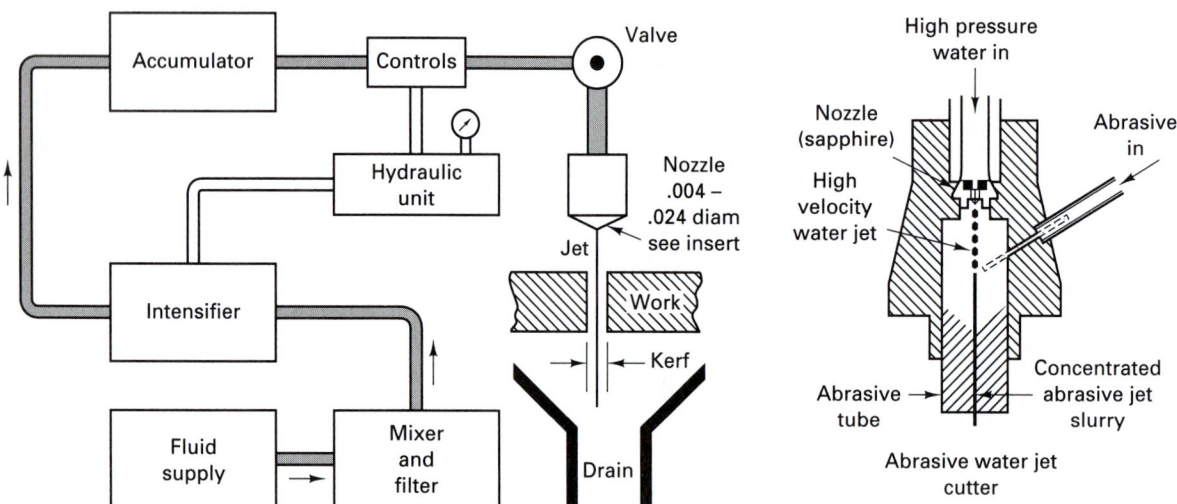

FIGURE 32-15   Schematic of hydrodynamic jet machining. The intensifier elevates the fluid to the desired nozzle pressure while the accumulator smooths out the pulses in the fluid jet. Schematic of an abrasive waterjet machining nozzle is shown on the right.

processing meats, baked goods, and frozen foods. The jet is typically 0.003 to 0.020 in. in diameter and exits the orifice at velocities up to 3000 ft/sec. This process has historically been used to cut soft nonmetalics such as acoustic tile, plastics, paperboard, asbestos (with no dust), leather, rubber, and fiberglass.

Abrasive Waterjet Machining

By adding a fine abrasive to the waterjet, a full range of ferrous and nonferrous metals have been machined. The majority of the metal-working applications require the addition of an abrasive to the waterjet stream. The abrasives are added after the stream has left the orifice (Figure 32-15b). Generally, the kerf of the cut is about 0.001 in. greater than the diameter of the jet and is not sensitive to dwell. Cutting rates vary from 20 in./min for acoustic tile, 50 in./min for epoxies, and 500 in./min for paper products (see Table 32-6).

Abrasive jet cutting adds these additional parameters to waterjet machining: abrasive material (density, hardness, shape), abrasive size or grit, flow rate (pounds per minute), abrasive feed method (pressurized or suction), types of nozzle and mixing chamber, abrasive orifice diameter size, and abrasive orifice or accelerator tube material.

Typical abrasive jet cutting systems operate under the following conditions: water pressure of 30,000 to 50,000 psi; water orifice sizes of 0.010 to 0.022 in. diameter; standoff distances 0.020 to 0.060 in.; abrasive grit sizes of 60, 80, 100, and 120; abrasive material of garnet, silica, or aluminum oxide. Flow rate of abrasive can be from $\frac{1}{2}$ to 3 lb/min.

The inside diameters of the abrasive nozzles or accelerator tubes are from 0.040 to 0.125 in. in diameter. These tubes are normally made of carbide. Abrasive waterjet cutting can be used to cut any material by using the appropriate abrasive, water-jet pressure, and feed rate. These processes are particularly good on composites, as the cutting rates are reasonable and they do not delaminate the layered material.

TABLE 32-6.   Typical Cutting Rates and Depths of Cut for Abrasive Waterjet Machining

Material	Depth of Cut (in.)	Cut Rate (in./min)
Aluminum	2.00	2.0–6.0
Chrome/vanadium 58R_C	0.69	2.0–3.5
Glass	0.75	24.0–40.0
Glass mirror	0.25	42.0–100.0
Graphite/epoxy	0.38	24.0–50.0
	0.44	20.0–45.0
	1.31	15.0–20.0
	2.00	5.0–7.0
Graphite/epoxy (Al honeycomb)	0.75	24.0–35.0
Hastelloy	0.06	19.0–26.0
HY-80	0.38	4.0–5.0
Inconel	1.50	1.0–1.5
	1.75 dia.	1.0–1.5
Kevlar	0.88	24.0–36.0
Stainless steel		
316	3.00 dia.	0.5–2.0
15-5 PH	2.50	0.5–1.0
15-5 PH	0.13	9.0–15.0
Titanium	1.00	1.5–2.5
Zircalloy	0.44	0.44–6.5

Process Parameters

Water pressure	40,000–50,000 psi
Water orifice	0.014 in. diameter
Abrasive flow	1.8 lb/min, 60-grit garnet
Water flow	approx. 1 gal/min
Standoff from part	0.030/0.050 in.

Source: 1985 SCTE Proceedings in Nontraditional Machining, ASM, Metals Park, Ohio, 1987.

Abrasive Flow Machining

Abrasive flow machining (AFM) uses a slurry of abrasive material flowing over or through a part, under pressure, to perform edge finishing, deburring, radiusing, polishing, or minor surface machining.   Aluminum oxide, silicon carbide, boron carbide, and diamonds are used as abrasives in grit sizes ranging from No. 8 to No. 700.   This process is useful for polishing and deburring inaccessible internal passageways and works for most materials.

The AFM process should be considered when workpieces have needed laborious hand finishing or have complex shapes or internal edges that are difficult to reach, or when bulk methods (tumbling, vibrating) are not satisfactory.   The process is effective in removing the recast layers produced by the EDM or laser beam machining discussed next.

■ 32.5 Thermal Processes

Electrical Discharge Machining

Electrical discharge machining (EDM) cuts metal by discharging electric current stored in a capacitor bank across a thin gap between the tool (cathode) and the workpiece (anode). Literally thousands of sparks per second are generated and each spark produces a tiny crater by melting and vaporization, thus eroding the shape of the tool into the workpiece.   The dielectric fluid (kerosene) flushes out the "chips" and confines the spark.   Of all the exotic

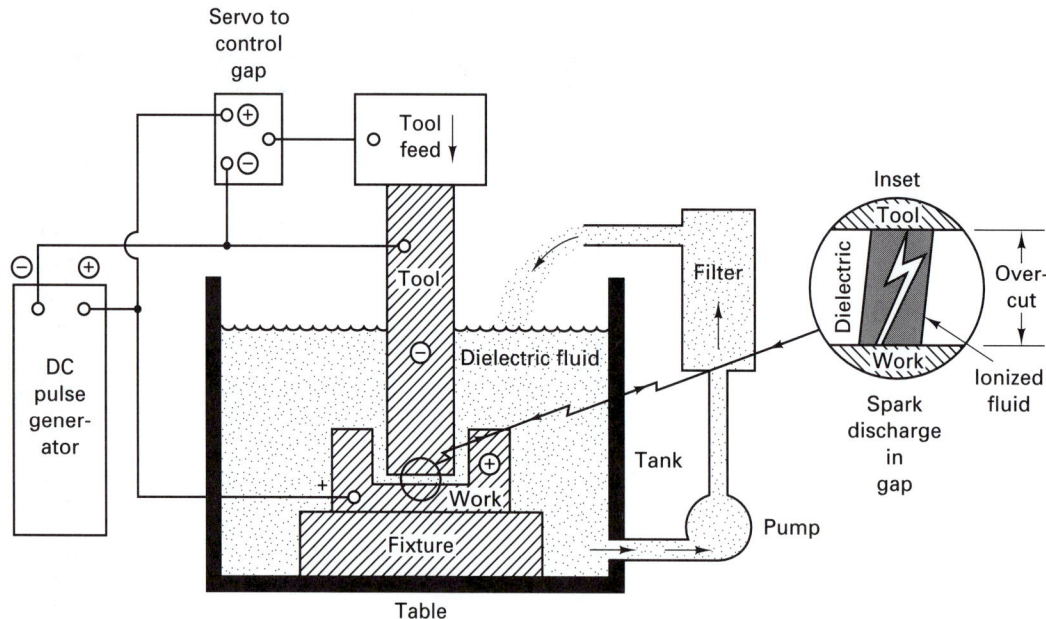

FIGURE 32-16 EDM or spark erosion machining of metal, using high-frequency spark discharges in a dielectric, between the shaped tool (cathode) and the work (anode). The table can make *X–Y* movements.

metalworking processes, none has gained greater industry-wide acceptance than EDM. Figure 32-16 shows a schematic of the process.

In EDM, each spark contains a discrete, measured, and controllable amount of energy; therefore, MRR and surface finish can be predicted while size is carefully controlled. The heat generated by the spark melts the metal, and the impact of the spark causes the metal to be ejected, possibly vaporized, and recast in the dielectric as spheres. Materials of any hardness can be cut as long as the material can conduct electricity. Any shape that can be cut into the tool can be reproduced in the workpiece. Spark reversals and unflushed chips cause sparking on the tool, producing unwanted tool wear and taper; therefore, for production runs tools are usually made in duplicate sets by numerical control machining. About 80 to 90% of EDM work is in the manufacture of tool and die sets for production casting, forging, stamping, and extrusions.

The absence of almost all mechanical forces makes it possible to machine fragile parts without distortion. In addition, fragile tools, even wires, can be used. The controllability, versatility, and accuracy of this method usually result in superior design flexibility. Significant cost reductions in the manufacture of tools and dies from steels of any hardness and carbides can be achieved.

EDM is a slow process compared to conventional methods and produces a matte surface finish composed of many small craters. Figure 32-17 shows a scanning electron micrograph of an EDM surface on top of a ground surface. Note the small sphere in the lower right corner attached to the surface. In EDM, surface finish varies widely as a function of the spark frequency, voltage, and current, parameters which, of course, also control the MRR.

The size of the cavity cut by the EDM tool will be larger than the tool. That is, the distance between the surface of the electrode and the surface of the workpiece represents

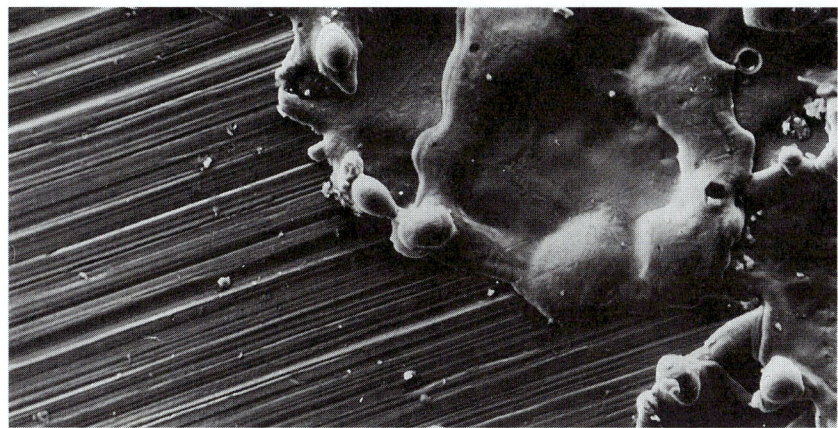

FIGURE 32-17   EDM surface on top of a ground surface in steel, spherodial nature of debris from the surface in evidence around the craters (750×).

the *overcut* and is equal to the length of the spark, which is essentially constant over all areas of the electrode, regardless of size or shape. Typical overcut values range from 0.0005 to 0.020 in. Overcut depends on the gap voltage plus the chip size, which varies with the amperage. EDM equipment manufacturers publish overcut charts for the different power supplies for their machines, but these values should be used mainly as guides for the tool designer. The dimensions of the electrode (the tool) are basically equal to the desired dimensions of the part less the overcut values.

While different materials are used for electrodes, graphite has emerged as the best EDM tool material. The choice depends on the application, considering how easily the electrode material can be machined, how fast it wears (this is called *spark erosion*), how fast it cuts, what kind of quality of finish it can produce, what type of power supply is being used, and how much the material costs. Copper, brass, copper–tungsten, aluminum, 70/30 zinc–tin, and other alloys are used for electrode material when graphite is not selected.

The dielectric fluid has four main functions: insulation between tool and work, spark conductor, coolant, and flushing medium. This fluid must ionize to provide a channel for the spark and deionize quickly to become an insulator. Polar compounds, such as glycerine water (90 : 10) with triethylene oil as an additive, have been shown to improve the MRR and decrease the tool wear when compared with traditional cutting fluids, such as kerosene.

Wire EDM is a special form of EDM, shown in Figure 32-18, wherein the electrode is a continuously moving conductive wire. The tensioned wire of copper, brass, or tungsten is used only once, traveling from a take-off spool to a take-up spool while being "guided" to produce a straight narrow kerf in plates up to 3 in. thick. The wire diameter ranges from 0.002 to 0.010 in. with positioning accuracy up to ±0.0002 in. in machines with NC or tracer control. The dielectric is usually deionized water. This process is widely used for the manufacture of punches, dies, and stripper plates, with modern machines capable of cutting die relief, intricate openings, tight radius contours, and corners routinely. See Figure 32-19 for examples of *wire EDM* products.

Advantages and Disadvantages of EDM.

Electrical discharge machining is applicable to all materials that are fairly good electrical conductors, including metals, alloys, and most carbides. The melting point, hardness, toughness, or brittleness of the material imposes no

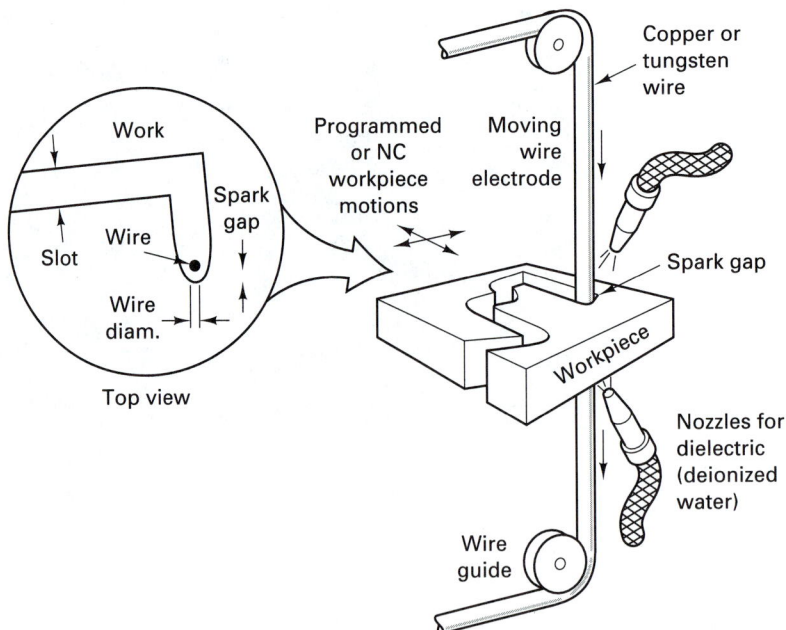

FIGURE 32-18 Schematic diagram of equipment for wire EDM using a moving wire electrode.

limitations. Thus, EDM provides a relatively simple method for making holes of any desired cross section in materials that are too hard or brittle to be machined by most other methods. Because the forces between the tool and the workpiece and virtually zero, very delicate work can be done. The process leaves no burrs on the edges. It is widely used to produce forming dies for sheet metal in the automotive industry.

On most materials, the process produces a thin, hard recast surface, which may be an advantage or a disadvantage, depending on the use. When the workpiece material is one that tends to be brittle at room temperatures, the surface may contain fine cracks caused by the thermally induced stresses. Consequently, some other finishing process often is used subsequent to EDM to remove a thin surface layer, particularly if the product will be used in a fatigue environment.

Electron Beam Machining

As a metals-processing tool, the electron beam is used mainly for welding (see Chapter 36), to some extent for surface hardening, and occasionally for cutting (mainly drilling). *Electron beam machining* (EBM) is a thermal NTM process that uses a beam of high-energy electrons focused on the workpiece to melt and vaporize metal. This micromachining process is performed in a vacuum chamber (10^{-5} mm/Hg). Magnetic lenses are used to focus the beam. Deflection coils control the position of the beam. The desired beam path can be programmed with a computer to produce any desired pattern in the work. The spot size diameters are on the order of 0.0005 to 0.001 in., and holes or narrow slits with depth-to-width ratios of 100:1 can be "machined" with great precision in a short time in any material. The interaction of the beam with the surface produces dangerous x-rays; therefore, shielding is necessary. The layer of recast material and the depth of heat damage is very small. For micromachining, processing speeds can exceed that of EDM or ECM.

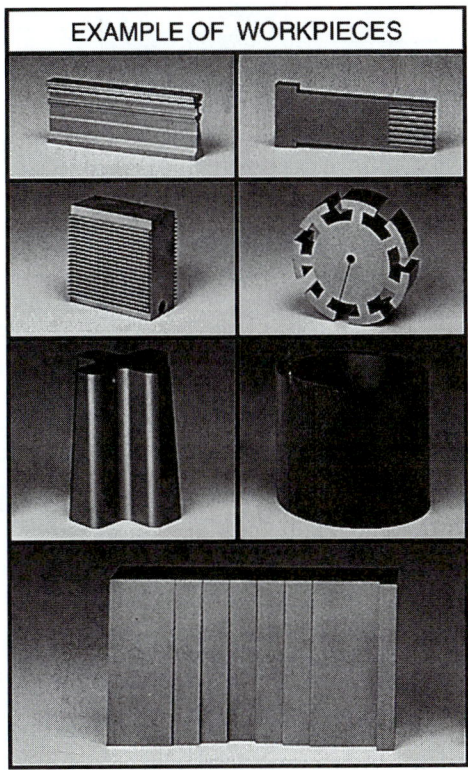

FIGURE 32-19   Examples of wire EDM workpieces made on NC machine (Hatachi).

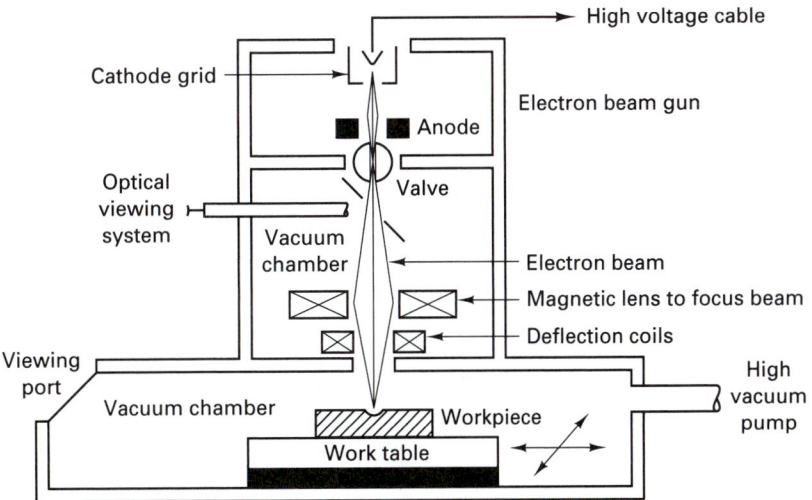

FIGURE 32-20   Electron beam machining uses high-energy electron beam (10^9 W/in^2) to melt and vaporize metal in 0.001-in.-diameter spots.

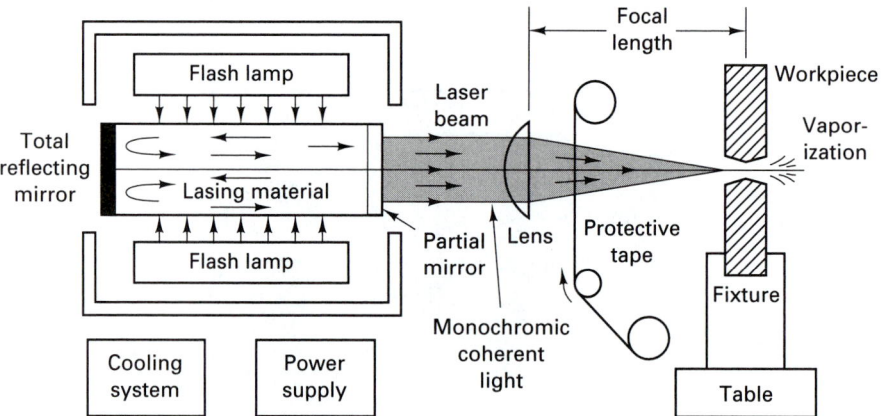

FIGURE 32-21 Schematic diagram of a laser beam machine, a thermal NTM process that can micromachine any material.

Typical tolerances are about 10% of the hole diameter or slot width. These machines require high voltages (50 to 200 kV) to accelerate the electrons to speeds of 0.5 to 0.8 the speed of light and should be operated by fully trained personnel. See Figure 32-20 for a schematic.

Laser Beam Machining

Laser beam machining (LBM) is a thermal NTM process that uses a laser to melt and vaporize materials. This process is also used for joining (welding, crazing, soldering leads in microelectronic circuits) and heat treating. See Chapters 5 and 36 for discussion. As shown in Figure 32-21, the beam can be focused down to 0.005 in. in diameter for drilling microholes through metals as thick as 0.100 in., but hole depth-to-diameter ratios of 10:1 are more typical. High-energy solid-state and gas lasers are needed, with the optical characteristics of the workpiece determining the wavelength of light energy that should be used. The range is 0.4880 μm for argon lasers up to 10.6 μm for carbon dioxide gas lasers. See Table 32-7 for a listing of commercially available lasers for material processing. The small beam divergence, high peak powers (in pulsed lasers), and single frequency provide power densities on the order of 10^5 to 10^{10} W/in^2 allowing holes 0.020 in. in diameter to be drilled in milliseconds with an accuracy of ±0.001 in. However, this is not a mass metal-removal process and is limited to rather thin stock (0.10 to 0.20 in.) as the cutting speed drops off rapidly with thickness. Hole geometry is irregular, and there will be a recast layer and a heat-affected zone which can be detrimental to material properties. High precision systems can hold line cuts to ±0.005 in. positional accuracy and some to ±0.001 in. per foot of travel. Lasers are used to cut, weld, heat-treat, and trim by varying the power density along with appropriate adjustments in output beam intensity, focus, and duration (Figure 32-22).

Off-the-shelf laser systems are now available with NC controls and are being used for applications ranging from cigarette paper cutting to drilling microholes in turbine engine blades. Protective materials are absolutely necessary when working around laser equipment because of the potential damage to eyesight from either direct or scattered laser light.

Plasma Arc Cutting

Plasma arc cutting (PAC) uses a superheated stream of electrically ionized gas to melt and remove material (Figure 32-23). The 20,000 to 50,000°F plasma is created inside a

TABLE 32-7. Commercial Lasers Available for Machining, Welding, and Trimming

Laser Type	Wave Length (μm)	Mode of Operation	Power (W)	Pulses per Second	Length of Time	Application	Comments
Argon	0.4880 0.5145	Repetitively pulsed	20 peak; 0.005 average	60	50 μs	Scribing thin films	Power low
Ruby	0.6943	Normal pulse	2×10^5 peak	Low (5–10)	0.2–7 ms	Large material removal in one pulse, drilling diamonds dies, spot welding	Often uneconomical
Nd–glass	1.06	Normal pulse	2×10^6 peak	Low (0.2)	0.5–10 ms	Large material removal in one pulse	Often uneconomical
Nd–YAG[a]	1.06	Continuous	1000	—	—	Welding	Compact: economical at low powers
	1.06	Repetitively Q-switched	3×10^5 peak; 30 average	1–24,000 300	50–250 ns 50 ns	Resistor trimming, electronic circuit fabrication	Compact and economical
	1.06	Normal pulsed	400	300	0.5 to 7 ms	Spot weld, drill	Very bulky at high powers
CO_2[b]	10.6	Continuous	15,000	—	—	Cutting organic materials, oxygen-assisted metal cutting	
	10.6	Repetitively Q-switched	75,000 peak; 1.5 average	400	50–200 ns	Resistor trimming	Bulky but economical
	10.6	Superpulsed	100 average	100	100 μs and up	Welding, hole production, cutting	Bulky but economical

Source: Modified from J. F. Ready, "Selecting a laser for material working," *Laser Focus*, Mar. 1970, p. 40.

[a]Neodynium–yttrium aluminum garnet.

[b]CO_2 plus He plus N_2 mixture.

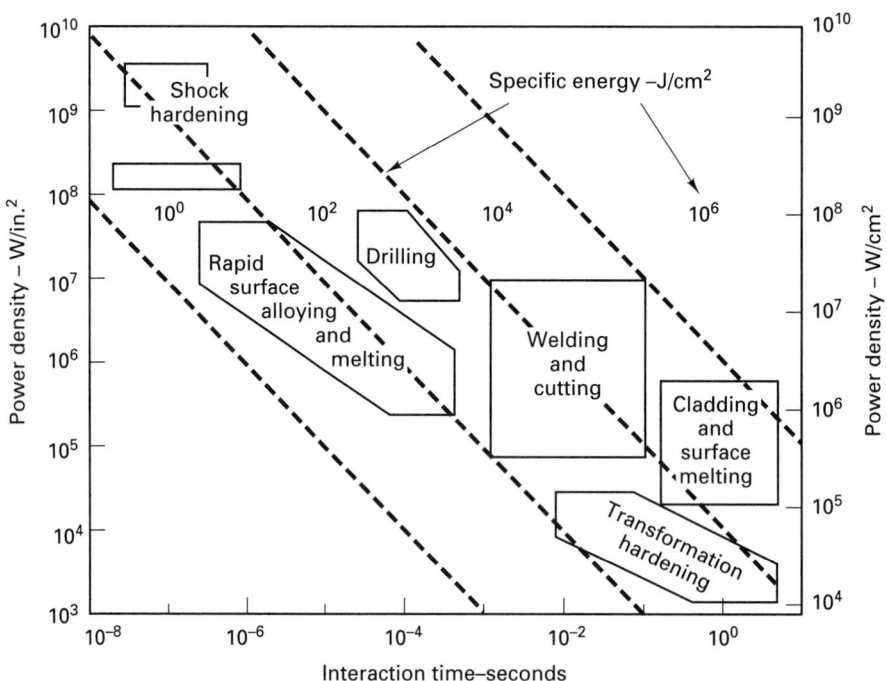

Source: "The role of laser metalworking in factory automation."
Westinghouse R & D center (Pittsburgh).

FIGURE 32-22   The power density input and the percent of surface heat absorbed
determines whether the laser cuts, welds, or heat treats. *(Courtesy of Westinghouse
R&D Center, Pittsburgh, Pa.)*

water-cooled nozzle by electrically ionizing a suitable gas such as nitrogen, hydrogen, argon, or mixtures of these gases.   The process can be used on almost any conductive
metal.   The plasma arc is a mixture of free electrons, positively charged ions, and neutral
atoms.   The arc is initiated in a confined gas-filled chamber by a high-frequency spark.
The high-voltage, direct-current power sustains the arc, which exits from the nozzle at
near sonic velocity.   The workpiece is electrically positive.   The high-velocity gases melt
and blow away the molten metal "chips."   Dual-flow torches use a secondary gas or water shield to assist in blowing the molten metal out of the kerf, giving a cleaner cut.   The
process may be performed under water using a large tank to hold the plates being cut.
The water assists in confining the arc and reducing smoke.   The main advantage of PAC
is speed.   Mild steel $\frac{1}{4}$ in. thick can be cut at 125 in./min.   Speed decreases with thickness.   Greater nozzle life and faster cutting speeds accompany the use of water-injection-
type torches.   Control of nozzle standoff from the workpiece is important.   One electrode
size can be used to machine a wide range of materials and thicknesses by suitable adjustments to the power level, gas type, gas flow rate, traverse speed, and flame angle.   PAC
is sometimes called plasma beam machining.

PAC can machine exotic metals at high rates.   Profile cutting of metals, particularly
of stainless steel and aluminum, has been the most prominent commercial application.
However, mild steel, alloy steel, titanium, bronze, and most metals can be cut cleanly and
rapidly.   Multiple-torch cuts are possible on programmed or tracer-controlled cutting tables
on plates up to 6 in. thick in stainless steel.   Smooth cuts free from contaminants are a

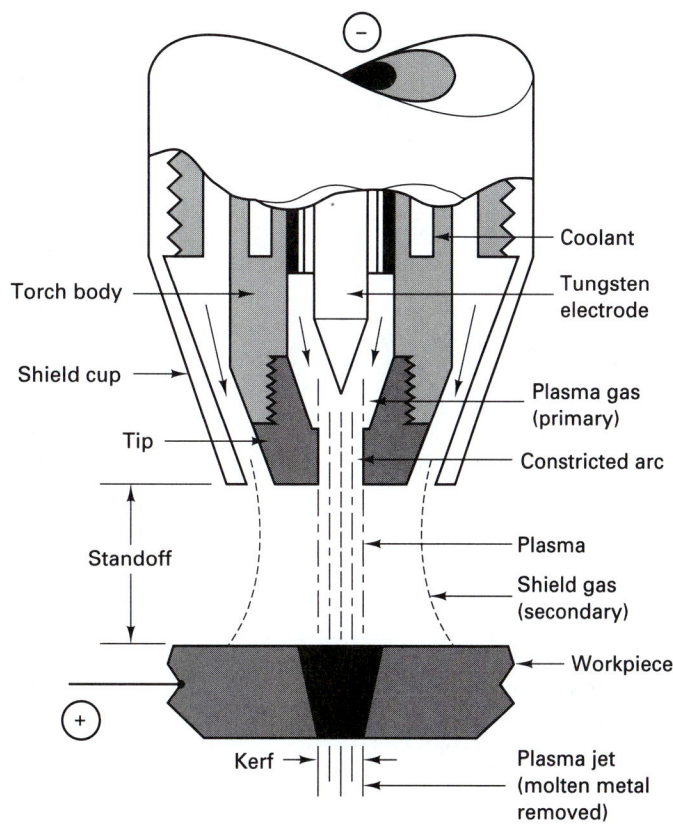

FIGURE 32-23 Plasma arc machining

PAC advantage. Well-attached dross on the underside of the cut can be a problem and there will be a heat-affected zone (HAZ). The depth of the HAZ is a function of the metal, its thickness, and the cutting speed. Surface heat treatment and metal joining are beginning to use the plasma torch. See Chapter 34 for more discussion on plasma arc processes.

■ KEY WORDS

abrasive flow machining
abrasive waterjet machining (AWM)
chemical blanking
chemical machining (CHM)
electrical discharge machining (EDM)
electrochemical grinding (ECG)
electrochemical machining (ECM)

electron beam machining (EBM)
electrochemical polishing
electrostream
etch factor
hydrodynamic machining
laser beam machining (LBM)
nontraditional machining (NTM)

photochemical machining
plasma arc cutting
thermochemical machining (TCM)
ultrasonic machining
waterjet machining
wire EDM

■ REVIEW QUESTIONS

1. What are the four basic categories of NTM processes?
2. Why might chipless machining processes have greater importance in the future?
3. How does the MRR for AWM compare to conventional metal cutting?
4. What are the steps in chemical machining using photosensitive resists?

5. Why is it preferable in chemical machining to apply the etchant by spraying instead of immersion?

6. What are the advantages of chemical blanking over regular blanking using punch and die methods?

7. Explain how multiple depths can be produced by chemical machining.

8. Would it be feasible to produce a groove 2 mm wide and 3 mm deep by chemical machining?

9. A drawing calls for making a groove 23 mm wide and 3 mm deep by chemical machining. Estimate the width of the opening in the maskant.

10. Could an ordinary steel weldment be chemically machined? Why or why not?

11. How would you produce a tapered section by chemical machining?

12. What is the principal application of thermochemical machining?

13. Is ECM related to chemical machining?

14. What effect does work-material hardness have on the metal-removal rate in ECM?

15. Explain the basic principle involved in electrochemical deburring.

16. What is the principal cause of tool wear in ECM?

17. Basically, what type of process is electrochemical grinding?

18. Would electrochemical grinding be a suitable process for sharpening ceramic tools? Why or why not? What about using ultrasonics?

19. Upon what factors does the metal-removal rate depend in ECM?

20. Why is the tool insulated in the ECM schematic?

21. In abrasive waterjet machining, what is the kerf?

22. In AWM, garnet is used in the abrasive flow. What is garnet?

23. In the AWM setup, what is the necessary function of the drain?

24. Is ultrasonic machining really a chipless process?

25. Where, in the ultrasonic stroke, is the acceleration and deceleration of the tool the greatest? Given that $F = ma$, what does this mean about the force given to the grits in the slurry?

26. What is the nature of the surface obtained by EDM?

27. What is the principal advantage of using a moving-wire electrode in EDM?

28. What effect would increasing the voltage have on the metal-removal rate in electrical discharge machining? Why?

29. If the metal from which a part is to be made is quite brittle and the part will be subjected to repeated tensile loads, would you select ECM or EDM for making it? Why?

30. If you had to make several holes in a large number of duplicate parts, would you prefer ECM, EDM, EBM, or LBM? Why?

31. What process would you recommend to make many small holes in a very hard alloy where the holes will be used for cooling and venting?

32. Why are the specific power values so large for LBM?

33. Explain (using a little physics and metallurgy) why the "chips" in thermal processes are often hollow spheres.

PART 6

JOINING PROCESSES

33 Gas Flame Processes: Welding, Cutting, and Straightening

34 Arc Processes: Welding and Cutting

35 Resistance Welding

36 Other Welding and Related Processes

37 Brazing and Soldering

38 Adhesive Bonding and Mechanical Fastening

39 Manufacturing Concerns in Welding and Joining

CHAPTER 33

GAS FLAME PROCESSES: WELDING, CUTTING, AND STRAIGHTENING

33.1 OVERVIEW OF WELDING PROCESSES

Classification of Welding Processes

33.2 OXYFUEL GAS WELDING

Oxyfuel Gas Welding Processes

Uses, Advantages, and Limitations

Pressure Gas Welding

33.3 OVERVIEW OF CUTTING PROCESSES

33.4 OXYGEN TORCH CUTTING

Processes

Fuel Gases for Oxyfuel Gas Cutting

Stack Cutting

Metal Powder Cutting and Chemical Flux Cutting

Underwater Torch Cutting

33.5 FLAME STRAIGHTENING

■ 33.1 OVERVIEW OF WELDING PROCESSES

Welding is a process by which two materials, usually metals, are permanently joined together by *coalescence*, which is induced by a combination of temperature, pressure, and metallurgical conditions. The particular combination of these variables can range from high temperature with no pressure to high pressure with no increase in temperature. Thus welding can be accomplished under a wide variety of conditions, and a number of welding processes have been developed. While welding has become a dominant joining process in manufacturing, the average person has little appreciation of its role or the large number of metal products that would have to be drastically modified, would be considerably more costly, or could not perform as efficiently if it were not available.

To obtain coalescence between two metals, there must be sufficient proximity and activity between the atoms of the pieces being joined to cause the formation of common metallic crystals. The ideal metallurgical bond, for which there would be no noticeable or detectable joint, would require (1) perfectly smooth, flat or matching surfaces; (2) clean surfaces, free from oxides, absorbed gases, grease, and other contaminants; (3) metals with no internal impurities; and (4) two metals that are both single crystals with identical crystallographic structure and orientation. While these conditions would be difficult to obtain under ideal laboratory conditions, they are virtually impossible to achieve in normal production. Consequently, the various joining methods have been designed to overcome or compensate for this inability. Surface roughness is overcome either by force, causing plastic deformation of the asperities, or by melting the two surfaces, so that fusion occurs. In solid-state welding, contaminated layers are removed by mechanical or chemical cleaning prior to welding, or by causing sufficient metal flow along the interface so that the impurities are squeezed out of the weld. In fusion welding, where a pool of molten metal exists, the contaminants are removed by the use of fluxing agents. When welding is performed in a vacuum the contaminants are removed much more easily, and coalescence is established with considerable ease. Thus, in outer space, mating parts may weld under extremely light loads, even when such was not intended.

The various welding processes differ not only in the way in which temperature and pressure are combined and achieved, but also in the attention that must be given to the cleanliness of the metal surfaces prior to welding and to possible oxidation or contamination during the welding process. If high temperatures are used, most metals react more rapidly with their surrounding environment. If actual melting occurs, serious modification of the metal may result. The metallurgical structure and quality of both the weld and adjacent material can be affected by the heating and cooling cycle of the welding process. These effects and their possible consequences should be considered when designing a product and selecting the joining process.

In summary, the production of a high-quality weld requires (1) a source of satisfactory heat and/or pressure, (2) a means of protecting or cleaning the metal, and (3) caution to avoid, or compensate for, harmful metallurgical effects.

Classification of Welding Processes

Whenever possible in this book we utilize the nomenclature of the American Welding Society, who has classified the various welding processes in the manner presented in Figure 33-1 and has assigned short letter symbols to facilitate their designation. These processes provide a variety of ways of achieving coalescence and make it possible to achieve effective and economical welds in nearly all metals and combinations of metals. As a result, welding has replaced other types of permanent fastening to such a degree that a large portion of manufactured products contain one or more welds.

FIGURE 33-1 Classification of common welding processes along with their AWS (American Welding Society) designations.

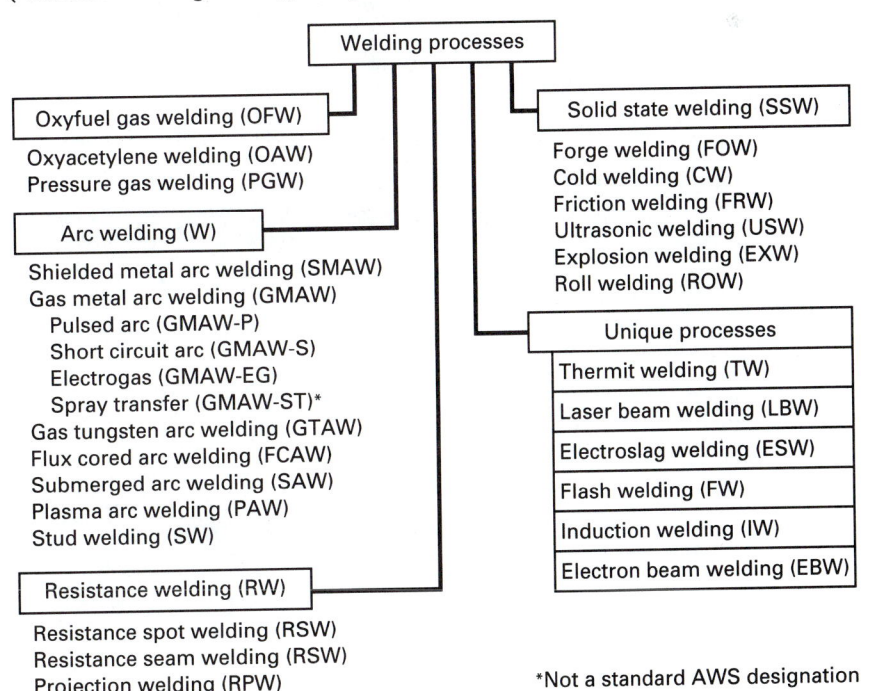

*Not a standard AWS designation

■ 33.2 Oxyfuel Gas Welding

Oxyfuel Gas Welding Processes

Oxyfuel gas welding (OFW) refers to a group of welding processes that use, as their heat source, the flame produced by the combustion of a fuel gas and oxygen. It was the development shortly after 1900, of a practical torch to burn acetylene and oxygen that brought welding out of the blacksmith's shop, demonstrated its potential, and started its development as a manufacturing process. While gas-flame welding has largely been replaced by other processes in the manufacturing environment, it is still popular for many applications because of its portability and the low capital investment required. Acetylene is still the principal fuel gas employed in the process.

The combustion of oxygen and *acetylene* (C_2H_2) by means of a welding torch of the type shown in Figure 33-2 produces a temperature of about 5850°F (3250°C) in a two-stage reaction. In the first stage, the oxygen and acetylene react to produce carbon monoxide and hydrogen:

$$C_2H_2 + O_2 \rightarrow 2CO + H_2$$

This reaction occurs near the tip of the torch. The second stage of the reaction involves the combustion of the CO and H_2 and occurs just beyond the first combustion zone. The specific reactions of stage 2 are

$$2CO + O_2 \rightarrow 2CO_2$$

$$H_2 + \tfrac{1}{2}O_2 \rightarrow H_2O$$

The oxygen for these secondary reactions is generally obtained from the surrounding atmosphere.

FIGURE 33-2   Typical oxyacetylene welding torch and cross-sectional schematic. *(Courtesy of Victor Equipment Company.)*

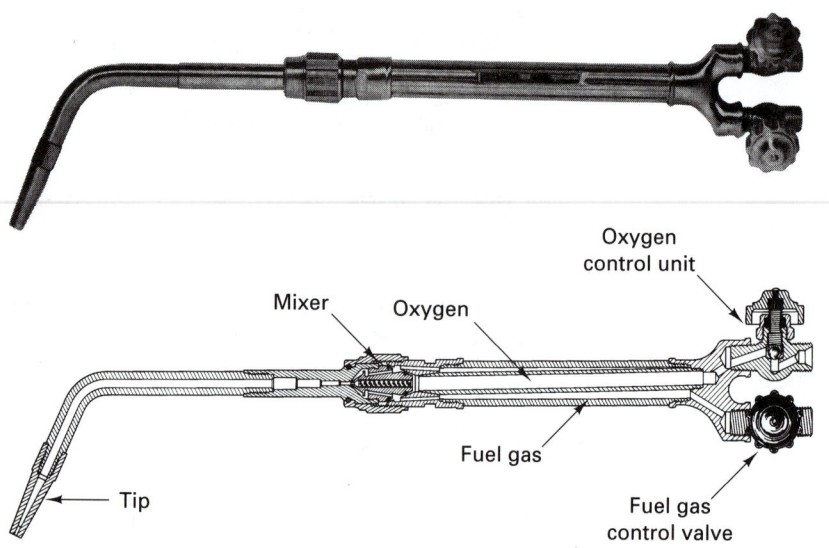

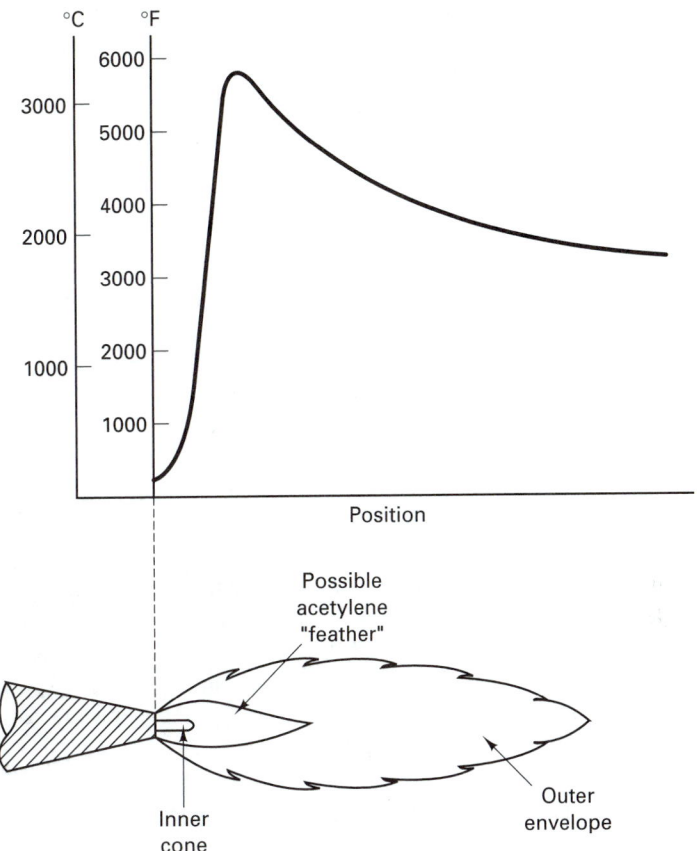

FIGURE 33-3 Typical oxyacetylene flame and the associated temperature distribution.

The two-stage combustion process produces a flame having two distinct regions. As shown in Figure 33-3, the maximum temperature occurs near the end of the inner cone, where the first stage of combustion is complete. Most welding should be performed with the torch positioned so that the point of maximum temperature is just above the metal being welded. The outer envelope of the flame serves to preheat the metal and, at the same time, provides shielding from oxidation, since oxygen from the surrounding air is used in the secondary combustion.

Three different types of flames can be obtained by varying the oxygen/acetylene (or oxygen/fuel gas) ratio. If the ratio is about 1:1 to 1.15:1, all reactions are carried to completion and a *neutral flame* is produced. Most welding is done with a neutral flame, since it will have the least chemical effect on the heated metal.

A higher ratio, such as 1.5:1, produces an *oxidizing flame*, hotter than the neutral flame (about 6000°F) but similar in appearance. Such flames are used when welding copper and copper alloys but are generally considered harmful when welding steel because the excess oxygen reacts with the carbon, decarburizing the region around the weld.

Excess fuel, on the other hand, produces a *carburizing flame*. The excess fuel decomposes to carbon and hydrogen, and the flame temperature is not as great (about 5550°F). Flames with a slight excess of fuel are reducing flames. No carburization occurs, but the metal is well protected from oxidation. Flames of this type are used in welding Monel (a nickel–copper alloy), high-carbon steels, and some alloy steels, and for applying some types of hard-facing material.

For welding purposes, the oxygen is usually supplied in relatively pure form from pressurized tanks, but, in rare cases, air can also be used. The acetylene is usually obtained in portable storage tanks that hold up to 300 ft³ (8.5 m³) at 250 psi (1.7 MPa) pressure. Because acetylene is not safe when stored as a gas at pressures above 15 psi (0.1 MPa), it is usually dissolved in acetone. The storage cylinders are filled with a porous filler such as balsa-wood chips or infusorial earth. Acetone is absorbed into the voids in the filler material and serves as a medium for dissolving the acetylene.

Stabilized methylacetylene propadiene, best known by the trade name of *MAPP* gas, has become a competitor of acetylene, particularly where portability is important. While flame temperature is slightly lower, the gas is more dense, thus providing more energy for a given volume. In addition, it can be stored in ordinary pressure tanks.

The pressures used in gas-flame welding range from 1 to 15 psi (7 to 105 MPa) and are controlled by pressure regulators on each tank. Because mixtures of acetylene and oxygen or air are highly explosive, precautions must be taken to avoid mixing the gases improperly or by accident. All acetylene fittings have left-hand threads, while those for oxygen are equipped with right-hand threads. This prevents improper connections.

The tip size (or orifice diameter) of the torch can be varied to control the shape of the inner cone and the flow rate of the gases. Larger tips permit greater flow of gases, resulting in greater heat input without the higher gas velocities that might blow the molten metal from the weld puddle. Larger torch tips are required for welding thicker metal, and operate with higher gas pressure.

Uses, Advantages, and Limitations

Almost all oxyfuel gas welding is *fusion welding*. The metals to be joined are simply melted where a weld is desired and no pressure is involved. Because a slight gap often exists between the pieces being joined, *filler metal* can be added in the form of a wire or rod. Welding rods come in standard sizes, with diameters from $\frac{1}{16}$ to $\frac{3}{8}$ in. (1.5 to 9.5 mm) and lengths from 24 to 36 in. (0.6 to 0.9 m). They are available in standard grades to provide specified minimum tensile strengths or in compositions that match the base metal. Figure 33-4 shows a schematic of oxyfuel gas welding with a consumable welding rod.

Fluxes may be used to clean the surfaces and remove contaminating oxide, thereby promoting the formation of a better bond. In addition, the gaseous shield produced by vaporizing flux can prevent oxidation during welding, and the slag produced by solidifying flux can protect the weld pool as it cools. Flux can be added as a powder, the welding rod can be dipped in a flux paste, or the rods can be precoated.

Good-quality welds can be obtained by the OFW processes if proper caution is exercised. The temperature of the work can be controlled easily. However, exposure of the heated and molten metal to the various gases in the flame and atmosphere makes it difficult to prevent contamination. Since the heat source is not concentrated, a large area of the metal is heated and distortion is likely to occur. Thus, in production applications, the flame welding processes have largely been replaced by arc welding. Nevertheless, flame welding is still quite common in field work, in maintenance and repairs, and in fabricating small quantities of specialized products. Table 33-1 summarizes the compatibility of oxyfuel welding with some common engineering materials.

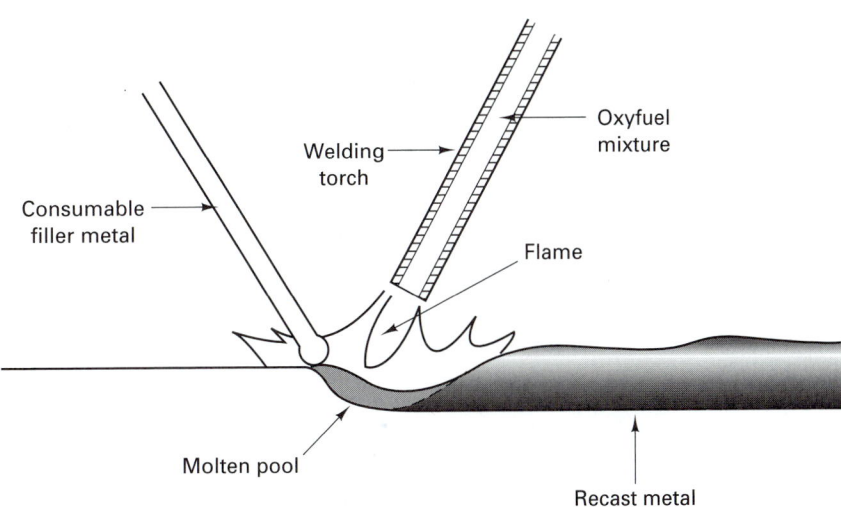

FIGURE 33-4   Schematic of oxyfuel gas welding with a consumable welding rod.

Material	Oxyfuel Welding Recommendation
Cast iron	Recommended with cast iron filler rods; braze welding recommended if there are no corrosion objections
Carbon and low-alloy steels	Recommended for low-carbon and low-alloy steels, using rods of the same material; more difficult for higher carbon
Stainless steel	Common for thinner material; more difficult for thicker
Aluminum and magnesium	Common for aluminum thinner than 1 in.; difficult for magnesium alloys
Copper and copper alloys	Common for most alloys; more difficult for some types of bronzes
Nickel and nickel alloys	Common for nickel, Monels, and Inconels
Titanium	Not recommended
Lead and zinc	Recommended
Thermoplastics, thermosets, and elastomers	Hot-gas welding used for thermoplastics, not used with thermosets and elastomers
Ceramics and glass	Seldom used with ceramics, but common with glass
Dissimilar metals	Difficult; best if melting points are within 50°F; concern for galvanic corrosion
Metals to nonmetals	Not recommended
Dissimilar nonmetals	Difficult

TABLE 33-1.   Engineering Materials and Their Compatibility with Oxyfuel Welding

Pressure Gas Welding

Pressure gas welding (PGW) is a process used to make butt joints between the ends of objects such as pipe and railroad rail. The ends are heated with a gas flame to a temperature below the melting point, and the soft metal is then forced together under considerable pressure. This process, therefore, is actually a form of solid-state welding.

■ 33.3 OVERVIEW OF CUTTING PROCESSES

For many years, metal sheets and plates have been cut by means of oxyfuel torches and electric arc equipment. Developed originally for use in salvage and repair work, then used to prepare plates for welding, these processes are now widely used to cut sheets and plates

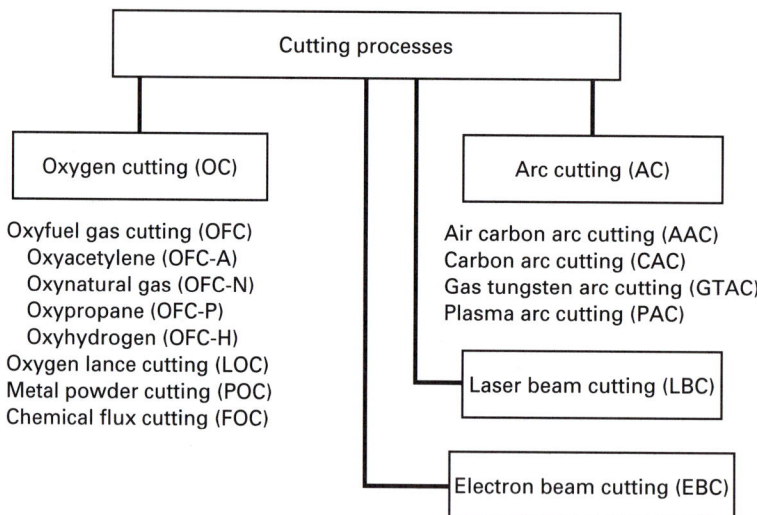

FIGURE 33-5 Classification of common cutting processes with their AWS (American Welding Society) designations.

into desired shapes for assembly and other processing operations. In recent years the development of laser and electron beam equipment has made possible the cutting of both metals and nonmetals at speeds of up to 1000 in./min (25 m/min). Accuracies of up to 0.01 in. (0.25 mm) are readily attainable and speeds of 50 in./min (1.3 m/min) are quite common. Figure 33-5 shows a chart of the commonly used thermal cutting processes with their AWS designations.

■ 33.4 OXYGEN TORCH CUTTING

Processes

By far the majority of all thermal cutting is done by *oxyfuel gas cutting* (OFC). In some cases, primarily when the metal being cut is nonferrous, the metal is merely melted by the flame of the oxyfuel gas torch and blown away to form a gap, or *kerf* (Figure 33-6). However, when ferrous metal is being cut, the process becomes one of rapid oxidation (burning) of iron at high temperatures according to the chemical equation:

$$3Fe + 2O_2 \rightarrow Fe_3O_4 + heat$$

Because this reaction does not occur until the metal is at approximately 1600°F (870°C), an oxyfuel flame is first used to raise the metal to the temperature at which burning will begin. Then a stream of pure oxygen is added to the torch (or the oxygen content of the oxyfuel mixture is increased) to oxidize the iron. The liquid iron oxide is then expelled from the joint by the kinetic energy of the oxygen-gas stream.

Theoretically, the heat of oxidation will be sufficient to keep the cut progressing, but additional heat is often supplied to compensate for losses to the atmosphere and the surrounding metal. If the workpiece is already hot, as when steel is cut in the mill immediately following hot processing, no supplemental heating is required, and a supply of oxygen through a small pipe is all that is needed to initiate and continue a cut. This is known as *oxygen lance cutting* (LOC). A temperature of about 2200°F (1200°C) is required for this procedure to be effective.

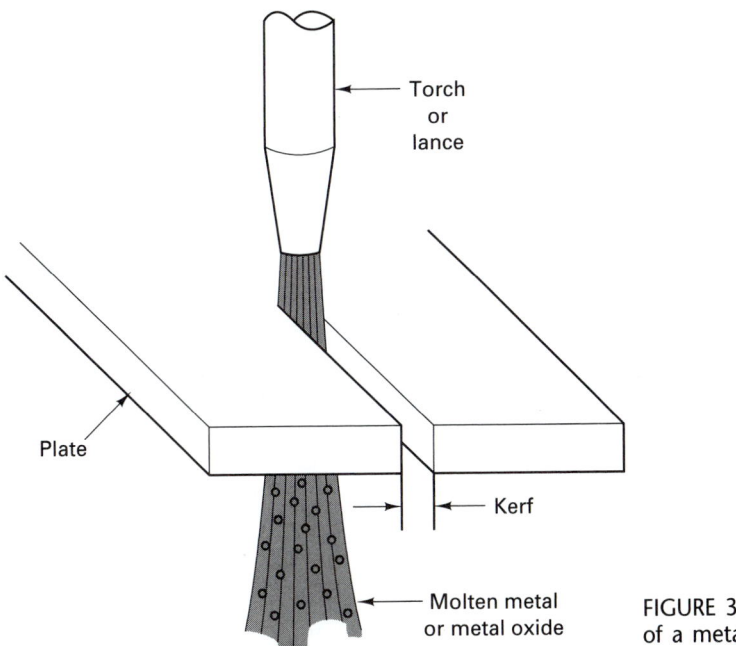

FIGURE 33-6   Flame cutting of a metal plate.

Oxyfuel gas cutting works best on metals that oxidize readily but do not have high thermal conductivities. Carbon and low-alloy steels can be readily cut, but cast irons and stainless steels contain oxidation-resistant ingredients and require special techniques. Aluminum and copper are difficult to cut.

Fuel Gases for Oxyfuel Gas Cutting

Acetylene is by far the most common fuel used in oxyfuel gas cutting, and the process is often referred to as *oxyacetylene cutting* (OFC-A). Figure 33-7 shows a typical cutting *torch*. The tip contains a circular array of small holes through which the oxygen–acetylene mixture is supplied to form the heating flame. A larger hole in the center supplies a stream of oxygen, controlled by a lever valve. The rapid flow of the cutting oxygen not only promotes the rapid oxidation but also serves to blow the formed oxides from the cut.

If the torch is adjusted and manipulated properly, a smooth cut can be produced, as shown at the top of Figure 33-8. As indicated in the remaining photos of that figure, high-quality cutting requires careful selection of the preheat condition, oxygen flow rate, and cutting speed. Oxygen purities over 99.5% are required for the most efficient cutting. A drop of purity to 98.5% will reduce cutting speed by 15%, increase oxygen consumption by 25%, and impair the quality of the cut.

Cutting torches are often manipulated manually. However, when the process is applied to manufacturing, the desired path is usually controlled by mechanical or programmable means. Many machines have been designed for special purposes, such as making straight cuts in flat stock, or cutting pipe. One of the more common applications is to prepare the edges of plates for subsequent welding. Figure 33-9 shows an example of a three-torch cut being used to produce a double-bevel edge configuration.

The marriage of computer numerically controlled (CNC) machines and cutting torches has also proven to be quite popular. This approach, along with the use of robot-mounted torches, provides great flexibility with good precision and control. Accuracies of ±0.015 in. (±0.40 mm) are possible, but ±0.03 to 0.04 in. (±0.75 to 1.0 mm) are more common.

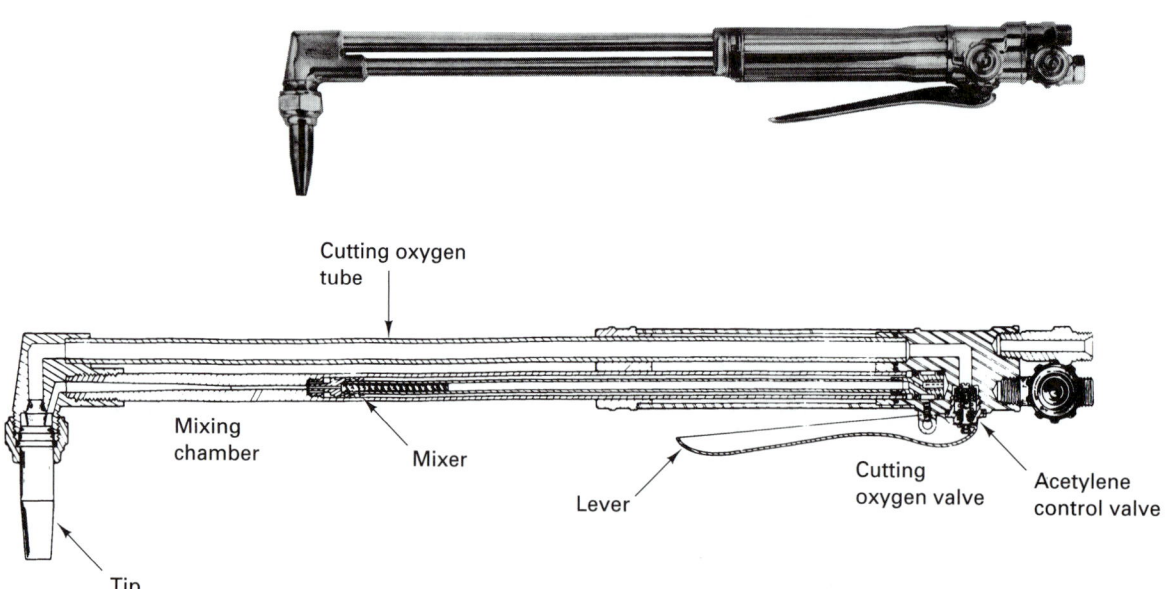

FIGURE 33-7 Typical oxyacetylene cutting torch and cross-sectional schematic. *(Courtesy of Victor Equipment Company.)*

Fuel gases other than acetylene can also used for oxyfuel gas cutting, the most common being natural gas (OFC-N) and propane (OFC-P). While their flame temperatures are lower than acetylene, their use is generally a matter of economics and gas availability. For certain special work, hydrogen can also be used (OFC-H).

Stack Cutting

When a modest number of duplicate parts are to be cut from thin sheet, but not enough to justify the cost of a blanking die, stack cutting may be the answer. The sheets should be flat, smooth, and free of scale, and they should be clamped together tightly so that there are no intervening gaps that could interrupt uniform oxidation or permit slag and molten metal to be entrapped. Obviously, the accuracy of stack cutting will be poorer than that obtained with a blanking die.

Metal Powder Cutting and Chemical Flux Cutting

When applying *thermal cutting* to the hard-to-cut materials, modified torch techniques may be required. *Metal powder cutting* (POC) injects iron powder into the flame to raise its cutting temperature. *Chemical flux cutting* (FOC) adds a fine stream of special flux to the cutting oxygen to increase the fluidity of the high-melting-point oxides. Both of these methods have largely been replaced by plasma arc cutting (PAC), which is discussed in Chapter 34. Laser and electron beam cutting is presented in Chapter 36.

Underwater Torch Cutting

The cutting of materials underwater presents a special challenge. Steel can be cut by use of a specially designed torch, such as the one shown in Figure 33-10. An auxiliary skirt surrounds the main tip and an additional set of gas passages conducts a flow of compressed air to provide secondary oxygen for the oxyacetylene flame and to expel water from the

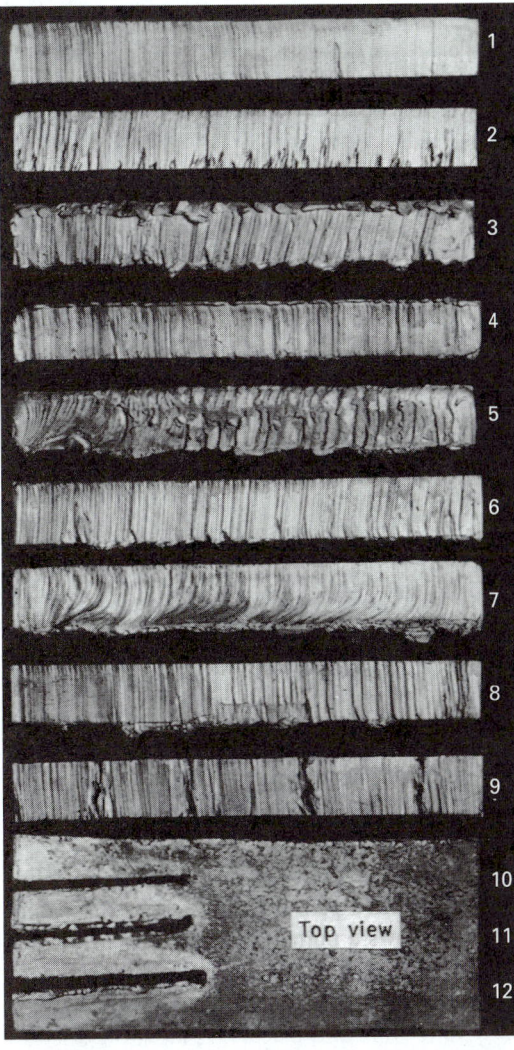

(1) Correct Procedure
Compare this cut in 1-in. plate with those below. The edge is square, the drag lines are vertical and not too pronounced.

(2) Preheat Flames Too Small
They are only about 1/8 in. long. Result: cutting speed was too slow, causing bad gouging effect at bottom,

(3) Preheat Flames Too Long
They are about 1/2 in. long. Result: surface has melted over, cut edge is irregular, and there is too much adhering slag.

(4) Oxygen Pressure Too Low
Result: top edge has melted over because of too slow cutting speed.

(5) Oxygen Pressure Too High
Nozzle size also too small. Result: entire control of the cut has been lost.

(6) Cutting Speed Too Slow
Result: irregularities of drag line are emphasized.

(7) Cutting Speed Too High
Result: a pronounced rake to the drag lines and irregularities on the cut edge.

(8) Blowpipe Travel Unsteady
Result: the cut edge is wavy and irregular.

(9) Lost Cut Not Properly Restarted
Result: bad gouges where cut was restarted.

(10) Good Kerf
Compare this good kerf (viewed from the top of the plate) with those below.

(11) Too Much Preheat
Nozzle also is too close to plate. Result: bad melting of the top edges.

(12) Too Little Preheat
Flames also are too far from the plate. Result: heat spread has opened up kerf at top. Kerf is tapered and too wide.

FIGURE 33-8   Cut edge of metal that has been cut properly and improperly by the oxyacetylene process. *(Courtesy of Linde Division, Union Carbide Corporation.)*

zone where the burning of metal occurs. The torch is either ignited in the usual manner before descent or by an electric spark device after being submerged. Acetylene gas is used for depths up to about 25 ft (7.5 m). For greater depths, hydrogen is used, since the environmental pressure is too great for the safe use of acetylene.

■ 33.5 FLAME STRAIGHTENING

Flame straightening is basically the creation of controlled, localized upsetting in order to straighten warped or buckled plates. The theory of the process can be illustrated by Figure 33-11. If a straight piece of metal is heated in a localized area, as indicated by the shaded area of the upper diagram, the metal on side *b* will be upset as it softens and tries to expand against the restraining cool metal. When the upset portion cools, it will contract and the resulting piece will be shorter on side *b*, forcing it to bend to the shape in the lower diagram.

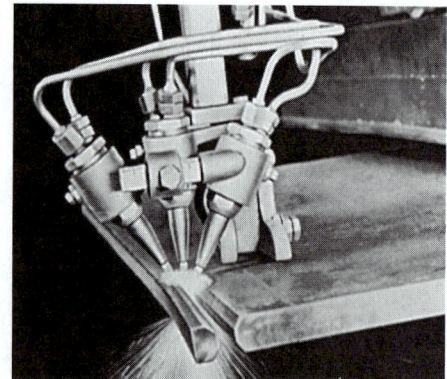

FIGURE 33-9 Plate edge being prepared for welding. The beveled shape is produced by three simultaneous oxyacetylene cuts. *(Courtesy of Linde Division, Union Carbide Corporation.)*

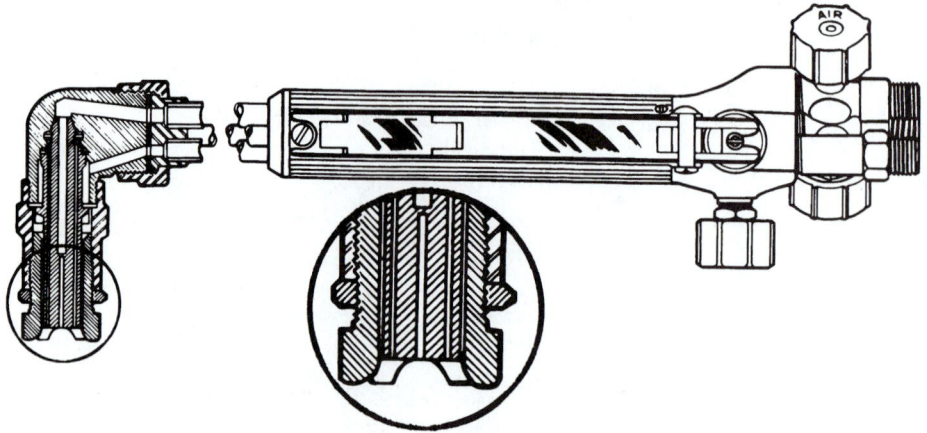

FIGURE 33-10 Underwater cutting torch. Note the extra set of gas openings in the nozzle to permit the flow of compressed air and the extra control valve. *(Courtesy of Bastian-Blessing Company.)*

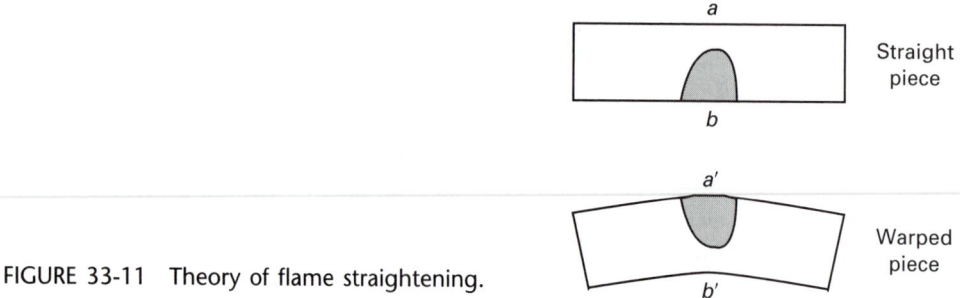

FIGURE 33-11 Theory of flame straightening.

If the starting material is bent or warped, as in the lower segment of Figure 33-9, the upper surface can be heated. *Upsetting* and subsequent thermal contraction will shorten the upper surface at a', bringing the plate back to a straight or flat configuration. Such a procedure can be used to repair structures that have been bent in an accident, such as automobile frames.

A similar process can be used to flatten metal plates that have become dished. Localized spots about 2 inches (50 mm) in diameter are quickly heated to the upsetting temperature so that the surrounding metal remains cool. Cool water is then sprayed on the plate and the contraction of the upset spot brings the buckle into an improved degree of flatness. To remove large buckles, the process may have to be repeated at several spots within the area.

Several cautions should be noted. When straightening steel, consideration should be given to the possible phase transformations that could occur during the heating and cooling. Since rapid cooling is used and martensite may form, a subsequent tempering operation may be required. In addition, one should also consider the residual stresses that are induced and their effect on subsequent cracking, stress-corrosion cracking, and other modes of failure. The effects of phase transformations and residual stresses are developed more fully in Chapter 39.

Flame straightening should not be attempted with thin material. For the process to work, the metal adjacent to the heated area must have sufficient rigidity to promote upsetting. If this is not the case, localized heating and cooling will simply transfer the buckle from one area to another.

■ KEY WORDS

acetylene	fusion welding	thermal cutting
carburizing flame	kerf	torch
coalescence	MAPP	upsetting
filler metal	neutral flame	welding
flame straightening	oxidizing flame	
flux	oxyfuel gas welding	

■ REVIEW QUESTIONS

1. What is welding?
2. What are the four conditions required for an "ideal metallurgical bond"?
3. What are some of the ways in which welding processes overcome or compensate for the inability to meet the ideal bond conditions?
4. What are some of the problems that might occur when high temperatures are used in welding?
5. Why does an oxyfuel gas welding torch usually have a flame with two distinct regions?
6. What is the location of the maximum temperature in an oxyacetylene flame?
7. What function or functions are served by the outer zone of the welding flame?
8. What three types of flames can be produced by varying the oxygen/fuel ratio?
9. What are some of the attractive features of MAPP gas?
10. Why might a welder want to change the tip size (or orifice diameter) in an oxyacetylene torch?
11. What are some of the processes that can be used to cut metal thermally?
12. In what way does the torch cutting of ferrous metals differ from cutting nonoxidizing metals?
13. Why might it be possible to cut cast steel strands continuously as they emerge from the casting operation using only an oxygen lance?
14. How does an oxyacetylene cutting torch differ from an oxyacetylene welding torch?
15. What are some of the ways in which cutting torches can be mechanically manipulated?
16. What modification must be incorporated into a cutting torch to permit it to cut metal underwater?
17. If a curved plate is to be straightened by flame straightening, should the heat be applied to the longer or shorter surface of the arc? Why?
18. Why does the flame straightening process not work for thin sheets of metal?

ARC PROCESSES: WELDING AND CUTTING

34.1 ARC WELDING
    Shielded Metal Arc Welding
    Gas Tungsten Arc Welding
      Gas Tungsten Arc Spot Welding
    Gas Metal Arc Welding
      Pulsed Arc Gas Metal Arc Welding
    Flux-Cored Arc Welding
    Submerged Arc Welding
      Bulk Welding
    Plasma Arc Welding
    Stud Welding
    Power Sources for Arc Welding

    Jigs and Positioners
34.2 ARC CUTTING
    Carbon Arc and Shielded Metal Arc Cutting
    Air Carbon Arc Cutting
    Oxygen Arc Cutting
    Gas Metal Arc Cutting
    Gas Tungsten Arc Cutting
    Plasma Arc Cutting
34.3 METALLURGICAL AND HEAT CONSIDERATIONS IN THERMAL CUTTING
Case Study: REPAIR OF A BICYCLE FRAME

■ 34.1 ARC WELDING

With the development of commercial electricity in the late nineteenth century, it was soon recognized that an *arc* between two *electrodes* was a concentrated heat source that could approach 7000°F (3900°C) in temperature. As early as 1881, various attempts were made to use such an arc as the heat source for fusion welding. Initially, carbon was selected as one electrode and the metal workpiece was the other, as illustrated in the basic electrical circuit of Figure 34-1. *Filler metal* was provided by a metallic wire or rod that was fed into the arc. As the process developed, the filler metal replaced carbon as the upper electrode. The metal wire not only carried the welding current, but as it melted in the arc, it also supplied the necessary filler.

    The results of these early efforts were extremely uncertain. Because of the instability of the arc, a great amount of skill was required to maintain it, and contamination and

FIGURE 34-1 Basic circuit for arc welding.

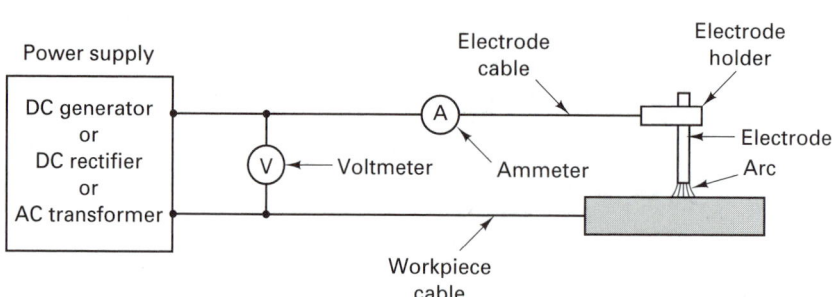

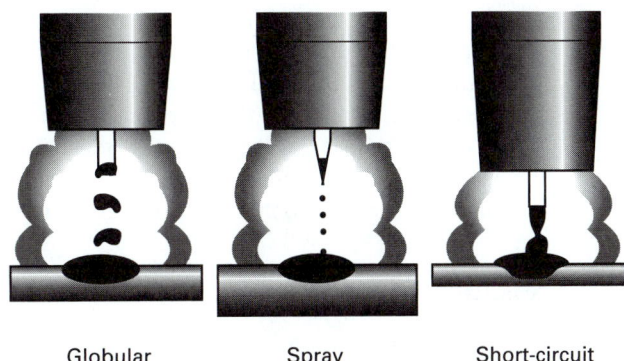

Globular   Spray   Short-circuit

FIGURE 34-2   Three modes of metal transfer during arc welding. *(Courtesy of Republic Steel Corporation.)*

oxidation of the weld resulted from the exposure of hot metal to the atmosphere. There was little or no understanding of the metallurgical effects and requirements of such a process. Consequently, the great potential was recognized, but very little use was made of the process until after World War I. Shielded metal electrodes were developed around 1920. These electrodes produced a stable arc by shielding it from the atmosphere, and provided a fluxing action to the molten pool. The major problems of arc welding were overcome, and the process began to expand rapidly.

All arc-welding processes employ the same basic circuit as depicted in Figure 34-1. If direct current is used and the work is made positive (the anode of the circuit), the condition is known as *straight polarity* (SPDC). When the work is negative and the electrode is positive, *reverse polarity* (RPDC) is employed. Since greater heat is liberated at the anode when bare electrodes are used, the SPDC conditions are generally preferred. Shielded electrodes may change the heat conditions, however, and produce better results when used in the reverse polarity mode. *Alternating current* is a popular alternative to the dc conditions.

In one large group of arc welding processes, the electrode is consumed (*consumable electrode processes*) and thus supplies the metal needed to fill the voids in the joint. Consumable electrodes have a melting temperature below the temperature of the arc. Small droplets are melted from the end of the electrode and pass to the workpiece. The size of these droplets varies greatly and the mechanism of the transfer depends on the type of electrode, welding current, and other process parameters. Figure 34-2 depicts metal transfer by the *globular*, *spray*, and *short-circuit transfer* modes. As the electrode melts, the arc length and the electrical resistance of the arc path will vary. To maintain a stable arc and satisfactory welding conditions, the electrode must be moved toward the work at a controlled rate. Manual arc welding is almost always performed with shielded (covered) electrodes. Continuous bare-metal wire can be used as the electrode in automatic or semiautomatic arc welding, but this is always in conjunction with a separate shielding and arc-stabilizing medium and automatic feed-controlling devices that maintain the proper arc length.

In the second family of arc welding processes, the electrode is made of tungsten, which is not consumed by the arc, except by relatively slow vaporization. In these *nonconsumable electrode processes*, a separate metal wire is used to supply the filler metal.

Because of the wide variety of processes available, arc welding has become a widely used means of joining material. Each process and application, however, requires the selection or specification of the welding voltage, welding current, arc polarity (straight polarity, reversed polarity, or alternating), arc length, welding speed (how fast the electrode is moved across the workpiece), arc atmosphere, electrode or filler material, and flux. Filler materials must be selected to match the base metal with respect to properties and/or alloy content (chemistry). For many of the processes, the quality of the weld also depends on the skill of the operator. Automation and robotics are reducing this dependence, but the selection and training of welding personnel are still of great importance.

Shielded Metal Arc Welding

Shielded metal arc welding (SMAW) is the most common of the arc welding processes because of its wide versatility and because it requires only low-cost equipment. The key to the process is the special electrode that consists of metal wire, usually from $\frac{1}{16}$ to $\frac{1}{4}$ in. in diameter and 9 to 14 in. in length. Around the wire is a bonded coating containing chemical components that add a number of desirable characteristics, including all or a number of the following:

1. Provide a protective atmosphere (a gas shield around the arc).
2. Stabilize the arc.
3. Act as a flux to remove impurities from the molten metal.
4. Provide a protective slag coating to accumulate impurities, prevent oxidation, and slow the cooling of the weld metal.
5. Reduce weld-metal spatter and increase the efficiency of deposition.
6. Add alloying elements.
7. Affect arc penetration (the depth of melting in the workpiece).
8. Influence the shape of the weld bead.
9. Add additional filler metal.

Coated electrodes are classified by the tensile strength of the deposited weld metal, the welding position in which they may be used, the preferred type of current and polarity (if direct current), and the type of covering. A four- or five-digit system of designation has been adopted by the American Welding Society and is presented in Figure 34-3. As an example, type E7016 is a low-alloy steel electrode that will provide a deposit with a minimum tensile strength of 70,000 psi in the non-stress-relieved condition; it can be used in all positions, with either alternating or reverse-polarity direct current; and it has a low-hydrogen plus potassium coating. To assist in identification, all electrodes are marked with colors in accordance with a standard established by the National Electrical Manufacturers Association. Electrode selection consists of determining the electrode coating, coating thickness, electrode composition, and electrode diameter.

A variety of electrode coatings have been developed. The cellulose and titania (rutile) coatings contain: SiO_2; TiO_2; small amounts of FeO, MgO, and Na_2O; and volatile matter. Upon decomposition, the volatile matter releases hydrogen, which may dissolve in the weld metal and lead to embrittlement or cracking in the joint. Low-hydrogen electrodes are available with various compositions designed to provide shielding without the emission of hydrogen. Since many electrode coatings can absorb moisture, and this is another source of undesirable hydrogen, the coated electrodes are often baked just prior to use.

To initiate a weld, the operator briefly touches the tip of the electrode to the workpiece and quickly raises it to a distance that will maintain a stable arc. The intense heat quickly melts the tip of the electrode wire, the coating, and a portion of the adjacent base metal.

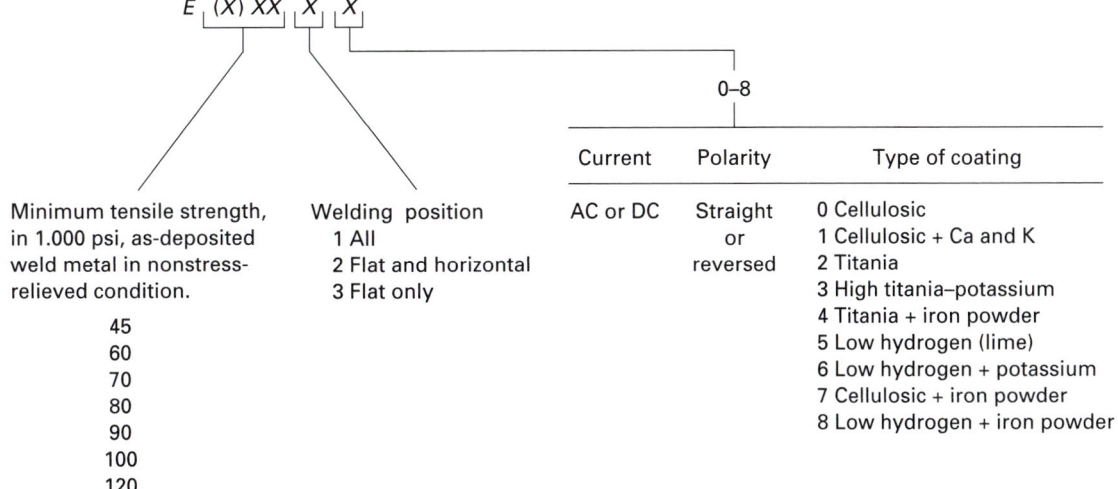

FIGURE 34-3   Designation system for arc-welding electrodes.

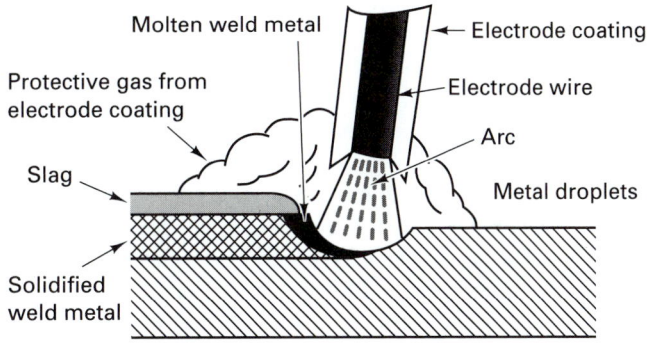

FIGURE 34-4   Schematic diagram of shielded metal arc welding (SMAW).   *(Courtesy of American Iron and Steel Institute, Washington, D.C.)*

As the coating on the electrode melts and vaporizes, it forms a protective atmosphere that stabilizes the arc and protects the molten and hot metal from contamination. Fluxing constituents unite with any impurities in the molten metal and float them to the surface to be entrapped in the slag coating that forms over the weld. This slag coating protects the cooling metal from oxidation and slows down the cooling rate to prevent the formation of hard, brittle structures. The slag is then easily chipped from the weld when it has cooled. Figure 34-4 provides a schematic of metal deposition from a shielded electrode.

Electrodes with iron powder in the coating can be used to significantly increase the amount of weld metal that can be deposited with a given size electrode wire and current. Alloy elements can be incorporated into the coating to adjust the chemistry of the weld. Other electrodes utilize coatings that are designed to melt more slowly than the filler wire.

If these *contact* or *drag electrodes* are tracked along the surface of the work, the center wire will be recessed by the proper length to maintain a stable arc.

Carbon steels, alloy steels, stainless steels, and cast irons are commonly welded by the shielded metal arc process, and welds can be made in all positions. Reverse-polarity dc is used to obtain deep penetration, with alternate modes being employed when welding thin sheet. The mode of metal transfer is either globular or short circuit, and the arc temperatures are rather low (9000°F or 5000°C). Typical welding voltages are 15 to 45 V with currents between 10 and 500 A.

Since electrical contact must be maintained with the center wire, most SMAW electrodes are finite-length "sticks." Length is limited since the current must be supplied near the arc or the electrode will tend to overheat and ruin the coating. Nevertheless, shielded metal arc welding is a simple and versatile process, requiring only a power supply, power cables, electrode holder, and a small variety of electrodes. Since the equipment is portable, and can even be powered by gasoline or diesel generators, it is a popular process in job shops and is used extensively in repair operations.

Gas Tungsten Arc Welding

Gas tungsten arc welding (GTAW), formerly known as TIG (tungsten inert-gas) welding, was one of the first developments away from the use of ordinary shielded electrodes. A tungsten electrode is positioned in a special holder through which inert gas (argon, helium, or a mixture of them) flows to provide a protective shield around both the arc and the pool of molten metal. Operating in this inert environment, the tungsten electrode (which is often alloyed with thorium or zirconium to provide better current-carrying and electron-emission characteristics) is not consumed at the temperatures of the arc. The arc length remains constant and the arc is stable and easy to maintain. Figure 34-5 shows a typical GTAW torch, with cables and passages for gas flow, power, and cooling water.

In applications where a close fit exists, no filler metal may be needed. When filler metal is required, it must be supplied as a separate wire, as illustrated in Figure 34-6. The filler metal is generally selected to match the chemistry of the metal being welded. Where high deposition rates are desired, a separate resistance heating circuit can be provided to preheat the filler wire. As shown in Figure 34-7, the deposition rate of heated wire can be several times that of a cold wire. The deposition rate can be further increased by oscillating the filler wire from side to side while making a weld pass. The hot-wire process is not practical when welding copper or aluminum, however, because of the low resistivities of the filler wire.

With skilled operators, gas tungsten arc welding can produce welds that are scarcely visible. In addition, the process produces very clean welds. Since no flux is employed, no special cleaning or slag removal is required. However, the surfaces to be welded must be clean and free of oil, grease, paint, and rust, because the inert gas does not provide any cleaning or fluxing action.

All metals and alloys can be welded by this process, and the use of inert gas makes it particularly attractive for the reactive metals, such as aluminum, magnesium, and titanium, as well as the high-temperature refractory metals. Maximum penetration is obtained with straight-polarity dc conditions, although ac may be specified to break up surface oxides (as when welding aluminum). Reverse polarity is rarely used because it tends to melt the tungsten electrode. Weld voltage is typically 20 to 40 V and weld current varies from less than 125 A for rpdc to 1000 A for spdc. A high-frequency, high-voltage alternating current is often superimposed on the regular ac or dc welding current to make it easier to start and maintain the arc.

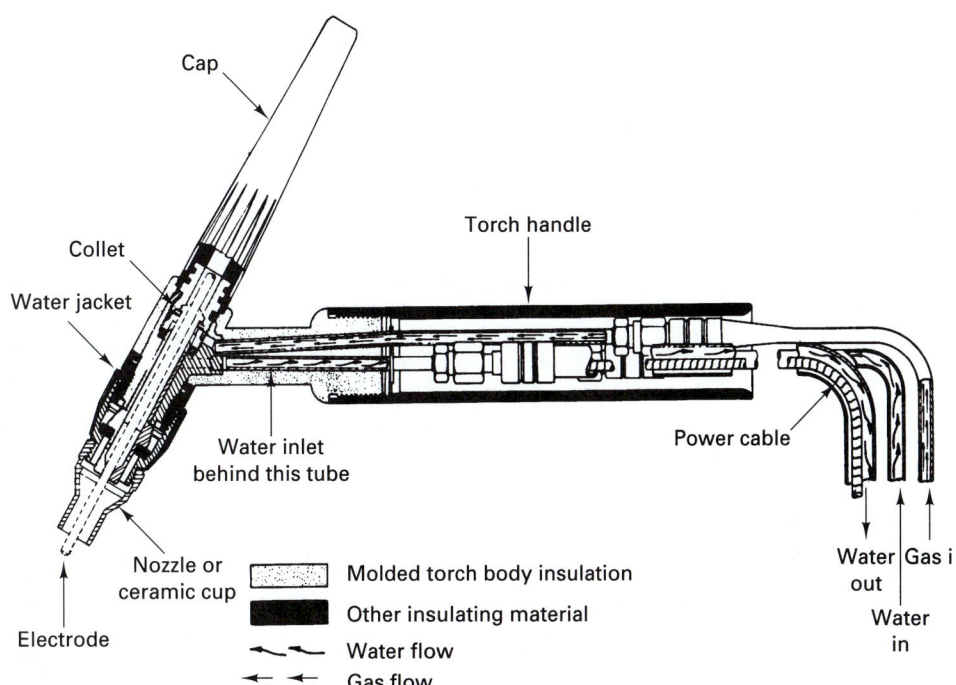

FIGURE 34-5   Welding torch used in nonconsumable-electrode, gas tungsten arc welding (GTAW), showing feed lines for power, cooling water, and inert gas flow. *(Courtesy of Linde Division, Union Carbide Corporation.)*

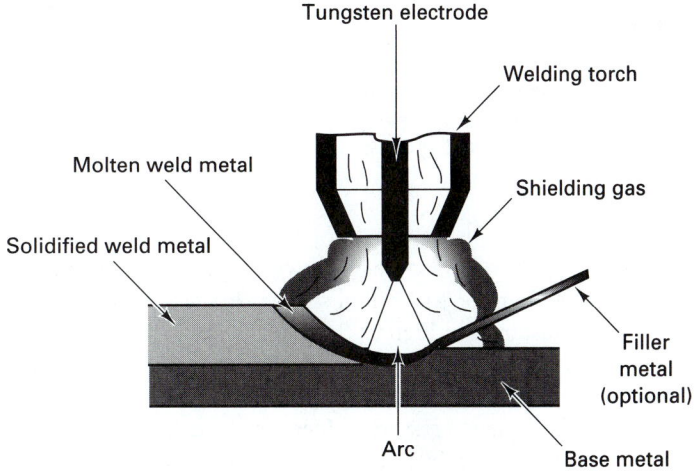

FIGURE 34-6   Schematic diagram of gas tungsten arc welding (GTAW). *(Courtesy of American Iron and Steel Institute, Washington, D.C.)*

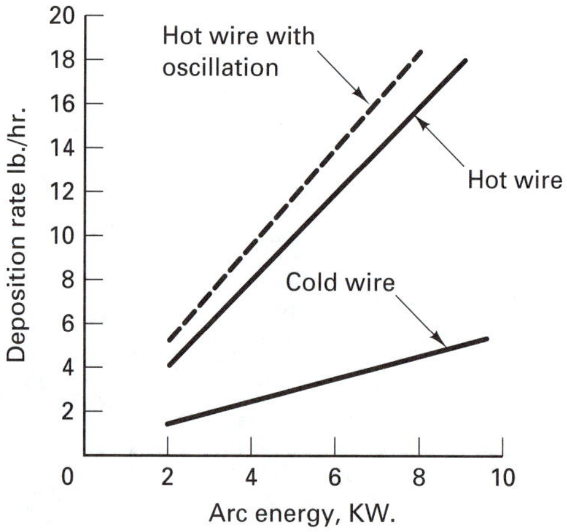

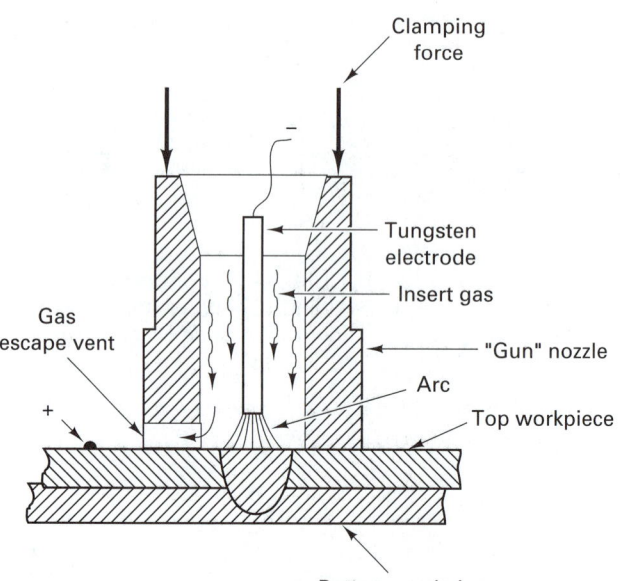

FIGURE 34-7 Comparison of the metal deposition rates in GTAW welding with cold, hot, and oscillating hot filler wire. *(Courtesy of Welding Journal.)*

FIGURE 34-8 Process schematic of spot welding by the inert-gas-shielded tungsten arc process.

Gas Tungsten Arc Spot Welding.

A variation of gas tungsten arc welding can be used to produce spot welds between two pieces of metal where access is limited to one side of the joint or where thin sheet is being attached to heavier material. The basic procedure is illustrated in Figures 34-8 and 34-9. A modified inert-gas tungsten arc gun is used with a vented nozzle on the end. The nozzle is pressed firmly against the material, holding the pieces in reasonably good contact. (The workpieces must be sufficiently rigid to sustain the contact pressure.) Inert gas, usually argon or helium, flows through the nozzle to provide a shielding atmosphere. Automatic controls advance the electrode to initiate the arc and then retract it to the correct distance for stabilized arcing. The duration of arcing is timed automatically to produce an acceptable spot weld. The depth and size of the weld nugget are controlled by the amperage, time, and type of shielding gas.

In arc spot welding, the weld nugget begins to form at the outside of one of the members being joined. With the more-standard resistance spot welding methods, the weld

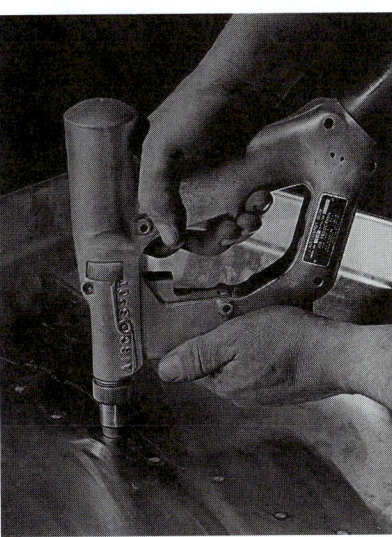

FIGURE 34-9 Making a spot weld by the inert-gas-shielded tungsten arc process. *(Courtesy of Air Reduction Company Inc.)*

nugget forms at the interface between the two members. Each technique has distinct advantages and disadvantages.

Gas Metal Arc Welding

Gas metal arc welding (GMAW), formerly known as MIG (metal inert-gas) welding, was a logical outgrowth of gas tungsten arc welding. The process is similar, but the arc is now maintained between the workpiece and an automatically fed bare-wire electrode. As shown in Figure 34-10, the consumable electrode provides the filler metal, so no additional feed is required. Argon, helium, and mixtures of the two can be used for welding virtually any metal but are used primarily with the nonferrous metals. When welding steel, some O_2 or CO_2 is usually added to improve the arc stability and reduce weld spatter. The cheaper CO_2 can be used alone when welding steel provided that a deoxidizing electrode wire is employed.

The specific shielding gases can have considerable effect on the nature of metal transfer from the electrode to the work and also influence the heat transfer behavior, penetration, and tendency for undercutting (weld pool extending laterally beneath the surface of the base metal). Several types of electronic controls can be used to alter the welding current, making it possible to control the mechanism of metal transfer, from drops, to spray, to short-circuiting drops. Some of these variations include *pulsed arc welding* (GMAW-P), *short-circuiting arc welding* (GMAW-S), and *spray transfer welding* (GMAW-ST). *Buried arc welding* (GMAW-B) is another variation in which carbon dioxide–rich gas is used and the arc is buried in its own crater.

Gas metal arc welding is fast and economical. There is no frequent change of electrodes as with the shielded metal arc process. No flux is required, and no slag forms over the weld. Thus multiple-pass welds can be made without the need for intermediate cleaning. The process can be readily automated, and the lightweight, compact welding unit lends itself to robotic manipulation. A reverse-polarity dc arc is generally used because of its deep penetration, spray transfer, and ability to produce smooth welds with good profile. Process variables include type of current, current magnitude, shielding gas, electrode diameter, electrode composition, electrode stickout (extension beyond the gun), welding speed,

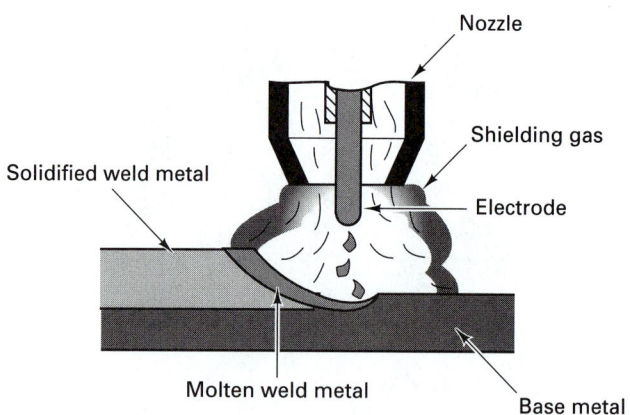

FIGURE 34-10   Schematic diagram of gas metal arc welding (GMAW). *(Courtesy of American Iron and Steel Institute, Washington, D.C.)*

welding voltage, and arc length. In a modification known as *advanced gas metal arc welding* (AGMAW), a second power source is used to preheat the filler wire before it emerges from the welding torch. With this modification, less arc heating is needed to produce a weld, so less base metal is melted, producing less dilution of the filler metal, and there is less penetration.

Pulsed Arc Gas Metal Arc Welding.   The benefits of heated filler metal are extended by the *pulsed arc* variation of gas metal arc welding (GMAW-P). A low welding current is first used to create a molten globule on the end of the filler wire. A burst of high current is then applied, which "explodes" the globule and transfers the metal across the arc in the form of a spray. By alternating the low and high currents at a rate of 60 to 120 times per second, the filler metal is transferred in a succession of rapid bursts, similar to the emissions of a rapidly squeezed aerosol atomizer. Because of the pulsed form of deposition, there is less heat input to the weld and the weld temperatures are reduced. Thinner material can be welded; distortion is reduced; workpiece discoloration is minimized; heat-sensitive parts can be welded; high-conductivity metals can be joined; electrode life is extended; electrode cooling techniques may not be required; and fine microstructures are produced in the weld pool. Welds can be made in all positions, and the use of pulsed power lowers spattering and improves the safety of the process. The high speed of the process is attractive for productivity and the energy or power required to produce a weld is lower than with other methods (reduced cost). Controls can be adjusted to alter the shape of the weld pool and vary the penetration.

Flux-Cored Arc Welding

Flux-cored arc welding (FCAW) utilizes a continuous tubular electrode wire filled with *flux* (Figure 34-11). In many ways the process is similar to shielded metal arc welding, but the electrode is now continuous and less bulky, since a binder is no longer required to hold the flux in place. When the arc is established, the vaporizing flux produces a protective atmosphere and also forms a slag layer over the weld pool that will require subsequent removal. Alloy additions (metal powders) can be blended into the flux to create a wide variety of filler metal chemistries.

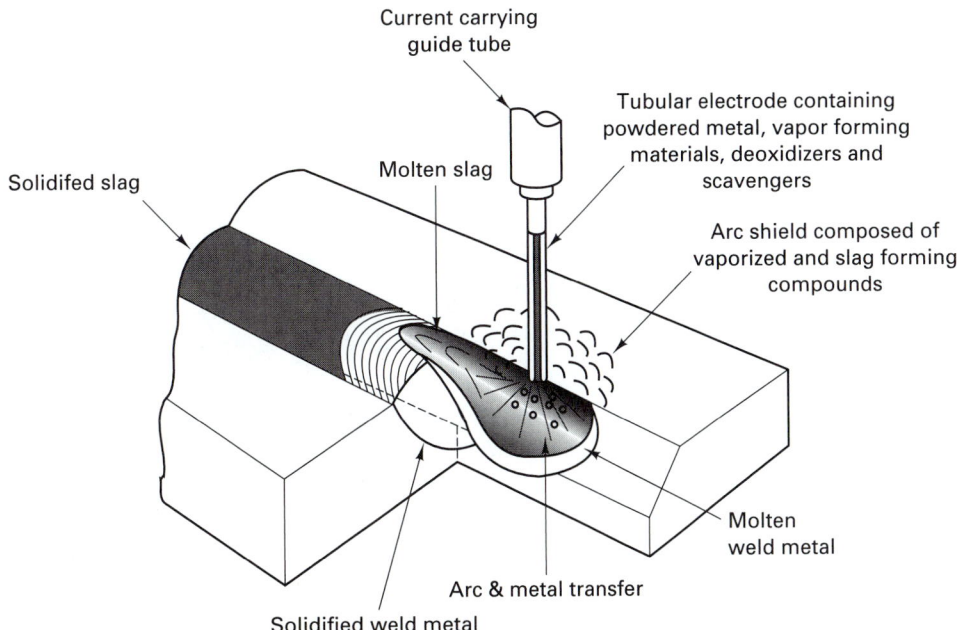

FIGURE 34-11   Schematic representation of the flux-cored arc welding process (FCAW). *(Courtesy of The American Welding Society, New York.)*

In other ways the flux-cored process is similar to gas metal arc welding. The continuous bare-wire electrode is fed automatically through a welding gun, with electrical contact being maintained through the exterior of the wire. Welds can be made in all positions, and high deposition rates are possible. For enhanced weld properties, an externally supplied shielding gas, generally CO_2, may be added, essentially combining the gas metal arc process with a flux-cored electrode. This process is used primarily for welding ferrous material, almost always with a reverse-polarity dc power source.

Submerged Arc Welding

In *submerged arc welding* (SAW), a thick layer of granular flux is deposited just ahead of a bare wire consumable electrode, and an arc is maintained beneath the blanket of flux with only a few small flames being visible. A portion of the flux melts and acts to remove impurities from the rather large pool of molten metal, while the unmelted excess provides additional shielding. The molten flux then solidifies into a glasslike covering over the weld. This layer, along with the flux that is not melted, provides a thermal insulation that slows the cooling of the weld metal and helps to produce soft, ductile welds. The solidified flux cracks loose from the weld upon cooling (due to the differential thermal contraction) and is easily removed. The unmelted flux is recovered by a vacuum system and reused. Figure 34-12 provides a schematic of the process.

Submerged arc welding is most suitable for making flat butt or fillet welds in low-carbon steel ($< 0.3\%$ carbon). With some preheat and postheat precautions, medium-carbon and alloy steels and some cast irons, stainless steels, copper alloys, and nickel alloys can be welded. The process is not recommended for high-carbon steels, tool steels, aluminum, magnesium, titanium, lead, or zinc. The reasons for this incompatibility are somewhat varied, including the unavailability of suitable fluxes, reactivity at high temperatures, and low sublimation temperatures.

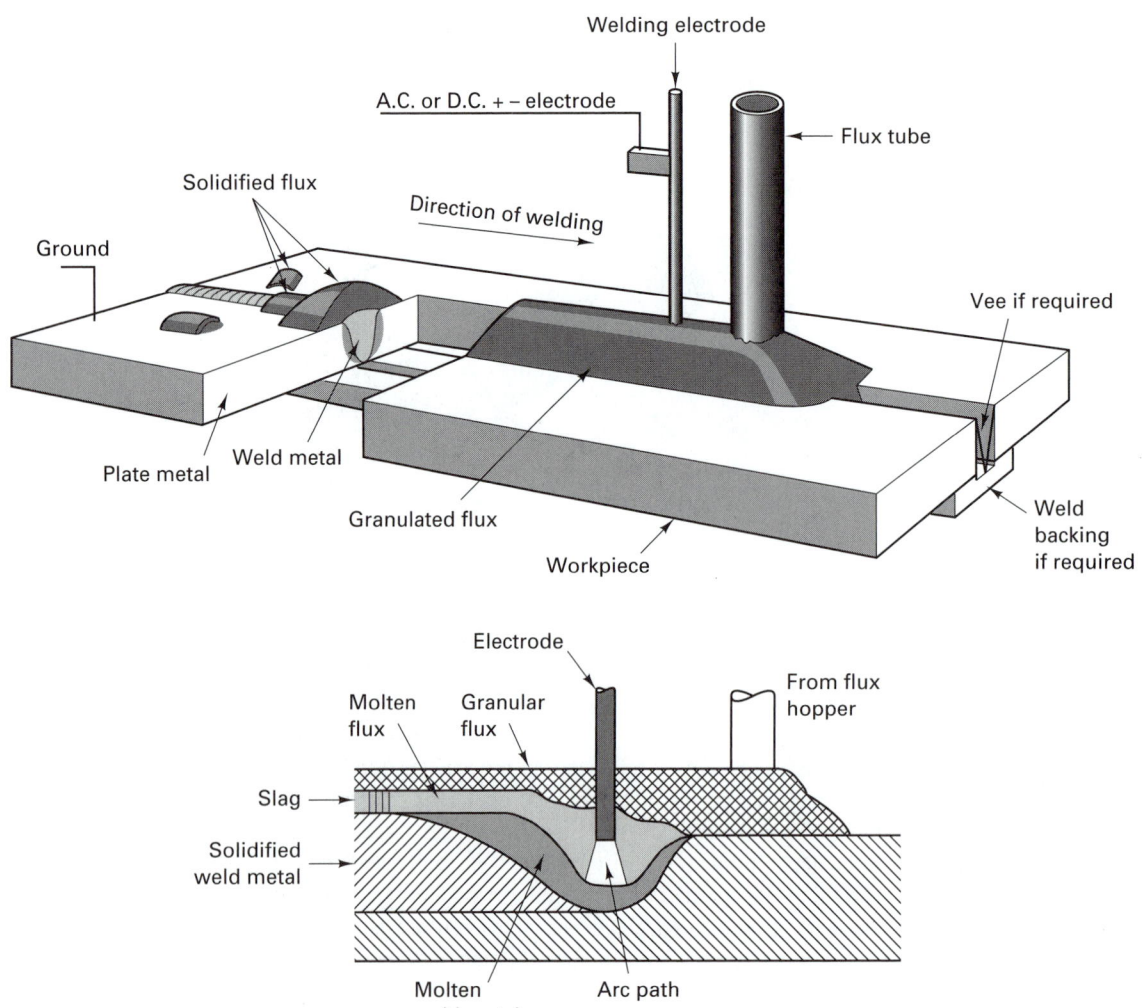

FIGURE 34-12 (*Top*) Basic features of the submerged arc welding process (SAW). *(Courtesy of Linde Division, Union Carbide Corporation.)* (*Bottom*) Cutaway schematic of submerged arc welding. *(Courtesy of American Iron and Steel Institute, Washington, D.C.)*

High welding speeds, high deposition rates, deep penetration, and high cleanliness (due to the flux action) are all characteristic of submerged arc welding. Welding speeds of 30 in./min in 1-in.-thick steel plate or 12 in./min in $1\frac{1}{2}$-in. plate are not uncommon. Single-pass welds $1\frac{1}{2}$ inches deep can be made, and almost any thickness of base metal can be joined. Because the metal is deposited in fewer passes than with alternative processes, there is less possibility of entrapped slag or voids, and weld quality is further enhanced. When even higher deposition rates are desired, multiple electrode wires can be employed.

Limitations to the process include the need for extensive flux handling, possible contamination of the flux by moisture (leading to porosity), the large volume of slag that must be removed, the high heat inputs that promote large-grain-size structures, and the slow cooling rate (which permits segregation and possible hot cracking). Welding is restricted to the horizontal position, since the flux and slag are held in place by gravity. In addition, chemical control is quite important, since the electrode material often comprises over 70% of the molten weld region.

The electrodes are generally classified by composition and are available in diameters from 0.045 in. to $\frac{3}{8}$ in. The larger electrodes carry higher currents and enable more rapid deposition, but penetration is shallower. Electrodes for welding alloy steels may take several forms: solid wire of the desired alloy, plain carbon steel with the alloy additions being incorporated into the flux, or tubular metal with the alloy additions in the hollow core. Various fluxes are also available and are selected for compatibility with the weld metal. All are designed to have low melting temperatures, good fluidity, and brittleness after cooling.

Submerged arc equipment may be semiautomatic, with the operator controlling the speed, or fully automatic. They may be portable (wherein the welder traverses over a stationary workpiece), or stationary (wherein the workpiece passes under the arc). Both ac and dc power can be used, with currents in the range of 600 to 2000 A.

Bulk Welding. In a modification of the submerged arc process known as *bulk welding*, iron powder is first deposited into the joint (ahead of the flux) as a means of increasing deposition rate. A single weld pass can then produce enough filler metal to be equivalent to seven or eight conventional submerged arc passes.

Plasma Arc Welding

In *plasma arc welding* (PAW) the arc is maintained between a nonconsumable electrode and either the welding gun (*nontransferred arc*) or the workpiece (*transferred arc*) (Figure 34-13). The nonconsumable tungsten electrode is set back within the "torch" in such a way as to force the arc to pass through a small-diameter nozzle. An inert gas (usually argon) is forced through this constricted arc, where it is heated to a high temperature and forms a *plasma*. The hot ionized gas then transfers its heat to the workpiece and melts the metal. When needed, filler metal can be fed into the plasma column, which is often surrounded by the separate flow of inert gas used to provide shielding for the weld pool.

Plasma arc welding is characterized by temperatures on the order of 30,000°F (16,500°C) and a high-energy concentration. This in turn offers fast welding speeds, narrow welds with deep penetration, a narrow *heat-affected zone*, reduced distortion, and a process that is insensitive to arc length. With a low-pressure plasma, the metal simply melts and flows into the joint. At higher pressures, a "keyhole" effect occurs in which the plasma gas creates a hole completely through the sheet (up to $\frac{1}{4}$ in. thick) that is surrounded by molten metal. As the torch is moved, liquid metal flows to fill the keyhole. If the gas pressure is increased even further, the molten metal is expelled from the region and the process becomes one of plasma cutting, which is discussed later in this chapter. Welds can be made in all positions, and nearly all metals and alloys can be welded.

Many plasma torches employ a small nontransferred arc within the torch to heat the orifice gas and ionize it. The ionized gas then forms a good conductive path for the main transferred arc. This dual-arc technique permits instant ignition of a low-current arc, which is more stable than that of an ordinary plasma torch. Separate dc power supplies are used for the pilot and main arcs.

Stud Welding

Stud welding (SW) is an arc-welding process used to attach studs or fasteners to a metal surface. A special gun is used, such as the one shown in Figure 34-14, into which the stud

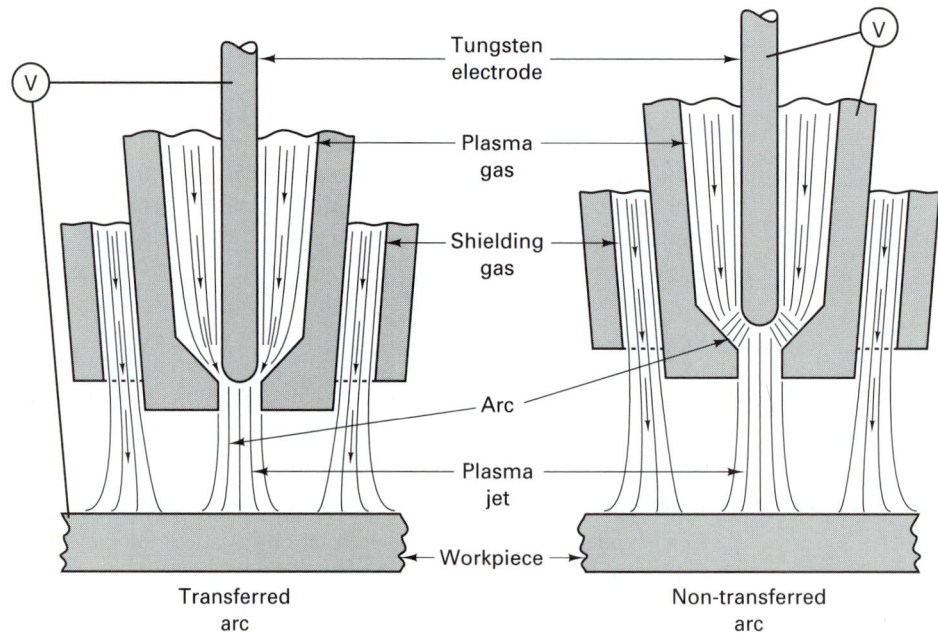

FIGURE 34-13 Two types of plasma arc torches. (*Left*) transferred arc; (*right*) nontransferred arc.

is inserted. The stud acts as an electrode and a dc arc is established between the end of the stud and the workpiece. After a small amount of metal is melted, the two pieces are brought together under light pressure and allowed to solidify. Automatic equipment controls the arc, its duration, and the application of pressure to the stud.

Figure 34-15 shows some of the wide variety of studs that are specially made for this process, often containing a recessed end that is filled with flux. A ceramic ferrule, such as the one shown in the center photo of Figure 34-15, is usually placed over the end of the stud before it is positioned in the gun. During the arc, the ferrule serves to concentrate the heat and isolate the hot metal from the atmosphere. It also confines the molten or softened metal and shapes it around the base of the stud, as shown in the right-hand photo of Figure 34-15. After the weld has cooled, the brittle ceramic can be broken free. Since burn-off or melting reduces the length of the stud, the original dimensions should be selected to compensate.

Stud welding requires almost no skill on the part of the operator. Once the stud and ferrule are placed in the gun and the gun positioned on the work, all the operator has to do is pull the trigger. The cycle is executed automatically and consumes less than 1 second. Thus the process is well suited to manufacturing and can be used to eliminate the drilling and tapping of many special holes. Production-type stud welders can produce over 1000 welds per hour.

Power Sources for Arc Welding

Arc welding requires a large electrical current, often in the range 150 to 1000 A. The load voltage is usually between 30 and 40 V, with the actual voltage across the arc varying from 12 to 30 V, depending on the arc length. Both dc and ac *power supplies* are available and

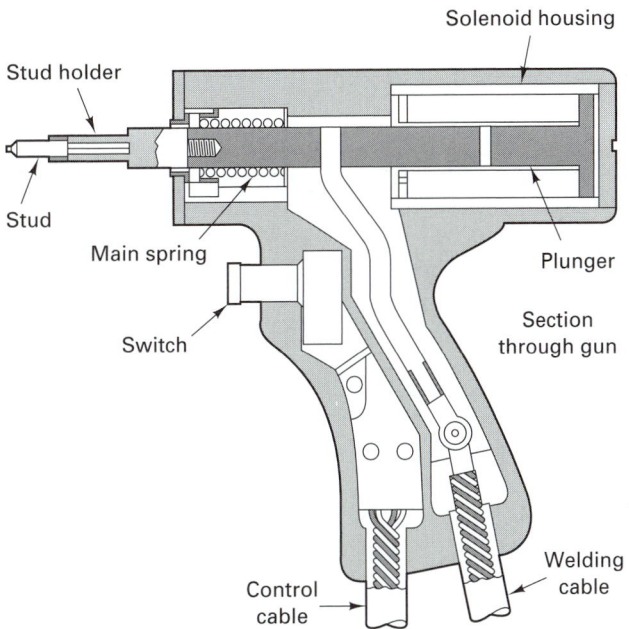

FIGURE 34-14 Schematic diagram of a stud welding gun.
(Courtesy of American Machinist.)

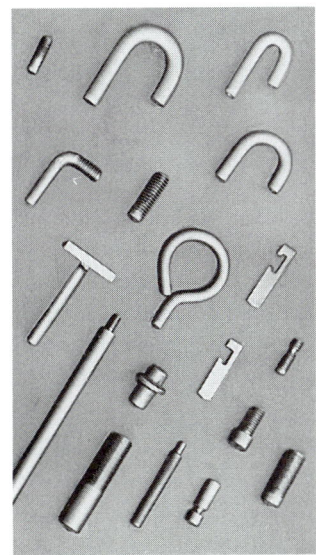
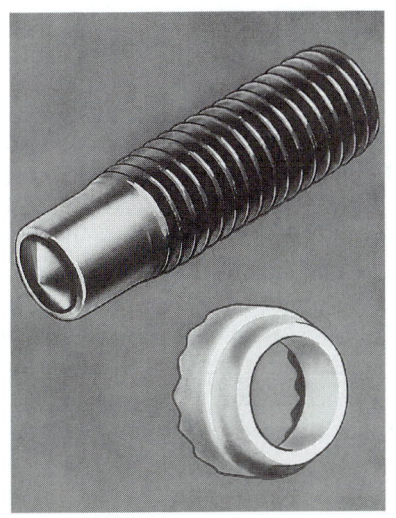

FIGURE 34-15 (*Left*) Types of studs used for stud welding. (*Center*) Stud and
ceramic ferrule. (*Right*) Stud after welding and a section through a welded stud.
(Courtesy of Nelson Stud Welding Co.)

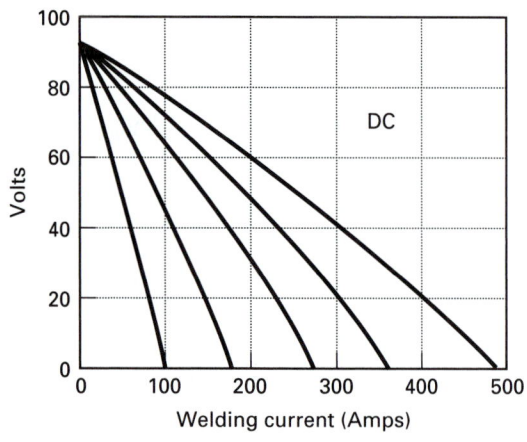

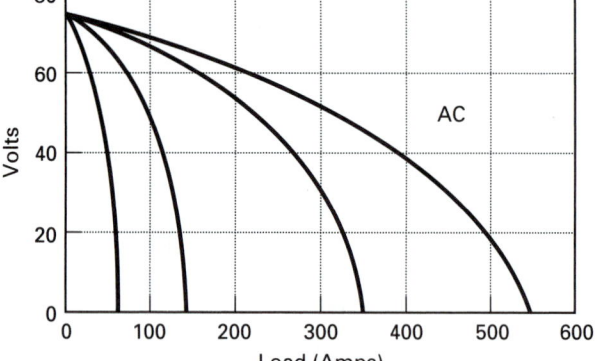

FIGURE 34-16 Drooping-voltage characteristics of typical arc-welding power supplies. (*Top*) Direct current; (*bottom*) Alternating current.

generally employ the "drooping voltage" characteristics shown in Figure 34-16. These characteristics are designed to minimize changes in welding current as the welding voltage fluctuates within the usual operating range.

In earlier years, most direct current for welding was provided by motor-generator sets. Today, however, solid-state transformer-rectifier machines, such as the one shown in Figure 34-17, are the most frequently used source of power. Operating on a three-phase electrical line, these machines can usually provide both ac and dc output. When welding is to be performed in remote locations, the gasoline-driven dc generators are still utilized.

If only ac welding is to be performed, relatively simple transformer-type power supplies can be used. These are usually single-phase devices with low power factors. When multiple machines are to be operated, as in a production-type shop, they are often connected to the various phases of a three-phase supply to help balance the load.

Jigs and Positioners

Jigs or fixtures (also called *positioners*) are frequently used to hold the work in production welding. By positioning and manipulating the workpiece, the welding operation can often be performed in a more favorable position. Parts can also be mounted on numerically controlled (NC) tables that manipulate and position the workpiece with respect to the welding tool.

FIGURE 34-17   Rectifier-type ac and dc welding power supply. *(Courtesy of Lincoln Electric Company.)*

◼ 34.2 ARC CUTTING

Virtually all metals can be cut by electric arc methods. Here the material is melted by the intense heat of the arc and then permitted, or forced, to flow away from the region of the slit or notch (*kerf*). Most of the processes are simply adaptations of the arc welding procedures discussed in this chapter.

Carbon Arc and Shielded Metal Arc Cutting

The *carbon arc cutting* (CAC) and *shielded metal arc cutting* (SMAC) methods use the arc from a carbon or shielded metal arc electrode to melt the metal, which is then removed from the cut by gravity or the force of the arc. Use is generally limited to small shops, garages, and homes where the equipment investment is small.

Air Carbon Arc Cutting

In *air carbon arc cutting* the arc is again maintained between a carbon electrode and the workpiece, but high-velocity jets of air are directed at the molten metal from holes in the electrode holder (Figure 34-18). While there is some oxidation, the primary function of the air is to blow the molten material from the cut. Thus the process can be used on metal that does not readily oxidize.

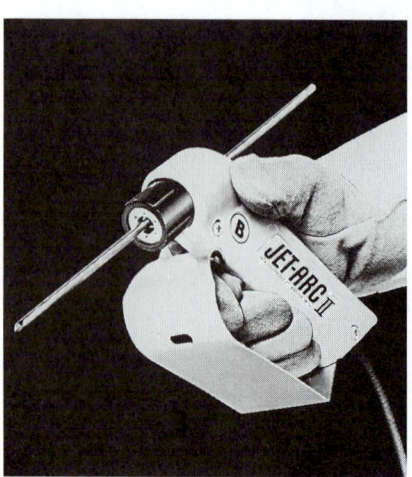

FIGURE 34-18   Gun used in the arc-air cutting process. Note the air holes surrounding the electrode in the holder. *(Courtesy of Jackson Products.)*

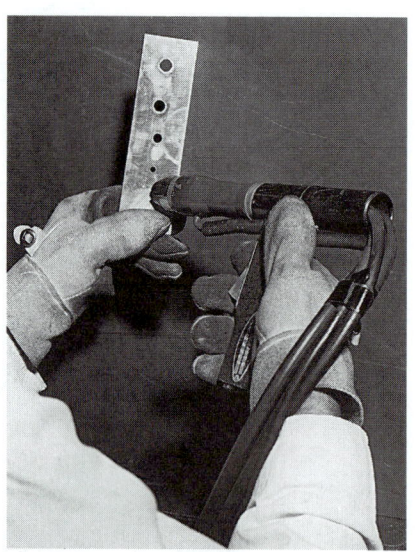

FIGURE 34-19   Making holes in sheet metal by the inert-gas arc process. *(Courtesy of Hobart Brothers Company.)*

Air carbon arc cutting is particularly effective for cutting cast iron and preparing steel plates for welding. Speeds up to 24 inches per minute are possible, but the process is quite noisy and hot metal particles tend to be blown out over a substantial area.

Oxygen Arc Cutting

In *oxygen arc cutting* (AOC), an electric arc and a stream of oxygen are employed to make the cut. The electrode is a coated ferrous-metal tube. The coated metal serves to establish a stable arc, while oxygen flows through the bore and is directed on the area of incandescence. In easily oxidizable metals, such as steel, the arc preheats the base metal, which then reacts with oxygen, is liquefied, and is expelled by the oxygen stream.

Gas Metal Arc Cutting

If the wire feed rate and other variables of gas metal arc welding (MIG welding) are adjusted so that the electrode penetrates completely through the workpiece, cutting rather than welding will occur. This is called *gas metal arc cutting* (GMAC). Wire feed rate controls the quality of the cut and the voltage determines the width of the slit or kerf.

Gas Tungsten Arc Cutting

The same basic circuit and shielding gas is used in *gas tungsten arc cutting* (GTAC) as in gas tungsten arc welding; however, a high-velocity jet of gas is added to expel the molten metal. Figure 34-19 shows a modification of gas tungsten arc spot welding being used to create holes up to $\frac{3}{8}$ in. in diameter in sheet metal.

Plasma Arc Cutting

The torches used in *plasma arc cutting* (PAC) produce the highest temperature available from any practical source. The two basic types of plasma torches were discussed earlier in this chapter. With the nontransferred-type torch, the arc column is completely within the nozzle, and a temperature of about 30,000°F (16,650°C) is obtained. With the transfer-type torch, the arc is maintained between the electrode and the workpiece, and the temperatures can be up to 60,000°F (33,300°C). Such high temperatures enable the cutting of virtually any material simply by melting it and blowing it from the cut.

Early efforts to employ this technique showed that the speed, versatility, and operating cost were far superior to those of the oxyfuel cutting methods. However, these early systems could not constrict the arc sufficiently to produce the quality of cut needed to meet the demands of manufacturing. Therefore, plasma arc cutting was generally limited to those materials that could not be cut by the oxidation-type cutting techniques. Radial impingement of water on the arc was found to produce the desired constriction and provides an intense, highly focused source of heat. Water-injected torches can now cut virtually any metal in any position. Magnetic fields have also been used to constrict the arc and produce ultrahigh-quality cuts without the need for water impingement.

Compared to the oxyfuel technique, plasma cutting is more economical (cost per cut is a fraction of oxyfuel), more versatile (can cut all metals as easily as mild steel), and much faster (typically, five to eight times faster than oxyfuel). Cutting speeds up to 300 in./min have been obtained in $\frac{1}{4}$-in. aluminum, and up to 100 in./min in $\frac{1}{2}$-in. steel. The combination of the extremely high temperature and jetlike action of the plasma produces narrow kerfs and remarkably smooth surfaces, nearly as smooth as can be obtained by sawing. Plasma cut surfaces are often within 2° of vertical, and surface oxidation is nearly eliminated by the cooling effect of the water spray. In addition, the heat-affected zone in the metal is only one-third to one-fourth as large as that produced by oxyfuel cutting. Heat-related distortion is virtually eliminated.

Integration with robots or CNC (computer numerically controlled) machines provides for fast, clean, and accurate cutting, as shown in Figure 34-20. Plasma arc torches have also been incorporated into punch presses to provide a manufacturing machine with outstanding flexibility in producing cut and punched products from a variety of materials. Transferred-type torches are usually used for cutting metals, while the nontransferred type are employed with the low-conductivity nonmetals.

FIGURE 34-20 Cutting sheet metal with a plasma torch. *(Courtesy of GTE Sylvania.)*

Inexpensive nitrogen is the primary gas used when cutting all types of metal. For cutting thick sections, such as 5 to 6 in. in thickness, an argon–hydrogen mixture is used to provide a deeper-penetrating arc.

■ 34.3 METALLURGICAL AND HEAT CONSIDERATIONS IN THERMAL CUTTING

The flame and arc processes expose materials to localized high temperatures, and can produce harmful metallurgical effects. If the cut edges will be subsequently welded, or if they will be removed by machining, there is little cause for concern. For other applications, consideration should be given to the effects of cutting heat and their interaction with the applied loads. In some cases, additional steps may be required to avoid or overcome harmful consequences.

For carbon steels with less than 0.25% carbon, thermal cutting does not produce serious metallurgical effects. However, in steels of higher carbon content, the metallurgical changes can be quite significant, and preheating and/or postheating may be required. For alloy steels, additional consideration must be given to the effects of the various alloy elements. Chromium, molybdenum, and tungsten can be particularly detrimental to cutting.

Because of the low rate of heat input, oxyacetylene cutting will produce a rather large heat-affected zone. The various arc cutting methods produce effects that are quite similar to those of arc welding. Plasma arc cutting is so rapid, and the heat is so localized, that the original properties of a metal are generally modified only within $\frac{1}{16}$ in. of the cut.

All of the thermal cutting processes produce some *residual stresses*, with the cut surface generally in tension. Except in the case of thin sheet, warping should not occur. However, if subsequent machining removes only a portion of the cut surface, or does not penetrate to a sufficient depth, the resulting imbalance in residual stresses can induce distortion. Therefore, if subsequent machining is to be done, it may be necessary to remove all cut surfaces to a substantial depth if dimensional stability is to be assured.

Thermal cutting can also introduce geometrical features into the product. All flame- or arc-cut edges are rough to varying degrees and thus contain notches that can act as stress raisers and reduce the endurance or fracture strength. If cut edges are to be subjected to high or repeated tensile stressing, the cut surface and the heat-affected zone should be removed by machining, or at least subjected to a stress-relief heat treatment.

■ KEY WORDS

arc	globular transfer	reverse polarity
arc cutting	heat-affected zone	shielded metal arc welding
arc welding	kerf	short-circuit transfer
bulk welding	nonconsumable electrode process	spray transfer
consumable electrode process	nontransferred arc	straight polarity
electrode	plasma	stud welding
filler metal	plasma arc welding	submerged arc welding
flux	power supply	transferred arc
flux-cored arc welding	pulsed arc	
gas tungsten arc welding	residual stresses	

■ REVIEW QUESTIONS

1. What sorts of problems plagued early attempts to develop arc welding?

2. What three basic types of current and polarity are used in arc welding?

3. What is the difference between a consumable and a nonconsumable electrode? For which processes does a filler metal have to be added by a separate mechanism?

4. What are the three types of metal transfer that can occur during arc welding?

5. What are some of the process variables that must be specified when setting up an arc welding process?

6. What are some of the functions of the electrode coatings in shielded metal arc welding?

7. How are welding electrodes commonly classified, and what information does the designation usually provide?

8. Why are shielded metal arc electrodes often baked just prior to welding?

9. What is the function of the slag coating that forms over a shielded metal arc weld?

10. What benefit can be obtained by placing iron powder in the coating of shielded metal arc electrodes that will be used to weld ferrous metals?

11. Why are shielded metal arc electrodes generally limited in length, making the process one of intermittent operation?

12. What are some of the commonly used shielding gases in the gas tungsten arc process?

13. What methods might be employed to increase the rate of filler metal deposition during gas tungsten arc welding?

14. What are some of the attractive features of gas tungsten arc welding?

15. What are some attractive features of gas tungsten arc spot welding?

16. Why might gas metal arc welding be preferred over shielded metal arc for production welding?

17. What are the primary process variables of gas metal arc welding?

18. Describe the metal transfer that occurs during pulsed arc gas metal arc welding.

19. What are some of the benefits that can be obtained by the reduced heating of the pulsed arc process?

20. What is the advantage of placing the flux in the center of an electrode (flux-cored arc welding) as opposed to the outside (shielded metal arc welding)?

21. What are some of the functions of the flux in submerged arc welding?

22. From a production viewpoint, what are some of the attractive characteristics of submerged arc welding? Major limitations?

23. What is the primary goal or objective in bulk welding?

24. How is the heating of the workpiece during plasma arc welding different from the other arc welding techniques?

25. What are some of the attractive features of plasma arc welding?

26. What is the primary difference between plasma arc welding and plasma arc cutting?

27. What is the special-use application of stud welding?

28. What is the function of the ceramic ferrule placed over the end of the stud in stud welding?

29. What is the purpose of the oxygen in oxygen arc cutting?

30. How are the gas metal arc welding and gas tungsten arc welding processes converted to cutting?

31. Why is plasma arc cutting an attractive way of cutting the high-melting-point metals?

32. What techniques can be used to constrict the arc in plasma arc cutting, thereby producing a narrower, more controlled cut?

33. What may make plasma arc cutting more attractive than the oxyfuel method?

34. Why is there little concern about heat-related effects on cut edges that will subsequently be welded?

35. How does the size of the heat-affected zone vary with the various cutting processes: oxyfuel, arc, and plasma?

36. Why might the residual stresses induced during cutting operations be objectionable?

37. Why might it be wise to machine away the cut edge and heat-affected zone of a stressed machine part?

*C*hapter 34 CASE STUDY

repair of a bicycle frame

The frame of a high-quality, light-weight bicycle has been fabricated from aluminum alloy tubing, where the various joints have been made by adhesively-bonding the tubes to connectors that incorporate either internal lugs or external sleeves. As the result of abuse, the frame has fractured near one of the joints. Seeking to recover use of the bicycle, the owner contacted a local auto body or muffler shop and requested that a repair be made using conventional gas tungsten arc welding. The repair seemed to be of good quality, but, shortly thereafter, the frame again broke. This time the fracture was adjacent to the repair weld and the characteristics of the break were different. While the first fracture was somewhat brittle in nature, the second appeared to be more ductile, with evidence of metal flow prior to fracture. Since the second fracture occurred in the unwelded tube material, the welder feels that the failure is not related to the weld and the material in the tubing must be defective.

1. If the material is cold-drawn aluminum tubing (i.e. strain hardened), explain what may have occurred during the repair. What is the probable cause of the second fracture? Was the weld in any way defective? Was the second failure related to the welding repair?

2. If the tubing had been strengthened by age hardening, could the same results have occurred? Explain.

3. Is there a better means of repairing the original fracture? What would have been your recommendation?

CHAPTER 35

RESISTANCE WELDING

35.1   THEORY OF RESISTANCE
       WELDING
       Heating
       Pressure
       Current Control
       Power Supply
35.2   RESISTANCE WELDING
       PROCESSES

       Resistance Spot Welding
          Spot Welding
          Equipment
          Spot-Weldable Metals
       Resistance Seam Welding
       Projection Welding
35.3   ADVANTAGES AND
       LIMITATIONS

■ 35.1 THEORY OF RESISTANCE WELDING

In *resistance welding*, both heat and pressure are used to induce *coalescence*. The heat is generated by the electrical resistance of the workpieces and the interface between them. The pressure is supplied externally and is varied throughout the weld cycle. A certain amount of pressure is applied initially to hold the workpieces in contact and thereby control the electrical resistance at the interface. When the proper temperature is attained, the pressure is increased to facilitate coalescence. Because pressure is utilized, coalescence occurs at a lower temperature than that required for oxyfuel gas or arc welding. In fact, melting of the base metal does not occur in many resistance welding operations. Resistance welding processes could well be considered as a form of solid-state welding, although they are not officially classified as such by the American Welding Society.

In some resistance welding processes, additional pressure is applied immediately after coalescence to provide a certain amount of forging action, with accompanying grain refinement. Additional heating can also be employed after welding to provide tempering and/or stress relief.

Usually, the required temperature can be attained and coalescence achieved in a few seconds or less. Consequently, resistance welding is a very rapid and economical process, extremely well suited to automated manufacturing. No filler metal is required, and the tight contact maintained between the piece excludes air and eliminates the need for fluxes or shielding gases.

Heating

The heat for resistance welding is obtained by passing a large electrical current through the workpieces for a short period of time. The amount of heat input can be determined by the basic relationship

$$H = I^2RT$$

where

H = total heat input in joules
I = current in amperes
R = electrical resistance of the circuit in ohms
T = length of time during which current is flowing in seconds.

It is important to note that the workpieces actually form part of the electrical circuit, as

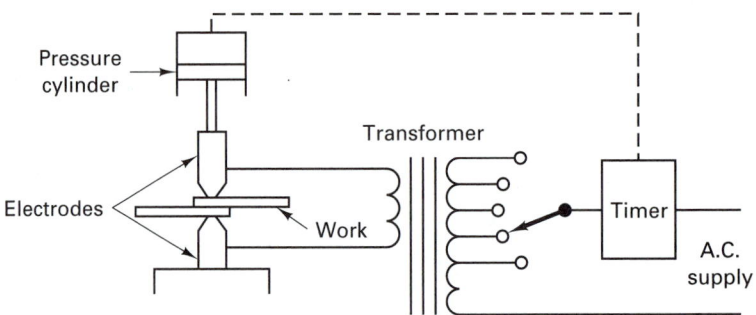

FIGURE 35-1 The fundamental resistance-welding circuit.

illustrated in Figure 35-1 and that the total resistance between the electrodes consists of three components:

1. Resistance of the workpieces
2. Contact resistance between the electrodes and the work
3. Resistance between the surfaces to be joined, known as the *faying surfaces*

Because it is desirable to have the maximum temperature occur where the weld is to be made, it is essential to keep resistances 1 and 2 as low as possible with respect to resistance 3. The resistance of the workpieces (1) is determined by the type and thickness of the metal. This is usually much less than the other two resistances, however, because of the larger area involved and the relatively high electrical conductivity of most metals. The resistance between the work and the electrodes (2) can be minimized by using electrode materials that are excellent electrical conductors, by controlling the shape and size of the electrodes, and by applying proper pressure. However, any change in the pressure between the electrodes and workpieces also changes the pressure between the faying surfaces. Therefore, only limited control of the electrode-to-work resistance can be obtained by variation in pressure.

Finally, the resistance between the faying surfaces (3) is a function of the quality (surface finish or roughness) of the surfaces; the presence of nonconductive scale, dirt, or other contaminants; the pressure; and the contact area. These factors must all be controlled to obtain uniform and consistent results.

As indicated in Figure 35-2, the objective of resistance welding is to bring both of the faying surfaces to the proper temperature simultaneously, while keeping the remaining material and the electrodes relatively cool. The electrodes are usually water cooled to keep their temperature low and thereby extend their useful life.

Pressure

Because the pressure in resistance welding promotes a forging action, resistance welds can be produced at lower temperatures than welds made by other processes. However, the control of both the magnitude and timing of the pressure is very important. If too little pressure is used, the contact resistance will be high and surface burning or pitting of the electrodes may result. If excessive pressure is applied, molten or softened metal may be expelled from between the faying surfaces or the softened workpiece may be indented by the electrodes. Ideally, moderate pressure should be applied to hold the workpieces in place and establish proper resistance at the interface prior to and during the passage of the

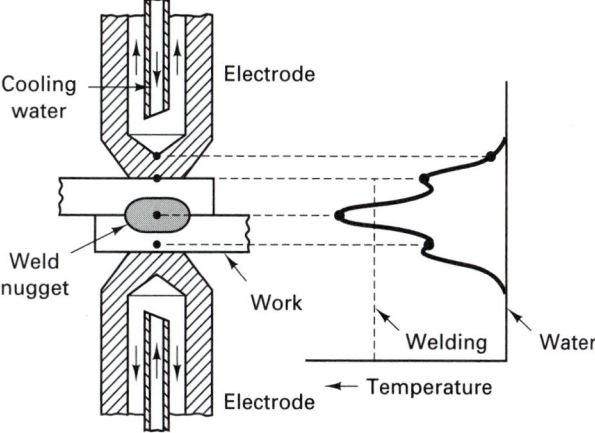

FIGURE 35-2   The desired temperature distribution across the electrodes and the workpieces in lap resistance welding.

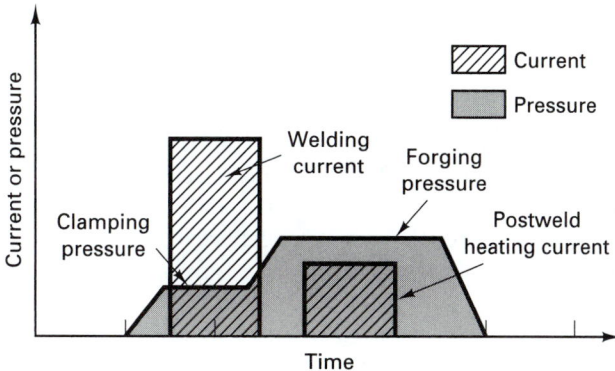

FIGURE 35-3   Typical current and pressure cycle for resistance welding. The cycle includes forging and postheating operations.

welding current. The pressure should then be increased considerably just as the proper welding heat is attained. This completes the coalescence and forges the weld to produce a fine-grained structure.

On small, foot-operated machines, only a single spring-controlled pressure is used. On larger, production-type welders, the pressure is generally applied through controllable air or hydraulic cylinders.

Current Control

With the surface conditions held constant and the pressure controlled, the temperature in resistance welding is then regulated by controlling the magnitude and duration of the welding current. Very precise and sophisticated controls are available for this purpose.

In large production-type welders, the magnitude, duration, and timing of both current and pressure can be programmed to follow specified cycles. Figure 35-3 shows a relatively simple cycle for a resistance weld that includes forging and postheating operations.

Power Supply

Since the overall resistance in the welding circuits can be quite low, high currents are generally required to produce a resistance weld. Power transformers convert the high-voltage,

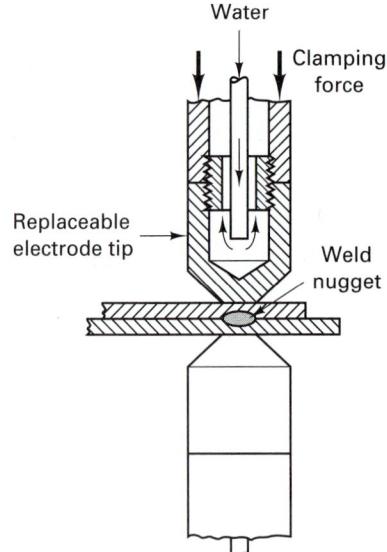

FIGURE 35-4 The arrangement of the electrodes and the work in spot welding, showing design for replaceable electrode tips.

low-current line power to the high-current (up to 100,000 A) low-voltage (0.5 to 10 V) power required for welding. While smaller machines may utilize single-phase circuitry, larger units generally operate on three-phase power. Many resistance welders use dc welding current, obtained through solid-state rectification of the three-phase power. These machines reduce the current demand per phase, give a balanced load, and produce excellent welds.

■ 35.2 RESISTANCE WELDING PROCESSES

Resistance Spot Welding

Resistance spot welding (RSW) is the simplest and most widely used form of resistance welding. As shown in Figure 35-4, overlapping sheets are positioned between water-cooled electrodes, which have reduced areas at the tips to produce welds that are usually from $\frac{1}{16}$ to $\frac{1}{2}$ in. (1.5 to 13 mm) in diameter. The electrodes close on the work, and the controlled cycle of pressure and current is applied to produce a weld at the metal interface. The electrodes then open and the work is removed.

A satisfactory spot weld, such as the one shown in Figure 35-5, consists of a *nugget* of coalesced metal formed between the faying surfaces. There should be little indentation of the metal under the electrodes. The strength of the weld should be such that in a tensile or tear test, the weld will remain intact and failure will occur in the heat-affected zone surrounding the nugget (Figure 35-6). If proper current density and timing, electrode shape, electrode pressure, and surface conditions are maintained, sound spot welds can be obtained with excellent consistency.

Spot-Welding Equipment. A variety of spot-welding equipment is available to meet the various needs of production operations. For light-production work where complex current-pressure cycles are not required, a simple *rocker-arm machine*, like that of Figure 35-7, is often used. The lower electrode arm is stationary, and the upper electrode, mounted on a pivot arm, is brought down into contact with the work by means of a spring-loaded foot

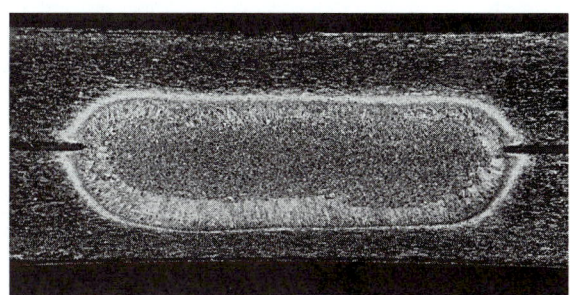

FIGURE 35-5 A spot-weld nugget between two sheets of 0.05-in. (1.3-mm) aluminum alloy. The nugget is not symmetrical because the radius of the upper electrode was greater than that of the lower electrode. *(Courtesy Locheed Aircraft Corporation.)*

FIGURE 35-6 Tear test of a satisfactory spot weld, showing how failure occurs outside of the weld.

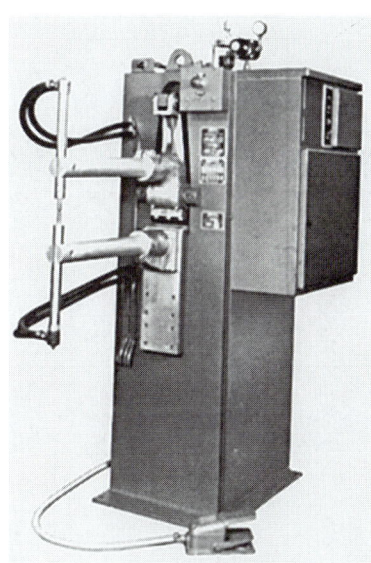

FIGURE 35-7 Foot-operated rocker-arm, spot-welding machine. *(Courtesy Sciaky Bros., Inc.)*

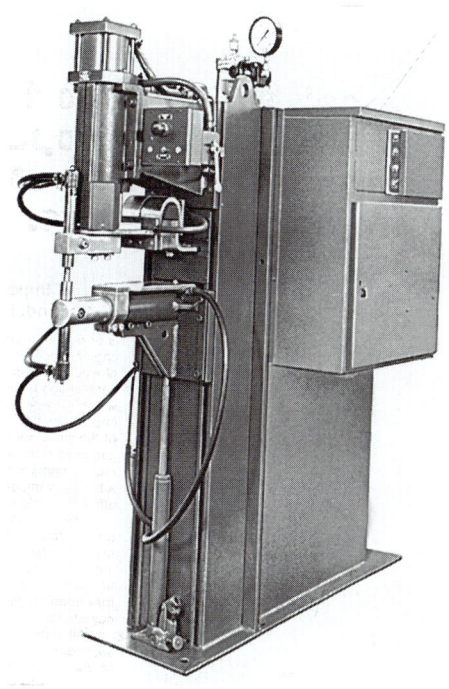

FIGURE 35-8 Single-phase, air-operated, press-type resistance welder with microprocessor control. *(Courtesy Sciaky Bros., Inc.)*

pedal. Rocker-arm machines are available with throat depths up to about 48 in. (1200 mm) and with transformer capacities up to 50 kVa. They are used primarily on steel.

Most large spot welders, and those used at high production rates, are of the *presstype* (Figure 35-8). On these machines, the movable electrode is controlled by an air or hydraulic cylinder, and complex pressure cycles can be programmed. Capacities up to 500 kVa with a 60-in. (1500-mm) throat depth are common. Special-purpose press-type welders can employ multiple welding heads to make up to 200 simultaneous spot welds in under 60 seconds.

The application of spot welding has been greatly extended through the use of *portable spot-welding guns.* These guns are connected to a stationary power supply and control unit by flexible air hoses, electrical cables, and water cooling lines. By permitting the welding process to be brought to the work, the portable guns extend the use of spot welding to applications where the work is too large to be positioned on a welding machine. These units can also be installed on industrial robots, which can be programmed to position the gun at various locations and produce spot welds in an automated fashion, without the need for an operator. Such a process offers a high degree of flexibility coupled with consistent quality and is now quite common in the automotive industry.

In the early 1980s, electronic advances enabled the development of welding transformers that would fit in the space occupied by the connecting cables, and integral transformer guns, or *transguns*, were introduced. By transforming the power immediately adjacent to the gun, these small, integrated units reduce power losses and operate with a noticeably enhanced efficiency. However, when accurate positioning is required in an articulated system like an industrial robot, the added weight of the integral transformer may be an offsetting disadvantage.

Spot-Weldable Metals. While steel is clearly the most common metal that is spot welded, one of the greatest advantages of the process is that virtually all of the commercial metals can be spot welded, and most of them can be spot welded to each other. In only a few cases do the welds tend to be brittle. Table 35-1 shows some of the combinations of metals that can be spot welded satisfactorily.

TABLE 35-1. Metal Combinations That Can Be Spot Welded

Metal	Aluminum	Brass	Copper	Galvanized Iron	Iron (Wrought)	Monel	Nichrome	Nickel	Nickel Silver	Steel	Tin Plate	Zinc
Aluminum	X										X	X
Brass		X	X	X	X	X	X	X	X	X	X	X
Copper		X	X	X	X	X	X	X	X	X	X	X
Galvanized iron		X	X	X	X	X	X	X	X	X	X	
Iron (wrought)		X	X	X	X	X	X	X	X	X	X	
Monel		X	X	X	X	X	X	X	X	X	X	
Nichrome		X	X	X	X	X	X	X	X	X	X	
Nickel		X	X	X	X	X	X	X	X	X	X	
Nickel silver		X	X	X	X	X	X	X	X	X	X	
Steel		X	X	X	X	X	X	X	X	X	X	
Tin plate	X	X	X	X	X	X	X	X	X	X	X	
Zinc	X	X	X									X

The majority of spot welding is performed to join wrought sheet, but other forms of metal can also be spot welded. Sheets can be spot welded to rolled shapes and steel castings, and some types of die castings can be welded without difficulty. Most metals require no special preparation, except to be sure that the surface is free of corrosion and is not badly pitted. For best results, aluminum and magnesium should be cleaned immediately

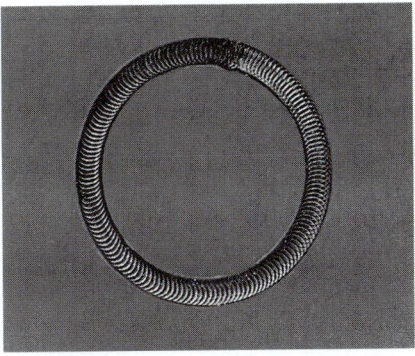

FIGURE 35-9   Seam welds made with overlapping spots of varied spacing.
(Courtesy Taylor-Winfield Corporation.)

prior to welding by mechanical or chemical techniques. Metals that have high electrical conductivity require clean surfaces to assure that the electrode-to-metal resistance is low enough for adequate temperature to be developed within the metal itself. Silver and copper are especially difficult to weld because of their high thermal conductivity. By using higher welding currents coupled with water cooling adjacent to the spot-welded area, adequate welding temperature can be obtained with the heat being restricted to the desired spot area.

When the two pieces being joined are of the same thickness, the practical limit for spot welding is about $\frac{1}{8}$ in. (3 mm) for each sheet. Different thicknesses of sheet can be joined, and thin pieces can be attached to material that is considerably thicker than $\frac{1}{8}$ in. When metals of different thickness or different conductivity are welded, however, a larger electrode or one with higher conductivity is often used against the thicker or higher-resistance material to assure that both pieces are brought to the desired temperature in a simultaneous fashion. Since spot welding is frequently used as an alternative to riveting, efforts are constantly being made to expand the useful range of the process. With special consideration, $\frac{1}{2}$-in. (12.7-mm)-thick steel plates have been joined successfully by spot welding.

Resistance Seam Welding

Resistance seam welds (RSEW) are made by two distinctly different processes. In one case the weld is made between overlapping sheets of metal, and the process is used to produce liquid- or gastight sheet metal vessels, such as gasoline tanks, automobile mufflers, and heat exchangers. The seam is actually a series of overlapping spot welds such as those in Figure 35-9. The basic equipment is the same as for spot welding, except that the electrodes are now in the form of rotating disks, as shown schematically in Figure 35-10 and in the actual unit of Figure 35-11. As the metal passes between the electrodes, timed pulses of current pass through it to form the overlapping welds. The timing of the welds and the movement of the work is controlled to assure that the welds overlap and the workpieces do not get too hot. The welding current is usually a bit higher than conventional spot welding, to compensate for the short circuit of the adjacent weld, and external cooling of the work, by air or water, is often employed. In a variation of the process, a continuous seam is produced by passing a continuous current through the rotating electrodes. A typical welding speed is about 60 in./min for thin sheet.

The second type of resistance seam welding is used to make butt welds between thicker metal plate. In this process the electrical resistance of the abutting metal(s) is used to generate heat, but high-frequency current (up to 450 kHz) is employed to restrict the

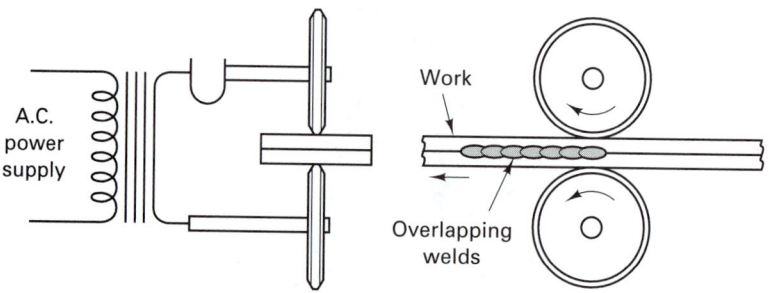

FIGURE 35-10   Schematic representation of seam welding.

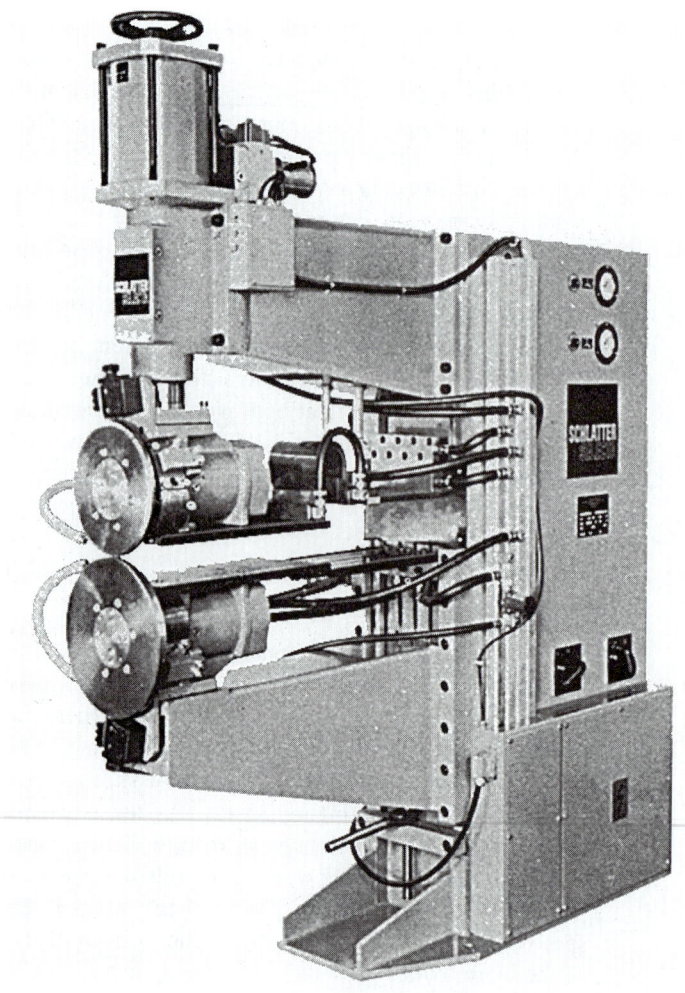

FIGURE 35-11   Typical commercial seam welder. *(Courtesy H. A. Schlatter AG.)*

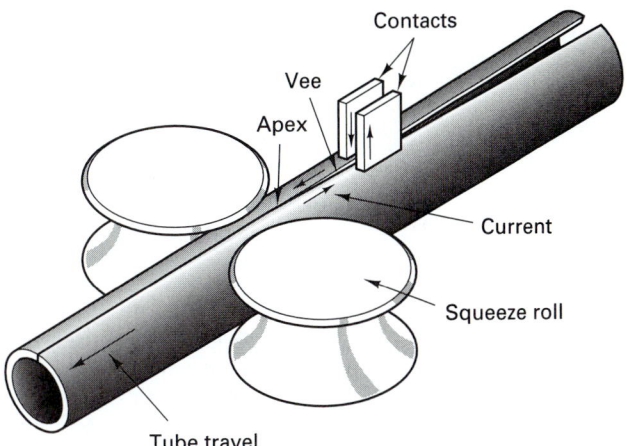

FIGURE 35-12   Using high-frequency ac current to produce
a resistance seam weld in butt-welded tubing. Arrows from
the contacts indicate the path of the high-frequency
current.

flow of current to the surfaces to be joined and their immediate surroundings. (*Note:* This
is similar to the results obtained by a parallel process, that of high-frequency induction
heating.) When they attain the desired temperature, the heated surfaces are pressed together to form the weld.

The most extensive use of resistance *butt welding* is probably in the manufacture of
pipe and tube, as illustrated in Figure 35-12, as well as simple structural shapes that can be
produced from plate. Material from 0.005 in. (0.13 mm) to more than $\frac{3}{4}$ in. (19 mm) in
thickness can be welded at speeds up to 250 ft (82 m) per minute. The combination of
high-frequency current and high welding speed produces a very narrow heat-affected zone.
Almost any type of metal can be welded, including dissimilar metals and high-conductivity
metals, such as aluminum and copper.

Projection Welding

Conventional spot welding is plagued by two significant limitations. Electrode condition
must be maintained continually, and only one spot weld is produced at a time. If additional strength is required, multiple welds must be made. *Projection welding* (RPW) provides a means of overcoming both of these limitations and has become an attractive means
of mass production.

The principle of projection welding is illustrated in Figure 35-13. A *dimple* is *embossed* into one of the workpieces at the location where a weld is desired. The workpieces
are then placed between large-area electrodes, and pressure and current are applied as in
spot welding. Because the current must flow through the points of contact, namely the
dimples, the heating is concentrated where the weld is desired. As the metal heats and becomes plastic, pressure causes the dimple to flatten and form a weld.

Because the projections are press-formed, they can often be produced during other
blanking and forming operations with virtually no additional cost. Another important advantage is that the dimples, or projections, can be made in almost any shape—such as round,
oval, or circular—to produce welds of shapes to suit various design purposes. They should
be designed, however, so that the weld forms outward from the center of the projection.

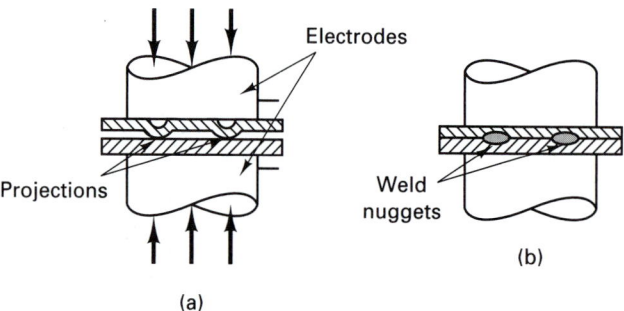

FIGURE 35-13 Principle of projection welding: prior to application of current and pressure (a) and after formation of the welds (b).

Several dimples can be incorporated into a sheet, and multiple welds can be made at one time. The number of projections is limited only by the ability of the machine to provide the required current and pressure. The attractiveness of the process is further enhanced by the fact that conventional spot-welding machines can be converted to projection welding simply by changing the size and shape of the electrodes. In addition, projection welding leaves no indentation mark on the free surface, a definite advantage over spot welding when good surface appearance is required.

In a variation of the process, bolts and nuts can be attached to other metal parts by projection welding. Contact is made at a projection that has been machined or forged onto the bolt or nut, current is applied, and the pieces are pressed together to form a weld.

■ 35.3 ADVANTAGES AND LIMITATIONS

The resistance welding processes have a number of distinct advantages that account for their wide use, particularly in mass production:

1. They are very rapid.
2. The equipment can often be fully automated.
3. They conserve material; no filler metal, shielding gases, or flux is required.
4. Skilled operators are not required.
5. Dissimilar metals can be easily joined.
6. A high degree of reliability and reproducibility can be achieved.

Resistance welding also has some limitations, the principal ones being:

1. The equipment has a high initial cost.
2. There are limitations to the type of joints that can be made (mostly *lap joints*).
3. Skilled maintenance personnel are required to service the control equipment.
4. For some materials, the surfaces must receive special preparation prior to welding.

The resistance welding processes are among the most common techniques for high-volume joining. However, the rapid heat inputs, short welding times, and rapid quenching by both the base metal and the electrodes can produce extremely high cooling rates in and around the weld. Martensite can form in steels containing more than 0.15% carbon. For these materials, some form of postweld heating is generally required to eliminate possible brittleness.

■ KEY WORDS

butt welding	lap joint	resistance spot welding
coalescence	nugget	transgun
embossed	projection welding	
faying surfaces	resistance seam welding	

■ REVIEW QUESTIONS

1. What are the two major roles of applied pressure in resistance welding?
2. Why might resistance welding be considered as a form of solid-state welding?
3. What are the three components that contribute to the total resistance between the electrodes?
4. What measures can be taken to reduce the resistance between the electrodes and the workpieces?
5. What are the possible consequences of too little pressure during the cycle? Too much pressure?
6. What is the ideal sequence for pressure application during resistance welding?
7. What is the simplest and most widely used form of resistance welding?
8. What is the typical size of a spot-weld nugget?
9. What are the two basic types of stationary spot-welding machines?
10. What is the major advantage of spot-welding guns?
11. What is a transgun? What is a disadvantage of a transgun when accurate positioning is required?
12. What is the most common metal that is spot welded?
13. What is the practical limit of the thicknesses of material that can be readily spot welded?
14. What design features can be altered to permit the joining of different thicknesses or different conductivity metals?
15. What is the difference between the seams produced by roll-spot welding and continuous seam welding?
16. Why is high-frequency ac current generally used for producing resistance butt-welded joints?
17. What two limitations of spot welding can be overcome by using the projection approach?
18. What limits the number of projection welds that can be formed in a single operation?
19. What are some of the attractive features of resistance welding when viewed from a manufacturing standpoint?
20. What are some of the primary limitations to the use of resistance welding?
21. What types of metallurgical problems might be encountered when spot welding medium- or high-carbon steels?

CHAPTER 36

OTHER WELDING
AND RELATED PROCESSES

36.1 INTRODUCTION
36.2 SOLID-STATE WELDING
      PROCESSES
      Forge Welding
      Forge-Seam Welding
      Cold Welding
      Roll Welding or Roll
         Bonding
      Friction Welding and Inertia
         Welding
      Ultrasonic Welding
      Diffusion Welding
      Explosive Welding
36.3 OTHER WELDING AND
      CUTTING PROCESSES
      Thermit Welding
      Electroslag Welding
      Electron Beam Welding
      Laser Beam Welding

      Laser Beam Cutting
      Laser Spot Welding
      Flash Welding
36.4 WELDING OF PLASTICS
36.5 WELDING-RELATED
      PROCESSES
      Surfacing
         Surfacing Materials
         Surfacing Methods and
            Applications
      Thermal Spray Coating or
         Metallizing
         Surface Preparation for
            Metallizing
         Characteristics and
            Applications of Sprayed
            Metals
Case Study: FIELD REPAIR TO A
             POWER TRANSFORMER
             CASE

■ 36.1 INTRODUCTION

As indicated in the lists of Figures 33-1 and 33-5, there are a number of welding and cutting processes that utilize heat sources other than oxyfuel flames, electric arcs, or electrical resistance. While some of these processes are quite old, others are among the newest in manufacturing. We begin with those processes that create a joint without melting of the workpiece or filler material—the *solid-state welding* processes.

■ 36.2 SOLID-STATE WELDING PROCESSES

Forge Welding

Being the most ancient of the welding processes, *forge welding* (FOW) has both historical and practical value, as it helps us to understand how and why the modern welding practices were developed. The armor makers of ancient times occupied positions of prominence in their society, largely because of their ability to join pieces of metal into a single, strong product. More recently, the village blacksmith was a master at the art of forge welding. With his hammer and anvil, coupled with skill and training, he could create a wide variety of useful shapes from metal.

Using a charcoal forge as a source of heat, the blacksmith heated the pieces to be welded to a practical forging temperature and then shaped the ends by hammering so that they could be fitted together without excessive thickness. The ends were then reheated and

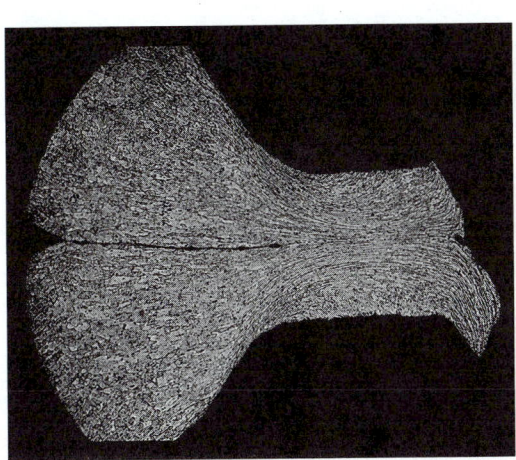
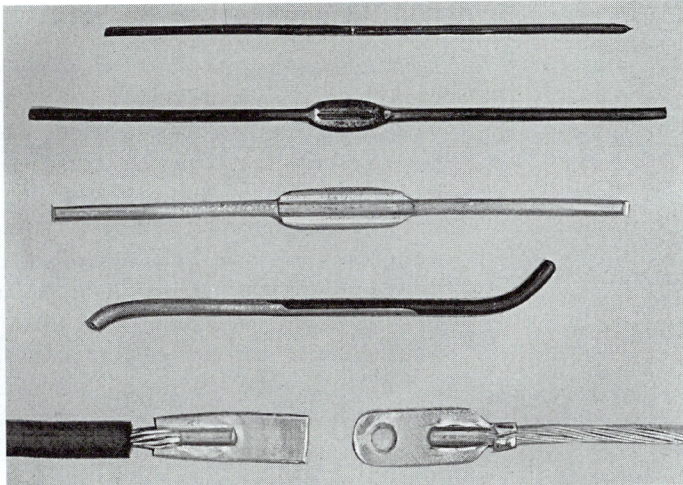

FIGURE 36-1   (*Left*) Section through a cold weld.   (*Right*) Small parts joined by cold welding.   (*Courtesy of Koldweld Corporation.*)

dipped into borax, which acts as a flux. Heating was continued until the blacksmith judged (by color) that the workpieces were at the proper temperature for welding. They were then withdrawn from the heat and struck either on the anvil or by the hammer to knock off any loose scale or impurities. The ends to be joined were then overlapped on the anvil and hammered to the degree necessary to produce an acceptable weld.

By a combination of heat and deformation, a competent blacksmith could produce joints that might be every bit as strong as the original metal. Because of the crudeness of the heat source, the uncertainty of temperature, and the difficulty in maintaining metal cleanliness, a great amount of skill was required and the results were highly variable. The quality of any deformation weld depends on the material temperature, surface cleanliness, and the amount of deformation induced by pressing or hammering.

Forge-Seam Welding

Although forge welding has largely been replaced by other joining methods, a large amount of *forge-seam welding* is used in the manufacture of pipe. As discussed in Chapter 18, a heated strip of steel is formed into a cylinder and the edges are welded together in either a lap or a butt configuration. Welding is the result of pressure and deformation as the metal is pulled through a conical welding bell or passed between welding rolls.

Cold Welding

Cold welding is a unique variation of forge welding that *uses no heating* but produces metallurgical bonds by means of cold plastic deformation. The surfaces to be joined are first cleaned and placed in contact. They are then subjected to localized pressures sufficient to cause about 30 to 50% localized cold working. A solid-state bond is produced, as shown in Figure 36-1. While some heating will occur due to the severe deformation, the primary factor in producing coalescence is the high localized pressure acting on newly formed surface material. The use of the process is generally confined to the joining of small parts made from soft, ductile material, such as the electrical connections shown in Figure 36-1.

Roll Welding or Roll Bonding

In the *roll welding* or *roll bonding* (ROW) process, two or more sheets of metal can be joined by passing them simultaneously through a rolling mill. As the materials are reduced in thickness, the length and/or width must increase to compensate. At the interface, the newly created uncontaminated surfaces are immediately exposed to the high pressures of rolling, and coalescence is produced. The process can be performed either hot or cold, and can be used to join either similar or dissimilar metals (as in Alclad aluminums or steel with a stainless steel *cladding*). The resulting bond strength can be quite good, as evidenced by the roll-bonded material used in the production of U.S. dimes and quarters.

By coating portions of one surface with a material that prevents bonding, the roll-bonding process can be used to produce sheets that are bonded, or not bonded, in selected regions. When these sheets are subsequently heated, the coating volatilizes and the no-bond regions expand to produce a flow path for gases or liquids. A common example of this technique is in the manufacture of refrigerator freezer panels, where inexpensive sheet metal is used to fabricate structural panels that also conduct the coolant, thereby eliminating the need for more-costly tubing.

Friction Welding and Inertia Welding

In *friction welding* (FRW) the heat required to produce the joint is generated by friction heating at the interface. The components to be joined are first prepared to have smooth, square-cut surfaces. As shown in Figure 36-2, one piece is then held stationary while the other is mounted in a motor-driven chuck or collet and rotated against it at high speed. A low contact pressure may be applied initially to permit cleaning of the surfaces by a burnishing action. This pressure is then increased, and contact friction quickly generates enough heat to raise the abutting surfaces to the welding temperature. As soon as this temperature is reached, rotation is stopped and the pressure is maintained or increased to complete the weld. The softened metal is squeezed out to form a flash and a "forged" structure

FIGURE 36-2 Sequence for making a friction weld. (a) Components with square surfaces are inserted into a machine where one part is rotated and the other is held stationary. (b) The components are pushed together with a low axial pressure to clean and prepare the surfaces. (c) The pressure is increased, causing an increase in temperature, softening, and possibly some melting. (d) Rotation is stopped and the pressure is increased rapidly, creating a forged joint with external flash.

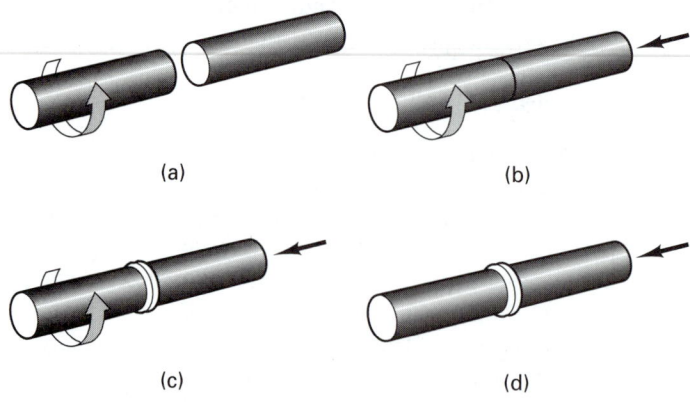

(a)　　　　(b)

(c)　　　　(d)

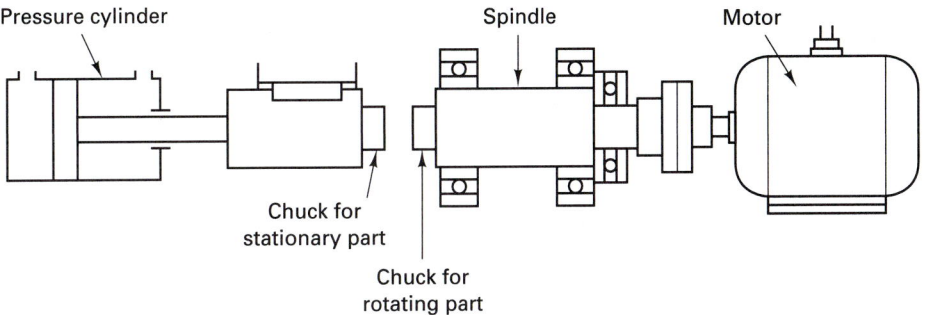

FIGURE 36-3 Schematic diagram of the equipment used for friction welding. *(Courtesy of Materials Engineering.)*

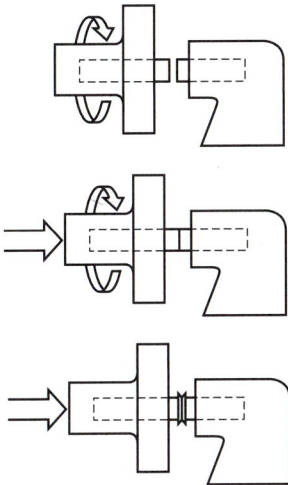

FIGURE 36-4 Schematic representation of the three steps in inertia welding.

is formed in the joint. If desired, the flash can be removed by subsequent machining. Friction welding has been used to join steel bars up to 4 in. in diameter and tubes with outer diameters up to 10 in. Figure 36-3 shows a schematic of the equipment required for friction welding.

Inertia welding is a modification of friction welding where the moving piece is attached to a rotating flywheel (Figure 36-4). The flywheel is brought to a specified rotational speed and is then separated from the driving motor. The rotating assembly is then pressed against the stationary member and the kinetic energy of the flywheel is converted into frictional heat. The weld is formed when the flywheel stops its motion and the pieces remain pressed together. Since the conditions of inertia welding are easily duplicated, welds of consistent quality can be produced, and the process can be readily automated. Figure 36-5 shows a plot of rotational speed (surface velocity), torque, and degree of upset throughout the inertia welding process.

In both friction and inertia welding, the time required to form a weld can be very short, often on the order of several seconds. Because of the short period of heating and the limited time for heat to flow away from the joint, the weld and heat-affected zones are usually very narrow. No material is melted, so the process is totally solid state. Oxides and other surface impurities on the joint faces tend to be displaced radially into the upset flash,

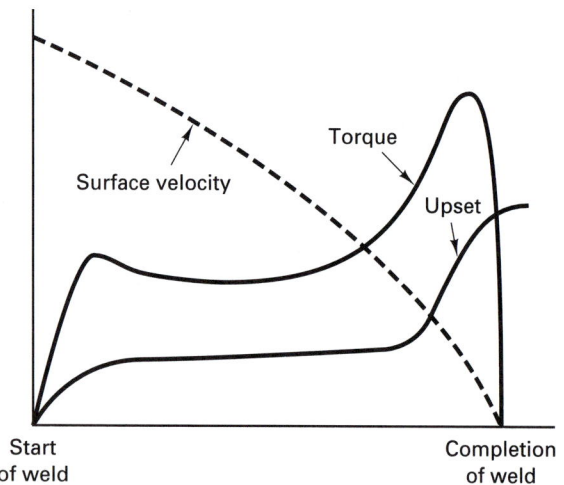

FIGURE 36-5 Relationship between surface velocity (speed), torque, and upset throughout the inertia welding process.

FIGURE 36-6 Some typical friction-welded parts. (*Left*) Impeller made by joining a chrome–moly steel shaft to a nickel–steel casting. (*Center*) Stud plate with two mild steel studs joined to a square plate. (*Right*) Tube component where a turned segment is joined to medium-carbon steel tubing. (*Courtesy of Newcor Bay City, Div. of Newcor, Inc.*)

which can be removed after welding if desired. Because virtually all of the energy is converted to heat, the process is very efficient and can be used to join a wide variety of metals or combinations of metals, including some not normally considered compatible, such as aluminum to steel. Since grain size is refined during the hot working, the strength of the weld is almost the same as the base metal. In addition, the processes are environmentally attractive since no smoke, fumes, or gases are generated, and no fluxes are required.

Unfortunately, the processes are restricted to joining round bars or tubes of the same size, or connecting bars or tubes to flat surfaces (joints where at least one of the pieces has circular symmetry). In addition, one of the components must be ductile when hot to permit deformation during the forging stage. Figure 36-6 shows some typical applications of friction welding.

Ultrasonic Welding

Ultrasonic welding (USW) is a solid-state process in which coalescence is produced by the localized application of high-frequency (10,000 to 200,000 Hz) shear vibrations to surfaces that are held together under rather light normal pressure. Although there is some increase in temperature at the *faying surfaces*, they generally do not exceed one-half of the melting point of the material (on an absolute temperature scale). Instead, it appears that the rapid reversals of stress along the contact interface facilitate coalescence by breaking up and dispersing the oxide films and surface contaminants, allowing clean metal to form a high-strength bond.

Figure 36-7 depicts the basic components of the ultrasonic welding process. The ultrasonic transducer is essentially the same as that employed in ultrasonic machining, depicted schematically in Figure 32-14. It is coupled to a force-sensitive system that contains a welding tip on one end. The pieces to be welded are placed between this tip and a

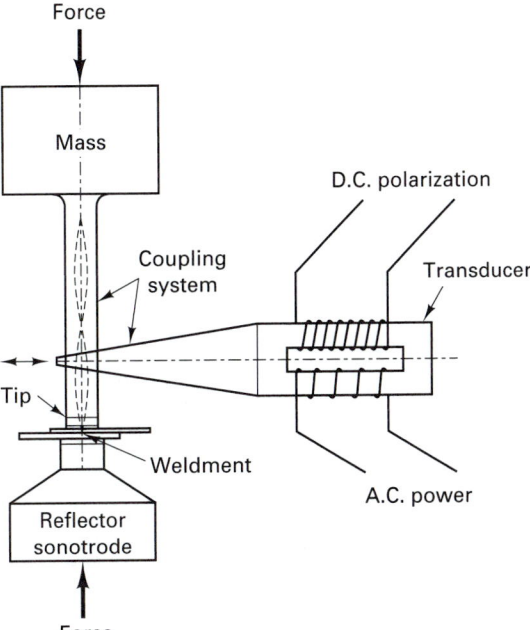

FIGURE 36-7 Schematic diagram of the equipment used in ultrasonic welding.

reflecting anvil, thereby concentrating the vibratory energy. Either stationary tips (for spot welds) or rotating disks (for seam welds) can be employed.

Ultrasonic welding is restricted to the lap joint welding of thin materials—sheet, foil, and wire—or the attaching of thin sheets to heavier structural members. The maximum thickness is about 0.1 in. (2.5 mm) for aluminum and 0.04 in. (1.0 mm) for harder metals. As indicated in Table 36-1, the process is particularly valuable because of the number of metals and dissimilar metal combinations that can be joined. It is even possible to bond metals to nonmetals, such as aluminum to ceramics or glass. Because the temperatures are low and no arcing or current flow is involved, the process can be applied to heat-sensitive electronic components. Intermetallic compounds seldom form and there is no contamination

TABLE 36-1. Metal Combinations Weldable by Ultrasonic Welding

Metal	Aluminum	Copper	Germanium	Gold	Molybdenum	Nickel	Platinum	Silicon	Steel	Zirconium
Aluminum	×	×	×	×	×	×	×	×	×	×
Copper		×		×		×	×		×	×
Germanium			×	×		×	×	×		
Gold				×		×	×	×		
Molybdenum					×	×			×	×
Nickel						×	×		×	×
Platinum							×		×	
Silicon										
Steel									×	×
Zirconium										×

of the weld or surrounding area. The equipment is simple and reliable, and only moderate skill is required of the operator. The required surface preparation is less than for most competing processes (such as resistance welding), and less energy is needed to produce a weld. Typical applications include joining the dissimilar metals in bimetallics, making microcircuit electrical contacts, welding refractory or reactive metals, bonding ultrathin metal, and encapsulating explosives or chemicals.

Diffusion Welding

Diffusion welding (DFW) *or diffusion bonding* occurs when properly prepared surfaces are maintained in contact under sufficient pressure and time at elevated temperature. In contrast to the deformation welding methods, plastic flow is limited and the principal bonding mechanism is atomic diffusion. A well-prepared interface can be viewed as a planar grain boundary with intervening voids and impurities. Under low pressure and elevated temperature, atomic diffusion will provide the necessary void shrinkage and grain boundary migration to form a metallurgical bond.

The quality of a diffusion weld depends on the surface condition of the materials, temperature, time at temperature, pressure, and the possible use of intermediate material layers, which can either promote diffusion or prevent the formation of undesirable intermetallic compounds. Some intermediate layers melt to form a temporary liquid that significantly accelerates the rate of atom movement.

Diffusion bonding is frequently used to join dissimilar metals and composite materials. Furnaces with inert or protective atmospheres can be used to produce high-quality joints with the reactive metals, such as titanium, beryllium, and zirconium, and the high-temperature refractory metals. Since the bonding process is quite slow, multiple parts are generally loaded into a furnace, or application is restricted to low-volume production.

Explosive Welding

Explosive welding (EXW) is used primarily for bonding sheets of corrosion-resistant metal to heavier plates of base metal (a *cladding* operation), particularly when large areas are involved. As shown in Figure 36-8, the bottom sheet or plate is positioned on a rigid base or anvil and the top sheet is inclined to it with a small open angle between the surfaces to be joined. An explosive material, usually in the form of a sheet, is placed on top of the two layers of metal and detonated in a progressive fashion, beginning where the surfaces touch. A compressive stress wave, on the order of hundreds of thousands of pounds per square inch (thousands of megapascals), sweeps across the surface of the plates. Surface films are liquefied or scarfed off the metals and are jetted out of the interface. The clean metal surfaces are then thrust together under high contact pressure. The result is a low-temperature weld with an interface configuration consisting of a series of interlocking ripples. The bond strength is quite high, and explosively clad plates can be subjected to a wide variety of subsequent processing, including further reduction in thickness by rolling. Because it is a solid-state welding processes, numerous combinations of dissimilar metals can be joined.

■ 36.3 OTHER WELDING AND CUTTING PROCESSES

Thermit Welding

The heating and coalescence in *thermit welding* (TW) is produced by the superheated molten metal and slag obtained from the chemical reaction between a metal oxide and a

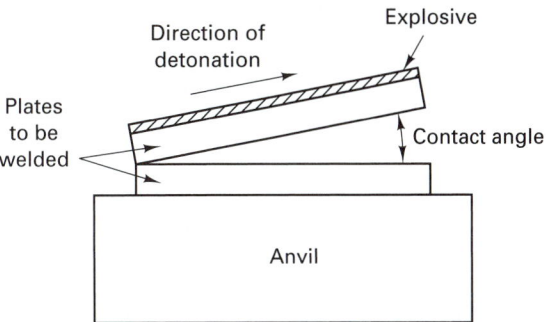

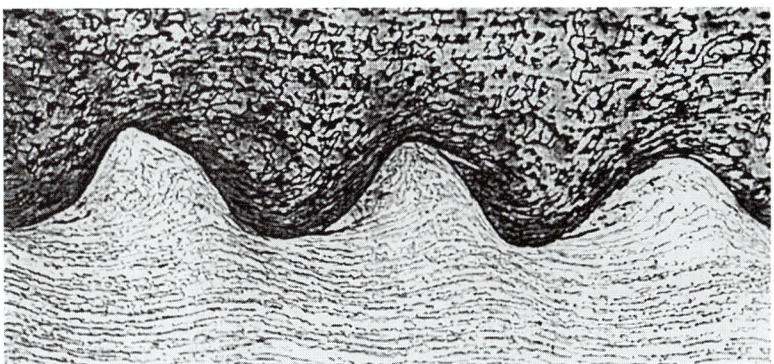

FIGURE 36-8   (*Top*) Schematic of the explosive welding process. (*Bottom*)
Explosive weld between mild steel (top) and stainless steel (bottom),
showing the characteristic wavy interface.

metallic reducing agent. The name *thermit* usually refers to a mechanical mixture of about
one part finely divided aluminum and three parts iron oxide. When this mixture is ignited
by a magnesium fuse (the ignition temperature is about 2100°F or 1150°C), it reacts according to the following chemical equation:

$$8Al + 3Fe_3O_4 \rightarrow 9Fe + 4Al_2O_3 + heat$$

A temperature of over 5000°F (2750°C) is produced in about 30 seconds, superheating the
molten iron, which then flows into a prepared joint, providing both heat and filler metal.
Runners and risers must be provided, as in a casting, to channel the molten metal and compensate for solidification shrinkage.

Copper, brass, and bronze can be welded using a starting mixture of copper oxide
and aluminum. Nickel, chromium, and manganese oxides have also been used in the thermit welding of the more exotic metals.

The thermit welding process is extremely old and has been replaced to a large degree
by alternative methods. Nevertheless, it is still effective and can be used to join thick sections of material, particularly in remote locations or where more-sophisticated welding
equipment is not available. One such application is the field repair of large steel castings
that have broken or cracked.

Electroslag Welding

Electroslag welding (ESW) (Figure 36-9) is a very effective process for welding thick sections of steel plate. There is no arc involved (except to start the weld), so the process is entirely different from submerged arc welding, and the electrical resistance of the metal being welded plays no part in producing the heat. Instead, the heat is derived from the passage of electrical current through a liquid slag. Resistance heating raises the temperature of the slag to around 3200°F (1760°C). The molten slag then melts the edges of the pieces that are being joined, as well as the continuously fed solid or flux-cored electrodes, which supply the filler metal. Multiple electrodes are often used to provide an adequate supply of filler and maintain the molten pool. Under normal operating conditions, there is a $2\frac{1}{2}$-in. (65-mm)-deep layer of molten slag, which serves to protect and cleanse the underlying $\frac{1}{2}$- to $\frac{3}{4}$-in. (12- to 20-mm)-deep pool of molten metal. These liquids are confined to the gap by means of sliding, water-cooled *molding plates* that are usually made of copper. As the weld metal solidifies at the bottom of the pool, the molding plates move upward at a rate that is typically between 0.5 and 1.5 in./min.

Since a vertical joint provides the easiest conditions for maintaining a deep slag bath, the process is used most frequently in this configuration. Circumferential joints can also be produced in large pipe by using special curved slag-holder plates and rotating the pipe to maintain the welding area in a vertical position.

Because large amounts of weld metal and heat can be supplied, electroslag welding is the best of all the welding processes for making welds in thick plates. The thickness of the plates can vary from $\frac{1}{2}$ to 36 in. (13 to 900 mm), and the length of the weld (amount of vertical travel) is almost unlimited. Edge preparation is minimal, requiring only squared edges separated by 1 to $1\frac{1}{2}$ in. (25 to 35 mm). Applications have included building construction, shipbuilding, machine manufacture, heavy pressure vessels, and the joining of large castings and forgings.

Solidification control is vitally important to obtaining a good electroslag weld, since slow cooling tends to produce a coarse grain structure. Cracking tendencies can be suppressed by adjusting the current, voltage, slag depth, number of electrodes, and electrode extension to produce a wide shallow pool of molten metal. A large heat-affected zone and extensive grain growth are also common features of the process. However, the long thermal cycle does serve to minimize residual stresses, distortion, and cracking in the heat-affected zone. If good fracture resistance is required, subsequent heat treatment of the welded structure may be necessary.

Electron Beam Welding

Electron beam welding (EBW) is a fusion welding process in which heating results from the impingement of a beam of high-velocity electrons on the metal to be welded. Originally developed for obtaining ultrahigh-purity welds in reactive and refractory metals, its unique qualities have led to use in a number of applications.

The electron optical system for the process is shown in Figure 36-10. A high-voltage current heats a tungsten filament to about 4000°F (2200°C), causing it to emit high-velocity electrons. By means of a control grid, accelerating anode, and focusing coils, these electrons are collected into a concentrated beam and directed onto the workpiece in a spot from $\frac{1}{32}$ to $\frac{1}{8}$ in. (0.8 to 3.2 mm) in diameter. The electron beam is an extremely intense heat source, capable of producing temperatures as high as several million degrees Fahrenheit. Unfortunately, electrons cannot travel well through air. To be effective as a heat source for welding, the beam must be generated and focused in a very high vacuum, typically at pressures of 1×10^{-4} mmHg (0.01 Pa) or less.

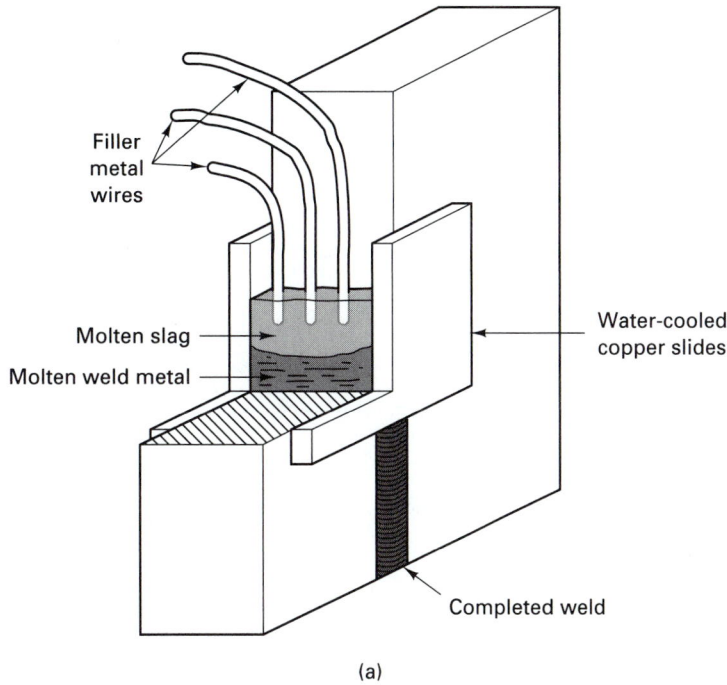

Filler metal wires

Molten slag

Molten weld metal

Water-cooled copper slides

Completed weld

(a)

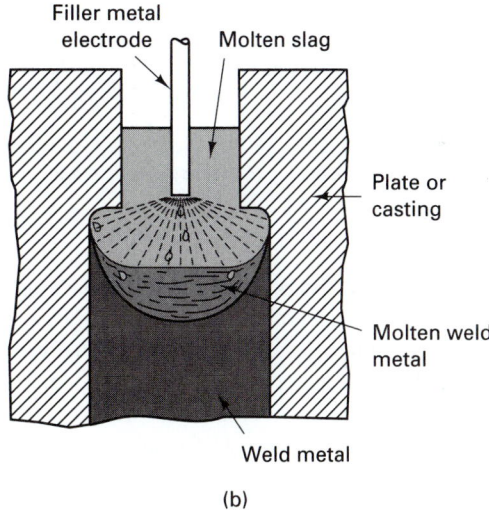

Filler metal electrode

Molten slag

Plate or casting

Molten weld metal

Weld metal

(b)

FIGURE 36-9   (*Top*) Arrangement of equipment and workpieces for making a vertical weld by the electroslag process. (*Bottom*) Cross section of an electroslag weld, looking through the water-cooled copper slide.

In many operations, the workpiece is also enclosed in the high-vacuum chamber and must be positioned and manipulated in this vacuum. When welding under these conditions, the vacuum assures degasification and decontamination of the molten weld metal, and very high quality welds are obtained. However, the size of the vacuum chamber tends to impose serious limitations on the size of the workpiece that can be accommodated, and

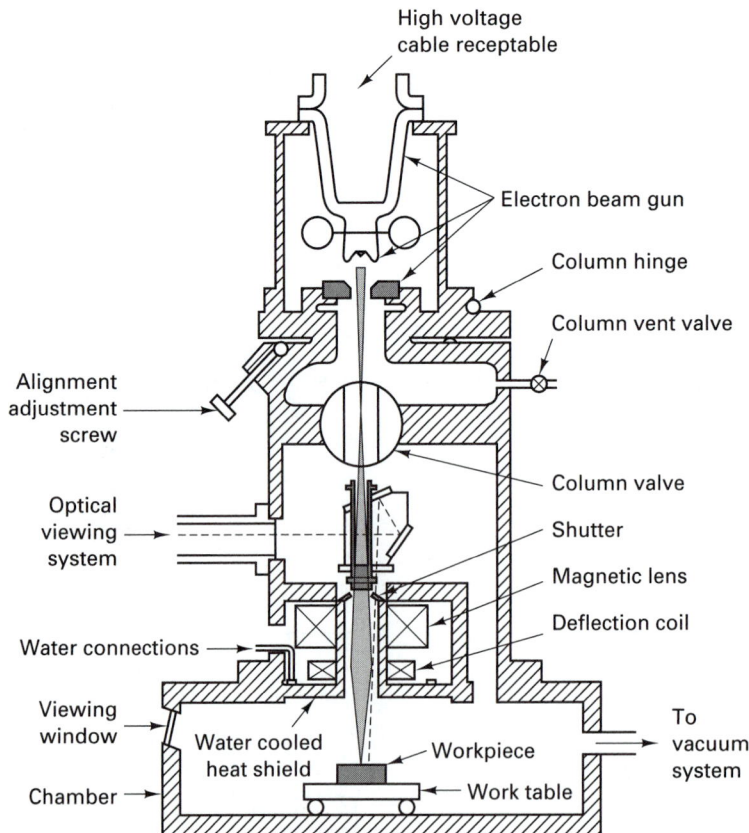

FIGURE 36-10 Schematic diagram of the electron beam welding process. *(Courtesy of American Machinist.)*

the need to break and reestablish the high vacuum as pieces are inserted and removed places a considerable restriction on productivity. As a consequence, electron beam welding machines have been developed that operate at pressures considerably higher than those required for beam generation. Some permit the workpiece to remain outside the vacuum chamber, the beam emerging through a small orifice in the vacuum chamber to strike the adjacent workpiece. High-capacity vacuum pumps are required to compensate for the leakage through the orifice. Although these machines offer more production freedom, they produce shallower, wider welds since the beam loses energy and diffuses as the pressure increases.

In general, two distinct ranges of voltage are employed in electron beam welding. High-voltage equipment employs 50 to 100 kilovolts and produces a smaller spot size and greater penetration than does the lower-voltage type, which uses from 10 to 30 kilovolts. Because of their high electron velocities, the high-voltage units emit considerable quantities of harmful X-rays and thus require expensive shielding and indirect viewing systems for observing the work. The X-rays produced by the low-voltage machines are sufficiently soft that they are absorbed by the walls of the vacuum chamber, and the parts can be viewed directly through viewing ports.

Almost any metal can be welded by the electron beam process, including those that are

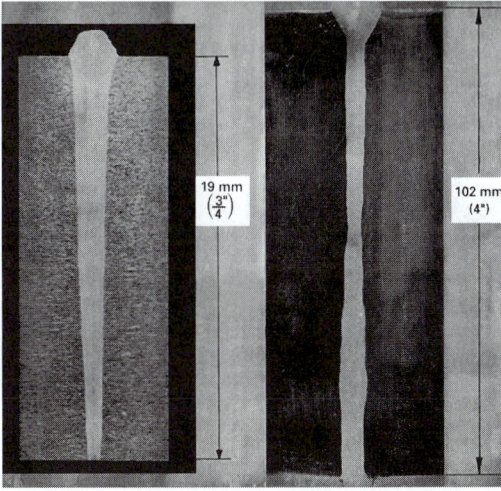

FIGURE 36-11   (*Left to right*) Electron beam welds in $\frac{3}{4}$-in.-thick 7079 aluminum, and 4-in.-thick stainless steel. (*Courtesy of Hamilton Standard*

difficult to weld by other methods, such as zirconium, beryllium, and tungsten. Dissimilar metals can be welded and the welds often exhibit a narrow profile and remarkable *penetrations* (Figure 36-11). The high power and heat concentrations can produce fusion zones with depth-to-width ratios of 25:1, coupled with low total heat input, low distortion, and a very narrow heat-affected zone. Heat-sensitive materials can be welded without damage to the base metal. High welding speeds are common; no shielding gas, flux, or filler metal is required; the process can be performed in all positions; and preheat or postheat is generally unnecessary.

On the negative side, the equipment is quite expensive and extensive joint preparation is required. Because of the deep and narrow weld profile, joints must be straight and precisely aligned over the entire length of the weld. Machining and fixturing tolerances are often quite demanding. The vacuum requirements tend to limit production rate and the size of the vacuum chamber may restrict the size of workpiece that can be welded.

The electron beam process is best employed where extremely high quality welds are required or where other processes will not produce the desired results. Nevertheless, its unique capabilities have resulted in its routine use in a number of applications, particularly in the automotive and aerospace industries.

Laser Beam Welding

As a heat source, laser beams can be used for welding, hole making, cutting, cladding, and heat treating. When used for *laser beam welding* (LBW) the focused laser beam usually provides power intensities in excess of 10 kW/cm^2. The high-intensity beam produces a very thin column of vaporized metal with a surrounding liquid pool. As the laser advances, the liquid flows into the channel to produce a weld with depth-to-width ratio generally greater than 4:1. The narrow weld pool solidifies quickly, producing a very thin heat-affected zone and little thermal distortion. Laser beam welding is most effective for simple fusion welds without filler metal (autogenous welds), but careful joint preparation is required to produce the narrow gap. Filler metal can be added if the gap is excessive, and inert-gas shielding may be used to protect the weld pool from oxidation.

The well-collimated beam of intense energy produces deep penetration welds that are similar to electron beam welds, but the laser beam technique offers several distinct advantages:

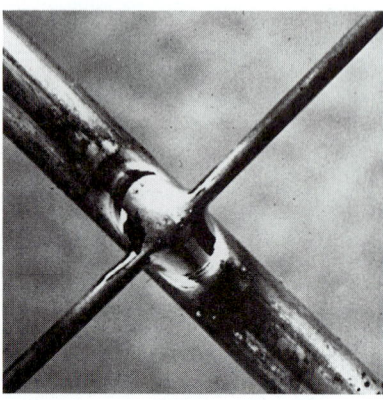 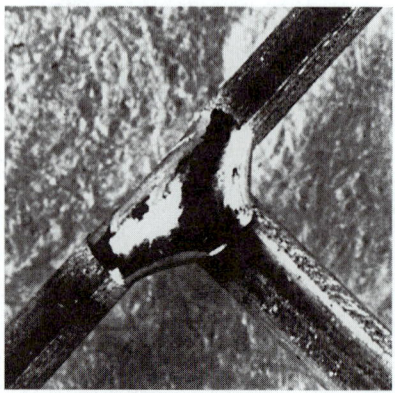 

FIGURE 36-12 *(Left)* Small electronic welds made by laser welding. *(Courtesy of Linde Division, Union Carbide Corporation.)*, *(Right)* Laser butt weld of 0.125-in. (3-mm) stainless steel, made at 60 in./min (1.5 m/min) with a 1250-W laser. *(Courtesy of Coherent, Inc.)*

1. The beam can be transmitted through air, so a vacuum environment is not required.
2. No X-rays are generated.
3. The laser beam is easily shaped, directed, and focused with both transmission and reflective optics (lenses and mirrors) and can be transmitted through fiber optic cables.
4. No direct contact is necessary to produce a weld, only optical accessibility. Welds can be made on materials that are encapsulated within transparent containers, such as components in a vacuum tube.

Because of the sharply focused beam and the short time (or high speed) of welding, laser welds are usually small, often less than 0.001 in. in width or diameter, and the total heat input is very low, often in the range 0.1 to 10 J. For this reason, laser welding has been adopted by the electronics industry for applications such as connecting wire leads to small electronic components. Lap, butt, tee, and cross-wire configurations can all be used. It is even possible to weld wires without removing the polyurethane insulation. The laser simply evaporates the insulation and completes the weld with the internal wire. Figure 36-12 shows several types of laser beam welds.

The equipment cost for a CO_2 or YAG (yttrium–aluminum–garnet) laser beam welding system is quite high, but this cost can be somewhat offset by the faster welding speeds, the ability to weld without filler metal, and low distortion, which enables a reduction in postweld straightening and machining. Caution should be used with such equipment, however, since reflected or scattered laser beams can be quite dangerous, even at great distances from the welding site. Eye protection is a must.

Laser Beam Cutting

Cutting small holes, narrow slots, and closely spaced patterns in a variety of materials is another widely used application of industrial lasers. *Laser beam cutting* (LBC) begins by "drilling" a hole through the material and then moving the beam in a programmed path. The

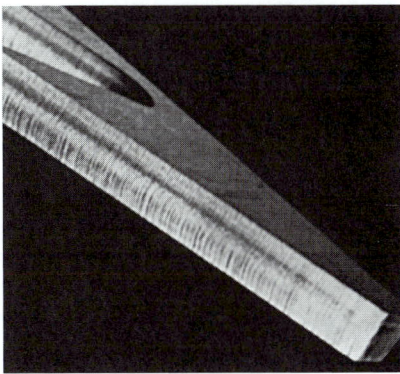

FIGURE 36-13   Surface of $\frac{1}{4}$-in. (6-mm)-thick carbon steel cut with a 1250-W laser at 70 in./min (1.8 m/min). *(Courtesy of Coherent, Inc.)*

intense heat from the laser is used to melt and/or evaporate the material being cut. A stream of *assist gas* is often used to blow the molten metal through the cut, cool the workpiece, and minimize the heat-affected zone. Oxygen is the usual gas for cutting mild steel, since it assists with the exothermic reaction and speeds the cut. Nitrogen or oxygen is used with stainless steel, nitrogen for aluminum, and, because of its high reactivity, titanium requires an inert gas, such as argon. Inert gases or air are used when cutting nonmetallics. Clean, accurate cuts are characteristic of the process, and the kerf and heat-affected zone are narrower than with any other thermal cutting process. No postcut finishing is required in many applications, even though the process does produce a thin recast surface. Figure 36-13 shows a laser cut of 0.25-in.-thick carbon steel, cut at 70 in./min with a 1250-W laser.

Both CO_2 and YAG lasers have been used in both continuous and pulsed modes. While cutting speed depends on the material being cut and its thickness, cutting is fastest in the continuous mode, and it is preferred for straight and mildly contoured cuts. The pulsed mode is preferred for thin materials and enables tight corners and intricate details to be cut without excessive burning. Metal plates up to 0.5 in. thick are routinely cut, as well as nonmetals with thicknesses up to 1 in. Cutting temperatures can be in excess of 20,000°F (11,000°C), and cutting speeds as high as 1000 in./min have been observed with some nonmetals.

The industrial lasers lend themselves to automation and robotic manipulation. Fiber optic systems can distribute a YAG-laser beam to multiple workstations in either a simultaneous or time-sharing fashion, and the individual stations can be as much as 300 feet from the laser. Through use of fiber-optic cables, laser energy can be piped directly to the end of a robot arm. This eliminates the need to mount and maneuver a heavy, bulky laser and enhances the speed and accuracy of both positioning and manipulation. Cutting, drilling, welding, and heat treating can all be performed with the same unit, and multiple axes of motion can be programmed for specific parts.

Lasers have also been mounted on traditional machine tools, such as CNC-type machines, or combined with traditional tools, such as punch presses, to produce extremely flexible hybrid equipment. Because no dedicated dies or tooling are required to produce a cut, and there is no setup time, the laser is an economical alternative to blanking or nibbling for prototype or short-run products or for materials that are difficult to cut by conventional methods, such as plastics, wood, and composites.

Laser cutting is achieving prominence in the cutting of composite materials. The more uniform the thermal characteristics of the components, the better the cut and the less thermal damage to the material. Kevlar-reinforced epoxy cuts easiest and gives a narrow heat-affected zone. Glass-reinforced epoxy is more difficult because of the greater thermal differences, and graphite-reinforced epoxy is even worse because of the high dissociation temperature and thermal conductivity of the graphite. By the time the graphite has absorbed sufficient cutting heat, the epoxy matrix has been decomposed to a significant depth. The use of lasers for machining is discussed further in Chapter 32.

Laser Spot Welding

Lasers have also been used to produce spot welds in a manner that offers unique advantages over the conventional resistance methods. A small clamping force is applied to assure contact of the workpieces and a fine-focused beam scans the area of the weld. Welding is performed in the keyhole mode, wherein the laser produces a small hole through the molten puddle. As the beam is moved, molten metal flows into the hole and solidifies, forming a fusion-type nugget.

Laser spot welding can be performed with access to only one side of the joint. It is a noncontact process and produces no indentations. No electrodes are involved, so electrode wear is no longer a production problem. Weld quality is independent of material resistance, surface resistance, and electrode condition, and no water cooling is required. The heat input is low, so the size of heat-affected zones is reduced. Speed of welding and strength of the resulting joint are comparable to resistance spot welds.

Flash Welding

In *flash welding* (FW), two pieces of metal are first secured in current-carrying grips and lightly touched together. An electric current may be passed through the joint to provide preheat (optional), after which the pieces are withdrawn slightly. An arc is then created across the gap (flashing), which melts the interface and expels the liquid and oxides. The pieces are then forced together under high pressure to upset the joint and form the weld. The electric current is turned off and the force is maintained until solidification is complete. After the product is removed from the machine, the upset portion can be removed by machining. Figure 36-14 shows a schematic of the flash welding process, including both the equipment and setup and the completed weld.

To produce a high-quality weld, it is important that the flashing action be continued long enough to melt the interface and also soften the adjacent metal. Sufficient plastic deformation must occur during the upsetting to enable the impurities and contaminants to be squeezed out into the *flash*.

Flash welding is usually employed for the butt welding of similar or dissimilar metals in solid or tubular shape. The equipment required is generally large and expensive, but excellent welds can be made at high production rates.

Percussion welding is a similar process where the heating is obtained by an arc produced by a rapid discharge of stored electrical energy, which is followed by a rapid application of force to expel the metal and produce the joint. In percussion welding, the arc duration is only 1 to 10 ms. While the heat is intense, it is also highly concentrated. Only a small amount of weld metal is produced, little or no upsetting occurs at the joint, and the heat-affected zone is quite small. Application is generally restricted to the butt welding of bar or tubing where heat damage is a major concern.

■ 36.4 WELDING OF PLASTICS

Mechanical fasteners, adhesives, and welding processes can all be employed to form joints between engineering plastics. Fasteners are quick and are suitable for most materials, but

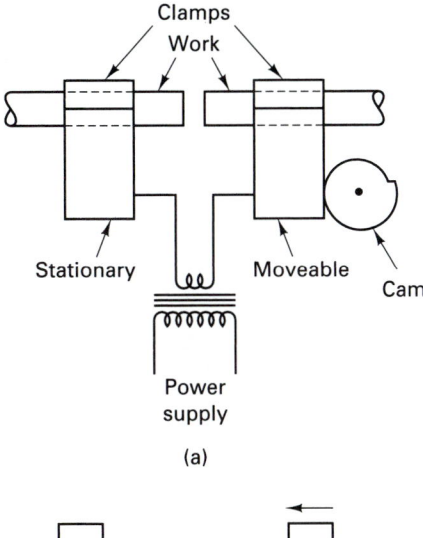

Clamps
Work

Stationary Moveable
Cam

Power
supply

(a)

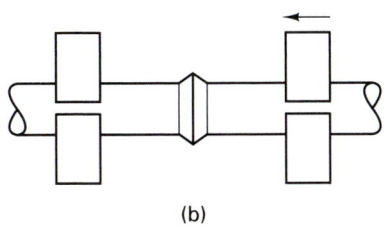

(b)

FIGURE 36-14 Schematic diagram of the flash welding process: (a) equipment and setup; (b) completed weld.

their use may be expensive, they generally do not provide leaktight joints, and the localized stresses may cause them to pull free of the polymeric material. Threaded metal inserts may have to be incorporated into the plastic components, further increasing the product cost. Adhesives can provide excellent properties and fully sound joints, but they are often difficult to handle and relatively slow to cure. In addition, considerable attention is required in the areas of joint preparation and surface cleanliness. In a modification of adhesive bonding, solvents can be used to soften surfaces, which are then pressed together to form a bond. Welding can be used to produce bonded joints with mechanical properties that approach those of the parent material. Unfortunately, only the *thermoplastic* polymers can be welded, since these materials can be melted or softened by heat without degradation. The thermosetting polymers do not soften with heat but tend only to char or burn.

Because the thermoplastics soften at such low temperatures, the heat required to weld these materials is significantly less than that required in the welding of metals. The processes used to weld plastics can be divided into two groups: (1) those that utilize mechanical movement and friction to generate heat, such as ultrasonic welding, friction welding, and vibration welding, and (2) those that involve external heat sources, such as hot-plate welding, hot-gas welding, and resistive and inductive implant welding.

Ultrasonic welding of plastics uses high-frequency mechanical vibrations to create the bond. Parts are held together and are subjected to ultrasonic vibrations (20 to 40 kH) perpendicular to the area of contact. The high-frequency stresses generate heat at the joint interface sufficient to produce a high-quality weld in a period of $\frac{1}{2}$ to $1\frac{1}{2}$ seconds. The process can be readily automated, but the tools are quite expensive and large production runs are generally required. In addition, welding is usually restricted to small components with weld lengths not to exceed a few inches or centimeters.

The *friction welding of plastics* (also called *spin welding*) is essentially the same as the friction welding of metals, but melting now occurs at the joint interface. High-quality

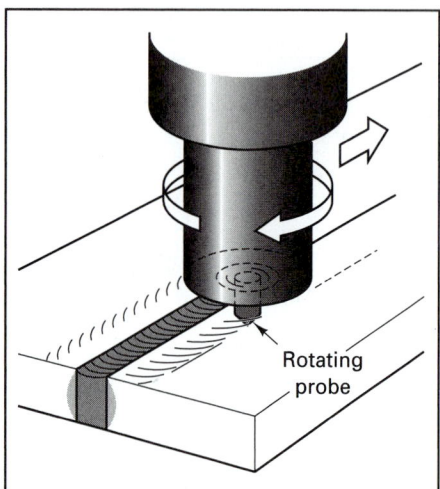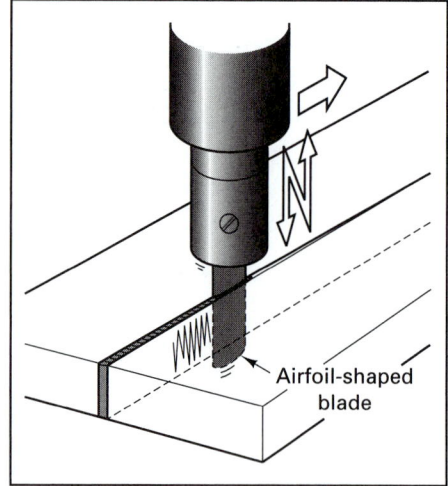

FIGURE 36-15 Friction stir welding using rotary and reciprocal motions to produce welds in plastics. The shoulder on the rotating probe provides additional friction heating to the top surface and prevents expulsion of the softened material from the joint. *(Courtesy of ASM International.)*

welds are produced with good reproducibility, and little end preparation is required. The major limitation is that at least one of the components must exhibit circular symmetry, and the axis of rotation must be perpendicular to the mating surface. Weld strengths vary from 50 to 95% of the parent material in bonds of the same plastic. Joints between dissimilar materials generally have poorer strengths.

In *vibration welding*, frictional heat is again generated by relative movement between the two parts, but the direction of movement is now parallel to the interface (as opposed to the ultrasonic case in which the direction is perpendicular), and the frequencies are considerably less, on the order of 100 to 240 Hz. Molten material is produced, the vibration is stopped, parts are aligned, and the weld region cools and solidifies. The entire process takes about 1 to 5 seconds. Long-length, complex joints can be produced at rather high production rates. Nearly all thermoplastics can be joined, independent of whether their prior processing was by injection molding, extrusion, blow molding, thermoforming, foaming, or stamping.

Butt welds have been made between plates of plastic (as well as plates of aluminum) using a friction process known as *friction stir welding*. As illustrated in Figure 36-15, the frictional heating is generated by a nonconsumable probe that is rotated or reciprocated between prepared edges which are being held together. The friction creates a plasticized region around the probe, which coalesces to form a solid-state bond as the probe traverses the joint. No filler material is required, distortion is low, and the process requires access to only one side of the joint.

Hot-plate welding is probably the simplest of the mass production techniques to join plastics. The parts to be joined are held in fixtures and then pressed against the opposite sides of an electrically heated tool. Contact is maintained until the surface has melted and the remaining material has softened to a specified distance from the interface. The parts separate, the tool is removed, and the two prepared surfaces are pressed together and allowed to cool. Contaminated surface material is usually displaced into a flash region.

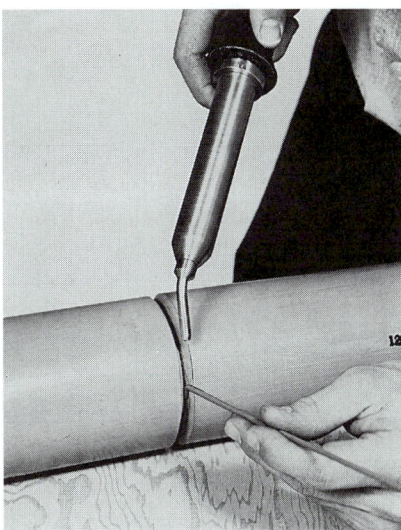

FIGURE 36-16   Using a hot-gas torch to make a weld in plastic pipe.

Weld times are comparatively slow, ranging from 10 seconds to several minutes. The joint strength is usually equal to that of the parent material, but the joint design is limited to a square-butt configuration. If the heated surface has a nonflat profile, shaped heating tools can be employed. Heated-tool welding can also be used to produce lap seam welds in flexible plastic sheets. Pressure is applied by rollers after the material has passed over a heater.

The *hot-gas welding of plastics* is similar to the oxyacetylene welding of metals. Compressed air, nitrogen, hydrogen, oxygen, or carbon dioxide is heated by an electric coil as it passes through a welding gun, such as the one shown in Figure 36-16. The hot-gas stream emerges from the gun at 400 to 570°F (200 to 300°C) and impinges on the joint area and selected filler material. V-groove or fillet welds are the most common joint configurations used with this process. Because the plastic does not melt and flow, filler material usually has to be added. As illustrated in Figure 36-16, thin rods of plastic material are heated simultaneously with the workpiece and are then forced into the softened joint area, providing both the filler material and the pressure needed to produce coalescence. This process is usually slow and the results are generally dependent on operator skill. Therefore, it seldom sees production application but is a popular process for the repair of thermoplastic materials.

In the *implant welding of plastics*, metal inserts are placed between the parts to be joined and are then heated by means of induction or resistance heating. (*Note:* The resistance method requires that wires be placed along the joint to carry current to the implants; this is not required for induction heating.) The thermoplastic material melts around the implants and flows to form a joint. Since a weld forms only in the vicinity of the implants, the process resembles spot welding and produces joints that are considerably weaker than those formed by processes that bond the entire contact area. When bonding is desired over larger areas, iron oxide paste or metal tapes can be used to concentrate the induction heating at the interface.

Still other processes to weld thermoplastics have been based on infrared or microwave heating. Laser welding has also been performed on plastics.

■ 36.5 WELDING-RELATED PROCESSES

Surfacing

Surfacing or *overlaying* is the process of depositing a layer of weld metal on the surface or edge of a base material of different composition. The usual objectives are to obtain improved resistance to wear, abrasion, heat, or chemical attack without having to make the entire piece from an expensive material, one that is difficult to fabricate, or one that would not possess the desired bulk properties. This process is often called *hard-facing*, since the deposited surfaces are generally harder than the base metal. This does not have to be true, however, for in some cases a softer metal (such as bronze) is applied to a harder base material.

Surfacing Materials. The materials most commonly used for surfacing include (1) carbon and low-alloy steels; (2) high-alloy steels and irons; (3) cobalt-based alloys; (4) nickel-based alloys, such as Monel, Nichrome, and Hastelloy; (5) copper-based alloys; (6) stainless steels; and (7) ceramic and refractory carbides, oxides, borides, silicides, and similar compounds.

Surfacing Methods and Applications. Since some of the base metal melts during the deposition, surfacing is a variation of fusion welding and can be performed by nearly all of the gas-flame or arc welding methods, including oxyfuel gas, shielded metal arc, gas metal arc, gas tungsten arc, submerged arc, and plasma arc. Arc welding is frequently used for the deposition of high-melting-point alloys. Submerged arc welding is used when large areas are to be surfaced or a large amount of surfacing material. The plasma arc process further extends the process capabilities because of its extreme temperatures. To obtain true fusion of the surfacing material, a transferred arc is used and the surfacing material is injected in the form of a powder. If a nontransferred arc is used, only a mechanical bond is produced, and the process becomes a form of metallizing. Lasers have also been used to perform hard-facing.

Thermal Spray Coating or Metallizing

The thermal spray processes offer a means of applying a coating of high-performance material (metals, alloys, ceramics, intermetallics, cermets, carbides, or even plastics) to more economical and more easily fabricated base metals. A wire or rod of the coating material is fed into a gas flame or arc, where it melts and becomes atomized by a stream of gas, such as argon, nitrogen, combustion gases, or compressed air. The gas stream propels the 0.0004- to 0.002-in. (10- to 50-μm)-diameter particles toward the target surface, where they impact ("splat"), cool, and bond. Very little heat is transferred to the substrate, whose peak temperatures generally range from 200 to 500°F (100 to 260°C). As a result, thermal spraying does not induce undesirable metallurgical changes or excessive distortion, and coatings can be applied to thin or delicate targets or to heat-sensitive materials such as plastics. The applied coating can range in thickness from 0.004 to 0.5 in. (0.1 to 12 mm).

Several of the thermal spray processes are adaptations of oxyfuel welding equipment. Figure 36-17 shows a schematic of an oxyacetylene metal spraying gun designed to utilize wire feed. The flame melts the wire and compressed air disintegrates the molten material and propels it to the workpiece. An alternative type of gun uses material in the form of powder, which is gravity or pressure fed into the flame, where it is melted and carried by the flame gas onto the target. The powder feed permits the deposition of material that would be difficult to fabricate into wire, such as cermets, oxides, and carbides. In addition,

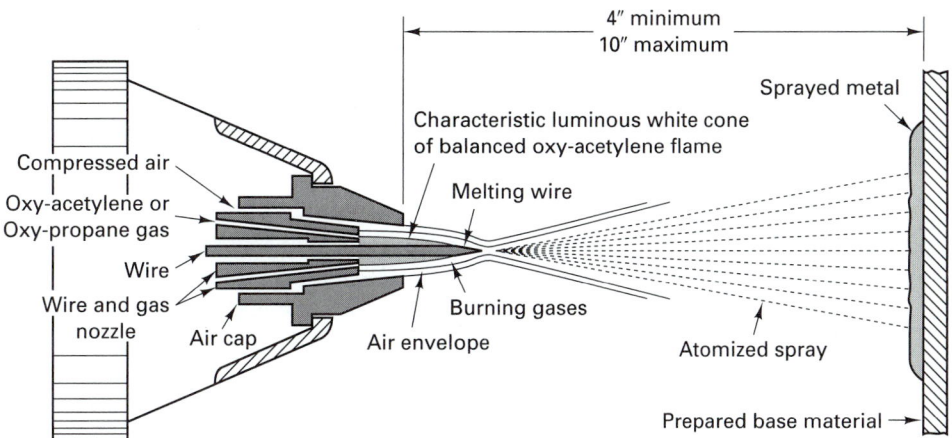

FIGURE 36-17 Schematic diagram of an oxyacetylene metal-spraying gun. *(Courtesy of METCO, Inc.)*

the droplet size is controlled by the powder, not by the factors that control atomization. The lower temperatures and lower particle velocities of the oxyfuel deposition methods result in coatings with high porosity and low cohesive strength.

In a process known as high-velocity oxyfuel (HVOF) spraying, the combustion flame is created in a high-pressure chamber and then exits through a small-diameter barrel to produce a supersonic stream of hot gas capable of heating and propelling the powder particles. The particles impact the substrate with high kinetic energy, producing dense, well-bonded coatings.

The simplest of the electric arc methods is probably wire arc or electric arc spraying. Two oppositely charged electrode wires are fed through the gun, meeting at the tip, where they form an arc. A stream of atomizing gas flows through the gun, stripping off the molten metal to produce a high-velocity spray. Since all of the input energy is used to melt the metal, this process is extremely energy efficient.

The most sophisticated of the metallizing techniques is the plasma spray process (Figure 36-18). A plasma-forming gas serves as both the heat source and propelling agent for the coating material. The particles attain high velocity, thereby producing dense, strongly bonded coatings. Since temperatures can reach 30,000°F (16,500°C), plasma spraying can be used to deposit materials with melting points as high as 6000°F (3300°C). Metals, alloys, ceramics, carbides, cermets, intermetallics, and plastic-based powders have all been deposited successfully. High-energy plasma spraying and vacuum plasma spraying are adaptations of the plasma technique.

While thermal spraying is similar to surfacing and is often applied for the same reasons, the coatings are usually thinner and the process is more suitable for irregular surfaces or heat-sensitive substrates. The deposition guns can be either hand-held or machine-driven. A stand-off distance of 6 to 10 in. (150 to 250 mm) is usually maintained between the spray nozzle and the workpiece. Table 36-2 compares the five basic methods of thermal spray deposition.

Surface Preparation for Metallizing. Unlike surfacing, metallizing does not melt the base metal. Since adhesion is entirely mechanical, it is essential that the base metal be prepared in a way that promotes good mechanical interlocking. This material must first be clean and free of dirt, moisture, oil, and other contaminates. The surface is then roughened

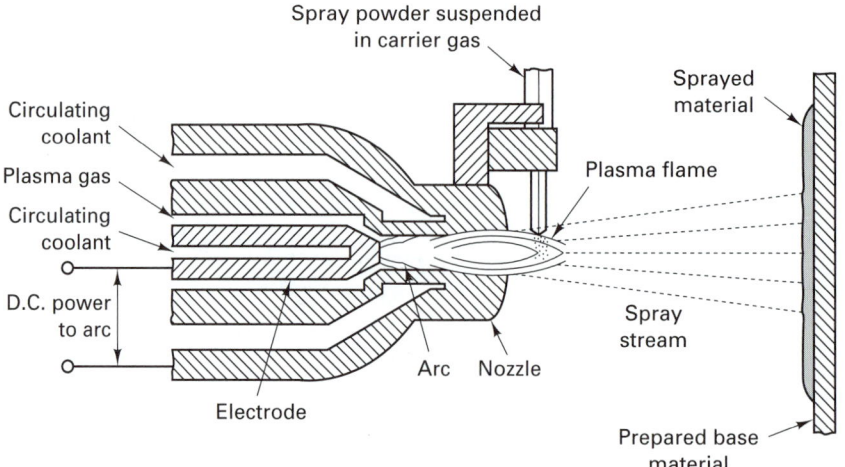

FIGURE 36-18   Schematic diagram of a plasma-arc spray gun. *(Courtesy of METCO, Inc.)*

TABLE 36-2.   Comparison of the Five Basic Thermal Spray Deposition Techniques

Method	Source	Heat		Deposited Materials	Particle Impact Velocity (ft/sec)	Adhesion Strength	Maximum Spray Rate (lb/hr)	Energy Required (kW/lb)
		Temperature (°F)						
Flame spray								
Wire	Oxyfuel	5,000		Metals	600	Medium	20	5–10
Powder	Oxyfuel	5,000		Metals, ceramics, plastics	100	Low	15	5–10
High-velocity oxyfuel (HVOF)	Oxyfuel	5,600		Metals, carbides	2000–3500	Very high	30	10¬90
Wire arc	DC arc	10,000		Metals only	800	High	35	0.1–0.2
Plasma spray	DC arc	10,000 to 30,000		All	800–4000	High to Very high	10–50	5–10

by one of a variety of methods to create minute crevices that can anchor the solidifying particles. Grit blasting with a sharp, abrasive grit is the most common technique. A surface roughness of 100 to 300 μin. is adequate for most applications.

Characteristics and Applications of Sprayed Metals. During deposition, the atomized, molten, and semimolten particles mix with air and then cool rapidly upon impact with the base metal. The deposited coatings consist of bonded particles that span a range of size, shape, and degree of melting. Some particles become oxidized and voids tend to become entrapped. As a result, the coatings are harder, more porous (0.1 to 15% porosity), and more brittle than the same material in its conventional wrought state. Thermal spray coatings add little, if any, additional strength to a part, since the strength of the porous coating is usually between one-third and one-half of its normal wrought strength. Applications, therefore, generally focus on the ability to provide resistance to heat, wear, erosion, and/or corrosion, or to restore worn parts to their original dimensions and specifications. In addition, the coating materials can be applied only where needed. Some typical applications include:

1. *Protective coatings.* Zinc and aluminum are sprayed on iron and steel to provide corrosion resistance, a process that may well extend the lifetime of bridges, buildings, and other infrastructure items. The interior surfaces of power boilers can be coated with high-chromium alloys to extend wall life through resisting both heat and corrosion.

2. *Building up worn surfaces.* Worn parts can often be salvaged by adding metal to the depleted regions. The repair and restoration of aircraft engine components is probably the largest single use of thermal spraying.

3. *Hard surfacing.* Although metal spraying should not be compared to hard-facing deposits that are applied by welding techniques, it can be useful when thin coatings are considered to be adequate. Typical applications might include automobile cylinder liners and piston rings, or thread guides in textile plants, and critical parts within pumps, bearings, and seals.

4. *Applying coatings of expensive metals.* Metal spraying provides a simple method for applying thin coatings of noble metals to surfaces where conventional plating would not be economical.

5. *Electrical properties.* Because metal can be sprayed on almost any surface, metal or nonmetal, it can be used to apply a conductive surface to an otherwise poor conductor or nonconductor. Copper, aluminum, or silver is frequently sprayed on glass or plastics for this purpose. Conversely, alumina (Al_2O_3) can be used to impart dielectric properties.

6. *Reflecting surfaces.* Aluminum, sprayed on the back of glass by a special fusion process, makes an excellent reflecting surface.

7. *Decorative effects.* One of the earliest and still important uses of metal spraying was to obtain decorative effects. Because sprayed metal can be treated in a variety of ways, such as buffed, wire brushed, or left in the as-sprayed condition, it is frequently specified for both manufactured products and architectural materials.

8. *Tailored surface characteristics.* Porous coatings of cobalt or titanium alloys, or certain ceramic materials, have been applied to medical implants to help promote adhesion and ingrowth of bone and tissue.

■ KEY WORDS

assist gas
cladding
cold welding
diffusion bonding
electron beam welding
electroslag welding
explosive welding
faying surfaces
flash
flash welding

forge welding
friction welding
hard facing
hot gas welding
hot plate welding
inertia welding
laser beam welding
metallizing
molding plates
penetration

percussion welding
roll bonding
solid-state welding
surfacing
thermit
thermoplastics
ultrasonic welding
vibration welding

■ REVIEW QUESTIONS

1. What were some of the limitations that made the forge welds of a blacksmith somewhat variable in terms of quality?
2. What is the primary feature that promotes coalescence in cold welding?
3. Describe how the roll bonding process can be used to fabricate products that contain pressure-tight fluid-flow channels that once required the use of metal tubing.
4. What is the source of heat in friction welding?
5. How does inertia welding differ from friction welding?
6. How are surface impurities removed in the friction and inertia welding processes?
7. What are some of the geometric limitations of friction and inertia welding?
8. What are some of the geometric limitations of ultrasonic welding?
9. What are some of the attractive features of ultrasonic welding?
10. What are the conditions necessary to produce high-quality diffusion welds?
11. If the interface of a weld is viewed in cross section, what is the distinctive geometric feature of an explosive weld?
12. In what ways is a thermit weld similar to the production of a casting?
13. What is the source of the welding heat in thermit welding?
14. For what types of applications might thermit welding be attractive?
15. What is the source of the welding heat in electroslag welding?
16. What are some of the various functions of the slag in electroslag welding?
17. Electroslag welding would be most attractive for the joining of what types of geometries and thicknesses?
18. Why is a high vacuum required in the electron beam chamber of an electron beam welding machine?
19. What types of production limitations are imposed by the high-vacuum requirements of electron beam welding? What compromises are made when welding is performed on pieces outside the vacuum chamber?
20. What are the major drawbacks of high-voltage electron beam welding equipment?
21. What are some of the attractive features of electron beam welding?
22. What are some of the ways in which laser beam welding is more attractive than electron beam welding?
23. Why is laser beam welding an attractive process for use on small electronic components?
24. What is the function of the "assist gas" in laser beam cutting?
25. What features enable industrial lasers to be integrated easily with robots and CNC machines?
26. What are some of the attractive features of laser spot welding?
27. In the flash welding process, why is it important to have a minimum duration of arcing and minimum amount of upsetting?
28. Why is it difficult to produce a weld between thermosetting polymers but not difficult with the thermoplastics?
29. What are some of the more-common techniques for welding thermoplastics? Identify the primary source of heat for each of these processes.
30. What are some of the primary methods by which surfacing materials can be deposited onto a metal substrate?
31. What are some of the techniques that can be used to apply a thermal spray coating?
32. How is thermal spraying similar to surfacing? How is it different?
33. Why is surface preparation such a critical feature of metallizing?

■ PROBLEMS

1. Many advanced engineering products, as well as composite materials, require the joining of dissimilar materials. Select several of the processes discussed in this chapter and investigate the capability of the process to join dissimilar materials and the associated limitations.
2. The processes described for the joining of plastics focused almost exclusively on the

thermoplastic polymers. What types of joining techniques could be applied to thermosetting

polymers? To elastomeric polymers? To ceramic materials?

*C*hapter 36 CASE STUDY

field repair to a power transformer case

Electric power transformers do not operate at 100% efficiency, and generally incorporate some means of cooling in their design. Large transformers are often submerged in oil-filled reservoirs, where the volume of oil provides a non-corrosive heat sink. In addition, it is not uncommon for additional features, such as horizontal cooling fins, to be added to the design to aid in dissipating heat from the reservoir.

Figure CS-36 shows the exterior of a large transformer that has been installed in a rural, somewhat-remote location. The reservoir housing has been constructed by welding 3/8- and 1/2-inch thick, low-carbon steel plates. While the transformer was in use, a service vehicle accidentally backed into the cooling-fin assembly, producing cracks in several of the fillet areas and a resulting loss of oil. Overheating occurred, and a repair is now necessary.

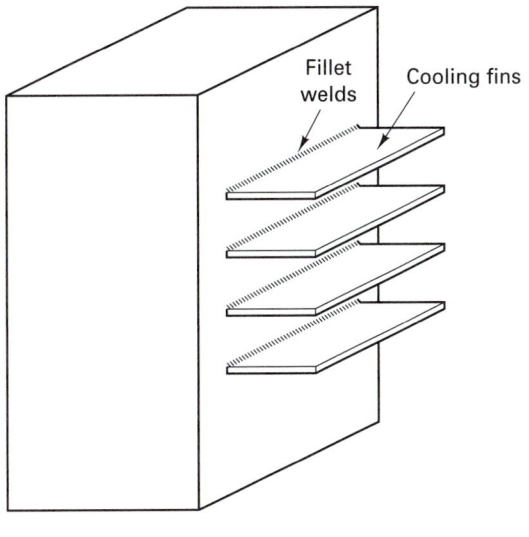

FIGURE CS-36

Because of the size of the transformer, some form of on-site repair is preferred. It is your job to determine the procedure and make the necessary arrangements.

1. Consider the full spectrum of welding processes and identify candidates that might be appropriate for this task.

2. For each of the candidate processes, identify its primary advantages and limitations.

3. Which of the candidate processes would you recommend? Why?

4. Describe the procedure that you would outline for such a repair. Are there any special concerns or precautions?

BRAZING AND SOLDERING

37.1	INTRODUCTION		Design of Brazed Joints
37.2	BRAZING		Braze Welding
	Nature and Strength of Brazed Joints	37.3	SOLDERING
			Solder Metals
	Brazing Metals		Soldering Fluxes
	Fluxes		Heating for Soldering
	Applying the Brazing Metal		Metals to Be Joined
	Heating Methods Used in Brazing		Design and Strength of Soldered Joints
	Flux Removal		Flux Removal
	Postbraze Operations and Inspection		Fluxless Soldering
		Case Study:	THE INDUSTRIAL DISPOSAL IMPELLER
	Fluxless Brazing		

■ 37.1 INTRODUCTION

There are many joining or assembly operations where welding may not be the best choice. Perhaps the heat of welding is objectionable, or the materials possess poor weldability, or welding is expensive. In such cases low-temperature joining methods may be preferred. These include brazing, soldering, adhesive joining, and the use of mechanical fasteners. In brazing and soldering, the metal surfaces are cleaned, the components assembled or fixtured, and a low-melting-point nonferrous metal is then melted, drawn into the space between the two solid surfaces by capillary action, and allowed to solidify. Adhesive bonding utilizes a nonmetallic filler material (often a polymerizable resin) to fill the space between the surfaces to be joined. Variations in surface finish and fit are more tolerable, since capillary action is not required, and oxides may not be a problem since the adhesive may actually adhere better to a tight oxide layer. Mechanical fasteners span a wide spectrum, including rivets, bolts, screws, staples, nails, and others. While some forms may be permanent, others offer the advantage of easy disassembly for service or replacement of components. In this chapter we discuss brazing and soldering. Adhesive bonding and mechanical fasteners are presented in Chapter 38.

■ 37.2 BRAZING

Brazing is the joining of metals through the use of heat and a filler metal whose melting temperature is above 840°F (450°C)* but below the melting point of the metals being joined. In comparison with welding, the brazing process is different in a number of ways:

1. The *composition* of the brazing alloy is significantly different from that of the base metal.
2. The *strength* of the brazing alloy is substantially lower than that of the base metal.
3. The *melting point* of the brazing alloy is lower than that of the base metal, so the base metal is not melted.

---

*The temperature is an arbitrary one, set to distinguish brazing from soldering.

4. Bonding requires *capillary action*, the specific flow being related to the viscosity of the liquid and the geometry of the joint.

Because of these differences, the brazing process has several distinct advantages:

1. Virtually all metals can be joined by some type of brazing metal.
2. The process is ideally suited for dissimilar metals, such as the joining of ferrous to nonferrous, or metals with widely different melting points.
3. Since less heating is required than for welding, the process can be performed quickly and economically.
4. The lower temperatures reduce problems associated with heat-affected zones, warping, or distortion. Thinner and more complex assemblies can be joined successfully.
5. Brazing is highly adaptable to automation and performs well when mass producing delicate assemblies. A strong permanent joint is formed.

A major disadvantage of brazing is that subsequent heating of the assembly can cause inadvertent melting of the braze metal, weakening or destroying the joint. Too often brazed joints fail when people apply heat in an attempt to straighten or repair damaged assemblies. Although this is certainly not the result of defective brazing, the consequences are still most unfortunate.

Another concern with brazed joints is their enhanced susceptibility to corrosion. Since the *filler metal* is of different composition from the materials being joined, the brazed joint is actually a localized galvanic corrosion cell. This effect can often be minimized by proper selection of the filler metal.

Nature and Strength of Brazed Joints

Just as in welding, brazing forms a strong metallurgical bond at the interfaces. The bonding is enhanced by clean surfaces, proper clearance, good wetting, and good fluidity. The resulting strength can be quite high, certainly higher than the strength of the brazing alloy and possibly higher than the strength of the metal being brazed. Attainment of this high strength, however, requires optimum processing and design.

Bond strength is a strong function of the *clearance* between the parts to be joined. If the joint is too tight, it may be difficult for the braze metal to flow into the gap (leaving unfilled voids), and flux may be unable to escape (remaining in locations that should be filled with braze material). There must be sufficient clearance so that the braze metal will wet the joint and flow into it under the force of capillary action. As the gap is increased beyond this optimum value, however, the joint strength decreases rapidly, dropping off to that of the braze metal itself. If the gap becomes too great, capillary forces may be insufficient to draw the material into the joint or hold it in place during solidification. Figure 37-1 shows the tensile strength of a butt-joint braze as a function of joint clearance.

Proper clearance varies considerably, depending primarily on the type of braze metal being used. The ideal clearance is usually between 0.0005 and 0.0015 in. (an "easy-slip" fit). A press fit can be acceptable if fluxes are not used and the surface roughness is sufficient to assure adequate flow of the filler metal into the joint. Clearances up to 0.003 in. can be accommodated with a more sluggish filler metal, such as nickel. However, when clearances range between 0.003 and 0.005 in., acceptable brazing becomes somewhat difficult, and joints with gaps in excess of 0.005 in. are almost impossible to braze. It should be noted that the cited dimensions are the clearances that should exist *at the temperature of the brazing process*. Any effects of thermal expansion should be compensated when

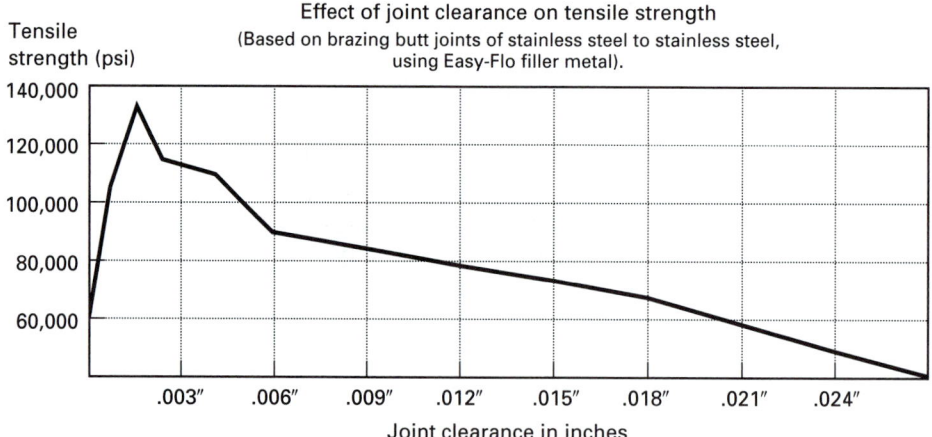

FIGURE 37-1 Typical variation of tensile strength with different joint clearances in a butt joint design. *(Courtesy of Handy & Harman).*

specifying the dimensions of the basic components.

Wettability is a strong function of the surface tensions between the braze metal and the base alloy. Generally, the wettability is good when the surfaces are clean and the two metals can form intermediate diffused alloys. Sometimes, the wettability can be improved, as is done when steel is tin plated to accept a lead–tin solder, or plated with nickel or copper to enhance brazing.

Fluidity is a measure of the flow characteristics of the molten braze metal and is a function of the metal, its temperature, surface cleanliness, and clearance.

Brazing Metals

Brazing materials should be selected on the basis of a variety of criteria, including compatibility with the base materials, brazing temperature restrictions, restrictions due to service or subsequent processing temperatures, the brazing process to be used, the joint design, anticipated service environment, desired appearance, desired mechanical properties (such as strength, ductility, and toughness), desired physical properties (such as electrical, magnetic, or thermal), and cost. In addition, the materials must be capable of flowing through small capillaries, "wetting" the joint surfaces, and partially alloying with the base metals. The most commonly used brazing metals are copper and copper alloys, silver and silver alloys, and aluminum alloys. Table 37-1 presents some common braze metal families, the metals they are used to join, and the typical brazing temperatures.

Copper is the basis for most of the commonly used brazing alloys. Unalloyed copper is used primarily for brazing steel and other high-melting-point materials, such as high-speed steel and tungsten carbide. Its use is confined almost exclusively to furnace operations conducted in a protective hydrogen atmosphere, in which the copper is extremely fluid and requires no flux. Its melting point is rather high (about 2000°F) and tight-fitting joints are required (gaps less than 0.003 in.).

Copper alloys were the earliest brazing materials. Copper–zinc alloys offer a lower melting point than that of pure copper, and are used extensively for brazing steel, cast irons, and copper. Copper–phosphorus alloys are used for the fluxless brazing of copper since the phosphorus can reduce the copper oxide film. These alloys should not be used

TABLE 37-1. Some Common Braze Metal Families, Metals They Are Used to Join, and Typical Brazing Temperatures

Braze Metal Family	Materials Commonly Joined	Typical Brazing Temperature (°F)
Aluminum–silicon	Aluminum alloys	1050–1150
Copper and copper alloys	Various ferrous metals as well as copper and nickel alloys and stainless steel	1700–2100
Copper–phosphorus	Copper and copper alloys	1300–1700
Silver alloys	Ferrous and nonferrous metals, except aluminum and magnesium	1150–1800
Precious metals (gold-based)	Iron, nickel, and cobalt alloys	1650–2000
Magnesium	Magnesium alloys	1100–1150
Nickel alloys	Stainless steel, nickel, and cobalt alloys	1700–2200

with ferrous or nickel-based materials, however, since they form brittle compounds with phosphorus and the resulting joints may be brittle. Manganese bronzes can also be used as brazing material.

Pure silver is used in brazing titanium. *Silver solders*, alloys of silver and copper with paladium, nickel, tin, or zinc, have brazing temperatures nearly 600°F below that of pure copper, and are used in joining steels, copper, brass, and nickel. Although these brazing alloys are quite expensive, such a small amount is required that the cost per joint is really quite low. The silver alloys are also used in brazing stainless steels. However, since the brazing temperatures are in the range of carbide precipitation (sensitization), only stabilized or low-carbon stainless steels should be brazed with these alloys if good corrosion resistance is required in the product.

Aluminum–silicon alloys, containing about 6 to 12% silicon, are used for brazing aluminum and aluminum alloys. By using a braze metal that is not greatly unlike the base metal, the possibility of galvanic corrosion is reduced. However, since these brazing alloys have melting points of about 1130°F (610°C), and the melting temperature of commonly brazed aluminum alloys such as 3003 is around 1290°F (670°C), control of the brazing temperature is critical. In brazing aluminum, proper fluxing action, surface cleaning, and/or the use of a controlled-atmosphere or vacuum environment is required to assure adequate flow of the braze metal.

Nickel- and cobalt-based alloys offer excellent corrosion- and heat-resistant properties. Gold and palladium alloys offer outstanding oxidation and corrosion resistance, as well as electrical and thermal conductivity. Magnesium alloys are used to braze magnesium.

Amorphous alloy brazing sheets have been produced by cooling metal at rates in excess of 1 million °C per second. The resulting metal foils are extremely thin, 0.0015 in. (0.04 mm) being typical, and exhibit excellent ductility and flexibility. Shaped inserts can be cut or stamped from the foil, inserted into the joint, and heated. Since the braze material is fully dense, no shrinkage or movement is observed during the brazing operation. A variety of brazing alloys are currently available in the form of amorphous foils. A nickel–chromium–iron–boron brazing alloy is used to braze alloys that are to be used at high temperatures. During the brazing operation, the boron in the brazing alloy diffuses into the base metal, raising the melting point of the remaining braze metal. The subsequent service temperature can then be above the melting point of the original braze alloy, and the braze material will not melt.

Fluxes

In a normal atmosphere, the heat required to melt the brazing alloy would also cause the formation of surface oxides that oppose the wetting of the surface and subsequent bonding. *Fluxes*, therefore, play a very important part in brazing by (1) dissolving oxides that may be on the surface prior to heating, (2) preventing the formation of oxides during heating, and (3) lowering the surface tension of the molten brazing metal and thus promoting its flow into the joint.

One of the primary factors affecting the quality and uniformity of brazed joints is cleanliness. Although fluxes will dissolve modest amounts of oxides, *they are not cleaners*. Before a flux is applied, dirt, grease, oil, rust, and heat-treat scale should be removed from the surfaces that are to be brazed. The less the flux has to do prior to heating, the more effective it will be during the brazing operation. Cast iron materials are not readily wettable because of the graphite. Consequently, before cast iron can be brazed, the graphite must be removed by etching.

Borax has been in common use as a brazing flux. *Fused borax* should be used because the water in ordinary borax causes bubbling when the flux is heated. Alcohol can be mixed with fused borax when a paste consistency is desired.

A number of modern fluxes are also available that have melting temperatures lower than borax and are somewhat more effective in removing oxidation. The particular flux should be selected for compatibility with the base metal being brazed and the particular process being used. Paste fluxes are utilized for furnace, induction, and dip brazing, and are usually applied by brushing. Either paste or powdered fluxes can be used with the torch brazing process. Application is usually done by dipping the heated end of the filler wire into the flux.

Fluxes for aluminum are usually mixtures of metallic halide salts, with sodium and potassium chlorides comprising 15 to 85% of the mixture. Activators such as fluorides and lithium compounds are also added. The aluminum brazing fluxes *do not* dissolve the surface oxide on aluminum.

Since most brazing fluxes are corrosive, the residue should be removed from the work immediately after brazing is completed. This is particularly important in the case of aluminum, where the chlorides are particularly detrimental. Considerable effort has been directed to developing fluxless procedures for brazing aluminum, as will be discussed later.

Applying the Brazing Metal

Brazing metal can be applied to joints in several ways. The oldest (and a common technique used when torch brazing) uses brazing metal in the form of a rod or wire. The joint area is first heated to a temperature high enough to melt the braze alloy and assure its remaining molten while flowing into the joint. The braze metal is then melted by the torch and capillary action draws it into the prepared gap.

This method of braze metal application requires considerable labor, and care is necessary to assure that the filler metal has flowed to the inner portions of the joint. To avoid these difficulties, the braze metal is often applied to the joint prior to heating, often in the form of wires, shims, powder, or preformed rings, washers, disks, or slugs. In cases where it can be done, rings or shims of braze metal are fitted into internal grooves in the joint before the parts are assembled. When this procedure is employed, the parts generally must be held together by press fits, riveting, staking, tack welding, or a jig, to maintain their proper alignment (Figure 37-2). Figure 37-3 shows various ways of applying braze metal wire, sheet, and foil. In such preloaded joints, care must also be exercised to assure that

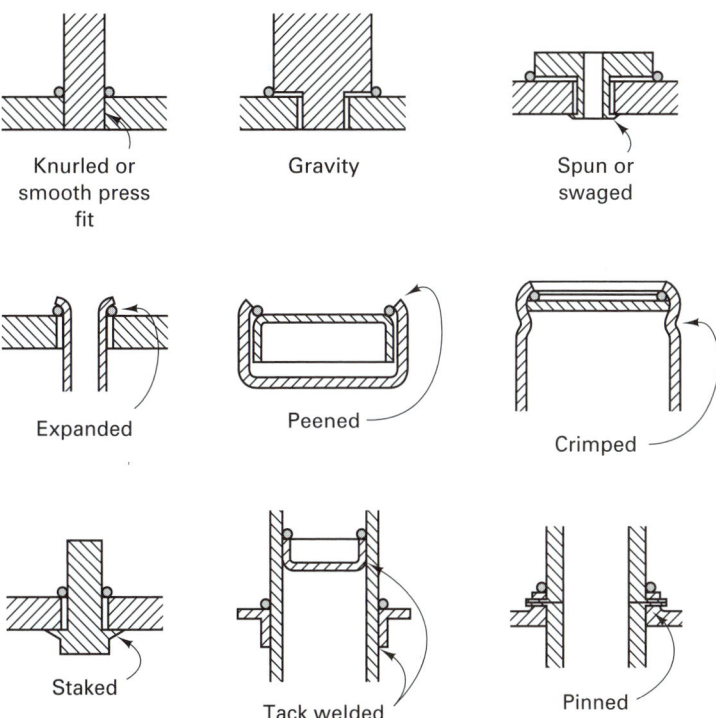

FIGURE 37-2 Methods of applying braze metal and positioning or fixturing various joints.

the filler metal is not drawn away from the intended surface by the capillary action of another surface of contact. Capillary action will always pull the molten braze metal into the smallest clearance, whether or not that was intended.

Another precaution that must be observed is that the flow of filler metal not be cut off by the absence of required clearances or by the presence of entrapped air. As the trapped gases expand upon heating, they can act to prevent the filler metal from flowing throughout the joint. Fillets and grooves within the joint can also act as reservoirs and trap the filler metal.

The least expensive approach to brazing is to design assemblies where all the components maintain a fixed position throughout the brazing cycle. When this is not possible, alignment and clearances can be maintained by tackwelding, riveting, staking, expanding or flaring, swaging, knurling, and dimpling. Shims, wires, ribbons, and screens can also be employed to assist in locating pieces or maintaining fit. When the components become more complex, special brazing jigs and *fixtures* are often used to hold the components during the heating. When these are used, however, it is usually necessary to provide springs that will compensate for thermal expansion, particularly when two or more dissimilar metals are being joined.

Another approach to the brazing process is to use sheets of clad metal where one or both of the surfaces are precoated with the brazing alloy. The desired bonds are made simply by placing the materials in contact and heating. By having the braze material already in place over the full area of contact, the joining operation does not have to rely on capillary action and metal flow. More complex assemblies can be produced than with conventional

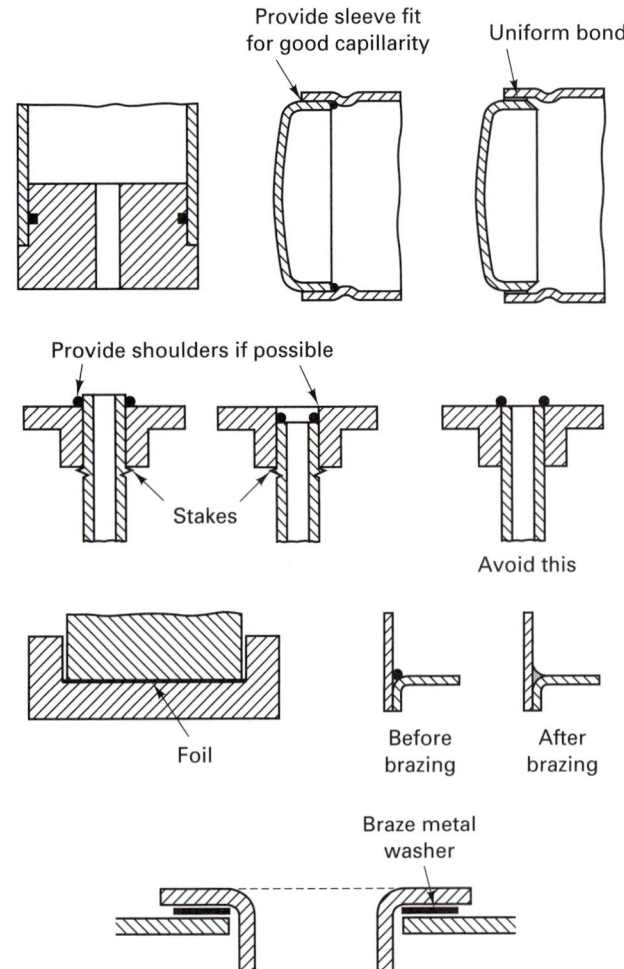

FIGURE 37-3 Techniques to apply brazing wire, foil, or sheet to assure proper flow into the joint.

methods, and the thickness of the braze material is precisely controlled to provide maximum strength to the joint.

Heating Methods Used in Brazing

A common source of heat for brazing is a gas-flame torch. In the *torch-brazing* procedure, oxyacetylene, oxyhydrogen, or other gas-flame combinations can be used. Most repair brazing is done in this manner because of its flexibility and simplicity, but the process is also widely used in production applications. Local heating permits the retention of most of the original material strength and permits large components to be joined with little or no distortion. The major drawbacks are the difficulty in controlling the temperature, maintaining uniformity of heating, and meeting the cost of skilled labor. A protective flux is required and the flux residue must be removed after brazing. In production-type torch brazing, specially shaped torches are often used to speed the heating and aid in reducing the amount of skill required.

Numerous brazing operations are performed in controlled-atmosphere or vacuum furnaces. In *furnace brazing*, the brazing metal must be preloaded into the work. If the work

FIGURE 37-4   Typical furnace-brazed assemblies. *(Courtesy of Pacific Metals Company.)*

is not of such a nature that its preassembly will maintain the parts in proper alignment, brazing jigs or fixtures must be used. Fortunately, assemblies that are to be furnace brazed can usually be designed so that jigs and fixtures are not needed; a light press fit will often suffice. Figure 37-4 shows a number of typical furnace-brazed assemblies.

Because excellent control of the furnace brazing temperatures can be obtained and no skilled labor is required, furnace brazing is particularly well suited for mass production operations. Either box- or continuous-type furnaces can be used, the latter being more suitable for mass production work. If the furnace atmosphere can reduce oxide films and prevent both the base and filler materials from oxidizing during the brazing operation, flux is not needed and the parts emerge clean and free of contaminants. When reactive materials are to be brazed, vacuum furnaces are often used.

A third form of heating is *salt-bath brazing*, where the parts are preheated and then dipped in a bath of molten salt that is maintained at a temperature slightly above the melting point of the brazing metal. This process offers several distinct advantages: (1) the work heats very rapidly because it is in complete contact with the heating medium, (2) the salt bath acts as a protective medium to prevent oxidation, and (3) thin pieces can easily be attached to thicker pieces without danger of overheating because the salt bath maintains a uniform temperature that is below the melting point of the parent metal. The latter feature makes the process well suited for brazing aluminum, where precise temperature control is required.

Here again, the parts must be held in jigs or fixtures (or be prefastened in some manner), and the brazing metal must be preloaded into the work. To assure that the bath remains at the desired temperature, its volume must be substantially larger than that of the assemblies to be brazed.

In *dip brazing*, the assemblies are immersed in a bath of molten brazing metal. The bath thus provides both the heat and braze metal for the joint. However, since the braze metal will usually coat the entire workpiece, it is a wasteful process and is usually employed only for small products.

Induction brazing utilizes high-frequency induction currents for heating. This process offers the following advantages, which account for its extensive use:

1. The heating is very rapid. Usually, only a few seconds are required for a complete cycle.
2. The operation can be made semiautomatic, so that only semiskilled labor is required.
3. Heating can be confined to the area of the joint through use of specially designed coils and short heating times. This minimizes softening and distortion and reduces problems associated with scale and discoloration.
4. Uniform results are easily obtained.
5. By making new, and relatively simple heating coils, a wide variety of work can be performed with a single power supply.

High-frequency power supplies are available in both large and small capacities at very modest cost. These are then coupled to a simple heating coil designed to fit around the joint. The heating coils are generally formed from copper tubing and are designed to carry a supply of cooling water. Although the filler material can be added to the joint manually after it is heated, the usual practice is to use preloaded joints to speed the operation and produce more-uniform bonds.

Some *resistance brazing* is also performed, in which the parts to be joined are pressed between two electrodes as a current is passed through. Unlike resistance welding, however, most of the resistance is provided by the electrodes, which are made of carbon or graphite. Thus most of the heating is by means of conduction from the hot electrodes. The resistance process is used primarily to braze electrical components, such as conductors, cable connectors, and similar devices. Equipment is generally an adaptation of conventional resistance welders.

Flux Removal

Although not all brazing fluxes are corrosive, most of them are. Thus, flux residues should be completely removed from the product. Since many of the commonly used fluxes are soluble in hot water, their removal is not very difficult. Immersion in a hot-water tank for a few minutes will usually give satisfactory results, provided that the water is sufficiently hot. In addition, flux removal is usually easier if the process is performed when the flux is still hot.

Blasting with grit or sand is another effective method of flux removal, but this procedure cannot be used if the surface finish is to be maintained. Fortunately, such drastic treatment is seldom necessary.

Postbraze Operations and Inspection

Postbraze operations often include heat treating, cleaning, and inspection. A visual examination is probably the simplest of the inspection techniques and is most effective when both sides of a brazed joint are accessible for examination. A proof test can be performed by subjecting the joint to loads in excess of those expected during service. Leak tests or pressure tests can assure gas- or liquid-tightness. Cracks and other flaws can be detected by dye-penetrant, magnetic particle, ultrasonic, or radiographic examination, as described in Chapter 11. Destructive forms of evaluation include peel tests, tension or shear tests, and metallographic examination.

Fluxless Brazing

Both the application and removal of brazing flux involves significant costs, particularly where complex joints and assemblies are involved. Consequently, a large amount of work has been devoted to the development of procedures where a flux is not required. Furnace atmospheres can often reduce existing oxides and prevent the formation of new ones. Thus a flux may not be necessary.

Much work in *fluxless brazing* has been directed to the brazing of aluminum. Because of its light weight and good thermal conductivity, aluminum is particularly attractive for applications such as automobile radiators where weight reduction is a desired objective. Several characteristics of aluminum make the metal particularly difficult to braze: its low melting point, the high galvanic potential, and the presence of a refractory oxide film. Successful fluxless brazing often requires vacuums up to 1×10^{-5} torr (0.0013 Pa). In addition, the aluminum must be cleaned and degreased carefully prior to brazing.

Design of Brazed Joints

Three types of brazed joints are used: *butt*, *scarf*, and *lap* or *shear*. Figure 37-5 shows a variety of braze joint designs for joining both flat and curved surfaces. Examples of good and poor joint design are presented in Figure 37-6. Because the basic strength of a brazed joint is generally less than that of the parent metals, the desired strength is often obtained by specifying sufficient joint area. Often, some type of lap joint is employed when maximum strength is required. If the joints are made very carefully, a lap of 1 to $1\frac{1}{4}$ times the

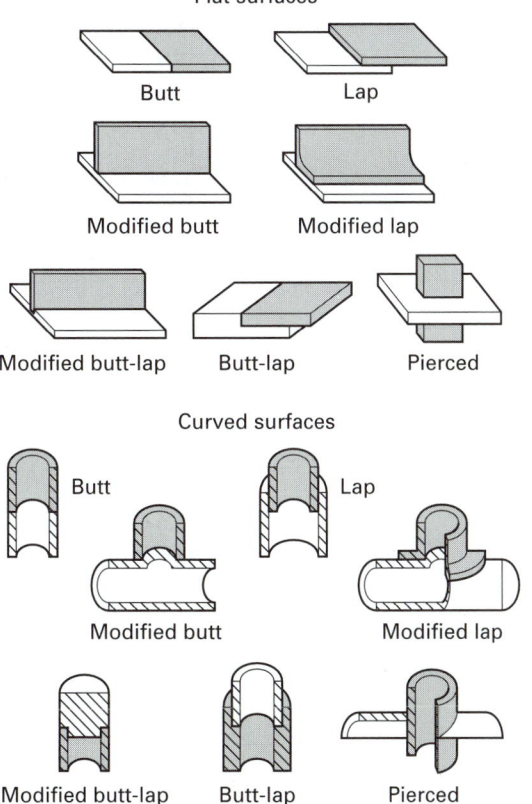

FIGURE 37-5   Some common designs of brazed joints for flat and curved surfaces. *(Adapted from* The Brazing Book, *Handy & Harman).*

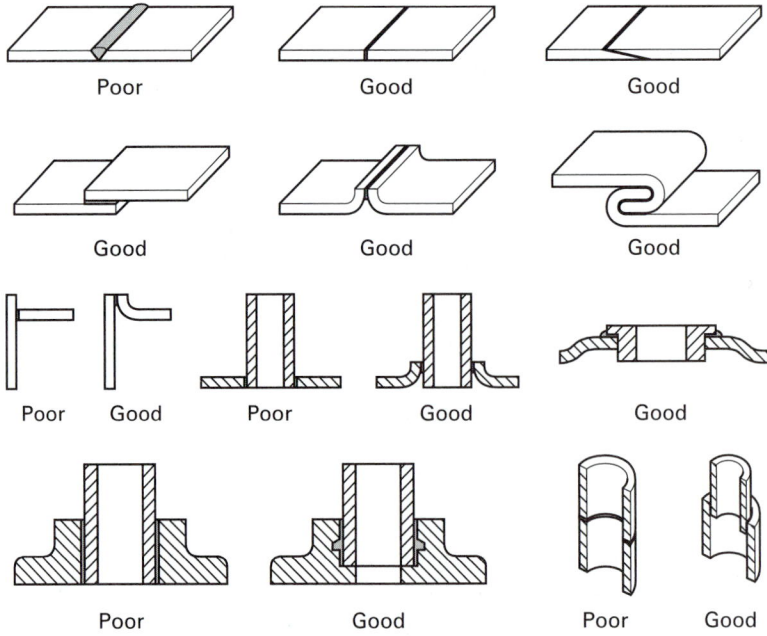

FIGURE 37-6 Examples of good and bad joint design for brazing.

thickness of the metal can develop strength as great as that of the parent metal. For joints that are made in routine production, however, it is best to use a lap equal to three times the material thickness. Table 37-2 summarizes the compatibility of various engineering materials with the brazing process.

TABLE 37-2. Engineering Materials and Their Compatibility with Brazing

Material	Brazing Recommendation
Cast iron	Somewhat difficult
Carbon and low-alloy steels	Recommended for low- and medium-carbon materials; difficult for high-carbon materials; seldom used for heat-treated alloy steels
Stainless steel	Recommended; Silver and nickel brazing alloys are preferred
Aluminum and magnesium	Common for aluminum alloys and some alloys of magnesium
Copper and copper alloys	Recommended for copper and high-copper brasses; somewhat variable with bronzes
Nickel and nickel alloys	Recommended
Titanium	Difficult, not recommended
Lead and zinc	Not recommended
Thermoplastics, thermosets, and elastomers	Not recommended
Ceramics and glass	Not recommended
Dissimilar metals	Recommended, but may be difficult, depending on degree of dissimilarity
Metals to nonmetals	Not recommended
Dissimilar nonmetals	Not recommended

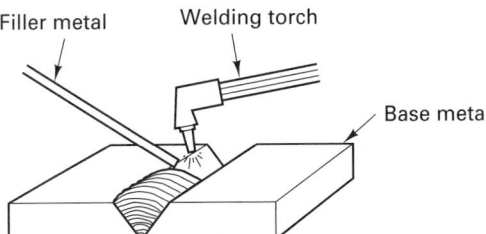

FIGURE 37-7   Schematic of the braze welding process.

Braze Welding

Braze welding differs from straight brazing in that capillary action is not required to distribute the filler metal. Here the molten filler is simply deposited by gravity, as in oxyacetylene gas welding. Because relatively low temperatures are required and warping is minimized, braze welding is very effective for the repair of steel products and ferrous castings. It is also attractive for joining cast irons since the low heat does not alter the graphite shape and the process does not require good wetting characteristics. Strength is determined by the braze metal being used and the amount applied. Considerable buildup may be required if full strength is to be restored to the repaired part.

Braze welding is almost always done with an oxyacetylene torch. The surfaces are first "tinned" with a thin coating of the brazing metal, and the remainder of the filler metal is then added. Figure 37-7 shows a schematic of braze welding.

■ 37.3 SOLDERING

By definition, *soldering* is a brazing type of operation where the filler metal has a melting temperature below 840°F (450°C). Effective soldering generally involves six important steps: (1) design of an acceptable solder joint, (2) selection of the correct solder for the job, (3) selection of the proper type of flux, (4) cleaning the surfaces to be joined, (5) application of flux, solder, and sufficient heat to allow the molten solder to fill the joint by capillary action and solidify, and (6) removal of the flux residue, if necessary.

Solder Metals

Because of their low cost and acceptable mechanical properties, most solders are alloys of lead and tin with the addition of a very small amount of antimony, usually less than 0.5%. The three most commonly used alloys contain 60, 50, and 40% tin and all melt below 465°F (240°C). Because tin is expensive, those alloys having higher proportions of tin are used only where their higher fluidity, higher strength, and lower melting temperature are desired. For *wiped* joints and for filling dents and seams, such as in automobile body work where little strength is required, solders containing only 10 to 20% tin are used.

Other soldering alloys are used for special purposes or where environmental or health concerns dictate the use of lead-free joints. Lead and lead compounds can be quite toxic. The use of lead-containing solders in drinking-water lines is now prohibited in the United States by federal legislation, and pending Environmental Protection Agency (EPA) regulations may extend this ban to other applications and industries. If substitute solders are to be acceptable, however, they should exhibit desirable characteristics in the areas of melting temperature, wettability, electrical and thermal conductivity, thermal-expansion coefficient, mechanical strength, ductility, creep resistance, thermal fatigue resistance, corrosion resistance, manufacturability, and cost. At present, none of the lead-free solders meet all of these requirements, and most are deficient in several areas. In addition, compatible fluxes must be chosen and the assembly methods may need to be modified.

Most alternative solders have been proposed from other eutectic alloy systems. Tin–antimony alloys are useful in electrical applications and have good strength and creep resistance. Their melting points are rather high, however. Bismuth alloys have very low melting points and good fluidity but suffer from poor wettability. Tin–indium alloys have been used to join metal to glass. They have very low melting points and good wettability but are quite expensive and can be somewhat brittle. Aluminum is often soldered with tin–zinc, cadmium–zinc, or aluminum–zinc alloys. Tin–silver and tin–gold also offer solder possibilities but are limited by both high melting points and high cost. Lead–silver or cadmium–silver alloys can be used for high-temperature service. Table 37-3 presents some of the more common solders, their melting properties, and typical applications.

TABLE 37-3. Some Common Solders and Their Properties

Composition (wt %)	Freezing Temperature (°F)			Applications
	Liquidus	Solidus	Range	
Lead–tin solders				
98 Pb–2 Sn	611	601	10	Side seams in three-piece can
90 Pb–10 Sn	576	514	62	Coating and joining metals
80 Pb–20 Sn	531	361	170	Filling and seaming auto bodies
70 Pb–30 Sn	491	361	130	Torch soldering
60 Pb–40 Sn	460	361	99	Wiping solder, radiator cores, heater units
50 Pb–50 Sn	421	361	60	General purpose
40 Pb–60 Sn	374	361	13	Electronic (low temperature)
Silver solders				
97.5 Pb–1 Sn–1.5 Ag	588	588	0	Higher-temperature service
36 Pb–62 Sn–2 Ag	372	354	18	Electrical
96 Sn–4 Ag	430	430	0	Electrical
Other alloys				
45 Pb–55 Bi	255	255	0	Low temperature
43 Sn–57 Bi	281	281	0	Low temperature
95 Sn–5 Sb	464	450	14	Electrical
50 Sn–50 In	257	243	14	Metal-to-glass
37.5 Pb–25 In–37.5 Sn	280	280	0	Low temperature

Soldering Fluxes

As in brazing, soldering requires that the metal surfaces must be clean and free of oxide so that the solder wets the surfaces and is drawn into the joint to produce an effective bond. Fluxes are used for this purpose, but it is essential that all dirt, oil, and grease be removed before the flux is applied. This precleaning or surface preparation can be performed by a variety of chemical or mechanical means, including solvent or alkaline degreasers, acid immersion (pickling), grit blasting, sanding, wire brushing, and other mechanical abrasion techniques.

Soldering fluxes further the cleaning and strip away surface oxides to expose the base metal. They are generally classified as *corrosive* and *noncorrosive*. A common noncorrosive flux is rosin (the residue after distilling turpentine) in alcohol. This is suitable for copper and brass and for tin-, cadmium-, and silver-plated surfaces, provided that the surfaces have been cleaned previously. Aniline phosphate is a more active noncorrosive flux, but it has limited use because it emits toxic gases when heated. It is suitable for use with copper and brass, aluminum, zinc, steel, and nickel.

The two most commonly used corrosive fluxes are muriatic acid and a mixture of zinc and ammonium chlorides. Acid fluxes are very active but are highly corrosive. Chloride fluxes are effective on aluminum, copper, brass, bronze, steel, and nickel, provided that no oil is present on the surface.

Heating for Soldering

The primary requirement for soldering is a source of sufficient heat and a means of transferring it to the metals being joined. Any method of heating that is suitable for brazing can be used for soldering, but furnace and salt-bath heating are seldom used. Dip soldering is used extensively for joining wire ends (particularly in electronics work), soldering automobile radiators, and tinning. Induction heating is used when large numbers of identical parts are to be soldered. Nevertheless, most soldering is still done with electric soldering irons or small torches. For low-melting-point solders, infrared heat sources can be employed.

In the various processes, the joints can be preloaded with solder, or the filler metal can be supplied from a wire. The particular method of heating usually dictates which procedure is used.

Metals to Be Joined

Table 37-4 summarizes the compatibility of the soldering processes with a variety of engineering materials. Copper, silver, gold, and tin-plated steels are all easily soldered. Aluminum has a strong, adherent oxide film that makes soldering difficult. Special fluxes and modified techniques may be required, but adequate joints are indeed possible, as shown by the number of aluminum radiators currently in automotive use.

TABLE 37-4. Engineering Materials and Their Compatibility with Soldering

Material	Soldering Recommendation
Cast iron	Seldom used since graphite and silicon inhibit bonding
Carbon and low-alloy steels	Difficult for low-carbon materials; seldom used for high-carbon materials
Stainless steel	Common for 300 series; difficult for 400 series
Aluminum and magnesium	Seldom used; however, special solders are available
Copper and copper alloys	Recommended for copper, brass, and bronze
Nickel and nickel alloys	Commonly performed using high-tin solders
Titanium	Seldom used
Lead and zinc	Recommended, but must use low-melting-temperature solders
Thermoplastics, thermosets, and elastomers	Not recommended
Ceramics and glass	Not recommended
Dissimilar metals	Recommended, but with consideration for galvanic corrosion
Metals to nonmetals	Not recommended
Dissimilar nonmetals	Not recommended

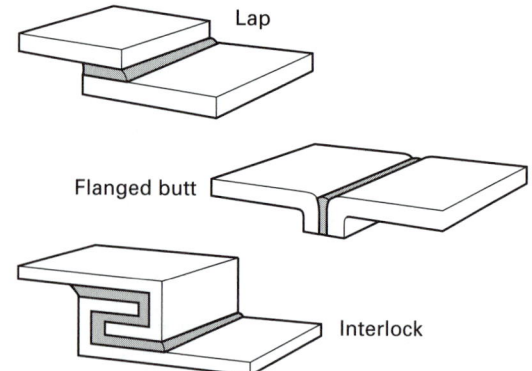

FIGURE 37-8 Some common designs for soldered joints. *(Courtesy of Lead Industries Association.)*

Design and Strength of Soldered Joints

Soldering can be used to join a wide variety of sizes, shapes, and thicknesses and is used extensively to provide electrical coupling or gas- or airtight seals. Soldered joints seldom develop shear strengths in excess of 250 psi (1.72 MPa), however. Consequently, if appreciable strength is required, soldered joints should not be used or some form of mechanical joint, such as a rolled-seam lock, should be made prior to soldering. Butt joints should never be used, and designs where peeling action is possible should be avoided. Figure 37-8 shows some of the more common solder joint designs: lap, flanged butt, and interlock.

As with brazing, there is an optimal clearance for best performance. For lap-type joints, a clearance of 0.003 to 0.005 in. provides for capillary flow of the solder, expulsion of the flux, and reasonable joint strength. In addition, the parts should be held firmly so that no movement can occur until the solder has cooled to well below the solidification temperature. Otherwise, the resulting joint may contain cracks and have very little strength.

Flux Removal

After soldering, the flux residues should generally be removed from the finished joints, either to prevent corrosion or for the sake of appearance. Flux removal is rarely difficult, provided that the type of solvent in the flux is known. Water-soluble fluxes can be removed with hot water and a brush. Alcohol will remove most rosin fluxes. However, when the flux contains some form of grease, as in most paste fluxes, a grease solvent must be used, followed by a hot water rinse. In many cases, solvents containing chlorofluorocarbons (CFCs) have been the cleaners of choice, but these chemicals have been implicated in the depletion of stratospheric ozone and should be avoided. Alternative means of flux removal can be used, or one might consider fluxless soldering.

Fluxless Soldering

At present, *fluxless soldering* techniques utilize controlled atmospheres (such as hydrogen plasma), thermomechanical surface activation (such as plasma gas impingement), or protective coatings that prevent oxides and enhance wetting. Additional success has been obtained with both laser and ultrasonic soldering.

■ KEY WORDS

braze welding
brazing
capillary action
clearance
dip brazing
filler metal

fixtures
fluidity
flux
fluxless brazing
fluxless soldering
furnace brazing

silver solder
soldering
torch brazing
wettability

■ REVIEW QUESTIONS

1. What are some of the possible low-temperature joining processes?
2. How does brazing differ from welding?
3. Why is brazing an appropriate method for joining dissimilar metals with widely different melting points?
4. Why are brazed joints more prone to corrosion problems than welded joints?
5. What features are necessary if the strength of a brazed joint is to equal or exceed the strength of the metal being brazed?
6. Why is it important to compensate for thermal expansion in setting the clearances for a brazing operation?
7. What are some important considerations when selecting a brazing alloy?
8. What are some of the most commonly used brazing metals?
9. What is a silver solder, and how can its cost be justified in a manufacturing application?
10. What are the three functions of a flux in brazing?
11. Why should flux residue be removed immediately following the brazing operation?
12. What are some necessary considerations when designing preloaded joints for brazing?
13. What are the primary limitations of the torch brazing process?
14. Why might reducing atmospheres or a vacuum be desirable in a furnace brazing operation?
15. What are some of the attractive features of salt-bath brazing?
16. Why is dip brazing usually restricted to use with small parts?
17. What are some of the attractive features of induction brazing?
18. What are some common postbraze operations?
19. What is the major advantage of fluxless brazing?
20. How can brazed joints be designed to compensate for the lower strength of the braze metal?
21. How does braze welding differ from brazing?
22. How does soldering differ from brazing?
23. What alloy system contains the majority of soldering alloys?
24. What are some of the potential difficulties in converting to a lead-free solder?
25. What are some of the techniques that can be used to clean surfaces in preparation for soldering?
26. What are some of the heat sources that can be used for soldering?
27. Why should soldering not be specified where appreciable strength is required in a joint?

■ PROBLEM

1. In brazing and soldering, joint clearance is a critical design parameter. What are some possible problems if the joint clearance is too small? Too large? Within the range of acceptable thicknesses, how do joint properties such as strength vary with thickness of the filler metal?

Chapter 37 CASE STUDY

the industrial disposal impeller

The impeller of a large industrial disposal unit is made of a circular segment of hot-rolled steel plate. Pieces of tungsten carbide are brazed into recesses in the plate with a copper-base brazing alloy as shown in Figure CS-37. In service, the impeller rotates at approximately 700 rpm, shredding the refuse from a pharmaceutical manufacturer (paper, bottles, chemicals, etc.) to a pulp. It is necessary to eliminate all identification of the medications before disposal is permitted off the premises of the manufacturer.

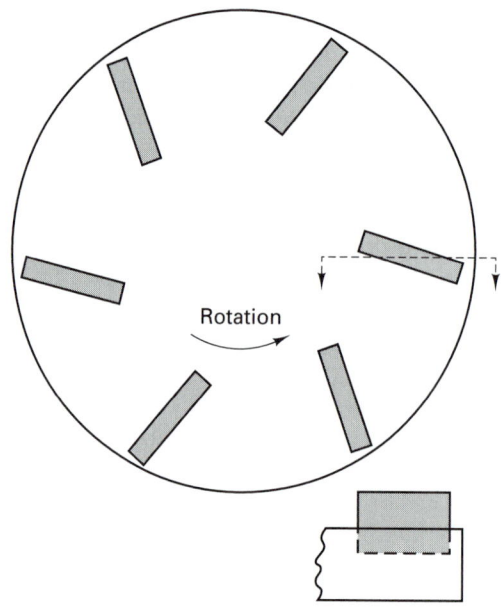

FIGURE CS-37

After five weeks of service, the impeller has failed. Several pieces of tungsten carbide have broken free and are chewing up the unit. Large craters have formed in the steel plate adjacent to the carbide inserts.

1. What do you suspect to be the cause of the problem?

2. How would you alter the design, materials or method of fabrication to prevent a recurrence?

CHAPTER 38

ADHESIVE BONDING AND MECHANICAL FASTENERS

38.1 ADHESIVE BONDING
Adhesive Materials and Their
Properties
Nonstructural and Special
Adhesives
Design Considerations
Advantages and Limitations

38.2 MECHANICAL
FASTENING
Introduction and Methods
Reasons for Selection
Manufacturing Concerns
Design and Selection
Case Study: GOLF CLUB HEADS

■ 38.1 ADHESIVE BONDING

The ideal *adhesive* bonds to any material, needs no surface preparation, cures rapidly, and maintains a high bond strength under all operating conditions. It also does not exist. Tremendous advances have been made, however, in the development of adhesives that are stronger, easier to use, less costly and more reliable than alternative methods of joining. From early applications, such as plywood and World War II aircraft component assembly, the use of structural adhesives (where the adhesive is a load-transmitting part of the product) has grown rapidly. Now, even such quality- and durability-conscious areas as the automotive and aircraft industries make extensive use of adhesive bonding. Adhesives in the automotive industry have advanced from the attaching of interior and exterior trim to the joining of major components, such as door, hood, and trunk assemblies. *Adhesive bonding* is the preferred means of assembly for sheet molding compound and reaction-injection-molded (RIM) panels. Since both metals and nonmetals can be bonded, adhesive bonding has grown significantly with the expanding role of plastics and composites.

Adhesive Materials and Their Properties

In adhesive bonding, a nonmetallic material is used to create a joint between two surfaces. The actual adhesives span a wide range of material types and forms, including *thermoplastic* resins, *thermosetting* resins, artificial *elastomers*, and even some ceramics. They can be applied as drops, beads, pellets, tapes, or coatings (films) and are available in the form of liquids, pastes, gels, and solids. *Curing* can be performed by the use of heat, radiation or light (photoinitiation), moisture, activators, catalysts, multiple-component reactions, or combinations thereof. Intended applications range from load-bearing (structural adhesives) to light-duty holding or fixturing, to sealing (the forming of liquid or gas-tight joints). With such a range of possibilities, the selection of the best adhesive for the task at hand can often be quite challenging.

Commonly used *structural adhesives* include epoxies, cyanoacrylates, anaerobics, acrylics, urethanes, silicones, high-temperature adhesives, and hot melts. For these materials, the bond can be stressed to a high percentage of its maximum load, for extended periods of time, without failure.

1. *Epoxies.* The thermosetting epoxies are the oldest, most common, and most diverse of the adhesive systems, and can be used to join most engineering materials,

1051

including metal, glass, and ceramic. They are strong, versatile adhesives that can be designed to offer high adhesion, good tensile and shear strength, high rigidity, creep resistance, easy curing with little shrinkage, and good tolerance to elevated temperatures. Various epoxies can be used over the temperature range -60 to $+500\,°F$. After curing at room temperature, shear strengths can be as high as 5000 to 10,000 psi (35 to 70 MPa).

Single-component epoxies use heat as the curing agent. Most, however, are two-component blends involving a resin and a curing agent, plus possible additives such as accelerators, plasticizers, and fillers that serve to enhance cure rate, flexibility, *peel* resistance, impact resistance, or other characteristics. Heat may again be required to drive or accelerate the cure.

Epoxy adhesives are limited by rather low peel strength and flexibility, and the bond strength can be sensitive to moisture and surface contamination. Epoxies are often brittle at low temperatures and the rate of curing is comparatively slow. Sufficient strength for structural applications is generally achieved in 8 to 12 hours, with full strength often requiring 2 to 7 days.

2. *Cyanoacrylates.* These are liquid monomers that polymerize when spread into a thin film between two surfaces. Trace amounts of moisture on the surfaces promote curing at amazing speeds, often as little as 2 seconds. Thus the cyanoacrylates offer a one-component adhesive system that cures at room temperature with no external impetus. Commonly known as *superglues*, this family of adhesives is now available in the form of liquids, gels, toughened versions designed to overcome brittleness, and even nonfrosting varieties.

The cyanoacrylates provide excellent tensile strength, fast curing and good shelf life, and adhere well to most commercial plastics. They are limited by their high cost, poor peel strength, and brittleness. Bond properties are poor at elevated temperature and effective curing requires good component fit (no or very small gaps).

3. *Anaerobics.* The anaerobics are one-component, thermosetting, polyester acrylics that remain liquid when exposed to air. When confined to small spaces and shut off from oxygen, as in a joint to be bonded, the polymer becomes unstable. In the presence of iron or copper, it polymerizes into a bonding-type resin without the need for elevated temperature. Additives can reduce odor, flammability, and toxicity and speed the curing operation. Slow-curing anaerobics require 6 to 24 hours to attain useful strength. With selected additives and heat, however, curing can be reduced to as little as 5 minutes.

The anaerobics are extremely versatile and can bond almost anything, including oily surfaces. The joints resist vibrations and offer good sealing to moisture and other environmental influences. Unfortunately, they are somewhat brittle and are limited to service temperatures below $300\,°F$.

4. *Acrylics.* The acrylic-based thermoplastic adhesives are relatively new, but offer good strength, toughness, and versatility, and are able to bond a variety of materials, including plastics, metals, ceramics, and composites, even through oily or dirty surfaces. Most involve application systems where a catalyst primer (curing agent) is applied to one of the surfaces to be joined and the adhesive is applied to the other. The pretreated parts can be stored separately for weeks without damage. Upon assembly, the components react to produce a strong thermoset bond at room temperature. Heat can often accelerate the curing, and at least one variety cures with ultraviolet light. In comparison to other varieties of adhesives, the acrylics offer strengths comparable to the epoxies, good resistance to water and humidity, and the added advantages of room temperature curing and a no-mix application system. Major limitations include low high-temperature strength, flammability, and an unpleasant odor when still uncured.

5. *Urethanes.* Urethane adhesives are a large and diverse family of polymers that are generally targeted for applications that involve temperatures under 150°F (65°C) and components that can undergo great elongation. Both one-part thermoplastic and two-part thermosetting systems are available. In general, they cure quickly to handling strength but are slow to reach the full-cure condition. Two minutes to handling with 24 hours to complete cure is common at room temperature.

Compared to the other structural adhesives, the urethanes offer good low-temperature adhesion coupled with good flexibility and toughness. They are somewhat moisture sensitive, degrade in many chemical environments, and can involve toxic components or curing products.

6. *Silicones.* The silicone thermosets cure from the moisture in the air or adsorbed moisture from the surfaces being joined. They form low-strength structural joints and are usually selected when considerable expansion and contraction is expected in the joint, flexibility is required (as in sheet metal parts), or good gasket or sealing properties are necessary. Metals, glass, paper, plastics, and rubbers can all be joined. The adhesives are relatively expensive and curing is rather slow, but the bonds that are produced can resist moisture, hot water, oxidation, and weathering, and retain their flexibility at low temperature.

7. *High-temperature adhesives.* When strength must be retained at temperatures over 500°F (290°C), high-temperature structural adhesives should be specified. These include epoxy phenolics, modified silicones or phenolics, polyamides, and some ceramics. High cost and long cure times are the primary limitations for these adhesives, which see primary application in the aerospace industry.

8. *Hot melts.* While generally not considered to be true structural adhesives, the *hot-melt adhesives* are being used increasingly to transmit loads, especially in composite material assemblies. They are thermoplastic resins which are solid at room temperature but melt abruptly when heated into the range of 200 to 300°F (100 to 150°C). They are generally applied as heated liquids and form a bond as the molten adhesive cools. Another method of application is to position the adhesive in the joint prior to operations such as the paint bake process in automobile manufacture. During the baking, the adhesive melts, flows into seams and crevices, and seals against the entry of corrosive moisture. The hot melts provide reasonable strength within minutes, but do soften and creep when exposed to elevated temperatures and become brittle when cold.

Table 38-1 presents a list of some popular structural adhesives along with their service and curing temperatures and expected strengths.

Nonstructural and Special Adhesives

There are a number of other types of adhesives whose limited load-bearing capabilities place them in a nonstructural classification. Nevertheless, they still play roles in manufacturing through a variety of uses, such as labeling and packaging. The hot-melt adhesives are often placed in this category but can be used for applications in both classifications. *Evaporative adhesives* use an organic solvent or water base, coupled with vinyls, acrylics, phenolics, polyurethanes, or various types of rubbers. Some commonly identified evaporative adhesives are rubber cements and floor waxes. *Pressure-sensitive adhesives* are

TABLE 38-1. Some Common Structural Adhesives, Their Cure Temperatures, Maximum Service Temperatures, and Strengths Under Various Types of Loadings

Adhesive Type	Cure Temperature (°F)	Service Temperature (°F)	Lap Shear Strength (psi at °F)[a]	Peel Strength at Room Temperature (lb/in.)
Butyral–phenolic	275 to 350	−60 to 175	1000 at 175 / 2500 at RT	10
Epoxy				
Room-temperature cure	60 to 90	−60 to 180	1500 at 180 / 2500 at RT	4
Elevated-temperature cure	200 to 350	−60 to 350	1500 at 350 / 2500 at RT	5
Epoxy–nylon	250 to 350	−420 to 180	2000 at 180 / 6000 at RT	70
Epoxy–phenolic	250 to 350	−420 to 500	1000 at 175 / 2500 at RT	10
Neoprene–phenolic	275 to 350	−60 to 180	1000 at 180 / 2000 at RT	15
Nitrile–phenolic	275 to 350	−60 to 250	2000 at 250 / 4000 at RT	60
Polyimide	550 to 650	−420 to 1000	1000 at 1000 / 2500 at RT	3
Urethane	75 to 250	−420 to 175	1000 at 175 / 2500 at RT	50

[a]RT, room temperature.

usually based on various rubbers, compounded with various additives, and bond at room temperature with a brief application of pressure. No cure is involved, and the tacky adhesive-coated surfaces require no activation by water, solvents, or heat. Peel-and-stick labels, cellophane tape, and Post-it notes are examples of this group of adhesives. *Delayed-tack adhesives* are similar to the pressure-sensitive systems, but are nontacky until activated by exposure to heat. They then remain tacky for several minutes to a few days to permit use or assembly.

While most adhesives are electrical and thermal insulators, *conductive adhesives* can be produced by incorporating selected fillers, such as silver, copper, or aluminum flakes or powder. Certain ceramic oxide fillers can be used to provide thermal conductivity coupled with electrical insulation.

Still another group of commercial adhesives are those designed to cure by exposure to radiation, such as visible, infrared, or ultraviolet light, microwaves, or electron beams. These *radiation-curing adhesives* offer rapid conversion from liquid to solid at room temperature and a curing mechanism that occurs throughout, rather than progressing from exposed surfaces (as with the competing low-temperature air or moisture cures). Current applications include a wide variety of dental amalgams that can fill cavities or seal surfaces while matching the color of the remaining tooth. In the manufacturing realm, heat-sensitive materials can be effectively bonded, and the rapid cure time significantly reduces the need for fixturing.

Design Considerations

The structural adhesives have been used for a wide range of applications in fields as diverse as automotive, aerospace, appliances, biomedical, electronics, construction, machinery, and sporting goods. Proper selection and use, however, requires consideration of a number of factors, including:

1. What materials are being joined? What are their porosity, hardness, and surface conditions? Will the thermal expansions or contractions be different?
2. How will the joined assembly be used? What type of joint is proposed, what will be the bond area, and what will be the applied stresses? How much strength is required? Will there be mechanical vibration, acoustical vibration, or impacts?
3. What temperatures might be required to effect the cure, and what temperatures might be encountered during service? Consideration should be given to highest temperature, lowest temperature, rates of temperature change, frequency of change, duration of exposure to extremes, the properties required at the various conditions, and differential expansions or contractions.
4. Will there be subsequent exposure to solvents, water or humidity, fuels or oils, light, ultraviolet radiation, acid solutions, or general weathering?
5. What is the desired level of flexibility or stiffness? How much toughness is required?
6. Over what length of time is stability desired? What portion of this time will be under load?
7. Is appearance important?
8. How will the adhesive be applied? What equipment, labor, and skill are required?
9. What will it cost?

Because of the large difference in bonding area, adhesive-bonded joints are often classified as either continuous surface or core to face. In *continuous-surface bonds*, both of the adhering surface areas are relatively large and are of the same size and shape. *Core-to-face bonds* have one adhered area that is very small compared to the other, as in the bonding of lightweight honeycomb-core structures to the face sheets.

Another major design consideration is the type of stress to which the joint will be subjected. As shown in Figure 38-1, applied stresses can subject the joint to tension (or compression), shear, cleavage, and peel. Most of the structural adhesives are significantly weaker in peel and cleavage than they are in shear or tension. Therefore, adhesively bonded joints should be designed so that as much of the stress as possible is in shear or tension, where all of the bonded area shares equally in bearing the load. Looking further, the shear strengths of common structural adhesives range from 2000 to 6000 psi (14 to 40 MPa) at room temperature, while the tensile strengths are only 600 to 1200 psi (4 to 8 MPa). Therefore, the best adhesive-bonded joints will be those that are designed to utilize the superior shear strengths. Additional stresses can be induced by creep, vibration and associated fatigue, thermal shock, and mechanical shocks. When vibration or shock loading is expected, the elastomeric adhesives can provide valuable damping and become quite attractive.

Figure 38-2 shows some commonly used joint designs and indicates their relative effectiveness. The butt joint is unsatisfactory because it offers only a minimum of bond surface area and little resistance to cleavage. Useful strength is generally obtained by increasing the bond area through the addition of straps or the conversion to some form of lap design. Figure 38-3 shows some designs recommended for corner and angle joints. Adhesives can also be used in combination with welding, brazing, or mechanical fasteners. Spot welds or rivets can provide additional strength or simply prevent movement of the components when the adhesive is not fully cured or is softened by exposure to elevated temperature.

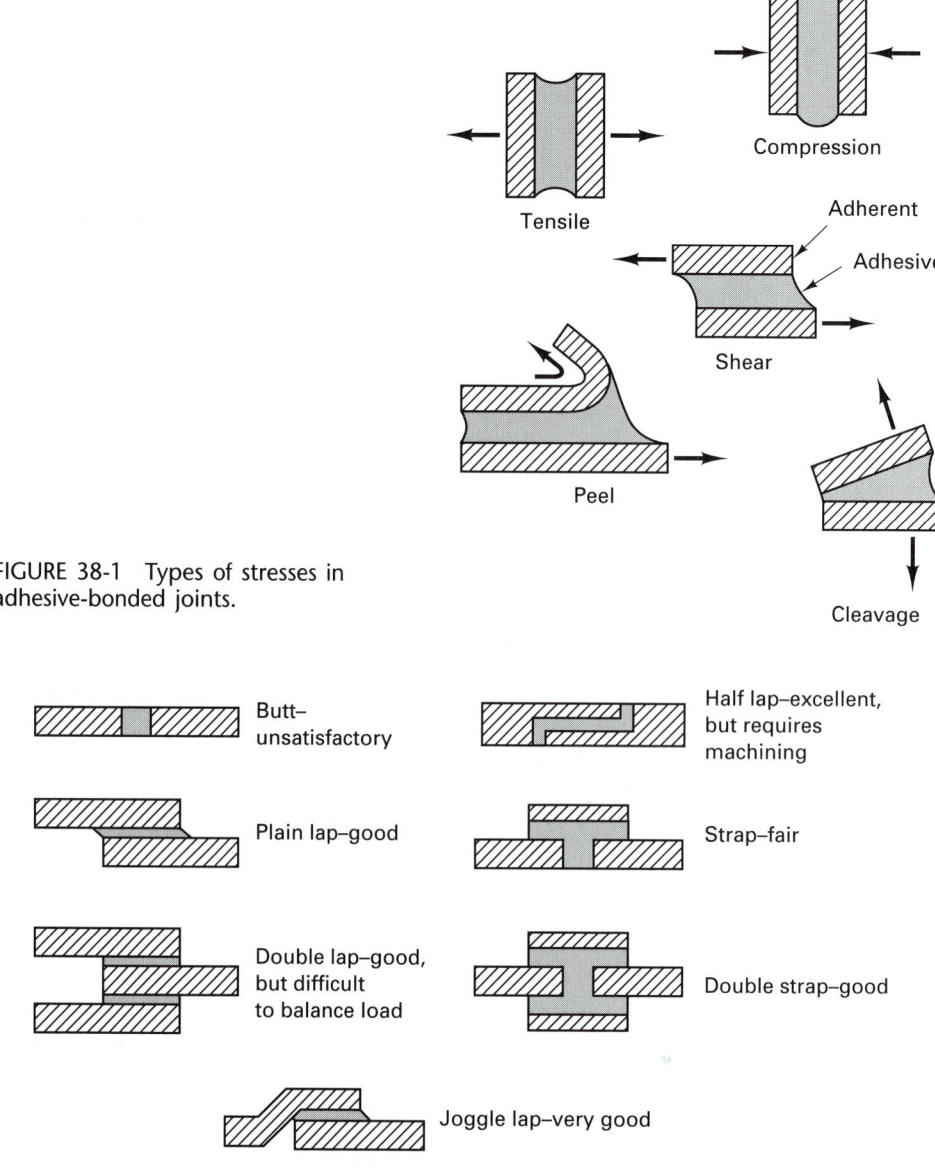

FIGURE 38-1 Types of stresses in adhesive-bonded joints.

FIGURE 38-2 Possible designs of adhesive-bonded joints and a rating of their performance in service.

To obtain satisfactory and consistent quality in adhesive-bonded joints, it is essential that the surfaces be prepared properly. Procedures vary widely but frequently involve cleaning of the surfaces to be joined. Contaminants and grease usually must be removed to assure adequate wetting of the surfaces by the adhesive. Chemical etching, steam cleaning, or abrasive techniques may be employed to further enhance wetting and bonding. While thick or loose oxide films are detrimental to adhesive bonding, a thin porous oxide can often provide surface roughness and enhance adhesion.

The destructive testing of adhesive joints, or the examination of joint failures, can reveal much about the effectiveness of the adhesive system. If failure occurs by separation

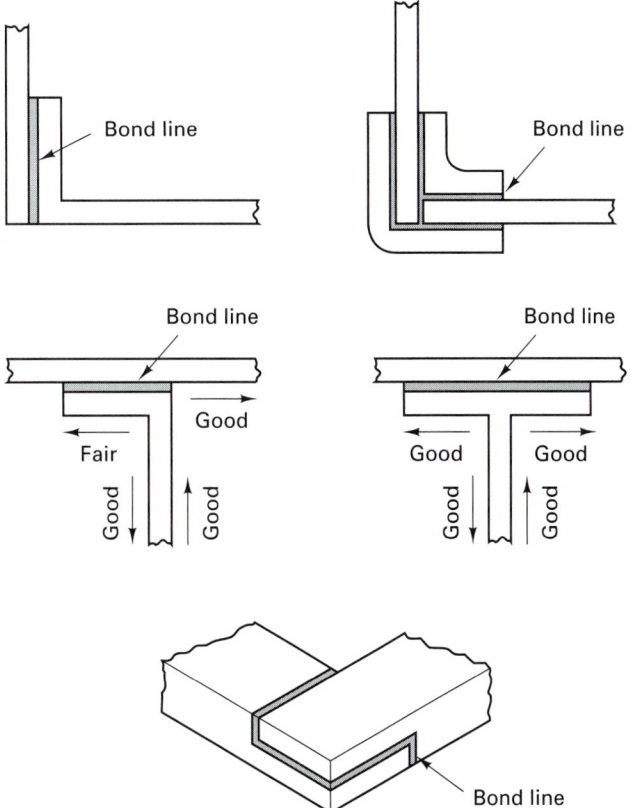

FIGURE 38-3 Adhesively bonded corner and angle joint designs.

at the adhesive–substrate interface, as shown in Figure 38-4a, it is indicative of a bonding or adhesion problem. If the failure lies entirely within the adhesive as in Figure 38-4b, the bonding with the substrate is adequate, but the strength of the adhesive may need to be enhanced. Finally, failure may occur within the substrate materials, as in Figure 38-4c. In this case the joint is good and failure is unrelated to the adhesive bonding operation.

Advantages and Limitations

Adhesive bonding has a number of obvious advantages. Almost any material or combination of materials can be joined in a wide variety of sizes, shapes, and thicknesses. For most adhesives, the curing temperatures are quite low, seldom exceeding 350°F (180°C). A substantial number cure at room temperature, or slightly above and can provide adequate strength for many applications. As a result, very thin or delicate materials such as foils can be joined to each other or to heavier sections. Heat-sensitive materials can be joined without damage and heat-affected zones are not present in the product. When joining dissimilar materials, the adhesive provides a bond that can tolerate the stresses of differential expansion and contraction.

Because adhesives bond the entire joint area, good load distribution and fatigue resistance are obtained and stress concentrations (such as those observed with screws, rivets, and spot welds) are avoided. Similarly, because of the large amount of contact area that can usually be obtained, the total joint strength compares favorably with that produced by

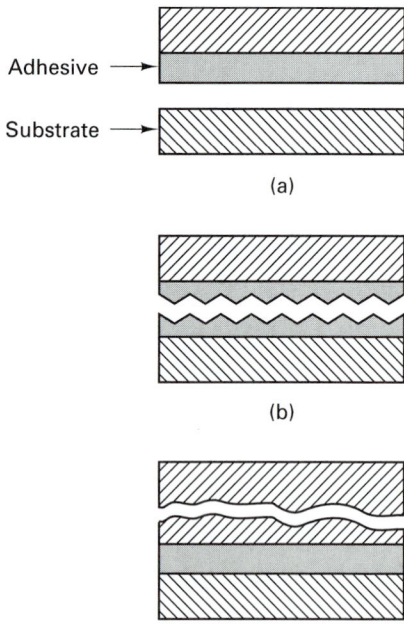

Adhesive

Substrate

(a)

(b)

(c)

FIGURE 38-4 Failure modes of adhesive joints: (a) adhesive failure; (b) cohesive failure within the adhesive; (c) cohesive failure within the substrate.

alternative methods of joining or attachment. (Shear strengths of industrial adhesives can exceed 3000 psi.) Additives can be incorporated to enhance strength, increase flexibility, or provide resistance to various environments.

Adhesives are generally inexpensive and frequently weigh less than the fasteners needed to produce a comparable-strength joint. In addition, the adhesive can also provide thermal and electrical insulation; act as a damper to noise, shock, and vibration; stop a propagating crack; and provide protection against galvanic corrosion when dissimilar metals are joined. By providing both a joint and a seal against moisture, gases, and fluids, adhesive-bonded assemblies often offer improved corrosion resistance throughout their useful lifetime. When used to bond polymers or polymer–matrix composites, the adhesive can be selected from the same family of materials to assure good compatibility.

From a manufacturing viewpoint, the formation of a joint does not require the flow of material, as in brazing and soldering. The bonding adhesive is applied directly to the surfaces, and the joint is then formed by the application of heat and/or pressure. Most adhesives can be applied quickly, and useful strengths are achieved in a short period of time. Some curing mechanisms take as little as 2 to 3 seconds! Surface preparation may be reduced since bonding can occur with an oxide film in place, and rough surfaces are actually beneficial because of the increased contact area. Tolerances are less critical since the adhesives are more forgiving than alternative methods of bonding. Smooth contours are obtainable, and no holes have to be made, as with rivets or bolts. These factors contribute to reduced manufacturing costs, which can be reduced further through the elimination of the mechanical fasteners and the absence of highly skilled labor. Bonding can often be achieved at locations that would prevent the access of many types of welding apparatus. Robotic dispensing systems can often be utilized.

The major disadvantages of adhesive bonding are:

1. There is no universal adhesive. Selection of the proper adhesive is often complicated by the wide variety of available options.

2. Most industrial adhesives are not stable above 350°F (180°C). Oxidation reactions are accelerated, thermoplastics can soften and melt, and thermosets decompose. While some adhesives can be used up to 500°F (260°C), elevated temperatures are usually a cause for concern.

3. High-strength adhesives are often brittle (poor impact properties). Resilient ones often creep. Some become brittle when exposed to low temperatures.

4. Long-term durability and life expectancy are difficult to predict.

5. Surface preparation and cleanliness, adhesive preparation, and curing can be critical if good and consistent results are to be obtained. Some adhesives are quite sensitive to the presence of grease, oil, or moisture on the surfaces to be joined. Surface roughness and wetting characteristics must be controlled.

6. Assembly times may be greater than for alternative methods, depending on the curing mechanism. Elevated temperatures may be required, as well as specialized fixtures.

7. It is difficult to determine the quality of an adhesive-bonded joint by traditional nondestructive techniques, although some inspection methods have been developed that give good results for certain types of joints.

8. Some adhesives contain objectionable chemicals or solvents, or produce them upon curing.

9. Many structural adhesives deteriorate under certain operating conditions. Environments that may be particularly hostile include ultraviolet light, ozone, acid rain (low pH), moisture, and salt. Thus durability and reliability of a joint over an extended service lifetime may be questioned.

10. Adhesively bonded joints cannot be readily disassembled.

Nevertheless, the extensive and successful use of adhesive bonding provides ample evidence that these limitations can be overcome if adequate quality-control procedures are adopted and followed.

One should note that while the unit area strengths of adhesives are relatively low, this is not a major disadvantage or limitation, since sufficient area can generally be provided in a properly designed joint.

■ 38.2 MECHANICAL FASTENING

Introduction and Methods

Assembly in manufacturing is often performed by some means of *mechanical fastening*, a classification that includes a wide variety of techniques and fasteners designed to suit the individual requirements of a multitude of joints and assemblies. Included within this family are integral fasteners, threaded discrete fasteners (which includes screws, bolts, studs, inserts, and other types), nonthreaded discrete fasteners (such as rivets, pins, retaining rings, staples, and wire stitches), special-purpose fasteners (such as the quick-release and tamper-resistant types), shrink and expansion fits, press fits, and others. Selection of the specific fastener or fastening method depends primarily on the materials to be joined, the function of the joint, strength and reliability requirements, weight limitations, dimensions of the components, and environmental conditions. Other considerations include cost, installation equipment and accessibility, appearance, and the need or desirability for disassembly. When disassembly and reassembly is desired (as for parts replacement, maintenance, or repair), threaded fasteners, snap-fits, or other fasteners that can be removed quickly and easily should be specified. Such fasteners should not have a tendency to loosen after installation, however. If disassembly is not necessary, permanent fasteners are often preferred, or threaded fasteners can be coupled with anaerobic adhesives which cure to full strength at room temperature.

Strength is imparted to the joint by mechanical interlocking and interference of the surfaces brought about by the clamping force; no fusion or adhesion of the surfaces is required. Fasteners and fastening processes should be selected to provide the desired strength and properties, reflecting the nature and magnitude of subsequent loading. Consideration should also be given to the presence of vibrations and/or cyclic stresses that would tend to promote loosening over a period of time. Weight considerations are significant in many applications, particularly aerospace and automotive. The need to withstand corrosive environments, operate at high or low temperatures, or face other severe conditions is often very influential when selecting fasteners or fastening processes.

The effectiveness of mechanical fasteners depends upon (1) the material of the fastener, (2) the fastener design (including the load-bearing area of the head), (3) hole preparation, and (4) the installation procedure. The desire is to achieve a uniform load transfer, a minimum of stress concentration, and uniformity of installation torque or interference fit. Various means are available for achieving these goals.

Integral fasteners are formed areas of a component that interfere or interlock with other components of the assembly and are most commonly found in sheet metal products. Examples include lanced or shear-formed tabs, extruded hole flanges, embossed protrusions, edge seams, and crimps. Figure 38-5 shows some of these techniques, all of which involve some form of metal shearing and/or forming. The common beverage can includes several of these joints: an edge seam to join the top of the can to the body (like Figure 38-5e) and an embossed protrusion that is subsequently flattened as a means of attaching the opener tab, as in Figure 38-5d.

Discrete fasteners (Figure 38-6) are separate pieces whose primary function is to join the primary components. These include bolts, nuts (with accessory washers, etc.), screws, nails, rivets, quick-release fasteners, staples, and wire stitches. Over 150 billion discrete fasteners are consumed annually in the United States, 27 billion by the automotive industry alone. The variety here is so immense that the major challenge is to select an appropriate fastener for the task at hand and, if possible, an optimum fastener. This task is made even more difficult by the inconsistent nomenclature and identification schemes. While some are classed or identified by the specific product or application, others are classed by the material from which they are made, their size, their shape, or primary operational features. (Discussion of the primary terms used in identifying discrete fasteners can be found in ANSI Standard B18.12.) On the positive side of the problem is the commercial availability of a wide range of standard and special types, sizes, materials, and strengths, such that an appropriate fastener can generally be found for most all joining needs. The fasteners are easy to install, easy to remove, and easy to replace. In addition, most standard varieties are interchangeable. Various finishes and coatings can be applied to withstand a multitude of service conditions.

Shrink and expansion fits form another major class of mechanical joining. Here a dimensional change is introduced to one or both of the components by heating or cooling (heating one part only, heating one and cooling the other, or cooling one). Assembly is then performed and a strong interference fit is established when temperature uniformity is restored. Joint strength is exceptionally high. In addition, the procedures can be used to produce a prestressed condition in a weak, low-cost material, often enabling it to replace a more-costly, stronger one. Similarly, a corrosion-resistant cladding or lining can be easily provided to a less-costly bulk material.

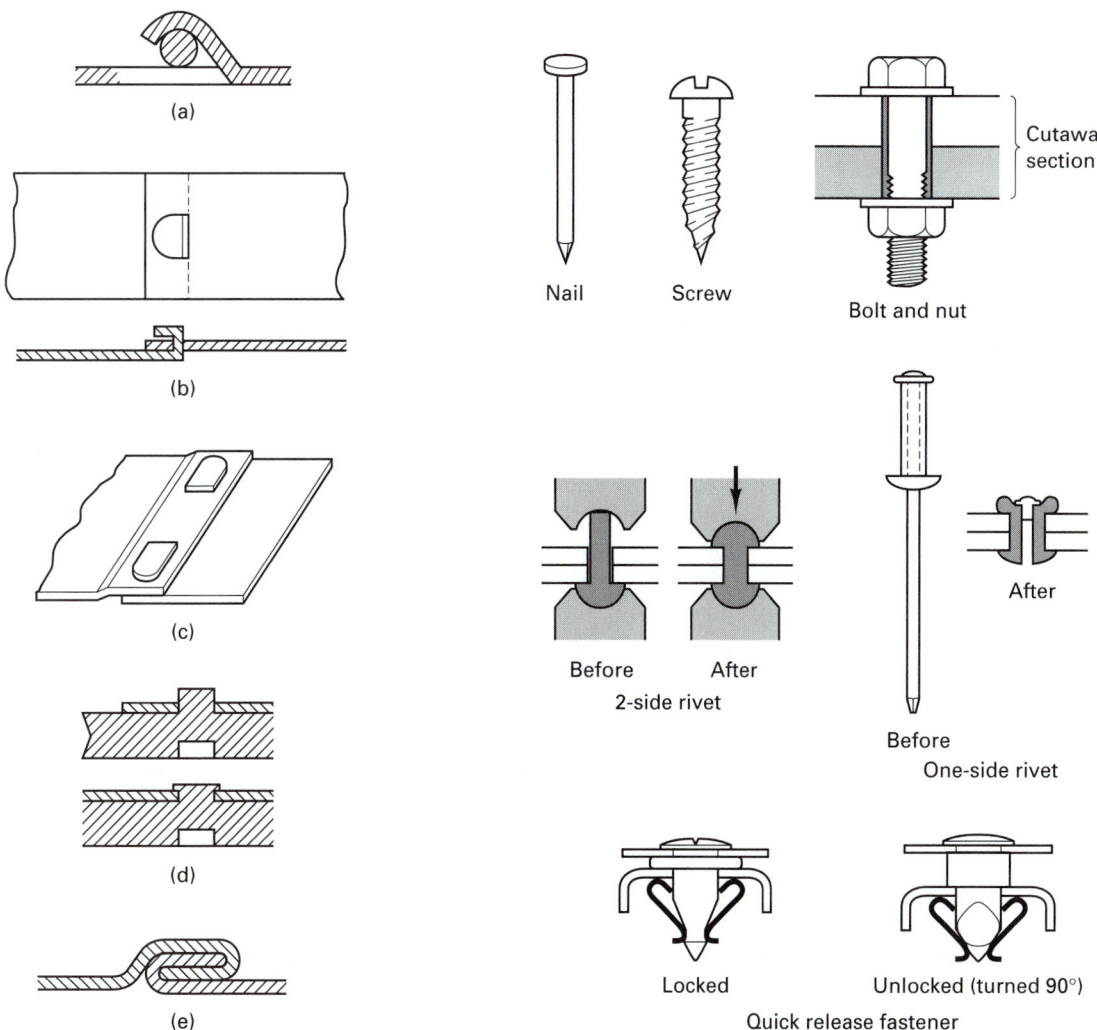

FIGURE 38-5   Several types of integral fasteners: (a) lanced tab to fasten wires or cables to sheet or plate; (b) and (c) assembly through folded tabs and slots for different types of loading; (d) use of a flattened embossed protrusion; (e) single-lock seam.

FIGURE 38-6   Various types of discrete fasteners, including a nail, screw, nut and bolt, two-side access supported rivet, one-side access blind rivet, and quick-release fastener.

Press fits are quite similar to shrink and expansion fits, with the same results being obtained by mechanical force instead of differential temperatures.

Reasons for Selection

Mechanical fastening offers a number of attractive features, among them being:

1. Ease of disassembly and reassembly. The threaded fasteners are noteworthy for this feature, and semipermanent fasteners (such as rivets) can be drilled out for a major disassembly.

2. The ability to join similar or different materials in a wide variety of sizes, shapes, and joint designs. Some joint designs, such as hinges and slides, permit limited motion between the components.

3. Low manufacturing cost. The fasteners are usually small formed components that cost little compared to the components being joined. They are readily available in a variety of mass-produced sizes.

4. Installation does not adversely affect the base materials as is often the case with techniques involving the application of heat and/or pressure.

5. Little or no surface preparation or cleaning is required.

Manufacturing Concerns

Many mechanical fasteners require that the components contain aligned holes. These holes can be produced in a variety of ways, including punching, drilling, electrical or chemical machining, and laser beams, as described elsewhere in this book. Each technique produces holes with characteristic surface finish, dimensional features, and properties. Castings, forgings, extrusions, and powder metallurgy components can be designed to include integral holes. Holemaking and proper positioning and alignment of the holes are major considerations in mechanical fastening.

Bolts, coupled with nuts, require access to both sides of the assembly during joining and require the added weight of the nut. Bolts can also be inserted into threaded (tapped) holes, thereby eliminating the nut and offering one-side assembly. Screws also offer this feature. When the bolt or screw is sufficiently hard, the fastener can actually cut its own threads, thereby eliminating the need for either a tapped hole or a nut. Stapling is a fast way of joining thin materials and does not require prior holemaking. Rivets offer good strength but are usually permanent or semipermanent. Snap fits utilize the elasticity of one of the components and require that the necessary elastic deformation occurs without fracture.

Design and Selection

The design and selection of a fastening method therefore requires numerous considerations, including the possible means of joint failure. When a product is assembled with fastened joints, the fasteners are extremely vulnerable sites. Mechanical joints generally fail because of oversight or lack of control in one of four areas: (1) the design of the fastener and the manufacturing techniques used to make it, (2) the material from which the fastener is made, (3) joint design, or (4) the means and details of installation. Fasteners may have insufficient strength or corrosion resistance or may be subject to stress-corrosion cracking or hydrogen embrittlement. Metal fasteners provide electrical conductivity between the components, and an inappropriate choice of fastener or component material can cause severe galvanic corrosion problems. Nonmetallic fasteners (such as threaded nylon, fiberglass, or graphite) can be used for low-strength applications where corrosion is a concern, but creep under load is a concern for these materials. Furthermore, mechanical fasteners join only at discrete points, so gases or liquids can enter the joint area and further aggravate conditions.

Many failures are the direct result of joint preparation and fastener installation. Approximately 90% of the cracks in aircraft structures originate at the fastener holes, and fatigue of fasteners is the largest single cause of fastener failure. Installation frequently imparts too much or too little preload (too tight or too loose). The joint surfaces may not be flat or parallel, and the area under the fastener head may be insufficient to bear the load. Vibration loosening enhances fastener fatigue. In addition, the details of joint design should consider stress distribution, since much of the load will be concentrated on the fasteners (in contrast to the previously discussed adhesive joints that distribute the load uniformly over the entire joint area).

Nearly all fastener failures can be avoided by proper design and fastener selection. Consideration should be given to the operating environment, required strength, and magnitude and frequency of vibration. The need for weight savings and the desirability for disassembly and reassembly will certainly influence decisions. Standard fasteners should be used whenever possible, and as little variety as necessary should be specified.

Fastener design should incorporate a shank-to-head fillet whenever possible. Rolled threads offer superior strength and corrosion-resistant coatings offer enhanced performance. Joint design should seek to avoid such features as offset or oversized holes. Proper installation and tightening are critical to good performance.

■ KEY WORDS

adhesive
adhesive bonding
anaerobics
continuous-surface bonds
core-to-face bonds
curing
discrete fasteners

elastomer
epoxies
hot-melt adhesives
integral fasteners
mechanical fastening
peel
press fit

shear
shrink and expansion fits
structural adhesive
thermoplastic
thermosetting

■ REVIEW QUESTIONS

1. What would be some of the characteristics of an ideal adhesive?
2. What is a structural adhesive?
3. What are some of the various ways in which structural adhesives can be cured?
4. What general magnitude of shear strength could be expected from a high-strength epoxy adhesive?
5. What are some possible additives to epoxy adhesives, and what properties might they contribute?
6. What promotes the curing of cyanoacrylates? Of anaerobics?
7. What are some attractive features of the acrylic adhesives when considered from a manufacturing viewpoint?
8. What features or characteristics might favor the selection of a silicone adhesive?
9. How can polymeric adhesives be made electrically or thermally conductive?
10. What are some of the temperature considerations that should be made in selecting an adhesive?
11. What are some of the environmental conditions for which exposure might reduce the performance or lifetime of a structural adhesive?
12. Why is it most desirable that adhesive joints be

designed so the adhesive is loaded in shear or compression?
13. What are some common techniques by which surfaces are prepared for adhesive bonding?
14. Why are the structural adhesives attractive when joining dissimilar metals or materials? Different sizes or thicknesses?
15. Why are adhesive joints not attractive for applications that involve exposure to elevated temperature?
16. In view of the relatively low strengths of the structural adhesives, how can adhesively bonded joints attain strengths comparable to those of other methods of joining?
17. What engineering properties or characteristics in addition to strength can be offered by the structural adhesives?
18. What are some of the manufacturing advantages of adhesives?
19. What features may limit the use of adhesives at elevated temperatures? At low temperatures?
20. What factors would influence the selection of a specific fastener or fastening method?
21. What types of fasteners should be selected if there is a need to disassemble and reassemble the product?

22. How do press fits differ from shrink or expansion fits? How are they similar?

23. How is the stress distribution different for adhesive joints and joints formed with discrete fasteners?

24. What are some of the common causes for failure of mechanically fastened joints?

25. From a manufacturing viewpoint, why is it desirable to use standard fasteners and minimize the variety of fasteners within a given product?

■ PROBLEMS

1. A contractor has installed aluminum siding on a house with steel nails. Use the galvanic series to evaluate the corrosion properties of this assembly. (*Note:* The aluminum is exposed to air, so the aluminum should be considered to be in its passive condition.) What do you expect will be the outcome of this fastener selection? Can you recommend a better alternative?

2. Mechanical fasteners are an attractive means of joining composite materials because they avoid exposing the composite to heat and/or high pressure. Assume that the composite is a polymer-based fiber-reinforced material with either uniaxial or woven fibers. For this particular system, what are some possible fastener-related problems? Consider joint preparation, assembly, and possible service failure.

Chapter 38 CASE STUDY

golf club heads

You are employed by a small manufacturer of sporting goods equipment, and your design team has recently proposed a new line of high-performance golf clubs. The club head is to be a "standard" AISI 431 martensitic stainless steel design with a metal insert incorporated into the striking face, as shown in Figure CS-38. The insert will be produced by powder metallurgy, and will consist of a copper-based alloy laced with particles of tungsten carbide. After a mild acid etching of the copper matrix, the carbide particles will protrude sufficiently to better grip the surface of the ball, imparting an enhanced amount of backspin to better control the "bite" upon landing. Since its purpose is to modify the striking face, the insert is rather thin, about 1/16- to 1/8-inch in thickness. It must be incorporated into the club face in a manner that does not dampen the impact or compromise the "feel" of the club.

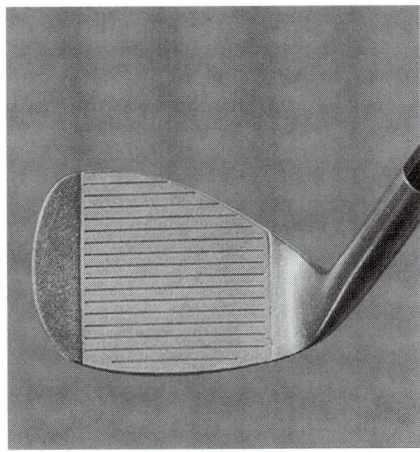

FIGURE CS-38

1. Your job is to devise the means of incorporating the insert into the club face. What are some possible means of joining or bonding the dissimilar metals? What are some of the advantages and limitations of each of your alternatives?

2. An additional joint occurs where the club head is attached to the shaft. If a graphite fiber reinforced epoxy is being proposed for the shaft, how might it be attached to the stainless steel head? If the shafts were metal, what other methods might be possible?

3. For production simplicity, it might be preferable to use the same joining procedure at all locations. In view of your recommendations above, does this appear to be possible? If the composite shaft were selected, what would you recommend?

MANUFACTURING CONCERNS IN WELDING AND JOINING

39.1	INTRODUCTION		Thermal-Induced Residual Stresses
39.2	TYPES OF FUSION WELDS AND TYPES OF JOINTS		Effects of Thermal Stresses
		39.5	WELDABILITY OR JOINABILITY
39.3	DESIGN CONSIDERATIONS		
		39.6	SUMMARY
39.4	HEAT EFFECTS	Case Study:	THE WELDED FRAME
	Welding Metallurgy		
	Thermal Effects in Brazing and Soldering		

■ 39.1 INTRODUCTION

Many of the problems that are inherent to welding and joining can be avoided by proper consideration of the particular characteristics and requirements of the process. Proper design of the joint is extremely critical. Selection of the specific process requires an understanding of the large number of available options, the variety of possible *joint configurations*, and the numerous variables that must be specified for each operation. Heating, melting, and resolidification can cause numerous problems, including drastic changes in the properties of both base and filler materials. Weld metal properties can be further changed by dilution of the filler by melted *base metal*, vaporization of various alloy elements, or a variety of gas–metal reactions.

Various types of weld defects can also be produced. These include cracks in a variety of forms, cavities (both gas and shrinkage), inclusions (slag, flux, and oxides), *incomplete fusion* between the weld and base metals, *incomplete penetration* (insufficient weld depth), unacceptable weld shape or contour, arc strikes, spatter, undesirable metallurgical changes (aging, grain growth, or transformations), and excessive *distortion*.

■ 39.2 TYPES OF FUSION WELDS AND TYPES OF JOINTS

There are four basic types of *fusion welds*, as illustrated in Figure 39-1. *Bead welds* (also called *surfacing welds*) require no edge preparation. However, because the weld is made directly onto a flat surface and the penetration depth is limited, bead welds are suitable only for joining thin sheets of metal, building up surfaces, and depositing hard facing (wear-resistant) materials.

Groove welds are used when full-thickness strength is desired on thicker material. Some sort of edge preparation is required to form a groove between the abutting edges. V, double V, U, and J (one-sided V) configurations are most common and are usually produced by oxyacetylene flame cutting. The specific type of groove usually depends on the thickness of the work, the welding process to be employed, and the position of the work. The objective is to obtain a sound weld throughout the full thickness with a minimum

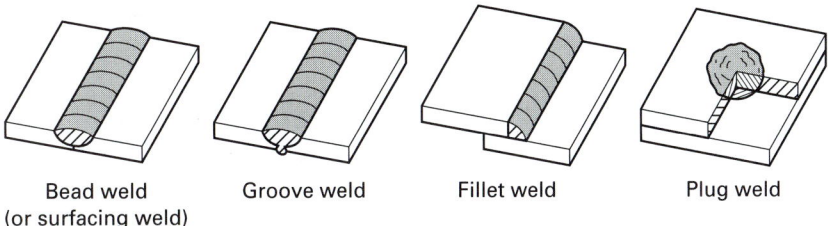

Bead weld   Groove weld   Fillet weld   Plug weld
(or surfacing weld)

FIGURE 39-1   Four basic types of fusion welds.

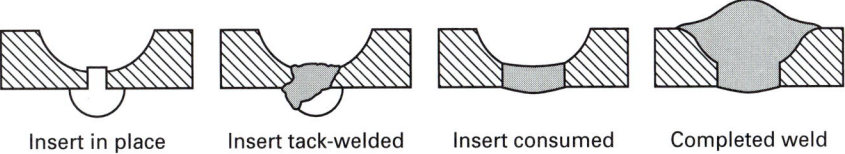

Insert in place   Insert tack-welded   Insert consumed   Completed weld

FIGURE 39-2   Use of a consumable backup insert in making fusion welds.
(Courtesy of Arcos Corporation.)

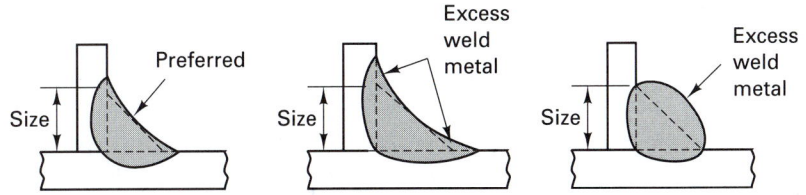

FIGURE 39-3   Preferred shape and the method of measuring the size of fillet
welds.

amount of *weld metal*. If possible, single-pass welding is preferred, but multiple passes may be required, depending on the thickness of the material and the welding process being used. As shown in Figure 39-2, special consumable inserts can be used to assure proper spacing between the mating edges and proper quality in the root pass. These inserts are particularly useful in pipeline welding and other applications where welding must be done from only one side of the work.

Fillet welds are used for tee, lap and corner joints. The size of the *fillet* is measured by the leg of the largest 45° right triangle that can be inscribed within the contour of the weld cross section. This is shown in Figure 39-3, which also depicts the proper shape for fillet welds to avoid excess metal deposition and reduce stress concentration. Fillet welds require no special edge preparation. They may be either continuous or intermittent (with spaces being left between short lengths of weld).

Plug welds are used to attach one part on top of another, replacing rivets or bolts. A hole is made in the top plate, and welding is started at the bottom of this hole.

Figure 39-4 shows the five basic types of joints that can be made with the use of bead, groove, and fillet welds, and Figure 39-5 shows several methods to make these joints. In selecting the type of weld joint to be used, the primary consideration should be the type of loading that will be applied. Too often, this basic principle is neglected, and a large proportion of what are erroneously called "welding failures" are actually the result of such

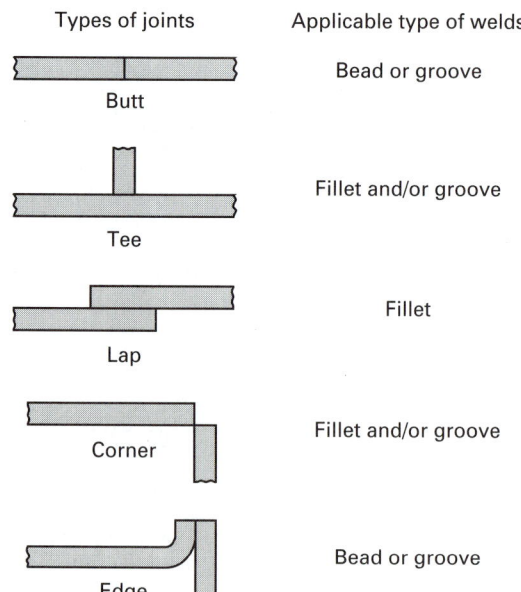

FIGURE 39-4 Basic types of fusion weld joints and the types of welds used in making them.

oversight. Cost and accessibility for welding are also important factors in joint selection but should be viewed as secondary to loading considerations. Cost is affected by the required edge preparation, the amount of weld metal that must be deposited, the type of process and equipment that must be used, and the speed and ease with which the welding can be accomplished. Accessibility will obviously have considerable influence on several of these factors.

■ 39.3 DESIGN CONSIDERATIONS

Welding is a unique process that cannot be substituted directly for other methods of fastening without proper consideration being given to its particular characteristics and requirements. Unfortunately, welding is so easy and convenient to use that these facts are often overlooked. Since the cause of many "welding failures" can often be traced to such negligence, it is important that proper design considerations be employed in all welding operations.

One very important fact is that welding produces *monolithic*, or one-piece structures. When two pieces are welded together, they become one piece. This can cause significant complications if not properly taken into account. For example, a crack in one piece of a multipiece structure may not be serious, because it will seldom progress beyond the single piece in which it occurs. In contrast, when a large structure, such as a ship hull, pipeline, storage tank, or pressure vessel, consists of many pieces welded together, a crack that starts in a single plate or weld can propagate for a great distance and cause complete failure. Obviously, such a fracture is not the fault of the welding process but is simply a reflection of the monolithic nature of the product.

Another factor that relates to design is that a given material in small pieces may not behave as it does in a large piece, like those often produced by welding. The importance of this fact is illustrated in Figure 39-6, which shows the relationship between energy absorption and temperature for the same steel when tested as a Charpy impact specimen and as a large, welded structure. In the form of a Charpy bar, the material exhibited ductile behavior and good energy absorption at temperatures down to 25°F (−4°C). When welded

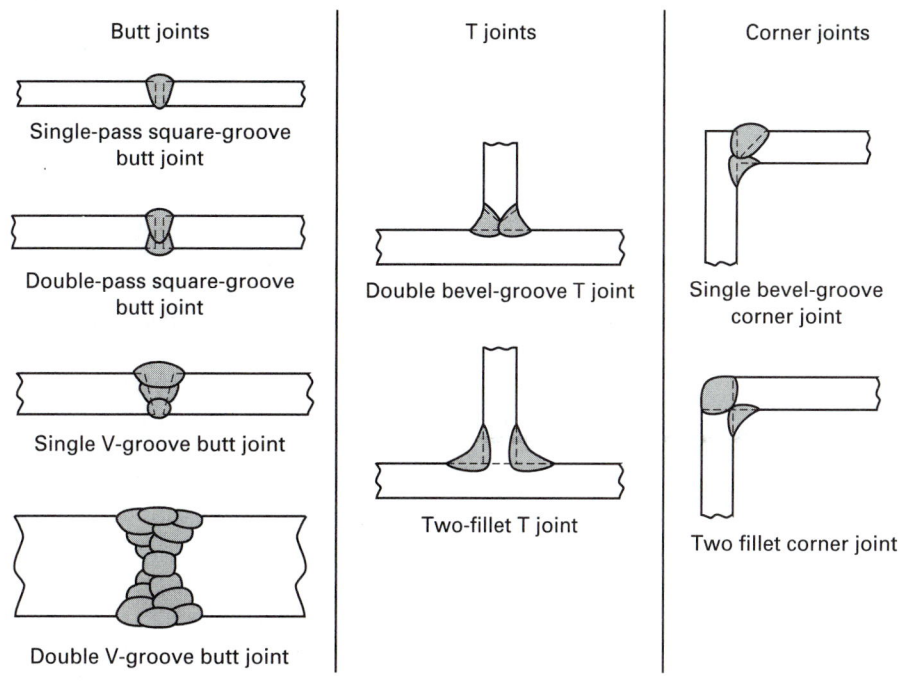

FIGURE 39-5 Various weld procedures used to form several common joints. *(Courtesy of Republic Steel Corporation.)*

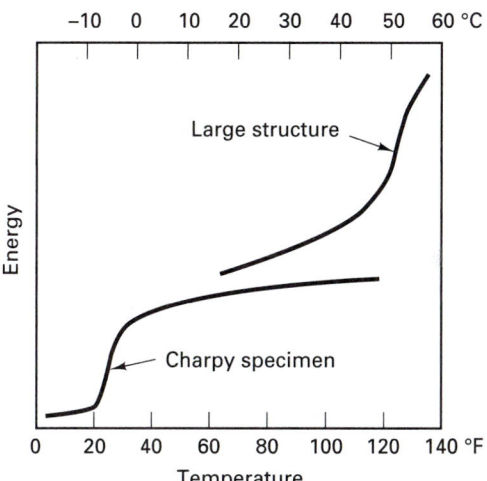

FIGURE 39-6 Effect of size on the transition temperature and energy-absorbing ability of a certain steel.

into a large structure, however, brittle behavior was observed at temperatures as high as 110°F (43°C). Thus the notch-ductility characteristics of steels used in large welded structures may be of great importance. More than one welded structure has failed because the designer failed to consider the effect of size.

Another common error is to make welded structures too rigid, thereby restricting their ability to redistribute high stresses and avoid failure. Considerable thought may be required to design structures and joints that provide sufficient flexibility. The multitude of successful welded structures, however, attests to the fact that such designs are indeed possible.

Accessibility of the joint for welding, welding position, component matchup, and the specific nature of the joint are other important considerations in welding design.

■ 39.4 HEAT EFFECTS

Welding Metallurgy

Heating and cooling are essential and integral components of all welding processes (except cold welding) and tend to produce metallurgical changes that are usually undesirable. In fusion welding, the heating is sufficient to produce some melting of the base metal and is then followed by rapid cooling. The thermal effects tend to be most pronounced for this type of welding, but also exist to a lesser degree in processes where the heating–cooling cycle is less severe. If the thermal effects are properly considered, the adverse results can be avoided and excellent service performance can be obtained. If they are overlooked, however, the results can be disastrous.

Because such a wide range of metals are welded and such a variety of processes are used, welding metallurgy is an extensive subject, and the material presented here serves only as an introduction. In fusion welding a pool of molten metal is created, with the molten metal coming from only the parent plate (*autogenous welding*), or a mixture of parent and filler material. Figure 39-7 shows a butt weld between a plate of material A and a plate of material B. A backing strip of material C is used with filler metal of material D. In this situation the molten pool is actually a complex alloy of all four materials. The molten material is held in place by a metal "mold" formed by the surrounding plate. Since the molten pool is usually small compared to the surrounding metal, fusion welding can often be viewed as *casting* a small amount of molten metal *into a metal mold*. The resultant structure and its properties can best be understood by first analyzing the casting and then considering the effects of the associated heat treatment on the adjacent base metal.

Figure 39-8 shows a typical microstructure produced by a fusion weld. In the center of the weld is a region composed of weld metal that has solidified from the molten state. The material in this weld pool, or *fusion zone*, is actually a mixture of parent metal and electrode or filler metal, with the ratio depending on the process used, the type of joint, and the edge preparation. Figure 39-9 compares two butt-weld designs where the weld pool in the upper design would contain a large percentage of base metal, and the weld pool in the lower design would be largely filler material. The metal in the fusion zone is cast material with a microstructure reflecting the cooling rate in the weld. Since their structures are different, this region cannot be expected to have the same properties and characteristics as those of the *wrought* material being welded. Adequate mechanical properties, therefore, can only be achieved by selecting filler rods or electrodes which have properties *in their*

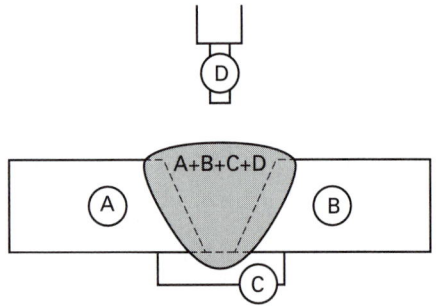

FIGURE 39-7 Schematic of a butt weld between a plate of metal A and a plate of metal B, with a backing plate of metal C and filler of metal D. The resulting weld nugget is a complex alloy of the four metals.

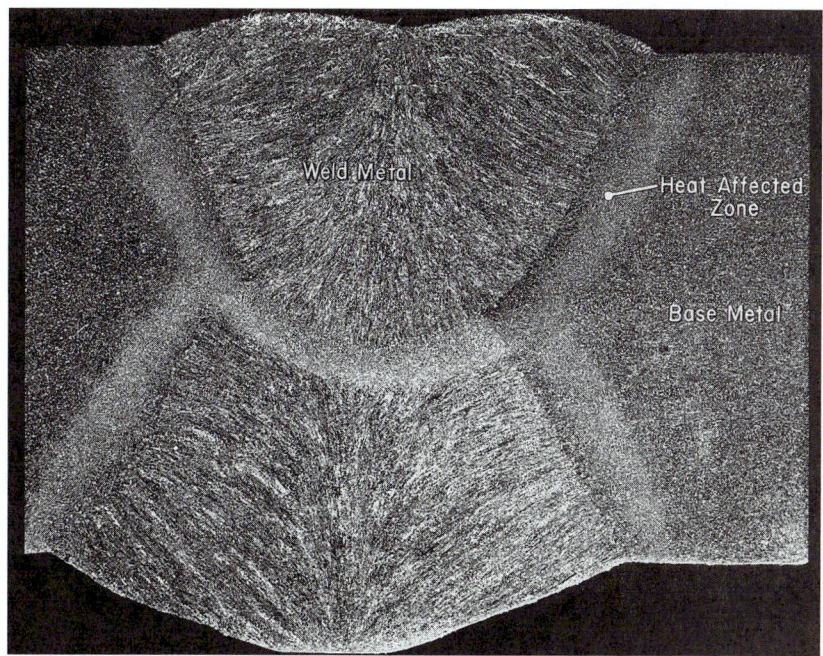

FIGURE 39-8   Grain structure and various zones in a fusion weld.

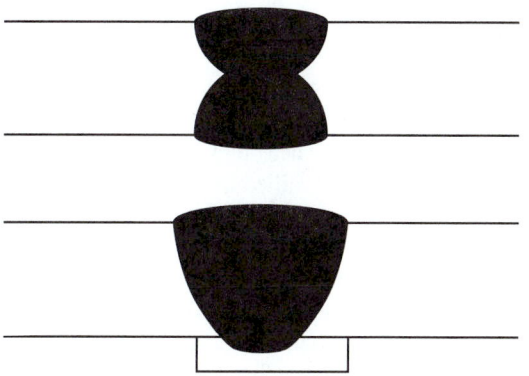

FIGURE 39-9   Comparison of two butt-weld designs. In the top weld, a large percentage of the weld pool is base metal. In the bottom weld, most of the weld pool is filler metal.

as-deposited condition that equal or exceed those of the wrought parent metal. It is not uncommon, therefore, for the filler metal to have a different chemistry than that of the metal being welded. The matching or exceeding of base metal strength, in the as-solidified condition, is the basis for several AWS (American Welding Society) specifications for electrodes and filler rods.

The grain structure in the fusion zone may be fine or coarse, equiaxed or dendritic, depending on the type and volume of weld metal and the rate of cooling. Most electrode and filler rod compositions tend to produce fine, equiaxed grains, but the volume of weld metal and variations in cooling rate can easily defeat these objectives.

The pools of molten metal created by fusion welding are prone to all of the problems and defects associated with metal casting, such as gas porosity, inclusions, blowholes, cracks, and shrinkage. Since the amount of molten metal is usually small compared to the

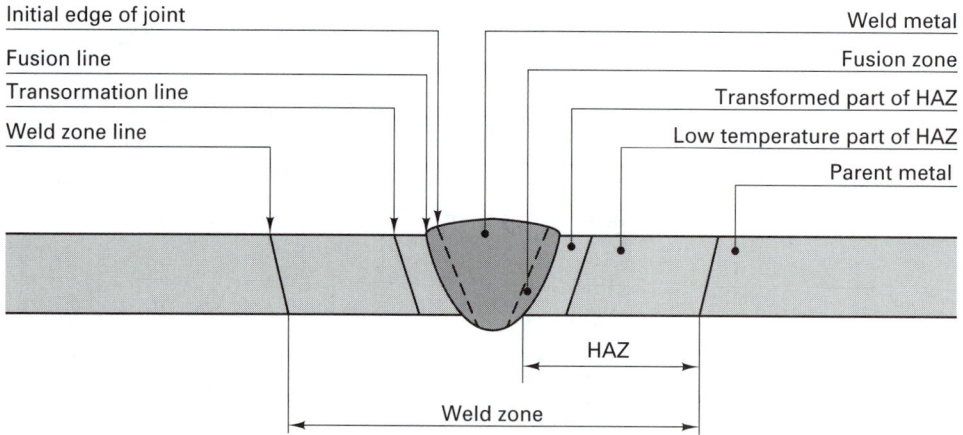

FIGURE 39-10 Schematic of a fusion weld in steel, presenting proper terminology for the various regions and interfaces. Part of the heat-affected zone has been heated above the transformation temperature. *(Courtesy of Sandvik AB.)*

total mass of the workpiece, rapid solidification and rapid cooling of the solidified metal are quite common. Associated with these conditions may be the entrapment of dissolved gases, chemical segregation, grain-size variation, grain shape problems, and orientation effects.

Adjacent to the fusion zone, and wholly within the base material, is the ever-present and generally undesirable *heat-affected zone* (HAZ). In this region the parent metal is not melted but is subjected to elevated temperatures for a brief period of time. Since the temperature and its duration vary widely with location, fusion welding might be more appropriately described as "a casting and an abnormal, widely varying, heat treatment." The adjacent metal may well experience sufficient heat to bring about structure and property changes, such as phase transformations, recrystallization, grain growth, precipitation or precipitate coarsening, embrittlement, or even cracking. The variation in thermal history produces a variety of microstructures and a range of properties. In steels, the structures can range from hard, brittle martensite all the way through coarse pearlite and ferrite.

Because of the altered structure, the heat-affected zone may no longer possess the desirable properties of the parent material, and since it was not molten, it cannot assume the properties of the solidified weld metal. Consequently, this is often the weakest area in the as-welded joint. Except where there are obvious defects in the weld deposit, most welding failures originate in the heat-affected zone. The heat-affected zone terminates where the base metal has experienced too little heat to be affected or altered by the welding process. Figure 39-10 presents a schematic of a fusion weld in a steel where part of the heat-affected zone has been heated above the transformation temperature. Standard terminology is presented for the various regions and interfaces.

Because of the melting, solidification, and exposure to a range of high temperatures, the structure and properties of welds can be extremely complex and varied. Associated problems, however, can often be reduced or eliminated. First, consideration should be given to the thermal characteristics of the various processes. Table 39-1 classifies some of the more common welding processes with regard to their *rate* of heat input. Processes with low rates of heat input (slow heating) tend to produce high total heat content within the metal, slow cooling rates, and large heat-affected zones. High-heat-input processes, on the other hand, have low total heats, fast cooling rates, and small heat-affected zones. The size of the heat-affected zone will also increase with increased starting temperature, decreased welding speed, increased thermal conductivity of the base metal, and a decrease in base metal thickness. Weld geometry is also important, with fillet welds producing smaller heat-affected zones than butt welds.

TABLE 39-1.   Classification of Common Welding Processes by Rate of Heat Input

Low rate of heat input
  Oxyfuel welding (OFW)
  Electroslag welding (ESW)
  Flash welding (FW)

Moderate rate of heat input
  Shielded metal arc welding (SMAW)
  Flux cored arc welding (FCAW)
  Gas metal arc welding (GMAW)
  Submerged arc welding (SAW)
  Gas tungsten arc welding (GTAW)

High rate of heat input
  Plasma arc welding (PAW)
  Electron beam welding (EBW)
  Laser welding (LBW)
  Spot and seam resistance welding (RW)
  Percussion welding

If the as-welded results are unacceptable, the entire piece might be heat treated after welding. Structure variation can be reduced or eliminated, but the results are limited to those structures that can be produced by heat treatment. The structures and properties associated with cold working, for example, could not be achieved. Additional problems may be encountered when one seeks to produce controlled heating and cooling (heat treatment) of the large, complex-shaped structures commonly produced by welding. Finally, furnaces, quench tanks, and related equipment may not be available to handle the full size of welded assemblies.

Another technique to reduce the variation in microstructure, or at least the sharpness of the variation, is to heat the base metal adjacent to the weld just prior to welding. This heating serves to reduce the cooling rate of both the weld deposit and the immediately adjacent metal in the heat-affected zone. The slower cooling rate results in the production of a more gradual change in microstructure and the elimination of a metallurgical stress raiser. Preheating is particularly important with the high-thermal-conductivity metals, such as copper and aluminum, where the cooling rate would otherwise be extremely rapid.

If the carbon content of plain carbon steels is greater than about 0.3%, the cooling rates encountered in normal welding may be sufficient to produce hard, untempered martensite and the associated loss of ductility. Alloy steels possess higher hardenability, so the likelihood of martensite formation is even greater with these materials. Therefore, special pre- and postwelding heat cycles may be required when welding these materials. For plain carbon steels, a preheat temperature of 200 to 400°F (100 to 200°C) is usually adequate. The weldable low-carbon, low-alloy steels are extremely attractive, therefore, because they can be welded without the necessity of *preheating* or *postheating*.

In processes where little or no melting occurs and there is considerable pressure applied to the heated metal (as in forge or resistance welding), the weld region will experience deformation and may assume a structure similar to that of wrought material.

While the discussion of metallurgical effects has largely focused on steels, it should be noted that other metals also exhibit heat-related changes. The exact effects of the heating and cooling associated with welding will depend on the specific transformations and structural changes that can occur within the alloy being joined.

Thermal Effects in Brazing and Soldering

In brazing and soldering, there is no melting of the base metal, but the joint still contains a solidified region and heat-affected portion of base material. For these processes, however, another thermal effect may be quite significant. The base and filler metals are usually of radically different chemistries, and elevated temperatures promote interdiffusion. Intermetallic phases can form at the interface and alter the properties of the joint. If present in small amounts, they can provide strength reinforcement. However, most *intermetallic compounds* are quite brittle, and too much intermetallic material can result in significant loss of both strength and ductility.

Thermal-Induced Residual Stresses

Another effect of the heating and cooling that accompanies welding is the introduction of *residual stresses*. In welding, these may be of two types and are most pronounced in fusion welding, where maximum heating occurs. Their effects can be observed in the form of dimensional changes, distortion, or cracking.

Residual welding stresses are the result of restraint to thermal expansion and contraction offered by the pieces being welded. Consider a rectangular bar of metal that is uniformly heated and cooled. When heated, the material expands and becomes larger in length, width, and thickness. Upon cooling, the material contracts and each dimension returns to its original value. Now clamp the ends of the bar in a vise so that lengthwise expansion cannot take place and repeat the thermal cycle. Upon heating, all of the expansion is restricted to the width and thickness, but the contraction upon cooling will occur uniformly. Thus the resulting rectangle will be shorter, thicker, and wider than the original specimen.

Let us now apply these principles to a weld between two plates, as illustrated in Figure 39-11. As the weld is made, the liquid region conforms to the shape of the "mold" and the adjacent plate material becomes hot and expands. Expansion perpendicular to the line of weldment can be absorbed by flow in the molten pool, but expansion parallel to the weld line tends to be restrained by the remaining plate material that has not been heated, and is therefore cooler and stronger. This resistance or restraining force can often be sufficient to induce plastic deformation of the heat-affected zone. The material is *upset*, responding to the thermal expansion by becoming thicker instead of longer.

FIGURE 39-11   Schematic of the residual stresses in a fusion weld.

FIGURE 39-12   Longitudinal residual stresses in a butt-welded joint.

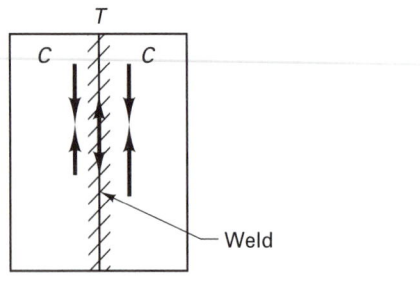

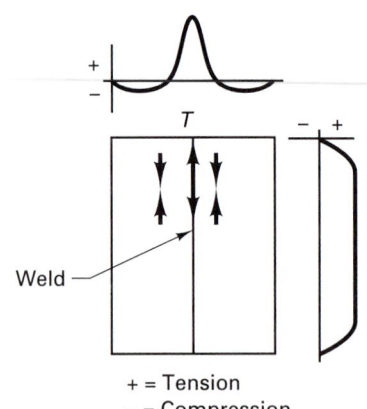

+ = Tension
– = Compression

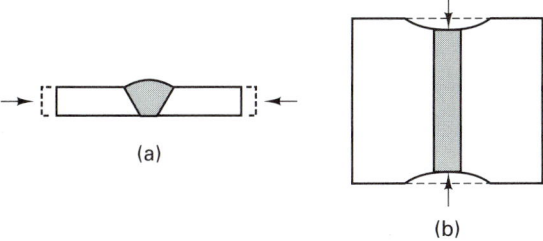

FIGURE 39-13   Shrinkage of a typical butt weld in the transverse (*left*) and longitudinal (*right*) directions.

After the *weld pool* solidifies, both the weld metal and adjacent heat-affected region cool and contract. This contraction is also resisted by the cooler, surrounding metal. The cooling region wants to contract, but is restrained and is forced to remain in a "stretched" condition, known as *residual tension* (region T). Similarly, this contracting region exerts forces that try to squeeze the adjacent material, producing regions of residual compression (regions C). While the net force is zero, in keeping with the basic laws of physics and mechanics, the localized variations can be substantial. Figure 39-12 depicts the residual stresses for such a weld. A high residual tension is observed in the weld metal, becoming compressive and then returning to zero as one moves away from the weld. The magnitude of the residual tension is relatively uniform along the weld, except at the ends, where the stresses can be relieved by a pulling in the edges.

Thermal contractions occur both parallel (longitudinal) and perpendicular (transverse) to the weld line, as shown in Figure 39-13. Generally, the transverse contractions are compensated by movement of the material being welded. The welded assembly simply becomes shorter than the positioned components were at the time of welding (Figure 39-13a). In the longitudinal direction, the assembly responds to the complex stresses of Figure 39-12 to produce the distorted shape of Figure 39-13b.

If the welded plates are restrained from movement, however, *additional stresses can be induced*. These are the second type of residual stresses and are known as *reaction stresses*. The magnitude of the reaction stresses is an inverse function of the length between the weld joint and the point of restraint. They can never exceed the yield strength of the parent metal (yielding would occur to relieve them) and are generally less than the stresses parallel to the weld.

Effects of Thermal Stresses

An obvious result of the thermal stresses induced by welding is a distortion or warping of the assembly as it responds to the thermal contraction. Figure 39-14 shows the distortion effects of several types of basic weldments. Unfortunately, no fixed rules can be given to avoiding warping because the specific conditions that cause it can be widely varied. The following suggestions, however, can help to reduce distortion problems.

Welds should be made with the least amount of weld metal necessary to form the joint. Faster welding speeds reduce the welding time and reduce the volume of metal that is heated. Welding sequences should be designed to use as few weld passes as possible and permit the base material to have a high freedom of movement. When constructing a multiweld assembly, it is beneficial to weld toward the point of greatest freedom, such as from the center to the edge.

The initial components can also be oriented out of position, so that the subsequent distortion will move them to the desired final shape. Another common procedure is to completely restrain the components during welding, thereby forcing some plastic flow in

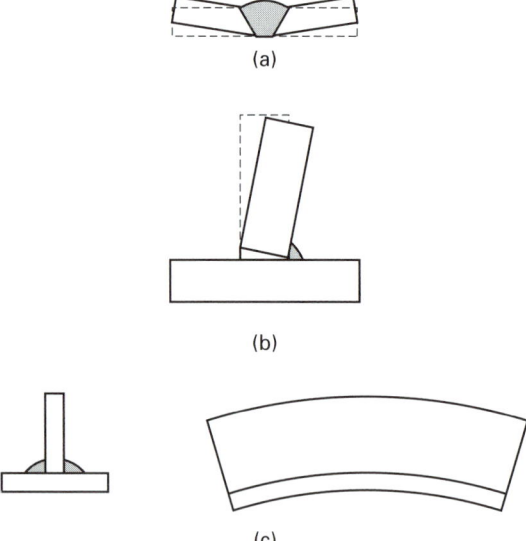

FIGURE 39-14 Distortions observed in several welding operations: (a) V-groove butt weld; (b) one-side fillet weld in a T-joint; (c) two-fillet weld T-joint with a high vertical web.

the material and/or the cooling weld metal. This procedure is used most effectively on small weldments, where the high reaction stresses are not likely to cause cracking. Still another procedure is to balance the resulting thermal stresses by depositing the weld metal in a specified pattern, such as short lengths or on alternating sides of a plate. Warping can also be reduced by the use of *peening*. As the weld bead surface is hammered with the peening tool, the metal is flattened and tries to spread. Being held back by the underlying material, the surface becomes compressed or squeezed. Surface rolling of the weld bead area can have the same effect. The compressive stresses induced by the deformation serve to offset the tensile stresses induced by welding.

Residual stresses should not have a harmful effect on the strength performance of weldments, *except in the presence of notches or in very rigid structures where no plastic flow can occur*. These two conditions should not exist in weldments *if they have been designed properly and proper workmanship has been employed*. Unfortunately, it is easy to inadvertently join heavy sections and produce rigid configurations that will not permit the small amounts of elastic or plastic movement required to reduce highly concentrated stresses. In addition, geometric notches, such as sharp interior corners, are often incorporated into welded structures. Other harmful "notches," such as gas pockets, rough beads, porosity, and arc "strikes" can also serve as initiation sites for weld failures, but these can be avoided by proper welding procedures, good workmanship, and adequate supervision and inspection.

The residual stresses of welding can cause additional warpage when subsequent machining removes metal and upsets the stress equilibrium balance. Consequently, weldments that are to undergo appreciable machining are frequently given a *stress relief* heat treatment prior to that operation.

While the reaction stresses can also contribute to distortion, they are most often associated with cracking during or immediately following the welding operation (as the weld is cooling). This cracking is most likely to occur when welds are made under conditions where there is great restraint to the shrinkage that occurs transverse to the direction of welding. When a multipass weld is being made, cracking tends to occur in the early beads where there is insufficient weld metal to withstand the shrinkage stresses. This can be quite serious if the resulting crack goes undetected and is not chipped out and repaired, or melted and rewelded during subsequent passes.

A recent study has determined that fracture, and the prevention of fracture through overdesign, costs the United States more than $120 billion per year. Since many of the structural failures occur at weldments, it is important that we take measures to reduce this possibility. Joints should be designed to keep restraint to a minimum, and the metals and alloys of the structure should be selected with welding in mind (more problems exist with higher-carbon steels, higher-alloy steels, and high-strength materials). In addition, fracture is more likely to occur with thicker materials.

Crack-prevention efforts can also be directed at maintaining the proper size and shape of the weld bead. While a concave fillet is desirable in machining, a concave weld profile has a greater tendency to crack upon cooling, since contraction actually increases the length of the surface. With a convex profile, the length of the surface will contract simultaneously with the volume, reducing the possibility of surface tension and cracking. Weld beads with high penetration (high depth/width ratio) are also more prone to cracking.

Other methods to suppress cracking focus on reducing the stresses by making the cooling more uniform, or relaxing them by promoting plasticity in the metals being welded. The metals to be welded may be preheated and additional heat may be applied between the welding passes to retard cooling. Some welding codes also require weldments to be subjected to a thermal stress relief after welding, but prior to use. Hydrogen dissolved in the molten weld metal can also induce cracking. Slower welding and cooling will allow any hydrogen to escape, and the use of low-hydrogen electrodes and low-moisture fluxes will reduce the likelihood of hydrogen being present.

■ 39.5 WELDABILITY OR JOINABILITY

It is important to note that not all joining processes are compatible with all engineering materials. The terms *weldability* and *joinability*, while denoting a material's ability to be welded or joined, are extremely nebulous. One process might produce excellent results when applied to a given material, while another may produce a dismal failure. Within a given process, the quality of results may vary greatly with variations in the process parameters, such as electrode material, shielding gases, welding speed, and cooling rate.

Table 39-2 shows the compatibility of the various joining processes with some significant types of engineering materials. In each case, the process is classified as recommended (R), commonly performed (C), performed with some difficulty (D), seldom used (S), and not used (N). It should be noted, however, that the classifications are generalizations, and exceptions often exist within both the family of materials (such as the various types of stainless steels) and the types of processes (arc welding here encompasses a large variety of specific processes). Nevertheless, the table can serve as a guideline to assist in process selection.

■ 39.6 SUMMARY

If the potential benefits of welding are to be obtained and harmful side effects are to be avoided, proper consideration should be given to the selection of the process, the design of the joint, and the effects of heating and cooling on both the weld and parent material. Parallel considerations apply to brazing and soldering operations, with additional attention to the effects of interdiffusion of the filler and base metals. Flame and arc cutting operations also involve localized heating and cooling, and they also experience altered structures in the heat-affected zone. Since many cut products undergo further welding or machining, however, these regions of undesirable structure may not be retained in the final product.

TABLE 39-2. Weldability and Joinability of Various Materials[a]

Material	Arc Welding	Oxyacetylene Welding	Electron Beam Welding	Resistance Welding	Brazing	Soldering	Adhesive Bonding
Cast iron	C	R	N	S	D	N	C
Carbon and low-alloy steel	R	R	C	R	R	D	C
Stainless steel	R	C	C	R	R	C	C
Aluminum and magnesium	C	C	C	C	C	S	R
Copper and copper alloys	C	C	C	C	R	R	C
Nickel and nickel alloys	R	C	C	R	R	C	C
Titanium	C	N	C	C	D	S	C
Lead and zinc	C	C	N	D	N	R	R
Thermoplastics	Heated tool R	Hot gas R	N	Induction C	N	N	C
Thermosets	N	N	N	N	N	N	C
Elastomers	N	N	N	N	N	N	R
Ceramics	N	S	C	N	N	N	R
Dissimilar metals	D	D	C	D	D/C	R	R

[a]C, commonly performed; R, recommended (easily performed with excellent results); D, difficult; N, not used; S, seldom used.

■ KEY WORDS

autogenous weld
base metal
bead weld (or surfacing weld)
distortion
fillet
fillet weld
fusion weld
fusion zone

groove weld
heat-affected zone
incomplete fusion
incomplete penetration
intermetallic compound
joint configuration
monolithic
peening

plug weld
postheating
preheating
residual stresses
stress relief
weldability
weld metal
weld pool

■ REVIEW QUESTIONS

1. Why might it be difficult to select the best procedure for making a given production-type weld?
2. What are some of the common types of weld defects?
3. What are the four basic types of fusion welds?
4. What types of weld joints commonly employ fillet welds?
5. What are some of the factors that influence the cost of making a weldment?
6. Why is it important to consider welded products as monolithic structures?
7. How does the fracture resistance of a metal vary with changes in material thickness?

8. How might excessive rigidity actually be a liability in a welded structure?
9. In what way is the weld pool segment of a fusion weld like a metal casting?
10. Why is it possible for the fusion zone to have a chemistry that is different from that of the filler metal?
11. Why is it not uncommon for the selected filler metal to have a chemical composition that is different from the material being welded?
12. What are some of the defects or problems that can occur in the molten metal region of a fusion weld?
13. Why do properties vary widely in most welding heat-affected zones?

14. What types of structure and property modifications can occur in welding heat-affected zones?

15. Why do most welding failures occur in the heat-affected zone?

16. Describe the cooling rates and size of the heat-affected zone for processes with low rates of heat input.

17. What are some of the difficulties or limitations encountered in heat treating large structures after welding?

18. What is the purpose of pre- and postheat in welding operations?

19. What additional thermal effect may produce adverse results in brazing and soldering operations?

20. What are the two types of residual stresses that occur in weldments, and what are their causes?

21. What is the typical nature of residual stresses that remain in the weld region, tension or compression? Based on your answer, what problems might you expect in this region?

22. How are reaction stresses affected by the distance between the weld and the point of fixed constraint?

23. What are some of the observed effects of the thermally induced residual stresses?

24. What are some of the techniques that can reduce the amount of distortion in a welded structure?

25. Why might a welded structure warp if the structure is machined after welding?

26. What are some of the techniques that can be employed to reduce the likelihood of cracking in a welded structure?

■ PROBLEMS

1. Two pieces of AISI 1025 steel are being shielded metal arc welded with E6012 electrodes. Some difficulty is being experienced with cracking in the weld beads and in the heat-affected zones. What possible corrective measures might you suggest?

2. A base for a special machine tool will weigh 1400 lb (635 kg) if made as a gray iron casting. Pattern cost would be $450 and the foundry has quoted a price of $0.60 per pound ($1.32 per kilogram) for making the casting. If the part is made as a weldment, it will require 800 lb (363 kg) of steel costing $0.14 per pound ($0.31 per kilogram). Cutting, edge preparation, and setup time will require 30 hours at a rate of $10 per hour for labor and overhead. Welding time will be 55 hours at an hourly rate of $9.50. Two hundred pounds (91 kg) of electrode will be required, costing $0.17 per pound ($0.37 per kilogram).

    (a) Which method of fabrication will be more economical if only one part is required?

    (b) What number of parts would be required for welding and casting to break even?

    (c) Since the base of the machine tool will now be made of steel rather than cast iron, what special property of the cast iron will be lost, and why might this be sufficient to dictate manufacturing process regardless of economics?

3. In the 1940s the hulls of freighters were constructed by riveting plates of steel together. When the defense efforts of World War II demanded accelerated production of freighters, construction of the hulls was converted to welding. The resulting "Liberty ships" proved quite successful, but also drew considerable attention when minor or moderate impacts produced cracks of lengths sufficient to scuttle the ship, often up to 50 ft or more. Since the only significant change had been the conversion from riveting to welding, the failures were blamed on the welding process. Is this a fair assessment? What do you think may have contributed to the problem? What evidence might you want to gather to support your beliefs?

*C*hapter 39 CASE STUDY

the welded frame

Figure CS-39 shows the proposed design for the primary frame of a home-made vehicle. The frame is to be constructed from hot-rolled, low carbon, box-channel material with mitre, butt and fillet welds at the twelve numbered joints. Due to the solidification shrinkage and subsequent thermal contraction of the joint material, welds are best made between unrestrained sections. If the structure is too rigid at the time of welding, the necessary dimensional changes are restrained and the associated residual stresses can lead to cracks or tears.

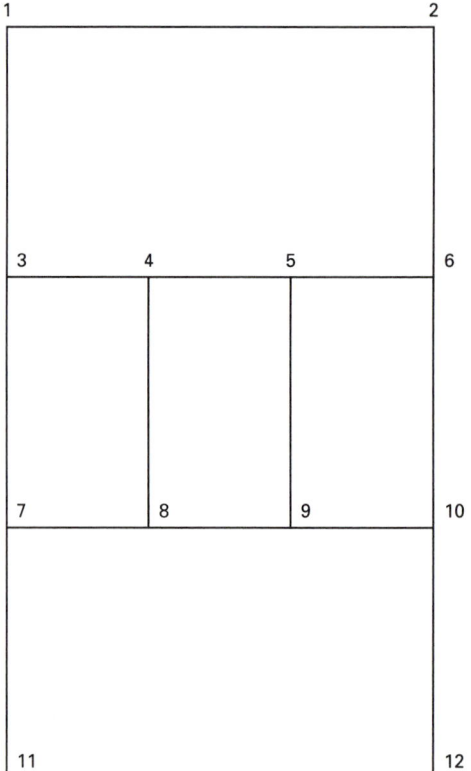

FIGURE CS-39

1. Consider the twelve welds in the proposed structure and recommend a welding sequence that will minimize the possibility of hot tears or cracks due to the welding of a restrained joint.

2. If your company is moving toward development of a computer-assisted design program, suggest one or more "rules" that may be programmed to aid in the selection of an acceptable weld sequence.

PART 7

PROCESSES AND TECHNIQUES RELATED TO MANUFACTURING

40 Surface Treatments and Finishing

41 Manufacturing Systems and Automation

42 Production Systems

43 Lean Production

CHAPTER 40

SURFACE TREATMENTS AND FINISHING

40.1	INTRODUCTION	
40.2	MECHANICAL CLEANING AND FINISHING	
	Abrasive Cleaning	
	Barrel Finishing or Tumbling	
	Vibratory Finishing	
	Media	
	Compounds	
	Summary of Mass-Finishing Methods	
	Belt Sanding	
	Wire Brushing	
	Buffing	
	Electropolishing	
40.3	CHEMICAL CLEANING	
	Alkaline Cleaning	
	Solvent Cleaning	
	Vapor Degreasing	
	Ultrasonic Cleaning	
	Acid Pickling	
40.4	BURR REMOVAL	
	Design to Facilitate or Eliminate Burr Removal	
40.5	COATINGS	
	Painting	

	Paint Application Methods	
	Drying	
	Hot-Dip Coatings	
	Chemical Conversion Coatings	
	Blackening and Coloring Metals	
	Electroplating	
	Anodizing	
	Electroless Plating	
	Electroless Composite Plating	
	Mechanical Plating	
	Porcelain Enameling	
40.6	VAPORIZED METAL COATINGS	
	Vacuum Metallizing	
	Sputtering	
	Chemical Vapor Deposition	
40.7	ION IMPLANTATION	
40.8	CLAD MATERIALS	
40.9	TEXTURED SURFACES	
40.10	COIL-COATED SHEETS	
40.11	EFFECTS OF SURFACE PROCESSING	
Case Study:	BURRS ON YO-GI'S COLLAR	

■ 40.1 INTRODUCTION

Most manufactured items go through a series of operations, including design, material selection, process selection, manufacturing, evaluation, and either acceptance or modification. If we focus on the manufacturing stage, we notice that most of the manufacturing processes that have been presented in earlier chapters are actually means of producing or altering the shape of a material. If the desired properties are obtained simultaneously with the desired shape, one is indeed fortunate. Quite often, property modification is required, and these changes must be induced in a way that retains shape. If the properties to be modified are bulk properties, such as strength or fracture resistance, heat treatment may be specified. Often, the properties that must be modified are surface properties, and the product undergoes one or more surface treatments.

The objectives of the surface modification processes can be quite varied. Some are designed to clean surfaces and remove the kinds of defects that occur during processing or handling (such as scratches, pores, burrs, fins, and blemishes). Others further improve or modify the products' appearance, providing features such as smoothness, texture, or color. Numerous techniques are available to improve resistance to wear or corrosion, or to reduce friction or adhesion to other materials. Scarce or costly materials can be conserved by making the interior of a product from a cheaper, more common material and then coating or plating the product surface.

As with all other processes, surface treatment requires time, labor, equipment, and material handling, and all of these have an associated cost. Efficiencies can be realized through process optimization and the integration of surface treatment into the entire manufacturing system. Design modifications can often facilitate automated or bulk finishing, eliminating the need for labor-intensive or single-part operations. Process selection should further consider the size of the part, the shape of the part, the quantity to be processed, the temperatures required for processing, the temperatures encountered during subsequent use, and any dimensional changes that might occur due to the surface treatment. Through knowledge of the available processes and their relative advantages and limitations, finishing costs can often be reduced or eliminated while maintaining or improving the quality of the product.

In addition to the above, the field of surface finishing has recently undergone significant change. Many chemicals that were once "standard" to the field, such as cyanide, cadmium, chromium, and chlorinated solvents, have now come under strict government regulation. Wastewater treatment and waste disposal have also become significant concerns. As a result, processes may have to be modified, or replacement processes may have to be used.

Because of their similarity to other processes, many surface finishing techniques have been presented elsewhere in the book. The *case hardening* techniques, both selective heating (flame, induction, and laser hardening) and altered surface chemistry (diffusion methods such as carburizing, nitriding, and carbonitriding), are presented in Chapter 5 as variations of heat treating. *Shot peening* and *roller burnishing* are presented in Chapter 19 as cold-working processes. *Roll bonding* and explosive bonding are discussed in Chapter 20 as means of producing laminar composites. *Hard facing* and metal spraying are included in Chapter 36 as adaptations of welding techniques. In the present chapter we focus on techniques for cleaning and surface preparation as well as the remaining methods of surface finishing or surface modification.

■ 40.2 MECHANICAL CLEANING AND FINISHING

Abrasive Cleaning

It is not uncommon for the various manufacturing processes to produce certain types of surface contamination. Sand from the molds and cores used in casting often adheres to product surfaces. Scale (metal oxide) can be produced whenever metal is processed at elevated temperatures. Oxides such as rust can form if material is stored between operations. These and other contaminants must be removed before decorative or protective surfaces can be produced. While vibratory shaking can be useful, some form of *abrasive cleaning* is usually required to remove the foreign material. In the most common method, sand, steel grit, metal shot, fine glass shot, or other form of abrasive is mechanically impelled against the surface to be cleaned. When sand is used, it should be clean, sharp-edged silica sand. Steel grit tends to clean more rapidly and generates much less dust, but is more expensive and less flexible.

FIGURE 40-1 Manual shot blasting. *(Courtesy of Norton Company.)*

When the parts are large, it may be easier to bring the cleaner to the part rather than the part to the cleaner. A common technique for such applications is *sand blasting* or *shot blasting*, where the abrasive particles are carried by a high-velocity blast of air emerging from a nozzle with about a $\frac{3}{8}$-in. opening. Air pressures between 60 and 100 psi are common when cleaning ferrous metals, and 10 to 60 psi is common for nonferrous. The abrasive may be sand or shot, or materials such as walnut shells, dry-ice pellets, or even baking soda. Pressurized water can also be used as a carrier medium.

When production quantities are large or the parts are small, the operation can be conducted in an enclosed hood, with the parts traveling past stationary nozzles. For large parts or small quantities, the blast may be directed by hand, as shown in Figure 40-1. Protective clothing and breathing apparatus must be provided and precautions taken to control the spread of the resulting dust. The process may even require a dedicated room or booth that is equipped with integrated air pollution control devices. Labor costs can also be extremely high.

From a manufacturing perspective, these processes are limited to surfaces that can be reached by the moving abrasive and cannot be used when sharp edges or corners must be maintained (since the abrasive tends to round the edges).

Barrel Finishing or Tumbling

Barrel finishing or *tumbling* is an effective means of finishing large numbers of small parts. In the Middle Ages, wooden casks were filled with abrasive stones and metal parts and were rolled about until the desired finish was obtained. Today, modifications of this technique can be used to deburr, radius, descale, remove rust, polish, brighten, surface-harden, or prepare parts for further finishing or assembly. The amount of stock removal can vary from as little as 0.0001 inch to as much as 0.005 in.

In the typical operation, the parts are loaded into a special barrel or drum until a predetermined level is reached. Occasionally, no other additions are made, and the parts are simply tumbled against one another. In most cases, however, an additional media of metal slugs or abrasives (such as sand, granite chips, slag, or ceramic pellets) is added. Rotation of the barrel causes the material to rise until gravity causes the uppermost layer to cascade

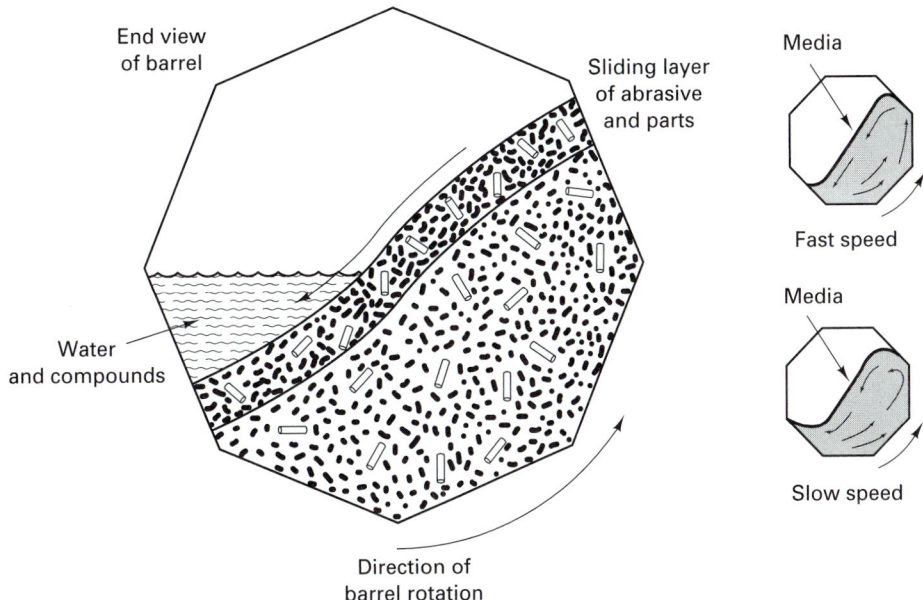

FIGURE 40-2 Schematic of the flow of material in tumbling or barrel finishing. The parts and media mass typically accounts for 50 to 60% of capacity.

downward in a "landslide" movement, as depicted in Figure 40-2. The sliding produces abrasive cutting that can effectively remove fins, flash, scale, and adhered sand. Since only a small portion of the load is exposed to the abrasive action, long times may be required to process the entire contents.

Increasing the speed of rotation adds centrifugal forces that cause the material to rise higher in the barrel. The enhanced action can often accelerate the process, provided that the speed is not so great as to destroy the cascading action and that the additional action does not damage the workpiece. By a suitable selection of abrasives, filler, barrel size, ratio of workpieces to abrasive, fill level and speed, a wide range of parts can be tumbled successfully. Delicate parts may have to be attached to racks within the barrel to reduce their movement while permitting the media to flow around them.

Natural and synthetic abrasives are available in a wide range of sizes and shapes, including those depicted in Figure 40-3, that enable the finishing of complex parts with irregular openings. The various media are often mixed in a given load, so that some will reach into all sections and corners to be cleaned.

Tumbling is usually done dry, but it can also be performed with an aqueous solution in the barrel. Chemical compounds can be added to the media to assist in cleaning, or descaling, or to provide features such as rust inhibition. Support equipment usually assists with loading and unloading the barrels, and with the separation of the workpieces from the abrasive media. The latter operation often uses mesh screens with selected size openings.

Barrel tumbling can be a very inexpensive way to finish large quantities of small parts and produce rounded edges and corners. Unfortunately, the abrasive action occurs on all surfaces and cannot be limited to selected areas. The cycle time is often long, and the process can be quite noisy.

In the *barrel burnishing* process, no cutting action is desired. Instead, the parts are tumbled against themselves, or with media such as steel balls, shot, rounded-end pins, or

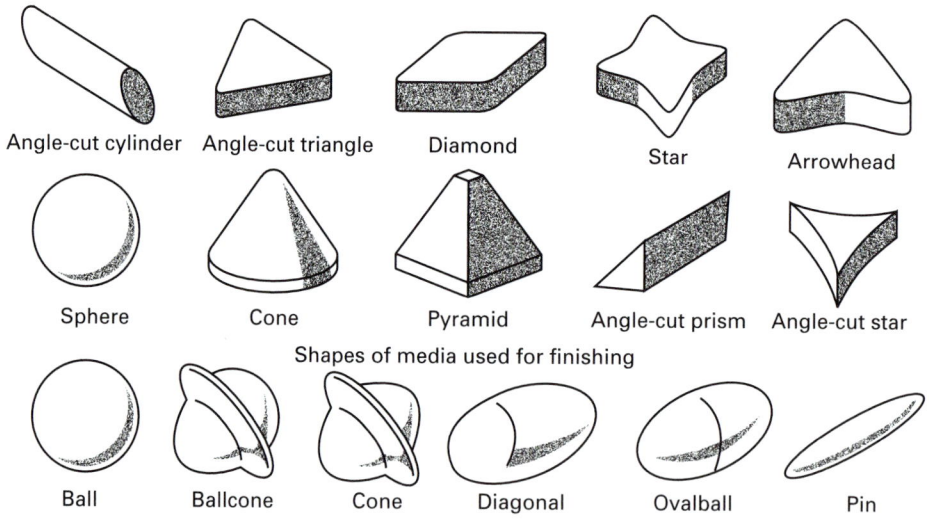

Shapes of media used for finishing

Steel media shapes used for burnishing

FIGURE 40-3 Synthetic abrasive media is available in a wide variety of sizes and shapes. Through proper selection, the media can be tailored to the product being cleaned.

ballcones. If the original material is free of visible scratches and pits, the combination of peening and rubbing will reduce minute irregularities and produce a smooth, uniform surface.

Barrel burnishing is normally done wet, using a solution of water and lubricating or cleaning agents, such as soap or cream of tartar. Because the rubbing action between the work and the media is very important, the barrel should not be loaded more than half full, and the volume ratio of media to work should be about 2:1, so the workpieces rub against the media and not each other. The speed of rotation should be set to maintain the cascading action and not fling the workpieces free of the tumbling mass.

Centrifugal barrel tumbling places the tumbling barrel at the end of a rotating arm. This adds centrifugal force to the weight of the parts in the barrel and can accelerate the process by as much as 25 to 50 times.

In *spindle finishing* the workpieces are attached to rotating shafts, and the assembly is immersed in media that is moving in a direction opposite to part rotation. This process is commonly applied to cylindrical parts and avoids the impingement of workpieces on one another. The abrasive action is accelerated, but time is required for fixturing and removal of the parts.

Vibratory Finishing

In contrast to the barrel processing, *vibratory finishing* is performed in open containers. As illustrated in Figure 40-4, tubs or bowls are loaded with workpieces and media and are vibrated at frequencies between 900 and 3600 cycles per minute. The specific frequency and amplitude are determined by the size, shape, weight, and material of the part, as well as the media and compound. Because the entire load is under constant agitation, cycle times are less than with barrel operations. The process is less noisy and is easily controlled and

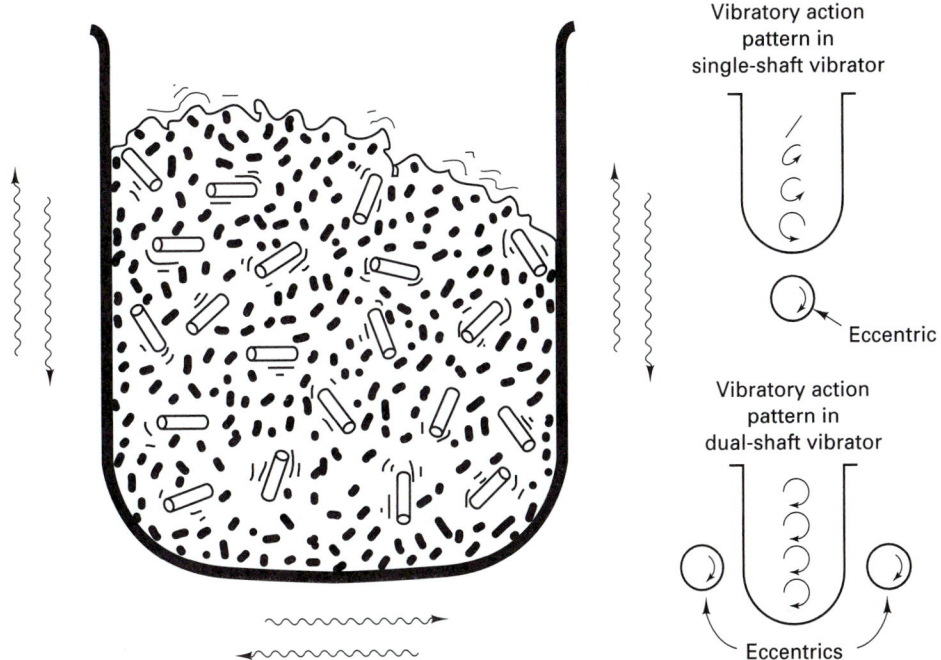

FIGURE 40-4   Schematic diagram of a vibratory finishing tub loaded with parts and media. The single eccentric shaft drive provides maximum motion at the bottom, which decreases as one moves upward. The dual shaft design produces more uniform motion of the tub, and reduces processing time.

automated. In addition, the open tubs allow for direct observation during the process, which can also deburr or smooth internal recesses or holes.

Media

The success of any of the mass-finishing processes depends greatly on *media* selection and the ratio of media to parts, as presented in Table 40-1. As developed in this table, a significant objective of the media may be to prevent the parts from impinging upon one another as it simultaneously cleans and finishes. Fillers, such as scrap punchings, minerals, leather scraps, and sawdust, are often added to provide additional bulk and cushioning.

Natural abrasives include slag, cinders, sand, corundum, granite chips, limestone, and hardwood shapes, such as pegs, cylinders, and cubes. Synthetic media typically contain 50 to 70 wt% of abrasives, such as alumina (Al_2O_3), emery, flint, and silicon carbide. This material is embedded in a matrix of ceramic, polyester, or resin plastic, which is softer than the abrasive, and erodes, allowing the exposed abrasive to perform the work. The synthetics are generally produced by some form of casting operation, so their sizes and shapes are consistent and reproducible (as opposed to the random sizes and shapes of the natural media). Steel media with no added abrasive is frequently specified for burnishing and light deburring.

Media selection should also be correlated with part geometry, since the abrasives should be able to contact all critical surfaces without becoming lodged in recesses or holes. This requirement has resulted in a wide variety of sizes and shapes, including those

TABLE 40-1.	Typical Media-to-Part Ratios for Mass Finishing
Media/Part Ratio by Volume	Typical Application
0:1	Part-on-part processing or burr removal without media
1:1	Produces very rough surfaces and is suitable for parts in which part-on-part damage is not a problem
2:1	Somewhat less severe part-on-part damage, but more action from less media
3:1	May be acceptable for very small parts and very small media. Part-on-part contact is likely on larger and heavier parts
4:1	In general, a good average ratio for many parts; a good ratio for evaluating a new deburring process
5:1	Better for nonferrous parts subject to part-on-part damage
6:1	Suitable for nonferrous parts, especially preplate surfaces on zinc parts with resin-bonded media
8:1	For improved preplate surfaces with resin-bonded media
10:1	Produces very fine finishes

Source: American Machinist, August 1983.

presented in Figure 40-3. The different abrasives, sizes, and shapes can be selected or combined to perform tasks ranging from light deburring with a very fine finish to heavy cutting with a rough surface.

Compounds

A variety of functions are performed by the *compounds* that are added in addition to the media and workpieces. These compounds can be liquid or dry, abrasive or nonabrasive, and acid, neutral, or alkaline. They are often designed to assist in deburring, burnishing, and abrasive cutting, as well as to provide cleaning, descaling, or corrosion inhibition.

In deburring and finishing, many small particles are abraded from both the media and the workpieces, and these must be suspended in the compound solution to prevent them from adhering to the parts. Deburring compounds also act to keep the parts and media clean and to inhibit corrosion. Burnishing compounds are often selected for their ability to develop desired colors and enhance brightness.

Cleaning compounds such as dilute acids and soaps are designed to remove excessive soils from both the parts and media and are often specified when the incoming materials contain heavy oil or grease. Corrosion inhibitors can be selected for both ferrous and nonferrous metals and are particularly important when steel media is being used.

Another function of the compounds may be to condition the water when aqueous solutions are being used. Consistent water quality, in terms of "hardness" and metal ion content, is important to ensure uniform and repeatable finishing results. Liquid compounds may also provide cooling to both the workpieces and the media.

Summary of Mass-Finishing Methods

The barrel and vibratory finishing processes are really quite simple and economical and can process large numbers of parts in a batch procedure. Soft, nonferrous parts can be finished in as little as 10 minutes, while the harder steels may require 2 hours or more. Sometimes the operations are sequenced, using progressively finer abrasives. Figure 40-5

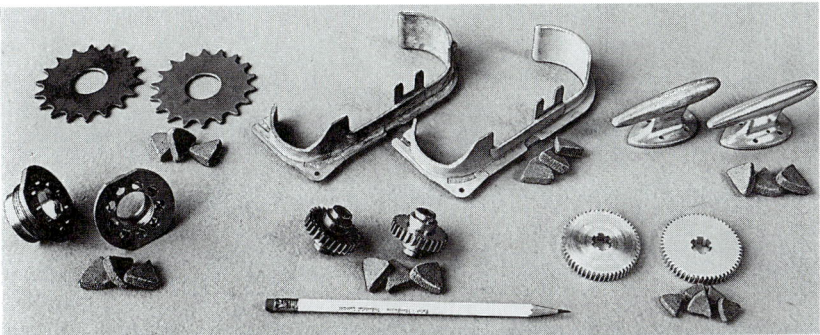

FIGURE 40-5   A variety of parts before and after barrel finishing with triangular-shaped media. *(Courtesy of Norton Company.)*

shows a variety of parts before and after the mass-finishing operation. The triangular abrasive is shown with each component.

Despite the high volume and apparent success, these processes may still be as much art as science. The key factors of workpiece, equipment, media, and compound are all interrelated, and the effect of changes can be quite complex. Media, equipment, and compounds are often selected by trial and error, with various approaches being tested until the desired result is achieved. Even then, maintenance of consistent results may still be difficult.

Belt Sanding

In the *belt sanding* operation, the workpieces are held against a moving abrasive belt until the desired degree of finish is obtained. Because of the movement of the belt, the resulting surface contains a series of parallel scratches with a texture set by the grit of the belt. When smooth surfaces are desired, a series of belts may be employed, with progressively finer grits.

The ideal geometry for belt sanding is a flat surface, for the belt can be passed over a flat table where the workpiece can be held firmly against it. Belt sanding is frequently a hand operation and is therefore quite labor intensive. Furthermore, it is difficult to apply when the geometry includes recesses or interior corners. As a result, belt sanding is usually employed when the number of parts is small and the geometry is relatively simple.

Wire Brushing

High-speed rotary *wire brushing* is sometimes used to clean surfaces and can also impart some small degree of material removal or smoothing. The resulting surface consists of a series of uniform curved scratches. For many applications, this may be an acceptable final finish. If not, the scratches can easily be removed by barrel finishing or buffing.

Wire brushing is often performed by hand application of a small workpiece to the brush, or the brush to a larger workpiece. Automatic machines can also be used where the parts are moved past a series of rotating brushes. In another modification the brushes are replaced with plastic or fiber wheels that are loaded with abrasive.

Buffing

Buffing is a polishing operation in which the workpiece is brought into contact with a revolving cloth wheel that has been charged with a fine abrasive, such as polishing rouge. The "wheels" are made of disks of linen, cotton, broadcloth, or canvas, and achieve the desired degree of firmness through the amount of stitching used to fasten the layers of cloth together. When the operation calls for very soft polishing or polishing into interior corners, the stitching may be totally omitted, the centrifugal force of the wheel rotation being sufficient to keep the layers in the proper position. Various types of polishing compounds are also available, with many consisting of ferric oxide particles in some form of binder or carrier.

The buffing operation is very similar to the lapping process that was discussed in Chapter 27. In buffing, however, the abrasive removes only minute amounts of metal from the workpiece. Fine scratch marks can be eliminated and oxide tarnish can be removed. A smooth, reflective surface is produced. When soft metals are buffed, a small amount of metal flow may occur, which further helps to reduce high spots and produce a high polish.

In manual buffing, the workpiece is held against the rotating wheel and manipulated to provide contact with all critical surfaces. Once again, the labor costs can be quite extensive. If the workpieces are not too complex, semiautomatic machines can be used, where the workpieces are held in fixtures and move past a series of individual buffing wheels. By designing the part with buffing in mind, good results can be obtained quite economically.

Electropolishing

Electropolishing is the reverse of electroplating (discussed later in this chapter) since material is removed from the surface rather than being deposited. A dc electrolytic circuit is constructed with the workpiece as the anode. As current is applied, material is stripped from the surface, with material removal occurring preferentially from any raised location. Unfortunately, it is not economical to remove more than about 0.001 in. of material from any surface. However, if the initial surface is sufficiently smooth (less than 8 μin. rms), and the grain size is small, the result will be a smooth polish with irregularities of less than 2 μin.—a mirrorlike finish.

Electropolishing was originally used to prepare metallurgical specimens for examination under the microscope. It was later adopted as a means of polishing stainless steel sheets and other stainless products. It is particularly useful for polishing irregular shapes that would be difficult to buff.

■ 40.3 CHEMICAL CLEANING

Chemical cleaning operations are effective means of removing oil, dirt, scale, or other foreign material that may adhere to the surface of a product, as a preparation for subsequent painting or plating. Because of environmental, health, and safety concerns, however, many processes that were once the industrial standard have now been eliminated or substantially modified. While the major concern with the mechanical methods has usually been airborne particles, the chemical methods often require the disposal of spent or contaminated solutions, and occasionally use hazardous, toxic, or environmentally unfriendly materials. Chlorofluorocarbons (CFCs) and carbon tetrachloride, for example, have been identified as ozone-depleting chemicals and have been phased out of commercial use. Process changes to comply with added regulations can significantly shift process economics. Manufacturers must now ask themselves if a part really has to be cleaned, what soils have to be removed, how clean the surfaces have to be, and how much they are willing to pay to accomplish that goal. Selection of the specific method will depend on the quantity of parts to be processed, part configuration, part material, desired surface finish, temperature of the process, and other variables.

Alkaline Cleaning

Alkaline cleaning is basically the "soap and water" approach to parts cleaning and is a commonly used method for removing a wide variety of soils (including oils, grease, wax, fine particles of metal, and dirt) from the surfaces of metals. The cleaners are usually complex solutions of alkaline salts, additives to enhance cleaning or surface modification, and surfactants or soaps that are selected to reduce surface tension and displace, emulsify, and disperse the insoluble soils. The actual cleaning occurs as a result of one or more of the following mechanisms: (1) saponification, the chemical reaction of fats and other organic compounds with the alkaline salts; (2) displacement, where soil particles are lifted from the surface; (3) dispersion or emulsification of insoluble liquids; and (4) dissolution of metal oxides.

Alkaline cleaners can be applied by immersion or spraying, and are usually heated to accelerate the cleaning action. The cleaning is then followed by a water rinse to remove all residue of the cleaning solution, as well as flush away some small amounts of remaining soil. A drying operation may also be required since the aqueous cleaners do not evaporate quickly, and some form of corrosion inhibitor (or rust preventer) may be required, depending on subsequent use.

Environmental issues relating to alkaline cleaning include (1) reducing or eliminating phosphate effluent, (2) reducing toxicity and increasing biodegradability, and (3) and recycling the cleaners to extend their life and reduce the volume of discard.

Solvent Cleaning

In *solvent cleaning*, oils, grease, fats, and other surface contaminants are removed by dissolving them in organic solvents derived from coal or petroleum, usually at room temperature. The common solvents include petroleum distillates (such as kerosene, naphtha, and mineral spirits), chlorinated hydrocarbons (such as methylene chloride and trichloroethylene), and liquids such as acetone, benzene, toluene, and the various alcohols. Small parts are generally cleaned by immersion, with or without assisting agitation, or by spraying (Figure 40-6). Products that are too large to immerse can be cleaned by spraying or wiping. The process is quite simple, and capital equipment costs are rather low. Drying is usually accomplished by simple evaporation.

Solvent cleaning is an attractive means of cleaning large parts, heat-sensitive products, materials that might react with alkaline solutions (such as aluminum, lead, and zinc), and products with organic contaminants (such as soldering flux or marking crayon). Virtually all common industrial metals can be cleaned, and the size and shape of the workpiece are rarely a limitation. Insoluble contaminants, such as metal oxides, sand, scale, and the inorganic fluxes used in welding, brazing, and soldering, cannot be removed by solvents. In addition, resoiling can occur as the solvent becomes contaminated. As a result, solvent cleaning is often used for preliminary cleaning.

Many of the common solvents have been restricted because of health, safety, and environmental concerns. Fire and excessive exposure are common hazards. Adequate ventilation is critical. Workers should use respiratory devices to prevent inhalation of vapors and wear protective clothing to minimize direct contact with skin. In addition, solvent wastes are often considered to be "hazardous materials" and may be subject to high disposal cost.

Vapor Degreasing

In *vapor degreasing*, the vapors of a chlorinated or fluorinated solvent are used to remove oil, grease, and wax from metal products. A nonflammable solvent, such as trichloroethylene, is

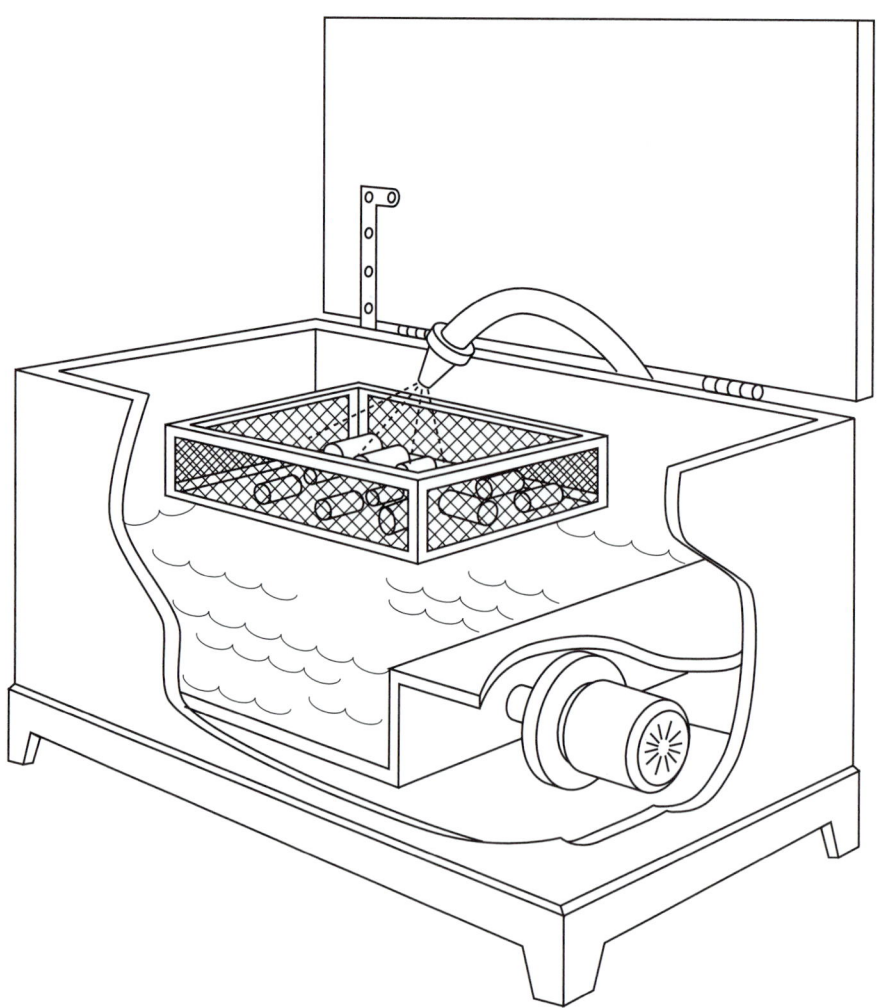

FIGURE 40-6   Solvent spray cleaning of small parts. The schematic shows the solvent reservoir, pump and spray nozzle, and the parts suspended in a wire mesh basket. *(Courtesy of ASM International.)*

heated to its boiling point, and the parts to be cleaned are suspended in its vapors. The vapor condenses on the work and washes the soluble contaminants back into the liquid solvent. Although the bath becomes dirty, the contaminants rarely volatilize at the boiling temperature of the solvent. Therefore, vapor degreasing tends to be more effective than cold solvent cleaning, since the surfaces always come into contact with clean solvent. Since the surfaces become heated by the condensing solvent, they dry almost instantly when they are withdrawn from the vapor.

Vapor degreasing is a rapid process that has almost no visible effect on the surface being cleaned. It can be applied to all common industrial metals, but the solvents may attack rubber, plastics, and organic dyes that might be present in product assemblies. A major limitation is the inability to remove insoluble soils, forcing the process to be coupled with another technique, such as mechanical or alkaline cleaning. Since hot solvent is present in the system, the process is often accelerated by coupling the vapor cleaning with an immersion or spray using the hot liquid.

Unfortunately, environmental issues have forced the almost complete demise of the process. While the vapor degreasing solvents are chemically stable, have low toxicity, are nonflammable, evaporate quickly, and can be recovered for reuse, they also have been identified as ozone-depleting compounds and have essentially been banned from use. It would be nice if there were a replacement solvent that could be used in the same process, or a replacement process that offered all of the qualities of a vapor degreaser, but these do not exist. Most manufacturers have been forced to convert to some form of water-based process using alkaline, neutral, or acid cleaners, or a process using chlorine-free, hydrocarbon-based solvents.

Ultrasonic Cleaning

When high-quality cleaning is required for small parts, *ultrasonic cleaning* may be preferred. Here the parts are suspended or placed in wire mesh baskets that are then immersed in a liquid cleaning bath, often a water-based detergent. The bath contains an ultrasonic transducer that operates at a frequency that causes cavitation in the liquid. The bubbles that form and implode provide the majority of the cleaning action, and if gross dirt, grease, and oil are removed prior to the immersion, excellent results can usually be obtained in 60 to 200 seconds. Because of the ability to use water-based solutions, ultrasonic cleaning has replaced many of the environmentally unfriendly solvent processes.

Acid Pickling

In the *acid pickling* process, metal parts are first cleaned to remove oils and other contaminants and then dipped into dilute acid solutions to remove oxides and dirt that are left on the surface by the previous processing operations. The most common solution is a 10% sulfuric acid bath at an elevated temperature between 150 and 185°F. Muriatic acid is also used, either cold or hot. As the temperature increases, the solutions can become more dilute.

After the parts are removed from the pickling bath, they should be rinsed to flush the acid residue from the surface and then dipped in an alkaline bath to prevent rusting. When it will not interfere with further processing, an immersion in a cold *milk of lime* solution is often used. Caution should be used to avoid overpickling, since the acid attack can result in a roughened surface.

■ 40.4 BURR REMOVAL

Burrs are the small, sometimes flexible projections of material that adhere to the edges of workpieces that are formed by cutting, punching, or grinding, like the exit-side burrs formed in the milled slot of Figure 40-7. Dimensionally, they are typically only 0.003 in. thick and 0.001 to 0.005 in. in height, but if not removed, they can lead to assembly failures, short circuits, injuries to workers, or even fatigue failures.

A number of different processes have been used for *burr removal*, including some discussed previously in this chapter and others presented as special types of machining. These include grinding, chamfering, barrel tumbling, vibratory finishing, centrifugal and spindle finishing, abrasive jet machining, water jet cutting, wire brushing, belt sanding, chemical machining, electropolishing, buffing, electrochemical machining, filing, ultrasonic machining, and abrasive machining.

Other methods may be quite specialized to the field of burr removal, such as thermal energy deburring. Here the parts are loaded into a chamber, which is then filled with a

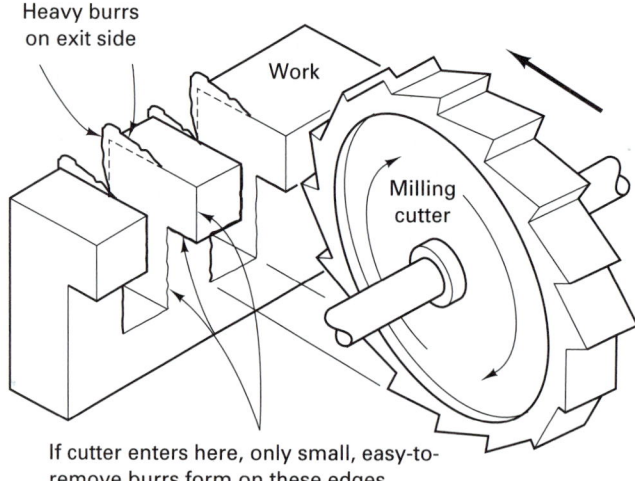

Heavy burrs
on exit side

Work

Milling
cutter

If cutter enters here, only small, easy-to-
remove burrs form on these edges

FIGURE 40-7 Schematic showing the formation of heavy burrs on the exit side of a milled slot. *(From L. X. Gillespie, American Machinist, November 1985.)*

combustible gas mixture. When the gas is ignited, the short-duration wavefront heats the small burrs to as much as 6000°F, while the remainder of the workpiece rarely exceeds 300°F. The burrs are vaporized in less than 20 ms, including those in inaccessible or difficult-to-reach locations. Since the process does not use abrasive media, there is no change to any of the product dimensions. The product surfaces are rarely affected by the generated heat, and the cycle (including loading and unloading) can be repeated as many as 100 times an hour. Unfortunately, there is a thin recast layer and heat-affected zone that forms where the burrs were removed. This region is usually less than 0.001 in. thick but may be objectionable in hardened steels and critically loaded parts.

Of all of the burr-removal methods, tumbling and vibratory finishing are usually the most economical, typically costing in the neighborhood of a few cents per part. Since most of the common methods also remove metal from exposed surfaces and produce a radius on all edges, it is important that the parts be designed for deburring. Table 40-2 provides a listing of the various deburring processes, as well as the edge radius, stock loss, and surface finish that would result from removal of a "typical burr" of 0.003 in. thickness.

Design to Facilitate or Eliminate Burr Removal

By knowing how and where burrs are likely to form, the engineer who designs parts may be able to make them easy to remove or even eliminate them. As shown in Figure 40-8, extra recesses or grooves can eliminate the need for deburring, since the burr produced by a cutoff tool or slot milling cutter will now lie below the surface. In this approach, one must determine whether it is cheaper to perform another machining operation (undercutting or grooving) or to remove the resulting burr.

Chamfers on sharp corners can also eliminate the need to deburr. The chamfering tool removes the large burrs formed by facing, turning, or boring and produces a relief for mating parts. The small burr formed during chamfering may be allowable or can easily be removed. Often, it may be preferable to give the manufacturer the freedom to use either a chamfer (produced by machining) or an edge radius (formed during the deburring operation) on all exposed corners or edges.

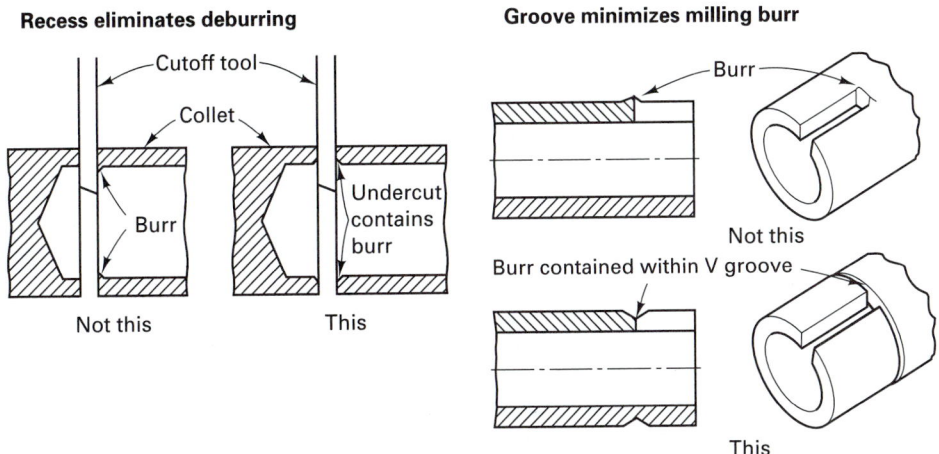

FIGURE 40-8 Designing extra recesses and grooves into a part may eliminate the need to deburr. *(From L. X. Gillespie,* American Machinist, *November 1985.)*

TABLE 40-2. Product Changes Associated with Removal of a Typical Burr[a]			
Process	Edge Radius (in.)	Stock Loss (in.)	Surface Finish (μin. AA[b])
Barrel tumbling	0.003–0.020	0–0.001	60–20
Vibratory deburring	0.003–0.020	0–0.001	70–35
Centrifugal barrel tumbling	0.003–0.020	0–0.001	70–20
Spindle finishing	0.003–0.020	0–0.001	70–20
Abrasive-jet deburring	0.003–0.010	0–0.002[c]	30–50
Water-jet deburring	0–0.005(p)		
Abrasive-flow deburring	0.001–0.020	0.001–0.005[d]	70–20
Chemical deburring	0–0.002	0–0.001	50–20
Ultrasonic deburring	0–0.002	0–0.001	20–15
Electrochemical deburring	0.002–0.010	0.001–0.003[e]	
Electropolish deburring	0–0.010	0.001–0.003[e]	30–15
Thermal-energy deburring	0.002–0.020	0	60–50
Power brushing	0.003–0.020	0–0.0005	
Power sanding	0.003–0.030[f]	0.0005–0.003	40–30
Mechanical deburring	0.003–0.060		
Manual deburring	0.002–0.015[g]		

[a]Based on a burr 0.08 mm (0.003 in.) thick and 0.13 mm (0.005 in.) high in steel. Thinner burrs can generally be removed much more rapidly. Values shown are typical. Stock-loss values are for overall thickness or diameter. Location A implies that loss occurs over external surfaces, B that loss occurs over all surfaces, and C that loss occurs only near edge. (*p*) indicates best estimate.

[b]Values shown indicate typical before and after measurements in a deburring cycle.

[c]Abrasive is assumed to contact all surfaces.

[d]Stock loss occurs only at surfaces over which medium flows.

[e]Some additional stray etching occurs on some surfaces.

[f]Flat sanding produces a small burr and no radius.

[g]Chamfer is generally produced with a small burr.

Source: L. X. Gillespie, *American Machinist*, November 1985.

■ 40.5 COATINGS

Each of the surface finishing methods previously presented has been a material removal process, designed to clean, smooth, or otherwise reduce the size of the part. A number of other techniques have been developed to add material to the surface of a part. If the material is deposited as a liquid or gas (or from a liquid or gas medium), the process is known as *coating*. If the added material is a solid during deposition, the process is known as *cladding*.

Painting

Paints and enamels are by far the most widely used finish on manufactured products, and a great variety are available to meet the wide range of product requirements. Most of today's commercial paints are synthetic organic compounds that contain pigments and dry by polymerization or by a combination of polymerization and adsorption of oxygen. Water is the most common carrying vehicle for the pigments. Heat can be used to accelerate the drying, but many of the synthetic paints and enamels will dry in less than an hour without the use of additional heat. The older oil-based materials have a long drying time and require excessive environmental protection measures. For these reasons they are seldom used in manufacturing applications.

Paints are used for a variety of reasons, usually to provide protection and decoration but also to fill or conceal surface irregularities, change the surface friction, or modify the light or heat absorption or radiation characteristics. Table 40-3 provides a list of some of the more commonly used organic finishes, along with their significant characteristics. *Nitrocellulose lacquers* consist of thermoplastic polymers dissolved in organic solvent. Although fast drying (by the evaporation of the solvent) and capable of producing very beautiful finishes, they are not sufficiently durable for most commercial applications. The *alkyds* are a general-purpose paint but are not adequate for hard-service applications. *Acrylic enamels* are widely used for automotive finishes and may require catalytic or oven curing. *Asphaltic paints*, solutions of asphalt in a solvent, are used extensively in the electrical industry, where resistance to corrosion is required and appearance is not of prime importance.

When considering a painted finish, the temptation is to focus on the outermost coat, to the exclusion of the underlayers. In reality, painting is a complex system that includes the substrate material, cleaning, and other pretreatments (such as anodizing, phosphating,

TABLE 40-3. Commonly Used Organic Finishes and Their Qualities

Material	Durability (Scale of 1–10)	Relative Cost (Scale of 1–10)	Characteristics
Nitrocellulose lacquers	1	2	Fast drying; low durability
Epoxy esters	1	2	Good chemical resistance
Alkyd–amine	2	1	Versatile; low adhesion
Acrylic lacquers	4	1.7	Good color retention; low adhesion
Acrylic enamels	4	1.3	Good color retention; tough; high baking temperature
Vinyl solutions	4	2	Flexible; good chemical resistance; low solids
Silicones	4–7	5	Good gloss retention; low flexibility
Fluoropolymers	10	10	Excellent durability; difficult to apply

and various conversion coatings), priming, and possible intermediate layers. The method of application is another integral feature to be considered.

Paint Application Methods

In manufacturing, almost all painting is done by one of four methods: *dipping*, *hand spraying*, *automatic spraying*, or *electrostatic spray finishing*. In most cases, at least two coats are required. The first (or *prime*) *coat* serves to (1) assure adhesion, (2) provide a leveling effect by filling in minor porosity and other surface blemishes, and (3) improve corrosion resistance, and thus prevent later coatings from being dislodged in service. These properties are less easily attainable in the more highly pigmented paints that are used in the final coats to promote color and appearance. When using multiple coats, however, it is important that the carrying vehicles for the final coats do not unduly soften the underlayers.

Dipping is a simple and economical means of paint application when all surfaces of the part are to be coated. The products can be manually immersed into a paint bath or passed through the bath while on or attached to a conveyor. Dipping is attractive for applying prime coats and for painting small parts where spray painting would result in a significant waste due to overspray. Conversely, the process is unattractive if only some of the surfaces require painting, or where a very thin, uniform coating would be adequate, as on automobile bodies. Other difficulties are associated with the tendency of paint to run, producing both a wavy surface and a final drop of paint attached to the lowest drip point. Good-quality dipping requires that the paint be stirred at all times and be of uniform viscosity.

Spray painting is probably the most widely used paint application process because of its versatility and the economy in the use of paint. In the conventional technique, the paint is atomized and transported by the flow of compressed air. In a variation known as *airless spraying*, mechanical pressure forces the paint through an orifice at pressures between 500 and 4500 psi. This provides sufficient velocity to produce atomization and also propel the particles to the workpiece. Because no air pressure is used for atomization, there is less spray loss (paint efficiency may be as high as 99%) and less generation of gaseous fumes.

Hand spraying is probably the most versatile means of application but can be quite costly in terms of labor and production time. When air or mechanical means provide the atomization, the workers must exercise considerable skill to obtain the proper coverage without allowing the paint to "run" or "drape." Only a very thin film can be deposited at one time, usually less than 0.001 in. As a result, several coats may be required with intervening time for drying.

One means of applying thicker layers in a single application is known as *hot spraying*. Special solvents are used which reduce the viscosity of the material when heated. Upon atomization, the faster evaporating solvents are removed, and the drop in temperature produces a more viscous, run-resistant material that can be deposited in thicker layers.

When producing large quantities of similar or identical parts, some form of automatic system is usually employed. The simplest automatic equipment consists of some form of parts conveyor that transports the parts past a series of stationary spray heads. While the concept is simple, the results may be unsatisfactory. A large amount of paint is wasted and it is difficult to get uniform coverage.

Industrial robots can be used to move the spray heads in a manner that mimics the movements of a human painter, maintaining uniform separation distance and minimizing waste. This is an excellent application for the robot, since a monotonous and repetitive process can be performed with consistent results. In addition, use of a robot removes the human from an unpleasant, and possibly unhealthy, environment.

Both manual and automatic spray painting can benefit from the use of *electrostatic deposition*. A dc electrostatic potential is applied between the atomizer and the workpiece. The atomized paint particles assume the same charge as the atomizer and are therefore repelled. The oppositely charged workpiece then attracts the particles, with the actual path of the particle being a combination of the kinetic trajectory and the electrostatic attraction. The higher the dc voltage, the greater the electrostatic attraction. Overspray can be reduced by as much as 60 to 80%, as well as the generation of airborne particles and other emissions. Unfortunately, part edges and holes receive a heavier coating than flat surfaces due to the concentration of electrostatic lines of force on any sharp edge. Recessed areas will receive a reduced amount of paint, and a manual touch-up may be required using conventional spray techniques. Despite these limitations, electrostatic spraying is an extremely attractive means of painting complex-shaped products where the geometry would tend to create large amounts of overspray.

In an electrostatic variation of airless spraying, the paint is fed onto the surface of a rapidly rotating cone or disk that is also one electrode of the electrostatic circuit. Centrifugal force causes the thin film of paint to flow toward the edge, where charged particles are spun off without the need for air assist. The particles are then attracted to the workpiece, which serves as the other electrode of the electrostatic circuit. Because of the effectiveness of the centrifugal force, paints can be used with high solids content, reducing the amount of volatile emissions and enabling a thicker layer to be deposited in a single application.

Powder coating is yet another variation of electrostatic spraying, but here the particles are solid rather than liquid. Several coats, such as primer and finish, can be applied and then followed by a single baking, in contrast to the baking after each coat that is required in the conventional spray processes. In addition, the overspray powder can often be collected and reused. While volatilized solvents are no longer a concern, operators must now address the possibility of powder explosion, as well as the health hazards of airborne particles.

Electrocoating or *electrodeposition* applies paint in a manner similar to the electroplating of metals. As shown schematically in Figure 40-9, the paint particles are suspended in an aqueous solution and are given an electrostatic charge by applying a dc voltage between the tank (cathode) and the workpiece (anode). As the electrically conductive workpiece enters and passes through the tank, the paint particles are attracted to it and deposit on the surface, creating a uniform, thin coating that is more than 90% resin and pigment. When the coating reaches a desired thickness, determined by the bath conditions, no more paint is deposited. The workpiece is then removed from the tank, rinsed in a water spray, and baked at a time and temperature that depends on the particular type of paint. Baking of 10 to 20 minutes at 375°F is somewhat typical.

Electrocoating combines the economy of ordinary dip painting with the ability to produce thinner, more uniform coatings. The process is particularly attractive for applying the prime coat to complex structures, such as automobile bodies, where good corrosion resistance is a requirement. Hard-to-reach areas and recesses can be effectively coated. Since the solvent is water, no fire hazard exists (as with the use of many solvents), and air and water pollution is reduced significantly. In addition, the process can be readily adapted to conveyor line production.

Drying

Most paints and enamels used in manufacturing require from 2 to 24 hours to dry at normal room temperature. This time can be reduced to between 10 minutes and 1 hour if the temperature can be raised to between 275 and 450°F. As a result, elevated temperature

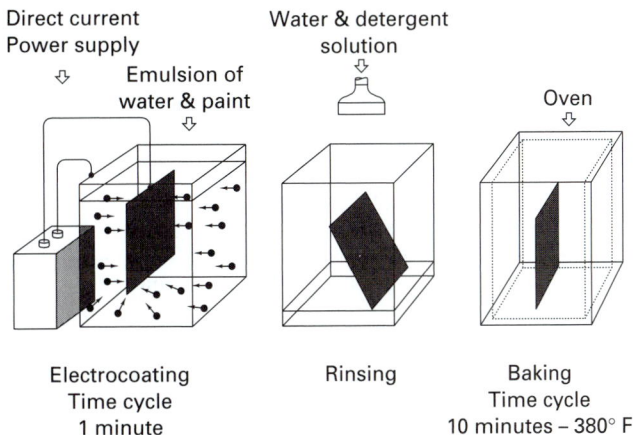

FIGURE 40-9 Basic steps in the electrocoating process.

drying is often preferred. Parts can be batch processed in ovens, or continuously passed through heated tunnels or under panels of infrared heat lamps.

Elevated-temperature drying is rarely a problem with metal parts, but other materials can be damaged by exposure to the moderate temperatures. For example, when wood is heated, the gases, moisture, and residual sap are expanded and driven to the surface beneath the hardening paint. Small bubbles tend to form that roughen the surface, or break, producing small holes in the paint.

Hot-Dip Coatings

Large quantities of metal products are given corrosion-resistant coatings by direct immersion into a bath of molten metal. The most common coating materials are zinc, tin, aluminum, and terne (an alloy of lead and tin).

Hot-dip galvanizing is the most widely used method of imparting corrosion resistance to steel. (The zinc acts as a sacrificial anode, protecting the underlying iron.) After the products, or sheets, have been cleaned to remove oil, grease, scale and rust, they are fluxed by dipping into a solution of zinc ammonium chloride and dried. Next, the article is completely immersed in a bath of molten zinc. The zinc and iron react metallurgically to produce a coating that consists of a series of zinc–iron compounds and a surface layer of nearly pure zinc.

The coating thickness is usually specified in terms of weight per unit area. Values between 0.5 and 3.0 oz/ft^2 are typical, with the specific value depending on the time of immersion and speed of withdrawal. Thinner layers can be produced by incorporating some form of air jet or mechanical wiping as the product is withdrawn. Since the corrosion resistance is provided through the sacrificial action of the zinc, the thin layers do not provide long-lasting protection. Extremely heavy coatings, on the other hand, may tend to crack and peel. The appearance of the coating can be varied through both the process conditions, as well as alloy additions of tin, antimony, lead, and aluminum. When the coatings are properly applied, bending or forming can often follow galvanizing without damage to the integrity of the coating.

The primary limitations to hot-dip galvanizing are the size of the product (which is limited to the size of the tank holding the molten zinc), and the "damage" that might occur when a metal is exposed to the temperatures of the molten material (approximately 850°F).

Tin coatings can also be applied by immersing in a bath of molten tin with a covering of flux material. Because of the high cost of tin and the relatively thick coatings applied by hot dipping, most tin coatings are now applied by electroplating. *Terne coating* utilizes an alloy of 15 to 20% tin and the remainder lead. This material is cheaper than tin and can provide satisfactory corrosion resistance for many applications.

Chemical Conversion Coatings

In *chemical conversion coating*, the surface of the metal is chemically treated to produce a nonmetallic, nonconductive surface that can impart a range of desirable properties. The most popular types of conversion coatings are chromate and phosphate.

Aluminum, magnesium, zinc, and copper (as well as cadmium and silver) can all be treated by a *chromate* conversion process that usually involves immersion in a chemical bath. The surface of the metal is converted into a layer of complex chromium compounds that can impart colors ranging from clear, through blue-bright, yellow, brown, olive drab, and black. Most of the films are soft and gelatinous when they are formed, but harden upon drying. They can be used to (1) impart exceptionally good corrosion resistance; (2) act as an intermediate bonding layer for paint, lacquer, or other organic finishes; or (3) provide specific colors by adding dyes to the coating when it is in its soft condition.

Phosphate coatings are formed by immersing metals (usually steel or zinc) in baths where metal phosphates (iron, zinc, and manganese phosphates are all common) have been dissolved in solutions of phosphoric acid. The resultant coatings can be used to precondition surfaces to receive and retain paint or enhance the subsequent bonding with rubber or plastic. In addition, phosphate coatings are usually rough and can provide an excellent surface for holding oils and lubricants. This feature can be used during manufacturing, where the coating holds the lubricants that assist in forming, or in the finished product, as with the black-color bolts and fasteners, whose corrosion resistance is provided by a phosphate layer impregnated with wax or oil.

Blackening or Coloring Metals

Many steel parts are treated to produce a black, iron oxide coating (Fe_3O_4), a lustrous surface that is resistant to rusting when handled. Since this type of oxide forms at elevated temperatures, the parts are usually heated in some form of special environment, such as spent carburizing compound or special blackening salts.

Chemical solutions can also be used to blacken, blue, and even "brown" steels. Brown, black, and blue colors can also be imparted to tin, zinc, cadmium, and aluminum through chemical bath immersions or wipes. The surfaces of copper and brass can be made to be black, blue, green, or brown, with a full range of tints in between.

Electroplating

Large quantities of metal *and plastic* parts are electroplated to produce a metal coating that imparts corrosion or wear resistance, improves appearance (through color or luster), or increases the overall dimensions. Virtually all commercial metals can be plated, including aluminum, copper, brass, steel, and zinc-based die castings. Plastics can be electroplated, provided that they are first coated with an electrically conductive material.

The most common platings are zinc, chromium, nickel, copper, tin, gold, platinum, and silver. The electrogalvanized zinc platings are thinner than the hot-dip coatings and can be produced without subjecting the base metal to the elevated temperatures of molten zinc. Nickel plating provides good corrosion resistance but is rather expensive and does

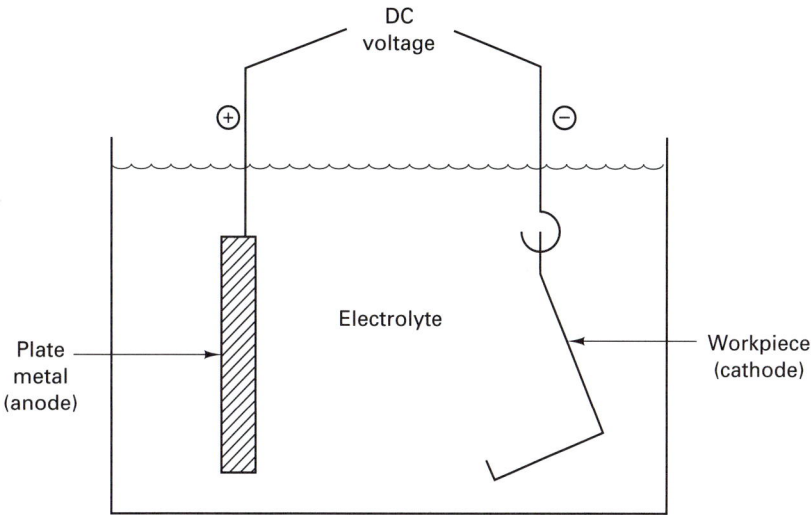

FIGURE 40-10 Basic circuit for an electroplating operation, showing the anode, cathode (workpiece), and electrolyte (conductive solution).

not retain its lustrous appearance. Consequently, when lustrous appearance is desired, a chromium plate is usually specified. Chromium is seldom used alone, however. An initial layer of copper produces a leveling effect and makes it possible to reduce the thickness of the nickel layer that typically follows to less than 0.0006 inch. The final layer of chromium then provides the attractive appearance. Gold, silver, and platinum platings are used in both the jewelry and electronics industries, where the thin layers impart the desired properties while conserving the precious metals.

Hard chromium plate, with Rockwell C hardnesses between 66 and 70, can be used to build up worn parts to larger dimensions and to coat tools and other products that need reduced surface friction and good resistance to both wear and corrosion. Hard chrome coatings are always applied directly to the base material and are usually much thicker than the decorative treatments, typically ranging from 0.003 to 0.010 in. thick. Even thicker layers are used in applications such as diesel cylinder liners. Since hard chrome plate does not have a leveling effect, defects or roughness in the base surface will be amplified. If smooth surfaces are desired, subsequent grinding and polishing may be necessary.

Figure 40-10 depicts the typical electroplating process. A dc voltage is applied between the parts to be plated (which is made the cathode) and an anode material that is either the metal to be plated or an inert electrode. Both of these components are immersed in a conductive electrolyte, which may also contain dissolved salts of the metal to be plated, as well as additions to increase or control conductivity. In response to the applied voltage, metal ions migrate to the cathode, lose their charge, and deposit on the surface. While the process is simple in its basic concept, the production of a high-quality plating requires selection and control of a number of variables, including the electrolyte and the concentrations of the various dissolved components, the temperature of the bath, and the electrical voltage and current. The interrelation of these features adds to the complexity and makes process control an extremely challenging problem.

The surfaces to be plated must also be prepared properly if satisfactory results are to be obtained. Pinholes, scratches, and other surface defects must be removed if a smooth,

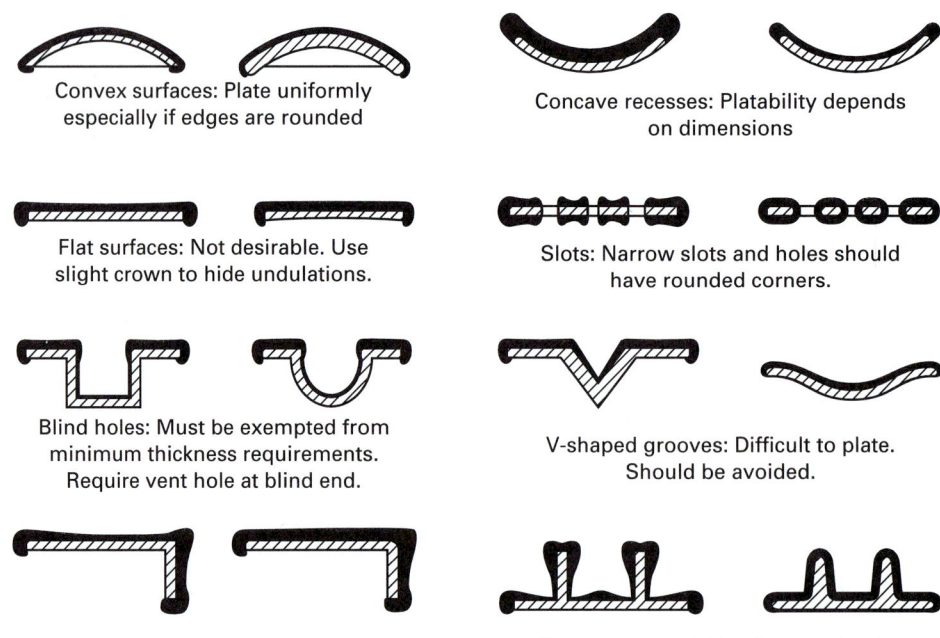

Convex surfaces: Plate uniformly
especially if edges are rounded

Concave recesses: Platability depends
on dimensions

Flat surfaces: Not desirable. Use
slight crown to hide undulations.

Slots: Narrow slots and holes should
have rounded corners.

Blind holes: Must be exempted from
minimum thickness requirements.
Require vent hole at blind end.

V-shaped grooves: Difficult to plate.
Should be avoided.

Sharply angled edges: Plating is thinner
in center areas. Round all areas.

Fins: Increase plating time and costs.
Reduce durability of finish.

FIGURE 40-11   Design recommendations for electroplating operations.

lustrous finish is desired. Combinations of degreasing, cleaning and pickling are used to assure a chemically clean surface, one to which the plating material can adhere.

As shown in Figure 40-11, the plated metal tends to be preferentially attracted to corners and protrusions. This makes it particularly difficult to apply a uniform plating to irregular shapes, especially ones containing recesses, corners, and edges. Design features can be incorporated to promote plating uniformity, and improved results can often be obtained through the use of multiple spaced anodes, or anodes whose shape resembles that of the workpiece.

Electroplating is frequently performed as a continuous process, where the individual parts to be plated are hung from conveyors. As they pass through the process, they are lowered into successive plating, washing, and fixing tanks. Ordinarily, only one type of workpiece is plated at a time, because the details of solutions, immersion times, and current densities are usually changed with changes in workpiece size and shape.

In the *electroforming* process, the coating becomes the final product. Metal is electroplated onto a mandrel (or mold) to a desired thickness and is then stripped free to produce small quantities of molds or other intricate-shaped sheet-type products.

Anodizing

In the *anodizing* process, which is somewhat the reverse of electroplating, a conversion-type coating can be developed on aluminum that can improve corrosion and wear resistance and impart a variety of decorative effects. If the workpiece is made the anode of an electrolytic cell, instead of a plating layer being deposited on the surface, a reaction progresses inward, increasing the thickness of the protective aluminum oxide layer that normally exists on aluminum. The product dimensions will increase, however, because the aluminum oxide coating occupies about twice the volume of the metal from which it formed.

The nature of the developed coating is controlled by the electrolyte. If the oxide coating is not soluble in the anodizing solution, it will grow until the resistance of the oxide prevents current from flowing. The resultant coating is thin, nonporous, and nonconducting, and finds use in a variety of electrical applications.

If the oxide coating is slightly soluble in the anodizing solution, dissolution competes with oxide growth, and a porous coating will be produced, where the pores provide for continued current flow to the metal surface. As the coating thickens, the growth rate decreases until it achieves steady state, where the growth rate is equal to the rate of dissolution. This condition is determined by the specific conditions of the process, including voltage, current density, electrolyte concentration, and electrolyte temperature. Sulfuric, chromic, oxalic, and phosphoric acids all produce electrolytes that dissolve oxide, with a sulfuric acid solution being the most common.

In a process variation known as *color anodizing*, a sulfuric acid bath is used to produce a layer of microscopically porous oxide that is transparent on pure aluminum and somewhat opaque on alloys. When this material is immersed in a dye solution, capillary action pulls the dye into the pores. The dye is then trapped in place by a sealing operation, usually performed simply by immersing the anodized metal in a bath of hot water. The aluminum oxide coating is converted to a monohydrate with accompanying increase in volume. The pores close and become resistant to further staining or the leaching out of the dye.

While most people are familiar with the variety of colors in aluminum athletic goods, such as softball bats, the actual applications range from giftware, through automotive trim, to architectural use. Aluminum can be made to look like gold, copper, or brass, or take on a variety of colors with a combined metallic luster that cannot be duplicated by other methods.

If PTFE (Teflon) is introduced into the pores, coatings can be produced that couple high hardness and low friction. The porous oxide layer can also be used to enhance the adhesion of an additional layer of material, such as paint, or carry lubricant during a subsequent forming operation. Since the coating is integral to the part, subsequent operations can often be performed without destroying its integrity or reducing its protective qualities.

Anodizing can also be performed on other metals, such as magnesium, and the process is similar to the passivation of stainless steel.

Electroless Plating

When using electroplating, it is almost impossible to obtain a uniform plating thickness on even moderately complex shapes, the platings cannot be applied to nonconductors, and a large amount of energy is required. For these reasons, a substantial effort has been directed toward the development of plating techniques that do not require an external source of electricity. These methods are known as *electroless*, or autocatalytic, *plating*. Considerable success has been achieved with nickel, but copper and cobalt, as well as some of the precious metals, can also be deposited.

In the electroless process, complex plating solutions (containing metal salts, reducing agents, complexing agents, pH adjusters, and stabilizers) are brought into contact with a substrate surface that acts as a catalyst or has been pretreated with catalytic material. The metallic ion in the plating solution is reduced to metal and deposits on the surface. Since the deposition is purely a chemical process, the coatings are uniform in thickness, independent of part geometry. Unfortunately, the rate of deposition is considerably slower than with electroplating.

Probably the most popular of the electroless coatings is electroless nickel, and various methods exist for its deposition using both acid and alkaline solutions. The coatings offer good corrosion resistance, as well as hardnesses between Rockwell C 49 and 55. In addition, the hardness can be increased further to as high as Rockwell C 80 by subsequent heat treatment.

Electroless Composite Plating

A very useful adaptation of the electroless process has been developed wherein minute particles are codeposited along with the electroless metal to produce composite material coatings. Finely divided, solid particles, with diameters between 1 and 10 μin, are added to the plating bath and deposit up to 50 vol% with the matrix. While it may appear that a large variety of materials could be codeposited, commercial applications have largely been limited to diamond, silicon carbide, aluminum oxide, and Teflon (PTFE).

Figure 40-12 shows a deposit of silicon carbide particles in a nickel alloy matrix, where the particles constitute about 25% by volume. The coating offers the same corrosion resistance as nickel, but the high hardness of the silicon carbide particles (about 4500 on the Vickers scale, where tungsten carbide is 1300 and hardened steel is about 900) contributes outstanding resistance to wear and abrasion. Since the deposition is electroless, the thickness of the coating is not affected by the shape of the part. Applications include the coating of plastic-molding dies, for use where the polymer resin contains significant amounts of abrasive filler.

Mechanical Plating

Mechanical plating, also known as peen plating or impact plating, is an adaptation of barrel finishing in which coatings are produced by cold-welding soft, malleable metal powder onto the substrate. Numerous small products are first cleaned and may be given a thin galvanic coating of either copper or tin. They are then placed in a tumbling barrel, along with a water slurry of the metal powder to be plated, glass or ceramic tumbling media, and chemical promoters or accelerators. The media particles peen the metal powder onto the surface, producing uniform thickness deposits (possibly a bit thinner on edges and thicker in recesses—the opposite of electroplating!). Any metal that can be made into fine powder can be deposited, but the best results are obtained for soft materials, such as cadmium, tin, and zinc. Since the material is deposited mechanically, the coatings can be layered or involve mixtures with bulk chemistries that would be chemically impossible due to solubility limits. The fact that the coatings are deposited at room temperature, and in an environment that does not induce hydrogen embrittlement, makes mechanical plating an attractive means of coating hardened steels.

Porcelain Enameling

Metals can also be coated with a variety of glassy, inorganic materials that impart resistance to corrosion and abrasion, decorative color, electrical insulation, or the ability to function in high-temperature environments. Multiple coats may be used, with the first or ground coat being selected to provide adhesion to the substrate and the cover coat to provide the surface characteristics. The material is usually applied in the form of a multicomponent suspension or slurry (by dipping or spraying), which is then dried and fired. An alternative dry process uses electrostatic spraying of powder and subsequent firing. During the firing operation, which

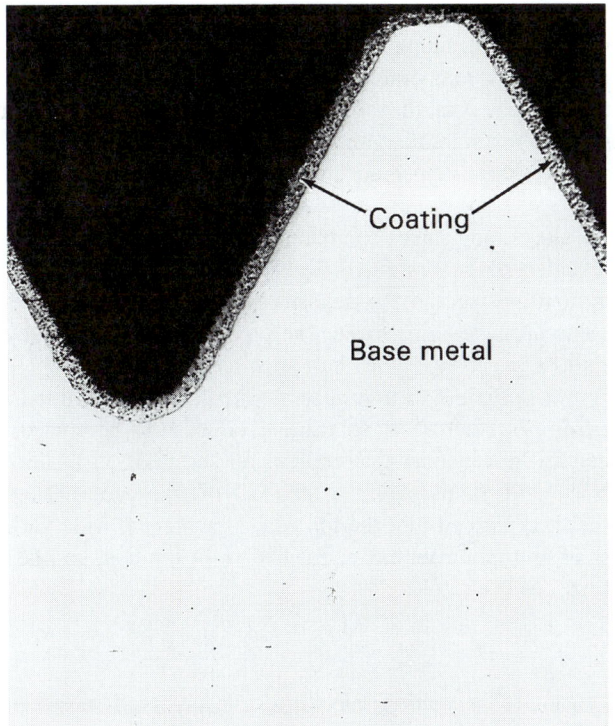
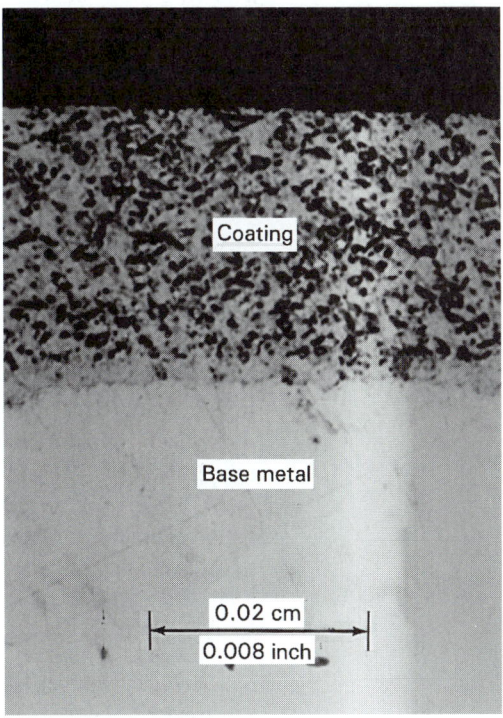

FIGURE 40-12   (*Left*) Photomicrograph of nickel carbide plating produced by electroless deposition. Notice the uniform thickness coating on the irregularly shaped product. (*Right*) High magnification cross section through the coating. (*Courtesy of Electro-Coatings Inc.*)

may require temperatures in the range of 800 to 1800°F, the coating materials melt, flow, and resolidify. Porcelain enamel is often found on the inner, perforated tubs of many washing machines and may be used to impart the decorative exterior on cookpots and frying pans.

■ 40.6 VAPORIZED METAL COATINGS

Vapor deposition processes can be classified into two main categories: *physical vapor deposition* (PVD) and *chemical vapor deposition* (CVD). While sometimes used as though it were a specific process, the term *PVD* applies to a group of processes in which the material to be deposited is carried physically to the surface of the workpiece. Vacuum metallizing and sputtering are key PVD processes, as are complex variations, such as ion plating. All are carried out in some form of vacuum, and most are line-of-sight processes in which the target surfaces must be positioned relative to the source. In contrast, the CVD processes deposit material through chemical reactions, and generally require significantly higher temperatures. Tool steels treated by CVD may have to be re-heat-treated, while most PVD processes can be conducted below normal tempering temperatures.

Vacuum Metallizing

Vacuum deposition or *vacuum metallizing* can be used to deposit thin films of metal or metal compounds on a wide variety of substrates, including glass, plastics, and metals.

The workpiece or multiple workpieces are placed in a low-pressure environment (usually less than one millionth of an atmosphere), along with a source of coating material. When the metal or compound is heated to high temperature, its vapor pressure exceeds that of its environment and it evaporates. The vapors then condense on cold surfaces within the chamber (i.e., the workpieces). The resultant coating is very thin (often as little as 1 µin.) and is generally used to provide decorative appearance, produce a reflective surface, or impart electrical properties.

Virtually all metals, most alloys, and many compounds can be deposited by vacuum metallizing, but several have emerged as commercially attractive. Aluminum is most widely used because it is easy to vaporize and the deposits provide good appearance and reflectivity. The thin layer that is produced also makes the process attractive for applying coatings of the precious metals to inexpensive substrates, such as plastic or steel.

Since the chamber vacuum is sufficient to minimize atomic collisions, emitted atoms travel in a straight line away from the source. If an entire product is to be coated, there must be provision to rotate the part in a manner that exposes all surfaces. In addition, the thin films usually lack durability and often require a top coat of transparent lacquer or resin. Dyes or pigments can be incorporated into the top coat to achieve a wide variety of metallic finishes. A thin layer of aluminum with a pigmented resin top coat can be made to look like copper, brass, bronze, or even gold.

Sputtering

Sputtering offers yet another means of depositing thin films onto prepared substrates. As illustrated in Figure 40-13, the substrate and the source material are both placed in a chamber that has been filled with low-pressure gas, usually argon. A high dc voltage is then applied between the source material (negative or cathode) and substrate (positive or anode). Since the applied potential exceeds the ionization energy of the gas, electrons emitted at the cathode strike the gas atoms, ionizing them to create a plasma. The positively charged gas ions are then accelerated toward the negatively charged source material. The impact energy is often sufficient to dislodge atoms from the surface, which then travel across the electrode gap and deposit on the substrate. Because of the kinetic energy of the impact and the cleanliness of the substrate (due in part to the partial vacuum), the adhesion is considerably better than

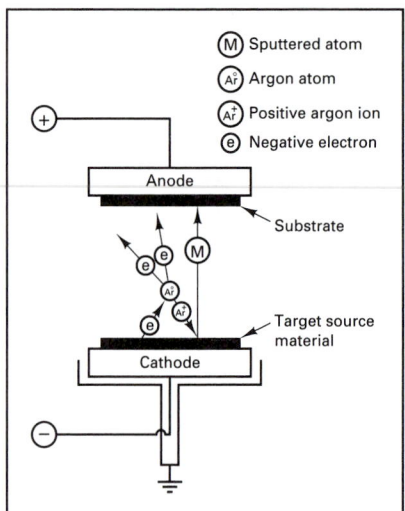

FIGURE 40-13 Schematic diagram of the sputtering process. The positively charged argon ions strike the negative cathode with sufficient energy to dislodge atoms, which then plate on the positively charged anode surfaces.

with ordinary deposition techniques. Applications are generally those where quality is a primary concern. Chromium has been sputtered onto the edges of razor blades to prevent corrosion, and the process is also used in the manufacture of solid-state devices and circuit boards. The primary limitation of the process is its relatively low deposition rate.

Chemical Vapor Deposition

A variety of pure metals, as well ceramic materials such as carbides, nitrides, borides, silicides, and oxides, can be deposited by the *chemical vapor deposition* (CVD) process, which is distinctly different from vacuum metallizing and sputtering. A metal halide or metal carbonyl vapor is brought into direct contact with a prepared and heated substrate (Figure 40-14). Deposition of the pure metal can then occur either by hydrogen reduction at a specified temperature or by direct pyrolytic decomposition at an even higher temperature. When the desire is to deposit a compound, the metal-containing vapor is accompanied by an additional reactive species, under controlled conditions of temperature, pressure, and composition such that the compound forms in preference to the pure metal. The reactions for the deposition of titanium carbide, titanium nitride, and aluminum oxide (all performed in a hydrogen environment) are as follows:

$$TiCl_4 + CH_4 \rightarrow TiC + 4HCl$$

$$TiCl_4 + N_2 + H_2 \rightarrow 2TiN + 4HCl$$

$$2AlCl_3 + 3H_2O \rightarrow Al_2O_3 + 6HCl$$

Refractory metals, alloys, and refractory compounds can be deposited as high-density coatings onto complex-shaped substrates. A substantial area of application is the hard coating of tools made from high-speed steels, tool and stainless steels, and cemented carbides. Coatings of titanium carbide, titanium nitride, aluminum oxide, silicon carbide, silicon nitride, and cubic boron nitride (and layered combinations thereof) have substantially extended tool life in harsh cutting conditions. They offer high hardness, high chemical stability, low friction, and antigalling characteristics. The metals most frequently deposited are those with high melting points, such as the refractory metals of tungsten, molybdenum, tantalum, and niobium. Since the coatings are produced by controlled chemical reactions, high purities can be achieved from somewhat impure reactants and deposits are produced on all surfaces, including through or blind holes. The primary limitation is the high temperature required to promote the reactions. While this is substantially below the melting points of the materials being deposited, it may still be sufficient to induce metallurgical change in the substrate or distortion of the part geometry.

Considerable growth has occurred in the area of diamond film deposition via chemical vapor deposition. Thin diamond films can be created that can be used to enhance the performance of existing products in the electronic, optical, and mechanical areas, and new products can be developed with previously unattainable properties. The coatings have hardnesses greater than the traditional carbides and nitrides, high chemical stability, low coefficient of sliding friction, optical transparency in the optical and infrared spectrums, high thermal conductivity, and high electrical resistivity. The specific properties are affected by the structure, thickness, and impurity levels of the film, which are influenced by the features of the deposition process. As with the other CVD processes, the application temperature is quite high, so the substrates are usually materials that can withstand high temperatures, such as metals and ceramics.

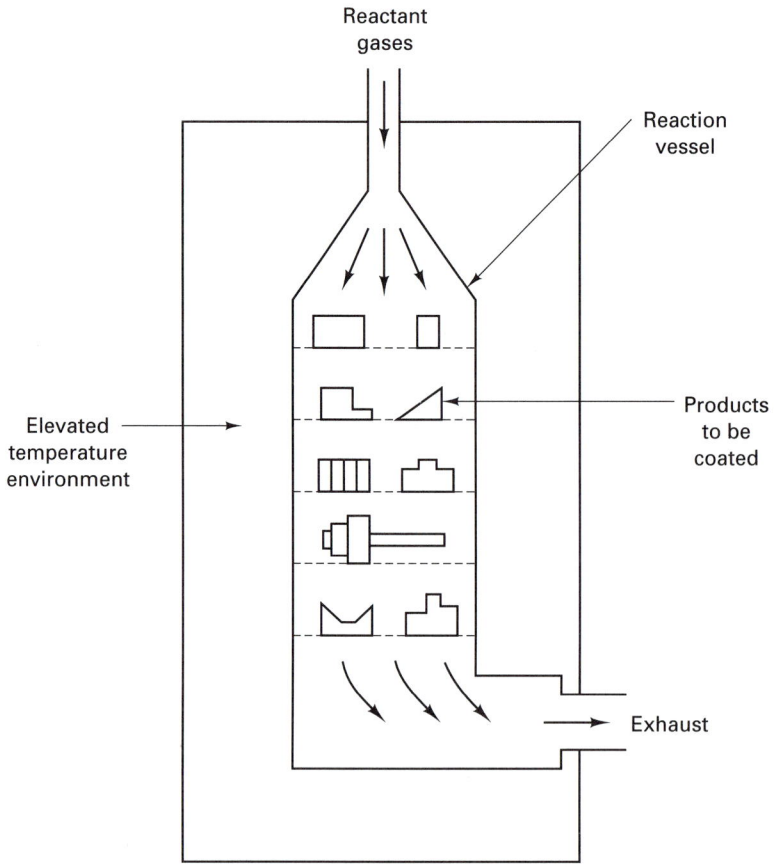

FIGURE 40-14 Schematic of the chemical vapor deposition (CVD) process. A controlled stream of reactant gases (carrier gas, hydrogen, and metal halide or metal carbonyl) is heated to reaction temperature in the presence of substrate surfaces. A coating of reaction product is deposited on the substrate surface.

■ 40.7 ION IMPLANTATION

Ion implantation, a technique originally developed to produce controlled doping of semiconductors, is now being used to modify the surface and subsurface chemistry (and properties) of a variety of products. As shown in Figure 40-15, atoms of a selected species are ionized and then passed down the evacuated column of a particle accelerator, toward a "target" that has been positioned at the end. Upon impact, the ions lose their kinetic energy through multiple collisions, eventually coming to rest in subsurface locations. The primary advantages are associated with the fact that it is a physical, rather than chemical or thermal process. Any ionizable species can be introduced into any target material, while maintaining temperatures below 300°F. The amount added is determined by the beam current and time of exposure and is limited only by an ultimate equilibrium with sputtering and not solubility limits. The beam can be positioned by electromagnetic controls, and the depth is a function of the acceleration voltage. Finally, the dimensions, surface finish and appearance of the target are unchanged by the process. Nitrogen ion implantation has been shown to outperform conventional nitriding, and metal ions, such as titanium and tantalum

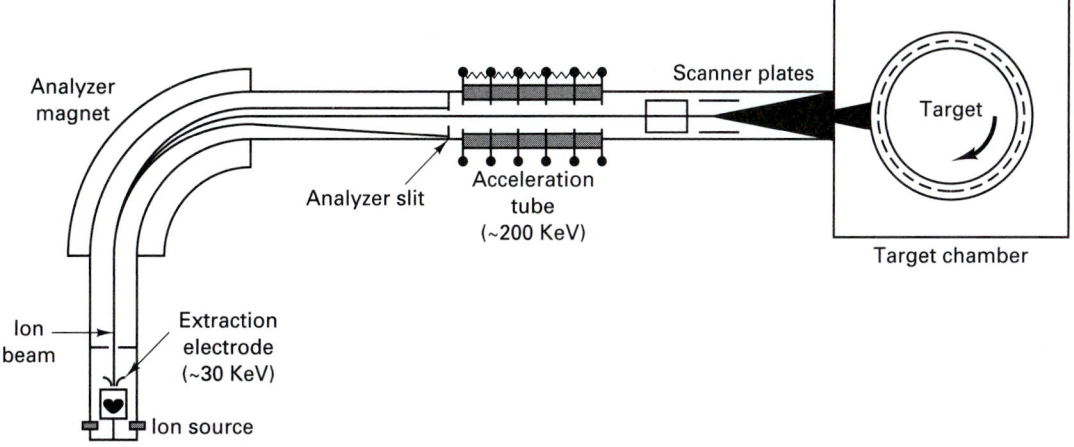

FIGURE 40-15 Schematic of the ion implantation process. Ions of a selected atomic species are separated by atomic mass and then accelerated toward a target surface. The entire process is performed in a high vacuum.

have been shown to impart desirable features to critical parts, such as orthopedic replacement joints and critical bearing assemblies. Unfortunately, the process is expensive for large areas and is limited to line-of-sight surfaces.

The capabilities of the process have been further extended by a technique known as *ion-beam mixing*. A thin film of metal or a compound is first deposited onto the target surface, and the energetic ion beam is then used to "mix" the deposited material into the surface of the product. The process is extremely flexible, offering a wide variety of substrate materials (metals, ceramics, polymers, and even glass), deposited layers, and implantation species, as well as specific process conditions.

■ 40.8 CLAD MATERIALS

Clad materials are actually a form of composite in which the components are joined as solids, using techniques such as roll bonding, explosive welding, and extrusion. The most common form is a laminate, where the surface layer provides properties such as corrosion resistance, wear resistance, electrical conductivity, thermal conductivity, or improved appearance, while the substrate layer provides strength or reduces overall cost. Alclad aluminum is a typical example. Here surface layers of weaker but more-corrosion-resistant single-phase aluminum alloys are applied to a base of high-strength but less-corrosion-resistant, age-hardenable material. Aluminum-clad steel meets the same objective but with a heavier substrate, and stainless steel can be used to clad steels, reducing the need for nickel and chromium alloy additions throughout.

Wires and rods can also be made as *claddings*. Here the surface layer often imparts conductivity, while the core provides strength or rigidity. Copper-clad steel rods are one example, and can be driven into the ground to provide electrical grounding for lightning rod systems.

■ 40.9 TEXTURED SURFACES

While technically not the result of a surface finishing process or operation, *textured surfaces* can be used to impart a number of desirable properties or characteristics. The types of textures that are often rolled onto the sheets used for refrigerator panels serve to conceal dirt, smudges, and fingerprints. Embossed or coined protrusions can enhance the grip of metal stair treads and walkways. Corrugations provide enhanced strength and rigidity. Still other textures can be used to modify the optical or acoustical characteristics of a material.

■ 40.10 COIL-COATED SHEETS

Traditionally, sheet metal components, such as appliance panels, have been fabricated from bare metal sheets. Parts are blanked and shaped by the traditional metal-forming operations, and the shaped panels are then finished on an individual basis. This requires individual handling and the painting or plating of geometries that contain holes, bends, and contours. In addition, there is the time required to harden, dry, or cure the applied surface finish.

An alternative approach is to apply the finish to the sheet material after rolling but before coiling. *Coatings* can be applied continuously to one or both sides of the material while it is in the form of a flat sheet. Thus the coiled material is effectively prefinished, and efforts need to be taken to protect the surface during the blanking and forming operations used to produce the final shape. Various paints have been applied successfully, as well as a full spectrum of metal coatings and platings. The sheared edges will not be coated, but if this feature can be tolerated, the additional measures to protect the surface may be an attractive alternative to the finishing of individual components.

■ 40.11 EFFECTS OF SURFACE PROCESSING

It is important to understand that the various manufacturing and surface finishing processes each impart distinct properties to the materials which will influence the performance of the product. The achievement of a satisfactory product obviously depends on a good design, high-quality manufacturing (including surface treatment), and proper assembly. The failure of parts in service, however, is usually the result of a combination of factors. A brief survey of features associated with surfaces and surface processing follows.

Each of the various machining processes produce characteristic surface textures (roughness, waviness, and lay) on the workpieces. In addition, the various processes tend to produce changes in the physical, mechanical, and metallurgical properties on or near the surfaces that are created. For the most part, these changes are limited to a depth of 0.005 to 0.050 inch below the surface. The effects can be beneficial or detrimental, depending on the process, material, and function of the product.

Machining processes (both chip forming and chipless) induce plastic deformation, and the cut surfaces are generally left with tensile residual stresses, microcracks, and a hardness that is different from the bulk material. Processes such as EDM and laser machining leave a layer of hard, recast metal on the surface that usually contains microcracks. Ground surfaces can have either residual tension or residual compression, depending on the mix between chip formation and plowing or rubbing during the grinding operation. If sufficient heat is generated, phase transformations can occur in the surface and subsurface regions.

Processes such as *roller burnishing* (described in Chapter 19) produce a smooth surface with compressive residual stresses. Shot peening and tumbling also impart residual compression to the surface. Welding processes produce tensile residual stresses as the deposited material shrinks upon cooling. Similar shrinkage occurs in castings, but the resulting stresses may be complex due to the variation of shrinkage or the lack of restraint. Tensile stresses on the surface can often be offset by a subsequent exposure to shot peening or tumbling.

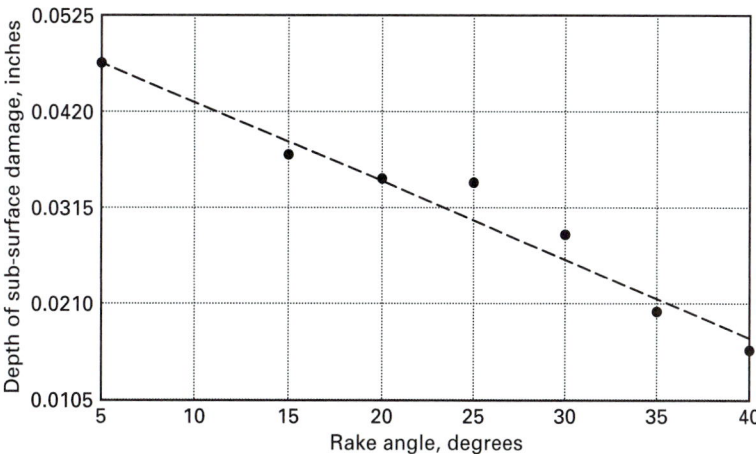

FIGURE 40-16   The depth of damage to the surface of a machined part increases with decreasing rake angle of the cutting tool.

In summary, the surface and subsurface regions of a material can be significantly altered due to (1) plastic strain or plastic deformation, (2) high temperatures, (3) differential expansions or contractions due to temperature changes or variations, and (4) chemical reactions.

To illustrate the complex nature of surface effects, consider Figure 40-16, which shows the depth of "surface damage" due to machining as a function of the rake angle of the tool. To increase the cutting speed (and thereby increase the rate of production), an engineer might change from a high-speed tool steel cutter with a large rake angle (such as 30°) to a carbide tool with a small rake (such as 5°). While the resulting surface finish may be similar, the depth of "surface damage" is doubled. Failures may occur in service, whereas previous parts had performed quite admirably.

Figure 40-17 shows how residual stresses couple with applied stresses to affect product performance. Suppose that a round beam has a load applied to it so that it is bent while rotating. At the top of the rotation, the surface is in tension, and at the bottom, it is in compression. The result is a condition of cyclic fatigue and the likelihood of a service life limited by fatigue failure. If the part is roller burnished or shot peened, the residual stress pattern of the middle figure is added to the applied stresses, producing the net pattern shown at the bottom. The net effect is a lowering of the peak tensile stress experienced by the surface and a related extension in fatigue life. The specific results will depend on the details of the process. For *shot peening*, the key variables include shot size, shot velocity, exposure time, distance between the nozzle and the surface, and the angle of impact.

Table 40-4 shows the results of rotating beam fatigue tests that were performed on "identical" specimens with eight different surface finishes. To assure a lifetime of 10^7 cycles, the same part can carry almost five times the load if its surface is prepared by ultrasonic machining rather than by EDM.

Figure 40-18 presents the results of another study in which specimens were prepared by milling and turning and then either polished, shot-peened, or roller-burnished. If an applied stress between 41,000 and 42,000 psi is experienced in a fatigue application, the difference in fatigue life between a milled specimen and one that has been milled and roller burnished is 610,000 cycles (90,000 cycles as opposed to 700,000 cycles). In essence, roller burnishing serves to induce a sevenfold extension to the fatigue life of the product. Similar results have been observed in the resistance to stress-corrosion cracking.

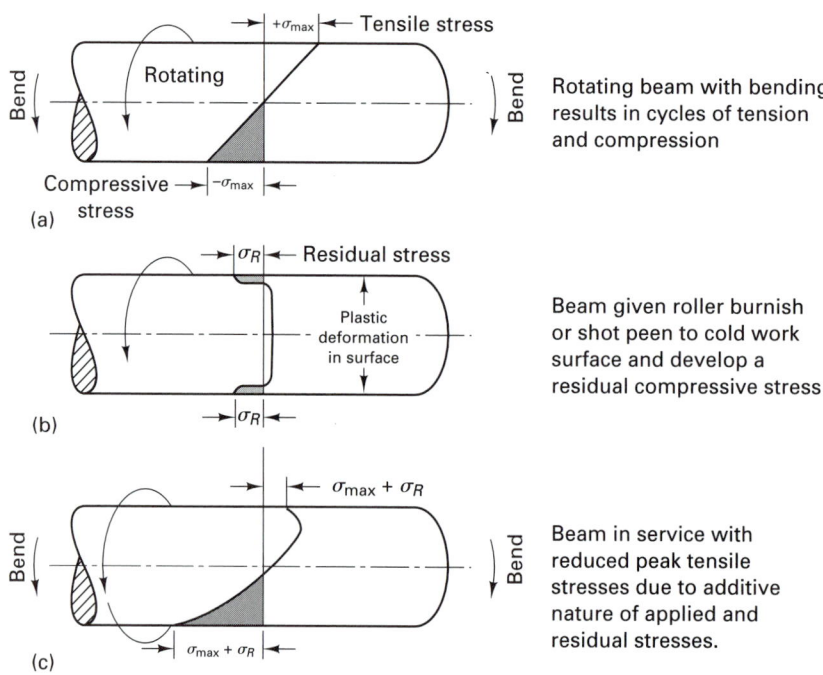

FIGURE 40-17   (*Top*) A cantilever-loaded (bent) rotating beam, showing the normal distribution of surface stresses (i.e., tension at the top and compression at the bottom). (*Center*) The residual stresses induced by roller burnishing or shot peening. (*Bottom*) Net stress pattern obtained when loading a surface-treated beam. The reduced magnitude of the tensile stresses contributes to increased fatigue life.

TABLE 40-4.   Fatigue Strength in Reverse Cantilever Bending of Ti–5 Al–2.5 Sn as a Function of Surface Generation Method for a Life of 10^7 Cycles

Surface Generation Method	Strength[a] (psi)
Ultrasonic	98,000
Slab mill	86,000
Chemical mill + vacuum anneal	77,000
Shot peen	76,000
As rolled and received	61,000
Chemical mill	59,000
Ground	52,000
Electric discharge machined	21,000

[a] Average values.

Source: Data from Rooney, WADC report.

As the data above shows, both the designer and the manufacturer need to be aware of the effects that manufacturing processes can have on the performance of a product. Maintaining the proper sequence of operations may be as important to the surface properties as the selection of the processes and control of the operating parameters.

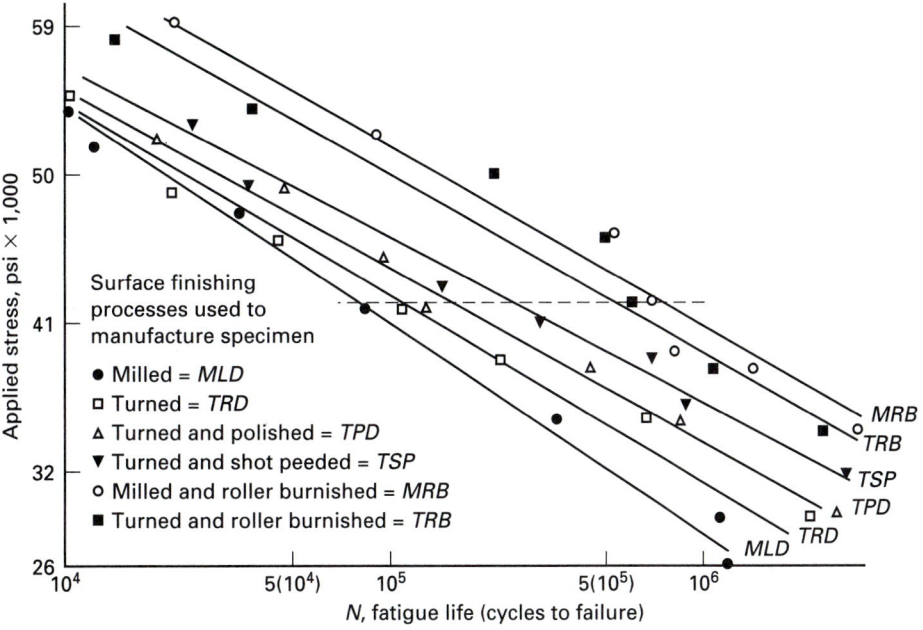

FIGURE 40-18   Fatigue life of rotating beam 2024-T4 aluminum specimens with a variety of surface finishing operations. Note the enhanced performance that can be achieved by shot peening and roller burnishing.

■ KEY WORDS

abrasive cleaning	compounds	physical vapor deposition
acid pickling	dipping	porcelain enameling
alkaline cleaning	electrocoating	powder coating
anodizing	electroforming	prime coat
barrel burnishing	electroless composite plating	residual stresses
barrel finishing	electroless plating	roll bonding
belt sanding	electroplating	roller burnishing
buffing	electropolishing	sand blasting
burr removal	electrostatic deposition	shot peening
case hardening	hard chromium plate	solvent cleaning
centrifugal barrel tumbling	hard facing	spray painting
chemical cleaning	hot-dip coating	sputtering
chemical conversion coating	ion implantation	textured surfaces
chemical vapor deposition	ion-beam mixing	tumbling
chromate	mechanical cleaning	ultrasonic cleaning
cladding	mechanical plating	vacuum metallizing
coating	media	vapor degreasing
coil-coated sheets	paint	vibratory finishing
color anodizing	phosphate	wire brushing

■ REVIEW QUESTIONS

1. What are some possible objectives of surface modification processes?
2. What are some of the factors that should be

considered when selecting a surface modification process?

3. What are some of the sources of foreign

material on the surface of manufactured products?

4. What are some of the common abrasive media used in blasting or abrasive cleaning operations?

5. What types of quantities and part sizes are most attractive for barrel finishing operations?

6. Describe why there might be an optimum fill level in barrel finishing; an optimum rotational speed.

7. How does barrel burnishing differ from abrasive barrel finishing?

8. Describe the primary differences between barrel finishing and vibratory finishing.

9. What are some of the attractive features of synthetic abrasive media?

10. What are some of the possible functions of the compounds that are used in abrasive finishing operations?

11. What is the ideal geometry for belt sanding? What types of geometric features would be difficult?

12. How is electropolishing different from electroplating?

13. What are some of the mechanisms of alkaline cleaning, and what types of soils can be removed?

14. What types of surface contaminants cannot be removed by solvent cleaning?

15. In view of its many attractive features, why has vapor degreasing become an unattractive process?

16. What is the primary type of surface contaminant removed by acid pickling?

17. What are burrs?

18. Describe the process of thermal energy deburring.

19. What is the difference between coating and cladding operations?

20. What are some of the reasons that paints may be specified for manufactured items?

21. What are some of the functions of a prime coat in a painting operation? What features are desired in the final coat?

22. What produces atomization and propulsion in airless spraying?

23. What features make industrial robots attractive for spray painting?

24. What are some of the attractive features of electrostatic spraying?

25. Why would it be difficult to apply electrostatic

spray painting to products made from wood or plastic?

26. What are some of the metal coatings that can be applied by the hot-dip process?

27. What are the two most common types of chemical conversion coatings?

28. How can nonconductive materials such as plastic be coated by electroplating?

29. What are the attractive properties of hard chrome plate?

30. What are some of the common process variables in an electroplating cell?

31. Why is it difficult to mix parts of differing size and shape in an automated electroplating system?

32. How is electroforming different from electroplating?

33. When anodizing aluminum, what features determine the thickness of the resulting oxide when the oxide is not soluble in the electrolyte? When it is partially soluble?

34. What produces the various colors in the color anodizing process?

35. What are some of the attractive features of electroless plating?

36. What types of particulate composites can be deposited by electroless plating?

37. What is mechanical plating?

38. What are some of the attractive properties of a porcelain enamel coating?

39. Describe the differences between physical vapor deposition (PVD) and chemical vapor deposition (CVD).

40. What are some of the common functions of a vacuum metallized coating?

41. Since the coating is deposited as a vapor, why is vacuum metallizing considered to be a "line-of-sight" process?

42. What features contribute to the excellent adhesion of sputtered coatings?

43. What types of materials can be deposited as coatings in the chemical vapor deposition process?

44. What are some of the attractive features of diamond-thin films?

45. What are some of the unique advantages of the ion implantation process?

46. How is ion-beam mixing different from ion implantation?

47. Describe the advantages and limitations of using coil-coated sheets in place of the painting of individual finished components.

48. What types of surface features or surface modification result from machining-type processes?

49. Why might processes that produce residual compressive stresses on product surfaces be attractive for mechanically loaded products?

■ PROBLEMS

1. Fishermen are among the most superstitious people in the world, and their superstitions affect the type of equipment that they use. As a result, hook manufacturers generally offer their products in a wide range of colors and finishes. Your company manufactures a range of hooks from AISI 1080 carbon steel wire, forming them to precision shape (eye, bends, barbs and point), and then heat treating them by a quench and temper treatment.

   Consider the size and shape of the product and the various properties that are required. The hooks must be strong enough to resist bending, but not so brittle that they might break. They must be corrosion-resistant to both fresh and salt water, and the desired appearance must be provided without fouling the point or the barbs. If the surface is applied before heat treatment, it must endure that process and maintain its appearance. If it is applied afterward, it cannot weaken or embrittle the hook.

   (a) Of the various surface modification processes, which ones might be attractive for such an application? (*Note:* Make sure that the process is appropriate! For example, barrel plating would probably produce a hopelessly snarled mass of wires!)

   (b) For each of the possible processes, describe the advantages, limitations, possible colors or finishes, and relative cost.

2. Select one or more of the following products (as directed) and recommend a surface treatment or coating. Consider the appropriateness of the technique to the size, shape, quantity, and material. Cite the specific features that make the recommended treatment the most attractive. What, if any, are the primary limitations or production concerns?

   (a) The exterior housing for the motor and drive unit of a chain saw that has been made as a magnesium alloy die casting.

   (b) Large quantities of steel bolts that are intended for use in outdoor construction.

   They have been fabricated from 4140 steel, have a shank diameter of $\frac{1}{2}$ in., and have been quenched and tempered to a final hardness of Rockwell C 45.

   (c) The handle of a household utility knife (retractable-blade razor-blade cutter) that has been made as a two-part zinc die casting.

   (d) The scoop portion of an inexpensive ice cream scoop that has been made as a zinc die casting.

   (e) A decorative handle for a kitchen cabinet that is made as a zinc die casting.

   (f) The exterior of an office filing cabinet that has been made from low-carbon steel sheet.

   (g) A high-quality combination wrench (open-end and box-end) that has been forged from 4147 steel bar stock.

   (h) Case of a moderately priced wristwatch that has been fabricated from yellow brass (to have a gold appearance).

   (i) Tubular frame of a lightweight bicycle that has been made from age-hardened aluminum.

   (j) The basket section of a grocery-store shopping cart that has been fabricated from welded steel mesh.

   (k) The exterior of an automobile muffler to be fabricated from steel sheet. Describe how the coating treatment might best be integrated into the fabrication sequence.

   (l) An inexpensive interior door knob that has been fabricated from deep-drawn cartridge brass sheet.

   (m) High-quality steel sockets for a socket-wrench set. These have been forged from AISI 4145 bar stock and subsequently heat treated by a quench and temper process.

   (n) Refrigerator door panels that have been fabricated from textured AISI 1008 steel sheet.

   (o) The interior and exterior surfaces of a 1000-gallon water storage tank that is fabricated by welding 5000 series aluminum plates. The water will be held at room temperature and is intended for human consumption.

(p) A standard office paper clip.

(q) A flashlight case that has been fabricated from deep-drawn yellow brass sheet.

(r) High-speed drill bits that have been fabricated from M1 tool steel.

(s) Injection-molded ABS plastic wheel covers for cars that are intended to look like chrome-plated metal.

(t) Inexpensive household scissors that have been cast from gray cast iron.

(u) The blade of a high-quality screwdriver that has been forged from AISI 1053 steel and quenched and tempered to Rockwell C 55.

(v) The exterior of high-quality, thick-walled cast aluminum cookware.

(w) A bathroom sink basin made from deep-drawn 1008 steel sheet.

(x) The body section of a child's toy wagon that has been deep-drawn from 1008 steel sheet.

Chapter 40 CASE STUDY

burrs on yo-gi's collar

The Yo-Gi Company requires 25,000 collars as shown in Figure CS-40. They are to be made from AISI 304 (18-8) stainless steel by machining on a screw machine using transfer and slotting attachments. The slotting process is creating large, thick burrs on the surface where the cutter exits the workpiece, as shown in sections A-A' (left). The part designer has recommended a redesigned part to eliminate the undercut and make the burrs more accessible—see view A-A (right). The burrs on the internal surfaces will remain, however, and are still inaccessible.

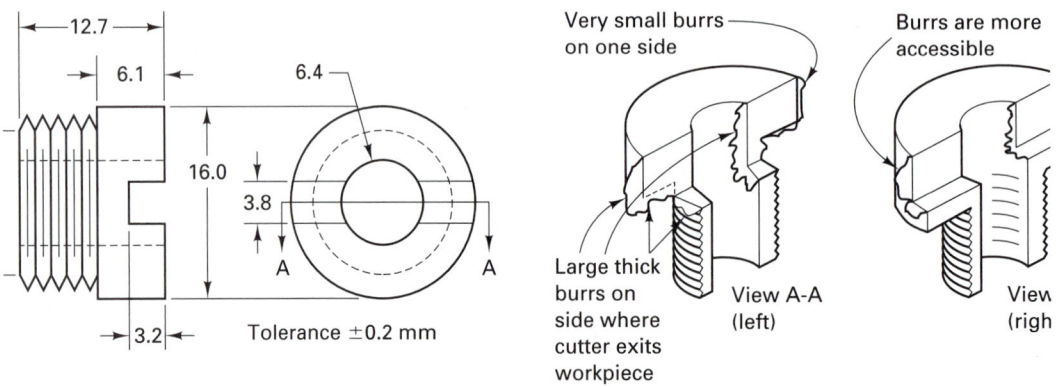

FIGURE CS-40

1. What are the methods that are commonly used in industry to remove burrs?

2. Outline the sequence of steps needed to produce this part on a screw machine from a 5/8-inch stainless steel bar in a way that would best serve to eliminate the burrs. It may be helpful to show the part progression.

3. Determine two other practical ways of making these collars. Estimate the cost of these alternative methods versus the screw machine method.

CHAPTER 41

MANUFACTURING SYSTEMS AND AUTOMATION

41.1	INTRODUCTION	Flow Shop
41.2	TRENDS IN MANUFACTURING SYSTEMS	Project Shop
		Continuous Processes
		Linked-Cell Manufacturing System
41.3	HOW THE FIRST MANUFACTURING SYSTEM EVOLVED	41.7 AUTOMATION
		Transfer Machines
41.4	SYSTEMS DEFINED	Flexible Manufacturing System
41.5	MANUFACTURING VERSUS PRODUCTION SYSTEMS	41.8 ROBOTICS
41.6	CLASSIFICATION OF MANUFACTURING SYSTEMS	Economics of Programmable Robots
		41.9 SUMMARY
	Job Shops	

■ 41.1 INTRODUCTION

In this chapter the difference between manufacturing processes and manufacturing systems are discussed. Up to this point, the book has emphasized processes and we have observed that in recent years, machine tools have become more powerful and versatile. Cutting tools have become more effective, capable of cutting at faster speeds with reduced cutting forces. Presses have become larger and tooling more sophisticated. However, in the typical factory, reducing costs by taking time out of the machining cycle (reducing machining time or tool change time) may do little to reduce overall manufacturing costs or make marked improvements in productivity. On an automatic screw machine producing a million-piece run, saving a couple of seconds per piece can be the difference between a healthy profit and a heavy loss. However, on a 50-piece lot, even a 10% reduction in machining time will have very little impact on cost for a part that takes 2 months to get through the plant. Table 41-1 compares time utilization in numerically controlled machining centers versus machining on conventional machine tools and machining in transfer machines. Observe that for machine tools used in lot production, only about one-fourth of the time, at best, is spent adding value (making chips). NC does reduce setup time, but considerable improvement can be made in the noncutting areas. Efficient utilization of the workers, materials, and machines is critical to a successful organization.

One of the reasons for this situation is the complexity of the typical manufacturing system, which requires elaborate schedules and long lead times in the planning. Superimposed on the inherent complexity are delays and disturbances caused by engineering design changes, vendor failures, material shortages, emergency orders, machine breakdowns, quality problems, and so on. The traditional way to cope with these factors has been to create a buffer at each machine to ensure that each machine has work available, but this, in turn, creates a large in-process inventory. Another strategy has been to use daily lists, colored

TABLE 41-1. Time Utilization of Conventional Equipment versus NC Machining Centers and Transfer Machines

	Percentage of Overall Machining Cycle		
Activity	Machining Center (Small to Medium-Sized Lots)	Conventional Machine (Small Lots)	Transfer Lines (Large Lots)
Metal cutting	23	20	50
Positioning, tool changing, or transfer	27	10	12
Gaging	8	15	1
Loading and unloading	10	15	—
Setup[a]	5	20	—
Waiting or idle[b]	14	13	22
Repair and technical problems	13	7	15

[a] Actual setup of a transfer line may take 6 to 24 months.

[b] Due to stockouts, personnel allowance, slow machine cycles, etc.

Source: After C. E. Carter, "Toward flexible automation," *Manufacturing Engineering*, Aug. 1982, p. 75.

tags, and people to expedite orders through the shop. Needless to say, this solution itself acts as a disturbance to the regularly scheduled work in progress. What is needed is an effective manufacturing system design that can easily be managed and organized to make sure that machines and people are utilized to the greatest possible extent. That is, more is to be gained through better overall management and planning of the manufacturing system with machines running at moderate speeds and feeds than operating at extremely fast but sporadic machining rates.

These final three chapters explain how manufacturing systems can be designed and integrated for low cost, superior quality, and on-time delivery. We begin with a discussion of the manufacturing systems, show how they are tied to production systems, and conclude with a detailed discussion of integrated manufacturing production systems.

■ 41.2 Trends in Manufacturing Systems

Significant changes are taking place in the design of manufacturing systems, fueled by the following trends:

1. The implementation of manufacturing cells as the first step in just-in-time manufacturing will increase.
2. Proliferation of the number and variety of products will continue, resulting in a decrease in quantities (lot size) as variety increases.
3. The customers desire for closer tolerances (more precision, better quality), reduced cost, and on-time delivery will continue to increase.
4. Variety in *materials*, including composite materials with widely diverse properties, will continue to increase, causing further proliferation of the manufacturing processes.
5. The cost of materials, including materials handling and energy, will continue to be a major part of the total product cost, while direct labor will be 5 to 10% of the total. That is, the number of direct laborers used in manufacturing will continue to decrease.
6. Product reliability will increase in response to the excessive number of product liability suits.
7. The time between design of a product and its manufacture will continue to decrease.

In general, a manufacturing system should be an integrated whole, composed of integrated subsystems, each of which interacts with the entire system. The system will have a number of objectives and its operation must optimize the whole. Optimizing pieces of the system (i.e., the processes or the subsystems) does not optimize the entire system. System operation requires information gathering and communication with decision-making processes that are integrated into the manufacturing system.

Each company will have many differences, resulting from differences in subsystem combinations, people, product design, and materials. From company to company, the differing interactions of the social, political, and business environments make each company unique, with its own set of problems. Clearly, there is a danger in grouping all the companies on a functional or departmental basis. In this chapter we extract some general principles. Let us begin with a historical perspective.

■ 41.3 How the First Manufacturing System Evolved

In the first industrial revolution, basic machine tools were invented and developed. With them came the first levels of mechanization and automation. Factories developed along with the manufacturing processes. Factories were needed to focus the resources (materials, workers, and processes) at the site where power was available. These early plants were arranged according to the kinds of machines, that is, functionally, because the machines were an extension of some human capability or human attribute. A machinist developed different skills from those of a leatherworker, an ironworker, or a foundry worker. The processes were divided according to the kinds of skills needed to operate the processes.

The early factories were placed beside rivers where water was used to turn waterwheels, which in turn drove overhead shafts that ran the length of the factory. A belt from the main shaft powered each machine. The grouping of similar machines that needed to run at about the same speed was logical and expedient. Steam engines and, later, electric motors replaced other types of machine power and greatly increased manufacturing system flexibility. However, the functional arrangement persisted and became known as the job shop. Thus the design of the first manufacturing system was based on functionality.

As product complexity increased and the factory grew larger, separate departments evolved for product design, accounting (bookkeeping), and sales. Later in the scientific management era of Taylor and Gilbreth,[*] departments for production planning, work scheduling, and methods improvement were added to what became the *production system*. The production system includes and services the manufacturing system. See Chapters 1 and 42. Because the production system was composed of functional areas to serve the functionally designed job shop, it naturally evolved with a functional structure.

■ 41.4 Systems Defined

The best definition of the concept *system* comes from Rubinstein's *Patterns of Problem Solving*[**] and is quite appropriate for manufacturing systems. The word *system* is used to define abstractly a relatively complex assembly (or arrangement) of physical elements characterized by *measurable parameters*. To model the system:

1. The system's boundaries or constraints must be defined.

---

[*] F. W. Taylor and Frank and Lillian Gilbreth are generally recognized as the founders of the industrial engineering profession.

[**] Moshe F. Rubinstein, *Patterns of Problem Solving*, Prentice Hall, Englewood Cliffs, N.J., 1975.

2. The system's behavior in response to excitations or disturbances from the environment must be predictable through its parameters.

Models are used to describe how the system works. Mathematical models for use in a control computer generally require a "theory" or equations that describe the system's boundaries and behavior through its input parameters. In short, if no theory exists, the model is not viable and the system is not controllable.

Manufacturing systems modeling and analysis are desirable but difficult because:

1. Objectives are difficult to define, particularly in systems that impact on social and political issues. Goals may conflict.
2. The data or information may be difficult to secure, inaccurate, conflicting, missing, or even too abundant to digest.
3. Relationships may be awkward to express in analytical terms, and interactions may be nonlinear; thus well-behaved functions embodied in many analysis tools do not apply.
4. System size may inhibit analysis due to implied time expenditures.
5. Systems are always dynamic and change during analysis. The environment can change the system, or vice versa.
6. All systems analysis will be subject to errors of omission and commission. Some of these will be related to breakdowns or delays in feedback elements because manufacturing systems include people in information loops.

Because of these difficulties, digital simulation is a widely used technique for manufacturing systems modeling and analysis as well as for systems design.

■ 41.5 MANUFACTURING VERSUS PRODUCTION SYSTEMS

As shown in Figure 41-1, inputs to the manufacturing system include materials, information, and energy. The system is a complex set of elements that includes machines (or machine tools), people, materials-handling equipment, and tooling. The materials are processed within the system and gain value as they are passed from machine to machine. Manufacturing system outputs may be either consumer goods or inputs to some other process (producer goods). Manufacturing systems are very interactive and dynamic. They involve many people who carry the process know-how. Change seems to be the only constant in this system.

Figure 41-1 gives a general definition for any manufacturing system. Observe that many of the inputs cannot be fully controlled (by management), and the effect of the disturbances must be counteracted by manipulating the controllable inputs or the system itself. Controlling material availability or predicting demand fluctuations may be difficult. The national economic climate can cause shifts in the business environment which can seriously change any of these inputs. In other words, in manufacturing systems, not all inputs are fully controllable.

The manufacturing systems themselves differ in structure or physical arrangement (design), as we shall see in the next section. However, all the manufacturing systems are serviced by a production system. Figure 41-2 shows this schematically. Because the oldest and most common manufacturing system is functionally organized, most production systems are functionally organized. Walls usually separate the people in these functional areas from all other areas. Breakdowns in communication links are common. Long lags in feedback loops result in manufacturing system problems.

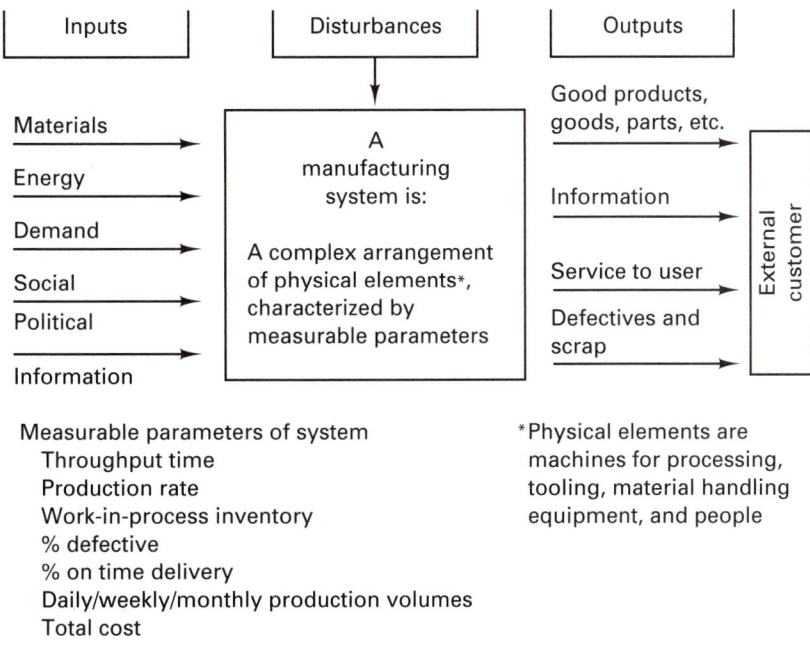

FIGURE 41-1   Generalized view of a manufacturing system with its input/output conditions.

■ 41.6 CLASSIFICATION OF MANUFACTURING SYSTEMS

Table 41-2 shows manufacturing industries which differ by the products that they make or assemble. Table 41-3 provides a partial list of service industries. Examination of all these industries reveals four kinds of classical manufacturing systems (or hybrid combinations thereof) and one new manufacturing system that is rapidly gaining acceptance in almost all of these industries. Table 41-4 lists examples of five types of manufacturing systems. More examples can be added to Tables 41-2 and 41-3. Neither list is meant to be complete. The classical systems here are the *job shop*, the *flow shop*, the *project shop*, and the *continuous process*. The *linked-cell* is a new kind of system. The *assembly line* is a form of the flow shop, and a system that has *batch flow* might also be added as another manufacturing system (MS). Figure 41-3 shows schematics of the four classical systems. *The linked-cell manufacturing system* is discussed later.

Job Shops

The *job shop*'s distinguishing feature is its functional design. In the job shop, a variety of products is manufactured, which results in small manufacturing lot sizes, often one of a kind. Job shop manufacturing is commonly done to specific customer order, but in truth, many job shops produce to fill finished-goods inventories. Because the plant must perform a wide variety of manufacturing processes, general-purpose production equipment is required. Workers must have relatively high skill levels to perform a range of different work assignments. Job shop products include space vehicles, aircraft, machine tools, special

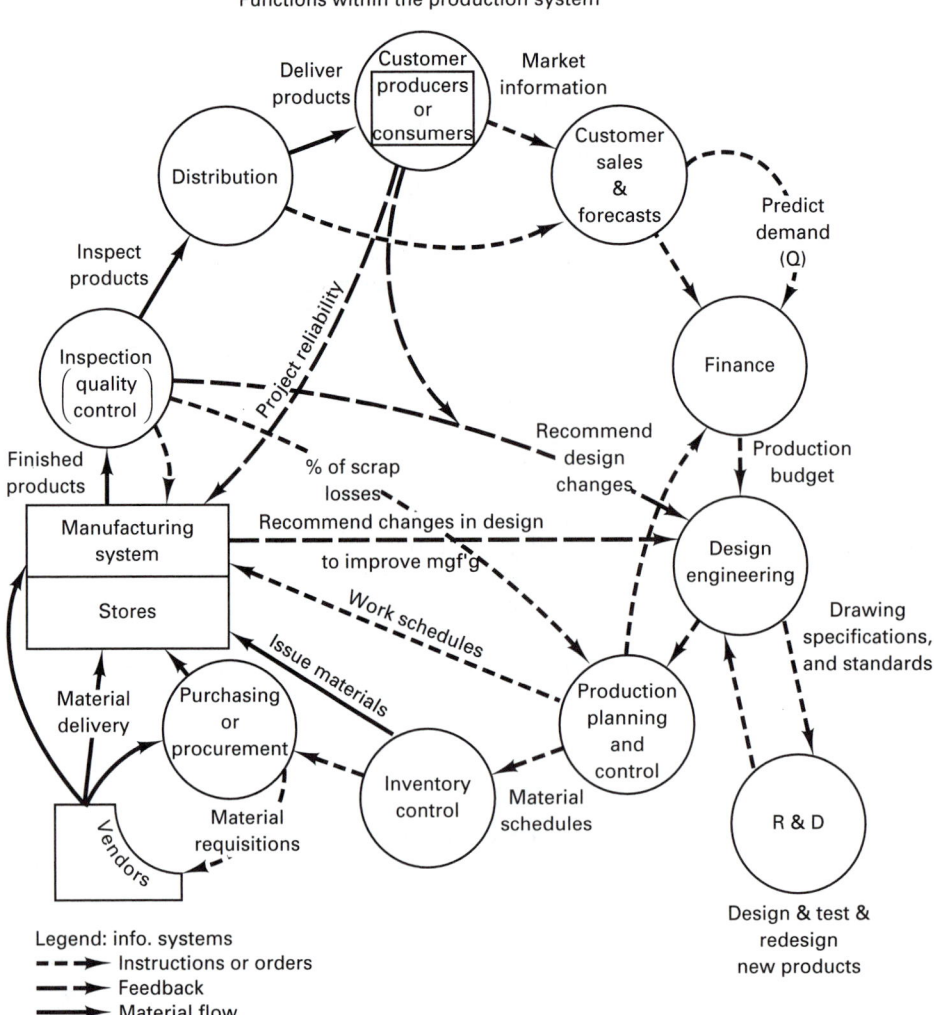

FIGURE 41-2 The production system includes the manufacturing system. The former is composed of many functional departments connected to one another by formal and informal information systems designed to service the manufacturing system that "makes the products."

tools, and equipment. Figure 41-4 depicts the functionally arranged job shop. Production machines are grouped according to the general type of manufacturing process. The lathes are in one department, drill presses in another, plastic molding is still another, and so on. The advantage of this layout is its ability to make a wide variety of products. Each different part requiring its own unique sequence of operations can be routed through the respective departments in the proper order. *Route sheets* are used as the production control device to control the movement of the material through the manufacturing system. Forklifts and handcarts are used to move materials from one machine to the next. As the company grows, the job shop evolves into a production job shop (PJS).

The production job shop becomes extremely difficult to manage as it grows, resulting

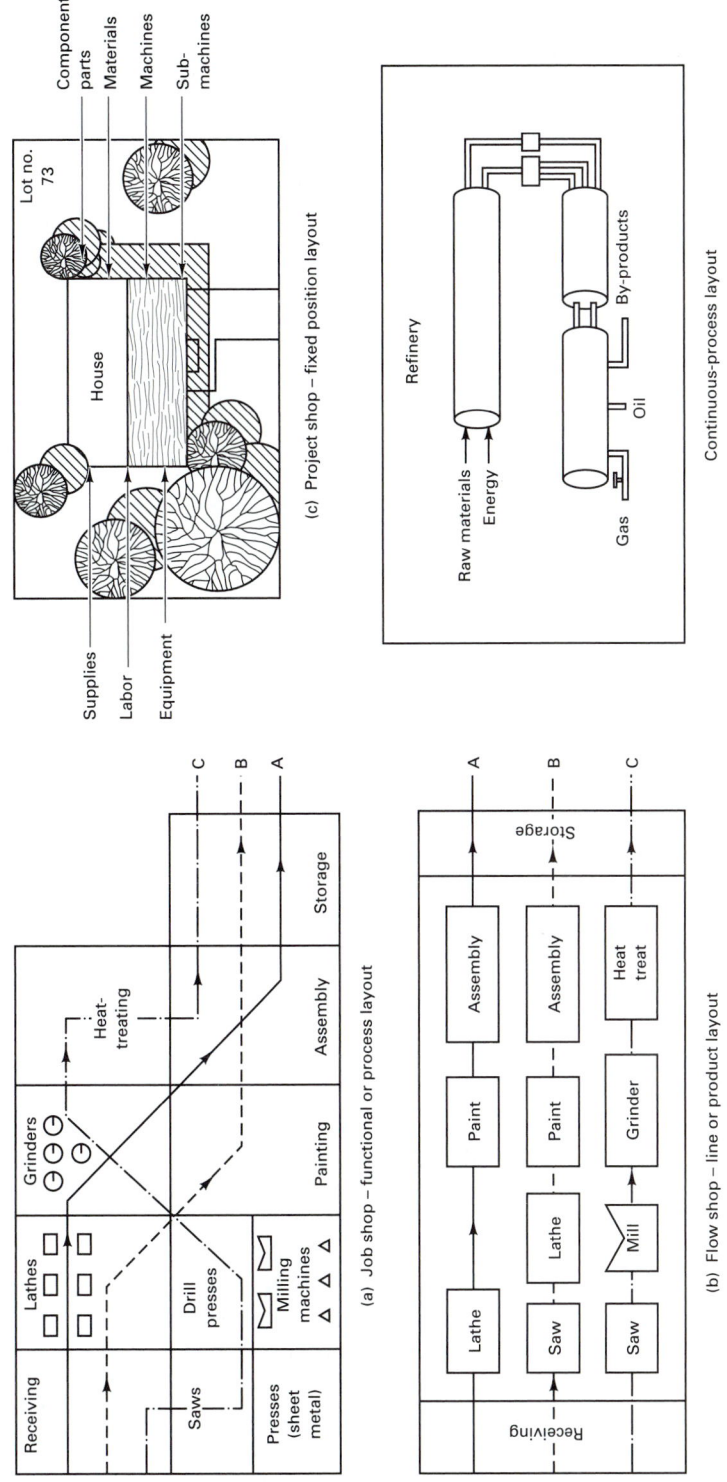

FIGURE 41-3 Four classical designs of manufacturing systems.

TABLE 41-2. Partial List of Producers, Consumer Goods, and Industries

Aerospace and airplanes	Foods (canned, dairy, meats, etc.)
Appliances	Footwear
Automotive (cars, trucks, vans, wagons, etc.)	Furniture
Beverages	Glass
Building supplies (hardware)	Hospital supplies
Cement and asphalt	Leather and fur goods
Ceramics	Machines
Chemicals and allied industries	Marine engineering
Clothing (garments)	Metals (steel, aluminum, etc.)
Construction	Natural resources (oil, coal, forest, pulp and paper)
Construction mats (brick, block, panels)	Publishing and printing (book, records, newspapers)
Drugs, soaps, cosmetics	Restaurants
Electrical and microelectronics	Retail (food, department store, etc.)
Energy (power, gas electric)	Ship building
Engineering	Textiles
Equipment and machinery (agricultural, construction	Tire and rubber
and electrical products, electronics, household products,	Tobacco
industrial machine tools, office equipment, computers,	Transportation vehicles (railroad, airline, truck, bus)
power generation)	Vehicles (bikes, cycles, ATVs, snowmobiles)

TABLE 41-3. Types of Service Industries

Advertising and marketing
Communication (telephone, computer networks)
Education
Entertainment (radio, TV, movies, plays)
Equipment and furniture rental
Financial (banks, investment companies, loan companies)
Health care
Insurance
Transportation and car rental
Travel (hotel, motel)

in long product throughput times and very large in-process inventory levels. In the job shop, parts spend 95% of the time waiting (delay) or being transported and only 5% of the time on the machine (Figure 41-4). Of the time during which parts are on the machine, only 20 to 30% involves machining, as was shown in Table 41-1. Thus the time spent actually adding value may only be 2 or 3% of the total time available.

The PJS manufacturing system builds large volumes of products but still builds in lots or batches, usually medium-sized lots of 50 to 200 units. The lots may be produced only once, or they may be produced at regular intervals. The purpose of batch production is often to satisfy continuous customer demand for an item. This system usually operates in the following manner. Because the production rate can exceed the customer demand rate, the shop builds an inventory of the item, then changes over the machines to produce other products to fill other orders. This involves tearing down the setups on many

TABLE 41-4. Types and Examples of Manufacturing Systems

Type of Manufacturing System	Examples	
	Service[a]	Product[b]
Job shop	Auto repair Hospital Restaurant University	Machine shop Metal fabrication Custom jewelry FMS
Flow shop or flow line	X-ray Cafeteria College registration Car wash	TV factory Auto assembly line
Cellular shop[c]	Fast-food restaurant (Kentucky Fried Chicken, Wendy's) Food court at mall 10-minute oil change	Family of "turned" parts Composite part families Design families Manned cells Robotic cells
Project shop	Producing a movie Broadway play TV show	Locomotive assembly Bridge construction House construction
Continuous process	Telephone company Phone company	Oil refinery Chemical plant

[a]Customer receives a service or perishable product.

[b]Products can be for customers or for other companies.

[c]New design of manufacturing system.

machines and resetting them for new products. When the stock of the first item becomes depleted, production is repeated to build the inventory again.

Some machine tools are designed for higher production rates. For example, automatic lathes capable of holding many cutting tools can have shorter processing times than engine lathes. The machine tools are often equipped with specially designed workholding devices, jigs, and fixtures, which increase process output rate, precision, accuracy, and repeatability.

Industrial equipment, furniture, textbooks, and components for many assembled consumer products (household appliances, lawn mowers, and so on) are made in production job shops. Such systems are called machine shops, foundries, plastic molding factories, and pressworking shops.

As much as 75% of all piece-part manufacturing is estimated to be in lot sizes of 50 pieces or fewer. Hence job shop production constitutes an important portion of total manufacturing activity.

Flow Shop

The flow shop has a product-oriented layout composed mainly of flow lines. When the volume gets very large, especially in an assembly line, this is called *mass production*. This kind of system can have (very) high production rates. Specialized equipment, dedicated to the manufacture of a particular product, is used. The entire plant may often be designed exclusively to produce the particular product or family of products, using special-purpose

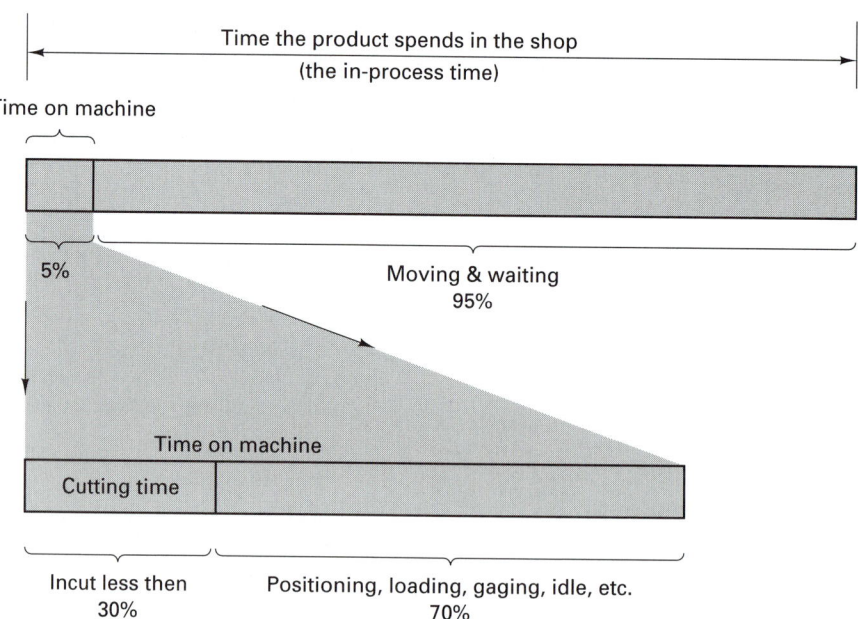

FIGURE 41-4 In the job shop, the parts spend most of the time waiting or moving and only 5% on the machine. *(Courtesy of C. F. Carter, Cincinnati Milacron.)*

rather than general-purpose equipment. The investment in the specialized machines and specialized tooling is high. Many production skills are transferred from the operator to the machines so that the manual labor skill level in a flow shop tends to be lower than in a production job shop. Items are made to "flow" through a sequence of operations by materials-handling devices (conveyors, moving belts, and transfer devices). The items move through the operations one at a time. The time the item spends at each station or location is fixed and equal (balanced).

Most factories are mixtures of the job shop with flow lines. Obviously, the demand for products can precipitate a shift from batch to high-volume production, and much of the production from these plants is consumed by that steady demand. Subassembly lines and final assembly lines are further extensions of the flow line, but the latter are usually much more labor intensive.

In the flow line manufacturing system, the processing and assembly facilities are arranged in accordance with the product's sequence of operations. Work stations or machines are arranged in line with only one work station of a type, except where duplicates are needed for balancing the time products take at each station. The line is organized by the processing sequence needed to make a single product or a regular mix of products. A hybrid form of the flow line produces batches of products moving through clusters of work stations or processes organized by product flow. In most cases the setup times to change from one product to another are long and often complicated.

Various approaches and techniques have been used to develop machine tools that would be highly effective in large-scale manufacturing. Their effectiveness was closely related to the degree to which the design of the products was standardized and the time over which no changes in the design were permitted. If a part or product is highly standardized

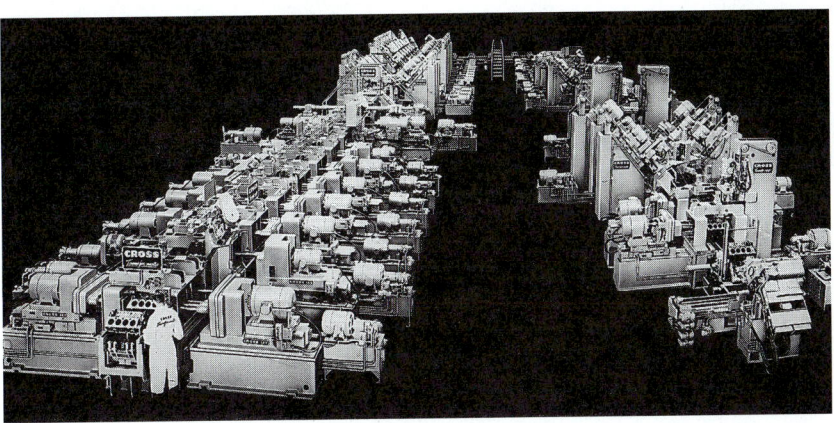

FIGURE 41-5   The transfer machine is an example of a flow line manufacturing system.   *(Courtesy of The Cross Company).*

and will be manufactured in large quantities, a machine that will produce the parts with a minimum of skilled labor can be developed. A completely tooled automatic screw machine is a good example for small parts.

On a much larger scale, Figure 41-5 shows an example of an automated transfer machine for the production of V-8 engine blocks at the rate of 100 per hour. There are five independent sections that perform 265 drilling, 6 milling, 21 boring, 56 reaming, 101 counterboring, 106 tapping, and 133 inspection operations. These are performed at 104 stations, including 1 loading, 53 machining, 36 visual inspection, and 1 unloading. Provision is made for banking parts between each section. However, such specialized machines are expensive to design and build and may not be capable of making any other products. Consequently, to be economical, such machines must be operated for considerable periods of time to spread the cost of the initial investment over many units. These machines and systems, although highly efficient, can be utilized only to make products in very large volume. Changes in design in the products are often avoided or delayed because it would be too costly to change the process or to scrap the machines.

However, as we have already noted, products manufactured to meet the demands of the free-economy, mass-consumption markets need to have changes in design for improved product performance as well as style changes. Therefore, hard automation systems need to be as flexible as possible while retaining the ability to mass-produce. This has led to the following developments in transfer lines.

First, the machines are constructed from basic building blocks or modular units that accomplish a function rather than produce a specific part. The production machine modules are combined to produce the desired system for making the product. These units are called self-contained power-head production machines. Such machines are then connected by automatic transfer devices to handle the material (i.e., the thing being produced), moving it automatically from one process to the next.

Second, the incorporation of *programmable logic controllers* (PLCs) and feedback control devices have made these machines more flexible. Modern PLCs have the functional sophistication to perform virtually any control task. These devices are rugged, reliable, easy to program, and economically competitive with any alternative control device, and they have replaced conventional hard-wired relay panels in many applications. Relay panels are hard to reprogram, whereas PLCs are very flexible. Relays have the advantage of being well understood by maintenance people and are invulnerable to electronic noise, but construction time is long and tedious. PLCs allow for mathematical algorithms to be included in the closed-loop control system and are being widely used for single-axis, point-to-point control as typically required in straight-line machining, robot handling, and robot-assembly applications. They do not at this time challenge computer numerical controls (CNCs) used on multiaxis contouring machines. However, PLCs are used for monitoring temperature, pressure, and voltage on such machines. PLCs are used on transfer lines to handle complex material movement problems, gaging automatic tool setting, on-line tool wear compensation, and automatic inspection, giving these systems flexibility that they never had before.

Third, the transfer line was combined with NC machines to form *flexible manufacturing systems* (FMSs). These systems are discussed further later in this chapter.

Project Shop

In the typical project manufacturing system, a product must remain in a fixed position or location during manufacturing because of its size and/or weight. The materials, machines, and people used in fabrication are brought to the site. Locomotive manufacturing, large aircraft assembly, and shipbuilding use fixed-position *layout*. Fixed-position *fabrication* is also used in construction jobs (buildings, bridges, and dams). As with the fixed-position layout, the product is large, and the construction equipment and workers must be moved to it. When the job is completed, the equipment is removed from the construction site. The project shop invariably has a job shop/flow shop manufacturing system making all the components for the large complex project and thus has a functionalized production system.

Continuous Processes

In the *continuous process*, the project physically flows. Oil refineries, chemical processing plants, and food-processing operations are examples. This system is sometimes called *flow production* when the manufacture of either complex single parts (such as a canning operation or automotive engine blocks) or assembled products (such as television sets) is described. However, these are not continuous processes but simply high-volume flow lines. In continuous processes, the products really do flow because they are liquids, gases, or powders. Continuous processes are the most efficient but least flexible kinds of manufacturing systems. They usually have the leanest, simplest production systems because these manufacturing systems are the easiest to control, having the least work-in-process (WIP).

Linked-Cell Manufacturing System

A cellular manufacturing system composed of linked cells is the newest manufacturing system. In cells, processes are grouped according to the sequence and operations needed to make a product. This arrangement is much like that of the flow shop but is designed for flexibility. The cell is designed in a *U-shape* so that the workers can move from machine to machine, loading and unloading parts. Figure 41-6 shows an example of a manned cell. The cell has one worker who can make a walking loop around the cell in 110 seconds. See the breakdown of time in the table to the right. The machines in the cell are all of the A(2) or higher level of automation, so that they can complete the desired processing untended, turning themselves off when done with a machining cycle. The operator comes to a machine, unloads a part, checks the part, loads a new part into the machine, and starts the machining cycle. The cell usually includes all the processing needed for a complete part or subassembly.

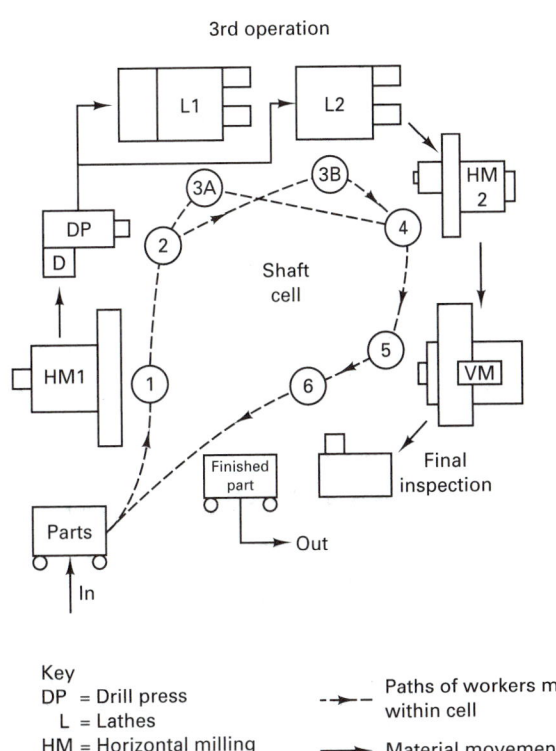

Work sequence	Name of operation	Time		
		Manual	Walking	Machine
1	Mill ends on work on HM (1)	12"	5"	30"
2	Drill hole on DP	15"	5"	20"
3A or 3B	Turn – bore on L1 or L2	13"	5" 8"	180"
4	Mill flats on HM2	12"	8" 5"	20"
5	Mill steps on VM	13"	7"	30"
6	Final inspect	10"	5"	
		75"	35"	280"

Cycle time = 75 sec + 35 sec = 110 seconds

Longest machining time = 180 seconds

Total machining time = 280 seconds

All machines in the cell are capable of running untended while the operator(s) are doing manual operations (unload, load, inspection, deburr) or walking from machine to machine. The time to change tools and workholders (perform setup) is not shown.

Key

DP = Drill press

L = Lathes

HM = Horizontal milling machine

VM = Vertial milling machine

- - -▸ - - Paths of workers moving within cell

───▸ Material movement paths

① Operation sequence

FIGURE 41-6 Example of a manned manufacturing cell with six machines and one multifunctional worker. The lathe operation is duplicated.

The machining time for the third operation is greater than the necessary cycle times. That is, 180 seconds > 110 seconds. Therefore, this operation is duplicated and the operator alternates between the two lathes, visiting each lathe every other trip he or she makes around the cell.

Cells are typically manned, but unmanned cells are beginning to emerge with a robot replacing the worker. A *robotic cell* design is shown in Figure 41-7 with one robot and four CNC machines. For the cell to operate autonomously, it must have adaptive control capability, that is, A(5) level of automation.

To form a linked-cell manufacturing system, the first step is to restructure portions of the job shop, converting it in stages into manned cells. At the same time, the flow shop parts are reconfigured into U-shaped cells as well (Figure 41-8). The flow shop manufacturing within the plant can be redesigned to make these systems operate like cells. To do this, the long setup times typical in flow lines must be vigorously attacked and reduced so that the flow lines can be changed quickly from making one product to making another. The need to line balance the flow line must be eliminated. This is accomplished by using standing, walking workers who are capable of operating multiples of operations. This makes them flexible and compatible with the cells designed to make piece parts and with

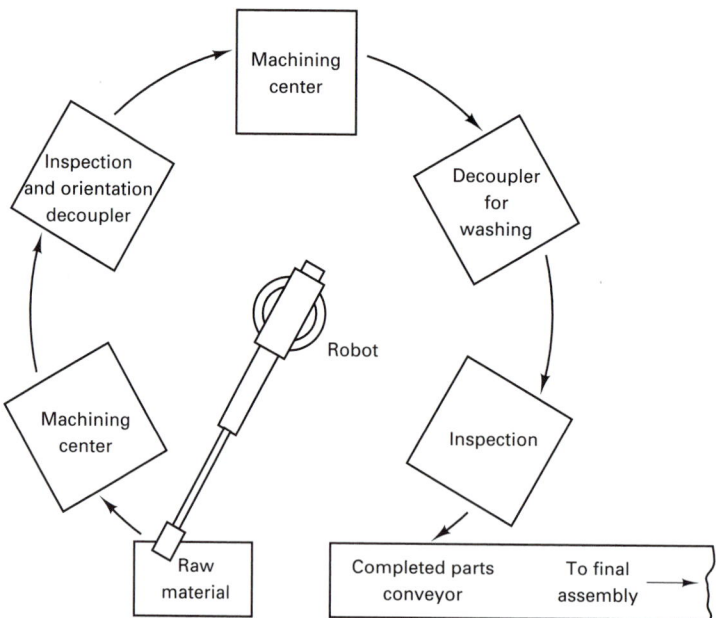

FIGURE 41-7 Example of U-shaped robotic cell. Decoupler checks part, reorients part for next machine, and acts as Kanban element in cell.

the subassembly lines and final assembly lines. Cells are designed to manufacture specific groups or families of parts. As described in Chapter 43, cells are linked either *directly* to each other or to subassembly points, or *indirectly* by the pull system of production control, called *kanban*.

Cells have many features that make them unique and different from other manufacturing systems. Parts move from machine to machine *one at a time* within the cell. For material processing, the *machines are typically A(2) or higher,* capable of completing a machining cycle initiated by a worker. The U shape puts the start and finish points of the cell next to each other. Every time the operator completes a walking trip around the cell, a part is completed. This time defines the cycle time (CT). The machining time (MT) for each machine needs only to be less than the time it takes for the operator to complete the walking trip around the cell. As shown in Figure 41-9, *machining processes are overlapping and need not be equal (balanced)* as long as no MT is greater than the CT. The cycle time is 150 seconds. The total machine time was $60 + 80 + 100 + 40 + 20 = 300$ seconds. A NC machining center capable of performing the five machining operations could easily replace the cell. However, the cycle time for a part would jump from 150 seconds to over 300 seconds because combining the processes into one machine prevents *overlapping of the machining times.* The CT for the cell can readily be altered by adding (a portion of) an additional worker to the cell (see Chapter 43). Figure 41-9b is a simple standard operations routine sheet, used to plan the manufacture of one of the parts in the family within the cell. The plan shows the relationships between the manual operations performed by the worker, the machining operations performed by the machine, and the time spent by the worker walking from machine to machine. The manual operations include loading and unloading the machine, checking quality, deburring, and taking chips out of fixtures. The cycle time is determined by the need of the system for the parts made by the cell.

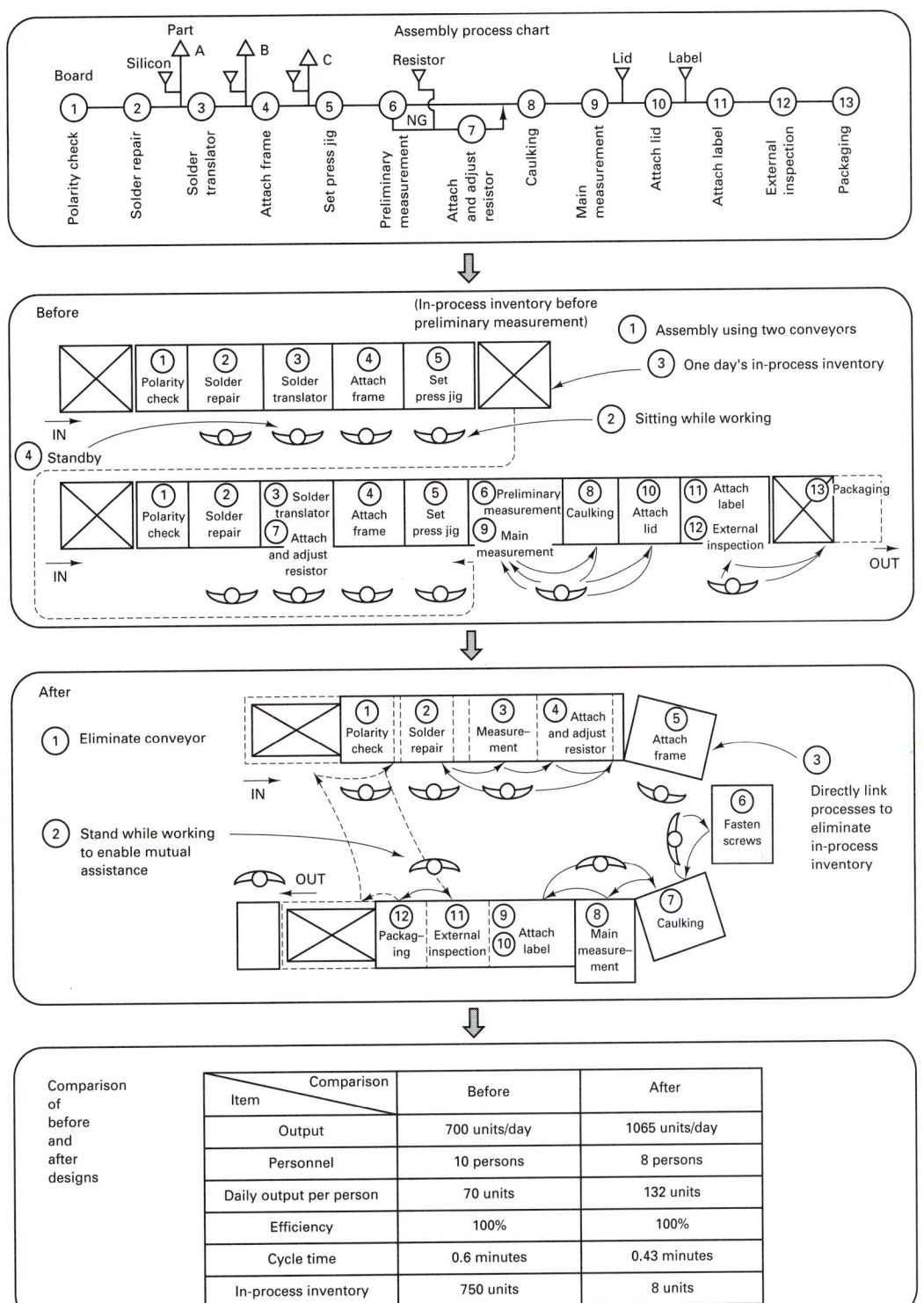

FIGURE 41-8   How a flow line can be reconfigured into a cell with standing, walking workers. *Source*: "One Piece Flow" by K. Sekine, Productivity Press.

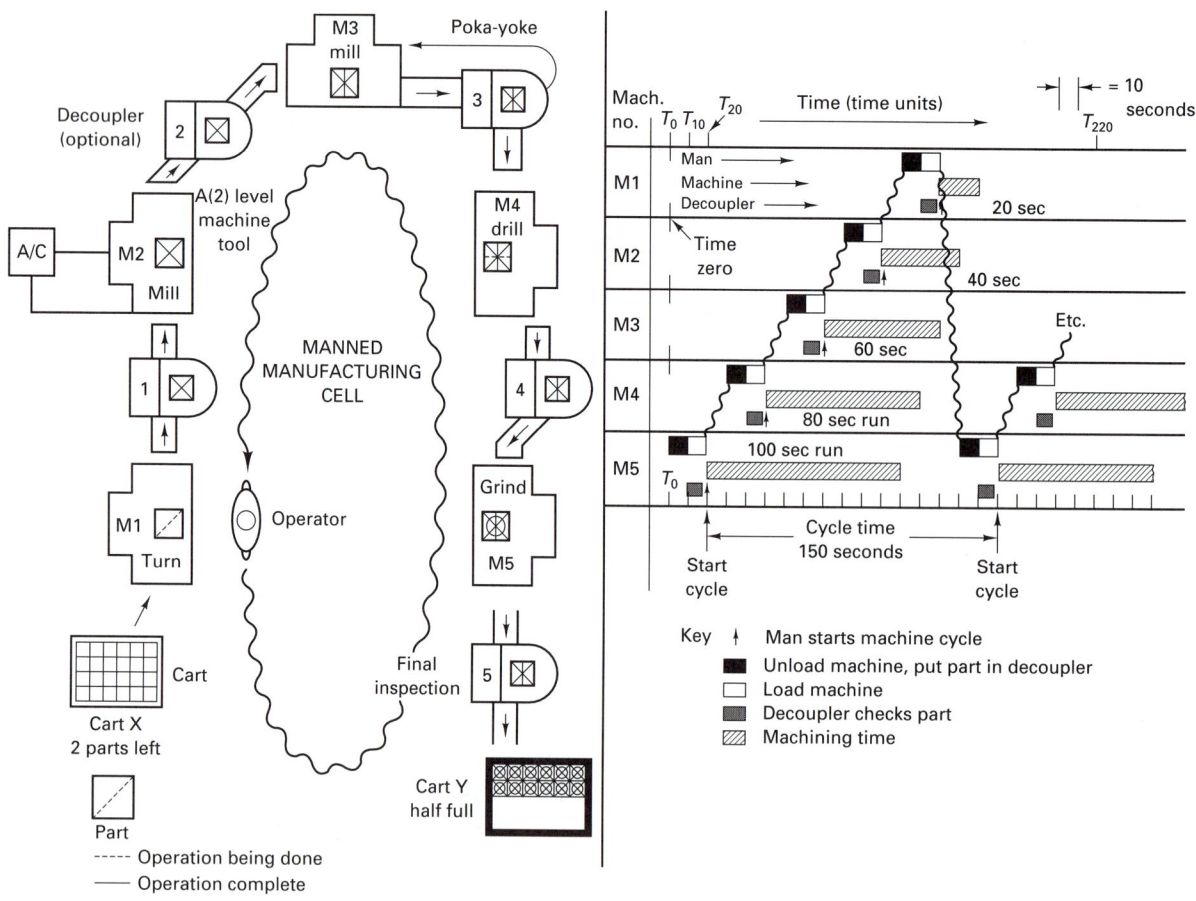

FIGURE 41-9 (a) Layout and (b) time bar diagram of a five-machine, U-shaped cell with decouplers, operator upstreaming, and variable machine times (MT).

Workers in the cells are multifunctional; each worker can operate more than one kind of process (or multiple versions of the same process) and also performs inspection and machine maintenance duties. Cells eliminate the job shop concept of one person/one machine and thereby greatly increase worker productivity and utilization. The restriction of the cell to a parts family makes reduction of setup in the cell possible. This is often called flexible fixturing. The general approach to setup reduction is discussed in Chapter 43.

In some cells *decouplers* are placed between the processes, operations, or machines to provide flexibility, part transportation, inspection for defect prevention (pokeyoka) and quality control, and process delay for the manufacturing cell. For example, referring again to Figure 41-9, notice the pokayoke arrangement for M3 mill and decoupler 3. The decoupler inspects the part for a critical dimension and feeds back adjustments to the machine to prevent the machine from making oversize parts as the milling cutter wears. A process delay decoupler would delay the part movement to allow the part to cool down, heat up, cure, or whatever is necessary for a period of time greater than the cycle time for the cell. Decouplers and flexible fixtures are vital parts of unmanned cells.

Table 41-5 provides a summary of the chief characteristics of the five basic manufacturing systems. Figure 41-10 summarizes the discussion on manufacturing systems by

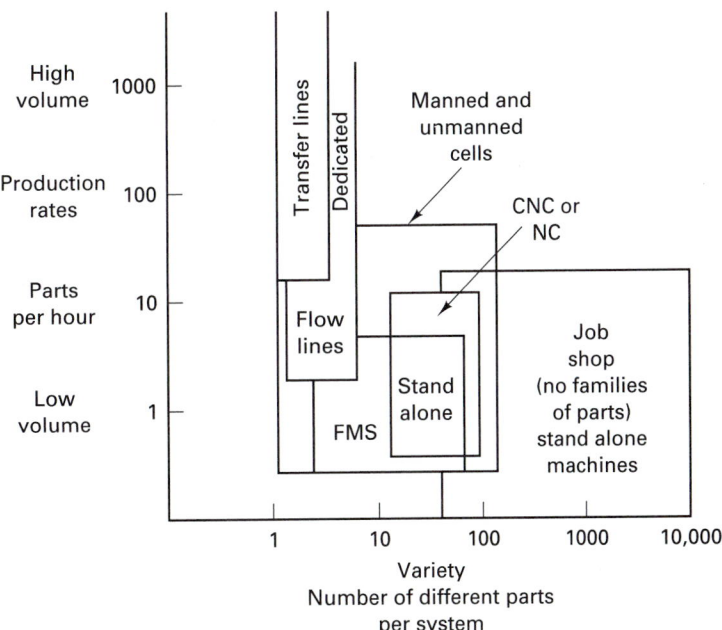

FIGURE 41-10   Part variety versus production rates for various manufacturing systems.

comparing different systems based on their production rate and product flexibility, that is, the number of different parts the system can handle. The project shop and continuous processes are not shown. The widely publicized Flexible Manufacturing System (FMS) is often operated like a job shop, wherein a random order of part movement is permitted and therefore scheduling of parts and machines within the FMS is necessary, just as it is in the job shop. This feature can make FMS designs difficult to link to the rest of manufacturing systems. Additional material on manufacturing cells and L-CMS can be found in Chapter 43.

Let us summarize the characteristics of manufacturing cells. The cell makes parts one at a time in a flexible design. Cell capacity (the cycle time) can be altered quickly to respond to changes in customer demand. The cycle time does not depend on the machining time.

Families of parts with similar designs, flexible workholding devices, and tool changers in programmable machines allow rapid changeover from one component to another. Rapid changeover means that quick or one-touch setup is employed, often like flipping a light switch. Significant inventory reductions between the cells is possible, and the inventory level can be directly controlled. Quality is controlled within the cell and the equipment within the cell is maintained routinely by the workers.

For robotic (unmanned) cells, the robot typically loads and unloads parts for one to five CNC machine tools, but this number may be increased if the robot can become mobile. A machining center can do the same sequence of steps but is not as flexible as a cell composed of multiple simple machines. Cellular layouts facilitate the integration of critical production functions while maintaining flexibility in producing superior-quality products.

TABLE 41-5. Characteristics of Basic Manufacturing Systems					
Characteristics	Job Shop	Flow Shop	Project Shop	Continuous Process	Linked Cells
Types of machines	Flexible, general purpose	Single purpose; single function	General purpose; mobile, manual automation	Specialized, high technology	Simple, customized single-cycle
Design of processes	Functional or process	Product flow layout	Project or fixed-position layout	Product	U-shaped, sequenced to flow for family of parts, overlapping
Setup time	Long, variable, frequent	Long and complex	Variable, every job	Rare and expensive	Short, frequent, one touch
Workers	Single functioned; highly skilled: one worker–one machine	One function; lower-skilled one worker–one machine	Specialized, highly skilled	Sill level varies	Multifunctional, respected
Inventories (WIP)	Large inventory to provide for large variety	Large to provide buffer storage	Variable, usually large	Very small	Small
Lot sizes	Small to medium	Large lot	Small lot	Very large	Small
Manufacturing lead time	Long, variable	Short, constant time	Long, variable lead	Very fast, constant	Short, constant

The product designer can easily see how parts are made in the cell since all the processes are together. Because quality-control techniques are also integrated into the cells, the designer knows exactly the cell's process capability. The designer can easily configure the future designs to be made in the cell. This is truly designing for manufacturing.

■ 41.7 AUTOMATION

The term *automation* has many definitions. Apparently, it was first used in the early 1950s to mean automatic handling of materials, particularly equipment used to unload and load stamping equipment. It has now become a general term referring to services performed, products manufactured and inspected, information handling, materials handling, and assembly, all done automatically (i.e., as an automatic operation).

In Chapter 1, Amber and Amber's *yardstick for automation* was presented. This classification is based on the concept that all work requires energy and information, and certain functions must be provided by worker or by machine. Whenever a machine replaces a human function or attribute, it is considered to have taken an "order" of automaticity. It today's factory, levels A(2), A(3), A(4), and A(5) are found, with occasional A(6) levels. This portion of the yardstick is given in greater detail in Figure 41-11. *Mechanization* refers to the first and second orders of automaticity, which includes semiautomatic machines. Virtually all of the machine tools described in previous chapters are A(2) machines. The machines used in manned manufacturing cells are basically A(2) or A(3) machines.

Automation as we know it today begins with the A(3) level. In recent years, this level has taken on two forms: *hard (or fixed-position) automation* and *soft (or flexible or programmable) automation*. Instructions to the machine, telling it what to do, how to do it, and when to do it are called the *program*. In hard automation transfer lines, the programming consists of cams, stops, slides, and hard-wired electronic circuits using relay logic. An example of

ORDER OF AUTOMATICITY	HUMAN ATTRIBUTE MECHANIZED	DISCUSSION	EXAMPLES

A (3)
 	
Automatic repeat cycle
 	
or
 	
Open loop control
	DILIGENCE
	
	Carries out routine instructions without aid of man
	
	Open end or nonfeedback
		All automatic machines
		  Loads, processes,
		    unloads, repeats–
		System assumed to be doing okay. Probability of malfunctions negligible
		Obeys fixed internal commands or external program
			• Record player with changer
			• Automatic screw mach.
			• Bottling machines
			• Clock works
			• Donut maker
			• Spot welder
			• Engine production lines
			• Casting lines
			• Newspaper printing machines
			• Transfer machines

A (4) Self-measuring and adjusting feedback or closed loop systems	JUDGEMENT Measures and compares result (output) to desired size or position (input) and adjusts to minimize any error	Self-adjusting devices Feedback from product position, size, velocity, etc. 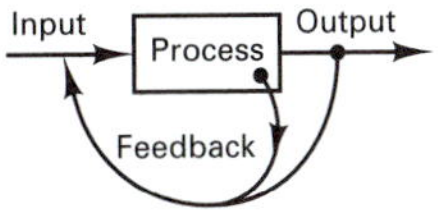 Multiple loops are possible	• Product control • Can filling • N/C machine tool with position control • Self-adjusting grinders • Windmills • Thermostats • Waterclock • Fly ball governor on steam engine
A (5) Adaptive or computer control: Automatic cognition	EVALUATION Evaluates multiple factors on process performance, evaluates and reconciles them Use mathematical algorithms	Process performance must be expressed as equation 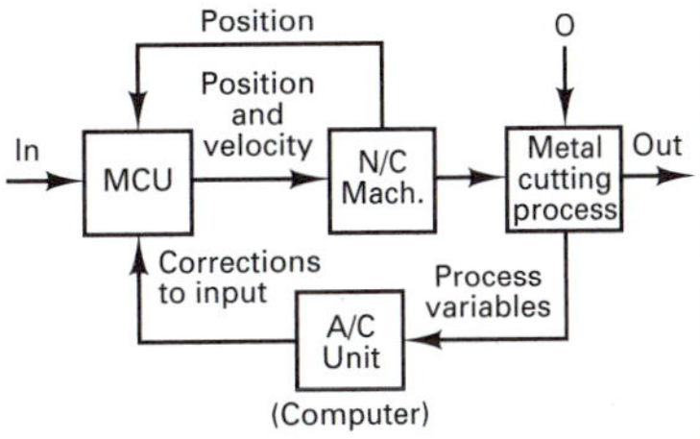 	N/C machine with A/C capability Maintaining pH level Turbine fuel control
A (6) Expert systems or limited self-programming	LEARNING BY EXPERIENCE	Subroutines are a form of limited self-programmming Trial and error sequencing Develops history of useage	Phone circuits Elevator dispatching

FIGURE 41-11   A portion of the Yardstick for Automation given in Chapter 1, highlighting levels A(3) and A(6).

this level of automation is the automatic screw machine. If the machine is programmed with a tape, a programmable controller (PC), a hand-held control box, microprocessor, or computers, control instructions are easily changed, with the software making the system or device much more adaptable.

Self-adjusting and measuring machines are of the A(4) level, replacing human adjustment and allowing these machines to be self correcting. This is commonly known as *feedback control* or *closed-loop operation*, meaning that information about the performance is fed back (or looped back) into the process. Examples of A(4) machines were given in Chapter 29. A simple A(4) level is one in which some parameter of the process is measured using a detection device (sensor). This information is fed back to a comparator, which makes comparisons with the desired level of operation. If the output and input are not equal, an error signal is created and the process adjusts to reduce the error. An automatic grinder that checks the diameter of a part and automatically repositions the grinding wheel to compensate for wheel wear would be an example of an A(4) machine.

The A(5) level, typified by *adaptive control* (AC), emulates human evaluation (refer to Figures 29-6 and 41-11). Basically, A(5) machines are capable of adapting the process itself so as to optimize it in some way. This level of automation requires that the system have a computer. Programmed into the computer are models (mathematical equations) that describe how this process or system behaves, how this behavior is bounded, and what aspect of the process or system is to be optimized. *This modeling obviously requires that the process be sufficiently well understood theoretically so that equations (models) can be written that describe how the real process works.* This level of automation has been achieved for continuous processes (oil refineries, for example) where the theory (of heat transfer and fluid dynamics) of the process is well understood and parameters are easier to measure. Unfortunately, the theory of metal forming and metal removal is less well understood and parameters are usually difficult to measure. These processes have resisted adequate theoretical modeling, and as a consequence there are very few A(5) machines on the shop floor. The basic elements of the adaptive control loop are:

1. *Identification:* measurements from the process itself, its output, or process inputs using sensors
2. *Decision analysis:* optimization of the process in the computer
3. *Modification:* signal to the controller to alter the inputs

To raise the automatic CNC grinder to the A(5) level, assume that it is a cylindrical center-type grinder on which part deflection as well as part size are measured. In this process, the cutting forces tend to deflect the part more as the grinding wheel gets farther from the centers. The adaptive control software program would have equations that relate deflection to grinding forces and infeed rates. The infeed would be decreased to reduce the force and minimize deflection. Notice that the overall system would still have to compensate for grit dulling and grit attrition, which will also alter the grinding forces. The A(5) level reflects deductive reasoning whereby particular outcomes are deduced from general principles.

An alternative approach is being evaluated in the research laboratories. The behavior of the process is described with linguistic terms. Fuzzy logic control statements replace closed-form mathematical models. The decision analysis element depends on a set of IF–THEN statements rather than specific equations.

The A(6) level reflects the beginnings of artificial intelligence (AI), in which the control software is infected with elements (subroutines) that permit some thinking on the part of the software. Few, if any such systems exist on the factory floor.

The A(6) level tries to *relate cause to effect.* Suppose that the effect was tool failure. At the A(5) level, the system would detect the increase in deflection due to increased forces (due to the tool's dulling) and reduce the feed to reduce the force. However, it might simultaneously try to increase the speed to maintain the MRR constant, which would increase the rate of tool wear. This was not the desired result. At the A(6) level, multiple factors are evaluated so that the system can recognize the need to change the tool rather than reduce the cutting feed. That is, the system learned by experience that feed reduction/speed increase was the wrong decision, that it was tool wear that caused force to increase. Such systems are often called *expert systems* when the software contains the collective experience of human experts or prior replications of the same process. The use

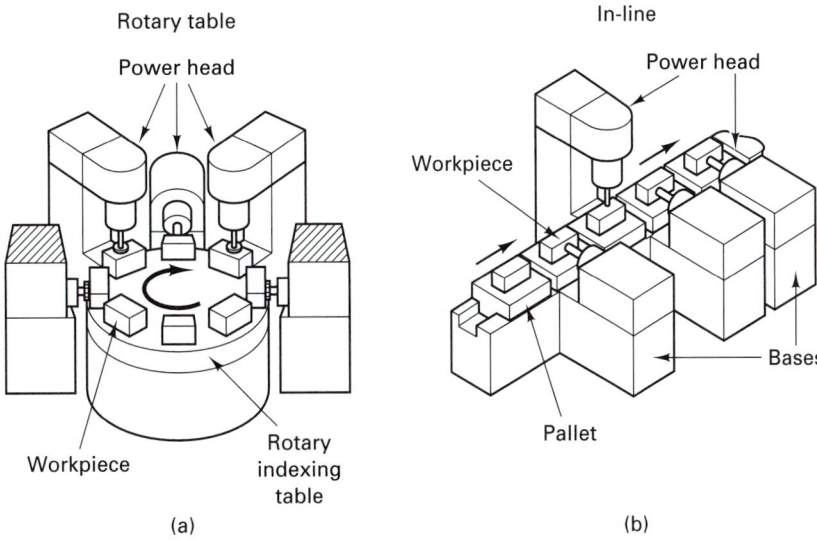

FIGURE 41-12 Examples of (a) rotary and (b) inline transfer units that can be constructed from components. *(Courtesy of Heald Machine Company.)*

of neural networks will allow the process to learn (i.e., the expert system will become more "intelligent" about the process).

The A(7) level reflects the next level of AI, whereby inductive reasoning is used. The system software can determine a general principle (the theory) based on the particular facts (the database) collected. NC and CNC machines are discussed in Chapter 29. Our next discussion concerns transfer lines and FMS.

Transfer Machines

A *transfer machine* is an automated flow line. Workpieces are automatically transferred from station to station, from one machine to another. Operations are performed sequentially. Ideally, work stations perform the operation(s) simultaneously on separate workpieces, with the number of parts equal to the number of stations. Each time the machine cycles, a part is completed. Figure 41-6 shows an example of a transfer line.

Transferring usually is accomplished by one or more of four methods. Frequently, the work is pulled along supporting rails by means of an endless chain that moves intermittently as required. In another method the work is pushed along continuous rails by air or hydraulic pistons. A third method, restricted to lighter workpieces, is to move them by an overhead chain conveyor, which may lift and deposit the work at the machining stations.

A fourth method often is employed when a relatively small number of operations, usually three to ten, are to be performed. The machining heads are arranged radially around a rotary indexing table, which contains fixtures in which the workpieces are mounted (Figure 41-12). The table movement may be continuous or intermittent. Face-milling operations sometimes are performed by moving the workpieces past one or more vertical-axis heads. Such circular configurations have the advantage of being compact and of permitting the workpieces to be loaded and unloaded at a single station without having to interrupt the machining.

Means must be provided for positioning the workpieces correctly as they are transferred to the various stations. One method is to attach the work to carrier pallets or fixtures that contain locating holes or points that mate with retracting pins or fingers at each work station. Excellent station-to-station precision is obtained as the fixtures thus are located and then clamped in the proper positions. However, carrier fixtures are costly. When it is possible they are eliminated and the workpiece transported between the machines on rails, which locate the parts by self-contained holes or surfaces. This procedure eliminates the labor required for fastening the workpieces to the carrier pallets, as well as the pallets.

In the design of large transfer machines, the matter of the geometric arrangement of the various production units must be considered. Whether or not transfer fixtures or pallets must be used is an important factor. These fixtures and pallets usually are quite heavy. Consequently, when they are used, a closed rectangular arrangement is often employed so that the fixtures are returned to the loading point automatically. If no fixtures or pallets are required, straight-line configurations can be employed. Whether pallets or fixtures must be used is dependent primarily on the degree of precision needed as well as on the size, rigidity, and design of the workpieces. If no transfer pallet or fixture is to be used, locating bosses or points must be designed or machined into the workpiece.

The matter of tool wear and replacement is of great importance when a large number of operations are incorporated in a single production unit. Tools must be replaced before they become worn and produce defective parts. Transfer machines often have more than 100 cutting tools. If the entire complex machine had to be shut down each time a single tool became dull and had to be replaced, overall productivity would be very low. This is avoided by designing the tooling so that certain groups have similar lives, monitoring tool thrust, and torque, and by shutting down the machine before the tooling has deteriorated. All the tools in the affected group can be changed so that repeated shutdowns are not necessary. If the transfer machine is equipped with ac drives, diagnostic feedback information of individual processes is easily accomplished (Figure 41-13). Programmable drives such as these replace feedboxes, limit switches, and hydraulic cylinders, and eliminate changing belts, pulleys, or gears to change feed rates and depths of cut and make the system much more flexible.

Methods have been developed for accurately presetting tools and for changing them rapidly. Tools can be changed in a few minutes, thereby reducing machine downtime. Increasingly, tools are preset in standard quick-change holders with excellent accuracy, often to within 0.0002 in. (0.005 mm).

In the use of multiple-station machines, the entire machine may have to be shut down when one or two stations become inoperative. This is usually avoided by arranging the individual units in groups, or sections with 10 to 12 stations per section, and providing for a small amount of buffer storage (*banking* of workpieces) between the sections. This permits production to continue on all remaining sections for a short time while one section is shut down for tool changing or repair.

The most significant problem deals with designing the line itself so that it operates efficiently as a whole. Processes are grouped together at stations. Transfer machines or, for that matter, any system wherein a number of processes are connected sequentially will require that the line be *balanced*. Line balancing means that the process time at each station must be the same, with the total nonproductive time for all other stations minimized. Theoretically, there will be one station that will have the longest time, and this station will control the cycle time for all the stations. Computer algorithms have been developed to deal with line balancing. When the line is the size of the one shown in

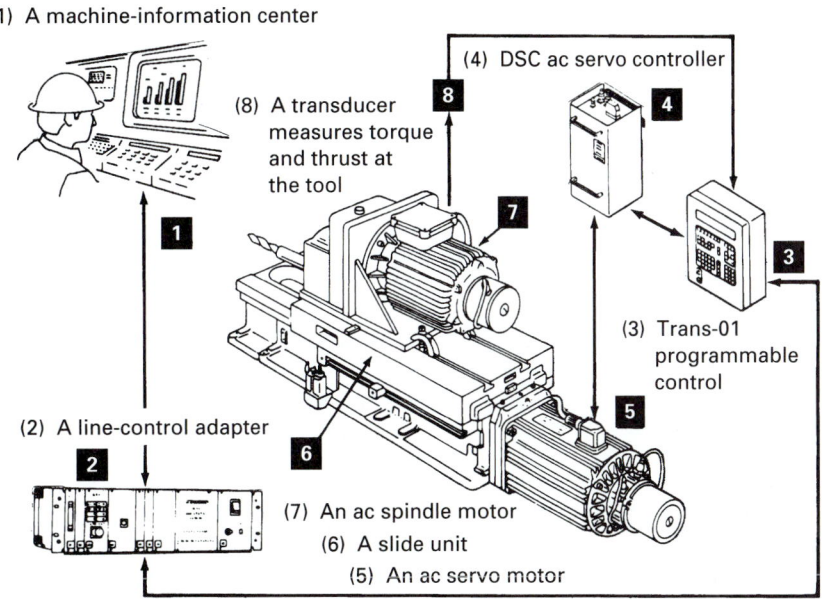

(1) A machine-information center

(8) A transducer measures torque and thrust at the tool

(4) DSC ac servo controller

(3) Trans-01 programmable control

(2) A line-control adapter

(7) An ac spindle motor

(6) A slide unit

(5) An ac servo motor

FIGURE 41-13   Distributed control systems monitor torque and thrust at the tool so that tool breakage is automatically prevented by reducing feed rate. *(From* American Machinist, *July 1985.)*

Figure 41-14, this becomes a rather complex problem. This line produces cylinder heads with 15 transfer machines, two assembly machines, part washers, gages, and inspection equipment.

Transfer machine 1, which has 30 stations, performs rough milling. Machines 2 and 3, with 23 and 25 stations, respectively, perform a variety of drill, plunge-mill, and spot-face operations. Valve throats are finished by 23-station machine 4, which also drills oil-feed holes. Transfer machine 6, with 31 stations, does rough and finish boring, drilling, reaming, and milling. This system produces engine blocks and cylinder heads at a rate of over 100,000 units per year (100 cylinder heads per hour).

In Figure 41-15 shows the design of the horizontal transfer machine that has CNC machines replacing slide units. In addition to CNC units with tool changers, this system features multiaxis fixture positioning, palletized automation, in-process gaging, size control, fault diagnostics, and excellent process capability.

Transfer machines are either A(3)-, A(4)-, or A(5)-level machines, depending on whether they have the built-in capacity for sensing when corrective action is required and how such corrections are made. Sensing and feedback control systems are essential requirements for the fourth level and all higher levels of automation.

Many machines have built-in *feedforward* devices. This means that the system takes information from the input side of the process rather than the output side of the process and uses that information to alter the process. For example, the temperature of the billet as it enters a hot-rolling or hot-extrusion process can be sensed and used as feedforward information to alter the process parameters. This would be an A(5) or adaptive control example. Sensors can be located in three positions, ahead of the process, in the process, or on the output side of the process. The feedforward concept can also be applied at the A(4)

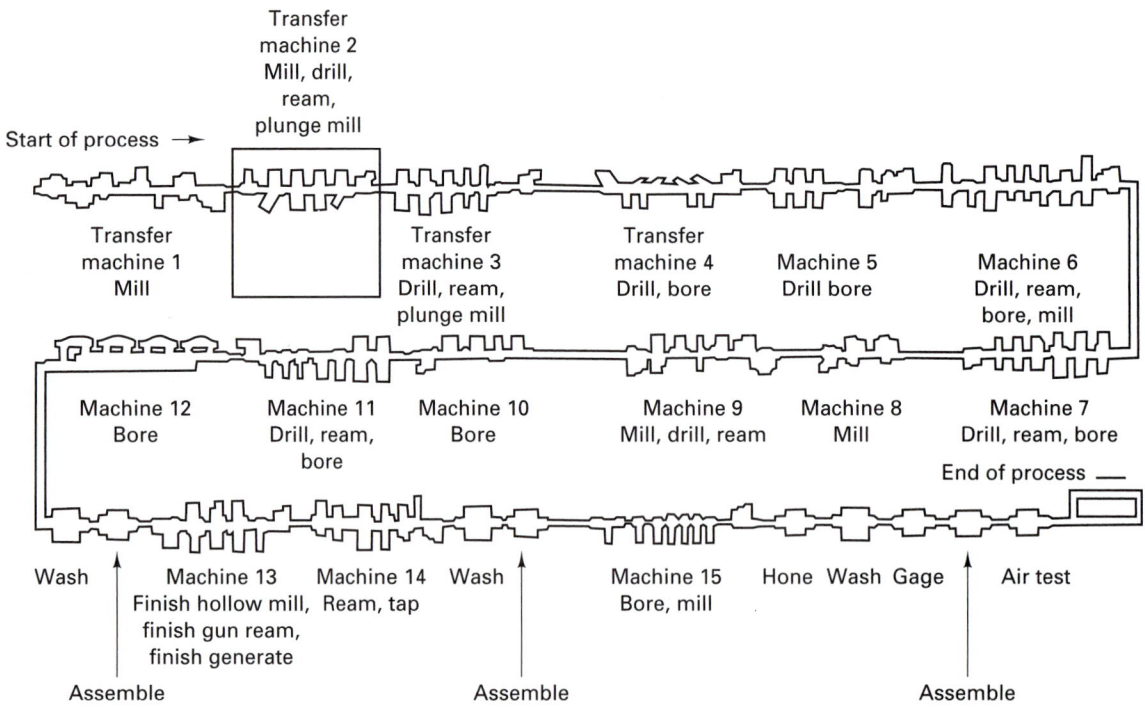

FIGURE 41-14 This megasystem at Ford is one of the largest flow lines in the world. It produces cylinder heads. *(From* American Machinist, *July 1985.)*

level. Suppose that you have a transfer line that processes two types of flywheel housings that are similar but require different machining operations. When the housings are fed into the machine, in mixed order, a sensing device contacts a distinguishing boss that is on one type but absent from the other.

The sensing and feedforward system then sets the proper tooling for that housing and omits operations that are required only for the other type.

It is common practice to equip transfer machines with automated inspection stations or probing heads that determine whether the operation was performed correctly and detect whether any tool breakage has occurred that might cause damage in subsequent operations. For example, after drilling, a hole is checked to make certain that it is clear prior to tapping.

Automation and transfer principles are also used very successfully for assembly operations. In addition to saving labor, automatic testing and inspection can be incorporated into such machines at as many points as desired. Such in-process inspection should be used to prevent defects from being made rather than finding defects after they are made. This assures superior quality. When defective assemblies are simply discovered and removed for rework or scrapped, the cause of the problem is not necessarily corrected.

The range of products now being completely or partially assembled automatically is very great. Figure 41-16 shows an automated assembly machine with a circular transfer table and a linear transfer line serviced by several robots, each performing a different task. Both tactile and visual sensors are needed.

In many cases, some manual operations are combined with some automatic operations. For example, one transfer machine for assembling steering knuckle, front-wheel

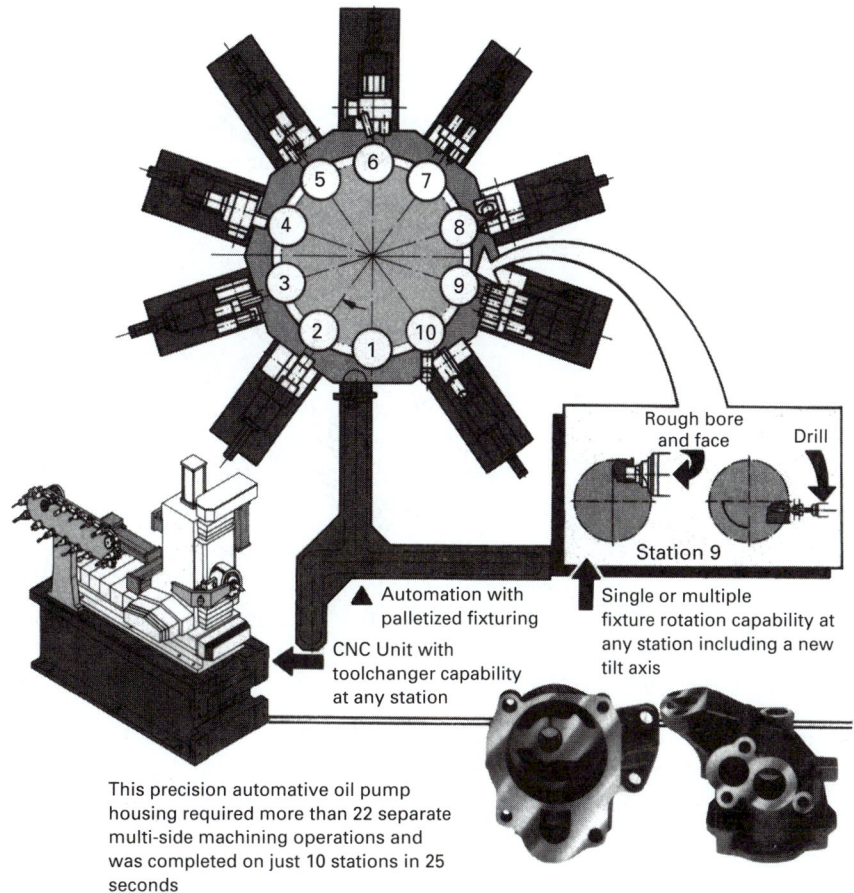

Rough bore
and face

Drill

Station 9

▲ Automation with
palletized fixturing

Single or multiple
fixture rotation capability at
any station including a new
tilt axis

CNC Unit with
toolchanger capability
at any station

This precision automative oil pump
housing required more than 22 separate
multi-side machining operations and
was completed on just 10 stations in 25
seconds

FIGURE 41-15 Horizontal, rotary transfer machine with CNC units in place of conventional slide units. *(Courtesy of Jestadt.)*

hub, and disk-brake assemblies has 16 automatic and 5 manual work stations. As with manufacturing operations, automatic assembly often can be greatly improved through proper part design.

Flexible Manufacturing System

The most publicized type of modern manufacturing systems is known as the *flexible manufacturing system* (FMS). The development of FMSs began in the United States in the 1960s. The idea was to combine the high reliability and productivity of the transfer line with the programmable flexibility of the NC machine in order to be able to produce a variety of parts. The first systems, called variable mission or flexible manufacturing systems, were built by Cincinnati Milacron and Sundstrand. In the late 1960s Sundstrand installed a system for machining aircraft speed drive housings that is still in use today. Overall, however, very few of these systems were sold until the late 1970s and early 1980s, when a worldwide FMS movement began. Table 41-6 shows the FMS installations and exports and imports for 15 countries. International trade in FMS is not significant. Despite all that has been written, there are fewer than 2000 systems in the world (less than 0.1% of the

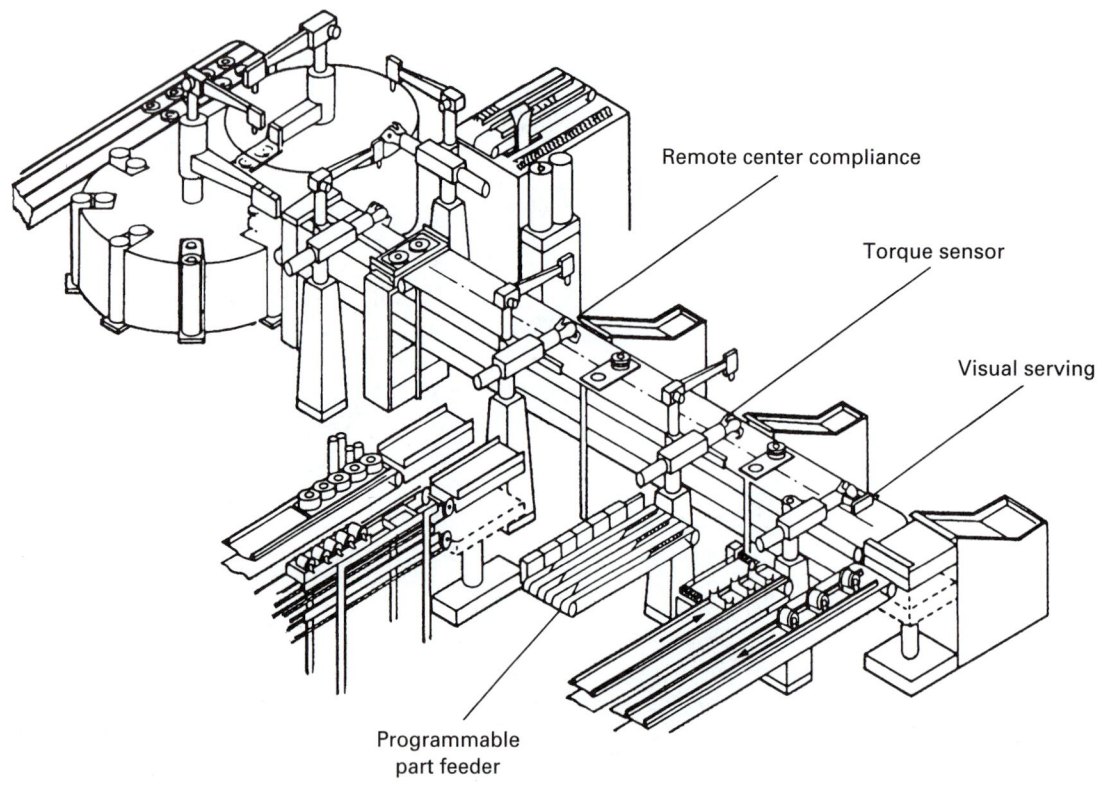

Remote center compliance

Torque sensor

Visual serving

Programmable
part feeder

FIGURE 41-16   Robotic assembly system using tactile and visual sensors. *(From Computers in America, Jan. 1984.)*

machine tool population).   There is also some evidence that the market for these large expensive systems became saturated around the mid-1980s.

The FMS differs from the cell in that the product can take random paths through the machines.   This system is fundamentally an automated conveyorized, computerized job shop.   The system is complex to schedule.   Because the machining time for different parts varies greatly, the FMS is difficult to link to an integrated system and often remains an island of expensive automation.

In about 64% of implementations the collection of components being machined consist of nonrotational (prismatic) parts such as crankcases and transmission housings, 24% are for rotational parts and the balance is a mixture of both types of parts.   The number of machine tools in an FMS varies from 2 to 10, as shown in Figure 41-17, with 3 or 4 being typical.   Annual production volumes for the systems are usually in the range 3000 to 10,000 parts, the number of different parts ranging from 2 to 20, with 8 being typical.   The lot sizes are typically 20 to 100 parts and the typical part has a machining time of about 30 minutes with a range of 6 to 90 minutes per part.   Each part typically needs two or three chuckings and needs about 40 operations.   Invariably, NC machining centers dominate older FMSs.   In recent years, CNC machine tools have been favored, leading to a considerable number of systems being operated under direct numerical control (DNC).   The machining centers always have tool changers.   To overcome the limitation of a single spindle, some systems are being built with head changers.   These are sometimes referred to as *modular machining centers.*

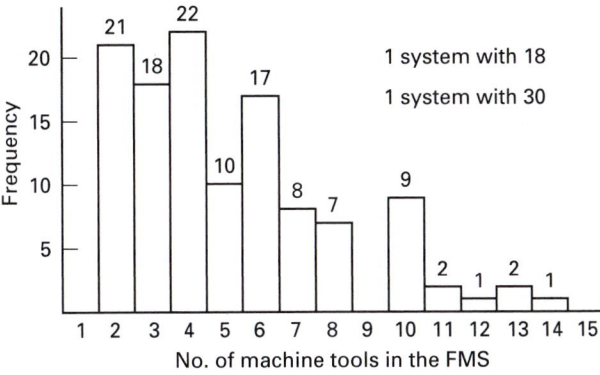

FIGURE 41-17   Histogram of the number of machine tools per FMS, reflecting systems installed worldwide. *(Data from U.S. Department of Commerce.)*

TABLE 41-6.   Number of Flexible Manufacturing Systems by Country in 1986

Country	Number of Systems	Comments
Japan	213	7 to 150 different parts per FMS; most systems home built; perhaps as many as 1000 FMSs in Japan by 1994.
United States	139	First FMS installation in 1963. FMS mainly in large companies. Cost is typically $1 million per machine. Perhaps as many as 1000 FMSs by 1994.
United Kingdom	97	Government sponsoring 20 to 50% of systems cost 5 to 120 parts/system. First installation in 1967.
West Germany	85	30% of FMS cost in material handling; 50 to 250 different parts per FMS.
France	72	Most installation in automotive, aerospace, and machine tool industries.
USSR	56	Pioneered GT: have developed standardized FMS for any turned part.
Italy	40	Extensive FMS installations in automotive industry.
Sweden	37	Most installations in automotive industry.
East Germany	30	First system installed in 1971; 14 to 200 different parts/FMS.
CSFR	23	
Bulgaria	15	
Finland	12	
Netherlands	8	
Switzerland	6	
Taiwan	5	
Others	42	
TOTAL	880	

Source: Data from R. U. Ayres, W. Haywood, and I. Tchijov, *Computer Integrated Manufacturing,* Vol. III, Chapman & Hall, London, 1992.

Other common features of FMSs, as shown in Figure 41-18, are *pallet changers*, underfloor conveyor systems for the collection of chips (not shown), and a conveyor system that delivers parts to the machine. This is also an expensive part of the system, as the conveyor systems are either powered rollers, mechanical pallet transfer conveyors, or, more recently, wire-guided carts [also called *automated guided vehicles* (AGVs)] operating on underground towlines or buried guidance cables (Figures 41-19 and 41-20). The carts are more flexible than the conveyors. The AGVs also serve to connect the islands of automation, operating between FMSs, replacing human guided vehicles (forklift trucks).

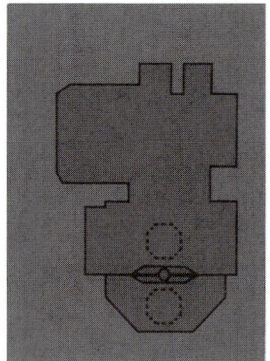

Machining center with
pattet changer and tool
changer

Machining
center
with automatic
material
handling
system

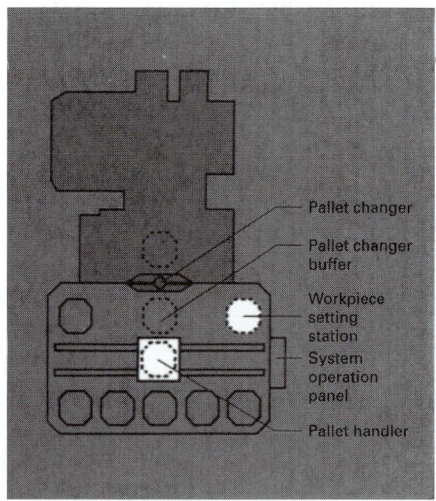

— Pallet changer

— Pallet changer
buffer

— Workpiece
setting
station

— System
operation
panel

— Pallet handler

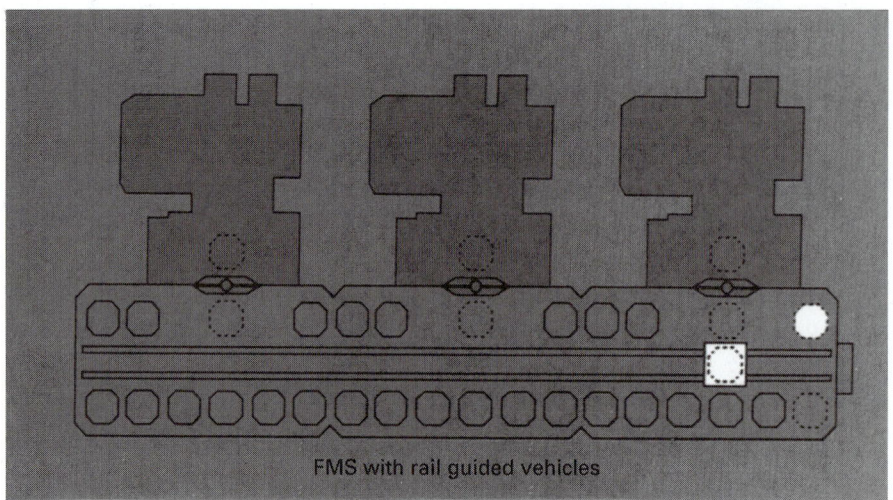

FMS with rail guided vehicles

FIGURE 41-18   The FMS often uses a pallet changer with the conveyor system to
deliver parts to the machine tool.

Pallets are a significant cost item for the FMS because the part must be accurately located in the pallet and the pallet accurately located in the machine. Since many pallets are required for each different component, a lot of pallets are needed and they typically represent anywhere from 15 to 20% of the total system cost. FMSs cost about $1 million per machine tool. Thus the seven-machine FMS shown in Figure 41-19 cost $6 million for hardware and software, with the transporter costing over $1 million.

The CNC machines receive programs as needed from a host minicomputer, which acts as a supervisory computer for the system, tracking the status of any particular machine in the system. In recent years, in-process inspection and automatic tool position correction for tool wear have been added features along with diagnostic routines to computer-monitor the condition of the machines. However, a common problem with these systems is the

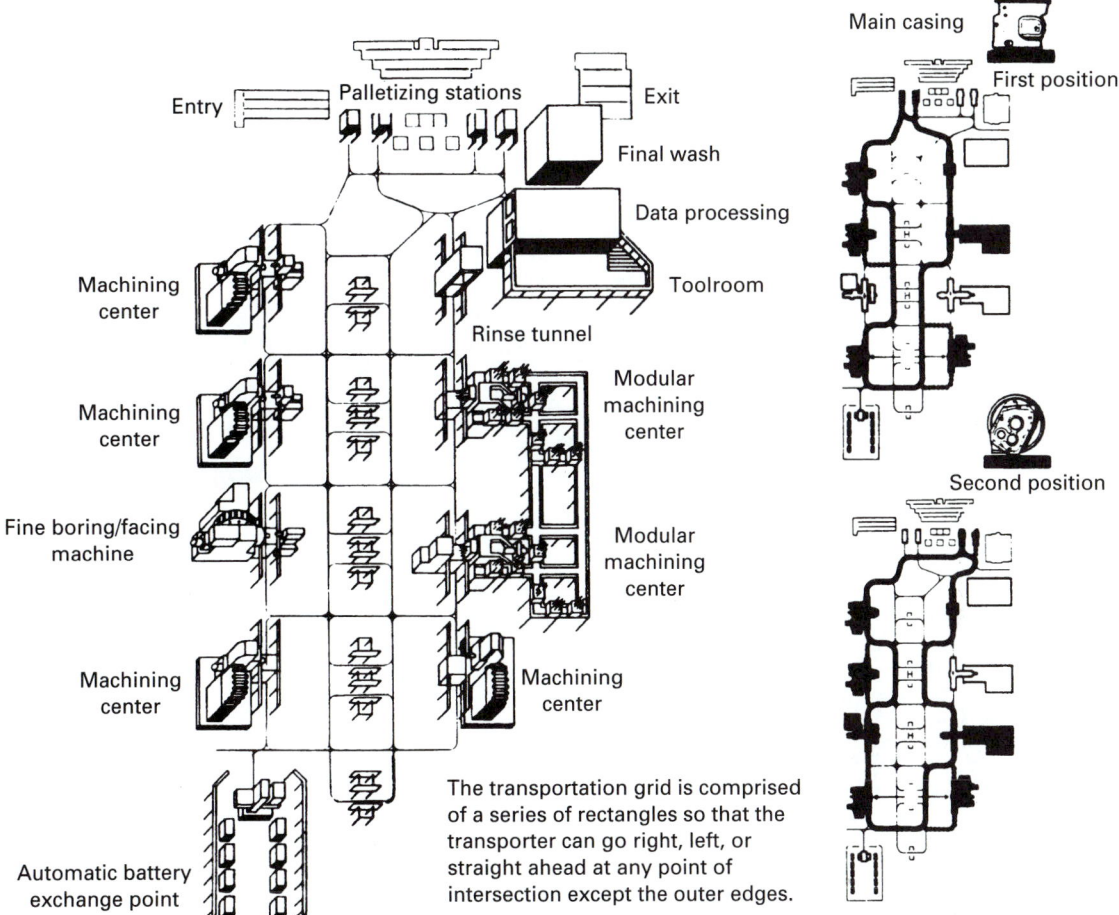

FIGURE 41-19 Renault's seven-machine FMS, with AGV transporter, routes different parts through different machines. *(From* Modern Machine Shop, *Apr. 1984.)*

monitoring of the tool condition and performance. Most installations also incur problems in the performance and reliability of the software and the control systems. It takes just as long to debug software as it does to debug hardware, and delays of 2 to 6 months in startup are not uncommon.

Figure 41-21 shows the levels of computer control in the FMS. (Note that operators are typically needed to load workpieces, unload finished parts, change worn tools, and perform equipment maintenance and repairs.) CNC and DNC functions can be incorporated into a single FMS. The system can usually monitor piece-part counts, tool changes, and machine utilization, with the computer also providing supervisory control of the production. The workpieces are launched randomly into the system, which identifies each part in the family and routes it to the proper machines. The systems generally display reduced manufacturing lead time, low in-process inventory, and high machine tool utilization, with reduced indirect and direct labor. The materials-handling system must be able to route any part to any machine in any order and provide each machine with a small queue of "banked parts" waiting to be processed so as to maximize machine utilization. Convenient access

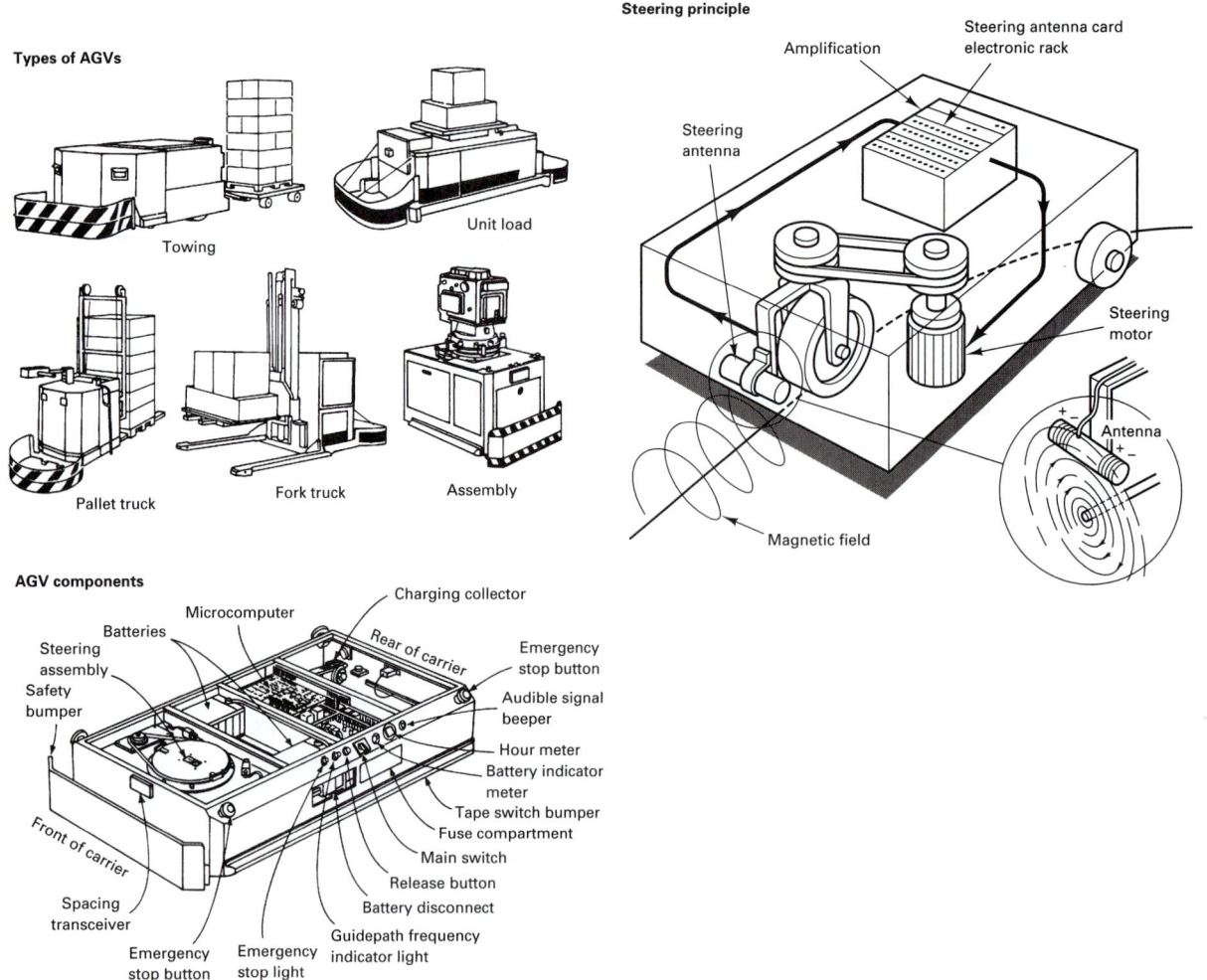

Types of AGVs

Towing

Unit load

Pallet truck

Fork truck

Assembly

Steering principle

Amplification

Steering antenna card electronic rack

Steering antenna

Steering motor

Antenna

Magnetic field

AGV components

Charging collector

Microcomputer

Batteries

Steering assembly

Safety bumper

Rear of carrier

Emergency stop button

Audible signal beeper

Hour meter

Battery indicator meter

Tape switch bumper

Fuse compartment

Main switch

Release button

Battery disconnect

Guidepath frequency indicator light

Front of carrier

Spacing transceiver

Emergency stop button

Emergency stop light

FIGURE 41-20 Automated guided vehicles can be used as a materials-handling system for automated assembly as well as in FMSs, delivering workpieces and tooling to the machines. One means of guidance is a buried-wire guidepath tracked by an antenna on the vehicle. The need for guidepath flexibility has inspired new guidance systems (infrared and inertial), adding further to the cost.

for loading and unloading parts, compatibility with the control system, and accessibility to the machine tools are other necessary design features for the materials-handling system.

The computer control for an FMS system has three levels. The master control monitors the entire system for tool failures or machine breakdowns, schedules the work, and routes the parts to the appropriate machine. The DNC computer distributes programs to the CNC machines and supervises their operations, selecting the required programs and transmitting them at the appropriate time. It also keeps track of the completion of the cutting programs and sends this information to the master computer. The bottom level of computer control is at the machines themselves.

It is difficult to design an FMS because it is, in fact, a very complex assembly of elements that must work together. Designing the FMS to be flexible is difficult. Many companies have found that between the time they ordered their system and had it installed and

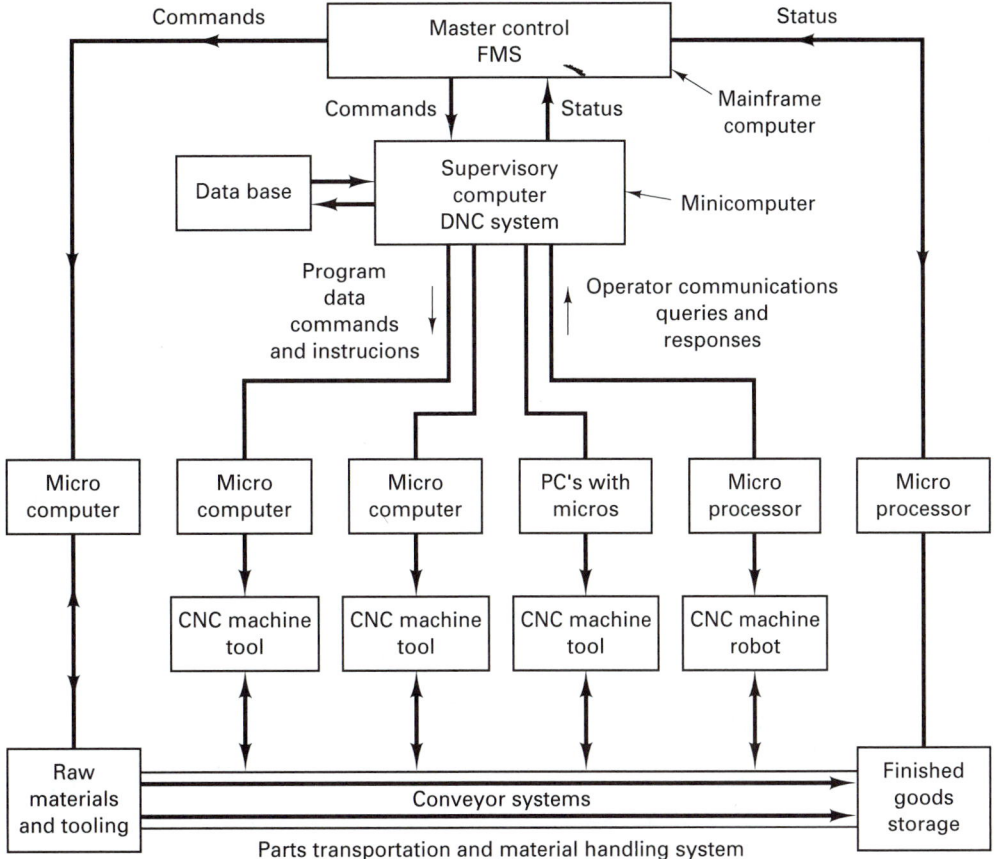

FIGURE 41-21   Three levels of computer control in a DNC system.

operational, design changes had eliminated a number of parts from the FMS. That is, the system was not as flexible as they thought. Figure 41-22 shows some more typical FMS designs.

The use of GT (group technology) to identify the initial family of parts around which the FMS is designed greatly improves the FMS design. One might say that FMSs were developed before their time, since they are being more readily accepted since GT has been used (at least conceptually) to identify families of parts.

As an FMS generally needs about three or four workers per shift to load and unload parts, change tools, and perform general maintenance, it cannot really be said to be self-operating. FMS systems are rarely left untended, as in third-shift operations. Other than the personnel doing the loading and unloading, the workers in the FMS are usually highly skilled and trained in NC and CNC. Most installations run fairly reliably (once they are debugged) over three shifts, with uptime ranging from 70 to 80%, and many are able to run one shift untended. Although the typical FMS installation may not be as flexible (in terms of the different parts it can make) as was thought to be when it was designed, it probably will yield the kinds of results shown in Table 41-7.

Table 41-8 gives the primary characteristics of FMSs. FMSs are, in fact, classic examples of supermachines. Such large expensive systems must be examined with careful and complete planning. It is important to remember that even though they are often

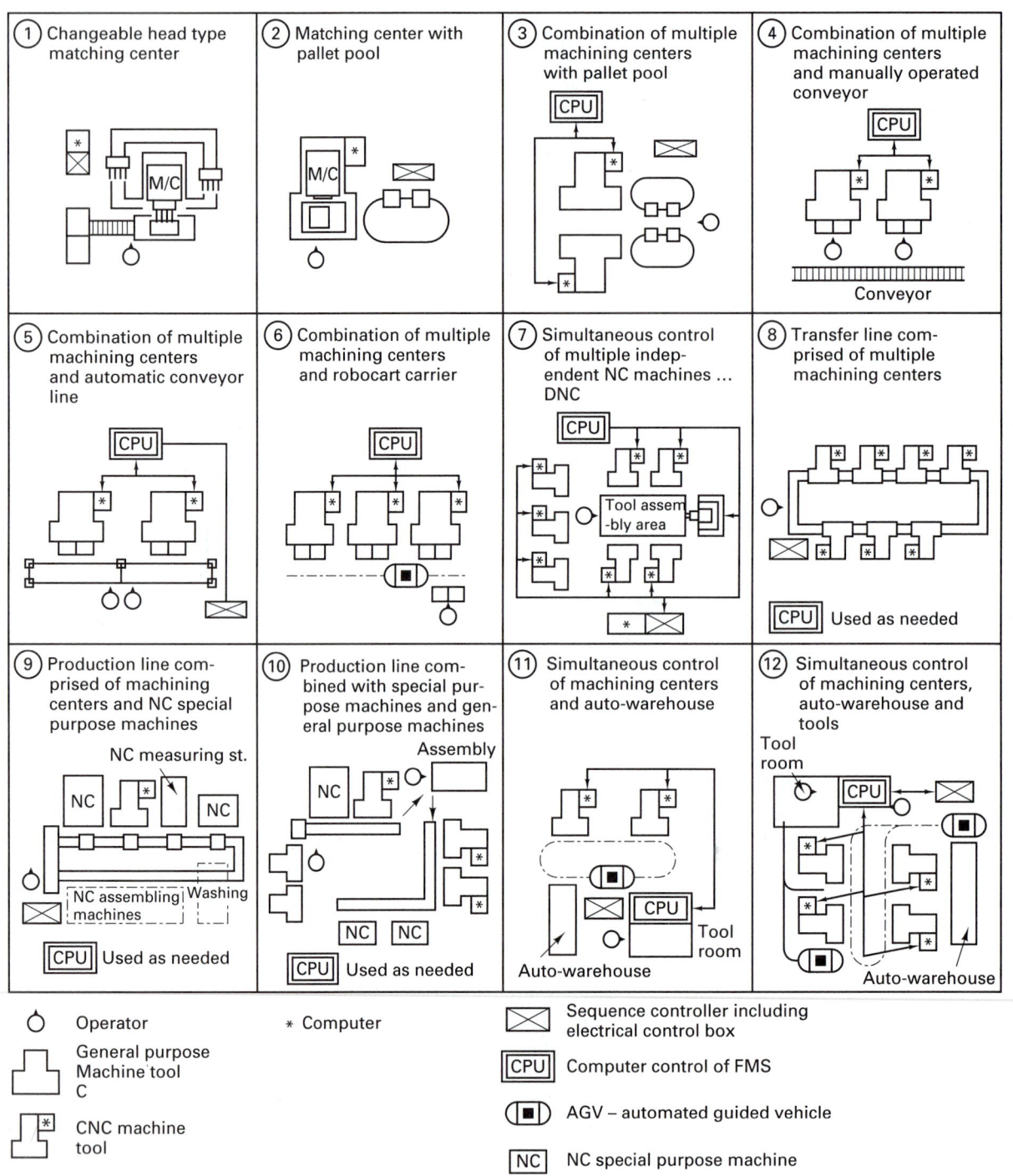

FIGURE 41-22   Examples of machining centers and FMS designs.

TABLE 41-7. What Users Expect from a Flexible Manufacturing System

Flexibility
  Small average batch size
  Variety in parts produced

Lower unit cost due to:
  Labor saving
  Reduced setup time
  Ability to accommodate model changes and design modification of parts
  Reduction of floor space
  Improvement in machine utilization (30% typical)
  Faster throughput times (increased output)

Product delivery
  Continuous check of production conditions
  Reduction of work-in-process (40% typical)
  Reduction of time to transfer products between processes
  Ability to diversify production parts
  Faster supply of spare or replacement parts
  Early detection of defects and system malfunctions

Improvement in quality
  Consistent process capability (machine accuracy and precision)
  Standardization
  Inprocess inspection of products
  First part right

marketed and sold as a "turnkey" installation (the buyer pays a lump sum and receives a system that can be turned on and run), this is only rarely possible with a system that has so many elements that must work together reliably. Taken in the context of integrated manufacturing systems, large FMSs restrict the flow of parts and are not really very flexible. The flexibility of the FMS requires variable speeds and cycles, numerical control, and a supervisory computer to coordinate cell operation. In the long run, smaller manned or unmanned cells may well be the better solution, in terms of system flexibility. Perhaps a better name for these systems is *variable mission or random path manufacturing systems.*

TABLE 41-8. Characteristics of Flexible Manufacturing Systems

Multiple machine tools (NC or CNC)
Automated materials-handling system
Computer control for system (DNC)
Multiple parts, medium-sized lots (200 to 10,000) with families of parts
Random sequencing of parts to machines (optional)
Automatic tool changing
Inprocess inspection
Parts washing (optional)
Automated storage and retrieval (optional)

■ 41.8 ROBOTICS

Robots are steel-collared workers. As defined by the Robot Institute of America, "A *robot* is a reprogrammable, multifunctional manipulator designed to handle material, parts, tools or specialized devices through variable programmed motions for the performance of a variety of tasks." The word *robot* was coined in 1921 by Karel Capek in his play *R.U.R.*

("Rossum's Universal Robots"). The term is derived from the Czech word for "worker." The principal inventor of the underlying control technology for robots was developed by George Devol, who worked for Remington-Rand in the 1950s. His patents were purchased by CONDEC, Inc., which developed the first commercial robot, called *Unimate.* Much of the work for the next 20 years was spearheaded by Joseph Engelberger, who came to be called the "father of robotics." Another famous author, Isaac Asimov, depicted robots in many of his stories and gave three laws that hold quite well for industrial robotic applications. Asimov's three laws of robotics were:

1. A robot may not injure a human being or, through inaction, allow a human being to be harmed. (Safety first.)
2. A robot must obey orders given by human beings except when that conflicts with the First Law. (A robot must be programmable.)
3. A robot must protect its own existence unless that conflicts with the First or Second Law. (Reliability.)

In considering the use of a robot, the following points should be considered. Anything that makes a job or task easy for the robot to do makes the job easy for a human being to do as well. The robot is a severely handicapped worker and is several orders of magnitude less flexible than a human being. Robots cannot think or solve problems on the plant floor. The big advantage of the robot is that it will do a job in an exact cycle time, whereas a human being often cannot. This is an important feature in cells.

For our purposes, if a machine is programmable, capable of automatic repeat cycles, and can perform manipulations in an industrial environment, it is an industrial robot.

All robots have the following basic components:

1. *Manipulators:* the mechanical unit, often called the "arm," that does the actual work of the robot. It is composed of mechanical linkages and joints with actuators to drive the mechanism directly or indirectly through gears, chains, or ball screws.
2. *Feedback devices:* transducers that sense the positions of various linkages and joints and transmit this information to the controllers in either digital or analog form [A(4)-level robots].
3. *End effector:* the "hand" or "gripper" portion of the robot, which attaches the end of the arm and performs the operations of the robot.
4. *Controller:* the brains of the system that direct the movements of the manipulator. In higher-level robots, computers are used for controllers. The functions of the controller are to initiate and terminate motion, store data for position and motion sequence, and interface with the "outside world," meaning other machines and human beings.
5. *Power supply:* electric, pneumatic, and hydraulic power supplies used to provide and regulate the energy needed for their manipulator's actuators.

The early robots were programmed using analog rather than digital control technology. The next generation of robots were more accurate and precise, with reliable electric (digital) controls. Many robot manufacturers made it hard to integrate their robots with other CNC equipment by retaining proprietary operating systems. Today, Japan leads the world in the implementation of robots into the factory. Many of the commercially available robots have one of the mechanical configurations shown in Figure 41-23. Cylindrical coordinate robots have a work envelope (shaded region) that is a portion of a sphere. Jointed-spherical coordinate robots have a jointed arm and a work envelope that approximates a portion of a

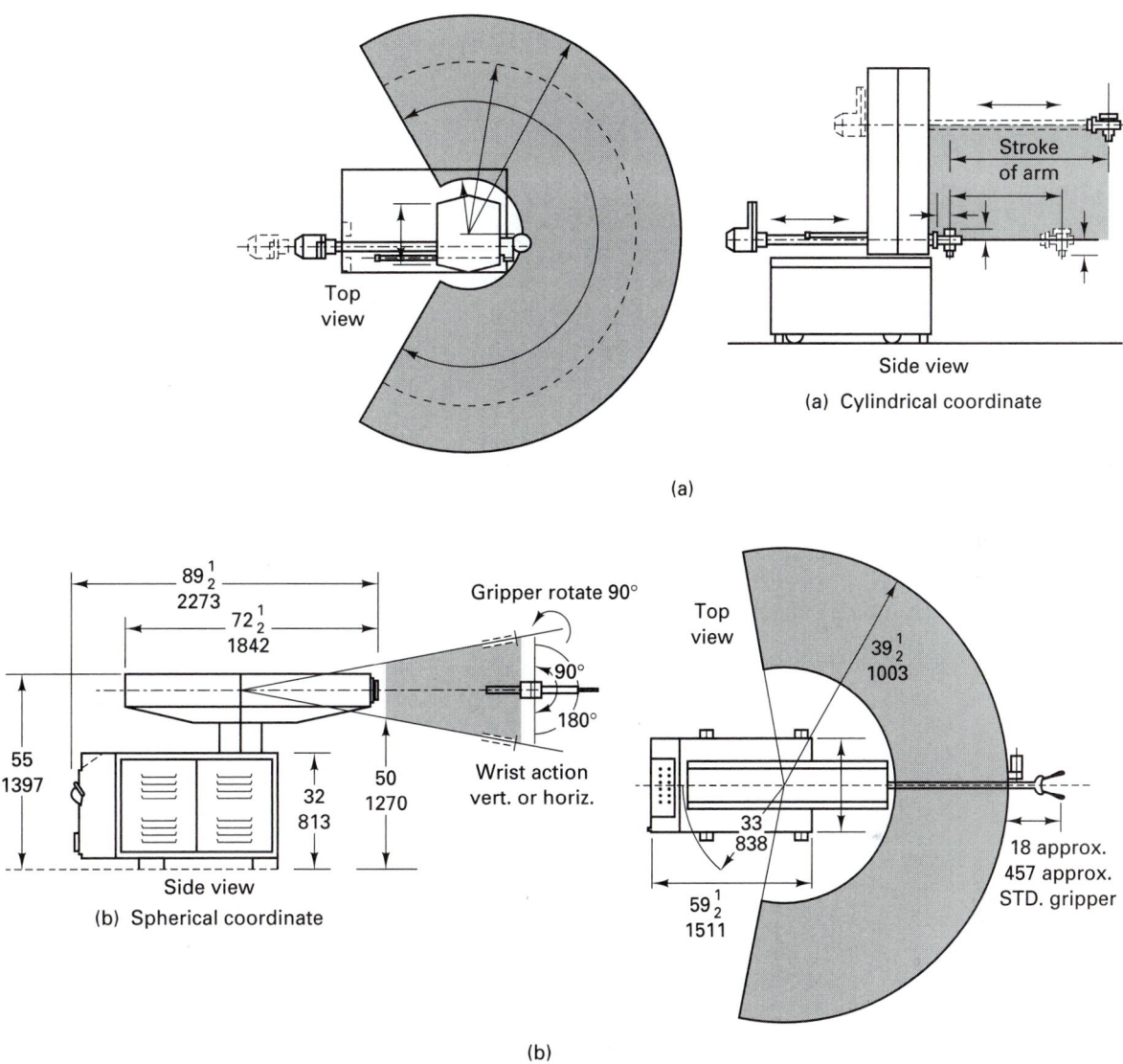

FIGURE 41-23 Work envelopes (shaded regions) for three typical industrial robot designs.

sphere. Rectangular coordinate robots (not shown) with a rectangular work envelope have been developed for high precision in assembly applications.

Figure 41-24 shows the six axes of motion typical for a jointed-arm robot. Robots may have two or three additional minor axes of motion at the end of the arm (commonly called the *wrist*). These three movements are *pitch* (vertical movement), *yaw* (horizontal motion), and *roll* (wrist rotation). The *hand* (or *gripper* or *end effector*), which is usually custom-made by the user, attaches to the wrist.

Industrial robots used in industry today fill three main functions: material handling, assembly, and materials processing. They have, for the most part, very primitive motor and intelligence capabilities, with most robots being A(3)-level machines. The sensory-interactive

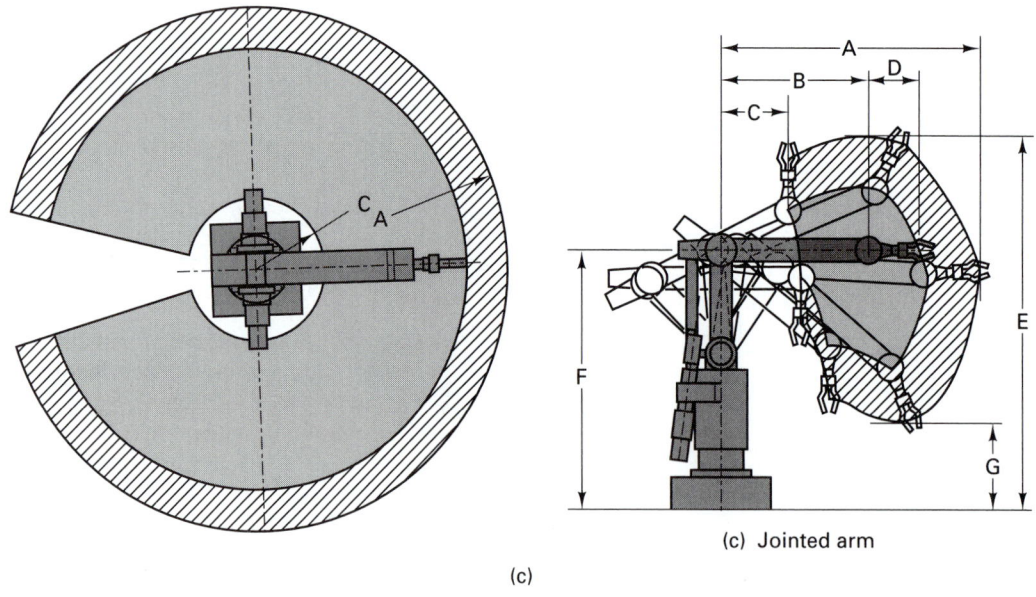

(c)

(c) Jointed arm

FIGURE 41-23 *(cont.)*

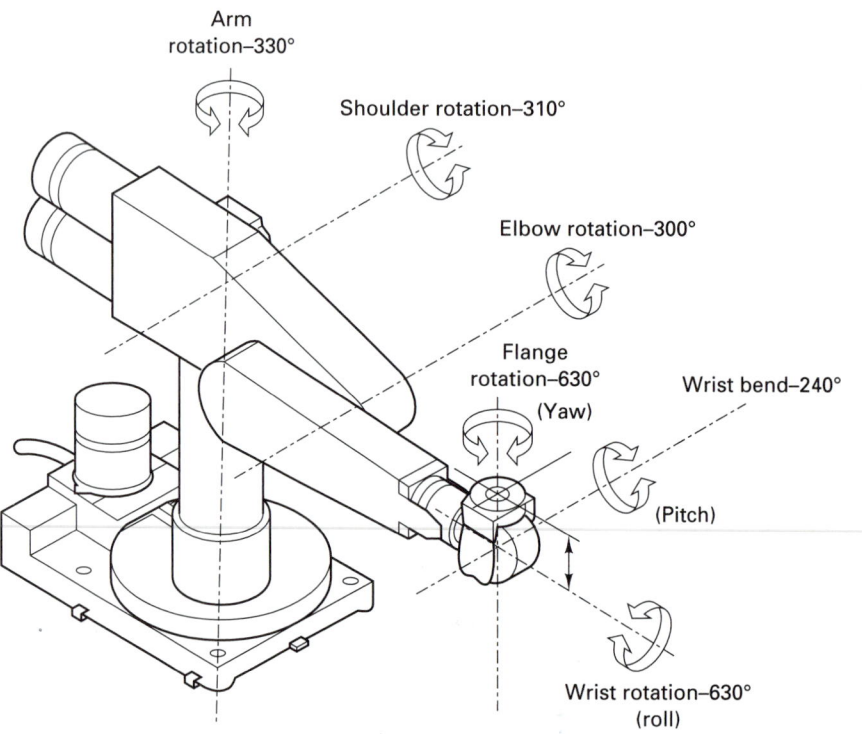

Arm
rotation–330°

Shoulder rotation–310°

Elbow rotation–300°

Flange
rotation–630°
(Yaw)

Wrist bend–240°

(Pitch)

Wrist rotation–630°
(roll)

FIGURE 41-24 The six axes of motion of a robot include roll, pitch, and yaw
movement of the wrist.

control, decision-making, strength-to-size, and artificial intelligence capabilities of robots are far inferior to those of human beings at this time. With regard to shortcomings in performance, the robot's major stumbling blocks are its accuracy and repeatability (i.e., process capability or dexterity), but robot capabilities are progressing steadily, with improvements in controls, ease of programming, operating speeds, and precision.

Robots are having a strong impact in the industrial environment, often doing those jobs that are hazardous, extremely tedious, and unpleasant. Robots perform well doing paint spraying, loading and unloading small forgings or die-casting machines, spot welding, and so on.

The A(3)-level robot, usually called a *pick-and-place machine*, is capable of performing only the simplest repeat-cycle movements, on a point-to-point basis, being controlled by an electronic or pneumatic control with manipulatory movement controlled by end stops.

All A(3) robots are usually small robots with relatively high speed movements, good repeatability (0.010 in.), and low cost. They are simple to program, operate, and maintain but have limited flexibility in terms of program capacity and positioning capability.

To raise the robot to an A(4)-level machine, sensory devices must be installed in the joints of the arm(s) to provide positional feedback and error signals to the servomechanisms, just as was the case for NC machines. The addition of an electronic memory and digital control circuitry allows this level of robot to be programmed by a human being guiding the robot through the desired operations and movements using a hand controller (Figure 41-25). The hand-held control box has rate-control buttons for each axis of motion of the robot arm. When the arm is in the desired position, the "record" or "program" button is pushed to enter that position or operation into the memory. This is similar to point-to-point NC machines, as the path of the robot arm movement is defined by selected endpoints when the program is played back. The electronic memory can usually store multiple programs and randomly access the required one, depending on the job to be done. This allows for a product mix to be handled without stopping to reprogram the machine. The addition of a computer, usually a minicomputer, makes it possible to program the robot to move its "hand" or "gripper" in straight lines or other geometric paths between given points, but the robot is still essentially a point-to-point machine.

There are three ways of controlling point-to-point motion independently: (1) sequential joint control, (2) noncoordinate joint control, and (3) terminally coordinate joint control. Point-to-point servo-controlled robots have the following common characteristics: high load capacity, large working range, and relatively easy programming, but the path followed by the manipulator during operation may not be the path followed during teaching.

To make an A(4) robot continuous-path, position and velocity data must be sampled on a time base rather than as discretely determined points in space. Due to the high rate of sampling, many spatial positions must be stored in the computer memory, thus requiring a mass storage system.

Continuous-path control techniques can be divided into three basic categories, based on how much information about the path is used in the motor control calculations (Figure 41-26). The conventional or *servo-control* approach uses no information about where the path goes in the future. The drive signals to the robot's motors are based on the past and present path tracking error. This is the control design used in most of today's industrial robots and process control systems. The second approach, called *preview control* or *feedforward control*, uses a tracking device to determine how the path is changing in addition to the past and present tracking error used by the servocontroller. The third type of path control is the *path planning* or *trajectory calculation* approach. Here the controller has available a complete description of the path the manipulator should follow from one point to

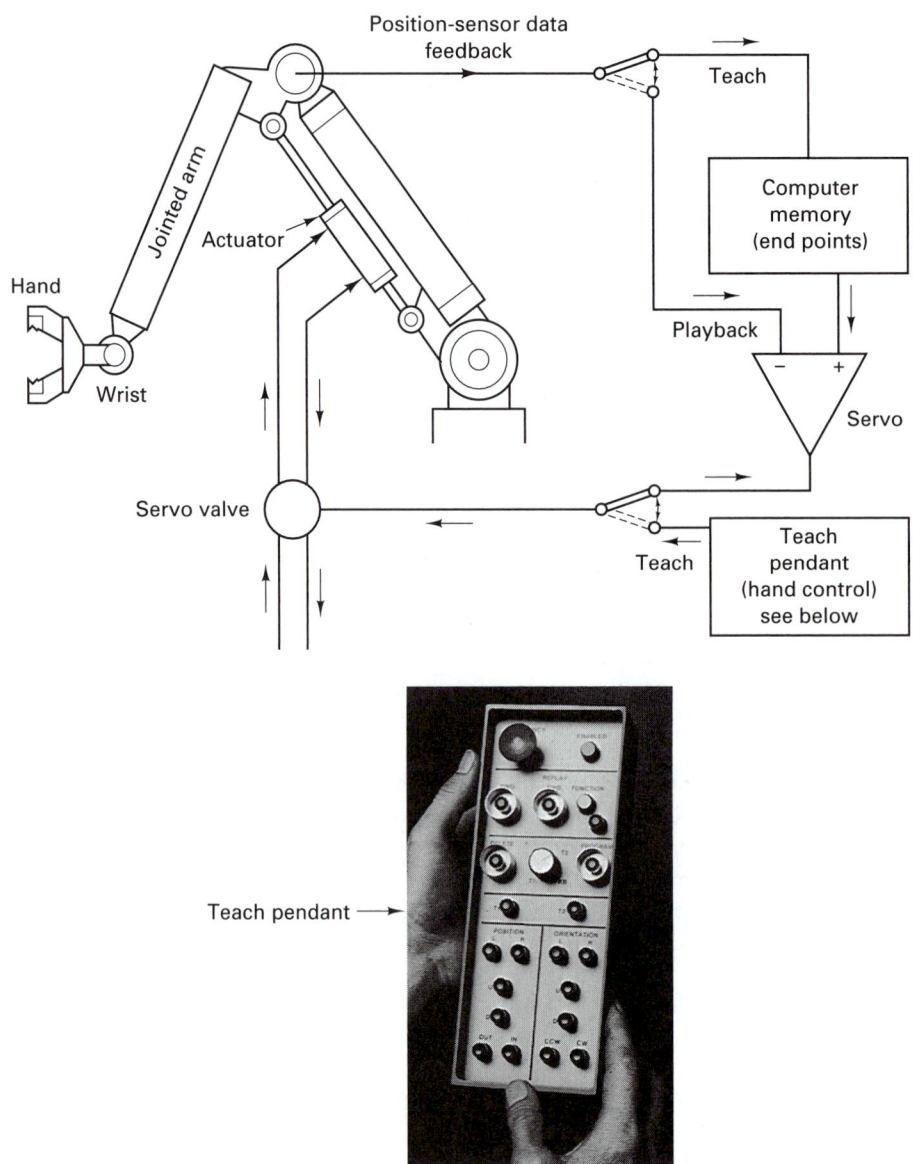

FIGURE 41-25   Manual programming of an industrial robot is accomplished by using a hand-held "teach pendant." The robot is guided through a sequence of operations. The successive positions of all the robotic joints are stored in an electronic memory. This establishes the endpoints of the motions of the arm. By switching from "teach" to "playback," the stored endpoints are replayed. *(Courtesy of Cincinnati Milacron, Inc.)*

another. Using a mathematical-physical model of the arm and its load, it precomputes an acceleration profile for every joint, predicting the nominal motor signals that should cause the arm to follow the desired path. This approach has been used for some robots to achieve highly accurate coordinate movements at high speed. Continuous-path machines tend to be smaller than point-to-point machines, with lighter load capacity, greater precision (±0.0020 to ±0.020 in.), and somewhat higher end-of-arm speed.

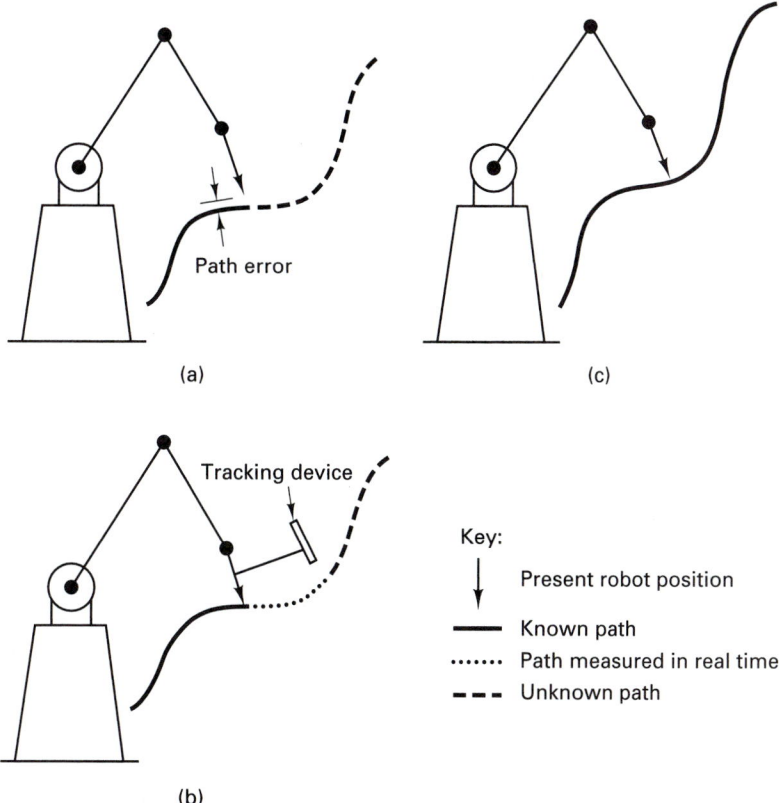

FIGURE 41-26   Continuous-path robots can use (a) basic servo control, (b) tracing or preview control, or (c) trajectory control to position the end effector.

Many of the industrial robots in use at this time are A(4) point-to-point machines. Most robots operate in systems wherein the items to be handled or processed are placed in precise locations with respect to the robots. Even robots with computer control that can follow a moving auto conveyor line while performing spot-welding operations have point-to-point feedback information, but this is satisfactory for most industrial applications.

To expand the capability of this handicapped worker made of steel, sensors are used to obtain information regarding position and component status. Tactile sensors provide information about force distributions in the joints and in the hand of the robot during manipulations. This information is then used to control movement rates. Visual sensors collect data on spatial dimensions by means of image recording and analysis. Visual sensors are used to identify workpieces; determine their position and orientation; check position, orientation, geometry, or speed of parts; determine the correct welding path or point, and so on.

To provide a robot with tactile or visual capability, powerful computers and sophisticated software are required, but this appears to be the most logical manner to raise the robot to the A(5) level of automation, at which it can adapt to variations in its environment. Vision systems can locate parts moving past a robot on a conveyor, identify those parts that should be removed from the conveyor and communicate this information to the robot. The robot tracks the moving part, orients its gripper, picks up the part, and moves it to the desired work station.

Table 41-9 represents a chronological listing of desirable attributes for robots. Table 41-10 provides some additional information on the current state of robotic vision and tactile sensing.

TABLE 41-9. Robot Attributes
Work space command with six infinitely controllable articulations between the robot base and the hand
Fast, "hands-on" programming
Local and remote program memory
Random program selection by external stimuli
Process capability (accuracy and repeatability) to meet application needs
Weight-handling capability (50 to 200 lb)
Intermixed point-to-point and path-following control
Synchronization with moving targets
Compatible computer interface for off-line programming
High reliability (at least 400-hour MTBF).
Vision capability
Tactile sensing
Multiple appendage hand–hand coordination
Computer-directed appendage trajectories
Mobility
Minimized spatial intrusion
Energy-conserving musculature
General-purpose hands
Worker–robot voice communication
Inherent safety (Asimov's laws of robotics)

Source: J. F. Engelberger, *The Industrial Robot*, Sept. 1979, with authors' modifications.

TABLE 41-10. Sensors Used on Robots and Typical Applications

Sensor Type	Design	Applications
Visual	Video pickup tubes (TV camera)	Position detection, part inspection
	Semiconductor sensors (lasers)	Parts detection, identification, sorting
	Fiber optics	Consistency testing (e.g., in manipulation, welding, and assembly)
		Guidance and control
Tactile	Feelers	Position detection
	Pin matrix	Tool monitoring (e.g., in casting, cleaning, grinding, manipulation, and assembly)
	Load cells (piezo, capacitance)	
	Conductive elastometers	Force sensing, part identification
	Silicon	Object recognition, pressure
Electrical (inductive capacitative)	Shunt (current determination)	Position detection
	Capacitor	Status determination (e.g., in manipulation and welding)
	Coil	

The current generation of robots is finding applications in the areas described below.

1. *Die casting.* In single- or multishift operations, custom or captive shops, robots unload machines, quench parts, operate trim presses, load inserts, ladle metal, and perform die lubrication. Die life is increased because die casting machines can be operated without breaks or shutdowns. Die temperatures remain stable and better controlled with uniform cycle times.

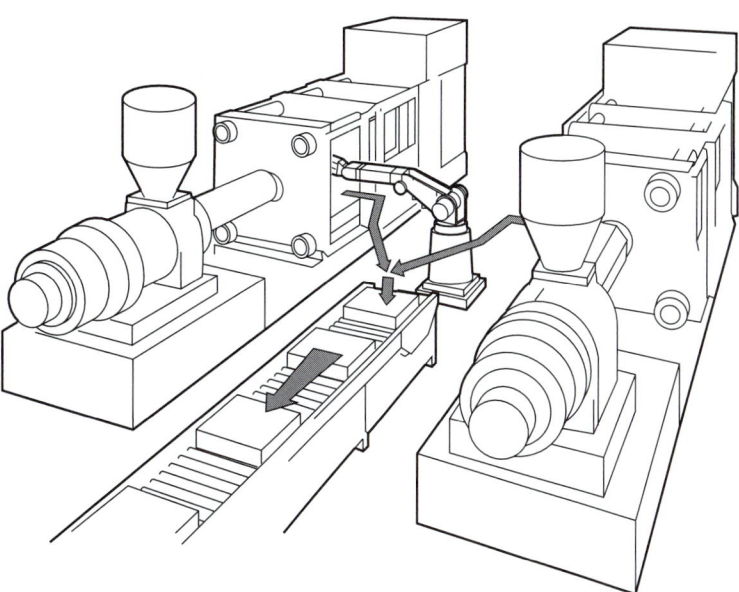

Unloading injection molders

FIGURE 41-27 Application of robots for materials handling, loading, and unloading two plastic injection molding machines. *(From Mert Corwin, A computer controlled robot for automotive manufacturing, Wolfsburg, West Germany, Sept. 12–15, 1977; courtesy of Cincinnati Milacron, Inc.)*

2. *Press transfer.* Robots in sheet metal press transfer lines guarantee consistent throughput shift after shift. Large and unwieldy parts can be handled at piece rates as high as 400 per hour, with no change in cycle time due to fatigue. Robots are adaptable for long- or short-run operations. Programming for new part sizes can be accomplished in minutes.

3. *Materials handling.* Strength, dexterity, and a versatile memory allow robots to pack goods in complex palletized arrays or to transfer workpieces to (and from) moving or indexing conveyors from machines. Savings are dramatic in these labor-intensive operations (Figure 41-27). Operating costs are reduced when robots feed forge presses and upsetters. They work continuously without fatigue or the need for relief in the hot, hostile environments commonly found in forging. Robots can easily manipulate the hot parts in the presses (Figure 41-28).

4. *Investment casting.* Scrap rates as high as 85% have been reduced to less than 5% when molds are produced by robots. The smooth, controlled motions of the robot provide consistent mold quality impossible to achieve manually.

5. *Material processing.* Product quality is improved and sustained with point-to-point, continuous-path robots in jobs such as routing, flame cutting, mold drying, polishing, and grinding. Once programmed, the robot will process each part with the same high quality. Figure 41-29 shows a robot cell with a vision system for drilling, tapping, buffing, filing, and deburring a family of plastic workpieces. The deburring program and the robot's hand are changed automatically when the robot retrieves a different workpiece. The input conveyor holds about a 3-hour supply of parts for the cell. The camera equipment is placed about 8 ft over the belt. When the controller

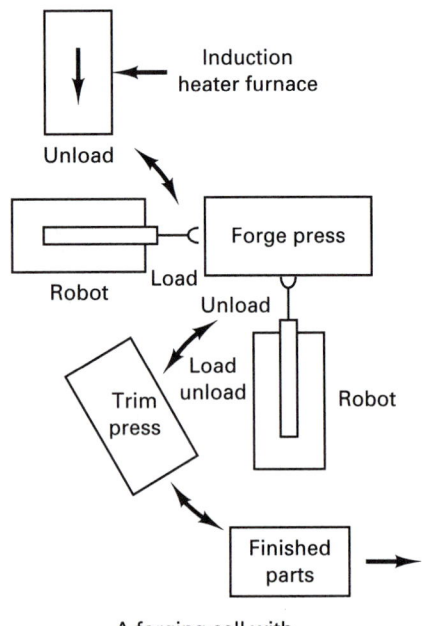

FIGURE 41-28 Layout of a forging cell with two robots for material handling.

A forging cell with two robots for material handling

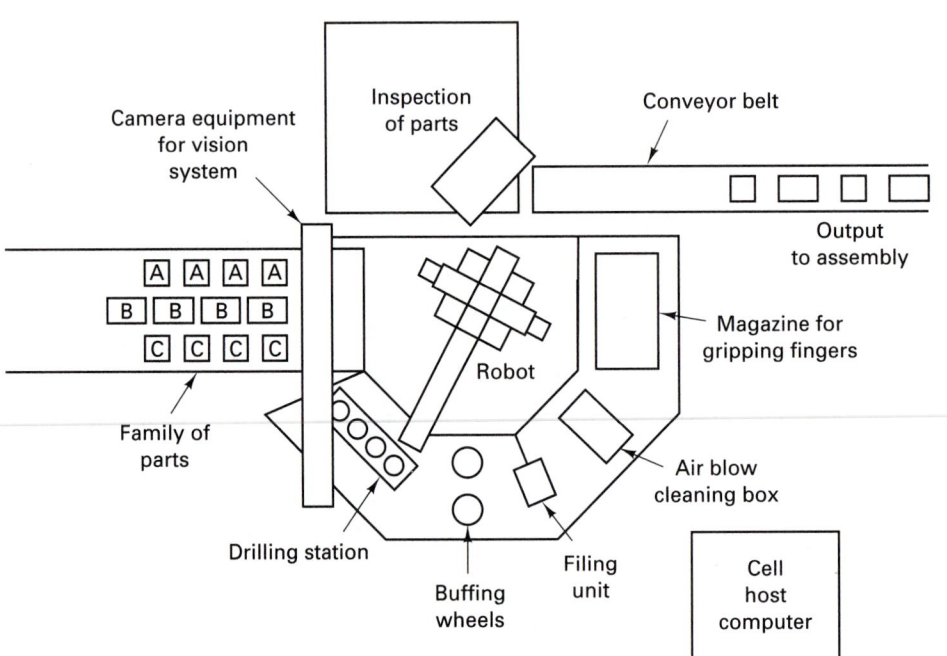

FIGURE 41-29 Robotic cell for a drilling, tapping, deburring family of parts.

receives a request for certain parts from assembly, the vision system identifies the raw material on the input side and the robot retrieves it and starts it through the cell.

6. *Welding.* Robots spot-weld cars and trucks for almost every major manufacturer in the world, with uniformity of spot location and weld integrity. With the addition of seam-tracking capability, robots can be used for arc welding, increasing arc time and freeing operators from hazardous environments, reducing the cost of worker protection, and improving the consistency of weld quality.

7. *Assembly.* The replacement of pneumatic and hydraulic systems with electric motors has improved the accuracy and precision of robots, permitting them to be more widely used in assembly. To perform assembly tasks using robots requires consideration of the entire system, from part presentation through joining, test, and inspection. Most important, the parts must be designed for robotic assembly. Flexible assembly, as this is now referred to, usually addresses midvolume products. As opposed to hard automation that uses special-purpose fixtures, part feeders, and work heads, flexible (or robotic) assembly uses general-purpose and programmable equipment and combinations of visual and tactile sensing so that a variety of parts can be assembled using the same equipment. This requires multiple degrees of freedom and general-purpose grippers which usually decrease the accuracy and precision of the robot, so a vision system is needed to compensate.

Most recent industrial assembly robots have been designed on the SCARA (selective compliance assembly robot arm) system (Figure 41-30). This design results in good positioning accuracy, high speeds, and lost cost for a robot with three or four axes, which is usually adequate for assembly tasks.

In summary, the integrating of robots into cellular arrangements of machine tools to process families of component parts where the robot performs tasks right along with one or more human beings is very efficient. The robot can provide part loading and unloading of machines grouped properly in a machining cell. Totally unmanned cells can facilitate maximum automation and productivity while maintaining programmable flexibility in producing small to medium-sized production lots of parts from compatible parts families. The robots can also change tools in the machines and, in the future, even the workholding devices, thereby adding more flexibility to the cell. These manufacturing cells help to achieve maximum machine tool utilization by greatly increasing the percentage of time the machines spend cutting, which in turn increases the output for the same investment. This is the name of the game in productivity. Examples of this concept are shown in Chapters 29 and 43.

Economics of Programmable Robots

Debugging hard automation systems can be costly and time consuming, taking an average of 12 months to bring a hard automation system on stream. If products change regularly, it is quite possible that the system will be obsolete before it becomes operational. Programmable automation generally takes less debugging and is less subject to obsolescence. As the robots become smarter, they will be able to replace more manual activities. As they become cheaper, it becomes more economical for them to replace hard automation types of functions as well as human labor.

Figure 41-31 shows a comparison of the cost of the steel-collar worker and that of the blue-collar worker in the U.S. automobile industry, beginning in 1970, up to 1992. In 1970, labor cost about $6.80 per hour. This figure rose to about $18 per hour. Throughout the same period, the cost of robot labor runs about $4.80 per hour. This figure includes

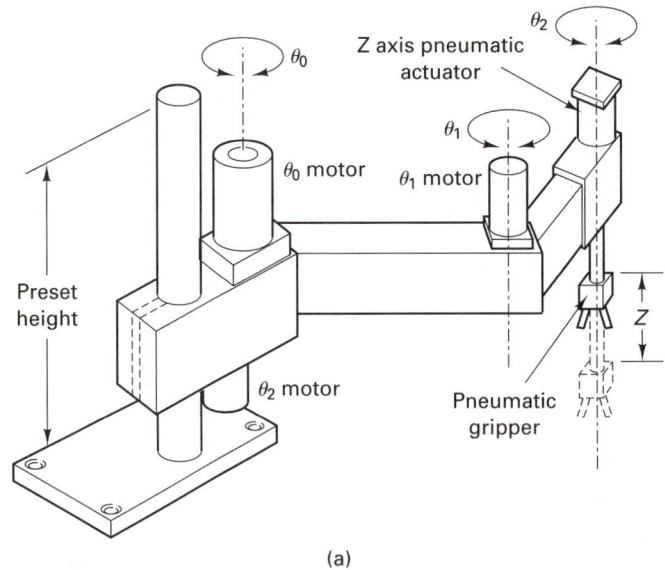

(a)

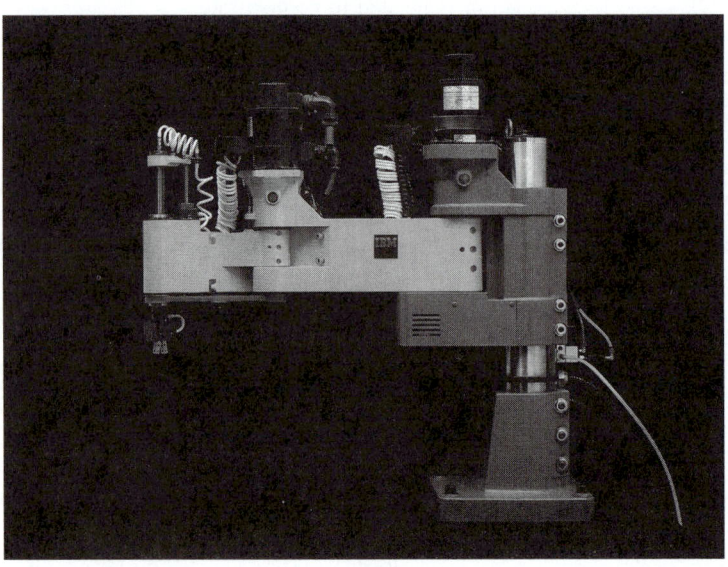

(b)

FIGURE 41-30 Example of a SCARA robot, a type of robot often used in electronic and assembly manufacturing because of its accuracy and precision.

capitalized cost for an 8-year life, maintenance, repair, installation, cost of power, and so on. This figure is conservative, since it is based on 32,000 hours of operating life, and many robots have exceeded 80,000 hours of in-plant service with uptimes exceeding 95%. Of course, economics is not the only reason to implement a robot. The loss of human capability and manufacturing flexibility in the system when a robot replaces a human being must be evaluated carefully.

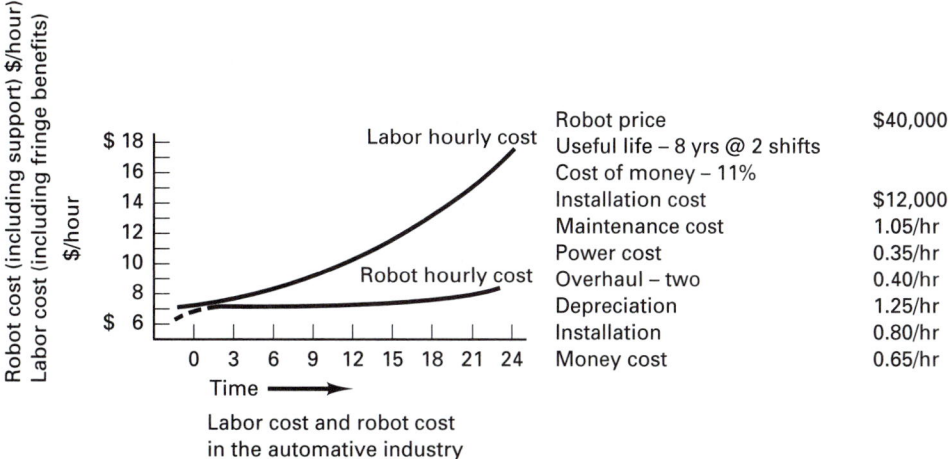

FIGURE 41-31   Cost per hour for a robot versus human labor in the automotive industry.

■ 41.9 SUMMARY

The robot and computers are seen as critical elements of advanced manufacturing technologies available for the next decade, which is being touted as the time when computer integrated manufacturing (CIM) will become a widespread reality. Technology abounds: computer-aided design, computer-aided manufacturing with NC, CNC, AC, and DNC, computer-aided testing and inspection (CATI), automatic assembly and warehousing, robots, and much more. But anyone can buy computers, robots, and other pieces of automation hardware and software. The secret to manufacturing success lies in the redesign of the manufacturing system so that it can achieve superior quality at low cost, with on-time delivery and still be flexible. This requires a visionary management team and a change in culture on the factory floor (i.e., an empowered and involved workforce). No better example of this can be found than at the Toyota Motor Company. Lead by their vice-president for manufacturing, Taiichi Ohno, who conceived, developed, and promoted Toyota's unique manufacturing system, this company has emerged as the world leader in car production. The Toyota system is unique and as revolutionary today as were the American armory system (job shops) or the Ford system for mass production in their day. It is significant that virtually every manufacturing system or technology cited in this chapter is practiced at Toyota. This new system is now being called *lean production* (to contrast it to mass production). Toyota does not use the word CIM because the computer is only a tool used in their system, a system that recognizes people as the most flexible element. This new system is discussed in Chapter 43. Students of manufacturing engineering are well advised to be knowledgeable of this unique system.

■ KEY WORDS

adaptive control
automated guided vehicle (AGV)
automation
continuous process
controller
decoupler
end effector
feedback device
feedforward

flexible manufacturing system (FMS)
flow shop
job shop
linked-cell manufacturing system
 (L-CMS)
manipulators
manufacturing system
mass production
measurable parameters

mechanization
pallet changer
production system
programmable logic controller (PLC)
project shop
robotic cell
route sheet
transfer machine
U-shaped cell

■ REVIEW QUESTIONS

1. Why does the elimination or reduction of setup time greatly improve the productivity of short-run or small-lot operations?

2. In Table 41-1, the transfer line gave 0% of time to loading/unloading and setup. Why?

3. What are four classical forms for manufacturing systems?

4. Can you give a service example of each of the four classical forms for manufacturing systems?

5. What is the new form of manufacturing system that has emerged in the last decade, and how does this new form differ from the previous classical forms?

6. What is meant by the statement, "The manufacturing system is usually a mixed or hybrid system"?

7. What are the trends that are driving companies toward small-lot production?

8. What is meant by the statement, "Processes proliferate"?

9. What is group technology, and how does it convert the functional job shop into a flexible cellular shop? Do you think that 100% of the job shop processes will be converted?

10. Discuss the statement, "Optimizing subsystems does not optimize the whole."

11. What is the definition of a manufacturing system? What are the physical elements?

12. One of the measurable parameters in the definition of a manufacturing system was throughput time. What is this parameter? How is it different from the reciprocal of the production rate?

13. What have been the benefits generally experienced by companies that have undergone conversions of their systems through GT?

14. How does the "do-nothing" alternative act as a

constraint on the implementation of a GT program?

15. Name examples of A(3) and A(4) systems that exist in the home. What about an A(5) or A(6) system?

16. Why is it so difficult to model manufacturing systems?

17. What is a route sheet, in which manufacturing system is it used, and for what purpose?

18. What is meant by the term *line balancing*?

19. Flexible manufacturing systems use CNC machines with AGVs, robots, or conveyors. What differentiates this system from a transfer line? That is, is the system shown in Figure 41-15 really a FMS rather than a rotary transfer machine?

20. Why do you think that transfer lines are using more programmable machines and PLCs?

21. In manned machining cells, why are the machines at least single-cycle automatics? What must be done it they are not?

22. How is it possible that the total machining time to make a part exceeds the cycle time for the part in the manufacturing cell? Recall from Table 41-1 that metal cutting represented about 23% of the cycle time in a machining center. Thus a part with a 4-minute cycle time would typically have about 1 minute of machining time. In the cell, a part with 4 minutes of machining time might have a 1-minute cycle time!

23. Define *flexibility* as a design criterion for the CMS.

24. Cells permit the product design function to be readily integrated (i.e., the designer knows beforehand the process capability of the set of processes that will make the part). How/why

does the designer know this for cells but not for functional job shops?

25. The A(6) level tries to relate causes to effect using AI. What is AI?

26. What is a feedforward device?

27. What are the main areas in which robots are utilized?

28. What are the basic components of all robots?

29. What are the common work envelopes for industrial robots?

30. How is positional feedback obtained in robots?

31. What are the primary differences between an instructional robot and an industrial robot? (See Chapter 43.)

32. What is digital simulation, and how is it used?

33. Compare a rotary transfer machine with an unmanned robotic cell.

34. Compare a human worker in a manned cell to the robotic worker in the unmanned cell. What are the advantages of one over the other?

35. In the analogy of manufacturing to football given in Section 1.2, what is the MPS equivalent of the kicking tee used for the placekicker on the football team?

36. What is tactile sensing? Name some examples of tactile sensing in the home.

37. The vast majority of robots are used in automobile fabrication for welding. What kind of welding is routinely done robotically?

CHAPTER 42

PRODUCTION SYSTEMS

42.1	INTRODUCTION	Quality Engineering
42.2	TYPICAL FUNCTIONAL AREAS	Procurement and Purchasing
	Marketing	Production Planning and Control
	Finance	Inventory Control
	Accounting	Inventory Models
	Personnel	Manufacturing Resource Planning
	Research and Development	
	Engineering	42.3 COMPUTER-INTEGRATED MANUFACTURING
	Product Design Engineering	
	Design Details Related to Casting	Computer-Aided Design
	Design Details Related to Forgings	Parts Classification and Coding for Group Technology
	Design Details Related to Machining	Computer-Aided Process Planning
	Manufacturing Engineering	Computer-Aided Manufacturing
	Basic Requirements for Process Planning	Manufacturing Automation Protocol
	Operations Sheets	42.4 SUMMARY
	Route Sheets	Case Study: UNDERGROUND STEAM LINE
	Quantity versus Process and Material Alternatives	
	Industrial Engineering	
	Plant Engineering	

■ 42.1 INTRODUCTION

In earlier chapters the basics of manufacturing processes and manufacturing systems were presented. The manufacturing processes make or assemble products, whereas the systems integrate the processes (with their operations) into selected sequences to produce entire products. We can detail and name the basic processes (casting, forming, machining, etc.) and the operations (transportation, inspection, storage). We can categorize the manufacturing systems (e.g., job shop, flow shop, linked cells), but we cannot yet clearly define the classical productions systems, even though many such systems exist. Classical production systems are mostly collections of subsystems devised to service the manufacturing systems. Their objectives are to aid the manufacturing systems in producing products to meet delivery schedules at a cost that maximizes profit.

In Section 1.2, the analogy of the university football program was presented. The athletic department was equated to the production system. Recall that the athletic department does many things for the football team (and other teams as well) to help it get ready to play (or produce). Thus, the production system (PS) serves the manufacturing system (MS) and the individual processes but does not actually make products. The people in the PS may design the product (even choose the color scheme), purchase materials and supplies, plan the implementation of the product into the manufacturing system, sell the products, forecast demand, maintain inventory, hire and fire workers, pay employees, and so on. In this chapter, the functions typically performed in a production system are examined.

■ 42.2 TYPICAL FUNCTIONAL AREAS

Production systems (PSs) or production operation systems serve and support the manufacturing processes and manufacturing systems by providing and transmitting information, energy, knowledge, skills, and services to the plant areas, to the company's vendors, and to its customers. Traditionally, a PS included the following areas (Figure 42-1):

- Marketing and sales department
- Finance (not shown)
- Accounting department
- Personnel department
- Research and development (not shown)
- Design engineering (product design)
- Purchasing department
- Production planning and scheduling
- Inventory control
- Quality control
- Plant engineering (maintenance)

In most companies, the people working in these areas are called *staff* or indirect labor to distinguish them from the *line* personnel of the manufacturing system. Although production systems have no standard design, they are usually arranged functionally. To connect the functional areas, informal lines of communication (information flows) are developed. Figure 42-2 shows the necessary communication links for a production job shop just in the areas of production control. This network for communication can become complex. Most managerial workers are in the PS, except for foremen, line supervisors, and manufacturing managers.

Often, the production worker views the services of quality and inventory control with distrust. The production worker may not understand how control charts monitor his process or how the materials requirement planning system controls the work-in-process (WIP). An adversarial relationship often develops between the people in the MS and the people in the PS. Computerizing this function merely complicates the problem. This is perhaps the most important difference between the lean production system discussed in Chapter 43 and the classical production system. That is, in the former, the key functions of the production system are infused into the manufacturing system. The critical control functions not only serve the manufacturing system but become an integral operational part of the making of the product. Briefly discussed also in Chapter 43 is the ongoing movement to restructure the production system into cross-functional teams, usually organized around product lines. In this chapter we discuss the more traditionally organized production system.

Marketing

The chief activities of *marketing* are forecasting sales, advertising, and estimating future demand for existing products. Selling the product is the primary interest of marketing. Promotional work, a highly specialized activity, involves advertising and customer relations. Customer service is a critical function for any manufacturing company. Thus marketing provides information and services concerning:

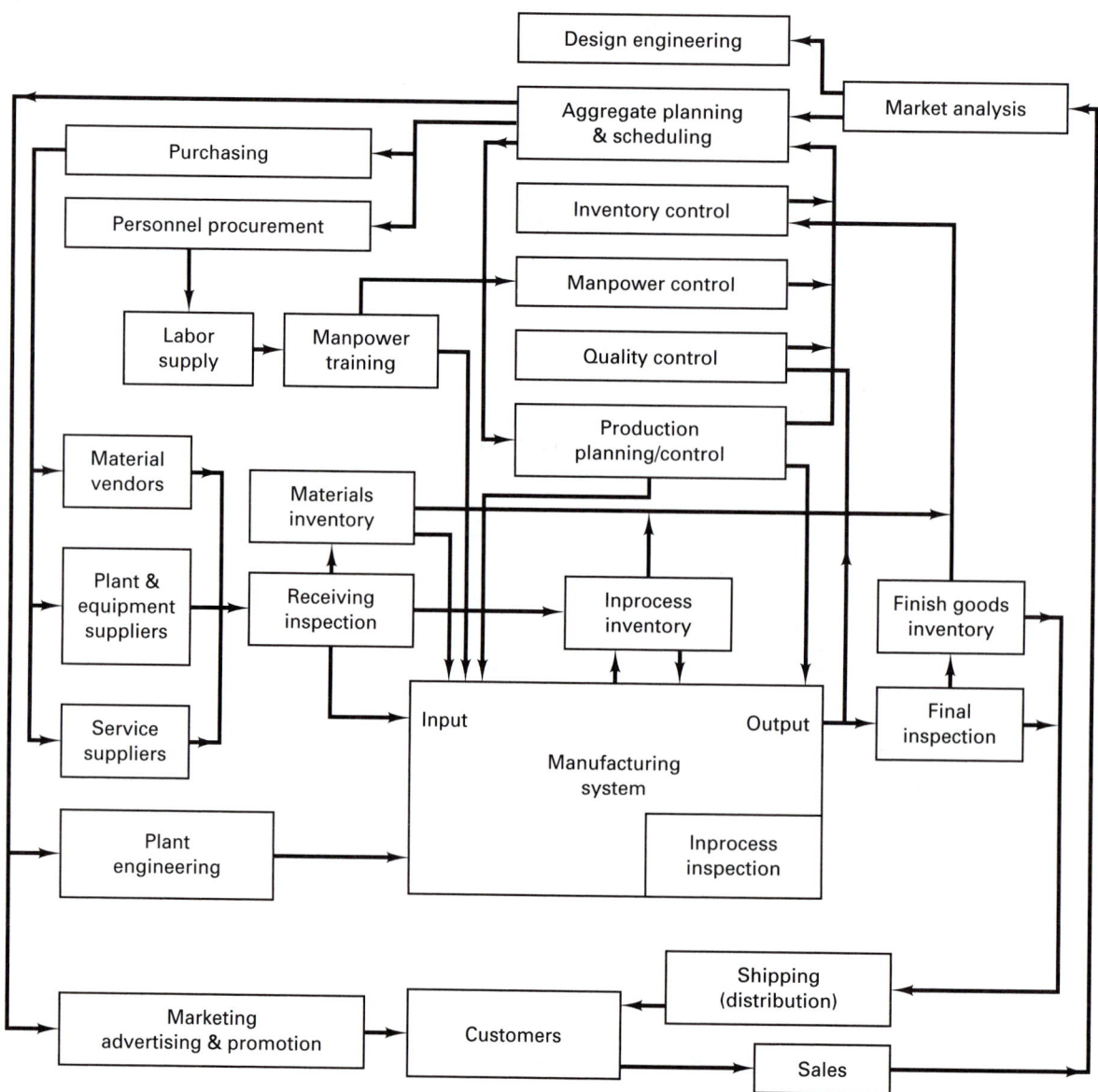

FIGURE 42-1 Classical production system showing inclusion of manufacturing system and major functional elements.

1. Sales forecast of future demand for existing products
2. Sales order data
3. Customer quality requirements
4. Customer reliability requirements
5. New products or modifications for existing products
6. Customer feedback on products
7. Customer service (repair or replace defective products)

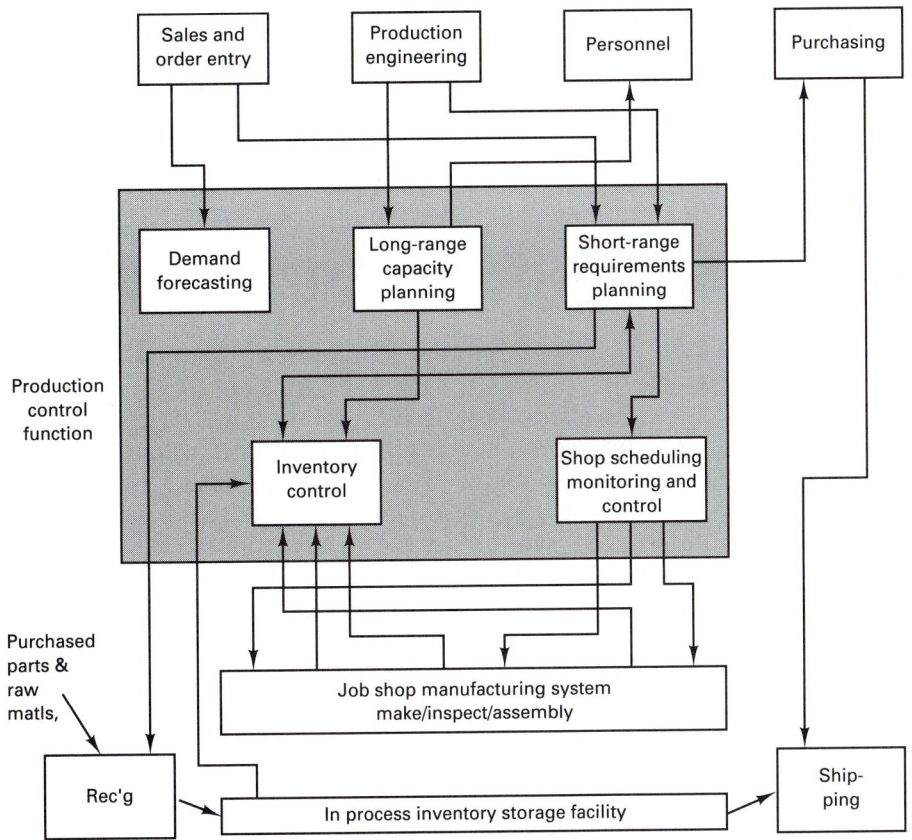

FIGURE 42-2 Flow of information in a production job shop manufacturing/production system, with emphasis on the production control.

There is no piece of information that is more vital to a company, or harder to come by accurately, than future demand. This information is required in order to plan effectively how much should be produced and to schedule that production when changes in demand are predicted. The faster the MS can respond to changes in product demand (for both existing and new products), the better, because the quick response reduces the need to develop accurate long-range forecasts. Short-range forecasts are more accurate than long-range forecasts.

Sales order information is central to *production planning and control.* Products are either made to stock (finished-goods inventory) or made to fill customer orders. Therefore, the orders determine how much, when, and what kinds of products or services must be produced.

Marketing develops information on new products or new uses for old products. This information usually goes to *research and development* or to product design engineers. Marketing also gathers customer feedback on existing products. The marketing department, which is in direct customer contact, gathers complaints about product performance and communicates them to design and/or manufacturing. Often, long-time users are the ones who identify product characteristics that create problems in its use. Clearly, customers want superior-quality products that give them reliable service. In this sense, quality and reliability are related, but they are functionally different. Quality is concerned with the prevention of defects and the conformance to specifications at the time the product is made or sold. Reliability is concerned with the performance of the product over time, while "in service" with the customer. In general, a superior-quality product is more reliable if the design is good. Failures of products that are well made but perform badly because of faulty design occur infrequently but are usually spectacular and newsworthy.

Finance

Financial functions involve management of the company's assets. For the production system, finance provides information and services concerning the following:

1. Internal capital financing
2. Budgeting
3. Investment analysis

Internal financing includes the review of budgets for operating sections, evaluation of proposed capital investments for production facilities, and preparation of financial statements such as balance sheets or profit-and-loss (or income-and-expense) statements.

Periodically, the manufacturing manager, as well as other managers, must submit budgets of expected financial requirements and expenditures to the finance department. The decisions made during budget preparation and the discussions on budget adjustments have a significant impact on the manufacturing system's operation. One of the strongest criticisms of the American system is that *decision makers know little about manufacturing processes or systems* and therefore make poor investment decisions. Few MBA programs have a course in manufacturing processes. Very few undergraduate business students take courses in manufacturing processes. Managers may not really know what business they are in when they do not understand what they do. However, American managers usually do have problem-solving and decision-making skills for handling investment alternatives which require knowledge of such concepts as rate of return, depreciation, sinking funds, payback periods, and compound interest. Managers must have the financial expertise to understand the very complex and constantly changing tax structure, tax regulations, and tax court decisions that affect the company's capital investment decisions.

Accounting

The *accounting* department maintains the company's financial records. Money is used to keep score, so to speak. Accounting also provides data needed for decision making. For the PS, accounting provides information and services on the following:

1. Cost accounting
2. Special reports
3. Data processing

The cost-accounting information indicates the level of efficiency of various departments and the cost of the products being manufactured. The unit-cost data (cost of materials, direct labor, and overhead) help the company to establish prices. Most American companies view this classical equation as follows:

$$\text{unit cost} + \text{profit} = \text{sales price}$$

If the unit cost goes up, then, to maintain profit, the sales price goes up. How many times have you heard this statement: "You'd better buy it now; next year it will cost you more"? Such a statement is an open invitation to those who view the equation this way:

$$\text{sales price} - \text{cost} = \text{profit}$$

The marketplace and the customer dictate the sales price. The only way to maintain or improve profit is to reduce the cost (unit cost)!

The purchasing (procurement) department uses manufacturing cost data in analyzing whether a product should be manufactured by the company (in-house) or purchased from a vendor (the classical make-or-buy decision). Accounting also produces special reports that monitor the status of the scrap and rework levels, raw-materials inventories, work-in-process inventories, finished-goods inventories, direct-labor hours and overtime, and so on. These reports provide quantitative measures of performance (measurable parameters) which can be compared with the original plans (estimates).

In large companies, the accounting department often controls the data-processing equipment. In companies that use computers for problem solving instead of for record keeping, data processing is a separate function.

Personnel

The *personnel* department represents workers, one of the key physical elements of the manufacturing system, and provides information and services concerning:

1. Recruitment
2. Training
3. Labor relations
4. Safety

Although the personnel department may not hire people directly, it assists the company managers by recruiting, screening, and testing potential employees for jobs in both the manufacturing system and the production system. It also handles the details of terminations and department transfers.

The personnel department can assist in training people. For example, in the area of safety, industrial accidents are both costly and disruptive to the workforce and production schedules. By working closely with the personnel department, management develops and institutes programs that can minimize safety problems. If the company has a union, the personnel department will handle labor relations, grievances, collective bargaining, and problems with the shop stewards and union officials.

Research and Development

Research and development (R&D) involves invention or discovery and innovation and their development in terms of achievable ends, such as new materials, products, processes, tools, and techniques. Many industries show the impact of R&D on their manufacturing systems. For example, for many years, the wood-products industry's manufacturing system produced only lumber products. In recent years R&D efforts have produced new products and processes for making plywood, particleboard (from wood chips), gardening mulch (from bark), laminated beams and panels, and chemicals (from wood).

The folks in the manufacturing system will work with R&D on ideas on the manufacture of new products and processes, and on the implementation of new process technology. Modern companies understand that unique process technology can provide a significant competitive edge in the marketplace. Often, R&D also provides ideas for product improvement and may answer questions on economical uses of by-products and waste products from manufacturing operations.

Engineering

Engineering functions are usually staff functions in the production system, providing information and services on the following:

1. Product design engineering or design
2. Manufacturing engineering
3. Industrial engineering
4. Plant engineering
5. Quality engineering

Product Design Engineering. In discussing design, the relationship of design to manufacturing must be reviewed. Through recognition of the systems approach, the design stage can save many dollars and much time later in manufacturing, inspection, assembly, packaging, and even distribution and marketing. Let us examine some of the simpler aspects of designing for manufacture or assembly, sometimes referred to as *producibility*.

Traditionally, a design will often develop in three phases, as outlined in Chapter 9. In the *conceptual* or *idea* phase, the designer conceives of an idea for a device that will accomplish some function. This stage establishes the functional requirements that must be met by the device. The functional requirements should be independent, according to Nam Suh's *Principles of Design.*[*]

In the second, *functional-design* stage, the product is designed so that it will achieve the functional requirements established in the conceptual stage. Often, more than one prototype will be made, suggesting alternative ways in which the functions can be met. At this stage, the designer is usually more concerned with materials than with processes and may ignore the fact that the designed configuration cannot be produced economically utilizing the material being considered.

The third phase of design is called *production design.* Although attention should also be given to the appearance of the product at this stage, particularly if sales appeal is important, the major emphasis is on providing a design that can be manufactured and assembled economically. The *design engineer* must, of course, know that certain manufacturing processes and operations exist that can manufacture the desired product. However, merely knowing that feasible processes exist is not sufficient. The designer must also know their limitations, relative costs, and process capabilities (accuracy, tolerance requirements, etc.) in order to *design for manufacture.* If maximum economy is to be achieved, the designer should be aware of the intimate relationship between design details and production operations.

It is extremely important that the relationship between manufacturing (including assembly) and design be given careful consideration throughout the design phase. Changes can be made for pennies in the design room that would cost hundreds or thousands of dollars to effect later in the factory. This type of consideration should be an integral and routine part of planning for manufacturing. Having the design engineers make a working prototype of each new model before any production drawings are made is one approach. If the model performs in accordance with the conceptual requirements, a second model is made, using, insofar as possible, the same manufacturing and assembly methods that will be used for production. Any changes that will permit easier and more economical production are incorporated into this second model. If the second model meets the functional requirements of the engineering design group, it is sent to the drafting room and production drawings are made. This practice eliminates details from products that are costly to produce and the need for a lot of design changes after a part has gone into production.

---

[*]Nam P. Suh, *Principles of Design*, Oxford University Press, N.Y., 1990.

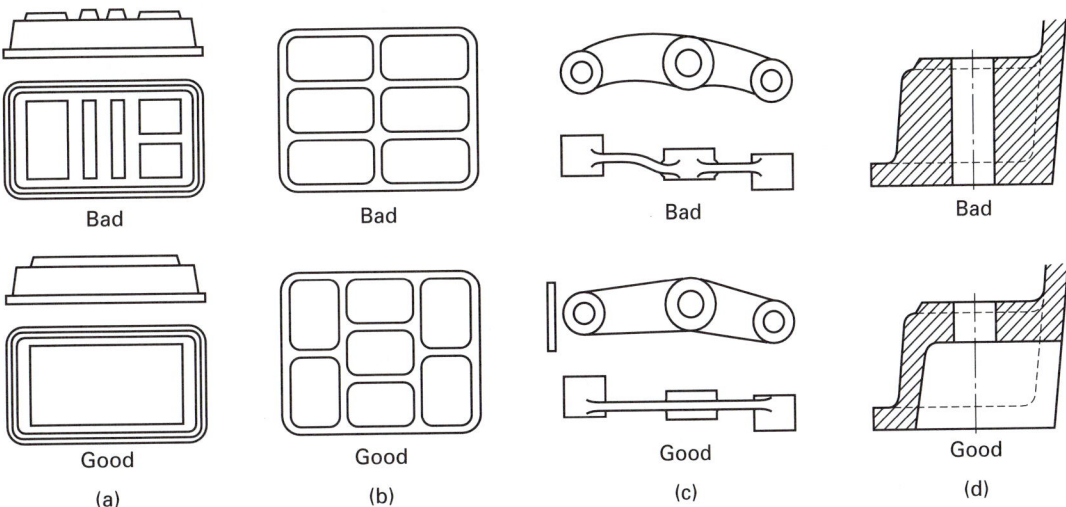

FIGURE 42-3 Good and bad details in casting design.

The designer plays a key role in determining what processes and equipment must be used to manufacture the product, although often indirectly. Clearly, one of the ways in which the designer can indirectly determine the process is through the selection of material. For example, the chassis for record players (stereos) are usually made from zinc and are die cast. Suppose, however, that the designer specifies a composite chassis to be made from fiber-reinforced plastic; then a form-molding process is needed instead of a die-casting machine. The designer specifies a particular joining process when he or she calls for a welded joint. These kinds of direct relationships are pretty obvious. However, other equipment and processes may be specified just as certainly in not so obvious ways.

One of the most common ways in which equipment and processes may be specified indirectly is through dimensional tolerances placed on a drawing. If a tolerance of 0.0002 in. is shown, a grinding operation may be specified just as definitely as if the word *grind* were placed on the drawing. Designers often fail to realize this fact and specify unnecessarily close tolerances; expensive and unnecessary operations result. On the other hand, important requirements may be ignored by the factory if dimensions and tolerances are indefinite. Designers should realize that the dimensions and tolerances they place on a drawing may have implications and results far beyond what they anticipated.

Design details are directly related to the processing that will be used, making the processing easy, difficult, or impossible and affecting the cost and/or quality. Some simple examples will help to amplify the discussion.

DESIGN DETAILS RELATED TO CASTING. Some design details that relate to casting are shown in Figure 42-3. These emphasize the basic fact that castings shrink during solidification and cooling and that difficulties may be experienced unless uniform sections are employed and adequate provision made to avoid excessive contraction stresses. Also, by remembering this principle, one often may reduce the weight and cost of castings, as illustrated. Figure 42-3c emphasizes that mold parting lines should be in a single plane, if practical. This also applies to forgings.

DESIGN DETAILS RELATED TO FORGINGS. When designing forgings, two primary factors should be kept in mind: (1) metal flow and (2) minimizing the number of operations and

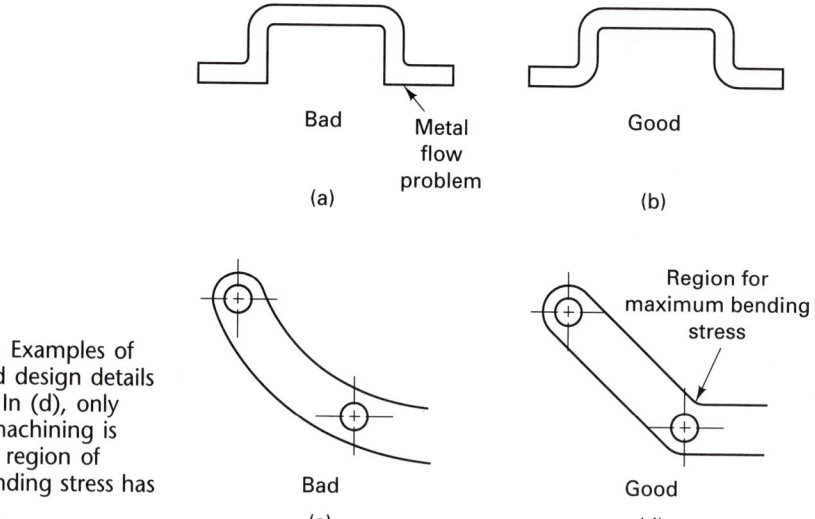

FIGURE 42-4 Examples of good and bad design details for forgings. In (d), only straight-line machining is needed but a region of maximum bending stress has been created.

dies. These two factors often are closely related. As illustrated in Figure 42-4a, the shape that would be very satisfactory for casting may be quite poor for forging, requiring more than one operation. This difficulty can be eliminated by the change shown in Figure 42-4b. Figure 42-4c and d illustrate how a shape that is easy to design may be costly to produce because of added difficulties in machining the forging die. Contouring is needed instead of straight-line machining. However, the change in design may create a problem in service (stress concentration at the bend with a hole). The designer must take these trade-offs into account.

DESIGN DETAILS RELATED TO MACHINING. The ways in which design details can affect machining are almost unlimited. Figure 42-5 illustrates just a few. It is far better for the designer to visualize how the workpiece will be machined and make minor modifications that will permit easy and economical machining than to force the manufacturing department to find some way to machine a needlessly bothersome detail, usually at excessive cost. Too often manufacturing will take for granted that the part has to be made as designed rather than take the trouble to contact the designer to ascertain whether some modification can be made that will facilitate machining.

Grinding may be made difficult, or virtually impossible, through poor design (Figure 42-6). With no provision for entry of the grinding wheel or its supporting arbor (top views), the required surfaces cannot be ground by ordinary procedures. A change in the design corrects the problem but may add another process. Another common deficiency in the design of parts that require the grinding of external cylindrical surfaces is the failure to provide any means to grip or hold the workpiece during grinding. This happens most frequently when only one or a few pieces are involved.

Through this kind of thinking *product design engineering* prepares the product design for the customer. If the product is proprietary, the manufacturer is responsible for developing and designing the product. The product design is documented with component drawings, specifications, and a bill of materials (Figure 42-7) that defines how many of each component go into the product. Initial designs may be based on information from R&D. Prototypes are often built for testing and demonstrating. Manufacturing engineering should be consulted on matters of producibility. In a further step sometimes referred

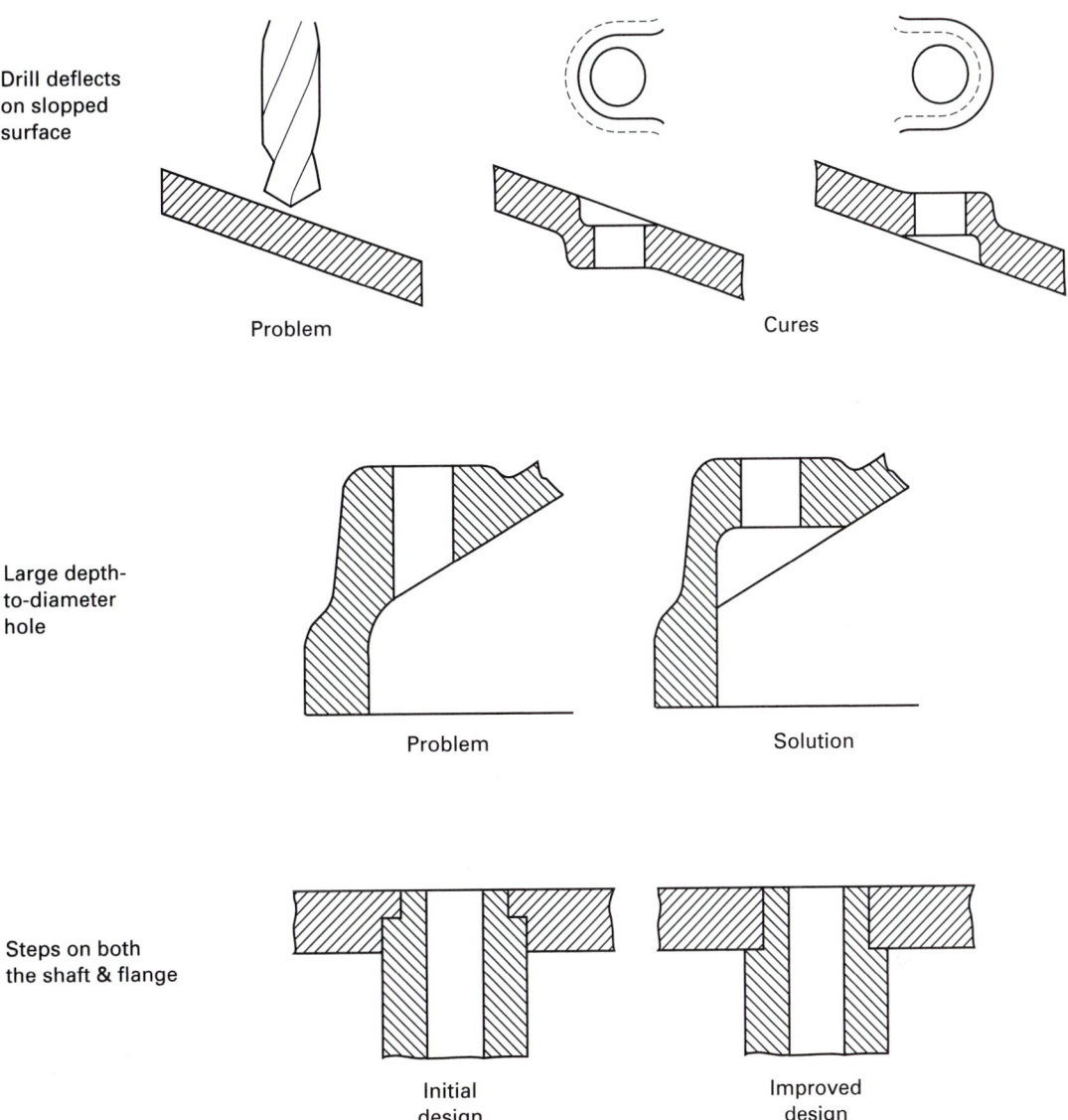

FIGURE 42-5   Examples of good and bad design details from the viewpoint of machining.

to as *value engineering*, engineers look for design changes that could reduce production costs and maintain quality or function. Manufacturing cost estimates are prepared at this point to help determine the market situation for the product.

Upon completion of the design and fabrication of the prototype, company management reviews the design and decides whether to manufacture the item. Engineering management must review and approve the product's design. Many companies call this an *engineering release*.

Corporate management must review and approve the product's general suitability. This second decision represents an authorization to produce the item. The design process

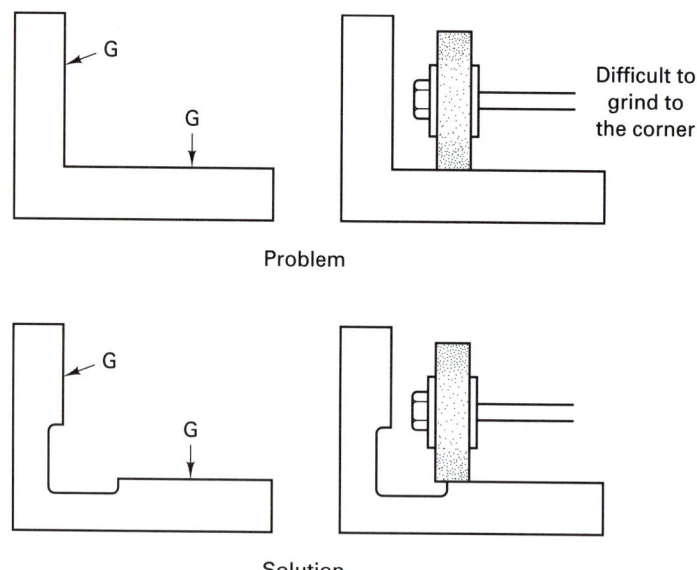

Problem

Solution

Difficult to grind to the corner

FIGURE 42-6   Method of providing proper clearance for grinding-wheel mounting and for overtravel.

discussed in Chapter 9 covered material selection factors, manufacturing considerations, materials substitutions, and product liability.

Manufacturing Engineering. Manufacturing engineers address the planning and management of all manufacturing processes and systems. Using the specifications, *manufacturing engineers* plan the manufacture of the product, determining which machines, workers, tools, and many other manufacturing system components should be used to meet quality and functional requirements. In the classical sense, manufacturing engineering includes responsibility for working with and giving advice to product designers on product producibility. Once production has started, engineering design changes are expensive. Manufacturing engineering may also design individual processes for machine tools (or modification), design tooling and specifications (workholding devices and cutting tools and dies), specify production processes and operations (called process planning), and solve processing problems on the plant floor as well.

In the job shop, *process planning* consists of determining the *sequence* of individual manufacturing processes and operations needed to produce the parts. The *route sheet* and the *operations sheet* are the documents that specify the process sequence through the job shop. The route sheet lists manufacturing operations and associated machine tools for each workpart (Figure 42-8). The route sheet travels with the parts, which move in batches (or lots) between the processes. When a cart of parts has to be moved from one point in the job shop to another, the route sheet provides routing (travel) information, telling the material handler which machine in which department the parts must go to next.

The operations sheet describes what machining or assembly operations are done to the parts at particular machines. See Figure 42-9, which details the operations performed on an engine lathe to make the part shown in Figure 42-10. Note that the details of speed, feed, and depth of cut are specified (actually, recommended, since the machinist may change them). Over time, many different people plan the same part; therefore, there can be many different process sequences and many different routes for the same or similar parts.

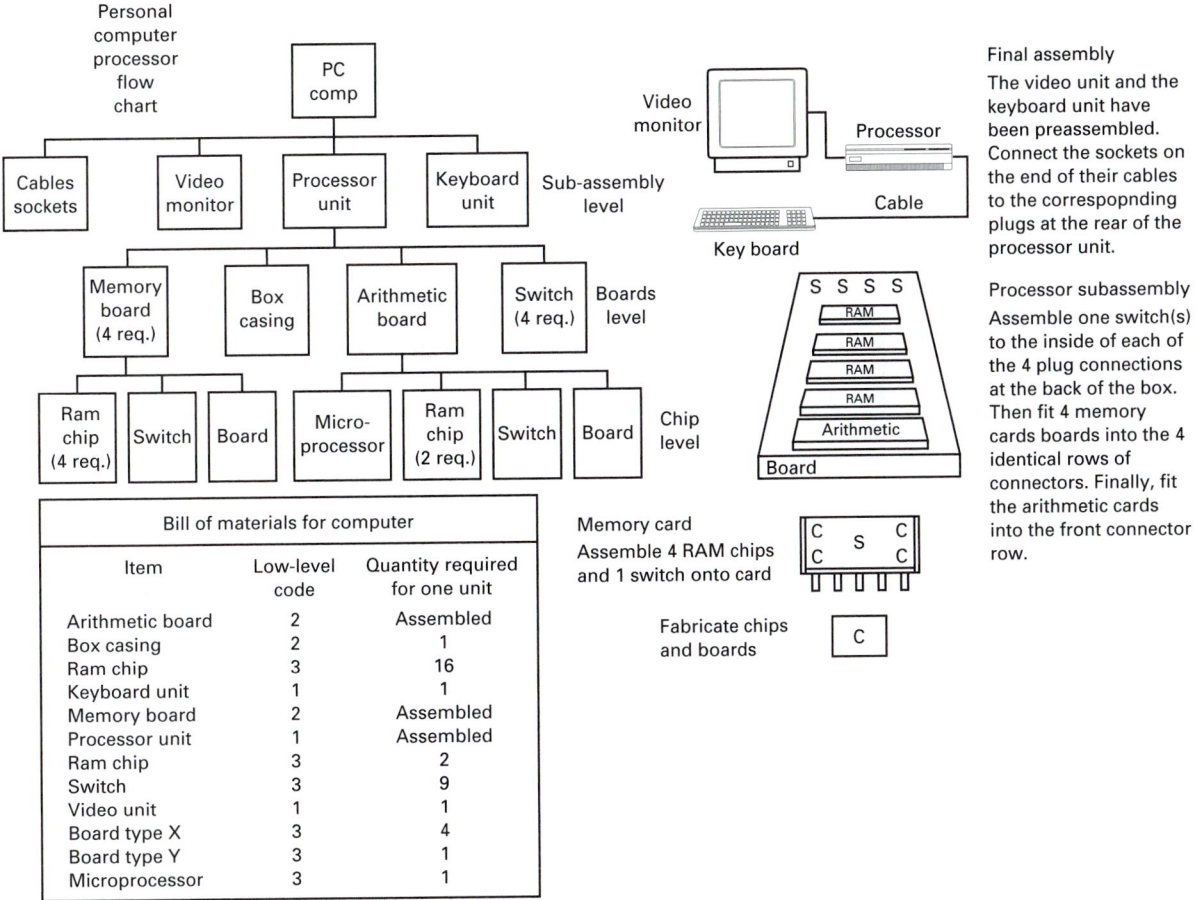

FIGURE 42-7   The process flow chart and the bill of materials (BOM) for a personal computer.

BASIC REQUIREMENTS FOR PROCESS PLANNING.   The first step in planning is to determine the basic job requirements that must be satisfied. These usually are determined by analysis of the drawings and the job orders. They involve consideration and determination of the following:

1. Size and shape of the geometric components of the workpiece
2. Tolerances, as applied by the designer
3. Material from which the part is to be made
4. Properties of material being machined
5. Number of pieces to be produced (see the section in this chapter on quantity versus process and case studies on economic analysis)

Such an analysis for the threaded shaft shown in Figure 42-10 would be as follows:

1. a. Two concentric and adjacent cylinders having diameters of 0.877/0.873 and 0.501/0.499, respectively, and lengths of 2 in. and $1\frac{1}{2}$ in.
   b. Three parallel plain surfaces forming the ends of the cylinders.

DARIC INDUSTRIES

ROUTING SHEET

| NAME OF PART | Punch | | PART NO. | 2 |
| QUANTITY | 1,000 | | MATERIAL | SAE 1040 |

OPERATION NUMBER	DESCRIPTION OF OPERATION	EQUIPMENT OR MACHINES	TOOLING
1	Turn, $\frac{5}{32}$, 0.125, and 0.249 diameters	J & L turret lathe	#642 box tool
2	Cut off to $1\frac{3}{32}$ length	"	#6 cutoff in cross turret
3	Mill $\frac{3}{16}$ radius	#1 Milwaukee	Special jaws in vise $\frac{3}{16}$ form cutter × 4″ D
4	Heat treat. 1,700° F for 30 minutes, oil quench	Atmosphere furnace	
5	Degrease	Vapor degreaser	
6	Check hardness	Rockwell tester	

FIGURE 42-8   Simple route sheet (or traveler) for making the punch shown in Figure 42-12. This part goes from a lathe to a mill to a furnace to vapor degreasing to final inspection. The lot size is 1000 parts.

    c. A $45°$ × $\frac{1}{8}$-in. bevel on the outer end of the $\frac{7}{8}$-in. cylinder.
    d. A $\frac{7}{8}$-in. NF-2 thread cut the entire length of the $\frac{7}{8}$-in. cylinder.
2. The tightest tolerance is 0.002 in., and the angular tolerance on the bevel is ±1°.
3. The material is AISI 1340 cold-rolled steel, BHN 200.
4. The job order calls for 25 parts.

    A number of conclusions regarding the processing can be drawn from this analysis. First, because concentric, external, cylindrical surfaces are involved, turning operations are required and the piece should be made on some type of lathe. Second, because 25 pieces are to be made, the use of an automatic screw machine would not be justified. As the company's

DARVIC INDUSTRIES

OPERATION SHEET

PART NAME: _____ Threaded Shaft 1340 Cold Rolled Steel _____   Part No. 7358-267-10

OPER. NO.	NAME OF OPERATION	MACH. TOOL	CUTTING TOOL	CUTTING SPEED		FEED ipr	DEPTH OF CUT Inches	REMARKS
				ft/min	rpm			
10	Face end of bar	Engine Lathe		120	458	Hand		Use 3-jaw Universal chuck
20	Center Drill End	"	Combination center drill		750	Hand		
30	Cut off to $3\frac{9}{16}$ length	"	Parting tool	120	458	Hand		To prevent chattering, keep overhang of work and tool at a minimum and feed steadily. Use lubricant.
40	Face to length	"	RH facing tool (small radius point)	120	458	Hand	(R) $\frac{1}{8}$ max. (F) .005	Before replacing part in 3-jaw chuck, scribe a line marking the $3\frac{1}{2}$ inch length.
50	Center Drill End	"	Combination center drill		750	Hand		
60	Place between centers, turn $\frac{.501}{.499}$ diameter, and face shoulder	"	RH turning tool (small radius point)	120 160	(R) 458 (F) 611	(R) .0089 (F) .0029	(R) .081 (3) (F) .007	
70	Remove and replace end for end and turn $\frac{.877}{.873}$ diameter	"	RH tools (R) (small radius point) (F) Round nose tool	120 160	(R) 458 (F) 611	(R) .0089 (F) .0029	(R) .057 (F) .005	
80	Produce 45°-chamfer	"	RH round nose tool	120	458	Hand	(R) $\frac{1}{8}$ max. (F) .005	
90	Cut $\frac{7}{8}$ - 14 NF-2 thread	"	Threading tool	60	208		(R) .004 (F) .001	(1) Swivel compound rest to 30 degrees. (2) Set tool with thread gage. (3) When tool touches outside diamter of work set cross slide to zero. (4) Depth of cut for roughing = .004. (5) Engage thread dial indicator on any line. (6) Depth of cut for finishing = .001 Use compound rest.
	Remove burrs and sharp edges	"	Hand file					

Date _____

FIGURE 42-9 Operations sheet for the threaded shaft shown in Figure 42-10.

NC lathe is in use, an engine lathe will be used. Third, because the maximum required diameter is approximately $\frac{7}{8}$ in., 1-in.-diameter cold-rolled stock will be satisfactory; it will provide about $\frac{1}{16}$ in. of material for rough and finish turning of the large diameter. From this information, the operations sheet(s) for a particular engine lathe is prepared.

OPERATIONS SHEETS. The *operations sheet* shown in Figure 42-9 lists, in sequence, the operations required for machining the threaded shaft shown in Figure 42-10. Typically, a single operation sheet lists the operations that are done in sequence on a single machine. However, the sheets may cover all the operations for a given part for a group of machines in a manned cell (see Chapter 43).

Operations sheets vary greatly as to details. The simpler types often list only the required operations and the machines to be used. Speeds and feeds may be left to the discretion of the operator, particularly when skilled workers and small quantities are involved.

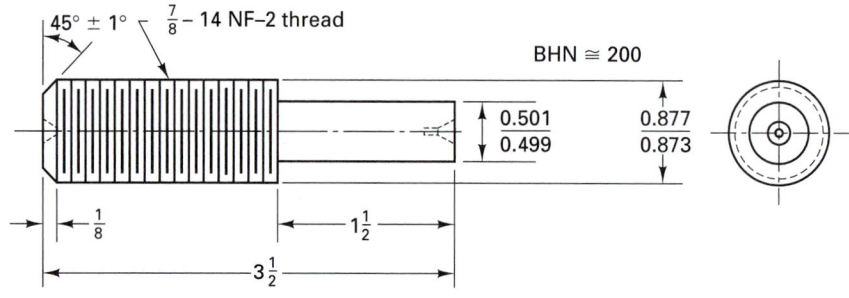

FIGURE 42-10   Threaded shaft to be manufactured in quantities of 25 units.

However, it is common practice for complete details to be given regarding tools, speeds, and often the time allowed for completing each operation. Such data are necessary if the work is to be done on NC machines, and experience has shown that these preplanning steps are advantageous when ordinary machine tools are used.

The selection of speeds and feeds required to manufacture the part may require referring to handbooks in which tables of the type shown in Figure 42-11 are given. For the threaded shaft, this table will give suggested values for the turning and facing operations for either high-speed-steel or carbide tools. Note that the tables are segregated by workpiece material (medium carbon alloy steels, wrought or cold worked) and then by process: in this case, turning. For these materials, additional tables for drilling and threading would have to be referenced. In Figure 42-9, notice that the depth of cut dictates the speed and feed selection and that rough and finish cuts are used. That is, operation 60 requires three roughing cuts and one finishing cut. Could you estimate or determine the cutting time for one of these roughing cuts?

While the operation sheet does not specifically say so, high-speed-steel tools are being used throughout. If the job were being done on an NC lathe, it is likely that all the cutting tools would be carbides. Notice also that the job as described for the engine lathe required two setups, a three-jaw chuck to get the part to length and produce the centers and a between-centers setup to complete the part between each setup. It was also necessary to stop the lathe to invert the part. Do you think that this part could be manufactured more efficiently in an NC lathe?

Data of the type shown in Figure 42-11 are currently being computerized to be compatible with computer-aided manufacturing and computer-aided process planning systems. Table 42-1 shows the most frequently reported system inputs and outputs for this kind of data.

As noted in Chapter 41, the cutting time, as computed from the machining parameters, represents only 20 to 30% of the total time needed to complete the part. Referring again to Figure 42-9, the operator will need to pick up the part after it is cut off at operation 30, stop the lathe, open up the chuck, take out the piece of metal in the chuck, scribe a line on the part marking the desired $3\frac{1}{2}$-in. length, place the part back in the chuck, change the cutoff tool to a facing tool, adjust the facing tool to the right height, and move the tool to the proper position for the desired facing cut in operation 40. All these operations take time and someone must estimate how much time is required for such noncutting operations if an accurate estimate of total time to make the part is to be obtained.

Turning, Single Point and Box Tools   MEDIUM-CARBON ALLOY STEELS, WROUGHT

MATERIAL	HARD-NESS	CONDITION	DEPTH OF CUT*	HIGH SPEED STEEL TOOL			CARBIDE TOOL						
							UNCOATED				COATED		
				SPEED	FEED	TOOL MATERIAL AISI	SPEED		FEED	TOOL MATERIAL GRADE C	SPEED	FEED	TOOL MATERIAL GRADE C
							BRAZED	INDEX-ABLE					
			in	fpm	ipr		fpm	fpm	ipr		fpm	ipr	
	Bhn		mm	m/min	mm/r	ISO	mm	m/min	mm/r	ISO	m/min	mm/r	ISO
5. ALLOY STEELS, WROUGHT (cont.)		Hot, Rolled, Annealed or Cold Drawn	.040	135	.007	M2, M3	375	500	.007	C 7	650	.007	CC 7
			.150	105	.015	M2, M3	300	400	.020	C 6	525	.015	CC 6
	175 to 225		.300	80	.020	M2, M3	240	315	.030	C 6	400	.020	CC 6
Medium Carbon			.625	65	.030	M2, M3	190	250	.040	–	–	–	–
			1	41	.18	S4, S5	115	150	.18	P10	200	.18	CP10
1340 4340 81B45			4	32	.40	S4, S5	90	120	.50	P20	150	.40	CP20
1345 50B40 8640			8	24	.50	S4, S5	73	95	.75	P30	120	.50	CP30
4042 50B44 8642			16	20	.75	S4, S5	58	76	1.0	–	–	–	–
4047 5046 8645		Annealed, Normalized, Cold Drawn or Quenched and Tempered	.040	115	.007	M2, M3	350	465	.007	C 7	600	.007	CC 7
4140 50B46 86B45	225 to 275		.150	90	.015	M2, M3	280	365	.020	C 6	475	.015	CC 6
4142 5140 8740			.300	70	.020	M2, M3	220	285	.030	C 6	375	.020	CC 6
4145 5145 8742			.625	55	.030	M2, M3	170	225	.040	–	–	–	–
4147 5147			1	35	.18	S4, S5	105	140	.18	P10	185	.18	CP10
			4	27	.40	S4, S5	85	110	.50	P20	145	.40	CP20
			8	21	.50	S4, S5	67	87	.75	P30	115	.50	CP30
			16	17	.75	S4, S5	52	69	1.0	–	–	–	–
			.040	90	.007	T15, M42+	330	440	.007	C 7	575	.007	CC 7
			.150	70	.015	T15, M42+	260	340	.015	C 6	450	.015	CC 6
			.300	55	.020	T15, M42+	200	270	.020	C 6	350	.020	CC 6
			.625	–	–	–	–	–	–	–	–	–	–
				27	.18	S9, S11+	100	135	.18	P10	175	.18	CP10
				21	.40	S9, S11+	79			P20		.40	CP20
					.50								CP30

FIGURE 42-11   Portion of table from the *Machining Data Handbook*, 3rd ed., which gives recommended speeds and feeds for a given material; broken down by operation, material hardness, and condition according to what tool material is being used at a given depth of cut. Such data are considered to be reliable, but their accuracy is not guaranteed. *(Courtesy of Metcut Research Associates, Inc.)*

TABLE 42-1.   Typical Inputs and Outputs for Computerized Machinability Data Systems in Use Today

Inputs	Outputs
Machine tool code	Recommended speed/feed
Cutting tool code	Production-time calculations
Material cost	Number and size of cuts
Operation code	Optimum speed/feed
Depth of cut	Cutting tool/grade
Material hardness	Machine tool

Source: Machining Briefs, Mar./Apr. 1982.

When an operation is machine-controlled, as in making a lathe cut of a certain length with power feed, the required time can be determined by simple mathematics. For example, if a cut is 10 in. in length and the turning speed is 200 rpm, with a feed of 0.005 in. per revolution, the cutting time required will be 10 minutes. Procedures are available for determining the time required for people-controlled machining elements, such as moving the carriage of a lathe by hand back from the end of one cut to the starting point of a following cut. Such determinations of time estimates are generally considered the job of the industrial engineer, and space does not permit us to cover this material in this book, but the

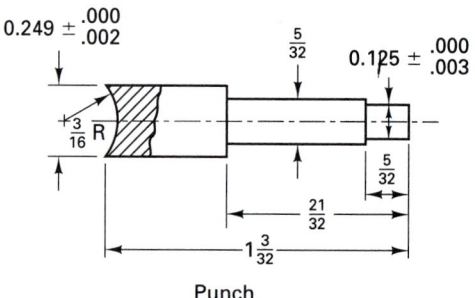

FIGURE 42-12 Part drawing of a punch.
See Figure 42-8 for the route sheet.

techniques are well established. Actual time studies, accumulated data from past operations, or some type of motion-time data, such as MTM (methods-time-measurement), can be employed for estimating such times. Each can provide accurate results that can be used for establishing standard times for use in planning. Various handbooks and books on machine-shop estimating contain tables of average times for a wide variety of elemental operations for use in estimating and setting standards. However, such data should be used with great caution, even for planning purposes, and they should never be used as a basis of wage payment. The conditions under which they were obtained may have been very different from those for which a standard is being set.

ROUTE SHEETS. *Route sheets* are very useful for general planning in the job shop even though they do not provide detailed information for the operation. The route sheet lists the processes that must be performed to produce the part, in their sequential order, and the machines or work stations and tooling that will be required for each operation. For example, Figure 42-12 shows the drawing of a small, round "punch," and Figure 42-8 is a routing sheet for making this part. Once the routing of a part has been determined, the detailed planning of each operation in the processing can then be done and the operations sheet prepared. The route sheet is also needed by *production planning,* another functional group in the PS that schedules the products (parts) through the machines and assigns the personnel to run the machines. Notice that the route sheet does not provide information on *when* (i.e., the timing) the parts are to be made.

QUANTITY VERSUS PROCESS AND MATERIAL ALTERNATIVES. Most processes are not equally suitable and economical for producing a range of quantities for a given product. Consequently, the quantity to be produced should be considered, and the design should be adjusted to the process that actually is to be used before the design is "finalized." As an example, consider the part shown in Figure 42-13. Assume that, functionally, brass, bronze, a heat-treated aluminum alloy, or ductile iron would be suitable materials. What material and process would be most economical if 10, 100, or 1000 parts were to be made?

If only 10 parts were to be made, contour sawing, followed by drilling the $\frac{3}{4}$-in. hole, would be very economical. The irregular surface would be difficult to machine unless a three-axis CNC contouring milling machine were available, and then the part could be machined out of a 2.5 in. × 3.0 in. × 2.0 in. block of metal. Casting would require the making of a pattern, which would be about as costly to produce as the part itself. It is unlikely that a suitable piece of ductile iron or heat-treated aluminum alloy would be readily available. Brass would be considerably cheaper than bronze. Therefore, brass, contour sawing, and drilling most likely would be the best combination. For 10 parts, the excess cost of

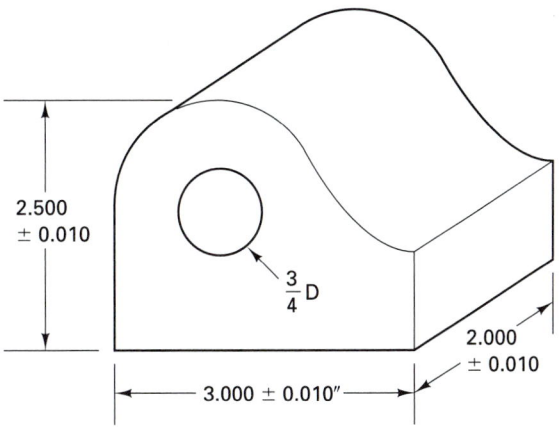

2.500
± 0.010

$\frac{3}{4}$ D

2.000
± 0.010

3.000 ± 0.010″

FIGURE 42-13   Part to be analyzed for production.

brass over ductile iron would not be great, and this combination would require no special consideration on the part of the designer.

For a quantity of 100 parts, an effort to minimize machining costs may be worthwhile. Casting might be the most economical process, followed by machining. Ductile iron would be cheaper than any of the other permissible materials. Although the design requirements for casting this simple shape would be minimal, the designer would want to consider them, particularly as to whether the hole should be cored.

For 1000 parts, entirely different solutions become feasible. The use of an aluminum extrusion, with the individual 2-in. units being sawed off, might be the most economical solution assuming that the time required to obtain a special extrusion die was not a factor. The hole in the part could be extruded, so this process is eliminated, but what are the trade-offs? What material would you recommend for the part now? How many feet of extrusion would be required, including sawing allowance? How much should be spent on the die so that the per-piece cost would not be great and probably would be more than offset by the savings in machining costs? If extrusion is used, the designer should make sure that any tolerances specified are well within commercial extrusion tolerances.

This simple example clearly illustrates how quantity can affect both material and process selection, and the selection of the process may require special considerations and design revisions on the part of the designer. Obviously, if the dimensional tolerances were changed, entirely different solutions might result. When more complex products are involved, these relations become more complicated, but they also are usually more important and require detailed consideration by the designer.

Other duties of the manufacturing engineer may include the responsible for the design of tools, jigs, and fixtures to produce the product. Just as the engineer who is concerned with manufacturing must understand the operation, functionality, and capability of machine tools but almost never designs them, similarly he or she should have a thorough understanding of the basic principles of jigs and fixtures so as to utilize them effectively. However, in most cases, the design of the tooling is left to the tool design specialists. In most companies, the manufacturing engineer makes recommendations on new machine tools, cutting tools, and other equipment.

After the product is in production, manufacturing problems invariably arise, and because of the functional design and operation of the existing system (the job shop), the manufacturing engineer has the responsibility for solving them. A typical scenario might entail poor-quality materials being received from a vendor and accepted (in error). Use of these poor-quality materials causes fixtures to work improperly, producing defective parts and resulting in components that cannot be assembled. Assembly workers, on incentive pay, will sacrifice quality for the sake of the piece rate, and the company ships defective products. Finding the cause of such problems is part of the manufacturing engineer's job and is sometimes called *troubleshooting* or "firefighting." The actual causes of problems may not be eliminated because of the pressure to keep on schedule. Since the causes are not eliminated, the defects keep coming back (in greater numbers) when material shortages occur in the system. The manufacturing engineer can find himself or herself responsible for a system that cannot possibly be controlled, because of its functional design.

Industrial Engineering. *Industrial engineers* are usually responsible for determining the level of workers, machines, and materials needed on the plant floor to turn the ideas developed in R&D, marketing, and procurement into real products. Industrial engineers look for the "better way" to produce products and services under uncertain conditions and constraints, such as the nature of the plant, materials, machines on hand, personnel, and available capital. The industrial engineering department is responsible for many elements of the MS and PS, including some that overlap with those of the manufacturing engineer. These include:

1. Production methods analysis (motion study)
2. Work measurement (time study and time standards)
3. Safety and ergonomics
4. Factory design (layout) and material handling
5. Quality engineering (may be a separate functional group)
6. Plant maintenance information

Part of the industrial engineering function is to determine, standardize, and analyze the methods used to produce particular products and services. Motion-study principles, videotapes and movies, SMED (see Chapter 43), and other techniques are used to determine how a product or service can best be produced and to develop efficient work methods. In other words, the sequence of activities as well as the machines, tools, and materials to be used must all be specified in the job shop environment.

After standardization of the job's content, information is obtained on how much time is required to do the job. This information is based on the time required for an average person to produce a given product or service, using average effort under normal working conditions. Time standards and studies are used to develop standard times for the standard methods. The method for a particular job can be analyzed using an *operations analysis sheet*.

Designing the plant layout and the associated materials-handling equipment falls to industrial engineering. Layouts that reduce manufacturing costs with minimized materials handling and inventory are fundamental to integrated manufacturing systems.

Plant Engineering. *Plant engineering*, another functional engineering group within the classical production system, is responsible for in-plant construction and *maintenance*, machine tool and equipment repair, heating and air-conditioning system maintenance, and for any other mechanical, hydraulic, or electric problems not necessarily related to the manufacturing system. For example, suppose that a new machine tool were being purchased. Plant engineers would be responsible for seeing to the installation of the machine.

Quality Engineering. *Quality engineering* is responsible for assuring that the quality of the product and its components meets the standards specified by the designer before, during,

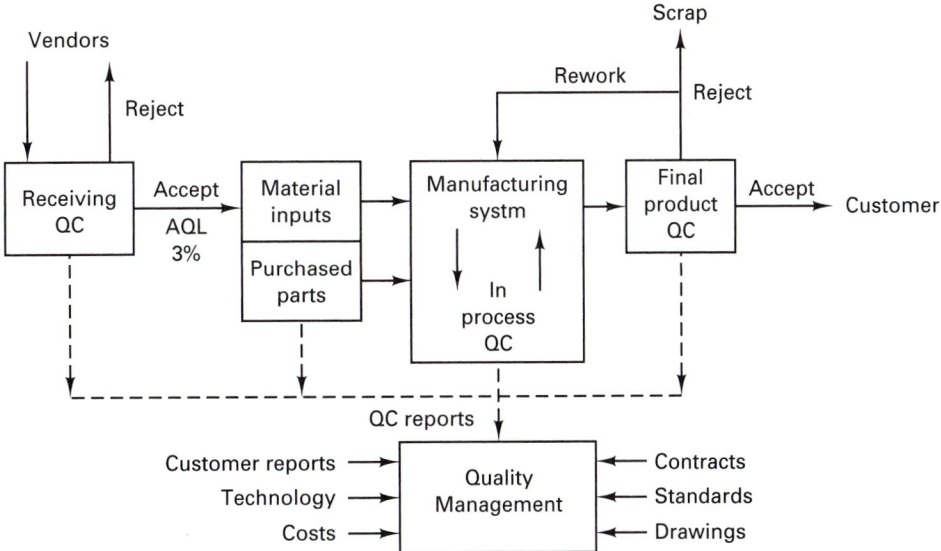

FIGURE 42-14   Classical quality control system.

and after manufacturing (Figure 42-14). In-process quality control inspections are performed at various points throughout the manufacturing system. In the classical system, materials and parts purchased from outside suppliers are inspected when they are received. The acceptable quality level (AQL) is traditionally around 2%, which means that the company is willing to accept 2% defectives from their vendors. Historically, many companies have believed that it costs too much to reduce defectives below this level. In recent years, goals of perfection and defect rates of parts per million have been the new way of life in the world of quality. Quality is designed into the product and the processes. Parts fabricated inside the company may be inspected many times during processing. Final inspection and testing of the finished product is performed to determine overall functional performance and appearance quality.

Procurement and Purchasing

The *procurement and purchasing* functions in a company involve primarily the acquisition of specified materials, equipment, services, and supplies of the proper quality, in the correct quantities, at the best prices, at the correct time. Many departments are involved in procurement, purchasing, manufacturing, marketing, finance, accounting, research and development, and engineering. For production systems, procurement provides information on vendors, prices, new products, and materials and determines the delivery schedule for purchased items.

Production Planning and Control

Production planning translates sales into forecasts by part number. The authority to manufacture the product is translated into a *master production schedule*, a key planning document specifying the products to be manufactured, the quantity to be produced, and the delivery date to the customer (Figure 42-15). The master schedule is converted into purchase orders

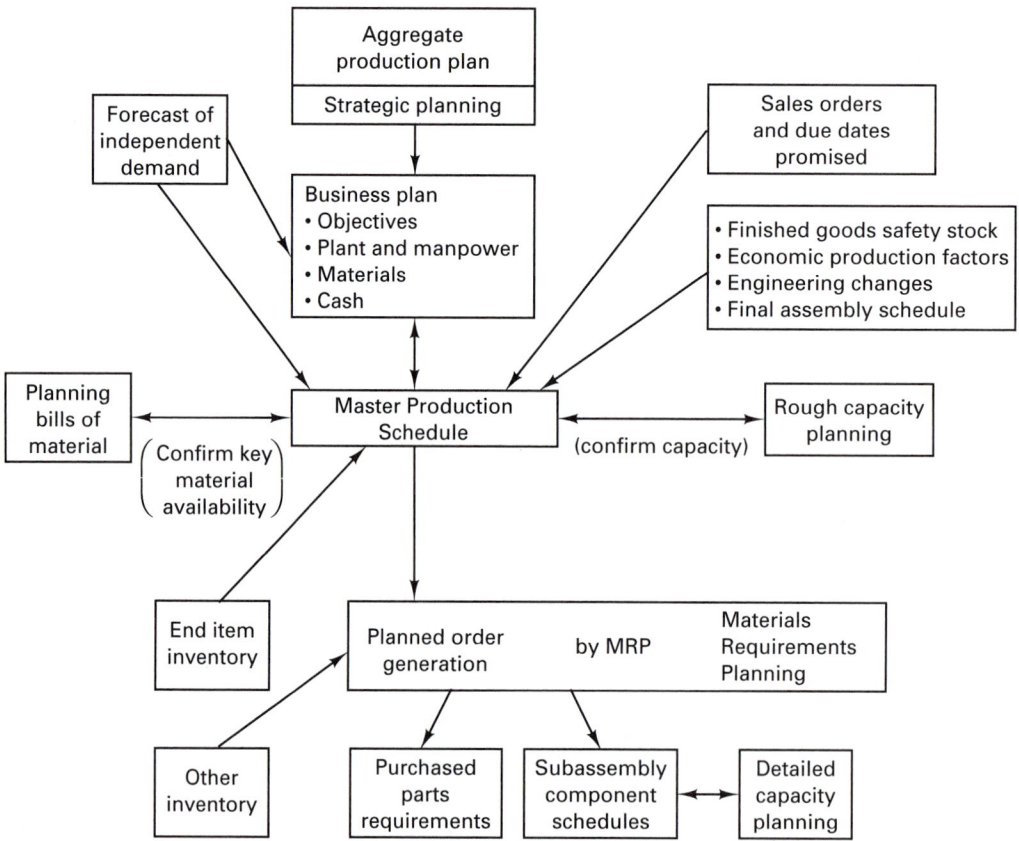

FIGURE 42-15   Master production schedule indicates what products to produce and when the products are needed.

for raw materials, orders for components from outside vendors, and production schedules for parts made in the manufacturing system. *Production control* develops the timing and coordination to ensure that delivery of the final product meets customer demand. Because of the complexity of the job shop MS, production is not controlled very well; therefore, many other control functions are needed.

The scheduling periods used in the master schedule are usually months. The master schedule must take into account the production capacity of the plant (how much can be built in a given period of time). The capacity of a job shop is tremendously variable and flexible and is not well controlled. Because of this characteristic, larger quantities of products are often requested in violation of the master schedule.

Based on the master schedule, individual components and subassemblies that make up each product are planned. Raw materials are ordered to make the various components. Purchased parts are ordered from vendors. Planning is a must if the components and assemblies are to arrive when needed, not months before or, even worse, days or weeks late. Many companies use a computer software technique called *material requirements planning* (MRP), which is discussed later.

The next task is *production scheduling*, in which start dates and due dates are assigned for the various components to be processed through the factory. Many factors make the scheduling job complex. The number of individual parts can be in the thousands. Each part seems to have its own individual process route through the plant. Parts are often routed through dozens of separate machines in many different departments. The number of machines in the shop is limited, and the machines are different, perform different operations, and have different features, capacities, and capabilities. In effect, the orders compete with each other for machines. In addition to these factors, parts become defective during the processing, cycle times vary, machines break down, and operators expand job times to fit the time available. All these factors destroy the validity of the planning schedule and require (in the classical system) huge amounts of resources (lots of people and paperwork) to manage all the exceptions. Chaos reigns supreme in the large job shop.

Dispatching is a production planning and control function requiring voluminous paperwork wherein individual orders comprised of order tickets, route sheets, part drawings, and job instructions are sent to the machine operators or foremen.

Expediters find lost or late materials by tracking the progress of the order against the production schedule. To speed up late orders, the expediter may rearrange the order-processing sequence for a certain machine, coax the foreman to tear down one setup so that another order can be run, or hand-carry parts from one department to the next just to keep production going. Obviously, the master schedule is disrupted. The size of the production control department and the number of expediters in a company is an informal measure of the level of chaos and inefficiency in the production control system.

Inventory Control

The life blood of a manufacturing system is its inventory. Inventory represents a major portion of a manufacturing facility's assets. For most companies, however, excess inventory represents idle investment dollars and wasted storage space. Even though the cost of carrying large inventories is substantial, many reasons are given for having them:

1. Fluctuation in demand and/or supply
2. Protection against process breakdowns
3. Replacement parts for lost batches or defective lots
4. Overproduction in anticipation of future demand
5. Protection from defective parts
6. Goods in transport
7. Just in case it is needed
8. Quantity purchasing

Inventory control governs finished goods, raw materials, purchased components, and work-in-process within the factory. The idea is to achieve a balance between too little inventory (with possible stockouts of raw materials) and too much inventory (with investments and storage space tied up).

The product, when finished, is either shipped directly to the customer or stocked in inventory. *Inventory control is* used to ensure that enough products of each type are available to satisfy customer demand. However, competing with this objective is the company's desire to minimize its financial investment in inventory. Inventory control interfaces with marketing and production control since coordination must exist between the various products' sales, production, and inventory levels. Although none of these three functions can operate effectively without information about what the others are doing, this information is often missing or out of date. Inventory control is often done by the production planning and control department.

The requirements for a total inventory control system are that it:

1. Analyze and plan inventory requirements.
2. Purchase raw materials and component parts in the amounts needed according to scheduled usage.
3. Receive and record the receipt of purchased materials.
4. Provide adequate facilities to store raw material, work-in-process, and finished-goods inventory.
5. Maintain accurate records of inventories on hand and on order.
6. Install realistic controls for materials in stores and for the issuance of materials, parts, and supplies.

A good inventory control system provides the correct quantity and quality of material at the correct time. This system also maintains accurate records/control of these materials.

Inventory Models. In 1915, Ford Harris and R. H. Wilson derived, independently, the *simple lot-size formula.* This model states (see Figure 42-16) that

$$\text{total annual cost} = \text{purchase cost} + \text{order cost} + \text{holding cost}$$

$$\text{TC} = RP + \frac{RC}{Q} + \frac{QiP}{2} \tag{42-1}$$

where

R = annual demand (units)
P = unit cost of an item
C = ordering cost per order
$H = iP$ = holding cost per unit per year
Q = lot size or order quantity (units)
i = annual holding cost as a fraction of unit cost

To find the value of Q that minimizes the total cost, use

$$\frac{d(\text{TC})}{dQ} = 0 - \frac{RC}{Q^2} + \frac{iP}{2} = 0$$

$$\frac{RC}{Q^2} = \frac{iP}{2}$$

$$Q = \sqrt{\frac{2RC}{iP}}$$

The model, called the *economic order quantity model,* has also been extended to cover economic production quantity by letting C equal the setup cost. The model assumes that the setup cost is fixed when, in fact, the setup cost can often be significantly reduced. Reduction in the setup cost results in a reduction in lot size. The best lot size is then the smallest lot size that permits a smoothly running process.

Many other models have been developed for the inventory process, and many models and systems have since evolved for stock or inventory replenishment, including the *reorder point model.* Figure 42-17 illustrates the relationship between lead time (L), order size (Q), safety stock (SS), expected demand rate (D), and the reorder point (ROP). The reorder point is based on how long it takes to obtain parts (lead time) and how many parts will be used up during this (lead) time.

One of the problems with the reorder point model is the dependence of the model on usage. The model shows that parts are used linearly. In fact, in batch processing, utilization

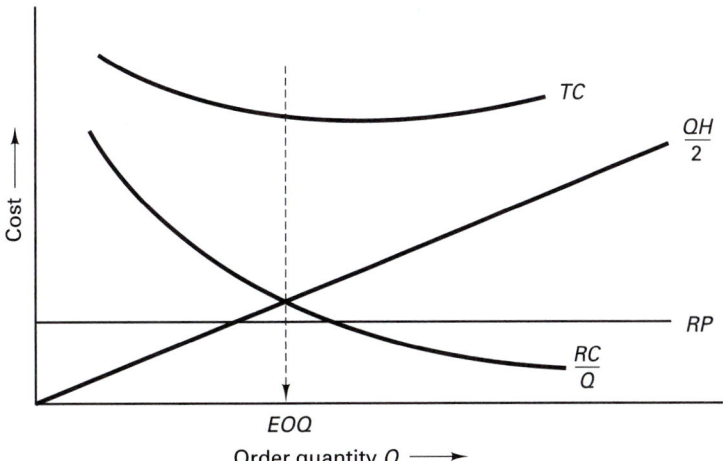

FIGURE 42-16   The economic order quantity (EOQ) model minimizes the total (annual) cost of inventory.

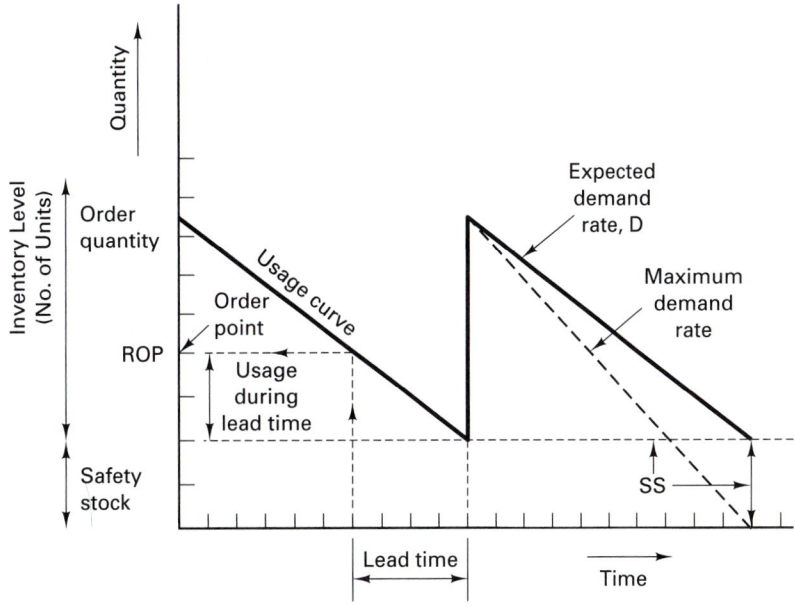

To find ROP
1. Establish lead time.
2. Back lead time off from order receipt time.
3. Go to usage curve.
4. Find order pont on vertical scale.

FIGURE 42-17   Classical inventory model for ROP (reorder point) methods.

is usually uneven, with large spikes or peaks. That is, 200 of an item may be used in week 2, zero in weeks 3 through 9, and 200 again in week 10. The reorder point model would cause the system to run out of parts in week 2, precipitating a rush order for parts that would remain on the shelf unnecessarily until week 10. The model also breaks down if the average utilization changes. To use the model effectively, demand must be constantly re-calculated.

The utilization of certain parts is dependent on the demand for other parts. For example, in the building of cars, the demand for tires depends on how many cars are built. While demand for cars must still be estimated, the demand for tires is known, based on the demand for cars.

These two types of demand are categorized as *independent* and *dependent*. The first must be "guessed" (forecasted). The second can be computed. The basis for these computations is a record of the relationship between the independent demand item (cars) and its dependent components, (tires, horn, windows.) This record is the *bill of materials* (see Figure 42-7).

Manufacturing Resource Planning.   To control inventory within the job shop, a computerized system called *material requirements planning* (mrp) was developed (Figure 42-18). Given a schedule showing the expected demand of independent demand items (a master production schedule) and given the relationship between independent and dependent demand items (bills of materials), mrp will calculate the quantities of dependent demand items needed and when they will be needed.

The potential value of this concept is significant. Suppose that a company has 100 finished goods items, 400 assemblies and subassemblies, and 1000 raw material items. Using statistical stock replenishment, it will need to forecast the average demand for 1500 items. Many of these items will have "lumpy" demand, which will cause the stockout/expedite/overstock cycle discussed earlier. With mrp, a *master production schedule* (MPS) for 100 items must be maintained; the other 1400 items will have their exact demand computed for every period.

Almost at the same time that mrp was developed, practitioners began to expand the concept's scope. Just as bills of materials could establish the usage of dependent materials, other records could be developed to tell the dependent requirements of labor hours, machine hours, capital, shipping containers—in fact any of the resources required to support the job shop MPS. Material requirements planning has become *manufacturing resource planning* (MRP). This system is responsible for final scheduling of production, dispatching, and releasing purchase orders. It will also maintain stock status, monitor output and scrap levels, and compare performance against the plan.

MRP and mrp are good *planning* techniques. There is no inherent capability to *control* and replan. Subsequent sophistication of MRP by adding feedback of actual results has led to *closed-loop manufacturing resource planning* (MRP II). Shop floor control and vendor control systems are added to the existing software so that revisions of dates and quantities will be taken into account in the next planning cycle.

The job shop is a complex dynamic manufacturing system. Inventory is its life blood. MRP is an attempt to control inventory using computers. Unfortunately, the MRP systems that have evolved are so complex that very few people in the companies that use them really understand them. Would you trust a system you do not understand? What is the value of an inventory control system that the majority of users do not comprehend?

What makes a good production/inventory control system? First, everyone who uses it must understand how it works. It must have accurate information and make accurate predictions or forecasts. The users of the system must act on the information that the

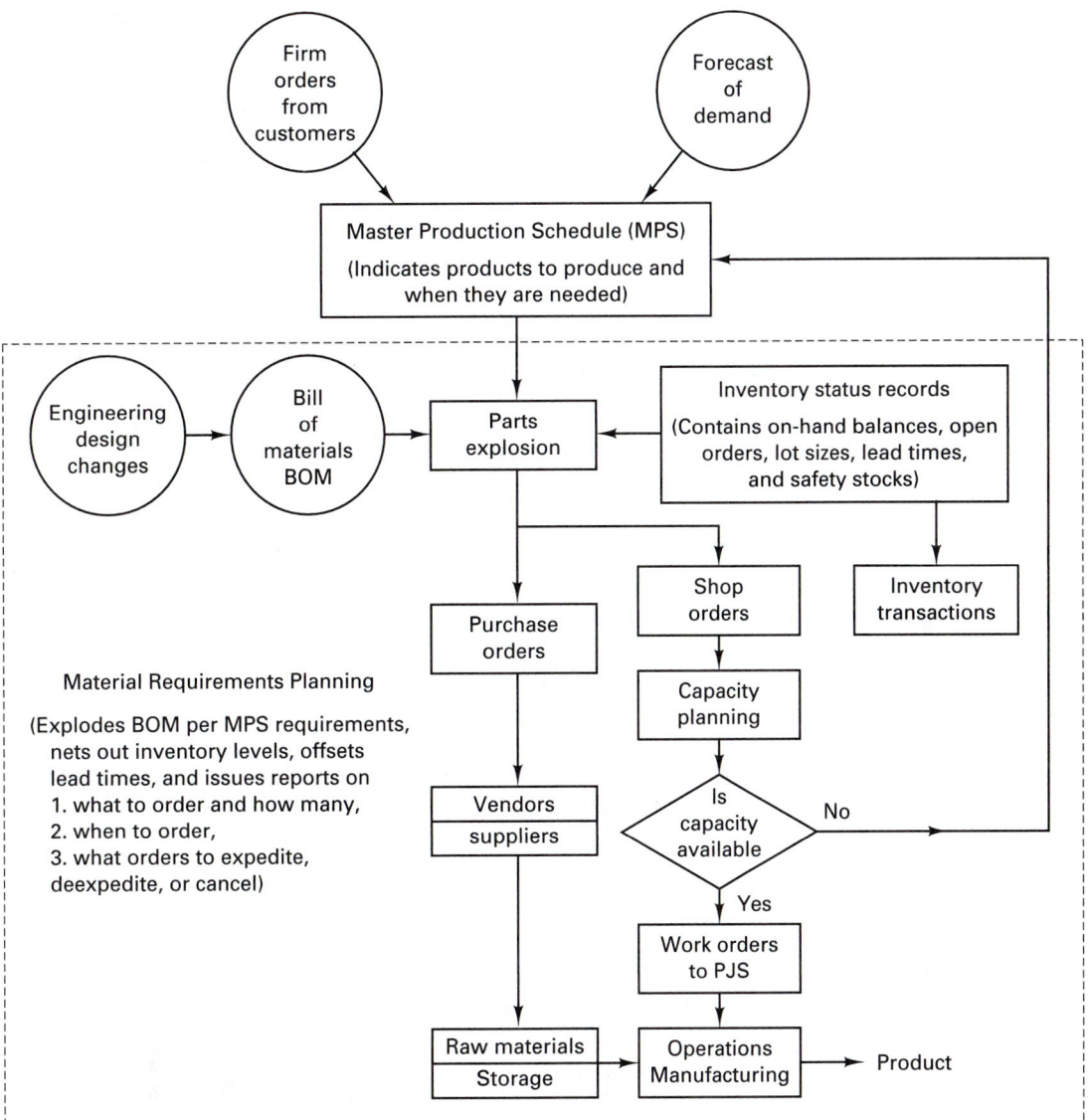

FIGURE 42-18 Material requirements planning is a computerized inventory system for the job shop.

system produces. In Chapter 43 a simple, remarkable system is described that has become known as the pull system (or the JIT kanban system). But a word of caution: This inventory/control system is not usable in the job shop. It is for a new MS called linked-cell MS, which can replace the job shop.

42.3 COMPUTER-INTEGRATED MANUFACTURING

A number of definitions have been developed for *computer-integrated manufacturing* (CIM). However, a CIM system is commonly thought of as an integrated system that encompasses all the activities in the production system from the planning and design of a product through the manufacturing system, including control. CIM is an attempt to combine existing computer technologies in order to manage and control the entire business. CIM is an approach that very few companies have adopted at this time, since surveys show that only 1 or 2% of U.S. manufacturing companies have approached full-scale use of FMS and CAD/CAM, let alone CIM systems.

As with traditional manufacturing approaches, the purpose of CIM is to transform product designs and materials into salable goods at a minimum cost in the shortest possible time. CIM begins with the design of a product (CAD) and ends with the manufacture of that product (CAM). With CIM, the customary split between the design and manufacturing functions is (supposed to be) eliminated.

CIM differs from the traditional job shop manufacturing system in the role the computer plays in the manufacturing process. Computer-integrated manufacturing systems are basically a network of computer systems tied together by a single integrated database. Using the information in the database, a CIM system can direct manufacturing activities, record results, and maintain accurate data. CIM is the computerization of design, manufacturing, distribution, and financial functions into one coherent system.

Figure 42-19 presents a block diagram illustrating the functions and their relationship in CIM. These functions are identical to those found in a traditional production (planning and control) system for a job shop MS. With the introduction of computers, changes have occurred in the organization and execution of production planning and control through the implementation of such systems as mrp, capacity planning, inventory management, shop floor control, and cost planning and control.

Engineering and manufacturing databases contain all the information needed to fabricate the components and assemble the products. As shown in Figure 42-19, the design engineering and process planning functions provide the inputs for the engineering and manufacturing database. This database includes all the data on the product generated during design, such as geometry data, parts lists, and material specifications. The bill of materials is shown separately, but it is a key part of the database. Figure 42-20 shows how the CAD/CAM database is related to the design and manufacturing activities. Included in the CAM is a CAPP (*Computer-aided process planning*) module, which acts as the interface between CAD and CAM.

Capacity planning is concerned with determining what labor and equipment capacity is required to meet the current master production schedule as well as the long-term future production needs of the firm. Capacity planning is typically performed in terms of labor and/or machine hours available. The master schedule is transformed into material and component requirements using mrp. These requirements are then compared with available plant capacity over the planning horizon. If the schedule is incompatible with capacity, adjustments must be made either in the master schedule or in plant capacity. The possibility of adjustments in the master schedule is indicated by the arrow in Figure 42-19 leading from capacity planning to the master schedule.

The term *shop floor control* in Figure 42-19 refers to a system for monitoring the status of manufacturing activities on the plant floor and reporting the status to management so that effective control can be exercised. The cost planning and control system consists of the database to determine expected costs to manufacture each of the firm's products. It also consists of the cost collection and analysis software to determine what the actual costs of manufacturing are and how these costs compare with the expected costs.

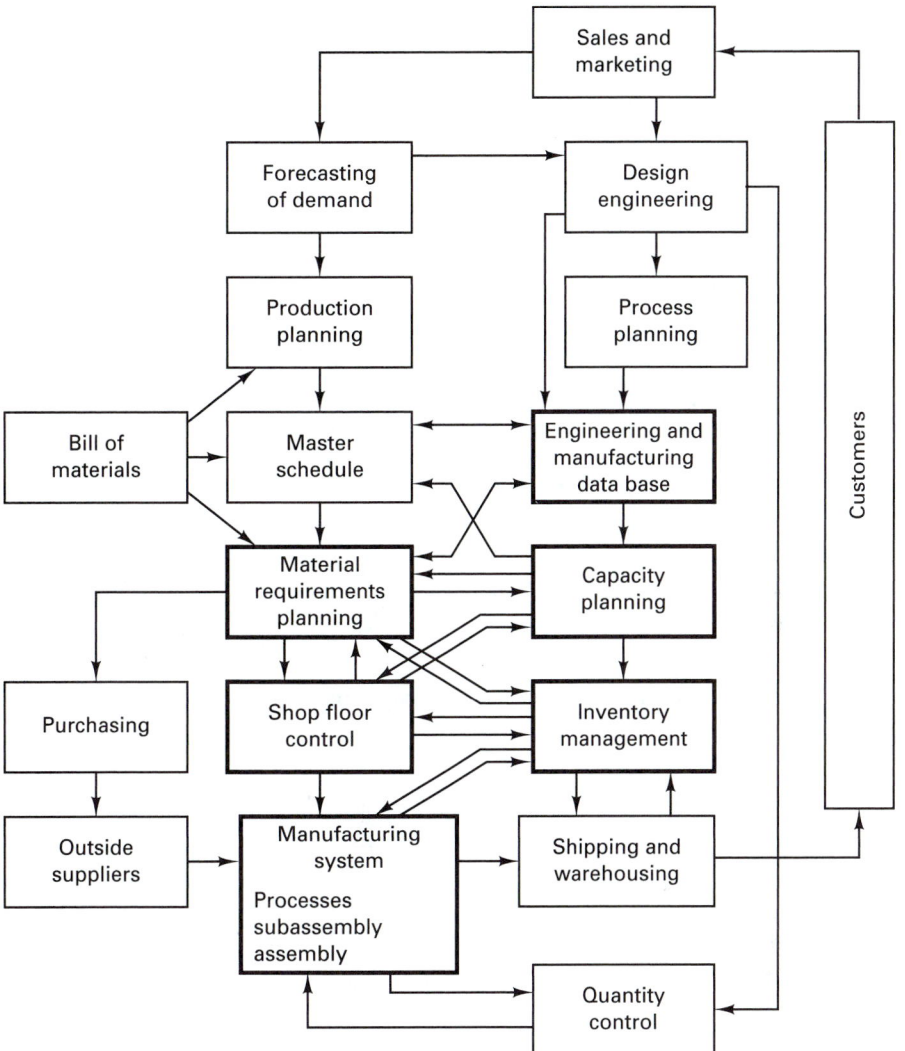

FIGURE 42-19 Cycles of activities in a computer-integrated manufacturing system.

Computer-Aided Design

A major element of a CIM system is a *computer-aided design* (CAD) system. CAD involves any type of design activity that makes use of the computer to develop, analyze, or modify an engineering design. The design-related tasks performed by a CAD system are:

- Geometric modeling
- Engineering analysis
- Design review and evaluation
- Automated drafting

Geometric modeling corresponds to the synthesis phase of the design process. To use geometric modeling the designer constructs the graphical image of the object on the CRT.

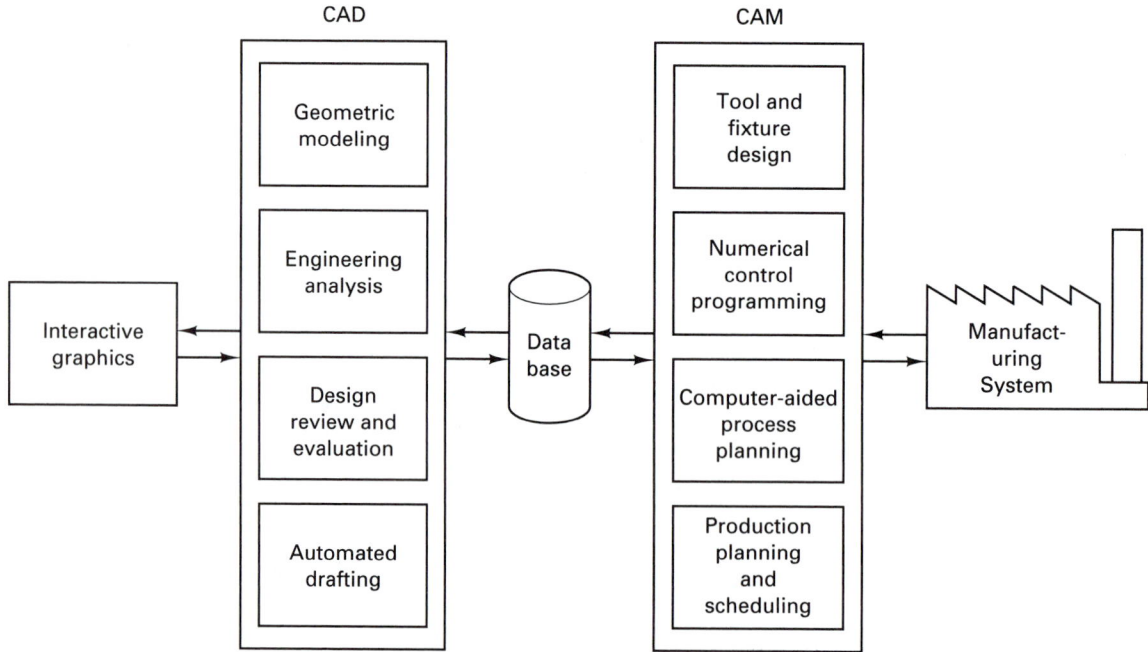

FIGURE 42-20   Database relationship to design and manufacturing.

The object can be represented using several different methods. The simplest systems use wire frames to represent the object. The wire frame geometric modeling is classified as:

- Two-dimensional representation used for flat objects
- Three-dimensional modeling of more complex geometries

Other enhancements to wire-frame geometric modeling are:

- Color graphics
- Dashed lines to portray rear edges that would be invisible from the front
- Removal of hidden lines
- Surface representation that makes the object appear solid to the viewer

The more advanced method of geometric modeling is solid modeling in three dimensions. This method uses solid geometry shapes called *primitives* to construct the object (Figure 42-21).

Engineering analysis is required in the formulation of any engineering design project. The analysis may involve stress calculations, finite element analysis for heat transfer computations, or the use of differential equations to describe the dynamic behavior of the system. Generally, commercially available general-purpose programs are used to perform these analyses.

Design review and evaluation techniques check the accuracy of the CAD design. Semiautomatic dimensioning and tolerancing routines that assign size specification to surfaces help to reduce the possibility of dimensioning errors. The designer can zoom in on the part design details for close scrutiny. Many systems have a layering feature that involves overlaying the geometric image of the final shape of a machined part on top of the

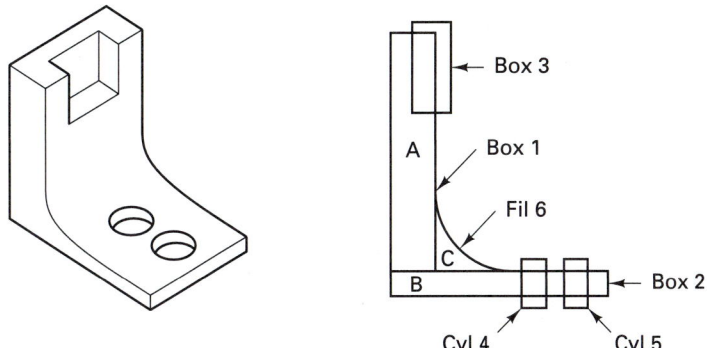

Part = Sum of Vols A, B, C
Volume A = Box 1 – Box 3
Volume B = Box 2 – Cyl 4 – Cyl 5
Volume C = Fil 6

FIGURE 42-21 The CAD (computer-aided design) part design is constructed by summing basic solid geometry shapes.

image of a rough part. Other features are interference checking and animation, which enhance the designer's visualization of the operation of the mechanism and help to ensure against interferences.

Automated drafting involves the creating of hard-copy engineering drawings from the CAD database. Typical features of CAD systems include automatic dimensioning, generation of crosshatched areas, scaling of drawings, ability to develop sectional views, enlarged views of particular part details, and the ability to rotate the part and perform transformations such as oblique, isometric, and perspective views.

Parts Classification and Coding for Group Technology

A group technology coding and classification system can also be developed using the CAD database. Parts coding involves assigning letters or numbers to parts to define their geometry or manufacturing process sequence. The codes are used to classify or group similar parts into families. Designers can use the classification and coding system to retrieve existing part designs rather than redesigning new parts.

Computer-Aided Process Planning

CAPP uses computer software to determine how a part is to be made. If GT is used, parts are grouped into part families according to how they are to be manufactured. For each part family, a standard process plan is established. That is, each part in the family is a variation of the same theme. The standard process plan is stored in computer files and then retrieved for new parts that belong to that family.

Figure 42-22 explains the CAPP process. The user initiates the procedure by entering the part code. The CAPP program then searches the part family file to determine whether a match exists. If the file contains an identical code number, the standard machine routing and operation sequence are retrieved for display to the user. The standard operation sequence is examined by the user to permit any necessary editing of the plan to make it compatible with the new part design. This is variant CAPP. After editing, the process

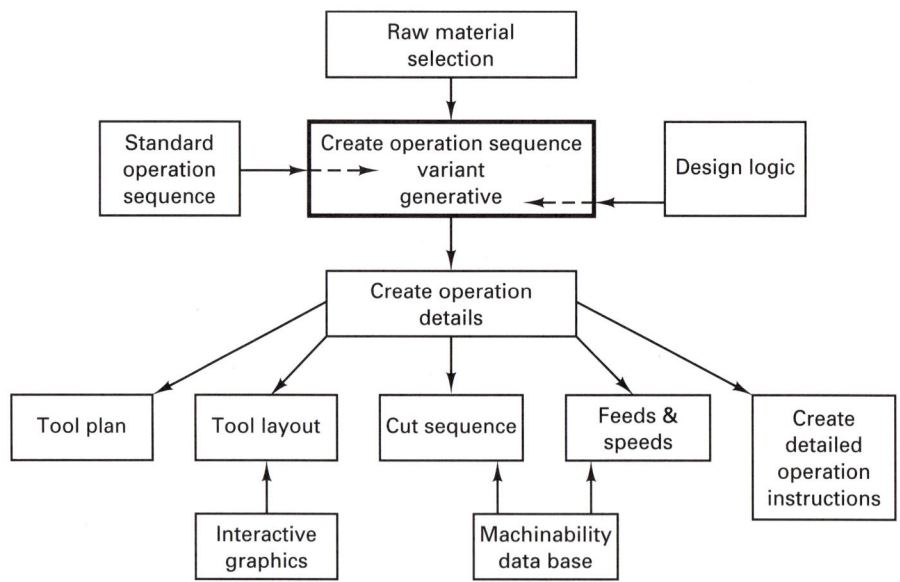

FIGURE 42-22 CAPP (computer-aided process planning) can be variant or generative.

plan formatter prepares the paper documents (the route sheet and operation sheets for the job shop).

If an exact match cannot be found, a new plan has to be *generated* based on design logic and the examination of geometry and tolerance information and then comparing these requirements to the machine capability. The software determines what processes can produce what surfaces and establishes the machining constraints. Once the process plan for a new part code number has been entered and verified, it becomes the standard process for future parts of the same classification. The process plan formatter may include software to compute machining conditions, layouts, and other detailed operation information for the job shop.

Computer-Aided Manufacturing

Another major element of CIM is *computer-aided manufacturing* (CAM). An important reason for using a CAD system is that it provides a database for manufacturing the product. However, not all CAD databases are compatible with manufacturing software. The tasks performed by a CAM system are:

- Numerical control or CNC programming
- Production planning and scheduling
- Tool and fixture design

Numerical control, discussed in Chapter 29, can use special computer languages. APT and COMPACT II are the two most common language-based computer-assisted programming systems used in industry today. These systems take the CAD data and adapt them to the particular machine control unit/machine tool combination used to make the part.

Manufacturing Automation Protocol. The CIM approach results in a vast collection of computers throughout the company. Because these computers are purchased at different times, they either cannot communicate with each other or require extensive (costly) custom interfaces. To deal with this problem, General Motors initiated MAP, an acronym for *manufacturing automation protocol*, which is a communication network specification defining a local area computer network—a network that offers reliable high-speed communication channels that are optimized for connecting different items of information-processing equipment within a limited geographic area. The major thrust of MAP is to provide a *local area network* (LAN) that is based on open rather than proprietary standards so that equipment from multiple vendors can be connected through the same network. To do this, each vendor's equipment must meet the universal MAP specification, a specification that most users feel is long overdue.

The network architecture created by most vendors in recent years has conformed to the International Standards Organization (ISO) model for open systems interconnection (OSI) (Table 42-2). This seven-layer model formalizes a developing consensus among communication network designers, provides the framework for layered networks, and introduces a uniform terminology for naming the various utilities involved. MAP uses the same ISO seven-layer reference model and further defines the protocol standards to be followed at each layer.

TABLE 42-2. International Standards Organization (ISO) Seven-Layer Model for Open-System Interconnects[a]	
ISO Protocol Layer	MAP 2.1 Protocol Selection
User program	MMFS/EIA 1393A
7. Application	ISO case kernel, FTAM
6. Presentation	Not used yet (null)
5. Session	ISO session kernel
4. Transport	ISO transport class 4
3. Network	ISO CLNS
2. Datalink	IEEE 802.2 link level, control class 1
1. Physical	IEEE 802.4, token access on broadband media

[a]Current MAP protocols selected for MAP version 2.1 shown.

The arrangement of hardware and software required for ISO seven-layer communications at each computer device is a MAP interface. When communications occur, each ISO layer (except any layer that is null) must pass the signals to the next so that the correct data are transferred from the coaxial cable to the computer—or from the computer to the coaxial cable, depending on whether the device is sending or receiving.

To aid understanding of the function performed at each layer, analogies have been made to sending a letter through the U.S. postal system. Table 42-3 defines the primary function of each ISO layer and a corresponding function of the postal system. The example relates to a message placed inside an envelope, the envelope addressed to a receiver, and mailed.

■ 42.4 SUMMARY

Regardless of all that has been written about CIM, the technology is not widespread. What has been called lean production appears to be more important to the future of manufacturing than to CIM. The evidence suggests that unless a company first adopts the basic tenets of lean production, the conversion to CIM is likely to fail.

In Chapter 43 this new approach to the factory of the future will be presented wherein many of the functions of the production system are integrated into the manufacturing system.

TABLE 42-3.	Comparison of Each ISO Layer to the U.S. Postal System Equivalent Function	
ISO Layer	ISO Function	U.S. Post Office Equivalent Function
7. Application	Provides all services directly comprehensible to application programs.	Message placed inside envelope, envelope addressed to receiver and sent on its way by mailing.
6. Presentation	Restructures data to/from standardized format used within the network.	Format and language of the letter, includes translation into proper language if required.
5. Session	Names/address translation access security, and synchronize/manage data.	Name, address, and zip code of both the receiver and the sender. The to and from.
4. Transport	Provides transparent, reliable data transfer from end node to end node.	Certified or registered mail providing verification to the sender that the letter arrived at the correct destination.
3. Network	Performs message routing for data transfer between nodes not in the same LAN.	The segment of the postal distribution system that transfers a letter outside the local postal system to a system in another city or country.
2. Datalink	Provides the means to establish, maintain and release logical datalinks between systems, transfer data frames between nodes in the same LAN, and detect and correct errors.	The segment of the postal distribution system that transfers a letter from the sender to a destination within the same postal system. (Either a receiver in the same system or to a distribution center for forwarding to a receiver in the system or to a distribution center for forwarding to a receiver in another system.)
1. Physical	Encodes and physically transfers messages between adjacent nodes.	Conveyance: postal worker, auto, truck, airplane, etc.

This requires that the job shop MS be replaced with a linked-cell MS. The functions of production control, inventory control, quality control, and machine tool maintenance are the first to be integrated.

■ KEY WORDS

accounting
computer-aided design (CAD)
computer-aided manufacturing
(CAM)
computer-aided process planning
(CAPP)
computer-integrated manufacturing
(CIM)
design for manufacture (DFM)
design for assembly (DFA)
dispatching
expediters

finance
industrial engineering
inventory control
local area network (LAN)
manufacturing automation protocol
(MAP)
manufacturing engineering
manufacturing resource planning
(MRP)
marketing
master production schedule (MPS)
material requirements planning

operations sheet
plant engineering
process planning
procurement
product design engineering
production system
purchasing
quality engineering
reorder point (ROP)
research and development (R&D)
route sheet

■ REVIEW QUESTIONS

1. How does a production system differ from a manufacturing system?
2. How is forecasting related to the manufacturing system and the production system as a whole?
3. Why is it so important for the designer to have intimate knowledge of the available manufacturing processes at the various stages of the design activity?

4. Explain how function dictates design with respect to the design of footwear. Use examples of different kinds of footwear (shoes, sandals, high heels, boots, etc.) to emphasize your points. For example, cowboy boots have pointed toes so that they slip into the stirrups easily and high heels to keep the foot in the stirrup.

5. Most companies, when computing or estimating costs for a job, will add in an overhead cost, often tying that cost to some direct cost, such as direct labor. What is included in this overhead cost?

6. How is this overhead cost, which if often given as 100 to 200% of direct labor cost, affected by better planning and scheduling?

7. How does the design of the product influence the design of the manufacturing system, including assembly and the production system?

8. Discuss this statement: "Software can be as costly to design and develop as hardware and will require long production runs to recover, even though these costs may be hidden in the overhead costs."

9. What is meant by this statement? "All processing schemes have a distinct technology and history and call for a distinct set of optimally suited materials."

10. If a designer called for a part to be made from cast steel, what processes has he eliminated from the possible list for casting this part?

11. Give an example of how the available cooking processes might limit what the chef can put on the hotel menu.

12. Give an example in design (detail) that might lead to a part failure or defect (see the Case Study).

13. In Figure 42-5, how would you have designed the shaft and flange pair if you were going to join these parts by friction welding?

14. Under what assembly conditions (again referring to Figure 42-5) might the initial design of the flange and shaft have been better than the improved design?

15. Can the part shown in Figure 42-10 be made more efficiently in an NC lathe? Why or why not?

16. What prevents the part shown in Figure 42-10 from being made by an automatic lathe?

17. Outline the manufacturing processes that you think might be needed to fabricate a razor blade and estimate what you think the blade itself actually costs. Razor blades have a final edge-cutting radius of 1000 Å or better. Did you determine the basic job requirements?

18. Figure 42-12 shows a drawing of a punch. What are the critical dimensions? Why are they critical? Why is overall length not critical? How is the part used?

19. Suppose that the part in Figure 42-10 were to be made in quantities of 250 rather than 25. What processes and machines would you have selected? Suppose that the quantity were 25,000; what then?

20. The part shown in Figure 42-13 is dimensioned incorrectly. There are missing dimensions. Redraw the part and put in the missing dimensions using a front view.

21. Assuming that the part shown in Figure 42-13 is made from aluminum, could you use extrusion and cutoff followed by hole drilling, considering the part tolerances. (*Hint:* What tolerances did you assign in Question 20?)

22. Find an example of a product or item the design of which has changed so that the item can be assembled automatically rather than manually. [This is called design for assembly (DFA).]

23. Manufacturing engineers create process plans. For the part shown in Figure 42-21, provide a process plan to make 1000 of these brackets. The brackets go into a metal bookcase and support the shelves.

24. Explain the difference between a route sheet and an operations sheet.

25. Find an example of an operations analysis sheet.

26. What does the study of ergonomics entail? (This was not discussed in Chapter 42.)

27. The IEs at the plant should assist the workers who are trying to reduce setup. What techniques are available to them for this effort?

28. What kinds of inspection systems have an AQL?

29. What is the difference between an expediter and a dispatcher in a job shop?

30. What is the difference between inventory control and production control?

31. What is the difference between dependent and independent demands?

32. How does the mrp use the BOM and the MPS to determine how many of an item should be ordered?

33. Define CIM in your own words.

34. A standard piece of CAD analysis is called FEA. What does a finite element analysis do?

35. What is the difference between the two types of CAPP?

36. What is MAP?

37. What is LAN?

Chapter 42 <u>CASE STUDY</u>

underground steam line

An underground steam line for a military base in Alaska, approximately 1.6 km (1 mile) in length, utilized the units depicted in Figure CS-42, each unit being 9.14 m (30 ft) in length. As shown, the steam line for each unit was enclosed in a cylindrical conduit made of sheet steel and weighed approximately 356 kg (785 lb). The conduit was to serve as the return drain line and to provide insulation. Each length of steam line was supported within the conduit by means of three U-shaped legs, made by cold-bending hot-rolled steel bar stock. See section A–A′ for a cross section. These legs were welded to the pipe about 2 m from each end.

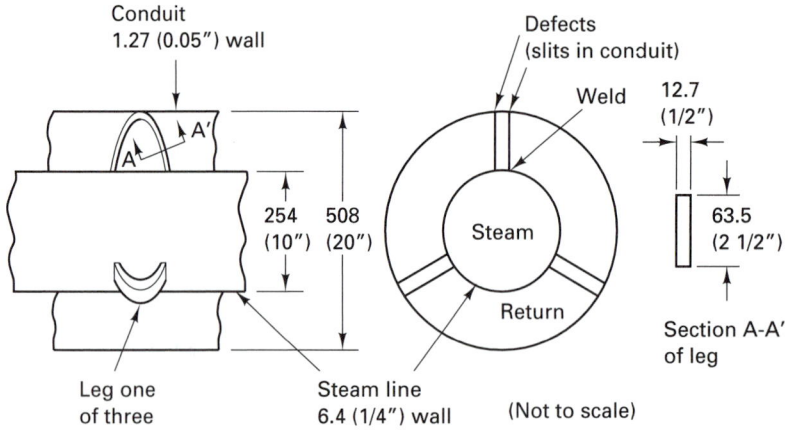

FIGURE CS-42 Schematic diagram showing method of supporting steam line in drainage conduit for an underground steam line.

The units were fabricated in California and were transported to Alaska by a sequence of truck, barge, and railroad. Because of an early winter, only one-half of the line was installed the first summer. Before work was resumed the following spring, someone decided to test the line and found that there were numerous holes in the conduit. The holes (slits in casing) were located where the edges of the U-shaped supporting legs contacted the outside conduit (casing) on the top of the casing. As the result of these holes, the project was delayed and a cost litigation ensued.

1. Why do you think these holes occurred?
2. Was it a design error? A fabrication error? A service error? A service environment error? What error(s) caused the failures? When did the defects actually develop?
3. What action would you recommend to avoid this problem in the future?
4. What should be done about the part of the line already installed?

CHAPTER 43

LEAN PRODUCTION: JIT MANUFACTURING SYSTEMS

44.1	INTRODUCTION	44.9	LINKING THE CELLS
44.2	TEN STEPS TO LEAN PRODUCTION	44.10	REDUCING THE WORK IN PROCESS
44.3	DESIGN OF THE LINKED-CELL FACTORY	44.11	VENDOR RELATIONSHIPS
44.4	HOW TO DESIGN MANUFACTURING CELLS	44.12	AUTONOMATION
		44.13	RESTRUCTURING THE REST OF THE BUSINESS
44.5	SETUP REDUCTION	44.14	BENEFITS OF CONVERSION
44.6	INTEGRATED QUALITY CONTROL	44.15	CONSTRAINTS TO CONVERSION
44.7	INTEGRATED PREVENTIVE MAINTENANCE	44.16	SUMMARY
44.8	LEVELING AND BALANCING THE MANUFACTURING SYSTEM	Case Study:	SNOWMOBILE ACCIDENT

■ 43.1 INTRODUCTION

Many people have been describing the factory of the future in current literature. There has been so much written about computers, computer-integrated manufacturing, FMS, and robotics in the recent years that it has become impossible to keep up with all the acronyms. Invariably, these articles describe a strategy in which one is automating, computerizing, or robotizing the existing manufacturing system, that is, the job shop. This approach has been defined as CIM, computer-integrated manufacturing. In the 1980s a strategy very different from CIM has emerged. This new strategy requires a systems-level change for the factory, a change that will affect every segment of the company, from accounting to shipping, that begins with the manufacturing system. This new approach is now being called *lean production* to contrast it with mass production. In this chapter the implementation strategy for lean production is boiled down into 10 steps. Through these steps, the critical control functions (discussed in Chapter 42 as part of the production system) are integrated into the manufacturing system. However, before embarking on this approach to the "factory with a future," some preliminary, preparatory steps must be discussed.

Integration of the production system functions into the manufacturing system requires commitment from top-level management and communication with everyone, particularly manufacturing. Total employee and union participation is absolutely necessary, but it is not usually the union leadership or the production workers who raise barriers to lean production. It is those in middle management who have the most to lose in this systems-level change.

The preliminary steps are as follows:

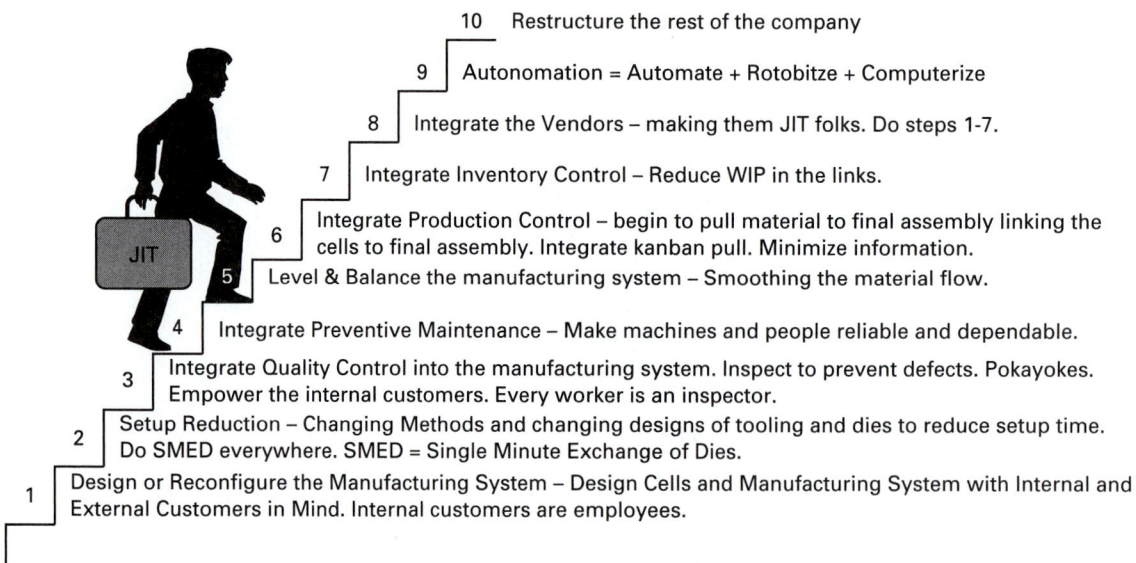

10   Restructure the rest of the company

9   Autonomation = Automate + Rotobitze + Computerize

8   Integrate the Vendors – making them JIT folks. Do steps 1-7.

7   Integrate Inventory Control – Reduce WIP in the links.

6   Integrate Production Control – begin to pull material to final assembly linking the cells to final assembly. Integrate kanban pull. Minimize information.

5   Level & Balance the manufacturing system – Smoothing the material flow.

4   Integrate Preventive Maintenance – Make machines and people reliable and dependable.

3   Integrate Quality Control into the manufacturing system. Inspect to prevent defects. Pokayokes. Empower the internal customers. Every worker is an inspector.

2   Setup Reduction – Changing Methods and changing designs of tooling and dies to reduce setup time. Do SMED everywhere. SMED = Single Minute Exchange of Dies.

1   Design or Reconfigure the Manufacturing System – Design Cells and Manufacturing System with Internal and External Customers in Mind. Internal customers are employees.

FIGURE 43-1   Ten steps to lean production, or how to design a linked-cell manufacturing system.

1. All levels in the plant, from the production workers to the president, must be educated in lean production philosophy and concepts.
2. Top management must be totally committed to this venture and everyone must be involved in the change.
3. Everyone in the plant must understand that cost, not price, determines profits; since the customers determine price and want superior quality and on-time delivery, customer satisfaction is the key.
4. Everyone must be committed to the elimination of waste. This is fundamental for getting lean.

■ 43.2 TEN STEPS TO LEAN PRODUCTION

Many companies have implemented lean production. Figure 43-1 outlines the 10 key steps to converting a factory from a job shop–flow shop manufacturing system to a true lean production system.

STEP 1. FORM U-SHAPED CELLS; RESTRUCTURE THE FACTORY FLOOR   In Chapter 41 the concept of cellular manufacturing was discussed. Cells replace the production job shop. The first task is to restructure and reorganize the basic manufacturing system into manufacturing cells that fabricate families of parts. This prepares the way for systematically creating a linked-cell system for one-piece movement of parts within cells and for small-lot movement between cells. Creating cells is the first step in designing a manufacturing system in which production and inventory control and quality control are integral parts.

STEP 2. RAPID EXCHANGE OF TOOLING AND DIES   Everyone on the plant floor must be taught how to reduce setup time using SMED (single-minute exchange of die). A setup reduction team assists production workers and foremen and demonstrates a project on the plant's worst setup problem. Reducing setup time is critical to reducing lot size.

STEP 3.   INTEGRATE QUALITY CONTROL   A multiprocess worker can run more than one kind of process. A multifunctional worker can do more than operate machines. He is also an inspector who understands process capability, quality control, and process improvement. In lean production, every worker has the responsibility and the authority to make the product right the first time and every time and the authority to stop the process when something is wrong. This integration of quality control into the manufacturing system markedly reduces defects while eliminating inspectors. Cells provide the natural environment for the integration of quality control. The fundamental idea is to inspect to prevent the defect from occurring.

STEP 4.   INTEGRATE PREVENTIVE MAINTENANCE   Making machines to operate reliably begins with the installation of an integrated preventive maintenance program, giving workers the training and tools to maintain equipment properly. The excess processing capacity obtained by reducing setup time allows operators to reduce the equipment speeds or feeds and to run processes at less than full capacity. Reducing pressure on workers and processes to produce a given quantity fosters in workers a drive to produce perfect quality.

Housekeeping rules:

1. A place for everything and everything in its place. Everything should be put away so it is ready to use the next time.
2. Each worker is responsible for cleanliness of workplace and equipment.

STEP 5.   LEVEL AND BALANCE   Level the entire manufacturing system by producing a mix of final assembly products in small lots. This is called *smoothing of production*. This will reduce the lumpiness of the demand for the component parts and subassemblies. Standardize the cycle time. Use a simplified and synchronized system to produce the proper number of components everyday, as needed. Begin at mixed-model final assembly and work backward through subassembly and cells. Each process, cell, and subassembly tries to match the build quantity and cycle time of final assembly. This is called *balancing*.

STEP 6.   LINK CELLS   Integration of production control is materialized by linking the cells, subassemblies, and final assembly elements, utilizing kanban. The structure of the manufacturing system now defines paths that parts can take through the plant. Begin by connecting the elements with kanban links. The need for route sheets is eliminated. The parts, that is, the in-process inventory, flow within the structure. All the linked cells, processes, subassemblies, and final assemblies start and stop together. They are synchronized. This is the integration of production control into the manufacturing system, forming a linked-cell manufacturing system.

STEP 7.   INTEGRATE INVENTORY CONTROL   The people on the plant floor directly control the inventory levels in their areas through the control of the kanban. This is the integration of the inventory control system to reduce systematically the *work-in-process*, (WIP). The reduction of WIP exposes problems that must be solved before inventory can be further reduced. The minimum level of WIP (which is actually the inventory *between* the cells and subassemblies) is determined by the percent defective, the reliability of the equipment, the setup time, and the transport distance to the next cell.

STEP 8.   INTEGRATE THE SUPPLIERS   Educate and encourage suppliers (vendors) to develop their own lean production system for superior quality, low cost, and rapid on-time delivery. They must be able to deliver material to the customer when it is needed and where it is needed without incoming inspection. The linked-cell network ultimately should include every supplier. Suppliers become a remote cell in the L-CMS.

After these eight steps have been completed, the manufacturing system has been redesigned and infused (integrated) with the critical production functions of quality control, inventory control, production control, and machine tool maintenance. Autonomation of the integrated system is next.

STEP 9. INSTITUTE AUTONOMATION Converting manned cells to unmanned cells is an evolutionary process, initiated by the need to solve problems in quality, reliability, or capacity (eliminate a bottleneck). It begins with mechanization of operations such as load, unload, inspect, and clamp and moves toward the automation of human thinking and automatic detection and correction of problems and defects. This is called *autonomation*, the autonomous control of quality and quantity.

STEP 10. RESTRUCTURE THE PRODUCTION SYSTEM Once the factory (the manufacturing system) has been restructured into a JIT manufacturing system and the critical control functions well integrated, the company will find it expedient to restructure the rest of the company. This will require removing the functionality of the various departments and forming teams, often along product lines. It will require the implementation of concurrent engineering teams to decrease the time needed to bring new products to market. This movement is gaining strength in many companies and is being called *Business Process Reengineering*; it is basically restructing the production system to be as waste free and efficient as the manufacturing system.

■ 43.3 DESIGN OF THE LINKED-CELL FACTORY

Figure 43-2 shows the layout of a *linked-cell* manufacturing system. In Chapter 41, the operation of a manned cell for shafts was described. This cell, designed for the manufacture of a family of parts, has the unique feature of *decoupling* the machine time (MT) from the cycle time (CT) while building products one at a time. This relaxes the line-balancing problem common to flow lines and transfer lines while greatly enhancing the flexibility. The *cycle time* is controlled by the time it takes the worker or workers to complete a walking loop through the cell or cells. Therefore, the cycle time can be altered by adding or deleting workers. The details of this system are shown in Figures 43-2 through 43-7 and Figure 41-6.

Beginning with Figure 43-2, the linked-cell factory is composed of cells to fabricate components, subassembly lines, and final assembly lines. Figure 43-3 indicates how the job shop–flow shop manufacturing system is reconfigured into manufacturing and assembly cells. Figure 43-4 shows the details of a hub cell described by means of a *standard operations routine sheet* (see also Figure 41-9). The design of the cell is shown at the lower left. Hubs are made by a four-station cold header, followed by three machining operations. The cell combines forming and machining. The third operation (finish milling) has a long MT, 70 seconds, compared to the needed cycle time of 55 seconds for the cell; therefore, this operation, the finish milling, is duplicated in the cell. The cold header can produce a part every 5 seconds. Its capacity production is much greater than needed by the cell, but it is *automated* to produce parts exactly at the rate the cell needs the parts and no faster. Overproduction will result in the need to store parts, transport parts to storage, retrieve the parts when needed, keep track of the parts (paperwork), and so on. All this requires people and costs money but adds no value. Like all the machines in the cell, the cold header is underutilized. The objective is to minimize the resources within the system that do not depreciate: direct material and labor.

As discussed in Chapter 41, there is no need to balance the MTs for the machines. It is necessary only that no MT be greater than the required CT. Since the MT for finish

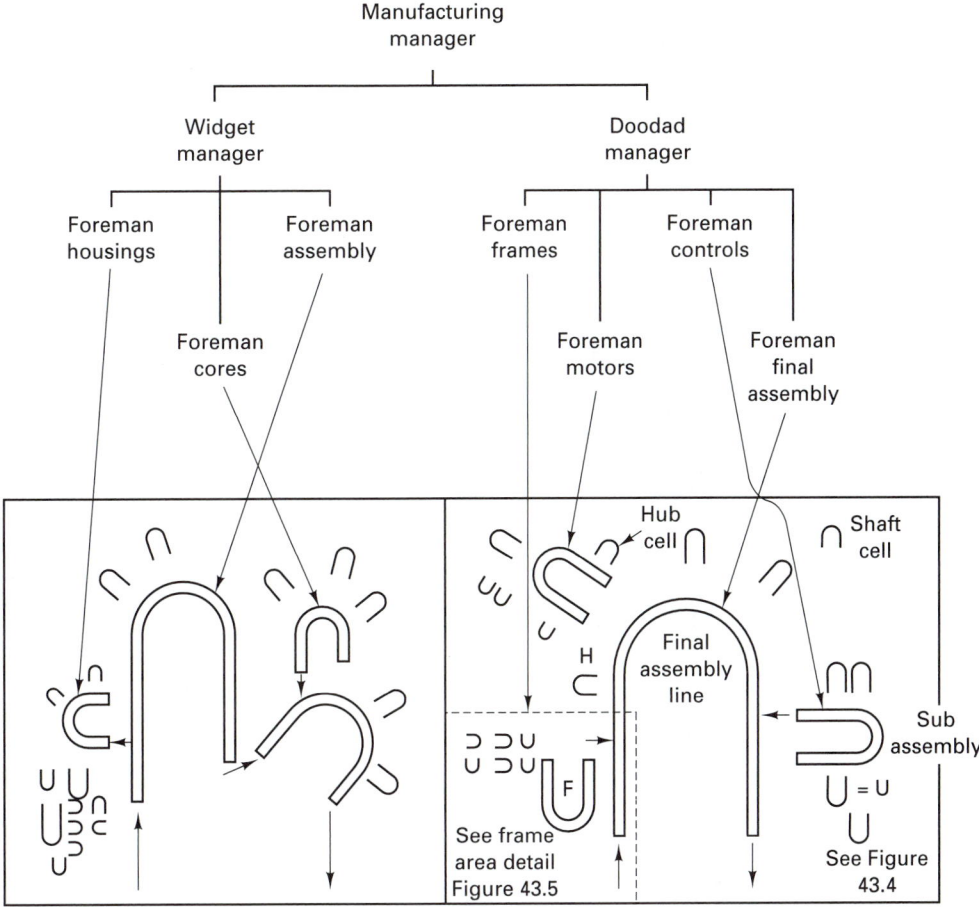

Linked cell manufacturing system

FIGURE 43-2   Plant layout for an integrated manufacturing system showing manufacturing cells, subassembly cells, and final assembly cells, all linked by kanban. See Figures 41-6, 43-4, 43-5, and 43-6 for details.

milling is greater than the cycle time (70 > 55), the finish-milling process is duplicated and the worker alternates machines on trips through the cell. The milling machines must do identical work. The cycle time is determined by the system's requirements since this cell, like all the others in the plant, is geared to produce parts as needed, when needed, by the subsequent process.

The machining speeds and feeds can be relaxed to extend the tool life of the cutting tools and reduce the wear and tear on the machines as long as the MT for a particular machine does not exceed CT for the cell. This increases the reliability of the process, reducing the probability of a breakdown. There is no mystery as to which process within the cell is the bottleneck, the machine with the longest MT. There is no need for any computer analysis to find the bottlenecks. Everyone in the manufacturing system can see and understand how the cell functions and therefore which process is the most likely to delay the cell's CT.

The cycle time is determined from the demand rate for the parts according to the following calculations:

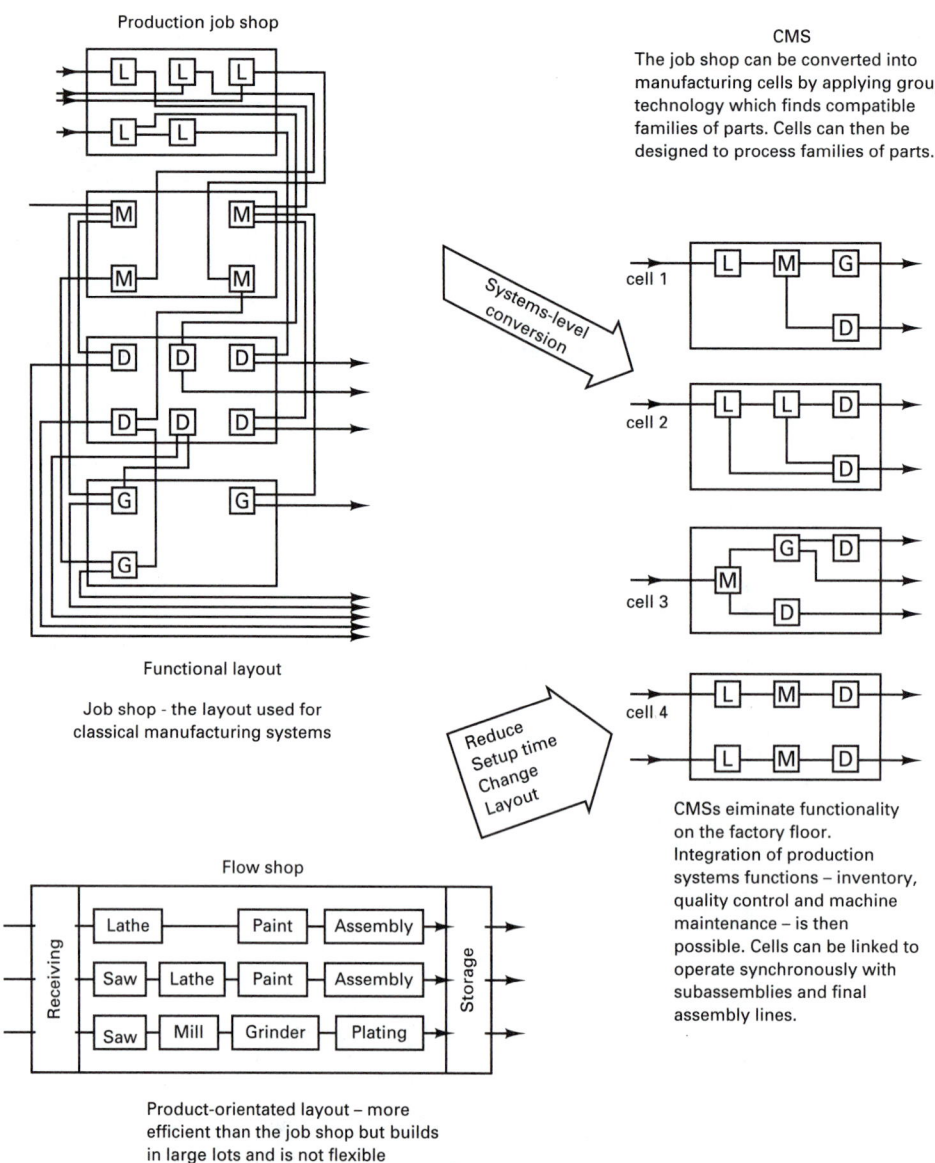

FIGURE 43-3 Two classical manufacturing systems in common use today, the job shop and the flow shop, require a systems-level conversion to reconfigure them into cells.

$$\frac{\text{daily demand}}{\text{for parts}} = \frac{\text{monthly demand (forecast plus customer orders)}}{\text{number of days in month}} \qquad (43\text{-}1)$$

$$\text{CT} = \frac{1}{\text{PR}} \qquad (43\text{-}2)$$

where

$$\text{PR} = \frac{\text{daily demand (parts)}}{\text{hours in day (hr)}}$$

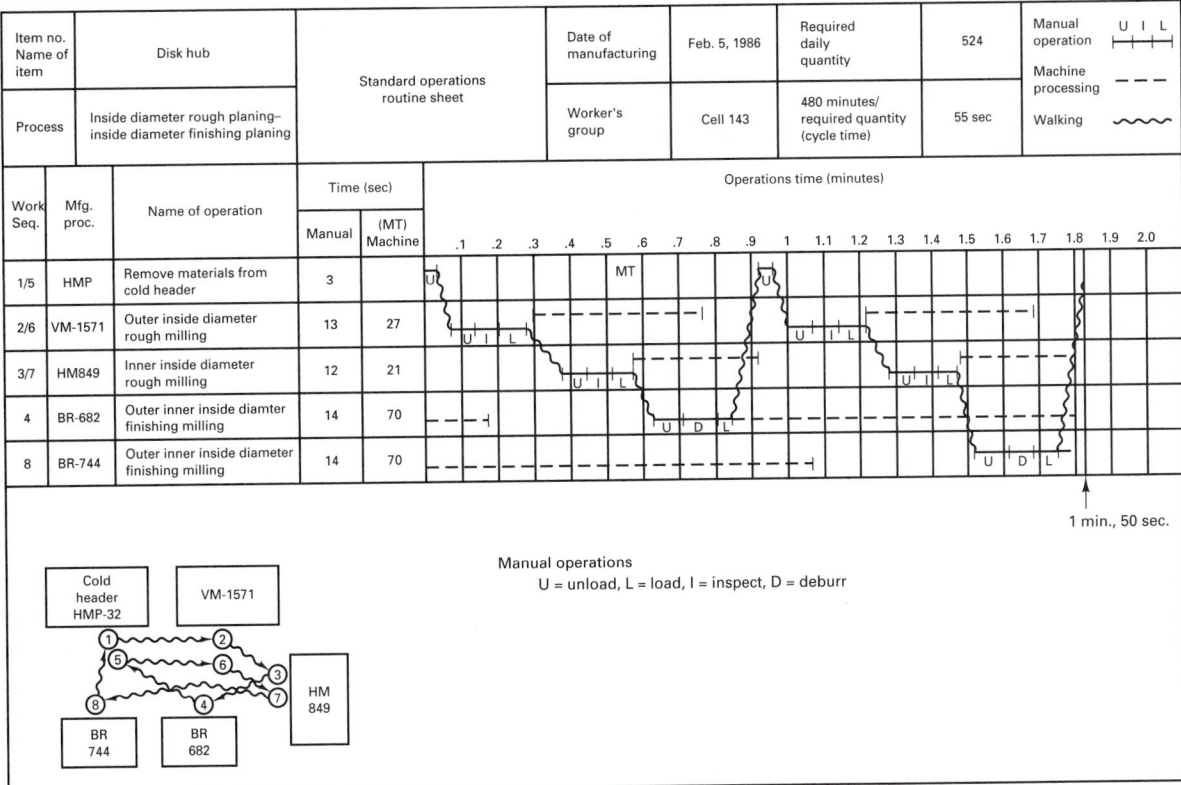

FIGURE 43-4   Example of a standard operations routine sheet for a manned cell. The hub manufacturing machine is a four-station cold header that makes parts as needed by the first milling machine (VM-1571).

You may think that this incredibly simple approach cannot possibly be the way in which JIT companies calculate cycle time, but life is simpler when the PJS has been eliminated and a linked-cell system has been installed.

Figure 43-5 shows the frame area of the plant in more detail. This area contains six component cells and one subassembly cell which directly feeds the mixed model final assembly line. These seven manned cells are "linked" by kanban (or linked directly) to another nearby cell, just as the subassembly line is directly linked to a "point of use" in the final assembly line. Cell *E* is directly linked to cell *F*. Cell *D* in this area is feeding parts to cell *H*, as shown in Figure 43-6, using kanban. Cell *H* withdraws parts from cell *D* as needed. In the linked-cell system, the work-in-process inventory is between the cells and is controlled by the kanban. The inventory within the cells is called the *stock-on-hand* (SOH). When the cell can make a very high percentage of perfect parts, have no machine breakdowns, and virtually no setups, the size of the kanban link can be reduced. It appears that the limit is three carts. See the discussion of step 7.

Figure 43-7 shows the frame area with two different allocations of workers. In the upper diagram, eight workers are tending the machines in seven cells. The cycle time was 120 seconds per unit. The next month a slower cycle time was needed because the demand decreased. Six workers were allocated to the seven cells; the result is a decrease in the production rate and an increase in the cycle time to 165 seconds. Notice that all the cells

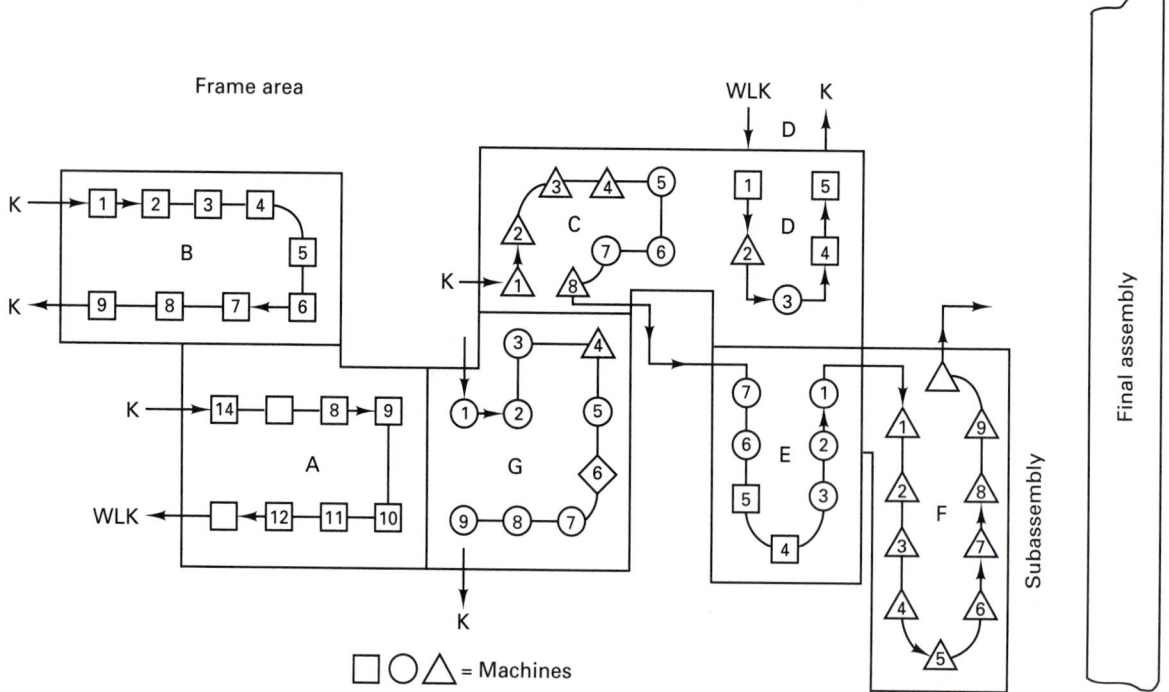

FIGURE 43-5   Seven manned cells showing the movement of parts. C and E are directly linked. See Figure 43-6 for linkage of cell D to cell H.

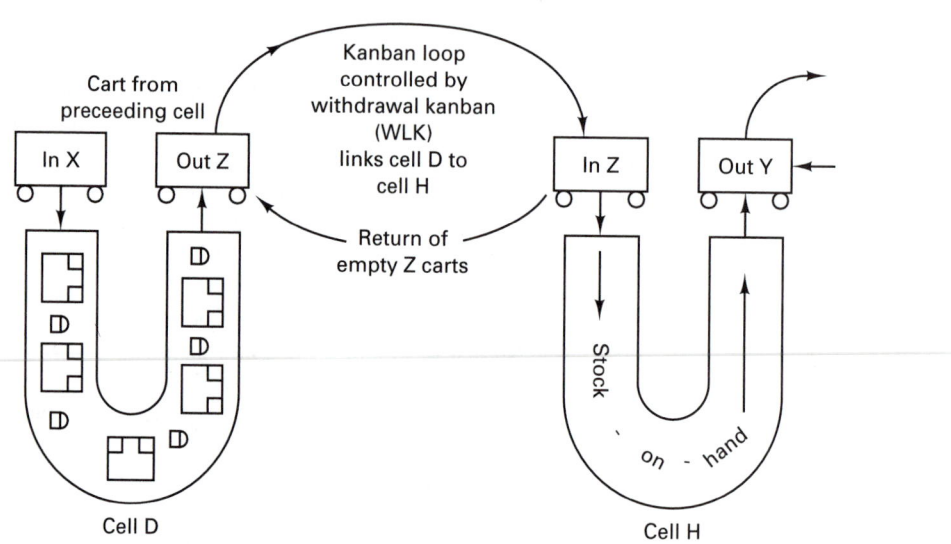

FIGURE 43-6   Kanban links the cells and controls the work-in-process inventory. The arrival of an empty cart at cell D is the signal for cell D to make a cart of parts for cell H.

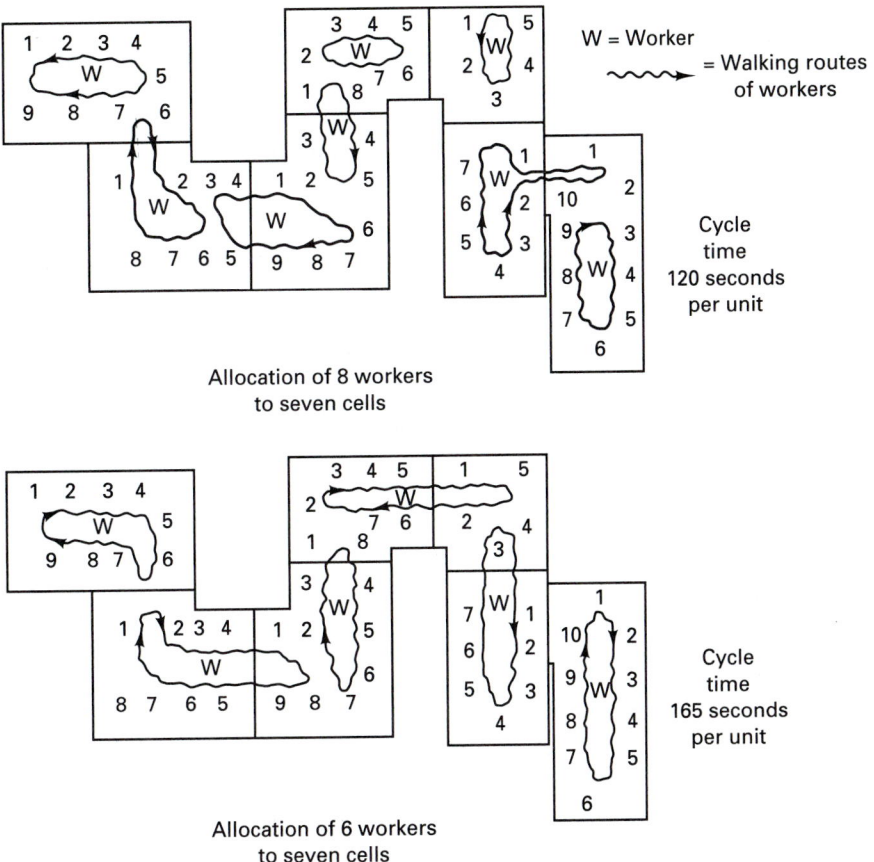

Allocation of 8 workers
to seven cells

Allocation of 6 workers
to seven cells

FIGURE 43-7   Number of workers is reduced as demand decreased and cycle time increased.

have the same cycle time in this area. For a cycle time of 120 seconds, each worker is tending 7 machines. For a cycle time of 165 seconds, each worker is tending 9 or 10 machines. The workers in these cells spend from 10 to 15 seconds at a machine and 5 seconds walking to the next machine. The frame foreman allocates the number of workers in the frame area, allocating the minimum number of workers to the area.

Workers can be added if a problem arises or if some cell continually falls behind. This is not a new concept. Fast-food restaurants such as Wendy's add workers to their cells when demand increases and remove workers when business is slow. Wendy's is an example of a CMS that produces hamburgers. The workers in the JIT factory are multifunctional, just like the workers at Wendy's. The machines at Wendy's are the A(2) level of automation. The fried potato and drink machines are loaded and started by the worker, but they shut themselves off automatically when the cycle is complete.

■ 43.4 How to Design Manufacturing Cells

Conversion of the functional system into a flexible, linked-cell system is a design task. Table 43-1 outlines the methods by which manufacturing cells can be formed. Most companies "design" their first cell by one of the trial-and-error techniques for expediency in

TABLE 43-1. How to Form Cellular Manufacturing Systems

JIT manufacturing systems are based on a linked-cell manufacturing system design. Knowing how to design the cells to be flexible is the key to successful manufacturing.

   I. Make tacit judgments based on axiomatic design principles.[a]
      A. Minimize function requirements (flexibility is chief criteria).
      B. Simplify the design of system.
      C. Minimize the design information in the product.
      D. Decouple those elements that are functionally coupled.

  II. Use group technology methodology.
      A. Production flow analysis finds families and defines cells.
      B. Coding/classification is more complete and expensive.
      C. Other GT methods, including eyeball or tacit judgments.
         1. Find the key machine, often a machining center, and declare all parts going to this machine a family. Move machines needed to complete all parts in a family around the key machine.
         2. Build the cell around a common set of components such as gears, splines, spindles, rotors, hubs, shafts, etc.
         3. Build the cell around a common set of processes, such as drill, bore, ream, keyset, chamfer holes.
         4. Build the cell around a set of parts that eliminates the longest (most time-consuming) element in setups between parts being made in the cell.

  III Simulation.

      A. Digital simulation of the system.
      B. Physical simulation of the system.
      C. Object-oriented and graphical simulation.

  IV. Pick a product or products.
      Design a linked-cell manufacturing system beginning with the final assembly line (convert FA to mixed model) and move backward through the subassembly to component parts and suppliers.

[a]Axiomatic design principles are presented in Nam P. Suh's book *Principles of Design*. Oxford University Press, N.Y., 1989.

gaining experience in cells. *Digital simulation* is gaining wider usage in designing and analyzing manufacturing systems with the advent of newer, more versatile languages.

Another technique being researched extensively is called *physical simulation*. This approach uses small robots and scaled-down versions of machine tools (minimachines) to emulate real-world systems. The small machines employ essentially the same minicomputers and software as the full-scale systems. In this way, the development of the software needed to integrate the machines and design of the cell can be done prior to installation of the full-scale system on the shop floor. Minimachine tool laboratories are ideal for providing hands-on instruction for students in manufacturing systems. Unmanned cells and FMSs can be simulated in the laboratory at quite reasonable cost. Generally speaking, the industrial robots and full-size machines are expensive and may be considered dangerous for student usage. *Instructional robots* make it possible for education systems and small businesses to gain hands-on experience with this technology. These robots cost but a fraction of the industrial versions but use essentially the same electronic controls with stepping motors or low-pressure hydraulics. The microprocessors in these machines use a small personal computer for program control. The trade-off here is one of scaling, as these robots will have much lower speed of response, lower weight-carrying capacity, and poorer process capability, and poorer reliability than their industrial counterparts but may cost only one-tenth of the industrial machine.

Group technology offers a systems solution to the reorganization of the functional system, restructuring the job shop into manufacturing cells. These conversions represent systems-level changes, which will create the potential for tremendous savings, but because

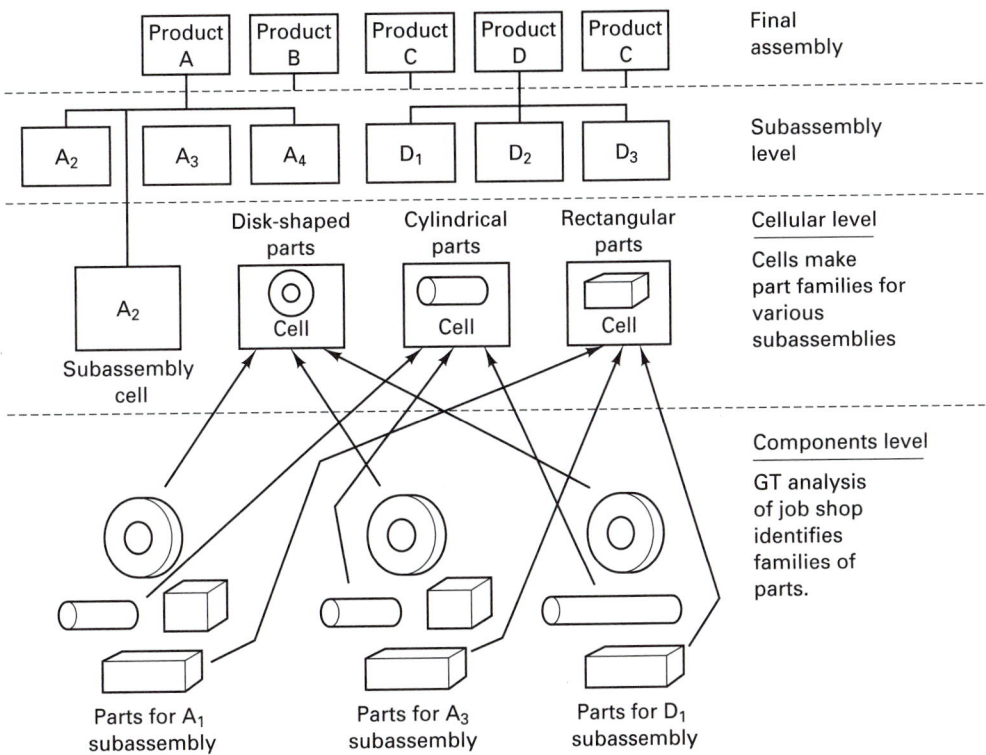

FIGURE 43-8   Short course on cell formation by group technology.

of the magnitude of the changes, careful planning and full cooperation from everyone involved are absolutely required.

The application of this concept to a manufacturing facility results in the grouping of units or components into families wherein the components have similar design or manufacturing sequences. Machines are then collected into groups or cells (machine cells) to process the family (Figure 43-8). By grouping similar components into families of parts, a group or set of processes can be collected together to make a family. This is a cell. Refer back to Figure 43-4. The order of machines in the cell now defines the manufacturing sequence.

The entire shop will not be able to convert into families immediately. Therefore, the total collection of manufacturing systems will be a mix, evolving toward a perfect linked-cell system over time. This will create scheduling problems, as in-process times for components made by cells will be vastly different from those made under traditional job shop conditions. However, as the volume in the functional area is decreased, the total system will become more efficient.

Judgment methods using axiomatic design principles are, of course, the easiest and least expensive, but also the least comprehensive. "Eyeball" techniques clearly work for restaurants but not in large job shops, where the number of components may approach 10,000 and the number of machines 300 to 500.

Production flow analysis (PFA) uses the information available on route sheets or cards. (See Figure 42-8 for an example.) The idea is to sort through all the components and group them by a matrix analysis, using product-routing information (Figure 43-9).

Job number	Machine Code Letter									
	A	B	C	D	E	F	G	H	I	J
1								X		
2		X	X							
3				X						
4							X	X		
5	X	X	X							
6									X	X
7	X		X							
8						X		X		
9									X	X
10				X	X					
11	X	X	X					X		
12						X	X			
13								X		
14				X	X					
15									X	X
16		X				X	X	X		
17										X
18	X	X								
19						X	X	X		
20					X					

A matrix of jobs (by number) and machine tools (by code letter) as found in the typical job shop.

Job number	Machine Code Letter									
	A	B	C	D	E	F	G	H	I	J
7	X		X							
11	X	X	X					X (exception)		
2		X	X							
5	X	X	X							
18	X	X								
14				X	X					
3				X						
10				X	X					
12						X	X			
4							X	X		
19						X	X	X		
16						X	X	X		
8						X		X		
1								X		
9					X (exception)				X	X
13								X		
6									X	X
15									X	X
17										X

Cell will have 3 machines F, G, H, for manufacture of 6 jobs.

A matrix rearranged to yield families of parts and associated groups of machines that can form a cell.

FIGURE 43-9   Schematic to explain the concept of production flow analysis. Matrix on the left shows routing information for specific job going to specific machines. Matrix on the right shows the jobs "collected" for families of parts made on a set of machines.

This method is more analytical than tacit judgment but not as comprehensive as coding/classification. PFA is a valuable tool in the systems reorganization problem. For example, it can be used as an up-front analysis, a sort of "before the fact" analysis that will yield some cost/benefit information. Decision makers would have some information on what the company could expect in terms of the percentage of their product that could be made by cellular methods, what would be a good "first cell" to undertake, what coding/classification system would work best for them, how much money they might have to invest in new equipment, and so on.

In short, PFA can greatly reduce the uncertainty in making the decision on reorganization. As part of this technique, an analysis of the flow of material in the entire factory is performed, laying the groundwork for the new linked-cell layout of the entire plant.

Many companies converting to a cellular system have used a *coding/classification* method. There are design codes, manufacturing codes, and codes that cover both design and manufacture.

Classification sorts items into classes or families based on their similarities. It uses a code to accomplish this goal. Coding is the assignment of symbols (letters or numbers or both) to specific component elements based on differences in shape, function, material, size, processes, and so on.

No attempt to review coding/classification (C/C) methods will be made here. C/C systems exist in bountiful number in published literature and from consulting firms.

Whatever C/C system is selected, it should be tailored to the particular company and should be as simple as possible so that everyone understands it. It is not necessary that old part numbers be discarded, but every component will have to be coded prior to the next step in the program, finding families of parts. This coding procedure will be costly and time consuming, but most companies that choose this conversion understand the necessity of performing this analysis.

The families of parts will not be all the same with regard to their material flow and therefore will require different designs (layouts). In some families, every part will go to every machine in exactly the same sequence, and no machine will be skipped and no back flow will be allowed. This is, of course, the purest form of a cellular system. Other families may require that some components skip some machines and that some machines be duplicated. However, back flow should be avoided.

The formation of families of parts leads to the design of cells, but cell design is by no means automatic. It is the critical step in the reorganization and must be carefully planned. Many companies begin with a pilot cell so that everyone can see how cells function. It will require time and effort to train the operators and they will need time to adjust to standing and walking. Nevertheless, the company should proceed with developing manned cells, not waiting until all the parts have been coded. Simply select a product or group of products that seems most logical. Only in this way will everyone learn how cells operate and how to reduce setup time on each machine. Machines will not be utilized 100%. Machine utilization rate usually improves but may not be what it was in the functional system. *The objective in manned cellular manufacturing is to utilize the people fully*, enlarging and enriching their jobs by allowing them to become multifunctional. In fact, one of the inherent results of the conversion to a cellular system is that the worker becomes multifunctional. That is, operators learn to operate many machines and/or perform tasks that include quality control, machine maintenance, and setup reduction. In unmanned cells and systems, the utilization of the equipment is more important because the most flexible element in the cell, the worker, has been removed and has been "replaced" by a robot.

The manned cellular system provides the worker with a natural environment for job enlargement. Greater job involvement enhances job enrichment possibilities and clearly provides an ideal arrangement for improving quality. Part quality can be checked between each step in the process.

The similarity in shape and processes needed in the family of parts allows setup time to be reduced or even eliminated. Being able to do the setup in less than 10 minutes is called *single-minute exchange of dies* (SMED), covered in Section 43.5.

■ 43.5 SETUP REDUCTION

The lean production approach to manufacturing demands that small lots be run. This is impossible to do if machine setups take hours to accomplish. The *economic order quantity* (EOQ) formula has been used in the United States to determine what quantity should run

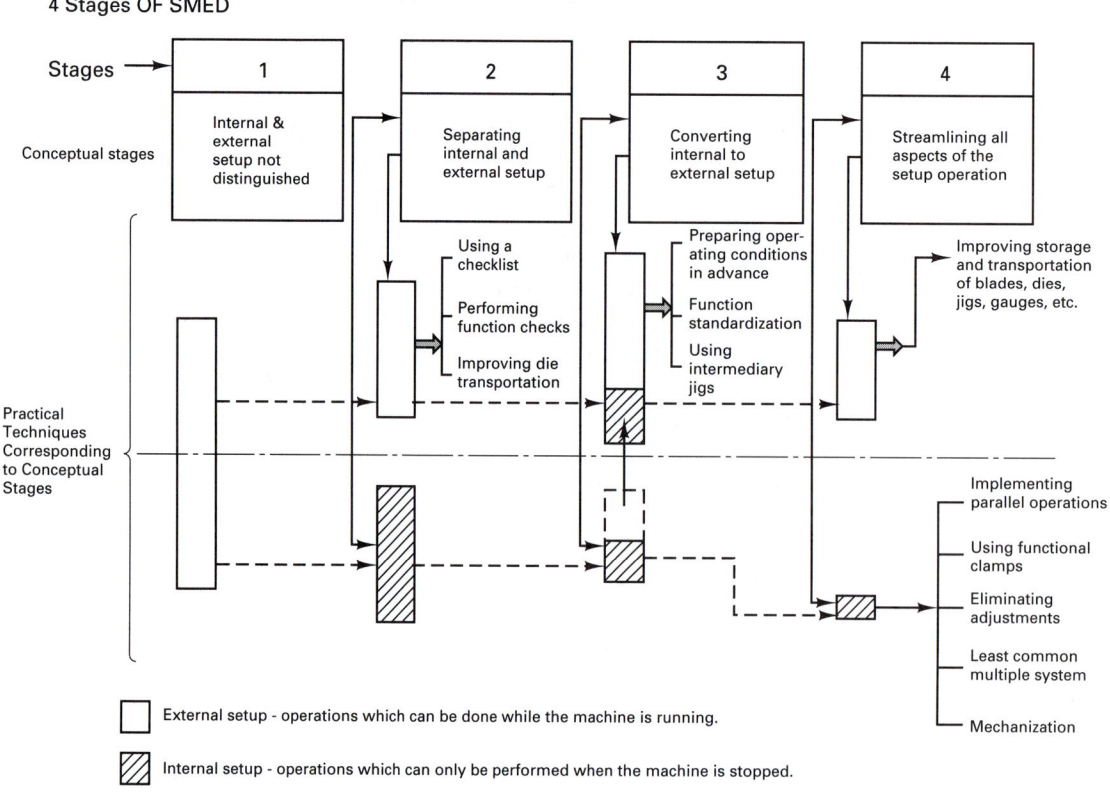

4 Stages OF SMED

Conceptual stages and pratical techniques of the SMED system
for the rapid exchange of tooling an dies, after Shingo

FIGURE 43-10 Conceptual stages of the SMED system for the rapid exchange of tooling and dies.

to cost-justify a long and costly setup time. The EOQ was a faulty suboptimal approach. Instead of accepting setup times as a large fixed number, we should have reduced setup times. This results in reduced lot sizes.

Successful setup reduction is easily achieved when approached from a methods engineering perspective. Much of the initial work in this area has been done by Shigeo Shingo and Taiichi Ohno. Large reductions can be achieved by applying time and motion studies and Shingo's SMED rules for rapid exchange of dies (Figure 43-10). Setup-time reduction occurs in four stages. The initial stage is to determine what currently is being done in the setup operation. The setup operation is usually videotaped and everyone concerned gets together and reviews the tape to determine the elemental steps in the setup. The next stage is to separate all setup activities into two categories, *internal* and *external*. Internal elements can be done only when the machine is not running; external elements can be done while the machine is running. This elemental division will usually shorten the lead time considerably. Stages 3 and 4 focus on reducing the internal time. The key here is for workers and setup people to learn how to reduce setup times and apply the simple principles and techniques. If a company must wait for the setup reduction team to examine every process, an integrated manufacturing system will never be achieved.

In the last stages of SMED, it may be necessary to invest capital to drive the setup times below 1 minute. Automatic positioning of dies, bolster plates on rollers, intermediate jigs, and duplicate workholders represent the typical kinds of hardware needed. The result is that 2-hour setup times can be reduced to under 5 minutes in relatively short order.

The initial objective here is to reduce the setup time until it is equal or less than to the cycle time (1 to 2 minutes). This will permit a significant initial reduction in lot size. After this, setup time reductions will result in further reductions in the lot size. The next goal is to get the setup time down to less than the time needed to load, unload, inspect, deburr, and so on at a machine. As shown in Table 43-2, when numerous processes are involved in cellular manufacturing of a family of parts, sequential setup changes are utilized. At the outset, the setup times may be long compared to the machining time (the run times). The setups are done sequentially. After setup, defect-free products should be made right from the start. The first part will be good. Ultimately, the ideal condition would be to eliminate setup between parts.

TABLE 43-2.   Example of Change from Part A to Part B for Four-Process Cell

Setup Change	Process			
	1	2	3	4
	A	A	A	A
	A	A	A	A
1	Setup change	A	A	A
2	B	Setup change	A	A
3	B	B	Setup change	A
4	B	B	B	Setup change
	B	B	B	B
	B	B	B	B

Figure 43-11 shows the drilling machine and vertical milling machine from the cell described in Figure 41-6. This cell was designed for a family of four parts. To reduce the setup time to a few seconds, or "one touch," the drilling machine was modified to eliminate setup as far as possible. The total cost to modify the drilling machine was around $300. This is an example of one-touch exchange of fixtures.

The table on the vertical mill was equipped with a digital readout device so that the four starting locations of the four machining operations were always the same. The fixtures for the four parts in the family are never removed from the oversize table. The first part out of these fixtures was always a good part and there is no time spent on fixture exchange. This is an example of no-touch exchange of fixtures.

Other setup reduction techniques include using group jigs (one jig to accommodate different parts, using adapters), training operators in rapid setup techniques, practicing rapid setups, and the intermediate jig concept. The *intermediate jig* is like the cassette for a VCR. The cassette is quickly loaded and locked in place. The tape inside every cassette can be different. To the machine (the tape player), every different fixture (tape) that is placed in the intermediate jig (cassette) looks the same. To the cutting tool (the tape head), every workholder looks different. Designing workholders so that they appear the same to the machine tool usually requires one to construct an intermediate jig or fixture plate to which the fixture itself is attached.

In summary, the savings in setup times are used to decrease the lot size and increase the frequency at which the lot is produced. The smaller the lot, the shorter the throughput

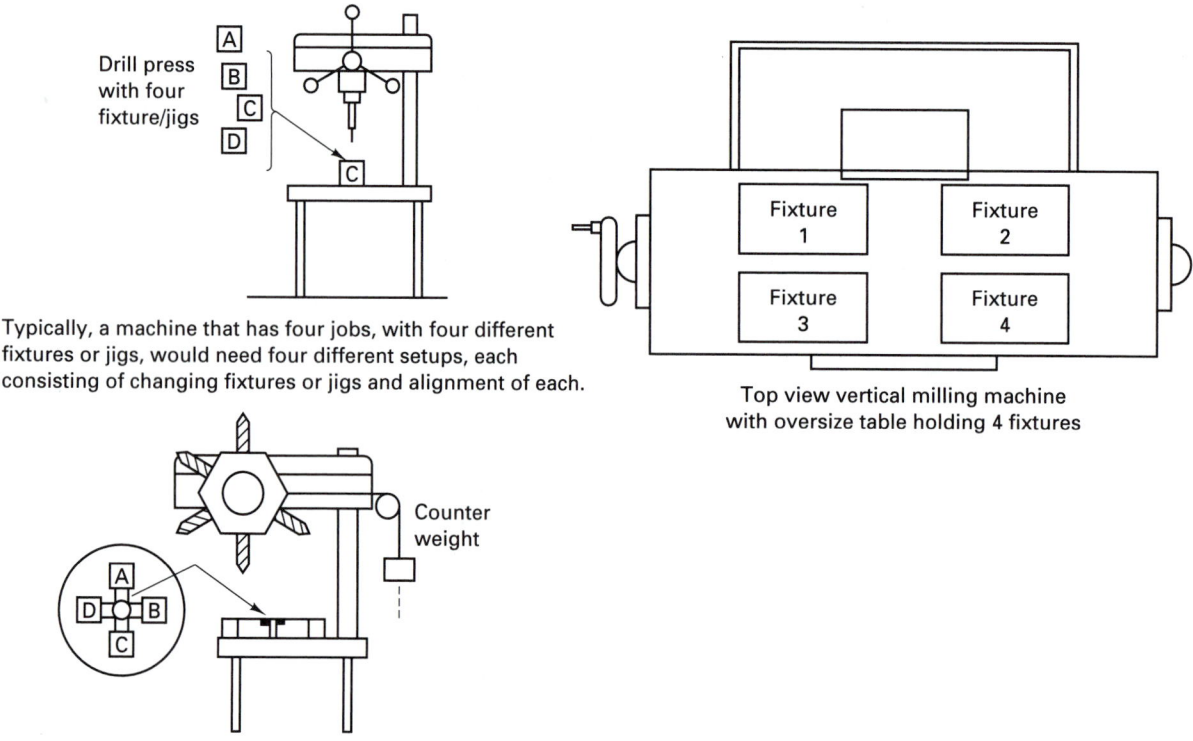

Drill press with four fixture/jigs

Typically, a machine that has four jobs, with four different fixtures or jigs, would need four different setups, each consisting of changing fixtures or jigs and alignment of each.

Top view vertical milling machine with oversize table holding 4 fixtures

Counter weight

With redesign, the four fixtures were mounted on a turntable and are permanently aligned to the spindle when locked in position. Turret replaces single spindle. Automatic feed replaces hand wheel.

FIGURE 43-11 Machines in cell process families of parts. Reducing the variety of parts coming to the machine permits one to modify the machine so that setup times can be reduced or eliminated.

time and the lead time. This makes the use of JIT principles feasible and makes the kanban inventory control system practical.

■ 43.6 INTEGRATED QUALITY CONTROL

When management and production workers trust each other, it is possible to implement an integrated quality control program. Japan was started on the road to superior quality with the visits of W. Edward Deming to Japan in the late 1940s and early 1950s. The Japanese were desperate to learn about quality, and statistical quality control techniques were readily accepted. They believed that everyone in America used statistical process control techniques. This of course was not the case. However, the Japanese did something that even those companies using SQC did not do. They taught the techniques and concepts to everyone, including top management and the production workers, who even had a quality journal on the subject. At Toyota, under the leadership of Taiichi Ohno, a new idea took hold, quite different in concept from our inspection philosophy. *Inspect to prevent* the defect from occurring rather than to *find* the defect after it has occurred. Ultimately, the concept of *autonomation* evolved (see Step 9).

The best approach is to make every worker an inspector and to give each person only one part to work on at a time, so that under no circumstance can a worker bury problems by working on alternative parts. Cells produce parts one at a time, just like assembly lines. In a nut shell, the idea is to "make one, check one, move one on." Pull cords are installed on the assembly lines to stop the lines if anything goes wrong. If workers find defective parts, if they cannot keep up with production, if production is going too fast according to the quantity needed for the day, or if a safety hazard is found, they are obligated to stop the lines. The problem is fixed immediately. Meanwhile, the other workers maintain their equipment, change tools, sweep the floor, or practice setups; but the line does not move until the problem is solved.

When every worker is responsible for quality, the number of inspectors on the plant floor is markedly reduced. Products that fail to conform to specification are immediately uncovered because they are used immediately.

Quite often, inspection devices are placed in the machines (inspection at the source) or in devices (called decouplers) between the machines, so the inspection is performed automatically. Again the idea is to prevent the defect from occurring rather than to inspect to find the defect after the part is made. Inspection by a machine instead of by a person is faster, easier, and more repeatable. This is called *in-process control inspection.*

Inspection becomes part of the production process and does not involve a separate location or person to perform it. Parts are 100% inspected by devices which either stop the process if a defect is found or correct the process before the defect can occur. The machine shuts off automatically when a problem arises. This prevents mass production of defective parts. The machine may also shut off automatically when the necessary parts have been made. This is part of inventory control and will be discussed later.

For manual work, another system for preventing defective work is called *Andon.* Andon is actually an electric light board which hangs high above the conveyor assembly lines so that everyone can see it. When a worker on the line needs help, he can turn on a yellow light. Nearby (multifunctional) workers who have finished their jobs within the allotted cycle time move to assist workers having problems. If the problem cannot be solved within the cycle time, a red light comes on and the line stops automatically until the problem is solved. In most cases the red lights go off within 10 seconds and the next cycle begins, a green light comes on, with all the processes beginning together. The name for this system is *Yo-i-don,* which literally means "ready, set, go." Such systems are built on teamwork and a cooperative spirit among the workers, fostered by a management philosophy based on harmony and trust.

Contrast this with the way things that often operate in the job shop. How long does it take to find a problem, to convince somebody that it is a problem, to get the problem solved, and to get the fix implemented? How many defective parts are produced in the meantime? Line shutdowns in JIT factories are encouraged to protect quality. Management must have confidence in the individual worker.

Another interesting technique, with which many American companies are already familiar, is *quality circles.* The Japanese call them *small group improvement activities* (SGIA) and they are a key part of company-wide quality control efforts in Japan (we call this *total quality management*). A quality circle is a group of employees who meet on a scheduled basis (daily, weekly) to discuss production and quality problems, to try to devise a solution to those problems, and to propose the solutions to their management. The group may be led by a foreman or by a production worker. It usually includes people from a given discipline or a given production area. Quality circles should be a natural entity within the manufacturing system and not artificially created. They are basically a technique to identify causes of problems and to generate ideas and suggestions to solve problems on the local level. In step 7 the means to identify the problems themselves is discussed.

■ 43.7 Integrated Preventive Maintenance

The multifunctional operators are trained to perform routine machine tool maintenance. Just adding lubricants (oiling the machine), checking for wear and tear, replacing damaged nuts and bolts, routinely changing and tightening belts and bolts, and listening for telltale whines and noises that signify impending failures can do wonders for machine tool reliability. The maintenance department must instruct the workers on how to do these things and help them prepare the routine check lists for machine maintenance. The workers are also responsible for keeping their areas of the plant clean and neat. Thus another function that is integrated into the MS is maintenance and housekeeping.

Naturally, the machines still need attention from the experts in the maintenance department, just as the airplane is taken out of service periodically for engine overhaul and maintenance. One alternative here is to switch to two 8-hour shifts separated by two 4-hour time blocks for machine maintenance, tooling changes, restocking, long setups, overtime, earlytime, and so on. This is called the 8-4-8-4 scheme. The main advantage that equipment has over people is that it can decrease variability, but it must be reliable and dependable. Smaller machines are simpler and easier to maintain and therefore are more reliable. Small machines in multiple copies add to the flexibility of the system as well. The linked-cell system permits certain machines in the cells to be slowed down and therefore, like the long-distance runner, to run farther and easier without breakdown. Many observers of the JIT manufacturing system come away with the feeling that the machines are "babied." In reality, they are being run at the pace needed to meet the demand.

True lean producers build and modify much of their manufacturing process technology. It is what makes them unique. In addition, they try to make or modify equipment when possible in multiple copies because two teams and two sets of equipment making the same products or families of parts are more flexible in the event of a machine failure and in terms of capacity. Thus *capacity is replicated in proven increments as demand grows.* Because the increment has an optimal design, this is an economic choice as well as having the security of dealing with a proven manufacturing process technology. Modifying existing equipment shortens the time needed to bring new technology on stream. Manufacturing in multiple versions of small-capacity machines retains the expertise and permits the company to keep improving and mistake-proofing the process. In contrast to this approach is the typical job shop, where a new supermachine would be purchased and installed when product demand increases. That is, many companies try to increase capacity by buying new, untried manufacturing technology which may take months, even years, to debug and make reliable.

■ 43.8 Leveling and Balancing the Manufacturing System

The steps outlined here are the amalgamated experience of many companies that have Americanized and implemented some version of the Toyota production system. Machine layout follows the flow of processes wherein products having common or similar processes are grouped together and quick conveyance between the processes is provided, along with the means to reduce setup time. The basic premise of the system is to produce the kind of units needed in the quantities needed at the time needed. The system, called *just-in-time* manufacturing (JIT), depends on *smoothing of the manufacturing system.* To eliminate variation or fluctuation in quantities in feeder processes, it is necessary to eliminate fluctuation in final assembly. This is called *leveling* the final assembly. Here is a simple example to show the basic idea.

Cycle time is determined as follows. Suppose that the forecast is for 240 cars per day and 480 production minutes are available (60 minutes × 8 hours/day). Thus cycle time = 2 minutes. Every 2 minutes a car rolls off the line. Suppose that the mix is as given in Table 43-3.

TABLE 43-3.   Example of Mixed Model Final Assembly Line That Determines the Cycle Time by Model

Q	Car Mix for Line Model	Cycle Time by Model (min)	Production Minutes by Model	Sequence (24 Cars)
50	Two-door coupe	9.6	100	TDC, TDF, TDF, FDS, FDW,
100	Two-door fastback	4.8	200	TDC, TDF, TDF, FDW, FDW,
25	Four-door sedan	18.2	50	TDC, TDF, TDF, FDS, FDW,
65	Four-door wagon	7.7	130	TDC, TDF, TDF, FDW, …
240	cars/8 hr	480 min/240 = 2 min per car		

The subprocesses that feed the two-door fastback are controlled by the cycle time for this model. Every 4.8 minutes the rear deck line will produce a rear hatch for the fastback version. Every 4.8 minutes two doors are made. Every car, regardless of model type, has an engine. Each engine needs four pistons; therefore, every 2.0 minutes, four pistons are produced. Parts and assemblies are produced in their minimum lot sizes and delivered to the next process, under the control of kanban. Let's look at a specific cell for this example.

Two disk hubs, made in cell 143, shown in Figure 43-4, are used in every car; the CT, then, for the cell is about 1 minute, actually 55 seconds. The extra 5 seconds per cycle provides for some spare capacity for making spare hubs for replacement parts. The cycle for a cell making parts exclusively for the wagon would be 7.7 minutes. The objective here is to make the same amount of product parts every day. Leveling the daily schedule takes the lumps out of the feeder processes. *Balancing is making the output from the cells equal to necessary demand for the parts downstream.* In summary, small lot sizes, made possible by setup reduction within the cells, single-unit conveyance within the cells, and standardized cycle times are the keys to accomplishing a smoothed manufacturing system. Every part, sequence of assembly operations, or subassembly has the same number of specified minutes. The minimum number of workers needed to produce 1 unit of output in the cycle time is used. This is called *leveling and balancing* the manufacturing system.

■ 43.9 LINKING THE CELLS

The cells are linked to each other by the pull system of production/inventory control called *kanban*. Kanban is a visual control system that is only good for lean production with its linked cells and its namesakes (Table 43-4); it is not good for the job shop. Figure 43-12 shows how the elements are linked together by kanban, thus providing control over the route that the parts must take (while doing away with the route sheet), control over the amount of material flowing between any two points, and information about when the parts will be needed. To accomplish this, there are two kinds of kanban: *withdrawal* (or conveyance) *kanban* (WLK) and *production-ordering kanban* (POK). One can think of the kanban as a loop connecting the output side of one cell with the input point of the next cell. The loop is filled with carts or containers that hold parts in specific numbers. Every cart has the same number of parts. Each cart has one WLK and one POK. Here is how kanban works.

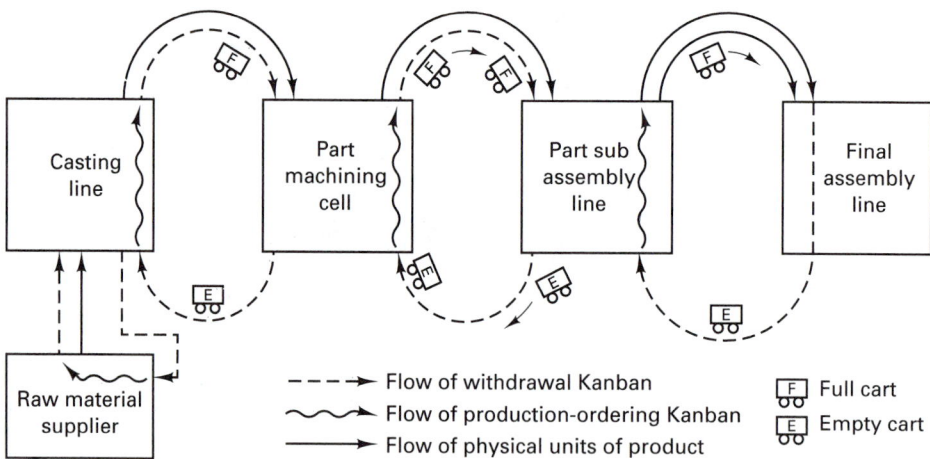

FIGURE 43-12   The processes are linked together by WLK kanban.

TABLE 43-4.   Other Names for Lean Manufacturing Systems
Just-in-time/total quality control, name coined by Dick Schonberger
ZIPS (zero inventory production system), Omark Industries
World class manufacturing, also by Dick Schonberger
MAN (material as needed), Harley-Davidson
MIPS (minimum inventory production system), Westinghouse
Ohno system, after Toyota's Taiichi Ohno, Mastermind of the System, many companies in Japan
Toyota production system, the "model" in reality or TPS
Stockless production, Hewlett Packard
Kanban, many companies in the United States and Japan
Modular manufacturing, apparel industries

Referring to Figure 43-12, the assembly line is being fed by the part subassembly line, which is fed by the part-machining cell. In kanban, the cell makes parts for the stations only when requested by means of a production-ordering kanban. That is, production is pulled, as needed, from the manufacturing cell. Suppose that a final assembly station uses up a container of parts. The empty container is moved back to the manufacturing cell by means of a WLK. The WLK is detached from the empty cart and attached to a full cart. The POK that was attached to the full cart is placed in the POK collection box. Thus an order is placed to replace the cart of parts that is being removed. The sequence in which the POKs are placed in the collection box dictates the order in which the lots will be made. The full cart with the WLK attached is taken to the assembly area. When assembly begins to use this cart of parts, the WLK is detached and placed in a collection box. The WLK can be used with an empty cart to withdraw more parts from the machining cell. The number of POKs equals the number of WLKs equals the number of carts. The carts hold a set number of parts. The exact amount of inventory at any place in the system is known:

$$\text{maximum inventory} = \text{number of carts} \times \text{number of parts in cart}$$

The same kind of loop is connecting the subassembly cell to the part-machining cell. All the other cells in linked-cell MS are similarly connected by the pull system for production control.

■ 43.10 REDUCING THE WORK-IN-PROCESS

Step 7 involves the integration of the inventory control. The inventory in the system is held in the links and is called the *work-in-process* (WIP). The WIP inventory has been analogized to the water in a river, as shown in Figure 43-13. A high river level is equivalent to a high level of inventory in the system. The high river level covers the rocks in the riverbed. Rocks are equivalent to problems. Lower the level of the river (inventory) and the rocks (problems) are exposed. This analogy is quite accurate. The problems receive immediate attention when exposed. When all the rocks are removed, the river can run very smoothly with very little water. However, if there is no water, the river has dried up. The notion of zero inventory is misleading. While zero defects is a proper objective, zero inventory is not possible. The idea is to minimize the necessary WIP between the cells. (Within the cell, parts are already handled one at a time, just as they are in assembly lines.)

The level of WIP between the stand-alone process, cells, subassembly, and assembly actually is controlled by the foremen in the various departments. The control is integrated and performed at the point of use. Here is how it works. Suppose that there are 10 carts in the link and that each cart holds 20 parts. The maximum inventory in this area is therefore 200 parts. The foreman goes to the stock area and picks up the kanban cards (one WLK, one POK), which puts one full cart of parts out of commission. The (maximum) inventory level is now 9×20, or 180 parts. The foreman waits until a problem appears. When it appears, the foreman immediately restores the kanban, which restores the inventory to its previous level. The cause of the problem may or may not be identified by the restoration of the inventory, but the condition is relaxed until a solution can be enacted. Once the problem is solved, the foreman repeats this procedure. If no other problems occur, the foreman then tries to drop the inventory to $8 \times 20 = 160$ parts. This procedure is repeated daily all over the plant. After a few months, the foreman in the frame area may be down to 5 carts of 20 parts. Over the weekend the system will be restored to 10 carts between the two points, but this time each cart will hold only 10 parts. If everything works smoothly, with the reduced WIP lot size, the foreman will then remove a cart to see what happens. More likely, set-up times will need to be reduced. In this way the inventory in the linked-cell system is continually reduced, exposing problems. The problems are solved one by one. The teams work on solving the exposed problems. This is how continuous improvement works in the JIT factory.

The minimum level of inventory that can be achieved is a function of the quality level, the probability of a machine breakdown, the length of the setups, the variability in the manual operations, the number of workers in the cell, parts shortages, the transportation distance, and so on. It appears that the minimum number of carts is three, and, of course, the minimum lot size is one. The significant point here is that inventory becomes a controllable independent variable rather than an uncontrollable variable dependent on cravings of the users of the manufacturing system for more inventory.

■ 43.11 VENDOR RELATIONSHIPS

The traditional purchasing department permits its vendors to make weekly/monthly/semiannual deliveries with long lead times, weeks/months are not uncommon. A large safety stock is kept just in case something goes wrong. Quantity variances are large and late and early deliveries are the norm. This situation leads to expediting.

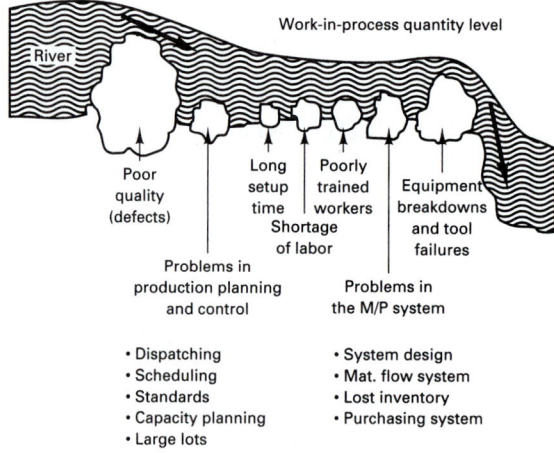

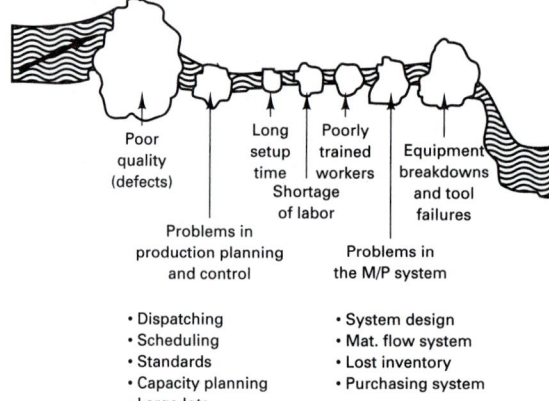

FIGURE 43-13 Lowering the work-in-process inventory causes problems in the system to be uncovered, just as lowering the level of a river causes the rocks to be uncovered.

As a hedge against vendor problems, multiple sources are developed. This may happen because one vendor cannot handle all the company's work. The purchasing department may claim that pitting one vendor against another gives the company a competitive advantage.

The lean manufacturing system does it differently. JIT purchasing is a program of continual long-term improvement. The buyer and vendor work together to reduce lead times, lot sizes, and inventory levels. Both companies become more competitive in the world marketplace.

In this environment, longer-term (18 to 24 month) flexible contracts are drawn up with 3 or 4 weeks' lead time at the outset. The buyer supplies updated forecasts every month which are good for 12 months, commits to long-term quantity, and perhaps even promises to buy out any excess materials. Exact delivery is specified by midmonth for the next month. Frequent communication between the buyer and the vendor is typical. Kanban controls the material movement between the vendor and the buyer. The vendor is a remote cell. Long-range forecasting for 6 months to 1 year is utilized. As soon as the buyer sees a change, the vendor is informed; this knowledge gives the vendor better visibility instead of a limited lead-time view. The vendor has *build schedule stability*, not "jerking" up and down of the build schedule.

The buyer moves toward fewer vendors, often going to local, sole sourcing. Frequent visits are made to the vendor by the buyer, who may supply engineering aid (quality, automation, setup reduction, packaging, and the like) to help the vendor become more knowledgeable on how to deliver, on time, the right quantity of parts that require no incoming inspection. The buyer and the seller must be willing to work together to solve problems.

The advantages of single sourcing are that resources can be focused on selecting, developing, and monitoring one source instead of many. When tooling dollars are concentrated in one source, there is a savings in tooling dollars. The higher volume should lead to lower costs. The vendor is more inclined to do special things for the buyer. The buyer and the vendor learn to trust each other. The quality is more consistent and easier to control and monitor.

■ 43.12 AUTONOMATION

Autonomation means the autonomous control of quality and quantity: Stop everything immediately when something goes wrong; control the quality at the source instead of using inspectors to find the problem that someone else may have created. The workers in the JIT factory inspect each other's work. Taiichi Ohno, former vice president of manufacturing for Toyota, was convinced that Toyota had to raise its quality to superior levels in order to penetrate the world automotive market. He wanted every worker to be personally responsible for the quality of the piece part or product that was produced. The need for automation simply reflects the gradual transition of the factory from manual to automated functions. Some people think of this as CIM. Others recognize that people are the most important (and flexible) asset in the company and see the computer as just another tool that is used in the process but is not the heart of the system. These companies are moving toward human integrated manufacturing (HIM), where a creative, motivated workforce is seen as the key to lean production.

In *robotic cells*, the microcomputers of the CNC machine tools and a robot are networked together with a cell host computer. It is difficult, if not impossible, to conceive of this kind of arrangement without resorting to some method that collects the work into compatible families. All the machines in the cell are programmable, and therefore this kind of automation is very flexible. See Figure 43-14 for an example of a robotic cell.

While it is an easy task to draw the boxes and connect lines in Figure 43-14, it is quite a different matter to realize all of the mechanical, electrical, and computer–engineering interfaces required to arrive at a fully unmanned, flexible, autonomous cell. In the real manufacturing environment, things are not as prescribed. The incoming work materials vary in geometry and specifications. The end effectors of the robot may not be able to accommodate such changes, or take note of random variations in the presentation of the components. Parts of the system will break, cutting tools will wear, and the quality control requirements will undoubtedly call for measuring more than one or two diameters.

A typical cell might have three different parts, with each part having six different sizes. To accommodate this family of 18 different parts may require that the process capability of robots be improved, that flexible grippers and workholders be designed, that controllers for existing machines be modified to allow supervision by a cell host computer, and that reliable cell control software be developed.

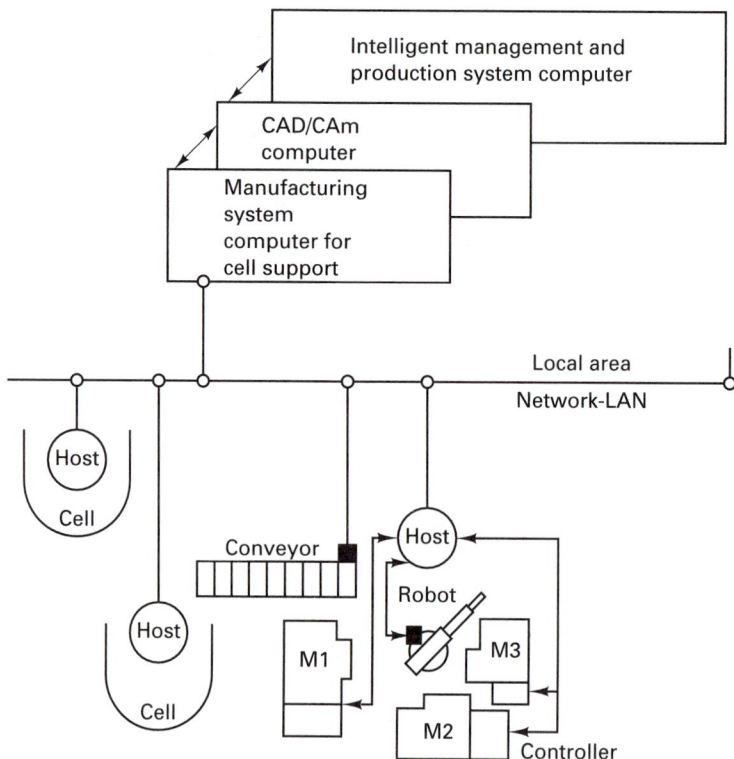

FIGURE 43-14 Manned cells evolve into unmanned cells. The cell host computer is linked directly to higher-level computers in the production system.

■ 43.13 RESTRUCTURING THE REST OF THE BUSINESS

Changing the manufacturing system will necessitate restructuring the *production system* and all its functions. In shifting from one type of system to another, the change will affect product design, tool design and engineering, production planning (scheduling) and control, inventories and their control, purchasing, quality control and inspection, and of course the production worker, the foreman, the supervisors, the middle managers, and so on, right up to top management. Such a conversion cannot take place overnight and must be viewed as a *long-term transformation* from one type of *production system* to another. This downsizing can be very traumatic for the business part of the company.

Step 10 recognizes the need for the rest of the company to reorganize (get lean). This effort often begins with building *product realization teams* designed to bring new products to the marketplace faster. In the automotive industry these are called *platform teams* and are an example of concurrent engineering. They are composed of people from design engineering, manufacturing, marketing, sales, finance, and so on. As the notion of team building spreads and the lean manufacturing system gets implemented, it is only natural that the production system will follow suit. Unfortunately, many companies are restructuring the business part of the company without having done the necessary steps 1 through 8 to get the manufacturing system lean and efficient. The reduction of the production system without simplifying and redesigning of the manufacturing system can lead to difficult times for the enterprise.

■ 43.14 BENEFITS OF CONVERSION

The conversion to lean manufacturing results in significant cost savings over a 2- to 3-year period. Specifically, manufacturing companies report significant reductions in raw materials and in-process inventories, setup costs, throughput times, direct labor, indirect labor, staff, overdue orders, tooling costs, quality costs, and the cost of bringing new designs on line.

However, this reorganization has a greater and immeasurable benefit. It prepares the way for CIM. The progression from the functional shop to the factory with linked cells and ultimately to robotic cells with computer control for the entire system must be accomplished in logical, economically justified steps, each building from the previous stage.

■ 43.15 CONSTRAINTS TO CONVERSION

Aside from the failure to recognize cells as a new form of manufacturing system, a major effort on the part of a business is required to undertake a conversion to lean manufacturing. The constraints on implementation are as follows:

1. *Systems changes are inherently difficult to implement.* Changing the *entire* manufacturing production system is a huge job.

2. *Companies spend freely for product innovation but not for process innovation.* It is easier to justify new hardware for the old manufacturing system than to rearrange the old hardware into a new manufacturing system (linked cells). However, anyone with capital can buy the newest equipment, often creating another island of automation.

3. *Fear of the unknown.* Decision making is choosing among the alternatives in the face of uncertainty. The greater the uncertainty, the more likely that the "do-nothing" alternatives will be selected. Converting to linked cells will free up additional capacity (setup time saved) and capital (funds not tied up in inventory). Such conversions will require expenditure of funds for equipment modifications, employee training (in quality, maintenance, and setup reduction), and so on. *The long-term payback equals a high-risk situation in the minds of the decision makers.*

4. *Faulty criteria for decision making.* Decisions should be based on the ability of the company to compete (quality, reliability, delivery time, flexibility for product change or volume change) rather than on output or cost alone.

5. The conversion to lean manufacturing represents a *real threat to staff and middle managers.* The functional tasks that they have been responsible for are being shifted and integrated into the manufacturing system. Also, the short-term life of the financially oriented middle managers is in conflict with the long-term nature of the program, which results in resistance to change in addition to the erosion of their functional empires.

6. *Lack of blue-collar involvement in the decision-making process of the company.* Getting the production workers involved in the decision-making process is in itself a significant change. The managers of the manufacturing system have had problems adjusting to this situation.

Clearly, education at all levels of the enterprise is needed to overcome these constraints. The attitudes of management and workers must change. Changing to linked cells requires an evolutionary, dynamic philosophy, but such conversions offer great potential for markedly improving the quality and productivity.

■ 43.16 SUMMARY

The factory with a future will require much higher levels of knowledge and more effective modes of information transfer about the quality and quantity of goods being manufactured.

The knowledge base of the factory worker must be increased to improve the productivity of the workers and ultimately the productivity of the company. Knowledge has a market value. Japan bought the technical knowledge it needed to build cars, electronics, machine tools, motorcycles, and even electron microscopes. The early Japanese transmission electron microscopes were virtually duplicates of the precision-made German instruments. A common thread that ties all the Japanese product areas together is high (sophisticated) technology. For example, the precision machining of the magnetic lens and the fabrication of high-voltage electronics were the key to the construction of high-quality electron microscopes.

Information clearly has value because people are willing to pay for it. Information also has cost because it costs something to produce the information. Thus the best manufacturing system is the one that minimizes the information needed to operate (control) the system.

The factory with a future will need superior information systems and people who can analyze, program, and otherwise deal with the information on the factory floor. The unique thing about knowledge (and information) is that, unlike energy, it does not follow the laws of conservation. *Knowledge is synergistic, breeding on itself.* Thus, as the factory worker becomes better educated and more knowledgeable about how the entire manufacturing production system works, the system will improve rapidly.

Manufacturing systems must be redesigned so that they are simpler and therefore easier to automate. By the next century, we will see significantly fewer workers on the plant floor. These workers will be far better educated and more productive. They will be involved in solving daily production problems, in working to improve the entire system, making decisions about how to improve their job, the processes, and the manufacturing system.

Many American industries are now undertaking massive educational programs to teach their employees about quality control, machine maintenance, setup reduction, and other elements of lean manufacturing. These educational steps are the key to improving productivity. This system is revolutionary and not just a passing fad. It represents the model for the third industrial revolution.

The attitude toward the production worker is critical. The production worker must not be viewed (by management) as a variable input but as a valuable asset. They must be respected. A job that a machine can do better than a human being should not be done by human beings—it is below their dignity. In the United States we believe in the value of human worth, but many workers have had to surrender these rights when they go on the factory floor. Workers are important as people. *People are America's greatest untapped resource.* Our production people can do much more than we are now giving them the opportunity to do. Management must give workers an opportunity to do more. The management system must provide all the workers with opportunities to display maximum abilities and to make contributions to improve their job, the products and the entire system. This must be practiced, not just preached. More employee training and education, at all levels, is absolutely necessary. The company must be prepared for large training costs in addition to large capital equipment costs. Workers need to be trained in quality control techniques and in setup reduction. All workers must be involved in the changes because there is no other way that the gains needed in quality and productivity can be achieved. Implementation is easy when the suggestions for changes needed to enhance quality or lower cost come from the workers. The step-by-step methodology that has been outlined will result in a company's becoming a factory with a future.

■ KEY WORDS

autonomation
business process reengineering (BPR)
coding/classification (C/C)
cycle time (CT)
decoupling
economic order quantity (EOQ)
group technology (GT)

intermediate jig
just-in-time manufacturing system
kanban
lean production
linked cell manufacturing system
product flow analysis (PFA)
production ordering kanban (POK)

single-minute exchange of die (SMED)
standard operations routine sheet
withdrawal kanban (WLK)
work-in-process (WIP)

■ REVIEW QUESTIONS

1. Why is setup reduction one of the first steps needed to convert current systems into a lean manufacturing system?
2. What is the difference between the Ford system and the Ohno system in terms of lot size?
3. In his book *Megatrends*, Naisbitt[*] argues that we are moving from a condition of forced technology toward high tech/high touch. What are high-touch aspects of cellular manufacturing?
4. What does the acronym SMED stand for?
5. In lean manufacturing, *integration* has a different meaning from the one you have heard or used previously. What is being integrated into what?
6. How is production control, a classical function of the production planning and control department, integrated into the manufacturing system?
7. What do we mean when we say a worker can operate multiple processes? Is this the same thing as being multifunctional? Are we all really multifunctional?
8. What is involved in the integration of quality control into the manufacturing system?
9. How does the strategy of less-than-full capacity loading or scheduling of the manufacturing system assist in improving quality?
10. Suppose that you are working in a cell within a

linked-cell JIT manufacturing system. How do you find out what you are to make and when you are to make it? (What to make and when to make it are the primary things you need to know.)
11. The effect of reducing the lot size is to take the lumps out of the manufacturing system's flow, making the flow smoother and more laminar. What other technique is used to level and balance the flow?
12. Perhaps the most powerful aspect of the linked-cell JIT system is its ability truly to control the level of the inventory. No system can operate without inventory. It is the life blood of the system, but how is the inventory level controlled and, in fact, continuously reduced in a JIT system?
13. What is the effect on the inventory in this or any manufacturing system of poor quality? Of unreliable machines? Of long setups?
14. What determines the minimum level of the inventory in a linked-cell JIT manufacturing system?
15. What is meant in stating that a vendor is just a remotely linked cell?
16. What is *autonomation*? (Note that this is not *automation*.)
17. Many factories have undertaken massive programs to computerize and automate their job shops. In fact, we have many examples wherein

---

[*] John Naisbitt, *Megatrends*, Warner Books, Inc., New York, 1982.

we have used the computer to improve a production system function. CAD is an excellent example of this. How is the CIM technique different from JIT?

18. How is cycle time determined for the cell?
19. How is the cycle time for the cell related to the number of workers in the cell?
20. What is concurrent engineering all about?
21. What is PFA?
22. What is an example of a code that you use every day which classifies you (i.e., tells the reader of the code something about you)?
23. What do we mean when we say that the worker is decoupled from the machine?
24. Why is a robotic cell not as flexible as a manned cell?
25. Is a robotic cell the same thing as a FMS? Explain why or why not. (*Hint:* Review material in Chapter 41.)

26. What is the difference between internal and external setup?
27. Explain the intermediate jig concept.
28. What is the key to implementing a successful setup reduction program?
29. How does the quality control approach described in step 3 differ from classical quality control?
30. Are Quality Circles an off-line QC method?
31. What are some other names for lean manufacturing?
32. How is work-in-process (WIP) or in-process inventory reduced?
33. Who is Taiichi Ohno?
34. How is JIT purchasing different from other purchasing strategies? That is, what is the goal of JIT purchasing?
35. What are some of the constraints on the implementation of lean production?

■ PROBLEM

1. Figure 43-4 shows the standard operations routine sheet for the machining of disk hubs. The machines for the cells are shown. The machine labeled "Hub cold header" cold forms the hubs, which then require three machining processes: two roughing processes and one finishing.
   (a) What is the cycle time for these parts?
   (b) What is the total machining time, exclusive of all delays, for one part?
   (c) Is the cell capable of meeting the required daily quantity?
   (d) Why does the worker visit BR744 every other trip or cycle through the cell?
   (e) What percentage of the cycle time is spent walking?
   (f) How could the walking time be reduced?

*C*hapter 43 CASE STUDY

snowmobile accident

Katrin B is suing the FlufferCat Snowmobile Company and LINUS Components for $750,000 over her friend's death. He was killed while racing his snowmobile through the woods in the upper peninsula of Michigan. Her lawyer claims that he was killed because a tie-rod broke, causing him to lose control and crash into a tree, breaking his neck. While it was impossible to determine whether the tie-rod broke before the crash or as a result of the crash, the following evidence has been put forth. The tie-rod was originally designed and made entirely out of low-carbon steel (heat treated by case hardening) in three pieces, as shown in Figure CS-43. These tie-rods were subcontracted by FlufferCat to LINUS Components. LINUS Components changed the material of the tie-rod bolts from steel to a heat-treated aluminum having the same UTS value as the steel. They did this because aluminum rods were easier to thread roll rather than to form by a thread-cutting operation. It was further found that threads on one of the tie-rod bolts were not as completely formed as they should have been. The sleeve of the tie-rod in question was split open and one of the tie-rod bolts was bent.

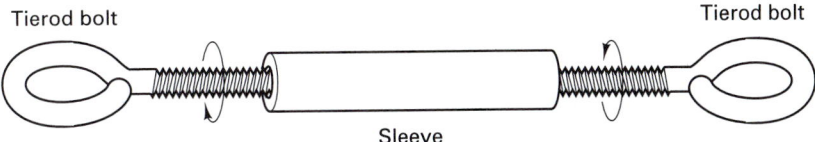

Tierod bolt   Tierod bolt

Sleeve

FIGURE CS-43   Assembled tie-rod for snowmobile.

Katrin's lawyer claimed that the tie-rod was not assembled properly. He claimed that one rod was screwed into the sleeve too far and the other not far enough, thereby giving it insufficient thread engagement. LINUS testified that these tie-rods are hand assembled and checked only for overall length and that such a misassembly was possible. Katrin's lawyer stated that the failure was due to a combination of material change, manufacturing error, and bad assembly, all combining to result in a failure of the tie-rod.

A design engineer for FlufferCat testified that the tie-rods were "way overdesigned" and would not fail even with slightly small threads or misassembly. FlufferCat's lawyer then claimed that the accident was caused by driver failure and that the tie-rod broke upon impact of the snowmobile with the tree. One of the men racing with Katrin's friend claimed that her friend's snowmobile had veered sharply just before he crashed, but under cross examination he admitted that they had all been drinking that night because it was so cold (he guessed 20° below). Since this accident had taken place over 5 years ago, he could not remember how much they had had to drink.

You are a member of the jury and have now been sequestered to decide if FlufferCat is guilty of negligence resulting in death. The rest of the jury, knowing you are an engineer, has asked for your opinion. What do you think? Who is really to blame for this accident? What actually caused the accident?

ADDITIONAL CASE STUDIES

The case studies in this section are provided as a supplement to those found at the end of the various chapters. Most combine material from multiple chapters. Several offer components that can be made by a wide variety of methods, and enable comparison of the relative merits of each approach. Others relate to product failures that might require a knowledge of materials, fabrication processes and the specific service environment.

- **A.** Heat-Treated Axle Shafts
- **B.** Handle and Body of a Large Rachet Wrench
- **C.** Diesel Engine Fuel Metering Lever
- **D.** Bevel Gear for a Riding Lawn Mower
- **E.** Rocker Pivot for an Electronic Scale
- **F.** Automobile Water Pump Impeller
- **G.** Flywheel for a High-Speed Computer Printer
- **H.** Broken Wire Cable
- **I.** Short-Lived Gear
- **J.** Broken Marine Engine Bearings
- **K.** Fire Extinguisher Pressure Gage

Appendix A CASE STUDY

heat-treated axle shafts

The high strength and fatigue requirements of automobile axle shafts generally require a heat-treated steel with a tempered martensite structure and a surface hardness of approximately R_c50. The shafts are approximately 35 mm ($1\frac{3}{8}$ in.) in diameter and 1.06 meters ($3\frac{1}{2}$ feet) in length, and distortion or warpage must be minimized. In the 1960s, these requirements were met by oil-quenching bars of medium-carbon alloy steels, such as 4140, 8640, and 5140. Now they are frequently manufactured from the less expensive plain-carbon steels, such as 1038 and 1040. How would you propose to heat-treat these lower-hardenability metals to the desired hardness and structure and still limit distortion?

A further cost saving can be obtained if a small segment of the axle shaft surface can serve as the inner race of a bearing. This, however, requires a surface hardness of R_c60 in this region. How would this additional requirement modify your material selection? Assuming that plain-carbon steel is still used, how might this region of high surface hardness be obtained?

Appendix B CASE STUDY

handle and body of a large rachet wrench

Figure CS-B shows the handle and body segment of a relatively large rachet wrench, such as those used with conventional socket sets. The design specifications require a material with a minimum yield strength of 50,000 psi and an elongation of at least 2% in all directions. Additional consideration should be given to weight minimization (because of the relatively large size of the wrench), corrosion resistance (due to storage and use environments), machinability (if finish machining will be required), and appearance.

1. Based on the size and shape of the product, describe several methods by which the component could be produced. For each method, briefly discuss its relative pros and cons.
2. What types of engineering materials might be able to meet the desired requirements? What would be the pros and cons of each general family?
3. For each of the shape generation methods in part 1, select an appropriate material from the alternatives discussed in part 2. (*Note:* Casting alloys should be matched with casting processes, high machinability alloys would be favored for cutting applications, etc.)
4. Which of the above alternatives do you feel would be the "best" solution to the problem? Why? For this system, outline the specific steps that would be necessary to produce the part from reasonable starting material.
5. For your proposed solution in part 4, would any additional heat treatment or surface treatment be required? If so, what would you recommend?
6. If a variation of this tool were to be marketed as a "safety tool" that could be used in areas of gas leaks where a spark could be fatal, how would you need to modify your previous recommendations? Discuss briefly.

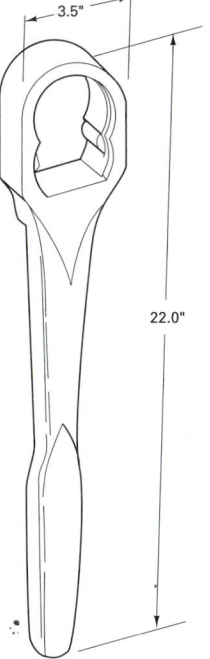

FIGURE CS-B

Appendix C CASE STUDY

diesel engine fuel metering lever

The component pictured in Figure CS-C is a fuel-metering device lever for a large diesel engine. The long dimension of the part is approximately $2\frac{1}{2}$ in. and the width is 7/8 in. The large hole through the center is 0.368 in. in diameter and the two small holes are to be tapped to accept 10–32 threaded bolts. A realistic production run would be for approximately 20,000 pieces.

1. Describe in a general way, three possible means of manufacturing the desired shape. For each, discuss briefly the pros and cons of the approach.
2. If the product were to require mechanical properties of a 55,000 psi tensile strength, a surface hardness of Rockwell B 55, and an elongation of 1%, what types of materials would seem appropriate for this part?
3. Briefly discuss the suitability of the materials proposed in Question 2 to the various processes proposed in Question 1. Are there any significant limitations or incompatibilities?
4. What do you feel would be the best material-and-process solution among those proposed above? For this solution, in what form would you purchase the starting material? Would heat treatment be required to achieve the necessary final properties?

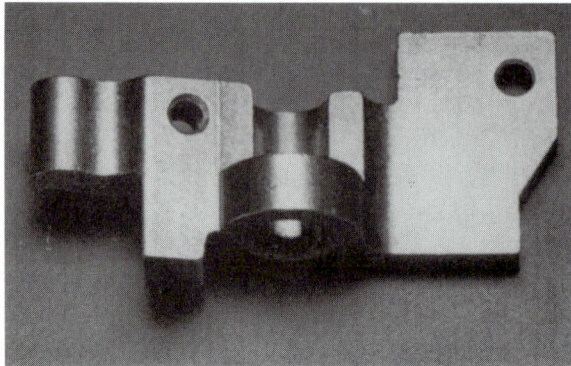

FIGURE CS-C

Appendix D CASE STUDY

bevel gear for a riding lawn mower

Figure CS-D is a bevel gear for the transaxle of a heavy-duty riding lawn mower. The gear has an outer diameter of 3.50 inches and a maximum thickness of 1/2-inch. At the root of the teeth, the thickness is approximately 3/16-inch. The minimum material properties have been estimated to be a yield strength of 125 ksi and a surface hardness of Rockwell C 25. The maximum operating temperature should be less than 250°F.

Since the mower has a multi-speed transmission, the transaxle gearing may be subject to sudden applications of load due to improper engagement of the clutch. The manufacturer has proposed a non-standard test in which a 10-pound weight is dropped onto the gear from a height of 15 inches. While this does not translate to a Charpy test value, it is clear that impact resistance will be an important property.

The gear will operate in an oil-filled enclosure so corrosion resistance need not be outstanding. A smooth surface finish is desirable (especially on the teeth), and the total production run has been placed between 25,000 and 50,000 units.

1. Based on the size and shape of the product, describe at least three reasonable ways in which the component could be produced. For each method, briefly discuss its relative pros and cons.
2. What types of engineering materials might be able to meet the desired requirements? What would be the pros and cons of each general family?
3. For each of the shape generation methods in part 1, select an appropriate material from the alternatives discussed in part 2. (*Note:* Casting alloys should be matched with casting processes, high machinability alloys would be favored for cutting applications, etc.).
4. Which of the above alternatives do you feel would be the "best" solution to the problem? Why? For this system, outline the specific steps that would be necessary to produce the part from reasonable starting material. Include any necessary heat treatments and/or surface treatments.
5. Your supervisor has asked you to evaluate the possibility of producing this part as a near-net-shape ferrous forging, with as-forged gear teeth. Investigate this process and determine if it would be a feasible approach. If so, would you recommend that the forging be performed cold, warm or hot? Would subsequent heat treatment or surface treatment be required? What concerns might be associated with these processes?

FIGURE CS-D

Appendix E <u>CASE STUDY</u>

rocker pivot for an electronic scale

Figure CS-E is a full-size photo of a pivot support for an extremely precise electronic scale. The two vertical holes are actually elongated slots, while the horizontal holes are round. All external surfaces are flat, including those that cannot be seen. All vertical surfaces are parallel to one another.

The part must have precise dimensions and be corrosion-resistant to a range of indoor (laboratory) environments. The only mechanical properties that were cited by the design team were a minimum yield strength of 25 ksi and a minimum tensile strength of 30 ksi. No hardness or impact specifications have been mentioned.

1. Based on the size and shape of the product, describe at least three reasonable ways in which the component could be produced. For each method, briefly discuss its relative pros and cons.
2. What types of engineering materials might be able to meet the desired requirements? What would be the pros and cons of each general family?
3. For each of the shape generation methods in part 1, select an appropriate material from the alternatives discussed in part 2. (*Note:* Casting alloys should be matched with casting processes, high machinability alloys would be favored for cutting applications, etc.).
4. Which of the above alternatives do you feel would be the "best" solution to the problem? Why? For this system, outline the specific steps that would be necessary to produce the part from reasonable starting material.
5. For your proposed solution in part 4, would any additional heat treatment or surface treatment be required? If so, what would you recommend?

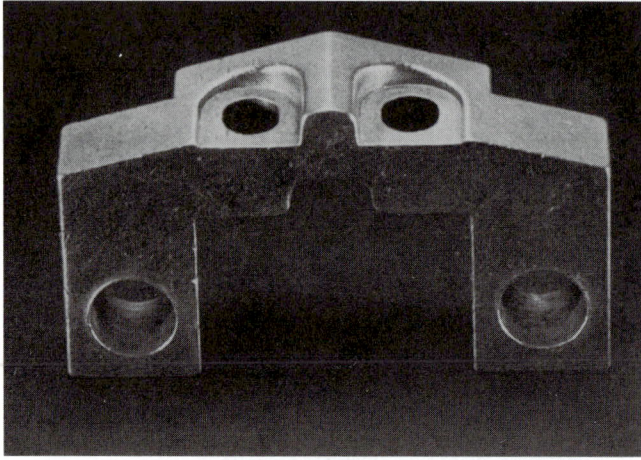

FIGURE CS-E

*A*ppendix F   CASE STUDY

automobile water pump impeller

Figure CS-F is a water pump impeller used by a major automotive parts manufacturer. The outer diameter of the component is 2.75 inches and the height is specified at 0.75 ± 0.005 inches. The I.D. of the center hole is 0.625 inches. The dimensions and positioning of the six blades is important from the viewpoint of vibration and balance. Smooth finish is desirable for good fluid flow. The operating temperature range has been estimated as −60°F to +300°F, and the mechanical property requirements have been given as 30,000 psi minimum tensile strength and a minimum elongation of only 0.5%. Since there is no metal-to-metal contact, high wear resistance is not necessary. The contact fluid should be a water/antifreeze mixture with corrosion-resistant additives. Production volume should be high, so a low total cost (material plus manufacturing) is a prime objective.

1. Based on the size and shape of the product, describe at least three reasonable ways in which the component could be produced. For each method, briefly discuss its relative pros and cons.
2. What types of engineering materials might be able to meet the desired requirements? What would be the pros and cons of each general family?
3. For each of the shape generation methods in part 1, select an appropriate material from the alternatives discussed in part 2. (*Note:* Casting alloys should be matched with casting processes, high machinability alloys would be favored for cutting applications, etc.).
4. Which of the above alternatives do you feel would be the "best" solution to the problem? Why? For this system, outline the specific steps that would be necessary to produce the part from reasonable starting material.
5. For your proposed solution in part 4, would any additional heat treatment or surface treatment be required? If so, what would you recommend?
6. Could this be a possible application for molded nylon, other polymer, or composite material? If so, what would be the candidate material? What would be the major differences in properties, performance and manufacturing process?

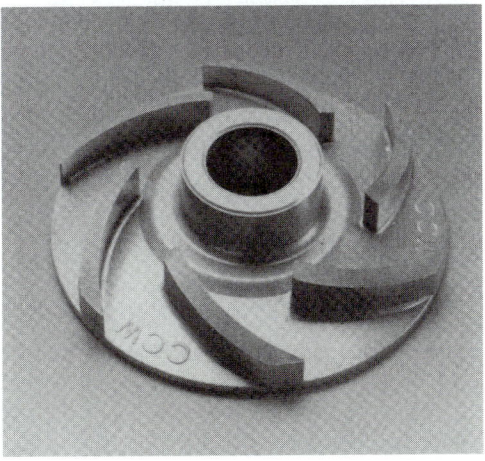

FIGURE CS-F

Appendix G <u>CASE STUDY</u>

flywheel for a high-speed computer printer

Figure CS-G shows a $5\frac{3}{4}$ inch diameter flywheel for use in a large, commercial-grade high-speed computer printer. The part is $3\frac{3}{4}$ inches thick to provide a total weight of approximately 12 pounds, assuming fabrication from a ferrous-based metal. The only mechanical requirement is a Rockwell B hardness equal to or greater than 50. The most restrictive requirement appears to be dimensional precision, since tooth, bore, hub and diameter dimensions must all be within 0.001 inch of specifications.

1. Based on the size and shape of the product, what are some possible means of producing the product? For each method, briefly discuss its relative pros and cons.
2. Assuming that a ferrous material will be used, select an appropriate material for each of the alternatives discussed in part 1. (*Note:* Casting alloys should be matched with casting processes, high machinability alloys would be favored for cutting applications, etc.).
3. Which of the above alternatives do you feel would be the "best" solution to the problem? Why? For this system, outline the specific steps that would be necessary to produce the part from reasonable starting material.
4. For your proposed solution in part 4, would any additional heat treatment or surface treatment be required? If so, what would you recommend?

FIGURE CS-G

Appendix H <u>CASE STUDY</u>

broken wire cable

Figure CS-H shows one end of a wire cable that broke in service, permitting the large boom of a crane to fall. Enlarged views of two typical wire ends are shown in the insets.

The company that manufactured the cable claimed that their product was of high quality and had simply been loaded beyond its rated capacity. An employee of the construction firm who was operating the crane was injured by the falling boom. He maintained that no overload had been applied and filed suit against the cable manufacturer, claiming that the cable was defective at that location.

Who do you feel was correct in this case? What evidence supports your position? From your knowledge of wire and cable manufacture, why is it unlikely that a wire cable would be defective at a specific location when new? What things could have happened to the cable while it was in use that could have contributed to the failure? Is it possible that the cable could have failed while under a load that was within its rated capacity? If so, who would have been at fault and responsible for the failure?

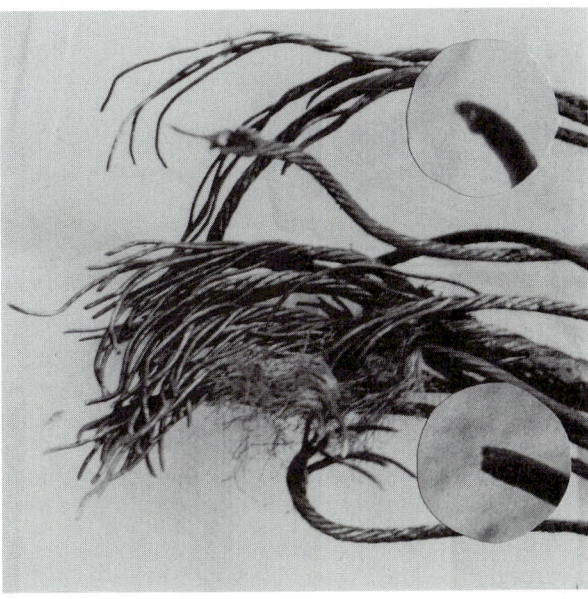

FIGURE CS-H

Appendix I CASE STUDY

short-lived gear

A 10-in. (250 mm) diameter gear has been fabricated from AISI 1080 steel. The gear blank has been hot-forged, air-cooled, and then full annealed in preparation for machining. Following finish machining, the gear teeth are flame-hardened and quenched to produce a surface hardness of Rockwell C 55. After a brief period of use, the teeth of the gear begin to deform and the gear fails to mesh properly.

1. What do you feel is the probable cause of the failure? How would you attempt to prove your hypothesis?
2. If you feel that the gear had been improperly manufactured, what change or changes in the manufacturing process would you recommend? How would your product be different from the one described above?

Appendix J CASE STUDY

broken marine engine bearings

The bearings on a small shipboard marine engine have been manufactured from AISI 52100 (bearing quality) steel that has been austenitized, quenched, and tempered to establish the desired final properties. Performance under normal operating conditions has been adequate for several months. After exposure to a period of subzero temperatures (a New England winter), the engine fails during its first operation. Tear-down reveals rather brittle cracks in the bearings. Micrometer measurements further reveal that the dimensions of the bearings are somewhat larger than manufacturer's specifications.

1. What do you suspect to be the cause of the failure? How would you go about proving your supposition? Who was at fault, the manufacturer of the bearings or the customer?
2. Assuming your conclusion in question 1 is correct, how would you attempt to prevent future occurrences of this type?

Appendix K CASE STUDY

fire extinguisher pressure gage

A s a materials engineer for the Fyre-Pro Extinguisher Company, you have recently been made aware of several potentially hazardous failures that have occurred in your company's products. Bourdon tube pressure gages (Figure CS-K) are used to monitor the internal pressure of your sodium bicarbonate (dry chemical) fire extinguishers. In several extinguishers, a longitudinal crack has formed along the axis of the bourdon tube, a curved tube of elliptical cross section that has been fabricated from phosphor bronze tubing. These cracks are particularly disturbing for several reasons. First, they allow the fire extinguisher to lose pressure and become inoperable. More significantly, however, the cracks allow the tube to deflect elastically in such a manner that the gage still indicates high internal pressure. Thus, while the extinguisher has actually lost all internal pressure and is useless, the reading on the gage would cause the owner to believe that he still had an operable fire-fighting device.

Your task is to determine the cause of these failures and suggest appropriate corrective measures.

1. What additional information would you like to have regarding the failed components, their fabrication history, and their service history? Why?
2. What might be some of the possible causes of these failures? What type of evidence would support each possibility? What types of additional tests or investigations might you propose?
3. Could these failures have occurred in "normal" use, or is it likely that some form of negligence, abuse, or misuse was involved?
4. What possible "corrective" or "preventative" measures might you suggest to prevent a recurrence?

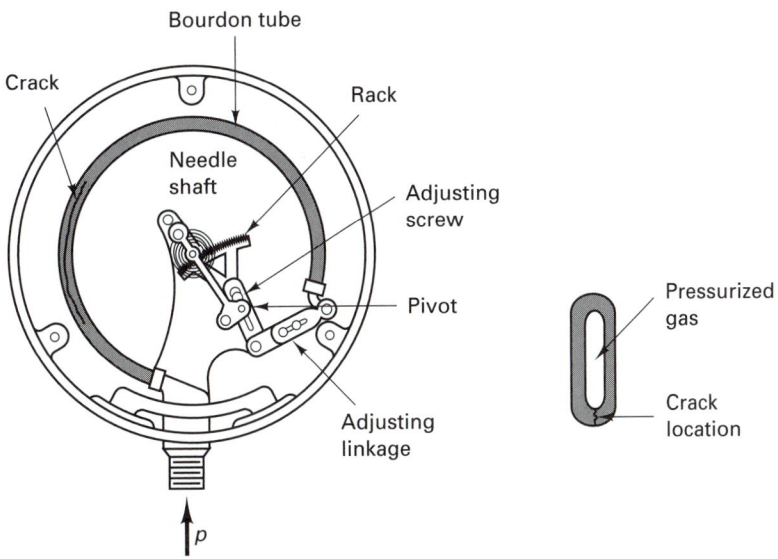

(a) Schematic of a bourdon-tube pressure gage   (b) Cross section of bourden tube

FIGURE CS-K

SELECTED REFERENCES FOR ADDITIONAL STUDY

Handbooks and General References

1. *Metals Handbook*, 10th Ed., ASM, Materials Park, Ohio.
   Vol. 1, Properties and Selection: Irons, Steels and High-Performance Alloys (1990)
   Vol. 2, Properties and Selection: Nonferrous Alloys and Special-Purpose Materials (1990)
   Vol. 3, Alloy Phase Diagrams (1992)
   Vol. 4, Heat Treating (1991)
   Vol. 5, Surface Engineering (1994)
   Vol. 6, Welding, Brazing and Soldering (1993)
   Vol. 18, Friction, Lubrication and Wear Technology (1992)

2. *Metals Handbook*, 9th Ed., ASM, Materials Park, Ohio.
   Vol. 1, Properties and Selection: Irons and Steels (1978)
   Vol. 2, Properties and Selection: Nonferrous Alloys and Pure Metals (1979)
   Vol. 3, Properties and Selection: Tool Materials and Special-Purpose Metals (1980)
   Vol. 4, Heat Treating (1981)
   Vol. 5, Surface Cleaning, Finishing, and Coating (1982)
   Vol. 6, Welding, Brazing, and Soldering (1983)
   Vol. 7, Powder Metallurgy (1984)
   Vol. 8, Mechanical Testing (1985)
   Vol. 9, Metallography and Microstructures (1985)
   Vol. 10, Materials Characterization (1986)
   Vol. 11, Failure Analysis and Prevention (1986)
   Vol. 12, Fractography (1987)
   Vol. 13, Corrosion (1987)
   Vol. 14, Forming and Forging (1988)
   Vol. 15, Casting (1988)
   Vol. 16, Machining (1988)
   Vol. 17, Nondestructive Evaluation & Quality Control (1989)

3. *Metals Handbook*, Desk Ed., ASM, Materials Park, Ohio (1984).

4. *Metals Handbook*, 8th Ed., ASM, Materials Park, Ohio.
   Vol. 1, Properties and Selection of Materials (1961)
   Vol. 2, Heat Treating, Cleaning and Finishing (1964)
   Vol. 3, Machining (1967)
   Vol. 4, Forming (1969)
   Vol. 5, Forging and Casting (1970)
   Vol. 6, Welding and Brazing (1971)
   Vol. 7, Atlas of Microstructures (1972)
   Vol. 8, Metallography, Structures and Phase Diagrams (1973)
   Vol. 9, Fractography and Atlas of Fractographs (1974)

Vol. 10, Failure Analysis and Prevention (1975)

Vol. 11, Nondestructive Inspection and Quality Control (1976)

5. *ASM Metals Reference Book*, Third Ed., ASM, Materials Park, Ohio (1993).

6. *ASM Engineered Materials Reference Book*, Second Ed., ASM, Materials Park, Ohio (1993).

7. *ASM Engineered Materials Handbook Series*, ASM, Materials Park, Ohio.

Vol. 1 Composites (1987)

Vol. 2 Engineering Plastics (1988)

Vol. 3 Adhesives and Sealants (1990)

Vol. 4 Ceramics and Glasses (1991)

8. *ASM Engineered Materials Handbook*, Desk Ed., ASM, Materials Park, Ohio (1995).

9. *ASM Specialty Handbook Series*, ASM, Materials Park, Ohio.

Aluminum and Aluminum Alloys (1993)

Stainless Steels (1994)

Tool Materials (1995)

10. *Tool and Manufacturing Engineers Handbook*, 4th Ed., Society of Manufacturing Engineers, Dearborn, Michigan.

Vol. 1, Machining (1983).

Vol. 2, Forming (1984).

Vol. 3, Materials, Finishing and Coating (1985).

Vol. 4, Quality Control and Assembly (1987).

Vol. 5, Manufacturing Management (1987).

Vol. 6, Design for Manufacturability (1992).

Vol. 7, Continuous Improvement (1994).

Vol. 8, Plastic Part Manufacturing (1996).

11. *SAE Handbook*, Part 1 - Materials (issued annually), Society of Automotive Engineers, Warrendale, PA.

12. "Materials Selector" Issue of *Materials Engineering* (issued annually), Penton/IPC Publications.

13. *Smithells Metals Reference Book*, 6th Ed., E. A Brandes, Butterworths, (1983).

14. *Materials Handbook*, 13th Ed., George S. Brady and Henry R. Clauser, McGraw-Hill (1995).

15. *Woldman's Engineering Alloys*, 8th Ed., J. Frick, ASM, Materials Park, Ohio (1994).

16. *Materials and Processes—Part A: Materials; Part B: Processes—*3rd Ed., James F. Young and Robert S. Shane (eds.), Marcel Dekker (1985).

17. *Production Handbook*, 4th Ed., John A. White (ed.), John Wiley and Sons, (1987).

18. *Production Processes: The Productivity Handbook*, 5th Ed., Roger W. Bolz (ed.), Industrial Press (1981).

19. *Manufacturing Processes Reference Guide*, R. H. Todd, D. K. Allen, and L. Alting, Industrial Press (1994).

20. *ASME Handbook*, McGraw-Hill, New York.

Vol. 1, Metals Engineering: Design

Vol. 2, Metals Properties

Vol. 3, Engineering Tables

Vol. 4, Metals Engineering: Processes

21. *ASTM Standards* (multiple volumes), published annually, ASTM, Philadelphia, PA.

22. *Source Book on Materials Selection*, Vols. 1 and 2, ASM, Materials Park, Ohio (1977).

23. *Source Book on Industrial Alloy and Engineering Data*, ASM, Materials Park, Ohio (1978).

24. *Military Standardization Handbook—*MIL-HDBK-5C - "Metallic Materials and Elements for Aerospace Vehicle Structures", U.S. Department of Defense (1976).

25. *CRC Handbook of Materials Science*, C. T. Lynch (ed.), CRC Press, Florida (1971).

Vol. 1—General Properties

Vol. 2—Metals, Composites and Refractory Materials

26. *Structural Alloys Handbook* (2 volumes), Mechanical Properties Data Center, Columbus, Ohio (1976).

27. *ASM Materials Engineering Dictionary*, J. R. Davis (ed.), ASM, Materials Park, Ohio (1992).

Basic and General Textbooks in Materials and Processes

28. *The Science and Engineering of Materials*, 3rd Ed., Donald R. Askeland, PWS (1994).

29. *Principles of Materials Science and Engineering*, 3rd Ed., W. F. Smith, McGraw-Hill (1996).

30. *Introduction to Materials Science for Engineers*, 4th. Ed., James F. Shackelford, Macmillan (1995).

31. *Materials Science and Engineering: An Introduction*, 3rd Ed., W. D. Callister, Wiley (1994).

32. *The Science and Design of Engineering Materials*, J. P. Schaffer, et.al., Richard D. Irwin Publications, Chicago (1995).

33. *Elements of Materials Science and Engineering*, 6th Ed., L. H. VanVlack, Addison Wesley, (1989).

34. *Materials Science and Engineering*, G. F. Carter and D. E. Paul, ASM (1991).

35. *Engineering Materials*, 5th Ed., Kenneth Budinski, Prentice Hall (1996).

36. *Physical Metallurgy Principles*, 3rd Ed., R. E. Reed-Hill and R. Abbaschian, PWS-Kent (1992).

37. *Structures and Properties of Engineering Alloys*, 2nd Ed., W. F Smith, McGraw-Hill, New York (1993).

38. *Manufacturing Engineering and Technology*, 3rd Ed., Serope Kalpakjian, Addison-Wesley (1995).

39. *Fundamentals of Modern Manufacturing*, Mikell P. Groover, Prentice Hall (1996).

40. *Manufacturing Engineering*, Kenneth C. Ludema, R. M. Caddell, and A. G. Atkins, Prentice Hall (1987)

41. *Manufacturing Engineering Processes*, Leo Alting, Marcel Dekker, New York (1982).

42. *Manufacturing Processes*, 8th Ed., B.H. Amsted, Phillip F. Ostwald and Myron L. Begeman, Wiley, New York (1987).

43. *Manufacturing Processes and Materials for Engineers*, 3rd Ed., L. E. Doyle, et.al., Prentice Hall, (1984).

44. *Processes and Materials of Manufacture*, 4th Ed., Roy A. Lindberg, Allyn and Bacon (1990).

45. *Introduction to Manufacturing Processes*, 2nd Ed., John A. Schey, McGraw-Hill (1987).

46. *Modern Manufacturing Process Engineering*, B. W. Niebel, A. B. Draper, and R. A. Wysk, McGraw-Hill (1989).

47. *Manufacturing Science*, A. Ghosh and A. K. Mallik, Ellis Harwood Ltd. (1986).

48. *Mechanical Metallurgy*, 3rd Ed., G. E. Dieter, McGraw-Hill, New York (1986).

49. *The Mechanical Behavior of Materials*, T. H. Courtney, McGraw-Hill (1990).

Ferrous Metals

50. *Engineering Properties of Steel*, ASM, Materials Park, Ohio (1982).

51. *The Making, Shaping and Treating of Steel*, 10th Ed., Association of Iron and Steel Engineers, (1985).

52. *Steels: Heat Treatment and Processing Principles*, G. Krauss, ASM, Materials Park, Ohio (1990).

53. *Steels: Metallurgy & Applications*, D.T. Llewellyn, Butterworth Heinemann (1992).

54. *Steel Selection: A Guide for Improving Performance and Profits*, R.F. Kern and M.E.Suess, Wiley-Interscience (1979).

55. *Worldwide Guide to Equivalent Irons and Steels*, 3rd. Ed., ASM, Materials Park, Ohio (1992).

56. *Modern Steels and Their Properties*, Bethlehem Steel Corporation, Bethlehem, PA (various editions).

57. *Nickel Alloy Steels Databook*, International Nickel.

58. *Steel Wire Handbook* (4 volumes), The Wire Association International, Guilford, CT (1965-1980).

59. *Handbook of Stainless Steels*, Donald Peckner and I. M. Bernstein (eds.), McGraw-Hill (1977).

60. *Stainless Steel*, R. A. Lula, ASM, Metals Park, Ohio (1985).

61. *Design Guidelines for the Selection and Use of Stainless Steel*, Designers Handbook Series, Specialty Steel Industry of the United States, Washington, D.C. (1992).

62. *Source Book on Stainless Steels*, ASM, Materials Park, Ohio (1976).

63. *Tool Steels*, 4th Ed., G. A. Roberts and R. A. Cary, ASM, Materials Park, Ohio (1980).

64. *Metallurgy and Heat Treatment of Tool Steels*, Robert Wilson, McGraw-Hill (1975).

65. *Properties and Selection of Tool Materials*, ASM, Materials Park, Ohio (1975).

66. *Cast Iron: Physical and Engineering Properties*, H. T. Angus, Butterworths (1976).

67. "Cast Irons", *Materials Engineering*, June, 1974.

Nonferrous Metals

68. *Worldwide Guide to Equivalent Nonferrous Metals and Alloys*, Second Ed., ASM, Materials Park, Ohio (1987).

69. *Nonferrous Wire Handbook* (2 volumes), The Wire Association International, Guilford, CT (1977-1981).

70. *Aluminum Standards and Data*, The Aluminum Association, Washington, D.C. (1990).

71. *Source Book on Selection and Fabrication of Aluminum Alloys*, ASM, Materials Park, Ohio, (1978).

72. *Aluminum Alloys: Structure and Properties*, L. F. Mondolfo, Butterworths, (1975).

73. *Aluminum*, ASM, Materials Park, Ohio.
     Vol. 1: Properties, Physical Metallurgy and Phase Diagrams (1967).
     Vol. 2: Design and Application
     Vol. 3: Fabrication and Finishing

74. *Aluminum: Properties and Physical Metallurgy*, John E. Hatch (ed.), ASM, Materials Park, Ohio (1984).

75. *Standards Handbook: Copper, Brass and Bronze* (7 volumes), Copper Development Association.

76. *Source Book on Copper and Copper Alloys*, ASM, Materials Park, Ohio (1979).

77. *Understanding Copper Alloys: The Manufacture and Use of Copper and Copper Alloy Sheet and Strip*, Olin Brass (1977).

78. *Materials Properties Handbook: Titanium Alloys*, R. Boyer, E.W. Collings and G. Welsch, ASM, Materials Park, Ohio (1994).

79. *Titanium Alloys Handbook*, Metals and Ceramics Information Center, Battelle Columbus Laboratories, Columbus, Ohio (1972).

80. *Titanium and Titanium Alloys Source Book*, ASM, Materials Park, Ohio (1982).

81. *Titanium: A Technical Guide*, M. Donachie (ed.), ASM, Materials Park, Ohio (1988).

82. "Titanium and Its Alloys", *Materials Engineering*, July, 1974.

83. *Engineering Properties of Zinc Alloys*, International Lead Zinc Research Organization, New York (1980).

Plastics, Composites, and Ceramics

84. *Composite Materials Handbook*, 2nd Ed., M. M. Schwartz, McGraw-Hill, New York (1995).

85. *Handbook of Composites*, George Lubin (ed.), Van Nostrand Reinhold, New York (1982).

86. *Engineers' Guide to Composite Materials*, John W. Weeton (Ed.), ASM, Materials Park, Ohio (1986).

87. *Fabrication of Composite Materials—Source Book*, M. M. Schwartz (ed.), ASM, Materials Park, Ohio (1985).

88. *Advanced Polymer Composites: Principles and Applications*, B. J. Jang, ASM, Materials Park, Ohio (1993).

89. *Fundamentals of Metal Matrix Composites*, S. Suresh (ed.), Butterworth-Heinemann (1993).

90. *Engineering Plastics and Composites*, 2nd Ed., W.A. Woishnis (ed.), ASM, Materials Park, Ohio (1993).

91. *Modern Plastics Encyclopedia*, published annually by Modern Plastics magazine, a McGraw-Hill publication.

92. *Handbook of Plastics, Elastomers and Composites*, 2nd Ed., C.A. Harper (Ed.), McGraw-Hill (1992).

93. *Polymer Processing: Analysis and Innovation*, ASME PED-Vol. 5, (1982).

94. *Plastics Process Engineering*, James, L. Throne, Marcel Dekker, New York (1979).

95. *Plastic Materials and Processes*, S. Schwartz and S. H. Goldman, Van Nostrand Reinhold, New York (1982).

96. *Plastic Part Technology*, E.A. Muccio, ASM, Materials Park, Ohio (1991).

97. *Plastics Technology Handbook*, 2nd Ed., M. Chanda and S. K. Roy (eds.), Marcel Dekker (1992).

98. *Plastics Processing Technology*, E.A. Muccio, ASM, Materials Park, Ohio (1994).

99. *Handbook of Thermoset Plastics*, S.H. Goodman, Noyes Publications (1986).

100. *Plastics Product Design Engineering Handbook*, 2nd Ed., Sidney Levy and J. H. DuBois, Chapman & Hall (1984).

101. *Plastic Blow Molding Handbook*, N. C. Lee (ed.), Van Nostrand Reinhold (1991).

102. *Structural Ceramics*, M. Schwartz, McGraw Hill (1994).

Elevated Temperature Applications

103. *Engineer's Guide to High Temperature Materials*, F. J. Clauss, Addison Wesley, Reading, Mass. (1969).

103. *Source Book on Materials for Elevated Temperature Applications*, ASM, Materials Park, Ohio (1979).

104. *Properties of Refractory Metals*, W. D. Wilkinson, Gordon and Breach (1969).

105. *High Temperature Property Data: Ferrous Alloys*, M. F. Rothman (ed.), ASM, Materials Park, Ohio (1987).

Electronic and Magnetic Applications

106. *Electronic Materials & Processes Handbook*, Second Ed., C. A. Harper and R. M. Sampson, McGraw-Hill (1994).

107. *The Magnetic Properties of Materials*, J. E. Thompson, CRC Press, Cleveland, Ohio (1968).

Design

108. *Engineering Design: A Materials and Processing Approach*, 2nd Ed., George Dieter, McGraw-Hill (1991).

109. *Handbook of Product Design for Manufacturing—A Practical Guide for Low-Cost Production*, James G. Bralla (ed.), McGraw-Hill, New York (1986).

110. *The Principles of Design*, Nam P. Suh, Oxford University Press (1990).

111. *Fundamentals of Tool Design*, 3rd Ed., David T. Reid (ed.), Society of Manufacturing Engineers, (1991).

112. *Tool Design*, 3rd Ed., C. Donaldson, G. LeCain, and V. C. Goold, Glencoe (1993).

113. *Atlas of Stress-Strain Curves*, H.E. Boyer (ed.), ASM, Materials Park, Ohio (1986).

114. *Atlas of Fatigue Curves*, Howard E. Boyer (ed.), ASM, Materials Park, Ohio (1986).

115. *Fatigue Data Book: Light Structural Alloys*, ASM, Materials Park, Ohio (1995).

116. *Atlas of Stress-Corrosion and Corrosion Fatigue Curves*, A. J. McEvily, ASM, Materials Park, Ohio (1990).

117. *Fracture Mechanics: Fundamentals and Applications*, T. L. Anderson, CRC Press (1991).

118. *Application of Fracture Mechanics for Selection of Metallic Structural Materials*, J. E. Campbell, et. al., ASM, Materials Park, Ohio (1982).

Casting

119. *Cast Metals Handbook*, 4th Ed., American Foundryman's Society (1957).

120. *Iron Castings Handbook*, Charles F. Walton, Iron Castings Society (1981).

121. *Gray and Ductile Iron Casting Handbook*, Gray and Ductile Iron Founder's Society, Cleveland, Ohio (1971).

122. *Steel Castings Handbook*, 6th Ed., Steel Founder's Society of America, Ohio (1995).

123. *Foundry Technology Sourcebook*, ASM and AFS, Materials Park, Ohio (1982).

124. *Investment Casting Handbook—1980*, Investment Casting Institute, Chicago, IL (1979).

125. *Principles of Metal Casting*, 2nd Ed., Heine, Loper and Rosenthal, McGraw-Hill (1967).

126. *The NFFS (Non-Ferrous Founders Society) Guide to Aluminum Casting Design: Sand and Permanent Mold*, NFFS (1994).

127. *Casting Kaiser Aluminum*, 3rd Ed., Kaiser Aluminum (1974).

128. *Copper, Brass and Bronze Castings: Their Structures, Properties, and Applications*, Non-Ferrous Founder's Society, Cleveland, Ohio (1961)—2 volumes.

129. *Copper Alloy Pressure Die Casting*, A. A. Machonis (ed.), International Copper Research Assoc., New York (1975).

Welding and Joining

130. *Welding Handbook*, 8th Ed., American Welding Society, New York (1987) - 3 volumes.

131. *Welding Handbook*, 7th Ed., American Welding Society, New York (1976-1984)—5 volumes.

132. *Welding Encyclopedia*, L. B. MacKenzie, 16th Ed. - revised and re-edited by T. B. Jefferson, Monticello Books, Morton Grove, IL (1968).

133. *Design of Weldments*, James F. Lincoln Foundation.

134. *Design of Welded Structures*, James F. Lincoln Foundation.

135. *Standard Handbook of Fastening and Joining*, 2nd Ed., R.O. Parmley, McGraw-Hill (1994).

136. *Metals Joining Manual*, M. M. Schwartz, McGraw-Hill (1979).

137. *Modern Welding Technology*, 2nd Ed., H. B. Cary, Prentice Hall, NJ (1989).

138. *Source Book on Electron Beam and Laser Welding*, ASM, Materials Park, Ohio (1980).

139. *Welding Kaiser Aluminum*, Kaiser Aluminum and Chemical Company, Oakland, CA (1967).

140. *Brazing Handbook*, American Welding Society, (1991).

141. *Source Book on Brazing and Brazing Technology*, ASM, Materials Park, Ohio (1980).

142. *Aluminum Brazing Handbook*, The Aluminum Association, Washington, D.C. (1974).

143. *Soldering Manual* 2nd Ed., American Welding Society (1978).

144. *Principles of Soldering and Brazing*, G. Humpston and D. M. Jacobson, ASM, Materials Park, Ohio (1993).

145. *The Basics of Soldering*, A. Rahn, John Wiley and Sons (1993).

146. *Solders and Soldering*, 3rd Ed., H. H. Manko, McGraw-Hill (1994).

147. *Aluminum Soldering Handbook*, The Aluminum Association, Washington, D.C.

148. *Handbook of Adhesives*, 2nd Ed., Irving Skiest (ed.), Van Nostrand Reinhold (1977).

149. *Adhesives Technology Handbook*, Arthur H. Landrock, Noyes Publications, (1985).

150. *Adhesives in Modern Manufacturing*, Society of Manufacturing Engineers, Dearborn, Mich. (1970).

151. *Joining of Composite Matrix Materials*, M. M. Schwartz , ASM, Materials Park, Ohio (1994).

152. *Ceramic Joining*, M. M. Schwartz, ASM, Materials Park, Ohio (1990).

Fabrication and Forming

153. *Handbook of Metal Forming*, Kurt Lange (ed.), McGraw-Hill, New York (1985).

154. *Handbook of Metalforming Processes*, B. Avitzur, Wiley-Interscience, New York (1983).

155. *Metal Forming: Fundamentals and Applications*, T. Altan, S-I Oh, and H. Gegel, ASM, Materials Park, Ohio (1983).

156. *Metalworking Science and Engineering*, Edward M. Mielnik, McGraw-Hill (1991).

157. *Handbook of Fabrication Processes*, O. D. Lascoe, ASM, Materials Park, Ohio (1988).

158. *Principles of Industrial Metalworking Processes*, G.W. Rowe, Arnold, London (1977).

159. *Finite-Element Plasticity and Metalforming Analysis*, G.W. Rowe, C.E.N. Sturgess, P. Hartley and I. Pillingen, Cambridge University Press (1991).

160. *Metalforming and the Finite-Element Method*, S. Kobayashi, S. Oh, and T. Altan, Oxford University Press (1989).

161. *Mechanics of Plastic Deformation in Metal Processing*, E. G. Thomsen, C. T. Yang, and S. Kobayashi, Macmillan (1965).

162. *Source Book on Cold Forming*, ASM, Materials Park, Ohio (1975).

163. *Mechanics of Sheet Metal Forming: Material Behavior and Deformation Analysis*, D. Koistinen, Plenum (1977).

164. *Techniques of Pressworking Sheet Metal*, 2nd Ed., D. F. Eary and E. A. Reed, Prentice Hall, (1974).

165. *Forging Industry Handbook*, 3rd Ed., T. G. Bryer (ed.), Forging Industry Association, Cleveland, Ohio and ASM (1985).

166. *Forging Equipment, Materials and Practice (MCIC-HB-03)*, Metals and Ceramics Information Center, Battelle Columbus Labs.

167. *Forging Materials and Practices*, A. M. Sabroff, et.al., Reinhold, N.Y. (1968).

168. *Open Die Forging Manual*, 3rd Ed., Forging Industry Association, Cleveland, Ohio (1982).

169. *Extrusion*, K. Laue and H. Stenger, ASM, Materials Park, Ohio (1981).

170. *Fine-Blanking: Practical Handbook*, Feintool A. G. Lyss, Switzerland (1972).

Powder Metallurgy

171. *Powder Metallurgy Science*, R. M. German, Metal Powder Industries Federation (1994).

172. *Powder Metallurgy, Principles and Applications*, F.V. Lenel, Metal Powder Industries Federation, Princeton, NJ (1980).

173. *Introduction to Powder Metallurgy*, J. S. Hirschorn, American Powder Met. Inst., New York (1969).

174. *Powder Metallurgy Processing: New Techniques and Analyses*, H. A. Kuhn and A. Lawley, Academic Press (1978).

175. *Source Book on Powder Metallurgy*, ASM, Materials Park, Ohio (1979).

176. *Handbook of Powder Metallurgy*, H. H. Hausner, Chemical Publishing Co. (1973).

177. *Powder Metallurgy Design Manual*, 2nd Ed., Metal Powder Industries Federation (1995).

178. *Powder Metallurgy Equipment Manual—3*, Samuel Bradbury (ed.), MPIF (1986).

179. *P/M Materials Standards and Specifications - MPIF Standard 35*, Metal Powder Industries Federation (1981).

180. *Powder Injection Molding*, R. M. German, Metal Powder Industries Federation (1990).

Machining

181. *Machining Data Handbook*, 3rd Ed., Machinability Data Center, Cincinnati, Ohio (1980)—2 volume set.

182. *McGraw-Hill Machining and Metalworking Handbook*, Ronald Walsh, McGraw-Hill (1994).

183. *Analysis of Material Removal Processes*, W. R. DeVries, Springer-Verlag (1992).

184. *Machining - Theory and Practice*, ASM, Materials Park, Ohio (1950).

185. *Design of Cutting Tools*, A. Bhattacharyya and Inyong Ham, Society of Manufacturing Engineers.

186. *Metal Cutting Principles*, M. C. Shaw, Oxford University Press (1984).

187. *Fundamentals of Metal Machining and Machine Tools*, G. Boothroyd and Winston Knight, Marcel Dekker (1989).

188. *Metal Cutting*, 2nd Ed., E. M. Trent, Butterworths (1984).

189. *Mechanics of Machining*, P. L. B. Oxley, Ellis Harwood Ltd. (1989).

190. *High Speed Machining*, ASME PED Vol. 12 (1984).

191. *Laser Machining*, G. Chryssolouris, Springer-Verlag (1991).

192. *Machining of Plastics*, Akira Kobayashi, Krieger Pub. Co. (1981).

Heat Treatment

193. *Heat Treater's Guide, Practices and Procedures for Irons and Steels*, 2nd Ed., ASM, Materials Park, Ohio (1995).

194. *Principles of the Heat Treatment Plain Carbon and Low Alloy Steels*, C. R. Brooks, ASM, Materials Park, OH (1995).

195. *Heat Treatment, Structure and Properties of Nonferrous Alloys*, Charles R. Brooks, ASM, Materials Park, Ohio (1982).

196. *Steel and Its Treatment*, Bofors Handbook, K. E. Thelning, Butterworths (1975).

197. *Principles of Heat Treatment of Steel*, G. Krauss, ASM, Materials Park, Ohio (1980).

198. *Principles of Heat Treatment*, M. A. Grossman and E. C. Bain, ASM, Materials Park, Ohio (1964).

199. *Practical Heat Treating*, Howard E. Boyer, ASM, Materials Park, Ohio (1984).

200. *The Heat Treating Source Book*, ASM, Materials Park, Ohio (1986).

201. *Source Book on Heat Treating*, ASM, Materials Park, Ohio (1975).
        Vol. 1: Materials and Processes
        Vol. 2: Production and Engineering Practices

202. *Heat Treatment in Fluidized Bed Furnaces*, R.W. Reynoldson, ASM, Materials Park, Ohio (1993).

203. *Atlas of Isothermal and Cooling Transformation Diagrams*, ASM, Materials Park, Ohio (1977).

204. *Atlas of Continuous Cooling Transformation Diagrams for Engineering Steels*, ASM, Materials Park, Ohio (1980).

205. *Atlas of Time-Temperature Diagrams for Irons and Steels*, G. VanderVoort (ed.), ASM, Materials Park, Ohio (1991).

206. *Atlas of Time-Temperature Diagrams for Nonferrous Alloys*, G. VanderVoort (ed.), ASM, Materials Park, Ohio (1991).

207. *Handbook of Quenchants and Quenching Technology*, G. E. Totten, C. E. Bates and N. A. Clinton, ASM, Materials Park, Ohio (1992).

Surfaces and Finishes

208. *Metal Finishing: Guidebook Directory*, Metals and Plastics Publications, Inc., Westwood, NJ (published annually).

209. *Surface Preparation and Finishes for Metals*, J. A. Murphy (ed.), Society for Manufacturing Engineers, Dearborn, Mich. (1971).

210. *Basic Metal Finishing*, J. A. Von Frauenhofer, Chemical Publishing (1976).

211. *Surface Finishing Systems*, George Rudski, ASM, Materials Park, Ohio (1983).

212. *Carburizing and Carbonitriding*, ASM, Materials Park, Ohio (1977).

213. *Source Book on Nitriding*, ASM, Materials Park, Ohio (1977).

214. *Electroplating Engineering Handbook*, 4th Ed., Lawrence Durney (ed.), Van Nostrand-Reinhold (1984).

215. *Properties of Electroplated Metals and Alloys*, 2nd Ed., Wm. H. Safranek, Amer. Electroplaters and Finishers Soc. (1986).

216. *Nickel and Chromium Plating*, 3rd Ed., J. K. Dennis and T. E. Such, Woodhead and ASM (1993).

217. *Electroless Nickel Plating*, Wolfgang Riedel, Finishing Publications Ltd. and ASM (1991).

218. *The Phosphating of Metals*, Second Ed., W. Rausch, Finishing Publications Ltd and ASM, (1990).

219. *Metallizing of Plastics: A Handbook of Theory and Practice*, R. Suchentruck, Finishing Publications Ltd and ASM, (1993).

220. *The Surface Treatment and Finishing of Aluminum and Its Alloys*, S. Wernick, R. Pinner and P. B. Sheasby, Finishing Publications Ltd. and ASM (1987).

Corrosion

221. *Handbook of Corrosion Data*, 2nd Ed., B. Craig and D. Anderson (eds.), ASM, Materials Park, Ohio (1995).

222. *NACE Corrosion Engineer's Reference Book*, 2nd Ed., R.S. Treseder (ed.), National Association of Corrosion Engineers (NACE) (1991).

223. *Corrosion Resistance Tables*, 4th Ed., P. A. Schweitzer, Marcel Dekker (1995).

224. *An Introduction to Metallic Corrosion*, 3rd Ed., Ulick R. Evans, Edward Arnold Ltd. and ASM (1981).

225. *Corrosion and Its Control—An Introduction to the Subject*, Atkinson and Van Droffelaar, National Association of Corrosion Engineers, Houston, TX (1982).

226. *Corrosion Engineering*, 3rd Ed., M. G. Fontana, McGraw-Hill, New York (1986).

227. *Corrosion and Corrosion Control*, 3rd Ed., H. H. Uhlig and R. Winston Revie, Wiley-Interscience, New York (1985).

228. *Corrosion, 2nd Ed.—Vol. 1: Corrosion of Metals and Alloys and Vol. 2: Corrosion Control*, L. L. Shrier (ed.).

229. *Material Selection for Corrosion Control*, S. I. Chawla and R. K. Gupta, ASM, Materials Park, Ohio (1993).

230. *Corrosion Prevention by Protective Coatings*, Charles G. Munger, National Association of Corrosion Engineers (1984).

231. *Corrosion in the Petrochemical Industry*, ASM, Materials Park, Ohio (1994).

232. *Marine Corrosion*, F. L. LaQue, Wiley-Interscience (1975).

233. *Marine Corrosion*, T. H. Rogers, Butterworth (1968).

234. *Guidelines for the Selection of Marine Materials*, International Nickel (1965).

235. *Corrosion of Metals in Marine Environments*, Metals and Ceramics Information Center, Report 86-50, (1986).

236. *High Temperature Corrosion of Engineering Alloys*, G.Y. Lai, ASM, Materials Park, Ohio (1990).

Testing, Inspection, and Quality Control

237. *The Testing of Engineering Materials*, 4th Ed., H. E. Davis, G. E. Troxell, and G. F. W. Hauck, McGraw-Hill, New York (1982).

238. *Nondestructive Testing Handbook*, 2nd Ed., (Multiple Volumes), American Society for Nondestructive Testing (1982-1985).

239. *Nondestructive Testing*, Louis Cartz, ASM, Materials Park, Ohio (1995).

240. *Metallography: Principles and Practice*, George VanderVoort, McGraw-Hill (1984).

241. *Handbook of Industrial Metrology*, Prentice Hall.

242. *Handbook of Dimensional Measurement*, 2nd Ed., F. T. Farago, Industrial Press (1982).

243. *ISO System of Limits and Fits, General Tolerances and Deviations*, American National Standards Institute.

244. *Introduction to Quality Engineering*, Genichi Taguchi, Kraus Int. Pub. (1986).

245. *Introduction to Statistical Quality Control*, 2nd Ed., D. C. Montgomery, John Wiley (1981).

246. *Statistical Quality Control*, 6th Ed, Eugene L. Grant and Richard S. Leavenworth, McGraw-Hill (1988).

247. *Guide to Quality Control*, Kaoru Ishikawa, Asian Prod. Organization (1984).

248. *Quality Control Handbook*, 4th Ed., J. M. Juran and F. M. Gigna, McGraw-Hill (1988).

249. *Quality, Productivity, and Competitive Position*, W. Edwards Deming, MIT Press (1982).

250. *Quality Control Source Book*, ASM, Materials Park, Ohio (1982).

251. *Quality Planning and Analysis*, 2nd Ed., Juran and Gryan, McGraw-Hill (1980).

252. *Statistical Quality Design and Control*, R. E. DeVor, T. Chang and J. W. Sutherland, Macmillan (1992).

253. *Taguchi Techniques for Quality Engineering*, P. J. Russ, McGraw-Hill (1988).

254. *Process Quality Control*, 2nd Ed., E. R. Ott and E. G. Schilling, McGraw-Hill (1990).

255. *Creating Quality*, William J. Kolarik, McGraw-Hill (1995).

256. *Quality Engineering in Production Systems*, G. Taguchi, E. A. ElSayed and T. Hsiang, McGraw-Hill (1989).

Failure Analysis and Product Liability

257. *Understanding How Components Fail*, Donald J. Wulpi, ASM, Materials Park, Ohio (1985).

258. *Case Histories in Failure Analysis*, ASM, Materials Park, Ohio (1979).

259. *Failure Analysis: The British Engine Technical Reports*, F. R. Hutchings and P. M. Unterweiser, ASM, Materials Park, Ohio (1981).

260. *Failure Analysis: Case Histories and Methodology*, F. K. Naumann, ASM, Materials Park, Ohio (1983).

261. *Handbook of Case Histories in Failure Analysis: Vols. 1 and 2*, K. A. Esaklul (ed.), ASM, Materials Park, Ohio (1992-93).

262. *Why Metals Fail*, R. D. Barer and B. F. Peters, Gordon and Breach, New York (1970).

263. *Analysis of Metallurgical Failures*, V. J. Conangelo and F. A. Heiser, Wiley-Interscience (1974).

264. *Metallurgical Failure Analysis*, C. R. Brooks and A. Choudhury, McGraw-Hill (1993).

265. *Source Book on Failure Analysis*, ASM, Materials Park, Ohio (1975).

266. *Failure of Materials in Mechanical Design*, J. A. Collins, Wiley-Interscience (1981).

267. *Stress Corrosion Cracking: Materials Performance and Evaluation*, Russell Jones (ed.), ASM, Materials Park, Ohio (1992).

268. *Metallography in Failure Analysis*, J. L. McCall and P. M. French, Plenum (1977).

269. *Fractography in Failure Analysis*, American Society for Testing and Materials, Philadelphia, PA (1978).

270. *Tool and Die Failures Source Book*, ASM, Materials Park, Ohio (1982).

271. *Residual Stresses and Fatigue in Metals*, J. O. Almen and P. H. Black, McGraw-Hill, New York (1963).

272. *Engineering Aspects of Product Liability*, V. J. Conangelo and P. A. Thornton, ASM, Materials Park, Ohio (1981).

273. *Product Liability and the Reasonably Safe Product*, A. S. Weinstein et.al., Wiley-Interscience, New York (1978).

Integrated Manufacturing Systems

274. *Computer Integrated Manufacturing*, Vols.1-4, R. U. Ayres, Chapman & Hall (1991-92).

275. *Design, Analysis and Control of Manufacturing Cells*, PED-Vol. 53, J T. Black, B. C. Jiang and G.J. Wiens, ASME (1991).

276. *The Design of the Factory with a Future*, J T. Black, McGraw- Hill (1991).

277. *Design and Analysis of Integrated Manufacturing Systems*, W. D. Compton, National Academy Press (1988).

278. *Automation, Production Systems and Computer-Integrated Manufacturing*, 2nd Ed., M. P. Groover, Prentice Hall (1987).

279. *Zero Inventories*, R.W. Hall, Dow Jones-Irwin (1983).

280. *Reinventing the Factory: Productivity Breakthroughs in Manufacturing Today*, R. L. Harmon and L. D. Peterson, The Free Press (1990).

281. *Dynamic Manufacturing—Creating the Learning Organization*, R.H. Hayes, S.C. Wheelwright and K.B. Clark, The Free Press (1988).

282. *Restoring Our Competitive Edge—Competing Through Manufacturing*, R. H. Hayes and S. C. Wheelwright, John Wiley (1984).

283. *Kanban and Just-In-Time at Toyota*, Japan Management Association, Trans. D. J. Lu, Productivity Press, Cambridge, MA (1986).

284. *JIT Factory Revolution*, H. Hirano and J T. Black, Productivity Press (1988).

285. *Just-in-Time Manufacturing Systems*, A. Satir (ed.), Elsevier (1991).

286. *The Shift to JIT*, I. Majima, Productivity Press (1992).

287. *Toyota Production System: An Integrated Approach to Just-in-Time*, 2nd Ed., Y. Monden, Industrial Engineering and Management Press, IIE (1993).

288. *TPM, Introduction to TPM: Total Productive Maintenance*, S. Nakajima, Productivity Press (1988).

289. *Toyota Production System: Beyond Large-Scale Production*, T. Ohno, Productivity Press (1988).

290. *Japanese Manufacturing Techniques: Nine Hidden Lessons in Simplicity*, R. J. Schonberger, The Free Press (1982).

291. *World Class Manufacturing*, R.J. Schonberger, The Free Press (1986).

292. *World Class Manufacturing Casebook: Implementing JIT and TOC*, R.J. Schonberger, The Free Press (1987).

293. *One-Piece Flow: Cell Design for Transforming the Production Process*, K. Sekine, Productivity Press (1990).

294. *A Revolution in Manufacturing: The SMED System*, S. Shingo, Productivity Press (1985).

295. *A Study of the Toyota Production System*, S. Shingo, Productivity Press (1989).

296. *Non-Stock Production: The Shingo System for Continuous Improvement*, S. Shingo, Productivity Press (1988).

297. *Zero Quality Control: Source Inspection and the Poka-Yoke System*, S. Shingo, Productivity Press (1986).

298. *The New Manufacturing Challenge*, K. Suzaki, The Free Press (1987).

299. *The Machine that Changed the World*, J. P. Womack, D. T. Jones and D. Roos, Harper Perennial (1991).

300. *Just-in-Time Manufacturing*, C. A. Voss, IFS Publications Ltd. (UK), Springer-Verlag (1987).

301. *Time-Based Manufacturing*, J.A. Bockerstette and R. L. Shell, McGraw-Hill (1992).

302. *Group Technology*, E. A. Arn, Springer-Verlag (1975).

303. *Group Technology*, Gallagher and Knight, Butterworths, London (1973).

304. *The Introduction of Group Technology*, J. L. Burbridge, Halsted Press, John Wiley (1975).

305. *Group Technology in the Engineering Industry*, J. L. Burbridge, Mechanical Engineering Publications Ltd., London (1979).

306. *Group Technology—Applications to Production Management*, I. Ham, K. Hitomi and T. Yoshida, Kluwer-Nijhoff Publishing Co., Boston (1985).

307. *Production Flow Analysis*, J. L. Burbridge, Oxford University Press (1989).

308. *Group Technology*, C. S. Snead, Van Nostrand Reinhold (1989).

Manufacturing Systems Analysis

309. *Stochastic Models of Manufacturing Systems*, J. A. Buzacott and J. G. Shanthikumar, Prentice Hall (1993).

310. *Modeling and Analysis of Manufacturing Systems*, R. G. Askin and C. R. Standridge, John Wiley (1993).

311. *Manufacturing Systems Engineering*, S. B. Gershwin, PTR, Prentice Hall (1994).

312. *Manufacturing Systems Design and Analysis*, 2nd Ed., B. Wu, Chapman and Hall (1994).

313. *Performance Modeling of Automated Manufacturing Systems*, N. Viswanadham and Y. Narahari, Prentice Hall (1992).

Manufacturing Engineering

314. *Manufacturing Engineering*, 2nd Ed., J. P. Tanner, Marcel Dekker (1991).

315. *Manufacturing Engineering*, 2nd Ed., D. T. Koenig, Taylor and Francis (1994).

316. *Expert Systems Applications in Engineering and Manufacturing*, A. B. Badiv, Prentice Hall (1992).

317. *Intelligent Manufacturing Systems*, A. Kusiak, Prentice Hall (1990).

318. *Concurrent Design of Products and Processes*, J. L. Nevins and D. E. Whitney, McGraw-Hill (1989).

319. *Manufacturing Intelligence*, P. K. Wright and D. A. Bourne, Addison-Wesley (1988).

320. *Manufacturing Systems Engineering*, K. Hitomi, Taylor and Francis Ltd. (1979).

Robotics

321. *Robotics for Engineers*, Y. Koren, McGraw-Hill (1985).

322. *Industrial Robots: Computer Interfacing and Control*, W. E. Snyder, Prentice Hall (1985).

323. *Industrial Robotics*, M. Groover, et. al., McGraw-Hill (1986).

324. *Industrial Robots and Robotics*, E. Kafrissen and M. Stephens, Reston (1984).

325. *Handbook of Industrial Robotics*, Nof (ed.), John Wiley (1985).

326. *Assembly with Robots*, T. Owen, Prentice Hall (1985).

327. *Robotics: Control Sensing, Vision and Intelligence*, K. S. Fu, R. C. Gonzalez and C. S. G. Lee, McGraw-Hill (1987).

328. *Introduction to Robot Technology*, P. Coiffet and M. Chizouze, McGraw-Hill (1993).

329. *Robotics and Manufacturing Automation*, 2nd Ed., C. R. Asfahl, John Wiley (1992).

Automation, CNC, FMS and CIM

330. *Handbook of Manufacturing Automation and Integration*, J. Stark (ed.), Auerbach (1989).

331. *Anatomy of Automation*, G. H. Amber and P. S. Amber, Prentice Hall (1962).

332. *Numerical Control and Computer Aided Manufacturing*, R. S. Pressman and J. E. Williams, John Wiley (1977).

333. *Principles of Numerical Control*, J. L. Childs, Industrial Press (1982).

334. *An Introduction to CNC Machining and Programming*, D. Gibbs and T. M. Crandell, Industrial Press (1991).

335. *Microcomputer Applications in Manufacturing*, A. G. Ulsoy and W. R. DeVries, John Wiley (1989).

336. *Computer Control of Machining and Processes*, J. G. Bollinger and N. A. Duffie, Addison Wesley (1988).

337. *NC Machine Programming and Software Design*, C-H Chang and M. A. Melkanoff, Prentice Hall (1989).

338. *Introduction to Computer Numerical Control*, J.V. Valentino and J. Goldenberg, Regents/Prentice Hall (1993).

339. *Flexible Manufacturing Systems Handbook*, Noyes Publishing (1984).

340. *The Design and Operation of FMS- Flexible Manufacturing Systems*, P. G. Ranky, IFS Publications Ltd. (UK) and North-Holland Publishing Company, New York (1983).

341. *Automatic Assembly*, G. Boothroyd, C. Poli and L. E. March, Marcel Dekker (1982).

342. *Computer-Integrated Design and Manufacturing*, D. D. Bedworth, M. R. Henderson and P. M. Wolfe, McGraw-Hill (1991).

Cost Estimating

343. *Manufacturing Cost Estimating Guide*, P. R. Ostwald, American Machinist (1982).

344. *Cost Estimator's Reference Manual*, R. D. Steward and R. W. Wyskida, John Wiley (1987).

345. *Engineering Cost Estimating*, 3rd Ed., P. E. Ostwald, Prentice Hall (1992).

346. *Realistic Cost Estimating for Manufacturing*, 2nd Ed., W. Winchell, Society of Manufacturing Engineers (1989).

INDEX

Abrasive cleaning, 1083
Abrasive flow machining (AFM), 955
Abrasive grain size, 787
Abrasive machining, 92, 783-813, 785
Abrasive waterjet, 787
Abrasive waterjet machining, 954
Abrasives, 86, 785
ABS, 200-201
Accounting, 1168
Accuracy, 257, 317, 662
Acetylene, 968
Acid pickling, 1093
Acoustic emission, 312
Acrylics , 200, 201, 1052
Acurad process, 405
Adaptive control, 852, 854, 1136
Addition polymerization, 195
Adhesive bonding, 21, 577, 1051
Adhesive joints, 1057
Admiralty metal, 171
Aerator, 371
Age hardening, 114-117, 123
Air furnace, 413
Air gage, 288
AISI-SAE Classification System, 151
Alclad, 177
Alkaline cleaning, 1091
Allotropic, 74
Allowance, 252, 256, 355
Alloy steels, 149, 152, 157
Alpha-ferrite, 100
Alumina (Al$_2$O$_3$), 209
Aluminum, 151, 356
Aluminum/Alloys, 174-81
Aluminum designation system, 176-180
Aluminum oxide, 2, 627, 786
Aluminum-lithium alloy, 181
Aluminum-silicon alloys, 1037
Amorphous structure, 73
Ampere, 246
Anaerobics, 1052
Angle bending, 511
Annealing, 111, 1102
Antioxident, 202
APT (Automatically programmed tools),
    848, 852, 868
Aramid, 217
Arc cutting, 993

Arc furnace, 414
Arc welding, 978
Artificial intelligence (AI), 28, 1137
Assembly, 18, 21
Assembly cell, 1131
Assignable causes, 328
Atomic bonds, 70
Atomic structure, 69
Attributes, 245
Attrition, 785
Ausforming, 132
Austempering, 131
Austenite, 99, 120
Austenitic stainless steel, 159-161
Autoclave, 582
Automated guided vehicles (AGV), 1143-46
Automation, 23, 1134, 1202
Autonomation, 23, 1202,
    fixed, 27
    programmable, 27
    yardstick for, 24
Autonomation, 1221

Babbitt, 189
Back rake angle, 599, 636
Bake-hardenable steel sheet, 157
Balance, 1201, 1216
Band-filing machine, 771
Bandsaw, 761, 764-69
Bar folder, 511
Barrel finishing, 1084
Belt sanding, 1089
Bending, 474, 509-15
Bevel gear, 912
Bevel protrator, 264, 282
Bickford drill point, 702
Bill of materials (BOM), 1175
Billit, 465-470
Binder, 431
Blanking, 521, 912
Blending, 424, 426
Blocking, 474
Bloom, 465-470
Blow molding, 558
Blowout, 504
Body-centered cubic metals, 73, 80, 100
Bonded product, 783
Bonding materials, 791

Borazon, 786
Boring, 654-57
Boring cutting tools, 680
Boring machines, 20, 672-82
Boron, 151, 578
Bottoming tap, 891-93
Brale, 49
Brass, 170, 356
Braze welding, 1045
Brazing, 21, 1034-45
Brazing thermal affects, 1074
Brinell hardness test, 46
Brittleness, 42
Broach nomenclature, 750
Broaching, 20, 597, 598, 615, 748-59,
    912, 916
Broaching machines, 756-59
Broad-hammer, 473
Bronze, 170-73
Brown & Sharpe, 673
BTA drill, 704-05, 710
Buffing, 1089
Built-up edge, 604, 632, 644
Bulk deformation process, 465
Burnishing, 509
Burr removal, 1093-95, 1116
Business process reengineering (BPR),
    1202
Butt welding, 1007, 1070

C-curve, 119
CAD-NC, 868
Calendering, 566
Candela, 246
Canning (in P/M), 433
Capillary action, 1035
Carbon equivalent in CI, 102
Carburizing, 134
Cast cobalt alloys, 624
Cast iron, 102, 356, 365
Casting, 18, 341-422, 557
Cause and effect diagram, 335
Cellulose acetate, 200-201
Cementite, 99
Center core drill, 702, 710
Center drill, 706
Centerless grinding, 802
Centrifugal barrel tumbling, 1086
Centrifugal casting, 408

Ceracon process, 434

Ceramic mold casting, 389-390

Ceramic-matrix composites (CMC), 220, 584

Ceramics, 207, 210, 572, 574

Cermented carbides, 624

Cermets, 211, 633

Chaplets, 386

Charpy test, 54

Check, 387

Chemical blanking, 941

Chemical cleaning, 1090

Chemical flux cutting, 974

Chemical machining, 21, 936, 941

Chemical NTM, 937

Chemical vapor deposition (CVD), 620, 629, 1005, 1107

Chill zone, 347

Chills, 355

Chip breaker groove, 625

Chip formation (see machining)

Chip thickness ratio, 602

Chisel end, 701

Choke, 351

Chromite, 209

Chromium, 150, 158

Chucks, 665, 689, 710

   collet, 832

   four jaw, 689

   Jacobs, 710

   quick-change, 709

   three jaw, 689

Chvorinsov's rule, 346

Circular saw, 764, 765

Cladding, 1012, 1016, 1109

Clamping, 817, 821, 832

Closed-loop control, 852

Cluster, 391

CNC grinder, 1136

CNC thread cutting, 889

Coalescence, 966

Coated carbide, 2

Coated electrode, 980

Coated product, 783, 809

Coating, 1096

Cobalt based alloys, 187, 190

Coding/classification, 1211

Coherency, 116

Cohesiveness, 371

Coining, 434, 507

Cold drawing, 528

Cold forging (heading), 499

Cold molding, 561

Cold rolling, 497

Cold shut, 350

Cold welding, 1011

Cold working, 84, 454, 457, 496-552

Cold-box core, 385

Cold-chamber die casting, 405

Cold-roll forming, 515

Collapsibility, 371, 374

Collet, 690

Coloring agent, 202

Columnar zone, 348

Combination set, 264

Compactability, 373

Compacted graphite iron, 107

Compacting, 424, 427

Composites, 214-223

Compound die, 527

Compounds, 1088

Compression molding, 559

Compression test, 52

Computed homography, 313

Computer-aided design (CAD), 30, 849, 1190-1191

Computer-aided manufacturing (CAM), 30, 852, 1190, 1194

Computer-aided process planning (CAPP), 30, 1190, 1193

Computer-integrated manufacturing (CIM), 1161, 1189

Computer numerical control (CNC), 850, 858

Computer-aided testing and inspection, 30

Conceptual design stages, 231

Condensation composites, 216, 577

Conform, 505

Constitutive relationships, 451

Consumable electrode, 979

Consumable-electrode remelting, 147

Consumer goods, 1

Continuous casting, 143-146, 411

Continuous chip, 604

Continuous cooling transformations, 123

Continuous extrusion, 505

Continuous path, 855

Continuous process, 9, 1128

Continuous surface bond, 1055

Contouring control, 855, 858-861

Control charts, 327, 331

Cooling curves, 92, 344

Coordinate measuring machine (CMM), 277

Cope, 343, 375

Copolymer, 195

Copper, 151, 168

Copper alloys, 1036

Copper alloys, 170-73

Core, 343, 386, 400, 406, 358

Core box, 343, 385

Core making, 383

Core print, 343, 344, 386

Core-oil process, 385

Core-to-face bond, 1055

Cored structure, 97, 144

Corrosion, 57

Corrosion resistance, 168, 174, 175, 186, 198

Corthias casting, 402

Corundum, 785

Counter-boring drill, 705

Counterboring, 717

Countersink, 706, 707, 717

Crack supression, 1077

Crack-arrestor, 455

Creep, 61

Creep feed grinding (CFG), 785, 798, 806

Crown gear, 912

Crucible furnace, 413

Crush dressing, 794

Crystal structure, 73

Cubic boron oxide (CBN), 786, 800

Cupola, 413

Cupronickels, 173

Curing, 1051

Cutoff, 521, 654

Cutter offset, 868

Cutting fluids, 647-49, 716, 798, 897

Cutting force, 604, 608, 610

Cutting speed, 590, 593, 633, 638, 642, 645, 699, 707, 728, 754, 774, 1179

Cutting time, 594, 656-58, 698, 727, 729, 751, 776, 895

Cutting tool, 13, 592, 682-87

   geometry, 636, 638

Cutting tool-materials

   carbides, 619, 684

   ceramics, 519, 627

   characteristics desired, 619

   coated carbides, 619, 627

   coated HSS, 629

   cubic boron nitride (CBN), 635

   diamond, 632

   high speed steel (HSS), 619, 622, 684

   properties, 621, 622, 627

   selection of 619, 620

Cyanoacrylates, 1052

Cycle time (CT), 1130, 1203, 1217

Cylindrical grinding, 801

Cylindricity, 259

Damping capacity, 46

Dead center, 687

Decouplers, 1133, 1202
Deep drawing, 535
Deep-hole drill, 704, 718
Degassification, 145
Delayed-tack adhesive, 1054
Delta, 99
Deoxidation, 145
Depth of cut line, 635
Depth of cut (DOC), 591, 655, 727, 784
Design engineer, 3, 30
Design engineering, 1170
Design for casting, 358, 1171
Design for forging, 1171
Design for machining, 1172
Design for manufacturing, 4, 1170
Design of P/M parts, 438
Designing for plastics, 568
Dezincification, 172
Dial indicator, 287
Diametral pitch, 908
Diamonds, 632, 785, 786
Die, 523
Die casting, 403-07, 1156
Die-filing machine, 770
Diffusion welding (DFW), 1016
Dinking, 521
Dipping, 571
Direct extrusion, 486
Direct numerical control (DNC), 849
Discontinuous chips, 604
Discrete fastener, 1060
Disk saw, 764
Dislocations, 80
Dispatching, 1185
Dispersion hardening 114, 116
Disposable inserts, 624
Distortion, 1076
Dog, 687, 692
Dog plate, 687
Doping, 87
Down milling, 729-731
Draft, 343, 356, 477
Draw bench, 528
Draw bending, 515
Drawing - hot, 490
Dressing, 793
Drill geometry, 697, 698, 702
Drill jig, 822-29
Drill press, 20, 711-16
Drilling problems, 720
Drop-hammer, 473
Dross, 348
Dry-sand core, 383
Dry-sand mold, 378

Dubbing, 700
Ductile cast iron, 106
Ductility, 42, 458
Ductility in copper, 168
Ductility - pressure-induced, 504
Dump core box, 384
Duplicators, 741-42
Duralumin, 177
Durometer hardness test, 51

Economic justification of jig, 840
Economic order quantity (EOQ), 1186
Economics of machining, 643, 652, 695, 700
Eddy-current testing, 310
Edging, 474
Eff-set process, 383
Elastic springback, 458
Elastomers, 205, 563, 571, 1051
Electric arc furnace, 142
Electrical discharge machining (EDM), 21, 955, 956, 1111
Electrical properties of metals, 86
Electrochemical machining (ECM), 21, 936, 947
Electrocoating, 1098
Electrode, 978
   designation system, 981
Electrodes, 1000
Electroforming, 545
Electroless plating, 1103
Electrolytic tough-pitch (ETP) copper, 168
Electromagnetic casting, 411
Electromagnetic forming, 542
Electron beam machining (EBM), 21, 958
Electron beam welding (EBW), 1018
Electron-beam hardening, 134
Electroplating, 1100
Electropolishing, 1090
Electroslag remelting process (ESR), 148
Electroslag welding (ESW), 1018
Electrostatic depositioning, 1098
Electrostatic workholder, 839
Emboss, 543, 1007
Emery, 785
End effector, 1150
End milling, 727
Endurance limit, 56
Engine lathe, 592, 665
Epoxies, 200-201, 1051
Equiaxed zone, 348
Equilibrium diagram 93-102
Equilibrium phase diagram, 91
Etch factor, 945
Ethyl cellulose, 200-201

Eutectic, 97
Eutectoid, 100
Evaporative adhesive, 1053
Exothermic hot top, 145
Expanded polystyrene casting, 393
Expansion fit, 1060
Expeditor, 1185
Expert systems, 28, 1135
Explosive bonding, 577
Explosive forming, 541
Explosive welding, (EXW), 1016
Extrusion, 465-470, 484-89, 534, 927
Face centered cubic metals, 73, 80, 100
Face milling, 727-30
Faceplate, 691, 692
Facing, 653, 657
Fatigue, 56
   effect of corrosion, 57
   relationship to surface finish, 1110
   testing, 56
Fatigue life, 1112, 1113
Faying surfaces, 1000
Fe_3C (see cementite)
Feed, 591, 593, 633, 638, 699, 707, 727, 748, 803, 1179
Feed force, 608
Feed rate, 595, 727
Feedback, 1136, 1150
Feedforward, 1139, 1153
Fellows gear shaper, 918-19
Ferrite, 99
Ferritic stainless steel, 159-161
Fiber reinforced composites, 217, 577
Fiber structure, 455, 501
Filament winding, 579
File test (for hardness), 51
Filing, 769-772
Filler, 202
Filler metal, 978, 1035
Fillet weld types, 1067-1069
Finance, 1168
Fine blanking, 519
Finishing (castings), 418
Finishing process, 22, 509, 1083
Fishbone diagram, 325
Fits, classes of, 252, 255
Fixture, 13, 817, 821, 829
Flame cutting, 973
Flame hardening, 133
Flame straightening, 975
Flanging, 517
Flash, 473
Flash welding (FW), 1024
Flask, 343, 375

Flaskless molding, 377

Flatness, 259

Flexforming, 538

Flexibility, 819, 1207

Flexible manufacturing system (FMS), 851, 1128, 1133, 1141, 1147, 1148

Floating plug, 530

Flow shop, 5, 9, 1126, 1140

Flow stress, 606

Flow turning, 533

Flow-reform process, 534

Fluidity, 350, 1036

Flurocarbons, 200-201

Flush-pin gage, 286

Flux, 348, 980, 987, 1038, 1046

Flux cored arc welding (FCAW), 986

Foam molding, 565

Follow rest, 691

Forge welding (FOW), 1010

Forged P/M, 435

Forging, 465-470, 472-484

Form milling, 913

Form turning, 654, 676, 685

Formability, 62

Forming process, 18, 448

    dependent variables, 449

    independent variables, 447

Four-facet drill, 702

Four-slide, 550

Fracture

    brittle, 83

    ductile, 83

    toughness, 63

Free-machining steels, 150-151, 156, 601

Freeze-drying, 91

Freezing range, 93, 345

Friability, 785

Friction, 452

Friction force, 604

Friction welding (FRW), 1012, 1025

Full annealing, 111

Full-mold process, 393-96, 565

Fullering, 474

Functional design stage, 231

Furnaces (for melting) 412-416

Fusion weld types, 1067

Fusion zone, 1070

G Ratio, 790

Gage blocks, 249

Gage capability, 261

Gaging, 245

Galvanizing, 183, 1099

Gang drill, 713

Garnets, 785

Gas flushing, 349

Gas metal arc welding (GMAW), 985

Gas porosity, 349

Gas tungsten arc spot welding, 984

Gas tungsten arc welding (GTAW), 982

Gating system 343, 351, 366

Gear finishing, 927

Gear inspection, 930

Gear manufacturing, 906-934

Gear shaving, 927

Gear terminology, 906, 907

Gear types, 906, 907, 910

Geometric tolerancing, 260, 280, 281

Glasses, 209, 210, 573

Go Not-Go gages, 283

Gooseneck, 404

Grade, 789

Grain boundaries, 76

Grain growth, 86

Grain size, 371, 374

Grain-size refinement, 114

Graphite, 190, 217, 385

Graphite in CI, 103

Gravite die casting, 400

Gray cast iron, 103

Green compact, 427

Green sand, 371, 377-79

Grinding , 598, 614, 785, 788, 798-800, 912

Grinding wheels, 795-96

Gripper, 1150

Group jig/fixture, 834

Group technology (GT), 9, 854, 1208

Growth, 76

Guerin process, 538

Gun drill, 704, 710

H-process, 377

Hacksaw, 761, 764

Hard chromium plating, 1101

Hardenability, 124-127

Hardness, 46

    relationship to shear stress, 607

    relationship to tensile strength, 51

Hastelloy X, 187

Heat (in machining), 611

Heat affected zone (HAZ), 1071-72

Heat checking, 404

    Heat treating, 18, 21, 110-138

    design concerns, 128-138

Helical gear, 910

Helix angle, 700

Hexagonal-close-packed metals, 75, 80

High strength light weight alloys, 174, 186

High temperature adhesive, 1053

High-energy-rate-forming (HERF), 540

High-strength low-alloy steels (HSLA), 154, 174

Histogram, 319

Hitchiner process, 403

Hobbing, 912, 920-921

Homogenization, 114

Honing, 785, 811

Hook's law, 39

Hop-dip coating, 1099

Hot drawing, 465-470

Hot hardness, 613

Hot isostatic pressing (HIP), 33, 623

Hot melt, 1053

Hot spots, 359

Hot tear, 374

Hot working, 84, 454, 464-92

Hot-box core, 385

Hot-chamber die casting, 404

Hot-compression molding, 560

Hot-isostatic pressing (HIP), 186

Hot-plate welding, 1026

Hubbing, 508

Hydroforming, 538

Hydrostatic extrusion, 502

Hypereutectoid steel, 100, 111

Hypoeutectoid steel, 101, 111

Illium, 187

Impact extrusion, 502

Impact grinding, 952

Impact tests, 54

Impregnation, 435

Impression-die forging, 472, 477

Incology, 188

Inconel, 601, 718, 187

Indexable insert drill, 709

Indirect extrusion, 487

Induction brazing, 1042

Induction furnace, 415

Induction hardening, 133

Industrial engineering, 4, 1182

Inertia welding, 1013

Infeed, 798

Infiltration, 436, 578

Injection molding, 431, 561-63, 584

Inoculation, 345

Insert tools, 624, 684

Inserts, 406

Insolubility, 95

Inspection, 13, 14, 244, 324, 327, 930, 1214

Integral fastener, 1060

Interchangeable parts, 25
Interference, 252, 273
Intermediate jig concept, 829, 1213
Intermetallic compounds, 97
Interpolation, 867
Inventory control, 1185, 1201
Investment casting, 390-393, 927, 1156
Involute curve, 908
Ion implantation, 1008
Ionitriding, 135
Iron carbide (see cementite)
Iron-carbon equilibrium diagram, 99
Ironing, 490, 542
Isostatic compaction, 430
Isothermal, 91
Isothermal forming, 460
Isothermal transformation (IT) diagram, 118
Isotropic, 216
Izod test, 54

Jig borer, 679
Jigs, 13, 817, 821, 822, 827-29, 992
Job, 11
Job shop, 5, 9, 1121, 1167, 1204
Jolting, 375
Jominy test, 124-126
Jump-cut teeth, 752
Just-in-time (JIT) manufacturing, 30, 1199

Kanban, 1201, 1217
Kelly-Bessemer process, 142
Kerf, 761, 972
Kevlar, 217, 218, 578
Killed (see deoxidized)
Knoop hardness test, 50
Knurling, 654, 660

Ladle metallurgy, 143
Laminar composites, 215, 577
Lamination, 577, 580
Lancing, 521
Lapping, 785, 813
Laser beam machining (LBM), 21, 960, 1022
Laser beam welding (LBW), 1021
Laser interferometer, 275
Laser spot welding, 1024
Laser-beam hardening, 133
Lathes, 665-67, 695
    automatic, 672-77, 695
    bench, 668
    designs, 665
    engine, 665
    screw machines, 672-676

sizes, 668
terminology, 665, 671
turret, 669
Lattice structure, 73
Lay pattern, 289
Lead, 881, 883
Lead screw, 887
Lead tin solder, 1046
Leak testing, 313
Lean production, 3, 1199-1225
Lean production - ten steps to, 1200
Least material condition (LMC), 259
Length, 245
Level, 1201, 1216
Limits, 255
Line balance, 1138
Linearity, 261
Linked cell manufacturing system (L-CMS), 1200, 1202-07
Linked-cell, 5, 9, 12, 1130, 1133
Liquid penetrant, 302
Liquid-metal forging, 408
Liquidus, 96, 352
Live center, 687
Location, 817, 820
Lost-form casting, 393, 565
Lost-wax process, 389
Low-force groove (LFG), 637
Low-pressure permanent mold (LPPM), 402
Low-stress grinding, 792
Lubricants, 202, 485
Lubrication, 452
Lucite, 200-201
Luders bands, 459

Machinability, 62, 646
Machine control unit (MCU), 855
Machine time (see cutting time), 1132, 1202
Machine tool, 8, 10, 25
Machine tools, 591, 665, 711-720, 737-42, 756, 764, 771, 777, 804, 847, 869, 912
Machining (ceramics), 575
Machining (plastics), 566
Machining, 18, 590-814
Machining allowance, 357
Machining center, 20, 27, 847, 857, 869
Machining surface damage, 1111
Machinist's rules, 264
Magnesium, 356
Magnesium/Alloys 182, 184, 356
Magnetic particle, 303
Magnetic workholder, 838
Magnification, 263

Malleable cast iron, 105
MAN, 1218
Managing steels, 158
Mandrel, 688
Manganese, 150
Manipulator, 1150
Mannesmann process, 491
Manufacturing automation protocol (MAP), 1195
Manufacturing cell, 12, 1128, 1132, 1208
Manufacturing cost, 3, 17
Manufacturing engineer, 4
Manufacturing engineering, 1174
Manufacturing process, 3, 8, 10, 16
Manufacturing system, 5, 7, 8, 1118-1149
    definition, 1121
    trends in, 1118
    types of, 1121, 1134
    volume vs variety, 1133
MAPP gas, 970
Marketing, 1165
Marquenching, 131
Martensite, 105, 119-122
Martensitic stainless steel, 159-161
Mass, 245
Master production schedule, 1184
Material processing by robots, 1157
Material processing families, 341, 1157
Material selection, 231-240
Materials 36-223
    physical 37, 64
Materials engineer, 4
Materials handling, 13, 1157
Materials requirement planning (MRP), 1184
Mats, 578
Maudsley, Henry, 25
Maximum material condition (MMC), 259
Measurement, 244
Mechanical fastening, 21, 1059
Mechanical NTM, 936, 939
Mechanical plating, 1104
Mechanization, 1134
Media, 1085-87
Melamines, 200-201
Mer, 196
Metal cores, 400
Metal cutting (see machining)
Metal forming, 446-461
Metal inert gas welding (MIG), 985
Metal molds, 400
Metal powder cutting (POC), 974
Metal powders, 425
Metal removal (see machining), 19

Metal removal rate (MRR), 594, 608, 656-58, 698, 727, 729, 751, 775
Metal spraying, 1029
Metal-matrix composites (MMC), 219, 584
Metallizing, 1028
Metric-english conversions, 248
Metrology, 261
Microalloyed steels, 155
Microchipping, 635
Microcrazing, 389
Microhardness test, 50
Micrometer caliper, 264
Milling, 20, 595, 598, 614, 659, 696-720, 912
Milling cutters, 732-37
Milling machine tools, 20, 737
Milling problems, 732
Misrun, 350
Modular fixturing, 833
Modulus of elasticity, 39, 174
Modulus of resilience, 41
Mold cavity, 342, 343
Molecular structure, 73
Molybdenum, 150
Monel, 186, 425
Mounted wheels, 808
MRP, 1188
Muller, 371
Multiple spindle, 713
Multislide, 550

NC tape control, 864
Nibbling, 521
Nichrome, 186
Nickel/alloys, 150, 158, 186, 187, 188
Nickel silver, 173
Niobiuum, 151
Nitriding, 135
No-bake molds, 380
Nodular cast iron, 106
Nominal, 320
Non-traditional machining (NTM), 935-963
Nondestructive testing, 23, 299-315
Nondestructive testing vs destructive testing, 300
Nonferrous metals and alloys, 167-190
Normal chance variations, 328
Normalizing, 112
Notching, 521
Nucleation, 76, 344
Nugget, 1002
Numerical control (NC), 9, 27, 846-875, 890, 957
Nylons, 200-201

Oblique machining, 601
Offset yield strength, 41
Ohno, Taiichi, 1218, 1221
Onset shear plane, 600
Open-die forging, 473
Open-hearth furnace, 142
Open-loop control, 24, 852
Operation, 13
Operations analysis sheet, 1182
Operations sheet, 1174
Optical comparator, 270
Orthogonal array, 325
Orthogonal machining, 599, 601
Osprey process, 434, 477
Overaged, 116
Overlaying, 1028
Oxyfuel gas welding, 971
Oxygen cutting, 972
Oxygen furnace, 142

Pack rolling, 469
Packaging, 13
Painting, 1096
Pallet, 1143
Parison, 558
Parsons, John, 848
Part program, NC, 863
Parting, 654
Parting line, 343, 356, 358
Patterns, 365-70, 388, 390, 343, 355
Pearlite, 101, 119
Peck drilling, 862
Peel stress, 1056
Peening, 509
Penetration, 350, 374
Percent elongation, 43, 154
Percussion welding, 1024
Perforating, 521
Peripheral milling, 726
Peritectic, 97
Permeability, 371, 432
Permenent mold casting, 399-420
Personnel, 1169
Phase, 90
Phase transformation strengthening, 115
Phenolics, 200-201
Phosphate coating, 1100
Photochemical machining (PCM), 941-42
Physical vapor deposition (PVD), 620, 630, 1105
Pickling, 459
Piercing, 49, 465, 521
Pig iron, 142
Pinion gear, 933

Pipe welding, 465, 491
Pitch, 750, 881
Pitch diameter (PD), 906
Plain-carbon steel, 148
Planer, 20, 778
Plant engineering, 1182
Plasma arc cutting (PAC), 994
Plasma arc machining, 962
Plasma arc welding (PAW), 989
Plasma jet machining, 21
Plasma spray forming, 546
Plaster mold casting, 387-89
Plastic deformation, 44, 78, 82
Plasticity, 446
Plasticizer, 202
Plastics, 194-205
Plate, 465-470
Plexiglas, 200-201
Plug gages, 283
Plug tap, 891
Plunge-cut grinding, 798
PM process (for cutting tools), 625
Pneumatic-mechanic forming, 541
Pocket milling, 862
Point-to-point control, 855, 858
Poly (ether ether ketone) (PEEK), 218, 558
Polybutadiene, 208
Polycarbonates, 200-201
Polycrystalline cubic boron nitride (PCBN), 635
Polycrystalline diamond (PCD), 633
Polyesters, 200-201
Polyethylenes, 200-201
Polymerization, 194-95, 563
Polymers, 194, 559
Polymorphic, 74
Polypropylene, 558
Polystyrenes, 200-201
Polyvinyl chloride (PVC), 558, 204
Porosity, 432
Positioner, 992
Postprocessor, 849, 850
Pouring cup, 343
Powder coating, 1098
Powder metallurgy (P/M), 423-442, 624-27, 927
Power (in cutting), 607
Precipitation hardness, 114-17
Precision, 257
Preform, 558, 560, 561
Prepreg, 578
Press brake, 511
Press fit, 1061
Press forging, 478

Presses, Metal forming, 479, 546-52
Pressure gas welding (PGW), 971
Pressure sensitive adhesive, 1053
Pressure-bag molding, 581
Pressure-temperature (P-T) diagram, 91
Preventative maintenance, 1201, 1216
Probes, 872-73
Process anneal, 112
Process capability, 317-26, 872
Process capability ratio, 320
Process planning, 1174
Procurement/Purchasing, 1183
Producer goods, 1
Producibility, 1170
Product life cycle, 17
Production control, 1167
Production flow analysis (PFA), 1209
Production job shop (PJS), 17
Production ordering kanban(POK), 1217
Production planning, 1165-67, 1175
Production system, 1120, 1122, 1164-96
Profilers, 741-42
Programmable logic controller (PLC), 1128
Progressive die set, 526
Project shop, 5, 9, 1128
Projection welding (RPW), 1007
Proportional limit, 39
Prototype, 231
Pull broach, 749
Pulsed arc gas metal arc welding, 986
Pultrusion, 579
Punch, 523
Push broach, 751

Quality circles, 335, 1215
Quality control, 317, 1201
Quality control - integrated, 1214
Quality engineering, 1182
Quantity vs process, 1180
Quartz, 785, 786
Quench, 115, 123
Quench media, 127
Quick-change setup, 831
Rack, 910
Racon drill point, 702
Radial drill, 713
Radial force, 608
Radiation curing adhesive, 1054
Radiography, 308
Radius gage, 286
Raker-tooth saw, 761
Ramming, hand, 374
Rapid-solidification process, 186
Rasp-cut file, 770

Reaction injection molding, 563
Reamer nomenclature, 721
Reaming, 660, 720-23
Recalescence, 345, 347
Reconditioning cutting tools, 643
Recrysallization, 84, 112, 454, 467
Refractoriness, 371
Rene 41, 80, 95, 187
Reorder point (ROP), 1186
Repeat accuracy, 263
Repressing, 434
Research & Development, 1167, 1169
Residual stress, 128, 1074, 1112
Resin-transfer molding, 582
Resistance welding, 999
Resistivity method, 313
Resolution, 263
Resultant force, 604
Retained austenite, 121
Rimming (in steel), 147
Ring gage, 284
Ring rolling, 470
Rise per tooth (RPT), 748, 752
Riser, 343, 353, 367
Riveting, 506
Robot, 419, 1149, 1156
Robotic cell, 1129, 1158, 1221
Rocker-arm machine, 1002
Rockwell hardness test, 48
Rod drawing, 528
Roll extrusion, 505
Roll finishing, 928
Roll forging, 483
Roll welding, 1012
Roller burnishing, 509, 1083, 1110, 1113
Rolling, 465-470, 912
Rotary gear shaper, 919
Rotary piercing, 491
Rotary swaging, 496
Rotational molding, 565
Rotor-cut teeth, 752
Roundness, 259
Route sheet, 1174
Roving, 578
Rubber, 206, 571
Rubber bag forming, 538
Rule of 10, 261, 262
Runner, 343, 351, 366, 367

S-N curve, 57
Sand blasting, 1084
Sand casting, 366, 367
Sand slinger, 377
Sand testing, 371

Saturated polymers, 194
Saw blade nomenclature, 762
Sawing, 20, 598, 615, 759-69
Scanning laser, 277
Schonberger, Richard, 1218
Screw thread, 880
Screw thread nomenclature, 881-84
Screw threads, types, 881-84
Seam welding (RSEW), 1005
Seaming, 515
Segmented chips, 604
Seleroscope test (for hardness), 51
Semicentrifugal casting, 408
Service production system, 7
Setup reduction, 1211
Shallow drawing, 535
Shaping, 20, 597, 598, 615, 772-77, 912, 917
Shaving, 521, 912
Shaw process, 389
Shear angle, 599, 602
Shear force, 604, 605
Shear front angle, 603
Shear strain, 603
Shear stress, 606, 607, 646
Shear velocity, 603
Shear-forming, 533
Shearing, 19, 518, 520
Sheet-forming process, 465-470
Shell drawing, 535
Shell molding, 380, 381
Shielded metal arc welding (SMAW), 980
Shop floor control, 1190
Shot blasting, 1084
Shot peening, 1083, 1111
Shrink fit, 1060
Shrink rules, 356
Shrinkage, 352
SI units, 247
Sialon,
Silica (S_1O_2), 209, 371
Silicon, 151
Silicon bronzes, 173
Silicon carbide, 622, 624, 633, 786
Silicone, 200-201, 1053
Silver solder, 1037, 1046
Sine bar, 283
Single minute exchange of dies (SMED), 1200, 1212
Sintering, 424, 432, 633
Sizing, 434, 506
Skelp, 491
Skin-dried mold, 378
Slab, 465-470
Slab milling, 726

Slag, 348

Slip, 79

Slip casting, 574

Slip jackel, 376

Slitting, 520

Slitting saw, 735

Slotter, 919

Slush casting, 401

Smoothing of production, 1201

Snagging, 785, 92

Snap gage, 284

Solder, 189

Soldering, 21, 1045

Solid coring drill, 705

Solid state welding, 1010

Solid-solution hardening, 114

Solidification, 96, 344
  shirinkage in, 144

Solidus, 96, 352

Solium Silicate - CO_2 Molding, 379

Solubility, 93

Solvent cleaning, 1091

Spade drill, 705, 707, 710

Spark erosion machining, 956

Spark-discharge forming, 541

Specific energy, 610, 646

Specific friction energy, 610

Specific horsepower (HP_s), 608, 609, 646

Specific shear energy, 610

Spheroidization, 113

Spindle finishing, 1086

Spinning, 531

Split core box, 385

Split nut, 887

Split point drill, 702

Spot facing, 717

Spot welding (RSW), 1002

Spray molding, 583

Springback, 458, 511

Sprue, 343, 351, 367, 36

Spur gear, 911

Sputtering, 1006

Squeeze casting, 408, 477

Squeezing, 375

Stability, 263

Stabilizer, 202

Stack cutting, 974

Stack molding, 377

Stagger tooth milling cutter, 735

Stainless steel, 158-161

Staking, 507

Standard operations routine sheet, 1202

Standards of measurement, 244

Station, 10

Statistical process control, 327

Steady rest, 691

Steel, 100, 142-64, 356

Steel-rule die, 526

Step, 748

Step drill, 707

Stiefel process, 491

Stocking cutter, 914

Stockless production, 1218

Storage, 13, 23

Straight line control, 855, 858

Straightening, 517

Straightness, 259

Strain, 37

Strain hardening, 45, 81, 114, 496

Strain rate, 603

Strain sensing, 313

Strength-to-weight, 198

Stress, 37

Stress relief, 1076

Stress-relief anneal, 112, 114

Stress-strain curve, 39, 44

Stresses in adhesive joints, 1056

Stretch forming, 534

Stretcher leveling, 517

Stretcher strains, 459

Stringers, 455

Strip, 465-470

Stripper plate, 23

Structural adhesive, 1051

Structural ceramics, 211

Structure, 789

Stud welding, 989

Stylus profile device, 292

Stylus tip radius, 292

Styrofoam, 201

Subland drill, 707

Submerged arc welding (SAW), 987

Subpress die, 524

Suh, Nam P., 1170

Sulfur, 150

Superabrasives, 209, 792

Superalloys, 187

Superfinishing, 812

Superglue, 1052

Superheat, 345

Supermicrometer, 268

Superplastic forming, 544

Surface finish (roughness)288-295, 422, 664, 746, 1083, 110

Surface finish blocks, 295

Surface finishing, 1083, 1110

Surface grinding, 784, 804

Surfacing, 1028

Swaging, 483

Swiss screw machine, 674

T-slot cutter, 737

Taguchi methods, 325

Tangential force, 604

Taper plug gage, 284

Taper tap, 891

Taper turning, 653, 661

Tapping problems, 896

Taylor's tool life model, 641

Teach pendant, 154

Teflon, 201

Temperature, 245

Temperature sensing, 313

Temperature-composition diagram, 91

Tempering, 112

Tensile test, 38, 40

Terpolymer, 195

Textured surfaces, 1109

Thermal fatigue, 404

Thermal NTM, 936, 940

Thermal spray coating, 1028

Thermit welding (TW), 1016

Thermochemical machining (TCM), 947

Thermoforming, 564

Thermoplastic, 197, 200, 558, 559, 1051

Thermosetting plastics, 197, 200, 559, 1051

Thread cutting, 880, 886-92

Thread cutting dies, 889

Thread designation, 885

Thread grinding, 899

Thread milling, 899

Thread plug gage, 284

Thread rolling, 880, 900-02

Thread tapping, 891

Three-two-one (3-2-1) principle, 820

Throwaway inserts, 624, 626, 684

Tie-lines, 96

Time, 245

Time-temperature-transformation (TTT)
  diagram, 118-123, 131, 132

Tin coating, 1100

Titanium/alloys, 151, 185

Titanium nitride coating, 897

Tolerance, 252, 256, 478

Tombstone, 848, 849

Tool materials-cermets, 632

Tool grinders, 807

Tool holders, 684, 685

Tool life, 640

Tool steels, 161-63, 620

Tool wear (in machining), 612, 630, 639

Tooling, 13

Toolmaker's flat, 274
Toolmaker's microscope, 270
Tooth set, 761
Torpedo, 561, 562
Total quality control, 335
Tow, 578
Toyota production system (TPS), 30
Tracer control, 742
Transfer machines, 1137, 1141
Transfer molding, 560
Treatments, 13, 18
Trepanning gun drill, 704-05, 710
Triple point, 91
Truing, 793
Tube drawing, 529
Tube sinking, 529
Tungsten, 151, 622-628
Tungsten inert gas welding (TIG), 982
Turning, 20, 594, 598, 614, 653-56, 677, 684
Turning centers, 870
Turret drill, 715
Turret lathe, 669, 686
Twist drill, 697, 705, 710

U-shaped cell, 1128, 1200
Udimet 500/700, 187
Ultimate tensile strength (UTS), 40, 154
Ultrasonic inspection, 305
Ultrasonic machining, 21, 952, 1093
Ultrasonic welding (USW), 1014, 1025
Undercooling, 344, 347
Universal dividing head, 744
Up milling, 729-31
Upset, 472

Upset forging, 479
Upsetting, 976
Urea-formaldehydes, 200-201
Urethane, 1053

V-process, 382
Vacuum arc remelting (VAR), 147
Vacuum chucks, 839
Vacuum induction melting (VAM), 148
Vacuum metallizing, 1105
Vacuum permanent mold casting, 403
Vacuum-bag molding, 581
Value engineering, 1173
Vanadium, 151
Vapor degreasing, 1091
Variability, 317, 325
Variables, 245
Vendor programs, 1219
Vendors, 1201
Vernier caliper, 264
Vibration welding, 1025
Vibratory finishing, 1086
Vickers hardness test, 49
Videograph, 600
Vinyls, 200-201
Vision systems, 275-78
Visual inspection, 301
Vitrification, 573
Vitrified bond, 791
Vixen-cut file, 770

Wall ironing, 490
Warm working, 454, 459
Waspaloy, 187

Waterglass, 379
Wave-set saw, 762
Waviness, 288
Weldability, 62, 1077
Welding, 21, 966-1031
Welding metallurgy, 1070
Welding-plastics, 1024-27
Wettability, 1036
White cast iron, 104
Whitney, Eli, 25
Whitworth, Joseph, 27
Wire brushing, 1089
Wire drawing, 530
Wire EDM, 957
Withdrawal kanban (WLK), 1217
Woodruff keyseater, 737
Work envelope, 1151
Work-hardening, 81
Work-in-process (WIP), 1219
Workability (see formability)
Workholding, 13, 591, 687, 710, 816-842
Workpiece, 15
Worm, 912

Yardstick for automation, 24, 1130, 1134, 1153
Yarns, 578
Yield strength, 40, 154
Yield-point runout, 459
Young's modulus, 39

Zemak, 3, 185
Zinc alloys, 183
Zircon, 371
Zirconia, 212